MULTIPLE ATOMIC WEIGHTS OF THE ELEMENTS

C_1	12.01	H_1	1.01	O_1	16.00
C_2	24.02	H_2	2.02	O_2	32.00
C_3	36.03	H_3	3.02	O_3	48.00
C_4	48.04	H_4	4.03	O_4	64.00
C_5	60.06	H_5	5.04	O_5	79.99
C_6	72.07	H_6	6.05	O_6	95.99
C_7	84.08	H_7	7.06	O_7	111.99
C_8	96.08	H_8	8.06	O_8	127.99
C_9	108.10	H_9	9.07	O_9	143.99
C_{10}	120.11	H_{10}	10.08	O_{10}	159.98

N_1	14.01	F_1	19.00	Cl_1	35.45
N_2	28.01	F_2	38.00	Cl_2	70.91
N_3	42.02	F_3	57.00	Cl_3	106.36
N_4	56.03	F_4	75.99	Cl_4	141.81
N_5	70.03	F_5	94.99	Cl_5	177.27

Br_1	79.91	I_1	126.90
Br_2	159.83	I_2	253.81
Br_3	239.75	I_3	380.71
Br_4	319.66	I_4	507.62
Br_5	399.58	I_5	634.52

Organic Chemistry

Organic Chemistry

Alan S. Wingrove
Towson State University

Robert L. Caret
Towson State University

Harper & Row, Publishers
New York

Cambridge
Hagerstown
Philadelphia
San Francisco

1817

London
Mexico City
São Paulo
Sydney

Sponsoring Editor: Malvina Wasserman
Project Editors: Eva Marie Strock and Penelope Schmukler
Text and Cover Designer: Nancy B. Benedict
Senior Production Manager: Kewal K. Sharma
Assistant Production Manager: Marian Hartsough
Compositor Syntax International Pte. Ltd.
Printer and Binder: R. R. Donnelley & Sons, Co.
Art Studio: J & R Services; Syntax International Pte. Ltd.

ORGANIC CHEMISTRY

Library of Congress Cataloging in Publication Data

Wingrove, Alan S., 1939–
    Organic chemistry.

    Includes bibliographical references and index.
    1. Chemistry, Organic.   I. Caret, Robert L.,
1947–   II. Title.
QD251.2.W55   547   81–845
ISBN 0–06–163400–X   AACR2

Permission for the publication herein of Sadtler Standard Spectra® has been granted, and all rights are reserved, by Sadtler Research Laboratories, division of Bio-Rad Laboratories, Inc.

To Betty and Virginia

Brief Contents

Preface xxi

1   Organic Chemistry 1

2   Atomic and Molecular Structure; Molecular Orbital Theory 17

3   Alkanes and Cycloalkanes: Structure, Properties, and Nomenclature 64

4   Alkanes and Cycloalkanes: Preparation and Reactions 121

5   Reaction Mechanisms 145

6   Stereochemistry and Stereoisomerism 184

7   Alkenes and Alkadienes: Structure and Properties 243

8   Alkenes and Alkadienes: Reactions; Natural and Synthetic Polymers 289

9   Alkyl Halides: Substitution and Elimination Reactions 365

10   Alcohols 409

11   Alkynes: Structure, Properties Nomenclature, Preparations, and Reactions 464

12   Spectroscopy 1 – Spectroscopic Methods: Infrared and Ultraviolet Spectroscopy 491

13   An Introduction to Benzene, Resonance Structures, and Aromaticity 522

14   Reactions of Aromatic Compounds: Electrophilic Aromatic Substitution 556

15   Spectroscopy 2 – Spectroscopic Methods: Nuclear Magnetic Resonance Spectroscopy 623

16   Aryl and Vinyl Halides 670

17   Phenols 695

18   Spectroscopy 3 – Spectroscopic Methods: Mass Spectrometry 721

19   Ethers, Epoxides, and Glycols 733

20   Aldehydes and Ketones: Preparation of the Carbon-Oxygen Double Bond 786

21   Aldehydes and Ketones: Reactions of the Carbon-Oxygen Double Bond 818

22   Carboxylic Acids and Dicarboxylic Acids 890

23   Derivatives of Carboxylic and Dicarboxylic Acids 939

24   Reactions of α Hydrogens in Carbonyl and β-Dicarbonyl Compounds: Condensation Reactions 1017

25   Lipids: Fats and Oils 1090

26   Amines 1104

27   Cycloaddition Reactions: Woodward-Hoffmann Rules 1178

28   Amino Acids, Peptides, and Proteins 1210

29   Carbohydrates 1260

Appendix: Reading References 1321

Index 1335

Detailed Contents

Preface   xxi

1   Organic Chemistry   1
   1.1   Introduction to Organic Chemistry   1
   1.2   Organic Versus Inorganic Chemistry   2
   1.3   Identification of Organic Compounds   6
   1.4   Classification of Organic Compounds; The Study of Organic
        Chemistry   10
   1.5   Study of Organic Chemistry   13

2   Atomic and Molecular Structure; Molecular Orbital Theory   17
   2.1   Atomic and Molecular Structure: An Historical Overview   17
   2.2   The Quantum Mechanical Atom: The Beginnings   18
   2.3   The Quantum Mechanical Atom: Wave Mechanics   20
   2.4   Molecular Structure; Molecular (Bond) Orbitals; Bond Length   24
   2.5   Covalent Bonds in Simple Molecules; The Sigma (σ) Bond   26
   2.6   Structure of Carbon and Methane; Hybridization   28
   2.7   Formation of More Complex Organic Molecules   33
   2.8   Structure of Reactive Intermediates in Organic Chemistry   41
   2.9   Hybridization in Boron, Nitrogen, and Oxygen   46
   2.10 Polarity and Electronegativity   50
   2.11 Atomic Orbitals: A More Detailed View   54
   2.12 Molecular Orbitals (MO's)   56
   2.13 Linear Combination of Atomic Orbitals (LCAO's)   56
   2.14 Molecular Orbitals and Bonding   57
   2.15 σ Versus π Molecular Orbitals   59
   2.16 Alkanes   60
   2.17 Alkenes and Alkynes   61

3   Alkanes and Cycloalkanes: Structure, Properties, and Nomenclature   64
   3.1   Methane   64
   3.2   Ethane   65
   3.3   Ethane and the Carbon-Carbon Single Bond; Conformational Analysis
        and Free Rotation   66
   3.4   Propane   71
   3.5   Butane; Structural Isomers   71
   3.6   Conformations of Butane   72
   3.7   Pentane   76
   3.8   Isomers of Higher Hydrocarbons; The Homologous Series   78
   3.9   System for Writing Structural Isomers   79
   3.10 Nomenclature: Common Names; IUPAC System   81

3.11 Physical Properties of Alkanes   92
3.12 General Structure and Nomenclature   95
3.13 Physical Properties of Cycloalkanes   97
3.14 Structural Properties of Unsubstituted Cycloalkanes   98
3.15 Evidence for Cycloalkane Stability: Heats of Combustion   106
3.16 Conformational Analysis of Monosubstituted Cyclohexanes   108
3.17 Disubstituted Cyclic Compounds; *cis-trans* Isomerism   113
3.18 Conformational Analysis for Disubstituted Cyclohexane Derivatives   116

4   Alkanes and Cycloalkanes: Preparation and Reactions   121
4.1   Industrial Sources   121
4.2   Alkanes from Alkyl Halides and Alkenes   124
4.3   Reactivity of Alkanes and Cycloalkanes   129
4.4   Halogenation; A Substitution Reaction   130
4.5   Examples of Halogenation of Alkanes   131
4.6   Free-Radical Halogenation: A Chain Reaction   134
4.7   Sulfuryl Chloride and the Halogenation of Alkanes   136
4.8   Nitration of Alkanes: Another Free-Radical Reaction   137
4.9   Combustion   138
4.10 Petroleum, Gasoline, Fuels, and Pollution   139
4.11 Air Pollution and Its Causes; Mechanism of Photochemical Smog
      Formation   140
4.12 Ozone and the Environment   141
4.13 Summary of Reactions   142

5   Reaction Mechanisms   145
5.1   What Is a Reaction Mechanism?   145
5.2   Chemical Reactions: A Graphical Representation   146
5.3   Thermodynamics   147
5.4   Energy of Activation   148
5.5   Transition State Versus Reaction Intermediate   149
5.6   Bond Breaking and Forming; Reversibility of Chemical Reactions   151
5.7   Reactive Intermediates   157
5.8   Kinetics Versus Thermodynamics   169
5.9   How To Study a Reaction Mechanism   171
5.10 Reaction Mechanisms: Product Distribution   171
5.11 Reaction Mechanisms: Intermediates   172
5.12 Reaction Mechanisms: Kinetics   173
5.13 Reaction Mechanisms: Catalysis   175
5.14 Reaction Mechanisms: Solvent Effects   175
5.15 Reaction Mechanisms: Substituent Effects   177
5.16 Nucleophilic Aliphatic Substitution: S_N2 177

6   Stereochemistry and Stereoisomerism   184
6.1   Optical Properties of Molecules; Optical Activity and the
      Polarimeter   184
6.2   Molecular Asymmetry and the Discovery of Enantiomers   188
6.3   Structures of Enantiomers; Requirements for Their Existence   190
6.4   Further Evidence for the Tetrahedral Carbon Atom; Symmetry and
      Stereochemistry   192
6.5   Interaction of Light with Matter; Enantiomers and Optical
      Activity   195

6.6   Properties of Enantiomers   197
6.7   Optical Purity; Racemic Compounds and Resolution   199
6.8   Designation of Configuration: *R* and *S* System   201
6.9   Reactions Involving Optically Active Compounds; Evidence for
       Free-Radical Structure and Nucleophilic Substitution   205
6.10 Another Way To Depict Structure of Enantiomers: Fischer
       Projection Formula   211
6.11 Stereoisomers of Molecules Containing Two Asymmetric Carbon
       Atoms   215
6.12 Stereoisomers of Compounds Containing Two *Analogous* Asymmetric
       Carbon Atoms   219
6.13 Determination of the Number of Stereoisomers   223
6.14 Properties of Diastereomers; Compounds That Contain Two
       Asymmetric Carbon Atoms   224
6.15 Examples of Biological Reactions Involving Optically Active
       Compounds   228
6.16 Reactions That Generate a Second Asymmetric Center in a
       Molecule   229
6.17 Stereoisomerism in Cyclic Compounds   232
6.18 Stereoisomerism in Cyclohexane Derivatives   235
6.19 Other Elements That Impart Asymmetry to Molecules   236

7   **Alkenes and Alkadienes: Structure and Properties   243**
7.1   Ethene: The Carbon-Carbon Double Bond   243
7.2   Propene   244
7.3   The Butenes; Geometric Isomerism   244
7.4   Criteria for the Existence of Geometric Isomers in Alkenes   246
7.5   Nomenclature of Alkenes   247
7.6   Cycloalkenes and Exocyclic Double Bonds: Nomenclature   251
7.7   Properties of Alkenes   252
7.8   Alkadienes: Nomenclature and Classification   254
7.9   Structure of Conjugated and Nonconjugated Dienes   256
7.10 Dehydrogenation of Alkanes   259
7.11 Dehydrohalogenation of Alkyl Halides; E2 Mechanism   261
7.12 Structure-Reactivity and Structure-Product Distribution Relationships
       in Dehydrohalogenation   264
7.13 Stereochemistry of the Dehydrohalogenation Reaction   266
7.14 Dehydration of Alcohols   270
7.15 Mechanism for Acid-Catalyzed Dehydration of Alcohols   271
7.16 Use of Carbocation Stability To Explain Reactivity in Dehydration
       Reactions   276
7.17 Product Distribution in Dehydration Reactions   277
7.18 Carbocation Rearrangements: 1,2-Hydride and 1,2-Alkyl Shifts   278
7.19 Dehalogenation of Dihalides   282
7.20 Preparation of Alkadienes   283
7.21 Summary of Reactions   283

8   **Alkenes and Alkadienes: Reactions; Natural and Synthetic Polymers   289**
8.1   Ionic Addition Reactions   289
8.2   Addition of Hydrogen Halides (Unsymmetrical Reagent) to Alkenes   290
8.3   Mechanism of Hydrogen Halide Addition to Alkenes   291
8.4   Further Evidence for Electrophilic Addition: Rearrangements   295

8.5   Hydration: Addition of Water (Unsymmetrical Reagent) to Alkenes   296
8.6   Geometry of Carbocations   298
8.7   Addition of Sulfuric Acid (Unsymmetrical Reagent) to Alkenes   300
8.8   Summary of Electrophilic Addition Reactions of Acidic Reagents; Markovnikov's Rule   301
8.9   Addition of Hydrogen Bromide to Alkenes in the Presence of Peroxides   302
8.10   Addition of Unsymmetrical Reagents to Alkadienes   304
8.11   Product Distribution in Addition of Unsymmetrical Reagents to Conjugated Alkadienes   308
8.12   Hydrogenation; Addition of Hydrogen (Symmetrical Reagent) to Alkenes   310
8.13   Hydrogenation of Alkadienes   313
8.14   Use of Catalytic Hydrogenation in Structure Determination   314
8.15   Heats of Hydrogenation; Evidence for Alkene Stability   315
8.16   Halogenation: Addition of Halogens (Symmetrical Reagent) to Alkenes   317
8.17   Mechanism and Stereochemistry of Bromine Addition   318
8.18   Stereochemistry of Bromine Addition to Open-Chain Alkenes   322
8.19   Chlorination: Addition of Chlorine to Alkenes   323
8.20   Halogenation of Alkadienes   324
8.21   Relative Reactivities of Alkenes in Addition Reactions   324
8.22   Addition of Hypohalous Acids (Unsymmetrical Reagent) to Alkenes; Formation of Halohydrins   326
8.23   Addition of Alkenes to Alkenes; Carbocation Reaction   327
8.24   Addition of Alkanes to Alkenes; Alkylation Reactions   329
8.25   Applications of Carbocation Chemistry to Natural Products   331
8.26   Carbocations: Summary   332
8.27   Reaction of Alkenes with Oxidizing Agents; cis-Hydroxylation and the Baeyer Test   332
8.28   Stereochemistry of Hydroxylation of Open-Chain Alkenes   335
8.29   Cycloaddition Reactions   335
8.30   Substitution Reactions of Alkenes; Allyl and Vinyl Radicals   336
8.31   Cleavage of Carbon-Carbon Double Bonds; Ozonolysis   339
8.32   Cleavage of Carbon-Carbon Double Bonds by Permanganate   342
8.33   Qualitative Analysis; Detection of Alkenes   344
8.34   Polymerization of Alkenes   345
8.35   Polymerization of Alkadienes; Rubber   348
8.36   Other Naturally Occurring Products That Contain Double Bonds   350
8.37   Summary of Reactions   355

9   Alkyl Halides: Substitution and Elimination Reactions   365
9.1   Structure, Nomenclature, and Preparation   365
9.2   Physical Properties   366
9.3   Nucleophilic Aliphatic Substitution   368
9.4   Nucleophilic Aliphatic Substitution; Effect of Alkyl Group Structure on Chemical Kinetics   368
9.5 S_N2 Reactions: Mechanism and Evidence   370
9.6   Effect of Alkyl Group Structure on S_N2 Reactivity   372
9.7   Effect of Nucleophiles on S_N2 Reactivity   375
9.8   Effect of Leaving Group on S_N2 Reactivity   378

9.9 S_N1 Reactions: Mechanism and Evidence 379
9.10 S_N1 Reactions: Structure-Reactivity Correlations 382
9.11 Effect of Leaving Group and Nucleophile on S_N1 Reactivity 385
9.12 Comparison of S_N1 and S_N2 Mechanisms 385
9.13 Effect of Solvent and Nucleophile Concentration on S_N1 and S_N2 Mechanisms 386
9.14 Special Reactivity of Allyl and Bicycloalkyl Halides 389
9.15 Competition Between Bimolecular Substitution and Elimination: S_N2 Versus E2 392
9.16 Competition Between Monomolecular Substitution and Elimination: S_N1 Versus E1 394
9.17 Elimination and Substitution 398
9.18 Qualitative Analysis; Classification of Alkyl Halides 399
9.19 Summary of Reactions of Alkyl Halides 401

10 Alcohols 409
10.1 Structure and Nomenclature 409
10.2 Physical Properties of Alcohols 413
10.3 Methods of Preparation of Alcohols: Survey 415
10.4 Hydroboration: Addition of Boron Hydrides to Double and Triple Bonds 416
10.5 Hydroboration: Synthetic Route to Alcohols; Stereochemistry of Addition 420
10.6 Hydroboration; Isomerization and Reduction Reactions 423
10.7 Industrially Important Alcohols 425
10.8 Reactions of Alcohols 427
10.9 Acidity of Alcohols; Formation of Metal Salts 428
10.10 Oxidation of Alcohols 430
10.11 Mechanism of Oxidation of Alcohols 435
10.12 Relationship Between Alcohols and Esters 438
10.13 Reaction of Alcohols with Hydrogen Halides 442
10.14 Mechanism of Reaction Between Hydrohalic Acids and Alcohols 444
10.15 Other Methods of Preparing Alkyl Halides from Alcohols 449
10.16 Qualitative Tests; Classification of Alcohols 453
10.17 Summary of Methods of Preparation of Alcohols 455
10.18 Summary of Reactions of Alcohols 456

11 Alkynes: Structure, Properties, Nomenclature, Preparations, and Reactions 464
11.1 Structure of the Carbon-Carbon Triple Bond 464
11.2 Nomenclature of Alkynes 465
11.3 Physical Properties 467
11.4 Preparation of Acetylene 467
11.5 Elimination Reactions: Dehydrohalogenation and Dehalogenation 468
11.6 Preparation of Alkynes by a Substitution Reaction 469
11.7 Alkynes As Carbon Acids; Terminal Alkynes 470
11.8 Reactions of Terminal Alkynes; Heavy Metal Salts 472
11.9 Reactions of Terminal Alkynes: Substitution Reactions and Synthesis of Alkynes 473
11.10 Addition Reactions; Hydrogenation 475
11.11 Addition of Halogens 477
11.12 Addition of Hydrogen Halides 477
11.13 Addition of Water 478

11.14 Hydroboration-Reduction Reactions   480
11.15 Addition of Alkynes to Alkadienes: Diels-Alder Reaction   480
11.16 Miscellaneous Addition Reactions of Alkynes   481
11.17 Cleavage Reactions of Alkynes   482
11.18 Qualitative Analysis; Detection of Alkynes   482
11.19 Summary of Methods of Preparation of Alkynes   482
11.20 Summary of Reactions of Alkynes   483

12   Spectroscopy 1–Spectroscopic Methods: Infrared and Ultraviolet
     Spectroscopy   491
12.1   Electromagnetic Radiation   492
12.2   Infrared Spectroscopy: What Is it?   495
12.3   Correlations Between IR Absorption and Structure   500
12.4   Interpretation of IR Spectra   502
12.5   UV Spectroscopy: What Is It?   502
12.6   UV Spectral Measurements   503
12.7   Effect of Structure on UV Absorption Spectra   505
12.8   Correlation Between UV Spectra and Structure   509
12.9   Use of UV Spectroscopy in Structure Identification   513

13   An Introduction to Benzene, Resonance Structures, and Aromaticity   522
13.1   Benzene: A Typical Aromatic Compound   522
13.2   Reactivity and Properties of Benzene   523
13.3   Electronic Structure of Benzene; Resonance   525
13.4   Molecular Orbital (MO) Theory   526
13.5   Representation of Benzene; Resonance and Delocalization Energy   528
13.6   Drawing Resonance Structures; Other Examples of Resonance   530
13.7   Aromaticity and Hückel's $(4n + 2)$ π-Electron Rule   534
13.8   Aromaticity and Molecular Orbital Theory   542
13.9   Nomenclature of Benzene Derivatives   545
13.10 Physical Properties of Aromatic Compounds   549
13.11 Source of Aromatic Compounds and Their Derivatives   550
13.12 Polynuclear Aromatic Hydrocarbons   551

14   Reactions of Aromatic Compounds: Electrophilic Aromatic Substitution   556
14.1   Reactions of Benzene: Electrophilic Aromatic Substitution   556
14.2   Why Substitution and Not Addition in Aromatic Compounds?   557
14.3   Nitration of Benzene: Monosubstitution   558
14.4   Halogenation of Benzene: Monosubstitution   561
14.5   Sulfonation of Benzene: Monosubstitution   562
14.6   Friedel-Crafts Alkylation: Monosubstitution   563
14.7   Other Methods of Alkylating Benzene   566
14.8   Introduction of a Second Group into a Monosubstituted Benzene
       Derivative: Disubstitution   568
14.9   General Factors in Reactivity   573
14.10 Orientation of Electrophilic Aromatic Substitution: Inductive Effect   574
14.11 Orientation and Reactivity Due to Resonance Effects   579
14.12 Orientation and Reactivity of Electrophilic Aromatic Substitution on
       Aryl Halides   583
14.13 Directing Effect of the Aryl Group   585
14.14 Disubstitution: Experimental Factors and Limitations   585
14.15 Synthetic Methods Involving Disubstituted Benzene Compounds   589

14.16 Synthesis of Aromatic Compounds Containing Three or More Substituents 590

14.17 Addition Reactions to the Aromatic Ring: Hydrogenation and Chlorination 592

14.18 Side-Chain Oxidation of Alkylbenzenes 593

14.19 Side-Chain Halogenation of Alkylbenzenes: Benzyl Radicals 594

14.20 Double Bonds in Side Chains on Aromatic Rings: Conjugation 598

14.21 Addition of Ionic Reagents to Alkenylbenzenes: The Benzyl Cation 599

14.22 Stable Carbocations 601

14.23 Thallium in Organic Synthesis 602

14.24 Qualitative Analysis: Detection of Aromatic Rings 604

14.25 Spectral Analysis of Aromatic Compounds 605

14.26 Summary of Reactions of Aromatic Compounds 609

15 **Spectroscopy 2 – Spectroscopic Methods: Nuclear Magnetic Resonance Spectroscopy 623**

15.1 Nuclear Magnetic Resonance Spectroscopy: What Is It? 623

15.2 NMR Spectrometer and Spectra 627

15.3 Interpretation of NMR Spectra 629

15.4 Identification of Different Kinds of Protons; Determination of the Number of Each Kind 629

15.5 Structure and Chemical Shift 632

15.6 Significance of Number of Peaks Associated with Each Kind of Proton: Spin-Spin Splitting and Coupling Constants 638

15.7 Summary of Interpreting NMR Spectra 648

15.8 NMR Spectra of Substituted Aromatic Compounds 650

15.9 NMR Spectra of Compounds That Contain Nonequivalent Geminal Hydrogens 652

15.10 NMR Spectra of Conformers 656

15.11 Effect of Other Atoms on NMR Spectra 658

15.12 Another Method for Analyzing Complex NMR Spectra: Spin Decoupling 660

16 **Aryl and Vinyl Halides 670**

16.1 Structure, Nomenclature, and Preparation 670

16.2 Physical Properties of Vinyl and Aryl Halides 672

16.3 Reactivity of Vinyl and Aryl Halides Toward Nucleophiles 672

16.4 Substitution Reactions on Aryl Halides: Nucleophilic Aromatic Substitution 672

16.5 Mechanism of Nucleophilic Aromatic Substitution 674

16.6 Nucleophilic Aromatic Substitution Mechanism: Structure-Reactivity Correlations 676

16.7 Comparison of Electrophilic and Nucleophilic Aromatic Substitution 681

16.8 Comparison of Aromatic and Aliphatic Nucleophilic Substitution 682

16.9 Dehydrohalogenation of Aryl Halides; Benzyne Formation 683

16.10 Polyhalogenated Hydrocarbons 688

16.11 Spectral Analysis of Alkyl and Aryl Halides 689

16.12 Summary of Reactions of Aryl Halides 691

17 **Phenols 695**

17.1 Structure and Nomenclature 695

17.2 Physical Properties 696

17.3   Preparation of Phenols   699
17.4   Phenols As Acids   700
17.5   Reactions of Phenols with Base   702
17.6   Electrophilic Aromatic Substitution Reactions on Phenols   704
17.7   Reactions of Phenoxide Ion   706
17.8   Oxidation Reactions of Phenols   706
17.9   Qualitative Analysis: Detection of Phenols   708
17.10 Spectral Analysis of Alcohols and Phenols   709
17.11 Summary of Methods of Preparation of Phenols   713
17.12 Summary of Reactions of Phenols   713

18   **Spectroscopy 3 — Spectroscopic Methods: Mass Spectrometry   721**
18.1   Mass Spectrometry: What Is It?   721
18.2   Interpretation of Mass Spectra   723

19   **Ethers, Epoxides, and Glycols   733**
19.1   Structure and Nomenclature of Ethers, Epoxides, and Glycols   733
19.2   Properties of Ethers   738
19.3   Uses and Occurrence of Ethers   740
19.4   Preparation of Ethers from Alcohols: Dialkyl Ethers   742
19.5   General Method of Ether Synthesis: Williamson Method   744
19.6   Variations in Williamson Ether Synthesis   748
19.7   Formation of Cyclic Ethers and Epoxides by the Williamson Method   751
19.8   C- Versus O-Alkylation of Phenoxide Ions   752
19.9   Reactions of Ethers: Cleavage   753
19.10 Preparation of Epoxides   756
19.11 Ring Opening of Epoxides by Acidic Reagents   757
19.12 Stereochemistry of Epoxide Hydrolysis: *anti*-Hydroxylation   760
19.13 Ring Opening of Epoxides by Basic Reagents   761
19.14 Orientation of Addition to Epoxides: Acidic Versus Basic Conditions   763
19.15 Biochemical Applications of Epoxide Ring Opening   765
19.16 Properties of Glycols   769
19.17 Methods of Preparation of Glycols   770
19.18 Reactions of Glycols   770
19.19 Spectral Analysis of Ethers   773
19.20 Summary of Methods of Preparation of Ethers   776
19.21 Summary of Reactions of Ethers   777
19.22 Summary of Methods of Preparation for Epoxides   777
19.23 Summary of Reactions of Epoxides   778
19.24 Summary of Reactions of Glycols   779

20   **Aldehydes and Ketones: Preparation of the Carbon-Oxygen Double Bond   786**
20.1   Structure of the Carbonyl Group   786
20.2   Nomenclature   787
20.3   Physical Properties of Aldehydes and Ketones   790
20.4   Methanal (Formaldehyde), Ethanal (Acetaldehyde), and Propanone (Acetone)   793
20.5   Preparation of Aldehydes: Summary   794
20.6   Other Methods for Preparing Aldehydes   795
20.7   Preparation of Ketones: Summary   797

20.8 Other Methods for Preparing Ketones   798
20.9 Spectral Analysis of Aldehydes and Ketones   805
20.10 Summary of Methods of Preparation of Aldehydes and Ketones   810

21 Aldehydes and Ketones: Reactions of the Carbon-Oxygen Double Bond   818
21.1 Reactions of Aldehydes and Ketones: Relative Reactivities   818
21.2 Nucleophilic Addition: General Mechanisms and Relative Reactivities of Aldehydes and Ketones   820
21.3 Stereochemistry of Addition to Carbonyl Compounds   824
21.4 Nucleophilic Addition of Hydrogen Cyanide: Cyanohydrins   824
21.5 Nucleophilic Addition of Sodium Bisulfite: Addition Compounds   826
21.6 Nucleophilic Addition of Alcohols: Hemiacetals and Acetals   828
21.7 Addition of Phenol to Formaldehyde: Condensation Polymerization Reactions   833
21.8 Nucleophilic Addition of Ammonia and Substituted Amines   835
21.9 Nucleophilic Addition of Water and Hydrogen Halides   839
21.10 Nucleophilic Addition of Grignard Reagents: Versatile Method for Preparing Alcohols   841
21.11 Application of Grignard Synthesis of Alcohols   846
21.12 Nucleophilic Addition of Organolithium Reagents: Alcohol Preparation   855
21.13 Reduction of the Carbonyl Group to Alcohol   856
21.14 Carbonyl to Methylene Reductions   861
21.15 Steroids   864
21.16 Qualitative Analysis: Detection of Aldehydes and Ketones   866
21.17 Alcohols As Key Intermediates in Aliphatic Chemistry   868
21.18 Summary of Reactions of Aldehydes and Ketones   871

22 Carboxylic Acids and Dicarboxylic Acids   890
22.1 Nomenclature of Carboxylic Acids   891
22.2 Nomenclature of Dicarboxylic Acids   894
22.3 Physical Properties of Carboxylic Acids and Dicarboxylic Acids   896
22.4 Methods of Preparation of Carboxylic Acids: Summary   897
22.5 Other Common Methods for Preparing Carboxylic Acids   898
22.6 Special Methods for Preparing Certain Common Carboxylic Acids and Dicarboxylic Acids   903
22.7 Prostaglandins   906
22.8 Reactions of Carboxylic Acids   908
22.9 Acidic Nature of Carboxylic Acids   908
22.10 Structure-Acidity Correlations: Aliphatic Carboxylic Acids   911
22.11 Aromatic Carboxylic Acids and Structure-Acidity Correlations   913
22.12 Dicarboxylic Acids and Structure-Acidity Correlations   915
22.13 Salt Formation of Carboxylic and Dicarboxylic Acids   916
22.14 Acidity of Various Hydroxyl-Containing Compounds; Use of Solubility in Separations   918
22.15 Resolution of Carboxylic Acids into Enantiomers   919
22.16 Reduction of Carboxylic Acids   921
22.17 Decarboxylation of Acids and Diacids   924
22.18 α Halogenation of Carboxylic Acids   925
22.19 Spectral Analysis of Carboxylic Acids   927
22.20 Summary of Methods of Preparation of Carboxylic Acids   929
22.21 Summary of Reactions of Carboxylic Acids   931

23 Derivatives of Carboxylic and Dicarboxylic Acids 939

23.1 Nomenclature 940
23.2 Physical Properties 944
23.3 Interconversion of Carboxylic Acid Derivatives and Nucleophilic Acyl Substitution 946
23.4 Preparation of Acid Chlorides 948
23.5 Reactions of Acid Chlorides 949
23.6 Preparation of Esters 954
23.7 Reactions of Esters 965
23.8 Preparation of Amides 975
23.9 Reactions of Amides 977
23.10 Occurrence and Use of Amides 979
23.11 Preparation of Anhydrides from Monocarboxylic Acids 980
23.12 Dicarboxylic Acids: Anhydride Formation Versus Decarboxylation 982
23.13 Reactions of Anhydrides 985
23.14 Use of Anhydrides in the Resolution of Alcohols 988
23.15 Preparation of Nitriles 990
23.16 Reactions of Nitriles 990
23.17 Summary of Interconversions of Carboxylic Acid Derivatives 991
23.18 Reactivity in Nucleophilic Acyl Substitution Reactions 992
23.19 Spectral Analysis of Acyl Compounds and Nitriles 995
23.20 Summary of Methods of Preparation of Carboxylic Acid Derivatives 1002
23.21 Summary of Reactions of Carboxylic Acid Derivatives 1005

24 Reactions of α Hydrogens in Carbonyl and β-Dicarbonyl Compounds: Condensation Reactions 1017

24.1 Carbon Acids; Formation and Structure of α Carbanions 1018
24.2 Evidence of α-Carbanion Formation: Isotopic Tracer Studies 1020
24.3 Stereochemical Evidence for Three-Dimensional Structure of Carbanions 1021
24.4 Base-Promoted Keto-Enol Equilibrium 1026
24.5 Base-Promoted Halogenation of Aldehydes and Ketones 1027
24.6 Halogenation of Methyl Ketones and Methyl Carbinols in Basic Solution: Haloform Test 1030
24.7 Acid-Catalyzed Reactions of Carbonyl Compounds 1033
24.8 Condensation of Aldehydes and Ketones: Aldol Condensation 1037
24.9 Crossed Aldol Condensations 1042
24.10 Applications of Aldol Condensations to Synthesis 1045
24.11 Reaction of Aldehydes That Contain No α Hydrogens: Cannizzaro Reaction 1047
24.12 Alkylation of Ketones: Enamine Synthesis 1048
24.13 Wittig Reaction: Use of Carbanions in Organic Synthesis 1051
24.14 Self-Condensation of Esters: Claisen Condensation 1053
24.15 Crossed Condensation of Esters 1057
24.16 Condensation of Esters with Ketones 1058
24.17 Condensation Reactions of Anhydrides: Perkin Condensation 1059
24.18 Cyano and Nitro Groups As Acidifying Substituents 1060
24.19 β-Dicarbonyl Compounds As Acids: Formation and Structure of Carbanions 1061
24.20 Malonic Ester Synthesis: Alkylation Reactions 1063

24.21 Acetoacetic Ester Synthesis: Alkylation   1068
24.22 Acylation of Malonic Ester and Acetoacetic Ester   1071
24.23 α,β-Unsaturated Carbonyl Compounds: Structure and Nucleophilic
       Addition Reactions   1072
24.24 Keto-Enol Tautomerism in Carbonyl and β-Dicarbonyl
       Compounds   1078
24.25 Summary of Reactions of α Hydrogens in Carbonyl and β-Dicarbonyl
       Compounds   1079

25 **Lipids: Fats and Oils   1090**
25.1 Glycerides   1090
25.2 Soaps and Detergents   1093
25.3 Unsaturated Glycerides   1097
25.4 Waxes   1099
25.5 Phospholipids: Micelles and Membranes   1100

26 **Amines   1104**
26.1 Structure and Nomenclature   1104
26.2 Physical Properties of Amines   1106
26.3 Basic Properties of Amines: Formation of Ammonium Salts   1108
26.4 Solubility Properties of Amines   1113
26.5 Heterocyclic Nitrogen Compounds   1114
26.6 Stereochemistry of Nitrogen   1119
26.7 Methods of Preparation of Amines   1121
26.8 Reactions of Amines Covered in Past Chapters   1129
26.9 Reactions of Amines Not Covered in Past Chapters   1136
26.10 Reactions of Selected Heterocyclic Compounds   1148
26.11 Naturally Occurring Amines and Amines of Medicinal, Environmental,
      and Industrial Interest   1152
26.12 Spectral Analysis of Amines   1161
26.13 Summary of Methods of Preparation of Amines   1164
26.14 Summary of Reactions of Amines   1166

27 **Cycloaddition Reactions: Woodward-Hoffmann Rules   1178**
27.1 Cycloaddition Reactions   1178
27.2 The Diels-Alder Reaction; A 4 + 2 Cycloaddition Reaction   1179
27.3 Bicyclic Compounds: Application of Diels-Alder Reaction   1181
27.4 Diels-Alder Reaction Using α,β-Unsaturated Carbonyl
      Compounds   1183
27.5 2 + 2 Cycloaddition Reactions   1187
27.6 Addition of Methylene to Alkenes: Cyclopropanes   1187
27.7 Utility of Wave Mechanics and Molecular Orbital Theory   1190
27.8 Conservation of Orbital Symmetry   1190
27.9 MO's: Mode of Attack   1191
27.10 Molecular Orbital Symmetry   1192
27.11 Applying MO Symmetry Rules to Organic Reaction Mechanisms   1193
27.12 Cycloaddition Reactions Revisited   1193
27.13 Generalized Selection Rules for Pericyclic Reactions   1199
27.14 Summary   1204

28 **Amino Acids, Peptides, and Proteins   1210**
28.1 α-Amino Acids: Structure, Properties, and Nomenclature   1210
28.2 Chemical and Physical Properties of α-Amino Acids   1212

28.3   Stereochemistry of α-Amino Acids   1216
28.4   Amino Acids, Peptides, and Proteins   1218
28.5   Synthesis of Peptides   1221
28.6   Automated Peptide Synthesis   1227
28.7   Synthesis of Amino Acids   1231
28.8   Synthesis of Proteins   1233
28.9   Structure of Proteins   1236
28.10 Factors Influencing Protein Structure   1238
28.11 Keratins   1244
28.12 Insulin   1245
28.13 Determination of Protein Structure   1248
28.14 Prosthetic Groups   1251
28.15 Chymotrypsin   1252

29   Carbohydrates   1260
29.1   Carbohydrates   1260
29.2   Photosynthesis   1261
29.3   Glucose Metabolism: Biosynthetic Pathways   1263
29.4   Stereochemistry of Carbohydrates   1263
29.5   D-Glucose: Mutarotation   1267
29.6   D-Glucose: Conformational Analysis   1275
29.7   Acetals and Ketals of Carbohydrates   1276
29.8   Nucleic Acids   1279
29.9   Reactions of Carbohydrates   1284
29.10 Biochemical Reactions Involving Carbohydrates   1296
29.11 Structure Elucidation in Carbohydrate Chemistry   1299
29.12 Biopolymers: Oligosaccharides and Polysaccharides   1305

Appendix: Reading References   1321
Index   1335

Preface

The inception, formulation, and eventual birth of a new textbook is an evolutionary process, beginning with a need by the author, a need for something better for the students. Upon beginning this project we felt that such a need existed, a text that was better organized and more coherent than currently available texts and written in a style students could both understand and find pleasant. This book is a result of our efforts to attain that goal.

Organic Chemistry is designed and written to be a comprehensive one-year text that can be used with equal facility in courses for chemistry majors, biological science students, and preprofessional students in the medical, dental, and pharmaceutical fields. The text provides a solid background in classical organic chemistry, and the numerous examples of organic compounds of current interest, concern, and/or biochemical importance provide special appeal to the students in the various disciplines.

The text is *written for the student,* in a clear and understandable style, and we made every effort to incorporate detailed explanations of the material so students can use the book on their own. Many published texts assume that students can read between the lines and understand the concepts being put forth; but this is not always true. The real key to studying and learning organic chemistry is a clear and coherent organization. The highlights of the organizational features of this text include:

Functional Group Approach

Organic Chemistry uses the functional group approach, which eliminates the needless repetition and discontinuity of thought that often arises in other texts. For example, we grouped alkanes and cycloalkanes (Chapters 3 and 4), alkenes and alkadienes (Chapters 7 and 8), and atomic and molecular structure (Chapter 2). Also, because the coverage of hydrocarbon chemistry has been reduced, other functional groups (alkyl halides in Chapter 9 and alcohols in Chapter 10) can be easily presented in the first semester. This provides a much more equitable distribution of the material between the two semesters than other texts.

Study Hints

Study hints and other suggestions for study methods that may be potentially useful in helping students succeed and avoid common pitfalls are presented throughout the text. Examples of these highlights, set off by color, include "How To Approach Organic Synthesis," "How To Study Organic Chemistry," "Flash Cards in Organic Chemistry," "How To Draw Resonance Structures," "Oxidation and Reduction in Organic Chemistry," "Reactive Intermediates," and "Electrophiles and Nucleophiles." A complete list appears in the Contents of the Study Guide.

Atomic and Molecular Structure

Atomic and molecular structure and molecular orbital theory are presented together in Chapter 2, and the various hybridizations of carbon are developed simultaneously in that chapter. The similarities and differences among these topics can thus be contrasted more readily, to avoid repetition and review in future chapters. For those instructors who choose not to discuss molecular orbital theory, we have separated the theory from the more qualitative discussions of bonding in Chapter 2. Thus, either the qualitative or/and the more theoretical approach to bonding can be taught since they appear in separate sections in the chapter.

Hydrocarbon Chemistry

Numerous modern texts dwell on hydrocarbon chemistry, often to the extent that the first-semester course seldom covers much beyond aromatic chemistry. We reduced the coverage of alkane chemistry, but we still hit the important highlights and introduce the basic concepts that evolve from them. This allows for the coverage of other functional groups, such as alkyl halides (Chapter 9) and alcohols (Chapter 10) in the first semester, permits greater flexibility in the second semester, and gives a better foundation for the laboratory portion of the course.

Mechanisms and Reactive Intermediates

Reaction theory and reaction mechanisms are first introduced and discussed together in Chapter 5, Reaction Mechanisms; nucleophilic aliphatic substitution is also introduced there. Product distribution, the effect of temperature and concentration, the use of stereochemistry, kinetics and isotope labels, and other experimental tools are discussed from the standpoint of explaining how reaction mechanisms are elucidated. The structure of carbocations, carbanions, carbon radicals, and carbenes are all presented in Chapter 2, and in Chapter 5 their relative stabilities, methods of formation, and reactions are contrasted. This material provides a strong foundation for the use of reaction mechanisms throughout the remainder of the text.

Spectroscopy

The spectroscopy material is presented in three separate chapters. Chapter 12 covers infrared and ultraviolet spectroscopy and follows the chapters on alkanes, alkenes, and alkynes; Chapter 15 covers nuclear magnetic resonance and follows aromatic chemistry; and Chapter 18 covers mass spectrometry and follows the aryl halides and phenols. This spreads out the material and gives students time to think about and use their newly gained knowledge of spectral techniques before delving into another technique. Once introduced, spectroscopy is integrated throughout all the subsequent chapters. This building-block approach provides a sound background in both the understanding and use of spectral methods.

Relevant Examples

Numerous relevant examples of organic compounds of bioorganic, medicinal, industrial, and environmental interest abound throughout the text, including sections on sweeteners, smog, soaps and detergents, natural products, prostaglandins, paints, and polymers. Complete chapters dealing with proteins (Chapter 28) and carbohydrates (Chapter 29) are also included. All these examples are grouped by area and listed in the Contents of the Study Guide.

Summaries

Many summaries of important material are included to help the student, such as summaries of K_a's, K_b's, spectral data, nomenclature, and bond energies. A summary of reactions follows each chapter; a complete reaction summary is compiled in the Study Guide.

Study Problems and Reading References

The text contains more than 900 problems of varying difficulty with over 3000 parts. About 50 percent of these problems are incorporated into the text as questions; about 50 percent are end-of-chapter study questions. Unlike most of the other texts in this field, this text's problems are written for the students – they are the type of problems instructors write for their own exams – and will provide a great deal of needed and useful practice. Reading references are presented in the appendix. Most of these references are from the *Journal of Chemical Education* and can be read and understood by the beginning student; selected textbooks are also listed whenever appropriate.

Study Guide and Answer Book

The Study Guide contains (1) a detailed answer for each problem, (2) a complete summary of nomenclature, (3) a complete summary of all reactions in the text, and (4) lists of relevant sections from the text by area (for example, industrial chemistry, bioorganic chemistry, medicinal chemistry, and environmental chemistry). The Study Guide will be very useful and invaluable to students as an aid in understanding and mastering organic chemistry.

Acknowledgments

A project of this magnitude is the result of the efforts of many individuals. We would like to thank the reviewers who read and criticized all or portions of the manuscript at the various stages of its development: Paul Barks (North Hennepin Community College, Minnesota), Paula Bruice (University of California, Santa Barbara), Robert Coley (Montgomery Community College, Maryland), Rasma Derrums (Miami-Dade Community College, Florida), Lloyd Dolby (University of Oregon, Eugene), Robert Gilman (Rochester Institute of Technology, New York), Kenneth Kemp (University of Nevada, Reno), Axtell Kramer, Jr.

(Meramec Community College, Missouri), Ronald Magid (University of Tennessee), John Mangravite (West Chester State College, Pennsylvania), Kenneth Marsi (California State University, Long Beach), James Mulvaney (University of Arizona, Tucson), Michael Ogliaruso (Virginia Polytechnic Institute), J. David Rawn (Towson State University, Maryland), Lanny Replogle (San Jose State University, California), Paul Robbins (Armstrong State College, Georgia), Richard Schowen (University of Kansas, Lawrence), Jan Simek (California Polytechnic Institute, California), Lawrence Singer (University of Southern California), Carl Snyder (University of Miami, Coral Gables), J. J. Topping (Towson State University, Maryland), and Gary Yost (Towson State University, Maryland).

We are also grateful to Leroy (Skip) Wade (Colorado State University) for twice reviewing the entire manuscript, for proofreading the galleys, for checking all the problems, and for his many helpful suggestions.

Thanks are due to the staff of Harper & Row who took a dream and made it a reality; in particular, Malvina (Mal) Wasserman, Chemistry Editor, Penny Schmukler and Eve Strock, Project Editors, and Lois Lombardo and Karen Judd, Managing Editors.

Our typists, Betty Caret, Kathy Mulqueen, Sharon Osenburg, and Nancy Scales, who made it all possible, deserve much credit.

And finally, we express our indebtedness to our students who gave us the reason for beginning, the courage to continue, and the fortitude to complete the project.

We are sincerely grateful to them all.

Alan S. Wingrove
Robert L. Caret

Organic Chemistry

1

Organic Chemistry

1.1   Introduction to Organic Chemistry

Until the early 1800s, **organic chemistry** referred to compounds that were derived from natural sources and living organisms. Ethanol and acetic acid had been obtained from the fermentation of grain or fruit juices, quinine had been isolated from the bark of the cinchona tree and was known to be an effective cure for malaria, and urea had been isolated from human urine. Oils and fats were also known. Coal and petroleum compounds came from fossil remains of once living sources. By contrast, inorganic compounds were those that could be obtained from minerals.

Because organic compounds seemed so closely linked with living processes, there appeared no hope of ever preparing them in the laboratory. The view was stated that a *vital force* was necessary for their formation. In 1828, the German chemist Wohler converted an inorganic compound, ammonium cyanate, into urea:

$$NH_4OCN \xrightarrow{\text{heat}} NH_2\overset{\displaystyle O}{\overset{\|}{C}}NH_2$$

Ammonium   Urea
cyanate

Following this discovery, many other organic compounds were synthesized in the laboratory, and finally the vital force concept was completely disproved.

Today, *organic chemistry is the study of carbon-containing compounds*. A distinction still exists between organic and inorganic chemistry, but the boundary between the two is often difficult to define. For example, the metal cyanides and metal carbonates are considered inorganic compounds even though they contain carbon. Some organic compounds are still isolated from living sources, but most are synthesized in the laboratory. Nevertheless, most proteins, nucleic acids, complex carbohydrates, and the essentials of living cells have defied laboratory synthesis because of their complexity. The field of biochemistry relates organic compounds to their mode of formation in living processes.

There are several million organic compounds known today, and thousands of new ones are prepared each year. It has been estimated that the number of known compounds is increasing at the rate of about 5% per year. More than 90% of the organic compounds have been synthesized in the laboratory, with the rest derived from living organisms (animals, plants, and fungi) or fossil remains (coal and petroleum). The number of possible organic compounds is virtually unlimited because of the unique electronic structure of the carbon atom.

TABLE 1.1   Examples of Organic Compounds

Type	Example(s)
Petroleum products and fuels	Gasoline, kerosene
Synthetic fibers	Dacron, nylon, Teflon, PVC (polyvinylchloride)
Agricultural chemicals	DDT, dieldrin, aldrin, malathion
Food additives and preservatives	MSG (monosodium glutamate)
Natural polymers	DNA, RNA, rubber
Drugs	Aspirin, penicillin
Refrigerants and aerosols	Freons (chlorofluorocarbons)
Proteins and enzymes	Hair, skin, amylase
Hormones	Insulin, adrenaline
Sugars and starches	Glucose, amylose
Fats and oils	Corn oil, vegetable oil, lard
Soaps and detergents	Dish and clothes detergents
Paints and resins	Oil-based and water-based paints
Explosives	TNT (trinitrotoluene)

We are surrounded by organic compounds. An idea of the far-reaching influence of organic chemistry can be seen from Table 1.1, which contains many, but certainly not all, of the organic compounds.

1.2   Organic Versus Inorganic Chemistry

Although we are aware of the existence of several million organic compounds, and the number of conceivable organic compounds is almost unlimited, only 75,000 to 100,000 inorganic compounds are known. Before looking at the detailed electron structure and bonding capabilities of the carbon atom, we should examine other differences between organic and inorganic compounds.

A. Physical Properties

With few notable exceptions (for example, carbon tetrachloride, CCl_4) organic compounds are combustible, whereas inorganic compounds are not. Indeed, this property can be used as an easy experimental method to determine whether a compound is organic or inorganic. The *flame test* involves putting a small portion of an unknown compound on the tip of a spatula and placing it in the flame of a burner. If the compound burns readily and leaves no residue, it is most likely organic. If the compound burns partially but leaves a residue, it may be the salt of an organic compound. If the sample is completely unaffected by heat, it is probably inorganic.

Organic compounds are gases, liquids, or solids (except for polymers) that melt below 400 °C. Contrast this with inorganic compounds, which are often ionic in nature and are solids that melt at very high temperatures. Most organic compounds are are not water-soluble unless they have polar groups [such as hydroxyl (—OH) or carboxyl (—COOH) groups] attached to them, whereas a greater percentage of inorganic compounds are water-soluble because of their ionic character.

B. Bonding

Many differences in chemical and physical properties can be explained in terms of bonding. **Ionic bonding** is most frequently encountered in inorganic chemistry, and it occurs as a result of the *transfer of electrons* from one atom to another. In general, metallic elements on the left side of the periodic table lose electrons to the more electronegative, nonmetallic elements on the far right side of the table. This gives rise to ionic salts, as exemplified by the formation of sodium chloride:

$$Na\cdot \longrightarrow Na^{\oplus} + e^{\ominus}$$

$$:\ddot{C}l\cdot + e^{\ominus} \longrightarrow :\ddot{C}l\overset{\ominus}{:}$$

$$\longrightarrow Na^{\oplus}:\ddot{C}l\overset{\ominus}{:}$$

After the transfer of electrons, the resulting ions have full outer shells and greater stability because their electron configurations resemble those of the *noble gases.* Electrostatic attraction between the positively charged sodium ion and the negatively charged chloride ion gives rise to the ionic bond.

On the other hand, carbon occupies a position in the center of the periodic table and has four valence electrons. It does not form stable ions. As we will discuss, however, there is evidence of various types of transient carbon ions in organic reactions. As with other elements, carbon would like to have a full octet of electrons in its outer shell, a feat it can accomplish by *sharing electrons with other atoms.* The resulting **covalent bond** is typical of most carbon compounds. In contrast to the ionic bond, where the atoms are held together by the electrostatic forces between the ions, the covalent bond utilizes electrostatic attractions between the nuclei of both atoms and the bonding electrons.

Following are five examples of covalent bonding in inorganic and organic compounds, using **Lewis electron structures,** where electrons are depicted by a dot ($\cdot$) or an ($\times$). Also shown are conventional structures, where a line (—) indicates one pair of bonding electrons. In the latter case, nonbonding electrons may or may not be included, depending on the concept being discussed.

$$H\cdot + H^{\times} \longrightarrow H\overset{\times}{:}H \qquad \text{or} \qquad H\!-\!H$$

$$:\ddot{C}l\cdot + {}^{\times}_{\times}\overset{\times\times}{C}l{}^{\times}_{\times} \longrightarrow :\ddot{C}l{}^{\times}_{\times}\overset{\times\times}{C}l{}^{\times}_{\times} \qquad \text{or} \qquad Cl\!-\!Cl$$

$$H^{\times} + \cdot\ddot{C}l: \longrightarrow H{}^{\times}_{\cdot}\ddot{C}l: \qquad \text{or} \qquad H\!-\!Cl$$

$$4H^{\times} + \cdot\overset{\cdot}{C}\cdot \longrightarrow \overset{\textstyle H}{\underset{\textstyle H}{H{}^{\times}_{\cdot}\overset{\times}{C}{}^{\times}_{\cdot}H}} \qquad \text{or} \qquad \overset{\textstyle H}{\underset{\textstyle H}{H\!-\!\overset{\textstyle |}{\underset{\textstyle |}{C}}\!-\!H}}$$

Methane

$$6H^{\times} + 2\cdot\overset{\cdot}{C}\cdot \longrightarrow \overset{\textstyle H\ \ H}{\underset{\textstyle H\ \ H}{H{}^{\times}_{\cdot}\overset{\times\times}{C}{}^{\times}_{\cdot}\overset{\times\times}{C}{}^{\times}_{\cdot}H}} \qquad \text{or} \qquad \overset{\textstyle H\ \ H}{\underset{\textstyle H\ \ H}{H\!-\!\overset{\textstyle |}{\underset{\textstyle |}{C}}\!-\!\overset{\textstyle |}{\underset{\textstyle |}{C}}\!-\!H}}$$

Ethane

In our study of organic chemistry we will encounter many types of covalent bonds. Their relative bond strengths and reactivities are governed largely by the atoms that are used to form them.

To help write Lewis electron structures for covalent molecules, first keep in mind that each nonmetallic atom in the first and second periods of the periodic table (hydrogen through neon) should satisfy the octet rule by having eight electrons about it (except for hydrogen, which has two electrons in its stable compounds). Second, account for all valence electrons that were associated with each atom originally. This can be done by considering the column in the periodic table where the element appears, since that number represents the number of valence electrons for the element. Third, remember that one, two, or three pairs of bonding electrons may be placed between two elements. Consider the Lewis electron structure for carbon dioxide, CO_2:

Element	$\dfrac{\text{Column in}}{\text{Periodic Table}} = \dfrac{\text{No. of Valence}}{\text{Electrons/Atom}}$	Total No. of Valence Electrons/Element
C	4	$4 \times 1 = 4$
O	6	$6 \times 2 = 12$
	Total no. of valence electrons in molecule	$= 16$

To write the electron dot structure for CO_2, there must be eight electrons around each element and a total of 16 valence electrons in the molecule. A reasonable Lewis electron structure can be written as

$$\ddot{\text{O}}::\text{C}::\ddot{\text{O}} \quad \text{or} \quad \ddot{\text{O}}{=}\text{C}{=}\ddot{\text{O}}$$

which has two pairs of bonding electrons between each pair of atoms; these pairs are referred to as **double bonds.**

This technique for writing Lewis structures may also be applied to positively and negatively charged ions. Compute the total number of valence electrons as before, and (1) if the ion is positively charged, decrease the total by the amount of the positive charge, and (2) if the ion is negative, increase the total by the amount of the negative charge. For example, in $NO_2^{\ominus}$, the total number of valence electrons is 1×5 (for N) $+ 2 \times 6$ (for O) $= 17$. The ion has a -1 charge, so add one more electron to 17, to give a grand total of 18 valence electrons for $NO_2^{\ominus}$. The Lewis structure is

$$\overset{\ominus}{\ddot{\text{O}}}:\ddot{\text{N}}::\ddot{\text{O}} \quad \text{or} \quad \overset{\ominus}{\ddot{\text{O}}}{-}\ddot{\text{N}}{=}\ddot{\text{O}}$$

We will encounter charged species, or ions, throughout this organic chemistry course. As the complexity of the molecules increases, it becomes more difficult to determine which atom(s), if any bear a charge.

Formal Charge

A simple and effective way to determine the charge on each atom in a molecule uses the following equation:

$$\begin{bmatrix}\text{Total number of electrons belonging} \\ \text{to atom in a molecule; } \textit{for purposes} \\ \textit{of determining charge only}\end{bmatrix} = \begin{bmatrix}\text{all nonbonded electrons} \\ \text{around atom}\end{bmatrix} + \frac{1}{2}\begin{bmatrix}\text{all bonded electrons} \\ \text{around atom}\end{bmatrix}$$

If the total number of electrons around an atom is greater than the number of valence electrons per atom (see Sec. 1.4), the atom is negatively charged: -1 (1 extra), -2 (2 extra), and so on. The opposite is true if the total is less than the number of valence electrons per atom. For example, using the equation just given, in methane each hydrogen atom $[1 = 0 + \frac{1}{2}(2)]$ is neutral, and the carbon atom $[4 = 0 + \frac{1}{2}(8)]$ is neutral. If we break a carbon-hydrogen bond in methane, two possible products are the methyl cation and the methyl anion. In the cation the carbon atom $[3 = 0 + \frac{1}{2}(6)]$ bears a +1 charge, and in the anion the carbon bears a −1 charge $[5 = 2 + \frac{1}{2}(6)]$. Work out the charges on the following species:

$$
\begin{array}{cccc}
\text{H} & \text{H} & \text{H} & \text{H} \\
\text{H:C:H} & \text{H:}\overset{..}{\text{C}}{}^{\oplus} & \text{H:}\overset{..}{\text{C}}\cdot & \text{H:}\overset{..}{\text{C}}{:}^{\ominus} \\
\text{H} & \text{H} & \text{H} & \text{H} \\
\text{Methane} & \text{Methyl cation} & \text{Methyl radical} & \text{Methyl anion}
\end{array}
$$

$$
\begin{array}{ccc}
\text{H:}^{\ominus} & \text{H}\cdot & \text{H}^{\oplus} \\
\text{Hydride ion} & \text{Hydrogen atom} & \text{Proton}
\end{array}
$$

Question 1.1

Write one reasonable Lewis electron dot structure for each of the following molecules or ions:

(a) $NO_2^{\oplus}$ (b) N_2 (c) SO_2 (d) SO_3
(e) CCl_4 (f) $H_3O^{\oplus}$ (g) $NH_4^{\oplus}$ (h) $NH_2^{\ominus}$

Question 1.2

Which atom bears the charge (where applicable) in the molecules or ions in Question 1.1? Explain.

That apparently unlimited numbers of organic compounds exist or can be prepared may be explained on the basis of the structure of the carbon atom itself. The covalent bonds that are formed between carbon and other atoms are very stable and strong. Thus, carbon can form bonds with other carbon atoms or with hydrogen, oxygen, nitrogen, the halogens, phosphorus, and sulfur, to name some of the more common elements found in organic compounds. Straight-chain, branched-chain, and cyclic compounds of carbon are common, and in theory, any possible organic compound that can be written on paper can be synthesized in the laboratory, provided that the octet rule is followed when writing the structures.

C. Reactivity

Most reactions of inorganic ions are very fast and usually quantitative. For example, as one learns in introductory chemistry, the neutralization reaction of an acid and a base is extremely rapid and can be used to determine quantitatively the amount of acid or base in an unknown mixture. Neutralization is a homogeneous reaction, but there are similar reactivity trends with inorganic ions that react to form a precipitate, that is, a heterogeneous reaction. The reaction between silver ion (from $AgNO_3$) and chloride ion to yield silver chloride is so slightly reversible that it is essentially irreversible and can be used to analyze a mixture for the amount of either $Ag^{\oplus}$ or $Cl^{\ominus}$ present.

$$H^{\oplus} + OH^{\ominus} \rightleftharpoons H_2O \qquad \text{Quantitative and}$$
$$Ag^{\oplus} + Cl^{\ominus} \rightleftharpoons AgCl \text{ (solid)} \qquad \text{very rapid reactions}$$

There are several driving forces behind these reactions. First, the reactions initially involve ions that are dissolved in aqueous solution, and because of the unlike charges on the species involved there is a great tendency for them to attract one another. Second, stable compounds result from these reactions: in neutralization, the resulting water molecule is only slightly dissociated, and silver chloride is a solid compound that is water-insoluble and removed from solution as a precipitate.

Most organic reactions are quite different from the reactions of inorganic ions or compounds. Except for some salts, the atoms of organic molecules are bonded predominantly through strong covalent bonds. They are neutral species and again, with the exception of some salts, do not dissociate into ions in solution. If we want to carry out a reaction between two different organic molecules, we usually must supply enough energy to break and then remake covalent bonds. To do this, energy is often supplied in the form of either heat (symbolized by delta Δ) or light (symbolized by hv, see Sec. 6.1). Such energy increases the number of collisions between the neutral molecules and thus increases the rate of the reaction. (For thermal energy, the reaction rate approximately doubles for each 10 °C rise in temperature.) For the most part, organic reactions are very slow and require special reaction conditions, whereas many inorganic reactions occur readily at room temperature. The rate of reaction between organic molecules is frequently increased by the addition of **catalysts.** The role of several of these catalysts is discussed in the upcoming chapters.

Organic reactions are seldom quantitative, and the yields of desired compounds are considerably less than those from inorganic reactions. If heat is used, both starting materials and products often decompose (with and without catalysts). Many organic reactions also are accompanied by the formation of undesirable side products. This reduces the yield of the desired compound and means that purification methods must be used in the laboratory; sometimes purification is the major problem, and losses of product often occur as a result of the mechanical processes involved. Yields in the 80 to 90% range are considered quite good for organic reactions.

1.3   Identification of Organic Compounds

Because there are several million known organic compounds and unlimited numbers of conceivable ones, it is impossible for us to know about all of them. How then can we approach organic chemistry without resorting to learning a compendium of facts about individual compounds? How can an unknown organic compound that is only one of millions be identified? On the surface it seems like looking for the proverbial needle in a haystack, but the key to studying and learning organic chemistry centers about a systematic approach to identifying compounds and a beautiful classification system for them. In this section we see how to approach organic chemistry!

Suppose an unknown organic compound is to be identified. The experimental approach that is used to elucidate its structure is outlined in the following four steps. This particular discussion presents procedures that can be carried out in the introductory organic laboratory.

A. Purification and Separation Techniques

Often organic compounds are obtained as mixtures and contain impurities in varying amounts. Before a compound can be identified, it must be obtained in pure form. There are four important methods for isolating and purifying compounds: **crystallization, distillation, extraction,** and **chromatography.** Each method uses physical or chemical properties of organic molecules. These techniques are the foundation of all laboratory work in organic chemistry.

An easy test that provides an indication of the purity of a solid is to find the **melting point.** Generally, a melting point range of 0.5 to 1 °C suggests that a compound is pure. When a solid is impure, crystallization is often used to purify it. Crystallization is accomplished by finding a solvent in which both the sample and the impurity it contains are mutually soluble but from which the sample can be selectively induced to crystallize, leaving the impurity in solution.

Similarly, finding the **boiling point** may be used to determine the purity of a liquid. Impure liquids are purified by several methods, the principal one being distillation, which makes use of differences in the boiling points of liquids.

Other methods may also be used, including extraction and chromatography. Additional information on these methods is found in the Reading References in the appendix.

B. Qualitative Elemental Analysis

As a prelude to determining the molecular formula and structure of a pure organic compound, the compound is analyzed to determine *what* elements are in it. In addition to carbon and hydrogen, several other elements found in organic compounds are the halogens, sulfur, oxygen, phosphorus, and nitrogen. Because the bonding in organic compounds is principally covalent, there is seldom a direct chemical means of analyzing the compounds. However, the covalent bonds may be ruptured by heating an organic compound with sodium metal, which converts the elements it contains into inorganic ions.

$$\text{Organic compound containing C, H, X, O, S, and N} \xrightarrow[\text{heat}]{\text{Na}} \xrightarrow{\text{hydrolysis}}$$
$$\text{Na}^{\oplus}, \text{X}^{\ominus}, \text{S}^{\ominus 2}, \text{CN}^{\ominus}, \text{CO}_2, \text{H}_2\text{O}$$

This process is called **sodium fusion.** Experimentally, it involves heating a small portion of an unknown compound with sodium metal and then carefully treating the fusion mixture with alcohol and water. This step destroys the excess sodium metal and dissolves the inorganic ions that were formed. The resulting aqueous solution is then analyzed to determine the inorganic ions in it.

The *halogens* (fluorine, chlorine, bromine, and iodine), if present, are converted into the corresponding halide ions, $\text{X}^{\ominus}$. (*Note:* The accepted designations for halogens and halide ions are X_2 and $\text{X}^{\ominus}$, respectively. These are used throughout this text.) A portion of the solution from the sodium fusion process is acidified with dilute nitric acid and boiled. The presence of halide is determined by adding silver nitrate solution and observing the formation of a precipitate of silver halide.

$$\text{Ag}^{\oplus} + \text{X}^{\ominus} \longrightarrow \text{AgX (solid)}$$

Although there are reliable techniques for distinguishing among the halides, some indication of the halide present can be determined by the color of the silver halide formed: silver chloride is *white* (and readily soluble in excess ammonia), silver bromide is *light yellow* (and somewhat soluble in excess ammonia), and silver iodide is *dark yellow* (and insoluble in excess ammonia). Fluoride cannot be detected by $\text{Ag}^{\oplus}$ because silver fluoride is water-soluble.

Sulfur, if present, is converted into sulfide ion, $\text{S}^{\ominus 2}$. When a second portion of the solution is acidified, the sulfur ion is converted to H_2S gas, which is allowed to come in contact with a strip of filter paper that has been saturated with a solution of lead acetate, $\text{Pb}(\text{CH}_3\text{CO}_2)_2$. Darkening of the filter paper caused by the formation of black lead sulfide indicates the presence of sulfur in the original compound.

$$S^{-2} \xrightarrow{H^{\oplus}} H_2S \text{ (gas)} \xrightarrow{Pb^{+2}} PbS \text{ (solid)}$$
Black

Nitrogen, if present, is converted into cyanide ion, $CN^{\ominus}$. After the acidity of a third portion of the solution has been adjusted and ferrous (Fe^{+2}) and ferric (Fe^{+3}) ions have been added, the cyanide ion is converted into ferroferricyanide, which is an intense blue precipitate and is called Prussian blue.

$$CN^{\ominus} \xrightarrow{Fe^{+2}} Fe(CN)_6^{-4} \xrightarrow[K^{\oplus}]{Fe^{+3}} KFeFe(CN)_6$$
Prussian blue

There are tests for carbon and hydrogen, but they are seldom used qualitatively since all organic compounds contain these elements. The quantitative determination of carbon, hydrogen, and oxygen is presented in Sec. 1.3C.

omit

Question 1.3

Silver ion, $Ag^{\oplus}$, reacts with sulfide ion, S^{-2}, to form Ag_2S, a black precipitate, and with cyanide ion, $CN^{\ominus}$, to form AgCN, a white precipitate. When testing for the presence of halide ion using the sodium fusion technique, the final test solution is acidified with dilute nitric acid and boiled before silver nitrate is added.

(*a*) Suggest why this must be done.
(*b*) What might happen if it were not done? (*Hint*: Look up the physical properties of hydrogen sulfide and hydrogen cyanide.)

C. Empirical Formulas, Molecular Weights, and Molecular Formulas

A compound's molecular weight and empirical formula are closely related in the determination of its molecular formula.

a. Empirical Formulas

*The **empirical formula** tells us only the simplest whole-number ratio of atoms in a molecule.* It does not tell us the molecular formula of that compound, although occasionally the molecular and empirical formulas coincide. For example, the empirical formula CH could correspond to any number of compounds and perhaps should be written as $(CH)_n$. With $n = 2$ we have the molecular formula C_2H_2, and with $n = 6$ we have C_6H_6. A molecular formula cannot be determined without the molecular weight of the compound being known.

Many of the methods used for the *quantitative* determination of these elements (as opposed to the *qualitative* analysis just described) are discussed extensively in most introductory chemistry courses. To review this material, consult one of the Reading References in the appendix.

Computation of Empirical Formula

Once the percentage (by weight) composition of a compound has been determined, the empirical formula can be computed. The following example reviews this method.

A certain organic compound was found by quantitative elemental analysis to contain 33.6% carbon, 5.6% hydrogen, and 49.6% chlorine. To find the compound's empirical formula, start by assuming that we have 100 grams (g) of the compound. If 33.6% of the compound is carbon, then it contains 33.6 g of carbon. Likewise, it contains 5.6 g of hydrogen and 49.6 g of chlorine. Since these weights do not total

100 g, the remainder of the compound is assumed to be oxygen (which cannot be measured directly), or 11.2 g of oxygen. Because the empirical formula represents the simplest whole-number ratio of atoms in the molecule, first determine the number of moles of each element in the 100-g sample by dividing the weight of each element by its atomic weight. Then the number of moles of each element is divided by the number of moles of that element present in the smallest amount to obtain whole-number ratios. (Sometimes this division must be repeated before whole numbers are obtained.) In this example, oxygen is present in the smallest molar amount, so the number of moles of each other element is divided by 0.7 to yield the smallest whole-number ratio.

Element	%	Conversion to Moles	Conversion to Smallest Whole-Number Ratio
C	33.6	$\dfrac{33.6 \text{ g}}{12 \text{ g/mole}} = 2.8$ moles	$\dfrac{2.8}{0.7} = 4$
H	5.6	$\dfrac{5.6 \text{ g}}{1.0 \text{ g/mole}} = 5.6$ moles	$\dfrac{5.6}{0.7} = 8$
Cl	49.6	$\dfrac{49.6 \text{ g}}{35.5 \text{ g/mole}} = 1.4$ moles	$\dfrac{1.4}{0.7} = 2$
O	11.2	$\dfrac{11.2 \text{ g}}{16 \text{ g/mole}} = 0.7$ mole	$\dfrac{0.7}{0.7} = 1$

The empirical formula is therefore $C_4H_8Cl_2O$.

b. Molecular Weights

Before the molecular formula can be determined, its **molecular weight** must be known. There are various ways to determine molecular weights; the more common are (1) **vapor density method** and (2) boiling point elevation (**ebullioscopic method**), freezing point depression (**cryoscopic method**), and vapor-pressure lowering (**osmotic pressure method**). These methods are discussed in most introductory chemistry courses.

The most recent and accurate and perhaps the fastest way to determine molecular weights utilizes the mass spectrometer and the method of **mass spectrometry.** Gas, liquid, or solid samples of organic compounds can be introduced into the mass spectrometer, which often yields the molecular weight in a matter of seconds. See Chap. 18 for a discussion of this method.

c. Molecular Formulas

We started by considering methods that are used to determine *what* elements are present in an unknown compound. We then reviewed ways to determine the empirical formula (that is, the ratio of the atoms present in a molecule) and the molecular weight (that is, the exact weight of 1 mole of a compound). There is now sufficient information to determine the **molecular formula.**

The molecular formula gives the number of each type of atom that the molecule is known to contain. It is always some multiple of the empirical formula, as illustrated by the following example.

Suppose a compound was found to have the empirical formula $C_6H_{11}Cl$ and a molecular weight of 237. The molecular weight corresponding to $C_6H_{11}Cl$ is 118.5. The molecular formula must be some multiple of the empirical formula, or $(C_6H_{11}Cl)_n$. We can write the equation

Weight corresponding to empirical formula $\times$ n = molecular weight

where n is the multiple of the empirical formula. In this case $(118.5) \times n = 237$, so $n = 2$. Thus, the molecular formula is $(C_6H_{11}Cl)_2$, or more correctly, $C_{12}H_{22}Cl_2$.

d. Spectroscopic Methods

In the past 20 to 30 yr, technological advances have led to the development of new instruments called *spectrometers* (or spectrophotometers), which produce a unique spectrum ("fingerprint") for a given compound. These instruments are discussed in Chaps. 12, 15, and 18.

1.4 Classification of Organic Compounds; The Study of Organic Chemistry

Now we know of the limitless possibilities for organic compounds and that the principal type of bonding is the covalent bond. And, we discussed some ways to purify organic compounds and to determine molecular formulas. However, we said nothing about how the atoms in a given compound are arranged, that is, the **molecular structure.**

For a given molecular formula we can often draw numerous possible structures. Consider the structures of all compounds with the molecular formula C_2H_6O:

$$
\begin{array}{cc}
\begin{array}{c}
\text{H}\ \ \text{H} \\
|\ \ \ | \\
\text{H}-\text{C}-\text{C}-\text{O}-\text{H} \\
|\ \ \ | \\
\text{H}\ \ \text{H}
\end{array}
& \text{and} &
\begin{array}{c}
\text{H}\ \ \ \ \ \text{H} \\
|\ \ \ \ \ | \\
\text{H}-\text{C}-\text{O}-\text{C}-\text{H} \\
|\ \ \ \ \ | \\
\text{H}\ \ \ \ \ \text{H}
\end{array} \\
(1) & & (2)
\end{array}
$$

In this simple case there are only two possible structures; note that compound (1) has a C—C—O—H linkage, whereas (2) has a C—O—C arrangement. These two structures represent an example of **structural isomerism;** *structural isomers have the same molecular formula but a different arrangement of atoms.* The word "isomer" comes from the Greek *isos,* meaning equal, and *meros,* meaning part. Be careful when drawing structural isomers: be certain each carbon atom has four pairs of bonding electrons attached to it to satisfy the octet rule. For other common atoms, the following bonding considerations must be adhered to:

Hydrogen: One pair of bonding electrons, —H

Oxygen and sulfur: Two pairs of bonding electrons and two pairs of nonbonding electrons, $-\overset{\cdot\cdot}{\underset{|}{\text{O}}}\!:$ and $-\overset{\cdot\cdot}{\underset{|}{\text{S}}}\!:$

Halogen: One pair of bonding electrons and three pairs of nonbonding electrons, $-\overset{\cdot\cdot}{\underset{\cdot\cdot}{\text{X}}}\!:$

Nitrogen: Three pairs of bonding electrons and one pair of nonbonding electrons, $-\overset{\cdot\cdot}{\underset{|}{\text{N}}}\!-$

These rules apply to most of the organic compounds we will encounter, but there are exceptions. Charged atoms do not adhere to these rules (H—$\overset{..}{\underset{..}{O}}$:$^{\ominus}$, H—$\overset{..}{\underset{|}{O}}{}^{\oplus}$—H,
H

and H:$^{\ominus}$, for example), nor do certain covalent molecules involving other than first- and second-period elements (IF_5, for example).

Let us return to the previous example and structural isomers (1) and (2). Since both isomers have the same molecular formula, is there a way of distinguishing between them and identifying which structure is which? To do this we must turn to the chemistry of organic compounds. These reactions and numerous others are discussed more fully in the following chapters; they are briefly presented here to give some idea of how the organic chemist approaches a seemingly complex question involving the elucidation of structure. The chemical reactions of the —O—H group in (1) are quite different from those of the C—O—C linkage in (2), even though both contain oxygen. One example is the difference in their reactivity toward sodium metal. Compound (1) reacts readily with sodium metal and liberates hydrogen gas, whereas (2) is completely unreactive toward it. This reactivity is perhaps not surprising

$$H-\overset{\overset{H}{|}}{\underset{\underset{H}{|}}{C}}-\overset{\overset{H}{|}}{\underset{\underset{H}{|}}{C}}-\overset{..}{\underset{..}{O}}-H + \text{Na metal} \xrightarrow[\text{temp.}]{\text{room}} H-\overset{\overset{H}{|}}{\underset{\underset{H}{|}}{C}}-\overset{\overset{H}{|}}{\underset{\underset{H}{|}}{C}}-\overset{..}{\underset{..}{O}}{:}^{\ominus} Na^{\oplus} + \frac{1}{2}H_2 \text{ (gas)}$$

(1)

$$H-\overset{\overset{H}{|}}{\underset{\underset{H}{|}}{C}}-\overset{..}{\underset{..}{O}}-\overset{\overset{H}{|}}{\underset{\underset{H}{|}}{C}}-H + \text{Na metal} \xrightarrow[\text{temp.}]{\text{room}} \text{no reaction}$$

(2)

when we consider how *explosively* sodium metal reacts with water:

$$H-O-H + \text{Na metal} \longrightarrow H-\overset{..}{\underset{..}{O}}{:}^{\ominus} Na^{\oplus} + \tfrac{1}{2}H_2 \text{ (gas)}$$

For both compound (1) and water this is a characteristic reaction of the —O—H group; in (1) a carbon-containing chain (that is, $H-\overset{\overset{H}{|}}{\underset{\underset{H}{|}}{C}}-\overset{\overset{H}{|}}{\underset{\underset{H}{|}}{C}}-$) is attached to the —O—H group, whereas in water a hydrogen is attached to it.

This example is an introduction to **functional groups** in organic chemistry. This term refers to specific arrangements of atoms or certain types of bonding in organic molecules. This functional-group concept is clarified further with the following two structures:

$$H-\overset{\overset{H}{|}}{\underset{\underset{H}{|}}{C}}-\overset{\overset{H}{|}}{\underset{\underset{H}{|}}{C}}-O-H \qquad \text{Functional group is } C-O-H$$

(1)

An alcohol: ethanol

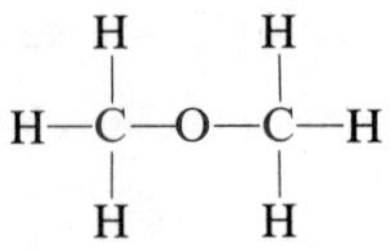

$$
\begin{array}{ccccc}
 & H & & H & \\
 & | & & | & \\
H- & C & -O- & C & -H \\
 & | & & | & \\
 & H & & H & \\
\end{array}
\qquad \text{Functional group is } C\!-\!O\!-\!C
$$

(2)

An ether: methyl ether

The functional group is all important in organic chemistry. In most cases the chemistry of organic compounds is dictated by the functional groups that are present and not by the number of carbon atoms they contain. That is, if there is an —O—H group in a molecule, it will react with sodium metal regardless of whether one carbon atom or hundreds of them are attached to it. Furthermore, the carbon atoms can be arranged in a straight chain (that is, in a continuous chain), a branched chain, or a cyclic structure, and the molecule will still have properties and undergo chemical reactions characteristic of the —O—H group.

TABLE 1.2   Common Functional Groups

Functional Group*	Name	Class of Organic Compound
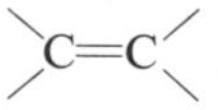 $\diagdown C = C \diagup$	Carbon-carbon double bond	Alkene
$-C\equiv C-$	Carbon-carbon triple bond	Alkyne
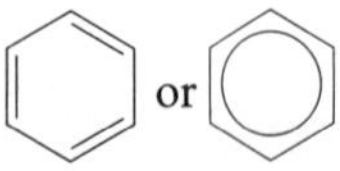 ⬡ or ⬡	Benzene ring	Aromatic
$-\overset{|}{\underset{|}{C}}-X$ (X = Cl, Br, F, I)	Halogen atom	Alkyl halide
$-\overset{|}{\underset{|}{C}}-OH$	Hydroxyl group	Alcohol
$-\overset{|}{\underset{|}{C}}-O-\overset{|}{\underset{|}{C}}-$	Alkoxy group	Ether
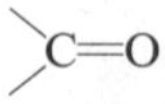 $\diagdown C = O$	Carbonyl group	Aldehyde or ketone
$-C = O$ $\quad|$ $\;OH$	Carboxyl group	Carboxylic acid
$-C = O$ $\quad|$ $\;G$ (G = Cl, OR, N$\diagup$, and others)	Acyl group	Carboxylic acid derivatives
$-\overset{|}{\underset{|}{C}}-N\diagup$	Amino group	Amine
$-C\equiv N$	Cyano group	Nitrile

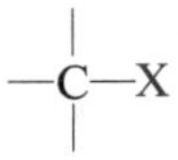

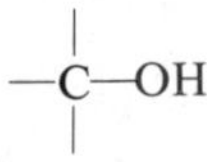

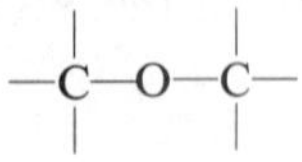

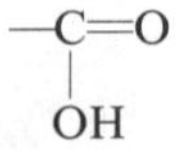

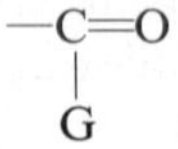

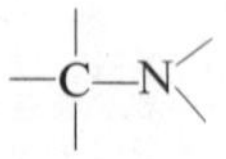

*Alkanes contain no functional group as such; they are saturated hydrocarbons (Chap. 3).

You soon will find that it is ill-advised, if not impossible, to try to learn the specific facts and reactions for each individual organic compound because there simply is no way that this can be done! Fortunately, a logical classification system for organic compounds and an organized and systematic method of studying organic chemistry are available. Molecules can be classified according to the functional group(s) they contain, and reactions can be studied by functional group (Table 1.2).

In this text we use this functional-group approach in presenting organic chemistry and study all the common groups.

1.5   Study of Organic Chemistry

During its early years organic chemistry was often viewed more as an art than a science, an art that centered around a large compendium of often seemingly isolated facts. The products from a given set of reactants (at specified reaction conditions) were known, but each reaction was treated more or less as independent of every other reaction. Organic chemistry resembled a cookbook—without theory, understanding, or even very much organization. Fortunately, there has been copious research with rewarding results: organic chemistry, as we know it today, is based largely on theory and understanding.

What does the organic chemist do? What is organic chemistry: the science? Thus far we glanced at the classification and systematic approach that can be used to identify an organic compound, but there is a lot more. We now survey briefly some highlights and mention several concepts that we will study in detail in this course.

A. Theories of Organic Chemistry

The underlying principles of organic chemistry are based on **structure**: the architecture and three-dimensional geometry of organic molecules. We will see how structure affects reactivity and how *structure-reactivity correlations* can be developed. Ideas and theories based on fact and observation will be developed about *how* organic reactions occur; this type of thinking is referred to as the study of mechanisms of organic reactions. This study is approached from the molecular and dynamic viewpoint; that is, we will consider reactions (insofar as possible) in terms of what happens to single molecules when bonds are made and broken as a chemical reaction occurs. For most reactions the considerable experimental evidence known in support of our theories and from which our theories arise will be presented and discussed.

B. Reactions of Organic Compounds

There are thousands of specific reactions in organic chemistry, but many of them fall into two general categories: **synthesis** and **degradation.**

In synthesis, the organic chemist may be concerned with ways to add more atoms to molecules. These molecules usually contain carbon but also may involve the halogens, sulfur, oxygen, nitrogen, or phosphorus. Sometimes a synthesis involves simply converting one functional group to another. Under certain conditions organic molecules may even rearrange into new molecules that have the same molecular formula. There may be an occasion when one wants to substitute one atom for another. With the help of theory and understanding (largely of reaction mechanisms) it is possible to limit the types of reactions that might be tried in the laboratory. Even though a reaction might be predicted to occur in theory, however, this does not necessarily mean that it can be done in the laboratory. The theories of reactions are invaluable in helping us draw analogies between known reactions and those we might write on paper or carry out in the laboratory. The following reaction illustrates a typical synthesis: the oxidation of an alcohol to an aldehyde represents the conversion of one functional group to another.

Synthesis:

$$
\underset{\text{Ethanol}}{\text{H}-\overset{\overset{\displaystyle H}{|}}{\underset{\underset{\displaystyle H}{|}}{C}}-\overset{\overset{\displaystyle H}{|}}{\underset{\underset{\displaystyle H}{|}}{C}}-\overset{..}{\underset{..}{O}}-H} \quad \xrightarrow[\substack{\text{oxidizing} \\ \text{agent}}]{[O]} \quad \underset{\substack{\text{Ethanal} \\ \text{(Acetaldehyde)}}}{\text{H}-\overset{\overset{\displaystyle H}{|}}{\underset{\underset{\displaystyle H}{|}}{C}}-\overset{\overset{\displaystyle :O:}{\|}}{C}-H}
$$

Degradation is a method for breaking bonds, thus splitting large, complex molecules into smaller fragments that may be identified by comparing them with known compounds. Degradation is often used to identify complex molecules, especially those occurring in natural sources, and also may be used synthetically to produce desired fragments that can be used to carry out a synthesis.

Degradation:

$$
\underset{\text{Propene}}{\text{H}-\overset{\overset{\displaystyle H}{|}}{\underset{\underset{\displaystyle H}{|}}{C}}-\overset{\overset{\displaystyle H}{|}}{C}=\overset{\overset{\displaystyle H}{|}}{C}-H} \quad \xrightarrow[\substack{\text{oxidizing} \\ \text{agent}}]{[O]} \quad \underset{\text{Ethanal}}{\text{H}-\overset{\overset{\displaystyle H}{|}}{\underset{\underset{\displaystyle H}{|}}{C}}-\overset{\overset{\displaystyle :O:}{\|}}{C}-H} \;+\; \underset{\substack{\text{Methanal} \\ \text{(Formaldehyde)}}}{\text{H}-\overset{\overset{\displaystyle :O:}{\|}}{C}-H}
$$

Structure determination is often difficult and challenging, frequently involving both synthesis and degradation. The proof of structure of a complex molecule may require the use of certain physical methods, such as X-ray crystallography and spectroscopy. Even after the structure has been determined, the final proof of structure comes from synthesizing the compound in the laboratory by using known reactions.

C. Applications of Organic Chemistry

Initially, our goal is to learn the theories and reactions of organic chemistry, but an ultimate goal is to apply these concepts to important compounds in nature. Most of the emphasis in this text is on learning these concepts. Once you do learn them, we will be able to discuss intelligently the chemistry of sugars, proteins and peptides, and natural products as well as the biochemistry of these compounds. For example, sugars involve cyclic structures, several types of functional groups, interesting aspects of molecular geometry (stereochemistry), and a whole host of organic reactions. One could not possibly hope to understand much about sugars, other than to memorize facts, until considerable exposure to and understanding of organic chemistry are achieved. A strong foundation in organic chemistry provides us with the background to discuss and understand the numerous examples of organic compounds we encounter daily: chemical sweeteners, smog, soaps and detergents, fats and oils, paints, drugs and so on—the list is endless.

About Studying Organic Chemistry

As a beginning student in organic chemistry you have just gotten your feet wet in some introductory material. Much of this may be review for you. Nonetheless, most of the material in Chap. 1 is developed further in subsequent chapters. Organic chemistry is cumulative; what you learn today is a basis for new material tomorrow. It is important that you attempt to master each concept before going to a new one. A fair number of students falter in organic chemistry for various reasons. Several ways you can do well, as well as some traps to avoid, are mentioned here. Read them—one or more will apply to you!

1. Keep up with your studying. Organic chemistry is cumulative and develops like a pyramid; new material depends on knowing the previous material. Organic chemistry is like a foreign language in some respects, for even though there is a systematic approach to the material, there are still many specific facts you must understand. Generally, students who fall behind seldom catch up; most often they fall further behind and ultimately give up. In short, do not fall behind; keep up with your studying from the very beginning. This means that you must start today!

2. Work study questions. The most important thing you can do to help ensure your success in and enjoyment of organic chemistry is to work study questions. In this text questions appear throughout the chapter as they apply to the particular sections that precede them. At the end of each chapter is a collection of study questions that emphasize the material of that chapter. The cumulative nature of organic chemistry means that you are responsible for all the previous material, which is reflected in the study questions. There are three main reasons why it is essential to work as many study questions as you can:

(*a*) *Working study questions allows you to find out what you do not know*. Basically, they provide a good medium for studying. Study questions are perhaps best done after you study *all* the material in that chapter and feel that you know it well.

(*b*) *Working study questions gives you added practice in working with names, structures, and reactions of organic compounds*. Often, the more times you write something on paper, the better engrained it becomes in your mind. You should study until the material is second nature to you.

(*c*) *Working study questions gives you confidence*. If you find that you can answer correctly the bulk of the study questions *on your own*, you have the right to feel that you have mastered the material.

First, work study questions as completely as possible on your own and without referring to solutions. If necessary, refer to the text or your lecture notes. Some students read the questions and then look at the answers—and they often recognize the answer as being "obvious" once they have seen it. Do not fall into this trap; you will only be kidding yourself because the most benefit results from diligent and independent study on your part.

3. Hints for more effective study. In addition to working study questions, there are three other study habits that may be useful to you.

(*a*) Reading (or at least scanning) the chapters before they are covered in lecture.

(*b*) Recopying lecture notes as soon as possible after lecture. This may even be done in conjunction with your text so that you can add items that help your understanding and that you may not have written in your original lecture notes.

(*c*) Studying with other students may also be helpful since you can often learn from each other. Later we present the effective study techniques of flash cards and reaction charts.

Study techniques are very much an individual matter; what some students find effective others do not. It is hoped that some of the suggestions help you. We started our venture into organic chemistry by assuming that you had a good background in general chemistry but knew nothing about organic chemistry. Your biggest jobs are to (1) keep up with the material and not fall behind and (2) work study problems and develop regular study habits.

Study Questions

Your instructor may assign selected questions of this set for you to study.

1.4 Define each of the following 12 terms, giving examples where appropriate:

(*a*) ionic bond	(*b*) covalent bond	(*c*) Lewis dot structure
(*d*) crystallization	(*e*) distillation	(*f*) sodium fusion

(*g*) formal charge   (*h*) empirical formula   (*i*) molecular formula
(*j*) molecular weight   (*k*) structural isomer   (*l*) functional group

1.5 Describe the sodium fusion process. How and why is it used to determine the presence of certain elements in an organic molecule? Write equations for the reactions that occur.

1.6 Keeping in mind the bonding rules in Secs. 1.2 and 1.4, write electron dot structures for the following molecules. Show all nonbonding valence (outer shell) electrons and all bonding valence (outer shell) electrons using dots. Indicate charges where important. (*Hint:* Some atoms may share more than one pair of electrons.)

(*a*) C_2H_2 (*b*) C_2H_4 (*c*) C_2H_6 (*d*) CH_3NH_2 (*e*) CHI_3 (*f*) CH_3CH_2OH
(*g*) NH_2OH (*h*) NH_4Br (*i*) CH_3OCH_3 (*j*) NaF (*k*) $BeCl_2$ (*l*) HCN

1.7 Write Lewis dot structures for the following molecules or ions. Indicate (where applicable) which atom bears the formal charge in each of the molecules.

(*a*) Br_2 (*b*) CO_3^{2-} (*c*) $CN^{\ominus}$ (*d*) H_2SO_4 (*e*) SO_2 (*f*) C_2H_4
(*g*) H_2O_2 (*h*) H_2SO_3 (*i*) C_3H_6 (*j*) C_5H_{12} (*k*) $C_2H_5^{\oplus}$

1.8 Indicate the formal charges, if any, on the atoms in red in each of the following.

$$\begin{array}{cccccc}
& & & & H & H \\
(a)\ H\!:\!\ddot{O}\!:\!H & (b)\ H\!:\!\ddot{O}\!:\!H & (c)\ H\!:\!\ddot{O}\!: & (d)\ H\!:\!\ddot{N}\!:\!H & (e)\ H\!:\!\ddot{N}\!:\!H \\
& & & & & H
\end{array}$$

$$\begin{array}{ccccc}
H & H & H & ,H & H \\
(f)\ H\!:\!\ddot{N}\!: & (g)\ H\!:\!\ddot{C}\!:\!H & (h)\ H\!:\!\ddot{C}\!:\!H & (i)\ H\!:\!\ddot{C}\!:\!H & (j)\ H\!:\!\dot{C}\!:\!H \\
& H
\end{array}$$

Take a good look at each of these structures for we will see them throughout the book.

omit **1.9** What is the percentage of each element in the following five compounds?

$CH_3CH_2CH_3$: propane, *bottled gas*

CH_3CHCH_3: isopropyl alcohol, *rubbing alcohol* (also called 2-propanol)
  |
  OH

NH_4OH: ammonium hydroxide, *household ammonia*

$NaOCl$:   sodium hypochlorite, *bleach*

$NaHCO_3$:   sodium bicarbonate, *baking soda*

*1.10 When 0.05 mole of hydrocarbon *H* was burned in the presence of oxygen, 4.4 g of carbon dioxide and 1.8 g of water were produced. Assuming the combustion was complete, what is the empirical formula of hydrocarbon *H*? The molecular formula? Draw the structure of a hydrocarbon that would provide the experimental data observed.

*1.11 Compound *A* consists of 60% carbon, 13.3% hydrogen, and 26.6% oxygen. What is its empirical formula? If 6 g of *A* are dissolved in 1,000 g of water, the freezing point of the water is lowered from 0° to $-0.186\,°C$. What is the molecular weight of the unknown? The molecular formula? Draw a plausible structure for a compound that would provide these experimental observations. (Hint: the molal freezing point depression constant for water is $-1.86\,°C$.)

*1.12 Ethylene glycol (common automobile antifreeze) consists of 38.7% carbon, 9.7% hydrogen, and 51.6% oxygen by weight. When 0.93 g of it is vaporized completely at 200° and at atmospheric pressure, 582 milliliters (ml) of vapor are formed. What is the molecular formula of ethylene glycol? Draw a plausible structure of this compound.

---

* See one of the general chemistry texts listed in the Reading References for this chapter in the appendix to review how to do these calculations.

2

Atomic and Molecular Structure; Molecular Orbital Theory

To initiate a discussion about the structure of carbon as it occurs in organic compounds, some salient facts about atomic structure and bonding are considered. Your course in general chemistry should have adequately provided the historical events that led to our present concept of atomic structure. (Much of the material in this section is review; for more details about atomic structure, consult your general chemistry text or a text listed in the Reading References in the appendix.)

We also trace the development of molecular orbital theory (MO theory) from the early origins of atomic theory to the present, beginning with J. J. Thomson's "plum pudding" picture of the atom and continuing with a review of Bohr's ideas and the contributions of Einstein and DeBroglie. Finally, the work of Erwin Schrödinger, who with others is credited with the development of quantum mechanics, is discussed.

Throughout the chapter our discussions are qualitative because the mathematics associated with the field of quantum mechanics are well beyond the scope of this text. We will see, however, that even a qualitative approach to the subject can yield a wealth of useful information to the enterprising chemist.

2.1   Atomic and Molecular Structure: An Historical Overview

The modern picture of atomic and molecular structure was developed during the early 1900s. New ideas and theories about matter and energy and their relationship to one another were being formulated. The proton, neutron, and electron had been recently discovered, and the atom had undergone a transformation from the plum pudding model put forward by J. J. Thomson in 1900 to the "quantized atom" introduced by Niels Bohr in 1912 (Fig. 2.1).

Bohr was among the first to postulate that the laws of the macroscopic world of classical physics could not be applied to the microscopic world of the atom. He envisioned that an entire new set of laws pertaining solely to the microscopic level would be needed to describe the atom and the particles it contains. Bohr viewed the atom as consisting of a positively charged nucleus surrounded by a sea of electrons moving symmetrically in circular orbits around the nucleus. In addition, Bohr felt the size of the nucleus was very small when compared with the sea of electrons, and essentially the entire mass of the atom was centered in the nucleus.

Bohr's picture of the atom was readily accepted because it explained many chemical phenomena that had been observed. However, the stationary orbit principle

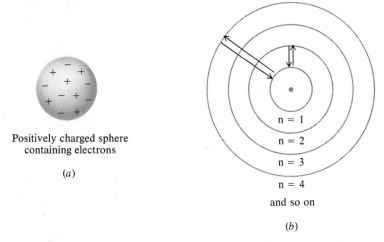

Positively charged sphere
containing electrons

(*a*)

n = 1
n = 2
n = 3
n = 4
and so on

(*b*)

FIGURE 2.1   (*a*) **Representation of J. J. Thompson's "plum pudding" picture of the atom. (*b*) The Bohr quantized atom: the promotion of electrons to higher energy states, with the concurrent absorption of energy shown in red—relaxation to lower states with the emission of energy in black.**

was against all the laws of classical physics, which predicted that an electron rotating in an orbit around a nucleus should gradually lose energy and eventually collapse into the nucleus. That atoms exist is sufficient proof that this does not occur. In the Bohr quantized atom then, an electron in a given orbit does not lose or gain energy. Energy is gained or lost only in transitions between the various orbits, and then only in discrete quantities called *quanta*. Although it may not appear so here, Bohr's postulates had their foundation in experimental observations. That the laws of classical physics should be ignored in the microscopic realm of the atom was becoming quite evident. All attempts to design a workable model using these laws failed.

Bohr's model for the atom was accepted for about 8 yr; then mounting evidence clearly indicated that another model was needed. Although the Bohr atom was adequate for explaining a one-electron atom like hydrogen, attempts at using this model to explain the observed data for more complex species failed. Also, the model did not present a clear picture of chemical bonding.

During this same period the laws of mechanics (classical physics) were also undergoing rapid change; the foundations of the *wave mechanical* or *quantum mechanical* atom were being laid.

2.2   The Quantum Mechanical Atom: The Beginnings

Several discoveries in the early 1900s led to the development of the *wave-particle nature of light*. In essence, this theory states that light behaves both like a particle and like a wave. The wave particle associated with light waves is termed the *photon* (see Sec. 6.1).

In 1924 DeBroglie published his doctoral dissertation titled "Investigations into Quantum Theory." DeBroglie reasoned, based on Einstein's work, that if one considered waves to be composed of particles, one could just as well assume that particles could exhibit wavelike behavior. A consequence of the theory is expressed by the following equation:

$$\lambda = \frac{h}{p}$$

where λ = wavelength of wave associated with given particle
$\qquad$ h = Planck's constant
$\qquad$ p = momentum of particle associated with wave (p = mass · velocity = $m \cdot v$)

This equation unequivocally relates the particle and wavelike nature of light; momentum is clearly a property one would relate to a particle, whereas wavelength could only be related to a wave. The wave-particle duality of nature was now firmly accepted for light and other forms of electromagnetic radiation as well as the electron (Fig. 2.2).

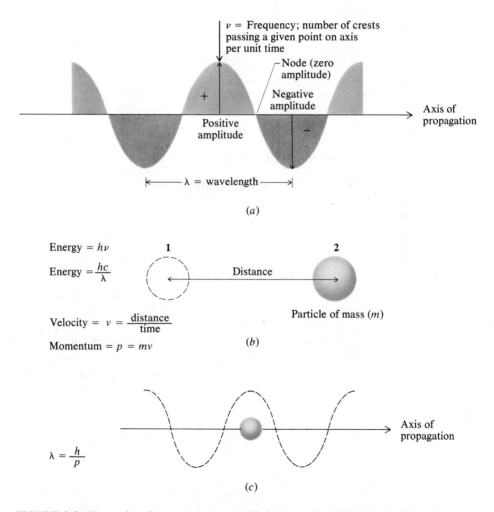

FIGURE 2.2   Properties of waves. (*a*) A wave with the properties of wavelength (λ) and frequency (v) indicated. The amplitude of the wave in what is defined as a positive ($+$) direction is called a *peak*, and that in the minus ($-$) direction is termed a *trough*. (*b*) A particle of mass *m* moving from point 1 to point 2 at velocity *v*. (*c*) DeBroglie's pictorial representation of a particle wave.

2.3   The Quantum Mechanical Atom: Wave Mechanics

Building on DeBroglie's postulates, Erwin Schrödinger in 1926 developed his theory of *wave mechanics*.[1] In this theory, electrons are described by a wave function, that is, a mathematical expression that accurately describes the motion of the electron in terms of the energy associated with it. This wave function is called *psi* (ψ).

The physical representation of these wave equations takes on new meaning when it is shown that psi can also be related to the probability of finding an electron at a given point in space. Although psi itself is not a measure of probability, it can be shown mathematically that ψ^2 (psi squared) is an accurate measure of the electron density as a function of position in space.[2] Throughout this chapter atomic orbitals are written as ψ and molecular orbitals as Ψ.

By plotting electron densities for various electrons as a function of position around a nucleus, one arrives at the familiar *orbitals* associated with the various electrons in an atom.

For example, electrons described by ψ_{1s}^2 have the greatest probability of being found in the ψ_{1s} (often written 1s) orbital. The energy of that electron is fixed (quantized) and is different from, for example, that of the electron in the ψ_{2s} (2s) orbital. Each electron in an atom can be described in terms of a given psi and has a certain discrete energy associated with it. Pictorial representations of these orbitals are in Figs. 2.3 and 2.4.

A. Geometry of Atomic Orbitals

There are four kinds of atomic orbitals—s, p, d, and f—that have different "shapes" depending on the energy of the electrons in them. These orbitals may also be written as ψ_s, ψ_p, ψ_d, and ψ_f. Both representations are used in this text.

A convenient method for depicting electrons in atomic orbitals shows the electron distribution in the form of a cloud. The pictorial representation implies a definite shape for orbitals, but remember that the electron cloud shows only the probability of finding an electron in a given region about the nucleus. Electron clouds show how the electronic charge is distributed in an orbital, but because electrons are in constant motion, we can speak of only the average negative charge per unit volume.

The lowest energy atomic orbital is the 1s (ψ_{1s}). It is *spherically symmetrical about the nucleus*, with its center at the nucleus. Its electron cloud is shown in Fig. 2.3. The boundaries are fuzzy and ill-defined. Because this figure represents only the probability of finding an electron, however, it tells us that the probability of finding

---

[1] The terms *wave mechanics* and *quantum mechanics* can, for all practical purposes, be used interchangeably. The development of modern wave mechanics, although most often associated with Schrödinger, was actually developed simultaneously by several individuals who used distinctly different mathematical approaches. The two approaches are called wave mechanics, referring to the wave nature of the electron, and quantum mechanics, referring to the quantization of energy in the atom. Among these individuals were Dirac, Heisenberg, Born, and Jordan. Unlike classical mechanics, quantum mechanics is a highly mathematical and abstract view of the atom. For the most part, we attempt to draw a physical picture of the atom as provided from wave mechanics without delving into the intricacies and abstractness of the mathematics.

[2] Although it can be expressed more concisely mathematically, the probability of finding a given electron in all space must have a finite value. We cannot, however, really discuss the probability of finding an electron at some fixed point; we are concerned with its probability of being found in some small region of space.

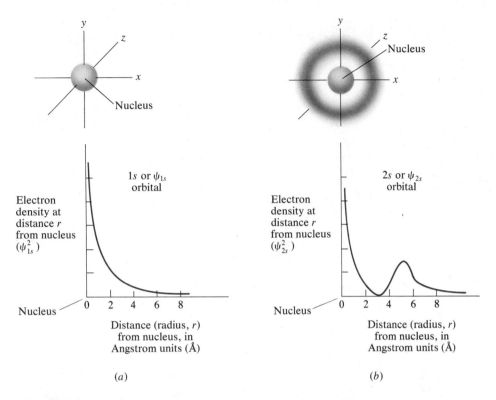

FIGURE 2.3 ψ_s Atomic orbitals. (*a*) Electron cloud associated with a ψ_{1s} atomic orbital, along with electron distribution curve. (*b*) Electron cloud associated with a ψ_{2s} atomic orbital, along with electron distribution curve.

an electron either at the nucleus or far from the nucleus (that is, completely outside the electron cloud) is very small. The electrons spend most of their time in the region indicated by the cloud but also have a finite probability of being found very near or very far from the nucleus. Because electron cloud drawings do not show accurately the electron distribution about the atom, electron density diagrams accompany them in Fig. 2.3.

The 2s (ψ_{2s}) atomic orbital is the next higher energy orbital and is also shown in Fig. 2.3. This orbital is also spherically symmetrical about the nucleus and is "larger" than the 1s orbital. It is of higher energy and lower stability than the 1s orbital because of the decreased electrostatic attraction between the 2s electrons and the nucleus of the atom caused by their being farther apart and separated from each other by the 1s electrons.

There are three 2p (ψ_{2p}) orbitals that have equal energy. In contrast to the 1s and 2s orbitals, which are spherically symmetrical about the nucleus, the 2p orbitals are *directional in character*, as shown in Fig. 2.4. They have shapes resembling dumbbells, with the nucleus at the center of each orbital. Each 2p orbital is perpendicular to another orbital; that is, one 2p orbital is perpendicular to the plane containing the other two. There are 90° angles between each two orbitals. To designate this three-dimensional character, the axes are labeled x, y, and z, and the orbitals are indicated by $2p_x$, $2p_y$, and $2p_z$ (or ψ_{2p_x}, ψ_{2p_y}, and ψ_{2p_z}). Electrons in these orbitals have the greatest probability of being found much farther from the nucleus than those in the 1s or 2s orbital.

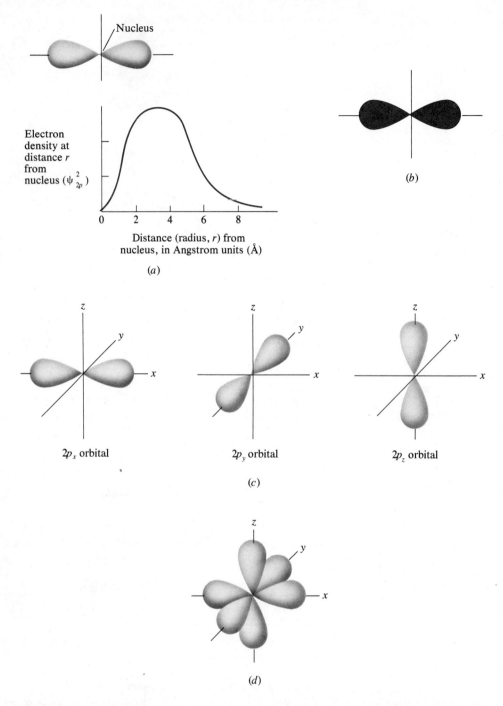

FIGURE 2.4 ψ_p Orbitals. (*a*) Three-dimensional shape and electron distribution curve of a single ψ_p orbital. (*b*) Cross-sectional (two-dimensional) drawing of a single ψ_p orbital. (*c*) Separate three-dimensional drawings of the three ψ_{2p} orbitals. (*d*) Composite drawing, showing all three ψ_{2p} orbitals.

B. Electronic Structure and Electron Configuration

There are several important rules governing the distribution of electrons in atoms, that is, the order for adding electrons to atoms (the *Aufbau principle* and *Hund's rule*) and the number of electrons permitted in each atomic orbital (the *Pauli exclusion principle*). The method for designating electronic structure combines these two principles to yield the *electron configuration* of atoms.

We start by viewing atomic structure from the standpoint of main shells and subshells. Historically, the main shells were called K, L, M, N, and so on. Each main shell contains subshells composed of the orbitals discussed previously. Today, the letter designation for main shells is seldom used; the letters have been replaced by numbers: K = 1, L = 2, M = 3, N = 4, and so forth. The numbers have the advantage that up to and including N = 4 they give the number of subshells in each main shell; that is, the first main shell (K = 1) has one subshell (composed of the 1s orbital), the second main shell (L = 2) has two subshells (composed of the 2s and 2p orbitals, respectively), the third main shell (M = 3) has three subshells (composed of the 3s, 3p, and 3d orbitals, respectively), and the fourth main shell (N = 4) has four subshells (composed of the 4s, 4p, 4d, and 4f orbitals, respectively).

The order of filling shells is stated by the Aufbau ("building up") principle and is a consequence of the energy levels of the various orbitals. The order is:

$$1s, 2s, 2p, 3s, 3p, 4s, 3d, 4p, 5s, 4d, 5p, 6s, 4f, 5d, 6p, 7s, 5f, 6d, 7p$$

$$\frac{\text{Order for filling subshells}}{\text{parallels increasing energy}} \longrightarrow$$

where the 1s orbital is closest to the nucleus and has the lowest energy, and the 7p orbitals are farthest from the nucleus and have the highest energy. *To provide the most stable configuration within a given set of orbitals of equal energy (for example, $2p_x$, $2p_y$, and $2p_z$), one electron each of identical (parallel) spin must be added to each orbital before a second electron can be added to any orbital.* This concept is called Hund's rule.

The last fundamental principle governing electronic structure that we use is the Pauli exclusion principle, which states that no more than two electrons can occupy a given atomic orbital and when there are two electrons present, they must have opposite spins, which are said to be paired.

We can summarize the principles of the electronic structure of atoms by writing electron configurations. This notation gives the main shell by number, the subshell (atomic orbital) by letter, and the number of electrons in each orbital as a superscript number. For example, the electron configuration of nitrogen (atomic number 7) is $1s^2 2s^2 2p^3$. For clarity, the three directional *p* orbitals can be indicated separately: $1s^2 2s^2 2p_x 2p_y 2p_z$ (the omission of a superscript implies one electron). Note that this obeys Hund's rule.

Table 2.1 summarizes the electronic structure for the first 12 elements in the periodic table. The electron spins are indicated by the arrows within a circle, and in accordance with the Pauli exclusion principle the spins are paired when the arrows are in opposite directions (for example, (↑↓)).

Question 2.1

Based on electronic structure, what similarities in chemical properties might you expect for the following three pairs of elements?

(*a*) Na and Li   (*b*) S and O   (*c*) Si and C

TABLE 2.1   Electronic Structure for the First 12 Atoms in the Periodic Table

Atom	Atomic Number	Main Levels: 1 — s	Main Levels: 2 — s	p_x	p_y	p_z	Main Levels: 3 — s	Electron Configuration
H	1	↑						$1s$
He	2	↑↓						$1s^2$
Li	3	↑↓	↑					$1s^2 2s$
Be	4	↑↓	↑↓					$1s^2 2s^2$
B	5	↑↓	↑↓	↑	○	○		$1s^2 2s^2 2p_x$
C	6	↑↓	↑↓	↑	↑	○		$1s^2 2s^2 2p_x 2p_y$
N	7	↑↓	↑↓	↑	↑	↑		$1s^2 2s^2 2p_x 2p_y 2p_z$
O	8	↑↓	↑↓	↑↓	↑	↑		$1s^2 2s^2 2p_x^2 2p_y 2p_z$
F	9	↑↓	↑↓	↑↓	↑↓	↑		$1s^2 2s^2 2p_x^2 2p_y^2 2p_z$
Ne	10	↑↓	↑↓	↑↓	↑↓	↑↓		$1s^2 2s^2 2p_x^2 2p_y^2 2p_z^2$
Na	11	↑↓	↑↓	↑↓	↑↓	↑↓	↑	$1s^2 2s^2 2p_x^2 2p_y^2 2p_z^2 3s$
Mg	12	↑↓	↑↓	↑↓	↑↓	↑↓	↑↓	$1s^2 2s^2 2p_x^2 2p_y^2 2p_z^2 3s^2$

Question 2.2

Based on electronic structure, what other elements might you expect to behave like fluorine? Why?

2.4   Molecular Structure; Molecular (Bond) Orbitals; Bond Length

Our major emphasis in studying bonding revolves about the covalent bond that is commonly encountered in organic compounds. However, as with the electronic structure of atoms and atomic orbitals, this is discussed only in a qualitative manner. Although we avoid the mathematical aspects of bonding, there are two qualitative concepts we should be aware of regarding bonding: (1) the pair of electrons used to form a covalent bond (one electron from one atom shared with one electron from the other atom) is *localized* between the nuclei of the two atoms involved, and (2) the shapes of the atomic orbitals on the original atoms are largely responsible for the *geometry* (three-dimensional shape) of the new molecule.

The localization of electrons in covalent bonds assumes that there is no inter-action of a given bond with any other bonds or atoms in the molecule. This is true for many covalent bonds, but there are some molecules in which bonding electrons interact with each other through adjacent bonds or even throughout an entire molecule. A typical example is the family of compounds related to benzene (called aromatic compounds); they are discussed in Chap. 13. We will discuss these molecules as the occasion arises but for now we consider bonding in molecules where there are no special electronic interactions.

The second premise of bonding involves the geometry and shapes of molecules and is generally more applicable to organic molecules. There are no rules that can be used universally to predict bond angles with precision; however, knowing the shapes (and angles) of atomic orbitals does give us an idea about the general shape of a molecule. These considerations permit reasonable predictions of bond angles, many of which agree well with experimental values.

When considering the formation of covalent bonds in a molecule, we often must use a great deal of imagination to envision the process of bond formation; indeed, most molecules could not possibly be made by the hypothetical processes we will consider.

A covalent bond is composed of a molecular orbital, called a **bond orbital** or a **molecular orbital.** *Atomic orbitals always refer to orbitals associated with individual atoms, and molecular orbitals are the orbitals that bond or link together two or more atoms.* Generally, formation of the covalent bond requires bringing together two atoms, each with an atomic orbital containing one electron.[1] In addition, these two atomic orbitals must be oriented so that they overlap to form the molecular (bond) orbital. A more stable covalent bond is formed in the process, with the two electrons now shared between the nuclei of the *two* atoms involved. The Pauli exclusion principle applies to molecular orbitals as well as individual atoms, so the electrons in the bond formation must have opposite spins and be paired.

What is the driving force for the formation of a covalent bond? Why do the bonds form? The answers to these questions come from a consideration of energy. When the atomic orbitals of two atoms overlap to yield a molecular bond orbital, the new arrangement of atoms is more stable than were the individual atoms themselves. The electrons in the covalent bond are attracted by the positively charged nuclei of two atoms, whereas in the individual atoms the electrons were attracted by a single atomic nucleus. *When a covalent bond is formed, energy is given off.* Conversely, *energy is required to break a covalent bond.* The energy required is called the **bond dissociation energy** and is symbolized by $\Delta H°$ or D (see Sec. 5.6). Bond dissociation energies are usually given in units of kilocalories per mole (kcal/mole) or kilojoules per mole (kJ/mole).

SI Units

Throughout this text we use those units most common to the literature of organic chemistry. However, the increasing use of the International System of Units (Systéme Internationale, abbreviated SI) by chemists, a system heretofore seldom used, requires that we, as chemists, be able to interchange units rapidly between the two forms. The common units and their conversion factors most likely to be useful to organic chemists are:

Quantity	Common Unit*	SI Unit	Conversion Factor
Length	Meter (m)	Meter (m)	
Mass	Kilogram (kg)	Kilogram (kg)	
Volume	Liter (L)	Cubic meter (m^3)	$1\ L = 10^{-3}\ m^3$
Temperature	Celsius (°C)	Kelvin (°K)	°K = °C + 273°
Energy	Calories (cal)	Joules (J)	1 cal = 4.184 J

* The units are often used in several of their equivalent metric forms, for example, 1 meter = 100 centimeters = 1000 millimeters, and so on.

---

[1] We see in Sec. 7.15, for example, how it is also possible to form a covalent bond by utilizing a non-bonding pair of electrons on one atom and a vacant atomic orbital on a second atom.

Bond lengths are another important physical property of molecules. Bond lengths in a given molecule may be estimated by comparing them with known molecules, calculating them theoretically, and measuring them experimentally with X-ray crystallography. Bond length is dictated largely by the type of orbitals (that is, $1s$, $2s$, $2p$, and so on) that overlap to form that bond. As shown in Figs. 2.4 and 2.5, there is considerable difference in the shapes of orbitals; that is, the $1s$ is closest to the nucleus, the $2s$ next, the $2p$ next, and so on. Examples of the bond lengths of some simple molecules are given in Sec. 2.5.

2.5   Covalent Bonds in Simple Molecules; The Sigma (σ) Bond

We now consider the formation of covalent bonds in some simple inorganic molecules as a prelude to extending these concepts to carbon-containing compounds.

A. The Hydrogen Molecule

The hydrogen molecule can be "made" by allowing the $1s$ orbitals (each containing one electron) of two atomic hydrogen atoms to overlap. This process is depicted in Fig. 2.5. Recall that imagination must be used when discussing the formation of molecules in terms of the atomic orbital and molecular orbital concept. The bond length in the hydrogen molecule is 0.74 angstroms (Å) ($1 \text{ Å} = 10^{-8}$ cm), which is the average distance between the two hydrogen nuclei. When this covalent bond is formed, two electrostatic interactions arise; the $1s$ electron of one hydrogen atom is attracted by the positive nucleus of both that atom and the other atom. At the same time, two positively charged hydrogen nuclei, which repel each other, are brought closer together. These forces of attraction and repulsion balance each other when the interatomic distance is 0.74 Å. The energy required to break this bond (bond dissociated energy for the hydrogen molecule) is 104 kcal/mole (435 kJ/mole).

$$\text{H:H} \longrightarrow \text{H·} + \text{H·} \qquad D_{\text{H—H}} \; 104 \text{ kcal/mole (435 kJ/mole)}$$

(In the symbol $D_{\text{H—H}}$ D represents the bond dissociation energy and the subscript H—H designates the bond in question.)

The hydrogen molecule would be expected to have a shape similar to the two hydrogen atoms from which it was made. It is cylindrically symmetrical about a straight line drawn through the two nuclei, as shown in Fig. 2.5. A cross section of the molecular orbital cut perpendicular to the bond axis is circular. When a molecular orbital is cylindrically symmetrical about a line passing through two nuclei, as is

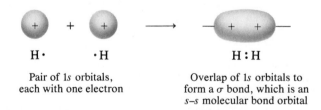

H·	·H	H:H
Pair of $1s$ orbitals, each with one electron		Overlap of $1s$ orbitals to form a σ bond, which is an s–s molecular bond orbital

FIGURE 2.5   Formation of hydrogen molecule, H—H. Bond formation by overlapping $1s$ orbitals. The + represents the nucleus.

the case for hydrogen, it is called a **sigma (σ) orbital** or a **σ bond.** Another designation for the covalent bond in hydrogen is *s-s molecular orbital* because it is made from two $1s$ atomic orbitals. The main shell number is omitted in this designation. Since hydrogen has two identical nuclei, the electrons are shared equally between them and the molecule is called *nonpolar;* that is, the electron density is distributed symmetrically about the two nuclei.

B. The Fluorine Molecule

The fluorine molecule, F_2, is formed by combining two fluorine atoms. The electron configuration of fluorine is $1s^2 2s^2 2p_x^2 2p_y^2 2p_z$, so there is one p orbital that is unfilled and contains one electron. The unfilled p orbital of one fluorine atom overlaps that of another fluorine atom, as shown in Fig. 2.6. The covalent bond formed in the fluorine molecule is also a σ bond and is called a *p-p molecular bond orbital* because it is formed by overlapping two $2p$ orbitals. As a result of the extremely high electronegativity (see Sec. 2.10) of the fluorine atom, the bulk of the electronic charge in the overlapping p orbitals is concentrated between the two nuclei, and the back lobes of those orbitals are considerably smaller than would be expected in a normal p orbital. Because the molecular orbital in the fluorine molecule is cylindrically symmetrical about a line passing through the two nuclei, it is a σ bond. The F—F bond length is 1.42 Å, and its bond dissociation energy, D_{F-F}, is about 37 kcal/mole (155 kJ/mole). Fluorine is a nonpolar molecule because the electrons are shared equally between the two identical fluorine nuclei.

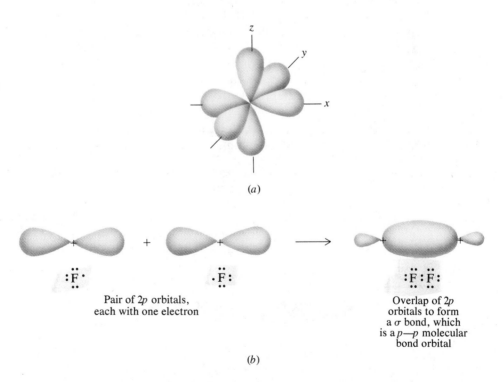

(a)

(b)

Pair of $2p$ orbitals,
each with one electron

Overlap of $2p$
orbitals to form
a σ bond, which
is a p—p molecular
bond orbital

FIGURE 2.6   Formation of the fluorine molecule, F—F. (*a*) **The fluorine atom with its three $2p$ orbitals.** (*b*) **Bond formation by overlapping the $2p$ orbitals of fluorine. The filled p_y and p_z orbitals have been omitted for clarity: the + represents the nucleus.**

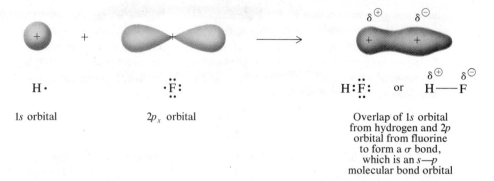

$$H \cdot$$

1s orbital

$$\cdot \ddot{F} \colon$$

$2p_x$ orbital

$$\overset{\delta \oplus}{H} \colon \overset{\delta \ominus}{\ddot{F}} \colon \quad \text{or} \quad \overset{\delta \oplus}{H} \!-\! \overset{\delta \ominus}{F}$$

Overlap of 1s orbital
from hydrogen and 2p
orbital from fluorine
to form a σ bond,
which is an s—p
molecular bond orbital

FIGURE 2.7   Formation of the hydrogen fluoride molecule, H—F. The electrons are shared unequally in H—F, with the greater concentration of electrons surrounding the fluorine atom. This charge distribution is indicated by $\delta^{\oplus}$ and $\delta^{\ominus}$.

C. The Hydrogen Fluoride Molecule

The hydrogen fluoride molecule, H—F, is "made" by overlapping the 1s atomic orbital of hydrogen with the 2p atomic orbital of fluorine, as shown in Fig. 2.7. The resulting σ bond can be called an s-p molecular orbital, which is cylindrically symmetrical about a line passing through the two nuclei. The bond dissociation energy is about 135 kcal/mole (565 kJ/mole) and the bond length is 0.91 Å.

The electrons in the σ bond in H—F are not shared equally between the two dissimilar atoms. Fluorine is considerably more electronegative than hydrogen (see Sec. 2.10), and as a result it has a greater attraction for electrons. The result is that the electrons "spend more of their time" *(have a higher probability of being found)* near fluorine so that hydrogen fluoride is a *polar* molecule; in other words, there is a higher concentration of electrons at the fluorine end of the molecule. This bond polarity can be designated by the symbols $\delta^{\oplus}$ *(delta plus,* meaning partial plus or partially positive) and $\delta^{\ominus}$ *(delta minus,* meaning partial minus or partially negative). Bond polarity and its relationship to electronegativity are discussed in Sec. 2.10.

2.6   Structure of Carbon and Methane; Hybridization

We now turn our attention to the electronic structure of carbon and the structure of the simplest organic molecule, methane, which has the molecular formula CH_4. Not only does methane have four hydrogen atoms attached to one carbon atom, but also all the H—C—H bond angles are equal (the value being 109°28′) and all C—H bond distances are equal (1.09 Å). If one were to attempt to derive the structure of methane using the type of arguments presented in Sec. 2.4, the following would result.

The electron configuration of carbon in its neutral ground state is

$$C \colon \; 1s^2 2s^2 2p_x 2p_y \quad \text{or} \quad \textcircled{\uparrow\downarrow} \; \textcircled{\uparrow\downarrow} \; \textcircled{\uparrow} \; \textcircled{\uparrow} \; \bigcirc$$

$$\underset{1s}{} \quad \underset{2s}{} \quad \underset{2p}{\underbrace{}}$$

Carbon has only two unpaired electrons, one in each of two 2p orbitals. If this atom were to combine with two hydrogen atoms, the molecule would have the formula CH_2, which is *not* the known formula CH_4 for methane.

How then can we account for four electrons in carbon being made available for bonding with four hydrogen atoms? We must develop an imaginary carbon atom;

we do so by considering several hypothetical steps involving electron structure. Once our new carbon atom is developed, hydrogen atoms will be added to it, forming methane. The hypothetical steps are considered from a qualitative standpoint and are fully supported by mathematical and theoretical calculations. Here also theory and fact are in agreement.

Start with the carbon atom in its neutral ground state and *promote one electron* from the $2s$ orbital to the $2p_z$ orbital. As might be expected, energy is required for this step since $2p$ orbitals are of higher energy than $2s$ orbitals. This unpairs all the outer-shell electrons and makes all four available for bonding to hydrogen.

Neutral ground state of carbon, C:

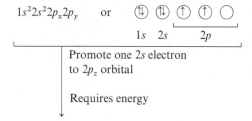

This new carbon atom can form one bond by an overlap between hydrogen and the $2s$ orbital and three bonds of another type by overlap with the $2p$ orbitals. This model would result in three directional bonds with the $2p$ orbitals at 90° angles to one another and another bond with the $2s$ orbital at some indeterminate position that would be closer to the nucleus. Since all the bonds in methane are equivalent, this is not a feasible explanation.

In addition to the promotion of an electron to make the $2s$ and $2p$ orbitals in carbon available for bonding, the orbitals must be hybridized or mixed together. **Hybridization** involves the combination of two or more different orbitals of the same main shell to form new orbitals, all of equal energy and equal shape. The electrons are relocated about the nucleus in hybridization, and this also requires energy. In carbon, the $2s$ orbital and the three $2p$ orbitals combine to form four hybrid equivalent sp^3 (pronounced s-p-three) hybrid atomic orbitals. (The term sp^3 refers to the fact that the hybrid orbitals are "made" from *one s* orbital and *three p* orbitals.) This process is shown in Fig. 2.8(a). Note that each sp^3 hybrid orbital (which can also be written ψ_{sp^3}) contains one electron and has the greatest concentration of electron charge in the larger lobe on one side of the nucleus and lesser charge in the smaller lobe on the opposite side. The sp^3 hybrid orbitals are much more directional than either the s or the p atomic orbitals from which they were derived.

Imaginary valence state of carbon, C:

$$1s^2 2s 2p_x 2p_y 2p_z \qquad \text{or}$$

1s | 2s   2p

Hybridize the s and the three p orbitals (shown in box)

Requires energy

Hybridized valence state of carbon, C:

$$1s^2 2(sp^3)2(sp^3)2(sp^3)2(sp^3) \qquad \text{or}$$

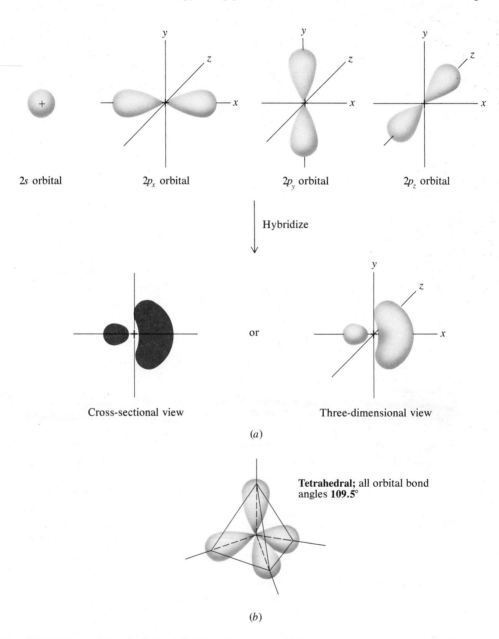

1s   four sp^3 hybrid
atomic orbitals

How do the sp^3 hybrid orbitals orient themselves around the carbon atom?
How can we account for energy being given off when methane is formed even though

2s orbital $2p_x$ orbital $2p_y$ orbital $2p_z$ orbital

Hybridize

Cross-sectional view   or   Three-dimensional view

(a)

Tetrahedral; all orbital bond
angles **109.5°**

(b)

FIGURE 2.8 ψ_{sp^3} or sp^3 **hybridization of carbon: the tetrahedral structure. (a) Hybridization
process and structure of four equivalent sp^3 hybrid atomic orbitals. Each orbital contains one electron.
The top four orbitals are all attached to the same carbon atom but are shown separately for clarity.
The $1s^2$ orbital has been omitted. (b) Three-dimensional structure with four sp^3 hybrid atomic orbitals
attached to a single carbon atom. Only the major lobes are shown.**

the electron promotion and hybridization processes require energy? These questions are closely related. Recall that electron promotion, followed by hybridization, yielded four identical sp^3 hybrid atomic orbitals (each containing one electron) about the carbon atom. To explain the orientation of these orbitals, consider the following simplification. Suppose we could put equal units of negative charge (for example, an electron) at the ends of four equal lengths of string that were all attached at a common point at the other end (for example, a carbon nucleus). How would these negative charges arrange themselves in space? Recalling that like charges repel one another, we predict that these negative charges would get as far away from one another as possible, and the only way they can do so is by occupying the four corners of a tetrahedron. To relate this to carbon, with four sp^3 hybrid orbitals, we predict that those four orbitals do likewise. Thus, the sp^3 hybridized carbon atom is **tetrahedral,** where the angles between all hybrid orbitals are equal to the tetrahedron angle of 109°28′, or approximately 109.5°. This structure is shown in Fig. 2.8(b). By forming the tetrahedral structure, the carbon atom has become *more* stable because the electrons are now farther apart (109.5° in the sp^3 carbon atom versus 90° in the p orbitals of the original carbon atom) and electrostatic interactions are decreased; this results in releasing energy.

The formation of the methane molecule itself provides further insight into bond stability and energy. First, recall that the sp^3 hybrid atomic orbitals are much more directional than either the s or p atomic orbitals. Now let us "construct" methane from a tetrahedral sp^3 carbon atom and four hydrogen atoms, as shown in Fig. 2.9. Because of their shape, the sp^3 atomic orbitals can overlap more extensively with hydrogen, resulting in much stronger covalent bonds. Bond formation also results in the liberation of energy. Even though electron promotion and hybridization

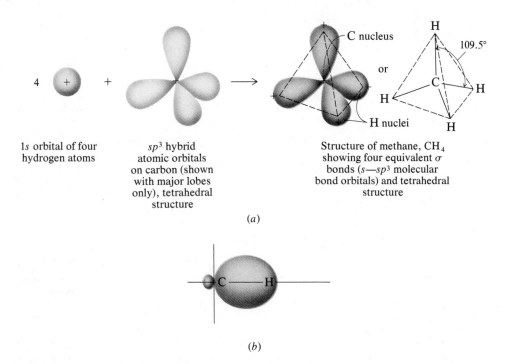

4   +   +   →

1s orbital of four sp^3 hybrid   Structure of methane, CH_4
hydrogen atoms   atomic orbitals   showing four equivalent σ
                                    on carbon (shown   bonds (s—sp^3 molecular
                                    with major lobes   bond orbitals) and tetrahedral
                                    only), tetrahedral   structure
                                    structure

(*a*)

(*b*)

FIGURE 2.9   Formation of the molecular orbitals in methane, CH_4 (small lobes omitted for clarity). (*b*) Schematic representation of a single C—H σ bond, with both the larger and smaller lobes shown.

both require some energy, the more stable tetrahedral structure that results and the formation of the strong covalent bonds evidently more than compensate for that energy requirement. The bond energy for each C—H bond in methane is about 104 kcal/mole (435 kJ/mole), and the total energy liberated in the formation of methane from its elements is about 416 kcal/mole (1741 kJ/mole).

When constructing the methane molecule, another important principle of molecular structure was invoked, namely, that the molecular structure resembles greatly the structure of the atomic orbitals used in forming the molecule (see Sec. 2.13). The tetrahedral structure of the sp^3 hybrid atomic orbitals in carbon suggests that methane should also be tetrahedral. This is true indeed, for all the H—C—H bond angles in methane are equal to 109.5°, the tetrahedral bond angle. The C—H bond distances in methane are 1.09 Å.

There are several ways to represent methane on paper, as shown by the following three structures:

$$\begin{array}{ccc}
\begin{array}{c} H \\ \cdot\cdot \\ H\!:\!\overset{\displaystyle \cdot\cdot}{\underset{\displaystyle \cdot\cdot}{C}}\!:\!H \\ H \end{array} &
\begin{array}{c} H \\ | \\ H\!-\!\overset{\displaystyle}{\underset{\displaystyle |}{C}}\!-\!H \\ H \end{array} &
CH_4
\end{array}$$

Lewis dot	Line	Condensed
structure	structure	structure
(1)	(2)	(3)

Methane

Structure (1) is a Lewis electron dot formula, where all valence electrons are shown as dots. Unless one is interested in these electrons—as in a particular reaction—this structure is seldom used. Structure (2) is similar to (1), but each pair of bonding electrons is replaced by a line that is understood to represent one pair of electrons. Structure (3) is called a *condensed* structure and is the type most frequently used in organic chemistry.

Unfortunately, structures (1) to (3) do not show the tetrahedral structure of the central carbon atom. When such representations are used, remember that the

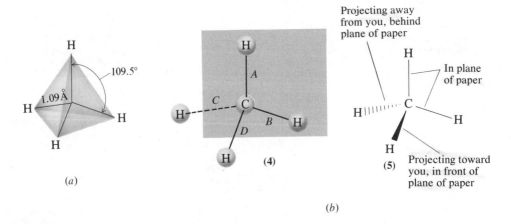

FIGURE 2.10   Representations of the three-dimensional structure of methane, CH$_4$. (a) Tetra-hedral methane structure. (b) Method for drawing three-dimensional structure of methane. Structure (4) is a drawing of a ball-and-stick model of methane, with the C—H bonds identified by letter and a plane drawn through the H—C—H bonds of bonds *A* and *B*. Structure (5) is the three-dimensional representation of structure (4).

real structure of methane is shown in the three-dimensional sketch in Fig. 2.10(a) or in the ball-and-stick model in Fig. 2.10(b). Because we are often interested in showing the three-dimensional nature of carbon compounds on paper, we can use the type of structures shown in Fig. 2.10(c). The convention for drawing structure (5) is to place any two C—H bonds in methane on the plane of the paper and indicate them with solid lines (C—H). Note that all the atoms in this position (H—C—H) of the molecule lie in the same plane. When this is done, one of the remaining C—H bonds sticks out in front of the plane and toward us; this bond is drawn as a wedge (C—◄H). The other C—H bond is behind the plane of the paper and away from us; this bond is indicated by a dashed wedge (C⋯⋯H). Molecular models are helpful for depicting three-dimensionality.

Thus far, only the sp^3 hybridized tetrahedral atomic orbitals for carbon have been discussed. In the following sections two other types of hybridized atomic orbitals for carbon are presented: (1) the sp^2 hybridized orbitals, which are *trigonal* (bond angles of 120°), and (2) the sp hybridized orbitals, which are *digonal* or *linear* (bond angles of 180°).

2.7 Formation of More Complex Organic Molecules

A. Ethane

The tetrahedral carbon atom accounts nicely for the structure of methane. We do not have to limit ourselves to adding only hydrogen atoms to it, however. Suppose, for example, we took a tetrahedral carbon atom and added only three hydrogen atoms to it, thereby leaving one sp^3 hybrid atomic orbital nonbonded. Bringing together two such species results in the formation of a carbon-carbon bond, as shown in Fig. 2.11. The resulting molecule, CH_3—CH_3, is called **ethane.** All the C—H bonds in it are s-sp^3 molecular orbitals, and the C—C bond is an sp^3-sp^3 molecular orbital because it is the result of overlapping two sp^3 hybrid atomic orbitals. All the bond angles are approximately 109.5°, and the C—C bond distance is 1.54 Å; the C—H bond distances are 1.10 Å.

Hydrogen atoms and other carbon-containing groups can be attached to an sp^3 tetrahedral carbon atom as well as other elements, such as oxygen, the halogens, sulfur, phosphorus, and so on. This versatility of carbon allows for the limitless number of organic compounds.

B. Ethene; The Carbon-Carbon Double Bond; The Pi (π) Bond

The simplest compound that contains a carbon-carbon double bond is **ethene,** which has the molecular formula C_2H_4 and the following molecular structure:

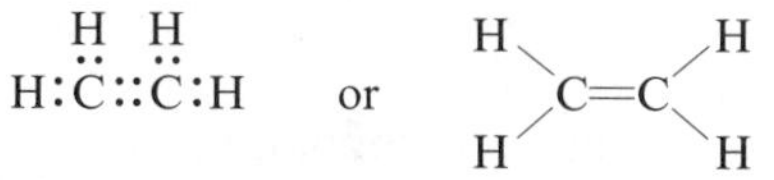

Ethene
(Ethylene)

Note that two pairs of electrons are shared between the two carbon atoms. Thus, the carbon atoms are joined by a **carbon-carbon double bond,** which is characteristic of the alkene family.

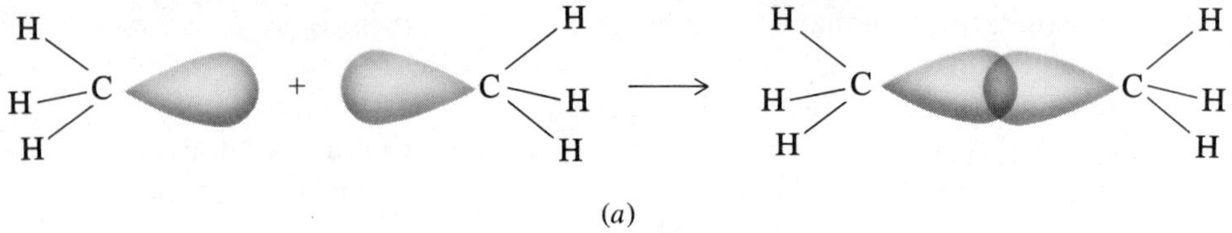

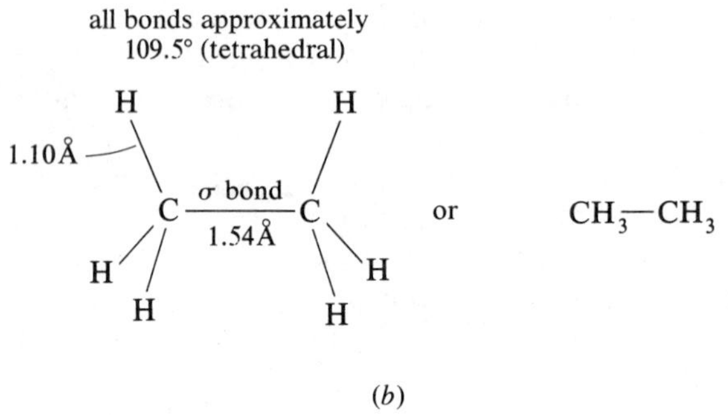

(a)

all bonds approximately
109.5° (tetrahedral)

(b)

FIGURE 2.11 **Hypothetical view of the formation of an ethane molecule, C_2H_6. (a) The left illustration shows carbon atoms with sp^3 hybrid orbitals (major lobes only, one electron each); the right illustration is the formation of the C—C σ bond (sp^3-sp^3 molecular bond orbital) in ethane. (b) Alternate ways of showing the molecular structure of ethane. Left structure shows all covalent bonds, using a line to show the shared electron pairs. Right structure is the "condensed" structure for the same molecule.**

Now let us look at a more detailed electronic structure of the carbon-carbon double bond. Quantum mechanics provides a theoretical, mathematical view of the structure, which appears to coincide and agree with all the chemical and physical properties of ethene and other alkenes. Envision the following hypothetical route for "making" this bond, starting with sp^2 hybridized carbon atoms. This type of carbon atom is formed by mixing (hybridizing) *one* 2s orbital with *two* 2p orbitals, which yields a carbon atom with *three equivalent* sp^2 (or ψ_{sp^2}) hybrid orbitals. To minimize electrostatic repulsion between the three sp^2 hybrid orbitals, they are oriented in an *equilateral planar* arrangement in the same plane as the carbon nucleus. This is often called **trigonal** geometry. This process is shown:

Neutral ground state of carbon, C:

$$1s^2 2s^2 2p_x 2p_y \quad \text{or} \quad \text{(diagram)}$$

1s   2s   2p

Promote electron from
2s to 2p orbital,
requires energy

Imaginary valence state of carbon, C:

$$1s^2 2s 2p_x 2p_y 2p_z \quad \text{or} \quad \text{(diagram)}$$

1s   2s   2p

Hybridization of the 2s orbital and *two* of the 2p orbitals produces three equivalent sp^2 hybrid orbitals and one unhybridized 2p orbital. The orbitals involved in hybridization are boxed in the following electronic structures:

Imaginary valence state of carbon, C:

$1s^2 \boxed{2s2p_x2p_y}2p_z$ or   ⟨↑↓⟩ ⟨↑⟩ ⟨↑⟩ ⟨↑⟩ ⟨↑⟩

                    $1s$ $2s$ $2p$

Hybridize $2s$ and *two*
$2p$ orbitals; requires energy

Hybridized valence state of carbon, C:

$1s^2 2(sp^2)2(sp^2)2(sp^2)2p$ or   ⟨↑↓⟩ ⟨↑⟩ ⟨↑⟩ ⟨↑⟩ ⟨↑⟩

                    $1s$ three sp^2 $2p$
                    hybrid
                    orbitals

The overlap of two of these carbon atoms and four hydrogen atoms forms the basic skeleton for ethene; the carbon-carbon bond is an sp^2-sp^2 molecular σ bond, and the four carbon-hydrogen bonds are s-sp^2 molecular σ bonds. Because this framework was made from carbon atoms with sp^2 hybrid atomic orbitals that were 120° apart, one might anticipate that the geometry at each carbon atom would reflect the trigonal structure, which is the case (Fig. 2.12).

Thus far we have considered only the sp^2 hybrid orbitals; what about the remaining p orbital on each carbon atom? Each p orbital has one lobe above and another below the plane defined by the carbon and hydrogen atoms involved in σ bonding; each orbital contains one electron. When the p orbital on one carbon atom overlaps in a parallel manner with the p orbital on the other carbon atom, the electrons pair up and form an additional covalent bond. Maximum overlap occurs when these two p orbitals are *parallel to one another*. The p orbitals are close enough to permit the electrons to be shared through space. The bond formed is called a **pi (π) bond.** The making of the ethene molecule is depicted in Fig. 2.12 (see also Sec. 2.17).

There are several noteworthy features of the double bond. It contains one strong σ bond, which results from the direct (cylindrically symmetrical) overlap of two sp^2 hybrid orbitals. It also contains one π bond, which results from the (parallel) overlap through space of the p orbitals. (Even though there is one electron cloud above and another below the plane of the atoms, these two clouds taken together represent one π bond.) The terms σ bond and π bond are used to differentiate these two types of bond.

There is less overlap between the p orbitals than between the *two sp^2* hybrid orbitals, so it is not surprising that the π bond is weaker than the σ bond. However, the total bond energy of the carbon-carbon double bond in ethene (163 kcal/mole, 682 kJ/mole) is greater than the bond energy of a carbon-carbon single bond (88 kcal/mole, 368 kJ/mole in ethane). Thus the energy associated with the π bond should be approximately 75 kcal/mole (163 kcal/mole $-$ 88 kcal/mole). The actual values are 95 kcal/mole (397 kJ/mole) and 68 kcal/mole (285 kJ/mole) for the σ and π bonds, respectively. The carbon-carbon double bond is stronger (requiring greater total energy to break) because of greater orbital overlap, so it is not surprising that the carbon-carbon double bond is shorter than the carbon-carbon single bond. The length of the carbon-carbon single bond in ethane is 1.54 Å, and the length of the carbon-carbon double bond in ethene is 1.34 Å.

Because of the trigonal nature of the sp^2 hybridized carbon atoms in the carbon-carbon double bond, one would expect the atoms attached to them to be coplanar with the two carbon atoms; that is, the ethene molecule should be flat. The bond angles should be 120° or very near thereto. As the following data indicate, this prediction is correct.

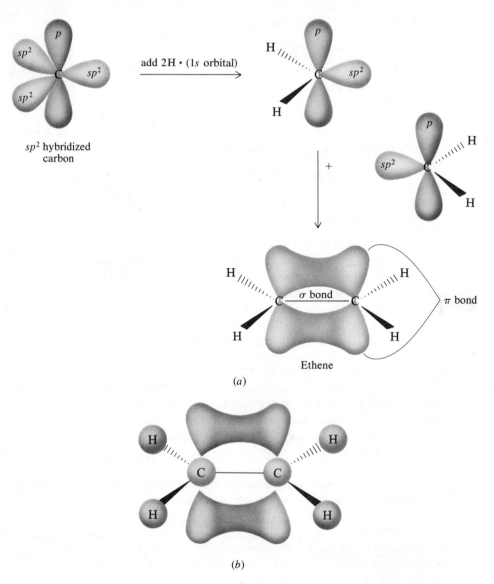

FIGURE 2.12   The ethene molecule, $CH_2{=}CH_2$. (*a*) Formation of ethene from sp^2 or ψ_{sp2} hybrid carbon atoms and hydrogen atoms. (See Sec. 2.17 and Fig. 2.29.) (*b*) Drawing of a molecular model of ethene, showing the planar nature of the molecule and the bond angles and the π bond.

Properties of Ethene and Ethane

Structure and bond angles:

Bonding:

Carbon-carbon	1 σ bond (sp^2-sp^2) + 1 π bond	1 σ bond (sp^3-sp^3)
Carbon-hydrogen	σ bonds (s-sp^2)	σ bonds (s-sp^3)

Bond distances:

Carbon-carbon	1.34 Å	1.54 Å
Carbon-hydrogen	1.09 Å	1.10 Å

Bond energies:

Carbon-carbon	σ bond = 95⎱ 163 kcal/mole π bond = 68⎰ (682 kJ/mole)	88 kcal/mole (368 kJ/mole)
Carbon-hydrogen	103 kcal/mole (431 kJ/mole)	98 kcal/mole (410 kJ/mole)

C. Ethyne; The Carbon-Carbon Triple Bond

Ethyne, C_2H_2, is the simplest member of the alkyne family. In this molecule three pairs of electrons are shared between the two carbon atoms; we refer to this as a carbon-carbon triple bond. To satisfy the octet rule, the following electronic structure is written for ethyne:

$$\text{H:C:::C:H} \qquad \text{or} \qquad \text{H—C≡C—H}$$

Ethyne
(Acetylene)

The triple bond is formed by allowing two sp hybridized carbon atoms to overlap. These orbitals are formed from the hybridization (mixing) of one s orbital and one p orbital, which leaves two directional p orbitals (each containing one electron) on each carbon to overlap with one another. The building up of the ethyne molecule is shown in Fig. 2.13.

The hybridization process results in two equivalent sp (or ψ_{sp}) hybrid orbitals on carbon, which lie on a straight line passing through the carbon nucleus. In this geometry they are a maximum distance apart to minimize the electrostatic repulsion between like charges. The hybrid orbitals are 180° apart. (In the same way, sp^3 hybrid orbitals are as far apart as possible in the tetrahedral geometry, and sp^2 hybrid orbitals are likewise in the trigonal geometry.) As we will see, the geometry of the ethyne molecule reflects the geometry of the sp hybrid orbitals from which it is made.

Figure 2.13 shows the overlap of two sp hybridized carbon atoms to form a carbon-carbon σ bond (sp-sp molecular bond orbital). Since each carbon atom contains two p orbitals perpendicular to one another and to the sp hybrid orbitals, theoretical calculations and experimental observation point to the fact that two pairs of p orbitals overlap to form two π bonds. The remaining two sp orbitals, one on each carbon, overlap with hydrogen atoms in the case of ethyne to form carbon-hydrogen σ bonds (s-sp molecular bond orbitals).

The two π bonds may be envisioned as follows: one is perpendicular to the plane of the paper, and the other is in the plane of the paper. It has been suggested, however, that these two π bonds are not completely independent but that they overlap to form a cylinderlike orbital surrounding the straight line defined by the H—C—C—H bond (Fig. 2.14).

The direct overlap of two sp hybrid orbitals produces one strong σ bond in ethyne, and the two π bonds are superimposed over the σ bond. As mentioned in Sec. 2.7B, π bonds involve less overlap than do σ bonds, and for this reason π bonds are weaker than the σ bond. However, the total bond energy of the carbon-carbon

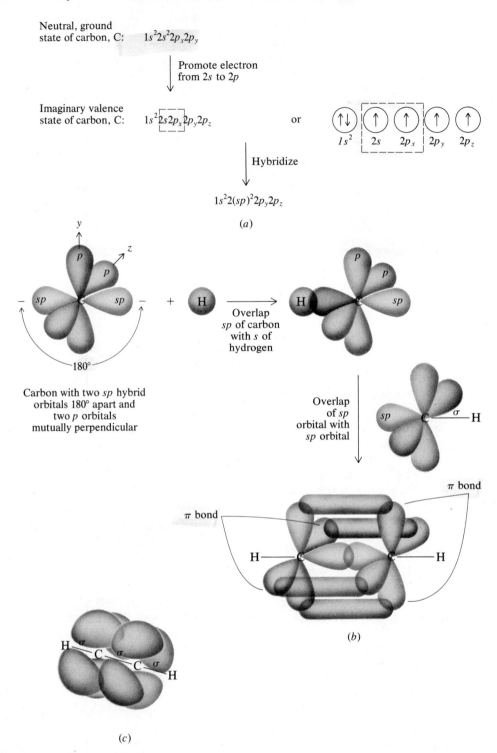

Neutral, ground
state of carbon, C: $1s^2 2s^2 2p_x 2p_y$

Promote electron
from $2s$ to $2p$

Imaginary valence
state of carbon, C: $1s^2 2s 2p_x 2p_y 2p_z$ or

$1s$ $2s$ $2p_x$ $2p_y$ $2p_z$

Hybridize

$1s^2 2(sp)^2 2p_y 2p_z$

(a)

Carbon with two sp hybrid
orbitals 180° apart and
two p orbitals
mutually perpendicular

Overlap
sp of carbon
with s of
hydrogen

Overlap
of sp
orbital with
sp orbital

π bond

π bond

(b)

(c)

FIGURE 2.13   "Making" the ethyne molecule. (a) Electronic changes in going from the neutral
carbon atom to the sp hybridized carbon. (b) Overlapping of the sp hybridized carbon atom first
with the s orbital of hydrogen and then with the sp hybrid orbital of another carbon atom. (c) Drawing
of the molecule showing two π bonds and the C—H and C—C σ bonds.

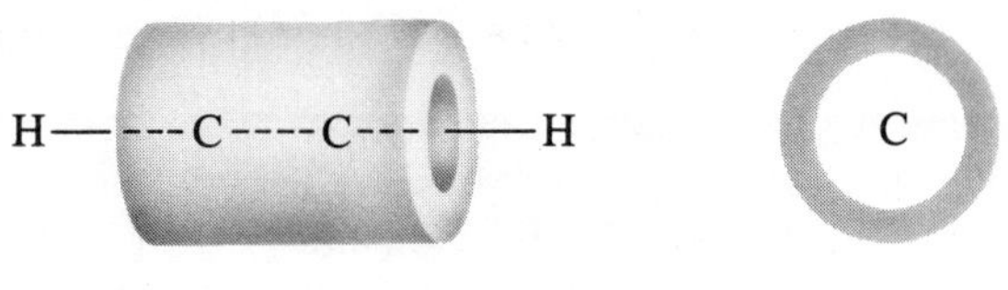

FIGURE 2.14   Carbon-carbon triple bond in ethyne, showing the cylindrical nature of the overlap between p orbitals forming the π MO. (*a*) Side view; (*b*) end view.

triple bond is about 230 kcal/mole (962 kJ/mole), so that it is stronger than the carbon-carbon double bond (163 kcal/mole, 682 kJ/mole, in ethene) and the carbon-carbon single bond (88 kcal/mole, 368 kJ/mole, in ethane). The triple bond is also the shortest of the three.

The pertinent physical data for the ethyne molecule are:

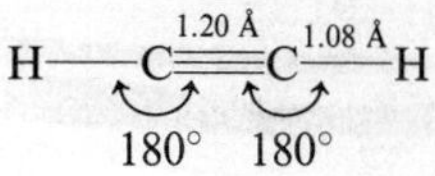

$$H\!\!-\!\!C\underset{180°}{\equiv}\overset{1.20\,\text{Å}}{C}\overset{1.08\,\text{Å}}{\underset{180°}{-}}H$$

Ethyne: a linear molecule

Since the sp hybrid orbitals from which it is made are linear, it is not surprising that the ethyne molecule is also linear. The C—C bond distance of 1.20 Å in ethyne is shorter than the distance in alkenes (1.34 Å in ethene) or in alkanes (1.54 Å in ethane). This is not surprising though since the triple bond is made up of two π bonds (which overlap through space and tend to pull the carbon atoms closer together), and the σ bond involves the overlapping of two sp hybrid orbitals that are relatively close to the carbon nucleus.

Percentage of s Character

Within a given main shell, an s orbital is always lower in energy (more stable) and less diffuse (more concentrated area of greatest probability) than a corresponding p orbital. As a simple extension of this, when comparing hybrid orbitals the percentage of s character can help explain many observed properties of that nucleus. An sp^3 hybrid orbital results from the combination of an s orbital and three p orbitals. The hybrid orbital is considered to be 25% s in character. Likewise, sp^2 orbitals are 33.3% s in character and sp orbitals are 50% s in character. The greater the percentage of s character, the lower in energy and more stable that orbital. Also, just as s orbitals are less diffuse than p orbitals, the greater the percentage of s character of a hybrid orbital, the less diffuse and closer to the nucleus it is. This concept is used in many instances to explain observed trends between structure and reactivity.

A carbon-carbon bond resulting from the overlap of two sp^2 orbitals would therefore be predicted to be shorter than one resulting from the overlap of two sp^3 orbitals. This is confirmed. For example,

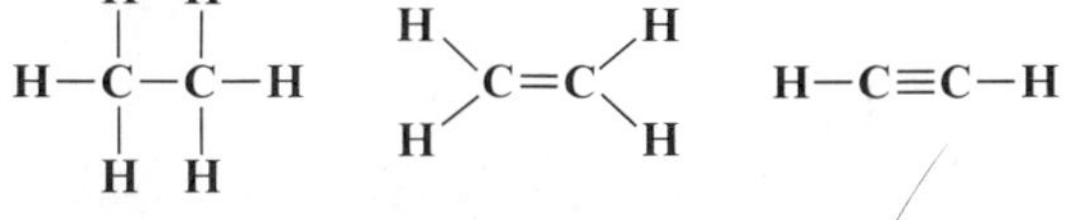

$$\underset{\displaystyle \overset{|}{H}\ \overset{|}{H}}{\overset{\displaystyle \overset{H}{|}\ \overset{H}{|}}{H\!-\!C\!-\!C\!-\!H}} \qquad \overset{H}{\underset{H}{>}}C\!=\!C\overset{H}{\underset{H}{<}} \qquad H\!-\!C\!\equiv\!C\!-\!H$$

C—C bond length:	1.54 Å	1.34 Å	1.20 Å
% s character:	25	33.3	50
Hybridization of C:	sp^3	sp^2	sp

D. The Carbonyl Group; The Carbon-Oxygen Double Bond

The carbon-oxygen double bond is analogous to the carbon-carbon double bond. The carbonyl-carbon is sp^2 hybridized; it contains three sp^2 hybrid orbitals lying 120° apart, with a p orbital perpendicular to the plane of the sp^2 orbitals. Two of the hybrid orbitals are involved in σ bonds with carbon or hydrogen atoms, and the third hybrid orbital is involved in bonding with an atom of oxygen, also through a σ bond. The electronic structure of the oxygen atom is indeed interesting. It possesses no three-dimensional character, so there is no way to determine experimentally the hybridization about oxygen. (Hybridization is most often determined by knowing the structure of the compounds that contain a particular atom, as, for example, carbon in methane, ethene, and so on.) Spectral studies on carbonyl compounds indicate that the electronic structure of oxygen is best treated as being unhybridized[1]; one p orbital of oxygen overlaps with an sp^2 hybrid orbital of carbon to form the σ bond, and a perpendicular p orbital of oxygen overlaps with the p orbital of carbon to form the π bond. The remaining s and p orbitals of oxygen contain the two unshared pairs of electrons. The following sketch shows the bonding and structure of the carbonyl group:

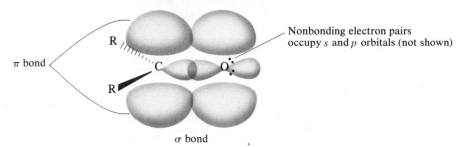

π bond

R

Nonbonding electron pairs occupy s and p orbitals (not shown)

σ bond

Thus, the C—O σ bond is an sp^2-p bond orbital, and the C—O π bond is the result of p-p overlap.

As we would expect based on our knowledge about the carbon-carbon double bond, the carbon-oxygen double bond and the two attached substituents (shown as R and R′) lie in a plane. The bond angles about the carbonyl-carbon are all very close to 120°:

120°

120° C=O

120°

π bond perpendicular
to plane of paper

Since the carbon-oxygen double bond is composed of atoms with differing electronegativities (carbon and oxygen), the electrons in the π cloud are shared unequally between them. The more electronegative oxygen atom has a greater affinity for electrons, so the electrons spend more time near oxygen than near carbon. The polarity of the carbon-oxygen double bond is conventionally indicated as follows:

$$\underset{R'}{\overset{R}{\diagdown}}C \overset{\delta\oplus}{=}\overset{\delta\ominus}{\underset{..}{\overset{..}{O}}}: \qquad \text{A } \textit{polar} \text{ bond: electrons shared unequally between C and O}$$

---

[1] The oxygen may also be sp^2 hybridized; there is no universal agreement on this point.

This polarity is invaluable in explaining many properties and reactions of aldehydes and ketones and carboxylic acids and their derivatives, families of compounds that contain the carbonyl group.

By comparison, the carbon-carbon double bond is nonpolar (unless, of course, it bears substituents that impart polar properties to the molecule); the bonding electrons are shared equally between the two identical carbon atoms.

The following data give average bond lengths and bond strengths and provide a comparison of the carbon-carbon and carbon-oxygen double bonds:

	$\diagdown C = C \diagup$ Bond	$\diagdown C = O$ Bond
Average bond length (Å)	1.34	1.22
Average bond strength (kcal/mole)	163	176 (in aldehydes) 179 (in ketones)

Thus, the carbon-oxygen bond is shorter and stronger than the carbon-carbon double bond.

2.8   Structure of Reactive Intermediates in Organic Chemistry

In our study of organic chemistry we will encounter several classes of intermediates, including free radicals, cations, and anions; these are referred to collectively as **reactive intermediates.**

A. The Methyl Radical

The **methyl radical** is a highly reactive intermediate. For now we simply present the salient features about its structure, and in Sec. 6.9 we present evidence—based on stereochemistry—to support the arguments given here. The methyl radical serves as a model for almost all free radicals of carbon.

The methyl radical is sp^2-hybridized and can be "made" from elemental carbon and hydrogen in a manner analogous to the construction of ethene in Sec. 2.7.

Each sp^2 hybrid orbital contains one electron, and there remains the $2p$ orbital that was unaffected by the hybridization process, which also contains one electron. The formation of this hybridized carbon atom is illustrated in Fig. 2.15.

The sp^2 hybrid orbitals are, of course, highly directional in space, providing a *trigonal* arrangement. The structure of this carbon atom is analogous to that of the carbon atom of ethene (see Sec. 2.7B). As we might expect, the sp^2 hybrid orbitals on carbon point to the three corners of a triangle to minimize electrostatic repulsion. Although energy was required to promote an electron and to hybridize the orbitals, more energy is liberated when these orbitals form bonds.

But where does the p orbital lie? The sp^2 hybrid orbitals are equivalent and trigonal, so predictably the p orbital assumes a location perpendicular to the plane of the sp^2 hybrid orbitals. This is also depicted in Fig. 2.15.

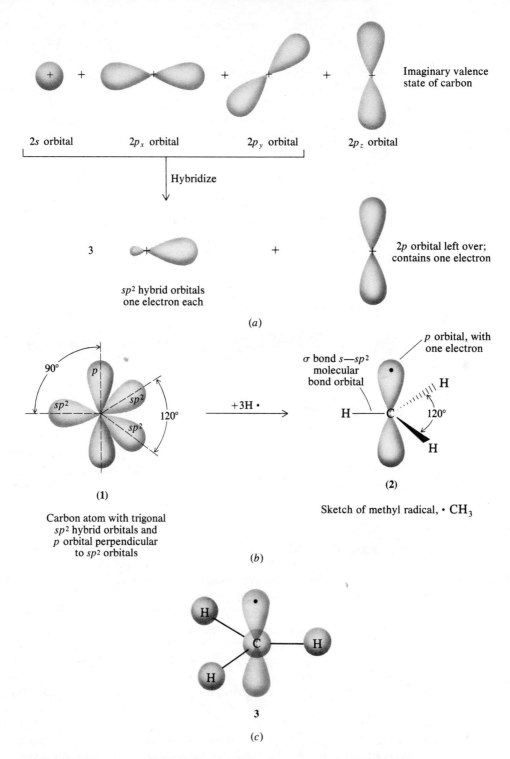

FIGURE 2.15 Formation of an sp^2 hybridized carbon atom and the structure of the methyl radical. (*a*) Hybridized sp^2 atomic orbitals of carbon. (*b*) Reaction showing formation of methyl radical from sp^2 hybridized carbon atom and hydrogen atoms. (*c*) Drawing of a molecular model showing geometry of the methyl radical.

The overlap of the sp^2 hybridized orbitals on carbon with the $1s$ orbitals of three hydrogen atoms yields three carbon-hydrogen σ bonds, which are called s-sp^2 molecular bond orbitals. We have now built up the methyl radical; its three-dimensional structure is shown in Fig. 2.15.

The theoretical treatment of the structure of radicals indicates that the radical can be planar or pyramidal, where the carbon is sp^3 hybridized and the odd electron occupies an sp^3 hybridized orbital in one corner of a tetrahedron.

Possible Structures for Methyl Radical

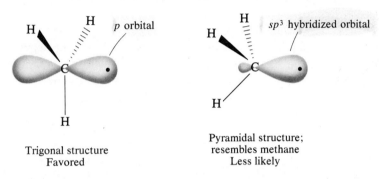

Trigonal structure
Favored

Pyramidal structure;
resembles methane
Less likely

The pyramidal structure resembles the tetrahedral structure of methane itself, which simply has a hydrogen atom abstracted to form the radical. Spectroscopic evidence indicates that the methyl radical is nearly flat. More compelling evidence yet for the trigonal structure comes from studies in stereochemistry—a most important tool for examining structure and one that is discussed in Chap. 6. However, in certain instances, for example the trifluoromethyl radical ($CF_3 \cdot$), a rapidly flipping (analogous to Fig. 2.17) pyramidal structure is preferred. Evidence is not so straightforward for more complicated radicals.

B. The Methyl Cation

The **methyl cation,**[1] a carbocation, is *planar* or *flat*. The carbon atom is sp^2 hybridized, and because there are only six electrons about it, there is a *vacant p orbital*. The structure of the carbocation is like that of the radical (see Sec. 2.8A), except that the radical has the odd electron occupying the p orbital that is perpendicular to the plane of atoms constituting the radical (Fig. 2.16).

Other evidence, which we are not prepared to discuss at this time, is in accord with this picture of the planar carbocation.

C. The Methyl Anion

On the basis of extensive investigations on the structure of **carbon anions,** usually called **carbanions,** it appears that the best description of their three-dimensional structure is the following. The carbon that bears the unshared electron pair and the negative charge is sp^3 hybridized, and the geometry about it is pyramidal. It is believed that carbanions undergo a rapid interconversion between two pyramidal forms, as

---

[1] The term *carbonium ion* (see Sec. 5.7B) is also used to designate these species.

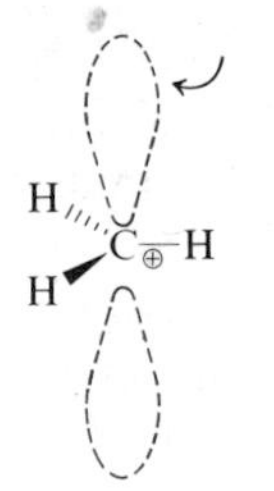

Vacant p orbital

$$H_{\prime\prime\prime\prime}\overset{\displaystyle{\mathrm{C}}}{\underset{\mathrm{H}}{\oplus}}\!\!-\!\mathrm{H}$$

FIGURE 2.16   The methyl cation, a carbocation.

shown in Fig. 2.17. This inversion is analogous to what happens when an umbrella is blown inside out by a strong wind, except that with carbanions one pyramidal form is converted to the other (and vice versa) very rapidly.

D. Methylene

Methylene has the formula $:CH_2$. Studies have shown that two types of methylene (also called **carbene**) occur, depending on their method of formation. The first type is called the *singlet methylene*, in which the two electrons are paired and therefore have opposite spin.

Singlet methylene, sp^2 hybridized:

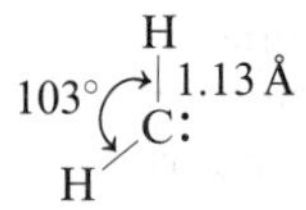

$$103° \overset{\displaystyle{\mathrm{H}}}{\underset{\mathrm{H}}{\bigg\langle}}\overset{| \; 1.13\,\text{Å}}{\mathrm{C}\colon}$$

Paired electrons

The other is called *triplet methylene*, in which the two electrons are unpaired.

Triplet methylene, sp hybridized:

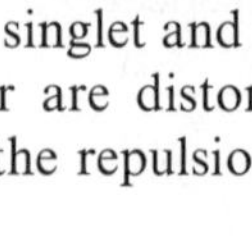

$$\sim 180°$$
$$\mathrm{H}\!-\!\overset{..}{\underset{..}{\mathrm{C}}}\!-\!\mathrm{H}$$
$$1.03\,\text{Å}$$

Unpaired electrons: a diradical

This species is a free radical or, more correctly, a *diradical*. As the data indicate, there is considerable difference in their geometries.

The bond angles in both the singlet and triplet methylene are not exactly 120° and 180°, respectively, but rather are distorted slightly from these values. This distortion is due in large part to the repulsions between the nonbonding electrons

FIGURE 2.17   The methyl anion, a carbanion. Inversion in simple, nonconjugated carbanions: pyramidal structure.

and the bonding electron pairs. There are also differences in the chemical reactivities of the two methylenes. Three-dimensional drawings are provided in the following summary section.

Summary of Structures of Reactive Intermediates

We have encountered the most important reactive intermediates in organic chemistry. They share one thing in common: They are highly reactive and often unstable, and their lifetimes are quite short. We now summarize the structures of these various intermediates. *Pay special attention to the net charge on each reactive species.*

Radicals
Radicals ($R_3C\cdot$) are planar, with a *p* orbital containing one electron perpendicular to the plane of the central carbon atom and the substituents attached to it (see Sec. 2.8A).

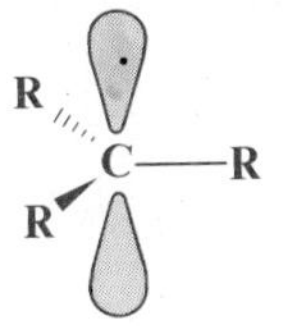

Carbonium Ions or Carbocations
Carbocations ($R_3C^{\oplus}$) are planarlike radicals. However, the perpendicular *p* orbital is empty (see Sec. 2.8B).

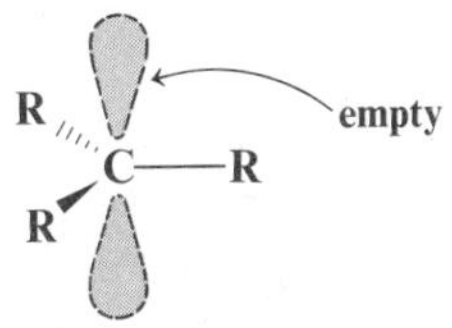

Carbenes
The geometry of carbenes ($R_2C:$) depends on whether they are singlet or triplet, as shown by the diagrams (see Sec. 27.6).

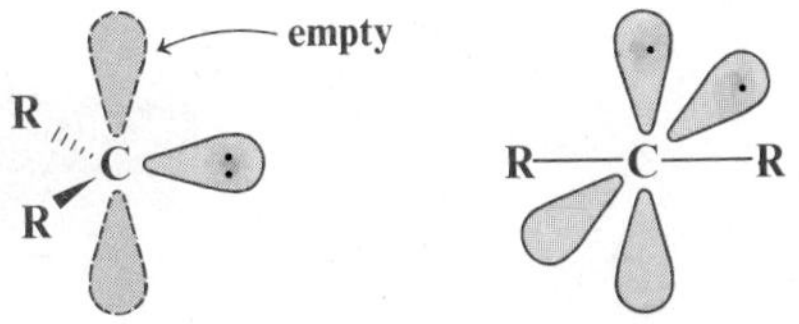

Singlet
carbene

Triplet
carbene

Carbanions
Simple carbanions consist of a pair of rapidly equilibrating pyramidal (sp^3 hybridized) species (see Sec. 2.8C).

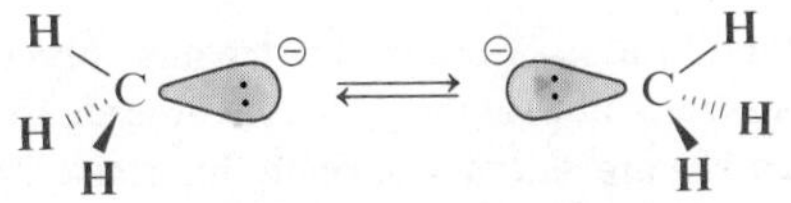

Inversion in simple, nonconjugated
carbanions: pyramidal structure

2.9   Hybridization in Boron, Nitrogen, and Oxygen

Hybridization of atoms other than carbon is also possible. For example, boron, nitrogen, and oxygen all exist in hybridized forms. One example of each of these is provided in this section.

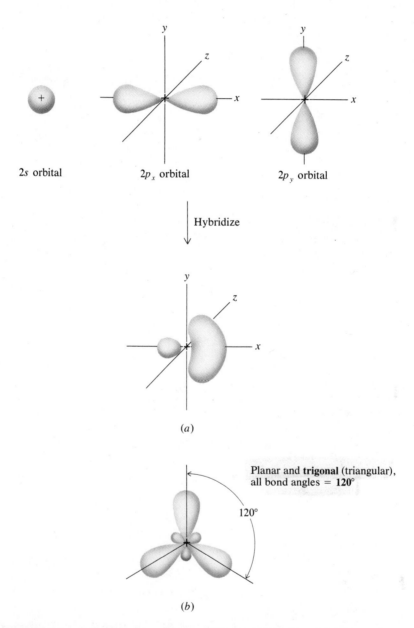

FIGURE 2.18 sp^2 Hybridization of boron; the trigonal structure. (*a*) The hybridization process and three-dimensional structure of a single sp^2 hybrid orbital. The top three orbitals are all attached to the same boron atom but are shown separately for clarity. The $1s^2$ orbital has been omitted. These orbitals hybridize to three equivalent sp^2 atomic orbitals (shown in three-dimensional perspective). Each orbital contains one electron. (*b*) Three-dimensional structure, with three sp^2 hybrid atomic orbitals attached to a boron atom. (Both large and small lobes are shown.)

A. Boron: Boron Trifluoride

Boron trifluoride, BF_3, contains covalent bonds in which all B—F bonds are equivalent. The conversion of elemental boron into a boron atom suitable for bonding (with fluorine or other atoms) is outlined in Figs. 2.18 and 2.19.

Question 2.3

Using arguments similar to those for carbon and boron, describe what electronic changes must occur when neutral ground-state beryllium, Be, is converted to beryllium hydride, BeH_2. BeH_2 contains two equivalent covalent bonds. What bond angles and shape would you predict for BeH_2? What atomic orbitals are used in bond formation?

B. Nitrogen: Ammonia

To consider the structure of ammonia, NH_3, it is necessary to look at the electronic structure of the central nitrogen atom, which has the electron configuration $1s^2 2s^2 2p_x 2p_y 2p_z$. In nitrogen, all three $2p$ orbitals are available for bonding and each has one electron that could accommodate one more electron by sharing through overlap with the $1s$ orbital of hydrogen. If this were to occur, the ammonia molecule would have its three N—H bonds perpendicular to one another (that is, bond angles of about 90°), which is the geometry of the p orbitals. Yet it has been found that the ammonia molecule is pyramidal, with H—N—H bond angles of about 107°.

The structure of ammonia has been studied theoretically, and the currently accepted and most plausible explanation for its structure involves hybridization of the nitrogen atom. The hybridization for nitrogen is similar in concept to that for carbon. In nitrogen we add the one additional valence electron it possesses to one of

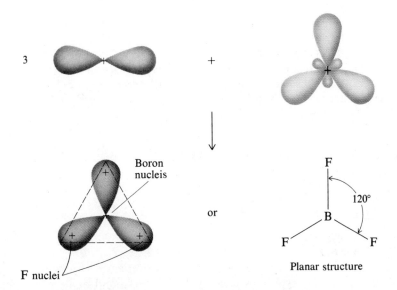

FIGURE 2.19 Formation of the boron trifluoride, BF_3, molecule. *Top:* Unfilled $2p$ orbital of three fluoride atoms and sp^2 hybrid atomic orbitals on boron (larger and smaller lobes both shown); trigonal structure. *Bottom:* Structure of BF_3, showing three equivalent B—F σ bonds (p–sp^2 molecular bond orbitals) and trigonal (planar) structure.

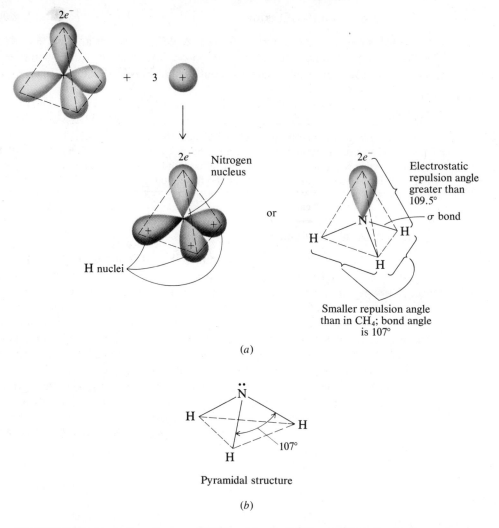

(a)

(b)

FIGURE 2.20   Formation of ammonia, NH_3 molecule. (a) *Top*: sp^3 hybrid orbitals on nitrogen (smaller lobes not shown), tetrahedral structure, and $1s$ orbitals of three hydrogen atoms. *Bottom*: Structure of NH_3, showing three equivalent N—H σ bonds (s-sp^3 molecular bond orbitals) and pyramidal structure. **(b)** Pyramidal structure of ammonia, with unshared pair of electrons shown.

the sp^3 hybrid orbitals *before* N—H bonds are made with the remaining three orbitals (Fig. 2.20).

C. Oxygen: Water

Finally, consider the electronic structure of water. The electron configuration of oxygen (neutral ground state) is: $1s^2 2s^2 2p_x^2 2p_y 2p_z$. Oxygen has two p orbitals available for bonding, but if we attempted to overlap the $1s$ orbitals of two hydrogen atoms, we would expect to have an H—O—H bond angle of 90° because the p orbitals are perpendicular to one another. In fact, the bond angle in water is about 104°, which is considerably greater than expected from using the p orbitals directly.

The structure of water has been studied theoretically and, like nitrogen in ammonia, sp^3 hybridization of oxygen seems to account best for its structure.

This sp^3 hybridization is the same type encountered for both carbon and nitrogen (Fig. 2.21).

About Hybridization

Before we temporarily leave the subject of hybridization, remember that we used it from a purely *qualitative* standpoint. We imagined some artificial ways to "make" atoms before constructing molecules and discussing molecular structure and geometry. We made mental models to aid our understanding. We cannot observe the excited valence states or hybridization states of atoms, and yet theory indicates that they are indeed involved in converting elements such as carbon, nitrogen, oxygen, and boron into molecules. Hybridization is the best explanation we have to account for the molecular structure of molecules that contain certain atoms. In future chapters we use hybridization frequently, and we will see that limitless numbers of organic compounds can be made from the three common types of hybridized carbon atom: sp^3, sp^2, and sp.

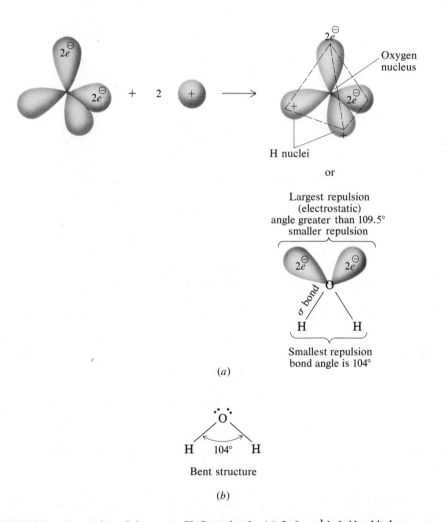

FIGURE 2.21 Formation of the water, H_2O, molecule. (*a*) *Left:* sp^3 hybrid orbitals on oxygen (smaller lobes not shown). Tetrahedral structure. *Middle:* 1s orbital of two hydrogen atoms. *Right:* Structure of H_2O, showing two equivalent O—H σ bonds (s-sp^3 molecular bond orbitals) and bent structure of water. (*b*) Bent structure of water, with unshared pairs of electrons shown.

See Sec. 6.19 for a discussion of hybridization in other than first- and second-period elements; see Reading Reference 1 for Chap. 1 in the appendix.

Reasoning by Analogy

Much of organic chemistry requires that you be able to understand material and to reason by analogy; that is, you need to apply what you have learned to new situations. To illustrate reasoning by analogy, suppose you were asked to tell the structure of dimethyl ether, CH_3—$\overset{..}{\underset{..}{O}}$—$CH_3$, and trimethylamine, $(CH_3)_3\overset{..}{N}$. We can think of these molecules as derived from or related to water and ammonia, respectively. We know that the bond angle is about 104° in water and about 107° in ammonia. With this information, we could guess the approximate bond angles in the compounds as follows:

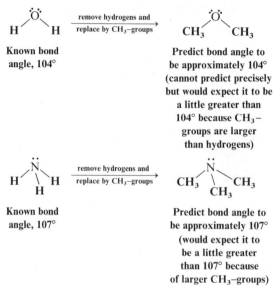

Known bond
angle, 104°

Predict bond angle to
be approximately 104°
(cannot predict precisely
but would expect it to be
a little greater than
104° because CH_3–
groups are larger
than hydrogens)

Known bond
angle, 107°

Predict bond angle to
be approximately 107°
(would expect it to
be a little greater
than 107° because
of larger CH_3–groups)

We will see a great deal of reasoning by analogy in our study of organic chemistry.

Question 2.4

Draw three-dimensional structures for the following five molecules. Show unshared pairs of electrons (if any) and indicate the approximate bond angles for all bonds. What atomic orbitals are used to make each molecular bond orbital? Label all σ bonds.

(*a*) CH_3—O—H   (*b*) $H_3O^{\oplus}$ (*c*) $NH_4^{\oplus}$ (*d*) CH_3—NH_2 (*e*) F—CH_2—CH_2—OH

2.10   Polarity and Electronegativity

The covalent bond is of paramount importance in organic chemistry, but until now we ignored for the most part bond polarity and the crucial role it plays in the reactivity and properties of organic compounds. Bond polarity is used to explain both *why* many reactions occur and many of the properties of molecules. We must therefore equip ourselves with a firm knowledge of this concept before proceeding further.

Although the two atoms involved in covalent bonding do share electrons, often the electrons are not shared equally between them. The key to understanding bond polarity involves the concept of electronegativity of atoms. The net electron affinity that the nucleus of each atom has for electrons is called **electronegativity.** Two major factors govern electronegativity: (1) the charge of the nucleus (kernel charge), and (2) the distance of outer-shell electrons from the nucleus. The larger the kernel charge, the greater the atom's affinity for electrons; the farther the outer-shell electrons from the nucleus, the less the atom's affinity for those electrons. The electronegativity scale devised by Linus Pauling is an arbitrary one based on bond energies; it is shown for selected elements in Table 2.2. Electronegativity increases in going from left to right and from bottom to top in the periodic table. Thus, fluorine, F, is the most electronegative element, and it very readily accepts an electron to complete its outer shell.

We encounter certain elements frequently in organic chemistry; thus you are advised to learn the following electronegativity order:

Order of decreasing electronegativity:

$$F > O > N = Cl > Br > C = S = I > H$$

Most	*Least*
electronegative	electronegative

Note that the electronegativity of carbon is very close to that of hydrogen, and there is still some debate about which element is more electronegative.

To predict the polarity of molecules using electronegativity, remember that we can consider only covalent bonds. We can use electronegativities for our predictions because they are all on the same relative scale. They represent the *relative abilities of atoms to attract electrons.* Covalent bonds may be classified into two major categories: nonpolar and polar.

TABLE 2.2   Electronegativities of Selected Elements

Electronegativity values are given in *italic* below each element.

Effective kernel charge (protons minus nonvalence electrons):	+1	+3	+4	+5	+6	+7	
	H	B	C	N	O	F	Increasing electronegativity
	2.1	*2.0*	*2.5*	*3.0*	*3.5*	*4.0*	Decreasing atomic size
			Si	P	S	Cl	
			1.8	*2.1*	*2.5*	*3.0*	
						Br	
						2.8	
						I	
						2.5	

Increasing kernel charge →

Increasing electronegativity →

1. A covalent bond is said to be **nonpolar** when the bonding electrons are shared equally between the two atoms involved. For example, the following molecules are nonpolar:

$$\text{H—H} \qquad :\overset{..}{\underset{..}{Cl}}—\overset{..}{\underset{..}{Cl}}: \qquad :\overset{..}{\underset{..}{F}}—\overset{..}{\underset{..}{F}}: \qquad \overset{..}{\underset{..}{O}}=\overset{..}{\underset{..}{O}} \qquad :N\equiv N:$$

2. A covalent bond is said to be **polar** when the bonding electrons are shared unequally between the atoms involved. The more electronegative element attracts electrons more strongly and, as a result, has the higher concentration of negative charge about it. The symbol $\nrightarrow$ by a bond indicates the direction of its polarity,

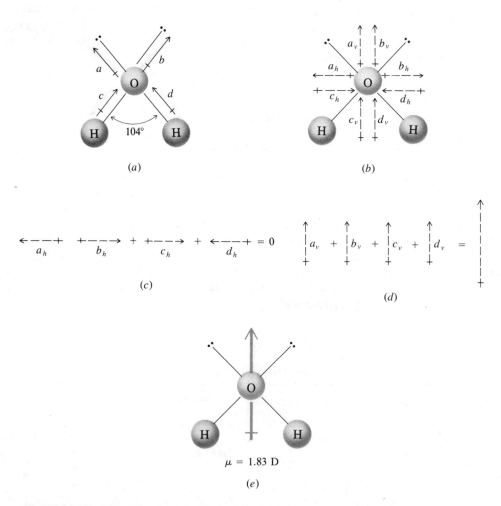

FIGURE 2.22   Schematic representation of the derivation of the net dipole moment of water. (*a*) Water molecule showing individual dipoles associated with each bond or unshared pair of electrons. Each dipole lettered for use in (*b*). (*b*) Each individual dipole in (*a*) resolved into horizontal (subscript *h*) and vertical (subscript *v*) vectorial components. Components are identified by letter to indicate which bond they came from in (*a*). (*c*) Algebraic addition of all individual horizontal vectors in (*b*). Magnitude and direction of a_h and b_h are equal and opposite, and the same is true of c_h and d_h. Thus the net horizontal dipole moment is zero. (*d*) Algebraic addition of all individual vertical vectorial components of (*b*). Magnitude and direction of net vertical dipole moment are indicated. (*e*) Diagram showing magnitude and direction of net dipole moment in the water molecule.

where one end of the symbol contains a + charge and indicates the positive end of the molecule, and the other (with the arrowhead) indicates the negative end. We can also indicate bond polarity by putting a $\delta^{\oplus}$ (meaning partial positive charge) over the electron-deficient atom (the less electronegative atom from which electrons are drawn) and a $\delta^{\ominus}$ over the electron-rich atom (the more electronegative atom attracting electrons). In a neutral molecule, *the sum of the $\delta^{\oplus}$'s must equal the sum of the $\delta^{\ominus}$'s.*

We can speak of the polarity of polyatomic molecules in terms of a *net dipole moment*, even though each individual covalent bond may have a dipole associated with it. We must consider the geometry of the molecule as well as the individual dipole moments associated with each bond to determine the direction of the (net) dipole moment in them.

Consider the water molecule. To determine the net dipole moment for water, we must resolve the individual dipole moments into their horizontal and vertical components using vector analysis. Determination of the vector sum of the individual vector components then yields the direction of the net dipole. This process is illustrated in Fig. 2.22. The presence of the unshared pairs of electrons often contributes largely to the net dipole moment of a molecule. Additional examples are provided in Fig. 2.23.

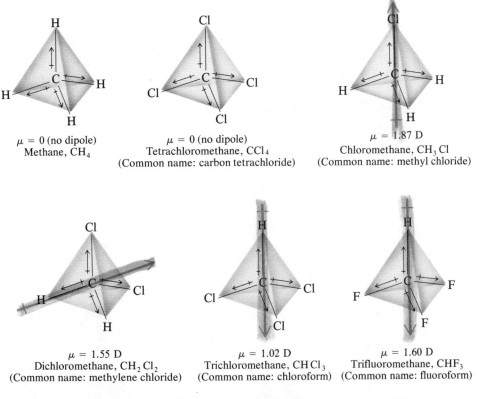

$\mu = 0$ (no dipole)
Methane, CH_4

$\mu = 0$ (no dipole)
Tetrachloromethane, CCl_4
(Common name: carbon tetrachloride)

$\mu = 1.87$ D
Chloromethane, CH_3Cl
(Common name: methyl chloride)

$\mu = 1.55$ D
Dichloromethane, CH_2Cl_2
(Common name: methylene chloride)

$\mu = 1.02$ D
Trichloromethane, $CHCl_3$
(Common name: chloroform)

$\mu = 1.60$ D
Trifluoromethane, CHF_3
(Common name: fluoroform)

FIGURE 2.23   Dipole moments of six selected organic molecules. Direction and magnitude of dipole moments and individual moments associated with each bond are shown. Dipole moments are usually given in Debye units (D), where the larger the value the greater the dipole moment of the molecule. Dipole moments commonly range between 0 and 2D.

2.11   Atomic Orbitals: A More Detailed View

A more complete representation of the *s* and *p* atomic orbitals presented in Figs. 2.3 and 2.4 is provided in Fig. 2.24. That for the *d* orbitals is given in Fig. 2.25.

Note that these orbitals have plus (+) and minus (−) signs associated with them. There are also points within these orbitals where there are nodes, nodal surfaces, and/or nodal planes. What is meant by these characterizations? If we refer to Fig. 2.2 for a moment, we can clarify these points. Remember that *the electron is a wave particle;* that is, it has both wave and particle-like properties. The + and − signs (often represented by two shades of color) associated with a given orbital are simply algebraic signs referring to the direction of the amplitude of the wave along its axis of propagation. These signs are completely arbitrary and are defined by the coordinate

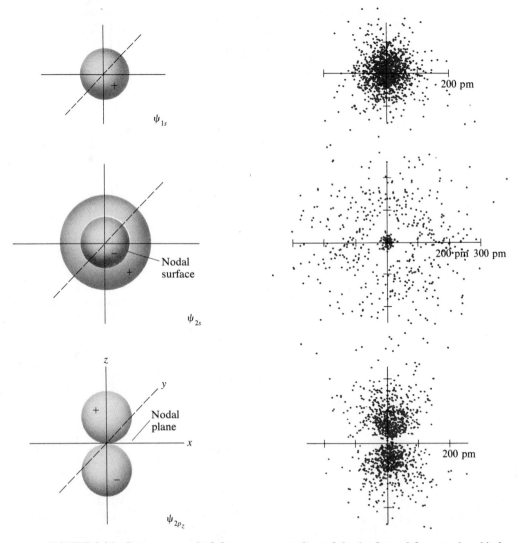

FIGURE 2.24   Structures on the left are representations of the 1s, 2s, and 2pₓ atomic orbitals. Those on the right are computer-generated plots of the position of a given electron versus time. (Computer-generated dot density diagrams of the 1s and 2s orbitals for hydrogen and lithium atoms from John W. Moore, William G. Davies, and Ronald W. Collins, *Chemistry*, McGraw-Hill Book Co., New York, 1978.)

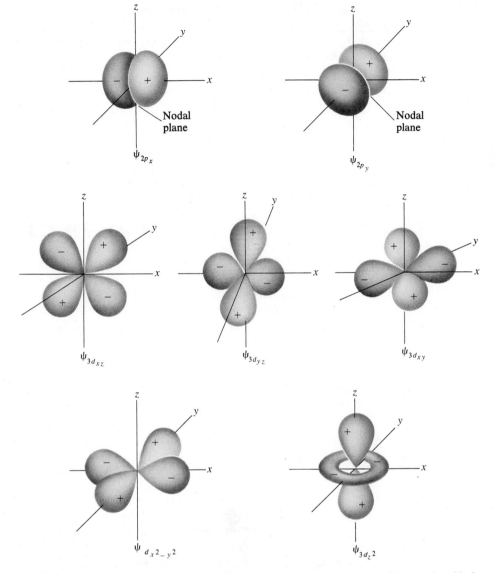

FIGURE 2.25   The structures of the $2p_x$, $2p_y$, $3d_{xz}$, $3d_{yz}$, $3d_{xy}$, $3d_{x^2-y^2}$, and $3d_{z^2}$ atomic orbitals. From John W. Moore, William G. Danes, and Ronald W. Collins, *Chemistry*, McGraw-Hill Book Co., New York, 1978.

system we choose as a reference. For example, one could define any amplitude directed away from us as − and toward us as +. We could just as easily and correctly define the opposite to be true. The sign of the wave is only important within a given coordinate system and may not be arbitrarily shifted from one system to another. One can get a good feeling of what these signs mean by thinking about the waves generated by dropping a small stone into a puddle of water. The waves can either reinforce or interfere with each other. If two identical waves are superimposed in such a way that the + amplitude of each is additive (crest of one superimposed on crest of the other), the amplitude is twice as large as each of the original waves and the waves are said to be in phase. Conversely, if the two waves are superimposed in such a way that they subtract from each other (crest of one superimposed on trough of the other), the amplitude of the resulting wave is zero and the waves are said to be out of phase.

The nodes, nodal surfaces, and nodal planes associated with a wave are points of zero amplitude. Since for orbitals these waves represent the probability of finding an electron in that region of space, the nodes represent regions of zero probability of finding a given electron.

2.12   Molecular Orbitals (MO's)

We were introduced to molecular orbitals in Sec. 2.4, where a qualitative approach to understanding molecular orbitals was presented. It was argued that the molecular orbital involved in bonding two nuclei would be expected to have a shape similar to the atomic orbitals from which it was made. How accurate is this prediction?

The basic assumption of what is referred to as **molecular orbital theory** is that an electron in a molecule can be described by a wave equation (psi, Ψ) just as we did for an electron in an atom. In the latter case the electron is in the electrostatic field of only one nucleus, whereas in the former the electron may be influenced simultaneously by the fields of many nuclei. The mathematics of these wave equations are so difficult that they are all but impossible to solve except for the simplest molecules. However, methods have been developed to solve these equations. They all involve making a variety of assumptions that simplify the mathematics; hence the solutions obtained using these methods are themselves really approximations. This is a small sacrifice considering the wealth of information one can obtain through the application of MO theory to chemical systems. In the following sections, we introduce several important uses and types of information one can obtain from MO theory.

2.13   Linear Combination of Atomic Orbitals (LCAO's)

The most widely used method for solving wave equations in molecular orbital theory is the **LCAO** method. LCAO stands for the *linear combination of atomic orbitals*.[1] In this method, molecular orbitals are constructed as combinations of the atomic orbitals involved in their formation. In other words, molecular orbitals can be described by a mathematical wave equation derived from each of the atomic wave equations of the atoms in question. Our original qualitative assumption concerning the similarity between molecular orbitals and the atomic orbitals involved in their formation can now be shown to be mathematically correct through the application of wave mechanics.

What do we mean by a linear combination of atomic orbitals? If an atomic orbital (ψ_A) combines with a second atomic orbital (ψ_B), a new molecular orbital (Ψ_{A-B}) is formed. ψ_A describes the electron in atom A, ψ_B describes the electron in atom B, and Ψ_{A-B} describes the electrons in the molecular orbital.

When two equivalent atomic orbitals combine, they always yield two molecular orbitals. One molecular orbital is called a **bonding orbital**; the other is referred to as an **antibonding orbital**. Antibonding orbitals are differentiated (symbolically) from bonding orbitals by the use of an asterisk, for example, Ψ^*.

---

[1] The term *linear* refers to the mathematical form of the equations we are using. Basically it means that all the terms in the equations are raised to the first power only. There are no terms raised to any higher power.

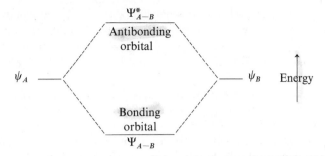

Electrons in a bonding orbital are of lower energy and hence are more stable than electrons in the original atomic orbitals. Electrons in antibonding orbitals, on the other hand, contain more energy and are therefore less stable than the original electrons in the atomic orbitals.

In terms of the phase of the orbitals involved, a bonding orbital is characterized by the positive overlap of two orbitals in phase with one another (and a high electron density in the area between the two nuclei). Antibonding orbitals are characterized by negative overlap, the combination of two orbitals out of phase, and a nodal plane in the region between the nuclei. For example:

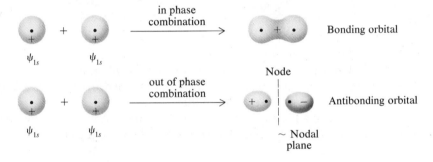

2.14   Molecular Orbitals and Bonding

The combination of two hydrogen atoms to form a hydrogen molecule can be illustrated using Lewis dot formulas:

$$H\cdot + H\cdot \longrightarrow H\!:\!H$$

The $1s$ atomic orbitals from each of the two hydrogen atoms, each containing one electron, combine to form the two molecular orbitals associated with the hydrogen molecule.

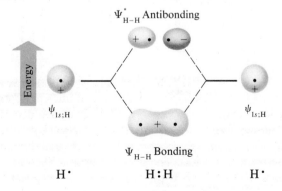

In taking a closer look at what happens to the two electrons we come once more to yet another very important piece of information that is derived from MO theory. Molecular orbitals obey all the laws or rules normally associated with atomic orbitals; that is, the Aufbau principle, Hund's rule, and the Pauli exclusion principle are all adhered to.

If we look at our atomic and molecular orbital diagram once more, we see that initially each atomic (ψ_{1s}) orbital of hydrogen contains one electron. If these two electrons are to be placed in the newly formed molecular orbitals of the hydrogen molecule in compliance with the rules, then both electrons should be placed in the lower energy (more stable) bonding orbital. This complies with the Aufbau principle and Hund's rule. In accordance with the Pauli exclusion principle, the two electrons must be of opposite spin, which is indicated by the direction of the two arrows used to symbolize the two electrons.

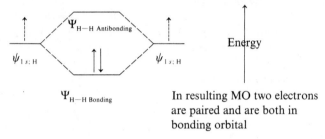

In resulting MO two electrons are paired and are both in bonding orbital

This electron configuration results in an overall lowering of the energy (stabilization) of the system and is therefore a favorable process in terms of energy. The formation of the hydrogen molecule from two hydrogen atoms should occur readily, and that is observed experimentally.

Finally, using a classical potential energy diagram, we can show how energy changes as a function of the internuclear distance between the two nuclei.

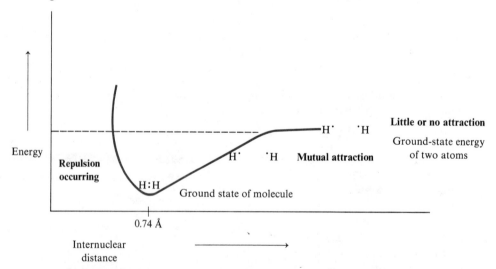

When the two nuclei are far apart (a large internuclear distance), there is little or no attraction between the positively charged nucleus of one atom and the negatively charged electrons in the other. As the internuclear distance decreases, however, there is a continual increase in the mutual attraction between the two atoms, which results in the lowering of the energy and hence the stabilization of the entire system. At some internuclear distance, an energy minimum is reached (in the case of H—H, this distance is 0.74 Å)—this is the most stable state (*ground state*) of the molecule.

If the two nuclei are forced closer together at this point, the effect of destabilization due to repulsion of the two nuclei outweighs the stabilization of attraction, and there again is an increase in the energy (destabilization) of the system.

2.15   σ Versus π Molecular Orbitals

In Sec. 2.4 we introduced the sigma molecular bond orbital (σ bond), and in Sec. 2.5 we introduced the pi molecular bond orbital (π bond). Both orbitals result from the combination (overlap) of atomic orbitals, but the ways the atomic orbitals overlap (and consequently the degrees of overlap) differ. σ molecular orbitals are cylindrically symmetrical about the bond axis that joins the two nuclei, whereas π molecular orbitals are formed from the parallel overlap of p orbitals. As we saw, the symmetry, bond strength, and length associated with a π bond are all quite different from those associated with a σ bond.

Examples of several molecular orbitals and the atomic orbitals from which they originate are given in Fig. 2.26. Note that the symbols associated with the molecular

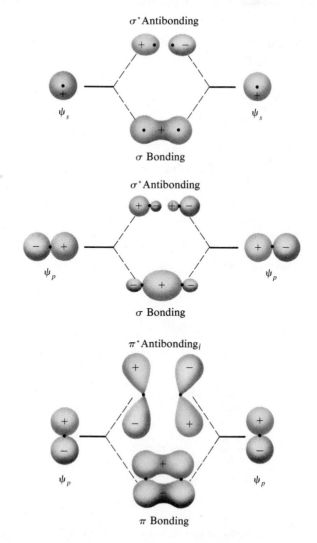

FIGURE 2.26   Bonding and antibonding MO's for σ and π bonds.

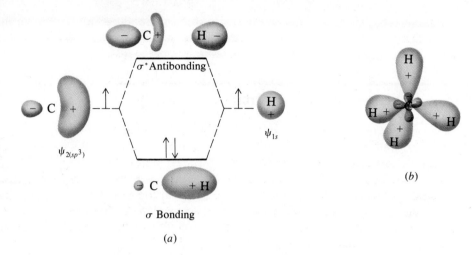

FIGURE 2.27 (*a*) **Combination of a carbon 2(sp^3) and hydrogen 1s atomic orbital to form a bonding (σ) and antibonding (σ*) MO. (*b*) Methane with four bonding MO's.**

orbitals are σ and π for the bonding molecular orbitals (Ψ_{molecule}) and σ* and π* for the antibonding molecular orbitals (Ψ*$_{\text{molecule}}$).

2.16   Alkanes

The structure of methane, and by analogy of the other alkanes, was presented in Sec. 2.6, and a quick review of that section may be useful at this point. An extension of that discussion in combination with the concepts regarding MO theory provides a more complete view of the bonding in these saturated hydrocarbons. This view is summarized in Fig. 2.27.

Each of the four sp^3 orbitals of carbon contains one electron that is involved in bonding with the s electron of a hydrogen atom. The combination of the $\psi_{2(sp^3)}$ orbital and the ψ_{1s} orbital results in the formation of both a bonding (σ) and an

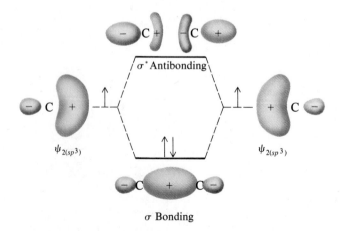

FIGURE 2.28   Combination of two carbon 2(sp^3) atomic orbitals to form a bonding and antibonding MO.

antibonding (σ^*) MO. The two electrons enter the σ bonding MO, resulting in the formation of the four carbon-hydrogen σ bonds of methane.

In the larger alkanes, in addition to carbon-hydrogen bonds there are carbon-carbon bonds. These result from the overlap of two $\psi_{2(sp^3)}$ orbitals, with the electrons again entering the more stable bonding MO (Fig. 2.28).

2.17   Alkenes and Alkynes

The molecular orbitals of a typical alkene were pictured in Fig. 2.26. The bonding molecular orbitals result from the overlap of the $2(sp^2)$ hybrid atomic orbitals of carbon with each other (to form a carbon-carbon σ bond) or with a hydrogen $1s$ atomic orbital (to form a carbon-hydrogen σ bond). In addition, parallel overlap of the two carbon $2p$ atomic orbitals provides the π component of the carbon-carbon double bond. The orbitals involved in this bonding pattern are presented in Fig. 2.29.

The bonding in the higher alkenes (including dienes) can be explained by extending the concepts provided in Sec. 2.7B. See also Secs. 13.4 and 13.6C and Question 13.1.

The bonding orbitals of the alkynes (the carbon-carbon triple bond) are presented in Fig. 2.30.

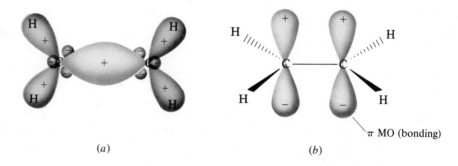

(a)   (b)

FIGURE 2.29   (a) Bonding σ MO's of ethene. (b) Bonding π MO of ethene.

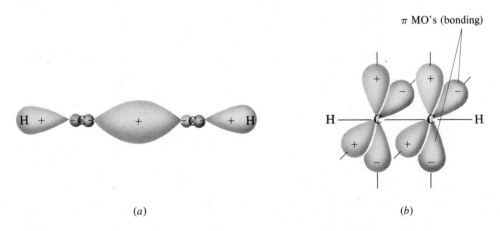

(a)   (b)

FIGURE 2.30   Bonding MO's (a) σ and (b) π for the ethyne molecule.

Study Questions

2.5 Briefly define each of the following.

(*a*) *s* orbital	(*b*) *p* orbital	(*c*) *sp*^{3} orbital
(*d*) *sp*^{2} orbital	(*e*) *sp* orbital	(*f*) atomic orbital
(*g*) molecular orbital	(*h*) σ bond	(*i*) π bond
(*j*) electron configuration	(*k*) Aufbau principle	(*l*) Hund's rule
(*m*) Pauli exclusion principle	(*n*) polar molecule	(*o*) nonpolar molecule
(*p*) hybridization	(*q*) tetrahedral	(*r*) trigonal
(*s*) pyramidal	(*t*) orbital overlap	(*u*) electronegativity
(*v*) dipole moment	(*w*) bond polarity	

2.6 Give the electron configuration (that is, $1s^2$, and so on) and draw diagrams for the electron distribution such as those shown in Table 2.1 for the following.

(*a*) Si   (*b*) P   (*c*) S   (*d*) Cl   (*e*) Ar
(*f*) Na$^{\oplus}$ (*g*) Be^{+2} (*h*) F$^{\ominus}$ (*i*) S^{-2}

2.7 Label the atomic orbitals involved in the bond designated by — in each of the following molecules (that is, give the molecular bond orbital designation). Describe the shapes of these molecules, and where applicable give the approximate bond angles that would be expected for them.

(*a*) H—Cl   (*b*) F—Br   (*c*) H$_3$C—H   (*d*) CH$_3$—CH$_3$
(*e*) (CH$_3$)$_2$N—CH$_3$ (*f*) H$_2$O$^{\oplus}$—H   (*g*) H$_2$B—H

2.8 In the following molecule, what atomic orbitals are used to form the numbered molecular bonds?

2.9 For each of the following eight bonds, indicate the atom that has the partial positive charge ($\delta^{\oplus}$) and the atom that has the partial negative charge ($\delta^{\ominus}$).

(*a*) C—H   (*b*) C—N   (*c*) C—Br   (*d*) C—S
(*e*) Si—Cl   (*f*) N—H   (*g*) S—O   (*h*) C—O

2.10 Suggest an explanation for the fact that the dipole moments of chloromethane, dichloromethane, and trichloromethane decrease as the number of chlorine atoms in these molecules increases.

2.11 Account for NF$_3$ having a dipole moment, whereas BF$_3$ has no dipole moment. When answering, draw structures of these molecules.

2.12 Arrange the following four molecules in order of *increasing* polarity; that is, list the least polar molecule first and the most polar molecule last.

CF$_4$ CH$_3$Cl   CH$_3$F   CH$_3$Br

2.13 Draw the three-dimensional structure of each of the following molecules. What is the hybridization of the central atom? What is the bond angle(s) in the molecule? If the molecule is polar, indicate the direction of the molecular dipole with a vector. If it is nonpolar, so indicate.

(*a*) BCl$_3$ (*b*) CCl$_4$ (*c*) HCl   (*d*) BeCl$_2$ (*e*) NF$_3$ (*f*) D$_2$O

2.14 What is the percentage of *s* character due to carbon of each of the bonds indicated in red in the following four structures?

—C≡N   —C=O   O=C=O   C≡O

2.15 Which bond in each of the following pairs is more polar?

(*a*) CH$_3$—Br   CH$_3$—F   (*b*) CH$_3$CH$_2$—NH$_2$ CH$_3$CH$_2$—OH
(*c*) H—I   H—OH   (*d*) Cl$_2$Al—Cl   Cl$_2$B—Cl
(*e*) H$_2$N—H   H$_2$P—H

2.16 (*a*) Sketch the structures of the CHF_3 and the BF_3 molecules and indicate the *approximate* bond angles in them. What atomic orbitals are involved in *each* of the bond orbitals in *each* of these molecules?

(*b*) Based on the structures you drew in (*a*), indicate the *net* dipole moments (if any) associated with each of these molecules. (You need show only the general direction of the net dipole moment.) Briefly explain.

2.17 Complete the following for the CF_2Cl_2 molecule.

(*a*) Draw an adequate representation to show the three-dimensional structure of the molecule. Indicate *approximate* bond angles.

(*b*) What atomic orbitals are used in forming the C—Cl and the C—F bond?

(*c*) Redraw the structural formula in (*a*) and include the individual dipole moments associated with *each* bond in the molecule. Then draw in the *net* dipole moment associated with the entire molecule.

2.18 Tetrafluorosilane, SiF_4, is a known compound. Draw the electron configuration of silicon's valence shell as indicated by the preceding structural formula. Explain what happens to the atomic orbitals and why it occurs. (*Hint:* Note the position of silicon versus carbon in the periodic table.)

2.19 It is well known that water can be protonated: $H_2\ddot{O}: + H^{\oplus} \rightleftharpoons H_3\overset{\oplus}{\ddot{O}}:$. This has been attributed to the interaction of one of the two unshared pairs of electrons on oxygen with the proton. $H_3\overset{\oplus}{\ddot{O}}:$ still has one unshared pair of electrons that could conceivably react with another proton to give $H_4O^{\oplus 2}$. Yet this species is not obtained. Briefly explain why.

2.20 Considering the π molecular orbitals only, draw the orbital picture of the nitrile group ($-C \equiv \ddot{N}$) and the carbon monoxide and carbon dioxide molecules.

~~omit~~ **2.21** The carboxyl, carbonyl, and acyl groups are drawn below. Draw the orbital picture of each species. Label all molecular orbitals as σ or π, and list the atomic orbitals from which each was derived.

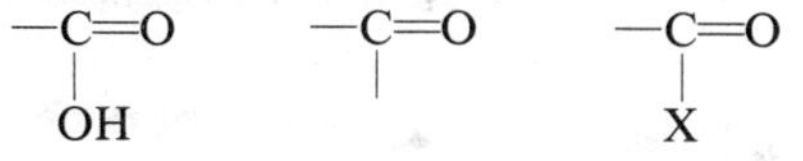

$$-C=O \qquad -C=O \qquad -C=O \qquad (X = Cl, Br, \text{ for example})$$
$$\ \ | \qquad\qquad\qquad\qquad\quad\ \ |$$
$$\ OH \qquad\qquad\qquad\qquad\ X$$

Carboxyl   Carbonyl   Acyl

3

Alkanes and Cycloalkanes: Structure, Properties, and Nomenclature

Organic compounds that contain only carbon and hydrogen are called **hydrocarbons**; this name is derived from *hydro*, referring to the hydrogen, and *carbon*, referring to the carbon they contain. Within this large family are two main subclassifications: **aliphatic** and **aromatic.** All aromatic compounds contain a unique structural feature: an *aromatic ring* (see Chap. 13). All hydrocarbons that do not contain an aromatic ring are classified as aliphatic compounds. Aliphatic hydrocarbons (from Greek *aleiphar* meaning fat or oil) can be *saturated* or *unsaturated* and are subdivided into families, which include alkanes, alkenes, alkynes, and so on, as shown. Saturated compounds contain only single bonds, whereas unsaturated molecules contain double bonds (alkenes) or triple bonds (alkynes).

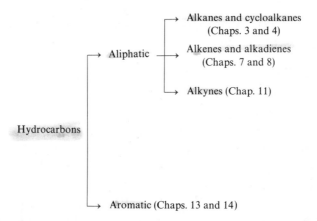

Our coverage of organic chemistry by families (functional groups) begins with the alkanes and the cycloalkanes in this chapter; the other hydrocarbons are discussed in subsequent chapters. We consider the structure, properties, and nomenclature of alkanes and cycloalkanes and turn to the methods of preparation and the reactions of these compounds in Chap. 4. These two families of compounds are similar, particularly in their reactions, so that they can be discussed together.

Alkanes: Structure, Properties, and Nomenclature

3.1   Methane

The tetrahedral structure of **methane, CH$_4$,** was discussed in Sec. 2.6, and it has been shown to be correct by physical methods (such as electron diffraction) and theoretical considerations (quantum mechanics). We start with a discussion of methane, the simplest alkane, and proceed through more complex alkanes until a systematic pattern becomes clear; then we can generalize certain aspects of the alkane family.

A. Properties

Methane is a colorless, odorless gas at room temperature. It has a melting point of $-182.6°$ and a boiling point of $-161.7°$. These properties are not surprising when we recall that methane contains only covalent bonds so that there is little attraction between the molecules; methane is a nonpolar molecule. Any attraction between the molecules is attributed to very weak van der Waals forces (see Sec. 3.11). Methane is insoluble in water (a very polar solvent) but soluble in a variety of non-polar organic solvents. Methane can be liquefied; liquid methane has a specific gravity of 0.424, one of the lowest of all organic compounds.

B. Occurrence

Methane is the major hydrocarbon constituent of *marsh gas*, formed by the *anaerobic* (meaning in the absence of oxygen) decay of plants, which involves the breakdown of complex organic molecules such as cellulose (a major constituent of plants and vegetation) by microorganisms. Highly flammable marsh gas is found in swamps and marshlands, where vegetation is covered by water, and often bubbles to the surface. Methane is perhaps better known as *natural gas*, which is found primarily with or near petroleum deposits. Methane can be separated from the other minor organic components of marsh gas (such as ethane, propane, butane, and isobutane) by fractional distillation, but natural gas is used as a fuel without purification for heating and cooking. Because it is odorless, it has odor added in the form of ethanethiol, CH_3CH_2SH, so that its presence can be detected easily. Methane is also present in coal mines, where it is known as *fire damp* because it explodes so easily.

3.2   Ethane

We discussed the molecular structure of ethane in Sec. 2.7. **Ethane** has the molecular formula C_2H_6 and the following structure:

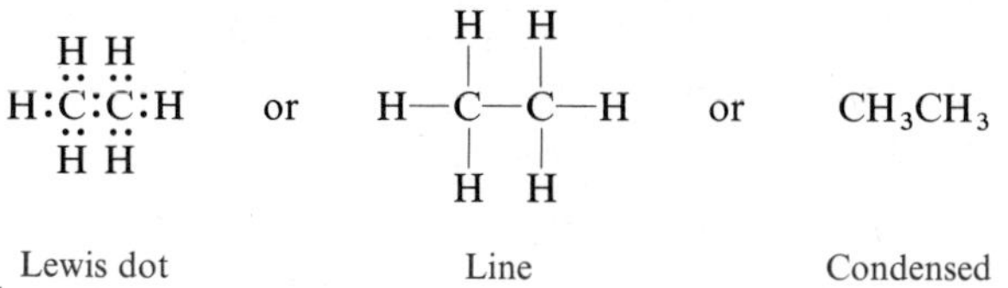

$$
\begin{array}{ccc}
\overset{\displaystyle \text{H H}}{\underset{\displaystyle \text{H H}}{\text{H:C:C:H}}} & \text{or} & \overset{\displaystyle \text{H H}}{\underset{\displaystyle \text{H H}}{\text{H—C—C—H}}} \quad \text{or} \quad CH_3CH_3
\end{array}
$$

Lewis dot	Line	Condensed
structure	structure	structure

Ethane

Physical experiments have shown that the carbon atoms in ethane are tetrahedral with bond angles of about 109.5°. The six C—H σ bonds have a length of 1.10 Å, and the C—C σ bond has a length of 1.54 Å. Indeed the C—C σ bonds in most alkanes have this bond length. The salient structural features of ethane are

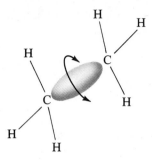

FIGURE 3.1   Carbon-carbon σ bond in ethane: a single bond with free rotation about the σ orbital.

summarized in the following structure:

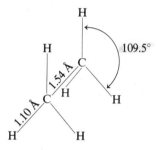

3.3   Ethane and the Carbon-Carbon Single Bond; Conformational Analysis and Free Rotation

The carbon-carbon bond in ethane is a single bond because it is composed of one pair of electrons in a bonding orbital. Its bond orbital (see Sec. 2.7) allows for free rotation. In other words, the orbital is cylindrically symmetrical (Fig. 3.1), and at first glance it appears that there is no obstacle to free rotation and that the carbon-hydrogen bonds on one *methyl group*[1] could have any orientation with respect to the carbon-hydrogen bonds on the other one. (Recall that a molecular orbital simply represents the region in a molecule where two bonding electrons have the greatest probability of being found; we can think of this orbital as being analogous to a frictionless "connecting" shaft between the two carbon atoms.)

To show the three-dimensional character of molecules, many structures for ethane, such as those in Fig. 3.2, can be drawn. Structures (1) and (2) are only two of an unlimited number of possible *spatial orientations* of the six hydrogen atoms. As they are drawn, the hydrogens attached to C-2 have been kept in the same position in (1) and (2), whereas the hydrogens on C-1 have been rotated 180° (by rotating the entire C-1 methyl group 180°).

There are several common ways to depict three-dimensional structures on paper, as shown in Figs. 3.1 and 3.2. The so-called **sawhorse structures** in Fig. 3.2(*b*), (*e*) bear a great resemblance to the ball-and-stick models of ethane, which are shown as structures in (*a*) and (*d*). The **Newman projection formulas** in Fig. 3.2(*c*), (*f*) are

---

[1] The *methyl group* is the methane molecule minus one hydrogen—the CH_3 group. The derivation of the name of this and other *alkyl groups* is given in Sec. 3.10.

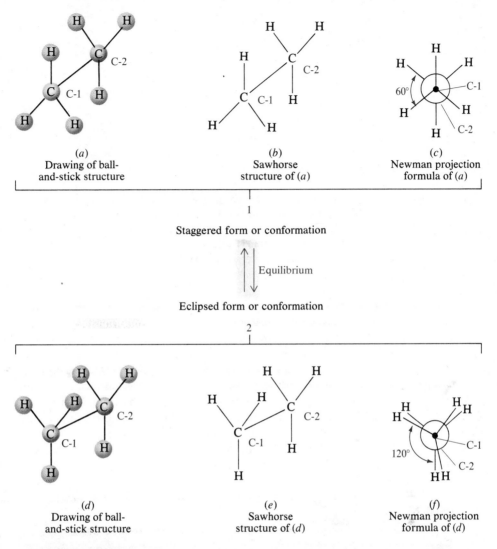

FIGURE 3.2 Methods for showing two possible conformations of ethane. (1) Staggered form or conformation; (2) eclipsed form or conformation.

especially useful for showing structures in simple molecules. The convention for drawing Newman projection formulas (Fig. 3.3) is as follows. Sight down a C—C single bond so that the two carbon atoms are superimposed on one another. Draw the atoms seen on the first carbon atom on the C—C bond and attach them to a common point (they will be 120° apart in the projection). Then draw the atoms seen on the second carbon atom (the atom directly behind the first one) and attach them about a circle (they too will be 120° apart in the projection). You will become more familiar with this convention by inspecting molecular models and drawing Newman projections from them.

Because of the electronic structure of the σ bond, we can say that there is free rotation about the carbon-carbon single bond. Thus the existence of structures (1) and (2) for ethane (Fig. 3.2) has been attributed to this concept of free rotation; they represent ethane and differ only in the three-dimensional location of the hydrogen atom in space. These arrangements of atoms, as well as an infinite number

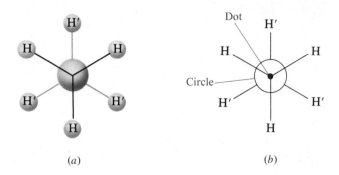

(a)   (b)

FIGURE 3.3   **Method for drawing the Newman projection formula for the staggered conformation of ethane. (a) End view of drawing of ball-and-stick structure of ethane. Hydrogens marked by H are attached to the first carbon atom we see; those marked by H′ are on the carbon atom behind the first one. (b) Newman projection formula as derived from the ball-and-stick structure in (a). Circle represents back carbon atom; dot represents front carbon atom.**

of others, exist and are converted from one to another through free rotation about the carbon-carbon single bond; we call such structures **conformations** or **conformers.** More generally, any two structures that differ only in their arrangement of atoms in space and are interconvertible by rotation about single bonds are conformers of one another. These two particular conformations are given the names **staggered** and **eclipsed** to refer to the three-dimensional arrangement of the atoms. For ethane, the staggered form or conformation is (1) in Fig. 3.2; note that each hydrogen on C-1 is as far as possible from each hydrogen on C-2. The eclipsed form or conformation (2) in Fig. 3.2 has these same hydrogen atoms (on C-1 with respect to C-2) as close to one another as possible. The end view of ethane in Fig. 3.2(f) reveals the hydrogens on the first carbon to be completely superimposed or eclipsed with those on the second carbon atom.

The myriad of conformations between the staggered and eclipsed forms are called **skew conformations.**

There is continuous rotation about the carbon-carbon single bond with a limitless number of possible conformations. As a result, all possible conformations are in equilibrium with one another. Each slight change in rotation about the single bond produces a new conformation. In most instances, conformational structures cannot be isolated as discrete, stable, and noninterconverting species because of the small amount of energy required to interconvert them. We can, however, speak of their *relative stabilities.* We expect the staggered conformation of ethane, (1) in Fig. 3.2, to be more stable than the eclipsed conformation, (2), and thus preferred because the electron clouds associated with the hydrogen atoms and the bonding electrons in the C—H σ bonds are a maximum distance apart in the staggered form. Electrostatic interactions (and perhaps to some extent *steric* interactions, that is, physical crowding of the atoms) are minimized in the staggered conformation; these interactions are shown in Fig. 3.4. In fact, more ethane molecules (at any instant) are in this staggered conformation than in any other possible conformation.

This suggests that our view of free rotation about carbon-carbon single bonds may not be completely correct, and indeed it is not. The *energy barrier to rotation* has been calculated to be about 3 kcal/mole (12.6 kJ/mole) in ethane; that is, this amount of energy is required to rotate the hydrogens on one methyl group from their most stable conformation—staggered—to their least stable conformation—eclipsed. Once the hydrogens have attained this least stable position, the methyl groups and

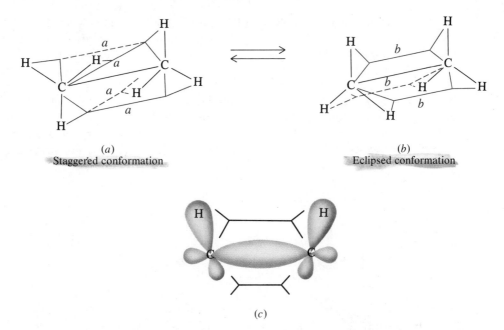

(a)
Staggered conformation

(b)
Eclipsed conformation

(c)

FIGURE 3.4 "Sawhorse" and electronic structures of the staggered and eclipsed conformations of ethane, showing electrostatic interactions between C—H bonds. In the staggered conformation (a), the C—H bonds are a maximum distance apart, and in the eclipsed conformation (b), the C—H bonds are a minimum distance apart; that is, distance $_a$ in (a) is much greater than distance $_b$ in (b). (c) Electrostatic repulsions between two eclipsed C—H bonds.

their attendant hydrogens are free to rotate in either the same or opposite direction back to the most stable conformation; when they do, up to 3 kcal/mole (12.6 kJ/mole) of energy is liberated, depending on the degree of rotation. The energy associated with the rotation of one atom or group of atoms past another atom or group of atoms (hydrogens past hydrogens in ethane) is called **torsional energy.** Torsional energy varies continuously with the degree of rotation of the methyl groups as shown in Fig. 3.5.

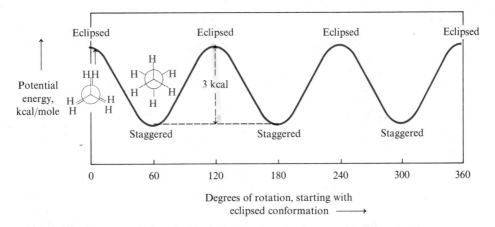

FIGURE 3.5 Graph showing the potential energy changes as a function of the degree of rotation about the carbon-carbon single bond in ethane.

The energy barrier of 3 kcal/mole (12.6 kJ/mole) is relatively small at room temperature, where the thermal energy of the surroundings exceeds this value; for the most part we may assume that free rotation exists about carbon-carbon single bonds. Ethane interconverts among its various conformers at a rate greater than 10^6 times per second at room temperature. It has been determined that the energy barrier must be greater than 20 to 25 kcal/mole (84 to 105 kJ/mole) to prevent thermal interconversions at room temperature, thereby rendering the respective conformers isolable. Even so, microwave spectroscopy studies on ethane gas at room temperature confirm our prediction that the preferred conformation is the most stable staggered form.

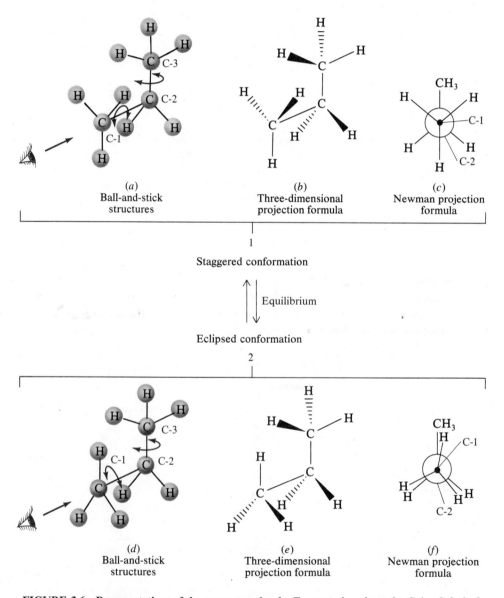

(a)
Ball-and-stick
structures

(b)
Three-dimensional
projection formula

(c)
Newman projection
formula

1

Staggered conformation

Equilibrium

Eclipsed conformation

2

(d)
Ball-and-stick
structures

(e)
Three-dimensional
projection formula

(f)
Newman projection
formula

FIGURE 3.6   Representations of the propane molecule. Free rotation about the C-1—C-2 single bonds in (a) and (d). Newman projection formulas (c) and (f) were drawn by sighting along the C-1—C-2 bonds in (a) and (d), respectively. (1) Staggered conformation; (2) eclipsed conformation.

3.4   Propane

Propane, with a molecular formula of C_3H_8, can be represented in one of the following ways:

H H H
H:C:C:C:H   or   H—C—C—C—H   or $CH_3CH_2CH_3$
H H H

Lewis dot   Line structure   Condensed
structure   structure

Propane

As the structures in Fig. 3.6 show, the C—C—C bond angle is about 109.5°. The preceding representations of the propane molecule do not show its bent nature. There is essentially free rotation about both carbon-carbon bonds in propane, and its staggered and eclipsed conformations are also shown in Fig. 3.6. In this case, a methyl group is brought near a hydrogen atom in the eclipsed conformation. We might predict this would introduce considerable added torsional strain into the molecule because the methyl group is larger and more bulky than hydrogen (compare with ethane in Sec. 3.3). However, the calculated energy barrier to rotation in propane is about 3.3 kcal/mole (13.8 kJ/mole), only slightly greater than the 3 kcal/mole (12.6 kJ/mole) for ethane.

Question 3.1

Draw a potential energy versus rotation diagram (similar to Fig. 3.5) for propane.

3.5   Butane; Structural Isomers

The next member in the alkane family is **butane,** with the molecular formula C_4H_{10}. Two possible structures can be constructed for compounds with this formula. Structure (1) is a **continuous-chain molecule** because all four carbon atoms form a continuous chain, whereas structure (2) is a **branched-chain molecule** because there is a three-carbon continuous chain with a carbon atom attached to the center atom. These structures are different because one cannot be converted to the other simply by rotation about a carbon-carbon bond. Further evidence that they are different comes from the types of bonds that exist. In (1), two carbon atoms are attached to three hydrogen atoms each and two carbon atoms to two hydrogens each. Structure (2) has three carbon atoms bonded to three hydrogens each and one carbon to one hydrogen. Examine these structures carefully to see the differences.

There is considerable evidence for these two butane structures. Each is a stable compound that can be isolated and kept indefinitely, and each has different physical properties. For example, (1) boils at −0.5° and melts at −135°, whereas (2) has a boiling point of −12° and a melting point of −159°. The two compounds have the same number of atoms but a different arrangement of atoms; that is, they have the same molecular formula (C_4H_{10}) but different structural formulas. The two structures of butane are said to be **structural isomers**[1]; they have an isomeric

---

[1] Structural isomers are sometimes called *constitutional isomers.*

$$
\begin{array}{cccc}
H & H & H & H \\
| & | & | & | \\
H-C-C-C-C-H \\
| & | & | & | \\
H & H & H & H
\end{array}
\qquad
\begin{array}{ccc}
H & H & H \\
| & | & | \\
H-C-C-C-H \\
| & | & | \\
H & & H \\
& H-C-H & \\
& | & \\
& H &
\end{array}
$$

or   or

$$CH_3-CH_2-CH_2-CH_3 \qquad CH_3-CH-CH_3$$
$$\qquad\qquad\qquad\qquad\qquad | $$
$$\qquad\qquad\qquad\qquad\qquad CH_3$$

(1)   (2)

Butane   Methylpropane
(n-butane)   (Isobutane)

relationship to one another. *Structural isomers* (see Chap. 6) *are two or more struc-tures that have the same molecular formula but are bonded together through a different bonding topology, that is, a different arrangement of which atoms are bonded to one another.* Unlike conformers, structural isomers can be interconverted only by breaking and reforming new covalent bonds.

We distinguish between the isomeric butanes by calling the continuous-chain compound (1) **butane** or **n-butane** and the branched-chain compound (2) **methyl-propane** or **isobutane**. The *n*-prefix of the former is the abbreviation for *normal* and is used to refer to any continuous-chain compound; whenever *n*- appears in a name, it is automatically pronounced "normal," so that *n*-butane is called "normal butane."

There is no equilibrium between structural isomers. Contrast this to conforma-tional structures (see Secs. 3.3 and 3.4), which are in equilibrium with one another and are interconverted simply by rotation about carbon-carbon single bonds, a process that requires 3 to 6 kcal/mole (12.6 to 25.2 kJ/mole) of energy. To convert butane (1) to methylpropane (2) or vice versa, we would have to break certain C—C and C—H bonds [which have bond energies ranging from 75 to 100 kcal/mole (314 to 418 kJ/mole), depending on the specific bonds] and then make certain new C—C and C—H bonds. Under ordinary conditions, nowhere near this amount of energy is available. (Structural isomers can be interconverted under drastic con-ditions of high temperature and pressure, but even then catalysts are often required; see Sec. 4.10).

3.6   Conformations of Butane

We know that the structural isomers of butane are stable, isolable, discrete compounds. Each can be examined in terms of its conformational structures, because there is still essentially free rotation about the carbon-carbon single bonds in both butane and isobutane.

The conformational picture for butane is complex. First, there is rotation about the C-1—C-2 and the C-3—C-4 bonds, where the carbon atoms are identified as follows:

$$\underset{1}{C}H_3-\underset{2}{C}H_2-\underset{3}{C}H_2-\underset{4}{C}H_3$$

Two of the many possible conformations (or conformers) available for butane by rotation about C-1—C-2 (or C-3—C-4) are shown in Fig. 3.7. Torsional strain

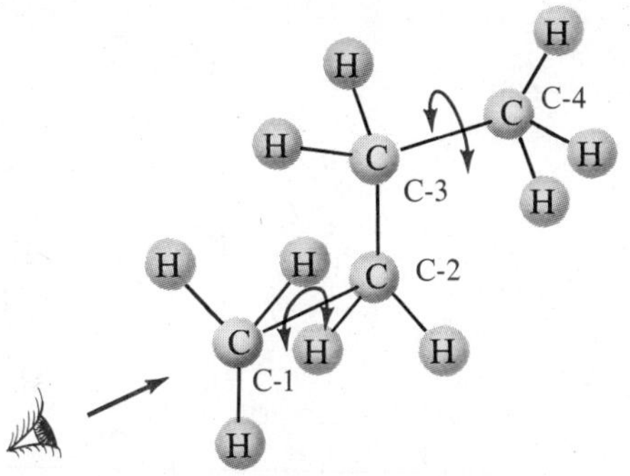

(a)
Drawing of ball-and-stick structure of
butane, showing possible rotation
about the C-1—C-2 and C-3—C-4 bonds

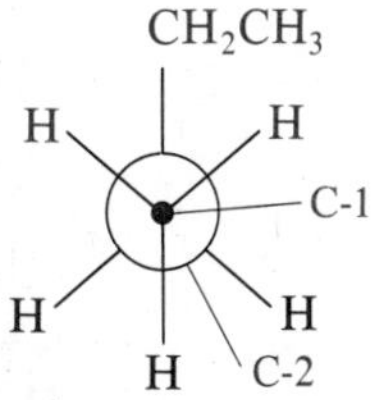

(b)
Newman projection formula
corresponding to structure (a)

Most Stable Conformation

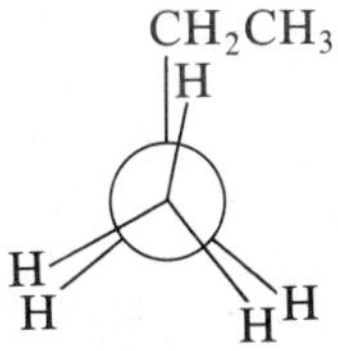

(c)
Obtained by rotation of C-1 in (b) by 60°, shown in Newman
projection formula

Least Stable Conformation

FIGURE 3.7   Most (a) and (b) and least stable (c) conformations of butane obtained by rotation about the C-1—C-2 bond. Identical conformations are obtained by rotation about the C-3—C-4 bond.

(see Sec. 3.4) is again encountered in these conformations. The most and least stable conformers are shown in the figure. The least stable conformation resulting from rotation about C-1—C-2 [Fig. 3.7(c)] involves torsional strain that is of the same order of magnitude as that for propane (see Sec. 3.4). Rotation about either of these two bonds gives identical conformations.

The story is different for rotation about the C-2—C-3 bond. The most and least stable conformations resulting from this rotation are shown in Fig. 3.8. The most stable conformation is the one in which the two methyl groups are as far from one another as possible, Fig. 3.8(1), and is called the **anti-conformation.** But this is only one of three possible staggered conformations; the other two, (3) and (4),

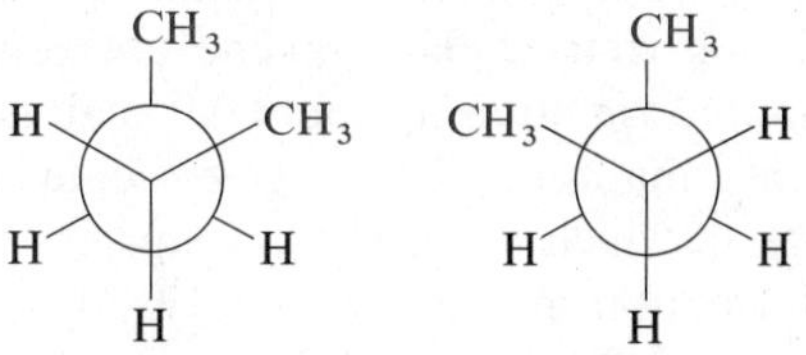

(3)   (4)

Gauche conformations of butane

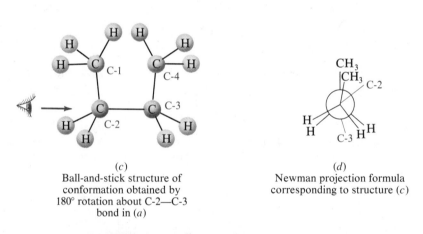

(a)
Ball-and-stick structure of
anti conformation of butane;
possible rotation about C-2—C-3
is indicated

(b)
Newman projection formula
corresponding to structure (a)

Most Stable (*anti*) Conformation of Butane

(c)
Ball-and-stick structure of
conformation obtained by
180° rotation about C-2—C-3
bond in (a)

(d)
Newman projection formula
corresponding to structure (c)

Least Stable (*eclipsed*) Conformation of Butane

FIGURE 3.8   Most stable (*anti*-) and least stable (eclipsed) conformations of butane obtained by rotation about the C-2—C-3 bond.

are called the **gauche conformations.** These are less stable than the *anti*-conformation because the bulky methyl groups are closer together. The least stable conformation is in Fig. 3.8(2), in which the two methyl groups are **eclipsed** with one another.

We might expect to find differences in energy associated with these conformations, and indeed this is true. The gauche conformations, (3) and (4), are less stable than the *anti*-conformation by about 0.9 kcal/mole (3.8 kJ/mole) (even though all are staggered conformations). It has been estimated that the least stable conformation, Fig. 3.8(c), (d), in which the methyl groups are closest together—is less stable than the *anti*-conformation in Fig. 3.8(a), (b) by 4.4 to 6.1 kcal/mole (18.4 to 25.5 kJ/mole). The energy of the molecule changes with rotation about the C-2—C-3 bond, as shown in Fig. 3.9.

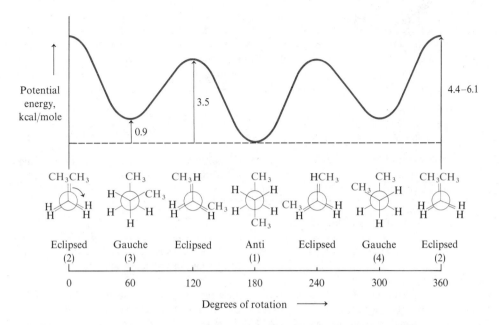

FIGURE 3.9 Graph showing the potential energy changes as a function of degree of rotation about the C-2—C-3 bond in butane. Rotation of C-2 occurs in the direction indicated on the Newman projection formula on the left.

We attributed the energy barrier to rotation to torsional strain (that is, electrostatic repulsion) for ethane (see Sec. 3.3) and propane (see Sec. 3.4). Rotation about the C-2—C-3 bond in butane, however, involves another factor in bringing the methyl groups closer together (Fig. 3.9). Both theoretical calculations and molecular models show that the methyl groups must be brought closer together than their size (as determined by the sum of their van der Waals radii) allows as they are rotated past one another. There is strong **steric repulsion** between the methyl groups. (Of course, there is still torsional strain as the two pairs of hydrogens rotate past one another at the same time as the two methyl groups pass one another.) There is enough thermal energy available at room temperature, however, to allow these methyl groups to rotate past one another so that rotation about the C—C single bond still occurs.

At equilibrium the various conformations of butane are rapidly interconverting because of free rotation, so they cannot be isolated. One can, however, speak of the relative stabilities of the various conformations. The following experimental data shed light on those for butane. At 14°, about 60% of the molecules have *anti*-conformation and about 40% have gauche conformation. There are virtually no conformational structures with the methyl groups eclipsed.

Question 3.2

(*a*) The torsional energy for ethane is 3 kcal/mole (12.6 kJ/mole), but this represents the energy to rotate *three* pairs of hydrogens past one another. What is the torsional energy for a single hydrogen-hydrogen interaction, assuming the energy associated with each of them is equivalent?

(*b*) Using the energy barriers to rotation of 3.3 kcal/mole (13.8 kJ/mole) for propane and 5.2 kcal/mole (21.8 kJ/mole) for the least stable form of butane, estimate the amount of energy associated with a methyl-hydrogen interaction and a methyl-methyl interaction.

Question 3.3

(*a*) Draw at least three different Newman projection formulas for the three following compounds by considering rotation only about the bond indicated by —. Insofar as possible, the Newman formulas should represent the following types of conformations: *skew*, *anti-*, *gauche*, *staggered*, and *eclipsed*. Label each appropriately.

$(CH_3)_2CH—CH_2CH_3$
$(CH_3)_2CH—C(CH_3)_3$
$(CH_3)_3C—C(CH_3)_3$

(*b*) Draw potential energy versus degree of rotation graphs (like Fig. 3.9) for each molecule. Using your answer to Question 3.2, estimate all the energy barriers to rotation that you can. Which conformations are the most and least stable?

Use of Molecular Models

The three-dimensional aspects of molecules, such as those involved in conformations, are not easy to depict on paper. Even photographs and drawings of ball-and-stick models do not do justice to the third dimension.

In our study of the architecture of molecules there is no substitute for using molecular models. You are encouraged to procure one of the sets of molecular models that are on the market—either ball-and-stick type or framework type. Use them frequently when you study the architecture of molecules. Of the material covered thus far, you should find them most useful in connection with conformations; you can examine the factors associated with the energy barrier to rotation and more easily draw the Newman projection formulas. Models, for example, would be extremely useful in working Question 3.3.

Question 3.4

Draw the conformations of isobutane corresponding to the energy maxima and minima as the molecule undergoes free rotation. Contrast them in terms of relative stability.

3.7   Pentane

Pentane, C_5H_{12}, is the next member of the alkane family. There are three and only three structural isomers for pentane, as shown in Table 3.1. As is typical of many structural isomers, these three compounds are stable and can be isolated; they have different boiling points (bp) and melting points (mp) as well as differences in other physical properties. Note that each structural isomer has the same molecular formula C_5H_{12}.

Writing structural isomers in condensed form, such as we did for the pentanes, is sometimes more difficult than it may seem. Although we want to write all possible structural isomers, we must avoid writing duplicate structures. We wrote the only three structural isomers for pentane. To show how one can be led astray, consider the following structures, all of which are identical as indicated by $\equiv$, a mathematical symbol for "identical to."

These structures are identical because they all have a continuous chain of five carbon atoms. Using free rotation about the carbon-carbon single bonds, we can "straighten out" the various bent bonds so that the carbon atoms lie in a straight line on paper. Remember, though, that all these carbon atoms are tetrahedral and that condensed structures do not depict shape; they are written simply for convenience.

TABLE 3.1   Melting and Boiling Points for Isomeric Pentanes and Hexanes*

Pentanes		
Structure and Name	bp, °C	mp, °C
CH$_3$CH$_2$CH$_2$CH$_2$CH$_3$ Pentane (*n*-Pentane)	36.1	−129.7
CH$_3$CH$_2$CHCH$_3$ $\quad\quad$ | $\quad\quad$ CH$_3$ Methylbutane (Isopentane)	27.9	−156.6
$\quad\quad$ CH$_3$ $\quad\quad$ | CH$_3$—C—CH$_3$ $\quad\quad$ | $\quad\quad$ CH$_3$ Dimethylpropane (Neopentane)	9.5	−16.6

Hexanes		
Structure and Name	bp, °C	mp, °C
CH$_3$CH$_2$CH$_2$CH$_2$CH$_2$CH$_3$ Hexane (*n*-Hexane)	68.7	−94.0
CH$_3$CH$_2$CH$_2$CHCH$_3$ $\quad\quad$ | $\quad\quad$ CH$_3$ 2-Methylpentane	60.3	−153.7
CH$_3$CH$_2$CHCH$_2$CH$_3$ $\quad\quad$ | $\quad\quad$ CH$_3$ 3-Methylpentane	63.3	−118.0
$\quad\quad$ CH$_3$ $\quad\quad$ | CH$_3$CH$_2$CCH$_3$ $\quad\quad$ | $\quad\quad$ CH$_3$ 2,2-Dimethylbutane	49.7	−98.2
CH$_3$CH—CHCH$_3$ $\quad$ | $\quad$ | $\quad$ CH$_3$ CH$_3$ 2,3-Dimethylbutane	58.0	−128.8

*   Common names in parentheses.

$$CH_3-CH_2-CH_2-CH_2-CH_3 \equiv CH_3-CH_2-CH_2-\overset{\displaystyle CH_3}{\underset{|}{CH_2}}$$

$$\equiv \overset{\displaystyle CH_3}{\underset{|}{CH_2}}-CH_2-\overset{\displaystyle CH_3}{\underset{|}{CH_2}} \equiv \underset{CH_3-CH_2}{\overset{|}{CH_2}}-CH_2-CH_3 \equiv CH_3-CH_2-\underset{CH_3-CH_2}{\overset{|}{CH_2}}$$

n-Pentane

There are numerous possible conformations of the isomeric pentanes because of the four carbon-carbon single bonds about which free rotation may occur simultaneously. Molecular models help us to see that many structures that look different on paper are really identical molecules.

More hints for writing structural isomers are in Sec. 3.9.

3.8   Isomers of Higher Hydrocarbons; The Homologous Series

We have been developing our discussion of the alkane family in a systematic fashion. We considered methane (CH_4), ethane (C_2H_6), propane (C_3H_8), butane (C_4H_{10}), and pentane (C_5H_{12}), each of which (after methane) contains one carbon atom and two hydrogen atoms more than the preceding compound; that is, each new compound contains the elements of an additional CH_2 unit. We may write the *general formula for the alkane family* as C_nH_{2n+2}, where $n = 1, 2, 3, \ldots$; note that

TABLE 3.2   Number of Theoretically Possible Structural Isomers for Alkane Family (C_nH_{2n+2})

Formula	Total No. of Isomers
CH_4	1
C_2H_6	1
C_3H_8	1
C_4H_{10}	2
C_5H_{12}	3
C_6H_{14}	5
C_7H_{16}	9
C_8H_{18}	18
C_9H_{20}	35
$C_{10}H_{22}$	75
$C_{11}H_{24}$	159
$C_{12}H_{26}$	355
$C_{13}H_{28}$	802
$C_{14}H_{30}$	1,858
$C_{15}H_{32}$	4,347
$C_{20}H_{42}$	366,319
$C_{30}H_{62}$	4,111,846,763
$C_{40}H_{82}$	62,491,178,805,831

each molecule mentioned conforms to this general formula. We will see that other families of organic compounds also have a general formula.

This general formula may be extended to apply to alkanes with a higher number of carbon atoms; for example, hexane ($n = 6$) is C_6H_{14}, heptane ($n = 7$) is C_7H_{16}, and so on. The structural isomers of the hexanes, which have the molecular formula C_6H_{14}, are shown in Table 3.1. As one increases the number of carbon atoms in alkanes, the possible number of structural isomers increases astronomically. There are three isomeric pentanes and five isomeric hexanes. There are nine isomeric heptanes (C_7H_{16}), all of which have been synthesized. Although there is no direct relationship between the carbon content and the number of structural isomers, a mathematical relationship has been established (see Reading Reference 1, Chap. 3, in the appendix); the results are tabulated in Table 3.2. All the possible alkanes from CH_4 to C_9H_{20} inclusive have been synthesized.

Table 3.2 shows how the numbers of carbon and hydrogen atoms increase systematically. Consider the sequence between CH_4 and $C_{15}H_{32}$ and note that as the number of carbon atoms increases by one, the number of hydrogens increases by two, preserving the ratio of C to H as n to $2n + 2$ as in the molecular formula C_nH_{2n+2}.

A series of compounds in which the structural formula of each differs from the immediately preceding one and the following one by a constant amount (usually CH_2) and in no other way is called a **homologous series.** Any compound in the series is called a **homolog** of the others. The *n*-alkane family (Table 3.2) is a typical example of a homologous series.

3.9   System for Writing Structural Isomers

Writing all the structural isomers corresponding to a particular molecular formula can be difficult, if not impossible, when approached randomly. A systematic approach is invaluable to help you write all the possible structures and avoid duplicating structures. Remember these important points:

1. All structural isomers must correspond to the molecular formula.
2. Avoid writing duplicate structures by checking a new one with all previous structures; this is best done by identifying the longest continuous chain, as shown for pentane in Sec. 3.7.

For example, suppose we want to write the structural isomers for heptane, C_7H_{16}. The following four systematic steps can be used.

1. Start with the continuous-chain structure of seven carbons, for which there is only one possible structure:

$$CH_3-CH_2-CH_2-CH_2-CH_2-CH_2-CH_3$$

Heptane

2. Consider next isomers with a six-carbon continuous chain, where a CH_3 group may be attached on one of the nonterminal carbon atoms to give branched-chain structures:

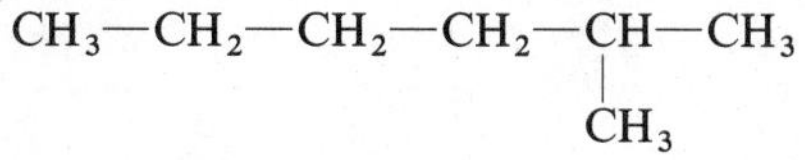

$$CH_3-CH_2-CH_2-CH_2-\underset{\underset{CH_3}{|}}{CH}-CH_3 \qquad CH_3-CH_2-CH_2-\underset{\underset{CH_3}{|}}{CH}-CH_2-CH_3$$

(1) 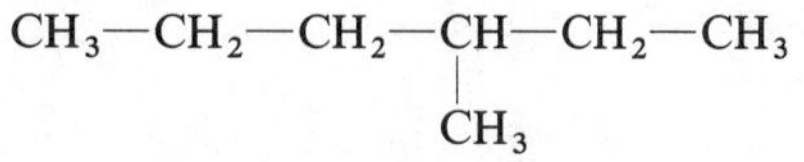   (2)

The following structures appear to differ from (1) and (2) but actually duplicate them. Be sure you understand why!

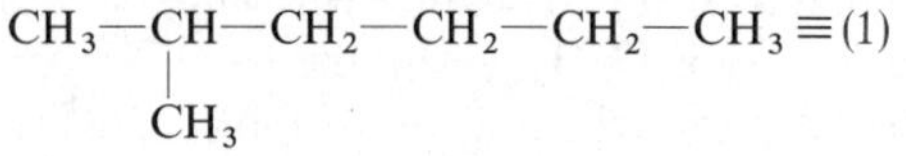

$$CH_3-CH-CH_2-CH_2-CH_2-CH_3 \equiv (1) \qquad \text{and}$$
$$\quad\quad | $$
$$\quad\quad CH_3$$

$$CH_3-CH_2-CH-CH_2-CH_2-CH_3 \equiv (2)$$
$$\quad\quad\quad\quad\quad | $$
$$\quad\quad\quad\quad\quad CH_3$$

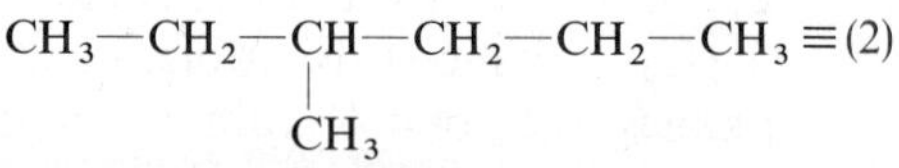

3. Next consider a five-carbon continuous chain, where the two additional carbon atoms can be attached either as two CH_3 groups or as one CH_3CH_2 group to the nonterminal carbon atoms. Of course, branched-chain structures are formed:

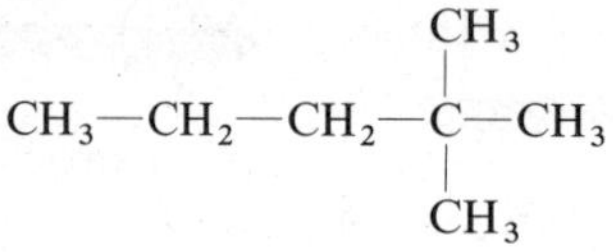

$$\qquad\qquad\qquad CH_3 \qquad\qquad\qquad\qquad CH_3$$
$$\qquad\qquad\qquad | \qquad\qquad\qquad\qquad\qquad |$$
$$CH_3-CH_2-CH_2-C-CH_3 \qquad CH_3-CH_2-C-CH_2-CH_3$$
$$\qquad\qquad\qquad | \qquad\qquad\qquad\qquad\qquad |$$
$$\qquad\qquad\qquad CH_3 \qquad\qquad\qquad\qquad CH_3$$

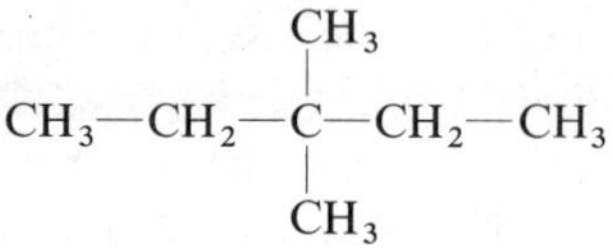

$$(3) \qquad\qquad\qquad\qquad\qquad (4)$$

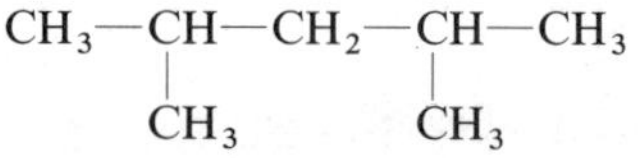

$$CH_3-CH-CH_2-CH-CH_3$$
$$\qquad | \qquad\qquad\quad |$$
$$\qquad CH_3 \qquad\qquad CH_3$$

$$(5)$$

$$\qquad\qquad CH_3$$
$$\qquad\qquad |$$
$$CH_3-CH_2-CH-CH-CH_3 \qquad CH_3-CH_2-CH-CH_2-CH_3$$
$$\qquad\qquad\qquad | \qquad\qquad\qquad\qquad\qquad |$$
$$\qquad\qquad\qquad CH_3 \qquad\qquad\qquad\qquad CH_2-CH_3$$

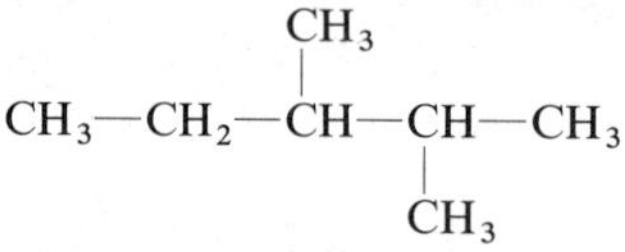

$$(6) \qquad\qquad\qquad\qquad\qquad (7)$$

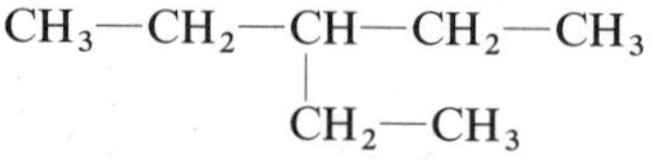

Some possible duplicate structures are:

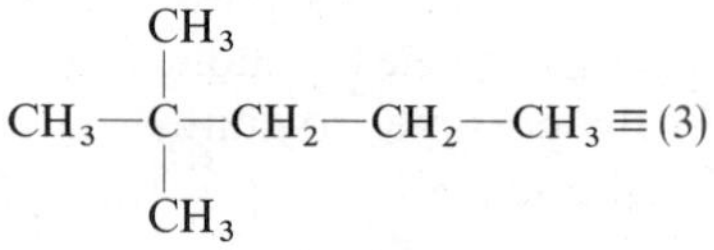

$$\qquad CH_3$$
$$\qquad |$$
$$CH_3-C-CH_2-CH_2-CH_3 \equiv (3) \qquad \text{and} \qquad CH_3-CH-CH_2-CH_2-CH_3 \equiv (2)$$
$$\qquad | \qquad\qquad\qquad\qquad\qquad\qquad\qquad\qquad\qquad |$$
$$\qquad CH_3 \qquad\qquad\qquad\qquad\qquad\qquad\qquad\qquad CH_2-CH_3$$

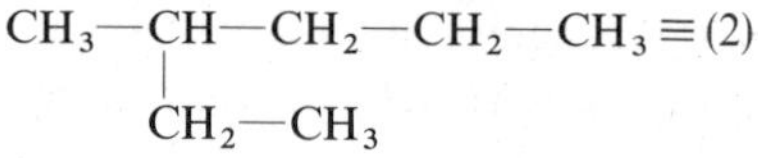

4. Finally, consider a four-carbon continuous chain, where the three additional carbon atoms can be attached either as three CH_3 groups or as one CH_3CH_2 group and one CH_3 group. There is only one branched-chain isomer of this type:

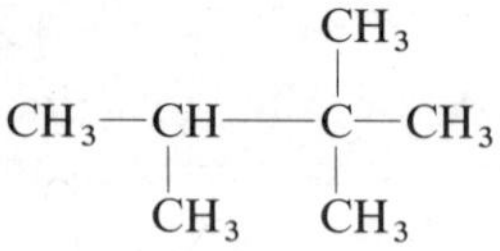

$$\qquad\qquad\qquad CH_3$$
$$\qquad\qquad\qquad |$$
$$CH_3-CH-\quad-C-CH_3$$
$$\qquad\quad | \qquad\quad |$$
$$\qquad\quad CH_3 \quad CH_3$$

$$(8)$$

Thus, there are nine structural isomers [n-heptane plus structures (1) to (8)] for heptane.

An alternate method for drawing structural isomers utilizes carbon skeletons, where hydrogens have been left off and "naked" carbon atoms are used. When this is done, care must be taken to go back and add the appropriate number of hydrogen atoms so that each carbon atom has four bonds about it. For example,

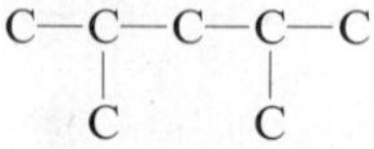

$$C-C-C-C-C \qquad \text{gives rise to} \qquad CH_3-CH-CH_2-CH-CH_3$$
$$\quad | \quad\quad | \qquad\qquad\qquad\qquad\qquad\qquad\quad | \qquad\qquad\quad |$$
$$\quad C \quad\quad C \qquad\qquad\qquad\qquad\qquad\qquad CH_3 \qquad\quad CH_3$$

Carbon skeleton   Correct condensed

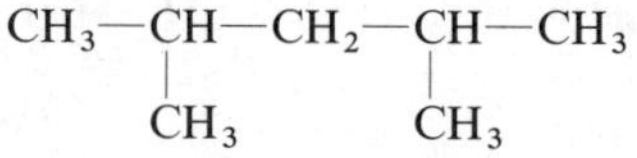

structure   structure

With either method, *always write in all the hydrogen atoms* and write the structures either in condensed form or with each bond shown by a line to indicate an electron pair.

Question 3.5

Which of the following four pairs of compounds represent identical structures? Which are structural isomers?

(a)
$$CH_3-CH_2 \diagdown$$
$$\qquad\qquad CH-CH_3 \quad \text{and} \quad CH_3-CH_2-CH-CH_2-CH_3$$
$$CH_3-CH_2 \diagup \qquad\qquad\qquad\qquad\qquad\qquad\qquad |$$
$$\qquad\qquad\qquad\qquad\qquad\qquad\qquad\qquad\qquad\qquad CH_3$$

(b)
$$\qquad\qquad CH_3-CH_2 \qquad\qquad\qquad CH_3 \quad CH_3$$
$$\qquad\qquad\qquad | \qquad\qquad\qquad\qquad\quad | \qquad |$$
$$CH_3-C-CH_3 \quad \text{and} \quad CH_3-CH-CH$$
$$\qquad\qquad\qquad | \qquad\qquad\qquad\qquad\qquad\qquad |$$
$$\qquad\qquad CH_3 \qquad\qquad\qquad\qquad\qquad\qquad CH_3$$

(c)
$$CH_3 \quad CH_3 \qquad\qquad\qquad\qquad CH_3$$
$$\quad | \qquad | \qquad\qquad\qquad\qquad\qquad |$$
$$CH-C-CH_3 \quad \text{and} \quad CH_3-C-CH-CH_3$$
$$\quad | \qquad | \qquad\qquad\qquad\qquad\quad | \qquad |$$
$$CH_3 \quad CH_3 \qquad\qquad\qquad\qquad CH_3 \quad CH_3$$

(d)
$$\qquad\qquad\qquad CH_2-CH_3 \qquad\qquad\qquad CH_3-CH_2-CH-CH_3$$
$$\qquad\qquad\qquad\quad | \qquad\qquad\qquad\qquad\qquad\qquad\qquad\qquad\quad |$$
$$CH_3-CH_2-CH-CH-CH_3 \quad \text{and} \quad CH_3-CH-CH-CH_3$$
$$\qquad\qquad\qquad CH_2-CH_3 \qquad\qquad\qquad\qquad\qquad\qquad\qquad CH_3$$

Question 3.6

Draw the structural isomers of: (*a*) all the octanes (C_8H_{18}) that have a five-carbon continuous chain, (*b*) the four chlorobutanes (C_4H_9Cl), and (*c*) the nine dichlorobutanes ($C_4H_8Cl_2$).

3.10 Nomenclature: Common Names; IUPAC System

We encountered the names of several organic compounds in our discussions about structure. Now we consider the system for naming all alkanes, which is the beginning of the internationally adopted method called the **IUPAC system** of nomenclature. The immense number of organic compounds necessitates that some system be used to name each compound unambiguously so that chemists worldwide can draw the same structure. This problem was first dealt with in 1892 at a meeting of the International Congress of Chemistry in Geneva, Switzerland. Periodic meetings have been held since to extend and modify the rules of nomenclature. The *Inter*national *U*nion of *P*ure and *A*pplied *C*hemistry (from which the initials IUPAC come), as it is now called, is still responsible for reexamining and updating nomenclature and is composed of chemists from the entire world. The IUPAC system is sometimes called the **Geneva system** because of the location of the original meetings.

The ideal system has not yet been found, but there is now considerable order for naming compounds. The situation is complicated by a number of **common names** for organic molecules that have been carried down through history; you should become familiar with some of these since they are frequently used today. You soon appreciate the need for systematic nomenclature, however, because it would be impossible to learn, for example, common names (if they were available) for the 75

TABLE 3.3   Molecular Structures and IUPAC System of Nomenclature for Alkane Family

Condensed Structure	Molecular Formula	IUPAC Name
CH_4	CH_4	Methane
CH_3CH_3	C_2H_6	Ethane
$CH_3CH_2CH_3$	C_3H_8	Propane
$CH_3(CH_2)_2CH_3$	C_4H_{10}	Butane
$CH_3(CH_2)_3CH_3$	C_5H_{12}	Pentane
$CH_3(CH_2)_4CH_3$	C_6H_{14}	Hexane
$CH_3(CH_2)_5CH_3$	C_7H_{16}	Heptane
$CH_3(CH_2)_6CH_3$	C_8H_{18}	Octane
$CH_3(CH_2)_7CH_3$	C_9H_{20}	Nonane
$CH_3(CH_2)_8CH_3$	$C_{10}H_{22}$	Decane
$CH_3(CH_2)_9CH_3$	$C_{11}H_{24}$	Undecane
$CH_3(CH_2)_{10}CH_3$	$C_{12}H_{26}$	Dodecane
$CH_3(CH_2)_{11}CH_3$	$C_{13}H_{28}$	Tridecane
$CH_3(CH_2)_{12}CH_3$	$C_{14}H_{30}$	Tetradecane
$CH_3(CH_2)_{13}CH_3$	$C_{15}H_{32}$	Pentadecane
$CH_3(CH_2)_{14}CH_3$	$C_{16}H_{34}$	Hexadecane
$CH_3(CH_2)_{15}CH_3$	$C_{17}H_{36}$	Heptadecane
$CH_3(CH_2)_{16}CH_3$	$C_{18}H_{38}$	Octadecane
$CH_3(CH_2)_{17}CH_3$	$C_{19}H_{40}$	Nonadecane
$CH_3(CH_2)_{18}CH_3$	$C_{20}H_{42}$	Eicosane
$CH_3(CH_2)_{19}CH_3$	$C_{21}H_{44}$	Heneicosane
$CH_3(CH_2)_{20}CH_3$	$C_{22}H_{46}$	Docosane
$CH_3(CH_2)_{21}CH_3$	$C_{23}H_{48}$	Tricosane
$CH_3(CH_2)_{28}CH_3$	$C_{30}H_{62}$	Triacontane
$CH_3(CH_2)_{29}CH_3$	$C_{31}H_{64}$	Hentriacontane
$CH_3(CH_2)_{30}CH_3$	$C_{32}H_{66}$	Dotriacontane
$CH_3(CH_2)_{38}CH_3$	$C_{40}H_{82}$	Tetracontane
$CH_3(CH_2)_{48}CH_3$	$C_{50}H_{102}$	Pentacontane
$CH_3(CH_2)_{58}CH_3$	$C_{60}H_{122}$	Hexacontane

structural isomers of $C_{10}H_{22}$! Certain common names are provided throughout this text, although the major emphasis is on the IUPAC system of nomenclature.

A. Names of Alkane Family

Start by considering the names of the continuous-chain hydrocarbons in Table 3.3. For the first four members of the alkane family (methane, ethane, propane, and butane), the names do not appear to follow any system except that they all end in -*ane*, which indicates that they are members of the alk*ane* family (that is, they fit the general formula C_nH_{2n+2}). For pentane and beyond, the names are derived by adding the Greek or Latin prefixes for the number of carbon atoms to the -*ane* ending. For example, *pent-*, *hex-*, *hept-*, *oct-*, *non-*, and *dec-* indicate 5, 6, 7, 8, 9, and 10 carbon atoms, respectively.

The naming of alkanes with more than 10 carbon atoms is also systematic. For example, the name for $C_{11}H_{24}$ is undecane, which is derived by adding *two* prefixes:

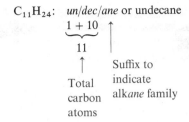

$C_{11}H_{24}$:   *un/dec/ane* or undecane

In a like manner, dodecane contains 2 (from *do*) plus 10 (from *dec*), or 12 carbon atoms. This method continues through nonadecane ($C_{19}H_{40}$).

The name for $C_{20}H_{42}$ is eicosane and for $C_{21}H_{44}$ it is heneicosane; these are irregular because the *ei* is dropped for alkanes $C_{22}H_{46}$ through $C_{29}H_{60}$. For these compounds, the names are derived by adding the prefixes: tricosane contains 3 (*tri*) plus 20 (*cos*), or 23 carbon atoms.

For alkanes with 30 or more carbon atoms, several prefixes are used to indicate numbers. Triacontane, $C_{30}H_{62}$, has 30 carbon atoms, tetracontane, $C_{40}H_{82}$, has 40 carbon atoms, and so on. The remainder of the names in Table 3.3 follow in the regular pattern that has been established.

The alkane names are of extreme importance, and you should memorize the names of at least the first 20. The names of most families of organic compounds are derived from the alkanes, and failure to learn them well means that you will be unable to name other compounds. As we will see, the names of other families involve changing the suffix.

B. Systematic (IUPAC) Approach to Nomenclature

The IUPAC system of nomenclature allows us to name simple as well as complex molecules. We start by giving the five steps (rules) for naming alkanes by the IUPAC system and follow with six examples to illustrate how to apply them.

Step 1. Find the longest continuous chain of carbon atoms in the molecule. The compound name is derived from the alkane corresponding to this longest chain, which is called the *parent* compound, and the various groups or atoms (called *substituents*) that have been substituted for hydrogen in it. If two continuous chains of equal length are found, choose the one with the larger number of substituents attached to it.

Step 2. Number the longest continuous chain from one end to the other, so that *the smallest numbers possible are used to designate the position of the substituents*. This may involve numbering the longest chain in both directions. If the first substituent from each end is the same number of carbons from the end, choose the name that gives the second (or third if the second is the same, and so on) substituent in the chain the lowest numerical designation.

Step 3. Identify the substituents that are attached to the parent compound and note the numbers of the carbon atoms to which they are attached.

Halogens: The following designations are used when the halogens appear as substituents in the compound:

F—,	fluoro
Cl—,	chloro
Br—,	bromo
I—,	iodo

Alkyl groups: A carbon-containing group that is obtained by removing one hydrogen atom from an alkane is called an **alkyl group,** which has the general formula C_nH_{2n+1}. These groups are named by dropping the *-ane* ending from the alkane name and replacing it by *-yl*; as shown in the following examples:

Formula and Name of Alkane	Formula and Name of Alkyl Group
CH_4, meth*ane* $\longrightarrow$	CH_3—, meth*yl*
C_2H_6, eth*ane* $\longrightarrow$	C_2H_5—, eth*yl*
C_3H_8, prop*ane* $\longrightarrow$	C_3H_7—, prop*yl*
$C_{10}H_{22}$, dec*ane* $\longrightarrow$	$C_{10}H_{21}$—, dec*yl*

There is only one structure for the methyl group (that is, CH_3—) and only one for the ethyl group (CH_3—CH_2—). When we consider alkyl groups that contain more than two carbon atoms, we have to account for isomerism. For example, there are two alkyl groups from propane that correspond to the formula C_3H_7—:

$$CH_3—CH_2—CH_3 \xrightarrow{\text{remove one H}} CH_3—CH_2—CH_2— \quad \text{Propyl group (or } n\text{-propyl group)}$$

$$\xrightarrow{\text{remove one H}} CH_3—CH—CH_3 \quad \text{Isopropyl group}$$

Removal of hydrogen from either end of propane gives the same alkyl group, which is called the **propyl (or *n*-propyl)** group. Removal of hydrogen from the central carbon atom gives the **isopropyl** group.

As the number of carbon atoms increases, the number of possible alkyl groups increases greatly. There are four alkyl groups with the formula C_4H_9— which are derived from butane.

$$CH_3—CH_2—CH_2—CH_3 \xrightarrow{\text{remove one H}} CH_3—CH_2—CH_2—CH_2— \quad \text{Butyl group (or } n\text{-butyl group)}$$

$$\xrightarrow{\text{remove one H}} CH_3—CH_2—CH—CH_3 \quad \textit{sec}\text{-Butyl group}$$

In the names given here, a continuous-chain (*normal*) structure with a hydrogen removed from one of the two equivalent terminal carbon atoms is written either without a prefix (for example, propyl) or with the prefix *n-* (for example, *n*-propyl). The prefix *sec-* stands for *secondary*; the *sec*-butyl group has a carbon atom with *two* other carbon-containing groups attached in addition to the bond that attaches it to the rest of the molecule. The prefix *tert-* stands for *tertiary*; in the *tert*-butyl group, the carbon bearing the bond for attachment to the rest of the molecule also contains three other carbon-containing groups. The prefix *iso* refers to branching at the next to last carbon on the chain. For example:

$$\begin{array}{cc}
\ce{CH3}\!\!\diagdown \\
\qquad\ce{CH-} \\
\ce{CH3}\!\!\diagup
\end{array}
\qquad
\begin{array}{c}
\ce{CH3}\!\!\diagdown \\
\qquad\ce{CH-CH2-} \\
\ce{CH3}\!\!\diagup
\end{array}$$

Isopropyl   Isobutyl

$$\begin{array}{c}
\ce{CH3}\!\!\diagdown \\
\qquad\ce{CH-CH2-CH2-} \\
\ce{CH3}\!\!\diagup
\end{array}
\qquad
\begin{array}{c}
\ce{CH3}\!\!\diagdown \\
\qquad\ce{CH-CH2-CH2-CH2-} \\
\ce{CH3}\!\!\diagup
\end{array}$$

Isopentyl   Isohexyl

By our definition of *sec-*, we would classify the isopropyl group as a secondary group. However, the term *isopropyl* is an exception to the general rule and must be memorized.

The various alkyl groups derived from propane and butane are used so commonly that you must memorize their names and structures. Beyond the isomeric butyl groups, the number of alkyl groups becomes very large and there is no prefix system for naming them.

Types of Carbon Atoms

The terms *methyl*, *primary* (*1°*), *secondary* (*2°*), and *tertiary* (*3°*) designate carbon atoms of a particular substitution pattern in organic chemistry. These designations are summarized here and discussed more fully in Sec. 4.1.

$$\ce{H-\underset{\displaystyle H}{\overset{\displaystyle H}{C}}-H} \qquad \ce{C-\underset{\displaystyle H}{\overset{\displaystyle H}{C}}-H} \qquad \ce{C-\underset{\displaystyle H}{\overset{\displaystyle H}{C}}-C} \qquad \ce{C-\underset{\displaystyle H}{\overset{\displaystyle C}{C}}-C}$$

Methyl   Primary carbon (1°)   Secondary carbon (2°)   Tertiary carbon (3°)

Question 3.7

Draw the structures of the eight alkyl groups that can be derived from the isomeric pentanes.

Step 4. If the same substituent occurs more than once, use the prefixes di, tri, tetra, and so forth to indicate how many of them are present. Precede them by a number (one number for each group present) to indicate which carbons in the parent chain they are attached to, for example, 2,2,3-trimethyl.

Step 5. If several substituents are present, arrange them either in alphabetical order or in order of increasing size or complexity. When alphabetizing the substituents, ignore prefixes that specify the number of a given type of substituent (di, tri, tetra, and so on) and any that are hyphenated (*n*-, *sec*-, *tert*-, and so forth). The prefixes *iso*, *neo*, and *cyclo* should not be ignored. For example, the letter to be used in alphabetizing the following substituents is shown in bold:

Di**m**ethyl-   **I**sopropyl-   **N**eopentyl-   *sec*-**B**utyl-

Alphabetical order is used in *Chemical Abstracts* and is the slightly preferred method; it is used in this text (listing in order of size, smallest to largest, is also correct). List all substituents, and preface each by a number to indicate where it is attached on the parent compound. Then add the name of the parent compound at the end. Omit the prefix *n*- from the parent name, since we have defined the parent compound as being the longest continuous chain in the molecule; to add *n*- would be redundant.

The following examples illustrate how to derive the IUPAC name from structure, as outlined by the preceding steps.

Example 1

Name:

$$CH_3{-}\overset{\overset{\displaystyle CH_3}{|}}{CH}{-}CH_2{-}CH_2{-}CH_3$$

Step 1. The longest continuous chain is boxed:

$$\boxed{CH_3{-}\overset{\overset{\displaystyle CH_3}{|}}{CH}{-}CH_2{-}CH_2{-}CH_3} \longleftarrow \text{Pentane (parent compound)}$$

The parent compound is a five-carbon continuous chain, so the parent name is *pentane*.
Step 2. Two possible ways of numbering the longest continuous chain are shown. The correct way has the substituent attached to C-2; the incorrect way has it attached to C-4.

$$CH_3{-}\overset{\overset{\displaystyle CH_3}{|}}{CH}{-}CH_2{-}CH_2{-}CH_3$$

Correct:	1	2	3	4	5
Incorrect:	5	4	3	2	1

Step 3. The substituent is the methyl group (CH_3) attached at C-2. It is called *2-methyl*.
Step 4. Not applicable.
Step 5. The parent name is *pentane* and there is a *2-methyl* substituent. Combine these to get the complete IUPAC name of **2-methylpentane.**

Note the (1) punctuation, (2) fact that *methyl* and *pentane* are run together, and (3) omission of *n*- before pentane.

Example 2

The following correct IUPAC names are derived as shown in Example 1. These are all monosubstituted compounds, and the longest continuous chain in each has been numbered correctly.

$$\overset{1}{CH_3}{-}\overset{2}{\underset{\underset{\displaystyle CH_3}{|}}{CH}}{-}\overset{3}{CH_3}$$

There is only one place where CH_3 group can be located—on C-2—and have a propane; if it were on C-1, we would have butane

IUPAC name: Methylpropane

$$CH_3-CH_2-\overset{2}{\underset{|}{CH}}-CH_3$$
$$\underset{Cl}{|}$$

with carbon numbering 4, 3, 2, 1 over $CH_3-CH_2-CH-CH_3$

IUPAC name: 2-Chlorobutane

$$CH_2-CH_3$$
$$\overset{4}{CH_2}-\overset{5}{CH}-\underset{|}{CH}-CH_3$$
$$\overset{3|}{CH_2}\quad \underset{6}{CH_2}-\overset{7}{CH_2}$$
$$\underset{2}{CH_2}-\underset{1}{CH_3}\quad \underset{8}{CH_2}-\underset{9}{CH_3}$$

IUPAC name: 5-*sec*-Butylnonane

Example 3

Name:

$$CH_3-CH_2 \qquad CH_3$$
$$CH-CH_2-\overset{|}{\underset{|}{C}}-CH_3$$
$$\underset{CH_3}{|} \qquad \underset{CH_3}{|}$$

Step 1. The longest continuous chain is boxed. We do *not* always take the longest horizontal chain (as previous examples may suggest) but must find the longest *continuous* chain:

$$CH_3-CH_2 \qquad CH_3$$
$$CH-CH_2-C-CH_3$$
$$CH_3 \qquad CH_3$$

The parent compound is a six-carbon continuous chain, so the parent name is *hexane*.
Step 2. Two ways to number the longest chain are shown. The correct way has the smaller substituent numbers:

Correct way: 2,2,   4

                             From two   From CH_3
                             CH_3 groups   group on
                             on C-2   C-4

Incorrect way: 3,5,5

$$\overset{6}{CH_3}-\overset{5}{CH_2} \qquad CH_3$$
$$\overset{4}{CH}-\overset{3}{CH_2}-\overset{2}{C}-\overset{1}{CH_3}$$
$$CH_3 \qquad CH_3$$

                      Correct

$$\overset{1}{CH_3}-\overset{2}{CH_2} \qquad CH_3$$
$$\overset{3}{CH}-\overset{4}{CH_2}-\overset{5}{C}-\overset{6}{CH_3}$$
$$CH_3 \qquad CH_3$$

                      Incorrect

Step 3. There are methyl groups at C-2, C-2, and C-4.
Step 4. Because there are three identical substituents (methyl groups) in the molecule, we call them *tri*methyl. To identify their points of attachment correctly, we use *2,2,4-trimethyl*. Note the punctuation—commas between numbers and a hyphen between number and name.
Step 5. Combine the parent name (*hexane*) and the names and positions of the substituents (*2,2,4-trimethyl*) to obtain the correct IUPAC name of **2,2,4-trimethylhexane.**

Example 4

Name:

$$CH_3-CH_2-CH_2-\overset{\overset{\displaystyle CH_3}{|}}{C}-CH_2-\overset{\overset{\displaystyle CH_3}{|}}{CH}-CH_2-CH_3$$

$$\overset{\overset{\displaystyle |}{CH}}{}-CH_2-\overset{\overset{\displaystyle |}{CH}}{}-CH_2-Cl$$

$$\overset{\overset{\displaystyle |}{CH_2}}{} \qquad\qquad \overset{\overset{\displaystyle |}{CH_3}}{}$$

$$\overset{\overset{\displaystyle |}{CH_3}}{}$$

Step 1. The longest continuous carbon chain is boxed. The parent name for a nine-carbon chain is *nonane*.

Step 2. The correct way to number the longest chain is shown (verify this as an exercise).

Steps 3, 4. We have methyl groups at positions 2, 5, and 7, which gives *2,5,7-trimethyl*, a chlorine atom at position 1, which is *1-chloro*, and an ethyl group on C-4, which is *4-ethyl*. Finally, we have a propyl group at position 5, which is *5-propyl*. Note that we must include prefixes such as *iso*, *sec-*, *tert-*, and *neo* when we name alkyl groups as *substituents*. In the case of the continuous-chain isomer, the *n-* is optional. Using propyl as an example, propyl (or *n*-propyl) ($CH_3CH_2CH_2-$) and isopropyl [$(CH_3)_2CH-$] have different points of attachment to the rest of the molecule; we indicate this by the correct prefix. Recall, however, that *we always omit n- from the parent names of the compound we are naming*.

Step 5. Combine the parent name (*nonane*) with the position-name combinations of the substituents (steps 3 and 4). We can arrange the substituents in alphabetical order (preferred) to yield the following correct IUPAC names. *In haloalkanes, the halogen always precedes the alkyl groups regardless of which convention is used.*

    Order of increasing size: **1-Chloro-2,5,7-trimethyl-4-ethyl-5-propylnonane**

    Alphabetical order (preferred):   **1-Chloro-4-ethyl-2,5,7-trimethyl-5-propylnonane**

Example 5

Draw the structure for 1-bromo-3,3-diethyl-4-isobutylnonane. Analyze the name completely and then break it apart in the following manner:

1-bromo-3,3-diethyl-4-isobutylnonane

$\uparrow$ $\quad$ $\nearrow$ $\uparrow$ $\quad$ $\uparrow$ $\qquad$ $\uparrow$

| Br on C-1 | 2 CH_3CH_2 Two CH_3CH_2- groups on C-3 | CH_3 $\backslash$ $CH-CH_2-$ CH_3 $/$ group on C-4 | Nine-carbon parent alkane |

Next draw a nine-carbon skelton (the parent chain) and number it:

$$\underset{C}{\overset{1}{}}-\underset{C}{\overset{2}{}}-\underset{C}{\overset{3}{}}-\underset{C}{\overset{4}{}}-\underset{C}{\overset{5}{}}-\underset{C}{\overset{6}{}}-\underset{C}{\overset{7}{}}-\underset{C}{\overset{8}{}}-\underset{C}{\overset{9}{}}$$

Then put the indicated substituents on the correct carbon atoms:

$$CH_2CH_3$$
$$\qquad\qquad |$$
$$Br-\overset{1}{C}-\overset{2}{C}-\overset{3}{C}\text{———}\overset{4}{C}-\overset{5}{C}-\overset{6}{C}-\overset{7}{C}-\overset{8}{C}-\overset{9}{C}$$
$$\qquad\qquad | \qquad\qquad |$$
$$\qquad CH_2CH_3 \quad CH_2$$
$$\qquad\qquad\qquad\qquad |$$
$$\qquad\qquad\qquad CH_3-CH-CH_3$$

Finally, put as many hydrogens as needed around each carbon atom to satisfy its valence requirement (that is, four bonds about each carbon atom).

The correct structure is

$$CH_2-CH_3$$
$$\qquad\qquad\qquad |$$
$$Br-CH_2-CH_2-C\text{———}CH-CH_2-CH_2-CH_2-CH_2-CH_3$$
$$\qquad\qquad\qquad | \qquad\quad |$$
$$\qquad\qquad CH_2-CH_3 \quad CH_2$$
$$\qquad\qquad\qquad\qquad\quad |$$
$$\qquad\qquad\qquad CH_3-CH-CH_3$$

Example 6

Name:

$$\underset{CH_3}{\overset{1}{}}-\underset{CH_2}{\overset{2}{}}-\underset{CH_2}{\overset{3}{}}-\underset{CH_2}{\overset{4}{}}-\underset{CH_2}{\overset{5}{}}-\underset{CH}{\overset{6}{}}-\underset{CH_2}{\overset{7}{}}-\underset{CH_2}{\overset{8}{}}-\underset{CH_2}{\overset{9}{}}-\underset{CH_2}{\overset{10}{}}-\underset{CH_2}{\overset{11}{}}-\underset{CH_2}{\overset{12}{}}-\underset{CH_3}{\overset{13}{}}$$
$$\qquad\qquad\qquad\qquad\qquad\qquad |$$
$$\qquad\qquad\qquad\qquad\qquad\quad CH_2$$
$$\qquad\qquad\qquad\qquad\qquad\qquad |$$
$$\qquad\qquad\qquad\qquad\quad Cl-C-CH_2CH_3$$
$$\qquad\qquad\qquad\qquad\qquad\qquad |$$
$$\qquad\qquad\qquad\qquad\qquad CH-CH_3$$
$$\qquad\qquad\qquad\qquad\qquad\qquad |$$
$$\qquad\qquad\qquad\qquad\qquad\quad CH_3$$

We have a 13-carbon continuous chain as the parent compound, which is *tridecane*. The side-chain substituent at C-6 is rather complex and one for which we have no simple name. First derive the name for it. The longest straight chain in it, which also includes the carbon attached to the main chain, contains four carbon atoms and will be named as a derivative of butane. Number this chain starting with C-1 at the point where it is attached to tridecane. Use numbers to indicate where the substituents are attached on this butane chain:

$$
\begin{array}{c}
\mid \\
^1CH_2 \\
\mid \\
Cl\!-\!^2C\!-\!CH_2CH_3 \\
\mid \\
^3CH\!-\!CH_3 \\
\mid \\
^4CH_3
\end{array}
$$

2-Chloro-2-ethyl-3-methylbutyl group

Note two things about this name: (1) the parent name, but*ane*, is changed to but*yl* because this is an alkyl side chain, and (2) the -1- has been omitted from in front of butyl because we defined the numbering system to start with that carbon atom and use the longest carbon chain within the side chain.

Now combine the name of the alkyl side chain with that of the parent compound to give the correct IUPAC name as **6-(2-chloro-2-ethyl-3-methylbutyl)tridecane.** Note that the entire name of the side-chain alkyl group is enclosed in parentheses; the numbers in parentheses refer to the longest carbon chain in that alkyl group. The number preceding the parentheses indicates that the alkyl group is attached to C-6 in tridecane.

Rules for naming more complex molecules can be found in the Nomenclature section of the Reading References for Chap. 3 in the appendix.

C. Prefixes *n-*, *iso*, *sec-*, *tert-*, and *neo*; Common Names of Alkanes

In the common system of nomenclature, the prefix *n-* is used whenever all the carbon atoms in a molecule form a continuous chain and also for unsubstituted alkanes:

$$CH_3(CH_2)_6CH_3$$

n-Octane

It can be used to indicate alkyl groups of any size as long as the carbon atoms are in a continuous chain and there is only one substituent attached on one of the terminal carbon atoms. Note also that in the common system the substituent name follows the name of the parent compound, which is just the opposite of the IUPAC system.

$$CH_3\!-\!I \qquad CH_3CH_2CH_2CH_2\!-\!Cl \qquad CH_3(CH_2)_7CH_2\!-\!Br$$

Methyl iodide   *n*-Butyl chloride   *n*-Nonyl bromide
(no *n-* needed)

The prefix *iso* indicates a $CH_3\!-\!\overset{|}{CH}\!-\!CH_3$ group in a molecule. We saw it in the isopropyl group. More generally, in the common system, *iso* can be used to name alkanes with six carbon atoms or less that contain this group. For example:

$$
\begin{array}{c}
CH_3 \\
\diagdown \\
CH\!-\!CH_3 \\
\diagup \\
CH_3
\end{array}
\qquad
\begin{array}{c}
CH_3 \\
\diagdown \\
CH\!-\!CH_2\!-\!CH_2\!-\!CH_3 \\
\diagup \\
CH_3
\end{array}
\qquad
\begin{array}{c}
\text{Note presence of} \\
CH_3 \\
\diagdown \\
CH\!- \\
\diagup \\
CH_3 \\
\text{shown by} \\
\text{dashed lines}
\end{array}
$$

Isobutane   Isohexane

This same prefix can be used to name alkyl groups so long as the point of attachment to another molecule is at the end opposite from where the $(CH_3)_2CH-$ group is attached:

$$CH_3 \diagdown$$
$$CH-CH_2-Cl \qquad Br-CH_2-CH_2-CH_2-CH \diagup^{CH_3}_{\diagdown CH_3}$$
$$CH_3 \diagup$$

Isobutyl chloride   Isohexyl bromide

If the branch or the substituent appears anywhere else in the molecule, this prefix cannot be used in the name. Note that the parent names (*but* in iso*but*yl chloride and *hex* in iso*hex*yl bromide) indicate the total number of carbons in those molecules. No hyphen is written between *iso* and the name of the alkyl group, but there is a hyphen after *n-*, *sec-*, and *tert-*.

The use of the *sec-* and *tert-* prefixes is outlined in Sec. 3.10B. The following examples further illustrate their use in the common system:

$$CH_3-\underset{\underset{Br}{|}}{CH}-CH_2-CH_3 \qquad CH_3-\underset{\underset{Cl}{|}}{\overset{\overset{CH_3}{|}}{C}}-CH_3 \qquad CH_3-\underset{\underset{OH}{|}}{\overset{\overset{CH_3}{|}}{C}}-CH_3$$

sec-Butyl bromide   *tert*-Butyl chloride   *tert*-Butyl alcohol

The *sec-* prefix is seldom used in naming hydrocarbons larger than butane. For example, examine *sec*-pentyl iodide, for which there are two possible structures. (Draw them.) The use of *sec-* in this instance is ambiguous and should be avoided.

Another prefix, one not often used, is *neo*, which indicates the presence of the

$$CH_3-\underset{\underset{CH_3}{|}}{\overset{\overset{CH_3}{|}}{C}}-CH_2-$$ grouping in a molecule. The following examples illustrate the

use of the prefix *neo* in the names of alkanes and monosubstituted alkyl compounds. No hyphen separates *neo* from the rest of the name.

$$CH_3-\underset{\underset{CH_3}{|}}{\overset{\overset{CH_3}{|}}{C}}-CH_3 \quad CH_3-\underset{\underset{CH_3}{|}}{\overset{\overset{CH_3}{|}}{C}}-CH_2-Cl \quad CH_3-\underset{\underset{CH_3}{|}}{\overset{\overset{CH_3}{|}}{C}}-CH_2-CH_3 \quad CH_3-\underset{\underset{CH_3}{|}}{\overset{\overset{CH_3}{|}}{C}}-CH_2-CH_2-Cl$$

Neopentane   Neopentyl chloride   Neohexane   Neohexyl chloride

Observe the characteristic group, $(CH_3)_3C$, in each molecule, as shown in the dashed lines. The *neo* prefix can be used only when there is a substituent on the carbon atom at the end opposite from this group.

Common Versus IUPAC Names

Confusion may arise between common and IUPAC names. For example, methyl chloride is a common name, and chloromethane is the IUPAC name for CH_3-Cl, yet methylpropane is an IUPAC name. How can an alkyl group (*methyl*, for example) be used in both systems? When all the carbons in a molecule are named as an alkyl group, a common name results. When simple carbon groupings are named as alkyl group substituents (that is, they are attached to a longer continuous chain, which is the parent compound), then they usually become part of the IUPAC system. The use of alkyl groups in common names is quite limited because those with five or more carbon atoms are usually too complex to have common names. Finally, be very careful not to mix the two systems because the result is names that are not correct in either, for example, 2-methylisobutane.

Question 3.8

Provide IUPAC names for the (*a*) four isomeric bromobutanes (C_4H_9Br), (*b*) three isomeric pentanes, (*c*) five isomeric hexanes, (*d*) nine isomeric heptanes, and (*e*) eight isomeric chloro-pentanes ($C_5H_{11}Cl$).

Question 3.9

Draw structures corresponding to the following names:

(*a*) 2,2-dimethylheptane   (*b*) 4-isopropyloctane
(*c*) hexadecane   (*d*) 2,2-dichloro-3,3-dimethylhexane
(*e*) 2,4-dibromo-3-(chloromethyl)pentane

3.11   Physical Properties of Alkanes

We now examine certain physical properties, such as boiling point, melting point, and density, for the alkane family. The pertinent data for selected alkanes are given in Table 3.4.

A. Boiling Points and van der Waals Forces

Alkane hydrocarbons are essentially nonpolar molecules. We anticipate no polarity associated with the carbon-carbon bonds and only slight polarity with the carbon-hydrogen bonds because of the small difference in electronegativity between carbon and hydrogen (see Sec. 2.10). As a consequence of the tetrahedral structure of carbon, any individual dipoles that might exist in alkanes cancel out.

There must be some type of interaction between alkane molecules because we know that alkanes liquefy and solidify. These forces are called **van der Waals forces.** We consider only the qualitative effects of van der Waals forces. The electrons in a covalent bond are in constant movement, and if the electron distribution were "frozen" at any moment, it would be *delocalized* (or dispersed) to some extent. It would behave like a small dipole if the electrons were nearer one atom than the other. A nearby molecule can behave similarly and have its own small dipole. When these two molecules are adjacent, their dipoles interact, with the positive end of one attracted by the negative end of the other. Also, a dipole in one molecule can affect the electron distribution in a neighboring molecule and *induce* a dipole to exist in the latter. The end result is the same: molecular attraction by **dipole-dipole interaction.** A rough diagram of this type of interaction is shown for an alkane:

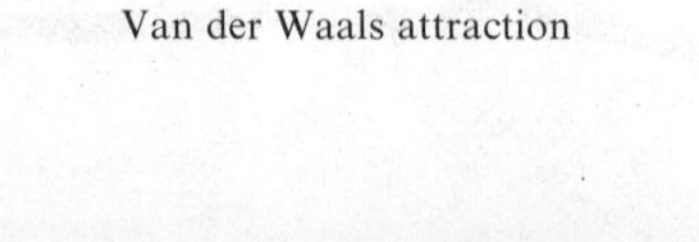

Van der Waals attraction

TABLE 3.4   Melting and Boiling Points and Densities for Selected _n_-Alkanes

Name	Formula	bp, °C	mp, °C	Density (g/cc) at 20°, d^{20}*	
Methane	CH_4	−161.7	−182.6	0.415	
Ethane	C_2H_6	−88.6	−172.0	0.572	Gases
Propane	C_3H_8	−42.2	−187.1	0.585	
_n_-Butane	C_4H_{10}	−0.5	−135.0	0.601	
_n_-Pentane	C_5H_{12}	36.1	−129.7	0.626	
_n_-Hexane	C_6H_{14}	68.7	−94.0	0.659	
_n_-Heptane	C_7H_{16}	98.4	−90.5	0.684	
_n_-Octane	C_8H_{18}	125.6	−56.8	0.703	
_n_-Nonane	C_9H_{20}	150.7	−53.7	0.718	
_n_-Decane	$C_{10}H_{22}$	174.0	−29.7	0.730	
_n_-Undecane	$C_{11}H_{24}$	195.8	−25.6	0.740	
_n_-Dodecane	$C_{12}H_{26}$	216.3	−9.6	0.749	Liquids
_n_-Tridecane	$C_{13}H_{28}$	230.0	−6.0	0.757	
_n_-Tetradecane	$C_{14}H_{30}$	251.0	5.5	0.764	
_n_-Pentadecane	$C_{15}H_{32}$	268.0	10.0	0.769	
_n_-Hexadecane	$C_{16}H_{34}$	280.0	18.1	0.775	
_n_-Heptadecane	$C_{17}H_{36}$	303.0	22.0	0.777	
_n_-Octadecane	$C_{18}H_{38}$	308.0	28.0	0.777	
_n_-Nonadecane	$C_{19}H_{40}$	330.0	32.0	0.778	
_n_-Eicosane	$C_{20}H_{42}$		36.4	0.778	
_n_-Heneicosane	$C_{21}H_{44}$		40.4	0.778	
_n_-Docosane	$C_{22}H_{46}$		44.4	0.778	
_n_-Tricosane	$C_{23}H_{48}$		47.4	0.780	
_n_-Tetracosane	$C_{24}H_{50}$		51.1	0.779	
_n_-Pentacosane	$C_{25}H_{52}$		53.3		Solids
_n_-Triacontane	$C_{30}H_{62}$		66.0		
_n_-Pentatriacontane	$C_{35}H_{72}$		74.6		
_n_-Tetracontane	$C_{40}H_{82}$		81.0		
_n_-Pentacontane	$C_{50}H_{102}$		92.0		
_n_-Hexacontane	$C_{60}H_{122}$		99.0		
_n_-Heptacontane	$C_{70}H_{142}$		105.0		

* Defined on p. 95, Sec. 3.11C.

The van der Waals forces are effective only over a very short range, so that interaction occurs only when molecules are very close to one another. In effect, it is a surface interaction. Remember, however, that the momentary dipoles and thus the dipole interactions are constantly changing; we see only the final result, which is a _weak net attraction_. Van der Waals forces have been used to explain the compressibility and liquefaction of inert gases such as helium and argon.

The homologous series of alkanes is an excellent example of the additive nature of van der Waals forces. As the data in Table 3.4 indicate, the boiling points of alkanes with more than four carbons increase by 20 to 30° for each additional —CH_2— group that is added to the molecule. The larger the molecule, the greater the surface area, and hence the greater the intermolecular van der Waals forces. This means that more energy is required to overcome those forces before the alkane molecules in the liquid phase can become completely separated and exist in gaseous

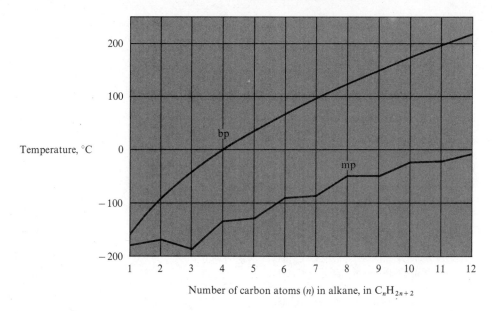

FIGURE 3.10   Plot of boiling point (bp) and melting point (mp) as a function of carbon content for alkanes.

forms. Boiling points give some indication of the relative amounts of energy required to overcome van der Waals forces.

Figure 3.10 shows a plot of boiling point versus carbon content for some alkanes. The boiling points of the structural isomers with a given carbon content are irregular, and their differences do not fit a pattern. Table 3.1 gives the boiling points for the isomeric pentanes and hexanes. Note that *the branched-chain isomers all have lower boiling points than the corresponding n-alkanes* and that the boiling point is lowered as the number of branches increases. These trends have been attributed to van der Waals forces, which are less effective as the shape of the molecule becomes more spherical and symmetrical through branching. Put another way, there are more sites where van der Waals forces can interact in a molecule like pentane than in a molecule like neopentane.

$$
\begin{array}{cc}
\text{Van der Waals} & \text{Van der Waals} \\
\text{interactions in pentane} & \text{interactions in neopentane}
\end{array}
$$

B. Melting Points

Tables 3.1 and 3.4 list the melting points of various alkanes. These are very useful identifying properties of organic molecules and bear no direct relationship to the boiling points; note that there is considerable variation from member to member and among structural isomers. The important general trend is that *the more symmetrical isomers tend to have higher melting points*. Melting point depends not only on the intermolecular forces but also on how well the molecules fit together in the crystalline form. For example, neopentane melts at $-16.6°$ and pentane melts at $-129.7°$. The complete symmetry in neopentane makes it look like a sphere and causes it to pack well in the crystalline lattice; thus it freezes (or melts) at a high temperature, which means that more energy is required to break apart its crystalline structure.

C. Solubility

Alkanes are insoluble in water and highly polar solvents. The adage "like dissolves like" applies to them for they are soluble in nonpolar organic solvents (such as cyclohexane) and in slightly polar solvents. They can be used as solvents themselves to dissolve other compounds of low polarity.

D. Density

The densities of the alkanes are listed in Table 3.4. Density is given in grams per cubic centimeter and is usually measured at $20°$. It is given the symbol d^{20}, where the superscript indicates the temperature at which the density is determined.

The densities start near 0.4 and increase steadily but gradually to a value of about 0.8, where they tend to level out. All alkanes, and indeed most organic compounds, are less dense than water. The addition of halogens to compounds increases their density greatly.

 Question 3.10

Without referring to Table 3.4, predict the approximate boiling point of dodecane, given that the boiling point of undecane is about $196°$.

Cycloalkanes: Structure, Properties, and Nomenclature

We examined the important structural aspects of alkanes in which the carbon atoms are attached to form open chains—both continuous and branched. These compounds have the common feature of possessing tetrahedral sp^3 hybridized carbon atoms and the general molecular formula C_nH_{2n+2}. We now turn to another family of compounds that also have sp^3 hybridized carbon atoms but with the carbon atoms arranged so that they form *rings*. They are called **cyclic** compounds or **cycloalkanes.** We consider their structure, properties, and nomenclature in this section; their chemistry is discussed in Chap. 4.

3.12   General Structure and Nomenclature

Cycloalkanes have the general molecular formula C_nH_{2n}. Their relationship to alkanes can be understood by realizing that the elements of two hydrogen atoms must be removed from an alkane to form a cycloalkane:

$$
\begin{array}{ccc}
\underset{\displaystyle CH_2 \quad \quad CH_2}{\overset{\displaystyle CH_2-CH_2}{\underset{|\quad\quad\quad\quad|}{CH_2-H \quad H-CH_2}}} & \xrightarrow[\text{hydrogen atoms}]{\text{removal of two}} & \underset{\displaystyle CH_2 \quad \quad CH_2}{\overset{\displaystyle CH_2-CH_2}{CH_2-CH_2}}
\end{array}
$$

$$
\begin{array}{ccc}
\text{Hexane} & & \text{Cyclohexane} \\
C_6H_{14} & \xrightarrow{\;-2H\;} & C_6H_{12}
\end{array}
$$

The result is a new carbon-carbon bond and a cyclic structure. Note, however, that hexane has all tetrahedral sp^3 hybridized carbon atoms and we might anticipate having tetrahedral sp^3 hybridized carbon atoms in cyclohexane, which indeed we do. Even though alkanes and cycloalkanes have different molecular formulas, C_nH_{2n+2} and C_nH_{2n}, respectively, they might be expected to have similar physical and chemical properties because of the tetrahedral carbon atom; we find this to be true also.

Cycloalkanes are named by placing the prefix **cyclo** in front of the name of the alkane that has the same number of carbon atoms. Some typical cyclic structures follow, both with all the carbon and hydrogen atoms shown and in the more conventional form as simple *line structures* (that is, a triangle to represent cyclopropane, a square for cyclobutane, a pentagon for cyclopentane, a hexagon for cyclohexane, and so forth).

Cyclopropane   Cyclobutane

Cyclopentane   Cyclohexane

When the line structures are used, it is assumed that one carbon and two hydrogens (that is, CH_2) are located at each corner unless substituents are attached.

Substituted cycloalkanes are usually named as a derivative of the cycloalkane itself. As with alkanes, cycloalkyl names can also be used. Examples of *monosubstituted cycloalkanes* are:

Bromocyclobutane
(Cyclobutyl bromide)

3-Cyclohexylhexane

This has been named as a derivative of hexane; cyclohex*ane* name for the six-membered ring has been changed to cyclohex*yl* and treated as an alkyl—in this case, cycloalkyl—substituent.

Disubstituted and other more highly substituted cycloalkanes are almost always named as a derivative of the cycloalkane. IUPAC nomenclature requires that the cyclic structure be numbered so that the substituents are given the lowest possible numbers. This approach is the same as for alkanes (see Sec. 3.10), as the following examples illustrate:

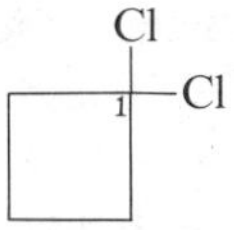

1,1-Dichloro-
cyclobutane

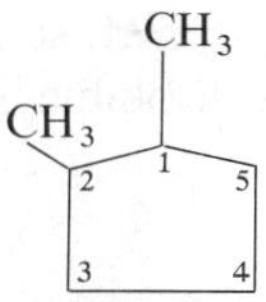

1,2-Dimethylcyclo-
pentane

and *not*

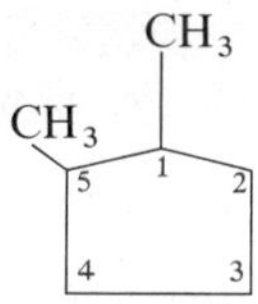

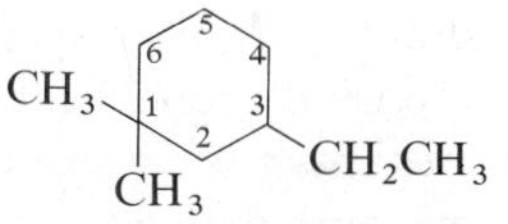

3-Ethyl-1,1-dimethylcyclohexane

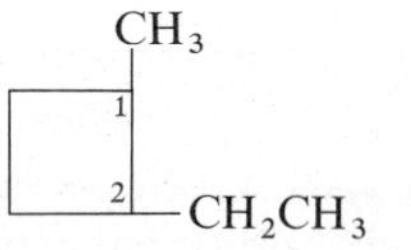

2-Ethyl-1-methylcyclobutane

Question 3.11

Draw structures corresponding to the following names:

(*a*) 1-bromo-3-chlorocyclobutane (*b*) 1-ethyl-2-isopropylcyclopropane
(*c*) cyclopentylcyclohexane (*d*) 1,2,4-tribromocyclohexane
(*e*) methylcyclononane (*f*) 1-ethyl-1-isobutylcyclopentane

3.13   Physical Properties of Cycloalkanes

Certain physical properties for the cycloalkanes are listed in Table 3.5. Compare them with those for the alkanes (Tables 3.1 and 3.4) and note that the boiling points, melting points, and densities are greater for the cycloalkanes than for alkanes that contain the same number of carbon atoms; for example, propane (bp −42.2, mp −187.1) versus cyclopropane (bp −33, mp −127) and hexane (bp 68.7, mp −94.0) versus cyclohexane (bp 81, mp 6.4). Symmetry must play an important role for the cycloalkanes as it does for the alkanes (see Sec. 3.11 and Table 3.1). Of course, unsubstituted cyclic hydrocarbons are highly symmetrical and should pack well in the crystalline lattice.

**TABLE 3.5   Melting and Boiling Points and Densities
for Selected Cycloalkanes**

Name	bp, °C	mp, °C	Density (g/cc) at 20° (as liquids)
Cyclopropane	−33	−127	0.688
Cyclobutane	11	−80	0.704
Cyclopentane	49	−94	0.746
Cyclohexane	81	6.4	0.778
Cycloheptane	117	−13	0.810
Cyclooctane	147	14	0.830
Methylcyclopentane	72	−142	0.749
Methylcyclohexane	100	−126	0.769

The cycloalkanes are mostly nonpolar, so they readily dissolve in nonpolar or weakly polar organic solvents but are insoluble in highly polar solvents (for example, water).

3.14   Structural Properties of Unsubstituted Cycloalkanes

In this section we discuss some structural features of cyclopropane, cyclobutane, cyclopentane, and cyclohexane. The important conclusions of the **Baeyer strain theory,** which was developed in 1885 and deals with the structure of cyclic compounds, are presented. Additional experimental evidence derived from the chemistry of these compounds will be presented in support of the Baeyer strain theory: (1) abnormal chemical reactivity (see Secs. 8.3, 8.5, 8.12, and 8.16) and (2) heats of combustion (see Sec. 3.15).

Start by recalling that open-chain alkanes have tetrahedral sp^3 hybridized carbon atoms. What changes in bond angles must occur in converting an alkane into the corresponding cycloalkane with the same number of carbon atoms in the ring? Let us see.

A. Cyclopropane

Cyclopropane is a three-membered ring and it must be triangular. Each of the three bond angles is 60°. To form this molecule from propane, we have to compress the bond angles from 109.5° to 60°, which causes cyclopropane to be *highly strained* and unstable compared with molecules in which the bond angles are 109.5°. In other words, many reactions in which the cyclopropane ring breaks apart to regain the tetrahedral structure about the carbon atoms are energetically favorable. Ring-opening reactions for cyclopropane are well known (see Secs. 8.3, 8.5, 8.12, and 8.16). Cyclopropane is the most strained cyclic compound, since its bond angles have been compressed by 49.5° (109.5° − 60°).

The structure of cyclopropane has been studied theoretically (using molecular orbital theory), and the bonding orbitals that form the carbon-carbon bonds are shown in Fig. 3.11. It is not surprising that this type of electronic structure exists because the carbon-carbon σ bonds are forced very close together in the strained-ring structure and they repel one another through electrostatic repulsion. It is more stable for the bonding orbitals to project outward from the molecule rather than to be close to one another. The bond is bent (often referred to as a "banana" bond) because the sp^3 hybridized orbitals from the individual carbon atoms do not directly overlap. This view is further supported by the observed H—C—H bond angles of 114°, which is greater than the expected tetrahedral angle of 109.5°.

Because the C—C molecular bond orbitals are not symmetrical about a line passing through the carbon nuclei, we speak of both the *internuclear* C—C—C bond angle, which is 60°, and the *interorbital* angle (the angle between the actual orbitals), which is 105°.

B. Cyclobutane

Cyclobutane, with a shape that resembles a square, has bond angles of about 90°. Its bonds have been compressed by about 19.5° (109.5° − 90°), and it is still quite strained but much less so than cyclopropane.

The picture of cyclobutane is incomplete. Theoretical considerations and experimental data indicate that some strain in the molecule can be relieved if it

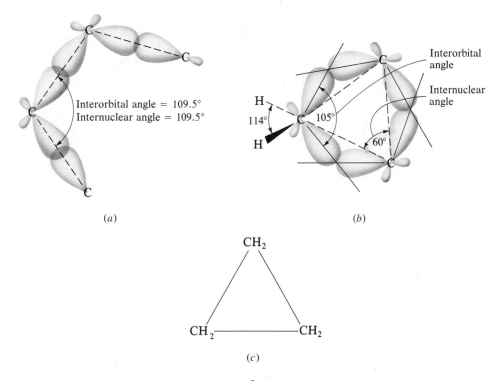

Interorbital angle = 109.5°
Internuclear angle = 109.5°

(a)

Interorbital angle

Internuclear angle

H
114°   105°
H
60°

(b)

CH₂

CH₂————CH₂

(c)

FIGURE 3.11   Orbital views for overlap of sp^3 hybridized orbitals used to form carbon-carbon bonds in (a) open-chain or large ring alkanes, (b) the strained cyclopropane molecule, and (c) two-dimensional representation. In (b) the bond orbitals are forced away from one another by electro-static repulsion.

adopts a "bent" envelope type of structure. The molecule is not planar but has the conformations shown in Fig. 3.12. The H—C—H bond angles have been found to be 112°, and the molecule appears to be bent by about 25°.

Later discussions will reveal the abnormal chemical reactivity of the strained cyclobutane molecule (see Sec. 8.12).

C. Cyclopentane

The next member in the cyclic series is **cyclopentane,** which is easily synthesized and quite stable. Assuming a completely planar structure, Baeyer originally reasoned that the internal C—C—C bond angles should equal those for a pentagon, or be 108°. This is very close to the tetrahedral angle of 109.5°, so it was thought that the cyclopentane molecule should be essentially strain-free. The Baeyer strain theory assumed that all the cycloalkanes were planar. Baeyer reasoned that the closer the C—C—C bond angles of the cycloalkanes were to 109.5°, the more stable the molecule. Thus he postulated that cyclopropane and cyclobutane should be unstable. As shown, his theory appeared to work quite well. Cyclopentane, which is quite stable chemically, further substantiated his theory.

The most stable conformation for cyclopentane, however, is not the planar structure, which would eclipse all the C—H bonds in the molecule. Recall, for example, the relative instability associated with the eclipsed conformation of ethane (see Sec. 3.3). Application of the torsional strain theory to cyclopentane suggests that

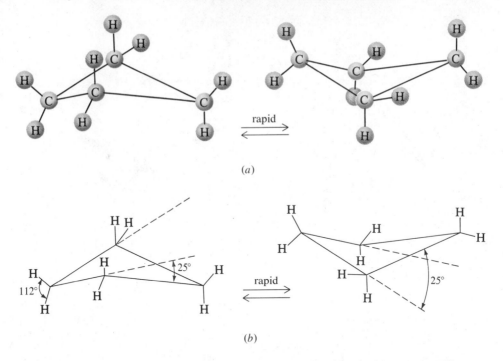

FIGURE 3.12   Conformations of cyclobutane. (*a*) Drawing of ball-and-stick structure; (*b*) paper representation of conformations, showing bond angles.

about 15 kcal/mole (63 kJ/mole) of torsional energy [five sets of torsional interactions × 3 kcal/mole (12.6 kJ/mole) per set] is associated with the planar cyclopentane conformation. This appears to be enough energy to distort the cyclopentane ring from the planar to a *puckered* conformation, where the C—C—C bond angles are about 105°; that is, some torsional energy is relieved even though more strained bonds are formed. The net strain in the puckered form is lower than in the planar form. This puckering moves rapidly around the ring due to the flexibility available to the molecule. The planar and puckered conformations of cyclopentane are shown in Fig. 3.13.

D. Cyclohexane

In the cycloalkanes discussed thus far we saw increasing stability and less strain in going from cyclopropane to cyclobutane to cyclopentane, with the latter being close to strain-free because its bond angles are nearly tetrahedral. These observations and explanations agree with the Baeyer strain theory. This is merely coincidental, however, since Baeyer's theory did not explain the puckering in cyclobutane and cyclopentane, which is an integral part of our modern view of the molecular structure of these two compounds.

The bond angles in **cyclohexane,** if it is planar, should be the same as those in a regular hexagon, namely 120°, which is somewhat larger than the normal 109.5° tetrahedral angle. Baeyer incorrectly proposed that cyclohexane should be a strained molecule because the bond angles had been *expanded* from 109.5° to 120°, whereas in cyclopropane and cyclobutane, which are also strained, they had been *compressed*. (Indeed, he incorrectly extended his theory to cycloheptane and other larger cycloalkanes in which the bond angles become larger and therefore should be even more unstable.)

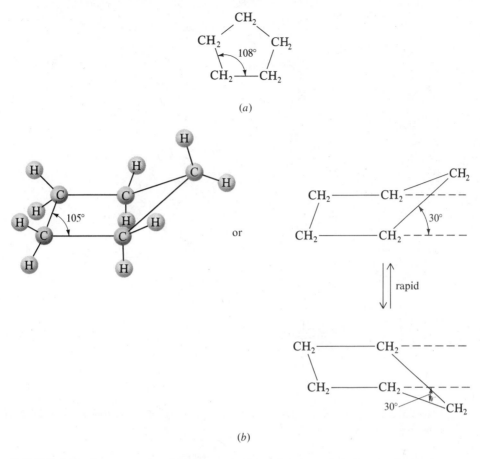

FIGURE 3.13 Structure of cyclopentane. (*a*) Planar representation of molecule (all hydrogens eclipsed); (*b*) puckered structure (hydrogens staggered), shown by ball-and-stick drawing and paper representation.

Cyclohexane, however, is readily synthesized and is a stable and well-known compound. It does not undergo any ring-opening reactions and does not appear to possess any unusual reactivity. It was at this point that the Baeyer strain theory broke down. Of course, he can not be faulted for this because the chemists in his time (*ca.* 1885) did not have the sophisticated instruments we now use to determine molecular structure and other physical properties of molecules. It is accepted today that cyclohexane is not a flat, planar molecule; the ring is puckered to such an extent that all C—C—C bond angles can be approximately 109.5° (they are actually 111.5°). Thus, we do not expect it to be unstable.

The study of the strain-free structures of cyclohexane represents another example of the field of **conformational analysis.** Our previous discussions (see Sec. 3.3, 3.4, and 3.6) of the conformers of ethane, propane, and butane are also part of this field of study. The six-membered ring is the simplest cycloalkane to study, both experimentally and theoretically.

We now examine the types of structures used to explain why it is strain-free. A conformation can be constructed for cyclohexane in which all the bond angles are 109.5° and all sets of bonds are in the staggered form to minimize torsional strain. This is shown in Fig. 3.14 and is called the **chair conformation** or **chair form** of cyclohexane because it resembles a lawn chair. Ball-and-stick drawings illustrate fairly

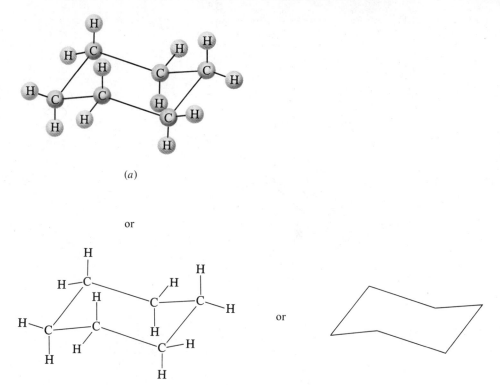

(a)

or

(b)

FIGURE 3.14   The chair conformation of cyclohexane. (*a*) **Ball-and-stick representation;** (*b*) **two paper representations.**

accurately the geometry of the molecule, and the paper representations are the common ways it is drawn. The use of molecular models is highly advised, for they enable you to see more details about the chair conformation and aid you considerably when you draw it.

   If we sight down each carbon-carbon bond in cyclohexane, we notice staggered bonds. These are analogous to those found in gauche butane (see Sec. 3.6) and are shown in Fig. 3.15. Thus, the chair conformation is free of angle strain and torsional strain. It is the most stable conformation of cyclohexane.

   Special terminology is used to identify the hydrogen atoms attached to the cyclohexane molecule. The six hydrogens that point radially outward from the ring (and are horizontally positioned) are called **equatorial** hydrogens. The three hydrogens that point vertically upward and the three that point vertically downward from the ring are called **axial** hydrogens. (To help remember these terms, think of the equatorial hydrogens as pointing toward the equator and the axial ones as pointing toward the axis of the earth.) These are abbreviated *e* for *equatorial* and *a* for *axial* in the following chair form:

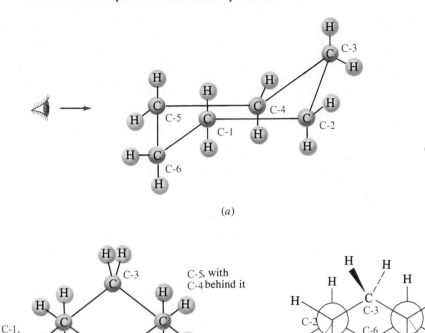

FIGURE 3.15 Derivation of Newman projection formula for the chair conformation of cyclohexane. Note in (*b*) and (*c*) that the staggered conformation is present. (*a*) Side view of ball-and-stick structure; (*b*) end view of ball-and-stick structure as viewed from direction shown in (*a*); (*c*) Newman projection formula corresponding to and drawn from end view (*b*).

Another conformation available to cyclohexane is the **boat conformation** or **boat form**, so called because it resembles the general shape of a boat. This conformation is obtained by starting with the chair conformation and "flipping" one end of the molecule in the manner shown in Fig. 3.16. This process involves some rotation about the carbon-carbon single bonds and results in a changed location of certain hydrogen atoms. (Molecular models are a great way to help you see this.) The boat form is much less stable than the chair form [an estimated 6.9 kcal/mole (28.9 kJ/mole) less stable] for two reasons. First, several bonds in the molecule are eclipsed with one another, as shown in the following Newman projection formula:

Boat cyclohexane: Newman projection

We would expect this conformation to possess considerable torsional energy (roughly twice that of a butane molecule in the eclipsed conformation). Second, steric

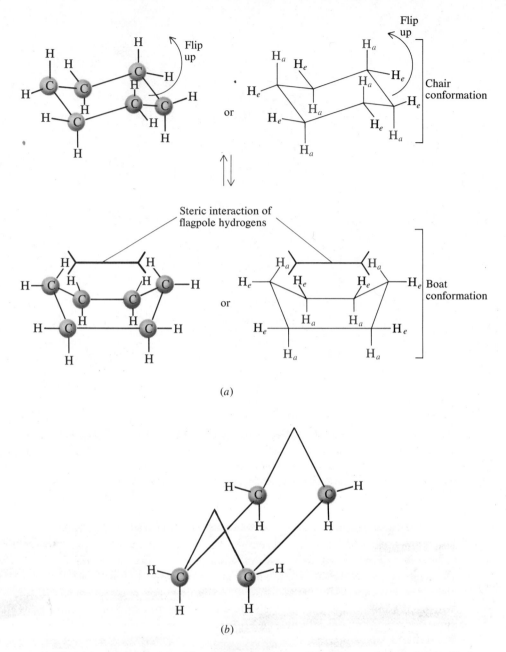

FIGURE 3.16   (a) Conversion of the chair conformation of cyclohexane into the less stable boat conformation, as shown from a side view. The boat conformation is less stable than the chair conformation by about 6.9 kcal/mole. Note "flagpole" hydrogens and their steric interaction. (b) End view of boat conformation; note the eclipsed bonds.

interactions of great magnitude are generated in the boat form when certain hydrogen atoms are brought near one another, as shown in Fig. 3.16. These crowded hydrogen atoms are called *flagpole hydrogens*. The boat form is so unstable relative to the chair conformation (and others mentioned later) that it is felt to be only a transition between the other more stable conformations; that is, the following type of equilibrium exists for cyclohexane:

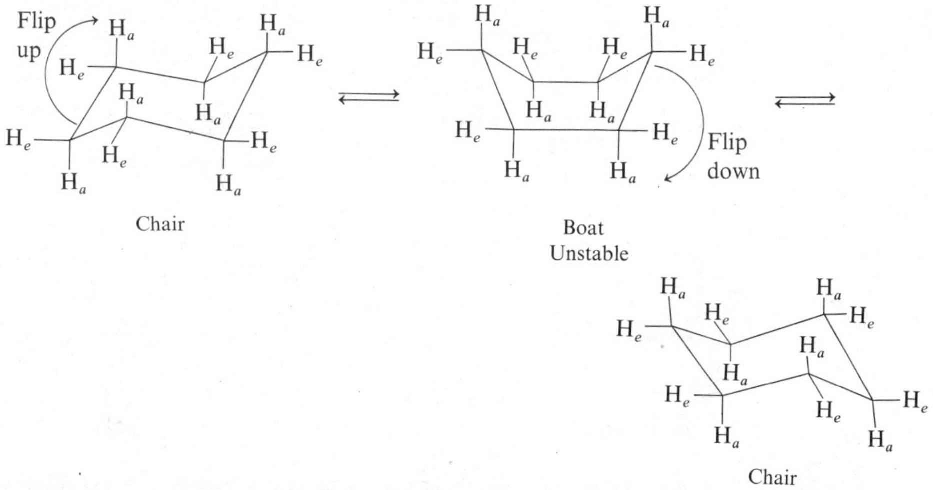

Chair

Boat
Unstable

Chair

We have considered only two possible conformations for cyclohexane, the stable chair form and the unstable boat form. As with all conformers, there are a myriad of intermediate conformational structures available to cyclohexane. Two of them are commonly referred to as the **half-chair form** and **twist form.** Theoretical calculations have estimated the following energies associated with these various structures.

Conformations of cyclohexane:

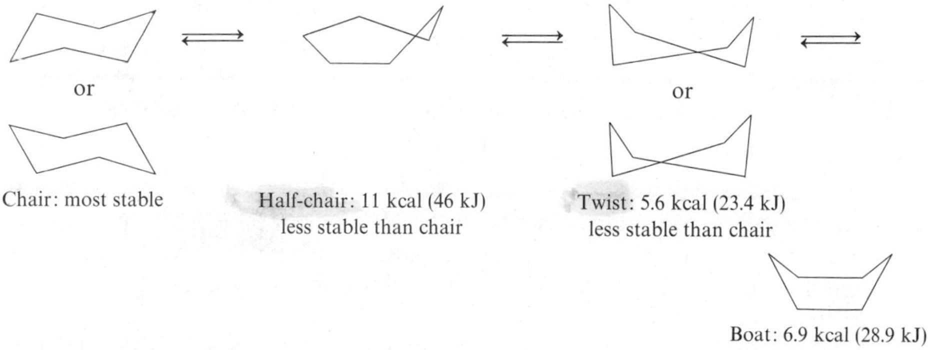

Chair: most stable   Half-chair: 11 kcal (46 kJ)   Twist: 5.6 kcal (23.4 kJ)
less stable than chair   less stable than chair

Boat: 6.9 kcal (28.9 kJ)
less stable than chair

The half-chair conformation is of very little importance at room temperature, where it has been estimated that there are about 10^8 chair forms for each half-chair form. Frequently, the chair and boat forms are the only conformational structures discussed for cyclohexane and its derivatives.

E. Cycloheptane and Larger Cycloalkanes

Cyclohexane is essentially strain-free as a result of its puckered conformation. This contradicted the Baeyer strain theory, which predicted that the larger cycloalkanes would be increasingly strained. However, the theory was correct up to a point; certain cycloalkanes with seven or more carbon atoms are somewhat strained but not nearly so much as cyclopropane and cyclobutane. Two additional factors are involved in the conformations of the larger rings: (1) steric hindrance between hydrogen atoms across the ring and (2) conformations where the torsional strain is intermediate between the staggered and the eclipsed form. The preferred conformations of these larger rings (seven-membered and larger) have minimum, torsional

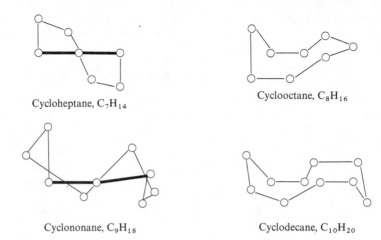

Cycloheptane, C_7H_{14} Cyclooctane, C_8H_{16}

Cyclononane, C_9H_{18} Cyclodecane, $C_{10}H_{20}$

FIGURE 3.17   Conformations of 7-, 8-, 9-, and 10-membered cycloalkanes, with hydrogen atoms omitted.

strain and steric hindrance, and they are always those of lowest energy. However, the relative stabilities of these conformations are not always so clean cut as they are for cyclohexane; sometimes there are several stable conformations of roughly equal energy. Ball-and-stick drawings for the stable conformations of cycloheptane, cyclooctane, cyclononane, and cyclodecane are shown in Fig. 3.17.

3.15   Evidence for Cycloalkane Stability; Heats of Combustion

When a cycloalkane or alkane is burned with an excess of oxygen, the products are carbon dioxide and water; this process is called **combustion.** (See also Sec. 4.9 for a discussion of the combustion of alkanes.) We now consider the cycloalkanes and how the heat associated with their combustion sheds light on their stability, that is, the strain in them.

Combustion liberates a certain amount of heat, which is dependent on the number of carbon atoms present. The amount of *heat liberated per carbon atom* can be computed by dividing the heat of combustion per mole by the number of carbon atoms present. The combustion of cyclohexane is an example:

$$C_6H_{12} + 9O_2 \longrightarrow 6CO_2 + 6H_2O + 944.48 \text{ kcal (3953 kJ)}$$

The heat of combustion per $-CH_2-$ group is therefore 944.48 kcal/6 $-CH_2-$ groups = 157.4 kcal (658.6 kJ). Because each $-CH_2-$ group gives the same products (CO_2 and water), the liberation of different amounts of energy *per* $-CH_2-$ group in various cycloalkanes must mean that those molecules contain different amounts of energy per $-CH_2-$ group.

The standard of comparison for the heats of combustion in cycloalkanes is the alkane, which can have no strain. For open-chain alkanes, each additional methylene group, $-CH_2-$, contributes almost 157.4 kcal/mole (658.6 kJ/mole) to the heat of combustion for the molecule. The heats of combustion for several cyclo-alkanes are listed in Table 3.6. This table also lists the total strain energy in kcal/mole

TABLE 3.6   Heats of Combustion and Strain Energy for Cycloalkanes

Cycloalkane	n in $(CH_2)_n$	Total Heat of Combustion, kcal/mole	Heat of Combustion/ CH_2 Group, kcal/mole	Total Strain Energy kcal/mole	Total Strain Energy kJ/mole
Cyclopropane	3	499.83	166.6	27.6	115.5
Cyclobutane	4	655.86	164.0	26.4	110.5
Cyclopentane	5	793.52	158.7	6.5	27.2
Cyclohexane	6	944.48	157.4	0.0	0.0
Cycloheptane	7	1,108.2	158.3	6.3	26.4
Cyclooctane	8	1,269.2	158.6	9.6	40.2
Cyclononane	9	1,429.5	158.8	11.2	46.9
Cyclodecane	10	1,586.0	158.6	12.0	50.2
Cycloundecane	11	1,742.4	158.4	11.0	46.0
Cyclododecane	12	1,891.2	157.6	2.4	10.0
Cyclotridecane	13	2,051.4	157.8	5.2	21.8
Cyclotetradecane	14	2,203.6	157.4	0.0	0.0
Cyclopentadecane	15	2,362.5	157.5	1.5	6.3
Cycloheptadecane	17	2,672.4	157.2	3.4	14.2

and kJ/mole, which is obtained by subtracting 157.4 kcal/mole (658.6 kJ/mole) (our standard) from the heat of combustion per —CH_2— group for each cycloalkane and then multiplying that difference by the total number of carbon atoms in the cycloalkane. For example, the strain energy per —CH_2— group in cyclopropane is 166.6 kcal/mole − 157.4 kcal/mole, or 9.2 kcal/mole (38.5 kJ/mole); since there are three —CH_2— groups in cyclopropane, the total strain energy is 9.2 kcal/mole × 3, or 26.7 kcal/mole (115 kJ/mole).

The data on heats of combustion provide a quantitative picture of the amounts of strain in cycloalkanes. Note that cyclohexane is strain-free. These data agree with theoretical calculations indicating that the strain order for the cycloalkanes is:

$$3 > 4 > 5 > 6 < 7 < 8 < 9 < 10 > 11 > 12$$

where these numbers represent n in $(CH_2)_n$. After cyclohexane, the strain increases to a maximum for cyclodecane and then decreases again. Very large cycloalkanes (for example, 15 or more carbon atoms per ring) have little or no strain. Also note that cyclopropane and cyclobutane have approximately equal stability. The instability of cyclopropane due to its greater C—C—C angle strain is equalized to some extent by stronger, more stable C—H bonds in the molecule.

The Baeyer strain theory is not applicable to cyclohexane and the higher-membered cycloalkanes because cyclohexane has tetrahedral bond angles, is puckered, and is strain-free. The bond angles in the higher cycloalkanes are puckered due to steric or torsional interactions and they have bond angles equal or nearly equal to the 109.5° tetrahedral angle. Baeyer argued that cycloheptane and higher cycloalkanes were strained because they were difficult to synthesize in the laboratory; that is, it was difficult to close the ring. These molecules are difficult to synthesize because ring closure requires that the two ends of the molecule come close enough together so that a C—C bond can form. In the larger, open-chain alkanes, the two ends are often quite far apart due to the free rotation about the single bonds. The *probability* of these ends coming close together and forming a cyclic structure is very greatly diminished when compared with shorter chains. Baeyer's assumption that strain was important is now known to be false.

3.16   Conformational Analysis of Monosubstituted Cyclohexanes

We now consider some monosubstituted cyclohexane derivatives and see what additional factors come into play in them. For cyclohexane itself, we showed two chair conformations, but they are identical because only hydrogen atoms are attached to the ring. If, however, we remove a hydrogen and replace it with a substituent (for example, a methyl group), we obtain two different chair conformations. These conformations as well as the boat conformer for methylcyclohexane are shown in Fig. 3.18.

As the ball-and-stick drawings show (bottom of Fig. 3.18), there is severe crowding between the axial methyl group and the other two axial hydrogens (that is, among the three axial bonds on the same side of the molecule). This type of interaction is called **1,3-diaxial interaction,** where 1,3 are the carbon atom numbers for the ring and thus the number of carbon atoms that separate the axial substituents. Crowding in the boat form is much greater than in cyclohexane itself, and although the boat form is probably an intermediate in the conversion of one chair form to another, it cannot be an important conformer.

We can analyze the 1,3-diaxial interaction for methylcyclohexane by considering the Newman projection formulas for the equatorial chair and the axial chair conformations. The Newman structures are drawn by sighting down the C-1—C-2 and the C-4—C-5 bonds. These interactions arise between the axial methyl group and the hydrogens pointing upward on C-3 and C-5. They are not present when the methyl group is equatorial. It has been estimated that the energy associated with each 1,3-diaxial methyl-hydrogen interaction is about 0.9 kcal/mole (3.8 kJ/mole).

Conformational structures:

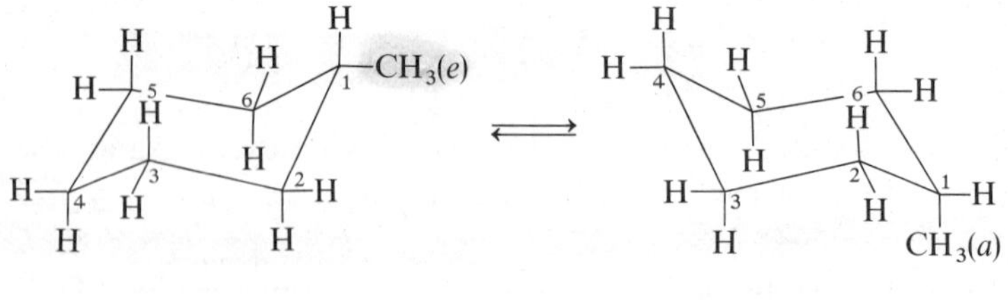

Newman projection formulas:

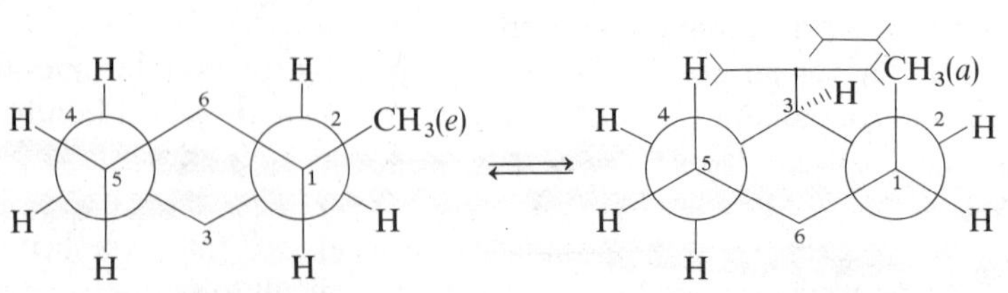

Methyl group is equatorial (*e*);   Methyl group is axial (*a*);
more stable conformation   less stable conformation

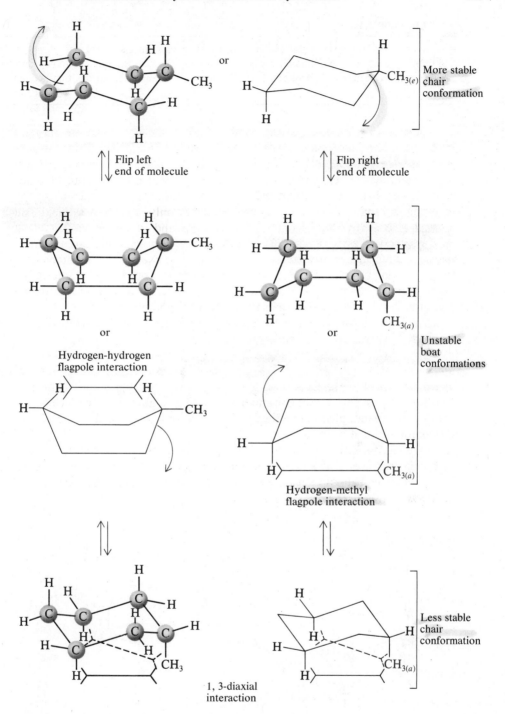

FIGURE 3.18   The two chair and two boat conformations for methylcyclohexane. Note the "flagpole" interactions in the boat conformation (there are two boat conformers, depending on which end of the molecule is flipped first).

Question 3.12

Note that the energies of the 1,3-diaxial methyl-hydrogen interaction of methylcyclohexane and the gauche interaction of butane (see Sec. 3.6) are both 0.9 kcal/mole (3.8 kJ/mole). Compare the Newman projections of these compounds, and point out any similarities that account for this.

In methylcyclohexane, as in cyclohexane, the ring flips rapidly from one chair form to another through the intermediate and very unstable boat form. This process is called **ring inversion.** With cyclohexane, the flipping occurs about 10^6 times a second, and the barrier to this process has been estimated to be about 11 kcal/mole (46 kJ/mole). Even at room temperature, there is more than enough energy for this to occur as rapidly as it does. In methylcyclohexane, there are "nonequivalent" chair conformations, where the methyl group can be axial or equatorial. Because the conformation with the equatorial group is so much more stable, it is not surprising that this conformer is preferred at equilibrium. The equilibrium constant $K_{eq} \approx 15$ for equatorial CH_3/axial CH_3; this corresponds to about 94% of the molecules existing in the equatorial CH_3 conformation. In terms of energy, the equatorial conformation is more stable than the axial conformation by about 1.7 kcal/mole. (7 kJ/mole.) [approximately 0.9 kcal/mole (3.8 kJ/mole) per 1,3-methyl-hydrogen axial-axial interaction in a cyclohexane ring].

In general, *a substituent (except for hydrogen) in the equatorial position produces the more stable and thus the preferred conformation:* this is largely because substituents are less crowded in the equatorial than in the axial position. Our arguments have been for the most part qualitative and based on the examination of molecular models. On the other hand, physical methods allow equilibrium constants to be determined for conformational isomers.

We can look further at the effect of substituents on the stability of conformational structures. The *tert*-butyl group, $(CH_3)_3C$—, is large and bulky. As shown by the following ball-and-stick drawings, there is considerable 1,3-diaxial interaction in *tert*-butylcyclohexane:

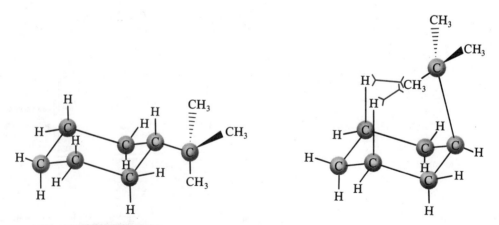

Left: **Equatorial *tert*-butylcyclohexane (preferred conformation)**; *right:* **axial *tert*-butylcyclohexane (1,3-diaxial interactions shown by ⟩—⟨).**

In terms of energy, the equatorial conformer is much more stable than the axial conformer, and the equilibrium constant is such that essentially all the *tert*-butyl groups are in the equatorial position.

TABLE 3.7   Energy Differences Showing Preference of Equatorial over Axial Positions in Monosubstituted Cyclohexanes*

Substituent	$\Delta G°$, Free-Energy Difference, kcal/mole[†]	K_{eq}	Relative Distribution, % eq:% ax
—CH$_3$	−1.7 (−7.1)	20	95:5
—CH$_2$CH$_3$	−1.8 (−7.5)	24	96:4
—CH(CH$_3$)$_2$	−2.1 (−8.8)	32	97:3
—C(CH$_3$)$_3$	Very large (> −5) (−20.9)	>1,000	>99.9: <0.1
—Cl, —Br, —I	−0.5 (−2)	2.4	71:29

* All values are approximate and taken at 25°.
[†] Values in parenthesis are kJ/mole.

It is not surprising that the magnitude of the 1,3-axial interactions varies with different substituents. For this reason, different monosubstituted cyclohexane derivatives exhibit different conformational preferences; that is, the energy difference between the axial and equatorial conformers can be larger or smaller depending on the substituent on the ring.

There is a direct relationship between this difference in energy, called the **free energy** (symbolized by $\Delta G°$),[1] and the equilibrium constant (K_{eq}) associated with a given equilibrium in solution. The relationship is:

$$\Delta G° = \text{difference in free energy} = -RT \ln K_{eq}$$

In this equation, R is the gas constant (0.00199 kcal/mole·°K) and T is the absolute temperature at which the equilibrium is measured (°K). The product of these two numbers and the natural logarithm of K_{eq} gives $\Delta G°$, the **free-energy difference** between the two conformers in kcal/mole.

The energy and equilibria data in Table 3.7 provide a quantitative picture for structure-conformation relationships. We can see that *the greater the free energy difference between the two conformers, the larger the K_{eq} associated with the axial/ equatorial equilibrium and the greater the preference of a particular group for occupying the equatorial position.* This is shown graphically for methylcyclohexane in Fig. 3.19.

Omit   **Question 3.13**

The equilibrium constant (K_{eq}) associated with the equatorial/axial equilibrium in cyclohexanol is approximately 5. What is the free-energy difference between the two conformers and the relative percentages of each at room temperature (25°)?

Question 3.14

(*a*) Draw the two chair conformers and a boat conformer for chlorocyclohexane.
(*b*) Of these three structures, which is the most stable and which is the least stable? Why?
(*c*) Do you think an axial chlorine atom would be more or less stable than an axial methyl group? Why?

[1] $\Delta G°$ is also called the **Gibbs standard free energy** (see Sec. 5.3).

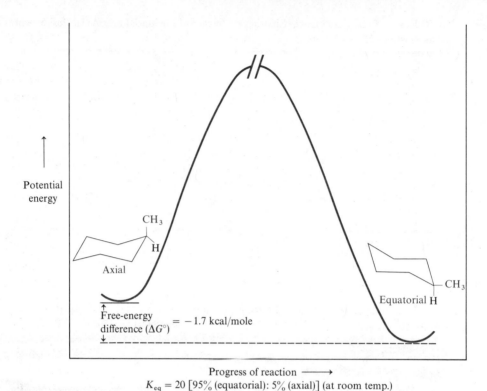

$K_{eq} = 20$ [95% (equatorial): 5% (axial)] (at room temp.)

FIGURE 3.19   Energy diagram illustrating the conformational equilibrium between equatorial and axial methylcyclohexane. This partial diagram depicts the two most stable conformers, the chair forms. The variety of conformational intermediates (e.g., boat conformers) involved in the equilibrium are omitted for clarity.

Drawing Conformational Structures for Cyclohexane Derivatives

We showed the transformation from one chair form to another occurring via the boat form, and we drew all three conformations. There is a simple procedure to follow, however, if we want to draw only the two chair forms.

Start by drawing one chair conformation and attach whatever number of atoms (substituents and/or hydrogens) you want to the carbon atoms. Then draw the carbon skeleton of the other chair conformation, which is derived by flipping one end of the molecule up and the opposite end down. The substituents on the first-drawn chair form can be drawn correctly on the other chair form as follows. Put the same substituents (or hydrogens) on the same carbon atoms, except that each substituent that was equatorial originally will be axial and each one that was axial originally will be equatorial on the second chair form. These equatorial → axial and axial → equatorial changes always take place when going from one chair form to another. Note, however, that this does not allow you to comment on the relative stabilities of the two conformations; this can be done only after they are correctly drawn.

For example:

Note also that certain bonds in the cyclohexane skeleton are parallel. This relationship can be used to your advantage in constructing accurate drawings. Draw the molecule in the order given in examples (1) to (5):

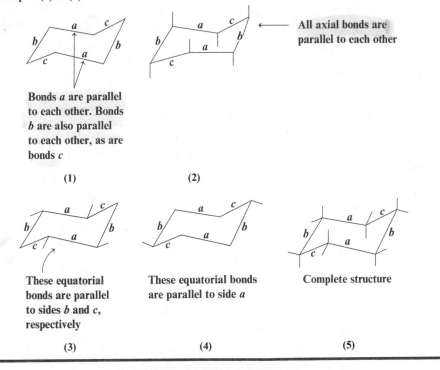

Bonds *a* are parallel to each other. Bonds *b* are also parallel to each other, as are bonds *c*

(1)

(2)

← All axial bonds are parallel to each other

These equatorial bonds are parallel to sides *b* and *c*, respectively

(3)

These equatorial bonds are parallel to side *a*

(4)

Complete structure

(5)

3.17 Disubstituted Cyclic Compounds; *Cis-Trans* Isomerism

We saw that there is only one structural isomer for monosubstituted cyclo-alkanes, although there are many possible conformations. The problem becomes more complex with disubstituted cyclic compounds, which we consider now.

There are two structural isomers for the dichlorocyclopropanes, 1,1-dichloro-rocyclopropane (1) and 1,2-dichlorocyclopropane (2):

(1)   (2)

When we examine structure (2) more closely, however, we find that the chlorine atoms may adopt two possible orientations in space. If they are both on the same side of the ring, we have what is called the *cis* isomer; both the chlorines may be either above or below the plane of the ring. (The term *cis* is Latin for "on this side.") If one chlorine is above and the other is below the plane of the ring, we have the *trans* isomer. (The term *trans* is Latin for "across.") The structures of these isomers are shown in Fig. 3.20.

The *cis* and *trans* isomers of 1,2-dichlorocyclopropane are stable and isolable compounds with their own unique physical properties. They cannot be interconverted without breaking and making bonds and are therefore not conformational isomers.

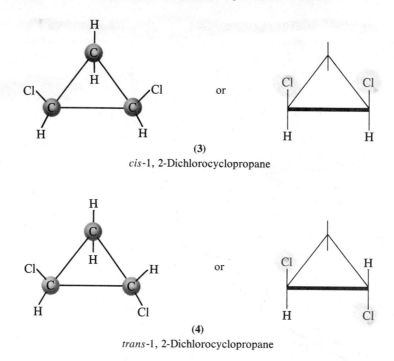

(3)

cis-1, 2-Dichlorocyclopropane

(4)

trans-1, 2-Dichlorocyclopropane

FIGURE 3.20   Ball-and-stick and perspective (paper) representations of (3) *cis* and (4) *trans* isomers of 1,2-dichlorocyclopropane.

Isomers like this are called **geometric isomers** or **cis-trans isomers.** [We return to geometric (*cis-trans*) isomers for a more complete look in Chap. 7.] This is a general phenomenon found in di- and polysubstituted cyclic compounds.

We must also modify our names for 1,2-dichlorocyclopropane to indicate the geometry of the chlorine atoms. The complete names are *cis*-1,2-dichlorocyclopropane (3) for the *cis* isomer where the chlorines are on the same side of the ring, and *trans*-1,2-dichlorocyclopropane (4) for the *trans* isomer where they are on opposite sides of the ring. Note that the prefix, *cis* or *trans*, comes at the beginning of the name.

We encounter an important new concept in *cis-trans* isomerism: restricted rotation about carbon-carbon single (σ) bonds in cyclic compounds. Compare the *cis* and *trans* structures with the corresponding open-chain propanes, where there is free rotation about the single bonds. In 1,2-dichloropropane, for example, we have many conformers or conformational structures, which are not isolable even though they have different spatial relationships; they are in equilibrium with one another.

Conformers; free rotation about carbon-carbon single bonds; equilibrium

If, however, we tie together the two ends of the molecule, which is what occurs in the cyclopropane structures, we no longer have free rotation about the single bonds. In cyclopropane molecules there is *no rotation* about the single bonds, though in the other cycloalkanes there is some twisting about bonds but never free rotation; hence, the term *restricted rotation* is used.

Cis-trans or *geometric isomerism;* no free rotation about carbon-carbon bonds; no equilibrium

cis (3)   *trans* (4)

There are many possible *cis* and *trans* compounds, as the following examples illustrate:

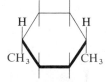

cis-1,3-Dimethylcyclobutane   *trans*-1,3-Dimethylcyclobutane

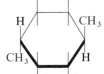

cis-1,3-Dibromocyclopentane   *trans*-1,3-Dibromocyclopentane

cis-1,4-Dimethylcyclohexane   *trans*-1,4-Dimethylcyclohexane

In these representations an attempt was made to show their perspective by darkening the bonds that project out of the plane of the paper toward the reader. Also, the cyclic structures are shown in their planar form only for convenience in drawing; we know that the four-, five-, and six-membered rings have various conformations. The use of vertical lines to attach substituents emphasizes their *cis* and *trans* relationship.

An interesting example of geometric isomerism occurs in benzene hexachloride, which is 1,2,3,4,5,6-hexachlorocyclohexane; this compound is prepared by treating benzene (see Sec. 14.17) with chlorine gas in the presence of ultraviolet light. One particular isomer of this compound is the active ingredient in the insecticide gammexane (lindane). It is called the γ isomer, which has been found to have the following structure:

γ-Benzene hexachloride (gammexane or lindane)

Note that several geometric isomers are possible for 1,2,3,4,5,6-hexachlorocyclo-hexane.

Question 3.15

Using planar structures throughout, draw all the structural and geometric isomers for the: (*a*) dibromocyclobutanes, (*b*) dimethylcyclopentanes, and (*c*) dimethylcyclohexanes. Name these compounds using the IUPAC system.

Question 3.16

Draw the structures of all cyclic compounds that have the formula C_5H_{10}. Include structural and geometric isomers where applicable.

3.18   Conformational Analysis for Disubstituted Cyclohexane Derivatives

We now return to the question of conformations that are available to disubstituted cyclohexane derivatives. For example, in *trans*-1,2-dimethylcyclohexane, there are two chair conformations:

(1)	(2)
trans-1,2-Dimethylcyclohexane	*trans*-1,2-Dimethylcyclohexane
Diequatorial methyl groups (more stable conformation)	*Diaxial methyl groups*

In conformation (1), the two methyl groups occupy equatorial positions, whereas in (2) they occupy axial positions. Indeed, in (1) the methyl groups may not appear to be on opposite sides of the ring even though they are. There is less crowding when both methyl groups are equatorial than when both are axial. Recall that the instability associated with an axial group (as compared with an equatorial group) has been attributed to 1,3-diaxial interactions. It is not surprising, therefore, that conformation (1) with diequatorial methyl groups is more stable than the diaxial conformation (2).

For *cis*-1,2-dimethylcyclohexane, there are also two chair conformations:

(3)	(4)
cis-1,2-Dimethylcyclohexane	*cis*-1,2-Dimethylcyclohexane
Axial-equatorial methyl groups	*Equatorial-axial methyl groups*

Equal Stabilities

These two conformations are of equal energy because each has one equatorial and one axial methyl group. (They are *mirror images*, which is another structural property of molecules we study in Chap. 6.) If we draw the planar structures of the *cis* and *trans* isomers of 1,2-dimethylcyclohexane, the methyl groups appear close together in the *cis* structure and far apart in the *trans* structure. However, molecular models of the more stable conformations of these isomers, that is, (1) and (3) or (4), show that the methyl groups are equal distances apart in both. Repulsion between methyl groups is not responsible for the differences in stability between the *cis* and *trans* isomers; it is the 1,3-diaxial interactions between an axial methyl group and the other two axial hydrogens that cause the difference. The *trans* isomer has been found to be more stable than the *cis* isomer by about 1.7 kcal/mole (7.1 kJ/mole).

For a given pair of disubstituted cyclohexane derivatives (*cis* and *trans* isomers), several steps must be followed to determine their relative stabilities. First, draw the two chair conformations for each isomer and determine which conformer of each pair is the more stable. Then compare the two more stable conformers to evaluate the relative stabilities of the *cis* and *trans* isomers.

In general, *the conformation with more substituents (other than hydrogen) in the equatorial position is more stable so long as the substituents are identical*. Glucose, the major constituent of starch and cellulose, is a six-membered ring that assumes the chair conformation and is structurally similar to cyclohexane (except that one —CH_2— group in cyclohexane has been replaced by —O— in glucose). There are two isomers of glucose (α and β), and the β isomer is the more stable because it has five equatorial groups, whereas the α isomer has four equatorial and one axial group.

β-D-Glucose

Five equatorial groups

α-D-Glucose

Four equatorial groups and one axial group

Determining the more stable conformation becomes more complex when different substituents are attached to a cyclohexane ring. In this case, we must use the differences in free energy, which show the preference of equatorial over axial positions and are listed in Table 3.7. The following example illustrates how to use these energies to predict the more stable conformation:

$\Delta G° = 1.7$ kcal/mole (for axial CH_3);
preferred conformation by
2.6 − 1.7, or 0.9 kcal/mole (3.8 kJ/mole)

$\Delta G° = 2.1$ kcal/mole [for axial $(CH_3)_2CH$—]
+ 0.5 kcal/mole (for axial—Cl)
= 2.6 kcal/mole (10.9 kJ/mole)

Calculations of this type are not overly accurate, but they do allow us to "guess" the preferred conformation. Recall that the energy differences in Table 3.7 already take into account both the 1,3-diaxial interactions and the torsional strain; that is, they represent the net energy required to change a substituent from the equatorial to the axial position. Thus, in the preceding example, we assume that the equatorial substituents are in their most stable position (zero relative energy), and we compute the energy associated with axial substituents only, neglecting the effects of, for example, 1,2-gauche interactions between the methyl and chloro groups.

Question 3.17

(*a*) Draw the chair conformations for *cis*-1,3-dimethylcyclohexane.
(*b*) Draw the chair conformations for *trans*-1,3-dimethylcyclohexane.
(*c*) Which is the more stable chair conformation for the *cis* isomer? For the *trans* isomer?

Question 3.18

Draw a chair conformation for a cyclohexane derivative that has an equatorial chlorine atom on C-1 and an axial chlorine atom on C-4. Is this the *cis* or the *trans* isomer?

omit

Question 3.19

Using the energy differences in Table 3.7, calculate the difference in energy between the two conformations of: (*a*) *cis*-4-ethylmethylcyclohexane, (*b*) *trans*-4-*tert*-butylmethylcyclohexane, and (*c*) the all-*cis* isomer of 1,2,4-trichlorocyclohexane (that is, all the chlorine atoms are on the same side of the ring).

Study Questions

3.20 Using examples where possible, define each of the following terms:

(*a*) dot, line, and condensed structures	(*b*) staggered conformation
(*c*) eclipsed conformation	(*d*) skew conformation
(*e*) gauche conformation	(*f*) *anti*-conformation
(*g*) free rotation	(*h*) conformers
(*i*) conformational structures	(*j*) Newman projection formulas
(*k*) three-dimensional projection formulas	(*l*) sawhorse structures
(*m*) torsional energy	(*n*) energy barrier to rotation
(*o*) homologous series	(*p*) homolog
(*q*) IUPAC system	(*r*) van der Waals forces
(*s*) induced dipole	(*t*) cycloalkanes
(*u*) conformational analysis	(*v*) chair conformation
(*w*) boat conformation	(*x*) combustion
(*y*) strain energy	(*z*) equatorial and axial substituents

3.21 Provide two acceptable names for the following two compounds, one of which must be the IUPAC name:

(*a*) $CH_3-CH-CH_2-CH_3$
                  $|$
                  CH_3

(*b*) CH_3
               $|$
     CH_3-C-CH_3
               $|$
               CH_3

3.22 Provide the IUPAC names for the following 13 compounds:

(*a*) Cl
               $|$
     $CH_3-CH-CH_2CH_2CH_2CH_3$

(*b*) CH_3
                    $|$
     $CH_3CH_2-CH-CH-CH_3$
                         $|$
                         CH_3

(*c*) $CH_3CH_2-CH-CH_2CH_2$
                  $|$ $|$
                  CH_3 CH_2
                             $|$
                             CH_3

(*d*) $CH_2CH_2CH_2CH_3$
                    $|$
     $CH_3CH_2-C-CH_2CH_2CH_3$
                    $|$
                    Br

$$\underset{\substack{\\ \text{CH}_3\text{CH}_2\text{CH}_2}}{(e)\ \ \text{CH}_3\text{CH}_2-\overset{\overset{\displaystyle \text{CH}_3}{|}}{\underset{}{\text{C}}}-\text{CH}_2-\overset{\overset{\displaystyle \text{CH}_3}{|}}{\underset{\underset{\displaystyle \text{CH}_3}{|}}{\text{CH}}}-\text{CH}_2}$$

(e) $\text{CH}_3\text{CH}_2-\overset{\text{CH}_3}{\underset{\text{CH}_3\text{CH}_2\text{CH}_2}{\text{C}}}-\text{CH}_2-\overset{\text{CH}_3}{\underset{\text{CH}_3}{\text{CH}}}-\text{CH}_2$

(f) $\text{CH}_3\text{CH}_2-\overset{}{\underset{\text{CH}_2-\text{Cl}}{\text{CH}}}-\text{CH}_2\text{CH}_2-\overset{}{\underset{\text{CH}_3}{\text{CH}}}-\text{Cl}$

(g) $\text{CH}_3-\overset{\text{CH}_3}{\underset{\text{CH}_3}{\text{C}}}-\text{CH}_2-\overset{\text{CH}_3}{\underset{\text{CH}_3}{\text{CH}}}-\overset{\text{CH}_3}{\underset{\text{CH}_3}{\text{C}}}-\text{CH}_3$

(h) $\text{CH}_3-\overset{}{\underset{\text{CH}_3}{\text{CH}}}-\text{CH}_2-\overset{\text{CH}_3\text{CHCH}_3}{\underset{}{\text{CH}}}-\overset{}{\underset{\text{CH}_3}{\text{CH}}}-\text{CH}_3$

(i) $(\text{CH}_3)_2\text{CHCH}_2\text{CH}_2\text{CH}(\text{CH}_2\text{CH}_3)_2$

(j) $\text{CH}_3-\overset{}{\underset{\text{CH}_2\text{CH}_3}{\text{CH}}}-\text{CH}_2-\overset{\text{CH}_2\text{CH}_3}{\underset{\text{CH}_2\text{CH}_3}{\text{CH}}}-\overset{}{\underset{}{\text{CH}}}-\text{CH}_3$

(k) $\text{CH}_3\text{CH}_2\text{CH}_2-\overset{}{\underset{\text{CH}_3\text{CH}_2-\text{CH}}{\text{CH}}}-\overset{\text{CH}_3}{\underset{\text{CH}_2\text{CH}_2\text{CH}_3}{\text{C}}}-\text{CH}_2\text{CH}_3$
$\overset{}{\underset{\text{CH}_2\text{CH}_3}{\text{CH}_2-\text{CH}-\text{CH}_3}}$

(l) $\text{CH}_3-\overset{}{\underset{\text{CH}_3}{\text{CH}}}-\overset{\text{CH}_2\text{CH}_2\text{CH}_3}{\underset{}{\text{CH}}}-\text{CH}_2\text{CH}_2-\overset{}{\underset{\text{CH}_3}{\text{CH}}}-\text{CH}_2\text{CH}_2-\overset{}{\underset{\text{CH}_2\text{CH}_3}{\text{CH}}}-\text{CH}_3$

(m) $\text{CH}_3-\overset{}{\underset{\text{CH}_3-\overset{}{\underset{\text{CH}_3}{\text{C}}}-\text{CH}_3}{\text{CH}}}-\text{CH}_2-\overset{\text{CH}_2-\overset{}{\underset{}{\text{CH}}}-\text{CH}_2\text{CH}_3}{\underset{\text{CH}_3}{\text{CH}}}-\text{CH}_2\text{CH}_3$

3.23 Which, if any, of the 13 structures in Study Question 3.22 contain the following alkyl groups?

(a) ethyl   (b) propyl   (c) isopropyl   (d) butyl
(e) isobutyl   (f) sec-butyl   (g) tert-butyl   (h) neopentyl

3.24 Draw structures corresponding to the following names (ignore conformations and draw cyclic structures carefully).

(a) 3-methylpentane   (b) 2,3-dimethylhexane
(c) octadecane   (d) 5-propyldecane
(e) 2-iodoeicosane   (f) 3-ethyl-3-methylheptane
(g) cis-1,3-diethylcyclopentane   (h) trans-1,2-dibromocyclobutane
(i) 2-cyclopentylpentane   (j) 4-ethyl-6-(2-ethyl-1,3-dimethylpentyl)dodecane
(k) 3,4,5-tribromooctane   (l) 4-tert-butylnonane
(m) 5-sec-butyl-5-isopropyldecane   (n) 2-chloro-2,4,4-trimethylpentane
(o) 2,4-dimethyl-4-ethyloctane   (p) cyclononane
(q) 4-(1,2,2-trichloroethyl)-3-ethyl-2,3,4-trimethylnonane
(r) 3-iodo-5-butyl-3,4-diethyldecane   (s) 2,5-dibromo-4-sec-butyl-5-ethylnonane

3.25 Draw at least three other structural isomers of $(\text{CH}_3)_2\text{CH}-\text{CH}(\text{CH}_3)_2$. Can structural isomers be isolated from one another? Why or why not?

3.26 Considering only rotation about the carbon-carbon bond shown by a line, draw the most stable conformation and the least stable conformation for $(\text{CH}_3)_2\text{CH}-\text{CH}(\text{CH}_3)_2$. Use Newman projection formulas. Can conformational structures (conformers) be isolated from one another? Why or why not?

3.27 We called $CH_3\!-\!\overset{|}{C}H\!-\!CH_2CH_3$ the *sec*-butyl group. What, if anything, is wrong with the name *sec*-pentyl?

3.28 How might you rationalize that *cis*-1,3-dimethylcyclobutane is more stable than the *trans*-isomer? (*Hint:* Consider the structure of cyclobutane in Fig. 3.12.)

3.29 (*a*) According to Table 3.7, the *tert*-butyl group has a *great* preference for occupying the equatorial position on a cyclohexane ring, and much more so than do ethyl and isopropyl groups. Why?

(*b*) *cis*-1,4-*Di-tert*-butylcyclohexane has been shown not to exist in the chair conformation (and certainly not in the boat form). Suggest a likely conformation for this compound.

3.30 Both the following compounds have the same molecular formula C_5H_{10}. Which is the more stable and why?

$$
\begin{array}{cc}
\overset{\displaystyle \diagup CH_2 \diagdown}{H_2C\!-\!CH\!-\!CH_2\!-\!CH_3} &
\overset{\displaystyle \overset{CH_2}{\diagup\ \ \diagdown}}{\underset{CH_2\!-\!-\!-\!-\!CH_2}{CH_2\qquad CH_2}}
\end{array}
$$

3.31 (*a*) Using Newman projection formulas, draw the *anti-* and *eclipsed* conformations for 1,2-dichloroethane. Which of these would you expect to be the more stable? Why?

(*b*) In a sense, butane could be considered a 1,2-disubstituted ethane in the same way as 1,2-dichloroethane, that is, $G\!-\!CH_2\!-\!CH_2\!-\!G$, where $G = Cl$ in the latter and $G = CH_3$ in the former. We discussed the conformations of butane (see Sec. 3.6). List the factors you think might be involved if we compared the relative stabilities of the *anti-* and the eclipsed conformations of 1,2-dichloroethane and of butane.

(*c*) We discussed the polarity associated with the carbon-chlorine bond which has a bond dipole of $\overset{\rightarrow}{C\!-\!Cl}$. How might these dipoles affect the stability of the eclipsed conformation of 1,2-dichloroethane? (*Hint:* In which conformation, *anti-* or eclipsed, would these dipoles have the greater interaction, that is, dipole-dipole interaction? In which would they have lesser? How would this type of interaction affect the stabilities of the conformations?)

(*d*) In the series 1,2-difluoroethane, 1,2-dichloroethane, 1,2-dibromoethane, and 1,2-diiodo-ethane, which compound would you expect to exhibit the greatest stability in the eclipsed conformation? Why?

3.32 Draw the two chair conformations for the γ isomer of benzene hexachloride (see Sec. 3.17). Which is more stable?

omit **3.33** The energy associated with a 1,3-diaxial methyl-hydrogen interaction has been estimated to be about 0.9 kcal/mole in methylcyclohexane where there are *two* methyl-hydrogen diaxial interactions. Thus, the conformational preference for an equatorial methyl group over an axial group is estimated to be 2 × 0.9, or 1.8 kcal/mole; the experimentally determined value (Table 3.7) is about 1.7 kcal/mole.

(*a*) Draw the two chair conformations for *cis*-1,3-dimethylcyclohexane.

(*b*) What type of diaxial interactions are involved in these conformations?

(*c*) Insofar as you can, predict which conformer is the more stable, and by what amount of energy it is more stable.

(*d*) Given that the energy difference between these two conformers is about 5.4 kcal/mole, how can you account for this?

omit **3.34** In Sec. 3.18 we studied the conformers for 1,2-dimethylcyclohexane. There are two *cis* conformers and two *trans* conformers.

(*a*) Using the value of 0.9 kcal/mole for each 1,3-diaxial methyl-hydrogen interaction and an additional 0.9 kcal/mole for each 1,2-diequatorial or 1,2-axial-equatorial methyl-methyl steric interaction, calculate the free-energy differences between each of these pairs of conformers.

(*b*) What do these energy differences mean in terms of the equilibrium constant for each pair and the relative percentages of each conformer in each pair?

(*c*) To determine the relative stabilities of the *cis* and *trans* conformations, compare the more stable *cis* conformer with the more stable *trans* conformer. What is the free-energy difference between these two species? (The experimentally measured value is 1.87 kcal/mole. If you did all the calculations correctly, your answer should be close to this.)

4

Alkanes and Cycloalkanes:
Preparation and Reactions

We studied the physical properties of the alkanes and cycloalkanes in Chap. 2, where our emphasis was on structure. In particular, interactions were explained in terms of the three-dimensional nature of these molecules, which results from the tetrahedral carbon atom in alkanes and cycloalkanes. Even though these two types of compounds have different molecular formulas, C_nH_{2n+2} for alkanes and C_nH_{2n} for cycloalkanes, their chemical reactions and the methods for their preparation show amazing similarity; only in a few cases do cycloalkanes react differently than alkanes.

We deferred discussion of the methods of preparation and reactions of alkanes and cycloalkanes until this time so they can be considered together. We stated that one of the greatest virtues of classifying organic compounds according to family is to allow us to see that particular sets of properties, chemical reactions, and methods of preparation are characteristic of a particular kind of structure. The characteristic structure of alkanes and cycloalkanes is the *tetrahedral carbon atom*.

4.1   Industrial Sources

Petroleum is the principal source of hydrocarbons in the United States. Petroleum products come from the fractional distillation of crude oil, which yields various fractions that are classified according to the range of their boiling points. The petroleum constituents of crude oil are listed in Table 4.1. As we learned in Sec. 3.11, the boiling points of the alkanes are related to their molecular weight, so it is not surprising that the distillation fractions are rich in certain alkanes.

Furthermore, the composition of the hydrocarbons in petroleum varies widely depending on the area of the country where the crude oil is obtained. In Pennsylvania the crude oil contains about 75% alkanes, 18% cycloalkanes, and 7% aromatic compounds. California crude oil contains about 35% alkanes, 54% cycloalkanes, and 11% aromatics. Most crude oil contains about the same amount of aromatics (7 to 11%), whereas the percentages of alkanes and cycloalkanes vary greatly.

The alkane fraction contains mostly continuous-chain compounds. There are virtually no alkenes (compounds that contain the carbon-carbon double bond) in petroleum.

The cycloalkane fraction contains mostly cyclopentane and cyclohexane and their alkyl derivatives. The petroleum industry refers to the cycloalkanes as *naphthenes*, a term not to be confused with the aromatic compound naphthalene (see Sec. 13.12), which is more familiar to us as mothballs.

TABLE 4.1   Components Obtained from Fractional Distillation of Crude Oil

Fraction	Carbon Content	Distillation Temperature, °C
Gas	C_1–C_4	Up to ~ 20
Petroleum ether	C_5–C_6	20–60
Ligroin (light naphtha)	C_6–C_7	60–100
Natural gasoline	C_6–C_{12} and cycloalkanes	50–200
Kerosene	C_{12}–C_{18} and aromatics	175–275
Gas oil (furnace and diesel oils)	Above C_{18}	Above 275
Lubricating oils	C_{20}–C_{30}, though mostly C_{26}–C_{30}	Nonvolatile liquids; can be removed by distillation at reduced pressures
Asphalt (petroleum coke)	Polycyclic structures	Residue; nonvolatile and nondistillable

The aromatic compounds are present in the smallest quantities. This fraction contains benzene, toluene, the isomeric dimethylbenzenes (xylenes), the isomeric trimethylbenzenes, naphthalene, and other polynuclear aromatic compounds (see Sec. 13.12). We discuss the structures of these aromatic compounds in Chaps. 13 and 14.

The lubricating oil fraction contains high molecular weight hydrocarbons that do not distill without decomposition at atmospheric pressure; they boil in excess of 400° at atmospheric pressure, which causes the carbon-carbon bonds to rupture and yield lower molecular weight compounds. However, they can be fractionally distilled (separated by boiling point differences) at reduced pressure to give light, medium, and heavy lubricating oils. Even though pure alkanes containing 20 or more carbon atoms are solids at room temperature, lubricating oils are liquids because they are *mixtures* of C_{20} to C_{30} hydrocarbons. Natural gas is the major source of methane and contains small amounts of ethane and propane.

The fractions obtained from the distillation of crude oil are indeed complex. A given sample of crude oil may contain 100 or more individual compounds, many of which have boiling points or other physical properties so similar that it is impractical to attempt to obtain them in pure form. The hydrocarbons obtained from petroleum are used mostly as fuels and oils, where the exact composition is relatively unimportant and purity is not necessary. We omit a discussion about gasoline here but take up this subject in relation to the combustion reactions (see Sec. 4.10).

Catalytic cracking, in which large hydrocarbons are broken down into small fragments under the influence of heat and catalysts, and *catalytic reforming*, in which both small and large fragments under very specific reaction conditions are reformed into other desired compounds, are valuable industrial methods used to prepare many hydrocarbons (see Sec. 4.10). Finally, *Fischer-Tropsch synthesis* (see Sec. 20.4A) is a convenient route to many industrially valuable organic chemicals, including alkanes.

Philosophy of Studying Chemical Reactions

We considered the structural properties of molecules, and we are now ready to study chemical reactions—reactions for the preparation of compounds and reactions of the compounds themselves, although by necessity these are closely interwoven.

We know that many alkanes and cycloalkanes are present in petroleum; however, petroleum chemistry is a specialized field and is more of industrial importance. Although certain hydrocarbons (methane, ethane, propane, butane, and the aromatic compounds) are isolated easily from crude oil in pure form, many are difficult to isolate in pure form and are prepared in the research laboratory. Some hydrocarbons are used so extensively that chemical companies have scaled up and modified laboratory methods to produce large amounts relatively inexpensively; yet many industrial processes use petroleum fractions as starting materials.

Our main concern in this text is the development of the theories and understanding of organic chemistry. Most reactions we discuss are normally done only in the laboratory, where relatively small quantities of pure compounds are desired. We may, for example, introduce reactions for the interconversion of two compounds that are available industrially; they may serve as models for the interconversion of other pairs of compounds that may not be commercially available. Basically, we develop our views about organic chemistry using simple examples that will allow us to extend our *reasoning by analogy* principle.

Throughout this text, many reactions are presented in a general way and followed by specific examples. When reactions are general for an entire family, we use a shorthand notation for the alkyl group, for which the symbol is R:

$$R = \text{any alkyl group}$$

TABLE 4.2 Classification System for Carbon and Hydrogen Atoms in a Molecule

General Structure	Structural Feature of Importance	Classification	Designation
H | R—C—H | H	Carbon atom bearing three hydrogen atoms and *one* other carbon atom (R group)	C: primary H: primary	C: 1° H: 1°
R′ | R—C—H | H	Carbon atom bearing two hydrogen atoms and *two* other carbon atoms (R groups)	C: secondary H: secondary	C: 2° H: 2°
R′ | R—C—H | R″	Carbon atom bearing one hydrogen atom and *three* other carbon atoms (R groups)	C: tertiary H: tertiary	C: 3° H: 3°
R′ | R—C—R‴ | R″	Carbon atom bearing no hydrogen atoms and *four* other carbon atoms (R groups) attached	C: quaternary	C: 4°

Thus, R—H represents methane, ethane, decane, tetradodecane, or, in general, any alkane hydro-carbon with any number of carbon atoms arranged in any way (that is, continuous or branched chain). R—X represents any alkyl halide, with limitless possible structures for R and with X being any halogen (F, Cl, Br, or I).

The use of the R *group* is so common that we indicate different R groups in the following manner:

$$R' \text{ or } R'' \text{ or } R''' = \text{ different alkyl groups}$$

$$\begin{array}{c} R' \\ | \\ R\text{—}C\text{—}R'' \\ | \\ R''' \end{array}$$

A compound with the general structure R—C—R'' has four different alkyl groups attached to one

carbon atom. Sometimes R^1, R^2, R^3, and R^4 are used for this purpose, so that $R^1R^2R^3R^4C$ means that four different alkyl groups are attached to carbon (see Table 4.2).

Organic reactions presented in this text are very seldom balanced. Our major concern is the (1) organic starting materials, (2) organic products, (3) inorganic reagents (if any) needed to effect the reaction, and (4) reaction conditions (temperature and pressure), where pertinent. For the most part, we indicate these features in our reactions. The laboratory chemist has to balance reactions to use the correct amount of each starting material.

Methods of Preparation of Alkanes and Cycloalkanes

4.2   Alkanes from Alkyl Halides and Alkenes

There are four common methods used to prepare alkanes, which we now discuss.

A. Coupling Reactions

When an alkyl halide reacts with lithium, the result is an organometallic species called an **alkyllithium reagent.** In the formation of the alkyllithium reagent, the lithium metal, Li·, partially donates its valence electron to the carbon to which it is attached; the reagent can be represented as $R\overset{\ominus}{:}\ Li^{\oplus}$. The C—Li bond is not com-pletely ionic, however, but rather highly polar. The following structure is perhaps a more accurate representation: $R^{\delta\ominus}\text{-----}Li^{\delta\oplus}$. The carbon atom(s) attached to a metal in an organometallic molecule are all **carbanions, $R\overset{..}{:}$**. The carbon-metal bond is very polar, with the highest electron density centered at carbon. The carbon behaves as a *carbon anion*, which is usually abbreviated as *carbanion*. Although the entire organometallic complex is involved in each reaction, we often write these reactions as involving free carbanions $(R\overset{\ominus}{:})$ for simplicity:

$$R\text{—}X \quad \xrightarrow{\text{Li}} \quad R\text{—Li}$$

Alkylhalide,   Alkyllithium
1°, 2°, or 3°

If the alkyllithium reagents react further with cuprous iodide (CuI), yet another organometallic reagent is formed—a **lithium dialkylcopper reagent:**

$$R\text{—Li} \quad \xrightarrow{\text{CuI}} \quad R_2\text{CuLi}$$

Lithium
dialkylcopper
reagent

The reaction of a lithium dialkylcopper reagent with an alkyl halide results in the formation of larger alkanes:

$$R_2CuLi + R'X \longrightarrow R-R'$$

Yields best when R' is
methyl or primary

Question 4.1

Beginning with the general alkyl halide R—X and terminating at the coupled alkane product (R—R'), determine the stoichiometry of the reactants in each reaction outlined previously.

The mechanism[1] appears to involve attack by the carbanion on the carbon that bears the halogen. As the electrons from the carbanion begin to attach to the carbon, the electrons in the carbon-halogen bond are repelled and the bond is broken; the halogen leaves as halide ion. This is probably an example of a **nucleophilic aliphatic substitution reaction,** which is discussed in Sec. 5.16.

$$R:\frown + R'-\ddot{X}: \longrightarrow R-R' + :\ddot{X}:^{\ominus}$$

Nucleophilic aliphatic substitution reaction

A **nucleophile (Nu:** $^{\ominus}$) (Greek *nucleo*, nucleus, and *phile*, loving) is any nucleus-loving species. They may be negatively charged (for example, $H\ddot{O}:^{\ominus}$, $:\ddot{Br}:^{\ominus}$, $H\ddot{S}:^{\ominus}$) or neutral (for example, $H-\ddot{O}-H$). See Sec. 9.7 for a more elaborate discussion of this species.

In the previous example the curved arrows ($\frown$) are used to keep track of electrons as a reaction proceeds from reactants to products. They show the electron flow in the course of bonds breaking and forming. For example, as the electron pair from the carbanion comes in to form a carbon-carbon bond (the alkane) between R and R', an electron pair leaves with the halide ion. Arrows are used frequently in organic chemistry simply as a device for keeping track of electrons; you might call it a bookkeeping device for electrons.

For example,

$$CH_3CH_2Br \xrightarrow{Li} CH_3CH_2Li \xrightarrow{CuI} \begin{matrix} CH_3CH_2 \\ \diagdown \\ CH_3CH_2 \diagup \end{matrix} CuLi$$

Bromoethane   Ethyllithium   Lithium diethylcopper

$$\begin{matrix} CH_3CH_2 \\ \diagdown \\ CH_3CH_2 \diagup \end{matrix} CuLi + CH_3Br \longrightarrow CH_3CH_2CH_3$$

Bromomethane   Propane

---

[1] A **mechanism** is the step-by-step process that we perceive as occurring as reactants are converted to products (see Sec. 5.1). When lithium dialkylcopper reagents are used as the nucleophiles, the nucleophilic aliphatic substitution mechanism is actually more complicated than we have shown. We simplified it for clarity.

Bromocyclohexane $\xrightarrow{\text{Li}}$ $\xrightarrow{\text{CuI}}$ Lithium dicyclohexylcopper

$\left(\langle\bigcirc\rangle\right)_2\text{CuLi} + CH_3CH_2Br \longrightarrow \langle\bigcirc\rangle-CH_2CH_3$

Bromoethane   Ethylcyclohexane

Question 4.2

(a) Prepare 2-methylbutane using the lithium dialkylcopper reaction.
(b) Which possible alkyl halides and organometallic reagents can be coupled to prepare this compound?

B. Cyclization Reactions

The action of a metal (usually Zn) on an alkyl dihalide results in *ring closure*, which is an intramolecular coupling reaction probably involving an organozinc intermediate. A bond is formed between two carbon atoms that are in the same molecule.

$$\begin{array}{c}CH_2{-}Cl \\ CH_2 \\ CH_2{-}Cl\end{array} \xrightarrow[\text{alcohol, heat}]{\text{Zn metal, NaI, aqueous}} \begin{array}{c}CH_2 \\ CH_2 \quad | \\ CH_2\end{array} \begin{array}{l}\text{New C}-\text{C} \\ \text{bond formed}\end{array}$$

1,3-Dichloropropane   Cyclopropane
                           80% yield

$$\begin{array}{c}CH_2 \\ CH_2 \quad CH_2{-}Br \\ CH_2 \quad CH_2{-}Br \\ CH_2\end{array} \xrightarrow{\text{as above}} \begin{array}{c}CH_2 \\ CH_2\end{array} \begin{array}{l}\text{New C}-\text{C} \\ \text{bond formed}\end{array}$$

1,6-Dibromohexane   Cyclohexane
                           44% yield

Zinc metal appears to work most efficiently. The best yields result from the synthesis of cyclopropane, whereas cyclization to give larger than six-membered rings is difficult and is accompanied by very low yields. The smaller rings are formed because the two ends of the chain are close enough to allow bond formation to occur. As the chains become longer, the probability of ring closure is greatly diminished.

It appears that an organozinc intermediate is involved and that the cycloalkane product results from an intramolecular displacement as illustrated:

$$\begin{array}{c}CH_2{-}Zn{-}X \\ CH_2 \\ CH_2{-}\ddot{X}:\end{array} \longrightarrow \begin{array}{c}CH_2 \\ CH_2 \quad | \\ CH_2\end{array} + Zn^{\oplus 2} + 2:\ddot{X}:^{\ominus}$$

This too is an example of a nucleophilic aliphatic substitution reaction. The sodium iodide catalyzes the reaction. The organozinc intermediate that forms exchanges chloride for iodide (X = Cl or I), which speeds up the reaction (see Sec. 9.8).

Question 4.3

The lithium organocopper reaction closely resembles the **Wurtz reaction,** in which 2 moles of alkyl halide react with excess sodium metal to give an alkane in which the carbon content is doubled:

$$2R-X \xrightarrow[\text{heat}]{Na^0} R-R$$

Alkyl halide,   Alkane
X = Br or I

How might the compound 2,7-dimethyloctane be obtained from an alkyl halide and the Wurtz reaction? Give structures and equations.

Question 4.4

Explain why the percentages indicated are expected for the reaction:

$$R-X + R'-X \xrightarrow[\text{heat}]{Na^0} R-R + R-R' + R'-R'$$

25%   50%   25%

Assume that equimolar amounts of R—X and R'—X were mixed together originally. (*Hint:* Only probabilities are involved in this question.)

Question 4.5

The Wurtz reaction is thought to involve an organosodium intermediate. Write the structure of this intermediate and compare it with the alkyllithium intermediate used in preparing the lithium organocopper reagent.

C. Zinc Metal Reduction

When an alkyl halide, R—X (X = Cl, Br, I, *but not F*), reacts with zinc metal under acidic conditions, it is converted into the corresponding alkane with no loss of carbon atoms and no change in the carbon skeleton. The general reaction, along with specific examples, follows:

$$R-X \xrightarrow[H^\oplus]{Zn\ metal} R-H$$

$$CH_3CH_2CH_2CH_2CH_2-Br \xrightarrow[H^\oplus]{Zn} CH_3CH_2CH_2CH_2CH_2-H$$

1-Bromopentane   Pentane

$$CH_3CH_2CH_2-\underset{\underset{Cl}{|}}{CH}-CH_3 \xrightarrow[H^\oplus]{Zn} CH_3CH_2CH_2-\underset{\underset{H}{|}}{CH}-CH_3$$

2-Chloropentane   Pentane

This reaction involves oxidation-reduction because the zinc metal is oxidized to $Zn^{\oplus 2}$ and the carbon is reduced. The mechanism of this reaction is not completely known.

D. Hydrogenation Reactions

One of the most important methods for preparing alkanes involves the *addition* of hydrogen to a carbon-carbon double bond:

$$\underset{}{\overset{}{C=C}} + H_2 \xrightarrow[\text{catalyst}]{Pt,\ Ni,\ or\ Pd} -\underset{\underset{H}{|}}{C}-\underset{\underset{H}{|}}{C}-$$

$$CH_3-CH=CH_2 + H_2 \xrightarrow{Ni} CH_3CH_2CH_3$$

Propene   Propane

The reaction is mentioned for the sake of completeness and because it is one of the most common methods used to prepare alkanes. We defer discussion of this topic to the chapter on alkenes (see Sec. 8.12).

Similarly, cyclohexanes can be prepared by *hydrogenating* (adding hydrogen to) aromatic compounds such as benzene, C_6H_6:

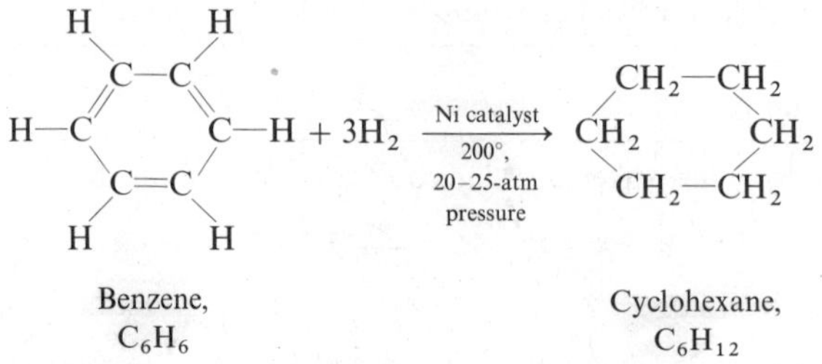

Benzene,
C_6H_6

Cyclohexane,
C_6H_{12}

This reaction is discussed in Sec. 14.17. It is useful in the preparation of cyclohexane and substituted cyclohexanes.

Both examples involve the conversion of an **unsaturated hydrocarbon** to an alkane or a cycloalkane. An unsaturated hydrocarbon does not contain the maximum number of hydrogen atoms that its carbon number allows for acyclic compounds (that is, C_nH_{2n+2}). By this definition, alkanes are **saturated hydrocarbons.** Cyclo-alkanes are also considered saturated because, other than the additional carbon-carbon single bond of the ring, they too contain the maximum number of hydrogen atoms their molecular formula allows. No general formula for the number of hydrogen atoms a cycloalkane may contain can be derived because this varies with the number of rings in each structure.

Oxidation States of Organic Compounds; Oxidation and Reduction

We mentioned that the conversion of an alkyl halide to an alkane is a reduction reaction. Just what is meant by oxidation and reduction of organic compounds? To answer this, consider the method for determining the oxidation state of carbon in organic molecules. The structure of the molecule must be drawn so that the three steps given here can be followed.

Step 1. Pick the carbon atom for which you wish to determine the oxidation number.
Step 2. Assign oxidation numbers to the atoms that are attached to the carbon atom in question in the following manner:

(a) Hydrogen (H) = +1, except in hydrides (for example, NaH) where H = −1
(b) Oxygen (O) = −2, except in peroxides (for example, H_2O_2) where O = −1
(c) Halogens (X = F, Cl, Br, I) = −1
(d) Hydroxyl groups (OH) = −1 (this can be deduced by remembering that H = +1 and O = −2, with a net result of −1 for the group that contains H and O)
(e) Other carbon-containing groups = 0

These values actually agree with our expectations based on bond polarity; that is, if we attach atoms that are different from carbon, we anticipate polarities of the following sort:

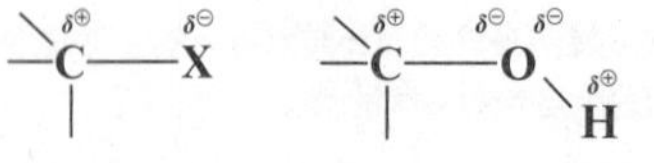

If we attach other carbon atoms, we expect the electrons to be shared equally, so that the other carbon-containing groups are assigned an oxidation number of 0.
Step 3. The sum of the oxidation numbers for all the atoms that are attached to the carbon atom plus the oxidation number of the carbon atom must equal the net charge on the molecule, which is usually zero.

In the following examples the circled carbon atom is the one for which we find the oxidation number.

$$C = -4 \qquad C = -3 \qquad C = -2 \qquad C = -1$$

$$C = 0 \qquad C = +1 \qquad C = 0 \qquad C = +1$$

$$C = +2 \qquad C = +2 \qquad C = +4 \qquad C = +4$$

Note the wide spectrum of oxidation states of carbon. Carbon in methane is in its most reduced form, with an oxidation state of -4; the most oxidized state of carbon appears in molecules such as CO_2, where it is $+4$.

We now return to the reduction of an alkyl halide to an alkane, where carbon has gained electrons. In the Zn metal reduction, the reducing agent is Zn^0 (metal), which donates electrons and is itself oxidized to Zn^{+2}.

Note that no carbon atom in the previous examples actually bears a formal charge. What is actually being determined by this bookkeeping device is the relative electron density of the various carbon atoms.

We will see numerous examples of oxidation and reduction in organic chemistry.

Reactions of Alkanes and Cycloalkanes

4.3   Reactivity of Alkanes and Cycloalkanes

All alkanes and the cycloalkanes with five-membered or larger rings are characteristically inert compounds. They are called **paraffins** (Latin *parum affinitas*, slight affinity) or **cycloparaffins**.

As there is very little polarity associated with these compounds and there are no nonbonded electrons in them, they do not generally react with concentrated acids and bases; for example, they are not affected by heating with concentrated sulfuric acid for extended periods of time. They are not affected by most strong oxidizing and reducing agents. They are truly inert to most reagents.

There is only a limited number of chemical reactions that can be carried out with alkanes and cycloalkanes, and most of these involve vigorous conditions. As we study these reactions, we find that we can understand why reactions occur and how variations in structure affect reactivity.

4.4   Halogenation: A Substitution Reaction

One of the few reactions of alkanes and cycloalkanes is their **halogenation.** This is a **substitution reaction** where the halogen is substituted for hydrogen. It may be shown in the most general form by

$$
\begin{array}{c}
R' \\
| \\
R-C-H + X_2 \\
| \\
R''
\end{array}
\xrightarrow[\text{or } h\nu]{\text{heat}}
\begin{array}{c}
R' \\
| \\
R-C-X + HX \\
| \\
R''
\end{array}
\qquad \text{Substitution of X for H}
$$

Alkane or cycloalkane,
R, R', or R'' = H or alkyl group

Usually a
mixture of
products

Reactivity: $F_2 \gg Cl_2 > Br_2 (> I_2)$

The most commonly used halogens (X_2) are chlorine, Cl_2, and bromine, Br_2. Iodine, I_2, does not react at all, and fluorine, F_2, is so reactive that the reaction is hard to control. Heat or ultraviolet light ($h\nu$) is usually required. As we will learn, this reaction is general for alkanes and cycloalkanes, and the reactivity depends on the structure of the hydrocarbon. The product distribution (that is, what products are formed) is also dependent on the hydrocarbon. The following are simple examples of this halogenation reaction in which only monosubstitution is considered:

$$CH_3-H + Br_2 \xrightarrow[h\nu]{125^\circ} CH_3-Br + HBr$$

Methane

Bromomethane
(Methyl bromide)

Cyclohexane

Chlorocyclohexane
(Cyclohexyl chloride)

Because the halogenation of alkanes and cycloalkanes often results in more than one product (that is, structural isomers) and even products that contain more than one halogen atom (that is, di-, tri-, and higher substituted compounds), the direct reaction of halogens with hydrocarbons is seldom used in the laboratory. It is, however, an important process in industry, where the various isomers can be separated by fractional distillation; usually all the products can be marketed. In this and the next chapter we cover the reaction in some detail because it provides an interesting example of how *reaction mechanisms* are deduced and used in organic chemistry. It also illustrates how we can use certain principles of reactivity to predict the product distribution for compounds that may never have been studied in the laboratory. Finally, we mention several other reagents that are used frequently in the laboratory for the halogenation of alkanes and cycloalkanes.

4.5 Examples of Halogenation of Alkanes

We now consider examples of the general halogenation reaction. With methane as the hydrocarbon, mixtures of methane and chlorine react vigorously when they are either heated to 300° or subjected to ultraviolet light at room temperature. Chlorine is substituted for hydrogen, and chloromethane (methyl chloride) is the first product formed:

$$CH_3—H + Cl—Cl \xrightarrow[\text{light, 25°}]{\text{300° or}} CH_3—Cl + H—Cl$$

Chloromethane   Hydrogen
(Methyl chloride)   chloride

Note, however, that chloromethane has three hydrogen atoms still attached to carbon, and on prolonged contact with chlorine gas, these hydrogens may be replaced by chlorine in a stepwise process:

$$CH_3Cl + Cl_2 \xrightarrow[\text{light, 25°}]{\text{300° or}} CH_2Cl_2 + HCl$$

Dichloromethane
(Methylene chloride)

$$\downarrow \text{Cl}_2, 300° \text{ or light, } 25°$$

$$CHCl_3 + HCl$$

Trichloromethane
(Chloroform)

$$\downarrow \text{Cl}_2, 300° \text{ or light, } 25°$$

$$CCl_4 + HCl$$

Tetrachloromethane
(Carbon tetrachloride)

A similar reaction occurs when bromine and methane are mixed, but more drastic conditions (125° and light) are required for bromination. Bromomethane is formed first, and then it may react further with bromine to yield di-, tri-, and tetra-bromo compounds.

As the preceding equations indicate, four products can be obtained from the halogenation of methane. We can control the reactions so that one of the four is the *major product*; however, the other three products are always formed to some extent. In the laboratory, this might be a problem, but industrially, where the halogenation of methane and other alkanes is widely used, all the products are marketed; they can be separated by fractional distillation, isolated in pure form, and sold. For example, trichloromethane has been used as an anesthetic and tetrachloromethane as a common cleaning agent that does not support combustion.

There are several variables one can manipulate to control a chemical reaction. The halogenation of methane (and other alkanes) can be controlled by varying the concentrations of the reactants, methane and halogen. We cannot form any dichloromethane until some chloromethane has been produced; in turn, trichloromethane must come from dichloromethane and tetrachloromethane must come from tri-chloromethane. To illustrate two extremes in the degree of substitution, consider the following. If a small amount of methane is mixed with a large excess—say, tenfold—of chlorine, we can be relatively certain that all the intermediate compounds (CH_3Cl, CH_2Cl_2, and $CHCl_3$) continue to react with the excess Cl_2 until the completely substituted CCl_4 compound forms. On the other hand, if we take a very large amount of methane—say, a tenfold excess—and a small amount of chlorine, it is much more likely for chlorine to react with a methane molecule than with a chloromethane molecule that has been formed.

If we start with equimolar amounts of chlorine and methane, we find quite a different picture. After the first few minutes of the reaction, chloromethane is formed and methane is disappearing. At some later time, there might be roughly equal amounts of chloromethane and methane, and CH_4 and CH_3Cl compete equally for chlorine; that is, CH_4 is converted to chloromethane while CH_3Cl is converted to CH_2Cl_2. Once formed, the dichloromethane reacts with chlorine to form tri-chloromethane, which can be converted in turn to tetrachloromethane. We might predict and do indeed find that a mixture of all four halogenated methane derivatives, as well as unreacted methane, is present when the reaction has stopped. The boiling points of these compounds are quite different (CH_4, $-161°$; CH_3Cl, $-24°$; CH_2Cl_2, $40°$; $CHCl_3$, $61°$; CCl_4, $77°$), and they can be separated by fractional distillation.

We will not burden ourselves with the particular experimental conditions (concentrations of reactants and temperatures) that are required to maximize the yields of these various mono-, di-, tri-, and tetrahalomethanes. You may assume that we can obtain any of them, but remember the experimental variations that are possible.

Let's now look at some other alkanes. When we allow ethane to react with bromine or chlorine, there is only one possible monosubstitution product:

$$CH_3—CH_3 \xrightarrow[\text{or } hv]{Cl_2, \text{ heat } (300°)} CH_3—CH_2—Cl$$

Ethane   Chloroethane
                                                    (Ethyl chloride)

$$\Big\downarrow \begin{smallmatrix} Br_2, \\ 125°, hv \end{smallmatrix}$$

$$CH_3—CH_2—Br$$

Bromoethane
(Ethyl bromide)

The halogenation of propane under typical free-radical halogenation conditions, however, results in two possible monosubstituted compounds. Furthermore, the product distribution from the chlorination of propane is different from that from bromination:

$$CH_3—CH_2—CH_3 \xrightarrow[25°]{Cl_2, hv} CH_3—CH_2—CH_2—Cl + CH_3—\overset{\displaystyle Cl}{\underset{\displaystyle |}{CH}}—CH_3$$

Propane   1-Chloropropane   2-Chloropropane
                           (n-Propyl chloride)   (Isopropyl chloride)
                                  45%   55%

$$\xrightarrow{Br_2, 125°, hv} CH_3—CH_2—CH_2—Br + CH_3—\overset{\displaystyle Br}{\underset{\displaystyle |}{CH}}—CH_3$$

                           1-Bromopropane   2-Bromopropane
                           (n-Propyl bromide)   (Isopropyl bromide)
                                  3%   97%

The number of possible structural isomers increases with increasing carbon content in the alkane. The halogenation of butane and 2-methylpropane is illustrated as follows:

$$CH_3-CH_2-CH_2-CH_3 \xrightarrow[25°]{Cl_2, hv} CH_3-CH_2-CH_2-CH_2-Cl + CH_3-CH_2-\overset{\overset{\displaystyle Cl}{|}}{C}H-CH_3$$

Butane	1-Chlorobutane	2-Chlorobutane
	(*n*-Butyl chloride)	(*sec*-Butyl chloride)
	28%	*72%*

$$\xrightarrow{Br_2, 125°, hv} CH_3-CH_2-CH_2-CH_2-Br + CH_3-CH_2-\overset{\overset{\displaystyle Br}{|}}{C}H-CH_3$$

1-Bromobutane	2-Bromobutane
(*n*-Butyl bromide)	(*sec*-Butyl bromide)
2%	*98%*

$$CH_3-\overset{\overset{\displaystyle CH_3}{|}}{C}H-CH_3 \xrightarrow[25°]{Cl_2, hv} CH_3-\overset{\overset{\displaystyle CH_3}{|}}{C}H-CH_2-Cl + CH_3-\overset{\overset{\displaystyle Cl}{|}}{\underset{\underset{\displaystyle CH_3}{|}}{C}}-CH_3$$

2-Methylpropane	1-Chloro-2-methylpropane	2-Chloro-2-methylpropane
(Isobutane)	(Isobutyl chloride)	(*tert*-Butyl chloride)
	64%	*36%*

$$\xrightarrow{Br_2, 125°, hv} CH_3-\overset{\overset{\displaystyle CH_3}{|}}{C}H-CH_2-Br + CH_3-\overset{\overset{\displaystyle Br}{|}}{\underset{\underset{\displaystyle CH_3}{|}}{C}}-CH_3$$

1-Bromo-2-methylpropane	2-Bromo-2-methylpropane
(Isobutyl bromide)	(*tert*-Butyl bromide)
1%	*99%*

The three pentane isomers can be halogenated, but the number of mono-substitution products greatly increases. Pentane yields three isomers, 2-methylbutane gives four isomers, and 2,2-dimethylpropane produces only a single compound.

It should now be clear that the halogenation of alkanes is a complex problem. As the previous yield data indicate, each new alkane gives a different product distribution—that is, different percentages of the monosubstitution products. It would not be enjoyable to burden ourselves with committing these percentages (and, of course, many more if we were to consider other alkanes) to memory, and fortunately, it is not necessary to do so! Once again the theories and understanding of organic chemistry come to our rescue and provide a method for explaining and determining both product distribution and selectivity in the halogenation of alkanes. A complete discussion of these theories is presented in Chap. 5.

Question 4.6

Draw the structural isomers of the: (*a*) monobromo derivatives of pentane, (*b*) monobromo derivatives of 2-methylbutane, and (*c*) monobromo derivatives of 2,2-dimethylpropane.

Question 4.7

There is only one monochloro derivative of cyclohexane, but there are eight monochloro derivatives of methylcyclohexane. Draw them. (*Hint:* Consider structural and geometric isomers.)

4.6   Free-Radical Halogenation: A Chain Reaction

All organic reactions occur through a mechanism. We can summarize the mechanism for the halogenation of a typical alkane such as methane as follows (chlorination is used in the example):

Step 1:

$$:\ddot{C}l:\ddot{C}l: \xrightarrow[\text{or } h\nu]{\text{heat}} :\ddot{C}l\cdot + :\ddot{C}l\cdot$$

Chlorine atoms

Step 2:

$$CH_3:H + :\ddot{C}l\cdot \longrightarrow CH_3\cdot + H:\ddot{C}l:$$

Methyl radical,
electron-deficient

Step 3:

$$CH_3\cdot + :\ddot{C}l:\ddot{C}l: \longrightarrow CH_3-Cl + :\ddot{C}l\cdot$$

Chloromethane
(product)

The mechanism is called **free-radical halogenation** or sometimes a **free-radical chain reaction,** and it is consistent with what is known experimentally. In this example, we know that heat or light is required to begin the reaction, but that once started (or *initiated*), the reaction proceeds for a time on its own (with the liberation of heat, an exothermic reaction) until the limiting reactant (the one that is used up first), CH_4 or Cl_2, is consumed.

In the previous equations, the curved arrows ($\smile$ and $\frown$) are again used to show electron flow, the directions in which the electrons move in the course of bonds breaking and forming. For example, in step 1, as the chlorine-chlorine bond breaks, one electron from the bond goes to each chlorine atom as it forms. Single-headed arrows ($\frown$) designate the flow of one electron, and full arrows ($\frown$) designate the flow of an electron pair.

If we look closely at the three steps in the reaction leading to the chlorination of methane, we find that they represent a **chain reaction.** The reaction starts by forming chlorine atoms (step 1), which are highly reactive. This is why heat or light is needed to break the weak Cl—Cl bond. The chlorine atoms are consumed when they react with methane to form the methyl radical and hydrogen chloride (step 2). However, they are regenerated when the methyl radical reacts with chlorine gas to form chloromethane and a new chlorine atom (step 3), which can react with another molecule of methane (or even CH_3Cl, once it is formed, and later with CH_2Cl_2 and $CHCl_3$) and thus start the reaction sequence (steps 2 and 3) over again. *A chain reaction is defined as one in which reactive intermediates are continually consumed and regenerated in a series of propagation steps.* This process can continue, theoretically at least, until one of the reactants is completely consumed.

No reaction can occur until some chlorine atoms are generated, and because energy is required for their formation, we call this *step 1* of the reaction the **chain-**

initiating step. Once the reaction is initiated, however, the **chain-propagating steps** (*steps 2 and 3*) continue the reaction through many cycles. The concentration of the chlorine atoms is always quite small in this sequence.

These three reactions produce a radical. The chlorination of methane stops by the recombination of these highly reactive radicals, which are present in very small concentrations:

Step 4:

$$:\ddot{C}l\cdot + \cdot\ddot{C}l: \longrightarrow :\ddot{C}l-\ddot{C}l:$$

or

$$CH_3\cdot + \cdot CH_3 \longrightarrow CH_3-CH_3$$
Ethane

or

$$CH_3\cdot + \cdot\ddot{C}l: \longrightarrow CH_3-\ddot{C}l:$$

These reactions represent all the possible combinations and permutations of the methyl radical and the chlorine atom. Note that the combination of two methyl radicals results in ethane, which is usually present in very small (trace) amounts in the halogenation of methane. Ethane can result only from a reaction involving methyl radicals, and its formation further supports their existence in this reaction.

Depending on the reaction conditions, the chain-propagating steps occur 100 to 10,000 times for each photon of light absorbed before the reaction is terminated (*step 4*), with the average being about 5,000 repeating cycles. This accounts for the yield of substitution product (chloromethane) based on the amount of energy supplied originally.

As in physics, where perpetual motion machines are unknown, chemical chain reactions also terminate, even though in theory they might continue for as long as reactants are present. The atoms and molecules lose energy as they collide with the walls of the container or with other molecules, and the atoms and radicals may be consumed by reacting with one another (*step 4*). Because their recombination uses up the chlorine atoms and methyl radicals, these reactions are called **chain-terminating steps.**

The steps are now summarized for the generalized halogenation of an alkane, R—H:

Chain-initiating step:

$$X_2 \xrightarrow{\text{slow}} 2:\ddot{X}\cdot \qquad (X_2 = Cl_2 \text{ or } Br_2)$$

Chain-propagating steps:

$$R-H + :\ddot{X}\cdot \xrightarrow[\text{determining step}]{\text{slowest, rate-}} R\cdot + H-X$$
$$R\cdot + X_2 \xrightarrow{\text{fast}} R-X + :\ddot{X}\cdot$$

These two steps repeat until termination

Chain-terminating steps:

$$:\ddot{X}\cdot + :\ddot{X}\cdot \longrightarrow X_2$$
$$R\cdot + :\ddot{X}\cdot \longrightarrow R-X$$
$$R\cdot + R\cdot \longrightarrow R-R$$

A carbon atom undergoes several changes during the halogenation reaction. We can summarize all these changes as in the following equation, which uses methane as a typical alkane:

| Reactant (tetrahedral, sp^3 hybridized carbon atom) | Activated complex or transition state (carbon developing radical character and becoming trigonal; C---H bond breaking and H---X bond forming) | Intermediate (trigonal, sp^2 hybridized carbon atom, with p orbital containing odd electron perpendicular to plane of CH_3 atoms) |

| Product (tetrahedral, sp^3 hybridized carbon atom) | Activated complex or transition state (carbon losing radical character and becoming tetrahedral again; X---X bond breaking and C---X bond forming) |

We should distinguish carefully between the terms **transition state**[1] and **intermediate.** Both are transient species. *A transition state structure is one in which bond making and/or bond breaking is occurring. An intermediate is a discrete structure that results from a chemical reaction carried out on a suitable precursor.* Reactions often have several transition states and intermediates associated with them. These definitions are elaborated on in Sec. 5.5.

We cannot see the transition states, nor can we see the changes in hybridization that have occurred. Remember that these numerous changes occur during the progress of the reaction, and we presented some types of evidence that have been amassed to suggest their intervention.

The mechanism of free-radical halogenation is discussed further in Chap. 5 (Reaction Mechanisms). We defer discussion of the energy changes involved in each step of the reaction until then.

4.7   Sulfuryl Chloride and Halogenation of Alkanes

Sulfuryl chloride, SO_2Cl_2, reacts with alkanes and cycloalkanes in the presence of peroxides, RO—OR. Benzoyl peroxide, C_6H_5—C—O—O—C—C_6H_5 is most often used as the peroxide because it catalyzes the reaction at 60 to 80°. The overall

---

[1] The term **activated complex** is also used to designate the transition state.

reaction is

$$R—H + Cl—\overset{\overset{O}{\|}}{\underset{\underset{O}{\|}}{S}}—Cl \xrightarrow[\text{60–80°}]{\text{trace of benzoyl peroxide}} R—Cl + SO_2 + HCl$$

Sulfuryl
chloride

This free-radical reaction is initiated by the decomposition of benzoyl peroxide into radicals, which in turn react with sulfuryl chloride to generate chlorine atoms:

Chain-initiating steps:

$$C_6H_5—\overset{\overset{O}{\|}}{C}—\overset{\cdot\cdot}{\underset{\cdot\cdot}{O}}\!:\!\overset{\cdot\cdot}{\underset{\cdot\cdot}{O}}—\overset{\overset{O}{\|}}{C}—C_6H_5 \xrightarrow{\text{60–80°}} 2C_6H_5—\overset{\overset{O}{\|}}{C}—\overset{\cdot\cdot}{\underset{\cdot\cdot}{O}}\!\cdot$$

$$C_6H_5—\overset{\overset{O}{\|}}{C}—\overset{\cdot\cdot}{\underset{\cdot\cdot}{O}}\!\cdot \longrightarrow C_6H_5\!\cdot + CO_2$$

$$\left.\begin{array}{c} C_6H_5—\overset{\overset{O}{\|}}{C}—\overset{\cdot\cdot}{\underset{\cdot\cdot}{O}}\!\cdot \\ \text{or} \\ C_6H_5\!\cdot \end{array}\right\} + Cl\!:\!\overset{\overset{O}{\|}}{\underset{\underset{O}{\|}}{S}}—Cl \longrightarrow \left.\begin{array}{c} C_6H_5—\overset{\overset{O}{\|}}{C}—\overset{\cdot\cdot}{\underset{\cdot\cdot}{O}}—Cl \\ \text{or} \\ C_6H_5—Cl \end{array}\right\} + \cdot\overset{\overset{O}{\|}}{\underset{\underset{O}{\|}}{S}}—Cl$$

$$\cdot\overset{\overset{O}{\|}}{\underset{\underset{O}{\|}}{S}}\!:\!\overset{\cdot\cdot}{\underset{\cdot\cdot}{Cl}}\!: \longrightarrow :\overset{\overset{O}{\|}}{\underset{\underset{O}{\|}}{S}} + :\overset{\cdot\cdot}{\underset{\cdot\cdot}{Cl}}\!\cdot$$

For example,

$$\underset{\text{2,3-Dimethylbutane}}{CH_3—\overset{\overset{CH_3}{|}}{\underset{\underset{H}{|}}{C}}—\overset{\overset{CH_3}{|}}{\underset{\underset{H}{|}}{C}}—CH_3} \xrightarrow[hv]{SO_2Cl_2} \underset{\substack{\text{1-Chloro-2,3-}\\\text{dimethylbutane}\\\textit{38\%}}}{CH_3—\overset{\overset{CH_3}{|}}{\underset{\underset{H}{|}}{C}}—\overset{\overset{CH_3}{|}}{\underset{\underset{H}{|}}{C}}—CH_2—Cl} + \underset{\substack{\text{2-Chloro-2,3-}\\\text{dimethylbutane}\\\textit{62\%}}}{CH_3—\overset{\overset{CH_3}{|}}{\underset{\underset{H}{|}}{C}}—\overset{\overset{CH_3}{|}}{\underset{\underset{Cl}{|}}{C}}—CH_3}$$

4.8   Nitration of Alkanes: Another Free-Radical Reaction

Another reaction of alkanes is **nitration,** in which a nitro group, —NO_2, is substituted for a hydrogen. This reaction requires exceptionally vigorous conditions and provides very small yields of product. With methane, the only product is nitromethane, CH_3—NO_2; this reaction is carried out with methane and concentrated nitric acid at 400 to 500° and under high pressure:

$$CH_4 + HO—NO_2 \xrightarrow[\text{pressure}]{\text{400–500°}} CH_3—NO_2 + H_2O$$

$$(HNO_3) \qquad\qquad \text{Nitromethane}$$

With higher alkanes, a mixture of products is obtained, some of which result from rupturing carbon-carbon bonds. For example, the nitration of ethane results in a mixture of nitromethane and nitroethane:

$$CH_3—CH_3 + HO—NO_2 \xrightarrow[\text{pressure}]{400-500°} CH_3—CH_2—NO_2 + CH_3—NO_2 + H_2O$$

$$\qquad\qquad\qquad\qquad\qquad\qquad\qquad\text{Nitroethane}\qquad\text{Nitromethane}$$

The mechanism for nitration has been postulated to occur by a free-radical route. It is most likely *initiated* by the thermal splitting of a few alkane molecules into free radicals:

$$CH_3CH_2{:}H \xrightarrow{400-500°} CH_3CH_2{\cdot} + H{\cdot}$$

and

$$CH_3{:}CH_3 \xrightarrow{400-500°} 2CH_3{\cdot}$$

Chain *propagation* probably occurs as follows:

$$CH_3CH_2{\cdot} + HO{:}NO_2 \longrightarrow CH_3CH_2—NO_2 + \quad H\ddot{O}{\cdot}$$

$$\qquad\qquad\qquad\qquad\qquad\qquad\qquad\qquad\qquad\text{Hydroxyl}$$
$$\qquad\qquad\qquad\qquad\qquad\qquad\qquad\qquad\qquad\text{radical}$$

$$CH_3CH_2{:}H + H\ddot{O}{\cdot} \longrightarrow CH_3CH_2{\cdot} + H_2\ddot{O}{:}$$

Chain *termination* involves the recombination of two free radicals.

Nitration of alkanes is exclusively a *commercial* process. As a result of the low yields and the mixture of isomers formed, aliphatic nitro compounds are seldom encountered. On the other hand, we will see in Chap. 14 that aromatic nitro compounds are easily prepared and widely used.

4.9   Combustion

One of the most important reactions of alkanes and cycloalkanes is their combustion in air to produce carbon dioxide and water. Combustion is accompanied by the liberation of heat. Because many hydrocarbons are obtained from natural gas and petroleum, they are inexpensive and very useful fuels. The heat produced by combustion is significant and is called the **heat of combustion.** For the burning of methane (natural gas), the combustion reaction

$$CH_4 + 2O_2 \longrightarrow CO_2 + 2H_2O$$

liberates 213 kcal/mole (890 kJ/mole) of methane. Propane liberates 526 kcal/mole (2,199 kJ/mole).

The inertness of hydrocarbons can be understood when we consider that energy, supplied by either a flame or a spark, is required to start a combustion reaction. The reaction then gives off heat, and that energy is usually enough to allow combustion to continue.

4.10   Petroleum, Gasoline, Fuels, and Pollution

The gasoline fraction from petroleum is called *straight-run* gasoline. It is a poor grade of gasoline and has poor performance as a fuel. Branched-chain alkanes are superior fuels to straight-chain compounds because they are more volatile, burn less rapidly in the cylinder, and thus reduce knocking. The alkenes (carbon-carbon double-bonded compounds) and aromatic compounds are also good fuels. Methods have been developed to convert hydrocarbons of higher and lower molecular weights than gasoline to that range, or to convert straight-chain hydrocarbons into branched ones. The branched-chain alkanes, alkenes, and aromatic compounds are obtained from petroleum hydrocarbons by *catalytic cracking* and *catalytic reforming* (see Secs. 4.1, 7.10, and 13.11). Branched-chain alkanes are prepared from alkenes and alkanes by alkylation reactions and from alkanes by *catalytic isomerization* (see Secs. 7.10, 8.23, and 8.24).

The antiknock quality of a fuel is measured by its *octane number*. Heptane is a very poor fuel that causes severe knocking in an automobile engine. It is given arbitrarily an octane number of 0. On the other hand, 2,2,4-trimethylpentane (which is *incorrectly* known as *isooctane* in the petroleum industry—a name still used because of the importance of this compound) is an excellent fuel and is given the arbitrary octane number of 100. Fuels are compared in a test engine with varying mixtures of heptane and 2,2,4-trimethylpentane. A gasoline that exhibits a performance equal to that of a mixture of 10 mole% heptane and 90 mole% 2,2,4-trimethylpentane has an octane rating of 90. Some hydrocarbons known today are better than 2,2,4-trimethylpentane, as shown in Table 4.3.

One way to improve the performance of gasoline is to add branched-chain alkanes that have a higher octane rating. Another is to add small amounts of tetraethyl lead, $(CH_3CH_2)_4Pb$, which yields *ethyl* gasoline. Tetraethyl lead, a highly toxic, colorless liquid, was discovered in 1922. On combustion, tetraethyl lead forms water, carbon dioxide, and lead oxide, which builds up and is harmful to the cylinder. 1,2-Dibromoethane, $BrCH_2CH_2Br$, is often added to ethyl gasoline, and it combines with the lead oxide to form volatile lead bromide, $PbBr_2$, which passes out with the exhaust fumes.

TABLE 4.3   Octane Numbers of Hydrocarbons

Compound	Octane No.
Heptane	0
2-Methylheptane	24
2-Methylpentane	71
Octane	−20
2-Methylbutane (*isopentane*)	90
2,2,4-Trimethylpentane (*isooctane,*	100
Benzene	101
Toluene	110
2,2,3-Trimethylpentane (*triptane*)	116
Cyclopentane	122
p-Xylene	128

The use of leaded gasolines has the disadvantage that toxic lead compounds pollute the atmosphere. The best solution seems to be to use highly branched alkanes, which can be prepared by isomerizing straight-chain alkanes.

4.11   Air Pollution and Its Causes; Mechanism of Photochemical Smog Formation

Air pollution is a serious problem today; it has been attributed largely to exhaust fumes from the internal combustion engine. Carbon monoxide and smog are perhaps the most vile air polluters. Although the formation of smog is complex and not completely understood, the major reactants appear to be branched-chain hydrocarbons (which are emitted from automobile exhausts because of incomplete combustion), sunlight, oxides of nitrogen (which are formed in the automobile engine), and oxygen.

We have some idea of how photochemical smog (smog produced by sunlight) is caused. The oxides of nitrogen are formed in the combustion engine, which takes in nitrogen and oxygen from the atmosphere and produces nitric oxide:[1]

$$N_2 + O_2 \longrightarrow 2NO$$

Nitric
oxide

The nitric oxide reacts readily with more oxygen from the atmosphere to form nitrogen dioxide, NO_2:

$$NO + \tfrac{1}{2}O_2 \longrightarrow NO_2$$

At this juncture, photochemical reactions are believed to take over. Smog production is initiated by an NO_2 molecule absorbing a photon of ultraviolet light from the sun. The activated NO_2 molecule, designated NO_2^*, then dissociates into NO and an atom of oxygen:

$$NO_2 + h\nu \longrightarrow NO_2^* \longrightarrow NO \;+\; :\overset{\cdot}{\underset{\cdot\cdot}{O}}\cdot$$

Oxygen atom
(contains two
unpaired electrons,
electron-deficient)

The oxygen atom is *very reactive* and behaves much like the halogen radical, $:\overset{\cdot\cdot}{\underset{\cdot\cdot}{X}}\cdot$, that we encountered in the free-radical halogenation of alkanes and cycloalkanes. The oxygen atom may, for example, react with molecular oxygen to form ozone, O_3, which is another pollutant in smog:

$$O_2 + :\overset{\cdot}{\underset{\cdot\cdot}{O}}\cdot \longrightarrow O_3$$

Ozone

Among other things, ozone attacks and cracks rubber and damages plants. The ozone thus formed may react with other NO molecules to generate more NO_2, which starts the cycle again.

$$O_3 + NO \longrightarrow NO_2 + O_2$$

---

[1] The nitrogen oxides that are emitted result in part from the combustion of nitrogen-containing organic compounds.

Some oxygen atoms react with nearly any kind of hydrocarbon to produce new organic free radicals:

$$\text{R—H} + :\overset{\cdot\cdot}{\underset{\cdot\cdot}{O}}\cdot \longrightarrow \text{R—}\overset{\cdot\cdot}{\underset{\cdot\cdot}{O}}\cdot$$

Hydrocarbon   Organic free radical
(very reactive)

These may in turn react with molecular oxygen to form even more reactive free radicals:

$$\text{R—}\overset{\cdot\cdot}{\underset{\cdot\cdot}{O}}\cdot + O_2 \longrightarrow \text{R—}O_3\cdot$$

Organic free radical
(exceedingly reactive)

This mixture of free radicals can react further with hydrocarbons to yield a variety of oxygen-containing molecules, many of which are undesirable. One particularly noxious eye irritant is peroxy acyl nitrate (*PAN*), $CH_3\overset{\displaystyle O}{\overset{\|}{—C}}\text{—O—O—NO}_2$, which is formed by the combination of free radicals with NO_2.

The steps involved in the production of photochemical smog are another example of a free-radical chain reaction. We can classify the various steps as follows:

Chain-initiating step:

$$N_2 + O_2 \longrightarrow 2NO$$

Chain-propagating steps:

$$NO + \tfrac{1}{2}O_2 \longrightarrow NO_2$$

$$NO_2 + h\nu \longrightarrow NO_2^* \longrightarrow NO + :\overset{\cdot}{\underset{\cdot\cdot}{O}}\cdot$$

$$O_2 + :\overset{\cdot}{\underset{\cdot\cdot}{O}}\cdot \longrightarrow O_3$$

$$O_3 + NO \longrightarrow NO_2 + O_2$$

The chain-terminating steps are difficult to specify, but they undoubtedly involve the reaction between hydrocarbon, R—H, and the oxygen atom, $:\overset{\cdot\cdot}{\underset{\cdot}{O}}\cdot$, to form products.

4.12   Ozone and the Environment

Ozone, O_3, plays a vital role in our daily lives. As mentioned in Sec. 4.11, it is a major constituent of photochemical smog. In this capacity, ozone is an oxidant and a primary source of atomic oxygen, which, as we saw, can participate in many undesirable reactions.

$$O_3 \xrightarrow[\text{light}]{\text{UV}} O_2 + \cdot\overset{\cdot\cdot}{\underset{\cdot\cdot}{O}}\cdot$$

In another capacity, however, ozone is necessary for life. The earth is constantly irradiated with ultraviolet (UV) light from the sun. The light is so intense that if it passes through our atmosphere unchanged, it could greatly increase the incidence of skin cancer and cataracts in human beings. Ozone, however, is present in the upper atmosphere and absorbs a significant portion of the ultraviolet radiation before it gets to the surface of the earth.

This property of ozone has led to several major confrontations between environmental and commercial interests. The first argument was over the supersonic transport (SST) plane. The commercial failure of these planes was caused, to a large extent, by the work of environmentalists who feared that the high concentrations of NO emitted from the exhaust of the plane would severely deplete the ozone layer in the atmosphere, where the following reaction is known to occur:

$$NO + O_3 \longrightarrow NO_2 + O_2$$

A new confrontation developed recently surrounding the use of Freons as propellants in aerosol sprays. Freons are a general class of chlorofluoromethanes; dichlorodifluoromethane, *Freon 12*, is the most common. Although the Freons are relatively unreactive, in the presence of UV light they can break down to give halogen atoms as one of the products. The halogen atoms can then react with ozone:

$$CX_4 \longrightarrow \cdot CX_3 + \ddot{\underset{\cdot\cdot}{X}}\cdot \qquad (X = Cl \text{ or } F)$$

$$\ddot{\underset{\cdot\cdot}{X}}\cdot + O_3 \longrightarrow [XO]\cdot + O_2$$

The overall result, as in the case of the SST, would be the eventual depletion of the ozone layer. Since the mid-1970s, other non-halogen-containing propellants have replaced the Freons.

4.13   Summary of Reactions

Methods of Preparation of Alkanes and Cycloalkanes

Coupling Reactions

1. Lithium dialkylcopper reagent:

$$R\!-\!X \xrightarrow{\text{Li}} R\!-\!Li \xrightarrow{\text{CuI}} R_2CuLi \qquad \text{Sec. 4.2A}$$

1°, 2°, or 3°
Alkyl halide

$$R_2CuLi \;+\; R'X \longrightarrow R\!-\!R'$$

1° Alkyl halide
or methyl halide

2. Cycloalkanes:

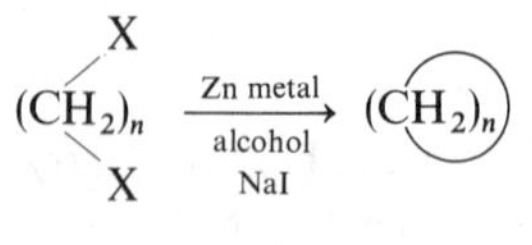

$$(CH_2)_n \xrightarrow[\substack{\text{alcohol} \\ \text{NaI}}]{\text{Zn metal}} (CH_2)_n \qquad \text{Sec. 4.2B}$$

$n = 3\text{–}6$

Reduction Reactions

1. Using metal and acid:

$$R\!-\!X \xrightarrow{\text{Zn, H}^\oplus} R\!-\!H \qquad \text{Section 4.2C}$$

2. Hydrogenation of alkenes:

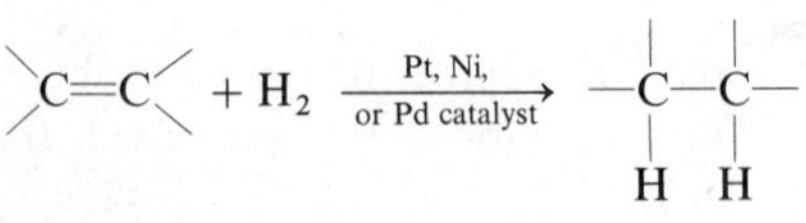

$$\underset{}{C\!=\!C} + H_2 \xrightarrow[\text{or Pd catalyst}]{\text{Pt, Ni,}} \underset{H\ \ H}{-\!C\!-\!C\!-} \qquad \begin{array}{l}\text{Introduced in Sec. 4.2D,} \\ \text{discussed in Sec. 8.12}\end{array}$$

Reactions of Alkanes and Cycloalkanes

Halogenation

1. $R—H + X_2 \xrightarrow[\text{light}]{\text{heat and/or}} R—X + HX$ General reaction: Sec. 5.16

   Reactivity

   X_2: $F_2 \gg Cl_2 > Br_2 (>I_2)$ Halogenation of alkanes: Sec. 4.5, Chap. 5

   H: $3° > 2° > 1° > CH_3—H$ Mechanism of halogenation: Sec. 4.6
   Methyl radical structure: Sec. 2.8

2. $R—H + SO_2Cl_2 \xrightarrow{\text{peroxides}} R—Cl + SO_2 + HCl$ Sec. 4.7

   ### Nitration

   $$R—H + HNO_3 \xrightarrow[\text{high pressure}]{400-500°} R—NO_2 + H_2O \qquad \text{Sec. 4.8}$$

   ### Combustion

   $$R—H + \text{excess } O_2 \longrightarrow CO_2 + H_2O + \text{heat} \qquad \text{Sec. 4.9}$$

Study Questions

4.8 Define or explain the following terms.

(*a*) R group
(*b*) Wurtz reaction (*Hint:* See Question 4.3.)
(*c*) oxidation state
(*d*) free-radical halogenation
(*e*) alkyl radicals
(*f*) halogen atoms
(*g*) chain-initiating step
(*h*) chain-propagating step
(*i*) chain-terminating step
(*j*) transition state versus intermediate
(*k*) activated complex
(*l*) sp^2 hybridized carbon
(*m*) product distribution
(*n*) type of carbon and hydrogen (for example, 1°, 2°, and so on)

4.9 Complete the following reactions by drawing the structures of all the organic products formed in each step. If no reaction occurs, write "no reaction."

(*a*) hexane + H_2SO_4(conc) $\longrightarrow$
(*b*) propane $\xrightarrow[hv]{Cl_2, 25°}$ monosubstitution

(*c*) isobutane $\xrightarrow[hv]{Cl_2, 25°}$ monosubstitution
(*d*) isobutane $\xrightarrow[hv]{Br_2, 125°}$ monosubstitution

(*e*) $CH_3(CH_2)_5CH_2—Br \xrightarrow[\text{heat}]{Na°}$ (*Hint:* For (*e*) to (*g*), see Question 4.3.)

(*f*) ⬡—$CH_2—Cl \xrightarrow[\text{heat}]{Na°}$
(*g*) $CH_3CH_2CH_2—Br + (CH_3)_2CH—Br \xrightarrow[\text{heat}]{Na°}$

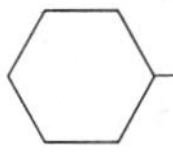

(2 moles)

(*h*) $Br—CH_2CH_2CH_2—Br \xrightarrow[\text{NaI}]{\text{Zn, alcohol}}$
(*i*) $Br—\underset{\underset{CH_3}{|}}{CH}—CH_2—\underset{\underset{CH_3}{|}}{CH}—Br \xrightarrow[\text{NaI}]{\text{Zn, alcohol}}$

(*j*) $C_8H_{18} + O_2$ (excess) $\longrightarrow$ complete combustion (balance this equation)

4.10 In the following three reactions, determine the oxidation state of the carbon atom in red. In which reactions, if any, has carbon been oxidized and in which reactions, if any, has it been reduced?

(*a*) $CH_4 + Cl_2 \xrightarrow[hv]{25°} CH_3—Cl + HCl$

(*b*) $(CH_3)_2CH_2 + Br_2 \xrightarrow[hv]{125°} (CH_3)_2CBr_2 + 2HBr$

(*c*) $CH_4 + 2O_2 \longrightarrow CO_2 + 2H_2O$ (combustion)

4.11 Illustrating with suitable drawings or sketches, describe the geometry of the ethyl radical $(CH_3CH_2\cdot)$. Include the shapes and orientation of the molecular bond orbitals and atomic orbitals (if any). Indicate all angles.

4.12 Consider the reaction:

$$CH_3-CH_2-CH_3 + Br_2 \xrightarrow[hv]{125°} CH_3-\underset{\underset{Br}{|}}{CH}-CH_3 + HBr$$

(**a**) Provide a *complete, stepwise* mechanism for this reaction. Clearly indicate and label the initiating step(s), the propagating step(s), and the terminating step(s). Show all steps, and indicate how your mechanism is a chain reaction.

(**b**) Describe what *changes* occur in the *hybridization* and *shape* of the carbon atom as the reaction proceeds (that is, during the time the C—H bond undergoing reaction is broken and the new C—Br bond is made). When describing the changes in shape, give approximate bond angles insofar as possible. Identify the transition state(s) and intermediate(s) that are encountered in this transformation.

4.13 Consider structural isomers of molecular formula C_6H_{14}.

(**a**) Which one reacts with chlorine and light to give two and only two monochloro products? Draw condensed structural formulas for it and the two products.

(**b**) Which isomer gives three and only three different monochloro products? Draw condensed structural formulas for it and the three products.

(**c**) Which isomer gives precisely four different monochloro products? Draw condensed structural formulas for it and the four products.

4.14 Complete the following reaction sequence by supplying structures for (*A*) to (*C*).

$$(A) \xrightarrow{\text{Li}} (B) \xrightarrow{\text{CuI}} (C) \xrightarrow{CH_3CH_2CH_2Br} CH_3CH_2CH_2CH(CH_3)_2$$

5

Reaction Mechanisms

A reaction mechanism is the step-by-step process that we perceive as occurring when reactants are converted to products. By considering two reactions, *free-radical substitution* (a familiar mechanism from Sec. 4.6) and *nucleophilic aliphatic substitution* (Sec. 4.2A), we present the terminology used by chemists in discussing mechanisms. We discuss the thermodynamics, kinetics, and geometric and structural changes associated with reactions and reaction mechanisms. Also, the common reactive intermediates of organic chemistry (free radicals, carbocations, and carbanions) are described in relation to their modes of formation, stability, and reactions. We see how chemists elucidate mechanisms and then how these mechanisms can be used by the enterprising chemist in organic synthesis.

5.1   What Is a Reaction Mechanism?

To understand what a reaction mechanism is, one must first know what a chemical reaction is. An astute description is provided by Professor B. Widom of Cornell University (*Science*, page 1555, 1965):

A chemical reaction is the result of a complex sequence of elementary molecular processes. Each molecule in a reaction mixture undergoes continual changes in its state—in its energy, geometry, and so on—most of these changes, perhaps, being minute and scarcely noticeable, others being sudden and violent as the result of extremely disruptive collisions. The processes are apparently aimless, each molecule being excited and de-excited with no obvious regularity. But when we view the reaction mixture as a macroscopic system the molecular turmoil is invisible, and what we see instead is a seemingly purposeful drive to equilibrium, rapid at first, then slower, until finally equilibrium is achieved and the chemical reaction is at an end.

A reaction mechanism is our view of this step-by-step drive to equilibrium. The mechanism shows which bonds are broken and formed and the order in which these changes occur. It is an effort to describe the changes in state, hybridization, carbon skeleton, functional group, and geometry that take place and the energy changes accompanying these processes. A reaction mechanism is, in a sense, a road map of the path followed by reactants in becoming products.

In our study of organic chemistry, the mechanisms associated with the majority of chemical transformations are presented and discussed. Although many mechanisms are well understood, others remain in large part a mystery.

5.2   Chemical Reactions: A Graphical Representation

A chemical reaction involves the conversion of reactants to products. Reactions may be *unimolecular* or *monomolecular* (involving one molecule) or *bimolecular* (involving two molecules); in only rare instances are more than two molecules involved. For the purpose of this discussion, we consider a generalized bimolecular reaction between two molecules, *A* and *B*:

$$A + B \longrightarrow C + D$$
Reactants   Products

How does this conversion occur? To begin, the two molecules come together through a collision of the two reacting species. If this collision is of sufficient energy

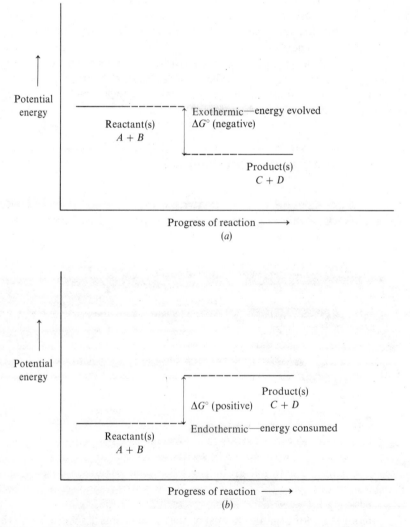

FIGURE 5.1   Partial graphical representations of (*a*) exothermic and (*b*) endothermic reactions for the generalized reaction $A + B \rightarrow C + D$.

and if the molecules are in a correct orientation with respect to one another, products are formed. There may be many unsuccessful attempts before a collision is successful and products result.

Regardless of the number of molecules involved in the collision or the actual chemical changes that occur (bond breaking or forming, conformational changes, changes in hybridization, and so on), the reaction may be described in terms of the energy changes in going from reactants to products. We can plot the change in free energy ($\Delta G°$) in progressing from reactants to products as a function of time. This is the same type of plot used in Sec. 3.16 when we discuss the free-energy changes that accompany conformational interconversions (which also may be viewed as a reactant converting to a product). Plots of a generalized **exothermic reaction** (in which energy is evolved in the course of the reaction) and an **endothermic reaction** (in which energy is consumed) are given in Fig. 5.1. In an exothermic reaction the products are of lower energy than the reactants, whereas the reverse is true of an endothermic reaction. A minus sign ($-$) is used to designate exothermic reactions and a plus sign ($+$) for endothermic reactions. The amounts of energy are commonly given in kilocalories per mole (kcal/mole) or kilojoules per mole (kJ/mole).

5.3 Thermodynamics

Before proceeding in our discussion of chemical reactions and the energy changes that accompany them, a discussion of the common thermodynamic parameters used is in order.

Thermodynamics is the branch of science dealing with changes in energy, often in view of the work performed by or on the system being studied. Many energy changes are possible in a given chemical reaction. Thus far, we have been concerned with the change in free energy ($\Delta G°$) (that is, potential energy) of a system. The total change in free energy, however, is a combination of two thermodynamic parameters, the enthalpy ($\Delta H°$) and entropy ($\Delta S°$) of the reaction, which are related by the following equation:

$$\Delta G° = \Delta H° - T\Delta S° \qquad (T = \text{absolute temperature, °K})$$

The enthalpy of a reaction ($\Delta H°$) is also called the *heat of reaction*. It is a measure of the amount of heat evolved (exothermic) or consumed (endothermic) in the course of a reaction and is given in kilocalories per mole or kilojoules per mole. The entropy ($\Delta S°$) is a measure of the randomness or order of the system. In going from a more ordered to a less ordered system, energy is released; the reverse is true in going from a less ordered to a more ordered system. In other words, nature (and chemical reactions) prefers a random, unordered system. The units associated with $\Delta S°$ are entropy units (eu), where 1 eu = 1 calorie per degree (4.184 joules per degree).

The more negative the $\Delta H°$ and the more positive the $\Delta S°$, the more negative the $\Delta G°$ and the more exothermic (favorable) the reaction. When discussing conformational changes (see Sec. 3.16), one discusses the total changes in free energy that occur ($\Delta G°$). However, when discussing most chemical reactions involving bond breaking and forming, one normally discusses changes in enthalpy ($\Delta H°$). In many organic reactions, the entropy change is most often quite small in relation to the change in enthalpy, and the following relationship holds:

$$\Delta G° \simeq \Delta H°$$

Do not make the rash assumption that this is always so, however. In some instances entropy may determine the success or failure of a transformation.

Before leaving this discussion on thermodynamics, refer to Sec. 3.16 and note that there is a direct relationship between the free-energy change ($\Delta G°$) and the equilibrium constant associated with that system. This relationship is given by

$$\Delta G° = -RT \ln K_{eq}$$

These thermodynamic parameters ($\Delta G°$, $\Delta H°$, $\Delta S°$, and others) are called *state functions*. They are absolute measures of the differences in energy between two points, and the path taken from one point to the other (reactants to products in these examples) does not change the value of the state function. The free-energy change is defined as the total difference in free energy between the reactants and products. It is a fixed value for a given system at a given temperature.

5.4   Energy of Activation

Because the products are at a lower energy state than the reactants in Fig. 5.1(a), there should be a driving force in the direction of the products and the reaction should proceed on its own. We might expect all exothermic reactions to proceed spontaneously, but in fact relatively few reactions occur spontaneously. More commonly, a required minimum amount of energy must be applied to the system to initiate the reaction regardless of whether the reaction is exothermic or endothermic. Once initiated, an exothermic reaction often produces sufficient energy to be self-sustaining, whereas an endothermic reaction stops if the applied energy is removed. This minimum amount of energy required to initiate a reaction is called the **energy of activation** and is given the symbol E_A. The units associated with the E_A are also kilocalories per mole or kilojoules per mole. The E_A is defined as the energy difference between the reactants and the highest point in potential energy along the reaction coordinate. This energy maximum is called the **activated complex.** It is not an intermediate isolable species but rather a *transition state* of high energy. The E_A is another thermodynamic state function.

The symbol $\Delta G^{\ddagger}$ is often used instead of E_A because E_A is simply a free-energy change—the change in free energy in going from reactants to the energy maximum along the reaction coordinate. There is also a $\Delta H^{\ddagger}$ and $\Delta S^{\ddagger}$ associated with the energy maximum. These parameters are related mathematically in the same way as the ground-state energy differences described in Sec. 5.3:

$$\Delta G^{\ddagger} = \Delta H^{\ddagger} - T\Delta S^{\ddagger}$$

The double dagger (‡) superscript distinguishes these state functions from ground-state energy changes.

Figure 5.2 is a complete energy diagram for the two generalized reactions in Fig. 5.1. The E_A is associated with a net gain of energy; that is, it is an energy-consuming process. In highly exothermic reactions E_A approaches 0 kcal/mole, and in highly endothermic reactions $E_A \approx \Delta H°$.

Question 5.1

Draw reaction coordinates for the two extreme types of reaction described in the preceding paragraph. Label the reactants, products, activated complexes (transition states), E_A's, and $\Delta H°$'s.

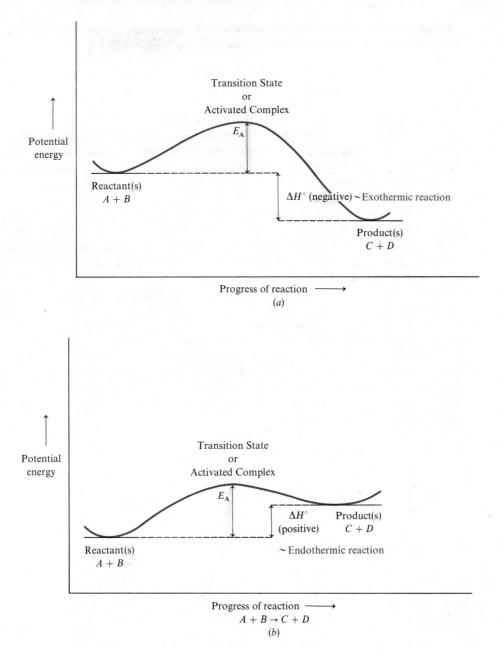

FIGURE 5.2   The energy profiles of a generalized (*a*) exothermic and (*b*) endothermic reaction, with the transition state E_A and $\Delta H°$ labeled.

5.5   Transition State Versus Reaction Intermediate

The conversion of reactants to products may occur in one concerted step:

$$A + B \xrightarrow[\text{activated complex}]{\text{through}} C + D$$

$$\text{Reactants} \qquad\qquad \text{Products}$$

with no isolable intermediates. Alternatively, the reaction may proceed through a series of two or more steps involving intermediates (see Sec. 4.6):

$$A + B \longrightarrow \text{intermediate} \longrightarrow C + D$$

$$\underset{\text{Reactants}}{\qquad\qquad} \qquad\qquad\qquad \underset{\text{Products}}{\qquad\qquad}$$

Examples of one-step reactions that do not involve any intermediates are given in Fig. 5.2(a) and (b). A reaction in which intermediates are produced is shown in Fig. 5.3. In the figure, the first step is an endothermic reaction and the second is exothermic. The reverse is also possible, or, alternatively, both steps could be exothermic or endothermic. The E_A and $\Delta H°$ values are all state functions and are independent of each other.

If the reaction involves more than two steps, it is plotted in exactly the same way. Each step goes from one energy minimum to a new energy minimum (reactants to product or intermediate) passing through an energy maximum, the transition state or activated complex (see Sec. 4.6), in each case. In all but the final step, the products are intermediates, which are (theoretically at least) isolable. Whether they are actually isolable depends on the reaction. In contrast, an activated complex is never isolable; it is a point along a reaction coordinate but does not exist for a finite period of time.

If a reaction proceeds through a series of steps to give a product, one step is usually much more difficult, that is, has a higher E_A, than any of the others. The rate of the overall reaction is dependent on this one slow step, the one with the highest energy barrier (E_A). This step is called the **rate-determining step.** In a one-step reaction, it is the only step. In the two-step reaction in Fig. 5.3, it is the first step ($E_A1 > E_A2$); once the intermediate is formed, there is sufficient energy available to convert it rapidly to product.

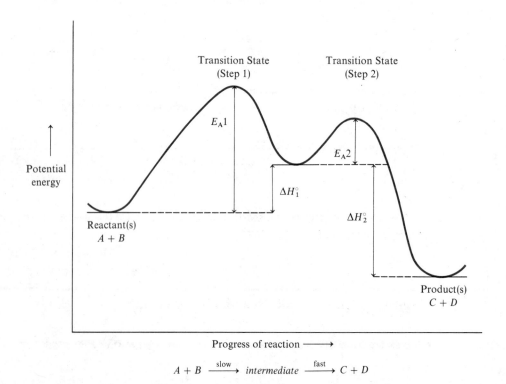

FIGURE 5.3   Energy profile of a typical two-step reaction.

5.6   Bond Breaking and Forming; Reversibility of Chemical Reactions

In reactions that involve the breaking and forming of bonds, the predominant energy changes result from the energy consumed in bond breaking or evolved in bond forming. The entropy changes in these reactions are usually negligible, so that $\Delta G° \approx \Delta H°$. Most mechanisms in this text involve the breaking and forming of bonds, both of which are directly related to the bond strength.

Bond strength is usually measured in kilocalories per mole or kilojoules per mole and is called the **bond dissociation energy.** It is simply the energy associated with that bond and is symbolized by either $\Delta H°$ or D. Bringing atoms together to form bonds results in greater stability and energy is evolved ($\Delta H°$ is negative); breaking a bond requires energy to move the atoms away from one another and energy is absorbed ($\Delta H°$ is positive). These concepts are illustrated by the following examples:

Bond making:

$$:\overset{..}{\underset{..}{Cl}}\cdot + :\overset{..}{\underset{..}{Cl}}\cdot \longrightarrow :\overset{..}{\underset{..}{Cl}}:\overset{..}{\underset{..}{Cl}}: \qquad$$ 58 kcal/mole of energy liberated
($\Delta H° = -58$ kcal/mole; -242.7 kJ/mole)

Bond breaking:

$$:\overset{..}{\underset{..}{Cl}}:\overset{..}{\underset{..}{Cl}}: \longrightarrow :\overset{..}{\underset{..}{Cl}}\cdot + :\overset{..}{\underset{..}{Cl}}\cdot \qquad$$ 58 kcal/mole of energy absorbed
($\Delta H° = +58$ kcal/mole; $+242.7$ kJ/mole)

The energies for the bond-making and bond-breaking processes are the same magnitude, as they must be because they are the reverse of one another. Most, though not all, chemical reactions are reversible; that is, reactants can convert to products, and products to reactants. The equilibrium lies in the direction of the more stable (lower energy) members of the reaction. All the thermodynamic parameters associated with either the forward or reverse reaction may be obtained from the reaction coordinate. For the preceding example, the reaction coordinate is given in Fig. 5.4.

Bond dissociation energies can be used to calculate the net energy changes that accompany the reactions of covalent bonds. Some selected bond dissociation energies are given in Table 5.1.

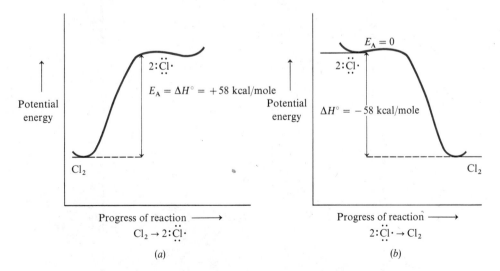

FIGURE 5.4   Bond (*a*) breaking and (*b*) forming in the chlorine molecule.

TABLE 5.1 Bond Dissociation Energies,* D, for Selected Molecules, in kcal/mole (kJ/mole)

Diatomic Molecules

H—H 104 (435.1)	H—F 135 (564.8)	H—Cl 103 (431)	H—Br 88 (368.2)	H—I 71 (297.1)
F—F 37 (154.8)	Cl—Cl 58 (242.7)	Br—Br 46 (150.6)	I—I 36 (150.6)	
O=O 119 (497.9)	N≡N 226 (945.6)			

Carbon-Hydrogen and Carbon-Halogen Bonds

CH_3—H 104 (435.1)	CH_3—F 108 (451.9)	CH_3—Cl 84 (351.5)	CH_3—Br 70 (292.9)	CH_3—I 53 (221.8)
CH_3CH_2—H 98 (410.0)	CH_3CH_2—F 106 (443.5)	CH_3CH_2—Cl 82 (343.1)	CH_3CH_2—Br 69 (288.7)	CH_3CH_2—I 54 (225.9)
$CH_3CH_2CH_2$—H 98 (410.0)		$CH_3CH_2CH_2$—Cl 82 (343.1)	$CH_3CH_2CH_2$—Br 69 (288.7)	
$(CH_3)_2CH$—H 94 (393.3)		$(CH_3)_2CH$—Cl 81 (338.9)	$(CH_3)_2CH$—Br 68 (284.5)	
$(CH_3)_3C$—H 91 (380.7)		$(CH_3)_3C$—Cl 79 (330.5)	$(CH_3)_3C$—Br 63 (263.6)	
$\begin{array}{c}CH_2-CH_2\\ \diagdown\\ CH_2-CH_2\end{array}\!\!CH$—H 94 (393.3)		$\begin{array}{c}CH_2-CH_2\\ \diagdown\\ CH_2-CH_2\end{array}\!\!CH$—Cl 73 (305.4)		

* Values given are for homolytic bond cleavage.

For simple molecules (for example, Cl_2) the bond dissociation energy is the energy needed to break the bond **homolytically,** with *one electron going to each atom.* It is also the amount of energy evolved when the bond is formed from the two radicals. For more complex molecules, such as methane, the bond dissociation energy is really an *average bond energy.* In methane, hydrogens can be removed in a stepwise fashion, each requiring its own characteristic energy.

	Bond Dissociation Energy, D, in kcal/mole (kJ/mole)
$CH_4 \longrightarrow \cdot CH_3 + H\cdot$	104 (435.1)
$\cdot CH_3 \longrightarrow \cdot \overset{\cdot}{C}H_2 + H\cdot$	105 (439.3)
$\cdot \overset{\cdot}{C}H_2 \longrightarrow \cdot \overset{\cdot}{C}H + H\cdot$	108 (451.9)
$\cdot \overset{\cdot}{C}H \longrightarrow \cdot \overset{\cdot}{\underset{\cdot}{C}}\cdot + H\cdot$	84 (351.5)
Overall reaction: $CH_4 \longrightarrow \cdot \overset{\cdot}{\underset{\cdot}{C}}\cdot + 4H\cdot$	401 (1677.8)

The average bond energy of a C—H bond in methane is obtained by summing the four bond dissociation energies and dividing by 4 because there are *four* C—H bonds in methane:

$$\text{Average C—H bond energy} = \frac{401 \text{ kcal/mole}}{4} \approx 100 \text{ kcal/mole (419.5 kJ/mole)}$$

We are concerned largely with reactions that require the breaking or making of specific bonds, and for this purpose, the bond dissociation energy, D, is most useful. A simple example of how these energies can be used involves the reaction between methane and bromine to give bromomethane (CH_3Br) and HBr.

$$CH_3—H + Br—Br \xrightarrow{\text{heat}} CH_3—Br + H—Br$$

If we consider only the bonds that change, we find that a CH_3—H bond and a Br—Br bond are broken. According to the values in Table 5.1, 104 and 46 kcal/mole, respectively, or a total of 150 kcal/mole (627.6 kJ/mole) is required. In the products, two new bonds are formed, CH_3—Br and H—Br. The energies associated with these bonds are 70 and 88 kcal/mole, respectively, and because these bonds are formed, a total of 70 + 88 or 158 kcal/mole (661.1 kJ/mole) is liberated. More energy is liberated when the products are formed (158 kcal/mole) than is required to break the bonds in the reactants (150 kcal/mole), so the *net* energy liberated by this reaction is 8 kcal/mole ($\Delta H° = -8$ kcal/mole or -33.5 kJ/mole). These results are summarized as follows:

Bonds broken	Bonds formed	Net energy liberated
$CH_3{-}H + Br{-}Br \longrightarrow$	$CH_3{-}Br + H{-}Br$	$\Delta H° = -8$ kcal/mole or -33.5 kJ/mole
104 + 46	70 + 88	
150 kcal/mole 627.6 kJ/mole *Required*	158 kcal/mole 661.1 kJ/mole *Liberated*	

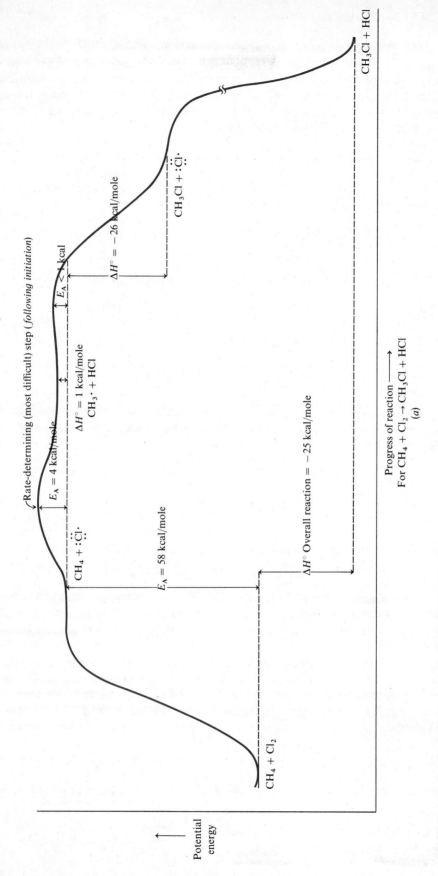

FIGURE 5.5   Graph of energy profile for chlorination of methane.

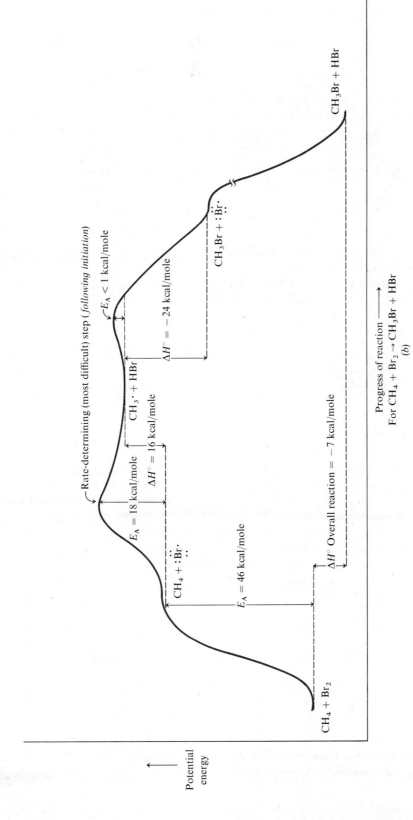

FIGURE 5.5 (*cont'd.*) Graph of energy profile for bromination of methane.

Contrasting the bromination and chlorination of methane, one gets the following:

Initiation:

$$Cl\!-\!Cl \longrightarrow 2:\ddot{\underset{..}{C}}l\cdot \qquad \Delta H^\circ = 58\ kcal/mole\ (242.7\ kJ/mole)$$

$$Br\!-\!Br \longrightarrow 2:\ddot{\underset{..}{B}}r\cdot \qquad \Delta H^\circ = 46\ kcal/mole\ (192.5\ kJ/mole)$$

When the chlorine or bromine atoms are formed, the next step is their attack on methane (see Sec. 4.6); the following E_A and ΔH° values apply to these reactions:

Propagation:

$$:\ddot{\underset{..}{C}}l\cdot + CH_3\!-\!H \longrightarrow CH_3\cdot + H\!-\!Cl \qquad E_A = 4\ kcal/mole\ (16.8\ kJ/mole)$$
$$\Delta H^\circ = 1\ kcal/mole\ (4.18\ kJ/mole)$$

$$:\ddot{\underset{..}{B}}r\cdot + CH_3\!-\!H \longrightarrow CH_3\cdot + H\!-\!Br \qquad E_A = 18\ kcal/mole\ (75.3\ kJ/mole)$$
$$\Delta H^\circ = 16\ kcal/mole\ (66.9\ kJ/mole)$$

The rate of hydrogen abstraction is faster in chlorination than in bromination because the E_A is lower for the former.

The final step in the halogenation reaction is the attack by the halogen on the methyl radical. This is a rapid, exothermic reaction for both bromine and chlorine. The energy of activation, E_A, is apparently quite small and has been estimated to be less than 1 kcal/mole (4.18 kJ/mole).

Propagation:

$$CH_3\cdot + Cl\!-\!Cl \longrightarrow CH_3\!-\!Cl + :\ddot{\underset{..}{C}}l\cdot \qquad E_A < 1\ kcal/mole\ (4.18\ kJ/mole)$$
$$\Delta H^\circ = -26\ kcal/mole\ (-108.8\ kJ/mole)$$

$$CH_3\cdot + Br\!-\!Br \longrightarrow CH_3\!-\!Br + :\ddot{\underset{..}{B}}r\cdot \qquad E_A < 1\ kcal/mole\ (4.18\ kJ/mole)$$
$$\Delta H^\circ = -24\ kcal/mole\ (-100.4\ kJ/mole)$$

The E_A and ΔH values associated with the steps in the bromination and chlorination reactions can be compared in graphs of the entire energy profile for both reactions. These are shown in Fig. 5.5(a), (b). The energy-profile diagrams make it easier to see why bromination is more difficult than chlorination: More energy is required to pass over the highest energy barrier (which is the rate-determining and most difficult step) in bromination than in chlorination.

Let us take a closer look at the rate-determining step in these two reactions, the step in which the halogen atom abstracts a hydrogen atom from methane to give the methyl radical:

First transition state,
rate-determining step

The amounts of bond breaking and bond making are not the same in chlorination and bromination. In general, when we consider similar reactions, *the higher the* E_A, *the later the transition state is reached in the reaction and the more the transition state resembles the final product.* Bromination has a higher E_A (18 kcal/mole; 75.3 kJ/mole) than chlorination (4 kcal/mole; 16.8 kJ/mole), so the transition state in bromination involves more breaking of the carbon-hydrogen bond than does chlorination. This can be translated into the following representations:

$$H-\overset{\overset{\displaystyle H}{|}}{\underset{\underset{\displaystyle H}{|}}{C}}{}^{\delta\cdot}-H-----{}^{\delta\cdot}Cl \qquad\qquad H-\overset{\overset{\displaystyle H}{|}}{\underset{\underset{\displaystyle H}{|}}{C}}{}^{\delta\cdot}----H--{}^{\delta\cdot}Br$$

Transition state
for chlorination
(reached early in reaction;
C—H bond hardly broken,
and carbon has little
radical character;
resembles reactants)

Transition state
for bromination
(reached late in reaction;
C—H bond fairly well broken,
and carbon has considerable
radical character;
resembles products)

We use transition state theory in the next section to explain the differences in selectivity in bromination versus chlorination.

Question 5.2

Using the values in Table 5.1, calculate the heats of reaction for the formation of the following three compounds from the corresponding hydrocarbons and halogens:
(*a*) ethyl bromide   (*b*) ethyl chloride   (*c*) isopropyl chloride

5.7   Reactive Intermediates

Intermediates are formed in all reactions other than concerted reactions. These intermediates, which correspond to energy valleys (energy minima) on the reaction coordinate, are not isolable in many instances. Reactive intermediates were first introduced in Chap. 2. The three classes of intermediates most commonly encountered in organic chemistry are **free radicals** ($R\cdot$), **carbocations** ($R^{\oplus}$, also called **carbonium ions,** see Sec. 5.7B), and **carbanions** ($R\overset{\ominus}{:}$). The geometry, hybridization, and orbital view of each of these species was presented in Sec. 2.8. We now look at several modes of formation of these intermediates as well as their relative stabilities and properties.

A. Free Radicals, R·

a. Stability of Free Radicals
Consider the energies required to form radicals via a direct homolytic bond cleavage:

$$R\overset{\frown}{:}H \xrightarrow[\text{bond cleavage}]{\text{homolytic}} R\cdot + H\cdot \qquad D = \Delta H^\circ = \text{bond dissociation energy}$$

As the following data (taken from Table 5.1) indicate, it is easier to form a tertiary radical than a secondary radical, and in turn it is easier to form a secondary radical than a primary radical. The methyl radical is the most difficult to form.

	Bond Dissociation Energy, D, kcal/mole (kJ/mole)
$CH_3-H \longrightarrow$ $CH_3\cdot$ $+ H\cdot$ Methyl radical	104 (435.1)
$CH_3-CH_2-H \longrightarrow CH_3-CH_2\cdot + H\cdot$ $1°H$ $1°$ Radical	98 (410.0)
$CH_3-CH_2-CH_2-H \longrightarrow CH_3-CH_2-CH_2\cdot + H\cdot$ $1°H$ $1°$ Radical	98 (410.0)
$CH_3-\overset{\displaystyle \mid}{\underset{\displaystyle H}{C}H}-CH_3 \longrightarrow CH_3-\overset{\displaystyle \cdot}{C}H-CH_3 + H\cdot$ $2°$ H $2°$ Radical	94 (393.3)
$CH_3-\overset{\displaystyle CH_3}{\underset{\displaystyle H}{\overset{\displaystyle \mid}{\underset{\displaystyle \mid}{C}}}}-CH_3 \longrightarrow CH_3-\overset{\displaystyle CH_3}{\underset{\displaystyle \cdot}{\overset{\displaystyle \mid}{C}}}-CH_3 + H\cdot$ $3°$ H $3°$ Radical	91 (380.7)

Various radicals and atoms are also formed in free-radical halogenation, in which the slow, rate-determining step is hydrogen abstraction:

$$R-H + :\overset{..}{\underset{..}{X}}\cdot \longrightarrow \quad R\cdot \; + \; H-X$$
$$\text{Alkyl}$$
$$\text{radical}$$

If we contrast thermal homolytic bond cleavage and free-radical formation in free-radical halogenation,

Thermal bond dissociation:

$$\overset{\frown}{R-H} \longrightarrow R\cdot + H\cdot$$
$$\text{Radical}$$

Hydrogen abstraction by halogen atom in free-radical halogenation reaction:

$$R-H + :\overset{..}{\underset{..}{X}}\cdot \longrightarrow R\cdot + H-X$$
$$\text{Radical}$$

on a *relative* basis, we see that the ease of removal of various types of hydrogens in halogenation parallels the relative ease of thermal dissociation; the order is:

Decreasing ease of removal of hydrogen
atoms to form carbon radicals:

$$3°\, H > 2°\, H > 1°\, H > CH_3-H$$

Another interesting relationship appears as a result of the intervention of free radicals. The bond dissociation energies given earlier apply to the formation of a given radical from the corresponding alkane. Each molecule has a different total energy, so we cannot speak of absolute energies for free radicals. The bond dissociation energies tell us, however, that it is more difficult to remove a hydrogen from methane to give the methyl radical than it is to remove a hydrogen from ethane to give an ethyl radical, and so forth. From a qualitative standpoint, the decreasing order of free-radical stability, *relative to the alkane from which each was formed*, is:

Decreasing order of
free-radical stability:

$$3° > 2° > 1° > CH_3\cdot$$

The ease of forming free radicals corresponds to their stability; that is, *the more stable the free radical, the more easily it is formed* (see Sec. 5.8).

We can now extend our discussion in Sec. 4.6 about the energetics of free-radical halogenation to the general case of alkanes, where various types (1°, 2°, 3°) of free radicals can be formed in a given molecule. Because the rate of halogenation is controlled by the slow, rate-determining step of hydrogen abstraction, it is reasonable that we consider the energies associated with this reaction. The following experimental data apply to the bromination of propane:

$$CH_3-CH_2-CH_3 \xrightarrow[-HBr]{:\ddot{B}r\cdot} CH_3-CH_2-CH_2\cdot \qquad E_A = 13\ \text{kcal/mole (54.4 kJ/mole)}$$

1° Radical $\Delta H° = 12\ \text{kcal/mole (50.2 kJ/mole)}$

$$\xrightarrow[:\ddot{B}r\cdot,\ -HBr]{} CH_3-CH-CH_3 \qquad E_A = 10\ \text{kcal/mole (41.8 kJ/mole)}$$

2° Radical $\Delta H° = 6\ \text{kcal/mole (25.1 kJ/mole)}$

These data are shown in Fig. 5.6.

The secondary radical is more stable than the primary radical even though both are formed from the same starting material (propane). Of greater importance is that the energies of activation are (1) quite large and (2) not very close together. Also, the energy of activation for the formation of the secondary radical is less than that for a primary radical. These data agree nicely with the reactivity and free-radical stability trends we observe. Now, let us consider product distribution.

b. Selectivity of Free Radicals
The *product distribution* (orientation) is governed by two factors:

1. *Relative reactivities* of the various types of hydrogens present in the molecule toward abstraction by the halogen atom $(:\ddot{X}\cdot)$
2. *Number* of each type of hydrogen in the molecule

We need to define the expression *types of hydrogen*. Fortunately, we have a system designed for this purpose based on the type of carbon atom to which the various hydrogen atoms are attached (see Sec. 4.1). Note the following example:

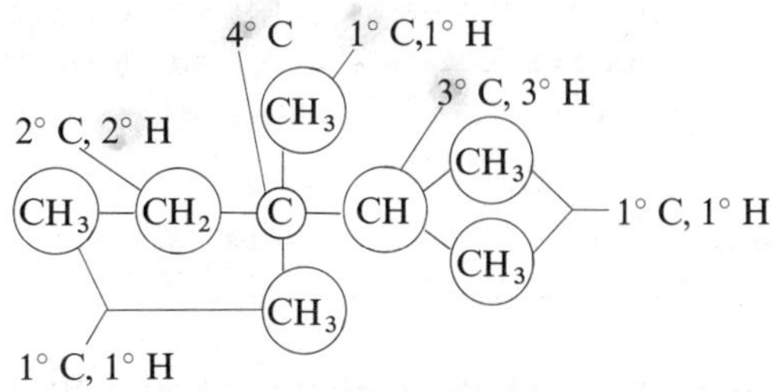

Now we return to the chlorination of propane and examine this reaction in terms of product distribution. Because free radicals—the propyl and isopropyl

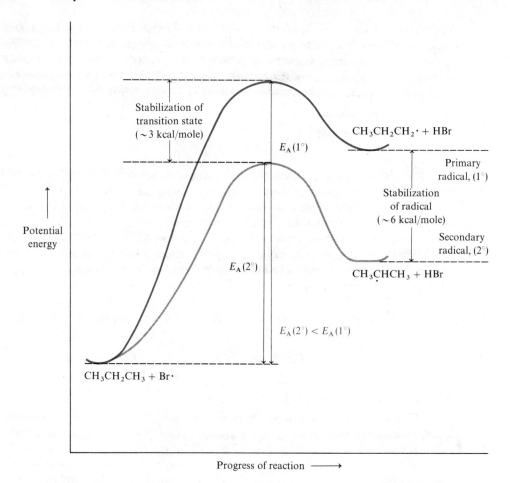

Potential energy

Stabilization of transition state (~ 3 kcal/mole)

$E_A(1°)$

$CH_3CH_2CH_2\cdot + HBr$

Primary radical, $(1°)$

Stabilization of radical (~ 6 kcal/mole)

Secondary radical, $(2°)$

$CH_3\underset{.}{C}HCH_3 + HBr$

$E_A(2°)$

$E_A(2°) < E_A(1°)$

$CH_3CH_2CH_3 + Br\cdot$

Progress of reaction $\longrightarrow$

FIGURE 5.6   Energy diagram comparing the progress of two competing reactions of propane with bromine atoms.

radicals—are involved as intermediates and hydrogen abstraction is the slow, rate-determining step we anticipate as soon as these radicals are formed, they are consumed by reaction with chlorine gas to form products. The product distribution (and thus the orientation of the chlorine atom in the chloropropanes) is dictated by the relative concentrations of the propyl and isopropyl radicals. *If* all eight hydrogens in propane were of equal bond strength, we would expect them to be equally easy to abstract when they collide with the chlorine atom. This would simply involve *probability:* attack on any of the six equivalent 1° hydrogens would give rise to the propyl radical, and attack on either of the two equivalent 2° hydrogens would give the isopropyl radical. Because the radicals form products immediately, we would predict, *on the basis of probabilities of attack alone,* that the product ratio would be

$$\frac{1\text{-Chloropropane}}{2\text{-Chloropropane}} = \frac{6}{2} = \frac{3}{1}$$

or 75% 1-chloropropane and 25% 2-chloropropane. (In the ratio, the 6 represents the six 1° hydrogens and the 2 represents the two 2° hydrogens.) However, these percentages do not correspond to the observed product distribution of 45% and 55%.

We assumed that it is equally probable that the chlorine atoms will attack any of the eight hydrogens in propane. Something else must also affect hydrogen abstraction, however, and this is the *relative reactivities* of the different types (1°, 2°, 3°) of hydrogen atoms in the molecule. Not all types of hydrogens have equal bond strength, and therefore they are not abstracted with equal ease. This is demonstrated by a simple experiment in which equimolar amounts of methane and ethane are allowed to react with a small amount of chlorine gas at 25° in the presence of light. When the products are analyzed, there are ~400 chloroethane molecules for each chloromethane molecule:

$$\underbrace{CH_4 + CH_3\!-\!CH_3}_{\text{Equimolar}} \xrightarrow[25°, \, h\nu]{Cl_2 \text{ (small amount)}} CH_3\!-\!Cl + \underset{1:400}{CH_3\!-\!CH_2\!-\!Cl}$$

The results must be corrected for the number of hydrogens in each molecule (four in methane and six in ethane), so 400 must be multiplied by $\frac{4}{6}$ or $\frac{2}{3}$; this reduces the relative reactivity of ethane to ~270 times that of methane. This type of experiment is often used to determine relative reactivities where two reactants are allowed to compete for the same reactive species (in this case, the chlorine atom). The product that is formed in larger quantity comes from the reactant that is more reactive. The reaction conditions in such an experiment are by necessity identical; this is especially useful in halogenation reactions where the reaction conditions are difficult to duplicate exactly. This technique is applied in later chapters.

Application of competition experiments to alkanes indicates the following general reactivity trends for chlorination:

Relative ease of hydrogen abstraction:

$$3° \, H > 2° \, H > 1° \, H \, (> CH_3\!-\!H)$$

Relative rates of hydrogen abstraction (per hydrogen atom):

$$\left. \begin{array}{l} 5.0 \text{ for } 3° \, H \\ 3.8 \text{ for } 2° \, H \\ 1.0 \text{ for } 1° \, H \end{array} \right\} \text{for chlorination with light at } 25°$$

These data were determined under identical conditions of temperature, pressure, and concentration. The reactivity of a primary hydrogen was arbitrarily set at 1.0, and the remaining data are relative. Thus it is 3.8 times easier to remove a secondary hydrogen than a primary one, and 5.0 times easier to abstract a tertiary hydrogen than a primary one. For chlorination reactions, the ease of hydrogen abstraction depends mostly on the particular type of hydrogen being removed and not on the overall structure of the alkane itself.

Now, let us apply this theory to the chlorination of propane. We must use two types of information: (1) the *number* of hydrogens that give the same product (that is, the probability factor) and (2) the *relative rates* of hydrogen abstraction. Propane has six identical 1° hydrogens and two identical 2° hydrogens. The rate of product formation is proportional to the rate of radical formation. We can calculate the rate of radical formation and thus determine the relative concentrations of products, which are proportional to the rate at which they are formed. For the reaction

$$CH_3\!-\!CH_2\!-\!CH_3 \xrightarrow[25°, \, h\nu]{Cl_2} CH_3\!-\!CH_2\!-\!CH_2\!-\!Cl + CH_3\!-\!\underset{\underset{Cl}{|}}{CH}\!-\!CH_3$$

1-Chloropropane   2-Chloropropane

we can set up and solve the following equation:

$$\frac{\text{1-Chloropropane}}{\text{2-Chloropropane}} = \frac{\text{rate of formation of 1-chloropropane}}{\text{rate of formation of 2-chloropropane}}$$

$$= \frac{\text{number of 1}° \text{ H} \times \text{relative reactivity of 1}° \text{ H}}{\text{number of 2}° \text{ H} \times \text{relative reactivity of 2}° \text{ H}}$$

$$= \frac{6 \times 1.0}{2 \times 3.8}$$

$$= \frac{6}{7.6}$$

Thus we predict the ratio of 1-chloropropane to 2-chloropropane to be $6 : 7.6$. These numbers can be converted to percentages:

$$\% \text{ 1-Chloropropane} = \frac{6}{6 + 7.6} \times 100 = 44.1\%$$

and

$$\% \text{ 2-Chloropropane} = \frac{7.6}{6 + 7.6} \times 100 = 55.9\%$$

These percentages are in very close agreement with experimental results, which, as we stated, are 45% 1-chloropropane and 55% 2-chloropropane.

Question 5.3

Calculate the product distribution for the isomeric alkyl chlorides that are obtained by allowing the following compounds to react with chlorine at 25° in the presence of light: (*a*) butane, (*b*) isopentane, (*c*) pentane, (*d*) 2,3-dimethylbutane, and (*e*) 2,2,4-tetramethylpentane.

From experiments the following general reactivity trends for bromination have been obtained:

Relative ease of hydrogen abstraction:

$$3° \text{ H} > 2° \text{ H} > 1° \text{ H} (>CH_3\!-\!H)$$

Relative rates of hydrogen abstraction (per hydrogen atom):

$$\left.\begin{array}{l} 1,600 \text{ for } 3° \text{ H} \\ 82 \text{ for } 2° \text{ H} \\ 1 \text{ for } 1° \text{ H} \end{array}\right\} \text{ for bromination with light at } 125°$$

Bromine is more selective than chlorine, but chlorination is faster. The greater selectivity can be explained by considering the transition states of the rate-determining steps of the two reactions. For chlorination, the transition state is *reactant-like* and for bromination, *product-like* (see Sec. 5.6). In bromination, therefore, the transition state has a great deal of free-radical character. The type of hydrogen reacting, and hence the type of radical forming, play a marked role in the rate of reaction and thus in the product distribution. This manifests itself in the greater observed selectivity mentioned earlier.

The structure-reactivity relationship governs radical formation, which in turn governs product distribution (orientation of halogen in product). Radical formation is related to radical stability and in turn to their energies of activation.

c. The Inductive Effect and Free-Radical Stability

We discuss radicals and their stability by explaining how electronic effects influence stability. We address ourselves to a most important concept, the **inductive effect,** which involves either electron donation or electron withdrawal through the σ bonds by adjacent groups.

Considerable chemical and physical evidence suggests that alkyl groups, relative to hydrogen atoms, release electrons to carbon. If, for example, an alkyl group is attached to an electron-deficient carbon, it donates electrons through the carbon-carbon σ bonds.

Now we apply this idea to radicals that are electron-deficient because the carbon atom has only seven valence electrons. If we attach groups that release (donate) electrons very slightly, as do the alkyl groups, then the carbon bearing the odd electron "feels" that it is starting to gain some negative electronic charge so that it resembles more the stable octet state. Remember that it comes nowhere close to gaining a full electron, but it does become more stable when the concentration of electrons on it is increased slightly by donation. For radical stability, we can envision the following, where the → indicates inductive electron donation from the R group to the radical:

$$
\begin{array}{cccc}
\overset{\displaystyle H}{\underset{\displaystyle H}{\text{H}-\text{C}\cdot}} &
\overset{\displaystyle H}{\underset{\displaystyle H}{\text{R}\rightarrow\text{C}\cdot}} &
\overset{\displaystyle R\downarrow}{\underset{\displaystyle H}{\text{R}\rightarrow\text{C}\cdot}} &
\overset{\displaystyle R\downarrow}{\underset{\displaystyle R\uparrow}{\text{R}\rightarrow\text{C}\cdot}} \\
\text{Methyl radical} & 1^\circ \text{ Radical} & 2^\circ \text{ Radical} & 3^\circ \text{ Radical}
\end{array}
$$

Increasing radical stability →

Electron donation: Increasing the number of alkyl (R) groups causes increasing radical stability

Generally, *greater numbers of alkyl substituents result in increased radical stability*. We say that alkyl groups are electron-releasing or electron-donating through the inductive effect.

All alkyl groups appear to have roughly equal electron-donating properties. Whether the R group attached is methyl or ethyl, for example, seems to be of minor importance in the overall donating ability of the group.

B. Carbocations, $R^{\oplus}$

The carbocation (or carbonium ion) is a carbon atom that bears a positive charge and has only six electrons (three bonding pairs) about it; it is *electron-deficient*.

Carbonium Ions Versus Carbocations

In a further attempt to render the nomenclature system of organic chemistry more consistent, the term *carbocation* was coined by Professor George Olah. The suffix *-onium* is used for species that are hypervalent (contain more bonds than usual) and positively charged.

$$
\begin{array}{cccc}
-\overset{|}{\underset{|}{\text{N}}}{}^{\oplus}- &
-\overset{\oplus}{\underset{|}{\overset{..}{\text{O}}}}- &
-\overset{\oplus}{\underset{|}{\overset{..}{\text{S}}}}- &
-\overset{|}{\underset{|}{\text{P}}}{}^{\oplus}- \\
\text{Ammonium} & \text{Oxonium} & \text{Sulfonium} & \text{Phosphonium}
\end{array}
$$

To be consistent, a carbonium ion should be pentavalent and positively charged. Indeed, species such as $CH_5^{\oplus}$ have been detected by mass spectrometry (see Chap. 18). To differentiate between this carbonium ion and the carbonium ion discussed routinely in aliphatic organic chemistry, Olah renamed the latter a *carbocation*.

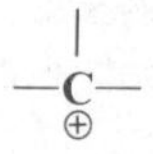

$$
\begin{array}{c}
| \\
-\underset{\oplus}{C}- \\
|
\end{array}
$$

Carbocation

Carbonium ion nomenclature, however, is well entrenched in organic literature. *In this text we use the preferred carbocation designation.* For example:

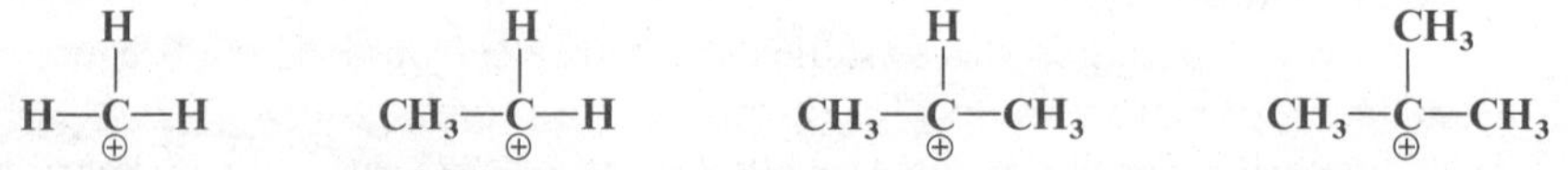

H	H	H	CH$_3$
$\underset{\oplus}{H-\underset{\|}{C}-H}$	$CH_3-\underset{\underset{\oplus}{\|}}{C}-H$	$CH_3-\underset{\underset{\oplus}{\|}}{C}-CH_3$	$CH_3-\underset{\underset{\oplus}{\|}}{C}-CH_3$
Carbonium ion or methyl cation	Methylcarbonium ion or ethyl cation *1° Carbocation*	Dimethylcarbonium ion or isopropyl cation *2° Carbocation*	Trimethylcarbonium ion or *tert*-butyl cation *3° Carbocation*

Carbocations result from a **heterolytic bond cleavage** (*both bonding electrons going with one atom*) of the following type:

$$R:\ddot{X}: \xrightarrow[\text{bond cleavage}]{\text{heterolytic}} R^{\oplus} + :\ddot{X}:^{\ominus}$$

Here, the $\Delta H°$ of the reaction is the heterolytic bond dissociation energy. The following values are provided for alkyl chlorides:

$CH_3\frown Cl \longrightarrow \overset{\oplus}{CH_3} + :\ddot{Cl}:^{\ominus}$ $\Delta H° = 227$ kcal/mole (949.8 kJ/mole)

Methyl
cation

$CH_3-CH_2\frown Cl \longrightarrow CH_3-\overset{\oplus}{CH_2} + :\ddot{Cl}:^{\ominus}$ $\Delta H° = 195$ kcal/mole (815.9 kJ/mole)

Ethyl cation or
methylcarbonium
ion: *a primary* (1°)
carbocation

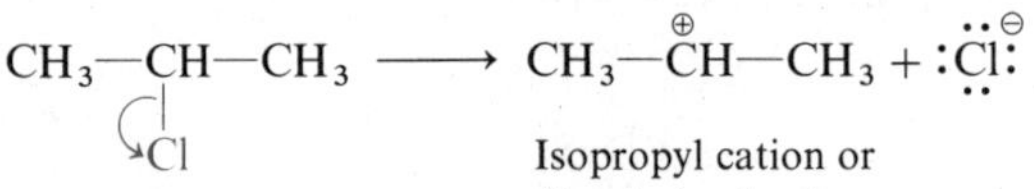

$CH_3-\underset{\underset{Cl}{\frown|}}{CH}-CH_3 \longrightarrow CH_3-\overset{\oplus}{CH}-CH_3 + :\ddot{Cl}:^{\ominus}$ $\Delta H° = 173$ kcal/mole (723.8 kJ/mole)

Isopropyl cation or
dimethylcarbonium
ion: *a secondary* (2°)
carbocation

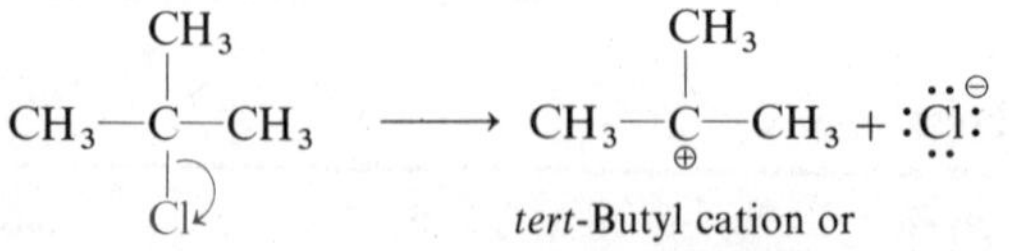

$$
CH_3-\underset{\underset{Cl}{\overset{\overset{\displaystyle CH_3}{|}}{\underset{\oplus}{C}}}}{}-CH_3 \longrightarrow CH_3-\underset{\oplus}{\overset{\overset{\displaystyle CH_3}{|}}{C}}-CH_3 + :\ddot{Cl}:^{\ominus}
$$ $\Delta H° = 157$ kcal/mole (656.9 kJ/mole)

tert-Butyl cation or
trimethylcarbonium
ion: *a tertiary*
(3°) *carbocation*

The values of $\Delta H°$ are thermal heterolytic bond dissociation energies determined in the gas phase. For comparison, it is also possible to generate carbocations in solution by a variety of methods, such as the following two:

Bond heterolysis by silver ion:

$$R\frown Cl + \underset{\frown}{Ag^{\oplus}} \longrightarrow R^{\oplus} + AgCl(s)$$

Addition of acids to alkenes:

$$\overset{}{\underset{}{C}}{=}\overset{}{\underset{}{C} + H^\oplus} \longrightarrow \ -\overset{|}{\underset{\oplus}{C}}-\overset{|}{\underset{H}{C}}-$$

Electrons flow to
form C—H bond

These methods and others are discussed in more detail in later chapters.

The previous data indicate that the order of ease for forming carbocations is:

Decreasing ease of formation
of carbocations:

$$3° > 2° > 1° > \overset{\oplus}{C}H_3$$

That is, 157 kcal/mole (656.9 kJ/mole) is required to form a *tertiary* ion—the *tert*-butyl cation from *tert*-butyl chloride—whereas 227 kcal/mole (949.8 kJ/mole) is required to form the methyl cation from methyl chloride. These are all very high-energy species, but relatively speaking, the methyl cation is more difficult to form by some 70 kcal/mole (292.9 kJ/mole) than the *tert*-butyl cation. The relative ease of forming carbocations should also parallel their relative stabilities because the more stable ion must be the one that is easier to form. Thus:

Decreasing stability
of carbocations:

$$3° > 2° > 1° > \overset{\oplus}{C}H_3$$

The electron-deficient radical (R·) is stabilized by substituents that donate electrons to the carbon atom bearing the odd electron. This effect is even more pronounced with positively charged systems, which are more stable when the charge is spread out over the rest of the molecule. Since the difference between the methyl cation and primary, secondary, and tertiary cations is the number of substituents that are attached to the electron-deficient carbon, the ions become more stable with additional alkyl groups. This is because *alkyl groups donate electrons through the inductive effect.* If electrons are donated to the positively charged carbon atom, then it becomes more stable (relative to the methyl cation).

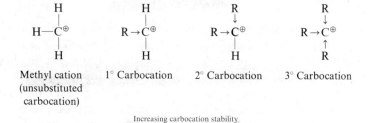

| Methyl cation (unsubstituted carbocation) | 1° Carbocation | 2° Carbocation | 3° Carbocation |

Increasing carbocation stability

Carbocations are very unstable and highly reactive. They normally exist for exceedingly short times (on the order of 10^{-9} sec). They undergo reactions in which a pair of electrons is provided to the positively charged carbon atom to complete its octet. One example is the reverse of the bond heterolysis shown earlier:

$$R^\oplus + :\overset{..}{\underset{..}{Cl}}:^\ominus \longrightarrow R:\overset{..}{\underset{..}{Cl}}:$$

Carbocations, in contrast to either free radicals or carbanions, often undergo *rearrangements* of their carbon skeletons in the course of a reaction. This occurs when a more stable carbocation results. For example,

$$CH_3-\underset{\underset{CH_3}{|}}{\overset{\overset{CH_3}{|}}{C}}-\overset{\overset{H}{|}}{\underset{\oplus}{C}}-CH_3 \xrightarrow{\text{rearranges to}} CH_3-\underset{\oplus}{\overset{\overset{CH_3}{|}}{C}}-\underset{\underset{CH_3}{|}}{\overset{\overset{H}{|}}{C}}-CH_3$$

2° Carbocation   3° Carbocation

In this reaction a secondary carbocation rearranges to a more stable tertiary carbocation, which then goes on to react further. When the carbon skeleton is rearranged, therefore, carbocations are the indicated intermediates. Carbocation rearrangements are discussed in detail in Sec. 7.18.

C. Carbanions, $R\overset{\ominus}{:}$

Carbanions are electron-rich moieties that result from a heterolytic bond cleavage in which both bonding electrons go with the carbon atom:

$$R\overset{\frown}{:}H \xrightarrow[\text{bond cleavage}]{\text{heterolytic}} R:\overset{\ominus}{} + H\overset{\oplus}{}$$

All that is required for the bond heterolysis is a strong enough base to abstract a proton from the alkane. The base used depends on the alkane undergoing reaction.

Organometallic compounds (see Sec. 4.2A) are similar to carbanions. An organolithium reagent, for example, may be depicted as

$$R^{\delta\ominus}\text{------}Li^{\delta\oplus}$$

in which the alkyl group is very much like a carbanion. This is true of many carbon-metal bonds.

From a series of equilibrium studies, the following order of carbanion stability has been determined.

Decreasing ease of formation of carbanions
and decreasing stability of carbanions:

$$\overset{\ominus}{:}CH_3 > 1° > 2° > 3°$$

This should not be surpising. We saw that alkyl groups donate electrons through the inductive effect. If electrons are donated to a negatively charged carbon, then that carbon becomes more negatively charged and less stable (relative to the methyl carbanion). The more alkyl groups, the less stable the carbanion.

Carbanions undergo a variety of reactions. For example, because they are strong bases, they react with almost any acid to abstract a proton. The reaction is reversible but is favored in the direction of the weak acid and weak base pair.

$$R:^{\ominus} + H:A \rightleftharpoons R:H + A:^{\ominus}$$
Strong   Strong   Weak   Weak
base   acid   acid   base

Acids and Bases

Acids and bases are commonly encountered in organic and inorganic chemistry. They are classified according to either the Lewis acid-base theory or the Brønsted-Lowry acid-base theory.

Lewis Acid-Base Theory

The free proton does not generally exist in solution; it reacts with molecules that contain at least one lone pair of electrons. In the Lewis definition, *a base is an electron-pair donor*, *and an acid is an electron-pair acceptor*, as illustrated by the following examples:

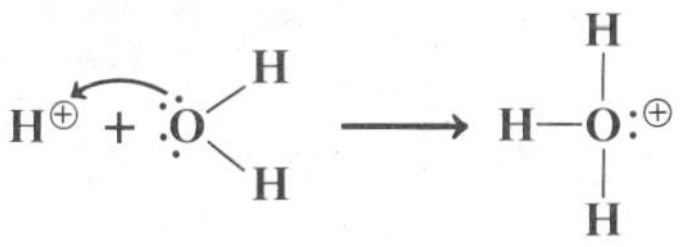

Lewis acid Lewis base

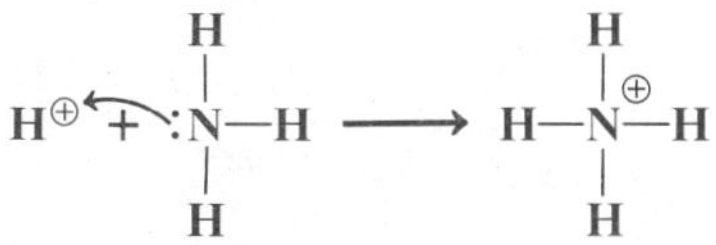

Lewis acid Lewis base

Lewis acids are *electron-deficient* and are either neutral or positively charged; for example:

$$H^{\oplus} \quad R^{\oplus} \text{ (a carbocation)} \quad BF_3 \quad AlCl_3 \quad FeCl_3 \quad ZnCl_2$$

Lewis bases are *electron-rich;* they must have at least one nonbonding pair of electrons, and they can be either neutral or negatively charged:

$$H_2\ddot{O}: \quad :NH_3 \quad :\ddot{\underset{..}{F}}:^{\ominus} \quad :\ddot{\underset{..}{Cl}}:^{\ominus} \quad H\ddot{\underset{..}{O}}:^{\ominus} \quad R:^{\ominus} \text{ (a carbanion)}$$

Typical reactions that involve the Lewis acid-base theory are further illustrated by the following equations:

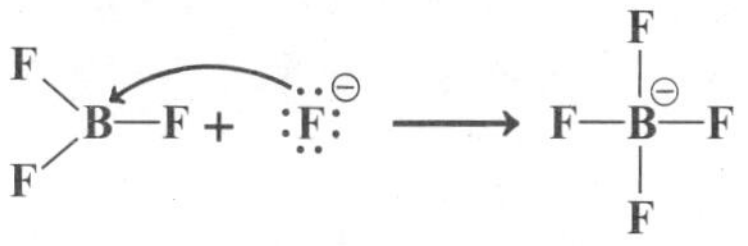

Lewis acid Lewis base

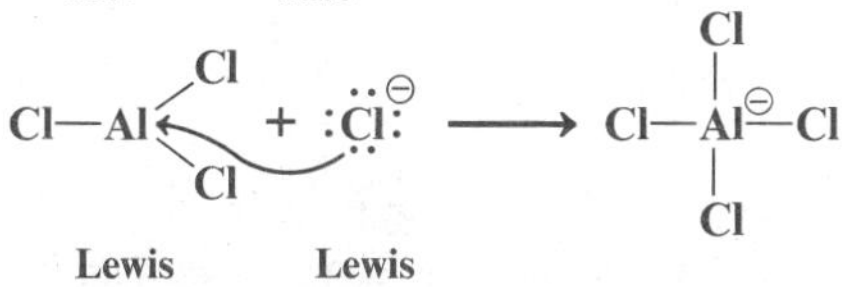

Lewis acid Lewis base

In the previous examples there is a net formal negative charge on the product. On the other hand, some reactions yield a neutral product, as in the following equation, which incorrectly utilizes the free proton for simplification:

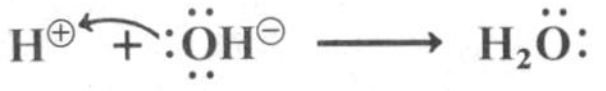

Lewis acid Lewis base

This reaction is better described in terms of the Brønsted-Lowry theory.

Brønsted-Lowry Theory

Protons do not exist as such under ordinary laboratory conditions, so most of their reactions involve a *proton transfer* from one species to another. This transfer is the Brønsted-Lowry acid-base theory. A *Brønsted-Lowry acid* is a proton donor, and a Brønsted-Lowry base is a proton acceptor.

Upon donating a proton, the acid becomes a conjugate base; upon accepting a proton, the original base becomes a conjugate acid, as the following examples illustrate:

$$H_3\overset{\oplus}{\overset{..}{O}}: + H\overset{..}{\underset{..}{O}}:^{\ominus} \longrightarrow H_2\overset{..}{\underset{..}{O}}: + H_2\overset{..}{\underset{..}{O}}:$$

  Acid   Base   Conjugate   Conjugate
                               base   acid

$$H-\overset{..}{\underset{..}{C}l}: + H_2\overset{..}{\underset{..}{O}}: \longrightarrow H_3\overset{\oplus}{\overset{..}{O}}: + :\overset{..}{\underset{..}{C}l}:^{\ominus}$$

  Acid   Base   Conjugate   Conjugate
                                acid   base

$$H_3\overset{\oplus}{\overset{..}{O}}: + \overset{..}{N}H_3 \longrightarrow H_2\overset{..}{\underset{..}{O}}: + NH_4^{\oplus}$$

  Acid   Base   Conjugate   Conjugate
                               base   acid

The next group of examples show the acid first, with the conjugate base in parenthesis:

$$H_2SO_4(HSO_4^{\ominus})\quad H_2S(HS^{\ominus})\quad HCN(CN^{\ominus})\quad CH_3OH(CH_3O^{\ominus})\quad CH_4(CH_3^{\ominus})$$

Equilibrium constants are used to determine how completely the proton is transferred from one species to another. For the generalized reaction between an acid, HA, and water, the following equilibrium expression and equilibrium constant is written:

$$HA + H_2\overset{..}{\underset{..}{O}}: \rightleftharpoons H_3\overset{..}{O}{}^{\oplus} + A:^{\ominus}$$

$$K_{eq}[H_2O] = K_a = \frac{[H_3O^{\oplus}][A^{\ominus}]}{[HA]} \qquad [H_2O] = 55.5\,M$$

This *acidity constant* can also be defined as $pK_a = -\log_{10}K_a$. From a quantitative determination of K_a or pK_a, the position of equilibrium can be determined, allowing us to determine which is the stronger acid and stronger base. The weaker acid and weaker base are those that appear on the side of the equilibrium equation which is favored at equilibrium. A large K_a means the acid is strong (equilibrium lies further to the right), and a small K_a means the acid is weak (equilibrium lies further to the left).

For a reaction to occur to any appreciable extent, a stronger acid must react with a stronger base to give a weaker acid and a weaker base, respectively:

$$HA \quad + \quad B:^{\ominus} \longrightarrow HB \quad + \quad A:^{\ominus}$$

Stronger   Stronger   Weaker   Weaker
 acid   base   acid   base

The reaction between a carbanion and an acid can now be interpreted in terms of Brønsted-Lowry theory, via the use of water as an acid and the carbanion as the base:

$$R:^{\ominus} + H-\overset{..}{\underset{..}{O}}H \longrightarrow R-H + H\overset{..}{\underset{..}{O}}:^{\ominus}$$

Strong   Strong   Weak   Weak
 base   acid   acid   base

This points to the need to compare the species involved. The carbanion ($R:^{\ominus}$) is a stronger base than hydroxide ion, and water is a stronger acid than the hydrocarbon, R—H.

We encounter numerous examples of acid-base theory throughout this text; refer to the Reading References in the appendix.

D. Summary

There are specific examples of the modes of formation and reactions of these intermediates in virtually every chapter of this text. We presented a broad overview of each species to show how each, though quite different in chemistry, can be discussed using the same concepts, for example, the role of the inductive effect on stability. All these concepts are built upon and expanded in future chapters.

Bond Polarity and Reactive Intermediates

In the study of chemical reactions, a knowledge of bond polarity often allows us to predict how charged ions might react with a particular bond. In the following example, G is more electronegative than carbon, $Nu\overset{\ominus}{:}$ is a negatively charged ion, and $E^{\oplus}$ is a positively charged one. $E^{\oplus}$ is a symbol for an electrophile (Greek *electro*, electron; *phile*, loving), an electron-loving reagent. Electrophiles are often positively charged, but need not be. Examples of electrophiles are $H^{\oplus}$ (the proton) and $R^{\oplus}$ (the carbocation). The other symbol used is $Nu\overset{\ominus}{:}$, the nucleophile (Greek *nucleo*, nucleus; *phile*, loving), a nucleus-loving species. Like the electrophile, it may be charged (negatively charged in this case) or neutral. Examples are $H\overset{..}{\underset{..}{O}}\overset{\ominus}{:}$ (hydroxide ion), $\overset{..}{\underset{..}{X}}\overset{\ominus}{:}$ (halide ion), and $R\overset{\ominus}{:}$ (the carbanion). The symbols $Nu\overset{\ominus}{:}$ and $E^{\oplus}$ are used throughout the text.

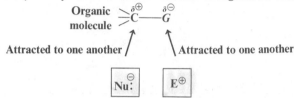

Knowing that *like charges repel* and *unlike charges attract*, we predict and find that the negatively charged ion reacts with the more positive atom in the organic molecule. By the same token, the electrophile $E^{\oplus}$, is attracted to the more electronegative atom, G, in the molecule. In the preceding example, the $Nu\overset{\ominus}{:}$ never reacts with G (that is, a $\ominus$ charge does not approach a $\delta^{\ominus}$ charge), and $E^{\oplus}$ *never* reacts with C.

5.8   Kinetics Versus Thermodynamics

In Sec. 5.7 there was the sentence: "The more stable the intermediate, the more easily it is formed." This could be expanded to: "The more stable the product, the more easily it is formed."

But is either statement true? Usually both are correct. The lower the energy of the product (or intermediate), the lower the E_A for its formation and the more quickly (easily) it forms because the rate of formation is directly related to the E_A. This is illustrated in Fig. 5.6, where one starting material, propane, may go to either the propyl or the isopropyl radical. In this instance the rule of thumb holds. The reaction that gives the more stable product has the lower E_A.

Another example is provided in Fig. 5.7. Here too the rule of thumb holds, assuming that the two different starting materials are of approximately the same energy. This is not a bad assumption as long as the molecules being compared are very similar. But remember that there may be exceptions. Fig. 5.8 shows a generalized view of what an exception might look like. The thermodynamic product is formed more slowly. (See Sec. 8.11 for a specific example where the kinetics and thermodynamics work to drive the reaction along two different paths.)

Reactions in which the product distribution is controlled by the relative energies of the products are said to be **thermodynamically controlled,** whereas those that are governed by the relative E_A values of the transition states are said to be **kinetically controlled.** Whether a reaction is kinetically or thermodynamically controlled depends

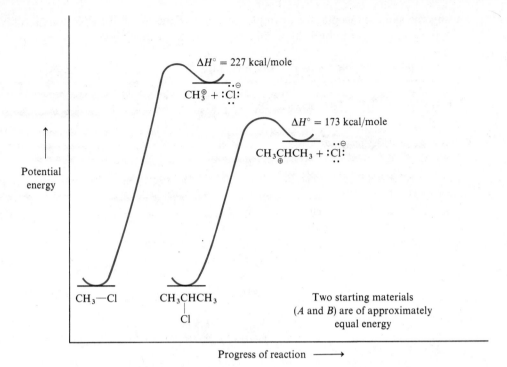

FIGURE 5.7   The formation of the methyl cation and the isopropyl cation, in which the more stable product, the isopropyl cation, is formed more easily.

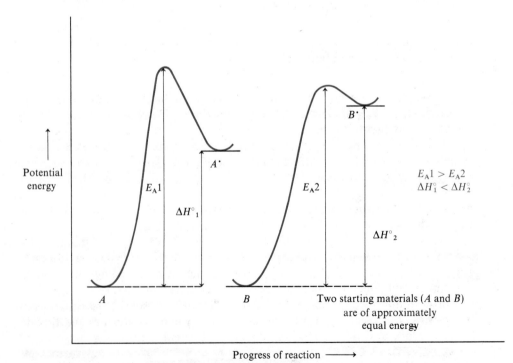

FIGURE 5.8   A generalized reaction in which the more stable product is not formed more easily because it has higher E_A.

on the particular thermodynamic parameters associated with it as well as the conditions under which it is run. (See Sec. 8.11 for an example of the role of reaction conditions in governing the reaction pathway.)

5.9   How to Study a Reaction Mechanism

A reaction mechanism should answer all our questions about a given reaction. It should show which bonds are broken, which are formed, and the order in which this is accomplished. It should explain any changes in stereochemistry, geometry, and hybridization. A reaction mechanism must explain **all** observations regarding a given chemical transformation.

In this section we introduce the tools used by chemists in elucidating reaction mechanisms. A more complete discussion of this topic is found in Reading Reference 1 for this chapter in the appendix. Our chemical background is too limited to go into any of these topics in great depth, but each is substantiated and expanded throughout the text.

The steps used to elucidate a reaction mechanism are:

Step 1. Products. Identify all the products formed in the reaction (see Sec. 5.10).
Step 2. Intermediates. Determine whether the reaction occurs through a one-step, concerted mechanism, or whether there is a series of two or more steps involving intermediates (see Sec. 5.11).
Step 3. Stereochemistry. Determine whether the three-dimensional geometry of the molecules changes in going from starting materials to products (discussed in Chap. 6).
Step 4. Kinetics. Measure the rate of the reaction (how fast it goes) and determine what factors affect this rate (see Sec. 5.12).
Step 5. Catalysis and solvent effects. Determine whether the reaction is catalyzed in any way (see Sec. 5.13) or whether there is a solvent effect.
Step 6. Substituent effects. Elucidate how the structure of each reactant affects the outcome of the reaction (see Sec. 5.15).
Step 7. Isotopically labeled compounds. Note whether the substitution of isotopes on specific positions of the reactant molecules affects the outcome or progress of the reaction in any way (see Secs. 10.11, 23.6A, and 23.6D).

Two of these steps, stereochemistry and isotopically labeled compounds, are discussed extensively in the text. Stereochemistry, which is the study of the three-dimensional arrangements of molecules in space, is presented in Chap. 6; isotopically labeled compounds are discussed principally in Secs. 10.11, 23.6A, and 23.6D. The remaining steps are discussed in turn in the following sections.

5.10   Reaction Mechanisms: Product Distribution

All products formed in the reaction under investigation must be identified. This step is essential. If you cannot account for all the materials, then there will always be some question as to whether or not the mechanism you propose is complete. By identifying all the products, you get a picture of what is occurring, start to know whether there is one mechanism or competing mechanisms, and build a foundation on which to plan your next move.

A reaction mechanism, **nucleophillic aliphatic substitution** (Sec. 4.2A), is discussed at this time. In its most general form, this substitution may be depicted by the equation:

$$R\!:\!L + Nu\!:\!^{\ominus} \rightarrow R\!:\!Nu + L\!:\!^{\ominus}$$

where
$$R = \text{methyl or primary alkyl group}$$
$$L\!:\!^{\ominus} = \text{leaving group}$$
$$Nu\!:\!^{\ominus} = \text{nucleophile}$$

As the reaction occurs, the leaving group, $L\!:\!^{\ominus}$, departs from carbon with the pair of bonding electrons, and the nucleophile is attached to carbon.

In this chapter we are concerned with one particular family of leaving groups and nucleophiles, the halides. A halogen has a great affinity for electrons because it is highly electronegative, so when a carbon-halogen bond breaks, it most often does so heterolytically with the bonding pair of electrons remaining with halogen. This happens in nucleophilic aliphatic substitution reactions on alkyl halides where a nucleophile is substituted for halogen:

$$R\!:\!\ddot{X}\!:\! + Nu\!:\!^{\ominus} \longrightarrow R\!:\!Nu \quad + \quad :\!\ddot{X}\!:\!^{\ominus}$$

Halide leaving
group

The nucleophiles include many organic and inorganic reagents (see Sec. 9.7). Because of the ready accessibility of many alkyl halides, a large number of organic molecules can be synthesized by this reaction.

The reaction used in this discussion is the formation of one alkyl halide from another alkyl halide:

$$R\!-\!\ddot{X}\!:\! + :\!\ddot{I}\!:\!^{\ominus} \longrightarrow R\!-\!\ddot{I}\!:\! + :\!\ddot{X}\!:\!^{\ominus}$$

In this reaction iodide is the nucleophile and halide (usually $:\!\ddot{Br}\!:\!^{\ominus}$ or $:\!\ddot{Cl}\!:\!^{\ominus}$) is the leaving group. A specific example is the conversion of bromomethane to iodomethane:

$$CH_3\!-\!\ddot{Br}\!:\! + :\!\ddot{I}\!:\!^{\ominus} \xrightarrow{\text{acetone}} CH_3\!-\!\ddot{I}\!:\! + :\!\ddot{Br}\!:\!^{\ominus}$$

The first step in elucidation is to isolate the products and identify them. In this example iodomethane and $:\!\ddot{Br}\!:\!^{\ominus}$ (as, for example, NaBr) are the only two products.

5.11   Reaction Mechanisms: Intermediates

After isolating and/or identifying the products of a given reaction, the next step is to determine whether the reaction occurs through a one-step, concerted mechanism or through a series of two or more steps involving intermediates.

Depending on the intermediate(s) involved (if any), it (they) may be:

1. Directly isolated.
2. Observed through the use of instruments (see Chaps. 12, 15, and 18).
3. Trapped by adding a compound it will react with. The product thus formed is isolated and the intermediate identified (see Sec. 16.9).

4. Independently synthesized, if stable, and allowed to undergo the reaction under investigation to see whether the same products result.

Intermediates may also be detected from indirect evidence. For example, if free radicals ($R^{\cdot}$) are involved, there is often a chain-reaction mechanism, which is easily recognizable (see Sec. 4.6). Alternatively, if carbocations ($R^{\oplus}$) are involved, the result is often rearrangements in the carbon skeleton of the starting material.

In our example, when 1-bromopropane is used, 1-iodopropane is the exclusive product (ignoring the NaBr). There is no rearrangement to produce, for example, 2-bromopropane. This is illustrated in the following proposed reaction scheme:

$$CH_3CH_2CH_2{-}Br \xrightarrow[\text{dissociation}]{\text{step 1}} CH_3CH_2\overset{\oplus}{C}H_2 \xrightarrow[\text{:I:} \ominus \text{attack}]{\text{step 2}} CH_3CH_2CH_2{-}I$$

1-Bromopropane (*n*-Propyl bromide) | $1°$ Carbocation | 1-Iodopropane (*n*-Propyl iodide)

rearrangement

Does not occur under the reaction conditions

$$CH_3\overset{\oplus}{C}HCH_3 \xrightarrow[\text{:I:} \ominus \text{attack}]{\text{step 3}} CH_3CHCH_3 \ (I)$$

$2°$ Carbocation | 2-Iodopropane (Isopropyl iodide)

2-Iodopropane is not formed under these reaction conditions. If carbocations were involved, some rearrangement to the more stable $2°$ carbocation would surely result. Since no 2-iodopropane is formed, this must not occur. Additionally, all attempts to isolate, trap, or observe any intermediates in this reaction prove fruitless. The reaction therefore must proceed through a one-step concerted mechanism.

Question 5.4

Here is a reaction of free radicals:

Radical | Nitrone | Nitrone adduct, stable free radical

The nitrone adduct can be observed, indirectly, by use of various instruments. How could you use this reaction as a tool in elucidating reaction mechanisms?

5.12 Reaction Mechanisms: Kinetics

The rate of a bimolecular chemical reaction depends on three factors: (1) the two reacting molecules must collide, (2) they must be in a correct orientation with respect to each other, and (3) the collision must be of some certain minimum energy. The E_A of the reaction governs the minimum amount of energy required, and concentration and temperature can be used to increase (or decrease) the number of collisions. The more collisions, the more likely there will be one of the correct orientation.

The rate of the reaction can be followed by monitoring the appearance of product or the disappearance of reactant. The rate is, after all, the number of molecules of reactant that are converted to product per unit of time. This field of study is called **kinetics.**

Mathematically, the rate of a chemical reaction can be expressed by the following general equation for the reaction $A + B + C \rightarrow$ products:

$$\text{Rate of reaction} = k_r[A][B][C]$$

where k_r = rate constant.

This expression relating the rate of reaction to the concentrations of the reactants is called the **rate law.** Rate laws must be determined experimentally, and the rate constant, k_r, varies depending on the reaction conditions (such as temperature and solvent). Concentrations are customarily expressed in moles per liter, so the rate of reaction is often expressed as a concentration change per unit of time (for example, moles per liter per second).

Furthermore, rate laws provide very important information about the reactants that are involved in the rate-determining step of a reaction. Many reactions proceed through more than one step, and when this is the case, the slowest step of the reaction (the rate-determining step) governs the overall reaction rate. In these cases, the other steps are normally faster, and as soon as the slow step occurs, the remaining steps of the reaction occur quite rapidly. Keep in mind, however, that the terms *slow* and *fast* are relative; they do not necessarily imply that the reaction itself is slow or fast.

The following generalized examples help illustrate the concept of reaction rates:

1. For the generalized reaction $A + B \rightarrow$ product(s),

$$\text{Rate} = k_r[A]$$

or

2. For the generalized reaction $A + B \rightarrow$ product(s),

$$\text{Rate} = k_r[A][B]$$

Valuable information concerning the mechanistic pathway in a given reaction can be determined from the rate law associated with that reaction. For example, in a reaction of type 1, changing the concentration of B (which could be present in the reaction, e.g., as the solvent) has no affect on the overall rate; B is not involved in the rate-determining step as evidenced by the rate expression. In contrast, if the concentration of B is doubled in a reaction of type 2, the overall rate doubles. Likewise, halving the concentration of B halves the rate. The effect of B is quite evident. Note that k_r is a constant; it does not vary with changes in concentration. Try using some real numbers to prove these points to yourself!

When the kinetics of the reaction between bromomethane and sodium iodide are measured, the rate expression derived from the experimental data is

$$CH_3\text{—}Br + :\overset{..}{\underset{..}{I}}:^{\ominus} \longrightarrow CH_3\text{—}I + :\overset{..}{\underset{..}{Br}}:^{\ominus}$$

$$\text{Rate} = k_r[CH_3Br]\left[:\overset{..}{\underset{..}{I}}:^{\ominus}\right]$$

The rate of the reaction depends on the concentration of both reactants; both are involved in the rate-determining step.

This is consistent with our previous findings. Because it appears that the reaction is a one-step process with no detectable intermediates, it follows that both reactants must be involved in this one step. The kinetic evidence provides added weight to the original findings.

5.13 Reaction Mechanisms: Catalysis

A very important question regarding any mechanism is whether or not it is catalyzed in any way. What is the effect of, for example, heat, light, acid strength (pH), or solvent? Does the reaction require the presence of acids, metals, or peroxides, for example? The answers to these questions provide valuable insight into the mechanism.

Examples of reactions that do require catalysts are free-radical substitution (see Sec. 4.6) and hydrogenation of alkenes (see Sec. 8.12). In the former, heat or light is necessary, and a metal is necessary for the latter.

Free-radical substitution:

$$R\!-\!H + X_2 \xrightarrow[\text{light}]{\text{heat or}} R\!-\!X + HX$$

Hydrogenation of alkenes:

$$\text{C}\!=\!\text{C} + H_2 \xrightarrow[\text{(Pt, Pd, or Ni)}]{\text{metal}} -\underset{H}{\overset{|}{\text{C}}}-\underset{H}{\overset{|}{\text{C}}}-$$

A catalyst speeds up a reaction by providing an alternate reaction pathway that involves but does not consume the catalyst. This alternate route has a lower E_A and is therefore a more rapid reaction. The preceding hydrogenation reaction, for example, does not occur unless one of the indicated metals (or some other catalyst) is added. With the metal catalyst, the reaction proceeds in a matter of minutes. The role of the catalyst in the hydrogenation mechanism is discussed in Sec. 8.12.

In nucleophilic aliphatic substitution reactions, catalysts are not usually needed, although there are some exceptions (see Sec. 10.13).

5.14 Reaction Mechanisms: Solvent Effects

The effect of the solvent is important to the success or failure of many chemical reactions. In nucleophilic aliphatic substitution, the solvent plays a key role. Before discussing this specific reaction, a discussion of solvents in general is in order.

A solvent may be either *polar* or *nonpolar* and either *protic* or *aprotic*. A protic solvent usually contains an —OH group; examples are water, alcohols (R—OH), and carboxylic acids (RCOOH). In a more literal sense, protic solvents are those that contain a moderately acidic proton. An aprotic solvent contains no acidic hydrogens and thus no —OH groups.

The polarity of a solvent is more difficult to define. A nonpolar one, of course, has no polarity associated with it at all; nonpolar solvents include hydrocarbons and certain symmetrically substituted haloalkanes, such as tetrachloromethane. A convenient scale for defining the polarity of a solvent is based on the *dielectric constant*, ε, which is a measure of its ability to insulate charges from one another.

Table 5.2 lists some common solvents along with their structures and dielectric constants. In general, the higher the dielectric constant, the greater the polarity of the solvent.

The changes that occur in nucleophilic substitution reactions are often subtle, but the rate of the reaction may be altered markedly by the choice of solvent. Protic

TABLE 5.2   Dielectric Constants of Certain Solvents

Solvent		Dielectric Constant, ε	Dipole Moment, μ
Name	Structure		
Protic solvents:			
Ethanoic acid (acetic acid)	$CH_3-\overset{\overset{\displaystyle O}{\|\|}}{C}-OH$	6	1.68
1-Butanol	$CH_3CH_2CH_2CH_2-OH$	8	1.60
Ethanol (ethyl alcohol)	CH_3CH_2-OH	26	1.66
Methanol (methyl alcohol)	CH_3-OH	32	2.87
Methanoic acid (formic acid)	$H-\overset{\overset{\displaystyle O}{\|\|}}{C}-OH$	58	1.82
Water	$H-OH$	81	1.84
Aprotic solvents:			
Cyclohexane		2	0
Benzene		2	0
Ethyl ether	$(CH_3CH_2)_2O$	4	1.15
Trichloromethane (chloroform)	$CHCl_3$	5	1.01
Propanone (acetone)	$(CH_3)_2C{=}O$	21	2.69
Hexamethylphosphoramide	$(CH_3)_2N-\overset{\overset{\displaystyle O}{\|\|}}{\underset{\underset{\displaystyle N(CH_3)_2}{\|}}{P}}-N(CH_3)_2$	30	5.54
Dimethylformamide	$H-\overset{\overset{\displaystyle O}{\|\|}}{C}-N(CH_3)_2$	38	3.86
Acetonitrile	$CH_3-C{\equiv}N$	38	3.44
Dimethyl sulfoxide	$CH_3-\overset{\overset{\displaystyle O}{\|\|}}{S}-CH_3$	48	3.90
Nitromethane	CH_3-NO_2	36	3.56

solvents (abbreviated HOS) solvate the nucleophile to a much greater extent than aprotic solvents because of hydrogen bonding interaction of the following type:

$$Nu:^{\ominus} \quad \overset{\delta^{\oplus}}{H}-\overset{\delta^{\ominus}}{O}-S$$

When this occurs, the unshared pairs of electrons on the nucleophile are tied up somewhat, thus reducing their effectiveness (that is, their nucleophilic character is reduced).

In practice, a nonpolar, aprotic solvent is difficult to work with because most inorganic salts are insoluble in it. Some of the more desirable solvents have a high dielectric constant and are aprotic, such as hexamethylphosphoramide, dimethyl-formamide, acetonitrile, dimethyl sulfoxide, and nitromethane (Table 5.2). The polarity of the solvent is mostly an experimental requirement so that the nucleophile and the organic compound are both present in a homogeneous reaction mixture. The reaction we are discussing is favored in polar, aprotic solvents.

5.15   Reaction Mechanisms: Substituent Effects

Once we have some idea of the mechanism, and perhaps even a rough guess of what it is, the next step is to gather sufficient evidence to substantiate it. This may be done by using compounds designed to provide us with answers to specific questions about the mechanism. These may be compounds with a specific carbon framework (skeleton) and/or various functional groups at specific positions in the molecule.

The conversion of an alkyl bromide to an alkyl iodide is an excellent example to illustrate this concept. For example, we could ask: How does the structure of the alkyl group affect the reactivity? Data indicate that there is a considerable difference in the reactivities of various alkyl groups in this reaction:

$$R—\overset{..}{\underset{..}{Br}}: + :\overset{..}{\underset{..}{I}}:^{\ominus} \longrightarrow R—\overset{..}{\underset{..}{I}}: + :\overset{..}{\underset{..}{Br}}:^{\ominus}$$

In one comparison, 1-bromopropane reacted 100 times faster than 2-bromopropane.

$$CH_3CH_2CH_2—\overset{..}{\underset{..}{Br}}: + :\overset{..}{\underset{..}{I}}:^{\ominus} \longrightarrow CH_3CH_2CH_2—\overset{..}{\underset{..}{I}}: + :\overset{..}{\underset{..}{Br}}:^{\ominus}$$

1° Carbon

$$\underset{\overset{|}{:\underset{..}{Br}}:}{CH_3CHCH_3} + :\overset{..}{\underset{..}{I}}:^{\ominus} \longrightarrow \underset{\overset{|}{:\underset{..}{I}}:}{CH_3CHCH_3} + :\overset{..}{\underset{..}{Br}}:^{\ominus}$$

2° Carbon

A *steric effect* is involved here. The more crowded the carbon bearing the leaving group, the slower the reaction (see Sec. 9.6). The reactivity order of the alkyl halides in this reaction is:

Decreasing order of reactivity
of R—X:

$$CH_3—X > 1° > 2° > 3°$$

Increasing the number of substituents on the carbon bearing the halogen makes the reaction more difficult.

5.16   Nucleophilic Aliphatic Substitution: S_N2

There is now sufficient evidence to develop a mechanism. The mechanism must explain all the experimentally determined results:

1. The reaction occurs in one step and does not, therefore, involve any intermediates.
2. The only products are alkyl iodide and sodium bromide.

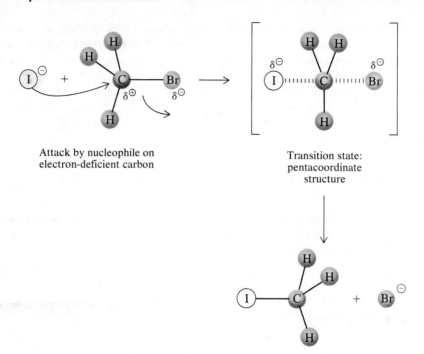

Attack by nucleophile on
electron-deficient carbon

Transition state:
pentacoordinate
structure

FIGURE 5.9   Nucleophilic substitution—the reaction that occurs between CH_3—Br and $:\overset{..}{\underset{..}{I}}:^{\ominus}$ to give CH_3—I and $:\overset{..}{\underset{..}{Br}}:^{\ominus}$.

3. The kinetics tell us that both alkyl halide and the iodide are involved in the rate-determining step.

All these are consistent with the mechanism illustrated in Fig. 5.9. Both alkyl bromide and iodide are involved in the transition state of a one-step reaction going from reactants to products. This is an example of a **bimolecular reaction** (see Sec. 9.5), a reaction involving two molecules.

4. The reaction is favored by polar, aprotic solvents—solvents that maximize the **nucleophilicity** (*nucleophilic* nature) of the nucleophile.

A good nucleophile is essential to a successful reaction. If the nucleophile does not attack the partially electron-deficient carbon, the transition state is never achieved and the reaction does not proceed.

5. Increasing the number of substituents on the carbon bearing the leaving group slows the reaction.

Increasing the number of substituents on the carbon being substituted makes it more difficult for the nucleophile, $:\overset{..}{\underset{..}{I}}:^{\ominus}$, to come within bonding distance of the carbon bearing the leaving group. In terms of energy, this means that the nucleophile must possess greater energy to overcome the nonbonding interactions and actually attack the carbon where substitution is to occur. The energies of activation, E_A values, provide an estimate of this barrier, as shown in the energy-profile diagrams in Figs. 5.10 and 5.11.

All the evidence is consistent with the mechanism in Fig. 5.9. This is called the **S_N2 mechanism.** The S stands for *substitution*, the N for *nucleophilic*, and the 2 for *bimolecular*; that is, in the rate-determining step, two molecules are involved in the transition state. The S_N2 reaction is discussed further in Chaps. 7 and 9.

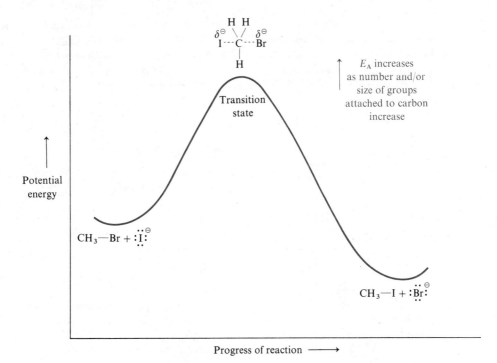

FIGURE 5.10   Energy-profile diagram showing energy changes in the reaction
CH_3—Br + :I:$^\ominus$ → CH_3—I + :Br:$^\ominus$.

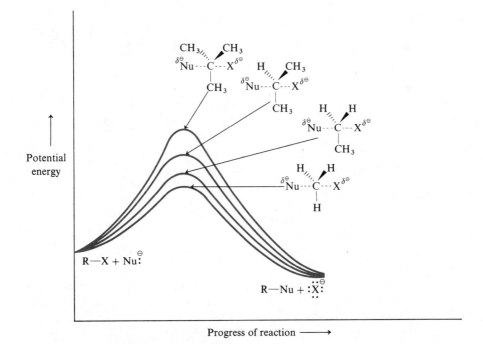

FIGURE 5.11   Energy-profile diagram of the nucleophilic displacement reaction of a series of alkyl halides by a nucleophile, Nu:$^\ominus$. The potential energy barriers are arbitrary but show that the energies of activation E_A increase in the sequence CH_3—X < CH_3CH_2—X < $(CH_3)_2$CH—X < $(CH_3)_3$C—X. The ground-state energies of the reactants have been arbitrarily drawn as equal to illustrate the relative differences in the E_A values.

The S_N2 reaction actually was introduced previously. The preparation of alkanes using lithium dialkylcopper reagents (see Sec. 4.2) probably involves this mechanism. The carbanionic carbon of the organometallic (the lithium dialkylcopper reagent) is the nucleophile and the halide is the leaving group. It should be clear now why the alkyl halide in the lithium dialkylcopper synthesis also must be methyl or primary.

$$R:\overset{\ominus}{} \quad + \quad R'\overset{\frown}{-}X \quad \xrightarrow{S_N2} \quad R—R' \; + \; :\overset{..}{\underset{..}{X}}:^{\ominus}$$

| Of the lithium dialkylcopper reagent | Methyl or primary alkyl halide | Alkane |

S_N2 Reaction

Study Questions

5.5 Define or explain the following 16 items.
(*a*) free-radical halogenation
(*b*) nucleophilic aliphatic substitution
(*c*) bond association energy, *D*
(*d*) *D* versus $\Delta H°$
(*e*) average bond energy
(*f*) exothermic versus endothermic reactions
(*g*) $\Delta H°$ versus E_A
(*h*) energy-profile diagram (reaction coordinate)
(*i*) transition state (activated complex)
(*j*) intermediate
(*k*) free-radical stability
(*l*) ease of formation of free radicals
(*m*) inductive effects
(*n*) kinetic versus thermodynamic control of reactions
(*o*) trapping of intermediates   (*p*) reaction mechanism

5.6 Arrange each of the following sets in order of decreasing stability:

(*a*)

$$CH_3CH_2-\overset{\overset{\displaystyle CH_3}{|}}{\underset{\displaystyle \cdot}{C}}-CH_3 \qquad CH_3-\overset{\overset{\displaystyle CH_3}{|}}{\underset{\underset{\displaystyle H}{|}}{C}}-CH_2\overset{\displaystyle \cdot}{C}H_2 \qquad CH_3-\overset{\overset{\displaystyle CH_3}{|}}{\underset{\underset{\displaystyle H}{|}}{C}}-\overset{\displaystyle \cdot}{C}HCH_3$$

(*b*)

$$\cdot\!\!\left\langle \bigcirc \right\rangle\!\!-CH_3 \qquad \left\langle \bigcirc \right\rangle\!\!-\overset{\displaystyle \cdot}{C}H_2 \qquad \left\langle \bigcirc \right\rangle\!\!\cdot\!\!-CH_3$$

(*c*)

$$CH_3\overset{\oplus}{C}HCH_3 \qquad \overset{\displaystyle CH_2^{\oplus}}{\bigcirc} \qquad CH_3(CH_2)_7\overset{\oplus}{C}H_2$$

(*d*)

$$(CH_3)_2\overset{\ominus}{\overset{..}{C}}H \qquad CH_3-\overset{\overset{\displaystyle CH_3}{|}}{\underset{\underset{\displaystyle CH_3}{|}}{C}}-\overset{\ominus}{\overset{..}{C}}H_2 \qquad CH_3-\overset{\overset{\displaystyle CH_2CH_3}{|}}{\underset{\underset{\displaystyle \ominus}{|}}{C}}-CH_3$$

5.7 The reaction $U + T = W$ is exothermic. Draw an energy diagram for it and clearly identify the (*a*) position of the reactants, (*b*) position of the products, (*c*) activation energy E_A, (*d*) heat of the reaction $\Delta H°$, and (*e*) position of the transition state.

5.8 Compounds *A* and *B* are in equilibrium.

$$A \rightleftharpoons B$$

For the reaction $A \to B$, the E_A is 15 kcal/mole (62.8 kJ/mole) and the $\Delta H°$ is 12 kcal/mole (50.2 kJ/mole). Draw the reaction coordinate for this equilibrium. What are the E_A and the $\Delta H°$ values for the reaction $B \to A$? What are the percentages of *A* and *B* in the reaction mixture at 25° (assume $\Delta G° \approx \Delta H°$)?

5.9 Consider the following reaction, which is the second step in free-radical halogenation:

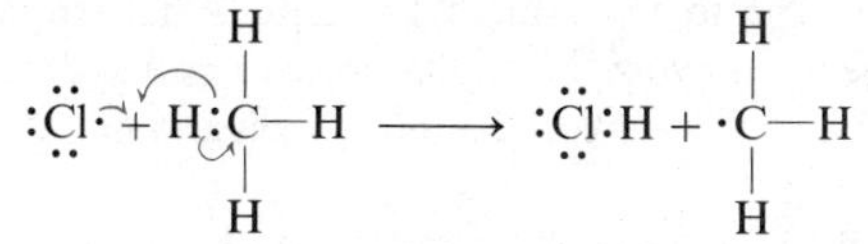

$$:\ddot{C}l\cdot + H:\underset{\underset{H}{|}}{\overset{\overset{H}{|}}{C}}-H \longrightarrow :\ddot{C}l:H + \cdot\underset{\underset{H}{|}}{\overset{\overset{H}{|}}{C}}-H$$

There is another pair of products possible. What are they? Which pair is more likely to form? (*Hint:* Determine this using heats of reactions; see Sec. 5.6.)

5.10 Calculate the product distribution for the isomeric alkyl chlorides that are obtained when the following compounds are allowed to react with chlorine gas at 25° in the presence of light:
(*a*) hexane   (*b*) 2,4-dimethylpentane
(*c*) 2,2,4-trimethylpentane   (*d*) 3,3-dimethylpentane

5.11 Draw all the structural isomers corresponding to the following formulas:
(*a*) $C_2H_3Cl_3$,   (*b*) $C_3H_6Cl_2$,   (*c*) $C_3H_5Cl_3$.

5.12 In the reaction

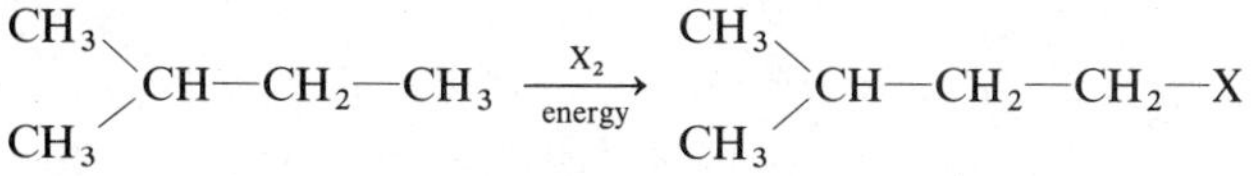

$$\underset{CH_3}{\overset{CH_3}{\diagdown}}CH-CH_2-CH_3 \xrightarrow[\text{energy}]{X_2} \underset{CH_3}{\overset{CH_3}{\diagdown}}CH-CH_2-CH_2-X$$

would you expect to obtain more than, less than, or the same amount of the indicated product from bromination ($X_2 = Br_2$) compared with chlorination ($X_2 = Cl_2$)? Briefly explain.

5.13 Consider the following reaction:

$$CH_3-CH_2-CH_2-Cl \xrightarrow[hv]{Cl_2, 25°} CH_3-CH_2-\underset{\underset{Cl}{|}}{CH}-Cl + CH_3-\underset{\underset{Cl}{|}}{CH}-CH_2-Cl$$

$$\underset{\underset{10\%}{(1)}}{} \qquad\qquad\qquad\qquad 45\%$$

$$+ Cl-CH_2-CH_2-CH_2-Cl$$
$$45\%$$

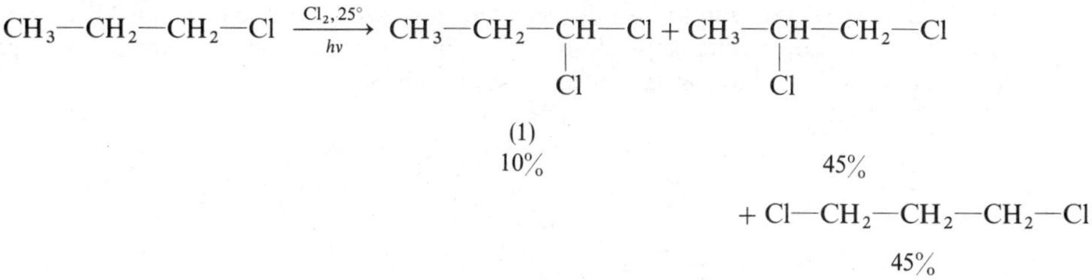

(*a*) What class of radicals (1°, 2°, or 3°) intervenes in the formation of the products? Draw their structures.

(*b*) Based on the discussion of radical stabilities in Sec. 5.7A, suggest an explanation for the relatively small amount (10%) of 1,1-dichloropropane, (1), that is formed.

5.14 (*a*) Briefly state the major factors that dictate the *product distribution* in the free-radical chlorination of alkanes (using light at room temperature). Answer qualitatively by giving the trends that are observed.

(*b*) How does the product distribution for the bromination of alkanes (at 125° with light) compare with that of chlorination, which you described in (*a*)? Give a qualitative comparison by indicating only the general trends.

(*c*) Suppose you wish to carry out the following reaction to obtain the greatest possible amount of the indicated product. In light of your general knowledge about the halogenation of alkanes, would you choose chlorination or bromination? Why?

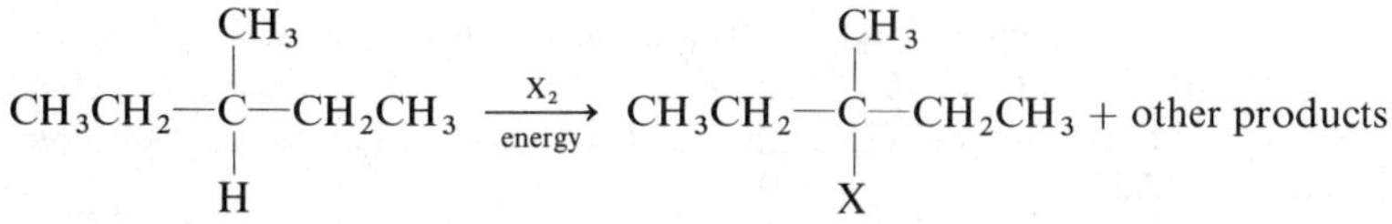

$$CH_3CH_2-\underset{\underset{H}{|}}{\overset{\overset{CH_3}{|}}{C}}-CH_2CH_3 \xrightarrow[\text{energy}]{X_2} CH_3CH_2-\underset{\underset{X}{|}}{\overset{\overset{CH_3}{|}}{C}}-CH_2CH_3 + \text{other products}$$

5.15 Compound *A*, a hydrocarbon, was brominated at 300° to yield two isomeric compounds, *B* and *C*. The reaction of 2 moles of *B* with 2 moles of Na metal yielded 2,3-dimethylbutane.
(*a*) What are the structures of compounds *A*, *B*, and *C*?
(*b*) Given the rate factors for bromination as 1,600:82:1 for 3°:2°:1°, what are the relative proportions of *B* and *C* formed?
(*c*) Give the structural formula and IUPAC name for the product of the reaction of 2 moles of *C* with 2 moles of Na.

(*d*) When *B* is treated with Li in anhydrous ether followed by the addition of CuI, compound *D* is formed. *D* reacts with *E* to give 2-methylpentane. What are the structures of *D* and *E*?

5.16 The reaction between hydrogen and fluorine to give hydrogen fluoride is: $H_2 + F_2 = 2HF$. This reaction occurs by a free-radical chain reaction and in a manner analogous to the halogenation of alkanes.

(*a*) Provide a complete, stepwise mechanism for this reaction and label the chain-initiating, chain-propagating, and chain-terminating steps. (*Hint:* See the bond energies in Table 5.1.)

(*b*) Using the bond energies in Table 5.1, calculate the heat of reaction, $\Delta H°$, for the *overall* reaction.

5.17 The following equations (shown in random order) are some possible free-radical reactions in the mechanism for the chlorination of methane:

$$H\cdot + :\overset{..}{\underset{..}{Cl}}\cdot \longrightarrow H\text{—}Cl \qquad\qquad H\cdot + H\cdot \longrightarrow H\text{—}H$$

$$CH_4 + :\overset{..}{\underset{..}{Cl}}\cdot \longrightarrow H\cdot + CH_3\text{—}Cl \qquad Cl\text{—}Cl \longrightarrow 2:\overset{..}{\underset{..}{Cl}}\cdot$$

$$:\overset{..}{\underset{..}{Cl}}\cdot + :\overset{..}{\underset{..}{Cl}}\cdot \longrightarrow Cl_2 \qquad\qquad H\cdot + Cl_2 \longrightarrow H\text{—}Cl + :\overset{..}{\underset{..}{Cl}}\cdot$$

(*a*) Arrange these equations into chain-initiating steps, chain-propagating steps, and chain-terminating steps.

(*b*) Using the bond energies in Table 5.1, calculate the heat of reaction, $\Delta H°$, for each chain-propagating step.

(*c*) Based on your results in (*b*), is this proposed mechanism a reasonable one? Briefly explain.

5.18 The chain-propagating steps for the free-radical chlorination of methane could perhaps have been written as follows instead of as those given in Sec. 4.6:

$$:\overset{..}{\underset{..}{Cl}}\cdot + CH_4 \longrightarrow CH_3Cl + H\cdot$$

and

$$H\cdot + Cl_2 \longrightarrow HCl + :\overset{..}{\underset{..}{Cl}}\cdot$$

(*a*) Calculate the $\Delta H°$ values for these two reactions using data from Table 5.1.

(*b*) Compare these values with the equations and their energies given in Sec. 5.6 and explain why the above cannot represent the path of the reaction.

5.19 (*a*) Using the bond energies in Table 5.1, calculate $\Delta H°$ for each of these two reactions:

$$:\overset{..}{\underset{..}{Cl}}\cdot + H\text{—}CH_3 \longrightarrow Cl\text{—}CH_3 + H\cdot$$

$$:\overset{..}{\underset{..}{Cl}}\cdot + H\text{—}CH_3 \longrightarrow \cdot CH_3 + H\text{—}Cl$$

(*b*) One of these reactions is the rate-determining step in the correct mechanism for the chlorination of methane. Which is it? Explain why, based on the $\Delta H°$ values you calculated in (*a*).

5.20 Consider the following reaction scheme. The conversion of *A* to *B* is *endothermic* by 30 kcal/mole (125.5 kJ/mole), and the conversions of *B* to *C* and *B* to *D* are *exothermic* by 35 and 40 kcal/mole (146.4 and 167.4 kJ/mole), respectively. The energies of activation (E_A) are as follows: *B* to *A*, 6 kcal/mole (25 kJ/mole); *B* to *C*, 4 kcal/mole (16.8 kJ/mole); and *B* to *D*, 2 kcal/mole (8.4 kJ/mole):

$$A \rightleftarrows B\underset{\searrow D}{\overset{\nearrow C}{}}$$

(*a*) Draw a completely labeled energy-profile diagram for this reaction scheme. Indicate the location of the transition states and calculate the heights of all the energy barriers that separate *A* to *D*.

(*b*) Indicate the rate-determining step(s) in the scheme.

5.21 In the following reaction,

$$R—Br + H\overset{..}{\underset{..}{O}}{:}^{\ominus} \longrightarrow R—OH + {:}\overset{..}{\underset{..}{Br}}{:}^{\ominus}$$

if R is a methyl or primary carbon, it is an S_N2 reaction. If, however, R is a tertiary carbon, the following different two-step mechanism is proposed:

$$R—\overset{..}{\underset{..}{Br}}{:} \xrightarrow{\text{dissociation}} R^{\oplus} + {:}\overset{..}{\underset{..}{Br}}{:}^{\ominus}$$

$$R^{\oplus} + H\overset{..}{\underset{..}{O}}{:}^{\ominus} \longrightarrow R—OH$$

Contrast the structural differences between these two mechanisms and explain why the two-step mechanism is more likely for tertiary halides (see Sec. 9.10).

5.22 Consider the following reaction of alkenes:

$$CH_3—CH{=}CH_2 \xrightarrow{\text{HCl}} CH_3\underset{\underset{Cl}{|}}{CH}CH_3 + CH_3CH_2CH_2—Cl$$

Major product

Explain why the major product should be expected. (See Sec. 5.7B for a hint.)

5.23 In the reaction

$$A \rightleftharpoons B \rightleftharpoons C$$

the following thermodynamic parameters have been determined:

$$A \longrightarrow B \quad E_A = 10 \text{ kcal/mole (41.9 kJ/mole)} \quad \Delta H° = 8 \text{ kcal/mole (33.5 kJ/mole)}$$

$$B \longrightarrow C \quad E_A = 4 \text{ kcal/mole (16.7 kJ/mole)} \quad \Delta H° = -5 \text{ kcal/mole } (-20.9 \text{ kJ/mole})$$

Draw the complete reaction coordinate for these two consecutive equilibria. Label all parts. Determine the thermodynamic parameters of the two reverse reactions.

6

Stereochemistry and Stereoisomerism

Any two nonidentical compounds with the same molecular formula are *isomers*. Several classes of isomers were presented in previous chapters: *structural* (see Secs. 3.5 and 3.9), *conformational* (see Secs. 3.3 and 3.16), and *geometric* (see Sec. 3.17).

We come now to another important type of isomerism, **stereoisomerism.** *Stereoisomers*, which include *enantiomers* and *diastereomers*, have different spatial arrangements of their atoms. This difference is due to an inherent lack of symmetry in the molecules themselves. Stereoisomers exhibit the same bonding topology (that is, they have the same atoms bonded to the same atoms), but as a result of the orientation of these atoms, any two stereoisomers are not superimposable; they are nonidentical, discrete compounds. To interconvert stereoisomers, bonds must be broken and formed. A breakdown of the various classes of isomers is given in Fig. 6.1. Note that geometric isomers (*cis-trans* isomers) are actually stereoisomers, a subclass of diastereomers.

Stereoisomerism is of great interest to us because of its significant applications in organic chemistry, particularly the study of reaction mechanisms. Stereochemistry and stereoisomers are often used to answer questions that cannot be solved by any other means. Most of this chapter is devoted to discussing stereoisomers and optical activity so that they may be used freely throughout the remainder of the text.

6.1   Optical Properties of Molecules; Optical Activity and the Polarimeter

Certain types of molecules *appear* to have identical physical properties in all respects but one. They have, for example, identical melting points, boiling points, and solubility properties; however, when *plane-polarized light* is passed through a solution of these compounds, the light is rotated in characteristic ways. Molecules with this property are said to exhibit *optical activity*.

Light is a form of electromagnetic radiation and, as such, exhibits properties of both a particle and a wave. The smallest quantized energy unit of electromagnetic radiation is called a *photon*. As with all waves, there is a characteristic wavelength (λ) and frequency (v) associated with the electromagnetic radiation. These are related to the overall energy of the radiation by the equation $E = hv$ or $E = hc/\lambda$, where h is Planck's constant and c is the speed of light (3×10^{10} cm/sec) (see Sec. 12.1).

White light can, therefore, be considered to consist of waves in motion. These waves possess a variety of wavelengths, each corresponding to a different color. The waves vibrate in all directions but are always perpendicular to the direction of

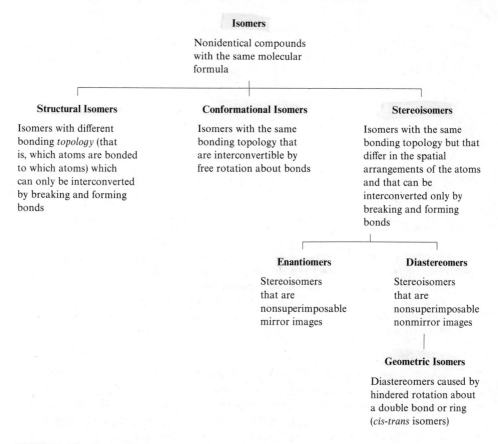

Isomers

Nonidentical compounds
with the same molecular
formula

Structural Isomers

Isomers with different
bonding *topology* (that
is, which atoms are bonded
to which atoms) which
can only be interconverted
by breaking and forming
bonds

Conformational Isomers

Isomers with the same
bonding topology that
are interconvertible by
free rotation about bonds

Stereoisomers

Isomers with the same
bonding topology but that
differ in the spatial
arrangements of the atoms
and that can be
interconverted only by
breaking and forming
bonds

Enantiomers

Stereoisomers
that are
nonsuperimposable
mirror images

Diastereomers

Stereoisomers
that are
nonsuperimposable
nonmirror images

Geometric Isomers

Diastereomers caused by
hindered rotation about
a double bond or ring
(*cis-trans* isomers)

FIGURE 6.1   Types of isomers.

the source. Special sources of light emit light of particular wavelengths; for example, a bulb filled with sodium vapor emits, among other wavelengths, a yellow light, and a mercury vapor light provides a silver color (but not white light). Sodium and mercury vapor lights are used for obtaining plane-polarized light and for studying the optical properties of molecules. The light from these sources can be separated into its various components by the use of filters to give single-color light called *monochromatic light.* Monochromatic light consists of waves of only one wavelength, but these waves still vibrate in all possible directions. Some of these wave properties of light are shown in Fig. 6.2.

   If a beam of monochromatic light is allowed to pass through a Nicol prism [which consists of properly oriented crystals of the mineral Iceland spar or calcite (calcium carbonate)], the emerging beam of light contains vibrations in only one direction or one plane. This is called *plane-polarized light* because the waves that are vibrating in all the other directions were filtered out and do not pass through the prism. This polarizing of light is analogous to that in Polaroid sunglasses, which reduce the intensity of light considerably. Polaroid glasses do not contain a Nicol prism, but they have an intricate grating of parallel lines made of a highly ordered arrangement of crystals that resemble venetian blinds; rotation of the glasses by 90° changes greatly the amount of light that passes through them. Polaroid lenses can be used in place of Nicol prisms.

   The instrument used to measure the optical activity or optical properties of molecules is called a **polarimeter.** It contains two Nicol prisms; one polarizes the

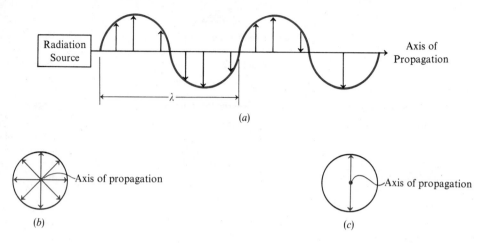

FIGURE 6.2   Representation of light as a wave motion. (*a*) Direction of propagation of light waves, showing wavelength λ and electric field vectors →. The axis of propagation shown is one of an infinite number emanating in all directions from the radiation source. (*b*) View from end of of axis of propagation shows ordinary light containing vibrations in many planes. (*c*) Polarized light containing vibrations in one plane, obtained by passing ordinary light through a Nicol prism or polarized filter.

light and is called the *polarizer* and the other is called the *analyzer*. The sample is placed in a tube between these two prisms. Many compounds do not affect the passage of plane-polarized light, so light emerges with no change in the direction of the plane of the light. When this occurs, the compound being studied is said to be **optically inactive.** If the plane of the polarized light is rotated, the substance is said to be **optically active.** Light passing through the optically active substance emerges vibrating in a different plane.

A better understanding of the polarimeter can be obtained from Fig. 6.3. As this figure illustrates, the Nicol prism analyzer can be rotated; as it is rotated, the intensity of the light increases to a maximum, then decreases to a minimum, and finally returns to its maximum intensity. To measure the angle (α) by which the light is rotated by a sample, the experimenter determines the position of the light maximum *without* the sample present; this is called the *zero point*. The sample is then inserted, and the light maximum is again determined. The difference between the zero point without the sample and the new angle is the **angle of rotation, α.** (Because we determine only a *net change* in rotation, the zero point does not have to be and often is not $0°$ on the scale marked on the analyzer.)

If the substance rotates light to the right, we call that particular substance the **dextrorotatory** (Latin *dexter*, right) or *d* **form** and indicate this by putting a + sign in front of the degrees of rotation. If light is rotated to the left, the substance is said to be the **levorotatory** (Latin *laevus*, left) or *l* **form;** a − sign is put in front of the degrees of rotation. (As an aid to using these terms, remember that both *l*eft and *l*evorotatory start with the letter *l*.)

The angle of rotation, α, measured by this procedure represents the *observed rotation* and is more correctly called α_{obs}, because this is what is actually observed in the laboratory. However, the degree of rotation depends on three factors: (1) the concentration of the compound in the sample tube, (2) the length of the sample tube, and (3) the temperature at which the measurement is taken. Optical rotation depends on the *number* of molecules that the light encounters in passing through the sample tube. It stands to reason that light traveling through a tube 20 decimeters

(a)

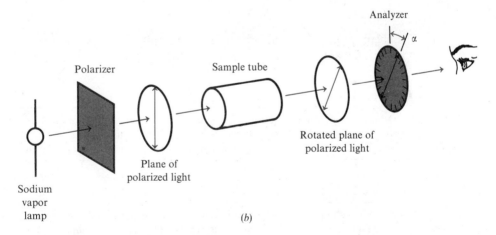

(b)

FIGURE 6.3   (a) Photograph of a polarimeter. (b) Schematic representation of a polarimeter.

(dm) long is affected by twice as many molecules as it is in a tube only 10 dm long; the same type of reasoning applies to comparing solutions with different concentrations.

If chemists around the world simply reported α_{obs} in the literature, rotations would have have no significance because of the dependence of rotation on concentration and the length of the sample tube. Chemical nomenclature has been standardized, and so has the method for reporting rotation. For this purpose, **specific rotation,** $[\alpha]$, is used and has been defined as the number of degrees of rotation based on unit concentration (1 gram/cubic centimeter, or 1 g/cc) and unit sample-tube length (1 dm). To calculate specific rotation, the following equation is used:

$$[\alpha] = \frac{\alpha_{obs}}{c \times l}$$

where α_{obs} = observed rotation

        l = sample-tube length (in dm)

       c = concentration of sample (in g/cc of solution)

Often a sample is dissolved in an organic solvent before its rotation is determined (this must be done for solids), and then it is customary to designate the solvent and concentration in reporting $[\alpha]$. Rotations are also a function of temperature and the wavelength of the light source. In the example

$$[\alpha]_D^{25°} = -4.275° \qquad (c\ 0.51,\ \text{chloroform})$$

the superscript 25° indicates the temperature and the subscript D indicates that the D line from a sodium lamp was used as the light source; the D line of sodium has a wavelength of 5,893 Å. The concentration (in this case 0.51 g/cc) and solvent are given in parentheses to the right.

If a sample is a pure liquid and enough of it is available, the rotation can be determined directly from the sample; we call such a sample *neat* (meaning pure and with no solvent). In this case, the concentration is the liquid's density, d, so that $[\alpha] = \alpha_{obs}/(d \times l)$. It is customary to designate that a neat liquid is used, as in the following example:

$$[\alpha]_D^{25°} = +47.2° \qquad (\text{neat})$$

The chemical and physical properties of optically active compounds are expanded in Secs. 6.9 and 6.16.

Question 6.1

An unknown compound weighing 4.2 g is dissolved in enough carbon tetrachloride to make a total volume of 250 cc. The observed rotation of this solution is $\alpha_{obs} = -2.25°$ in a 25-cm cell using the sodium D line. Calculate and designate properly $[\alpha]$ for this compound.

Question 6.2

A solution of a compound in a solvent is placed in a polarimeter and exhibits no apparent rotation of plane-polarized light. However, it is conceivable that this compound is optically active but the observed rotation is 360° or some multiple thereof. Suggest how diluting the sample and measuring the rotation at other concentrations can resolve this problem. Be specific as to *how* you would interpret the data.

6.2   Molecular Asymmetry and Discovery of Enantiomers

Until the middle of the nineteenth century, very few compounds were known to exhibit optical activity. As early as 1815, the optical interaction of light with molecules was discovered by the physicist Jean-Baptiste Biot. However, the foundation of stereochemistry was laid in 1848 when Louis Pasteur observed that the salts obtained from the residue in wine kegs could be crystallized and separated into two forms. Pasteur's work involved physically separating these two types of crystals by using a magnifying glass and tweezers. He obtained two piles of crystals, one containing identical "right-handed" crystals and the other "left-handed" crystals. These types were nonsuperimposable *mirror images*; that is, they resembled one another in the same way one's left and right hands do, as shown in Fig. 6.4.

The right-handed crystals were dissolved in water and observed in the polarimeter to rotate light in one direction, whereas the left-handed crystals rotated light in the opposite direction. Both sets of crystals were *optically active!* More significant yet was his observation that *the rotations were of equal magnitude* (for solutions of equal concentration) *but opposite in direction.* For example, if the rotation for one set of crystals was $[\alpha]_D^{20°} = +4.7°$, that for the other set of crystals would be $[\alpha]_D^{20°} = -4.7°$. Furthermore, when the rotation of the original mixture of crystals was tested, *no net rotation* was observed. It appeared that there were equal amounts of both types of crystals in the original mixture. After separation, the two types of crystals had optical properties, and when they were mixed together in equal amounts, again no net rotation was observed. Pasteur suggested that on the microscopic level the three-dimensional arrangements of atoms in these two types of crystals (that is, these two optically active forms) were different. He postulated that the atoms bore a mirror-image relationship to each other like the crystals themselves on the macroscopic level. This discovery was truly exciting, but Pasteur did not propose a model to explain the nature of the asymmetry (lack of symmetry) that allows for the construction of the two separate mirror-image forms.

There was considerable fortuity involved in Pasteur's discovery. As is often the case in scientific discoveries, he happened to be in the right place at the right time. Very few compounds crystallize so that equal amounts of right-handed and left-handed crystals form with each being uncontaminated by the other. Pasteur exploited his discoveries to the fullest, and his work is indicative of his genius. In Sec. 6.14 we discuss the three-dimensional structures of the salts of tartaric acid that Pasteur studied.

In the fall of 1874, the Dutch physical chemist van't Hoff and the French chemist LeBel independently proposed an explanation for the asymmetry of molecules—that is, for the existence of left-handed and right-handed structures. They observed that in the known cases of compounds with optical properties, the molecules contain at least one carbon atom that has four *different* atoms or groups of atoms attached to

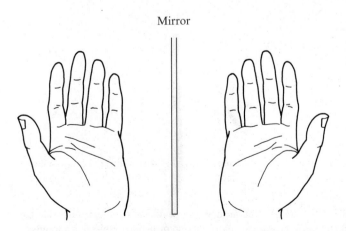

FIGURE 6.4   Mirror-image relationship of the left and right hands.

it. They pointed out that if the four bonds about this carbon atom are directed toward the four corners of a tetrahedron, the resulting molecule would *lack* symmetry. In the three-dimensional sense, they illustrated that two mirror-image forms of the molecule can be constructed and that these two forms would be nonsuperimposable and thus nonidentical. Two molecular structures that bear such a mirror-image relationship are called **enantiomers.**

We know from Chap. 2 that carbon is tetrahedral, so the conclusions of van't Hoff and LeBel may not be surprising to us. However, these researchers in 1874 did not have modern electronic equipment for studying molecular structure; quantum mechanics was completely unknown, and indeed very little was known about atomic and molecular structure.

6.3   Structures of Enantiomers; Requirements for Their Existence

Let us first consider a general molecule, *Cabcd*, where C is a carbon atom and *a*, *b*, *c*, and *d* represent different atoms or groups of atoms. If a molecular model corresponding to this molecule is held in front of a mirror, the structure in the mirror is the mirror image of the original three-dimensional structure. These two structures are shown in Fig. 6.5, as are the three-dimensional projection formulas corresponding to them.

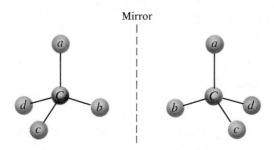

Mirror

Nonsuperimposable mirror images: enantiomers

(*a*)

Mirror

Nonsuperimposable mirror images: enantiomers

(*b*)

FIGURE 6.5   Representation of the three-dimensional structures for the molecule *Cabcd* using (*a*) ball-and-stick structures and (*b*) three-dimensional projection formulas.

But are these two structures enantiomers of one another or are they identical molecules? To answer this, we state the general criteria that must be satisfied to have enantiomers:

1. The two structures must be mirror images of one another.
2. The two structures must not be superimposable on one another.

The structures in Fig. 6.5 are indeed mirror images, and if molecular models are constructed, they are *not* superimposable on one another. There is no manner or means of superimposing them, bond for bond and atom for atom, without breaking and remaking certain bonds. (*Suggestion:* Make models and convince yourself that this is true!) Since the two criteria are met, these two structures represent a pair of enantiomers. *They differ from one another only in the three-dimensional (spatial) arrangement of the four groups about the carbon atom.*

A carbon atom that bears four different atoms or groups of atoms is called an **asymmetric carbon atom.** This may be a misleading designation because the carbon atom per se is not asymmetric; it means that a carbon atom is bonded to four different atoms or groups of atoms and that molecules of this type lack symmetry of any sort (see Sec. 6.4). We say that such molecules are *asymmetric* (Latin *a*, without), that is, without symmetry.

The following two pairs of enantiomers are drawn using three-dimensional projection formulas, and the asymmetric carbon atom is starred (*) for identification. Note that these compounds have only one asymmetric carbon atom.

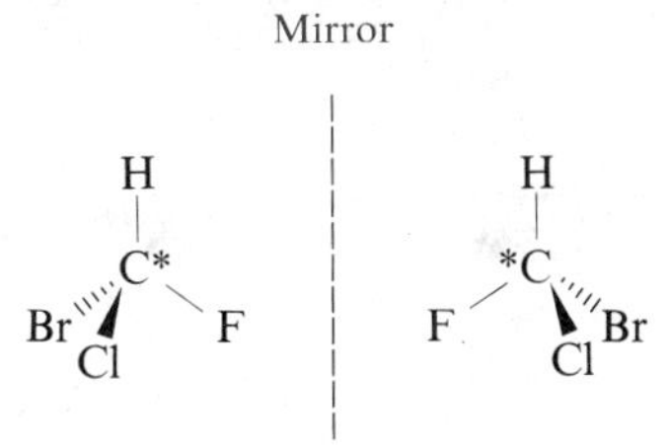

Bromochlorofluoromethane
Nonsuperimposable mirror images: enantiomers

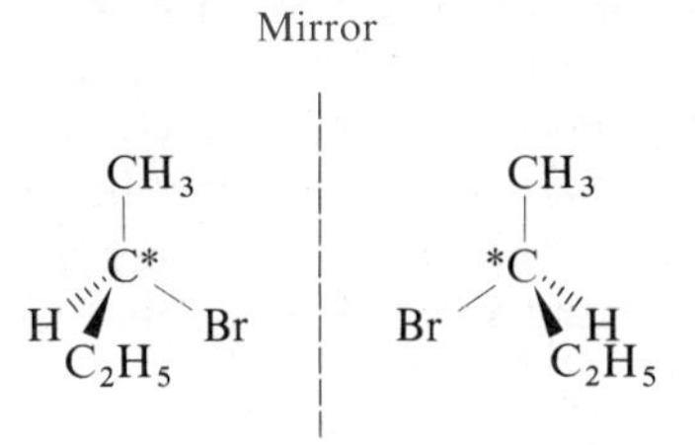

2-Bromobutane
Nonsuperimposable mirror images: enantiomers

The requirement that enantiomers be nonsuperimposable mirror images is quite general. Also, if a molecule can be constructed in two enantiomeric (mirror image) forms, it can exhibit optical activity. There are several factors that allow molecules to be enantiomers. One is the presence of *one* asymmetric carbon atom, as discussed.

The only reliable way to determine whether or not two stereoisomers are non-superimposable mirror images is to construct (or draw) their structures and then compare them. In the special case where there is only one asymmetric carbon atom in a molecule, the asymmetric carbon atom should first be identified. For example,

$$a—\overset{\overset{\displaystyle c}{|}}{\underset{\underset{\displaystyle d}{|}}{C^*}}—b$$

Molecules that have more than one asymmetric carbon atom are discussed in Sec. 6.11.

6.4   Further Evidence for the Tetrahedral Carbon Atom; Symmetry and Stereochemistry

Van't Hoff and LeBel used their concept for the existence of enantiomers to explain why simple compounds with two or more identical groups attached to a single carbon atom could never be separated into enantiomers. For example, only one isomer corresponding to CH_3Cl was known, and the same was true for compounds such as $CHCl_3$ despite exhaustive attempts to find other stereoisomers. For disubstituted methane derivatives, CH_2ab (where $a = b \neq H$ or $a \neq b \neq H$), as for example CH_2Cl_2 or CH_2ClF, again only one isomer was known.

Though there are several structures possible for carbon-containing compounds [see, for example, (1), (2), and (3)], only the tetrahedral carbon atom (3) accounts for all the known facts.

(1)   (2)   (3)
Square planar   Pyramidal   Tetrahedral

Build models of these compounds to prove this for yourself.

Deductive reasoning of this type led van't Hoff and Lebel to discard the planar and pyramidal structures for carbon compounds, thus leaving only the tetrahedral model. At the time they reported their findings it might have been argued that other isomers did exist but had eluded detection because of inadequate experimental techniques. We now know from numerous types of physical evidence (X-ray crystallography, spectroscopy, and electron diffraction methods) that these compounds are indeed tetrahedral.

Important Terms in Stereochemistry

The terminology associated with stereochemistry can often be very confusing. Let us take a moment to define some of the most common terms and see how they are used and what they mean. We must first have a system for discussing the molecular geometry of the compounds with which we are concerned. Fortunately, a system of rules, called symmetry elements, has been developed and can be used in this regard. We present some of their more important characteristics and uses at this time.

Symmetry elements describe molecules in terms of the various types of symmetry they exhibit. Compounds with similar symmetry properties are therefore organized into several symmetry groups. Before discussing these groups in detail, consider three of the more important symmetry elements.

Axis of Symmetry

A molecule contains an axis of symmetry if, by rotation about the axis passing through it, the molecule can be converted into a second molecule identical in all respects including the three-dimensional arrangement of the atoms in space. The axis is called a C_n axis, where $360°/n$ is the number of degrees of rotation necessary to convert the molecule into its identical twin.

All molecules contain an infinite number of C_1 axes. If you rotate anything $360°$ ($360°/1 = 360°$) about an axis passing through it in any direction, the result is the same species in the same three-dimensional orientation it exhibited originally.

For example:

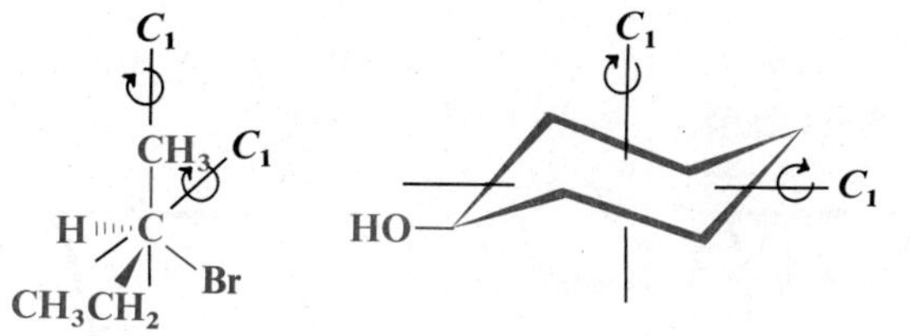

In addition, we encountered many different molecules that contain at least one C_2 or C_3 axis requiring $180°$ ($360°/2$) and $120°$ ($360°/3$) of rotation for interconversion; four of these are illustrated:

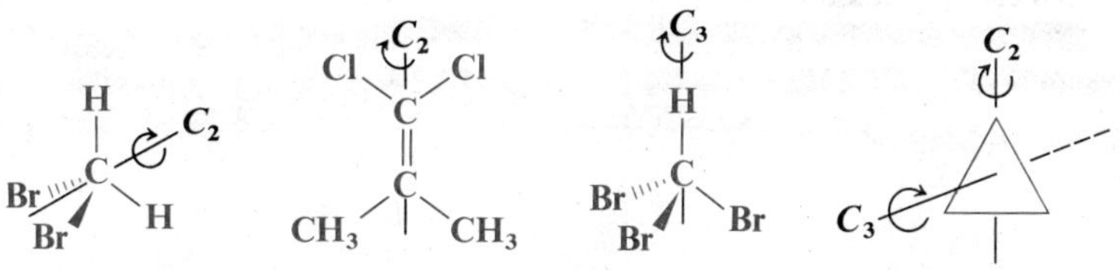

To be sure you understand the concept of an axis of symmetry, look at one more example. *cis*-1,2-Dibromoethene contains a C_2 axis that lies in the plane of the page and passes through the molecule (which is also in the plane of the page) midway between the two bromine atoms bisecting the carbon-carbon double bond.

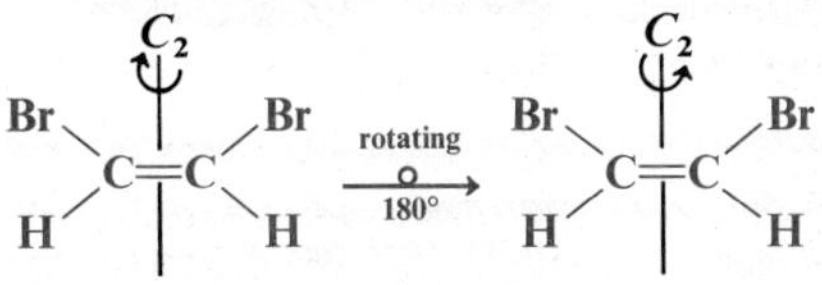

A second C_2 axis can be drawn that also bisects the carbon-carbon double bond and is perpendicular to the first. *This second C_2 axis is not a true axis of symmetry, however.* When the molecule is rotated $180°$ about this second axis, the molecule generated, although identical with the first, differs in its three-dimensional arrangement of atoms in space. In the original molecule the bromine atoms point upward to the top of the page, but in the molecule generated by rotation they point downward to the bottom of the page. The second C_2 axis is, by definition, not allowed.

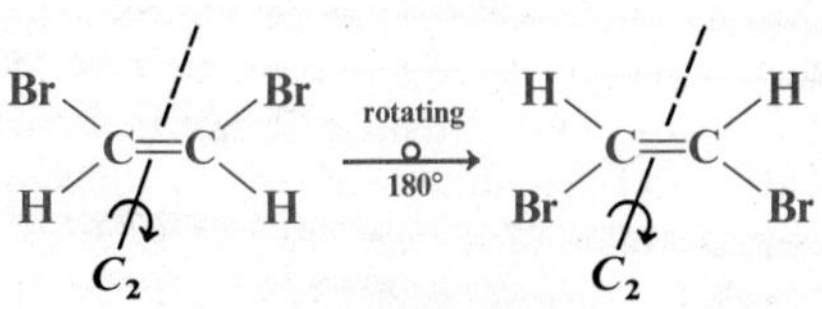

Not a true C_2 axis

Plane of Symmetry

A molecule is said to contain a plane of symmetry if a plane can be placed through the molecule, bisecting it in such a way that the half of the molecule on one side of the plane is the mirror image of the half on the other side. A plane of symmetry is often called a sigma plane (σ plane) and is represented as C_s symmetry. Examples of molecules that contain one or more C_s planes of symmetry are drawn here.

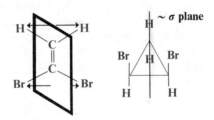

Dichloromethane, CH_2Cl_2, has *two* planes of symmetry, and bromochloromethane, CH_2BrCl, has *one* plane of symmetry; these are illustrated as follows:

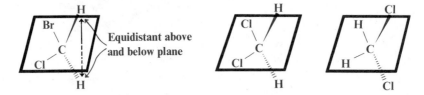

Center of Symmetry

If all the atoms or groups of atoms in a molecule can be interconverted through a central point to give the same molecule in the original spatial arrangement, the molecule is said to contain a center of symmetry. The symbolic designation of a center of symmetry is C_i. An examples is:

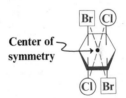

The C_i symmetry element is also called a *center of inversion*.

Now that we know these symmetry elements, let us take a closer look at some terms.

1. *Optical activity:* When we say a compound is optically active, or exhibits optical activity, we mean that the compound rotates plane-polarized light when placed in a polarimeter. To determine whether a given compound is optically active, you must put it in a polarimeter and see whether the light is rotated.

2. *Dissymmetry or chirality:* A molecule that is dissymmetric or chiral has a nonsuperimposable mirror image; an enantiomer. This may also be defined by using the symmetry groups mentioned earlier. If a molecule (or object) is dissymmetric or chiral, it does not contain any σ planes or centers of symmetry (no C_s or C_i symmetry)[1]; that is, a σ plane cannot be drawn through the molecule and the molecule cannot be interconverted through a center of symmetry. The word *dissymmetric* simply refers to a lack of some symmetry, or in this case C_s or C_i symmetry. A compound that is dissymmetric or chiral (a synonym) has the inherent ability to rotate plane-polarized light. The key word is *ability*; the compound may or may not rotate plane-polarized light when placed in a polaimeter, but it does have the inherent ability to do so. Enantiomers always have the inherent ability to rotate light, but as observed by Pasteur, an equal mixture of two enantiomers does not rotate light.

3. *Asymmetry:* A compound that is asymmetric, or exhibits asymmetry, belongs to a second symmetry group that contains those compounds lacking all but C_1 symmetry. Contrast this to dissymmetry, which refers to the lack of C_s or C_i symmetry only. A dissymmetric molecule may contain, for example, a C_2 axis, but an asymmetric molecule may not. Asymmetric molecules usually contain one or more asymmetric carbon atoms. Asymmetric compounds are just a subdivision of dissymmetric compounds. All compounds that are asymmetric are also dissymmetric,

---

[1] There are exceptions to this general statement, but they are few in number and beyond the scope of this text.

but the reverse is not necessarily true. For example, the coil of wire drawn here is dissymmetric, but not asymmetric.

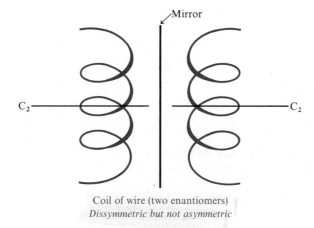

Coil of wire (two enantiomers)
Dissymmetric but not asymmetric

Additional examples of dissymmetric and/or asymmetric molecules are given throughout the chapter.

6.5   Interaction of Light with Matter; Enantiomers and Optical Activity

We are now able to present the following general rules:

1. All dissymmetric or chiral molecules exist as one or more discrete pairs of enantiomers.
2. All dissymmetric molecules are capable of exhibiting optical activity.

Some molecules exist in more than two enantiomeric forms. These always exist as discrete pairs, with one member of the pair rotating light in a clockwise (dextrorotatory) direction and the other member rotating light an equal magnitude in a counterclockwise direction (levorotatory). Also, even though a compound is dissymmetric, it may not exhibit optical activity. As pointed out in Sec. 6.4, if an equal mixture of two enantiomers is placed in a polarimeter, no rotation is observed. The ability to rotate light is there, but the rotation is not perceived. Several examples of enantiomers are given in the following paragraphs.

A. Molecules with One Asymmetric Carbon Atom; Asymmetric Molecules

An important naturally occurring compound, lactic acid, (1), is found to be dextrorotatory (rotates light to the right; rotation has a positive value) when isolated from animal muscle tissue. Lactic acid contains an asymmetric carbon atom.

$$CH_3-\overset{\overset{\displaystyle OH}{|}}{\underset{\underset{\displaystyle H}{|}}{C^*}}-COOH$$

(1)

The two enantiomers of this compound are:

Mirror

$$CH_3 \underset{H}{\overset{COOH}{\underset{|}{C^*}}} OH \quad \bigg| \quad HO \underset{H}{\overset{HOOC}{\underset{|}{^*C}}} CH_3$$

Lactic acid

Nonsuperimposable mirror images: enantiomers

Note that *we cannot determine which enantiomer is dextrorotatory and which is levorotatory*. To do this, we would have to know the absolute configuration of each enantiomer (see Sec. 6.8) and how it relates to the observed rotation. Refer to these structures after you have read this section.

When identifying an asymmetric carbon atom, be certain to look at each *entire* group that is bonded to a given carbon atom. For the following compound

$$CH_3-CH_2-\overset{H}{\underset{Cl}{C^*}}-CH_2-CH_2-CH_3 \equiv C_2H_5-\overset{H}{\underset{Cl}{C^*}}-C_3H_7\text{-}n$$

we might say (incorrectly) that there is no asymmetric carbon atom because two identical CH_2 groups are attached to a carbon atom. However, we must look further and note that the starred carbon atom has *four* different groups: one hydrogen, one chloro, one ethyl, and one propyl. This molecule does exist in two enantiomeric forms. Note also that the entire alkyl group (for example, ethyl or propyl) is treated as one distinct R group attached to the asymmetric center. As long as the groups attached to the chiral center (the asymmetric carbon atom) are not dissymmetric themselves, they are their own mirror images and may be treated as an R group. 3-Chlorohexane, drawn earlier, may also be simplified to

$$R-\overset{H}{\underset{Cl}{C^*}}-R'$$

See also 2-bromobutane drawn in Sec. 6.3.

B. Dissymmetric Molecules That Contain No Chiral Centers (No Asymmetric Carbon Atoms)

Exposure to this class of stereoisomers in this text is quite limited, but it is important to see at least an example. 1,3-Dichloroallene (see Sec. 7.8 for a discussion of allenes) is given here. Be sure to build models of the two enantiomers to get a clear picture of the dissymmetry inherent in molecules of this type.

Mirror

$$\underset{Cl}{\overset{H}{\diagdown}}C=C=C\underset{H}{\overset{Cl}{\diagup}} \quad \bigg| \quad \underset{H}{\overset{Cl}{\diagup}}C=C=C\underset{Cl}{\overset{H}{\diagdown}}$$

1,3-Dichloroallene

Nonsuperimposable mirror images: enantiomers

Question 6.3

What symmetry elements are present in the two preceding examples?

Question 6.4

Build a model of allene, $CH_2{=}C{=}CH_2$. What is the hybridization of each of the three carbons?

Question 6.5

Identify all the asymmetric carbon atoms in the compounds shown in Question 3.22 *except* for compounds (*i*) and (*j*).

6.6 Properties of Enantiomers

Although enantiomers appear to have identical physical properties (boiling point, melting point, density, and so on) and to differ only in their interaction with plane-polarized light, this is not a complete picture. Light is actually composed of two chiral (dissymmetric) components. To simplify, we use a single wave of plane-polarized, monochromatic light in our discussion.

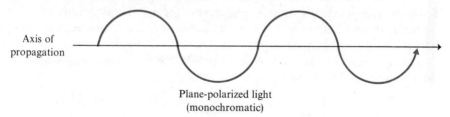

Axis of propagation

Plane-polarized light
(monochromatic)

This light wave actually has two components, a *right-handed circularly polarized* (RHCP) component and a *left-handed circularly polarized* (LHCP) component.

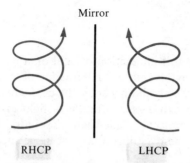

Mirror

RHCP   LHCP

Right- and left-handed circularly
polarized components of
plane polarized light,
an enantiomeric relationship

The plane-polarized wave normally drawn is simply a vector sum of these two components along the axis of propagation. That is, the sum of these two circularly polarized components gives the light wave we are used to seeing. Since each component in a light wave is chiral, the light itself is chiral.

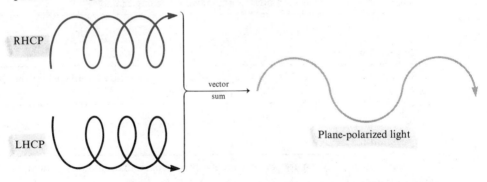

A light beam consists of two chiral components of electromagnetic radiation that interact with the electrons associated with the atoms and bonds in a molecule. This can cause the electrons to be perturbed to some extent. (It is unlikely that enough energy is available to promote electrons to an excited state so that light is not absorbed by the molecules.) Put another way, the electric field associated with the radiation causes the electrons in the molecule to be polarized slightly. This changes slightly the direction of the oscillating electric field as it passes near each atom or group of atoms. As might be expected, each atom or group of atoms has a different electron density associated with it, so that different atoms rotate light to different extents and in different directions.

If the molecule is achiral (not chiral), the light is rotated in one direction by one atom (or group of atoms), but because of the inherent symmetry of the molecule, when that same light passes near a like atom (or group of atoms) it is rotated in the reverse direction by an equal amount. These two individual and equal but opposite rotations cancel one another so that there is no net rotation and no optical activity. The following sketch illustrates this.

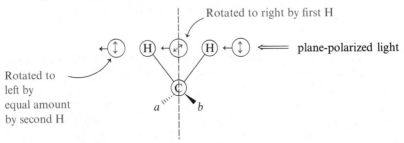

We considered this interaction of light on a molecular basis (that is, with one single molecule). Probably only a small fraction of molecules are oriented in solution in just this way, so we should consider the case when many molecules are present. One molecule might exhibit net rotation of light in one direction, but there most certainly is another molecule somewhere in solution that rotates it in an equal but opposite direction. The overall result is the same: no net rotation and thus optical inactivity.

If the molecule is chiral, then the light is rotated in a certain direction by one atom or group of atoms and, because of the inherent lack of symmetry in the molecule, in different but indeterminate directions by other atoms or groups of atoms in the molecule. These electromagnetic interactions (rotations) do not average to zero (that is, they do not completely cancel one another). The result is a *net* rotation of light in one direction or another and thus optical activity.

The interaction of light with a single asymmetric molecule produces minute amounts of rotation. However, there are typically on the order of 10^{21} molecules in a sample tube, so we measure the sum of these rotations.

Light is chiral, and any one member of a pair of enantiomers interacts differently with this probe. *In general, enantiomers have identical physical and chemical properties when interacting or reacting with achiral probes or compounds, but they react (or interact) differently with chiral probes or compounds.*

A molecule's optical activity is a characteristic and unique physical property. For example, lactic acid [(1) in Sec. 6.5] can be separated into dextrorotatory and levorotatory enantiomers, both of which have a melting point of $53°$ and are infinitely soluble in water. However, the dextrorotatory enantiomer has $[\alpha]_D^{25°} = +3.8°$ and its mirror image isomer (the levorotatory enantiomer) has $[\alpha]_D^{25°} = -3.8°$.

There is no predictable relationship between sign of rotation and configuration. Configuration refers to the *absolute* position of each substituent in relation to the other substituents in three-dimensional space (see Sec. 6.8). We cannot say that all compounds with a positive rotation are of one configuration, and the same for compounds with negative rotation. On the other hand, if the configuration has been related to the sign of rotation by independent means, then if we know the sign of rotation, we always know the configuration. The symbols d and l are synonymous with the $+$ and $-$ signs of rotation, respectively.

To illustrate this, consider the following compounds with their specific rotations indicated; all have the same configuration about carbon and differ only in the substituents that are attached. Note that there is great variation in the magnitude and sign of rotation.

$[\alpha]_D^{25°} +3.3°$ $-8.2°$ $-75.5°$ $-66.4°$

Enantiomers are stable, isolable compounds that differ from one another in their three-dimensional spatial arrangement. *Enantiomers cannot be interconverted under ordinary conditions,* because to do so requires that bonds be broken and new bonds remade.

Not in equilibrium

When a chemical reaction is carried out on an enantiomer such that, for example, the d form is converted to the l form or vice versa, it is called **inversion of configuration.** Simply put, one enantiomer is converted to its mirror image.

6.7   Optical Purity; Racemic Compounds and Resolution

Although many compounds, like lactic acid [(1) in Sec. 6.5], can be obtained from natural resources as a single enantiomer, this is not always true in the laboratory. More frequently, compounds that have been synthesized are in optically inactive form; that is, they contain equal amounts of the d and l forms.

If a pure enantiomer is needed, it must be separated from a mixture of enantiomers. This process of separating enantiomers from one another is called **resolution.** We do not discuss resolution until we have more chemical reactions at our disposal; see, for example, Secs. 23.14 and 26.6.

A 50:50 mixture of the *d* and *l* forms of any chiral compound is *optically inactive* and is called a **racemic mixture.** The *d* and *l* forms have rotations that are equal in magnitude but opposite in direction, so they exactly cancel one another. A racemic mixture is often identified by placing ($\pm$) in front of the compound name, whereas ($+$) identifies the dextrorotatory form and ($-$) the levorotatory form. For example, racemic lactic acid is ($\pm$)-lactic acid, and the optically active enantiomers are ($+$)-lactic acid and ($-$)-lactic acid.

If a racemic mixture can be separated (that is, resolved) into its *d* and *l* components so that each enantiomer is uncontaminated with the other one, then we say that each enantiomer is **optically pure.** On the other hand, if the resolution is incomplete, some samples of the compound are richer in one enantiomer than the other; the *net* result is that this sample is *optically active* but not optically pure. For example, optically pure lactic acid has $[\alpha]_D^{25°} = +3.8°$ for the *d* form and $[\alpha]_D^{25°} = -3.8°$ for the *l* form. Lactic acid with $[\alpha]_D^{25°} = +1.1°$ is optically active but not optically pure; this sample contains more of the *d* form than of the *l* form, so the net rotation is positive.

In practice, resolution must be repeated several times to reach optical purity, and the researcher is fortunate to end up with a 5 to 10% yield of a pure enantiomer; the theoretical maximum of 50% of each enantiomer is seldom achieved. See Sec. 23.14 for more detail.

Here is a summary of several of these relationships and terms:

Compound with one asymmetric carbon atom:

Racemic mixture	$\xrightarrow[\text{(complete separation)}]{\text{resolution}}$	*d* **form**	+	*l* **form**
Contains 50:50 mixture of *d* and *l* enantiomers: ($\pm$) form *Optically inactive*		Dextrorotatory (rotates light to the right), ($+$) isomer *Optically pure* *50% Yield*		Levorotatory (rotates light to the left), ($-$) isomer *Optically pure* *50% Yield*

For example:

$$CH_3CH_2-\overset{H}{\underset{OH}{\overset{|}{\underset{|}{C^*}}}}-CH_3 \quad \xrightarrow{\text{resolution}}$$

2-Butanol
racemic mixture,
$[\alpha]_D^{25°} = 0°$

$+13.9°$ $-13.9°$

$$CH_3-\overset{H}{\underset{NH_2}{\overset{|}{\underset{|}{C^*}}}}-COOH \quad \xrightarrow{\text{resolution}}$$

Alanine (amino acid)
racemic mixture,
$[\alpha]_D^{25°} = 0°$

$-3.1°$ $+3.1°$

6.8 Designation of Configuration: *R* and *S* System

Recall from Sec. 6.6 that there is no predictable relationship between the sign of rotation and the configuration of enantiomers unless they have been related to one another by independent means. Is there a method to designate the configuration about asymmetric carbon atoms, that is, the position each substituent bears in relation to the other substituents in three-dimensional space? The answer is yes. The method used today was proposed by R. S. Cahn, Sir C. K. Ingold, and V. Prelog and uses the prefixes *R* and *S* in front of the compound name to designate the *absolute configuration* of the substituents.

A. Step 1: Classification of Substituents by Priority

The first step involves the classification of various atoms or groups of atoms in *order of priority*.

1. *Atoms.* We first look at the four *atoms* that are attached to the asymmetric carbon atom and arrange them in decreasing order according to atomic number; the higher the atomic number, the higher the priority. This provides the following order of *decreasing priority*:

I, Br, Cl, S, P, F, O, N, C, H, unshared electron pair

Highest   Lowest
priority   priority

Isotopes of an element (which have the same atomic number) are arranged by decreasing mass. For example, if hydrogen ($_1^1H$) and deuterium, D or hydrogen of mass 2 ($_1^2H$), are present, D has the higher priority.

2. *Groups of atoms.* If two or more of the *first atoms* (those attached to the asymmetric carbon atom) have the same atomic number (that is, are identical), we turn to the *second group of atoms* attached to the first atom and compare their atomic numbers. For example, compare —CH_2CH_3 with —CH_2Cl. Both first atoms are carbon and thus identical, so we turn to the —Cl and —CH_3 groups. Since Cl in —CH_2—Cl has a higher atomic number than does C in —CH_2—CH_3, —CH_2—Cl has higher priority than —CH_2CH_3.

A higher atomic number always takes precedence in determining the priority even though different numbers of substituents may be attached. For example, the —CH_2Cl group has priority over the —$C(CH_3)_3$ group, even though the latter has 3C and the former has 2H + Cl; this is because Cl has a higher atomic number than does C. It makes no difference that C in —$C(CH_3)_3$ has *three* groups (other than hydrogen) attached to it, whereas C in —CH_2Cl has only *one*.

This selection process is continued to the third, fourth, and so on sets of atoms (working away from the asymmetric carbon atom) until a unique priority order is determined.

3. *Multiple bonds.* Doubly and triply bonded atoms are considered equivalent to two and three atoms in the second and third groups, respectively. For example:

$$
\begin{array}{ccc}
& & \text{C} \quad \text{C} \\
& & | \quad\quad | \\
-\text{C}=\text{CR}_2 & \text{is equivalent to} & -\text{C}-\text{C}-\text{R} \\
| & & | \quad\quad | \\
\text{H} & & \text{H} \quad \text{R}
\end{array}
$$

$$-C\equiv C-R \quad \text{is equivalent to} \quad \overset{\overset{\text{C}\ \text{C}}{|\ \ |}}{-\underset{\underset{\text{C}\ \text{C}}{|\ \ |}}{C}-C-R}$$

$$-\underset{\underset{\text{H}}{|}}{C}=O \quad \text{is equivalent to} \quad -\underset{\underset{\text{H}}{|}}{\overset{\overset{\text{O}\ \text{C}}{|\ \ |}}{C}}-C-O$$

The atoms in color do not actually exist but are used as part of the defined system in assigning priorities to the attached groups. For these three groups, the order of decreasing priority is:

$$-\underset{\underset{\text{H}}{|}}{C}=O \qquad -C\equiv C-R \qquad -\underset{\underset{\text{H}}{|}}{C}=CR_2$$

Highest   Lowest
priority   priority

If we compare $-\underset{\underset{\text{H}}{|}}{C}=O$ (which is equivalent to $2O + 1H$ attached to the indicated carbon) with $-CH_2-OH$ (which contains $1O + 2H$), the former has higher priority because we are comparing $2O + 1H$ for it with $1O + 2H$ for the latter; $2O$'s have greater priority than $1O$.

The phenyl group, C_6H_5-, is treated as though it is present as one of the Kekulé structures (see Chap. 13):

is equivalent to

4. *Summary.* The common atoms and some common functional groups are shown. Each group is arranged in *decreasing order of priority*.

Atoms:

I, Br, Cl, S, P, F, O, N, C, H, unshared electrons

Groups:

a. $-C(CH_3)_3, -CH(CH_3)_2, -CH_2CH_3, -CH_3$

b. $-\overset{\overset{\text{O}}{||}}{C}-X, -\overset{\overset{\text{O}}{||}}{C}-OR, -\overset{\overset{\text{O}}{||}}{C}-OH, -\overset{\overset{\text{O}}{||}}{C}-NH_2, -\overset{\overset{\text{O}}{||}}{C}-R, -\overset{\overset{\text{O}}{||}}{C}-H$

c. $-C\equiv N, -C_6H_5, -C\equiv C-H, -CH=CH_2$

Highest priority   Lowest priority

Although you may be unfamiliar with many of these functional groups, they are presented here for completeness and can be used as they arise in later chapters.

B. Step 2: *R* and *S* Designation Based on Priorities

The three-dimensional structure of the molecule must be visualized so that the group of *lowest* priority is as far as possible from the sight of the viewer. When this is done, we see the arrangement of the remaining atoms or groups of atoms pointing out of the molecule toward us, and designate the configuration based on the relative priorities of those atoms or groups.

The **R configuration** exists when the sequence of the other three groups (in *decreasing* order of priority) is viewed as being in the **clockwise** (right-handed) direction; the symbol *R* stands for the Latin word *rectus*, meaning right. [As an aid to remembering this symbol, keep in mind that *R* (or *r*) is the first letter in *right*.]

The **S configuration** exists when the sequence obtained is viewed as being in the **counterclockwise** (left-handed) direction. The symbol *S* stands for the Latin word *sinister*, meaning left.

C. Examples of *R* and *S* Configuration

Now we consider some simple examples to illustrate how the priority system and the *R* and *S* designation can be applied.

Example 1

What is the *R* and *S* designation for the enantiomers of bromochlorofluoromethane?

$$Br{-}\overset{\displaystyle Cl}{\underset{\displaystyle H}{C^*}}{-}F$$

Using the priority numbers for atoms, we arrive at the following sequence:

Atom:

	—Br	—Cl	—F	—H

Priority:

	1	2	3	4
	Highest			Lowest

We now construct the two enantiomers, with the lowest priority atom—hydrogen, H, in this case—as far from the eye as possible.

Enantiomers of bromochlorofluoromethane:

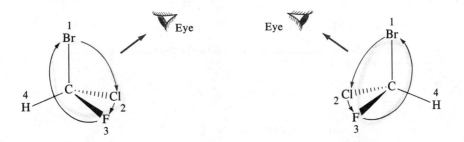

R-Bromochlorofluoromethane

Clockwise rotation in going from highest to lowest priority, *R* configuration

S-Bromochlorofluoromethane

Counterclockwise rotation in going from highest to lowest priority, *S* configuration

Example 2

What is the R and S designation for the enantiomers of 2-chlorobutane?

$$CH_3{-}CH_2{-}\overset{\displaystyle Cl}{\underset{\displaystyle H}{\overset{|}{\underset{|}{C^*}}}}{-}CH_3$$

We apply the rules for assigning priorities and arrive at the following sequence for the groups attached to the asymmetric carbon atom (C-2):

Group:

$$-Cl \qquad -CH_2-CH_3 \qquad -CH_3 \qquad -H$$

Priority:

1	2	3	4
Highest			Lowest

Hydrogen has the lowest priority and is farthest from the eye of the viewer in the following three-dimensional projection formulas.

Enantiomers of 2-chlorobutane:

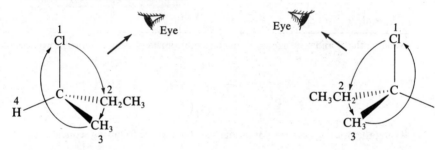

R-2-Chlorobutane

Clockwise sequence,
R configuration

S-2-Chlorobutane

Counterclockwise sequence,
S configuration

Example 3

Draw the configuration for R-1,2-dibromo-3-methylbutane.

$$Br{-}CH_2{-}\overset{\displaystyle Br}{\underset{\displaystyle H}{\overset{|}{\underset{|}{C^*}}}}{-}\overset{\displaystyle CH_3}{\underset{\displaystyle CH_3}{CH}}$$

We identify the substituents attached to the asymmetric carbon atom and then assign priority numbers:

Group:

$$-Br \qquad -CH_2-Br \qquad -\overset{\displaystyle CH_3}{\underset{\displaystyle CH_3}{CH}} \qquad -H$$

Priority:

1	2	3	4
Highest			Lowest

The —CH_2—Br has a higher priority than —$CH(CH_3)_2$ because the atom with the higher atomic number (in this case, bromine) takes precedence over any substituents with a lower atomic number. See step 1 in this section. We can now arrange the groups in clockwise order with the lowest priority atom, hydrogen, farthest from our eye, in the following manner.

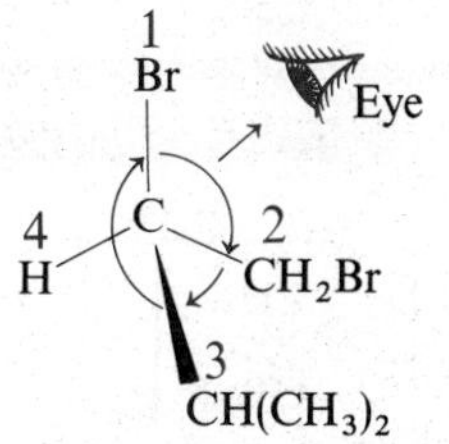

R-1,2-Dibromo-3-methylbutane

If we know the precise configuration of a stereoisomer as well as its rotation, we can combine the R and S designation and the sign of rotation in the compound name, and thus provide complete information about a particular stereoisomer. For example, S-1-chloro-2-methylbutane has a positive rotation, so its complete name is S-(+)-1-chloro-2-methylbutane; its enantiomer is R-(−)-1-chloro-2-methylbutane.

Question 6.6

Draw three-dimensional projection formulas for the enantiomers (if any) of the following five compounds, and designate each as being the R or S configuration:

(*a*) 1-bromo-1-fluoroethane   (*b*) 1,2-dichloropropane
(*c*) 2,2-dichloropropane   (*d*) 2-butanol, CH_3—CH—CH_2CH_3
                                              |
                                              OH

(*e*) alanine (a naturally occurring amino acid), CH_3—CH—COOH
                                                          |
                                                          NH_2

Question 6.7

Draw three-dimensional projection formulas for the following molecules:

(*a*) R-lactic acid, CH_3—CH—COOH   (*b*) S-2-bromopentane
                         |
                         OH

(*c*) R-3-chloro-1-pentene, CH_3—CH_2—CH—CH=CH_2 (*d*) S-3-chloro-3-methylhexane
                                          |
                                          Cl

A second system of absolute configurations known as the D/L system is presented in Chap. 28.3. We delay discussion of it until then.

6.9   Reactions That Involve Optically Active Compounds; Evidence for Free-Radical Structure and Nucleophilic Substitution

Stereochemistry and the optical properties of certain molecules are powerful tools for elucidating reaction mechanisms. For now we discuss three types of reactions of optically active compounds: (1) reactions that do *not* involve the asymmetric carbon atom, (2) reactions that take place at the asymmetric carbon atom, and (3) reactions that generate an asymmetric carbon atom.

A. Reactions That Do Not Involve the Asymmetric Carbon Atom

Some reactions involving organic molecules do not affect the asymmetric carbon atom, for example, the free-radical chlorination of optically active 2-chloro-butane (1).

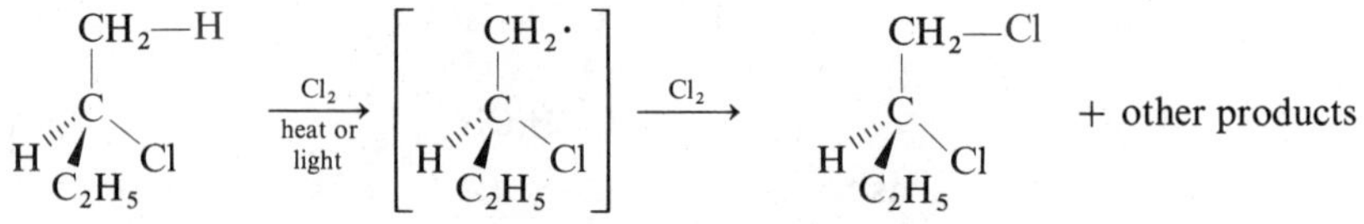

(1)	Radical	(2)
(−)-2-Chlorobutane	intermediate	1,2-Dichlorobutane
Optically active		*Optically active*

Although other structural isomers are formed, consider only 1,2-dichlorobutane (2), which in isolation from the other products of the reaction is found to be optically active. This is not unreasonable because the reaction carried out on (1) does not affect the hydrogen attached to the asymmetric carbon. Thus, the configuration at the asymmetric carbon atom is unchanged; H on C-1 is simply replaced with a Cl atom, which has proceeded through the —CH$_2$· radical. We say that this reaction has occurred with **retention of configuration** because there are no changes at the asymmetric carbon atom itself. Note that the configuration of (1) is R and that of (2) is S because of the new priorities of the groups in (2). The retention of configuration is relative (all groups are in the same positions relative to one another before and after the reaction) but not necessarily absolute.

As a second example of this type of reaction, consider the following conversion, which involves the reduction of an alkene (3) to an alkane (4):

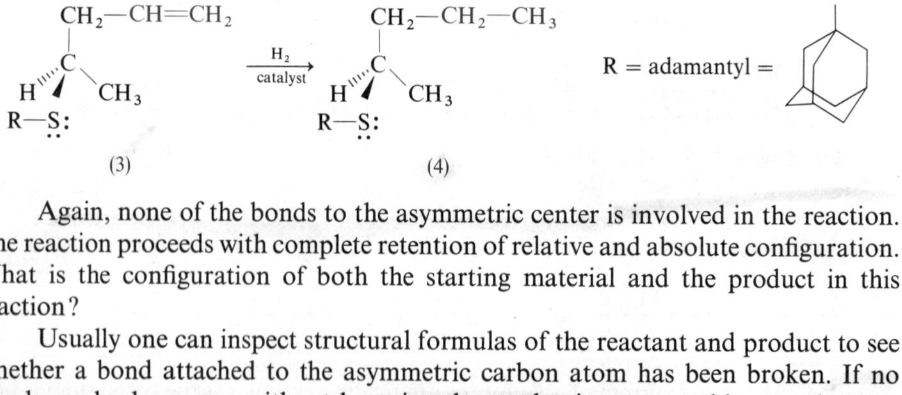

Again, none of the bonds to the asymmetric center is involved in the reaction. The reaction proceeds with complete retention of relative and absolute configuration. What is the configuration of both the starting material and the product in this reaction?

Usually one can inspect structural formulas of the reactant and product to see whether a bond attached to the asymmetric carbon atom has been broken. If no bonds are broken, even without knowing the mechanism, we could guess the configuration at the asymmetric carbon. In general, reactions that do not involve breaking bonds attached to an asymmetric carbon atom occur with retention of configuration. (As illustrated in Question 6.10, some optically active compounds form optically inactive products even though the asymmetric center is unaffected.)

B. Reactions That Involve an Asymmetric Carbon Atom

a. Free-Radical Substitution

Consider the free-radical chlorination of *optically active* (−)-1-chloro-2-methylbutane (5). One particular product, 1,2-dichloro-2-methylbutane (6), is obtained from this reaction as an *optically inactive* compound even though it still contains an asymmetric carbon atom.

$$\underset{\substack{(5)\\ (-)\text{-1-Chloro-2-}\\ \text{methylbutane}\\ \textit{Optically active}}}{\overset{\displaystyle CH_3}{\underset{\displaystyle H}{C_2H_5-\overset{|}{\underset{|}{C^*}}-CH_2-Cl}}} \xrightarrow[\text{or light}]{Cl_2,\ heat} \underset{\substack{(6)\\ 1,2\text{-Dichloro-2-methylbutane}\\ \textit{Optically inactive}}}{\overset{\displaystyle CH_3}{\underset{\displaystyle Cl}{C_2H_5-\overset{|}{\underset{|}{C^*}}-CH_2-Cl}}} + \text{other products}$$

To understand this reaction, consider what happens. A C—H bond attached to the asymmetric carbon atom in (5) is broken, and a new C—Cl bond at the same carbon atom in (6) is formed. Recall from Sec. 4.6 that the carbon radical is usually a planar species. We now have at our disposal the tools necessary to support partially the planar structure of radicals.

We evaluate this reaction using three-dimensional projection formulas as shown in Fig. 6.6. However, close examination of the planar free radical indicates that both faces appear to be identical, and if this is true, there is no reason to expect that molecular chlorine prefers to attack one side over the other. As Fig. 6.6 shows, attack from one side produces one stereoisomer and attack from the opposite side produces its mirror image. These two structures are nonsuperimposable mirror images and thus enantiomers. Because there is a 50:50 chance of chlorine attacking from one side or the other, the two enantiomers are formed in a 50:50 ratio. As we saw, equal mixtures of enantiomers produce an optically inactive (racemic) mixture.

If a reaction occurs at an asymmetric carbon atom in an optically active molecule and gives racemic (optically inactive) products, we say the reaction has occurred with **racemization.** In the preceding example, therefore, the reaction has occurred with *complete* racemization. In contrast, many reactions occur with *partial* racemization; that is, there is some net retention or inversion of configuration.

Other compounds like (5) have been studied and the same type of observations recorded. Although we cannot see the free radical, several pieces of evidence prove its existence (see Sec. 2.8), and its three-dimensional structure has been reinforced by stereochemistry.

An alternative proposal for the structure of the free radical is that it might be *pyramidal*; however, our stereochemical observations tend to disprove this suggestion. If the radical were pyramidal, we anticipate that the entering chloro group would occupy the same location that the departing hydrogen did, so that optically active starting material would produce optically active product. This incorrect model for the free radical is illustrated using (−)-1-chloro-2-methylbutane (5) as the example.

Incorrect model for free-radical structure:

(5)

Pyramidal structure
of free radical;
does not exist

(6)

Retained configuration,
same configuration as (5);
would be optically active

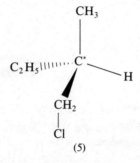

R-(−)-1-Chloro-2-methylbutane

Optically active

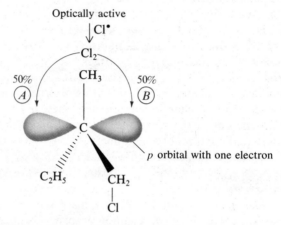

Planar free radical in *p* orbital perpendicular
to vertical plane defined by carbon atoms

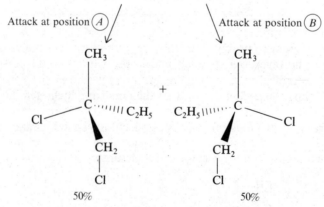

Equal mixture of enantiomers: optically inactive
(±)-1, 2-dichloro-2-methylbutane (6)

FIGURE 6.6   Free-radical chlorination at an asymmetric carbon atom. The entering chlorine atom can become attached to either face of the planar free radical, with 50% probability of attack at position Ⓐ and 50% probability of attack at position Ⓑ. This produces a 50:50 mixture of enantiomers that is racemic and optically inactive.

If the free radical were pyramidal, it would resemble the structure of the starting compound (5) and would in turn react with molecular chlorine to yield optically active compound (6). Using this model, (6) would be formed with retention of configuration and would be optically active. The results contradict this.

It has been suggested that free radicals are pyramidal but rapidly interconvert between two pyramidal structures before they react with chlorine. This would involve radicals that are enantiomers of one another and on reaction with chlorine, they would produce a racemic mixture of product. This model is discounted for several reasons. First, it assumes that interconversion of radicals is much faster than the reaction between a radical and molecular chlorine; this seems unlikely considering the great reactivity of a radical. Second, equilibration between the two pyramidal radicals would have to occur completely so that there would be precisely a 50:50 mixture of the enantiomeric radicals. That this should occur time and time again in all molecules is unlikely.

Even though quantum-mechanical calculations indicate that planar and pyramidal free radicals are of roughly equivalent energy, spectroscopic and stereochemical studies both indicate that *free radicals exist in planar form*, with the odd electron occupying the *p* orbital perpendicular to the plane defined by the carbon bearing it and the attached substituents.

Question 6.8

Starting from (−)-1-chloro-2-methylbutane (5), draw the three-dimensional projection formulas for the enantiomeric free radicals that would be formed *if they were pyramidal*. Then show how they would lead to a racemic product on reaction with molecular chlorine.

Question 6.9

Suppose the reaction pictured in Fig. 6.6 involved a rapidly interconverting pyramidal intermediate. What stereochemistry do you think would be associated with this type of reaction? Would configuration of product be retained (same as starting material) or inverted (mirror image of starting material), or would racemization occur. Why?

Question 6.10

Suppose optically active 1,2-dibromopropane is allowed to react with bromine in the presence of light at 125°.

(*a*) Draw the structural isomers of all the tribromopropanes, $C_3H_5Br_3$, that would be formed.
(*b*) For each structural isomer formed, describe the stereochemistry you would expect it to exhibit (that is, draw three-dimensional projection formulas for the stereoisomers). Which, if any, of these stereoisomers would be optically active, and why? Explain your answer based on the structure of the free radical.
(*c*) If the optically active 1,2-dibromopropane used were of the *R* configuration, specify whether each of the stereoisomers produced would be of the *R* or the *S* configuration.

b. Nucleophilic Aliphatic Substitution, S_N2

The S_N2, nucleophilic aliphatic substitution, mechanism was presented in Sec. 5.16. Consider the course of this mechanism when optically active compounds are used. The reaction of S-2-bromobutane with iodide ion is drawn here:

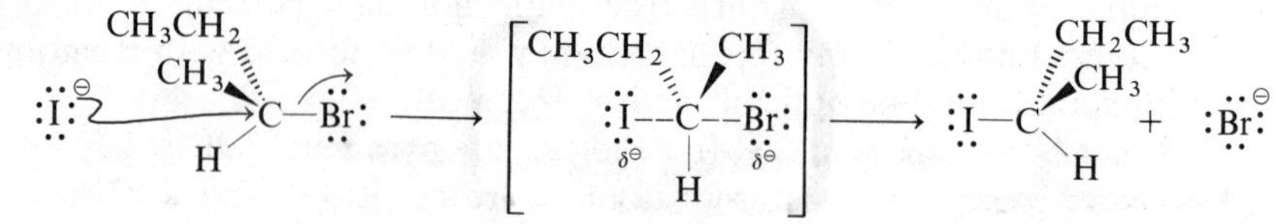

S-2-Bromobutane	Pentacoordinate	R-2-Iodobutane
Optically active	transition state	*Optically active*

We know that this mechanism involves attack by the nucleophile ($: \overset{\cdot\cdot}{\underset{\cdot\cdot}{I}} :^{\ominus}$) on the carbon bearing the leaving group ($: \overset{\cdot\cdot}{\underset{\cdot\cdot}{Br}} :^{\ominus}$). The iodide attacks the carbon to which the bromine is attached in a *backside* manner. The reaction is concerted, and in one step the carbon-iodine bond forms and the carbon-bromine bond breaks. The pentacoordinate transition state accompanying this reaction is also drawn. As a result of the mechanistic pathway of this reaction, the product has exactly the *opposite relative configuration* (and in this case, absolute configuration) from the starting material. The reaction proceeds with **complete inversion of relative configuration.** (See Sec. 9.5 for another example of this mechanism using stereochemical probes.)

Question 6.11

If the S-2-bromobutane used in the previous reaction were 100% optically pure, what would be the optical purity of the resulting product? How would the results differ if we started with S-2-bromobutane that is only 60% optically pure?

C. Reactions That Generate an Asymmetric Carbon Atom

We now turn from consideration of reactions at an asymmetric carbon to reactions that generate an asymmetric carbon atom in a molecule that previously possessed none. In the monochlorination of pentane, for example, there are three structural isomers, but only 2-chloropentane has an asymmetric carbon atom. On isolation and purification, however, this compound is *optically inactive*.

$$CH_3-CH_2-CH_2-CH_2-CH_3 \xrightarrow[h\nu]{Cl_2} CH_3-CH_2-CH_2-CH_2-CH_2-Cl$$

1-Chloropentane

$$+ CH_3-CH_2-CH_2-\overset{Cl}{\underset{}{\overset{|}{\overset{*}{C}H}}}-CH_3 + CH_3-CH_2-\underset{\underset{Cl}{|}}{CH}-CH_2-CH_3$$

2-Chloropentane   3-Chloropentane

Based on what we know about the mechanism of free-radical halogenation and the structure of the intervening free radical, this should not be surprising. The planar free radical can be attacked from either side by chlorine with equal probability and thus produce a 50:50 mixture of the two enantiomers of 2-chloropentane. This is depicted in Fig. 6.7.

Recall from Sec. 6.6 that enantiomers have identical physical properties (with respect to achiral probes), such as boiling and melting points, solubilities, and densities, so that they cannot be separated by fractional distillation. Thus, 2-chloropentane is obtained only as a racemic mixture that is optically inactive. Later we

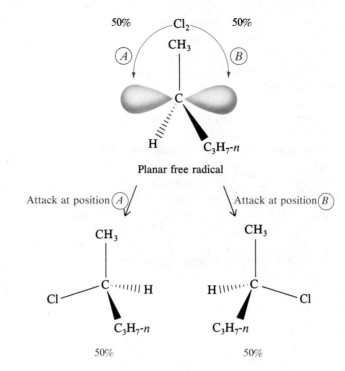

Planar free radical

Equal mixture of enantiomers: optically inactive

FIGURE 6.7 Free-radical chlorination of the 2-pentyl radical, which is derived from hydrogen abstraction at C—2 on pentane. The entering chlorine atom can become attached to either face of the planar free radical, with 50% probability of attack at position Ⓐ and 50% probability of attack at position Ⓑ. This produces a 50:50 mixture of enantiomers that is racemic and optically inactive. (See also Fig. 6.6.)

will study special methods used to resolve enantiomers, which is one way to confirm the formation of enantiomers in reactions of this type.

This leads us to a very important general conclusion: *Optically inactive reactants[1] (even though they may have an asymmetric carbon atom) always yield optically inactive products (even though an asymmetric carbon atom may be generated during the reaction).* The chlorination of pentane demonstrates the validity of this statement for the general case where we start with a reactant that contains no asymmetric center and obtain a product in which one has been formed.

6.10 Alternate Way to Depict Structure of Enantiomers: Fischer Projection Formula

We used *three-dimensional projection formulas* to show how groups are arranged about an asymmetric carbon atom. Although these are not difficult to draw for simple molecules, they can be with more complex molecules—those with two or more

---

[1] Reactants here include all chemical reactants and nonchemical factors, such as catalysts, light, and electric or magnetic fields that may be used in a chemical reaction.

asymmetric carbon atoms. Carbohydrates (see Chap. 29), for example, contain as many as five asymmetric carbon atoms.

The convention presented here was devised by Emil Fischer around 1900 when his famous structure elucidation work on sugars was made public. This new type of structure is no substitute for using molecular models or even for drawing three-dimensional projection formulas, but it is a simpler representation of the stereo-chemistry of compounds that contain one or more asymmetric carbon atoms, and it can be drawn much faster!

The convention for drawing the Fischer representation of molecules that contain one asymmetric carbon is as follows. A cross is drawn, and the four groups that are attached to the asymmetric carbon atom are placed on the four ends of the cross. The point where the horizontal and vertical lines cross is the asymmetric carbon atom. The horizontal lines represent the bonds that come out of the plane of the page toward us, and the vertical lines represent the bonds going behind the plane of the page away from us. The resulting structure appears two-dimensional and is called the **Fischer projection formula.**

Consider some examples. Fig. 6.8 shows the ball-and-stick drawings, three-dimensional projection formulas, and Fischer projection formulas for the enantiomers of 1-bromo-1-chloroethane:

$$CH_3 - \overset{\overset{\displaystyle H}{|}}{\underset{\underset{\displaystyle Br}{|}}{C^*}} - Cl$$

In effect, Fischer projection formulas are two-dimensional representations of molecules. As we look straight at the ball-and-stick models in Fig. 6.8, the Cl and Br atoms "appear" to be connected by a straight *horizontal* line, and the CH_3 and H atoms "appear" to be connected by a straight *vertical* line. We *draw what we see* to get Fischer projection formulas of these enantiomers, and these formulas can be thought of as two-dimensional photographs of the ball-and-stick drawings.

Fischer Projection Formulas

A Fischer projection is always drawn as a cross. The asymmetric carbon is not actually drawn but is the point of intersection of the horizontal and vertical lines. Fischer projections are always drawn this way and labeled as being such. Unless a drawing is so labeled, it is not a Fischer projection. For example,

$$\begin{array}{c} CH_3 \\ | \\ H \!-\!\!\!-\!\!\!-\!\!\! Br \\ | \\ CH_2CH_3 \end{array} \qquad\qquad \begin{array}{c} CH_3 \\ | \\ H - C - Br \\ | \\ CH_2CH_3 \end{array}$$

Fischer projection	Not a Fischer projection
Drawn correctly and labeled	*Drawn incorrectly*

The use of Fischer projection formulas to represent three-dimensional structures in two dimensions requires great caution. They must be used only for molecules with asymmetric carbon atoms. These two-dimensional structures allow us to show the three-dimensional, *absolute positions* of the atoms with respect to one another. We can slide these structures around on the plane of a paper and rotate them end for end (in 180° multiples, but not 90° at a time). A structure may not be removed

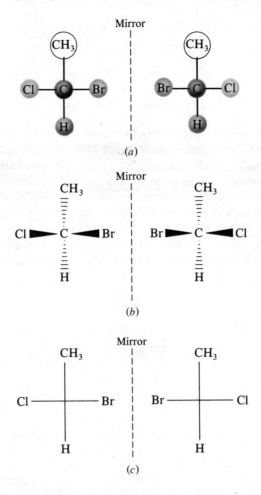

FIGURE 6.8   Methods for depicting the two enantiomers of 1-bromo-1-chloroethane. (*a***) Ball-and-stick structures; (***b***) three-dimensional projection formulas; (***c***) Fischer projection formulas. Note the mirror image relationship of each pair of enantiomers.**

from the plane of the paper and flipped over, however, because then it would represent a new three-dimensional structure, that of the other enantiomer. *This is not allowed*. These comments are illustrated by the following two examples for 1-bromo-1-chloroethane (taken from Fig. 6.8).

1. *Rotation.* Rotation is best envisioned by placing a pin at the center of the molecule (indicated by the heavy dot) and *rotating 180°* about that point. Note that the positions of all groups change: The positions of the CH_3 and H groups and the Cl and Br groups are interchanged.

One enantiomer   Same enantiomer

2. *Flipping.* Flipping is not allowed. The positions of the CH_3 and H groups are the same, and only the positions of the Cl and Br are interconverted.

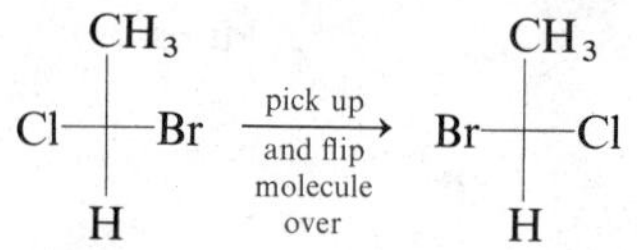

$$\underset{\text{One}\atop\text{enantiomer}}{\overset{CH_3}{\underset{H}{Cl\!-\!\!\!-\!\!\!-\!Br}}} \quad \xrightarrow[\substack{\text{molecule}\\\text{over}}]{\substack{\text{pick up}\\\text{and flip}}} \quad \underset{\text{Other}\atop\text{enantiomer}}{\overset{CH_3}{\underset{H}{Br\!-\!\!\!-\!\!\!-\!Cl}}}$$

Drawing Fischer projection formulas requires a systematic, organized approach. A good procedure is to take any two substituents attached to the asymmetric carbon atom and place them on the vertical line. Then place the other two substituents on the horizontal line to form one enantiomer. The other enantiomer can be drawn simply by interchanging the positions of the two substituents on the horizontal line. Using lactic acid, $CH_3CH(OH)COOH$, as an example, place the CH_3 and COOH groups at the ends of the vertical line. In structure (1) place the H on the left and the OH on the right on the horizontal line, and in structure (2) keep the substituents on the vertical line the same and interchange the positions of H and OH

Mirror

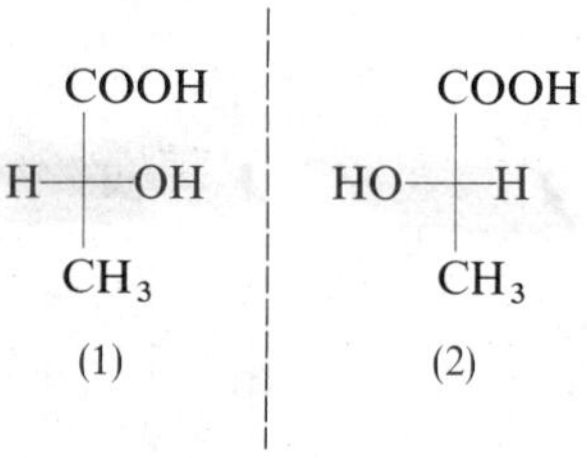

$$\overset{COOH}{\underset{CH_3}{H\!-\!\!\!-\!\!\!-\!OH}} \quad\Big|\quad \overset{COOH}{\underset{CH_3}{HO\!-\!\!\!-\!\!\!-\!H}}$$

(1)   (2)

Nonsuperimposable mirror images: enantiomers

Drawing Fischer projection formulas for compounds with a single asymmetric carbon is even easier than the method shown earlier. By simply drawing the structure of one enantiomer and then drawing its mirror image, we get the other enantiomer. Drawing stereoisomers for compounds with more than one asymmetric carbon atom is more difficult, and the preceding system outlined is essential.

To show what is likely to happen when one tries to draw Fischer projection formulas in a random fashion, consider the following structures—all of which represent the *same* enantiomer:

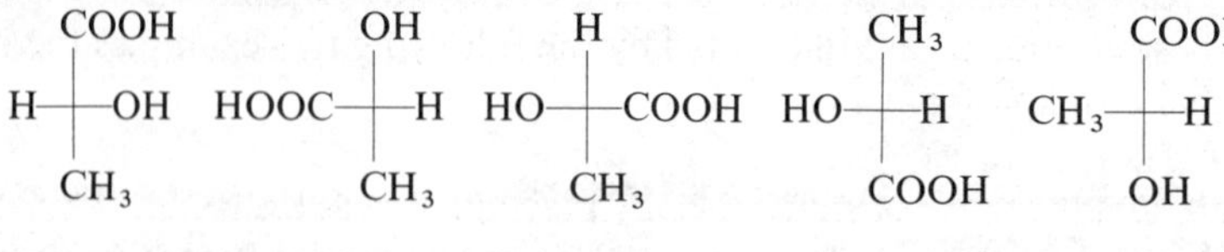

$$\overset{COOH}{\underset{CH_3}{H\!-\!\!\!-\!OH}} \quad \overset{OH}{\underset{CH_3}{HOOC\!-\!\!\!-\!H}} \quad \overset{H}{\underset{CH_3}{HO\!-\!\!\!-\!COOH}} \quad \overset{CH_3}{\underset{COOH}{HO\!-\!\!\!-\!H}} \quad \overset{COOH}{\underset{OH}{CH_3\!-\!\!\!-\!H}} \quad \text{and so on}$$

Make a molecular model for lactic acid and convince yourself that all these structures do indeed represent the same molecule.

Question 6.12

Draw three-dimensional and Fischer projection formulas for the enantiomers of the following molecules:

(*a*) 2-bromobutane

(*b*) 2-bromopropanoic acid, $CH_3\!-\!\underset{\underset{Br}{|}}{CH}\!-\!COOH$

(*c*) 1-chloro-2-methylbutane

6.11   Stereoisomers of Molecules That Contain Two Asymmetric Carbon Atoms

Molecules with two asymmetric carbon atoms may be classified in two ways: (1) those that contain *analogous* asymmetric carbon atoms and (2) those that contain *different* asymmetric carbon atoms. What do we mean when we say that a molecule has two different or two analogous asymmetric carbon atoms? Each of the two asymmetric carbon atoms has four different substituents attached to it, but if the two sets of four substituents are identical (ignoring the absolute configuration of the carbons), then the asymmetric carbon atoms are analogous. If, on the other hand, one or more substituents are different between the two sets, then the asymmetric carbon atoms are different. To illustrate this, consider the following molecules.

Different asymmetric carbon atoms:

$$CH_3-\overset{Cl}{\underset{H}{\overset{|}{\underset{|}{C^*}}}}-\overset{Br}{\underset{H}{\overset{|}{\underset{|}{C^*}}}}-Cl$$

(1)
1-Bromo-1,2-dichloropropane

$\overset{*}{C}$-1 contains: $-H, -Cl, -Br, -CH-CH_3$
                                          $\underset{\quad}{\overset{|}{Cl}}$

$\overset{*}{C}$-2 contains: $-H, -Cl, -CH_3, -CH-Cl$
                                          $\underset{\quad}{\overset{|}{Br}}$

Analogous asymmetric carbon atoms:

$$CH_3-\overset{H}{\underset{Br}{\overset{|}{\underset{|}{C^*}}}}-\overset{H}{\underset{Br}{\overset{|}{\underset{|}{C^*}}}}-CH_3$$

2,3-Dibromobutane

$\overset{*}{C}$-2 and $\overset{*}{C}$-3 *both* contain the same substituents:

$-H, -CH_3, -Br, -CH-CH_3$
                       $\underset{\quad}{\overset{|}{Br}}$

A. Drawing Stereoisomers with Two Different Asymmetric Centers

Consider the stereoisomers associated with a molecule that contains two different asymmetric carbon atoms, for example, 1-bromo-1,2-dichloropropane (1). How many possible stereoisomers exist for this molecule, and what are their structures?

As a convenient point of departure, we start by constructing a molecular model of one stereoisomer and then make its mirror image, as shown:

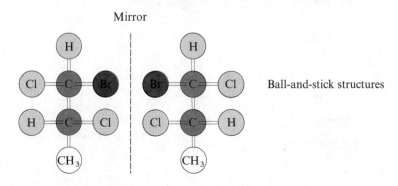

Ball-and-stick structures

$$
\begin{array}{ccc}
\text{H} & & \text{H} \\
| & & | \\
\text{Cl}\!\!\blacktriangleright\!\!\text{C}\!\!\blacktriangleleft\!\!\text{Br} & & \text{Br}\!\!\blacktriangleright\!\!\text{C}\!\!\blacktriangleleft\!\!\text{Cl} \\
\text{H}\!\!\blacktriangleright\!\!\text{C}\!\!\blacktriangleleft\!\!\text{Cl} & & \text{Cl}\!\!\blacktriangleright\!\!\text{C}\!\!\blacktriangleleft\!\!\text{H} \\
| & & | \\
\text{CH}_3 & & \text{CH}_3
\end{array}
$$

Three-dimensional projection formulas

Mirror

$$
\begin{array}{ccc}
\text{H} & & \text{H} \\
| & & | \\
\text{Cl}\!\!-\!\!\!-\!\!\text{Br} & & \text{Br}\!\!-\!\!\!-\!\!\text{Cl} \\
\text{H}\!\!-\!\!\!-\!\!\text{Cl} & & \text{Cl}\!\!-\!\!\!-\!\!\text{H} \\
| & & | \\
\text{CH}_3 & & \text{CH}_3 \\
(2) & & (3)
\end{array}
$$

Fischer projection formulas

*Nonsuperimposable mirror
images: enantiomers*

Molecular models demonstrate that there is no possible way to superimpose these two mirror images on one another. There is free rotation about the carbon-carbon bonds, but still these two structures cannot be superimposed on one another, so they are not conformational structures. Just as our general definition in Sec. 6.5 dictates, these represent a *pair of enantiomers*. Regardless of how many asymmetric carbon atoms a molecule contains, this definition holds; all enantiomers exist as a discrete pair, one (+) and one (−).

Further examination reveals that there is yet another stereoisomer for 1-bromo-1,2-dichloropropane (1), as shown by structure (4) that follows. Comparison of structure (4) with (2) and (3) indicates that it is not a mirror image of either of them and is also not superimposable on either of them. Using models, we can make the mirror image of (4) and note that this new stereoisomer (5) is not superimposable on (4). Also, (5) is not a mirror image of or superimposable on either (2) or (3). [As in (2) and (3), we can rotate about the carbon-carbon bond in (4) and (5) to obtain other conformations that are still not superimposable on one another or on (2) and (3).]

Mirror

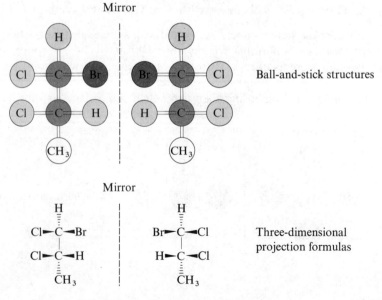

Ball-and-stick structures

Mirror

$$
\begin{array}{ccc}
\text{H} & & \text{H} \\
| & & | \\
\text{Cl}\!\!\blacktriangleright\!\!\text{C}\!\!\blacktriangleleft\!\!\text{Br} & & \text{Br}\!\!\blacktriangleright\!\!\text{C}\!\!\blacktriangleleft\!\!\text{Cl} \\
\text{Cl}\!\!\blacktriangleright\!\!\text{C}\!\!\blacktriangleleft\!\!\text{H} & & \text{H}\!\!\blacktriangleright\!\!\text{C}\!\!\blacktriangleleft\!\!\text{Cl} \\
| & & | \\
\text{CH}_3 & & \text{CH}_3
\end{array}
$$

Three-dimensional projection formulas

```
                      Mirror
      H                 |         H
      |                 |         |
 Cl———Br                |    Br———Cl            Fischer projection formulas
      |                 |         |
 Cl———H                 |    H————Cl
      |                 |         |
     CH₃                |        CH₃
      (4)                        (5)
         Nonsuperimposable mirror
            images: enantiomers
```

There are no more stereoisomers that can be made for (1), for which we have two pairs of enantiomers, (2) and (3), and (4) and (5). These molecules differ from one another only in the three-dimensional arrangement of atoms about the two asymmetric carbon atoms. We know that these molecules all have the same structural formulas (that is, they are all 1-bromo-1,2-dichloropropanes) and we know the relationship between (2) and (3) and between (4) and (5) (they are two pairs of enantiomers), but now how is (2) related to (4) or (5) and (3) related to (4) and (5)? They are called **diastereomers** of one another. *If two stereoisomers are not enantiomers of one another, then they are diastereomers.* Diastereomers are stereoisomers that do not bear a mirror-image relationship to each other. These relationships are summarized as follows:

Enantiomers:

(2) and (3), (4) and (5)

Diastereomers:

(2) and (4), (2) and (5), (3) and (4), (3) and (5)

Molecular models may not be available when we wish to draw all the possible stereoisomers for compounds such as (1). Then the systematic method for drawing Fischer projection formulas from Sec. 6.10 is indispensable. As with compounds that contain a single asymmetric carbon atom, keep as many atoms and substituents in the same position as possible, and then vary the positions of the other substituents to end up with all the possible stereoisomers—but no more. We must show all possible combinations and permutations of configurations for each asymmetric carbon atom.

Start by taking one substituent on each of the two asymmetric carbon atoms and placing them on the vertical line; carbon-containing substituents that are part of the longest continuous chain are used if possible. Using compound (1) as an example, we have:

```
        H
        |
     ———+———
        |
     ———+———
        |
       CH₃
```

Then place the remaining two substituents of one asymmetric carbon atom on the Fischer formula in one configuration. Draw two identical projection formulas, keeping the configuration on the first carbon atom the same in both, and place the two remaining substituents on the other asymmetric carbon on the projection formulas so that one is on the right and the other on the left in one formula and vice versa in the other.

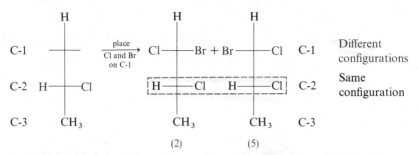

(2)   (5)

Note that the configurations at C-1 are different, whereas the configuration at C-2 is the same in both formulas.

Finally, draw two more Fischer projection formulas in which the configuration that was fixed before is reversed; that is, draw the mirror image of (2) and of (5). Thus, for compound (1) the following results:

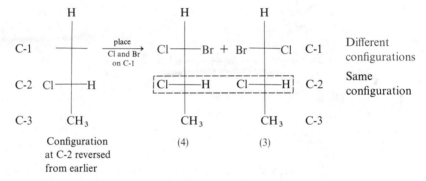

Configuration   (4)   (3)
at C-2 reversed
from earlier

Note that the configurations at C-1 are different, but the configurations at C-2 are the same in both formulas.

The four stereoisomers drawn here are the same four that were derived by constructing molecular models and drawing the structures that correspond to them. These stereoisomers have been identified by the same structure numbers [(2) to (5)] used previously. Furthermore, *all* the Fischer projection formulas we drew using the systematic approach have the H on C-1 and the CH_3 group in the same position.

As they are drawn, the Fischer projection formulas appear to be rigid structures. But remember that free rotation is still occurring about the carbon-carbon single bonds and that these structures simply represent the real three-dimensional configurations in two dimensions. To illustrate how a random approach can lead to chaos, the following Fischer projection formulas are all equivalent and result from free rotation about the C-1—C-2 bond:

Ball and stick structures:

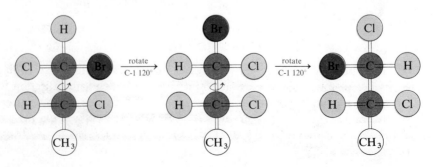

Fisher projection formulas:

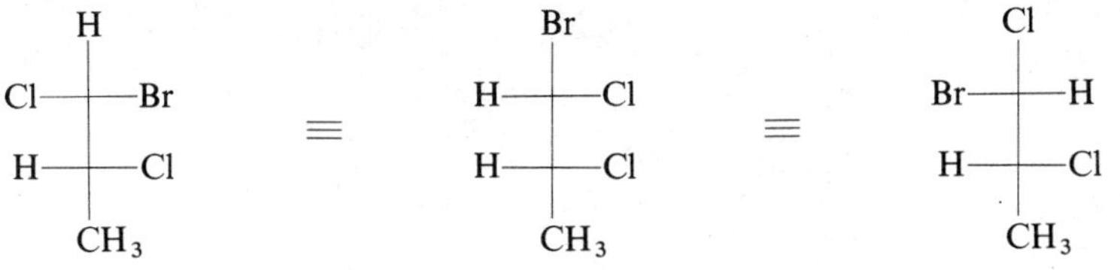

All represent stereoisomer (2)

B. The *R* and *S* System

We can also use the *R* and *S* designation for configurations of molecules that contain more than one asymmetric center. We indicate by carbon atom number the asymmetric center we are so designating. The configuration about each asymmetric carbon atom is determined independently of the other asymmetric carbons present. For compound (1)

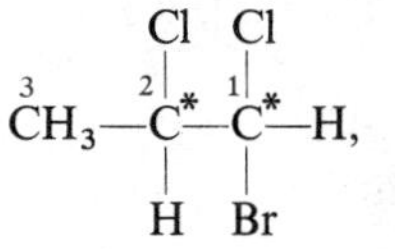

the substituents around asymmetric C-1 are, in order of *decreasing* priority: Br—, Cl—, CH_3—CH—, H—. Likewise, the order of *decreasing* priority for the sub-
$\quad\quad\quad\quad\quad$|
$\quad\quad\quad\quad\quad$Cl
stituents attached to asymmetric C-2 are: Cl—, H—C—, CH_3—, H—. Using these
$\quad\quad\quad\quad\quad\quad\quad\quad\quad\quad\quad\quad$|
$\quad\quad\quad\quad\quad\quad\quad\quad\quad\quad\quad\quad$Br
priorities, we obtain the following designations for the four configurational stereo-isomers of (1):

Structure (2): (1*S*,2*R*)-1-bromo-1,2-dichloropropane
Structure (3): (1*R*,2*S*)-1-bromo-1,2-dichloropropane
Structure (4): (1*S*,2*S*)-1-bromo-1,2-dichloropropane
Structure (5): (1*R*,2*R*)-1-bromo-1,2-dichloropropane

As a check, (2) and (3) are enantiomers of one another, and the configurations are opposite at both asymmetric carbon atoms, as indicated by the *R* and *S* designations. Likewise, (4) and (5) are enantiomers, and their *R* and *S* designations reflect this mirror-image relationship.

6.12 Stereoisomers of Compounds That Contain Two *Analogous* Asymmetric Carbon Atoms

A. Drawing Stereoisomers with Two Analogous Asymmetric Centers

A typical example of a compound that contains two analogous asymmetric carbon atoms is 2,3-dibromobutane (1).

Using molecular models, construct the various configurational stereoisomers for this compound. Start, for example, by drawing (2) and its mirror image (3). Careful examination of these two structures shows that they are nonsuperimposable, and since they also are mirror images, they represent a pair of enantiomers.

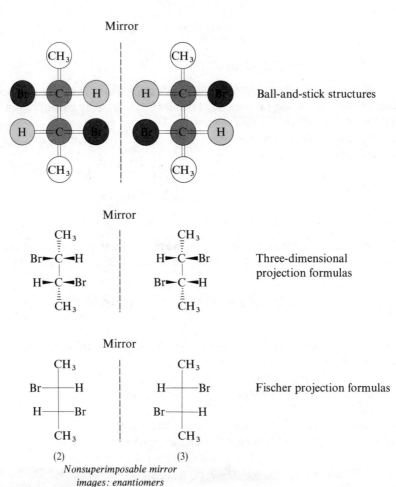

(1)

2,3-Dibromobutane

Ball-and-stick structures

Three-dimensional
projection formulas

Fischer projection formulas

(2)   (3)

*Nonsuperimposable mirror
images: enantiomers*

   Another stereoisomer (4) and its mirror image (5) can be constructed. Here, however, (4) and (5) are completely superimposable on one another if we rotate either structure end for end by 180°. (Recall from Sec. 6.10 that Fischer projections cannot be lifted out of the plane of the paper but can be rotated end for end by keeping the relative positions of all substituents the same.) Thus (4) and (5) are identical structures in every respect; they represent two ways to draw the same three-dimensional structure. This structure has no nonsuperimposable mirror image and therefore does not possess an enantiomer. It is *optically inactive* and is called a **meso compound.** A *meso* compound contains two or more chiral centers, yet it is superimposable on its mirror image; it is *achiral* and does not exhibit optical activity.

   A *meso* compound and its mirror image are superimposable and thus identical structures. But, if they are superimposable, where is the plane of symmetry? In the structure of (4) [or of (5), which is identical with (4)], a plane bisects the molecule equidistant between C-2 and C-3, splitting the molecule into two equal parts. These two parts are mirror images of one another.

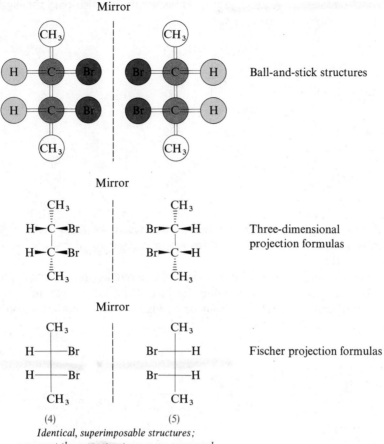

Ball-and-stick structures

Three-dimensional projection formulas

Fischer projection formulas

(4)　　　　　　(5)

Identical, superimposable structures;
represent the same structure: meso compound
(4) ≡ (5)

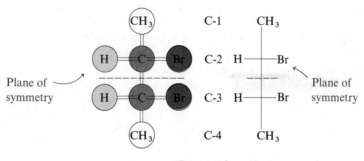

meso Compound

Since half the molecule is the mirror image of the other half, it is not surprising that *meso* compounds are optically inactive, because the optical activity of one half is exactly canceled by that of the other half. Here the halves have been "tied" together by a chemical bond. This is analogous to a racemic mixture in which the two enantiomers exist as discrete molecules, but because they are present in equimolar amounts, their rotations also exactly cancel one another and the mixture is optically inactive.

We can recognize *meso* compounds when we (1) have compounds that contain two or more asymmetric carbon atoms, and (2) find that a particular stereoisomer contains a plane of symmetry. In structure (4) the plane of symmetry passes through a bond between two carbon atoms. However, it can also pass through an atom or group of atoms, as is the case with

$$CH_3$$

H———*———Br

---HCH---   Plane of
                symmetry
H———*———Br

$$CH_3$$

(6)

meso

Note also that two asymmetric carbon atoms do not have to be adjacent to one another. Asymmetric carbon atoms are often separated by one or more carbon atoms, as they are in (6).

As with compounds that contain two different asymmetric centers, a systematic approach can be used for drawing the Fischer projection formulas for the stereo-isomers of compound (1). Without going into as much detail, the following stereo-isomers can be drawn:

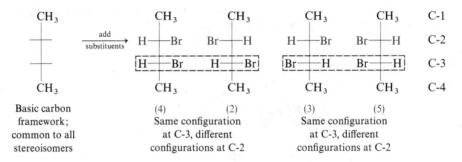

CH_3		CH_3	CH_3	CH_3	CH_3	C-1

Basic carbon framework; common to all stereoisomers

(4)   (2)   (3)   (5)
Same configuration   Same configuration
at C-3, different   at C-3, different
configurations at C-2   configurations at C-2

Extreme care must be exercised in using this systematic method. Compare each stereoisomer with every other one to identify which are enantiomers and to determine whether or not a *meso* compound exists.

The *three* (not four) stereoisomers for 2,3-dibromobutane may be summarized as follows:

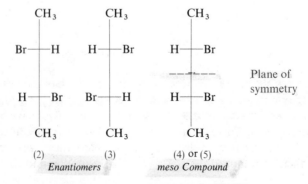

(2)   (3)   (4) or (5)
Enantiomers   *meso Compound*

There are one pair of enantiomers, (2) and (3), and one *meso* compound, (4) or (5). But how are (2) and (4) and (3) and (4) related? They are *diastereomers* because they are all 2,3-dibromobutanes but are not enantiomers of one another.

B. The *R* and *S* System

The configurations of compounds (2) to (4) may be specified using the *R* and *S* designation. Structure (2) is called (2*R*, 3*R*)-2,3-dibromobutane and (3) is (2*S*, 3*S*) -2,3-dibromobutane. (Because these are analogous asymmetric carbon atoms, the numbers preceding the *R*'s and *S*'s can be omitted and the naming is still correct.) Structure (4) is called (2*R*,3*S*)-2,3-dibromobutane, which is correct because the molecule contains equal halves that are mirror images.

Question 6.13

For each of the following compounds, indicate the asymmetric carbon atoms with a *, and identify each as being analogous or different asymmetric centers. Then draw Fischer projection formulas for all the stereoisomers and indicate pairs of enantiomers, *meso* compounds, and at least two pairs of diastereomers. Which structures are optically active?

(*a*) $BrCH_2-CH-CH-CH_2Br$:
$\qquad\qquad\quad |\quad\ |$
$\qquad\qquad\quad Br\ \ Br$

(*b*) $CH_3-CH-CH-CH_2Br$
$\qquad\qquad\quad |\quad\ |$
$\qquad\qquad\quad Br\ \ Br$

(*c*) $CH_3-CH-CH_2$
$\qquad\qquad |\quad\ \ |$
$\qquad\qquad F\quad F$

(*d*) $CH_3-CH-CH_2CH_2-CH-CH_3$
$\qquad\qquad\quad |\qquad\qquad\quad |$
$\qquad\qquad\quad Cl\qquad\qquad\quad Br$

(*e*) $CH_3-CH_2-\overset{\displaystyle CH_3}{\underset{\displaystyle Cl}{C}}-\overset{\displaystyle CH_3}{\underset{\displaystyle Cl}{C}}-CH_2-CH_3$

(*f*) $CH_3-CH_2-\overset{\displaystyle CH_3}{\underset{\displaystyle Cl}{C}}-\underset{\displaystyle Cl}{CH}-CH_3$

Question 6.14

For the stereoisomers drawn in Question 6.13, specify the *R* and *S* configurations.

6.13   Determining the Number of Stereoisomers

We saw how to draw stereoisomers for compounds that contain two analogous or two different asymmetric carbon atoms and that there are three stereoisomers for the former and four for the latter type of compound. Is there a way to determine the maximum number of stereoisomers one can expect for compounds containing asymmetric carbon atoms? Yes, the formula for determining this is:

$$\textit{Maximum number of stereoisomers} = 2^n$$

where $\qquad\qquad n = $ number of assymetric carbon atoms in molecule

With one asymmetric carbon atom ($n = 1$), 2^1 or 2 stereoisomers are possible, which is one pair of enantiomers. With two asymmetric carbon atoms ($n = 2$), 2^2 or 4 stereoisomers are possible. But earlier examples show that there are four stereoisomers when the two asymmetric carbon atoms in a molecule are substituted differently but only three stereoisomers when a molecule contains two analogously substituted asymmetric carbon atoms. The point is that the formula tells us the *maximum* number of stereoisomers; there can never be more, but there may be less. The number of structures will be maximum when a molecule contains different asymmetric carbon atoms, but when there are two or more analogous asymmetric carbons, the number will be less. Drawing the stereoisomers allows you to determine the precise number that can exist.

Compounds with more than two asymmetric carbon atoms exist in a number of stereoisomeric forms. For example, a six-carbon sugar has four different asymmetric carbon atoms and thus there are a maximum of 2^4 or 16 stereoisomers possible! One such sugar, D-glucose, has the structure:

$$
\begin{array}{c}
\text{CHO} \\
\text{H} \overset{*}{-\!\!\!|\!\!\!-} \text{OH} \\
\text{HO} \overset{*}{-\!\!\!|\!\!\!-} \text{H} \\
\text{H} \overset{*}{-\!\!\!|\!\!\!-} \text{OH} \\
\text{H} \overset{*}{-\!\!\!|\!\!\!-} \text{OH} \\
\text{CH}_2\text{OH}
\end{array}
$$

D-Glucose

We discuss its structure, along with those of many other sugars, in Chap. 29.

Question 6.15

Indicate the *maximum* number of stereoisomers expected for the following compounds:

(a) $\text{CH}_3\text{—CH—CH—CH—CH}_3$
$\qquad\qquad\;\; |\quad\;\; |\quad\;\; |$
$\qquad\qquad\;\, \text{OH OH OH}$

(b) $\text{HO—CH}_2\text{—CH—CH—CH—CH}_3$
$\qquad\qquad\qquad\quad |\quad\;\; |\quad\;\; |$
$\qquad\qquad\qquad\;\;\, \text{OH OH OH}$

(c) $\text{HO—CH}_2\text{—CH—CH—CH—CH—COOH}$
$\qquad\qquad\qquad\;\; |\quad\;\; |\quad\;\; |\quad\;\; |$
$\qquad\qquad\qquad\; \text{OH OH OH OH}$

(d) $\text{HO—CH}_2\text{—CH—CH—CH—CH—CH—COOH}$
$\qquad\qquad\qquad\;\; |\quad\;\; |\quad\;\; |\quad\;\; |\quad\;\; |$
$\qquad\qquad\qquad\; \text{OH OH OH OH OH}$

6.14   Properties of Diastereomers; Compounds That Contain Two Asymmetric Carbon Atoms

Whereas enantiomers, in the absence of any other chiral influence, exhibit identical chemical and physical properties, diastereomers have different physical properties (such as boiling and melting points, densities, and so on) and chemical properties, and only by chance do they ever have identical signs and magnitudes of rotation; see Figs. 6.9(a) to (d). They contain identical functional groups, but often their chemical properties are not identical. For example, they usually all react with a given reagent, but the *rate* at which a reaction occurs is different from one diastereomer to another.

Differences in the physical properties of diastereomers usually allow them to be separated from one another by classical methods. Liquids can be separated by fractional distillation if the boiling points are different, solids can be separated by crystallization if the melting points and solubility properties are different, and sometimes even chromatography can be used for separating because of differences in polarity.

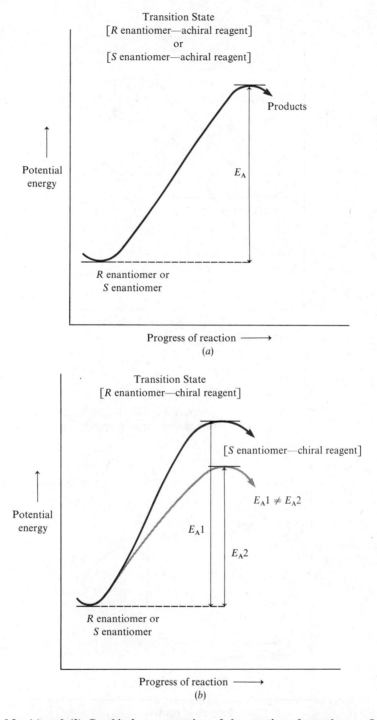

FIGURE 6.9 (*a*) and (*b*) Graphical representation of the reaction of enantiomers. In (*a*) the enantiomers are reacting with achiral reagents: the two enantiomers are of identical energy. On reaction with an achiral reagent, they result in transition states that are also mirror images (enantiomeric) of each other, of equal energy, and with the same E_A values. There is no difference in their chemical reactivities. In (*b*) the enantiomers are reacting with chiral reagents: when two enantiomers of identical energy react with the same chiral reagent, two transition states are possible. These transition states are diastereomeric, of unequal energy, and have different E_A values. There is a difference in the chemical reactivity of the two enantiomers.

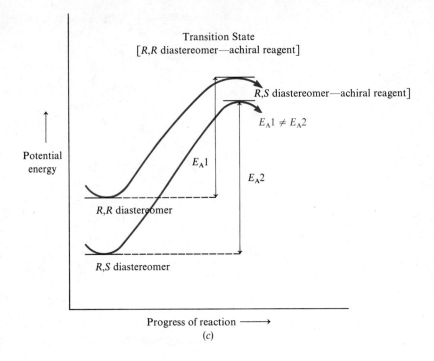

$$
(c)
$$

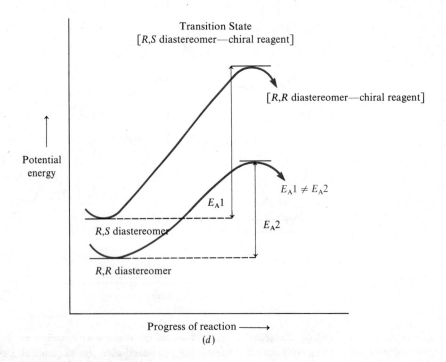

$$
(d)
$$

FIGURE 6.9 (*cont'd.*) (*c*) **and** (*d*) **Graphical representation of the reaction of diastereomers. The diastereomers are reacting with** (*c*) **achiral or** (*d*) **chiral reagents: any two diastereomers are of unequal energy (except in rare instances by coincidence). These general examples of any pair of diastereomers are completely arbitrary. When a pair of diastereomers reacts with either a chiral or achiral reagent, diastereomeric transition states result. There is always (again, except by co-incidence) a difference in the chemical reactivities of two diastereomers.**

Consider some examples that illustrate how enantiomers and diastereomers differ. For tartaric acid, HOOC—$\overset{*}{C}$H—$\overset{*}{C}$H—COOH, which has two analogous
                                          | |
                                          OH   OH
asymmetric centers, the stereoisomers can be written as follows:

Mirror

COOH	COOH		COOH
HO——H	H——OH		H——OH
			———————— Plane of symmetry
H——OH	HO——H		H——OH
COOH	COOH		COOH
(−)-Tartaric acid (l form)	(+)-Tartaric acid (d form)		meso-Tartaric acid (meso form)

Enantiomers

mp	170°	170°	140°
$[\alpha]_D^{25°}$	−12.7°	+12.7°	0
Solubility in water (g/100 ml)	140	140	125

The physical properties and optical rotations are given below each structure. Based on what is known about the properties of enantiomers and diastereomers, it is not surprising that the *d* and *l* forms have equal but opposite rotations and the same melting points and solubilities in water. As anticipated, the *meso* form, which is a diastereomer of both the *d* and *l* forms, has a different melting point, its solubility in water is quite different, and it shows no optical activity.

The tartaric acid molecule, which is an example of a family of compounds called *hydroxy acids*, is interesting from another standpoint. This is the compound that Pasteur separated in his historic discovery of enantiomers (see Sec. 6.2). Although enantiomers are inseparable most of the time, that was (fortunately) a rare exception because the two enantiomers crystallized so that they looked different and could be separated physically using tweezers and a magnifying glass. Actually, Pasteur separated the sodium ammonium salts of *d*- and *l*-tartaric acid,

$$Na^{\oplus \ominus}OOC—CH—CH—COO^{\ominus}NH_4^{\oplus}.$$
$$\qquad\qquad |\quad\ |$$
$$\qquad\quad OH\ \ OH$$

This may now look like there are two different asymmetric carbon atoms (as a result of the $Na^{\oplus}$ on one end and $NH_4^{\oplus}$ on the other), but when the salts are dissolved in water (to measure the optical activity) ionization occurs to give the anions which interact with polarized light and rotate it.

$COO^{\ominus}$	$COO^{\ominus}$
HO——H	H——OH
H——OH	HO——H
$COO^{\ominus}$	$COO^{\ominus}$

Enantiomers

6.15   Biological Reactions Involving Optically Active Compounds

Much information was presented about enantiomers and diastereomers, especially definitions and terminology. One important aspect of the chemistry of these molecules yet to be mentioned is their important applications in biological chemistry. We mention only a few here.

A. Cholesterol

To illustrate how many stereoisomers can exist for a given compound, look at cholesterol. Cholesterol is thought to be linked to heart attacks and other heart problems in humans, and it is found in fairly high concentrations in eggs, dairy products such as cheese, and red meat.

Cholesterol contains the eight asymmetric carbon atoms indicated by * in the following diagram. There are 2^8 or 256 possible stereoisomeric forms for cholesterol, but the naturally occurring compound has the structure shown here in three dimensions.

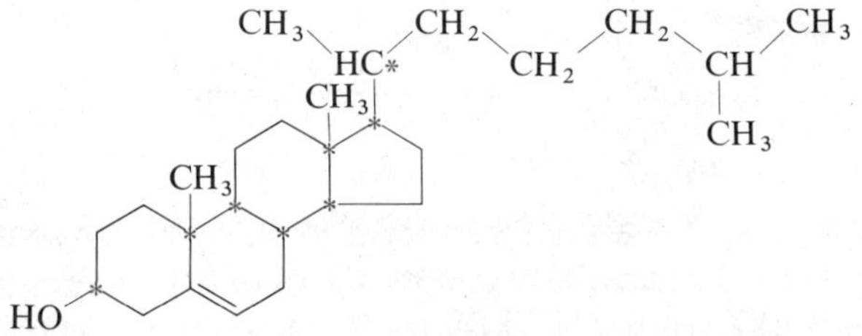

Cholesterol
Planar drawing

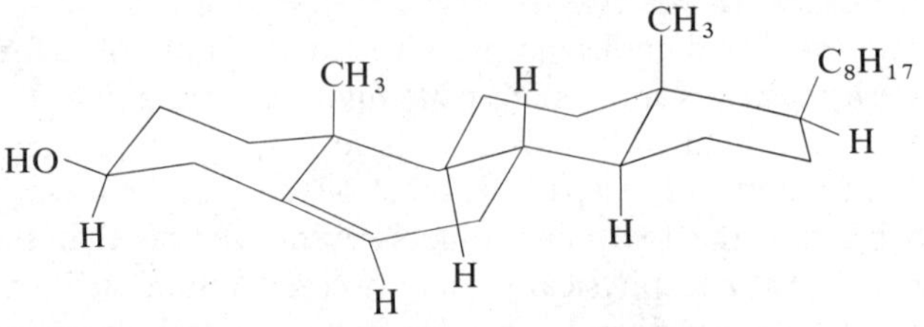

Three-dimensional structure

B. Penicillin

Penicillin is produced by the penicillin mold, *Penicillum glaucum*. This mold reacts preferentially with (+)-ammonium tartrate and only after all the (+) isomer has been consumed, does the mold start to utilize the (−) form. If the reaction is stopped at the appropriate time, (−)-ammonium tartrate can be isolated. This represents a way to *resolve* enantiomers (see Sec. 6.7) to obtain at least one enantiomer. It is also an example of biological selection.

C. Adrenaline

Naturally occurring *adrenaline* rotates light counterclockwise and is about 20 times more active biologically than its enantiomer. Only this form is generated in the human system. Synthetic adrenaline contains a mixture of the enantiomeric forms, but when it is administered to a person, only the proper form is used; its enantiomer is nontoxic and is unaffected by and does not affect the human system. On the other hand, cases are known of one enantiomer being a useful medication but its enantiomer toxic; separation in such cases must be completed before the compound is used as a medicine (see Sec. 23.14).

Adrenaline

D. Carvone

We mentioned that the physical properties of two enantiomers are identical in the absence of a chiral influence. It is possible in several cases, however, to distinguish between enantiomers on the basis of their odor. This is true for ($+$)- and ($-$)-carvone, which have the structure given here. The ($+$) isomer is associated with oil of caraway and has the associated odor. ($-$)-Carvone, on the other hand, is associated with oil of spearmint and has a strong mint odor. One can detect these two isomers probably because the odor receptors in the nose are also chiral. When one molecule of carvone interacts with the receptors, a stereoisomer is formed. When the other isomer interacts with the same receptors, a second stereoisomer (a diasteromer of the first) is formed. These two diastereomers have different physical properties and in this case can be distinguished by their odors. (Draw the two enantiomers.) (For further discussion of this phenomenon, see the Reading References for this chapter in the appendix.)

Carvone

We encounter numerous optically active natural products throughout this text, for example, carbohydrates (sugars, see Chap. 29) and amino acids and proteins (see Chap. 28). We will find, for example, that only one configuration of amino acids is produced and consumed in the body and that enzymes play a key role in dictating why this occurs.

6.16   Reactions That Generate a Second Asymmetric Center in a Molecule

We discussed stereochemical reactions that (1) do not affect the asymmetric carbon atom, (2) generate a single asymmetric center from a molecule originally containing none, and (3) occur at an asymmetric center (see Sec. 6.9). We now include a reaction that generates a second asymmetric center in an optically active molecule.

Consider, for example, the introduction of a bromine atom into an optically active sample of 2-chlorobutane. At first we limit our discussion to one of the many isomers formed in this reaction—2-bromo-3-chlorobutane:

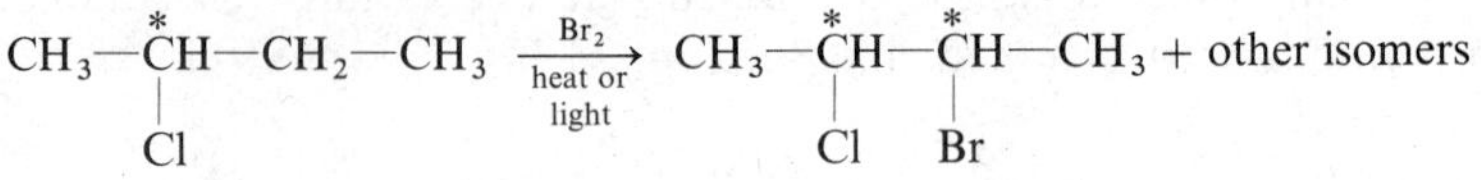

$$CH_3\overset{*}{-}\!\!\underset{\underset{Cl}{|}}{CH}\!\!-CH_2-CH_3 \xrightarrow[\substack{\text{heat or}\\ \text{light}}]{Br_2} CH_3\overset{*}{-}\!\!\underset{\underset{Cl}{|}}{CH}\!\!\overset{*}{-}\!\!\underset{\underset{Br}{|}}{CH}\!\!-CH_3 + \text{other isomers}$$

2-Chlorobutane   2-Bromo-3-chlorobutane
Optically active   Optically active

In this reaction a second asymmetric carbon atom is formed in a molecule that already contains one asymmetric carbon atom. In the generation of the second asymmetric center, none of the bonds to the original asymmetric atom is involved; the carbon that contains the chlorine atom has the same relative configuration in the product that it had in the reactant. What can be said about the stereochemistry of the new asymmetric center? To answer this, recall from Sec. 6.9 that a planar free radical intervenes in the free-radical bromination reaction. The stereochemical fate of this reaction, beginning with the *R* enantiomer of 2-chlorobutane, is as follows:

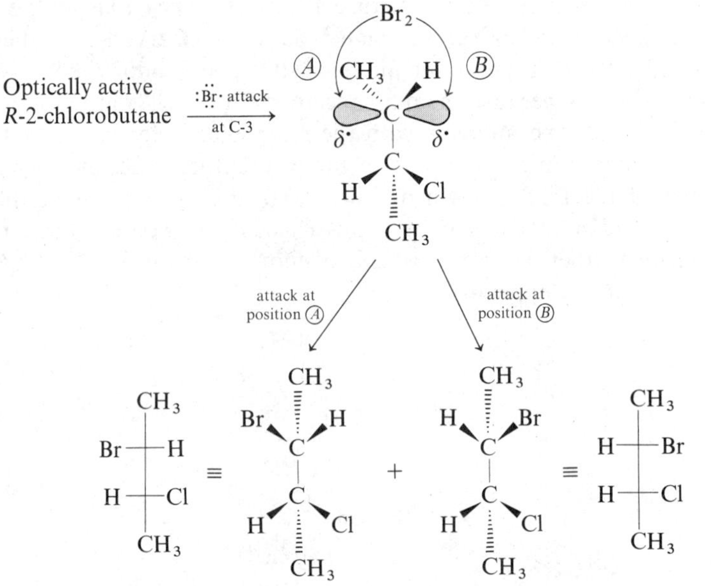

(2R,3R)-2-Bromo-3-chlorobutane   (2S,3R)-2-Bromo-3-chlorobutane
Optically active   Optically active

Diastereomers

Two stereoisomers result from substitution at C-3, and both are optically active. They are (2R,3R)-2-bromo-3-chlorobutane and (2S,3R)-2-bromo-3-chloro-butane, two diastereomers. Because they are diastereomers, the energy of activation for the formation of one need not equal that of the other, and unequal amounts of the two diastereomers are quite likely. The synthesis of diastereomers in unequal proportions, which is quite common when one or more asymmetric centers are already present in the molecule, is called **asymmetric induction**. Reactions of this type, in which one of several diastereomeric products predominates, are called **stereoselective reactions**. Several other examples of asymmetric induction and stereoselective reactions are presented in Secs. 8.17, 8.18, and 8.27. These concepts are particularly important to our understanding of many biochemical reaction pathways.

Now we look at what would happen if we performed the identical reaction outlined earlier on *S*-2-chlorobutane.

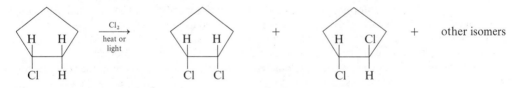

We get the other two possible diastereomers—(2*R*,3*S*)- and (2*S*,3*S*)-3-bromo-3-chlorobutane. These products are also diastereomeric to each other and form in unequal proportions. The proportion of the (2*S*,3*S*) isomer in this reaction equals the proportion of the (2*R*,3*R*) isomer in the previous reaction. The same is true of the (2*S*,3*R*) and (2*R*,3*S*) pairs of enantiomers produced in the two reactions. Consider this one step further by allowing a racemic mixture of equal parts *R*- and *S*-2-chlorobutane to react. A reaction mixture is produced that is also racemic. This mixture consists of equal percentages of the enantiomers in each of the two pairs of enantiomers formed. This is consistent with our observations in Sec. 6.9 that optically inactive reactants always yield optically inactive products in the absence of any other chiral influence. In this case, racemic reactants yield racemic products.

As another example of free-radical substitution involving stereoselectivity, consider the following reaction:

This reaction results in the formation of diastereomeric products (geometric isomers in this example) with one of the possible diastereomers favored over the others. This is in line with our definition of stereoselectivity. Note, however, that the entire reaction mixture is not optically active. The *cis* isomer is a *meso* compound, and the *trans* isomer is a racemic mixture.

Question 6.16

In the preceding example, a molecule with no asymmetric centers is converted in one step to a molecule with two asymmetric centers. Recalling the general discussion in Sec. 6.14, discuss the stereochemical implications of this reaction. Would you expect an equal mixture of the two products? If so, why? If not, why not?

Question 6.17

The following four compounds are subjected to free-radical chlorination (chlorine in the presence of heat or light) so that one additional chlorine atom is introduced into each molecule. The products are separated by a suitable laboratory technique (that is, fractional distillation or recrystallization) so that each is obtained in pure form. Draw the Fischer projection formulas for *all* the products that would be formed in the original reaction, and then indicate how many pure products would be obtained on separation. For each product so obtained, indicate whether it would be optically active or optically inactive.

(*a*) optically active 2-bromobutane (use either enantiomer)
(*b*) racemic 2-bromobutane
(*c*) optically active 1-chloro-2-methylbutane (use either enantiomer)
(*d*) optically active 2-chloropentane (use either enantiomer)

Question 6.18

Answer Question 6.17 for the following four compounds and configurations. Indicate the configuration, *R* or *S*, for *each* product.

(*a*) *R*-2-bromobutane   (*b*) racemic 2-bromobutane
(*c*) *S*-1-chloro-2-methylbutane   (*d*) *S*-2-chloropentane

Question 6.19

When 2-chlorobutane is reacted with Cl_2 and heat and the 2,3-dichlorobutane is isolated, the two possible diastereomers are formed in a ratio of 29%:71% chiral isomer:*meso* isomer. The chiral enantiomeric pair is racemic. It was asserted in Sec. 6.16 that the generation of a second asymmetric center in free-radical chlorination yields unequal mixtures of diastereomers, in this case, a 29%:71% ratio. On the other hand, the generation of a single asymmetric center forms an *equal* mixture of enantiomers, as in the conversion of butane to 2-chlorobutane in Sec. 6.9. These facts seemingly contradict one another, and yet they are correct.

(*a*) Draw the Newman projection formulas for the conformations of the *sec*-butyl radical, CH_3—$\dot{C}H$—C_2H_5; keep in mind that the radical is planar.
(*b*) On the basis of those conformations, what is the most likely stereoisomer of each product?
(*c*) Examine these conformations carefully and determine what stereochemical relationship they have to one another. (That is, are they enantiomers or diastereomers?)
(*d*) Using them, explain why a 50:50 mixture of enantiomers of 2-chlorobutane should be expected, and thus why the results stated earlier do not contradict each other.
(*e*) Perform a similar breakdown and analysis of the mechanism involved in the conversion of 2-chlorobutane to 2,3-dichlorobutane.

6.17   Stereoisomerism in Cyclic Compounds

Cyclic compounds exist in *cis* and *trans* forms (see Sec. 3.17). Looking more closely at these compounds, we see that stereoisomerism is also important when they contain asymmetric centers.

Consider cyclic compounds that contain a single asymmetric center. 1,1,2-Trichlorocyclopropane (1) contains an asymmetric center and thus should be capable of being resolved into its enantiomers.

On the other hand, a simple monosubstituted cyclic compound, such as chlorocyclopropane (2), has no asymmetric center, possesses a plane of symmetry, and is achiral.

Mirror

Cl   Cl   ┊   Cl   Cl
      *   ┊   *
Cl   H   ┊   H   Cl

(1)
1,1,2-Trichlorocyclopropane
Nonsuperimposable mirror images: enantiomers

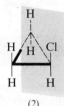

(2)
Chlorocyclopropane
Contains plane of symmetry

Disubstituted cyclic compounds can exist as *cis* and *trans* isomers. If we look at a two-dimensional representation of a disubstituted compound, we see that it contains two analogous asymmetric carbons. For example, consider 1,2-dichloro-cyclopropane (3).

Cl   *   *   Cl

(3)
1,2-Dichlorocyclopropane
Contains two analogous asymmetric carbon atoms

Now we construct three-dimensional structures for the *cis* and *trans* isomers to determine whether or not mirror images exist, and if so, whether they are super-imposable. For *trans*-1,2-dichlorocyclopropane, the following results:

Mirror

H   Cl   ┊   Cl   H
Cl   H   ┊   H   Cl

trans-1,2-Dichlorocyclopropane, *trans* (3)
Nonsuperimposable mirror images: enantiomers

These two enantiomers are resolvable.
    For *cis*-1,2-dichlorocyclopropane, we get

Mirror

H   H   ┊   H   H
Cl   Cl   ┊   Cl   Cl

cis-1,2-Dichlorocyclopropane, *cis* (3)
Superimposable mirror images: identical structures

Even though there are two asymmetric carbon atoms in this structure, it is optically inactive because the mirror images are superimposable on the original compound; they represent the same structure. This could be predicted because *cis* (3) contains a plane of symmetry and must be achiral.

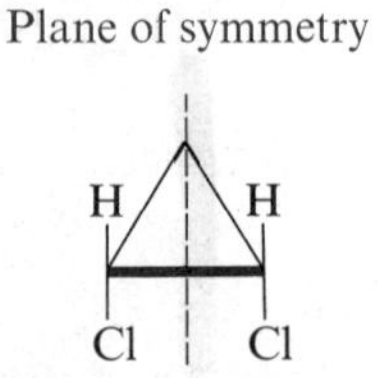

Plane of symmetry

cis-1,2-Dichlorocyclopropane, *cis* (3)
Contains plane of symmetry: meso compound

Because the *trans* isomer has two nonsuperimposable, mirror-image structures that are enantiomers, either of them is a diastereomer of the *meso*, *cis* isomer.

Similar approaches can be used to draw all the stereoisomers of each structural isomer of the cyclobutane and cyclopentane compounds. When drawing these stereoisomers, we often assume they are planar because they are nearly so. Before looking at the cyclohexane derivatives in Sec. 6.18, which present some special problems, consider a particular cyclobutane isomer, *trans*-1,3-dimethylcyclobutane, which has the following mirror-image structures:

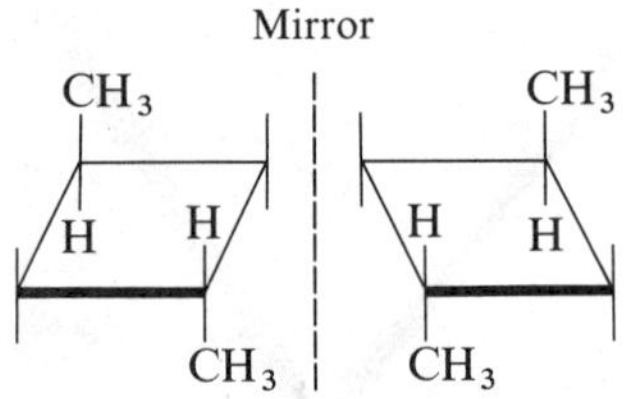

Mirror

trans-1,3-Dimethylcyclobutane
Superimposable mirror images: identical structures (achiral compound)

It is much less obvious that these are superimposable mirror images than for the previous compounds studied. By simply rotating, these two structures can be superimposed on one another. Further proof that this is an achiral compound comes from the observation that it possesses a plane of symmetry.

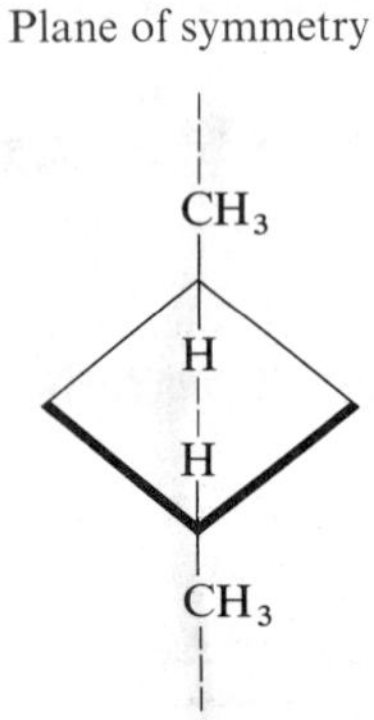

Plane of symmetry

trans-1,3-Dimethylcyclobutane
Contains plane of symmetry: achiral compound

Question 6.20

Draw all possible structural isomers for the following four compounds. For each structural isomer, draw the stereoisomers. Identify pairs of enantiomers and *meso* compounds, and tell which stereoisomers will be optically active and which will be inactive.

(*a*) dimethylcyclopropanes   (*b*) dibromocyclobutanes

(*c*) dimethylcyclopentanes   (*d*) dimethylcyclobutanes

6.18   Stereoisomerism in Cyclohexane Derivatives

We know from Secs. 3.16 and 3.17 that cyclohexane derivatives exist in numerous conformations that are in equilibrium with one another. As a consequence, we must consider not only *cis-trans* isomerism but also conformations when discussing the stereoisomers and optical properties of these molecules.

Let us consider *trans*-1,2-dichlorocyclohexane and draw its more stable chair conformation. The chlorine atoms are both equatorial. Now we draw the mirror image of this conformation.

Mirror

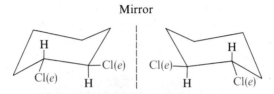

trans-1,2-Dichlorocyclohexane
Nonsuperimposable and noninterconvertible
mirror images: resolvable racemic mixture of enantiomers

These mirror images are not superimposable. They are also noninterconvertible, because if we flip either of them into the other, less stable chair conformation, the two chlorine atoms are moved from the equatorial to the axial position. *trans*-1,2-Dichlorocyclohexane should be resolvable into two enantiomers, each of which should be optically active. Additionally, some diaxial conformation is present and it also exists as a pair of nonsuperimposable, *noninterconvertible* enantiomers, although relatively little of this conformation exists at equilibrium.

Now we draw a chair conformation of *cis*-1,2-dichlorocyclohexane and construct its mirror image. These two mirror images are nonsuperimposable and thus enantiomers.

Mirror

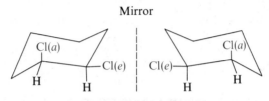

cis-1,2-Dichlorocyclohexane
Nonsuperimposable but convertible mirror
images: nonresolvable racemic mixture of enantiomers

The chlorine atoms are equatorial-axial, and on flipping either chair form, the other one is obtained; these two conformations are *interconvertible* by fllipping. As a result, they are really conformational enantiomers that are readily interconvertible. At room temperature, there is enough energy to permit equilibration between these two enantiomers. They represent a nonresolvable racemic mixture of enantiomers that cannot be separated. In a sense, this is like a *meso* compound because its structural properties render it optically inactive. This is because the enantiomers are non-resolvable.

The existence and interconvertibility of the chair conformations of cyclohexane derivatives often complicate the problem of predicting optical properties (enantiomers, diastereomers, and the like). It is strongly suggested that you construct models to help understand this discussion. Conventional, planar structures are useful in predicting the stereoisomerism in cyclohexane derivatives, as the following examples illustrate:

Mirror

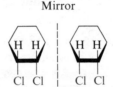

cis-1,2-Dichlorocyclohexane
Superimposable mirror images: identical structures

Mirror

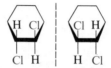

trans-1,2-Dichlorocyclohexane
Nonsuperimposable mirror images: enantiomers

Question 6.21

Draw the more stable chair conformation for each of the following four compounds, and then draw its mirror image. Which compound can be resolved into a pair of enantiomers, and which is incapable of resolution? Which, if any, contains a plane of symmetry?

(*a*) *cis*-1,3-dibromocyclohexane   (*b*) *trans*-1,3-dibromocyclohexane
(*c*) *cis*-1,4-dimethylcyclohexane   (*d*) *trans*-1,4-dimethylcyclohexane

Question 6.22

Draw the mirror images of the more stable chair conformations of the following six compounds. Indicate *meso* compounds and pairs of enantiomers. Are any of these compounds *meso*-like, that is, do they exist as a nonresolvable racemic mixture?

(*a*) *cis*-2-chlorobromocyclohexane   (*b*) *trans*-2-chlorobromocyclohexane
(*c*) *cis*-3-chlorobromocyclohexane   (*d*) *trans*-3-chlorobromocyclohexane
(*e*) *cis*-4-chlorobromocyclohexane   (*f*) *trans*-4-chlorobromocyclohexane

6.19   Other Elements That Impart Asymmetry to Molecules

The organic chemist is mostly interested in carbon-containing compounds, and we devoted most of our attention thus far to asymmetry imparted to molecules by the tetrahedral carbon atom. There are several other common elements that impart asymmetry to molecules, however, and we survey them briefly here.

A. Nitrogen

Amines are the bases of organic chemistry, and the simplest amine is ammonia, NH_3 (see Chap. 26). Replacing the hydrogens in ammonia with carbon-containing groups (that is, $R' \neq R'' \neq R'''$) gives a molecule that appears to be asymmetric because of the pyramidal structure of nitrogen. However, such molecules have never been resolved because of the rapid interconversions (millions per second) between the two enantiomeric forms.

Mirror

$$R'' \text{---} N \quad \mid \quad N \text{---} R''$$

Nonsuperimposable mirror images of a tertiary nitrogen compound: nonresolvable enantiomers

However, if the carbon-containing groups attached to nitrogen are held rigidly in place, for example, when nitrogen is in a three-membered cyclic structure or is at a bridgehead position, then such an arrangement possesses asymmetry and the enantiomers are resolvable. For example,

Asymmetric nitrogen

Tröger's base

2-Methyl-*N*-bromoaziridine

Asymmetric nitrogen

Asymmetric nitrogen

6,6-Dimethyl-2-quinuclidone

The stereochemical integrity of the aziridines and other cyclic compounds containing nitrogen as part of the ring (nitrogen heterocyclics, see Sec. 26.5, is discussed further in Sec. 26.6).

Quaternary ammonium salts, on the other hand, have been resolved because of the tetrahedral structure of the substituted ammonium ion; that is, the following compound can be thought of as a derivative of the ammonium ion, $NH_4^{\oplus}$, where all four hydrogens have been replaced by four nonidentical carbon-containing groups (see Sec. 26.3).

Mirror

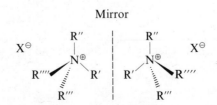

Nonsuperimposable mirror images of a quaternary nitrogen compound: enantiomers

B. Phosphorus

Unlike nitrogen, phosphines and other phosphorus compounds are configurationally stable (do not undergo pyramidal inversion at phosphorus) and are isolable in optically active forms.

$$R—\overset{R''}{\underset{..}{\overset{|}{P^*}}}—R' \equiv \quad R^{''''}\overset{P}{\underset{R''}{\diagup}}R' \qquad R—\overset{R''}{\underset{O}{\overset{|}{P^*}}}—R'$$

Phosphines   Phosphine oxides
(pyramidal)   (tetrahedral)

C. Sulfur

Sulfonium salts (1) and sulfoxides (2) have been prepared and resolved into their enantiomeric forms even though they both contain an unshared pair of electrons. In these compounds, the sulfur atom maintains a rather rigid geometry and does not interconvert between the two enantiomeric forms.

$$R''—\overset{R'}{\underset{\overset{\ominus}{X}\quad R}{\overset{\oplus\,|}{S^*}}}: \qquad O{=}\overset{R'}{\underset{R}{\overset{|}{S^*}}}:$$

(1)   (2)
Sulfonium salt   Sulfoxide
(pyramidal)   (pyramidal)

D. Silicon and Germanium

A limited number of silicon- and germanium-containing compounds have been prepared and resolved into their enantiomers. Because both elements are in the same column of the periodic table as carbon, however, it is reasonable that they exhibit the tetrahedral structure and thus exist as enantiomers.

$$R'''{—}\overset{R''}{\underset{R}{\overset{|}{\overset{*}{Si}}}}{—}R' \qquad R'''{—}\overset{R''}{\underset{R}{\overset{|}{\overset{*}{Ge}}}}{—}R'$$

Asymmetric silicon   Asymmetric germanium
compound   compound
(tetrahedral)   (tetrahedral)

E. Isotopes

Enantiomers have even been synthesized and resolved that owe their asymmetry solely to different isotopes.

$$\bigcirc{-}^{12}CH_{2*}{-}\overset{O}{\underset{..}{\overset{||}{S}}}{-}^{13}CH_2{-}\bigcirc \qquad CH_3{-}\bigcirc{-}\overset{^{16}O}{\underset{^{18}O}{\overset{||}{\underset{||}{S^*}}}}{-}CH_3 \qquad CH_3CH_2{-}\overset{H}{\underset{D}{\overset{|}{\overset{*}{C}}}}{-}CH_3$$

Question 6.23

Draw the two enantiomers and assign the absolute configuration of each for the compounds listed in Sec. 6.19.

Study Questions

6.24 Define each of the following terms, or indicate what is meant by them.

(*a*) optical activity
(*b*) plane-polarized light
(*c*) polarimeter
(*d*) observed rotation
(*e*) specific rotation
(*f*) three-dimensional projection formula
(*g*) Fischer projection formula
(*h*) asymmetric molecule
(*i*) asymmetric carbon atom
(*j*) plane of symmetry
(*k*) mirror-image molecules
(*l*) superimposable mirror images
(*m*) nonsuperimposable mirror images
(*n*) enantiomers
(*o*) diastereomers
(*p*) *meso* compound
(*q*) *d* and *l*
(*r*) + and −
(*s*) *R* and *S*
(*t*) racemic mixture
(*u*) retention of configuration
(*v*) inversion of configuration
(*w*) racemization
(*x*) chiral (chirality, achiral)
(*y*) dissymmetric
(*z*) analogous asymmetric carbon atoms
(*a′*) different asymmetric carbon atoms
(*b′*) dextrorotatory and levorotatory
(*c′*) symmetry (C_s, C_i, C_n)

6.25 Compare the physical and optical properties of enantiomers and diastereomers. Given a mixture of enantiomers, can they be separated from one another by physical means (distillation, crystallization, or chromatography)?

6.26 Compare the physical and optical properties of enantiomers and conformers. Can either or both of these two types of stereoisomers be separated from one another? Why or why not?

6.27 For *d*-tartaric acid, $[\alpha]_D^{20°} = +12.0°$ in water. Suppose we have a solution that has 2 g/100 ml of this compound in water at 20° and is in a 2-dm polarimeter tube.

(*a*) What would α_{obs} be for this sample?

(*b*) What would $[\alpha]_D^{20°}$ be for this sample?

6.28 The specific rotation of (+)-2-butanol is $[\alpha]_D^{25°} = +5.76°$. Calculate the specific rotation expected for a mixture containing 80% (+)-2-butanol and 20% (−)-2-butanol.

6.29 Describe conclusive chemical evidence used to show that the saturated carbon atom in alkanes and other compounds is not flat, with 90° bond angles between the molecular bond orbitals.

6.30 Consider the following five pairs of stereoisomers. For each pair, state whether they are identical structures, enantiomers, or diastereomers. Which structures are optically active? Which are optically inactive because they are *meso* compounds?

6.31 1,2-Cyclopropanedicarboxylic acid (1) exists in two isomeric forms. One form, compound *A*, has a melting point of 130°, and the other form, compound *B*, melts at 175°. Compound *B* can be resolved into two optically active forms, both of which are optically active. Compound *A* is nonresolvable.

(a) Draw the structures of compounds *A* and *B*.

(b) State as much as you can about the optical activity of the forms derived from *B*.

(c) What might you deduce about the optical properties of *A*?

$$HOOC \qquad COOH$$

(1)

6.32 Do you think that the compound CH_3—CH—CD_3 could exist in enantiomeric forms?
$$\underset{Br}{|}$$
Why or why not?

6.33 Draw Fischer projection formulas for all the stereoisomers of the following compounds. Which are optically active, assuming that they are separated from all other stereoisomers? Which are *meso* compounds? In each part, label pairs of enantiomers.

(a) CH_3—CH—CH—CH_3
$$\underset{Cl}{|} \quad \underset{Br}{|}$$

(b) CH_3—CH—CH—CH_2CH_3
$$\underset{Br}{|} \quad \underset{Br}{|}$$

(c) CH_3CH_2—CH—CH—CH_2CH_3
$$\underset{CH_3}{|} \quad \underset{CH_3}{|}$$

(d) $HOCH_2$—CH—CH—CH—CH_2OH
$$\underset{OH}{|} \quad \underset{OH}{|} \quad \underset{OH}{|}$$

(e) CH_3—CH—CH_2—CH_2—CH—CH_3
$$\underset{Cl}{|} \qquad \qquad \underset{Cl}{|}$$

(f) CH_3—CH—CH—CH—CH_2Br
$$\underset{Br}{|} \quad \underset{Br}{|} \quad \underset{Br}{|}$$

6.34 For compounds **(a)**, **(b)**, **(c)**, and **(e)** in Question 6.33, indicate the configurations using the *R* and *S* convention.

6.35 Draw an enantiomer (if any exists) of each of the following four structures and, where possible, draw at least one diastereomer. Which, if any, of the structures would rotate polarized light?

(a)
$$\begin{array}{c} CH_2Cl \\ D{-\!\!\!|\!\!\!-}Cl \\ CH_3 \end{array}$$

(b)
$$\begin{array}{c} CH_3 \\ Cl{-\!\!\!|\!\!\!-}H \\ H{-\!\!\!|\!\!\!-}Cl \\ CH_3 \end{array}$$

(c)
$$OH \qquad OH$$

(d)
$$\begin{array}{c} CH_3 \\ H{-\!\!\!|\!\!\!-}H \\ Br{-\!\!\!|\!\!\!-}H \\ CH_3 \end{array}$$

6.36 Using planar rings, draw the possible geometric (*cis-trans*) isomers of the following four cyclic compounds. By inspection, deduce which of them are *meso* compounds and which exist enantiomeric form. Draw the enantiomers where possible.

(a)
$$CH_3 \qquad OH$$

(b)
$$\begin{array}{c} OH \\ CH_3 \qquad CH_3 \end{array}$$

(c)
$$\begin{array}{c} OH \\ CH_3 \qquad CH_2CH_3 \end{array}$$

(d)
$$CH_3 \qquad CH_2CH_3$$

6.37 Draw three-dimensional projection formulas for the following compounds:

(a) R configuration of CHClBrI

(b) S configuration of
$$CH_3CH_2CH_2-\underset{\underset{Cl}{|}}{\overset{\overset{CH_3}{|}}{C}}-CH_2CH_3$$

(c) (3R,4S)-3,4-dichlorohexane

6.38 Provide complete configurational names, including the R and S designation, for the following compounds.

(a)
$$D\!-\!\!\!\underset{\underset{Cl}{\vdots}}{\overset{\overset{CH_3}{\vdots}}{C}}\!\!\!-\!H$$

(b)
$$H\!-\!\!\!\underset{\underset{CH_2CH_3}{\vdots}}{\overset{\overset{CH_3}{\vdots}}{C}}\!\!\!-\!Cl$$

6.39 In the following two reactions, suppose the stereoisomers indicated are used as reactants. Are they optically active or optically inactive? Are the products optically active or optically inactive? Why?

(a)
$$\underset{\underset{C_2H_5}{}}{\overset{\overset{H}{|}}{\underset{CH_3}{\diagdown}}\!\!\!\overset{}{C}\!\!\!\diagdown CH_2\!-\!Cl}} \xrightarrow[heat]{Na}$$ Draw the three-dimensional structure of this product, which has the formula $C_{10}H_{22}$. (*Hint:* See Sec. 4.2.)

(b)
```
   H—C=O                    COOH
H ──┼── OH       HNO₃    H ──┼── OH
H ──┼── OH      ──────►  H ──┼── OH
H ──┼── OH               H ──┼── OH
   CH₂OH                    COOH
```

6.40 What is the lowest molecular weight alkane compound that has an asymmetric carbon (not due to isotopes) atom? Draw three-dimensional and Fischer projection formulas for the optical isomers of the compound(s).

6.41 Optically active 1,2-dibromo-1-chloroethane (1), shown in Fischer projection form, is heated with chlorine gas and all products with the formula $C_2H_2Br_2Cl_2$ are isolated from the reaction.

```
        CH₂Br
   H ────┼──── Cl
        Br
```
(1)

(a) Draw the structure of a product that is optically inactive because it does not contain an asymmetric center.

(b) Draw the Fischer projection structure of a product that is optically inactive because it is a *meso* compound.

(c) Draw the Fischer projection structure of a product that is optically active, and state why it is optically active.

6.42 Two unlabeled alkanes, A and B, both have the formula C_4H_{10}. To help identify them, some chlorination studies are carried out. Compound A gives two monochloro derivatives, C and D, each with the formula C_4H_9Cl. On further treatment with chlorine, C gives four dichloro products and D gives four dichloro products. The dichloro products have the formula $C_4H_8Cl_2$. Independently, it is found that compound D can be prepared in *optically active* form. Compound B gives two monochloro derivatives, E and F. On further treatment with chlorine, E gives only one dichloro compound, whereas F gives three dichloro compounds.

On the basis of this information, draw the structures of compounds A to F. Also, draw the structures of all the dichloro compounds that arise from C to F. *Give your reasoning!* Indicate what the observations teach you about the structures of the lettered compounds.

6.43 The following reaction is carried out starting with *optically active* compound (1).

Is product (2) optically active or optically inactive? Briefly explain your answer by drawing suitable *three-dimensional* structures of reactant, intermediate(s), and product.

6.44 Which of the following are chiral? Achiral? Explain. Build models to help answer this question.

(*a*) $CH_3CH{=}C{=}CHCH_3$ (*b*) $CH_3CH_2CH{=}C{=}CH_2$

(*c*)

(*d*)

(*e*)

7

Alkenes and Alkadienes: Structure and Properties

In our discussions of alkanes and cycloalkanes we mentioned the **carbon-carbon double bond,** which characterizes another important functional group in the hydrocarbon family. Compounds that contain this functional group are called **alkenes.** They are also known as *olefins*, which is derived from an old name for ethylene (an alkene, C_2H_4), namely, *olefiant gas* (oil-forming gas). The general formula for alkenes is C_nH_{2n}, which means that each member has two fewer hydrogen atoms than does the corresponding alkane (formula C_nH_{2n+2}). Because alkenes do not contain the maximum number of hydrogen atoms that their carbon number allows, they are called **unsaturated hydrocarbons,** in contrast to the alkanes, which are called **saturated hydrocarbons.** The cycloalkanes also have a molecular formula of C_nH_{2n}. Although alkenes and cycloalkanes have the same general formula, they can be distinguished by their chemical reactivity and other physical properties; several of these differences are presented in this chapter and the next.

Compounds that contain two carbon-carbon double bonds are called **alkadienes.** Because their properties, structure, and reactivity are similar to those of alkenes, we discuss alkenes and alkadienes together. This chapter has two major subdivisions: (1) structure, properties, and nomenclature and (2) methods of preparation. The reactions of alkenes and alkadienes are discussed in Chap. 8.

Structure, Properties, and Nomenclature for Alkenes and Alkadienes

7.1 Ethene; The Carbon-Carbon Double Bond

The simplest compound that contains the carbon-carbon double bond is ethene, which was introduced in Sec. 2.7, where a detailed view of the carbon-carbon double bond was presented. These features are summarized in Fig. 7.1.

Our view of the electronic structure of the carbon-carbon double bond suggests not only that the π bond is weaker than the σ bond but also that the electrons in it are much more exposed than those in a σ bond. In many ways, the π bond is analogous to an unshared pair of electrons, which accounts for many reactions of this functional group. There is also restricted rotation about the carbon-carbon double bond because of the π bond (see Sec. 7.3), whereas free rotation is possible in open-chain alkanes.

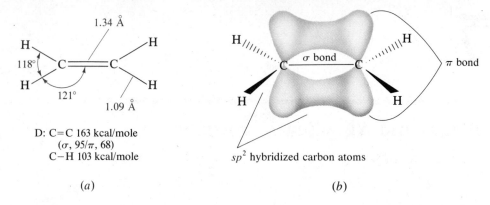

D: C=C 163 kcal/mole
   (σ, 95/π, 68)
   C—H 103 kcal/mole

sp^2 hybridized carbon atoms

(a)   (b)

FIGURE 7.1   Ethene and the carbon-carbon double bond. (a) Structural formula; (b) σ and π bonds.

7.2   Propene

Two compounds have the molecular formula C_3H_6: cyclopropane and **propene.** Propene is a member of the alkene family and may be represented by

or   or CH_3—CH=CH_2

Propene

Propene (C_3H_6) can be distinguished from cyclopropane (also C_3H_6) on the basis of chemical reactivity (see Chap. 8). The double bond in propene has the same electronic structure as in ethene; it is composed of an sp^2-sp^2 σ bond and a π bond. The carbon-carbon single bond is an sp^2-sp^3 σ bond.

7.3   The Butenes; Geometric Isomerism

The next members of the alkene family are the **butenes, C_4H_8.** The structures of the possible isomers that have this formula and contain a carbon-carbon double bond are the following:

CH_2=CH—CH_2—CH_3 CH_3—CH=CH—CH_3 $\begin{matrix} CH_3 \\ \diagdown \\ \diagup \\ CH_3 \end{matrix} C$=$CH_2$

1-Butene   2-Butene   Methylpropene

Two of these structures (1-butene and 2-butene) have a four-carbon continuous chain and differ only in the position of the double bond, whereas the third (methylpropene) has a branched chain. (For now do not worry about the nomenclature of these compounds; it is discussed in Sec. 7.5.)

When butenes are studied in the laboratory, four isomeric compounds can be isolated. They have boiling points of approximately $-7°$, $-6°$, $+1°$, and $+4°$. On catalytic hydrogenation (the addition of the elements of hydrogen to a carbon-carbon double bond, see Sec. 8.12), three of the compounds yield butane and the fourth yields isobutane (methylpropane). Further studies indicate that the compounds with boiling points of $+1°$ and $+4°$ have the double bond in the 2 position; that is, they are 2-butenes, CH_3—CH=CH—CH_3. How can we explain existence of four alkenes having the formula C_4H_8, or more specifically, the existence of *two* 2-butenes?

The planar structure for ethene (see Sec. 7.1) requires that all its atoms lie in the same plane. If molecular models are made for 2-butene, two distinct three-dimensional structures can be constructed, as shown in Fig. 7.2. In one structure the two methyl groups are on the same side of the molecule, and in the other they are on opposite sides. These two structures are distinctly different. They represent another example of **geometric isomerism,** which also exists in cyclic compounds (see Sec. 3.17). The isomer with the two methyl groups on the same side of the double bond is the *cis* isomer, and the one with the methyl groups opposite one another (that is, across the double bond) is the *trans* isomer. To designate the three-dimensional geometry of these isomers, the prefixes *cis* and *trans* are added to the names; thus, (1) is *cis*-2-butene and (2) is *trans*-2-butene. Physical methods are commonly used to distinguish between *cis* and *trans* isomers (see Sec. 7.7). It has been found that *cis*-2-butene is the isomer with a boiling point of $+4°$, and *trans*-2-butene boils at $+1°$.

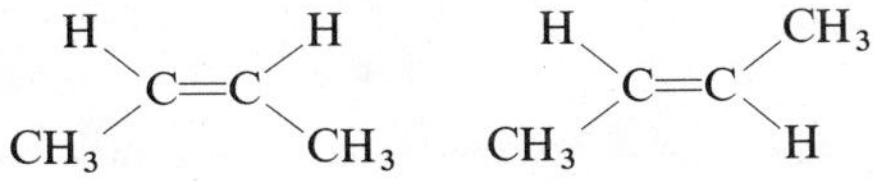

(1)	(2)
cis-2-Butene	*trans*-2-Butene
(bp $+4°$)	(bp $+1°$)

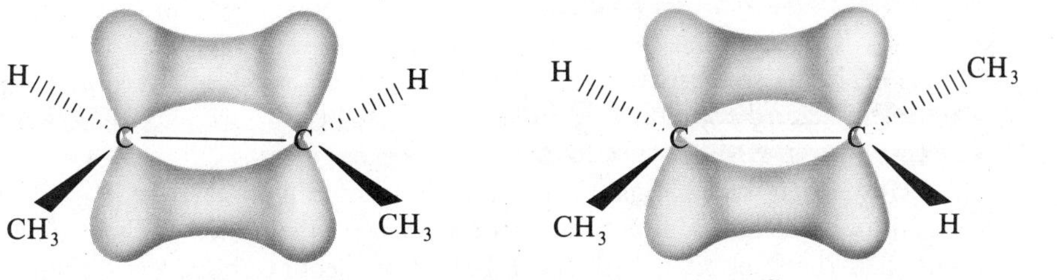

(a)

Geometric isomers

(b)

FIGURE 7.2 Structures of the isomeric 2-butenes (a) using three-dimensional drawing and (b) showing suitable planar representations. In (1) the methyl groups are on the same side of the double bond, and in (2) the methyl groups are on opposite sides. (The infinite number of possible arrangements caused by free rotation about the carbon-carbon single bonds has been ignored.)

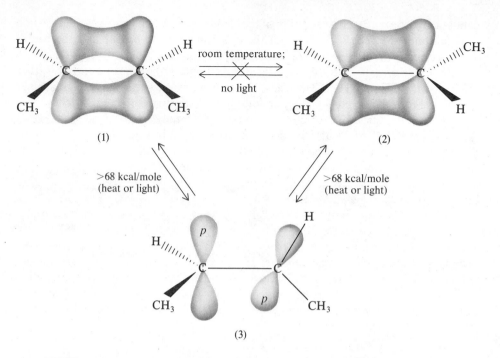

FIGURE 7.3   Interconversion of the two isomeric 2-butenes. Rotation about the carbon-carbon double bond prevents overlap of the p orbitals and thus breaks the π bond, as shown by (3). Once (3) is formed, it can re-form (1) and (2) to give a mixture of these two isomeric butenes.

The *cis* and *trans* isomers of alkenes are not interconvertible under ordinary conditions because of the π bond, which has a bond strength of 68 kcal/mole (284.5 kJ/mole). This amount of energy is available only at high temperature (or with ultraviolet light), so these two isomers exist as stable compounds at room temperature. There is **restricted rotation** about the carbon-carbon double bond, which accounts for the geometric isomers (often called *cis-trans* isomers).

How, then, can geometric isomers be interconverted? If energy in excess of 68 kcal/mole (284.5 kJ/mole) the π bond energy, is supplied by heat or ultraviolet light, then the π bond is broken and rotation occurs about the carbon-carbon σ bond. The p orbitals that make up the double bond are moved to a point where they are perpendicular to one another so that overlap can no longer occur. This process is shown in Fig. 7.3. Regardless of which isomer we start with, rotation is possible once the π bond is broken, and a mixture of *cis* and *trans* isomers is formed.

7.4   Criteria for Existence of Geometric Isomers in Alkenes

Several general features indicate whether or not an alkene can exist as a pair of geometric isomers. There is hindered rotation about any carbon-carbon double bond. However, *geometric isomers exist only if both the carbon atoms in the carbon-carbon double bond contain two different groups.* The following structures illustrate this, where *a*, *b*, *c*, and *d* represent different groups.

More specifically, the following simple alkenes do *not* exist as geometric isomers. The identical groups on one carbon atom are shown in red.

$$a \diagdown C=C \diagup b \qquad a \diagdown C=C \diagup c \qquad a \diagdown C=C \diagup c \qquad b \diagdown C=C \diagup c$$
$$b \diagup \quad \diagdown a \qquad b \diagup \quad \diagdown d \qquad a \diagup \quad \diagdown d \qquad b \diagup \quad \diagdown c$$

No geometric isomerism

$$b \diagdown C=C \diagup b \qquad a \diagdown C=C \diagup d$$
$$a \diagup \quad \diagdown a \qquad b \diagup \quad \diagdown c$$

Geometric   Geometric
isomers   isomers

$$H \diagdown C=C \diagup H \qquad H \diagdown C=C \diagup H \qquad H \diagdown C=C \diagup CH_3$$
$$H \diagup \quad \diagdown CH_3 \qquad H \diagup \quad \diagdown CH_2CH_3 \qquad H \diagup \quad \diagdown CH_3$$

Propene   1-Butene   Methylpropene

Remember the terms *structural*, *conformational*, and *stereoisomers*. Geometric isomers are part of that latter class.

From all compounds that have the formula C_4H_8, there are five stable structural isomers: 1-butene, 2-butene, methylpropene, cyclobutane, and methylcyclopropane. The first three have double bonds and the latter two are cyclic compounds, but all fit the general formula C_nH_{2n}. In addition, for 2-butene there are two geometric isomers: *cis*-2-butene and *trans*-2-butene, which can be isolated.

Question 7.1

Which of the following four compounds, if any, can exist as geometric isomers?

(*a*) $(CH_3)_2C=CHCH_3$ (*b*) $CH_3CH_2C=CHCH_3$
                                                          $|$
                                                        CH_3

(*c*) $CH_3CH_2C=CHCH_2CH_3$ (*d*) $CH_3CH_2CH=CHCH_2CH_3$
              $|$
            CH_3

For those that exist as geometric isomers, draw their structures to show accurately the bond angles about the carbon-carbon double bond.

7.5   Nomenclature of Alkenes

A. Common Names

Simple alkenes have carried some common names with them through the years. *Ethylene* is a common name for $CH_2=CH_2$. *Propylene* and *butylene* are common names for the three- and four-carbon alkenes, respectively. Sometimes the isomers with a given carbon content are referred to collectively, such as the *amylenes* (pentylenes, five-carbon molecules) and *hexylenes* (six-carbon molecules).

Substituted alkenes can be named as derivatives of a substituted ethylene. For example,

$$CH_3 \diagdown C=C \diagup CH_3 \qquad F \diagdown C=C \diagup F \qquad CH_2=CHCl$$
$$CH_3 \diagup \quad \diagdown CH_3 \qquad F \diagup \quad \diagdown F$$

Tetramethylethylene   Tetrafluoroethylene   Chloroethylene
                           (used in manufacture   or
                                of Teflon)   vinyl chloride

B. Systematic (IUPAC) Nomenclature

The rules for naming alkenes resemble those for alkanes (see Sec. 3.10), but with the following modifications:

1. Select the longest continuous carbon chain *that contains the double bond*, and use this as the parent compound.
2. Number this longest continuous chain so that the double bond has the lowest possible number, regardless of what the number(s) of the other substituent(s) are.
3. Take the alkane name corresponding to the longest continuous carbon chain and change the -*ane* ending to **-ene.** To indicate the position of the carbon-carbon double bond, add the number of the lower numbered carbon atom of the carbon-carbon double bond in front of the alkene name.
4. Side chains and substituents are named as usual, and their position is indicated by number to show to which carbon atom they are attached.

The following examples illustrate the naming of open-chain alkenes:

$$CH_2{=}CH_2 \qquad CH_2{=}CH{-}CH_3 \qquad \overset{1}{C}H_2{=}\overset{2}{C}H{-}\overset{3}{C}H_2{-}\overset{4}{C}H_3 \qquad \overset{1}{C}H_3{-}\overset{2}{C}H{=}\overset{3}{C}H{-}\overset{4}{C}H_3$$

Ethene   Propene   1-Butene   2-Butene
                                    (*not* 3-butene)   (*cis* and *trans*)

$$\overset{3}{C}H_3{-}\overset{2}{C}H{=}\overset{1}{C}H{-}Cl \qquad\qquad \overset{5}{C}H_3{-}\overset{4}{C}H{-}\overset{3}{\underset{\underset{CH_3}{|}}{\overset{\overset{CH_3}{|}}{C}}}H{-}\overset{2}{C}H{=}\overset{1}{C}H_2$$

1-Chloropropene   3,4-Dimethyl-1-pentene
(*cis* and *trans*)
(*not* 3-chloro-2-propene)

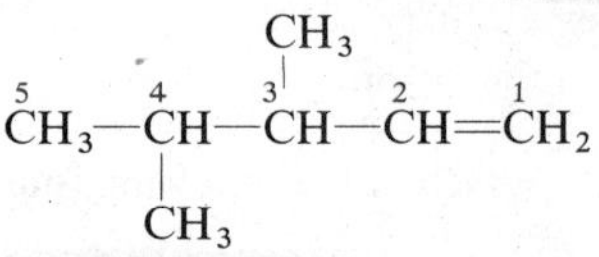

$$\overset{2}{C}H{-}\overset{1}{C}H_3$$
$$\overset{3}{\|}$$
$$CH$$
$$CH_3{-}CH_2{-}CH_2{-}\overset{4}{|}\overset{4}{C}H{-}\overset{5}{C}H_2{-}\overset{6}{C}H_2{-}\overset{7}{C}H_2{-}\overset{8}{C}H_2{-}\overset{9}{C}H_3$$

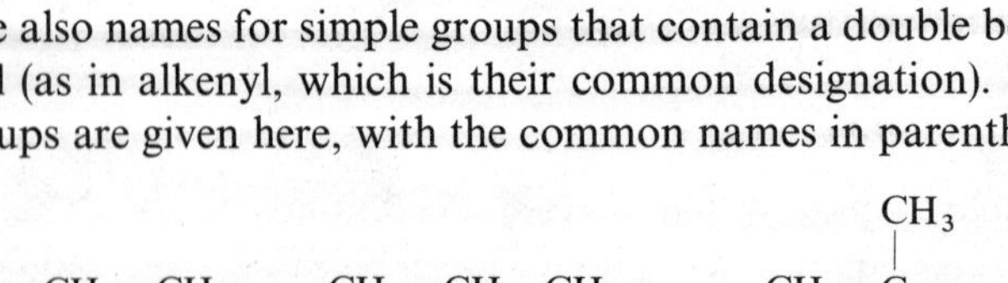

4-Propyl-2-nonene
(*cis* and *trans*)

There are also names for simple groups that contain a double bond; they have the suffix **-enyl** (as in alkenyl, which is their common designation). The names for three such groups are given here, with the common names in parentheses:

$$CH_2{=}CH{-} \qquad CH_2{=}CH{-}CH_2{-} \qquad CH_2{=}\underset{\underset{CH_3}{|}}{C}{-}$$

Ethenyl   2-Propenyl   1-Methylethenyl
(Vinyl)   (Allyl)   (Isopropenyl)

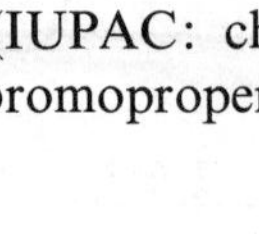

Thus, $CH_2{=}CH{-}Cl$ can be called *vinyl chloride* (IUPAC: chloroethene) and $CH_2{=}CH{-}CH_2{-}Br$ is *allyl bromide* (IUPAC: 3-bromopropene). Vinyl cyclopropane (IUPAC: ethenylcyclopropane) is

$$CH_2{=}CH{-}\triangleleft$$

In the IUPAC system the alkenyl groups are numbered starting with the carbon atom that bonds the group to another substituent. Two examples of alkenyl groups are:

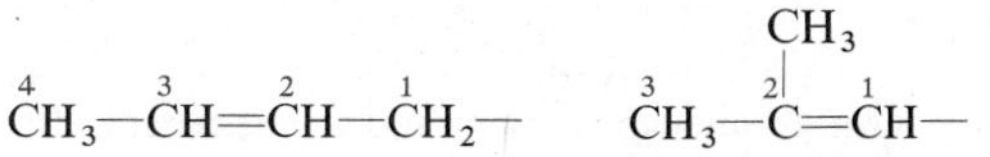

$$\underset{4}{CH_3}-\underset{3}{CH}=\underset{2}{CH}-\underset{1}{CH_2}-$$

$$\underset{3}{CH_3}-\overset{\overset{CH_3}{|}}{\underset{2}{C}}=\underset{1}{CH}-$$

(*cis* or *trans*)-2-Butenyl   2-Methyl-1-propenyl

The convention for naming *cis* and *trans* isomers was introduced briefly in Sec. 7.3 with relatively simple alkenes. Those conventions are extended as illustrated here.

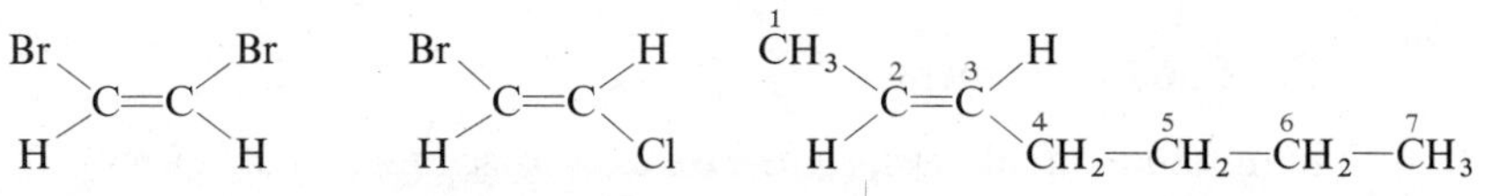

cis-1,2-Dibromoethene   *trans*-1-Bromo-2-chloroethene   *trans*-2-Heptene

When the alkene is more complex, it is often difficult to determine the correct convention to apply. Generally the geometric configuration is that of the *longest continuous chain of carbon atoms that contains the carbon-carbon double bond*. Thus, in compound (1) two ethyl groups are *cis* to one another, and a propyl and an isopropyl group are *cis* to one another. Also an ethyl and a propyl group are *trans* to one another, and an isopropyl and an ethyl group are *trans* to one another. Without some suitable convention, placement of the prefix *cis* or *trans* in the complete name has no significance in terms of the geometry about the double bond. The compound is properly named by using the longest continuous chain, which is numbered and shown in red, and the geometry of this chain is *cis*. The substituents are added by indicating the carbon atoms to which they are attached.

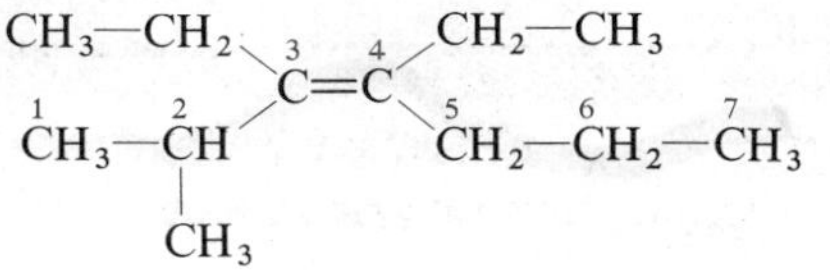

(1)
3,4-Diethyl-2-methyl-*cis*-3-heptene

Question 7.2

Draw the structural isomers and, where applicable, the structures of the geometric isomers for the following four compounds. Provide the IUPAC name for each. (Ignore the possibility of enantiomers.)

(*a*) dichloroethylenes, $C_2H_2Cl_2$ (*b*) pentenes
(*c*) bromopropenes, C_3H_5Br (*d*) dibromopropylenes, $C_3H_4Br_2$

Question 7.3

The IUPAC name for isobutylene, $(CH_3)_2C=CH_2$, is *methylpropene*, where the positions of the methyl group and the double bond are omitted. Why is methylpropene a satisfactory name?

Question 7.4

Draw the geometric isomers for the following compound. (*Hint:* There are four of them.) Ignore enantiomers.

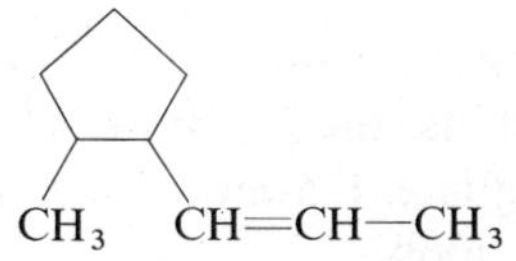

$$CH_3 \qquad CH=CH-CH_3$$

Question 7.5

Considering both geometric isomers and enantiomers, draw the stereoisomers expected for this compound:

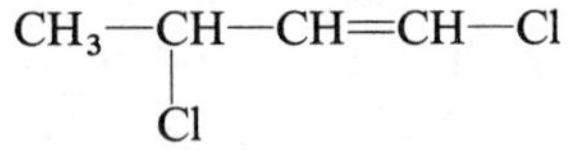

$$CH_3-CH-CH=CH-Cl$$
$$\mid$$
$$Cl$$

C. *Z* and *E* Isomers

In the study of alkenes, some instances arise in which it is very difficult to name a compound unambiguously. Would you, for example, name the following compound *cis*- or *trans*-2-bromo-1-chloro-1-iodopropene? Rules must be set up for naming all

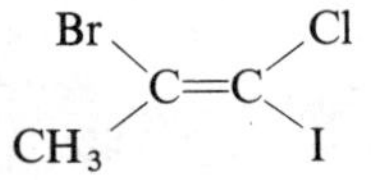

organic compounds, and the *cis* and *trans* system is much too ambiguous for certain alkenes. Another system has been designed using the symbols *Z* (German *zusammen*, together) and *E* (German *entgegen*, opposed to). The system has the following rules:

1. Compare the two groups on one carbon of the carbon-carbon double bond.
2. Assign the two groups priorities using the Cahn-Ingold-Prelog rules for *R* and *S* configuration in Sec. 6.8.
3. Repeat steps 1 and 2 for the second carbon of the carbon-carbon double bond.
4. If the two groups of highest priority are on the same side of the double bond, we have the *Z* isomer. If they are on opposite sides, we have the *E* isomer. (As an aid to distinguishing *Z* and *E*, remember that the word *zusammen* contains the elements of the word *same*, thus referring to the substituents being on the same side.)

The following three examples help clarify the use of these rules. The group of highest priority on each carbon is in red.

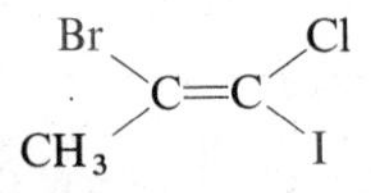

E-2-Bromo-1-chloro-1-iodopropene 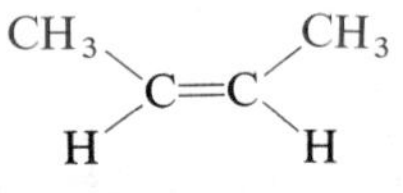   *Z*-2-Butene 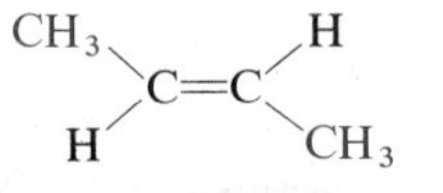   *E*-2-Butene

Question 7.6

A chemist prepared the oxime of 2-hexanone but found that two isomers formed, both of which were oximes. Explain. (*Hint:* The nitrogen atom in an oxime is sp^2 hybridized.)

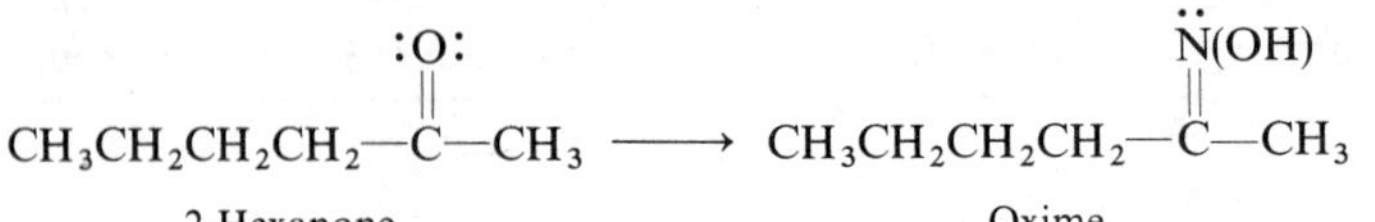

2-Hexanone   Oxime
two isomers

As with all the common names, the *cis-trans* system has been around for so long and is so widely used that the Z/E system is used primarily in examples in which the older system is inadequate. However, we must be aware of both systems and we use them throughout the book.

7.6   Cycloalkenes and Exocyclic Double Bonds: Nomenclature

Cyclic compounds that contain a double bond in the ring are called **cycloalkenes** and the double bond is called an **endocyclic double bond.** This type of compound is named systematically (IUPAC) in a method analogous to that for open-chain alkenes, except that *the numbering always starts at one carbon atom of the double bond.* It continues around the ring through the double bond in such a way as to keep the substituent numbers as small as possible. In the following examples, note that the position of the carbon-carbon double bond does not appear in the name because by convention it is always at C-1.

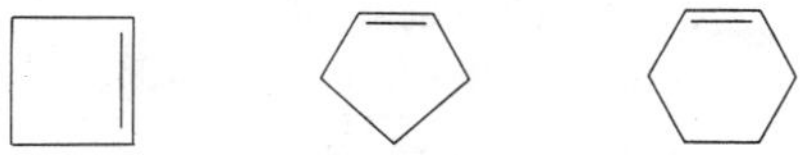

Cyclobutene   Cyclopentene   Cyclohexene

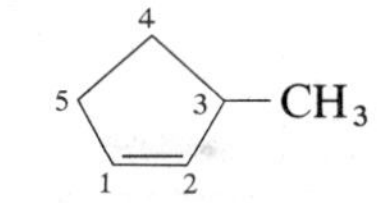

3-Methylcyclopentene
(*not* 5-methylcyclopentene)

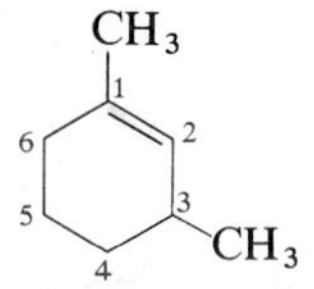

1,3-Dimethylcyclohexene
(*not* 1,5-dimethylcyclohexene)

Note that the geometry of the double bond in the smaller cycloalkenes must always be *cis.* Build models to prove this to yourself.

Some cyclic compounds contain a carbon-carbon double bond that is **exocyclic**—that is, it involves one of the ring carbons but is not contained in the ring. When this is the case, the $=CH_2$ group is called the **methylene group** (generally speaking, the elements of $-CH_2-$ are called *methylene*). For example,

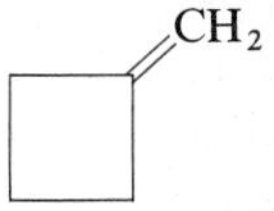

Methylenecyclobutane 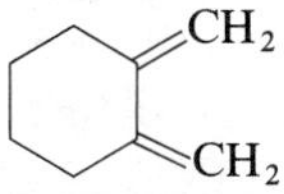   1,2-Dimethylenecyclohexane

Question 7.7

Draw structures for and name the structural and geometric (if any) isomers for the following compounds; neglect enantiomers:

(*a*) all cycloalkenes with the formula C_5H_8
(*b*) all dimethylcyclohexenes that contain at least one methyl group on the carbon-carbon double bond.

Question 7.8

Build molecular models of the cycloalkenes containing three, four, five, six, seven, and eight carbon atoms. Are both *cis* and *trans* isomers possible for any of these molecules? Why or why not?

TABLE 7.1   Physical Properties of Selected Alkenes*

Name	Structure	mp, °C	bp, °C	Density (g/cc) at 20°
Ethene	$CH_2\!\!=\!\!CH_2$	−169.4	−102.4	0.610
Propene	$CH_2\!\!=\!\!CH\!\!-\!\!CH_3$	−185.0	−47.7	0.610
1-Butene	$CH_2\!\!=\!\!CHCH_2CH_3$		−6.5	0.643
Methylpropene	$CH_2\!\!=\!\!C(CH_3)_2$	−140.7	−6.6	0.627
1-Pentene	$CH_2\!\!=\!\!CH(CH_2)_2CH_3$		30.1	0.643
2-Methyl-1-butene	$CH_2\!\!=\!\!C\diagdown_{CH_2CH_3}^{CH_3}$		31.0	0.650
3-Methyl-1-butene	$CH_2\!\!=\!\!CHCH(CH_3)_2$		20.1	0.634
1-Hexene	$CH_2\!\!=\!\!CH(CH_2)_3CH_3$	−138.0	64.0	0.675
1-Heptene	$CH_2\!\!=\!\!CH(CH_2)_4CH_3$	−119.0	93.0	0.698
1-Octene	$CH_2\!\!=\!\!CH(CH_2)_5CH_3$		123.0	0.716

* Properties of the *cis* and *trans* isomers of 2-butene and 2-pentene are in Table 7.2.

7.7   Properties of Alkenes

Alkene hydrocarbons have many physical properties similar to those of alkanes. They are water-insoluble but quite soluble in nonpolar organic solvents such as benzene, alkanes, chloroform, and carbon tetrachloride. Certain physical properties of alkenes are collected in Tables 7.1 and 7.2. Note that the alkenes have boiling points close to those of the corresponding alkanes, and the boiling points increase 20° to 30° for each additional CH_2 (methylene) group. They are less dense than water.

In general, the *trans* isomers of alkenes are more stable than the *cis* isomers. This is often attributed to steric hindrance. For example, in the 2-butenes the methyl groups are far apart in the *trans* isomer but much closer in the *cis* isomer. The larger the substituents attached to a double bond, the greater the stability of the *trans*

TABLE 7.2   Physical Properties and Heats of Combustion for Selected Geometric Isomers of Alkene Hydrocarbons

Name	Structure	mp, °C	bp, °C	Heat of Combustion ($\Delta H°$), kcal/mole (kJ/mole)
cis-2-Butene	} $CH_3\!\!-\!\!CH\!\!=\!\!CH\!\!-\!\!CH_3$	−139	3.7	−606.4 (−2537.2)
trans-2-Butene		−106	0.9	−605.4 (−2533.0)
cis-2-Pentene	} $CH_3\!\!-\!\!CH_2\!\!-\!\!CH\!\!=\!\!CH\!\!-\!\!CH_3$	−150	37.7	−752.5 (−3148.5)
trans-2-Pentene		−140	36.4	−751.7 (−3145.1)

isomer over that of the *cis* isomer. Predicting alkene stability becomes more difficult when a halogen is present or there are three or more substituents on a double bond.

Let us return to the question of stability for simple alkenes. As we saw in Sec. 3.15, heats of combustion are useful in predicting the stability of certain cyclic systems, and they can also help identify the geometric isomers of certain alkenes. In general, the lower the heat of combustion, the greater the stability (this parallels cycloalkane trends), and as Table 7.2 shows, the *trans* isomers liberate less heat than do the corresponding *cis* isomers. Furthermore, the *cis* isomers have higher boiling points and lower melting points than the *trans* isomers.

An interesting difference in physical properties between alkanes and alkenes is their polarities. Whereas alkanes are relatively nonpolar, certain alkenes are polar because of the forced geometry of the double bond. Since alkyl groups release electrons through the inductive effect, it is not surprising that propene and 1-butene both have a small dipole moment of the magnitude and direction shown here:

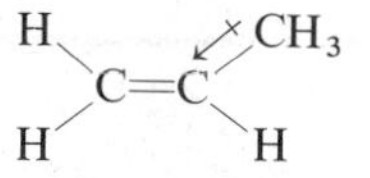

Net dipole: $\mu = 0.36$ D 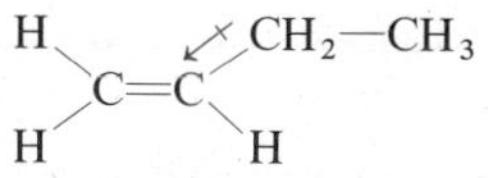   Net dipole: $\mu = 0.37$ D

(Recall that the electrons are shared equally between the two carbon atoms of a carbon-carbon double bond, so there is no dipole associated with that bond per se.)

In a substituted alkene, such as 2-butene, consider each isomer separately. For example, *trans*-2-butene has no dipole moment because the identical methyl groups have individual bond dipole moments that cancel out. On the other hand, the individual bond dipole moments in *cis*-2-butene sum to produce a small net dipole as shown:

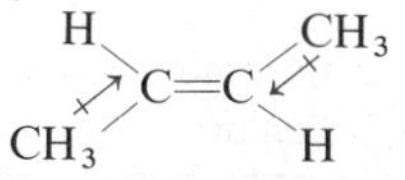

trans-2-Butene   *cis*-2-Butene
Dipoles cancel, $\mu = 0$ Net dipole: μ is small
bp $+0.9°$ bp $+3.7°$
mp $-106°$ mp $-139°$

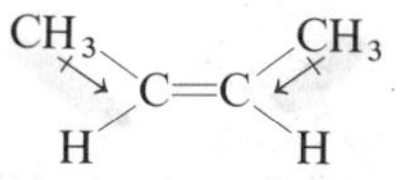

The dipole moment of *cis*-2-butene has not been measured. However, that it is slightly polar may be deduced because the *cis* isomer has a slightly higher boiling point than does the *trans* isomer.

The 1,2-dihaloethenes are another interesting example of how dipole moments and boiling and melting points can be used to help distinguish between *cis* and *trans* isomers (*Z* and *E* isomers). For the general compound, the *trans* isomer has no net dipole moment because the two individual C—X bond dipoles cancel one another. The *cis* isomer, on the other hand, has a net dipole in the direction indicated because the two individual bond dipoles are additive in that direction.

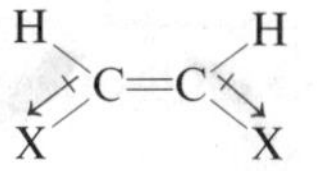

cis-1,2-Dihaloethene   *trans*-1,2-Dihaloethene
Net dipole   No net dipole

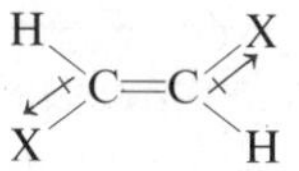

Table 7.3 compares the dipole moments and boiling and melting points of the two geometric isomers of the 1,2-dihaloethenes.

TABLE 7.3 Dipole Moments and Boiling and Melting Points for *cis* and *trans* Isomers of 1,2-Dihaloethenes

Name	Dipole Moment, μ	mp, °C	bp, °C
cis-1,2-Dichloroethene	1.85	−80	60
trans-1,2-Dichloroethene	0	−50	47
cis-1,2-Dibromoethene	1.35	−53	112
trans-1,2-Dibromoethene	0	−6	108
cis-1,2-Diiodoethene	0.75	−14	188
trans-1,2-Diiodoethene	0	+72	190

Question 7.9

Compare the stabilities of *cis*- and *trans*-2-butene with the corresponding *cis* and *trans* isomers of 1,2-di-*tert*-butylethene. (Give systematic names for the latter two compounds.)

Question 7.10

Indicate the direction of the net dipole moment, if any, in each of the following four compounds:

(*a*) *trans*-1-bromo-2-chloroethene   (*b*) 2,3-dibromo-*cis*-2-butene
(*c*) *trans*-1,2-dichloro-1-propene   (*d*) 1,1-dichloro-2,2-difluoroethene

Question 7.11

Predict where *cis*-1,2-difluoroethene would fall in the dipole moments listed in Table 7.3, and briefly justify your answer.

Question 7.12

Resolve the individual bond dipole moments for the *cis*- and *trans*-1,2-dihaloethenes into their horizontal and vertical components using the method shown in Fig. 2.22. Then show that the *trans* isomer would be expected to have no net dipole, whereas the *cis* isomer would. (*Hint:* Resolve the bond dipoles so that one component is perpendicular to the double bond and the other is parallel to it.)

7.8   Alkadienes: Nomenclature and Classification

A. Nomenclature

Compounds that contain two or more double bonds are called alkadienes, alkatrienes, alkatetraenes, and so on, depending on the number of double bonds they contain. The suffixes *di-*, *tri-*, *tetra-*, and so forth are added to the name, and the location of each double bond is specified by the appropriate number. The IUPAC system requires that the double bonds be given the lowest possible numbers, as for the alkenes. The following examples illustrate this system of nomenclature, both in open-chain and in cyclic compounds containing two or more double bonds.

$\overset{1}{C}H_2{=}\overset{2}{C}H{-}\overset{3}{C}H{=}\overset{4}{C}H_2$ $\overset{1}{C}H_2{=}\overset{2}{C}H{-}\overset{3}{C}H_2{-}\overset{4}{C}H{=}\overset{5}{C}H_2$ $\overset{1}{C}H_2{=}\overset{2}{C}H{-}\overset{3}{C}H{=}\overset{4}{C}H{-}\overset{5}{C}H_3$

  1,3-Butadiene   1,4-Pentadiene   1,3-Pentadiene

$\overset{1}{C}H_2{=}\overset{2}{C}H{-}\overset{3}{C}H{=}\overset{4}{C}H{-}\overset{5}{C}H{=}\overset{6}{C}H_2$ $\overset{1}{C}H_2{=}\overset{2}{C}{-}\overset{3}{C}{=}\overset{4}{C}H_2$ with CH$_3$ CH$_3$ substituents

  1,3,5-Hexatriene   2,3-Dimethyl-1,3-butadiene   1,3-Cyclopentadiene

    1,3-Cyclohexadiene   1,4-Cyclohexadiene

B. Classification of Dienes

Dienes are classified according to the positions of the double bonds. A compound that contains *alternate* double and single bonds is called a **conjugated diene.**

$$\ce{C=C-C=C}$$

General structure for a
conjugated diene

When the two double bonds are separated by at least one saturated carbon atom, the compound is said to be an **isolated diene.**

$$\ce{C=C-[C]_n-C=C}$$

$n \geq 1$
General structure for an
isolated diene

The final class consists of compounds in which one carbon atom bears *two* carbon-carbon double bonds and is called a **cumulated diene.** This class of compounds is much less important because they are difficult to prepare. They are called *allenes* (see Sec. 6.5).

$$\ce{C=C=C}$$

General structure for a
cumulated diene

For example, $CH_2{=}C{=}CH_2$, is *allene* or *propadiene*, and $CH_2{=}C{=}CH{-}CH_3$ is *1,2-butadiene.*

C. Geometric Isomers in Dienes

Geometric isomers are possible in compounds with the carbon-carbon double bond. Each carbon-carbon double bond can contribute one pair of *cis* and *trans* isomers, *if each carbon atom is attached to two different substituents.* For example, there are three geometric isomers for 2,4-hexadiene:

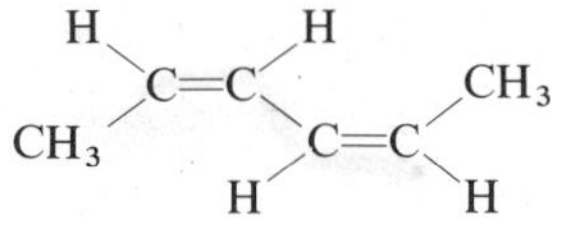

cis,cis-2,4-Hexadiene
or Z,Z-2,4-hexadiene

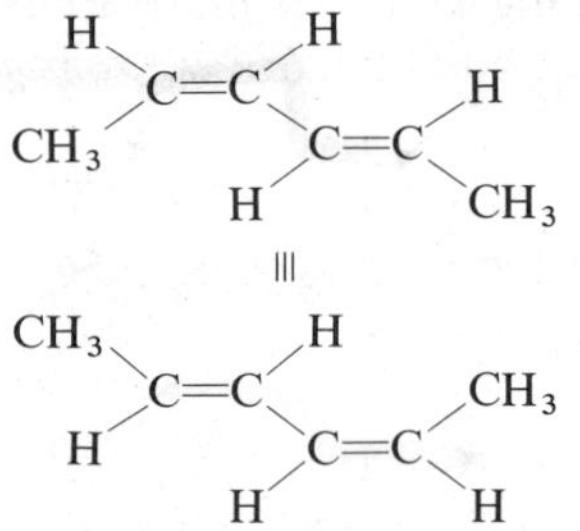

cis,trans-2,4-Hexadiene
or E,Z-2,4-hexadiene

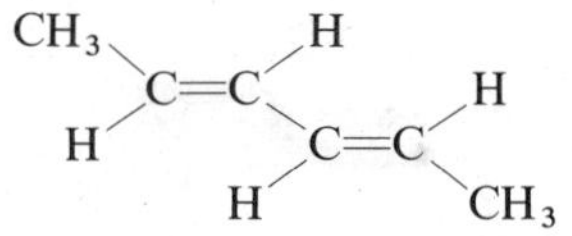

trans,trans-2,4-Hexadiene
or E,E-2,4-hexadiene

As with alkenes, each carbon-carbon double bond must be examined carefully to determine whether or not each carbon atom is attached to identical or different substituents. Then all combinations and permutations of *cis* and *trans* isomers about each double bond must be drawn.

Question 7.13

Draw the geometric isomers (if any) for each of the following compounds, and name them using the *cis/trans* and *Z/E* systems:

(*a*) 1,3-pentadiene   (*b*) 1,5-hexadiene   (*c*) 1,4-hexadiene
(*d*) 1,4-dichloro-1,3-butadiene   (*e*) 3-methyl-1,3-pentadiene

7.9   Structure of Conjugated and Nonconjugated Dienes

In Sec. 8.10 we will see that the *chemical reactivity* of conjugated dienes is quite different from that of either isolated or cumulated dienes or of alkenes. Consider the structure of a typical conjugated diene, 1,3-butadiene. There are two planar geometries for this molecule, the *s-cis* and *s-trans* structures, which result from free rotation about the carbon-carbon single bond. These terms indicate the positions of the two double relative to the single (*s*) bond. These two forms *cannot* be isolated under ordinary conditions because they are conformations; they are simply the favored conformations.

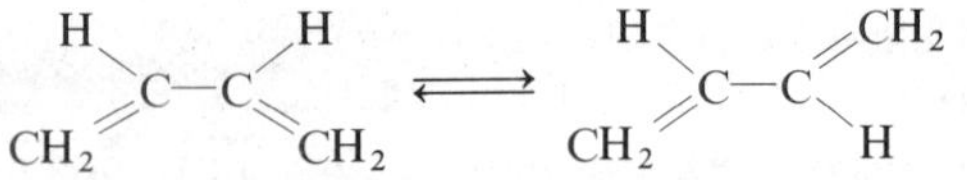

s-cis   *s-trans*

Conformations of 1,3-butadiene

Redrawing these structures to show the molecular orbitals that form each double bond, as in Fig. 7.4, illustrates that all four *p* orbitals are parallel. The overlap of two *p* orbitals forms the π bond, but why couldn't there be some overlap between the *p* orbitals on the carbon atoms that constitute the single bond? The answer comes from experimental evidence, which indicates that there is some overlap; that is, the carbon-carbon single bond does have some double-bond or π-bond character.

Heat of hydrogenation refers to the energy liberated in the reaction:

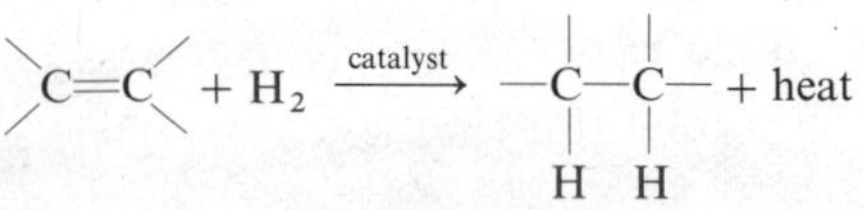

$$C=C + H_2 \xrightarrow{\text{catalyst}} -\underset{\underset{H}{|}}{C}-\underset{\underset{H}{|}}{C}- + \text{heat}$$

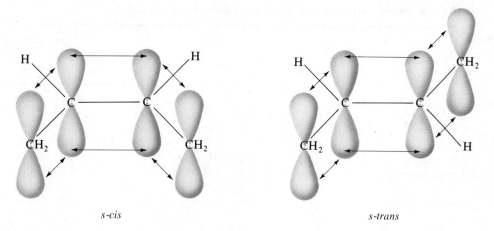

s-cis   s-trans

FIGURE 7.4 Orbital view of the two preferred conformations of 1,3-butadiene. Overlap of p orbitals to form π bonds shown by arrows.

We discuss this reaction in more detail in Sec. 8.12. The heats of hydrogenation for selected alkenes and alkadienes are given in Table 7.4. They show that there is interaction between the two π bonds in a conjugated diene.

Consider the hydrogenation of 1-pentene and 1,4-pentadiene, both of which give pentane as the product:

$$CH_2{=}CH{-}CH_2{-}CH_2{-}CH_3 + H_2 \xrightarrow{\text{catalyst}}$$

1-Pentene

$$CH_3{-}CH_2{-}CH_2{-}CH_2{-}CH_3 \qquad \Delta H° = -30.1 \text{ kcal/mole} (-125.9 \text{ kJ/mole})$$

Pentane

$$CH_2{=}CH{-}CH_2{-}CH{=}CH_2 + 2H_2 \xrightarrow{\text{catalyst}}$$

1,4-Pentadiene

$$CH_3{-}CH_2{-}CH_2{-}CH_2{-}CH_3 \qquad \Delta H° = -60.8 \text{ kcal/mole} (-254.4 \text{ kJ/mole})$$

Pentane

In 1,4-pentadiene the two double bonds are both terminal, and in 1-pentene the double bond is also terminal (*terminal* means that it is at the end of the carbon chain). If the two double bonds in the diene are independent of one another (that is, isolated) and if they do not interact with one another in any way, then we would predict that the heat of hydrogenation for 1,4-pentadiene is twice that of 1-pentene. (Compounds with more than one double bond are expected to have a heat of hydrogenation equal to the sum of the heats of hydrogenation for each individual double bond.) In this case, 2×30.1 or 60.2 kcal/mole (252 kJ/mole) (predicted) is close to the observed value of 60.8 kcal/mole (254.4 kJ/mole). Using this same reasoning, the predicted value for 1,5-hexadiene is 60.2 kcal/mole (251.9 kJ/mole), and the experimentally observed value of 60.5 kcal/mole (253 kJ/mole) is close to it. We study more about the effect of alkene structure on heats of hydrogenation in Sec. 8.15.

Now let us look at 1-butene and 1,3-butadiene, both of which on hydrogenation give the same compound, butane. Applying the principle of additivity of heats of hydrogenation, the predicted value for 1,3-butadiene is twice that of 1-butene, or $2 \times 30.3 = 60.6$ kcal/mole (253.6 kJ/mole). However, the observed value is 57.1 kcal/mole (238.9 kJ/mole), which is about 3 kcal/mole (13 kJ/mole) lower than

TABLE 7.4   Heats of Hydrogenation for Selected Alkenes and Alkadienes

Name	Structure	Heat of Hydrogenation ($-\Delta H°$), kcal/mole (kJ/mole)
1-Butene	$CH_2{=}CH{-}CH_2{-}CH_3$	30.3 (126.8)
1,3-Butadiene	$CH_2{=}CH{-}CH{=}CH_2$	57.1 (238.9)
1-Pentene	$CH_2{=}CH{-}CH_2{-}CH_2{-}CH_3$	30.1 (125.9)
1,3-Pentadiene	$CH_2{=}CH{-}CH{=}CH{-}CH_3$	54.1 (226.4)
1,4-Pentadiene	$CH_2{=}CH{-}CH_2{-}CH{=}CH_2$	60.8 (254.4)
1-Hexene	$CH_2{=}CH{-}CH_2{-}CH_2{-}CH_2{-}CH_3$	30.1 (125.9)
1,5-Hexadiene	$CH_2{=}CH{-}CH_2{-}CH_2{-}CH{=}CH_2$	60.5 (253.1)
Cyclohexene		28.6 (119.6)
1,3-Cyclohexadiene		55.4 (231.8)
2-Methyl-1,3-butadiene	$CH_2{=}\underset{\underset{CH_3}{\textstyle\vert}}{C}{-}CH{=}CH_2$	53.4 (223.4)
2,3-Dimethyl-1,3-butadiene	$CH_2{=}\underset{\underset{CH_3}{\textstyle\vert}}{C}{-}\underset{\underset{CH_3}{\textstyle\vert}}{C}{=}CH_2$	53.9 (225.5)
Propadiene (*allene*)	$CH_2{=}C{=}CH_2$	71.1 (297.5)

predicted. Thus, from heat of hydrogenation data, 1,3-butadiene is about 3 kcal/mole (13 kJ/mole) more stable than expected.

This finding is fairly common for conjugated dienes. For example, the predicted heat of hydrogenation for 1,3-pentadiene is about 58 kcal/mole (242.7 kJ/mole) and the observed value is about 54 kcal/mole (226 kJ/mole) (see Question 7.15).

An even more direct comparison of the stabilities of dienes involves 1,3-pentadiene [heat of hydrogenation = 54.1 kcal/mole (226.4 kJ/mole)] and 1,4-pentadiene [heat of hydrogenation = 60.8 kcal/mole (254.4 kJ/mole)]. Both these molecules consume 2 moles of hydrogen and give pentane as the product. These data indicate that 1,3-pentadiene is about 6.7 kcal/mole (28 kJ/mole) more stable than 1,4-pentadiene because it evolves that amount less energy.

In general, then, conjugated dienes (which have alternating double and single bonds) are more stable than isolated double bonds. The additional stability of conjugated dienes is attributed to the **delocalization** of the π electrons in the two adjacent double bonds. When all the *p* orbitals are parallel, as they are in *s-cis-* and *s-trans-*1,3-butadiene, delocalization is possible. The bonding in 1,3-butadiene may be represented either by the orbital picture in (Fig. 7.4) or by the following, which shows some weak π bonding superimposed on the carbon-carbon single bond.

Delocalization occurs (in these and other molecules encountered later) when electrons are "spread out" over a larger area of the molecule through interaction with other orbitals within the molecule. Geometries must be favorable for this to occur, and we refer only to the electron distribution within the molecule; the atoms that make up the molecule do not change their location or arrangement (see Secs. 8.30 and 13.6 and Question 13.19).

Weak π bonding   Strong
π bonding

H   H   H CH_2

C⚌C   C⚌C

CH_2 CH_2 CH_2 H

Strong π bonding   Weak π bonding

s-cis-1,3-Butadiene   *s-trans*-1,3-Butadiene

Question 7.14

Using the heats of hydrogenation for 1-pentene and *cis*-2-pentene of 30.1 and 28.6 kcal/mole, respectively, calculate the heat of hydrogenation predicted for *cis*-1,3-pentadiene. Compare this result with the observed value of 54.1 kcal/mole.

Question 7.15

Draw the *s-cis* and *s-trans* conformations for both *cis*- and *trans*-1,3-pentadiene. Comment on the relative stabilities of the two conformations for each geometric isomer. Which do you think would be favored?

Methods for Preparing Alkenes and Alkadienes

In Secs. 7.10 and 7.11 some common methods used to prepare alkenes and alkadienes are discussed. Although certain simple alkenes are available from petroleum (see Sec. 4.10), most laboratory preparations involve the **elimination** of two atoms or substituents that are attached on adjacent carbon atoms.

Elimination reaction:

$$—\underset{A}{\overset{|}{C}}—\underset{B}{\overset{|}{C}}— \longrightarrow \quad \overset{}{>}C{=}C\overset{}{<} + A—B$$

where *A—B* may be H—H, H—X, X—X, H—OH, or $H—HSO_4$ as well as others.

There are some special methods for preparing alkenes, which are mentioned in this and future chapters.

7.10   Dehydrogenation of Alkanes

The first reaction we discuss involves the loss (elimination) of hydrogen atoms from adjacent carbon atoms and is called **dehydrogenation** (*de* means taking away, so this term means "taking away hydrogen").

$$—\underset{H}{\overset{|}{C}}—\underset{H}{\overset{|}{C}}— \xrightarrow[\text{heat}]{\text{catalyst}} \quad >C{=}C< + H—H$$

Alkane   Alkene

For example, the passage of ethane gas through a tube heated to between 500° and 700° removes the elements of hydrogen and forms ethene.

$$CH_3-CH_3 \xrightarrow{500-700°} CH_2=CH_2 + H_2$$

Ethane   Ethene

This is **pyrolysis** (Greek *pyros*, fire; *lysis*, a loosing), and in industrial plants it is called "**cracking**." The pyrolysis of propane gives propene, ethene, methane, and hydrogen:

$$CH_3-CH_2-CH_3 \xrightarrow{500-700°} CH_3-CH=CH_2 + CH_2=CH_2 + CH_4 + H_2$$

Propane   Propene   Ethene   Methane

Thermal cracking of more complex alkanes gives a complex mixture of products, but ethene is often the major one.

The reaction is speeded up by certain catalysts, and **catalytic cracking** involves passing hydrocarbon vapors over a solid catalyst of SiO_2 and Al_2O_3 at about 450 to 500° and 15 to 30 pounds per square inch (lb/sq in.) pressure.

Thermal cracking is an example of a free-radical chain reaction, which we studied in Sec. 4.6. As heat is applied to hydrocarbons, C—H and C—C bonds are ruptured; the bond energies in Table 5.1 indicate that the C—C bonds are weaker and thus more easily broken. Possible mechanisms for the cracking of propane that account for the observed products are described next.

The reaction is *initiated* by breaking some C—C and C—H bonds, for example:

$$CH_3-CH_2-CH_3 \begin{cases} CH_3CH_2CH_2\cdot + H\cdot \\ CH_3CH_2\cdot + CH_3\cdot \end{cases}$$

The free radicals and hydrogen atoms that are formed then *propagate* the chain reaction by forming a new free radical for each one consumed.

$$CH_3-CH_2-CH_3 + H\cdot \longrightarrow CH_3-CH_2-CH_2\cdot \text{ or } CH_3-\overset{..}{C}H-CH_3 + H_2$$

$$CH_3-CH_2-CH_3 + CH_3\cdot \longrightarrow CH_3-CH_2-CH_2\cdot \text{ or } CH_3-\overset{..}{C}H-CH_3 + CH_4$$

(or $CH_3CH_2\cdot$)   (or CH_3-CH_3)

$$CH_3\overset{\frown}{:}CH_2\overset{\frown}{-}CH_2\cdot \longrightarrow CH_3\cdot + CH_2=CH_2$$

$$H\overset{\frown}{:}CH_2\overset{\frown}{-}CH_2\cdot \longrightarrow H\cdot + CH_2=CH_2$$

$$CH_3-\overset{\overset{\displaystyle H}{|}}{C}H\overset{\frown}{-}CH_2\cdot \longrightarrow CH_3-CH=CH_2 + H\cdot$$

and so on. Other reactions also involve the attack of radicals on hydrocarbons.

Thermal cracking reactions are often used to convert petroleum fractions that boil at higher temperatures than the gasoline fraction (see Sec. 4.10) into hydrocarbons containing eight or fewer carbon atoms.

Cracking is indeed a complex process that forms many products in proportions that depend on temperature and pressure as well as on the type of catalyst used. It is *not* a convenient laboratory process!

Question 7.16

The pyrolysis of butane produces, among others, the following products: hydrogen, ethene, ethane, methane, propene, propane, 1-butene, 2-butene, and 1,3-butadiene. Write a stepwise mechanism to account for the formation of these products, identifying the initiating, propagating, and terminating steps.

7.11   Dehydrohalogenation of Alkyl Halides; E2 Mechanism

The *elimination* of the elements of hydrogen halide, H—X, can be accomplished by treating an alkyl halide with a suitable base. The base used most often is potassium hydroxide dissolved in alcohol (ethyl alcohol or ethanol, C_2H_5OH). This reaction is referred to as **dehydrohalogenation** (this word can be broken into several familiar parts: *de-*, taking away, *-hydro-*, hydrogen, and *-halogen-*, X; thus, *dehydrohalogenation* means literally "taking away H—X").

A.   General Dehydrohalogenation Reaction

$$-\overset{\underset{|}{|}}{C}-\overset{\underset{|}{|}}{C}- \;+\; OH^{\ominus}\,(+K^{\oplus}) \xrightarrow[\text{solvent}]{\text{alcohol}} \;\;>\!C\!=\!C\!<\; +\; X^{\ominus}\;+\;HOH\,(+K^{\oplus})$$

$$\overset{\underset{\text{H}}{}\;\underset{\text{X}}{}}{}$$

Alkyl   Alkene
halide

Because the potassium ion, $K^{\oplus}$, plays no important role in dehydrohalogenation, it remains unchanged. Other strong bases, such as NaOH, could be used as well, but potassium hydroxide has greater solubility in alcohol. The alcohol is a good solvent for both the alkyl halide and the product (alkene). There is no limitation on the nature of the halide, although alkyl fluorides are seldom used because they are difficult to obtain. Alkyl iodides are used less frequently because they are more expensive than the other halides.

The general reaction does not show how the structure of the alkyl halide affects the structure of the alkenes produced. In the following examples, only one possible alkene is formed, since each alkyl halide is a primary (terminal) halide.

$$CH_3-CH_2-Cl \xrightarrow[\text{alcohol}]{\text{KOH}} CH_2\!=\!CH_2$$
Chloroethane   Ethene

$$CH_3-CH_2-CH_2-Br$$
1-Bromopropane

or

$$CH_3-\overset{\underset{|}{Br}}{CH}-CH_3 \xrightarrow[\text{alcohol}]{\text{KOH}} CH_3-CH\!=\!CH_2$$
                                                        Propene

2-Bromopropane

$$CH_3-CH_2-CH_2-CH_2-Br \xrightarrow[\text{alcohol}]{\text{KOH}} CH_3-CH_2-CH\!=\!CH_2$$
1-Bromobutane   1-Butene

On the other hand, with other secondary alkyl halides, two possible alkenes can be formed because a molecule of H—X can be eliminated in two possible ways (see top of page 262).

As the reaction indicates, 2-butene is the major product (81%). This complies with the general trend, which is discussed further in Sec. 7.12. Complications can occur because the hydroxide ion not only can cause elimination but also can displace the halide and become attached to carbon instead. This reaction is discussed in Sec. 9.15. However, under the conditions for elimination indicated—namely, KOH in alcohol—dehydrohalogenation is the major reaction and the only reaction we consider at this time.

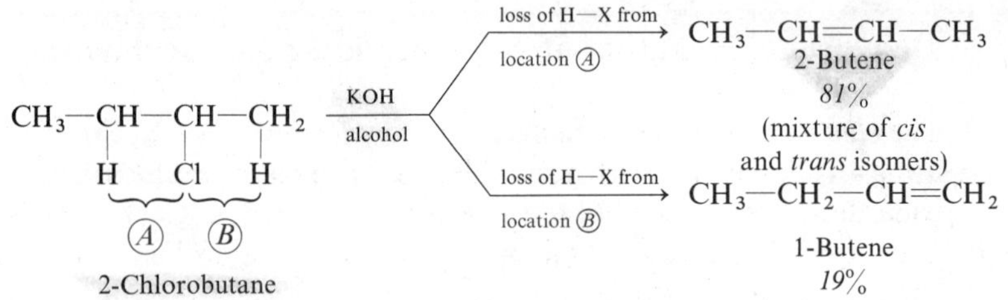

$$CH_3-CH-CH-CH_2 \xrightarrow[\text{alcohol}]{\text{KOH}}$$

with H and Cl and H below (A and B)

2-Chlorobutane

loss of H—X from location (A) → $CH_3-CH=CH-CH_3$
2-Butene
81%
(mixture of *cis*
and *trans* isomers)

loss of H—X from location (B) → $CH_3-CH_2-CH=CH_2$
1-Butene
19%

B. Mechanism of Dehydrohalogenation

The study of *kinetics* (reaction rates) is indispensable to any mechanistic investigation. Many elimination reactions of H—X are second order overall; that is, the rate of reaction is proportional to the concentrations of $OH^{\ominus}$ and of the alkyl halide, R—X. The rate law for the reaction is:

$$\text{Rate of reaction} = k_2[OH^{\ominus}][R-X]$$

where k_2 = rate (proportionality) constant.

This means that the reaction involves one molecule of $OH^{\ominus}$ and one molecule of R—X in the rate-determining step. Doubling the concentration of either base or alkyl halide doubles the rate of reaction, and doubling both of them increases the rate by a factor of four. The *slow, rate-determining step* has been shown to be the attack by $OH^{\ominus}$ on a hydrogen atom attached to the carbon atom that is adjacent to the one bearing the halogen atom (the carbon bearing the hydrogen is the β carbon atom and the one to which the halogen is attached is the α carbon atom).

Why can a hydrogen attached to carbon be removed by a base like hydroxide ion? To answer this, recall that halogen is a very electronegative element, which exerts a strong electron-withdrawing inductive effect on the carbon atom to which it is attached. This, in turn, causes electrons in nearby σ bonds to be withdrawn to some extent by the inductive effect. Consider the following structures:

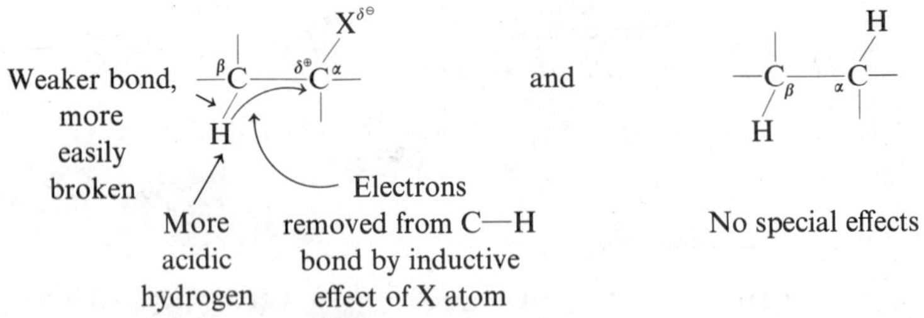

Weaker bond, more easily broken $^\beta C \xrightarrow{\delta^{\oplus}} C^{\alpha} - X^{\delta\ominus}$ and

More acidic hydrogen   Electrons removed from C—H bond by inductive effect of X atom

No special effects

In the structure that contains the halogen atom, the C—X bond is polar, with a $\delta^{\oplus}$ charge on the α carbon. This carbon, in turn, withdraws electrons to some extent from the C—H bond on the β carbon atom, which weakens that bond and makes hydrogen the easier to abstract. The hydrogen is more acidic than it would be without the presence of the halogen atom. The hydrogen is abstracted with the concerted formation of the double bond.

Base-promoted elimination reactions of this sort are called **E2 reactions,** where E stands for *elimination* and 2 for *bimolecular* because the rate of the reaction depends on two molecules, R—X and $OH^{\ominus}$. The mechanism, which is consistent with all known facts, may be depicted as follows, where the base removes the hydrogen on the β carbon atom as a proton and forms water. Simultaneously, the pair of electrons that originally held that hydrogen forms the π bond and the halogen departs with its bonding electrons as a halide ion.

$$-\overset{\beta}{\underset{\underset{HO:^{\ominus}}{\overset{|}{\underset{}{}}}}{C}} - \overset{\alpha}{\underset{\underset{}{\overset{|}{\underset{H}{}}}}{C}} - \quad\longrightarrow\quad \overset{}{\underset{}{}}C{=}C\overset{}{\underset{}{}} + :\overset{..}{\underset{..}{X}}:^{\ominus} + H_2\overset{..}{\underset{..}{O}}:$$

The curved arrows indicate the electron shifts. *Electrons always flow to a more positive (electron-withdrawing or electron-deficient) center.* In this reaction, the arrows show the electrons flowing from the hydroxide ion (which is rich in electrons originally and has a negative charge) to the highly electronegative halogen atom.

A more complete and accurate picture of this mechanism follows in which various bonds are broken or made as indicated; conventionally, partially broken or made bonds are shown by dotted lines. This is called a **concerted reaction** because all the bond breaking and making occur simultaneously in one step.

Concerted mechanism for E2 reaction:

$$
\begin{array}{c}
\text{(reactant structure)} \longrightarrow
\left[
\begin{array}{c}
\text{Bonds being} \\
\text{broken} \\
\text{Transition} \\
\text{state}
\end{array}
\right]
\end{array}
$$

Transition
state

$$\overset{}{\underset{}{}}C{=}C\overset{}{\underset{}{}} + :\overset{..}{\underset{..}{X}}:^{\ominus} + H_2\overset{..}{\underset{.}{O}}$$

This reaction is also referred to as **_anti_-elimination,** where *anti-* refers to the orientation of the leaving groups (in this case the H and X, which are 180° apart in an *anti*-conformation) and *not* to the geometry of the product. The stereochemistry of this elimination reaction is discussed further in Sec. 7.13.

The chemical kinetics (that is, the rate law) indicate a bimolecular reaction and demand that the transition state for this reaction involve one molecule of alkyl halide and one molecule of hydroxide ion, $OH^{\ominus}$. The preceding transition state complies with this. The negative charge is greatly dispersed in this reaction, and it is only reasonable that the $\delta^{\ominus}$ charges be as far apart as possible, as they are in the transition state. Furthermore, the transition state contains a partially formed carbon-carbon double bond and resembles the product.

There are several driving forces for this reaction. The first, which we mentioned, is the electron-withdrawing inductive effect of the halogen atom, which makes the β hydrogen susceptible to attack by base. Another is the formation of water, which results when hydroxide ion attacks the C—H bond, and the concurrent formation of a π bond. Finally, the departing halide ion is highly solvated by the alcohol molecules, ROH (Fig. 7.5). Solvation is essential to a facile reaction and in many instances may control the reaction pathway.

We discuss further evidence for this mechanism in Sec. 7.12 and 7.13. We see in Sec. 9.15 that the E2 reaction often competes with the S_N2 reaction, a mechanism introduced in Sec. 4.2A.

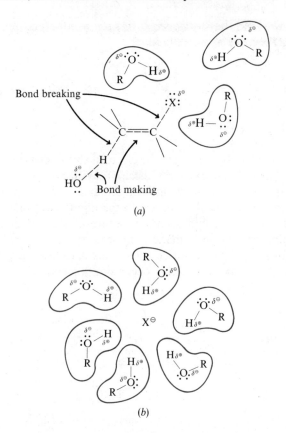

FIGURE 7.5   (*a*) The carbon-halogen bond breaking in dehydrohalogenation, showing the role of the solvent. (*b*) The halide ion solvated by the polar alcohol solvent.

7.12   Structure-Reactivity and Structure-Product Distribution Relationships in Dehydrohalogenation

We noted in Sec. 7.11 that the dehydrohalogenation of a secondary alkyl halide, in which either of the two adjacent hydrogen atoms can be removed, yields two products. We now examine the effect of the structure of the alkyl halide on (1) its reactivity toward elimination and (2) the product distribution of the isomeric alkenes formed.

Some experimental results were formulated into a useful generalization (at best, a rule of thumb) in 1875 by the Russian chemist Alexander Saytzeff. The **Saytzeff rule** states that in reactions involving the loss of hydrogen halide from an alkyl halide, the hydrogen is lost preferentially from the more highly branched carbon atom; that is, 3° > 2° > 1°. (As we will see in Sec. 7.17, this rule also applies to the dehydration of alcohols to form alkenes.) The reason this occurs is related to the *relative stabilities* of the alkenes formed in a given elimination reaction. Using heats of hydrogenation, the relative stability of alkenes has been found to be the following:

Decreasing stability of alkenes:

$$R_2C{=}CR_2 > R_2C{=}CHR > R_2C{=}CH_2 \sim RCH{=}CHR$$

$$> RCH{=}CH_2 \; ({>}CH_2{=}CH_2)$$

Generally, *the preferred product in dehydrohalogenation elimination reactions is the alkene that bears the greater number of alkyl groups on the carbon-carbon double bond;* the *structure* of the alkyl groups may also play a role, but the *number* of substituents is more important.

Since Saytzeff's time, transition-state theory has developed to the point where it can be applied to dehydrohalogenation. In this example, thermodynamic control governs the favored reaction pathway. *The more stable alkene is formed more readily.* The transition state for this reaction (see Sec. 7.11) has considerable double-bond character, so it is not surprising that the stability of the competing transition states parallels the stability of the products.

In the dehydrohalogenation of 2-bromobutane, removal of any one of the three hydrogens on C-1 yields 1-butene, and removal of either hydrogen on C-3 yields 2-butene. Based solely on the probability of hydrogen removal, one would *predict* that the ratio of 1-butene to 2-butene would be 3:2. Yet 2-butene is the major product because it is more stable and is formed faster.

Three examples illustrate *the more highly substituted alkene is the major product in the dehydrohalogenation reaction.*

$$CH_3-CH_2-\underset{\underset{\displaystyle Br}{|}}{C}H-CH_3 \xrightarrow[\text{alcohol}]{\text{KOH}} CH_3-CH_2-CH{=}CH_2 + CH_3-CH{=}CH-CH_3$$

2-Bromobutane   1-Butene   2-Butene

(sec-butyl bromide)   *20%*   (mixture of *cis* and *trans* isomers) *80%*

$$CH_3-CH_2-CH_2-\underset{\underset{\displaystyle Cl}{|}}{C}H-CH_3 \xrightarrow[\text{alcohol}]{\text{KOH}} CH_3-CH_2-CH_2-CH{=}CH_2$$

2-Chloropentane   1-Pentene   *30%*

$$+ CH_3-CH_2-CH{=}CH-CH_3$$

2-Pentene (mixture of *cis* and *trans* isomers) *70%*

$$CH_3-CH_2-\underset{\underset{\displaystyle Cl}{|}}{\overset{\overset{\displaystyle CH_3}{|}}{C}}-CH_2-CH_3 \xrightarrow[\text{alcohol}]{\text{KOH}} CH_3-CH{=}\underset{\underset{\displaystyle}{}}{\overset{\overset{\displaystyle CH_3}{|}}{C}}-CH_2-CH_3 + CH_3-CH_2-\overset{\overset{\displaystyle CH_2}{||}}{C}-CH_2-Cl$$

3-Chloro-3-methylpentane   3-Methyl-2-pentene (*cis* and *trans*) (contains *three* alkyl substituents attached to double bond) *Major product*   2-Ethyl-1-butene (contains *two* alkyl substituents attached to double bond) *Minor product*

The generalization regarding product distribution does not allow us to predict exact product ratios; we can state only which alkene is the major product. When *cis* and *trans* isomers are formed, predicting which geometric isomer predominates can only be done when the product is simple (see Sec. 7.7).

Regarding the effect of the structure of the alkyl halide on *reactivity*, two factors appear to play an important role. Consider, for example, the following as representative of primary, secondary, and tertiary alkyl halides:

$$CH_3-CH_2-CH_2-CH_2-Br \longrightarrow CH_3-CH_2-CH=CH_2$$

1°

$$CH_3-CH_2-\underset{\underset{Br}{|}}{CH}-CH_3 \longrightarrow CH_3-CH=CH-CH_3 + CH_3-CH_2-CH=CH_2$$

2°   (cis and trans)

$$CH_3-\underset{\underset{Br}{\overset{\overset{CH_3}{|}}{C}}}{|}-CH_3 \longrightarrow \underset{CH_3}{\overset{CH_3}{>}}C=CH_2$$

3°

The amount of branching increases in going from 1° to 2° to 3°, and in these examples this increases the number of hydrogens that can be attacked by base. In the 1° compound there are two such hydrogens (shown in red), in the 2° halide there are five such hydrogens, and in the 3° alkyl bromide there are nine. The first factor, then, is the *probability* of hydrogen removal. The second factor is alkene stability, and as we recall, the more stable the alkene, the more rapidly it is formed. In the preceding example, the more highly substituted alkene is formed from the 3° compound. The amount of alkene substitution is less from the 2° halide and even less from the 1° halide. These factors combine to provide the order of reactivity in terms of the alkyl halide, R—X, as:

Order of decreasing reactivity of R—X in dehydrohalogenation reactions:

3° R—X > 2° R—X > 1° R—X

Question 7.17

For each of the following alkyl halides, indicate all the alkenes that would be formed on dehydrohalogenation. In each case, which product (if any) would be formed in the largest amount and why?

(*a*) 1-chlorohexane   (*b*) 2-chlorohexane
(*c*) 2-chloro-2-methylhexane   (*d*) 3-chloro-2-methylhexane
(*e*) 2-chloro-2,3-dimethylpentane   (*f*) 1-chloro-2-methylbutane
(*g*) 3-chloropentane   (*h*) 3-chloro-3,4-dimethylhexane

(*i*) ⬡—CH₂—I   (*j*) ⬡—CH₃ (with Cl)

7.13   Stereochemistry of Dehydrohalogenation Reaction

One additional piece of information concerning the mechanism of the base-promoted dehydrohalogenation of alkyl halides—the E2 reaction—comes from the stereochemistry of the reaction. We indicated in Sec. 7.12 that there is *anti*-elimination (the two groups or atoms being eliminated are 180° apart in an *anti*-confromation). Now let us see the evidence for this preference.

A. Open-Chain Compounds

Consider first the general reaction in both the Newman projection and the three-dimensional projection formulas.

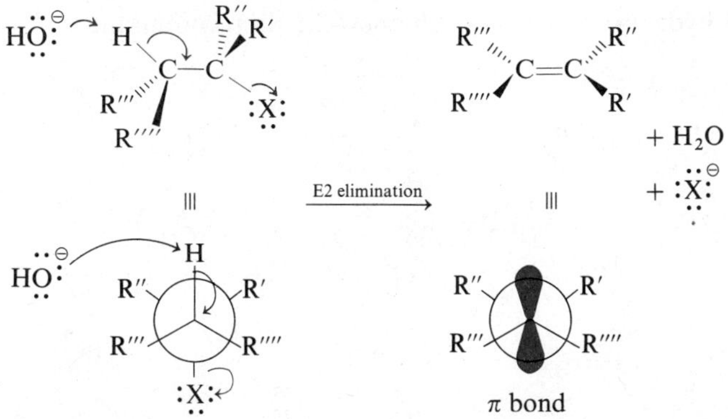

Elimination is easy when the molecule can rotate about the carbon-carbon single bond to allow for *anti*-elimination. The rate of elimination is fastest in conformations in which large groups are not forced close together.

In the following examples the various stereoisomers of 2,3-dibromobutane are allowed to react with hydroxide ion in alcohol. From *meso*-2,3-dibromobutane (1) we would predict that *anit*-elimination gives 2-bromo-*cis*-2-butene (2), and this is the observed product. On the other hand, either enantiomer, *d*-2,3-dibromobutane or *l*-2,3-dibromobutane (3), should give the same product, 2-bromo-*trans*-2-butene (4), and this is observed.

Dehydrobromination of *meso*-2,3-dibromobutane:

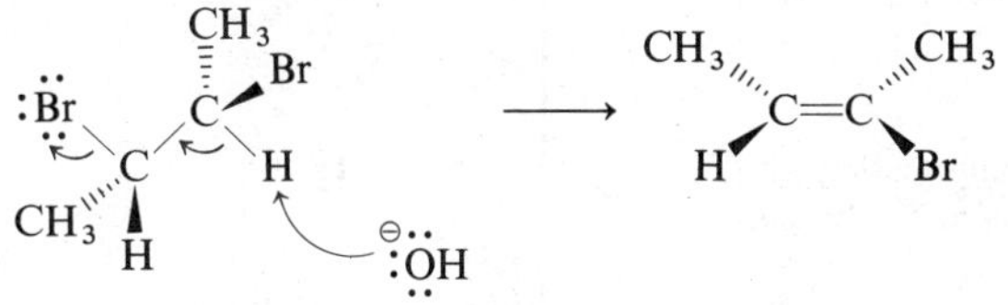

(1)
meso

(2)
cis

or

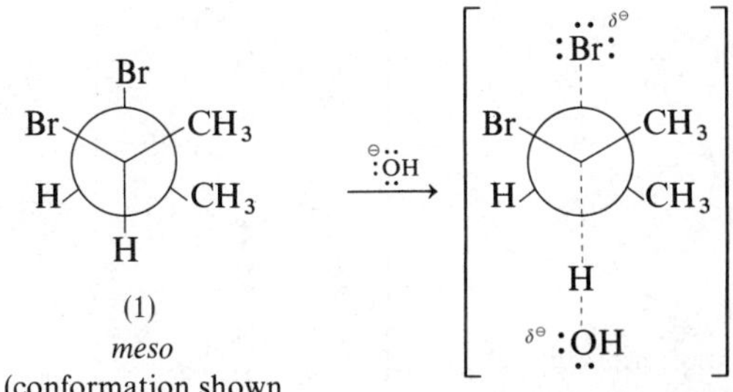

(1)
meso
(conformation shown
with leaving groups,
—H and —Br, *anti* to
one another)

Transition state
(shows bond breaking and
making; two methyl groups
brought close together)

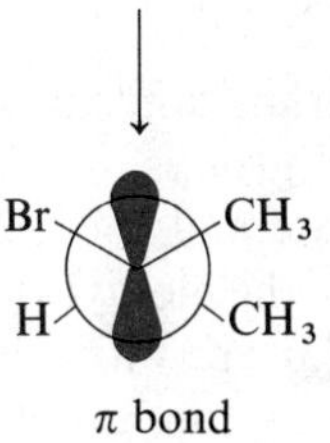

π bond

(2)
cis

Dehydrobromination of *d*- and *l*-2,3-dibromobutane:

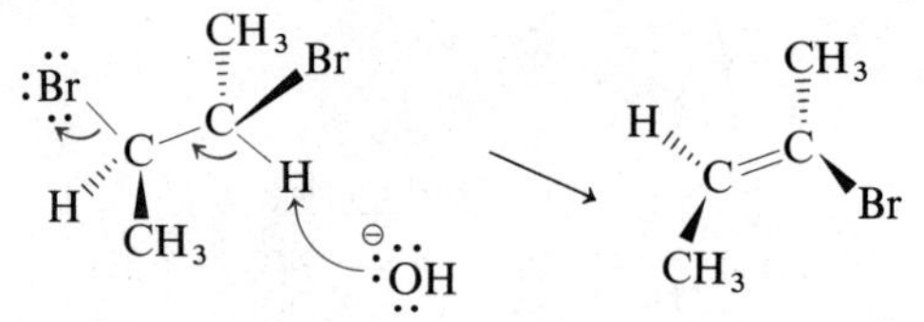

(3a)

(4)

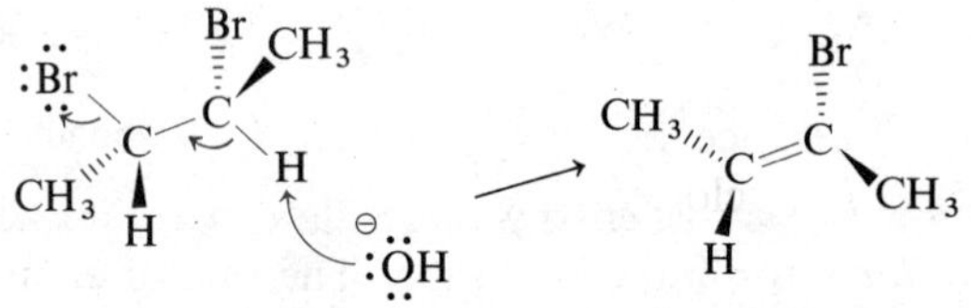

(3b)

(4)

Enantiomers: *d* and
l forms

Identical compounds

or

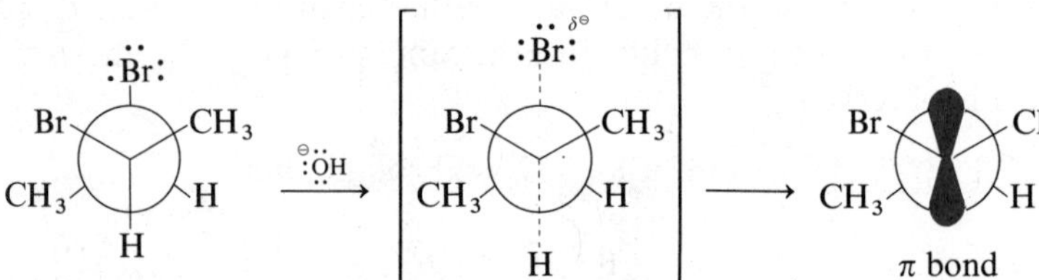

(3a)

Conformation with
—H and —Br *anti-*

Transition state
(shows bond breaking
and making; methyl
groups far apart)

π bond

(4)

trans

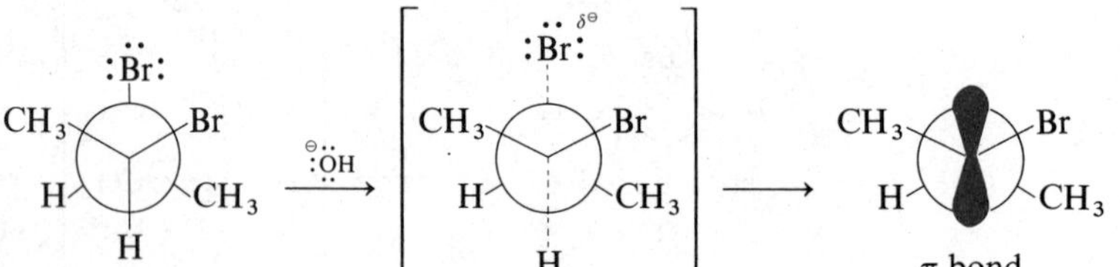

(3b)

Enantiomer of (3a)
Conformation with
—H and —Br *anti-*

Transition state
(shows bond breaking
and making; methyl
groups far apart)

π bond

(4)

trans

(same as structure above)

In the transition state and in the product from the *meso*-dibromide (1), the two bulky methyl groups are brought close together (eclipsed). On the other hand, the transition state and the product from the enantiomeric (*d* and *l*) dibromides, (3a) and (3b), have the methyl groups far apart. It might be expected that the enantiomeric dibromides (3) are more reactive than the *meso*-dibromide (1) because the steric interactions are minimized in both the transition state and the product, and this is true. Recall from Sec. 6.14 that enantiomers react at the same rate, whereas diastereomers react at different rates. Thus, because the *meso-* and *d,l*-dibromides, (1) and (3), are diastereomers of one another, it is not surprising that they react at different rates. Transition-state theory and structures of products allow us to predict relative reactivities.

It would be difficult to account for the stereochemistry shown for the dehydrohalogenation reaction in the preceding example *if anti-elimination did not occur*. This supports the notion that the leaving groups must be as far apart as possible.

B. Cyclic Compounds

Additional evidence regarding the *anti*-elimination mechanism comes from cyclic systems. In the cyclohexane series, the substituents being eliminated must bear the *anti*-relationship (*trans*-diaxial) for E2 reactions to occur at a reasonable rate.

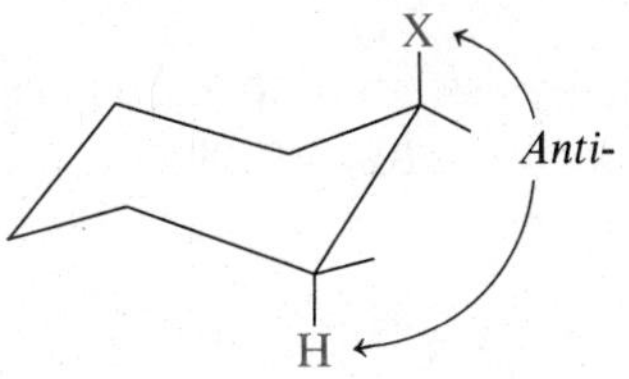

Dehydrohalogenation can occur

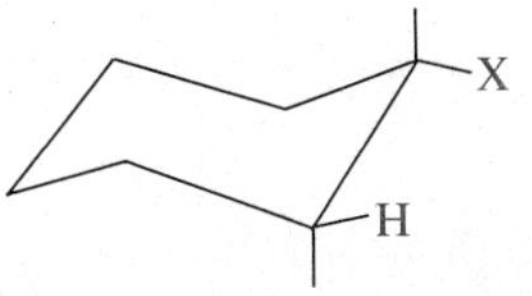

Dehydrohalogenation does not occur *or* occurs very slowly

Recall from Sec. 3.16 that a *tert*-butyl group on a cyclohexane ring occupies an equatorial position almost exclusively. Using *cis-* and *trans-4-tert*-butyl-1-chlorocyclohexane as examples, the *cis* isomer undergoes rapid bimolecular (E2) elimination, whereas the *trans* isomer undergoes various types of reactions that we have not yet studied in preference to E2 elimination because the E2 pathway is energetically unfavored. This results from the conformational constraints imparted to the system by the bulky *tert*-butyl group, as shown here.

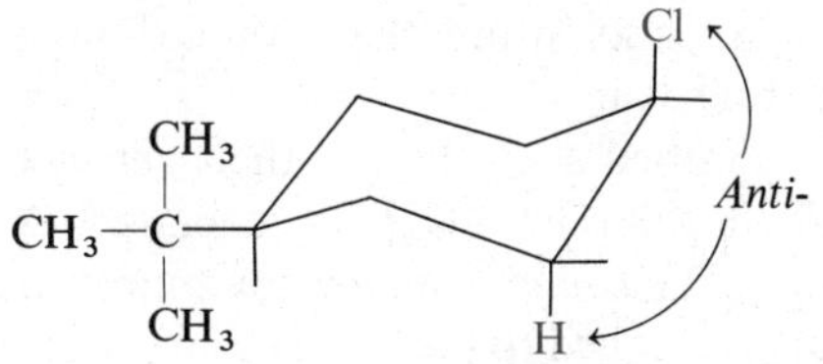

cis-4-tert-Butyl-1-chlorocyclohexane
Undergoes E2 reaction

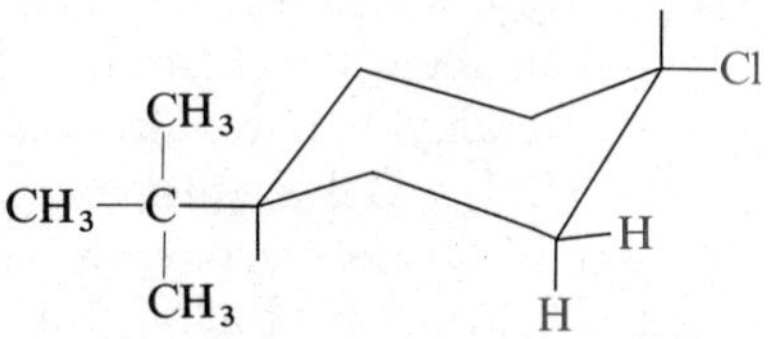

trans-4-tert-Butyl-1-chlorocyclohexane
Undergoes E2 reaction only with great difficulty and very slowly

Question 7.18

Using the principles of E2 elimination, account for the following two reactions. In particular, explain the exclusive formation of 2-menthene from menthyl chloride in the second reaction even though the double bond in it is less substituted than that in 3-menthene.

Neomenthyl chloride   2-Menthene   3-Menthene

25%   *75%*

2-Menthene

Only elimination product

Menthyl chloride

7.14 Dehydration of Alcohols

We now turn to another important laboratory preparation of alkenes, **dehydration** (which literally means "taking away hydrate or water"). This reaction requires a catalyst and heat.

Alcohol   Alkene

Decreasing ease of dehydration:

$$3° \ ROH > 2° \ ROH > 1° \ ROH$$

The dehydration reaction is usually carried out by heating the alcohol with sulfuric acid (H_2SO_4) or phosphoric acid (H_3PO_4). Dehydration can also be effected by passing alcohol vapor over heated alumina (Al_2O_3), where alumina serves as a Lewis acid (see Sec. 5.7).

We need to learn a little more about another family of organic compounds, the **alcohols.** Their general formula is R—OH, where R represents any alkyl group. The —OH group in an alcohol has many chemical properties analogous to those of water. An alcohol may, for example, be protonated in acidic solution to form $R—OH_2^{\oplus}$, which is similar to the protonation of water to give $H—OH_2^{\oplus}$ or $H_3O^{\oplus}$. We study this and other reactions in this and future chapters, although the main discussion about the chemistry of alcohols is in Chap. 10.

Like alkyl halides, alcohols are classified according to their general structure. Thus, $RCH_2—OH$ is a primary (1°) alcohol, $R_2CH—OH$ is a secondary (2°) alcohol, and $R_3C—OH$ is a tertiary (3°) alcohol. The *common names* for simple alcohols are derived by naming the alkyl group that is attached to the —OH group and adding the word *alcohol*. For example,

$$CH_3—OH \quad\quad CH_3—CH_2—CH_2—OH \quad\quad CH_3—CH_2—\underset{\underset{OH}{|}}{CH}—CH_3 \quad\quad CH_3—\underset{\underset{OH}{\overset{\overset{CH_3}{|}}{C}}}{}—CH_3$$

Methanol   1-Propanol   2-Butanol   2-Methyl-2-propanol

(methyl   (*n*-propyl alcohol)   (*sec*-butyl alcohol)   (*tert*-butyl

alcohol)   *1°*   *2°*   alcohol)

1°   *3°*

The *systematic nomenclature* of alcohols is discussed in detail in Sec. 10.1.

Following are seven typical dehydration reactions:

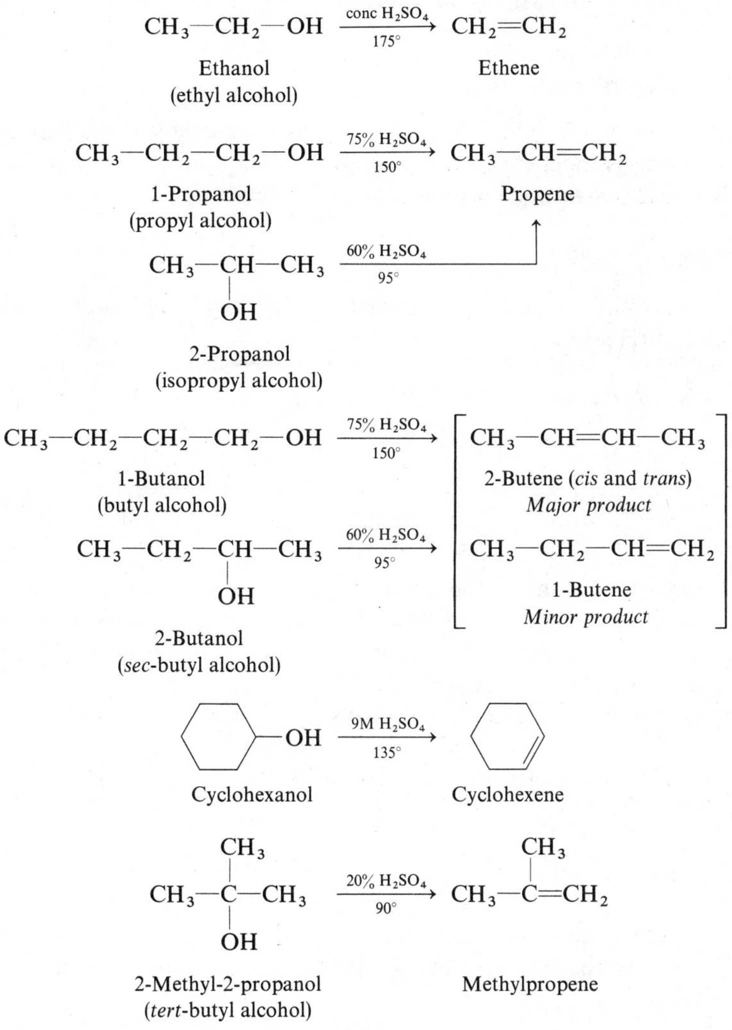

$$CH_3-CH_2-OH \xrightarrow[175°]{conc\ H_2SO_4} CH_2=CH_2$$

Ethanol
(ethyl alcohol)
Ethene

$$CH_3-CH_2-CH_2-OH \xrightarrow[150°]{75\%\ H_2SO_4} CH_3-CH=CH_2$$

1-Propanol
(propyl alcohol)
Propene

$$CH_3-\underset{\underset{OH}{|}}{CH}-CH_3 \xrightarrow[95°]{60\%\ H_2SO_4}$$

2-Propanol
(isopropyl alcohol)

$$CH_3-CH_2-CH_2-CH_2-OH \xrightarrow[150°]{75\%\ H_2SO_4} \left[\begin{array}{c} CH_3-CH=CH-CH_3 \\ \text{2-Butene (}cis\text{ and }trans\text{)} \\ \textit{Major product} \\ \\ CH_3-CH_2-CH=CH_2 \\ \text{1-Butene} \\ \textit{Minor product} \end{array} \right.$$

1-Butanol
(butyl alcohol)

$$CH_3-CH_2-\underset{\underset{OH}{|}}{CH}-CH_3 \xrightarrow[95°]{60\%\ H_2SO_4}$$

2-Butanol
(*sec*-butyl alcohol)

Cyclohexanol $\xrightarrow[135°]{9M\ H_2SO_4}$ Cyclohexene

$$CH_3-\underset{\underset{OH}{\overset{\overset{\displaystyle CH_3}{|}}{C}}}{}-CH_3 \xrightarrow[90°]{20\%\ H_2SO_4} CH_3-\underset{\overset{\displaystyle CH_3}{|}}{C}=CH_2$$

2-Methyl-2-propanol
(*tert*-butyl alcohol)
Methylpropene

Careful examination of these examples shows that milder conditions (acid concentration *and* temperature) are required to dehydrate a 3° alcohol than a 2° alcohol and that the 1° alcohols are the most difficult to dehydrate. 2-Propanol and 1-propanol yield the same product, propene, but this is not surprising because only one alkene can be formed. However, that both 1-butanol and 2-butanol yield a mixture of two alkenes (1-butene and 2-butene) *is* surprising; we discuss this seemingly unusual reaction in Sec. 7.17.

When two isomeric alkenes are formed, the isomer bearing the greater number of alkyl substituents is the major product. This is the same trend we encountered in the dehydrohalogenation reactions (see Sec. 7.11).

7.15   Mechanism for Acid-Catalyzed Dehydration of Alcohols

In this section we show the currently accepted mechanism for the dehydration of a general alcohol. This reaction is *acid-catalyzed* and acid is necessary to the reaction. Kinetic studies on these reactions show that the rate of reaction is proportional only to the concentration of alcohol:

$$\text{Rate of reaction} = k[\text{R—OH}]$$

where k = rate (proportionality) constant.[1]

In other words, the rate of reaction increases directly with increasing alcohol concentration and is independent of acid concentration, provided that *some* acid is present. Our mechanism must account for these facts.

A. Step 1

The first step of the dehydration reaction is the protonation of the alcohol by the acid catalyst (H_2SO_4 or H_3PO_4):

Alcohol	Proton (electron-deficient)	Protonated alcohol: oxonium ion

The curved arrow shows the donation of one pair of electrons on oxygen to the highly electron-deficient proton. The oxygen atom thus acquires a *formal positive charge*, because it shares one of its nonbonding pairs of electrons with hydrogen in forming a new covalent hydrogen-oxygen bond.

This reaction is analogous to the protonation of water:

Hydronium
ion

The protonated alcohol is called an **oxonium ion,** which resembles *hydronium ion*, $H_3O^\oplus$; because the reactions are so similar, it is not surprising that the names are similar also.

B. Step 2

The second step of the reaction is the loss of the water molecule from the oxonium ion:

Oxonium ion	Carbocation (electron-deficient carbon atom bears positive charge and contains only six electrons)	

---

[1] The rate expression for this reaction is actually

$$\text{Rate of reaction} = k[\text{H}^\oplus][\text{R—OH}]$$

which shows that the reaction rate is dependent on the concentration of acid, $H^\oplus$. The hydrogen ion concentration, however, remains constant throughout the reaction. The ion is a catalyst and is not consumed as the reaction progresses. Because both k, the rate constant, and $[H^\oplus]$ are constants, they are written as a combined constant, where $k = k[H^\oplus]$. The reaction is called **pseudo first order,** that is, it exhibits first-order kinetics but is not really first order overall.

This forms a new electron-deficient species, the **carbocation,** which is a carbon atom that bears a positive charge and contains only six electrons (three bonding pairs). It is said to be *electron-deficient.*

That this reaction occurs can be rationalized on the basis of the positively charged (electropositive) oxygen atom in the oxonium ion. A positive charge in a molecule is expected to attract electrons in the bonds attached to it. Here, the positively charged oxygen pulls one pair of electrons away from carbon and toward itself, as indicated by the curved arrow. In breaking this bond, a new positively charged species, the carbocation, is formed and water is liberated as a neutral molecule.

$$R\text{—}\overset{\oplus}{\underset{H}{\overset{H}{O}}}: \longrightarrow \left[\overset{\delta\oplus}{R}\text{----}\overset{\delta\oplus}{\underset{H}{\overset{H}{O}}}: \right] \longrightarrow R^{\oplus} + H_2\overset{..}{\underset{..}{O}}:$$

Reactant	Transition state	Products
(positive charge on oxygen)	(carbon-oxygen bond broken; both carbon and oxygen have partial positive charges)	(positive charge on carbon)

Alternatively, it might be argued that the positively charged oxygen in the oxonium ion attracts the electrons in the bonds that attach the hydrogens to it. They do, but if a bonding pair of electrons is attracted away from hydrogen, the starting alcohol and a proton are regenerated. Thus, two equilibria can be written:

$$-\overset{|}{\underset{H}{C}}\text{—}\overset{|}{\underset{:\overset{..}{O}H}{C}}\text{—} + H^{\oplus} \Longleftrightarrow -\overset{|}{\underset{H}{C}}\text{—}\overset{|}{\underset{\overset{\oplus}{\underset{H}{O}}\overset{..}{}H}{C}}\text{—} \Longleftrightarrow -\overset{|}{\underset{H}{C}}\text{—}\overset{|}{\underset{\oplus}{C}}\text{—} + H_2\overset{..}{O}:$$

Reactants	Oxonium ion	Carbocation

The steps have been simplified to some extent for clarity. For example, protonation of the alcohol does not necessarily involve $H^{\oplus}$ but is probably the transfer of a proton from an acid such as H_2SO_4 to an alcohol:

$$R\text{—}\overset{..}{\underset{H}{O}}: + H\text{—}\overset{..}{\underset{..}{O}}\text{—}SO_3H \Longleftrightarrow R\text{—}\overset{\oplus}{\underset{H}{\overset{H}{O}}}: + HSO_4^{\ominus}$$

Base	Acid	Oxonium ion

This reaction is an example of a Lowry-Brønsted acid-base reaction (see Sec. 5.7); the alcohol is the proton acceptor (the base) and sulfuric acid is the proton donor (the acid).

C. Step 3

The final step of the dehydration reaction leading to alkene is the loss of a hydrogen atom adjacent to the carbocation, which is ejected as a proton:

$$-\overset{|}{\underset{H}{C}}\text{—}\overset{|}{\underset{\oplus}{C}}\text{—} \Longleftrightarrow \overset{}{\underset{}{C}}{=}\overset{}{\underset{}{C}} + H^{\oplus}$$

Carbocation	Alkene

As pointed out in previous sections, a proton is seldom lost as $H^{\oplus}$. Rather, solvent or some other base in solution pulls the acidic proton off the (in this example) carbocation. The reaction in this example might involve H_2O or $HSO_4^{\ominus}$ acting as bases to give $H_3O^{\oplus}$ or H_2SO_4. This is a very important concept and should be kept in mind when writing mechanisms. For the sake of simplifying the mechanism, the base is often omitted in the drawing. Regardless of whether it is actually drawn or not, its role should not be overlooked.

The carbocation needs a pair of electrons so that it can complete its octet, which it accomplishes by attracting the electrons in the adjacent C—H bond. The curved arrow shows the movement of one pair of electrons away from hydrogen and toward the positively charged carbon atom. The result is a carbon-carbon double bond and liberation of a proton.

The three-dimensional structure of the carbocation was discussed in Sec. 2.8.

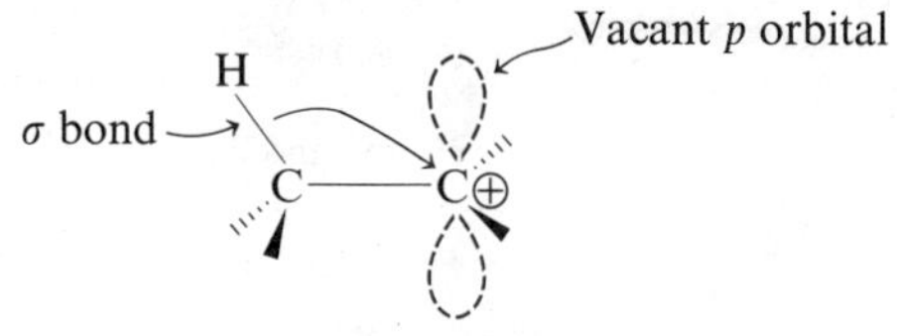

Carbocation

1° Carbocations

Although there is direct evidence for the formation of both 3° and 2° carbocations as *intermediates* in organic synthesis carried out in solution, there is virtually no evidence that 1° carbocations exist as free intermediates under normal reaction conditions in solution. Many reactions in which 1° carbocations are intermediates that go on to react in a subsequent step probably occur in a one-step, concerted process. We use the convention of depicting free carbocations in reactions that involve these intermediates for simplification. If a given reaction proceeds by way of a 1° carbocation, however, the caveat given here should be kept in mind. For example,

Two steps:

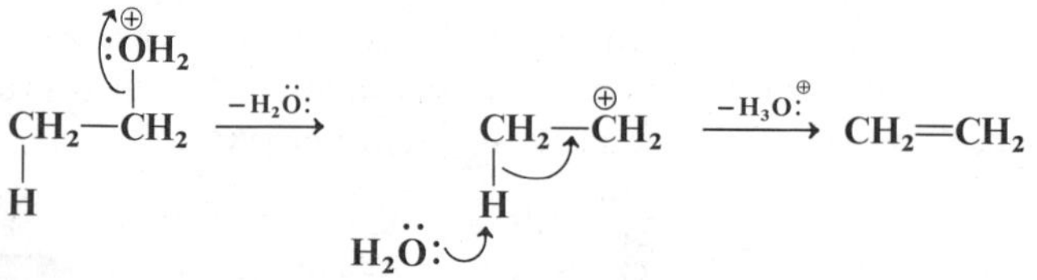

1° Carbocation

One step:

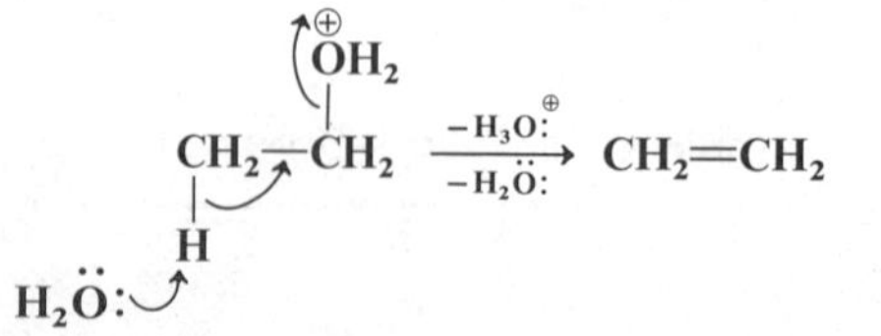

Preferred mechanism in solution

Let us summarize the three steps in the dehydration mechanism.

Step 1:

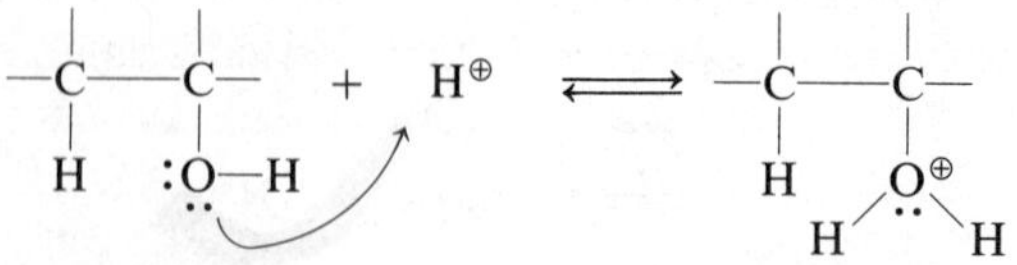

Alcohol Oxonium ion

Equilibrium lies to right

Step 2:

Equilibrium lies far to left (slow, rate-determining step)

Step 3:

Equilibrium lies far to right

Do these steps account for the known facts about dehydration? The reaction is acid-catalyzed; that is, the acid is not consumed in the course of the reaction. Step 1 consumes a proton but step 3 regenerates it. Furthermore, no alkene can be formed until the carbocation is formed in step 2. Alcohols, like water, are readily protonated, so there is always a large concentration of oxonium ions in solution. In strong acid, the equilibrium is essentially shifted quantitatively to the right—all the alcohol molecules are protonated. Thus, the slow rate-determining step is step 2, the rate of which is proportional to the concentration of oxonium ions, which in turn is proportional to the concentration of starting alcohol. Because the overall rate of reaction is proportional only to the concentration of alcohol and is independent of acid concentration, the previous steps are consistent with the known facts.

Because the rate of dehydration depends only on the concentration of alcohol, this reaction is called an **E1 reaction,** where E stands for *elimination* and 1 for *monomolecular* or *unimolecular*. Recall from Secs. 7.11 and 7.12 that base-promoted dehydrohalogenation is an E2 reaction because its rate of reaction depends on the concentration of alkyl halide *and* base; that is, it is a bimolecular reaction.

All three steps have been shown as equilibria because each is reversible. The equilibrium of step 1 lies far to the right because alcohols are readily protonated by acid (as mentioned in the previous paragraph). Step 2 lies far to the left because it involves the formation of the unstable and highly reactive electron-deficient carbocation. Once the carbocation is formed, however, it rapidly loses a proton (step 3), so that the equilibrium involved in this step lies far to the right.

The hydrogen ion catalyst plays the important role of weakening the carbon-oxygen bond in the alcohol so that it can be broken more easily. This is because the positive charge on oxygen in the oxonium ion pulls bonding electrons away from carbon.

Strong bond, difficult to break	Weakened bond, easier to break
Alcohol	Oxonium ion

Several experimental techniques are used to shift the equilibrium to favor alkene production. Using concentrated sulfuric acid or phosphoric acid (both of which are strong dehydrating agents; that is, they "tie" up water as $H_3O^\oplus$) helps remove water as it is formed. Also, alkenes are usually rather volatile and can be removed from the reaction mixture by distillation as soon as they are formed.

There is additional proof that the steps in the dehydration reaction are equilibria. We will see later that alkenes can be converted to alcohols if the reaction conditions are changed.

Use of Arrows in Mechanisms

As mentioned several times, curved arrows do *not* show how electrons are likely to move but are only a "bookkeeping" device. The convention used is the following.

1. Place the tail of the arrow near the source of electrons.
2. Draw the arrow so that the head points toward the atom or groups of atoms that *attract* electrons; this is usually an electronegative element or a positively charged (electron-deficient) center.
3. Sometimes a series of electron changes occur so that several arrows are used; these show the stepwise electron transfer from one atom to the next or from one bond to another.

Let us review briefly the previous reactions where arrows were used, and see that they have been drawn in accord with the preceding conventions.
In dehydrohalogenation:

Electronegative atom that *attracts* electrons

Electron source

Note *net* flow of electrons *from* electron source *to* electronegative atom, which has affinity for electrons

In dehydration:

Positively charged (electron-deficient) atom that attracts electrons

σ bond electron source

Note *net* flow of electrons to positive charge, which pulls pair of electrons away from H and leaves it as a proton ($H^{\oplus}$), which is one of the products

We use curved arrows to keep track of electrons throughout this text.

7.16   Use of Carbocation Stability to Explain Reactivity in Dehydration Reactions

In Sec. 7.15 carbocations were invoked in the acid-catalyzed E1 dehydration reaction. With this in mind, consider the dehydration mechanism and the effect of varying the substitution of the alcohol on the rate of the reaction. Does the reaction exhibit a *substituent effect* (see Sec. 5.15)?

When the experiments are actually performed, one finds that the rate of the reaction follows the order $3° > 2° > 1°$. The reactivity order can be explained on the basis of carbocation stability. Assuming that the starting alcohols are all of roughly comparable stability (equal energy), the following order of reactivity versus carbocation stability is found:

Increasing reactivity order:

Primary alcohol		Secondary alcohol		Tertiary alcohol
RCH_2—OH	<	R_2CH—OH	<	R_3C—OH

$\downarrow$ H$^\oplus$, heat $\downarrow$ H$^\oplus$, heat $\downarrow$ H$^\oplus$, heat

$$R\overset{\oplus}{C}H_2 \quad < \quad R_2\overset{\oplus}{C}H \quad < \quad R_3C^\oplus$$

Primary carbocation	Secondary carbocation	Tertiary carbocation

Because the tertiary carbocation is the most stable, it is easiest to form; that is, the energy of activation, E_A, for its formation is the lowest of the three alcohols. The secondary carbocation is the next easiest to form, so the secondary alcohol is less reactive than the tertiary alcohol. Finally, the primary carbocation is even less stable, so the primary alcohol is even less reactive. Here, then, the reactivity of alcohols parallels the stability of the carbocations they produce. The more stable the carbocation, the easier it is to form.

7.17   Product Distribution in Dehydration Reactions

The *Saytzeff rule* used to predict the product distribution in the base-promoted dehydrohalogenation reactions of alkyl halides (see Sec. 7.12) also applies to the dehydration of alcohols.

The relative stability of substituted alkenes is unaffected by the method of their formation. *In dehydration, the more highly substituted alkene (that is, the double bond bearing the greater number of alkyl substituents) is the preferred product.* Thus, in the dehydration of 2-butanol, the major product is 2-butene.

$$CH_3-CH_2-\underset{\underset{OH}{|}}{CH}-CH_3 \xrightarrow[\text{heat}]{\text{conc } H_2SO_4} CH_3-CH=CH-CH_3 + CH_3-CH_2-CH=CH_2$$
$$\text{2-Butene}$$

2-Butanol (*sec*-butyl alcohol)	(*cis* and *trans*) (double bond contains two alkyl groups) *Major product*	1-Butene (double bond contains one alkyl group) *Minor product*

In the following case, where three alkenes can be formed, the most highly substituted product is the preferred one, and all that can be said about the other two is that, taken together, they are the minor products.

Transition-state theory complies with the expectation that the more stable alkene is the one that is formed faster and thus is the major product. In this case, the transition state for proton loss has considerable double-bond character and thus resembles product (see Sec. 5.6).

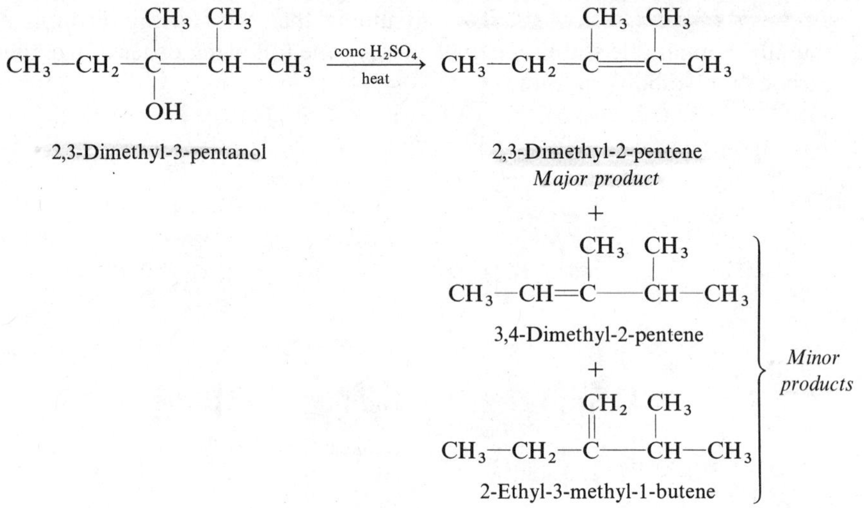

$$CH_3-CH_2-\underset{\underset{OH}{|}}{\overset{\overset{CH_3}{|}}{C}}-\underset{\overset{|}{CH_3}}{CH}-CH_3 \xrightarrow[\text{heat}]{\text{conc } H_2SO_4} CH_3-CH_2-\underset{}{\overset{\overset{CH_3}{|}}{C}}=\underset{}{\overset{\overset{CH_3}{|}}{C}}-CH_3$$

2,3-Dimethyl-3-pentanol

2,3-Dimethyl-2-pentene
Major product

+

$$CH_3-CH=\underset{}{\overset{\overset{CH_3}{|}}{C}}-\underset{\overset{|}{CH_3}}{CH}-CH_3$$

3,4-Dimethyl-2-pentene

+

$$CH_3-CH_2-\underset{}{\overset{\overset{CH_2}{||}}{C}}-\underset{\overset{|}{CH_3}}{CH}-CH_3$$

2-Ethyl-3-methyl-1-butene

} *Minor products*

7.18   Carbocation Rearrangements: 1,2-Hydride and 1,2-Alkyl Shifts

Carbocations explain the observed products from the acid-catalyzed dehydration of certain alcohols. For example:

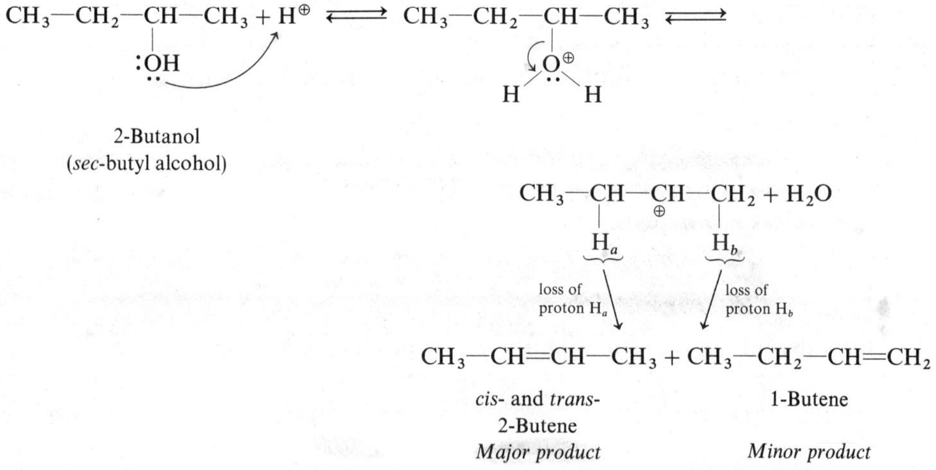

$$CH_3-CH_2-\underset{\underset{:OH}{|}}{CH}-CH_3 + H^{\oplus} \rightleftharpoons CH_3-CH_2-\underset{\underset{\overset{O^{\oplus}}{\underset{H \quad H}{}}}{|}}{CH}-CH_3 \rightleftharpoons$$

2-Butanol
(*sec*-butyl alcohol)

$$CH_3-CH-\underset{\oplus}{CH}-CH_2 + H_2O$$
$$\overset{H_a}{\underbrace{\qquad}} \qquad \overset{H_b}{\underbrace{\qquad}}$$

loss of proton H_a ↙   ↘ loss of proton H_b

$$CH_3-CH=CH-CH_3 + CH_3-CH_2-CH=CH_2$$

cis- and *trans*-
2-Butene
Major product

1-Butene

Minor product

How can we explain the formation of a mixture of 1-butene and 2-butene from 1-butanol, $CH_3-CH_2-CH_2-CH_2-OH$, which gives the same products as does 2-butanol? If a carbocation is involved, we expect 1-butanol to give the *n*-butyl cation, which then loses a proton to give *only* 1-butene.

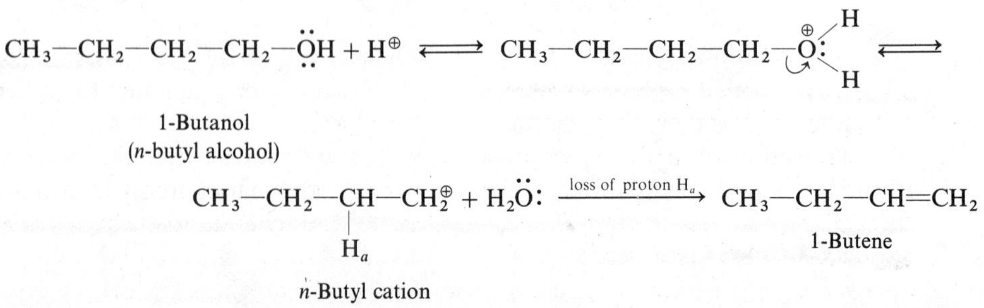

$$CH_3-CH_2-CH_2-CH_2-\overset{..}{\underset{..}{O}}H + H^{\oplus} \rightleftharpoons CH_3-CH_2-CH_2-CH_2-\overset{\oplus}{\underset{H}{\overset{H}{O:}}} \rightleftharpoons$$

1-Butanol
(*n*-butyl alcohol)

$$CH_3-CH_2-\underset{\underset{H_a}{|}}{CH}-CH_2^{\oplus} + H_2\overset{..}{O}: \xrightarrow{\text{loss of proton } H_a} CH_3-CH_2-CH=CH_2$$

1-Butene

n-Butyl cation

Carbocation theory accounts for the products of the dehydration of 2-butanol. But what about 1-butanol? We know from Sec. 5.7 that the order of carbocation stability is: $3° > 2° > 1°$. If the *n*-butyl cation rearranges to give the more stable secondary carbocation—the *sec*-butyl cation—then proton loss gives the two observed products (there are two products obtained from the dehydration of both 1- and 2-butanol):

$$\overset{4}{CH_3}-\overset{3}{CH_2}-\overset{2}{CH_2}-\overset{1}{CH_2^{\oplus}} \xrightarrow[\substack{\text{two hydrogens on}\\\text{C-2 moving to C-1}}]{\text{rearranges by one of}} \overset{4}{CH_3}-\overset{3}{CH_2}-\underset{\oplus}{\overset{2}{CH}}-\overset{1}{CH_2}-H$$

1° Carbocation   More stable 2° carbocation

$$\downarrow$$

$$CH_3-CH{=}CH-CH_3$$
$$+$$
$$CH_3-CH_2-CH{=}CH_2$$

That a more stable carbocation forms should not surprise us, but how does the rearrangement occur? The modern view involves the migration of a hydrogen ion or (as we will see in the examples that follow) an alkyl group from an adjacent carbon atom to the electron-deficient (positive) carbon atom. When this migration occurs, the hydrogen ion or alkyl group takes the pair of bonding electrons with it, thereby leaving a new carbocation. This is called a **1,2 shift** or **1,2 migration** because the migration occurs between one carbon atom and an adjacent carbon atom. Note that the numbers *1,2* have nothing to do with the nomenclature of the compound.

When hydrogen migrates with its pair of electrons, the migrating species is $H{:}^{\ominus}$, which is the *hydride ion*; this is called a **hydride shift** or a **hydride migration.** When an alkyl group migrates with its electron pair, $R{:}^{\ominus}$ is the migrating species, and this is called an **alkyl shift** or **alkyl migration.** In general, the following changes occur:

1,2-Hydride shift:

$$-\overset{\beta}{C}\!-\!\overset{\alpha}{\underset{\oplus}{C}}- \longrightarrow \left[-\overset{|}{C}\underset{H}{-}\overset{|}{C}- \right]^{\oplus} \longrightarrow -\overset{|}{\underset{\oplus}{C}}-\overset{|}{\underset{H}{C}}-$$

(arrow shows movement   Transition
of electrons away   state
from C_β toward C_α)

1,2-Alkyl shift:

$$-\overset{\beta}{C}\!-\!\overset{\alpha}{\underset{\oplus}{C}}- \longrightarrow \left[-\overset{|}{C}\underset{R}{-}\overset{|}{C}- \right]^{\oplus} \longrightarrow -\overset{|}{\underset{\oplus}{C}}-\overset{|}{\underset{R}{C}}-$$

(arrow shows movement   Transition
of electrons away   state
from C_β toward C_α)

In terms of bond orbitals, the 1,2 shift may be envisioned as the attraction by the carbocation of the electrons in the bond that holds the hydrogen atom or alkyl group on the carbon adjacent to the carbocation. Migration occurs *if* the new carbocation that is formed as a result of the 1,2 shift is more stable (that is, *if* it contains a greater number of alkyl groups) than the starting carbocation.

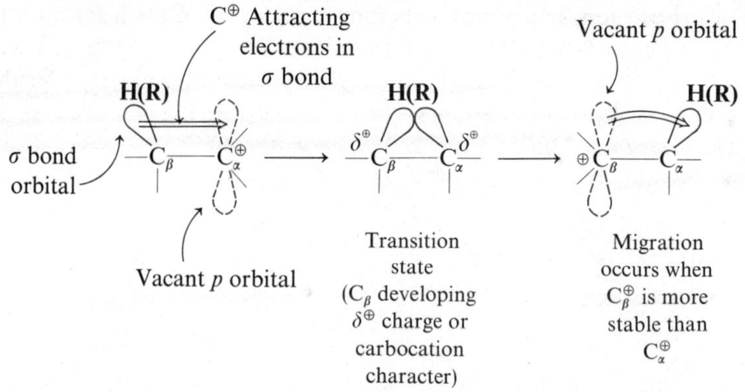

Rearrangement takes place when a 1,2-hydride or a 1,2-alkyl shift produces a new, more stable (that is, more highly substituted) carbocation. In the dehydration of alcohols, a carbocation is formed first. If it can rearrange to a new, more stable carbocation, it does so, and the new ion then loses a proton to yield one or more alkenes. If no rearrangement is possible, then the original carbocation simply loses a proton to give one or more alkenes.

Now consider some dehydration reactions where 1,2 shifts occur, keeping in mind that alkyl shifts always produce a new arrangement of carbon atoms in the carbon skeleton. In the following examples, the starting materials and *observed* products are shown on the first line and the mechanism is shown below; the steps involving the protonation of the alcohol and the loss of the water molecule are omitted. The symbols $\sim H\overset{\ominus}{:}$ and $\sim R\overset{\ominus}{:}$ mean that hydrogen and the alkyl group *migrate with their bonding electrons.*

Example 1

Hydride shift

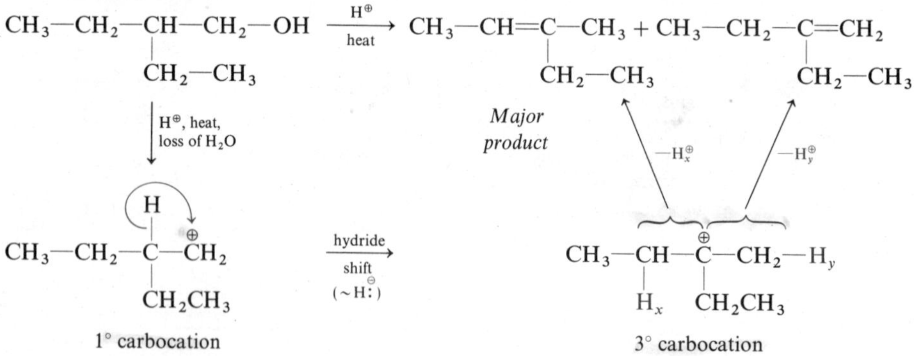

In this example the carbon adjacent to the carbocation contains hydrogen *and* an alkyl group (the ethyl group, $-CH_2-CH_3$). *If* the alkyl group migrates (as it sometimes does), the resulting carbocation is

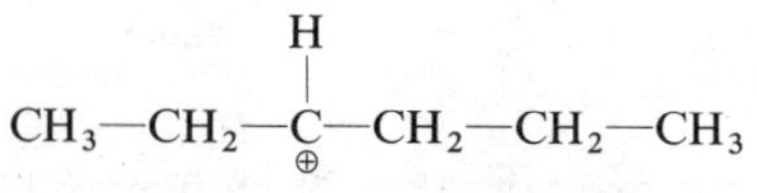

which is a *secondary* carbocation. That *no* hexene derivatives are formed indicates that the hydride shift not only is preferred but also occurs exclusively to give the even more stable *tertiary* carbocation.

Example 2

Alkyl shift

$$CH_3-\overset{\overset{\displaystyle CH_3}{|}}{\underset{\underset{\displaystyle CH_3}{|}}{C}}-CH_2-OH \xrightarrow[\text{heat}]{H^{\oplus}} CH_2=\overset{\overset{\displaystyle CH_3}{|}}{C}-CH_2-CH_3 + CH_3-\overset{\overset{\displaystyle CH_3}{|}}{C}=CH-CH_3$$

2-Methyl-1-butene   2-Methyl-2-butene

Minor product   *Major product*

↓ H⊕, heat, loss of H₂O

$$CH_3-\overset{\overset{\displaystyle CH_3}{|}}{\underset{\underset{\displaystyle (CH_3)}{|}}{\overset{\oplus}{C}}}-\overset{\oplus}{C}H_2 \xrightarrow[\substack{\text{migration} \\ (\sim CH_3\overset{\ominus}{:})}]{\text{methyl}} CH_3-\overset{\overset{\displaystyle CH_3}{|}}{\underset{\oplus}{C}}-CH_2-CH_3$$

1° carbocation   3° carbocation

In this example the carbon adjacent to the carbocation contains three methyl groups, any of which can migrate to the primary carbocation and thus form the much more stable tertiary carbocation. Note that a new carbon skeleton is obtained.

It is also possible for the loss of water and the migration (in this example of $CH_3\overset{\ominus}{:}$) to occur simultaneously. There is very little experimental evidence to substantiate the existence of primary carbocations in solution. For this reason, the one-step, concerted route from the oxonium ion directly to the secondary or tertiary carbocation may be the preferred pathway in solution.

An alternate mechanism:

$$CH_3-\overset{\overset{\displaystyle CH_3}{|}}{\underset{\underset{\displaystyle CH_3}{|}}{C}}-CH_2-\overset{\oplus}{O}H_2 \xrightarrow[\substack{\text{water and} \\ \text{migration} \\ \text{of } CH_3\overset{\ominus}{:}}]{\substack{\text{simultaneous} \\ \text{loss of}}} CH_3-\overset{\overset{\displaystyle CH_3}{|}}{\underset{\oplus}{C}}-CH_2-CH_3 \xrightarrow{-H^{\oplus}} \text{products}$$

Oxonium ion   3° carbocation

Question 7.19

Outline the complete, stepwise mechanisms for the dehydration of the following four alcohols:

(*a*) 1-propanol

(*b*) cyclohexanol, ⬡—OH

(*c*) $CH_3-\overset{\overset{\displaystyle CH_3}{|}}{C}H-\overset{\overset{\displaystyle OH}{|}}{C}H-CH_3$

(*d*) $(CH_3)_3C-\overset{\overset{\displaystyle OH}{|}}{C}H-CH_3$

Question 7.20

Which, if any, of the alcohols in Question 7.19 yield more than one alkene? For those that do, indicate which alkene would be expected to form in the greater (greatest) amount.

Question 7.21

Which alcohol in Question 7.19 would you expect to react the slowest? Why?

Question 7.22

Provide mechanistic explanations for the following reaction:

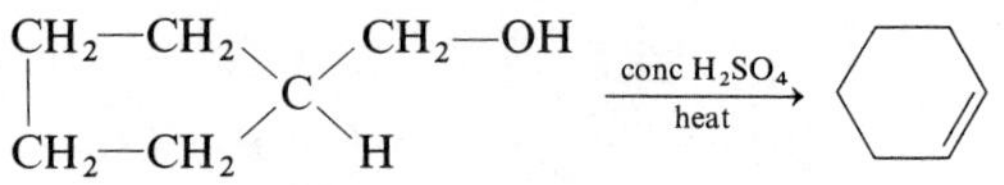

$$\text{CH}_2\text{—CH}_2 \quad \text{CH}_2\text{—OH}$$
$$\text{CH}_2\text{—CH}_2 \quad C \quad H \quad \xrightarrow[\text{heat}]{\text{conc H}_2\text{SO}_4} \quad \bigcirc$$

7.19   Dehalogenation of Dihalides

Alkyl dihalides can be converted to alkenes by reacting with zinc metal in alcohol (solvent):

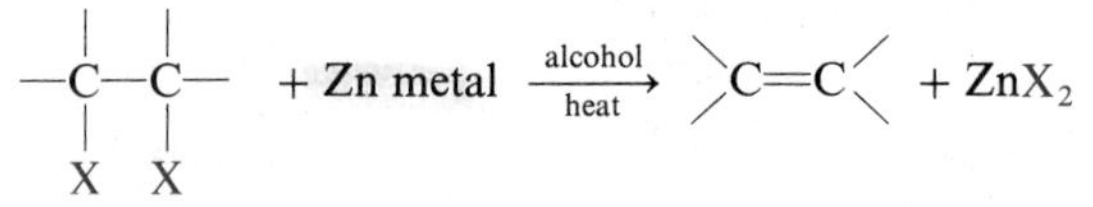

$$\underset{\substack{X \quad X \\ \text{Alkyl dihalide}}}{-\overset{|}{\underset{|}{C}}-\overset{|}{\underset{|}{C}}-} \ + \text{Zn metal} \xrightarrow[\text{heat}]{\text{alcohol}} \underset{\text{Alkene}}{>\!C\!=\!C\!<} \ + \text{ZnX}_2$$

Dihalides with the preceding structure are called *vicinal* (Latin *vicinalis*, neighboring) *dihalides* or neighboring halides.

This method of alkene preparation is not generally useful because most vicinal dihalides are prepared from alkenes (Sec. 8.18). However, the removal of bromine is stereospecific, and stereochemistry allows us to understand the mechanism of the dehalogenation reaction.

In the same way that dehydrohalogenation is a bimolecular *anti*-elimination reaction (see Sec. 7.11), so is dehalogenation. Using debromination as an example, either of the two enantiomers of *d,l*-2,3-dibromobutane yields *cis*-2-butene as the product.

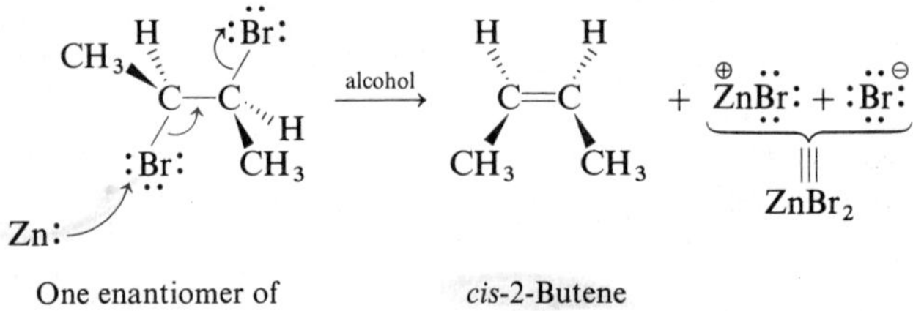

One enantiomer of   *cis*-2-Butene
d, l-2, 3-dibromobutane
(note anti-relationship of
two bromine atoms)

The transition state for this reaction is believed to be similar to that for dehydrohalogenation (see Sec. 7.11):

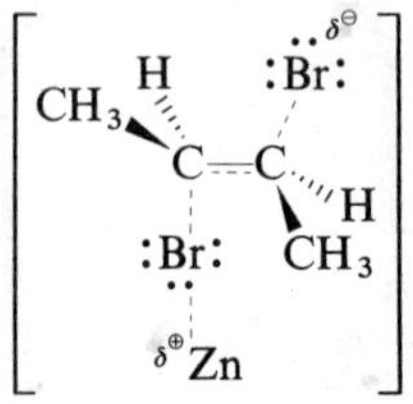

Transition state

Note also that oxidation and reduction occur in dehalogenation. The vicinal dihalide is reduced while zinc metal is oxidized.

We see in Sec. 8.16 how these dihalides are prepared from alkenes.

Question 7.23

Draw the three-dimensional projection formula for *meso*-2,3-dibromobutane and deduce the stereochemistry (*cis* or *trans*) of the 2-butene that is formed.

7.20 Preparation of Alkadienes

Substituted dienes are generally more difficult to prepare than 1,3-butadiene. However, the principles we studied so far can be extended to dienes. In particular, the dehydration of alcohols that contain two —OH groups (and are called *diols*) yield dienes:

$$CH_2\text{—}CH_2\text{—}CH_2\text{—}CH_2 \xrightarrow[\text{heat}]{H^\oplus} CH_2\text{=}CH\text{—}CH\text{=}CH_2$$
$$\quad| \qquad\qquad\qquad |$$
$$OH \qquad\qquad\qquad OH$$

1,4-Butanediol   1,3-Butadiene

$$\begin{array}{cc} CH_3 & CH_3 \\ | & | \\ CH_3\text{—}C\text{——}C\text{—}CH_3 \\ | & | \\ OH & OH \end{array} \xrightarrow[\text{heat}]{H^\oplus} \begin{array}{cc} CH_3 & CH_3 \\ | & | \\ CH_2\text{=}C\text{——}C\text{=}CH_2 \end{array}$$

2,3-Dimethyl-2,3-   2,3-Dimethyl-1,3-butadiene
butanediol

The dehydration of diols often leads to undesired side products.

Industrially, 1,3-butadiene is prepared from butane by cracking (see Sec. 7.10):

$$CH_3\text{—}CH_2\text{—}CH_2\text{—}CH_3 \xrightarrow[\text{catalyst}]{\text{heat}} CH_3\text{—}CH_2\text{—}CH\text{=}CH_2 + CH_3\text{—}CH\text{=}CH\text{—}CH_3 + H_2$$

$$\downarrow \begin{array}{l}\text{heat,}\\\text{catalyst}\end{array}$$

$$CH_2\text{=}CH\text{—}CH\text{=}CH_2 + H_2$$

1,3-Butadiene

7.21 Summary of Reactions

A. Preparation of Alkenes

1. Dehydrohalogenation of alkyl halides

$$\begin{array}{cc} | & | \\ \text{—}C\text{—}C\text{—} \\ | & | \\ H & X \end{array} \xrightarrow[\text{alcohol}]{KOH} \;\; {\Large \diagdown}C\text{=}C{\Large \diagup}$$

Alkyl halide   Alkene

Reactivity of R—X: $3° > 2° > 1°$
Product distribution: Most highly substituted alkene is *major* product

General reaction and mechanism:
   see Sec. 7.11
Structure-reactivity and
   structure-reactivity
   relationship: Sec. 7.12
Stereochemistry—*anti*- elimination:
   Sec. 7.13

2. Dehydration of alcohols

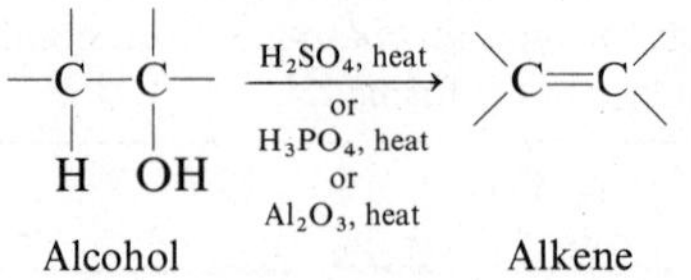

Alcohol Alkene

Reactivity of R—OH: 3° > 2° > 1°
Product distribution: Most highly substituted alkene is *major* product

General reaction: Sec. 7.14
Mechanism: Sec. 7.15
Carbocation stability: Sec. 5.7
Carbocation stability and reactivity: Sec. 7.16
Product distribution: Sec. 7.17
Rearrangements: Sec. 7.18

3. Dehalogenation of vicinal dihalides

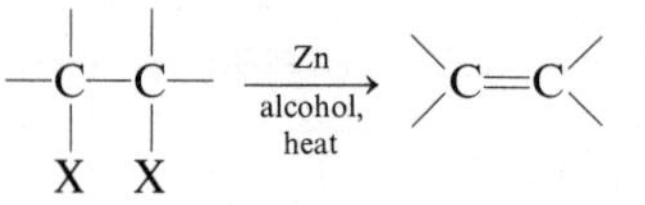

Vicinal dihalide Alkene

General reaction and mechanism: Sec. 7.19

B. Preparation of Alkadienes

1. Dehydration

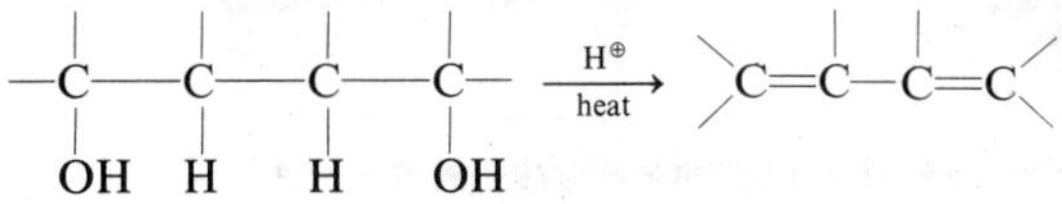

Dialcohol (diol) Diene

Reaction and examples: Sec. 7.20

2. Dehydrogenation (cracking)

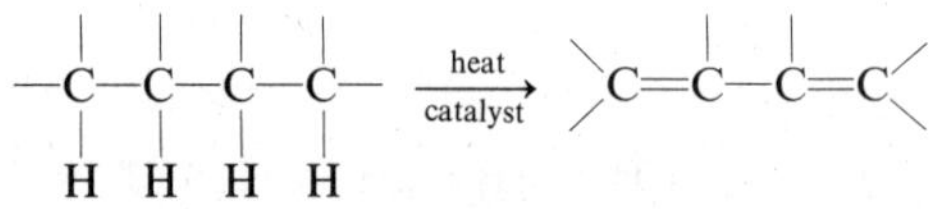

Reaction and examples: Sec. 7.20

Study Questions

7.24 Define and, where possible, give examples of each of the following terms:
(*a*) E1 reaction (*b*) E2 reaction (*c*) dehydration (*d*) dehydrohalogenation
(*e*) dehalogenation (*f*) π bond (*g*) conjugated diene (*h*) isolated diene
(*i*) 1,2-hydride shift (*j*) 1,2-alkyl shift (*k*) geometric isomers (*l*) *cis* isomer
(*m*) *trans* isomer (*n*) carbocation stability (*o*) electron-donating group
(*p*) *s-cis* (*q*) *s-trans* (*r*) *anti*-elimination

7.25 Briefly and concisely indicate what is meant by the term *geometric isomerism.* Why do geometric isomers exist in stable forms (and thus can be isolated), whereas conformational isomers do not (and thus cannot be isolated)?

7.26 Draw the structures corresponding to the following nine names. Which molecules would you expect to exist as geometric isomers? For those that do, draw the structures of the geometric isomers.
(*a*) 1-chloro-2-methyl-2-butene (*b*) 2-methyl-2-butene (*c*) 1-bromo-1-chloro-1-propene
(*d*) 2-octene (*e*) 1-octene (*f*) 1,1,2-tribromoethene
(*g*) cyclopentene (*h*) 2-methyl-2-hexene (*i*) 3-methyl-2-hexene

7.27 Provide IUPAC names for the following compounds.

(a) CH_3—CH_2—CH_2—$\overset{\overset{\displaystyle CH_3}{|}}{CH}$—$CH$=$CH_2$ (b) CH_3—$\overset{\overset{\displaystyle }{|}}{\underset{\underset{\displaystyle CH_3}{|}}{CH}}$—$CH$=$\overset{\overset{\displaystyle CH_3}{\diagup}}{\underset{\underset{\displaystyle CH_3}{\diagdown}}{C}}$

(c) CH_3—CH_2—$\overset{\overset{\displaystyle CH_3}{|}}{CH}$—$CH_2$—$\overset{\overset{\displaystyle CH_3}{|}}{C}$=$\overset{\underset{\underset{\displaystyle CH_2—CH_2—CH_2—CH_3}{|}}{}}{C}$—$CH_3$ (d) $(CH_3CH_2)_2C$=$C(CH_3)_2$

(e) CH_3—$\overset{\overset{\displaystyle CH_3}{|}}{\underset{\underset{\displaystyle CH_3}{|}}{C}}$—$CH_2$—$\overset{\overset{\displaystyle CH_3}{|}}{CH}$—$CH_2$—$\overset{\overset{\displaystyle CH_2—CH_3}{|}}{\underset{\underset{\displaystyle H—C=CH—CH_3}{|}}{C}}$—$CH_2$—$CH_2$—$CH_2$—$CH_3$ (ignore stereochemistry)

(f) CH_2=$\overset{\underset{\underset{\displaystyle Cl}{|}}{}}{C}$—$\overset{\overset{\overset{\displaystyle }{||}}{}}{\underset{\underset{\displaystyle CH_2}{}}{C}}$—$CH_2$—$CH_3$ (g) $\overset{\overset{\displaystyle CH_3}{\diagup}}{\underset{\underset{\displaystyle CH_3}{\diagdown}}{C}}\overset{CH_3}{\underset{\underset{\underset{\displaystyle CH_3}{}}{C=C}}{\diagdown}}\overset{H}{\diagup}\overset{}{\underset{CH_3}{}}$ (h) $\overset{\overset{\displaystyle CH_3—CH_2}{\diagdown}}{\underset{\underset{\displaystyle Br}{|}}{C}}$=$\overset{\overset{\displaystyle CH_3}{\diagup}}{\underset{\underset{\displaystyle Br}{|}}{C}}$

(i) $\overset{\overset{\displaystyle CH_3—CH_2—CH_2—CH_2}{\diagdown}}{\underset{\underset{\displaystyle CH_3—CH_2—CH_2}{\diagup}}{C}}$=$\overset{\overset{\displaystyle CH_2—CH_3}{\diagup}}{\underset{\underset{\displaystyle CH_3}{\diagdown}}{C}}$ (j) [cyclohexene with Cl, Cl, CH_3]   (k) [cyclopentene with CH_2CH_3, CH_3]

7.28 Draw structures corresponding to the following eight names.
(a) 4-methyl-1-heptene   (b) 4,4-dimethyl-7,7-dichloro-1-decene
(c) 3-methylcyclopentene   (d) 3-ethyl-2-methylcyclohexene
(e) 3-methyl-1,3-cyclohexadiene   (f) 3-ethyl-2-methyl-1,3-butadiene
(g) *cis*-3,4-dibromo-2-pentene   (h) *trans*-1,6-dibromo-3-hexene

7.29 Sketch a three-dimensional picture of the structure of ethene. Clearly label the carbon-carbon σ bond and the carbon-carbon π bond. What are the *approximate* bond angles in this molecule, and what is the hybridization at each carbon atom?

7.30 Draw a suitable three-dimensional projection formula for the following two molecules. Indicate approximate bond angles throughout. What atomic orbitals are used to form *each* bond orbital?
(a) CH_2=CH—CH_2—OH (b) CH_3—CH=CH—CH=CH_2 (*cis* isomer)

7.31 Indicate the products formed from dehydration of the following alcohols. Which product in each case is formed in the *major* amount, and why?

(a) CH_3—CH_2—CH_2—CH_2—CH_2—OH (b) CH_3—CH_2—$\overset{\overset{\displaystyle OH}{|}}{\underset{\underset{\displaystyle CH_3}{|}}{C}}$—$CH_3$

(c) CH_3—$\overset{\overset{\displaystyle OH}{|}}{\underset{\underset{\displaystyle CH_3}{|}}{C}}$—$\overset{\overset{\displaystyle }{}}{\underset{\underset{\displaystyle CH_3}{|}}{CH}}$—$CH_3$ (d) [cyclopentane with OH and CH_3]   (e) CH_3—$\overset{\overset{\displaystyle }{}}{\underset{\underset{\displaystyle CH_3}{|}}{CH}}$—$\overset{\overset{\displaystyle }{}}{\underset{\underset{\displaystyle CH_3}{|}}{CH}}$—$CH_2$—$OH$

7.32 Describe the *changes* in hybridization that occur in (a) dehydration of an alcohol and (b) dehydrobromination of an alkyl bromide.

7.33 In each of the three pairs of alkyl halides, which is the more reactive toward potassium hydroxide in alcohol? Why?

(*a*) $CH_3-CH_2-CH_2-Br$ or $(CH_3)_2CH-Br$

(*b*) $CH_3-\underset{\underset{CH_3}{|}}{CH}-CH_2-Cl$ or $CH_3-CH_2-\underset{\underset{CH_3}{|}}{CH}-Cl$

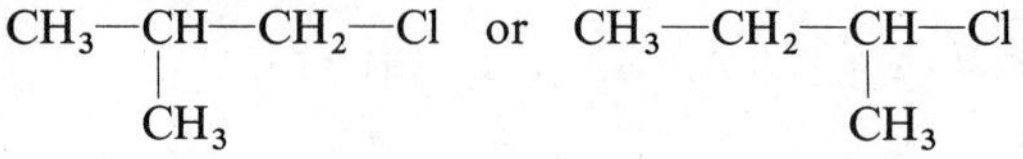

(*c*) $CH_3-CH_2-\underset{\underset{CH_3}{|}}{CH}-\overset{\overset{Br}{|}}{CH}-CH_3$ or $CH_3-CH_2-\underset{\underset{CH_3}{|}}{\overset{\overset{Br}{|}}{C}}-CH_2-CH_3$

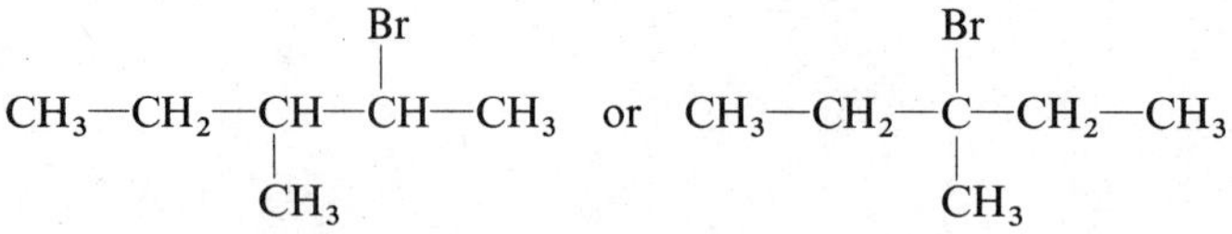

7.34 Give the product(s) of dehydrohalogenation of the following five alkyl halides. In cases where more than one product is formed, indicate which alkene is the major product. Why?

(*a*) $CH_3-\underset{\underset{CH_3}{|}}{CH}-\underset{\underset{CH_3}{|}}{CH}-CH_2-CH_2-Br$ (*b*) $CH_3-\underset{\underset{CH_3}{|}}{CH}-\underset{\underset{CH_3}{|}}{CH}-\underset{\underset{Br}{|}}{CH}-CH_3$

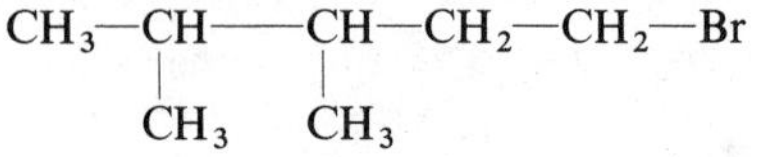

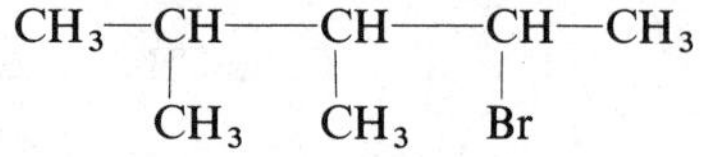

(*c*) 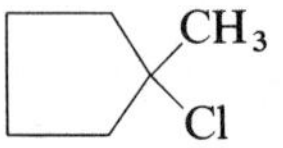   (*d*) 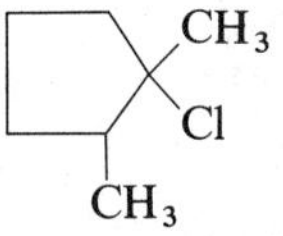   (*e*) $CH_3-CH_2-\underset{\underset{CH_3}{|}}{CH}-\underset{\underset{CH_3}{|}}{CH}-Br$

(*f*) $CH_3-CH_2-\underset{\underset{CH_3}{|}}{\overset{\overset{Cl}{|}}{C}}-CH_2-CH_2-CH_3$

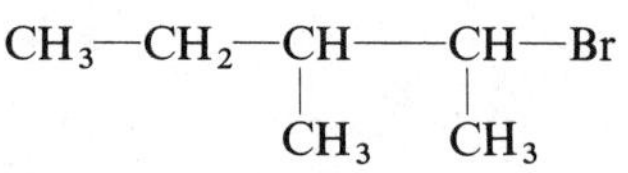

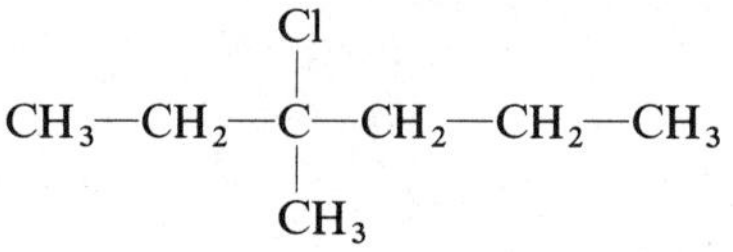

7.35 In the following four pairs of alkenes, which is the more stable alkene and why?

(*a*) $CH_3-CH_2-CH{=}CH_2$ or $(CH_3-CH_2)_2C{=}CH_2$

(*b*) *cis*-2-pentene   or   *trans*-2-pentene

(*c*) $CH_3CH_2-\underset{\underset{CH_3}{|}}{C}{=}CH_2$ or $CH_3-CH_2-\underset{\underset{CH_3}{|}}{C}{=}C\overset{\diagup CH_3}{\diagdown CH_3}$

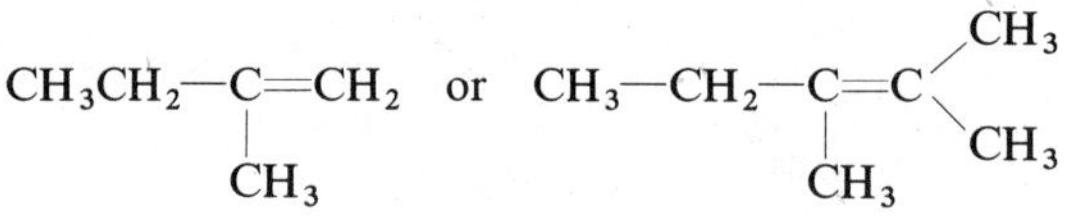

(*d*)   or

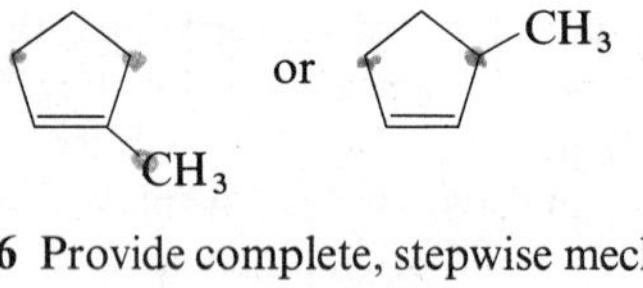

7.36 Provide complete, stepwise mechanisms for the following reactions. Show all intermediates.

(*a*) $CH_3-CH_2-\underset{\underset{OH}{|}}{\overset{\overset{CH_3}{|}}{C}}-CH_2-CH_3 \xrightarrow[\text{heat}]{H^{\oplus}} C_6H_{12}$ (two products, ignoring geometric isomers; draw their structures)

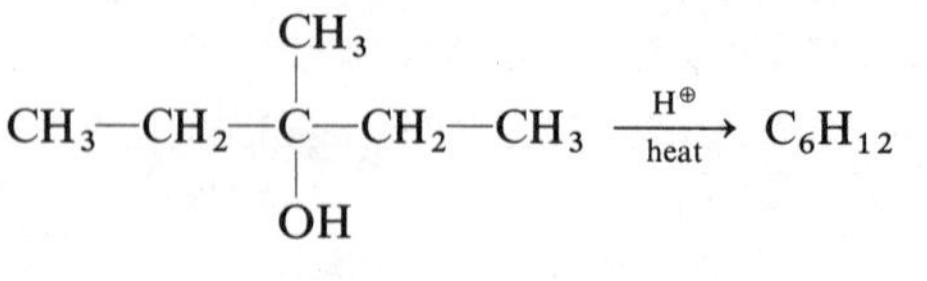

(*b*) $CH_3-CH_2-\underset{\underset{CH_3}{|}}{CH}-CH_2-OH \xrightarrow[\text{heat}]{H^{\oplus}} \underset{H}{\overset{CH_3}{\diagup}}C{=}C\underset{\diagdown CH_3}{\overset{\diagup CH_3}{}}$

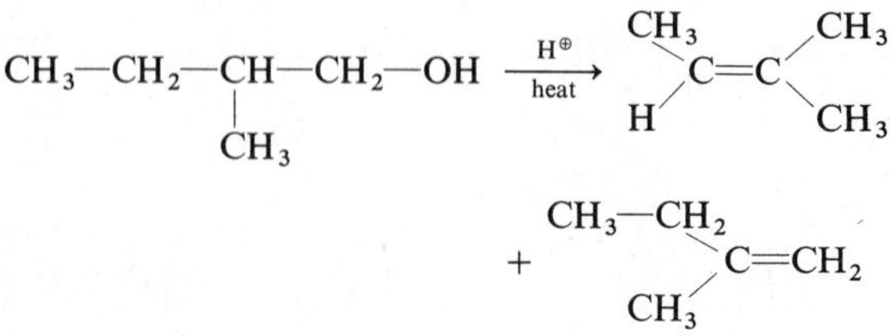

$+ \quad \underset{CH_3}{\overset{CH_3-CH_2}{\diagdown}}C{=}CH_2$ Which is the major product? Why?

(*c*) $CH_3-\underset{\underset{CH_3}{|}}{\overset{\overset{CH_3}{|}}{C}}-CH_2-OH \xrightarrow[\text{heat}]{H^{\oplus}} C_5H_{10}$ (two products; draw their structures)

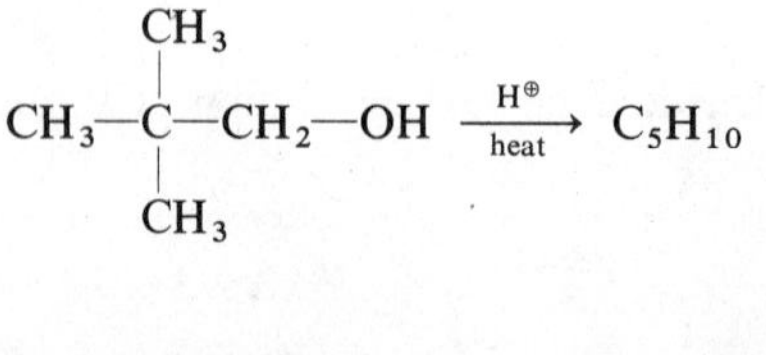

(d) CH_3—$\overset{\overset{\displaystyle CH_3}{|}}{\underset{\underset{\displaystyle CH_3 \ \ OH}{|\ \ \ \ |}}{C}}$—CH—$CH_3$ $\xrightarrow[\text{heat}]{H^{\oplus}}$ CH_3—$\overset{\overset{\displaystyle CH_3}{|}}{\underset{\underset{\displaystyle CH_3}{|}}{C}}$—CH=$CH_2$ + CH_3—$\overset{\overset{\displaystyle CH_3 \ \ CH_3}{|\ \ \ \ |}}{C}$=C—$CH_3$

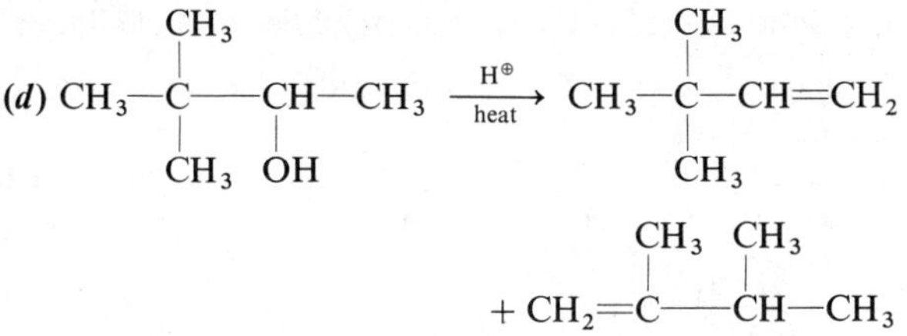

$+ CH_2$=$\overset{\overset{\displaystyle CH_3 \ \ CH_3}{|\ \ \ \ |}}{C}$—CH—$CH_3$ \qquad Which is the major product? Why?

(e) CH_3—$\overset{\overset{\displaystyle CH_3}{|}}{CH}$—$\overset{\overset{\displaystyle CH_3}{|}}{\underset{\underset{\displaystyle OH}{|}}{CH}}$—$\overset{\overset{\displaystyle CH_3}{|}}{\underset{\underset{\displaystyle CH_3}{|}}{C}}$—$CH_3$ $\xrightarrow[\text{heat}]{H^{\oplus}}$ CH_3—$\overset{\overset{\displaystyle CH_3}{|}}{C}$=CH—$\overset{\overset{\displaystyle CH_3}{|}}{\underset{\underset{\displaystyle CH_3}{|}}{C}}$—$CH_3$

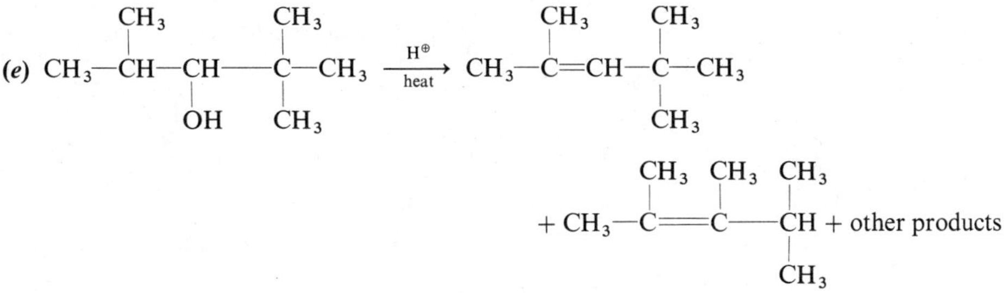

$+ CH_3$—$\overset{\overset{\displaystyle CH_3 \ \ CH_3 \ \ CH_3}{|\ \ \ \ |\ \ \ \ |}}{C}$=C—$\underset{\underset{\displaystyle CH_3}{|}}{CH}$ + other products

(f) CH_3—CH_2—$\underset{\underset{\displaystyle Br}{|}}{CH}$—$CH_2$—$CH_2$—$CH_3$ $\xrightarrow[\text{alcohol}]{KOH}$ CH_3—CH=CH—CH_2—CH_2—CH_3

$+ CH_3$—CH_2—CH=CH—CH_2—CH_3

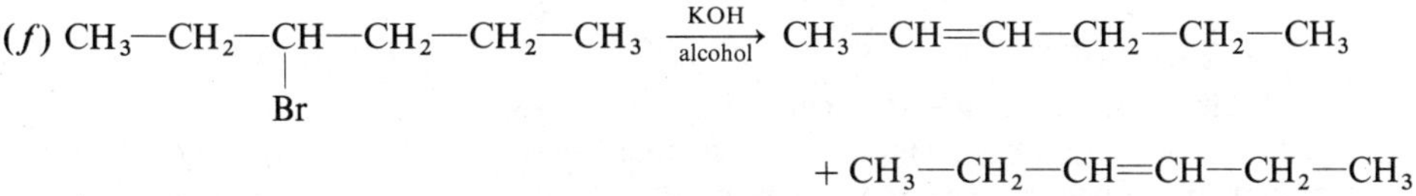

7.37 Treatment of alkyl halide (1) with a strong base (KOH in alcohol) yields only 2-methyl-1-butene as the product. Treatment of alcohol (2), however, with acid and heat yields 2-methyl-2-butene as the major product.
(a) Show a mechanism for reactions (1) and (2) to account for the observed products.

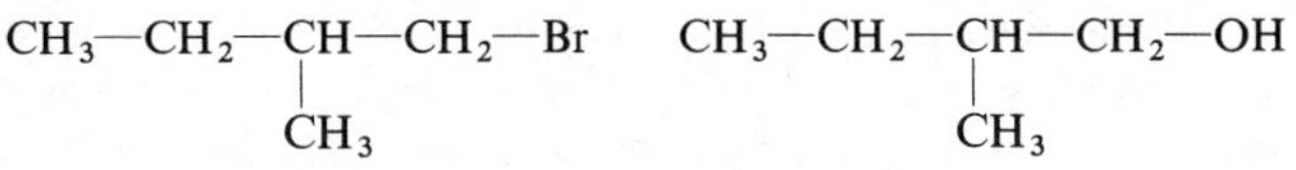

CH_3—CH_2—$\underset{\underset{\displaystyle CH_3}{|}}{CH}$—$CH_2$—Br \qquad CH_3—CH_2—$\underset{\underset{\displaystyle CH_3}{|}}{CH}$—$CH_2$—OH

\qquad (1) \qquad\qquad\qquad\qquad (2)

(b) Based on these results, which method, *dehydrohalogenation* or *dehydration*, is preferred for preparing 2-methyl-1-butene?
7.38 Draw the structure of a bromoalkane (alkyl bromide) that will: (a) give only 1-pentene on dehydrobromination; (b) give only 2-pentene on dehydrobromination.
7.39 Suggest a reason why the acid-catalyzed dehydration of 1,4-cyclohexanediol (1) gives almost exclusively 1,3-cyclohexadiene (2) and very little 1,4-cyclohexadiene (3).

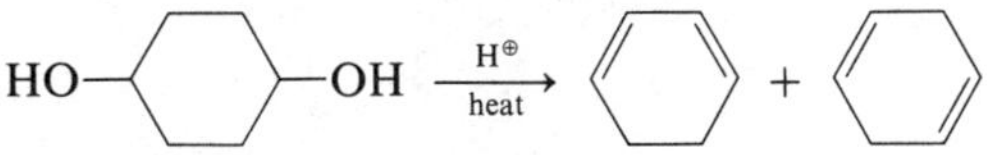

HO—⬡—OH $\xrightarrow[\text{heat}]{H^{\oplus}}$ ⬡ + ⬡

\qquad (1) \qquad\qquad (2) \qquad (3)

\qquad\qquad\qquad *Major*

7.40 Provide a reasonable explanation for the following observed results:

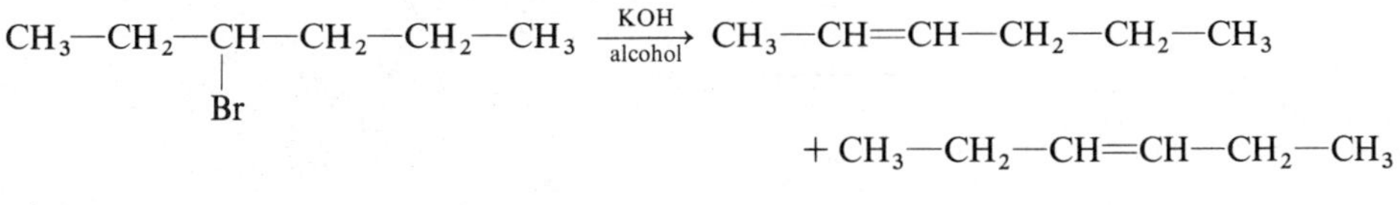

CH_3—CH_2—$\underset{\underset{\displaystyle Br}{|}}{CH}$—$CH_2$—$CH_2$—$CH_3$ $\xrightarrow[\text{alcohol}]{KOH}$ CH_3—CH=CH—CH_2—CH_2—CH_3

$+ CH_3$—CH_2—CH=CH—CH_2—CH_3

whereas

CH_3—CH_2—$\underset{\underset{\displaystyle Br}{|}}{CH}$—$CH_2$—CH=$CH_2$ $\xrightarrow[\text{alcohol}]{KOH}$ CH_3—CH_2—CH=CH—CH=CH_2

\qquad\qquad\qquad\qquad\qquad\qquad Only product

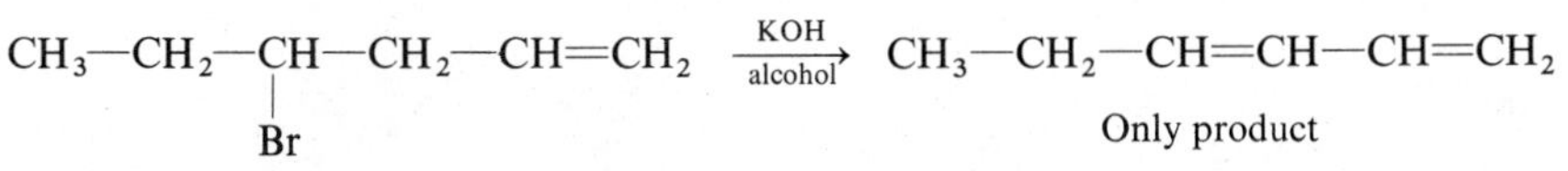

7.41 We saw that electron-donating groups (those that release electrons inductively) stabilize carbocations, $R^{\oplus}$, and radicals, $R\cdot$, because both these species are electron-deficient. Using the principles that explain these observations, predict the relative stabilities of the following carbanions: $\overset{\ominus}{RCH_2:}$, $\overset{\ominus}{R_2CH:}$, $\overset{\ominus}{R_3C:}$, and $\overset{\ominus}{CH_3:}$, where R is an electron-donating group (for example, alkyl group). Justify your prediction.

7.42 Give the geometry of the indicated product of the base-promoted elimination of the following compounds. Which reaction would you expect to be faster? Briefly explain why.

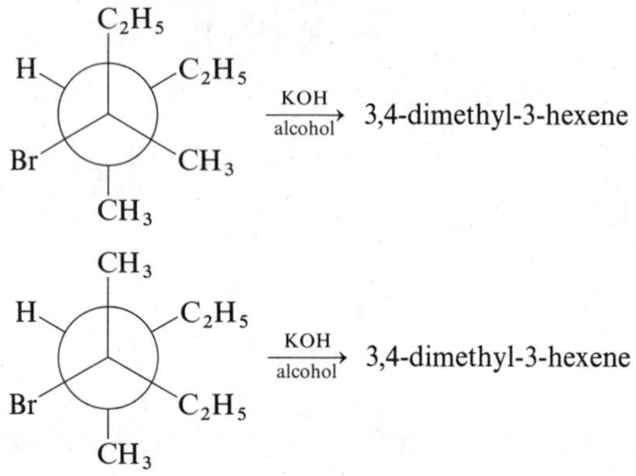

C_2H_5
H
C_2H_5
Br
CH_3
CH_3
KOH
alcohol
3,4-dimethyl-3-hexene
CH_3
H
C_2H_5
Br
C_2H_5
CH_3
KOH
alcohol
3,4-dimethyl-3-hexene

7.43 The E2 reaction was presented in Sec. 7.11 as being a concerted *anti*-elimination in which all bond breaking and making occur simultaneously. An alternative explanation has been proposed involving a stepwise mechanism:

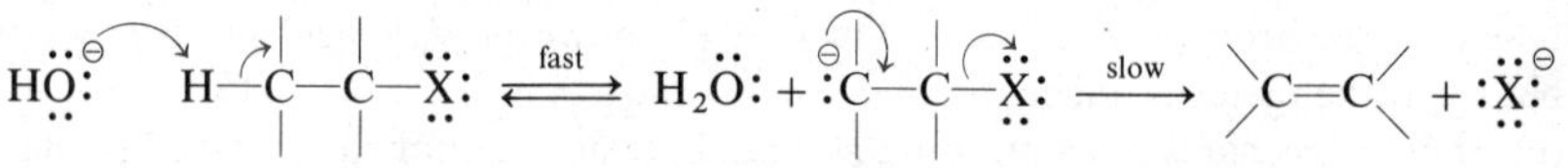

fast
slow

(***a***) Would the rate of reaction for this mechanism be first order in both hydroxide ion and alkyl halide, that is, rate of reaction = $k_2[OH^{\ominus}]$ [alkyl halide]? Why or why not?

(***b***) This mechanism has been disproved by experiments in which $OH^{\ominus}$ and H_2O are replaced with $OD^{\ominus}$ and D_2O (D = deuterium). What results would you expect with deuterium present if the reaction is E2, as previously described, or if it occurs as shown here (that is, an E1 type of reaction)? How would your results support the E2 mechanism and disprove the two-step E1 mechanism?

8

Alkenes and Alkadienes: Reactions; Natural and Synthetic Polymers

In this chapter we examine the various reactions of alkenes—reactions that are characteristic of the carbon-carbon double bond or that occur because the double bond is present in the molecule. Three general types of reactions are taken up: **addition, substitution,** and **cleavage.** Of the three types, addition reactions predominate.

As the discussion progresses, sometimes it is not only desirable but also essential to compare the reactions of alkenes with those of alkadienes (compounds containing two carbon-carbon double bonds). In several instances we see that the reactions of the carbon-carbon double bond are similar to those of two strained cycloalkanes: cyclopropane and cyclobutane.

The chapter concludes by examining polymerization reactions and applications of alkene and alkadiene chemistry to naturally occurring compounds.

8.1 Ionic Addition Reactions

The general equation for the addition of a reagent A—B to a carbon-carbon double bond is:

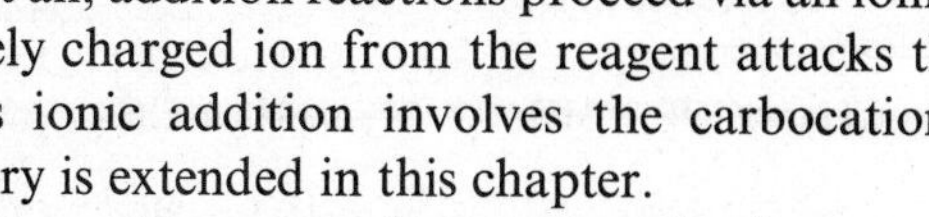

$$>C=C< \; + \; A-B \quad \longrightarrow \quad -\underset{A}{\overset{|}{C}}-\underset{B}{\overset{|}{C}}-$$

Addition: π bond broken

where A—B represents a variety of reagents, such as H—H, H—X, X—X, and H—OH.

Most, but not all, addition reactions proceed via an ionic, stepwise mechanism, whereby a positively charged ion from the reagent attacks the π bond and forms a carbocation. Thus ionic addition involves the carbocation theory developed in Chap. 7. This theory is extended in this chapter.

Because there are two ways the reagent A—B can add to a carbon-carbon double bond—either as shown here or with A and B reversed—we should also consider the *orientation of addition.*

It is not surprising that the carbon-carbon double bond reacts with a positively charged ion. The π bond is electron-rich and the electrons in it are generally exposed and open to attack. Also, the electrons in the π bond are quite polarizable (see Sec. 9.7). They are readily available to attack by an electron-deficient (positively charged) ion, an *electrophile*. Reactions that occur *via* this ionic mechanism are called **electrophilic addition reactions** because the first step is the attack of the electrophile on the π cloud, resulting in the addition of the electrophile to the electron-rich double bond; a carbocation is produced, as the following illustrates:

$$E^{\oplus}$$

Electrophile:
is electron-
deficient,
attracts
electrons from
polarizable
π cloud

σ bond

π Cloud is electron-
rich; is easily
attacked by $E^{\oplus}$,
provides source
of electrons

Adds to give more
stable carbocation:
π bond is broken

Two general classes of reagents add to alkenes: (1) those in which two identical atoms add (**symmetrical reagents,** such as X_2) and (2) those in which two different atoms or groups of atoms add (**unsymmetrical reagents,** such as H—X, HOH, H_2SO_4). We consider unsymmetrical reagents first because they allow us to develop principles that are used in the discussion of symmetrical reagents.

8.2   Addition of Hydrogen Halides (Unsymmetrical Reagent) to Alkenes

The general reaction for the addition of hydrogen halide, H—X, to a carbon-carbon double bond is:

General reaction for addition of H—X:

$$\text{C=C} + \text{H—X} \xrightarrow[\text{(solvent)}]{\text{CCl}_4} \text{—C—C—}$$
$$\qquad\qquad\qquad\qquad\qquad\quad \text{H} \quad \text{X}$$

Alkene   Alkyl halide

where X=Cl, Br, I

This reaction is normally carried out in an inert solvent such as carbon tetrachloride, and the hydrogen halide is used in its anhydrous gaseous form. Also, highly purified reactants and solvents are needed for a good reaction.

When the alkene is symmetrical (that is, it contains identical substituents on each carbon atom of the double bond), only one possible product can be obtained. For example:

$$CH_2{=}CH_2 + HBr \xrightarrow{CCl_4} CH_2{-}CH_2$$
$$\qquad\qquad\qquad\qquad\qquad\quad \text{H} \qquad \text{Br}$$

Ethene   Bromoethane

Cyclohexene   + HCl $\xrightarrow{CCl_4}$ Chlorocyclohexane

$$CH_3 \diagdown \atop H \diagup C=C \diagup CH_3 \atop \diagdown H \quad + \text{HI} \xrightarrow{\text{CCl}_4} \quad CH_3-\underset{\underset{H}{|}}{CH}-\underset{\underset{I}{|}}{CH}-CH_3$$

cis-2-Butene   2-Iodobutane
(or *trans*-)   (*Product racemic; see Ques. 8.2*)

Unsymmetrical alkenes could yield two products. For example, the reaction between 1-butene and hydrogen bromide could produce both 1-bromobutane and 2-bromobutane, but 2-bromobutane is the only observed product.

$$CH_3-CH_2-\underset{\underset{H}{\uparrow}}{CH}=\underset{\underset{Br}{\uparrow}}{CH_2} \xrightarrow{\text{CCl}_4} CH_3-CH_2-\underset{\underset{H}{|}}{CH}-\underset{\underset{Br}{|}}{CH_2}$$

1-Bromobutane
Not formed to any appreciable extent

$$CH_3-CH_2-\underset{\underset{Br}{\uparrow}}{CH}=\underset{\underset{H}{\uparrow}}{CH_2} \xrightarrow{\text{CCl}_4} CH_3-CH_2-\underset{\underset{Br}{|}}{CH}-\underset{\underset{H}{|}}{CH_2}$$

2-Bromobutane
Predominant product

The addition of hydrogen chloride to methylpropene could also give two products, 1-chloro-2-methylpropane and 2-chloro-2-methylpropane. The only product isolated, however, is 2-chloro-2-methylpropane.

$$CH_3 \diagdown \atop CH_3 \diagup C=CH_2 \atop \underset{H-Cl}{\uparrow \ \uparrow} \quad \xcancel{\xrightarrow{\text{CCl}_4}} \quad CH_3 \diagdown \atop CH_3 \diagup C-CH_2 \atop \underset{H \quad Cl}{|\ \ |}$$

1-Chloro-2-methylpropane
*Not formed to any
appreciable extent*

$$CH_3 \diagdown \atop CH_3 \diagup C=CH_2 \atop \underset{Cl-H}{\uparrow \ \uparrow} \quad \xrightarrow{\text{CCl}_4} \quad CH_3 \diagdown \atop CH_3 \diagup C-CH_2 \atop \underset{Cl \quad H}{|\ \ |}$$

2-Chloro-2-methylpropane
Predominant product

Careful examination of the predominant products in the previous examples reveals that the halide ends up on the more highly substituted carbon atom. Now we consider the mechanism of addition and use it to explain why only one product is formed.

8.3   Mechanism of Hydrogen Halide Addition to Alkenes

In the introduction to the reactions of alkenes in Sec. 8.1, a general mechanism for the addition of various reagents to the carbon-carbon double bond was given. Now let us apply it to the addition of hydrogen halides. The reaction is normally done with anhydrous H—X in an inert organic solvent; under these conditions, H—X reacts directly with the alkene to form a carbocation, which then reacts with the nucleophile to give the observed product.

Step 1:

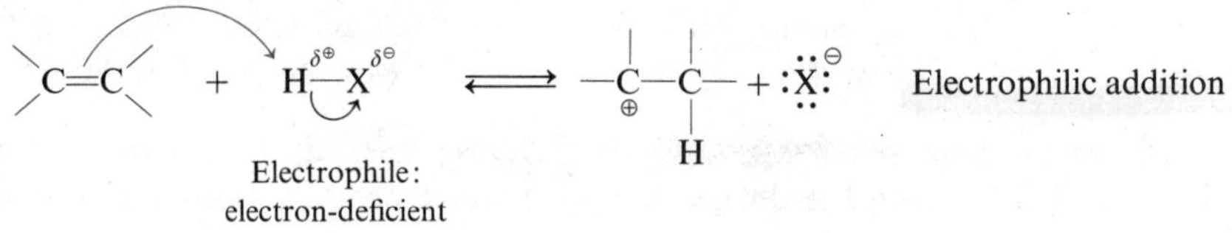

Electrophilic addition

Electrophile:
electron-deficient

Step 2:

Reaction between carbocation and nucleophile

Nucleophile:
electron-rich

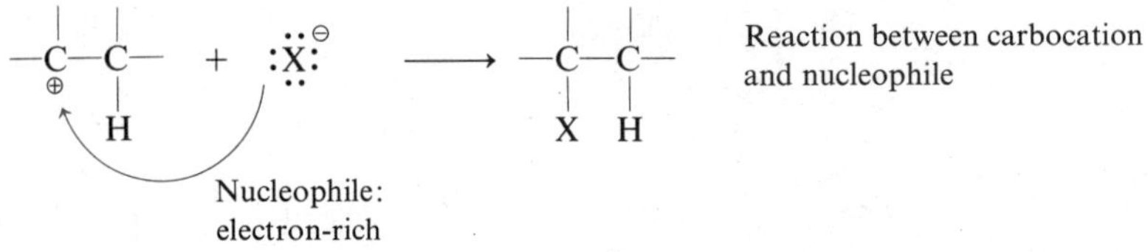

In step 1 the electron-deficient proton $H^{\oplus}$ is the electrophile that seeks electrons and finds them in the π bond. This forms a carbocation, which is also electron-deficient and *very unstable and reactive*. The carbocation finds electrons on the electron-rich halide ion, a *nucleophile*. These two ions then react to form a carbon-halogen bond.

The addition of H—X to the carbon-carbon double bond is called electrophilic addition because the positive part, $H^{\oplus}$, which is the electrophile, adds *first*, and then the nucleophilic halide ion, $:\overset{..}{\underset{..}{X}}:^{\ominus}$, adds to the carbocation that is formed.

The reactions of symmetrical alkenes with hydrogen halides can now be interpreted in terms of carbocation intermediates and electrophilic addition:

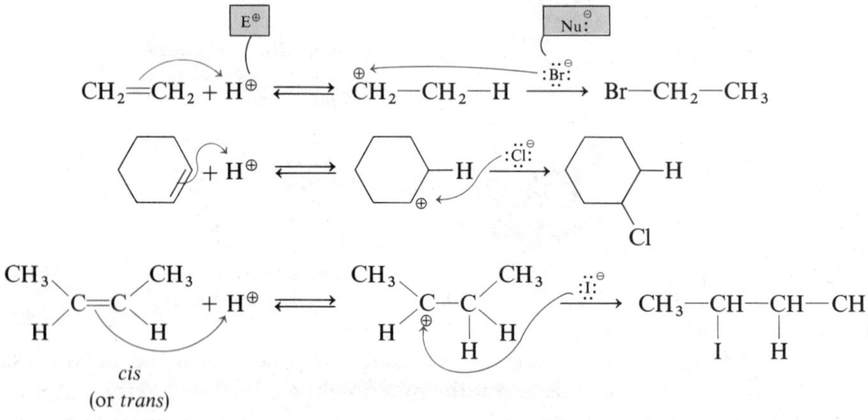

We now have another reaction characteristic of carbocations, which may:

1. Lose a proton to give an alkene;
2. Rearrange to give a more stable carbocation;
3. React with an electron-rich species, a nucleophile.

Further support for the two-step mechanism involving carbocations comes from the addition of hydrogen halides to unsymmetrical alkenes. Consider the reaction between 1-butene and HBr. The proton can be added in one of two ways to give either a primary carbocation, the *n*-butyl cation, or a more stable secondary carbocation, the *sec*-butyl cation. Because the order of carbocation stability is

$3° > 2° > 1° > \overset{\oplus}{CH_3}$, the preferred mode of addition is the one that gives the *sec*-butyl cation. This ion then reacts immediately with the bromide ion to form the observed product, 2-bromobutane.

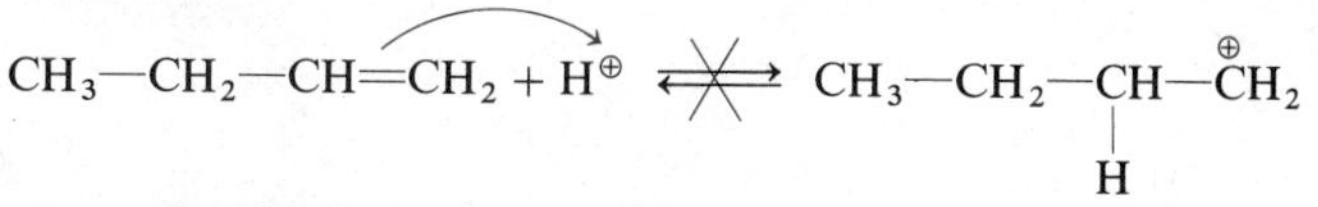

$$CH_3—CH_2—CH\!=\!CH_2 + H^\oplus \; \overset{\times}{\rightleftharpoons} \; CH_3—CH_2—\overset{\oplus}{CH}—CH_2$$
$$\underset{H}{|}$$

(*Note:* When tail of arrow is moving *away* from C-1 or C-2, it signifies that π electrons are taken *away* from that particular atom. Carbon that ends up with three bonds to it has had the electrons removed.)

1° carbocation
Not formed

$$CH_3—CH_2—CH\!=\!CH_2 + H^\oplus \; \rightleftharpoons \; CH_3—CH_2—\overset{\oplus}{CH}—CH_2$$
$$\underset{H}{|}$$

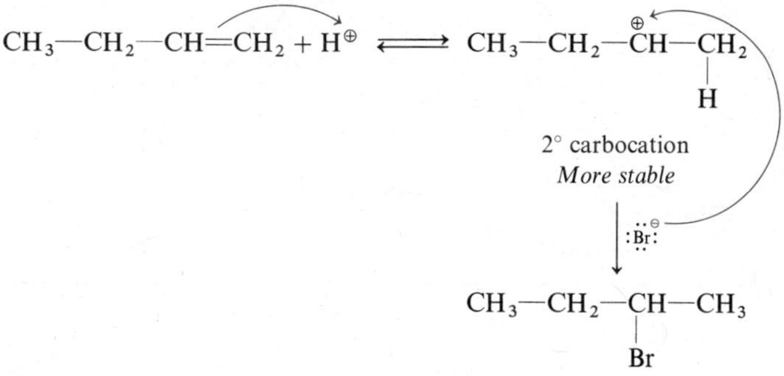

2° carbocation
More stable

$$:\!\overset{..}{\underset{..}{Br}}\!:^{\ominus}$$

$$CH_3—CH_2—CH—CH_3$$
$$\underset{Br}{|}$$

Observed product

The energy-profile diagram in Fig. 8.1 shows the two possible modes of addition of the proton to 1-butene. Not only is the secondary carbocation more stable, but the energy of activation, E_A, for its formation is less than that for the primary ion, so it is also formed faster. Thus, formation of the secondary carbocation is greatly favored.

Analogously, the addition of hydrogen chloride to methylpropene is viewed as follows:

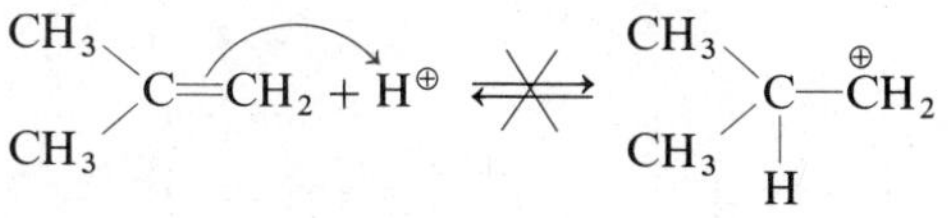

$$\underset{CH_3}{\overset{CH_3}{\diagdown\!\!\diagup}}C\!=\!CH_2 + H^\oplus \; \overset{\times}{\rightleftharpoons} \; \underset{CH_3}{\overset{CH_3}{\diagdown\!\!\diagup}}\overset{\oplus}{C}—CH_2$$
$$\underset{H}{|}$$

1° carbocation
Not formed

$$\underset{CH_3}{\overset{CH_3}{\diagdown\!\!\diagup}}C\!=\!CH_2 + H^\oplus \; \rightleftharpoons \; \underset{CH_3}{\overset{CH_3}{\diagdown\!\!\diagup}}\overset{\oplus}{C}—CH_2$$
$$\underset{H}{|}$$

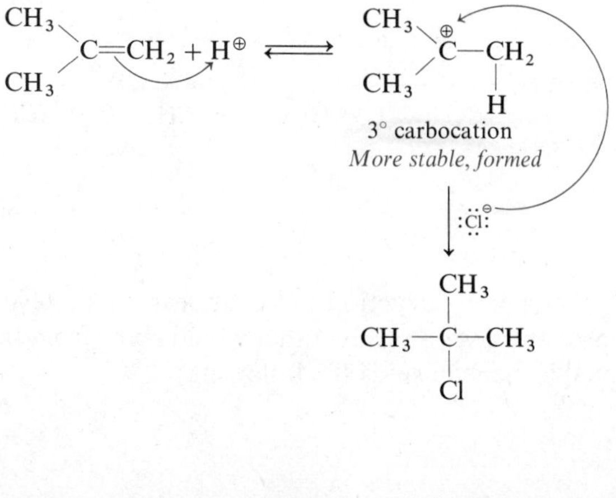

3° carbocation
More stable, formed

$$:\!\overset{..}{\underset{..}{Cl}}\!:^{\ominus}$$

$$CH_3—\overset{\overset{\displaystyle CH_3}{|}}{\underset{\underset{\displaystyle Cl}{|}}{C}}—CH_3$$

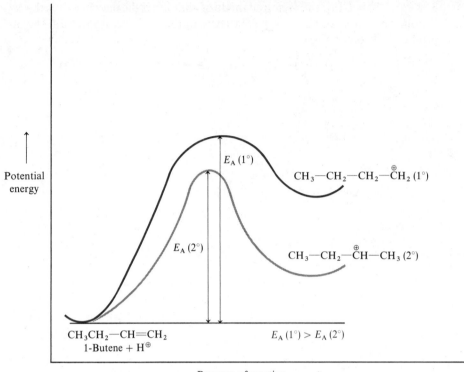

Figure 8.1   Energy-profile diagram showing the reaction between a proton and 1-butene to give two possible carbocations: the primary *n*-butyl cation and the *sec*-butyl cation. The more stable secondary ion is formed faster.

Cyclopropanes and substituted cyclopropanes also react with hydrogen halides to give open-chain products. This abnormal reactivity has been attributed to the great amount of strain in the three-membered ring; the larger cycloalkanes do not so react.

$$\begin{array}{c} CH_2 \\ | \\ CH_2 \end{array}\!\!\!\!> CH_2 + HBr \longrightarrow \underset{\underset{H}{|}}{CH_2}-CH_2-\underset{\underset{Br}{|}}{CH_2}$$

1-Bromopropane

$$\begin{array}{c} CH_2 \\ | \\ CH_2 \end{array}\!\!\!\!> CH-CH_3 + HI \longrightarrow \underset{\underset{H}{|}}{CH_2}-CH_2-\underset{\underset{I}{|}}{CH}-CH_3$$

2-Iodobutane

These ring-opening reactions also appear to involve carbocations because methylcyclopropane gives a product that would come from the more stable carbocation; a reasonable mechanism is the following:

$$\underset{CH_2}{\overset{CH_2}{\diagdown}}CH-CH_3 + H^{\oplus} \longrightarrow CH_2-CH_2-\overset{\oplus}{\underset{H}{C}H}-CH_3$$

E$^{\oplus}$

2° carbocation

$\overset{\ominus}{\overset{\cdot\cdot}{\underset{\cdot\cdot}{I}}}$:

Nu:$^{\ominus}$

$$CH_2-CH_2-\underset{I}{\overset{|}{C}H}-CH_3$$
$$\overset{|}{H}$$

2-Iodobutane
Observed product

The other possible carbocations that might result by breaking the ring carbon-carbon bonds are primary and so are not formed.

$$\underset{\overset{\oplus}{CH_2}}{\overset{CH_3}{\diagdown}}CH-CH_3 \quad or \quad \underset{CH_3}{\overset{\overset{\oplus}{CH_2}}{\diagdown}}CH-CH_3$$

This abnormal ring opening of cyclopropane led Baeyer to set forth his strain theory (see Sec. 3.15).

Question 8.1

Write mechanisms to show how the following addition reactions occur, paying particular attention to the structure of the product. There is only one product from each reaction.

(*a*) 2-methyl-2-butene + HBr   (*b*) [cyclopentene with CH₃] + HCl

(*c*) 2-methyl-1-butene + HI   (*d*) 2,3-dimethyl-1-butene + HCl
(*e*) 1,1-dimethylcyclopropane + HBr

Question 8.2

The addition of hydrogen iodide, HI, to 2-pentene gives *racemic* 2-iodopentane as the major product plus a lesser amount of 3-iodopentane. Provide mechanisms for the formation of these products, and deduce what must be true about the relative stabilities of the two carbocations that are formed.

8.4   Further Evidence for Electrophilic Addition: Rearrangements

The intervention of carbocations in the addition of hydrogen halides to alkenes is supported further by rearrangements occurring in certain compounds (see Sec. 7.18). For example, the addition of HCl to 3-methyl-1-butene gives two products:

$$CH_3-\underset{\overset{|}{CH_3}}{\overset{|}{C}H}-CH=CH_2 \xrightarrow[CCl_4]{HCl} CH_3-\underset{\overset{|}{CH_3}}{\overset{|}{C}H}-\underset{Cl}{\overset{|}{C}H}-CH_3 + CH_3-\underset{Cl}{\overset{\overset{|}{CH_3}}{\overset{|}{C}}}-CH_2-CH_3$$

3-Methyl-1-butene   3-Chloro-2-methylbutane   2-Chloro-2-methylbutane

The formation of these two products is accounted for nicely by carbocations. First, addition of the proton to (1) gives the more stable secondary carbocation (2). Then, (2) may do two things: react with chloride ion to form one product and rearrange to form a new, more stable tertiary carbocation (3), which reacts with chloride ion to form the second product. These steps are shown here. The rearrangement occurs through a hydride shift. Note that a methyl shift would result in the formation of a secondary carbocation that has no added stability; thus, only a hydride shift occurs. (Draw this pathway.)

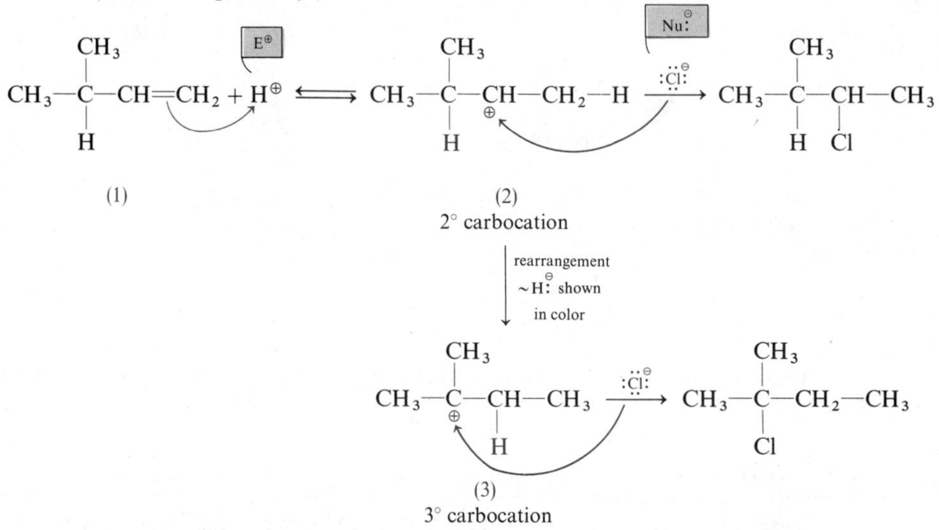

$$CH_3-\overset{\underset{|}{CH_3}}{\underset{|}{C}}-CH=CH_2 + H^\oplus \rightleftharpoons CH_3-\overset{\underset{|}{CH_3}}{\underset{|}{C}}-\overset{\oplus}{CH}-CH_2-H \xrightarrow{:\overset{..}{\underset{..}{Cl}}:^\ominus} CH_3-\overset{\underset{|}{CH_3}}{\underset{|}{C}}-\overset{\underset{|}{Cl}}{CH}-CH_3$$

(1)

(2)

2° carbocation

rearrangement
~H:$^\ominus$ shown
in color

$$CH_3-\overset{\underset{|}{CH_3}}{\underset{|}{\overset{\oplus}{C}}}-\overset{\underset{|}{H}}{CH}-CH_3 \xrightarrow{:\overset{..}{\underset{..}{Cl}}:^\ominus} CH_3-\overset{\underset{|}{CH_3}}{\underset{|}{C}}-CH_2-CH_3$$

(3)

3° carbocation

Question 8.3

The addition of hydrogen bromide to 3,3-dimethyl-1-butene in carbon tetrachloride solvent yields two alkyl bromides. Derive the structures of these compounds by considering the mechanism of the addition. Show all steps in the mechanism you write.

8.5   Hydration: Addition of Water (Unsymmetrical Reagent) to Alkenes

In Sec. 7.14 we saw how alcohols can be dehydrated (that is, water removed) by strong, concentrated acids:

Dehydration:

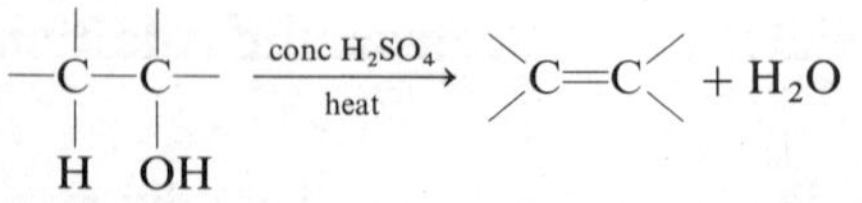

$$-\overset{\underset{|}{H}}{\underset{|}{C}}-\overset{\underset{|}{OH}}{\underset{|}{C}}- \xrightarrow[\text{heat}]{\text{conc } H_2SO_4} C=C + H_2O$$

Alcohol   Alkene

On the other hand, the elements of water can be added to an alkene in the presence of dilute, aqueous acid when there is a large excess of water present:

Hydration:

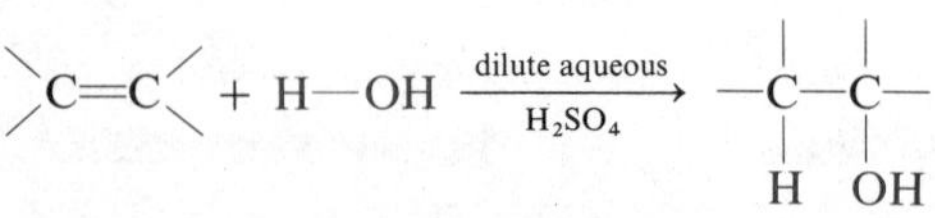

$$C=C + H-OH \xrightarrow[H_2SO_4]{\text{dilute aqueous}} -\overset{\underset{|}{H}}{\underset{|}{C}}-\overset{\underset{|}{OH}}{\underset{|}{C}}-$$

Hydration does not occur, however, unless dilute mineral acid is added; this strongly suggests that the acid is needed for accomplishing the electrophilic addition of water. Three typical examples of the hydration of alkenes are:

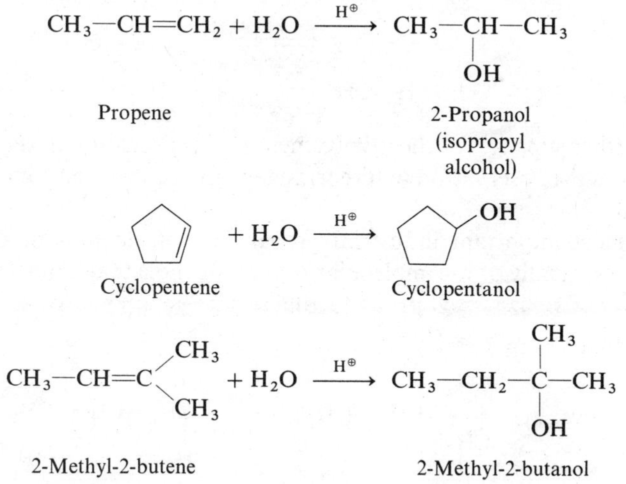

$$CH_3—CH=CH_2 + H_2O \xrightarrow{H^{\oplus}} CH_3—CH—CH_3$$
$$\underset{OH}{|}$$

Propene

2-Propanol
(isopropyl
alcohol)

Cyclopentene   Cyclopentanol

$$CH_3—CH=C\begin{smallmatrix}CH_3\\CH_3\end{smallmatrix} + H_2O \xrightarrow{H^{\oplus}} CH_3—CH_2—\underset{\underset{OH}{|}}{\overset{\overset{CH_3}{|}}{C}}—CH_3$$

2-Methyl-2-butene   2-Methyl-2-butanol

Another important observation about hydration is that the reaction is *acid-catalyzed*. A general mechanism consistent with these facts is the following:

Step 1: Proton addition to give carbocation

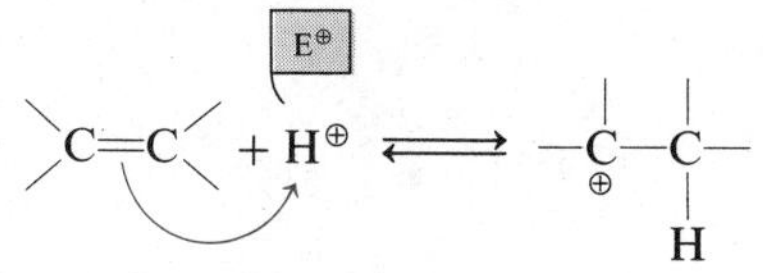

Step 2: Reaction of carbon with water, a nucleophile

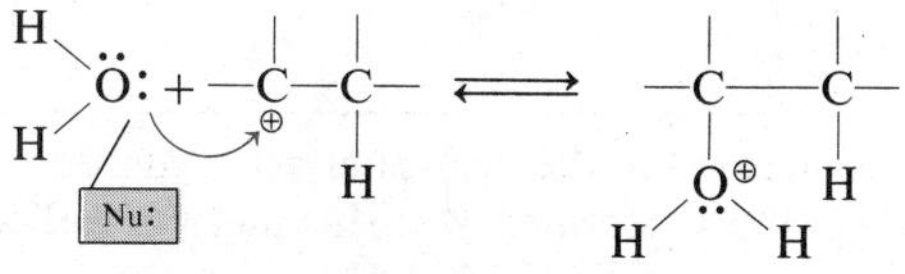

Step 3: Loss of proton to give alcohol

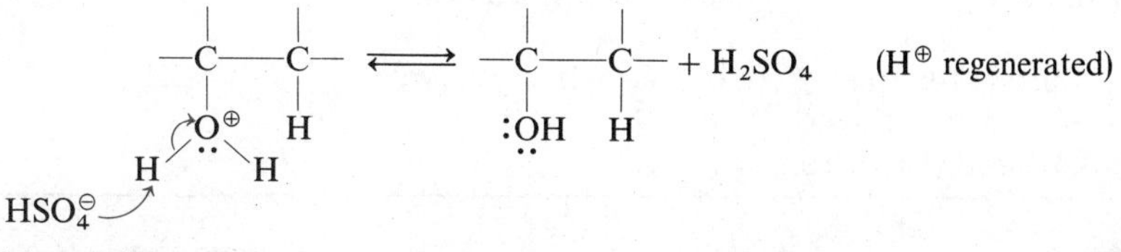

This mechanism accounts for the acid-catalyzed aspect of the reaction, because a proton is consumed in step 1 and regenerated in step 3 (as H_2SO_4). More important, though, is that each step is shown as reversible, and for good reason: *This hydration mechanism is just the reverse of that for dehydration.* If a reaction is truly reversible, then it and its reverse must follow the same individual steps but in opposite directions. Such a reaction is said to obey the **principle of microscopic reversibility.** Therefore, examine carefully the dehydration and hydration mechanisms and see that they are just the reverse of each other.

What are the differences between hydration and dehydration if their mechanisms are identical? Simply, the reaction conditions are different. In dehydration the strong, concentrated acid removes water as it is formed and usually the alkene is volatile enough so that it too is removed as it is formed. In hydration an excess of water is present and every effort is made to keep the alkene in the reaction vessel. Hydration and dehydration may be summarized as follows:

$$
\begin{array}{c}
\text{Dehydration} \\
-\overset{|}{\underset{|}{C}}-\overset{|}{\underset{|}{C}}- \underset{\xrightarrow{\ H^{\oplus},\ H_2O\ (\text{excess})\ }}{\overset{\text{conc } H_2SO_4,\ \text{heat}}{\rightleftarrows}} \quad \overset{}{\underset{}{C}}=\overset{}{\underset{}{C}} \\
\overset{}{\underset{\text{H}\quad\text{OH}}{}} \qquad \text{Hydration}
\end{array}
$$

Further support for the involvement of a carbocation in the hydration reaction comes from the observation that rearrangement occurs when a more stable cation can be formed.

A most important industrial preparation of alcohols involves the hydration reaction, especially of low molecular weight alcohols that come from simple alkenes. The highly strained cyclopropane molecule also opens when treated with water under acidic conditions:

$$
\begin{array}{ccc}
\overset{\text{CH}_2}{\underset{\text{CH}_2}{\diagdown}}\!\!\!>\!\text{CH}_2 \;+\; H_2O & \xrightarrow{\;H^{\oplus}\;} & \text{CH}_2-\text{CH}_2-\text{CH}_2 \\
 & & \quad\;\;|\qquad\qquad\;| \\
 & & \quad\;\;\text{H}\qquad\quad\;\;\text{OH}
\end{array}
$$

Cyclopropane   1-Propanol
(*n*-propyl alcohol)

$$
\begin{array}{ccc}
\overset{\text{CH}_2}{\underset{\text{CH}_2}{\diagdown}}\!\!\!>\!\text{CH}-\text{CH}_3 \;+\; H_2O & \xrightarrow{\;H^{\oplus}\;} & \text{CH}_2-\text{CH}_2-\text{CH}-\text{CH}_3 \\
 & & \quad\;\;|\qquad\qquad\;\;| \\
 & & \quad\;\;\text{H}\qquad\quad\;\;\;\text{OH}
\end{array}
$$

Methylcyclopropane   2-Butanol
(*sec*-butyl alcohol)

Carbocations are believed to be involved here also.

Question 8.4

Write the mechanisms for the acid-catalyzed hydration of each of the following five alkenes, and show the product(s) formed. Keep in mind that carbocation rearrangements are also possible.

(*a*) propene   (*b*) 2-methyl-2-butene

(*c*) 3-methyl-1-butene   (*d*) methylenecyclopentane, ⬠=CH₂

(*e*) 3,3-dimethyl-1-butene

8.6   Geometry of Carbocations

Carbocations, though highly reactive and very unstable, have been studied stereochemically to deduce their geometry. The sp^2 hybridized carbocation was introduced in Sec. 2.8. We now see some evidence that led to this structural assignment.

Optically active compounds are often used in determining the structure of highly reactive intermediates that cannot be isolated. From our study of carbocations we know that they can be generated by treating an alcohol with acid and that, once generated, they react with water to form an alcohol. When an optically active alcohol is used, however, it loses its optical activity on heating with aqueous acid but is unaffected by heating with aqueous base. Let us explain these results.

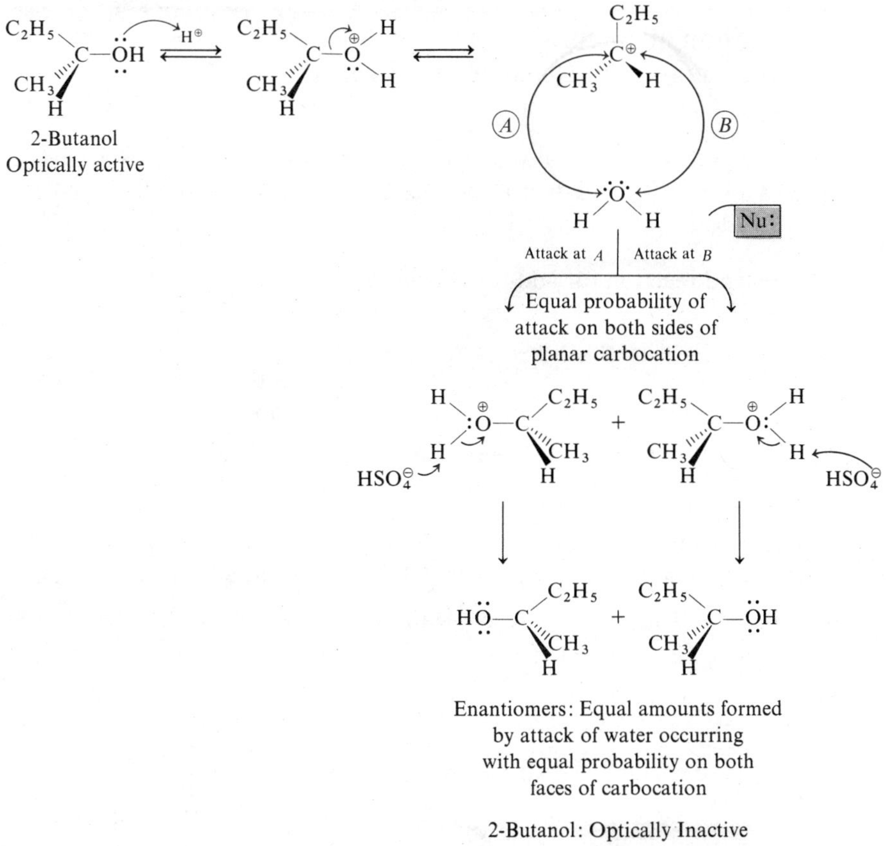

FIGURE 8.2 **Reaction involving a carbocation intermediate. The reaction occurs when 2-butanol is heated with aqueous acid.**

A readily available, optically active alcohol is 2-butanol; in the presence of acid, it can form the *sec*-butyl cation. When this reaction is carried out in aqueous solution, the carbocation can react with water to regenerate 2-butanol, which on isolation is found to have lost optical activity (the extent of the activity loss depends on the acid concentration and reaction time). The key to this reaction is the optically inactive alcohol, which must consist of an equal mixture of enantiomers. Because we know that a carbocation is the intermediate in this reaction, it is reasonable to look to this species in explaining the observed results. Our job then is to interpret these results and deduce a plausible structure for the incipient intermediate. One explanation for the observed racemization is that the carbocation is planar and is attacked by water with equal probability on both faces. These reactions are outlined in Fig. 8.2.

Question 8.5

What stereochemistry would you predict from heating either pure *cis*- or pure *trans*-3-methylcyclohexanol (1) with aqueous acid under conditions where no elimination occurs? Briefly explain.

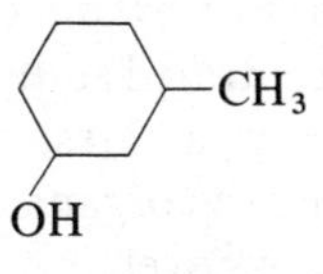

(1)

8.7   Addition of Sulfuric Acid (Unsymmetrical Reagent) to Alkenes

When alkenes are allowed to react with *cold, concentrated sulfuric acid*, the elements of H_2SO_4 add to the double bond and form alkyl hydrogen sulfates, $R-O-SO_3H$:

General reaction for addition of sulfuric acid:

$$\underset{\text{Alkene}}{\overset{}{>}C=C\overset{}{<}} + \underset{\substack{\text{Sulfuric acid} \\ \text{(conc)}}}{H-O-SO_3H} \xrightarrow{\text{cold}} \underset{\text{Alkyl hydrogen sulfate}}{-\overset{|}{\underset{|}{C}}-\overset{|}{\underset{|}{C}}-}$$

This reaction is another example of electrophilic addition, where the proton from sulfuric acid ($H_2SO_4 \longrightarrow H^{\oplus} + HSO_4^{\ominus}$) first attacks the double bond to give a carbocation. This ion then reacts with bisulfate ion, $HSO_4^{\ominus} \equiv :\overset{\ominus}{\underset{..}{O}}-SO_3H$, which is the nucleophile.

$$>C=C< + H^{\oplus} \longrightarrow -\overset{|}{\underset{|}{C}}-\overset{|}{\underset{\oplus}{C}}- \quad \overset{\ominus}{:}\overset{..}{O}-SO_3H \longrightarrow -\overset{|}{\underset{|}{C}}-\overset{|}{\underset{|}{C}}-$$

As expected, this reaction is also susceptible to rearrangements.

After the alkyl hydrogen sulfate is formed, it can be converted to the corresponding alcohol by adding excess water and heating. In a sense, an alkyl hydrogen sulfate is an ester of sulfuric acid, which is hydrolyzed by water—a reaction quite typical of esters (see Chap. 23). Alkyl hydrogen sulfates per se have little use in organic chemistry; however, they are used as an intermediate in the industrial conversion of alkenes to alcohols. Two typical examples are:

$$\underset{\text{Ethene}}{CH_2{=}CH_2} \xrightarrow{\text{conc } H_2SO_4} \underset{\text{Ethyl hydrogen sulfate}}{CH_3-CH_2-O-SO_3H} \xrightarrow[\text{heat}]{H_2O} \underset{\text{Ethanol}}{CH_3-CH_2-OH} + H_2SO_4$$

$$\underset{\text{Propene}}{CH_3-CH{=}CH_2} \xrightarrow[\text{cold}]{80\% \ H_2SO_4} \underset{\substack{\text{Isopropyl} \\ \text{hydrogen sulfate}}}{CH_3-\underset{\underset{O-SO_3H}{|}}{CH}-CH_3} \xrightarrow[\text{heat}]{H_2O} \underset{\text{2-Propanol}}{CH_3-\underset{\underset{OH}{|}}{CH}-CH_3} + H_2SO_4$$

The research chemist often uses this reaction in the laboratory because it is a good way to remove unwanted alkenes (sometimes formed as a side product) from nonbasic organic compounds. For example, alkanes and alkyl halides do not react with and are insoluble in concentrated sulfuric acid, whereas alkenes readily dissolve in it to form the alkyl hydrogen sulfates. Thus they can be removed from alkanes and/or alkyl halides by treating the organic mixture with cold, concentrated sulfuric acid. Usually the alkene is not recovered.

8.8 Summary of Electrophilic Addition Reactions of Acidic Reagents; Markovnikov's Rule

In the past several sections we discussed the addition of various unsymmetrical reagents to the carbon-carbon double bond: H—X, H_2O, and H_2SO_4. These reactions have three things in common: (1) all involve an acidic medium and thus require the presence of the proton, (2) they are similar mechanistically because all involve the addition of the proton to the carbon-carbon double bond (attack is at the π bond) to form the more stable carbocation, and (3) the carbocation then reacts with an electron-rich nucleophile. A summary of these reactions is given:

Alkyl halide

Alkyl hydrogen sulfate

Alcohol

That unsymmetrical reagents add to alkenes has been known for a long time, and the Russian chemist Markovnikov set forth a generalization in 1905 concerning the orientation of addition. He based his "rule" on known reactions. The **Markovnikov rule** states that, under normal addition conditions, *the more positive part of the reagent adds to the carbon bearing the greater number of hydrogen atoms originally, whereas the negative part of the reagent adds to the other carbon atom of the double bond*. Reexamination of the examples in Secs. 8.2 to 8.6 reveals that many additions do obey the Markovnikov rule; so far, these reactions are limited to reagents that contain protons.

At best, however, Markovnikov's rule is a rule of thumb; it is often ambiguous and cannot be used to predict the orientation of addition to some alkenes. For example, Markovnikov's rule does not take into account molecular rearrangements (see Sec. 8.4).

Sometimes special structural features in molecules affect the orientation of addition. For example, the ionic addition of HBr to bromoethene, $CH_2{=}CH{—}Br$, yields 1,1-dibromoethane, $CH_3{—}CHBr_2$, and hence "obeys" the rule. The structures of the two possible carbocations formed from proton addition to the double bond are:

$$Br{—}\overset{\oplus}{C}H{—}CH_3 \quad \text{and} \quad Br{—}CH_2{—}\overset{\oplus}{C}H_2$$

(1)   (2)

More stable   Less stable

At first glance we would predict (1) to be less stable than (2) because of the attachment of the electronegative bromine atom to the carbon bearing the positive charge in (1). Because it withdraws electrons, bromine would be expected to destabilize carbocation (1):

$$Br \leftarrow \overset{\oplus}{C}H—CH_3$$
(1)

Electron-withdrawing ability of bromine destabilizes carbocation to some extent

That 1,1-dibromoethane is the important product of this reaction indicates that the opposite is true; (1) is more stable than (2). The effect of an attached bromine on the stability of a positively charged center is discussed further in Sec. 14.12.

As we saw, orientation of addition and product distribution are well accounted for by carbocation theory. Using well-understood mechanisms and carbocation stabilities will never fail you.

Question 8.6

The addition of hydrogen bromide to 3,3,3-trifluoropropene, $CH_2\!=\!CH—CF_3$, yields 1,1,1-trifluoro-3-bromopropane, $Br—CH_2—CH_2—CF_3$, as the predominant product. Rationalize mechanistically this result, which is anti-Markovnikov addition.

8.9   Addition of Hydrogen Bromide to Alkenes in the Presence of Peroxides

The reaction of unsymmetrical reagents with alkenes occurs via a two-step electrophilic addition and involves ions (electrophiles, $E^{\oplus}$, and nucleophiles, $\overset{\ominus}{Nu\!:}$). A different type of reaction occurs between hydrogen bromide and alkenes if peroxides (R_2O_2) are present. Under typical ionic conditions (for example, carbon tetrachloride solvent), hydrogen bromide is added in accordance with carbocation theory, but *if peroxides are present, the orientation of addition is just the reverse.* The former is said to be *Markovnikov addition* and the latter is **anti-Markovnikov addition** because of the different orientations. For example,

$$CH_3—CH\!=\!CH_2 + HBr$$

Propene

$$\xrightarrow{CCl_4 \text{ solvent}} \underset{\substack{| \quad\quad |\\ Br \quad\; H}}{CH_3—CH—CH_2}$$

2-Bromopropane

$$\xrightarrow[RO—OR]{peroxides} \underset{\substack{| \quad\quad |\\ H \quad\; Br}}{CH_3—CH—CH_2}$$

1-Bromopropane

The differences in these reactions are ascribed to differences in the highly reactive species that intervene. In ionic additions, carbocations are formed because the proton attacks the π cloud. When peroxides are added, however, the mechanism changes from an ionic one to one involving free radicals. There is considerable evidence for the existence of free radicals.

To account for the addition of the elements of hydrogen and bromine to a carbon-carbon double bond, the following steps have been set forth.

Chain-initiating steps:

Step 1:

$$R—\overset{\cdot\cdot}{\underset{\cdot\cdot}{O}}\overset{\cdot\cdot}{\underset{\cdot\cdot}{O}}—R \longrightarrow 2R—\overset{\cdot\cdot}{\underset{\cdot\cdot}{O}}\cdot$$

Alkoxy radical

Step 2:

$$R—\overset{\cdot\cdot}{\underset{\cdot\cdot}{O}}\cdot + H\overset{\cdot\cdot}{\underset{\cdot\cdot}{Br}}\colon \longrightarrow R—\overset{\cdot\cdot}{\underset{\cdot\cdot}{O}}\colon\!H \quad + \quad \colon\!\overset{\cdot\cdot}{\underset{\cdot\cdot}{Br}}\cdot$$

Bromine atom:
electron-deficient,
free radical

Chain-propagating steps:

Step 3:

$$\colon\!\overset{\cdot\cdot}{\underset{\cdot\cdot}{Br}}\cdot + \underset{}{>}C\!\!=\!\!C\underset{}{<} \longrightarrow \begin{array}{c} | \quad | \\ —C—C— \\ | \quad \cdot \\ Br \end{array}$$

Electron-deficient
bromine atom attacks
π cloud of double bond

Attack occurs to give more
stable free radical

Step 4:

$$\begin{array}{c} | \quad | \\ —C—C\cdot \\ | \\ Br \end{array} + H\overset{\cdot\cdot}{\underset{\cdot\cdot}{Br}}\colon \longrightarrow \begin{array}{c} | \quad | \\ —C—C— \\ | \quad | \\ Br \quad H \end{array} \quad + \quad \colon\!\overset{\cdot\cdot}{\underset{\cdot\cdot}{Br}}\cdot$$

Attack of electron-
deficient carbon
radical on HBr

Is regenerated
and starts back
in step 3

The reaction is initiated by the decomposition of the peroxide, R—O—O—R, into alkoxy radicals, R—$\overset{\cdot\cdot}{O}\cdot$, which attack hydrogen bromide to form an alcohol, R—OH, and bromine atoms, $\colon\!\overset{\cdot\cdot}{\underset{\cdot\cdot}{Br}}\cdot$. This is another method for generating bromine atoms, which were first encountered in the free-radical halogenation of alkanes and were generated by the thermal decomposition of elemental bromine: $Br_2 \rightarrow 2\colon\!\overset{\cdot\cdot}{\underset{\cdot\cdot}{Br}}\cdot$.

The chain-propagating steps involve the attack of the electron-deficient bromine atom on the electron-rich π cloud of the double bond. This occurs so as to yield the more stable free radical. (See later discussion about orientation.) The new carbon radical is electron-deficient and finds an electron by attacking the hydrogen atom in another molecule of hydrogen bromide. This forms the alkyl halide and regenerates a bromine atom, which in turn reacts with another molecule of alkene. Thus the reaction is said to be initiated by peroxides; note that it is indeed a chain reaction. The chain is terminated by the reaction of radicals with themselves.

Question 8.7

Draw the possible termination steps for the reaction of HBr/R_2O_2 with propene.

To understand the orientation of addition, recall that the order of radical stability is as follows (see Sec. 5.7):

Decreasing stability of carbon radicals:

$$3° > 2° > 1° > \cdot CH_3$$

When HBr is added to propene, the bromine atom can add to the double bond in two possible ways:

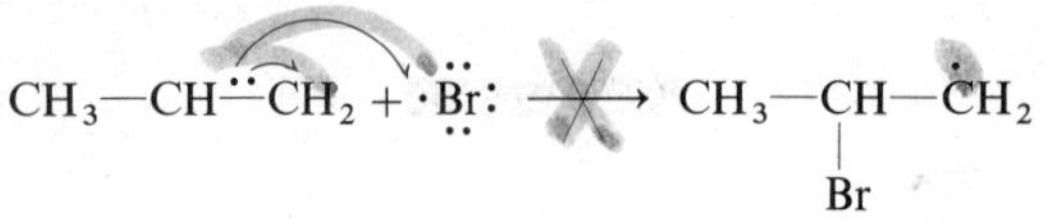

$$CH_3-CH\!\cdot\!\cdot CH_2 + \cdot\ddot{B}r: \;\;\xrightarrow{\;\;\times\;\;}\;\; CH_3-CH-\dot{C}H_2$$
$$\underset{Br}{|}$$

1° free radical
*Less stable and not
formed*

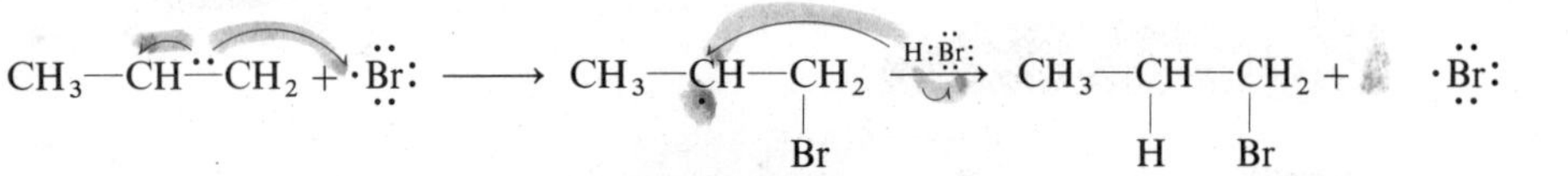

$$CH_3-\dot{C}H\!\cdot\!\cdot CH_2 + \cdot\ddot{B}r: \;\longrightarrow\; CH_3-\dot{C}H-CH_2 \;\xrightarrow{\;H:\ddot{B}r:\;}\; CH_3-CH-CH_2 + \;\cdot\ddot{B}r:$$
$$\underset{Br}{|} \qquad\qquad\qquad\qquad \underset{H}{|}\;\;\underset{Br}{|}$$

2° free radical   Observed   Regenerated
More stable and   product
favored

The addition of free radicals to a carbon-carbon double bond occurs so as to form the more stable intermediate carbon free radical. With propene, the bromine atom adds to the double bond to give the more stable secondary free radical as the favored intermediate, which in turn reacts with HBr to form product.

The peroxide-initiated addition of hydrogen bromide to alkenes gives anti-Markovnikov orientation; that is, the hydrogen ends up on the carbon bearing the lesser number of hydrogen atoms initially and the bromine atom ends up on the other carbon atom of the double bond.

The principles in ionic and free-radical addition are the same. The first step gives the more stable intermediate (carbocation or radical), which reacts further to give product. In ionic addition hydrogen adds first as a proton, and in free-radical addition bromine adds first as a bromine atom.

Question 8.8

Give the products expected when the following alkenes are allowed to react with HBr in the presence of peroxides:

(*a*) 1-butene   (*b*) methylpropene   (*c*) 1-methylcyclohexene
(*d*) *cis*-2-butene   (*e*) 2-methyl-2-butene

Question 8.9

In the presence of peroxides, chloroform ($CHCl_3$) reacts with propene to give 1,1,1-trichlorobutane, $CH_3-CH_2-CH_2-CCl_3$. This reaction is believed to occur in a manner analogous to the free-radical addition of hydrogen bromide to alkenes.

(*a*) Write a complete mechanism for this reaction.
(*b*) What product(s) are expected if in the above reaction the propene were replaced by methylpropene? Why?

8.10   Addition of Unsymmetrical Reagents to Alkadienes

In alkadienes we might expect electrophilic addition to occur at one or both of the double bonds, depending on the amount of acidic reagent present. This expectation is correct, but depending on the type of alkadiene (isolated or conjugated) used, different products are observed.

With an isolated diene, such as 1,4-pentadiene, the expected products are obtained because each double bond is completely independent of the other. For example:

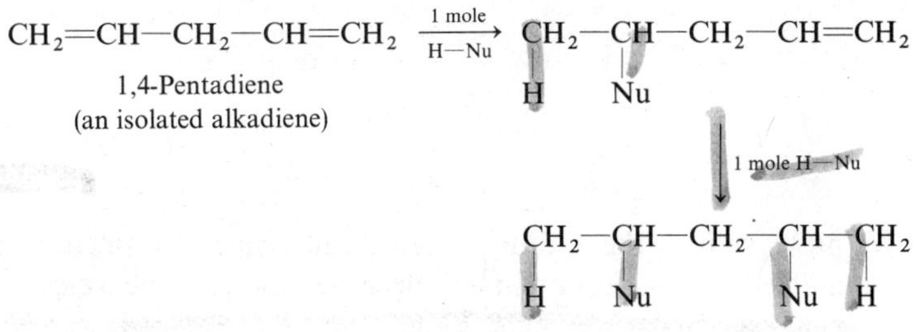

$$CH_2{=}CH{-}CH_2{-}CH{=}CH_2 \xrightarrow[H-Nu]{1\ mole} \underset{\underset{H\quad Nu}{\big|\quad\big|}}{CH_2{-}CH{-}CH_2{-}CH{=}CH_2}$$

1,4-Pentadiene
(an isolated alkadiene)

$$\Big\downarrow 1\ mole\ H{-}Nu$$

$$\underset{\underset{H\quad Nu\qquad Nu\quad H}{\big|\quad\big|\qquad\big|\quad\big|}}{CH_2{-}CH{-}CH_2{-}CH{-}CH_2}$$

where Nu = X, OH, OSO$_3$H, for examples

The addition of hydrogen chloride to 1,4-pentadiene is typical:

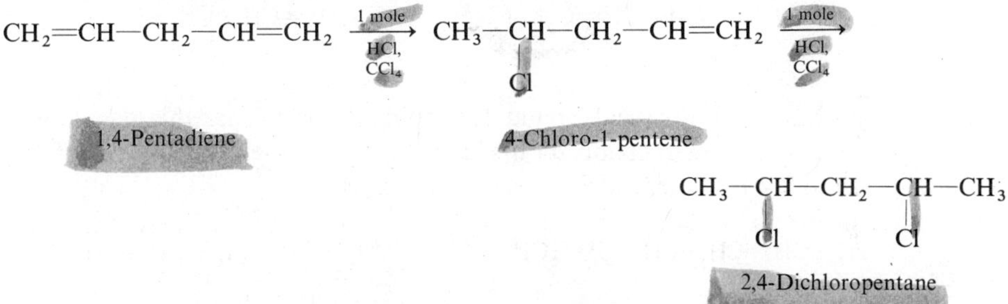

$$CH_2{=}CH{-}CH_2{-}CH{=}CH_2 \xrightarrow[HCl,\ CCl_4]{1\ mole} \underset{\underset{Cl}{\big|}}{CH_3{-}CH{-}CH_2{-}CH{=}CH_2} \xrightarrow[HCl,\ CCl_4]{1\ mole}$$

1,4-Pentadiene   4-Chloro-1-pentene

$$\underset{\underset{Cl\qquad\qquad Cl}{\big|\qquad\qquad\big|}}{CH_3{-}CH{-}CH_2{-}CH{-}CH_3}$$

2,4-Dichloropentane

Note that the orientation of addition is explained on the basis of carbocation stability, where the more stable cation (a 2° ion) is formed and then reacts with chloride ion, which is the nucleophile (Nu:$^{\ominus}$). These steps occur twice, resulting in Markovnikov addition to both carbon-carbon double bonds.

When this same type of reaction is carried out with a *conjugated* alkadiene, such as 1,3-butadiene, two products are also obtained when an acidic reagent is added; these products are shown here:

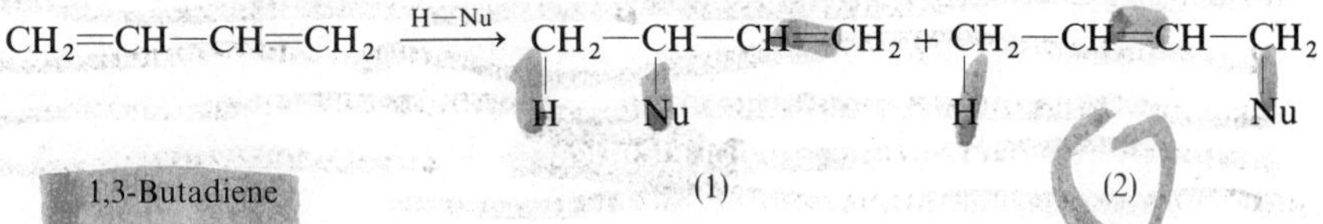

$$CH_2{=}CH{-}CH{=}CH_2 \xrightarrow{H-Nu} \underset{\underset{H\quad Nu}{\big|\quad\big|}}{CH_2{-}CH{-}CH{=}CH_2} + \underset{\underset{H\qquad Nu}{\big|\qquad\big|}}{CH_2{-}CH{=}CH{-}CH_2}$$

1,3-Butadiene   (1)   (2)

where Nu = X, OH, OSO$_3$H, for example

The formation of (1) is expected, but compound (2) has undergone several bonding changes; the terminal double bond is no longer present and there is a double bond between the two central carbon atoms. A specific example is the electrophilic addition of hydrogen bromide to a conjugated diene:

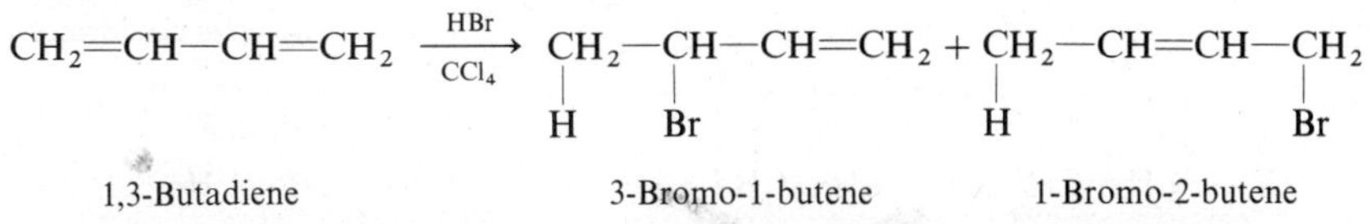

$$CH_2{=}CH{-}CH{=}CH_2 \xrightarrow[CCl_4]{HBr} \underset{\underset{H\quad Br}{\big|\quad\big|}}{CH_2{-}CH{-}CH{=}CH_2} + \underset{\underset{H\qquad Br}{\big|\qquad\big|}}{CH_2{-}CH{=}CH{-}CH_2}$$

1,3-Butadiene   3-Bromo-1-butene   1-Bromo-2-butene

Other addition reactions, for example, halogenation (see Sec. 8.20) and hydrogenation (see Sec. 8.13), also yield two products. In general, *in the addition reactions of conjugated dienes, a reagent adds across the adjacent atoms that constitute a double bond* (**1,2 addition**) *and also to the carbon atoms that constitute the two ends of the conjugated system* (**1,4 addition**). (*Note:* The numbers 1,2 and 1,4 do *not* necessarily refer to the numbering of the alkadiene used for naming it; these numbers refer to the four carbon atoms that constitute the conjugated system.) Frequently, the major product comes from 1,4 addition.

$$
\overset{1}{C}=\overset{2}{C}-\overset{3}{C}=\overset{4}{C} + E^{\oplus}\ Nu:^{\ominus} \longrightarrow -\overset{1}{\underset{E}{C}}-\overset{2}{\underset{Nu}{C}}-\overset{3}{C}=\overset{4}{C} + -\overset{1}{\underset{E}{C}}-\overset{2}{C}=\overset{3}{C}-\overset{4}{\underset{Nu}{C}}-
$$

1,2 Addition   1,4 Addition

We now examine the addition of an acidic reagent, HBr, to conjugated dienes and account for the products obtained. Because this is a typical electrophilic addition reaction, the first step is the attack of the proton on one of the double bonds to give the more stable carbocation. In 1,3-butadiene there is no doubt that the more stable ion is the secondary cation.

$$
CH_2{=}CH{-}CH{=}CH_2 + H^{\oplus} \longrightarrow CH_2{=}CH{-}\overset{\oplus}{C}H{-}\underset{H}{C}H_2
$$

But how do both the 1,2- and 1,4-products form from this one cation? It is easier to answer this question by first looking at another example; 2,4-hexadiene. Addition to 2,4-hexadiene also gives two products:

$$
CH_3{-}CH{=}CH{-}CH{=}CH{-}CH_3 \xrightarrow{HBr}
$$

$$
\boxed{CH_3{-}\underset{H}{CH}{-}\underset{Br}{CH}{-}CH{=}CH{-}CH_3}
$$

4-Bromo-2-hexene

$$
+ CH_3{-}\underset{Br}{CH}{-}\underset{H}{CH}{-}CH{=}CH{-}CH_3 + CH_3{-}\underset{H}{CH}{-}CH{=}CH{-}\underset{Br}{CH}{-}CH_3
$$

5-Bromo-2-hexene   2-Bromo-3-hexene
Not formed   *Two products formed*

Here, however, there are two possible ways for the proton to add to one of the double bonds, and each possibility produces a 2° carbocation. The production of 4-bromo-2-hexene as the only 1,2-addition product (5-bromo-2-hexene is not formed) suggests that these two carbocations are not identical and that the one with the positive charge adjacent to the remaining double bond is favored.

$$
CH_3{-}CH{=}CH{-}CH{=}CH{-}CH_3 + H^{\oplus} \xcancel{\longrightarrow} CH_3{-}CH{=}CH{-}CH{-}\overset{\oplus}{C}H{-}CH_3
$$

2,4-Hexadiene   H

Would lead to
5-bromo-2-hexene
which is not formed

or

$$
CH_3{-}CH{=}CH{-}CH{=}CH{-}CH_3 + H^{\oplus} \longrightarrow CH_3{-}CH{=}CH{-}\overset{\oplus}{C}H{-}\underset{H}{C}H{-}CH_3
$$

(1)
Favored

Careful study of carbocation (1) shows that it is not an ordinary secondary carbocation; the positive charge *is* adjacent to the other double bond. Viewed in terms of the bonding orbitals involved, the picture is:

$$CH_3-C...C...C^{\oplus}...CH_2-CH_3$$ (with H atoms)

(1)   Vacant p orbital

The positive charge attracts electrons and the π cloud next to it has a high concentration of electrons, so these electrons are drawn toward the positive charge. When this happens, the carbon atom in the double bond from which electrons have been withdrawn becomes positive. This may look like an endless process with a pair of π electrons shifting back and forth between carbon atoms:

$$CH_3-CH=CH-\overset{\oplus}{CH}-\underset{\underset{H}{|}}{CH}-CH_3 \longleftrightarrow CH_3-\overset{\oplus}{CH}-CH=CH-\underset{\underset{H}{|}}{CH}-CH_3$$

(1)   (2)

These are both reasonable electronic structures; the only differences between them are the position of the double bond and the location of the positive charge. No atoms have changed location; they differ only in electron distribution. Structures (1) and (2) are **resonance structures;** resonance is indicated by a double-headed arrow ($\longleftrightarrow$). Neither (1) nor (2) adequately represents the entire molecule (see Sec. 13.6), but each represents some characteristics of the molecule. The molecule is best described as a combination (sum) of all the resonance structures; this is called a *resonance hybrid*. The hybrid is best represented by *one equivalent structure* in which the positive charge is spread out (*delocalized*) over several carbon atoms:

$$CH_3-\overset{\delta\oplus}{CH}\text{---}CH\text{---}\overset{\delta\oplus}{CH}-\underset{\underset{H}{|}}{CH}-CH_3$$

The bonding π molecular orbital involves all three carbon atoms (see Sec. 13.6D). The electron deficiency is greatest at the two ends of the orbital.

$$CH_3\underset{\delta\oplus}{-}CH-CH-CH\underset{\delta\oplus}{-}CH_2CH_3$$

Delocalized allyl cation

Because of the distribution of positive charge in the π cloud, the two carbon atoms indicated previously have partial positive charge; each can react with a nucleophile, such as bromide ion, to produce the two products that are observed.

attack at Ⓐ

$$CH_3-CH=CH-\underset{\underset{Br}{|}}{CH}-\underset{\underset{H}{|}}{CH}-CH_3$$

4-Bromo-2-hexene
1,2-Addition product

$$CH_3-\overset{\delta\oplus}{CH}\text{---}CH\text{---}\overset{\delta\oplus}{CH}-\underset{\underset{H}{|}}{CH}-CH_3$$

Ⓑ   Ⓐ

$:\overset{..}{\underset{..}{Br}}:^{\ominus}$

$Nu:^{\ominus}$

attack at Ⓑ

$$CH_3-\underset{\underset{Br}{|}}{CH}-CH=CH-\underset{\underset{H}{|}}{CH}-CH_3$$

2-Bromo-3-hexene
1,4-Addition product

Thus, structural and electronic theory explain why it is reasonable that two products form when an acidic reagent is allowed to react with a conjugated diene.

Getting back to our original question, we now see that the cation that results from the addition of a proton to 1,3-butadiene is also a delocalized cation:

$$CH_2=CH-CH-CH_3 \longleftrightarrow CH_2-CH=CH-CH_3 \equiv \overset{\delta\oplus}{CH_2}\cdots CH\cdots \overset{\delta\oplus}{CH}-CH_2$$

Hybrid
electrons delocalized

Nucleophilic attack at the two $\delta\oplus$ carbons gives the two observed products: 3-bromo-1-butene and 1-bromo-2-butene.

Several important new principles are involved in this type of reaction. The first concerns the unusual stability of carbocations that are adjacent to a π bond. Regardless of how the positive charge is formed, this type of ion, called an **allyl cation**,

$$-\overset{}{\underset{\oplus}{C}}-C=C\diagdown$$

Allyl cation

is shown by physical measurements to be as stable as, and in some cases more stable than, a tertiary cation, depending on the groups attached.

Decreasing order of carbocation stability:

$$\text{Allyl} \geq 3° > 2° > 1° > \overset{\oplus}{CH_3}$$

Second, we saw one way to form the allyl cation and examined its electronic structure on the basis of the orbitals involved and the mutual attraction between positive charge and the electrons in the π cloud. The positive charge is spread out (delocalized) over three carbon atoms. *Spreading out a charge generally results in a more stable electronic configuration.*

Question 8.10

Predict the products of addition of 1 mole of hydrogen chloride to

(*a*) 1,3-pentadiene   (*b*) 1,5-hexadiene   (*c*) 1,3-butadiene

8.11   Product Distribution in Addition of Unsymmetrical Reagents to Conjugated Alkadienes

The addition of acidic reagents such as HBr to conjugated dienes gives two products. The proportions of the 1,2- and 1,4-addition products vary with reaction temperature. For example, the addition of HBr to 1,3-butadiene at $-80°$ yields a mixture containing 80% 3-bromo-1-butene (1,2 addition) and 20% 1-bromo-2-butene (1,4 addition). When this same reaction is carried out at 40°, the product distribution is 80% of the 1,4-addition product and only 20% of the 1,2-addition product.

$$CH_2=CH-CH=CH_2 + HBr \longrightarrow \underset{\overset{|}{H}\;\;\overset{|}{Br}}{CH_2-CH-CH=CH_2} + \underset{\overset{|}{H}\;\;\;\;\;\;\;\;\overset{|}{Br}}{CH_2-CH=CH-CH_2}$$

	1,2 Addition	1,4 Addition
At $-80°$:	80%	20%
At $+40°$:	20%	80%

Heating (at 40°) either the 1,2- or the 1,4-addition product yields the same 20:80 mixture of 1,2 and 1,4 products, respectively; yet each compound is stable at temperatures of −80° or less.

These facts have been interpreted in terms of an equilibrium that must exist between the two compounds at temperatures greater than −80°. Furthermore, because 1,2 addition predominates at low temperatures (for example, −80°), it has been suggested that the 1,2-addition product is formed more quickly than the 1,4-addition product. Yet the 1,4-addition product must be more stable since it is preferred at higher temperatures. This is reasonable because the double bond in the 1,4 product is more highly substituted than in the 1,2 product; we saw that more highly substituted alkenes are more stable and thus favored.

This competitive 1,2- and 1,4-addition reaction, which is typical of conjugated dienes in their reaction with various reagents, is an example of **kinetic versus thermodynamic control** (see Sec. 5.8). In this example at lower temperatures, chemical kinetics (that is, the rates at which the two reactions occur) dictate which product is preferred. The reaction that occurs the fastest produces the preferred product, which in this case is the 1,2 addition product. We say that this is *kinetic control*. At higher temperatures, however, the more stable product is preferred. This reaction is *thermodynamically controlled* because the energies associated with the products dictate which is preferred.

The essential differences between kinetic and thermodynamic control may be translated into the energy-profile diagrams shown in Fig. 8.3. The differences in the product stabilities are shown by the heats of reaction ($\Delta H°$). That $\Delta H°$ for the formation of the 1,4 product is more negative than for the 1,2 product means that the 1,4 product is more stable; this is, in effect, the thermodynamic control over the reaction. The importance of kinetic control is readily seen by comparing the energies of activation, E_A, for the formations of 1,2 product and 1,4 product; the E_A for 1,2 addition is less than that for 1,4 addition. Energies of activation represent the amounts

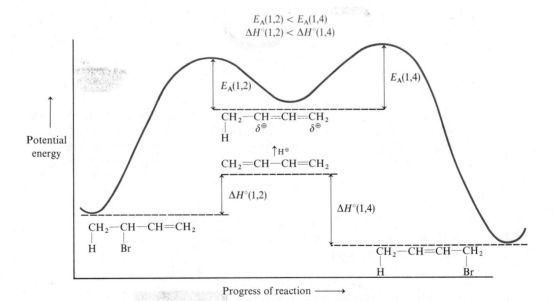

FIGURE 8.3   Energy-profile diagram showing progress of the reaction between 1,3-butadiene and HBr. Only the product-producing step (carbocation to product) is diagrammed to scale. The reactant (1,3-butadiene) is on the chart for comparison of the enthalpies of the two reactions.

of energy that must be supplied to convert starting material to product. It is reasonable that 1,2 addition is preferred over 1,4 addition at low temperatures where there is less energy available and the path of least resistance is followed. As the temperature increases, there is enough energy to easily overcome the energy of activation for 1,2 or 1,4 addition. Then the reaction can occur by either 1,2 or 1,4 addition, but it prefers the route that yields the more stable product (that is, the route that has a more negative $\Delta H°$ and liberates the most energy).

Mechanistically, these results require two explanations. First, the initial reaction of 1,3-butadiene with HBr involves the allyl cation, which reacts with bromide ion to form the 1,2 and the 1,4-addition products. The product distribution depends on the reaction temperature, and the energies of activation dictate that temperature dependence.

$$CH_2=CH-CH=CH_2 \xrightarrow{H^{\oplus}} CH_2-\overset{\delta\oplus}{CH}\text{---}CH\text{---}\overset{\delta\oplus}{CH}_2 \longrightarrow 1,2\ addition + 1,4\ addition$$

with H below the second carbon and $:\overset{\ominus}{Br}:$ below.

Allyl cation

Second, the conversion of 1,2 or 1,4 product (or a mixture of the two) to the 20:80 equilibrium mixture of 1,2 and 1,4 product (which is accomplished by heating) is more difficult to explain. However, as we will see in Sec. 9.11, allyl halides are particularly susceptible to ionization in which the carbon-halogen bond is broken to give the relatively stable allyl cation. This is the reverse reaction of bromide ion attack on the allyl cation. This appears to occur when the 1,2 or 1,4 product is heated, and the allyl cation then rereacts with bromide ion to give a mixture of products. The ionization and recombination sequence is believed to occur until the equilibrium mixture of products is obtained. The equilibrium mixture is constant at a given temperature and is dependent on the energy difference between the two possible products.

$$CH_2-CH-CH=CH_2 \underset{\text{ionization}}{\rightleftharpoons} CH_2-\overset{\delta\oplus}{CH}\text{---}CH\text{---}\overset{\delta\oplus}{CH}_2 \rightleftharpoons CH_2-CH=CH-CH_2$$

| | | |
| H | Br | |

1,2 Product   Allyl cation   1,4 Product
                                                   Favored

The influence of kinetic and thermodynamic controls of reactions depends on the specific reactants and conditions used. Therefore, care must be exercised in using reaction rates to predict the major product of a reaction. Sometimes, as in the case here, the less stable product is formed faster, and the more stable product is formed slower.

8.12   Hydrogenation: Addition of Hydrogen (Symmetrical Reagent) to Alkenes

The addition of hydrogen to alkenes, called **hydrogenation,** is simple because there is no problem about the orientation of addition. The reaction is not ionic, however, but instead appears to involve the addition of hydrogen atoms for reasons we will soon see.

Hydrogenation:

$$\text{>C=C<} + \text{H—H} \xrightarrow[\text{(Pt, Pd, or Ni)}]{\text{catalyst}} \text{—C—C—}$$

Alkene   Alkane

Hydrogenation requires the presence of a metal catalyst, such as platinum (Pt), palladium (Pd), or nickel (Ni), without which the reaction proceeds at a negligible rate even at elevated temperatures. Even with a catalyst, heat is often needed. In general, adding a catalyst to a reaction lowers the energy of activation, E_A, and frequently alters the reaction mechanism.

Three typical examples of the catalytic hydrogenation of alkenes follow:

$$CH_2\text{=}CH_2 \xrightarrow{H_2,\ Pd} CH_3\text{—}CH_3$$

Ethene   Ethane

$$\begin{array}{c}CH_3 \\ \\ H\end{array}C\text{=}C\begin{array}{c}CH_3 \\ \\ H\end{array} \xrightarrow{H_2,\ Pt} CH_3\text{—}CH_2\text{—}CH_2\text{—}CH_3$$

cis-2-Butene   Butane

$$\begin{array}{c}CH_3 \\ \\ H\end{array}C\text{=}C\begin{array}{c}H \\ \\ CH_3\end{array} \xrightarrow{H_2,\ Pt}$$

trans-2-Butene

Hydrogenation of geometric isomers gives the same alkane (see 2-butene above), and because there is free rotation about the new σ bond in the product, these reactions tell nothing about the mechanism of addition. Use of a cyclic alkene, on the other hand, yields quite meaningful results. The addition of hydrogen to a carbon-carbon double bond is *syn* addition, as illustrated by the hydrogenation of 1,2-dimethyl-cyclohexene:

CH$_3$ CH$_3$ H$_3$C CH$_3$

1,2-Dimethylcyclohexene   cis-1,2-Dimethylcyclohexane

That *only* the *cis* isomer of 1,2-dimethylcyclohexane is formed shows that both hydrogen atoms must have added to the *same side* of the carbon-carbon double bond.

The hydrogenation reaction is very exothermic, but mixtures of an alkene and hydrogen gas do not react at normal temperatures unless there is a metal catalyst. Although the precise mechanism of hydrogenation is still being studied, it is believed that the metallic catalyst adsorbs hydrogen gas. The metal provides electrons to hydrogen and forms metal-hydrogen bonds, thus causing the hydrogen gas to dissociate into atoms. As the alkene approaches the surface of the metal catalyst, it encounters the adsorbed hydrogen atoms, which add to the double bond in a *syn* fashion as shown in Fig. 8.4.

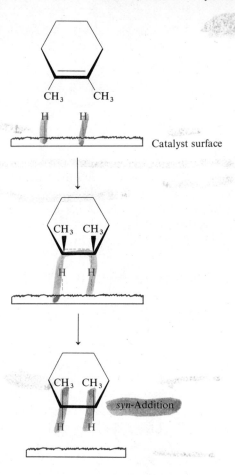

FIGURE 8.4 Stereochemistry of catalytic hydrogenation: *syn* addition.

Hydrogen readily reduces most double and triple bonds, such as $\diagdown C{=}O$, $-C{=}N-$, $-C{\equiv}C-$, and $-C{\equiv}N$. This reaction is one of the most general and valuable tools available to the organic chemist.

Cycloalkanes are not affected by catalytic hydrogenation, *except for cyclopropane and cyclobutane*, which, as we recall, are highly strained molecules. These two compounds undergo ring opening to form open-chain alkanes:

$$\triangle \xrightarrow[80°]{H_2,\,Ni} CH_3{-}CH_2{-}CH_3$$

Cyclopropane   Propane

$$\square \xrightarrow[200°]{H_2,\,Ni} CH_3{-}CH_2{-}CH_2{-}CH_3$$

Cyclobutane   Butane

Note that the opening of cyclobutane requires more drastic conditions (200°) than does cyclopropane (80°). This is expected because the latter is considerably more strained. Also, both compounds are less reactive than most alkenes.

Question 8.11

In this section hydrogenation was referred to as a *reduction* reaction. In the following two reactions, determine the oxidation states of carbon in the double bond and in the corresponding alkane (see the end of Sec. 4.2), and then explain why reduction occurs on the basis of oxidation changes on carbon.

$$CH_2{=}CH_2 + H_2 \xrightarrow{Pt} CH_3{-}CH_3$$

$$(CH_3)_2C{=}C(CH_3)_2 + H_2 \xrightarrow{Pt} (CH_3)_2CH{-}CH(CH_3)_2$$

Question 8.12

Deuterium gas, D_2, is an isotope of H_2 but has molecular weight 4. In the presence of a metal catalyst, D_2 also adds to double bonds. Suppose D_2 is allowed to react with *cis*-2-butene to give 2,3-dideuteriobutane, $CH_3{-}CHD{-}CHD{-}CH_3$.

(*a*) Using three-dimensional projection structures, draw the product that is formed.
(*b*) If *trans*-2-butene were used, what is the three-dimensional projection formula for the product?
(*c*) How, if at all, do the products from *cis*-2-butene and *trans*-2-butene differ in stereochemistry?

8.13   Hydrogenation of Alkadienes

The hydrogenation of alkadienes, like that of alkenes, requires a metal catalyst. Two possible alkenes form as a result of 1,2 and 1,4 addition (see Sec. 8.11). When there is an excess of hydrogen gas, both double bonds are reduced. These reactions are illustrated using 1,3-butadiene as an example.

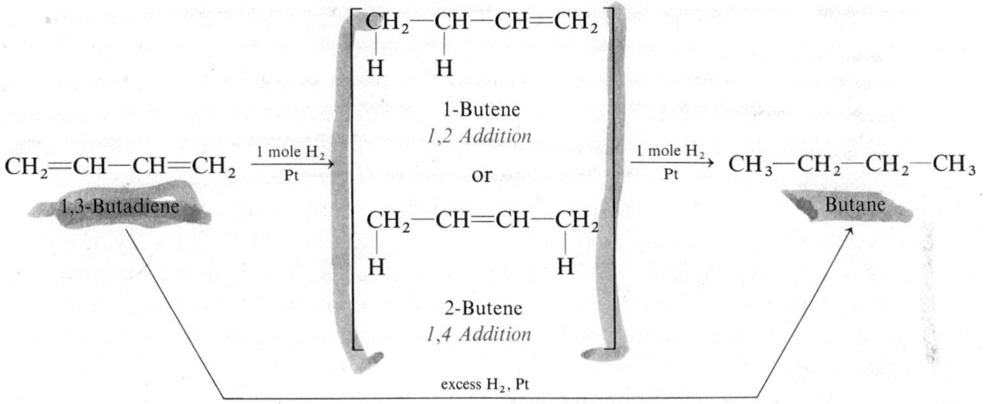

The mechanism of the 1,2 and 1,4 additions of hydrogen is much less well understood than is the ionic addition of unsymmetrical reagents because hydrogenation is a surface reaction and most likely involves the transfer of hydrogen atoms from the catalyst surface to the double bond. One possible mechanism involves an *allyl radical,* which is formed by adding a hydrogen atom to the terminal carbon atom. As we see in Sec. 8.30, allyl radicals, like allyl cations, are unusually stable.

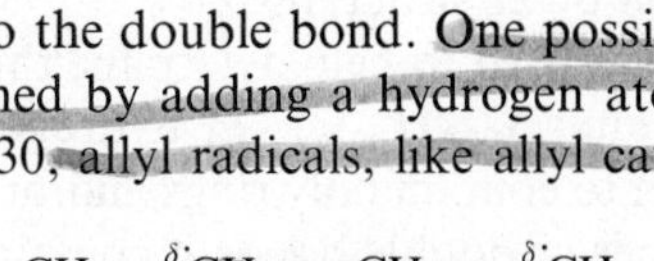

$$CH_2{-}\overset{\delta\cdot}{C}H\!\!=\!\!=\!\!CH\cdots\cdots\overset{\delta\cdot}{C}H_2$$
$$\underset{H}{|}$$

A second hydrogen atom can attack either carbon atom in the allyl radical that bears partial radical character, thus giving 1,2 and 1,4 addition.

$$CH_2=CH-CH=CH_2 + \cdot H \longrightarrow \overset{\delta\cdot}{C}H_2\text{-----}CH\text{-----}\overset{\delta\cdot}{C}H-CH_2 \quad \text{allyl radical}$$

on catalyst
surface

or

Ⓐ   Ⓑ

H

H·

on catalyst surface

attack at Ⓐ   attack at Ⓑ

$$CH_2-CH=CH-CH_2 \qquad\qquad CH_2=CH-CH-CH_2$$

H   H   H   H

1,4 Addition   1,2 Addition

8.14   Use of Catalytic Hydrogenation in Structure Determination

Catalytic hydrogenation has considerable analytic application. It is, for example, possible to ascertain the number of double bonds in a compound by accurately measuring the number of moles of hydrogen consumed per mole of compound. Experimentally, this is done by either measuring the volume of hydrogen that is taken up or by determining the pressure drop in the system.

To determine the number of double bonds (or triple bonds, see Chap. 11) and/or rings in an unknown hydrocarbon, the molecular formula of the unknown is first determined. Then the compound is hydrogenated and the number of moles of hydrogen absorbed per mole of unknown is determined, or alternatively, the formula of the hydrogenated compound is determined. Because each carbon-carbon double bond consumes 1 mole equivalent of hydrogen gas, we can compute the number of double (or triple) bonds and rings that the molecule contains.

For example, an unknown compound has the molecular formula C_6H_{10}. *If* this compound were completely saturated, its formula would be C_nH_{2n+2} or C_6H_{14}. Thus, C_6H_{10} is *deficient* 2 moles of hydrogen (that is, four hydrogen atoms), so it may contain two double bonds or one ring and one double bond (or one triple bond, $-C\equiv C-$). This is sometimes called the **index of hydrogen deficiency** or **degree of unsaturation.**

Now suppose the unknown were hydrogenated to give a new compound with the formula C_6H_{12}, which could not be hydrogenated further. This implies that the original unknown contained one double bond that reacted with hydrogen and one ring that was unreactive toward hydrogen. One possible structure for the unknown is cyclohexene. (Suggest other structures.)

Hydrogenation was used to help determine the structure of vitamin A_1, which has the molecular formula $C_{20}H_{30}O$. On hydrogenation, it consumes 5 moles of hydrogen and is known to contain only rings and/or double bonds. These data allow us to deduce the number of double bonds it contains. A compound with 20 carbon atoms contains $2n + 2$ or 42 hydrogen atoms if it is completely saturated. Upon hydrogenation, vitamin A_1 consumes 5 moles of H_2 and produces a compound with

40 hydrogen atoms, or two less than complete saturation. Thus vitamin A_1 must contain *five* double bonds and *one* ring. The structure of vitamin A_1 is given here:

Vitamin A_1

Question 8.13

The naturally occurring product *β-carotene* has the formula $C_{40}H_{56}$. On catalytic hydrogenation, it gives a compound with the molecular formula $C_{40}H_{78}$.

(*a*) How many moles of hydrogen did *β*-carotene consume?
(*b*) How many double bonds and how many rings does it contain?

omit ## 8.15   Heats of Hydrogenation; Evidence for Alkene Stability

There are several ways to determine the relative stabilities of some organic compounds. For example, heats of combustion were used to determine cycloalkane stability in Sec. 3.15 because they provide a measure of the *total* energy in organic compounds. Heats of combustion were also used in Sec. 7.7 to determine the relative stabilities of the *cis* and *trans* isomers of alkenes.

We now see how the heats of hydrogenation are used to support the relative stabilities of alkenes that were categorically asserted in Sec. 7.12. They also reinforce the relative stabilities of geometric isomers determined by combustion data (see previous paragraph).

The amount of energy liberated by the catalytic hydrogenation of a double bond is called the **heat of hydrogenation.** The approximate energy liberated in the hydrogenation of a double bond may be computed as follows, where the *average* bond energy for the C—H bond is taken to be 101 kcal/mole (422.6 kJ/mole):

$$\underset{\substack{\text{68 kcal/mole}\\ \text{(only } \pi \text{ bond}\\ \text{broken)}}}{\text{C}=\text{C}} + \underset{\text{104 kcal/mole}}{\text{H}-\text{H}} \xrightarrow{\text{Pt}} \underset{2 \times 101 \text{ kcal/mole}}{-\text{C}-\text{C}-} \qquad \Delta H° = -30 \text{ kcal/mole} (-125.5 \text{ kJ/mole})$$

The heats of hydrogenation for selected alkenes are shown in Table 8.1.

In general, if two alkenes liberate different amounts of energy on hydrogenation, the one that liberates less is the more stable because it contained less energy originally. For example, ethene liberates 32.8 kcal/mole (137.2 kJ/mole), whereas monosubstituted ethenes (propene, 1-butene, 1-pentene, and 1-hexene) liberate 30.1 to 30.3 kcal/mole (125.9 to 126.8 kJ/mole). Additional comparisons are seen in the

TABLE 8.1   Heats of Hydrogenation of Selected Alkenes

Name	Structure	Heat of Hydrogenation $(-\Delta H°)$ kcal/mole (kJ/mole)	General Structure
Ethene (ethylene)	$CH_2{=}CH_2$	32.8 (137.2)	
Propene (propylene)	$CH_3{-}CH{=}CH_2$	30.1 (125.9)	
1-Butene	$CH_3CH_2{-}CH{=}CH_2$	30.3 (126.8)	
1-Pentene	$CH_3CH_2CH_2{-}CH{=}CH_2$	30.1 (125.9)	
1-Hexene	$CH_3CH_2CH_2CH_2{-}CH{=}CH_2$	30.1 (125.9)	$R{-}CH{=}CH_2$
3-Methyl-1-butene	$CH_2{=}CH{-}\overset{\overset{\displaystyle CH_3}{\vert}}{CH}{-}CH_3$	30.3 (126.8)	
3,3-Dimethyl-1-butene	$CH_2{=}CH{-}\overset{\overset{\displaystyle CH_3}{\vert}}{\underset{\underset{\displaystyle CH_3}{\vert}}{C}}{-}CH_3$	30.3 (126.8)	
trans-2-Butene	$\overset{CH_3}{\underset{H}{}}{>}C{=}C{<}\overset{H}{\underset{CH_3}{}}$	27.6 (115.5)	
cis-2-Butene	$\overset{CH_3}{\underset{H}{}}{>}C{=}C{<}\overset{CH_3}{\underset{H}{}}$	28.6 (119.7)	$R{-}CH{=}CH{-}R$
Methylpropene (isobutylene)	$\overset{CH_3}{\underset{CH_3}{}}{>}C{=}CH_2$	28.0 (117.2)	$\overset{R}{\underset{R}{}}{>}C{=}CH_2$
2-Methyl-2-butene	$CH_3{-}\overset{\overset{\displaystyle CH_3}{\vert}}{C}{=}CH{-}CH_3$	26.9 (112.5)	$\overset{R}{\underset{R}{}}{>}C{=}CH{-}R$
2,3-Dimethyl-2-butene	$\overset{CH_3}{\underset{CH_3}{}}{>}C{=}C{<}\overset{CH_3}{\underset{CH_3}{}}$	26.6 (111.3)	$\overset{R}{\underset{R}{}}{>}C{=}C{<}\overset{R}{\underset{R}{}}$

following:

$CH_3(CH_2)_3{-}CH{=}CH_2$ $CH_3{-}CH{=}CH{-}CH_3$

Monosubstituted   Disubstituted
ethene   ethene
30.1 kcal/mole   *cis*, 28.6; *trans*, 27.6 kcal/mole
(125.9 kJ/mole)   (119.7; 115.4 kJ/mole)

$CH_3{-}\overset{\overset{\displaystyle CH_3}{\vert}}{C}{=}CH{-}CH_3$ $CH_3{-}\overset{\overset{\displaystyle CH_3}{\vert}}{C}{=}\overset{\overset{\displaystyle CH_3}{\vert}}{C}{-}CH_3$

Trisubstituted   Tetrasubstituted
ethene   ethene
26.9 kcal/mole   26.6 kcal/mole
(112.5 kJ/mole)   (111.3 kJ/mole)

These data, along with others in Table 8.1, indicate the following stability order for substituted alkenes:

Decreasing order of alkene stability:

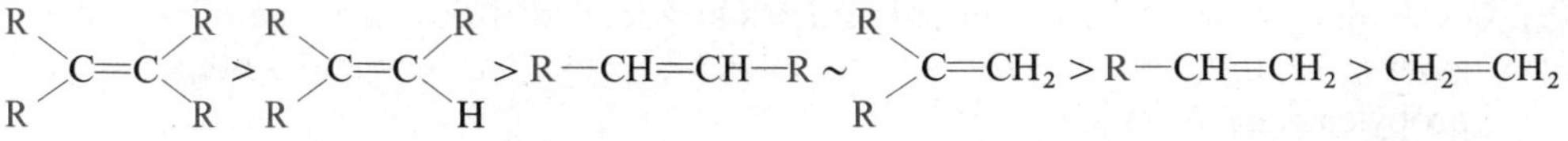

cis and *trans*

We discussed and used this order in Secs. 7.12 and 7.16 on the product distribution in dehydrohalogenation and dehydration, which are the two most common methods for preparing alkenes in the laboratory.

The heat of hydrogenation of *trans*-2-butene (27.6 kcal/mole; 115.5 kJ/mole) is less than that of *cis*-2-butene (28.6 kcal/mole; 119.7 kJ/mole), thus indicating that the *trans* isomer is more stable than the *cis* isomer. This trend was also observed from the heats of combustion in Sec. 7.7.

8.16 Halogenation: Addition of Halogens (Symmetrical Reagent) to Alkenes

Alkenes react readily with bromine or chlorine to give *addition* products, whereas iodine does not react and fluorine is so reactive that it not only adds to the double bond but also reacts with the carbon-hydrogen bonds in the molecule to give substitution products.

General reaction for halogenation:

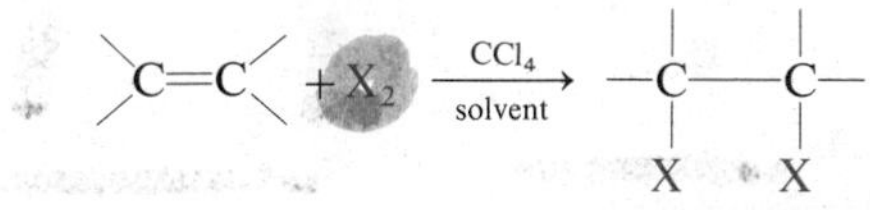

Alkene — Vicinal dihalide

where $X_2 = Cl_2, Br_2$

An inert organic solvent is usually used in this reaction; carbon tetrachloride is often chosen. The reaction is remarkably fast (almost instantaneous at room temperature) and is often used as a *qualitative test* for the presence of the carbon-carbon double bond. For example, a solution of bromine in carbon tetrachloride is reddish brown, and if it is added to an alkene, the bromine reacts and the characteristic bromine color disappears. Alkanes do not react under these conditions. For example:

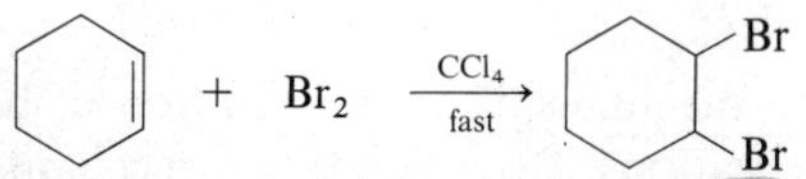

Cyclohexene *Colorless* — *Reddish brown* — *Colorless*

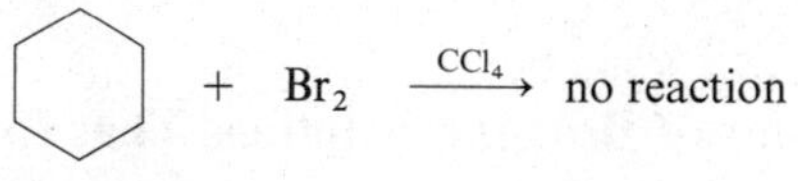

Cyclohexane

No color change, solution remains reddish brown

Halogenation of alkenes is the best way to prepare *vicinal dihalides*. The reverse of this addition reaction is dehalogenation (see Sec. 7.19). The starting material for dehalogenation usually comes from the halogenation of alkenes, and for this reason dehalogenation is seldom used for the preparation of alkenes. In the following two examples, common names in parentheses are derived from the name of the alkene from which the product is formed; for example, propylene bromide is prepared from propylene and bromine.

$$CH_2{=}CH_2 + Cl_2 \xrightarrow{CCl_4} \underset{\underset{Cl}{|}}{CH_2}{-}\underset{\underset{Cl}{|}}{CH_2}$$

Ethene   1,2-Dichloroethane
(ethylene)   (ethylene chloride)

$$CH_2{=}CH{-}CH_3 + Br_2 \xrightarrow{CCl_4} \underset{\underset{Br}{|}}{CH_2}{-}\underset{\underset{Br}{|}}{CH}{-}CH_3$$

Propene   1,2-Dibromopropane
(propylene)   (propylene bromide)

Note that the product no longer contains the double bond even though the common name might imply otherwise.

Cyclopropane also undergoes an addition reaction with bromine or chlorine. The highly strained ring is opened to give an open-chain dihalide, although a Lewis-acid catalyst such as $AlCl_3$ is needed to speed up the reaction.

$$\begin{matrix}CH_2 \\ | \quad \searrow CH_2 \\ CH_2 \end{matrix} \xrightarrow[\substack{CCl_4, \\ AlCl_3}]{Br_2} \underset{\underset{Br}{|}}{CH_2}{-}CH_2{-}\underset{\underset{Br}{|}}{CH_2}$$

Cyclopropane   1,3-Dibromopropane
                  Major product

Ring opening does not occur with larger cycloalkanes; even with cyclopropane, it is much slower than the corresponding addition reaction of alkenes. This is one way cyclopropane appears to resemble alkenes, which, as we saw, undergo a variety of addition reactions.

8.17   Mechanism and Stereochemistry of Bromine Addition

The accepted mechanism for the addition of bromine to alkenes is based on experimental observations. Recalling that a mechanism must be consistent with all the observed facts that have been gathered, we now look at some facts for bromination. (There are some slight differences for chlorination; see Sec. 8.19.)

A.   Reactions of Bromine Solutions That Contain Inorganic Salts

When alkenes are allowed to react with bromine that contains added inorganic salts, the following are observed:

$$CH_2{=}CH_2 \begin{cases} \xrightarrow[\text{aqueous}]{Br_2,\ Na^\oplus I^\ominus} & \underset{\underset{\text{1,2-Dibromoethane}}{Br\quad Br}}{CH_2{-}CH_2} + \underset{\underset{\text{1-Bromo-2-iodoethane}}{Br\quad I}}{CH_2{-}CH_2} \\[3em] \xrightarrow[\text{aqueous}]{Br_2,\ Na^\oplus Cl^\ominus} & \underset{Br\quad Br}{CH_2{-}CH_2} + \underset{\underset{\text{1-Bromo-2-chloroethane}}{Br\quad Cl}}{CH_2{-}CH_2} \end{cases}$$

$$CH_3{-}CH{=}CH_2 \xrightarrow[\text{aqueous}]{Br_2,\ Na^\oplus Cl^\ominus} \underset{\underset{\text{1,2-Dibromopropane}}{Br\quad Br}}{CH_3{-}CH{-}CH_2} + \underset{\underset{\substack{\text{1-Bromo-2-}\\\text{chloropropane}\\\textit{Major}\\\textit{chlorine-}\\\textit{containing}\\\textit{compound}}}{Cl\quad Br}}{CH_3{-}CH{-}CH_2} + \underset{\underset{\substack{\text{2-Bromo-1-chloropropane}\\\textit{Minor chlorine-containing}\\\textit{compound}}}{Br\quad Cl}}{CH_3{-}CH{-}CH_2}$$

These reactions are normally done by dissolving bromine and sodium halides in water and adding the alkene. Several products are obtained from each reaction, and in addition, all the reactions are accompanied by the formation of a **halohydrin,** for example, $X{-}CH_2{-}CH_2{-}OH$. We discuss how this is formed in Sec. 8.22, but for now we focus on the halogenation reactions.

B. Stereochemistry

The reaction between bromine and ethene gives a product of which the stereo-chemistry of addition cannot be determined because of free rotation about the carbon-carbon single bond in the product.

$$CH_2{=}CH_2 + Br_2 \xrightarrow{CCl_4} \underset{Br\qquad Br}{CH_2 \curvearrowright CH_2}$$

Ethene

Stereochemistry of addition
indeterminate due to
free rotation about
carbon-carbon single bond

If an alkene is selected so that the addition product has no free rotation, however, then the stereochemistry of addition can be learned. One alkene so used is cyclopentene, from which the product of bromine addition is *only trans*-1,2-dibromocyclopentane.

Cyclopentene   *trans*-1,2-Dibromocyclopentane

It is important that no *cis* isomer is formed. The same results are obtained from cyclohexene. Because of this evidence, the reaction is said to occur with *anti*-addition. Contrast this with catalytic hydrogenation, which occurs by *syn* addition (see Sec. 8.12).

C. Mechanism of Bromination: *anti*-Addition

The following mechanism is consistent with the stereochemistry and formation of multiple products when inorganic salts are present:

(1)
Bromonium ion
Intermediate

Transition state
π Cloud polarizes Br_2

step 3

attack at C_α attack at C_β

where X = Br, Cl, I

Steps 1 and 2 occur because, when the two molecules collide, the electron-rich π bond in the alkene *polarizes* the bromine molecule by forcing electrons away from the bromine atoms nearest the π bond, as shown in the transition state. Even though the electrons are shared equally between the identical bromine atoms in the bromine molecule, which is nonpolar, this electron distribution is affected when the molecule is brought into another electron environment—in this case, that of the π cloud. As the polarization occurs, the bromine atom nearest the alkene becomes partially positive (electron-deficient) and is attracted to it; this breaks the π bond and produces a *bridged ion*, the **bromonium ion** (1), in which the bromine atom shares one of its nonbonding electron pairs with carbon and is positively charged itself. The negative bromide ion is also produced by these steps. When there are identical substituents on the double bond, the *bromonium ion is symmetrical;* that is, the carbon atoms that originally constituted the double bond share positive charge equally. It has been suggested that several electronic structures (*resonance structures*) that place some positive charge on carbon can be written for (1):

Resonance hybrid (average
of all contributing
structures); carbon bears
some positive charge, some
carbocation character

Step 3 shows the attack by the bromide ion (formed as a product of attack by bromine on the starting alkene), but there are two carbon atoms where attack can occur. Attack at either carbon atom in (1) occurs from the side remote from the positively charged bromine atom and produces two products, both having the bromine atoms *trans* (*anti*) to one another.

Now we apply this mechanism to the examples at the beginning of this section. First, the formation of multiple products when Br_2 is allowed to react with an alkene in the presence of $Na^\oplus X^\ominus$ ($Na^\oplus Cl^\ominus$ or $Na^\oplus I^\ominus$) is accounted for by invoking the bridged bromonium ion:

$$CH_2\!=\!CH_2 \xrightarrow{\ Br_2\ } CH_2\!\!-\!\!CH_2 + :\overset{..}{\underset{..}{Br}}:^{\ominus} \qquad \text{Formation of bromonium ion}$$

$$\underset{:\overset{\oplus}{Br}:}{}$$

$$CH_2\!\!-\!\!CH_2 \longrightarrow Br\!-\!CH_2\!-\!CH_2\!-\!Br$$

Bromide ion and halide ion compete for bromonium ion

$$CH_2\!\!-\!\!CH_2 \longrightarrow X\!-\!CH_2\!-\!CH_2\!-\!Br$$

where $X = Cl, I$

According to this picture, once the bromonium ion is formed, it can be attacked by either bromide ion ($Br^{\ominus}$) or halide ion ($Cl^{\ominus}$ or $I^{\ominus}$, depending on which sodium halide is added). These two anions *compete* for the bromonium ion and thus form two products; the product distribution depends on the concentration of added salt.

The stereochemistry is also accounted for by the bridged bromonium ion:

d,l-trans-1,2-Dibromocyclopentane
Racemic: equal mixture of enantiomers

In this reaction, two *trans*-dibromo products are formed. On careful examination, we see that they are a mixture of enantiomers. There is equal probability of attack at either C_α or C_β, so a 50:50 mixture of enantiomers is formed and the product is racemic.

On the other hand, consider the stereochemistry if the bridged bromonium ion were not involved. The reaction between bromine and cyclopentene could, alternatively, produce a free carbocation that is planar (see Sec. 8.6). This ion could then be attacked from either the top or bottom side of the ring, and a mixture of *cis*- and *trans*-1,2-dibromocyclopentane would result.

That no *cis* isomer is observed is strong evidence that a free carbocation is *not* involved. The bridged bromonium ion discussed earlier seems to be the best explanation for all the observed results.

Finally, the orientation of addition to unsymmetrical alkenes fits nicely into the bromonium ion picture. Because the bromonium ion places some carbocation character on the two carbons to which bromine is attached, the carbon atom that contains the greater number of alkyl substituents bears the greater positive charge and thus is more likely to be attacked by the nucleophilic halide ion. For example:

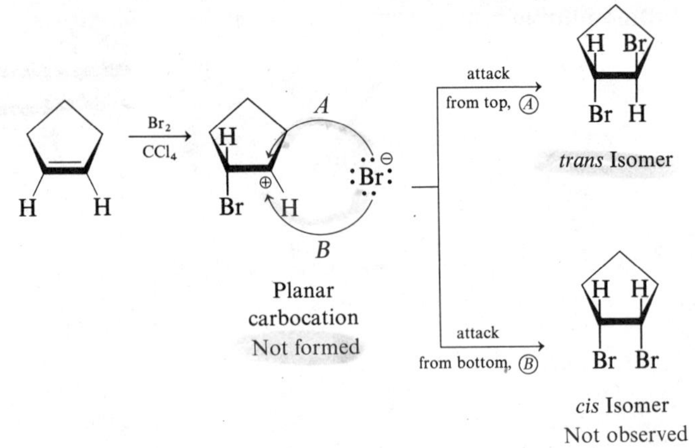

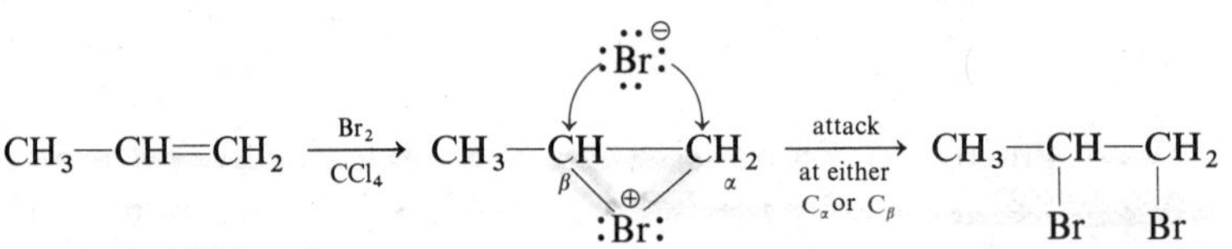

When the same reaction is carried out with added sodium halide, say NaCl, the 1,2-dibromopropane is formed by the preceding mechanism, but the major chlorine-containing compound is 1-bromo-2-chloropropane. The bromonium ion is believed to be involved, but in terms of possible electronic structures, the one with the partially positive charge on C_β is greatly favored over the structure with the charge on C_α. Attack of the nucleophilic chloride ion occurs almost exclusively at this position to give 1-bromo-2-chloropropane.

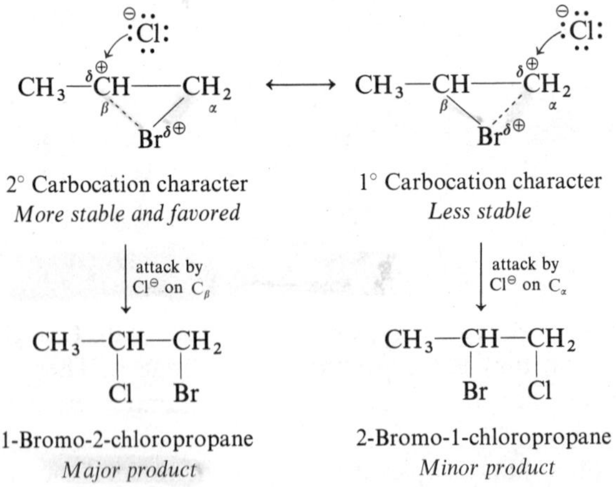

---

omit **8.18   Stereochemistry of Bromine Addition to Open-Chain Alkenes**

Cycloalkenes are only one piece of evidence collected to support the **stereospecific**[1] *anti*-addition of bromine. Certain open-chain alkenes are also used to prove

---

[1] A **stereospecific reaction** is one in which reactants that are diastereomerically different preferentially produce one of several diastereomeric products.

this mode of addition. For example, *trans*-2-butene reacts with bromine in carbon tetrachloride to yield *meso*-2,3-dibromobutane only, whereas *cis*-2-butene gives a racemic mixture of *d,l*-2,3-dibromobutane:

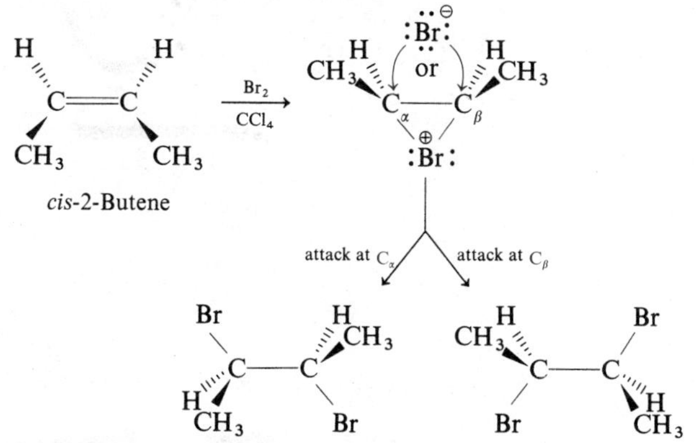

cis-2-Butene

attack at C_α attack at C_β

Nonsuperimposable mirror images: enantiomers
Racemic mixture

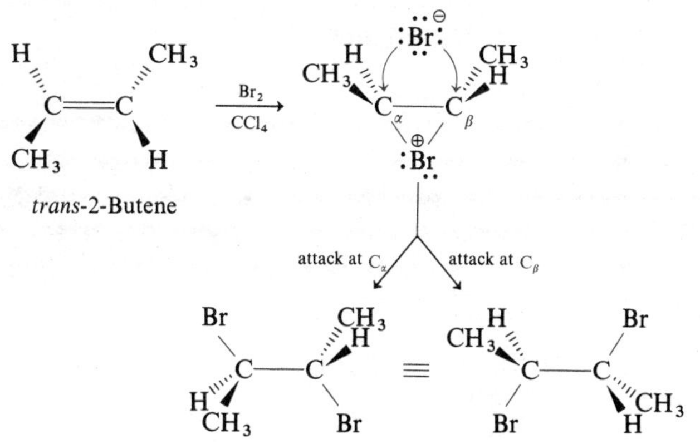

trans-2-Butene

attack at C_α attack at C_β

meso Compound: two identical structures

That the product from *trans*-2-butene is *meso* can be seen by rotating one of the central carbon atoms by 180° and noting the plane of symmetry.

Question 8.14

(*a*) Predict the stereochemistry in the products resulting from bromine addition to *trans*-2-pentene.
(*b*) Predict the stereochemistry from bromine addition to *cis*-2-pentene.

8.19   Chlorination: Addition of Chlorine to Alkenes

In previous sections the stereospecific *anti*-addition of bromine to alkenes was illustrated and documented by experimental fact. In comparison, although chlorination is stereospecific in certain instances, it often is not and produces a mixture of *syn* and *anti*-addition products with alkenes where such stereochemistry can be determined. This is not completely unreasonable because chlorine is much more electronegative than bromine and is thus less likely to bear positive charge, as it must in the bridged chloronium ion. It has been suggested that chlorination is much more likely to occur via a free carbocation:

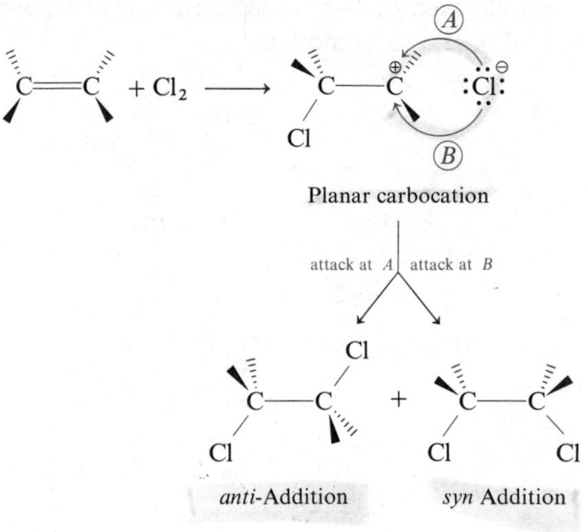

Planar carbocation

attack at *A*   attack at *B*

anti-Addition   *syn* Addition

8.20   Halogenation of Alkadienes

Halogenation of alkadienes occurs in two steps. The addition of halogen in carbon tetrachloride produces the 1,2- and 1,4-addition products, which on reaction with more halogen give the tetrahalo derivative. The 1,2- and 1,4-addition products are believed to be formed as a result of the allyl cation rather than the bridged ion invoked in the halogenation of alkenes. For example, the chlorination of 1,3-butadiene is shown here:

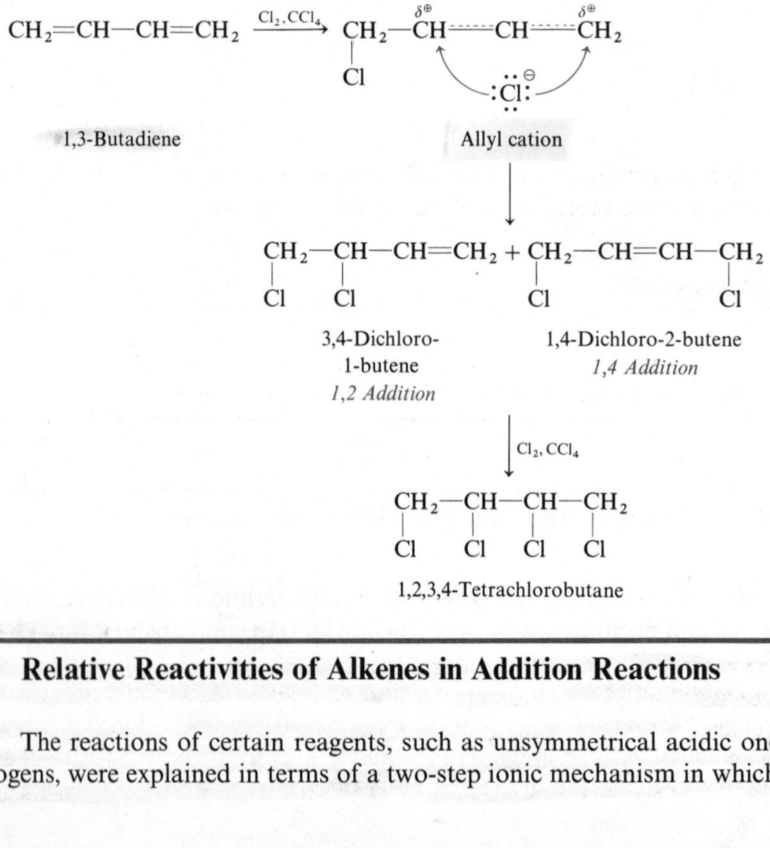

1,3-Butadiene   Allyl cation

3,4-Dichloro-   1,4-Dichloro-2-butene
1-butene   *1,4 Addition*
1,2 Addition

Cl_2, CCl_4

$$CH_2-CH-CH-CH_2$$
$$ClClClCl$$

1,2,3,4-Tetrachlorobutane

Light 8.21   Relative Reactivities of Alkenes in Addition Reactions

The reactions of certain reagents, such as unsymmetrical acidic ones and the halogens, were explained in terms of a two-step ionic mechanism in which the more

TABLE 8.2   Relative Rates of Addition of Bromine to Selected Alkenes

Alkene	Relative Rate of Bromine Addition (in CH_2Cl_2 Solution) at $-78°$*
$(CH_3)_2C=C(CH_3)_2$	14.0
$(CH_3)_2C=CH_2$	5.5
$CH_3CH=CH_2$	2.0
$CH_2=CH_2$ (standard)	1.0 (arbitrarily chosen standard)
$BrCH=CH_2$	Very slow

*Rate of addition of bromine to ethene set at 1.0.

positive part of the reagent adds to the double bond to give the more stable carbocation. We can now explain why some alkenes are more reactive than others. Two important factors are involved: (1) the nature of the substituents attached to the double bond and (2) the number of substituents so attached.

First consider the relative reactivities of certain alkenes toward bromine addition, as shown in Table 8.2. As these data show, the more highly substituted the alkene, the more reactive it is; our standard is ethene, which has its rate of reaction arbitrarily set at 1.0. However, further examination indicates that the presence of an electron-withdrawing group, the bromine atom in vinyl bromide ($CH_2=CH-Br$), markedly slows the reaction to less than the rate for addition to ethene.

There are two ways to view the effect of structure on reactivity. One is to consider the relative "concentration" of electrons in the π cloud that the electrophile attacks. The greater the electron density, the greater the affinity of the positively charged electrophile for that electron source. Substituents that donate electrons inductively to the double bond (such as alkyl groups) increase reactivity; substituents that withdraw electrons inductively (for example, the halogens) decrease reactivity toward addition. These comparisons are relative to ethene, our standard of reference. Thus, we have

$$\underset{CH_3}{\overset{CH_3}{>}}C=C\underset{CH_3}{\overset{CH_3}{<}} \quad \underset{CH_3}{\overset{CH_3}{>}}C=CH_2 \quad CH_3{\rightarrow}CH=CH_2 \quad \underset{\text{Standard}}{CH_2=CH_2} \quad Br{\leftarrow}CH=CH_2$$

$\longleftarrow$ Increasing density of electrons in π cloud

and increasing reactivity toward electrophile, $E^{\oplus}$

The second approach to understanding the reactivity of substituted alkenes involves the relative stabilities of the intermediate carbocations that result from attack by the electrophile. The following list is more complete because it shows the alkenes as well as the cations that result from electrophilic attack.

Decreasing order of reactivity of alkenes toward ionic addition:

$$R_2C=CR_2 > R_2C=CHR > R_2C=CH_2 \gtrless RCH=CHR > RCH=CH_2 > CH_2=CH_2 > X-CH=CH_2$$

$$\downarrow E^{\oplus} \qquad \downarrow E^{\oplus} \qquad \downarrow E^{\oplus} \qquad \downarrow E^{\oplus} \qquad \downarrow E^{\oplus} \qquad \downarrow E^{\oplus} \qquad \downarrow E^{\oplus}$$

$$\overset{\oplus}{R_2C}-\underset{E}{CR_2} \simeq \overset{\oplus}{R_2C}-\underset{E}{CHR} \simeq \overset{\oplus}{R_2C}-\underset{E}{CH_2} > \overset{\oplus}{RCH}-\underset{E}{CHR} \simeq \overset{\oplus}{RCH}-\underset{E}{CH_2} > \overset{\oplus}{CH_2}-\underset{E}{CH_2} > X-\overset{\oplus}{CH}-\underset{E}{CH_2}$$

Approximate order of decreasing carbocation stability $\longrightarrow$

The more stable carbocation should be formed faster, so the decreasing order of reactivity roughly parallels the order of carbocation stability.

These two arguments, (1) the relative affinities of the electrophile for the π clouds in various substituted alkene reactants and (2) the relative stabilities of the intermediate carbocations that are formed by electrophilic attack on the double bond, complement one another and lead to the same conclusions regarding the effect of structure on reactivity.

8.22   Addition of Hypohalous Acids (Unsymmetrical Reagent) to Alkenes; Formation of Halohydrins

When aqueous solutions of chlorine or bromine react with alkenes, two addition products are normally formed. One is the *vicinal dihalide* and the other contains a halogen on one carbon atom and a hydroxy group (—OH) on the other—a type of compound commonly referred to as a **halohydrin.**

General reaction:

$$
\begin{array}{c}
\diagup \\
C=C \\
\diagup
\end{array}
+ X_2 + H_2O \longrightarrow
\begin{array}{cc}
| & | \\
-C-C- \\
| & | \\
X & X
\end{array}
+
\begin{array}{cc}
| & | \\
-C-C- \\
| & | \\
X & OH
\end{array}
$$

Alkene	Vicinal dihalide	Halohydrin

For example:

$$CH_2{=}CH_2 \xrightarrow{Cl_2,\ H_2O}
\begin{array}{cc}
CH_2{-}CH_2 \\
| \quad | \\
Cl \quad Cl
\end{array}
\ +\
\begin{array}{cc}
CH_2{-}CH_2 \\
| \quad | \\
Cl \quad OH
\end{array}$$

Ethene	1,2-Dichloroethane (ethylene chloride)	2-Chloroethanol (ethylene chlorohydrin)

$$CH_3{-}CH{=}CH_2 \xrightarrow{Br_2,\ H_2O}
\begin{array}{ccc}
CH_3{-}CH{-}CH_2 \\
| \quad | \\
Br \quad Br
\end{array}
\ +\
\begin{array}{ccc}
CH_3{-}CH{-}CH_2 \\
| \quad | \\
OH \quad Br
\end{array}$$

Propene	1,2-Dibromopropane (propylene bromide)	1-Bromo-2-propanol (propylene bromohydrin)

Even though the halohydrin formation involves the elements of an unsymmetrical reagent, hypohalous acid (HO—X), the reaction mechanism most likely involves the bridged bromonium ion in the case of bromine. Once formed, the bridged ion reacts with water (a typical nucleophile) to give an oxonium ion, which loses a proton to give halohydrin, or with bromide ion to give the vicinal dihalide.

The nucleophiles, water or bromide ion, compete with one another for attack on the bromonium ion. The more substituted carbon atom in the bridged ion is attacked preferentially (see Sec. 8.17).

Although the mechanism for the formation of the chlorohydrin of ethene could involve a bridged chloronium ion, chlorine addition to alkenes seems to invoke a free carbocation (see Sec. 8.19). These mechanisms account for both the orientation of the addition and the formation of multiple products. Recall from Sec. 8.17 that halohydrins are also formed in the reactions of bromine that contains inorganic salts with alkenes. In those reactions, water is the solvent used to dissolve the sodium halides.

| 1,2-Dibromopropane | 1-Bromo-2-propanol |
| (vicinal dibromide) | (bromohydrin) |

An alternative explanation for halohydrin formation invokes the formation of hypohalous acid, HO—X, which occurs when a halogen is dissolved in water; the elemental halogen is both oxidized and reduced, and one of the products is HOX:

$$X_2 + H_2O \longrightarrow HOX + H^\oplus + :\overset{..}{\underset{..}{X}}:^\ominus$$

The oxidation state of halogen in hypohalous acid is $+1$, which is reasonable because oxygen is more electronegative than, for example, chlorine. In HOX, then, we have $H\overset{..}{\underset{..}{O}}{}^{\delta\ominus}—\overset{..}{\underset{..}{X}}{}^{\delta\oplus}:$, which adds to the double bond in accordance with Markovnikov's rule; that is, the positive halogen $(:\overset{\oplus}{\underset{..}{X}})$ adds to the carbon bearing the greater number of hydrogens, and the negative hydroxyl group $(:\overset{..}{\underset{..}{O}}H^\ominus)$ adds to the other carbon atom. In either instance, the carbocation theory is the simplest explanation for halohydrin formation and for the formation of multiple products.

Question 8.15

Predict the products formed when a solution of bromine in methanol (CH_3OH) is allowed to react with propene. (*Hint:* Keep in mind the similarity in structure between water, $H_2\overset{..}{\underset{..}{O}}:$, and $CH_3—\overset{..}{\underset{|}{\underset{H}{O}}}:$, particularly with regard to their nucleophilic character.)

8.23   Addition of Alkenes to Alkenes: Carbocation Reaction

Upon treatment with sulfuric acid or phosphoric acid, propene is converted into a mixture of two alkenes, each containing six carbons and having the molecular formula C_6H_{12}. Catalytic hydrogenation of the alkene mixture gives a single alkane, 2-methylpentane. The two alkenes must therefore have the same carbon framework, so they differ only in the position of the double bond. The two alkenes have the following structures:

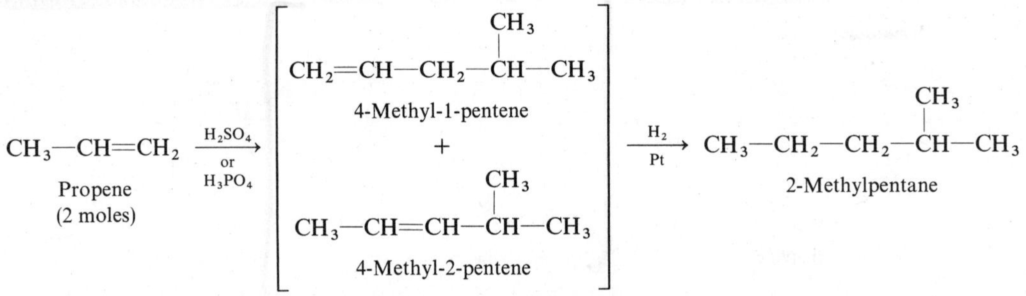

$$CH_3-CH=CH_2 \xrightarrow[\text{or}\atop H_3PO_4]{H_2SO_4} \left[\begin{array}{c} \overset{\displaystyle CH_3}{\underset{|}{CH_2=CH-CH_2-CH-CH_3}} \\ \text{4-Methyl-1-pentene} \\ + \\ \overset{\displaystyle CH_3}{\underset{|}{CH_3-CH=CH-CH-CH_3}} \\ \text{4-Methyl-2-pentene} \end{array} \right] \xrightarrow[\text{Pt}]{H_2} \overset{\displaystyle CH_3}{\underset{|}{CH_3-CH_2-CH_2-CH-CH_3}}$$

Propene
(2 moles)

2-Methylpentane

The reaction conditions must be carefully controlled because (as we will see) the addition can continue and produce compounds that contain more carbon atoms. (Note that 4-methyl-2-pentene can exist as *cis-trans* isomers; our earlier reference to two alkenes means they differ in the position of the double bond and does not mean they are geometric isomers.)

This particular reaction is called **dimerization** because the products contain precisely twice the carbon and hydrogen content of the starting compound, propene. The products are called **dimers** because they contain two (*di-*) parts (*-mer*) of propene. By analogy, a trimer contains three parts, a tetramer four parts, and so on.

One requirement for dimerization is that there be a small amount of acid catalyst, so the reaction probably involves carbocations. Step 1 is the addition of a proton to propene to give the more stable secondary carbocation:

Step 1:

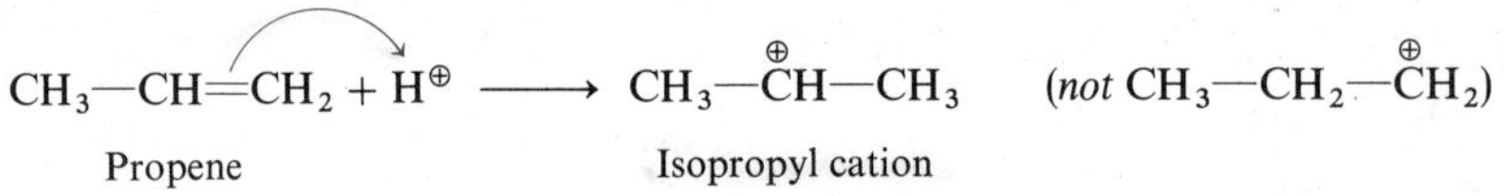

$$CH_3-CH{=}CH_2 + H^\oplus \longrightarrow CH_3-\overset{\oplus}{CH}-CH_3 \quad (not\ CH_3-CH_2-\overset{\oplus}{CH}_2)$$

Propene   Isopropyl cation

The carbocation, once formed, may complete its octet by reacting with various nucleophiles, such as $:\!\overset{..}{\underset{..}{X}}\!:^\ominus$, $:\!\overset{..}{\underset{..}{O}}H^\ominus$, $HSO_4^\ominus$, and H_2O. Under the conditions of dimerization, however, there are virtually no nucleophiles of the type mentioned earlier. (Recall that the acid is present only in catalytic amounts.) Also remember that the π cloud in the propene molecule is an excellent source of electrons, and it is reasonable that the carbocation (an electrophile) attacks those electrons and forms a new carbon-carbon bond. The orientation of the addition of the carbocation to the second molecule of propene should, of course, occur to produce the more stable cation (Markovnikov's addition). This is exactly what happens. Now the two three-carbon molecules are united to give the six-carbon dimer.

Step 2:

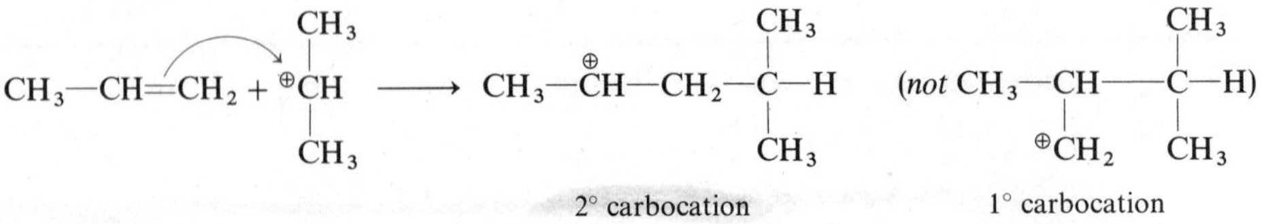

$$CH_3-CH{=}CH_2 + {}^\oplus\overset{\displaystyle CH_3}{\underset{\displaystyle CH_3}{\underset{|}{\overset{|}{CH}}}} \longrightarrow \overset{\displaystyle CH_3}{\underset{|}{CH_3-\overset{\oplus}{CH}-CH_2-\overset{|}{\underset{|}{C}}-H}} \quad (not\ CH_3-\overset{\displaystyle CH_3}{\underset{\displaystyle {}^\oplus CH_2}{\underset{|}{\overset{|}{CH}}}}-\overset{\displaystyle CH_3}{\underset{|}{\overset{|}{C}}}-H)$$

2° carbocation   1° carbocation

The new secondary carbocation can in principle undergo a variety of reactions. It could react with another molecule of propene to give a trimer. Or it could simply lose a proton to form alkenes, which it can do in either of the ways shown in step 3 to give the alkenes that are obtained from the dimerization of propene.

Step 3:

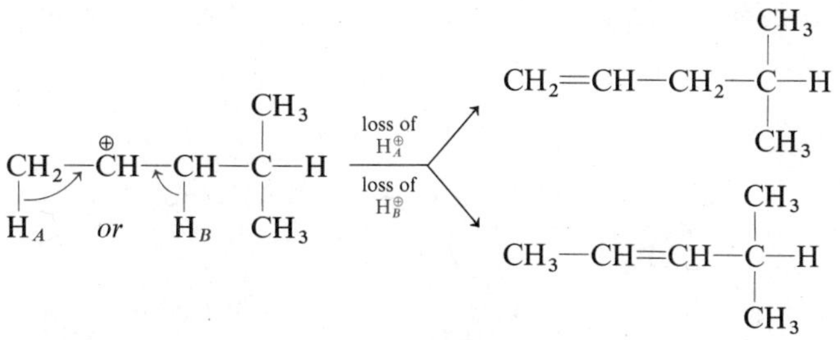

Reaction conditions must be controlled very carefully to ensure that only dimerization occurs. In Sec. 8.34 we see that certain conditions lead to polymerization and polymers (many units). The conditions needed for favoring dimerization, trimerization, or polymerization must be determined experimentally.

In dimerization, the carbocation is an electrophile, $E^{\oplus}$, reacting in the same way as, for example, a proton, $H^{\oplus}$; both are electron-deficient.

Question 8.16

The dimerization of methylpropene, $(CH_3)_2C{=}CH_2$, in the presence of an acidic catalyst gives a mixture of two alkenes, which on catalytic hydrogenation yield 2,2,4-trimethylpentane (which is frequently called *isooctane* and used as a standard in the octane rating system; see Sec. 4.10). Provide a complete, stepwise mechanism to account for the formation of the mixture of alkenes. (Draw their structures on the basis of your mechanism.)

8.24 Addition of Alkanes to Alkenes: Alkylation Reactions

An important reaction in the petroleum industry involves the addition of an alkane to an alkene. Frequently the starting alkane and alkene boil below the boiling point of gasoline, but they combine to give a new, highly branched alkane with a higher boiling point and an increased octane rating. Adding an alkane to an alkene is called **alkylation.** Strong acids, such as *concentrated* sulfuric acid or hydrofluoric acid, are used as catalysts.

As an example, consider the reaction between methylpropane and ethene:

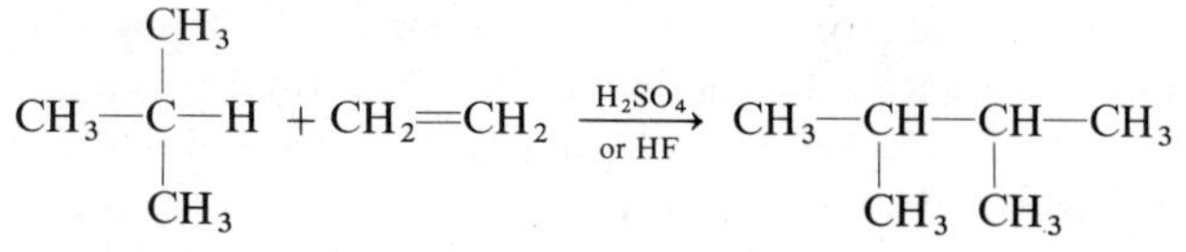

Methylpropane Ethene 2,3-Dimethylbutane

The formation of 2,3-dimethylbutane can be explained in terms of a carbocation mechanism. The first step is the addition of a proton to ethene:

Step 1:

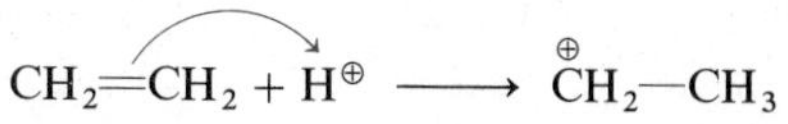

Ethyl cation

Because the ethyl cation is highly reactive (it is a primary carbocation), it abstracts a hydride ion (a hydrogen atom with its bonding pair of electrons) from methylpropane to form the much more stable tertiary cation and ethane.

Step 2:

$$CH_3-\underset{\underset{CH_3}{|}}{\overset{\overset{CH_3}{|}}{C}}-H + \overset{\oplus}{C}H_2-CH_3 \xrightarrow[\substack{(H:) \\ \text{transfer}}]{\text{hydride}} CH_3-\underset{\underset{CH_3}{|}}{\overset{\overset{CH_3}{|}}{C}}{}^{\oplus} + CH_3-CH_3$$

Ethyl   tert-Butyl cation
cation

The *tert*-butyl cation then reacts with a molecule of ethene to give a new primary carbocation (step 3).

Step 3:

$$CH_3-\underset{\underset{CH_3}{|}}{\overset{\overset{CH_3}{|}}{C}}{}^{\oplus} + CH_2{=}CH_2 \longrightarrow CH_3-\underset{\underset{CH_3}{|}}{\overset{\overset{CH_3}{|}}{C}}-CH_2-\overset{\oplus}{C}H_2$$

This new carbocation can undergo several 1,2 shifts to form first a secondary and then a tertiary carbocation, as shown in step 4.

Step 4:

$$CH_3-\underset{\underset{CH_3}{|}}{\overset{\overset{CH_3}{|}}{C}}{-}\underset{H}{\overset{}{C}}H{-}\overset{\oplus}{C}H_2 \xrightarrow[\substack{\text{shift} \\ (\sim H:^{\ominus})}]{\text{hydride}} CH_3-\underset{\underset{CH_3}{|}}{\overset{\overset{CH_3}{|}}{C}}{-}\overset{\oplus}{C}H{-}CH_3$$

1° carbocation   2° carbocation

$$\downarrow \substack{\text{methyl shift} \\ (\sim :CH_3)}$$

$$CH_3-\underset{\oplus}{\overset{\overset{CH_3}{|}}{C}}{-}\overset{\overset{CH_3}{|}}{C}H{-}CH_3$$

3° carbocation

This is an interesting sequence of rearrangements, and clearly the driving force is the formation of carbocations of increasing stability.

The carbon skeleton in the 3° carbocation in step 4 is identical with that in the observed product, 2,3-dimethylbutane. The final product is formed when this 3° cation abstracts a hydride ion from another molecule of methylpropane to form a *tert*-butyl cation, which may then react with more ethene and start steps 3 and 4 over.

Step 5:

$$CH_3-\underset{\underset{CH_3}{|}}{\overset{\overset{CH_3}{|}}{C}}-H + CH_3-\underset{\oplus}{\overset{\overset{CH_3}{|}}{C}}{-}\overset{\overset{CH_3}{|}}{C}H{-}CH_3 \xrightarrow[\substack{(H:) \\ \text{transfer}}]{\text{hydride}} CH_3-\underset{\underset{CH_3}{|}}{\overset{\overset{CH_3}{|}}{C}}{}^{\oplus} + CH_3-\overset{\overset{CH_3}{|}}{C}H{-}\overset{\overset{CH_3}{|}}{C}H{-}CH_3$$

2,3-Dimethylbutane
Observed product

The 3° carbocation formed continues the reaction sequence.

8.25 Applications of Carbocation Chemistry to Natural Products

There are many interesting and important examples of carbocation chemistry in natural products and polymer chemistry (see Sec. 8.34). We saw, for example, that it is possible for a carbocation to react with a double bond. This also happens when the double bond is elsewhere in the molecule containing the carbocation and occurs often when either a five- or six-membered ring may form. (The formation of smaller rings is unlikely because they are highly strained, and rings with more than six members seldom form because it is unlikely that the carbocation will come close enough to the double bond to allow reaction to occur.) For example,

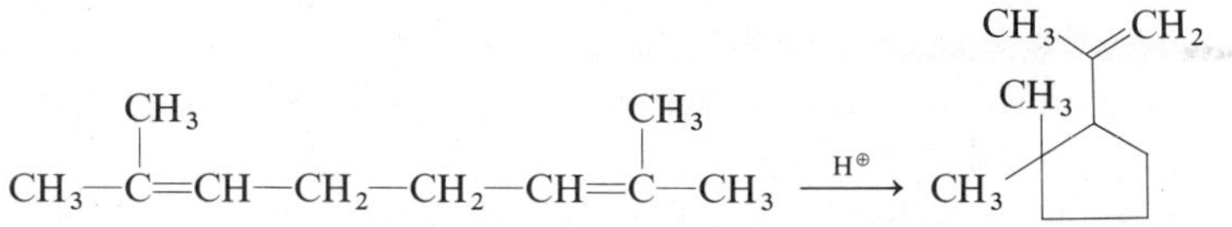
CH3
CH3—C=CH—CH2—CH2—CH=C—CH3
CH3
H⊕
CH3
CH2
CH3
CH3

A more careful examination of the changes reveals that the following carbocation reactions intervene. (In this example, a more convenient notation is used; a short line represents a methyl group, and the intersection of two or more lines signifies that a carbon atom is present at that intersection unless otherwise indicated. For example, 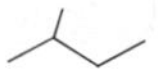 is 2-methylbutane. This convention is used rarely in this text and only when complex molecules are drawn.)

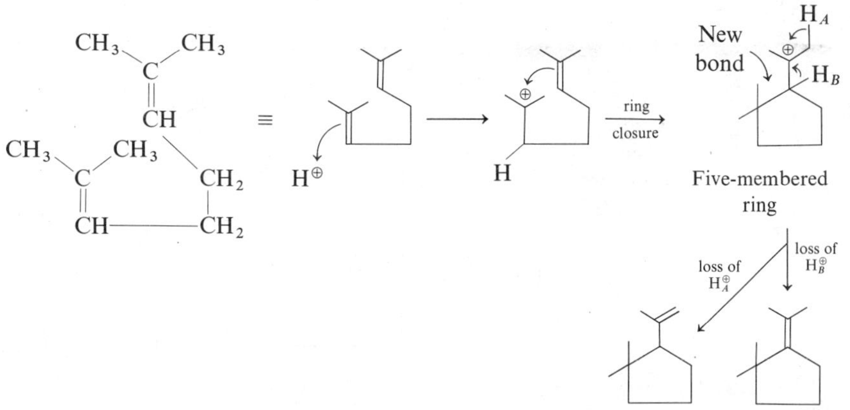
CH3
CH3
C
CH
CH3
CH3
C
CH2
CH
CH2
≡
H⊕
H
ring
closure
New
bond
HA
HB
Five-membered
ring
loss of
H⊕A
loss of
H⊕B

Another example involves the formation of a carbocation from an alcohol via the conversion of *farnesol* to *bisabolene*, both of which are members of a large family of naturally occurring compounds—the **terpenoid family** (see Sec. 8.35). The reaction and its mechanism are shown here. Note that the reaction results in the formation of a six-membered ring.

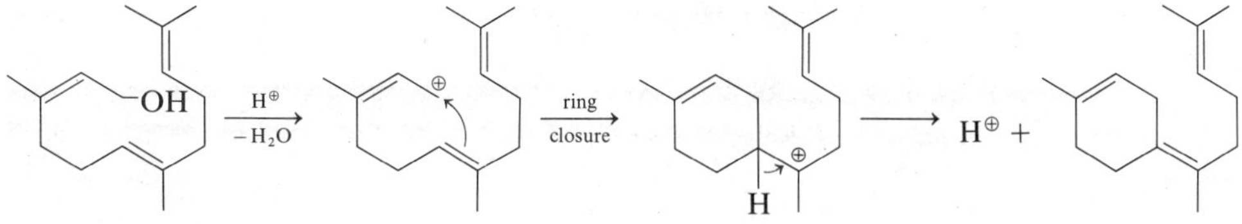
OH
H⊕
−H2O
ring
closure
H
H⊕ +

Farnesol Bisabolene

We will encounter other examples where principles of organic chemistry apply to natural products.

8.26   Carbocations: Summary

It now seems desirable to summarize the important facts about carbocations:

1. *Methods of formation.* Carbocations may be formed
(*a*) From alcohols in acid solution (see Sec. 7.15).
(*b*) By adding an electrophile ($E^{\oplus}$, often a proton) to a double bond (see Sec. 8.3); orientation of addition is such that the more stable carbocation is formed (see Sec. 8.3).
2. *Stability of carbocations.* The order of decreasing stability of carbocations is (see Sec. 8.10):

$$\text{allyl} \approx 3° > 2° > 1° > \overset{\oplus}{C}H_3$$

3. *Reactions of carbocations.* Carbocations may:
(*a*) Lose a proton to form alkenes (see Sec. 7.15); the most highly substituted alkene is the major product (see Sec. 7.17).
(*b*) Rearrange to a more stable carbocation by a 1,2-hydride or 1,2-alkyl shift (see Sec. 7.18).
(*c*) React with a nucleophile, $\overset{\ominus}{Nu}$:, which is often an inorganic ion that is neutral or negatively charged and has at least one unshared pair of electrons (see Sec. 8.3).
(*d*) Add to an alkene to form a new carbocation (see Sec. 8.23).
(*e*) Abstract a hydride ion from an alkane to form a new alkane and a new carbocation (usually more stable) (see Sec. 8.24).
Note: The carbocations formed in reactions (*b*), (*d*), and (*e*) may then undergo any of reactions (*a*) to (*e*).
4. *Structure of carbocations.* Carbocations have a planar geometry (see Secs. 2.8 and 8.6).

Question 8.17

Indicate two different reactions for the formation of this structure:

$$CH_3-\underset{\underset{H}{|}}{\overset{\overset{CH_3}{|}}{C}}-\overset{\oplus}{C}H-CH_3$$

Question 8.18

Indicate at least four different reactions that the carbocation shown in Question 8.17 may undergo.

8.27   Reaction of Alkenes with Oxidizing Agents; *cis* Hydroxylation and Baeyer Test

The reaction between an alkene and certain oxidizing agents results in a product in which the elements of two hydroxyl (—OH) groups are added across the double bond:

$$\overset{}{\underset{}{>}}C=C\overset{}{\underset{}{<}} \xrightarrow[\text{or } OsO_4 + H_2O]{KMnO_4 + H_2O} -\underset{\underset{OH}{|}}{C}-\underset{\underset{OH}{|}}{C}-$$

Alkene   Glycol

The product is called a **glycol** (or **diol,** di-alcohol), and the reaction is often called **hydroxylation** because hydroxyl groups are added to the alkene. Two common reagents that accomplish this transformation are (1) aqueous potassium permanganate, $KMnO_4$, often containing dilute base, and (2) osmium tetroxide, OsO_4, in the presence of water. The latter reaction requires a second step, the addition of a reducing agent, usually Na_2SO_3. Other oxidizing agents such as peroxy acids, RCO_2OH (see Sec. 19.17), can also be used to oxidize alkenes to diols. Some examples of this hydroxylation reaction are:

$$CH_2{=}CH_2 + KMnO_4 \xrightarrow[OH^\ominus]{H_2O} \underset{\overset{|}{OH}\quad\overset{|}{OH}}{CH_2{-}CH_2} + MnO_2$$

Ethene   *Purple*   1,2-Ethanediol   *Brown-black*
                               (ethylene glycol)   *precipitate*

$$CH_3{-}CH{=}CH_2 \xrightarrow[\text{2. }Na_2SO_3]{\text{1. }OsO_4,\ H_2O} \underset{\overset{|}{OH}\quad\overset{|}{OH}}{CH_3{-}CH{-}CH_2} + OsO_3$$

Propene   1,2-Propanediol
                               (propylene glycol)

Cyclopentene   1,2-Cyclopentanediol
                               (cyclopentene glycol)

The common names of glycols are derived from the name of the alkene with the suffix *glycol* added as a separate word. Glycols can be named by indicating the positions of attachment of the two —OH groups in front of the parent alkane name and adding the suffix *-diol* to the parent name. Both types of names are indicated in the preceding examples.

Extensive studies have been carried out on the mechanism of the hydroxylation reaction. Stereochemistry plays an important role in elucidating the mechanism. For example, the reaction between cyclopentene and either potassium permanganate or osmium tetroxide yields only *cis*-1,2-cyclopentanediol.

Cyclopentene   *cis*-1,2-Cyclopentanediol

Clearly, the stereochemistry of the addition is revealed by this result and is often called *syn* or *cis* hydroxylation. (The similar reaction between a peroxy acid and an alkene yields only the *trans* isomer. The mechanism is much different and is discussed in Chap. 19.) Also, hydroxylation is an oxidation-reduction reaction. For example, manganese is reduced from its +7 state in $KMnO_4$ to +4 in MnO_2, and osmium is reduced from its +8 state in OsO_4 to +6 in OsO_3. Thus, our mechanism must account for both the stereochemistry and the oxidation-reduction aspects of this reaction.

Using potassium permanganate as an example, the currently accepted mechanism involves the addition of the tetrahedral permanganate ion across the double bond (the π cloud is attacked), and the resulting cyclic intermediate is hydrolyzed by dilute base (or water) to give the glycol and manganese dioxide. The hydrolysis reaction involves only the manganese-oxygen bonds and not the carbon-oxygen bonds. The geometry of the permanganate ion requires that its addition occur so that both oxygens are on the same side of the double bond, thus giving *syn* addition. The mechanism and stereochemistry are illustrated for the hydroxylation of cyclopentene.

Tetrahedral   Cyclic intermediate   *cis*.Hydroxylation

As fast as manganese (V) ion is formed, it is oxidized by permanganate ion, $MnO_4^{\ominus}$, to manganate ion, $MnO_4^{\ominus 2}$ [which contains manganese(VI)]. This unstable ion decomposes to give permanganate ion and manganese dioxide:

$$\left[\overset{V}{MnO_4}{}^{\ominus 3}\right] + \overset{VII}{MnO_4}{}^{\ominus 1} \longrightarrow \left[\overset{VI}{MnO_4}{}^{\ominus 2}\right] \longrightarrow \overset{VII}{MnO_4}{}^{\ominus 1} + \overset{IV}{MnO_2}$$

Unstable   Unstable   Brown-black
                                                                   precipitate

The reaction involving osmium tetroxide also involves a cyclic intermediate from addition to the double bond, followed by reduction to give the corresponding *cis*-glycol:

syn Addition

The disadvantage of using osmium tetroxide is that it is very toxic and costly.

Experimentally, the reaction between alkenes and aqueous potassium permanganate is very fast and easy to carry out, and it is often used to help determine the presence of the double bond[1] in unknown molecules. This reaction is the basis of the **Baeyer test**. If a reaction has occurred, the characteristic purple of the permanganate ion disappears and is replaced by a brown-black precipitate of manganese dioxide, MnO_2. The detection of functional groups is discussed in Sec. 8.33.

In contrast to alkenes, the strained cyclopropane molecule does not react with oxidizing agents such as potassium permanganate.

Question 8.19

Compare the stereochemistry of *syn* hydroxylation with that of the addition of bromine to cyclopentene. Why do the stereochemistries differ?

---

[1] Potassium permanganate also reacts with triple bonds, certain alcohols, and aldehydes. Thus a positive reaction with $KMnO_4$ is *not* conclusive evidence for the double bond.

Omit **8.28   Stereochemistry of Hydroxylation of Open-Chain Alkenes**

The stereochemistry of hydroxylation is also *syn* with certain open-chain alkenes. For example, *cis*-2-butene gives *meso*-2,3-butanediol, whereas *trans*-2-butene gives *d,l*-2,3-butanediol.

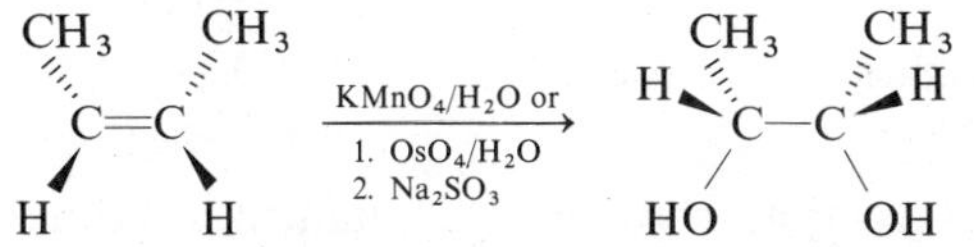

cis-2-Butene   *meso*-2,3-Butanediol

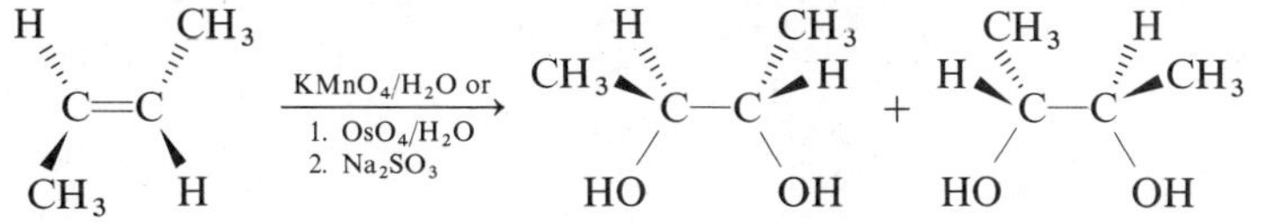

trans-2-Butene   *d,l*-2,3-Butanediol
Enantiomers: Optically inactive racemic mixture

It may not be immediately apparent why the racemic mixture of *d,l*-2,3-butanediol is formed, but the two stereoisomers arise from the equally probable attack by permanganate or osmium tetroxide on the top or bottom side of the double bond. (*Problem:* Work this out using molecular models; this requires rotating one of the stereoisomers by 180° to see the mirror-image relationship.)

Question 8.20

Draw the three-dimensional structures for the stereoisomers that result from the hydroxylation of *cis*-2-pentene and of *trans*-2-pentene.

Question 8.21

The hydroxylation of 1-butene yields 1,2-butanediol

$$CH_3-CH_2-\underset{\underset{OH}{|}}{CH}-\underset{\underset{OH}{|}}{CH_2}$$

which is optically inactive even though it has an asymmetric center (where is it?). Explain this result.

Omit **8.29   Cycloaddition Reactions**

One of the more useful synthetic methods available to the organic chemist is the reaction between a conjugated diene and an alkene. In this reaction the alkene is sometimes referred to as a **dienophile** (Greek *philos*, loving; thus "diene-loving"). There are many variations for this reaction, which is called the **Diels-Alder reaction** in honor of its developers, Otto Diels and Kurt Alder. When an alkadiene and an alkene are allowed to react, new carbon-carbon bonds form to give a cyclohexene derivative. There is overall 1,4 addition of the dienophile to the diene, as shown by the general formation:

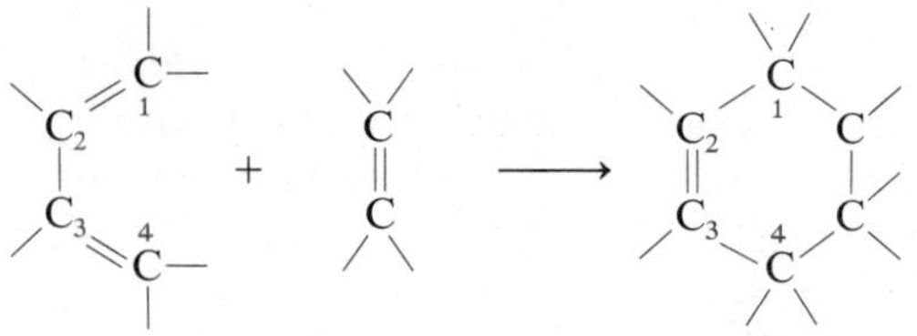

| Conjugated diene | Dienophile | Adduct *1,4 Addition* |

For example, the reaction between 1,3-butadiene and ethene would give cyclohexene:

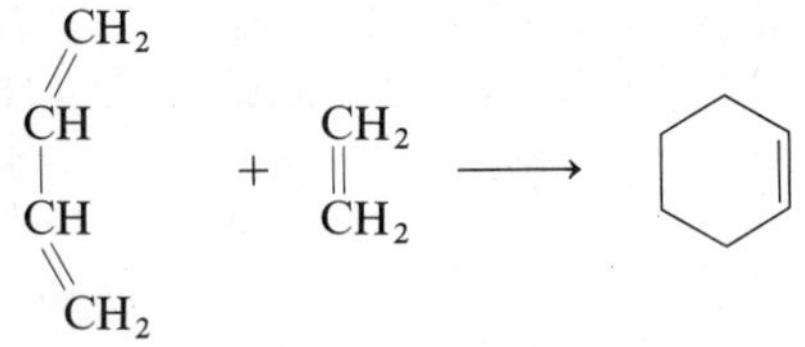

1,3-Butadiene   Ethene   Cyclohexene

Considerable research has been devoted to elucidating the mechanism of the Diels-Alder reaction and other similar reactions. These are called **cycloaddition** reactions. They are discussed in detail in Chap. 27.

Another example of a cycloaddition reaction involves the reaction of an alkene with the reactive intermediate, methylene, $CH_2:$, to produce cyclopropane. This reaction is shown here and is also discussed in Chap. 27.

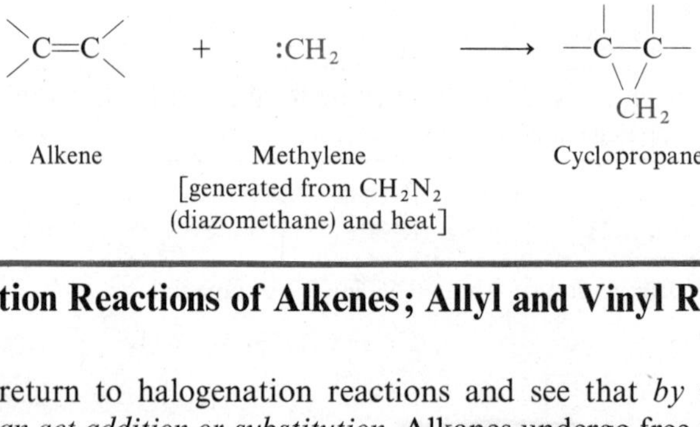

Alkene   Methylene   Cyclopropane
              [generated from CH_2N_2
              (diazomethane) and heat]

8.30   Substitution Reactions of Alkenes; Allyl and Vinyl Radicals

We now return to halogenation reactions and see that *by varying reaction conditions, we can get addition or substitution.* Alkanes undergo free-radical substitution on heating with bromine or chlorine. On the other hand, alkenes undergo ionic (electrophilic) addition when they react with bromine or chlorine in carbon tetrachloride solution.

Now consider molecules that exhibit the properties of being an alkene and an alkane at the same time. Propene, for example, can undergo addition or substitution.

$$CH_2{=}CH{-}CH_3 + Cl_2 \xrightarrow{CCl_4} \underset{\underset{Cl}{|} \quad \underset{Cl}{|}}{CH_2{-}CH{-}CH_3} \qquad \text{Ionic \textbf{addition} in solution}$$

Propene   1,2-Dichloropropane
                          (propylene chloride)

$$CH_2{=}CH{-}CH_3 + Cl_2 \xrightarrow{400°} CH_2{=}CH{-}CH_2{-}Cl + HCl \qquad \text{Free-radical \textbf{substitution} in the gas phase}$$

Propene   3-Chloro-1-propene
                          (allyl chloride)

The free-radical substitution reaction is primarily of industrial importance. In the laboratory, where more complicated reactants are involved, mixtures of addition and substitution products are often obtained and alternate routes are used to effect substitution. For example, the compound **N-bromosuccinimide** (abbreviated NBS) is a very convenient laboratory source for introducing a bromine atom at the allylic position in alkenes.

Cyclohexene   N-Bromosuccinimide   3-Bromocyclohexene   Succinimide
              (NBS)   *80%*

The reaction is enhanced by the presence of ultraviolet light or traces of peroxides, so it is clear that radicals are involved. Because traces of HBr are present, it has been suggested that this acid reacts with NBS to form bromine at a slow, but constant rate: HBr + N-bromosuccinimide $\longrightarrow$ Br$_2$ + succinimide. There are several alkanelike hydrogens (that is, those not attached to the double bond) in cyclohexene, yet substitution occurs predominantly on the carbon adjacent to the double bond. Also there is no substitution for the hydrogen attached to the double bond.

To explain the selectivity that halogen atoms, $:\overset{..}{\underset{..}{X}}\cdot$, exhibit for certain hydrogens in alkenes, the free-radical halogenation of propene is used as an example. The abstraction of the hydrogen on the carbon adjacent to the double bond gives the very stable **allyl radical** (see Sec. 8.10). This in turn reacts with halogen to give 3-halo-1-propene (often called an *allyl halide* because CH$_2$=CH—CH$_2$— is the allyl group).

$$CH_2=CH-\underset{\underset{H}{|}}{CH_2} + :\overset{..}{\underset{..}{X}}\cdot \longrightarrow CH_2=CH-\overset{\cdot}{C}H_2 + H-X$$

Allyl radical

$$CH_2=CH-\overset{\cdot}{C}H_2 + X_2 \longrightarrow CH_2=CH-CH_2-X + :\overset{..}{\underset{..}{X}}\cdot$$

Allyl halide

The reason for the unusual stability of the allyl radical is best seen by examining its orbital picture:

Allyl radical

The *p* orbital that the odd electron (the radical) occupies can be parallel to the *p* orbitals of the π bond. When this occurs, a new molecular orbital results and the odd electron is *delocalized* (spread out) through the rest of the molecule by overlapping with the π bond. This is another example of *resonance*, where two reasonable electronic structures for the allyl radical can be drawn *without* shifting the position of any atom in the molecule (see Sec. 13.6). Neither structure accurately represents the electron distribution in the radical, which is best shown by the delocalized structure:

$$\left[CH_2\!\!=\!\!CH\!\!-\!\!\dot{C}H_2 \longleftrightarrow \dot{C}H_2\!\!-\!\!CH\!\!=\!\!CH_2 \right]$$

Contributing resonance structures
for allyl radical

Best represented by $\overset{\delta\cdot}{C}H_2\!\!=\!\!CH\!\!=\!\!\overset{\delta\cdot}{C}H_2$

Allyl radicals are often more stable than even tertiary (3°) radicals. This explains why, for example, substitution occurs almost exclusively at the carbon adjacent to the double bond in cyclohexene, in preference to the other carbon atoms that also contain secondary hydrogens. The hydrogens adjacent to the double bond are more easily abstracted in free-radical substitution reactions because of the formation of the very stable allyl radical.

Question 8.22

(*a*) Give the structure(s) of the major monosubstitution products formed from the reaction between bromine and alkenes (1) and (2) (or equally well, between them and *N*-bromosuccinimide).

$$(CH_3)_2C\!\!=\!\!CHCH_3 \qquad (CH_3)_2C\!\!=\!\!CH\!\!-\!\!CH_2\!\!-\!\!CH_2\!\!-\!\!CH_3$$

(1)   (2)

(*b*) Compound (2) in (*a*) is 96 times more reactive toward monosubstitution than is compound (1). Suggest a reason for this reactivity difference. (*Hint:* Consider the different types of allylic hydrogens in these two compounds.)

Free-radical substitution for the hydrogens attached to the carbons bearing the double bond can be induced to occur but only with great difficulty. Higher reaction temperatures are normally required, but this substitution appears to occur by the same mechanism as does substitution in alkanes. For example:

$$CH_2\!\!=\!\!CH_2 \xrightarrow{\;:\ddot{C}l\cdot\;} CH_2\!\!=\!\!\dot{C}H \xrightarrow{\;Cl_2\;} CH_2\!\!=\!\!CH\!\!-\!\!Cl$$

Ethene   Vinyl radical   Chloroethene
                  Very unstable   (vinyl chloride)

(The *vinyl* group is $CH_2\!\!=\!\!CH\!\!-\!\!.$)

However, all available evidence indicates that abstraction of a *vinyl hydrogen* (that is, a hydrogen from a double bond) is exceedingly difficult and happens *only* if there are no more reactive hydrogens in the molecule. This is supported by the free-radical chlorination of propene giving 3-chloro-1-propene, $CH_2\!\!=\!\!CHCH_2Cl$, and *no* other monochlorinated product.

The relative reactivity sequence for hydrogen abstraction and thus for free-radical stability (because the two are parallel) can be extended to include these two groups.

Vinylic   Allylic hydrogen
hydrogens

Decreasing ease of hydrogen abstraction:

Allylic > 3° > 2° > 1° > CH_3—H > vinylic

Decreasing stability of free radicals:

$$\text{Allyl} > 3° > 2° > 1° > \cdot CH_3 > \text{vinyl}$$

The free-radical substitution reactions are of great industrial importance, though considerably less useful to the researcher.

Question 8.23

How might you account for the major product being as shown in the following reaction?

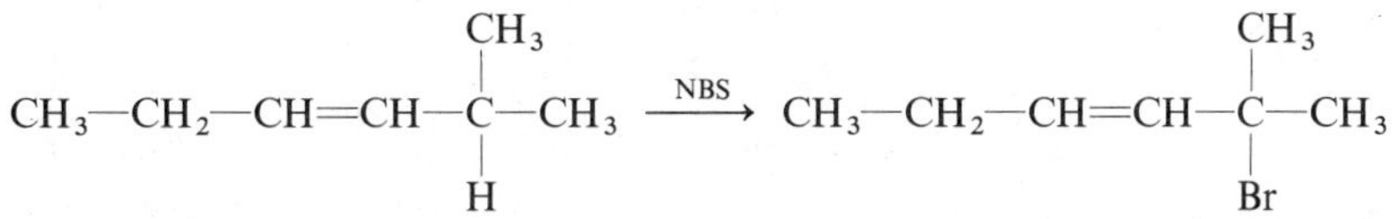

$$CH_3-CH_2-CH=CH-\overset{\overset{\displaystyle CH_3}{|}}{\underset{\underset{\displaystyle H}{|}}{C}}-CH_3 \xrightarrow{\text{NBS}} CH_3-CH_2-CH=CH-\overset{\overset{\displaystyle CH_3}{|}}{\underset{\underset{\displaystyle Br}{|}}{C}}-CH_3$$

8.31 Cleavage of Carbon-Carbon Double Bonds; Ozonolysis

We now turn to the third general type of reaction: **cleavage.** In general, it is difficult to break carbon-carbon bonds without affecting other parts of the molecule, but one way this is done utilizes *ozone* (O_3) and is called **ozonolysis.** The reaction usually occurs in two steps: treatment of the alkene with ozone followed by the addition of water and zinc.

General reaction for ozonolysis:

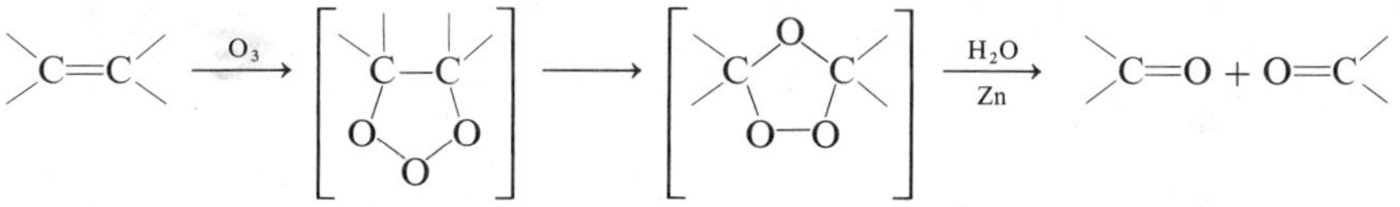

| Alkene | Molozonide *Unstable* | Ozonide | Observed cleavage products (aldehydes and ketones) |

The net effect is cleavage of the double bond and the formation of two new products; note that the carbon-carbon double bond is replaced by *two* carbon-oxygen double bonds, $\diagup C=O$.

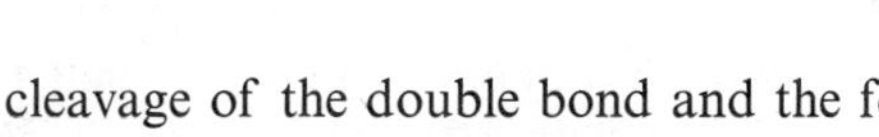

$$\diagup C \not{=} C \diagdown \xrightarrow[\text{2. H}_2\text{O, Zn}]{\text{1. O}_3} \diagup C=O + O=C \diagdown$$

The products of cleavage by ozone are aldehydes and/or ketones, depending on the substituents:

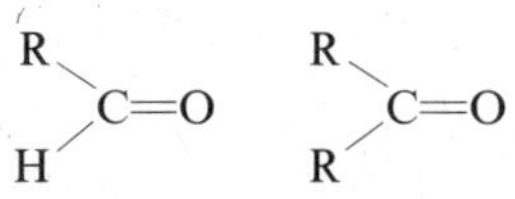

Three examples of cleavage by ozonolysis are shown here. The two separate steps are indicated without drawing the structure of the intermediate ozonide; this shorthand notation means that the alkene is treated first with ozone and then with water and zinc metal.

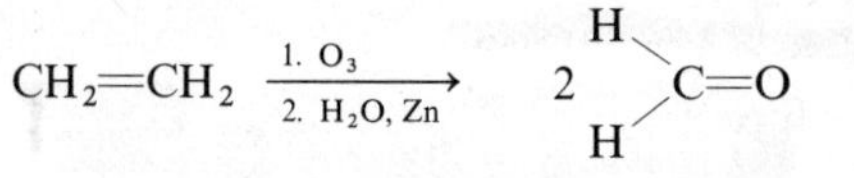

$$CH_2{=}CH_2 \xrightarrow[\text{2. } H_2O, Zn]{\text{1. } O_3} \quad 2 \quad \underset{H}{\overset{H}{\diagdown}}C{=}O$$

Ethene

Aldehyde: methanal
(formaldehyde)

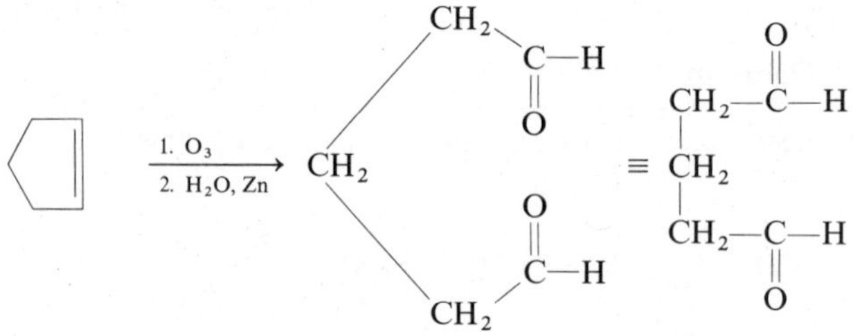

Cyclopentene

Dialdehyde: pentanedial

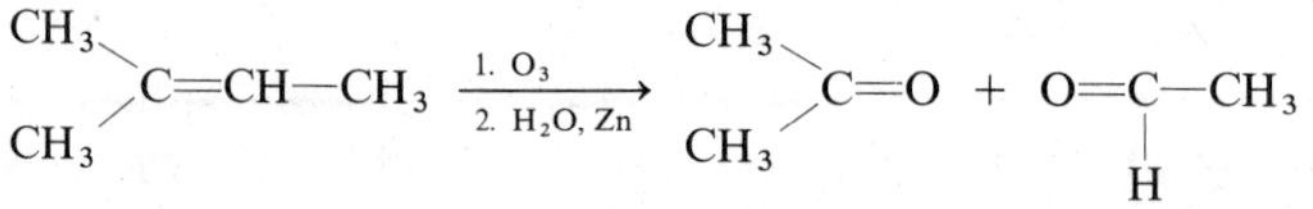

2-Methyl-2-butene

Ketone: Aldehyde: ethanal
propanone   (acetaldehyde)
(acetone)

The chief use of ozonolysis is in proof of structure, where the cleavage products are often simpler compounds that can be isolated and identified. For example, a compound that has the molecular formula C_6H_{12} decolorizes bromine in carbon tetrachloride and reacts with ozone followed by water and zinc, to give two compounds:

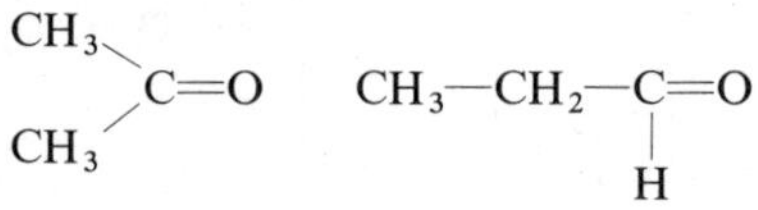

Ketone: Aldehyde:
2-propanone   propanal

The appearance of 2-propanone indicates that the grouping $\underset{CH_3}{\overset{CH_3}{\diagdown}}C{=}$ was present, and likewise the formation of propanal implies that the $CH_3{-}CH_2{-}CH{=}$ grouping was also there. Putting these two fragments together, we arrive at the conclusion that the structure of the unknown alkene is:

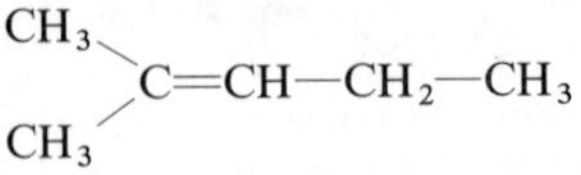

2-Methyl-2-pentene

Question 8.24

Draw the products formed from the ozonolysis of the four butene (C_4H_8) isomers? How does the structure of the products allow you to deduce which isomer is which? Does this method provide a unique structural assignment for all four isomers? If not, what is the ambiguity?

The mechanism of ozonolysis has been studied extensively, and although it is not elucidated with certainty, a considerable body of information now sheds light on it. Ozone, O_3, is a highly reactive form of oxygen produced by passing molecular oxygen between the electrodes of a discharge tube:

$$3O_2 \xrightarrow[\text{discharge}]{\text{electric}} 2O_3$$

Oxygen   Ozone

Two equivalent electronic structures that are resonance structures of one another can be drawn for ozone; the molecule also has *formal charges* (as indicated on the resonance structures) so that each oxygen can accommodate eight electrons to satisfy the octet rule.

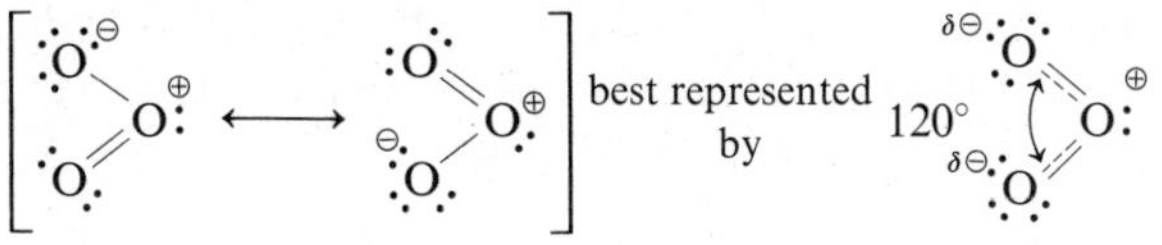

Because of the polarity of ozone, the initial step in ozonolysis is believed to be *addition*, where the positively charged oxygen is an electrophile and attacks the electron-rich double bond. A one-step ring closure provides the primary ozonide, which is called *molozonide*.

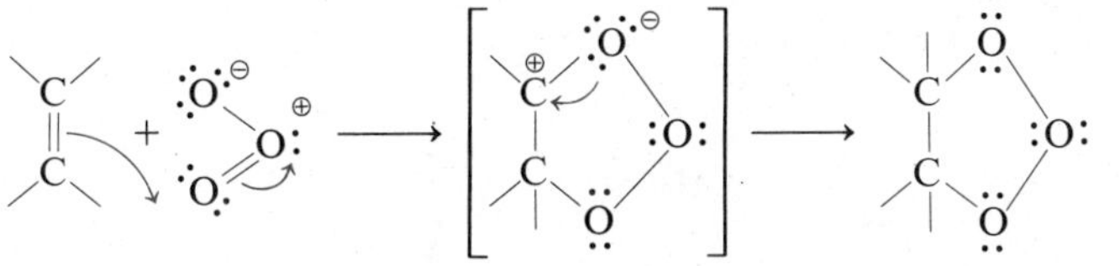

Molozonide
(a primary ozonide)

The molozonide rearranges immediately to an ozonide. The ozonides are unstable and sometimes decompose explosively; they are seldom isolated and are usually treated further without purification.

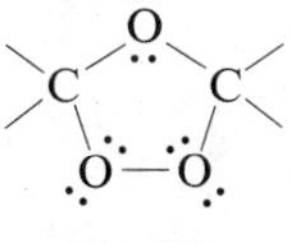

Ozonide

The ozonide may be decomposed in several ways. The most common is hydrolysis with hot water, which gives the cleavage products and hydrogen peroxide. A reducing agent, such as zinc metal dust, removes the hydrogen peroxide or other peroxide-containing oxidizing agents that may form; decomposition is usually done by adding both water and zinc metal.

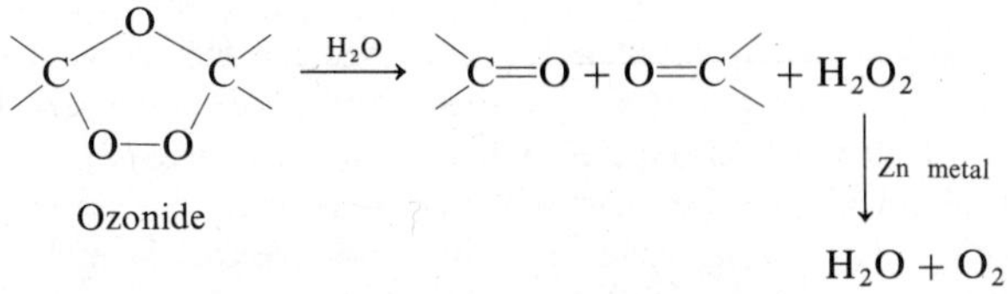

The other common method of decomposition is catalytic hydrogenation:

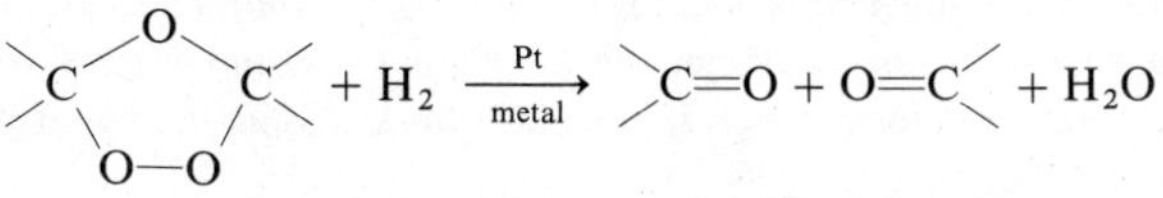

Care must be taken in running this reaction because hydrogen can also react with aldehydes or ketones to give alcohols (see Sec. 21.13).

8.32   Cleavage of Carbon-Carbon Double Bonds by Permanganate

Alkenes are also cleaved by the action of permanganate under vigorous conditions. The products are usually ketones and carboxylic acids, except when there is a terminal carbon-carbon double bond ($=CH_2$), which produces carbon dioxide. This reaction is believed to involve the corresponding glycol, which is formed first (see discussion of the Baeyer test in Sec. 8.27) and then further oxidized. The general reaction is:

$$\underset{\substack{\text{Glycol} \\ \textit{Not isolated}}}{} \longrightarrow \text{acids} + \text{ketones} + \text{carbon dioxide}$$

Two specific examples are:

Methylpropene   Ketone: 2-propanone

2-Methyl-2-pentene   Ketone: 2-propanone   Carboxylic acid salt: potassium propanoate

Studying Reactions

One key to success in working synthesis problems is knowing the reactions to the point that you do not have to think about them; they should become so familiar that you can consider various possible steps in syntheses without writing them down. There are, however, several study devices that may help you to master organic chemistry. One is an outline showing the interconversion of various functional groups, and the other is flash cards. Both are mentioned here, but remember these are only suggestions. Use the study techniques that work best for you.

(1) Reaction outlines. Many reactions we presented are incorporated into Chart 8.1, although certain details (such as reactivities and stereochemistry) are left out. *This chart is not complete* and is only a guide. If you find this type of outline useful, make one of your own and then extend and expand it as new reactions are encountered in future chapters.

(2) Flash cards. As increasing numbers of reactions are encountered, you may find it useful to consider flash cards as a study aid. Because study habits vary, flash cards are *optional*. Commercial flash cards for organic chemistry are available at most bookstores, and although they contain most reactions presented in this text, the cards must be sorted continuously to find the desired ones.

If you want to try this technique, obtain blank cards and prepare your own. Like all else, you must keep up in making flash cards; if you get behind, it is unlikely that they will benefit you. They

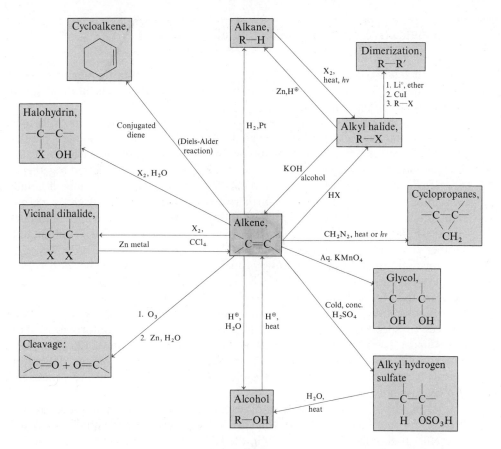

CHART 8.1 Outline of most of the functional groups and their reactions appearing in Chaps. 1 to 8.

are perhaps most useful for studying reactions. The following samples indicate some types of information that can be put on flash cards.

Sample 1: Reactant and product given; reagent and so on missing

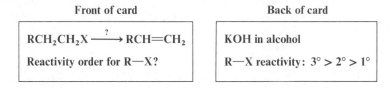

Sample 2: Reactant, reagent, and so on given; product(s) missing

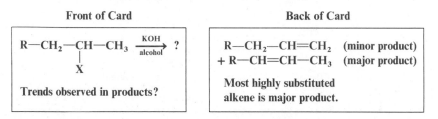

If you decide to make flash cards, remember to make them so they are *most useful to you*; you can easily put more or less on each card than is suggested by these samples.

8.33   Qualitative Analysis; Detection of Alkenes

One fascinating aspect of organic chemistry is **qualitative analysis,** which involves detecting various functional groups within a molecule. The ultimate goal is to identify the compound completely.

Useful information obtained during the identification of a compound comes from elemental analysis (see Sec. 1.3) and determination of the compound's physical properties, such as boiling point, melting point, density, and refractive index. Tables of thousands of organic compounds are available, but they are collected together by *functional group*. Because many compounds have identical boiling points (if liquids) or melting points (if solids), their further separation by functional group permits an unknown compound to be pinpointed as one of several possible compounds fairly quickly. Final identification is often done by converting the unknown compound to another compound called a *derivative*. Tables of organic compounds arranged by functional group and then by boiling or melting point usually list the properties of the derivatives. Qualitative analysis is purely a laboratory technique, but some reactions used for it should be discussed in textbooks.

In this day of elaborate electronic instruments, much laboratory work is replaced by spectroscopic methods that yield large amounts of information about compounds. The more common methods are discussed in Chaps. 12, 15, and 18. Nonetheless, qualitative analysis is an integral part of many organic laboratory courses because it aids learning the chemistry of the various functional groups and also because it is fun!

Using the fairly limited number of reactions we studied, we can discuss qualitative tests for the alkenes. But first, what are the requirements for a qualitative test to be useful? It should be done quickly in the laboratory, thereby excluding reactions that involve product analysis and the attendant separation and purification of products. For example, ozonolysis is a good technique for proof of structure, but it is a poor qualitative test for alkenes because the aldehyde and ketone products have to be isolated and identified. For the most part, qualitative tests should involve reactions that occur rapidly and involve a visible reaction, such as color change, formation of a second layer, or gas evolution. Elemental analysis is another example of a test in which the presence or absence of precipitates under various conditions allows one to draw a tentative conclusion about the presence or absence of the halogens, sulfur, and nitrogen.

Tentative conclusions about the presence or absence of a carbon-carbon double bond can be deduced from two simple qualitative tests: (1) bromine in carbon tetrachloride and (2) the Baeyer test (aqueous potassium permanganate). Both reactions are fast and involve readily observable changes:

$$\text{C=C} \ + \ \text{Br}_2 \ \xrightarrow{\text{CCl}_4} \ -\underset{\text{Br}}{\overset{|}{\text{C}}}-\underset{\text{Br}}{\overset{|}{\text{C}}}-$$

Alkene	Red-brown	Vicinal dibromide
Colorless		*Colorless*

$$\text{C=C} \ + \ \text{MnO}_4^{\ominus} \ \xrightarrow[\text{OH}^{\ominus}]{\text{H}_2\text{O}} \ -\underset{\text{OH}}{\overset{|}{\text{C}}}-\underset{\text{OH}}{\overset{|}{\text{C}}}- \ + \ \text{MnO}_2$$

Alkene	Purple	Glycol	Brown-black
Colorless		*Colorless*	*precipitate*

Great caution should be used in interpreting all qualitative tests because many reagents give positive results with several functional groups.

Polymerization Reactions and Naturally Occurring Polymers

Alkenes and alkadienes play a very important role in many natural and synthetic polymers. There are two general types of polymerization reactions: (1) **addition**— molecules join *without* the loss of any atoms, and (2) **condensation**—molecules combine with the expulsion of simple inorganic molecules (often water). In the remainder of this chapter we consider some natural and synthetic addition polymers of alkenes and alkadienes.

8.34   Polymerization of Alkenes

Polymerization involves the combination of small, simple alkenes called **monomers** (Greek *mono*, one, and *meros*, part; or "one part") to form large molecules called **polymers** (Greek *poly*, many; or "many parts"). The general reaction is:

$$n \; \overset{\diagdown}{\diagup}C{=}C\overset{\diagup}{\diagdown} \quad \xrightarrow{\text{catalyst}} \quad \sim\!\!\!\sim\!\!\!\sim\left[\begin{matrix} | & | \\ C{-}C \\ | & | \end{matrix}\right]_n\!\!\!\sim\!\!\!\sim\!\!\!\sim$$

Monomer   Polymer
(*n* = large number)

Typically, hundreds of monomers combine to form polymers. A familiar example of a polymer is *polyethylene*, which is formed by heating ethylene (the monomer) under pressure and in the presence of oxygen.

$$n\mathrm{CH_2{=}CH_2} \quad \xrightarrow[\text{pressure}]{\mathrm{O_2, heat}} \quad \sim\!\!\!\sim\!\!\!\sim\left[\mathrm{CH_2{-}CH_2}\right]_n\!\!\!\sim\!\!\!\sim\!\!\!\sim$$

Polyethylene

Polyethylene is a plasticlike material used as a packaging film and in electrical insulators and other household articles. The length of the polymer, that is, the number of carbon atoms in each polymer unit, depends on the reaction conditions (temperature, pressure, concentration, and nature of the catalyst). Typically a unit contains 700 to 800 ethylene molecules and has a molecular weight of 20,000 to 25,000.

Other useful polymers obtained from industrial production include Saran, polyvinyl chloride, Teflon, and Orlon.

$$n\mathrm{CH_2{=}C}\overset{\diagup \mathrm{Cl}}{\diagdown_{\mathrm{Cl}}} \quad \longrightarrow \quad \sim\!\!\!\sim\!\!\!\sim\left[\mathrm{CH_2{-}\overset{\overset{\textstyle Cl}{|}}{\underset{\underset{\textstyle Cl}{|}}{C}}}\right]_n\!\!\!\sim\!\!\!\sim\!\!\!\sim \qquad \text{Saran[1] (packaging film)}$$

1,1-Dichloroethylene   Polydichloroethylene
Monomer   *Polymer*

---

[1] Saran is actually a **copolymer**; *it contains two different monomeric units.* The two monomers are 1,1-dichloroethylene and vinyl chloride in about a 1:1 ratio. The two monomers are randomly copolymerized. Using the two structures given here for reference, draw a more accurate structure for Saran.

$$n\text{CH}_2=\text{CH}-\text{Cl} \longrightarrow \left[\text{CH}_2-\underset{\underset{\text{Cl}}{|}}{\text{CH}}\right]_n \qquad \text{PVC (food packaging, plastic pipe, records)}$$

Vinyl chloride   Polyvinyl chloride
Monomer   *Polymer*

$$n\;\underset{\text{F}}{\overset{\text{F}}{\diagup}}\text{C}=\text{C}\underset{\text{F}}{\overset{\text{F}}{\diagdown}} \longrightarrow \left[\underset{\underset{\text{F}}{|}}{\overset{\overset{\text{F}}{|}}{\text{C}}}-\underset{\underset{\text{F}}{|}}{\overset{\overset{\text{F}}{|}}{\text{C}}}\right]_n \qquad \text{Teflon}$$

Tetrafluoroethylene   Polytetrafluoroethylene
Monomer   *Polymer*

$$n\text{CH}_2=\text{CH}-\text{C}\equiv\text{N} \longrightarrow \left[\text{CH}_2-\underset{\underset{\text{C}\equiv\text{N}}{|}}{\text{CH}}\right]_n \qquad \text{Orlon}$$

Acrylonitrile   Polyacrylonitrile
Monomer   *Polymer*

Polymerization reactions are catalyzed by acids (*acidic* or *cationic conditions*), bases (*basic or anionic conditions*), and radicals (*free-radical polymerization*). Most industrial reactions involve free-radical polymerization. The acid-catalyzed (cationic) polymerization of ethylene is discussed first. This reaction starts with the addition of a proton to a molecule of ethylene, and the resulting ethyl cation reacts with another molecule of ethylene (the addition of a carbocation to an alkene, see Sec. 8.23); this process continues to attach a very large number of ethylene monomer units together as shown:

$$\text{H}^{\oplus} + \text{CH}_2=\text{CH}_2 \longrightarrow \text{CH}_3-\overset{\oplus}{\text{C}}\text{H}_2 \xrightarrow{\text{CH}_2=\text{CH}_2} \text{CH}_3-\text{CH}_2-\text{CH}_2-\overset{\oplus}{\text{C}}\text{H}_2 \xrightarrow{\text{CH}_2=\text{CH}_2}$$

$$\text{CH}_3-\text{CH}_2-\text{CH}_2-\text{CH}_2-\text{CH}_2-\overset{\oplus}{\text{C}}\text{H}_2 \xrightarrow[\text{times}]{\text{repeat} \atop \text{many}} \text{CH}_3-\text{CH}_2-\left[\text{CH}_2-\text{CH}_2\right]_n\text{CH}_2-\overset{\oplus}{\text{C}}\text{H}_2$$

The reaction terminates when the carbocation either loses a proton or reacts with a nucleophile. This polymerization reaction is favored by a *high ethylene concentration*, and only a trace of acid catalyst (typically a mineral acid) is added. Yields of high molecular weight polymers are usually quite low. For this reason the Ziegler-Natta method (discussed later) is used in this polymerization.

The free-radical polymerization is conceptually similar to cationic polymerization. A free-radical initiator (usually a peroxide, RO—OR) decomposes on heating to give a free radical (the alkoxy radical, RÖ·, when a peroxide is used). The free radical then *adds* to the carbon-carbon double bond in much the same way that halogen radicals add to alkenes in the free-radical addition of hydrogen bromide (see Sec. 8.9). The new free radical adds to another molecule of alkene, and so the process continues as shown:

$$\text{R}-\overset{..}{\underset{..}{\text{O}}}{:}\overset{..}{\underset{..}{\text{O}}}-\text{R} \xrightarrow{\text{heat}} 2\text{R}\overset{..}{\underset{..}{\text{O}}}\cdot$$

Peroxide   Alkoxy radical

$$\text{R}\overset{..}{\underset{..}{\text{O}}}\cdot + \text{CH}_2\overset{..}{\underset{..}{}}\text{CH}_2 \longrightarrow \text{RO}-\text{CH}_2-\overset{.}{\text{C}}\text{H}_2 \xrightarrow{\text{CH}_2\overset{..}{\underset{..}{}}\text{CH}_2}$$

$$\text{RO}-\text{CH}_2-\text{CH}_2-\text{CH}_2-\text{CH}_2 \xrightarrow[\text{times}]{\text{repeat} \atop \text{many}} \text{RO}-\text{CH}_2-\text{CH}_2-\left[\text{CH}_2-\text{CH}_2\right]_n\text{CH}_2-\overset{.}{\text{C}}\text{H}_2$$

The reaction terminates when the free radical comes in contact with impurities or when two radicals come together. In contrast, free-radical inhibitors are often added to increase the shelf life of chemicals that have a tendency to polymerize.

Another interesting aspect of free-radical polymerization reactions is that branching occurs under certain conditions (frequently high temperatures). Chain branching may be viewed as follows:

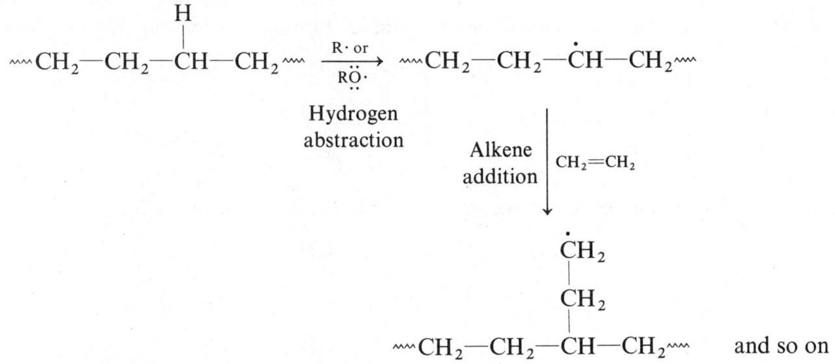

This branching is not a possible side reaction in ionic addition.

Exciting advances in polymer chemistry were announced in 1953 and later by Karl Ziegler and Giulio Natta, who illustrated how *stereoregular* polymers can be prepared. Various catalysts, such as the aluminum alkyls (AlR_3) in the presence of metal halides like titanium tetrachloride ($TiCl_4$), were developed for this purpose, and Ziegler and Natta were awarded the Nobel Prize in chemistry in 1963 for their work.

In the presence of various catalysts, propylene can be polymerized into polypropylene to give three principal forms of polymer:

1. *Isotactic*, where the configuration at the branching carbon is such that the methyl groups are all on the same side. This polymer is highly crystalline and forms strong fibers.
2. *Syndiotactic*, where the methyl groups alternate in a regular fashion along the chain.
3. *Atactic*, where the methyl groups are randomly disposed along the polymer. This polymer is rubbery and shows elastic properties.

Fig. 8.5 shows these three types of polymers.

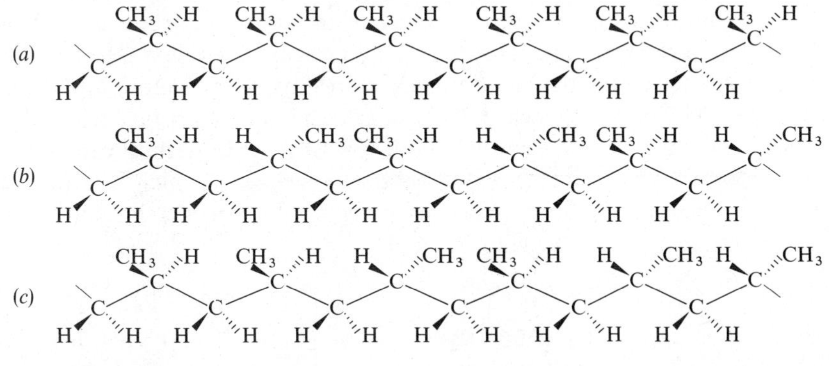

FIGURE 8.5 Structures of the stereospecific polymers of propene (propylene): (a) isotactic, (b) syndiotactic, (c) atactic.

The precise role the catalyst plays is not well understood. However, through much experimental work, there are procedures available to produce any of these three types of polymers.

Question 8.25

Provide a mechanistic rationale for both cationic and free-radical polymerization of isobutylene forming a polymer of the structure (and this is the *only* way they are linked together):

$$\left[\text{CH}_2{-}\underset{\underset{\text{CH}_3}{|}}{\overset{\overset{\text{CH}_3}{|}}{\text{C}}} \right]_n$$

rather than one in which there are some of the following types of linkages:

$$\text{CH}_2{-}\underset{\underset{\text{CH}_3}{|}}{\overset{\overset{\text{CH}_3}{|}}{\text{C}}}{-}\underset{\underset{\text{CH}_3}{|}}{\overset{\overset{\text{CH}_3}{|}}{\text{C}}}{-}\text{CH}_2$$

8.35   Polymerization of Alkadienes; Rubber

Conjugated alkadienes, like alkenes, undergo polymerization under conditions of free-radical catalysis. Natural rubber, for example, is the result of the polymerization of 2-methyl-1,3-butadiene, which is commonly called *isoprene*. Natural rubber has all *cis* orientations.

2-Methyl-1,3-butadiene
(isoprene)

Natural rubber
(*cis*-polyisoprene)

The *cis* relationship refers to the geometry of the atoms that constitute the continuous chain; that is, the —CH$_2$— groups are on the same side of the double bond. Natural rubber, of course, comes from the rubber tree, which does not synthesize it as described here. However, isoprene can be polymerized in the laboratory by some of the metal alkyl catalysts developed by Ziegler and Natta (or by metallic lithium) to form a polymer that possesses the characteristic *cis* configuration of rubber and has properties nearly identical with those of natural rubber.

Another important synthetic rubber is *neoprene*, which was one of the first successful synthetic rubbers and is prepared by the polymerization of 2-chloro-1,3-butadiene (*common name:* chloroprene).

2-Chloro-1,3-butadiene
(chloroprene)

Polychloroprene
(Neoprene)

Neoprene has properties similar to those of crude rubber, and on vulcanization, it produces a rubberlike material that has special uses.

In the mechanism for the polymerization of alkadienes, free-radical initiators start the reaction, which occurs predominantly by 1,4 addition. This results in the breaking of certain π bonds and the forming of new ones. This reaction is viewed as probably occurring by the following stepwise route, as illustrated for 1,3-butadiene:

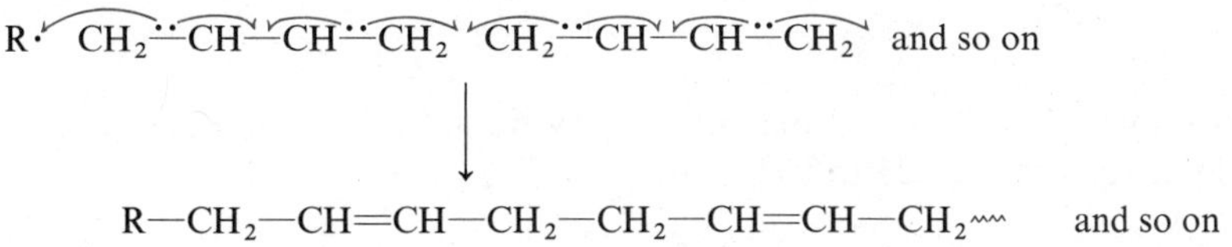

$$R \cdot \quad CH_2 \cdots CH \cdots CH \cdots CH_2 \quad CH_2 \cdots CH \cdots CH \cdots CH_2 \quad \text{and so on}$$

$$\downarrow$$

$$R—CH_2—CH=CH—CH_2—CH_2—CH=CH—CH_2 \sim \quad \text{and so on}$$

Other important synthetic polymers have been synthesized in recent years. For example, during World War II a practical rubber substitute was developed after natural rubber supplies were cut off. This synthetic material, known as GR-S (Government Rubber Styrene) or Buna-S, was prepared by polymerizing an alkene (styrene, phenylethene) with an alkadiene (1,3-butadiene) to give a product used in tire manufacturing. This type of reaction is called *copolymerization* because it occurs between two different compounds that contain double bonds. The partial structure of GR-S is

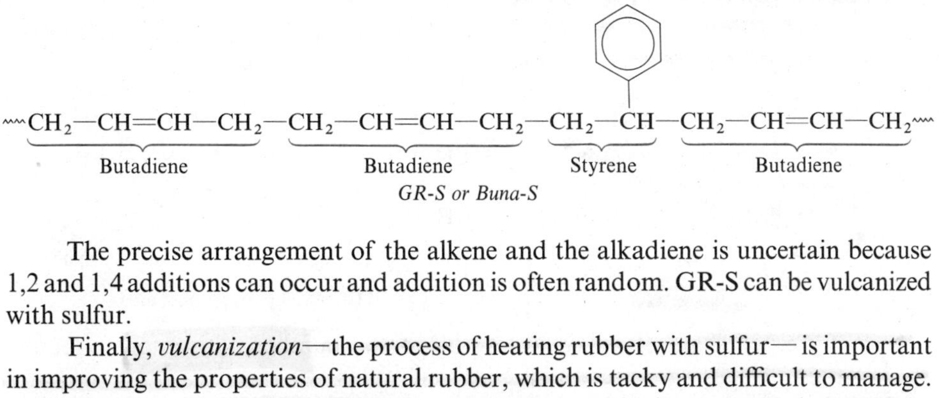

$$\sim CH_2—CH=CH—CH_2—CH_2—CH=CH—CH_2—CH_2—CH—CH_2—CH=CH—CH_2 \sim$$

| Butadiene | Butadiene | Styrene | Butadiene |

GR-S or Buna-S

The precise arrangement of the alkene and the alkadiene is uncertain because 1,2 and 1,4 additions can occur and addition is often random. GR-S can be vulcanized with sulfur.

Finally, *vulcanization*—the process of heating rubber with sulfur—is important in improving the properties of natural rubber, which is tacky and difficult to manage. Vulcanization causes the individual polymer units (which are long chains) to be hooked together by sulfur bridges; this is referred to as cross linking and in simplified form appears to resemble the following structure:

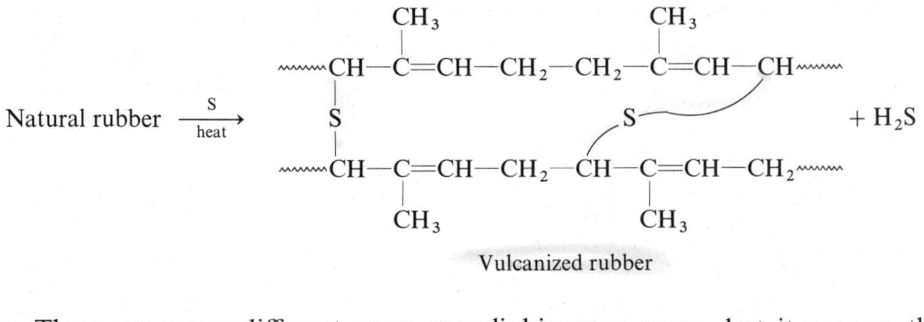

$$\text{Natural rubber} \xrightarrow[\text{heat}]{S}$$

$$\begin{array}{ccc} & CH_3 & & CH_3 \\ & | & & | \\ \sim CH—C=CH—CH_2—CH_2—C=CH—CH \sim & \\ & | & & \\ & S & & S \\ & | & & \\ \sim CH—C=CH—CH_2—CH—C=CH—CH_2 \sim & & +H_2S \\ & | & & | \\ & CH_3 & & CH_3 \end{array}$$

Vulcanized rubber

There are many different ways cross-linking can occur, but it appears that radicals are involved; in the preceding structure, the allylic hydrogens adjacent to the double bonds are replaced by sulfur linkages. There is evidence that even some carbon-carbon double bonds are broken to form cross links; some, but not all, of the unsaturation disappears.

Vulcanization often requires 5 to 8% by weight of sulfur, and special hard rubbers require as much as 25 to 50% sulfur.

8.36   Other Naturally Occurring Products That Contain Double Bonds

A host of natural products, called **terpenes,** are found in plants. In 1877 O. Wallach (University of Gottingen) recognized that terpenes can be divided mentally into five-carbon *isoprene* (2-methyl-1,3-butadiene) units, and in 1910 Wallach received the Nobel Prize in chemistry for his elegant work. This is known as the **isoprene rule.** He observed that terpenes can be constructed mentally by combining isoprene units together in a *head-to-tail* manner. The actual intermediate involved in the formation of terpenes is *isopentenyl pyrophosphate.*

Isoprene   Isopentenyl pyrophosphate

Terpenes are classified according to the number of isoprene subunits they contain. *Monoterpenes* contain 2 isoprene units (10 carbon atoms), *sesquiterpenes* contain 3 units (15 carbon atoms), *diterpenes* contain 4 units (20 carbon atoms), *triterpenes* contain 6 units (30 carbon atoms), and *tetraterpenes* contain 8 isoprene units (40 carbon atoms).

The isoprene rule is illustrated as follows for two simple monoterpenes. Note that this rule refers only to the carbon framework and not to the location of double bonds or other atoms (such as oxygen) in the terpenes.

Myrcene   Limonene

The following are typical examples of some classes of terpenes. Note the stereospecificity in these molecules in regard to *cis* and *trans* double bonds.

A. Monoterpenes

These are hydrocarbons with the molecular formula $C_{10}H_{16}$ and may be one of four general classes: (1) open-chain molecule with three double bonds, (2) cyclic structure with two double bonds, (3) bicyclic structure with one double bond, and (4) tricyclic structure. Unlike the first three, the fourth type is not found in nature. Many monoterpenes contain oxygen. The presence of oxygen or other functional groups often alters the structure of the terpene so that the general classes listed here may not hold.

Geraniol	Carvone	Camphor
(geranium oil)	(caraway oil)	(from Formosan camphor tree)

B. Sesquiterpenes

Their general formula is $C_{15}H_{24}$, and they occur in nature as open-chain, monocyclic, and bicyclic compounds.

Nerolidol	Zingiberene	β-Selinene
(oil of neroli)	(oil of ginger)	(oil of celery)

C. Diterpenes

Naturally occurring diterpenes contain 20 carbon atoms and exist in open-chain, cyclic, bicyclic, and tricyclic structures. One of the most important is vitamin A_1, which is derived from oxidation of β-carotene at the center of the molecule. Vitamin A_1 is essential for good night vision.

Vitamin A_1

D. Triterpenes

These 30-carbon compounds exist in a variety of structural types. For example, the open-chain hydrocarbon *squalene* can be cyclized under acidic conditions to give *lanosterol*. This is a particularly interesting cyclization reaction because it involves carbocations and several 1,2 shifts; its mechanism is discussed in Sec. 19.15.

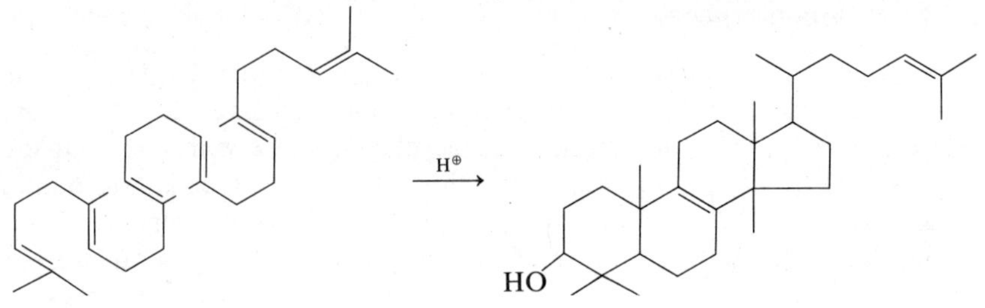

Squalene   Lanosterol

E. Tetraterpenes

These are the red to yellow pigments in plants and are called *carotenes*. For example, *lycopene* is an open-chain compound responsible for the red color in watermelons and tomatoes.

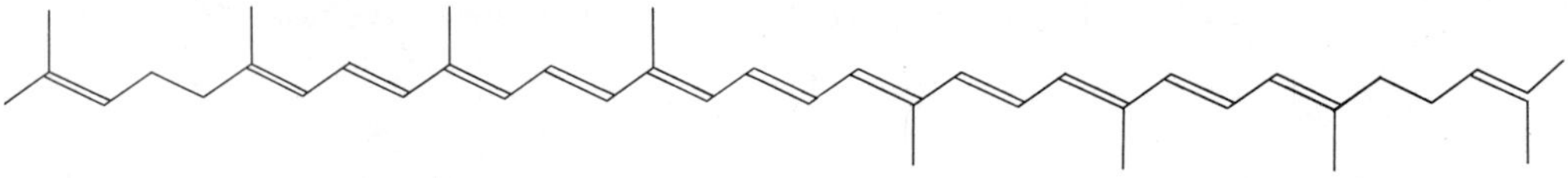

Lycopene

β-Carotene is a yellow tetraterpene found in carrots, tomatoes, and spinach. When this molecule is cleaved in the position indicated, two molecules of vitamin A are produced.

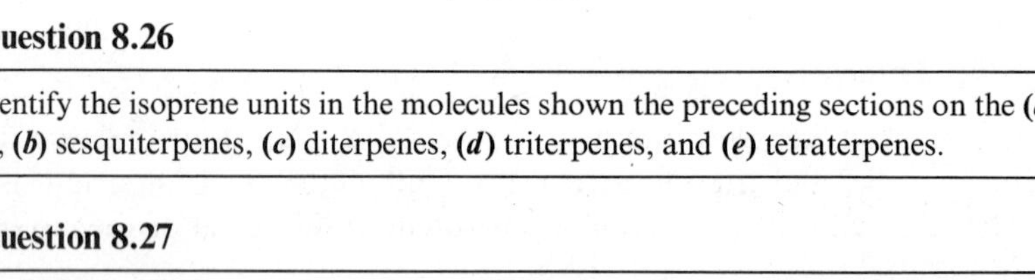

Cleavage here gives vitamin A

β-Carotene

Question 8.26

Identify the isoprene units in the molecules shown the preceding sections on the (*a*) mono-terpenes, (*b*) sesquiterpenes, (*c*) diterpenes, (*d*) triterpenes, and (*e*) tetraterpenes.

Question 8.27

There is symmetry in β-carotene, which on cleavage yields two molecules of vitamin A. Is there symmetry in lycopene? If so, where?

Multistep Synthesis

One of the organic chemist's main interests is the laboratory synthesis of new compounds. Sometimes the goal is simply to prepare a new compound for a mechanistic study, or it may be that a structure has been determined using degradation procedures (like ozonolysis, see Sec. 8.31, or permanganate oxidation, see Sec. 8.32) and the job now is to synthesize the same compound from known starting materials to confirm the structure assignment. Compounds are also synthesized be-cause of their potential usefulness as a medicine, insecticide, soap, detergent, and so on. Whatever the need, seldom does a one-step reaction convert a commercially available compound into the desired one. Thus, several steps must be used; such a procedure is called a multistep synthesis, which is a sequence of reactions that convert a known starting material to a new compound. Only practice with problems can make you proficient at devising a good multistep synthesis on paper, which is what a chemist must do before going into the laboratory. *In this section, you are shown only the approach to working synthesis problems* along with some things you should consider when working this type of problem, with the hope that they will get you started. You must master and understand the following.

1. *Know the chemical reactions and methods of preparation of various functional groups.* There is no way you can combine chemical reactions in a logical sequence so that one compound is converted to a new one, unless you know all the reactions well and do not have to stop and think about them.

2. *Determine whether the carbon content is increasing, decreasing, or unchanged in the desired transformation.*

(*a*) If the carbon content is *unchanged*, think of the reactions of functional groups because you usually use some combination of those reactions. Sometimes the number of carbon atoms is unchanged but the carbon skeleton is different; if this is the case, think about rearrangements (that is, 1,2-hydride and 1,2-alkyl shifts; see Sec. 7.18).

(*b*) If the carbon content *increases*, think of the reactions that form new carbon-carbon bonds. For example:

Wurtz reaction (see Sec. 4.2):

$$2R—X \xrightarrow[\text{heat}]{Na°} R—R \qquad \text{Doubles carbon content}$$

Symmetrical
alkane

Lithium dialkylcopper reaction (see Sec. 4.2):

$$R_2CuLi + R'X \longrightarrow R—R'$$

Addition of carbocations to alkenes (see Sec. 8.23):

$$R—OH \xrightarrow[\text{heat}]{H^\oplus}$$

or

Alkene $\xrightarrow{H^\oplus}$

$$R^\oplus \xrightarrow{\overset{\diagup}{C}=\overset{\diagdown}{C}} R—\overset{|}{\underset{|}{C}}—\overset{\oplus}{\underset{|}{C}}— \longrightarrow \text{product} \qquad \text{(proton loss, and so on)}$$

This list will grow rapidly, but these represent the methods you now have for making carbon-carbon bonds.

(*c*) If the carbon content *decreases*, think of reactions that break carbon-carbon bonds. The only reaction you had that is practical in the laboratory is ozonolysis (see Sec. 8.31) or permanganate cleavage (see Sec. 8.32). (The cracking of alkanes causes degradation, so this is hardly a good laboratory method.)

3. *Work backward.* Consider all the reactions that could be used to obtain the final desired product from any intermediate compounds. Then consider the reactions that could be used to synthesize the intermediate compound. Work backward in this fashion until finally the desired starting material is reached.

4. *Choose a sequence of reactions that give good yields of pure products.* A series of reactions is not desirable if each produces a number of isomers or undesired by-products. This means that the yields are likely to be poor and also that the isomers have to be separated in the laboratory—which is time-consuming and often ineffective. Where possible, the sequence should contain the *minimum* number of steps.

Often more than one acceptable route can be devised, but only as long as good yields and pure products are obtained. The characteristic reactions of the various functional groups play an important role in synthetic schemes. For example, a double bond is hydrogenated regardless of whether there are 2 or 100 carbon atoms in the molecule; this is characteristic of the alkene functional group. When showing reactions in multistep syntheses, draw the structures of each intermediate compound. Equations are normally not balanced, but the necessary reagents and critical reaction conditions are always included; for example:

$$CH_3—CH_2—Br \xrightarrow[\text{alcohol}]{KOH} CH_2=CH_2 \xrightarrow{H^\oplus \ H_2O} CH_3—CH_2—OH$$

Now consider some sample multistep syntheses and see the general method of attack that can be applied to them.

Example 1

Convert

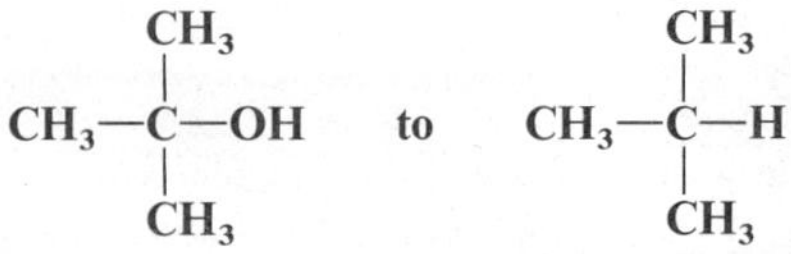

$$CH_3-\underset{\underset{CH_3}{|}}{\overset{\overset{CH_3}{|}}{C}}-OH \quad \text{to} \quad CH_3-\underset{\underset{CH_3}{|}}{\overset{\overset{CH_3}{|}}{C}}-H$$

First note that the carbon content is unchanged; that is, there are four carbon atoms in both starting material and product.

Now we ask: What are *all* the possible ways to prepare isobutane from *any* intermediate compound? Thus far, we have only the following reactions:

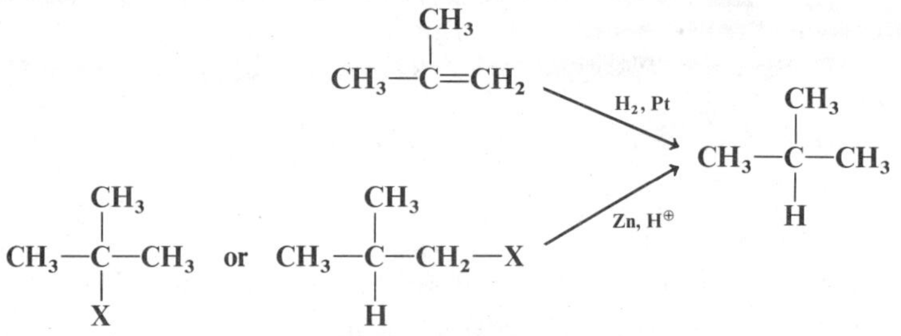

$$CH_3-\underset{}{\overset{\overset{CH_3}{|}}{C}}{=}CH_2$$

$$\xrightarrow{\text{H}_2,\text{ Pt}}$$

$$CH_3-\underset{\underset{X}{|}}{\overset{\overset{CH_3}{|}}{C}}-CH_3 \quad \text{or} \quad CH_3-\underset{\underset{H}{|}}{\overset{\overset{CH_3}{|}}{C}}-CH_2-X$$

$$\xrightarrow{\text{Zn, H}^{\oplus}}$$

$$CH_3-\underset{\underset{H}{|}}{\overset{\overset{CH_3}{|}}{C}}-CH_3$$

Clearly neither compound is the one with which we must start the synthesis, so we continue to work backward. Can either compound be prepared from another intermediate compound? Yes, as follows:

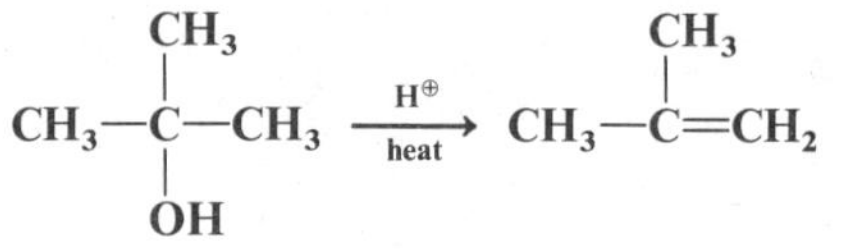

$$CH_3-\underset{\underset{OH}{|}}{\overset{\overset{CH_3}{|}}{C}}-CH_3 \xrightarrow[\text{heat}]{\text{H}^{\oplus}} CH_3-\underset{}{\overset{\overset{CH_3}{|}}{C}}{=}CH_2 \qquad (1)$$

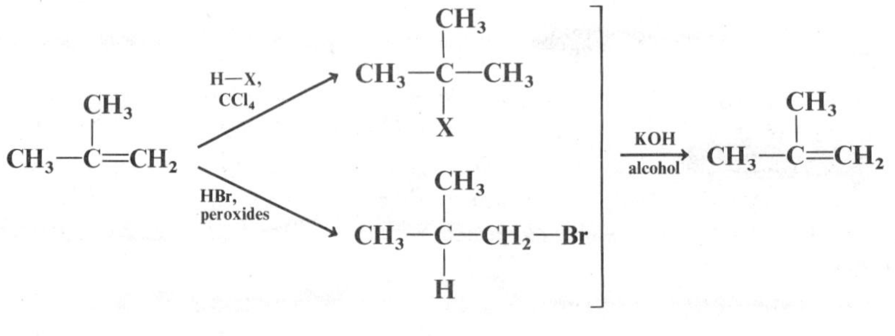

and

$$CH_3-\overset{\overset{CH_3}{|}}{C}{=}CH_2$$

$$\xrightarrow[\text{CCl}_4]{\text{H}-\text{X,}} CH_3-\underset{\underset{X}{|}}{\overset{\overset{CH_3}{|}}{C}}-CH_3$$

$$\xrightarrow[\text{peroxides}]{\text{HBr,}} CH_3-\underset{\underset{H}{|}}{\overset{\overset{CH_3}{|}}{C}}-CH_2-Br$$

$$\left.\right] \xrightarrow[\text{alcohol}]{\text{KOH}} CH_3-\overset{\overset{CH_3}{|}}{C}{=}CH_2 \qquad (2)$$

or

$$CH_3-\underset{\underset{H}{|}}{\overset{\overset{CH_3}{|}}{C}}-CH_3 \xrightarrow[\substack{\text{heat,}\\ hv}]{\text{X}_2} CH_3-\underset{\underset{H}{|}}{\overset{\overset{CH_3}{|}}{C}}-CH_2-X + CH_3-\underset{\underset{X}{|}}{\overset{\overset{CH_3}{|}}{C}}-CH_3 \qquad (3)$$

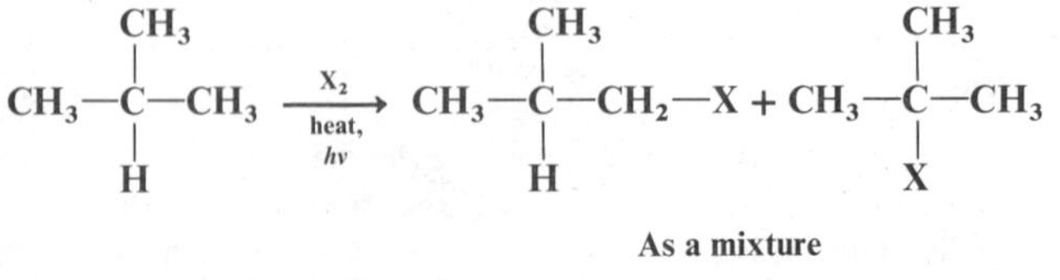

As a mixture

Reaction (1) does convert what happens to be the starting compound to an alkene and is the method of choice. But it is instructive to mention why the other two reactions are less desirable. Reaction (2) starts with an alkene, which on reaction with H-X under ionic conditions (carbon tetrachloride solvent) gives one alkyl halide that could be used in the final step of the synthesis; allowing this same alkene to react with HBr under free-radical conditions (peroxides) gives the other alkyl halide that could also be used in the final step. Either alkyl halide could be dehydrohalogenated to give the alkene that might be used in the last step. But reaction (2) starts and ends with the same product, and reaction (3) is of no use because it starts with the compound we want to prepare. Reaction (1) is the only logical pathway. Combining these, we obtain the following multistep synthesis, which is the answer to this problem:

$$CH_3-\underset{\underset{OH}{|}}{\overset{\overset{CH_3}{|}}{C}}-CH_3 \xrightarrow[heat]{H^\oplus} CH_3-\overset{\overset{CH_3}{|}}{C}=CH_2 \xrightarrow[Pt]{H_2} CH_3-\underset{\underset{H}{|}}{\overset{\overset{CH_3}{|}}{C}}-CH_3$$

Indicated   Desired
starting material   product

Example 2

Convert

$$\underset{CH_3}{\overset{CH_3}{>}}CH-CH_3 \quad to \quad CH_3-\underset{\underset{OH}{|}}{\overset{\overset{CH_3}{|}}{C}}-\underset{\underset{OH}{|}}{CH_2}$$

In this example we present only the final answer. Follow the example through carefully and consider alternative possibilities where they occur; try to figure out what they are and why they were eliminated.

$$\underset{CH_3}{\overset{CH_3}{>}}CH-CH_3 \xrightarrow[\underset{hv}{heat,}]{X_2} \left[\begin{array}{c} \underset{CH_3}{\overset{CH_3}{>}}CH-CH_2-X \\ + \\ CH_3-\underset{\underset{X}{|}}{\overset{\overset{CH_3}{|}}{C}}-CH_3 \end{array} \right] \xrightarrow[alcohol]{KOH}$$

$$\underset{CH_3}{\overset{CH_3}{>}}C=CH_2 \xrightarrow[H_2O]{KMnO_4} CH_3-\underset{\underset{OH}{|}}{\overset{\overset{CH_3}{|}}{C}}-\underset{\underset{OH}{|}}{CH_2}$$

In conclusion, keep in mind that these examples are presented to show you *how to approach multistep synthesis problems*. The best way to solve problems of this type is by working the Study Questions at the end of the chapter. After repeated contact with them, many students find multistep synthesis problems *fun*, and they are a good way to put together the many reactions that are presented about different families of compounds.

8.37   Summary of Reactions

A. Addition Reactions of Alkenes

$$\overset{}{>}C=C\overset{}{<} \xrightarrow[CCl_4]{HX} -\overset{|}{C}-\overset{|}{C}-$$

Alkene   H X

Alkyl halide

Orientation: Sec. 8.2
Mechanism: Sec. 8.3
Carbocation structure: Sec. 2.8
Rearrangement: Sec. 8.4

$$\overset{}{>}C=C\overset{}{<} \xrightarrow[H^\oplus]{H_2O} -\overset{|}{C}-\overset{|}{C}-$$

Alkene   H OH

Alcohol

Mechanism and orientation: Sec. 8.5

$$\text{C}=\text{C} \xrightarrow[\substack{\text{(cold,}\\ \text{conc)}}]{\text{H}_2\text{SO}_4} \overset{|\ \ \ \ |}{-\text{C}-\text{C}-} \xrightarrow[\text{heat}]{\text{H}_2\text{O}} \overset{|\ \ \ \ |}{-\text{C}-\text{C}-}$$

Alkene   H O—SO$_3$H   H OH

Mechanism and orientation:
Sec. 8.7

Alkyl hydrogen
sulfate

$$\text{C}=\text{C} \xrightarrow[\text{peroxides}]{\text{H—Br}} \overset{|\ \ \ \ |}{-\text{C}-\text{C}-}$$

Alkene   Br H

Mechanism and orientation
(reverse addition): Sec. 8.9

Alkyl bromide

$$\text{C}=\text{C} \xrightarrow[\text{Pt, Ni, Pd}]{\text{H}_2} \overset{|\ \ \ \ |}{-\text{C}-\text{C}-}$$

Alkene   H H

Mechanism and stereochemistry: Sec. 8.12

Alkane

$$\text{C}=\text{C} \xrightarrow[\text{(methylene)}]{\text{CH}_2\text{N}_2,\ \text{heat}} \text{C} - \text{C}$$

Alkene   CH$_2$

Secs. 8.29 and 27.6

Cyclopropane

$$\text{C}=\text{C} \xrightarrow[\text{CCl}_4]{\text{Br}_2} \overset{|\ \ \ \ |}{-\text{C}-\text{C}-}$$

Alkene   Br Br

Stereochemistry and mechanism:
     Secs. 8.16 and 8.17
Stereochemistry for open-chain alkenes:
     Sec. 8.18

Vicinal dibromide
(*anti*-addition)

$$\text{C}=\text{C} \xrightarrow[\text{CCl}_4]{\text{Cl}_2} \overset{|\ \ \ \ |}{-\text{C}-\text{C}-}$$

Alkene   Cl Cl

Mechanism and stereochemistry:
     Sec. 8.19

Vicinal dichloride

$$\text{C}=\text{C} \xrightarrow[\text{H}_2\text{O}]{\text{X}_2} \overset{|\ \ \ \ |}{-\text{C}-\text{C}-}$$

Alkene   X OH

Orientation and mechanism: Sec. 8.22

Halohydrin

$$\text{C}=\text{C} \xrightarrow[\text{H}^{\oplus}]{\text{C}=\text{C}} \overset{|\ \ \ \ |}{-\text{C}-\text{C}-}\text{C}=\text{C}$$

Alkene   Alkene

Dimerization: Sec. 8.23
Orientation and mechanism: Sec. 8.23
In natural products: Sec. 8.25

$$\text{C}=\text{C} \xrightarrow{\text{R—H}} \overset{|\ \ \ \ |}{-\text{C}-\text{C}-}$$

Alkene   R H

Orientation and mechanism: Sec. 8.24

Alkane

$$\text{C}=\text{C} \xrightarrow[\substack{\text{1. OsO}_4\\ \text{2. Na}_2\text{SO}_3}]{\substack{\text{KMnO}_4,\ \text{OH}^{\ominus},\ \text{H}_2\text{O}\\ \text{or}}} \overset{|\ \ \ \ |}{-\text{C}-\text{C}-}$$

Alkene   OH OH

Mechanism and stereochemistry:
     Sec. 8.27
Stereochemistry of addition
     to open-chain alkenes: Sec. 8.28

Glycol
(*syn* addition)

B. Substitution Reactions of Alkenes

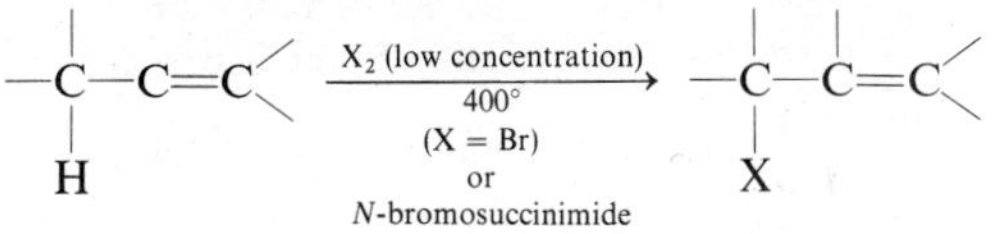

$$\underset{\underset{H}{|}}{-C}-C=C\diagup \xrightarrow[\substack{400° \\ (X = Br) \\ \text{or} \\ N\text{-bromosuccinimide}}]{X_2 \text{ (low concentration)}} \underset{\underset{X}{|}}{-C}-C=C\diagdown$$ Mechanism: Sec. 8.30

C. Cleavage Reactions of Alkenes

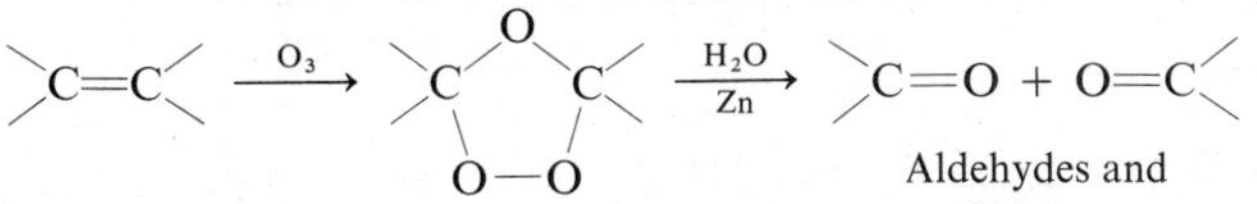

$$\diagup C=C\diagdown \xrightarrow{O_3} \text{(ozonide)} \xrightarrow[Zn]{H_2O} \diagup C=O + O=C\diagdown$$ General reaction: Sec. 8.31
Aldehydes and ketones   Mechanism: Sec. 8.31

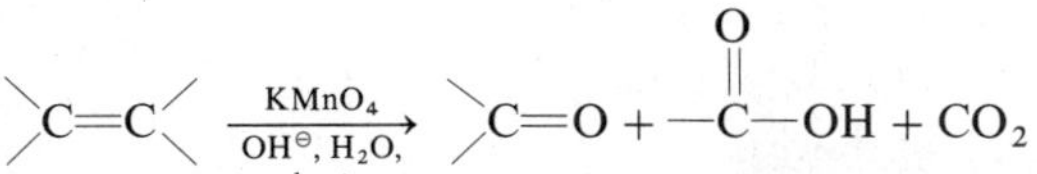

$$\diagup C=C\diagdown \xrightarrow[\substack{\text{heat}}]{\substack{KMnO_4 \\ OH^\ominus, H_2O,}} \diagup C=O + \underset{\underset{}{\overset{\overset{O}{\|}}{}}}{-C}-OH + CO_2$$ Sec. 8.32
Ketones and/or acids and/or carbon dioxide

D. Addition Reactions of Alkadienes

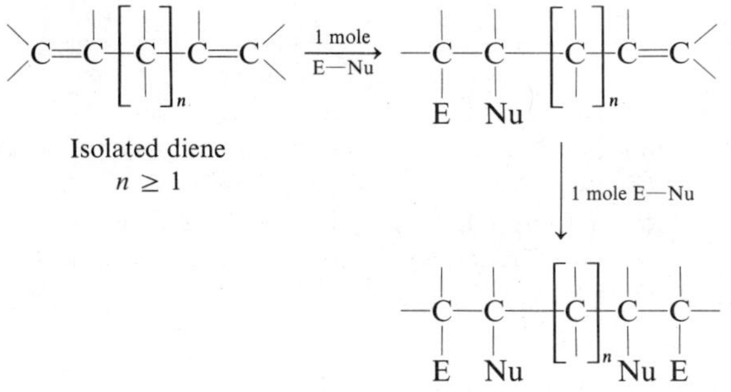

Isolated diene
$n \geq 1$

$$\xrightarrow{\substack{1 \text{ mole} \\ E-Nu}}$$

Sec. 8.10

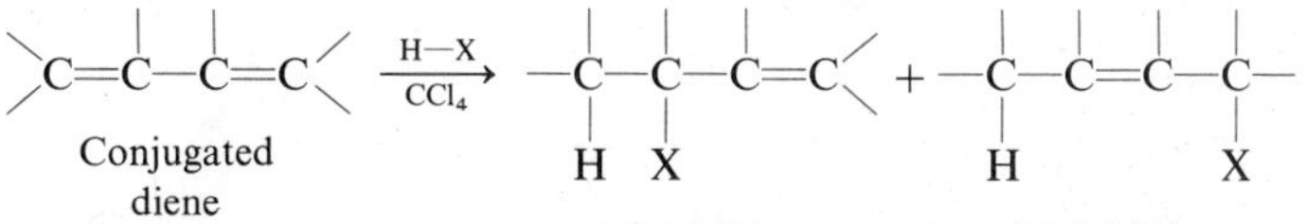

1 mole E—Nu

Conjugated diene

$$\xrightarrow[CCl_4]{H-X}$$ 1,2 Addition   1,4 Addition

Mechanism, orientation, product distribution: Sec. 8.11

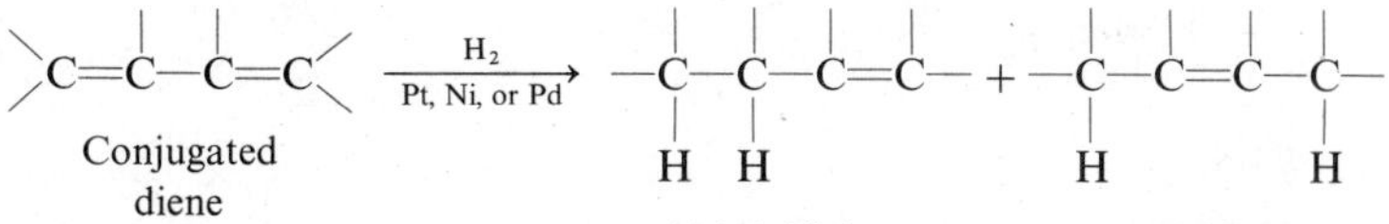

Conjugated diene

$$\xrightarrow[Pt, Ni, \text{ or } Pd]{H_2}$$ 1,2 Addition   1,4 Addition

Sec. 8.13

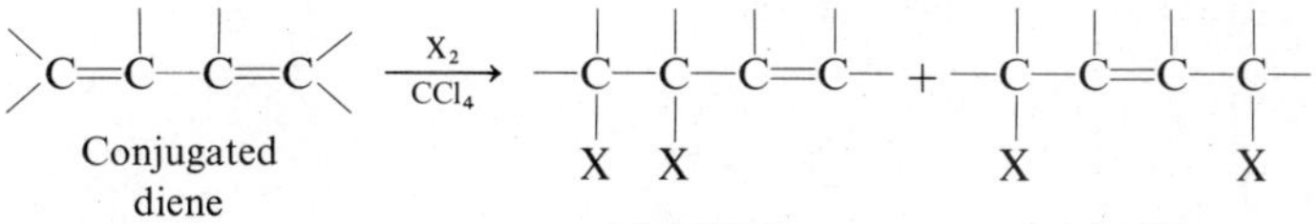

Conjugated diene

$$\xrightarrow[CCl_4]{X_2}$$ 1,2 Addition   1,4 Addition

Sec. 8.20

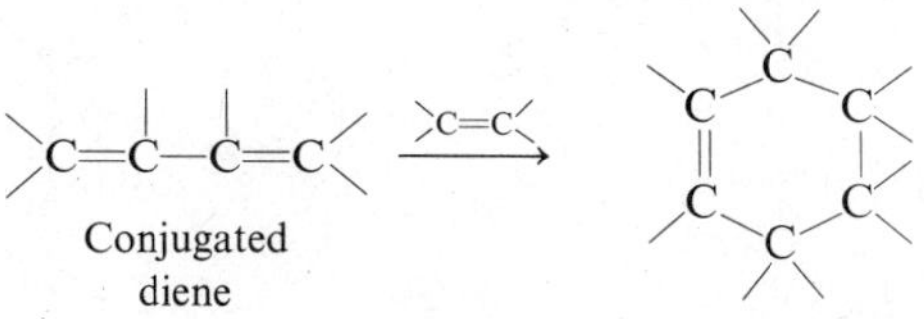

Conjugated diene

$$\xrightarrow{}$$

Diels-Alder reaction

Secs. 8.29 and 27.2

E. Addition Reactions of Cyclopropane

△ $\xrightarrow[\text{CCl}_4]{\text{HX}}$ $CH_3-CH_2-CH_2-X$ Sec. 8.3

Cyclopropane

△ $\xrightarrow[\text{H}^\oplus]{\text{H}_2\text{O}}$ $CH_3-CH_2-CH_2-OH$ Sec. 8.5

Cyclopropane

△ $\xrightarrow[80°]{\text{H}_2,\ \text{Pt}}$ $CH_3-CH_2-CH_3$ Sec. 8.12

Cyclopropane

△ $\xrightarrow[\text{CCl}_4]{\text{X}_2}$ $\underset{X}{CH_2}-CH_2-\underset{X}{CH_2}$ Sec. 8.16

Cyclopropane

F. Addition Reactions of Cyclobutane

▢ $\xrightarrow[200°]{\text{H}_2,\ \text{Pt}}$ $CH_3-CH_2-CH_2-CH_3$ Sec. 8.12

Cyclobutane

Study Questions

8.28 Define and, where possible, give examples of the following 15 terms:

(*a*) Baeyer test	(*b*) electrophile	(*c*) nucleophile
(*d*) monomer	(*e*) polymer	(*f*) addition polymerization
(*g*) terpene	(*h*) isoprene rule	(*i*) ozonolysis
(*j*) *syn* hydroxylation	(*k*) *anti*-bromination	(*l*) *syn* hydrogenation
(*m*) natural rubber	(*n*) bromonium ion	(*o*) electrophilic addition

8.29 Identify numbered [that is, (1), (2), (3)] intermediates or organic products by drawing their structure(s).

(*a*) $CH_3-\underset{\underset{CH_3}{|}}{C}=CH-CH_3 + H_2SO_4$ (cold, conc) $\longrightarrow$ (1) $\xrightarrow[\text{heat}]{\text{H}_2\text{O}}$ (2)

(*b*) $CH_3-CH=CH_2 + HBr \xrightarrow{\text{CCl}_4}$ (3)

(*c*) $CH_3-CH=CH_2 + HBr \xrightarrow{\text{peroxides}}$ (4)

(*d*) $CH_3-\underset{\underset{H}{|}}{\overset{\overset{CH_3}{|}}{C}}-\underset{\underset{CH_3}{|}}{\overset{\overset{CH_3}{|}}{C}}-Cl \xrightarrow[\text{alcohol}]{\text{KOH}}$ (5)

(*e*) $Br-CH_2-CH_2-CH_2-CH_2-Br + Zn$ metal $\longrightarrow$ (6)

(*f*) $CH_3-CH_2-\underset{\underset{CH_3}{|}}{CH}-CH_2-Br \xrightarrow[\text{heat}]{\text{Na metal}}$ (7)

(*g*) $CH_3-\underset{\underset{H}{|}}{\overset{\overset{CH_3}{|}}{C}}-\underset{\underset{H}{|}}{\overset{\overset{CH_3}{|}}{C}}-CH_3 + Br_2 \xrightarrow[\text{hv}]{\text{heat}}$ monosubstitution (8)

(*h*)   ▷$-CH=CH_2 + H_2 \xrightarrow[\text{heat}]{\text{Pt}}$ (9)

(*i*)   ⬡$=CH_2 + H_2O \xrightarrow[\text{(trace)}]{\text{H}^\oplus}$ (10)

(*j*) (CH₃)₂C cyclopropane + HCl $\xrightarrow[\text{(solvent)}]{\text{CCl}_4}$ (11)

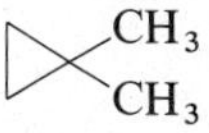

(*k*) $CH_3-CH_2-CH=CH_2$ + *N*-bromosuccinimide $\longrightarrow$ (12)

(*l*) octahydronaphthalene $\xrightarrow[\text{2. H}_2\text{O, Zn}]{\text{1. O}_3}$ (13)

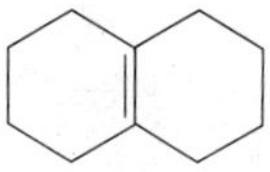

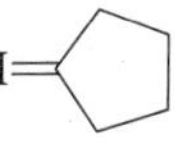

(*m*) $CH_3-CH=$cyclopentane $\xrightarrow[\text{2. H}_2\text{O, Zn}]{\text{1. O}_3}$ (14) +(15)

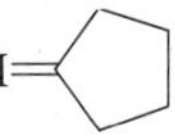

(*n*) $CH_3-CH=$cyclopentane $\xrightarrow[\substack{\text{:ÖH}^\ominus\text{, H}_2\text{O,} \\ \text{heat}}]{\text{KMnO}_4}$ (16) +(17)

(*o*) (+)-*sec*-Butyl alcohol $\xrightarrow[\substack{\text{H}_2\text{O,} \\ \text{heat}}]{\text{KOH}}$ (18)

(*p*) $CH_3-CH=CH-CH=CH-CH_3$ + Cl_2 $\xrightarrow{\text{CCl}_4}$ (19) +(20)

(*q*) *cis*-$CH_3-CH=CH-CH_3$ $\xrightarrow[\text{H}_2\text{O, heat}]{\text{KMnO}_4, \text{HÖ:}^\ominus}$ (21) +(22)

(*r*) $CH_3-\underset{\underset{CH_3}{|}}{C}=CH_2$ $\xrightarrow[\text{2. H}_2\text{O, Zn}]{\text{1. O}_3}$ (23) +(24)

(*s*) $(CH_3)_2C=CH_2$ $\xrightarrow[\substack{\text{:ÖH}^\ominus\text{, H}_2\text{O} \\ \text{(cold)}}]{\text{KMnO}_4}$ (25) +(26)

(*t*) $(CH_3)_2C=CH_2$ $\xrightarrow[\text{2. Na}_2\text{SO}_3]{\text{1. OsO}_4, \text{H}_2\text{O}}$ (27)

(*u*) $CH_3-CH_2-\underset{\underset{Br}{|}}{CH}-\underset{\underset{Br}{|}}{CH_2}$ $\xrightarrow[\text{alcohol}]{\text{Zn}}$ (28) $\xrightarrow[\text{H}^\oplus \text{ (trace)}]{\text{H}_2\text{O}}$ (29)

(*v*) cyclohexadiene + Br_2 $\xrightarrow{\text{CCl}_4}$ (30) +(31)

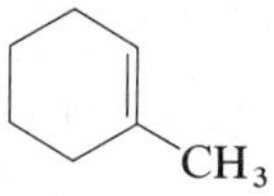

(*w*) methylcyclohexene + HBr $\xrightarrow{\text{peroxides}}$ (32)

(*x*) cyclopentene + NBS $\xrightarrow{\text{CCl}_4}$ (33)

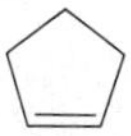

(*y*) $CH_3-\underset{\underset{OH}{|}}{CH}-CH(CH_3)_2$ $\xrightarrow[\text{heat}]{\text{H}^\oplus}$ (34) $\xrightarrow[\text{2. H}_2\text{O, Zn}]{\text{1. O}_3}$ (35)

(*z*) $(CH_3)_2C=CH_2$ $\xrightarrow[\text{CCl}_4]{\text{HBr}}$ (36) $\xrightarrow[\text{alcohol}]{\text{KOH}}$ (37) $\xrightarrow[\text{H}_2\text{O}]{\text{Cl}_2}$ (38)

(*aa*) $CH_2=CH-CH=CH_2$ + HBr $\xrightarrow{\text{CCl}_4}$ (39) +(40)

(*bb*) cyclopentene + Br_2 $\xrightarrow[\text{(solvent)}]{\text{CH}_3\text{O}-\text{H}}$ (41) +(42)

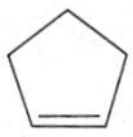

(*cc*) $CH_3-\underset{\underset{Br}{|}}{CH}-CH_2-\underset{\underset{Br}{|}}{CH}-CH_3$ $\xrightarrow[\text{heat}]{\text{Zn}}$ (43)

8.30 Complete the following chart by supplying the missing reagents or products, *which are indicated by number*. Any needed organic or inorganic reagents may be used; indicate approximate reaction conditions where necessary. No more than two steps are needed for each set of reagents.

$$CH_3-CH-CH_2-Cl$$
$$\quad\quad|$$
$$\quad\quad OH$$

(1)

(7)
1. O_3
2. H_2O, Zn

$$CH_3-CH=CH_2 \quad\xrightarrow{(6)}\quad CH_3-CH-CH_2$$
$$\quad\quad\quad\quad\quad\quad\quad\quad\quad\quad\quad\quad\quad | \quad\quad |$$
$$\quad\quad\quad\quad\quad\quad\quad\quad\quad\quad\quad\quad\quad Cl \quad Cl$$

(2)

(3)

(4)

(5)

$$CH_3-CH-CH_3$$
$$\quad\quad|$$
$$\quad\quad OH$$

$$CH_3-CH_2-CH_2-Cl \quad\quad CH_2-CH=CH_2$$
$$\quad\quad\quad\quad\quad\quad\quad\quad\quad\quad\quad\quad\quad\quad\quad |$$
$$\quad\quad\quad\quad\quad\quad\quad\quad\quad\quad\quad\quad\quad\quad\quad Cl$$

$$CH_3-CH=CH_2$$

8.31 Show how hexane could be prepared from each of the following three compounds:
(*a*) 1-bromopropane   (*b*) 2-chlorohexane   (*c*) 1-hexene

8.32 Indicate the hybridization of carbon involved in the C—H bond in each of the following:
(*a*) ethene   (*b*) methyl cation   (*c*) cyclohexane   (*d*) methyl radical

8.33 Describe the *changes* in hybridization that occur in these three reactions:
(*a*) hydrogenation of an alkene,
(*b*) addition of hydrogen bromide to an alkene, and
(*c*) rearrangement of a carbocation to a more stable ion.

8.34 Provide a three-dimensional structure that shows clearly the geometry of the methyl cation, $CH_3^\oplus$. Label the atomic orbitals involved in the bonding and indicate the approximate bond angles.

8.35 There are two alkene structures with the molecular formula C_6H_{12} that would be named methylpentenes by the IUPAC nomenclature and for which there are *cis* and *trans* isomers. Write unambiguous configurational formulas for the four compounds. Give complete IUPAC names for each structure, including *cis-trans* designations.

8.36 (*a*) Draw structural (or condensed structural) formulas for the five isomeric cyclic hydrocarbons with molecular formula C_5H_{10} (ignore stereoisomers).
(*b*) Which of the compounds in (*a*) will react with Br_2 in CCl_4 at ordinary temperatures (no light)? Explain why these compounds react, whereas the others do not, and give the equation(s) for the reaction(s) that occur.

8.37 Indicate the starting materials required to form the carbocation if it was desired to start with an (*a*) alcohol and (*b*) alkene.

$$CH_3$$
$$\quad\quad|$$
$$CH_3-C-\overset{\oplus}{C}H-CH_3$$
$$\quad\quad|$$
$$\quad\quad CH_3$$

8.38 In each of these eight reactions, indicate the products and the stereochemistry observed. Account for the stereochemistry on the basis of currently accepted mechanisms.

(*a*) [cyclohexene structure] $+ Br_2 \xrightarrow{CCl_4}$

(*b*) *trans*-$CH_3-CH_2-CH=CH-CH_2-CH_3 + Br_2 \xrightarrow{CCl_4}$

(*c*) [cyclohexene structure] $+ KMnO_4 \xrightarrow[OH^\ominus]{H_2O}$

(*d*) *cis*-$CH_3-CH_2-CH=CH-CH_2-CH_3 + KMnO_4 \xrightarrow[OH^\ominus]{H_2O}$

(*e*) *cis*-$CH_3-CH=CH-CH_3 + H_2O \xrightarrow{H^\oplus}$

(*f*) + Br$_2$ $\xrightarrow{CCl_4}$

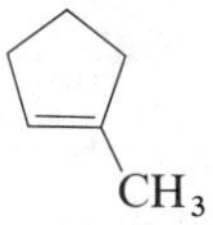

(*g*) + H$_2$ $\xrightarrow{Pt}$

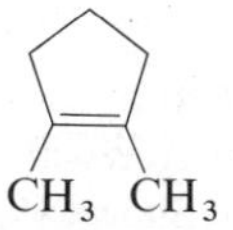

(*h*) *trans*-CH$_3$—CH$_2$—CH=CH—CH$_2$—CH$_3$ + H$_2$ $\xrightarrow{Pt}$

8.39 Provide a mechanistic explanation for the following observation. The ionic addition of HCl to 2-methyl-1,3-butadiene gives 3-chloro-3-methyl-1-butene and 1-chloro-3-methyl-2-butene but not 3-chloro-2-methyl-1-butene or 1-chloro-2-methyl-2-butene.

8.40 When cyclopentene is allowed to react with chlorine gas dissolved in water, products (1) and (2) (and their enantiomers) are observed. Explain this result using mechanisms.

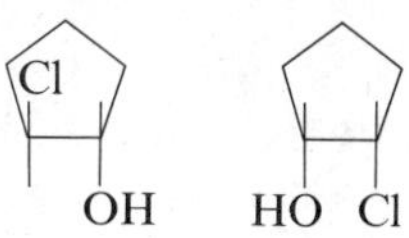

(1)   (2)

8.41 When 1-butene or *cis*-2-butene or *trans*-2-butene is treated with a trace of acid, each of these three alkenes gives a mixture that contains all three, and the composition of the mixture is the same regardless of which starting alkene was used. Explain these results, assuming that no polymerization or addition occurs. Use mechanisms.

8.42 Provide *complete, stepwise mechanisms* for *both* the following reactions, and use these mechanisms to explain the differences in orientation of the addition of HBr. Show all intermediates.

$$(CH_3)_2C=CH_2 + HBr \xrightarrow[\text{solvent}]{CCl_4} (CH_3)_3C-Br$$

$$(CH_3)_2C=CH_2 + HBr \xrightarrow[\text{(RO—OR)}]{\text{peroxides}} (CH_3)_2CH-CH_2-Br$$

8.43 Draw the structure of the carbocation that results from the addition of H$^\oplus$ to CH$_2$=CH—CH=CH$_2$, and briefly discuss the *electronic structure* of the resulting ion.

8.44 Provide complete, stepwise mechanisms for these six reactions. Show all intermediates.

(*a*) $\xrightarrow{\text{Br}_2}{\text{NaI}}$ (Also give products)

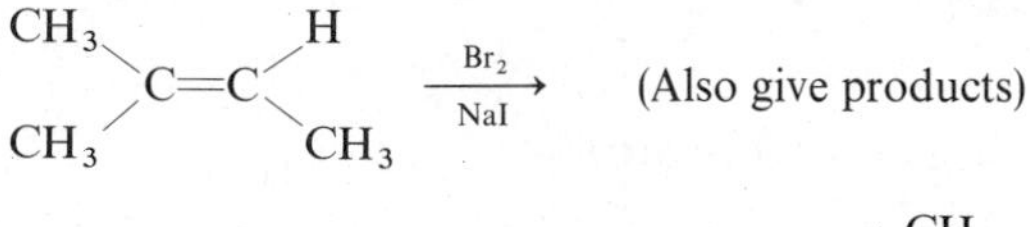

(*b*) $\xrightarrow[\text{H}_2\text{O}]{\text{H}^\oplus \text{ (trace)}}$ CH$_3$—CH$_2$—$\overset{\overset{\displaystyle CH_3}{|}}{\underset{\underset{\displaystyle CH_3}{|}}{C}}$—$\overset{\overset{\displaystyle }{}}{\underset{\underset{\displaystyle CH_3}{|}}{CH}}$—$\overset{\overset{\displaystyle CH_3}{|}}{\underset{\underset{\displaystyle CH_3}{|}}{C}}$—OH

(*c*) $\xrightarrow{\text{Br}_2 \text{ in}}{\text{CH}_3\text{OH}}$ +   (Indicate stereochemistry of products)

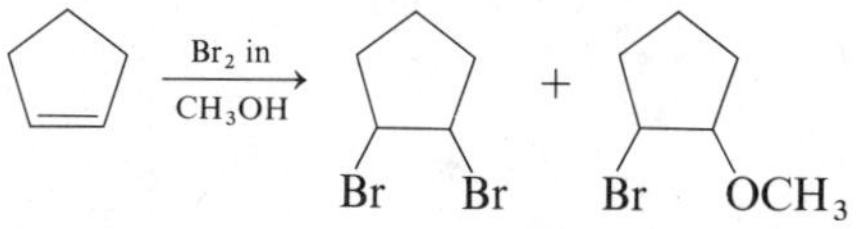

(*d*) CH$_3$—CH—CH=CH$_2$ $\xrightarrow[\text{CCl}_4]{\text{HCl}}$ C$_5$H$_{11}$Cl   (Two products; draw their structures)
$\qquad\quad$ |
$\qquad\;\;$ CH$_3$

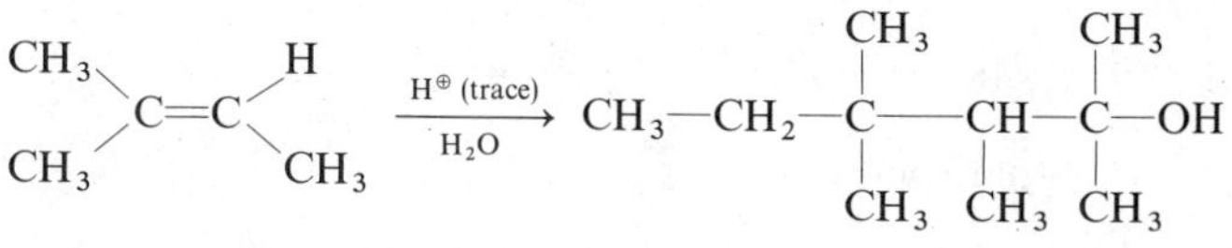

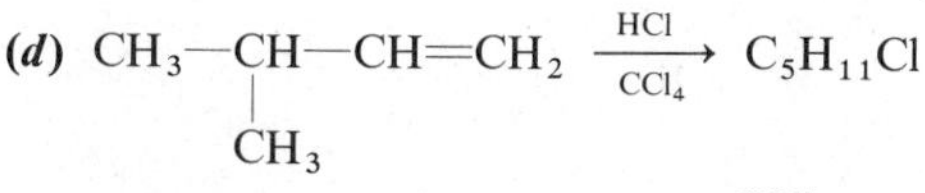

(*e*) CH$_3$—CH—CH$_2$—CH$_2$—Cl $\xrightarrow[\text{alcohol}]{\text{KOH}}$ 3-methyl-1-butene
$\qquad\quad$ |
$\qquad\;\;$ CH$_3$

(*f*) CH$_3$—$\overset{\overset{\displaystyle CH_3}{|}}{\underset{\underset{\displaystyle H}{|}}{C}}$—CH=CH—$\overset{\overset{\displaystyle CH_3}{|}}{\underset{\underset{\displaystyle H}{|}}{C}}$—CH$_3$ + HBr $\xrightarrow{CCl_4}$ C$_8$H$_{17}$Br   (Two products; draw their structures)

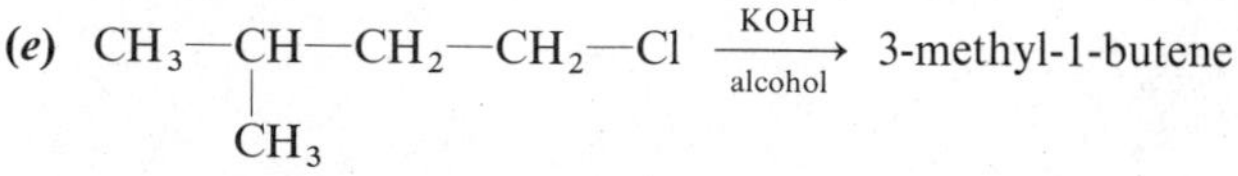

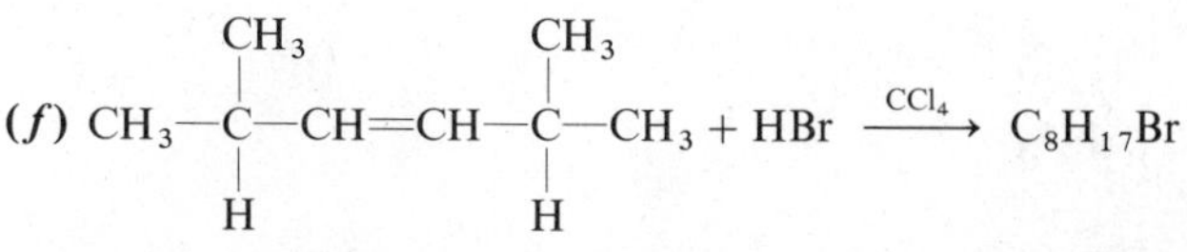

8.45 The dimerization of propene (see Sec. 8.23) yields two products, 4-methyl-1-pentene and 4-methyl-2-pentene.
(*a*) How might ozonolysis be used to distinguish between these two compounds?
(*b*) Does ozonolysis allow one to differentiate *cis* and *trans* isomers, such as those for 4-methyl-2-pentene?

8.46 The ionic addition of hydrogen bromide (carbon tetrachloride solvent) to an alkene, such as cyclohexene, gives a single organic product under *anhydrous conditions*. When moisture is present, however, another product is also formed. What is this likely to be? Show the mechanism for its formation.

8.47 Predict the major product obtained when the following inorganic reagents are allowed to react with propene under ionic conditions (anhydrous carbon tetrachloride solvent):
(*a*) FBr   (*b*) ICl   (*c*) ClBr

8.48 Provide a reasonable explanation, based on mechanisms, for the observed orientation of HCl addition (ionic conditions) in this reaction:

$$\underset{\underset{CH_3}{|}}{\overset{\overset{CH}{|}}{CH_3-\overset{\oplus}{N}}}-CH=CH_2 + HCl \xrightarrow{CCl_4} \underset{\underset{CH_3 \quad H \quad Cl}{|\qquad|\quad\ |}}{\overset{\overset{CH}{|}}{CH_3-\overset{\oplus}{N}}}-CH-CH_2$$

8.49 Devise suitable laboratory syntheses (that is, multistep syntheses) to show how the following transformations can be carried out in good yield. You may use any needed inorganic reagents or organic solvents (for example, ether, carbon tetrachloride) but no other source of carbon.

(*a*) $\underset{\underset{CH_3}{|}}{CH_3-CH}-CH_2-CH_2-Br \longrightarrow \underset{\underset{CH_3}{|}}{CH_3-CH}-CH_2-CH_3$ (Do by *two* different methods)

(*b*) Cyclohexane $\longrightarrow$ cyclohexene

(*c*) $CH_3-CH_2-CH_2-OH \longrightarrow (CH_3)_2CH-Cl$

(*d*) Cyclopentane $\longrightarrow$ *trans*-1,2-dibromocyclopentane

(*e*) Propane $\longrightarrow CH_3-CH_2-CH_2-Br$

(*f*) 1-Chlorobutane $\longrightarrow$ 2-iodobutane

(*g*) 1-Chlorobutane $\longrightarrow$ 3,4-dimethylhexane

(*h*) 〈 〉—OH $\longrightarrow$ 〈H Br〉   (*trans* isomer)
Br H

(*i*) Propene $\longrightarrow$ 1,2,3-tribromopropane

(*j*) $\underset{\underset{CH_3}{|}}{\overset{\overset{CH_3}{|}}{CH_3-C}}-OH \longrightarrow$ 1-bromo-2-methylpropane

(*k*) $\underset{\underset{CH_3}{|}}{CH_3-CH}-CH_2-OH \longrightarrow \underset{\underset{OH}{|}}{\overset{\overset{CH_2-OH}{|}}{CH_3-C}}-CH_3$

(*l*) 〈 〉—OH $\longrightarrow$ 〈 〉   (*cis* isomer)
HO OH

(*m*) 1-Bromopropane $\longrightarrow$ CH_3—$\underset{\underset{OH}{|}}{CH}$—$\underset{\underset{Cl}{|}}{CH_2}$

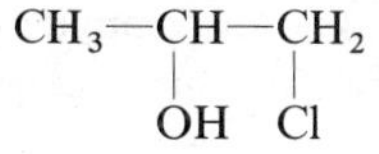

(*n*) Propane $\longrightarrow$ CH_2=CH—CH_2—Br

(*o*) $\longrightarrow$

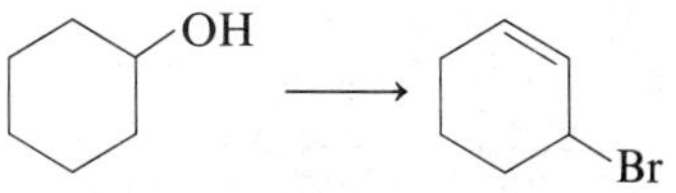

(*p*) Propene $\longrightarrow$ CH_2—CH—CH_2
       with Cl, OH, OH substituents: $\underset{\underset{Cl}{|}}{CH_2}$—$\underset{\underset{OH}{|}}{CH}$—$\underset{\underset{OH}{|}}{CH_2}$

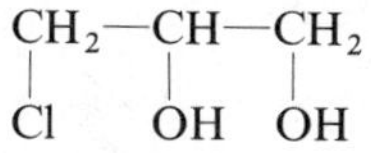

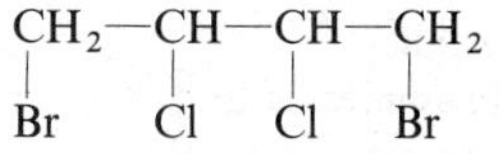

(*q*) CH_2=CH—CH=CH_2 $\longrightarrow$ $\underset{\underset{Br}{|}}{CH_2}$—$\underset{\underset{Cl}{|}}{CH}$—$\underset{\underset{Cl}{|}}{CH}$—$\underset{\underset{Br}{|}}{CH_2}$

(*r*) CH_3—$\underset{\underset{CH_3}{|}}{\overset{\overset{CH_3}{|}}{C}}$—$\underset{\underset{OH}{|}}{CH}$—$CH_3$ $\longrightarrow$ CH_3—$\underset{\underset{CH_3}{|}}{CH}$—$\underset{\underset{CH_3}{|}}{CH}$—$CH_3$ (*Hint:* Carbon skeleton is rearranged without changing carbon content)

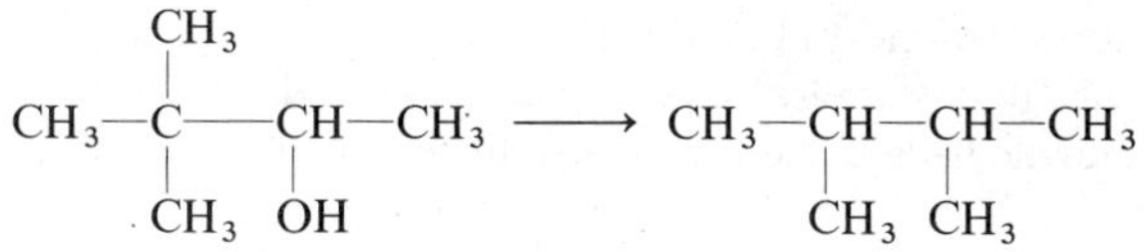

(*s*) Cyclohexene $\longrightarrow$

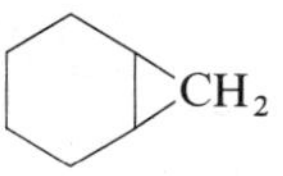

(*t*) —Br $\longrightarrow$ (*Hint:* Cyclohexene is an intermediate)

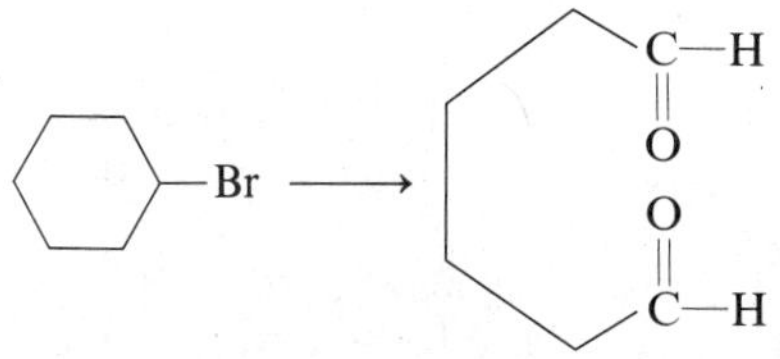

(*u*) $(CH_3)_2\underset{\underset{Br}{|}}{C}$—$\underset{\underset{Br}{|}}{CH_2}$ $\longrightarrow$ $(CH_3)_3CH$

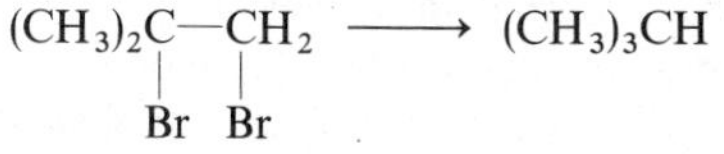

8.50 Starting with 1-methylcyclohexene (1), show how each of the following four compounds can be prepared in the laboratory. If your synthesis produces a particular stereochemistry in the product, indicate (by drawing or words) what it is. Some conversions may require more than one step.

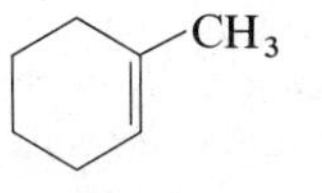

(1)
1-Methylcyclohexene

(*a*) 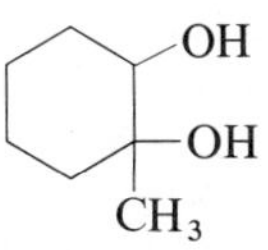   (*b*) 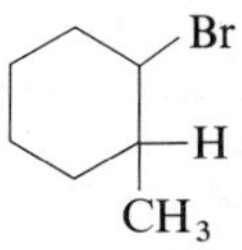   (*c*) 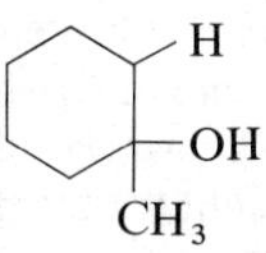   (*d*)

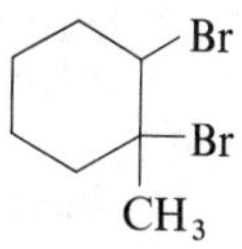

8.51 What alkene or alkenes yield the following products on ozonolysis?

(*a*) H_2C=O + CH_3—$\underset{\underset{H}{|}}{C}$=$O$ (*b*) 2 moles of =O

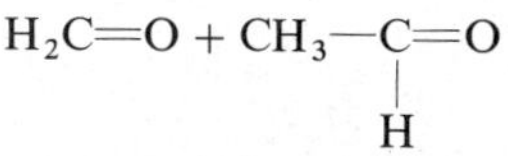

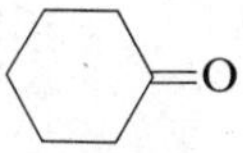

(*c*) 2 moles of CH_3—$\underset{\underset{H}{|}}{C}$=$O$ (*d*) H_2C=O + CH_3—$\underset{\underset{O}{||}}{C}$—$\underset{\underset{O}{||}}{C}$—$CH_3$ + CH_3—$\underset{\underset{H}{|}}{C}$=$O$

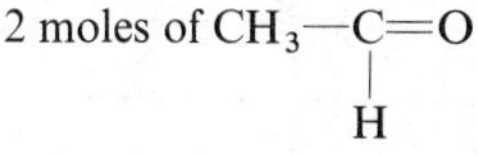

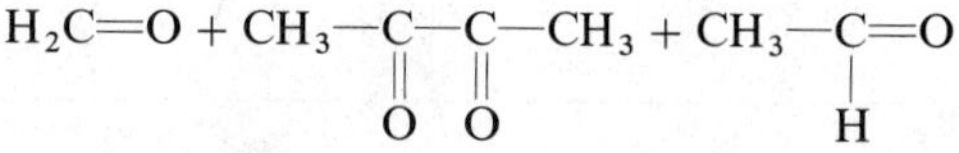

8.52 On catalytic hydrogenation, an unknown compound A with the formula C_6H_{10} absorbs only 1 mole of hydrogen. Ozonolysis of A yields just one product with the structure shown here. What is the structure of A?

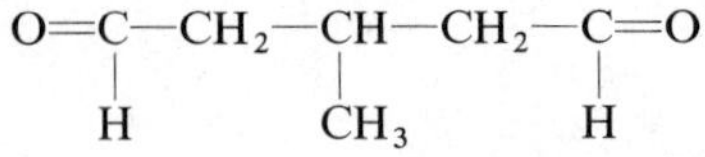

$$O=C-CH_2-CH-CH_2-C=O$$
$$\quad\;\; |\qquad\qquad |\qquad\qquad |$$
$$\quad\;\; H\qquad\qquad CH_3\qquad\; H$$

8.53 Compound A, C_4H_9Br, on treatment with alcoholic KOH, gives two isomeric compounds, B and C, with the formula C_4H_8. B and C react readily with Br_2 in CCl_4. On ozonolysis followed by hydrolysis, B gives only one product ($CH_3C=O$), whereas C gives two different products.

$$|$$
$$H$$

Identify compounds A, B, and C by drawing the structural formula for each. Make clear what each of the observations means.

8.54 Two isomeric alkyl bromides, A and B, $C_5H_{11}Br$, yield the following results in the laboratory. A, on treatment with alcoholic potassium hydroxide, gives two new compounds C and D, C_5H_{10}. C, on treatment with ozone and then with Zn/H_2O, gives compounds (1) and (2). B, on treatment with alcoholic potassium hydroxide, gives compounds D and E, C_5H_{10}. Compounds C, D, and E, on catalytic hydrogenation, all give the same compound F, C_5H_{12}.

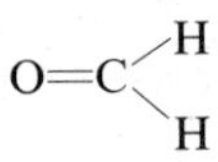

$$O=C{\overset{H}{\underset{H}{\diagdown}}} \qquad\qquad CH_3-\overset{\overset{\displaystyle CH_3}{|}}{CH}-\underset{\underset{\displaystyle H}{|}}{C}=O$$

(1) 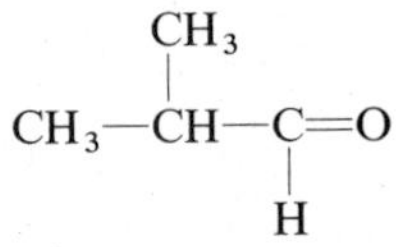   (2)

Deduce the structures of compounds A to F. Ignore the possibility of optical or geometric isomers (that is, none of the lettered compounds are optical isomers or geometric isomers of one another).

8.55 A well-known aviation fuel A, C_8H_{18}, is prepared by allowing B, C_4H_8, and C, C_4H_{10}, to react at $0°$ in the presence of concentrated sulfuric acid. When B is treated with concentrated sulfuric acid at $80°$, it is converted into a mixture of two isomeric alkenes, D and E, each of which has the molecular formula C_8H_{16}. Hydrogenation (catalytic) of either D or E produces compound A. Upon ozonolysis, D and E yield acetone, $(CH_3)_2C=O$, and formaldehyde, $CH_2=O$, respectively, as the most volatile products. Catalytic hydrogenation of B produces C. Deduce and draw the structures of compounds A to E on the basis of the information given.

8.56 When 1,5-dimethyl-1,5-cyclooctadiene (1) is dissolved in concentrated sulfuric acid, two hydrocarbons, A and B, are formed (along with copious amounts of tar). A and B are isomeric with (1), and all three compounds have the (same) molecular formula $C_{10}H_{16}$.

(a) How many moles of unsaturation are indicated by the molecular formula?

(b) Both A and B take up 1 mole of hydrogen on catalytic hydrogenation and produce the same hydrocarbon C with the molecular formula $C_{10}H_{18}$. How many double bonds appear to be present in A and B? How many rings?

(c) Treatment of A with ozone and zinc metal and then with water produces formaldehyde, $H_2C=O$, and compound D, whereas similar treatment of B gives only compound E. Give a structure for A and B consistent with the results of the ozonolysis reaction.

(d) Draw the structure of C, which is the hydrogenation product of either A or B.

(e) Based on your knowledge of the reactions of sulfuric acid with alkenes, propose a reasonable mechanism to account for the production of A and B from (1).

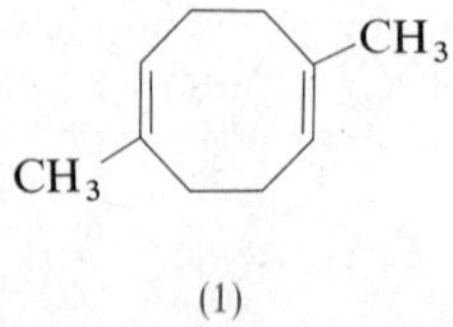

(1) 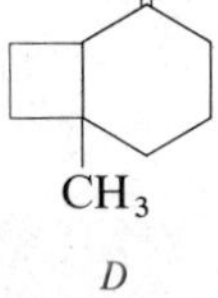 D 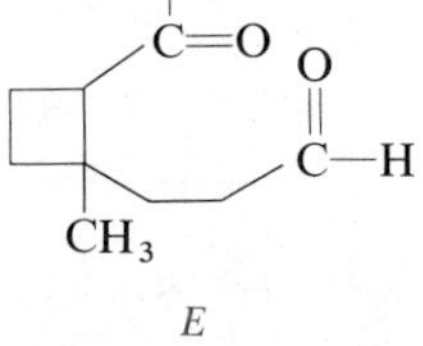 E

9

Alkyl Halides: Substitution and Elimination Reactions

Alkyl halides, R—X, were discussed in many of the previous chapters. We learned how to name them, saw several ways to prepare them, and studied several reactions of this family of compounds. In this chapter we delve further into their reactions, specifically **nucleophilic aliphatic substitution reactions** (introduced in Chap. 5) and competing **elimination reactions** (introduced in Chap. 7).

Two other closely related families of compounds are aryl halides and vinyl halides. Because aromatic compounds have not been discussed, we do not treat certain aspects of their chemistry here. There is a great difference in the reactivities of these two classes of compounds as compared with the alkyl halides, however; these differences are presented and compared in Chap. 16.

9.1 Structure, Nomenclature, and Preparation

The structure and nomenclature of the alkyl halides, R—X, were integral parts of earlier material in this text and are not reiterated here (see Sec. 3.10).

The most important method of preparing alkyl halides is from the corresponding alcohols. This is discussed in detail as a reaction of alcohols in Secs. 10.13 and 10.15. We already saw, however, several other methods of preparing these compounds. These methods are summarized here along with references to the appropriate text sections. In addition, we discuss a few miscellaneous reactions that result in the formation of alkyl halides in Chap. 11 and in Chap. 14

1. *Free-radical halogenation of alkanes and alkyl side chains*

(a) $R—H + X_2 \xrightarrow[\text{light}]{\text{heat and/or}} R—X + H—X$ Sec. 4.4

   Alkane   Alkyl
                     halide

Limitation: usually gives several products; bromination more selective than chlorination

(b) $R—H + SO_2Cl_2 \xrightarrow{\text{peroxides}} R—Cl + HCl + SO_2$ Sec 4.7

   Alkane   Alkyl
                     chloride

Limitation: usually gives several products

(c) $\diagdown$C=C—C— + X$_2$ $\xrightarrow{\text{heat}}$ $\diagdown$C=C—C— + H—X   Sec. 8.30
            | |
            H   X

      Alkene   Allyl halide
                                  Predominant product

Limitation: usually gives several products

(d) $\diagdown$C=C—C— + NBS $\xrightarrow{\text{CCl}_4}$ $\diagdown$C=C—C— + HBr   Sec. 8.30
            | |
            H   Br

      Alkene   Allyl bromide
                             Allyl substitution predominates

2. *Hydrogen halide addition to alkenes*

$\diagup$C=C$\diagdown$ $\xrightarrow[\text{CCl}_4]{\text{H—X}}$ H—C—C—   Sec. 8.2
H | |
                                       H X

                                  Markovnikov
                                   addition

Limitation: reaction subject to rearrangement

$\xrightarrow[\text{peroxides}]{\text{HBr}}$ H—C—C—   Sec. 8.9
                                | |
                                Br H

                              Anti-Markovnikov
                                  addition

Limitation: reaction works only with HBr

3. *Halogen addition to alkenes*

$\diagdown$C=C$\diagup$ $\xrightarrow[\text{CCl}_4]{\text{X}_2}$ —C—C—   Sec. 8.16
                                       | |
                                       X X

      Alkene   Vicinal dihalide

4. *Halide exchange: nucleophilic aliphatic substitution*

$$R—\ddot{\underset{\cdot\cdot}{X}}\colon + \colon\!\ddot{\underset{\cdot\cdot}{X}}\colon^{\ominus} \longrightarrow R—\ddot{\underset{\cdot\cdot}{X}}\colon + \colon\!\ddot{\underset{\cdot\cdot}{X}}\colon^{\ominus}$$ Sec. 5.9

Limitation: R- must be primary or methyl

9.2   Physical Properties

The physical properties of selected alkyl halides are listed in Table 9.1. In a series of alkyl halides that have the same alkyl group structure, the bromides boil at higher temperatures than the corresponding chlorides and the iodides boil at higher temperatures than the bromides. The boiling point also rises with increasing carbon content. The alkyl chlorides boil at temperatures greater than the alkanes from which they are prepared, but alkyl chlorides and alkanes of comparable weight boil at roughly the same temperature.

TABLE 9.1 Physical Properties of Selected Alkyl Halides

Name*	Structure	Chloride bp, °C	Chloride Density†	Bromide bp, °C	Bromide Density†	Iodide bp, °C‡	Iodide Density†
Halomethane (methyl)	CH_3—X	−24	0.920	5	1.732	42	2.279
Haloethane (ethyl)	CH_3CH_2—X	13	0.910	38	1.430	72	1.933
1-Halopropane (n-propyl)	$CH_3CH_2CH_2$—X	46	0.890	71	1.353	102	1.747
2-Halopropane (isopropyl)	$(CH_3)_2CH$—X	37	0.860	60	1.310	89	1.702
1-Halobutane (n-butyl)	$CH_3(CH_2)_3$—X	78	0.884	102	1.275	130	1.617
1-Halo-2-methylpropane (isobutyl)	$(CH_3)_2CHCH_2$—X	69	0.875	91	1.263	120	1.609
2-Halobutane (sec-butyl)	CH_3CH_2—CHX—CH_3	68	0.871	91	1.261	119	1.595
2-Halo-2-methylpropane (tert-butyl)	$(CH_3)_3C$—X	51	0.851	73	1.222	100d‡	
1-Halopentane (n-pentyl)	$CH_3(CH_2)_4$—X	108	0.883	130	1.223	157	1.517
1-Halohexane (n-hexyl)	$CH_3(CH_2)_5$—X	134	0.882	156	1.173	180	1.441
Halocyclohexane (cyclohexyl)	cyclo-C_6H_{11}—X	143	1.000	165			
3-Halopropene (allyl)	CH_2=CH—CH_2—X	45	0.938	71	1.398	102	1.848
Dihalomethane	CH_2X_2	40	1.336	99	2.490	180d‡	3.325
Trihalomethane	CHX_3	61	1.489	151	2.89	Solid	
Tetrahalomethane	CX_4	77	1.597	189	3.424	Solid	

* Common name in parentheses if two names are given.
† Grams per cubic centimeter at 20°.
‡ d = decomposes on boiling.

Alkyl halides are generally water-insoluble in spite of their polar nature. Alkyl chlorides are slightly less dense than water, and the bromides and iodides are denser than water.

Alkyl halides are insoluble in and inert toward concentrated sulfuric acid, as are alkanes. As a result, alkyl halides and alkanes can be purified by extracting them with H_2SO_4, which removes alkenes, alkynes, alcohols, and so on, because they react with sulfuric acid.

Question 9.1

An organic mixture contains cyclohexane, cyclohexyl bromide, cyclohexene, and cyclohexanol. Write equations for the reactions that occur when this mixture is treated with cold, concentrated sulfuric acid.

9.3 Nucleophilic Aliphatic Substitution

The principal reaction of the alkyl halides is **nucleophilic aliphatic substitution** (see Sec. 5.16). In its most general form, this can be depicted by the equation

$$R:L + Nu \overset{\ominus}{:} \longrightarrow R:Nu + L \overset{\ominus}{:}$$

where R *must* be an alkyl group
$L: =$ leaving group
$Nu: =$ nucleophile

For reaction to occur, the leaving group, $L:$, departs carbon with the pair of bonding electrons. The nucleophile, $Nu:$, must contain at least one unshared pair of electrons and must be either neutral or negatively charged (see Sec. 9.7 for a review of nucleophiles). The preceding equation is general, so no attempt is made to assign formal charges to the leaving group and the nucleophile.

One of the most commonly used families of leaving groups is the halides (see Sec. 9.8). This is illustrated here:

$$R:\overset{..}{\underset{..}{X}}: + Nu\overset{\ominus}{:} \longrightarrow R:Nu + :\overset{..}{\underset{..}{X}}:^{\ominus}$$

Several factors govern nucleophilic substitution. The *structure of the alkyl group and the nature of the leaving group and the nucleophile* are three major variables that control the mechanism of substitution and the formation of side products.

Nucleophiles may be divided into two general categories: (1) those that are relatively nonbasic (such as halide ions) and (2) those that also possess some degree of basicity (such as hydroxide ion). In our study of nucleophilic substitution at a saturated carbon atom, we will find that the major side reaction is **elimination** and that it becomes increasingly important as the basicity of the nucleophile increases. Furthermore, changes in the structure of the alkyl group, the nucleophile, or the reaction conditions (concentration, temperature, and solvent) often cause either substitution or elimination to be the major reaction.

9.4 Nucleophilic Aliphatic Substitution; Effect of Alkyl Group Structure on Chemical Kinetics

One very convenient method for examining the effect of structure on reactivity is chemical kinetics, in which the rate of chemical reaction is studied as a function of the concentration of the various reactants (see Sec. 5.12). In the reaction of an alkyl halide with, for example, hydroxide ion, $\overset{\ominus..}{:OH}$, or halide ion, $:\overset{..}{\underset{..}{X}}:$ (typical nucleophiles), the rate law changes as the structure of the alkyl halide is changed from primary to tertiary. Let us consider some specific examples.

The reaction between methyl iodide and hydroxide ion occurs with substitution of the $\overset{\ominus..}{:OH}$ ion for $:\overset{..}{\underset{..}{I}}:^{\ominus}$:

$$CH_3 \overset{..}{\underset{..}{-I}}: + \underset{\boxed{Nu\overset{\ominus}{:}}}{\overset{\ominus..}{:OH}} \xrightarrow[\text{ethanol}]{H_2O} CH_3 \overset{..}{\underset{..}{-O}}H + \underset{\boxed{L\overset{\ominus}{:}}}{:\overset{..}{\underset{..}{I}}:^{\ominus}}$$

Methyl   Methanol
iodide

This reaction is usually carried out in the laboratory by dissolving a base, such as sodium hydroxide, in a mixture of ethanol and water used as the solvent; without the added ethanol, the reaction is heterogeneous because methyl iodide is insoluble in water.

The chemical kinetics of the reaction between methyl iodide and hydroxide ion have been determined, and the rate of reaction is proportional to the concentrations of methyl iodide and hydroxide ions. The reaction exhibits second-order kinetics, and this relationship may be written mathematically as:

$$\text{Rate of reaction} = k_2[CH_3\!-\!I][:\!\overset{..}{O}H^{\ominus}]$$

Chemically this means that the rate of the reaction is doubled as the concentration of *either* $CH_3\!-\!I$ or $:\overset{..}{O}H^{\ominus}$ is doubled. Doubling the concentration of both species results in the reaction proceeding four times faster. Reducing the concentration of either $CH_3\!-\!I$ or $:\overset{..}{O}H^{\ominus}$ by a factor of two results in the reaction going half as fast. The rate constant k_2 has a unique value for each set of reaction conditions—namely, temperature and solvent. The rate law expresses the dependence of the rate of the reaction on concentration. This reaction obeys *second-order kinetics* and is called a *bimolecular reaction*. The significance of the chemical kinetics are seen when we study the mechanism of this reaction in Sec. 9.5.

The same results are obtained when, for example, methyl bromide reacts with iodide ion (see Sec. 5.12). The kinetics of this halide exchange reaction are also second order.

$$CH_3\!-\!\overset{..}{\underset{..}{Br}}: + \;\; \overset{\boxed{Nu:}}{:\overset{..}{I}:^{\ominus}} \;\;\longrightarrow\;\; CH_3\!-\!\overset{..}{\underset{..}{I}}: + :\overset{\boxed{L:}}{\underset{..}{Br}}:^{\ominus}$$

When the primary halide, methyl iodide or bromide, however, is replaced by *tert*-butyl iodide and the chemical kinetics are redetermined, the rate of the reaction depends *only* on the concentration of *tert*-butyl iodide.

$$\underset{\underset{CH_3}{|}}{\overset{\overset{CH_3}{|}}{CH_3\!-\!C\!-\!\overset{..}{\underset{..}{I}}:}} \;\; + \text{Nu}:^{\ominus} \;\;\longrightarrow\;\; \underset{\underset{CH_3}{|}}{\overset{\overset{CH_3}{|}}{CH_3\!-\!C\!-\!Nu}} + :\overset{..}{\underset{..}{I}}:^{\ominus}$$

2-Iodo-2-methylpropane
(tert-butyl iodide)

where $\text{Nu}:^{\ominus} = H\overset{..}{\underset{..}{O}}:^{\ominus}$ or $:\overset{..}{\underset{..}{Br}}:^{\ominus}$, for example

In this case, the rate law is

$$\text{Rate of reaction} = k_1[(CH_3)_3C\!-\!I]$$

Here, doubling the concentration of the alkyl iodide doubles the rate of the reaction, but doubling or changing in any way the concentration of the nucleophile $\text{Nu}:^{\ominus}$ has no effect on the rate. This reaction obeys *first-order kinetics* and is called a *monomolecular* (or unimolecular) reaction. The kinetics are important in deducing the mechanism of this reaction also (see Sec. 9.10).

In practice, the reaction between a tertiary halide and hydroxide ion is also carried out in a mixed solvent, such as ethanol and water, so that a homogeneous reaction mixture may be used. The concentration of base is normally very small

(0.05 M or less). For reasons that become clear soon, only a small fraction of the tertiary halide is converted into the corresponding tertiary alcohol (*tert*-butyl alcohol) because of the overwhelming importance of elimination to form an alkene (methylpropene). When a halide exchange reaction is run (see Sec. 5.9), acetone is the solvent of choice, and the yields of the substitution products are much better than in the previous example.

Note that the rate laws are the same regardless of the halogen attached to the alkyl group. The kinetics are governed by the structure of the alkyl group, not the leaving group. Even though the rate laws are the same, however, the rate constants are different for different halogens; for example, an alkyl chloride reacts at a different rate than an alkyl bromide.

9.5   S$_N$2 Reactions: Mechanism and Evidence

The bimolecular nature of the reaction between methyl iodide and hydroxide ion suggests that these two species come together and are both involved in the transition state that intervenes between the reactants and products, with the overall reaction being

$$CH_3-\overset{..}{\underset{..}{I}}: + :\overset{..}{\underset{..}{O}}H^{\ominus} \longrightarrow CH_3-\overset{..}{\underset{..}{O}}H + :\overset{..}{\underset{..}{I}}:^{\ominus}$$

Recalling that the carbon-iodine bond, C—I, is polar because of the electronegativity of iodine, it is not surprising that this nucleophilic substitution reaction involves attack of $:\overset{..}{\underset{..}{O}}H^{\ominus}$, the nucleophile, on the partially positive carbon atom. This occurs through a collision between the two molecules. Of course, the collision must be of some minimum energy, E_A, and the orientation must be correct. If these criteria are met, a successful reaction results. Experimental evidence (see later) shows that nucleophilic attack occurs from the *backside* of the methyl iodide molecule, as shown by the following three-dimensional projection formulas:

| Nucleophilic backside attack | Transition state [Like ($\delta^{\ominus}$) charges as far apart as possible] | Inversion of configuration at carbon |

Figure 9.1 shows this same reaction depicted by drawings of molecular models. In the transition state, the two partial negative charges are as far apart as possible to minimize the electrostatic repulsions between like charges. The reaction is *concerted*; the C—I bond is broken at the same time the C—OH bond is made, as indicated by the dashed lines. The electron pair is transferred from oxygen on the nucleophile (H$\overset{..}{\underset{..}{O}}:^{\ominus}$) to carbon, and the bonding pair that held carbon and iodine together is transferred to the electronegative iodine atom that leaves as the iodide ion,

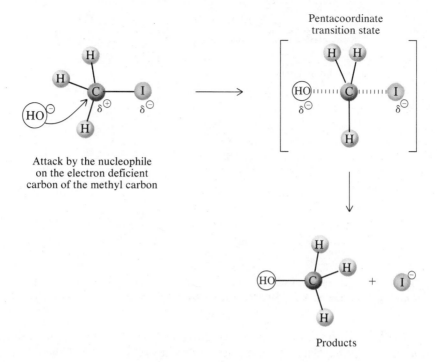

FIGURE 9.1   The S$_N$2 reaction between methyl iodide and hydroxide ion.

$:\overset{\cdot\cdot}{\underset{\cdot\cdot}{I}}:^{\ominus}$ (the leaving group, $\overset{\ominus}{L:}$). The transition state is highly unstable because its structure is *pentacovalentlike* and carbon does not normally bear five bonds. The iodine and the —OH group are a maximum distance apart, and the three hydrogens attached to carbon lie in the same plane as carbon, with the hydrogens directed to the three corners of a triangle. This is an example of an S$_N$2 reaction (see Sec. 5.16). In this case, an alkyl halide is converted to an alcohol.

A. Stereochemical Evidence for S$_N$2 Mechanism

Some of the strongest and most compelling evidence in support of the S$_N$2 mechanism comes from stereochemistry. For example, optically pure 2-bromobutane can be obtained, and if this alkyl bromide is treated with hydroxide ion under conditions where an S$_N$2 reaction occurs (that is, where second-order kinetics are followed), then optically pure 2-butanol is obtained as the product. But the 2-butanol has exactly the opposite configuration of the starting bromide. (The relative configurations of starting material and product are determined by independent means.) This reaction is shown in Fig. 9.2.

*When the product of a reaction has the opposite configuration of the corresponding starting material, then the reaction has proceeded with **inversion of configuration.*** The configuration of each 2-bromobutane molecule was inverted in the preceding reaction. Although we are unable to examine the stereochemistry of the S$_N$2 reaction with methyl iodide, compounds with an asymmetric carbon atom provide an elegant method for determining the mechanism of the S$_N$2 reaction, which proceeds with complete inversion of configuration.

The transition state for the inversion mechanism (Fig. 9.2) shows the three-dimensional location of the methyl, ethyl, and hydrogen groups, which are inverted

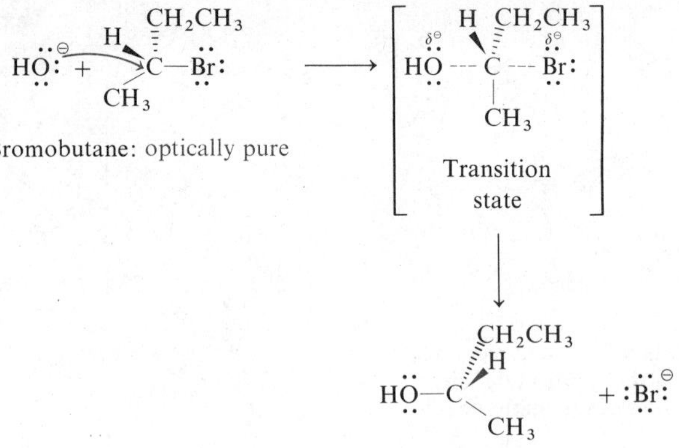

2-Bromobutane: optically pure

2-Butanol optically pure
100% Inverted Configuration
Relative to Optically Pure
2-Bromobutane

FIGURE 9.2 Stereochemistry of the S_N2 reaction between optically active 2-bromobutane and hydroxide ion, showing 100% inverted configuration that occurs via backside attack.

as they pass through the pentacovalent transition state. This process has been compared to an umbrella turning inside out in a strong wind. Thus there is little doubt that the S_N2 reaction occurs by backside attack of the nucleophile on the carbon bearing the leaving group.

Other stereochemical evidence comes from cyclic compounds. For example, when *trans*-3-bromomethylcyclopentane (1) reacts with hydroxide ion under S_N2 conditions, the exclusive substitution product is the corresponding *cis*-alcohol (2).

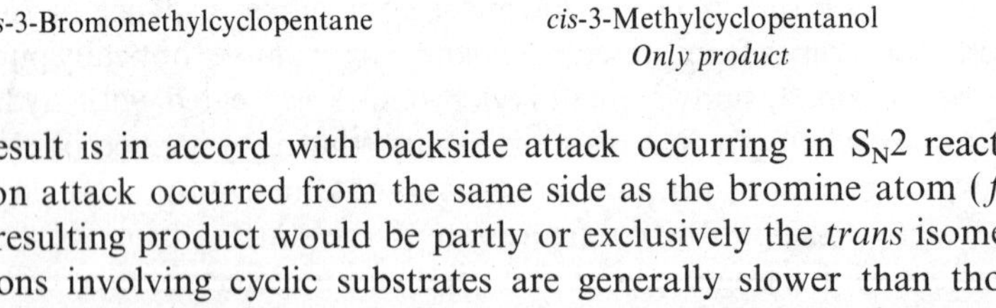

(1)
trans-3-Bromomethylcyclopentane

(2)
cis-3-Methylcyclopentanol
Only product

This result is in accord with backside attack occurring in S_N2 reactions. If hydroxide ion attack occurred from the same side as the bromine atom (*frontside attack*), the resulting product would be partly or exclusively the *trans* isomer.

Reactions involving cyclic substrates are generally slower than those with acyclic substrates. This is due to steric strain between the incoming nucleophile and substituents on the substrate skeleton and the added strain of attaining the required pentacoordinate transition-state geometry the ring imparts on the system.

9.6 Effect of Alkyl Group Structure on S_N2 Reactivity

How does the structure of the alkyl group affect S_N2 reactivity? As we saw in Sec. 9.5, the transition state for the S_N2 reaction is pentacovalent and the —OH

**TABLE 9.2   Relative Rates of S_N2 Reactions for the Reaction
$R\!-\!\ddot{Br}\!: + :\overset{\ominus}{\ddot{I}}\!: \longrightarrow R\!-\!\ddot{I}\!: + :\overset{\ominus}{\ddot{Br}}\!:$ Showing Effect of α Substitution**

Alkyl Group (R)		Relative Rate of Reaction*[†]
Name	Structure	
Methyl	$\overset{\alpha}{CH_3}\!-$ (no α substituents)	200,000
Ethyl	$CH_3\!-\!\overset{\alpha}{CH_2}\!-$ (1 α substituent)	1,000
Isopropyl	$\begin{array}{c} CH_3 \\ \diagdown \\ CH_3 \end{array}\!\!\!\overset{\alpha}{CH}\!-$ (2 α substituents)	12
tert-Butyl	$CH_3\!-\!\overset{\overset{\displaystyle CH_3}{\mid}}{\underset{\underset{\displaystyle CH_3}{\mid}}{C^{\alpha}}}\!-$ (3 α substituents)	1 (standard)

* The relationship

$$E_{A\,\text{reaction}\,a} - E_{A\,\text{reaction}\,b} = 1.37 \log \frac{\text{rate reaction } a}{\text{rate reaction } b}$$

(where a and b refer to any two reactions compared) shows that although there is a large difference between the rates of methyl and t-butyl, the difference in the energies of activation of these two reactions is relatively small, on the order of 7 kcal/mole (29.3 kJ/mole).
[†] Rate of tert-butyl bromide arbitrarily set at 1.0.

group (the nucleophile) and the —Br atom (the leaving group) both have partial negative charges. The central carbon atom is not charged to any appreciable extent, so the attachment of electron-donating or electron-withdrawing groups to that carbon atom should have little effect on the molecule's reactivity. (We know from studies on carbocations that these types of groups affect the stability of these charged ions.) Yet the data in Table 9.2 indicate that there is a considerable difference in the reactivities of various alkyl groups in the S_N2 reaction:

$$R\!-\!\ddot{Br}\!: + :\overset{\ominus}{\ddot{I}}\!: \longrightarrow R\!-\!\ddot{I}\!: + :\overset{\ominus}{\ddot{Br}}\!:$$

In this example we convert one alkyl halide to a new alkyl halide via an S_N2 reaction. This is an example of the halide exchange reaction introduced in Sec. 5.9. Here, one halide ion acts as the nucleophile and the second halide ion as the leaving group. The structural change of the alkyl groups shown in Table 9.2 involves increasing the degree of substitution on the carbon bearing the bromine atom (called the α carbon); this is called α substitution.

These data suggest that the decreased S_N2 reactivity is due to steric factors (see Sec. 5.15). As the number of substituents attached to the reacting center increases, it becomes increasingly more difficult for a nucleophile to enter the backside of the molecule (backside attack). The steric hindrance is seen more vividly in drawings of molecular models of these compounds in Fig. 9.3. Thus, *the reactivity of compounds in S_N2 reactions is attributed to **steric factors*** (and not electronic factors).

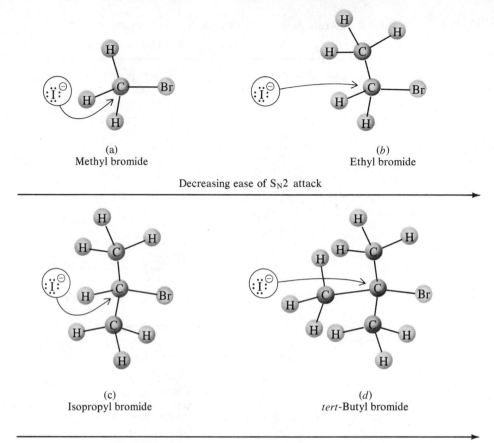

FIGURE 9.3   Three-dimensional models of (*a*) methyl bromide, (*b*) ethyl bromide, (*c*) isopropyl bromide, and (*d*) *tert*-butyl bromide showing the effect of increased steric hindrance on the backside attack by iodide ion $:\ddot{\text{I}}:^{\ominus}$ (the nucleophile).

From this discussion, the reactivity order of alkyl halides in S_N2 reactions is:

Decreasing order of S_N2 reactivity of R—X:

$$CH_3—X > 1° > 2° > 3°$$

Put another way, increasing the number of substituents on the α carbon makes it more difficult for the nucleophile, $:\ddot{\text{I}}:^{\ominus}$, to come within bonding distance of the carbon bearing the leaving group. This means that the nucleophile must possess enough energy to overcome the nonbonding interactions and actually attack the carbon where substitution is to occur. The energies of activation, E_A, provide an estimate of this barrier, as shown in the energy-profile diagram in Fig. 5.11.

Further data are available concerning the steric effects on S_N2 reactivity. Increasing the number of substituents on the carbon adjacent to the one bearing the leaving group (the *β-carbon atom*) also slows the rate of bimolecular nucleophilic substitution, as indicated in Table 9.3. In these compounds, the leaving group is always attached to a —CH_2— group, so the alkyl halide is always primary. This illustrates again that care must be exercised in making sweeping generalizations about relative reactivities. In some cases the decision is clear cut, whereas in others it may not be.

TABLE 9.3   Relative Rates of S_N2 Reaction for Reaction

$$R\!-\!\overset{..}{\underset{..}{Br}}\!: + CH_3CH_2\overset{..}{\underset{..}{O}}\!:^{\ominus} \longrightarrow R\!-\!\overset{..}{\underset{..}{O}}CH_2CH_3 + :\overset{..}{\underset{..}{Br}}\!:^{\ominus} \text{ Showing Effect of } \beta \text{ Substitution*}$$

Alkyl Group (R)		Relative Rate of Reaction[†]
Name	Structure	
Ethyl	$\overset{\beta}{CH_3}\!-\!CH_2\!-$ (no β substituents)	500,000
n-Propyl	$CH_3\!-\!\overset{\beta}{CH_2}\!-\!CH_2\!-$ (1 β substituent)	28,000
Isobutyl	$\begin{array}{c}CH_3\diagdown \\ \qquad\overset{\beta}{CH}\!-\!CH_2\!- \\ CH_3\diagup\end{array}$ (2 β substituents)	4,000
Neopentyl	$\begin{array}{c}CH_3 \\ \mid \\ CH_3\!-\!\overset{\beta}{C}\!-\!CH_2\!- \\ \mid \\ CH_3\end{array}$ (3 β substituents)	1 (standard)

* Note that the nucleophile in these examples is an alkoxide ion (the ethoxide ion) and the product is an ether.
† Rate of neopentyl bromide arbitrarily set at 1.0.

9.7   Effect of Nucleophiles on S_N2 Reactivity

A nucleophile is a base in two senses: (1) the Brønsted-Lowry sense, in which a base is a proton acceptor, and (2) the Lewis sense, in which a base is a substance that donates a pair of electrons. It follows, therefore, that nucleophilic substitution involves the transfer of a pair of electrons from the nucleophile to a carbon atom; the nucleophile reacts as a Lewis base.

There is a reasonable, although not exact, correlation between the basicity of an ion or molecule and its ability to serve as a nucleophile. In general, the stronger bases are the better nucleophiles (referring, of course, to nucleophilic substitution at carbon). Examples of strong bases are the hydroxide ion, $:\overset{..}{O}H$, derived from the weak acid water, and the alkoxide ion, $R\overset{..}{O}:^{\ominus}$, derived from the weak acid alcohol (ROH).

Nucleophiles that are weakly basic are H_2O, ROH (alcohols), cyanide ion, $^{\ominus}:C\!\equiv\!N:$, and the iodide ion, $:\overset{..}{\underset{..}{I}}:^{\ominus}$. We will see, however, that stronger bases are more effective in producing side reactions (predominantly elimination).

We are now primarily concerned with the halides, however. The order of nucleophilic reactivities of the halide ions in nucleophilic aliphatic substitution is:

Decreasing order of nucleophilic reactivity (in protic solvents):

$$:\overset{..}{\underset{..}{I}}:^{\ominus} > :\overset{..}{\underset{..}{Br}}:^{\ominus} > :\overset{..}{\underset{..}{Cl}}:^{\ominus} > :\overset{..}{\underset{..}{F}}:^{\ominus}$$

Best	Poorest
nucleophile	nucleophile

This order is the reverse of the basicity order of the halide ions ($:\ddot{\overset{..}{F}}:^{\ominus}$ > $:\ddot{\overset{..}{Cl}}:^{\ominus}$ > $:\ddot{\overset{..}{Br}}:^{\ominus}$ > $:\ddot{\overset{..}{I}}:^{\ominus}$). Thus the correlation between basicity and nucleophilicity does not hold for the halides. What are the reasons for the unexpected reactivity order?

Two important factors have been proposed to explain the halide ion reactivities in nucleophilic substitution reactions: (1) polarizability and (2) solvation. The reactivity order ($:\ddot{\overset{..}{I}}:^{\ominus}$ > $:\ddot{\overset{..}{Br}}:^{\ominus}$ > $:\ddot{\overset{..}{Cl}}:^{\ominus}$ > $:\ddot{\overset{..}{F}}:^{\ominus}$) parallels atom size because of the number of outer-shell electrons and the size of the outer orbitals. The larger atoms are more **polarizable** than smaller ones, which means the electron clouds are more easily distorted in the larger atoms because they are less tightly held by the nucleus. As the electron clouds of the reacting molecules are distorted or polarized, very weak instantaneous dipoles are formed. Attraction between the oppositely charged ends of these dipoles draws the molecules together. We can envision partial bond formation at relatively large distances as the electron cloud of the nucleophile is "pulled" toward the carbon atom where substitution is to occur. The activation energy of the substitution is decreased when easily polarizable nucleophiles (that is, those with electron clouds that are easily distorted) are used.

The second factor is **solvation,** which is the extent to which a nucleophilic species is surrounded by solvent molecules. Because many nucleophilic substitution reactions are carried out in *protic* solvents (see Sec. 9.13), the nucleophiles dissolved in them are surrounded by solvent molecules, as shown by the example of $:\ddot{\overset{..}{X}}:^{\ominus}$ dissolved in water:

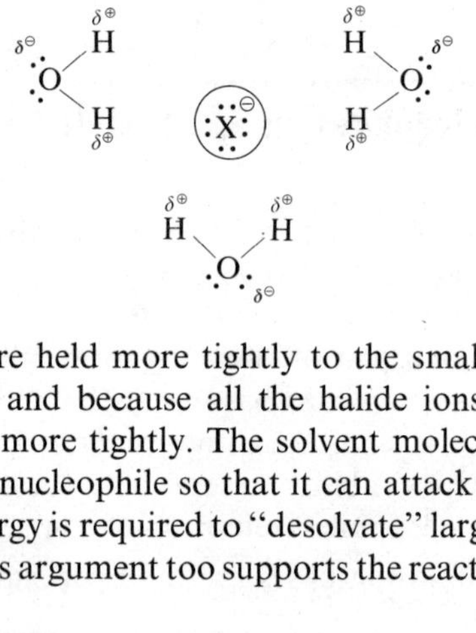

The water dipoles are held more tightly to the smaller halide ions that have higher charge/size ratios, and because all the halide ions have a -1 charge, the smaller ions hold solvent more tightly. The solvent molecules must be partially or wholly removed from the nucleophile so that it can attack the carbon where substitution is to occur. Less energy is required to "desolvate" larger ions because they hold the solvent less tightly. This argument too supports the reactivity order of the halogen nucleophiles.

To illustrate the dramatic effect that the nature of the nucleophile has on S_N2 reactivity, various nucleophilic substitution reactions have been carried out on methyl bromide, with the general reaction:

$$CH_3-\ddot{\overset{..}{Br}}: + Nu:^{\ominus} \longrightarrow CH_3-Nu + :\ddot{\overset{..}{Br}}:^{\ominus}$$

The *relative* rates of substitution for this reaction are given in Table 9.4, where the standard for comparison is the reaction with water, the poorest nucleophile of those studied. (The relative rates are calculated from actual rates determined under identical conditions.) Note that $H\ddot{\overset{..}{O}}:^{\ominus}$ is a much better nucleophile than is HOH.

The following gives the approximate S_N2 reactivity order of several nucleophiles. This order is reasonably valid for reactions carried out in *protic* solvents. Nucleophilic reactivity is affected by solvent, so care must be exercised in using it. (The listing contains several groups you have not studied; they are included for completeness.)

TABLE 9.4   Relative Rates of Nucleophilic Substitution for Reaction

$$CH_3-\overset{..}{\underset{..}{Br}}: + Nu:^{\ominus} \longrightarrow CH_3-Nu + :\overset{..}{\underset{..}{Br}}:^{\ominus}$$

Nucleophile, Nu:	Relative Rates of Reaction*
$H_2\overset{..}{O}:$	1.0 (standard)
$:\overset{..}{\underset{..}{Cl}}:^{\ominus}$	1.1×10^3
$:\overset{..}{\underset{..}{Br}}:^{\ominus}$	8×10^3
$H\overset{..}{\underset{..}{O}}:^{\ominus}$	1.8×10^4
$:\overset{..}{\underset{..}{I}}:^{\ominus}$	1.1×10^5

* Rate of reaction with water set at 1.0.

Decreasing order of nucleophilic reactivity in protic solvents:

$$H\overset{..}{\underset{..}{S}}:^{\ominus} > R\overset{..}{\underset{..}{S}}:^{\ominus} > :C\equiv N:^{\ominus} > :\overset{..}{\underset{..}{I}}:^{\ominus} > \overset{..}{N}H_3 > H\overset{..}{\underset{..}{O}}:^{\ominus} > N_3^{\ominus} > :\overset{..}{\underset{..}{Br}}:^{\ominus} > R\overset{..}{\underset{..}{O}}:^{\ominus} > :\overset{..}{\underset{..}{Cl}}:^{\ominus} > :\overset{..}{\underset{..}{F}}:^{\ominus} > H_2\overset{..}{O}:$$

Factors That Favor Nucleophilicity

1. *Negative charge:* Anions (bases) are stronger than their corresponding conjugate acids.

$$H_2\overset{..}{\underset{}{N}}:^{\ominus} > H_3N: \qquad H\overset{..}{\underset{..}{O}}:^{\ominus} > H_2\overset{..}{O}:$$

$$\underset{R-C-\overset{..}{\underset{..}{O}}:^{\ominus}}{\overset{O}{\overset{\|}{}}} > \underset{R-C-\overset{..}{O}H}{\overset{O}{\overset{\|}{}}} \qquad R\overset{..}{\underset{..}{O}}:^{\ominus} > R\overset{..}{O}H$$

2. *Basicity:* Within a given *family*, nucleophilicity increases with *polarizability not basicity*, although this is often the result of the reaction conditions as explained in Sec. 9.13.

$$:\overset{..}{\underset{..}{I}}:^{\ominus} > :\overset{..}{\underset{..}{Br}}:^{\ominus} > :\overset{..}{\underset{..}{Cl}}:^{\ominus} > :\overset{..}{\underset{..}{F}}:^{\ominus}$$

$$R\overset{..}{\underset{..}{Se}}:^{\ominus} > R\overset{..}{\underset{..}{S}}:^{\ominus} > R\overset{..}{\underset{..}{O}}:^{\ominus}$$

Within a given *row*, nucleophilicity increases with *basicity*.

$$H_2\overset{..}{\underset{}{N}}:^{\ominus} > R\overset{..}{\underset{..}{O}}:^{\ominus} > H\overset{..}{\underset{..}{O}}:^{\ominus} > :\overset{..}{\underset{..}{F}}:^{\ominus}$$

3. *Size:* The bulkier the nucleophile, the lower its nucleophilicity.

$$CH_3CH_2\overset{..}{N}H_2 > (CH_3CH_2)_2\overset{..}{N}H > (CH_3CH_2)_3N:$$

$$CH_3CH_2\overset{..}{\underset{..}{O}}:^{\ominus} > (CH_3)_2CH\overset{..}{\underset{..}{O}}:^{\ominus} > (CH_3)_3C\overset{..}{\underset{..}{O}}:^{\ominus}$$

All these factors are strongly influenced by the particular conditions (for example, solvent) under which the reaction is run and should therefore be applied with caution in making predictions.

Based on the discussions in this section, it might appear that S_N2 reactions depend exclusively on the nature of the nucleophile. In Sec. 9.8, however, we discuss the relative abilities of certain groups to serve as leaving groups, and in Sec. 9.13 we expand on the important role the solvent often plays in nucleophilic substitution reactions. (See Sec. 5.14 also.)

9.8   Effect of Leaving Group on S_N2 Reactivity

The reactivity of various leaving groups in S_N2 reactions is quite surprising. We saw in Sec. 9.7 that iodide ion is a good nucleophile, yet it is also the best leaving group of the halides. The decreasing order of reactivities of alkyl halides toward the same nucleophile is as follows:

Decreasing order of reactivity of R—X toward Nu:

$$R—I > R—Br > R—Cl \gg R—F$$

The same feature that makes the iodide ion a good nucleophile—polarizability—also makes it a good leaving group. It forms a relatively weak, loose covalent bond with carbon because it is so large; its facile deformation means that the C—I bond can be lengthened and thus broken more easily. The carbon-iodine bond energy is relatively low (only 53 kcal/mole; 221.8 kJ/mole) compared with that of the other carbon-halogen bonds.

The general trend in leaving group reactivity is that *the less basic groups are the better leaving groups.* Two factors govern the ease with which a leaving group departs. One is the nature of the nucleophile, which helps "push" the leaving group off carbon as the former donates electrons to that carbon. The second is the ability of the solvent to "pull" the leaving group off by solvating it. These factors are closely allied with the strength of the bond between carbon and the leaving group. We find, for example, that the hydroxyl group (—OH) attached to carbon is a miserable leaving group per se, and yet it is a good leaving group when protonated to give the oxonium ion:

$$R—\overset{..}{\underset{..}{O}}H \quad + H^{\oplus} \rightleftharpoons R—\overset{\oplus}{\underset{..}{O}} \overset{H}{\underset{H}{\diagdown}}$$

Poor leaving   Good leaving
group   group

Carbon and hydrogen virtually never leave with a bonding electron pair, because the hydride ion, $H\!:^{\ominus}$, and the carbanion, $—\overset{|}{\underset{|}{C}}\!:^{\ominus}$, are exceptionally strong bases.

Carbon and hydrogen are not very polarizable atoms, and they form very strong bonds with other carbon atoms.

$$R\!:\!H \qquad R\!:\!\overset{|}{\underset{|}{C}}—$$

Very poor leaving groups

A more comprehensive list of leaving groups is as follows:

Order of decreasing ease of displacement from a saturated carbon atom:

$$p\text{-Br}—C_6H_4—SO_2\overset{..}{\underset{..}{O}}— > p\text{-CH}_3—C_6H_4—SO_2\overset{..}{\underset{..}{O}}— > :\!\overset{..}{\underset{..}{I}}— > :\!\overset{..}{\underset{..}{Br}}— > :\!\overset{..}{\underset{..}{Cl}}— > :\!\overset{..}{\underset{..}{F}}—$$

$$\approx —\overset{..}{\underset{..}{O}}Ac > —\overset{\oplus}{N}R_3 > —\overset{..}{\underset{..}{O}}R \approx —\overset{..}{\underset{..}{O}}H > —\overset{..}{N}H_2$$

This list also includes some groups you have not studied for the sake of completeness and future reference. The tosylates, $p\text{-CH}_3—C_6H_4—SO_2O—$, fall into this category. They are *very good* leaving groups and their use is discussed in Sec. 19.6.

9.9   S$_N$1 Reactions: Mechanism and Evidence

We now come to the second example of nucleophilic substitution mentioned in Sec. 9.3, namely, the reaction

$$
\underset{\underset{\text{CH}_3}{|}}{\overset{\overset{\text{CH}_3}{|}}{\text{CH}_3-\text{C}-\ddot{\text{I}}:}} + \text{Nu}:^{\ominus} \longrightarrow \underset{\underset{\text{CH}_3}{|}}{\overset{\overset{\text{CH}_3}{|}}{\text{CH}_3-\text{C}-\text{Nu}}} + :\ddot{\text{I}}:^{\ominus}
$$

where Nu$:^{\ominus}$ = H$\ddot{\text{O}}:^{\ominus}$ or $:\ddot{\text{B}}\text{r}:^{\ominus}$, for example

The rate of this reaction depends only on the concentration of the *tert*-butyl iodide:

$$\text{Rate of reaction} = k_1[(\text{CH}_3)_3\text{C}-\text{I}]$$

Because the rate of this reaction does *not* depend on the concentration of the nucleophile, the mechanism must depend on some reaction involving only the tertiary alkyl iodide. Based on the chemical kinetics and other evidence we soon discuss, the mechanism of this reaction has been shown to occur in two steps:

Step 1. Ionization of the alkyl iodide; the rate-determining step

$$
\underset{\underset{\text{CH}_3}{|}}{\overset{\overset{\text{CH}_3}{|}}{\text{CH}_3-\text{C}-\ddot{\text{I}}:}} \overset{\text{slow}}{\rightleftharpoons} \underset{\underset{\text{CH}_3}{|}}{\overset{\overset{\text{CH}_3}{|}}{\text{CH}_3-\text{C}^{\oplus}}} + :\ddot{\text{I}}:^{\ominus}
$$

	tert-Butyl cation	Iodide leaves with its bonding pair of electrons

Step 2. Reaction of the tertiary carbocation with the nucleophile; fast reaction

$$
\underset{\underset{\text{CH}_3}{|}}{\overset{\overset{\text{CH}_3}{|}}{\text{CH}_3-\text{C}^{\oplus}}} + \text{Nu}:^{\ominus} \overset{\text{fast}}{\longrightarrow} \underset{\underset{\text{CH}_3}{|}}{\overset{\overset{\text{CH}_3}{|}}{\text{CH}_3-\text{C}-\text{Nu}}}
$$

One noteworthy feature of this two-step mechanism is that an old friend, the relatively stable *tert*-butyl cation, is the product of the first step of the reaction, and from what we know about the stability of 3° carbocations, it is not surprising that one forms. Another feature is that the reaction between the tertiary carbocation and the nucleophile is very fast and does not occur until the carbocation is generated. This accounts nicely for the absence of the nucleophile in the rate law for the reaction—it is not involved in the rate-determining step.

This reaction is called an **S$_N$1 reaction,** where the S stands for *substitution*, the N for *nucleophilic*, and the 1 for *monomolecular*. The rate of this reaction depends only on how fast the alkyl halide ionizes to the *tert*-butyl cation. The energy profile diagram in Fig. 9.4 shows that the greatest amount of energy (the energy of activation, E_A) is required for the formation of the carbocation, which then reacts readily with the nucleophile to give the product. As we see later, a major side reaction of the S$_N$1 reaction is elimination (alkene formation).

Now we look at some evidence that supports this two-step mechanism and the intervention of carbocations. In addition to structure-reactivity correlations discussed

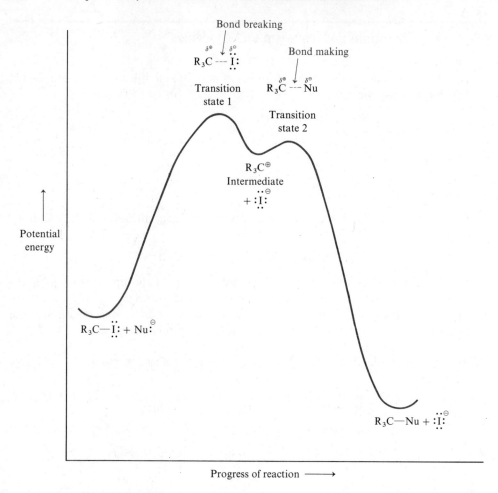

FIGURE 9.4   Energy-profile diagram for the S_N1 reaction between *tert*-butyl iodide, symbolized by $R_3C—I$, and a nucleophile, $Nu:^{\ominus}$.

in Sec. 9.10, some of the most compelling evidence comes from stereochemical studies and molecular rearrangements.

A. Stereochemistry of S_N1 Reaction

The hydrolysis (reaction with water) of *optically active* 1-chloro-1-phenylethane (note that the benzene ring, when treated as a substituent, is called the **phenyl** group; see Sec. 13.9), is carried out under conditions that favor the S_N1 mechanism. When the product is isolated, it exhibits only slight optical activity; the configuration of the predominant isomer is also inverted.

$$\text{1-Chloro-1-phenylethane} + H_2O \xrightarrow[\text{reaction}]{S_N1} \text{1-Phenylethanol}$$

1-Chloro-1-phenylethane
(α-phenethyl chloride)
Optically pure

1-Phenylethanol
(α-phenethyl alcohol)
*Slightly optically active,
inverted configuration*

Because only 2% optical activity is observed in the product, it has been concluded that this S_N1 reaction occurs with predominant **racemization.** We start with one enantiomer of the alkyl halide, but the reaction forms a mixture of enantiomers, the *d* and *l* forms of 1-phenylethanol, with the enantiomer that has a configuration opposite that of the starting alkyl chloride predominating.

To account for these results, the intervention of a planar carbocation, the 1-phenylethyl cation, is invoked. As we know from Sec. 2.8, the carbocation is planar and can be attacked from either side of its face by a nucleophile. The mechanism of the S_N1 solvolysis of 1-chloro-1-phenylethane is shown in Fig. 9.5, along with the planar carbocation and its reaction with the nucleophile, water. That racemization is not complete is attributed to the leaving group, the chloride ion ($:\overset{\cdot\cdot}{\underset{\cdot\cdot}{Cl}}:^{\ominus}$), which stays in the vicinity of the carbocation after the carbon-chlorine bond is broken. The chloride ion, in effect, shields the carbocation partially from attack by water on that side, thus producing slightly more backside attack by water and thus slightly more of the alcohol with inverted configuration. There is no doubt that racemization is quite extensive, however, thus supporting the carbocation mechanism.

In general, **S_N1 reactions occur with a predominance of racemization,** in contrast to the S_N2 reaction, which has the stereochemistry of complete inversion. Even so, care must be used in making sweeping generalizations about the stereochemistry of the S_N1 reaction. Different solvents give different results, and some compounds possess special structural features that favor net retention, complete racemization,

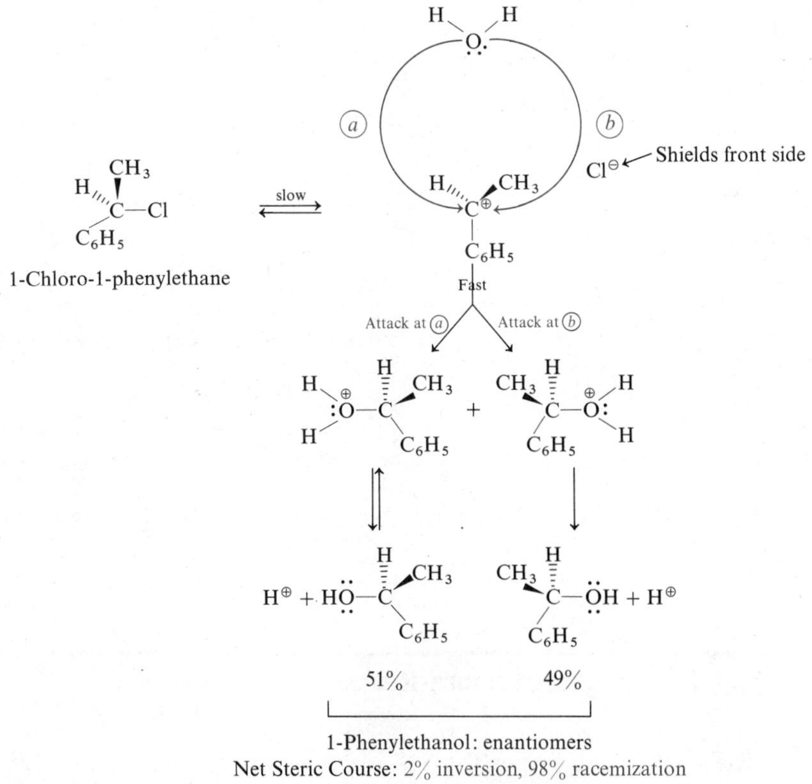

1-Phenylethanol: enantiomers
Net Steric Course: 2% inversion, 98% racemization

FIGURE 9.5 Stereochemistry of the hydrolysis of optically active 1-chloro-1-phenylethane showing formation of predominantly optically inactive (racemic) product.

or net inversion of configuration. Some very elegant mechanistic work demonstrates that alkyl halides often exist in ion pairs—that is, $R^{\oplus}X^{\ominus}$—in solvolysis reactions.

B. Molecular Rearrangements

If indeed carbocations are involved in S_N1 reactions, then molecular rearrangements should occur when possible (see Sec. 5.7). We do not anticipate any rearrangements in S_N2 reactions, however, because bond breaking and making occur simultaneously. As the following example illustrates, extensive rearrangement occurs with S_N1 conditions but no rearrangement is observed when S_N2 conditions are used.

$$CH_3-\overset{\overset{\displaystyle CH_3}{|}}{\underset{\underset{\displaystyle CH_3}{|}}{C}}-CH_2-OH$$

2,2-Dimethyl-1-propanol
(neopentyl alcohol)
No rearrangement

$\overset{\ominus}{:}\overset{..}{O}H,$
H_2O (S_N2 conditions)

$$CH_3-\overset{\overset{\displaystyle CH_3}{|}}{\underset{\underset{\displaystyle CH_3}{|}}{C}}-CH_2-I$$

1-Iodo-2,2-dimethylpropane
(neopentyl
iodide)

$H_2\overset{..}{O}:$ (S_N1 conditions)

$$\left[:\overset{..}{\underset{..}{I}}:^{\ominus} + CH_3-\overset{\overset{\displaystyle CH_3}{|}}{\underset{\underset{\displaystyle (CH_3)}{|}}{C}}-\overset{\oplus}{C}H_2 \right] \xrightarrow[\text{migrates}]{\sim CH_3^{\ominus}:} CH_3-\overset{\overset{\displaystyle CH_3}{|}}{\underset{\oplus}{C}}-CH_2-CH_3$$

1° carbocation

3° carbocation

$H_2\overset{..}{O}:,$
$-H^{\oplus}$

$-H^{\oplus}$

$$CH_3-\overset{\overset{\displaystyle CH_3}{|}}{\underset{\underset{\displaystyle OH}{|}}{C}}-CH_2-CH_3$$

2-Methyl-2-butanol
(*tert*-pentyl alcohol)

$$CH_3-\overset{\overset{\displaystyle CH_3}{|}}{C}=CH-CH_3$$

2-Methyl-2-butene

Compounds such as neopentyl iodide react very slowly by both S_N1 and S_N2 mechanisms.

9.10 S_N1 Reactions: Structure-Reactivity Correlations

The relative reactivities for S_N2 reactions are $CH_3-X > 1° > 2° > 3°$, and the reason for this is steric factors. Similar reactivity correlations were determined for S_N1 reactions, again as a function of the structure of the attached alkyl group. The data in Table 9.5 show the reactivities of some alkyl bromides in the hydrolysis

TABLE 9.5   Relative Rates of Reactivity of Alkyl Bromides in S$_N$1 Hydrolysis Reaction R—$\overset{..}{\underset{..}{Br}}$: + H$_2$$\overset{..}{O}$: $\longrightarrow$ R—$\overset{..}{\underset{..}{O}}$H + H—$\overset{..}{\underset{..}{Br}}$:

Name and Structure of Alkyl Bromide	Class of Alkyl Group	Relative Rate of Reaction*
Methyl bromide, CH$_3$—Br	CH$_3$—	1.0
Ethyl bromide, CH$_3$CH$_2$—Br	1°	1.6
Isopropyl bromide, (CH$_3$)$_2$CH—Br	2°	32
tert-Butyl bromide, (CH$_3$)$_3$C—Br	3°	10^7

* Rate of CH$_3$—Br set at 1.0.

reaction (that is, the reaction of alkyl bromide with water). These data indicate the following S$_N$1 reactivities of alkyl halides:

Decreasing order of S$_N$1 reactivity of R—X:

$$3° > 2° > 1° > CH_3—X$$

Is this reactivity order expected on the basis of the mechanism set forth for the the S$_N$1 reaction in Sec. 9.9? Indeed it is, for the alkyl group structure in the alkyl halide parallels the carbocation stability (see Sec. 5.7).

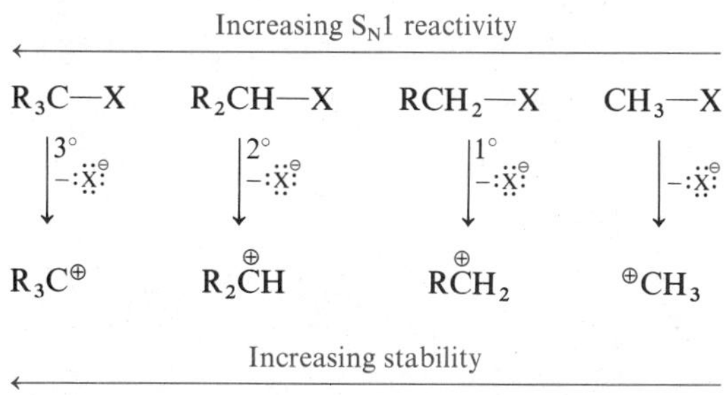

In general, **S$_N$1 reactivity is governed by electronic effects.** Substituents that donate electrons increase the S$_N$1 reactivity of an alkyl halide, whereas those that withdraw electrons decrease it.

As with S$_N$2 reactions, care must be exercised in generalizing the reactivities of alkyl halides by class of alkyl group (that is, 1°, 2°, or 3°). Increasing the size of the alkyl groups in a tertiary alkyl halide greatly affects the rate of reaction. For example, compare the relative rates of hydrolysis of these two alkyl halides:

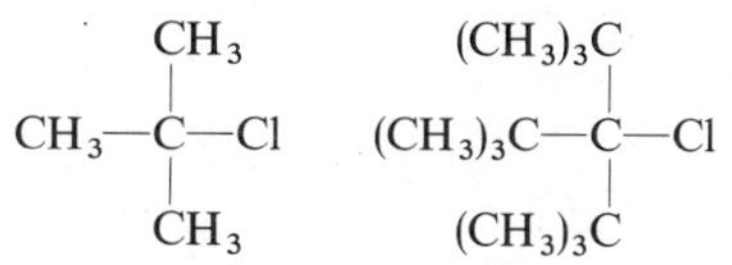

Even though both compounds are tertiary chlorides, the more highly substituted one reacts much faster. This is attributed to the relief of steric strain in the formation of the planar carbocation; that is, there is a bond angle change from $\sim 110°$ in the alkyl halide to $120°$ in the carbocation.

Four factors provide the driving force for reaction by the S_N1 mechanism: (1) formation of a relatively stable tertiary carbocation, (2) relief of steric strain in going from alkyl halide to the carbocation, (3) assistance by solvent in pulling off the leaving group, and (4) removal of halide ion from solution, thereby shifting the equilibrium in the desired direction. We discussed the first two factors, but the latter two deserve further mention.

Many S_N1 reactions are carried out in a polar solvent, commonly water or a mixture of water and some other organic solvent such as methanol or ethanol. Water is very polar, and the positive end of its dipole surrounds the halide ion and aids the ionization process by helping "pull" the halide ion off. Once liberated, the halide ion is highly solvated by water molecules, as is the carbocation; the carbocation, however, is solvated by the negative end of the water molecule dipole. This solvation effect is shown in Fig. 9.6. When we write the formula of a carbocation as $R^\oplus$, it is really an oversimplified representation of the actual structure because the ion is highly solvated under most conditions.

Carbocation formation is expedited by the addition of certain inorganic ions that react with and remove the halide ion from solution. The addition of, for example, silver ion greatly eases the ionization of the tertiary halide because of the formation of the insoluble silver halide salt, which precipitates out. Removal of the halide ion from solution causes the equilibrium of the ionization reaction to shift to the right, thus producing more carbocation.

$$R_3C{-}X + Ag^\oplus \; \rightleftharpoons \; \overset{\delta\oplus}{R_3C}{-}{-}\overset{\delta\ominus}{-}X{-}{-}\overset{\delta\oplus}{-}Ag \; \longrightarrow \; R_3C^\oplus + \quad AgX$$

$$\text{Precipitate}$$

This reaction is the basis for a common method of distinguishing primary, secondary, and tertiary alkyl halides (see Sec. 9.18).

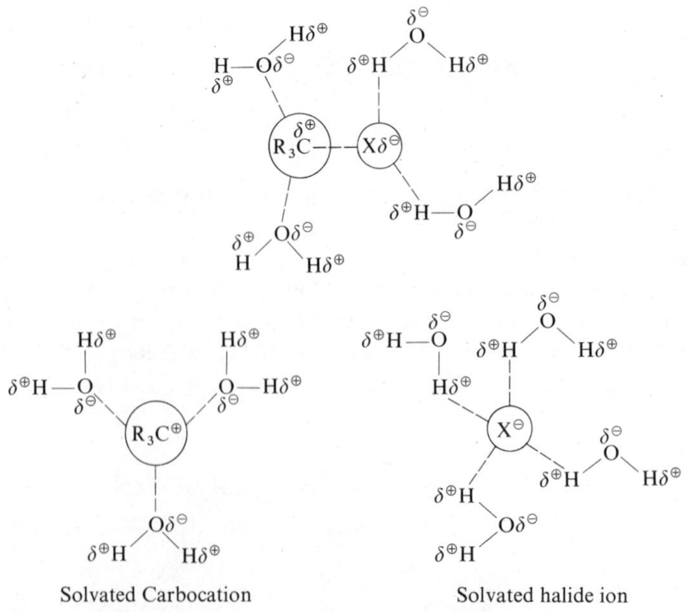

Solvated Carbocation   Solvated halide ion

FIGURE 9.6 An approximate picture of the role polar solvents (e.g., water and alcohol) play in assisting the ionization of an alkyl halide.

Question 9.2

Arrange the following four compounds in order of increasing S$_N$1 reactivity:

$$\underset{\underset{\displaystyle C(CH_3)_3}{|}}{\overset{\overset{\displaystyle C(CH_3)_3}{|}}{(CH_3)_3C-C-Br}} \qquad (CH_3)_3C-Br \qquad \underset{\underset{\displaystyle CH_3}{|}}{\overset{\overset{\displaystyle CH(CH_3)_2}{|}}{CH_3-C-Br}} \qquad \underset{\underset{\displaystyle CH_2CH_3}{|}}{\overset{\overset{\displaystyle C(CH_3)_3}{|}}{CH_3CH_2-C-Br}}$$

9.11 Effect of Leaving Group and Nucleophile on S$_N$1 Reactivity

The relative abilities of various groups to leave carbon in S$_N$1 reactions are roughly the same as those for S$_N$2 reactions (see Sec. 9.8). The net result in both reactions is the same: the leaving group departs with the pair of bonding electrons. Under identical conditions, there should be no difference in leaving group abilities in S$_N$1 and S$_N$2 reactions.

Reflection on the mechanism of the S$_N$1 reaction brings us to the conclusion that *there is little dependence of the rate of reaction on the nucleophile.* The limiting step of the reaction is ionization and carbocation production; the latter is highly reactive and reacts rapidly with a nucleophile. As we see in Sec. 9.12, however, the nucleophile and its concentration often play an important role in determining whether a particular reaction occurs by the S$_N$1 or S$_N$2 mechanism. In Sec. 9.13 the effect of solvent on these two types of mechanisms is discussed.

9.12 Comparison of S$_N$1 and S$_N$2 Mechanisms

The important similarities and differences for the S$_N$1 and S$_N$2 mechanisms are summarized in Table 9.6.

The reactivity sequences for alkyl halides are not so simple as they appear, however. With primary and methyl halides, nucleophilic substitution occurs via the S$_N$2 mechanism. With tertiary halides, the S$_N$1 mechanism is important and very little (if any) substitution occurs via the S$_N$2 mechanism. With secondary alkyl halides,

TABLE 9.6 Summary of Factors in S$_N$1 and S$_N$2 Reactions

Factor	S$_N$1 Reaction	S$_N$2 Reaction
Chemical kinetics (molecularity of reaction)	Monomolecular	Bimolecular
Rate law	Rate = k_1[R—X]	Rate = k_2[R—X][Nu:]
Stereochemistry	Predominant racemization	Complete inversion of configuration
Reactivity of alkyl halides	3° > 2° > 1° > CH$_3$—X	CH$_3$—X > 1° > 2° > 3°
Special features	Rearrangements *may* occur	No rearrangements are observed

it is difficult (if not impossible) to predict which mechanism is important; most often, substitution on 2° alkyl halides occurs by both S_N1 and S_N2 mechanisms. Without considerable experimental evidence, which includes information about reaction conditions (concentration, solvent, and so on), it is simply not possible to predict whether a given secondary alkyl halide is going to react via the S_N1 or the S_N2 mechanism.

Optically pure 2-octyl brosylate, for example, may be converted to 2-octanol by reaction with aqueous dioxane. The octanol thus produced, however, is only 77% optically pure and predominantly of inverted absolute configuration. These results can be rationalized by invoking two different mechanisms—S_N1 and S_N2. Evidently the reaction proceeds predominantly via an S_N2 pathway, but the partial racemization can be explained only by an S_N1 route.

$$CH_3CH_2CH_2CH_2CH_2CH_2CHCH_3 \xrightarrow[\text{dioxane}]{H_2\ddot{O}:}$$
$$\overset{|}{O\text{---}Bs}$$

2-Octyl brosylate
100% optically pure

$$CH_3CH_2CH_2CH_2CH_2CH_2CHCH_3 + Bs\text{---}\overset{..}{\underset{..}{O}}\overset{\ominus}{:} + H^{\oplus}$$
$$\overset{|}{O\text{---}H}$$

2-Octanol
77% optically pure, primarily
of inverted configuration

where

$$-O\text{---}Bs = -O\text{---}\overset{\overset{O}{\parallel}}{\underset{\underset{O}{\parallel}}{S}}\text{---}\bigcirc\text{---}Br$$

Brosylate group
(*good* leaving group, see Sec. 19.6)

In Sec. 9.13 we see how certain experimental conditions may be varied to favor either the S_N1 or S_N2 reaction path.

9.13   Effect of Solvent and Nucleophile Concentration on S_N1 and S_N2 Mechanisms

The solvent plays a very important role in nucleophilic substitution reactions (see Sec. 5.14).

A. S_N1 Reactions

S_N1 reactions are very sensitive to solvent because their transition state involves charge separation and ultimately ionization of the carbon-halogen bond:

$$R_3C\text{---}\overset{..}{\underset{..}{X}}: \longrightarrow [R_3\overset{\delta\oplus}{C}\text{-----}\overset{\delta\ominus}{\underset{..}{X}}:] \longrightarrow R_3C^{\oplus} + :\overset{..}{\underset{..}{X}}:^{\ominus} \longrightarrow \text{Products}$$

Polar transition
state (favored by
polar solvents)

We discussed the effect of water on this reaction in Sec. 9.10 and Fig. 9.6. Because of its high dielectric constant ($\varepsilon = 81$, from Table 5.2), we might expect water to be an exceptionally good solvent for the S_N1 reaction. In fact it is poor. Most alkyl halides have very little solubility in water, so reactions between these two reactants are heterogeneous (two phases). As a result, mixtures of water and organic solvents are used for S_N1 reactions. For example, the hydrolysis of *tert*-butyl chloride has been carried out in pure ethanol and in mixtures of ethanol and water. In a 50:50 mixture of ethanol and water, *tert*-butyl chloride reacts approximately 10,000 times faster than in pure ethanol.

In general, **S_N1 reactions are favored in polar, protic solvents.** A polar solvent favors the ionization of the C—X bond in the slow, rate-determining step. A protic solvent, such as water, alcohol, or acid, is necessary to assist the ionization step by the ion-dipole interactions with the carbocation and the halide ion (Fig. 9.6). Whereas the rate of hydrolysis of *tert*-butyl chloride is enhanced by polar, protic solvents (see previous paragraph), the reaction between *tert*-butyl chloride and iodide ion occurs very slowly in anhydrous acetone ($\varepsilon = 21$), which is a moderately polar aprotic solvent.

B. S_N2 Reactions

S_N2 reactions are less susceptible to changes in solvent than are S_N1 reactions. In the former, both the transition state and the reactants have a net charge of -1:

$$\text{H}\overset{..}{\underset{..}{\text{O}}}\text{:}^{\ominus} + \text{R—}\overset{..}{\underset{..}{\text{X}}}\text{:} \longrightarrow [\text{H}\overset{\delta\ominus}{\overset{..}{\text{O}}}\text{---R---}\overset{\delta\ominus}{\overset{..}{\underset{..}{\text{X}}}}] \longrightarrow \text{products}$$

Transition state
*Same net charge
as reactants*

No new charges are created in the transition state, and the negative charge is less concentrated in the transition state than in the reactants because it is dispersed over a larger area of the molecule. S_N2 reactions occur a little more slowly in polar than in nonpolar solvents. Protic solvents solvate the nucleophile to a much greater extent than do aprotic solvents (see Secs. 5.14 and 9.13). However, the polarity of the solvent allows the nucleophile and organic compound to be present together in a homogeneous reaction mixture. In general, **S_N2 reactions are favored in polar, aprotic solvents.**

$$\text{Nu:}^{}\text{----}\overset{\delta\oplus}{\text{H}}\text{—}\overset{\delta\ominus}{\overset{..}{\underset{..}{\text{O}}}}\text{—Solvent}$$

Solvation of the nucleophile
by protic solvents

Indeed, the effect of changing the solvent is often reflected in the reactivity order of nucleophiles in S_N2 reactions. *In protic solvents*, such as water and alcohols, polarizability plays a major role in determining nucleophilicity. The decreasing order of nucleophilicity is: $:\overset{..}{\underset{..}{\text{I}}}:^{\ominus} > :\overset{..}{\underset{..}{\text{Br}}}:^{\ominus} > :\overset{..}{\underset{..}{\text{Cl}}}:^{\ominus} > :\overset{..}{\underset{..}{\text{F}}}:^{\ominus}$. *In aprotic solvents*, such as dimethyl sulfoxide, acetone, and dimethylformamide, in which the anion is less highly solvated, the order may be reversed ($:\overset{..}{\underset{..}{\text{F}}}:^{\ominus} > :\overset{..}{\underset{..}{\text{Cl}}}:^{\ominus} > :\overset{..}{\underset{..}{\text{Br}}}:^{\ominus} > :\overset{..}{\underset{..}{\text{I}}}:^{\ominus}$), thus reflecting the true order of base strength in this series. It should be emphasized that the order of nucleophile reactivity given in Sec. 9.7 is for protic solvents; when aprotic solvents are used, that order is changed. The precise change depends on the solvent used, and it is beyond the scope of this text to list them.

C. Effect of Nucleophile Concentration

The concentration of nucleophile plays some role in dictating the mechanism of nucleophilic aliphatic substitution. The rate of an S_N1 reaction is for the most part independent of the concentration of the nucleophile, although S_N1 reactions are favored by weaker nucleophiles (why?). On the other hand, the S_N2 reaction is favored when the concentration and strength of nucleophile is fairly high.

D. Summary

The purpose of this section is to give you some idea of the complexity of nucleophilic substitution in terms of solvent effects. So many factors are involved that it is impossible (and really not desirable) to provide "recipes" for various S_N1 and S_N2 reactions. These reactions have been the subject of many working years of research, yet much is not understood about them. The general trends involved are important, however, especially the effect of alkyl group structure on the type of mechanism. For most purposes, it is sufficient to indicate that a reaction of a secondary alkyl halide is either S_N1 or S_N2, but remember that the experimental conditions must be chosen to favor one or the other.

Question 9.3

Arrange the following four compounds in order of increasing reactivity toward iodide ion in acetone:

$(CH_3)_3C—Br$ $CH_3CH_2CH_2—Br$ $(CH_3)_2CHCH_2—Br$ $CH_3—\overset{\displaystyle |}{\underset{\displaystyle Br}{CH}}—CH_2CH_3$

Question 9.4

Considering only substitution, arrange the compounds in Question 9.3 in order of increasing reactivity toward hydroxide ion in a solvent such as 60% ethanol–40% water.

Question 9.5

Optically active 2-bromooctane is allowed to react under the following sets of conditions: (1) sodium hydroxide dissolved in dimethyl sulfoxide and (2) sodium hydroxide dissolved in 60% ethanol–40% water. From each reaction the corresponding alcohol

$$(2\text{-octanol}, CH_3—CHOH—CH_2CH_2CH_2CH_2CH_2CH_3)$$

is isolated. What could you predict about the optical activity of the product from each reaction? Which conditions favor S_N1 reactivity? S_N2 reactivity?

Question 9.6

An important method for preparing amines involves the reaction between an alkyl halide and ammonia (see Sec. 26.7), which occurs via an S_N2 mechanism when a primary alkyl halide is used:

$$H_3\overset{..}{N}: + R—CH_2—\overset{..}{\underset{..}{X}}: \longrightarrow \left[H_3\overset{\delta\oplus}{N}----\underset{\displaystyle R}{CH_2}----\overset{\delta\ominus}{X} \right] \longrightarrow H_3\overset{\oplus}{N}—\underset{\displaystyle R}{CH_2} + :\overset{..}{\underset{..}{X}}:^{\ominus}$$

Based on the trends of solvent effect on S_N2 reactions, do you think a polar solvent would speed up this reaction or have virtually no effect on it? Briefly explain.

9.14 Special Reactivity of Allyl and Bicycloalkyl Halides

As part of our general picture of the reactivity of alkyl halides in nucleophilic substitution reactions, we showed that primary halides react very slowly under S_N1 conditions. We now study a primary alkyl halide that not only reacts very rapidly under S_N1 conditions but also reacts more rapidly than do typical tertiary alkyl halides. Finally, we learn about a tertiary alkyl halide that is completely unreactive in S_N1 and S_N2 reactions.

A. Allyl Halide Reactivity

Chemical kinetic measurements indicate that allyl halides are more reactive than tertiary alkyl halides in S_N1 reactions; that is,

Order of decreasing reactivity of R—X in S_N1 reactions:

$$\text{Allyl} > 3° > 2° > 1° > CH_3-X$$

This order parallels the order of carbocation stability discussed in Sec. 5.7. Allyl halides, $CH_2\!\!=\!\!CH-CH_2-X$, exhibit enhanced S_N1 reactivity because of the formation of the relatively stable allyl carbocation, which is resonance-stabilized:

$$CH_2\!\!=\!\!CH-CH_2-\overset{..}{\underset{..}{Cl}}: \xrightarrow{\text{slow}} CH_2\!\!=\!\!CH-\overset{\oplus}{C}H_2 \longleftrightarrow \overset{\oplus}{C}H_2-CH\!\!=\!\!CH_2 + :\overset{..}{\underset{..}{Cl}}:^{\ominus}$$

3-Chloro-1-propene
(allyl chloride)

$$\overset{\delta\oplus}{C}H_2\!=\!\!=\!\!=\!\!CH\!=\!\!=\!\!=\!\overset{\delta\oplus}{C}H_2$$

Resonance hybrid

Support for the allyl carbocation comes from the hydrolysis of 1-chloro-2-butene (1) and 3-chloro-1-butene (2), which give a mixture of the same two alcohols under S_N1 conditions.

$$CH_3-CH\!\!=\!\!CH-CH_2-\overset{..}{\underset{..}{Cl}}: \qquad CH_3-CH-CH\!\!=\!\!CH_2$$
$$\overset{|}{:\overset{..}{\underset{..}{Cl}}:}$$

(1)   (2)
1-Chloro-2-butene   3-Chloro-1-butene

$$\searrow \underset{\text{slow}}{-:\overset{..}{\underset{..}{Cl}}:^{\ominus}} \qquad \swarrow \underset{\text{slow}}{-:\overset{..}{\underset{..}{Cl}}:^{\ominus}}$$

$$CH_3-CH\!\!=\!\!CH-\overset{\oplus}{C}H_2 \longleftrightarrow CH_3-\overset{\oplus}{C}H-CH\!\!=\!\!CH_2$$

Common resonance-stabilized carbocation

$$CH_3-\overset{\delta\oplus}{C}H\!=\!\!=\!\!=CH\!=\!\!=\!\!=\overset{\delta\oplus}{C}H_2$$

$$:\overset{..}{O}:$$
$$-H^{\oplus} \underset{\text{fast}}{\diagup} \quad H \qquad H \quad \underset{\text{fast}}{\diagdown} -H^{\oplus}$$

$$CH_3-CH-CH\!\!=\!\!CH_2 \qquad\qquad CH_3-CH\!\!=\!\!CH-CH_2-OH$$
$$\overset{|}{OH}$$

(3)   (4)
60% from (1)   40% from (1)
38% from (2)   62% from (2)

Certain reaction conditions permit both S_N1 and S_N2 reactions to occur, sometimes simultaneously; when this happens, (1) and (2) give different product mixtures. For example:

$$CH_3-CH=CH-CH_2-Cl$$

H_2O
(S_N1 conditions)

H_2O
(S_N2 conditions)

(3) + (4)   Mostly (4)

B. Bicycloalkyl Halides

The requirement for S_N2 reaction is that the backside of the alkyl halide be relatively unhindered, and for S_N1 reaction that the carbocation be able to be planar. What happens when neither requirement can be satisfied? Certain bicyclic molecules contain structural features that hinder the backside from nucleophilic attack and at the same time constrain the carbocation in a geometry such that it cannot become planar.

a. Nomenclature of Bicyclic Compounds

Before discussing these reactions we must digress and mention the nomenclature used for many of the products, which are called **bicyclic** compounds. As the name implies, they have two (*bi-*) rings. The following molecule has the common name *norbornane*:

Norbornane

On careful examination, the molecule contains two five-membered rings and one six-membered ring. The carbon atoms in red are common to those three rings, and they are called the *bridgehead carbon atoms*. Systematic naming requires the following: (1) Numbering starts at a bridgehead carbon and proceeds around the largest ring to the second bridgehead carbon atom, and then around the second largest ring back to the original bridgehead so as to give substituents and double bonds the lowest possible numbers. (2) Any remaining bridgehead carbon atoms are numbered last. (3) The parent name of the compound is derived from the alkane name that corresponds to the *total* number of carbon atoms in the rings (that is, excluding substituents). (4) The prefix *bicyclo* is added to the alkane name. (5) Between the *bicyclo* prefix and the parent name is a series of numbers enclosed in brackets and separated by periods; these numbers indicate the number of carbon atoms in each "bridge" connecting the bridgehead carbon atoms. The largest number comes first, then the next largest, and so on. The following examples illustrate the nomenclature for four bicyclic compounds.

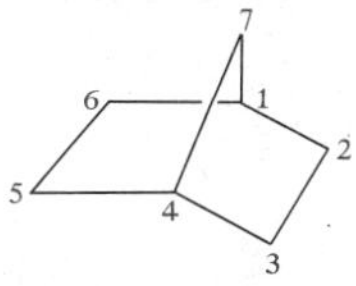

Bicyclo[2.2.1]heptane
(norbornane)

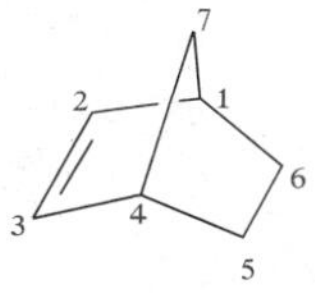

Bicyclo[2.2.1]-
2-heptene
(norbornene)

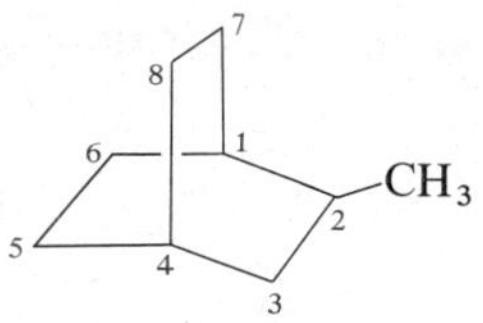

2-Methylbicyclo[2.2.2]octane

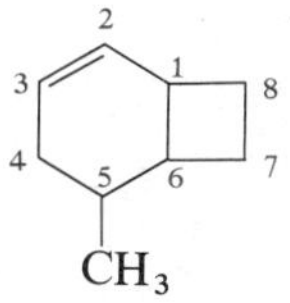

5-Methylbicyclo[4.2.0]-2-octene

b. Reactions

Each preceding compound contains one or more carbons that hinder backside attack in an S_N2 reaction. Let us look at one specific example.

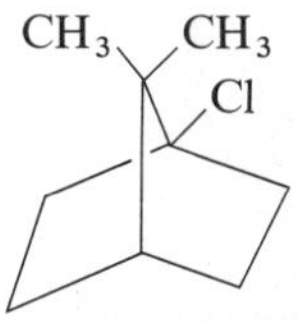

1-Chloro-7,7-dimethylbicyclo[2.21]heptane
(apocamphyl chloride)

This compound is unreactive when heated with concentrated aqueous sodium hydroxide solution for many hours, and on prolonged boiling with alcoholic silver nitrate, no silver chloride is observed. The molecular model of this compound and the corresponding carbocation (if it were formed) are shown in Fig. 9.7. Note that the carbocation is *not planar*, and study of this compound provides elegant evidence for the planarity requirement for the formation of carbocations. Note also the steric hindrance from the backside of the C—Cl bond, which prohibits S_N2 reaction.

Bicycloalkyl halides are generally unreactive when the halogen is attached to the *bridgehead carbon atom*. Attachment of halogen elsewhere in the molecule gives

(a)

(b)

FIGURE 9.7   Structural drawings of the structure of (a) apocamphyl chloride and (b) nonplanar carbocation that would be produced by S_N1 reaction (dissociation of the C—Cl bond) of apocamphyl chloride.

a compound that is reactive by either the S_N1 or the S_N2 mechanism. For example:

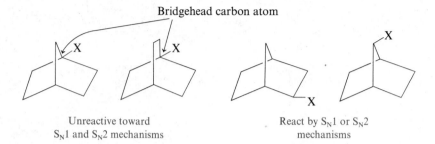

Bridgehead carbon atom

Unreactive toward   React by S_N1 or S_N2
S_N1 and S_N2 mechanisms   mechanisms

Question 9.7

Assuming that some slight amount of reaction occurs when the following compounds are heated with alcoholic silver nitrate (S_N1 conditions), which compound would react more rapidly? Why?

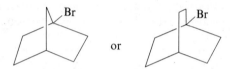

9.15   Competition Between Bimolecular Substitution and Elimination: S_N2 Versus E2

We emphasized the importance of nucleophilic aliphatic substitution, but another important reaction of alkyl halides is their ability to lose hydrogen halide (dehydrohalogenation, see Sec. 7.11) to give alkenes. In some cases substitution occurs, and in others elimination. What factors favor substitution or elimination?

With a typical primary alkyl halide, like ethyl iodide, there can be two different bimolecular reactions: substitution by the S_N2 mechanism and elimination by the E2 mechanism:

$$CH_3-CH_2-I$$
Ethyl iodide

$\xrightarrow[\text{alcohol}]{\text{conc KOH}} CH_2=CH_2$ Elimination *E2 reaction*

$\xrightarrow[\text{alcohol}]{\text{dilute :}\ddot{O}H^{\ominus},\ H_2O} CH_3-CH_2-OH$ Substitution *S_N2 mechanism*

Both elimination and substitution require the presence of hydroxide ion, yet the former is favored when the solvent is alcohol and the latter is favored when a mixture of water and alcohol is the solvent.

By way of review, the E2 elimination reaction occurs by a concerted mechanism in which base attacks a β hydrogen and there is simultaneous creation of the carbon-

carbon double bond and departure of the halogen as the halide ion (see Sec. 7.11):

$$\underset{\underset{\underset{HO:^{\ominus}}{H}}{CH_2-CH_2}}{\overset{\ddot{I}:}{}} \longrightarrow CH_2=CH_2 + :\ddot{I}:^{\ominus} + H_2O$$

Thus, in the general case, there are two possible competing reactions, one giving substitution and the other giving elimination, when one reactant can serve *both* as a nucleophile and as a base. Hydroxide ion, $:\ddot{O}H^{\ominus}$, is often encountered in dehydrohalogenation.

1° Alkyl halide

$:\ddot{O}H$ attack gives S_N2 substitution:

$:\ddot{O}H$ attack gives E2 elimination: $C=CH_2$

Certain ions have the dual capacity of being good nucleophiles for substitution and good bases for elimination, so it is desirable to indicate the trends observed for such ions. Strong bases are required to bring about bimolecular (E2) elimination, and their effectiveness roughly parallels their base strength; the order of decreasing base strength is:

$$H_2\ddot{N}:^{\ominus} > CH_3CH_2\ddot{O}:^{\ominus} > H\ddot{O}:^{\ominus} > :C\equiv N:^{\ominus} > CH_3-\overset{\overset{O}{\|}}{C}-\ddot{O}:^{\ominus}$$

Amide ion	Ethoxide ion	Hydroxide ion	Cyanide ion	Acetate ion

Stronger bases	Weaker bases
Favor elimination	*Favor substitution*

This listing also represents roughly the order of decreasing nucleophilicity (that is, the ability of the ions to serve as nucleophiles). Finally, a variety of very weakly basic substances, such as methanol (CH_3OH), water, and halide ion, have considerable nucleophilic character but are so weakly basic that they cause little, if any, elimination.

In summary, the strong bases listed earlier are very reactive toward hydrogen abstraction and thus favor elimination (E2). The weak bases are intermediate in relation to promoting elimination and giving rise to substitution (that is, they compete in S_N2 and E2 reactions). With some of these species, the major mode of reaction can be shifted from substitution to elimination, or vice versa, by changing the solvent and the reaction conditions. For example, elimination is favored over substitution at higher temperature, which is one variable at our disposal.

The effect of changing solvent is seen vividly in the reaction between ethyl iodide and hydroxide ion. When the solvent contains a mixture of ethanol and water, the predominant mode of reaction is substitution (although it is accompanied by a small amount of elimination). When the solvent is pure ethanol, elimination is the exclusive reaction, because of the equilibrium between ethanol and hydroxide ion, to form a small amount of ethoxide ion, which is an even stronger base than hydroxide ion:

$$CH_3CH_2\overset{..}{\underset{..}{O}}{-}H + :\overset{..}{\underset{..}{O}}H^{\ominus} \;\rightleftharpoons\; CH_3CH_2\overset{..}{\underset{..}{O}}:^{\ominus} + H_2\overset{..}{\underset{..}{O}}:$$

Ethoxide ion
Stronger base

The effect of the structure of the alkyl group on S_N2 and E2 reactions is as follows:

Increasing S_N2 reactivity
$\longleftarrow$

RCH₂X   R₂CHX   R₃CX

$\longrightarrow$
Increasing E2 reactivity

The S_N2 reactivity sequence was discussed in Sec. 9.6, and the reasons for the E2 reactivity sequence were presented in Sec. 7.12. Remember that at least one β hydrogen is required for E2 elimination.

Question 9.8

Arrange the following four compounds in order of (*a*) decreasing S_N2 reactivity and (*b*) decreasing E2 reactivity:

$(CH_3)_2CH{-}Cl$ $CH_3CH_2CH_2CH_2{-}Cl$ $CH_3{-}\underset{\underset{Cl}{|}}{CH}{-}CH_2CH_3$ $CH_3CH_2{-}\overset{\overset{CH_3}{|}}{\underset{\underset{Cl}{|}}{C}}{-}CH_3$

9.16 Competition Between Monomolecular Substitution and Elimination: S_N1 Versus E1

The **E1 reaction** (monomolecular elimination, see Sec. 7.15) is a two-step reaction in which the first step is the formation of a carbocation and the second is the rapid loss of a proton to form an alkene. We learned that tertiary alkyl halides dissociate into tertiary carbocations and halide ions. We also know that carbocations lose protons, usually to some basic substance. This, then, is another mechanism by which the elements of H—X can be removed from an alkyl halide to give an alkene:

E1 Mechanism for Dehydrohalogenation of an Alkyl Halide

Step 1. *Ionization of carbon-halogen bond, slow rate-determining step*

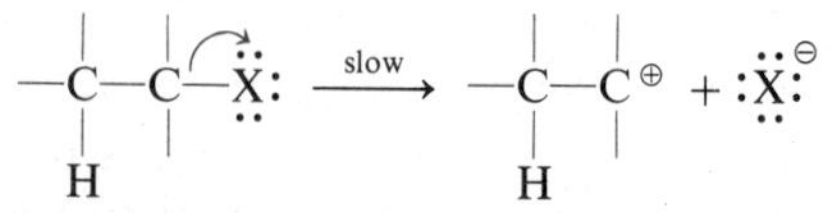

$$-\overset{|}{\underset{|}{C}}-\overset{|}{\underset{\underset{H}{|}}{C}}\overset{\frown}{-}\overset{..}{\underset{..}{X}}: \;\xrightarrow{\text{slow}}\; -\overset{|}{\underset{|}{C}}-\overset{|}{\underset{\underset{H}{|}}{C}}{}^{\oplus} + :\overset{..}{\underset{..}{X}}:^{\ominus}$$

3° Alkyl halide

Step 2. *Loss of proton to basic substance B:, fast reaction*

$$-\overset{|}{\underset{|}{C}}-\overset{|}{\underset{H}{C}}{}^{\oplus} \xrightarrow{\text{fast}} \overset{\diagup}{\underset{\diagdown}{C}}{=}\overset{\diagup}{\underset{\diagdown}{C}} + \text{B:H}^{\oplus}$$

$$\text{B:}\nearrow \qquad\qquad\qquad \text{Alkene}$$

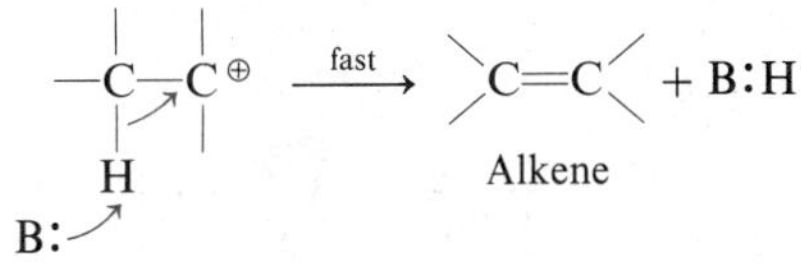

There is considerable evidence for the El mechanism. We first encountered it in the dehydration of alcohols in Sec. 7.14, in which the El mechanism operates for primary, secondary, and tertiary alcohols. This dehydration reaction follows first-order kinetics and is first order in $[\text{ROH}]$.

Tertiary alkyl halides obey first-order kinetics in both S_N1 and El mechanisms where the rate depends only on the concentration of alkyl halide, $[R_3C—X]$. If our ideas about the mechanisms of the S_N1 and El reactions are correct (both involve the same carbocation intermediate), then the S_N1 and E1 reactivities of alkyl halides should be parallel because the rate-determining (slow) step in each is the ionization of the alkyl halide. In accord with these expectations, the reactivity order of alkyl halides is as follows:

Increasing S_N1 reactivity $\longrightarrow$

$RCH_2X \qquad R_2CHX \qquad R_3CX$

$\longrightarrow$ Increasing E1 reactivity

Neither the elimination reaction nor the substitution reaction can occur until the carbocation is formed. El elimination reactions are also accompanied by molecular rearrangements (when they can occur) of the same type observed in S_N1 reactions. These factors point to the intervention of a common carbocation intermediate. The following examples illustrate the similarity of the S_N1 and El mechanisms.

Example 1

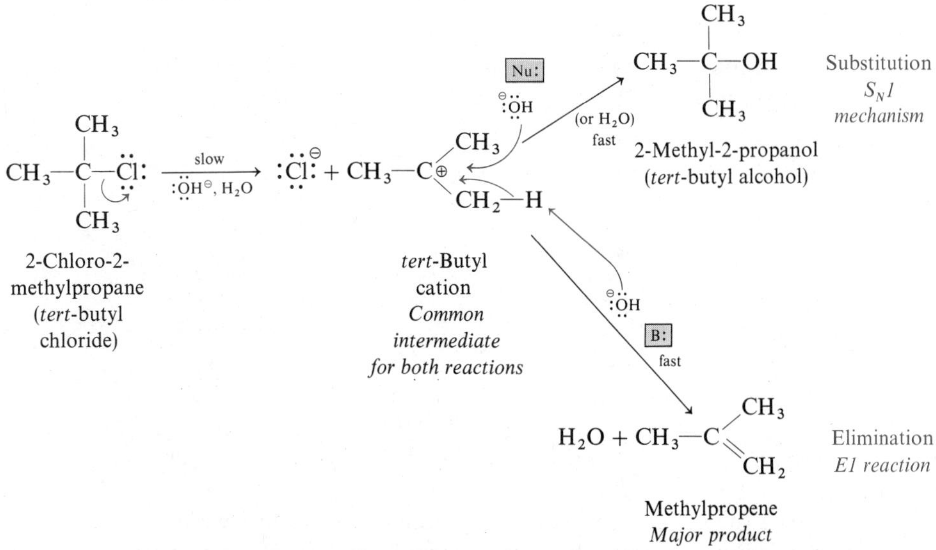

Example 2

(see also Sec. 5.11)

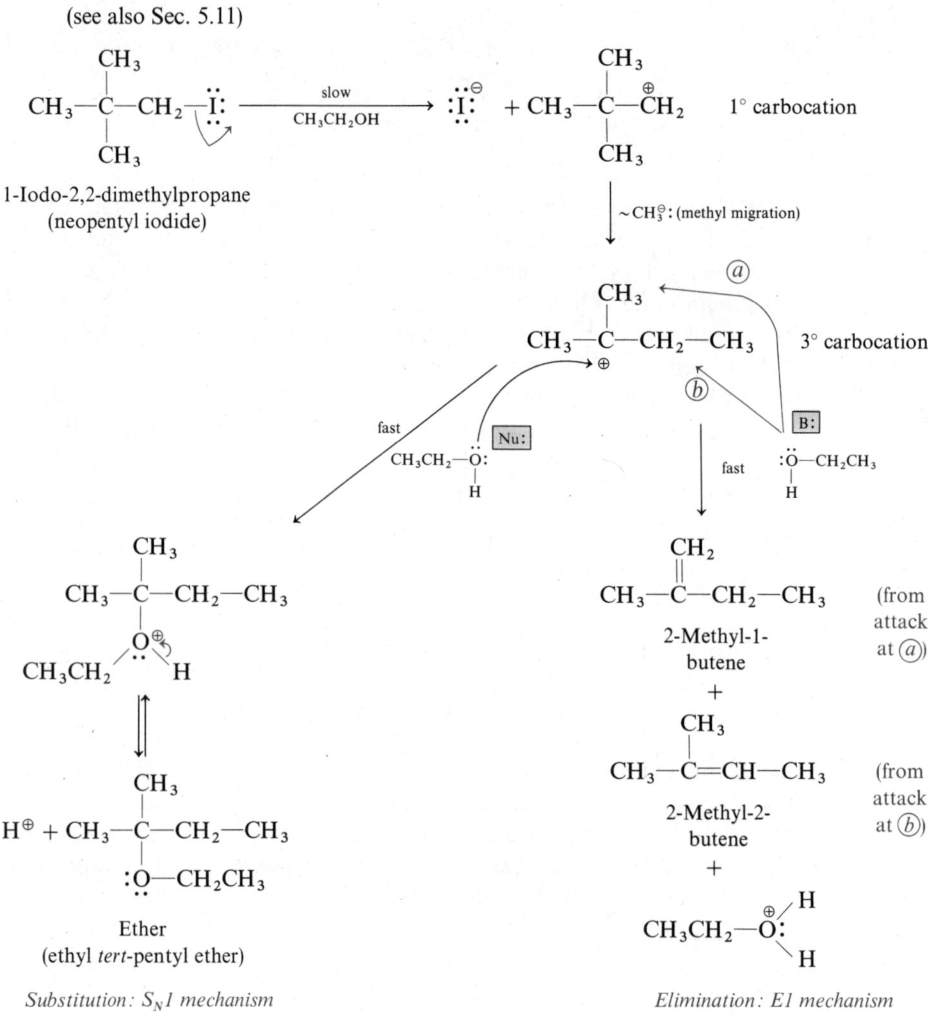

$$CH_3-\overset{\overset{CH_3}{|}}{\underset{\overset{|}{CH_3}}{C}}-CH_2-\ddot{\overset{..}{\underset{..}{I}}}: \xrightarrow[\text{CH}_3\text{CH}_2\text{OH}]{\text{slow}} :\overset{..}{\underset{..}{I}}:^{\ominus} + CH_3-\overset{\overset{CH_3}{|}}{\underset{\overset{|}{CH_3}}{C}}-\overset{\oplus}{CH_2} \quad 1° \text{ carbocation}$$

1-Iodo-2,2-dimethylpropane
(neopentyl iodide)

$\sim CH_3^{\ominus}$: (methyl migration)

$$CH_3-\overset{\overset{CH_3}{|}}{\underset{\oplus}{C}}-CH_2-CH_3 \quad 3° \text{ carbocation}$$

ⓐ

ⓑ

fast
$CH_3CH_2-\ddot{O}:$
H

Nu:

B:

fast

:Ö—CH₂CH₃
H

$$CH_3-\overset{\overset{CH_3}{|}}{\underset{\overset{|}{\underset{CH_3CH_2}{O^{\oplus}}\diagdown H}}{C}}-CH_2-CH_3$$

$$\Updownarrow$$

$$H^{\oplus} + CH_3-\overset{\overset{CH_3}{|}}{\underset{\underset{:\ddot{O}-CH_2CH_3}{|}}{C}}-CH_2-CH_3$$

Ether
(ethyl *tert*-pentyl ether)

Substitution: S_N1 mechanism

$$CH_3-\overset{\overset{CH_2}{\|}}{C}-CH_2-CH_3$$

2-Methyl-1-
butene

(from
attack
at ⓐ)

+

$$CH_3-\overset{\overset{CH_3}{|}}{C}=CH-CH_3$$

2-Methyl-2-
butene

(from
attack
at ⓑ)

+

$$CH_3CH_2-\overset{\oplus}{\underset{\diagdown H}{\ddot{O}}}\diagup^{H}$$

Elimination: E1 mechanism

In summary, the competition between S_N1 and E1 reactions is shown by the following scheme:

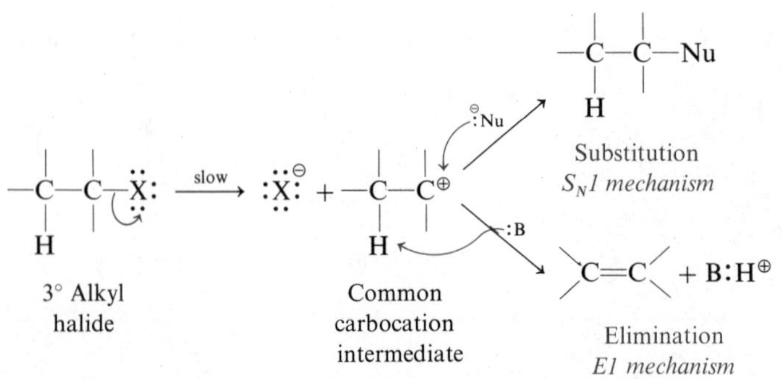

$$-\overset{|}{\underset{|}{C}}-\overset{|}{\underset{H}{C}}-Nu$$

Substitution
S_N1 mechanism

$$-\overset{|}{\underset{|}{C}}-\overset{|}{\underset{H}{C}}-\overset{..}{\underset{..}{X}}: \xrightarrow{\text{slow}} :\overset{..}{\underset{..}{X}}:^{\ominus} + -\overset{|}{\underset{|}{C}}-\overset{\oplus}{\underset{H}{C}}$$

$^{\ominus}$:Nu

:B

$$\diagup C=C\diagdown + B:H^{\oplus}$$

Elimination
E1 mechanism

3° Alkyl
halide

Common
carbocation
intermediate

Note the common intermediate—the carbocation—which can react with the nucleophile Nu: to give substitution or with the base B: to give elimination; sometimes Nu: and B: are the same species, as when water or ethanol, CH_3CH_2OH, is used as the solvent without added base.

Because of the competition between S_N1 and $E1$ reactions, we again can change experimental conditions to favor one type of reaction or the other. The rate-determining step in both reactions is ionization of the carbon-halogen bond, so the solvent must be reasonably polar (that is, have a dielectric constant, ε, greater than 20). The S_N1 reaction is favored by a solvent which has nucleophilic character (that is, one or more unshared pairs of electrons) but is a very weak base to minimize elimination. Two such solvents are water and alcohol. The $E1$ reaction is favored by the presence of a weak or strong base (see Sec. 9.15) dissolved in a polar solvent; the common bases are hydroxide, $:\overset{..}{\underset{..}{O}}H^{\ominus}$, alkoxide, $R\overset{..}{\underset{..}{O}}:^{\ominus}$, and amide, $:\overset{..}{N}H_2^{\ominus}$.

As an example of the S_N1 reaction, a tertiary alkyl halide can be converted to the corresponding alcohol by allowing it to react with water (*hydrolysis*) in the presence of a very weak base, such as carbonate, $CO_3^{\ominus 2}$, or bicarbonate, $HCO_3^{\ominus}$, which removes hydrogen ion as it is formed. Under these conditions the mechanism for the reaction between *tert*-butyl chloride and water is:

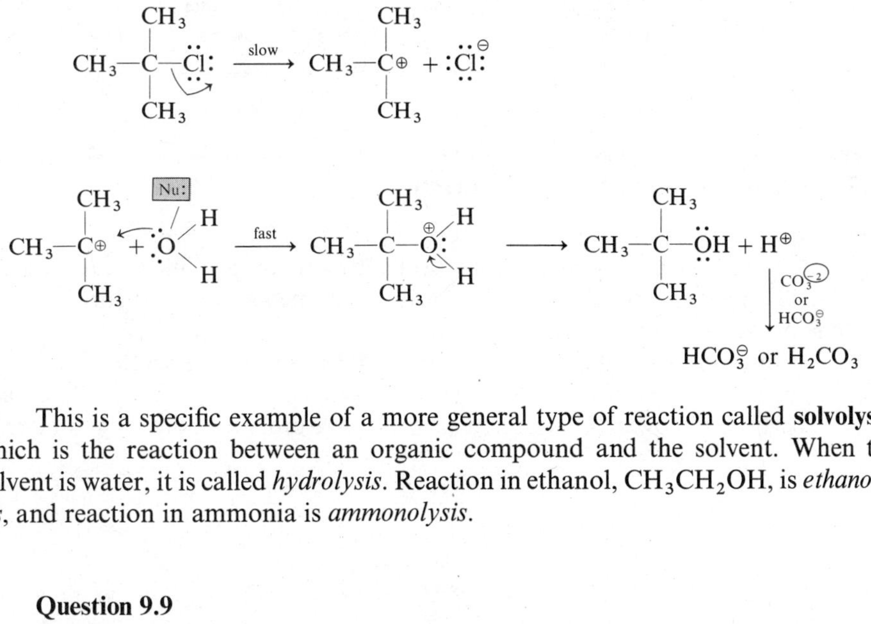

This is a specific example of a more general type of reaction called **solvolysis,** which is the reaction between an organic compound and the solvent. When the solvent is water, it is called *hydrolysis*. Reaction in ethanol, CH_3CH_2OH, is *ethanolysis*, and reaction in ammonia is *ammonolysis*.

Question 9.9

Arrange these three compounds in order of (*a*) increasing S_N1 reactivity and (*b*) increasing E1 reactivity:

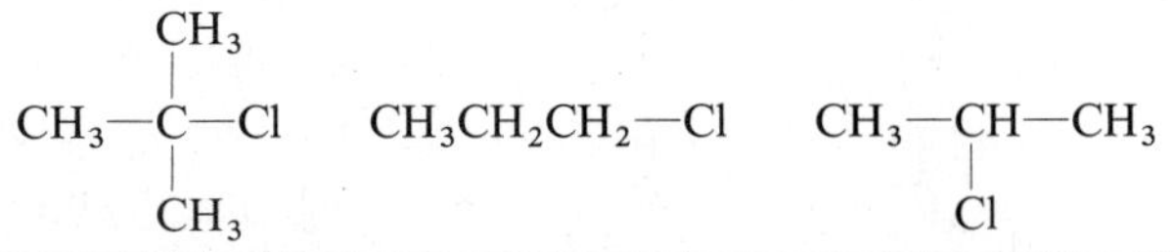

$$CH_3-\underset{\underset{\textstyle CH_3}{|}}{\overset{\overset{\textstyle CH_3}{|}}{C}}-Cl \qquad CH_3CH_2CH_2-Cl \qquad CH_3-\underset{\underset{\textstyle Cl}{|}}{CH}-CH_3$$

Question 9.10

Considering only E2 mechanisms, which compound in each of the following pairs would be expected to react more rapidly? Briefly explain.

(*a*) $(CH_3)_2CH-Cl$ or $(CH_3)_3C-Cl$ (*b*) $(CH_3)_2CHCH_2-Cl$ or $(CH_3)_3C-Cl$

Question 9.11

(*a*) Provide a mechanistic rationale for the following reaction. (*Hint:* Draw conformations.)

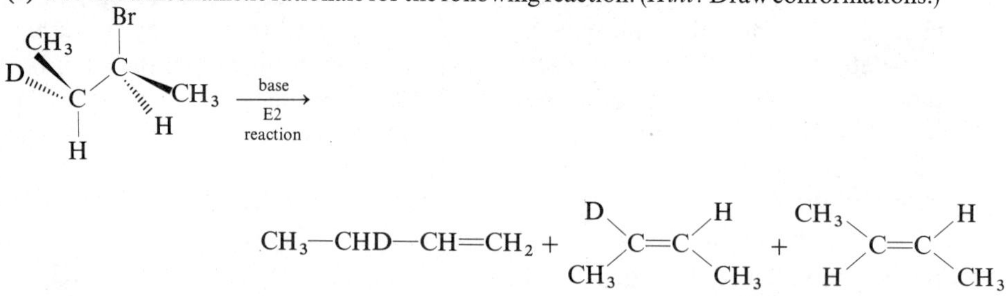

$$CH_3—CHD—CH=CH_2 \;+\; \underset{CH_3}{\overset{D}{>}}C=C\underset{CH_3}{\overset{H}{<}} \;+\; \underset{H}{\overset{CH_3}{>}}C=C\underset{CH_3}{\overset{H}{<}}$$

(*b*) Of the three products, which would you anticipate being formed in the greatest amount? Why?

(*c*) What significance do you attach to the fact that *neither* of the following 2-butenes was produced? (*Hint:* Think in terms of conformations.)

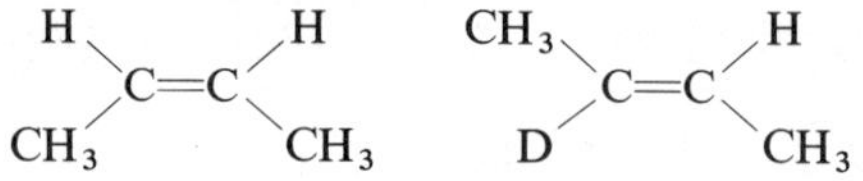

$$\underset{CH_3}{\overset{H}{>}}C=C\underset{CH_3}{\overset{H}{<}} \qquad \underset{D}{\overset{CH_3}{>}}C=C\underset{CH_3}{\overset{H}{<}}$$

(No D present)

9.17   Elimination and Substitution

We know that elimination and substitution compete in the reactions of alkyl halides. The biggest problem is to devise experimental methods for favoring either one or the other for laboratory work.

Regardless of whether the E1 or the E2 mechanism operates in the dehydrohalogenation of alkyl halides, the end result is that *alkenes are formed.* The reactivities of alkyl halides toward elimination have the same order for both E1 and E2 reactions:

Increasing E1 reactivity
⟶

$$RCH_2X \qquad R_2CHX \qquad R_3CX$$

Increasing E2 reactivity
⟶

Thus, when we want to carry out a dehydrohalogenation reaction, we use a polar solvent with a strong base. Three common solvent-base systems are: (1) concentrated KOH dissolved in alcohol; (2) sodium alkoxide, $Na^{\oplus} \;^{\ominus}\!:\!\overset{..}{\underset{..}{O}}R$, dissolved in alcohol, ROH; and (3) sodium alkoxide dissolved in an aprotic solvent, such as dimethyl sulfoxide, $(CH_3)_2SO$. With such strong bases, elimination is rapid and there is little need to worry about possible molecular rearrangements.

Reaction conditions that favor substitution are more difficult to obtain, especially when a tertiary alkyl halide is used. The experimental factors that favor S_N2 reactions were presented in the latter part of Sec. 9.15 and those favoring S_N1 reactions in the latter part of Sec. 9.16. Remember, however, that the greatest amount of crossover between mechanisms occurs in secondary alkyl halides, and in this family of compounds one can most easily favor either monomolecular or bimolecular substitution. Substitution is also favored by low reaction temperatures. For

example,

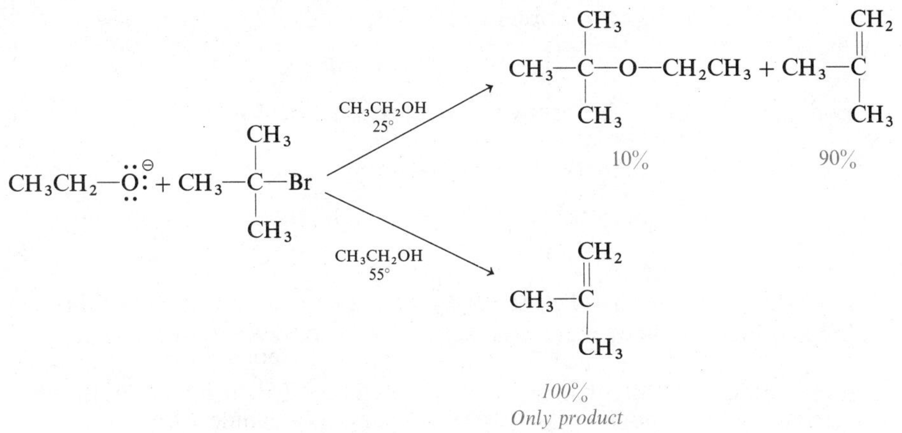

A careful selection of conditions can often completely change the observed product distribution:

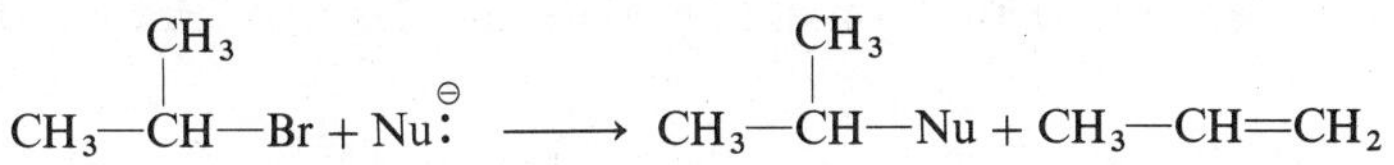

$$Nu\colon^\ominus = CH_3CH_2\ddot{\overset{..}{O}}\colon^\ominus \qquad\qquad 25\% \qquad\qquad 75\%$$
$$\text{(as } CH_3CH_2O^\ominus Na^\oplus/CH_3CH_2OH)$$
$$Nu\colon^\ominus = \colon\overset{..}{\underset{..}{Cl}}\colon^\ominus \qquad\qquad 100\% \qquad\qquad 0\%$$
$$\text{(as } n\text{-Bu}_4N^\oplus Cl^\ominus/\text{acetone)}$$

9.18   Qualitative Analysis; Classification of Alkyl Halides

The first hint that an unknown compound may be an alkyl halide is the detection of halogen in a sodium fusion test. Alkyl halides that contain no other functional groups, such as —OH, or a double or triple bond are insoluble in cold, concentrated sulfuric acid and do not react with reagents that detect the presence of these groups.

Two simple classification tests can detect an alkyl halide and determine whether it is primary, secondary, or tertiary. One test uses the alcoholic silver nitrate reagent and the other involves sodium iodide dissolved in acetone. Both one-step tests are *quick*, and if reaction occurs, it can be observed *visually* in the laboratory.

The **alcoholic silver nitrate reagent** involves an S_N1 reaction (see Sec. 9.10), in which the silver ion assists ionization by reacting with the halide and forming the insoluble silver halide salt, which is readily visible:

$$R—X + Ag^\oplus \xrightarrow[\text{water}]{\text{alcohol}} R^\oplus + AgX(s)$$

Precipitate

AgCl, white

AgBr, light yellow

AgI, dark yellow

The color of the silver halide that forms gives some indication of the halide present, although is not always reliable.

The reactivity trends of the alkyl halides are as follows:

Decreasing reactivity order of R—X toward alcoholic $AgNO_3$:

$$R—I > R—Br > R—Cl$$

For a given halogen, the reactivity sequence as a function of alkyl group structure is:

Decreasing reactivity of R—X toward alcoholic $AgNO_3$ (for a given X):

$$Allyl > R_3C—X > R_2CH—X > RCH_2—X > CH_3—X$$
$$3° \qquad 2° \qquad 1°$$

Note that this order parallels the S_N1 reactivity and carbocation stability sequences. Most primary halides react *very* slowly, if at all, even when heated. Anomalous results occur with allyl halides, $-\overset{|}{C}=\overset{|}{C}-CH_2—X$, which react mostly by the S_N1 mechanism even though they may be primary alkyl halides (see Sec. 9.14).

The **sodium iodide in acetone reagent** involves the S_N2 substitution of iodide for other halogens (see Sec. 5.9). Sodium iodide is soluble in acetone, $(CH_3)_2C=O$ (a polar aprotic solvent), whereas sodium bromide and sodium chloride are insoluble in it. This test involves the formation of a precipitate of NaBr or NaCl in the following reaction:

$$R—\overset{..}{\underset{..}{X}}: + Na^{\oplus} + :\overset{..}{\underset{..}{I}}:^{\ominus} \xrightarrow[\text{anhydrous}]{\text{acetone}} R—\overset{..}{\underset{..}{I}}: + Na^{\oplus}:\overset{..}{\underset{..}{X}}:^{\ominus}$$
$$\text{Precipitate}$$

where X = Cl, Br

Alkyl bromides are more reactive than alkyl chlorides, and the reactivity order as a function of alkyl group structure is:

Order of decreasing reactivity of R—X (X = Cl, Br) toward NaI in acetone:

$$CH_3—X > RCH_2—X > R_2CH—X > R_3C—X$$

This order also parallels the S_N2 reactivity sequence for alkyl halides. Tertiary halides react *very* slowly, if at all, even when heated. The test is necessarily limited to chlorides and bromides.

The results of these two tests must be interpreted with caution because of differences in the reactivities of chlorides, bromides, and iodides. Also, allyl halides that are primary react positively with both reagents. In the laboratory, these two tests are normally run against "standards," so that the rates of reaction with primary, secondary, and tertiary alkyl groups containing different halogens can be compared with an "unknown" alkyl halide.

Aryl halides, Ar—X, and vinyl halides, $-\overset{|}{C}=\overset{|}{C}-X$, do *not* react with either test reagent.

Question 9.12

Indicate what simple test you would perform and what you would observe in the laboratory to distinguish between the following pairs of compounds.

(*a*) $CH_3CH_2CH_2—Br$ and $(CH_3)_2CH—Br$

(*b*) $CH_3CH_2CH_2—\underset{\underset{Br}{|}}{CH}—CH_3$ and $CH_3CH_2CH_2—\underset{\underset{Cl}{|}}{CH}—CH_3$

(c)

[structure: cyclohexane ring with CH$_3$ and Cl substituents] and [structure: cyclohexane ring]—CH$_2$—Cl

(d) CH$_3$—CH=CH—CH$_2$—Br and CH$_2$=CH—CH$_2$CH$_2$—Br

Question 9.13

Suggest a sequence of simple chemical tests that would allow you to label bottles containing the following five compounds, assuming that each bottle contains only one pure compound: cyclohexane, cyclohexene, cyclohexyl bromide, *n*-butyl bromide, and *tert*-butyl chloride. What would you observe in each test?

9.19 Summary of Reactions of Alkyl Halides

See Sec. 9.1 for a review of preparations covered in previous chapters.

A. Nucleophilic Substitution

$$R—\overset{..}{\underset{..}{X}}: + Nu:^{\ominus} \longrightarrow R—Nu + :\overset{..}{\underset{..}{X}}:^{\ominus} \qquad \text{Secs. 5.9 and 9.3}$$

Effect of R-group structure on reactivity,
  general discussion (Sec. 9.4)
S$_N$2 (bimolecular) reactions
  Mechanism and stereochemistry (*inversion*
    of configuration) (Secs. 5.9 and 9.5)
  Effect of alkyl group structure on
    reactivity (Sec. 9.6)

$$R—X: \qquad CH_3X > 1° > 2° > 3°$$

Effect of nucleophile (Sec. 9.7)

$$:\overset{..}{\underset{..}{I}}:^{\ominus} > :\overset{..}{\underset{..}{Br}}:^{\ominus} > :\overset{..}{\underset{..}{Cl}}:^{\ominus} > :\overset{..}{\underset{..}{F}}:^{\ominus} \qquad \text{(in protic solvents)}$$

Effect of leaving group (Sec. 9.8)

$$R—I > R—Br > R—Cl \gg R—F$$

S$_N$1 (monomolecular) reactions
  Mechanism and stereochemistry (*racemization*), molecular rearrangements
    (Sec. 9.9)
  Effect of alkyl group structure on reactivity (Sec. 9.10)

$$R—X: \qquad Allyl > 3° > 2° > 1° > CH_3—X$$

Effect of leaving group (Sec. 9.11)

$$R—I > R—Br > R—Cl \gg R—F$$

Effect of nucleophile (Sec. 9.11)
  Nature of nucleophile (Sec. 9.11)
  Concentration of nucleophile, S$_N$1 reaction favored by low concentration
    (Sec. 9.13)

S_N1 versus S_N2 reactions
  Comparison (Sec. 9.12)
  Effect of solvent and nucleophile (Secs. 5.14 and 9.13)
Reaction of allyl and bicycloalkyl halides (Sec. 9.14)

$$R—X:\quad \text{Allyl} > 3° > 2° > 1° > CH_3—X$$

Bicycloalkyl halides (halide at bridgehead) unreactive

Competition between S_N2 and E2 reactions
  Effect of solvent, nucleophile, and basicity of nucleophile (Sec. 9.15)

Increasing S_N2 reactivity
$$\overleftarrow{}$$
$$RCH_2X \qquad R_2CHX \qquad R_3CX$$
$$\overrightarrow{}$$
Increasing E2 reactivity

Competition between S_N1 and E1 reactions (Sec. 9.16)
  E1 mechanism for elimination
  Both S_N1 and E1 involve common carbocation

Increasing S_N1 reactivity
$$\overrightarrow{}$$
$$RCH_2X \qquad R_2CHX \qquad R_3CX$$
$$\overrightarrow{}$$
Increasing E1 reactivity

Elimination and substitution, experimental factors favoring one or the other
  (Sec. 9.17)
Qualitative tests for alkyl halide classification (Sec. 9.18)
  Reactivity toward alcoholic $AgNO_3$

$$R—I > R—Br > R—Cl$$

$$R—X:\quad \text{Allyl} > 3° > 2° > 1° > CH_3—X$$

  Reactivity toward NaI in acetone

$$R—Br > R—Cl$$

$$R—X:\quad CH_3—X > 1° > 2° > 3°$$

Aryl, vinyl, and bicycloalkyl (X on bridgehead) are unreactive

B. Elimination

Dehydrohalogenation (Sec. 7.11), E1 and E2

$$-\overset{|}{\underset{H}{C}}-\overset{|}{\underset{X}{C}}- \xrightarrow{\text{base}} \enspace {>}C{=}C{<}$$

Dehalogenation (Sec. 7.19)

$$-\overset{|}{\underset{X}{C}}-\overset{|}{\underset{X}{C}}- \xrightarrow[\text{alcohol}]{\text{Zn}} \enspace {>}C{=}C{<}$$

C. Coupling

$$R_2CuLi \quad + \quad R'X \quad \longrightarrow \quad R—R' \quad \text{(Sec. 4.2)}$$

Lithium dialkylcopper   1° Alkyl
reagent   halide
or
methyl halide

$$R—X \xrightarrow[\text{heat}]{\text{Na metal}} R—R \quad \text{Wurtz reaction (Sec. 4.2)}$$

2 moles

Study Questions

9.14 Indicate how *vinyl chloride*, CH_2=CH—Cl, can be prepared from the following using any other inorganic reagents that may be needed:

(*a*) ethene   (*b*) 1,2-dichloroethane   (*c*) 1,1-dichloroethane

9.15 Indicate how the following conversions can be carried out in good yield. Use no source of carbon other than that indicated. Any needed inorganic reagents or organic solvents may be used. (*Hint:* See the Study Hint on Multistep Synthesis, end of Chap. 8)

(*a*) $(CH_3)_2CH—OH \longrightarrow CH_3—CH=CH_2$

(*b*) $(CH_3)_2CH—OH \longrightarrow Cl—CH_2—CH=CH_2$

(*c*) [cyclohexyl]—OH $\longrightarrow$ [3-bromocyclohexene with Br]

(*d*) CH_2=CH—CH$_3$ $\longrightarrow$ CH_2=CH—CH$_2$—I

(*e*) $(CH_3)_4C \longrightarrow (CH_3)_3CCH_2—Br$

9.16 Supply the missing reactants or products as required to complete the following five equations. The items to be inserted are indicated by number, for example, (1). If no reaction occurs, so indicate.

(*a*) $CH_3(CH_2)_3CH_2—Cl \xrightarrow{(1)} CH_3(CH_2)_3CH_2—OH$

$\downarrow (2)$

$CH_3(CH_2)_2CH=CH_2 \xrightarrow{(3)} CH_3(CH_2)_2\overset{\displaystyle Br}{\underset{\displaystyle |}{CH}}—CH_3 \xrightarrow[H^\oplus]{Zn} (4)$

(*b*) $(CH_3)_2CH—Br \xrightarrow[\text{acetone}]{\text{NaI}} (5)$

(*c*) CH_2=CH—CH$_3$ $\xrightarrow{(6)} CH_2$=CH—CH$_2$—Cl $\xrightarrow[\text{heat}]{H_2O} (7)$

(*d*) $CH_3CH_2CH_2OH \xrightarrow[\text{heat}]{H^\oplus} (8)$

(*e*) $CH_3—\overset{\displaystyle CH_3}{\underset{\displaystyle CH_3}{\overset{\displaystyle |}{\underset{\displaystyle |}{C}}}}—Br \xrightarrow{(9)} CH_3—\overset{\displaystyle CH_3}{\underset{\displaystyle CH_3}{\overset{\displaystyle |}{\underset{\displaystyle |}{C}}}}—OH$

$\downarrow (10)$

$(CH_3)_2C=CH_2 \xrightarrow{(11)} (CH_3)_2\overset{\displaystyle OH}{\underset{\displaystyle |}{C}}—\overset{\displaystyle OH}{\underset{\displaystyle |}{C}}H_2$

$\underset{\text{(12)}}{\swarrow} {\scriptstyle \begin{array}{l}1.\ O_3\\2.\ Zn,\ H_2O\end{array}}$ $\downarrow {\scriptstyle KMnO_4,\ :\overset{..}{O}H^\ominus,\ H_2O,}\atop{\text{heat}}$

(13)

9.17 Predict whether each of the following reactions would be successful if carried out under conditions with the indicated type of mechanism. Briefly explain.

(a) CH_3—$\overset{\overset{\displaystyle CH_3}{|}}{\underset{\underset{\displaystyle CH_3}{|}}{C}}$—Br + :ÖH $\xrightarrow{S_N2}$ CH_3—$\overset{\overset{\displaystyle CH_3}{|}}{\underset{\underset{\displaystyle CH_3}{|}}{C}}$—OH + :Br:

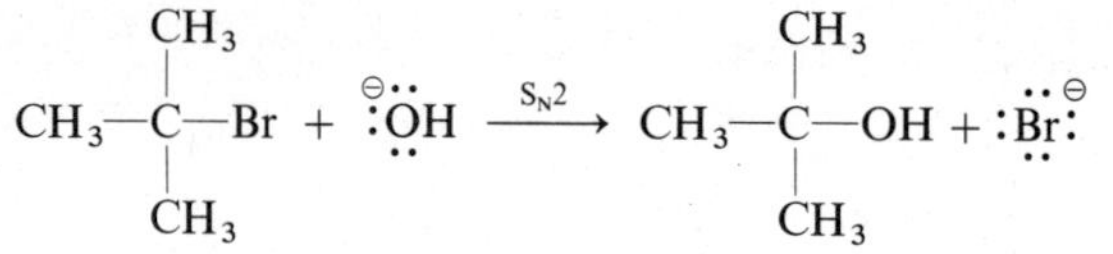

(b) + :C≡N: $\xrightarrow{S_N2}$ + :Cl:

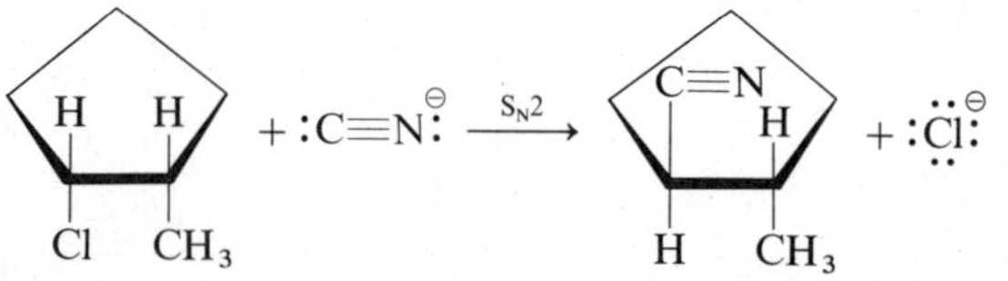

(c) $CH_3CH_2CH_2$—Br + CH_3OH $\xrightarrow{S_N1}$ $CH_3CH_2CH_2$—O—CH_3 + $H^\oplus$ + :Br:

(d) + :ÖH $\xrightarrow{S_N1}$ + :Cl:

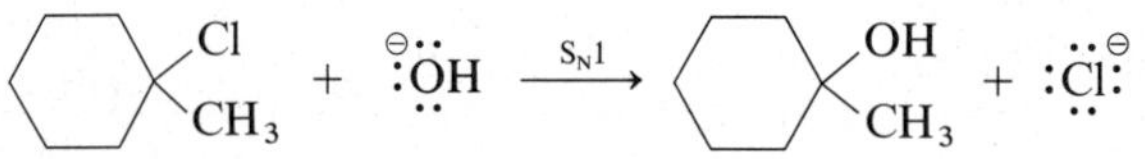

(e) $CH_3CH_2CH_2$—OH + :Cl: $\xrightarrow{S_N2}$ $CH_3CH_2CH_2$—Cl + :ÖH

9.18 Outline briefly the mechanistic *differences* between (a) S_N1 and S_N2 reactions and (b) E1 and E2 reactions.

9.19 In each of the following reactions, two possible products, *A* and *B*, are shown. Provide a complete, stepwise mechanism to show how product *A* is formed, and on the basis of your understanding of mechanisms, explain why product *A* is formed in preference to (but not necessarily to the exclusion of) product *B*.

(a) $\underset{Br}{\overset{H}{\diagdown}}C=C\underset{H}{\overset{CH_2CH_2-Br}{\diagup}}$ $\xrightarrow[\text{organic solvent}]{CH_3\ddot{O}:Na^\oplus}$

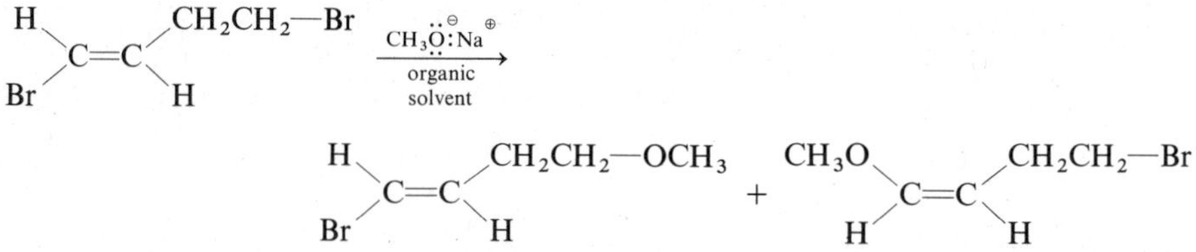

$\underset{Br}{\overset{H}{\diagdown}}C=C\underset{H}{\overset{CH_2CH_2-OCH_3}{\diagup}}$ + $\underset{H}{\overset{CH_3O}{\diagdown}}C=C\underset{H}{\overset{CH_2CH_2-Br}{\diagup}}$

AB

(b) CH_3—$\overset{\underset{\underset{\displaystyle CH_3}{|}}{||}}{C}$=CH—$CH_2$—Br $\xrightarrow[\text{heat}]{H_2O}$ CH_3—$\overset{\overset{\displaystyle OH}{|}}{\underset{\underset{\displaystyle CH_3}{|}}{C}}$—CH=$CH_2$ + CH_3—$\overset{\underset{\underset{\displaystyle CH_3}{|}}{||}}{C}$=CH—$CH_2$—OH

AB

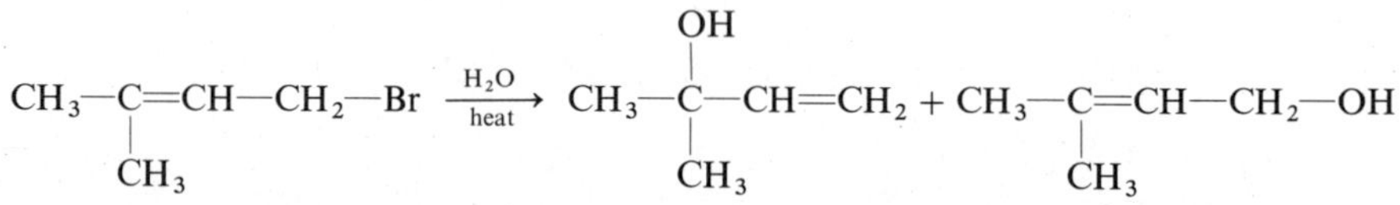

9.20 When *optically active* 2-iodooctane is allowed to stand in a solution containing acetone and sodium iodide, it gradually loses its optical activity. Why?

9.21 Provide a mechanistic interpretation for the following reactivities obtained from the hydrolysis of the alkyl chlorides under S_N1 conditions. The rates are relative to 1-chlorobutane.

$CH_3CH_2CH_2CH_2$—Cl$$$CH_3CH_2$—Ö—$CH_2$—Cl$$$CH_3$—Ö—$CH_2CH_2$—Cl

Relative rates
of reaction:1$$$1 \times 10^9$$$0.2

9.22 Of the following compounds:

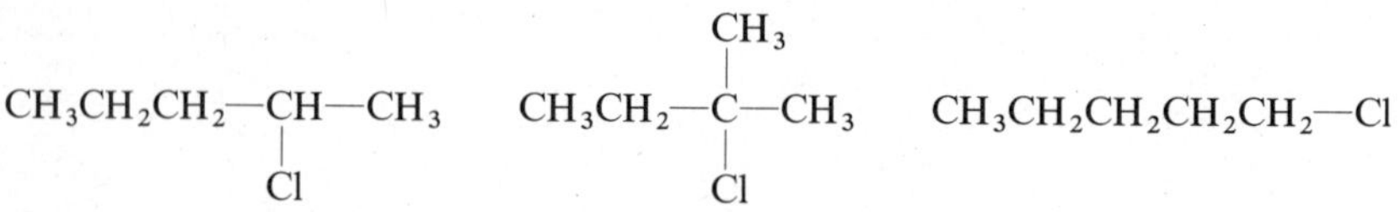

$CH_3CH_2CH_2$—$\overset{\underset{\underset{\displaystyle Cl}{|}}{|}}{C}H$—$CH_3$$$$CH_3CH_2$—$\overset{\overset{\displaystyle CH_3}{|}}{\underset{\underset{\displaystyle Cl}{|}}{C}}$—$CH_3$$$$CH_3CH_2CH_2CH_2CH_2$—Cl

ABC

$$CH_3CH_2—CH—CH_2CH_3$$
$$\underset{Cl}{|}$$

D

which one is the most reactive toward:
(*a*) S_N2 displacement   (*b*) S_N1 displacement   (*c*) E2 elimination   (*d*) E1 elimination

9.23 Consider the following compounds and answer the questions.

$$CH_3CH_2—\underset{\underset{H}{|}}{\overset{\overset{Br}{|}}{C}}—CH_3 \qquad CH_3CH_2—\underset{\underset{Br}{|}}{\overset{\overset{CH_3}{|}}{C}}—CH_3 \qquad CH_3—\underset{\underset{Br}{|}}{\overset{\overset{CH_3}{|}}{CH}}—CH—CH_3$$

A   *B*   *C*

$$Br—CH_2CH_2—\underset{\underset{}{|}}{\overset{\overset{CH_3}{|}}{CH}}—CH_3 \qquad CH_3CH_2—\underset{\underset{}{|}}{\overset{\overset{CH_3}{|}}{CH}}—CH_2—Br$$

D   *E*

(*a*) Arrange the compounds in order of *increasing* S_N2 reactivity toward $OH^\ominus$ (that is, list the least reactive compound first and the most reactive last). If you think two or more compounds are of roughly comparable reactivity, so indicate.

(*b*) Arrange the compounds in order of *increasing* S_N1 reactivity toward $OH^\ominus$ (that is, list the least reactive compound first and the most reactive last). If you think two or more compounds are of roughly comparable reactivity, so indicate.

(*c*) Which, if any, of the compounds is likely to react by *both* S_N1 and S_N2 mechanisms?

(*d*) Using one of the compounds you would expect to react almost exclusively with $:\overset{..}{\underset{..}{O}}H^\ominus$ via a S_N2 mechanism, write the complete mechanism for it. Show all steps and intermediates and indicate relative rates of reaction where important.

(*e*) Which, if any, of the compounds is *most* likely to undergo an E1 elimination reaction?

(*f*) Using one of the compounds you would expect to react almost exclusively with $OH^\ominus$ via a S_N1 mechanism, write a complete mechanism for it. Show all steps and intermediates and indicate relative rates of reaction where important.

9.24 Each of the following statements is true of one or more of the mechanisms S_N1, S_N2, E1, and E2. After each statement, indicate any and all mechanisms for which the statement is valid concerning *alkyl halides*. Use the abbreviations S_N1, S_N2, E1, and E2.

(*a*) involves carbocation intermediates   (*b*) follows first-order kinetics
(*c*) follows second-order kinetics   (*d*) reactivity of R—X: $3° > 2° > 1°$
(*e*) reactivity of R—X: $1° > 2° > 3°$ (*f*) rearrangements possible
(*g*) results in dehydrohalogenation   (*h*) substitution occurs with racemization
(*i*) substitution occurs with inversion   (*j*) involves *anti*-elimination
(*k*) occurs via a concerted mechanism   (*l*) involves backside attack
(*m*) contains five groups around carbon in the transition state
(*n*) results in racemization on substitution

9.25 Using real compounds with five or fewer carbon atoms (do *not* use R, R′, and so on), give an example of an alkyl halide that on treatment with hydroxide ion would: (*a*) give more alkene than does isopropyl iodide, and (*b*) give less alkene than does isopropyl iodide.

9.26 The reaction between *optically active* 2-bromobutane and silver nitrate is a S_N1 reaction, from which the substitution product is 2-butyl nitrate,

$$CH_3—CH—CH_2CH_3$$
$$\underset{ONO_2}{|}$$

When the silver nitrate is replaced by sodium iodide, the substitution product is 2-iodobutane, which is formed by a S_N2 reaction. Using three-dimensional structures, provide complete, stepwise mechanisms to illustrate these reactions; pay special attention to the stereochemistry of the product. (*Hint:* One resonance structure for the nitrate ion is the following.)

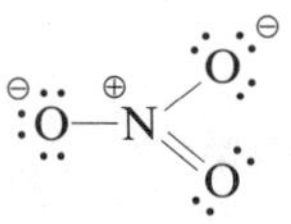

$$\overset{\ominus}{:}\overset{..}{O}—\overset{\oplus}{N}\overset{:\overset{..}{O}:^{\ominus}}{\underset{:\overset{..}{O}\cdot}{}}$$

9.27 *Glycerol* (1) is an important industrial commodity and the precursor of trinitroglycerine. It is usually prepared industrially starting with propene, as shown by the following sequence of reactions:

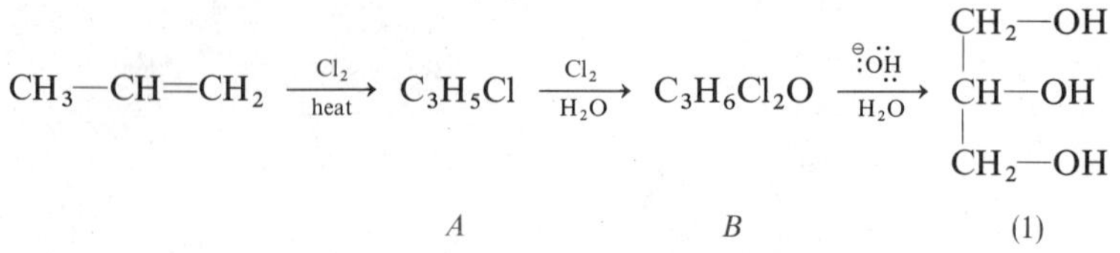

$$CH_3—CH=CH_2 \xrightarrow[\text{heat}]{Cl_2} C_3H_5Cl \xrightarrow[H_2O]{Cl_2} C_3H_6Cl_2O \xrightarrow[H_2O]{\ominus:\overset{..}{O}\overset{..}{H}} \begin{matrix} CH_2—OH \\ | \\ CH—OH \\ | \\ CH_2—OH \end{matrix}$$

$$\qquad\qquad\qquad\qquad\qquad A \qquad\qquad\qquad B \qquad\qquad\qquad (1)$$

Trace the above reactions by drawing the structures of *A* and *B*. What type of reaction is involved in each step of this process?

9.28 Consider the reaction between isopropyl iodide, $(CH_3)_2CH—I$, and hydroxide ion in aqueous solution, from which isopropyl alcohol, $(CH_3)_2CH—OH$, is the major product. Some propene is also formed.

(*a*) Write a mechanism for the production of isopropyl alcohol, *assuming* that the reaction is completely S_N2. Show all steps as well as intermediates.

(*b*) Write a mechanism for the production of isopropyl alcohol, *assuming* that the reaction is completely S_N1. Show all steps as well as intermediates.

(*c*) Write one possible mechanism to account for the formation of the propene as a side product. Show all steps and intermediates. Is the mechanism you have written for an E1 or an E2 reaction?

9.29 Consider the reaction between *optically active* 2-bromobutane and hydroxide ion to give 2-butanol, $CH_3—CHOH—CH_2CH_3$.

(*a*) When this reaction is carried out under one set of conditions, the 2-butanol is *optically active*. Provide a complete mechanism to explain this result. What, if anything, can be said about the relationship between the configuration of the starting alkyl halide and the product?

(*b*) When this reaction is carried out under different conditions, the 2-butanol is *optically inactive*. Provide a complete mechanism to explain this result.

9.30 On treatment with aqueous sodium hydroxide solution, *trans*-4-bromocyclohexanol gives only *cis*-1,4-cyclohexanediol (2). On the other hand, *trans*-4-bromo-4-methylcyclohexanol (3), on treatment with aqueous sodium hydroxide solution, gives a mixture of *cis*- and *trans*-1-methyl-1,4-cyclohexanediol, *cis* (4) and *trans* (4). Provide reasonable, stepwise mechanisms to explain these results. (The structures of the compounds are shown here and the results are summarized in equation form.)

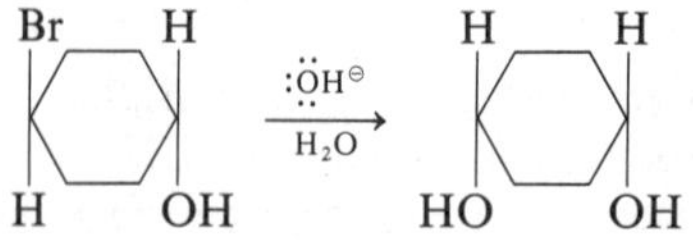

(1)   (2)
Only product

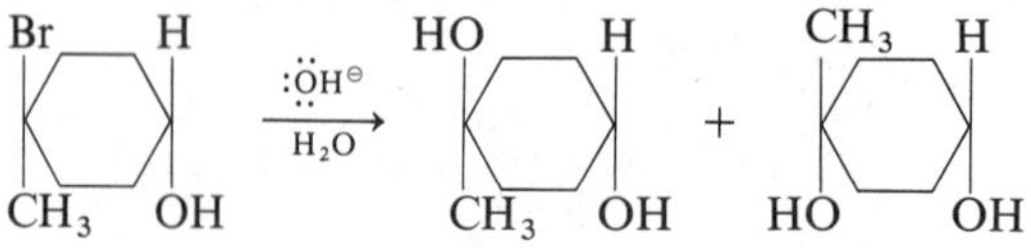

(3)   *trans* (4)   *cis* (4)

9.31 When *optically pure* 2-chlorobutane (1) is allowed to react with sodium methoxide in methanol under certain conditions, the following products are obtained:

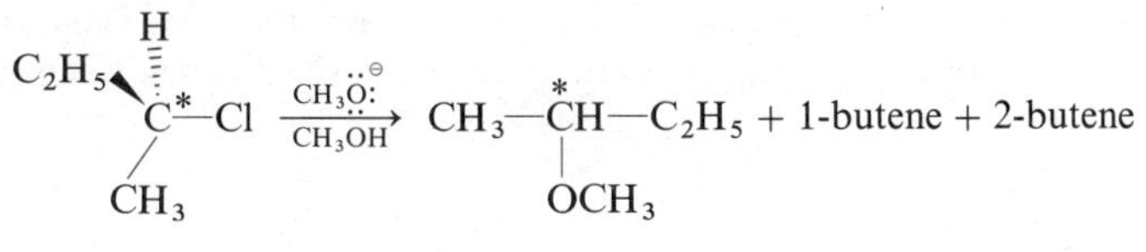

$$\underset{\substack{| \\ CH_3 \\ (1)}}{\overset{\substack{H \\ | \\ C_2H_5 \diagdown \\ C\overset{*}{}-Cl}}{}} \xrightarrow[CH_3OH]{CH_3\ddot{O}:^{\ominus}} \underset{\substack{| \\ OCH_3 \\ (2)}}{CH_3-\overset{*}{C}H-C_2H_5} + \text{1-butene} + \text{2-butene}$$

On isolation, ether (2) was found to be *partially optically active* and of *inverted configuration* [inverted relative to the configuration of (1)]. (*Hint:* The expression "partially optically active" means that some of the ether is racemic.)

(*a*) Provide a mechanistic interpretation of the reactions leading to the formation of ether (2). Use three-dimensional structures throughout. Show all steps and all intermediates or transition states, and be sure that your mechanisms account for the stereochemistry. Identify your mechanisms as being S_N1 or S_N2.

(*b*) Provide a mechanism to show how 1-butene is formed from (1) via an E2 mechanism. Show all steps and intermediates or transition states.

9.32 Consider the following reactions carried out on *optically active* 2-chlorobutane:

Reaction I:

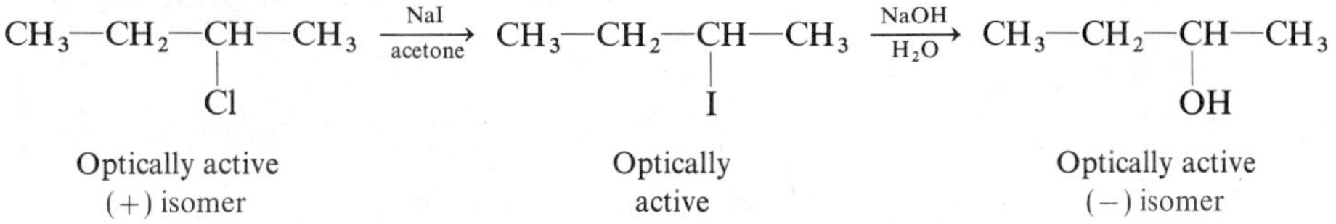

$$\underset{\substack{| \\ Cl}}{CH_3-CH_2-CH-CH_3} \xrightarrow[acetone]{NaI} \underset{\substack{| \\ I}}{CH_3-CH_2-CH-CH_3} \xrightarrow[H_2O]{NaOH} \underset{\substack{| \\ OH}}{CH_3-CH_2-CH-CH_3}$$

<table>
<tr><td>Optically active
(+) isomer</td><td>Optically
active</td><td>Optically active
(−) isomer</td></tr>
</table>

Reaction II:

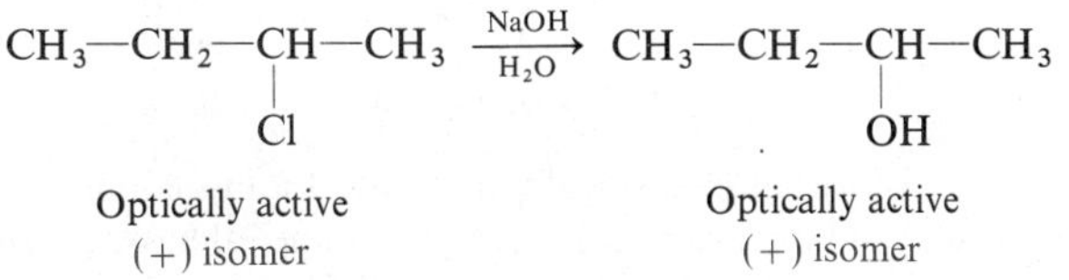

$$\underset{\substack{| \\ Cl}}{CH_3-CH_2-CH-CH_3} \xrightarrow[H_2O]{NaOH} \underset{\substack{| \\ OH}}{CH_3-CH_2-CH-CH_3}$$

<table>
<tr><td>Optically active
(+) isomer</td><td>Optically active
(+) isomer</td></tr>
</table>

The *rotations* of the 2-butanol obtained from these two reactions are *equal in magnitude* but *opposite in sign*. Note that the signs of rotation are indicated. Give a mechanistic explanation of these results, using three-dimensional structures throughout. Show important intermediates and account for the stereochemical results.

9.33 When 2-chloro-3-methylbutane (1) is heated with dilute aqueous base, three products are formed.

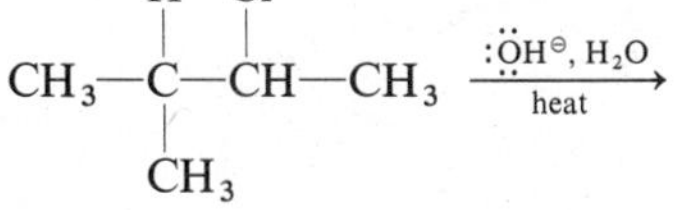

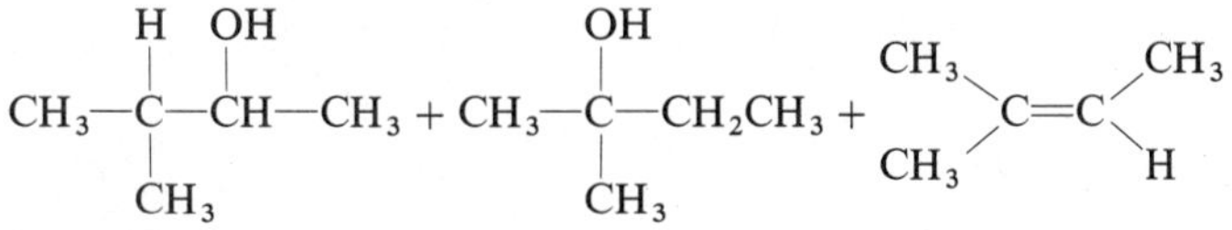

$$\underset{\substack{| \\ CH_3 \\ (1)}}{\overset{\substack{H \quad Cl \\ | \quad | \\ CH_3-C-CH-CH_3}}{}} \xrightarrow[heat]{:\ddot{O}H^{\ominus},\, H_2O}$$

$$\underset{\substack{| \\ CH_3 \\ (2)}}{\overset{\substack{H \quad OH \\ | \quad | \\ CH_3-C-CH-CH_3}}{}} + \underset{\substack{| \\ CH_3 \\ (3)}}{\overset{\substack{OH \\ | \\ CH_3-C-CH_2CH_3}}{}} + \underset{(4)}{\overset{CH_3}{\underset{CH_3}{}}\diagup \overset{}{C}=\overset{}{C}\diagdown \overset{CH_3}{\underset{H}{}}}$$

Provide the following mechanisms. Show all steps and intermediates and indicate relative rates of reaction where important. Identify each mechanism as being S_N1, S_N2, E1, or E2.

(*a*) Two different mechanisms to show how (2) can be obtained from compound (1)

(*b*) A reasonable mechanism to show how (3) can be obtained from compound (1)

(*c*) A reasonable mechanism to show how (4) can be obtained from compound (1)

9.34 *Optically active* bromide (1), with the absolute configuration shown, is allowed to react with sodium ethoxide, $CH_3CH_2\ddot{O}:^{\ominus} Na^{\oplus}$, in ethanol, CH_3CH_2OH, from which a mixture of S_N2

displacement product and E2 elimination product are obtained. Provide complete, stepwise mechanisms for these two reactions; also, show the transition states. Keep in mind the geometric requirements for elimination, and indicate the stereochemistry expected in both the S_N2 and the E2 product.

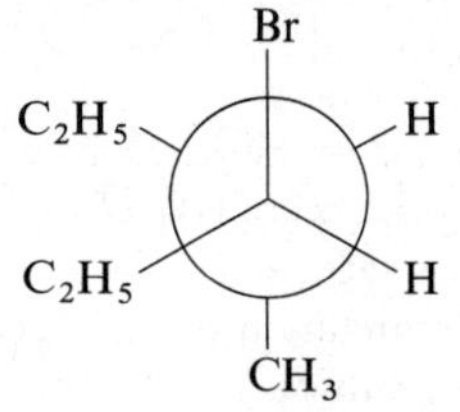

(1)

9.35 Provide mechanistic explanations for the following observation: 2-Bromooctane reacts with water three times faster than 2-chlorooctane. The products in both cases are 2-octanol and 2-octene. The alkene/alcohol ratio in the mixture of products is the *same* from both the bromo and the chloro compound, even though the alkyl halides react at different rates.

9.36 The reaction between an alkyl halide and sodium metal (*the Wurtz reaction*, see Sec. 4.2) has been suggested to occur via two steps:

$$R\text{—}X + 2Na° \longrightarrow R\!:\!Na^{\ominus}\overset{\oplus}{} + Na^{\oplus} + :\!\ddot{X}\!:^{\ominus}$$

$$R\text{—}X + R\!:\!Na^{\ominus}\overset{\oplus}{} \longrightarrow R\text{—}R + Na^{\oplus} + :\!\ddot{X}\!:^{\ominus}$$

Coupling
product

Overall reaction: $2R\text{—}X + 2Na° \longrightarrow R\text{—}R + 2Na^{\oplus} + 2:\!\ddot{X}\!:^{\ominus}$

(**a**) Interpret this reaction mechanistically in terms of a nucleophilic substitution reaction. Are there any other analogous types of reactions you studied that might support your mechanism?
(**b**) The Wurtz reaction works best with primary alkyl halides, gives some product with secondary alkyl halides, and fails completely with tertiary alkyl halides. Based on your knowledge of substitution and elimination reactions, discuss the possible reasons for the effect of alkyl group structure on this reaction. What do you think are the major side reaction(s) involved with the Wurtz reaction?

10

Alcohols

The next functional group to be studied is the **hydroxyl group,** —OH, found in two families of compounds, **alcohols** and **phenols.** This chapter presents the properties, preparation, and reactions of alcohols. In Chap. 17 phenols are discussed, and it is there that their properties and reactions are compared and contrasted with those of alcohols.

$$R—OH$$

Alcohol

The great importance of alcohols in organic chemistry will become apparent in this chapter. They are the backbone in the synthesis of aliphatic compounds because the —OH group can be converted, either directly or indirectly, to many other functional groups.

10.1   Structure and Nomenclature

The attachment of an —OH group to a saturated carbon atom results in the family of compounds called *alcohols*, R—OH. The classification of alcohols depends on the structure of the alkyl group to which the —OH group is attached. For example, RCH_2—OH is a *primary* (*1°*) alcohol, R_2CH—OH is a *secondary* (*2°*) alcohol, and R_3C—OH is a *tertiary* (*3°*) alcohol. The R group may contain other functional groups, but the requirement that the —OH group be attached to a saturated carbon atom is rigid.

There may be double bond(s) in the molecule, as in *allyl alcohol,*

$$CH_2=CH_2—CH_2—OH$$

but if the —OH group is attached directly to a double bond, a *vinyl alcohol* results (vinyl alcohols are usually unstable, see Sec. 24.4).

In most instances, a carbon atom cannot bear two hydroxyl groups because this functional group, called a ***gem*-diol** (Latin *geminus*, twin; see Sec. 21.9), is unstable and loses water to give either an aldehyde or a ketone.

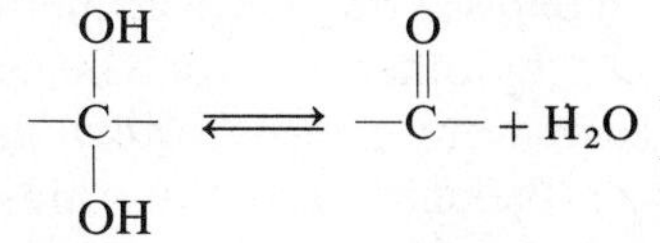

$$
\begin{array}{ccc}
\overset{\displaystyle OH}{\underset{\displaystyle OH}{-\overset{|}{\underset{|}{C}}-}} & \rightleftharpoons & \overset{\displaystyle O}{-\overset{\|}{C}-} + H_2O
\end{array}
$$

gem-diol   Aldehyde or
Unstable   ketone

Conversely, water cannot be added to most aldehydes and ketones. As we will learn, there must be special structural features in a *gem*-diol for it to be stable and isolable, as for example in chloral hydrate, CCl_3—$CH(OH)_2$. In Chap. 17 we see that aromatic rings containing the —OH group (that is, phenols) are stable even though the —OH is attached to an sp^2 carbon atom.

It is possible to have two hydroxyl groups (or more) in a molecule, provided they are not on the same carbon. These dihydroxy compounds are called **diols.** 1,2-Diols, for example, HO—CH_2CH_2—OH, are also known as *vic*-diols or vicinal diols (Latin *vicinus*, near; see Sec. 19.1). *Vic*-diols are also called **glycols** (see Sec. 10.7D).

Three naming systems are used for alcohols: Sec. 10.1A, common names, Sec. 10.1B, carbinol names, and Sec. 10.1C, IUPAC (systematic) names.

A. Common Names

The use of common names is limited to fairly simple alcohols. It involves the name of the alkyl group followed by the word *alcohol*, and thus is limited by the availability of common names for alkyl groups. For example:

CH_3—OH CH_3CH_2—OH

| Methyl alcohol | Ethyl alcohol | Isopropyl alcohol | *tert*-Butyl alcohol | Cyclohexyl alcohol |

$$CH_3(CH_2)_7OH$$

| Isopentyl alcohol | *n*-Octyl alcohol | β-Bromoethyl alcohol |

An older common name for a five-carbon alkyl group is *amyl*, which is often used in place of *pentyl*. For example:

$$CH_3CH_2CH_2CH_2CH_2\text{—OH} \qquad CH_3CH_2\text{—}\underset{\underset{OH}{|}}{\overset{\overset{CH_3}{|}}{C}}\text{—}CH_3$$

| *n*-Amyl alcohol or *n*-pentyl alcohol | *tert*-Amyl alcohol or *tert*-pentyl alcohol |

B. Carbinol Names

An older system for naming alcohols is the carbinol system, and because these names are still used occasionally, some passing familiarity with them is warranted. This system considers alcohols as derived from methyl alcohol in which the hydrogens on CH_3OH are replaced by other groups. The substituents, except for hydrogen, bonded to the C—OH unit are named as alkyl groups and followed by the suffix *-carbinol*. The names of the substituents are most often arranged in alphabetical order (not considering *di*, *tri*, and so on), and the complete name appears as one word with all substituents and the suffix run together. Two examples are:

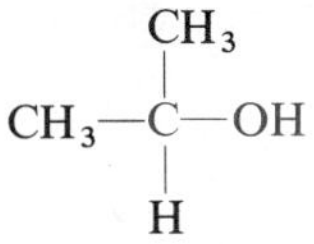

$$CH_3CH_2CH_2-OH$$

Dimethylcarbinol 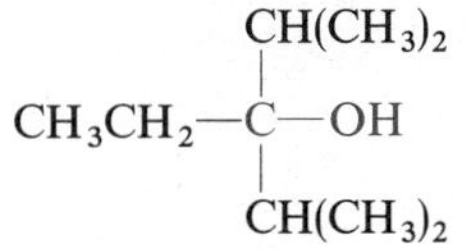   Ethyldiisopropylcarbinol

C. IUPAC (Systematic) Names

The IUPAC method of nomenclature is the most generally useful and versatile for naming alcohols. The general approach for using it is as follows:

1. The longest continuous chain of carbon atoms to which the —OH group is bonded is chosen as the parent compound. The alkane name corresponding to the longest chain of carbon atoms is the parent name of the alcohol, and the -*e* ending is replaced by **-ol**. Thus, the characteristic ending for an alcohol is -*ol* as in the general name *alkanol*.

2. The longest chain is numbered so that the —OH group is attached to the carbon bearing the lowest possible number. The —OH group assumes *priority* in numbering over double and triple bonds.

3. The position of the —OH group usually precedes the alkanol name, unless double or triple bonds are present, and then their position must also be indicated (see the following examples). The positions of the remaining substituents are indicated by number in the usual fashion.

The following six examples illustrate alcohol nomenclature by these IUPAC rules:

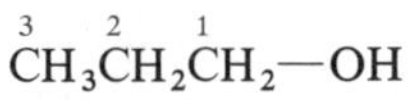

1-Propanol 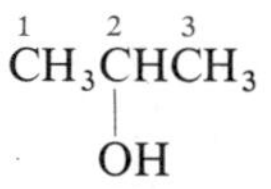   2-Propanol 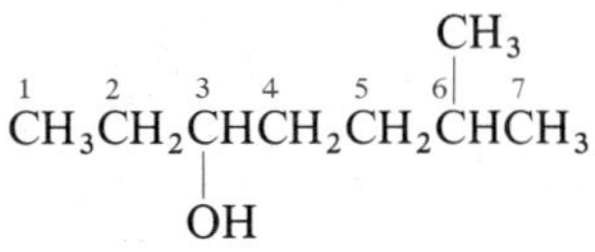   6-Methyl-3-heptanol

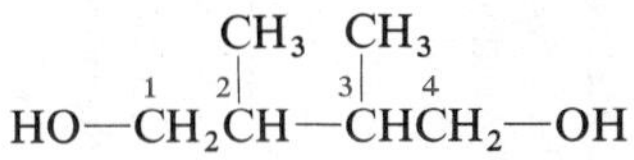

2,3-Dimethyl-1,4-butanediol 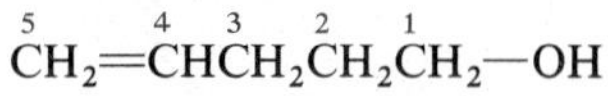   4-Penten-1-ol

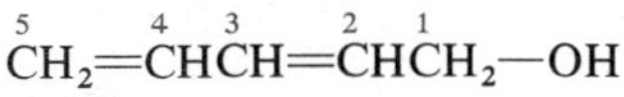

2,4-Pentadien-1-ol

Question 10.1

Draw the structures of the isomeric alcohols that have the following molecular formulas and give the IUPAC name for each:

(*a*) $C_4H_{10}O$ (provide two other acceptable names for each compound)   (*b*) $C_5H_{12}O$
(*c*) $C_6H_{14}O$

TABLE 10.1 Physical Properties of Selected Alcohols

Common Name	IUPAC Name	Formula	mp, °C	bp, °C	Density at 20°, g/cc	Solubility, g/100 g H$_2$O
Methyl alcohol	Methanol	CH$_3$OH	−97	64.7	0.792	∞
Ethyl alcohol	Ethanol	CH$_3$CH$_2$OH	−114	78.3	0.789	∞
n-Propyl alcohol	1-Propanol	CH$_3$CH$_2$CH$_2$OH	−126	97.2	0.804	∞
n-Butyl alcohol	1-Butanol	CH$_3$(CH$_2$)$_2$CH$_2$OH	−90	117.7	0.810	7.8
n-Pentyl alcohol	1-Pentanol	CH$_3$(CH$_2$)$_3$CH$_2$OH	−78.5	138	0.817	2.3
n-Hexyl alcohol	1-Hexanol	CH$_3$(CH$_2$)$_4$CH$_2$OH	−52	155.8	0.820	0.6
n-Heptyl alcohol	1-Heptanol	CH$_3$(CH$_2$)$_5$CH$_2$OH	−34	176	0.822	0.2
n-Octyl alcohol	1-Octanol	CH$_3$(CH$_2$)$_6$CH$_2$OH	−16	194	0.827	0.053
n-Decyl alcohol	1-Decanol	CH$_3$(CH$_2$)$_8$CH$_2$OH	6	233	0.829	
Isopropyl alcohol	2-Propanol	(CH$_3$)$_2$CHOH	−88.5	82.3	0.786	∞
Isobutyl alcohol	2-Methyl-1-propanol	(CH$_3$)$_2$CHCH$_2$OH	−108	107.9	0.802	10
sec-Butyl alcohol	2-Butanol	CH$_3$CH$_2$CHOHCH$_3$	−114	99.5	0.808	12.7
tert-Butyl alcohol	2-Methyl-2-propanol	(CH$_3$)$_3$COH	25	82.5	0.789	∞
Isopentyl alcohol	3-Methyl-1-butanol	(CH$_3$)$_2$CHCH$_2$CH$_2$OH	−117	131.5	0.812	3
tert-Pentyl alcohol	2-Methyl-2-butanol	(CH$_3$)$_2$COHCH$_2$CH$_3$	−12	101.8	0.809	12.9
Neopentyl alcohol	2,2-Dimethyl-1-propanol	(CH$_3$)$_3$CCH$_2$OH	52	113	0.812	
Cyclopentyl alcohol	Cyclopentanol	cyclo-C$_5$H$_9$OH		140	0.949	
Cyclohexyl alcohol	Cyclohexanol	cyclo-C$_6$H$_{11}$OH	−24	161.5	0.962	
Allyl alcohol	2-Propen-1-ol	CH$_2$=CH—CH$_2$OH	−129	97	0.855	∞
Benzyl alcohol	Phenylmethanol	C$_6$H$_5$CH$_2$OH	−15.3	205.4	1.046	4
Diphenylmethyl alcohol (diphenylcarbinol)	Diphenylmethanol	(C$_6$H$_5$)$_2$CHOH	69	298		0.05
Triphenylmethyl alcohol (triphenylcarbinol)	Triphenylmethanol	(C$_6$H$_5$)$_3$COH	162.5			

10.2   Physical Properties of Alcohols

Alcohols are our first example of a family of compounds that possess considerable polarity because of the presence of the $—O^{\delta\ominus}—H^{\delta\oplus}$ bond. This polarity is reflected in their physical properties: boiling point, melting point, and solubility in other solvents. Recall that hydrocarbons (alkanes, alkenes, alkynes, cycloalkanes) are essentially nonpolar and exhibit very low solubility in polar solvents such as water. Alkyl halides and certain geometric isomers of the alkenes have dipole moments, yet these compounds show little to no solubility in polar solvents, and they are all soluble in nonpolar solvents. Alcohols exhibit solubility characteristics radically different from any other family we have studied.

Table 10.1 lists several alcohols and their physical properties. The *n*-alcohols show a rather constant increase in boiling point as the number of carbon atoms increases; like alkanes, the boiling point increases by $\sim 20°$ for each added $—CH_2—$ group in the homologous series. Carbon chain branching causes a decrease in boiling point; for example, *n*-butyl alcohol boils at 118°, isobutyl alcohol at 108°, and *tert*-butyl alcohol at 83°. These are the same trends observed with alkanes.

Comparison of the boiling points of alcohols with those of the alkanes from which they may be derived or with those of alkanes of similar molecular weights indicates that the alcohols boil at much higher temperatures. For example, *n*-butyl alcohol (mol wt 74) boils at 118°, whereas *n*-pentane (mol wt 72) boils at 36°. This phenomenon is similar to that for water, ammonia, and hydrogen fluoride, which all have boiling points considerably higher than would be expected for compounds with similar molecular weights. Table 10.2 shows the boiling points of various families of inorganic compounds as a function of molecular weight.

The importance of **intermolecular hydrogen bonding** is seen in the alcohols, as well as in water, ammonia, and hydrogen fluoride. Alcohol molecules tend to associate in the liquid phase in the following manner:

Hydrogen bonding in alcohols (or water when R=H)

Hydrogen bonding is represented by dashed lines; it is a very weak type of bonding between the hydrogen of one alcohol molecule and the unshared pair of electrons on the oxygen atom in another. Hydrogen bond energies are estimated to be in the range of 4 to 8 kcal/mole (16 to 33 kJ/mole), and these values should be compared with 50 to 100 kcal/mole (200 to 400 kJ/mole) for most covalent bonds. As a result, extra energy is required to overcome the hydrogen bonding to change an alcohol from the liquid to the gaseous phase. This energy is supplied thermally, as evidenced by the increased boiling points of alcohols (and other molecules with hydrogen bonding) over compounds in which no hydrogen bonding is possible. The hydrogen bonding interaction in ammonia and hydrogen fluoride may be depicted as follows:

TABLE 10.2   Boiling Points and Molecular Weights of Simple Molecules

Compound and Formula	Mol Wt	bp, °C	Comment
Column IV compounds:			
Methane, CH_4	16	-161.5	No hydrogen bonding
Silane, SiH_4	32	-111.8	
Column V compounds:			
Ammonia, NH_3	17	-33.4	Abnormally high due to hydrogen bonding
Phosphine, PH_3	34	-87.7	
Column VI compounds:			
Water, H_2O	18	100.0	Abnormally high due to hydrogen bonding
Hydrogen sulfide, H_2S	34	-85.5	
Hydrogen selenide, H_2Se	81	-41.5	
Hydrogen telluride, H_2Te	129.6	-2.2	
Column VII compounds:			
Hydrogen fluoride, HF	20	19.5	Abnormally high due to hydrogen bonding
Hydrogen chloride, HCl	36.5	-84.9	
Hydrogen bromide, HBr	80.9	-67.0	
Hydrogen iodide, HI	127.9	-35.4	

Hydrogen bonding is important between molecules that contain fluorine (F), nitrogen (N), and oxygen (O); at least one hydrogen must be bonded to one of these electronegative atoms. The molecular attraction occurs between the partially positive hydrogen and the unshared pair(s) of electrons on F, N, or O. As the trends in Table 10.2 indicate, this effect drops off very rapidly as one goes down a given column in the periodic table, that is, F to Cl to Br to I, and N to P, and O to S to Se to Te. Indeed, hydrogen bonding occurs with oxygen, nitrogen, and fluorine because these elements are highly electronegative and their negative charge is highly concentrated because they are so small.

There are interesting trends in the solubilities of alcohols in various solvents. The lower molecular weight alcohols (three carbons or fewer) are infinitely soluble in water. As Table 10.1 shows, the water solubility of alcohols decreases with increasing molecular weight, largely because increasing carbon content means that the alcohol has more hydrocarbon-like character and thus properties more characteristic of the hydrocarbon alkyl group than of the $-OH$ group. The water solubility of alcohols is attributed to hydrogen bonding between ROH and HOH, which is reasonable because of their similar structures.

Conversely, low molecular weight alcohols are not very soluble in hydrocarbon solvents. For example, methanol is only partially soluble in *n*-octane, whereas the higher alcohols are completely miscible in it because of their increased hydrocarbon content.

Some indication of the relative polarities of alcohols can be derived from their dielectric constants (ε) and dipole moments. For example, water is one of the most polar protic solvents known ($\varepsilon = 81$, see Table 5.2); methanol, ethanol, and 1-butanol are arranged in decreasing order of polarity, with dielectric constants of 32, 26, and 8, respectively (see Table 5.2).

Alcohols comply with the old adage "like dissolves like," and yet, as we saw, this property is limited by certain structural features, most of which involve the carbon content of an alcohol.

Question 10.2

Explain the following.

(a) There are two isomeric compounds with the molecular formula C_2H_6O: CH_3CH_2—OH and CH_3—O—CH_3. The former boils at 78° and the latter at $-24°$.

(b) *Dimethylamine*, $(CH_3)_2NH$, has a molecular weight of 45 and boils at 7.4°. *Trimethylamine*, $(CH_3)_3N$, on the other hand, has a higher molecular weight (59) and yet boils at a lower temperature (3.5°).

Question 10.3

It has been suggested that chloroform-methanol mixtures involve hydrogen bonding. Explain how this hydrogen bonding might occur. Do you think it would be greater than, the same as, or less than what might be observed in fluoroform (CHF_3)-methanol mixtures? In iodoform (CHI_3)-methanol mixtures? Why?

10.3 Methods of Preparation of Alcohols: Survey

We studied several important methods used for preparing alcohols; these are summarized here:

1. *Acid-catalyzed hydration of alkenes*

$$\text{>C=C<} + H_2O \xrightarrow{H^\oplus} \underset{\substack{| \quad | \\ H \quad OH}}{-\text{C}-\text{C}-} \qquad \text{Sec. 8.5}$$

Characteristics: Obeys Markovnikov's rule of addition.

Rearrangements of carbon skeleton may result.

Gives secondary and tertiary alcohols, except for the hydration of ethene, which gives ethanol (a primary alcohol).

2. *Hydrolysis of alkyl halides*

Primary: $RCH_2X \xrightarrow[H_2O]{:\overset{..}{O}H^\ominus} RCH_2OH$ S_N2 reaction, Sec. 9.15

Secondary: $R_2CHX \xrightarrow[H_2O]{:\overset{..}{O}H^\ominus} R_2CHOH$ S_N1 or S_N2 reaction, Sec. 9.13

Tertiary: $R_3CX \xrightarrow{H_2O} R_3COH$ S_N1 reaction, Sec. 9.16

Characteristics: Primary and secondary alkyl halides usually require hydroxide, whereas tertiary do not.

Major side reaction is elimination (alkene formation), especially with 3° alkyl halides.

3. *Addition of hypohalous acid to alkenes*

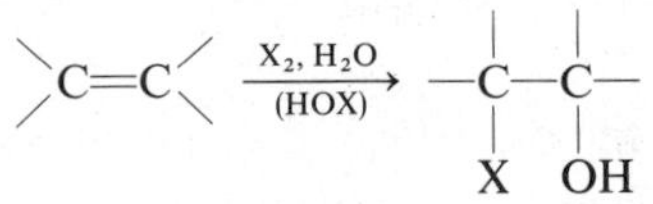

$$\text{C}{=}\text{C} \xrightarrow[\text{(HOX)}]{X_2,\ H_2O} \quad -\overset{\displaystyle |}{\underset{\displaystyle X}{\text{C}}}-\overset{\displaystyle |}{\underset{\displaystyle OH}{\text{C}}}- \qquad \text{Sec. 8.22}$$

Characteristics: Obeys Markovnikov's rule of addition.

Produces only secondary or tertiary alcohols, except for addition to ethene, which gives $X-CH_2CH_2-OH$ (a primary alcohol).

In Sec. 10.4 we study another method for obtaining alcohols from alkenes: *hydroboration.* However, two of the most important general methods for the preparation of alcohols are explained in later chapters:

4. Reaction between Grignard reagents and aldehydes or ketones (see Sec. 21.10) and esters (see Sec. 23.7)
5. Reduction of aldehydes or ketones (see Sec. 21.13), esters (see Sec. 23.7), and carboxylic acids (see Sec. 22.16)

10.4   Hydroboration: Addition of Boron Hydrides to Double and Triple Bonds

As a prelude to another method for obtaining alcohols, we consider the addition of the inorganic reagent *borane* (BH_3), a boron hydride, to double and triple bonds. More specifically, the addition reactions involve *diborane*, $(BH_3)_2$, which is the dimer of BH_3. There is considerable evidence concerning the structures of borane and diborane, which are shown here. The monomer exists in tetrahydrofuran, in which a Lewis acid-Lewis base complex is formed between these two molecules; tetrahydrofuran is the Lewis base and donates a pair of electrons to boron, which is the Lewis acid because it accepts that electron pair.

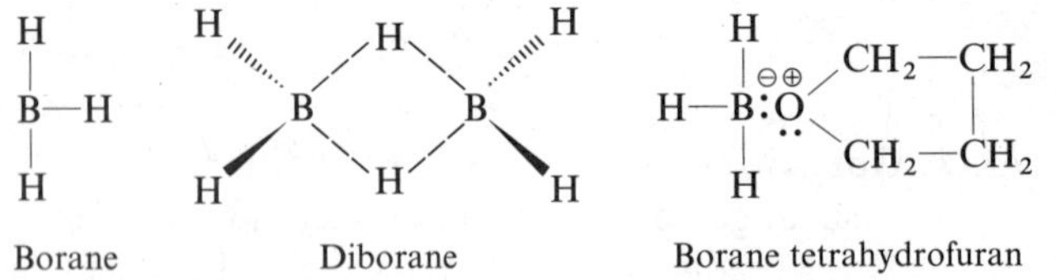

Borane   Diborane   Borane tetrahydrofuran
                                           A molecular complex

The structure of diborane is interesting because it resembles ethene in many respects. The two boron-hydrogen-boron bridges that hold the two BH_3 molecules together are somewhat analogous to the π cloud of the double bond. As shown in the structures, the B—H—B bridges are in the plane of the page, and each boron molecule contains two additional hydrogens that protrude in front of and behind the plane of the page. The π bond is superimposed on the σ bond in the carbon-carbon double bond, which when taken together is a very strong bond. In contrast, the hydrogen bridges between the two BH_3 molecules are quite weak and easily broken, as we will see later in this section when we discuss the mechanism of BH_3 addition to double and triple bonds. Boranes are a versatile class of reagents but must be treated with care. Borane itself is toxic; it is flammable in air and reacts explosively with water.

When allowed to react with alkenes, diborane gives *addition;* the overall reaction with ethene is:

$$6 \ CH_2{=}CH_2 + (BH_3)_2 \longrightarrow 2 \ (CH_3CH_2)_3B$$

Ethene   Diborane   Triethylboron

This reaction is believed to occur in three steps, where the intermediate addition products continue to react with alkene as long as a hydrogen is attached to boron:

$$CH_2{=}CH_2 + BH_3 \longrightarrow CH_3CH_2{-}BH_2 \xrightarrow{CH_2=CH_2} (CH_3CH_2)_2B{-}H$$

$$\downarrow {\scriptstyle CH_2=CH_2}$$

$$(CH_3CH_2)_3B$$

Triethylboron

This reaction is called **hydroboration,** which involves the addition of hydrogen and boron across the carbon-carbon double bond.

Diborane is usually prepared by the reaction of sodium borohydride, $NaBH_4$, with boron trifluoride, BF_3. The common ether solvents used are tetrahydrofuran and diglyme:

$$\begin{array}{c} CH_2{-}CH_2 \\ | \quad\quad | \\ CH_2 \quad CH_2 \\ \diagdown \quad \diagup \\ O \end{array}$$

Tetrahydrofuran

$$CH_3{-}O{-}CH_2CH_2{-}O{-}CH_2CH_2{-}O{-}CH_3$$

Diglyme

$$3 \ NaBH_4 + 4 \ BF_3 \xrightarrow[\text{solvent}]{\text{ether}} 2 \ (BH_3)_2 + 3 \ NaBF_4$$

Boron trifluoride is a gas, but when dissolved in diethyl ether it forms a molecular complex $(CH_3CH_2)_2O^{\oplus}{-}^{\ominus}BF_3$, which is a solution that is easy to handle in the laboratory. Because of the high reactivity of diborane, hydroboration reactions are usually carried out by mixing together sodium borohydride, boron trifluoride-ether solution (called *boron trifluoride etherate*), and the alkene. The boron addition compounds themselves are seldom isolated but are allowed to react further without purification; we study some important reactions of the alkylboron compounds in Sec. 10.5.

Much is known about the mechanism of the diborane addition reaction. For example, the **orientation of addition** becomes clear when diborane is allowed to react with an unsymmetrical alkene:

$$6 \ CH_3CH_2{-}CH{=}CH_2 + (BH_3)_2 \longrightarrow 2 \ (CH_3CH_2CH_2CH_2)_3B$$

1-Butene   Diborane   Tri-*n*-butylboron

This orientation of addition has been explained in the following way. The monomeric borane is presumably the reactive species formed by the reaction between diborane and the ether solvent, and it is probably solvated by ether:

$$(BH_3)_2 + 2 \ R_2\overset{..}{O}: \rightleftharpoons 2 \ R_2\overset{..}{\underset{\oplus}{O}}{:}\underset{\ominus}{BH_3}$$

Diborane Ether   Borane-ether
(solvent)   complex

In a simplified manner, addition is viewed as involving the electrophilic attack of the electron-deficient boron atom in BH_3 on the π electrons of the carbon-carbon double bond. At the same time, one hydrogen on BH_3 is transferred to the other carbon atom of the double bond. As we will see in Sec. 10.5, stereochemical results indicate that this addition reaction is *concerted*, and the carbon-boron and carbon-hydrogen bonds are made at the same time the carbon-carbon double bond is broken. A four-centered transition state is involved, and one of the carbon atoms that originally constituted the double bond develops *carbocation character*. This is illustrated for 1-butene:

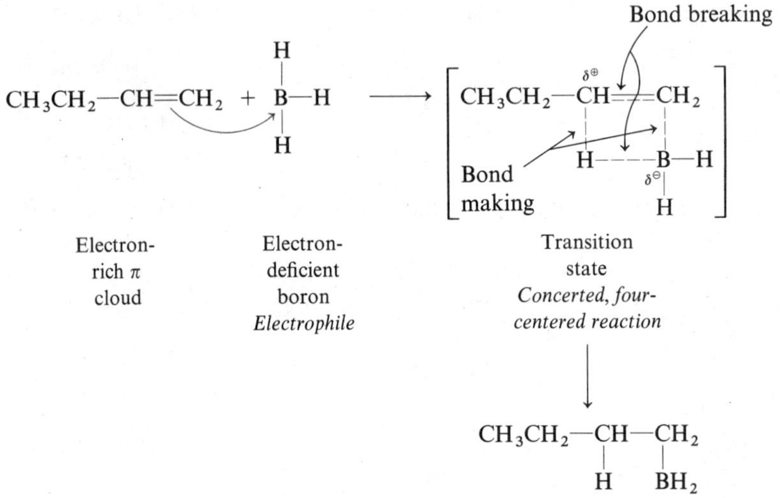

The intermediate then goes on to react with 2 more moles of 1-butene to give tri-*n*-butylboron, $(CH_3CH_2CH_2CH_2)_3B$.

The key to understanding the orientation of addition is the structure of the transition state that possesses carbocation character. There are two possible ways borane can add to 1-butene, so we should look at the structures of the possible addition products:

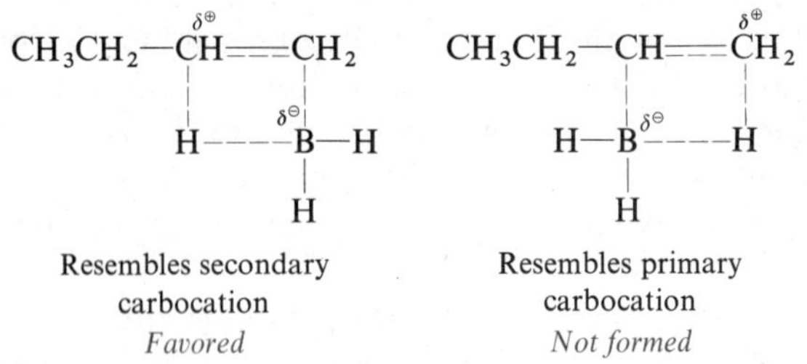

Our knowledge about carbocation stability allows us to predict that addition of BH_3 favors the formation of the transition state that possesses secondary and not primary carbocation character.

The addition of BH_3 to alkenes greatly resembles the addition of unsymmetrical acidic reagents (such as H_2O and H—X), with the major difference being the details of the addition. Acidic reagents add in *two steps*, the first being attack of the electrophile on the double bond and the second being the addition of the nucleophile to the resulting carbocation. On the other hand, *BH_3 addition occurs in one step*—a concerted reaction—in which boron is the electrophile and the nucleophile is the hydride ion, H:$^{\ominus}$. Here, however, the transfer is *internal*, because the B—H bond is broken and hydrogen is transferred to the partially positive charged carbon atom. The following two equations show these analogies:

Electrophilic addition of acidic reagents, H—Nu:

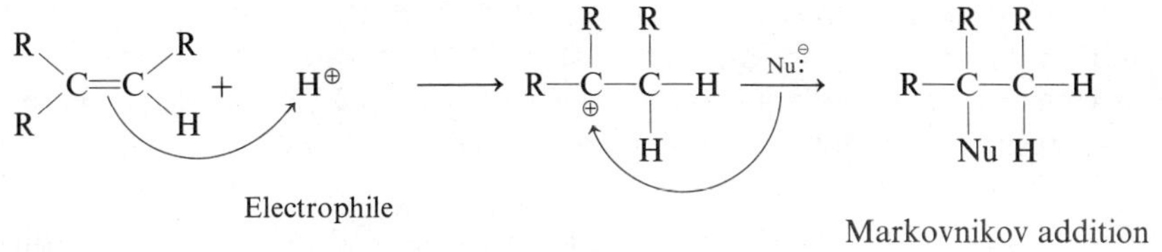

Electrophilic addition of borane, BH_3:

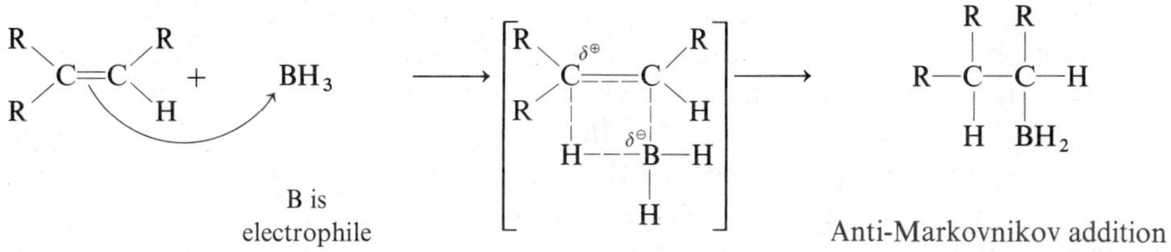

The rule of thumb regarding electrophilic addition of acidic reagents to alkenes is Markovnikov's rule (see Sec. 8.8), which states that the hydrogen goes to the carbon bearing the greater number of hydrogens originally. The addition of borane to alkenes is an example of anti-Markovnikov addition, where the hydrogen on boron becomes attached to the carbon bearing the smaller number of hydrogens originally.

In addition to the stereochemical evidence supporting the mechanism of borane addition to alkenes (see Sec. 10.5), it is interesting that, under the normal conditions used in hydroboration, *borane addition occurs without any molecular rearrangement*. Even though the transition state involves carbocation character, internal hydride transfer from boron to partially positive carbon occurs before there can be any rearrangement. For example, the addition of BH_3 to 3-methyl-1-butene gives only one unrearranged product, whereas the addition of HCl (ionic conditions) to the same alkene results in extensive rearrangement (see Sec. 8.4).

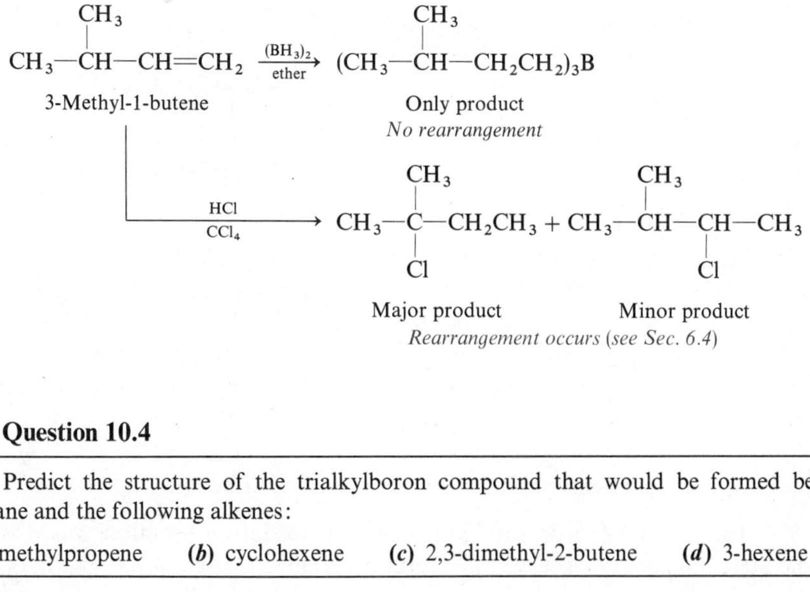

Question 10.4

Predict the structure of the trialkylboron compound that would be formed between diborane and the following alkenes:

(*a*) 2-methylpropene   (*b*) cyclohexene   (*c*) 2,3-dimethyl-2-butene   (*d*) 3-hexene

10.5   Hydroboration: Synthetic Route to Alcohols; Stereochemistry of Addition

This section presents some important synthetic applications of the borane addition reaction in organic chemistry. Hydroboration of alkenes, followed by treatment of the trialkylboron adduct with hydrogen peroxide in the presence of base, results in the formation of alcohols. For example:

$$CH_2{=}CH_2 + (BH_3)_2 \longrightarrow (CH_3CH_2)_3B \xrightarrow[:\ddot{O}H^{\ominus}]{H_2O_2} CH_3CH_2{-}OH + B(OH)_3$$

   Ethene   Diborane   Triethylboron   Ethanol   Boric acid

This sequence is referred to as **hydroboration-oxidation** and is a two-step reaction. In reality, the hydroboration step is carried out in the manner described in Sec. 10.4; the intermediate trialkylboron is not isolated or purified but is treated directly with hydrogen peroxide and base to produce ethyl alcohol and boric acid.

The main driving force for this cleavage reaction is the affinity of boron for electrons and thus for Lewis bases. Boron forms a stronger bond with oxygen than with carbon. The nonbonding orbitals on oxygen overlap with the vacant p orbital on boron, thereby satisfying to some extent boron's desire for electrons.

Three other examples of the usefulness of hydroboration-oxidation follow; in each case, the *net* result is the addition of the elements of water, H—OH, across a carbon-carbon double bond, but in the anti-Markovnikov sense.

$$CH_3CH_2{-}CH{=}CH_2 \xrightarrow[\text{ether}]{(BH_3)_2} \xrightarrow[:\ddot{O}H^{\ominus}]{H_2O_2} CH_3CH_2CH_2CH_2{-}OH$$

   1-Butene   1-Butanol
                                                              1°

      Cyclohexylethene   2-Cyclohexylethanol
                                                              1°

$$\underset{\text{2-Methyl-2-butene}}{CH_3{-}\overset{\overset{\displaystyle CH_3}{|}}{C}{=}CH{-}CH_3} \xrightarrow[\text{ether}]{(BH_3)_2} \xrightarrow[:\ddot{O}H^{\ominus}]{H_2O_2} \underset{\text{3-Methyl-2-butanol}}{CH_3{-}\overset{\overset{\displaystyle CH_3}{|}}{CH}{-}\underset{\underset{\displaystyle OH}{|}}{CH}{-}CH_3}$$

                                                              2°

This method provides mostly primary and secondary alcohols, whereas the acid-catalyzed hydration of alkenes yields mostly secondary and tertiary alcohols. Accordingly, hydroboration-oxidation is exceedingly useful for obtaining alcohols that cannot be produced by other common methods. It is not accompanied by molecular rearrangements, whereas the acid-catalyzed hydration of alkenes often is.

A. Stereochemistry of Hydroboration-Oxidation

The mechanism of the hydroboration-oxidation reaction was deduced from the stereochemistry observed for certain alkenes. For example, the hydroboration-oxidation of 1-methylcyclopentene gives only *trans*-2-methylcyclopentanol:

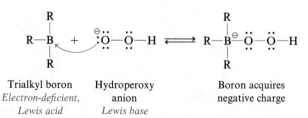

1-Methylcyclopentene   *trans*-2-Methylcyclopentanol

Because of the orientation of hydroboration and the lack of molecular rearrangement during its addition to an alkene, diborane must add in the *syn* fashion. Therefore, *the oxidation reaction must occur with retention of configuration*, as shown by the sequence:

1-Methylcylopentene

Transition state
syn Addition

Stops at
dialkylboron

trans-2-Methylcyclopentanol

To account for the stereospecific oxidation reaction, the following mechanism is proposed, where R_3B is any trialkyl boron compound.

Step 1. Reversible reaction between hydroxide ion and hydrogen peroxide

Hydrogen peroxide   Hydroperoxy anion

The equilibrium shown here is similar to that between hydroxide ion and an alcohol.

Step 2. Attack of hydroperoxy anion on trialkyl boron compound

Trialkyl boron   Hydroperoxy   Boron acquires
Electron-deficient,   anion   negative charge
Lewis acid   *Lewis base*

This reaction results in the electron-deficient boron obtaining another pair of electrons to complete its octet. The hydroperoxy anion $H—\overset{..}{\underset{..}{O}}—\overset{..}{\underset{..}{O}}:^{\ominus}$ is a Lewis base (electron-pair donor) and boron in R_3B is a Lewis acid (electron-pair acceptor).

Step 3. Migration of alkyl group from boron to oxygen with retention of configuration and loss of hydroxide ion

$$R—\overset{\overset{\displaystyle R}{|}}{\underset{\underset{\displaystyle R}{|}}{B}}{}^{\ominus}—\overset{..}{\underset{..}{O}}—\overset{..}{\underset{..}{O}}—H \longrightarrow R—\overset{\overset{\displaystyle R}{|}}{B}—\overset{..}{\underset{..}{O}}—R + :\overset{..}{\underset{..}{O}}H^{\ominus}$$

Alkyl group
migrates with
electron pair

This reaction appears to involve the migration of the alkyl group, R—, from boron to oxygen—a process that occurs with *complete retention of configuration*. The alkyl group migrates with its bonding pair of electrons in the same way alkyl groups migrate in carbocation rearrangements. It is accompanied by the displacement of hydroxide ion.

Repeating steps 2 and 3 twice more results in the formation of a trialkoxyboron compound, $(RO)_3B$, which is hydrolyzed in basic solution to give boric acid, $B(OH)_3$, and 3 moles of alcohol, ROH. This reaction probably occurs by the following route; because it does not involve rupturing the carbon-oxygen bond, the overall stereochemistry of the oxidation reaction is *net* retention of configuration.

$$RO—\overset{\overset{\displaystyle OR}{|}}{\underset{\underset{\displaystyle OR}{|}}{B}} + :\overset{..}{O}H^{\ominus} \longrightarrow RO—\overset{\overset{\displaystyle OR}{|}}{B}—\overset{..}{\underset{..}{O}}H + R\overset{..}{\underset{..}{O}}:^{\ominus}$$

Trialkoxyboron   Alkoxide ion

$$R\overset{..}{\underset{..}{O}}:^{\ominus} + H_2\overset{..}{\underset{..}{O}}: \rightleftharpoons R\overset{..}{\underset{..}{O}}H + H\overset{..}{\underset{..}{O}}:^{\ominus}$$

Product

then

$$RO—\overset{\overset{\displaystyle OR}{|}}{B}—OH \xrightarrow[\text{above steps twice}]{\text{repeat}} 2\ ROH + B(OH)_3$$

Thus, diborane adds to a carbon-carbon double bond in a *syn* fashion in a concerted, four-center reaction.

Additional evidence concerning the stereochemistry of diborane addition comes from open-chain alkenes. For example, the addition of deuterated diborane, $(BD_3)_2$, gives the following results:

trans-2-Butene   3-Deuterio-2-butanol
                                   (+ other enantiomer)

In summary, the hydroboration-oxidation reaction is useful for preparing a host of primary alcohols that cannot be obtained by other means, and it also permits the synthesis of certain stereoisomers from both cyclic and noncyclic alkenes.

Question 10.5

Compare the products obtained in the reaction between *trans*-2-butene and *cis*-2-butene with $(BD_3)_2$ followed by H_2O_2 and $H\overset{..}{\underset{..}{O}}{:}^{\ominus}$. Pay particular attention to the stereochemistry of the products.

Question 10.6

Give the products that result from hydroboration-oxidation of the following alkenes:

(*a*) 3,3-dimethyl-1-butene (*b*) methylenecyclohexane
(*c*) 2-methyl-2-hexene (*d*) 1-octene

Question 10.7

(*a*) Compare the product(s) obtained by acid-catalyzed hydration (draw their structures) of the alkenes in Question 10.6 with those obtained from hydroboration-oxidation.
(*b*) Do the same for the examples given in this section.

Question 10.8

Starting with any needed alkanes or alkenes, devise syntheses for the following three compounds. Your method should produce each compound in good yield and free of isomers:

(*a*) 1-butanol (*b*) 2-butanol (*c*) 2-methyl-1-propanol

10.6 Hydroboration; Isomerization and Reduction Reactions

Hydroboration-oxidation of alkenes (see Sec. 10.5) yields alcohols, but several other important reactions also may follow. One is the conversion of the alkylboron intermediate to an alkane, which occurs when the alkylboron is heated with acid:

$$\left(-\overset{|}{\underset{|}{C}}-\overset{|}{\underset{|}{C}}-\right)_3 B \xrightarrow[\text{heat}]{H^{\oplus},\, H_2O} -\overset{|}{\underset{|}{C}}-\overset{|}{\underset{|}{C}}-H + B(OH)_3$$

Trialkylboron   Alkane   Boric
                                               acid

Based on the starting alkene, this is a reduction reaction. Anhydrous acetic acid, CH_3COOH, or anhydrous propanoic acid, CH_3CH_2COOH, can also be used, and then the boron ends up as a boron triester. The following example illustrates this method:

$$CH_3CH_2\!-\!CH\!=\!CH_2 \xrightarrow{(BH_3)_2} (CH_3CH_2CH_2CH_2)_3B \xrightarrow[\text{(acetic acid)}]{\overset{\overset{\text{O}}{\|}}{CH_3C\!-\!OH}}$$

1-Butene   Tri-*n*-butylboron

$$CH_3CH_2CH_2CH_3 + B(O\overset{\overset{\text{O}}{\|}}{C}CH_3)_3$$

n-Butane   Boron
                               triacetate

The hydroboration portion of the hydroboration-reduction reaction sequence occurs through the normal stereospecific pathway, but what about the reduction portion? Experiments show that this reaction proceeds with complete retention of relative configuration at the site of reduction, the carbon-boron bond. A six-membered transition state explains this quite well.

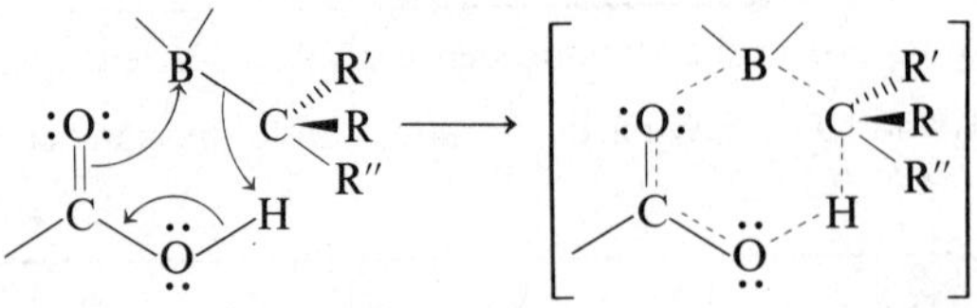

Reduction occurring with retention of configuration

Six-membered transition state

This transition state also accounts for carboxylic acids being much more efficient than mineral acids in effecting the reduction.

At elevated temperatures, alkylboron compounds undergo isomerization. They form alkenes and reform alkylboron compounds in a reversible reaction, and ultimately the boron migrates along the carbon chain until the least substituted alkylboron compound is formed. Either oxidation (H_2O_2, $:\ddot{O}H^{\ominus}$) or reduction ($H^{\oplus}$, H_2O, heat) of the final alkylboron compound yields an alcohol or alkane, respectively. For example:

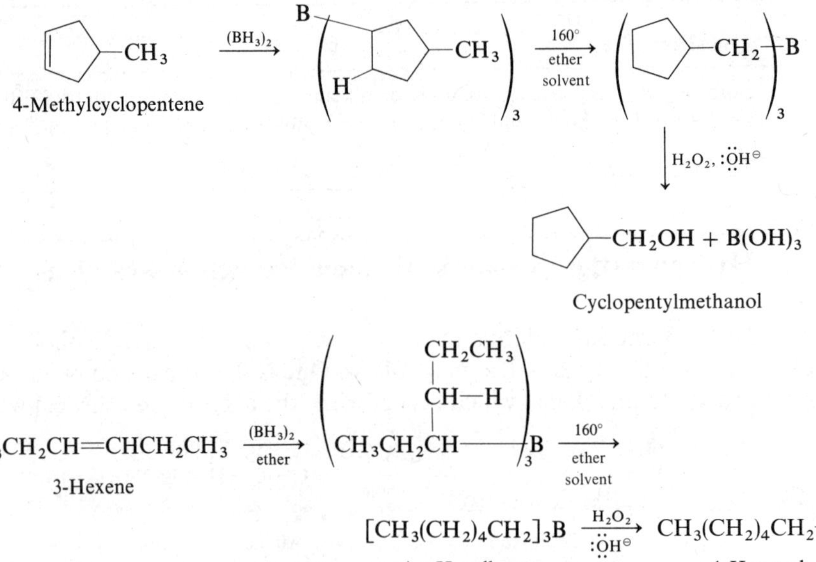
4-Methylcyclopentene
(BH₃)₂
160°
ether
solvent
H₂O₂, :ÖH⊖
CH₂OH + B(OH)₃
Cyclopentylmethanol

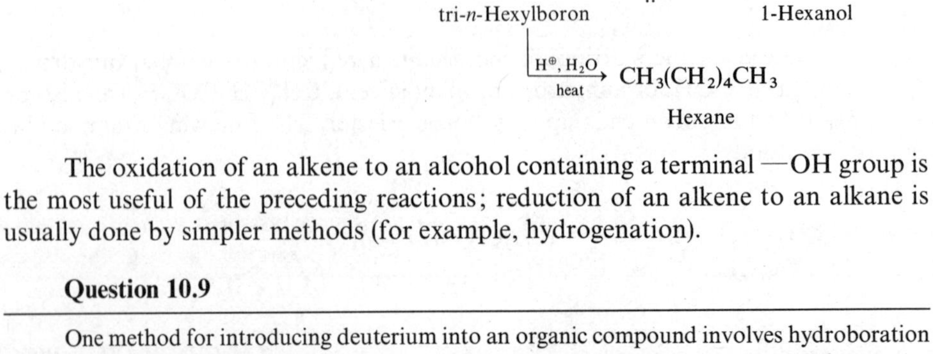
CH₃CH₂CH=CHCH₂CH₃
3-Hexene
(BH₃)₂
ether
160°
ether
solvent
[CH₃(CH₂)₄CH₂]₃B
tri-n-Hexylboron
H₂O₂
:ÖH⊖
CH₃(CH₂)₄CH₂—OH
1-Hexanol
H⊕, H₂O
heat
CH₃(CH₂)₄CH₃
Hexane

The oxidation of an alkene to an alcohol containing a terminal —OH group is the most useful of the preceding reactions; reduction of an alkene to an alkane is usually done by simpler methods (for example, hydrogenation).

Question 10.9

One method for introducing deuterium into an organic compound involves hydroboration using deuterated diborane, $(BD_3)_2$, followed by treatment with acid and water. What monodeuterated product is formed from the following alkenes if this method is used?

(*a*) 1-hexene (*b*) 1-methylcyclohexene (*c*) $(CH_3)_2C{=}CHCH_3$

Question 10.10

Give the product that would be obtained by treating each of these three compounds with diborane, by heating the alkylboron compounds to 160°, and finally by treating the alkylboron compounds so formed with hydrogen peroxide and base.

(a) (b) (c) $(CH_3)_3CCH_2CH=CH-CH_3$

10.7 Industrially Important Alcohols

Several alcohols are of great industrial importance, and they are useful to the organic chemist as *solvents* and as *reactants*. Methanol and ethanol are the two most commonly encountered alcohols. Special mention is made of them here, followed by some general comments about the industrial sources of other alcohols.

A. Methanol

Methanol, or methyl alcohol, CH_3OH, is commonly prepared from inorganic compounds. The hydrogenation of carbon monoxide can be carried out in the presence of a metal catalyst at high temperature and pressure:

$$CO + 2 H_2 \xrightarrow[400°, 200 \text{ atm}]{ZnO-CrO_3} CH_3OH$$

The yield of methanol from this process is excellent.

Methanol is highly *poisonous*. It causes blindness and ultimately death if consumed internally or allowed to come in prolonged contact with the skin. Breathing its vapors is also harmful.

B. Ethanol

Ethanol, or ethyl alcohol, CH_3CH_2OH, is the alcohol in alcoholic beverages. Because of its great utility as a solvent in organic chemistry and as a starting material for many syntheses, ethanol is one of the most readily available and frequently used alcohols.

Ethanol is prepared industrially by the hydration of ethene, which is readily available from petroleum sources. Ethene and steam are allowed to react at elevated temperatures in the presence of phosphoric acid catalyst:

$$CH_2=CH_2 + H_2O \text{ (steam)} \xrightarrow[300-350°]{H_3PO_4} CH_3CH_2OH$$

Another method for obtaining ethanol is the *fermentation of carbohydrates*. Carbohydrates are present in many natural products, including grain, starch, and sugar. The term grain alcohol is synonymous with ethanol and refers to the source of the alcohol. The fermentation of various other natural products, such as rye, corn, and grapes, gives rise to alcoholic beverages, and the sugar content in these compounds is responsible for alcohol formation. The type of beverage produced (scotch, bourbon, rosé wine, burgundy wine, and so on) depends on how the fermentation is carried out. Three factors involved are (1) whether the carbon dioxide is allowed to escape, (2) whether the beverage is distilled, and (3) whether other substances are added.

Pure ethanol is obtained only with moderate difficulty. The fermentation process in the production of liquors or wines very seldom gives an alcohol content in excess of 18% by volume (36 proof). Industrial processes give ethanol that is up to 95.6% pure, with the remainder being water. Distillation of an aqueous solution of ethanol gives a 95.6:4.4 (volume percent) mixture of ethanol and water, which boils at 78.15°. Regardless of the ratio of ethanol to water in the original mixture, the distillate always contains 95.6% ethanol and 4.4% water; it is called an **azeotropic mixture.**[1] Even though it is a mixture, it has a constant boiling point.

Pure ethanol (100% pure) can be obtained by removing the remaining water from 95.6% ethanol. This can be accomplished by chemical means (for example, by treatment with magnesium turnings, which react with the water and form magnesium hydroxide and hydrogen gas) or by distillation with benzene. The benzene forms an azeotropic mixture with ethanol and water (7.5% water, 74% benzene, and 18.5% ethanol), which boils at 65°, lower than the boiling point of the ethanol-water azeotrope. Thus, this azeotropic mixture distills first and leaves behind pure ethanol, which is collected by distillation.

Absolute ethanol obtained by treatment of 95.6% ethanol with benzene is unfit for human consumption because the benzene is toxic. Because absolute alcohol is frequently needed for chemical and medical research, it is made undrinkable by the addition of certain denaturants, such as benzene or methanol, before being released to the user. This is called **denatured alcohol,** which is absolute ethanol that cannot be consumed internally.

C. Other Monofunctional Alcohols

Most other low molecular weight alcohols (those with as many as four carbon atoms) are obtained by hydration of suitable alkenes, which are available from petroleum. For example, isopropyl alcohol comes from propylene, and *tert*-butyl alcohol comes from isobutylene. Because the orientation of hydration is such that the hydrogen ends up on the carbon bearing the greater number of hydrogens originally (the Markovnikov rule), hydration produces only secondary and tertiary alcohols (except for the hydration of ethylene, which gives ethanol, a primary alcohol).

Other important organic compounds are often synthesized from these lower molecular weight alcohols. Acetone, $(CH_3)_2C{=}O$, a highly useful organic solvent and reagent, is prepared industrially from the oxidation of isopropyl alcohol, $(CH_3)_2CHOH$, which in turn comes from the hydration of propylene. Isopropyl alcohol is commonly used as rubbing alcohol and a deicing liquid.

D. Polyfunctional Alcohols

Polyfunctional alcohols contain two or more hydroxyl groups. Two of the most common ones are *ethylene glycol* and *glycerol*, also known as *glycerine*.

$$
\begin{array}{ccc}
\text{CH}_2\text{—CH}_2 & \text{CH}_2\text{—CH—CH}_3 & \text{CH}_2\text{—CH—CH}_2 \\
|\quad\ \ | & |\quad\ \ | & |\quad\ \ |\quad\ \ | \\
\text{OH}\ \ \text{OH} & \text{OH}\ \ \text{OH} & \text{OH}\ \ \text{OH}\ \ \text{OH}
\end{array}
$$

1,2-Ethanediol   1,2-Propanediol   1,2,3-Propanetriol
(ethylene glycol)   (propylene glycol)   (glycerol or glycerine)

---

[1] An **azeotrope** is a solution of definite composition having a constant minimum (or maximum) boiling point lower (or higher) than the boiling point of any of its components.

Diols (dihydroxy compounds) are called **glycols** (Greek *glykys*, sweet) because many are sweet-tasting compounds; the same is true of polyhydroxy compounds (polyols) such as glycerol.

Although methanol was once the principal constituent of automobile antifreeze, it is now replaced by the higher boiling and more efficient ethylene glycol. Ethylene glycol is also used extensively in the production of Dacron (see Sec. 23.7) and Mylar film, which is used in recording and computer magnetic tapes. It is obtained commercially from ethylene by one of the following two routes:

$$CH_2{=}CH_2 \xrightarrow[\text{(HOCl)}]{Cl_2,\, H_2O} HOCH_2CH_2Cl \xrightarrow[S_N2]{:\ddot{O}H^\ominus,\, H_2O} HOCH_2CH_2OH$$

Ethylene   2-Chloroethanol   1,2-Ethanediol
                   (ethylene   (ethylene glycol)
                   chlorohydrin)

$$CH_2{=}CH_2 \xrightarrow[\text{heat}]{O_2,\, Ag\ metal} \underset{O}{CH_2{-}CH_2} \xrightarrow[H_2O]{:\ddot{O}H^\ominus,} HOCH_2CH_2OH$$

Ethylene   Ethylene oxide

The first reaction involves reactions we have studied: addition of HOCl to a double bond (see Sec. 8.22) and nucleophilic substitution of hydroxide ion on an alkyl halide (see Sec. 9.12 ff). The second route involves reactions we will discuss in Sec. 19.10. Treatment of ethylene with cold, dilute, aqueous potassium permanganate (the Baeyer test, see Sec. 8.32) also produces this glycol.

Glycerol or glycerine was originally obtained as one of the products of saponification of soap (see Sec. 25.2), but it is now prepared commercially from propylene. Glycerol retains moisture quite well and is used in making candy, skin lotions, and other pharmaceuticals. It is also frequently used as an aid in pushing glass through rubber corks.

10.8   Reactions of Alcohols

There are two ways in which the —OH group of an alcohol is affected by other reagents. One reaction is the rupture of the oxygen-hydrogen bond, and the other is the rupture of the carbon-oxygen bond.

General Reactions of Alcohols

1. Rupture of oxygen-hydrogen bond, acidity of alcohols

$$RO{\ddag}H \longrightarrow \text{metal salts or esters or aldehydes, ketones, acids}$$

2. Rupture of carbon-oxygen bond, substitution and elimination

$$R{\ddag}OH \longrightarrow R{-}G \quad \text{or} \quad \text{alkene}$$

                          Substitution   Elimination
                                         (dehydration)

where G = various groups

In the following sections we examine these two types of reactions in detail.

Reactions Involving the Oxygen-Hydrogen (RO—H) Bond of Alcohols

10.9   Acidity of Alcohols; Formation of Metal Salts

Before discussing the acidity of alcohols, let us recall what we know about water: It is weakly acidic (dissociation constant, $K_a = 1.98 \times 10^{-16}$; $pK_a = 15.7$) and it reacts readily with group I and group II metals. For example, water reacts *violently* with sodium metal to liberate hydrogen gas:

$$2\,H_2O + 2\,Na^0 \longrightarrow 2\,Na^{\oplus} + 2\,{:}\overset{..}{\underset{..}{O}}H^{\ominus} + H_2$$

Water   Hydroxide
                               ion

By analogy with water, we would expect the structurally similar alcohol molecule to react in the same manner because the two differ only in that one of the hydrogens in the former is replaced by a carbon-containing group in the latter:

$$H\overset{\overset{\displaystyle :\overset{..}{O}:}{\diagup}}{}\!\!\diagdown H \xrightarrow[\text{H by R}]{\text{replacement of}} R\overset{\overset{\displaystyle :\overset{..}{O}:}{\diagup}}{}\!\!\diagdown H$$

Water   Alcohol

Accordingly, alcohols are expected to and in fact do react with group I and group II metals, and also with certain group III metals, but at a much slower rate. The general reaction is:

$$RO{-}H + M^0 \longrightarrow \underset{\substack{\text{or }M^{\oplus 2}\\ \text{or }M^{\oplus 3}}}{M^{\oplus} +} \quad R\overset{..}{\underset{..}{O}}{:}^{\ominus} + H_2$$

Alcohol   Alkoxide
                                    ion

This is an oxidation-reduction reaction because the metal is oxidized (that is, it loses electrons and becomes positively charged: $M \rightarrow M^{\oplus} + e^{\ominus}$) and hydrogen is reduced (it gains electrons and forms the neutral hydrogen gas molecule: $2H^{\oplus} + 2e^{\ominus} \rightarrow H_2$). Three reactions of alcohols with metals are:

$$CH_3O{-}H \xrightarrow{\;Na^\circ\;} CH_3\overset{..}{\underset{..}{O}}{:}^{\ominus}\,Na^{\oplus} + H_2$$

Methanol   Sodium
(methyl   methoxide
alcohol)

$$CH_3CH_2O{-}H \xrightarrow{\;Mg^\circ\;} (CH_3CH_2\overset{..}{\underset{..}{O}}{:}^{\ominus})_2Mg^{\oplus 2} + H_2$$

Ethanol   Magnesium
(ethyl   ethoxide
alcohol)

$$\underset{\underset{\displaystyle CH_3}{|}}{\overset{\overset{\displaystyle CH_3}{|}}{CH_3{-}C{-}O{-}H}} \xrightarrow{\;K^\circ\;} \underset{\underset{\displaystyle CH_3}{|}}{\overset{\overset{\displaystyle CH_3}{|}}{CH_3{-}C{-}\overset{..}{\underset{..}{O}}{:}^{\ominus}\,K^{\oplus}}} + H_2$$

2-Methyl-2-propanol   Potassium-2-methyl-2-propoxide
(*tert*-butyl   (potassium
alcohol)   *tert*-butoxide)

The general name of a metal salt is *metal alkoxide*, where *metal* is the name of the metallic cation and *alk-* is the name of the alkyl group less the *-yl* ending. For example, $CH_3\overset{\cdot\cdot}{\underset{\cdot\cdot}{O}}\colon^{\ominus}$ is the *methoxide* ion, where *meth-* is derived from *methyl* (the name for the CH_3 group). The name of $CH_3\overset{\cdot\cdot}{\underset{\cdot\cdot}{O}}\colon^{\ominus} K^{\oplus}$ is *potassium methoxide*. (This naming system is more general than it may at first appear, because for $^{\ominus}\colon\overset{\cdot\cdot}{\underset{\cdot\cdot}{O}}H$, *hydroxide* is a contraction of hydrogen oxide, where *hydro-* refers to hydrogen.)

The reactivity of alcohols toward various metals is closely related to their acidity. Because the reaction of a metal with an alcohol is an oxidation-reduction reaction, it is not surprising that the availability of protons is a factor in determining the relative rates at which reaction occurs. Toward a given metal, the relative reactivities of various alcohols are:

Decreasing order of reactivities of alcohols toward group I, II, or III metals:

$$CH_3OH > 1° > 2° > 3°$$

As we will see, the relative acidities of alcohols parallel this sequence.

The acidities of the alcohols are best understood by considering the alkoxide anions that are formed as the proton is donated. The electron-donating effect of the alkyl group substituents increases the concentration of electrons on the oxygen bearing the negative charge, and as we might expect, it becomes increasingly more difficult to place negative charge (during the process of forming the alkoxide ion) on oxygens that already bear some increased partial negative charge:

Decreasing alkoxide ion stability;
decreasing ease of forming alkoxide
⟶

Thus, not only are primary alcohols more acidic than secondary or tertiary alcohols, but the corresponding alkoxide ions are also more stable. It appears that acidity is directly related to alkoxide ion stability. This trend agrees with experimental observation. Alcohols that contain only alkyl groups typically have dissociation constants in the range $K_a = 10^{-16}$ to 10^{-18}, depending on their structure. *tert*-Butyl alcohol is, for example, a weaker acid than methyl alcohol, and in their reactivities toward potassium metal, methyl alcohol reacts rapidly (almost explosively), whereas *tert*-butyl alcohol reacts very slowly—about 24 hr being required for complete reaction. Keep in mind that this explanation is true on a *relative basis*, comparing the various classes of alcohols.

Class of alcohol: CH_3—OH   1°   2°   3°
Decreasing acidity of alcohols ⟶

These alkoxide bases are all stronger than hydroxide, and as a result, they are better at promoting dehydrohalogenation reactions (elimination of alkyl halides to produce alkenes). They are normally prepared by dissolving a weighed amount of metal in anhydrous alcohol, thus giving the alkoxide anion dissolved in the corresponding alcohol.

Several metal alkoxides are commercially available, including sodium methoxide (often called *sodium methylate*) and sodium, potassium, rubidium, and cesium *tert*-butoxides. The *tert*-butoxides are exceptionally strong bases, especially when dissolved in a polar, aprotic solvent such as dimethyl sulfoxide, $(CH_3)_2SO$. This base-solvent combination is often used in elimination reactions because it produces a fast reaction with a minimum of substitution; for example:

$$CH_3CH_2CH_2CH_2\!-\!Br \xrightarrow[\text{dimethyl sulfoxide}]{(CH_3)_3C\overset{..}{\underset{..}{O}}{:}^{\ominus} K^{\oplus}} CH_3CH_2\!-\!CH\!=\!CH_2$$

1-Bromobutane
(*n*-butyl bromide)

1-Butene

Because of their basicity, the alkoxide bases react readily with water to form the corresponding alcohol and hydroxide ion:

$$R\overset{..}{\underset{..}{O}}{:}^{\ominus} + H_2\overset{..}{O}{:} \; \rightleftharpoons \; R\overset{..}{O}H + H\overset{..}{\underset{..}{O}}{:}^{\ominus}$$

Although this is a reversible reaction, the equilibrium lies far to the right.

More recent measurements of the acidities of various alcohols relative to water in the *gas phase* yield interesting (though perhaps puzzling) results. The solution pK_a values of a series of common alcohols are listed here:

Substrate	pK_a (solution)
H_2O	15.7
CH_3OH	16
CH_3CH_2OH	18
$CH_3CHOHCH_3$	18
$(CH_3)_3COH$	19

In the gas phase, however, the reverse trend is observed; that is, *tert*-butyl alcohol is the strongest acid and water the weakest. This leads to the belief that solvation effects in solution play the principal role in determining the acidity of a given alcohol in solution and that any argument involving the polar inductive effect of the alkyl groups, though it appears to give the correct result, works merely through a fortunate coincidence (see Sec. 5.7 for a general discussion of acidities).

10.10 Oxidation of Alcohols

A very important interrelationship between two functional groups—alcohols and the carbonyl group—starts with the oxidation of alcohols. In its most general form, this oxidation reaction is:

$$\underset{\text{Alcohol}}{-\overset{\overset{\displaystyle O-H}{|}}{\underset{\underset{\displaystyle H}{|}}{C}}-} \xrightarrow{[O]} -\overset{\overset{\displaystyle O}{\|}}{C}-$$

Alcohol

where the two hydrogens are removed and the $H—\overset{|}{\underset{|}{C}}—O—H$ linkage is converted into the carbonyl group, $>C=O$. The symbol [O] indicates oxidation and stands for any reagent that is capable of supplying oxygen to an organic molecule. Oxidation in organic chemistry refers to the loss of electrons by carbon and the elevation of its oxidation state to a more positive value.

The precise details of the oxidation and the structure of the functional group obtained depend on the structure of the alcohol and sometimes on the reaction conditions. These various situations are discussed according to the type of alcohol used (primary, secondary, or tertiary). Some oxidation reactions also provide a useful means of classifying primary, secondary, and tertiary alcohols (see Sec. 10.18).

A. Oxidation of Primary Alcohols

The oxidation of primary alcohols, $RCH_2—OH$, gives either an aldehyde, $R—\overset{O}{\overset{||}{C}}—H$, or a carboxylic acid, $R—\overset{O}{\overset{||}{C}}—OH$, depending on the reaction conditions. When the alcohol is heated with copper metal at 200 to 300° or reacted with the CrO_3-pyridine complex,[1] the sole organic product is the corresponding aldehyde:

$$R—CH_2—OH \xrightarrow[\substack{or \\ CrO_3\text{-pyridine}}]{Cu° \\ 200-300°} R—\overset{O}{\overset{||}{C}}—H$$

Primary alcohol   Aldehyde
Only organic product

In the former reaction the copper metal is a catalyst. This is a dehydrogenation reaction because hydrogen gas is the other product. This method is used relatively seldom in the laboratory because of the conditions required and the experimental difficulty in carrying it out; CrO_3-pyridine is preferred. For example, methanol yields methanal and ethanol gives ethanal:

$$CH_3OH \xrightarrow[\substack{or \\ CrO_3\text{-pyridine}}]{Cu° \\ 200-300°} H—\overset{O}{\overset{||}{C}}—H$$

Methanol   Methanal
(methyl   (formaldehyde)
alcohol)

$$CH_3CH_2OH \xrightarrow[\substack{or \\ CrO_3\text{-pyridine}}]{Cu° \\ 200-300°} CH_3—\overset{O}{\overset{||}{C}}—H$$

Ethanol   Ethanal
(ethyl   (acetaldehyde)
alcohol)

Chromic acid, H_2CrO_4, is another common oxidizing agent that can be used to convert primary alcohols to aldehydes. Chromic acid is itself unstable for long periods of time, and as a result, it is usually prepared in solution by treating potassium dichromate, $K_2Cr_2O_7$, with either sulfuric or acetic acid. It can also be prepared

---

[1] Pyridine is a common organic solvent (see Sec. 13.7C).

by dissolving chromic anhydride, CrO_3, in water under acidic conditions. The reactions are:

$$K_2Cr_2O_7 \xrightarrow{H_2SO_4} [H_2Cr_2O_7] \xrightarrow{H_2O} 2H_2CrO_4$$

Potassium   Unstable   Chromic acid
dichromate

$$CrO_3 + H_2O \longrightarrow H_2CrO_4$$

Chromium
trioxide

In this text we always indicate the required reactants (H_2SO_4, $K_2Cr_2O_7$), but remember that the active oxidizing agent is chromic acid.

This method has the potential disadvantage that it also oxidizes aldehydes to carboxylic acids, especially if the aldehyde is allowed to remain in contact with the chromic acid for a period of time.

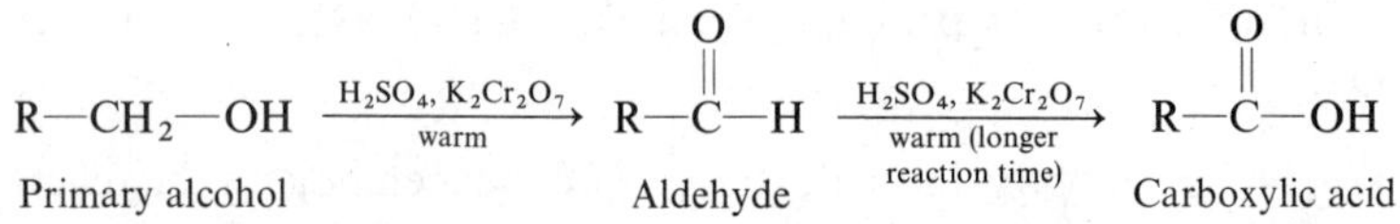

$$R-CH_2-OH \xrightarrow[\text{warm}]{H_2SO_4,\ K_2Cr_2O_7} R-\overset{\overset{\displaystyle O}{\|}}{C}-H \xrightarrow[\substack{\text{warm (longer)}\\ \text{reaction time)}}]{H_2SO_4,\ K_2Cr_2O_7} R-\overset{\overset{\displaystyle O}{\|}}{C}-OH$$

Primary alcohol   Aldehyde   Carboxylic acid

The chromium-containing product of this reaction is Cr^{+3}, so chromium is reduced from its $+6$ oxidation state in H_2CrO_4 to $+3$ in Cr^{+3}. The mechanism of chromic acid oxidation is presented in Sec. 10.11. For example, the following oxidation reactions of primary alcohols occur:

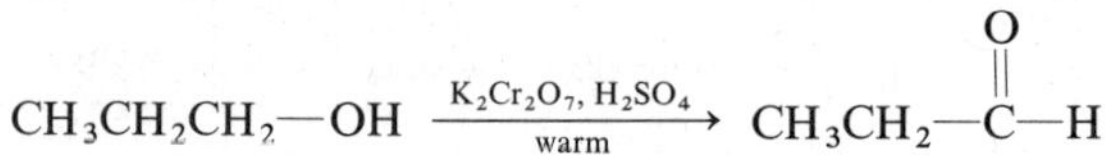

$$CH_3CH_2CH_2-OH \xrightarrow[\text{warm}]{K_2Cr_2O_7,\ H_2SO_4} CH_3CH_2-\overset{\overset{\displaystyle O}{\|}}{C}-H$$

1-Propanol   Propanal
(*n*-propyl alcohol)   (propionaldehyde)

$$\begin{matrix} CH_3 \\ \ \ \ \ \ \diagdown \\ \ \ \ \ \ \ \ \ \ CH-CH_2-OH \\ \ \ \ \ \diagup \\ CH_3 \end{matrix} \xrightarrow[\text{warm}]{K_2Cr_2O_7,\ H_2SO_4} \begin{matrix} CH_3 \\ \ \ \ \ \ \diagdown \ \ \ \ \ \ \overset{\displaystyle O}{\|} \\ \ \ \ \ \ \ \ \ \ CH-C-H \\ \ \ \ \ \diagup \\ CH_3 \end{matrix}$$

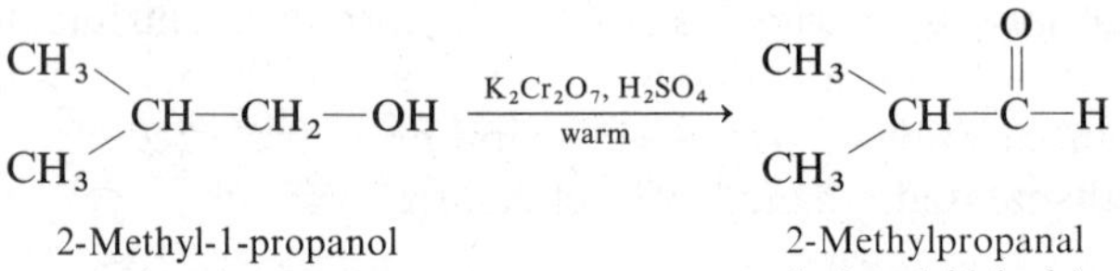

2-Methyl-1-propanol   2-Methylpropanal
(isobutyl alcohol)   (isobutylaldehyde)
(bp, 107–108°)   (bp, 65°)

Experimentally, however, the selective oxidation of alcohols to aldehydes can be carried out fairly easily by removing the aldehyde product by distillation as it is formed. This can be done because aldehydes boil at lower temperatures than the corresponding alcohols. For example, in the preparation of 2-methylpropanal (illustrated here), the aldehyde boils at 65° and the alcohol from which it is prepared boils at 107 to 108°, so the aldehyde is easily removed from the reaction mixture by fractional distillation. Also, alcohols are usually inexpensive starting materials, and further oxidation of the aldehyde to the acid can be minimized by adding chromic acid in small portions to the alcohol, thus ensuring that there is no excess oxidizing agent.

For compounds that are sensitive to acid, the CrO_3-pyridine complex can be used successfully to oxidize both primary and secondary alcohols to aldehydes and ketones, respectively. If an aldehyde is the desired product, care must be taken not to oxidize too vigorously or else the aldehyde is oxidized to a carboxylic acid.

Neutral or basic potassium permanganate is a third common oxidizing agent, but it is much more powerful than chromic acid. Although there is evidence that an aldehyde is an intermediate in the permanganate oxidation, it is oxidized so rapidly that the only observed product is the corresponding carboxylic acid. The carboxylic acid is obtained as a salt in the strongly basic solution but is readily converted to the free acid by the addition of mineral acid. The product is shown as the free acid for simplicity, which is common practice in *paper chemistry*. This too is an oxidation-reduction reaction because manganese is reduced from its $+7$ oxidation state in permanganate, $MnO_4^{\ominus}$, to $+4$ in manganese dioxide, MnO_2.

$$R—CH_2—OH \xrightarrow[\substack{:\ddot{O}H^{\ominus}, \\ H_2O}]{KMnO_4} \left[\begin{array}{c} O \\ \parallel \\ R—C—H \end{array} \right] \xrightarrow[\substack{:\ddot{O}H^{\ominus}, \\ H_2O}]{KMnO_4} \begin{array}{c} O \\ \parallel \\ R—C—OH \end{array}$$

Primary alcohol / Aldehyde *Not isolated* / Carboxylic acid

As a result, *permanganate is not a useful reagent for preparing aldehydes*, yet it is one of the best reagents for converting a primary alcohol to a carboxylic acid.

The oxidation reactions of primary alcohols may be summarized as follows:

$$R—CH_2—OH \xrightarrow[\substack{\text{or } CrO_3, H_2O, \text{ warm} \\ \text{or } CrO_3\text{-pyridine}}]{\substack{Cu^{\circ}, 200-300^{\circ} \\ \text{or } K_2Cr_2O_7, H_2SO_4, \text{ warm}}} \begin{array}{c} O \\ \parallel \\ R—C—H \end{array} \xrightarrow[\substack{\text{or } CrO_3, H_2O, \text{ heat} \\ \text{or } KMnO_4, :\ddot{O}H^{\ominus}, \text{ heat}}]{\substack{K_2Cr_2O_7, H_2SO_4, \text{ heat} \\ \text{or } CrO_3, H_2O, \text{ heat}}} \begin{array}{c} O \\ \parallel \\ R—C—OH \end{array}$$

Primary alcohol / Aldehyde / Carboxylic acid

$$\xrightarrow{KMnO_4, :\ddot{O}H^{\ominus}, \text{ heat}}$$

B. Oxidation of Secondary Alcohols

The oxidation of secondary alcohols is more straightforward because the only possible product is a ketone that contains the same number of carbon atoms. The four reagents, copper metal and heat, chromic acid, CrO_3-pyridine, and basic potassium permanganate can be used, although the latter are used most frequently in the laboratory. The general reaction is:

$$\begin{array}{c} O—H \\ | \\ R—C—R' \\ | \\ H \end{array} \xrightarrow[\substack{\text{or } KMnO_4, :\ddot{O}H^{\ominus} \\ \text{or } CrO_3\text{-pyridine}}]{\substack{Cu^{\circ}, 200-300^{\circ} \\ \text{or } K_2Cr_2O_7, H_2SO_4}} \begin{array}{c} O \\ \parallel \\ R—C—R' \end{array} \xrightarrow[\substack{\text{vigorous} \\ \text{conditions}}]{\substack{K_2Cr_2O_7, H_2SO_4 \\ \text{or } KMnO_4, :\ddot{O}H^{\ominus}}} \text{decomposition of ketone}$$

Secondary alcohol / Ketone

The ketone is stable and there is no further reaction unless very vigorous conditions are provided, which result in the ketone being broken down into smaller molecules. Two specific examples of the oxidation of secondary alcohols are:

$$\begin{array}{c} OH \\ | \\ CH_3—CH—CH_2CH_3 \end{array} \xrightarrow[H_2SO_4]{K_2Cr_2O_7} \begin{array}{c} O \\ \parallel \\ CH_3—C—CH_2CH_3 \end{array}$$

2-Butanol / 2-Butanone
(*sec*-butyl alcohol)

$$\text{Cyclohexanol} \xrightarrow[:\ddot{O}H^{\ominus}, H_2O]{KMnO_4} \text{Cyclohexanone}$$

C. Oxidation of Tertiary Alcohols

As we saw, the oxidation of primary and secondary alcohols results in two hydrogens being removed from the H—C—O—H moiety. When a compound does not contain a hydrogen bonded to the carbon bearing the —OH group, we do not expect an oxidation reaction to occur. This is indeed the case with tertiary alcohols, where the tertiary carbon atom is unable to accommodate the carbon-oxygen double bond without breaking a carbon-carbon bond. *Under basic conditions*, *tertiary alcohols do not undergo oxidation.* Under acidic conditions, they are readily dehydrated to an alkene, which is oxidized with breakdown of the original carbon skeleton into smaller molecules.

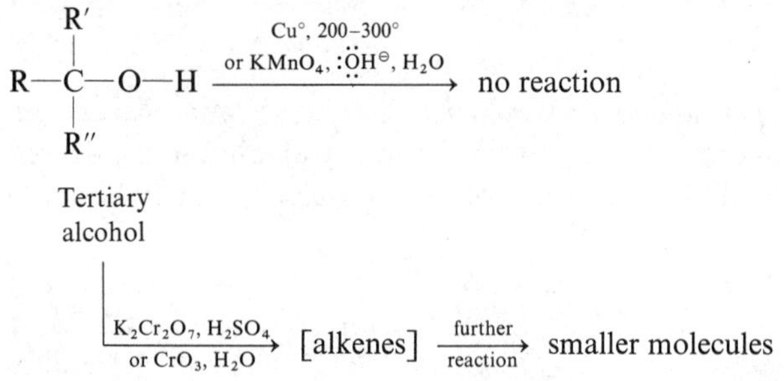

$$
\begin{array}{c}
R' \\
| \\
R-C-O-H \\
| \\
R''
\end{array}
\xrightarrow[\text{or KMnO}_4, \ :\ddot{O}H^{\ominus}, \text{H}_2\text{O}]{\text{Cu}^\circ, \ 200\text{--}300^\circ}
\text{no reaction}
$$

Tertiary
alcohol

$$
\xrightarrow[\text{or CrO}_3, \text{H}_2\text{O}]{\text{K}_2\text{Cr}_2\text{O}_7, \text{H}_2\text{SO}_4} [\text{alkenes}] \xrightarrow[\text{reaction}]{\text{further}} \text{smaller molecules}
$$

Question 10.11

Indicate the *changes* in the oxidation state of carbon that occur in each step of the four following transformations:

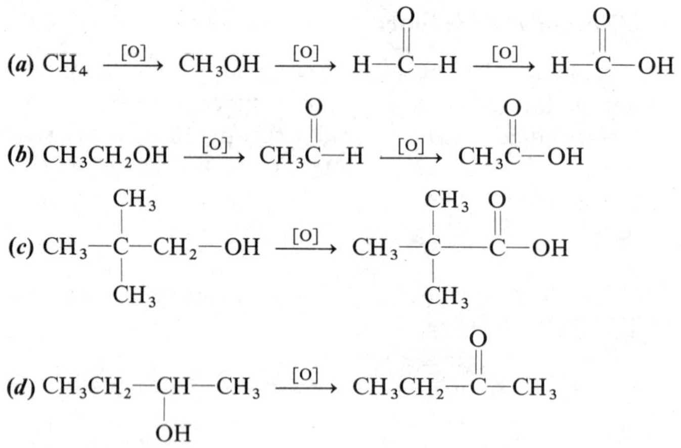

(*a*) $CH_4 \xrightarrow{[O]} CH_3OH \xrightarrow{[O]} \overset{\displaystyle O}{\underset{\displaystyle \|}{H-C-H}} \xrightarrow{[O]} \overset{\displaystyle O}{\underset{\displaystyle \|}{H-C-OH}}$

(*b*) $CH_3CH_2OH \xrightarrow{[O]} \overset{\displaystyle O}{\underset{\displaystyle \|}{CH_3C-H}} \xrightarrow{[O]} \overset{\displaystyle O}{\underset{\displaystyle \|}{CH_3C-OH}}$

(*c*) $\begin{array}{c} CH_3 \\ | \\ CH_3-C-CH_2-OH \\ | \\ CH_3 \end{array} \xrightarrow{[O]} \begin{array}{c} CH_3 \ \ O \\ | \ \ \| \\ CH_3-C \ \ \ C-OH \\ | \\ CH_3 \end{array}$

(*d*) $\begin{array}{c} CH_3CH_2-CH-CH_3 \\ | \\ OH \end{array} \xrightarrow{[O]} \begin{array}{c} O \\ \| \\ CH_3CH_2-C-CH_3 \end{array}$

Question 10.12

Give the structure(s) of the products obtained when each of the following alcohols is allowed to react with (1) copper metal at 300°, (2) potassium dichromate and sulfuric acid under mild conditions, (3) potassium dichromate and sulfuric acid under more vigorous conditions, and (4) basic potassium permanganate solution.

(*a*) 1-propanol   (*b*) 2-propanol   (*c*) cyclohexanol   (*d*) 3-methyl-3-hexanol

10.11　Mechanism of Oxidation of Alcohols

The mechanism of the oxidation of alcohols by chromic acid and permanganate has been studied extensively, but several aspects remain in doubt. The problem in studying these and most oxidation-reduction processes is that the transfer of electrons cannot be followed, and even *how* they are transferred at all is sometimes a mystery. (We use stereochemistry and isotopes for the mechanistic study of many reactions; it is indeed too bad that we cannot label or see electrons, for instance by having the electrons in one compound colored red and those in another blue!) In this section the currently accepted alternatives to the oxidation mechanism are presented.

The oxidation of isopropyl alcohol to acetone by chromic acid has been studied extensively. The initial step is believed to be the reversible formation of an ester between chromic acid and isopropyl alcohol. The intermediate chromate ester is highly unstable and is not normally isolated.

2-Propanol (isopropyl alcohol)　　Chromic acid　　Isopropyl hydrogen chromate *Ester*

Once formed, isopropyl hydrogen chromate may decompose in one of two possible ways. One (*route 1*) involves protonation of the ester, followed by abstraction of an α hydrogen by base. In this case water is likely to be the base, and the elimination, which greatly resembles the E2 mechanism, forms H_2CrO_3. The alcohol is oxidized and chromium is reduced from its $+6$ oxidation state in H_2CrO_4 to the unstable $+4$ oxidation state in H_2CrO_3. Chromium(VI) and chromium(IV) react rapidly to product chromium(V), HCr^VO_3. Then 2 moles of HCr^VO_3 react with 2 more moles of alcohol to form 2 moles of ketone and chromium(III), which is the stable oxidation state of chromium that is *observed* as a product.

Route 1: E2 type of elimination:

Then

$$H_2Cr^{IV}O_3 + H_2Cr^{VI}O_4 \longrightarrow 2HCr^VO_3 + H_2O$$

and

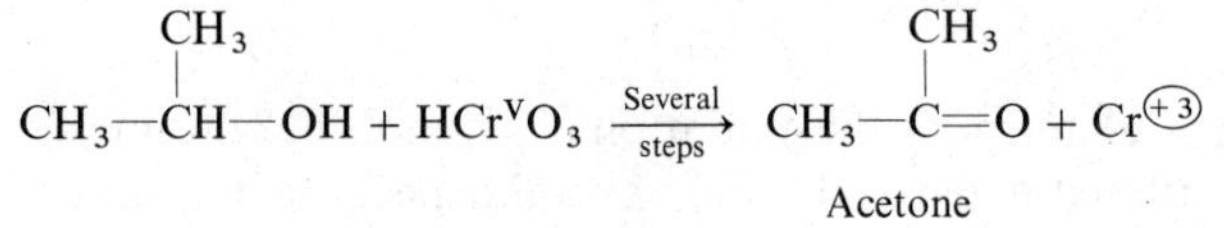

$$
\underset{\text{CH}_3\text{—CH—OH}}{\overset{\text{CH}_3}{|}} + \text{HCr}^\text{V}\text{O}_3 \xrightarrow[\text{steps}]{\text{Several}} \underset{\underset{\text{Acetone}}{\text{CH}_3\text{—C}=\text{O}}}{\overset{\text{CH}_3}{|}} + \text{Cr}^{\oplus 3}
$$

The second way the chromate ester may decompose is via the cyclic mechanism (*route 2*). Whereas route 1 uses an external base (for example, H_2O), the electrons attached to oxygen in the ester are used as the base in route 2. Here also $H_2Cr^\text{IV}O_3$ is formed, and it reacts further with $H_2Cr^\text{VI}O_4$ as before (*route 1*).

Route 2: Cyclic (concerted) mechanism for oxidation:

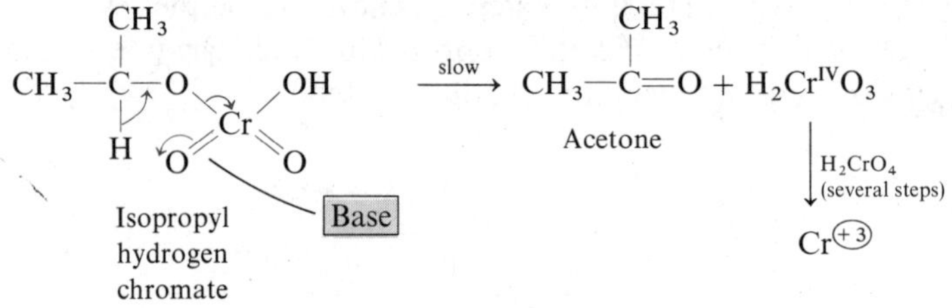

Some of the most compelling evidence for either route of the preceding mechanism comes from studying **deuterium isotope effects.**

Kinetic Isotope Effects

If two molecules containing a different isotope of an element at a given position (for example ^{2}H in one molecule and ^{1}H in the other) are allowed to react under identical conditions, the two molecules often react at different rates. This applies to reactions that are reversible (that is, equilibrium is involved) or irreversible (no equilibrium). The differences in the reactivity or in the position of equilibrium, as the case may be, are called isotope effects and are most often observed if one of the bonds broken in a reaction involves the isotope. This is referred to as a *primary isotope effect.*

Secondary isotope effects, which involve isotope substitution on a bond not actually broken in a reaction, are also observable, although they are often significantly more difficult to detect.

For now we concern ourselves with reactivity differences that result when hydrogen is replaced by deuterium on a bond that is undergoing reaction; this is a *primary hydrogen isotope effect.* If a reaction involves the breaking of a carbon-hydrogen or an oxygen-hydrogen bond, for example, then replacement of hydrogen by deuterium in a molecule results in the same bond-breaking reaction taking place at a slower rate.

A closer look at what occurs as energy is applied to a molecule, resulting in the eventual breaking of a bond, provides a much clearer picture of the origin of isotope effects. Directing our attention to the bond that is broken (the same bond the isotope is substituted on), we can envision the following. The bond is constantly undergoing vibration, and the energy of the vibration involves numerous quantized and allowed values. As energy is added to the molecule, the energy is absorbed and the frequency (energy) of the vibration increases. Although molecules are found in virtually all allowed vibrational energy levels, as energy is added a much higher percentage is found in higher levels. As more and more energy is added, this process continues until the frequency (energy) of the vibration is greater than the bond dissociation energy $\Delta H°$, at which time the bond breaks. Both masses of atoms involved in the bond and the nature of the bond itself (bond strength) affect the various vibrational allowed energy levels of the molecule. The greater the mass of the atoms, the greater the energy required to break the bond. Since deuterium is twice as heavy as hydrogen, when it is substituted for hydrogen on a given bond, that bond is more difficult to break—an isotope effect is observed.

Most often, differences in rates of reaction (that is, chemical kinetics) are used to evaluate the isotope effect, and these types of measurements are called kinetic isotope effects. Consider the rate differences of breaking carbon-hydrogen and carbon-deuterium bonds:[1]

$$-\overset{|}{\underset{|}{C}}-H + W \quad \xrightarrow{k_H} \quad \left[-\overset{|}{\underset{|}{C}}---H----W \right] \quad \longrightarrow \quad -\overset{|}{\underset{|}{C}}\cdot + (W - H)^{\cdot}$$

Transition state

$$-\overset{|}{\underset{|}{C}}-D + W \quad \xrightarrow{k_D} \quad \left[-\overset{|}{\underset{|}{C}}---D----W \right] \quad \longrightarrow \quad -\overset{|}{\underset{|}{C}}\cdot + (W - D)^{\cdot}$$

Transition state

In these equations the *rates* of removal of hydrogen and deuterium are k_H and k_D, respectively. At room temperature the *ratio* of these rate constants, that is, k_H/k_D, is on the order of 5–7:1; in other words, the rate of removal of hydrogen is 5 to 7 times faster than the rate of removal of deuterium (though lower values are often observed) *if there is an isotope effect*. The energy profiles for these two reactions are given in Fig. 10.1. On the other hand, if there is no difference in the rates of removal of hydrogen and of deuterium ($k_H/k_D = 1$), there is no isotope effect; hydrogen or deuterium removal must not be involved in the rate-determining step (or more correctly in the transition state of the rate-determining step). The kinetic isotope effect has been studied from a theoretical standpoint and supported by numerous experimental investigations.

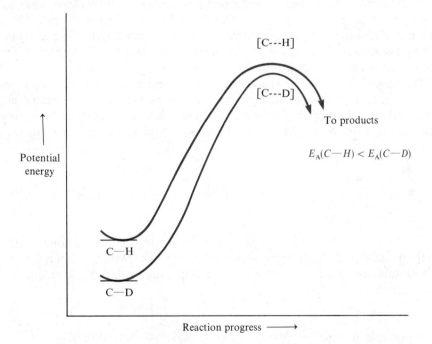

FIGURE 10.1 Energy profile for the breaking of a C—H versus a C—D bond in a generalized reaction in which an isotope effect is observed. The C—H and C—D are the same bond in the molecule being investigated. In one reaction H is attached to the molecule, and in a separate reaction D is attached. The results of the two reactions are then compared. Note that the difference in energy (ΔE) between the two molecules is greater in the ground state than in the transition state. (See Sec. 12.2, and Reading References for Chap. 12 in the Appendix, for an explanation.)

When chromic acid is used in the oxidation of CH_3—CDOH—CH_3 and CH_3—CHOH—CH_3, the deuterated compound reacts about six times slower, so

---

[1] Homolytic bond cleavage shown; heterolytic also possible.

that $k_H/k_D \sim 6$; this shows that the removal of the α hydrogen is involved in the slow, rate-determining step. Hydrogen removal is involved in both routes 1 and 2, so both decomposition mechanisms are in accord with this result.

Another possiblity arises from the observation that electron-donating groups on the carbon bearing the α-hydrogen result in an increased rate of reaction. This is not what we would expect if hydrogen is removed as a proton, as it is in routes 1 and 2. Thus, it has been suggested that hydrogen is lost as a hydride ion to $HCrO_3^{\oplus}$.

Route 3:

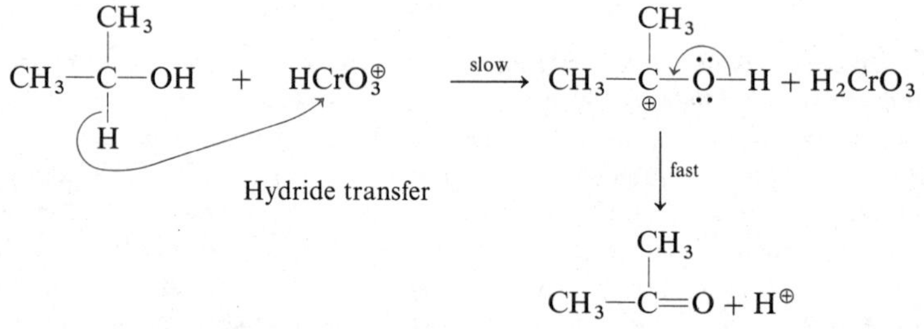

$$
\begin{array}{c}
CH_3 \\
| \\
CH_3-C-OH \;+\; HCrO_3^{\oplus} \xrightarrow{\text{slow}}
\end{array}
\qquad
\begin{array}{c}
CH_3 \\
| \\
CH_3-C-\overset{\oplus}{\underset{..}{O}}-H \;+\; H_2CrO_3 \\
\end{array}
$$

Hydride transfer

$$\downarrow \text{fast}$$

$$
\begin{array}{c}
CH_3 \\
| \\
CH_3-C{=}O \;+\; H^{\oplus}
\end{array}
$$

This mechanism is also compatible with the observed deuterium isotope effect.

The chromic acid oxidation mechanism is not clear cut, but at least it has been narrowed down to several reasonable possibilities. The intervention of the chromate ester is considered to be more likely, so route 1 or 2 is favored. Other frequently used oxidizing agents have been studied in less detail and are less well supported. Only brief mention is made of the mechanism of the permanganate oxidation.

Permanganate oxidation of alcohols can be accomplished in acidic or basic solution, but in acid, the permanganate ion is converted to permanganic acid, $HMnO_4$, which is dangerously explosive. In basic solution, equilibrium between the alcohol and hydroxide ion gives some alkoxide ion, which is an intermediate in the oxidation reaction.

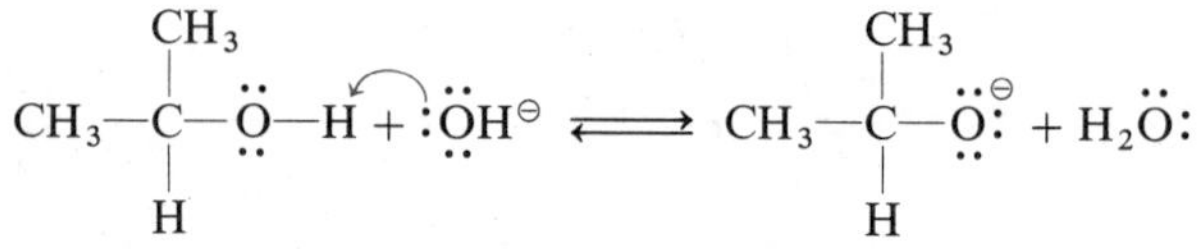

$$
\begin{array}{c}
CH_3 \\
| \\
CH_3-C-\overset{..}{\underset{..}{O}}-H \;+\; :\overset{..}{\underset{..}{O}}H^{\ominus} \\
| \\
H
\end{array}
\rightleftharpoons
\begin{array}{c}
CH_3 \\
| \\
CH_3-C-\overset{..}{\underset{..}{O}}{:}^{\ominus} \;+\; H_2\overset{..}{\underset{..}{O}}{:} \\
| \\
H
\end{array}
$$

The slow, rate-determining step in the basic permanganate oxidation is believed to be the transfer of the hydride ion to permanganate; the $HMnO_4^{\ominus 2}$ that is formed is highly unstable and disproportionates to give MnO_2 and $MnO_4^{\ominus}$. These reactions are shown here:

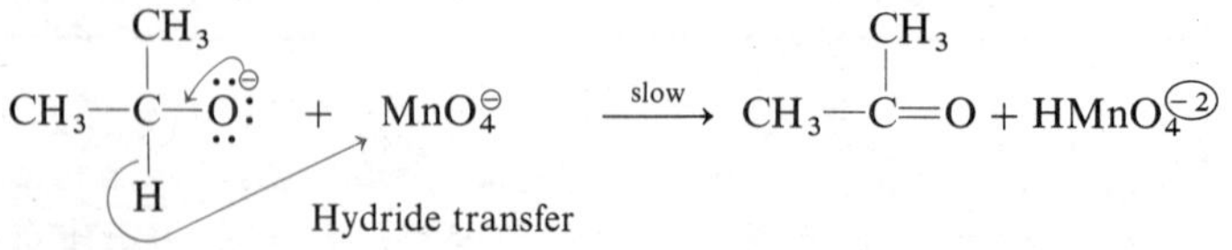

$$
\begin{array}{c}
CH_3 \\
| \\
CH_3-C-\overset{..}{\underset{..}{O}}{:}^{\ominus} \;+\; MnO_4^{\ominus} \xrightarrow{\text{slow}} \\
| \\
H
\end{array}
\qquad
\begin{array}{c}
CH_3 \\
| \\
CH_3-C{=}O \;+\; HMnO_4^{\ominus 2}
\end{array}
$$

Hydride transfer

and

$$
2HMnO_4^{\overset{V}{\ominus 2}} \xrightarrow{\text{fast}} \overset{IV}{MnO_2} + \overset{VII}{MnO_4^{\ominus}} + 2H\overset{..}{\underset{..}{O}}{:}^{\ominus} + e^{\ominus}
$$

10.12   Relationship Between Alcohols and Esters

A host of esters are derived from inorganic acids and organic alcohols. Our first contact with them was in Sec. 8.7 in the reaction between an alkene and cold, concentrated sulfuric acid to give an alkyl hydrogen sulfate ester:

$$R-CH\!=\!CH_2 + HO-\underset{\underset{O}{\|}}{\overset{\overset{O}{\|}}{S}}-OH \xrightarrow{\text{cold}} R-\underset{CH_3}{\overset{}{CH}}-O-\underset{\underset{O}{\|}}{\overset{\overset{O}{\|}}{S}}-OH \qquad \text{Obeys Markovnikov rule}$$

Alkene   Sulfuric acid   Alkyl hydrogen sulfate *Ester*

Our intent in this section is simply to survey the different esters that can be obtained from an alcohol and various inorganic acids. *An ester is typified by the R—O—G linkage*, where G represents the central atom in an inorganic acid; for example, G = N in HNO_3, S in H_2SO_4, and P in H_3PO_4.

The reactions discussed are similar in that they all involve an inorganic acid and an alcohol that unite to form an ester; the general equation is:

$$R-O-H \;+\; H-O-G \longrightarrow R-O-G + H_2O$$

Alcohol   Various inorganic acids ($HONO_2$, $HOSO_3H$, $HOPO_3H_2$)   Ester

Ester formation usually involves the loss of a molecule of water. Because the oxygen-hydrogen bond in the alcohol is often broken, ester formation is considered along with reactions of the O—H bond.

A. Esters of Carboxylic Acids

The most important family of esters is derived from a carboxylic acid and an alcohol. Many esters occur in natural products (for example, fats and oils; see Sec. 25.1). We discuss esters in great detail in Chap. 23. Their general structure is

$$\underset{\text{Ester}}{R-\overset{\overset{O}{\|}}{C}-O-R'} \quad \text{derived from} \quad \underset{\substack{\text{Carboxylic} \\ \text{acid}}}{R-\overset{\overset{O}{\|}}{C}-O-H} \quad \text{and} \quad \underset{\text{Alcohol}}{H-O-R'}$$

B. Esters of Sulfuric Acid

Esters of sulfuric acid are prepared by several routes, one of which was mentioned earlier (the addition of cold, concentrated sulfuric acid to an alkene). Cold, concentrated sulfuric acid also reacts with an alcohol to give first an *alkyl hydrogen sulfate*, $ROSO_3H$, and with excess alcohol to give an *alkyl sulfate*, $ROSO_2OR$. These reactions are reversible because both esters are hydrolyzed in excess water.

The hydrolysis of alkyl hydrogen sulfates, which are obtained from the addition of sulfuric acid to alkenes, is an important industrial preparation of alcohols.

Certain detergents are prepared by the sulfonation of an alcohol. For example, the reduction of fats and oils (see Sec. 25.2) produces various continuous-chain normal alcohols with an even number of carbon atoms (usually 12, 14, 16, 18, and 20). Lauryl alcohol, or 1-dodecanol, reacts with cold, concentrated sulfuric acid to give the corresponding lauryl hydrogen sulfate, which on neutralization with sodium hydroxide gives sodium lauryl sulfate. This product is a commercially available detergent.

$$ROH + HO\overset{\overset{\textstyle O}{\|}}{\underset{\underset{\textstyle O}{\|}}{S}}OH \underset{+H_2O}{\overset{-H_2O}{\rightleftharpoons}} RO\overset{\overset{\textstyle O}{\|}}{\underset{\underset{\textstyle O}{\|}}{S}}O-H \underset{+H_2O}{\overset{\overset{ROH}{-H_2O}}{\rightleftharpoons}} RO\overset{\overset{\textstyle O}{\|}}{\underset{\underset{\textstyle O}{\|}}{S}}OR$$

Alcohol

Sulfuric acid   Alkyl hydrogen sulfate *Ester*   Alkyl sulfate *Ester*

↑ cold, conc H_2SO_4

Alkene

$$CH_3(CH_2)_{11}OH \xrightarrow[H_2SO_4]{cold, conc} CH_3(CH_2)_{11}-O-SO_3H \xrightarrow{NaOH} CH_3(CH_2)_{11}-O-SO_3^{\ominus}Na^{\oplus}$$

Lauryl alcohol   Lauryl hydrogen sulfate   Sodium lauryl sulfate *Detergent*

Sulfate esters should not be confused with sulfonic acids, $R-SO_3H$, which contain a carbon-sulfur bond. Sulfate esters contain the relatively weak oxygen-sulfur bond, which easily undergoes hydrolysis.

$$R-O-\overset{\overset{\textstyle O}{\|}}{\underset{\underset{\textstyle O}{\|}}{S}}-OH \qquad\qquad R-\overset{\overset{\textstyle O}{\|}}{\underset{\underset{\textstyle O}{\|}}{S}}-OH$$

Sulfate ester   Sulfonic acid
Easily hydrolyzed   *Not subject to hydrolysis*
to ROH + H₂SO₄

C. Esters of Nitric Acid

The reaction between cold, concentrated nitric acid and an alcohol produces *nitrate esters*, $RONO_2$. Often sulfuric acid is a catalyst for the reaction. One of the most common organic compounds that can be classified as a nitrate ester is nitroglycerine; its formation is shown in the following:

$$\begin{array}{l} CH_2-OH \\ | \\ CH-OH + 3HO-NO_2 \\ | \\ CH_2-OH \end{array} \quad \underset{\underset{H_2O}{\xleftarrow{\hspace{1cm}}}}{\overset{H_2SO_4}{\underset{10-20°}{\xrightarrow{\hspace{1cm}}}}} \quad \begin{array}{l} CH_2-O-NO_2 \\ | \\ CH-O-NO_2 \\ | \\ CH_2-O-NO_2 \end{array}$$

Glycerine   Nitric acid   Trinitroglycerine

Nitroglycerine contains three nitrate ester groups, and like sulfate esters, they can be hydrolyzed back into nitric acid and glycerol. An important use of nitroglycerine is in the treatment of persons with heart disease; nitroglycerine dilates blood vessels and alleviates arterial tension. It was also used extensively as a military explosive. It is highly sensitive to shock, and decomposes explosively to nitrogen, carbon dioxide, water, and oxygen in the ratio 6:12:10:1, respectively, per *four* molecules of nitroglycerine. Not only is the reaction highly exothermic, but that it liberates seven molecules of gaseous product per molecule of nitroglycerine explains the great explosive force.

An important distinction between nitrate esters and nitro compounds is that in the former the carbon-oxygen-nitrogen bond in $R-O-NO_2$ is readily hydrolyzed, whereas in the latter the carbon-nitrogen bond in $R-NO_2$ prohibits this reaction from occurring.

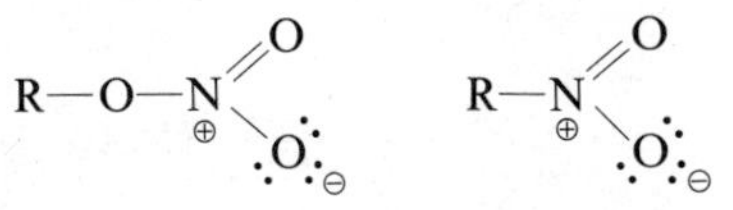

$$R-O-\overset{\oplus}{N}\overset{\displaystyle O}{\underset{\displaystyle \overset{..}{\underset{..}{O}}:_\ominus}{\big\|}} \qquad\qquad R-\overset{\oplus}{N}\overset{\displaystyle O}{\underset{\displaystyle \overset{..}{\underset{..}{O}}:_\ominus}{\big\|}}$$

Nitrate ester	Nitro compound
Easily hydrolyzed	*Not subject*
to ROH + HONO₂	*to hydrolysis*

D. Esters of Phosphoric Acid

The reaction between an alcohol and phosphoric acid produces various phosphate esters depending on the amount of alcohol used. The various steps are shown here:

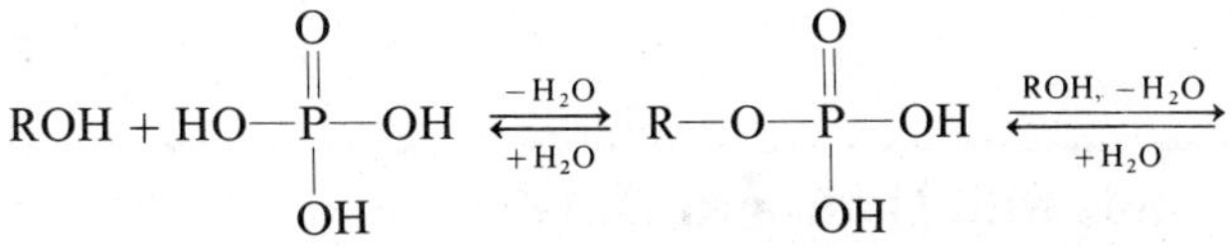

$$ROH + HO-\overset{\displaystyle O}{\underset{\displaystyle OH}{\overset{\|}{P}}}-OH \underset{+H_2O}{\overset{-H_2O}{\rightleftharpoons}} R-O-\overset{\displaystyle O}{\underset{\displaystyle OH}{\overset{\|}{P}}}-OH \underset{+H_2O}{\overset{ROH,\ -H_2O}{\rightleftharpoons}}$$

Alkyl
dihydrogen
phosphate

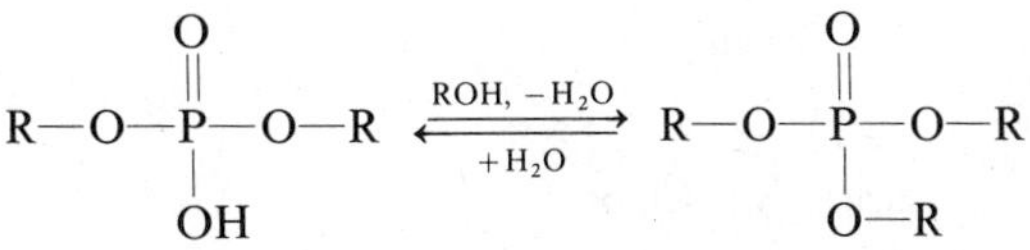

$$R-O-\overset{\displaystyle O}{\underset{\displaystyle OH}{\overset{\|}{P}}}-O-R \underset{+H_2O}{\overset{ROH,\ -H_2O}{\rightleftharpoons}} R-O-\overset{\displaystyle O}{\underset{\displaystyle O-R}{\overset{\|}{P}}}-O-R$$

Dialkyl	Trialkyl
hydrogen	phosphate
phosphate	

Phosphate esters are readily hydrolyzed under acidic or basic conditions into phosphoric acid and alcohol, as indicated.

Some of the most important phosphate esters come from natural products that have various phosphate linkages. One example is adenosine triphosphate (ATP), which is discussed in more detail in Sec. 29.10. The structure of the phosphate linkage of ATP is shown here, with R— representing the adenosine moiety; stepwise removal of the phosphate produces adenosine diphosphate (ADP) and finally adenosine monophosphate (AMP), which are also shown.

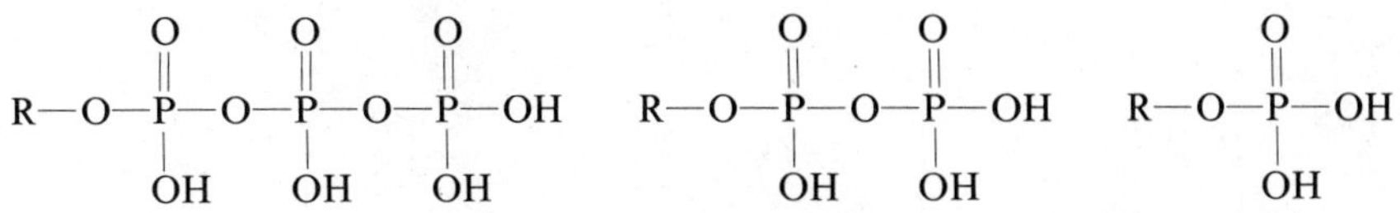

$$R-O-\overset{\displaystyle O}{\underset{\displaystyle OH}{\overset{\|}{P}}}-O-\overset{\displaystyle O}{\underset{\displaystyle OH}{\overset{\|}{P}}}-O-\overset{\displaystyle O}{\underset{\displaystyle OH}{\overset{\|}{P}}}-OH \qquad R-O-\overset{\displaystyle O}{\underset{\displaystyle OH}{\overset{\|}{P}}}-O-\overset{\displaystyle O}{\underset{\displaystyle OH}{\overset{\|}{P}}}-OH \qquad R-O-\overset{\displaystyle O}{\underset{\displaystyle OH}{\overset{\|}{P}}}-OH$$

Adenosine triphosphate	Adenosine	Adenosine
(ATP)	diphosphate	monophosphate
	(ADP)	(AMP)

Reactions Involving the Carbon-Oxygen (R—OH) Bond of Alcohols

In Sec. 7.14 we discussed a very common reaction that involves the rupture of the carbon-oxygen bond in an alcohol in dehydration. Recall that this reaction is acid-catalyzed and subject to rearrangements of the incipient carbocation:

$$-\overset{|}{\underset{\underset{H}{|}}{C}}-\overset{|}{\underset{\underset{OH}{|}}{C}}- \xrightarrow[\text{heat}]{H^{\oplus}} \hspace{0.3em} {>}C{=}C{<} + H_2O$$

Alcohol reactivity:

$$\text{Allyl} > 3° > 2° > 1°$$

Orientation:

Most highly substituted alkene is produced

In the following sections we discuss another very important reaction of alcohol in which the —OH group is replaced by a halogen, —X.

10.13   Reaction of Alcohols with Hydrogen Halides

An alcohol may be converted in one step into an alkyl halide, as shown in the following general reaction:

$$R{-}OH + H{-}X \longrightarrow R{-}X + H_2O \qquad \text{substitution reaction}$$

where X = Cl, Br, I

Alkyl bromides can be prepared from the corresponding alcohols by heating them with concentrated hydrobromic acid, HBr, to which sulfuric acid is sometimes added. Another useful method involves preparing HBr *in situ* (meaning *in solution*) by mixing sulfuric acid and sodium bromide in the presence of the alcohol and heating the resulting mixture. Another alternative often used for secondary alcohols is to pass HBr gas through the heated alcohol. Because of elimination, aqueous concentrated sulfuric acid cannot be used with secondary or tertiary alcohols (see Sec. 9.12). These two examples are illustrative:

$$CH_3CH_2CH_2CH_2{-}OH \xrightarrow[\text{reflux}]{\text{NaBr, aq } H_2SO_4} CH_3CH_2CH_2CH_2{-}Br$$

1-Butanol   1-Bromobutane
(*n*-butyl alcohol)   (*n*-butyl bromide)
85–90%

$$\text{[cyclopentyl]}{-}OH \xrightarrow[\text{gas, heat}]{\text{anhydrous HBr}} \text{[cyclopentyl]}{-}Br$$

Cyclopentanol   Bromocyclopentane
(cyclopentyl   (cyclopentyl
alcohol)   bromide)

Alkyl iodides are prepared in a manner similar to that for alkyl bromides. For example:

$$CH_3CH_2CH_2{-}OH \xrightarrow[\text{heat}]{\text{HI, H}_2\text{SO}_4} CH_3CH_2CH_2{-}I$$

1-Propanol   1-Iodopropane
(*n*-propyl alcohol)   (*n*-propyl iodide)

$$CH_3CH_2CH_2{-}\underset{\underset{OH}{|}}{CH}{-}CH_3 \xrightarrow[\text{heat}]{\text{conc HI}} CH_3CH_2CH_2{-}\underset{\underset{I}{|}}{CH}{-}CH_3$$

2-Pentanol   2-Iodopentane

Alkyl chlorides are less easy to prepare because only the more reactive alcohols are converted to alkyl chlorides on treatment with hydrochloric acid or dry hydrogen chloride gas. However, a mixture of hydrogen chloride and zinc chloride, $ZnCl_2$, is usually effective in converting the less reactive (usually primary) alcohols to the corresponding alkyl chlorides. For example:

$$CH_3CH_2CH_2CH_2{-}OH \xrightarrow[\text{reflux}]{\text{HCl, ZnCl}_2} CH_3CH_2CH_2CH_2{-}Cl$$

1-Butanol   1-Chlorobutane
(*n*-butyl alcohol)   (*n*-butyl chloride)

$$CH_3CH_2{-}\underset{\underset{OH}{|}}{CH}{-}CH_3 \xrightarrow[\text{reflux}]{\text{HCl, ZnCl}_2} CH_3CH_2{-}\underset{\underset{Cl}{|}}{CH}{-}CH_3$$

2-Butanol   2-Chlorobutane
(*sec*-butyl alcohol)   (*sec*-butyl chloride)

The effect of the zinc salt is believed to involve a rapid equilibrium between the zinc chloride and a zinc chloride–alcohol complex:

$$R{-}\overset{..}{\underset{\underset{H}{|}}{O}}{:} \; + \; ZnCl_2 \;\rightleftharpoons\; \left[R^{\underline{\delta}\oplus}{-}{-}{-}\overset{..}{\underset{\underset{H}{|}}{O}}^{\underline{\delta}\ominus}{-}{-}^{\delta\ominus}ZnCl_2 \right] \xrightarrow{H^\oplus} R^\oplus + ZnCl_2 + H_2O$$

Lewis base   Lewis acid

Here it is the complex that actually undergoes dissociation to the carbocation. The zinc complexes with the alcohol and polarizes the carbon-oxygen bond, thereby weakening it, and at the same time provides a better leaving group in the $(ZnCl_2OH)^\ominus$ complex. This complex goes on to give zinc chloride and water.

The HCl-$ZnCl_2$ mixture is called the *Lucas reagent* and provides a convenient method for classifying alcohols as primary, secondary, or tertiary (see Sec. 10.16). Many tertiary alcohols are converted to their respective tertiary alkyl chlorides when treated with concentrated mineral acid. For example:

$$CH_3{-}\overset{\overset{CH_3}{|}}{\underset{\underset{OH}{|}}{C}}{-}CH_2CH_3 \xrightarrow[25^\circ]{\text{conc HCl}} CH_3{-}\overset{\overset{CH_3}{|}}{\underset{\underset{Cl}{|}}{C}}{-}CH_2CH_3$$

2-Methyl-2-butanol   2-Methyl-2-chlorobutane
(*tert*-pentyl alcohol;   (*tert*-pentyl chloride;
tert-amyl alcohol)   *tert*-amyl chloride)

Alkyl fluorides are not prepared directly from alcohols. Instead, the alcohol is converted to the alkyl chloride and then to the alkyl fluoride using the halide exchange reaction discussed in Sec. 5.10.

10.14   Mechanism of Reaction Between Hydrohalic Acids and Alcohols

We start our presentation of the mechanisms of alkyl halide formation from an alcohol by considering some known facts about the reaction:

1. *Acid catalyst.* The reaction requires the presence of protons. A mixture of sodium halide and alcohol never produces an alkyl halide regardless of the reaction time or temperature. Addition of a strong acid such as sulfuric acid causes the reaction to occur very rapidly.
2. *Rearrangements.* In certain alcohols, molecular rearrangements occur.
3. *Stereochemistry.* There is inversion of configuration with primary and some secondary alcohols, whereas other secondary and all tertiary alcohols give racemic alkyl halides.
4. *Reactivity.* The reactivity order for the reaction of alcohols with H—X is:

$$Allyl > 3° > 2° > 1° < CH_3OH$$

The mechanism of alkyl halide formation from an alcohol must account for all these facts. As we will see, the mechanism is largely dictated by the structure of the alcohol (primary, secondary, or tertiary) and two different mechanisms must be invoked.

All these reactions have one feature in common, however; acid is required to protonate the alcohol.

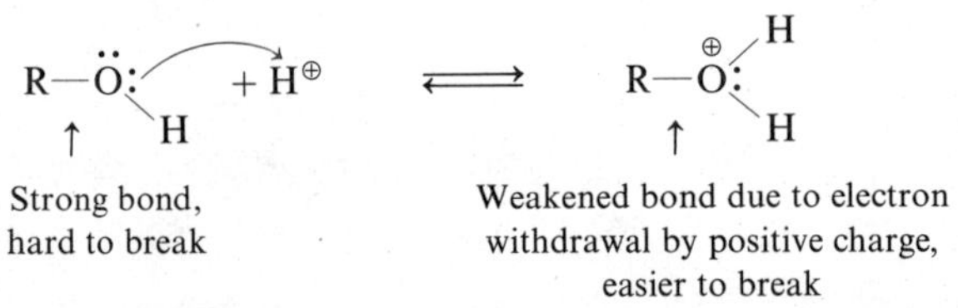

$$R—\overset{..}{\underset{\uparrow\ \ H}{O}} \quad + H^{\oplus} \quad \rightleftharpoons \quad R—\overset{\oplus}{\underset{\uparrow\ \ H}{O}}\diagup^{H}$$

| Strong bond, hard to break | Weakened bond due to electron withdrawal by positive charge, easier to break |

This equilibrium occurs constantly in acidic solution. Protonation weakens the carbon-oxygen bond sufficiently so that substitution may ultimately occur.

A. Mechanism for Methanol and Most Primary (1°) Alcohols: S_N2 Reaction

The preparation of primary alkyl and methyl halides from the corresponding alcohols involves the S_N2 mechanism shown here.

Step 1. Protonation of alcohol; an equilibrium reaction

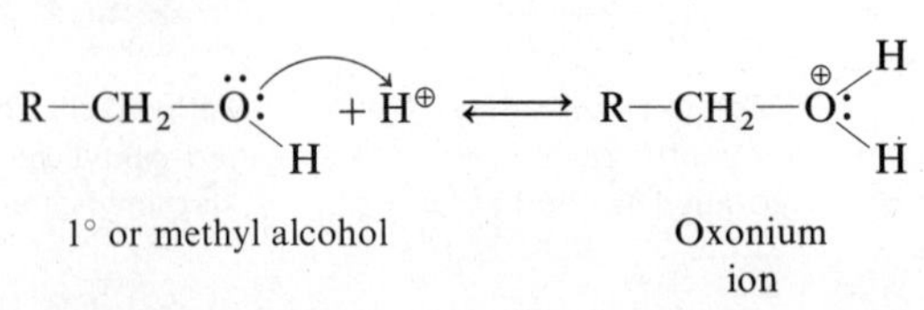

$$R—CH_2—\overset{..}{\underset{H}{O}} \quad + H^{\oplus} \rightleftharpoons R—CH_2—\overset{\oplus}{\underset{H}{O}}\diagup^{H}$$

| 1° or methyl alcohol | Oxonium ion |

Step 2. Substitution of halide ion, $:\overset{..}{\underset{..}{X}}:^{\ominus}$, for H_2O; a concerted, one-step reaction

$$:\overset{..}{\underset{..}{X}}:^{\ominus} + \underset{R}{CH_2}-\overset{\oplus}{\underset{|}{O}}\overset{H}{\underset{H}{\diagdown}} \longrightarrow \left[\overset{\delta\ominus}{:\overset{..}{X}}---\underset{R}{CH_2}---\overset{\delta\oplus}{\overset{..}{O}}\overset{H}{\underset{H}{\diagdown}} \right] \longrightarrow :\overset{..}{X}-\underset{R}{CH_2} + :\overset{..}{O}\overset{H}{\underset{H}{\diagdown}}$$

| Reactants | Transition state | Products |

This substitution reaction is very similar to the S_N2 reaction between alkyl halides and various nucleophiles described in Sec. 9.3. Some major differences between these two S_N2 reactions are indicated in the following equations:

Alkyl halide + nucleophile:

$$Nu:^{\ominus} + \underset{R}{CH_2}-\overset{..}{\underset{..}{X}}: \longrightarrow \left[\overset{\delta\ominus}{Nu}---\underset{R}{CH_2}---\overset{\delta\ominus}{\overset{..}{X}}: \right] \longrightarrow Nu-\underset{R}{CH_2} + :\overset{..}{\underset{..}{X}}:^{\ominus}$$

| Negatively charged | Neutral | Negatively charged transition state | Neutral | Negatively charged |

Protonated alcohol + halide:

$$:\overset{..}{\underset{..}{X}}:^{\ominus} + \underset{R}{CH_2}-\overset{\oplus}{\overset{..}{O}}H_2 \longrightarrow \left[\overset{\delta\ominus}{:\overset{..}{X}}---\underset{R}{CH_2}---\overset{\delta\oplus}{\overset{..}{O}}H_2 \right] \longrightarrow :\overset{..}{X}-\underset{R}{CH_2} + H_2\overset{..}{O}:$$

| Negatively charged | Positively charged | Neutral transition state (highly polar) | Neutral | Neutral |

Note especially the *charges* on the reactants, transition states, and products. Substitution on halides results in the liberation of the negatively charged ion, whereas the neutral water molecule departs in the process of converting an alcohol to an alkyl halide. There is conservation of charge.

These two nucleophilic substitution reactions are clearly similar because both involve a nucleophile and a pentacovalent-like transition state. The major difference is that *protonation, a reversible reaction, precedes nucleophilic substitution on an alcohol.*

What evidence supports the mechanism for the S_N2 reaction on protonated primary alcohols? Perhaps we should ask what alternative mechanism might be used instead; then the S_N1 mechanism comes to mind. Because the S_N1 reaction involves carbocations, we look for evidence that implies their presence. Two important facts exclude carbocation formation and thus the S_N1 mechanism in the displacement reaction on protonated methyl and primary alcohols: (1) lack of rearrangement and (2) occurrence of complete inversion of configuration with optically active alcohols. Let us look at these further.

No rearrangement of the carbon skeleton occurs when a typical primary alcohol is allowed to react with H—X. For example, *n*-butyl alcohol gives *only n*-butyl bromide and *no sec*-butyl bromide. The following equations show mechanistically how a carbocation intermediate is excluded, starting with the protonated alcohol.

Observed:

$$CH_3CH_2CH_2\overset{\oplus}{\underset{|}{C}}H_2-\overset{..}{O}\overset{H}{\underset{H}{\diagdown}} \xrightarrow[\text{reaction}]{\overset{:\overset{..}{\underset{..}{Br}}:^{\ominus}}{S_N2}} CH_3CH_2CH_2CH_2-\overset{..}{\underset{..}{Br}}: + H_2\overset{..}{O}:$$

| Protonated 1-butanol (*n*-butyl alcohol) | 1-Bromobutane (*n*-butyl bromide) |

Not observed:

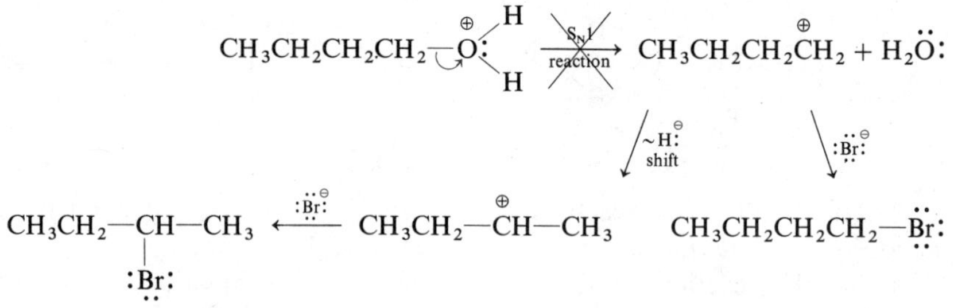

$$CH_3CH_2CH_2CH_2-\overset{\oplus}{\underset{H}{\overset{H}{O}}} \xrightarrow[\text{reaction}]{S_N1} CH_3CH_2CH_2\overset{\oplus}{C}H_2 + H_2\ddot{O}:$$

$$\swarrow \underset{\text{shift}}{\sim H:^{\ominus}} \qquad \searrow :\ddot{\underset{..}{Br}}:^{\ominus}$$

$$CH_3CH_2-\overset{|}{\underset{:Br:}{C}H}-CH_3 \xleftarrow{:\ddot{Br}:^{\ominus}} CH_3CH_2-\overset{\oplus}{C}H-CH_3 \qquad CH_3CH_2CH_2CH_2-\ddot{\underset{..}{Br}}:$$

2-Bromobutane
(*sec*-butyl bromide)
Not formed

More stable 2° ion

Stereochemistry is difficult to study with primary alcohols because it *appears* that there is no way to have an optically active primary alcohol in which the —OH group is attached to the asymmetric center. A primary carbon atom can be made asymmetric, however, by replacing one hydrogen atom by deuterium. For example, 1-butanol-1-*d* (1) has been prepared in optically active form and converted to 1-bromo-1-deuteriobutane. The reaction occurs with *complete inversion of configuration* so that the S_N2 mechanism must be occurring. The steric course of this reaction is:

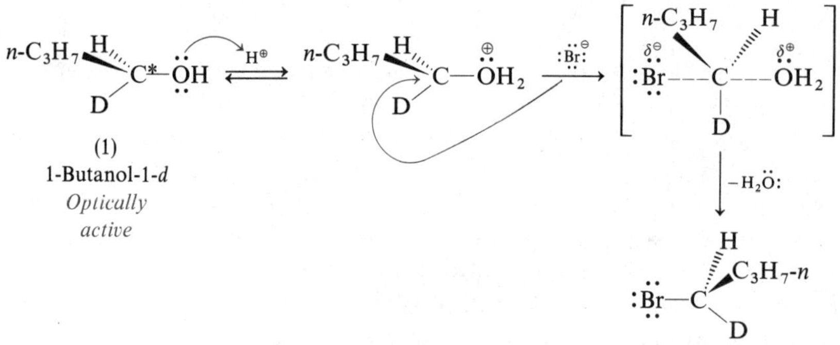

(1)
1-Butanol-1-*d*
*Optically
active*

1-Bromo-1-deuteriobutane
Inversion of configuration

B. Mechanism for Secondary (2°) and Tertiary (3°) Alcohols: S_N1 Reaction

The conversion of most secondary and all tertiary alcohols to their corresponding alkyl halides occurs via the S_N1 mechanism shown here. As with primary alcohols, the first step is the protonation of the alcohol in an equilibrium reaction.

Step 1. Protonation of alcohol, an equilibrium reaction

Secondary alcohol:

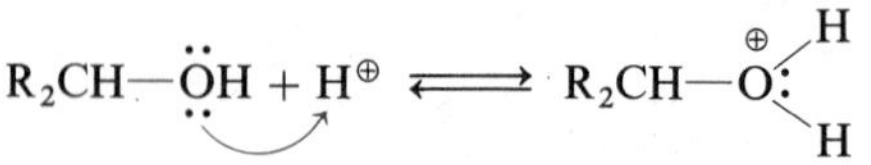

$$R_2CH-\ddot{O}H + H^{\oplus} \rightleftharpoons R_2CH-\overset{\oplus}{\underset{H}{\overset{H}{O}}}:$$

Tertiary alcohol:

$$R_3C-\ddot{O}H + H^{\oplus} \rightleftharpoons R_3C-\overset{\oplus}{\underset{H}{\overset{H}{O}}}:$$

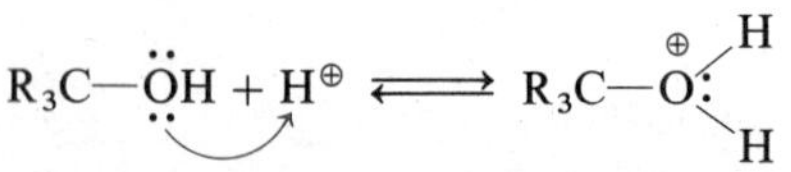

Steps 2 and 3. Loss of water to give carbocation and attack by halide ion

Secondary alcohol:

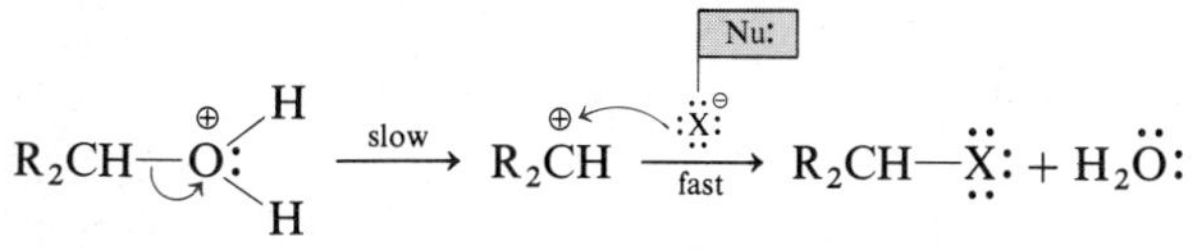

$$R_2CH\!-\!\overset{\oplus}{O}\!\!\underset{H}{\overset{H}{\diagup}} \xrightarrow{\text{slow}} R_2\overset{\oplus}{CH} \xrightarrow[\text{fast}]{:\overset{..}{\underset{..}{X}}:^{\ominus}} R_2CH\!-\!\overset{..}{\underset{..}{X}}: + H_2\overset{..}{\underset{..}{O}}:$$

Tertiary alcohol:

$$R_3C\!-\!\overset{\oplus}{O}\!\!\underset{H}{\overset{H}{\diagup}} \xrightarrow{\text{slow}} R_3\overset{\oplus}{C} \xrightarrow[\text{fast}]{:\overset{..}{\underset{..}{X}}:^{\ominus}} R_3C\!-\!\overset{..}{\underset{..}{X}}: + H_2\overset{..}{\underset{..}{O}}:$$

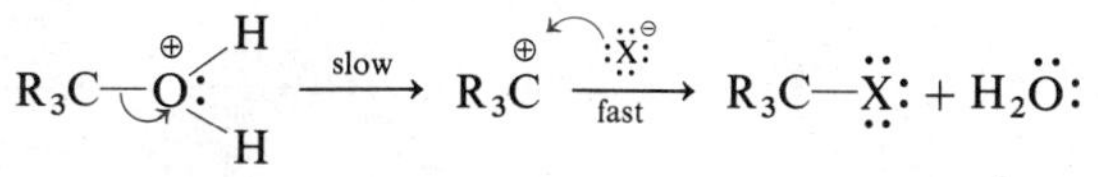

This reaction is very similar to the S_N1 reaction of alkyl halides with various nucleophiles. With alcohols, the leaving group is the neutral water molecule, whereas the halide ion, $:\overset{..}{\underset{..}{X}}:^{\ominus}$, departs in nucleophilic substitution on alkyl halides. The driving forces for the departure of these leaving groups are similar; halide ion leaves because of its highly electronegative character, and water leaves because the positive charge on the oxonium ion, $R\!-\!\overset{\oplus}{O}H_2$, "pulls" electrons toward oxygen and away from carbon. *The main difference between substitution on alkyl halides and substitution on alcohols is that the latter requires prior protonation to give the oxonium ion before reaction occurs.* The following reactions illustrate the similarity in leaving groups.

Alkyl halides:

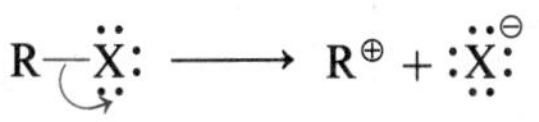

$$R\!-\!\overset{..}{\underset{..}{X}}: \longrightarrow R^{\oplus} + :\overset{..}{\underset{..}{X}}:^{\ominus}$$

Alcohols:

$$R\!-\!\overset{\oplus}{O}\!\!\underset{H}{\overset{H}{\diagup}} \longrightarrow R^{\oplus} + H_2\overset{..}{\underset{.}{O}}:$$

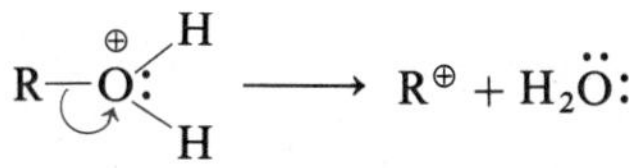

There is considerable evidence to support the carbocation mechanism for most secondary and all tertiary alcohols. The reactivity order of alcohols ($3° > 2° > 1° < CH_3OH$) parallels the stability order for carbocations, except for the methyl cation. The relative instability of primary carbocations causes primary alcohols to react mostly by the S_N2 mechanism, and the methyl cation is even more unstable so that methyl alcohol reacts entirely by the S_N2 mechanism.

The molecular rearrangements of certain secondary and tertiary alcohols are explained conveniently by a carbocation mechanism. Rearrangement sometimes produces an alkyl halide in which the location of the halide differs from that of the original OH group, and sometimes it produces rearrangement of the carbon skeleton.

Example 1

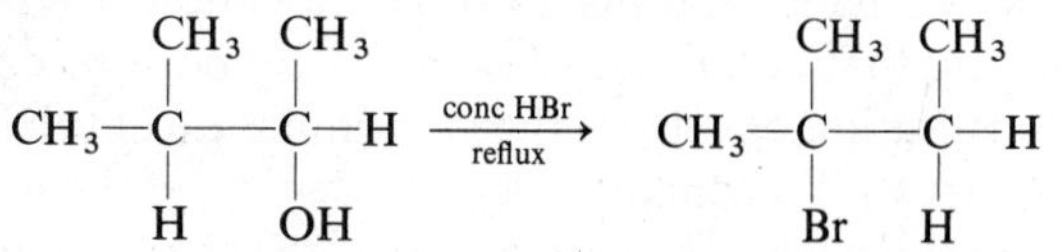

$$\underset{\text{3-Methyl-2-butanol}}{CH_3\!-\!\underset{\underset{H}{|}}{\overset{\overset{CH_3}{|}}{C}}\!-\!\underset{\underset{OH}{|}}{\overset{\overset{CH_3}{|}}{C}}\!-\!H} \xrightarrow[\text{reflux}]{\text{conc HBr}} \underset{\substack{\text{2-Bromo-2-methylbutane} \\ \textit{Only product}}}{CH_3\!-\!\underset{\underset{Br}{|}}{\overset{\overset{CH_3}{|}}{C}}\!-\!\underset{\underset{H}{|}}{\overset{\overset{CH_3}{|}}{C}}\!-\!H}$$

Mechanism:

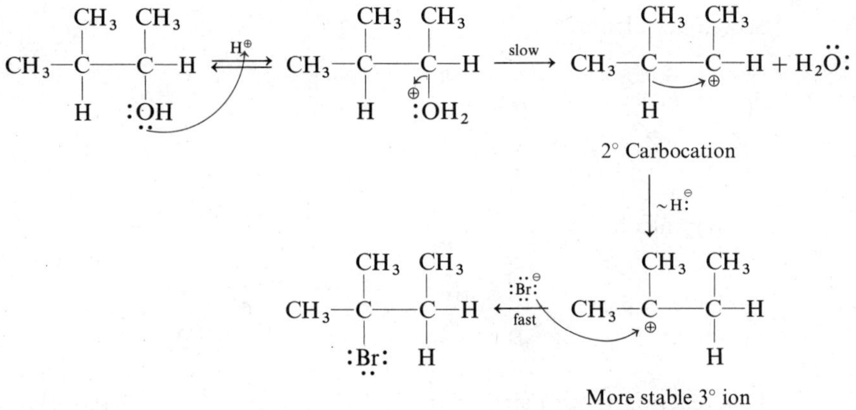

2° Carbocation

More stable 3° ion

Example 2

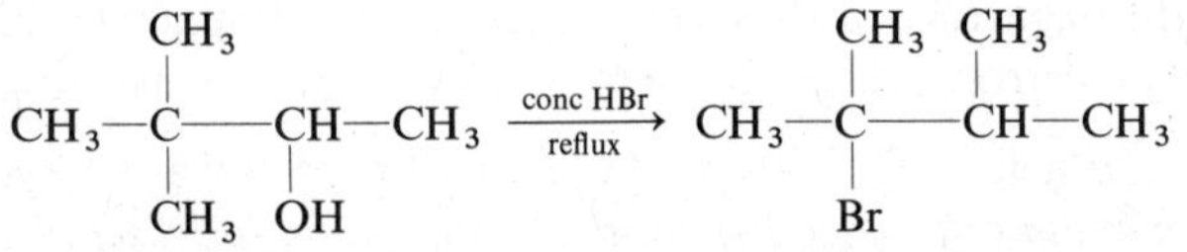

3,3-Dimethyl-2-butanol   2-Bromo-2,3-dimethylbutane

Mechanism:

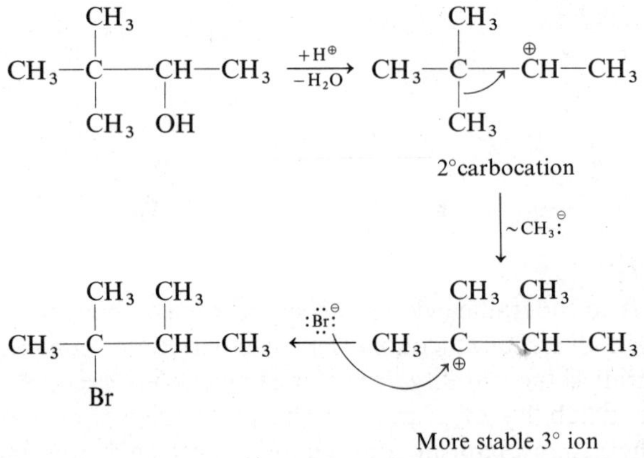

2°carbocation

More stable 3° ion

C. Summary

In general, alcohols tend to react with hydrohalic acids by the S_N1 mechanism, so the reaction is more likely to involve carbocations. The crossover point in the alkyl group structure occurs somewhere between primary and secondary; some primary alcohols, most secondary alcohols, and all tertiary alcohols react by the S_N1 mechanism. Methanol and most primary alcohols react by the S_N2 mechanism. It is not surprising that certain alcohols react by both mechanisms; the analogous nucleophilic substitution on alkyl halides is similar except that the crossover between S_N1 and S_N2 occurs with secondary alkyl halides. As in the reactions of alkyl halides, experimental conditions may be varied so that the conversion of a secondary alcohol to a corresponding alkyl halide occurs predominantly by either the S_N1 or S_N2 mechanism.

Substitution Versus Elimination of Alcohols

In connection with our discussion of the substitution reaction on primary alcohols, the curious student may question the validity of the S_N2 mechanism upon recalling that the dehydration of primary alcohols occurs via a carbocation (the E1 reaction, see Sec. 7.15).

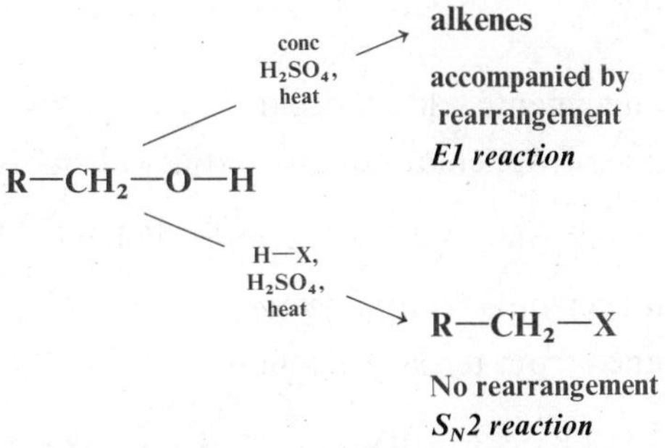

$$R{-}CH_2{-}O{-}H$$

conc H₂SO₄, heat → alkenes
accompanied by rearrangement
E1 reaction

H—X, H₂SO₄, heat → $R{-}CH_2{-}X$
No rearrangement
S_N2 reaction

The answer to this apparent discrepancy comes from differences in reaction conditions. Hot, concentrated sulfuric acid is used for dehydration, where the protonated alcohol has little choice but to lose water because there is no good nucleophilic species present for a reaction. This also means that a highly reactive carbocation, once formed, has little choice but to rearrange and/or lose a proton to form alkenes. On the other hand, conditions that favor alkyl halide formation include a large concentration of the relatively nucleophilic halide ion, so that nucleophilic substitution occurs preferentially. Even so, the preparation of alkyl halides from alcohols is often accompanied by some elimination. The alkenes can be removed by "washing" the alkyl halide with cold, concentrated sulfuric acid, which reacts with and removes the alkenes and the unreacted alcohol (see Sec. 10.12).

Question 10.13

When neopentyl alcohol, $(CH_3)_3CCH_2{-}OH$, is heated with HCl-ZnCl₂, it is converted to 2-chloro-2-methylbutane; no alkyl halide with the structure $(CH_3)_3CCH_2{-}Cl$ is produced.

(*a*) Provide a mechanism to explain the formation of the observed product.
(*b*) Neopentyl alcohol is a primary alcohol, but judging from the above results it apparently does not undergo an S_N2 reaction. Based on your general knowledge about S_N2 reactions, why is this so? (*Hint:* See Secs. 5.11 and 9.13.)

Question 10.14

Consider the following reaction:

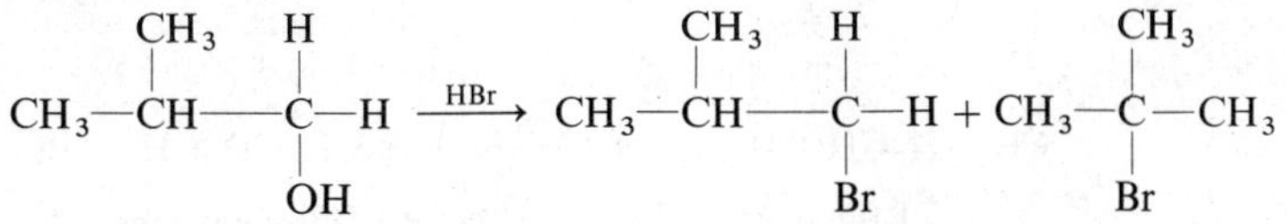

$$CH_3{-}\overset{\overset{\displaystyle CH_3}{|}}{CH}{-}\overset{\overset{\displaystyle H}{|}}{\underset{\underset{\displaystyle OH}{|}}{C}}{-}H \xrightarrow{\text{HBr}} CH_3{-}\overset{\overset{\displaystyle CH_3}{|}}{CH}{-}\overset{\overset{\displaystyle H}{|}}{\underset{\underset{\displaystyle Br}{|}}{C}}{-}H + CH_3{-}\overset{\overset{\displaystyle CH_3}{|}}{\underset{\underset{\displaystyle Br}{|}}{C}}{-}CH_3$$

Provide a mechanistic interpretation for this result, being careful to explain why this 1° alcohol yields two different alkyl bromides.

10.15   Other Methods for Preparing Alkyl Halides from Alcohols

Other common laboratory methods are used for converting alcohols to the corresponding alkyl halides. Their equations are shown in general form and then discussed in detail.

A. Using Phosphorus Halides

$$3 \text{ R—OH} + \text{PX}_3 \xrightarrow{\text{heat}} 3 \text{ R—X} + \text{P(OH)}_3$$

Phosphorus   Orthophosphorus
trihalide   acid

where X = Cl, Br

Features: Rearrangements seldom occur.

Poor yields obtained from chlorides and tertiary alcohols.

$$6 \text{ R—OH} + 2 \text{ P} + 3 \text{ I}_2 \xrightarrow{\text{heat}} 6 \text{ R—I} + 2 \text{ P(OH)}_3$$

Features: Rearrangements seldom occur.

Poor yields obtained from tertiary alcohols.

$$\text{R—OH} + \text{PX}_5 \xrightarrow{\text{heat}} \text{R—X} + \text{POX}_3 + \text{HX}$$

Phosphorus   Phosphorus
pentahalide   oxyhalide

where X = Cl, Br

Features: Rearrangements seldom occur.

Poor yields obtained from chlorides and tertiary alcohols.

These reactions are similar because they all involve some form of phosphorus halide acting on an alcohol. Phosphorus tribromide and trichloride, PBr_3 and PCl_3, are commercially available, whereas phosphorus triiodide, PI_3, is too unstable. However, it is prepared in situ by mixing alcohol, elemental red phosphorus, and iodine. The reaction between phosphorus and iodine is: $2\text{P} + 3\text{I}_2 \rightarrow 2\text{PI}_3$, but the reactants are normally indicated as P and I_2 when writing the reaction. Heating the resulting mixture gives a good yield of alkyl iodide when primary and secondary alcohols are used. Phosphorus pentachloride, PCl_5, is commercially available, but phosphorus pentabromide, PBr_5, is prepared in situ from PBr_3 and Br_2.

All the phosphorus halide reactions work poorly with tertiary alcohols but give good yields with most primary and secondary alcohols. That most secondary alcohols can be converted to the corresponding alkyl halides without rearrangement makes the reaction quite useful. Four examples follow:

$$\text{CH}_3\text{CH}_2\text{—OH} \xrightarrow[\text{heat}]{\text{P,I}_2} \text{CH}_3\text{CH}_2\text{—I}$$

Ethanol   Iodoethane
(ethyl alcohol)   (ethyl iodide)

$$\text{CH}_3\text{CH}_2\text{CH}_2\text{CH}_2\text{—OH} \xrightarrow[\text{heat}]{\text{PBr}_3} \text{CH}_3\text{CH}_2\text{CH}_2\text{CH}_2\text{—Br}$$

1-Butanol   1-Bromobutane
(n-butyl alcohol)   (n-butyl bromide)

⬡—OH $\xrightarrow{\text{PBr}_3}$ ⬡—Br

Cyclohexanol   Bromocyclohexane
(cyclohexyl alcohol)   (cyclohexyl bromide)

$$\text{CH}_3\text{CH}_2\text{—}\underset{\underset{\text{OH}}{|}}{\text{CH}}\text{—CH}_3 \xrightarrow[25°]{\text{PBr}_3} \text{CH}_3\text{CH}_2\text{—}\underset{\underset{\text{Br}}{|}}{\text{CH}}\text{—CH}_3$$

2-Butanol   2-Bromobutane
(sec-butyl alcohol)   (sec-butyl bromide)

The mechanism of some of these reactions has been studied and is believed to involve the initial formation of a phosphite ester, which is then attacked by halide ion. The currently accepted mechanism is shown here:

Step 1. Formation of phosphite ester

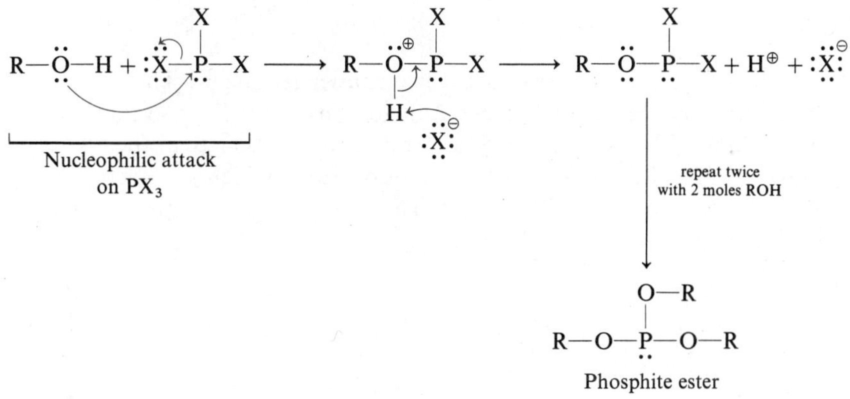

$$R\!-\!\ddot{O}\!-\!H + :\ddot{X}\!-\!\overset{\overset{\displaystyle X}{|}}{P}\!-\!X \longrightarrow R\!-\!\overset{\oplus}{\ddot{O}}\!-\!\overset{\overset{\displaystyle X}{|}}{P}\!-\!X \longrightarrow R\!-\!\ddot{O}\!-\!\overset{\overset{\displaystyle X}{|}}{P}\!-\!X + H^{\oplus} + :\ddot{X}\!:^{\ominus}$$

Nucleophilic attack on PX_3

repeat twice with 2 moles ROH

$$R\!-\!O\!-\!\overset{\overset{\displaystyle O-R}{|}}{P}\!-\!O\!-\!R$$

Phosphite ester

Step 2. Attack of halide ion on phosphite ester

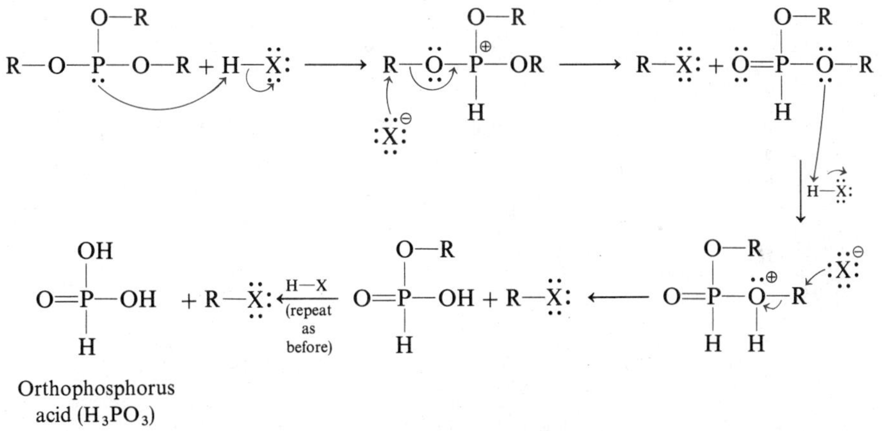

$$R\!-\!O\!-\!\overset{\overset{\displaystyle O-R}{|}}{P}\!-\!O\!-\!R + H\!-\!\ddot{X}\!: \longrightarrow R\!-\!\overset{\oplus}{\ddot{O}}\!-\!\overset{\overset{\displaystyle O-R}{|}}{\underset{\underset{\displaystyle H}{|}}{P}}\!-\!OR \longrightarrow R\!-\!\ddot{X}\!: + \ddot{O}\!=\!\overset{\overset{\displaystyle O-R}{|}}{\underset{\underset{\displaystyle H}{|}}{P}}\!-\!\ddot{O}\!-\!R$$

$$O\!=\!\overset{\overset{\displaystyle OH}{|}}{\underset{\underset{\displaystyle H}{|}}{P}}\!-\!OH + R\!-\!\ddot{X}\!: \xleftarrow[\substack{\text{(repeat}\\ \text{as}\\ \text{before)}}]{H-X} O\!=\!\overset{\overset{\displaystyle O-R}{|}}{\underset{\underset{\displaystyle H}{|}}{P}}\!-\!OH + R\!-\!\ddot{X}\!: \longleftarrow O\!=\!\overset{\overset{\displaystyle O-R}{|}}{\underset{\underset{\displaystyle H}{|}}{P}}\!-\!\overset{\oplus}{\ddot{O}}\!-\!R$$

Orthophosphorus acid (H_3PO_3)

One driving force for this reaction is the great affinity of phosphorus for oxygen, as evidenced by the reaction between alcohol and PX_3 to form the trialkylphosphite, $(RO)_3P:$. The conversion of the phosphite to orthophosphorus acid and alkyl halide is accomplished by a series of three S_N2 reactions, in which halide attacks the alkyl group and displaces the moiety containing phosphorus and oxygen. Phosphine oxide groups are better leaving groups than is the hydroxyl group, and careful examination of the formulas reveals that the leaving group in each case contains phosphorus and oxygen.

The possibility that substitution occurs via an S_N1 mechanism has not been excluded. The rare occurrence of rearrangements suggests that carbocations are not formed, however. Also, alkyl chlorides are produced in poor yields compared with the bromides and iodides, and this supports the bimolecular displacement reaction, for we recall that iodide and bromide ions are better nucleophiles than is chloride ion in S_N2 reactions (the nature of the nucleophile plays a much less important role in the S_N1 reaction).

B. Using Thionyl Chloride

Thionyl chloride, $SOCl_2$, is often used to prepare alkyl chlorides from alcohols, especially because phosphorus trichloride gives poor yields. The reaction is:

$$R\!-\!OH + SOCl_2 \xrightarrow{\text{heat}} R\!-\!Cl + SO_2 + HCl$$

Thionyl
chloride

This reaction has several experimental advantages. The first is that thionyl chloride is a liquid with a relatively low boiling point ($79°$) so that excess reagent is easily removed by distillation. The two products, other than the alkyl chloride, are sulfur dioxide, SO_2, and hydrogen chloride, HCl, which are liberated as gases.

The mechanism of this reaction has been studied in some detail, and it involves the intermediate chlorosulfite ester (1).

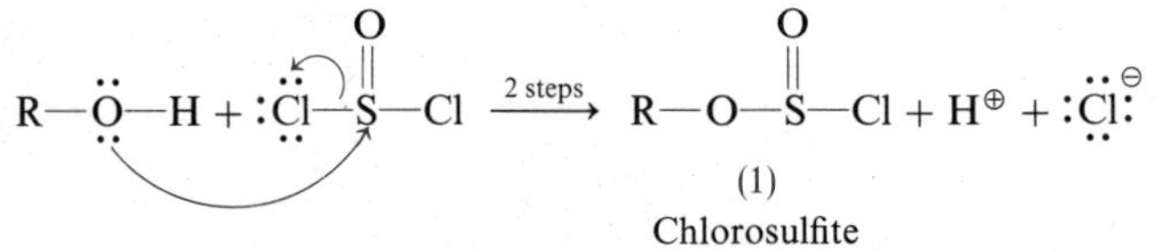

(1)
Chlorosulfite
ester

The chlorosulfite ester can decompose by either the S_N1 or S_N2 mechanism as follows:

S_N1 mechanism:

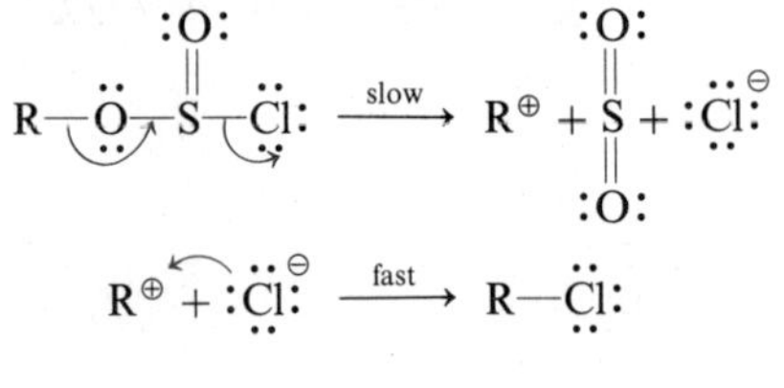

S_N2 mechanism:

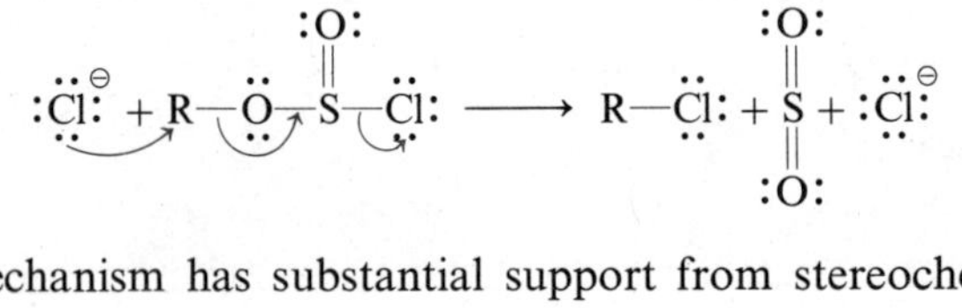

A third mechanism has substantial support from stereochemical studies. It involves an *internal reaction* in which the chlorine on the chlorosulfite ester group is transferred to carbon on the same side that the group originally occupied. A cyclic transition state is proposed to explain this intramolecular reaction.

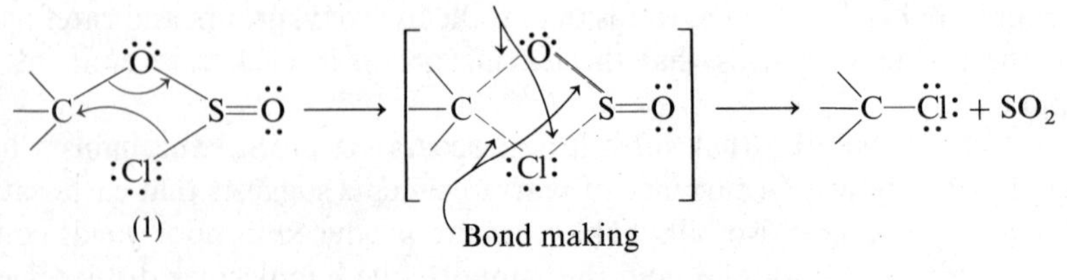

Bond breaking

Bond making

(1)

Cyclic transition state
(Cl attaches to carbon on
the same side as the
original —OSOCl group)

This mechanism is called $S_N i$ for *internal nucleophilic substitution*. It is highly dependent on the nature of the solvent, which we do not discuss in detail here. The stereochemical evidence comes from allowing an optically active secondary alcohol, such as 2-octanol, to react with thionyl chloride. Studies of the product show that the reaction occurs with *net retention of configuration* but that some racemization does occur possibly due to competing $S_N 1$ and $S_N 2$ mechanisms.

An alternative mechanism proposed also explains the high degree of stereo-specificity observed in the $S_N i$ reaction. In contrast to the cyclic transition state diagrammed earlier, this mechanism involves an intermediate species resulting from the breaking of the oxygen-carbon bond of the chlorosulfite ester to yield a *tight ion pair*. This tight ion pair then undergoes a rapid reaction to produce the alkyl halide of retained configuration and sulfur dioxide.

There is experimental evidence favoring both mechanisms, but the latter is the most widely accepted today. In particular, the intervention of the tight ion pair provides a good explanation of the observed (partial) racemization that accompanies the reaction.

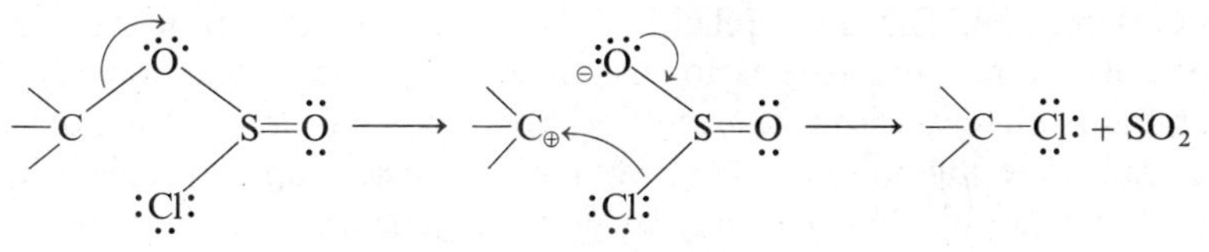

Tight ion pair

Question 10.15

Give the structure of the alcohol and the reagents that would be required to produce each of the following alkyl halides in good yield and without contamination by side products:

(*a*) 2-chloro-3-methylbutane   (*b*) *tert*-butyl iodide
(*c*) isohexyl iodide   (*d*) *n*-decyl chloride
(*e*) *cis*-4-bromomethylcyclohexane (two methods from two different alcohols)

Question 10.16

Explain mechanistically the following experimental observation:

$$CH_2{=}CH{-}CD_2{-}OH \xrightarrow[\text{ether}]{\text{SOCl}_2} CH_2{=}CH{-}CD_2{-}Cl + CD_2{=}CH{-}CH_2{-}Cl$$

Major   Minor

10.16   Qualitative Tests; Classification of Alcohols

One simple laboratory test that may be used to *suggest* the presence of an alcohol (or other organic molecules that contain hydrogens of comparable or greater acidity) is the reaction of sodium metal with an unknown compound. If the sodium dissolves with the liberation of a gas, there is at least a possibility that the compound is an alcohol. Other functional groups, such as phenols and carboxylic acids, also react with sodium.

Two important reactions studied in this chapter provide the backbone for the classification of alcohols according to alkyl group structure. These are the Lucas test and chromic acid test. Although results from these tests often reinforce one another, we discuss them separately.

The Lucas test is often used to distinguish primary, secondary, and tertiary alcohols, and the reagent is a mixture of concentrated hydrochloric acid and anhydrous zinc chloride. When allowed to react with alcohols, the Lucas reagent converts them to the corresponding alkyl chlorides, and the *rate* of the reaction provides an indication of alkyl group structure. Primary alcohols give no appreciable reaction, secondary alcohols react more rapidly, and tertiary alcohols react very rapidly. For this test to be observed in the laboratory, the alcohol must be soluble in the Lucas reagent because the alkyl chloride is not. The formation of a *second layer* or an *emulsion* is a *visual* indication that a reaction is occurring.

Primary: RCH_2—OH + HCl $\xrightarrow{ZnCl_2}$ no reaction

Secondary: R_2CH—OH + HCl $\xrightarrow{ZnCl_2}$ R_2CH—Cl + H_2O moderately fast

Tertiary: R_3C—OH + HCl $\xrightarrow{ZnCl_2}$ R_3C—Cl + H_2O very fast

Heating to 50 or 60° may be required before the reaction occurs in a reasonable period of time. The most reliable results are obtained from the Lucas test when controls are run against unknowns; that is, the rate of reaction of an unknown is compared directly with rates resulting from known primary, secondary, and tertiary alcohols. The solubility of the alcohol in the Lucas reagent places severe limitations on the type of alcohol that may be used. In general, monofunctional alcohols with six or fewer carbon atoms and almost all polyfunctional alcohols can be used.

The chromic acid in acetone test is based on the oxidation reactions that primary and secondary alcohols and aldehydes undergo. The reagent is prepared by dissolving chromic anhydride, CrO_3, in sulfuric acid, with acetone, $(CH_3)_2C=O$, used as the solvent. This reagent oxidizes primary and secondary alcohols, and as it does, there is a distinctive color change. Tertiary alcohols or ketones are unaffected by the reagent. The chromic acid reagent is orange, and a positive test is indicated when this color is replaced by a green or blue-green precipitate or emulsion.

Primary: RCH_2—OH $\xrightarrow{H_2CrO_4}$ $R-\overset{\displaystyle O}{\overset{\|}{C}}-H$ $\xrightarrow{H_2CrO_4}$ $R-\overset{\displaystyle O}{\overset{\|}{C}}-OH$

Secondary: R_2CH—OH $\xrightarrow{H_2CrO_4}$ $R-\overset{\displaystyle O}{\overset{\|}{C}}-R$ $\xrightarrow{H_2CrO_4}$ no further reaction

Tertiary: R_3C—OH $\xrightarrow{H_2CrO_4}$ no visible reaction

Primary and secondary alcohols and aldehydes give a positive reaction in 5 sec or less, whereas tertiary alcohols and ketones are unaffected by the test. Aldehydes can be distinguished from alcohols by other qualitative tests (see Sec. 21.16) as well. As with the Lucas test, it is helpful to run the chromic acid test on knowns and unknowns at the same time; slight differences in reaction time or in the color of the residue can indicate differences in the alkyl group structure.

Some other functional group tests give the following five results with alcohols:

1. They do not decolorize bromine in carbon tetrachloride.
2. They dissolve in concentrated sulfuric acid.
3. They normally give a negative Baeyer test (dilute, *cold* $KMnO_4$ in water). There are exceptions, however. Also, other functional groups (such as aldehydes) react positively with this reagent, so a positive Baeyer test must be regarded with reservation.

4. They give a negative test with $Ag(NH_3)_2^{\oplus}$ or $Cu(NH_3)_2^{\oplus}$ (the test for a terminal alkyne, see Sec. 11.18).

5. They give a negative chloroform-aluminum trichloride test (the test for the aromatic ring, see Sec. 14.24).

Question 10.17

Indicate what single laboratory test you would perform to distinguish between the following pairs of compounds. State the observations you would expect to see as a positive test.

(*a*) *sec*-butyl alcohol and *tert*-butyl alcohol
(*b*) cyclohexanol and *n*-hexyl alcohol
(*c*) allyl alcohol (CH_2=CH—CH_2—OH) and *n*-propyl alcohol
(*d*) isopentyl alcohol and isopropyl alcohol

10.17   Summary of Methods of Preparation of Alcohols

1. Acid-catalyzed hydration of alkenes (Secs. 8.5 and 10.3)

$$\text{C=C} + H_2O \xrightarrow{H^{\oplus}} -\overset{|}{\underset{H}{C}}-\overset{|}{\underset{OH}{C}}-$$

2. Hydrolysis of alkyl halides (Secs. 9.13, 9.15, 9.16, and 10.3)

$$R-X \xrightarrow[\text{or } H_2O]{H_2O,\,:\ddot{O}H^{\ominus}} R-OH$$

3. Hypohalous acid addition to alkenes (Secs. 8.22 and 10.3)

$$\text{C=C} \xrightarrow[\text{(HOX)}]{X_2,\,H_2O} -\overset{|}{\underset{X}{C}}-\overset{|}{\underset{OH}{C}}-$$

4. Hydroboration-oxidation of alkenes (Secs. 10.4 and 10.5)
(*a*) Hydroboration

$$\text{C=C} \xrightarrow{(BH_3)_2} \left(-\overset{|}{\underset{H}{C}}-\overset{|}{C}-\right)_3 B$$

Stereospecific *syn* addition
orientation (anti-Markovnikov,
B goes to carbon bearing greater
number of hydrogens)

(*b*) Oxidation

$$\left(-\overset{|}{\underset{H}{C}}-\overset{|}{C}-\right)_3 B \xrightarrow[:\ddot{O}H^{\ominus}]{H_2O_2} -\overset{|}{\underset{H}{C}}-\overset{|}{\underset{OH}{C}}-$$

Occurs with retention of configuration

5. Hydroboration-reduction of alkenes and alkynes (Sec. 10.6)

$$\overset{}{\underset{}{C}}=\overset{}{\underset{}{C} } \xrightarrow{(BH_3)_2} \left(-\overset{|}{\underset{H}{C}}-\overset{|}{\underset{|}{C}}- \right)_3 B \xrightarrow[H_2O]{H^\oplus} -\overset{|}{\underset{H}{C}}-\overset{|}{\underset{H}{C}}-$$

Alkene   Alkane

6. Hydroboration-isomerization (Sec. 10.6)

$$\left(-\overset{|}{\underset{H}{C}}-\overset{|}{\underset{|}{C}}- \right)_3 B \xrightarrow[\substack{\text{ether} \\ \text{solvent}}]{160°} \text{terminal alkyl boron compound}$$

Least substituted,
least hindered isomer

oxidation
or reduction

alcohol or alkane

10.18   Summary of Reactions of Alcohols

A. Reactions Involving Oxygen-Hydrogen (RO—H) Bond

1. Reaction with active metals (Sec. 10.9)

$$RO-H + M^0 \longrightarrow M^\oplus + \;\; R\ddot{O}\!:^\ominus \;\; + H_2$$

or $M^{\oplus 2}$ Alkoxide
or $M^{\oplus 3}$ ion

M^0 = group I (Na, K, Li, and so on) metal
      or group II (Mg, Ca, Ba, and so forth) metal
      or group III (Al, and so on) metal

ROH reactivity:

$$CH_3OH > 1° \; ROH > 2° \; ROH > 3° \; ROH$$

ROH acidity:

$$CH_3OH > 1° \; ROH > 2° \; ROH > 3° \; ROH$$

2. Oxidation reactions
(a) Primary alcohols (Sec. 10.10A)

$$R-CH_2-OH \xrightarrow[\substack{\text{warm} \\ \text{or } CrO_3\text{-pyridine} \\ \text{or } CrO_3, H_2O, \text{ warm}}]{\substack{Cu, \text{ heat} \\ \text{or } K_2Cr_2O_7, H_2SO_4,}} R-\overset{O}{\overset{||}{C}}-H \xrightarrow[\substack{\text{or } CrO_3, H_2O, \text{ heat} \\ \text{or } KMnO_4, HO^\ominus, \text{ heat}}]{\substack{K_2Cr_2O_7, \\ H_2SO_4, \text{ heat}}} R-\overset{O}{\overset{||}{C}}-OH$$

Aldehyde   Carboxylic acid

$$KMnO_4, OH^\ominus, \text{ heat}$$

(b) Secondary alcohols (Sec. 10.10B)

$$R-\overset{OH}{\underset{H}{\overset{|}{\underset{|}{C}}}}-R' \xrightarrow[\substack{\text{or } K_2Cr_2O_7, H_2SO_4 \\ \text{or } KMnO_4, OH^\ominus \\ \text{or } CrO_3\text{-pyridine}}]{Cu°, \text{ heat}} R-\overset{O}{\overset{||}{C}}-R'$$

Ketone

(*c*) Tertiary alcohols (Sec. 10.10C)

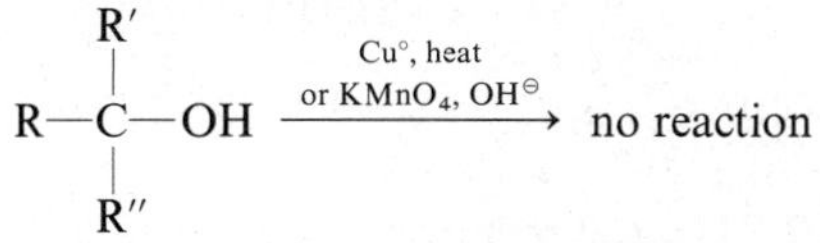

$$\underset{\underset{R''}{|}}{\overset{\overset{R'}{|}}{R-C-OH}} \quad \xrightarrow[\text{or KMnO}_4,\ \text{OH}^\ominus]{\text{Cu}^\circ,\ \text{heat}} \quad \text{no reaction}$$

K$_2$Cr$_2$O$_7$ and H$_2$SO$_4$ oxidize 3° alcohols capable of being dehydrated

(*d*) Mechanism of oxidation reaction (Sec. 10.11)

3. Formation of esters from alcohols (Sec. 10.12)
(*a*) Carboxylic acid esters

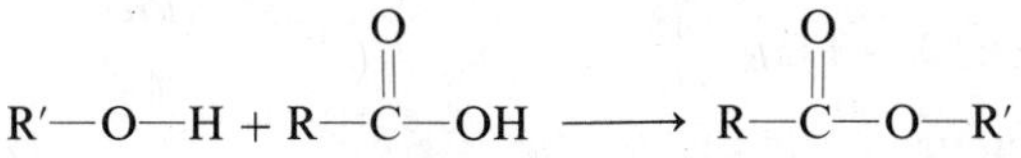

$$R'-O-H + R-\overset{\overset{\displaystyle O}{\|}}{C}-OH \longrightarrow R-\overset{\overset{\displaystyle O}{\|}}{C}-O-R'$$

$$\underset{\text{acid}}{\text{Carboxylic}} \qquad\qquad \underset{\text{acid ester}}{\text{Carboxylic}}$$

(*b*) Sulfate esters

$$R-O-H + HO-SO_3H \xrightarrow{\text{cold}} R-O-SO_3H$$

$$\underset{\text{acid}}{\text{Sulfuric}} \qquad\qquad \underset{\text{sulfate ester}}{\text{Alkyl hydrogen}}$$

$$\downarrow \text{ROH}$$

$$R-O-SO_2-O-R$$

$$\underset{\text{ester}}{\text{Dialkyl sulfate}}$$

(*c*) Nitrate esters

$$R-O-H + HO-NO_2 \xrightarrow[10-20^\circ]{\text{H}_2\text{SO}_4} R-O-NO_2$$

$$\underset{\text{acid}}{\text{Nitric}} \qquad\qquad \underset{\text{ester}}{\text{Alkyl nitrate}}$$

(*d*) Phosphate esters

$$R-O-H + HO-PO_3H_2 \longrightarrow R-O-PO_3H_2 \xrightarrow{\text{ROH}}$$

$$\underset{\text{acid}}{\text{Phosphoric}} \qquad \underset{\text{phosphate ester}}{\text{Alkyl dihydrogen}}$$

$$R-O-\overset{\overset{\displaystyle O}{\|}}{\underset{\underset{\displaystyle OH}{|}}{P}}-O-R \xrightarrow{\text{ROH}} R-O-\overset{\overset{\displaystyle O}{\|}}{\underset{\underset{\displaystyle O-R}{|}}{P}}-O-R$$

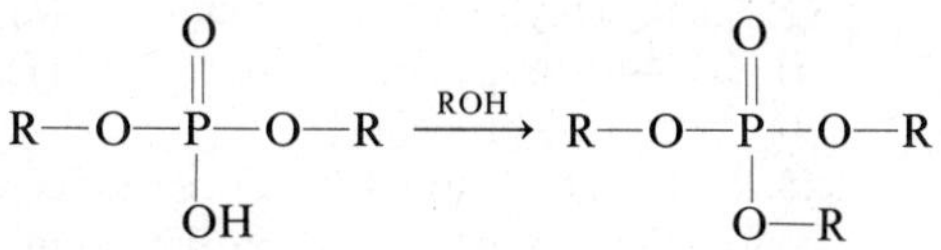

$$\underset{\text{phosphate ester}}{\text{Dialkyl hydrogen}} \qquad\qquad \underset{\text{ester}}{\text{Trialkyl phosphate}}$$

B. Reactions Involving Carbon-Oxygen (R—OH) Bond

1. Reaction with hydrogen halides (Secs. 10.13 and 10.14)

$$R-OH + HX \longrightarrow R-X + H_2O$$

where X = Cl, Br, I

Reactivity:

$$\text{Allyl} > 3° > 2° > 1° < CH_3OH$$

Rearrangements may occur.

Mechanism and stereochemistry:

Methanol and most 1° ROH: S_N2 reaction

2° and 3° ROH: S_N1 reaction

2. Reaction with phosphorus halides (Sec. 10.15A)

$$ROH \xrightarrow{PX_3} RX + P(OH)_3$$

$$ROH \xrightarrow{P, I_2} RI + P(OH)_3$$ Poor yields with 3° ROH, rearrangements seldom occur

$$ROH \xrightarrow{PX_5} RX + POX_3$$

where X = Cl, Br (poor with Cl)

3. Reaction with thionyl chloride (Sec. 10.15B)

$$ROH \xrightarrow{SOCl_2} RCl + SO_2 + HCl$$

Study Questions

10.18 Name the following nine compounds by the IUPAC system.

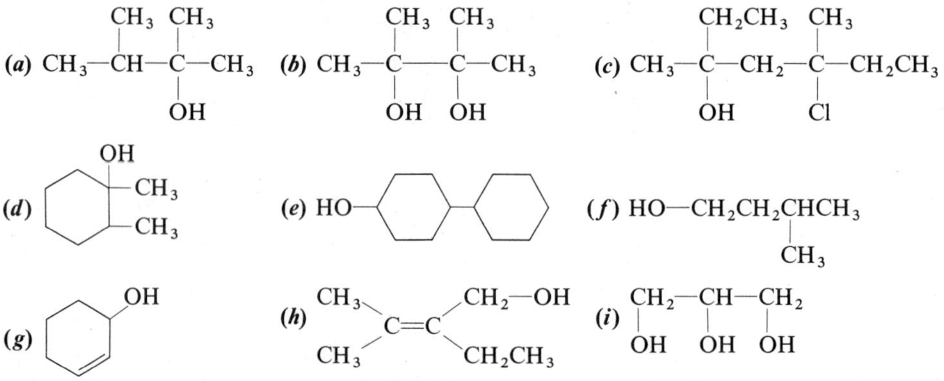

(a) $CH_3-\underset{}{CH}-\underset{}{C}-CH_3$ (b) $CH_3-\underset{}{C}-\underset{}{C}-CH_3$ (c) $CH_3-\underset{}{C}-CH_2-\underset{}{C}-CH_2CH_3$

(d) cyclohexanol with CH_3, CH_3, OH

(e) $HO-$ (bicyclohexyl)

(f) $HO-CH_2CH_2CHCH_3$ with CH_3

(g) cyclohexenol with OH

(h) $\underset{CH_3}{\overset{CH_3}{>}}C=C\underset{CH_2CH_3}{\overset{CH_2-OH}{<}}$

(i) $\underset{OH}{CH_2}-\underset{OH}{CH}-\underset{OH}{CH_2}$

10.19 Provide structures to correspond to the following 11 names.

(a) 2-methyl-2-pentanol   (b) 2,3-dimethyl-2-butanol

(c) cis-1,3-cyclobutanediol   (d) tert-pentyl alcohol

(e) isoheptyl alcohol   (f) diethylcarbinol

(g) methylethyl-tert-butylcarbinol   (h) cyclopentylcarbinol

(i) 2,2,4-trimethyl-3-pentanol   (j) trans-1,4-cyclohexanediol

(k) 1-chloro-2-propanol

10.20 Indicate the products or reactants, as required, to complete the following reactions. The items to be supplied are indicated by number, for example, (1). If no reaction occurs, so indicate.

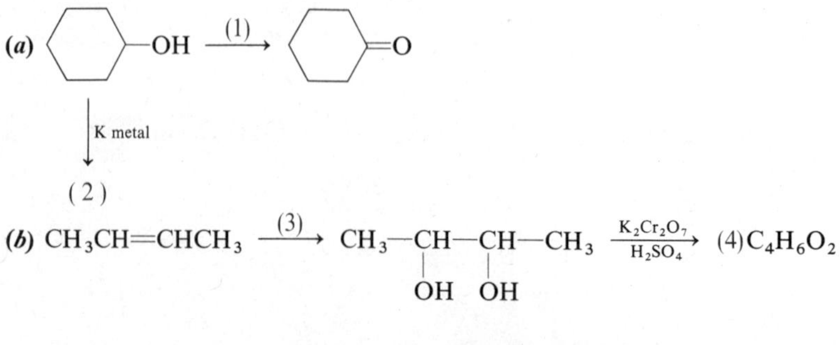

(a) (cyclohexyl)$-OH \xrightarrow{(1)}$ (cyclohexyl)$=O$

      $\downarrow$ K metal

    (2)

(b) $CH_3CH=CHCH_3 \xrightarrow{(3)} CH_3-\underset{OH}{CH}-\underset{OH}{CH}-CH_3 \xrightarrow[H_2SO_4]{K_2Cr_2O_7} (4)C_4H_6O_2$

(c) $CH_3CH_2CH_2CH_2$—OH $\xrightarrow{(5)}$ $CH_3CH_2CH_2CH_2$—Br $\xrightarrow[\text{alcohol}]{\text{KOH}}$ (6)

$\downarrow$(8) $\searrow$(9) $\searrow$(7) $(CH_3)_2CuLi$

$CH_3CH_2CH_2$—$\overset{\overset{\displaystyle O}{\|}}{C}$—H $CH_3CH_2CH_2$—$\overset{\overset{\displaystyle O}{\|}}{C}$—OH

(d) CH_3—$\overset{\overset{\displaystyle OH}{|}}{CH}$—$CH_3$ $\xrightarrow[\text{acetic acid}]{CrO_3}$ (10)

$\downarrow$ P, I$_2$

(11)

(e) (12) $\xleftarrow[\text{acetic acid}]{CrO_3}$ $CH_3CH_2CH_2$—OH $\xrightarrow[\text{heat}]{Cu^\circ}$ (16)

$\swarrow$ KMnO$_4$, :$\ddot{O}$H$^\ominus$, strong heat $\downarrow$ KMnO$_4$, :$\ddot{O}$H$^\ominus$, warm $\searrow$ K$_2$Cr$_2$O$_7$, H$_2$SO$_4$, warm

(13)   (14)   (15)

(f) CH_3CH_2—$\overset{\overset{\displaystyle OH}{|}}{CH}$—$CH_3$ $\xrightarrow{SOCl_2}$ (17)

$\downarrow$ Mg   (18) $\searrow$

(19) CH_3CH_2—$\overset{\overset{\displaystyle I}{|}}{CH}$—$CH_3$ $\xrightarrow{(20)}$ CH_3CH_2—$\overset{\overset{\displaystyle OH}{|}}{CH}$—$CH_3$

(g) CH_3—CH=CH$_2$ $\xrightarrow{(21)}$ Cl—CH$_2$—CH=CH$_2$ $\xrightarrow[\substack{KMnO_4, \\ H_2O}]{\text{cold}}$ (24)

$\swarrow$(22) $\downarrow$ 1. O$_3$ 2. H$_2$O, Zn

HO—CH$_2$—CH=CH$_2$ (23)

(h) [cyclohexene] $\xrightarrow[\text{CHCl}_3]{\text{NBS}}$ (25)

$\searrow$ Br$_2$, H$_2$O   (26)

$\downarrow$ Br$_2$, CCl$_4$

(27)

(i) CH_3—$\overset{\overset{\displaystyle CH_3}{|}}{C}$=CH$_2$ $\xrightarrow{(28)}$ CH_3—$\overset{\overset{\displaystyle CH_3}{|}}{\underset{\underset{\displaystyle OH}{|}}{C}}$—CH$_3$ $\xrightarrow{(29)}$ CH_3—$\overset{\overset{\displaystyle CH_3}{|}}{\underset{\underset{\displaystyle :\overset{\ominus}{\ddot{O}}:\, K^\oplus}{|}}{C}}$—CH$_3$

$\downarrow$ (BH$_3$)$_2$

(30) $\xrightarrow{H_2O_2,\ OH^\ominus}$ (31)

(j) [cyclopentene] $\xrightarrow{(32)}$ HO—[cyclopentane]—HO $\xrightarrow[\substack{H_2SO_4, \\ \text{mild}}]{K_2Cr_2O_7}$ (33)

10.21 Arrange each of the following four series of compounds in order of *increasing* reactivity toward the Lucas reagent, HCl-ZnCl$_2$. If two or more compounds are expected to have roughly comparable reactivity, so indicate.

(a) $(CH_3)_2CHOH$ $(CH_3)_3COH$ CH_3CH_2—$\overset{\overset{\displaystyle }{}}{\underset{\underset{\displaystyle OH}{|}}{CH}}$—$CH_3$ $CH_3(CH_2)_4OH$

(b) $(CH_3)_3COH$ CH_3CH_2OH CH_3OH $CH_3CH_2CH_2-\underset{\underset{OH}{|}}{CH}-CH_3$

(c)

(d) $CH_2{=}CH-CH_2-OH$ $CH_3CH_2CH_2-OH$ $(CH_3)_2\underset{\underset{CH_2CH_3}{|}}{C}-OH$ $(CH_3CH_2)_2CH-OH$

10.22 Arrange each of the following series of compounds in order of *decreasing* acidity. In cases where you expect two or more compounds to have roughly the same acidity, so indicate.
(a) compounds listed in (d) Study Question 10.21
(b) Cl_2CHCH_2OH $ClCH_2CH_2OH$ CH_3CH_2OH Cl_3CCH_2OH F_3CCH_2OH
(c) *tert*-butyl alcohol   triethylcarbinol   diethylcarbinol   *n*-propyl alcohol

10.23 Draw the structures of compounds *A* to *E* in the following. Indicate the reactions that occur and how they allow you to deduce the structures of the unknown compounds.
(a) Compound *A*, $C_4H_{10}O$, reacts with sodium metal with the liberation of hydrogen gas. *A* is unaffected by chromic acid in acetone.
(b) Compound *B*, $C_4H_{10}O$, reacts with sodium metal and liberates hydrogen gas. *B* reacts *very rapidly* with HCl-$ZnCl_2$.
(c) Compound *C*, $C_7H_{14}O$, reacts with warm chromic acid to give a single compound *D*, $C_7H_{12}O$. When *C* is treated with concentrated sulfuric acid in the presence of heat, compound *E*, C_7H_{12}, is produced, and no other isomeric compounds are observed. On treatment with hot, basic potassium permanganate solution, *E* gives compound (1).

$$HO-\overset{\overset{O}{\|}}{C}-CH_2CH_2-\overset{\overset{CH_3}{|}}{CH}-CH_2-\overset{\overset{O}{\|}}{C}-OH$$

(1)

10.24 Indicate how the following 16 transformations can be carried out in good yield, using the starting materials shown as the only source of carbon. Inorganic reagents and organic solvents may be used freely.
(a) 1-Propanol $\longrightarrow$ *n*-propyl iodide   (b) 2-Propanol $\longrightarrow$ *n*-propyl iodide
(c) 1-Propanol $\longrightarrow$ 2-propanol   (d) 2-Propanol $\longrightarrow$ 1-propanol
(e) *n*-Butyl alcohol $\longrightarrow$ 1-butene   (f) *n*-Pentyl alcohol $\longrightarrow$ 1,2-dichloropentane

(g)

(h)

(i) 1-Butanol $\longrightarrow$ 1,2-butanediol   (j) 1-Butanol $\longrightarrow$ 1-butene
(k) 1-Pentanol $\longrightarrow$ *n*-pentane (by a route *not* involving an alkene)
(l) 1-Pentanol $\longrightarrow$ *n*-pentane (by another route)
(m) 1-Pentanol $\longrightarrow$ 1-pentene
(n) 1-Pentanol + cyclopentane $\longrightarrow$ *n*-pentylcyclopentane

(o) $CH_3-\underset{\underset{CH_3}{|}}{CH}-OH \longrightarrow CH_3-\underset{\underset{CH_3}{|}}{CH}-\underset{\underset{CH_3}{|}}{CH}-CH_3$

(p) 1-Pentanol $\longrightarrow$ 4,5-dimethyloctane

10.25 Indicate how the following conversions can be accomplished, paying particular attention to the stereochemistry where it is indicated. Use any needed organic or inorganic reagents.

(a)

cis-2-Methylcyclopentanol

(b) (+)-2-Butanol $\longrightarrow$ CH$_3$—$\overset{\displaystyle O}{\overset{\|}{C}}$—CH$_2CH_3$

(c) Optically active 2-butanol $\longrightarrow$ 2-iodobutane, which is optically active but has a configuration opposite that of the starting alcohol

(d) Cyclopentanol $\longrightarrow$ cyclopentyl bromide

(e) Cyclopentanol $\longrightarrow$ cyclopentyl fluoride

(f) Cyclopentanol $\longrightarrow$ cyclopentyl mercaptan, —SH

(g) Cyclopentanol $\longrightarrow$ cyclopentyl nitrile, —CN

10.26 Compounds *A* to *D* have roughly the same molecular weight.

$$CH_3CH_2-\overset{\overset{\displaystyle |}{\underset{\displaystyle |}{OH}}}{CH}-CH_2CH_2CH_3 \qquad CH_3CH_2CH_2CH_2CH_2CH_2CH_3$$

A   *B*

$$CH_3CH_2CH_2-\overset{\overset{\displaystyle CH_3}{|}}{\underset{\underset{\displaystyle OH}{|}}{C}}-CH_3 \qquad CH_3CH_2CH_2CH_2CH_2CH_2OH$$

C   *D*

(a) Without reference to tables of physical properties, arrange these compounds in decreasing order of boiling point; that is, list the compound with the highest boiling point first. Briefly explain the reasoning used to derive this order.

(b) Which compounds are soluble in cold, concentrated sulfuric acid? Why?

(c) Which compounds will react immediately with HCl-ZnCl$_2$? Give the structure(s) of the product(s).

(d) Which compounds will be unreactive toward HCl-ZnCl$_2$? Why?

10.27 Explain each of the following, using structural formulas and equations as needed.

(a) Ethanol, CH$_3$CH$_2$OH (mol wt 46) is soluble in water, whereas propane (mol wt 44) is insoluble in water.

(b) Ethanol (bp 78°) boils at a much higher temperature than propane (bp −42°).

(c) The dipole moment of water is 1.89 D, whereas the dipole moment of methanol, CH$_3$OH, is 1.69 D.

10.28 When *optically active* 2-bromobutane (1) is allowed to react with sodium hydroxide under certain conditions, *optically active* 2-butanol is formed. When the resulting 2-butanol is treated with concentrated HBr, *optically active* 2-bromobutane is regenerated. The final product has the *same sign of rotation* and the *same magnitude of rotation* as did the starting 2-bromobutane.

Provide a complete mechanistic interpretation of these results using three-dimensional structures throughout. *Show all steps and all intermediates* and, where important, indicate relative rates of reaction.

10.29 Assume (as is true most of the time) that primary and secondary alkyl bromides undergo complete inversion of configuration (S_N2 reaction) when treated with aqueous hydroxide ion. Consider the following reaction sequence, and then explain how these data allow you to deduce whether the conversion of (1) to (2) occurs with complete retention or complete inversion of configuration. Provide a complete mechanism, showing all steps and intermediates, that is consistent with your answer.

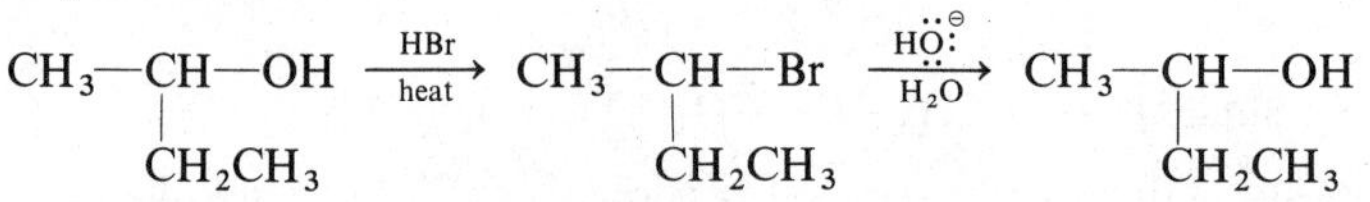

$$CH_3-CH-OH \xrightarrow[\text{heat}]{HBr} CH_3-CH-Br \xrightarrow[\text{H}_2\text{O}]{\overset{\cdot\cdot\ominus}{\text{HO}:}} CH_3-CH-OH$$
$$\quad\quad | \quad\quad\quad\quad\quad\quad\quad\quad | \quad\quad\quad\quad\quad\quad\quad\quad |$$
$$\quad CH_2CH_3 \quad\quad\quad\quad\quad CH_2CH_3 \quad\quad\quad\quad\quad CH_2CH_3$$

$$\quad\quad (1) \quad\quad\quad\quad\quad\quad\quad (2)$$
$$\quad\text{2-Butanol} \quad\quad\quad\text{2-Bromobutane} \quad\quad\quad\text{2-Butanol}$$
$$\quad [\alpha]_D^{25} +23° \quad\quad\quad [\alpha]_D^{25} -10° \quad\quad\quad [\alpha]_D^{25} +23°$$

10.30 For each of the following eight pairs of compounds, state a simple chemical test that would help you distinguish between the members of each pair. Indicate what you would observe as a positive test in each case.
(*a*) cyclohexanol and cyclohexane
(*b*) cyclohexanol and cyclohexene
(*c*) bromomethane and *tert*-butyl bromide
(*d*) $CH_2{=}CH-CH_2CH_2OH$ and $CH_3-CH{=}CH-CH_2-OH$

(*e*) —CH_2OH and —OH

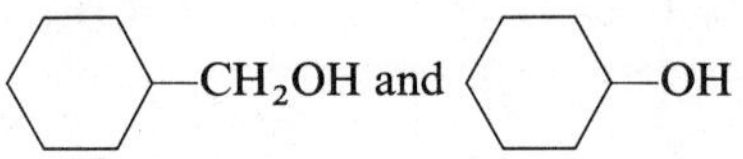

(*f*) *sec*-butyl alcohol and isobutyl alcohol
(*g*) *n*-butyl alcohol and *n*-butyl bromide
(*h*) $CH_3CH_2CH_2CH_2-OH$ and $CH_3CH{=}CH-CH_2-OH$

10.31 Provide complete, stepwise mechanisms for these four reactions, being careful to show all intermediates.

$$\quad\quad\quad CH_3 \quad\quad\quad\quad\quad\quad\quad\quad\quad\quad CH_3$$
$$\quad\quad\quad | \quad\quad\quad\quad\quad\quad\quad\quad\quad\quad\quad |$$
$$(a)\ CH_3-C-CH-CH_3 \xrightarrow[\text{heat}]{\text{conc HBr}} CH_3-C-CH_2-CH_3$$
$$\quad\quad\quad | \quad | \quad\quad\quad\quad\quad\quad\quad\quad\quad |$$
$$\quad\quad\quad H \quad OH \quad\quad\quad\quad\quad\quad\quad\quad Br$$

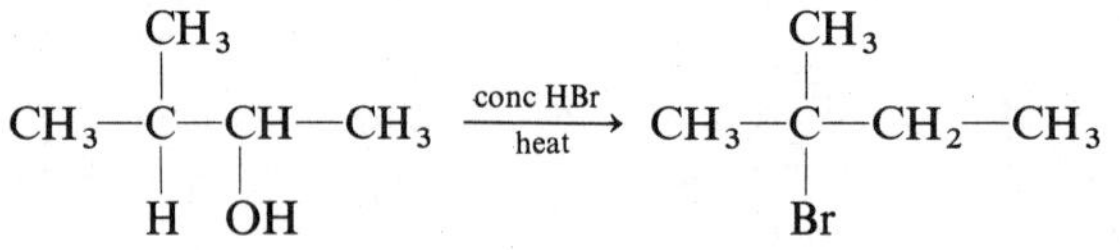

Is this an S_N1 or an S_N2 reaction? Explain the basis for your decision.

$$\quad CH_3 \quad\quad\quad\quad CH_3$$
$$\quad\searrow \quad\quad\quad\quad\quad |$$
$$(b)\quad\quad CH-CH-C-CH_3 \xrightarrow[\text{heat}]{\text{conc}\atop\text{HCl,}}$$
$$\quad\nearrow \quad\quad | \quad |$$
$$\quad CH_3 \quad\quad OH \quad CH_3$$

$$\quad CH_3 \quad\quad\quad CH_3 \quad\quad\quad\quad CH_3 \quad\quad CH_3 \quad\quad\quad\quad CH_3 \quad\quad CH_3$$
$$\quad\searrow \quad\quad\quad | \quad\quad\quad\quad\quad | \quad\quad\quad | \quad\quad\quad\quad\quad | \quad\quad\quad |$$
$$\quad CH-CH-C-CH_3 + CH_3-C-CH_2-C-CH_3 + CH_3-C-CH-CH-CH_3$$
$$\quad\nearrow \quad\quad | \quad | \quad\quad\quad\quad\quad | \quad\quad\quad | \quad\quad\quad\quad\quad | \quad\quad |$$
$$\quad CH_3 \quad\quad Cl \quad CH_3 \quad\quad\quad\quad Cl \quad\quad CH_3 \quad\quad\quad\quad Cl \quad CH_3$$

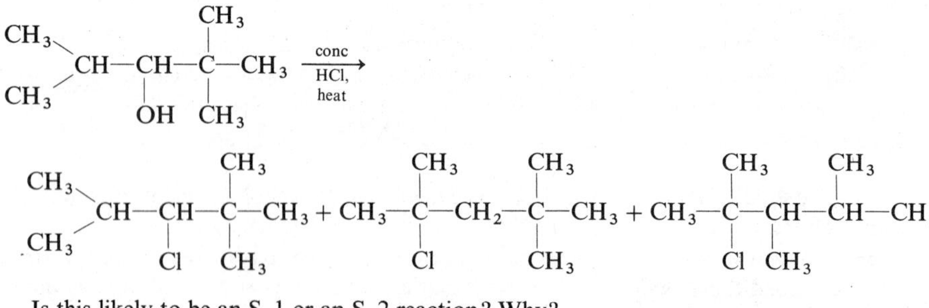

Is this likely to be an S_N1 or an S_N2 reaction? Why?

$$(c)\quad \text{[cyclopentane ring]} \xrightarrow[\text{heat}]{H_2SO_4} \text{[cyclohexene ring]}$$
$$\quad\quad\quad CH_2OH$$

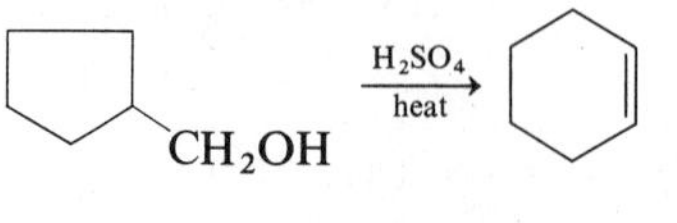

$$\quad\quad\quad CH_3 \quad\quad\quad\quad\quad\quad\quad\quad\quad\quad CH_3$$
$$\quad\quad\quad | \quad\quad\quad\quad\quad\quad\quad\quad\quad\quad\quad |$$
$$(d)\ CH_3-C-CH_2OH \xrightarrow{\text{conc HCl}} CH_3-C-CH_2CH_3$$
$$\quad\quad\quad | \quad\quad\quad\quad\quad\quad\quad\quad\quad\quad\quad |$$
$$\quad\quad\quad CH_3 \quad\quad\quad\quad\quad\quad\quad\quad\quad\quad Cl$$

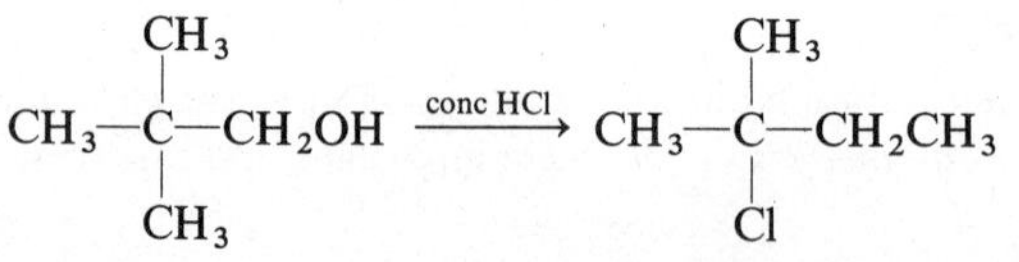

10.32 In which of the following pairs of reactions would you expect to find an isotope effect? Briefly explain.

(a) $^{12}CH_3$—O—H + Na → $^{12}CH_3O^{\ominus}$ + $\frac{1}{2}H_2$ and $^{14}CH_3$—O—H + Na → $^{14}CH_3O^{\ominus}$ + $\frac{1}{2}H_2$

(b) $CH_3{}^{18}O$—H + Na → $CH_3{}^{18}O^{\ominus}$ + $\frac{1}{2}H_2$ and $CH_3{}^{16}O$—H + Na → $CH_3{}^{16}O^{\ominus}$ + $\frac{1}{2}H_2$

(c) CD_3—CH_2—Br $\xrightarrow[\text{alcohol}]{\text{KOH}}$ CD_2=CH_2 and CH_3—CH_2—Br $\xrightarrow[\text{alcohol}]{\text{KOH}}$ CH_2=CH_2

(d) CH_3—CD_2—Br $\xrightarrow[\text{alcohol}]{\text{KOH}}$ CH_2=CD_2 and CH_3—CH_2—Br $\xrightarrow[\text{alcohol}]{\text{KOH}}$ CH_2=CH_2

11

Alkynes: Structure, Properties, Nomenclature, Preparations, and Reactions

We now turn our attention to another family of hydrocarbons, the **alkynes,** which have as their characteristic functional group the **carbon-carbon triple bond.** Alkynes have the general molecular formula C_nH_{2n-2}. Like the alkenes, they are referred to as *unsaturated hydrocarbons*. An alkyne contains four fewer hydrogen atoms than an alkane of equal carbon number; an alkene contains two fewer hydrogens than the corresponding alkane. Consider, for example, the following straight-chain hydrocarbons.

Alkanes, C_nH_{2n+2}	Alkenes, C_nH_{2n}	Alkynes, C_nH_{2n-2}
C_4H_{10}	C_4H_8	C_4H_6
$CH_3CH_2CH_2CH_3$	$CH_3CH_2CH=CH_2$	$CH_3CH_2C\equiv CH$
Butane	1-Butene	1-Butyne
	or	or
	$CH_3CH=CHCH_3$	$CH_3C\equiv CCH_3$
	2-Butene	2-Butyne

In this chapter we investigate the structure of the carbon-carbon triple bond and consider both the methods of preparation and the reactions of this functional group. In addition, we often contrast these reactions to those of the other hydrocarbon families we encountered: alkanes, alkenes, cycloalkanes, and alkadienes.

Structure, Nomenclature, and Properties

11.1 Structure of the Carbon-Carbon Triple Bond

The detailed electronic structure of the triple bond has been investigated by quantum mechanics, which gives a view consistent with experimental results. The triple bond is formed by two overlapping *sp* hybridized carbon atoms (see Sec. 2.17). The geometry of the acetylene molecule shown in Fig. 11.1 reflects the geometry of the *sp* hybrid orbitals from which it is made.

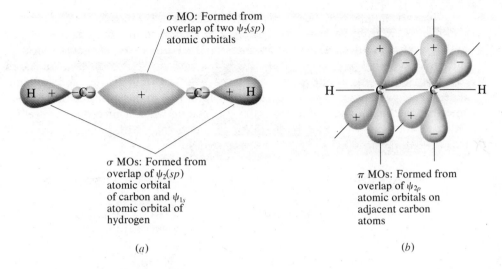

σ MO: Formed from
overlap of two $\psi_2(sp)$
atomic orbitals

σ MOs: Formed from
overlap of $\psi_2(sp)$
atomic orbital
of carbon and ψ_{1s}
atomic orbital of
hydrogen

π MOs: Formed from
overlap of ψ_{2p}
atomic orbitals on
adjacent carbon
atoms

(a)   (b)

FIGURE 11.1   The bonding MO's of acetylene: (a) σ and (b) π.

11.2   Nomenclature of Alkynes

As with alkanes and alkenes, two types of nomenclature are used in naming alkynes: the common name and the IUPAC method.

A. Common Names

Alkynes can be named as a derivative of the simplest alkyne, acetylene. Replacement of one or both of the hydrogen atoms by alkyl groups gives a compound that is named by placing the alkyl group names in front of the parent acetylene name. For example:

$$H-C\equiv C-CH_3 \qquad CH_3-C\equiv C-CH_3 \qquad CH_3CH_2-C\equiv C-CH_2CH_2CH_2CH_3$$

Methylacetylene   Dimethylacetylene   Ethyl-*n*-butylacetylene

B. Systematic (IUPAC) Nomenclature

Systematic nomenclature requires that the characteristic **-yne** ending be used. The naming of alkynes follows the same general rules given for alkenes in Sec. 3.10, except that the *-ene* ending for alkenes is replaced by the *-yne* ending. The parent compound must contain the triple bond in the longest continuous carbon chain, which is numbered so that the triple bond has the *lowest* number. Substituents are indicated by number, as they are in alkenes. For example:

$$\overset{1}{C}H_3-\overset{2}{C}\equiv\overset{3}{C}-\overset{4}{C}H_2-\overset{5}{C}H_3 \qquad\qquad \overset{7}{C}H_3-\overset{6}{C}H-\overset{5}{C}H_2-\overset{4}{C}\equiv\overset{}{C}-\overset{3}{C}H-\overset{2}{C}H_2-\overset{1}{C}H_3$$

CH₃ above position 3; CH₂—CH₃ (8,9) below position 6.

2-Pentyne   3,7-Dimethyl-4-nonyne

$$H-\overset{1}{C}\equiv\overset{2}{C}-\overset{3}{C}H_2-\overset{4}{C}\equiv\overset{5}{C}-\overset{6}{C}H_3$$

1,4-Hexadiyne

The only new problem involves the systematic naming of molecules that contain *both double and triple carbon-carbon bonds*. The longest continuous chain is chosen so that it contains all (or the maximum number, if all cannot be included) the double and triple bonds in the molecule. The chain is numbered to obtain the smaller number combinations for the multiple bond positions. The two endings, *-ene* and *-yne*, are added to the parent name, with *-ene* coming before *-yne*. In the examples that follow, note that the final *e* is dropped from the *-ene* ending when it precedes a vowel (in this case, the *y* in *-yne*). Where two different numberings with the same priorities are possible, the double bond is given the lower number.

$$\overset{6}{C}H_3-\overset{5}{C}H=\overset{4}{C}H-\overset{3}{C}H_2-\overset{2}{C}\equiv\overset{1}{C}-H \qquad \overset{1}{C}H_2=\overset{2}{C}H-\overset{3}{C}H_2-\overset{4}{C}\equiv\overset{5}{C}H$$

4-Hexen-1-yne   1-Penten-4-yne
(not 2-hexen-5-yne,   (not 4-penten-1-yne, double
1 before 2)   bond before triple bond)

$$\overset{7}{C}H_3-\overset{6}{C}H=\overset{5}{C}-\overset{4}{C}=\overset{3}{C}H-\overset{2}{C}\equiv\overset{1}{C}-H$$
$$\qquad\qquad | \qquad |$$
$$\qquad\quad CH_3 \quad CH_3$$

4,5-Dimethyl-3,5-heptadien-1-yne

Question 11.1

Draw all the structural isomers of alkynes that have the formula C_5H_8 and name them by the IUPAC system.

TABLE 11.1   Physical Properties of Selected Alkynes

Name	Structure	mp, °C	bp, °C	Density at 20°, g/cc
Ethyne (acetylene)	$HC\equiv CH$	-82	-75	0.618
Propyne	$CH_3-C\equiv C-H$	-101.5	-23.3	0.671
1-Butyne	$CH_3-CH_2-C\equiv C-H$	-122.5	8	0.668
2-Butyne	$CH_3-C\equiv C-CH_3$	-28	27	0.694
1-Pentyne	$CH_3-CH_2-CH_2-C\equiv C-H$	-98	40	0.695
2-Pentyne	$CH_3-CH_2-C\equiv C-CH_3$	-101	55.5	0.713
3-Methyl-1-butyne	$CH_3-\underset{\underset{CH_3}{\vert}}{CH}-C\equiv C-H$		28	0.665
1-Hexyne	$CH_3(CH_2)_3-C\equiv C-H$	-124	71	0.720
2-Hexyne	$CH_3-CH_2-CH_2-C\equiv C-CH_3$	-92	84	0.731
3-Hexyne	$CH_3-CH_2-C\equiv C-CH_2-CH_3$	-51	82	0.726
3,3-Dimethyl-1-butyne	$CH_3-\underset{\underset{CH_3}{\vert}}{\overset{\overset{CH_3}{\vert}}{C}}-C\equiv C-H$	-81	38	0.669
1-Heptyne	$CH_3(CH_2)_4-C\equiv C-H$	-80	100	0.733
1-Octyne	$CH_3(CH_2)_5-C\equiv C-H$	-70	126	0.748
1-Nonyne	$CH_3(CH_2)_6-C\equiv C-H$	-65	151	0.763
1-Decyne	$CH_3(CH_2)_7-C\equiv C-H$	-36	182	0.770

Question 11.2

Draw all the structural isomers of alkynes that have the formula C_6H_{10} and name them by the IUPAC system.

11.3   Physical Properties

Alkynes are generally nonpolar compounds with physical properties similar to those of the alkanes, cycloalkanes, alkenes, and alkadienes. They are insoluble in water but are usually soluble in most nonpolar organic solvents. In general, they boil at somewhat higher temperatures than do alkanes and alkenes that contain the same number of carbon atoms. Some physical properties are listed in Table 11.1.

Methods of Preparation

11.4   Preparation of Acetylene

Acetylene is an important industrial compound. It is used in oxygen-acetylene torches and in miners' lights. Acetylene (which is called *ethyne* in the IUPAC system of nomenclature) is readily prepared by the action of water on *calcium carbide*, CaC_2:

$$CaC_2 + 2H_2O \longrightarrow H-C\equiv C-H + Ca^{\oplus 2} + 2OH^{\ominus}$$

Calcium
carbide

Ethyne
(acetylene)

This is an interesting reaction because the electronic structure of calcium carbide (shown as follows) reveals that the carbide ion, $C_2^{\ominus 2}$, is a *dicarbanion*, which abstracts protons from water. In this case, $C_2^{\ominus 2}$ is a strong base and water serves as the acid.

Calcium carbide:
$$Ca^{\oplus 2} \overset{\ominus}{:}C\equiv C\overset{\ominus}{:}$$

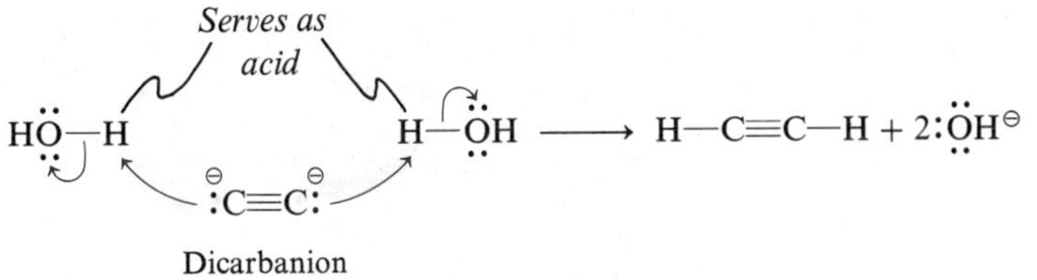

Serves as acid

$$H\ddot{O}-H \qquad H-\ddot{O}H \longrightarrow H-C\equiv C-H + 2:\ddot{O}H^{\ominus}$$

$$\overset{\ominus}{:}C\equiv C\overset{\ominus}{:}$$

Dicarbanion
Strong base

The precursor to acetylene, calcium carbide, is obtained from cheap raw materials:

$$CaCO_3 \overset{heat}{\longrightarrow} CaO + CO_2$$

Limestone

and then

$$3C + CaO \overset{2000°}{\longrightarrow} CaC_2 + CO$$

Coke

Acetylene is a valuable starting material for many organic compounds. It provides important alkenes and dienes for polymerization reactions that produce plastics and synthetic rubbers.

11.5   Elimination Reactions: Dehydrohalogenation and Dehalogenation

A general approach for preparing alkynes involves a double elimination reaction in which an alkene intervenes as an intermediate:

General reaction:

$$
\underset{A\quad B}{\overset{A\quad B}{-\!\overset{|}{C}\!-\!\overset{|}{C}\!-}}\ \xrightarrow{-AB}\ \underset{}{\overset{A\quad B}{-C\!=\!C-}}\ \xrightarrow{-AB}\ -C\!\equiv\!C-
$$

Alkene   Alkyne

A. Dehydrohalogenation of Vicinal Dihalides

A practical laboratory method for preparing alkynes is outlined here, starting from an alkene:

$$
\underset{H}{\overset{}{\diagdown}}C\!=\!C\underset{}{\overset{}{\diagup}}_{H}\ \xrightarrow[\text{CCl}_4]{X_2}\ \overset{X\quad X}{\underset{H\quad H}{-\!\overset{|}{C}\!-\!\overset{|}{C}\!-}}\ \xrightarrow[\text{alcohol}]{\text{KOH}}\ \underset{H}{\overset{}{\diagdown}}C\!=\!C\overset{X}{\underset{}{\diagdown}}\ \xrightarrow[\substack{\text{(sodium}\\ \text{amide)}}]{\text{NaNH}_2}\ -C\!\equiv\!C-
$$

Alkene   Vicinal   Vinyl halide   Alkyne
                    dihalide   *Stable and*
                                      less reactive

The dehydrohalogenation normally must be done in two steps. Treatment of the vicinal dihalide with potassium hydroxide in alcohol removes one molecule of hydrogen halide and produces a *vinyl halide*, which is considerably less reactive than an alkyl halide. (We learned of the difficulty in abstracting a vinyl hydrogen by a radical in Sec. 8.30.) The removal of the second molecule of H—X requires a strong base; sodium amide, $\overset{\oplus}{Na}\overset{\ominus}{:}\overset{\cdot\cdot}{N}H_2$ (often called *sodamide*), is often used. The preparation and properties of sodium amide are discussed in Sec. 11.7.

As an example, 2-butene can be converted to 2-butyne by the sequence:

$$
CH_3\!-\!CH\!=\!CH\!-\!CH_3 \xrightarrow[\text{CCl}_4]{\text{Br}_2} \overset{Br\quad Br}{\underset{H\quad H}{CH_3\!-\!\overset{|}{C}\!-\!\overset{|}{C}\!-\!CH_3}} \xrightarrow[\text{alcohol}]{\text{KOH}} \overset{H}{\underset{Br}{CH_3\!-\!\overset{|}{C}\!=\!C\!-\!CH_3}}
$$

2-Butene   2,3-Dibromobutane   2-Bromo-2-butene
(*cis* and *trans*)   (*cis* and *trans*)

$$\Big\downarrow \text{NaNH}_2$$

$$CH_3\!-\!C\!\equiv\!C\!-\!CH_3$$
2-Butyne

B. Dehalogenation of Tetrahalides

In this reaction 2 moles of halogen are removed by the action of zinc metal in alcohol on a suitable tetrahalide.

General reaction:

$$
\begin{array}{c}
\overset{X}{\underset{X}{\mid}}\ \overset{X}{\underset{X}{\mid}} \\
-\!\!\overset{\mid}{\underset{\mid}{C}}\!\!-\!\!\overset{\mid}{\underset{\mid}{C}}\!\!-
\end{array}
\xrightarrow[\substack{\text{alcohol,}\\ \text{heat}}]{\text{Zn}}
\left[
\begin{array}{c}
\overset{X}{\mid}\ \overset{X}{\mid} \\
-\overset{\mid}{C}\!=\!\overset{\mid}{C}-
\end{array}
\right]
\xrightarrow[\substack{\text{alcohol,}\\ \text{heat}}]{\text{Zn}}
-C\!\equiv\!C- + ZnX_2
$$

$+\ ZnX_2$ under the bracket; "Alkyne" under the product.

For example:

$$
CH_3\!-\!\overset{\overset{\displaystyle Br}{\mid}}{\underset{\underset{\displaystyle Br}{\mid}}{C}}\!-\!\overset{\overset{\displaystyle Br}{\mid}}{\underset{\underset{\displaystyle Br}{\mid}}{C}}\!-\!CH_3
\xrightarrow[\substack{\text{alcohol,}\\ \text{heat}}]{\text{Zn}}
CH_3\!-\!C\!\equiv\!C\!-\!CH_3
$$

2,2,3,3-Tetrabromobutane   2-Butyne

An alkene intermediate is involved, and this reaction is analogous to the dehalogenation of a vicinal dihalide (see Sec. 7.19):

$$
\begin{array}{c}
\overset{\mid}{\underset{X}{\underset{\mid}{C}}}\!-\!\overset{\mid}{\underset{X}{\underset{\mid}{C}}} \\
\end{array}
\xrightarrow[\substack{\text{alcohol,}\\ \text{heat}}]{\text{Zn}}
\ \ \overset{}{>}\!C\!=\!C\!\overset{}{<} + ZnX_2
$$

Vicinal dihalide   Alkene

Like the previous reaction for alkenes, the dehalogenation of tetrahalides is of relatively little use because the starting material is obtained from an alkyne originally (see Sec. 11.11).

11.6   Preparation of Alkynes by a Substitution Reaction

One of the best ways to prepare substituted alkynes is by a substitution reaction of the following type:

$$
-C\!\equiv\!C\!:^{\ominus} + R\!-\!X \longrightarrow -C\!\equiv\!C\!-\!R + :\ddot{X}\!:^{\ominus} \qquad \text{Substitution}
$$

Alkynyl anion   New C—C bond formed

where the **alkynyl anion,** $-C\!\equiv\!C\!:^{\ominus}$, displaces the halide as $:\ddot{X}\!:^{\ominus}$ and forms a new carbon-carbon bond. Discussion of this reaction is deferred to Sec. 11.9, where we investigate the reactions leading to the formation of the alkynyl anion as well as its reaction with alkyl halides.

Reactions of Alkynes

Alkynes undergo two types of reactions. One is characteristic of the triple bond, which undergoes addition in two steps:

$$
-C\!\equiv\!C- \xrightarrow[A-B]{1\ \text{mole}}
\begin{array}{c}
\overset{A}{\mid}\ \overset{B}{\mid} \\
-C\!=\!C-
\end{array}
\xrightarrow[A-B]{1\ \text{mole}}
\begin{array}{c}
\overset{A}{\mid}\ \overset{B}{\mid} \\
-\overset{\mid}{C}\!-\!\overset{\mid}{C}- \\
\underset{A}{\mid}\ \underset{B}{\mid}
\end{array}
\qquad \text{Addition}
$$

The second reaction is characteristic of a triple bond that bears *at least one hydrogen atom*. This type of alkyne, which includes ethyne, H—C≡C—H, and more generally, R—C≡C—H, is called a **terminal alkyne** because the triple bond must be at the end of a carbon chain. The terminal hydrogen on a alkyne is moderately acidic and thus can be removed by a strong base:

$$-C≡C-H + B:^{\ominus} \longrightarrow -C≡C:^{\ominus} + B:H \qquad \text{Proton removal}$$

Base   Alkynyl anion

The resulting alkynyl anion, a *carbanion*, undergoes several important types of reactions, which are discussed in Sec. 11.9.

11.7   Alkynes As Carbon Acids; Terminal Alkynes

The terminal hydrogen of an alkyne, —C≡C—H, is relatively acidic compared with hydrogens in alkanes (that is, hydrogens attached to sp^3 hybridized carbons) or in alkenes (hydrogens attached to sp^2 hybridized carbons). As a result of extensive investigations on the acidity of carbon-hydrogen bonds, the order of decreasing acidity is found to be:

Decreasing acidity of hydrocarbons[1]:

$$-C≡C-H \quad > \quad \underset{H}{>}C=C< \quad > \quad -\underset{|}{\overset{|}{C}}-\underset{|}{\overset{|}{C}}-H$$

$$K_a \approx 10^{-25} \qquad K_a \approx 10^{-35} \qquad K_a < 10^{-45}$$

One consequence of the moderate acidity of the hydrogen atom on a terminal triple bond is that it can be removed by several reagents. A terminal alkyne such as ethyne, for example, reacts with sodium metal to form hydrogen gas and sodium acetylide (a salt):

$$H-C≡C-H + Na \text{ metal} \longrightarrow H-C≡C:^{\ominus} Na^{\oplus} + \tfrac{1}{2}H_2 \text{ (gas)}$$

Ethyne   Acetylide ion
(acetylene)   *Carbanion*

This is an oxidation-reduction reaction, in which sodium metal is oxidized to $Na^{\oplus}$ and hydrogen is reduced from $H^{\oplus}$ to H_2. It is not surprising that this type of reaction occurs, because sodium reacts *explosively* with the acidic hydrogens of water to form sodium hydroxide and hydrogen gas:

$$H_2\ddot{O}: + Na \rightarrow Na^{\oplus}:\ddot{O}H^{\ominus} + \tfrac{1}{2}H_2$$

Another common way for removing the hydrogen is to use *sodium amide* or *sodamide* (which is a contraction for *sod*ium *amide*), $Na^{\oplus\ominus}:\ddot{N}H_2$. Sodium metal reacts with pure liquid ammonia (bp −33°) to form sodium amide, a reaction that involves oxidation-reduction:

$$\ddot{N}H_3 \text{ (liquid)} + Na \xrightarrow[\text{catalyst}]{Fe^{(+3)}} Na^{\oplus\ominus}:\ddot{N}H_2 + \tfrac{1}{2}H_2$$

Sodium amide

---

[1] For a discussion of K_a's, see Sec. 5.7.

The addition of acetylene to sodium amide results in the formation of the acetylide ion and ammonia:

$$H-C\equiv C-H \ + \ :NH_2^\ominus \ \rightleftharpoons \ H-C\equiv C:^\ominus \ + \ H-\ddot{N}H_2$$

Acid	Base,	Acetylide	Ammonia
Stronger acid	amide ion	ion	*Weaker acid*
$(K_a \approx 10^{-25})$	*Stronger base*	*Weaker base*	$(K_a \approx 10^{-34})$

This is a typical acid-base reaction in which the amide ion $:\ddot{N}H_2^\ominus$ as the base pulls the hydrogen from acetylene as a proton and forms ammonia. Put another way, $:\ddot{N}H_2^\ominus$ has a greater affinity for hydrogen (a proton) than does carbon. It can also be said that a terminal alkyne is a stronger acid than ammonia, because the former donates a proton to the amide ion, $:\ddot{N}H_2^\ominus$.

Further evidence regarding the reactivity of the acetylide ion comes from its reaction with water. It abstracts a proton from water and becomes a terminal alkyne again:

$$H-C\equiv C:^\ominus \ Na^\oplus \ + \ H-OH \ \rightleftharpoons \ H-C\equiv C-H \ + \ :\ddot{O}H^\ominus \ + \ Na^\oplus$$

Base	Acid	Weaker acid	Weaker base
Stronger base	*Stronger acid*	*Weaker acid*	*Weaker base*
	$(K_a \approx 10^{-15})$	$(K_a \approx 10^{-25})$	

In this case the acetylide ion serves as a base and water as an acid. The acetylide ion has a greater affinity for protons than does water or, alternatively, water is a stronger acid than a terminal alkyne.

We now know the position of a terminal alkyne relative to water and ammonia in terms of acidity. The order of acidity is: *water > acetylene > ammonia*. Keep in mind, however, that ammonia here means pure ammonia and not the familiar aqueous solutions of ammonia (that is, ammonium hydroxide) we have in the laboratory.

To explain the acidity of terminal alkynes, one must look at the electronic structure of the conjugate base, the anion, that is formed. The pair of electrons on carbon in the anion are closer to the carbon nucleus because they occupy a *sp* hybrid orbital (see Sec. 2.7), which has 50% *s* character. This means that the electrons in this orbital are closer to carbon in the terminal alkyne than they are, for example, in an alkene (33.3% *s* character in *sp*² hybrid orbitals) or in an alkane (25% *s* character in *sp*³ hybrid orbitals). (See Sec. 2.7 for a discussion of % *s* character versus hybridization.) This results in a more stable electronic configuration and therefore a more stable anion. For example,

	R—H	R—CH=CH₂	R—C≡C—H
	Alkane	Alkene	Alkyne
~K_a	10^{-45}	10^{-35}	10^{-25}
s character	25%	33.3%	50%

The following picture is obtained by evaluating the electronic structures of anionic bases in alkanes, alkenes, and alkynes:

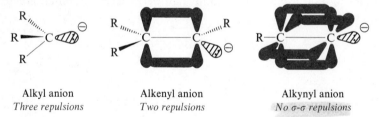

Alkyl anion	Alkenyl anion	Alkynyl anion
Three repulsions	*Two repulsions*	*No σ-σ repulsions*

In the alkyl anion, the electron pair of the anion is $\sim 109.5°$ from the *three* other σ bonds attached to carbon, and there are *three* electrostatic repulsions between the lone pair and the electrons in the σ bonds. In the alkenyl anion, the carbanion is $120°$ from the *two* nearest σ bonds, and there are *two* repulsions that are farther apart than in the alkyl carbanion. The carbanion is $180°$ from the other σ bond in the alkynyl anion, so electrostatic repulsions are minimized even further. Basically, then, an alkynyl anion has less electrostatic repulsion than does the carbanion attached to a double bond and still less than a carbanion attached to an sp^3 hybridized carbon (that is, an alkyl carbanion). There is some electrostatic repulsion between the electron pair and the π clouds in a double or triple bond, but this is less important because π bonds are formed from the overlap of p orbitals, which are fairly far from the nucleus [see, for example, Fig. 2.7(c)]. The point is that the repulsion between a carbanion and a π bond is much less than that between a σ bond and a carbanion. (Electrostatic forces depend on the inverse square of the distance between charges, so even small changes in distance result in enormous changes in electrostatic effects, both repulsive and attractive.)

Question 11.3

Based on the relative acidities of hydrocarbons, what, if anything, do you think would happen when an alkyne such as propyne is allowed to come in contact with *n*-propyl magnesium bromide, $CH_3—CH_2—CH_2 \overset{\ominus}{:} [MgBr]^{\oplus}$? Briefly explain.

11.8   Reactions of Terminal Alkynes; Heavy Metal Salts

Because of the acidity of a terminal alkyne, it undergoes certain characteristic reactions. The alkynyl anion, $R—C{\equiv}C\overset{\ominus}{:}$, reacts with several metallic cations, $Ag^{\oplus}$ and $Cu^{\oplus}$ (*cuprous*) for example, to form insoluble heavy metal salts. These reactions are normally carried out in aqueous ammonia (ammonium hydroxide); under these conditions the silver and cuprous ions are present as the silver ammonia complex, $Ag(NH_3)_2^{\oplus}$, and the cuprous ammonia complex, $Cu(NH_3)_2^{\oplus}$, respectively. The reactions that occur are:

$$R—C{\equiv}C—H + Ag(NH_3)_2^{\oplus} \longrightarrow R—C{\equiv}C—Ag(s) + NH_3 + NH_4^{\oplus}$$

R = H or alkyl   Silver alkynide
Precipitate
(light color)

and

$$R—C{\equiv}C—H + Cu(NH_3)_2^{\oplus} \longrightarrow R—C{\equiv}C—Cu(s) + NH_3 + NH_4^{\oplus}$$

R = H or alkyl   Cuprous alkynide
Precipitate
(brick red)

The ammonia in the test reagent is a weak base, but it is apparently strong enough to establish a slight equilibrium between the terminal alkyne and the corresponding anion:

$$R—C{\equiv}C—H + \overset{..}{N}H_3 \rightleftharpoons R—C{\equiv}C\overset{\ominus}{:} + NH_4^{\oplus}$$

$\downarrow$ $Ag^{\oplus}$ or $Cu^{\oplus}$

$$R—C{\equiv}C—Ag \quad or \quad R—C{\equiv}C—Cu$$

Precipitate

As soon as the anion is formed, it reacts with $Cu^{\oplus}$ or $Ag^{\oplus}$ to form the insoluble precipitate of the heavy metal salt, and this provides the driving force for the reaction. The silver and cuprous salts have substantial covalent character in the carbon-metal bond and are not true ionic salts like the corresponding sodium and potassium salts.

The starting alkyne can be regenerated from the heavy metal salt with the addition of mineral acid, which is usually hydrogen chloride because the liberated silver ion reacts with the chloride ion to form insoluble silver chloride:

$$R-C\equiv C-Ag + H^{\oplus} + :\overset{..}{\underset{..}{Cl}}:^{\ominus} \longrightarrow R-C\equiv C-H + \underset{\text{Precipitate}}{AgCl}$$

Because heavy metal salts are often explosive when dry, they should be destroyed while still wet.

The formation of heavy metal salts is used to distinguish *terminal* from *nonterminal* alkynes (those *not* having the triple bond at the end of the chain). The former gives a precipitate, whereas the latter does not, so the results are readily observed in the laboratory. This is a good qualitative test for a terminal alkyne; for example:

$$\underset{\substack{\text{1-Hexyne} \\ \textit{Terminal alkyne}}}{CH_3-(CH_2)_3-C\equiv C-H} \xrightarrow[\text{or } Cu(NH_3)_2^{\oplus}]{Ag(NH_3)_2^{\oplus}} \underset{\text{Precipitate}}{CH_3-(CH_2)_3-C\equiv C-Ag \text{ (or Cu)}}$$

whereas

$$\underset{\substack{\text{3-Hexyne} \\ \textit{Nonterminal alkyne}}}{CH_3-CH_2-C\equiv C-CH_2-CH_3} \xrightarrow[\text{or } Cu(NH_3)_2^{\oplus}]{Ag(NH_3)_2^{\oplus}} \text{no reaction}$$

Question 11.4

Describe a *laboratory* procedure by which a mixture of 1-hexyne and 3-hexyne can be separated so that each is obtained in pure form and is uncontaminated with the other.

11.9 Reactions of Terminal Alkynes: Substitution Reactions and Synthesis of Alkynes

The sodium salts of terminal alkynes are used in the synthesis of mono- and disubstituted acetylene derivatives. The general reaction, starting from acetylene, is:

$$H-C\equiv C-H \xrightarrow[NaNH_2]{Na \text{ or}} H-C\equiv C:\overset{\ominus}{}Na^{\oplus} \xrightarrow{R-X} H-C\equiv C-R \xrightarrow[NaNH_2]{Na \text{ or}} Na^{\oplus}\overset{\ominus}{:}C\equiv C-R$$

$$\underset{\substack{R-X \text{ must} \\ \text{be } 1° \text{ or} \\ \text{methyl}}}{} \qquad \underset{\substack{R'-X \\ \text{must be } 1° \\ \text{or methyl}}}{} \Big\downarrow R-X$$

$$R'-C\equiv C-R$$

The basis for this reaction is nucleophilic aliphatic substitution. Careful examination reveals that the carbanion, $R-C\equiv\overset{\ominus}{C}:$, serves as a nucleophile and displaces halide from R—X and R'—X. This is an example of the S_N2 mechanism (see Sec. 9.3).

Nucleophilic aliphatic substitution, S_N2 mechanism:

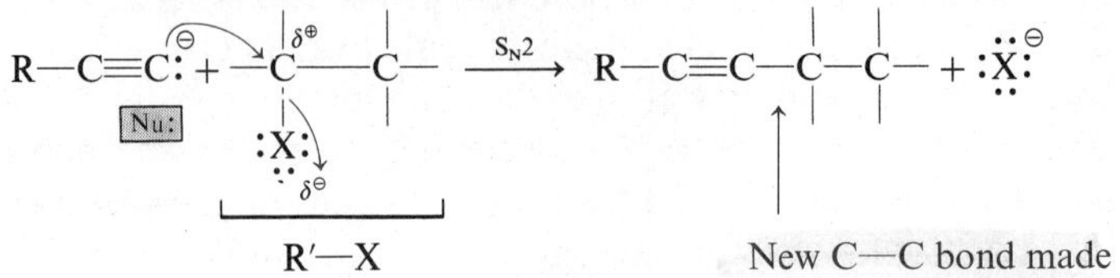

Now consider a reaction introduced in Chap. 7. We know from studies on the preparation of alkenes from alkyl halides that base-promoted E2 eliminations occur:

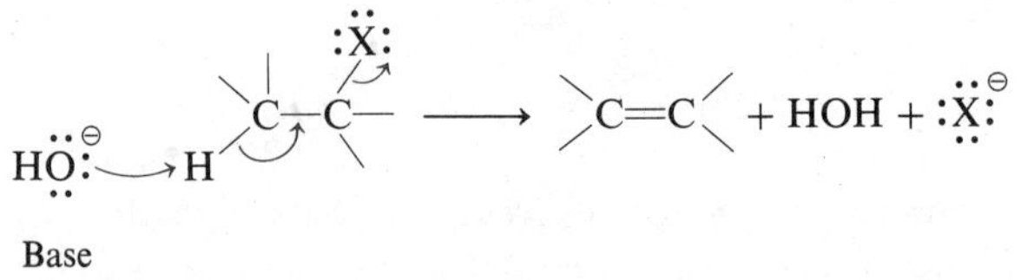

The $R-C\equiv C:^{\ominus}$ carbanion is also a base and, as such, can cause elimination by attacking a hydrogen attached to the β carbon atom in the alkyl halide. By analogy to the preceding reaction, we have:

E2 Elimination (dehydrohalogenation):

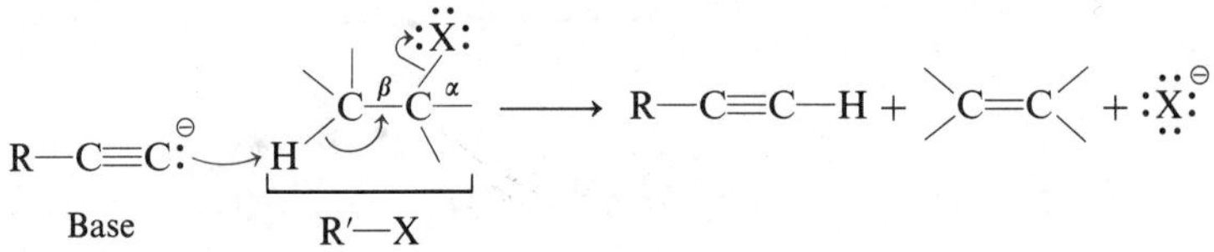

Thus, substitution can occur on the alkyl halide to give a new alkyne, and there is also the possibility of elimination, which converts the alkyl halide to an alkene and the alkynide carbanion, $R-C\equiv C:^{\ominus}$, to the corresponding alkyne, $R-C\equiv C-H$. It is possible to have *both* substitution and elimination when an alkyl halide reacts with an alkynide carbanion, but which factors favor substitution and which favor elimination?

In Sec. 7.12 we saw that ease of elimination is a function of the structure of the alkyl halide; the reactivity order is:

$$3° \text{ RX} > 2° \text{ RX} > 1° \text{ RX}$$

It should not be surprising that this same order applies for eliminations caused by the base $R-C\equiv C:^{\ominus}$. In Sec. 9.4 we also learned that the *reactivity order for nucleophilic substitution* is:

$$CH_3X > 1° \text{ RX} > 2° \text{ RX} > 3° \text{ RX}$$

We find, therefore, that primary or methyl aliphatic halides react smoothly with alkynide carbanions ($R-C\equiv C:^{\ominus}$) to give good yields of a new alkyne (that is, *substitution* occurs). Some secondary alkyl halides give fair yields accompanied by considerable elimination, and tertiary halides give elimination exclusively. *For the most part, only reactions that involve primary or methyl alkyl halides should be used for substitution.*

Armed with these details about substitution and elimination reactions of the alkynide carbanion, let us look at an actual example of how ethyne can be converted to 2-methyl-4-nonyne. Note that two different alkyl groups are introduced into a molecule containing a triple bond.

$$H-C\equiv C-H \xrightarrow[\text{NaNH}_2]{\text{Na or}} H-C\equiv C:\overset{\ominus}{}\overset{\oplus}{\text{Na}} \xrightarrow{\overset{\text{CH}_3}{|}} H-C\equiv C-CH_2CHCH_3$$

Ethyne   Sodium   1-Bromo-2-methylpropane $H-C\equiv C-CH_2CHCH_3$
(acetylene)   acetylide   *1° Halide*
                                                                S_N2

with 1-Bromo-2-methylpropane labeled Br—CH₂CHCH₃ with CH₃ substituent, and the far right product showing CH₃ branch.

Na or
NaNH₂

$$CH_3CH_2CH_2CH_2-C\equiv C-CH_2\overset{\text{CH}_3}{\underset{|}{C}}HCH_3 \xleftarrow{CH_3CH_2CH_2CH_2-Cl} \overset{\oplus}{\text{Na}}\ :\overset{\ominus}{}C\equiv C-CH_2\overset{\text{CH}_3}{\underset{|}{C}}HCH_3$$

2-Methyl-4-nonyne   1-Chlorobutane
Desired product   *1° Halide*
                                                              S_N2

The utility of this reaction can be appreciated more after studying some reactions of the triple bond that convert it to a variety of other functional groups. Of great importance is that *this reaction is a method for making carbon-carbon bonds.* Starting from ethyne, we can introduce virtually any number of carbon atoms into a molecule. The study questions on synthesis at the end of the chapter will acquaint you with some of these possibilities.

11.10   Addition Reactions; Hydrogenation

We now turn to reactions that involve the triple bond itself. The triple bond undergoes a variety of addition reactions in the same way as alkenes, and indeed careful control of concentration and reaction conditions often permits the alkenes to be isolated.

Catalytic hydrogenation in the presence of platinum, palladium, or nickel results in the addition of 2 moles of hydrogen and the formation of the corresponding alkane. The general reaction is:

$$R-C\equiv C-R \xrightarrow[\text{Pt, Ni, or Pd}]{\text{1 mole H}_2} \left[\overset{R}{\underset{H}{}}C=C\overset{R}{\underset{H}{}} \right] \xrightarrow[\text{Pt, Ni, or Pd}]{\text{1 mole H}_2} R-\overset{H}{\underset{H}{C}}-\overset{H}{\underset{H}{C}}-R$$

Alkyne   Alkene   Alkane

For example:

$$CH_3-CH_2-C\equiv C-H$$
1-Butyne

2 moles H₂
Pt

2 moles H₂
Pt

$$CH_3-CH_2-CH_2-CH_3$$
Butane

$$CH_3-C\equiv C-CH_3$$
2-Butyne

Several special metal catalysts allow catalytic hydrogenation to stop at the alkene stage. The following catalysts are commonly used: (1) palladium metal coated with carbon (Pd/C), which is commonly called *Lindlar's catalyst;* (2) palladium metal coated with calcium carbonate (Pd/CaCO₃); and (3) a nickel-boron (Ni/B) catalyst, which is called the *P-2 catalyst.* With these catalysts the reaction stops at the alkene stage and produces predominantly the *cis* isomer.

The *trans*-alkene is the predominant product when the triple bond is reduced with sodium metal or lithium metal in liquid (pure) ammonia. The mechanism of this reaction is not well understood but probably involves electron transfer through free-radical intermediates and results in the overall oxidation of sodium (or lithium) and reduction of the alkyne. These two reductions and their stereochemistry are depicted here:

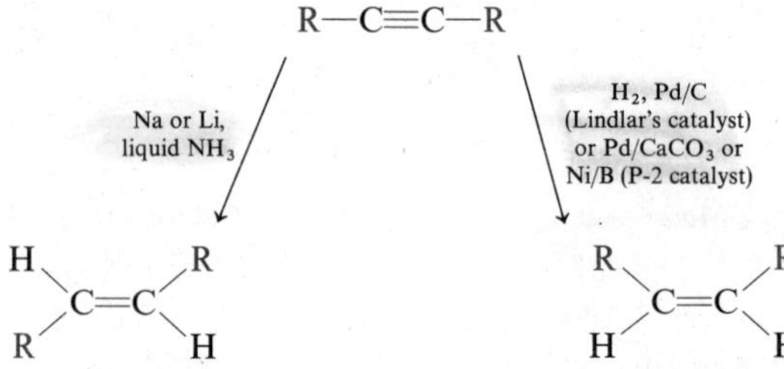

$$R-C\equiv C-R$$

Na or Li, liquid NH_3

H_2, Pd/C (Lindlar's catalyst) or $Pd/CaCO_3$ or Ni/B (P-2 catalyst)

$$\underset{R}{\overset{H}{>}}C=C\underset{H}{\overset{R}{<}}$$

trans Isomer

$$\underset{H}{\overset{R}{>}}C=C\underset{H}{\overset{R}{<}}$$

cis Isomer

For example:

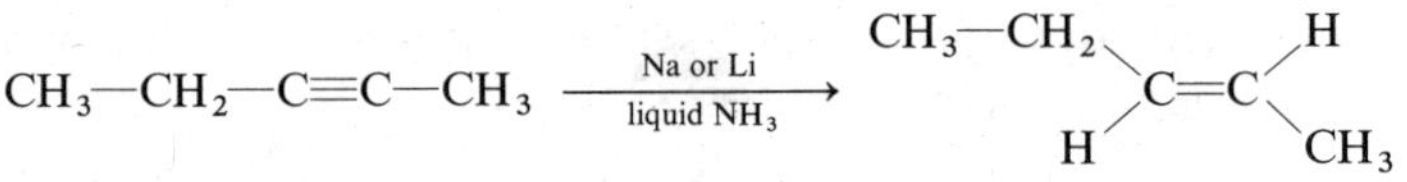

$$CH_3-CH_2-C\equiv C-CH_3 \xrightarrow[\text{liquid } NH_3]{\text{Na or Li}}$$

$$\underset{H}{\overset{CH_3-CH_2}{>}}C=C\underset{CH_3}{\overset{H}{<}}$$

2-Pentyne

trans-2-Pentene
Predominant isomer

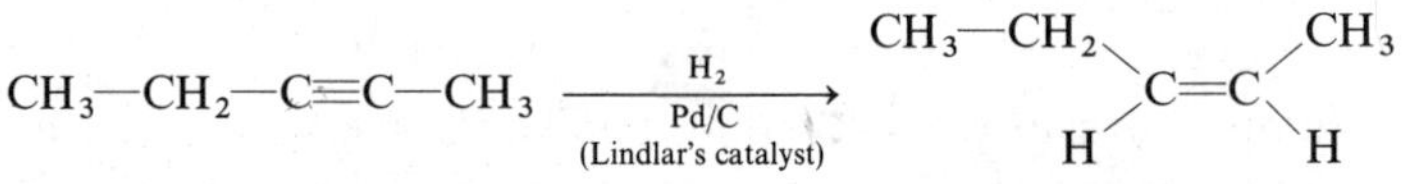

$$CH_3-CH_2-C\equiv C-CH_3 \xrightarrow[\text{(Lindlar's catalyst)}]{\underset{\text{Pd/C}}{H_2}}$$

$$\underset{H}{\overset{CH_3-CH_2}{>}}C=C\underset{H}{\overset{CH_3}{<}}$$

cis-2-Pentene
>95% pure

That *syn* addition of hydrogen in the catalytic hydrogenation process occurs should not be surprising. Recall from Sec. 8.12 that *syn* addition occurs in the hydrogenation of alkenes, and the mechanism of the hydrogenation of alkynes is believed to be similar; that is, the alkyne "sits" on the catalyst surface and hydrogen is transferred to the triple bond as shown in Fig. 11.2. (To help remember which reagents give *cis*-alkenes and which give *trans*-alkenes, keep in mind the mechanism for catalytic hydrogenation, which must give the *cis* isomer.)

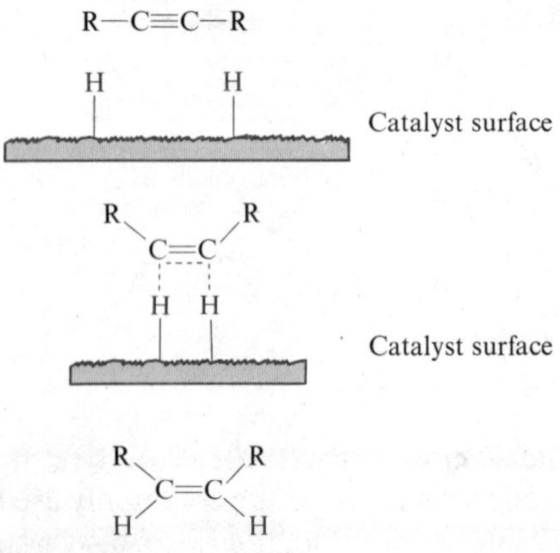

$$R-C\equiv C-R$$

H   H

Catalyst surface

$$\underset{H \quad H}{\overset{R}{>}}C=C\overset{R}{<}$$

Catalyst surface

$$\underset{H}{\overset{R}{>}}C=C\underset{H}{\overset{R}{<}}$$

syn-Addition

FIGURE 11.2   *Syn* hydrogenation of an alkyne to give a *cis*-alkene.

Question 11.5

Outline a sequence of reactions to convert *trans*-2-butene to:

(*a*) 2-butyne   (*b*) *cis*-2-butene

11.11   Addition of Halogens

The addition of halogens to alkynes follows much the same route as the addition to alkenes except that the alkene intermediate intervenes. The addition appears to involve carbocations for chlorine addition and bromonium ions for bromine addition, as evidenced by the predominance of *trans*-dibromide in the alkene intermediate. The general reaction is:

$$R-C{\equiv}C-R \xrightarrow[CCl_4]{1\ mole\ X_2} \underset{\substack{Mostly \\ trans\ when \\ X = Br}}{\overset{R}{\underset{X}{>}}C{=}C\overset{X}{\underset{R}{<}}} \xrightarrow[CCl_4]{1\ mole\ X_2} R-\underset{\underset{X}{|}}{\overset{\overset{X}{|}}{C}}-\underset{\underset{X}{|}}{\overset{\overset{X}{|}}{C}}-R$$

where $X_2 = Cl_2, Br_2$

For example:

$$CH_3-C{\equiv}C-CH_3 \xrightarrow[CCl_4]{1\ mole\ Br_2} \overset{Br}{\underset{CH_3}{>}}C{=}C\overset{CH_3}{\underset{Br}{<}} \xrightarrow[CCl_4]{1\ mole\ Br_2} CH_3-\underset{\underset{Br}{|}}{\overset{\overset{Br}{|}}{C}}-\underset{\underset{Br}{|}}{\overset{\overset{Br}{|}}{C}}-CH_3$$

2-Butyne   2,3-Dibromo-2-butene   2,2,3,3-Tetrabromobutane
                    (predominantly *trans*)

Because tetrahalides are made from alkynes, the usefulness of the dehalogenation reaction is very limited (see Sec. 7.19). Note that this reaction and the reaction of alkynes with hydrogen halides covered in Sec. 11.12 provide two new routes to alkyl halides.

11.12   Addition of Hydrogen Halides

The ionic addition of hydrogen halides is a typical electrophilic addition reaction that occurs in two steps, as the following general reaction illustrates:

$$R-C{\equiv}C-R \xrightarrow[CCl_4]{1\ mole\ H-X} R-\underset{\underset{H}{|}}{C}{=}\underset{\underset{X}{|}}{C}-R \xrightarrow[CCl_4]{1\ mole\ H-X} R-\underset{\underset{H}{|}}{\overset{\overset{H}{|}}{C}}-\underset{\underset{X}{|}}{\overset{\overset{X}{|}}{C}}-R$$

where $H-X = HCl, HBr, HI$

Halides with this structure are called *geminal* (Latin *geminus*, twin) *dihalides*, or twin halides. Note that once 1 mole of H—X is added, the second mole adds in accordance with Markovnikov's rule (see Sec. 8.8), so that both hydrogens end up on the same carbon and the two halogens end up on the other carbon atom. Because of the stepwise nature of the reaction, it is possible to add 1 mole of one hydrogen halide and then add 1 mole of a different hydrogen halide; for example:

$$
CH_3-C\equiv C-H \xrightarrow[CCl_4]{\text{1 mole HBr}} CH_3-\underset{\underset{Br}{|}}{C}=\underset{\underset{H}{|}}{C}-H \xrightarrow[CCl_4]{\text{1 mole HBr}} CH_3-\underset{\underset{Br}{|}}{\overset{\overset{Br}{|}}{C}}-\underset{\underset{H}{|}}{\overset{\overset{H}{|}}{C}}-H
$$

| Propyne | 2-Bromopropene | 2,2-Dibromopropane |

$$
CH_3-C\equiv C-CH_3 \xrightarrow[CCl_4]{\text{1 mole HCl}} CH_3-\underset{\underset{Cl}{|}}{C}=\underset{\underset{H}{|}}{C}-CH_3 \xrightarrow[CCl_4]{\text{1 mole HI}} CH_3-\underset{\underset{Cl}{|}}{\overset{\overset{I}{|}}{C}}-\underset{\underset{H}{|}}{\overset{\overset{H}{|}}{C}}-CH_3
$$

| 2-Butyne | 2-Chloro-2-butene | 2-Chloro-2-iodobutane |

These addition reactions involve carbocations in much the same way that similar reactions of alkenes do; the positive reagent adds to give the more stable carbocation. The orientation of addition of HBr to propyne in the preceding reaction also obeys Markovnikov's rule.

The addition of unsymmetrical reagents to alkynes should be easy to master because of the formation of the intermediate alkenes, which were studied in Chap. 8.

11.13 Addition of Water

The addition of water to alkynes presents some new aspects of organic chemistry. Note that *only* 1 mole of water adds, whereas the other reagents discussed add either *1 or 2 moles* depending on the reaction conditions. Water adds in accordance with Markovnikov's rule to give as the first intermediate a vinyl alcohol, but this alcohol is very unstable and undergoes rearrangement to form an aldehyde or ketone (see Chap. 20), which contains the carbonyl group, $\ce{>C=O}$. The general reaction is:

$$
-C\equiv C- + H-OH \xrightarrow[HgSO_4]{H_2SO_4} \left[\underset{H}{\overset{}{>}}C=C\overset{O-H}{\underset{}{<}} \right] \rightleftharpoons -\underset{\underset{H}{|}}{\overset{\overset{H}{|}}{C}}-\overset{\overset{O}{\|}}{C}-
$$

| Vinyl alcohol | Aldehyde or ketone |
| *Unstable* | |

The addition of water to ethyne is of great industrial importance because this is a major method for producing ethanal (acetaldehyde). Ethanal can be oxidized to ethanoic acid (acetic acid), which is an important organic compound. These reactions are:

$$
H-C\equiv C-H + H_2O \xrightarrow[HgSO_4]{H_2SO_4} \left[\underset{H}{\overset{H}{>}}C=C\overset{O-H}{\underset{H}{<}} \right] \rightleftharpoons
$$

Vinyl alcohol
Unstable

$$
H-\underset{\underset{H}{|}}{\overset{\overset{H}{|}}{C}}-\overset{\overset{O}{\|}}{C}-H \xrightarrow{\text{oxidation}} CH_3-\overset{\overset{O}{\|}}{C}-OH
$$

| Ethanal | Ethanoic acid |
| (acetaldehyde) | (acetic acid) |

The conversion of a vinyl alcohol, a functional group that is often called an *enol* (a name derived from the functional groups present—the carbon-carbon double bond, *-ene*, and the —OH group, *alcohol* or *-ol*—which are contracted to give the word *enol*), to a carbonyl-containing or keto compound is an equilibrium reaction called **keto-enol tautomerism.** However, the equilibrium greatly favors the keto structure. As soon as the enol is formed, it undergoes rearrangement and produces the more stable keto compound:

Enol structure	Keto structure
	Favored

Under the acidic conditions of the reaction, the enol is readily protonated to give the resonance-stabilized carbocation as shown in the following equation. Loss of a proton from the carbocation gives the keto form.

Enol	Resonance structures	Keto

In the enol, a stronger acid (O—H) is converted into a weaker acid (C—H) of the keto form. It is not surprising that the equilibrium favors the keto form. This phenomenon is called **tautomerism.** The *two different structures* which are in equilibrium with one another are called **tautomers.** Because the keto and the enol forms are involved in the equilibrium, this is called **keto-enol tautomerism.**

Before leaving this addition reaction, mention should be made of the reactants used. Water and acid (H_2SO_4) are required to effect the electrophilic addition reaction to the triple bond. The reaction is catalyzed by certain metallic salts, such as mercuric ion (Hg^{+2}), cuprous ion ($Cu^{\oplus}$), and nickel ion (Ni^{+2}). Of these, mercuric ion is used most frequently, and it is believed to form a complex with the triple bond. This complex formation aids electron withdrawal from the triple bond and the carbons become slightly electron-deficient so that the addition of water is facilitated.

R—C≡C—R +	Hg^{+2}	R—C≡C—R
Lewis base	Lewis acid	Mercuric ion
		complex

Many types of tautomerism are known. One common type, which is presented in more detail when we study the carbohydrates in Chap. 29, is called *ring-chain*

Aldose sugar	Chain form	Ring form

tautomerism. Considering the previous example, the logic of the name should be evident. Equilibrium to the ring form is generally favored.

11.14 Hydroboration-Reduction Reactions

As we saw in Secs. 10.4 and 10.6, hydroboration is a versatile synthetic tool. Although it is used principally in the synthesis of alcohols, the hydroboration-reduction of alkenes provides an easy route to alkanes.

This reaction has several other powerful uses in synthesis. For example, alkynes react with diborane faster than do alkenes, and if equimolar amounts of alkyne and diborane are mixed, the reaction stops at the alkene stage. As with alkenes, borane adds stereospecifically to alkynes in the *syn* fashion, and mild hydrolysis of the resulting vinylboron compound gives a good yield of the *cis*-alkene.

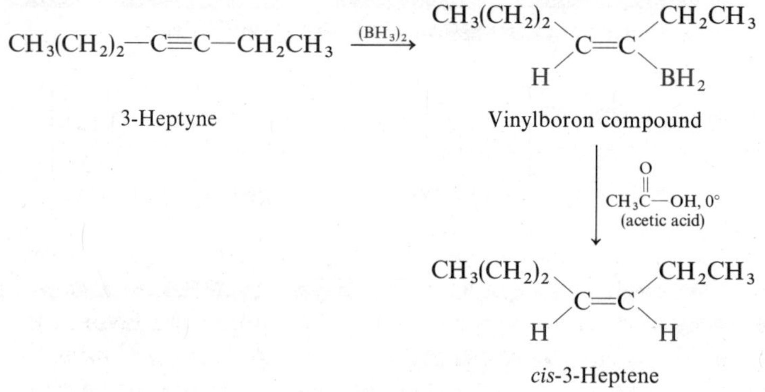
CH3(CH2)2—C≡C—CH2CH3
(BH3)2
CH3(CH2)2
CH2CH3
C=C
H
BH2
O
CH3C—OH, 0°
(acetic acid)
CH3(CH2)2
CH2CH3
C=C
H
H
3-Heptyne; Vinylboron compound; *cis*-3-Heptene 95%

11.15 Addition of Alkynes to Alkadienes: Diels-Alder Reaction

The Diels-Alder reaction (see Secs. 8.29 and 27.2) occurs when alkenes or alkynes are reacted with dienes in the presence of heat. For example, acetylene and 1,3-butadiene react at high temperature and under pressure to give 1,4-cyclohexadiene:

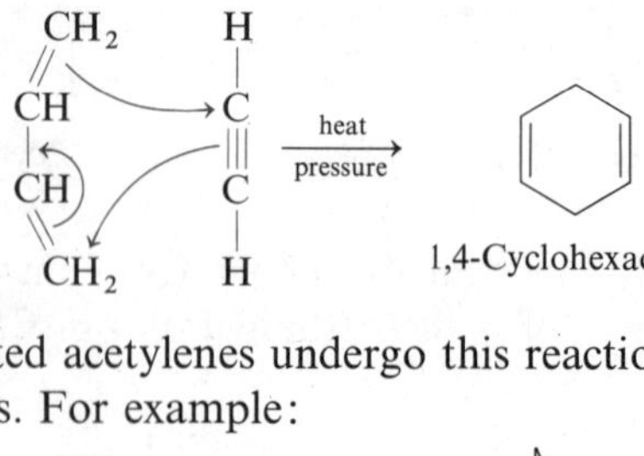
CH2
CH
CH
CH2
H
C
C
H
heat
pressure
1,4-Cyclohexadiene

Mono- and disubstituted acetylenes undergo this reaction as well, with both open-chain and cyclic dienes. For example:

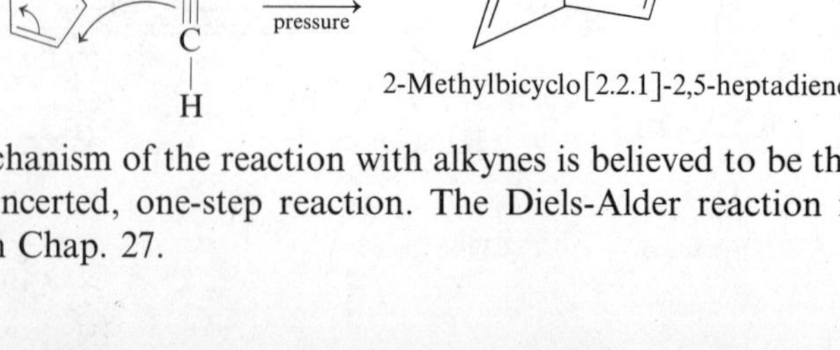
CH3
C
C
H
heat
pressure
CH3
2-Methylbicyclo[2.2.1]-2,5-heptadiene

The mechanism of the reaction with alkynes is believed to be the same as with alkenes: a concerted, one-step reaction. The Diels-Alder reaction is discussed in more detail in Chap. 27.

11.16   Miscellaneous Addition Reactions of Alkynes

omit
(notat)

Several important addition reactions of alkynes are of great use industrially because they provide compounds used in polymerization reactions. The following examples are illustrative.

A. Dimerization of Acetylene

Acetylene dimerizes when it is passed into an aqueous solution of cuprous chloride ($CuCl \equiv Cu_2Cl_2$) and ammonium chloride:

$$H-C\equiv C-H \xrightarrow[NH_4Cl]{CuCl} H-C\equiv C-CH=CH_2$$

2 moles   1-Buten-3-yne
                                (vinylacetylene)

This reaction may be thought of as being the *addition* of a C—H bond from one molecule of acetylene across the triple bond of a second molecule:

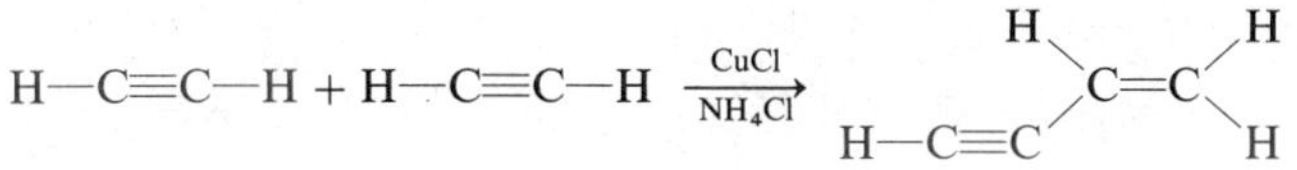

$$H-C\equiv C-H + H-C\equiv C-H \xrightarrow[NH_4Cl]{CuCl}$$

The product of this reaction can be subjected to selective catalytic hydrogenation, which reduces the triple bond to a double bond and produces 1,3-butadiene; iron metal is often used as the catalyst:

$$H-C\equiv C-CH=CH_2 \xrightarrow[Fe]{H_2} CH_2=CH-CH=CH_2$$

1-Buten-3-yne   1,3-Butadiene

B. Addition of Alcohols

Alcohols react with acetylene in the presence of a basic catalyst such as potassium hydroxide to form *vinyl alkyl ethers:*

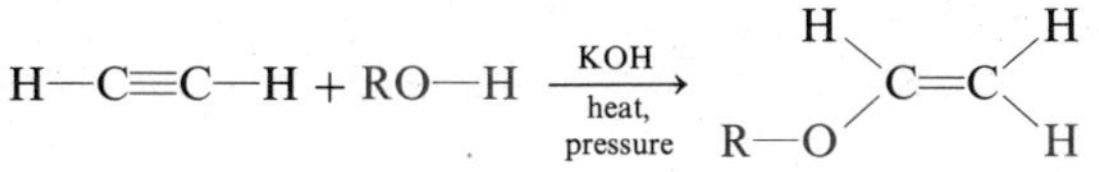

$$H-C\equiv C-H + RO-H \xrightarrow[\substack{heat, \\ pressure}]{KOH}$$

Vinyl alkyl ether

The vinyl ether is very reactive in polymerization reactions.

C. Addition of Hydrogen Cyanide

Hydrogen cyanide adds to acetylene in the presence of hydrogen chloride and cuprous chloride to give acrylonitrile (also called *vinyl cyanide*):

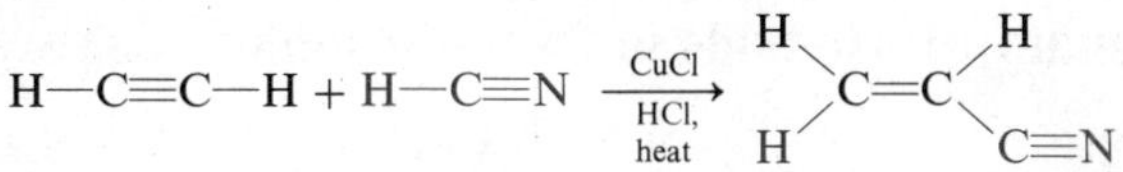

$$H-C\equiv C-H + H-C\equiv N \xrightarrow[\substack{HCl, \\ heat}]{CuCl}$$

Hydrogen   Acrylonitrile
cyanide   (vinyl cyanide)

The polymerization of acrylonitrile gives Orlon (see Sec. 8.34).

11.17   Cleavage Reactions of Alkynes

omit

Carbon-carbon triple bonds react with potassium permanganate under basic conditions, and the products are carboxylic acids:

$$CH_3—CH_2—CH_2—C\equiv C—CH_3 \xrightarrow[\substack{:\ddot{O}H^{\ominus}, H_2O \\ heat}]{KMnO_4}$$

2-Hexyne

$$CH_3—CH_2—CH_2—\overset{\overset{\displaystyle O}{\|}}{C}—\overset{..}{\underset{..}{O}}{:}^{\ominus} K^{\oplus} + K^{\oplus\ominus}{:}\overset{..}{\underset{..}{O}}—\overset{\overset{\displaystyle O}{\|}}{C}—CH_3$$

Carboxylic acid salts

The reaction has the same visible results that occur when permanganate reacts with alkenes; the purple of the permanganate ion is discharged and replaced by a brown-black precipitate of manganese dioxide.

Ozonolysis of alkynes also results in cleavage of the triple bond and the formation of carboxylic acids, whereas the same reaction with alkenes gives aldehydes and ketones; for example:

$$CH_3—CH_2—CH_2—C\equiv C—CH_3 \xrightarrow[2.\ H_2O,\ Zn]{1.\ O_3} CH_3—CH_2—CH_2—\overset{\overset{\displaystyle O}{\|}}{C}—OH + HO—\overset{\overset{\displaystyle O}{\|}}{C}—CH_3$$

2-Hexyne   Butanoic acid   Ethanoic acid
Carboxylic acids

11.18   Qualitative Analysis; Detection of Alkynes

We discussed several reactions characteristic of the triple bond. A terminal alkyne can be distinguished from a disubstituted alkyne on the basis of the reaction the former undergoes with silver ammonia complex or cuprous ammonia complex:

$$R—C\equiv C—H \xrightarrow[or\ Cu(NH_3)_2^{\oplus}]{Ag(NH_3)_2^{\oplus}} R—C\equiv C—Ag\ (or\ Cu)$$

Metal salt precipitate

$$R—C\equiv C—R \xrightarrow[or\ Cu(NH_3)_2^{\oplus}]{Ag(NH_3)_2^{\oplus}} no\ reaction$$

Alkenes and alkynes, on the other hand, react with bromine in carbon tetrachloride, although alkynes react more slowly than alkenes. They both react with an aqueous solution of basic potassium permanganate (Baeyer test).

Quantitative hydrogenation is frequently used to determine the degree of unsaturation (see Sec. 8.12), and degradation of the starting unsaturated compound by ozone or permanganate is used to complete the identification.

11.19   Summary of Methods of Preparation of Alkynes

1. Hydrolysis of calcium carbide: acetylene (Sec. 11.4)

$$CaC_2 \xrightarrow{H_2O} H—C\equiv C—H$$

2. Dehydrohalogenation of vicinal dihalides (Sec. 11.5A)

$$R-\underset{\underset{H}{|}}{\overset{\overset{X}{|}}{C}}-\underset{\underset{H}{|}}{\overset{\overset{X}{|}}{C}}-R \xrightarrow[\text{alcohol}]{\text{KOH}} \left[\underset{H}{\overset{R}{>}}C=C\underset{R}{\overset{X}{<}} \right] \xrightarrow{\text{NaNH}_2} R-C\equiv C-R$$

3. Dehalogenation of tetrahalides (Sec. 11.5B)

$$R-\underset{\underset{X}{|}}{\overset{\overset{X}{|}}{C}}-\underset{\underset{X}{|}}{\overset{\overset{X}{|}}{C}}-R \xrightarrow[\substack{\text{alcohol,}\\\text{heat}}]{\text{Zn}} R-C\equiv C-R$$

4. Substitution reactions of alkynyl anions on alkyl halides (Secs. 11.6 and 11.9)

$$H-C\equiv C-H \xrightarrow[\text{NaNH}_2]{\text{Na or}} H-C\equiv \overset{\ominus}{C}\!:\!\overset{\oplus}{\text{Na}} \xrightarrow{R-X} H-C\equiv C-R \xrightarrow[\text{NaNH}_2]{\text{Na or}}$$

$$\overset{\oplus\ominus}{\text{Na}}\!:\!C\equiv C-R \xrightarrow{R'-X} R'-C\equiv C-R$$

R and R′ must be primary or methyl.

11.20   Summary of Reactions of Alkynes

A. Alkynes As Acids; Terminal Alkynes

1. Removal of acidic hydrogen (Sec. 11.7)

$$R-C\equiv C-H \xrightarrow[\text{NaNH}_2]{\text{Na or}} R-C\equiv \overset{\ominus}{C}\!:\!\overset{\oplus}{\text{Na}}$$

$$(\text{Na} + \text{NH}_3 \xrightarrow[\text{catalyst}]{\text{Fe}^{+3}} \text{NaNH}_2 + \tfrac{1}{2}\text{H}_2)$$

Only for terminal alkynes

2. Formation of heavy metal salts (Sec. 11.8)

$$R-C\equiv C-H \begin{cases} \xrightarrow{\text{Ag(NH}_3)_2^{\oplus}} R-C\equiv C-Ag \xrightarrow{H^{\oplus}} R-C\equiv C-H \\ \quad\quad\quad\text{Light-colored} \\ \quad\quad\quad\text{precipitate} \\ \\ \xrightarrow{\text{Cu(NH}_3)_2^{\oplus}} R-C\equiv C-Cu \xrightarrow{H^{\oplus}} R-C\equiv C-H \\ \quad\quad\quad\text{Brick-red} \\ \quad\quad\quad\text{precipitate} \end{cases}$$

B. Addition Reactions of Triple Bonds

$$R-C\equiv C-R' \xrightarrow[\text{Pt, Ni, Pd}]{\text{1 mole H}_2} R-CH=CH-R' \xrightarrow[\text{Pt, Ni, Pd}]{\text{1 mole H}_2} R-CH_2-CH_2-R' \quad \text{Sec. 11.10}$$

$$R-C\equiv C-R' \xrightarrow[\text{Pd/C or Pd/CaCO}_3 \text{ or Ni/B}]{\text{H}_2} \underset{H}{\overset{R}{>}}C=C\underset{H}{\overset{R'}{<}} \quad \text{Sec. 11.10}$$

cis-isomer

$$R-C\equiv C-R' \xrightarrow[\text{liquid NH}_3]{\text{Na or Li}} \quad \underset{R}{\overset{H}{\big\backslash}}C=C\underset{H}{\overset{R'}{\big/}} \qquad \text{Sec. 11.10}$$

Predominantly
trans isomer

$$R-C\equiv C-R' \xrightarrow[\text{CCl}_4]{\text{1 mole X}_2} \quad \underset{X}{\overset{R}{\big\backslash}}C=C\underset{R'}{\overset{X}{\big/}} \xrightarrow[\text{CCl}_4]{\text{1 mole X}_2} \quad R-\overset{\overset{X}{|}}{\underset{\underset{X}{|}}{C}}-\overset{\overset{X}{|}}{\underset{\underset{X}{|}}{C}}-R' \qquad \text{Sec. 11.11}$$

where $X_2 = Cl_2$, Br_2 (mostly *trans addition*)

$$R-C\equiv C-R' \xrightarrow[\text{CCl}_4]{\text{1 mole H—X}} R-\overset{\overset{|}{C}}{\underset{\underset{H}{|}}{C}}=\overset{\overset{|}{C}}{\underset{\underset{X}{|}}{C}}-R' \xrightarrow[\text{CCl}_4]{\text{1 mole H—X}} R-\overset{\overset{H}{|}}{\underset{\underset{H}{|}}{C}}-\overset{\overset{X}{|}}{\underset{\underset{X}{|}}{C}}-R' \qquad \text{Sec. 11.12}$$

where H—X=HCl, HBr, HI

Orientation: Obeys Markovnikov's rule

$$R-C\equiv C-R' \xrightarrow[\text{HgSO}_4]{\text{H}_2\text{O, H}_2\text{SO}_4} \left[R-\overset{\overset{|}{C}}{\underset{\underset{H}{|}}{C}}=\overset{\overset{|}{C}}{\underset{\underset{OH}{|}}{C}}-R' \right] \rightleftharpoons R-\overset{\overset{|}{C}}{\underset{\underset{H}{|}}{C}}-\overset{\overset{|}{C}}{\underset{\underset{O}{||}}{C}}-R' \qquad \text{Sec. 11.13}$$

Unstable

Orientation: Obeys Markovnikov's rule

$$R-C\equiv C-R' \xrightarrow{\text{(BH}_3)_2} \quad \underset{H}{\overset{R}{\big\backslash}}C=C\underset{BH_2}{\overset{R'}{\big/}} \xrightarrow[0°]{\text{CH}_3\text{COOH}} \quad \underset{H}{\overset{R}{\big\backslash}}C=C\underset{H}{\overset{R'}{\big/}} \qquad \text{Sec. 11.14}$$

Stereochemistry: *syn* addition

$$H-C\equiv C-H \xrightarrow[\substack{\text{heat} \\ \text{(Diels-Alder reaction)}}]{\text{CH}_2=\text{CH—CH}=\text{CH}_2} \quad \bigcirc \qquad \text{Secs. 11.15 and 27.2}$$

$$H-C\equiv C-H \xrightarrow[\text{CuCl, NH}_4\text{Cl}]{\text{H—C}\equiv\text{C—H}} H-C\equiv C-CH=CH_2 \qquad \text{Sec. 11.16A}$$

$$H-C\equiv C-H \xrightarrow[\substack{\text{KOH, heat,} \\ \text{pressure}}]{\text{RO—H}} RO-CH=CH_2 \qquad \text{Sec. 11.16B}$$

$$H-C\equiv C-H \xrightarrow[\text{HCl, CuCl, heat}]{\text{H—C}\equiv\text{N}} CH_2=CH-C\equiv N \qquad \text{Sec. 11.16C}$$

Omit

C. Cleavage of Triple Bond

1. By potassium permanganate (Sec. 11.17)

$$R-C\equiv C-R' \xrightarrow[\substack{\text{H}_2\text{O, OH}^\ominus, \\ \text{heat OH}^\ominus}]{\text{KMnO}_4} R-\overset{\overset{O}{||}}{C}-\overset{..}{\underset{..}{O}}{:}K^{\oplus} \; ^\ominus + K^{\oplus}\; ^\ominus:\overset{..}{\underset{..}{O}}-\overset{\overset{O}{||}}{C}-R'$$

Carboxylic acid salts

2. By ozone (Sec. 11.17)

$$R-C\equiv C-R' \xrightarrow[\text{2. H}_2\text{O, Zn}]{\text{1. O}_3} R-\overset{\overset{O}{||}}{C}-OH + HO-\overset{\overset{O}{||}}{C}-R'$$

Carboxylic acids

Study Questions

11.6 Using electronic configurations (for example, $1s$, $2s$) describe the *hypothetical changes* in electronic structure that occur when the carbon atom (neutral, ground state) is transformed into the carbon atom in acetylene. Also,

(a) Indicate the shape of acetylene by giving bond angles, and tell how the electronic structure of the carbon atoms in it accounts for its shape.

(b) State the atomic orbitals used in forming the molecular bond orbitals in the molecule.

(c) Identify and properly label π and/or σ bonds in the molecule.

(d) Draw the antibonding orbitals of the acetylene molecule that correspond to the bonding orbitals in Fig. 11.1.

11.7 Indicate clearly the difference between a carbon-carbon σ bond and a carbon-carbon π bond. To illustrate this, draw a structure of propyne in which the hybrid atomic orbitals of each carbon atom are represented and properly labeled. The overlap of atomic orbitals to form carbon-carbon bond orbitals should be shown. The nature of each carbon-carbon bond should be labeled appropriately as σ or π.

11.8 Consider the allene molecule, $CH_2{=}C{=}CH_2$.

(a) Designate the hybridization of *each* carbon atom in the molecule.

(b) Make an unambiguous, *three-dimensional drawing* of the molecule, showing clearly the geometry of the molecular orbitals in the following manner:

(1) Use straight lines for the σ orbitals and indicate the angles between them.

(2) For the π orbitals, show the p atomic orbitals and then show how they overlap to form the π molecular orbitals. Indicate shapes and spatial relationships.

(c) Explain how 2,3-pentadiene, $CH_3{-}CH{=}C{=}CH{-}CH_3$, can be resolved into optically active enantiomers, even though it has no asymmetric carbon atom. (*Hint:* Recall from Sec. 6.5 the general criteria for the existence of enantiomers, and figure this out with the aid of molecular models.)

(d) On the basis of your answers to **(a)** and **(b)**, deduce the structure of the carbon dioxide molecule, $O{=}C{=}O$.

11.9 Give IUPAC names corresponding to the following structures:

(a) $CH_3{-}CH_2{-}CH_2{-}CH_2{-}C{\equiv}C{-}H$

(b) $CH_3{-}\overset{\displaystyle |}{\underset{\displaystyle CH_3}{CH}}{-}CH_2{-}CH_2{-}C{\equiv}C{-}CH_3$

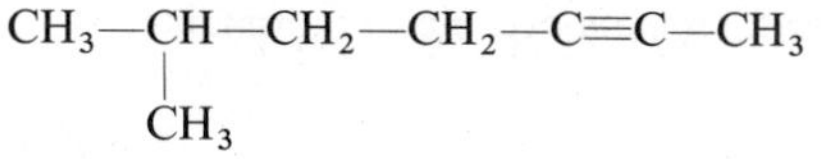

(c) $H{-}C{\equiv}C{-}\overset{\displaystyle CH_3}{\overset{\displaystyle |}{CH}}{-}\overset{\displaystyle CH_3}{\overset{\displaystyle |}{\underset{\displaystyle |}{\underset{\displaystyle CH_3}{C}}}}{-}CH_3$

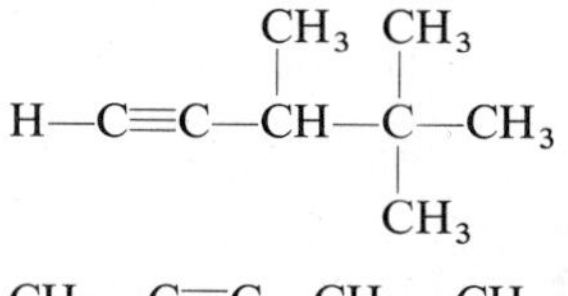

(d) $CH_3{-}C{\equiv}C{-}CH_2{-}CH_2{-}C{\equiv}C{-}CH_3$

(e) $CH_3{-}\overset{\displaystyle CH_3}{\overset{\displaystyle |}{\underset{\displaystyle |}{\underset{\displaystyle CH_3}{C}}}}{-}C{\equiv}C{-}CH_2{-}CH_3$ Give two acceptable names for this compound, one of which must be IUPAC.

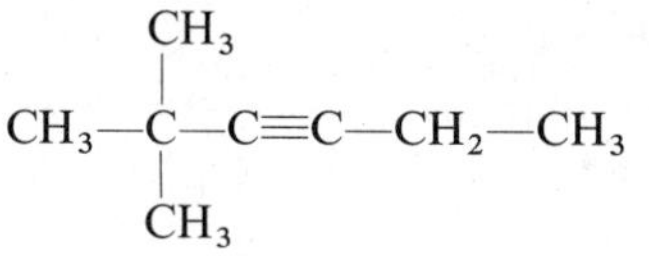

(f) $CH_3{-}\overset{\displaystyle |}{\underset{\displaystyle \overset{\displaystyle CH}{\underset{\displaystyle ||}{\underset{\displaystyle CH_2}{}}}}{CH}}{-}CH_2{-}C{\equiv}C{-}CH_2{-}\overset{\displaystyle |}{\underset{\displaystyle \overset{\displaystyle CH_2}{\underset{\displaystyle |}{\underset{\displaystyle CH_3}{}}}}{CH}}{-}CH_3$

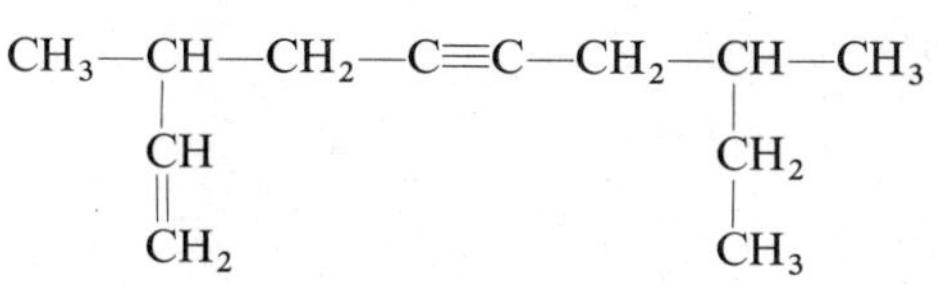

11.10 Give the structures corresponding to the following names:

(a) 2-heptyne   **(b)** di-*tert*-butylacetylene   **(c)** 2,4-octadiyne

(d) 2-methyl-5-isopropyl-3-octyne   **(e)** 2,2,5,5-tetramethyl-3-hexyne   **(f)** cyclodecyne

11.11 Draw the two possible vinyl alcohols that could be formed by addition of water (in the presence of H_2SO_4 and $HgSO_4$) to propyne, and then draw the keto structures derived from each vinyl alcohol. When this reaction is carried out in the laboratory, the only compound isolated is acetone,

$$CH_3-\overset{\overset{\displaystyle O}{\|}}{C}-CH_3$$

(*a*) What can you conclude about the *orientation* of the addition of water to the triple bond (that is, does it obey Markovnikov's rule)?

(*b*) What product would you expect to obtain from the hydration of 1-butyne and 2-butyne?

11.12 For the following molecule, give the *approximate* bond angles for the lettered bonds and the hybridization of the lettered carbon atoms. The bond angles are shown in an arbitrary fashion and do not necessarily represent the correct angle.

11.13 (*a*) Draw structural formulas for all *open-chain* compounds that have the molecular formula C_4H_6, and name each of them.

(*b*) Which of the compounds in (*a*) will react with silver ammonia complex, $Ag(NH_3)_2^{\oplus}$, to give a precipitate? Write the equation for the reaction that occurs, and briefly explain why it occurs.

11.14 Draw structures for and name all isomeric alkynes that have the molecular formula C_8H_{14} and contain a five-carbon continuous chain (that is, all C_8H_{14} compounds named as substituted pentynes).

11.15 (*a*) Write structural formulas for the vinyl alcohols with the molecular formula C_4H_8O, and then write the structural formulas for the actual compounds you would isolate if you tried to synthesize the vinyl alcohols.

(*b*) In light of your answer to (*a*), explain what the following terms mean: tautomerism, tautomers, keto structure, enol structure.

11.16 Complete the following 16 reactions by supplying the missing organic product(s), the missing reactant(s), and the missing reagent(s), as needed. The numbers in parentheses [that is, (1), (2), and so on] represent the missing items. Use any organic or inorganic reagents you want.

(*a*) (1) + (2) $\longrightarrow$ H—C≡C—H

(*b*) $CH_3-C{\equiv}C-H \xrightarrow{(3)} CH_3-\overset{\overset{\displaystyle Br}{|}}{\underset{\underset{\displaystyle Br}{|}}{C}}-CH_3$

(*c*) $CH_3-CH_2-\overset{\overset{\displaystyle Br}{|}}{\underset{\underset{\displaystyle Br}{|}}{C}}-\overset{\overset{\displaystyle Br}{|}}{\underset{\underset{\displaystyle Br}{|}}{C}}-CH_3 \xrightarrow[\substack{\text{alcohol,}\\ \text{heat}}]{\text{Zn metal}}$ (4) $CH_3-CH_2-C{\equiv}C-CH_3$

(*d*) $CH_3-C{\equiv}C-H \xrightarrow[Pt]{2\text{ M}(5)\,H_2} CH_3-CH_2-CH_3$

(*e*) $CH_3-CH_2-C{\equiv}C-CH_3 \xrightarrow[Ni/B]{(6)\,H_2}$ $\underset{CH_3-CH_2}{\overset{H}{\diagdown}}C{=}C\underset{CH_3}{\overset{H}{\diagup}}$

cis Isomer

(*f*) $CH_3-C{\equiv}C-H \xrightarrow{(7)} CH_3-\overset{\overset{\displaystyle O}{\|}}{C}-CH_3$

(*g*) $CH_3-C{\equiv}C-H \xrightarrow[\text{2. 1 mole HBr, CCl}_4]{\text{1. 1 mole HCl, CCl}_4}$ (8) $CH_3-\overset{\overset{\displaystyle Br}{|}}{\underset{\underset{\displaystyle Cl}{|}}{C}}-\overset{\overset{\displaystyle H}{|}}{\underset{\underset{\displaystyle H}{|}}{C}}-H$

(*h*) (9), $C_{17}H_{30} \xrightarrow{H_2,\ Pt}$ (10), $C_{17}H_{32} \xrightarrow{H_2,\ Pt}$ cycloheptadecane (*i*) $CH_3-C\equiv C-CH_3$ $\xrightarrow[\substack{2. \text{ 1 mole HBr,} \\ CCl_4}]{1. \text{ Na, NH}_3}$ (11) $\xrightarrow{\text{Li}}$ (12) $\xrightarrow{H_3O^{\oplus}}$ (13)

(*j*) $CH_3-CH_2-CH_2-Br$ $\xrightarrow{(14)}$ $CH_3-CH=CH_2$ $\xrightarrow[CCl_4]{Br_2}$ (15) $\xrightarrow[\text{2. NaNH}_2]{1. \text{ KOH, alcohol}}$ (16)

(*k*) $H-C\equiv C-H$ $\xrightarrow{(17)}$ $H-C\equiv C:\overset{\oplus}{\underset{\ominus}{Na}}$ $\xrightarrow{(18)}$ $H-C\equiv C-CH_2-CH_2-CH_3$

$\downarrow Ag(NH_3)_2^{\oplus}$

$H-C\equiv C-CH_2-CH_2-CH_3 \xleftarrow{(20)}$ (19)

(*l*) $CH_3-CH_2-C\equiv C-CH_3$ $\xrightarrow[\text{2. H}_2O, Zn]{1. O_3}$ (21)

(*m*) $CH_3-C\equiv C:\overset{\oplus}{\underset{\ominus}{Na}} + CH_3-\overset{\overset{\displaystyle CH_3}{|}}{\underset{\underset{\displaystyle CH_3}{|}}{C}}-Br \longrightarrow$ (22) $CH_3-C\equiv C-\overset{\overset{\displaystyle CH_3}{|}}{\underset{\underset{\displaystyle CH_3}{|}}{C}}-CH_3 + NaBr$

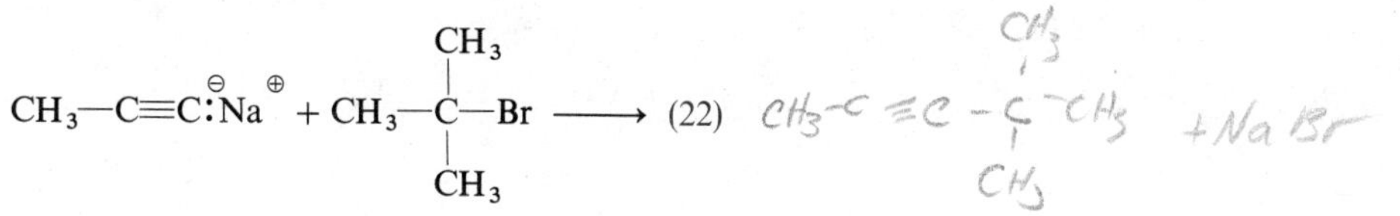

(*n*) $CH_3-C\equiv C:\overset{\oplus}{\underset{\ominus}{Na}} + H_2O \longrightarrow$ (23)

(*o*) $CH_3-C\equiv C-CH_3$ $\xrightarrow{(24)}$ 2 moles $CH_3-\overset{\overset{\displaystyle O}{||}}{C}-OH$

(*p*) $CH_3-C\equiv C-CH_3$ $\xrightarrow{(27)}$ $\underset{H}{\overset{CH_3}{\diagdown}}C=C\underset{CH_3}{\overset{H}{\diagup}}$ $\xrightarrow{(28)}$ 2 moles $CH_3-\overset{\overset{\displaystyle O}{||}}{C}-OH$

trans Isomer

(25)$\downarrow$

$\underset{H}{\overset{CH_3}{\diagdown}}C=C\underset{H}{\overset{CH_3}{\diagup}}$ $\xrightarrow{(26)}$ 2 moles $CH_3-\overset{\overset{\displaystyle O}{||}}{C}-H$

cis Isomer

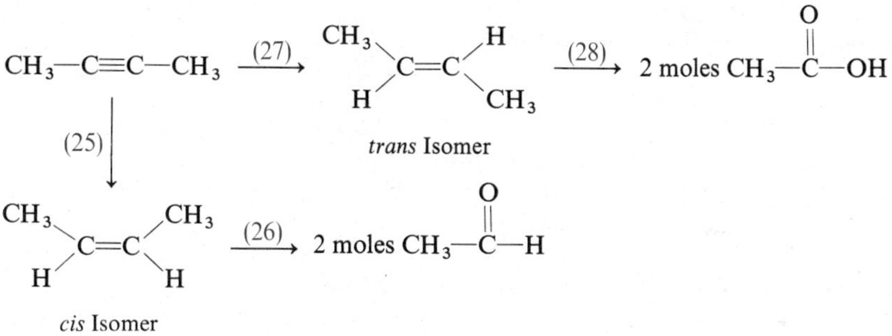

11.17 Using calcium carbide and methane as the *only* sources of carbon, outline methods for preparing each of the following five alkynes. Use any other needed reagents. (Note from this problem that you can prepare compounds with three-, four-, five-, and six-carbon atoms, where the triple bond is in different locations in some of the compounds. This should emphasize how substitution reactions can be used to build up organic molecules.)

(*a*) $CH_3-C\equiv C-H$ (*b*) $CH_3-CH_2-C\equiv C-H$

(*c*) $CH_3-C\equiv C-CH_3$ (*d*) $CH_3-CH_2-C\equiv C-CH_3$

(*e*) $CH_3-CH_2-C\equiv C-CH_2-CH_3$

11.18 (*a*) As discussed in Sec. 10.5, diborane adds to triple bonds in a *syn* fashion. Suppose the following sequence of reactions is carried out on 2-butyne, and the observed product is not the corresponding alcohol but instead is the ketone shown here. Explain these results.

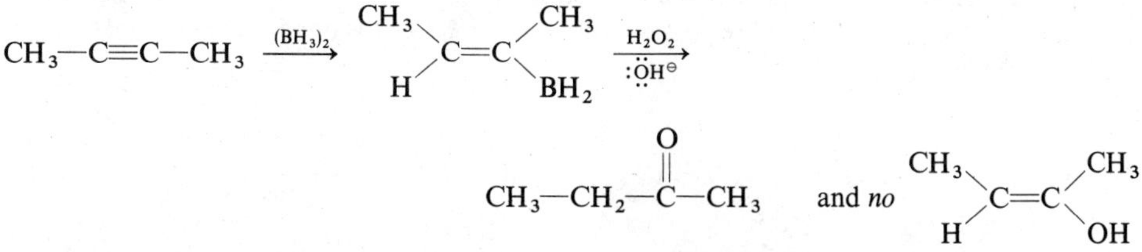

$CH_3-C\equiv C-CH_3$ $\xrightarrow{(BH_3)_2}$ $\underset{H}{\overset{CH_3}{\diagdown}}C=C\underset{BH_2}{\overset{CH_3}{\diagup}}$ $\xrightarrow[:\overset{..}{O}H^{\ominus}]{H_2O_2}$

$CH_3-CH_2-\overset{\overset{\displaystyle O}{||}}{C}-CH_3$ and *no* $\underset{H}{\overset{CH_3}{\diagdown}}C=C\underset{OH}{\overset{CH_3}{\diagup}}$

(*b*) How, if at all, would the results be affected if 2-pentyne were used in place of 2-butyne?

(*c*) What product(s) would be formed if 1-butyne were used in place of 2-butyne?

(*d*) Compare this hydroboration-oxidation reaction with acid-catalyzed hydration ($HgSO_4$, H_2SO_4, H_2O), using 1-butyne as the starting alkyne in both cases.

11.19 Devise laboratory syntheses (that is, multistep syntheses) that accomplish the following 13 transformations in good yield. Use any needed inorganic reagents and specify approximate reaction conditions where important. Use no source of carbon other than that given in the starting material.

(*a*) $CH_3-CH_2-Br \longrightarrow CH_3-CHBr_2$

(*b*) $CH_3-CH_2-Br \longrightarrow CH_3-CH_2-C\equiv C-CH_2-CH_3$

(*c*) $CH_3-\overset{\displaystyle |}{\underset{\displaystyle Br}{C}}H-CH_3 \longrightarrow CH_3-\overset{\displaystyle Br}{\underset{\displaystyle I}{\overset{\displaystyle |}{\underset{\displaystyle |}{C}}}}-CH_3$

(*d*) $CH_3-CH_2-CH_2-Cl$ and acetylene $\longrightarrow$

$$CH_3-CH_2-CH_2-CH=CH-CH_2-CH_2-CH_3$$
cis Isomer

(*e*) Acetylene $\longrightarrow CH_3-CH_2-\overset{\displaystyle |}{\underset{\displaystyle Cl}{C}}H-\overset{\displaystyle |}{\underset{\displaystyle Cl}{C}}H-CH_2-CH_3$

(*f*) $CH_3-CH_2-CH_2-OH \longrightarrow CH_3-C\equiv C-H$

(*g*) Acetylene $\longrightarrow CH_3CH_2CH_2\overset{\displaystyle Cl}{\underset{\displaystyle Cl}{\overset{\displaystyle |}{\underset{\displaystyle |}{C}}}}-\overset{\displaystyle Cl}{\underset{\displaystyle Cl}{\overset{\displaystyle |}{\underset{\displaystyle |}{C}}}}-H$

(*h*) Acetylene $+ CH_3CH_2Br \longrightarrow CH_3-CH_2-\overset{\displaystyle Cl}{\underset{\displaystyle Cl}{\overset{\displaystyle |}{\underset{\displaystyle |}{C}}}}-CH_3$

(*i*) Acetylene $\longrightarrow CH_2=CH-Br$

(*j*) $CH_3-\overset{\displaystyle |}{\underset{\displaystyle OH}{C}}H-CH_3 \longrightarrow CH_3-\overset{\displaystyle |}{\underset{\displaystyle Cl}{C}}=CH_2$

(*k*) $CH_3-CH=CH_2 \longrightarrow CH_3-CH_2-CH_2-CH=CH-CH_3$
trans Isomer

(*l*) $CH_3-\overset{\displaystyle CH_3}{\underset{\displaystyle CH_3}{\overset{\displaystyle |}{\underset{\displaystyle |}{C}}}}-C\equiv C-H \longrightarrow CH_3-\overset{\displaystyle CH_3}{\underset{\displaystyle Cl}{\overset{\displaystyle |}{\underset{\displaystyle |}{C}}}}\text{---}\overset{\displaystyle CH_3}{\overset{\displaystyle |}{C}}H-CH_3$ (*Hint:* A rearrangement occurs on adding HCl/CCl_4 to a certain intermediate)

11.20 Section 11.3 stated that the enol structure is less stable than the keto structure, which implies that the keto structure possesses less energy. Using the bond energies given here, compute the $\Delta H°$ for the reaction:

$$\underset{R'}{\overset{R}{>}}C=C\underset{R''}{\overset{O-H}{<}} \rightleftharpoons R'-\overset{\displaystyle R}{\underset{\displaystyle H}{\overset{\displaystyle |}{\underset{\displaystyle |}{C}}}}-\overset{\displaystyle O}{\overset{\displaystyle ||}{C}}-R''$$

where the alkyl groups are the same in both structures. [Bond energies: $C-H = 104$, $O-H = 111$, $C-C = 163$ (σ, 95; π, 68), and $C=O = 179$ (σ, 108; π, 71) kcal/mole.] Based on the $\Delta H°$ value you computed, is the enol more or less stable than the keto structure? By how much (in terms of energy)?

11.21 Indicate simple laboratory tests that could be used to distinguish between the members in each of the following pairs of compounds. Indicate what you would *observe*, and explain how the results allow for differentiating between the compounds.

(*a*) propane and propene
(*b*) 1-butyne and 2-butyne
(*c*) 1-hexyne and cyclohexene
(*d*) 2-hexyne and cyclohexane
(*e*) 1,3-butadiene and 1-butyne
(*f*) 1-butene and 1-butyne
(*g*) 3-bromo-1-butene and 1-butene
(*h*) cyclopropane and 2-butyne
(*i*) cyclopropane and 1,3-butadiene

11.22 Unknown compound A has the formula C_6H_{10} and is *optically active*. A gives a precipitate when treated with silver ammonia complex, $Ag(NH_3)_2^{\oplus}$. On catalytic reduction (hydrogen in the presence of platinum), A gives compound B, C_6H_{14}. However, hydrogenation of A in the presence of nickel-boron catalyst gives C, C_6H_{12}. On treatment first with ozone and then with water and zinc metal, C gives compounds D, CH_2O, and E, $C_5H_{10}O$.

On the basis of the information given, draw the structures of compounds A to E. Indicate what the various tests tell you, and how the results lead to your conclusion regarding the structures of the various compounds. Be sure to account for the optical activity in A.

11.23 An unknown compound A, $C_{11}H_{14}$, was analyzed to determine its structure. When A is treated with hydrogen gas in the presence of Pt metal, it gives compound B, $C_{11}H_{22}$, which is identified as *cis*-4-propyl-1-ethylcyclohexane. A is unaffected by $Ag(NH_3)_2^{\oplus}$. A is also unaffected by H_2 gas in the presence of Ni/B catalyst. When A is treated with ozone and then with Zn and water, the following compounds are isolated from 1 mole of A:

1 mole   1 mole   2 moles

(*a*) On the basis of the observations just given, draw the structures of compounds A and B. Indicate briefly how you used *each* observation to help you solve this problem, and give the logic and reasoning you used to arrive at these structures.

(*b*) Is the structure you proposed for A unique (that is, is there one and only one structure consistent with these data)? If not, state what ambiguity exists, and draw a structure for one other possibility.

11.24 An unknown compound A, C_6H_{12}, reacts rapidly with hydrogen chloride in carbon tetrachloride to form compound B, $C_6H_{13}Cl$. When B is treated with KOH in alcohol, compound C is formed; C is isomeric with but not identical with A. Reaction of C with excess bromine in carbon tetrachloride, followed by treatment with KOH in alcohol and then with excess sodium amide ($NaNH_2$), gives compound D, C_6H_{10}. When C is treated with dilute acid, it is isomerized to a third isomer E having the formula C_6H_{12}. When E is treated with ozone and then with water and zinc metal, compounds (1) and (2) are obtained. When E is treated with dilute, basic $KMnO_4$ at room temperature, compound F results.

(1)   (2)

Give the structures of compounds A to F, and explain your reasoning.

11.25 Compound A, having the formula C_7H_{10}, is found to be *optically active*. To determine its structure, the following results are obtained from laboratory experiments.

1. A decolorizes solutions of bromine in carbon tetrachloride and of aqueous potassium permanganate.

2. A, when treated with $Cu(NH_3)_2^{\oplus}$, gives no precipitate.

3. On treatment with hydrogen gas in the presence of Pt catalyst, A gives B, C_7H_{16}. B is *optically active*.

4. On treatment with hydrogen gas in the presence of Ni/B catalyst, A gives C, C_7H_{12}. C is also *optically active*.

5. C, when treated with ozone and then with water and zinc metal, gives *equimolar* amounts of the following products:

$$\underset{H}{\overset{H}{\diagdown}}C{=}O + CH_3{-}\underset{\underset{H}{|}}{C}{=}O + O{=}\underset{\underset{H}{|}}{C}{-}\underset{\underset{CH_3}{|}}{CH}{-}\underset{\underset{H}{|}}{C}{=}O$$

On the basis of these observations, draw the structures for compounds A to C. Indicate briefly what the various observations teach you, and give the logic and reasoning you used in arriving at these structures. Be sure to account for the optical activity of the various compounds.

11.26 Unknown compound A, with molecular formula C_5H_8O, gives a precipitate when treated with ammoniacal silver nitrate. A is hydrogenated over a platinum catalyst to give compound B, $C_5H_{12}O$, which does not react with ammoniacal silver nitrate solution or with bromine in carbon tetrachloride. When heated with a trace of concentrated sulfuric acid, B produces two isomeric compounds C and D, C_5H_{10}. Both these compounds rapidly decolorize a solution of bromine in carbon tetrachloride, and both give 2-methylbutane on treatment with hydrogen gas in the presence of palladium. C, on treatment with ozone and then with water and zinc metal, gives acetone $(CH_3{-}\underset{\underset{O}{\|}}{C}{-}CH_3)$ and acetaldehyde $(CH_3{-}\underset{\underset{O}{\|}}{C}{-}H)$.

(a) How many moles of unsaturation are present in A, and what structural feature is suggested by the ammoniacal silver nitrate test?

(b) On the basis of the information given, deduce and draw the structures of compounds A to D. Give equations for the reactions that occur, and rationalize the structure assignments you make.

12

Spectroscopy 1—Spectroscopic Methods: Infrared and Ultraviolet Spectroscopy

Before the development of modern instruments, chemists were faced with the nearly impossible task of analyzing products from reactions by "classical" methods, that is time-consuming laboratory techniques of functional group identification leading ultimately to the complete determination of the structure of compounds. Even then, small amounts of products often went undetected because refined laboratory techniques were lacking.

Amazing strides in technology and equipment development have taken place in the past 20 to 30 yr. Today's chemist is able to separate an impure mixture into its components by using various types of chromatography (see Sec. 1.3). Then a vast amount of information about an isolated unknown compound can be gained in several hours by the use of instruments; only small amounts of sample [for example, less than 10 milligrams (mg)] are often required.

The main goal of this chapter and subsequent chapters on spectroscopy is to provide information about how modern instruments can be used to aid in the identification of organic compounds. The measurements are taken on instruments called **spectrometers** or **spectrophotometers,** which record certain facts on sheets of paper called **spectrograms** or **spectrums,** thus providing a ready reference to other scientists. This aspect of chemical study is called **spectroscopy,** of which four common methods are: (1) *infrared (IR) spectroscopy*, (2) *nuclear magnetic resonance (NMR) spectroscopy*, (3) *ultraviolet (UV) spectroscopy*, and (4) *mass spectrometry (MS)*.[1] In addition, several special techniques, such as optical rotatory dispersion (ORD) and electron spin resonance (ESR), are used to study molecules that have special optical and free-radical properties, respectively.

Of the methods listed, we focus our attention on IR and NMR spectroscopy because these methods are more generally used. Ultraviolet spectroscopy can be used only for molecules that contain multiple bonds. Although MS is a highly useful technique for structure determination, the data recorded on a mass spectrum are often complex and difficult to interpret. Mass spectrometry does, however, allow for an easy determination of the molecular weight of an unknown compound, and today it replaces all classical methods for doing so (see Sec. 18.2). Because of the potential importance of UV and MS, we examine both methods in a cursory sense.

One reason IR and NMR spectroscopy have come into vogue is that spectrometers are now available at reasonable prices. These highly complex electronic instruments are routinely available in research laboratories and even in many

---

[1] Although the term *spectroscopy* is often associated with this technique, *spectrometry* is the more correct name and is used throughout this text. Spectroscopy deals with examining the spectrum of light or radiation from any source; spectrometry deals with the examination of spectra from any source.

undergraduate organic laboratories. A brief introduction to the operation of these instruments is presented here.

12.1   Electromagnetic Radiation

First let us look at absorption spectra in general. Electromagnetic radiation (see Sec. 6.1), of which light is a subclassification, is usually described by either its *frequency*, v, which is the number of waves per second, or its *wavelength*, λ, which is the distance between waves, generally given in centimeters (cm). The product of frequency and wavelength is the speed of light, c, which is a constant: $c = \lambda v$. Electromagnetic radiation is composed of individual units (which are wave particles) of energy, called *quanta;* their energy is related by the equation: $E = hv$. Because $c = \lambda v$, this equation can be rearranged to: $E = hc/\lambda$, where h is Planck's constant. This equation states simply that energy is proportional to the frequency of the radiation and inversely proportional to the wavelength. A more convenient method for expressing energy uses kilocalories per mole, for which the following equations exist:

$$E = hv = \frac{hc}{\lambda} = \frac{28.6}{\lambda} \text{ kcal/mole or } \frac{119.7}{\lambda} \text{ kJ/mole} \quad \text{when } \lambda \text{ is expressed in microns } (\mu)$$

$$= \frac{28,600}{\lambda} \text{ kcal/mole or } \frac{119,662}{\lambda} \text{ kJ/mole} \quad \text{when } \lambda \text{ is expressed in nanometers (nm) or millimicrons (m}\mu)$$

The wavelength of light is usually expressed in **microns** (μ), with units of micrometers. Units of **millimicrons** (mμ) are also frequently used, as are their modern equivalent **nanometers** (nm). (This text uses nm instead of mμ for expressing frequency.)

TABLE 12.1   Regions of Electromagnetic Radiation Spectrum

Region	Wavelength, λ	Frequency, v, in Wave Numbers, cm^{-1}	Energy, kcal/mole (kJ/mole)
Cosmic rays	5×10^{-5} nm		
Gamma (γ) rays	10^{-3}–0.14 nm		
X rays	0.01–15 nm		
Far ultraviolet	15–200 nm	666,667–50,000	1,907–143 (7979.8–598.3)
Near ultraviolet*	200–400 nm	50,000–20,000	143–71.5 (598.3–299.2)
Visible	400–800 nm	25,000–12,500	71.5–35.7 (299.2–149.4)
Near infrared	0.8–2.5 μ	12,500–400	35.7–11.4 (149.4–47.7)
Vibrational infrared*	2.5–25 μ	4,000–400	11.4–1.14 (47.7–4.8)
Far infrared	0.025–0.5 mm	400–200	1.14–0.57 (4.8–2.4)
Microwave radar	0.5–300 mm	200–0.033	0.57–9.4 $\times 10^{-5}$ (2.4–2.2 $\times 10^{-4}$)
Various radio frequencies	0.3–10^9 m		

* Used most commonly in organic chemistry.

The following are the conversion factors for units:

$$1\ \mu = 10^{-6}\ m = 10^{-3}\ mm$$

$$1\ m\mu = 1\ nm = 10^{-9}\ m = 10^{-6}\ mm = 10^{-3}\ \mu$$

The spectrum of electromagnetic radiation is shown in Table 12.1 and Fig. 12.1. The visible region represents just a small fraction of the spectrum. In the visible region, the lower end corresponds to violet light at ~400 nm, and the long end at ~800 nm corresponds to red light; the various colors associated with visible light are shown in Table 12.2.

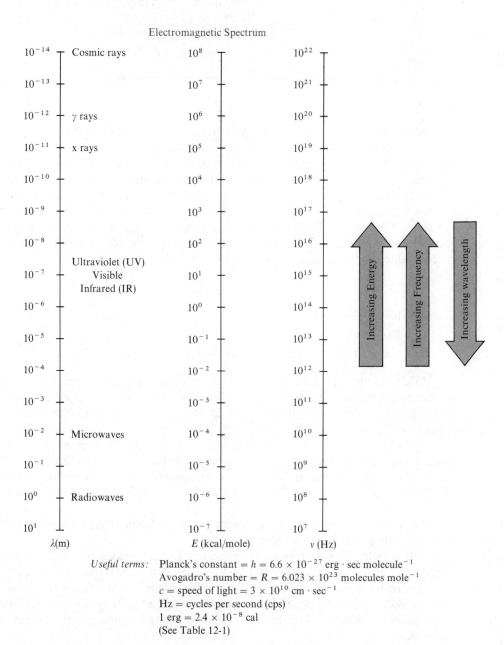

Electromagnetic Spectrum

Useful terms: Planck's constant $= h = 6.6 \times 10^{-27}$ erg · sec molecule^{-1}
Avogadro's number $= R = 6.023 \times 10^{23}$ molecules mole^{-1}
$c =$ speed of light $= 3 \times 10^{10}$ cm · sec^{-1}
Hz = cycles per second (cps)
1 erg $= 2.4 \times 10^{-8}$ cal
(See Table 12-1)

FIGURE 12.1   The electromagnetic spectrum.

TABLE 12.2   Relationship Between Absorption and Color in Visible Region

Wavelength Absorbed, nm	Color of Light Absorbed	Apparent Visible Color
400	Violet	Yellow
450	Blue	Orange
500	Blue-green	Red
530	Yellow-green	Violet
550	Yellow	Blue
600	Orange-red	Green-blue
700	Red	Green

Molecules are constantly undergoing molecular motions that are classified as vibrational motions (atoms within the molecule undergo various types of vibration with respect to one another) or rotational motions (atoms rotate about one or more axes which bisect the molecule). These motions are quantized and may be of only certain discrete allowed values. In addition, it is possible for a molecule, through various arrangements of electrons in molecular orbitals (see Sec. 2.12), to exist in a variety of electronic states that are also quantized. Transitions between these energy states is possible when energy equal to the energy difference between any two states is added to the system:

$$\Delta E = E_i - E_f$$

where E_i = initial energy state

E_f = final energy state

ΔE = change in energy in going from state i to f

When the energy source is removed, the molecule may either continue to react or relax to the original lower energy state. If relaxation occurs, energy equal to ΔE is released, often in the form of heat or light. In IR and UV spectroscopy the relaxation process predominates.

With UV light there is sufficient energy to cause transitions between the electronic energy levels in molecules, thus providing the basis of UV spectroscopy. This occurs when there are "mobile" electrons as in nonbonding pairs of electrons or in π bonds and π clouds, but normally not within σ bonds, where the electrons are tightly bound.

In the IR region there is enough energy to measure the stretching and bending of bonds, thus resulting in IR spectroscopy.

The very low energy region corresponding to microwaves is used to provide information about bond lengths and bond angles, but microwave spectroscopy is seldom used to gain information for the proof of molecular structure. It is usually used on molecules that have a known gross structure and for which more detailed information is desired. Rotational energy changes may be measured by microwave spectroscopy.

With this brief introduction to spectroscopy, we begin more detailed discussions. We are interested in learning as much as possible about the structure of organic molecules. The ideal goal is to use these simple spectral methods, which require small amounts of sample and do not result in their destruction (except for MS), to ascertain what functional group(s) are present in addition to determining the precise structure.

Infrared Spectroscopy

12.2 Infrared Spectroscopy: What Is It?

Infrared spectra yield a large body of information about the structural details of organic compounds. *Infrared spectroscopy allows the detection and identification of virtually all functional groups* because they show specific and characteristic absorptions in the infrared region. Of the spectral methods we study, IR is the only one that allows for *direct* determination of the functional group.

An IR spectrum is unique for a given compound, so IR spectroscopy is an excellent method for confirming the structure of an unknown compound, provided that the same compound is known from other sources. It is highly unlikely that any two organic compounds possess identical IR spectra in both the position and intensity of the peaks. An example of the use of IR spectroscopy is in the synthesis of organic compounds. If a known natural product is synthesized in the laboratory and if its IR spectrum is completely identical with that from a natural source, then the two compounds are identical and the synthesis has been successful.

Infrared spectra are complex and often difficult to interpret completely. Infrared spectroscopy is largely empirical, and one must be able to recognize and identify certain characteristic absorption peaks of various functional groups. The best way to *learn* spectroscopy (IR and others) is by looking at and interpreting numerous spectra. In our study of IR, we present the basic principles and then see how to interpret spectra from the standpoint of functional group identification. Emphasis is given to those functional groups that have been presented. In future chapters, the characteristic IR absorption spectra of new functional groups are indicated and discussed.

Infrared spectroscopy measures the changes in the *stretching* and *bending* vibrations that occur when a molecule absorbs electromagnetic energy in the IR region of the electromagnetic spectrum (Fig. 12.2).

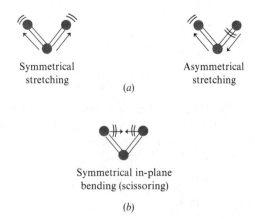

Symmetrical
stretching

(*a*)

Asymmetrical
stretching

Symmetrical in-plane
bending (scissoring)

(*b*)

FIGURE 12.2 (*a*) **Stretching and** (*b*) **bending vibrational modes of a group of atoms,** XY_2.

The possible vibrational motions of a molecule depend on two things: (1) the masses of the atoms that form a given bond and (2) the nature of the bond itself, that is, its bond strength.

The energies associated with the vibrational modes of molecules (Fig. 12.2 and Table 12.1) may be illustrated (*approximately*) by a system of balls and springs that also move with various stretching and bending vibrations. Atoms (or balls) with greater masses move with less frequency, and the stronger the bond (or spring), the more rapid the movement. A mathematical relationship from classical physics can be derived relating the frequency of vibration of a given bond to the bond strength. This relationship is given here:

$$v = \frac{1}{2\pi c} \sqrt{\frac{k}{m^*}}$$

where v = frequency of absorption in cm^{-1}

c = speed of light (2.99×10^{10} cm/sec)

k = force constant of bond

m^* = reduced mass of two atoms constituting bond; that is, $m^* = (m_1 + m_2)/m_1 m_2$, where m_1 and m_2 are masses of two atoms

The relationship states that the frequency of vibration is directly proportional to the square root of the force constant of the bond, k. The force constant is particular to and characteristic of a given bond. It is another physical constant, for example, like a melting point. In the preceding model, the force constant may be likened to the stiffness and strength of the spring. Also, the frequency is inversely proportional to the square root of the reduced mass, m^*, of the system. The greater the mass, the lower the frequency of absorption. Likewise, the stronger the bond (k), the greater the frequency of absorption. With this in mind, consider Table 12.3 and the vibrational frequencies of the C—C, C=C, and C≡C bonds. If our ideas about bond strength are correct, there should be some correlation between IR absorption and the nature of the chemical bond. For example, we asserted that the decreasing energy of carbon-carbon bonds is: triple > double > single. The stronger the bond, the greater the amount of energy required to *stretch* it. Because wave numbers are proportional to energy, they give some indication of bond strength. From Table 12.3, we find that such a correlation exists.

Infrared wave numbers:

$\sim 2,200$ cm^{-1} $\sim 1,650$ cm^{-1} $\sim 1,200$ cm^{-1}

Type of bond:

C≡C   C=C   C—C

<u>Increasing bond strength</u>
←

But look further. What do we know about a carbon-hydrogen bond compared with a carbon-carbon bond? Bond dissociation energies indicate that a C—H bond [104 kcal/mole (435 kJ/mole) in methane] is not much stronger than a C—C bond [88 kcal/mole (368.2 kJ/mole) in ethane]. Yet the stretching frequencies for these two bonds are:

Infrared wave numbers:

$\sim 3,000$ cm^{-1} $\sim 1,200$ cm^{-1}

TABLE 12.3   IR Absorption Frequencies of Functional Groups

Compound Type	Bond Type	Wave Number (Frequency), $\tilde{\nu}$, cm^{-1} (Intensity in Parentheses)	Wavelength, λ, μ
Alkane	—C—H	2,850–2,970 (s)* 1,340–1,470 (s)	3.37–3.51 6.8–7.5
Alkane	—C—C—	1,100–1,300 (w)	7.7–9.1
Alkane	—C—D	~2,200 (s)	4.55
Alkene	=C—H	3,010–3,095 (m) 675–995 (s)	3.24–3.32 10.1–14.8
Alkene	C=C	1,610–1,680 (v)	5.95–6.2
Alkyne	≡C—H	3,200–3,300 (s)	3.0–3.1
Alkyne	—C≡C—	2,100–2,260 (v)	4.4–4.8
Aromatic	Ar—H	3,010–3,100 (m) 690–900 (s)	3.24–3.33 11.1–14.5
Aromatic ring	C=C	1,500–1,600 (v)	6.25–6.67
Monomeric alcohol, phenol	—O—H	3,590–3,650 (v)	2.74–2.79
Hydrogen-bonded alcohol, phenol	—O—H	3,200–3,600 (v)	2.8–3.1
Monomeric carboxylic acid	—O—H	3,500–3,650 (m)	2.79–2.86
Hydrogen-bonded carboxylic acid	—O—H	2,500–3,000 (v, broad)	3.7–4.0
Amine, amide	—N—H	3,300–3,500 (m)	2.79–3.1
Amine, amide	—C—N	1,180–1,360 (s)	7.4–8.5
Nitrile	—C≡N	2,210–2,280 (s)	4.4–4.8
Alcohol, ester, carboxylic acid	—C—O—	1,050–1,300 (s)	7.7–9.5
Aldehyde, ketone, carboxylic acid, ester	C=O	1,690–1,760 (s)	5.7–5.9
Nitro compound	—NO$_2$	1,500–1,570 (s) 1,300–1,370 (s)	6.4–6.67 7.3–7.7

* s = strong, m = moderate, w = weak, v = variable.

Type of bond:

$$C—H \qquad\qquad C—C$$

Increasing bond strength

This also agrees with our expectations. The mass of the C—H bond is much less than that of the C—C bond, and this accounts well for the observed frequencies. Work this out mathematically, assuming the force constants for the two bonds are approximately equal.

Question 12.1

If the stretching frequency of a C—H bond in a molecule is measured to be 2,900 cm^{-1}, and the H atom is replaced with one of its isotopes, tritium, T ($_1^3$H), what would be the stretching frequency of the C—T bond? Do the same for the C—D bond.

The frequency of movement depends not only on the bond strength and the atoms involved but also on the environment and the entire molecule. If we hit a ball-and-spring system, the motion of the entire system increases as well as that of a given ball. The same thing happens in molecules; when the electromagnetic radiation in IR radiation strikes a bond, the amplitude of the vibration of that bond increases. But there is one important difference between the ball-and-spring model and bonds in molecules: all possible energies change the amplitude of a ball attached to a spring, whereas a bond in a molecule is affected (that is, its energy level is raised) only when IR radiation with a certain energy impinges on the bond. Put another way, the energies associated with bonds are quantized in much the same way that electrons in atoms are quantized; only certain energies affect certain bonds. At the instant the "correct" energy hits a bond, the amplitude of a particular vibration increases suddenly and by a certain amount; the change is not so gradual as increasing the impact on a ball-and-spring system. The IR spectrophotometer changes the frequency (and thus the energy) of the IR beam continuously, and when the energy of the beam passing through the sample is exactly equal to the energy needed to bend or stretch a certain bond, the sample absorbs that energy. When the transmitted beam of radiation is absorbed, the amount of radiation passing through the sample greatly decreases. The IR spectrum then records the change in beam intensity as a function of wavelength on graph paper.

To understand *how* a molecule interacts with electromagnetic radiation in the IR region, we must first see how an IR spectrometer operates. Fig. 12.3 is a schematic diagram of a typical double-beam IR spectrophotometer. It is called a "double-beam" instrument because two beams of radiation are generated, one of which passes through a cell containing the sample dissolved in a solvent and the other passes through a cell containing only the solvent. More details of the operation of the spectrophotometer appear beneath Fig. 12.3, but special mention should be made of the cells that contain the sample. The sample cells and the various optical parts (such as prisms) of the IR spectrophotometer are made of fused rock salt (NaCl) or other inorganic salts. As a result, great care must be exercised in handling the cells, and contact with moisture must be avoided at all times; the cells are often stored in desiccators and dried with inert gases. These special optics are used because IR radiation does not pass through ordinary glass.

Because the energy associated with the IR region is relatively low [1 to 11 kcal/mole (4 to 46 kJ/mole), Table 12.1], it does not permanently change the molecule

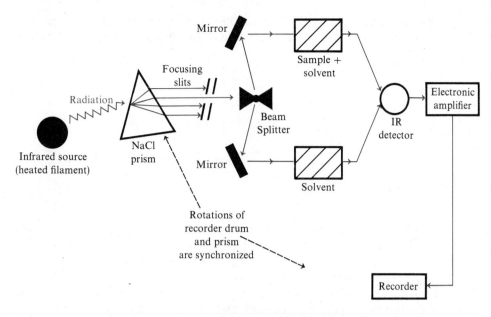

FIGURE 12.3   Simplied view of the recording double-beam infrared (IR) spectrophotometer, indicating the various components. The IR source emits a continuum of wavelengths. The radiation passes through a prism, which separates the wavelengths. The prism is then rotated to let radiation of increasing wavelength pass through the slits. This gives a single wavelength (monochromatic) beam that is split into two beams and focused by the mirrors; one beam passes through the compound being studied (often dissolved in a solvent), and the other beam passes through solvent alone. The intensities of the two beams are determined by the IR detector, which combines them to produce net transmittance. (The intensity of the light passing through the solvent cell is subtracted from the intensity of the light passing through the solvent plus compound cell because the solvent itself absorbs IR radiation. The net amount of transmittance is then caused by the compound itself.) This transmittance signal is recorded on graph paper as a function of wavelength. The rotations of the prism and recording drum are synchronized so that wavelength (frequency) and transmittance are correlated on the graph paper with what is actually going through the sample at that time.

(that is, no bonds are broken). The light energy of the radiation is absorbed by the molecule but is then very quickly liberated as heat.

Although absorbance could be plotted, most infrared instruments plot the *percent transmittance* ($\%T$), which is the ratio of the intensity of the light passing through the sample (I) to the intensity of the light striking the sample (I_0) times 100. The $\%T$ is plotted versus the wavelength (or frequency) of radiation. This relationship is:

$$\%T = \frac{I}{I_0} \times 100$$

Infrared spectra are obtained by irradiation of the sample with light from the IR region of the electromagnetic spectrum. This is the 5,000 to 500-cm^{-1} region. The units most commonly used are *reciprocal centimeters*, also called *wave numbers*. The wave number associated with a given wavelength is obtained by taking the reciprocal of the wavelength in centimeters. It tells the number of wavelengths of light contained in 1 cm.

Because the energy-frequency relationship for electromagnetic radiation is $E = h\nu$ [where h = Planck's constant = 6.62×10^{-27} erg/sec, and ν is the frequency of the radiation in cycles per second or hertz (Hz)], the frequency of the radiation

$(\tilde{v})$ can be expressed in cm^{-1} by dividing v by the speed of light, c (in cm/sec). The wavelength in microns (μ) of the radiation can be obtained by dividing 10,000 by the frequency $(\tilde{v})$ in cm^{-1}.

v = frequency in cycles per second or hertz (Hz)

$\tilde{v}$ = frequency in cm^{-1}

λ = wavelength in microns (μ)

$$\lambda = \frac{10,000}{\tilde{v}}$$

$$\tilde{v} = \frac{v}{3 \times 10^{10}}$$

The positions of absorption bands in IR spectra are commonly expressed in either microns (μ) or wave numbers (cm^{-1}). Both methods have virtue. The wave-number scale has the advantage that it is directly proportional to energy (2.85 cal = 1 cm^{-1}). The micron scale has the advantage that smaller numbers are involved (typically 2 to 15 μ), which are probably easier to master, but microns are inversely proportional to energy. Because both methods are commonly used, the IR spectra in this text contain both scales.

12.3   Correlations Between IR Absorption and Structure

In general, IR absorptions between 5,000 and 1,250 cm^{-1} are attributed to vibrational stretching modes of various functional groups. The correlation between the bond being affected and IR radiation is given in Table 12.3, which includes for completeness many functional groups we have not studied.

Careful analysis of the 5,000 to 1,250-cm^{-1} (2 to 8 μ) region gives considerable information about the functional groups in a molecule. The presence or absence of a given functional group can often be deduced from this analysis with the aid of the information given in Table 12.3. Remember, however, that the presence of certain other functional groups in the molecule or even attached to a given functional group may affect its location in the IR spectrum and thus lead one to conclude its absence incorrectly.

Absorption peaks observed between 1,250 and 500 cm^{-1} (8 to 20 μ) of the IR spectrum are usually a result of a complex combination of vibration and rotation excitation of the entire molecule. Much of this absorption is due to bending, which requires less energy than does stretching; there are numerous directions in which slight bending may occur, thus giving rise to numerous IR absorptions. The spectrum in the 1,250 to 500-cm^{-1} (8 to 20 μ) region is characteristic and unique for each compound and is often called the **fingerprint region.** There is little likelihood that two molecules have the same absorptions (location and intensity) in the fingerprint region, although they may be quite similar in the functional group region (5,000 to 1,250-cm^{-1} or 2 to 8 μ).

We cannot possibly discuss all the structural features revealed by IR spectroscopy, but this discussion should be sufficient to get you started. Consult the comprehensive textbooks listed in the Reading References for this chapter in the appendix for a more detailed survey of this topic. Expertise in spectroscopy is gained only by experience in examining and interpreting spectra.

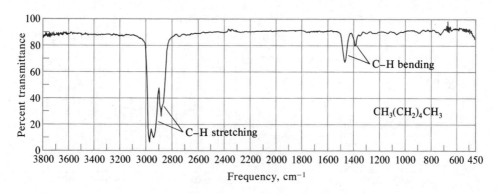

FIGURE 12.4 IR spectrum of *n*-hexane. [Data from *Gases and Vapors: High Resolution Infrared Spectra*, © Sadtler Research Laboratories, Division of Bio-Rad Laboratories, Inc. (1972)]

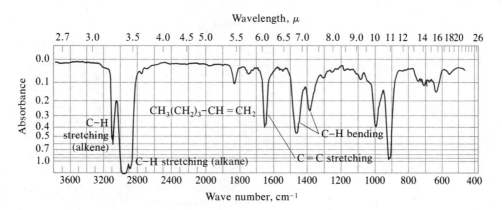

FIGURE 12.5 IR spectrum of 1-hexene. (Data from *Indexes to Evaluated Infrared Reference Spectra*, copyright 1975, The Coblentz Society, Norwalk CT)

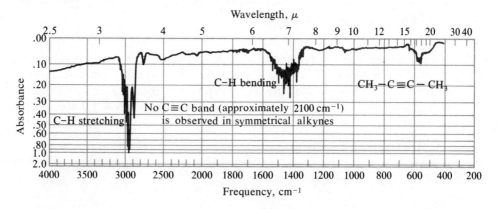

FIGURE 12.6 IR spectrum of 2-butyne. [Data from *The Sadtler Handbook of Infrared Spectra*, W. W. Simmons (ed.), © Sadtler Research Laboratories, Division of Bio-Rad Laboratories, Inc. (1978)]

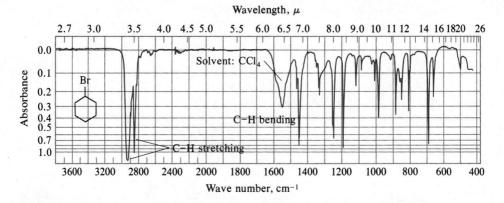

FIGURE 12.7   IR spectrum of bromocyclohexane. (Data from *Indexes to Evaluated Infrared Reference Spectra*, copyright 1975, The Coblentz Society, Norwalk CT)

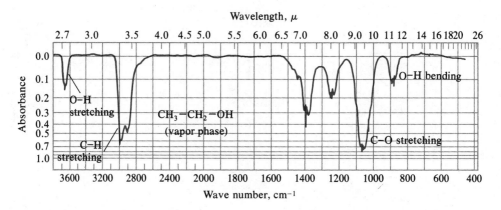

FIGURE 12.8   IR spectrum of ethanol. (Data from *Indexes to Evaluated Infrared Reference Spectra*, copyright 1975, The Coblentz Society, Norwalk CT)

12.4   Interpretation of IR Spectra

As a means of extending our background in IR spectroscopy, several typical IR spectra are presented in Figs. 12.4 to 12.8. These contain functional groups we studied, and the peaks are identified as far as possible. Often it is impossible to identify each and every peak, especially in the fingerprint region below 1,250 cm^{-1} (8 μ).

Ultraviolet Spectroscopy

12.5   UV Spectroscopy: What Is It?

In contrast to IR spectroscopy, which measures the stretching and bending of covalent bonds, UV spectroscopy is an instrumental method for measuring *electronic transitions* within molecules. As Table 12.1 indicates, the UV and visible regions of the electromagnetic spectrum are adjacent to one another. Depending on the molecule being studied, absorptions may occur in either or both the UV and visible regions. Usually UV spectroscopy involves absorption of energy that is not visible to the naked eye.

Electric discharge on inert gases results in electrons being elevated to higher atomic orbitals. When the electrons return to their ground state, energy is emitted in the form of visible light. Each inert gas produces its own unique color in electric discharge tubes. Ultraviolet spectroscopy involves basically the same thing—electronic changes between energy levels—but this time the electronic changes are brought about by the application of UV light and generally involve either unshared pairs of electrons or electrons in π bonds. Ultraviolet spectra normally cover the region from 200 to 400 nm, where 1 nm $= 10^{-9}$ m. Before discussing the types of molecules usually studied in the UV region, let us first look at the technique of UV spectroscopy.

12.6   UV Spectral Measurements

The principles of a UV spectrophotometer are similar to those of an IR spectrophotometer (see Sec. 12.2 and Fig. 12.3), where the sample under study is dissolved in a suitable organic solvent and placed in a cell, while pure solvent is placed in another cell. Typical solvents for UV spectra are methanol, ethanol, hexane, and water, which are transparent to UV radiation. The light source is usually a hydrogen lamp, and the optics and cells are made of quartz because most other clear materials absorb UV energy. The light passes through both cells, and the net amount of energy absorbed is recorded on a chart paper in the manner described momentarily.

Several factors govern the amount of incident light or energy *absorbed* by a sample. One of these is the structure of the compound being studied. Another is the number of molecules in the sample, which depends on the concentration of the compound and the length of the cell that is used. The relationship among absorbance, concentration, and cell length may be expressed by the following equation, which is known as the *Beer-Lambert law*:

$$\text{Absorbance} = \log \frac{I_0}{I} = kcl$$

where $I_0 =$ intensity of incident light (at a certain wavelength)

$I =$ intensity of light transmitted through sample (same wavelength)

$c =$ concentration of sample, e.g., in moles per liter

$l =$ length of cell, e.g., in cm

$k =$ absorption coefficient (proportionality constant)

The UV spectrophotometer plots *absorbance* (A) as a function of the wavelength of incident radiation. When concentration is given in molarity and path length in centimeters, the absorption coefficient k is called the *molar extinction coefficient*, symbolized by ε. The preceding equation now becomes

$$A = \varepsilon cl$$

Absorbance, concentration (c), and cell length (l) are known, so the molar extinction coefficient (ε) can be calculated; it is valuable information that can be used in connection with UV spectra. Spectra are often redrawn by plotting ε against λ, expressed

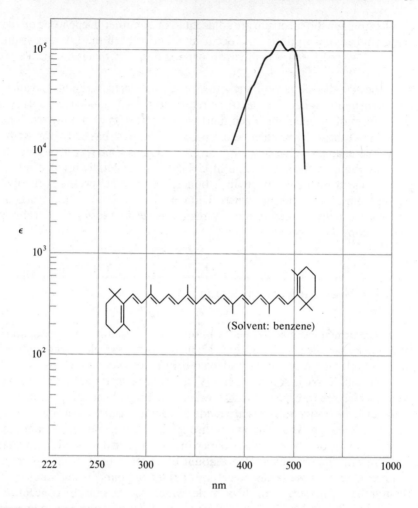

FIGURE 12.9 **UV spectrum of β-carotene, a yellow plant pigment. (Data from the *DMS UV Atlas*, 1966–1971, Plenum Press, New York)**

in either nm or Å. The molar extinction coefficient typically ranges from 10 to 100,000, and often log ε is used in the UV plot. A typical UV spectral plot is shown in Fig. 12.9.

It is not uncommon for organic compounds to have two or more absorption maxima in the UV region. Typically, UV data are reported as $\lambda_{max}^{ethanol}$ 232 nm (log ε = 4.18), where the solvent is specified and log ε is the molar extinction coefficient at the wavelength where the UV maximum (λ_{max}) occurs. Note also the broad nature of the band, which is common in UV spectroscopy. The energy put into the sample is sufficient to allow a large number of different electronic transitions to occur between the many different allowed vibrational and rotational energy levels of the molecule.[1] Although each transition is quantized, the ΔE values are so close that complete resolution in the solution phase is impossible. A broad absorption band results, composed of all these overlapped transitions. In the gas phase with relatively simple molecules it is possible to resolve these bands (Fig. 12.10).

---

[1] For a more thorough discussion, see Reading References 1 to 4 for this chapter in the appendix.

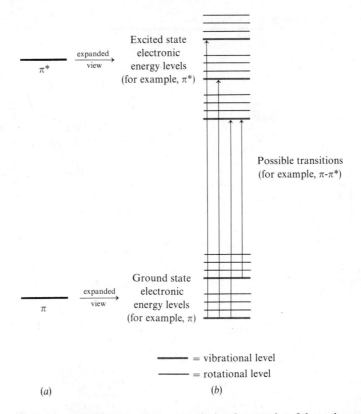

FIGURE 12.10 (a) Typical energy-level diagram relating the energies of the various energy states of a molecule. (b) Expanded view of (a) showing the different vibrational, rotational, and electronic energy levels of each state.

12.7 Effect of Structure on UV Absorption Spectra

With the groundwork for some terms associated with UV spectroscopy laid, we can now study the effect of structure on UV spectra. The basic type of change that can be measured in UV spectroscopy is *electronic transitions*, in which *energy* is required to promote electrons to a higher energy level. It is this absorption of energy that is measured and recorded in UV spectroscopy. The amount of energy required for electronic excitation varies as the nature of the chemical bond and the molecular structure change.

A. σ Bonds

The electrons in a σ bond are very tightly bound between atomic nuclei, and they can be excited to a higher energy state only by the very short wavelengths in the far UV region. The excitation of the valence electrons in the C—C and C—H σ bonds (as in alkanes) requires radiation of 130 nm or less (220 kcal/mole; 920.5 kJ/mole). The raising of these electrons to a higher energy state is called σ → σ* *transition* because the electrons are excited from a σ bonding orbital to a σ* anti-bonding orbital (see Sec. 2.13). These transitions require significant energy and are

Relative ΔE of the common UV transitions

$\sigma - \sigma^* > n - \sigma^* > \pi - \pi^* > n - \pi^*$

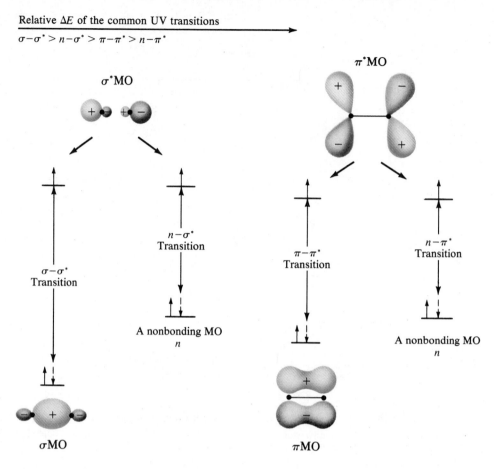

FIGURE 12.11   Possible UV transitions diagrammed in relation to the energy required for the transition to occur.

seldom observed. This transition is diagrammed relative to several other possible UV transitions in Fig. 12.11.

B. π Bonds

In contrast to the electrons that constitute σ bonds, the bonding electrons in π bonds are less tightly bound and more mobile. As a result, it is easier to excite them to higher (and less stable) energy levels, and these electronic transitions can be observed in the UV region with most instruments. These transitions are called $\pi \rightarrow \pi^*$ *transitions* because electrons undergo transitions between the π bonding and π^* antibonding orbitals.

Examples of functional groups that exhibit $\pi \rightarrow \pi^*$ transitions are C=C (alkenes), C≡C (alkynes), and C=O (carbonyl groups). Using a carbon-carbon double bond, a transition is shown in Fig. 12.11. Ethene, for example, exhibits a $\lambda_{max} = 163$ nm, and the energy associated with the $\pi \rightarrow \pi^*$ transitions for other alkenes, alkynes, and carbonyl compounds is of the same order of magnitude. In general, the UV spectra of compounds containing just one carbon-carbon multiple bond are difficult to measure because they absorb in the far UV region.

C. Excitation of Unshared Electron Pairs

Another type of electronic excitation involves unshared pairs of electrons on atoms that are part of a multiple bond. For example, one nonbonding electron on

TABLE 12.4   UV Absorption Frequencies of Functional Groups

Compound Type	Example	Wavelength, λ, nm	Extinction Coefficient* (approximate)
Hydrocarbons:			
Alkanes	CH_3—CH_3	135	7,000 (vapor phase)
Alkenes	CH_2=CH_2	163	15,000 (vapor phase)
	(cyclohexene)	182	7,500
Alkynes	$CH\equiv CH$	173	6,000 (vapor phase)
	$CH_3C\equiv C(CH_2)_4CH_3$	196	2,000
Polyenes	CH_2=CH—CH=CH_2	217	21,000
(conjugated)			
	(cyclohexadiene)	256	8,000
	CH_3CH=$CHCH$=CH_2	223 (*cis*)	22,600
		223.5 (*trans*)	23,000
Aromatics			
	X = H	203	7,500
	X = CH_3	206	7,000 (alcohol)
	X = CH=CH_2	244	12,000 (alcohol)
	X = NO_2	252	10,000
	X = CN	224	13,000 (water)
Alcohols:			
Aliphatic	CH_3OH	183	150
Aromatic	⟨⟩—OH	210.5	6,200 (water)
Halides:			
Aliphatic	CH_3Cl	173	200
	CH_3I	259	400
Aromatic	⟨⟩—Cl	209.5	7,400 (water)
	⟨⟩—Br	210	7,900 (water)
Ethers:			
Aliphatic	CH_3CH_2—O—CH_2CH_3	171	4,000 (vapor phase)
Aromatic	⟨⟩—OCH_3	217	6,400 (water)
Amines:			
Aliphatic	$(CH_3)_3N$	199	4,000
Aromatic	⟨⟩—NH_2	230	8,600 (water)

TABLE 12.4   (*continued*)

Compound Type	Example	Wavelength, λ, nm	Extinction Coefficient* (approximate)
Acids and derivatives:			
Aliphatic	$CH_3-\overset{O}{\overset{\|}{C}}-OH$	204	60 (water)
	$CH_3-\overset{O}{\overset{\|}{C}}-OCH_2CH_3$	211	57 (alcohol)
	$CH_3-\overset{O}{\overset{\|}{C}}-Cl$	220	100
	$CH_3-\overset{O}{\overset{\|}{C}}-NH_2$	178	9,500
	$CH_3-\overset{O}{\overset{\|}{C}}-O-\overset{O}{\overset{\|}{C}}-CH_3$	225	47
Aromatic	⬡$-\overset{O}{\overset{\|}{C}}-OH$	230	10,000 (water)
Aldehydes and ketones:			
Aliphatic	$CH_3-\overset{O}{\overset{\|}{C}}-CH_3$	279	15
	$CH_3-\overset{O}{\overset{\|}{C}}-H$	293	12
	$CH_3-\overset{O}{\overset{\|}{C}}-CH_2CH_3$	279	16
Aromatic	⬡$-\overset{O}{\overset{\|}{C}}-H$	249.5	11,500 (water)
	⬡$-\overset{O}{\overset{\|}{C}}-CH_3$	245.5	10,000 (water)

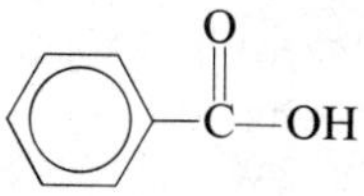

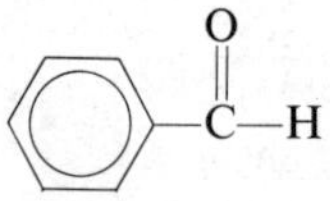

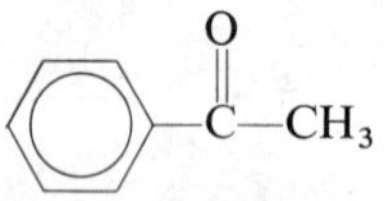

* All in hydrocarbon solvents unless otherwise specified.

oxygen in the carbonyl group, $C=\overset{..}{\underset{..}{O}}$, can be shifted to a higher, unstable energy state by the application of energy. This type of transition is called an $n \rightarrow \pi^*$ *transition*, where n refers to the nonbonding electron pairs, in this example, on oxygen. Simple aldehydes and ketones exhibit an absorption maximum at ~ 280 nm because of the $n \rightarrow \pi^*$ excitation. Whereas the extinction coefficients of the $\pi \rightarrow \pi^*$ transitions are quite large ($\varepsilon \approx 10,000$), those of the $n \rightarrow \pi^*$ transitions are usually low ($\varepsilon \approx 10$ to 100). The $n \rightarrow \sigma^*$ transitions are also possible but are of less importance and use in structure determination. Table 12.4 is a summary of common UV spectral parameters.

D. Effect of Conjugation on UV Spectra

One tangible application of UV spectroscopy is in the detection and identification of conjugated systems, such as C=C—C=C and C=C—C=O, where the $\pi \to \pi^*$ and $n \to \pi^*$ transitions occur at lower energies and higher wavelengths and can be measured in the UV region. 1,3-Butadiene, CH_2=CH—CH=CH_2, for example, has λ_{max} at 217 nm, whereas for 2-butene λ_{max} is at 210 nm. In general, increasing the number of substituents attached to conjugated systems or adding more conjugated double bonds increases the absorption maximum.

Because of the relationship between wavelength (λ) and energy ($E = hc/\lambda$), the absorption maximum, λ_{max}, indicates the difference between the ground-state energy and the excited-state energy of a molecule. The conjugated systems of π electrons are delocalized (see Sec. 8.10), and judging from UV spectral data, they appear to be more stable in the excited state than a corresponding nonconjugated system would be. This is supported by UV data on conjugated alkenes, where 1,3-butadiene has λ_{max} at 217 nm, 1,3,5-hexatriene (one added C=C) has λ_{max} at 258 nm, and 1,3,5,7-octatetraene (one more added C=C) has λ_{max} at 286 nm. As the amount of conjugation increases, the ΔE between the ground and excited states decreases; the excited state is more stable relative to the ground state.

Additional examples of structure-UV spectra correlations are given in Sec. 12.8.

12.8   Correlation Between UV Spectra and Structure

The correlations between molecular structure and UV spectra that are used commonly today are the result of much experimental data. From these data the organic chemist has developed a series of rules for computing wavelength maxima for compounds that absorb in the UV region.

Before we present these rules, it should be pointed out that various molecules have discrete energy levels that result in the absorption maxima. We have no way to draw intelligently electronic structures for the excited states of most *complex* organic molecules. The exact electronic structures are known for simple molecules for which the various antibonding (π^*) energy levels have been determined. Of great importance, however, is the *additivity principle*, which governs absorption maxima. As mentioned, the energy difference between the ground state and excited state becomes smaller as the number of conjugated bonds in the molecule increases; as the energy difference between levels decreases, the amount of energy required for electronic excitation also decreases, thus resulting in the absorption maxima appearing at longer wavelengths (larger λ).

A. Terminology of UV Spectroscopy

The following definitions are used commonly in UV spectroscopy:

Chromophore. An unsaturated group (or collection of groups) that is responsible for UV absorption; it literally means "color bearing," although many colorless compounds produce UV spectra.

Auxochrome. A substituent (usually saturated) that alters the wavelength and molar extinction coefficient of the absorption maximum when it is attached to a chromophore.

Bathochromic shift. The shift of the absorption maximum to longer wavelength (larger λ) caused by either substitution onto the chromophore (by an auxochrome) or a change in the solvent.

Hypsochromic shift. The shift to shorter wavelength (smaller λ) as a result of substitution or solvent change.

Hyperchromic effect. An increase in the molar extinction coefficient (ε).

Hypochromic effect. A decrease in the molar extinction coefficient.

B. Rules for Predicting Absorption Maxima for Conjugated Systems

Many empirical values have been tabulated for the absorption maxima of conjugated systems that contain several different substituents. An example given here uses the Woodward rules developed by Dr. R. B. Woodward of Harvard University. Consult the Reading References for this chapter in the appendix for detailed information. The intent of our discussion on UV spectroscopy is to provide only the most general rules governing the additivity principle and the effect that added substituents have on λ_{max}. The starting point of these rules is the following two chromophores, with their indicated λ_{max}:

Conjugated diene
$\lambda = 217$ nm   Conjugated unsaturated ketone
$\lambda = 215$ nm

The addition of the following *auxochromes* to these chromophores results in the indicated changes:

1. Extra conjugated double bond (C=C): increase of λ_{max} by 30 nm.

2. Substitution of alkyl group (R—), halogen (X—), or alkoxy group (—OR) for hydrogen on a conjugated *diene*: increase of λ_{max} by 5 nm.

3. Existence of the diene or unsaturated ketone in the *s-cis* conformation: λ_{max} increased by ~ 40 nm more than that of the *s-trans* conformation. (See Ques. 12.12.)

s-cis Diene   *s-trans* Diene

4. Presence of a conjugated *exocyclic* double bond (see Sec. 7.6): increase of λ_{max} by 5 nm.

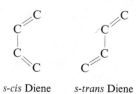

C. Examples

For now our major concern is the conjugated dienes. The following three examples illustrate how these general rules may be applied.

Example 1

$$CH_3 \quad CH_3$$
$$CH_2{=}C{-}C{=}CH_2$$

2,3-Dimethyl-1,3-butadiene

Calculated: $\lambda_{max} = 217 + (2 \times 5) = 227$ nm
Observed: $\lambda_{max} = 226$ nm

The two added methyl groups on C-2 and C-3 of the basic conjugated diene system result in a $5 + 5$ or 10-nm increase in λ_{max}; the observed absorption maximum is close to that calculated for 2,3-dimethyl-1,3-butadiene.

Example 2

$$CH_3{-}CH{=}CH{-}CH{=}CH_2$$

1,3-Pentadiene

Calculated: $\lambda_{max} = 217 + 5 = 222$ nm
Observed: $\lambda_{max} = 223$ nm ($\varepsilon = 22,600$) for *cis* isomer
                     $= 223.5$ nm ($\varepsilon = 23,000$) for *trans* isomer

The methyl group added onto the conjugated butadiene system increases λ_{max} by about 5 nm. The observed λ_{max} is close to the calculated value. Note that the *trans* isomer absorbs at a longer wavelength and with a larger molar extinction coefficient. This is generally true of *cis* and *trans* isomers, and UV spectroscopy is one method that can be used to confirm the structure of geometric isomers.

Example 3

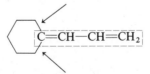

Calculated: $\lambda_{max} = 217 + (2 \times 5) + 5 = 232$ nm
Observed: $\lambda_{max} = 236.5$ nm

The two alkyl groups (shown by arrows) add 2×5 nm to the λ_{max}, and at the same time the double bond attached to cyclohexane is an *exo* double bond, so an additional 5 nm must be added.

As the additivity rules indicate, the presence of more and more conjugated double bonds increases λ_{max} (see Table 12.5). For example, β-*carotene* (see Sec. 8.36) is the pigment in carrots, tomatoes, and spinach, and because it contains 11 conjugated double bonds, it absorbs in the visible region, with λ_{max} at 450, 485, and

TABLE 12.5   UV Absorption Frequencies of Seven α,β Unsaturated Compounds

Compound	Wavelength, λ_{max}, nm	Extinction Coefficient* (approximate)
$CH_2{=}CH{-}\overset{\overset{\displaystyle O}{\|}}{C}{-}OH$	200	1,000
$CH_3CH{=}CH{-}\overset{\overset{\displaystyle O}{\|}}{C}{-}OH$	205 (*trans*) 205.5 (*cis*)	14,000 13,500
$\langle hexane \rangle{=}CH{-}\overset{\overset{\displaystyle O}{\|}}{C}{-}OH$	220	14,000
$CH_3{-}\overset{\overset{\displaystyle O}{\|}}{C}{-}CH{=}CH_2$	213	7,000
$CH_3CH{=}CH{-}\overset{\overset{\displaystyle O}{\|}}{C}{-}CH_3$	224	10,000
cyclohexenone	225	10,000
$HC{\equiv}C{-}\overset{\overset{\displaystyle O}{\|}}{C}{-}(CH_2)_2CH_3$	214	4,500

* All measured in alcohol solutions.

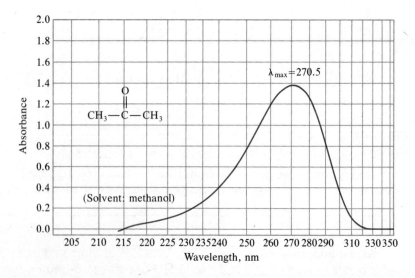

FIGURE 12.12   UV spectrum of acetone. [Data from *The Sadtler Handbook of Ultraviolet Spectra*, W. W. Simmons (ed.), © Sadtler Research Laboratories, Division of Bio-Rad Laboratories, Inc. (1979)]

525 nm. *Lycopene* (Fig. 12.15), also containing 11 conjugated double bonds (see Sec. 8.36), is red and is found in watermelons and tomatoes; its UV absorption maxima occur at 477, 507, and 548 nm.

12.9 UV Spectroscopy in Structure Identification

One principal use of UV spectroscopy is to identify structure when the position of a double bond is uncertain on the basis of either IR or NMR spectra. The requirements for using UV spectroscopy are considerably more restrictive than those for the other two instrumental methods, yet there are many examples where the structure of an organic molecule can be determined only by UV spectroscopy.

As an example, the following two compounds can be distinguished on the basis of their UV spectra:

Compound (1) contains a conjugated diene, which shows an absorption maximum at approximately 232 nm: the calculated value is also 232; 217 (for basic diene system) + 2 × 5 (for two alkyl substituents) + 5 nm (for exocyclic double bond). Compound (2) absorbs near 175 nm, the value for a single double bond, because its two double bonds are *not* conjugated.

Sometimes the possible compounds all contain conjugated dienes but in different positions. By knowing the possible structures and using the additivity rules, the structure of the unknown compound can often be determined on the basis of its UV spectra.

Examples of UV spectra are in Figs. 12.12 to 12.15.

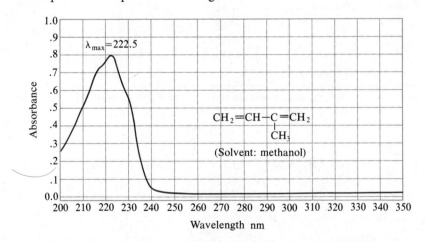

FIGURE 12.13 UV spectrum of isoprene. [Data from *The Sadtler Handbook of Ultraviolet Spectra*, W. W. Simmons (ed.), © Sadtler Research Laboratories, Division of Bio-Rad Laboratories, Inc. (1979)]

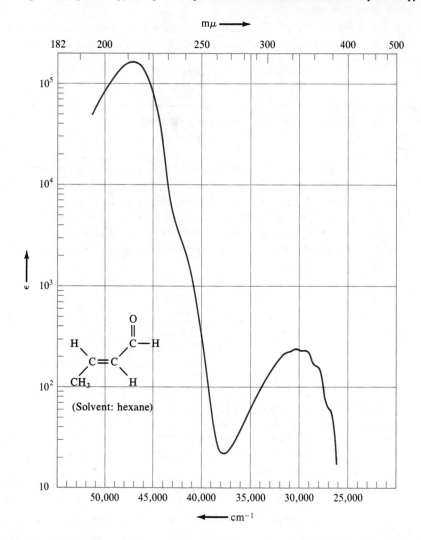

FIGURE 12.14   UV spectrum of crotonaldehyde. (Data from the *DMS UV Atlas*, 1966–1971, Plenum Press, New York)

Study Questions

12.2 For each of these six compounds, indicate one major IR absorption band that should differentiate it from the others.

(*a*) ⬡

(*b*) $CH_3CH_2CH_2CH_2CH_3$

(*c*) $CH_3—\overset{\overset{\displaystyle CH_3}{|}}{\underset{\underset{\displaystyle OH}{|}}{C}}—CH_3$

(*d*) Br—⬡

(*e*) $CH_2{=}CH{-}$⬡

(*f*) $CH_3—\overset{\overset{\displaystyle O}{\|}}{C}—$⬠

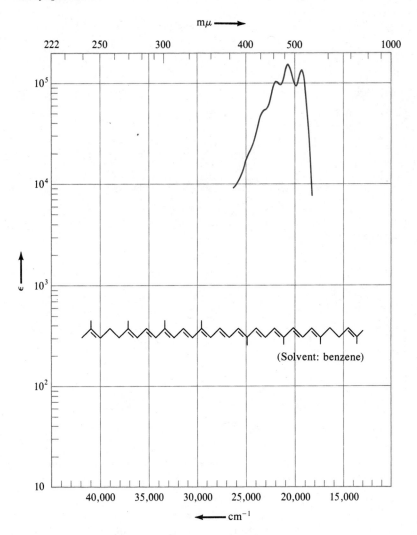

FIGURE 12.15 **UV spectrum of lycopene. (Data from the *DMS UV Atlas*, 1966–1971, Plenum Press, New York)**

12.3 Consider the following reaction:

OH

$\xrightarrow[\text{heat}]{9\ M\ H_2SO_4}$ + H_2O

bp 161°   bp 85°

(*a*) The reaction is reversible. Explain *briefly* how one could use Le Chatelier's principle in the laboratory to provide a very high yield of the desired cyclohexene product.

(*b*) Which of the following IR spectra is of cyclohexanol? Which is of cyclohexene? Label at least one major peak in each spectrum that led you to your conclusion.

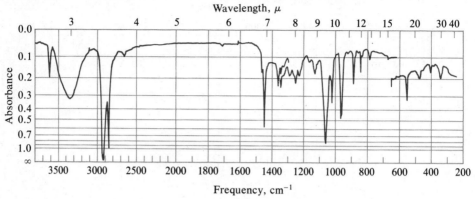

PROBLEM 12.3 Data from *Indexes to Evaluated Infrared Reference Spectra*, copyright 1975, The Coblentz Society, Norwalk CT.

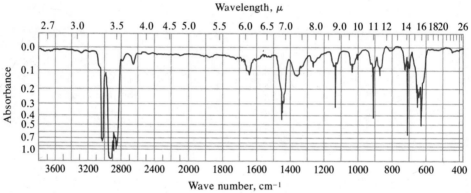

PROBLEM 12.3 Data from *Indexes to Evaluated Infrared Reference Spectra*, copyright 1975, The Coblentz Society, Norwalk CT.

12.4 These IR spectra are of compounds *A*, *B*, and *C*.

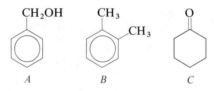

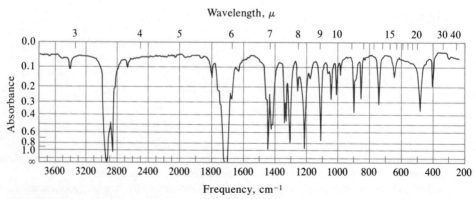

PROBLEM 12.4 Data from *Indexes to Evaluated Infrared Reference Spectra*, copyright 1975, The Coblentz Society, Norwalk CT.

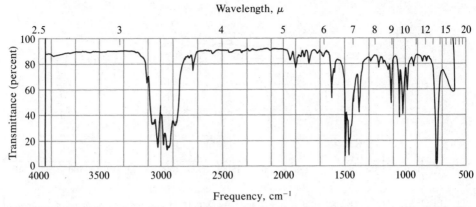

PROBLEM 12.4   Data from *Indexes to Evaluated Infrared Reference Spectra*, copyright 1975, The Coblentz Society, Norwalk CT.

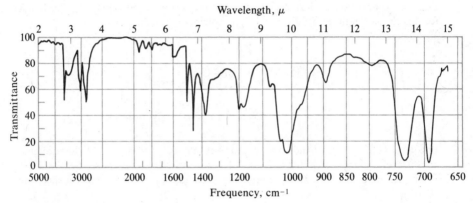

PROBLEM 12.4   Data from *Indexes to Evaluated Infrared Reference Spectra*, copyright 1975, The Coblentz Society, Norwalk CT.

Which represents *A*? Which represents *B*? Which represents *C*? Label the major peaks *on each spectrum* that led to your conclusion.

12.5 Compounds *A* and *B* are both C_5 hydrocarbons. The IR spectra of these two compounds are given here, as are their structural formulas. Match the correct formula with the correct spectrum.

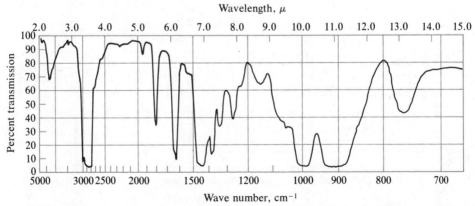

PROBLEM 12.5   IR spectrum of compound (*A*). [Data from *The Sadtler Handbook of Infrared Spectra*, W. W. Simmons (ed.), © Sadtler Research Laboratories, Division of Bio-Rad Laboratories, Inc. (1978)]

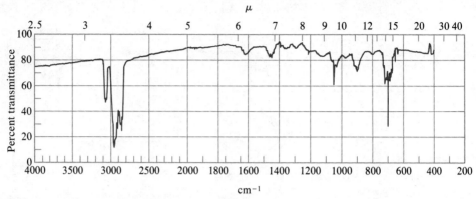

PROBLEM 12.5   IR spectrum of compound (B). [Data from *The Sadtler Handbook of Infrared Spectra*, W. W. Simmons (ed.), © Sadtler Research Laboratories, Division of Bio-Rad Laboratories, Inc. (1978)]

Explain those portions of the spectrum that led to your conclusion. (*Hint*: The Reading References for this chapter in the appendix may help in solving this problem.)

$$CH_3CH_2CH_2CH=CH_2$$

1-Pentene   Cyclopentene

12.6 Several absorption bands may appear between 650 and 850 cm^{-1} that are characteristic of benzene, a monosubstituted benzene, or a disubstituted benzene (see Sec. 14.25). Spectra I, II, and III are partial spectra of compounds *A*, *B*, and *C*. Match the correct spectrum with the compound.

A   *B*   *C*

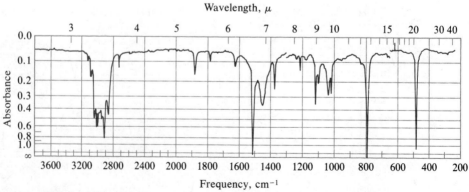

PROBLEM 12.6   IR spectrum of compound (I). (Data from *Indexes to Evaluated Infrared Reference Spectra*, copyright 1975, The Coblentz Society, Norwalk CT)

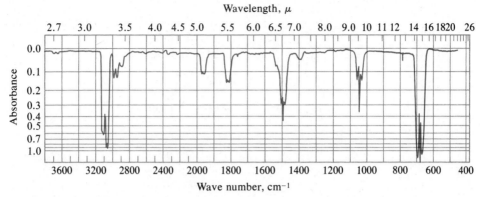

PROBLEM 12.6   IR spectrum of compound (II). (Data from *Indexes to Evaluated Infrared Reference Spectra*, copyright 1975, The Coblentz Society, Norwalk CT)

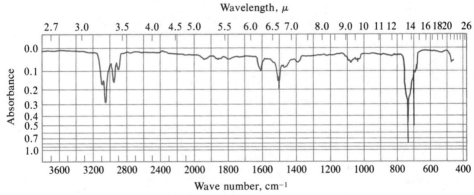

PROBLEM 12.6   IR spectrum of compound (III). (Data from *Indexes to Evaluated Infrared Reference Spectra*, copyright 1975, The Coblentz Society, Norwalk CT)

12.7 Compound A decolorizes Br_2 in CCl_4. When compound A reacts with ozone followed by treatment with Zn and water, 1 mole of compound B, C_2H_4O, and 1 mole of compound C, C_4H_8O, are produced. Compound A has a molecular weight of 84 and consists of approximately 86% C and 14% H. The IR spectra of B and C are given here. From this information, deduce plausible structures for A, B, and C.

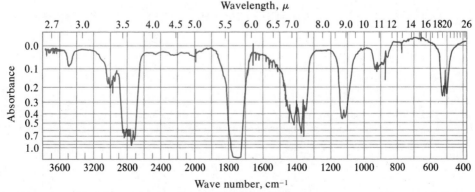

PROBLEM 12.7   IR spectrum of compound (B). (Data from *Indexes to Evaluated Infrared Reference Spectra*, copyright 1975, The Coblentz Society, Norwalk CT)

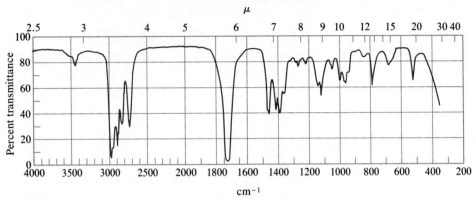

PROBLEM 12.7   IR spectrum of compound (*C*). [Data from *The Sadtler Handbook of Infrared Spectra*, W. W. Simmons (ed.), © Sadtler Research Laboratories, Division of Bio-Rad Laboratories, Inc. (1978)]

12.8 Indicate how UV spectroscopy can be used to distinguish between the members of each of the following pairs of compounds.

(*a*) CH_2=CH—CH_2—CH=CH_2 and CH_3—CH=CH—CH=CH_2

(*b*)   [structures: cyclohexene with =CH₂ and CH₃ substituents]   and   [cyclohexane with two =CH₂ groups]

12.9   Predict the UV absorption maxima of the following four compounds.

(*a*) CH_2=$\overset{\displaystyle CH_3}{\underset{\displaystyle CH_3}{C}}$—$C$=$CH$—$CH_3$

(*b*)   [ring structure with CH₃ and C(CH₃)₂]

(*c*) CH_3—CH=CH—CH=$\overset{\displaystyle CH_3}{C}$—$CH$=$CH$—$CH_3$

(*d*) CH_3—CH=CH—CH_2—CH=CH—CH=CH—CH_3

12.10   The absorption spectrum for crotonaldehyde (CH_3CH=CH—CH=O) is shown in Fig. 12.14. If the sample were run in a 1-cm cell at a concentration of 1×10^{-5} *M* in ethanol, calculate the observed absorbance (*A*) for the compound. If the same solution were run in a 0.25-cm cell, what would be the observed absorbance (*A*)?

12.11   Compounds *A* to *D* have λ_{max} values of 177, 217, 232, and 277 nm. Which compound goes with which maximum?

CH_2=CH—CH=CH—CH=CH—CH=CH_2 $CH_3CH_2CH_2CH_2CH_2CH_2CH$=$CH_2$

A   *B*

CH_2=CH—CH=CH_2 [cyclohexane ring]=CH—CH=CH_2

C   *D*

12.12   When a conjugated diene is part of a cyclic six-membered ring, that is, cyclohexadiene, a λ_{max} of 253 nm is used as a base value for all calculations. Predict the λ_{max} of each of the following compounds.

(*a*)   [ring with CH₃]   (*b*)   [ring]—CH=CH_2 (*c*)   [ring with CH₃ CH₃]—CH=CH_2

12.13 Predict the λ_{max} of each of the following six chromophores.

(a)

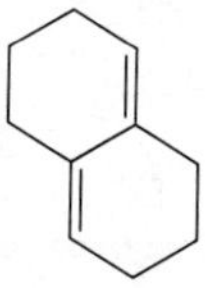

(b)

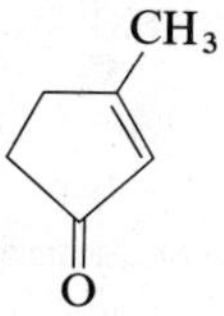

(c)

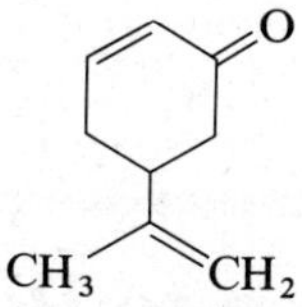

(d)

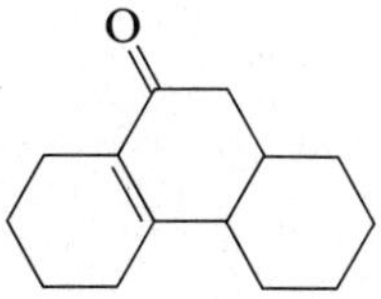

(e)

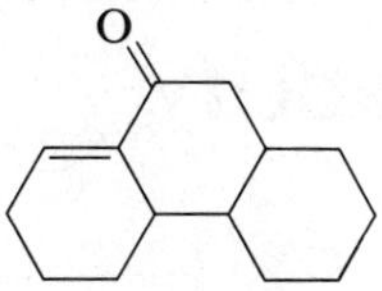

(f)

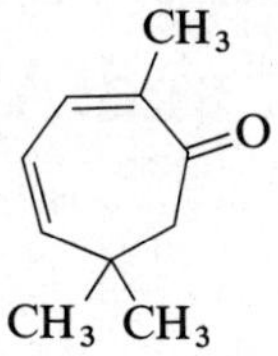

13

Introduction to Benzene, Resonance Structures, and Aromaticity

We concentrated thus far on the broad class of hydrocarbons called *aliphatic compounds*. We now begin a discussion of the second broad class of hydrocarbons, the **aromatic compounds.** This classification is somewhat artificial because compounds that contain an **aromatic ring** may also contain substituents that are classified as aliphatic. As we will see, reaction conditions often dictate whether the aliphatic or the aromatic part of such molecules undergoes reaction.

The family of aromatic compounds includes the parent compound, **benzene** and other compounds that resemble benzene in chemical behavior. This chemical behavior is attributed to the unique electronic structure of the aromatic compounds. The pleasant aroma of many members of this class is responsible for the term aromatic.

To see what is meant by the term *aromatic compound*, we start with the simplest aromatic compound, benzene, and discuss its salient structural and chemical features.

13.1   Benzene: A Typical Aromatic Compound

Elemental analysis and quantitative molecular formula determination show that benzene has the formula C_6H_6. This compound has a high degree of *unsaturation*; when compared with C_6H_{14} (the completely saturated six-carbon alkane), it is deficient by 4 moles of hydrogen. It must therefore have some combination of double bonds, triple bonds, and/or rings. Many molecules can be drawn to satisfy the formula C_6H_6; eight are shown:

$$CH_3-C\equiv C-C\equiv C-CH_3 \qquad CH_2=CH-C\equiv C-CH=CH_2$$

(5)   (6)   (7)

$$HC\equiv C-CH=CH-CH=CH_2$$

(8)

Proposed structures often can be supported or refuted on the basis of experimental fact. Hydrogenation and bromination data and independent synthesis can be used to discard certain of the possible structures for benzene.

Several of these structures can be excluded because catalytic hydrogenation of benzene results in only 3 moles of hydrogen being absorbed; cyclohexane, C_6H_{12}, is produced. No further hydrogen is absorbed, even under vigorous conditions of high temperature and pressure. This fact alone excludes structures (5) to (8); (5) would give a cyclopentane derivative and (6) to (8) would give *n*-hexane. This leaves only structures (1) to (4) as reasonable alternatives. However, bromination of benzene gives *only one* monobromo derivative, C_6H_5Br, thus excluding (2) which would give two possible monobromo compounds, and (4), which would give three. (*Question:* What are the structures of the latter five compounds?) It seems apparent that benzene is a symmetrical molecule and that substitution of any one hydrogen by bromine is equivalent to substitution of any of the remaining five hydrogens. In addition, compounds (2) to (4) have been synthesized by independent means in recent years and are different from benzene. Of structures (1) to (8), it appears that (1) is the most satisfactory.

Benzene is often represented as two equivalent structures called *Kekulé structures*, (1) and (1'):

(1)   (1')
Kekulé structures

These structures are named after August Kekulé, who proposed them in 1865. But is structure (1) or structure (1') the "correct" structure for benzene? The answer, as we will see, is that neither is completely correct.

13.2   Reactivity and Properties of Benzene

Benzene is a particularly stable compound. It can, for example, withstand sustained heat and high pressure and maintain its structural identity. Benzene, as drawn in (1) and (1'), contains alternating single and double bonds. We could call either structure *1,3,5-cyclohexatriene.* We know that cyclohexene, 1,3-cyclohexadiene, and 1,4-cyclohexadiene undergo rapid electrophilic addition reactions with such reagents as bromine in carbon tetrachloride.

Cyclohexene $\xrightarrow[\text{fast}]{\text{Br}_2,\ \text{CCl}_4}$ 1,2-Dibromocyclohexane   Addition

1,3-Cyclohexadiene $\xrightarrow[\text{fast}]{\text{Br}_2\ \text{(excess)},\ \text{CCl}_4}$ 1,2,3,4-Tetrabromocyclohexane   Addition

1,4-Cyclohexadiene $\xrightarrow[\text{fast}]{\text{Br}_2\ \text{(excess)},\ \text{CCl}_4}$ 1,2,4,5-Tetrabromocyclohexane   Addition

If we extend this analogy, we might expect that the carbon-carbon double bonds in benzene would also react rapidly with bromine in carbon tetrachloride to give an addition product. Yet when this reaction is attempted, nothing happens:

or $\xrightarrow[\text{CCl}_4]{\text{Br}_2}$ no reaction

Benzene

Similar results occur with other reagents that typically react with alkenes and alkadienes. For example, aqueous potassium permanganate reacts with cyclohexene and the cyclohexadienes, but not with benzene; the same is true of hydrogen halides. Cyclohexene and the cyclohexadienes undergo catalytic hydrogenation at room temperature and atmospheric pressure, whereas benzene requires elevated temperatures ($\sim 100°$) and high pressure ($\sim 1{,}000$ lb/sq in.) for hydrogenation to occur.

On the other hand, when benzene reacts with bromine in the presence of a catalyst, such as $FeBr_3$, a reaction does occur at room temperature; the products are a compound with the formula C_6H_5Br and hydrogen bromide:

$$C_6H_6 + Br_2 \xrightarrow{FeBr_3} C_6H_5{-}Br + H{-}Br$$

Benzene   Bromobenzene

Substitution occurs; bromine is substituted for hydrogen on the ring. This is in contrast to cyclohexene and the cyclohexadienes, where addition reactions occur.

The chemical reactivity and great stability suggest that there are indeed no normal double bonds in benzene. It is never called 1,3,5-cyclohexatriene, in part because it does not undergo reactions typical of compounds containing the double bond.

Before discussing the electronic structure of benzene, some other known facts must be reconciled. The carbon-carbon bond lengths are 1.39 Å in benzene, as compared with the 1.34-Å length of most carbon-carbon double bonds and the 1.54-Å length of most carbon-carbon single bonds. We are dealing with a bond intermediate in length between the carbon-carbon single and double bonds. There is only one "type" of carbon-carbon bond in benzene because all carbon-carbon bonds have the same bond length. If benzene were actually one of the Kekulé structures, it would have alternating double and single carbon-carbon bonds. Also, the carbon-carbon bond angles in benzene are 120° and the molecule is completely planar (flat); see Figs. 13.1 and 13.2.

13.3 Electronic Structure of Benzene; Resonance

During the past 20 to 30 yr structure theory has developed to the point that it satisfactorily explains the experimental facts known for benzene. Quantum mechanics deals with the structure of benzene from a mathematical point of view, and the results derived from it are in accord with the modern structural view.

The orbital picture of benzene reveals a great deal about its electronic structure. The carbon atoms in benzene are sp^2 hybridized, resulting in a planar molecule. If we take six sp^2 hybrid carbon atoms and use them to form a six-membered ring, we get the structure shown in Fig. 13.1 after adding the six hydrogen atoms. Each of the six carbon atoms contains one p orbital perpendicular to the sp^2 hybrid orbitals, and each p-orbital contains one electron. Rather than each pair of p orbitals overlapping to form a separate π bond, *all six p orbitals overlap with one another to form a π cloud,* as shown in Fig. 13.2.

The electrons that form the π cloud are no longer associated with just one or two carbon atoms as they are in the π bond in alkenes; this delocalization of electrons imparts greater stability to the molecule by dispersing the charge (the electrons) over a much larger area. The amount of added stability is called the **delocalization energy,** which is discussed in Sec. 13.5.

The physical data for benzene agree well with this molecular view. The pertinent data for benzene, which have been determined from X-ray and electron diffraction

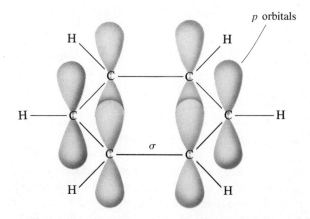

FIGURE 13.1   Orbital picture of the benzene molecule, showing the σ bonds and the p orbitals (overlap of p orbitals not shown).

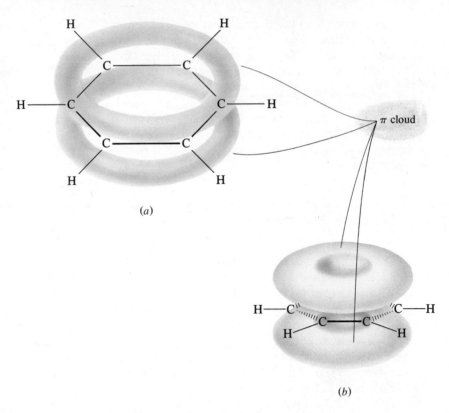

FIGURE 13.2   Benzene molecule showing the π clouds that result from overlap of the p orbitals shown in Fig. 13.1: (a) perspective view; (b) side view.

methods, are summarized as follows:

Benzene (π cloud omitted)

13.4   Molecular Orbital (MO) Theory

We learned in Sec. 2.14 that molecular orbitals obey the Aufbau principle, Hund's rule, and the Pauli exclusion principle. How can we explain that the π molecular orbital of benzene contains six π electrons when all atomic and molecular orbitals can contain a maximum of two? Actually, when the six atomic p orbitals of benzene combine to form the π molecular orbitals of benzene, this results in six molecular orbitals (see Sec. 2.13), which are shown in Fig. 13.3.

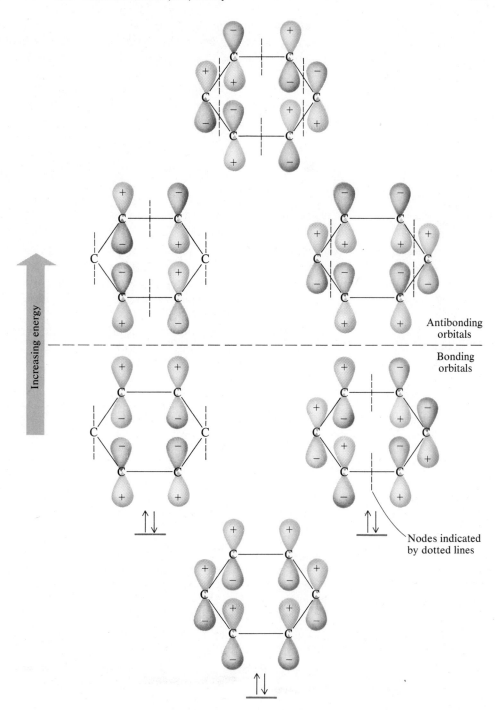

FIGURE 13.3   MO's of the π system of benzene.

How do we know what these molecular orbitals look like? What about their relative stabilities? The answers come from the mathematical analysis of this system by quantum mechanics. There are several generalizations we can use in our analysis of benzene and (as we will see) other aromatic systems. One is summarized in the next paragraph.

When working with an extended π molecular orbital system, it is possible to arrange the orbitals in order of increasing energy by using the rule of thumb that the energy of an orbital increases along with the number of nodes (points of zero amplitude) where the positive lobe of one orbital overlaps with the negative lobe of another orbital (see Sec. 2.11). The two lobes cancel out. Note, for example, that the two molecular orbitals associated with the π-electron system of ethene contain zero and one node, respectively:

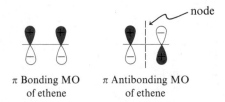

π Bonding MO π Antibonding MO
  of ethene   of ethene

Extending this concept to the benzene MO's in Fig. 13.3, we see that these too are arranged according to the number of nodes in each structure. The six π electrons enter the three bonding molecular orbitals, resulting in a stable bonding structure. The structures in Figs. 13.1 and 13.2 are compilations of these three bonding molecular orbitals.

In Sec. 13.8 there is a simplified approximation that allows us to determine routinely the numbers and relative energies of the various molecular orbitals associated with the π-electron system of an aromatic compound.

Question 13.1

Using this general rule, draw, in order of increasing energy, the molecular orbitals associated with the π-electron system of the allyl free radical. Show the nodes and indicate which orbitals contain electrons, and the number of electrons (see Sec. 13.6C).

13.5   Representation of Benzene; Resonance and Delocalization Energy

We now come to the problem of how best to represent the electronic structure of benzene on paper. The best possible representation is the molecular orbital view presented in Sec. 13.4. We also said that the two Kekulé structures in Sec. 13.1 represent benzene, but how can this be?

Reexamination of these two Kekulé structures, (1) and (1'), reveals that both satisfy the molecular formula C_6H_6. They both also have the same carbon and hydrogen framework; the only difference between (1) and (1') is the location of the electrons. *When two or more electronic structures differ only in the location of the bonding electrons, they are said to be* **resonance structures;** resonance structures are indicated by a double-headed arrow ($\leftrightarrow$). The following resonance structures

(1)   (1')

are best represented by this resonance hybrid:

(2)

Even though we drew two electronic structures, (1) and (1′), for benzene, keep in mind that *neither of these structures exists.* No benzene molecules contain alternating double and single bonds in the ring; all benzene molecules are the same. These are simply two ways to represent benzene *on paper.* As we saw in previous sections, the benzene molecule has the π electrons spread out or *delocalized* over the entire molecule. The real molecule is a **hybrid** (mixture) of all the resonance structures you can draw for the compound (see Sec. 13.6). The molecule does not exist as any one resonance form, but rather as an average of them. Benzene is best represented by structure (2). The circle indicates that the π electrons are spread out equally over the entire molecule. Henceforth benzene will usually be represented by a hexagon with a circle in it; it is assumed that there is a C—H unit at every corner of the hexagon unless another substituent is drawn in place of hydrogen. [Be careful not to confuse a hexagon with a circle in it—the representation for benzene—with a plain hexagon (*no circle*), which represents cyclohexane.]

Heats of combustion and hydrogenation were used to determine alkene stability in Sec. 8.15, and they can be applied to benzene. For example, the heat of hydrogenation of cyclohexene is 28.6 kcal/mole (119.7 kJ/mole):

$\Delta H° = -28.6$ kcal/mole (-119.7 kJ/mole)

Cyclohexene   Cyclohexane

Using cyclohexene as a reference, *if* benzene contained three isolated double bonds, *then* its heat of hydrogenation should be three times that of cyclohexene or -85.8 kcal/mole (-359 kJ/mole) for the hypothetical 1,3,5-cyclohexatriene molecule. The heat of hydrogenation for benzene determined by experiment is -49.8 kcal/mole (-208.4 kJ/mole), which is 36 kcal/mole (150.6 kJ/mole) *less* energy than predicted by our alkene model:

$\Delta H° = -49.8$ kcal/mole (-208.4 kJ/mole)
[predicted: 3×-28.6 or -85.8 kcal/mole (-359 kJ/mole]

    This difference means that benzene contains 36 kcal/mole (150.6 kJ/mole) less energy than it would be expected to have if it were 1,3,5-cyclohexatriene. Put another way, benzene is more stable than expected by 36 kcal/mole (105.6 kJ/mole). This energy is called the *delocalization energy,* and it represents the stabilization obtained through delocalization of the π electrons over the entire benzene molecule, as shown in Fig. 13.2. (This energy is also called *resonance energy.* Because the electrons are really delocalized over the entire carbon framework in benzene—and in other aromatic molecules or ions—the energy is more accurately called delocalization

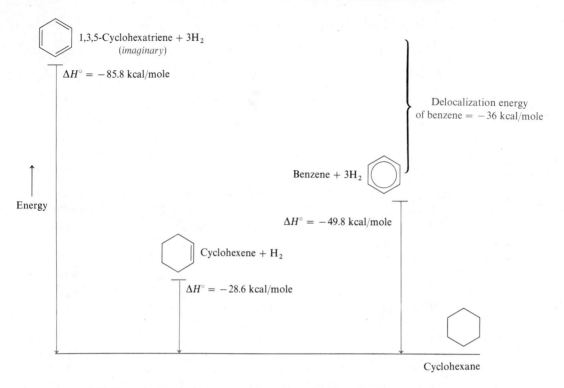

FIGURE 13.4   Diagram representing the delocalization energy of benzene.

energy.) Figure 13.4 depicts the delocalization energy of benzene compared with that of the hypothetical 1,3,5-cyclohexatriene molecule.

Heat of combustion data provide similar results for benzene, from which the delocalization energy is estimated to be ~ 42 kcal/mole (175.7 kJ/mole).

Question 13.2

The heat of hydrogenation of 1,3-cyclohexadiene is 55.4 kcal/mole (231.8 kJ/mole). Using the value of 28.6 kcal/mole (119.7 kJ/mole) for the heat of hydrogenation of cyclohexene, is 1,3-cyclohexadiene more or less stable than you would predict and by how much? To what do you attribute this difference? Cite examples from previous discussions to support your statement.

13.6   Drawing Resonance Structures; Other Examples of Resonance

A. Resonance Structures

What are the criteria for the existence of resonance structures, and how and where may they be drawn? We encountered several examples of resonance in previous chapters, and we now point out four common features that resonance structures must possess.

1. They are electronic structures that do not involve any shift or movement of the *atoms* within the molecule or ion. If a molecule or ion can be represented by two or more *viable electronic structures* that differ only in the arrangement of the electrons, then there is **resonance.**

2. All resonance structures for a given molecule or ion must contain the *same number of paired electrons*.

3. If the electronic structures are of *comparable stability*, then resonance is more important than if structures with varying degrees of stability are involved.

4. In general, the *greater the number* of resonance structures, the greater the degree of delocalization and the *more stable* the molecule or ion.

B. Resonance Structures of Organic and Inorganic Molecules

We now apply these rules of thumb to several simple organic and inorganic molecules. For benzene, we get the following:

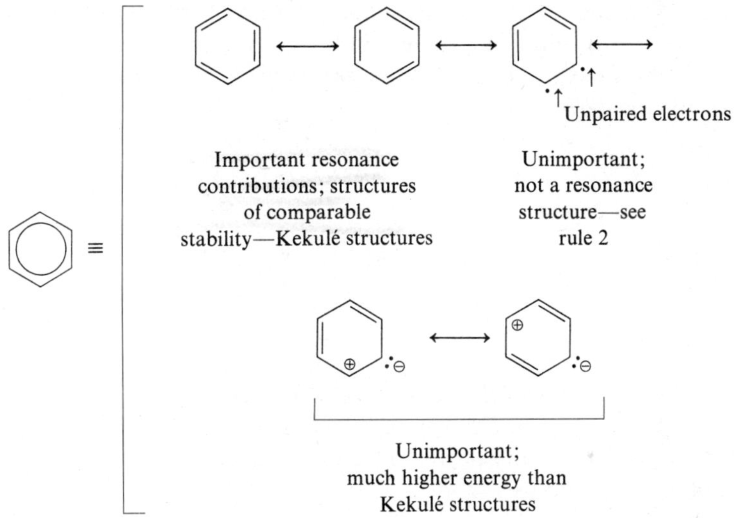

Bonds between different atoms are often polarized because of differences in electronegativity. Polarity may also be described by resonance structures. For example, the following resonance structures can be drawn for hydrogen fluoride:

$$[\text{H}\!-\!\ddot{\text{F}}\!: \quad \longleftrightarrow \quad \text{H}^{\oplus} \; :\!\ddot{\text{F}}\!:^{\ominus}] \equiv \overset{\delta\oplus}{\text{H}}\!-\!\!-\!\overset{\delta\ominus}{\text{F}}$$

Note that the dipolar structure is represented formally as having no H—F bond.

In addition to the preceding two structures, there is a third resonance structure possible for H—F:

$$\text{H} :^{\ominus} \ddot{\text{F}} :^{\oplus}$$

Why is it not included in our discussion? This question illustrates a very important point concerning possible resonance structures—they must be *chemically logical*. It is not logical for the H—F bond to be polar with a negative charge on hydrogen and a positive charge on the much more electronegative fluorine atom.

Chloromethane is polar and may be depicted by the following resonance structures:

$$[\text{CH}_3\!-\!\ddot{\text{Cl}}\!: \quad \longleftrightarrow \quad \text{CH}_3^{\oplus} \; :\!\ddot{\text{Cl}}\!:^{\ominus}] \equiv \overset{\delta\oplus}{\text{CH}_3}\!-\!\overset{\delta\ominus}{\text{Cl}}$$

Rule 1 must be applied with some knowledge of the structure and properties of the molecules involved. For example, it is possible to draw several resonance structures for methane, but they are ignored because they contribute little, if anything, to its stability:

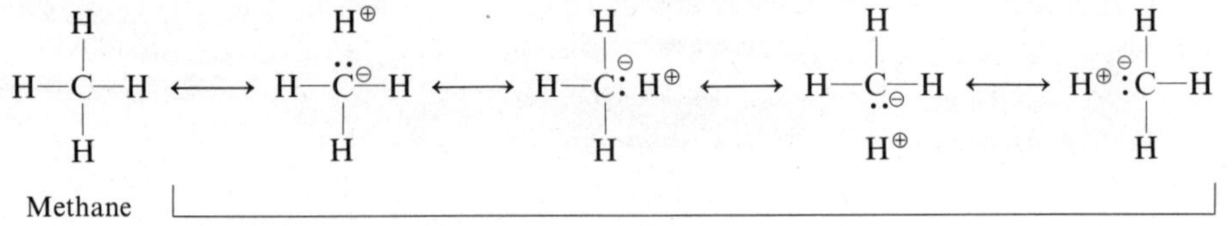

Methane

Negligible and minor contribution

Many other organic ions and molecules cannot be described accurately by a single conventional electronic structure. The same is true for certain inorganic ions. For example, the carbonate ion, $CO_3^{\ominus 2}$, is difficult to depict so that it contains a total of 24 valence electrons and yet each carbon and oxygen atom has an octet of valence electrons. Physical evidence indicates that the carbonate ion is flat and that all the carbon-oxygen bond lengths are equal. Three electronic structures can be drawn for the carbonate ion:

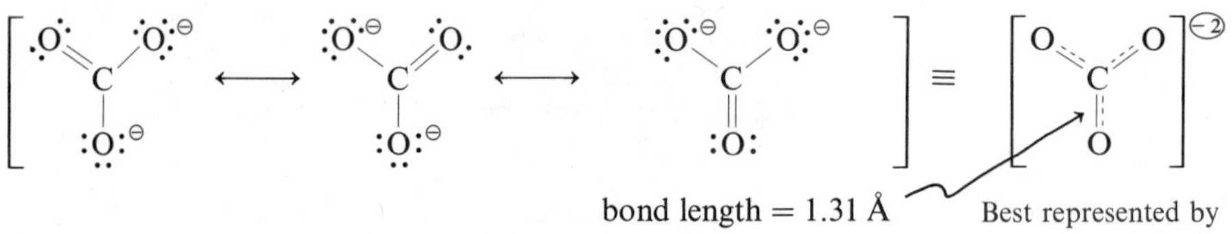

bond length = 1.31 Å

Best represented by resonance hybrid

The orbital view of this ion places six electrons in the π orbitals that are spread out (delocalized) over the entire molecule. This accounts well for the bond equivalency as well as for the stability of the carbonate ion. Thus, π bonding occurs between all pairs of adjacent atoms.

Question 13.3

How would you explain that although the carbon-oxygen bonds are equal in carbonate ion, $CO_3^{\ominus 2}$, in carbonic acid, H_2CO_3, there is one carbon-oxygen double bond and two carbon-oxygen single bonds?

Question 13.4

Draw resonance structures for these six ions or molecules:

(a) SO_2 (b) SO_3 (c) $NO_2^{\ominus}$ (d) $NO_2^{\oplus}$ (e) $NO_3^{\ominus}$ (f) O_3

Question 13.5

Draw the resonance structures for the nitro group in nitromethane, $CH_3—NO_2$. Do they account for the two nitrogen-oxygen bonds being equivalent? Briefly explain.

C. Allyl Free Radical and Allyl Cation

The stability of the allyl radical (see Sec. 8.30) and the allyl cation (see Sec. 8.10), which are redrawn to emphasize their similarity to previous examples, is best explained by the concept of delocalization.

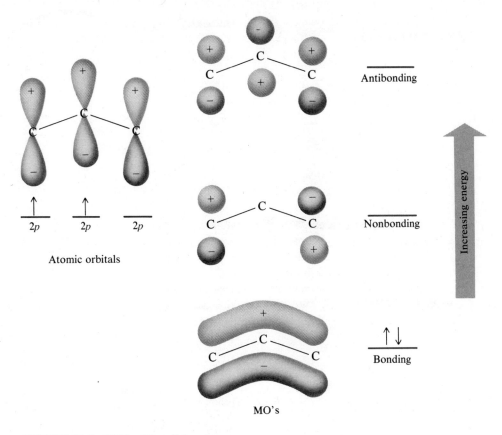

FIGURE 13.5   MO's of the allyl system.

Allyl radical (Sec. 8.30):

Best represented
by resonance hybrid

Allyl cation (Sec. 8.10)

Best represented
by resonance hybrid

The molecular orbitals for the allyl cation are shown in Fig. 13.5. Note that both π electrons are in the bonding molecular orbital of the π system.

D. Hyperconjugation

Before leaving resonance temporarily, we should mention the application of resonance theory to carbocations in general, as discussed in Sec. 5.11. The stabilizing effect of alkyl groups attached to a carbocation lies in the apparent ability of the C—H (or C—C) σ bonds adjacent to the electron-deficient carbon atom to contribute electrons to the carbon bearing the positive charge through a type of electron delo-calization. Using the *tert*-butyl cation as an example, this can be envisioned as follows:

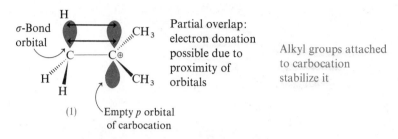

As the number of alkyl groups attached to a positively charged carbon atom increases, the number of possible interactions like that shown in structure (1) increases. A similar type of electron donation is possible when a C—C bond replaces the C—H bonds we used in our example.

The type of interaction shown in (1) cannot occur when hydrogens are attached directly to the positively charged carbon because the electrons in that σ bond are pointing away from the charged carbon atom; there can be no overlap through space.

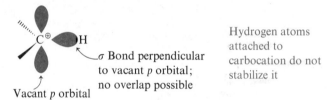

The type of interaction represented by structure (1) is often called **hypercon-jugation** or *no-bond resonance*. The name refers to the delocalization of electrons occurring through space and not directly through the σ or π bonds; refer to structure (1).

13.7   Aromaticity and Hückel's (4n + 2) π-Electron Rule

Among the unique qualities attributed to benzene, a typical aromatic com-pound, is its unreactivity compared with the alkenes and alkynes. As a result, *benzene and compounds related to it are said to be aromatic or at least to have aromatic character.* More simply, the delocalization of electrons in benzene is a quality often referred to as **aromaticity.**

Aromaticity is not limited to benzene but is generally associated with certain molecules that have four common features:

1. They are cyclic structures containing some number of π bonds.
2. They are planar structures, or at least very close to planar.
3. Each atom in the ring must be sp^2 hybridized (in some cases they may be sp hybridized).
4. Delocalization of the π electrons can occur.

Three signs suggest that a molecule may possess aromatic character:

1. Unusual stability, as evidenced by a lack of reactivity toward reagents that normally react rapidly with their noncyclic counterparts.
2. Thermochemical measurements of delocalization energy by heats of hydrogenation or heats of combustion, both of which indicate a molecule of much greater stability than one would predict based on noncyclic models.
3. The use of physical measurements, such as nuclear magnetic resonance (NMR); this method is particularly useful for organic ions. (See Sec. 16.11D.)

Theoretical molecular orbital calculations (presented by Hückel in 1931) predicted that monocyclic molecules that contain cyclic clouds of delocalized π electrons exhibit aromatic character when there are $(4n + 2)$ π electrons in the molecule, where n is any integer. Cyclic compounds that have 2, 6, 10, 14, ..., $(4n + 2)$ delocalized π electrons are aromatic. Since that time, more refined calculations predict that monocyclic compounds that have $4n$ π electrons are less stable than the noncyclic analogues. Those that have 4, 8, 12, 16, ..., $(4n)$ delocalized π electrons are not aromatic and are called **antiaromatic** because of their high degree of instability.

Benzene satisfies all the criteria for aromaticity. It is cyclic and planar, each carbon atom in the ring is sp^2 hybridized, and the ring contains six π electrons, a Hückel number. Benzene would be correctly predicted to be aromatic.

A. Cyclopentadienyl Anion

The cyclopentadienyl anion $C_5H_5^{\ominus}$, is prepared by allowing cyclopentadiene to react with either sodium or potassium metal. Cyclopentadiene is a relatively strong carbon acid so that one of the hydrogens is easily removed; oxidation (of Na → Na$^{\oplus}$) and reduction (of H$^{\oplus}$ → H$_2$) occur:

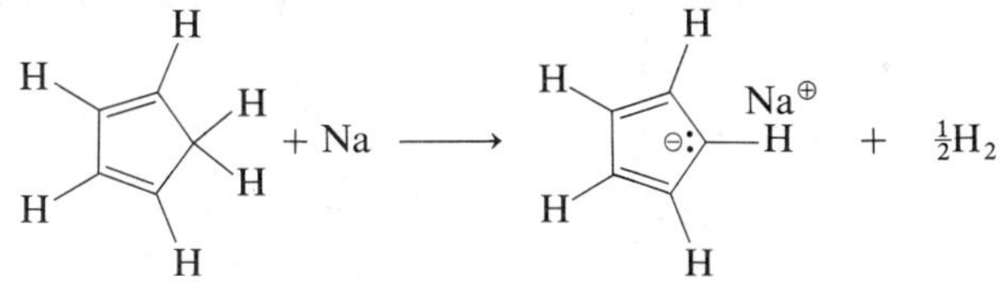

Cyclopentadiene   Cyclopentadienyl anion

The anion contains six π electrons and thus satisfies Hückel's $(4n + 2)$ π-electron rule for aromaticity for $n = 1$. The system is also cyclic and its geometry (that is, it contains five sp^2 hybridized carbon atoms) allows orbital overlap and electron delocalization to occur. Several resonance structures may be written for this anion; they are summarized best by the hybrid shown here:

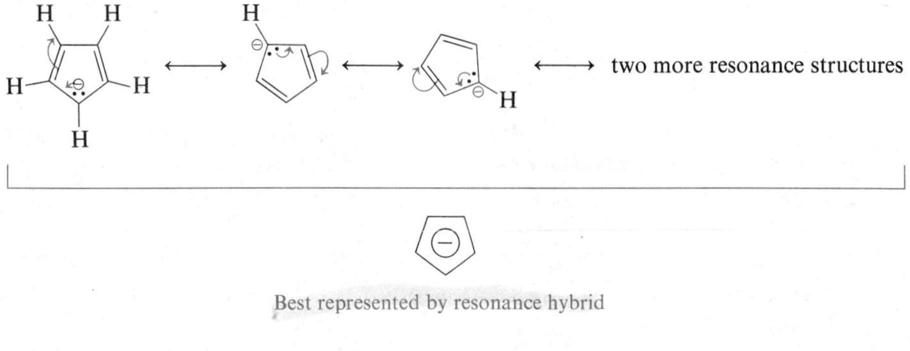

Best represented by resonance hybrid

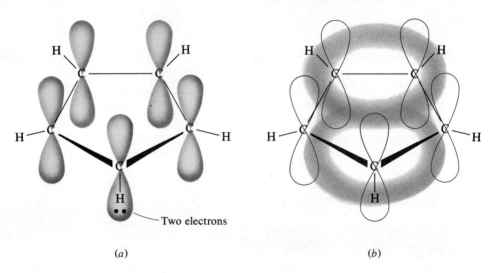

(a)   (b)

FIGURE 13.6   Orbital picture of the stable cyclopentadienyl anion showing the (*a*) four *p* orbitals, each containing one electron, and the fifth *p* orbital containing two electrons, and (*b*) overlap of the *p* orbitals to form the delocalized π-electron clouds.

From the molecular orbital viewpoint, this anion can be represented best by delocalized π-electron clouds above and below the planar ring as in Fig. 13.6. In this ion, the overwhelming stability associated with the electron delocalization must overcome the energy that is required to compress the sp^2 carbon bond angles from 120° to 108° (the pentagon bond angle).

Cyclopentadienyl anions also react with ferrous ion, Fe^{+2}, in the ratio of 2:1, respectively, to form a novel compound called **ferrocene.** This is one of many compounds with similar structure in which one or more organic ions are held in place by inorganic metal ions.

2 ⬠⊖ + Fe^{+2} ⟷ Fe^{+2}

Cyclopentadienyl   Ferrocene
anion   Stable molecule

The carbon-carbon bonds in ferrocene are all of equal length, thus supporting the delocalized structure we described for the anion from which it was prepared.

B. Cycloheptatrienyl Cation

The *anion* of cyclopentadiene is unusually stable, and the *cation* of cycloheptatriene is also unusually stable. The cation can be prepared by allowing the hydrocarbon to react with a carbocation salt; the one often used is the stable triphenylmethyl cation fluoroborate salt, $(C_6H_5)_3C^{\oplus}BF_4^{\ominus}$, which we discuss in Sec. 14.22 and represent here simply as $R^{\oplus}BF_4^{\ominus}$. The carbocation, a Lewis acid, abstracts a hydride ion from cycloheptatriene, a Lewis base, to give a new, more stable carbocation, the **cycloheptatrienyl cation:**

Cycloheptatriene   Carbocation   (1)
Stable   Cycloheptatrienyl
fluoroborate
More stable
carbocation

This may appear to be a peculiar reaction, but as we saw in Sec. 9.9 with other carbocations, a less stable ion can abstract a hydride ion from a hydrocarbon to form a more stable ion. The fluoroborate salt (1) is water-soluble, and when either bromide ion or iodide ion (as NaBr or NaI) is added, the insoluble cycloheptatrienyl halide salt precipitates out:

(1)   Cycloheptatrienyl
Water-soluble   iodide
Water-insoluble

We mention this because the cycloheptatrienyl cation—often called the **tropylium ion**—was first made around 1900 but was discarded in the water used to extract the reaction. It was not until 1954 that the tropylium salt was isolated and found to be unusually stable—that is, to obey Hückel's $(4n + 2)$ rule.

The cycloheptatrienyl cation is unusually stable because it contains six π electrons that overlap to form a π-electron cloud. We can write resonance structures to indicate the various ways the electrons are distributed in the molecule, but we usually write a resonance hybrid that best shows the electronic structure from an orbital point of view. These structures are as follows:

Four more resonance structures
(Draw them)

Best represented by
resonance hybrid

C. Heterocyclic Compounds

A very important family of organic compounds contains **hetero** atoms—nitrogen, oxygen, sulfur, and others—in place of carbon in cyclic structures; consequently, they are called **heterocyclic compounds.** We briefly discuss their structure because many of them are aromatic compounds, but we defer their chemistry until Chaps. 19 and 26.

Pyridine is a common heterocyclic compound that has aromatic character. It is analogous to benzene except that nitrogen replaces one of the carbon atoms in the ring; there is no hydrogen atom attached to the nitrogen in pyridine. There is an electron pair on nitrogen, which is in a p orbital parallel to the π cloud and interacts with it through orbital overlap. Nitrogen also contains an sp^2 orbital that contains a pair of electrons and is perpendicular to the aromatic π-electron cloud and does not overlap with it. The orbital picture of pyridine is shown here:

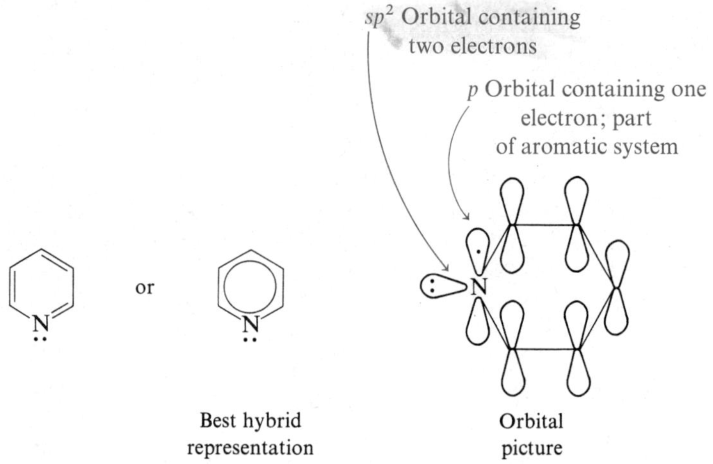

sp^2 Orbital containing two electrons

p Orbital containing one electron; part of aromatic system

Best hybrid representation

Orbital picture

Pyridine

The complete orbital diagram for pyridine resembles that for benzene. It has three bonding and three antibonding orbitals, and the six π electrons completely fill the bonding orbitals.

Pyrrole is a five-membered heterocyclic compound that contains six π electrons (two double bonds and a pair of nonbonded electrons on nitrogen). There is a nitrogen atom in the ring in place of carbon. The nitrogen has a hydrogen attached to it, and it also bears one pair of nonbonding electrons. The electron pair occupies a p orbital and is delocalized by resonance into the π-electron cloud, as shown here:

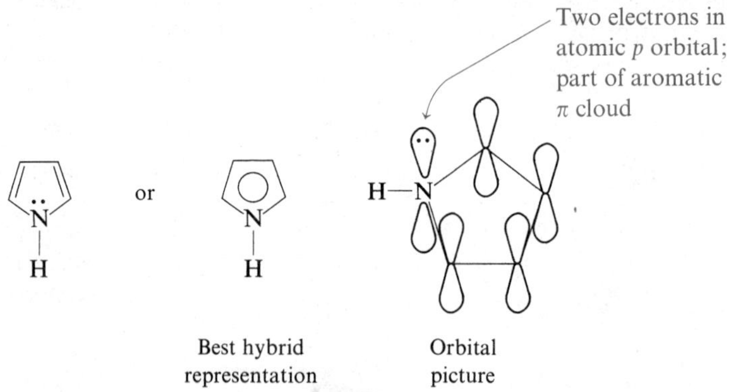

Two electrons in atomic p orbital; part of aromatic π cloud

Best hybrid representation

Orbital picture

Pyrrole

Question 13.6

We know that nitrogen normally exhibits a pyramidal geometry arising from sp^3 hybridization. Why is nitrogen sp^2 hybridized in pyrrole and pyridine; that is, what is the driving force behind this change in hybridization?

Furan and thiophene are similar to pyrrole in that they are both five-membered ring compounds, the former containing oxygen and the latter containing sulfur in the ring. These compounds also contain six π electrons and thus are aromatic.

The orbital picture of furan is shown, and that of thiophene is anologous except sulfur replaces oxygen in the ring.

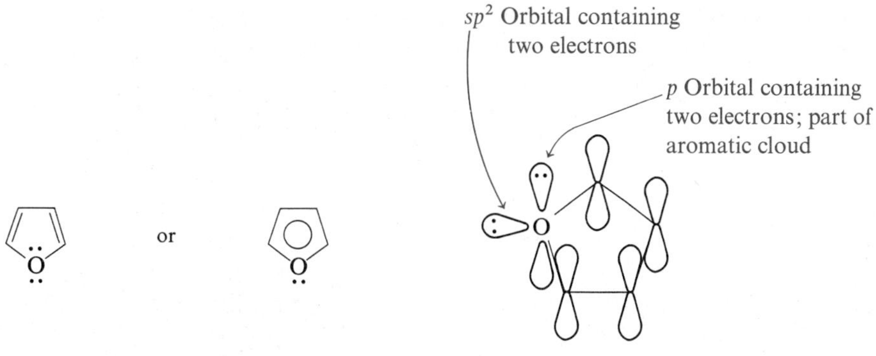

sp² Orbital containing
two electrons

p Orbital containing
two electrons; part of
aromatic cloud

or

Best hybrid
representation

Orbital picture

Furan

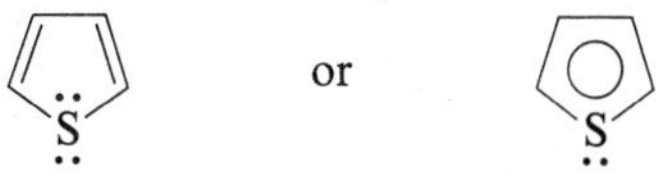

or

Best hybrid
representation

Thiophene

In both furan and thiophene one of the nonbonding pairs of electrons occupies a *p* orbital and overlaps with the other *p* orbitals to form the π cloud. The other nonbonding pair of electrons occupies an *sp²* orbital that sticks out from the ring and is roughly perpendicular to the *p* orbitals.

D. Summary

We surveyed several aromatic systems, ranging from cations (like the cyclo-heptatrienyl cation) to anions (like the cyclopentadienyl anion) to neutral molecules (like benzene and the heterocyclic compounds). These are all unusually stable molecules or ions compared with analogous open-chain compounds. The "magic" number of π electrons required for aromaticity is dictated by Hückel's (4*n* + 2) π-electron rule.

We started our discussion about stability based on resonance, but *the existence of many resonance structures does not necessarily mean that a compound or ion is unusually stable*. The number of π electrons is the major criterion for aromaticity. For example, the cyclopentadienyl cation, $C_5H_5^{\oplus}$, can be represented by five resonance structures, yet this ion is of high energy and is antiaromatic (unusually unstable) because it contains four π electrons:

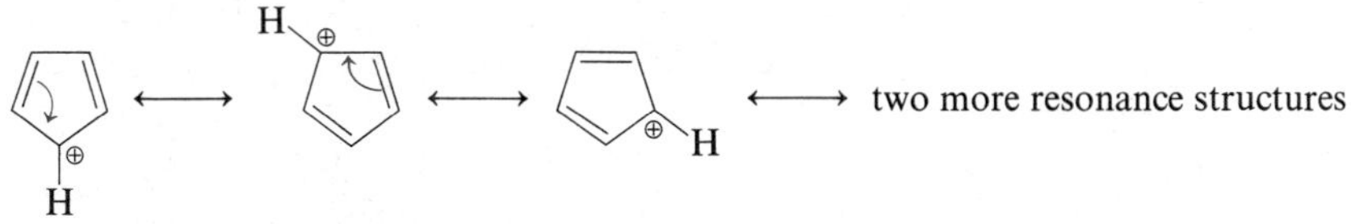

two more resonance structures

On the other hand, we drew the same number of resonance structures for the cyclopentadienyl anion, which is unusually stable not because of the resonance structures but because it contains six π electrons.

We mentioned that planarity is essential in the structures of aromatic compounds and ions. As an example of the research carried on in the field of aromaticity, consider the 14-membered ring compound, [14]-annulene, which has been shown to be aromatic on the basis of NMR (see Sec. 16.11).

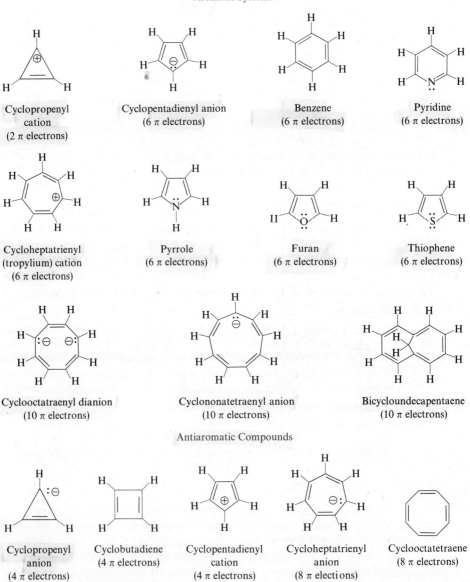

[14]-Annulene
14 π electrons: aromatic

It is not immediately obvious that the π electrons in [14]-annulene interact to form a π cloud, but all evidence available suggests they do. The chemist is interested

Aromatic Systems

Cyclopropenyl
cation
(2 π electrons)

Cyclopentadienyl anion
(6 π electrons)

Benzene
(6 π electrons)

Pyridine
(6 π electrons)

Cycloheptatrienyl
(tropylium) cation
(6 π electrons)

Pyrrole
(6 π electrons)

Furan
(6 π electrons)

Thiophene
(6 π electrons)

Cyclooctatraenyl dianion
(10 π electrons)

Cyclononatetraenyl anion
(10 π electrons)

Bicycloundecapentaene
(10 π electrons)

Antiaromatic Compounds

Cyclopropenyl
anion
(4 π electrons)

Cyclobutadiene
(4 π electrons)

Cyclopentadienyl
cation
(4 π electrons)

Cycloheptatrienyl
anion
(8 π elections)

Cyclooctatetraene
(8 π electrons)

FIGURE 13.7   Monocyclic molecules and ions that are classified as aromatic [(4n + 2) rule] or antiaromatic [(4n) rule].

in learning what limitations of structure and geometry govern aromaticity, and the validity of the ($4n + 2$) rule is supported by numerous examples (Fig. 13.7).

Question 13.7

Predict which of the following six compounds would be expected to exhibit aromaticity. Briefly explain. *Note:* Treat unpaired electrons the same way you treat paired electrons.

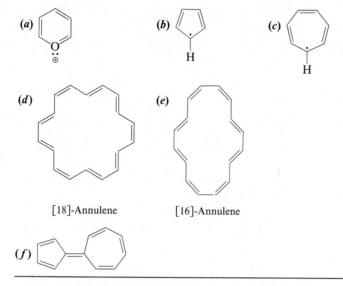

(a)

(b)

(c)

(d)

(e)

[18]-Annulene   [16]-Annulene

(f)

Question 13.8

Draw the resonance structures for the following:

(a) cycloheptatrienyl anion, $C_7H_7^{\ominus}$ (b) pyridine   (c) furan

Comment on the aromaticity associated with each.

Question 13.9

For each of the following nine pairs of structures, decide whether they represent resonance structures of one another. If not, why not? (You have not encountered some of these functional groups but you should be able to reason by knowing and understanding the concepts of resonance.)

(a) $CH_2{=}CH{-}CH{=}CH_2$ and $\overset{\oplus}{C}H_2{-}CH{=}CH{-}CH_2{:}^{\ominus}$

(b) $CH_3{-}\underset{\underset{H}{|}}{C}{=}O$ and $CH_2{=}\underset{\underset{H}{|}}{C}{-}OH$ (c) $CH_3{-}\underset{\underset{H}{|}}{C}{=}O$ and $CH_3{-}\underset{\underset{H}{|}}{\overset{\oplus}{C}}{-}\overset{..}{\underset{..}{O}}{:}^{\ominus}$

(d) $CH_3{-}C\overset{\displaystyle O}{\underset{\underset{..}{\overset{..}{O}}{:}^{\ominus}}{\diagup\!\!\diagdown}}$ and $CH_3{-}C\overset{\overset{..}{\overset{..}{O}}{:}^{\ominus}}{\underset{O}{\diagup\!\!\diagdown}}$ (e) and

(f) $CH_2{=}\underset{\underset{CH_3}{|}}{\overset{\overset{O{-}H}{|}}{C}}$ and $CH_3{-}\underset{\overset{\|}{O}}{C}{-}CH_3$ (g) $CH_3{-}\overset{\oplus}{C}H_2$ and $\underset{\overset{\oplus}{H}}{CH_2{=}CH_2}$

(h) $CH_3{-}\underset{\overset{..}{..}}{S}{-}CH_3$ and $CH_3{-}\underset{\overset{..}{..}}{S}{=}CH_2$ (i) $\underset{\ominus}{CH_2}{-}\overset{\overset{\|}{O}}{C}{-}H$ and $CH_2{=}C{-}H$ with $:\overset{..}{\underset{..}{O}}{:}^{\ominus}$

omit **13.8   Aromaticity and Molecular Orbital Theory**

Why is the Hückel number $(4n + 2)$ so important? The answer comes from molecular orbital theory. The molecular orbitals associated with the π-electron system of any simple aromatic compound can be generated in the following manner:

1. Draw the ring so that the vertex of one ring atom is pointed toward the bottom of the page and below all other atoms in the ring, which are symmetrically distributed above it. We draw the benzene ring in the following way:

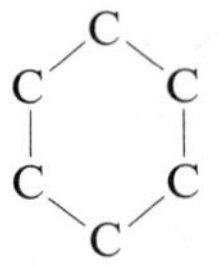

2. Each atom in the ring now represents an energy level associated with a given molecular orbital. In the case of benzene, where the molecular π system is composed of six atomic p orbitals, we wind up with the six molecular orbitals drawn here:

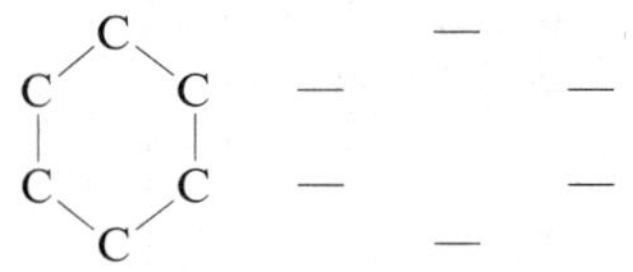

3. If the system is composed of an even number of ring atoms, the molecular orbitals drawn in this way will be divided to provide an equal number of bonding and anti-bonding orbitals. These are symmetrically dispersed about a line drawn through the center of the cyclic system.

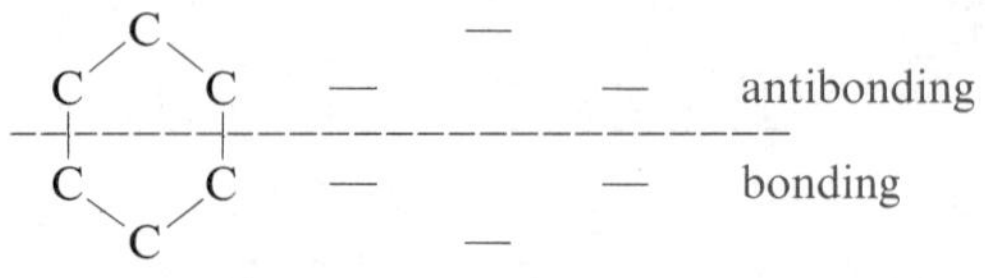

4. If the system contains an odd number of ring atoms, then there will be an unequal number of bonding and antibonding orbitals. They again are distributed about a line drawn horizontally through the center of the cyclic system. This is shown for the cyclopentadienyl system.

Let us now draw the molecular orbitals associated with several aromatic systems we encountered previously.

A. Benzene

The benzene molecular orbitals are shown in Fig. 13.8. Note that *all the bonding orbitals are filled.* This means that the electronic configuration results in maximum overlap (maximum bonding) and hence greatest stability. For benzene, any number less than the Hückel number of π electrons (in this case 6) results in less overlap and less stability, whereas any number greater results in electrons entering antibonding

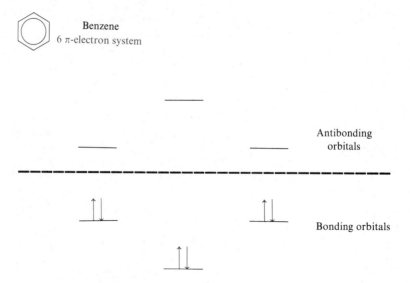

FIGURE 13.8   MO's associated with the π-electron system of benzene: an aromatic system.

orbitals that in turn cause an increase in energy and destabilization of the system. This characteristic is common to all aromatic compounds.

B. Cyclopentadienyl Anion

The cyclopentadienyl anion is also aromatic (see Sec. 13.7). The complete molecular orbital diagram for this ion is shown in Fig. 13.9. Note that the bonding orbitals are completely filled.

C. Cyclopropenyl System

The cyclopropenyl cation (Figs. 13.10 and 13.11) is also aromatic, but the cyclopropenyl anion (Fig. 13.12) is extremely unstable. The delocalization of the π

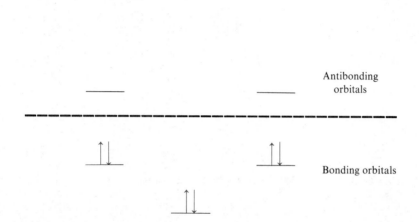

FIGURE 13.9   MO's associated with the π-electron system of the cyclopentadienyl anion: an aromatic system.

 Cyclopropenyl cation:
2π-electron system

FIGURE 13.10   MO's associated with the π-electron system of the cyclopropenyl cation: an aromatic system.

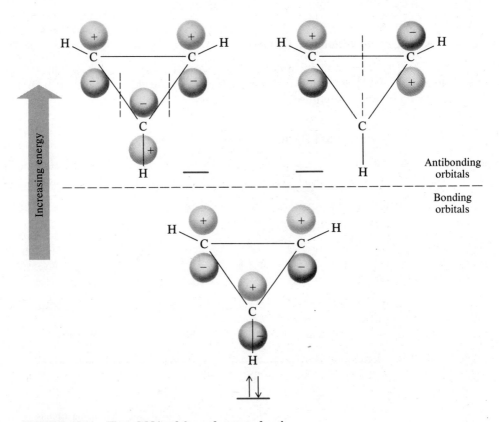

FIGURE 13.11   The π MO's of the cyclopropenyl cation.

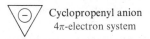

Cyclopropenyl anion
4π-electron system

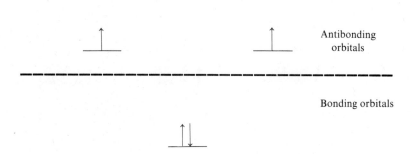

FIGURE 13.12 **MO's associated with the π-electron system of the cyclopropenyl anion: an antiaromatic system.**

electrons over the cyclic system in this case (and in analogous cases) results in the generation of molecular orbitals that when occupied result in an extremely unstable electronic configuration. Systems of this type are referred to as **antiaromatic systems.**

This can be determined in a qualitative sense simply by viewing the orbital diagrams in Figs. 13.10 and 13.12. In the former the bonding orbitals of the molecule are completely filled; the molecule is definitely aromatic. But in the latter electrons are also found in the antibonding orbitals. In fact, two different antibonding orbitals but only one bonding orbital are occupied. Antiaromatic compounds contain $4n$ π electrons. The delocalization of electrons in antiaromatic systems results in an increase in energy and thus in a decrease in stability over the localized electronic configuration. *Antiaromatic compounds have either unfilled bonding orbitals or electrons in antibonding orbitals.*

13.9   Nomenclature of Benzene Derivatives

Before we study the reactions of benzene and its derivatives, we must learn the nomenclature of aromatic compounds. We consider monosubstituted, disubstituted, and polysubstituted benzene derivatives in that order.

A. Monosubstituted Benzenes

Many monosubstituted benzene compounds are named by placing the name of the substituent in front of the word *-benzene.* Some typical examples are:

Cl — Chlorobenzene

Br — Bromobenzene

NO_2 — Nitrobenzene

CH_2-CH_3 — Ethylbenzene

$CH_3-\overset{\displaystyle CH_3}{\underset{\displaystyle CH_3}{C}}$ — *tert*-Butylbenzene

These names are fairly easy to remember because they contain the name of the substituent.

Certain other monosubstituted benzene compounds have names that bear no resemblance to the substituents they contain. The student must learn the following names.

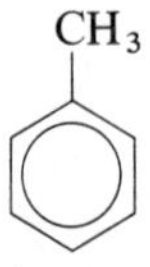

Toluene
(*not* methylbenzene)

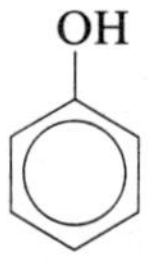

Phenol
(*not* hydroxybenzene)

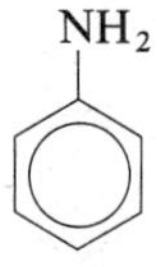

Aniline
(*not* aminobenzene)

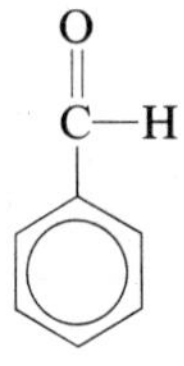

Benzaldehyde

Anisole
(*not* methoxybenzene)

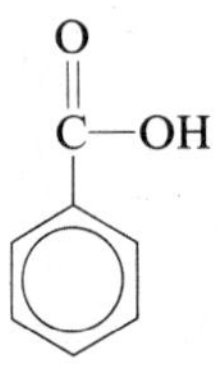

Benzoic acid

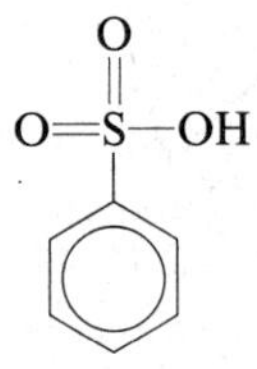

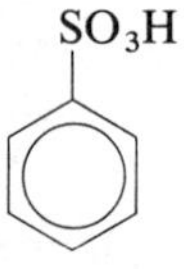

Benzenesulfonic acid

Sometimes an aromatic ring is attached to a fairly large and complex substituent that does not possess a simple alkyl group name. To circumvent this problem, benzene with a hydrogen removed is called the *phenyl* group:

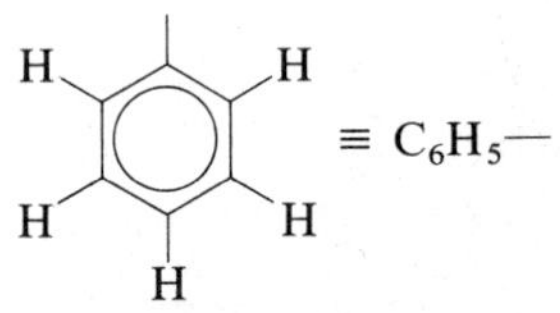

$\equiv C_6H_5-$

Phenyl group

This is a special name because the convention used for alkyl groups (that is, take the alkane name, drop the -*ane* ending, and add -*yl*) does not apply. Because the word *phenyl* bears no resemblance to the name of the parent compound, benzene, it must be learned. It can be used in nomenclature as follows:

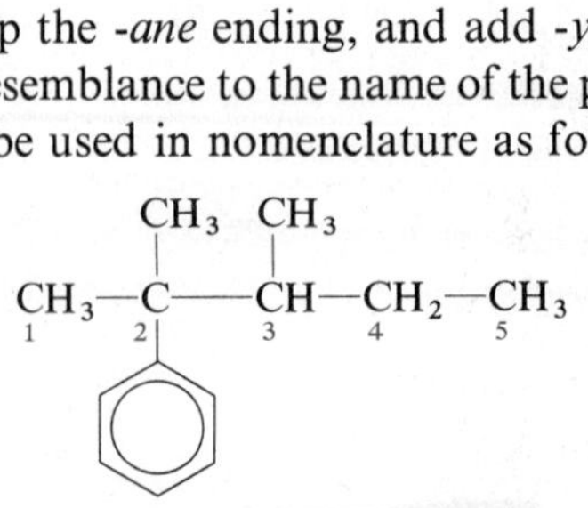

2,3-Dimethyl-2-phenylpentane

Another group containing the phenyl group is the *benzyl group*, which is discussed fully in Secs. 14.19 and 14.21. Examples of its use in the common system of nomenclature are given here.

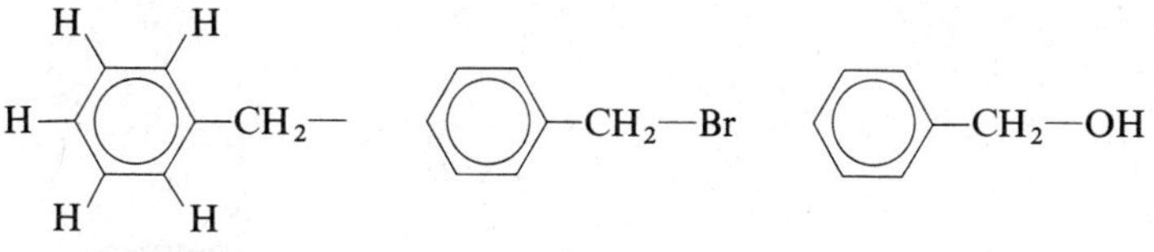

Benzyl group   Benzyl bromide   Benzyl alcohol

Benzyl bromide is also called α-*bromotoluene*, which is a common name.

B. Disubstituted Benzenes

There are three possible ways to attach two substituents to a benzene ring. We must name the substituents and also indicate their positions relative to one another. The three possible structural isomers are commonly designated by the prefixes *ortho* (Greek *orthos*, regular), *meta* (*meta*, between), and *para* (*para*, beside), which are abbreviated *o-*, *m-*, and *p-*, respectively. For a benzene ring that contains two chlorine atoms, we have:

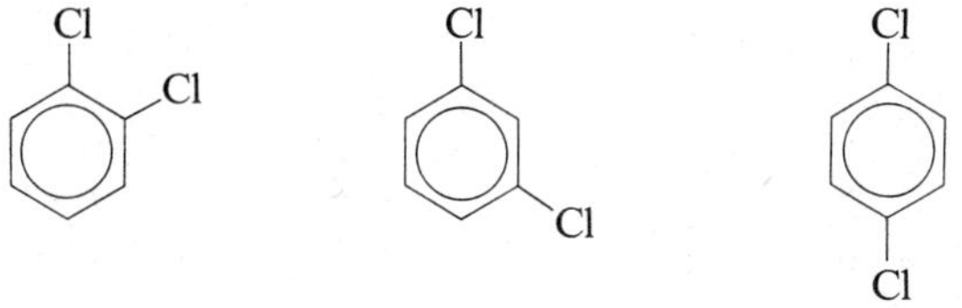

o-Dichlorobenzene   m-Dichlorobenzene   p-Dichlorobenzene
(*ortho* isomer)   (*meta* isomer)   (*para* isomer)

Because of the planar nature of benzene, the ring can be rotated about the center and we still have the same compound; that is, it is the relative location of the substituents that is important. For example, the following are all *o*-dichlorobenzene:

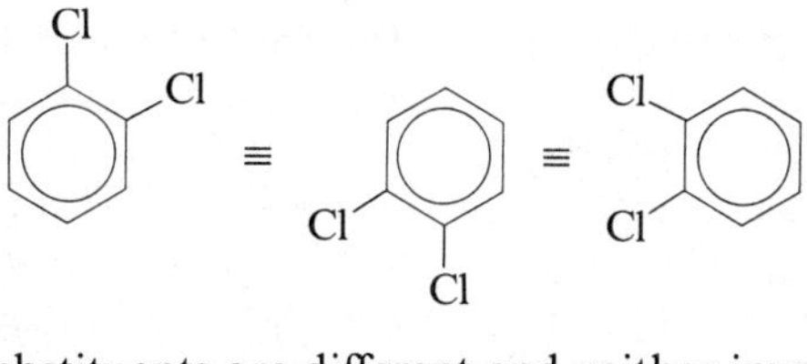

When the two substituents are different and neither imparts a special name to benzene (for example, an —OH group or —NH$_2$ group on benzene results in the special names *phenol* and *aniline*, respectively), then the names of the two substituents are inserted between the position prefix (*o-*, *m-*, or *p-*), in alphabetical order, and the word -*benzene;* for example:

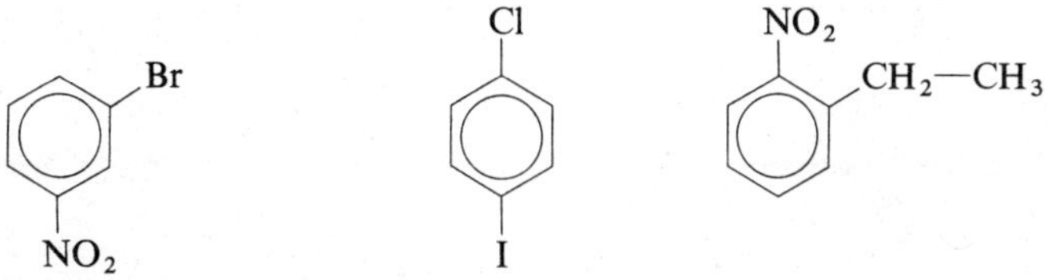

m-Bromonitrobenzene   *p*-Chloroiodobenzene   *o*-Ethylnitrobenzene

When one or both of the two substituents attached to benzene yield a special name then the compound is named as a derivative using a special name. In the following examples, the part of the molecule that has a special name is outlined in the box.

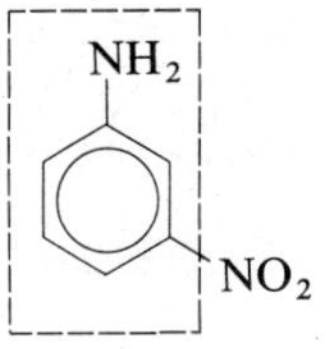

m-Nitroaniline

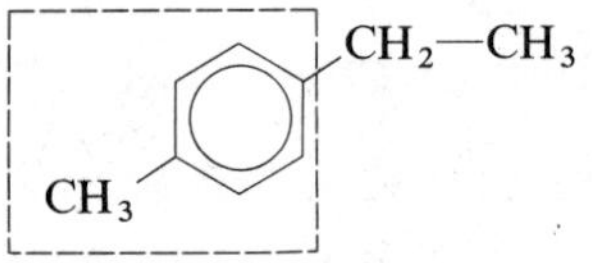

p-Ethyltoluene

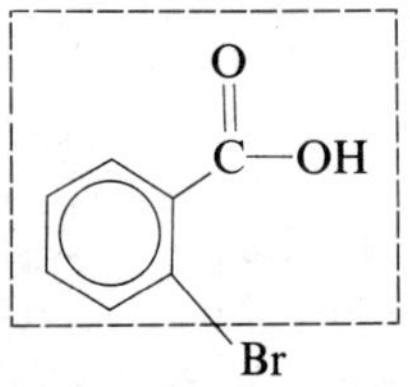

o-Bromobenzoic acid

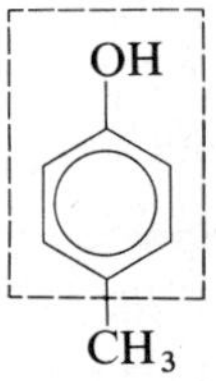

p-Methylphenol or *p*-hydroxytoluene
(*p*-cresol)

There are also some common names that are accepted for disubstituted benzene compounds, for example:

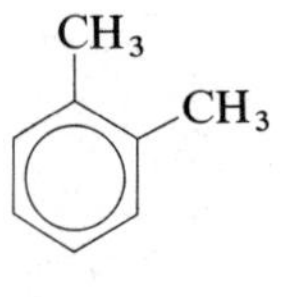

o-Xylene 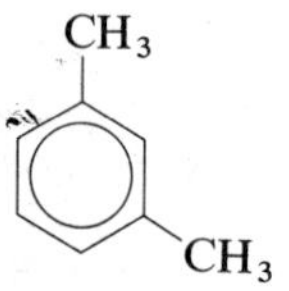   *m*-Xylene 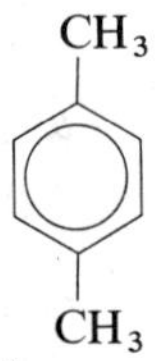   *p*-Xylene 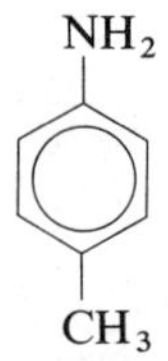   *p*-Toluidine 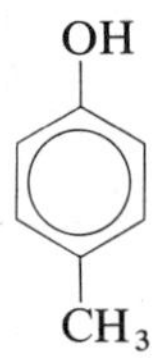   *p*-Cresol

C. Tri- and Polysubstituted Benzenes

When there are more than two substituents on the benzene ring, it is numbered to indicate their positions and to give the smallest substituent numbers. The substituents are listed alphabetically.

When one substituent attached to the ring imparts a common name to benzene, then the compound is named as a derivative of that substituent and numbering is always done so that substituent is attached to C-1. In the complete name, the number 1 is always omitted. Finally, when all the substituents are the same, each is numbered, as in 1,2,3,4-tetrabromobenzene.

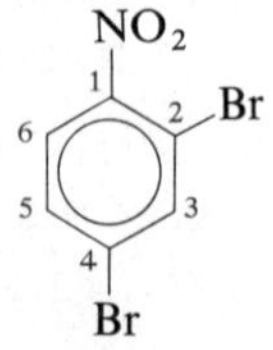

2,4-Dibromonitrobenzene 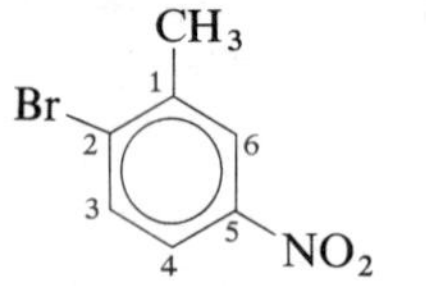   2-Bromo-5-nitrotoluene 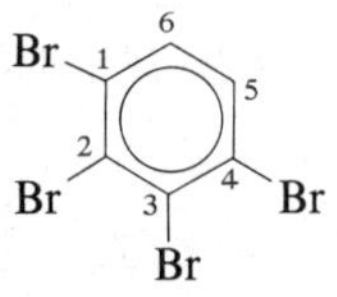   1,2,3,4-Tetrabromobenzene

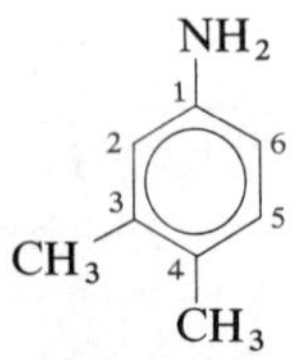

3,4-Dimethylaniline 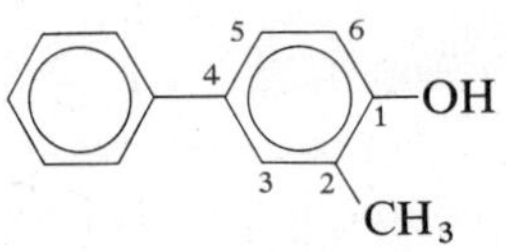   2-Methyl-4-phenylphenol

D. Special Names for Benzene-Containing Compounds

In addition to the names for aromatic compounds studied thus far, there are other benzene-containing molecules with nomenclature that deserve special mention. When two phenyl rings are attached, the resulting molecule is called **biphenyl:**

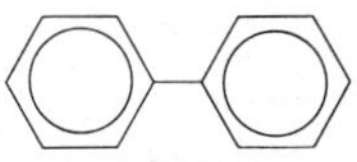

Biphenyl

Aromatic compounds can also be named as derivatives of the alkane to which they are attached:

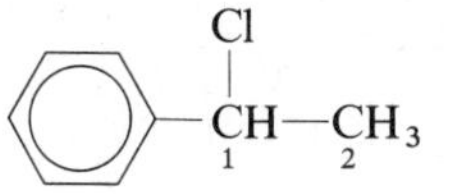

1-Chloro-1-phenylethane
(α-phenethyl chloride)

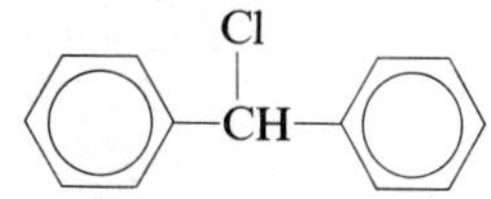

Chlorodiphenylmethane

Aromatic compounds containing a double or triple bond in an attached side chain are also given common names, as in the following examples:

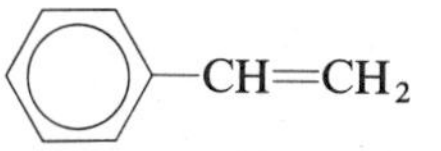

Phenylethene
(styrene or vinylbenzene)

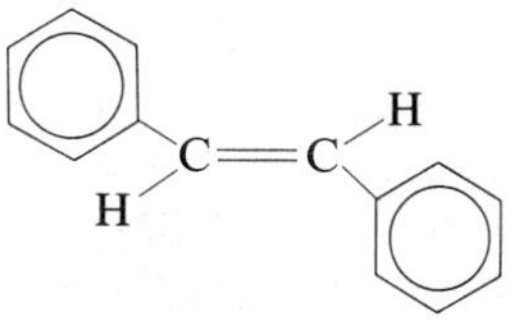

trans-1,2-Diphenylethene
(*trans*-stilbene)

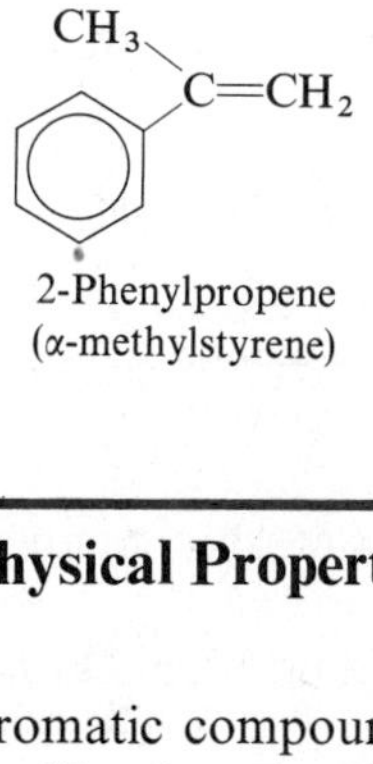

2-Phenylpropene
(α-methylstyrene)

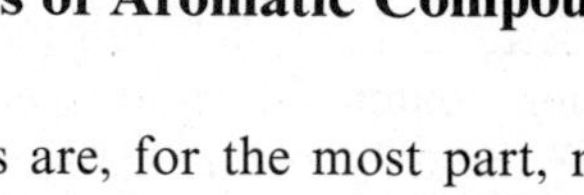

3-Phenylpropene
(allylbenzene)

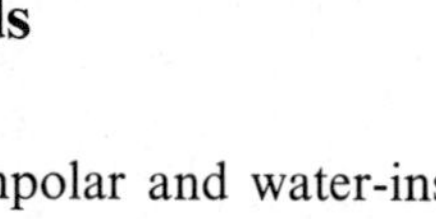

Phenylethyne
(phenylacetylene)

13.10   Physical Properties of Aromatic Compounds

Aromatic compounds are, for the most part, nonpolar and water-insoluble. They are like the other hydrocarbons we studied because they are soluble in many organic solvents, especially nonpolar ones. Some compounds and their physical properties are listed in Table 13.1.

Some interesting trends can be observed in the aromatic compounds. The most notable is the effect of structure on melting point; note that the *p* isomer generally has a higher melting point than either the *o* or the *m* isomer. This is attributed to the symmetrical structure of the *p* isomer, thus allowing it to fit into the solid crystalline lattice (that is, crystal packing) better than either the *o* or *m* isomer. To illustrate this, *p*-xylene freezes at $+13°$, whereas *o*- and *m*-xylenes remain liquid at temperatures well below 0°. On the other hand, the xylenes boil at temperatures fairly close to one another. The symmetrical molecule, 1,2,4,5-tetramethylbenzene (*common name:*

TABLE 13.1   Physical Constants of Selected Aromatic Hydrocarbons

Name	Formula	mp, °C	bp, °C	Density at 20°, g/cc
Benzene	C_6H_6	5.4	80.1	0.879
Toluene	$C_6H_5CH_3$	−93	110.6	0.866
o-Xylene	1,2-$(CH_3)_2C_6H_4$	−28	144	0.880
m-Xylene	1,3-$(CH_3)_2C_6H_4$	−54	139	0.864
p-Xylene	1,4-$(CH_3)_2C_6H_4$	13	138	0.861
Hemimellitene	1,2,3-$(CH_3)_3C_6H_3$	−25	176	0.895
Pseudocumene	1,2,4-$(CH_3)_3C_6H_3$	−44	169	0.876
Mesitylene	1,3,5-$(CH_3)_3C_6H_3$	−57	165	0.864
Prehnitene	1,2,3,4-$(CH_3)_4C_6H_2$	−4	205	0.902
Isodurene	1,2,3,5-$(CH_3)_4C_6H_2$	−24	196	0.891
Durene	1,2,4,5-$(CH_3)_4C_6H_2$	80	195	0.838
Pentamethylbenzene	$C_6H(CH_3)_5$	53	231	0.917
Hexamethylbenzene	$C_6(CH_3)_6$	166	265	
Ethylbenzene	$C_6H_5CH_2CH_3$	−93	136	0.867
Propylbenzene	$C_6H_5CH_2CH_2CH_3$	−99	159.5	0.862
Cumene	$C_6H_5CH(CH_3)_2$	−96	152	0.862
Butylbenzene	$C_6H_5CH_2CH_2CH_2CH_3$	−81	180	0.860
tert-Butylbenzene	$C_6H_5C(CH_3)_3$	−58	168	0.865
p-Cymene	*p*-$CH_3C_6H_4CH(CH_3)_2$	−73.5	177	0.857
1,3,5-Triethylbenzene	1,3,5-$(CH_3CH_2)_3C_6H_3$	−66.5	215	0.862
Hexaethylbenzene	$C_6(CH_2CH_3)_6$	129	305	0.831
Styrene	$C_6H_5CH{=}CH_2$	−31	146	0.908
Allylbenzene	$C_6H_5CH_2CH{=}CH_2$		156	0.893
trans-Stilbene	*trans*-$C_6H_5CH{=}CHC_6H_5$	124	307	0.971
cis-Stilbene	*cis*-$C_6H_5CH{=}CHC_6H_5$	6		
Diphenylmethane	$(C_6H_5)_2CH_2$	27	265	1.006
Triphenylmethane	$(C_6H_5)_3CH$	94	359	1.014
Biphenyl	$C_6H_5C_6H_5$	70.5	255	1.990
Phenylacetylene	$C_6H_5C{\equiv}CH$	−45	142	0.930
Diphenylacetylene	$C_6H_5C{\equiv}CC_6H_5$	63	300	0.966

durene) melts at $+80°$, whereas its isomers are liquids at room temperature. Hexamethylbenzene is a symmetrical, high-melting solid (mp 166°).

Increasing the carbon content increases the boiling point. For example, the xylenes on the average boil about 29° higher than toluene, and the tri- and tetramethyl benzenes are divided into groups with boiling points that are about 30° apart. As with the other hydrocarbons, boiling point depends greatly on molecular weight, whereas melting point depends on symmetry (or lack of it) in the molecules.

13.11   Source of Aromatic Hydrocarbons and Their Derivatives

Many aromatic compounds are commercially available at low cost. The research chemist never has to prepare simple aromatic hydrocarbons like benzene, toluene, of the xylenes. Petroleum provides aromatic compounds as well as aliphatic molecules that are converted into aromatic compounds by **aromatization,** more commonly called **reforming, hydroforming,** and **platforming** by industrial concerns. For example, the C_6, C_7, and C_8 alkanes from petroleum can be dehydrogenated to produce benzene, toluene, and the xylenes. The catalyst is often platinum supported on aluminum oxide (Al_2O_3), and the alkanes are passed over the catalyst at 500 to 700° and 300 to 700 lb/sq in. pressure; the following reactions occur (in addition to cracking, see Sec. 4.10, and other side reactions):

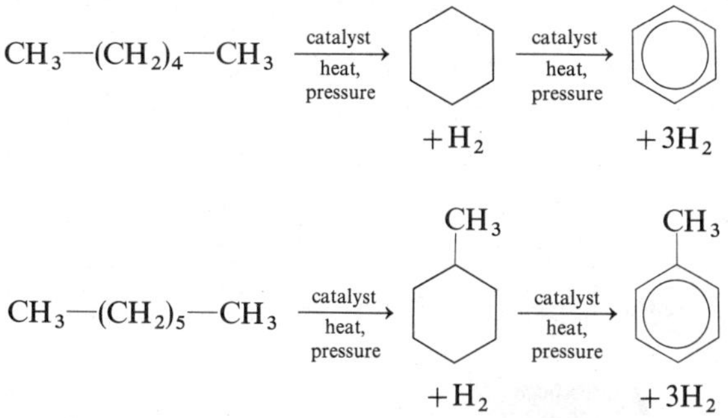

$$CH_3\text{---}(CH_2)_4\text{---}CH_3 \xrightarrow[\substack{heat, \\ pressure}]{catalyst} \bigcirc \xrightarrow[\substack{heat, \\ pressure}]{catalyst} \bigcirc$$
$$+H_2 \qquad +3H_2$$

$$CH_3\text{---}(CH_2)_5\text{---}CH_3 \xrightarrow[\substack{heat, \\ pressure}]{catalyst} \overset{CH_3}{\bigcirc} \xrightarrow[\substack{heat, \\ pressure}]{catalyst} \overset{CH_3}{\bigcirc}$$
$$+H_2 \qquad +3H_2$$

Coal also provides several aromatic compounds. When heated at temperatures of 800 to 1,000° in the near absence of air for 15 to 20 hr, coal decomposes. The volatile materials are driven off and condensed. The gas that remains is called *coal gas*, and the liquid that condenses is called *coal oil*. Coal oil is fractionally distilled to yield a variety of aromatic compounds.

Typically, petroleum yields only hydrocarbons, whereas coal oil distillation gives (in addition to benzene, toluene, and the xylenes) phenols, cresols, and some polynuclear aromatic hydrocarbons, for example, naphthalene, anthracene, and phenanthrene, which are discussed later. Coal oil also gives heterocyclic bases such as pyridine and quinoline (see Chap. 26).

13.12 Polynuclear Aromatic Hydrocarbons

The **polynuclear aromatic hydrocarbons (PAHs)** are a family of aromatic hydrocarbons composed of polycyclic benzenoid structures, that is, compounds composed of two or more benzene rings that are joined through a common edge (a common C—C bond). Rings joined in this way are called *fused-ring systems.*

The three most common PAHs are naphthalene ($C_{10}H_8$, moth balls), anthracene ($C_{14}H_{10}$), and phenanthrene ($C_{14}H_{10}$). The structural formulas and orbital pictures of these compounds are given in Fig. 13.13.

These compounds, as well as their higher homologs, are classified as aromatic because they behave chemically like aromatic compounds. That is, they are resistant to oxidation, reduction, and any type of addition. Generally they are more stable than one would predict based on aliphatic models (for example, cyclohexene). They are, however, less resistant than benzene to the reactions listed previously. For example, the reaction of anthracene or phenanthrene with chromic acid results in oxidation and the formation of the respective 9,10-quinones (diketones), which are cyclic and conjugated. Here reaction does not result in a complete destruction of the aromatic nature of the starting material. Benzene, on the other hand, does not react under these conditions.

The principal reaction of PAHs is electrophilic aromatic substitution. Heats of hydrogenation, heats of combustion, and NMR indicate that these compounds

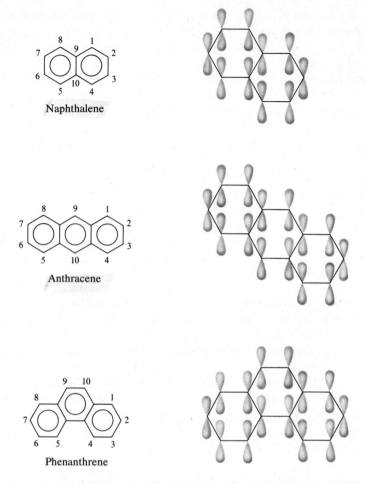

FIGURE 13.13   Structural formula and MO picture of naphthalene, anthracene, and phenanthrene.

are indeed aromatic. This is not surprising because these compounds are cyclic, contain $(4n + 2)$ π electrons, and are planar, and every carbon in the ring is sp^2 hybridized. Delocaliztion of electrons is possible.

We will encounter PAH compounds in future chapters. There is a great deal of interest in these compounds, particularly the larger homologs, because they have been shown to be *carcinogenic* (cause cancer) and are found in nature. They are probably formed primarily as products of natural and industrial combustion. These three homologs, for example, are found in cigarette smoke:

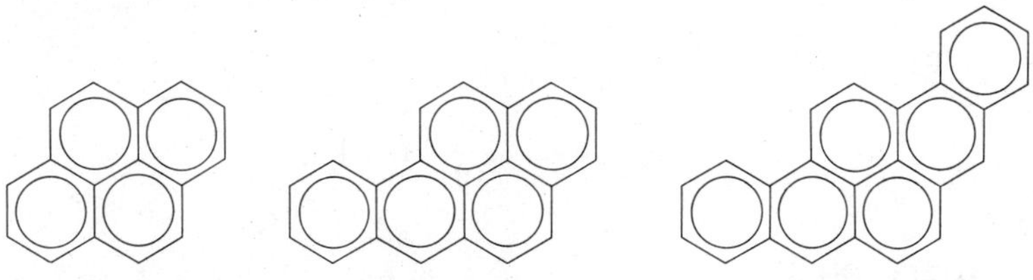

| Pyrene | Benzopyrene | Dibenzopyrene |

In contrast to benzene, the PAHs listed thus far all bear more than one type of hydrogen. If they are allowed to react under conditions of electrophilic aromatic substitution, they yield several monosubstituted isomers. In general, the position in a given PAH that is substituted can be controlled by reaction conditions. Changes in the temperature, solvent, or catalyst often yield quantitative amounts of one isomer in preference to the others. In a very general sense, naphthalene reacts preferentially at the α position (1 position), whereas anthracene and phenanthrene both react most often at the 9,10 positions. Examples are given here. Consult the Reading References for this chapter in the Appendix for additional details concerning orientation and reactivity of PAHs in electrophilic aromatic substitution reactions.

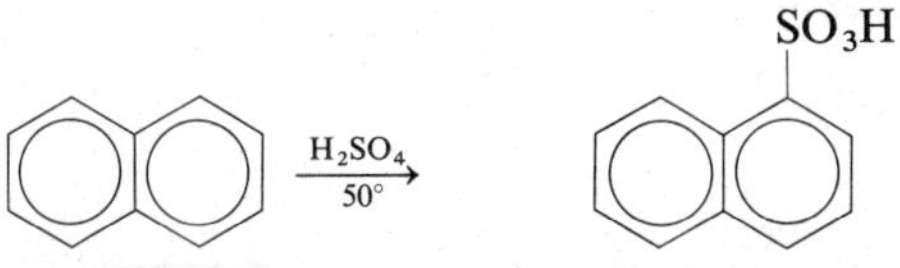

Naphthalene $\xrightarrow[50°]{H_2SO_4}$ 1- or α-Naphthalenesulfonic acid

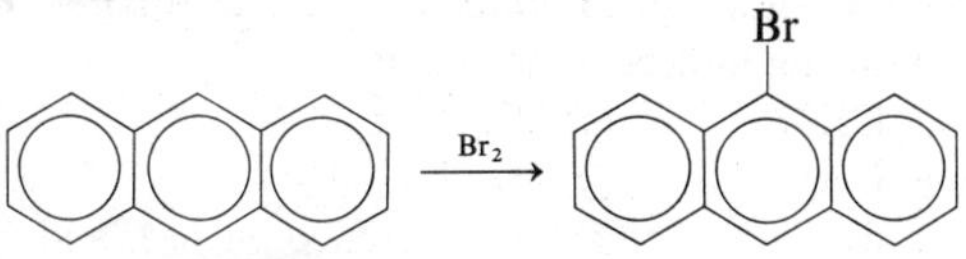

Anthracene $\xrightarrow{Br_2}$ 9-Bromoanthracene

$\xrightarrow[H_2SO_4]{HNO_3}$ 9-Nitrophenanthrene + other isomers

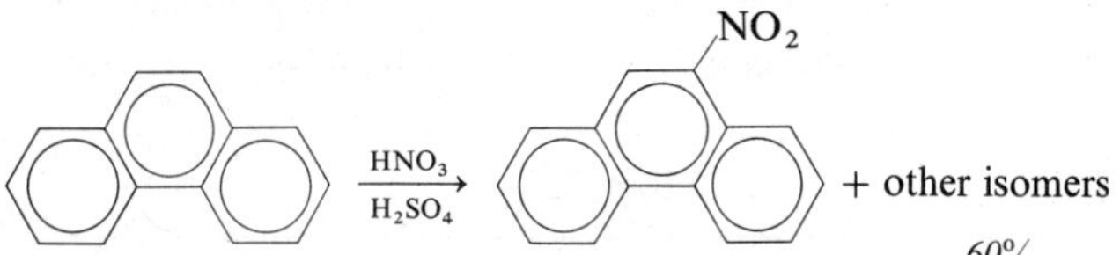

40%   60%

For more detail regarding the reactions of this class of hydrocarbons, see Sec. 19.18.

Study Questions

13.10 Name the following eight compounds.

(*a*) 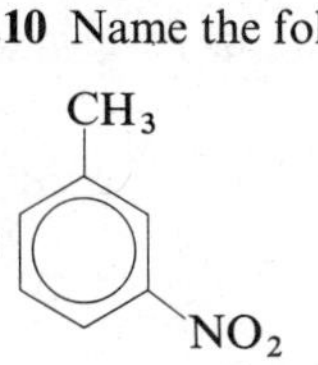   (*b*) 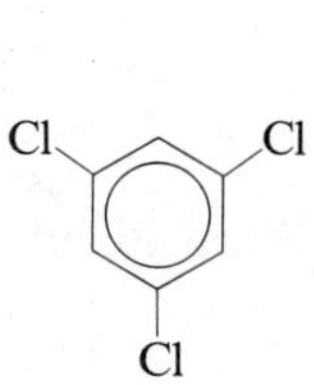   (*c*)

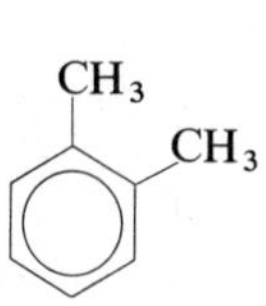

(d) $\langle\bigcirc\rangle$—CH_2—$\overset{\overset{\displaystyle CH_3}{|}}{CH}$—$CH_3$

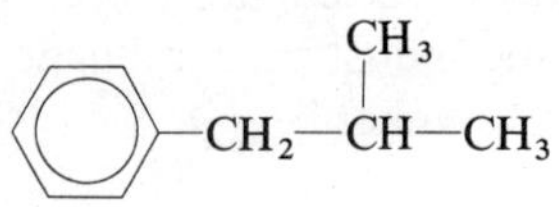

(give two names)

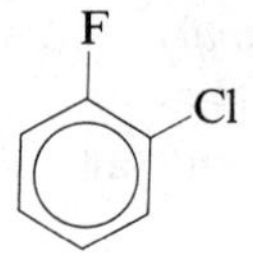

(e) [structure: benzene ring with F and Cl]

(f) [structure: benzene ring with NH_2, Cl, and CH_3]

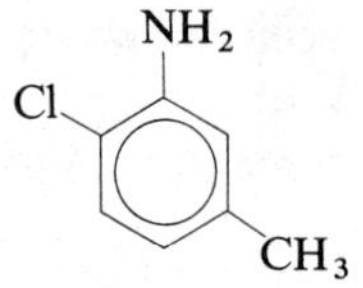

(g) $(C_6H_5)_3C$—CH_3

(h) [structure: benzene ring with CH_3—$\overset{\overset{\displaystyle CH_3}{|}}{C}$—$CH_3$, NO_2, and NO_2]

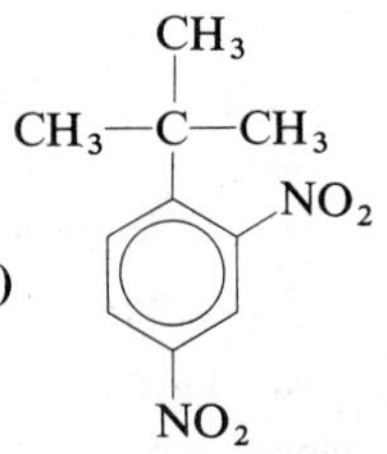

13.11 Draw structures to correspond to each of the following names.
(a) 2-chloro-2-phenylpentane
(b) m-bromotoluene
(c) α,p-dibromotoluene (*Hint:* Think carefully. What are the structures of p-bromotoluene and α-bromotoluene?)
(d) o-nitrobenzenesulfonic acid
(e) cyclohexylbenzene
(f) o-nitrophenol
(g) 3,4,5-tribromoaniline
(h) 2,4-dinitrochlorobenzene
(i) sec-butylbenzene
(j) diphenylmethane
(k) biphenyl
(l) 2,4-dibromo-5-nitrotoluene
(m) 2,4-diethylaniline

13.12 Draw the structures of all the aromatic compounds (that is, those containing the benzene ring) that have the following molecular formulas:
(a) C_7H_7Br (b) C_8H_{10} (c) $C_6H_4Br_2$
(d) $C_6H_3Br_3$ (e) $C_6H_2Br_4$ (f) $C_{10}H_{14}$

13.13 Column A lists five compounds we have encountered, and Column B contains various statements, some of which properly describe some properties of those compounds. In the blank provided by each compound in column A, place the *letters* of any or all statements in column B that properly describe or apply to that compound. Each compound has one or more correct statements.

Column A

Column B

(1) _____ Ethene
(2) _____ Ethyne
(3) _____ Cyclopropane
(4) _____ Benzene
(5) _____ Cyclohexane

A. Contains sp^2 carbon
B. Contains sp carbon
C. Contains sp^3 carbon
D. Contains only one π bond
E. Contains only two π bonds
F. Has $109\frac{1}{2}°$ bond angle
G. Has 120° bond angle
H. Has 180° bond angle
I. Undergoes free-radical substitution when heated with chlorine gas
J. Reacts with bromine in carbon tetrachloride (25 to 80°, no catalyst)
K. Reacts with $Ag(NH_3)_2^{\oplus}$

13.14 Which of the following seven structures are aromatic? Explain?

13.15 How could 1,3,5,7-cyclononatetraene be converted into an aromatic substance? Indicate the reactions that would be used, and draw the structures of the starting material and the product.

13.16 3,4-Dibromocyclobutene (1) has been prepared, and it is the potential precursor to an interesting hydrocarbon having the formula C_4H_4.

(*a*) What reagent would be used to try to convert (1) into C_4H_4? What is the structure of C_4H_4?

(*b*) When this conversion is attempted, C_4H_4 is not produced. Can you suggest a reason for this?

(1)

13.17 The following allyl carbocations can be prepared by addition of a proton to the corresponding alkadiene. However, ion (1) is *much more stable* than ion (2). Write resonance structures for these two carbocations, and discuss their role in accounting for the greater stability of ion (1).

(1)   (2)

13.18 Referring to Sec. 13.8, contrast the electronic configurations of the allyl radical, allyl cation, and allyl anion. Do the same for the cyclopropenyl and cyclopentadienyl systems.

13.19 Draw the molecular orbitals of the 1,3-butadiene molecule. (*Hint:* See Question 13.1.) Indicate all nodes in the orbitals with a dashed line.

14

Reactions of Aromatic Compounds: Electrophilic Aromatic Substitution

In Chap. 13 we defined aromaticity and listed the requirements necessary for a compound to have this character. We now look in more detail at the chemical behavior of the aromatic compounds by contrasting their reactivity and behavior with those of the familiar aliphatic compounds.

Benzene, as well as the other aromatic systems, contains a rich source of electrons—the π-electron cloud. As in the alkenes and alkynes, the electrons in the π cloud are not tightly bound and are therefore relatively accessible to attack by electron-deficient species—**electrophiles, $E^{\oplus}$**. In addition, the aromatic system has a very stable arrangement of electrons, so compounds that are aromatic tend to undergo reactions that allow them to remain aromatic. The principal reaction of aromatic compounds is **electrophilic aromatic substitution,** which does indeed preserve the aromatic integrity of the system.

When discussing the properties of aromatic systems we used the simplest member of the family, benzene. We continue this approach by presenting the various reactions of this family of hydrocarbons using benzene as our model.

Electrophilic Aromatic Substitution

14.1   Reactions of Benzene: Electrophilic Aromatic Substitution

Most reactions of benzene involve substitution rather than addition. This is attributed to the stability of the aromatic ring system, which would be destroyed by addition reactions.

The benzene ring has an electron-rich π cloud that is accessible to attack by electrophiles, $E^{\oplus}$. This is not surprising because electrophiles also attack the π bond in alkenes and alkynes in the first step of addition reactions.

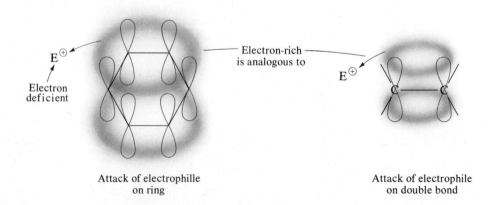

Attack of electrophille
on ring

Electron-rich
is analogous to

Attack of electrophile
on double bond

We start by summarizing the steps in the reactions that lead to substitution on benzene. These reactions are examples of *electrophilic aromatic substitution*. (This term is derived from *electrophiles* attacking an *aromatic* ring and giving rise to *substitution*.) In these reactions, an electrophile ($E^{\oplus}$) is substituted for a proton ($H^{\oplus}$).

Step 1. Generation of the electrophile: equilibrium reaction

$$E\!-\!Nu \underset{}{\overset{catalyst}{\rightleftharpoons}} E^{\oplus} + \overset{\ominus}{:}Nu$$

Step 2. Attack of the electrophile on the aromatic nucleus: slow, rate-determining step

Step 3. Loss of proton to a proton-accepting species (for example, $Nu\overset{\ominus}{:}$) to give substitution product: fast reaction

In the slow, rate-determining step (step 2), the aromatic nucleus (benzene) and the electrophile ($E^{\oplus}$) come together to form a new bond between the species. As a result of this *bimolecular* attack, electrophilic aromatic substitution is often called an S_E2 reaction, where S stands for *substitution*, E for *electrophilic*, and 2 for the *bimolecular* nature of the reaction.

In the next few sections we discuss various **monosubstitution** reactions of benzene. After monosubstitution, we consider putting a second, or even a third or fourth, substituent on the ring and see that other factors—reactivity and orientation—play an important role in electrophilic aromatic substitution reactions.

14.2   Why Substitution and Not Addition in Aromatic Compounds?

The curious student may ask this question. As we said it is not surprising that an electrophile attacks the electron-rich π cloud on the aromatic ring. But why does the anion that is often formed as a side product in the generation of the electrophile not react with the positively charged ring to give addition rather than the ring losing a proton to give substitution? The reason can be seen from energy considerations, as illustrated by the following examples:

Addition:

Product more stable
than reactant

Substitution:

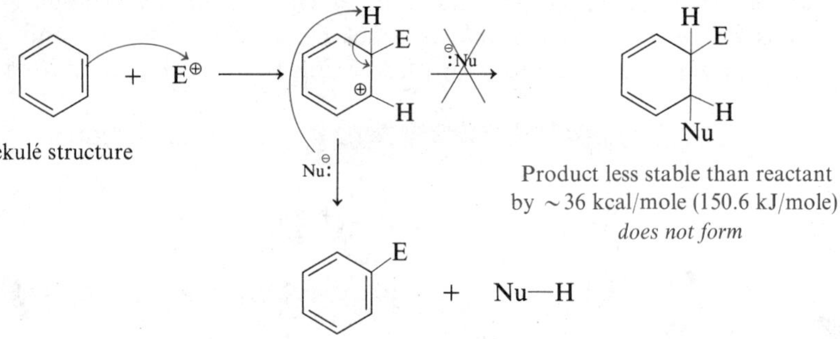

Kekulé structure

Product less stable than reactant
by ~36 kcal/mole (150.6 kJ/mole);
does not form

Product more stable
than reactant

In the preceding typical addition reaction of an alkene, the product is more stable than the reactant and thus is favored. On the other hand, in the addition reaction with benzene, the resulting product would be considerably *less* stable than the reactant because it would involve destroying the π cloud; because the delocalization energy of benzene is 36 kcal/mole (150.6 kJ/mole), destroying the π cloud would require ~36 kcal/mole (150.6 kJ/mole) of energy. This does not occur because loss of the proton regenerates the aromatic system and gives a product that is *more stable than the reactant.*

14.3 Nitration of Benzene: Monosubstitution

When allowed to react with a mixture of concentrated nitric acid and concentrated sulfuric acid at temperatures between 50 and 60°, benzene is converted smoothly into nitrobenzene. Temperatures in excess of 60° often cause disubstitution (see Sec. 14.15A).

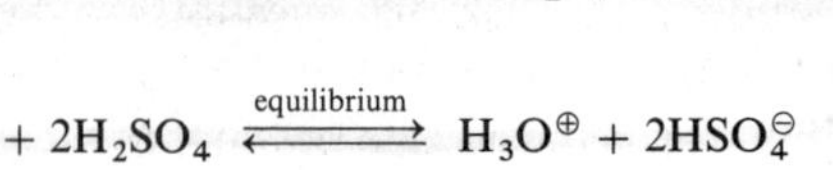

Nitrobenzene

The currently accepted mechanism for nitration involves the following three steps. These are discussed in more detail in the subsequent paragraph.

Step 1. Generation of nitronium ion. $NO_2^\oplus$ (the electrophile)

$$HO-NO_2 + 2H_2SO_4 \underset{\text{equilibrium}}{\rightleftharpoons} H_3O^\oplus + 2HSO_4^\ominus + NO_2^\oplus \boxed{E^\oplus}$$

Nitronium
ion

Step 2. Attack of nitronium ion on benzene

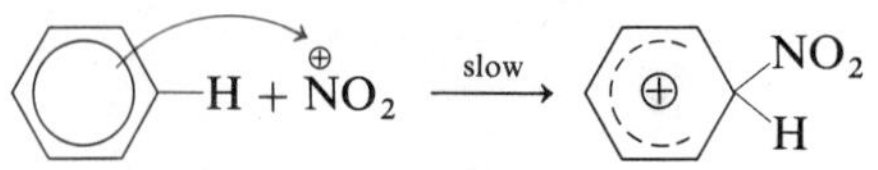

Step 3. Loss of proton to give nitrobenzene

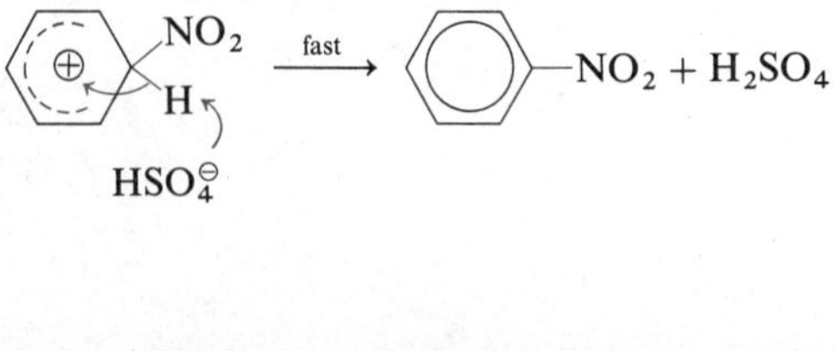

The reaction between concentrated sulfuric and nitric acids (step 1) results in the formation of the **nitronium ion**, $^{\oplus}NO_2$, which is an electron-deficient species that attacks the π cloud of the benzene ring. There is considerable evidence that the nitronium ion is indeed the nitrogen-containing species that permits the electrophilic substitution reaction to occur. For example, a salt such as nitronium perchlorate, $NO_2^{\oplus}ClO_4^{\ominus}$ (which is a stable, isolable compound), reacts with benzene to give nitrobenzene even when no sulfuric or nitric acid is present. The formation of the nitronium ion from nitric and sulfuric acids has been visualized as occurring in the following manner:

$$H^{\oplus} + H-\overset{..}{\underset{..}{O}}-NO_2 \rightleftharpoons \overset{H}{\underset{H}{\searrow}}\overset{..}{\underset{\oplus}{O}}-NO_2 \rightleftharpoons H_2\overset{..}{O}: + {}^{\oplus}NO_2$$

From   Nitric acid   Nitronium
H_2SO_4 ion

This reaction is analogous to the protonation of an alcohol, followed by loss of water to give a carbocation:

$$H^{\oplus} + H-\overset{..}{\underset{..}{O}}-R \rightleftharpoons H_2\overset{\oplus}{O}-R \rightleftharpoons H_2\overset{..}{O}: + R^{\oplus}$$

The second step (step 2) is the attack of the electron-deficient nitronium ion on the π cloud of benzene. The nitronium ion is electron deficient and "finds" a rich source of electrons on benzene. It attaches to one of the six equivalent carbon atoms via a covalent bond. The resulting species is a carbocation with a positive charge on the ring formed as a result of benzene "donating" a pair of electrons to the electrophile, $^{\oplus}NO_2$. The structure of the carbocation has been simplified somewhat because this *intermediate* ion formally contains a tetrahedral carbon atom and should be depicted as follows:

often depicted as

For simplicity of drawing, we depict it as shown in step 2.

The electronic structure of the intermediate carbocation deserves further mention. If we revert to the Kekulé-type structure (which we know to be incorrect) for illustration, we can redraw the attack of the electrophile on benzene in terms of its reaction with one of the double bonds. Three resonance structures can be drawn for this carbocation, and because they are equally stable, the carbocation is best represented by the hybrid in which the positive charge is distributed in the ring:

Kekulé structure

Best represented by

or

Most common
representation

These resonance structures do tell us, however, that electrons are withdrawn more strongly from the two positions *ortho* and the one position *para* to the entering nitro group than from the two *meta* positions.

The final step of the reaction (step 3) is the loss of a proton from the intermediate carbocation to form nitrobenzene. As the carbon-hydrogen bond is broken, the hydrogen-oxygen bond of H_2SO_4 forms. From studies of carbocations we know that one of their reactions is proton loss to give a double bond. In the case of electrophilic substitution, the driving force for proton loss is the regeneration of the very stable aromatic electronic system that was disturbed by the attack of the nitronium ion on the π cloud in step 2.

The energy changes that occur during the reaction most likely involve several transition states and intermediates as shown in Fig. 14.1.

The reaction that has the greater energy of activation (E_A) is the slow step, so the formation of the carbocation (*transition state 1*) is the slow, rate-determining step. The products are more stable than the reactants, so that nitration is an exothermic reaction. The energy-profile diagram for nitration (Fig. 14.1) is similar to that for all electrophilic aromatic substitution reactions. Refer to Sec. 5.6 for a discussion of multistep reactions and the energy changes that accompany them.

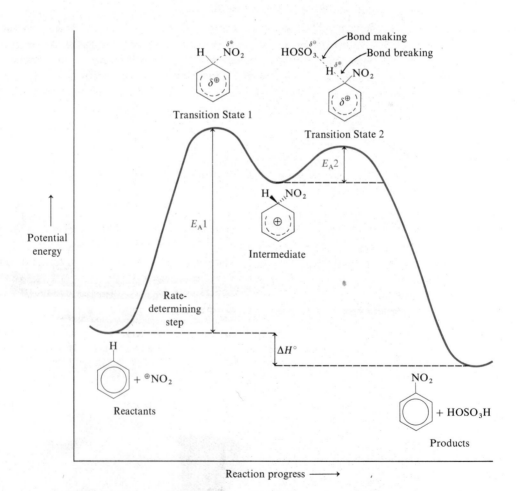

FIGURE 14.1   Energy-profile diagram of the reaction between the nitronium ion (generated from nitric and sulfuric acids) and benzene.

14.4 Halogenation of Benzene: Monosubstitution

When benzene is allowed to react with either bromine or chlorine in the absence of a catalyst, the reaction, if any, is exceedingly slow. Yet when iron metal or ferric halide is added in catalytic amounts, the reaction proceeds smoothly and forms the halobenzene; hydrogen halide is also liberated. The bromination reaction is commonly carried out by adding some "iron tacks" to the reaction mixture. The iron metal reacts with bromine to form some ferric bromide catalyst. Certain other Lewis acids such as AlX_3 also are effective catalysts.

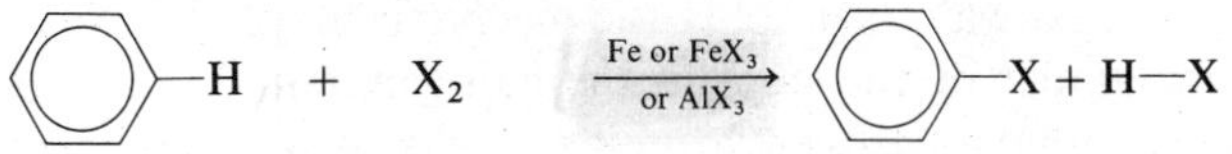

$X_2 = Cl_2, Br_2$

$X = Cl$: Chlorobenzene
$X = Br$: Bromobenzene

The currently accepted mechanism for halogenation agrees with the general mechanism for electrophilic aromatic substitution. A typical mechanism for bromination is shown here:

Step 1. Generation of electrophile, $Br^\oplus$: reversible reaction

$$FeBr_3 + Br_2 \underset{}{\overset{equilibrium}{\rightleftharpoons}} FeBr_4^\ominus + :\overset{..}{\underset{..}{Br}}{}^\oplus$$

Step 2. Attack of $Br^\oplus$ on benzene ring

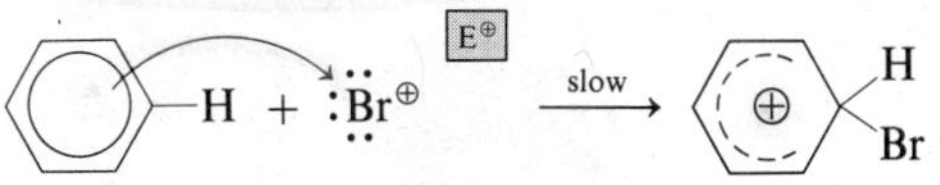

Step 3. Loss of proton to form bromobenzene

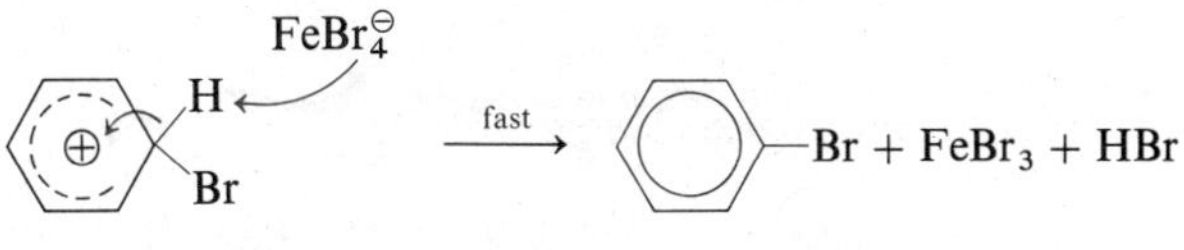

The first step of the reaction involves a new and important type of reaction. The use of a catalyst-like ferric bromide, $FeBr_3$, has been interpreted as responsible for the generation of the electrophilic species, $:\overset{..}{\underset{..}{Br}}{}^\oplus$ (or $:\overset{..}{\underset{..}{Cl}}{}^\oplus$ in the case of Cl_2 and $FeCl_3$):

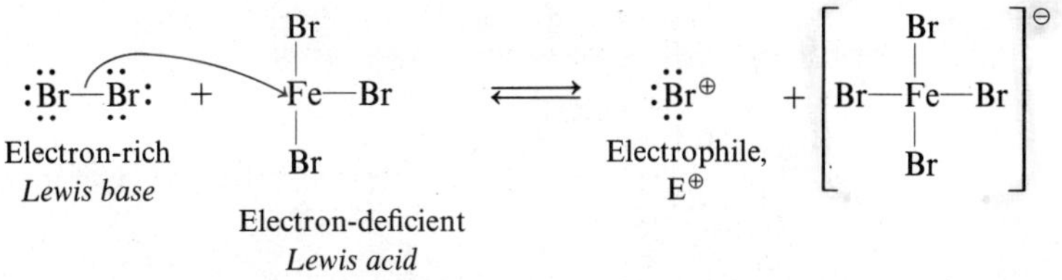

This is another acid-base reaction, this time in the Lewis sense (see Sec. 5.7). The iron atom in $FeBr_3$ is electron-deficient; it may complete its octet by accepting a pair of electrons from a bromine atom and it then acquires a formal negative charge. The net effect of this Lewis acid-Lewis base reaction is the formation of an electrophilic species, the $:\overset{..}{\underset{..}{Br}}{}^\oplus$ ion, which is electron-deficient and attacks the aromatic ring.

Kinetic data indicate that the rate of halogenation depends not only on the concentration of halogen and aromatic ring but also on the $FeBr_3$ concentration. The current view is that either an ion pair

$$:\overset{\cdot\cdot}{\underset{\cdot\cdot}{Br}}{}^{\oplus}FeBr_4^{\ominus}$$

or a molecular complex with ionic character

$$\overset{\delta\oplus}{Br}----\overset{}{Br}----\overset{\delta\ominus}{FeBr_3}$$

is involved in the reaction. In either case, though, the electron-deficient bromine atom attacks the ring. Even at that, it is doubtful that any free $:\overset{\cdot\cdot}{\underset{\cdot\cdot}{Br}}{}^{\oplus}$ (or $:\overset{\cdot\cdot}{\underset{\cdot\cdot}{Cl}}{}^{\oplus}$) exists. We simply write it that way to simplify the mechanism.

The remainder of the steps in the halogenation reaction are straightforward. Note, however, that hydrogen bromide is liberated in the final step and that the catalyst, ferric bromide, is regenerated; it can rereact with bromine and benzene, so in this sense it is a catalyst.

The same steps and comments apply for chlorination.

14.5   Sulfonation of Benzene: Monosubstitution

The sulfonation of benzene is often carried out with *fuming* sulfuric acid (sulfuric acid containing sulfur trioxide, SO_3), although concentrated sulfuric acid ($95\%\ H_2SO_4$, $5\%\ H_2O$) is used for "activated" rings (the meaning of the term *activated* rings is explained in Sec. 14.9).

$$\text{C}_6\text{H}_5-H \xrightarrow[\substack{(H_2SO_4 + SO_3) \\ 25°}]{\text{fuming } H_2SO_4} \text{C}_6\text{H}_5-SO_3^{\ominus} + H^{\oplus}$$

The mechanism for the sulfonation of benzene (and other aromatic compounds) is as follows:

Step 1. Generation of the electrophile, SO_3

$$2H_2SO_4 \xrightleftharpoons{\text{equilibrium}} H_3O^{\oplus} + HSO_4^{\ominus} + SO_3 \boxed{E^{\oplus}}$$

Step 2. Attack of SO_3 on benzene ring

Step 3. Loss of proton to form the benzenesulfonate ion

There seems little doubt that the electrophile is the neutral sulfur trioxide, SO_3, molecule:

With fuming sulfuric acid as the sulfonating agent, SO_3 is present, but even with concentrated sulfuric acid, the electrophile can be produced as shown in step 1. Chemical kinetics show that the rate of sulfonation depends on the concentrations of benzene (or aromatic ring) and sulfur trioxide, and not sulfuric acid.

In sulfonation we start with two neutral molecules, benzene and SO_3, and after attack (step 2), the aromatic ring is positively charged and the $-SO_3^{\ominus}$ group is negatively charged. In this sense, SO_3 has served as a Lewis acid (an electron-pair acceptor) and benzene as a Lewis base (an electron-pair donor). (Compare this reaction with attack by $\overset{..}{\underset{..}{X}}{}^{\oplus}$ or $^{\oplus}NO_2$ in the halogenation or nitration of benzene.)

The final step of sulfonation is proton loss, which probably involves the transfer of a proton to a proton acceptor (Lowry-Brønsted base) such as bisulfate ion, $HSO_4^{\ominus}$. The product is shown as the benzenesulfonate ion because benzenesulfonic acid is strong (comparable to sulfuric acid) and highly dissociated.

If the reaction is carried out in the presence of a large excess of *deuteriosulfuric acid* at approximately 100°, the formation of the deuterated product takes place along with the sulfonation process:

Deuteriobenzene   Hexadeuteriobenzene

The products probably result from the direct attack on the ring by $D^{\oplus}$ followed by loss of $H^{\oplus}$. (Draw the mechanism for this substitution reaction.)

14.6   Friedel-Crafts Alkylation: Monosubstitution

Named after its discoverers, the Friedel-Crafts reaction is important for introducing alkyl groups onto an aromatic ring. It involves an *alkyl* halide and a Lewis acid catalyst, usually aluminum trichloride, $AlCl_3$:

R = alkyl   Alkylbenzene

*Fails with aryl
and vinyl halides*

Other catalysts, for example, FeX_3, BF_3, and HF, can be used.

The Friedel-Crafts alkylation reaction appears to follow the general pattern for electrophilic aromatic substitution:

Step 1. Generation of the electrophile, $R^{\oplus}$

Lewis
acid

Step 2. Attack of $R^\oplus$ on benzene ring

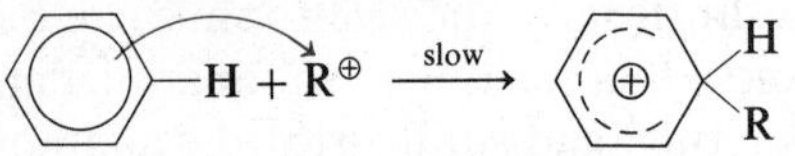

Step 3. Loss of proton to form the alkyl benzene

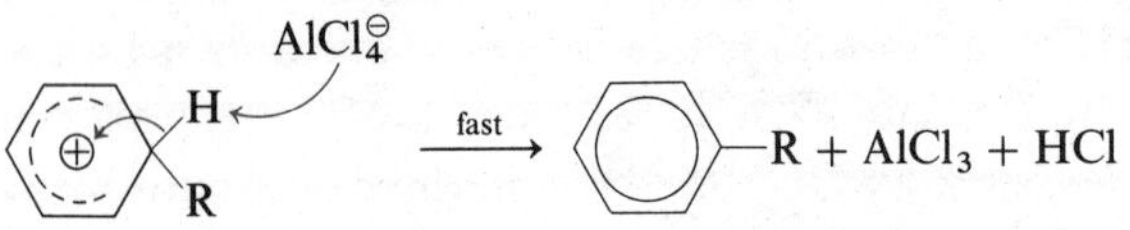

The formation of the electrophile (step 1) has been oversimplified to some extent because evidence suggests that the alkyl halide (for example, an alkyl chloride) reacts with aluminum chloride to form a complex or ion pair of the type shown here:

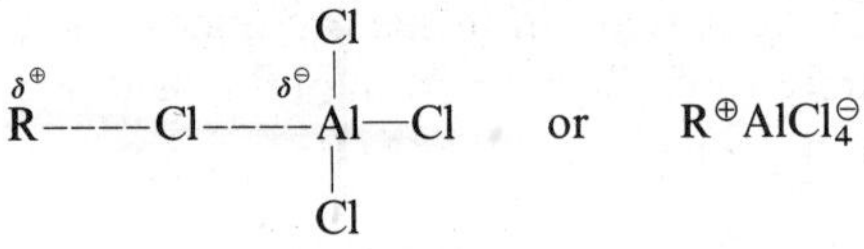

$$\overset{\delta\oplus}{R}----Cl----\overset{\overset{\displaystyle Cl}{|}}{\underset{\underset{\displaystyle Cl}{|}}{Al}}\overset{\delta\ominus}{}—Cl \qquad or \qquad R^{\oplus}AlCl_4^{\ominus}$$

Undissociated   Ion pair
complex

Aluminum chloride polarizes the carbon-halogen bond in the alkyl chloride and generates a partial positive charge on carbon; this entire complex then attacks the benzene ring, but the more positive part—the partially positive carbon atom— actually attacks the electron-rich π cloud.

The other catalysts are also Lewis acids (electron-pair acceptors). Hydrogen fluoride is a rather special case, and it helps to polarize the halogen on the alkyl halide by forming a **hydrogen bond** between the two molecules:

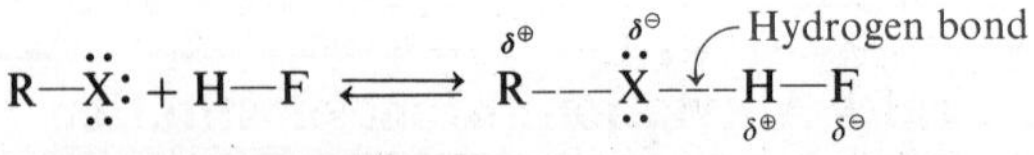

$$R—\overset{..}{\underset{..}{X}}: + H—F \rightleftharpoons \overset{\delta\oplus}{R}---\overset{\delta\ominus}{\underset{..}{X}}---\overset{\delta\oplus}{H}—\overset{\delta\ominus}{F}$$

Hydrogen bond

We now have a new method for producing carbocations: *the reaction between an alkyl halide and a Lewis acid, such as AlCl$_3$*. We also have a new reaction of carbocations: *they may attack and alkylate the electron-rich aromatic ring.*

If indeed carbocations are involved in this reaction, is there other evidence to support their intervention? The answer is unquestionably yes! Consider, for example, the following alkylation reaction in which rearrangement occurs. The reaction between neopentyl chloride and benzene with AlCl$_3$ as the catalyst gives 2-methyl-2-phenylbutane:

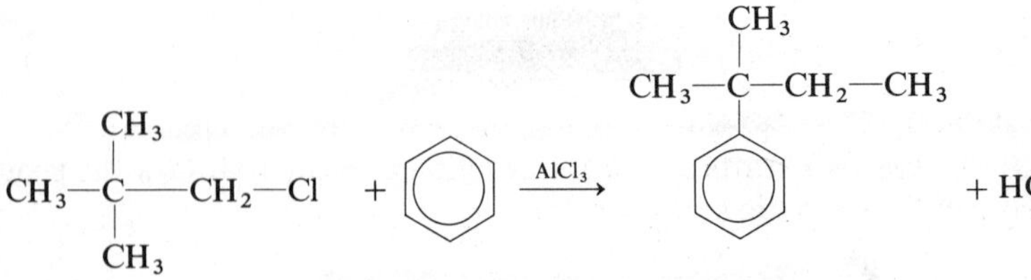

2,2-Dimethyl-1-chloropropane
(neopentyl chloride)

2-Methyl-2-phenylbutane
Major product

In this reaction a primary carbocation is formed first and then rearranges by methyl migration to form a more stable tertiary carbocation. This may occur in two steps (through a *free* 1° carbocation) or in one step (directly through the *carbocation-like* transition state). The final 3° carbocation then attacks the benzene ring and, on loss of a proton, gives product.

Partial mechanism:

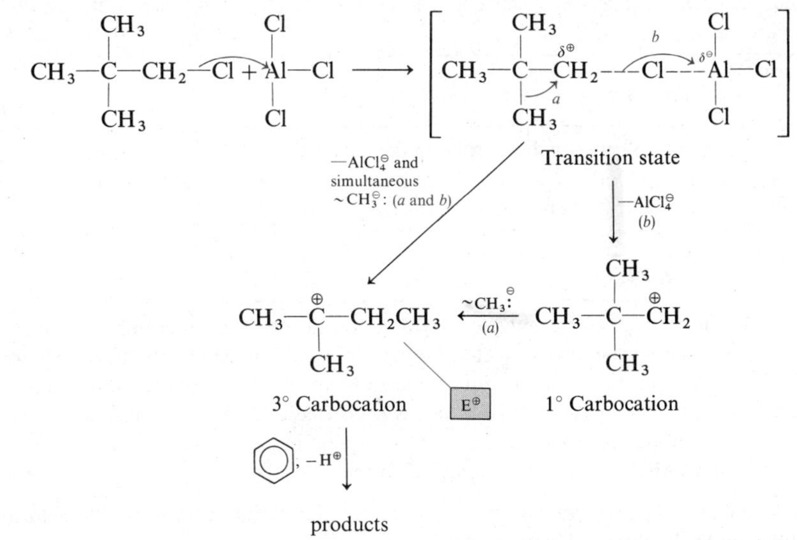

$$CH_3-\overset{\overset{\displaystyle CH_3}{|}}{\underset{\underset{\displaystyle CH_3}{|}}{C}}-CH_2-Cl + \overset{\overset{\displaystyle Cl}{|}}{\underset{\underset{\displaystyle Cl}{|}}{Al}}-Cl \longrightarrow \left[CH_3-\overset{\overset{\displaystyle CH_3}{|}}{\underset{\underset{\displaystyle CH_3}{|}}{C}}\overset{\delta\oplus}{\underset{a}{-}}CH_2\cdots\overset{b}{Cl}\cdots\overset{\delta\ominus}{\underset{\underset{\displaystyle Cl}{|}}{\overset{\overset{\displaystyle Cl}{|}}{Al}}}-Cl \right]$$

$-AlCl_4^{\ominus}$ and
simultaneous
$\sim CH_3^{\ominus}$: (a and b)

Transition state

$-AlCl_4^{\ominus}$
(b)

$$CH_3-\overset{\overset{\displaystyle\oplus}{|}}{\underset{\underset{\displaystyle CH_3}{|}}{C}}-CH_2CH_3 \quad\xleftarrow[(a)]{\sim CH_3^{\ominus}:}\quad CH_3-\overset{\overset{\displaystyle CH_3}{|}}{\underset{\underset{\displaystyle CH_3}{|}}{C}}-\overset{\oplus}{C}H_2$$

3° Carbocation $\boxed{E^{\oplus}}$ 1° Carbocation

$\bigcirc$, $-H^{\oplus}$ ↓

products

The picture of the Friedel-Crafts alkylation reaction is not completely clear. It has been the subject of many investigations and is still being studied. To provide some idea of how seemingly subtle changes can affect the reaction, a ferric chloride catalyst often gives alkylation *without rearrangement*, whereas the same reaction using $AlCl_3$ results in extensive rearrangement. Changing reaction conditions (solvent, temperature, concentration, and catalyst) often produces varying amounts of rearrangement. Also, attachment of an alkyl group to the ring makes the ring more susceptible to further reaction to give di- and polysubstituted products (see Sec. 14.15B).

Friedel-Crafts Alkylation; Rearrangements

As a further complication, rearrangements of the side chain may also occur after it is attached to the ring. (Evidence indicates that the alkylation reaction is not reversed; that is, the sequence of dealkylation, rearrangement of the carbocation, and realkylation does not occur.) There is direct rearrangement of the alkyl benzene after it is formed. For example, 2-methyl-2-phenylbutane (formed in a previous example) rearranges to 2-methyl-3-phenylbutane, presumably by the following mechanism:

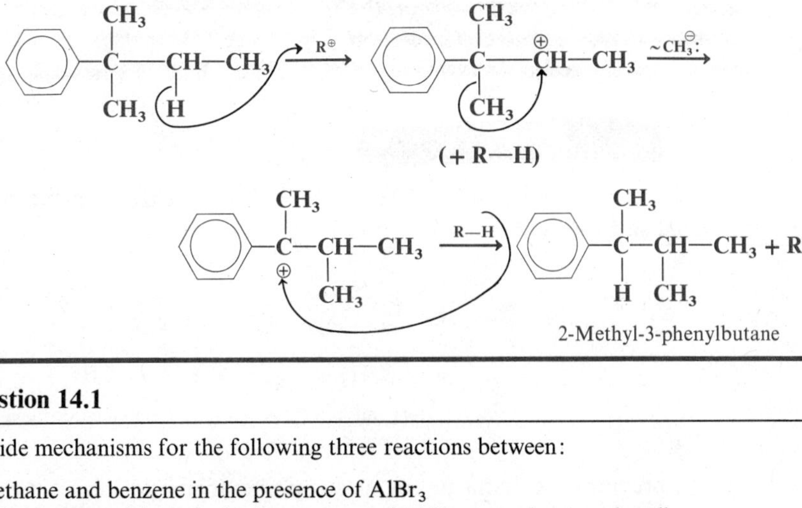

$$\bigcirc-\overset{\overset{\displaystyle CH_3}{|}}{\underset{\underset{\displaystyle CH_3}{|}}{C}}-CH-CH_3 \xrightarrow{R^{\oplus}} \bigcirc-\overset{\overset{\displaystyle CH_3}{|}}{\underset{\underset{\displaystyle CH_3}{|}}{C}}-\overset{\oplus}{C}H-CH_3 \xrightarrow{\sim CH_3^{\ominus}:}$$

$(+ R-H)$

$$\bigcirc-\overset{\overset{\displaystyle CH_3}{|}}{\underset{\oplus}{C}}-\overset{}{\underset{\underset{\displaystyle CH_3}{|}}{CH}}-CH_3 \xrightarrow{R-H} \bigcirc-\overset{\overset{\displaystyle CH_3}{|}}{\underset{\underset{\displaystyle H}{|}}{C}}-\overset{}{\underset{\underset{\displaystyle CH_3}{|}}{CH}}-CH_3 + R^{\oplus}$$

2-Methyl-3-phenylbutane

Question 14.1

Provide mechanisms for the following three reactions between:

(*a*) bromoethane and benzene in the presence of $AlBr_3$
(*b*) isobutyl chloride and benzene in the presence of $AlCl_3$ to produce *tert*-butylbenzene

(*c*) *n*-butyl chloride and benzene in the presence of AlCl₃ to give a mixture of two alkyl benzenes, *n*-butylbenzene and *sec*-butylbenzene (the ratio of products is 34:66, but you do not have to explain it).

Question 14.2

The reaction between dichloromethane, CH_2Cl_2, and an *excess* of benzene in the presence of aluminum trichloride produces predominantly diphenylmethane, $C_6H_5—CH_2—C_6H_5$. Provide a reasonable sequence of mechanisms to explain this.

Question 14.3

An interesting experiment sheds light on the nature of the intermediate that forms when an alkyl halide comes in contact with aluminum trichloride. The Friedel-Crafts alkylation of benzene by *tert*-butyl chloride is carried out in the presence of AlCl₃ that contains trace amounts of radioactive chlorine, ^{36}Cl. The HCl that is produced and the AlCl₃ remaining in the reaction mixture are radioactive to the extent that complete interchange of ^{36}Cl has occurred; that is, both HCl and AlCl₃ contain radioactive chlorine.

(*a*) What can you conclude about the possibility of chlorine from *tert*-butyl chloride becoming equivalent to the three chlorines in AlCl₃?
(*b*) Does this result support or refute the formation of the *tert*-butyl cation and $AlCl_4^{\ominus}$? If it refutes it, what alternative reaction do you envision between *tert*-butyl chloride and AlCl₃ containing ^{36}Cl?

Question 14.4

Based on your knowledge about the structure of carbocations, what would you expect to observe when an optically active alkyl halide, such as *sec*-butyl chloride, reacts with benzene in the presence of AlCl₃? Provide a mechanism, using three-dimensional projection formulas, to explain your prediction.

14.7   Other Methods of Alkylating Benzene

The Friedel-Crafts reaction involves the attack of either a free carbocation or a partially positive carbon atom on the electron-rich π cloud of the benzene ring. The reaction evolves about carbocations. There are other ways to form carbocations, however, so there should be other types of reactions that can be used for alkylation.

A. Alkylation by Alcohols

The reaction between *tert*-butyl alcohol and benzene in the presence of concentrated sulfuric acid yields *tert*-butylbenzene:

tert-Butyl
alcohol   *tert*-Butylbenzene

In previous mechanisms for electrophilic aromatic substitution, the first step was a single equilibrium leading to the generation of the electrophile. Here, however, there are several equilibrium steps:

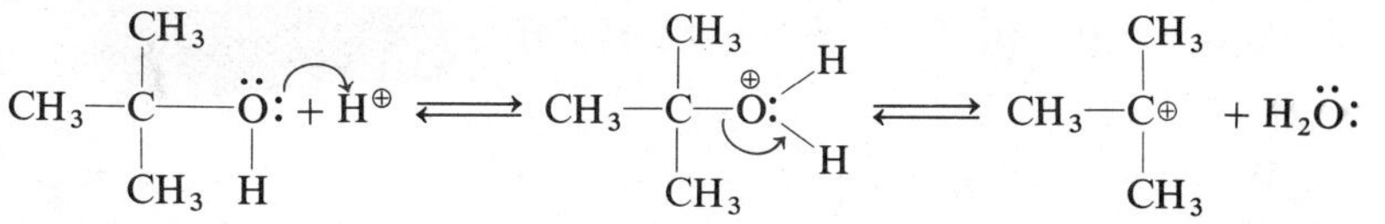

The *tert*-butyl cation, which is the electrophile, then reacts with benzene to form *tert*-butylbenzene:

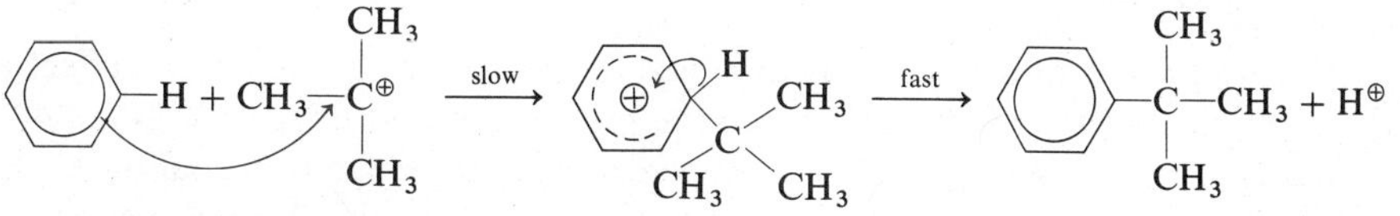

B. Alkylation by Alkenes

When an alkene is allowed to come in contact with benzene in the presence of an acid catalyst, alkylation occurs. For example, the reaction between propene and benzene in the presence of sulfuric acid produces isopropylbenzene:

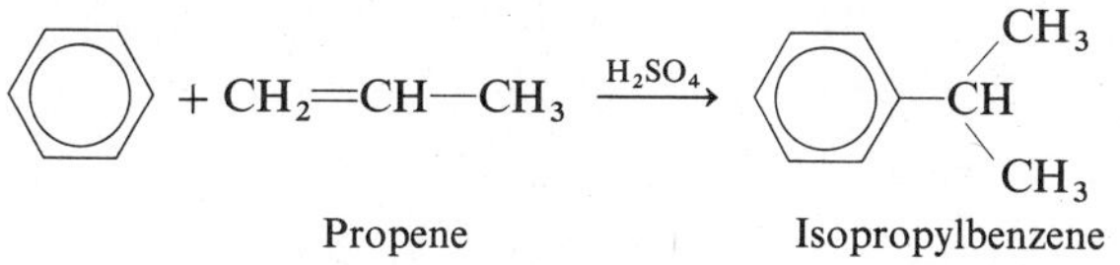

Propene
(propylene)

Isopropylbenzene
(cumene)

The carbocation is generated by the addition of a proton to propene, which in turn reacts with benzene in the usual manner:

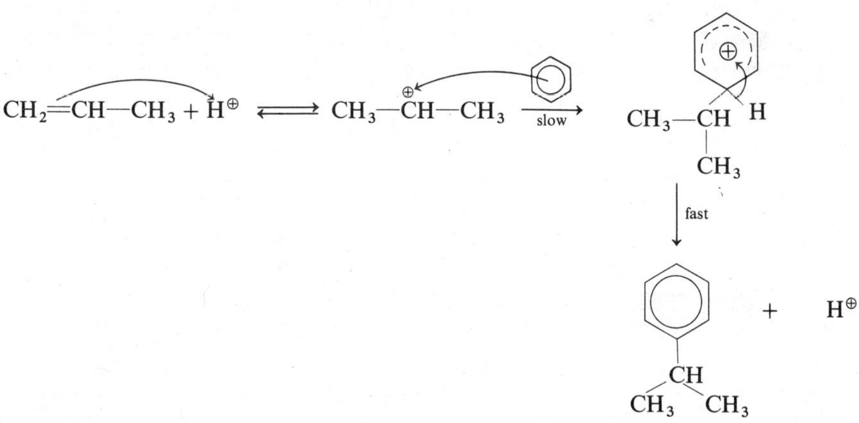

Question 14.5

The reaction between *n*-butyl chloride and benzene in the presence of $AlCl_3$ gives a mixture of *n*-butylbenzene and *sec*-butylbenzene. However, the reaction between *n*-butyl alcohol ($CH_3CH_2CH_2CH_2OH$) and benzene in the presence of concentrated sulfuric acid gives almost entirely *sec*-butylbenzene. From these results, what can you conclude regarding the nature of the intermediates in these two reactions?

Question 14.6

Provide a complete mechanism to account for the production of 2-methyl-2-phenylbutane from benzene and 3-methyl-1-butanol, $(CH_3)_2CHCH_2CH_2$—OH, in the presence of concentrated sulfuric acid.

14.8   Introduction of a Second Group into a Monosubstituted Benzene Derivative: Disubstitution

We considered several common reactions that introduce a single substituent onto the benzene molecule. Suppose, however, we have a monosubstituted benzene derivative and want to introduce a second substituent onto the ring. Where does it go? What factors govern disubstitution (or even polysubstitution, which we discuss in Sec. 14.15)?

If the substituent already attached to the aromatic ring had no effect at all on where an entering group goes, then the product should have 40% *ortho*, 40% *meta*, and 20% *para* disubstitution. These percentages come from probabilities; there are two equivalent *ortho* positions, two equivalent *meta* positions, and one *para* position where an entering group can attach (that is, the ratio $o:m:p = 2:2:1$ or $40:40:20$).

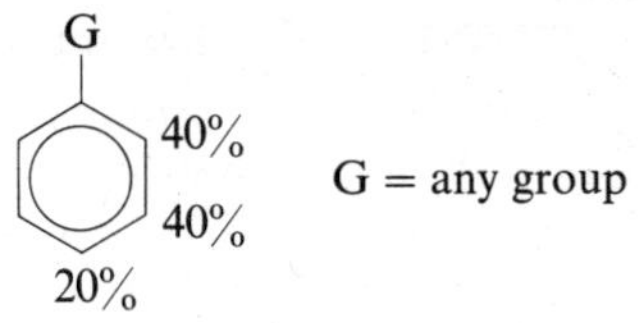

Predicted positions and
amounts of disubstitution
based on probabilities

This assumes that all *ortho*, *meta*, and *para* positions are equally reactive.

As the following examples illustrate, disubstitution does not follow strict probabilities, and different substituents orient an incoming group to different positions. The nitration of toluene yields mostly *ortho*- and *para*-nitrotoluene, whereas the same reaction with nitrobenzene produces mostly *meta*-dinitrobenzene:

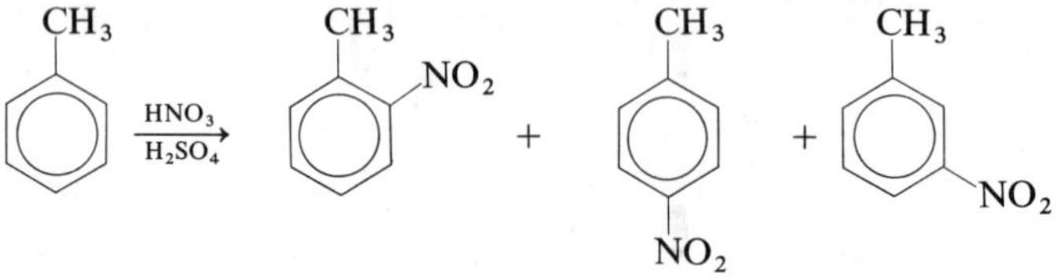

| Toluene | *o*-Nitrotoluene *57%* | *p*-Nitrotoluene *40%* | *m*-Nitrotoluene *3%* |

Major products   *Minor product*

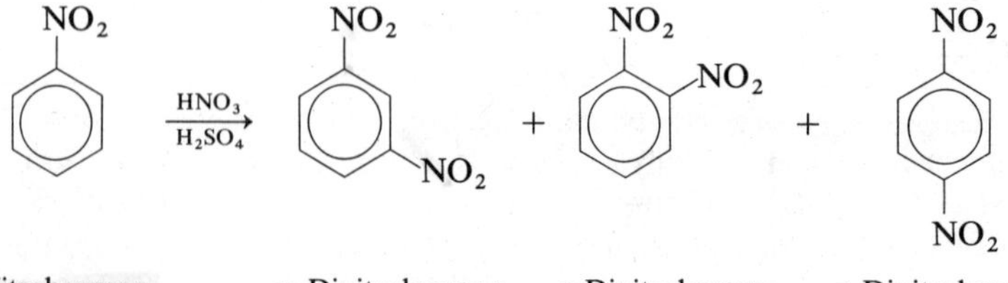

| Nitrobenzene | *m*-Dinitrobenzene *93%* | *o*-Dinitrobenzene *6%* | *p*-Dinitrobenzene *1%* |

Major product   *Minor products*

Thus, the methyl group differs from the nitro group in the way it orients an incoming nitro group. These examples do not show the whole picture because there is a tremendous difference in the rates at which the second nitro group enters toluene and nitrobenzene. Toluene is 2.4×10^8 times more reactive than nitrobenzene!

There are two important factors to consider: (1) the **orientation** of an incoming group and (2) the **reactivity** of the monosubstituted benzene compound toward an incoming group. *Both orientation and reactivity depend on the group that is initially present in the molecule*, and for the most part they depend much less on the nature of the incoming group, which is an electrophile.

A group that orients an incoming group mostly into the *ortho* and *para* positions is called an ***ortho, para* director,** whereas one that gives mostly *meta* substitution is called a ***meta* director.**

We start our discussion about disubstitution reactions with the data in Table 14.1, which indicate the orientation and reactivity in the nitration of certain monosubstituted benzene compounds.

TABLE 14.1 Orientation and Reactivity for Nitration of Monosubstituted Benzene Compounds

Substituent, G	Relative Reactivity	% (G, ortho-NO₂)	% (G, meta-NO₂)	% (G, para-NO₂)	Ratio ortho + para/ meta
—OH	Very fast	55	Trace	45	100/0
—NH—C(=O)—CH₃	Fast	19	1	80	99/1
—CH₃	24	57	3	40	97/3
—C(CH₃)₃	16	12	8	80	92/8
—CH₂Cl	0.3	32	16	52	84/16
—F	0.03	12	Trace	88	100/0
—Cl	0.03	30	1	69	99/1
—Br	0.03	37	1	62	99/1
—I		38	2	60	98/2
—H (Benzene)	1.0	Standard for relative reactivities			
—NO₂	$\sim 10^{-7}$	6	93	1	7/93
—C(=O)—OC₂H₅	3×10^{-4}	28	68	4	32/68
—N(CH₃)₃⁺	Slow	0	89	11	11/89
—C(=O)—OH	Slow	19	80	1	20/80
—SO₃H	Slow	21	72	7	28/72
—CF₃	Slow	0	100	0	0/100

The reactivities are relative to the rate of nitration of benzene, which is arbitrarily set at 1.0 and is the standard of comparison. Relative reactivity may be defined as:

$$\text{Relative reactivity} = \frac{\text{rate of substitution on } C_6H_5\text{—G}}{\text{rate of substitution on } C_6H_6 \text{ (benzene)}}$$

As the data indicate, there is great variation in the reactivities of these compounds. Contrast phenol and nitrobenzene. Although the numbers are not available for nitration, the bromination of C_6H_5—OH is $\sim 10^{16}$ times faster than the bromination of C_6H_5—NO_2, and these two groups represent the extremes in reactivity in electrophilic substitution reactions.

The orientation data for the various substituents are also instructive. The top half of Table 14.1 (G = —OH to —I) lists substituents that give predominantly *ortho,para* substitution and thus are *ortho,para* directors. The substituents below G = —H give predominantly *meta* substitution and are *meta* directors.

The table does not include all substituents commonly encountered in organic chemistry because all the nitration data are not available. However, from various electrophilic substitution reactions (which include a fairly large number of reactions), the general trends are given here. The following list contains the (1) predominant *orientation* that each group imparts to an incoming electrophile and (2) *reactivity* that group imparts to an aromatic ring *relative to benzene*.

Orientation-reactivity properties of various substituents in electrophilic aromatic substitution:

Decreasing order of reactivity:

$$\text{—NH}_2, \text{—NHR}, \text{—NR}_2 \qquad \text{—NH} \overset{\overset{\displaystyle O}{\|}}{-} \text{C—CH}_3, \qquad\qquad \text{—C}_6\text{H}_5,$$

—OH	>	—OR	>	—R (—CH₃, —C₂H₅, and so on)	>

| Strongly activating | | Moderately activating | | Weakly activating | |

ortho, para Directors

$$\text{—NO}_2, \text{—}\overset{\oplus}{\text{N}}\text{R}_3, \text{—C}\equiv\text{N},$$

Benzene	>	—X (—F, —Cl, —Br, —I)	>	$-\overset{\overset{\displaystyle O}{\|}}{C}\text{—OH}, -\overset{\overset{\displaystyle O}{\|}}{C}\text{—H}, -\overset{\overset{\displaystyle O}{\|}}{C}\text{—R},$

$$\text{—SO}_3\text{H}, \text{—CF}_3$$

Benzene	Weakly deactivating	Strongly deactivating
Standard of comparison	*ortho, para* Directors	*meta* Directors

This classification of these various substituents should be learned for future use. Although some functional groups have not been studied yet, this listing is provided for completeness at this point. The listing indicates the reactivity of the substituents as *activating or deactivating relative to benzene*. Note that the halogens are weakly deactivating and yet are *ortho, para* directors. All the *meta* directors are strongly deactivating.

The *meta* directors deserve special mention because in each the atom attached directly to the aromatic ring either is partially positive or bears a full positive *formal charge*. To illustrate this, the *meta* directors are redrawn as follows and the polarities of some are indicated by partial charges. Although we have not studied many of these functional groups, knowing the electronegativities allows the polarities of various bonds to be indicated. For example, in the carbon-oxygen double bond, the polarity is $\overset{\delta\oplus}{C}=\overset{\cdot\cdot}{\overset{\delta\ominus}{O}}$ because oxygen is more electronegative than carbon. In other groups the central atom bears a full *formal* positive charge ($+1$ in the nitro group and $+1$ in the amino group). Some groups have no *net* charge and yet have a formal positive charge because the formal charges are internal. For example, in the nitro group, there is a $+1$ charge on nitrogen and a -1 charge on oxygen, so the nitro group as a unit has no charge. On the other hand, $-\overset{\oplus}{N}R_3$ is not neutral because it has a $+1$ charge; when it is present on the aromatic ring, some anion must be present to account for neutrality.

Groups bearing full positive formal charge:

$$\underset{(-NO_2)}{\overset{:O:}{\underset{\oplus}{-N}}\diagdown_{\overset{\cdot\cdot}{\underset{\cdot\cdot}{O}}:^{\ominus}}} \qquad \underset{(-\overset{\oplus}{N}R_3)}{\overset{R}{\underset{R}{\overset{|}{-\overset{\oplus}{N}-R}}}}$$

Groups bearing partial positive charge:

$$\overset{\delta\oplus}{-C}\overset{\cdot\cdot}{\equiv}\overset{\delta\ominus}{N} \qquad \overset{:O:^{\delta\ominus}}{\underset{\delta\oplus}{-C}}-\overset{\cdot\cdot}{\underset{\cdot\cdot}{O}}-H \qquad \overset{:O:^{\delta\ominus}}{\underset{\delta\oplus}{-C}}-H \qquad \overset{:O:^{\delta\ominus}}{\underset{\delta\oplus}{-C}}-R \qquad \underset{(-CF_3)}{\overset{F^{\delta\ominus}}{\underset{F_{\delta\ominus}}{\overset{\uparrow}{\underset{\delta\oplus}{C}\rightarrow F^{\delta\ominus}}}}} \qquad \underset{(-SO_3H)}{\overset{:O:^{\delta\ominus}}{\underset{:O:_{\delta\ominus}}{-S}}\overset{\delta\ominus}{\underset{\parallel}{-}}\overset{\cdot\cdot}{O}-H}$$

Table 14.1 shows that the substituent already present on benzene dictates the position of an entering group, with the orientation being either *ortho* and *para* or *meta*. For any given substituent, however, the product distribution varies depending on the nature of the electrophile. Table 14.2 shows the *ortho/para* ratio for different entering groups on different monosubstituted benzenes. When the sum of the *ortho* and *para* products is *not* equal to 100, the difference represents the amount of *meta* substitution. Note that these substituents are predominantly *ortho, para* directors.

TABLE 14.2   Dependence of Disubstitution Product Distribution on Substituent on Ring

Compound	Substituent in C_6H_5—G	Ratio of *ortho/para* isomers when following groups are introduced into compounds			
		—Cl	—Br	—NO$_2$	—SO$_3$H
Toluene	—CH$_3$		40:60	57:40	32:62
Chlorobenzene	—Cl	39:55	11:86	30:69	0:100
Bromobenzene	—Br	45:53	13:85	37:62	0:100
Phenol	—OH	50:50	10:90	55:45	

Several other factors, such as reaction temperature and steric hindrance, play a role in dictating the *ortho/para* ratio. Without going into great detail, we consider the effect that the size (bulk) of the substituent already on the ring has on product distribution. It is reasonable that larger substituents "shield" to some extent the *ortho* position from attack by an electrophile, and when this occurs, more *para* substitution takes place. In the following sequence of alkyl benzenes, the alkyl groups increase in size, and the *ortho/para* ratios are indicated for nitration:

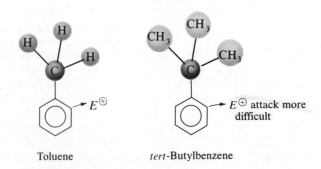

	Toluene	Ethylbenzene	Isopropylbenzene	*tert*-Butylbenzene
Ortho/para ratio in nitration:	*1.57*	*0.93*	*0.48*	*0.22*

In this sequence, as the number of methyl groups increases, the amount of steric hindrance at the *ortho* position increases. The E_A for this particular pathway is thereby raised because the reaction is more difficult. Molecular model drawings of the two extremes, toluene and *tert*-butylbenzene, are shown in Fig. 14.2 and demonstrate the increased steric hindrance of the *ortho* position toward attack by an electrophile.

Steric factors are even involved in the attacking electrophile. For example, the chlorination of chlorobenzene ($:\ddot{C}l^{\oplus}$ is the electrophile) gives a mixture of dichlorobenzenes in which the *ortho/para* ratio is about 0.7. On the other hand, nitration of chlorobenzene (the electrophile is the more bulky nitronium ion, $^{\oplus}NO_2$) gives an *ortho/para* ratio of about 0.4.

Some facts in this section were presented to give an idea of the complexity of disubstitution. The most important features, however, are the orientation and reactivity of the various substituents commonly encountered in organic chemistry. The trends are clear, but the precise details of predicting the *relative* amounts of *ortho*, *meta*, and *para* substitution are not available, and there seems little reason to memorize a lot of numbers. It is important to understand the theory behind orientation. We now consider *why* one group is an *ortho*, *para* director and another is a *meta* director, and *why* one group activates the ring toward further substitution whereas another deactivates it.

FIGURE 14.2 **Use of drawings of molecular models to illustrate steric hindrance, which makes attack by electrophile E$^+$ more difficult.**

14.9   General Factors in Reactivity

There are two ways to look at the factors that govern the reactivity of aromatic rings toward an electrophile. The first involves the electron concentration in the aromatic ring in its ground state before attack by the electrophile. The greater the electron density (electron concentration), the more strongly the molecule attracts an electrophile. Also, the polarizability of the electron cloud (see Sec. 9.7) plays an important role. The more easily the electron cloud is polarized, the more easily reaction with the electrophile occurs.

The second approach regarding reactivity comes from comparing the relative stabilities of the carbocation intermediates formed in electrophilic aromatic substitution.

How are these factors influenced by groups attached to the aromatic ring? In a general sense, groups attached to the aromatic ring have the effect of either donating or withdrawing electrons from the ring. The effects are summarized here:

Ground state:

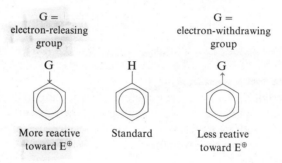

Intermediate (*para* attack shown):

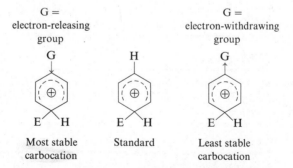

The electron-releasing groups increase the electron density in the aromatic ring in the ground state, making the ring more reactive toward electrophilic attack. Also, once the electrophile has attacked, the electron-releasing group stabilizes the resulting intermediate carbocation. Transition-state theory states that the more stable the intermediate, the more rapidly it is formed (Sec. 5.8). The most stable carbocation is the one to which electrons are donated by the G group. The least stable is the one from which electrons are withdrawn by the G group.

Hence, the relative order of carbocation stability and the relative electron concentration on the aromatic ring before the reaction are two ways of looking at the same trend; either approach predicts the same conclusion. *Substituents that donate (or release) electrons to the aromatic ring activate it for further substitution. Substituents that withdraw electrons from the aromatic ring deactivate it for further substitution.* The following classification reiterates the reactivity each group imparts to an aromatic ring:

Electron-donating (electron-releasing) groups (all are *ortho, para* directors):

$$-\ddot{N}H_2, -\ddot{N}HR, -\ddot{N}R_2 \qquad -\ddot{O}H, -\ddot{O}R$$

$$-C_6H_5, -R \qquad CH_3-\overset{\overset{\displaystyle :O:}{\|}}{C}-\ddot{N}HR$$

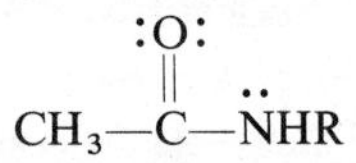

Electron-withdrawing groups (all are *meta* directors, except the halogens, —X, which are *ortho, para* directors):

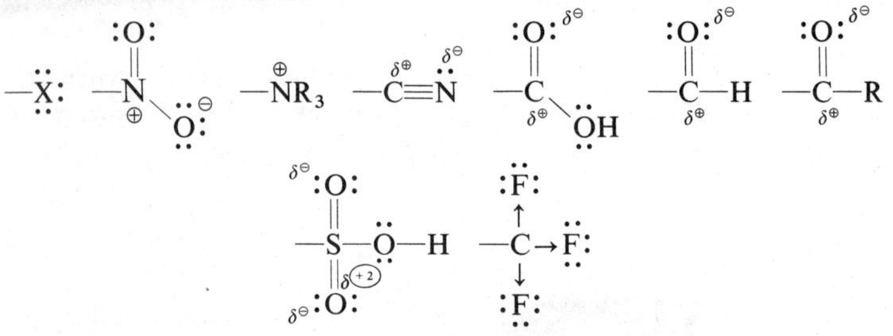

Note that all the deactivating groups (*except for halogen*) have either a partial or full positive charge on the atom that is attached to the ring, and the positive charge withdraws electrons from the ring.

To present a more complete picture of activation and deactivation in electrophilic aromatic substitution, we next discuss how each group acts as an electron-releasing or withdrawing group. We also consider the orientation these various substituents impart to the aromatic ring.

Question 14.7

Which of these two molecules, $C_6H_5-CF_3$ or $C_6H_5-CCl_3$, would be more reactive toward electrophilic aromatic substitution? Briefly explain.

Question 14.8

Arrange the following four compounds in order of *increasing* reactivity toward ring bromination: $C_6H_5-CH_3$, $C_6H_5-CH_2Cl$, $C_6H_5-CHCl_2$, and $C_6H_5-CCl_3$. Briefly explain the basis for your choice.

14.10   Orientation of Electrophilic Aromatic Substitution: Inductive Effect

In this section we discuss the theory of orientation for *substituents that either donate or withdraw electrons via the* **inductive effect.** We consider (1) alkyl (R—) substituents and (2) all *meta* directors. The special factors of other activating substituents are discussed in Secs. 14.11 and 14.12.

One convenient way to approach the question of orientation is to consider the electronic structures of the possible carbocations that are formed in each reaction. With no prior knowledge about orientation, we can gain a good idea about this subject by considering attack first at the *ortho* position, then at the *meta*, and finally at the *para* position and asking whether there is any special stability or instability associated with these various carbocations.

A. Orientation Obtained with Alkyl (R—) Substituents

As an example of alkyl substituents, which are activating, *ortho, para* directors, consider electrophilic substitution of toluene. The carbocations formed from *ortho, meta,* and *para* attack by an electrophile, $E^{\oplus}$, are shown here. It is convenient to use the Kekulé-type structures, not because they actually exist but because they depict more accurately the ring carbon atoms that bear positive charge.

ortho Attack:

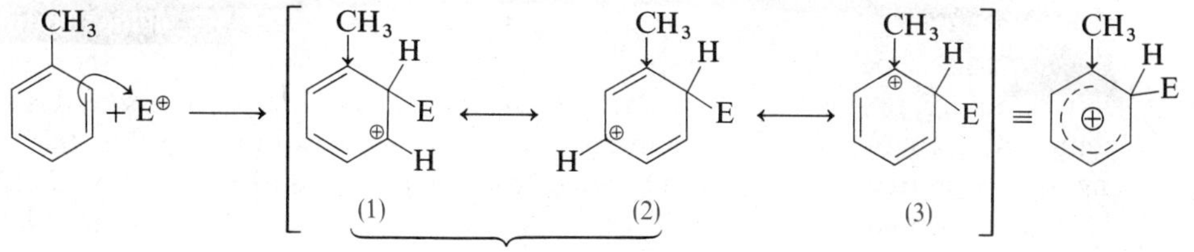

Positive charge on carbons bearing hydrogen
Comparable but less stable

Positive charge on carbon bearing electron-donating substituent
More stable

Best representation

(1)   (2)   (3)

meta Attack:

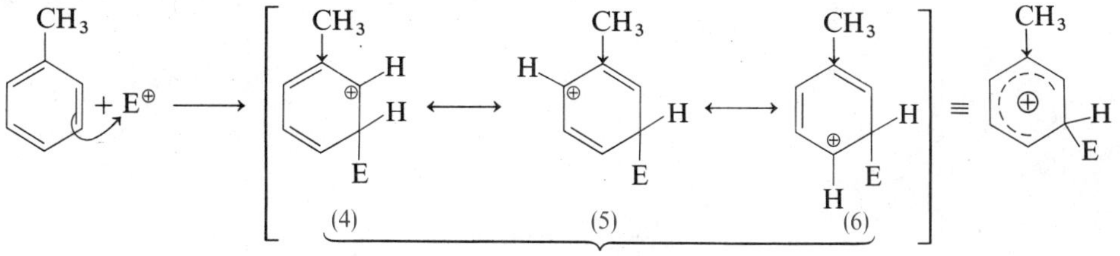

Positive charge on carbons bearing hydrogen; no special stability or instability
All of comparable stability

Best representation

(4)   (5)   (6)

para Attack:

Positive charge on carbon bearing hydrogen
Comparable but less stability

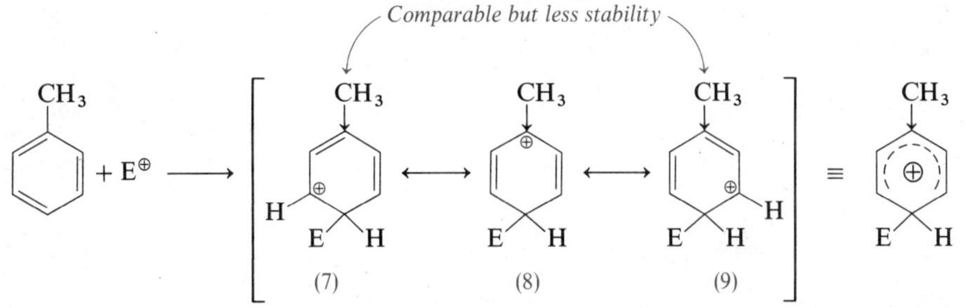

Positive charge on carbon bearing electron-donating substituent
More stable

Best representation

(7)   (8)   (9)

Of these electronic structures, structure (3) of *ortho* attack should be more stable than structures (1) and (2) because of the electron-donating methyl group on the carbon bearing a positive charge in (3). For *meta* attack, all the electronic structures, (4) to (6), are of comparable stability and none is especially stable or unstable.

Comparing *ortho* and *meta* attack, it is not surprising that the electrophile prefers to attach itself to the *ortho* position because then stabilization of the positive charge in the cation is possible. The analogous argument applies when comparing *meta* with *para* attack. In the latter, electronic structure (8) is particularly stable compared with (7) and (9); as before, there is nothing special about *meta* attack, so attack at the *para* position is preferred over the *meta* position. Thus, we predict that *ortho* and *para* attack are favored over *meta* attack, which is indeed the case.

The added stability due to the interaction of the electron-releasing methyl group with the positive charge on the ring is properly referred to as **delocalization of charge.** This is perhaps best explained by transition-state theory; the energies of activation, E_A, for the transition states for *ortho* and *para* attack are less (and *roughly comparable to one another*) than that for *meta* attack. Furthermore, the intermediates resulting from *ortho* and *para* attack are more stable than for *meta* attack. These energy relationships are shown on the energy-profile diagram in Fig. 14.3.

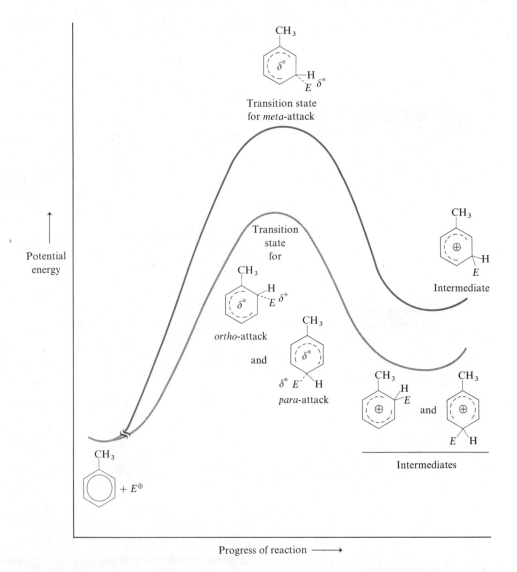

FIGURE 14.3   Energy-profile diagram showing the transition states and intermediates for the first step of electrophilic attack on toluene.

B. Orientation Obtained with *meta* Directors

We use a similar approach to deduce the orienting effect of substituents that deactivate the ring because they contain a partial or full positive formal charge on the atom attached to the ring. For illustration we consider possible *ortho*, *meta*, and *para* attack by an electrophile on benzenesulfonic acid, for which the following resonance structures may be drawn.

ortho Attack:

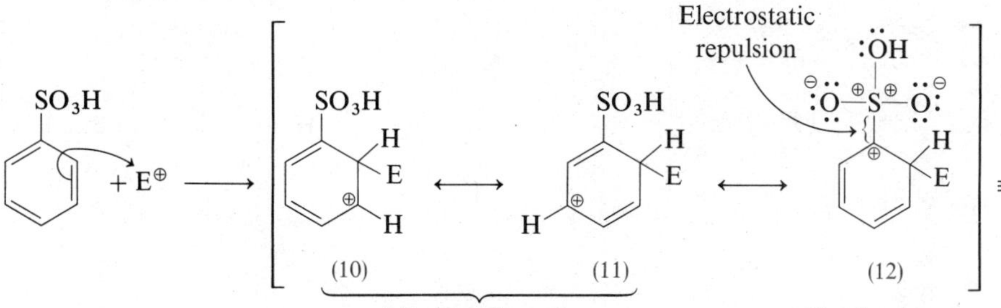

(10)   (11)   (12)

Positive charge on
carbon bearing hydrogen
Comparable stability

Positive charge
on carbon
adjacent to
positive charge
on sulfur
*Particularly
unstable*

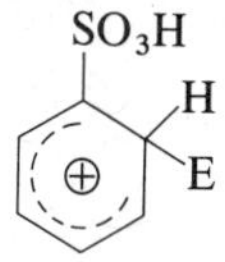

Best
representation

meta Attack:

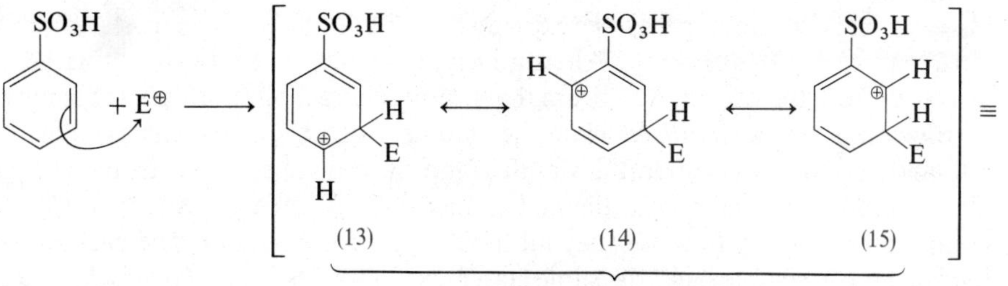

(13)   (14)   (15)

Positive charge on carbons
bearing hydrogen; no special
stability or instability
All of comparable stability

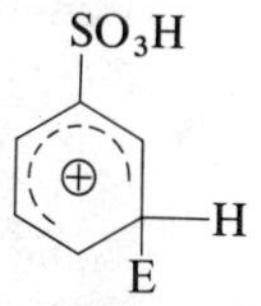

Best
representation

para Attack:

Positive charge on
carbons bearing hydrogen;
no special stability or
instability
Comparable stability

Electrostatic
repulsion

(16)   (17)   (18)

Positive charge
on carbon
adjacent to
positive charge
on sulfur
*Particularly
unstable*

Best
representation

In *ortho* attack structure (12) is particularly unstable because the carbon bearing the positive charge is adjacent to the sulfur atom that bears formal positive charges. This instability can be explained in terms of the electrostatic repulsion between like positive charges. (Alternatively, we could say that sulfur is withdrawing electrons from the carbon that bears the positive charge, but both say the same thing: Carbocations are *destabilized* when they are near other positive charges.) Structures (10) and (11) are of comparable stability and have no special electronic effects that either stabilize or destabilize them. For *meta* attack, structures (13) to (15) are of comparable stability because they all have a positive charge on the carbon bearing hydrogen. For *para* attack, the same statements apply as for *ortho* attack. Electronic structure (17) is particularly unstable because it has positive charges on carbon and sulfur adjacent to one another. Structures (16) and (18) are of comparable stability but more stable than (17) because they contain no special electronic features that either stabilize or destabilize them. Thus, we predict that electrophilic substitution occurs via the route that provides the most stable intermediates, and in this case, *meta* attack should be predominant; indeed, this fits the facts.

In electrophilic substitution reactions involving *meta* directors, *meta* attack occurs because *ortho* and *para* attack give intermediate carbocations that are *less stable* than those resulting from *meta* attack. Simply put, *meta* attack is the best of three evils. The energy of activation, E_A, for the formation of the transition state for *meta* attack is less than that for either *ortho* or *para* attack.

Writing Resonance Structures

Our discussion about orientation has revolved about the details of electronic structures for the intermediate carbocations. The use of resonance structures allows us to pinpoint carbon atoms that have increased positive charge, and, depending on the substituents attached, we can deduce whether *ortho,para* or *meta* substitution is most likely. Resonance structures do not exist as such, but remember that the introduction of an electrophile onto an aromatic ring results in *increased positive charge on the two carbons ortho to the entering group and the one para to it.* Using benzene as an example, the following resonance structures may be drawn, where the location of the positive charge is emphasized by arrows:

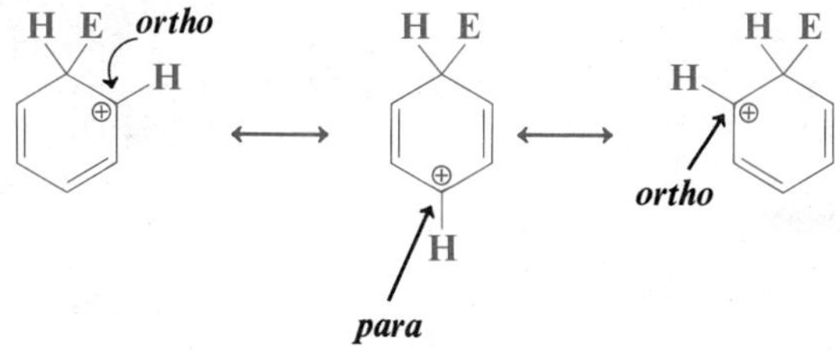

The same type of structures arise when a substituent is present on the ring already. We must evaluate each structure individually and see whether there is special *stability* or *instability*.

14.11   Orientation and Reactivity Due to Resonance Effects

We now come to the interesting question of why the following substituents are all very strongly activating and are *ortho*, *para* directors.

$$-\overset{\cdot\cdot}{\text{N}}\text{H}_2 \quad -\overset{\cdot\cdot}{\text{N}}\text{HR} \quad -\overset{\cdot\cdot}{\text{N}}\text{R}_2 \quad -\overset{\cdot\cdot}{\text{N}}\text{H}-\overset{\overset{\displaystyle O}{\|}}{\text{C}}-\text{CH}_3 \quad -\overset{\cdot\cdot}{\underset{\cdot\cdot}{\text{O}}}-\text{H} \quad -\overset{\cdot\cdot}{\underset{\cdot\cdot}{\text{O}}}-\text{R}$$

Attachment of nitrogen or oxygen to the aromatic ring should deactivate the ring because both of these elements are more electronegative than carbon (see Sec. 2.10) and thus should withdraw electrons by the inductive effect. Why, then, do they so strongly activate the ring toward electrophilic aromatic substitution? Let us examine the electronic structures of the possible intermediate carbocations that are formed when an electrophile attacks an aromatic ring bearing, for example, an —OH group. As before, we consider *ortho*, *meta*, and *para* attack.

ortho Attack:

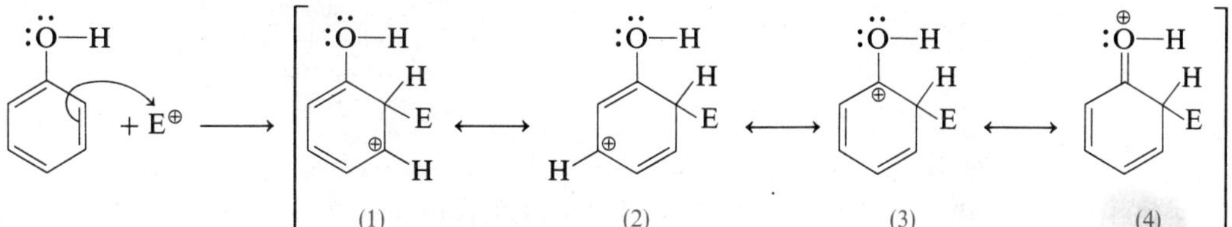

(1) (2)	(3)	(4)
Positive charge on carbons bearing hydrogens *No special stability or instability*	Positive charge adjacent to electron-rich oxygen atom *Unstable*	Additional resonance structure showing delocalization of positive charge due to unshared electron pairs on oxygen

meta Attack:

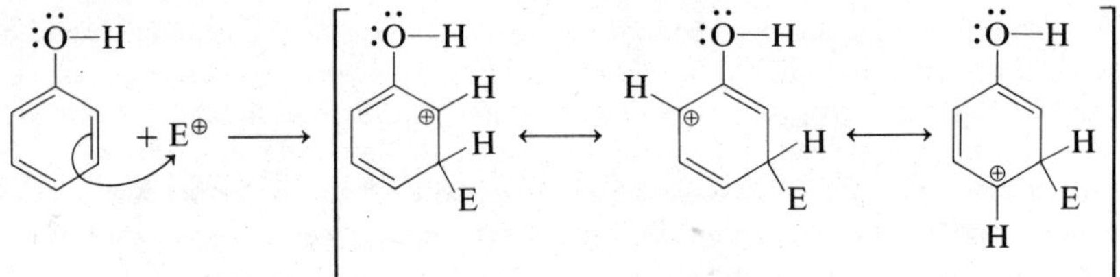

Positive charge
on carbons bearing hydrogens
No special stability
or instability

para Attack:

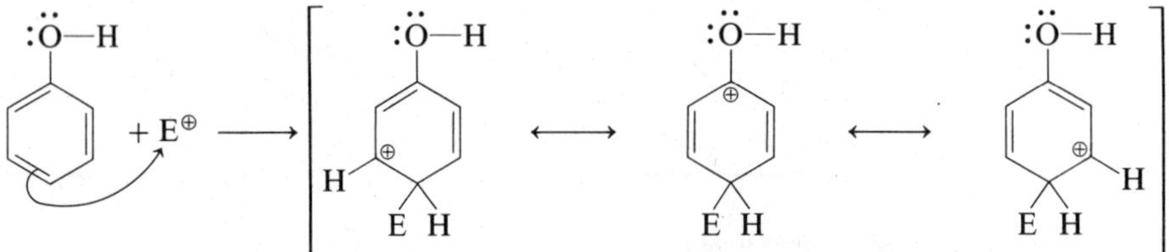

(8)	(9)	(10)
No special stability or instability	Positive charge adjacent to electron-rich oxygen *Unstable*	*No special stability or instability*

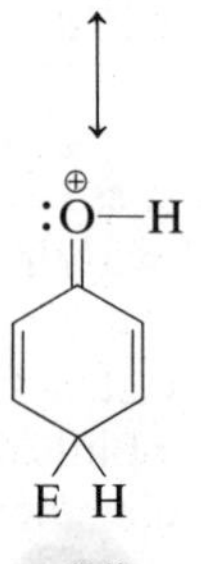

(11)

Additional resonance structure
showing delocalization of
positive charge due to unshared
electron pairs on oxygen

    There is no special stability or instability associated with carbocation structures (5) to (7) resulting from *meta* attack. However, careful examination of the electronic structures for *ortho* and *para* attack reveals that structure (4) for *ortho* attack and structure (11) for *para* attack are particularly stable, but why? Carbocations (3) and (8) have an incomplete octet of electrons, whereas after donation of a pair of electrons by oxygen, *each* atom in structures (4) and (11) has a full octet of electrons.

    Note that oxygen acquires a positive charge in the donation, thus resulting in the electrons and the concomitant positive charge being spread out (delocalized) over a larger area of the molecule (because the resonance hybrid is an average of all the resonance structures involved). Also, oxygen (and for that matter, nitrogen in the other groups that behave similarly to the —OH group) readily donates electrons.

(3)	(4)	(9)	(11)
Carbon has six electrons	Each atom has eight electrons	Carbon has six electrons	Each atom has eight electrons

For example, both water and ammonia are protonated in acidic solution because they can donate one unshared pair of electrons to a proton:

$$H_2\overset{..}{\overset{.}{O}}: + H^\oplus \rightleftharpoons H\overset{\oplus}{\underset{\overset{|}{H}}{\overset{..}{O}}}H$$

Water   Hydronium ion

$$H_3N: + H^\oplus \rightleftharpoons H\overset{\overset{H}{|}}{\underset{\overset{|}{H}}{N^\oplus}}H$$

Ammonia   Ammonium ion

The interaction of an oxygen adjacent to a carbon bearing a positive charge is referred to as the **resonance effect.** This type of interaction is understandable when the orbital view of either the starting compound or the intermediate carbocation is examined:

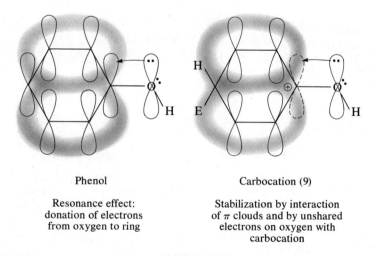

Phenol   Carbocation (9)

Resonance effect:
donation of electrons
from oxygen to ring

Stabilization by interaction
of π clouds and by unshared
electrons on oxygen with
carbocation

In these two structures, electron donation occurs through overlap of a *p* orbital on oxygen with the π cloud of the ring. This increases the electron density (electron concentration) in the ring and makes it more susceptible to attack by an electrophile. In the intermediate carbocation, the same type of interaction exists, but here the electrons of a nonbonded pair on oxygen are donated to the positively charged carbocation. Because stabilizing interactions of this type are not available for *meta* attack, the transition states and intermediates are of lower energy for *ortho, para* attack; the energy of activation, E_A, of the transition state in their forma-tion is lower. Thus, *ortho, para* attack is favored.

A priori there is no way to predict whether the inductive effect or the resonance effect will predominate or whether the two effects are of roughly equal importance. For example, in the case of phenol, the inductive effect of oxygen is electron with-drawal from carbon. In the preceding example, oxygen is more electronegative, and its nucleus withdraws electrons from the adjacent carbon and thus from the ring; this alone would deactivate the ring. On the other hand, electron donation through resonance would activate the ring.

We must turn to experimental facts to assess the relative importance of these two effects. Because phenol (and the other strongly activating *ortho, para* directors) is much more reactive than benzene and gives *ortho, para* substitution almost ex-clusively, the resonance effect is overwhelmingly the more important and the inductive effect has little influence if any.

What about the rest of the strongly activating *ortho, para* directors? They all behave similarly because *they all have at least one unshared pair of electrons* on the atom attached to the aromatic ring, and they all behave like phenol.

When each compound is attacked by an electrophile in the *ortho* or *para* position, the result is especially stable ions involving the interaction of one unshared pair of electrons with the ring (shown here for *para* attack):

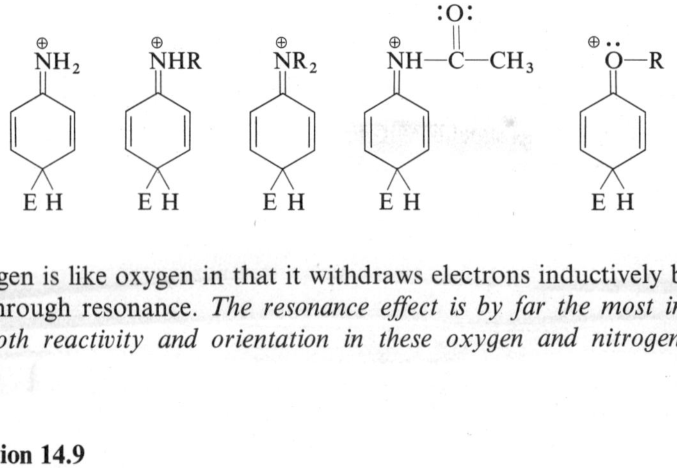

Nitrogen is like oxygen in that it withdraws electrons inductively but donates electrons through resonance. *The resonance effect is by far the most important in dictating both reactivity and orientation in these oxygen and nitrogen-containing compounds.*

Question 14.9

On the basis of electronic theory, explain the following observations:

(*a*) The $-\ddot{N}(CH_3)_2$ group is an *ortho, para* director and strongly activates the ring, whereas the $-\overset{\oplus}{N}(CH_3)_3$ group is a *meta* director and deactivates the ring toward electrophilic substitution.
(*b*) The $-NO_2$ group is a *meta* director, and its presence on the ring makes electrophilic sub-stitution more difficult as compared with the $-\ddot{N}H_2$ group, which is an *ortho, para* director.

Question 14.10

Draw resonance structures for the carbocations that result when each of the following five compounds is subjected to monobromination. Consider *ortho, meta,* and *para* attack and draw structures for all three possibilities. Then, on the basis of electronic theory, deduce whether these groups are *ortho, para* directors or *meta* directors. Are these compounds more or less reactive than benzene?

(*a*) aniline, $C_6H_5-NH_2$

(*b*) the anilinium ion, $C_6H_5-\overset{\oplus}{N}H_3$

(c) acetanilide, C_6H_5—NH—C—CH_3 (d) benzaldehyde, C_6H_5—C—H

with O double bonds as drawn

(e) acetophenone, C_6H_5—C—CH_3

14.12   Orientation and Reactivity of Electrophilic Aromatic Substitution on Aryl Halides

The aryl halides, Ar—X, such as chloro- and bromobenzene, are unique in terms of electrophilic aromatic substitution. A halogen atom *deactivates* the ring toward further substitution compared with benzene, and yet it is an *ortho, para* director. This does not seem consistent with our previous discussions about the inductive and resonance effects, but it is. Let us see why.

The presence of the strongly electronegative halogen atom—for example, chlorine—on the ring should cause electron withdrawal through the inductive effect. The inductive effect leads to a decreased affinity for electrophilic attack on the ring and decreased stability of the carbocation that results. The net effect is that the halobenzenes are less reactive than benzene.

To explain the orientation, we again look at the electronic structures of the intermediate carbocations that result from electrophilic attack at the *ortho*, *meta*, and *para* positions.

ortho Attack:

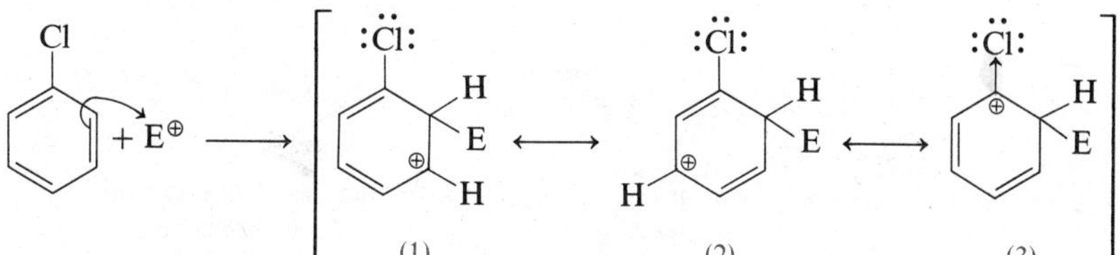

(1)   (2)   (3)

No specially stable or
unstable structures
Comparable stability

Carbocation
bearing electron-
withdrawing
chlorine atom
*Particularly
unstable*

meta Attack:

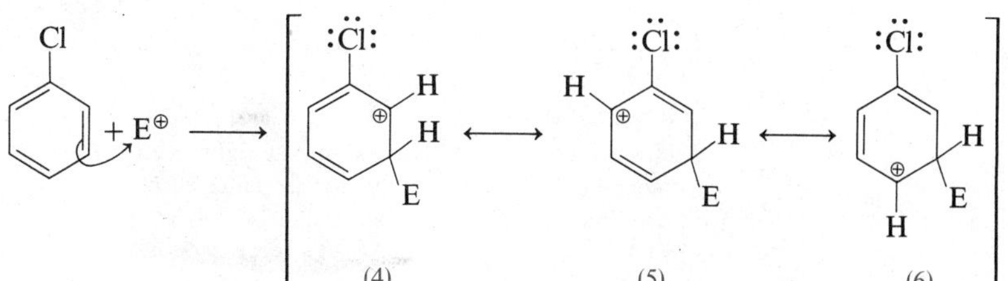

(4)   (5)   (6)

No specially stable
or unstable structures
All of comparable stability

para Attack:

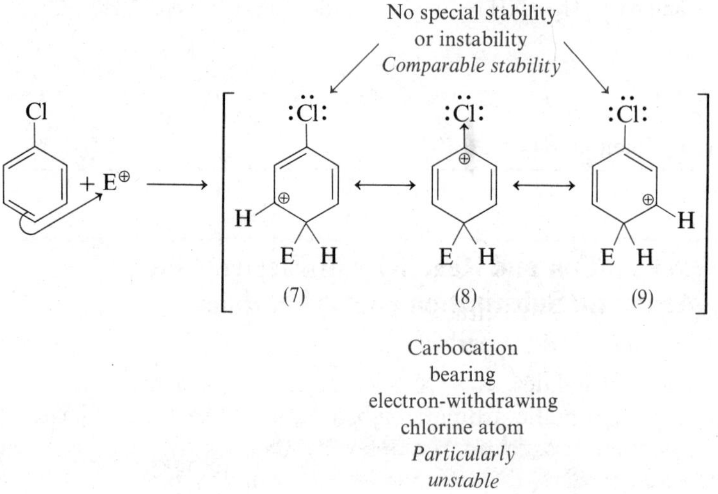

No special stability
or instability
Comparable stability

Carbocation
bearing
electron-withdrawing
chlorine atom
*Particularly
unstable*

If we use previous arguments about orientation, we would predict that halogens, such as chlorine, are *meta* directors, because *ortho* and *para* attack each produces one particularly unstable structure, (3) and (8), respectively. However, look at these two structures more closely! Chlorine contains three unshared pairs of nonbonding electrons, and unshared electron pairs adjacent to an aromatic ring can donate electrons through the resonance effect. This possibility means that we can write additional electronic (resonance) structures for *ortho* and *para* attack:

ortho Attack:

(3)
Six electrons on
carbon and
electron withdrawal
by Cl
Unstable

(10)
Each atom has eight electrons
due to resonance effect;
positive charge delocalized
Much greater stability

para Attack:

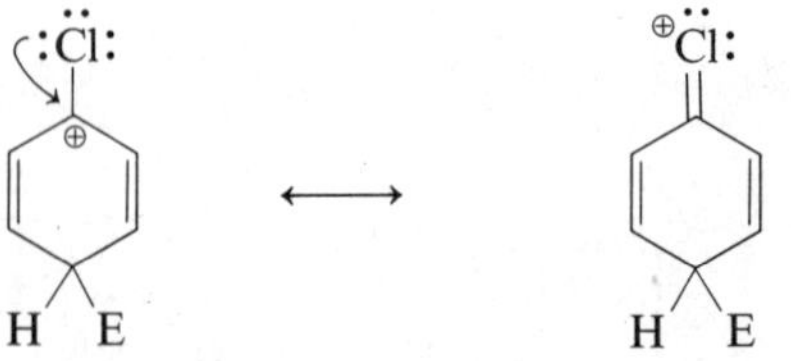

(8)
Six electrons on
carbon and
electron withdrawal
by Cl
Unstable

(11)
Each atom has eight electrons
due to resonance effect;
positive charge delocalized
Much greater stability

In structures (10) and (11), all atoms (except hydrogen) have a complete octet of electrons as a result of resonance donation, so they possess more stability than structures (3) and (8), respectively. This results in chlorine bearing some positive charge, which is therefore spread out (delocalized) over a larger area of the molecule.

As before, there is no a priori way to predict the relative importance of these two effects. We must turn to experimental facts, which indicate that the halobenzenes are deactivated but that an incoming group is oriented *ortho* and *para*. From this, we conclude that the inductive withdrawal and the resonance donation effects are of comparable importance in dictating the orientation and reactivity of the halogen atom attached to an aromatic ring.

The inductive effect is largely responsible for dictating reactivity and the resonance effect is largely responsible for dictating orientation in reactions of aryl halides. In some substituents the resonance effect completely outweighs the inductive effect, as in the —NH_2 and —OH substituents (see Sec. 14.11). With the halogens, however, these effects are much more nearly equal in magnitude.

14.13 Directing Effect of the Aryl Group

The aryl group, Ar-, is an *ortho*, *para* director and a weak activating group.

The Symbol Ar-

In the same way that we use the symbol R- to represent any alkyl group, we commonly use Ar- to represent *any aryl group* in which other substituents are bonded to the aromatic ring. This general symbol can be used to represent the simplest aromatic ring (phenyl, although Ph- is sometimes used in this case) or one containing any number of substituents.

A closer look at the influence this group has on the aromatic ring to which it is attached indicates that the aryl group is an electron-withdrawing group by the inductive effect but an electron-releasing group through resonance. The resonance effect is more important in dictating the chemistry of this moiety. The orbital view illustrating electron donation through resonance is shown here for the ground state:

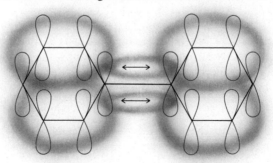

Electron donation through
resonance of an aryl group

14.14 Disubstitution: Experimental Factors and Limitations

We now briefly survey reactions of monosubstituted benzene derivatives to learn what experimental conditions are required and what limitations, if any, exist.

A. Reactions of Nitro Compounds

The nitro group deactivates the ring toward further substitution and also orients incoming groups to the *meta* position. Because of the deactivation more vigorous reaction conditions must be used to introduce a second group into the molecule containing the nitro group. For example,

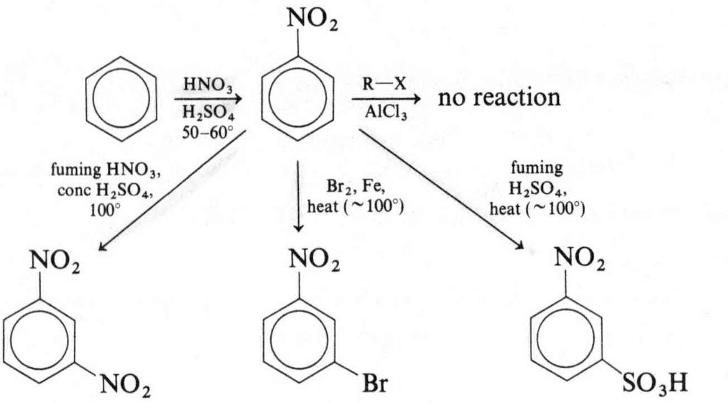

| *m*-Dinitrobenzene | *m*-Bromonitrobenzene | *m*-Nitrobenzenesulfonic acid |

The nitration, bromination (or more generally, halogenation), and sulfonation of nitrobenzene occur in the normal fashion; however, it is impossible to carry out a Friedel-Crafts reaction on nitrobenzene. Even more generally, *Friedel-Crafts reactions cannot be carried out on compounds containing a meta director or with aromatic rings containing the* $-\ddot{O}H$, $-\ddot{N}H_2$, $-\ddot{N}HR$, *and* $-\ddot{N}R_2$ *groups.*[1] The *meta* directors not only deactivate the ring but also react with the AlX_3 catalyst:

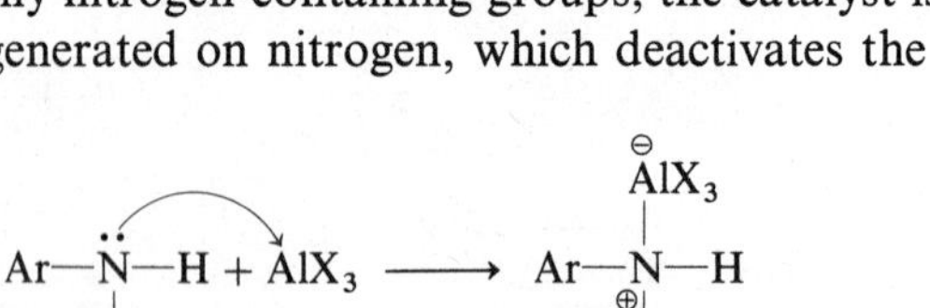

This results in extensive electron withdrawal and removal of the catalyst from the reaction. In many nitrogen-containing groups, the catalyst is also tied up and a positive charge is generated on nitrogen, which deactivates the ring; for example:

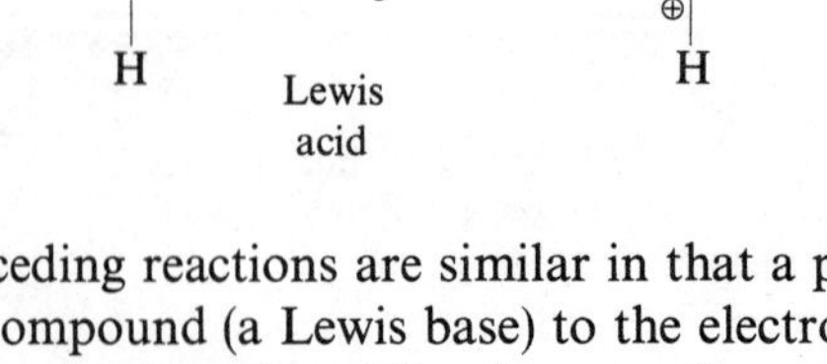

Both the preceding reactions are similar in that a pair of electrons is donated from the organic compound (a Lewis base) to the electron-deficient AlX_3 molecule (a Lewis acid). Note the formal positive charges adjacent to the aryl group.

---

[1] Actually, Friedel-Crafts reactions may sometimes be run with one of these groups present in the molecule. Yields are very low, however, and the reactions should be used only if no other synthetic pathway is available.

Because the amine group is particularly susceptible to a variety of interfering reactions (for example, acid-base and oxidation), the amine is often converted to an *amide* prior to synthesis. The amide does not undergo the side reactions, and the free amine can be regenerated whenever desired.

Aniline   Acetanilide   *p*-Acetamido toluene

p-Toluidine

The *acetamido* group is an *ortho, para* director but less activating than the amino group. It is also a much weaker Lewis base. Yields may still be low, but they are much better than those obtained in the presence of the free —N̈H₂ group. (See Sec. 26.8A for a more detailed discussion.)

B. Reactions of Alkylbenzenes

Alkylbenzenes are more reactive than benzene, and the alkyl group orients an incoming electrophile into the *ortho* and *para* positions. The reactivity is enhanced when there is one alkyl group, and a problem encountered in the Friedel-Crafts alkylation reaction is the formation of disubstituted alkylbenzenes:

Toluene   *o*-Xylene   *p*-Xylene

Sulfonation, nitration, and halogenation of toluene (or of any alkylbenzene, Ar—R) occur in a straightforward manner, although reaction conditions are milder (usually temperatures are lower) because of the activating effect of the methyl group.

An interesting reaction sequence involving alkylation and then sulfonation is the commercial preparation of detergents, which have the general formula R—Ar—SO₃⊖Na⊕. One of the first detergents made industrially involved the alkylation of benzene with tetrapropylene (which comes from the controlled polymerization of propylene, see Sec. 8.34), followed by sulfonation and conversion of the resulting sulfonic acid to its sodium salt:

$$CH_3\text{—}CH\text{—}CH_2\text{—}CH\text{—}CH_2\text{—}CH\text{—}CH_2\text{—}CH\text{=}CH_2 +$$
$$\quad\quad |CH_3 \quad\quad |CH_3 \quad\quad |CH_3$$

Tetrapropylene

$\xrightarrow[\text{AlCl}_3\text{—HCl}]{\text{HF or}}$

$C_{10}H_{21}\text{—}CH\text{—}CH_3$ $\xrightarrow{H_2SO_4}$ $C_{10}H_{21}\text{—}CH\text{—}CH_3$ (SO$_3$H) $\xrightarrow{Na_2CO_3}$

(+ some *ortho* isomer)

SO$_3^{\ominus}$Na$^{\oplus}$

$$CH_3\text{—}CH\text{—}CH_2\text{—}CH\text{—}CH_2\text{—}CH\text{—}CH_2\text{—}CH\text{—}CH_3$$
$$\quad |CH_3 \quad\quad |CH_3 \quad\quad |CH_3$$

Detergent

This detergent, although effective as a cleansing agent, is not affected by microorganisms and therefore was not destroyed after being discharged into ground-waters and rivers. After considerable research it was found that detergents containing straight-chain alkyl groups are degraded by microorganisms, that is, are *biodegradable*. Since 1965 most manufacturers have incorporated these nonbranched alkyl groups into detergents, and the starting alkenes are often made from polymerization of ethylene. Side chains containing 10 to 14 carbons appear to be degraded most effectively.

Question 14.11

(*a*) Propose a reasonable mechanism for the AlCl$_3$/HCl-catalyzed isomerization of *o*-xylene to *m*-xylene, which occurs at 80°. (*Hint:* A dealkylation-realkylation reaction is involved.)
(*b*) Which product, *o*-xylene or *m*-xylene, is thermodynamically more stable? Why?

C. Reactions of Benzenesulfonic Acids

Benzenesulfonic acids can be nitrated and halogenated, and a second sulfonic acid group can be introduced into the molecule. Because the —SO$_3$H group is a *meta* director, however, the Friedel-Crafts reactions cannot be carried out on aryl-sulfonic acids. The sulfonic acid group deactivates the ring toward further substitution.

D. Reactions of Halobenzenes

The halogen atom deactivates an aromatic ring but orients an incoming group into the *ortho* and *para* positions. Halobenzenes can be sulfonated, alkylated, and nitrated under normal conditions.

14.15 Synthetic Methods Involving Disubstituted Benzene Compounds

We can now develop simple synthetic schemes leading to certain disubstituted aromatic compounds. The important feature to keep in mind when synthesizing disubstituted compounds is which group should be introduced first because that group directs the second group into the *ortho* and *para* positions or the *meta* position. Usually an *ortho*, *para* mixture can be separated into its pure components because the *para* isomer normally has a higher freezing point and is less soluble in many solvents than is the *ortho* isomer.

To illustrate how changing the *order* of substitution can yield different products, the following outline shows the synthesis of the three isomeric chloronitrobenzenes:

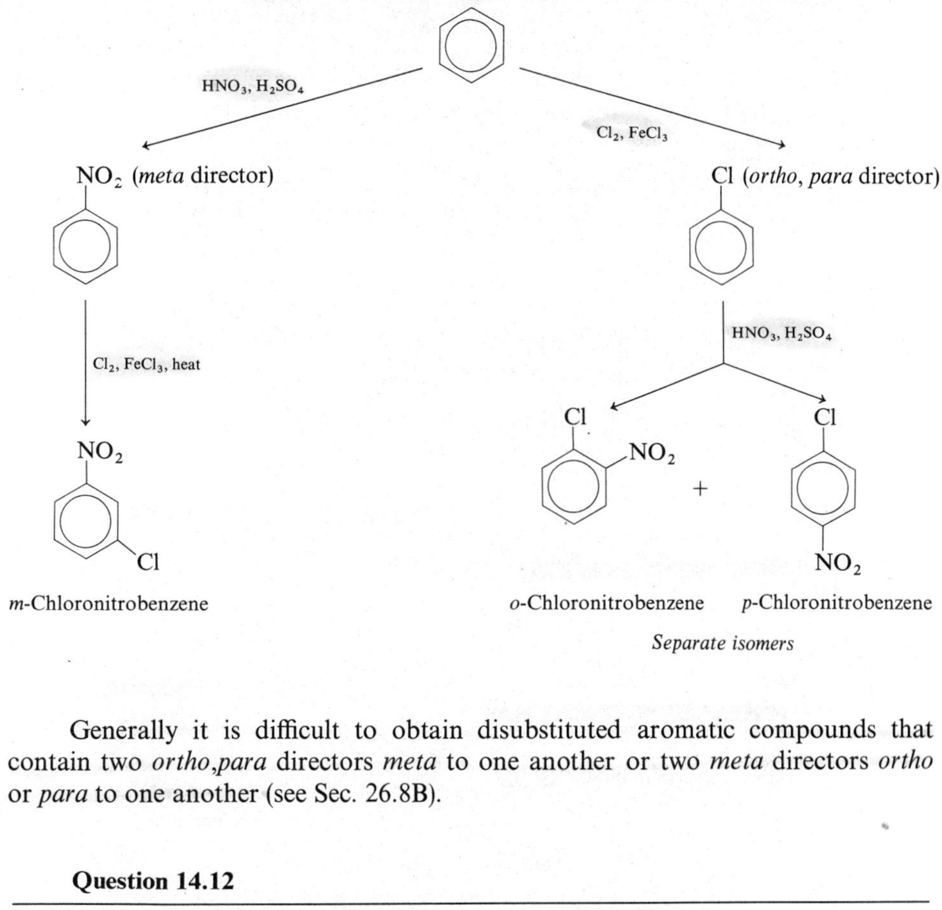

Generally it is difficult to obtain disubstituted aromatic compounds that contain two *ortho,para* directors *meta* to one another or two *meta* directors *ortho* or *para* to one another (see Sec. 26.8B).

Question 14.12

Devise suitable reaction sequences whereby each of the following compounds can be prepared from benzene in good yield. Assume *ortho* and *para* isomers can be separated from one another, and use any other reagents you desire.

(*a*) *p*-bromoisopropylbenzene   (*b*) *o*-chlorotoluene
(*c*) *p*-bromochlorobenzene   (*d*) *m*-nitrobenzenesulfonic acid
(*e*) *o*-ethyltoluene   (*f*) *p*-*tert*-butylbenzenesulfonic acid
(*g*) *o*-xylene   (*h*) *m*-xylene

14.16   Synthesis of Aromatic Compounds Containing Three or More Substituents

The synthesis of compounds with three or more substituents requires that we consider both *orientation* and *reactivity* effects of the groups already on the ring. To make a reasonable evaluation of where an entering group is most likely to go, consider the following general approach.

1. *The substituents that are already on the ring are responsible for the orientation of the incoming group.* Orientation does not depend much on the nature of the entering electrophile.

2. *In some cases the substituents already present on the ring orient an incoming group into the same position that one of them already occupies.*

3. *When there are two or more substituents with different activating powers, the one with the greatest activating power usually dictates the orientation of the entering group.* The relative activating powers of various substituents are given in Sec. 14.8. Care must be used in making sweeping generalizations when comparing some of these substituents. Consider the following examples in which the position of attack is shown by a red arrow:

Both groups in (1) and (2) direct to the same position.

The two compounds that contain two aromatic rings (called *biphenyls*) are instructive examples; in (4) the phenyl ring is activated by both another phenyl ring and the —NH—COCH$_3$ group, and in (5) one phenyl ring is deactivated by the nitro group, thus causing substitution to occur on the other ring.

On the other hand, when the groups are close together in activation order, the results are less clear cut. There is so little difference between a methyl group and a halogen that a mixture of products is obtained on further substitution:

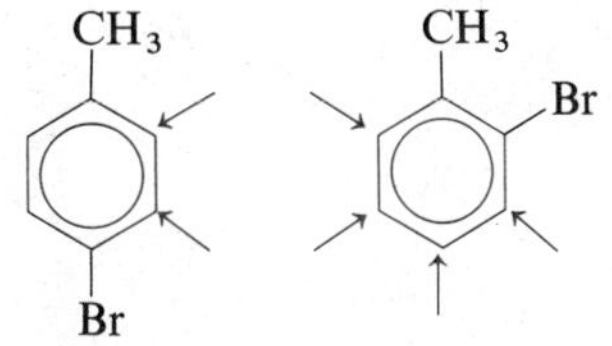

4. *When two substituents are meta to one another, there is little substitution on the carbon between them.* This property is attributed to the steric hindrance of the groups present, thus hindering the attack of the electrophile in that position. In the following two examples, the major products are indicated by a solid arrow and those formed in minor amounts (usually less than 10%) by a dashed arrow:

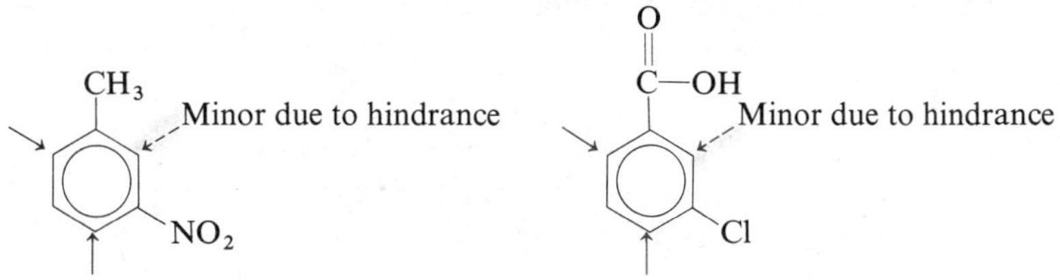

Question 14.13

Suppose each of the following seven compounds were nitrated under conditions that cause mononitration. At what position or positions would *major* substitution occur? How many isomers would be produced in significant amounts in each case?

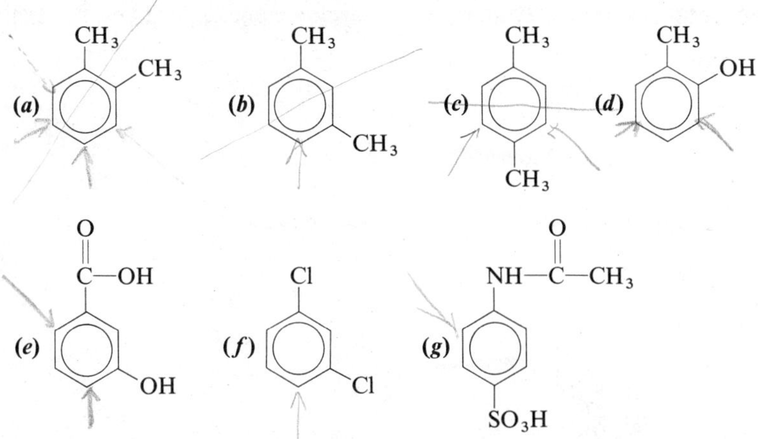

Question 14.14

Devise suitable routes to carry out the following syntheses in the laboratory. Assume that *ortho* and *para* isomers can be separated from a mixture containing both of them. Start from benzene and use any other needed reagents.

(a) 2,4,6-trinitrotoluene (*TNT*, an explosive)   (b)

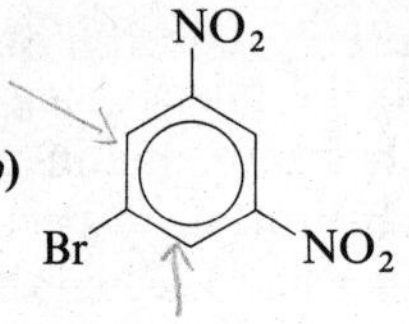

(c) 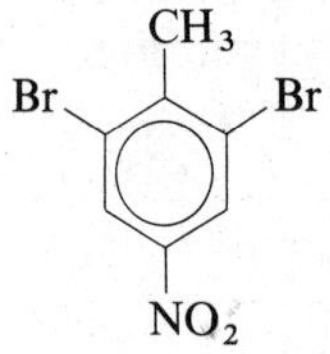   (d) 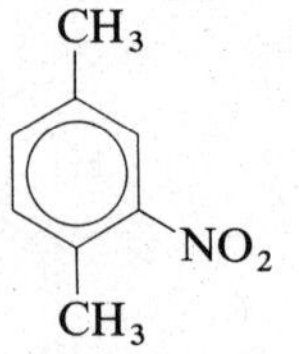   (e)

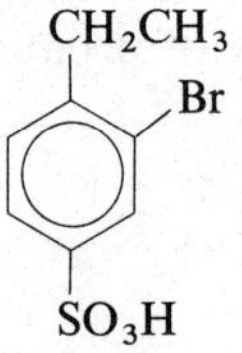

Miscellaneous Reactions of Aromatic Compounds

In the remainder of this chapter, we present reactions with aromatic compounds that do *not* involve electrophilic aromatic substitution on the ring. For the most part they involve the alkyl side chains.

14.17   Addition Reactions to Aromatic Ring: Hydrogenation and Chlorination

Although it is quite unreactive toward addition, the aromatic ring can be hydrogenated and chlorinated under drastic conditions. The hydrogenation reaction is the same one discussed in Sec. 13.5 in connection with the quantitative determination of the delocalization energy of benzene. When there are substituents on the ring, hydrogenation still occurs and produces cyclohexane derivatives, for example:

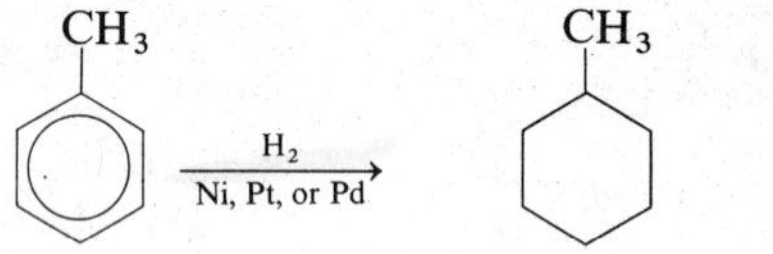

Toluene   Methylcyclohexane

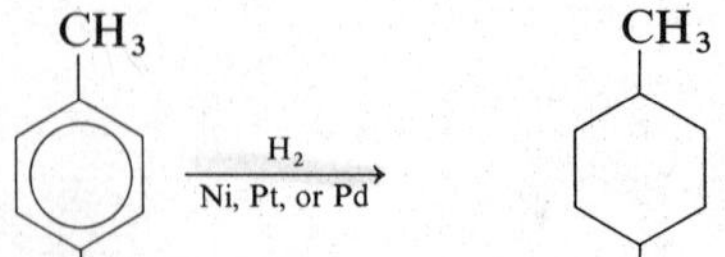

p-Xylene   1,4-Dimethylcyclohexane

The reaction of chlorine with benzene in the presence of light or at high temperatures and pressures adds 3 moles of chlorine per mole of benzene:

Benzene   Benzenehexachloride (BHC)

Several stereoisomers are formed in this reaction, but perhaps the most important is the γ isomer; it is an insecticide and has been given the names *gammexane*
and *lindane* (see Sec. 3.17). This reaction is believed to involve free-radical addition.

14.18   Side-Chain Oxidation of Alkylbenzenes

The oxidation of the alkyl side chain in alkylbenzenes (Ar—R) is accomplished
by either (1) $KMnO_4$ in aqueous base or (2) chromic acid, H_2CrO_4, formed from the
reaction between potassium dichromate ($K_2Cr_2O_7$) and sulfuric acid. Both reactions
require vigorous heating. In the simplest case, the methyl group in toluene is oxidized
to the carboxylic acid group, —COOH:

Toluene   Benzoic acid

Other substituents seldom affect this oxidation reaction, but when there are
several methyl groups, all are oxidized in an analogous manner.

m-Nitrotoluene   *m*-Nitrobenzoic acid

o-Xylene   Phthalic acid

The oxidation reaction has considerable use in synthesis because it is a good
method for preparing various aromatic carboxylic acids. On the other hand, oxidation

can also be used to help identify the position of alkyl groups that may be attached to the aromatic ring. It is interesting that alkyl groups containing more than one carbon atom are oxidized until only one carbon remains, namely the carbon in the —COOH group attached to the ring; for example:

n-Butylbenzene   Benzoic acid

p-Isopropyltoluene   Terephthalic acid

The mechanisms of chromic acid or permanganate oxidation of alkylbenzenes are not well understood, but a limited number of facts point to the intervention of alkyl radicals.

14.19   Side-Chain Halogenation of Alkylbenzenes: Benzyl Radicals

Alkylbenzenes, Ar—R, contain both an aromatic ring and an alkyl side chain. The ring undergoes electrophilic substitution with ionic reagents and alkanes undergo free-radical substitution reactions. Thus it is not surprising that alkylbenzenes undergo both reactions depending on the reaction conditions used. For example, toluene reacts with chlorine in the presence of ferric chloride to give ring chlorination, whereas with chlorine in the presence of light there is substitution on the methyl group.

o-Chlorotoluene   *p*-Chlorotoluene   Toluene   (Chloromethyl)benzene
(benzyl chloride or
α-chlorotoluene)

Results of electrophilic   *Result of free-*
aromatic substitution   *radical halogenation*
on CH$_3$ group

(The derivation of the alternate name α-chlorotoluene for benzyl chloride is explained later in this section.)

The two types of reaction of toluene with chlorine are different because of the kind of reactive chlorine intermediates produced under the different reaction conditions. In electrophilic aromatic substitution, the electrophile $:\overset{..}{\underset{..}{Cl}}{}^{\oplus}$ is generated and then attacks the ring. Under free-radical conditions, chlorine atoms, $:\overset{..}{\underset{..}{Cl}}\cdot$, are generated and attack the hydrogens on the methyl group; free-radical substitution is the result.

Treatment of toluene with an excess of chlorine in the presence of heat or light ultimately yields a product in which all three hydrogens on the methyl group are substituted by chlorine; this product is formed as a result of stepwise substitutions of chlorine for hydrogen on toluene:

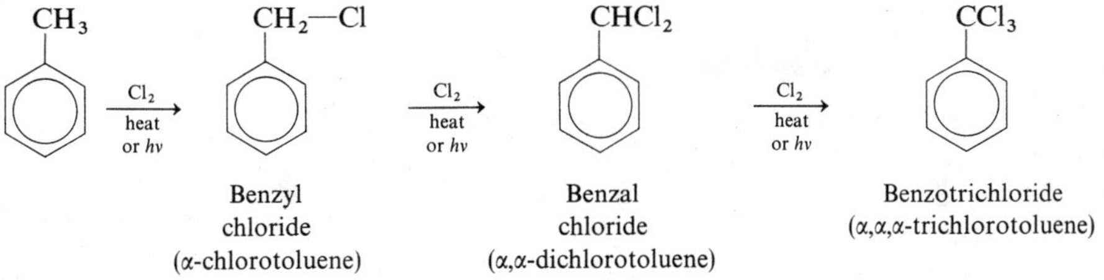

| | Benzyl chloride (α-chlorotoluene) | Benzal chloride (α,α-dichlorotoluene) | Benzotrichloride (α,α,α-trichlorotoluene) |

The orientation of an incoming halogen in the side chain becomes more interesting as the number of carbons in it increases. The free-radical halogenation of ethylbenzene, for example, occurs in such a way that the entering halogen atom is on the carbon adjacent to the aromatic ring (the α-carbon atom):

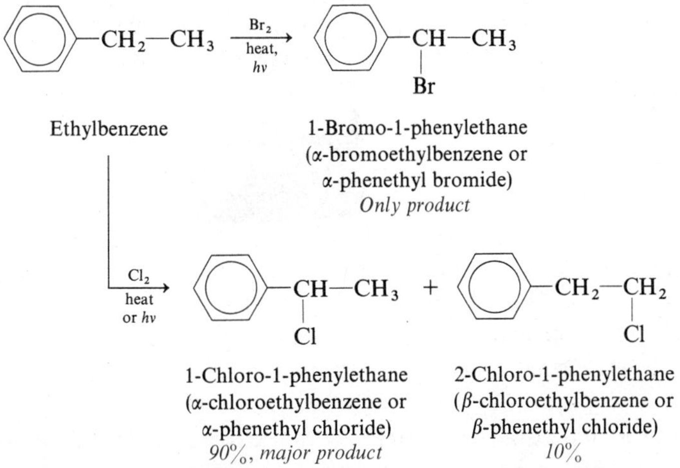

Note that the common names in parentheses use the Greek letters α, β, γ, δ, and so on to designate the position of the substituent on the side chain attached to the aromatic ring, starting with α for the carbon adjacent to the aromatic ring.

The experimental results indicate that the hydrogens on the α-carbon atom have unusual reactivity and that they are removed in preference to the other hydrogens on the side chain. These α hydrogens are called **benzylic hydrogens**:

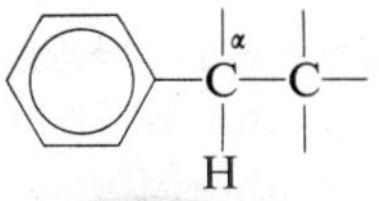

Benzylic hydrogen
(α *Hydrogen*)

In the mechanism of free-radical substitution, the following steps are involved in the halogenation of ethylbenzene:

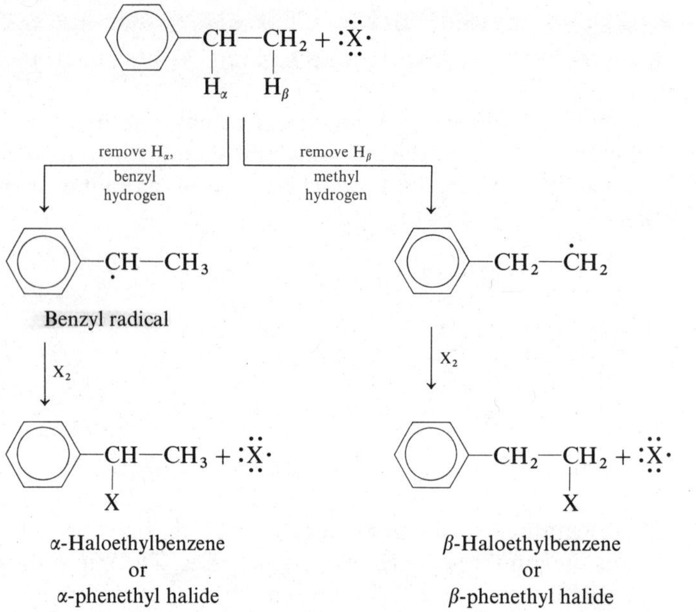

So far we considered empirical data collected from experimental results. What is the real driving force behind the greatly enhanced reactivity of the benzyl hydrogens? To answer this, we must look at some known facts about the relative stability of a benzyl radical:

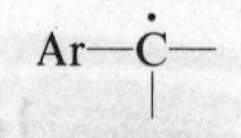

$$Ar—\overset{\centerdot}{C}—$$

benzyl radical

The bond dissociation energy for breaking the tertiary hydrogen in isobutane, $(CH_3)_3C—H$, is 91 kcal/mole (380.7 kJ/mole) (see Table 5.1) to give the *tert*-butyl radical. By comparison, the energy required for formation of the benzyl radical, $C_6H_5—\overset{\centerdot}{C}H_2$, from toluene is 79 kcal/mole (330.5 kJ/mole), and that of the allyl radical (the 3-propenyl radical, $CH_2{=}CH—\overset{\centerdot}{C}H_2$) from propene is 77 kcal/mole (322.2 kJ/mole). Thus benzyl and allyl radicals are easier to form and of comparable stability, and both are more stable than the *tert*-butyl radical. With this information we can extend the series of radical stabilities from Sec. 5.7 as follows:

Decreasing stability of free radicals and decreasing ease of hydrogen atom abstraction from alkanes:

$$Allyl \approx benzyl > 3° > 2° > 1° > \cdot CH_3$$

The unsual stability of the benzyl radical is attributed to delocalization of the odd electron on the α carbon into the π cloud of the aromatic ring system. In the orbital view of this radical, the odd electron occupies a *p* orbital on an sp^2 hybridized carbon atom that overlaps with the six *p* orbitals making up the π cloud; the following picture shows this delocalization:

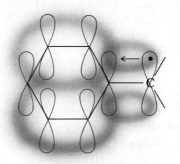

Benzyl radical
(all *p* orbitals contain
seven electrons; delocalization
of radical electron stabilizes system)

The delocalization of the odd electron can be represented on paper by the following resonance structures, and the stabilization of the benzyl radical can be referred to as *resonance* stabilization.

Best
representation

This electronic view of benzyl radicals is consistent with what is known about alkyl radicals that intervene in free-radical halogenation. It is not surprising that the free-radical halogenation of alkyl side chains on aromatic rings occurs with predominant substitution on the α carbon for chlorination (that is, 90% α- and 10% β-chloroethylbenzene) and with exclusive substitution on the α carbon in bromination.

Question 14.15

Based on your knowledge of the effects of various substituents on electrophilic aromatic substitution reactions, predict the relative stabilities of the following *para*-substituted benzyl radicals:

$$G-\!\!\!\bigcirc\!\!\!-\dot{C}H_2$$

where G = OCH_3, Cl, NO_2

Question 14.16

Arrange these four aromatic compounds in order of *increasing* reactivity in free-radical bromination and briefly justify your choice.

$$CH_3O-\!\!\!\langle\bigcirc\rangle\!\!\!-CH_3 \qquad \langle\bigcirc\rangle\!\!\!-CH_3 \qquad Cl-\!\!\!\langle\bigcirc\rangle\!\!\!-CH_3 \qquad O_2N-\!\!\!\langle\bigcirc\rangle\!\!\!-CH_3$$

14.20   Double Bonds in Side Chains on Aromatic Rings: Conjugation

The presence of a carbon-carbon double bond in the side chain of an aromatic compound characterizes a general hydrocarbon family called **alkenylbenzenes;** two common examples are styrene and α-methylstyrene:

$$CH=\!CH_2 \qquad\qquad \overset{\displaystyle CH_3}{\underset{\displaystyle}{\diagdown}}C=\!CH_2$$

Phenylethene   2-Phenylpropene
(styrene)   (α-methylstyrene)

Styrene is frequently used in polymerization reactions (see Sec. 8.34) and is prepared industrially by catalytic dehydrogenation of ethylbenzene; ethylbenzene is prepared via a Friedel-Crafts alkylation reaction between ethene and benzene, both of which are cheap industrial compounds.

$$\langle\bigcirc\rangle + CH_2\!=\!CH_2 \xrightarrow{\;H^{\oplus}\;} \langle\bigcirc\rangle\!-CH_2\!-CH_3 \xrightarrow[\text{heat}]{\text{catalyst}} \langle\bigcirc\rangle\!-CH=\!CH_2 + H_2$$

Ethylbenzene   Styrene

In the laboratory, either dehydration or dehydrohalogenation is the preferred method. Some interesting trends in the orientation of the double bond are seen from these two types of reactions. For example, the dehydrohalogenation of 1-phenyl-2-bromobutane and the dehydration of the alcohol 1-phenyl-2-butanol both give a single alkene as product, even though the double bonds in both possible products contain the same number of substituents. The Saytzeff rule (see Sec. 7.12) predicts that neither should be greatly preferred.

$$\langle\bigcirc\rangle\!-CH_2\!-\!\underset{\underset{\displaystyle Br}{|}}{CH}\!-CH_2\!-CH_3 \xrightarrow[\substack{\text{alcohol,}\\\text{heat}}]{KOH} \langle\bigcirc\rangle\!-CH=\!CH-CH_2\!-CH_3$$

1-Phenyl-2-bromobutane   1-Phenyl-1-butene (*cis* and *trans*)
                                         Only product

+

$$\langle\bigcirc\rangle\!-CH_2\!-\!\underset{\underset{\displaystyle OH}{|}}{CH}\!-CH_2\!-CH_3 \xrightarrow[\text{heat}]{H^{\oplus}} \langle\bigcirc\rangle\!-CH_2\!-CH=\!CH-CH_3$$

1-Phenyl-2-butanol   1-Phenyl-2-butene
                                 Not observed

This and similar observations indicate that the presence of a double bond *adjacent* to an aromatic ring characterizes an especially stable molecule. The double bond is said to be **conjugated** with the aromatic ring:

Conjugated system
Unusually stable

The reason for this unusual stability is best seen from the orbital picture of a conjugated system:

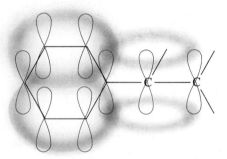

Conjugated system
(all *p* orbitals overlap with one another)

Delocalization of electrons through overlap is important in the unusual stability of other systems as well (for example, radicals). The added stability of a conjugated system explains the orientation of the previous elimination reactions.

Question 14.17

Use equations to illustrate each of the following:

(*a*) Two different alcohols that could be used to prepare styrene
(*b*) Two different alkyl halides that could be used to prepare styrene
(*c*) An alkyl halide that could be used to prepare α-methylstyrene

14.21   Addition of Ionic Reagents to Alkenylbenzenes: The Benzyl Cation

The ionic addition of unsymmetrical reagents to alkenes first gives the more stable carbocation. This is followed by the union of the resulting ion with a nucleophile. Recall that the addition of an acidic reagent to a conjugated diene system occurs to give the more stable carbocation (see Sec. 8.11):

Conjugated diene   Allyl cation

In this case the carbocation is stabilized because of the proximity of the positive charge to the adjacent electron-rich π bond.

It is not surprising that this same important principle governs the addition of unsymmetrical reagents, such as HX and HOH, to a conjugated alkenylbenzene. Consider the addition of H_2O to 1-phenyl-1-propene in the presence of an acidic catalyst; the reactive intermediate is a benzyl cation, in which the positive charge is as close as possible to the electron-rich π cloud of the aromatic ring:

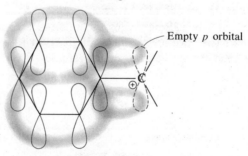

2° Carbocation
(less stable
than benzyl
cation)
Not formed

2° Carbocation
(also benzyl cation)
Formed

1-Phenyl-1-propanol

The ionic addition of the general reagent E—Nu follows this route, where the more positive part of the reagent ($E^{\oplus}$, the electrophile) adds to give the more stable benzyl cation, followed by the addition of the nucleophile, $Nu\!:^{\ominus}$

$$Ar—C\!\!=\!\!C\diagdown + E^{\oplus} \longrightarrow Ar—\underset{\oplus}{C}—\underset{E}{C}— \xrightarrow{\;:Nu^{\ominus}\;} Ar—\underset{Nu}{C}—\underset{E}{C}—$$

Benzyl cation

The orbital view of the benzyl cation sheds light on its stability, where the *empty p* orbital of the carbocation can accept electrons through overlap with the electron-rich π cloud of the aromatic ring:

Empty *p* orbital

Benzyl cation
(positive charge delocalized over
ring and thus stabilized)

Resonance can also be used to represent this stabilization effect. The following electronic structures indicate that the positive charge is more concentrated on the *ortho* and *para* positions of the ring:

Resonance hybrid,
best representation

Experimental measurements on the formation of the benzyl cation, $C_6H_5{-}\overset{\oplus}{C}H_2$, from toluene indicate that it is *slightly* more stable than a tertiary carbocation or an allyl cation. It may be included in the carbocation stability sequence of Sec. 5.7 as follows:

Decreasing stability of carbocations:

$$\text{Benzyl} \approx \text{allyl} \approx 3° > 2° > 1° > \overset{\oplus}{C}H_3$$

Question 14.18

Based on your knowledge of the *net* activating or deactivating properties of various substituents, predict the relative stabilities of the following five benzyl cations; briefly justify your choice.

Question 14.19

Based on your knowledge of the stabilities of free radicals, predict the orientation of the peroxide-catalyzed addition of hydrogen bromide to 1-phenyl-1-propene, $C_6H_5{-}CH{=}CH{-}CH_3$. Provide a complete, stepwise mechanism for this reaction. How does the orientation of HBr addition under these conditions compare with HBr addition under ionic conditions, that is, in CCl_4 solvent? Why?

14.22   Stable Carbocations

Several known carbocations are stable and can be isolated. Perhaps the most notable is the triphenylmethyl cation, which can be prepared in the laboratory by allowing triphenylmethanol to react with fluoroboric acid in acetic anhydride as a solvent:

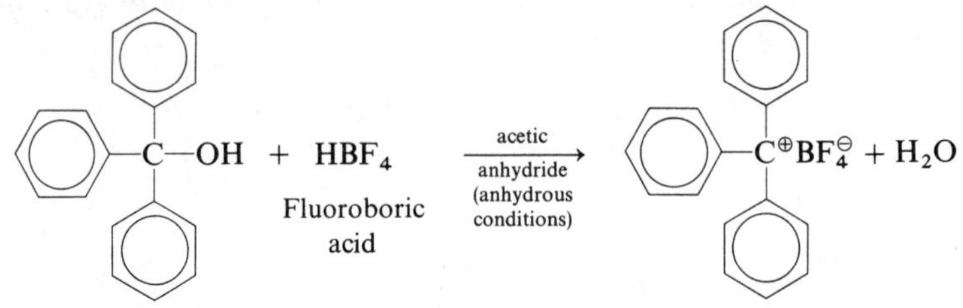

Triphenylmethanol
White

Triphenylmethyl fluoroborate
Stable, isolable,
bright orange

The triphenylmethyl fluoroborate salt is bright orange and is obtained in crystalline form when anhydrous ether is added to the reaction mixture in which it is prepared. Fluoroboric acid is a strong acid, and its reaction with an alcohol is typical of the reaction of any acid and alcohol (see Sec. 7.15). In this case, however, the carbocation cannot rearrange and cannot lose a proton, so it remains as a carbocation.

The triphenylmethyl cation is unusually stable because the positive charge is delocalized and thus stabilized by interaction with the three electron-rich π clouds that are adjacent to it.

The triphenylmethyl cation also reacts with cycloheptatriene to form the cycloheptatrienyl cation (tropylium ion) by abstraction of a hydride ion from the hydrocarbon; the triphenylmethyl cation is converted to triphenylmethane in the process. The cycloheptatrienyl cation (an aromatic ion) is more stable than the triphenylmethyl cation, thus providing the driving force for the reaction (see Sec. 13.7B).

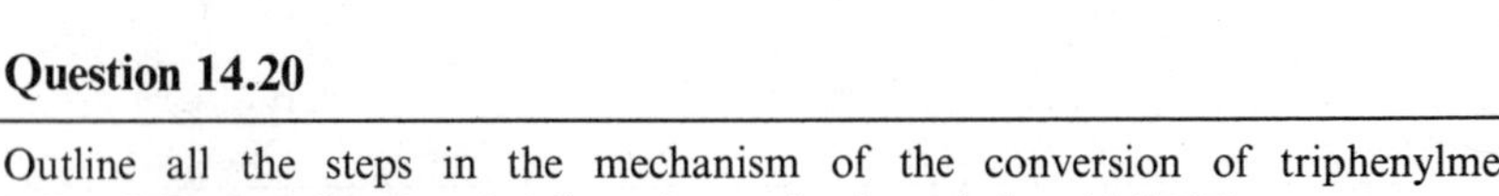

Cycloheptatriene

Triphenylmethyl
fluoroborate

Cycloheptatrienyl
cation
More stable
ion

Triphenylmethane

Question 14.20

Outline all the steps in the mechanism of the conversion of triphenylmethanol, $(C_6H_5)_3C$—OH, to triphenylmethyl fluoroborate by fluorobric acid, HBF_4.

Question 14.21

Triphenylmethyl fluoroborate is a stable carbocation and can be kept in the absence of moisture. What reaction do you think occurs when it comes in contact with moisture and is thereby destroyed? Provide a mechanism to show what happens.

14.23   Thallium in Organic Synthesis

In recent years a new and highly versatile reagent, thallium trifluoroacetate $[Tl(OOCCF_3)_3]$, has emerged for use in the substitution of one or more positions in aromatic systems.

Although thallium was discovered in 1861, not until recent years has the reagent been routinely used in organic synthesis. The lack of work with thallium in the past was partly due to its toxicity. Most of the credit for the work in this area goes to Drs. Edward C. Taylor of the Princeton University and A. McKillop of the University of East Anglia. Although several different thallium reagents have been designed and investigated by these individuals, we limit this discussion to thallium trifluoroacetate (TTFA).

TTFA is prepared by the reaction of thallium oxide and trifluoroacetic acid:

$$Tl_2O_3 \; + \; 3 \; CF_3\overset{\displaystyle O}{\overset{\|}{C}}\!-\!OH \; \longrightarrow \; 2 \; Tl(OCCF_3)_3 + 3 \; H_2O$$

Thallium oxide	Trifluoroacetic acid	Thallium trifluoroacetate (TTFA)

When TTFA reacts with aromatic compounds, an organothallium reagent is

$$CF_3\overset{\displaystyle O}{\overset{\|}{C}}\!-\!O \qquad O\!-\!\overset{\displaystyle O}{\overset{\|}{C}}CF_3$$

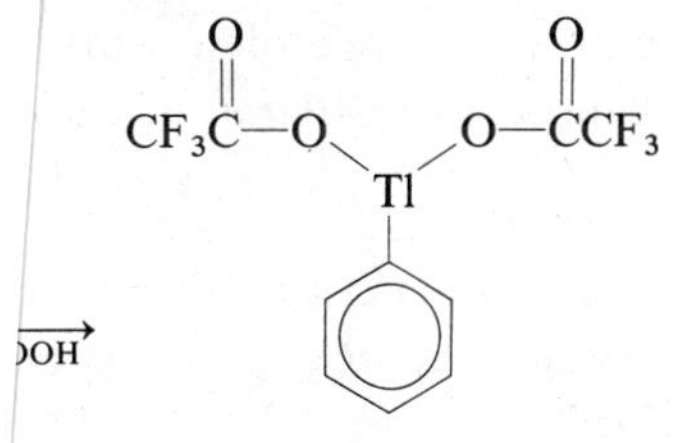

Organothallium reagent

hen react (in situ) with a variety of reagents ted aromatic compounds in high yield:

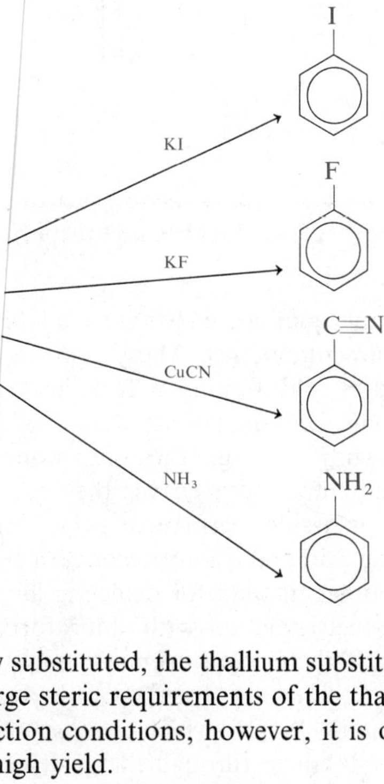

If the aromatic system is already substituted, the thallium substitutes generally at the *para* position because of the large steric requirements of the thallium electrophile. By slight variations in the reaction conditions, however, it is often possible to prepare one isomer selectively in high yield.

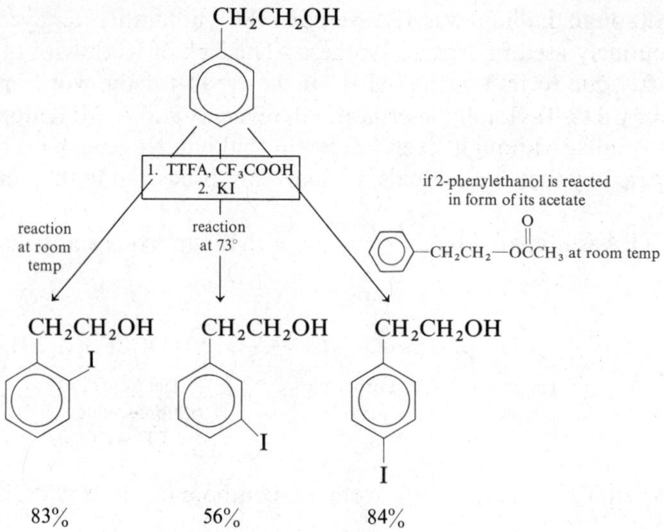

Also, the presence of substituents capable of complexing with the TTFA on the ring can lead to *ortho* attack exclusively. This is true for aryl carboxylic acids

$$
\begin{array}{ccc}
& O & & O \\
& \| & & \| \\
(Ar&-C-OH), & \text{esters (Ar}&-C-OR), & \text{ethers (Ar}-O-R), & \text{and alcohols (Ar}-OH),
\end{array}
$$

(Ar—C—OH), esters (Ar—C—OR), ethers (Ar—O—R), and alcohols (Ar—OH), to name a few. The orientation is probably caused by a complexation of the type depicted here:

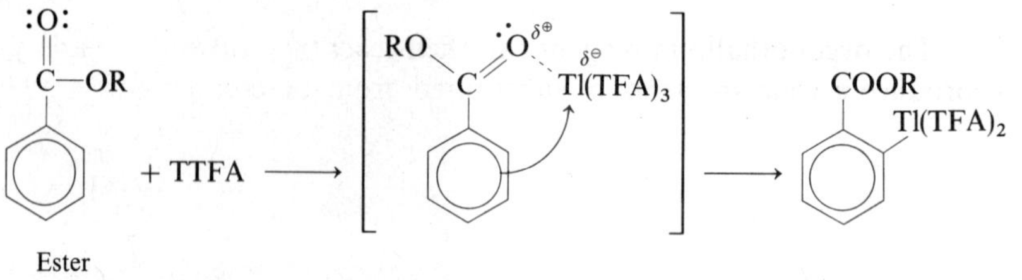

14.24   Qualitative Analysis: Detection of Aromatic Rings

Simple qualitative tests are available for rapid detection of the aromatic nucleus in organic compounds. Many aromatic compounds are readily sulfonated when dissolved in cold, fuming sulfuric acid, and the resulting sulfonic acids are soluble in the acidic solution. On the other hand, alkenes and alkynes also dissolve in fuming sulfuric acid, although these functional groups are readily distinguishable from aromatic compounds because they decolorize bromine in carbon tetrachloride or aqueous potassium permanganate solution. Alkanes and alkyl halides are insoluble in sulfuric acid and do not react with Br_2 or aqueous $KMnO_4$.

Another method for detecting alkylbenzenes and other aromatic compounds involves their reaction with chloroform in the presence of anhydrous aluminum chloride. If there is an aromatic nucleus, the solution's color varies from red to orange. A series of Friedel-Crafts reactions ultimately produces triarylmethyl carbocations, $Ar_3C^{\oplus}$, which are responsible for the colors. The positive charge on carbon is stabilized through charge delocalization of the type described for the stable triphenylmethyl cation in Sec. 14.22. The Friedel-Crafts reactions believed to occur are the following:

$$CHCl_3 + AlCl_3 \longrightarrow {}^{\oplus}CHCl_2 + AlCl_4^{\ominus}$$

$$\downarrow \text{Ar—H}$$

$$\left[Ar \underset{CHCl_2}{\overset{H}{<}} \right]^{\oplus} \longrightarrow Ar\text{—}CHCl_2 + H^{\oplus}$$

$$\downarrow \text{reacts with two more moles of Ar—H}$$

$$Ar_3C\text{—H}$$

Triarylmethane, Ar_3C—H, is converted to the triarylmethyl cation that is responsible for the color. This conversion may well involve a hydride-ion transfer from triarylmethane to a diarylmethyl cation:

$$Ar_3C\text{—H} + Ar_2\overset{\oplus}{C}H \longrightarrow Ar_3C^{\oplus} + Ar_2CH_2$$

Less stable	More stable
2° ion	3° ion

The diarylmethyl cation can be formed from the reaction between diarylmethyl chloride, Ar_2CH—Cl, which is an intermediate in the Friedel-Crafts reaction, and $AlCl_3$:

$$Ar_2CH\text{—Cl} + AlCl_3 \longrightarrow Ar_2\overset{\oplus}{C}H + AlCl_4^{\ominus}$$

The color of the solution gives some indication of the nature of the aromatic compound. Benzene and its homologs, as well as aryl halides, give orange to red solutions. Other aromatic compounds (studied later) also react; biphenyl and phenanthrene give purple solutions, napthalene gives a blue solution, and anthracene gives green.

14.25   Spectral Analysis of Aromatic Compounds

The IR and UV spectra of a variety of aromatic hydrocarbons were presented in Chap. 12. Of all the chromophores encountered in organic chemistry, the aromatic ring is the most studied.

A. Infrared Spectroscopy

The IR spectra of aromatic compounds normally contain peaks of moderate intensity at 3,010 to 3,100 cm^{-1} and sharp peaks at 675 to 995 cm^{-1} due to C—H stretching and bending vibrations, respectively. Peaks of variable intensity are observed at 1,500 to 1,600 cm^{-1}, which is the C=C bond stretching frequency (see Table 12.3). Also, weak bands, called *overtone* bands, resulting from less probable vibrational transitions are observed in the 1,650 to 2,000-cm^{-1} region of the spectrum. These overtones, though quite weak, are easily picked out because there are so few other absorptions in this region. The overtone bands are extremely helpful in elucidating the structure of a substituted aromatic benzene ring because varying substituted carbon skeletons have very characteristic patterns. However, these patterns appear similar when seen on routine instruments and must be used with caution. The overtone bands coupled with the absorptions observed in the 675 to 995-cm^{-1} region give a reasonably clear picture of the substitution pattern of aromatic compounds. The IR spectra of o-, m-, and p-xylene in Fig. 14.4 illustrate these fundamental absorptions.

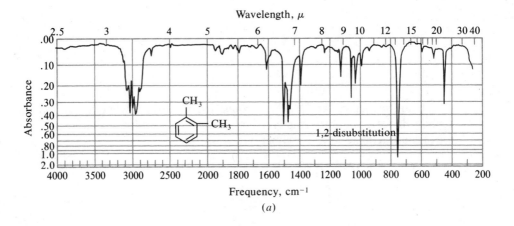

(a)

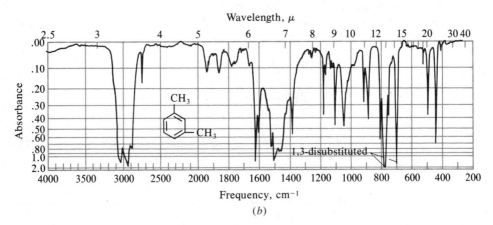

(b)

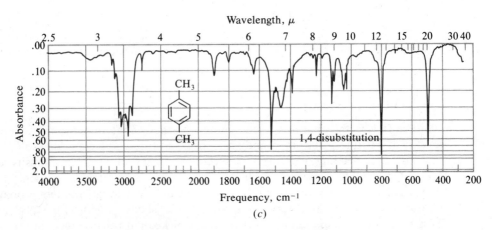

(c)

FIGURE 14.4   The IR spectra of (a) *ortho*-xylene, (b) *meta*-xylene, and (c) *para*-xylene. Primary absorption bands are labeled contrasting the similarities and differences caused by the various substitution patterns.

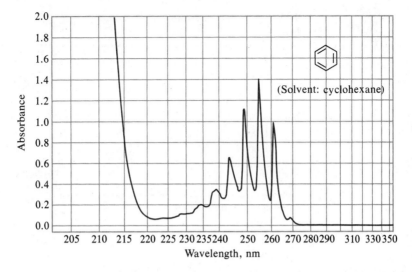

FIGURE 14.5   UV spectrum of benzene. [Data from *Sadtler Handbook of Ultraviolet Spectra*, W. W. Simons, © Sadtler Research Laboratories, Division of Bio-Rad Laboratories, Inc. (1979).]

ठ‍मित

B. Ultraviolet Spectroscopy

The $\pi \rightarrow \pi^*$ transitions of the π electrons of the benzene and substituted benzene chromophores are observed in the 200 to 260-nm region of the spectrum. A very intense band caused by these transitions appears in the region around 180 nm. The extinction coefficient (ε) associated with this band is about five times larger than that of the 200-nm band, which has an extinction coefficient of around 10,000. The UV spectrum of benzene, which is typical of an aromatic compound, is shown in Fig. 14.5.

The substitution of the benzene skeleton with another group causes shifts in all the observed bands. These shifts result from changes in the electronic energy levels associated with the molecule and may be explained in terms of the inductive, resonance, and steric effects of the attached groups. In general, alkyl groups tend to cause slight shifts to longer wavelength; other substitutents, particularly polar substituents, also cause quite large shifts in the direction of longer wavelength. See Table 12.4 for examples.

C. Spectroscopy and Structure Elucidation

Polynuclear aromatic hydrocarbons (PAHs) were introduced in Sec. 13.12. These compounds are aromatic, but, depending on the number of rings involved and the bonds involved in the fusion of these rings, they contain π electron clouds of varying size and geometry. Depending on the arrangements of the aromatic nuclei that make up the PAH, particular shifts of the fundamental aromatic absorption bands can be predicted.

In all PAHs the three common absorption bands are shifted to longer wavelengths with respect to benzene (Table 14.3). Also, the more rings, the greater the shift. However, the magnitude of these shifts is significantly different when the ring system is linear, anthracene for example, from when it is angular, phenanthrene for example. The angular arrangement results in less shift than does the linear arrangement. The UV spectra of anthracene and phenanthrene are provided in Fig. 14.6.

TABLE 14.3   UV Absorptions of Aromatic Hydrocarbons

Compound	Wavelength of Absorption, λ, nm*		
Benzene	184	204	255
Naphthalene	221	275	311
Anthracene	252	376	. . .
Phenanthrene	251	292	350

* All in hydrocarbon solvents.

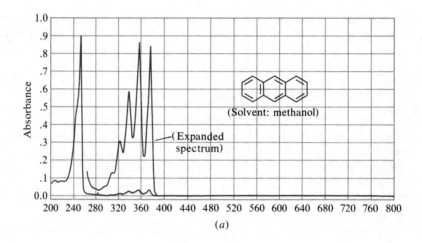

(a)

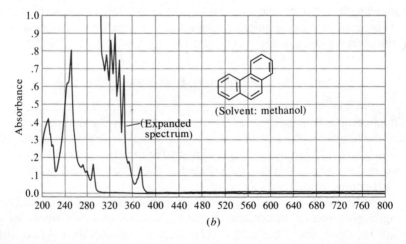

(b)

FIGURE 14.6   UV spectrum of (a) anthracene and (b) phenanthrene. [Data from *Sadtler Handbook of Ultraviolet Spectra*, W. W. Simons, ed. © Sadtler Research Laboratories, Division of Bio-Rad Laboratories, Inc. (1980).]

14.26 Summary of Reactions of Aromatic Compounds

A. Electrophilic Aromatic Substitution Reactions

a. General Reaction

$$Ar-H + E^{\oplus} \longrightarrow Ar-E + H^{\oplus}$$

Mechanism and discussion (Secs. 14.1 and 14.2)

b. Monosubstitution Reactions

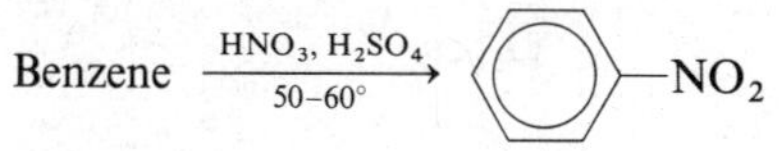

Benzene $\xrightarrow[50-60°]{HNO_3, H_2SO_4}$ ⟨benzene ring⟩—NO$_2$ Mechanism (Sec. 14.3)

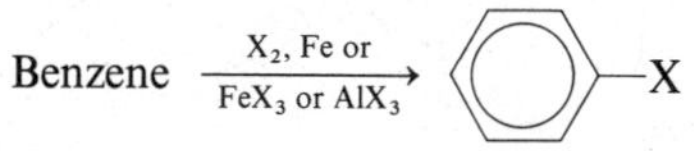

Nitrobenzene

Benzene $\xrightarrow[FeX_3 \text{ or } AlX_3]{X_2, Fe \text{ or}}$ ⟨benzene ring⟩—X   Mechanism (Sec. 14.4)

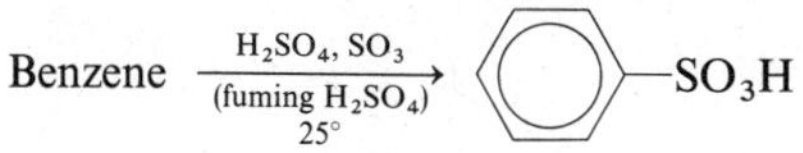

Halobenzene
(X = Cl, Br)

Benzene $\xrightarrow[\substack{(fuming\ H_2SO_4) \\ 25°}]{H_2SO_4,\ SO_3}$ ⟨benzene ring⟩—SO$_3$H   Mechanism (Sec. 14.5)

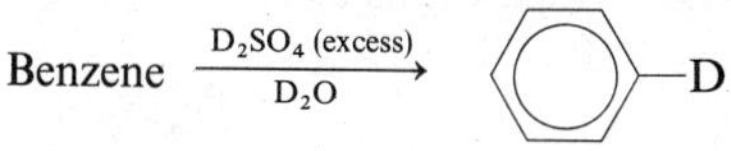

Benzenesulfonic
acid

Benzene $\xrightarrow[D_2O]{D_2SO_4 \text{ (excess)}}$ ⟨benzene ring⟩—D   Sec. 14.5

Deuteriobenzene

Benzene $\xrightarrow[\substack{R = alkyl,\ not \\ aryl}]{R-X,\ AlCl_3}$ ⟨benzene ring⟩—R   Mechanism and rearrangements in alkyl groups (Sec. 14.6)

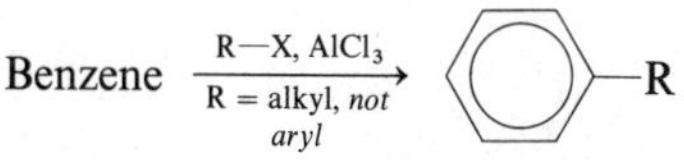

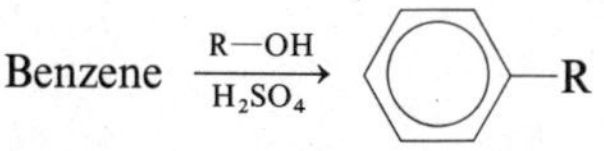

Alkylbenzene
Rearrangements may occur

Benzene $\xrightarrow[H_2SO_4]{R-OH}$ ⟨benzene ring⟩—R   Mechanism (Sec. 14.7A)

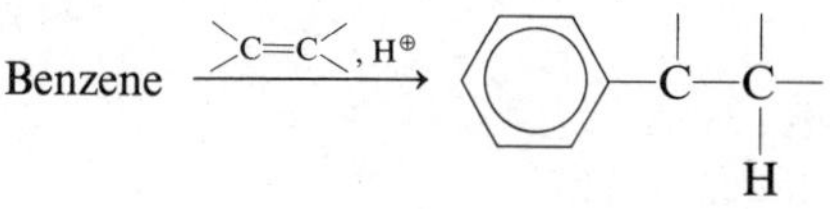

Benzene $\xrightarrow{\substack{>C=C<,\ H^{\oplus}}}$ ⟨benzene ring⟩—$\overset{\displaystyle |}{\underset{\displaystyle |}{C}}-\overset{\displaystyle |}{\underset{\displaystyle H}{C}}$—   Mechanism (Sec. 14.7B)

c. Disubstitution Reactions: General

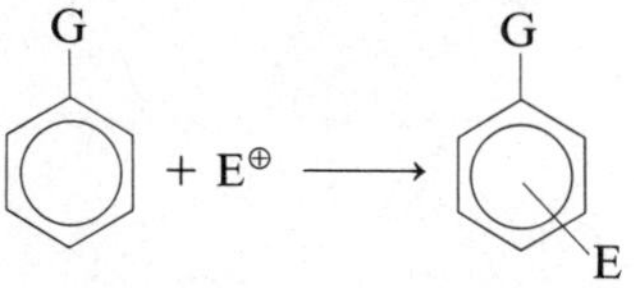

⟨benzene ring with G⟩ + E$^{\oplus}$ ⟶ ⟨benzene ring with G and E⟩

General discussion of orientation and reactivity (Sec. 14.8)
Factors affecting reactivity (Sec. 14.9)
Orientation and reactivity (Secs. 14.10A and 14.11)

d. Disubstitution Reactions: Examples

1. 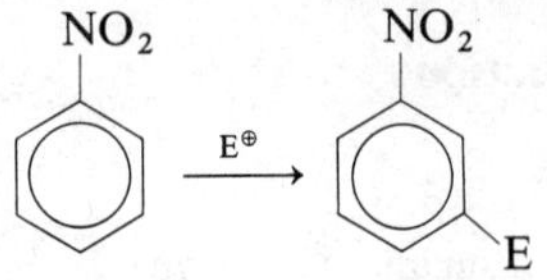   Sec. 14.14A

where E = NO_2, X, SO_3H. Cannot alkylate (Friedel-Crafts reaction)

2. 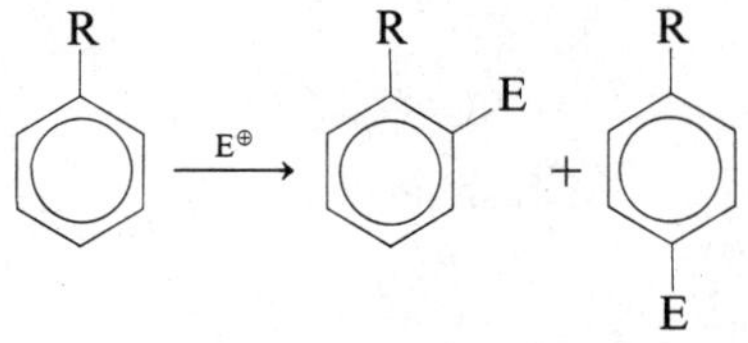   Sec. 14.14B

ortho   *para*

where E = R, SO_3H, NO_2, X

3. 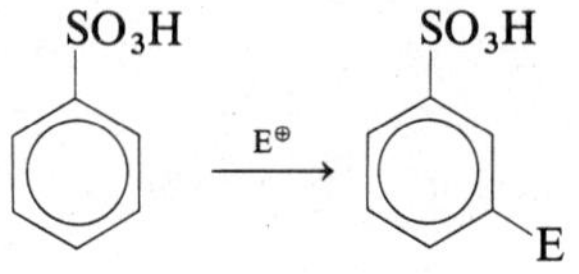   Sec. 14.14C

meta

where E = NO_2, X, SO_3H. Cannot alkylate (Friedel-Crafts reaction)

4. 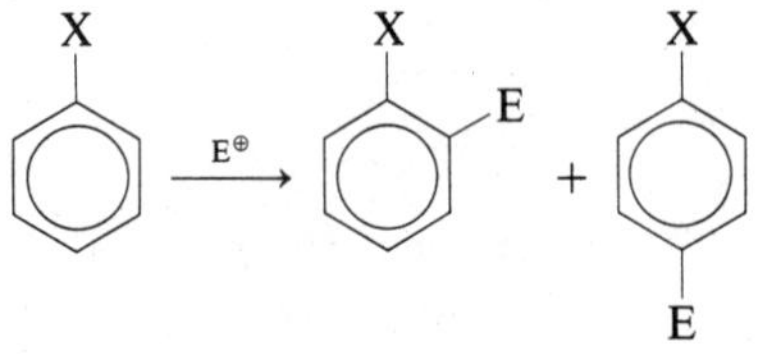   Sec. 14.14D

ortho   *para*

where E = R, SO_3H, NO_2, X

5. Applications to synthesis (Sec. 14.15)

e. Trisubstitution Reactions
Applications to synthesis (Sec. 14.16)

B. Addition Reactions of Aromatic Ring

a. Hydrogenation (Reduction)

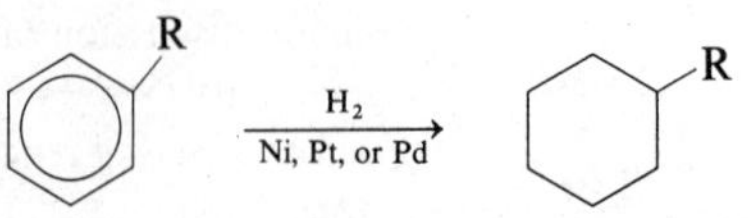

Sec. 14.17

Alkylbenzene   Alkylcyclohexane

b. Chlorination

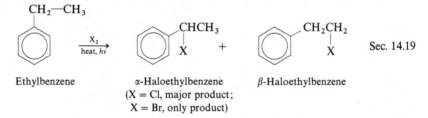

Benzene $\xrightarrow[\text{pressure}]{\text{Cl}_2}$ Benzenehexachloride   Sec. 14.17

C. Oxidation of Alkyl Side Chains

Alkylbenzene $\xrightarrow[\text{or K}_2\text{Cr}_2\text{O}_7, \text{H}_2\text{SO}_4, \text{heat}]{\text{KMnO}_4, \text{OH}^\ominus, \text{H}_2\text{O}, \text{heat}}$ Benzoic acid $+$ CO_2 Sec. 14.18

D. Side-Chain Halogenation

Ethylbenzene (CH_2—CH_3) $\xrightarrow[\text{heat, }hv]{X_2}$ α-Haloethylbenzene (X = Cl, major product; X = Br, only product) $+$ β-Haloethylbenzene   Sec. 14.19

E. Conjugated Alkenylbenzenes

a. Preparation

—$CH_2CHCH_2CH_3$ (X) $\xrightarrow[\text{heat}]{\text{KOH}\atop\text{alcohol}}$ —$CH=CHCH_2CH_3$ Sec. 14.20

only product

—$CH_2CHCH_2CH_3$ (OH) $\xrightarrow{\text{H}^\oplus, \text{heat}}$

b. Addition Reactions

—$CH=CHCH_3$ $+$ H—X $\xrightarrow{\text{CCl}_4}$ —$CHCH_2CH_3$ (X)   Sec. 14.21

1-Phenyl-1-propene   1-Halo-1-phenylpropane

F. Stable Carbocations

$(C_6H_5)_3C$—OH $\xrightarrow{\text{HBF}_4}$ $(C_6H_5)_3C^\oplus BF_4^\ominus$ Sec. 14.22

Triphenylmethanol   Triphenylmethyl fluoroborate
White   *Orange*

G. Organothallium Reagents

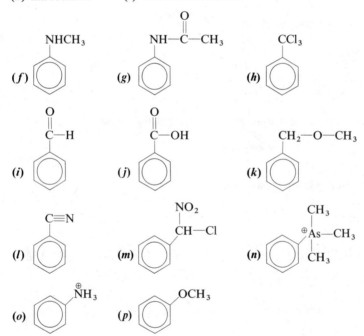

where G = —I, —F, —CN, –NH$_2$. Dependent on temperature and solvent

Study Questions

14.22 Suppose each of the following 16 compounds is monochlorinated on the ring. Give the structure(s) of the principal product(s) expected from each of them, and indicate whether this chlorination reaction would be faster than, slower than, or equal to the rate of chlorination of benzene itself.

(*a*) bromobenzene   (*b*) *n*-propylbenzene   (*c*) aniline
(*d*) nitrobenzene   (*e*) benzenesulfonic acid

(*f*) NHCH$_3$ (*g*) NH—C(=O)—CH$_3$ (*h*) CCl$_3$

(*i*) C(=O)—H   (*j*) C(=O)—OH   (*k*) CH$_2$—O—CH$_3$

(*l*) C≡N   (*m*) NO$_2$ / CH—Cl   (*n*) CH$_3$ / ⊕As—CH$_3$ / CH$_3$

(*o*) ⊕NH$_3$ (*p*) OCH$_3$

14.23 Draw the structure(s) of the principal product(s) that would be obtained by introducing a chlorine atom (by electrophilic aromatic substitution) into the following compounds.

(*a*) *m*-dinitrobenzene   (*b*) *p*-methylaniline   (*c*) *p-tert*-butyltoluene
(*d*) *o*-nitrotoluene   (*e*) *m*-nitrotoluene   (*f*) *m*-nitrophenol
(*g*) *o*-xylene (*o*-dimethylbenzene)   (*h*) *m*-xylene
(*i*) *p*-xylene

(*j*) NH—C(=O)—CH$_3$ / NO$_2$ (*k*) OH / CH$_3$

(l)

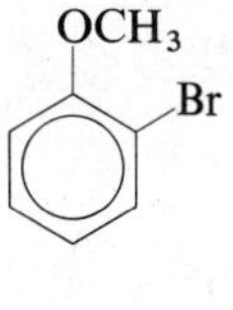

OCH₃
Br

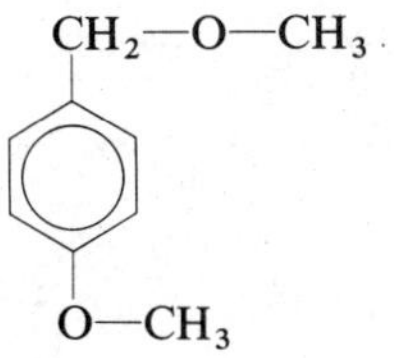

(m)

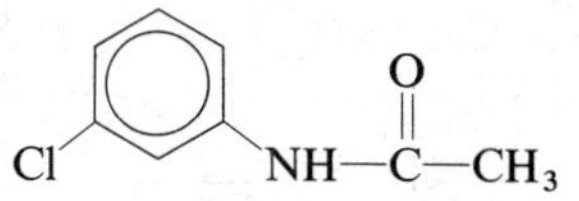

Cl
NH—C—CH₃
O

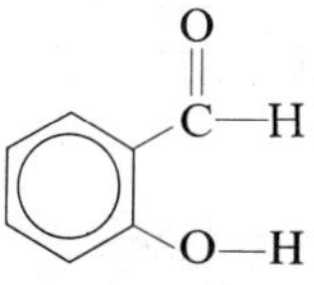

(n)

CH₂—O—CH₃

O—CH₃

(o)

C—H
O
O—H

(p)

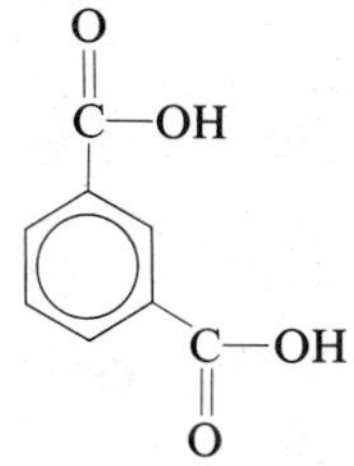

C—OH
O

C—OH
O

(q)

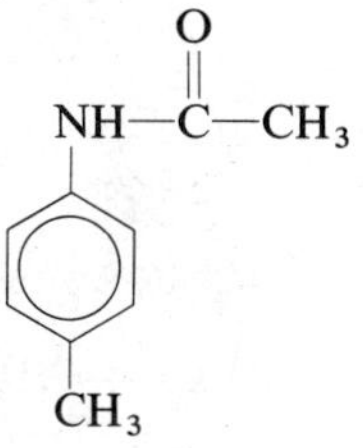

NH—C—CH₃
O

CH₃

(r) CH₃—⬡—⬡

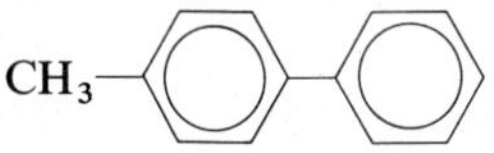

(s) ⬡—⬡

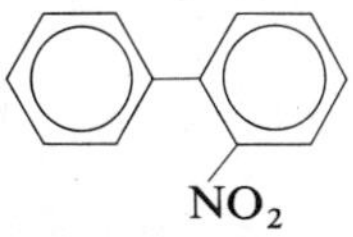

NO₂

(t) ⬡—CH₂—O—⬡

(u)

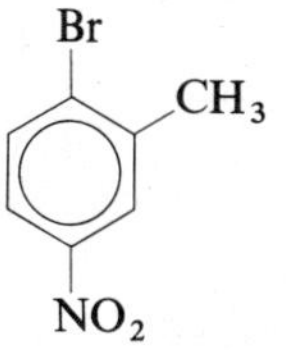

Br
CH₃

NO₂

(v)

Br
CH₃

O₂N

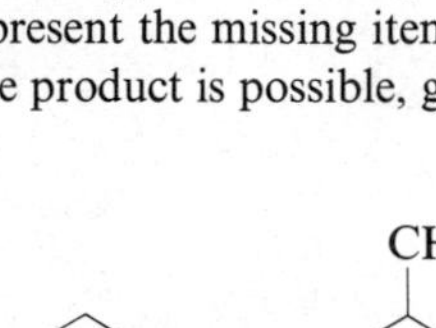

14.24 Complete the following 14 reactions by supplying the missing organic product(s), missing reactant(s), and missing reagent(s) as needed. The numbers in parentheses [for example, (1)] represent the missing items. Use any organic or inorganic reagents you want. When more than one product is possible, give the structures of all major products.

(a) ⬡ $\xrightarrow{(1)}$

CH₂CH₃
⬡ $\xrightarrow[\text{H}_2\text{SO}_4]{\text{HNO}_3}$ (2), monosubstitution

$\downarrow$ K₂Cr₂O₇, H₂SO₄, heat

(3)

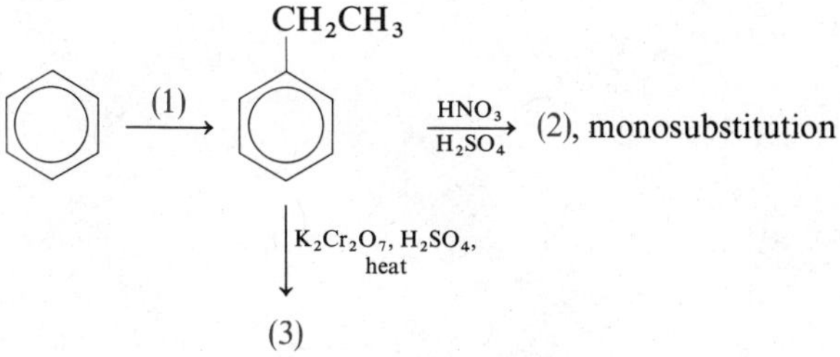

(*b*)

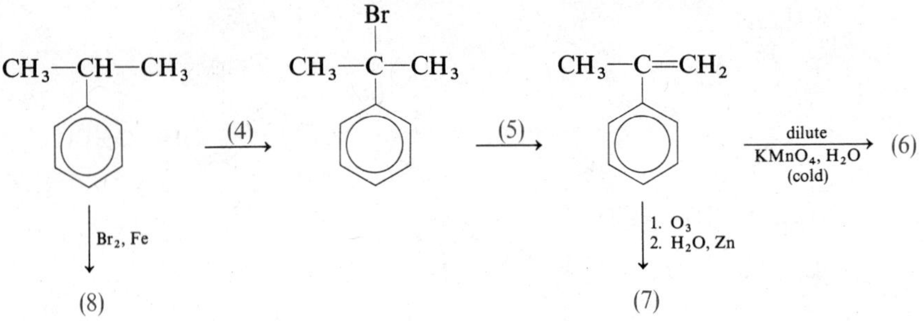
Br
CH3—CH—CH3
(4)
CH3—C—CH3
(5)
CH3—C=CH2
dilute
KMnO4, H2O
(cold)
(6)
Br2, Fe
(8)
1. O3
2. H2O, Zn
(7)

(*c*)

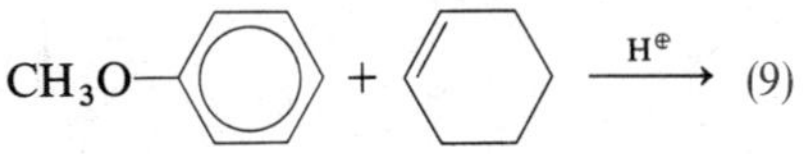
CH3O
+
H⊕
(9)

(*d*)

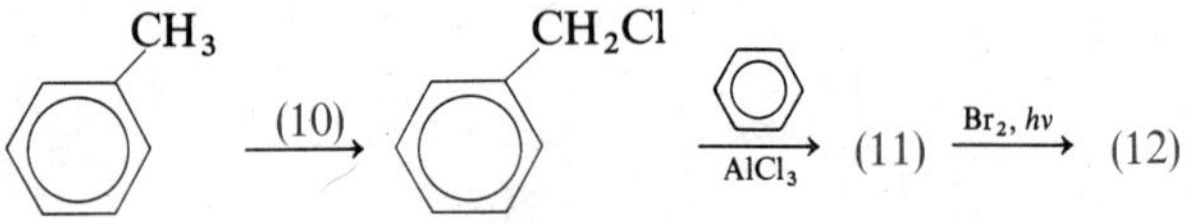
CH3
(10)
CH2Cl
AlCl3
(11)
Br2, hν
(12)

(*e*)

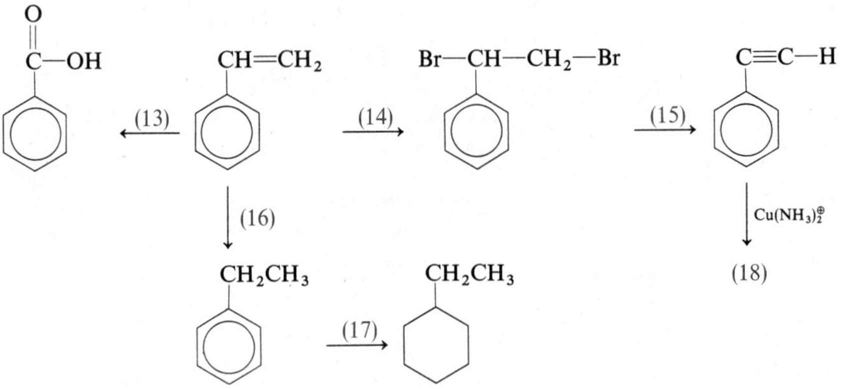
O
C—OH
(13)
CH=CH2
(14)
Br—CH—CH2—Br
(15)
C≡C—H
(16)
Cu(NH3)2⊕
(18)
CH2CH3
(17)
CH2CH3

(*f*)

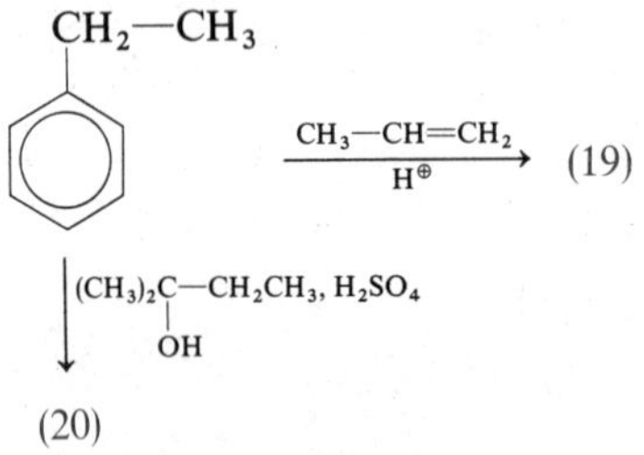
CH2—CH3
CH3—CH=CH2
H⊕
(19)
(CH3)2C—CH2CH3, H2SO4
OH
(20)

(*g*)

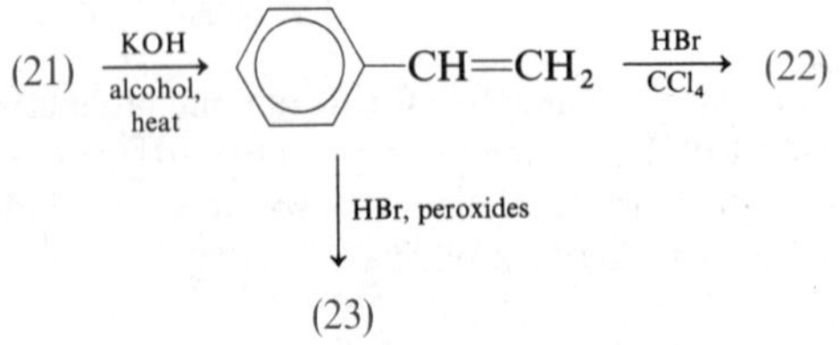
(21)
KOH
alcohol,
heat
CH=CH2
HBr
CCl4
(22)
HBr, peroxides
(23)

(*h*)

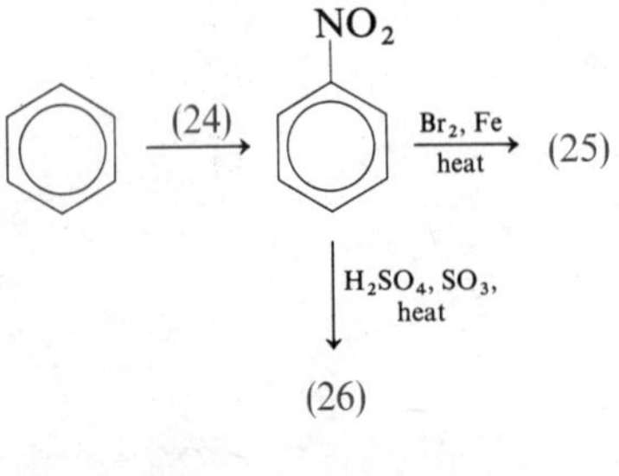
(24)
NO2
Br2, Fe
heat
(25)
H2SO4, SO3,
heat
(26)

(*i*) (27) $\xleftarrow[\substack{\text{liquid} \\ \text{NH}_3}]{\text{Li}}$ Ph—C≡C—CH$_3$ $\xrightarrow{(28)}$

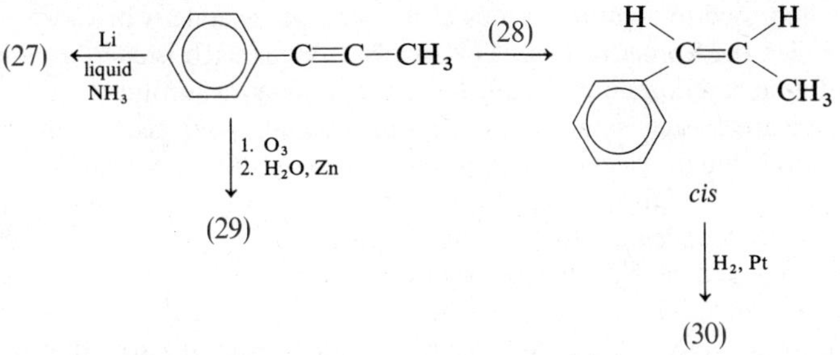

cis

$\downarrow$ 1. O$_3$
2. H$_2$O, Zn

(29)

$\downarrow$ H$_2$, Pt

(30)

(*j*) Ph—CH$_2$—CH$_2$—OH $\xrightarrow{(31)}$ Ph—CH=CH$_2$ $\xrightarrow[\text{CCl}_4]{\text{HCl}}$ (32)

$\downarrow$ KMnO$_4$, OH$^{\ominus}$
H$_2$O, heat

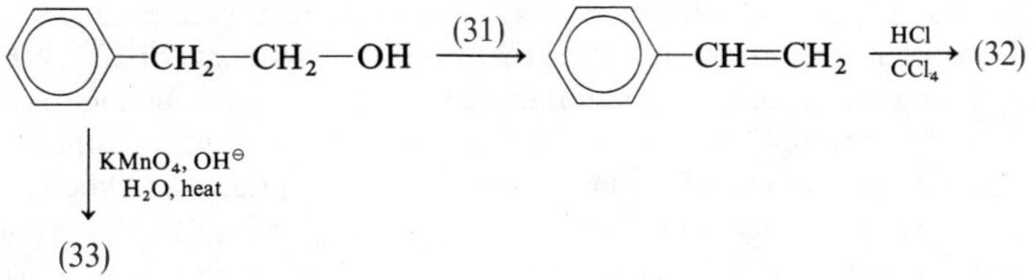

(33)

(*k*) Cl-Ph $\xrightarrow[\text{AlCl}_3, 0°]{\text{CH}_3\text{Cl}}$ (34)

$\downarrow$ (35)

SO$_3$H / Cl $\xrightarrow[\text{H}_2\text{O}]{\text{NaOH}}$ (36)

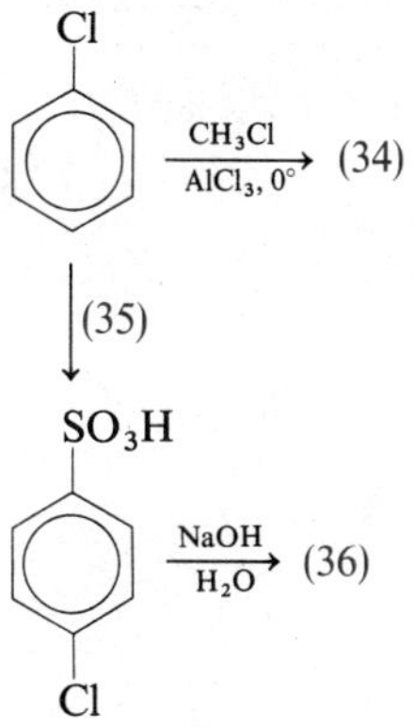

(*l*) CH$_3$-Ph $\xrightarrow[\text{SO}_3]{\text{H}_2\text{SO}_4}$ (37) + CH$_3$-Ph-SO$_3$H $\xrightarrow{(38)}$ CH$_3$-Ph(Br)-SO$_3$H $\xrightarrow[\text{heat}]{\text{KMnO}_4, \text{aq.}}$ (39)

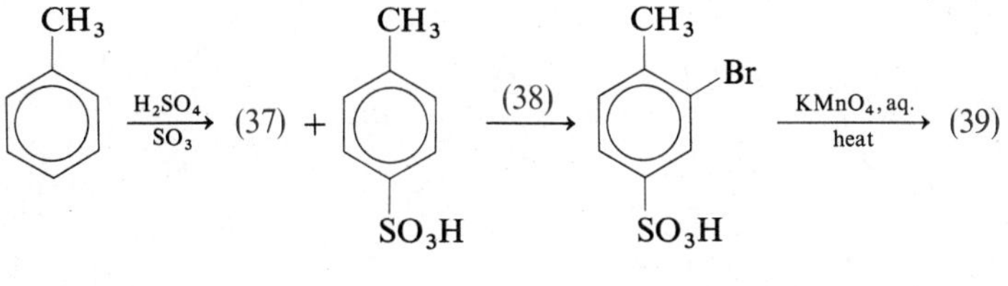

(*m*) CH$_3$-Ph $\xrightarrow[\text{AlCl}_3]{\text{CH}_3\text{CH}_2\text{CH}_2\text{—Cl}}$ (40)

$\downarrow$ Br$_2$, heat, $h\nu$

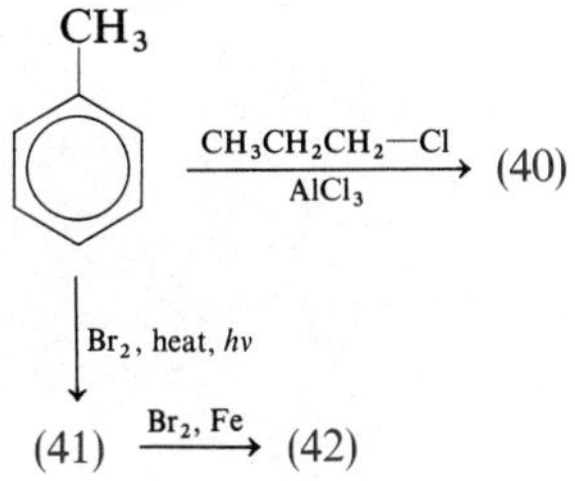

(41) $\xrightarrow{\text{Br}_2, \text{Fe}}$ (42)

(*n*) Ph—C(=O)CH$_3$ $\xrightarrow[\text{TTFA}]{\text{TFAA}}$ (43) $\xrightarrow{\text{CuCN}}$ (44)

$\downarrow$ KF

(45)

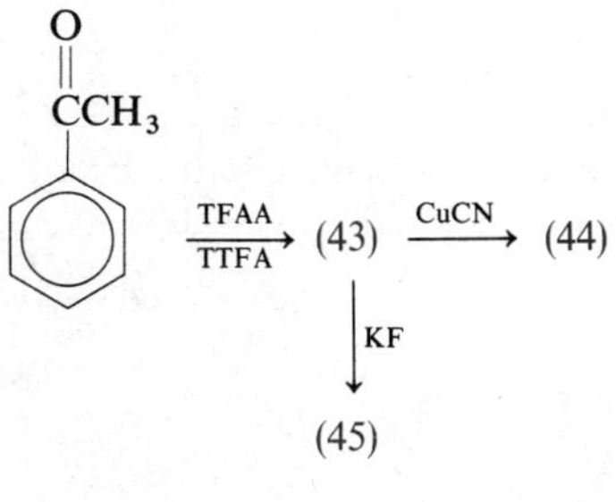

14.25 An early method used to identify a series of isomeric benzene derivatives was developed by Körner and is called the **Körner method.** This involves taking each isomer, carrying out an electrophilic substitution reaction on it so as to introduce one new substituent, and then determining *how many* derivatives each isomer yields. This method takes into account the production of all possible derivatives; but the yield is unimportant because only the number of new derivatives is of interest. For example, groups that orient *ortho* and *para* may give a small amount of *meta* substitution as well, thus producing a total of three derivatives.

(a) There are three isomeric dimethylbenzenes (*xylenes*). One isomer, on nitration, gives precisely two mononitro derivatives, another isomer gives one mononitro derivative, and the third isomer gives three mononitro derivatives. Based on this information, draw the structure of the three xylenes as well as the structure(s) of the mononitro derivative(s) that each produces.

(b) There are three isomeric tribromobenzenes that have melting points of 120°, 87°, and 44°. Mononitration of these three compounds produces one, two, and three mononitro derivatives, respectively. On the basis of this information, draw the structures of the three isomeric tribromobenzenes as well as the structure(s) of the mononitro derivative(s) that each produces.

14.26 Devise practical laboratory procedures for synthesizing each of the following 19 compounds from *benzene* and any other needed organic or inorganic reagents. Assume that pure *ortho* and *para* isomers can be isolated from a mixture containing both of them. Show all reagents and reaction conditions where pertinent, but do not balance equations.

(a) *o*-bromotoluene   **(b)** *p*-bromochlorobenzene   **(c)** benzoic acid
(d) *p*-isopropyltoluene   **(e)** 3,5-dinitrochlorobenzene

(f) O_2N—⟨benzene⟩—CH_2Br **(g)** Br—⟨benzene⟩—$\overset{\displaystyle O}{\overset{\|}{C}}$—$OH$

(h) ⟨benzene with Br⟩—$\overset{\displaystyle O}{\overset{\|}{C}}$—$OH$ **(i)** C_6H_5—$CH{=}CH_2$

(j) ⟨cyclohexane⟩—CH_2CH_3 **(k)** Br—⟨benzene⟩—NO_2 (with Br)

(l) C_6H_5—$C{\equiv}C$—H **(m)** Br—⟨benzene⟩—Br (with Br)

(n) HO—$\overset{\displaystyle O}{\overset{\|}{C}}$—⟨benzene⟩—$\overset{\displaystyle O}{\overset{\|}{C}}$—$OH$ **(o)** C_6H_5—CH_2—CH_2—Br

(p) Br—⟨benzene⟩—$CH{=}CH_2$ **(q)** C_6H_5—$\overset{\displaystyle }{\underset{\displaystyle CH_3}{C}}{=}CH_2$

(r) ⟨benzene with SO_3H⟩—$\overset{\displaystyle Br}{\underset{\displaystyle CH_3}{C}}$—$CH_3$ **(s)** iodobenzene

14.27 Indicate how the following conversions can be carried out. Use the indicated organic compounds as the only source of carbon (except for organic solvents, such as CCl_4 and ether) as well as any inorganic reagents you want.

(a) Benzene + ethane $\longrightarrow$ C_6H_5—$\overset{\displaystyle Cl}{\underset{\displaystyle }{CH}}$—$CH_3$

(b) m-Xylene $\longrightarrow$

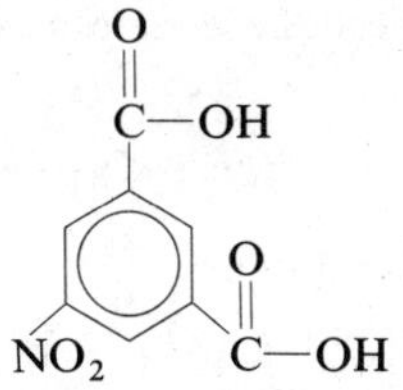

(c) Ethylbenzene $\longrightarrow$

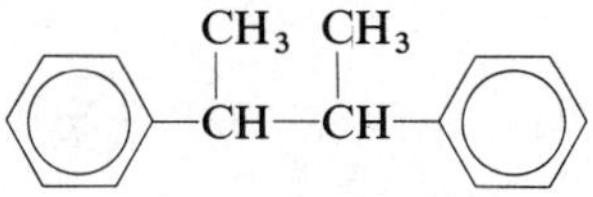

(d) Benzene + methane $\longrightarrow$

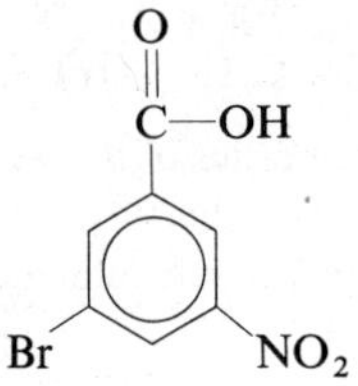

(e) Toluene $\longrightarrow$

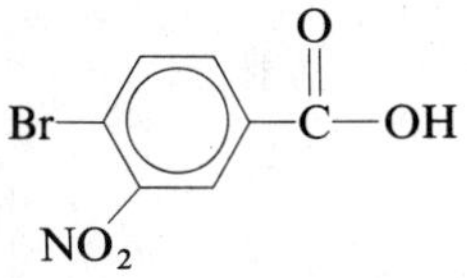

(f) Toluene $\longrightarrow$

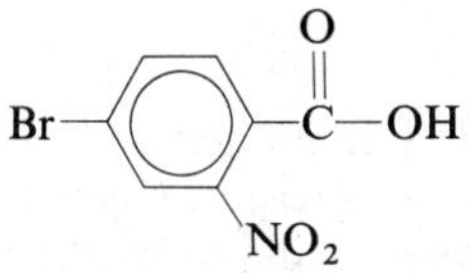

(g) Anisole $\longrightarrow$

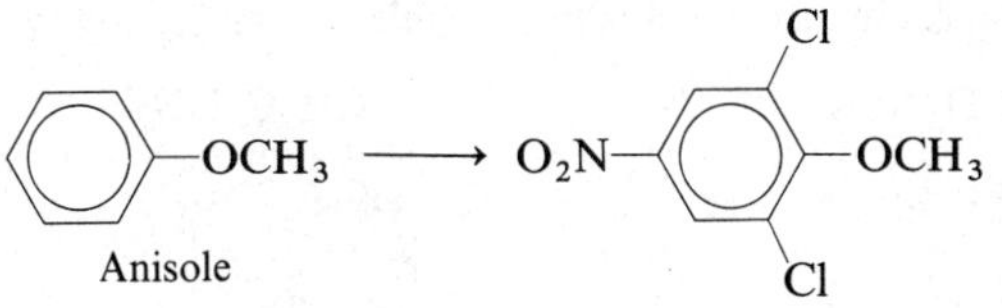

(h) Acetylene + toluene $\longrightarrow$ $C_6H_5-CH_2-C\equiv C-H$

(i) Acetylene + toluene $\longrightarrow$ $CH_2=C$ (with CH_2-phenyl and Cl)

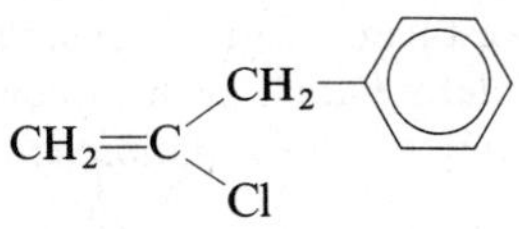

(j) Toluene $\longrightarrow$

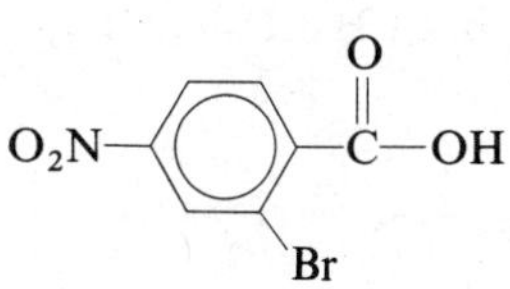

(k) Benzene + CH_4 $\longrightarrow$

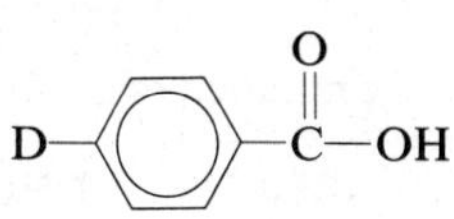

14.28 Following are 15 common substituents found on aromatic rings. Classify each in one or more of the following six categories, as applicable:

1. *ortho, para* Director
2. *meta* Director
3. Releases electrons mostly by the inductive effect
4. Withdraws electrons mostly by the inductive effect
5. Releases electrons mostly by the resonance effect

6. Withdraws electrons by the inductive effect and donates electrons by the resonance effect, both to about the same extent

(*a*) —NO_2 (*b*) —SO_3H (*c*) —O—CH_3 (*d*) —Br

(*e*) —CH_2—CH_3 (*f*) C_6H_5—   (*g*) —$\overset{\overset{\text{O}}{\|}}{C}$—H   (*h*) —C≡N

(*i*) —O—H   (*j*) —NH_2 (*k*) —$\overset{\oplus}{N}H_3$ (*l*) —$\overset{\overset{\text{O}}{\|}}{C}$—OH

(*m*) —CCl_3 (*n*) —NH—$\overset{\overset{\text{O}}{\|}}{C}$—$CH_3$ (*o*) —Cl

14.29 Arrange these four compounds in order of *decreasing reactivity* toward ring bromination. List the most reactive compound first and the least reactive last. Briefly justify your choice in terms of how inductive and/or resonance effects influence reactivity in electrophilic substitution reactions.

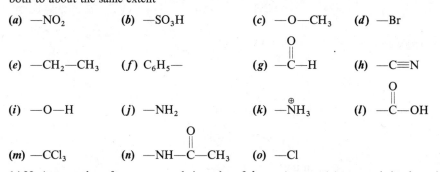

14.30 (*a*) Arrange the following four compounds in order of *increasing* reactivity toward electrophilic aromatic substitution (that is, list the least reactive compound first and the most reactive last). Justify your answer, and *for each compound* indicate whether the inductive effect (electron donation or withdrawal) and/or the resonance effect is responsible for the reactivity.

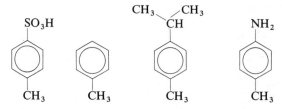

(*b*) Which compound would yield the highest percentage of the *meta* isomer (monosubstitution) and which would yield the lowest percentage of the *meta* isomer? Briefly explain.

14.31 Provide a mechanistic interpretation of the following results.

(*a*) The amount of *meta* substitution in the nitration of the following compounds is as indicated:

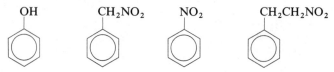

% *meta*:   100   88   19   5

(*b*) The amount of *meta* substitution in the nitration of the following compounds is as indicated:

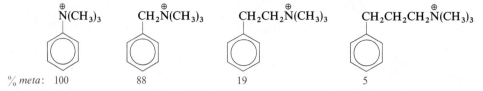

% *meta*:   4   14   34   64

14.32 Arrange the two series of compounds in Study Question 14.31 in order of *decreasing* reactivity toward electrophilic aromatic substitution; that is, list the most reactive compound first. Briefly justify your answer.

14.33 Benzene does not react with hypobromous acid, HOBr, but when extra mineral acid is added, bromobenzene is produced. Suggest a reasonable mechanism for this reaction. (*Hint:* Reason by analogy with the nitration of benzene.)

14.34 Consider the possibility of *ortho, para* attack and of *meta* attack by an electrophile, $E^\oplus$, in the following two compounds. Write the resonance structures for the intermediate carbocations that result, and on the basis of them, explain the orienting effects of the substituents attached to the benzene ring.

(*a*) $C_6H_5\overset{..}{-}N=O$

Nitrosobenzene

(*b*) $C_6H_5-CH=CH_2$

Styrene

14.35 Treatment of benzene with any of the following three reagents yields *tert*-butylbenzene. Provide complete, stepwise mechanisms to show how the reaction occurs in each case. What is the common intermediate in this reaction?

(*a*) *tert*-butyl bromide + $AlCl_3$
(*b*) *tert*-butyl alcohol $[(CH_3)_3COH]$ + conc H_2SO_4
(*c*) 2-methylpropene (isobutylene) + conc H_2SO_4

14.36 Provide a complete, stepwise mechanism to explain the following observed reaction. Show all intermediates.

14.37 The reaction between 1 mole of benzene and 1 mole of dichloromethane in the presence of $AlCl_3$ gives compound A, C_7H_7Cl. When A reacts with excess benzene in the presence of $AlCl_3$, compound B, $C_{13}H_{12}$, is formed. B is also produced when an excess of benzene reacts with dichloromethane and $AlCl_3$. Provide reasonable mechanisms to show what reactions are occurring, and draw the structures of compounds A and B.

14.38 When compound (1) is treated with $AlCl_3$ at room temperature, a new compound with the molecular formula $C_{10}H_{12}$ is formed. What is the likely structure of the new compound? (*Hint:* Friedel-Crafts reaction)

(1)

14.39 The reaction between benzene and *n*-butyl chloride in the presence of $AlCl_3$ gives a mixture of *n*-butylbenzene and *sec*-butylbenzene. The same reaction with *sec*-butyl chloride instead of *n*-butyl chloride gives only *sec*-butylbenzene.

(*a*) Provide a mechanistic interpretation for these results. Why does *sec*-butyl chloride yield *no* *n*-butylbenzene when it reacts with benzene?

(*b*) Suppose *optically active sec*-butyl chloride was used in the reaction. Would you expect the *sec*-butylbenzene from this reaction to be optically active or not? Why?

14.40 Predict the products that would be formed when each of the following alkyl halides is treated with alcoholic potassium hydroxide. Where two or more products can be formed, indicate which would be formed in the greater (greatest) amount.

(*a*) $C_6H_5-CH_2-CH_2-CH_2-Br$

(*b*) $C_6H_5-CH_2-\underset{\underset{Br}{|}}{CH}-CH_3$

(*c*) $CH_3-\underset{\underset{C_6H_5}{|}}{CH}-\overset{\overset{Cl}{|}}{CH}-CH_3$

(*d*) $CH_3-\overset{\overset{Cl}{|}}{CH}-CH_2-\underset{\underset{C_6H_5}{|}}{CH_2}$

(e) $CH_3-\overset{\overset{\displaystyle Br}{|}}{\underset{\underset{\displaystyle C_6H_5}{|}}{C}}-CH_2-CH_3$ (f) $CH_3-CH_2-\overset{\overset{\displaystyle CH_3}{|}}{\underset{\underset{\displaystyle Br}{|}}{C}}-\overset{}{\underset{\underset{\displaystyle C_6H_5}{|}}{CH}}-CH_3$

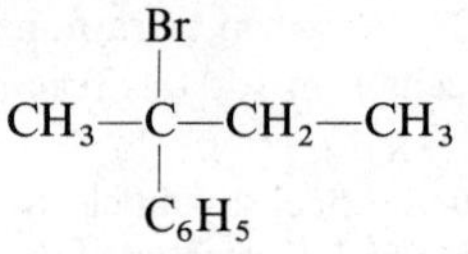

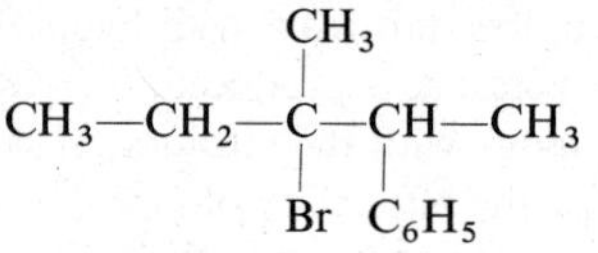

14.41 (a) Treatment of optically active α-iodoethylbenzene (1) with sodium methoxide $(CH_3\overset{..}{\underset{..}{O}}\text{:}^{\ominus}\ Na^{\oplus})$ in methanol (CH_3OH) gives an optically active ether (2). Propose a reasonable mechanism to account for this result, and clearly show the intermediate(s) involved. What is the absolute configuration of (2)? Use three-dimensional projection formulas throughout.

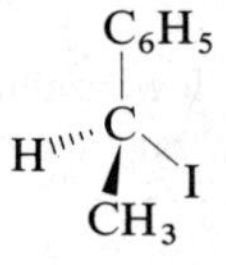

$$H^{\text{\tiny\textbackslash\textbackslash\textbackslash}}\overset{\overset{\displaystyle C_6H_5}{|}}{\underset{\underset{\displaystyle CH_3}{|}}{C}}\diagdown I \qquad\qquad C_6H_5-\overset{}{\underset{\underset{\displaystyle O-CH_3}{|}}{CH}}-CH_3$$

$$(1) \qquad\qquad\qquad (2)$$

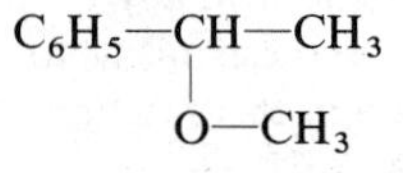

(b) When the reaction in **(a)** is carried out in methanol (without sodium methoxide), (2) is also obtained. However, (2) possesses considerably *less* optical activity than it did in the preceding reaction. Provide a mechanistic explanation of this observation, again using three-dimensional projection formulas.

14.42 Arrange the following five benzyl chlorides in order of *increasing* S_N1 reactivity; that is, list the least reactive first and the most reactive last. Explain the basis for your decision.

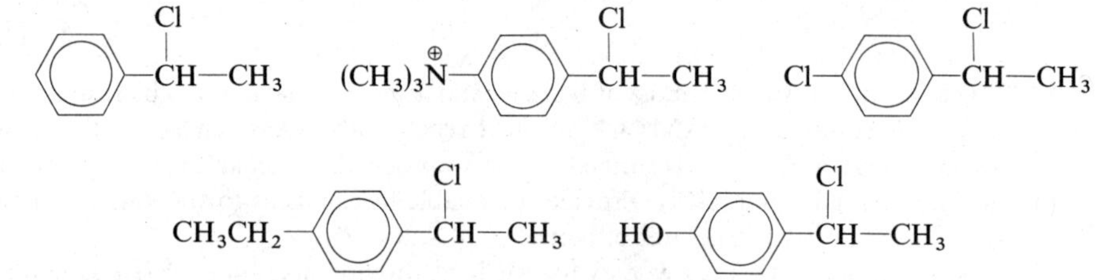

14.43 *N*-Bromosuccinimide reacts preferentially with the allyl hydrogens in alkenes (see Sec. 8.30). Reasoning by analogy, where do you think the major substitution would occur when *N*-bromo-succinimide reacts with *n*-propylbenzene? Briefly explain.

14.44 Indicate simple chemical tests that could distinguish between members of each of the following pairs of compounds. Indicate what you would observe in each case.

(a) cyclohexene and benzene
(b) cyclohexane and benzene
(c) $C_6H_5-C\equiv C-CH_3$ and $C_6H_5-CH_2-C\equiv C-H$
(d) benzene and 1,3,5-hexatriene
(e) $C_6H_5-CH=CH_2$ and ethylbenzene
(f) chlorobenzene and nitrobenzene

14.45 When toluene labeled with *tritium* (^3_1H or T) in an *ortho* position is nitrated, half the *ortho* hydrogens in the resulting *o*-nitrotoluene are labeled:

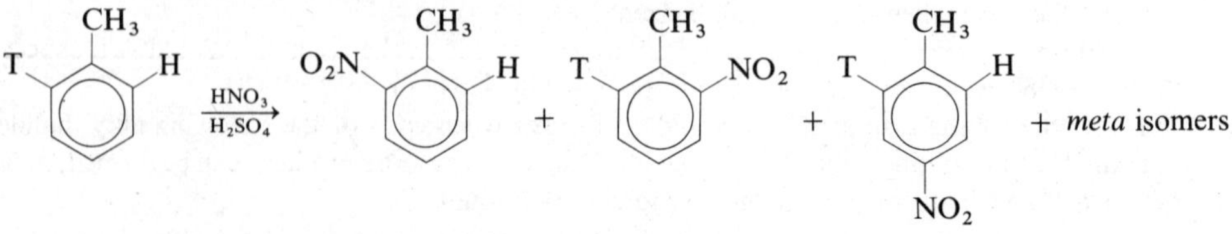

(a) Are these results consistent with the mechanism for nitration presented in Sec. 14.3?
(b) What, if anything, does this experiment with tritium indicate about the equivalence of the *ortho* positions? Explain.

14.46 An *aromatic* compound *A*, C_8H_{10}, on treatment with chlorine in the presence of heat and light, gives two isomeric compounds, *B* and *C*, each having the formula C_8H_9Cl. *B* is present in large amounts, but only a small amount of *C* is formed.

When *A* is treated with chlorine in the presence of ferric chloride, three isomeric compounds, *D*, *E*, and *F*, result, all with the formula C_8H_9Cl. *D* and *E* are formed in large amounts, whereas only a trace of *F* is observed.

Identify the lettered compounds *A* to *F* by drawing the structural formula for each. *Give your reasoning*, and indicate clearly what each observation means in helping deduce the structures of the compounds.

14.47 Compound (1), tropolone, is the simplest member of the family of compounds known as tropolones. X-ray crystallographic study of (1) indicates that all the C—C bonds are of nearly equal length (approximately 1.40 Å). Tropolone also undergoes nitration with dilute HNO_3 to give nitrotropolone, with the nitro group substituting predominantly at the 5 position. How can you explain these observations?

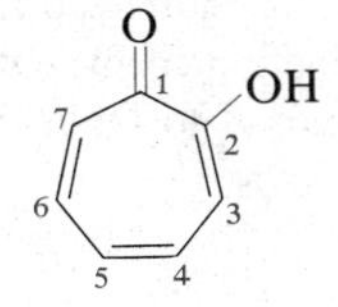

(1)
Tropolone

14.48 Identify the hydrocarbons in each of the following.
(*a*) Hydrocarbon *A*, C_8H_{10}, gives the following compound on oxidation with basic potassium permanganate under vigorous conditions:

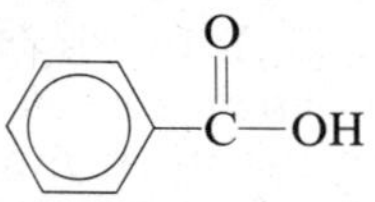

(*b*) Hydrocarbon *B*, C_8H_{10}, gives the following compound on oxidation with basic potassium permanganate under vigorous conditions:

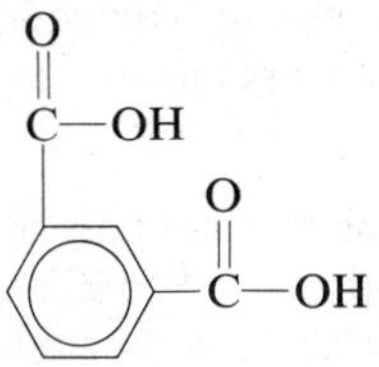

(*c*) Hydrocarbon *C*, C_9H_{10}, gives the same compound as does hydrocarbon *B* in (*b*) on vigorous oxidation.
(*d*) Hydrocarbons *D*, *E*, and *F* are isomeric with one another and have the molecular formula $C_{10}H_{14}$. On oxidation with hot chromic acid (H_2CrO_4) derived from sulfuric acid and potassium dichromate, all three yield the following compound:

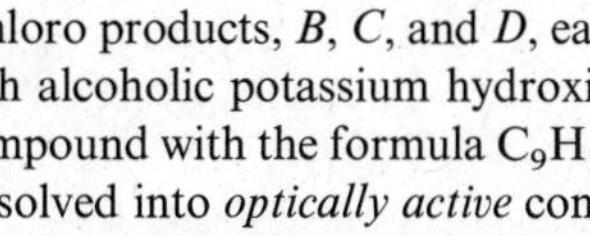

(*e*) Hydrocarbon *G*, $C_{10}H_{12}$, gives the same product as does *D*, *E*, and *F* in (*d*) on vigorous oxidation. Is there a unique structure for *G*? If not, draw other structures(s) that are consistent with the information provided.
14.49 Compound *A*, C_9H_{12}, is *aromatic*. Treatment of *A* with chlorine gas in the presence of light gives three isomeric monochloro products, *B*, *C*, and *D*, each having the formula $C_9H_{11}Cl$. Both *B* and *C*, on treatment with alcoholic potassium hydroxide, give the same compound *E*, C_9H_{10}; on the other hand, *no* compound with the formula C_9H_{10} can be obtained from *D* under the same conditions. *C* can be resolved into *optically active* compounds that are enantiomers of one another, whereas *B* and *D* cannot. Treatment of *A* with chlorine in the presence of ferric chloride gives *two and only two isomeric* monochloro products, *F* and *G*, each having the formula $C_9H_{11}Cl$. *No traces* of any other isomers can be detected.

(*a*) On the basis of these observations, draw the structures of compounds *A* to *G*. Indicate briefly how you used *each* observation to help you solve this problem.

(*b*) Is the structure you proposed for *A* unique? (That is, is there one and only one structure for *A* consistent with these data?) If not, state what ambiguity exists and draw a structure for one other possibility.

14.50 Compounds *A* and *B* are isomeric with one another, each having the molecular formula $C_{15}H_{14}$. Both *A* and *B* decolorize bromine in carbon tetrachloride and give a brown precipitate with cold, aqueous potassium permanganate. The addition of HBr (carbon tetrachloride solvent) to *A* gives a mixture of two bromo compounds, *C* and *D*; on dehydrohalogenation with alcoholic potassium hydroxide, both *C* and *D* give compound *B*. Vigorous oxidation of either *A* or *B* gives two new compounds, benzoic acid (1) and phthalic acid (2). The UV spectra of *A* and *B* are virtually identical. Both have an absorption band for the $\pi \rightarrow \pi^*$ transition at ~290 nm (283 nm for *A* and 295 nm for *B*). The extinction coefficient for both is in the range of 10,000 to 20,000. Deduce the structures of compounds *A* to *D*; briefly explain.

(1)   (2)

14.51 An *aromatic* compound *A*, C_8H_8BrCl, is *optically active*. To elucidate its structure, the following results are obtained from some laboratory experiments.

On treatment with alcoholic KOH, *A* gives compound *B*, C_8H_7Br, which is optically *inactive*. *B*, when subjected to catalytic hydrogenation ($Pt + H_2$) under mild conditions, gives compound *C*, C_8H_9Br. (*Note:* This hydrogenation did not affect the aromatic ring.) *C* is also optically *inactive*. When treated with bromine in the presence of $FeBr_3$, *C* gives two and *only two* isomeric dibromo compounds, *D* and *E*, each with the formula $C_8H_8Br_2$. On vigorous oxidation (basic potassium permanganate and heat), *A* gives compound *F*,

$C_7H_5O_2Br$, which contains one and only one carboxylic acid ($-\overset{O}{\overset{\|}{C}}-OH$) group.

Identify the lettered compounds *A* to *F* by drawing structural formulas for each. *Give your reasoning*, and trace the various reactions described. Indicate what the observations tell you, and explain how they allow you to deduce the structures of the lettered compounds. Indicate the source of the optical activity in *A*.

15

Spectroscopy 2—Spectroscopic Methods: Nuclear Magnetic Resonance Spectroscopy

Infrared (IR) spectroscopy came into vogue in the early 1950s when a variety of IR spectrophotometers became available at such economic prices that all research and many teaching laboratories were able to obtain them. Nuclear magnetic resonance (NMR) spectroscopy was developed in the early 1960s, and now most research laboratories have NMR spectrometers. NMR spectroscopy is a very useful technique for identifying and analyzing organic compounds.

This extremely important experimental technique is based on the magnetic property *nuclear spin*, which is exhibited by many atomic nuclei: ^1H, ^{15}N, ^{13}C, ^{19}F, ^{17}O, ^{31}P, for example—that is, nuclei with odd mass numbers. Nuclei with even mass numbers and odd atomic numbers, such as ^2H, ^{10}B, and ^{14}N, possess other types of magnetic properties, whereas elements with even mass numbers and even atomic numbers, such as ^{12}C, ^{16}O, and ^{32}S, have no magnetic properties. As we will see, atomic nuclei have nuclear spin and thus magnetic properties, which are important in NMR. For now we confine ourselves to *hydrogen, ^1H*. In NMR discussions, the hydrogen nucleus is often referred to simply as a *hydrogen* or a *proton*, but these are misnomers because the hydrogen atoms in organic compounds are covalently bonded to other atoms and are neither the free atomic hydrogen nor a proton, $H^\oplus$.

15.1 Nuclear Magnetic Resonance Spectroscopy: What Is It?

The hydrogen nucleus, which is representative of the other nuclei that have magnetic properties, has a uniform distribution of charge. As a result of spinning about an axis, the hydrogen nucleus generates a small *nuclear magnetic moment* along the axis of spin and behaves like a little bar magnet. The rotation and magnetic moment of the hydrogen nucleus can be represented as a vector ($\rightarrow$) as shown in Fig. 15.1.

When the hydrogen nucleus is placed in an external magnetic field of strength H_0, the small magnetic dipole moment generated by the spin of the nucleus becomes oriented either *with* the applied field (that is, in the same direction) or *against* it (in the opposite direction). The allowed energy levels are quantized in the presence of the external field. Regardless of the direction of orientation, the nuclear magnetic dipole is not aligned precisely parallel with the applied magnetic field, but it traces out a circular path about the axis defined by the external magnetic field H_0; this phenomenon is called *precession*. (Precession of nuclear magnetic moments is

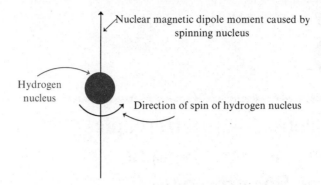

FIGURE 15.1 Magnetic and spin properties of the hydrogen nucleus in the absence of any external effects.

analogous to a spinning gyroscope that is not oriented parallel to the earth's gravitational field.) The state in which the nuclear magnetic dipole of the nucleus is oriented *with* the applied field is more stable because energy must be absorbed to flip the direction of nuclear spin. These two possible energy states are shown in Fig. 15.2. The difference in the energies of these two states is small, and at ordinary temperatures there is only a small excess of protons in the more stable state.

The frequency of the precession ω_0 is directly proportional to the strength of the applied external magnetic field H_0:

$$\omega_0 = \gamma_H H_0$$

where γ_H = the *gyromagnetic ratio* for the *proton* (a constant, different for each nuclei). Thus, ω_0 increases as H_0 increases.

Protons in a strong, uniform magnetic field are precessing about the axis of the applied magnetic field. Now suppose an electromagnetic frequency is applied to the

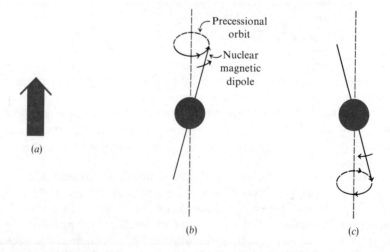

FIGURE 15.2 The two energy states of the hydrogen nucleus that are available to it in an applied external magnetic field H_0 (*a*). Precessional orbit is shown (*b*) with applied magnetic field: more stable; (*c*) against applied magnetic field: less stable.

sample in such a way that its magnetic component H_1 is at right angles to the applied magnetic field H_0 and is rotating with the precessing proton. An oscillator coil may be used to generate this electromagnetic frequency, which when placed at right angles to the applied magnetic field as in Fig. 15.3 generates the linear oscillating magnetic field H_1. This type of linear oscillating magnetic field is equivalent to two components that rotate in opposite directions. The one rotating in the same direction as the precession of the nuclear magnetic dipole of the proton is important. At a constant external magnetic field H_0, an increase in the oscillator frequency, v, causes an increase in the angular velocity of the rotating magnetic field H_1. When the angular velocity of H_1 is equal to the precessional frequency, ω_0, of the proton, the two are said to be in *resonance*. At this point the nucleus absorbs the energy and flips to its higher energy state. This energy absorption is picked up by electronic detectors and recorded on graph paper as a peak. The reversal of the direction of the nuclear magnetic dipole (hydrogen spin) is also shown on Fig. 15.3.

Before discussing the NMR spectrometer itself, let us briefly examine more of the theory of its operation. The frequency of the rotation of the magnetic component can be varied by changing the frequency of the oscillator used to produce the oscillating magnetic field H_1. The energy required to flip the nucleus also depends on the strength of the applied magnetic field H_0. The following equations provide these relationships:

Energy due to applied magnetic field:

$$\Delta E = \frac{h\gamma_H H_0}{2\pi}$$

where $\gamma_H H_0 = \omega_0$

Energy due to oscillator:

$$\Delta E = hv$$

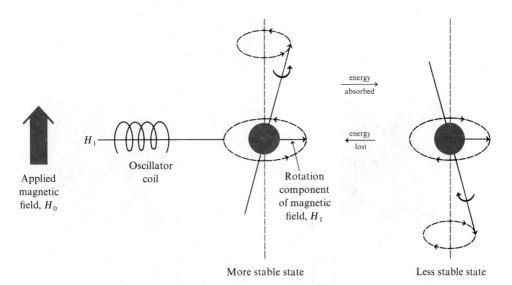

FIGURE 15.3   The interconversion of the two states of a hydrogen nucleus in an applied external magnetic field H_0. System of nuclei is subjected to a second, oscillating magnetic field H_1, which is generated by the oscillator coil and is at right angles to H_0. Flipping occurs when the angular velocity of the rotation component of H_1 is equal to the precessional frequency ω_0 of the proton.

Because the energies, ΔE, must be equal for flipping to occur, we have

$$h\nu = \frac{h\gamma_H H_0}{2\pi} \qquad \text{or} \qquad \nu = \frac{\gamma_H H_0}{2\pi}$$

where ΔE = energy difference between two spin states

γ_H = gyromagnetic ratio (a constant)

h = Planck's constant

H_0 = applied magnetic field strength at nucleus

ν = frequency of resonance absorption as produced by oscillator

This final equation indicates that stronger applied magnetic fields, H_0, require higher oscillator frequencies, ν. This is completely reasonable because stronger magnetic fields hold the nuclear magnetic dipole more tightly in its more stable state, so that more energy (greater oscillator frequency) is required to flip the nucleus (that is, reverse its direction).

Experimentally, when the applied magnetic field H_0 is 14,092 gauss (G), the frequency of the oscillator ν must be 60 megacycles per second (Mcps) to cause resonance of the proton. (Any other desired combination in the same ratio of H_0 to ν works.) One megacycle is 10^6 cycles, and the designation hertz (Hz) is often used for cycles per second (cps); thus the term megahertz (MHz) is equivalent to Mcps, and both mean 1 million cycles per second. In this text we use the more contemporary notation Hz and MHz. Frequencies in the range of 60 MHz occur in the radio-frequency (rf) region of the electromagnetic spectrum, so an rf oscillator is used in NMR spectrometers.

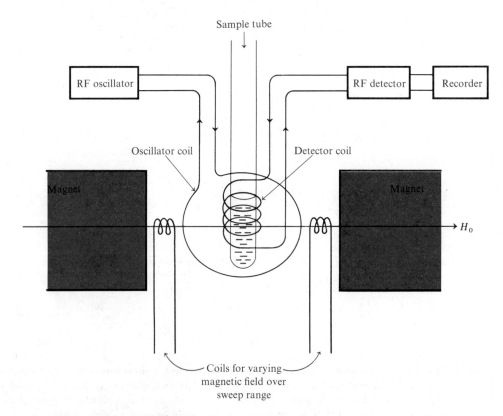

FIGURE 15.4   Schematic diagram of the essential components of an NMR spectrometer.

Two variables affect the resonance of the proton—the strength of the magnetic field, H_0, and the frequency of the rf oscillator, v—so the condition of resonance (spin flipping) for the proton can be achieved by keeping H_0 constant and varying v, or by maintaining v at a constant value and varying H_0. The latter approach is more practical from the standpoint of instrument design and is used commonly in modern commercial NMR spectrometers.

A simplified schematic diagram of the modern NMR spectrometer is shown in Fig. 15.4. In addition to the large magnets, small coils are used to vary the magnetic field over the sweep range of the NMR spectrum.

The magnetic field H_0 and/or the rf frequency v must be changed when studying other nuclei that have magnetic moments. Keep in mind, therefore, that the conditions described apply for protons only.

15.2   NMR Spectrometer and Spectra

If all the protons in an organic molecule were identical, they would absorb energy under identical conditions of magnetic field and frequency, and NMR would be of little or no practical use. However, organic compounds contain hydrogen atoms that are in different molecular environments. The electrons in the bonds between, for example, carbon and hydrogen, hydrogen and oxygen, or hydrogen and nitrogen place an "electron shield" around the nucleus. This causes the *effective* magnetic field (due to the applied field H_0) at the nucleus to differ from one nucleus to another in the same molecule. When protons are in different electronic environments (as dictated by the molecular structure of the molecule), they are shielded to varying extents. This causes protons in organic molecules to show resonance (absorb energy when the spins flip) at different energies. This resonance measurement of nuclear magnetic dipoles (spin flipping) is the basis for NMR spectroscopy, and the spectrum obtained from NMR is highly dependent on the structure of the particular molecule being studied. For this reason, NMR is a valuable tool for studying structure.

The NMR spectrum of 1,3-dichloropropane is shown in Fig. 15.5; it is exactly what is obtained from the graph recorded by the NMR spectrometer but reduced in size. The strength of the applied magnetic field H_0 *increases* in going from left to right; this is called the *upfield direction*. Because the frequency of the rf oscillator is held constant, the energy required to produce spin flipping results from varying the magnetic field H_0; that the magnetic field increases in going from left to right on the spectrum means that the energy required for the flipping must also increase in going from left to right. Typically, most protons in organic molecules can be detected in the range of 0 Hz to 600 to 700 Hz, the scale shown in the spectrum.

Fig. 15.5 reveals two additional interesting features. First, there are two groups of peaks, one centered at 132 Hz and another at 222 Hz. Second, one group consists of five peaks and the other of three peaks. This indicates that not all hydrogens in the molecule are equivalent. But, how can they be identified and how can we account for both their location and the number of peaks? Before we consider these topics, we must acquaint ourselves with certain definitions relating to NMR spectroscopy.

The positions of the peaks in the NMR spectra of most compounds are measured relative to the position of the peak produced by a standard compound. For most measurements, the standard is *tetramethylsilane*, $(CH_3)_4Si$, abbreviated as TMS. This standard is used because its 12 equivalent hydrogens give a sharp single

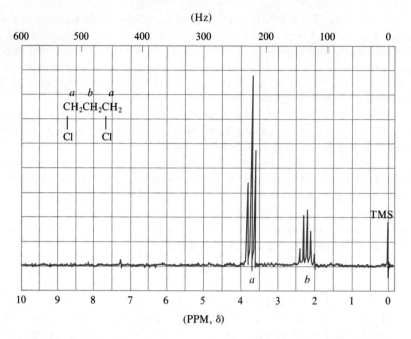

FIGURE 15.5 NMR spectrum of 1,3-dichloropropane. Note labeling of various features of the spectrum. [Data from *The Aldrich Library of NMR Spectra*, C. J. Pouchert and J. R. Campbell (eds.), copyright 1974, Aldrich Chemical Co., Milwaukee]

peak, which occurs at a field higher than the absorptions of protons in most organic compounds. The values of 132 and 222 Hz in the spectrum of 1,3-dichloropropane are called **chemical shifts**, which are measured *relative* to the TMS peak and are *downfield* from it. These chemical-shift values represent the two types of hydrogen nuclei in the compound. All these measurements are taken when the rf oscillator is set at a constant frequency, commonly 60 MHz. The standard TMS is usually used internally by adding a small amount of it (commonly 1%) to the sample tube containing the compound and organic solvent.

In general, the chemical shift of a hydrogen is dependent on the frequency of the rf oscillator. But, rather than giving chemical-shift values in hertz (which would vary depending on the strength of the magnetic field and radio frequency being used), a more convenient method for expressing chemical shifts requires their conversion to a *frequency-independent* scale—the **delta (δ) scale**. Division of the chemical shift (in hertz) by the frequency of the oscillator (also in hertz) followed by multiplication of the result by 10^6 gives δ (a unitless scale):

$$\delta = \frac{\text{chemical shift (Hz)}}{\text{oscillator frequency (Hz)}} \times 10^6$$

In this scale, TMS has $\delta = 0$, and peaks downfield from TMS have *positive* δ values; those upfield from TMS have *negative* values of δ.

The two groups of peaks in 1,3-dichloropropane (Fig. 15.5) are converted to the δ scale as follows:

$$\text{Group a:} \qquad \frac{222 \text{ Hz}}{60 \times 10^6 \text{ Hz}} \times 10^6 = 3.7\,\delta$$

$$\text{Group b:} \qquad \frac{132 \text{ Hz}}{60 \times 10^6 \text{ Hz}} \times 10^6 = 2.2\,\delta$$

Something interesting happens to these calculations when the chemical shift is divided simply by the oscillator frequency (without multiplying by 10^6). For group a, the result is $(132)/(60 \times 10^6)$ or 2.2×10^{-6}, or 2.2 ppm (parts per million), which is another common way to express chemical shift; multiplication of 2.2×10^{-6} by 10^6 gives δ, which is a unitless scale.

Another standard scale is the **tau (τ) scale,** which is also independent of frequency. This system defines the TMS peak as occurring at 10 τ. To convert from δ to τ, the chemical shift as measured on the δ scale is subtracted from 10:

$$\tau = 10 - \delta$$

Both the τ and δ scales are used commonly, so the organic chemist must be familiar with both.

NMR spectra are usually determined for compounds dissolved in some suitable organic solvent. One necessary requirement for the solvent is that it contain no hydrogens that would appear in the spectrum and possibly obscure the compound's absorption peaks. Carbon tetrachloride, CCl_4, is often used for moderately nonpolar compounds. When a compound is insoluble in it, however, other solvents must be used. To circumvent the requirement that they contain no hydrogens, deuterated solvents are used with great success; some common ones are deuterochloroform, $CDCl_3$, hexadeuteroacetone, $(CD_3)_2C{=}O$, and hexadeutero dimethyl sulfoxide $(CD_3)_2SO$. Recall from Sec. 15.1 that ^2H (deuterium, D) has a nuclear magnetic dipole, so it should absorb energy and appear as a peak in the NMR spectrum. It does this only under different conditions of applied field strength and oscillator frequency, however, and so does not interfere with the proton magnetic resonance spectra.

15.3   Interpretation of NMR Spectra

A wealth of information can be gained from correctly interpreting the NMR spectrum of a compound. In the next sections we see how to interpret the following aspects of a spectrum, with the final goal being the identification of unknown compounds:

1. *Identification of different kinds of protons* and *determination of the number of each different kind present* (Sec. 15.4)
2. *Structure and chemical shift* (Sec. 15.5)
3. *Significance of the number of peaks associated with each different kind of proton: spin-spin splitting* (Sec. 15.6)

Part of this discussion refers to the spectrum of 1,3-dichloropropane in Fig. 15.5, and additional illustrative examples are presented to emphasize some important concepts in spectra interpretation.

15.4   Identification of Kinds of Protons; Determination of the Number of Each Kind

A. Proton Equivalence

Examining the structure of a molecule and determining how many types or kinds of hydrogens are present are little different from classifying hydrogens as primary, secondary, or tertiary. In methane, CH_4, all four hydrogens are equivalent because all bond distances and bond angles are equal and all four hydrogens are in identical environments; the same is true for ethane, CH_3—CH_3. In propane, CH_3—CH_2—CH_3, there are six equivalent primary hydrogens and two equivalent secondary hydrogens.

NMR spectroscopy makes this same distinction because *all the hydrogens in the same molecular environment absorb at the same position on the NMR scale.* If there are different sets of protons, then there are different sets of peaks on the NMR spectrum. Protons in the same molecular environment are said to be *equivalent*. For the most part, we speak of *nonequivalent sets of protons* that *give rise to different NMR chemical shifts.* Several examples of proton equivalency are shown here; the equivalent sets of protons are labeled with identical letters:

$$CH_3-CH_2-Br \qquad CH_3-\overset{\overset{\displaystyle Br}{|}}{C}H-Br \qquad Br-CH_2-CH_2-Br$$
$$\quad b \qquad a \qquad\qquad b \qquad a \qquad\qquad\quad a \qquad\quad a$$

Bromoethane	1,1-Dibromoethane	1,2-Dibromoethane
Two sets of protons,	*Two sets of protons,*	*One set of protons,*
two NMR groups	*two NMR groups*	*one NMR group*

$$CH_3-CH_2-CH_2-NO_2 \qquad CH_3-\overset{\overset{\displaystyle NO_2}{|}}{C}H-CH_3$$
$$\quad c \qquad\; b \qquad\; a \qquad\qquad\quad a \qquad\; b \qquad a$$

1-Nitropropane	2-Nitropropane
Three sets of protons,	*Two sets of protons,*
three NMR groups	*two NMR groups*

In the NMR spectrum of 1,3-dichloropropane in Fig. 15.5, two sets or groups of peaks are some distance removed from one another (that is, they are centered at different chemical shifts). Does this agree with the concept of equivalence of protons? Yes, because the two protons on both C-1 and C-3 are equivalent, and the two protons on C-2 are equivalent (but different from the other four protons):

$$Cl-\overset{\overset{\displaystyle H_a}{|}}{\underset{\underset{\displaystyle H_a}{|}}{C}}-\overset{\overset{\displaystyle H_b}{|}}{\underset{\underset{\displaystyle H_b}{|}}{C}}-\overset{\overset{\displaystyle H_a}{|}}{\underset{\underset{\displaystyle H_a}{|}}{C}}-Cl \quad \text{or} \quad Cl-CH_2-CH_2-CH_2-Cl$$
$$\qquad\qquad\qquad\qquad\qquad\qquad\qquad a \qquad\; b \qquad\; a$$

The identical protons are labeled H_a and H_b in this structure. All protons labeled H_a are equivalent because they are all attached to a carbon that bears a —Cl and a —CH_2—CH_2—Cl group. In the same way, all H_b protons are equivalent because the carbon to which they are attached contains two identical —CH_2—Cl groups. Thus, these two sets of equivalent protons appear in the NMR spectrum as two groups of peaks.

B. Proton Counting

NMR tells us not only which protons are equivalent but also *how many of each different kind of proton are in the molecule.* Because the intensity of the absorption at any applied field strength (at constant frequency) is proportional to the number of protons that are absorbing energy, the number of each kind of proton is represented by the area under the relevant absorption peak. *The area under each peak is directly proportional to the number of protons that produces that peak.* The areas under peaks can be determined by cutting them out of the graph paper and weighing the amount of cut-out paper. If any single set of peaks is known to represent a certain number of protons, then the number of protons represented by each other set of peaks can be computed. More frequently (but with less accuracy), the areas under the NMR peaks are measured electronically and recorded directly on the graph paper in the form of a horizontal line tracing that rises in steps as each peak is passed over. The total rise of this line represents the total number of protons in the compound. The number of protons represented in each peak (as shown by a one-step rise in the integration line) can be estimated if any one set of protons is known, because the ratio of the areas under the peaks is equivalent to the ratio of the numbers of protons that caused those peaks. This process is referred to as **peak integration.**

The NMR spectrum of 1,3-dichloropropane in Fig. 15.5 is redrawn in Fig. 15.6 with the peak integrations shown. By measuring the height of each step (either with a millimeter ruler or by counting the number of squares on the crosshatched NMR chart paper), we get the *ratio* of the two types of protons that are in the molecule. This calculation follows:

Calculation: 1,3-Dichloropropane contains six hydrogens (6H's)

Peak a: 2.5 grids

Peak b: 1.2 grids

Total   3.7 grids

$$\frac{3.7 \text{ grids}}{6 \text{ H's}} = 0.6 \text{ grids/H}$$

therefore, $\dfrac{2.5 \text{ grids}}{0.6 \text{ grids/H}} = 4 \text{ H's}$

$$\frac{1.2 \text{ grids}}{0.6 \text{ grids/H}} = 2 \text{ H's}$$

Because the accuracy of electronic integration is of the order of $\pm 5\%$, we seldom obtain whole-integer ratios. However, it is usually clear what the nearest

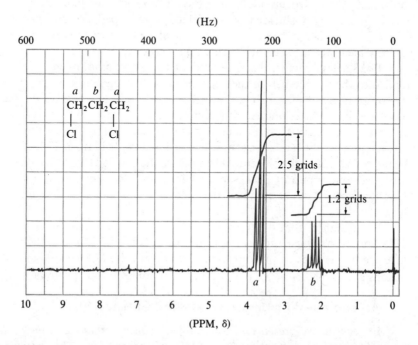

FIGURE 15.6   NMR spectrum of 1,3-dichloropropane, showing integration markings. Integration markings of spectra in this book have been omitted for clarity and replaced with a number indicating the relative number of hydrogens comprising the peak, for example, (3) = 3H's [Data from *The Aldrich Library of NMR Spectra*, C. J. Pouchert and J. R. Campbell (eds.), copyright 1974, Aldrich Chemical Co., Milwaukee]

whole-number ratio should be. For example, if the ratio 1.05:1.99:4.07 is obtained from integration measurements, then the proton ratio is 1:2:4. It must be emphasized that integration gives only *proton ratios*. Frequently, however, the chemist has a good idea about the structure of a compound. Then if one set of protons can be identified and if the number of them present in the compound is known, the peak ratios can be used to determine the absolute number of protons represented by each group of peaks. Also, if the molecular formula is known, the number of protons in each equivalent group can often be determined.

We know that different sets of equivalent protons appear in different places in the NMR spectrum and that we can determine the ratios of these sets of equivalent protons by peak areas. We considered only "straightforward" compounds, however, in which there are no special structural features such as asymmetric carbon atoms or *cis* and *trans* isomers; these other types of compounds are discussed in Sec. 15.6.

15.5   Structure and Chemical Shift

We now ask *why* certain sets of protons appear at one place on the NMR spectrum, whereas others appear elsewhere. More simply, what is the relationship between chemical shift and structure? In the beginning of Sec. 15.2, we suggested that different molecular environments produce differing electronic environments around hydrogen nuclei. The externally applied magnetic field H_0 is uniform over the entire molecule, so it cannot make different protons nonequivalent. On the other hand, the magnetic field that is induced by the movement of electrons (which are also charged and have a magnetic moment associated with them) over the molecule (for example, those in σ bonds) is not uniform, and thus causes protons to be nonequivalent. The hydrogen nuclei near these magnetic fields induced by the molecule may be more highly shielded or deshielded; they experience an *effective* magnetic field H_0 that is either *less* or *greater* than that of the externally applied magnetic field H_0 due to the induced field either opposing or reinforcing the externally applied field. For shielded nuclei, this means that stronger external magnetic fields must be applied to the sample before the spins of those hydrogen nuclei flip and appear on the spectrum as they absorb energy. The reverse effect is *deshielding*. Actually, all nuclei are shielded to some extent with respect to the external field; it is the degree of shielding that varies from nucleus to nucleus. The terms *shielding* and *deshielding* are therefore used in a relative sense; the farther downfield from TMS a peak appears the more *deshielded*, the closer to TMS the less deshielded (the more *shielded*). When more than one group of peaks appears in an NMR spectrum, this simply reflects the differences in the strength of the induced magnetic field at different points in the molecule.

The distances that different groups of absorption peaks are shifted *relative to TMS* are the *chemical shifts* we referred to in Sec. 15.2. For 1,3-dichloropropane, there are two groups of protons, one centered at 2.2 δ (7.8 τ) corresponding to two protons and the other at 3.7 δ (6.3 τ) corresponding to four protons.

A. Electron-Withdrawing Groups and Chemical Shift

In general, *the presence of electron-withdrawing groups in a molecule causes a downfield chemical shift* (relative to TMS) of the protons. The inductive effect is largely responsible for this, because inductive withdrawal of electrons away from hydrogens causes them to be *deshielded* (that is, the induced magnetic field is decreased, so less externally applied energy is required to flip the spin of those protons). Does this prediction agree with the observed NMR spectrum of 1,3-dichloroporpane?

To answer this, we compare the chemical shifts of 1,3-dichloropropane with those of propane:

$$2.2\ \delta\ (7.8\ \tau) \qquad\qquad 1.43\ \delta\ (8.57\ \tau)$$

$$Cl\!-\!CH_2\!-\!CH_2\!-\!CH_2\!-\!Cl \qquad CH_3\!-\!CH_2\!-\!CH_3$$

$$3.7\ \delta\ (6.3\ \tau) \qquad\qquad 0.9\ \delta\ (9.1\ \tau)$$

1,3-Dichloropropane   Propane

Note that the attachment of chlorine atoms, which withdraw electrons strongly through the inductive effect, in place of the hydrogens on propane causes a great downfield shift, from $0.9\ \delta\ (9.1\ \tau)$ in propane to $3.7\ \delta\ (6.3\ \tau)$ in 1,3-dichloropropane. The chlorine atoms also have some electron-withdrawing effect on the central hydrogens, which are shifted downfield from $1.43\ \delta\ (8.57\ \tau)$ in propane to $2.2\ \delta\ (7.8\ \tau)$ in 1,3-dichloropropane; this effect, however, is not nearly so great as when the chlorine atoms are attached directly to a carbon bearing hydrogens.

Further support for the deshielding effect of the halogens is seen in the following sequence of compounds, which differ only in the halogen attached to the methyl group:

$$CH_3\!-\!F \qquad CH_3\!-\!Cl \qquad CH_3\!-\!Br \qquad CH_3\!-\!I$$

$4.26\ \delta$	$3.0\ \delta$	$2.82\ \delta$	$2.16\ \delta$
$(5.74\ \tau)$	$(7.0\ \tau)$	$(7.18\ \tau)$	$(7.84\ \tau)$

Decreasing chemical shift
from TMS

The decreasing chemical shift parallels the decreasing electronegativity (that is, the decreasing electron-withdrawing effect) of the halogens.

The attachment of increasing numbers of halogens to the same carbon atom also causes the chemical shift to increase (that is, larger δ values), as the following show:

$$
\begin{array}{cccc}
Cl & Cl & H & H \\
| & | & | & | \\
Cl\!-\!C\!-\!H & Cl\!-\!C\!-\!H & Cl\!-\!C\!-\!H & H\!-\!C\!-\!H \\
| & | & | & | \\
Cl & H & H & H
\end{array}
$$

$7.24\ \delta$	$5.28\ \delta$	$3.0\ \delta$	$0.9\ \delta$
$(2.76\ \tau)$	$(4.72\ \tau)$	$(7.0\ \tau)$	$(9.1\ \tau)$

The electronegativity of atoms played an important role in the choice of tetramethylsilane, $(CH_3)_4Si$, as the standard for comparison in NMR spectroscopy. The hydrogens in TMS are all identical (equivalent), and the presence of silicon, which has a very low electronegativity, means that these hydrogens have a great deal of shielding and appear far upfield (at $0.00\ \delta$, or $10.0\ \tau$, by convention). By comparison, the hydrogens in neopentane, $(CH_3)_4C$, appear at $0.94\ \delta\ (9.06\ \tau)$; they are shifted downfield from TMS because of the more electronegative (than silicon) central carbon atom.

Another way of looking at the effect of electron-withdrawing atoms on chemical shift is as follows. When electrons are withdrawn from a hydrogen by substituents with increasing electron-withdrawing ability (that is, through covalent σ bonds), that hydrogen becomes increasingly positive and thus more like a true proton, $H^{\oplus}$. In cases where electron withdrawal is complete or nearly complete, as in certain organic (carboxylic) acids, the proton appears far downfield and typically at a δ value greater than 10.

We have considered chemical shift from a qualitative viewpoint to help us predict the positions of chemical shift in simple compounds (see Table 15.1).

TABLE 15.1 Chemical Shifts at 60 MHz of Hydrogens Attached to Various Functional Groups

Functional Group; hydrogen type shown as H	Chemical Shift, ppm	
	δ	τ
TMS, $(CH_3)_4Si$	0	10.0
Cyclopropane	0–0.4	9.6–9.9
Alkanes:		
RCH_3	0.9	9.1
R_2CH_2	1.3	8.7
R_3CH	1.5	8.5
Alkenes:		
$-C{=}C-H$ (vinyl)	4.6–5.9	4.1–5.4
$-C{=}C-CH_3$ (allyl)	1.7	8.3
Alkynes:		
$-C{\equiv}C-H$	2–3	7–8
$-C{\equiv}C-CH_3$	1.8	8.2
Aromatic:		
$Ar-H$	6–8.5	1.5–4
$Ar-C-H$ (benzyl)	2.2–3	7–7.8
Fluorides, $F-C-H$	4–4.5	5.5–6
Chlorides:		
$Cl-C-H$	3–4	6–7
$Cl-C-H$ (with Cl)	5.8	4.2
Bromides, $Br-C-H$	2.5–4	6–7.5
Iodides, $I-C-H$	2–4	6–8
Nitroalkanes, O_2N-C-H	4.2–4.6	5.4–5.8

TABLE 15.1 (*continued*)

Functional Group; hydrogen type shown as H	Chemical Shift, ppm δ	τ
Alcohols, ethers:		
HO—C—H	3.4–4	6–6.6
RO—C—H	3.3–4	6–6.7
Acetals: —O—C(H)—O—	5.3	4.7
Esters:		
R—C(=O)—O—C—H	3.7–4.1	5.9–6.3
RO—C(=O)—C—H	2–2.6	7.4–8
Carboxylic acids:		
HO—C(=O)—C—H	2–2.6	7.4–8
R—C(=O)—O—H	10.5–12	−2–(−0.5)
Aldehydes, ketones:		
R—C(=O)—C—H	2–2.7	7.3–8
Aldehydes:		
R—C(=O)—H	9–10	0–1
Amides:		
R—C(=O)—N—H	5–8	2–5
Alcohols: R—O—H	4.5–9	1–5.5
Phenols: Ar—O—H	4–12	−2–6
Amines: R—NH_2	1–5	5–9

B. Ring Currents: Shielding and Deshielding

The chemical shifts of protons attached to triply bonded atoms and to aromatic rings (that is, aromatic protons) are much more varied than might be expected. For example, protons on benzene appear at $7.37\ \delta$ ($2.63\ \tau$), whereas those on acetylene absorb at $\sim 2.9\ \delta$ ($7.1\ \tau$). The reasons for these differences are not obvious but, as with other proton chemical shifts, these protons are affected by the degree of shielding.

Induced magnetic fields are responsible for the chemical shifts in aromatic compounds and in triply bonded groups. The π electrons that circulate in the π cloud of the aromatic ring and in the triple bond induce a magnetic field to occur in the molecules when an external magnetic field is applied; this induced electron circulation is called a **ring current.** This induced field is either in the same direction as the external field H_0 or in the opposite direction.

If the ring current is in the same direction, the protons in its region are *deshielded* because the proton nuclei "feel" a magnetic field from two sources: the ring current and the external field. Thus less external energy is required for proton resonance, and those protons absorb at lower field strengths than otherwise expected.

If the ring current is in the opposite direction of the applied field, then it cancels out part of the applied field. Hydrogen nuclei under these conditions "feel" an effective magnetic field that is less than the applied field H_0, so more external energy is required for absorption. These protons are said to be *shielded* and appear at higher field strengths than otherwise predicted.

The benzene and acetylene molecules demonstrate two extremes of ring currents. In benzene the effect of the ring current on the protons reinforces the external magnetic field and the aromatic hydrogens appear at high δ values (low τ values), namely $7.37\ \delta$ ($2.63\ \tau$). This is shown in Fig. 15.7(*a*). On the other hand, the effect of the ring current on the protons in acetylene opposes the external field, and acetylenic protons appear upfield at $2.9\ \delta$ ($7.1\ \tau$) [Fig. 15.7(*b*)].

The following two comparisons illustrate this shielding and deshielding effect:

Cyclohexene	Benzene	Acetylene
$5.57\ \delta$	$7.37\ \delta$	$2.9\ \delta$
$(4.43\ \tau)$	$(2.63\ \tau)$	$(7.1\ \tau)$
	Deshielded	*Shielded*

1-Methylcyclohexene	Toluene
$1.68\ \delta$	$2.32\ \delta$
$(8.32\ \tau)$	$(7.68\ \tau)$
	Deshielded

The aromatic hydrogens on benzene are deshielded and the hydrogens on acetylene are shielded, as evidenced by the chemical shift of these protons compared with that of cyclohexene (which has a hydrogen attached to a carbon-carbon double bond). The deshielding effect induced by the ring current in the benzene ring system extends beyond the hydrogens attached to the aromatic ring, as shown by the chemical shift of the methyl protons on toluene, which appear downfield from those on the

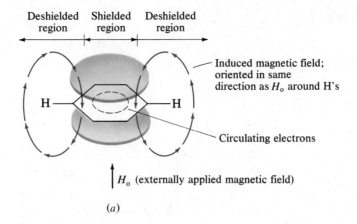

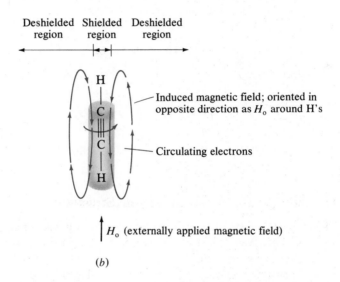

FIGURE 15.7 **(a) Ring current induced in benzene, resulting in deshielding of aromatic protons. (b) Ring current induced in acetylene, resulting in shielding of protons.**

methyl group attached to the double bond in 1-methylcyclohexene. The deshielding effect is less pronounced because these hydrogens are farther from the ring current.

C. Ring Currents and Aromaticity

The idea of ring currents provides the organic chemist with compelling evidence of aromaticity in compounds that obey Hückel's $(4n + 2)$ π-electron rule (see Sec. 13.7). Deshielding occurs in molecules that are aromatic [that is, they contain $(4n + 2)$ π electrons], and the aromatic protons are therefore shifted downfield. Compounds that are not aromatic (according to the $4n + 2$ rule) do not exhibit any downfield aromatic shift. Certain large-ring compounds obey Hückel's rule and contain some protons that are shielded and others that are deshielded. The two examples here illustrate the application of ring currents and NMR spectroscopy to aromaticity.

Aromatic:

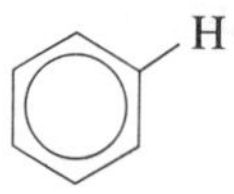

Benzene: 6 π electrons
7.37 δ (2.63 τ)
Deshielded

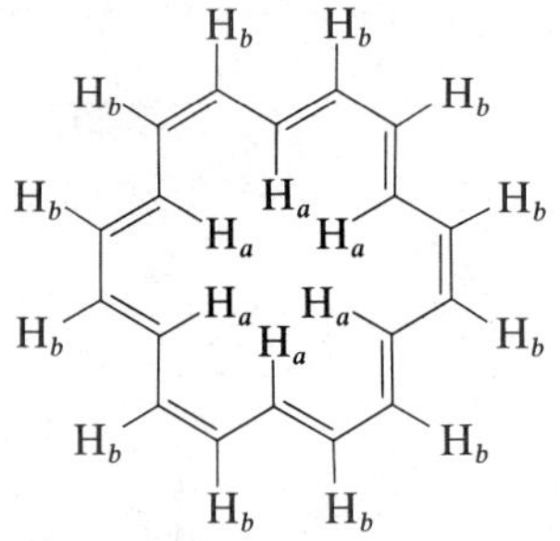

[18]-Annulene: 18 π electrons
6 *inside hydrogens* (H_a): $-1.9\ \delta$ (11.9 τ)
*Highly shielded due to being
inside ring current*

12 *outside hydrogens* (H_b):
8.8 δ (1.2 τ) *Deshielded
due to being outside ring current*

Nonaromatic:

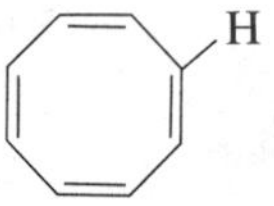

Cyclooctatetraene: 8 π electrons
5.69 δ (4.31 τ)

D. Chemical Shift Correlations with Structure

The chemical shifts of hydrogens attached to a given functional group are for the most part relatively constant; only special structural features in a molecule cause great variations to occur. The chemical shifts often provide valuable information about both the structure and the functional groups present in an unknown molecule. Extensive tables are available that correlate structure with chemical shift, and a partial listing of common functional groups is given in Table 15.1. This listing contains many functional groups not yet studied; future chapters provide more details about the NMR chemical shifts of these functional groups.

15.6   Significance of Number of Peaks Associated with Each Kind of Proton: Spin-Spin Splitting and Coupling Constants

The NMR spectroscopy studied thus far has largely dealt with chemical shifts and intensities of proton absorptions. The spectra are even more complex and informative (as seen in Fig. 15.5), however, because of additional interactions between the atoms in a molecule.

A. Spin-Spin Splitting

The existence of several peaks within each group of absorptions in the NMR spectrum of 1,3-dichloropropane (Fig. 15.5) is caused by **spin-spin splitting** or **coupling,** which results from the magnetic interactions between nonequivalent magnetic hydrogen nuclei in the molecule. This phenomenon is observed in the vast majority of organic molecules. The coupling of one hydrogen with another equivalent one that is attached to the same carbon atom or a neighboring carbon atom is not observed. In addition, coupling of protons that are more than two carbon atoms from one another, though possible, is uncommon. *As a general rule, splitting occurs only between neighboring protons* (*usually within three bonds of each other*) *that have different chemical shifts* (*the protons are therefore nonequivalent*).

These nonequivalent protons may be on adjacent carbons or on the same carbon. This discussion is concerned solely with protons on adjacent carbon atoms. In the case of 1,3-dichloropropane, the spins of the H_a protons couple with those of the H_b protons, and vice versa; the result of this coupling is *spin-spin splitting* (Fig. 15.5):

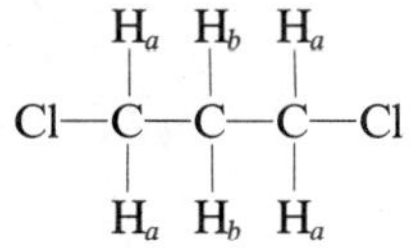

$$\begin{array}{ccccc}
 & H_a & H_b & H_a & \\
 & | & | & | & \\
Cl-&C-&C-&C-&Cl \\
 & | & | & | & \\
 & H_a & H_b & H_a &
\end{array}$$

a. Protons on Adjacent Carbon Atoms

Before considering the splitting in other real compounds, let us see what causes it. Splitting is the result of one set of equivalent protons (those absorbing energy and appearing as a peak on the spectrum) "feeling" different magnetic field environments because of nonequivalent protons on the adjacent carbon atom(s). When equivalent protons are absorbing energy, they are at the instant prior to absorbing the energy all aligned with the externally applied magnetic field H_0 where the nuclear magnetic dipoles of the other protons in the molecule are aligned, some with the field and some against it. By means of the following three general examples, let us explain spin-spin splitting in terms of how the nonabsorbing protons (as a result of their magnetic fields) affect the "effective" magnetic field that the absorbing protons "feel."

Example 1   Two tertiary protons on adjacent carbon atoms

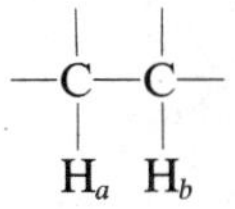

$$\begin{array}{cc}
 | & | \\
-C-&C- \\
 | & | \\
H_a & H_b
\end{array} \qquad \text{where } \delta_{H_a} > \delta_{H_b}$$

1. Effect of H_b on H_a when H_a absorbs energy. Suppose a sufficient external magnetic field is applied so that H_a absorbs energy. How does H_b affect the magnetic field that H_a feels?

The nuclear magnetic dipole of the hydrogen nucleus H_b may be aligned in two allowed directions, indicated by the arrows ↑ (with the field) and ↓ (against the field). If the H_b spin orientation is ↓, it opposes H_0, and H_a feels an *effective* magnetic field that is *less* than the applied field, so that more external energy must be supplied for resonance to occur. This may be justified qualitatively by considering the nature of a magnetic field, which has a north (N) and south (S) pole. Thus, the strength of the applied magnetic field H_0 is somewhat canceled when the spin of H_b is ↓, so that a greater magnetic field must be applied. The following diagram helps clarify this concept:

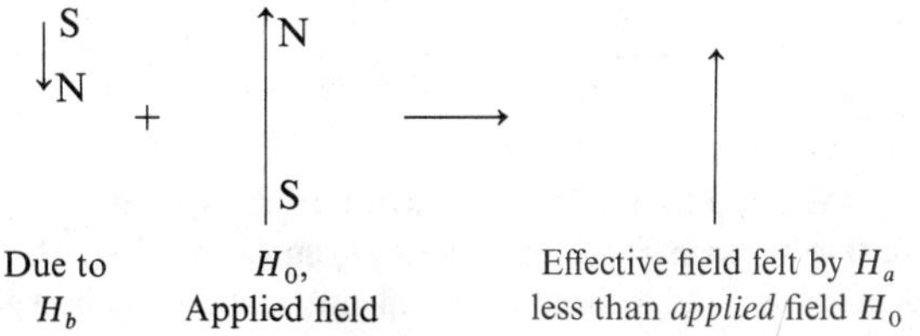

$$\begin{array}{ccccc}
\downarrow S & & \uparrow N & & \\
\downarrow N & + & & \longrightarrow & \uparrow \\
& & S & &
\end{array}$$

| Due to H_b | H_0, Applied field | Effective field felt by H_a less than *applied* field H_0 |

This gives one H_a peak at higher field strength.

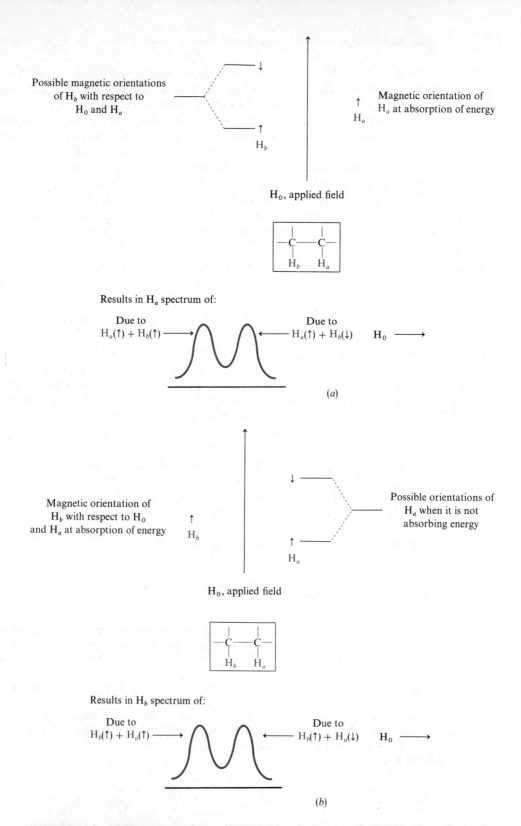

Possible magnetic orientations of H_b with respect to H_0 and H_a

H_b

Magnetic orientation of H_a at absorption of energy

H_a

H_0, applied field

Results in H_a spectrum of:

Due to
$H_a(\uparrow) + H_b(\uparrow)$

Due to
$H_a(\uparrow) + H_b(\downarrow)$

H_0

(a)

Magnetic orientation of H_b with respect to H_0 and H_a at absorption of energy

H_b

Possible orientations of H_a when it is not absorbing energy

H_a

H_0, applied field

Results in H_b spectrum of:

Due to
$H_b(\uparrow) + H_a(\uparrow)$

Due to
$H_b(\uparrow) + H_a(\downarrow)$

H_0

(b)

FIGURE 15.8   NMR spectrum of two adjacent tertiary hydrogens, H_a and H_b, where H_a absorbs at higher chemical shifts than H_b. (a) Effect of H_b on H_a when H_a absorbs energy; (b) effect of H_a on H_b when H_b absorbs; (c) idealized NMR spectrum showing absorptions of H_a and H_b, both of which are split into doublets of equal intensity.

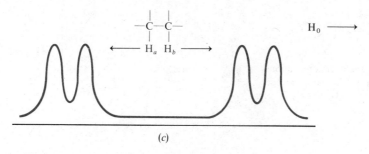

(c)

FIGURE 15.8 (*continued*)

If the H_b spin orientation is ↑, it reinforces H_0, and H_a feels an *effective* magnetic field that is *greater* than the applied field. Less external energy must be applied for resonance of H_a to occur; this gives another peak at lower field strength.

As a result of H_b, *two peaks (a doublet) are produced* for H_a, as shown in Fig. 15.8(*a*).

2. Effect of H_a on H_b when H_b absorbs energy. This is very much like the preceding situation, except that now the spin of H_b is aligned with the applied field H_0 as energy is absorbed. There are two equally probable orientations of the spin of the nonabsorbing H_a proton. The interaction of these two orientations with that of H_b also gives rise to a doublet (two peaks) in the absorption spectrum for H_b. This is shown in Figure 15.8(*b*).

3. Composite spectrum. Although we considered protons H_a and H_b separately, in the NMR spectrum for the entire compound, we see the absorptions caused by both of them but at different chemical shifts. Each appears as a doublet; that is, there is a pair of doublets as shown in Fig. 15.8(*c*). The so-called *center of gravity* of each doublet represents the δ (or τ) value reported for the chemical shift of each proton. The amount of splitting that has occurred is discussed later in this section.

Example 2   One tertiary proton and two secondary protons on adjacent carbon atoms

---

$$-\underset{\underset{H_a}{|}}{\overset{|}{C}}-\underset{\underset{H_b}{|}}{\overset{|}{C}}-H_b \qquad \text{where } \delta_{H_a} > \delta_{H_b}$$

1. Effect of H_a on H_b when H_b absorbs energy. We are again interested in learning how one H_a proton affects the absorption of two *equivalent* H_b protons. This is identical with part 2 of Example 1, because each of the two equivalent H_b protons is aligned with the applied magnetic field at the point where energy is absorbed, and each individually feels the same effective magnetic field that just one proton felt in Example 1. When there are *two* protons absorbing energy, however, the *intensity* of the absorbing peaks [a doublet in this case, Fig. 15.9(b)] is twice as great as when there is just one proton.

2. Effect of H_b on H_a when H_a absorbs energy. When H_a absorbs energy, there are two equivalent H_b protons and their spins may be aligned in all possible combinations of direction with (↑) or against (↓) the applied field. These possible alignments are shown in Fig. 15.9(*a*). Note that the combinations ↑↓ and ↓↑ are of identical energy, and because the spins in each are in opposite directions, each results in H_a feeling an effective magnetic field identical with the externally applied field H_0.

Of the four possible ways that the spins of the two equivalent H_b protons can be aligned, *one* (↓↓) is at higher energy, *two* (↑↓ and ↓↑) are at equal energies, and *one* (↑↑) is at lower energy than the applied field. This results in three peaks (a *triplet*), with peak intensities 1:2:1. These values result from the number of ways the spins of the two nonabsorbing H_b protons

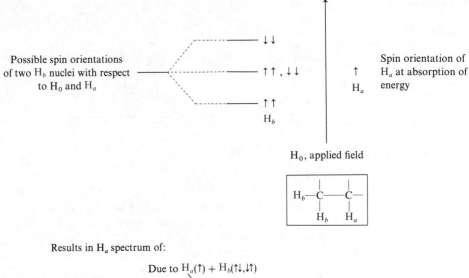

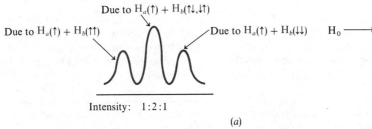

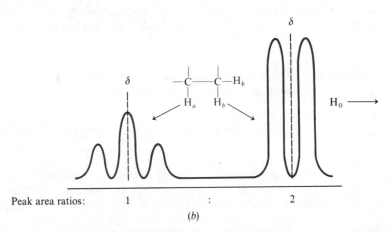

FIGURE 15.9   NMR spectrum of a compound with one tertiary hydrogen, H_a, adjacent to two equivalent secondary hydrogens, H_b, where H_a absorbs at higher chemical shift than H_b. (*a*) Effect of H_b on H_a when H_a absorbs energy; (*b*) NMR spectrum showing absorptions of H_a and H_b, where H_b appears as a doublet and H_a as a triplet.

can be aligned with the proton being observed. When two (or more) alignments are of equal energy, they result in increased intensity of the peak. The center of the triplet is the chemical shift reported for H_a.

3. Composite spectrum. Again we considered the H_a and two equivalent H_b protons separately. When the entire NMR spectrum is taken [Fig. 15.9(*b*)], both groups of peaks are observed but at different chemical shifts. There is an upfield doublet caused by the H_b protons, and each peak in the doublet is of equal intensity. There is a downfield triplet due to the H_a proton, and the three peaks within the triplet have the intensity ratio 1:2:1. As far as *total* peak areas are concerned, the ratio of the doublet to the triplet is 2:1 because there are two H_b protons and one H_a proton.

Example 3 One tertiary proton and three primary protons on adjacent carbon atoms

$$\overset{\displaystyle H_b}{\underset{\displaystyle H_a \quad H_b}{-\overset{|}{C}-\overset{|}{C}-H_b}} \qquad \text{where } \delta_{H_a} > \delta_{H_b}$$

1. Effect of H_a on three equivalent H_b protons when they absorb energy. When the H_b protons absorb energy, as in Examples 1 and 2, H_a may exist in two possible spin orientations, ↑ and ↓, and because all the H_b protons are equivalent, each individually feels the same effective magnetic field. Thus, when the H_b protons absorb energy, they produce a *doublet;* the total area under the doublet now corresponds to three hydrogens.

2. Effect of three equivalent H_b protons on H_a when it absorbs energy. When H_a starts to absorb energy from the externally applied field H_0, there are three equivalent H_b protons on the adjacent carbon atom; their spins may be aligned with one another in the possible combinations shown in Fig. 15.10(a) and they interact with H_a. Note that some spin combinations are of identical

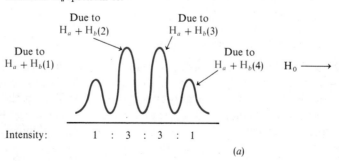

(a)

FIGURE 15.10 Idealized NMR spectrum of a compound with one tertiary hydrogen, H_a, adjacent to three equivalent H_b protons, where H_a absorbs at a higher chemical shift than H_b. (a) Effect of H_b on H_a when H_a absorbs energy; (b) NMR spectrum showing absorptions of H_a and H_b, where H_b appears as a doublet and H_a as a quartet.

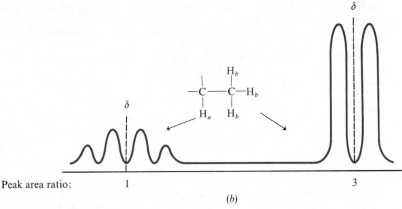

Peak area ratio: 1   3

(b)

FIGURE 15.10 (*continued*)

energy and that increasing the number of spins in the direction ↓ increases the energy (see part 1 in Example 1 for explanation).

Of the eight possible ways the spins of the three equivalent H_b protons can be combined, one (↓↓↓) is at the highest energy and one (↑↑↑) is at the lowest energy; there are two sets of three combinations each at intermediate energies. These spin combinations give rise to four peaks (a *quartet*) with intensity ratios 1:3:3:1. (As in Example 2, the peak intensities are proportional to the *number* of the spin combinations in each energy level.) The center of the quartet is taken to represent the chemical shift reported for H_a.

3. Composite spectrum. The composite spectrum in Fig. 15.10(*b*) shows the quartet further downfield for H_a and the doublet further upfield for the three equivalent H_b protons. The total peak area ratio of the quartet to the doublet is 3:1 because there are three H_b protons and one H_a proton.

These three general examples show how *spin-spin coupling* causes NMR signals to be split and to appear as two or more peaks as a result of the interaction between magnetically nonequivalent hydrogen nuclei in a given molecule. This coupling probably occurs both through space and through bonds. We considered only interactions between vicinal hydrogens (those on adjacent carbon atoms) thus far; geminal hydrogens are discussed at the end of this section as well as in Sec. 15.9. To summarize the results in the preceding examples, a tertiary hydrogen is split into a doublet by one adjacent hydrogen, into a triplet by two equivalent adjacent hydrogens, and into a quartet by three equivalent adjacent hydrogens.

The spin-spin splitting trend is general. The number of peaks (N) into which a proton signal may be split can be calculated by the formula:

$$N = n + 1$$

where n = number of equivalent vicinal protons

Note that this formula does *not* contain the number of equivalent protons that are responsible for the signal because they govern only the intensity of the signal. It is primarily the number of neighboring protons that dictates the number of peaks into which a proton signal is split. The number of peaks observed for a given proton (or group of equivalent protons) is often referred to as *multiplicity*, and the peak is referred to as a doublet, triplet, quartet, pentet, and so on.

B. Coupling Constants Due to Spin-Spin Splitting

Several other important features are observed in spin-spin splitting of proton absorptions. One is the magnitude of the spin-spin splitting constants between hydrogen nuclei, which is normally measured in cycles per second (hertz); these values are called **coupling constants, J**. Furthermore, when adjacent protons couple and cause splitting, the magnitude of the separation (that is, J) of the peaks in one signal for one group of equivalent hydrogens is identical with that in the signal for the adjacent group of equivalent hydrogens. This coupling between the two groups of protons must be the same because the protons are the same distance apart and are coupling through identical space and bonds.

Before discussing the splitting constants for protons in various functional groups, let us return to the example of 1,3-dichloropropane in Fig. 15.6 and examine the splitting and coupling constants observed for it.

$$
\begin{array}{c c c}
H_a & H_b & H_a \\
| & | & | \\
Cl-C_1-C_2-C_3-Cl \\
| & | & | \\
H_a & H_b & H_a
\end{array}
$$

Note that a *pentet* of peaks is centered at 2.2 δ (7.8 τ) because of the hydrogens (H_b) on C-2. These two identical hydrogens are surrounded by four identical hydrogens, H_a (two on C-1 and two on C-3). We would predict that the four H_a's would couple with the H_b's and cause ($n + 1$) or ($4 + 1$) peaks to be produced in the NMR absorption spectrum of H_b. The four equivalent H_a's are coupled with the two equivalent H_b's, which cause the H_a absorption to be split into a triplet ($2 + 1$ peaks). Both splittings are observed in the NMR spectrum. Furthermore, we stated that the coupling constants, J, should be identical in the spectra of both H_a and H_b protons. Fig. 15.6 reveals that $J_{ab} = 7$ Hz, where the subscripts a and b identify the hydrogens which are coupled in 1,3-dichloropropane. This is as it should be because the H_a protons couple with and split the H_b protons, and vice versa.

The NMR spectra of other typical organic molecules are shown in Figs. 15.11 to 15.13, which illustrate the type of information that can be obtained from NMR.

Fig. 15.11 shows the NMR spectrum of *1,1,2-trichloroethane* with the following features:

$$
\begin{array}{c}
H_a \\
| \quad b \\
Cl-C-CH_2-Cl \\
| \\
Cl
\end{array}
$$

The tertiary hydrogen, H_a, appears farther downfield than the two equivalent secondary hydrogens, H_b, because H_a is bonded to the carbon bearing two halogen atoms, and the two H_b protons are attached to the carbon that has one electronegative chlorine attached to it. Proton H_a is split into a *triplet* ($2 + 1$ peaks, intensities $1:2:1$) because of the two adjacent, equivalent H_b protons. The two equivalent H_b protons are split into a *doublet* ($1 + 1$ peaks, intensities $1:1$) because they are adjacent to a single H_a proton. The splitting constants are $J_{ab} \approx 6$ Hz, and the integrated peak areas are in the ratio 2:1 for $H_b:H_a$.

Fig. 15.12 shows the NMR spectrum of *2-nitropropane* with the following features:

$$
\begin{array}{c}
H_b \\
a \quad | \quad a \\
CH_3-C-CH_3 \\
| \\
NO_2
\end{array}
$$

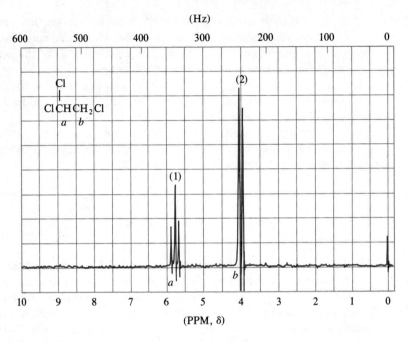

FIGURE 15.11   NMR spectrum of 1,1,2-trichloroethane, with TMS as the internal standard. [Data from *The Aldrich Library of NMR Spectra*, C. J. Pouchert and J. R. Campbell (eds.), copyright 1974, Aldrich Chemical Co., Milwaukee]

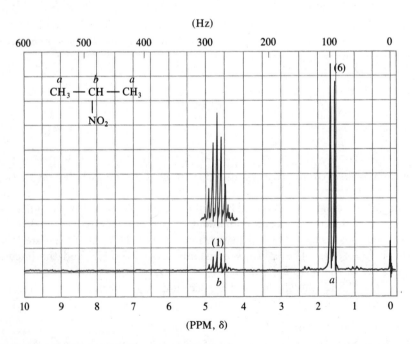

FIGURE 15.12   NMR spectrum of 2-nitropropane, with TMS as the internal standard. [Data from *The Aldrich Library of NMR Spectra*, C. J. Pouchert and J. R. Campbell (eds.), copyright 1974, Aldrich Chemical Co., Milwaukee]

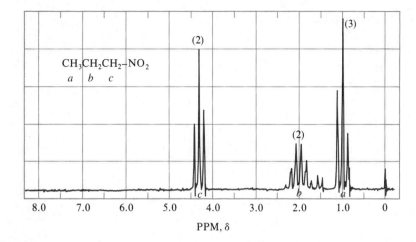

FIGURE 15.13   NMR spectrum of 1-nitropropane, with TMS as the internal standard. [Data from the *Sadtler Standard Reference Spectra* (*NMR*), © Sadtler Research Laboratories, Division of Bio-Rad Laboratories, Inc. (1965–1974)]

The six methyl hydrogens, H_a, appear farther upfield than the single tertiary hydrogen, H_b, because the latter is bonded to a carbon atom bearing the electron-withdrawing nitro (—NO_2) group. The H_a protons are split into a *doublet* (1 + 1 peaks, intensities 1:1) because they are adjacent to a single proton, H_b. The one H_b proton is split into a *septet*, which is rather difficult to see because the two peaks at the ends are very small; the septet results from the coupling of the six equivalent methyl protons (H_a) with H_b to give seven (6 + 1) peaks. The spin-spin coupling constants are all identical, with $J_{ab} \approx 7$ Hz. The integrated peak areas are in the ratio 6:1 for $H_a : H_b$.

Fig. 15.13 shows the NMR spectrum of *1-nitropropane*, which has the following features.

$$\overset{a}{CH_3}—\overset{b}{CH_2}—\overset{c}{CH_2}—NO_2$$

This molecule contains three sets of nonequivalent hydrogens, which appear at three different places on the NMR spectrum. The methyl protons H_a, which are farthest from the electron-withdrawing nitro group, appear farthest upfield. Next come the two equivalent H_b protons, which are closer to the nitro group; the electron-withdrawing effect causes a downfield shift. Finally, the H_c protons, which are on the carbon bearing the nitro group, are the farthest downfield. The methyl protons H_a are split into a *triplet* because they are adjacent to the two equivalent H_b protons. The two equivalent H_b protons appear as a *sextet* because they are coupled with five adjacent protons (three H_a protons and two H_c protons). (See Question 15.19.) Finally, the two equivalent H_c protons appear as a triplet because they are coupled with and split by the two equivalent H_b protons. The coupling constants are roughly the same for all the protons in this molecule, so that $J_{ab} \approx J_{bc} \approx 7$ Hz. The peak areas for these three types of protons are in the ratio 3:2:2 for $H_a : H_b : H_c$.

The *magnitude* of the spin-spin splitting constants, J, often provides valuable information about the structure of an organic molecule. Some typical hydrogen-

hydrogen coupling constants, J_{HH}, are:

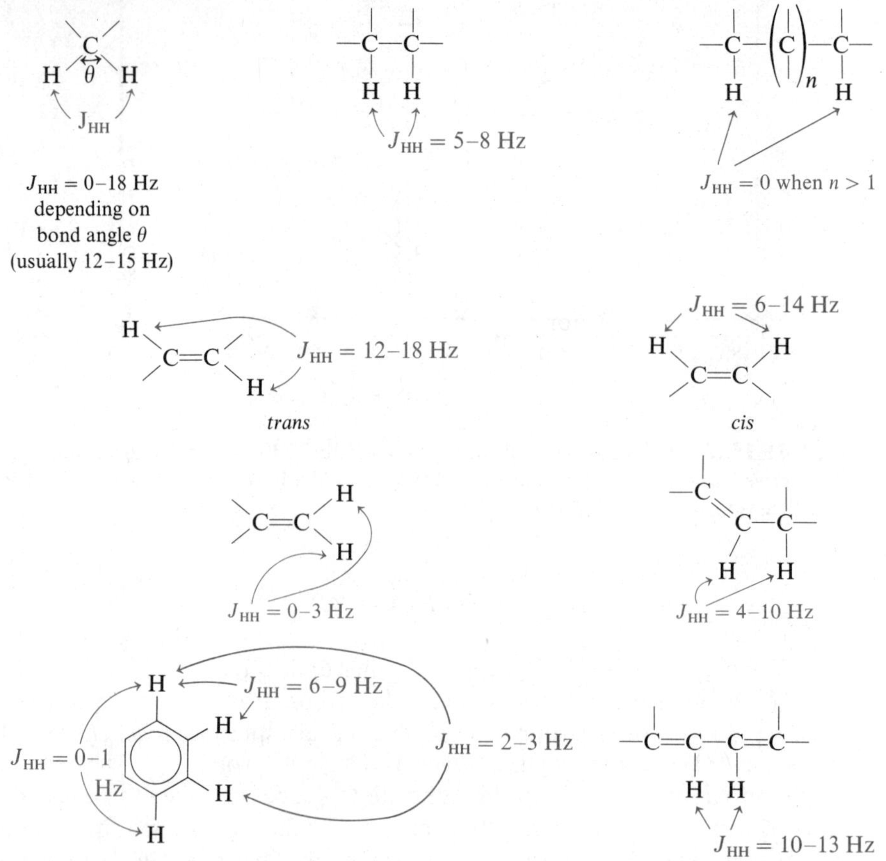

$J_{HH} = 0-18$ Hz
depending on
bond angle θ
(usually 12–15 Hz)

$J_{HH} = 5-8$ Hz

$J_{HH} = 0$ when $n > 1$

$J_{HH} = 12-18$ Hz

trans

$J_{HH} = 6-14$ Hz

cis

$J_{HH} = 0-3$ Hz

$J_{HH} = 4-10$ Hz

$J_{HH} = 6-9$ Hz

$J_{HH} = 2-3$ Hz

$J_{HH} = 0-1$ Hz

$J_{HH} = 10-13$ Hz

In this listing of coupling constants, it may seem surprising that two hydrogens attached to the same carbon atom have a splitting constant (couple with one another); as with vicinal protons, they do so only when those two hydrogens are nonequivalent (see Sec. 15.9).

Splitting patterns are not always easy to analyze because peaks often overlap and add together, even though they result from absorption by different groups of protons. Remember the basic principles involved in NMR spectroscopy and the basic types of information that can be obtained from NMR spectra. It is only with practice that familiarity is gained and complex spectra can be interpreted.

15.7   Summary of Interpreting NMR Spectra

Before moving to more complex examples of NMR spectroscopy, including its applications to chemical problems, let us summarize the important types of information that can be gained from an NMR spectrum.

1. From the *chemical shift* of each peak or group of split peaks, one can obtain some idea of the *type of functional group* to which the hydrogen(s) is attached. Often the spectrum must be considered as a whole, so that the chemical shifts can be evaluated relative to those in the rest of the spectrum.

2. From the *multiplicity of the spin-spin splitting pattern* (that is, the number of peaks associated with each nonequivalent proton or group of protons), information can be gained about the number of hydrogen atoms on the adjacent carbon atom (that is, the nearest hydrogen neighbors). Note, however, that *splitting occurs between adjacent hydrogens only when they are nonequivalent;* there is, for example, no splitting in Br—CH₂—CH₂—Br.

3. From the *peak area integration*, knowledge can be obtained about the relative numbers of each type of hydrogen in the molecule. The peak areas even play a role in the spin-spin splitting analysis, because the intensities of the peaks in a given multiplicity often vary from one splitting pattern to another.

The NMR peaks for a given compound are usually reported in the following manner:

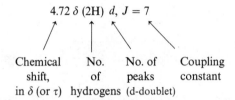

$$4.72 \; \delta \; (2H) \; d, \, J = 7$$

Chemical shift, in δ (or τ)	No. of hydrogens	No. of peaks (d-doublet)	Coupling constant

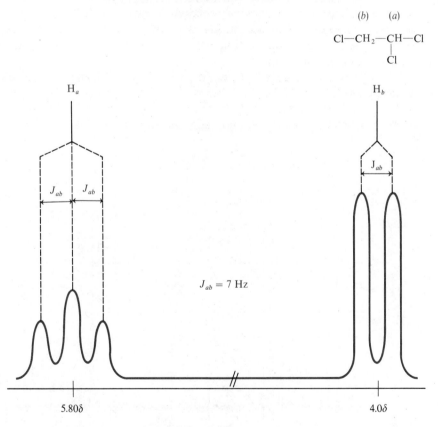

FIGURE 15.14   Idealized NMR spectrum, as illustrated for 1,1,2-trichloroethane,

$Cl_2CH—CH_2Cl$.

Peak intensities within each group are as they should be theoretically. Actual spectrum of the compound is shown in Fig. 15.11.

The *number of peaks* uses the following common abbreviations: $s =$ singlet, $d =$ doublet, $t =$ triplet, $q =$ quartet, $m =$ multiplet (that is, many nondistinguishable peaks). Beyond the use of q for quartet, the number of split peaks is normally written out (for example, pentet, sextet, septet, and so on).

Finally, many multiplets are not symmetrical looking, even though theory predicts that they should be. For example, in the NMR spectrum for 1,1,2-trichloroethane in Fig. 15.11, the triplet and doublet should look like the peaks shown in Fig. 15.14, with intensities of 1:2:1 and 1:1, respectively. In the spectrum of the actual compound in Fig. 15.11, however, the inner peaks (that is, the right-hand peak of the downfield triplet and the left-hand peak of the upfield doublet) are taller than the outer peaks. This is often observed in NMR spectra unless the coupling constant J within the signal is much smaller than the chemical shift differences between signals—the various sets of split peaks have chemical shifts that are far apart.

15.8 NMR Spectra of Substituted Aromatic Compounds

The NMR spectra of various substituted aromatic compounds are instructive. First, peaks are attributed to the substituent(s) if they contain protons. Second, certain substituents affect the chemical shift of the remaining protons that are attached to the aromatic ring. In this section, we will see some typical NMR spectra of substituted aromatic compounds.

The NMR spectrum of toluene, shown in Fig. 15.15, contains two absorption peaks, one centered at $2.32\ \delta$ ($7.68\ \tau$) corresponding to three protons and one centered at $7.17\ \delta$ ($2.83\ \tau$) corresponding to five protons. Both peaks appear as singlets, although in many aromatic compounds the ring protons have different chemical

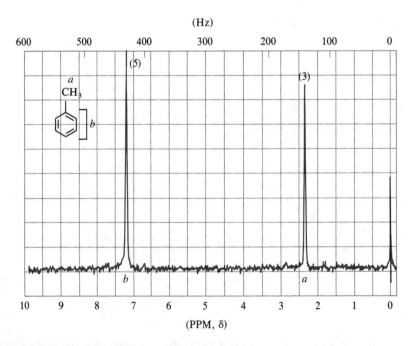

FIGURE 15.15 NMR spectrum of toluene, with TMS as the standard. [Data from *The Aldrich Library of NMR Spectra*, C. J. Pouchert and J. R. Campbell (eds.) copyright 1974, Aldrich Chemical Co., Milwaukee]

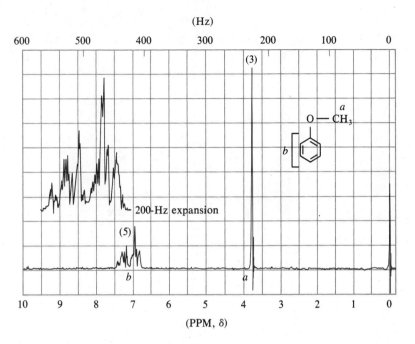

FIGURE 15.16   NMR spectrum of anisole. [Data from *The Aldrich Library of NMR Spectra*, C. J. Pouchert and J. R. Campbell (eds.), copyright 1974, Aldrich Chemical Co., Milwaukee]

shifts and appear as a split peak. In toluene, the *ortho*, *meta*, and *para* hydrogens all have about the same chemical shift and appear as a singlet because they are nearly equivalent. The methyl protons are close enough to the ring current of the aromatic ring to be deshielded slightly, so they appear a little more downfield than otherwise expected for aliphatic hydrogens (see Sec. 15.5B). Furthermore, the methyl protons are not split because there are no adjacent hydrogens; the nearest hydrogen is attached to the *ortho* position of the ring and is thus more than two carbon atoms (four bonds) removed from the methyl protons.

The NMR spectrum of anisole is shown in Fig. 15.16. In accord with the general chemical shift pattern (Table 15.1), the protons on the —O—CH$_3$ group are de-shielded and shifted downfield. Also, the absorption in the aromatic region ($\sim 7\,\delta$ or $3\,\tau$) shows that there is a multiplet of peaks. This chemical shift pattern is often observed for hydrogens that are *ortho* to substituents. In this example, the protons that are *ortho* to the methoxy group, —OCH$_3$, are found at lower field (higher δ values) than the other aromatic protons.

As a final example, consider the NMR spectrum of *p*-nitrochlorobenzene in Fig. 15.17, which contains only aromatic protons. Here the effect of electron-withdrawing groups is very dramatic because as we know from electrophilic aromatic substitution, the nitro group is very much more deactivating and electron-withdrawing than the chloro group. In this NMR spectrum, the nitro group greatly deshields the two protons *ortho* to it, thus shifting them downfield. The chloro group is slightly deactivating, and it causes the two protons *ortho* to it to be shifted downfield but to a lesser extent than the nitro group. The four peaks in the aromatic region are identified on the spectrum. Note also that the coupling constants between the two pairs of *ortho* hydrogens are identical, which further supports this assignment. As is typical of sets of peaks with chemical shifts that are close together, the two outer peaks are smaller than the two inner peaks, even though each pair of peaks taken together integrates to be equal to two protons.

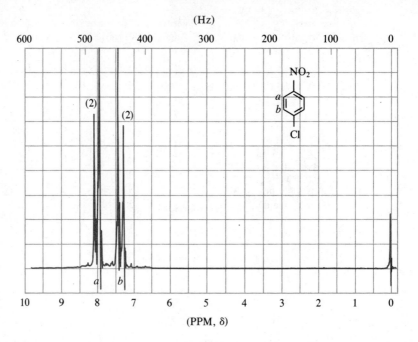

FIGURE 15.17   NMR spectrum of *para*-nitrochlorobenzene, with TMS as the internal standard. [Data from *The Aldrich Library of NMR Spectra*, C. J. Pouchert and J. R. Campbell (eds.), copyright 1974, Aldrich Chemical Co., Milwaukee]

15.9   NMR Spectra of Compounds That Contain Nonequivalent Geminal Hydrogens

We explain many NMR absorption spectra on the basis of proton equivalence and nonequivalence, yet the underlying principle of equivalence is based on the molecular structure of the compound.

Equivalent Versus Nonequivalent Protons

Examination of the protons in ethylbenzene leads to a fairly straightforward determination of which hydrogens are equivalent and which are nonequivalent. For example, the methyl hydrogens (*a*), methylene hydrogens (*b*), and the *ortho* (*c*), *meta* (*d*), and *para* (*e*) hydrogens are five non-equivalent sets of protons that should each give a discrete peak in the NMR spectrum. The various types of hydrogens in several other examples are assigned here:

A simple chemical test to determine the equivalency or nonequivalency of two given protons is outlined as follows:

1. Build (or draw) two three-dimensional models of the molecule being investigated.
2. Take one of these models and replace one of the two hydrogens you are comparing with some other group (—G).
3. Take the second model and replace the second hydrogen (the one you are comparing to the hydrogen replaced in 2) with the same G group.
4. Compare the two molecules.
(*a*) If they are superimposable, the two hydrogens are equivalent.
(*b*) If the two molecules are enantiomers of each other, the two hydrogens bear a mirror-image relationship to each other and may be treated as equivalent hydrogens for purposes of discussing the NMR spectrum. This type of hydrogen is called *enantiotopic*.
(*c*) If the two molecules are diastereomers of each other, then the hydrogens (*diastereotopic*) are nonequivalent.

Example 1

Consider the two methylene hydrogens in ethylbenzene:

1. Build two identical models (using Fischer projections, for example):

$$
\begin{array}{cc}
C_6H_5 & C_6H_5 \\
H\!\!-\!\!\!\!+\!\!\!\!-\!\!H & H\!\!-\!\!\!\!+\!\!\!\!-\!\!H \\
CH_3 & CH_3
\end{array}
$$

2. Replace each of the two hydrogens being compared with a G group:

$$
\begin{array}{cc}
C_6H_5 & C_6H_5 \\
H\!\!-\!\!\!\!+\!\!\!\!-\!\!G & G\!\!-\!\!\!\!+\!\!\!\!-\!\!H \\
CH_3 & CH_3
\end{array}
$$

3. Comparing the two, we find that they are mirror images. The two hydrogens are therefore enantiotopic and, for NMR purposes, may be treated as equivalent.

Example 2

Consider the two geminal hydrogens of bromochloromethane:

1. Build two identical models:

$$
\begin{array}{cc}
Br & Br \\
H\!\!-\!\!\!\!+\!\!\!\!-\!\!H & H\!\!-\!\!\!\!+\!\!\!\!-\!\!H \\
Cl & Cl
\end{array}
$$

2. Replace each of the two hydrogens being compared with a G group:

$$
\begin{array}{cc}
Br & Br \\
H\!\!-\!\!\!\!+\!\!\!\!-\!\!G & G\!\!-\!\!\!\!+\!\!\!\!-\!\!H \\
Cl & Cl
\end{array}
$$

3. Comparing the two, we find that they are enantiomers. The hydrogens, therefore, bear a mirror-image relationship to each other (they are enantiotopic) and, for purposes of NMR, may be treated as equivalent.

Example 3

Consider the two geminal hydrogens of bromoethene:

1. Build two identical models:

$$
\begin{array}{cc}
\text{H} \diagdown \text{C} \diagup \text{H} & \text{H} \diagdown \text{C} \diagup \text{H} \\
\text{C} \diagdown & \text{C} \diagup \\
\text{H} \diagup \text{C} \diagdown \text{Br} & \text{H} \diagup \text{C} \diagdown \text{Br}
\end{array}
$$

2. Replace each of the two hydrogens being compared with a G group:

$$
\begin{array}{cc}
\text{G} \diagdown \text{C} \diagup \text{H} & \text{H} \diagdown \text{C} \diagup \text{G} \\
\text{C} \diagdown & \text{C} \diagup \\
\text{H} \diagup \text{C} \diagdown \text{Br} & \text{H} \diagup \text{C} \diagdown \text{Br}
\end{array}
$$

3. Comparing the two resulting molecules, we find that they are diastereomers. The two hydrogens are therefore nonequivalent.

The presence of an asymmetric carbon atom in a molecule results in an amazingly large number of NMR signals. For example, the NMR spectrum of 2-nitrobutane in Fig. 15.18 shows a very complex multiplet centered at $\sim 1.9\ \delta$ $(8.1\ \tau)$, which corresponds to two hydrogens; yet the initial examination of the structure of the molecule suggests that those *two* protons should appear as a pentet $(4 + 1$ peaks) in the spectrum. The explanation of this phenomenon lies in *molecular nonequivalence*.

$$
\overset{a}{\text{CH}_3}-\overset{b}{\text{CH}_2}-\overset{c}{\overset{*}{\text{CH}}}-\overset{d}{\text{CH}_3}
$$
$$
\underset{\text{NO}_2}{|}
$$

2-Nitrobutane

This compound has an asymmetric carbon atom, and when we examine its possible conformational structures (Fig. 15.19), we find that the hydrogens on the *b*

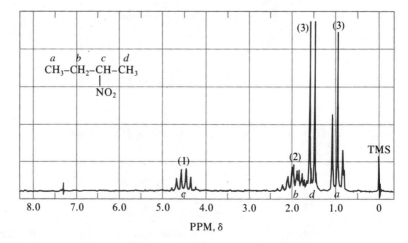

FIGURE 15.18 NMR spectrum of 2-nitrobutane, with TMS as the internal standard. (Data from the *Varian High Resolution NMR Spectra Catalog*, Varian Associates, Palo Alto, CA.)

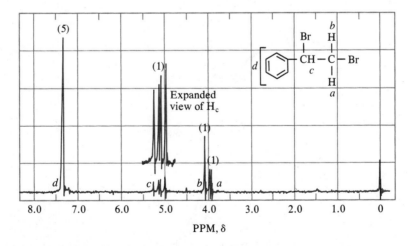

FIGURE 15.19   Staggered conformations (shown in Newman projection formulas) for an enantiomer of 2-nitrobutane, showing the nonequivalence of the H_b protons.

carbon atoms adjacent to the asymmetric center (that is, the vicinal hydrogens) are
always in different electronic environments even though there is free rotation about
the carbon-carbon bonds. The hydrogens near the electron-withdrawing nitro group
feel a different shielding effect than those farther from it, and regardless of the conformation, there will always be one H_b hydrogen nearer the nitro group. As we see in
Sec. 15.10, conformational stereoisomers do not exhibit any unusual effects in the
NMR spectrum as long as the hydrogens are equivalent. It is, however, the nonequivalence of the hydrogens (they are diastereomeric because of the asymmetric
carbon) in the three-dimensional sense that gives rise to the complex splitting pattern
of the H_b protons in 2-nitrobutane.

   It is interesting that the presence of any asymmetric carbon atom in compounds
causes this effect; it is not limited to specific compounds like 2-nitrobutane. Also,
the effect is observed on both the racemic mixture of enantiomers and the optically
pure compounds since there is no difference in the NMR spectra of a mixture of
enantiomers and each of the pure enantiomers themselves.

   Another example of the nonequivalence of protons is shown in Fig. 15.20,
which depicts the NMR spectrum of 1,2-dibromo-1-phenylethane. This spectrum
is simple enough so that the various peaks can be identified. The splitting caused by
the various protons is shown in the figure. Note also that the absorption peaks at
approximately 4.0 δ (6.0 τ) result from both H_a and H_b.

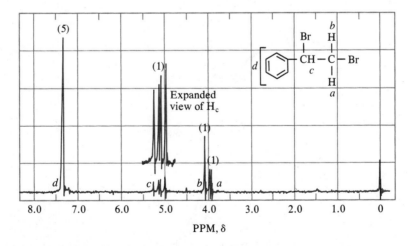

FIGURE 15.20   NMR spectrum of 1,2-dibromo-1-phenylethane, showing nonequivalence of pro-
tons. Various peaks and splitting are identified by reference to the structure drawn on the spectrum.
[Data from the *Sadtler Standard Reference Spectra (NMR)*, © Sadtler Research Laboratories,
Division of Bio-Rad Laboratories, Inc. (1965–1974)]

15.10   NMR Spectra of Conformers

Because of the importance of the molecular environment, it is not surprising that geometry plays an important role in the splitting patterns and coupling constants of protons. Considerable theoretical work has been done on this subject, and the results indicate that the coupling constant J should depend on how much the two hydrogens are removed from one another. The experimental results support the theoretical work, from which the following conclusions are drawn.

The coupling constant J varies with the **dihedral angle,** ω, which is the angle that separates two vicinal hydrogens:

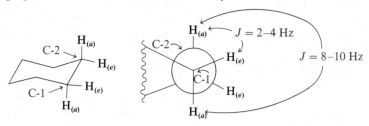

The variation of coupling constant with dihedral angle is shown in the graph in Fig. 15.21.

Particularly instructive examples of the dihedral angle-coupling constant relationship come from cyclohexane and substituted cyclohexanes. For example, in cyclohexane the following relationships exist between equatorial and axial hydrogens; Newman projection formulas are used for clarity:

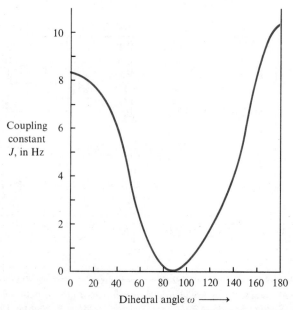

FIGURE 15.21   **Effect of dihedral angle ω on splitting constant J between two vicinal hydrogens.**

When the NMR spectrum of cyclohexane is determined, however, it consists of a single peak and there is no coupling. This is because cyclohexane is rapidly interconverting between chair conformations, with intervening twist-boat and boat forms. This interconversion occurs so rapidly at room temperature that the equatorial and axial hydrogens cannot be "seen" separately by NMR. Instead, the spectrum that is recorded represents the average population of the two chair forms, where an axial hydrogen in one form is equatorial in the other, and vice versa. One can use the analogy of NMR being like a camera with a shutter speed too slow to view the rapidly converting conformers; therefore an average picture is obtained.

When the temperature of a cyclohexane sample is lowered, however, the rate of interconversion between chair forms is slowed down to the point that the equatorial and axial hydrogens can be seen as discrete peaks in the NMR spectrum. For example, the single peak observed at room temperature is separated into two peaks of equal intensity at $-100°$, one of which is attributed to the equatorial hydrogens and the other to the axial hydrogens. Under these conditions, ring flipping still occurs but at a slow rate. Methods of this sort also enable us to estimate energies of activation for conformational ring flipping and indeed NMR provides an almost unique way to get such measurements.

From studies on 4-substituted *tert*-butylcyclohexane derivatives, we know that the bulky *tert*-butyl group occupies the equatorial position almost exclusively (see Sec. 3.16). In NMR spectroscopy, these types of compounds have been used to determine that an equatorial hydrogen generally absorbs downfield (greater δ value, smaller τ value) when compared with an axial hydrogen in the same position. As with many findings in organic chemistry, this is a good rule of thumb but several exceptions have been found.

Application of the preceding principle has been reported for compounds such as bromocyclohexane, with an NMR spectrum that contains a peak at 4.16δ (5.84τ) downfield from the absorption peaks of the rest of the hydrogens in the molecule. This peak is assigned to the hydrogen attached to the carbon bearing bromine. On cooling, this single peak separates into two new peaks, one at 3.97δ (6.03τ) and one at 4.64δ (5.36τ). The latter is attributed to the equatorial hydrogen (axial bromine); furthermore, the peaks contain unequal areas and are in the ratio 4.6: 1, respectively (but the two taken together still represent one hydrogen in bromo-cyclohexane). From these data, it is concluded that equatorial bromocyclohexane is preferred over axial bromocyclohexane by a ratio of 82:18 at $-75°$.

82%   18%
Conformations of bromocyclohexane

Conformations also exist in noncyclic compounds, in which free rotation about carbon-carbon bonds may occur. As we know, conformations are in equilibrium with one another. At ordinary temperatures, where rotation is rapid, the NMR instrument is unable to distinguish between conformers. It records only what it sees on the average. To cite a specific example, consider isopropyl bromide, which gives two sets of peaks in the NMR, one a doublet corresponding to the six methyl hydrogens and one a septet (6 + 1 peaks) corresponding to the hydrogen attached to carbon bearing the bromine atom. If we examine the following conformers, which result from rotation about one of the carbon-carbon bonds,

Conformations of isopropyl bromide

we see that each of the three methyl hydrogens (H, H′, and H″) is in a different molecular environment; in theory, these should appear as three different hydrogens in the NMR spectrum. The NMR records the presence of all three hydrogens, but because they are in equilibrium with one another, it records the average environment of the three of them. The resulting signal is just what would be expected—a doublet (1 + 1 peaks).

These instances should be compared with the NMR spectrum obtained for a nonequivalent environment, like the one in compounds containing an asymmetric center (see Sec. 15.9).

15.11   Effect of Other Atoms on NMR Spectra

We considered only hydrogen-hydrogen splitting and spin coupling thus far. Although these are important in NMR spectroscopy, several other atoms produce interesting results in NMR spectra and provide useful information about the structure of molecules. We examine deuterium, fluorine, and carbon-13 as representative atoms.

Deuterium, ^2H or D, which is a heavy isotope of hydrogen, has been used considerably in NMR spectroscopy for two reasons. First, it is fairly easy to introduce deuterium into a molecule. Second, the presence of deuterium in a molecule is not detected in the proton NMR spectrum. Deuterium has a much smaller magnetic dipole moment than hydrogen, and because of this it absorbs at different field strengths. Even though it broadens the peak of a neighboring proton, it does not split it because it couples only slightly with the proton.

The effect of deuterium on NMR spectra can best be seen by a simple example in which deuterium replaces the methyl hydrogens in, for example, ethyl bromide. The following schematic representations of NMR spectra show the changes that occur.

Note that the multiplicity of the —CH_2— peaks changes from a quartet (3 + 1 peaks) to a triplet (2 + 1 peaks) to a doublet (1 + 1 peaks) and finally to a singlet as the hydrogens on the —CH_3 group are replaced by D. In Br—CH_2—CD_3 no splitting occurs between the two remaining hydrogens and the neighboring methyl, which is now completely deuterated.

Another example of how information can be gained about the conformational structure of cyclohexane comes from deuterium labeling. For example, $C_6D_{11}H$ (1) has been prepared and examined by proton NMR spectroscopy. At room temperature, a single peak is observed for the one remaining hydrogen. As the temperature is lowered, the singlet separates into a doublet of equal areas. These two peaks are separated by 29 Hz and are attributed to equatorial and axial hydrogens in cyclohexane.

1

As before, the rate of interconversion between conformers is very fast at room temperature (it has been estimated that flipping occurs about 1 million times per second!), but at low temperature the interconversion is much slower (estimated to be once every 10 sec at $-100°$). Under these conditions, the NMR spectrometer is able to detect and record the presence of protons that are in different molecular environments and thus have different chemical shifts. (*Question:* Which proton, the equatorial or the axial one, is farthest downfield?)

The presence of fluorine (^{19}F) in organic molecules does, on the other hand, interact with and split adjacent protons. Fluorine-19 has magnetic properties, and it is possible to measure the NMR spectra of fluorine (the same is true of deuterium), although this requires the use of different rf frequencies and applied field strengths. In the proton NMR spectra, fluorine like deuterium per se cannot be detected, but its presence can be seen by observing how it couples with and splits adjacent protons. For example, the proton NMR spectrum of 2,2,2-trifluoroethanol,

$$CF_3—CH_2—O—H$$

shows a sharp singlet at 3.38 δ (6.62 τ), which is due to the proton on the —OH group, and a quartet centered at 3.93 δ (6.07 τ) corresponding to the two —CH_2—protons being split by the three fluorine atoms on the —CF_3 group. The splitting constant for the hydrogen-fluorine interactions in $J_{HF} = 9$ Hz. Because fluorine does not absorb in the proton NMR spectrum under these conditions, these are the only peaks observed for 2,2,2-trifluoroethanol.

Much work has recently been done using the carbon-13 nucleus. In carbon-13 NMR (CMR) spectroscopy, the instrument is designed to observe the naturally occurring carbon-13 isotopes in the sample (samples can be synthesized selectively to include carbon-13 at various position also). Carbon-13 is the isotope found in one of every 10,000 carbon atoms. When the NMR instrument is set up to observe this particular nucleus, only the carbon-13 nuclei give rise to peaks. This is the same effect observed in the proton NMR spectra of compounds containing fluorine or deuterium. The carbon-13 nuclei are, however, coupled with the hydrogen nuclei, and a highly complex spectrum often results. To render the spectrum more easily interpretable, the protons are irradiated and decoupled (see Sec. 15.12). The spectrum thus produced consists of a series of singlets—one for each different type of carbon present in the molecule. Fig. 15.22 shows the completely interpreted CMR spectrum of dimethyl-*p*-tolylsulfonium tetrafluoroborate, a sulfonium salt.

Although CMR spectra are much more difficult to obtain than the corresponding proton NMR (PMR) spectra because of the much lower natural abundance of the

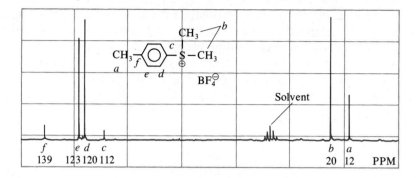

FIGURE 15.22   Carbon 13 NMR spectrum of dimethyl-*para*-tolylsulfonium tetrafluoroborate.

carbon-13 nucleus that often requires much larger samples, the large shift differences between various carbon atoms makes the technique extremely well suited to structure elucidation. Seldom do peaks of two different atoms overlap, which is often the case in PMR spectroscopy. Although chemical shifts in PMR generally range over 10 ppm, there is a 200-ppm range in CMR. This allows for much easier interpretation and peak assignment, eventually leading to complete structure determination.

15.12   Another Method for Analyzing Complex NMR Spectra: Spin Decoupling

The introduction of deuterium into an organic molecule often simplifies the NMR spectrum so that overlapping peaks can be identified more easily. Another technique, called **spin decoupling** (or *double resonance*), is considerably simpler from the experimental viewpoint.

It is possible to *saturate* a sample with a second rf signal at the same time the regular rf signal is applied. The regular rf signal is responsible for measuring the NMR spectrum of a given set of protons, and while this measurement is taking place, the second rf signal is set at the frequency at which the adjacent protons absorb, thus saturating this signal.[1] In this way, the protons that are detected on the NMR spectrum do not couple with and are not split by the adjacent protons.

A simple example to illustrate this concept is the NMR spectrum of ethanol, CH_3—CH_2—O—H, which consists of a 1:2:1 triplet centered at 1.22 δ (8.78 τ) due to the CH_3— group, a 1:2:2:1 quartet centered at 3.70 δ (6.30 τ) due to the —CH_2— group, and a singlet at 5.3 δ (4.7 τ) due to the proton on the —OH group. Now, if a second rf signal is applied to this sample at the energy corresponding to the —CH_2— absorption and if at the same time the NMR spectrum is recorded between 1 and 1.5 δ (8.5 and 9 τ), a *singlet* appears at 1.22 δ (8.78 τ) due to the CH_3— absorption because the adjacent protons are no longer able to couple with and split the protons on the CH_3— group (Fig. 15.23).

Note also that we do not observe coupling between the —OH and the adjacent —CH_2— even though the protons are only three bonds apart. This is common with nuclei that have acidic protons. In solution these protons often undergo rapid

---

[1]When a large amount of rf energy is applied to a sample in an NMR spectrometer, an equilibrium state results in which the nuclei are distributed equally between the two allowed states, ↑ and ↓. Although the nuclei are rapidly equilibrating between these two states, there is no net absorption of energy. The overall effect is that the protons are nonexistent with respect to the instrument. They are not observable and do not couple with neighboring protons.

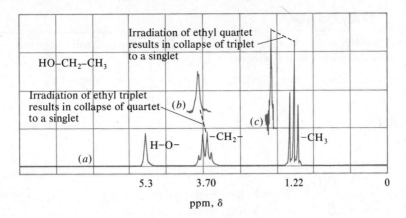

FIGURE 15.23   (*a*) NMR spectrum of ethanol and (*b*) partially decoupled NMR spectrum of ethanol showing result of irradiation at the ethyl triplet. (*c*) Result of irradiation at the ethyl quartet. The irradiated peaks have been shifted to give a clearer view of the results of the decoupling.

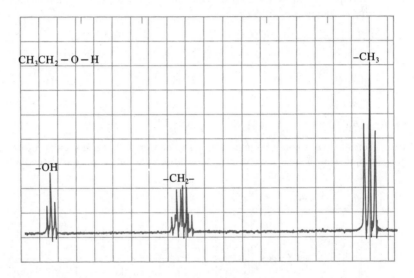

FIGURE 15.24   NMR spectrum of pure, dry ethanol, showing coupling between —OH and —CH$_2$—. (Data from L. M. Jackman and S. Sternhell, *Applications of Nuclear Magnetic Resonance Spectroscopy in Organic Chemistry*, 2d ed., p. 16, 1969, Pergamon Press, New York)

exchange with traces of, for example, water; that is:

$$R—CH_2—\overset{..}{\underset{..}{O}}—H + H_2O \rightleftharpoons H—\overset{\overset{\displaystyle H}{|}}{\underset{\oplus}{O}}—H + R—CH_2—\overset{..}{\underset{..}{O}}{:}^{\ominus}$$

At one instant, for example, the —CH$_2$— group attached to oxygen "sees" the —OH proton, but in the next instant it is gone. If exchange is too rapid for the NMR to "see," then no coupling is observed. This is what was seen in the preceding case. If however, *dry* ethanol is run on the NMR as a pure liquid, exchange is slowed down considerably however, and coupling may be observed (Fig. 15.24).

Question 15.1

Why is the exchange process slowed down when pure dry alcohol is used?

Study Questions

15.2 In each of the following nine molecules, identify the equivalent hydrogens by assigning a letter to identical sets of hydrogens.

(*a*) CH_3-CH_2-I (*b*) $Cl-CH_2-CH_2-Cl$ (*c*) $CH_3-CH-CH_2-CH_3$
$$\qquad\qquad\qquad\qquad\qquad\qquad\qquad\qquad\qquad\qquad\qquad CH_3$$

(*d*) CH_3-CHCl_2 (*e*) $CH_3-CH_2-CH_2-Cl$ (*f*) $CH_3-CH-CH_3$
$$\qquad\qquad\qquad\qquad\qquad\qquad\qquad\qquad\qquad\qquad\qquad Cl$$

(*g*) ⬡ $\overset{CH_3}{\underset{H}{}}$ (*h*) $Cl-CH_2-CH_2-CH_2-Cl$ (*i*) $Cl-CH_2-\overset{CH_3}{\underset{|}{CH}}-CH_2-Cl$

15.3 For the compounds listed in Study Question 15.2, indicate the *approximate* positions of the chemical shifts you would expect to observe in their proton NMR spectra.

15.4 For each of the following eight compounds, indicate the *relative* chemical shifts you would expect for the various hydrogens. Also indicate the multiplicity (that is, the number of peaks) that would be expected for each set of identical protons, as well as the relative peak areas for each group of identical hydrogens.

(*a*) $BrCH_2CH_2Br$ (*b*) CH_3CHBr_2 (*c*) $BrCH_2CHBr_2$

(*d*) CH_3CH_2Br (*e*) ⬡$-CH_2CH_3$ (*f*) ⬡$-\overset{CH_3}{\underset{CH_3}{C}}-H$

(*g*) $(C_6H_5)_2CHCH_3$ (*h*) $(CH_3)_2CHCH_2NO_2$

15.5 There are two isomeric compounds, *A* and *B*, each with the molecular formula $C_3H_3O_2Cl$; both decolorize bromine in carbon tetrachloride and are soluble in aqueous sodium bicarbonate.

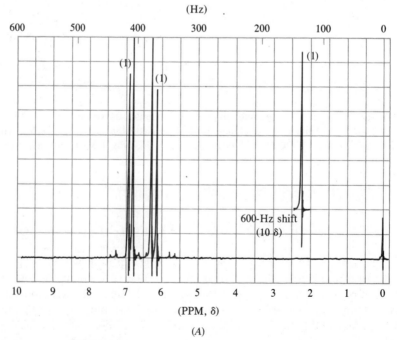

(*A*)

PROBLEM 15.5   NMR spectrum of compound (*A*). [Data from *The Aldrich Library of NMR* Spectra, C. J. Pouchert and J. R. Campbell (eds.), copyright 1974, Aldrich Chemical Co., Milwaukee]

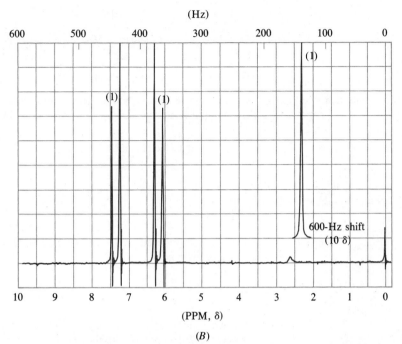

(Hz)

(PPM, δ)

(B)

PROBLEM 15.5   NMR spectrum of compound (B). [Data from *The Aldrich Library of NMR Spectra*, C. J. Pouchert and J. R. Campbell (eds.), copyright 1974, Aldrich Chemical Co., Milwaukee]

The NMR spectra of compounds *A* and *B* are shown here. On the basis of these spectra, identify *A* and *B*.

15.6 Draw sketches of the proton NMR spectrum you would anticipate observing at 60 MHz and relative to TMS for each of the following eight compounds. Indicate the relative positions of each group of protons, and show spin-spin splitting patterns insofar as possible. Also indicate the integrated peak areas expected for each group of protons. Keep in mind that F has magnetic properties and splits adjacent hydrogens.

(*a*) CH_3CH_2Br (*b*) CH_3CH_2F (*c*) CH_3CD_2F (*d*) $CH_3CD_2CH_2Cl$

(*e*) $(CH_3)_2CHCH_2Cl$ (*f*) $(CH_3)_3CH$ (*g*) $(CH_3)_3CD$ (*h*) $F_2CHCHBr_2$

15.7 Using the chemical test described in Sec. 15.9, determine whether each of the following pairs of hydrogens (indicated in red) are equivalent or bear an enantiomeric or diastereomeric relationship to each other.

(*a*)

(*b*) $H-\overset{\overset{\displaystyle Cl}{|}}{\underset{\underset{\displaystyle Cl}{|}}{C}}-H$

(*c*) $H-\overset{\overset{\displaystyle Cl}{|}}{\underset{\underset{\displaystyle Br}{|}}{C}}-H$

(*d*)

(*e*) $CH_3-\overset{\overset{\displaystyle H}{|}}{\underset{\underset{\displaystyle H}{|}}{C}}-Br$

(*f*)

(*g*) $Cl-\overset{\overset{\displaystyle H}{|}}{\underset{\underset{\displaystyle H}{|}}{C}}-\overset{\overset{\displaystyle H}{|}}{\underset{\underset{\displaystyle Cl}{|}}{C}}-CH_3$

15.8 For each of the following seven compounds, indicate the total number of peaks you would predict in the proton NMR spectrum. Also indicate the relative chemical shift you would predict and the multiplicity expected.

(*a*) $CH_3\underset{\underset{\displaystyle Cl}{|}}{C}HCH_3$ (*b*) $CH_3CH_2CH_2Cl$ (*c*) $CH_3-\overset{\overset{\displaystyle CH_3}{|}}{C}=CH_2$

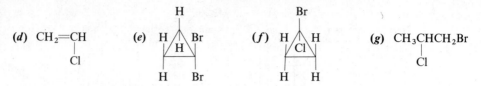

(*d*) $CH_2=CH$
            |
            Cl

(*e*)

(*f*)

(*g*) CH_3CHCH_2Br
                    |
                    Cl

15.9 Deuterium oxide (D_2O) is a common solvent in NMR spectroscopy. When using D_2O, however, TMS is not used as the internal reference standard but is replaced by sodium 4,4-dimethyl-4-silapentane sulfonate (1).

$$(CH_3)_3Si—CH_2CH_2CH_2—SO_3^{\ominus}Na^{\oplus}$$

(1)

Explain why this substitution is made.

15.10 Explain the following observations.

CH_3
|
$(CH_2)_6$ ◄——— Chemical shift of broad singlet due to six pairs of
| methylene hydrogens is 1.27 δ.
CH_3

$CH_2—(CH_2)_2$
|
$(CH_2)_2$ ◄——— Chemical shift of these two pairs of methylene hydrogens
| is only 0.6 δ. Compounds of this general structure
$CH_2—(CH_2)_2$ (aromatic rings joined to each other by two positions
                    that are nonadjacent) are called *cyclophanes*.

15.11 A sample of propane gas is chlorinated and four dichloro products, *A*, *B*, *C*, and *D*, with the formula $C_3H_6Cl_2$ are isolated. Each of these dichloro products is chlorinated again, and on analysis of the resulting trichloro products (formula $C_3H_5Cl_3$), the following results are obtained:

1. Compound *A* gives only one trichloro product.
2. Compound *B* gives two isomeric trichloro products.
3. Compounds *C* and *D* each gives three trichloro products.

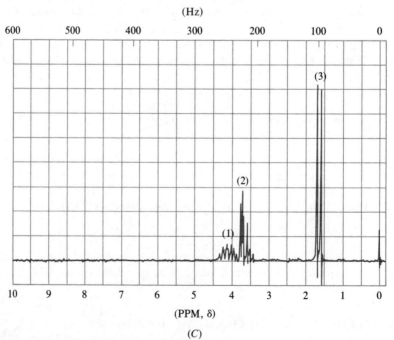

(*C*)

PROBLEM 15.11   NMR spectrum of compound (*C*). [Data from *The Aldrich Library of NMR Spectra*, C. J. Pouchert and J. R. Campbell (eds.), copyright 1974, Aldrich Chemical Co., Milwaukee]

The NMR spectrum of compound C is given. On the basis of these observations, draw the correct structures for A to D, as well as those of the trichloro products. *Briefly* give your reasoning.

15.12 Consider the following reaction:

$$A\ (C_4H_{10}) \xrightarrow[\text{heat}]{Br_2} B \text{ and } C \text{ (both } C_4H_9Br)$$

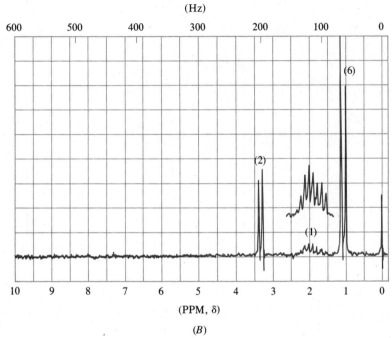

(B)

PROBLEM 15.12 NMR spectrum of compound (B). [Data from *The Aldrich Library of NMR Spectra*, C. J. Pouchert and J. R. Campbell (eds.), copyright 1974, Aldrich Chemical Co., Milwaukee]

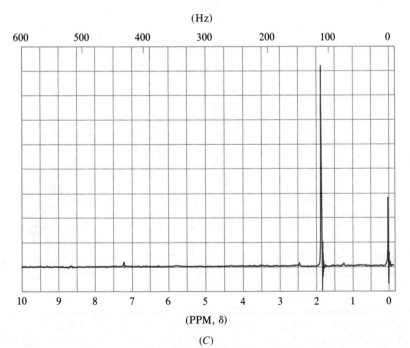

(C)

PROBLEM 15.12 NMR spectrum of compound (C) [Data from *The Aldrich Library of NMR Spectra*, C. J. Pouchert and J. R. Campbell (eds.), copyright 1974, Aldrich Chemical Co., Milwaukee]

The NMR spectra of B and C are given here. What are the structures of B and C? Draw complete structural formulas for each and provide the IUPAC name.

15.13 An *aromatic* compound A, C_8H_{10}, on treatment with chlorine in the presence of heat and light, gives two isomeric compounds, B and C, each having the formula C_8H_9Cl. B is present in large amounts, but only a small amount of C is formed. Identify compounds A and B from the NMR spectra provided here. What is the structure of C?

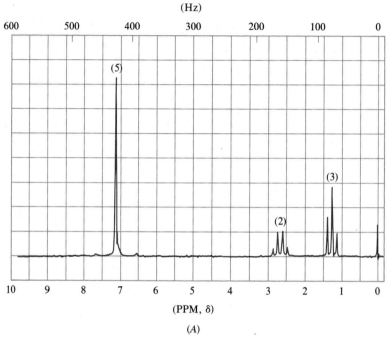

(A)

PROBLEM 15.13   NMR spectrum of compound (A). [Data from *The Aldrich Library of NMR Spectra*, C. J. Pouchert and J. R. Campbell (eds.), copyright 1974, Aldrich Chemical Co., Milwaukee]

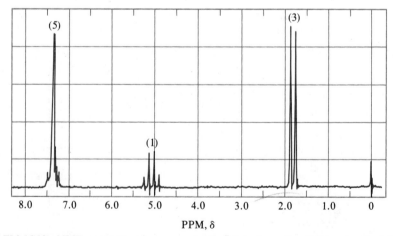

PPM, δ

PROBLEM 15.13   NMR spectrum of compound (B). [Data from the *Sadtler Standard Reference Spectra (NMR)*, © Sadtler Research Laboratories, Division of Bio-Rad Laboratories, Inc. (1965–1974)]

15.14 When neopentyl alcohol, $(CH_3)_3CCH_2$—OH, is heated with concentrated sulfuric acid, it is slowly converted into a mixture of two alkenes, each with the formula C_5H_{10}. Analysis of the mixture shows that there is 85% of one alkene and 15% of the other. The NMR spectrum of the predominant isomer is provided here. What are the structures of these two alkenes?

Propose a reasonable, stepwise mechanism to show how they are formed. Which alkene is the major product, and why?

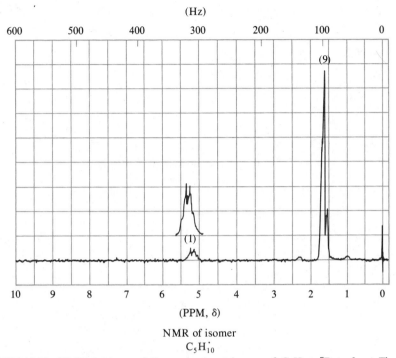

(Hz)

NMR of isomer
C_5H_{10}

PROBLEM 15.14 NMR spectrum of the predominant isomer of C_5H_{10}. [Data from *The Aldrich Library of NMR Spectra*, C. J. Pouchert and J. R. Campbell (eds.), copyright 1974, Aldrich Chemical Co., Milwaukee]

15.15 Optically active α-phenethyl bromide (1-bromo-1-phenylethane) reacts with aqueous sodium hydroxide and two products, *A* and *B*, are isolated from the reaction mixture. Neither compound exhibits any optical activity. *A* is converted back to the starting material by reaction with PBr_3; the regenerated starting material is racemic.

The following data are available for compound *B*:

1. Reacts with Br_2/CCl_4.
2. Reacts with cold, concentrated $KMnO_4$.
3. IR: strong band at 3,000 to 3,100 cm^{-1}; weak band at 1,640 to 1,680 cm^{-1}.
4. NMR:
   7.25 δ (5H) *bs*
   6.7 δ (1H) *q* (or two doublets)
   5.4 δ (2H) *q* (or two doublets)
5. When *B* is reacted with concentrated sulfuric acid and then heated in water, compound *A* ([α] = 0°) is obtained.

What is the structure of compound *A*? Compound *B*? What are the names of the mechanisms involved in each of these transformations (that is, 1-bromo-1-phenylethane → *A* → *B*)? Draw the mechanisms for each of these reactions.

15.16 Two isomeric compounds, *A* and *B*, have the formula $C_2H_4Br_2$. In the NMR spectrum of *A*, there is a singlet at 3.63 δ (6.37 τ) as the only absorption, and in the NMR spectrum of *B*, there is a *doublet* centered at 2.45 δ (7.55 τ) and a *quartet* centered at 5.85 δ (4.15 τ). What are the structures of *A* and *B*?

15.17 Compound *A* has a boiling point of ~80° ±4°. It reacts with sodium metal and, when oxidized with warm chromic acid, gives *B*, C_3H_6O. The NMR spectrum of *A* and the IR spectrum of *B* are shown here. Give the structures of *A* and *B*.

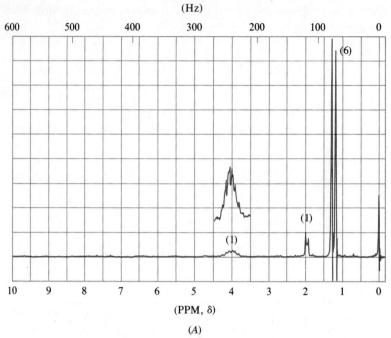

(A)

PROBLEM 15.17   NMR spectrum of unknown (*A*). [Data from *The Aldrich Library of NMR Spectra*, C. J. Pouchert and J. R. Campbell (eds.), copyright 1974, Aldrich Chemical Co., Milwaukee]

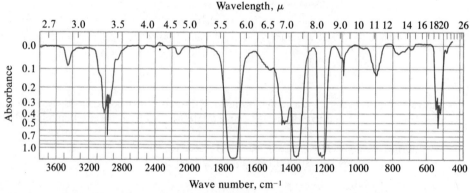

PROBLEM 15.17   IR spectrum of unknown (*B*). [Data from *Indexes to Evaluated Infrared Reference Spectra*, copyright 1975, The Coblentz Society, Norwalk CT]

15.18 The NMR spectrum of each of the following two compounds is reasonably complex and quite difficult to interpret. Focusing on one particular aspect of the spectra, the coupling between the pairs of diastereotopic hydrogens labeled H_a and H_b, explain how knowing the magnitude of this coupling constant enables one to determine which compound is being discussed. Predict a coupling constant ($J_{H_aH_b}$) for each compound.

15.19 Although the rule of thumb we use for determining the multiplicity expected for a given peak in the NMR spectrum works well for most simple organic compounds, there are many instances in which the $n + 1$ splitting rule does not fully explain experimentally determined spectra. If, for example, a given group of protons is split by two (or more) different groups of adjacent protons, the following modifications to the rule should be applied.

1. If the two groups of protons coupling to the protons observed have approximately the same coupling constant, they may be treated as one group of $n + 1$ protons.

2. If, however, the coupling between one group and the protons observed is quite different than that between the observed protons and the second group, they must be treated separately. The total number of peaks to be expected can be determined by application of the following equation:

$$N = (n_a + 1)(n_b + 1) \cdots (n_x + 1)$$

where $n_a, n_b, \ldots, n_x$ refer to the number of various groups of protons coupled to the protons being observed.

The following splitting patterns are observed for the peaks indicated in red. Explain these observations and draw that portion of the NMR spectra described.

(*a*) Br—CH_2—CH_2—CH_2—Cl   Pentet
       a   *b*   *c*

$$\overset{\displaystyle O}{\underset{}{\|}}$$

(*b*) CH_3—CH=CH—$\overset{O}{\overset{\|}{C}}$—H   Two doublets
   a   *b*   *c*

15.20 Treatment of an alkyl halide with antimony pentahalide (an extremely strong Lewis acid that has a strong affinity for halide ion) results in the formation of the corresponding hexahaloantimonate salt. The general equation is

$$R—X + SbX_5 \longrightarrow R^{\oplus}SbX_6^{\ominus}$$

This reaction was used successfully by Professor G. Olah of the University of Southern California to show the existence of a stable carbocation. The ion he generated was the *tert*-butyl hexafluoroantimonate ion, formed by treatment of *tert*-butyl fluoride with antimony pentafluoride. What would you predict to be the difference in the NMR spectra of *tert*-butyl fluoride and the *tert*-butyl cation? Explain how this information can be used as confirmation of a free, stable carbocation species.

16

Aryl and Vinyl Halides

The substitution reactions of alkyl halides provide conclusive evidence that not all carbon-halogen bonds are identical. Some are broken more easily than others. In contrast to alkyl halides, the aryl halides, Ar—X, are quite unreactive toward nucleophilic substitution. This chapter examines some of the chemistry of the aryl halides so that they can be compared with the alkyl halides. We also mention vinyl halides, $-\overset{|}{C}=\overset{|}{C}-X$, which are especially unreactive toward substitution of the halogen by a nucleophile. Alkynyl halides, $-C\equiv C-X$, are difficult to prepare, seldom encountered, and of limited interest. More can be learned about this family from the Reading References for this chapter in the appendix.

16.1 Structure, Nomenclature, and Preparation

The nomenclature of aryl halides was discussed in detail in Sec. 13.9. One important method of preparation of aryl halides is direct electrophilic aromatic substitution (see Sec. 14.4), and another method, presented in Sec. 14.23, involves organothallium reagents. Diazonium salts, Ar—$N_2^{\oplus}$, provide yet another convenient route (see Sec. 26.8).

The structure of aryl halides, Ar—X, and vinyl halides, $-\overset{|}{C}=\overset{|}{C}-X$, is of particular interest because the nature of the carbon-halogen bond in these molecules is largely responsible for their lack of reactivity toward nucleophilic substitution. The carbon-halogen bond lengths in aryl and vinyl halides are shorter than the length in alkyl halides. The C—Cl bond lengths in chlorobenzene and chloroethene (vinyl chloride), CH_2=CH—Cl, are close to 1.69 Å, as compared with C—Cl bond lengths in many alkyl chlorides that are in the range of 1.78 to 1.81 Å. The C—Br bond length in bromobenzene and bromoethene (vinyl bromide), CH_2=CH—Br, is 1.86 Å, whereas it ranges from 1.91 to 1.93 Å in most alkyl bromides. Thus, the carbon-halogen bonds in aryl and vinyl halides are usually short, thus suggesting that there may be some double bond character in the carbon-halogen bond. This is reasonable since the unshared pairs of electrons on halogen in aryl and vinyl halides can donate electrons to the π cloud via *resonance* (see Sec. 14.12). This is seen in the orientation effect that halogen imparts to the aromatic ring in electrophilic aromatic substitution. Although we examined resonance in connection with the carbocation intermediates that result from electrophilic attack on the aromatic ring, resonance structures can also be drawn for neutral molecules. The following structures are possible for an aryl halide:

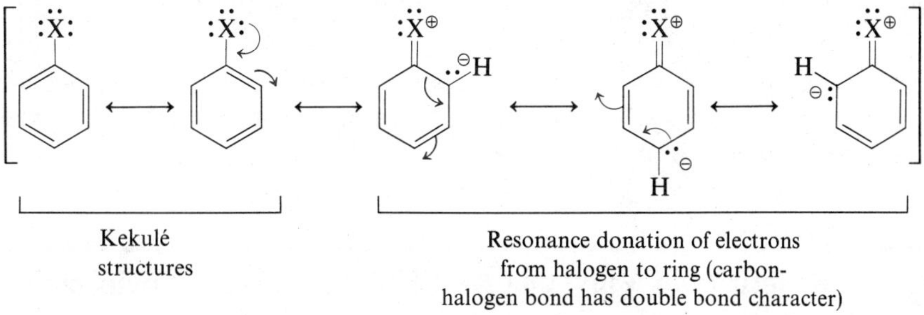

| Kekulé structures | Resonance donation of electrons from halogen to ring (carbon-halogen bond has double bond character) |

Aryl halide

The resonance interaction between a halogen and the carbon-carbon double bond can also explain the bond-sharing effect in vinyl halides:

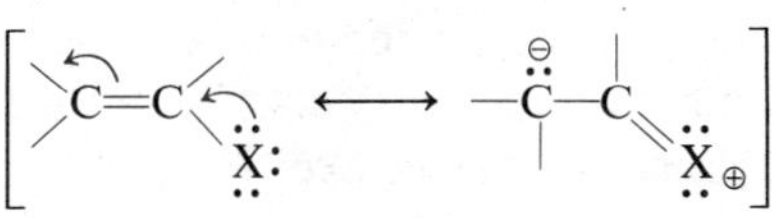

Vinyl halide

Changing the hybridization of the carbon to which the halogen is attached also may account for the C—X bond shortening. In alkyl halides the molecular bond orbital is sp^3-p for the C—X bond; in vinyl and aryl halides it is sp^2-p for the C—X bond. The higher percentage of s character in the bond orbitals accounts for this bond shortening (see Sec. 2.7).

TABLE 16.1   Physical Properties of Selected Vinyl and Aryl Halides

Compound	bp, °C	mp, °C	Density at 20°, g/cc
Aryl compounds:			
Fluorobenzene	85	−42	1.024
Chlorobenzene	132	−45	1.106
Bromobenzene	155	−31	1.499
Iodobenzene	189	−31	1.832
o-Chlorotoluene	159	−31	1.082
m-Chlorotoluene	162	−34	1.073
p-Chlorotoluene	162	−48	1.070
o-Dichlorobenzene	180	−17	1.305
m-Dichlorobenzene	173	−24	1.288
p-Dichlorobenzene	175	52	1.248
o-Nitrochlorobenzene	245	32	1.368
m-Nitrochlorobenzene	236	48	1.343
p-Nitrochlorobenzene	239	83	1.298
Vinyl compounds:			
Chloroethene (vinyl chloride)	−160	−14	0.912
Bromoethene (vinyl bromide)	−138	16	1.493

16.2   Physical Properties of Vinyl and Aryl Halides

The physical properties of some selected vinyl and aryl halides are given in Table 16.1. Like alkyl halides (see Sec. 9.2), vinyl and aryl halides are water-insoluble and often more dense than water.

16.3   Reactivity of Vinyl and Aryl Halides Toward Nucleophiles

Many vinyl and aryl halides are completely *unreactive* toward a wide variety of nucleophiles. For example, vinyl chloride or chlorobenzene can be boiled for many hours with alcoholic silver nitrate solution with no trace of silver chloride formed. These same compounds can be heated with aqueous hydroxide ion for a long time with no evidence of substitution. They also do not undergo Friedel-Crafts reactions with other aromatic compounds (see Sec. 14.6).

As we will see in Sec. 16.4, nucleophilic substitution on aryl halides *does* occur, but only under vigorous reaction conditions or when special structural features are present.

16.4   Substitution Reactions on Aryl Halides:
Nucleophilic Aromatic Substitution

The reactions we now study involve the attack by a nucleophile on an aromatic ring, giving rise to substitution of that nucleophile for a leaving group:

$$Ar-L + Nu\!:^{\ominus} \longrightarrow Ar-Nu + L\!:^{\ominus}$$

where L = leaving group

Nu$:^{\ominus}$ = various nucleophiles

Accordingly, this reaction is called **nucleophilic aromatic substitution.**

Nucleophilic substitution occurs readily on alkyl halides, which makes the reaction useful in synthesis because of the availability of numerous alkyl halides. In contrast to alkyl compounds, aryl halides must be forced to undergo nucleophilic substitution. A limited number of nucleophilic substitution reactions are known for vinyl compounds and they are seldom used.

Unsubstituted aryl halides are normally unreactive toward hydroxide ion, but under high temperature and pressure, nucleophilic substitution of —OH for —X can be forced. This is one industrial method for the preparation of phenol,

$$C_6H_5—OH$$

for example:

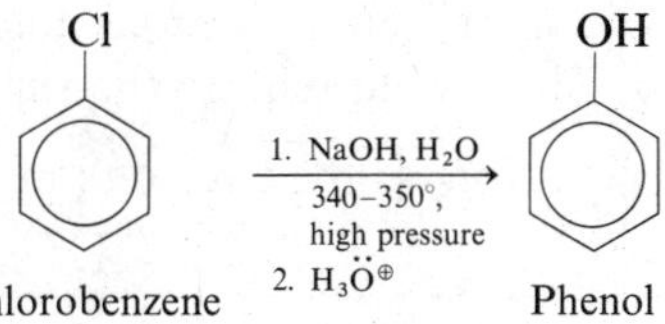

Chlorobenzene

1. NaOH, H$_2$O
340–350°,
high pressure
2. H$_3$Ö$^\oplus$

Phenol

Dow process

Other nucleophiles, such as RÖ:$^\ominus$, N̈H$_3$, and :C≡N:$^\ominus$, require similarly vigorous conditions to displace —X from an aryl halide.

Structure-reactivity correlations for this displacement reaction show that the presence of strong electron-withdrawing groups on positions *ortho* and *para* to the halogen greatly increase the rate of displacement. The following three examples illustrate that milder reaction conditions are required as more *ortho* and *para* substituents are added:

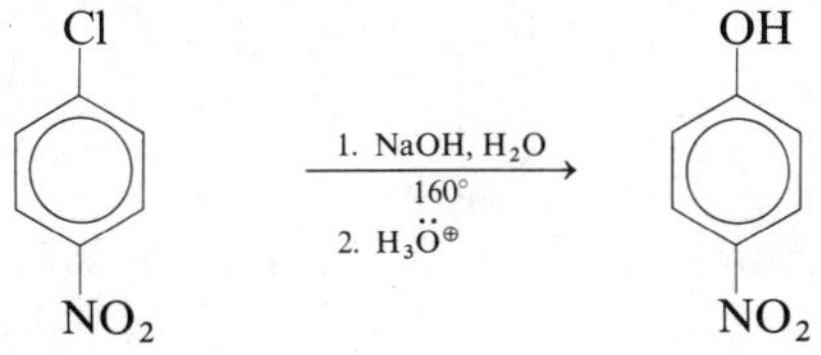

p-Nitrochlorobenzene

1. NaOH, H$_2$O
160°
2. H$_3$Ö$^\oplus$

p-Nitrophenol

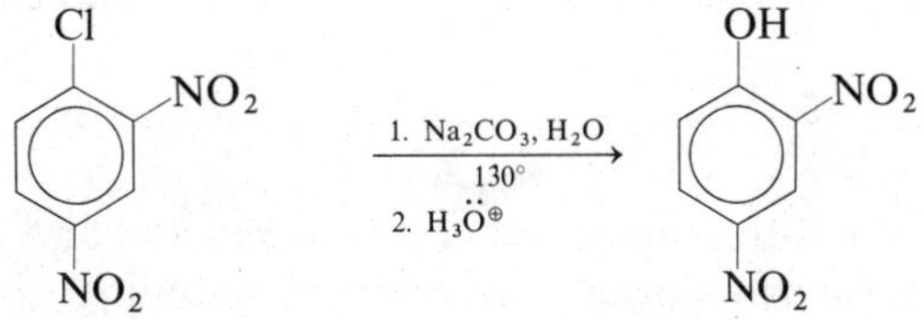

2,4-Dinitrochlorobenzene

1. Na$_2$CO$_3$, H$_2$O
130°
2. H$_3$Ö$^\oplus$

2,4-Dinitrophenol

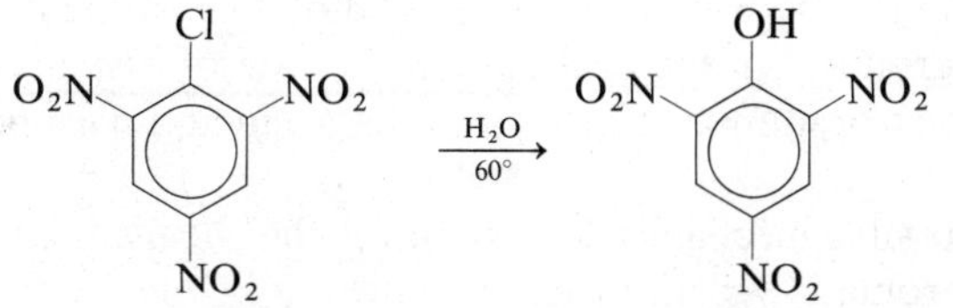

2,4,6-Trinitrochlorobenzene
(picryl chloride)

H$_2$O
60°

2,4,6-Trinitrophenol
(picric acid)

On the other hand, electron-withdrawing substituents that are *meta* to the halogen on the aromatic ring have little effect on reactivity.

Other common electron-withdrawing groups have similar effects. The following groups are the *meta* directors of electrophilic aromatic substitution:

$$-SO_3H \qquad -COOH \qquad -C\equiv N \qquad -\overset{\oplus}{N}R_3 \qquad \underset{O}{-\overset{\|}{C}-R} \qquad \underset{O}{-\overset{\|}{C}-H}$$

Other substituents in addition to halogen can be displaced by nucleophilic aromatic substitution; the most notable is the sulfonate group on $Ar—SO_3^{\ominus}$. We discuss this reaction in Sec. 17.3. Unlike electrophilic aromatic substitution, in which an electrophile displaces $H^{\oplus}$, hydrogen is seldom displaced by nucleophilic aromatic substitution reactions. This would require the rupture of a strong carbon-hydrogen bond with hydrogen departing as the hydride ion, $H\overset{\ominus}{:}$, a poor leaving group.

16.5   Mechanism of Nucleophilic Aromatic Substitution

The currently accepted mechanism for nucleophilic aromatic substitution involves two steps:

Step 1. Attack of nucleophile on aromatic ring: slow, rate-determining step

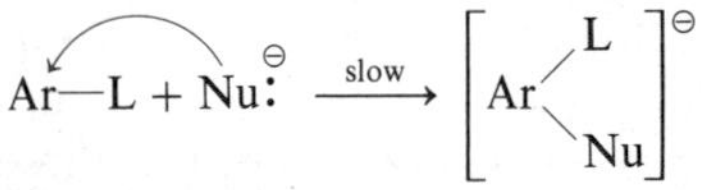

$$Ar—L + Nu\overset{\ominus}{:} \xrightarrow{\text{slow}} \left[Ar\overset{L}{\underset{Nu}{<}} \right]^{\ominus}$$

Step 2. Departure of leaving group, giving product: fast step

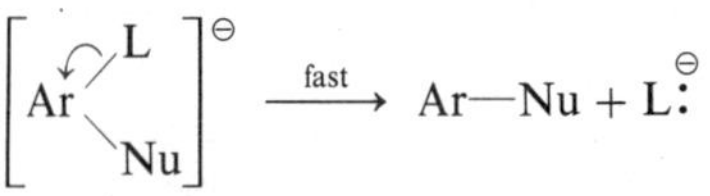

$$\left[Ar\overset{L}{\underset{Nu}{<}} \right]^{\ominus} \xrightarrow{\text{fast}} Ar—Nu + L\overset{\ominus}{:}$$

The following three facts support this mechanism for the displacement reaction on aryl halides:

1. The reaction follows second-order kinetics; that is, the rate of the reaction depends on the concentrations of the aryl halide, $Ar—X$, and the nucleophile, $Nu\overset{\ominus}{:}$

$$\text{Rate of reaction} = k_2[Ar—X][Nu\overset{\ominus}{:}]$$

2. The rates of displacement on aryl *bromides* and aryl *chlorides* are about the same and both are much slower than the corresponding displacement reaction on aryl *fluorides*. All three react much more quickly than the *iodides*.
3. The rate of displacement is significantly enhanced by the presence of *ortho* and *para* substituents on aryl halides when those substituents are strong electron-withdrawing groups. On the other hand, the rate of displacement is slowed by the presence of electron-donating groups in the *ortho* and *para* positions.

Three possible mechanisms account for the chemical kinetics. By analogy with electrophilic aromatic substitution, these are:

1. Single-step mechanism in which attack by the nucleophile is accompanied by simultaneous departure of the leaving group (halide); this is analogous to the S_N2 mechanism for alkyl halides.

2. Two-step mechanism, in which the first slow step is the attack of the nucleophile on the aromatic ring to give an intermediate, which then loses the leaving group in a fast reaction.

3. Two-step mechanism in which the first step is the attack of the nucleophile on the ring as a fast reaction, followed by departure of the leaving group in a slow reaction.

Sequence 2 best accounts for the experimental evidence for nucleophilic aromatic displacement reactions.

Let us look at a possible explanation for the inertness of aryl halides toward nucleophilic displacement. It is not surprising that a nucleophile experiences great difficulty in coming close to the electron-rich π cloud of the aromatic ring because the nucleophile has at least one unshared pair of electrons and is often negatively charged; that is, there is considerable electrostatic repulsion between the like charges of the nucleophile and the π cloud:

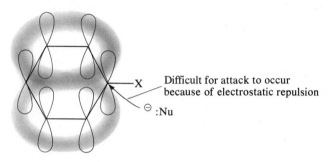

(By comparison, attack of the aromatic ring by an electrophile, $E^\oplus$, is relatively easy because of attraction between unlike charges.)

Some compelling evidence for the two-step mechanism shown at the beginning of this section comes from the rates of departure of various halides. The order of decreasing bond dissociation energies of carbon-halogen bonds is: C—F > C—Cl > C—Br > C—I. That the rates of nucleophilic substitution on aryl halides are about the same for aryl chlorides and aryl bromides indicates that *the breaking of the carbon-halogen bond is not the rate-determining step;* if it were, then aryl bromides should be much more reactive than aryl chlorides. This excludes the one-step, concerted mechanism, (1), and two-step mechanism, (3), because the slow step in each is the breaking of the carbon-halogen bond. Note also that the reactivity order of aryl halides is the reverse of that for alkyl halides (R—I > R—Br > R—Cl > R—F, see Sec. 9.8).

It is surprising that aryl fluorides are the most reactive aryl halides in nucleophilic substitution. Reasons have been offered for this. Because the rate-determining step occurs before cleavage of the aryl-halogen bond, the fluorine atom must affect the rate of nucleophilic attack. It is a very small atom compared with chlorine or bromine and thus provides a less hindered approach for the nucleophile to attack the carbon atom to which fluorine is bonded. It is also the most electronegative of the halogens, thus providing a greater partial positive charge on the carbon where attack is to occur, facilitating nucleophilic attack. The stronger C—F bond may indeed break more slowly than the other carbon-halogen bonds, but compared with the rate of initial nucleophilic attack, all carbon-halogen bonds may break very rapidly. There is as yet no way to test these possibilities. Based on what we *observe*, we can say with certainty only that the slow, rate-determining step is nucleophilic attack on the carbon bearing the halogen atom.

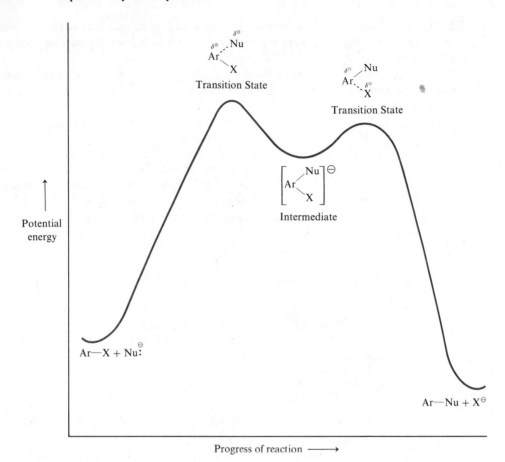

FIGURE 16.1   Energy profile diagram for the two-step bimolecular nucleophilic aromatic substitution reaction Ar—X + Nu:$^{\ominus}$ → Ar—Nu + X$^{\ominus}$.

Studying the rates of reaction with different leaving groups (that is, different halides) is somewhat analogous to studying the isotope effect in the oxidation of alcohols (see Sec. 10.11), where we studied the reactivity differences in the breaking of R—H and R—D bonds. There are more variables in the aryl halide series, including mass, atomic size, and electronegativity of the various halogens, but the experimental results demonstrate that the reaction occurs by the proposed two-step mechanism.

The energy-profile diagram for the two-step mechanism is shown in Fig. 16.1, and its similarity with Fig. 14.1 for electrophilic aromatic substitution should be noted.

16.6   Nucleophilic Aromatic Substitution Mechanism: Structure-Reactivity Correlations

Now let us reconstruct the currently accepted mechanism and add the supporting explanations. The first step, attack of the nucleophile on the carbon bearing the leaving group, can be viewed as follows for the general aryl halide, Ar—X:

The intermediate carbanion that is formed may be represented by the following resonance structures to emphasize the *ring* carbon atoms that bear the greatest amount of negative charge:

The best representation of this intermediate shows the negative charge spread over the entire ring. Note that the carbon bearing the halogen in the aryl halide is trigonal (sp^2 hybridization) and the same carbon atom in the intermediate carbanion is tetrahedral (sp^3 hybridization). Consequently, the intermediate should be drawn to depict the tetrahedral nature of that carbon atom, but the planar drawing is usually used for convenience.

The carbanion intermediate also accounts for the structure-reactivity correlations observed in nucleophilic aromatic substitution. As with electrophilic aromatic substitution (see Sec. 14.9), there are two explanations of the reactivity of aryl halides containing substituents on the ring. The presence of electron-donating groups increases the "concentration" of electrons in the π cloud, thus making it less susceptible to nucleophilic attack, whereas electron-withdrawing groups decrease the electron density in the π cloud and make it more susceptible to attack by a nucleophile. We can also consider the stability of the carbanion that results from the attack of the nucleophile. The carbanion with electron-donating substituents attached is less stable than the one with electron-withdrawing substituents. Our standard of comparison is an unsubstituted aryl halide. These effects are illustrated in Fig. 16.2. Note that the substituents that activate a ring in electrophilic aromatic substitution deactivate it in nucleophilic aromatic substitution, and vice versa. These effects can best be understood by considering the resonance structures of the possible carbanions that intervene in nucleophilic aromatic substitution. The structures which result from attack on an aryl halide containing a nitro group are shown:

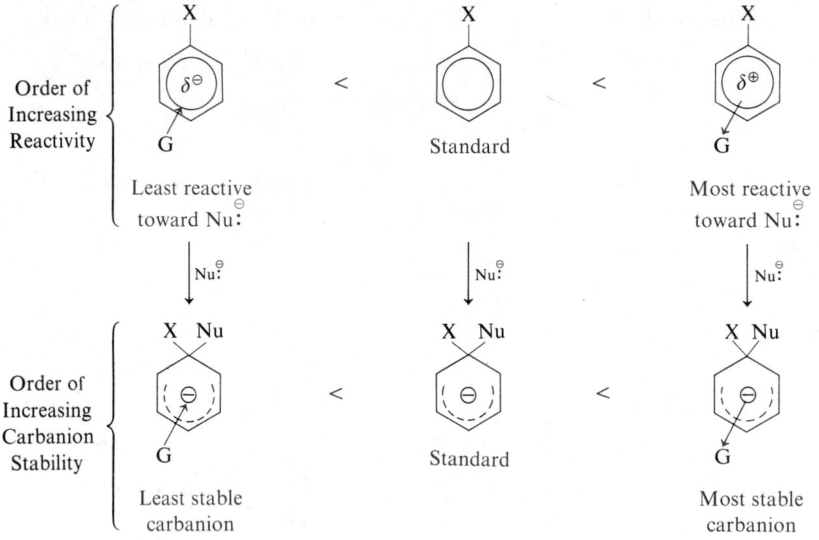

G = electron-donating groups: —NH₂, —NR₂, —OH, —OR, —Ar, —R

G = electron-withdrawing groups:

$$-\overset{\oplus}{N}R_3, -NO_2, -SO_3H, -C\equiv N, -X, -\overset{\overset{\displaystyle O}{\|}}{C}-OH, -\overset{\overset{\displaystyle O}{\|}}{C}-H, -\overset{\overset{\displaystyle O}{\|}}{C}-R$$

FIGURE 16.2   Effects of substituents on nucleophilic aromatic substitution.

ortho Substitution:

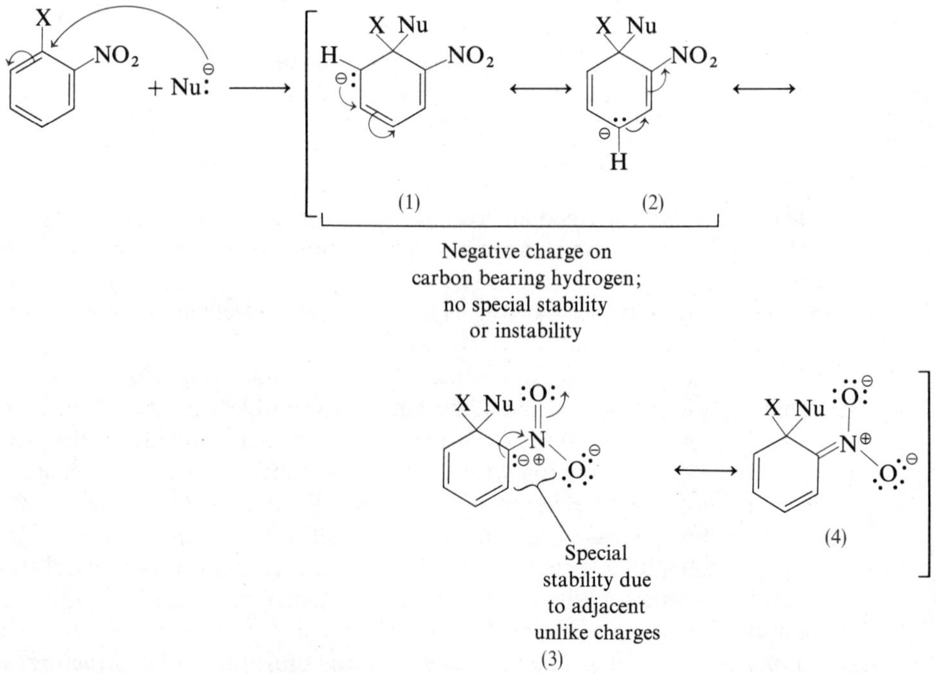

meta Substitution:

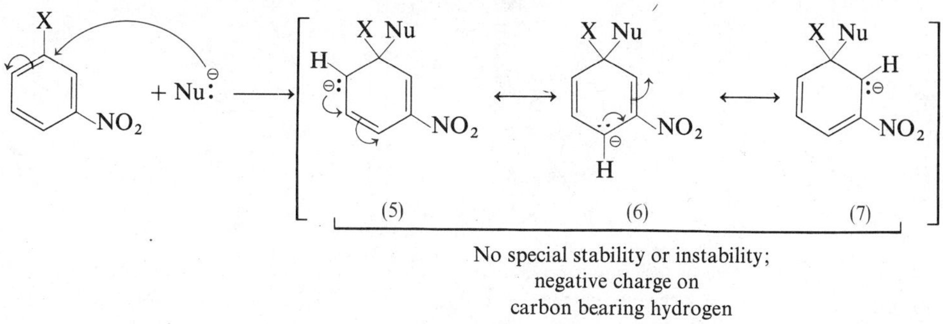

(5)   (6)   (7)

No special stability or instability;
negative charge on
carbon bearing hydrogen

para Substitution:

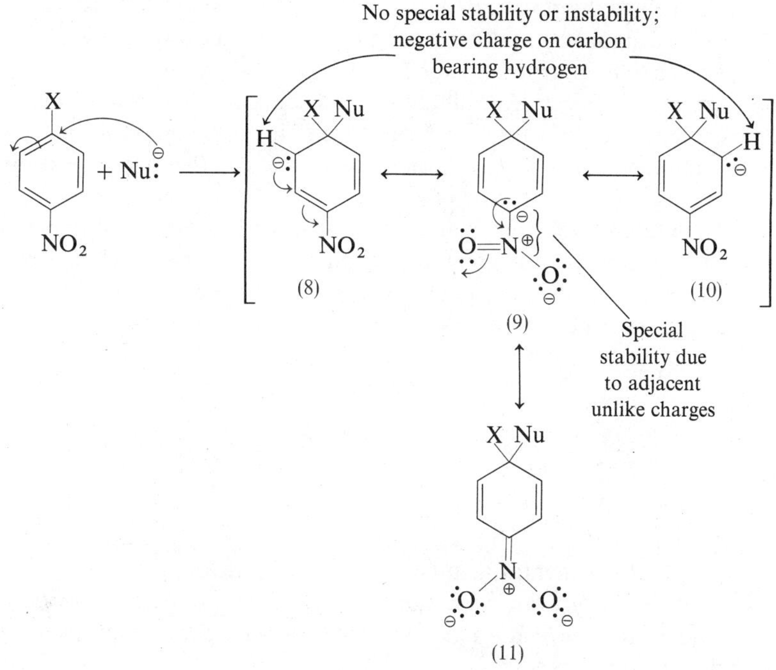

No special stability or instability;
negative charge on carbon
bearing hydrogen

(8)   (9)   (10)

Special
stability due
to adjacent
unlike charges

(11)

The resonance structures show that (1) and (2) are of comparable stability for *ortho* substitution, whereas structure (3) possesses special stability because positive and negative charges are on adjacent nitrogen and carbon atoms, respectively, and allow for another resonance form (4) to contribute. For *meta* substitution structures (5) to (7) are of comparable stability and none is especially stable or unstable. For *para* substitution structures (8) and (10) are of comparable stability, whereas (9) is especially stable because of unlike charges on adjacent atoms resulting in the contributing structure (11). Thus, it is not surprising that electron-withdrawing groups, like $-NO_2$, on the *ortho* and *para* positions enhance the stability of the carbanion intermediate in nucleophilic substitution.

Note the nature of the two structures that are especially stable. For structures (3) and (9) we can write one additional resonance structure each showing the interaction between positive and negative charges. Structure (4) results from (3) and structure (11) results from (9). The additional structures, (9) and (11), show that negative charge is delocalized over a larger portion of the molecule, and they are thus resonance-stabilizing structures. They are redrawn here for emphasis.

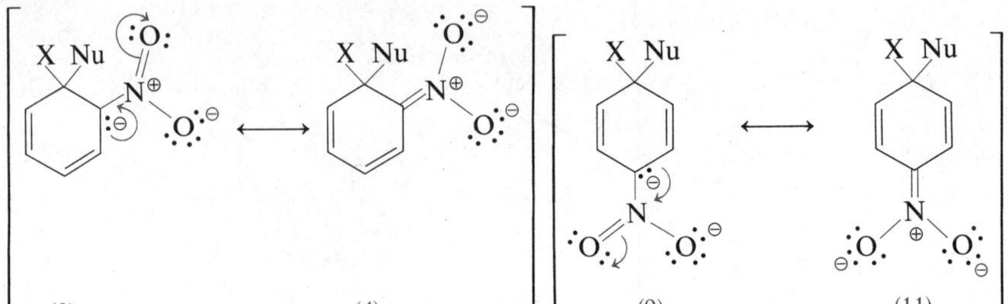

| (3) | (4) | (9) | (11) |

ortho Attack,
resonance stabilization of
negative charge

para Attack,
resonance stabilization of
negative charge

Thus we see again that spreading charge (positive or negative) over a larger area of the molecule usually results in added stability of an intermediate.

It is reasonable that other strong electron-withdrawing groups (the *meta* directors of electrophilic aromatic substitution) behave similarly through the inductive and/or resonance effects. Recall that they all contain either formal positive or partially positive charges on the atom adjacent to the ring, so they stabilize a negative charge in those electronic structures in which the charge is on the carbon bearing the electron-withdrawing groups.

The presence of more than one electron-withdrawing group *ortho* and/or *para* to the halide provides for greater delocalization of negative charge because nucleophilic aromatic substitution places increased negative charge on the positions *ortho* and *para* to the center of attack:

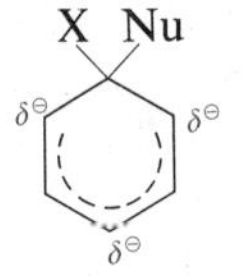

As the experimental results in Sec. 16.4 indicate, the structure-reactivity correlation is: *Increasing the number of electron-withdrawing substituents on the positions ortho and para to a halide increases the ease with which they undergo nucleophilic substitution.*

Question 16.1

Which compound in each of the following four pairs would be more reactive toward displacement of the halide by hydroxide ion? Briefly explain.

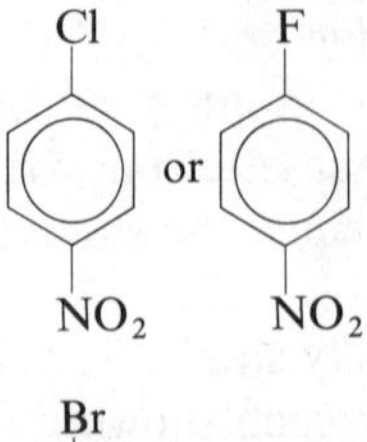

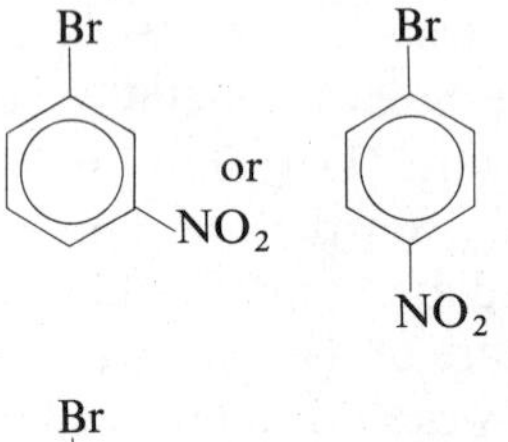

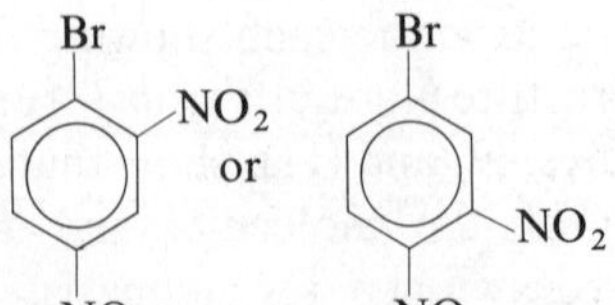

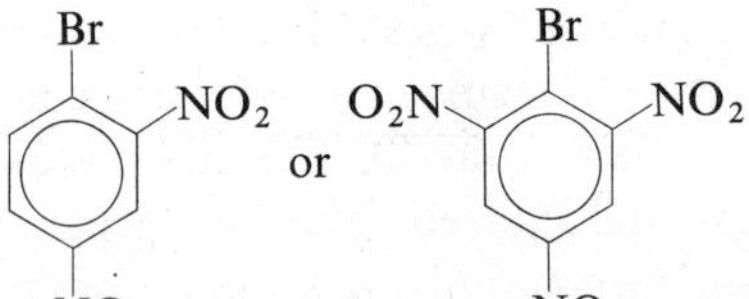

Question 16.2

Which chlorine atom would be displaced more readily in this molecule? Why?

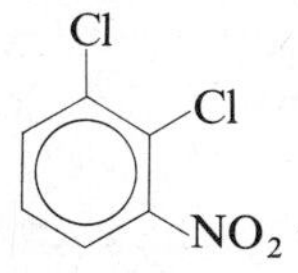

Question 16.3

Draw resonance structures for nucleophilic displacement by hydroxide ion in each of the following molecules, and use them to explain the reactivity order indicated.

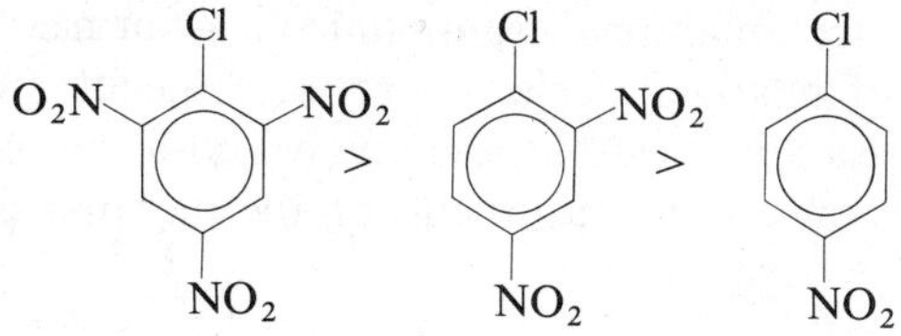

Question 16.4

In this section we considered the reaction between aryl halides and a negatively charged nucleophile. On the other hand, certain neutral nucleophiles, such as water, undergo substitution with some aryl halides; for example:

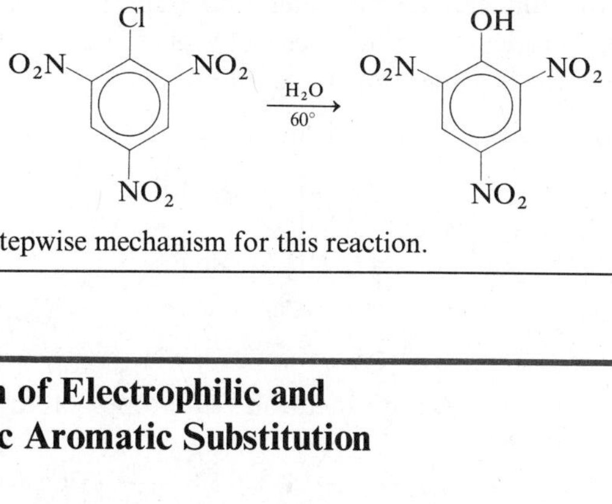

Propose a complete, stepwise mechanism for this reaction.

16.7 Comparison of Electrophilic and Nucleophilic Aromatic Substitution

There are similarities as well as differences between electrophilic and nucleophilic aromatic substitution reactions. The mechanisms for these two reactions are shown here.

Electrophilic aromatic substitution:

$$Ar-H + E^\oplus \xrightarrow{\text{slow}} \left[Ar \begin{matrix} H \\ \diagdown \\ E \end{matrix} \right]^\oplus \xrightarrow{\text{fast}} Ar-E + H^\oplus$$

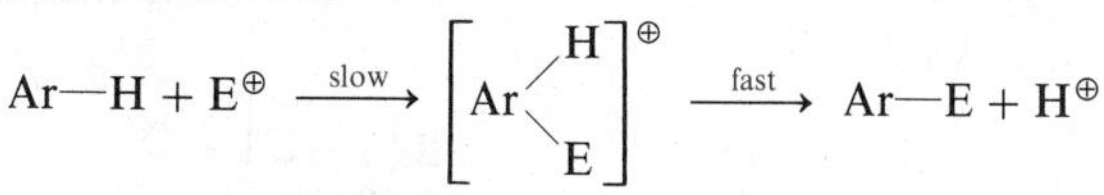

Positively charged intermediate

Nucleophilic aromatic substitution:

$$Ar-L + Nu\colon^\ominus \xrightarrow{\text{slow}} \left[Ar \begin{matrix} L \\ \diagdown \\ Nu \end{matrix} \right]^\ominus \xrightarrow{\text{fast}} Ar-Nu + L\colon^\ominus$$

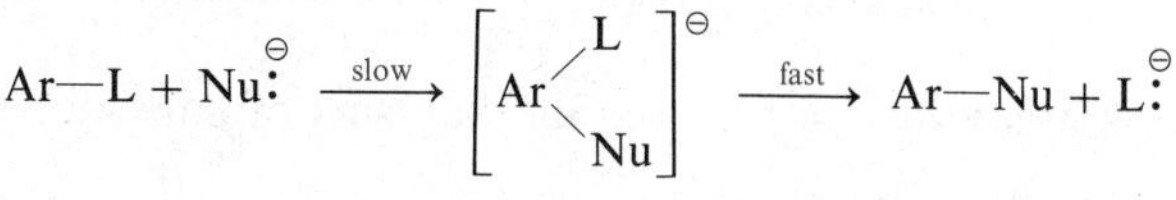

Negatively charged intermediate

Both mechanisms occur in two steps: the first is the slow, rate-determining attack of the electrophile or nucleophile on the aromatic ring, and the second step is the fast loss of a proton or halide ion (leaving group). Electrophilic substitution involves displacement of hydrogen as a proton, whereas nucleophilic substitution involves displacement of some highly electronegative atom (usually halogen) or group of atoms.

The intermediate in electrophilic aromatic substitution is positively charged, so that electron-donating groups attached to the ring enhance reactivity whereas electron-withdrawing groups suppress it. The intermediate in nucleophilic aromatic substitution is negatively charged, so that electron-donating groups suppress reactivity and electron-withdrawing groups enhance reactivity.

Of these two reactions, electrophilic aromatic substitution is much more useful because it introduces functional groups onto an aromatic ring. Nucleophilic aromatic substitution is of relatively little use because special structural features must be present before reaction occurs; these involve both the nature of the leaving group and the presence of certain substituents on the ring in *ortho* or *para* positions.

16.8 Comparison of Aromatic and Aliphatic Nucleophilic Substitution

Nucleophilic substitution on aromatic rings occurs in two steps. In contrast, aliphatic substitution can be in either one step (S_N2 reaction) or two steps (S_N1 reaction). The considerable difference between the types of intermediates in these reactions is outlined in general form as follows:

Nucleophilic aromatic substitution:

$$\text{Ar—L} + \text{Nu:}^{\ominus} \xrightarrow{\text{slow}} \left[\text{Ar} \begin{matrix} \diagup \text{L} \\ \diagdown \text{Nu} \end{matrix} \right]^{\ominus} \xrightarrow{\text{fast}} \text{Ar—Nu} + \text{L:}^{\ominus}$$

Full negative
charge on ring

Nucleophilic aliphatic substitution: monomolecular (S_N1 reaction):

$$\text{R—L} \xrightarrow{\text{slow}} \text{R}^{\oplus} + \text{L:}^{\ominus} \xrightarrow[\text{fast}]{\text{Nu:}^{\ominus}} \text{R—Nu}$$

Full positive
charge on carbon

where R = 2° or 3° alkyl group

Nucleophilic aliphatic substitution: bimolecular (S_N2 reaction):

$$\text{R—L} + \text{Nu:}^{\ominus} \xrightarrow{\text{slow}} \left[\overset{\delta\ominus}{\text{Nu}}\text{----R----}\overset{\delta\ominus}{\text{L}} \right] \xrightarrow{\text{fast}} \text{Nu—R} + \text{L:}^{\ominus}$$

Little to no
charge on carbon

where R = methyl, 1° and 2° alkyl groups

There is great variation in the mechanisms of substitution reactions, but the net result is the same: displacement of some leaving group, $\text{L:}^{\ominus}$, by some nucleophile, $\text{Nu:}^{\ominus}$ or Nu:.

Question 16.5

For each of the preceding nucleophilic substitution reactions, indicate:

(a) The stereochemistry, where applicable
(b) Whether an *intermediate* or a *transition state* intervenes between starting material and product
(c) The *changes* in hybridization and geometry that carbon undergoes in being converted from starting material to product

16.9   Dehydrohalogenation of Aryl Halides; Benzyne Formation

The evidence we examined strongly supports nucleophilic displacement on aryl halides, but one requirement is that electron-withdrawing groups be present on the ring. With unsubstituted aryl halides or those containing electron-donating groups, substitution is very difficult to accomplish, and strong bases, and/or vigorous conditions of temperature and pressure are required. As a result of some elegant research work by Professor John D. Roberts (California Institute of Technology), there is compelling evidence for a completely different type of mechanism in many of these reactions. Let us examine this work.

Our emphasis has been on displacement reactions involving moderately weak bases, such as hydroxide ion. Recall, for example, that chlorobenzene, C_6H_5—Cl, can be converted to phenol, C_6H_5—OH, by aqueous hydroxide at 340 to 350° and high pressure. Stronger bases can also be used to accomplish nucleophilic displacement reactions, however, and when this is done the product distribution can be quite different than it is in the previous bimolecular mechanism. For example, the amide ion, $:\overset{..}{N}H_2^{\ominus}$, reacts with bromobenzene in liquid ammonia at $-33°$ to give aniline:

Bromobenzene   Aniline

When this same reaction is carried out with o-chlorotoluene, m-chlorotoluene, and p-chlorotoluene, products are obtained in which the amino group occupies a position on the ring that is different from that of the original chlorine atom.

o-Chlorotoluene   o-Aminotoluene   m-Aminotoluene
                              or   or
                         o-Toluidine   m-Toluidine
                            45%   55%

m-Chlorotoluene   52%   40%   p-Aminotoluene
                                                                 or
                                                             p-Toluidine
                                                                8%

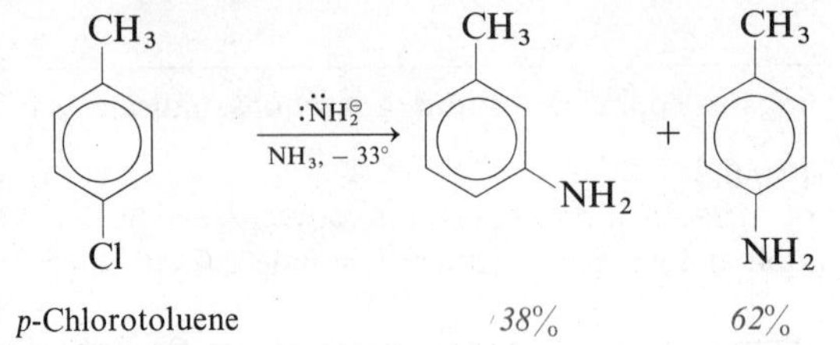

p-Chlorotoluene   38%   62%

The same type of reaction occurs when, for example, *p*-chlorotoluene is heated with concentrated aqueous sodium hydroxide:

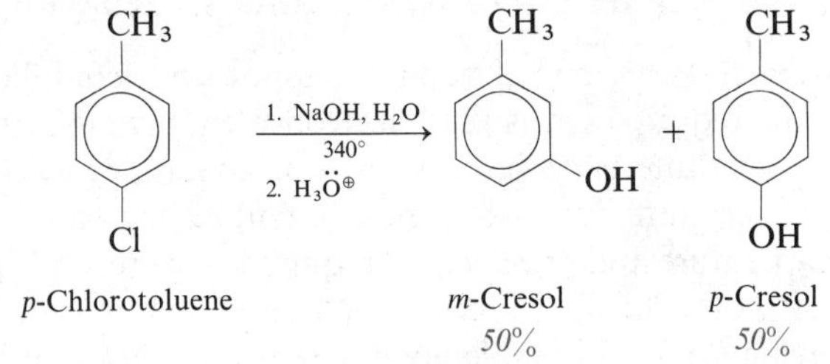

p-Chlorotoluene   *m*-Cresol   *p*-Cresol
                           50%   50%

These results and others indicate the intervention of a highly reactive intermediate that contains a *carbon-carbon triple bond* in the aromatic ring. This intermediate, called **benzyne** or **dehydrobenzene,** results from the elimination of the elements of H—X from the aryl halide. The formation of benzyne may be envisioned as occurring in one of the following two ways:

1. Concerted bimolecular elimination (E2) reaction:

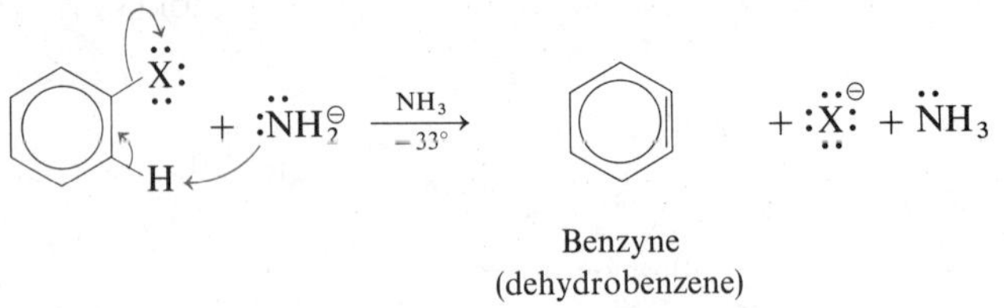

Benzyne
(dehydrobenzene)

where X = Br, Cl, I

This reaction is analogous to the E2 elimination reaction that occurs when an alkyl halide is treated with base (see Sec. 7.11). This mechanism is supported by the observation of an isotope effect in aryl bromides and chlorides that contain deuterium. For example, bromobenzene containing deuterium *ortho* to the bromine atom reacts more slowly than bromobenzene. The carbon-hydrogen or carbon-deuterium bond is broken in the rate-determining step, and deuterium slows down the reaction (see the discussion on isotope effects in Sec. 10.11).

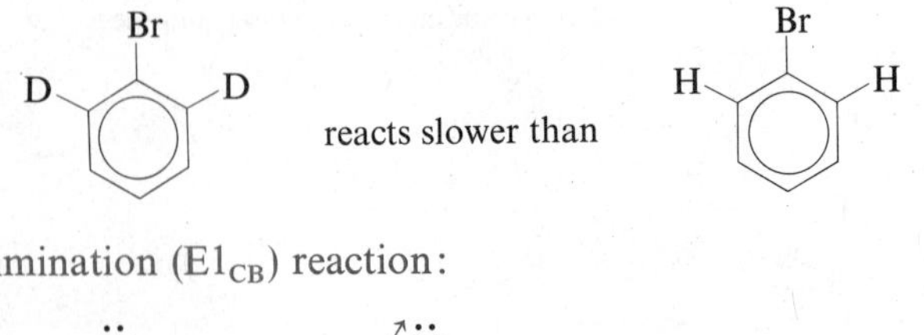

reacts slower than

2. Two-step elimination (E1$_{\text{CB}}$) reaction:

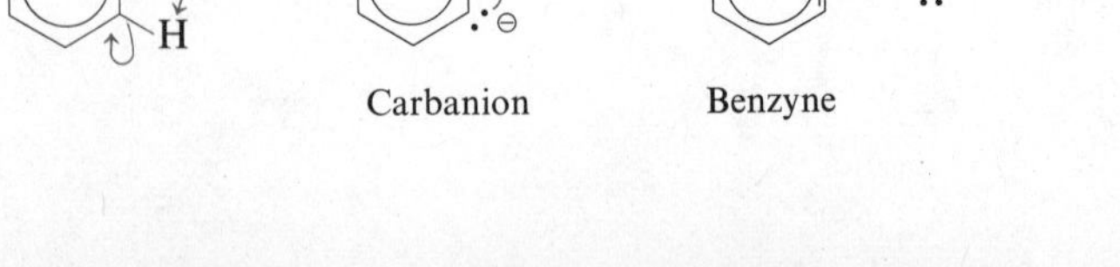

Carbanion   Benzyne

This mechanism is supported by the observation that fluorobenzene, when allowed to react with deuterated amide ion, $:\ddot{N}D_2^{\ominus}$, in deuterated ammonia, $\ddot{N}D_3$, contains deuterium when the unreacted starting material is analyzed. This suggests that the hydrogens *ortho* to fluorine are removed and replaced in a rapid reversible reaction, and deuterium is incorporated into the molecule. This also supports the observation that the slow, rate-determining process in benzyne formation from fluorobenzene is the loss of fluoride ion. The deuterium incorporation results are shown in the following scheme:

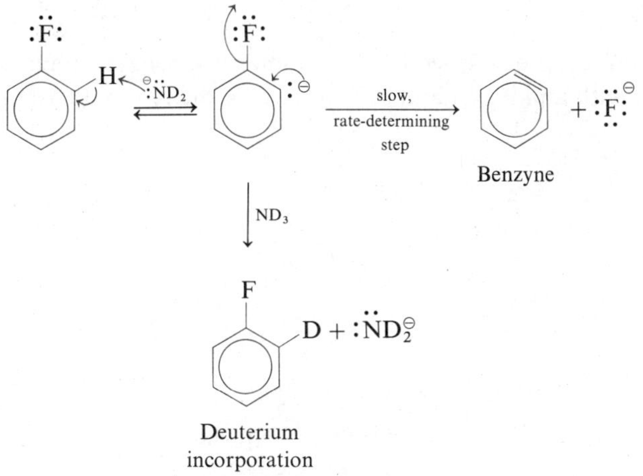

Benzyne

Deuterium
incorporation

This reaction is often called an **E1$_{CB}$ reaction** (E for elimination, 1 for mono-molecular, and CB for conjugate base) because the slow, rate-determining step is the loss of fluoride ion from the conjugate base of the original aryl halide.

The mechanism involved in the formation of benzyne, then, appears to differ depending on the precursors.

The benzyne intermediate is highly reactive. It reacts indiscriminately with solvent or base. For example, in the reaction between bromobenzene and amide ion in liquid ammonia, the addition of ammonia probably occurs in the following manner:

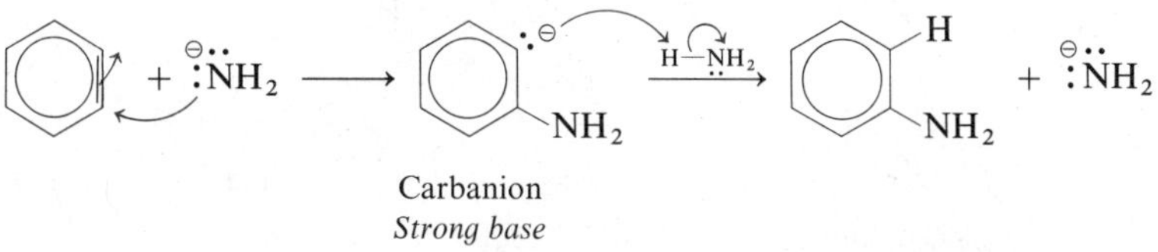

Carbanion
Strong base

Alternatively, the addition of solvent to benzyne may be a concerted process in which carbon-nitrogen and carbon-hydrogen bonds are formed simultaneously:

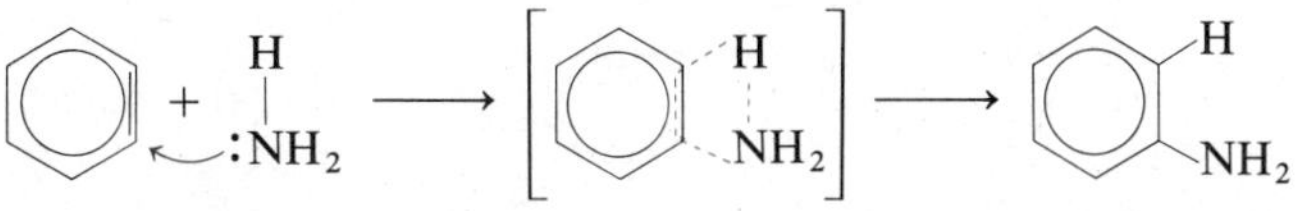

There are other examples of this type of reaction, such as hydroboration (see Sec. 10.4). However, there is no conclusive evidence in benzyne chemistry to confirm either the one-step, concerted addition or the two-step addition of solvent or base.

The overall reaction in which an aryl halide is converted to aniline is referred to as an **elimination-addition** reaction; the elements of HBr are eliminated and the elements of NH_3 are added. Support for the elimination-addition mechanism comes from reactions in which the carbon bearing the halogen is labeled with carbon 14 (^{14}C indicated by a * in the following structures).

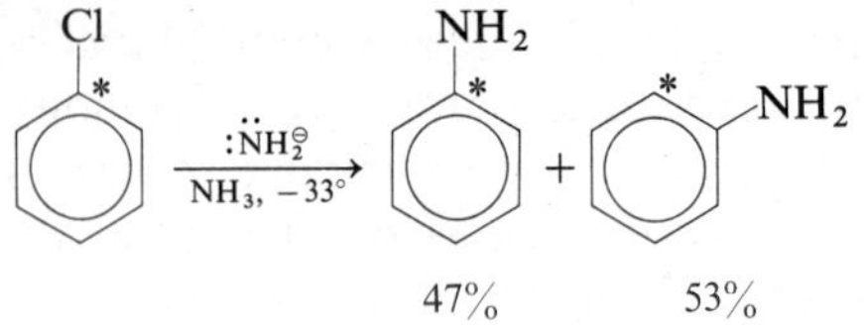

$$47\% \qquad 53\%$$

where * = carbon 14

This result is explained by the symmetrical benzyne intermediate, which reacts randomly with amide ion and solvent to give the two observed products:

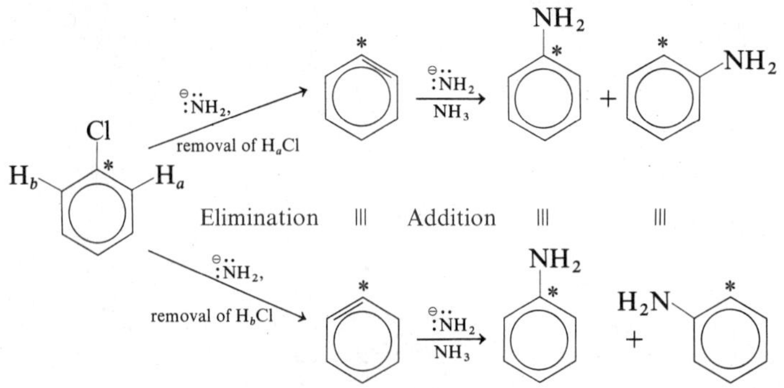

The reactions of the chlorotoluenes presented at the beginning of this section are explained by the "benzyne" intermediate (perhaps more correctly called *toluyne*):

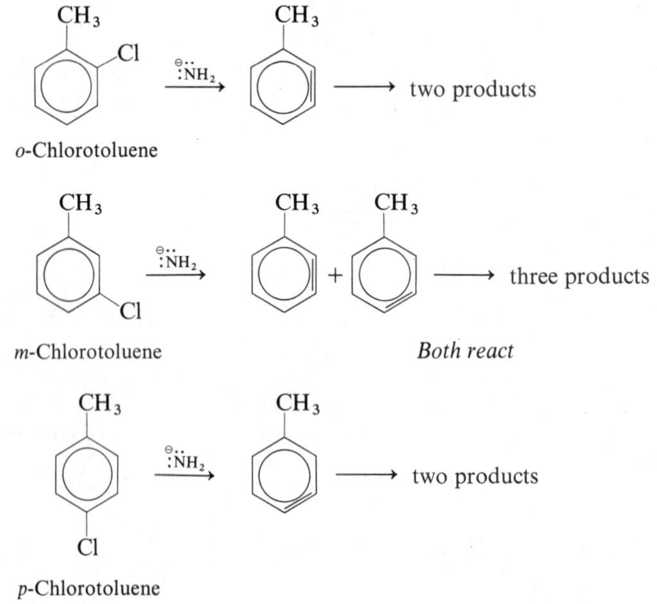

o-Chlorotoluene

m-Chlorotoluene   *Both react*

p-Chlorotoluene

Note that *m*-chlorotoluene gives two possible benzyne intermediates, each of which reacts with solvent and base to give a product.

The benzyne molecule contains a "triple bond" between two adjacent carbon atoms, one of which originally contained hydrogen and the other originally contained the halogen. More correctly, though, benzyne consists of two overlapping sp^2 hybrid atomic orbitals as shown:

Orbital picture of benzyne:
overlapping sp^2 hybrid orbitals

The sp^2 hybrid orbitals are parallel to the plane of the ring and perpendicular to the p orbitals that constitute the π cloud. Because of the geometry of the aromatic ring, however, the sp^2 hybrid orbitals are not parallel but instead point away from one another. This overlap is not strong as it is in, for example, a normal π bond.

The formation of a benzyne type of intermediate is favored when the amide ion, $:\overset{..}{N}H_2^{\ominus}$, is used as the base. Because of its strongly basic nature, it can abstract hydrogens from the aromatic ring (as protons) to form ammonia as one product. The halide ion, $:\overset{..}{\underset{..}{X}}:^{\ominus}$, departs, thus forming benzyne. Note that there must be a

hydrogen *ortho* to the halogen for benzyne to be formed. Its formation is favored with aryl halides that contain electron-donating substituents.

On the other hand, the presence of electron-withdrawing substituents on the ring causes the bimolecular nucleophilic aromatic substitution reaction to be favored.

As shown, the benzyne intermediate rapidly reacts with the solvent and/or base to give products. It is possible, however, to trap or capture the benzyne by means of the Diels-Alder reaction (Sec. 8.29). When benzyne is formed in the presence of a diene, the benzyne reacts as the dienophile to give the Diels-Alder product. A diene often used for this purpose is anthracene, which provides the structurally interesting molecule triptycene:

| Benzyne | Anthracene | Triptycene |
| (dienophile) | (diene) | |

Question 16.6

Suggest why the following compounds do not react with the amide ion in liquid ammonia.

(*Hint:* Steric factors, although present, are not responsible for their lack of reactivity.)

Question 16.7

Suggest an explanation for the difference between the following reactions (* = carbon 14):

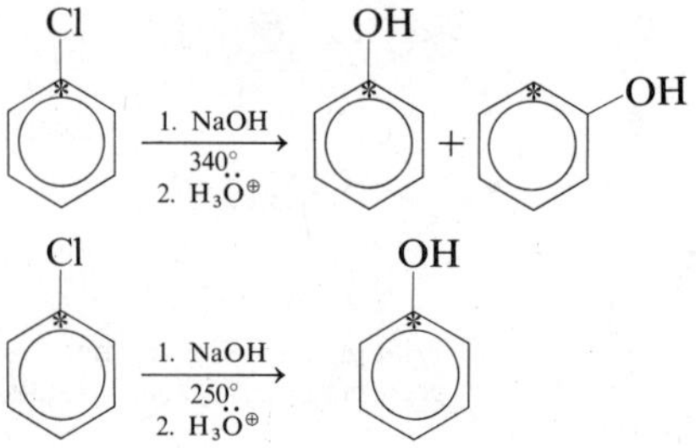

Only product

16.10 Polyhalogenated Hydrocarbons

It is almost impossible to pick up a newspaper or magazine without reading about a member of the family of hydrocarbons called polyhalogenated hydrocarbons. Unfortunately, however, these articles are usually warnings about undesirable properties. Do the abbreviations DDT, PCB, PBB, and BHC seem familiar? The structures of these and other polyhalogenated hydrocarbons follow along with a short synopsis of the use of each.

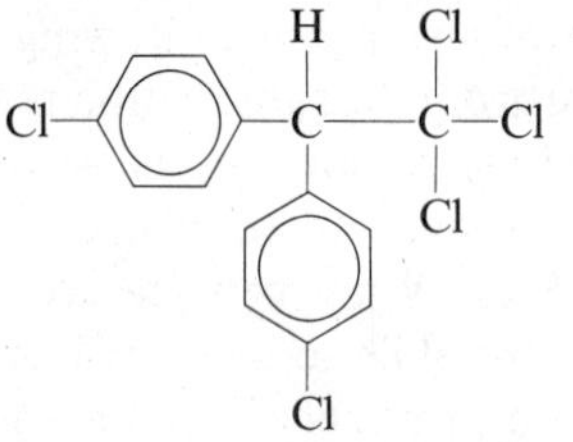

Use: Most versatile insecticide of twentieth century. Wide spectrum of uses, including against the tsetse fly (sleeping sickness), *Anopheles* mosquito (malaria), and body louse (typhus).

DDT: Dichlorodiphenyltrichloroethane (misnomer)

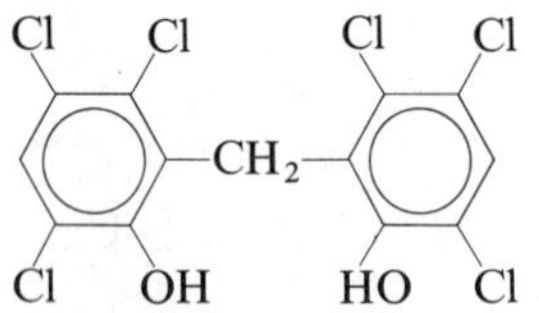

Use: Antiseptic. Used in germicidal soaps and face creams. Highly versatile and effective.

Hexachlorophene

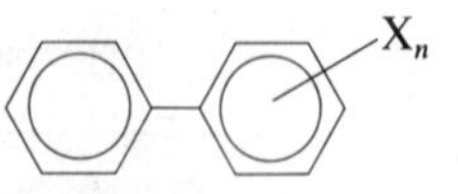

Use: Plasticizer, electrical insulators.

PCBs and PBBs: Polychloro(or bromo)biphenyls

where X = Cl or Br, *n* = 5 to 6 normally; all halogens may be in one ring or spread out between two rings

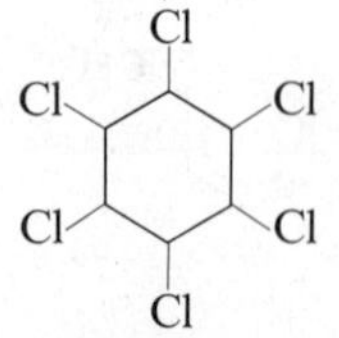

Use: Insecticide (see Secs. 3.17 and 14.17).

BHC: Benzenehexachloride; lindane

Other common insecticides are

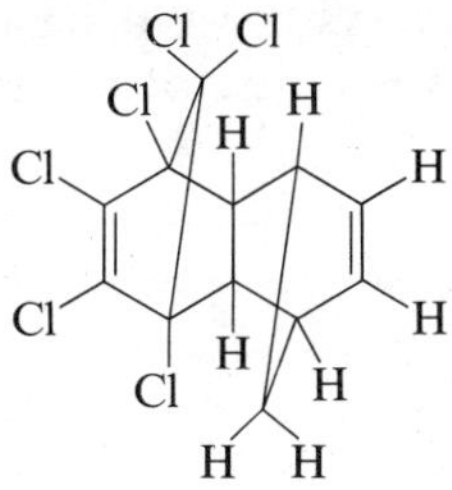

Aldrin 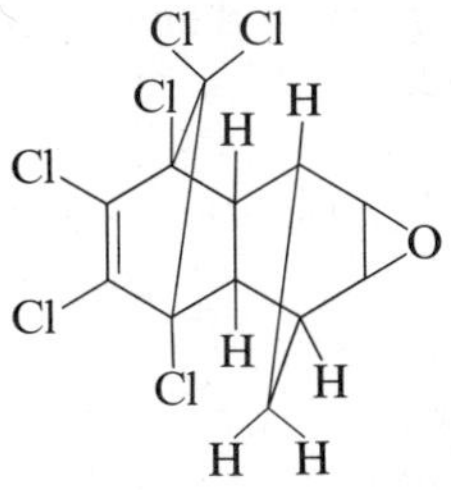   Dieldrin 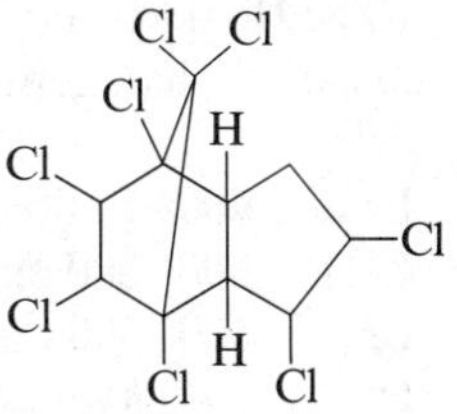   Chlordane

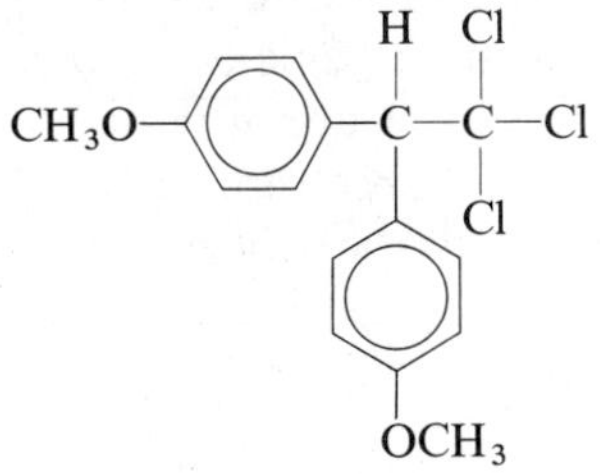

Methoxychlor 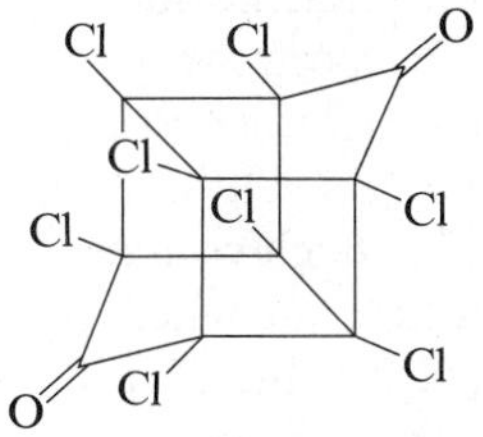   Kepone

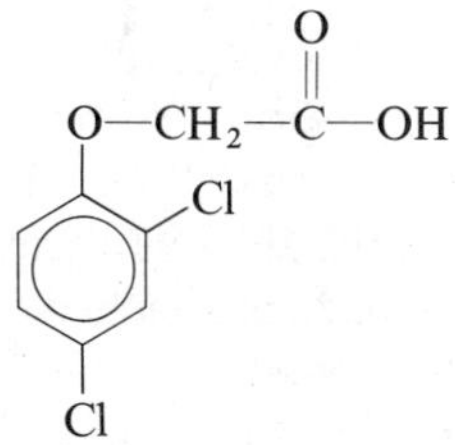

2,4-Dichlorophenoxyacetic
acid (see Sec. 13.3)
2,4-D

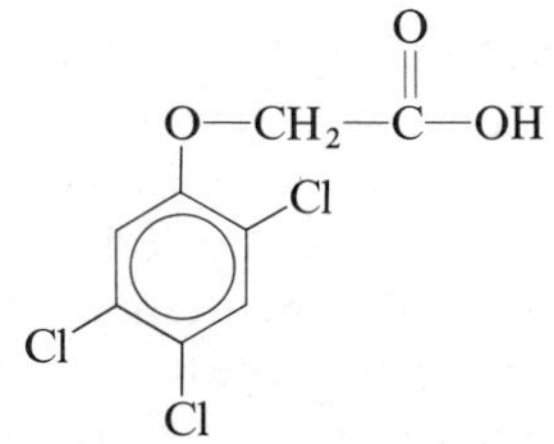

2,4,5-Trichlorophenoxyacetic acid
2,4,5-T

    All these compounds are nonpolar and as such are quite soluble in the nonpolar tissue (fatty acid tissue) of the body. In addition, these compounds are extremely stable; that is, they do not undergo chemical or biochemical degradation at any reasonable rate. This allows for the constant accumulation of these materials in the environment and in the body. They have been shown to cause death (DDT), result in deterioration of the central nervous system (DDT), be carcinogenic (hexachloro-phene), and cause birth defects (PBB). Compounds such as hexachlorophene, which until a very few years ago was found in many hand and face soaps, now are either totally banned or at least severely regulated.

    In the case of DDT and other insecticides, these adverse side effects are amplified many times by biological magnification. For example, once the DDT from crops enters the food chain its concentration along the cycle constantly increases. Micro-scopic organisms such as plankton ingest it; these in turn are ingested by other fish or mammals (or fish to mammals), which are eventually eaten by human beings.

16.11 Spectral Analysis of Alkyl and Aryl Halides

    Alkyl and aryl halides exhibit the characteristic spectral absorptions of the aromatic ring (see Sec. 14.25) and the carbon-carbon double bond (see Chap. 12).

A. Infrared Spectroscopy

The IR spectra of alkyl and aryl halides contain few bands that can be assigned to the carbon-halogen bond. The absorption bands for these bonds are as follows:

C—F:	$1{,}400-1{,}000 \text{ cm}^{-1}$	$(7.14-10 \ \mu)$
C—Cl:	$800-600 \text{ cm}^{-1}$	$(12.5-16.6 \ \mu)$
C—Br:	$600-500 \text{ cm}^{-1}$	$(16.6-20 \ \mu)$
C—I:	$\sim 500 \text{ cm}^{-1}$	$(20 \ \mu)$

Generally these are strong bands, but with varying locations in the spectra.

Carbon tetrachloride, CCl_4, and chloroform, $CHCl_3$, are commonly used as solvents in IR spectroscopy because many organic compounds are soluble in them and they have relatively simple IR spectra. Carbon tetrachloride contains a single type of bond (C—Cl) and gives a simpler spectrum than chloroform. By use of a reference cell, which contains pure solvent, the absorption caused by the solvent is largely canceled out when determining the spectrum of a compound dissolved in the solvent. See Table 12.2 for more detail.

B. Ultraviolet Spectroscopy

The alkyl fluorides and chlorides (monosubstituted) show no absorption in the UV spectral region (any absorptions are below 180 nm). The alkyl bromides and iodides, however, often exhibit a λ_{max} between 200 and 250 nm, but the absorption is weak (small ε). Methyl iodide, for example, absorbs at 259 nm and has an extinction coefficient of 400 (in hexane). When a halogen is attached to a double bond, conjugation occurs, shifting the normal absorption band to a longer wavelength. For example, the substitution of a halogen on the butadiene skeleton shifts the observed absorption band from 217 nm to 222 nm (see Sec. 12.7), and bromobenzene has a λ_{max} at 210 nm versus 203 nm for benzene.

C. Nuclear Magnetic Resonance Spectroscopy

The NMR spectra of some alkyl and aryl halides were presented and discussed in Chap. 15. Because of its electronegativity, a halogen causes a great downfield shift of the hydrogens on the carbon bearing it. This effect varies with the electronegativity of the attached group and its proximity to the nuclei being observed (see Sec. 15.5).

D. Spectroscopy and Structure Elucidation

In Sec. 13.7D the annulenes were introduced. An annulene is any monocyclic compound that can be drawn in a manner (resonance form) in which double and single bonds alternate. The total number of carbons making up the backbone of the ring is used in naming the annulene. [14]-Annulene is drawn here:

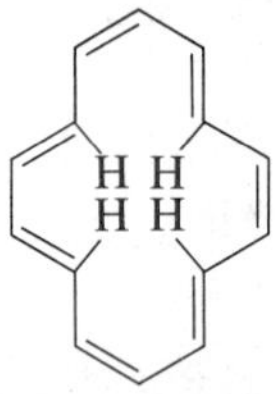

[14]-Annulene

[14]-Annulene is a 14 π-electron system. Because 14 is a $(4n + 2)$ number and the molecule fulfills all the other requirements associated with aromaticity, [14]-annulene should be aromatic. One piece of evidence that it is indeed aromatic is its NMR spectrum. The four inner protons absorb at $\sim -2\,\delta$. They are shielded far upfield from normal aromatic hydrogens ($\delta \approx 7$). This can be explained if the annulene is aromatic; then the four hydrogens, because of their position within the ring, feel the effect of the aromatic ring current. In particular, they are in the shielding portion of the internal magnetic field of the ring. Only the aromatic character of the annulene explains the remarkable upfield shift of these protons. The bromoannulene drawn below exhibits a quite different effect. In this case the π cloud is composed of p orbitals on sp and sp^2 hybridized carbon atoms. The 12 π-electron system is non-aromatic (see Sec. 13.8). The ring current effect of the π cloud is reversed. The protons in the outer portion of the cloud are shielded, and those within are deshielded. The inner proton shown below in red absorbs at 16.4 δ, well downfield from the normal aromatic region.

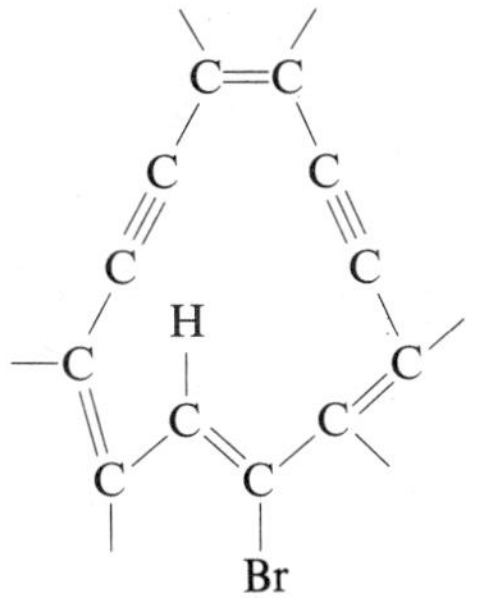

16.12   Summary of Reactions of Aryl Halides

A. Nucleophilic Aromatic Substitution

$$Ar\!-\!\ddot{X}\!: + Nu\!:^{\ominus} \longrightarrow Ar\!-\!Nu + :\ddot{X}\!:^{\ominus}$$

General discussion (Secs. 16.3 and 16.4)
Mechanism of substitution (Sec. 16.5)
Structure-reactivity correlations (Sec. 16.6)
Electron-withdrawing substituents in *ortho* and *para* positions enhance reactivity
Electron-donating substituents suppress reactivity

B. Benzyne Formation: Dehydrohalogenation of Aryl Halides

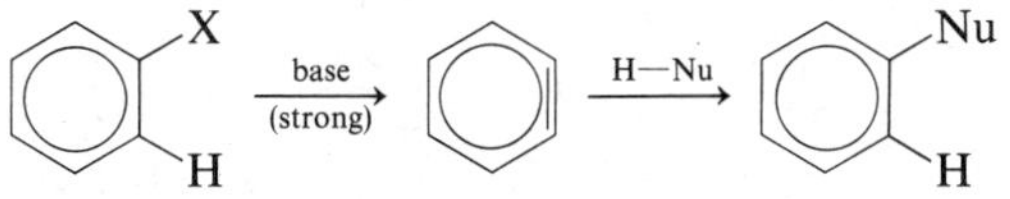

Elimination   Benzyne   Addition

Discussion of reaction and mechanism (Sec. 16.9)

Study Questions

16.8 Indicate how *vinyl chloride*, $CH_2\!=\!CH\!-\!Cl$, can be prepared from the following carbon compounds, using any other inorganic reagents that may be needed.
(*a*) ethene   (*b*) ethyne   (*c*) 1,2-dichloroethane   (*d*) 1,1-dichloroethane

16.9 Supply the missing reactants or products, as required, to complete the following eight equations. The items to be inserted are indicated by number [for example, (1)]. If no reaction occurs, so indicate.

(a) benzene $\xrightarrow{(1)}$ toluene $\xrightarrow{(2)}$ 4-chlorotoluene

(b) benzene $\xrightarrow{(3)}$ iodobenzene

(c) benzene $\xrightarrow{(4)}$ 4-bromobenzoic acid

(d) benzene $\xrightarrow{(5)}$ 4-bromonitrobenzene

(e) 3,4-dimethyl-1-chlorobenzene with CH₃ groups $\xrightarrow[\text{NH}_3, -33°]{\ominus\ddot{N}H_2}$ (6)

(f) 2-chloro-1,3-dimethylbenzene $\xrightarrow[\text{NH}_3, -33°]{\ominus\ddot{N}H_2}$ (7)

(g) 4-chloro-3-nitrotoluene $\xrightarrow[\text{heat}]{\ddot{O}H^{\ominus}, \text{H}_2\text{O}}$ (8)

(h) (9) $\xrightarrow[\text{NH}_3, -33°]{\ominus\ddot{N}H_2}$ 3-ethylaniline + 2-ethylaniline

16.10 Outline briefly the mechanistic *differences* between these three pairs of reactions: **(a)** S_N1 and S_N2 reactions, **(b)** E1 and E2 reactions, **(c)** nucleophilic aliphatic substitution and nucleophilic aromatic substitution.

16.11 Does nucleophilic aromatic substitution most closely resemble the S_N1 or the S_N2 reaction? In what way are they similar? Different?

16.12 Provide a mechanism to explain the following two experimental observations.

(a) O_2N—(2-bromo-1-bromobenzene)—Br $\xrightarrow[\text{H}_2\text{O, heat}]{\ddot{H}\ddot{O}:^{\ominus}}$ O_2N—(2-bromophenol)—OH

(b)   F—(4-methoxyphenyl) + 2 (phenyl)—Li $\xrightarrow[\text{hydrolysis}]{\text{followed by}}$ 2-methoxybiphenyl + 3-methoxybiphenyl

16.13 Two isomeric compounds, *A* and *B*, have the formula $C_2H_4Br_2$. In the NMR spectrum of *A* there is a singlet at $3.63\ \delta$ $(6.37\ \tau)$ as the only absorption, and in the NMR spectrum of *B* there is a *doublet* centered at $2.45\ \delta$ $(7.55\ \tau)$ and a *quartet* centered at $5.85\ \delta$ $(4.15\ \tau)$. What are the structures of *A* and *B*?

16.14 Compound *A* reacts with allyl bromide (in the presence of $AlCl_3$) to give compound *B*. *B* reacts with H_2/Pt to give *C*, $C_{10}H_{14}$. The NMR spectrum of *A* is given here. The UV spectrum of *B* exhibits a maximum at 245 nm for the primary $\pi \rightarrow \pi^*$ transition band. *A* also reacts with bromine and light at room temperature to give a very strong lacrimator. Give structures for *A* to *C*.

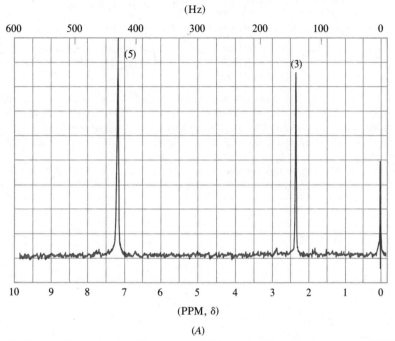

(A)

PROBLEM 16.14 NMR spectrum of unknown (*A*). [Data from *The Aldrich Library of NMR Spectra*, C. J. Pouchert and J. R. Campbell (eds.), copyright 1974, Aldrich Chemical Co., Milwaukee]

16.15 The nitroso group, $-\overset{..}{N}=\overset{..}{O}:$, is an *ortho,para* director in electrophilic aromatic substitution reactions, and its presence *ortho* or *para* to a suitable leaving group results in enhanced nucleophilic aromatic substitution reactivity. Such a case is shown by the following equation:

Explain mechanistically why this reaction occurs, paying special attention to the resonance structures (draw them) of the intermediate. *Note:* In this reaction, the observed products are really

$$O=N-\!\!\!\!\bigcirc\!\!\!\!-\overset{..}{\underset{..}{O}}:^{\ominus} + (CH_3)_2NH$$

16.16 When compound *A* reacts under the following conditions, compounds *B* and/or *C* are formed:

$$A\ (C_7H_7Cl)$$

a $\xrightarrow[\text{900 lb/sq in.}]{\overset{..}{N}H_3,\ 200°} B\ (C_7H_9N)$

b $\xrightarrow{:\overset{..}{N}H_2^{\ominus},\ \text{liquid NH}_3} C\ (C_7H_9N) + B$

Compound A is a low-boiling liquid (bp 162°), which is insoluble in all aqueous solvents (acids and bases) and exhibits the following NMR and IR spectra:

NMR: 7.15 δ (4H) m IR: strong absorption bands at 3,050 cm^{-1},
            2.32 δ (3H) s 2,950 cm^{-1}, and 810 cm^{-1}

Compounds B and C are isomers of each other and exhibit similar NMR and IR spectra:

NMR: 6.78 δ (4H) m IR: strong bands at $\sim$3,500 cm^{-1},
            3.35 δ (2H) s $\sim$3,010 cm^{-1}, and $\sim$2,900 cm^{-1}
            2.28 δ (3H) s

Both B and C react with benzenesulfonyl chloride and base to give a clear solution. When the solution is acidified, an oil precipitates (*Hint:* see Sec. 26.8A, a). If anthracene is added to the reaction mixture under conditions a, the anthracene is isolated unreacted in quantitative yield. When this is done under conditions b, methyltriptycene ($C_{21}H_{16}$) is isolated.

(**a**) What are the structures and names of compounds A to C?
(**b**) What is the name of the mechanism involved in A going to B? Draw it out in detail.
(**c**) What is the mechanism in A going to B and C?

17

Phenols

We now study the family of compounds called **phenols,** with the general formula Ar—OH. The hydroxyl group in phenols is attached to the aromatic ring, whereas in alcohols it is bonded to an alkyl group. In addition to examining the properties and methods of preparation of phenols, we will contrast the reactivity of this family of compounds with that of alcohols. Phenols are more acidic than alcohols, but most important, the —OH group bonded to an aromatic ring is not susceptible to substitution as it is when it is bonded to an alkyl group. There are some similarities between alcohols and phenols, but for the most part these two families are quite different.

17.1 Structure and Nomenclature

In the IUPAC system, phenols are often named by considering compounds containing the —OH group on the ring as derivatives of the parent compound, phenol. Sometimes common names must take precedence; then the —OH group is named as a substituent and is given the name *hydroxy*.

Phenol   *m*-Bromophenol   3,4-Dinitrophenol   *o*-Hydroxybenzenesulfonic acid

Care must be used not to confuse the name *phenyl*, which stands for C_6H_5—, with *phenol*, the name of the compound C_6H_5—OH; for example:

p-Phenylphenol

Other compounds have common names that are used so often that they must be learned; for example:

p-Cresol   Catechol   Resorcinol   Hydroquinone

As a family, phenols are active bactericides. Phenol dissolved in water is commonly called *carbolic acid*, which is used as an antiseptic. Many other phenolic compounds also have been synthesized and found useful as antiseptics. For example, *hexylresorcinol* is used in throat lozenges; *hexachlorophene* (see Sec. 16.10) is frequently found in soaps, deodorants, and toothpaste; and *o-phenylphenol* is in many antiseptics.

Hexylresorcinol   Hexachlorophene   o-Phenylphenol

Some natural products are substituted phenols; for example, the oil responsible for mint flavor is *thymol*, and *carvacrol* has the flavor of summer savory. Both compounds are used medically as fungicides and bactericides; as bactericides, they are more effective, less toxic, and more pleasant smelling than phenol or cresols.

Carvacrol   Thymol

In Chap. 19 on ethers, we find some naturally occurring products that are derivatives of phenols.

17.2   Physical Properties

The physical properties of selected phenols are given in Table 17.1. The parent compound, phenol, is more water-soluble than the other phenols. This is attributed to hydrogen bonding between water and the —OH group, a property similar to that found in alcohols (see Sec. 10.2). The hydrogen bonding between phenols and water is envisioned as follows:

TABLE 17.1   Physical Properties of Selected Phenols

Name and Formula	mp, °C	bp, °C	Solubility, in g/100 g H_2O at 25°	Dissociation Constant, K_a
Phenol, C_6H_5OH	43	181	9.3	1.28×10^{-10}
o-Cresol, o-CH_3—C_6H_4—OH	30	191	2.5	0.65×10^{-10}
m-Cresol, m-CH_3—C_6H_4—OH	11	201	2.5	0.98×10^{-10}
p-Cresol, p-CH_3—C_6H_4—OH	35.5	201	2.3	0.67×10^{-10}
o-Chlorophenol, o-Cl—C_6H_4—OH	8	176	2.8	77×10^{-10}
m-Chlorophenol, m-Cl—C_6H_4—OH	29	214	2.6	17×10^{-10}
p-Chlorophenol, p-Cl—C_6H_4—OH	37	217	2.8	6.5×10^{-10}
p-Fluorophenol, p-F—C_6H_4—OH	48	185		1.1×10^{-10}
p-Bromophenol, p-Br—C_6H_4—OH	64	236	1.4	5.6×10^{-10}
p-Iodophenol, p-I—C_6H_4—OH	94			6.3×10^{-10}
o-Aminophenol, o-NH_2—C_6H_4—OH	174			5×10^{-10}
m-Aminophenol, m-NH_2—C_6H_4—OH	123		2.6	69×10^{-10}
p-Aminophenol, p-NH_2—C_6H_4—OH	186			7×10^{-10}
o-Nitrophenol, o-NO_2—C_6H_4—OH	44.5	214	0.2	600×10^{-10}
m-Nitrophenol, m-NO_2—C_6H_4—OH	96		1.4	50×10^{-10}
p-Nitrophenol, p-NO_2—C_6H_4—OH	114		1.7	680×10^{-10}
2,4-Dinitrophenol, 2,4-$(NO_2)_2$—C_6H_3—OH	113			$100,000 \times 10^{-10}$
2,4,6-Trinitrophenol, 2,4,6-$(NO_2)_3$—C_6H_2—OH	122		1.4	Very large (0.6)
2,4,6-Trichlorophenol, 2,4,6-Cl_3—C_6H_2—OH	69	244		260×10^{-10}
2,4,6-Tribromophenol, 2,4,6-Br_3—C_6H_2—OH	95			
Catechol, 1,2-$(OH)_2$—C_6H_4	105	245	45	4×10^{-10}
Resorcinol, 1,3-$(OH)_2$—C_6H_4	110	281	123	4×10^{-10}
Hydroquinone, 1,4-$(OH)_2$—C_6H_4	170	286	8	1×10^{-10}

and so on

Note that increasing the number of —OH groups on the aromatic ring greatly increases the water-solubility, as evidenced by the solubility data for catechol, resorcinol, and hydroquinone.

It is especially informative to compare certain physical properties of phenol with those of the corresponding saturated cyclic alcohol, cyclohexanol. The boiling points of cyclohexanol and phenol are 161° and 181° and the freezing points are 26° and 43°, respectively. Thus, phenol boils and melts at higher temperatures, presumably because of more intermolecular hydrogen bonding. Cyclohexanol is soluble in water to the extent of 3.6 g/100 g of water, and under the same conditions, the solubility of phenol is 9.3 g/100 g of water. Hydrogen bonding is apparently greater in water solutions of phenol than in aqueous cyclohexanol solutions.

Pure phenol and substituted phenols are often colorless, crystalline solids. However, phenols undergo rapid oxidation to colored organic compounds (see

Sec. 17.8), and as a result many phenols are pink or brown because of the impurities formed from oxidation.

Several important principles are involved in the physical properties of the isomeric nitrophenols. *o*-Nitrophenol is a low-melting solid that can be distilled without difficulty. It also has a very low solubility in water compared with the *meta* and *para* isomers, which suggests that it does not undergo intermolecular hydrogen bonding to any appreciable extent with either like molecules or water. Indeed, spectral studies show that **internal hydrogen bonding** occurs between the two neighboring polar groups—the —OH and —NO$_2$ groups; the following structure depicts this interaction, which is often referred to as **intramolecular hydrogen bonding**:

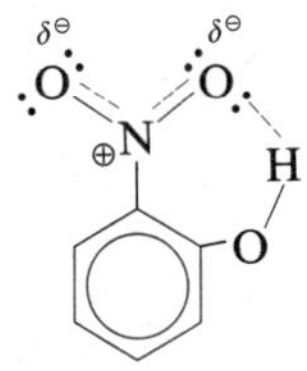

o-Nitrophenol

Internal hydrogen bonding

As a result of the internal hydrogen bonding, *o*-nitrophenol does not hydrogen bond so strongly with solvent or other molecules, thus reducing its water-solubility and lowering its boiling and melting points.

On the other hand, *m*- and *p*-nitrophenol, because of geometric constraints, have no choice but to hydrogen bond with either the solvent or one another (intermolecular hydrogen bonding). This bonding holds the molecules together to a much greater extent, so that more energy must be supplied to overcome the strong hydrogen bonds; thus they melt and boil at much higher temperatures. Indeed, at atmospheric pressure, *p*-nitrophenol undergoes decomposition on heating.

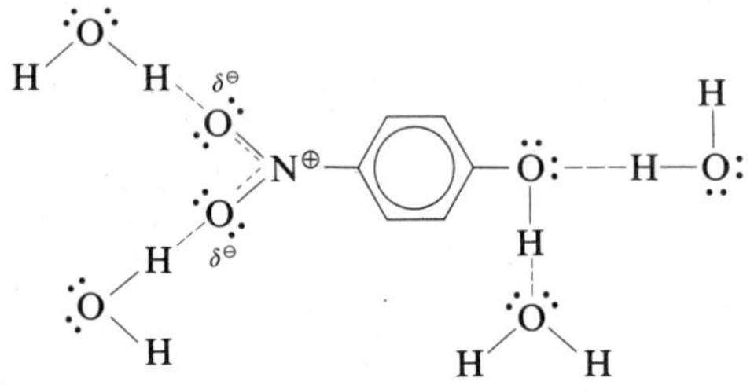

p-Nitrophenol dissolved in water

Hydrogen bonding with water

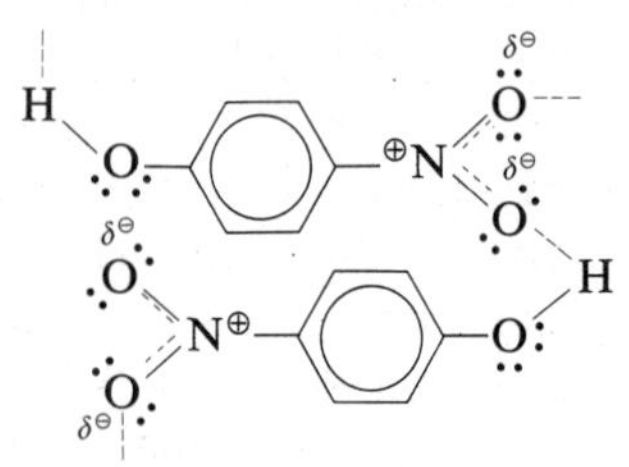

Pure *p*-nitrophenol

Intermolecular hydrogen bonding between molecules

17.3   Preparation of Phenols

A. From Halobenzenes

Several common industrial methods are available for the preparation of phenol. One of the most important, the Dow process developed by the Dow Chemical Company, involves the reaction between chlorobenzene and aqueous sodium hydroxide at elevated temperature and pressure:

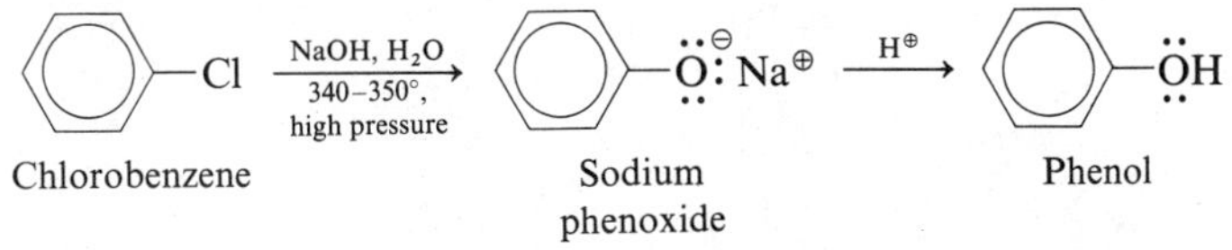

The mechanism of this reaction was discussed in detail in Sec. 16.4 and 16.5 as being representative of nucleophilic aromatic substitution. For reasons discussed in Sec. 17.5, the Dow process produces the sodium salt of phenol, and the reaction mixture must be acidified with a strong mineral acid, such as HCl or H_2SO_4, to obtain phenol.

B. From Arylsulfonic Acids

Another industrial method involves fusing benzenesulfonic acid with solid sodium hydroxide. This also produces the sodium salt of phenol, and phenol is liberated on acidification.

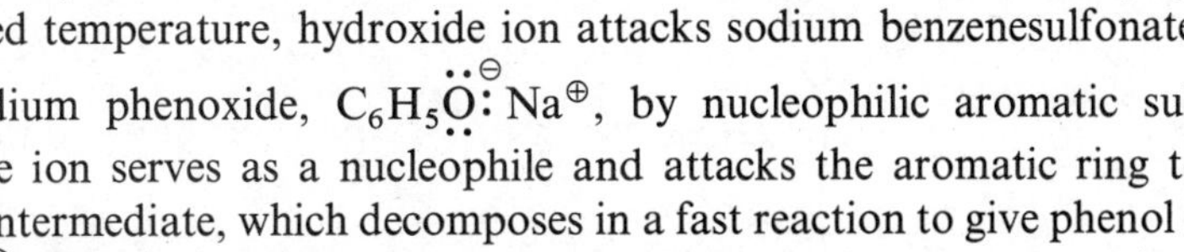

In this reaction the sodium hydroxide first reacts with benzenesulfonic acid, a strong acid, and converts the latter to sodium benzenesulfonate, $C_6H_5SO_3^\ominus Na^\oplus$. At elevated temperature, hydroxide ion attacks sodium benzenesulfonate and produces sodium phenoxide, $C_6H_5\overset{..}{\underset{..}{O}}{}^\ominus Na^\oplus$, by nucleophilic aromatic substitution. Hydroxide ion serves as a nucleophile and attacks the aromatic ring to give the unstable intermediate, which decomposes in a fast reaction to give phenol and sulfite ion, $SO_3^{\ominus 2}$, the leaving group. Under the conditions of the reaction, phenoxide ion is obtained as the product. This is a suitable laboratory method for preparing certain substituted phenols, but other functional groups on the ring (such as —NO_2 and —X) give rise to undesirable side reactions under the drastic conditions of fusion with sodium hydroxide.

C. From Aromatic Amines

The most important laboratory preparation of phenols involves the hydrolysis of the corresponding diazonium salt, which is discussed in Sec. 26.8.

ō m̄ ̄τ   **D. Through Organothallium Reagents**

Another versatile method for the laboratory synthesis of phenols was recently developed. This method involves the synthesis of the organothallium reagents introduced in Chap. 14. The organothallium reagent is prepared as described in Sec. 14.23 by the reaction of benzene or a substituted benzene with thallium trifluoroacetate in trifluoroacetic acid. The arylthallium reagent formed by this procedure is then treated with lead tetraacetate and triphenylphosphine [$(C_6H_5)_3P$]. The ester thus formed is hydrolyzed with base to give the phenol after neutralization.

The specificity exhibited by the thallation reactions and the dependence of this specificity on experimental conditions (see Sec. 14.23) can be used by the resourceful chemist to provide optimum yields of many phenols that would often be more difficult and costly to prepare with classical synthetic procedures.

17.4   Phenols As Acids

Phenols and alcohols undergo similar reactions involving the rupture of the oxygen-hydrogen bond of the —OH group. Whereas most aliphatic alcohols are quite weak acids (weaker than, for example, water, as explained in Sec. 10.9), phenols are more acidic. Indeed, an aqueous solution of phenol is acidic to litmus paper.

$$R\ddot{O}-H \rightleftharpoons R\ddot{O}:^{\ominus} + H^{\oplus} \qquad K_a = 10^{-16} \text{ to } 10^{-18}$$

Alkoxide
ion

but

$$K_a = 1.28 \times 10^{-10}$$

Phenoxide
ion

Two factors may be responsible for the increased acidity of phenols over alcohols. The first is the relative stabilities of the anions, the conjugate bases, which are formed on ionization. But perhaps the most important reason is the effect of solvation (see Sec. 10.9).

In Sec. 10.9 we attributed the decreased acidity of alcohols over water to electron donation by the alkyl group substituents, which destabilizes the alkoxide anion formed in the reaction. Extending this argument, let us examine the electronic effect of the phenoxide anion, $C_6H_5\ddot{O}:^{\ominus}$, which forms on dissociation of phenol. Several resonance structures can be drawn for the phenoxide ion, which show how the negative charge is distributed on the ring carbon atoms:

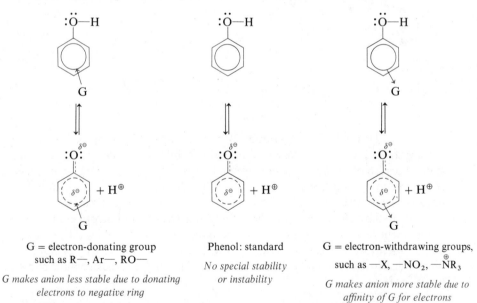

Kekulé
structures

Best represented by the resonance hybrid

Phenoxide ion

In contrast to the alkoxide ion and hydroxide ion in which the negative charge is concentrated on oxygen, the phenoxide ion has the negative charge distributed over a much larger area of the molecule; this resonance stabilization results in greater stability of phenoxide and hence it is the weakest base of the three.

Structure-acidity correlations for a series of substituted phenols are in accord with the notions presented earlier regarding phenol acidity. Ring substituents that withdraw electrons stabilize the negative charge placed in the ring after dissociation occurs. Substituents that release electrons to the ring decrease the acidity because the resulting substituted phenoxide ion is less stable. The following illustrates this *relative* effect, compared with phenol as the standard:

G = electron-donating group
such as R—, Ar—, RO—

*G makes anion less stable due to donating
electrons to negative ring*

Phenol: standard

*No special stability
or instability*

G = electron-withdrawing groups,
such as —X, —NO$_2$, —$\overset{\oplus}{N}$R$_3$

*G makes anion more stable due to
affinity of G for electrons*

The dissociation constants, K_a, listed in Table 17.1 for various substituted phenols, support this theory of structure-acidity relationship. For example, the dissociation constants for the series of *para*-substituted phenols, *p*-G—C$_6$H$_4$—OH, show that *p*-cresol (G = CH$_3$) is less acidic ($K_a = 0.67 \times 10^{-10}$) than phenol ($K_a = 1.28 \times 10^{-10}$), which is reasonable because methyl donates electrons by the inductive effect. On the other hand, *p*-chlorophenol, *p*-bromophenol, and *p*-iodophenol have

K_a values of 6.5×10^{-10}, 5.6×10^{-10}, and 6.3×10^{-10}, respectively, and are more acidic than phenol because of the electron-withdrawing effect of the halogens. p-Nitrophenol is a stronger acid yet ($K_a = 680 \times 10^{-10}$) because of the strong electron-withdrawing effect of the —NO$_2$ group.

For certain substituents like nitro, "resonance stabilization" of the anion occurs in addition to the powerful inductive electron-withdrawing effect attributed to the nitro group. For example, the resonance structures for the p-nitrophenoxide ion (*neglecting the Kekulé structures*) are:

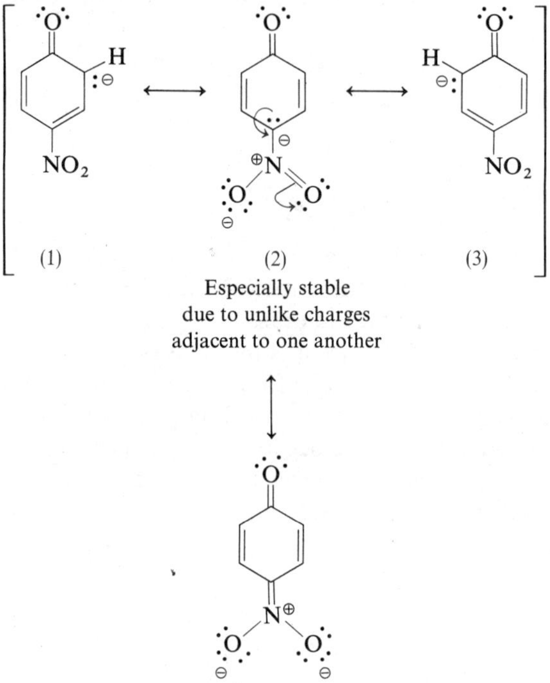

(1)   (2)   (3)

Especially stable
due to unlike charges
adjacent to one another

(4)
Resonance-stabilized structure,
negative charge delocalized
over larger area of molecule

Resonance structures (1) and (3) have comparable stability because in each the negative charge is on the carbon bearing a hydrogen. Structure (2) is especially stable because of the adjacent positive and negative charges on nitrogen and carbon, respectively; an additional resonance structure (4) can be drawn to show how that negative charge can be delocalized over a larger area of the molecule. This extra resonance stabilization explains the exceptional acidity of p-nitrophenol over phenol. Attachment of nitro groups *ortho* and *para* to the —OH group on the ring increases acidity the most for reasons analogous to those that explain why *ortho*- and *para*-nitro-substituted aryl halides are more reactive in nucleophilic aromatic substitution (see Sec. 16.6).

17.5   Reaction of Phenols with Base

An important consequence of the acidity of phenols is their reaction with certain bases, a reaction not exhibited by alcohols. As Table 17.1 indicates, most

substituted phenols have low water-solubility, yet treatment of a phenol with dilute
sodium hydroxide solution converts it into its water-soluble sodium salt:

$$\text{Ar}\ddot{\text{O}}\text{—H} \; \underset{\text{H}^{\oplus}}{\overset{:\ddot{\text{O}}\text{H}^{\ominus}}{\rightleftarrows}} \; \text{Ar}\ddot{\text{O}}\text{:}^{\ominus} \; + \; \text{H}_2\text{O}$$

A substituted phenol   A phenoxide salt
Water-insoluble   Water-soluble

The reaction is reversible; the addition of excess mineral acid to the water-soluble
salt converts it back into the phenol.

This reaction was alluded to several times in the discussion of methods for
preparing phenols in Sec. 17.3. The action of a strong base on chlorobenzene or
benzenesulfonic acid involves nucleophilic aromatic substitution, in which $:\ddot{\text{O}}\text{H}^{\ominus}$
first displaces $:\ddot{\text{C}}\text{l}:^{\ominus}$ or $\text{SO}_3^{\ominus 2}$, respectively. As soon as it is formed, the resulting
phenol is *rapidly* converted to its sodium salt because the acidic phenol molecule
reacts with base: $\text{C}_6\text{H}_5\text{O—H} + \text{Na}^{\oplus}:\ddot{\text{O}}\text{H}^{\ominus} \rightarrow \text{C}_6\text{H}_5\ddot{\text{O}}:^{\ominus} \text{Na}^{\oplus} + \text{H}_2\text{O}$. From the
main reaction, sodium phenoxide is obtained. To obtain the free phenol, the reaction
mixture must be acidified with mineral acid.

Other strong bases (for example, KOH) could be used to remove the proton
from phenol; however, most other common inorganic bases, such as sodium car-
bonate (Na_2CO_3) and sodium bicarbonate (NaHCO_3), are not basic enough to do so.

After we study the acidic properties of carboxylic acids in Sec. 22.9, we discuss
some practical and quite useful applications of the solubility of acids with varying
strength in different bases.

We can now use the solubility properties of phenols to separate them from
other families of compounds. For example, suppose we have a mixture of phenol
and chlorobenzene, which are soluble in one another and form a homogeneous
mixture. If we wish to separate these two aromatic compounds, we can do so by
extracting the mixture with several portions of aqueous sodium hydroxide solution.
The sodium hydroxide converts phenol to its water-soluble sodium salt, which is
present in the aqueous layer; the chlorobenzene remains as a second layer. Acidification of the aqueous solution regenerates the phenol, thus completing the separation.
The following scheme shows this separation:

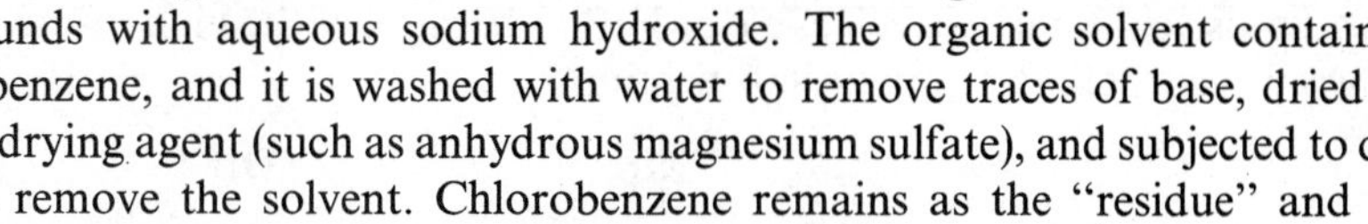

Organic mixture   organic layer → chlorobenzene, $\text{C}_6\text{H}_5\text{Cl}$
Phenol, $\text{C}_6\text{H}_5\text{OH}$ NaOH
Chlorobenzene, $\text{C}_6\text{H}_5\text{Cl}$ H_2O aqueous layer $\xrightarrow[\text{H}_2\text{O}]{\text{H}^{\oplus}}$ phenol, $\text{C}_6\text{H}_5\text{OH}$
$\text{C}_6\text{H}_5\ddot{\text{O}}:^{\ominus} \text{Na}^{\oplus}$

This separation would normally be done in the laboratory by dissolving the
original mixture in some organic solvent (for example, ether or dichloromethane) in
which both compounds are soluble and then extracting the solvent containing the
compounds with aqueous sodium hydroxide. The organic solvent contains only
chlorobenzene, and it is washed with water to remove traces of base, dried over a
typical drying agent (such as anhydrous magnesium sulfate), and subjected to distillation to remove the solvent. Chlorobenzene remains as the "residue" and can be
purified further by simple distillation. The aqueous extract containing sodium
phenoxide is acidified to regenerate phenol, and it is usually removed from the
aqueous solution by extraction with ether or dichloromethane. This organic layer is
then washed with water and dried, and removal of the volatile organic solvent leaves
phenol. It too can be obtained in a purer form by distillation.

Question 17.1

Suggest an explanation for phenol being only slightly soluble in aqueous sodium bicarbonate, whereas 2,4,6-trinitrophenol is readily soluble in this base. (*Hint:* Look at the K_a values in Table 17.1.)

17.6   Electrophilic Aromatic Substitution Reactions on Phenols

The presence of the —OH group on an aromatic ring causes it to be highly activated toward electrophilic aromatic substitution, as explained in Sec. 14.11. Not only does the hydroxyl group activate the ring, but it also is an *ortho,para* director. Several groups closely related to the —OH group have the following order of decreasing reactivity toward electrophilic aromatic substitution:

$$Ar-\overset{..}{\underset{..}{O}}:^{\ominus} \, > \, Ar-\overset{..}{\underset{..}{O}}H \, > \, Ar-\overset{..}{\underset{..}{O}}-R$$

In Sec. 17.5 we saw how ArOH can be converted to $Ar\overset{..}{\underset{..}{O}}:^{\ominus}$, and in Chap. 19 we will learn how to convert Ar—O—H (a phenol) into $Ar-\overset{..}{\underset{..}{O}}-R$ (an alkyl aryl ether). All these substitutents possess unshared pairs of electrons on the oxygen adjacent to the ring, which activate because of the resonance effect. We now survey several salient features of halogenation, nitration, sulfonation, and alkylation of phenol. In particular, note the much milder reaction conditions required with such activated aromatic compounds.

A. Halogenation

Phenol can be chlorinated at elevated temperatures (40 to 150°) with either no solvent or a nonpolar solvent. A mixture of *o*- and *p*-chlorophenol is produced, although the *p* isomer predominates.

Phenol   *o*-Chlorophenol   *p*-Chlorophenol
                                                 Predominant product

The monobromination of phenol requires the use of a nonpolar solvent (for example, carbon disulfide, CS_2) and low temperature ($\sim 5°$), and the *para* isomer is again the predominant product. Note that this reaction does *not* require the presence of a Lewis acid catalyst, such as Fe or $FeBr_3$, because of the high reactivity of phenol toward electrophilic aromatic substitution.

o-Bromophenol   *p*-Bromophenol
                      Predominant product

In this reaction CS_2 is the chemical formula for *carbon disulfide*, a noxious-smelling and low-boiling solvent. (*Question:* By analogy to carbon dioxide, what geometry do you expect for carbon disulfide?)

When a nonpolar solvent is replaced by a polar one, quite different results are obtained in the bromination of phenol. A common polar solvent is water, and when *aqueous* bromine (often called *bromine water:* Br_2, H_2O) reacts with phenol, tri-substitution occurs. The 2,4,6-tribromophenol forms *immediately* and is a white precipitate. Here also no Lewis acid catalyst is required.

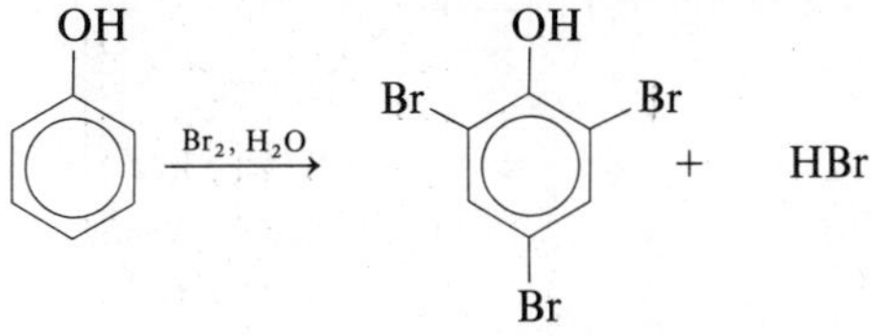

2,4,6-Tribromophenol
Very fast

This reaction is the basis for a simple qualitative test for the presence of a phenol. The test is so sensitive that it can detect the presence of phenol in concentrations as low as one part per 100,000 ($\sim 1 \times 10^{-4}$ M).

B. Nitration

Phenol can be nitrated with dilute nitric acid at room temperature, and the *ortho* isomer is the predominant product. The overall yield is quite low because of the competing oxidation of phenol by nitric acid (see Sec. 17.8).

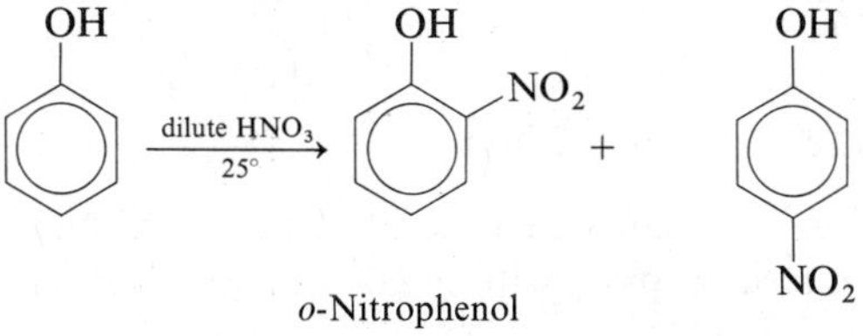

o-Nitrophenol
Predominant product

p-Nitrophenol

C. Sulfonation

Phenol can be sulfonated, but experimental conditions dictate which mono-sulfonic acid isomer is produced. For example, phenol and concentrated sulfuric acid at 25° provides an equal mixture of *p*- and *o*-hydroxybenzenesulfonic acid, but at 100° with dilute sulfuric acid the *para* isomer predominates:

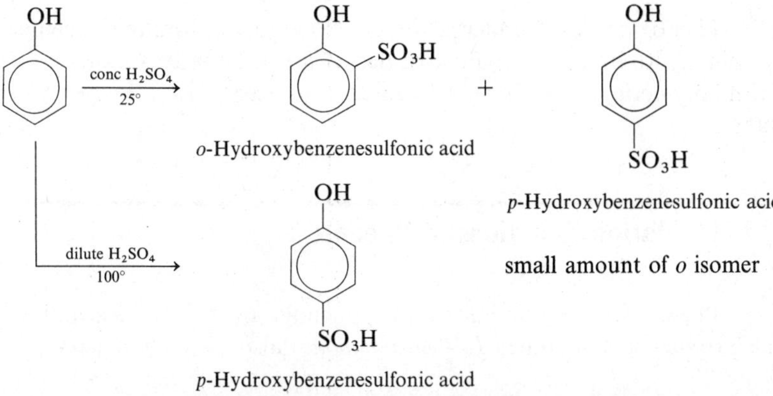

o-Hydroxybenzenesulfonic acid

p-Hydroxybenzenesulfonic acid

small amount of *o* isomer

p-Hydroxybenzenesulfonic acid

D. Friedel-Crafts Alkylation

Like other aromatic compounds, phenols can be alkylated, but alkylation often occurs so rapidly that *di-* and *tri*alkylated products are formed. Careful control of the reaction, usually by a "weak" Lewis acid catalyst like HF, does give mono-substitution; for example:

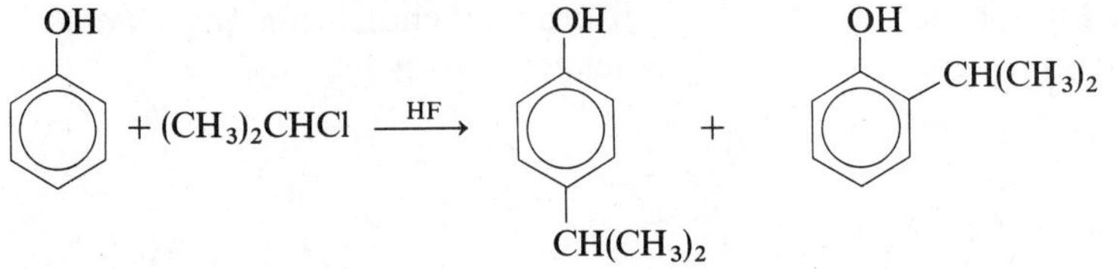

Phenol   + (CH₃)₂CHCl   →   *p*-Isopropylphenol   +   *o*-Isopropylphenol
                                   Major product

17.7   Reactions of Phenoxide Ions

Although phenols are highly reactive in electrophilic aromatic substitution reactions, phenoxide ions are even more reactive because of the partial negative charge in the aromatic ring:

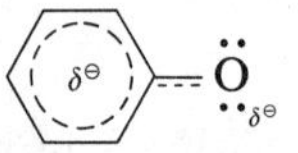

However, the number of reactions that can be carried out on phenoxide ions is severely limited by the requirement of most electrophilic substitution reactions for acidic conditions.

A. Bromine Water Reaction

One reaction that can be carried out in basic solution uses bromine water, which reacts even more rapidly with phenoxide ion than with phenol itself:

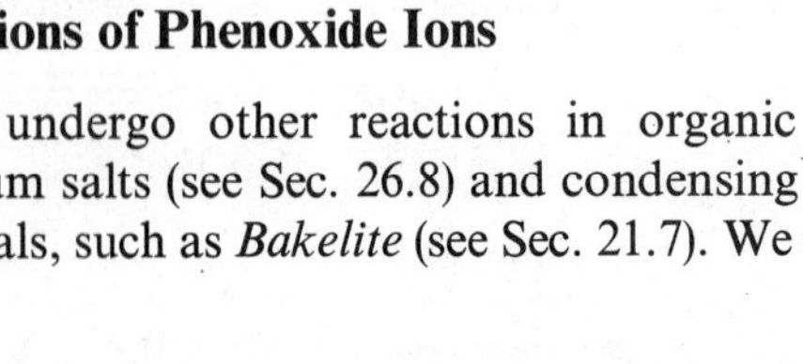

B. Other Reactions of Phenoxide Ions

Phenoxide ions undergo other reactions in organic chemistry, including coupling with diazonium salts (see Sec. 26.8) and condensing with formaldehyde to form polymeric materials, such as *Bakelite* (see Sec. 21.7). We discuss these reactions later.

17.8   Oxidation Reactions of Phenols

Phenol and certain substituted phenols are highly susceptible to oxidation; the product is sometimes *1,4-benzoquinone* (also called *p*-benzoquinone or simply quinone) or a substituted 1,4-benzoquinone. Phenol itself is readily oxidized by chromic acid in acetic acid to 1,4-benzoquinone; the chromic acid (H_2CrO_4) is prepared by dissolving chromium trioxide, CrO_3, in aqueous acetic acid:

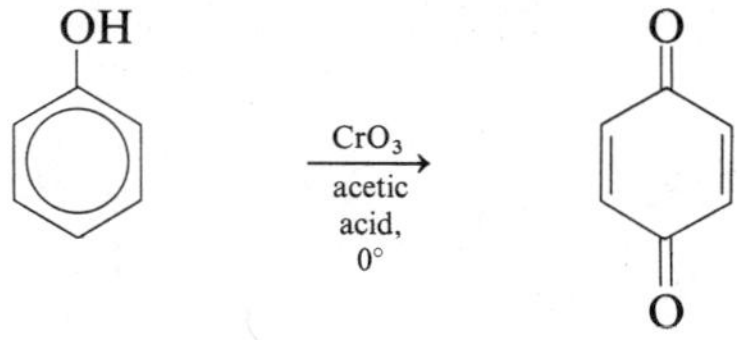

Phenol

1-4-Benzoquinone
or *p*-benzoquinone
or quinone
Yellow

A variety of oxidizing agents accomplish this reaction, and even the oxygen in the air is sufficient to cause some oxidation of phenol when it is hot. The 1,4-benzoquinone product is yellow, and substituted benzoquinones are yellow, orange, or red. Their facile formation is responsible for the appearance of colored "impurities" in most samples of phenols.

The presence of strong electron-donating groups on phenol makes oxidation more rapid. For example, *p*-aminophenol is oxidized with the very mild oxidizing agent, ferric ion:

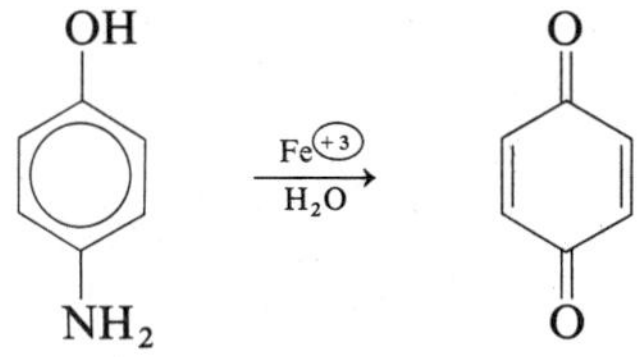

p-Aminophenol

1,4-Benzoquinone

Certain substituents *para* to the —OH group are eliminated during the oxidation reaction, although there must be more vigorous conditions for this to occur.

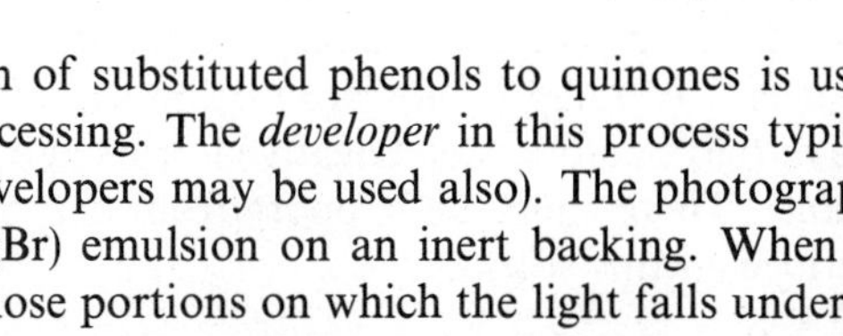

2,4-Dimethylphenol

2-Methyl-1,4-benzoquinone
(toluquinone)

The oxidation of substituted phenols to quinones is used routinely in black and white film processing. The *developer* in this process typically is hydroquinone (although other developers may be used also). The photographic film consists of a silver bromide (AgBr) emulsion on an inert backing. When the silver bromide is exposed to light, those portions on which the light falls undergo a change in crystal structure, producing a more reactive activated silver bromide (AgBr*). The activated silver bromide reacts more rapidly with hydroquinone than does the nonactivated material. The developer is oxidized to a quinone and the silver bromide is reduced to silver, which precipitates on the emulsion. The unactivated silver bromide can then be removed with a *fixer* (usually sodium thiosulfate), leaving the precipitated silver on the emulsion; this is the photographic negative.

$$\text{Hydroquinone} + 2AgBr^* + 2H\ddot{O}{:}^{\ominus} \longrightarrow \text{1,4-Benzoquinone} + 2Ag^0 + 2H_2O + 2{:}\ddot{B}r{:}^{\ominus}$$

Hydroquinone
(developer)

1,4-Benzoquinone

17.9   Qualitative Analysis: Detection of Phenols

Several rapid chemical tests described in this section are useful in determining whether an unknown compound is a phenol.

A. Solubility

Phenols, like many organic compounds, are insoluble or only slightly soluble in water. An unknown that is water-insoluble, soluble in dilute sodium hydroxide solution, but insoluble in both sodium carbonate and sodium bicarbonate solution is *often*, but not *always*, a phenol. The solubility test gives a preliminary indication of a phenol, but more conclusive evidence comes from the other tests described here.

B. Bromine Water

As discussed in Sec. 17.6A, phenols and substituted phenols react very readily with bromine water, (Br_2, H_2O). The formation of an immediate precipitate, accompanied by the disappearance of the characteristic bromine color, strongly suggests a phenol. Other compounds such as substituted anilines, $ArNH_2$, behave similarly, but as we see in Sec. 17.5, they can be differentiated from phenols by solubility tests.

C. Ferric Chloride Test

Because of the facile oxidation of phenols to quinones (see Sec. 17.8), aqueous ferric chloride, $FeCl_3$, can indicate their presence. Most phenols and enols react with ferric chloride to give colored complexes and/or oxidation products. The nature of the colored complexes is still unknown, and the ferric chloride test is also sensitive to the solvent. Although color production is typical of phenols and enols, many of them do *not* give colors, so a negative ferric chloride test cannot exclude a phenolic compound. There must be support from the bromine water and ceric nitrate tests described in this section.

D. Ceric Nitrate Test

Alcohols and phenols replace nitrate ions in ceric ammonium nitrate (1), and when this exchange occurs, it is accompanied by a color change:

$$(NH_4^{\oplus})_2Ce(NO_3)_6^{\ominus 2} + ROH \longrightarrow (NH_4^{\oplus})_2Ce(OR)(NO_3)_5^{\ominus 2} + H^{\oplus} + NO_3^{\ominus}$$

(1)	Alcohol	*Red* when R = alkyl
Ceric ammonium nitrate	or phenol	
Yellow		*Brown to greenish brown* when R = aryl

As indicated, the colors observed depend on whether an alcohol or a phenol is used in the test. Aromatic amines are oxidized by the reagent and give a color close to that obtained from a phenol. Caution must therefore be exercised in interpreting results from this test.

17.10 Spectral Analysis of Alcohols and Phenols

A. Infrared Spectroscopy

The IR spectrum of the hydroxyl group, —OH, in alcohols reveals a great deal about their structure. Two extremes provide limits on hydrogen bonding; in the pure liquid alcohol, hydrogen bonding is maximized because the molecules are in direct contact with each other, and in the vapor state, hydrogen bonding in alcohols is essentially nonexistent because of the large distances between molecules. For example, in the gas phase, ethanol exhibits a strong, *sharp* absorption peak at 3,700 cm^{-1} (2.7 μ) due to *O—H stretching*. On the other hand, ethanol dissolved in carbon tetrachloride (a nonpolar solvent) gives a very weak band at 3,640 cm^{-1} (2.747 μ) and another strong, broad absorption peak at 3,350 cm^{-1} (2.98 to 2.99 μ) due to hydrogen-bonded O—H stretching. As the concentration of ethanol in CCl$_4$ increases, so does the intensity of the 3,350-cm^{-1} peak. The IR spectra of ethanol in the vapor phase and neat are shown in Fig. 17.1.

In summary, the *non-hydrogen-bonded* (free) O—H stretching appears as a strong, broad peak at 3,640 to 3,700 cm^{-1} ($\sim$2.7 μ). Hydrogen bonding normally causes a shift in the absorption peak by about 300 cm^{-1} toward lower frequency.

Carbon-oxygen stretching is also observed in the IR spectrum of alcohols, and the exact location depends largely on the class of alcohol being studied. C—O stretching normally appears as a strong, broad band with the following correlations between frequency and structure: 1° ROH, $\sim$1,050 cm^{-1} (9.52 μ); 2° ROH, $\sim$1,100 cm^{-1} (9.09 μ); and 3° ROH, $\sim$1,160 cm^{-1} (8.62 μ).

As might be expected, phenols resemble alcohols in their O—H stretching frequencies in the IR spectra, which occur as a strong, broad peak at 3,200 to 3,600 cm^{-1} (3.125 to 2.77 μ) for phenols. The C—O stretching frequency in phenols occurs as a strong, broad peak at about 1,230 cm^{-1} (8.13 μ), whereas that in alcohols occurs between 1,050 cm^{-1} (9.52 μ) and 1,200 cm^{-1} (8.33 μ), as discussed in Chap. 12. The IR spectrum of phenol is given in Fig. 17.2.

B. Ultraviolet Spectroscopy

The —O—H group is an auxochrome. It does not absorb in the UV portion of the electromagnetic spectrum, but it does affect the shift of any chromophore to which it is attached. This is observed, for example, in the positions of the primary absorption bands of benzene and phenol, which are 203 and 210.5 nm, respectively. A shift to longer wavelength is observed, as expected, on substitution by the hydroxyl group (refer to Table 12.4).

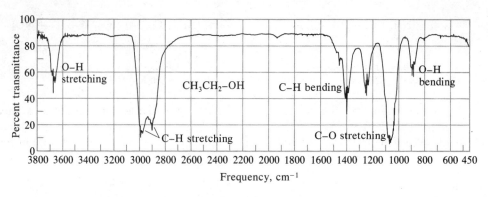

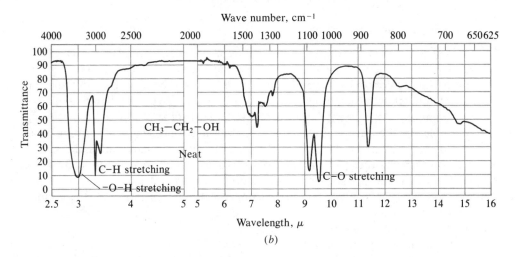

FIGURE 17.1   IR spectra of ethanol in (*a*) the vapor phase and (*b*) neat. [(*a*) Data from *Gases and Vapors: High Resolution Infrared Spectra*, © Sadtler Research Laboratories, Division of Bio-Rad Laboratories, Inc. (1972); (*b*) data from *The Aldrich Library of Infrared Spectra*, 2d ed., C. J. Pouchert (ed.), copyright 1975, Aldrich Chemical Co., Milwaukee]

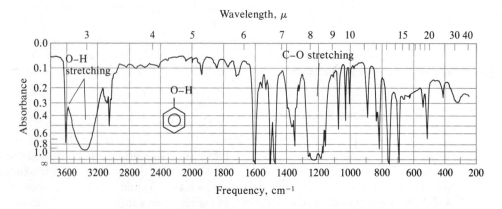

FIGURE 17.2   IR spectrum of phenol. (Data from *Indexes to Evaluated Infrared Reference Spectra*, copyright 1975, The Coblentz Society, Norwalk CT)

C. Nuclear Magnetic Resonance Spectroscopy

The NMR chemical shift of the O—H proton is variable because it depends on the degree of hydrogen bonding in the sample. It also depends on the acidity of the —OH group, that is, the degree of proton character the hydrogen possesses. Typically, the O—H proton resonance in pure alcohols (no solvent) occurs at 4 to 5 δ, relative to TMS, and in phenols at 4 to 7 δ. As the sample is diluted by solvent, hydrogen bonding decreases, and the hydroxyl proton absorption may be shifted upfield by as much as 3 ppm compared with the pure alcohol in which hydrogen bonding is maximized. Attachment of substituents may cause the O—H bond to become more or less acidic (see Sec. 10.9); the more acidic the alcohol, the further downfield the O—H proton resonance. If intramolecular hydrogen bonding occurs, the proton may be shifted downfield between 6 to 12 δ. The NMR spectrum of ethanol dissolved in carbon tetrachloride is shown in Fig. 17.3.

Because of trace amounts of water, acid or base, the hydroxyl proton absorption occurs as a broad singlet in the NMR spectrum; it does not enter into spin-spin splitting with adjacent hydrogens attached to carbon, and conversely adjacent carbon hydrogens are not split by the hydroxyl proton. This is explained in terms of a rapid exchange of protons between alcohol molecules. This exchange is so rapid that the hydroxyl proton is never attached to a given oxygen long enough to "see" adjacent protons. The following type of equilibrium is present:

$$-\overset{|}{\underset{H_1}{C}}-O-H_\alpha \;+\; -\overset{|}{\underset{H_2}{C}}-O-H_\beta \;\rightleftharpoons\; -\overset{|}{\underset{H_1}{C}}-O-H_\beta \;+\; -\overset{|}{\underset{H_2}{C}}-O-H_\alpha$$

The equilibrium is catalyzed by traces of acid or base.

The lack of splitting by hydroxyl protons is a consequence of the time scale of the NMR spectrometer, which records only an average of what it "sees" (the camera shutter speed analogy). For example, the proton attached to oxygen at a given instant may have a $\uparrow$ spin, and when that proton is replaced by another through rapid exchange, there is a 50% probability that the new proton has the reverse spin ($\downarrow$). This exchange occurs much faster than the NMR spectrometer can record it, so the recorded spectrum is an "average" of the hydroxyl proton's structure. The adjacent protons behave as though they are not coupled with the hydroxyl proton.

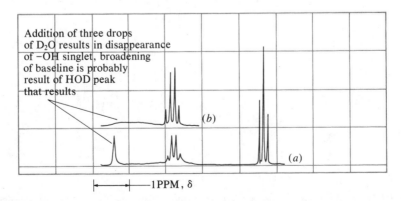

FIGURE 17.3   (a) NMR spectrum of ethanol. (b) Hydroxyl peak disappears upon treatment of sample with D_2O.

Proton Exchange

Employing the technique we originally used to explain coupling and looking at the H_1 protons in the equilibrium equation, the following energy states result:

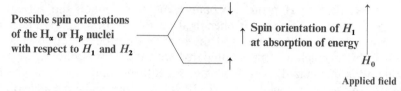

Possible spin orientations of the H_α or H_β nuclei with respect to H_1 and H_2

Spin orientation of H_1 at absorption of energy

H_0

Applied field

When H_1 and H_α or H_β are permanently covalently bound to each other, the NMR spectrometer "sees" two states of different energy ($\uparrow\uparrow$ and $\downarrow\uparrow$), each giving rise to an absorption peak resulting in a doublet. If the protons are unlike, however, as in the preceding equilibrium, the two states are constantly interchanging rapidly—faster than the NMR spectrometer can see. From the viewpoint of the instrument, one state is observed—an average of the previous two. One peak results and no coupling. The same rationale can be applied to viewing either H_α or H_β with respect to H_1 and H_2.

On the other hand, if the alcohol is highly purified and anhydrous, then the rate of exchange is greatly decreased (because of the absence of acid, in this case water, to catalyze the equilibrium) and spin-spin splitting occurs in accordance with the $(n + 1)$ peak rule. In effect, exchange is now so slow that the NMR spectrometer sees both individual states. This effect is also observed with strongly hydrogen-bonded samples, which show the rate of exchange. The NMR spectrum of pure, dry ethanol in Fig. 17.4 illustrates the former effect.

The NMR absorption caused by the OH peak is easily detected. As pointed out in Sec. 15.11, *deuterium* does not appear in the proton NMR spectrum. Because of the acidity of alcohols (see Sec. 10.9), the O—H proton can be exchanged for deuterium by shaking the NMR sample (and solvent) with a small amount of *deuterium oxide*, D_2O; the following exchange occurs:

$$R—O—H + D_2O \rightleftharpoons R—O—D + HOD$$

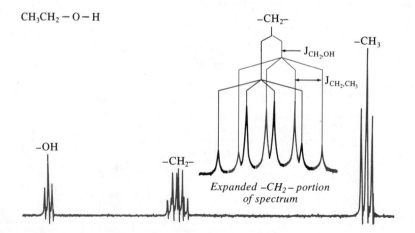

FIGURE 17.4   High-resolution NMR spectrum of pure, anhydrous ethanol, relative to TMS as the internal standard. (Data from L. M. Jackman and S. Sternhell, *Applications of Nuclear Magnetic Resonance Spectroscopy in Organic Chemistry*, 2d ed., p. 16, 1969, Pergamon Press, New York)

Comparison of the NMR spectrum before and after shaking with D_2O reveals the absorption caused by the O—H peak, *as evidenced by its disappearance* after the D_2O treatment (Fig. 17.3). A simultaneous appearance of a peak at approximately 5 δ, caused by the HOD formed during the exchange, adds additional support to the assignment.

17.11   Summary of Methods of Preparation of Phenols

1. From aryl halides

$$Ar—X \xrightarrow[\substack{H_2O, \\ heat, \\ pressure}]{:\ddot{O}H^{\ominus}} Ar\ddot{O}:^{\ominus} \xrightarrow{H^{\oplus}} ArOH \qquad Sec.\ 17.3\ (see\ also\ Secs.\ 16.4\ and\ 16.5)$$

2. From aryl sulfonic acids

$$ArSO_3H \xrightarrow[fuse]{NaOH} Ar\ddot{O}:^{\ominus} Na^{\oplus} \xrightarrow{H^{\oplus}} ArOH \qquad Sec.\ 17.3$$

3. Through organothallium reagents

$$Ar—H \xrightarrow[CF_3COOH]{Tl(OOCCF_3)_3} Ar—Tl(OOCCF_3)_2 \xrightarrow[(C_6H_5)_3P]{Pb(OOCCH_3)_4}$$

$$Ar—OOCCF_3 \xrightarrow[2.\ H^{\oplus}]{1.\ H\ddot{O}:^{\ominus}} Ar—OH \qquad Sec.\ 17.3$$

4. From aromatic amines

To be discussed in Chap. 26   Sec. 26.8

17.12   Summary of Reactions of Phenols

1. Reaction with base

$$Ar\ddot{O}H \xrightleftharpoons[H^{\oplus}]{:\ddot{O}H^{\ominus}} Ar\ddot{O}:^{\ominus} \qquad Secs.\ 17.4\ and\ 17.5$$

Must use strong base: NaOH

2. Electrophilic aromatic substitution on phenols
Decreasing order of reactivity:

$$Ar\ddot{O}:^{\ominus} > Ar\ddot{O}H > Ar\ddot{O}—R \qquad Sec.\ 17.6$$

Above groups are highly activating *ortho, para* directors

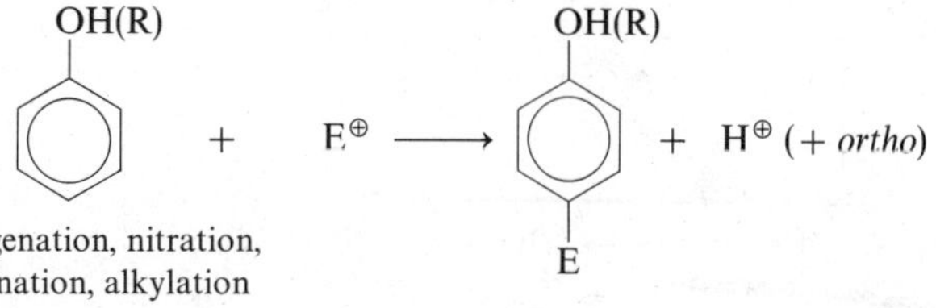

Halogenation, nitration,
sulfonation, alkylation

3. Electrophilic aromatic substitution on phenoxide ion with bromine water

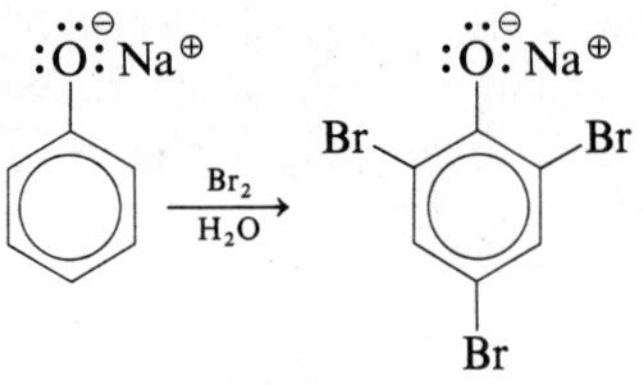

Sec. 17.7A

Very fast reaction

4. Oxidation

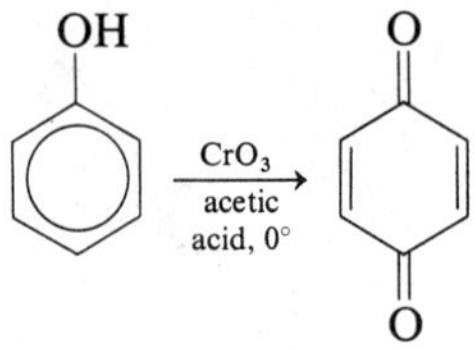

Sec. 17.8

Certain *para* substituents
(—CH$_3$, —NH$_2$) are removed
on oxidation

Study Questions

17.2 Provide structures to correspond to the following eight names.

(*a*) *m*-cresol   (*b*) catechol   (*c*) resorcinol
(*d*) hydroquinone   (*e*) 2,4-dibromophenol   (*f*) 2,4,6-trinitrophenol
(*g*) 4-methylcatechol   (*h*) α-phenyl-2,4-dihydroxytoluene

17.3 Name each of the following compounds by the IUPAC system.

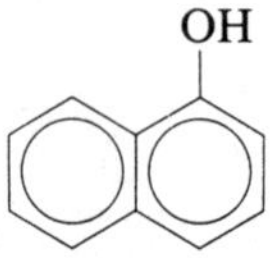

17.4 Supply the missing reagents or products, as needed, to complete the following 17 reactions. The items to be supplied are indicated by number [for example, (1)]. If no reaction is expected, so indicate.

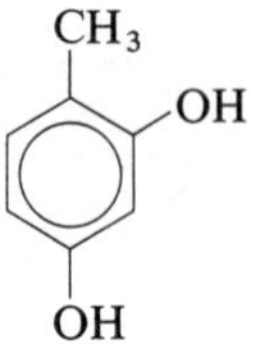

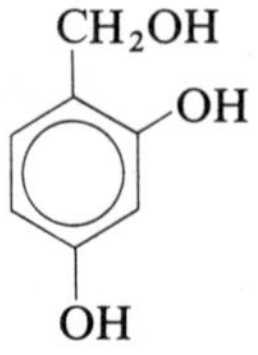

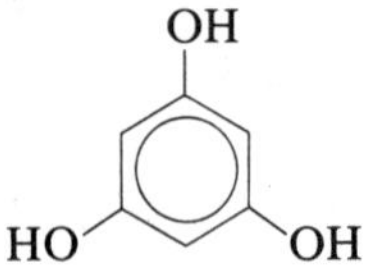

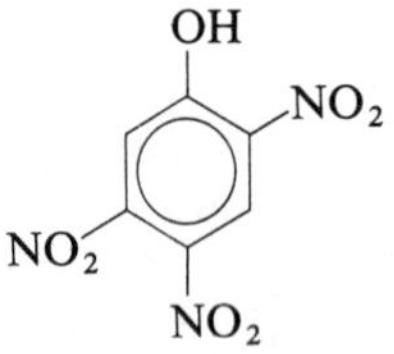

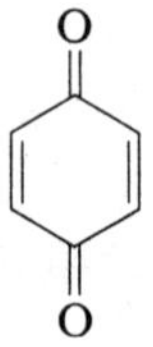

(*a*) Toluene $\xrightarrow{\text{H}_2\text{SO}_4}$ (1) and (2) $\xrightarrow[\text{fusion}]{\text{NaOH}}$ (3) and (4)

Monosubstitution

(*b*) HO—⟨⟩ $\xrightarrow[\text{heat}]{\text{PBr}_3}$ (5)

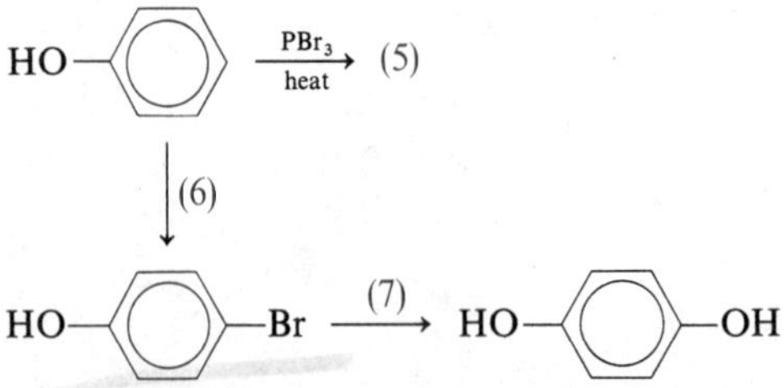

(c) HO—⟨benzene⟩ $\xrightarrow{Br_2(excess),\ H_2O}$ (8)

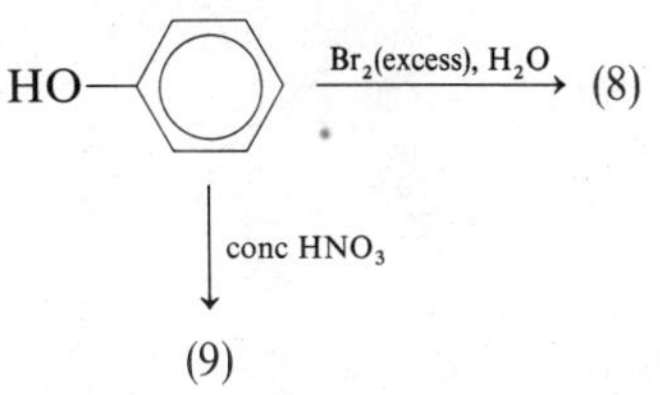

$\downarrow$ conc HNO_3

(9)

(d) Benzene $\xrightarrow{(10)}$ ⟨benzene⟩—CH(CH$_3$)$_2$ $\xrightarrow{(11)}$ ⟨benzene⟩—COOH

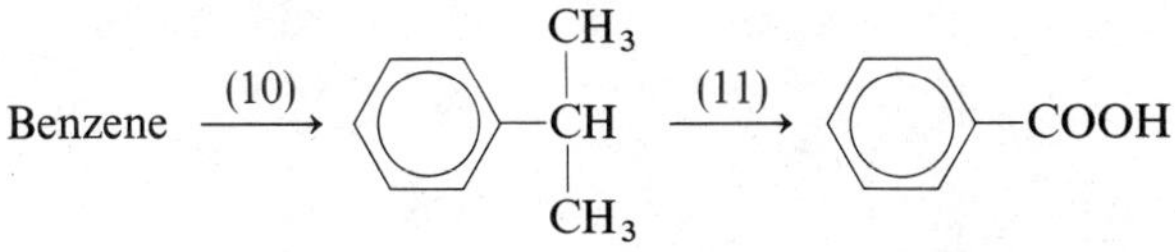

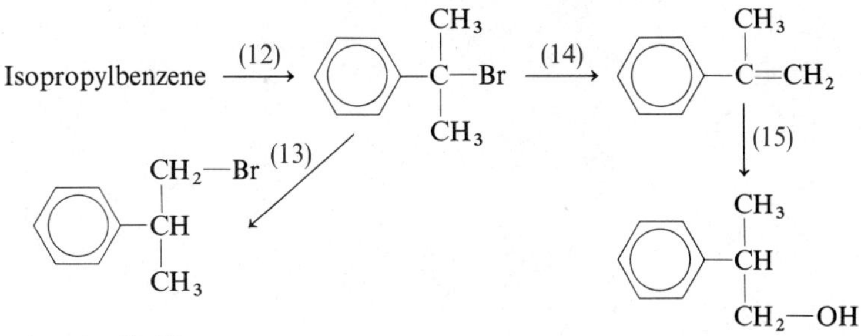

(e) Isopropylbenzene $\xrightarrow{(12)}$ ⟨benzene⟩—C(CH$_3$)$_2$—Br $\xrightarrow{(14)}$ ⟨benzene⟩—C(CH$_3$)=CH$_2$

$\xrightarrow{(13)}$ ⟨benzene⟩—CH(CH$_3$)—CH$_2$—Br

$\downarrow$(15)

⟨benzene⟩—CH(CH$_3$)—CH$_2$—OH

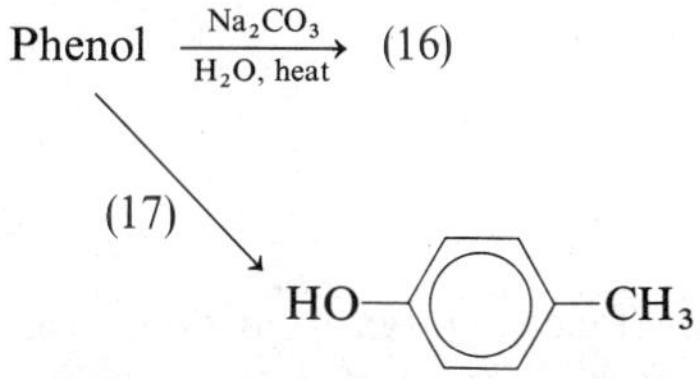

(f) Phenol $\xrightarrow[H_2O,\ heat]{Na_2CO_3}$ (16)

(17)

HO—⟨benzene⟩—CH$_3$

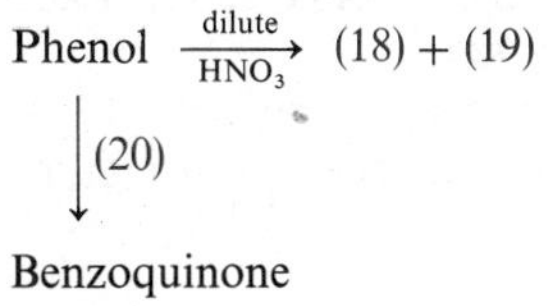

(g) Phenol $\xrightarrow[HNO_3]{dilute}$ (18) + (19)

$\downarrow$(20)

Benzoquinone

(h) ⟨benzene with OH and CH$_2$OH⟩ $\xrightarrow[heat]{PBr_3}$ (22)

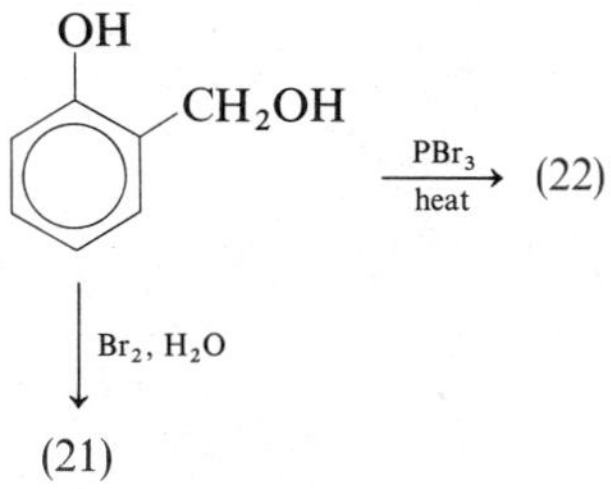

$\downarrow$ Br$_2$, H$_2$O

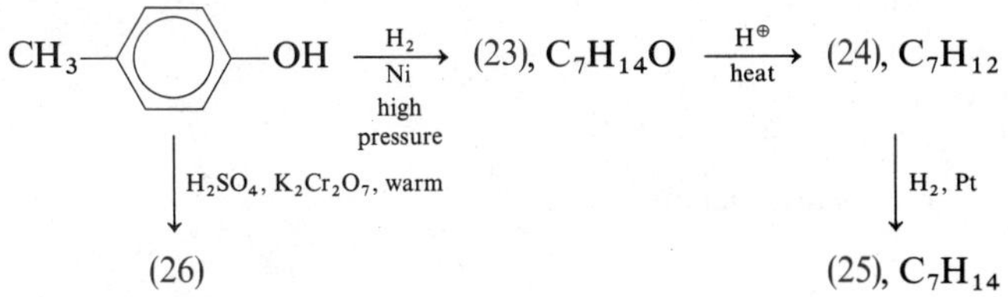

(21)

(i) CH$_3$—⟨benzene⟩—OH $\xrightarrow[\substack{Ni\\high\\pressure}]{H_2}$ (23), $C_7H_{14}O$ $\xrightarrow[heat]{H^{\oplus}}$ (24), C_7H_{12}

$\downarrow$ H$_2$SO$_4$, K$_2$Cr$_2$O$_7$, warm

(26)

$\downarrow$ H$_2$, Pt

(25), C_7H_{14}

(j) (27), C_7H_8O $\xrightarrow{(28)}$ O=⟨cyclohexadiene⟩=O

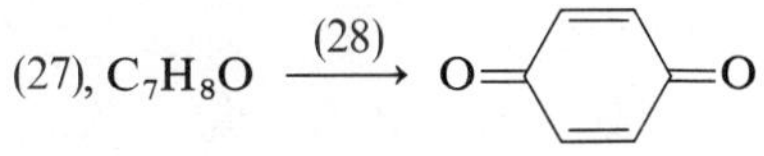

(k) ⟨benzene with OH and CH$_3$⟩ $\xrightarrow[H_2O]{NaOH}$ (29)

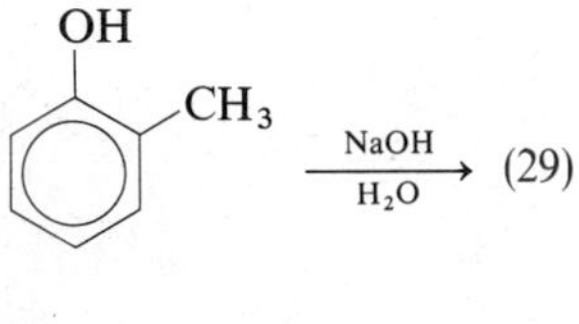

(*l*) CH$_3$—[phenol ring with OH, CH$_3$]—CH$_3$ $\xrightarrow{(30)}$ CH$_2$—[phenol ring with OH]—CH$_2$ with Br groups (31)...

(*l*) CH$_3$ \[benzene with OH\] CH$_3$ $\xrightarrow{(30)}$ \[benzene with OH, CH$_2$Br, CH$_2$Br\]

(*m*) \[naphthalene\] $\xrightarrow[\text{CF}_3\text{COOH}]{\text{Tl(OOCCF}_3)_3}$ $\xrightarrow[\text{2. HO:}^{\ominus}]{\text{1. Pb(OOCCH}_3)_4, (\text{C}_6\text{H}_5)_3\text{P}}$ (31)

(*n*) \[anthracene with OH top and CH$_2$OH bottom\] $\xrightarrow[\text{H}_2\text{O}]{\text{NaOH}}$ (32)

(*o*) \[benzene\]—$\underset{\text{H}}{\overset{\text{CH}_3}{\text{C}}}$—$\underset{\text{H}}{\overset{\text{CH}_3}{\text{C}}}$—OH $\xrightarrow[\text{heat}]{\text{H}_2\text{SO}_4}$ (33) $\xrightarrow{(\text{BH}_3)_2 \text{ ether}}$ (34) $\xrightarrow[\text{HO:}^{\ominus}]{\text{H}_2\text{O}_2}$ (35)

(*p*) \[benzene\]—CH$_2$OH $\xrightarrow[200-300°]{\text{Cu}^0}$ (36)

(*q*) \[benzene\]—$\overset{\text{OH}}{\text{CH}}$—CH$_3$ $\xrightarrow[200-300°]{\text{Cu}^0}$ (37)

17.5 Devise suitable methods for accomplishing the following transformations, using the indicated starting materials as the only sources of carbon. Other organic solvents and inorganic reagents may be used freely.

(*a*) Benzenesulfonic acid $\longrightarrow$ phenol   (*b*) Chlorobenzene $\longrightarrow$ phenol

(*c*) Phenol $\longrightarrow$ anisole (see Sec. 19.5)   (*d*) Benzenesulfonic acid $\longrightarrow$ *m*-nitrophenol

(*e*) Benzenesulfonic acid $\longrightarrow$ *p*-bromophenol

(*f*) Chlorobenzene $\longrightarrow$ 2,4-dinitrophenol

(*g*) Phenol $\longrightarrow$ \[cyclohexane\]=O (cyclohexanone)

(*h*) Toluene $\longrightarrow$ HO—\[benzene\]—CH$_2$OH

(*i*) Toluene $\longrightarrow$ HO—\[benzene with Br\]—CH$_3$

(*j*) Benzene $\longrightarrow$ O=\[ring\]=O

(*k*) Phenol $\longrightarrow$ 2,4,6-trimethylphenol

(*l*) Toluene $\longrightarrow$ \[benzene with OH, Br, CH$_3$, Br\]

(*m*) Benzene $\longrightarrow$ *p-tert*-butylphenol

(*n*) Benzyl alcohol $\longrightarrow$ α,*p*-dibromotoluene

(*o*)   Benzenesulfonic acid $\longrightarrow$

OH

CH$_3$

SO$_3$H

(*p*)   Toluene $\longrightarrow$

CH$_2$OH

Br   Br

OH

17.6 (*a*) Draw the structural isomers for all aromatic compounds with the molecular formula $C_8H_{10}O$ which contain a benzene ring.

(*b*) One of the isomers in (*a*), *A*, does not react with bromine water. On treatment with hot, concentrated sulfuric acid, *A* gives *B*, C_8H_8, which decolorizes bromine in carbon tetrachloride. *A* cannot be resolved into optically active enantiomers. The UV spectrum of *B* is given here along with the NMR spectrum of *B*. What are the structures of *A* and *B*?

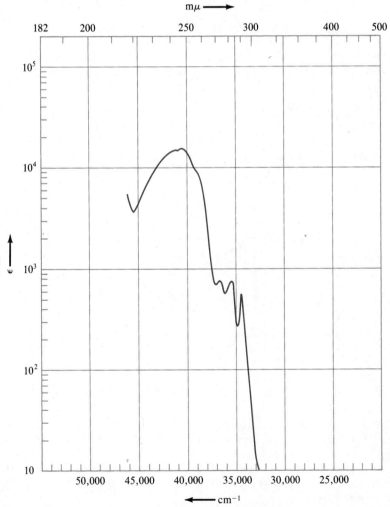

PROBLEM 17.6   UV spectrum of unknown compound (*B*). (Data from the *DMS UV Atlas*, 1966–1971, Plenum Press, New York)

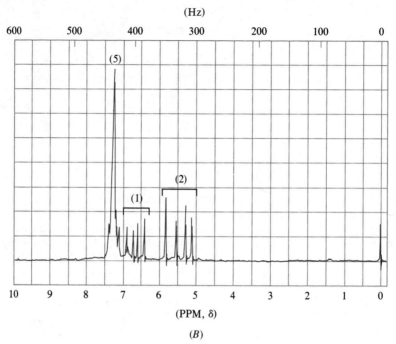

PROBLEM 17.6   NMR spectrum of unknown compound (*B*). [Data from *The Aldrich Library of NMR Spectra*, C. J. Pouchert and J. R. Campbell (eds.), copyright 1974, Aldrich Chemical Co., Milwaukee]

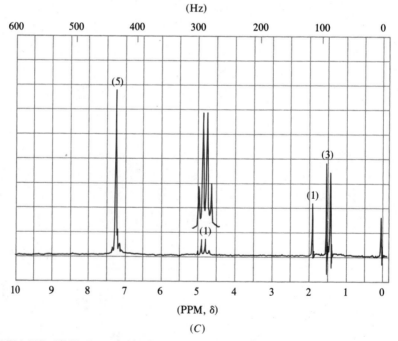

PROBLEM 17.6   NMR spectrum of unknown compound (*C*). [Data from *The Aldrich Library of NMR Spectra* C. J. Pouchert and J. R. Campbell (eds.), copyright 1974, Aldrich Chemical Co., Milwaukee]

(*c*) Another of the isomers in (*a*), *C*, gives exactly the same reactions as does *A*. However, *C* can be resolved into optically active enantiomers. What is the structure of *C*? The NMR of *C* is given here.

(*d*) Another isomer, *D*, gives a positive test with ferric chloride, and on treatment with bromine water, *D* reacts immediately to form *E*, $C_8H_8Br_2O$. What are the structures of *D* and *E*? Is the structure of *D* unique? If not, what other alternative structure(s) fits the facts?

(*e*) Another isomer, *F*, reacts similarly to *D* except that on treatment with bromine water, *F* gives *G*, $C_8H_7Br_3O$. What are the structures of *F* and *G*? Is the structure of *F* unique? If not, what other alternative structure(s) fits the facts?

17.7 List the following in order of their reactivity with the Lucas reagent; explain the observed reactivity.

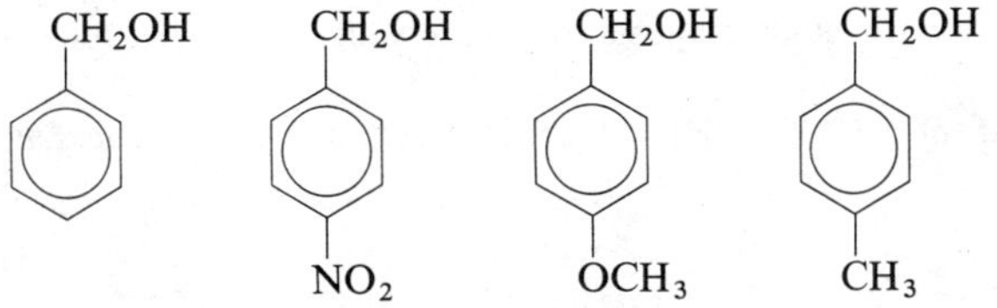

17.8 Given highly purified, anhydrous samples of the following three alcohols, predict the splitting pattern(s) you would expect to observe in the NMR spectra: (*a*) methyl alcohol, (*b*) isopropyl alcohol, (*c*) *tert*-butyl alcohol.

17.9 What simple laboratory tests (chemical or spectral) could be used to differentiate between each member of the following pairs of compounds? Indicate what you would expect to observe.

(*a*) phenol and toluene   (*b*) phenol and 2,4,6-trimethylphenol
(*c*) phenol and 2,4,6-trinitrophenol   (*d*) phenol and *p*-hydroxybenzyl alcohol
(*e*) phenol and cyclohexanol

17.10 The acid-catalyzed hydration of ethyne (acetylene) does not give vinyl alcohol,

$$CH_2=CH-OH;$$

the observed product is ethanal (acetaldehyde), $CH_3-CH=O$. In an analogous manner, consideration of one of the Kekulé structures of phenol might lead one to believe that it is the *enol* of the corresponding ketone, yet it has been proved conclusively that phenol contains none of the keto compound. In other words, the following equilibrium lies *far* to the left:

Enol form of phenol   Keto form of phenol

Suggest an explanation for this seemingly contradictory phenomenon.

17.11 The dissociation constant, K_a, for phenol is 1.28×10^{-10}. Consider the substituted phenols shown here and identify each with its correct K_a from the list of possible K_a values given. Briefly explain.

Possible phenols:

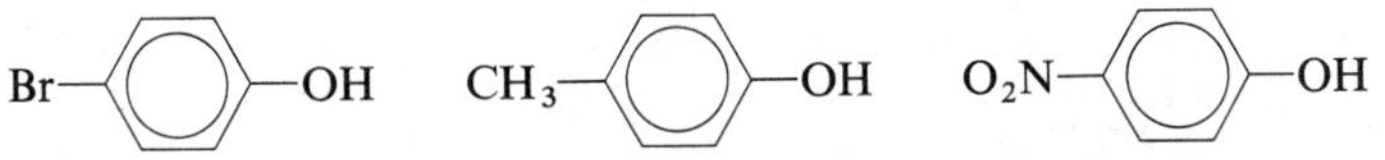

Possible K_a values: 6.8×10^{-8} 5.6×10^{-10} 6.7×10^{-11}

17.12 The dissociation constants of the following compounds are *approximately* as indicated:

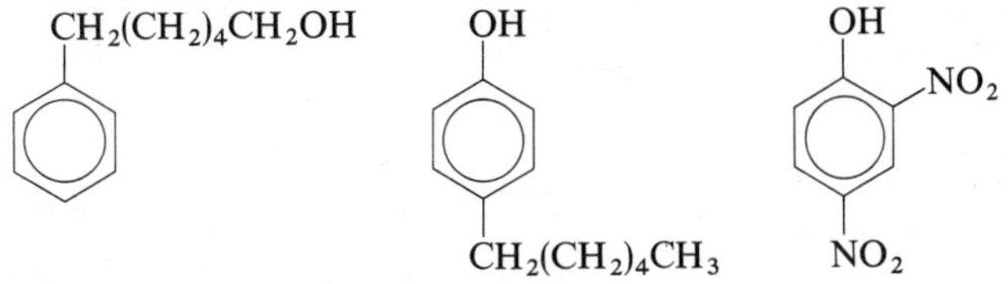

$K_a \approx 10^{-17}$ $K_a \approx 10^{-11}$ $K_a \approx 10^{-4}$

All these compounds are insoluble or slightly soluble in water but completely soluble in ether. The sodium salts of the phenols are completely soluble in water. Suppose that you had at your disposal only the following bases, in addition to water and ether:

1. Na_2CO_3, the salt of bicarbonate, $HCO_3^{\ominus}$; $K_a \approx 10^{-11}$
2. $NaHCO_3$, the salt of carbonic acid, H_2CO_3; $K_a \approx 10^{-7}$

3. NaOH, the salt of water; $K_w = 10^{-14}$, $K_a \approx 10^{-15}$
4. sodium acetate, NaOAc, the salt of acetic acid, HOAc; $K_a \approx 10^{-5}$

Using only these reagents, or some combination thereof, explain how the three organic compounds shown can be separated from one another by means of simple extraction. Write equations for the reactions involved in the separation. (*Hint:* Use the acidities of the organic compounds and those of the inorganic reagents to deduce which basic substance will dissolve each organic compound.)

17.13 Treatment of (1) (G = H) with concentrated hydrochloric acid affords (2) (G = H) in good yield:

(1)   (2)

(*a*) Provide a complete, stepwise mechanism for this reaction and show all intermediates and resonance forms (if any). What general type of reaction is this?

(*b*) Which of the two alcohols, (1) when G = NO$_2$ or (1) when G = OCH$_3$, would be expected to react faster in the preceding reaction? Explain in terms of the reaction mechanism.

(*c*) Suppose the reaction described in (*b*) was studied when the groups are *meta* rather than *para*. Would this change in structure have any effect on your answer to (*b*)? If so, would the *meta* isomers react faster or slower than the corresponding *para* isomers? Briefly explain.

18

Spectroscopy 3—Spectroscopic Methods: Mass Spectrometry

Infrared, ultraviolet, and nuclear magnetic resonance spectroscopy have as their goal the identification and characterization of the structures of organic molecules. These three methods share the feature of not destroying the sample being investigated.

Another important instrumental method is *mass spectrometry* (MS),[1] which has been developed extensively during the past few years. Mass spectrometry is an excellent method for determining molecular weights. Indeed, MS has almost completely replaced many classical methods of determining molecular weight (see Sec. 1.3) because it is *much faster* and *much more accurate*, although the sample is destroyed in the process. High-resolution mass spectrometers give molecular weights typically accurate to ± 0.001 unified atomic mass units (amu).

18.1   Mass Spectrometry: What Is It?

The *mass spectrometer*, shown in Fig. 18.1, is a complex electronic instrument that bombards a minute sample of an organic compound with high-energy electrons. This causes an electron to be knocked out of the molecule (usually a nonbonding electron or an electron from a π bond or π cloud, if the molecule contains one of these features, because it is less tightly bound to the molecule; an electron from a σ bond is much more difficult to remove). The resulting species is an *ion radical*, which is positively charged. The **parent ion,** $P^{\oplus}$ (actually a radical cation $P^{\oplus}_{\bullet}$, written as $P^{\oplus}$ for simplification), has the same weight as the starting molecule (less the negligible weight of the lost electron) and it usually can be detected by the spectrometer. The parent ion under further bombardment breaks down into smaller ion radicals, carbocations, and neutral molecules, and all the new positively charged ions can be detected. Thus, the spectrum derived from a mass spectrometer registers the masses of many different fragments, hence the name *mass spectrum* (literally, a spectrum of different masses). Because of the structure of the mass spectrometer, only particles that are *positively* charged are shown on the spectrum.

The general equation for the bombardment of a molecule, M, with electrons to produce the positively charged parent ion, $P^{\oplus}$, is:

$$\text{M} \;+\; e^{\ominus} \;\longrightarrow\; P^{\oplus} \;+\; 2e^{\ominus}$$

Molecule   Parent
                               ion radical

---

[1] See the introduction to Chap. 12.

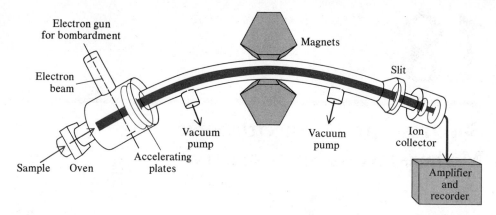

FIGURE 18.1 Simplified schematic diagram of the mass spectrometer. The sample is vaporized and introduced into an evacuated chamber, where it is bombarded with high-energy electrons (70 eV, 1,600 kcal/mole) that strip an electron from the molecule. The resulting positive ions are attracted to and accelerated by the negative accelerating plates into a bent chamber that is surrounded by a magnetic field perpendicular to the plane of the paper. Varying the strength of the magnetic field changes the radius of the curvature of the different ions of differing molecular weights going through the chamber, so that the spectrum of ions is gradually swept past the detector slit. The spectrum is recorded on graph paper as a plot of mass (actually mass/charge ratio) versus intensity. Neutral molecules are not detected and are removed from the system by several vacuum pumps.

Each positive ion produced either directly or by fragmentation of the original molecule has a unique *mass/charge* ratio, called the *m/e ratio*, which is recorded on graph paper. Most ions have a single positive charge ($e = +1$), so that the *m/e* ratio is the actual mass of the ion being detected.

The parent ion ($P^{\oplus}$) directly provides the molecular weight of the sample. It can usually be detected, although in some cases, it produces a peak of very low intensity.

The mass spectrum of molecules is not quite so simple as depicted here for two major reasons. The parent ion breaks apart and often gives many fragments, which appear as numerous peaks in the mass spectrum. Also, most elements exist as a mixture of stable isotopes.

The common elements in organic molecules exist in several stable isotopic forms, as Table 18.1 indicates. The percentages are derived by setting the mass of

TABLE 18.1   Natural Abundance of Heavy Isotopes for Common Elements

Element	Abundance (%) of Lowest Weight Isotope	Relative Abundance (%) of Other Isotopes	
Carbon	100 ^{12}C	1.08 ^{13}C	
Hydrogen	100 ^{1}H	0.016 ^{2}H	
Nitrogen	100 ^{14}N	0.38 ^{15}N	
Oxygen	100 ^{16}O	0.04 ^{17}O	0.20 ^{18}O
Sulfur	100 ^{32}S	0.78 ^{33}S	4.40 ^{34}S
Chlorine	100 ^{35}Cl	32.5 ^{37}Cl	
Bromine	100 ^{79}Br	98.0 ^{81}Br	

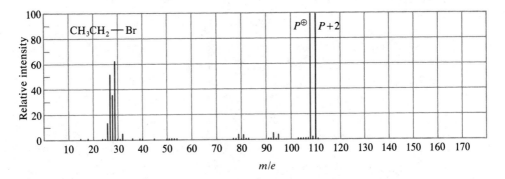

FIGURE 18.2   Graphic representation of the mass spectrum of ethyl bromide. [Data from the *EPA/NIH Mass Spectral Data Base*, **S. R. Heller** and **G. W. A. Milne** (eds.), vols. 1–4, copyright 1978, U.S. Dept. of Commerce]

the lowest weight isotope equal to 100% and computing the percentages of the other isotopes relative to it.

Mass spectra often contain peaks of significant intensity that are attributed to the presence of isotopes. Two of these are called the $P + 1$ and $P + 2$ peaks, where P is the mass of the parent ion ($P^{\oplus}$) and 1 and 2 represent mass increases due to the heavier isotopes of 1 or 2 mass units. For example, if a compound contains a single chlorine atom, then about 32.5 of the molecules contain ^{37}Cl for each 100 that contain ^{35}Cl (see Table 18.1). This molecule gives rise to a $P + 2$ peak that is 32.5% as intense as the parent ($P^{\oplus}$) peak, where the latter is caused by ^{35}Cl and the former by ^{37}Cl. By the same token, when a compound contains one carbon atom, then 1.08% of all the molecules contain ^{13}C (see Table 18.1), and the mass spectrum of this compound shows a $P^{\oplus}$ peak and a $P + 1$ peak one mass unit higher with about 1.08% the intensity of the $P^{\oplus}$ peak. The same distribution appears in the other *fragments* that come from the parent ion (see Question 18.2). The mass spectrum of ethyl bromide (bromine has two naturally occurring isotopes, Table 18.1) is presented in Fig. 18.2. Note particularly the two very intense peaks at m/e ratios 108 and 110; these are the $P^{\oplus}$ and $P + 2$ peaks, respectively.

The interpretation of mass spectra is made difficult because of the large number of fragments that can be formed from the parent ion. Yet the same concepts we found so successful in understanding many organic reactions—namely, mechanisms—are equally useful in understanding the fragmentation pattern and identifying the fragments obtained from organic molecules in the mass spectrometer. For example, fragmentation occurs preferentially at bonds that, on breaking, give relatively stable carbocations (allyl $\approx$ benzyl $> 3° > 2° > 1° > \overset{\oplus}{C}H_3$). Fragmentations that produce simple neutral molecules, such as water, ammonia, nitrogen, carbon monoxide, ethene (ethylene), and ethyne (acetylene), also have a greater tendency to occur. Several examples of these fragmentation patterns are presented in Sec. 18.2.

18.2   Interpretation of Mass Spectra

Fig. 18.3 shows the *graphic representation* of the mass spectrum of toluene. The spectrum that comes from the mass spectrometer is a long strip of paper (6 to 12 in. wide) containing spikes (which resemble the sharp peak observed for an NMR singlet), the height of which represents intensity. The paper is calibrated in mass units

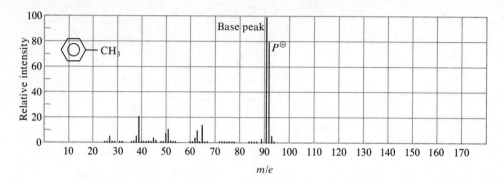

FIGURE 18.3 Graphic representation of the mass spectrum of toluene.[Data from the *EPA/NIH Mass Spectral Data Base*, S. R. Heller and G. W. A. Milne (eds.), vols. 1–4, copyright 1978, U.S. Dept. of Commerce]

(usually through the use of internal reference standards). The graphic representation is constructed by plotting mass/charge ratio versus relative abundance, or percentage of base peak, where the base peak is the *most intense peak* in the spectrum. The base peak is arbitrarily assigned 100% abundance. As shown in Fig. 18.3, the base peak is shown at 100% relative abundance at m/e ratio 91, the parent peak ($P^{\oplus}$) at m/e 92, and the $P + 1$ and $P + 2$ peaks at m/e 93 and 94, respectively.

The fragmentation of toluene in the mass spectrometer is explained in terms of the formation of a stable carbocation. There is a characteristically strong peak at m/e ratio 91 in the mass spectrum of toluene and other alkyl-substituted benzene. This suggests that the ion radical forms first, followed by breaking of the C—H bond shown here. The product is not the benzyl cation as we might expect but instead an even more stable carbocation, the tropylium cation (see Sec. 13.7B):

It has been suggested that the benzyl cation forms first, but it is also likely that the tropylium cation forms directly. Once formed, the tropylium cation can lose a molecule of ethyne to form $C_5H_5^{\oplus}$, m/e 65:

The mass spectrum of 3,3-dimethylheptane is shown in Fig. 18.4. Note that the parent peak is missing at m/e 128 because of the extensive and rapid fragmentation

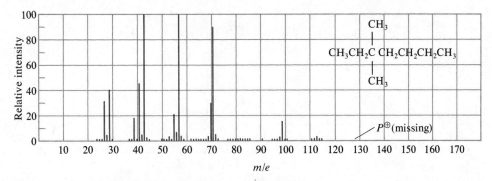

FIGURE 18.4   Graphic representation of the mass spectrum of 3,3-dimethylheptane. [Data from the *EPA/NIH Mass Spectral Data Base*, S. R. Heller and G. W. A. Milne (eds.), vols. 1–4, copyright 1978, U.S. Dept. of Commerce]

of the parent ion radical once it is formed. This example shows how fragmentation readily occurs to give stable tertiary carbocations. The following scheme illustrates this behavior, and the various important fragments shown here can be identified in the spectrum in Fig. 18.4.

$$\left[CH_3{-}CH_2{-}\overset{\overset{\displaystyle CH_3}{|}}{\underset{\underset{\displaystyle CH_3}{|}}{C}}{-}CH_2{-}CH_2{-}CH_2{-}CH_3 \right]^{\overset{\cdot}{\oplus}}$$

Ⓐ   Ⓑ   Ⓒ

Ion radical of
3,3-dimethylheptane

cleavage
at Ⓐ

cleavage
at Ⓑ

cleavage
at Ⓒ

$CH_3{-}CH_2^{\oplus}$

m/e 29

and

$\overset{\overset{\displaystyle CH_3}{|}}{\underset{\underset{\displaystyle CH_3}{|}}{\overset{\oplus}{C}}}{-}CH_2{-}CH_2{-}CH_2{-}CH_3$

m/e 99

$CH_3^{\oplus}$

m/e 15

and

$CH_3{-}CH_2{-}\overset{\overset{\displaystyle\oplus}{}}{\underset{\underset{\displaystyle CH_3}{|}}{C}}{-}CH_2{-}CH_2{-}CH_2{-}CH_3$

m/e 113

$CH_3{-}CH_2{-}\overset{\overset{\displaystyle CH_3}{|}}{\underset{\underset{\displaystyle CH_3}{|}}{\overset{\oplus}{C}}}$

m/e 71

and

$CH_3{-}CH_2{-}CH_2{-}CH_2^{\oplus}$

m/e 57

Although not all peaks in Fig. 18.4 are assigned, there are several noteworthy features. The peak at *m/e* 113 is of low intensity because its formation also produces the highly unstable $CH_3^{\oplus}$ ion; this fragmentation is not very likely to occur. The high-intensity peak at *m/e* 43 is due to $C_3H_7^{\oplus}$, which can result from further fragmentation of some larger fragments or combinations of smaller fragments. Surrounding each peak with relatively high intensity are smaller peaks. Many of those

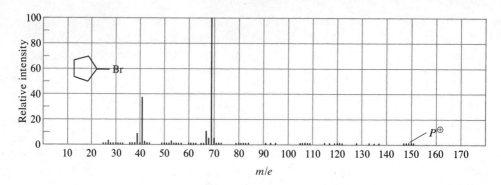

FIGURE 18.5   Mass spectrum of cyclopentyl bromide. [Data from the *EPA/NIH Mass Spectral Data Base*, S. R. Heller and G. W. A. Milne (eds.), vols. 1–4, copyright 1978, U.S. Dept. of Commerce]

with one more mass unit are caused by the presence of isotopes, that is, are $P + 1$ peaks, whereas many of those with two less mass units are due to alkenyl cations, those that have lost the elements of hydrogen.

Several more interesting spectra are shown in Figs. 18.5 to 18.7.

The intention of this section on mass spectroscopy is to introduce the basic underlying principles and the type of information that can be gleaned from mass

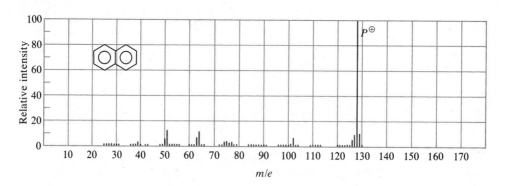

FIGURE 18.6   Mass spectrum of naphthalene. [Data from the *EPA/NIH Mass Spectral Data Base*, S. R. Heller and G. W. A. Milne (eds.), vols. 1–4, copyright 1978, U.S. Dept. of Commerce]

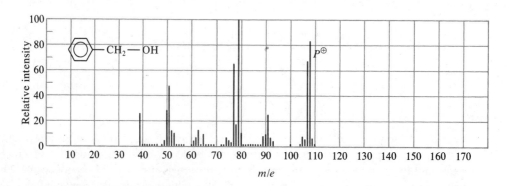

FIGURE 18.7   Mass spectrum of benzyl alcohol. [Data from the *EPA/NIH Mass Spectral Data Base*, S. R. Heller and G. W. A. Milne (eds.), vols. 1–4, copyright 1978, U.S. Dept. of Commerce]

spectra. For further information and additional examples, consult the Reading References for this chapter in the appendix. However, as with the other spectroscopic (or spectrometric) methods, a brief summary of common fragmentation patterns is given in future chapters for the various functional groups being studied.

Study Questions

18.1 The mass spectrum given here is for one of the following two isomers:

$$CH_3$$
$$|$$
$$CH_3CHCH_2CH_3 \qquad CH_3-\underset{\underset{CH_3}{|}}{\overset{\overset{CH_3}{|}}{C}}-CH_3$$

Isopentane   Neopentane

Which isomer corresponds to the mass spectrum? Explain your reasoning.

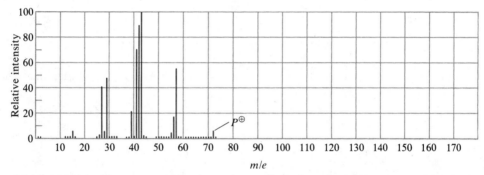

PROBLEM 18.1   Mass spectrum of unknown isomer. [Data from the *EPA/NIH Mass Spectral Data Base*, S. R. Heller and G. W. A. Milne (eds.), vols. 1–4, copyright 1978, U.S. Dept. of Commerce]

18.2 In Sec. 18.1 we learned that the relative percentage of the $P+1$ peak is primarily determined by the number of carbons in the parent ion. Based on this fact, what will be the relative magnitude of the $P+1$ peak of each of the following five compounds?

(*a*) $CH_3CH_2CH_2CH_2CH_2CH_3$ (*b*) ⬡

(*c*) ⬡ OH   (*d*) ⬡ CH_3 (*e*) ⬡ $CH_2CH_2CH_3$

In reality, the relative intensity of the $P+1$ peak is caused by the natural abundance of ^{13}C versus ^{12}C *and* the natural abundance of ^2H and ^{15}N and so on. The intensity of the $P+1$ peak can be determined by application of the following equation:

$$\text{Intensity of } P+1 = \text{(no. of carbon atoms)(natural abundance of } ^{13}C)$$
$$+ \text{(no. of hydrogen atoms)(natural abundance of } ^2H)$$
$$+ \text{(no. of nitrogen atoms)(natural abundance of } ^{15}N)$$

Using this formula, determine the relative intensity of the $P+1$ peaks for each of the preceding five compounds. The natural abundances are in units of percent (%).

18.3 Compound *A*, C_7H_8, reacts with bromine and heat to give compound *B*. The mass spectrum of *B* is given here. Completely interpret the spectrum and provide structures for *A* and *B*.

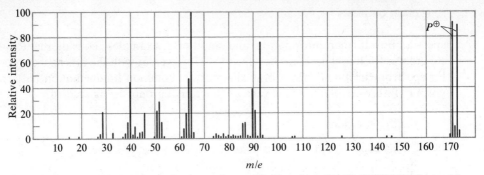

PROBLEM 18.3   Mass spectrum of unknown isomer (*B*). [Data from the *EPA/NIH Mass Spectral Data Base*, S. R. Heller and G. W. A. Milne (eds.), vols. 1–4, copyright 1978, U.S. Dept. of Commerce]

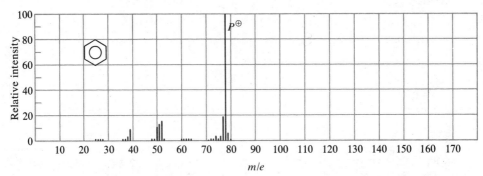

PROBLEM 18.4   Mass spectrum of benzene. [Data from the *EPA/NIH Mass Spectral Data Base*, S. R. Heller and G. W. A. Milne (eds.), vols. 1–4, copyright 1978, U.S. Dept. of Commerce]

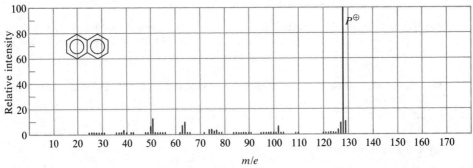

PROBLEM 18.4   Mass spectrum of naphthalene. [Data from the *EPA/NIH Mass Spectral Data Base*, S. R. Heller and G. W. A. Milne (eds.), vols. 1–4, copyright 1978, U.S. Dept. of Commerce]

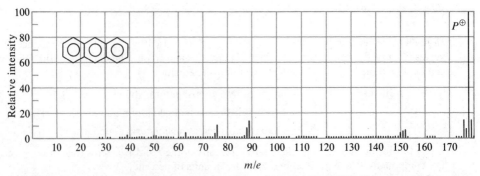

PROBLEM 18.4   Mass spectrum of anthracene. [Data from the *EPA/NIH Mass Spectral Data Base*, S. R. Heller and G. W. A. Milne (eds.), vols. 1–4, copyright 1978, U.S. Dept. of Commerce]

18.4 The mass spectra of benzene, naphthalene, and anthracene are given here. Completely interpret these spectra and determine the common features exhibited by these aromatic systems.

18.5 Compound A, C_3H_4, reacts with sodamide to give B, C_3H_3Na. B reacts with C, C_7H_7Br, to give D, $C_{10}H_{10}$. D is reduced with 1 mole of H_2/Ni—B to give E, $C_{10}H_{12}$. Ozonolysis of E gives 1 mole of F, C_2H_4O, plus G, C_8H_8O. The NMR, IR, and mass spectra of F and G are given here. Determine structures for A to G.

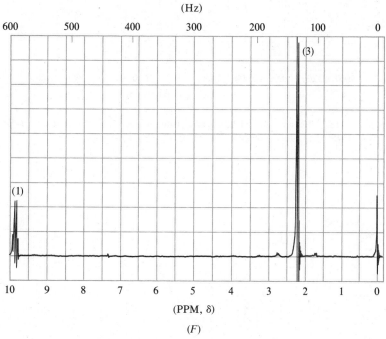

(F)

PROBLEM 18.5   NMR spectrum of (F). [Data from *The Aldrich Library of NMR Spectra*, C. J. Pouchert and J. R. Campbell (eds.), copyright 1974, Aldrich Chemical Co., Milwaukee]

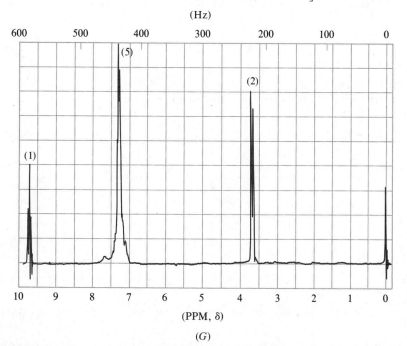

(G)

PROBLEM 18.5   NMR spectrum of (G). [Data from *The Aldrich Library of NMR Spectra*, C. J. Pouchert and J. R. Campbell (eds.), copyright 1974, Aldrich Chemical Co., Milwaukee]

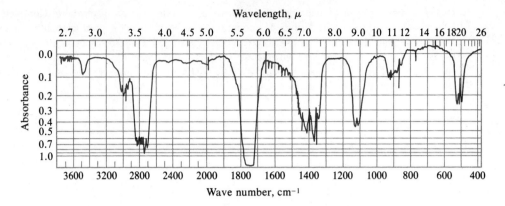

PROBLEM 18.5   IR spectrum of (*F*). [Data from *Indexes to Evaluated Infrared Reference Spectra*, copyright 1975. The Coblentz Society, Norwalk CT]

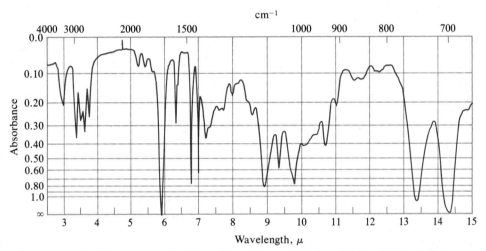

PROBLEM 18.5   IR spectrum of (*G*). [Data from *Indexes to Evaluated Infrared Reference Spectra*, copyright 1975. The Coblentz Society, Norwalk CT]

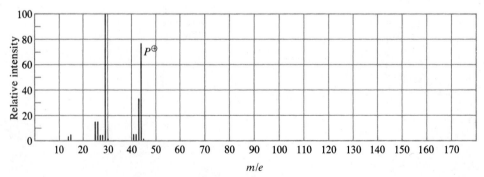

PROBLEM 18.5   Mass spectrum of (*F*). [Data from the *EPA/NIH Mass Spectral Data Base*, S. R. Heller and G. W. A. Milne (eds.), vols. 1–4, copyright 1978, U.S. Dept. of Commerce]

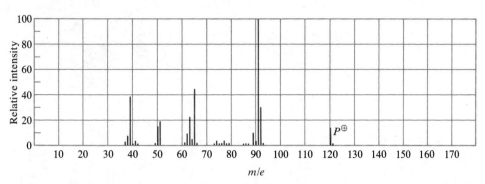

PROBLEM 18.5   Mass spectrum of (*G*). [Data from the *EPA/NIH Mass Spectral Data Base*, S. R. Heller and G. W. A. Milne (eds.), vols. 1–4, copyright 1978, U.S. Dept. of Commerce]

18.6 The following two mass spectra are of ethylbenzene. Spectrum 1 was run at a much higher ionization voltage than spectrum 2. Explain the differences in the spectra and the pros and cons of running one versus the other.

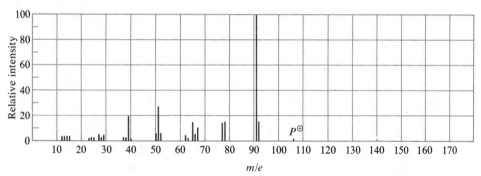

PROBLEM 18.6   1: Mass spectrum of ethylbenzene. [Data from the *EPA/NIH Mass Spectral Data Base*, S. R. Heller and G. W. A. Milne (eds.), vols. 1–4, copyright 1978, U.S. Dept. of Commerce]

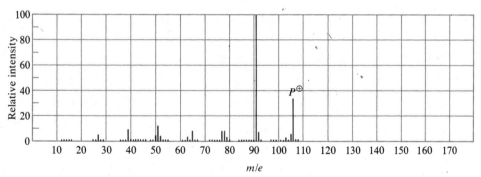

PROBLEM 18.6   2: Mass spectrum of ethylbenzene. [Data from the *EPA/NIH Mass Spectral Data Base*, S. R. Heller and G. W. A. Milne (eds.), vols. 1–4, copyright 1978, U.S. Dept. of Commerce]

18.7 Compound A, C_8H_9Br, reacts with alcoholic potassium hydroxide to give B, C_8H_8. B decolorizes Br_2/CCl_4 and permanganate solution. Reaction of B with HBr and peroxides gives A. Ozonolysis of B gives 1 mole of C, C_7H_6O, and 1 mole of D, CH_2O. Reduction of C with $LiAlH_4$ (see Sec. 21.13B) gives E, C_7H_8O. When B is heated in the presence of a trace of benzoyl peroxide (a free-radical initiator), a large molecular weight polymer is obtained. The NMR, IR, and mass spectra of B are given. What are the structures of A to E?

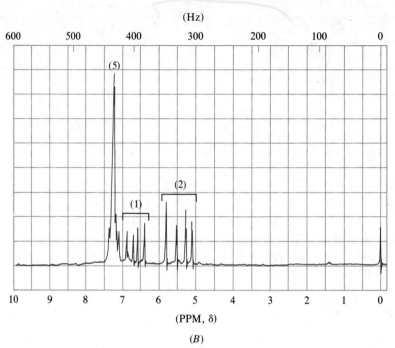

PROBLEM 18.7   NMR spectrum of compound (*B*). [Data from *The Aldrich Library of NMR Spectra*, C. J. Pouchert and J. R. Campbell (eds.), copyright 1974, Aldrich Chemical Co., Milwaukee]

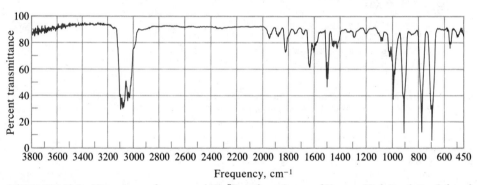

PROBLEM 18.7   IR spectrum of compound (*B*). [Data from *Gases and Vapors: High Resolution Infrared Spectra*, © Sadtler Research Laboratories, Division of Bio-Rad Laboratories, Inc. (1972)]

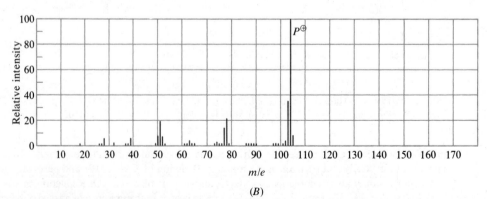

PROBLEM 18.7   Mass spectrum of compound (*B*). [Data from the *EPA/NIH Mass Spectral Data Base*, S. R. Heller and G. W. A. Milne (eds.), vols. 1–4, copyright 1978, U.S. Dept. of Commerce]

19

Ethers, Epoxides, and Glycols

In this chapter we discuss the **ether** family of compounds, which is characterized by the **C—O—C** linkage. We pay particular attention to the three-membered cyclic ethers, the **epoxides,** which are much more reactive toward many reagents than are ethers in general. In previous chapters we mentioned **glycols,** but we now cover them in greater detail because synthetically they are closely related to epoxides. The inter-relationship of ethers, epoxides, and glycols makes it desirable to include them all in this single chapter.

$$R{-}O{-}R \qquad Ar{-}O{-}R \qquad Ar{-}O{-}Ar \qquad \overset{\diagup}{\underset{\diagdown}{C}}{-}\overset{\diagup}{\underset{O}{C}} \qquad {-}\overset{|}{\underset{OH}{C}}{-}\overset{|}{\underset{OH}{C}}{-}$$

Epoxide   Glycol

Ethers

19.1 Structure and Nomenclature of Ethers, Epoxides, and Glycols

A. Ethers

We can consider ethers as derivatives of water in much the same way we did for alcohols. Replacement of both hydrogens on H_2O by carbon-containing groups gives rise to the ether family:

$$\underset{\text{Water}}{H{\diagup}\overset{\ddot{\cdot}\ddot{O}\ddot{\cdot}}{\diagdown}H} \xrightarrow[\text{of one H}]{\text{replacement}} \underset{\substack{\text{R: alcohol} \\ \text{Ar: phenol}}}{H{\diagup}\overset{\ddot{\cdot}\ddot{O}\ddot{\cdot}}{\diagdown}R(Ar)} \xrightarrow[\text{of second H}]{\text{replacement}} \underset{\text{Ether}}{(Ar)R{\diagup}\overset{\ddot{\cdot}\ddot{O}\ddot{\cdot}}{\diagdown}R(Ar)}$$

The substitution of alkyl and aryl groups for hydrogen is not always simple, as we learned in discussing the preparation of alcohols. We will see how ethers can be prepared in later sections of this chapter.

All ethers contain the C—O—C linkage, and several subclassifications of them depend on whether alkyl or aryl groups are present:

R—O—R   Ar—O—R   Ar—O—Ar

Dialkyl   Alkyl   Diaryl
ether   aryl ether   ether

Ethers that contain two identical alkyl or aryl groups are called *symmetrical ethers*, and those that contain two different groups (alkyl and/or aryl) are referred to as *unsymmetrical ethers*.

a. Common Names

Ethers are named by taking the names of the two alkyl or aryl group substituents attached to oxygen and following them with the word *ether*. For symmetrical ethers, the name of the alkyl or aryl group is often used instead of *di*alkyl or *di*aryl; use of the prefix *di-* is optional.

$$CH_3—O—CH_3 \qquad CH_3CH_2CH_2CH_2—O—CH_2CH_2CH_2CH_3$$

Methyl ether   *n*-Butyl ether   Phenyl ether
(or dimethyl ether)   (or di-*n*-butyl ether)   (or diphenyl ether)

The term *ether* used alone usually refers to ethyl ether (or diethyl ether), $(CH_3CH_2)_2O$, because of its frequent use in chemistry. Unsymmetrical ethers require the names of both groups:

$$CH_3—O—\overset{\overset{\displaystyle CH_3}{|}}{CH}—CH_2CH_3 \qquad CH_3—O—CH=CH_2 \qquad CH_3—O—$$

Methyl *sec*-butyl   Methyl vinyl   Methyl phenyl
ether   ether   ether
                                              (or, IUPAC, anisole)

Several ether solvents other than diethyl ether are frequently encountered and go by their common names:

$$CH_3—O—CH_2CH_2—O—CH_3 \qquad CH_3—O—CH_2CH_2—O—CH_2CH_2—O—CH_3$$

Glyme   Diglyme
(acronym for ethylene   (acronym for diethylene
glycol dimethyl ether)   glycol dimethyl ether)

b. IUPAC Nomenclature

Common names are used more extensively than IUPAC names for ethers (especially simple ones). The IUPAC system for naming ethers considers them as derivatives of the longest carbon-containing alkyl group attached to oxygen, and the —OR group is considered as a substituent that replaces hydrogen. The —OR group is called *alkoxy* ($CH_3O—$ is *methoxy*, $CH_3CH_2CH_2CH_2O—$ is *butoxy* or *n-butoxy*, and so on), and the —OAr group is called *aryloxy* ($C_6H_5O—$ is *phenyloxy*, but often *phenoxy* for convenience in pronunciation). The following examples are illustrative:

$$CH_3O—\overset{2}{\underset{}{C}}\overset{3}{H}\overset{}{CH_2}\overset{4}{CH_3}$$

with $\overset{1}{CH_3}$ substituent

2-Methoxybutane   *p*-Chlorophenoxyethane   1,2-Diethoxyethane

$$CH_3CH_2\overset{3}{C}HCH_2CH_3$$

3-Propoxypentane   1-Methoxy-1-phenylhexane

One version of the IUPAC nomenclature allows fairly complex ethers to be named, especially those that cannot be named easily by the previous approach. This variation uses **oxa** to indicate the presence of oxygen. The oxygen atom is made a part of the longest continuous chain and a number is assigned to it as though it were a methylene group, $-CH_2-$. The position of oxygen is indicated by number and is followed by *oxa* as a prefix added to the alkane name; for example:

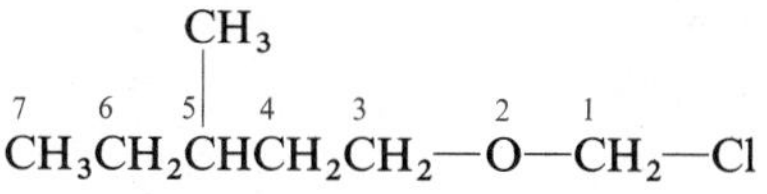

$$\overset{7}{C}H_3\overset{6}{C}H_2\overset{5}{\underset{\underset{\displaystyle CH_3}{|}}{C}}H\overset{4}{C}H_2\overset{3}{C}H_2-\overset{2}{O}-\overset{1}{C}H_2-Cl$$

1-Chloro-5-methyl-2-oxaheptane

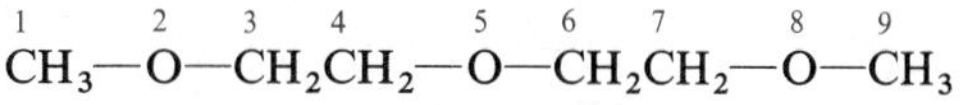

$$\overset{1}{C}H_3-\overset{2}{O}-\overset{3}{C}H_2\overset{4}{C}H_2-\overset{5}{O}-\overset{6}{C}H_2\overset{7}{C}H_2-\overset{8}{O}-\overset{9}{C}H_3$$

2,5,8-Trioxanonane

Question 19.1

Draw the structures of all the isomeric ethers with the molecular formula $C_5H_{12}O$, and provide common and IUPAC names for each isomer.

Question 19.2

Draw the structures of all the isomeric ethers with the formula $C_6H_{14}O$.

B. Cyclic Ethers

When oxygen is part of a cyclic ring system, the result is a cyclic ether. In the most general form, this family of compounds is characterized by the formula:

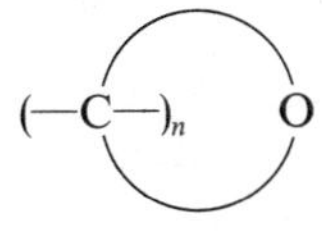

$$(-C-)_n \qquad O$$

Cyclic ethers
$n \geq 2$

Compounds with three-, four, and five-membered rings are shown here along with their common names:

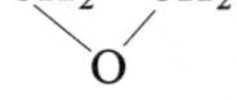

$$\begin{array}{ccc}
CH_2-CH_2 & CH_2-CH_2 & CH_2-CH_2 \\
\diagdown O \diagup & | \qquad | & | \qquad | \\
 & CH_2-O & CH_2 \quad CH_2 \\
 & & \diagdown O \diagup
\end{array}$$

Ethylene   Trimethylene   Tetrahydrofuran
oxide   oxide

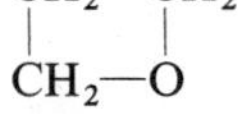

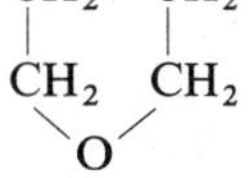

Another common cyclic ether is 1,4-dioxane; the name is a contraction of 1,4-dioxacyclohexane, the IUPAC name for the molecule (see IUPAC nomenclature in Sec. 19.1A,b).

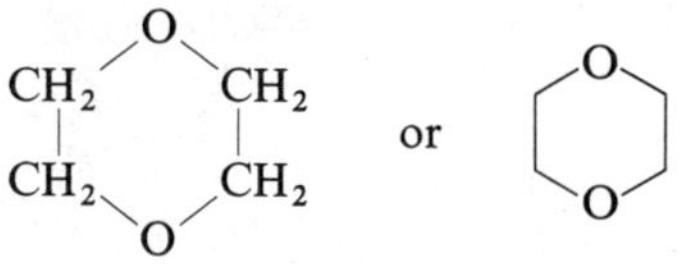

$$\begin{array}{c}
\overset{\displaystyle O}{CH_2 \diagup \diagdown CH_2} \\
| \qquad\qquad | \\
CH_2 \diagdown \diagup CH_2 \\
O
\end{array} \qquad \text{or}$$

1,4-Dioxane or "dioxane"
(IUPAC: 1,4-dioxacyclohexane)

1,4-Dioxane, often called *dioxane*, is used frequently as a solvent in organic reactions.

Cyclic ethers can also be unsaturated, as is the case with furan, which is an aromatic compound (see Sec. 13.7):

$$CH\!\!-\!\!CH$$
$$CH \qquad CH \quad \text{or}$$
$$\diagdown O \diagup$$

Furan

C. Epoxides

The three-membered cyclic ether, ethylene oxide, is the simplest compound in the family of *epoxides*. Epoxides play an especially important role in organic chemistry. As we shall see, they are unusually reactive toward many electrophilic and nucleophilic reagents.

Ethylene oxide also goes by the name *oxirane*, which indicates that the three-membered ring structure contains oxygen. Epoxides can be named by considering them as substituted oxiranes; the names and positions of the substituents serve as prefixes to the oxirane name, where the ring oxygen is numbered 1 in the ring.

IUPAC Nomenclature of Heterocyclic Compounds

IUPAC nomenclature has been extended to the heterocyclic compounds discussed in this and previous chapters. By combining the appropriate suffix and prefix, any heterocyclic can be named unambiguously. The prefixes given here are added to the suffix of the appropriate heteroatom.[1] The suffix for each heteroatom is also given here. Although these names appear straightforward, the common names in these cases are so well entrenched in chemical terminology that the IUPAC names are rarely seen. Their use is increasing, however, particularly in research publications.

$$
\begin{array}{cccc}
\underset{G}{\triangledown} & \underset{G}{\diamondsuit} & \underset{G}{\pentagon} & \underset{G}{\hexagon} \\
\textit{-irane} & \textit{-etane} & \textit{-olane} & \textit{-ane}
\end{array}
$$

$$G = O \qquad \text{Prefix:} \quad ox\text{-}$$
$$ S \qquad\qquad\qquad thi\text{-}$$
$$ P \qquad\qquad\qquad phosph\text{-}$$

Examples:

$$
\begin{array}{ccc}
\underset{O}{\triangledown} & \underset{S}{\diamondsuit} & \underset{P}{\pentagon} \\
\text{Oxirane} & \text{Thietane} & \text{Phospholane}
\end{array}
$$

Epoxides may also be named by calling the oxygen bridge *epoxy* and then indicating the numbers of the carbon atoms to which oxygen is attached; here, however, the parent name of the compound is derived from the longest continuous chain of carbon atoms that bears the epoxy group.

Because epoxides are usually made from the corresponding alkenes (see Sec.

$$\diagup C\!\!=\!\!C \diagdown \xrightarrow[\text{reagents}]{\text{various}} \diagup C\!\!-\!\!C \diagdown$$
$$ \diagdown O \diagup$$

Alkene Epoxide

19.10), they are often named by taking the alkene name and adding the word *oxide*.

---

[1] For the nitrogen heterocyclics, see Chap. 26.

The following examples illustrate these three methods for naming epoxides (common names are in parentheses):

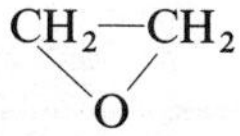

$$CH_2\!-\!CH_2 \quad CH_3\!-\!CH\!-\!CH_2 \quad CH_3\!-\!\underset{\underset{O}{|}}{\overset{CH_3}{\overset{|}{C}}}\!-\!CH_2$$

Oxirane
(ethylene oxide or
epoxyethane)

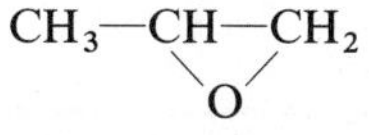

Methyloxirane
(propylene oxide or
1,2-epoxypropane)

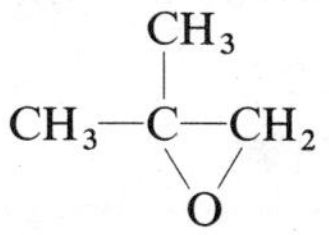

2,2-Dimethyloxirane
(isobutylene oxide or
1,2-epoxy-2-methylpropane)

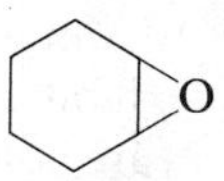

7-Oxabicyclo [4.1.0] heptane
(cyclohexene oxide or
1,2-epoxycyclohexane)

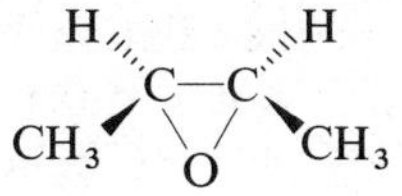

cis-2,3-Dimethyloxirane
(cis-2-butene oxide or
cis-2,3-epoxybutane)

D. Glycols

We had passing contact with glycols in Sec. 8.32 because they are formed in the Baeyer test. They were also discussed in Secs. 10.1 and 10.7. Glycols are compounds that contain two hydroxyl groups on adjacent carbons:

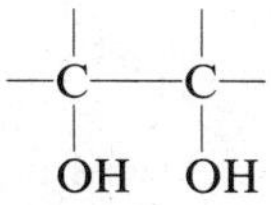

$$-\underset{\underset{OH}{|}}{C}\!-\!\underset{\underset{OH}{|}}{C}-$$

Vicinal diol or glycol

gem-Diols—that is, compounds that have two hydroxyl groups on the same carbon atom—are *usually unstable*; they lose water to form an aldehyde or ketone:

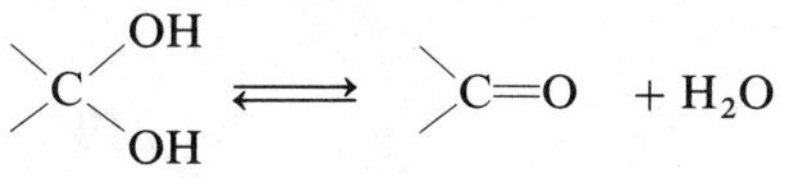

$$\overset{/}{\underset{\diagdown}{C}}\underset{OH}{\overset{OH}{\diagdown}} \quad \rightleftharpoons \quad \overset{\diagdown}{\underset{/}{}}C\!=\!O \quad + H_2O$$

gem-Diol
Unstable

Aldehyde or
ketone

As the following five examples illustrate, glycols may be named by the IUPAC system and some simpler molecules have common names as well.

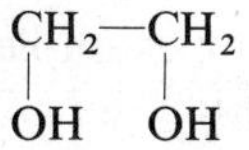

$$CH_2\!-\!CH_2 \qquad CH_3\!-\!\underset{\underset{OH}{|}}{\overset{\overset{CH_3}{|}}{C}}\!-\!CH_2 \qquad CH_2CH_2CH_2CH_2$$
$$\underset{OH}{|} \quad \underset{OH}{|} \qquad\qquad \underset{OH}{|}\quad\underset{OH}{|} \qquad\qquad \underset{OH}{|}\qquad\qquad\underset{OH}{|}$$

1,2-Ethanediol
(ethylene glycol)

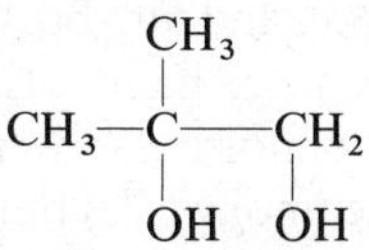

2-Methyl-1,2-propanediol
(isobutylene glycol)

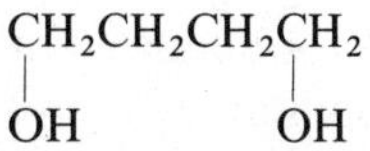

1,4-Butanediol
(tetramethylene glycol)

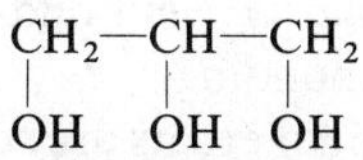

$$CH_2\!-\!CH\!-\!CH_2$$
$$\underset{OH}{|}\quad\underset{OH}{|}\quad\underset{OH}{|}$$

1,2,3-Propanetriol
(glycerol)

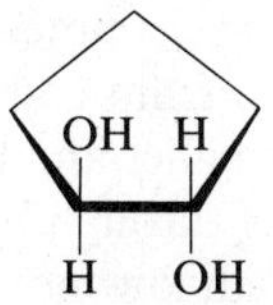

trans-1,2-Cyclopentanediol

Because glycols can be obtained from the corresponding alkenes, they are often given common names which combine the name of the alkene with the word *glycol*. *Ethylene glycol* and *isobutylene glycol*, shown earlier, are examples of this system.

Ethers

19.2   Properties of Ethers

A. Physical Properties

Because of the relationship of ethers to water indicated at the beginning of this chapter, it is not surprising that they exhibit structural similarities to water, especially in the C—O—C bond angle. Ethers are polar and possess dipole moments, and the bond angle about oxygen is near 110°:

~110°

It is suggested that the bond angle in ethers is greater than the 104.5° angle in water because of steric repulsion between the R groups.

The following data indicate the weakly polar nature of an ether molecule, as compared to water or methyl alcohol:

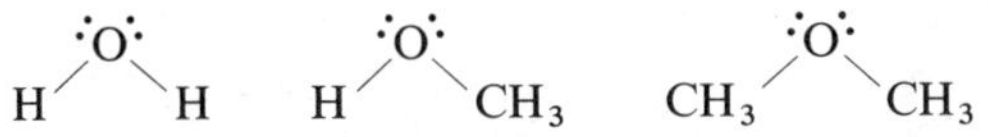

Water	Methyl alcohol	Methyl ether
bp 100°	bp 65°	bp −24°
mp 0°	mp −98°	mp −140°
$\mu = 1.89$ D	$\mu = 1.69$ D	$\mu = 1.29$ D

The polarity of methyl ether is attributed to the unshared pairs of electrons on oxygen as well as to the electron-releasing effect of the methyl groups. The net dipole moment is as shown:

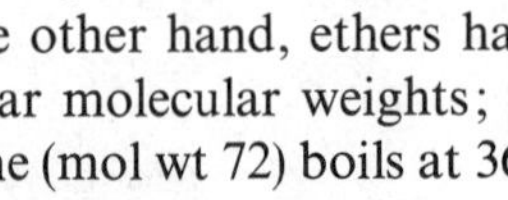

Ethers boil at much lower temperatures than do alcohols that have the same molecular formula. For example, ethyl ether boils at 34.5°, whereas *n*-butyl alcohol, *sec*-butyl alcohol, isobutyl alcohol, and *tert*-butyl alcohol boil at 117.7°, 99.5°, 107.9°, and 82.5°, respectively. On the other hand, ethers have boiling points comparable to those of alkanes with similar molecular weights; for example, ethyl ether (mol wt 70) boils at 34.5° and pentane (mol wt 72) boils at 36.1°. The difference in physical properties between ethers and alcohols of equivalent molecular weights is attributed to the lack of hydrogen bonding in ethers. Even though ethers are polar molecules, as pure compounds in the liquid form they are unable to develop appreciable inter-molecular forces between molecules; the "bulkiness" of the attached R groups prevents the molecules from coming close together (Fig. 19.1). On the other hand, alcohols contain the smaller and highly polar O—H bond, which can associate with neighboring molecules through hydrogen bonding.

On the other hand, low molecular weight ethers are somewhat soluble in water. It is suggested that ether dissolves in water for much the same reason that an alcohol

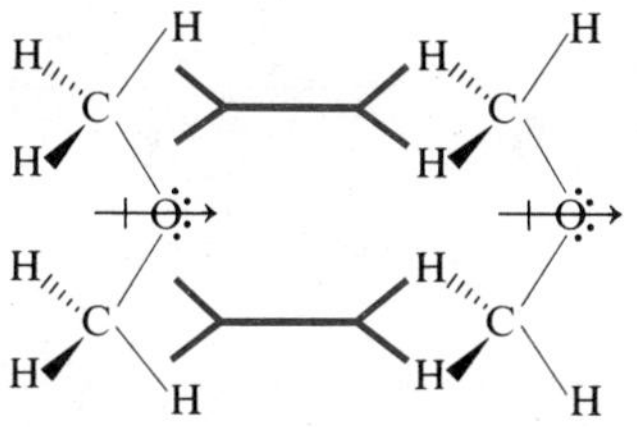

FIGURE 19.1 Structure of methyl ether, showing the inability of various molecules to come close enough together to allow intermolecular attractive forces between molecular dipoles to become important.

does, namely hydrogen bonding between the polar water molecule and the unshared pairs of electrons on oxygen in ROH and ROR:

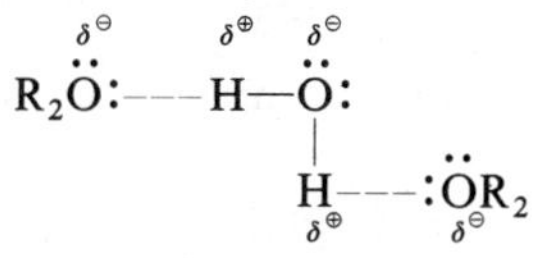

Ether dissolved in water
Hydrogen bonding

Table 19.1 gives the physical properties of selected ethers.

TABLE 19.1 Physical Properties of Selected Ethers

Name	Structure	mp, °C	bp, °C	Density at 20°, g/cc
Methyl ether	$(CH_3)_2O$	−140	−24.9	0.661
Methyl ethyl ether	$CH_3CH_2OCH_3$		7.9	0.725
Ethyl ether	$(CH_3CH_2)_2O$	−116	34.5	0.714
n-Propyl ether	$(CH_3CH_2CH_2)_2O$	−122	90.5	0.736
Isopropyl ether	$[(CH_3)_2CH]_2O$	−60	68	0.735
n-Butyl ether	$(CH_3CH_2CH_2CH_2)_2O$	−95	141	0.768
n-Pentyl ether (n-amyl ether)	$[CH_3(CH_2)_4]_2O$	−69	188	0.775
Vinyl ether	$(CH_2{=}CH)_2O$		39	0.773
Allyl ether	$(CH_2{=}CHCH_2)_2O$		94	0.826
Tetrahydrofuran	$\begin{array}{c} CH_2{-}CH_2 \\ \mid \qquad\quad \rangle O \\ CH_2{-}CH_2 \end{array}$	−108	65.4	0.888
1,4-Dioxane	$\begin{array}{c} CH_2{-}CH_2 \\ O\langle \qquad\quad \rangle O \\ CH_2{-}CH_2 \end{array}$	11	101	1.034
Anisole (methyl phenyl ether)	$C_6H_5OCH_3$	−37.3	154	0.994
Phenetole (ethyl phenyl ether)	$C_6H_5OCH_2CH_3$	−33	172	0.970
Diphenyl ether	$(C_6H_5)_2O$	27	259	1.072

Question 19.3

Suggest an explanation for the dipole moment of ethyl ether being 1.18 D, whereas that for methyl ether is 1.29 D.

B. Basic Properties

The unshared pairs of electrons on ethers impart basic properties to them. They are considered bases in the Lewis sense, because they can donate unshared electron pairs to Lewis acids. This important property of ethers is very useful in organic synthesis as summarized here:

1. Ethers solvate Grignard reagents and thus render them soluble in the solvent (see Sec. 21.10).

2. Ethers form stable molecular complexes with BF_3, for example, $R_2\overset{\cdot\cdot}{\overset{\oplus\ominus}{O}:BF_3}$ (see Sec. 10.4).

3. Hydroboration reactions are often carried out in ether solution, which breaks down diborane, $(BH_3)_2$, into borane: $(BH_3)_2 + 2R_2\overset{\cdot\cdot}{O}: \rightleftarrows 2R_2\overset{\cdot\cdot}{\overset{\oplus\ominus}{O}:BH_3}$ (see Sec. 10.2).

Later in this chapter we discuss another reaction—the cleavage of ethers in acidic solution—which involves protonation of the unshared electron pairs; here also ether serves as a base (a proton acceptor) in the Lowry-Brønsted sense (see Sec. 19.9):

$$R_2\overset{\cdot\cdot}{O}: + \ H^\oplus \ \rightleftarrows \ R_2\overset{\oplus}{\overset{\cdot\cdot}{O}:H}$$

$$\text{Base} \qquad \text{Acid} \qquad\qquad \text{Oxonium ion}$$

19.3   Uses and Occurrence of Ethers

The most important ether, **ethyl ether,** $(CH_3CH_2)_2O$, or simply *ether*, is a solvent in many organic reactions, including the Grignard reaction. It is frequently used as an extraction solvent for removing an organic product from an aqueous solvent, but the solubility of ether in water (7.5%) results in large losses of solvent during extraction. Ether has another advantage in that its low boiling point (34.5°) allows it to be removed easily from other less volatile organic compounds. On the other hand, ether is highly flammable and thus presents tremendous potential fire danger. *Ether should never be heated with an open flame* or in an open system on a hot plate. Ether vapor is denser than air and it tends to "creep" across laboratory benches and floors to other places where it can be ignited. When ethers are stored for long periods, they may undergo air oxidation to form extremely dangerous, explosive peroxides. Precautions should be taken to assure that ethers are free of peroxides before use. A good, safe procedure is to date all ether containers when they are first opened.

Ether has been used since 1842 as an anesthesic during surgery, but it must be administered with great care because of its flammability and volatility. The possible generation of static electricity must be avoided. *Divinyl ether,* $(CH_2{=}CH)_2O$, is often used in a like manner, but because it is about seven times more potent than ether, care must be exercised to avoid overdoses of it. Divinyl ether often goes by the more common name vinethene.

Tetrahydrofuran (THF), a five-membered cyclic ether, $(CH_2)_4O$, is also an important solvent for organic reactions, largely because it is miscible with water yet can be recovered from an aqueous solution by distillation. It is less volatile than ether and is the solvent of choice for preparing the Grignard reagent from chlorobenzene (see Sec. 21.10).

Several naturally occurring phenolic ethers occur in plants, and they are recognized by their flavors and scents. Four examples are:

Vanillin (vanilla)   Eugenol (cloves)   Isoeugenol (nutmeg)   Safrole (sassafras)

Vitamin E is also a phenolic ether. Its role in metabolism is unclear, but it appears to retard the oxidation of fats and oils and is also reported to be an aphrodisiac.

Vitamin E

Several other alkyl aryl ethers are produced commercially for agricultural uses. Two are:

1,1,1-Trichloro-2,2-di(p-methoxyphenyl)ethane (methoxychlor)
Insecticide

2,4-Dichlorophenoxyethanoic acid (2,4-D)
Herbicide

One component that plays a large role in the narcotic activity of *marijuana* is tetrahydrocannabinol (THC). The dried flowers of the *Indian hemp* plant produce a tobaccolike material called *hashish* or *marijuana*. Although marijuana is not listed as a habit-forming or addictive drug, its use and sale are illegal. THC is a cyclic phenolic ether with the structure:

Tetrahydrocannabinol
(THC)

Question 19.4

Olivetol [3,5-dihydroxy-1-(1-pentyl)benzene] can be converted to THC by an acid-catalyzed cyclization with the terpene *citral*. Locate the two isoprene units of citral and the olivetol skeleton in the preceding THC structure. Can you draw citral and olivetol from these? Attempt to draw the cyclization mechanism.

19.4   Preparation of Ethers from Alcohols: Dialkyl Ethers

Lower molecular weight ethers are most often prepared from the corresponding alcohols, but the following general reaction is limited to the preparation of *symmetrical* ethers:

$$2\ R\!-\!O\!-\!H \quad \xrightarrow[140°]{H_2SO_4} \quad R\!-\!O\!-\!R\ +\ H_2O$$

Excess   Symmetrical
R = primary   ether
or secondary
R may not be aryl

For reasons discussed later, this reaction is poor for tertiary alcohols. Also, careful control of the temperature is important for minimizing side reactions.

Neither diaryl ethers nor alkyl aryl ethers can be prepared from the corresponding phenols by this method. Attempts to prepare aryl ethers by this route result in sulfonation of the aromatic ring. Methods for preparing alkyl aryl ethers and diaryl ethers are discussed in Sec. 19.6.

A. Mechanism of Dialkyl Ether Formation

Two substitution mechanisms explain how dialkyl ethers are formed; primary alcohols react by one mechanism and secondary and tertiary alcohols by another. In the general reaction a molecule of water is lost during ether formation, and we must account for this in the mechanisms. Ether formation requires the presence of acid, and as in the conversion of an alcohol to an alkyl halide, acid protonates the alcohol before substitution can occur:

$$R\ddot{O}H + H^{\oplus} \rightleftharpoons R\!-\!\overset{\oplus}{\underset{H}{O}}\!\overset{H}{\vdots}$$

Oxonium
ion

Methanol and primary alcohols appear to react by the S_N2 mechanism because there usually is no molecular rearrangement in converting a primary alcohol to the corresponding ether. (This parallels the trend observed in converting a primary alcohol to the corresponding alkyl halide; see Sec. 10.13.) Nucleophilic attack by the alcohol on a molecule of protonated alcohol gives the ether; an alcohol molecule is substituted for a water molecule. This typical S_N2 mechanism is

$$R-CH_2-\overset{..}{\underset{H}{\overset{Nu:}{\overset{..}{O}}}} + R\overset{\oplus}{\underset{H}{CH_2-\overset{/H}{\overset{..}{O}}}} \xrightarrow{\text{slow}} \left[R-CH_2-\overset{\delta\oplus}{\underset{H}{O}}---\overset{}{\underset{R}{CH_2}}---\overset{\delta\oplus}{O}\overset{H}{\underset{H}{\diagdown}} \right]$$

1° Alcohol	Protonated 1° alcohol		Transition state

$$R-CH_2-\overset{..}{\underset{..}{O}}-CH_2-R + H^{\oplus} \rightleftharpoons R-CH_2-\overset{\oplus}{\underset{\underset{H}{|}\,\underset{R}{|}}{O}}-CH_2 + H_2\overset{..}{\underset{..}{O}}:$$

Symmetrical ether	Protonated ether

Secondary and tertiary alcohols, on the other hand, appear to react via the S_N1 mechanism; molecular rearrangements are observed. The following mechanism for a secondary alcohol could occur equally well for a tertiary alcohol. As before, the alcohol is protonated in a prior equilibrium reaction before undergoing substitution:

$$R_2CH-\overset{..}{\underset{..}{O}}H + H^{\oplus} \rightleftharpoons R_2CH-\overset{\oplus}{\overset{..}{O}}\overset{/H}{\underset{H}{\diagdown}}$$

and then

$$R_2CH-\overset{\oplus}{\overset{..}{O}}\overset{/H}{\underset{H}{\diagdown}} \xrightarrow{\text{slow}} R_2\overset{\oplus}{CH} + H_2\overset{..}{\overset{..}{O}}:$$

$$\Big\downarrow \text{fast} \quad R_2CH-\overset{..}{O}H$$

$$R_2CH-\overset{..}{\underset{..}{O}}-CHR_2 + H^{\oplus} \rightleftharpoons R_2CH-\overset{\oplus}{\underset{\underset{H}{|}}{\overset{..}{O}}}-CHR_2$$

Symmetrical ether

B. Competition Between Ether Formation and Elimination

One major difficulty with using the acid-catalyzed reaction on alcohols to form ethers is the competition between nucleophilic substitution to give an ether and elimination to give an alkene or a mixture of alkenes. Also, if an S_N1 or E1 mechanism is involved, the carbocation intermediate may react with other nucleophiles in the solution (for example, bisulfate ion) to give undesired side products. Indeed, when tertiary alcohols are used, polymers often result. Carbocation rearrangements are also possible.

C. Experimental Control of Ether Formation Versus Elimination

Because of the possible competition between ether formation and elimination, what does the organic chemist do to favor one product or the other? The experimental conditions, usually reaction temperature and concentration of reactants, can be varied.

Elimination is favored at higher temperatures and when the concentration of the catalyst is increased. For example, ethanol is dehydrated to ethene at about 180°:

$$CH_3CH_2OH \xrightarrow[180°]{\text{conc } H_2SO_4} CH_2{=}CH_2$$

Ethanol   Ethene

Polymerization is minimized by removing the ethene as soon as it is produced.

Ether formation is favored at lower temperatures when there is an excess of alcohol. Recall that ether formation by the S_N1 *or* S_N2 mechanism requires that an unprotonated molecule of alcohol serve as a nucleophile and attack either a carbocation (S_N1 reaction) or a protonated alcohol molecule (S_N2 mechanism). Thus it is imperative to use an excess of alcohol. The reaction is normally carried out by placing some concentrated sulfuric acid in a reaction vessel and continually adding alcohol to it to ensure that there is an excess. The use of excess alcohol is usually no problem because most low molecular weight alcohols are commercially available at low cost. For example, *n*-butyl ether is prepared from excess 1-butanol at about 140°:

$$CH_3CH_2CH_2CH_2OH \xrightarrow[140°]{\text{conc } H_2SO_4} (CH_3CH_2CH_2CH_2)_2O$$

1-Butanol   *n*-Butyl ether
(*n*-butyl alcohol)
Excess

Question 19.5

Suppose you want to prepare propyl *n*-butyl ether. Suggest why the reaction between 1-propanol and 1-butanol in the presence of concentrated sulfuric acid is not a feasible route. (This is generally true for the attempted preparation of unsymmetrical ethers, ROR′.) What products are formed?

Question 19.6

What alcohols might be used to prepare the following ethers?

(*a*) isobutyl ether   (*b*) benzyl ether $(C_6H_5CH_2)_2O$
(*c*) $(ClCH_2CH_2)_2O$ (*d*) tetrahydrofuran

19.5   General Method for Ether Synthesis: Williamson Method

The direct conversion of an alcohol to the corresponding ether has two fairly stringent limitations. First, it yields only symmetrical ethers containing alkyl groups, and second, it produces molecular rearrangements with certain secondary alcohols.

A more convenient and general method for ether preparation is the **Williamson synthesis,** which involves the reaction between an alkyl halide and an alkoxide or a phenoxide ion:

$$R{-}\overset{..}{\underset{..}{O}}{:}^{\ominus} \; + \; R'{-}\overset{..}{\underset{..}{X}}{:} \longrightarrow R{-}\overset{..}{\underset{..}{O}}{-}R' \; + \; {:}\overset{..}{\underset{..}{X}}{:}^{\ominus}$$

Alkoxide ion   Alkyl halide   Dialkyl ether

where R and R′ may be different; R′ must be methyl, primary, or secondary

$$Ar{-}\overset{..}{\underset{..}{O}}{:}^{\ominus} + R'{-}\overset{..}{\underset{..}{X}}{:} \longrightarrow Ar{-}\overset{..}{\underset{..}{O}}{-}R' + {:}\overset{..}{\underset{..}{X}}{:}^{\ominus}$$

Phenoxide ion   Alkyl aryl ether

where R′ must be methyl, primary, or secondary

These are typical nucleophilic substitution reactions that occur via the S_N2 mechanism:

$$R—\overset{..}{\underset{..}{O}}{:}^{\ominus}$$

or $$\overset{|}{\underset{|}{C}}—X{:}$$ or

$$Ar—\overset{..}{\underset{..}{O}}{:}^{\ominus}$$

$$\boxed{Nu{:}^{\ominus}}$$

S_N2 Mechanism

$$R—\overset{..}{\underset{..}{O}}—\overset{|}{\underset{|}{C}}—$$

or

$$Ar—\overset{..}{\underset{..}{O}}—\overset{|}{\underset{|}{C}}—$$

The nucleophiles are $R—\overset{..}{\underset{..}{O}}{:}^{\ominus}$ and $Ar—\overset{..}{\underset{..}{O}}{:}^{\ominus}$.

Because the Williamson ether synthesis occurs by the S_N2 mechanism, it is not surprising that structural limitations must be placed on the alkyl halide. Alkoxide ion and phenoxide ion are good nucleophiles, but they are also strong bases, so elimination is an important side reaction. Recall from Sec. 9.15 and 9.17 that the reactivity order of alkyl halides toward base-promoted elimination (E1 and E2) is $3° \ R_3CX > 2° \ R_2CHX > 1° \ RCH_2X$. In practice, tertiary alkyl halides undergo elimination exclusively so that *the Williamson method is limited to methyl, primary, and secondary alkyl halides.*

Experimentally, the alcohol is often used as a solvent as well as to form the alkoxide ion. The corresponding alkoxide ion can be prepared in situ by carefully adding a weighed amount of sodium metal to the anhydrous solvent.

The Williamson method has the distinct advantage of allowing for the preparation of unsymmetrical dialkyl ethers and alkyl aryl ethers. Even better, the starting reagents can often be derived from readily available alcohols or phenols. The following two examples illustrate how *dialkyl ethers* can be synthesized.

Example 1

Synthesize ethyl *n*-butyl ether from any desired alcohols.

This compound can be prepared by either of the following two routes. Particular note should be made of the conversion of the alcohol to the corresponding alkoxide ion.

$$CH_3CH_2OH \xrightarrow[\text{metal}]{Na} CH_3CH_2\overset{..}{\underset{..}{O}}{:}^{\ominus}Na^{\oplus}$$

Ethanol   Ethoxide   mix
(ethyl alcohol)   ion   together

$$CH_3CH_2CH_2CH_2OH \xrightarrow{PBr_3} CH_3CH_2CH_2CH_2Br$$

1-Butanol   1-Bromobutane
(*n*-butyl alcohol)   (*n*-butyl bromide)

$$CH_3CH_2—O—CH_2CH_2CH_2CH_3$$

1-Ethoxybutane
(ethyl *n*-butyl ether)

or

$$CH_3CH_2OH \xrightarrow[\text{heat}]{P, I_2} CH_3CH_2I$$

Iodoethane   mix
(ethyl iodide)   together

$$CH_3CH_2CH_2CH_2OH \xrightarrow[\text{metal}]{Na} CH_3CH_2CH_2CH_2\overset{..}{\underset{..}{O}}{:}^{\ominus}Na^{\oplus}$$

n-Butoxide ion

$$CH_3CH_2—O—CH_2CH_2CH_2CH_3$$

Example 2

Synthesize *n*-butyl *tert*-butyl ether from any desired alcohols.

Superficially, it may appear that this compound can be prepared by either of the following two routes:

$$CH_3CH_2CH_2CH_2OH \xrightarrow[\text{heat}]{\overset{\text{NaBr}}{\text{H}_2\text{SO}_4,}} CH_3CH_2CH_2CH_2Br$$

1-Butanol
(*n*-butyl alcohol)

1-Bromobutane
(*n*-butyl bromide)

$$CH_3 \overset{\overset{\displaystyle CH_3}{|}}{\underset{\underset{\displaystyle CH_3}{|}}{C}} OH \xrightarrow[\text{metal}]{K} CH_3 \overset{\overset{\displaystyle CH_3}{|}}{\underset{\underset{\displaystyle CH_3}{|}}{C}} \overset{..}{\overset{\ominus}{O}} :K \overset{\oplus}{}$$

2-Methyl-2-propanol
(*tert*-butyl alcohol)

tert-Butoxide ion

mix together

$$CH_3 \overset{\overset{\displaystyle CH_3}{|}}{\underset{\underset{\displaystyle CH_3}{|}}{C}} O-CH_2CH_2CH_2CH_3$$

2,2-Dimethyl-3-oxaheptane
(*n*-butyl *tert*-butyl ether)
Obtained in good yield

or

$$CH_3CH_2CH_2CH_2OH \xrightarrow[\text{metal}]{Na} CH_3CH_2CH_2CH_2\overset{..}{\underset{..}{O}}:\overset{\ominus}{} Na \overset{\oplus}{}$$

n-Butoxide ion

$$CH_3 \overset{\overset{\displaystyle CH_3}{|}}{\underset{\underset{\displaystyle CH_3}{|}}{C}} OH \xrightarrow[\text{HBr}]{\text{conc}} CH_3 \overset{\overset{\displaystyle CH_3}{|}}{\underset{\underset{\displaystyle CH_3}{|}}{C}} Br$$

2-Bromo-2-methylpropane
(*tert*-butyl bromide)

mix together

$$CH_3 \overset{\overset{\displaystyle CH_3}{|}}{C}=CH_2 + CH_3CH_2CH_2CH_2OH + :\overset{..}{\underset{..}{Br}}:\overset{\ominus}{}$$

2-Methylpropene
(isobutylene)

No n-butyl tert-butyl ether produced

The first route proceeds as expected; *tert*-butoxide reacts smoothly with *n*-butyl bromide, a primary alkyl halide, to give the desired ether with a minimum of side product (elimination). The second route, however, involves the reaction between *n*-butoxide and *tert*-butyl bromide, a tertiary alkyl halide, which undergoes elimination exclusively.

Mechanistic considerations must also be used in evaluating possible routes to *alkyl aryl ethers*, which are prepared as an adjunct of the Williamson method. However, only one possible combination of reagents works: A phenoxide ion must be allowed to react with an alkyl halide. This is feasible from a mechanistic viewpoint because Ar—$\overset{..}{\underset{..}{O}}$:$\overset{\ominus}{}$ serves as a nucleophile and displaces a halide ion from an alkyl halide.

S_N2 Mechanism:

$$Ar-\overset{..}{\underset{..}{O}}:^{\ominus} + \overset{|}{\underset{|}{C}}-\overset{..}{\underset{..}{X}}: \longrightarrow Ar-\overset{..}{\underset{..}{O}}-\overset{|}{\underset{|}{C}}-$$

Phenoxide   Methyl   Alkyl aryl
ion   1° or 2°   ether
           Alkyl halide

Consider, on the other hand, the reaction between an alkoxide ion and an aryl halide. Knowing the extreme difficulty of carrying out nucleophilic displacement reactions on aryl halides from Secs. 16.5 and 16.8, it is not surprising that this route most often fails; at best, reaction occurs with great difficulty.

Nucleophilic aromatic substitution:

$$R-\overset{..}{\underset{..}{O}}:^{\ominus} + Ar-\overset{..}{\underset{..}{X}}: \xrightarrow{\text{slow}} \left[Ar \overset{\overset{..}{\underset{..}{O}}-R}{\underset{\underset{..}{X}:}{}} \right]^{\ominus} \xrightarrow{\text{fast}} Ar-\overset{..}{\underset{..}{O}}-R + :\overset{..}{\underset{..}{X}}:^{\ominus}$$

Alkoxide   Aryl   Alkyl
ion   halide   aryl ether

Occurs with great difficulty

As in the preparation of dialkyl ethers, alkyl aryl ethers are obtainable from phenols and alcohols that are commercially available or can be synthesized in the laboratory. With these limitations firmly in mind, consider the method that might be used to prepare *n*-butyl phenyl ether:

Phenol $\xrightarrow{\text{NaOH}}$ Phenoxide ion

$CH_3CH_2CH_2CH_2OH \xrightarrow{P, I_2} CH_3CH_2CH_2CH_2I$
1-Butanol (*n*-butyl alcohol)   1-Iodobutane (*n*-butyl iodide)

mix together → —O—$CH_2CH_2CH_2CH_3$
1-Phenoxybutane (*n*-butyl phenyl ether)

The alternative—allowing bromobenzene to react with *n*-butoxide ion—fails completely unless very vigorous conditions are used.

The synthesis of *diaryl ethers* poses a much more challenging problem; the only way they can be made is by allowing an aryloxide ion to react with an aryl halide.

Nucleophilic aromatic substitution:

$$Ar-\overset{..}{\underset{..}{O}}:^{\ominus} + Ar'-\overset{..}{\underset{..}{X}}: \longrightarrow Ar-O-Ar' + :\overset{..}{\underset{..}{X}}:^{\ominus}$$

Question 19.7

Starting with any alcohols or phenol or toluene, provide the best possible synthesis for the following. Choose a method that minimizes side reactions.

(*a*) *n*-butyl isobutyl ether
(*b*) *n*-propyl isopropyl ether
(*c*) methyl phenyl ether (anisole)

(*d*) *p*-tolyl *n*-propyl ether, CH_3—$\langle\text{benzene ring}\rangle$—O—$CH_2CH_2CH_3$

(*e*) cyclopentyl *tert*-butyl ether
(*f*) optically active 2-methoxybutane

Question 19.8

Phenyl ether, C_6H_5—O—C_6H_5, is formed as a side product in the commercial preparation of phenol from the reaction between chlorobenzene and aqueous sodium hydroxide at high temperature and pressure (the Dow process). Provide a complete mechanism to show this.

omit

19.6   Variations in Williamson Ether Synthesis

A. Using Tosylates

Although the normal approach for ether synthesis makes use of alkyl halides, another functional group closely related to alcohols can be employed. *Tosylates*, ROTs, which are prepared from alcohols and *p*-toluenesulfonyl chloride in the presence of base, also react with alkoxide ions and yield ethers.

The correct name of the sulfonate ester (see Secs. 10.12 and 23.6 for a discussion of esters) is *alkyl p-toluenesulfonate*, but *p*-toluenesulfonate is abbreviated as **tosylate** and often symbolized as —**OTs.** In the laboratory, tosylates are *actually* synthesized from an alcohol and *p*-toluenesulfonyl chloride (1) in the presence of a weak base,[1] B:. The reaction is:

$$R-O-H \ + \ Cl-\overset{\overset{O}{\|}}{\underset{\underset{O}{\|}}{S}}-\!\!\!\langle\bigcirc\rangle\!\!-CH_3 \xrightarrow[\text{(base)}]{B:} R-O-\overset{\overset{O}{\|}}{\underset{\underset{O}{\|}}{S}}-\!\!\!\langle\bigcirc\rangle\!\!-CH_3 + :\overset{\ominus}{\underset{..}{Cl}}: + \overset{\oplus}{B}:H$$

R may not be tertiary	(1) *p*-Toluenesulfonyl chloride (*abbreviated* *TsCl*)	Alkyl *p*-toluenesulfonate (*tosylate, abbreviated R—OTs*)

(For now, this reaction must be accepted as fact; it is discussed in more detail in Sec. 23.6.) Tosylate formation involves breaking only the oxygen-hydrogen bond in the alcohol, so it occurs with *complete retention of configuration* at the carbon of the alcohol.

When tosylates react with various nucleophiles, the entire tosylate group departs. The equation for this general displacement reaction is:

$$\overset{\ominus}{Nu}: + R-\overset{..}{\underset{..}{O}}-\overset{\overset{O}{\|}}{\underset{\underset{O}{\|}}{S}}-\!\!\!\langle\bigcirc\rangle\!\!-CH_3 \longrightarrow Nu-R + :\overset{\ominus}{\underset{..}{O}}-\overset{\overset{O}{\|}}{\underset{\underset{O}{\|}}{S}}-\!\!\!\langle\bigcirc\rangle\!\!-CH_3$$

$$\underbrace{\qquad\qquad\qquad}_{\equiv\ —OTs} \qquad\qquad \underbrace{\qquad\qquad\qquad}_{\equiv\ Ts\overset{..\ominus}{\underset{..}{O}}:}$$

Leaving group

---

[1] One of the most commonly used weak bases in this reaction is *pyridine*, C_5H_5N, which reacts with and removes protons from the reaction mixture as they are produced; the reaction is:

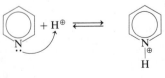

Pyridine *Base*	Pyridinium ion

This type of reaction is discussed in Sec. 26.5.

Tosylates are readily prepared in good yield from most primary and secondary alcohols; however, they usually cannot be prepared from tertiary alcohols.

Symmetrical or unsymmetrical alkyl or alkyl aryl ethers are obtained by this method, as illustrated by the synthesis of *n*-butyl ether:

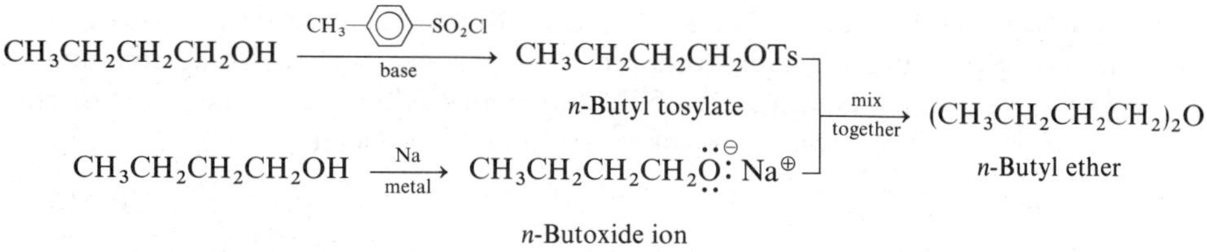

$$CH_3CH_2CH_2CH_2OH \xrightarrow[\text{base}]{CH_3-\bigcirc-SO_2Cl} CH_3CH_2CH_2CH_2OTs$$

n-Butyl tosylate

$$CH_3CH_2CH_2CH_2OH \xrightarrow[\text{metal}]{Na} CH_3CH_2CH_2CH_2\ddot{O}:^{\ominus} Na^{\oplus}$$

n-Butoxide ion

$$\xrightarrow[\text{together}]{\text{mix}} (CH_3CH_2CH_2CH_2)_2O$$

n-Butyl ether

Using tosylates for ether formation has several advantages over alkyl halides. First, tosylates are usually easy to obtain in good yield. Second, tosylates are more reactive than alkyl bromides toward nucleophilic substitution, and they give less elimination with alkoxide ion than do alkyl halides.

Tosylates are also used to allow the stereospecific displacement of an —OH group by various nucleophiles. That tosylates have the same configuration as the alcohol from which they were prepared makes them very useful for stereospecific reactions. For example, optically active 2-octanol can be converted to optically active 2-octyl tosylate, which on reaction with a variety of nucleophiles under S_N2 conditions gives other optically active 2-octyl compounds. In the reaction shown here, 2-octyl tosylate reacts with bromide ion under S_N2 conditions (solvents such as acetone or dimethyl formamide are frequently chosen for this purpose) to give optically active 2-bromooctane. Tosylate formation occurs with retention of configuration, and the S_N2 reaction occurs with complete inversion of configuration, so that this two-step reaction results in complete inversion of configuration (comparing, of course, the configurations of 2-octanol and 2-bromooctane).

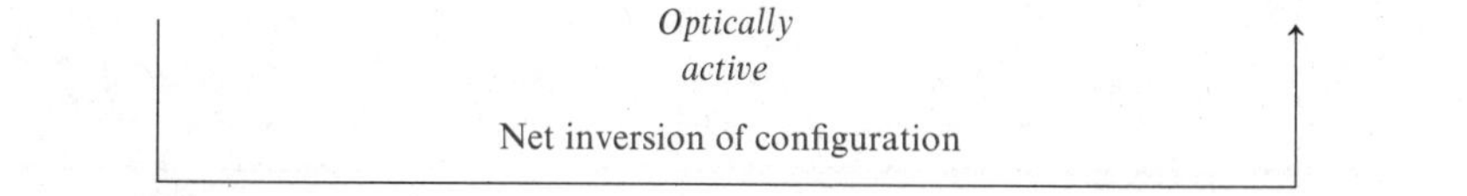

2-Octanol
Optically active

2-Octyl tosylate
Optically active

2-Bromooctane
Optically active

Net inversion of configuration

Tosylates can also be used to introduce functional groups into a molecule that are difficult to introduce by other routes. For example, an alcohol cannot be converted directly into an alkyl fluoride, but treatment of a tosylate with fluoride ion allows this to be accomplished in good yield and without rearrangement; for example:

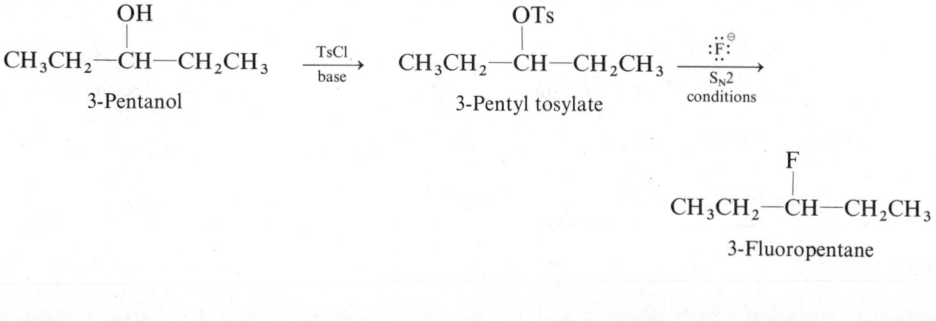

$$\underset{\text{3-Pentanol}}{CH_3CH_2-\overset{\overset{\displaystyle OH}{|}}{CH}-CH_2CH_3} \xrightarrow[\text{base}]{TsCl} \underset{\text{3-Pentyl tosylate}}{CH_3CH_2-\overset{\overset{\displaystyle OTs}{|}}{CH}-CH_2CH_3} \xrightarrow[\text{conditions}]{\ddot{:}\ddot{F}\ddot{:}^{\ominus} \\ S_N2}$$

$$\underset{\text{3-Fluoropentane}}{CH_3CH_2-\overset{\overset{\displaystyle F}{|}}{CH}-CH_2CH_3}$$

Treatment of tosylates with bisulfide ion, $H\ddot{S}\colon^{\ominus}$, gives mercaptans, R—SH, and with cyanide ion, $\colon\!C\!\equiv\!N\colon^{\ominus}$, gives nitriles, $R\!-\!C\!\equiv\!N$, and so on.

Another advantage of this reaction is the absence of molecular rearrangements. For example, treatment of 3-methyl-2-butanol with concentrated HCl produces only 2-chloro-2-methylbutane (see Sec. 10.13). However, conversion of this alcohol to the corresponding tosylate, followed by treatment with chloride ion under S_N2 conditions, affords a high yield of 2-chloro-3-methylbutane. None of the rearranged isomer is observed. This is shown in the following scheme:

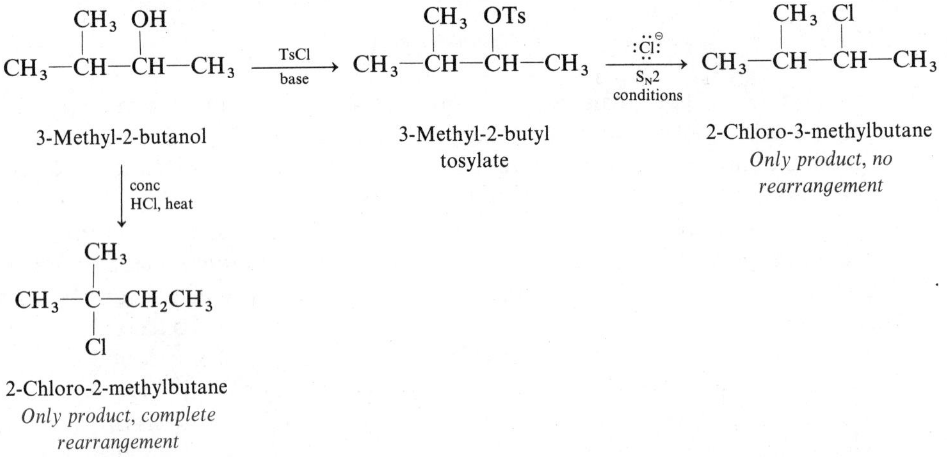

3-Methyl-2-butanol

3-Methyl-2-butyl tosylate

2-Chloro-3-methylbutane
Only product, no rearrangement

2-Chloro-2-methylbutane
Only product, complete rearrangement

Sulfonates as Leaving Groups

Many synthetic investigations use tosylates as the leaving group because of the ease with which they are obtained from readily available alcohols. Like alkyl halides, secondary tosylates often dissociate into carbocations under S_N1 conditions (polar, protic solvents, such as formic and acetic acid and methanol and ethanol, are used to favor carbocation formation). Certain other aryl sulfonate esters, such as the two shown here, can be prepared from alcohols, and they behave analogously to tosylates.

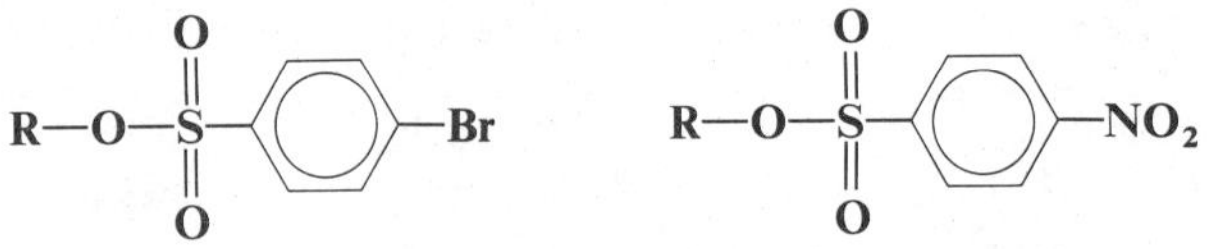

Alkyl *p*-bromobenzenesulfonate
"*Brosylate*," *abbreviated R—OBs*

Alkyl *p*-nitrobenzenesulfonate
"*Nosylate*," *abbreviated R—ONs*

Question 19.9

Based on your knowledge of the electron-withdrawing and electron-donating abilities of various substituents attached to an aromatic ring, which of the three—tosylate, brosylate, or nosylate—do you think would be the *best* leaving group and which the poorest leaving group?

Question 19.10

Trace out these two reactions by drawing suitable three-dimensional structural formulas:

(*a*) *trans*-4-Methylcyclohexanol $\xrightarrow[\text{base}]{\text{TsCl}}$ tosylate $\xrightarrow[S_N2]{CH_3\ddot{O}\colon^{\ominus}}$ product

(*b*) Optically active 2-butanol $\xrightarrow[\text{base}]{\text{TsCl}}$ tosylate $\xrightarrow[S_N2]{CH_3\ddot{O}\colon^{\ominus}}$ product

Question 19.11

It is well known that a primary alcohol, such as 1-butanol, reacts readily with HBr in the presence of concentrated sulfuric acid to form the corresponding alkyl bromide. Yet when the same reaction is carried out with H_2S or HF or HCN, no substitution product is obtained. Explain why this is so, taking into account that mercaptans (R—SH), alkyl fluoride (R—F), and alkyl cyanides (R—CN) are prepared from tosylates in the manner described in this section.

B. Using Dimethyl Sulfate

A specialized variation of the previous tosylate reaction is the use of dimethyl sulfate, $CH_3OSO_2OCH_3$ or $(CH_3O)_2SO_2$, which is the dimethyl ester of sulfuric acid (see Sec. 10.12). This molecule is especially reactive toward nucleophilic substances, and when alkoxide or phenoxide ions are used the corresponding methyl ethers are produced:

$$2R\ddot{O}\colon^{\ominus} (\text{or } 2Ar\ddot{O}\colon^{\ominus}) + CH_3 - \overset{\overset{O}{\parallel}}{\underset{\underset{O}{\parallel}}{\ddot{O} - S}} - O - CH_3 \longrightarrow 2R\ddot{O}CH_3 (\text{or } 2Ar\ddot{O}CH_3) + SO_4^{\ominus 2}$$

Dimethyl sulfate

Mechanistically, this reaction occurs in a manner analogous to the displacement reactions of tosylates (see Sec. 19.6A), except that both methyl groups are amenable to substitution.

19.7   Formation of Cyclic Ethers and Epoxides by the Williamson Method

The synthesis of certain cyclic ethers can be accomplished by the Williamson method. The reaction is novel in that it is intramolecular; that is, the alkoxide ion is generated in the molecule containing a halogen. A hydroxy alkyl halide reacts with aqueous sodium hydroxide to generate some alkoxide ion, which displaces the halide ion by an S_N2 mechanism; for example:

4-Chloro-1-butanol

Oxolane
(tetrahydrofuran)

Epoxides can also be synthesized by this type of reaction. For example, oxirane (ethylene oxide) can be obtained from ethene via the following route, which starts with the addition of aqueous chlorine (the addition of the elements of hypochlorous acid, HOCl) across the double bond (see Sec. 8.22):

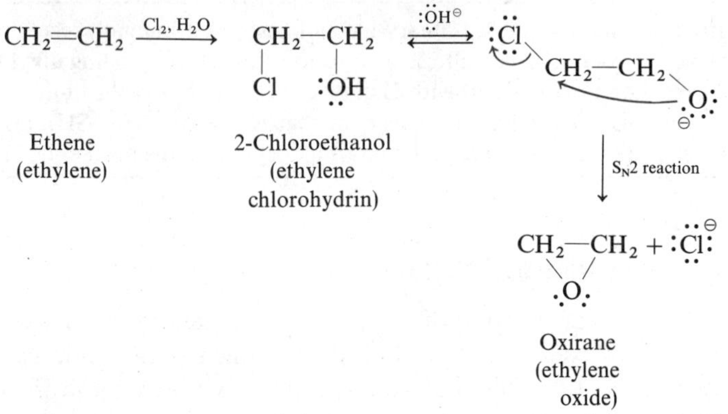

Ethene
(ethylene)

2-Chloroethanol
(ethylene
chlorohydrin)

S_N2 reaction

Oxirane
(ethylene
oxide)

In an analogous manner, cyclohexene can be converted to cyclohexene oxide (1). This conversion involves the *anti*-addition of aqueous chlorine to cyclohexene, and the backside S_N2 attack of the alkoxide ion formed from the chlorohydrin:

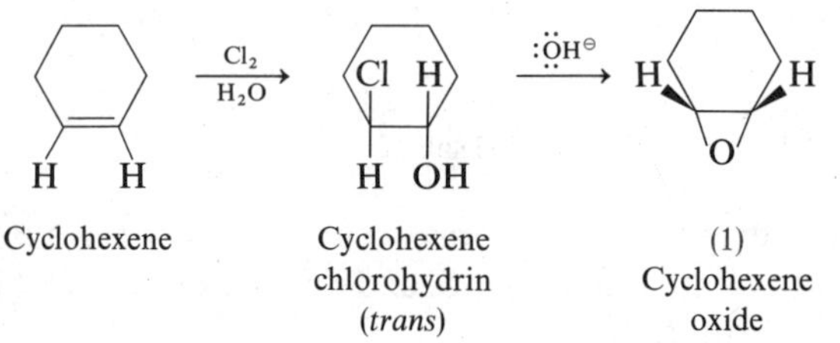

Cyclohexene

Cyclohexene
chlorohydrin
(*trans*)

(1)
Cyclohexene
oxide

Other routes for producing epoxides are described in Sec. 19.10.

19.8   C- Versus O-Alkylation of Phenoxide Ions

The discussion about the preparation of alkyl aryl ethers in Secs. 19.5 and 19.6 is fairly accurate for most alkyl halides. However, the reaction can be changed so that the alkyl group becomes attached directly to the ring (**C-alkylation**) rather than to oxygen (**O-alkylation**). C-alkylation is favored with particularly reactive alkyl halides, such as allyl halides, and by nonpolar solvents or solvents that promote strong hydrogen bonding (protic solvents).

On the other hand, O-alkylation is favored by highly polar, aprotic solvents, even with reactive alkyl halides. The following reaction illustrates C- and O-alkylation:

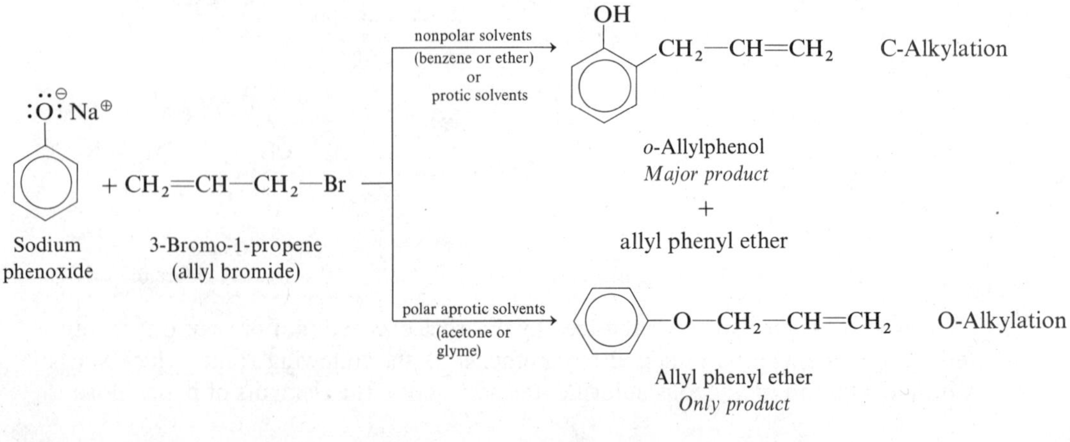

nonpolar solvents
(benzene or ether)
or
protic solvents

OH

CH_2—CH=CH_2 C-Alkylation

o-Allylphenol
Major product

+

allyl phenyl ether

Sodium
phenoxide

3-Bromo-1-propene
(allyl bromide)

polar aprotic solvents
(acetone or
glyme)

O—CH_2—CH=CH_2 O-Alkylation

Allyl phenyl ether
Only product

The mechanism of the formation of *o*-allylphenol from phenoxide ion is as follows:

Phenoxide
ion, resonance structures

C-alkylation competes with O-alkylation, so that appreciable amounts of both allyl phenyl ether and *o*-allylphenol are obtained from sodium phenoxide and allyl bromide in a nonpolar solvent.

Question 19.12

In the **Reimer-Tiemann reaction**, phenol reacts with $CHCl_3$ in aqueous $\overset{\cdot\cdot}{HO}\overset{\ominus}{:}$; an *o*-hydroxybenzaldehyde results:

Phenol   *o*-Hydroxybenzaldehyde

In the first step of the reaction, Cl_2C: (dichlorocarbene) is produced, it then reacts with phenoxide ion to give C-alkylation. Write a mechanism to explain the observed product.

19.9   Reactions of Ethers: Cleavage

There are relatively few reactions of ethers because the C—O—C linkage is inert toward many reagents. Epoxides are more reactive than the other types of ethers and are discussed in Secs 19.12 and 19.13. Ethers are, however, cleaved by concentrated solutions of hydrohalic acids under fairly drastic reaction conditions:

$$R—O—R' \xrightarrow[\text{several steps}]{\text{excess HX, heat}} R—X + R'—X$$

$$Ar—O—R \xrightarrow[\text{several steps}]{\text{excess HX, heat}} Ar—OH + R—X \quad \text{but } no \text{ Ar—X}$$

The reactivity order of HX in cleavage reactions is HI > HBr > HCl, and usually HI or HBr is used.

The mechanism of the reaction has been investigated in some detail, and the need for a high concentration of acid suggests that the first step involves protonation of the ether. This is not surprising because alcohols and other oxygen-containing compounds are readily protonated in acidic solution.

The protonated ether can then undergo cleavage through either the S_N1 or the S_N2 mechanism, depending on the structure of the alkyl groups attached. Consider, for example, the cleavage of methyl *tert*-butyl ether by hydrogen iodide:

$$CH_3\!-\!O\!-\!\underset{\underset{CH_3}{|}}{\overset{\overset{CH_3}{|}}{C}}\!-\!CH_3 \xrightarrow[130-150°]{\text{conc HI}} CH_3\!-\!I \;+\; (CH_3)_3C\!-\!I \;+\; H_2O$$

Methyl *tert*-butyl ether	Iodomethane (methyl iodide)	2-Iodo-2-methylpropane (*tert*-butyl iodide)

Because this ether contains both a primary and a tertiary alkyl group, its cleavage may occur by one or both of the following mechanisms. It is likely that both mechanisms operate simultaneously.

Step 1. Protonation of ether, a reversible reaction

$$CH_3\!-\!\ddot{\underset{..}{O}}\!-\!C(CH_3)_3 + H^\oplus \;\rightleftharpoons\; CH_3\!-\!\overset{\oplus}{\underset{\underset{H}{|}}{O}}\!-\!C(CH_3)_3$$

Step 2. Reaction by S_N2 mechanism: iodide ion attack on methyl group

$$:\!\overset{..}{\underset{..}{I}}\!:^\ominus + CH_3\!-\!\overset{\oplus}{\underset{\underset{H}{|}}{O}}\!-\!C(CH_3)_3 \longrightarrow CH_3\!-\!\overset{..}{\underset{..}{I}}\!: + H\overset{..}{\underset{..}{O}}\!-\!C(CH_3)_3$$

$$\Big\downarrow \text{HI, } S_N1$$

$$I\!-\!C(CH_3)_3$$

Step 2 is similar to the reaction between protonated methanol, $CH_3\overset{\oplus}{\underset{..}{O}}H_2$, and iodide ion, $:\!\overset{..}{\underset{..}{I}}\!:^\ominus$. In the preceding case the leaving group is *tert*-butyl alcohol rather than H_2O. When *tert*-butyl alcohol is formed, it is converted rapidly into *tert*-butyl iodide by an S_N1 mechanism that is typical for tertiary alcohols (see Sec. 10.14). Alternatively, ether cleavage may occur as follows:

Alternate Step 2. Reaction by S_N1 mechanism: carbocation formation and attack by iodide ion

$$CH_3\!-\!\overset{\oplus}{\underset{\underset{H}{|}}{O}}\!-\!C(CH_3)_3 \xrightarrow{\text{slow}} CH_3\!-\!\overset{..}{\underset{..}{O}}H \;+\; \overset{\oplus}{C}(CH_3)_3$$

$$\Big\downarrow \text{HI, } S_N2 \qquad\qquad\qquad \Big\downarrow :\!\overset{..}{\underset{..}{I}}\!:^\ominus \text{, fast, } S_N1$$

$$CH_3\!-\!I \qquad\qquad\qquad I\!-\!C(CH_3)_3$$

Several lines of evidence support the mechanism for ether cleavage, one of which comes from allowing the reaction to go partially to completion and analyzing the reaction mixture. Often a mixture of an alcohol and an alkyl halide is detected in addition to unreacted ether.

In light of the mechanism for ether cleavage and our knowledge about the lack of reactivity of phenols toward substitution of a nucleophile for the —OH group (see Sec. 16.5), it is not surprising that *alkyl aryl ethers* give only the phenol and an alkyl halide. No aryl halide is formed. For example:

$$\text{C}_6\text{H}_5\text{—O—CH}_2\text{CH}_3 \xrightarrow[150°]{\text{conc HI}} \text{C}_6\text{H}_5\text{—OH} + \text{CH}_3\text{CH}_2\text{I}$$

Ethyl phenyl ether   Phenol   Iodoethane
(phenetole)   (ethyl iodide)

$$\Big\downarrow \text{conc HI,}\ 150°$$

no reaction

Finally, *diaryl ethers* are completely inert to cleavage by acid because of the great difficulty in rupturing the carbon-oxygen bonds attached to the aromatic ring:

$$\text{Ar—O—Ar}' \xrightarrow{\text{HX, heat}} \text{ no reaction}$$

Question 19.13

The acid-catalyzed cleavage of several different ethers is carried out and the following is the numbers of moles of product *per mole* of starting ether. What is the structure of the ether in each case? (*Hint:* Consider also cyclic ethers and ethers that contain more than one ether linkage per molecule.)

(*a*) 2 moles of isopropyl iodide
(*b*) 1 mole of n-propyl bromide and 1 mole of isobutyl bromide
(*c*) 2 moles of ethyl bromide and 1 mole of 1,2-dibromoethane
(*d*) 1 mole of 1,4-dibromobutane
(*e*) 2 moles of 1,2-diiodoethane

(*f*) 1 mole of

(*g*) 1 mole of phenol and 1 mole of benzyl bromide

Question 19.14

Suppose the following sequence of reactions is carried out in the laboratory. What stereochemical results do you think would be obtained in each step, and what would be the net stereochemical relationship (retention, racemization, or inversion) between 1-butanol-1-*d* and 1-bromobutane-1-*d*? Explain mechanistically.

$$\text{CH}_3\text{CH}_2\text{CH}_2\overset{\text{H}}{\underset{\text{D}}{\overset{|}{\underset{|}{\text{—}\overset{*}{\text{C}}\text{—}}}}}\text{OH} \xrightarrow[\text{heat}]{\text{H}_2\text{SO}_4} (\text{CH}_3\text{CH}_2\text{CH}_2\text{—CHD})_2\text{O} \xrightarrow[\text{heat}]{\text{HBr}} \text{CH}_3\text{CH}_2\text{CH}_2\overset{\text{D}}{\underset{}{\overset{|}{\text{—CH—Br}}}}$$

1-Butanol-1-*d*   1-Bromobutane-1-*d*
Optically active

Question 19.15

How, if at all, would your answer to Question 19.14 be affected if you started with optically active α-phenylethyl alcohol, C_6H_5—$\overset{*}{C}HOH$—CH_3, instead of 1-butanol-1-d? Be specific.

Epoxides

We now discuss further the preparation and chemistry of compounds containing the oxirane ring, which are commonly referred to as *epoxides*. The nomenclature of this family of compounds was discussed in Sec. 19.1C.

19.10   Preparation of Epoxides

One common method for preparing epoxides involves the reaction of an alkene with aqueous chlorine, followed by treatment with base (see Sec. 19.7). The general reaction is:

Alkene   Chlorohydrin   Epoxide

Oxirane (ethylene oxide), the simplest epoxide, is prepared commercially by allowing ethene to react with oxygen under carefully controlled conditions; a metallic catalyst, such as silver, is usually used:

$$CH_2{=}CH_2 \xrightarrow[300°]{O_2,\ Ag\ metal} CH_2{-}\!\!\!\overset{\displaystyle\diagdown O \diagup}{}\!\!\!{-}CH_2$$

Ethene   Oxirane
(ethylene)   (ethylene oxide)

Most laboratory syntheses of epoxides make use of **peroxyacids,** however, which are derivatives of carboxylic acids. Peroxybenzoic acid and *m*-chloroperoxybenzoic acid are commonly used because they are stable solids.

Peroxy   Peroxybenzoic acid   *m*-Chloroperoxybenzoic acid
acid

In another sense, peroxy acids can be considered as derivatives of hydrogen peroxide, H—O—O—H, in which one hydrogen has been replaced by the Ar—C— group. The reaction between a peroxy acid and an alkene produces the corresponding epoxide; for example:

Cyclohexene	Peroxybenzoic acid	Cyclohexene oxide	Benzoic acid

1-Hexene   *m*-Chloroperoxy-
                              benzoic acid

1,2-Epoxyhexane   *m*-Chlorobenzoic
                             acid

The —OH of the peracid is electrophilic and attacks the double bond of the alkene to give *syn* addition of the epoxide oxygen:

For example,

Cyclopentene   Cyclopentene oxide

Epoxides are usually simple to prepare in the laboratory. The peroxy acid and the alkene are mixed together in chloroform solution and allowed to react at room temperature for a short time. The epoxide is usually formed in high yield and is easily isolated in pure form.

19.11   Ring Opening of Epoxides by Acidic Reagents

A. General Reaction

Because of the strain involved in the three-membered epoxide ring, it is not surprising that epoxides are highly susceptible to ring opening. Their 60° internuclear bond angles resemble those in cyclopropane (see Sec. 3.14), so that the C—O—C bond angle is greatly compressed compared with the average bond angle of 110° in noncyclic ethers. The lack of direct overlap of the atomic orbitals in epoxides also means that the bonds are weaker and more easily broken.

In this and the next several sections, we examine various ring-opening reactions of epoxides. We start with reactions involving acidic reagents. Ring opening occurs between an epoxide and the acidic reagent, H—Nu:

$$
\begin{array}{c}
-\overset{|}{C}-\overset{|}{C}- + H-Nu \xrightarrow{\ H^{\oplus}\ } -\overset{|}{\underset{OH}{C}}-\overset{|}{\underset{Nu}{C}}- \\
\diagdown O \diagup
\end{array}
$$

The mechanism for this general reaction is shown along with comments that illustrate its similarity to other mechanisms we have studied.

Step 1. Protonation of oxygen

$$
-\overset{|}{C}-\overset{|}{C}- + H^{\oplus} \rightleftharpoons -\overset{|}{C}\underset{\overset{..}{O}}{\underset{\overset{|}{H}}{\oplus}}\overset{|}{C}-
$$

Oxonium ion

This is analogous to the protonation of other ethers and alcohols, but an epoxide is more readily protonated because the unshared electron pairs on oxygen extrude from the ring and are less crowded and more susceptible to attack.

Step 2. Attack of nucleophile on oxonium ion

$$
-\overset{}{C}\underset{\underset{\overset{|}{H}}{\overset{..}{O}}}{\oplus}\overset{}{C}- + Nu\overset{\ominus}{:} \longrightarrow -\overset{}{\underset{:OH}{C}}-\overset{\overset{Nu}{|}}{\underset{|}{C}}-
$$

The oxonium ion is structurally similar to the bromonium ion (see Sec. 8.17), which is formed initially in the ionic addition of bromine to an alkene:

$$
-\overset{}{C}\underset{\overset{\oplus}{\underset{.\,Br\,.}{}}}{}\overset{}{C}-
$$

Bromonium ion

$$
-\overset{}{C}\underset{\underset{\overset{|}{H}}{\overset{..}{O}}}{\oplus}\overset{}{C}- \longleftrightarrow -\overset{}{C}\underset{\underset{\overset{|}{H}}{:\overset{..}{O}}}{\oplus}\overset{}{C}- \longleftrightarrow -\overset{}{C}\underset{\underset{\overset{|}{H}}{\overset{..}{O}:}}{\oplus}\overset{}{C}- \equiv \overset{\delta\oplus}{-}\overset{}{C}\underset{\underset{\overset{|}{H}}{\overset{..}{O}_{\delta\oplus}}}{}\overset{\delta\oplus}{C}-
$$

Resonance structures showing charge distribution   Resonance hybrid

The nucleophile may attack either C_α or C_β to give product:

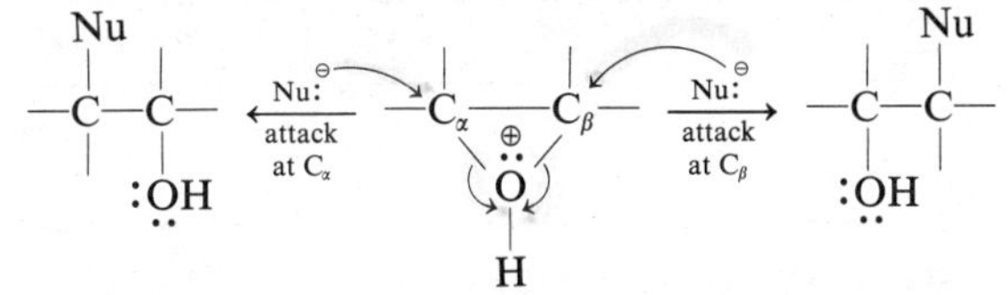

In Sec. 19.13 we study evidence that supports this mode of nucleophilic attack; several general examples are given here:

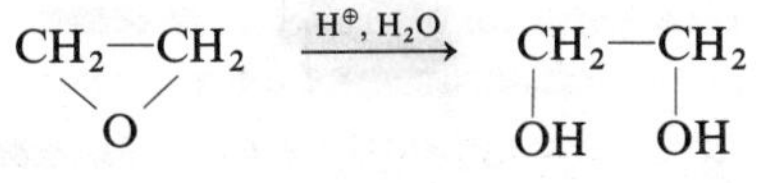

$$CH_2\!\!-\!\!CH_2 \xrightarrow{H^\oplus, H_2O} CH_2\!\!-\!\!CH_2$$

Oxirane   1,2-Ethanediol
(ethylene oxide)   (ethylene glycol)

$$CH_2\!\!-\!\!CH_2 \xrightarrow[\text{(or ArOH)}]{H^\oplus, ROH} CH_2\!\!-\!\!CH_2$$

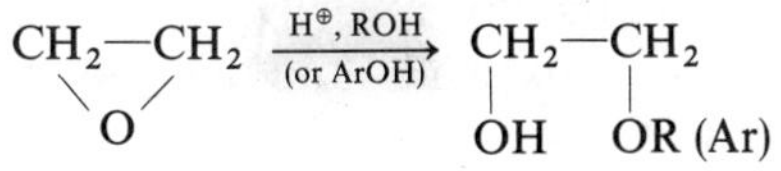

Alkoxy alcohol

$$CH_2\!\!-\!\!CH_2 + HX \xrightarrow[\text{solvent}]{\text{inert}} CH_2\!\!-\!\!CH_2$$

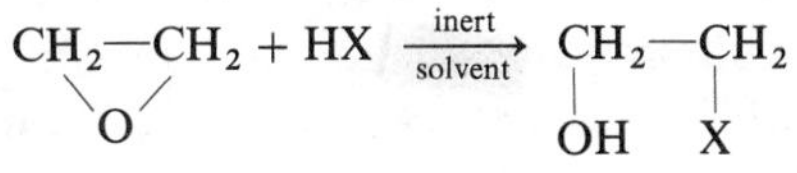

Halohydrin

Question 19.16

The reaction between ethylene oxide and hydrogen bromide gas in an inert solvent gives only 2-bromoethanol, but when it is carried out in aqueous hydrogen bromide, some ethylene glycol is also formed. Propose a mechanism for the formation of the ethylene glycol, and in light of your answer, defend the use of an inert solvent and anhydrous conditions for the preparation of halohydrins from ethylene oxide.

Question 19.17

By varying the experimental conditions, ethylene oxide can be induced to undergo different types of reactions, two of which are indicated here. Provide a reasonable mechanism for each reaction.

(a) Ethylene oxide $\xrightarrow[H_2O]{H^\oplus}$ ethylene glycol $\xrightarrow[H^\oplus]{\text{ethylene oxide}}$

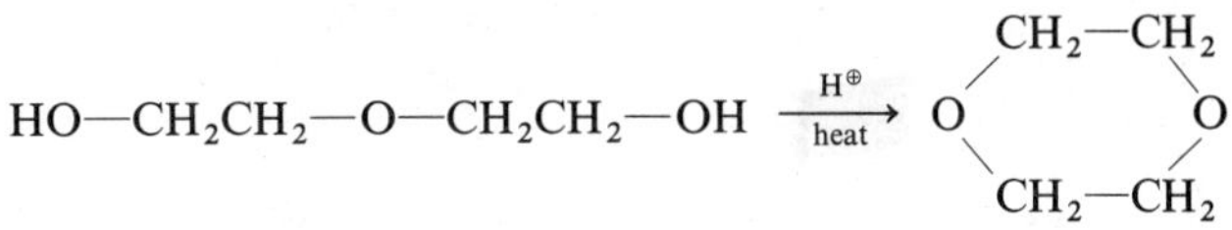

$$HO\!\!-\!\!CH_2CH_2\!\!-\!\!O\!\!-\!\!CH_2CH_2\!\!-\!\!OH \xrightarrow[\text{heat}]{H^\oplus}$$

Diethylene glycol   1,4-Dioxane

(b) Ethylene oxide $\xrightarrow[H^\oplus]{CH_3OH} CH_3\!\!-\!\!O\!\!-\!\!CH_2CH_2\!\!-\!\!OH \xrightarrow[H^\oplus]{\text{ethylene oxide}}$

$$CH_3\!\!-\!\!O\!\!-\!\!CH_2CH_2\!\!-\!\!O\!\!-\!\!CH_2CH_2\!\!-\!\!OH$$

Methyl carbitol

Question 19.18

Based on the reactions outlined in Question 19.17, suggest suitable synthetic routes for the following, starting with readily available organic compounds.

(a) $CH_3—O—CH_2CH_2—O—CH_3$, *glyme*
(b) $CH_3—O—CH_2CH_2—O—CH_2CH_2—O—CH_3$, *diglyme*
(c) $C_6H_5—O—CH_2CH_2—O—CH_2CH_2—O—C_6H_5$, *diphenyl carbitol* (a common name)

19.12   Stereochemistry of Epoxide Hydrolysis: *anti*-Hydroxylation

The mechanism for the hydrolysis of an epoxide described in Sec. 19.12 invokes a cyclic oxonium ion, which is attacked by a variety of nucleophiles. There is a considerable body of evidence supporting the involvement of an oxonium ion, some of which comes from stereochemical studies.

Cyclic systems are especially amenable to study because the products can usually be separated from one another fairly easily. For example, cyclopentene is converted to cyclopentene oxide by peroxyacids, and the oxide linkage must be attached to the carbon atoms on the same side of the ring.

If our ideas about an oxonium ion being formed from an epoxide are correct, then cyclopentene oxide should form the corresponding oxonium ion on treatment with acid. On reaction with water, the oxonium ion should open to give the two hydroxyl groups *trans* to one another. This expectation is borne out by experimental results. The mechanism and stereochemistry of epoxide hydrolysis are viewed as follows:

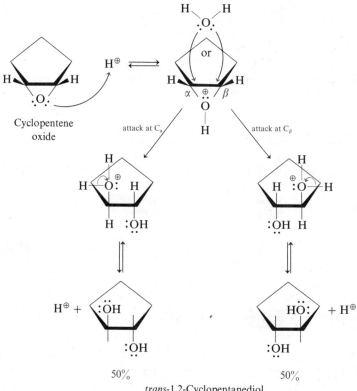

trans-1,2-Cyclopentanediol

Racemic mixture:
equal mixture
of enantiomers

In this mechanism, there is an equal (50:50) probability that the nucleophile, water, will attack either of the carbon atoms that constitute the oxonium ion from the top side. This gives an equal mixture of enantiomers that represents a racemic mixture. (This mechanism is analogous to the reaction of various nucleophiles with the bromonium ion, see Sec. 8.18.)

Thus, the sequence cycloalkene → cycloalkene oxide → 1,2-cycloalkanediol results in the *anti*-addition of the two hydroxyl groups and is referred to as ***anti*-hydroxylation.** On the other hand, recall from Sec. 8.32 that the reaction between aqueous potassium permanganate or osmium tetroxide gives *syn* hydroxylation.

Stereochemical studies on open-chain compounds support the *anti*-hydroxylation mechanism presented here (see Question 19.19).

Question 19.19

The following two reactions have been observed:

$$\text{\textit{trans}-2-Butene} \longrightarrow \text{\textit{trans}-2,3-epoxybutane} \longrightarrow \text{2,3-butanediol}$$

The product from this reaction is optically inactive and no optically active enantiomers can ever be obtained from it.

$$\text{\textit{cis}-2-Butene} \longrightarrow \text{\textit{cis}-2,3-epoxybutane} \longrightarrow \text{2,3-butanediol}$$

The product from this reaction is optically inactive, but it can be resolved into two optically active enantiomers.

Given that the epoxidation reaction occurs with retention of configuration (that is, *trans*-2-butene gives *trans*-2,3-epoxybutane, and so on), show how the data presented here support the *anti*-hydroxylation reaction. Use three-dimensional projection formulas throughout. Briefly explain.

Question 19.20

How, if at all, would the stereochemical results observed in the products in Question 19.19 differ if *cis*- and *trans*-2-pentene were used in place of the 2-butenes? Be specific.

19.13   Ring Opening of Epoxides by Basic Reagents

It is not surprising that epoxides are opened by acidic reagents, because they are a member of the ether family, and we know from Sec. 19.9 that ethers are cleaved by hydrohalic acids. On the other hand, cyclic and open-chain ethers, *except for epoxides*, are generally unaffected by basic reagents. The major driving force for epoxide ring opening by basic reagents is the relief of the strain that is present. The generalized mechanism is:

| Basic nucleophile | Epoxide | Basic alkoxide ion | Addition of elements of HNu |

| Addition | | Protonation |

In this section we examine some different types of basic reagents that typically react with epoxides and produce ring opening, for example:

$$CH_2 \underset{\underset{\ddot{\cdot}\ddot{O}\ddot{\cdot}}{\diagdown\diagup}}{\text{------}} CH_2 \xrightarrow{\ddot{O}H^{\ominus}, H_2O} \underset{\underset{OH}{|}}{CH_2} \text{------} \underset{\underset{OH}{|}}{CH_2}$$

Ethylene oxide   Ethylene glycol

As the following mechanism illustrates, this base-catalyzed reaction starts with the attack of hydroxide ion (a good nucleophile) on the epoxide ring. Ring opening generates an alkoxide ion that is more basic than hydroxide ion. It abstracts a proton from water to form the glycol, and hydroxide ion is regenerated.

Step 1. Attack of hydroxide ion on epoxide ring

$$HO\overset{\ominus}{\ddot{\cdot}} + \underset{\underset{\underset{\delta^\ominus}{\ddot{O}}}{\diagdown\diagup}}{\overset{\delta^\oplus}{CH_2}} \text{------} CH_2 \longrightarrow HO\text{---}\underset{\underset{:\ddot{O}:}{|}}{CH_2CH_2}^{\ominus}$$

Step 2. Reaction of alkoxide ion with water to form glycol and regenerate hydroxide ion; acid-base reaction

$$HO\text{---}\underset{\underset{:\ddot{O}:}{|}}{CH_2CH_2} + H\text{---}\ddot{O}H \underset{}{\overset{fast}{\rightleftharpoons}} HO\text{---}\underset{\underset{:OH}{|}}{CH_2CH_2} + :\ddot{O}H^{\ominus}$$

Strong	Weak	Weaker	Weaker
base	acid	acid	base

Two other example of epoxide ring opening under basic conditions are given here:

$$CH_3CH_2\ddot{O}\overset{\ominus}{\cdot}\ Na^{\oplus} + CH_2\underset{\underset{O}{\diagdown\diagup}}{\text{------}}CH_2 \xrightarrow{CH_3CH_2OH} \underset{\underset{OH}{|}}{CH_2}\text{------}\underset{\underset{OCH_2CH_3}{|}}{CH_2}$$

Ethoxide   2-Ethoxyethanol
ion

$$\langle\!\!\!\bigcirc\!\!\!\rangle\text{---}\ddot{O}\overset{\ominus}{\cdot}\ Na^{\oplus} + CH_2\underset{\underset{O}{\diagdown\diagup}}{\text{------}}CH_2 \xrightarrow{H_2O} \langle\!\!\!\bigcirc\!\!\!\rangle\text{---}O\text{---}CH_2CH_2OH$$

Phenoxide   2-Phenoxyethanol
ion

The base-catalyzed ring-opening reactions of epoxides are often used in the manufacture of polymers. For example, when a large excess of ethylene oxide is treated with a trace amount of hydroxide ion, a small amount of ethylene glycoxide ion forms first:

$$CH_2\underset{\underset{\ddot{\cdot}\ddot{O}\ddot{\cdot}}{\diagdown\diagup}}{\text{------}}CH_2 \xrightarrow{:\ddot{O}H^{\ominus}} \overset{\ominus\cdot\cdot}{:O}\ \ CH_2CH_2\text{---}OH$$

Ethylene glycoxide ion

Because there is a large excess of ethylene oxide and virtually no protons are available for protonation, the ethylene glycoxide ion opens other ethylene oxide molecules in a chain reaction to form a polymer. The following shows how a small portion of the polymer is formed:

$$HO-CH_2CH_2O^{\ominus} \quad CH_2-CH_2 \quad CH_2-CH_2 \quad CH_2-CH_2 \quad CH_2-CH_2 \quad \text{and so on} \longrightarrow$$
$$O \qquad O \qquad O \qquad O$$

$$HO-CH_2CH_2-O\left[CH_2CH_2O\right]_n CH_2CH_2-OH$$

Poly(ethylene glycol) polymer

The polymers of ethylene oxide range from viscous liquids to waxlike solids depending on the molecular weight (20,000 and up). The higher molecular weight polymers are reacted with diisocyanates to form tough elastomeric solids called *carbowaxes* that are often used in the manufacture of foam rubber and as stationary phases in chromatography.

Question 19.21

Draw the general structure of the polymer derived from glycerol (1,2,3-propanetriol) and ethylene oxide.

Question 19.22

Most polymers derived from epoxides are *water-soluble*, even those with molecular weights in the millions. This is remarkable because virtually all high molecular weight monomeric organic compounds and most polymers are completely insoluble in water. Suggest a reason for this water-solubility.

The conversion of epoxides to alcohols via Grignard reagents is covered in Sec. 21.10.

19.14 Orientation of Addition to Epoxides: Acidic Versus Basic Conditions

omit

In Sec. 9.13 we considered only simple epoxides in which there is no problem with orientation. With unsymmetrical epoxides, however, nucleophilic attack can occur at either carbon on the epoxide ring; as we will see, the orientation of addition depends on the reaction conditions. For example, the elements of ethanol add to epoxides under both acidic and basic conditions with the following results:

Acid catalyzed:

$$\begin{array}{c} CH_3 \\ | \\ CH_3-C-CH_2 \\ \diagdown / \\ O \end{array} + CH_3CH_2OH \xrightarrow{H^{\oplus}} \begin{array}{c} CH_3 \\ | \\ CH_3-C-CH_2-OH \\ | \\ OCH_2CH_3 \end{array}$$

1,2-Epoxy-2-methylpropane
(isobutylene oxide)

2-Ethoxy-2-methyl-1-propanol
Major product

Base catalyzed:

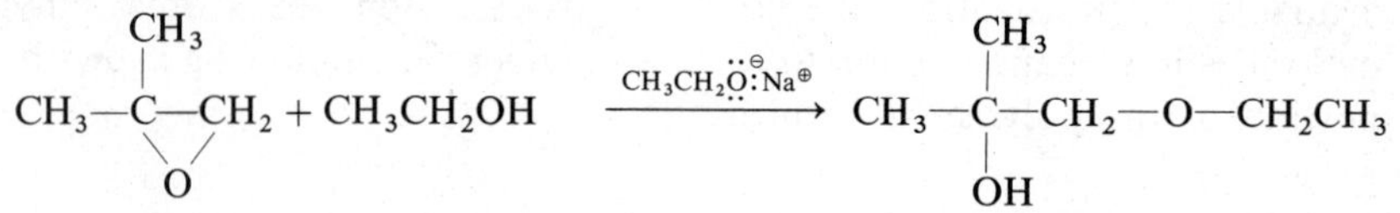

$$CH_3-\overset{\overset{\displaystyle CH_3}{|}}{\underset{\underset{\displaystyle O}{\diagup}}{C}}CH_2 + CH_3CH_2OH \quad\xrightarrow{CH_3CH_2\ddot{\overset{\ominus}{O}}:Na^{\oplus}}\quad CH_3-\overset{\overset{\displaystyle CH_3}{|}}{\underset{\underset{\displaystyle OH}{|}}{C}}-CH_2-O-CH_2CH_3$$

1-Ethoxy-2-methyl-2-propanol
Major product

To understand the orientation in these reactions, we must consider their mechanisms. In the acid-catalyzed addition reaction, the epoxide is protonated and then the nucleophilic ethanol molecule attacks the ring carbon atoms. But which one does it attack? The positive charge on oxygen attracts electrons from the adjacent σ bonds, thus placing a partial positive charge on the two carbon atoms that constitute the ring system. However, the more highly substituted carbon atom bears positive charge better than the less substituted carbon atom, as shown by the following structures:

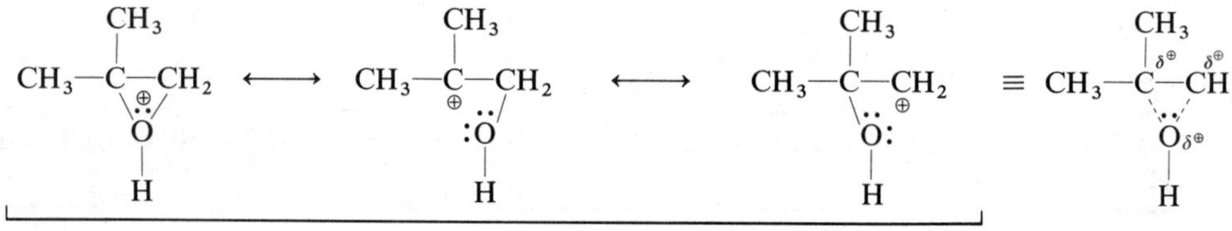

| Protonated epoxide; charge on oxygen | Partial positive charge on tertiary carbon *More stable and favored* | Partial positive charge on primary carbon *Not favored* | Resonance hybrid |

Resonance structures

The nucleophile quite naturally attacks preferentially the carbon bearing the greater positive charge to give the observed product:

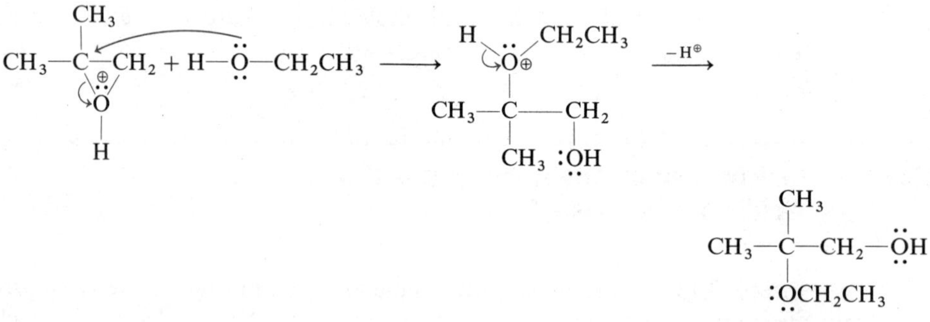

In general, *in the addition of unsymmetrical acidic reagents to epoxides, the nucleophile ends up on the more highly substituted carbon atom.* Even though this addition is mostly an S_N2 reaction, the *carbocation character* of the transition state dictates the orientation.

The addition of basic reagents to epoxides occurs by the S_N2 mechanism. The slight polarity of the carbon-oxygen bonds causes the initial attack by the nucleophile. As with other S_N2 reactions involving basic nucleophiles, the opening of the epoxide ring is dictated by steric factors. Here, however, the nucleophile can attack at two positions and it chooses the *least hindered* one.

In general, *epoxide ring opening by basic reagents occurs with the nucleophile ending up on the least substituted carbon atom.*

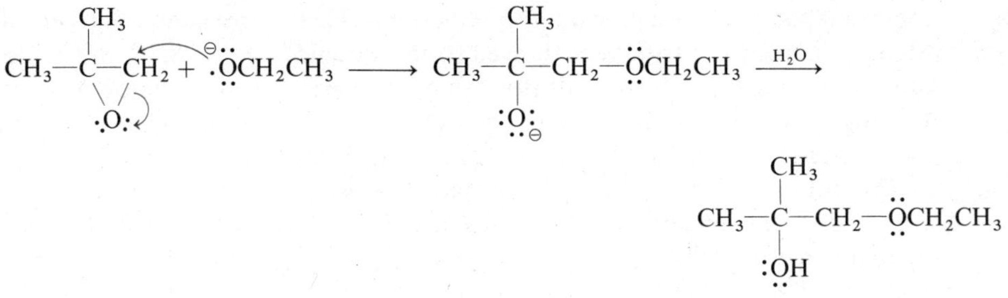

The orientation trends of epoxide ring opening presented here are useful for predicting the *major* product. Two isomeric products may be formed and the product ratio is often dictated by both experimental conditions and structural features of the starting epoxide.

Question 19.23

What is the major product from the reaction between propylene oxide and each of the following five reagents?

(a) aqueous sodium phenoxide, $C_6H_5\ddot{O}\!:^{\ominus} Na^{\oplus}$

(b) aqueous hydrogen bromide

(c) anhydrous hydrogen bromide (HBr gas)

(d) methyl magnesium bromide, followed by hydrolysis (see Sec. 21.10)

(e) potassium isopropoxide in isopropyl alcohol

Question 19.24

Indicate the expected orientation resulting from the addition of the following three reagents to methylenecyclohexane oxide (1).

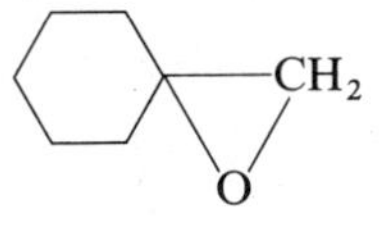

(1)

(a) anhydrous hydrogen chloride   (b) sodium ethoxide in ethanol

Question 19.25

One possible way to determine the orientation of the addition of water to epoxides makes use of isotopic tracers, namely oxygen 18 (^{18}O). What results would you expect from the reaction between propylene oxide and these two reagents? Explain mechanistically.

(a) $H_2^{18}O$ in the presence of $H^{\oplus}$ (b) $H_2^{18}O$ in the presence of $^{18}\!:\!\ddot{O}H^{\ominus}$

19.15   Biochemical Applications of Epoxide Ring Opening

A. Cholesterol

One interesting example of epoxide ring opening that involves considerable carbocation chemistry is the conversion of *squalene* to *lanosterol*. Lanosterol is important because it is a precursor to *cholesterol*. These three compounds are closely related; squalene is a triterpene derived from isoprene (see Sec. 8.36) and is found in large quantities in many animals (for example, shark liver oil). Biological processes in living systems (animals) make use of enzymes to convert squalene into lanosterol and thence into cholesterol. We discuss principles that are known *from laboratory studies* to be involved in the transformation of squalene to lanosterol.

Let us start by considering the following structures of squalene (1), lanosterol (2), and cholesterol (3):

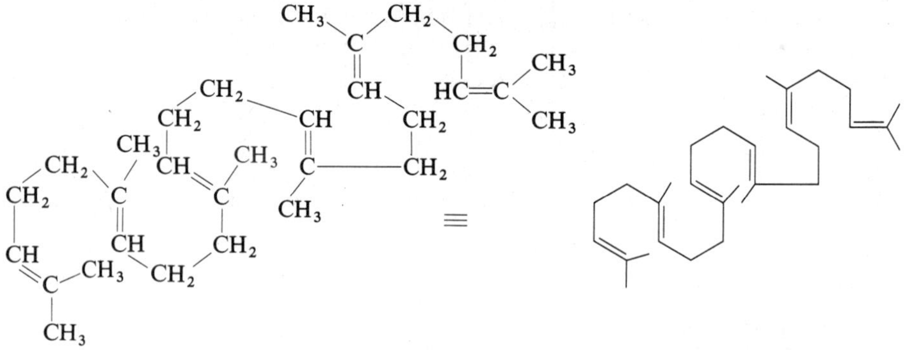

Structure showing
all atoms

More conventional
"line" structure

(1)
Squalene

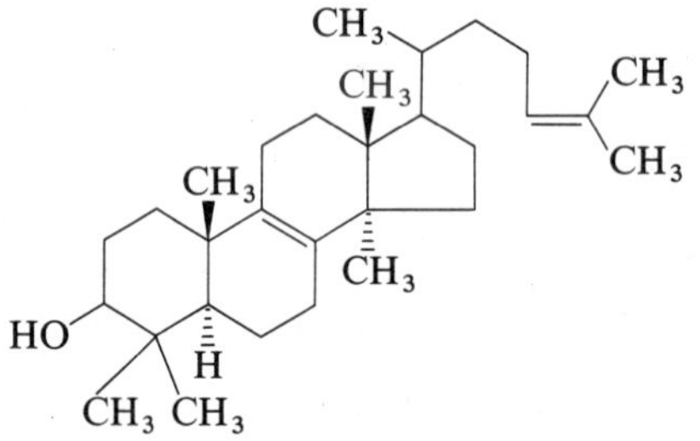

(2)
Lanosterol

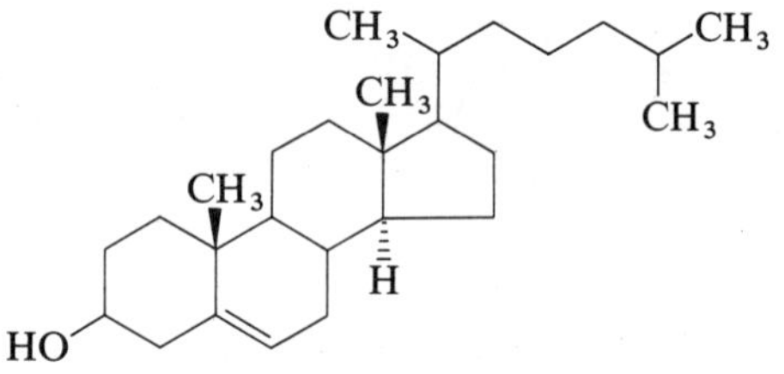

(3)
Cholesterol

Structures (1) to (3) are shown partially in the more common "line" type formulas to simplify drawing them. Aside from lanosterol being cyclic and squalene being an open-chain, highly unsaturated compound, these two compounds have great structural similarity; the major difference is in the location of the one methyl group in each compound shown in **red.** Also, lanosterol contains a hydroxyl group, whereas squalene does not. Lanosterol and cholesterol possess the same ring structure, although the latter contains three less methyl groups and one less double bond, with the other double bond located in a different position.

The conversion of squalene to lanosterol, which *looks* foreboding, is actually straightforward and easily explained by the principles of organic chemistry we have studied. (In animals, of course, the transformation is induced by enzymes.) The currently accepted mechanism, shown here, has been substantiated in the laboratory:

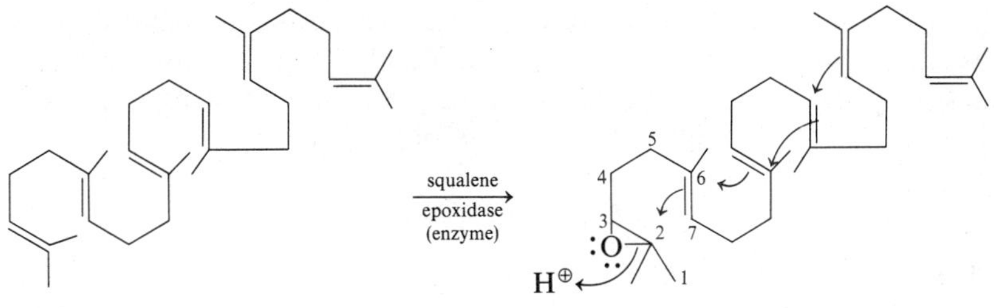

(1)
Squalene

squalene
epoxidase
(enzyme)

(4)
Squalene 2,3-epoxide

squalene oxide
cyclase (enzyme)

lost as H⊕

⟶ Lanosterol
(2)

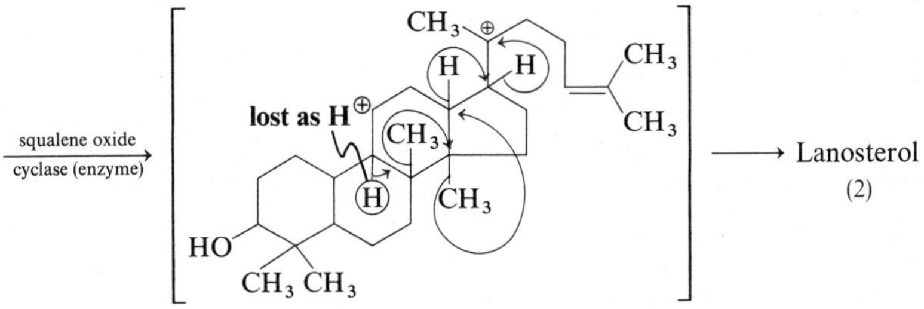

(5)
Intermediate

The first step in the reaction is the conversion of squalene to squalene 2,3-epoxide (4) by an enzyme; although there are several double bonds in squalene, the enzyme epoxidizes only the one indicated. On treatment with another enzyme, the expoxide in squalene 2,3-epoxide (4) is protonated, and at the same time a sequence of reactions occurs throughout the molecule. Pay special attention to the following reactions:

1. One of the double bonds in (4) attacks the protonated epoxide and forms a new carbon-carbon bond. Cyclization occurs (a six-membered ring is the result) because the double bond is in the same molecule; this is analogous to the reaction of an alkene with a carbocation (see Sec. 8.23) except that here the electron-deficient center is the protonated epoxide ring system.

2. The attack of the double bond on the epoxide ring leaves a new carbocation at C-6 in (4). Once this ion forms, there is a succession of attacks of double bonds on carbocations (these attacks probably occur in a concerted manner), as shown by the arrows in (4). In simple terms, a net flow of electrons occurs throughout the molecule toward a positive center—in this case, the protonated epoxide. It is surprising that all these reactions occur, but molecular models show that all the orbitals are near enough to one another to facilitate bond formation. Note that *four* new rings are formed, one of which is five-membered and the other three six-membered. Our knowledge about the relative stability of five- and six-membered rings certainly makes this reasonable.

3. Once the ring-closure reactions occur to give intermediate (5), a series of methyl and hydride migrations take place as shown. These 1,2 shifts seem to be a "cascading" effect, where the hydrogen nearest the carbocation shifts first and forms a new carbocation. This is followed by a series of shifts (hydrogen, methyl, and methyl), which likely occur stepwise but very rapidly. The indicated hydrogen (in **bold**) is lost as a proton. Individually, these 1,2 shifts are the same type we studied in carbocation chemistry, and this particular example illustrates how several migrations can occur within a given molecule.

Keep in mind that lanosterol and cholesterol are produced in living systems (usually animals), which take simple molecules such as water and carbon dioxide and build them into very complex molecules. The following sequence summarizes the relationship of isoprene to lanosterol:

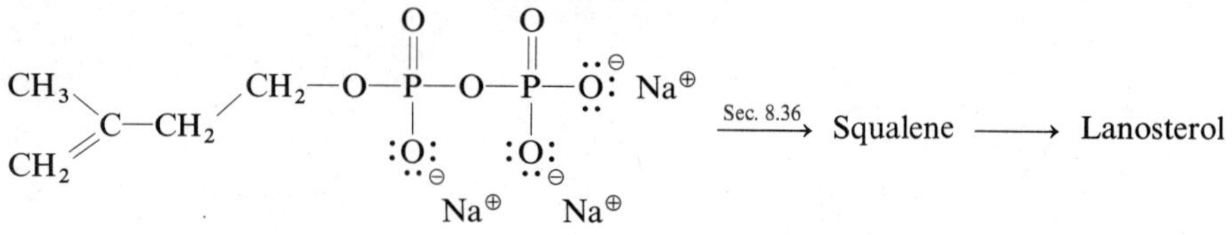

Isopentenyl
pyrophosphate

Many other complex ring structures observed in natural products are derived from isoprene (actually, isopentenyl pyrophosphate), as shown by the "isoprene rule" in Sec. 8.36.

The transformation of lanosterol (2) to cholesterol (3) is considerably more complex and little is known about it.

B. Polynuclear Aromatic Hydrocarbons and Cancer

In Sec. 13.12 we discussed polynuclear aromatic hydrocarbons (PAH) and mentioned that they are carcinogenic. Mechanistically it appears that their carcinogenicity stems from their ability to become covalently attached to the nucleic acids (DNA and RNA) in the cells. Since DNA and RNA are the templates for protein synthesis in the body, the PAH, by covalently bonding to these templates, effectively short circuits the normal metabolic pathways and cancerous cells result.

The covalent bonding between the nucleic acid and the PAH involves an epoxide intermediate. Enzymes oxidize the PAH to epoxides. The epoxide then undergoes ring opening by a nitrogen nucleophile on the nucleic acid. The nucleic acids contain many nucleophilic sites that can react this way (see Sec. 29.8).

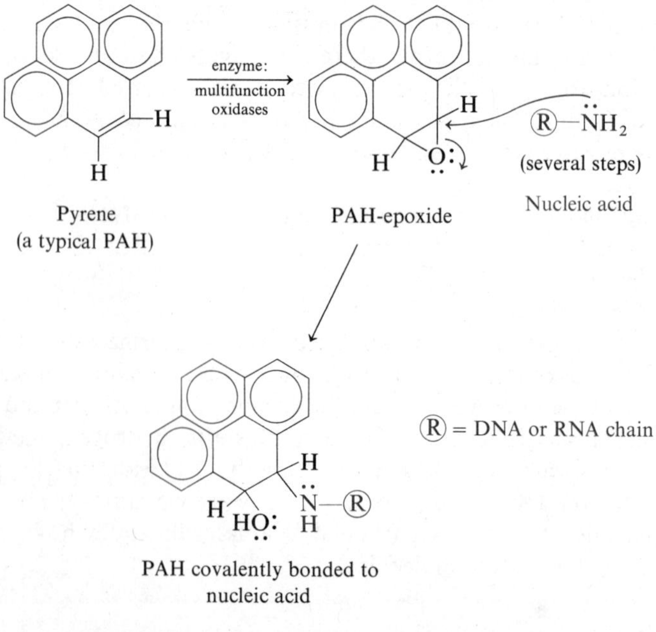

Pyrene
(a typical PAH)

PAH-epoxide

Nucleic acid

(R) = DNA or RNA chain

PAH covalently bonded to
nucleic acid

Glycols

19.16   Properties of Glycols

The physical properties of selected glycols and polyhydroxy compounds are listed in Table 19.2. Note that they are all high-boiling compounds and very soluble in water. For example, ethylene glycol, CH_2OHCH_2OH, is used in antifreeze because of its high boiling point, solubility in water, and low freezing point. It is effective in raising the boiling point of water in radiators, which reduces overheating

TABLE 19.2 Physical Properties of Selected Glycols and Polyhydroxy Compounds

Name	Structure	mp, °C	bp, °C	Solubility, g/100 g water
1,2-Ethanediol (ethylene glycol)	$HOCH_2$—CH_2OH	−16	197	∞
1,2-Propanediol (propylene glycol)	CH_3—$CHOH$—CH_2OH		187	∞
1,3-Propanediol	$HOCH_2$—CH_2—CH_2OH		215	∞
1,2,3-Propanetriol (glycerol)	$HOCH_2$—$CHOH$—CH_2OH	18	290	∞
1,2-Butanediol	CH_3CH_2—$CHOH$—CH_2OH		192	Slight
1,4-Butanediol	$HOCH_2$—CH_2CH_2—CH_2OH	17	231	∞
Pentaerythritol	$HOCH_2$—$\overset{\displaystyle CH_2OH}{\underset{\displaystyle CH_2OH}{C}}$—$CH_2OH$		260	6.2
cis-1,2-Cyclopentanediol		30		
trans-1, 2-Cyclopentanediol		55		
cis-1,2-Cyclohexanediol		98		
trans-1,2-Cyclohexanediol		104		

in the summer, and it lowers the freezing point of water to protect the radiator from freezing and cracking in the winter.

19.17   Methods of Preparation of Glycols

We studied several of the most common methods for preparing glycols from alkenes. These methods, with references to previous discussions, are given here.

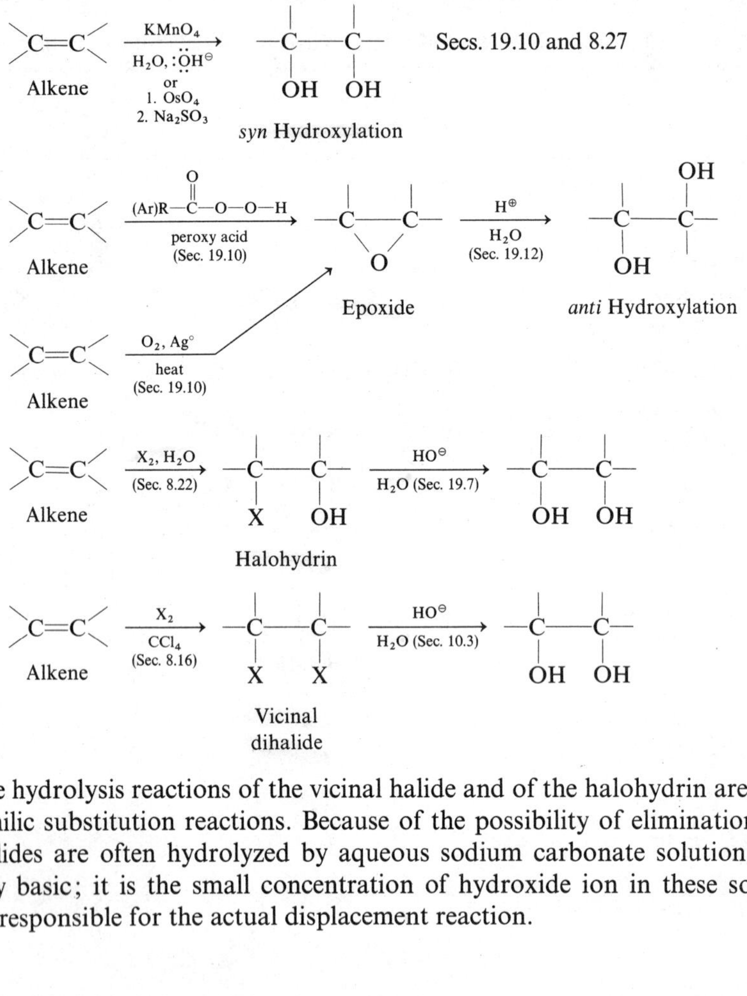

syn Hydroxylation

Epoxide   *anti* Hydroxylation

Halohydrin

Vicinal
dihalide

The hydrolysis reactions of the vicinal halide and of the halohydrin are typical nucleophilic substitution reactions. Because of the possibility of elimination, these alkyl halides are often hydrolyzed by aqueous sodium carbonate solution, which is weakly basic; it is the small concentration of hydroxide ion in these solutions which is responsible for the actual displacement reaction.

19.18   Reactions of Glycols

Many reactions of glycols are typical of the hydroxyl group (—OH) and are therefore similar to the reactions of alcohols (see Chap. 10). However, reactions that require the presence of two hydroxyl groups on adjacent carbon atoms are unique for glycols. In this section we discuss *periodic acid oxidation*.

Glycols, α-hydroxy aldehydes, α-hydroxy ketones, and α,β-dicarbonyl compounds are oxidized by periodic acid, HIO_4. The products shown in the following equations are formed in addition to iodic acid, HIO_3. Note that a carbon-carbon bond is broken in each. The mechanism is given in Sec. 29.9A,e.

omit

Glycols:

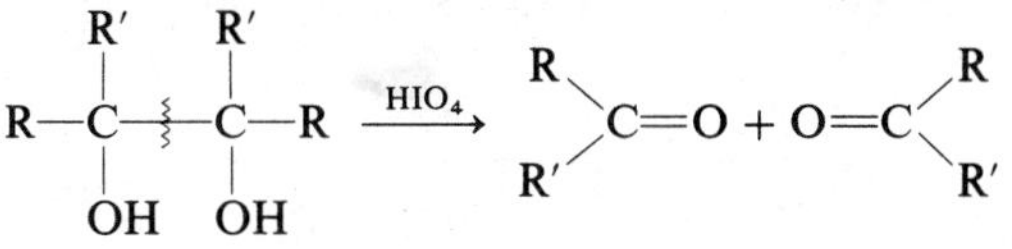

$$R-\underset{\underset{OH}{|}}{\overset{\overset{R'}{|}}{C}}\overset{\overset{R'}{|}}{\underset{\underset{OH}{|}}{C}}-R \xrightarrow{HIO_4} \underset{R'}{\overset{R}{>}}C=O + O=C\underset{R'}{\overset{R}{<}}$$

Aldehydes or ketones,
depending on R and R'

but

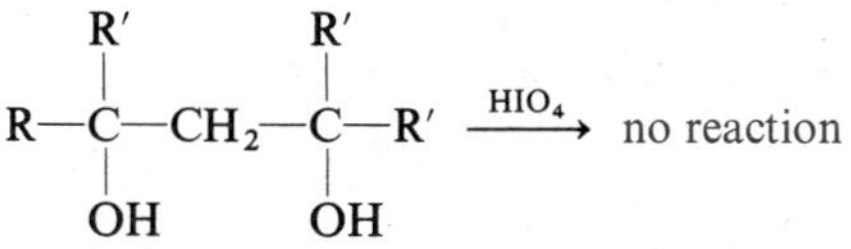

$$R-\underset{\underset{OH}{|}}{\overset{\overset{R'}{|}}{C}}-CH_2-\underset{\underset{OH}{|}}{\overset{\overset{R'}{|}}{C}}-R' \xrightarrow{HIO_4} \text{no reaction}$$

α-Hydroxy aldehydes and ketones:

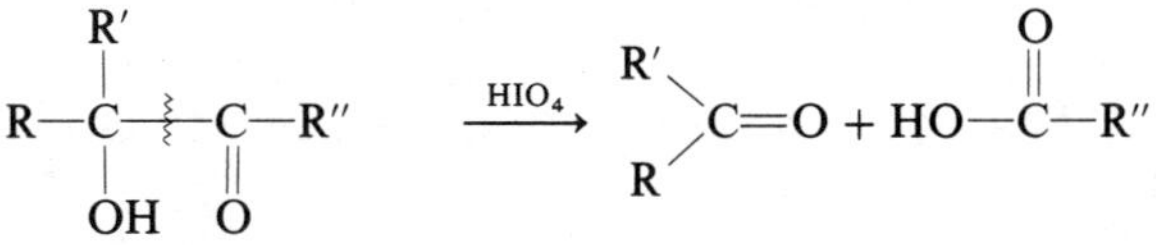

$$R-\underset{\underset{OH}{|}}{\overset{\overset{R'}{|}}{C}}\overset{\overset{O}{\parallel}}{C}-R'' \xrightarrow{HIO_4} \underset{R}{\overset{R'}{>}}C=O + HO-\overset{\overset{O}{\parallel}}{C}-R''$$

R″ = *H*, α-hydroxy aldehyde   Aldehyde   Carboxylic
R″ = alkyl or aryl group,   or   acid
     α-hydroxy ketone   ketone

α,β-Dicarbonyl compounds:

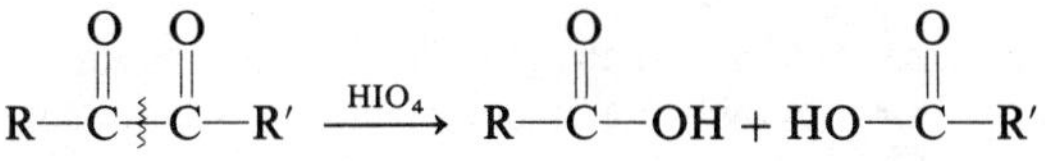

$$R-\overset{\overset{O}{\parallel}}{C}\overset{\overset{O}{\parallel}}{C}-R' \xrightarrow{HIO_4} R-\overset{\overset{O}{\parallel}}{C}-OH + HO-\overset{\overset{O}{\parallel}}{C}-R'$$

R and R' may   Carboxylic acids
be H, alkyl, or
aryl

Periodic acid oxidation has considerable use in structure determination be-cause each carbon-carbon bond that is cleaved consumes 1 mole of HIO_4.

Note these two special features of this oxidation reaction:

1. Two hydroxyl groups, or one hydroxyl group and one carbonyl group, or two carbonyl groups must be on *adjacent* carbon atoms for reaction to occur.
2. Periodic acid is a very mild oxidizing agent compared with permanganate ion or chromic acid. For example, if an aldehyde is produced that does not have an α-hydroxyl group, no further reaction occurs; that is, the aldehyde group is *not* oxidized to a carboxylic acid group by excess HIO_4, as it would be with a stronger oxidizing agent.

We discuss this reaction in more detail in connection with the chemistry and structure of carbohydrates (which are polyhydroxy compounds, see Chap. 29), but for now let us examine the effect of having additional hydroxyl groups in the molecule. For example, when compound (1) reacts with periodic acid, a stepwise sequence of reactions occurs:

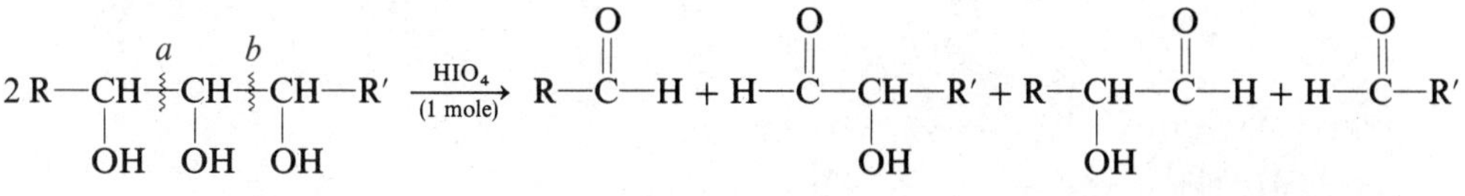

$$2\,R-\underset{\underset{OH}{|}}{\overset{a}{CH}}\underset{\underset{OH}{|}}{\overset{b}{CH}}\underset{\underset{OH}{|}}{CH}-R' \xrightarrow[\text{(1 mole)}]{HIO_4} R-\overset{\overset{O}{\parallel}}{C}-H + H-\overset{\overset{O}{\parallel}}{C}-\underset{\underset{OH}{|}}{CH}-R' + R-\underset{\underset{OH}{|}}{CH}-\overset{\overset{O}{\parallel}}{C}-H + H-\overset{\overset{O}{\parallel}}{C}-R'$$

(1)   From cleavage at *a*   From cleavage at *b*

After the first cleavage reaction occurs at either position *a* or *b*, an α-hydroxy aldehyde is produced which undergoes further cleavage with periodic acid:

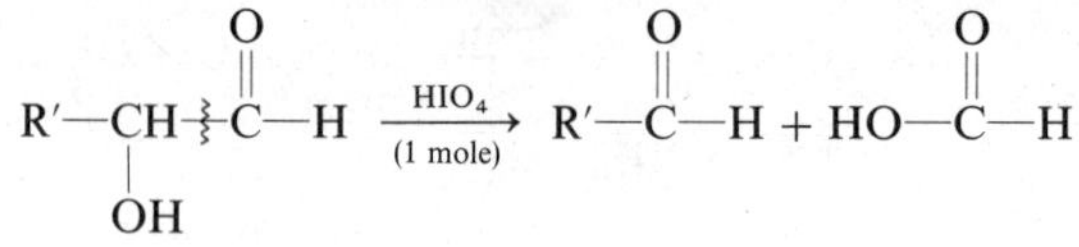

$$R'-CH \overset{\overset{\displaystyle O}{\|}}{\underset{\underset{\displaystyle OH}{|}}{C}}-H \xrightarrow[\text{(1 mole)}]{HIO_4} R'-\overset{\overset{\displaystyle O}{\|}}{C}-H + HO-\overset{\overset{\displaystyle O}{\|}}{C}-H$$

and

$$R-CH \overset{\overset{\displaystyle O}{\|}}{\underset{\underset{\displaystyle OH}{|}}{C}}-H \xrightarrow[\text{(1 mole)}]{HIO_4} R-\overset{\overset{\displaystyle O}{\|}}{C}-H + HO-\overset{\overset{\displaystyle O}{\|}}{C}-H$$

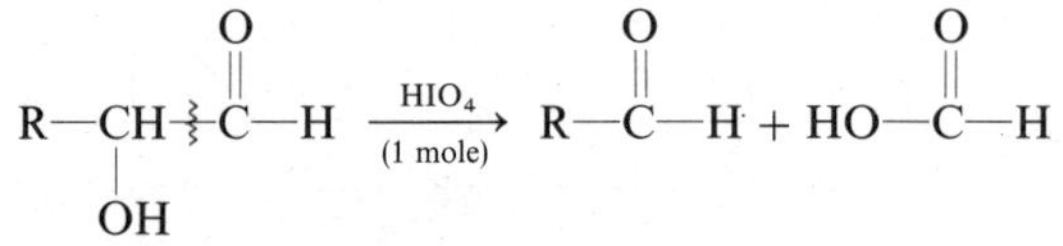

Regardless of the order in which the reactions occur, the overall reaction with excess HIO_4 is:

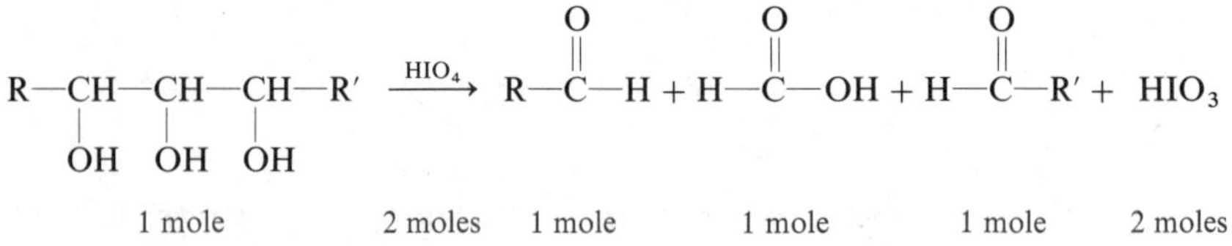

$$R-\underset{\underset{\displaystyle OH}{|}}{CH}-\underset{\underset{\displaystyle OH}{|}}{CH}-\underset{\underset{\displaystyle OH}{|}}{CH}-R' \xrightarrow{HIO_4} R-\overset{\overset{\displaystyle O}{\|}}{C}-H + H-\overset{\overset{\displaystyle O}{\|}}{C}-OH + H-\overset{\overset{\displaystyle O}{\|}}{C}-R' + HIO_3$$

| 1 mole | 2 moles | 1 mole | 1 mole | 1 mole | 2 moles |

Note that 2 moles of HIO_4 are consumed.

Question 19.26

The periodic acid reaction with the glycols, α-hydroxy aldehydes, or ketones is an oxidation-reduction reaction. For the following reaction, indicate what changes in oxidation state occur.

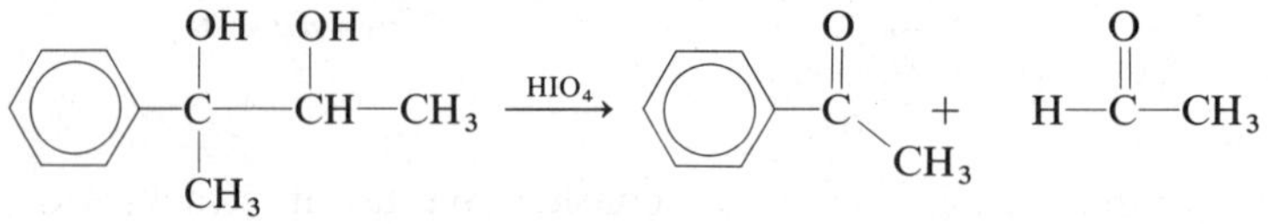

2-Phenyl-2,3-butanediol   Acetophenone   Ethanal (acetaldehyde)

What is the oxidizing agent? The reducing agent?

Question 19.27

Indicate the cleavage products (if any) that are formed when each of the following seven compounds is allowed to react with excess HIO_4. Also state how many moles of periodic acid are consumed by each compound.

(*a*) ethylene glycol
(*b*) glycerol
(*c*) 2,4-pentanediol
(*d*) 1,2,3,4-butanetetraol

(*e*) $CH_3O-CH_2-\underset{\underset{\displaystyle OH}{|}}{CH}-\underset{\underset{\displaystyle OH}{|}}{CH}-OCH_3$

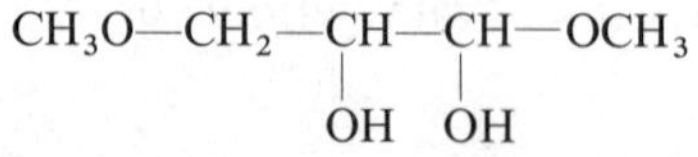

(*f*) $HOCH_2-(CHOH)_4-\overset{\overset{\displaystyle O}{\|}}{C}-H$

(*g*) $\underset{\underset{\displaystyle OH}{|}}{CH_2}-\underset{\underset{\displaystyle OH}{|}}{CH}-\underset{\underset{\displaystyle OH}{|}}{CH}-\underset{\underset{\displaystyle OH}{|}}{CH}-\overset{\overset{\displaystyle O}{\|}}{C}-CH_2OH$

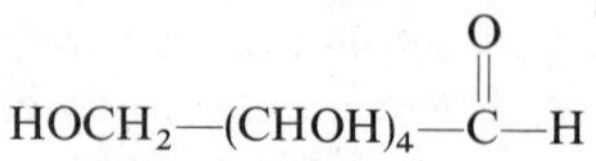

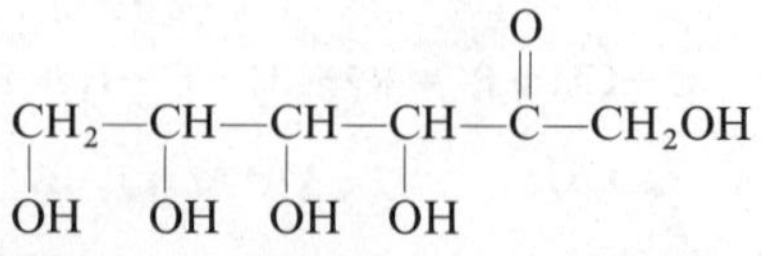

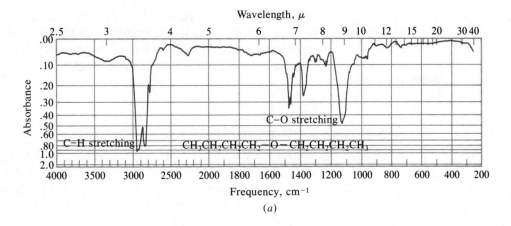

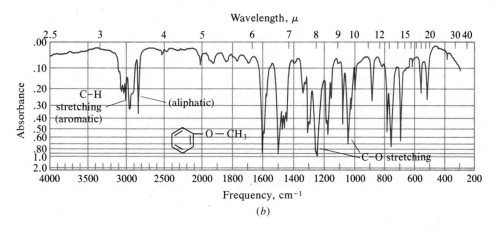

FIGURE 19.2   IR spectra of (*a*) n-butyl ether and (*b*) anisole (methyl phenyl ether). [Data from *The Sadtler Handbook of Infrared Spectra*, W. W. Simmons (ed.), © Sadtler Research Laboratories, Division of Bio-Rad Laboratories, Inc. (1978)]

19.19   Spectral Analysis of Ethers

A. Infrared Spectroscopy

The carbon-oxygen-carbon linkage absorbs in the 1,050 to 1,300-cm^{-1} region of the infrared. Because both alcohols and ethers contain this linkage, the main difference in their IR spectra is that the former show characteristic absorption due to the O—H group and the latter do not.

The carbon-oxygen stretching occurs as a strong, broad band at 1,050 to 1,150 cm^{-1} (9.52 to 8.7 μ) in alkyl ethers, and at 1,200 to 1,275 cm^{-1} (8.33 to 7.84 μ) in aryl and vinyl ethers. The latter also often have a weaker band at 1,020 to 1,080 cm^{-1} (9.80 to 9.26 μ). Fig. 19.2 shows the IR spectra of a typical alkyl and aryl ether.

B. Ultraviolet Spectroscopy

The RO— group, like the HO— group, is an auxochrome. It absorbs only in the region below 185 nm in the ultraviolet. Ethers and alcohols are commonly used as solvents in UV spectroscopy for this reason. The RO— group contains nonbonding electrons on oxygen, and therefore it can interact with chromophores to which it is attached. The effect again is analogous to that of the alcohols; that is, the RO— group shifts the absorptions of the chromophores to which it is attached to longer wavelengths.

C. Nuclear Magnetic Resonance Analysis

The presence of the electronegative oxygen atom in ethers causes the adjacent protons to be deshielded and to appear downfield. Typically, these adjacent hydrogens produce peaks at 3.2 to 4.5 δ (5.5 to 6.8 τ). [This effect is felt also on the hydrogens β to the oxygen where the peaks appear at 1.3 to 1.6 δ (8.4 to 8.7 τ).] The NMR spectrum of anisole is shown in Fig. 15.16.

D. Mass Spectrometry

When ethers are subjected to the electron beam in the ionization chamber of the mass spectrometer, the primary fragmentation that occurs results in the cleavage of the carbon-carbon bond of the carbons α and β to the oxygen. An example of a typical (partial) fragmentation pattern for *sec*-butyl ethyl ether is shown here. The molecular ion, though usually present, is not relatively abundant.

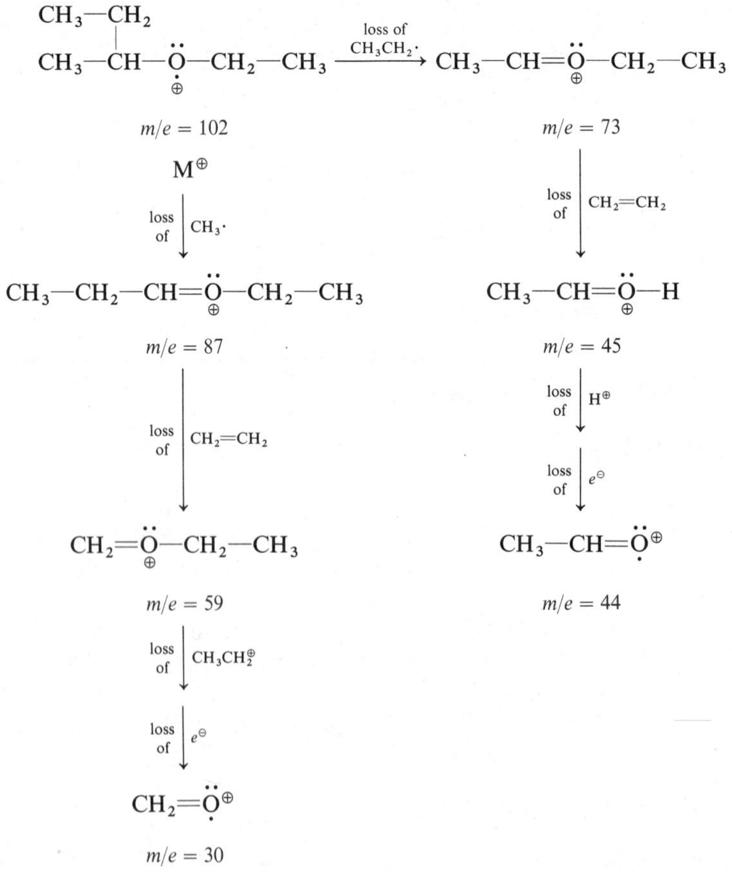

ᵒᵐⁱᵗ **E. Spectroscopy and Structure Elucidation**

The NMR spectrum of styrene oxide is shown in Fig. 19.3. The similarities in the chemical environment of the *ortho*, *meta*, and *para* hydrogens of the aromatic ring result in a broad singlet at 7 α.

To aid in the elucidation of structure from such spectra, a new family of reagents has been developed, called **shift reagents.** These reagents are generally Lewis acids containing a metal cationic center; when added to a compound containing a basic center (for example, O, N), they complex with that center; for example,

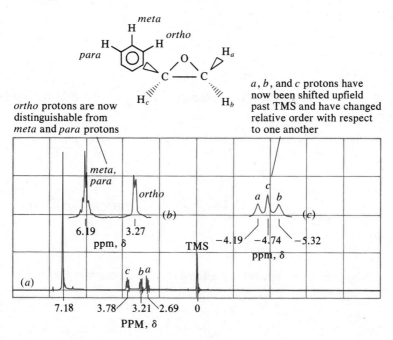

The NMR spectrum of the ether with the added shift reagent is an average of the spectrum of the free ether and the complex, which are in rapid equilibrium. The complex directly affects the observed spectrum. The magnetic field associated with the metal ion, which is a paramagnetic moiety, causes marked changes in the observed shifts of the protons in the substrate—hence the name *shift reagent.*

Some shift reagents cause upfield shifts of the substrate protons, whereas others cause downfield shifts. In all cases, the magnitude of the shift is a function of the distance from the shift reagent to the protons being observed. The closer they are, the greater the effect. Fig. 19.3 shows the spectrum of styrene oxide with added shift reagent.

FIGURE 19.3 NMR spectrum of styrene oxide in CCl₄ (*a*) without shift reagent and (*b*) and (*c*) with Pr(DPM)₃ added.

Two of the more common shift reagents used are lanthanide complexes of europium (Eu) and praseodymium (Pr); for example, *tris*-(dipivaloylmethanato)-praseodymium, $Pr(DPM)_3$:

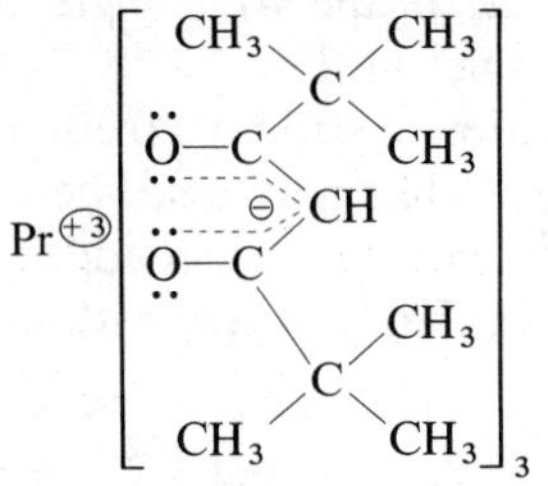

$$Pr(DPM)_3$$

19.20   Summary of Methods of Preparation of Ethers

1. Dehydration of primary and secondary alcohols

$$2ROH \xrightarrow[140°]{H_2SO_4} ROR + H_2O$$

Excess
$R \neq$ aryl

Mechanism (Sec. 19.4A)
Side reactions (Sec. 19.4B)
Experimental control of reaction
(Sec. 19.4C)

2. Reaction of alkoxide or phenoxide ions with alkyl halides: Williamson method (Sec. 19.5)

$$ROH \xrightarrow[metal]{Na} R\ddot{O}{:}^{\ominus} \xrightarrow[(S_N2)]{R'X} ROR'$$

Alkoxide   R' must be
ion   methyl
1° or 2°

$$ArOH \underset{}{\overset{{:}\ddot{O}H^{\ominus}}{\rightleftarrows}} Ar\ddot{O}{:}^{\ominus} \xrightarrow[(S_N2)]{R'X} ArOR'$$

Phenoxide   R' must be
ion   methyl
1° or 2°

3. Reaction of alkoxide and phenoxide ions with tosylates and dimethyl sulfate (Sec. 19.6)

$$R\ddot{O}{:}^{\ominus} + R'{-}OTs \longrightarrow ROR' + Ts\ddot{O}{:}^{\ominus}$$

Tosylate

$$2R\ddot{O}{:}^{\ominus} (2Ar\ddot{O}{:}^{\ominus}) + CH_3O{-}\overset{\displaystyle O}{\underset{\displaystyle O}{\overset{\|}{\underset{\|}{S}}}}{-}OCH_3 \longrightarrow 2ROCH_3 \ (2ArOCH_3) + SO_4^{\ominus 2}$$

Dimethyl sulfate

4. Cyclic ethers and epoxides via the Williamson method (Sec. 19.7)

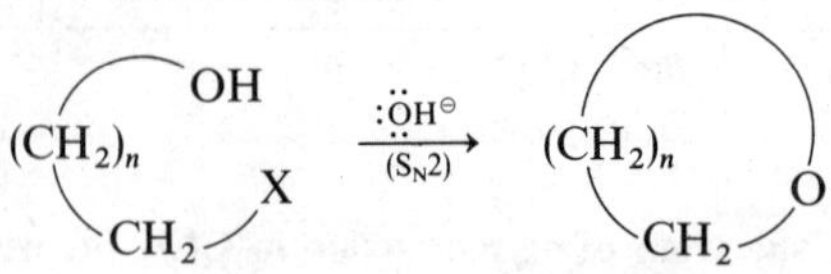

$n = 1, 2, 3, 4$

Cyclic ether
(or epoxide when $n = 1$)

5. Phenoxide ions and allyl halides: C- Versus O-Alkylation (Sec. 19.8)

19.21   Summary of Reactions of Ethers

Cleavage by acid (HX) (Sec. 19.9):

$$ROR' + HX \xrightarrow{\text{heat}} RX + R'X$$

Reactivity of HX:

$$HI > HBr > HCl$$

$$ArOR + HX \xrightarrow{\text{heat}} ArOH + RX$$

$$ArOAr' + HX \xrightarrow{\text{heat}} \text{no reaction}$$

19.22   Summary of Methods of Preparation for Epoxides

1. From alkenes via chlorohydrin

Alkene   Chlorohydrin   Epoxide

2. From alkenes on treatment with peroxy acids

3. Special preparation of ethylene oxide

$$CH_2=CH_2 + O_2 \xrightarrow[300°]{\text{Ag metal}} CH_2-CH_2 \quad \text{Sec. 19.10}$$

Ethylene   Ethylene
                                 oxide

19.23   Summary of Reactions of Epoxides

A. Ring Opening by Acidic Reagents

General:

$$-\underset{\underset{\displaystyle O}{\diagdown\diagup}}{C}-\underset{}{C}- \; + \; H-Nu \; \xrightarrow{\;H^{\oplus}\;} \; -\underset{\underset{\displaystyle OH}{|}}{C}-\underset{\underset{\displaystyle Nu}{|}}{C}- \qquad \text{Sec. 19.11A}$$

Stereochemistry:

$$-\underset{\underset{\displaystyle O}{\diagdown\diagup}}{C}-\underset{}{C}- \; + \; H-Nu \; \xrightarrow{\;H^{\oplus}\;} \; -\underset{}{\overset{\overset{\displaystyle Nu}{|}}{C}}-\underset{\underset{\displaystyle OH}{|}}{C}- \qquad \text{Sec. 19.12}$$

anti addition

With H_2O, *anti*-hydroxylation occurs.

Orientation:

$$R-\underset{\underset{\displaystyle O}{\diagdown\diagup}}{\overset{\overset{\displaystyle R}{|}}{C}}-CH_2 \; + \; H-Nu \; \xrightarrow{\;H^{\oplus}\;} \; R-\underset{\underset{\displaystyle Nu}{|}}{\overset{\overset{\displaystyle R}{|}}{C}}-\underset{\underset{\displaystyle OH}{|}}{CH_2} \qquad \text{Sec. 19.14}$$

Major product

Nucleophile ends up on more highly substituted carbon.

1. Reaction with water

$$-\underset{\underset{\displaystyle O}{\diagdown\diagup}}{C}-\underset{}{C}- \; + \; H_2O \; \xrightarrow{\;H^{\oplus}\;} \; -\underset{\underset{\displaystyle OH}{|}}{C}-\underset{\underset{\displaystyle OH}{|}}{C}- \qquad \text{Sec. 19.11B}$$

2. Reaction with alcohol or phenol

$$-\underset{\underset{\displaystyle O}{\diagdown\diagup}}{C}-\underset{}{C}- \; + \; ROH\,(ArOH) \; \xrightarrow{\;H^{\oplus}\;} \; -\underset{\underset{\displaystyle (Ar)RO}{|}}{C}-\underset{\underset{\displaystyle OH}{|}}{C}- \qquad \text{Sec. 19.11C}$$

3. Reaction with hydrohalic acid

$$-\underset{\underset{\displaystyle O}{\diagdown\diagup}}{C}-\underset{}{C}- \; + \; HX \; \xrightarrow{\;\text{anhydrous}\;} \; -\underset{\underset{\displaystyle X}{|}}{C}-\underset{\underset{\displaystyle OH}{|}}{C}- \qquad \text{Sec. 19.11D}$$

B. Ring Opening by Basic Reagents

General:

$$-\underset{\underset{\displaystyle O}{\diagdown\diagup}}{C}-\underset{}{C}- \; + \; Nu\overset{\ominus}{:} \; \longrightarrow \; -\underset{\underset{\displaystyle Nu}{|}}{C}-\underset{\underset{\displaystyle :\overset{\ominus}{O}:}{|}}{C}- \; \xrightarrow{\;H^{\oplus}\;} \; -\underset{\underset{\displaystyle Nu}{|}}{C}-\underset{\underset{\displaystyle OH}{|}}{C}- \qquad \text{Sec. 19.13}$$

Epoxide

Orientation:

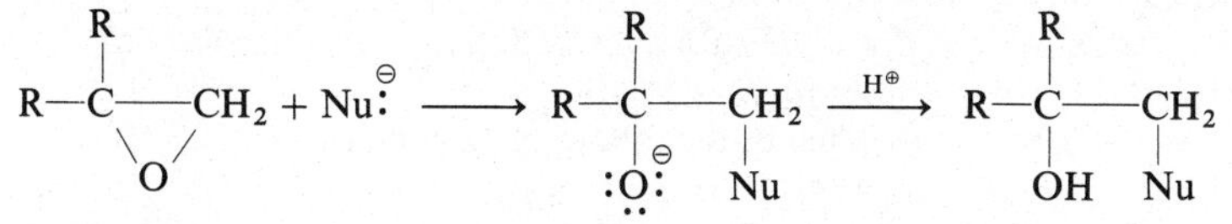

Sec. 19.14

Epoxide

Nucleophile ends up on less highly substituted carbon.

Stereochemistry:

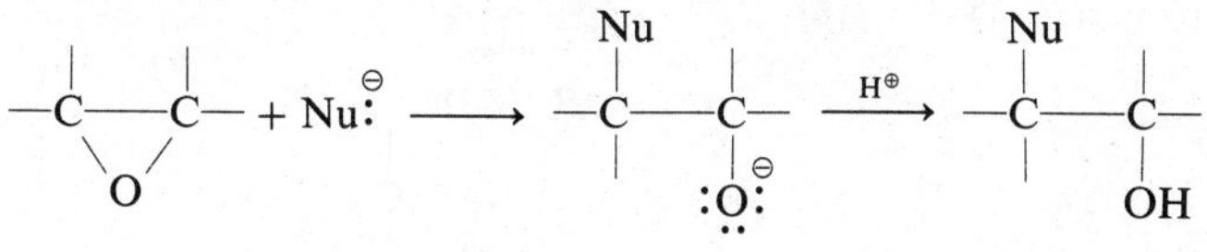

Sec. 19.13

anti addition

1. Reaction with water

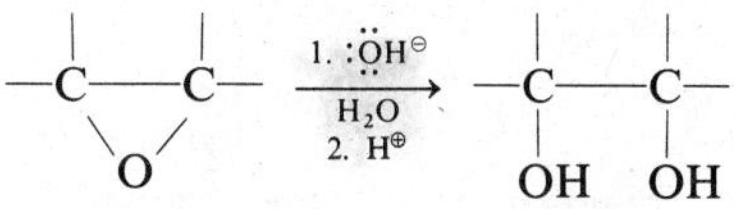

Sec. 19.13A

Epoxide Glycol

2. Reaction with alcohol or phenol

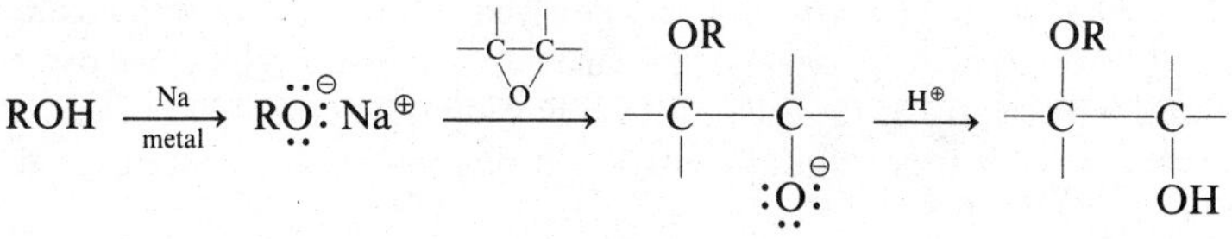

Sec. 19.13B

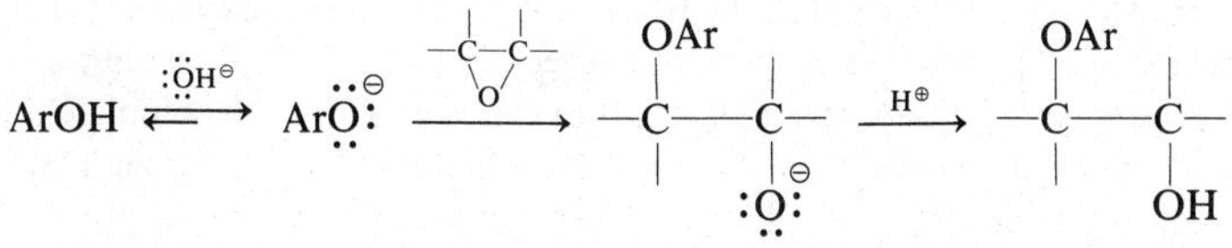

Sec. 19.13B

19.24 Summary of Reactions of Glycols

(*Note:* A summary of the methods of preparing glycols is given in Sec. 19.17.)

Oxidation by periodic acid:

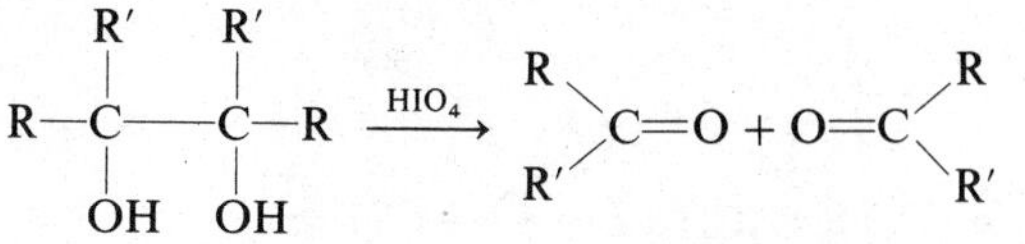

Sec. 19.18

R and R′ may be alkyl or aryl groups or hydrogen

Aldehydes or ketones, depending on R and R′

Study Questions

19.28 Draw and name all the isomeric compounds that contain an aromatic ring and have the molecular formula C_7H_8O.

19.29 Draw and name all the isomeric ethers that have the molecular formula $C_4H_{10}O$.

19.30 Ignoring stereoisomers, draw and name the cyclic ethers and epoxides with the molecular formula $C_5H_{10}O$.

19.31 Ignoring stereoisomers, draw the structures of the isomeric glycols with the molecular formula $C_4H_{10}O_2$.

19.32 Provide suitable names for each of the following 12 compounds.

(*a*) $[(CH_3)_2CHCH_2]_2O$ (*b*) $C_6H_5—O—CH_3$

(*c*) $C_6H_5CH_2—O—CH_2CH_3$ (*d*) $CH_2=CH—O—CH_3$

(*e*) CH_3O—⬡—SO_3H (*f*) $(CH_3)_3C—O—CH(CH_3)_2$

(*g*) $CH_3CH_2CHCH_2CH_2OH$
                  |
                  OCH_3

(*h*) $C_6H_5—CH—CH_2$
              | |
             OH   OH

(*i*) $C_6H_5—CH{-}{-}CH_2$
                   \ /
                    O

(*j*) $HOCH_2CH_2CH_2CH_2OH$

(*k*) ⬭
     OH   OH

(*l*)   O
           /   \
        CH_3 CH_3

19.33 Draw structures that correspond to the following 15 names.

(*a*) *n*-propyl ether   (*b*) benzyl ether   (*c*) phenyl ether
(*d*) methyl vinyl ether   (*e*) ethyl allyl ether   (*f*) phenyl allyl ether
(*g*) 1,1-dimethoxycyclohexane   (*h*) anisole   (*i*) ethyl *p*-nitrophenyl ether
(*j*) propylene glycol   (*k*) isobutylene glycol   (*l*) 2,2-dimethyloxirane
(*m*) *trans*-2,3-diphenyloxirane (*n*) 4-oxadecane   (*o*) cyclobutene oxide

19.34 Outline a suitable synthesis for each of the following, using only phenol and alcohols of four carbon atoms or less as the starting materials. Common organic solvents and inorganic reagents may be used.

(*a*) ethyl *n*-propyl ether   (*b*) *n*-butyl phenyl ether   (*c*) *n*-butyl cyclohexyl ether
(*d*) benzyl ethyl ether   (*e*) *n*-butyl isobutyl ether   (*f*) *n*-propyl *tert*-butyl ether

19.35 Supply the missing reactants, reagents, or products, as required, to complete the following 19 equations. The items to be supplied are indicated by number [for example, (1)].

(*a*) Cyclopentene oxide $\xrightarrow[\text{CH}_3\text{OH}]{\text{CH}_3\ddot{\text{O}}\!\!{}^{\ominus}}$ (1)   ⬡^{OH} $_{O\text{-}CH_3}$

$(CH_3)_3 C-\overset{\ominus}{\underset{..}{O}}\!:N^{\oplus}$

(*b*) (2) + (3) $\longrightarrow$ $CH_3—\overset{\displaystyle CH_3}{\underset{\displaystyle CH_3}{\overset{|}{\underset{|}{C}}}}—O—CH_2—$⬡

⬡$-CH_2-Br$

(*c*) $CH_3\ddot{\underset{..}{O}}\!{}^{\ominus} + (CH_3)_3CBr \longrightarrow$ (4) $CH_3OH + CH_3\overset{\displaystyle CH_3}{\underset{|}{C}}=CH_2$

(*d*) $(CH_3)_3C\ddot{\underset{..}{O}}\!{}^{\ominus} + CH_3I \longrightarrow$ (5) $(CH_3)_3-C-O-CH_3$

(*e*) Ethylene oxide $\xrightarrow[\text{H}_2\text{O}]{\overset{\oplus}{(6)}}$ $HOCH_2CH_2OH$ $\xrightarrow[\substack{H_2SO_4,\\ \text{heat}}]{\text{NaBr}}$ (7) $Br-CH_2-CH_2-Br$

(*f*) $CH_3—\overset{\text{OH}\quad\text{OH}}{\underset{}{\overset{|\quad\ \ |}{CH—CH}}}—CH_3 \xrightarrow{\text{HIO}_4}$ (8) $2\ CH_3-\overset{\displaystyle O}{\overset{\|}{C}}-H$

(*g*) $CH_2=CH_2 \xrightarrow{(9)} HOCH_2CH_2Cl \xrightarrow{(10)} CH_2=CHCl$

$\downarrow{(11)}$

$(ClCH_2CH_2)_2O \xrightarrow[\text{alcohol}]{\text{KOH}}$ (12) $CH_2=CH-O-CH=CH_2$

(*h*) $C_6H_5-CH=CH_2$ $\xrightarrow{Ar-\overset{\overset{\displaystyle O}{\|}}{C}-O-OH}$ (13)

$\downarrow H_3\ddot{O}^{\oplus}$ $\overset{\text{1. }(BH_3)_2}{\underset{\text{2. }:\ddot{O}H^{\ominus},\ H_2O_2}{\searrow}}$

(14)   (15)

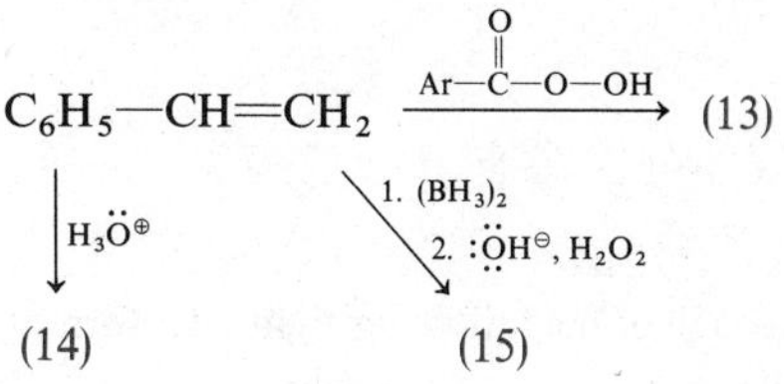

(*i*) $(CH_3)_2CHCH_2-O-C(CH_3)_3$ $\xrightarrow[\text{heat}]{\text{conc HI}}$ (16)

(*j*) $C_6H_5-O-C(CH_3)_3$ $\xrightarrow[\text{heat}]{\text{conc HI}}$ (17)

(*k*) Cl—⟨○⟩—OH $\xrightarrow[\text{Na OH}]{(18)}$ Cl—⟨○⟩—$\ddot{O}:^{\ominus}$ $\xrightarrow{(19)}$ Cl—⟨○⟩—$O-CH_2CH_2CH_3$

$\downarrow (CH_3O)_2SO_2$

(20)

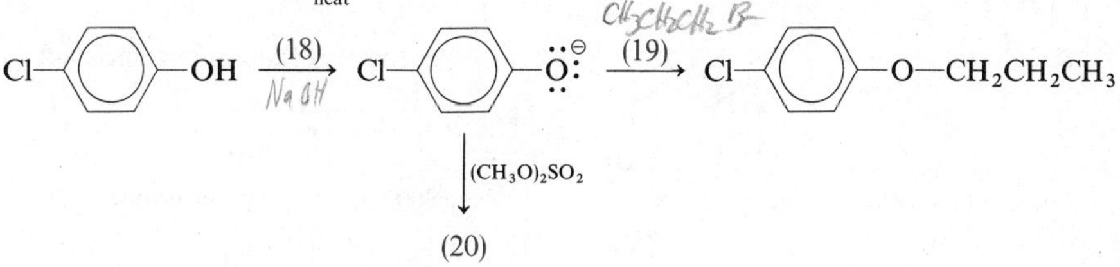

(*l*) $HO(CH_2)_4OH$ $\xrightarrow{PBr_3}$ (21)

(*m*) $C_6H_5-CH\overset{\displaystyle O}{\underset{}{\diagdown\diagup}}CH_2$ $\xrightarrow{C_2H_5OH,\ H^{\oplus}}$ (22)

$\downarrow\ \overset{C_2H_5\ddot{O}:^{\ominus}Na^{\oplus},}{C_2H_5\ddot{O}H}$

(23) $\xrightarrow[\text{heat}]{H_2SO_4}$ (24) $\xrightarrow{H_2,\ Pt}$ (25)

(*n*) $CH_2\overset{\displaystyle O}{\underset{}{\diagdown\diagup}}CH_2$ $\xrightarrow[H_2O]{H\ddot{O}:^{\ominus}}$ (26) $\xrightarrow{HIO_4}$ (27)

$\downarrow (28)$

$HOCH_2CH_2Br$ $\xrightarrow{(29)}$ $CH_2\overset{}{\underset{O}{\diagdown\diagup}}CH_2$

$\downarrow H\ddot{O}:^{\ominus}$

(30)

(*o*) $(CH_3)_2CHCH_2OH$ $\xrightarrow{H_2SO_4 \atop 140°}$ (31)

$\downarrow$ Na metal

(32) $\xrightarrow{(33)}$ $(CH_3)_2CHCH_2-O-CH_2CH_3$

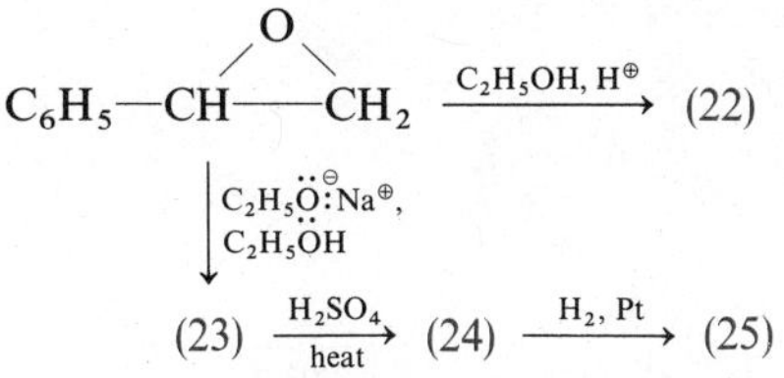

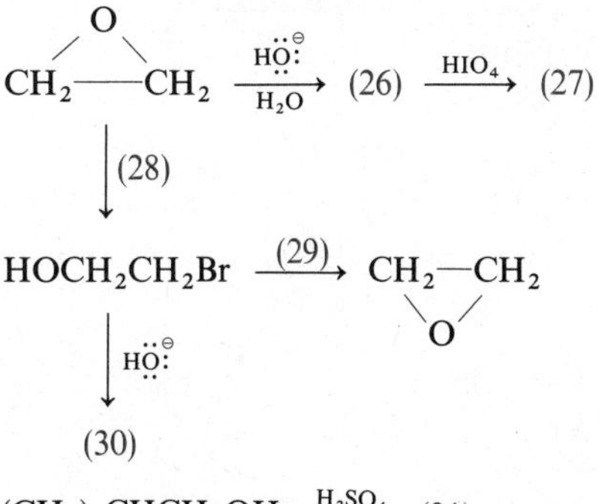

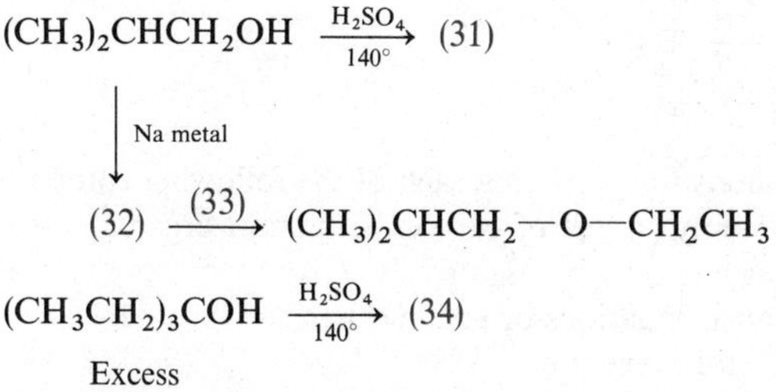

(*p*) $(CH_3CH_2)_3COH$ $\xrightarrow{H_2SO_4 \atop 140°}$ (34)
      Excess

(*q*) $C_6H_5\ddot{O}:^{\ominus}Na^{\oplus} + CH_2=CH-CH_2-Br$ $\xrightarrow[\text{solvent}]{\text{polar}}$ (35) $\xrightarrow{H_2,\ Pt}$ (36)

$\downarrow\ CH_2\overset{}{\underset{O}{\diagdown\diagup}}CH_2$

(37)

(*r*) $\overset{H\text{\tiny//}}{\underset{C_6H_5}{}}C\overset{\displaystyle O}{\underset{}{\diagdown\diagup}}C\overset{C_6H_5}{\underset{H}{}}$ $\xrightarrow{H^{\oplus},\ H_2O}$ (38) (Give three-dimensional structure)

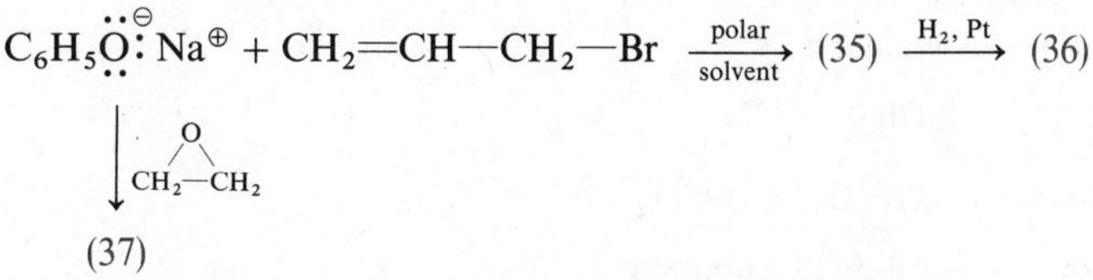

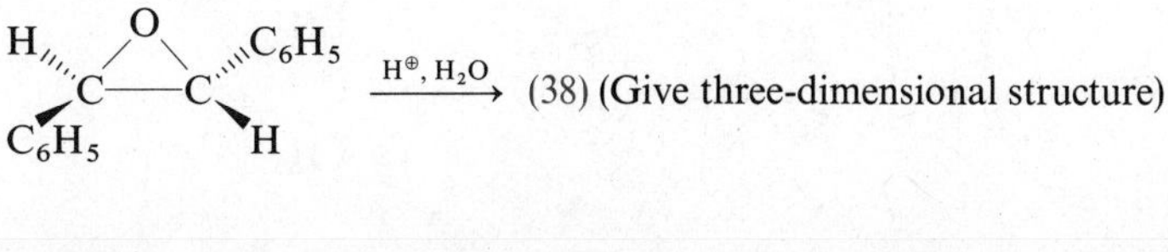

(s) [structure: benzene ring with OCH$_3$ substituent] $\xrightarrow[\text{H}_2\text{SO}_4]{\text{HNO}_3}$ (39), C$_7$H$_7$NO$_3$ $\xrightarrow[\text{heat}]{\text{conc HBr}}$ (40)

19.36 Devise suitable methods for carrying out each of the following eight transformations using any other needed organic or inorganic reagents.

(a) Ethylbenzene $\longrightarrow$ C$_6$H$_5$—CH—CH$_2$ (with OH groups on both carbons)

(b) Cyclohexene $\longrightarrow$ [cyclohexane ring with OH and OCH$_3$]   (Stereochemistry not important)

(c) Cyclohexene $\longrightarrow$ [cyclohexane ring with OH and O—C$_6$H$_5$]   (Stereochemistry not important)

(d) CH$_3$C≡CH $\longrightarrow$ [epoxide structure with H, H on top carbons and CH$_3$, CH$_2$CH$_3$ below]

cis Isomer

(e) Chlorobenzene $\longrightarrow$ O$_2$N—[benzene ring]—O—C(CH$_3$)$_3$ [with three CH$_3$ groups]

(f) Benzene $\longrightarrow$ [benzene ring with OCH$_3$ and Br substituents]

(g) [cyclopentane ring]=CH$_2$ $\longrightarrow$ [cyclopentane ring with C(OH) and CH$_2$OCH$_3$]

(h) [cyclopentane ring]=CH$_2$ $\longrightarrow$ [cyclopentane ring with C(OCH$_3$) and CH$_2$OH]

19.37 Devise suitable synthetic methods for preparing each of the following compounds, using the given starting material(s) as the only source of carbon. Common organic solvents (ether, CCl$_4$, and so on) and inorganic reagents may be used freely. Give reactants and approximate reaction conditions, but do not balance equations or give mechanisms.

(a) Cyclohexane $\longrightarrow$ *trans*-1,2-cyclohexanediol

(b) Cyclohexane $\longrightarrow$ *cis*-1,2-cyclohexanediol

(c) Di-*n*-butyl ether $\longrightarrow$ *n*-octane

(d) Propylene $\longrightarrow$ CH$_2$=C(CH$_3$)—O—C(CH$_3$)=CH$_2$

(e) Tetrahydrofuran $\longrightarrow$ 1,4-dichlorobutane

(f) Toluene + C$_3$ compounds $\longrightarrow$ CH$_3$—[benzene ring]—O—CH(CH$_3$)$_2$

(*g*) 1,4-Butanediol $\longrightarrow$ CH$_2$—CH—CH——CH$_2$
 $\qquad\qquad\qquad\qquad\qquad$ O $\qquad$ O

(*h*) ⬡—CH$_2$OH $\longrightarrow$ ⬡⟨CH$_2$ / O⟩

(*i*) Cyclohexene $\longrightarrow$ 1,6-hexanediol

19.38 Arrange these three compounds in order of increasing boiling point, and briefly justify your choice.

$$CH_3CH_2—\underset{\underset{OH}{|}}{CH}—CH_2CH_2CH_3 \qquad CH_3CH_2CH_2CH_2CH_2CH_2CH_3$$

$$CH_3CH_2CH_2—O—CH_2CH_2CH_3$$

19.39 Diethyl ether is slightly soluble in water but completely soluble in concentrated hydrochloric acid. Explain.

19.40 *tert*-Butyl phenyl ether is difficult to obtain and is not produced when sodium phenoxide reacts with *tert*-butyl bromide. (What are the products from this reaction?) However, when bromobenzene reacts with potassium *tert*-butoxide in *tert*-butyl alcohol, *tert*-butyl phenyl ether is produced in fair yield. Suggest two possible mechanisms to explain its formation, and then suggest an experiment that could be used to show that the reaction does *not* occur by nucleophilic aromatic substitution (which it does not).

19.41 Compare the reactivities of ethylene oxide and trimethylene oxide, $(CH_2)_3O$, in ring-opening reactions. (*Hint:* Some analogies can be made between these two compounds and cyclopropane and cyclobutane; the latter two compounds were discussed in Chap. 3.)

19.42 When propylene oxide is treated with pure, anhydrous HBr under polar conditions, two possible products might be formed as shown. Explain mechanistically which one is preferred (that is, formed in major amounts).

$$CH_2——CH—CH_3 \xrightarrow[\substack{polar \\ solvent}]{pure\ HBr} \underset{\underset{OH}{|}}{CH_2}—\underset{\underset{Br}{|}}{CH}—CH_3 + \underset{\underset{Br}{|}}{CH_2}—\underset{\underset{OH}{|}}{CH}—CH_3$$

19.43 Provide reasonable mechanisms for the following five reactions; show all steps and intermediates.

(*a*) $(CH_3CH_2CH_2CH_2)_2O \xrightarrow[\substack{HBr \\ heat}]{conc} 2CH_3CH_2CH_2CH_2Br$

(*b*) $C_6H_5—O—C(CH_3)_3 \xrightarrow[heat]{conc\ HI} C_6H_5OH + (CH_3)_3CI$

(*c*) $C_6H_5—O—CH_3 \xrightarrow[heat]{conc\ HBr} C_6H_5OH + CH_3Br$

(*d*) ⬡⟨OH / OH⟩ + CH$_2$I$_2$ $\xrightarrow[H_2O]{:\ddot{O}H^\ominus}$ ⬡⟨O / O⟩CH$_2$

(*e*) $CH_3CH_2OH + (CH_3)_2C{=}CH_2 \xrightarrow[\substack{heat, \\ pressure}]{H^\oplus} CH_3CH_2—O—C(CH_3)_3$

19.44 Given a sample of *optically active* 2-butanol, suggest methods to convert it into optically active *sec*-butyl methyl ether with absolute configuration (*a*) the same as that of the starting alcohol and (*b*) opposite that of the starting alcohol.

19.45 In the mass spectrometer, the fragmentation of any compound that is capable of generating the benzyl cation does so. Why?

⬡—$\overset{\oplus}{C}H_2$

Benzyl cation

Each of the following two compounds exhibits a reasonably intense peak at m/e ratio 91. This is not the benzyl cation. It is an aromatic molecule; what is it? How does it result (mechanistically)?

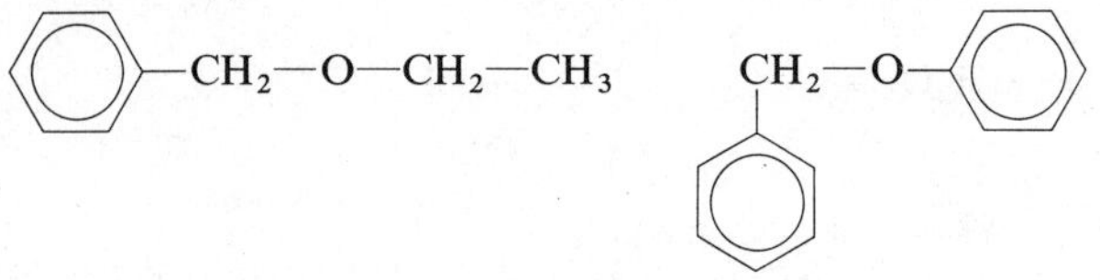

$$\langle\bigcirc\rangle\text{—CH}_2\text{—O—CH}_2\text{—CH}_3 \qquad \text{CH}_2\text{—O—}\langle\bigcirc\rangle$$

19.46 When certain alkyl halides react with alcohols, the products are ethers and the reaction is called *alcoholysis*. Explain mechanistically why *tert*-butyl bromide in isopropyl alcohol produces a reasonable amount of isopropyl *tert*-butyl ether, whereas *n*-butyl bromide under the same conditions does not react at all. How would you expect allyl bromide, $\text{CH}_2\text{=CH—CH}_2\text{—Br}$, to behave under these conditions? Why?

19.47 For each of the following 10 pairs of compounds, indicate a simple chemical test that could be used to distinguish one member from the other. What would you *observe* in each case as positive evidence that a reaction had occurred?

(*a*) cyclopentane and tetrahydrofuran
(*b*) cyclohexyl methyl ether and anisole
(*c*) *n*-butyl bromide and *n*-butyl ether
(*d*) *n*-propyl ether and allyl ether
(*e*) cyclohexyl methyl ether and cyclohexene
(*f*) propylene glycol and 1,3-propanediol
(*g*) ethylene oxide and ethyl ether
(*h*) ethylene glycol and allyl chloride
(*i*) allyl alcohol and glycerol
(*j*) 1,2-butanediol and 1,3-butanediol

19.48 When anisole is fragmented, the mass spectrum contains a prominent peak at m/e 65. The same peak is observed in the mass spectrum of phenol. This transient species is nonaromatic. What is it? Draw a mechanism illustrating the formation of this species from both anisole and phenol.

19.49 *p*-Toluenesulfonate esters (tosylates, see Sec. 19.6), like alkyl halides, undergo both substitution and elimination, depending on the nature of the nucleophile. Suppose ethyl tosylate, $\text{CH}_3\text{CH}_2\text{OTs}$, reacts with hydroxide ion and both ethanol and ethylene are produced from the reaction. Provide complete mechanisms to show the formation of both these products. Do you think this elimination reaction is E1 or E2? Why?

19.50 Consider the following reaction sequence:

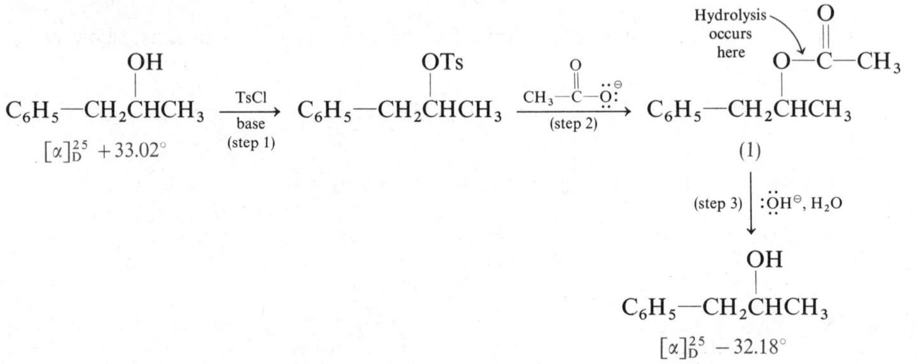

$$\begin{array}{c}\overset{\text{OH}}{\underset{|}{}} \\ \text{C}_6\text{H}_5\text{—CH}_2\text{CHCH}_3 \\ [\alpha]_D^{25} \; +33.02° \end{array} \xrightarrow[\substack{\text{base} \\ \text{(step 1)}}]{\text{TsCl}} \begin{array}{c}\overset{\text{OTs}}{\underset{|}{}} \\ \text{C}_6\text{H}_5\text{—CH}_2\text{CHCH}_3 \end{array} \xrightarrow[\text{(step 2)}]{\underset{}{\text{CH}_3\text{—C—O:}^{\ominus}}} \begin{array}{c}\overset{}{\underset{}{}} \\ \text{C}_6\text{H}_5\text{—CH}_2\text{CHCH}_3 \\ (1) \end{array}$$

Hydrolysis occurs here $\searrow$ $\overset{O}{\overset{||}{O\text{—C—CH}_3}}$

$$\text{(step 3)} \Big\downarrow :\overset{..}{\text{O}}\text{H}^{\ominus}, \text{H}_2\text{O}$$

$$\begin{array}{c}\overset{\text{OH}}{\underset{|}{}} \\ \text{C}_6\text{H}_5\text{—CH}_2\text{CHCH}_3 \\ [\alpha]_D^{25} \; -32.18° \end{array}$$

The carbon-oxygen bond indicated by an arrow in compound (1) is broken in step 3. With this and your knowledge about the stereochemistry of tosylate formation (step 1), what can you deduce about the stereochemistry and thus the mechanism of step 2? Draw the mechanism of this substitution reaction, paying special attention to the nucleophilic properties of the acetate ion,

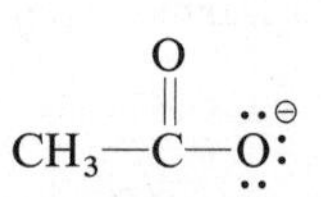

$$\text{CH}_3\text{—}\overset{\overset{\textstyle O}{||}}{\text{C}}\text{—}\overset{..}{\underset{..}{\text{O}}}\text{:}^{\ominus}$$

19.51 Compound *A*, $\text{C}_{11}\text{H}_{16}\text{O}$, gives two new compounds, *B*, $\text{C}_4\text{H}_9\text{Br}$, and *C*, $\text{C}_7\text{H}_8\text{O}$, on heating with concentrated HBr for an extended time. *B* reacts at a moderate rate with both sodium iodide in acetone and alcoholic silver nitrate solution. The NMR spectrum of *B* is shown.

C gives positive results on the ferric chloride test, and on treatment with aqueous bromine solution, *C* gives *D*, $C_7H_5Br_3O$. Draw the structures of compounds *A* to *D*, and explain the reactions described.

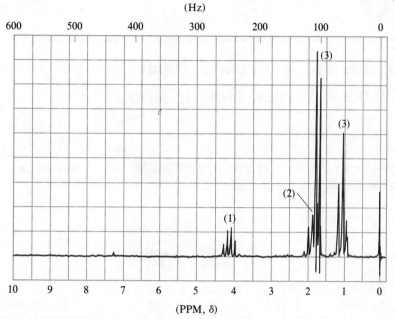

PROBLEM 19.51 NMR spectrum of unknown (*B*). [Data from *The Aldrich Library of NMR Spectra*, C. J. Pouchert and J. R. Campbell (eds.), copyright 1974, Aldrich Chemical Co., Milwaukee]

20

Aldehydes and Ketones: Preparation of the Carbon-Oxygen Double Bond

In this chapter we study the **carbonyl group, C=O,** and two new families of organic compounds, **aldehydes** and **ketones:**

$$\underset{\substack{\text{Aldehyde}}}{R-\overset{\displaystyle\overset{O}{\|}}{C}-H} \qquad \underset{\substack{\text{Ketone}}}{R-\overset{\displaystyle\overset{O}{\|}}{C}-R'}$$

R and R' may be alkyl or aryl

Aldehydes and ketones are similar in their molecular structures and their reactions with many reagents. Several general methods can be used to prepare either of them, and whether one obtains aldehyde or ketone depends on the structure of the starting material and/or the reaction conditions. On the other hand, some methods provide only aldehydes and others yield only ketones. In Chap. 21 we study the reactions of aldehydes and ketones *as they involve the carbon-oxygen double bond itself*.

20.1   Structure of Carbonyl Group

The carbon-oxygen double bond (see Sec. 2.7) is analogous in some respects to the carbon-carbon double bond. The carbonyl-carbon is sp^2 hybridized, with three sp^2 hybrid orbitals lying 120° apart and a p orbital perpendicular to the plane of the sp^2 orbitals. The oxygen is perhaps best treated as being unhybridized; one p orbital on oxygen overlaps with an sp^2 hybrid orbital of carbon to form the σ bond, and a perpendicular p orbital on oxygen overlaps with the p orbital of carbon to form the π bond. The remaining s and p orbitals on oxygen contain the two unshared pairs of electrons. The following sketch shows the bonding and structure of the carbonyl group:

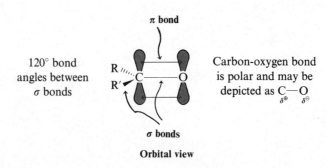

120° bond angles between σ bonds

Carbon-oxygen bond is polar and may be depicted as $\underset{\delta\oplus}{C}-\underset{\delta\ominus}{O}$

Orbital view

20.2 Nomenclature

A. Aldehydes

a. Common Names

The names of simple aldehydes are derived from the common names of the corresponding carboxylic acids (see Chap. 22). The characteristic -*ic acid* ending in the common name of the carboxylic acid is dropped and replaced by -*aldehyde*. For example, the common name for the one-carbon carboxylic acid, HCOOH, is formic acid, so the name of the corresponding aldehyde, HCHO, is *formaldehyde*. Similarly, CH_3CHO is called *acetaldehyde* and is derived from acetic acid, CH_3COOH. The common names for various simple aldehydes are listed in Table 20.1; the first five continuous-chain aldehydes are as follows:

$$\underset{\text{Formaldehyde}}{H\overset{\displaystyle O}{\overset{\|}{-}C}-H} \qquad \underset{\text{Acetaldehyde}}{CH_3-\overset{\displaystyle O}{\overset{\|}{C}}-H} \qquad \underset{\text{Propionaldehyde}}{CH_3CH_2-\overset{\displaystyle O}{\overset{\|}{C}}-H}$$

$$\underset{\text{Butyraldehyde}}{CH_3CH_2CH_2-\overset{\displaystyle O}{\overset{\|}{C}}-H} \qquad \underset{\text{Valeraldehyde}}{CH_3CH_2CH_2CH_2-\overset{\displaystyle O}{\overset{\|}{C}}-H}$$

Substituted aldehydes can be named as derivatives of the parent, continuous-chain aldehyde. The parent compound must contain the aldehyde group, and the carbon atoms on the carbon chain are assigned Greek letters in the following manner:

$$\overset{\varepsilon}{C}-\overset{\delta}{C}-\overset{\gamma}{C}-\overset{\beta}{C}-\overset{\alpha}{C}-\underset{\underset{\displaystyle H}{|}}{C}=O$$

The names and Greek-letter designations of the substituents indicate their location on the longest chain. For example:

Phenylacetaldehyde
(α is omitted because
this is a unique name)

β-Methylbutyraldehyde

α,β-Dimethylvaleraldehyde

The "aldehyde group," $-\overset{\displaystyle O}{\overset{\|}{C}}-H$, can be called *formyl*, which must be used in the naming of certain compounds; for example:

4-Formylbenzenesulfonic acid

3-Formylcyclohexene

b. IUPAC Names

The IUPAC system of naming aldehydes uses the parent alkane name corresponding to the longest continuous chain containing the aldehyde group. The final -**e** of the alkane name is dropped and replaced by -**al,** the characteristic ending for an aldehyde. The numbering of the longest continuous chain starts with C-1 as the carbon bearing the aldehyde group. Substitutents are indicated by name and position number. The following examples illustrate this system; some common names are indicated in parentheses:

$$H-\overset{\overset{\textstyle O}{\|}}{C}-H \qquad CH_3-\overset{\overset{\textstyle O}{\|}}{C}-H \qquad CH_3CH_2CH_2CH_2-\overset{\overset{\textstyle O}{\|}}{C}-H \qquad CH_3(CH_2)_8-\overset{\overset{\textstyle O}{\|}}{C}-H$$

Methanal   Ethanal   Pentanal   Decanal
(formaldehyde)   (acetaldehyde)   (valeraldehyde)

Benzaldehyde   o-Bromobenzaldehyde   o-Hydroxybenzaldehyde
(salicylaldehyde)

$$\underset{5}{CH_3CH_2}-\underset{4}{CH}=\underset{3}{CH}-\underset{2}{}\overset{\overset{\textstyle O}{\|}}{\underset{1}{C}}-H \qquad \underset{5}{CH_3CH_2}\underset{4}{CH_2}-\underset{3}{}\underset{2}{CH}-\overset{\overset{\textstyle O}{\|}}{\underset{1}{C}}-H$$

2-Pentenal   2-Ethylpentanal
(α-ethylvaleraldehyde)

$$\underset{4}{CH_3}-\underset{3}{\overset{\overset{\textstyle \text{(phenyl)}}{|}}{\underset{\underset{\textstyle CH_3}{|}}{C}}}-\underset{2}{CH_2}-\overset{\overset{\textstyle O}{\|}}{\underset{1}{C}}-H$$

3-Methyl-3-phenylbutanal
(β-methyl-β-phenylbutyraldehyde)

Be *very careful not* to intermix the naming systems. *Common names require the use of Greek-letter prefixes, and IUPAC names require numbers.* Note also that Greek letters start on the carbon adjacent to the aldehyde group, whereas IUPAC numbering starts *with* the aldehyde carbon.

B. Ketones

a. Common Names

Ketones can be named by using the alkyl or aryl group names of the substituents that are attached to the carbonyl group and following them by the word *ketone*. There is also a simple, common name for the simplest ketone, CH_3COCH_3, *acetone*. For example:

$$CH_3-\overset{\overset{\textstyle O}{\|}}{C}-CH_3 \qquad CH_3CH_2-\overset{\overset{\textstyle O}{\|}}{C}-CH_3$$

Dimethyl ketone   Methyl ethyl ketone
or
acetone

$$CH_2=CH-\overset{\overset{\textstyle O}{\|}}{C}-CH_3 \qquad \overset{\overset{\textstyle O}{\|}}{C}-CH_2CH_2CH_2CH_3$$

Methyl vinyl ketone   Butyl phenyl ketone

Ketones that have the grouping $C_6H_5-\overset{\overset{\textstyle O}{\|}}{C}-R(-Ar)$ are designated by the term *phenone*, which is preceded by the name of the carboxylic acid, RCOOH or ArCOOH, from which the ketone is derived; the *-ic acid* (or *-oic acid*) ending is dropped from the carboxylic acid name and replaced by *-o-* before the *-phenone* ending is attached. The following three examples illustrate this, and the names of the carboxylic acids (which we have not studied) are included:

Acetophenone

derived from $HO-\overset{\overset{\textstyle O}{\|}}{C}-CH_3$, acetic acid

Benzophenone

derived from $HO-\overset{\overset{\textstyle O}{\|}}{C}-$, benzoic acid

Propiophenone

derived from $HO-\overset{\overset{\textstyle O}{\|}}{C}-CH_2CH_3$, propionic acid

b. IUPAC Names

The longest chain that contains the keto group is used as the parent compound in the IUPAC system. Ketone names are derived from the corresponding alkane name by dropping the **-e** ending and replacing it with **-one**. The substituents are identified by name and position number, and the longest chain is numbered so that the carbon-oxygen double bond is given the lowest possible number; for example:

$CH_3-\overset{\overset{\textstyle O}{\|}}{C}-CH_3$

Propanone

$CH_3CH_2-\overset{\overset{\textstyle O}{\|}}{C}-CH_3$

Butanone

$\underset{5\quad\quad 4}{CH_3CH_2}-\underset{3}{\overset{\overset{\textstyle O}{\|}}{C}}-\underset{2\quad\quad 1}{CH_2CH_3}$

3-Pentanone

$\underset{5}{CH_3}-\underset{4}{\overset{\overset{\textstyle O}{\|}}{C}}-\underset{3}{CH_2}-\underset{2}{\overset{\overset{\textstyle O}{\|}}{C}}-\underset{1}{CH_3}$

2,4-Pentanedione

$\underset{4}{CH_2}=\underset{3}{CH}-\underset{2}{\overset{\overset{\textstyle O}{\|}}{C}}-\underset{1}{CH_3}$

3-Buten-2-one

$\underset{1}{CH_3}-\underset{2}{\overset{\overset{\textstyle CH_3}{|}}{C}}-\underset{3}{CH_2}-\underset{4}{\overset{\overset{\textstyle O}{\|}}{C}}-\underset{5\quad 6\quad 7}{CH_2CH_2CH_3}$

2-Methyl-2-phenyl-4-heptanone

1,4-Cyclohexanedione

2,2-Dimethylcyclohexanone

The word *oxo* is sometimes used as a prefix for the carbonyl group; for example:

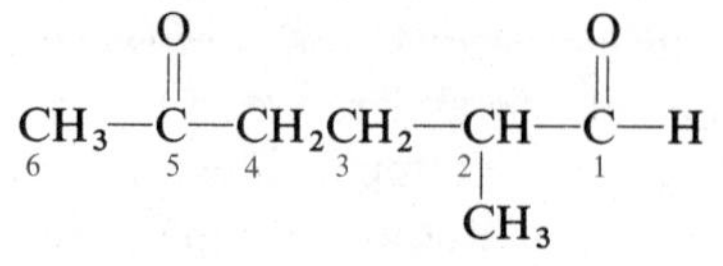

$$CH_3-\underset{6}{\overset{\overset{\displaystyle O}{\|}}{C}}-\underset{5}{CH_2}\underset{4}{CH_2}-\underset{\underset{\displaystyle CH_3}{2}}{CH}-\underset{1}{\overset{\overset{\displaystyle O}{\|}}{C}}-H$$

2-Methyl-5-oxohexanal

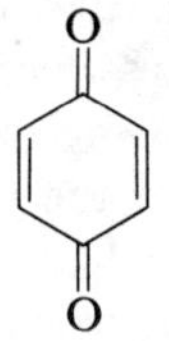

3,6-Dioxo-1,4-cyclohexadiene
(1,4-benzoquinone)

Substituted benzophenones can be named by numbering the two aromatic rings, using numbers (1, 2, and so on) for one ring and primed numbers (1', 2', and so forth) for the other ring; for example:

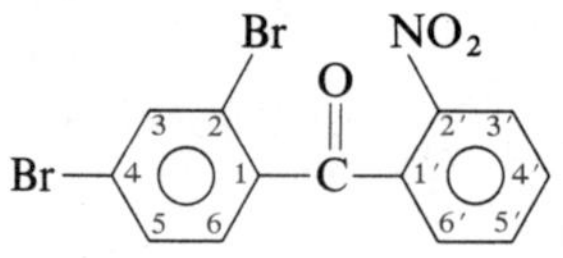

2,4-Dibromo-2'-nitrobenzophenone

Question 20.1

Draw the structural isomers for the following, giving names as required:

(*a*) All aldehydes having the molecular formula C_4H_8O, common and IUPAC names
(*b*) All aldehydes having the molecular formula $C_5H_{10}O$, common and IUPAC names
(*c*) All ketones having the molecular formula $C_5H_{10}O$, common and IUPAC names
(*d*) All aldehydes and ketones having the molecular formula $C_6H_{12}O$, IUPAC names
(*e*) All aromatic aldehydes and ketones having the molecular formula C_8H_8O, provide a suitable name
(*f*) All aromatic aldehydes and ketones having the molecular formula $C_9H_{10}O$, provide a suitable name.

Question 20.2

We saw how different families of organic compounds can have the same general molecular formula. For example, C_nH_{2n+2} can represent only an alkane, whereas C_nH_{2n} can represent either an alkene or a cycloalkane (which we know to have quite different physical and chemical properties). For each of the following five general formulas, indicate the possible type of bonding that may be present. Consider only those functional groups studied thus far.

(*a*) C_nH_{2n-2} (*b*) C_nH_{2n-4} (*c*) $C_nH_{2n+2}O$ (*d*) $C_nH_{2n}O$ (*e*) $C_nH_{2n-2}O$

20.3   Physical Properties of Aldehydes and Ketones

Selected physical properties of some typical aldehydes and ketones are listed in Tables 20.1 and 20.2, respectively.

In general, aldehydes and ketones boil at higher temperatures than ethers and at lower temperatures than alcohols that have the same carbon content. They also boil at higher temperatures than alkanes and alkenes of comparable molecular weight.

TABLE 20.1   Physical Properties of Selected Aldehydes

Common Name	IUPAC Name	Structure	mp, °C	bp, °C	Density at 20°, g/cc
Formaldehyde	Methanal	CH_2O	−92	−21	0.815
Acetaldehyde	Ethanal	CH_3CHO	−123	21	0.781
Propionaldehyde	Propanal	CH_3CH_2CHO	−81	49	0.807
n-Butyraldehyde	Butanal	$CH_3(CH_2)_2CHO$	−97	75	0.817
Isobutyraldehyde	2-Methylpropanal	$(CH_3)_2CHCHO$	−66	61	0.794
n-Valeraldehyde	Pentanal	$CH_3(CH_2)_3CHO$	−91	103	0.819
Isovaleraldehyde	3-Methylbutanal	$(CH_3)_2CHCH_2CHO$	−51	93	0.803
Caproaldehyde	Hexanal	$CH_3(CH_2)_4CHO$		129	0.834
n-Heptaldehyde	Heptanal	$CH_3(CH_2)_5CHO$	−45	155	0.850
Acrolein	Propenal	$CH_2{=}CH{-}CHO$	−88	53	0.841
Crotonaldehyde	2-Butenal	$CH_3{-}CH{=}CH{-}CHO$	−77	104	0.859
	Benzaldehyde	C_6H_5CHO	−56	179	1.046
	o-Tolualdehyde	$o\text{-}CH_3{-}C_6H_4{-}CHO$		196	1.039
	m-Tolualdehyde	$m\text{-}CH_3{-}C_6H_4{-}CHO$		199	1.019
	p-Tolualdehyde	$p\text{-}CH_3{-}C_6H_4{-}CHO$		205	1.019
Salicylaldehyde	o-Hydroxybenzaldehyde	$o\text{-}HO{-}C_6H_4{-}CHO$	2	197	1.146
	p-Hydroxybenzaldehyde	$p\text{-}HO{-}C_6H_4{-}CHO$	116		1.129

TABLE 20.2   Physical Properties of Selected Ketones

Common Name	IUPAC Name	Structure	mp, °C	bp, °C	Density at 20°, g/cc
Acetone	Propanone	CH_3COCH_3	−95	56	0.792
Methyl ethyl ketone	Butanone	$CH_3COCH_2CH_3$	−86	80	0.805
Methyl n-propyl ketone	2-Pentanone	$CH_3COCH_2CH_2CH_3$	−78	102	
Diethyl ketone	3-Pentanone	$CH_3CH_2COCH_2CH_3$	−41	101	0.814
	2-Hexanone	$CH_3CO(CH_2)_3CH_3$	−57	127	0.830
	3-Hexanone	$CH_3CH_2COCH_2CH_2CH_3$		124	0.818
Di-n-propyl ketone	4-Heptanone	$(CH_3CH_2CH_2)_2CO$	−34	144	0.821
Chloroacetone	Chloropropanone	$ClCH_2COCH_3$	−45	119	1.162
Mesityl oxide	4-Methyl-3-penten-2-one	$(CH_3)_2C{=}CHCOCH_3$	−59	131	0.863
Methyl vinyl ketone	1-Buten-3-one	$CH_2{=}CHCOCH_3$		80	0.863
	Cyclohexanone	⬡=O		157	0.948
Methyl phenyl ketone	Acetophenone	$C_6H_5COCH_3$	21	202	1.024
Ethyl phenyl ketone	Propiophenone	$C_6H_5COCH_2CH_3$	21	218	
Diphenyl ketone	Benzophenone	$C_6H_5COC_6H_5$	48	305	1.083

For example, compare the boiling points of the following five compounds:

$$CH_3CH_2CH_2CH_3 \qquad CH_3—O—CH_2CH_3 \qquad CH_3CH_2—\overset{\displaystyle O}{\overset{\|}{C}}—H$$

Butane	Methoxyethane	Propanal
(*n*-butane)	(methyl ethyl ether)	(propionaldehyde)
bp −0.5°	bp 7.9°	bp 49°
mol wt 58	mol wt 60	mol wt 58

$$CH_3—\overset{\displaystyle O}{\overset{\|}{C}}—CH_3 \qquad CH_3CH_2CH_2—OH$$

Propanone	1-Propanol
(acetone)	(*n*-propyl alcohol)
bp 56°	bp 97.2°
mol wt 58	mol wt 60

The unusually high boiling point of 1-propanol is due to hydrogen bonding in the liquid state, and the boiling point of butane is unusually low because of the complete absence of polar effects and hydrogen bonding. The comparison of the boiling points of propanone, propanal, and methyl ethyl ether is particularly interesting. These three compounds contain the same number of carbon and oxygen atoms and have molecular weights that are quite similar. Yet the ether boils at a much lower temperature than the other two compounds. Why? An ether, as we recall from Sec. 19.2, contains polar carbon-oxygen bonds that are "hidden" beneath adjacent hydrogen atoms or substituents. As a consequence, ether molecules are unable to associate very much in the liquid phase through intermolecular dipole interactions; the molecules cannot get close enough for these short-range forces to be effective and important (Fig. 19.1). On the other hand, the planar and polar carbonyl group in aldehydes and ketones permits them to associate with neighboring molecules in the liquid phase through dipole-dipole interactions (Fig. 20.1). Thus aldehydes and ketones boil at higher temperatures than ethers.

Lower molecular weight aldehydes and ketones show appreciable solubility in water because of the hydrogen bonding between water and the polar carbonyl group:

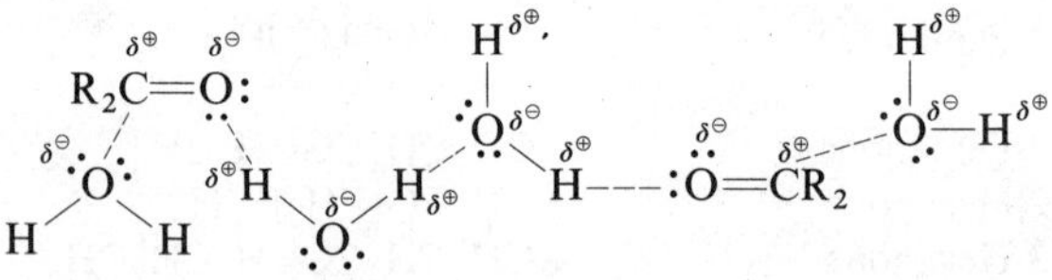

For example, propanone is completely miscible in water, whereas methyl ethyl ether (C_3H_8O) is soluble only to the extent of about 7.5 g/100 g of water at room temperature. The differences in the polarity of these two functional groups are responsible for their solubilities.

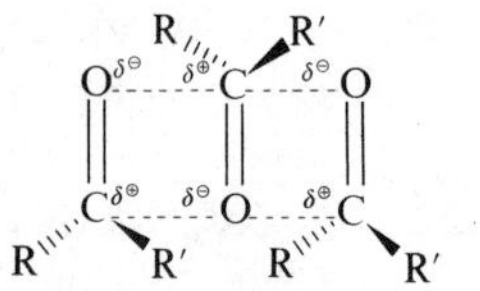

FIGURE 20.1   Intermolecular dipole-dipole interaction that occurs for aldehydes and ketones in the liquid phase.

Dipole moments are in accord with the polarity the carbonyl group imparts to organic molecules. As shown by the following examples, increasing the number of electron-releasing alkyl (methyl) groups attached to the carbonyl group results in increased polarity of the molecule:

| Methanal (formaldehyde) 2.34 D | Ethanal (acetaldehyde) 2.70 D | Propanone (acetone) 2.88 D |

Alcohols, ethers, and even dihalo compounds (which contain the highly polar carbon-halogen bond) are less polar than aldehydes and ketones; for example:

| Difluoromethane (methylene fluoride) 1.93 D | Methanol (methyl alcohol) 1.69 D | Methoxymethane (dimethyl ether) 1.29 D |

CH_3-O-H CH_3-O-CH_3

Question 20.3

Explain why salicylaldehyde (*o*-hydroxybenzaldehyde) has a much lower melting point than *p*-hydroxybenzaldehyde (2° for the former versus 116° for the latter) and salicyladehyde steam distills whereas *p*-hydroxybenzaldehyde does not.

20.4 Methanal (Formaldehyde), Ethanal (Acetaldehyde), and Propanone (Acetone)

Many industrially important aldehydes and ketones are prepared using the **Fischer-Tropsch synthesis,** which is outlined here:

$$CO + H_2 \xrightarrow[\text{coke}]{\text{steam}} \text{aldehydes and ketones}$$

Carbon monoxide and hydrogen are allowed to react while being heated by hot steam. The coke serves as a source of carbon monoxide. Carbonyl compounds are the principal products when the reaction is run at high pressures, whereas hydrocarbons predominate in low-pressure reactions.

Three important industrial carbonyl compounds are discussed in this section.

A. Methanal (Formaldehyde)

Methanal (formaldehyde), the simplest member of the aldehyde family, is a gas (bp $-21°$) and is readily soluble in water, where it forms the corresponding hydrate, $HO-CH_2-OH$:

$$CH_2=O + H_2O \rightleftharpoons HO-CH_2-OH$$

Methanal (formaldehyde)   Methanediol

This is a reversible reaction, and attempted isolation of the hydrate (called methanediol) yields only formaldehyde. An aqueous solution of formaldehyde is called *Formalin*. Formaldehyde readily polymerizes to give two solid compounds: *paraformaldehyde* and the cyclic trimer *trioxane*. If dry, pure formaldehyde is needed for a reaction, it is usually obtained by heating one of the solid compounds:

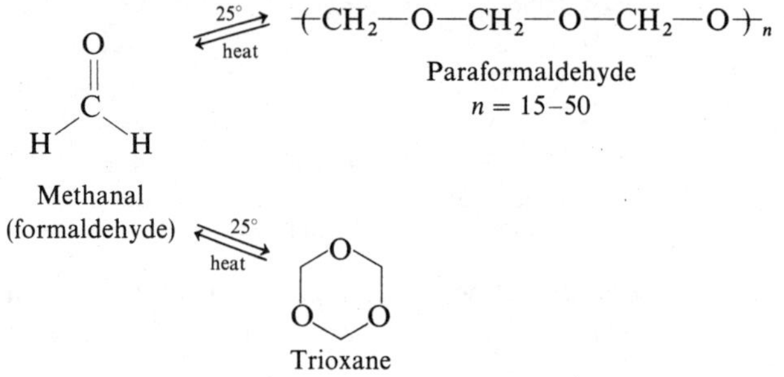

$$CH_2-O-CH_2-O-CH_2-O\}_n$$

Paraformaldehyde

$n = 15-50$

Methanal
(formaldehyde)

Trioxane

Formaldehyde is produced commercially by heating methanol with air in the presence of a copper metal catalyst at high temperatures.

B. Ethanal (Acetaldehyde)

Ethanal (acetaldehyde), $CH_3-\overset{O}{\underset{||}{C}}-H$ (bp 21°), is an important precursor to ethanoic acid (acetic acid). It readily forms the trimer, *paraldehyde*, which boils at 125°. Paraldehyde regenerates pure ethanal on heating with a trace of acid:

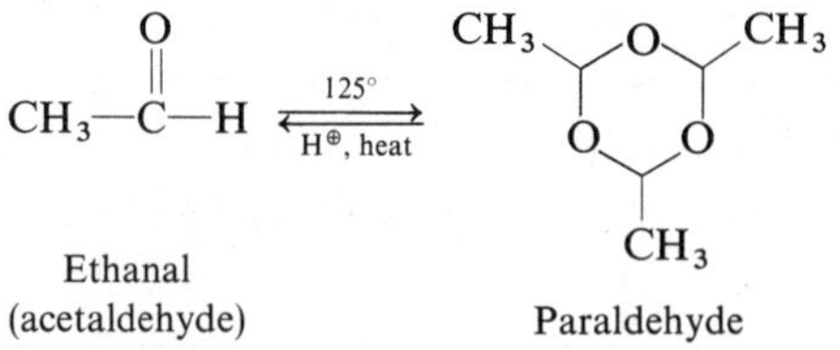

Ethanal
(acetaldehyde)

Paraldehyde

Acetaldehyde is prepared industrially by heating ethanol with copper metal at ~250° (see Sec. 10.10) or by treating ethyne (acetylene) with aqueous sulfuric acid in the presence of mercuric sulfate ($HgSO_4$) catalyst (see Sec. 11.13).

Other, higher molecular weight aldehydes do not spontaneously polymerize and very seldom form trimers.

C. Propanone (Acetone)

Propanone (acetone), $CH_3-\overset{O}{\underset{||}{C}}-CH_3$, is an important and versatile solvent for organic reactions, and it is usually obtained by the air oxidation of 2-propanol or by heating 2-propanol with copper metal at 250°. 2-Propanol is obtained from the acid-catalyzed hydration of propene, a major product of the petroleum industry.

20.5   Preparation of Aldehydes: Summary

We encountered several common methods for preparing aldehydes; these methods are now reviewed.

A. Oxidation of Primary Alcohols (Sec. 10.10)

One of the most important methods for preparing an aldehyde is the controlled oxidation of the corresponding primary alcohol. Chromic acid oxidation is the most common *laboratory* method.

$$R-CH_2-OH \quad \xrightarrow[\text{warm}]{\begin{array}{c}\text{Cu, 200-300}^\circ \\ K_2Cr_2O_7, H_2SO_4 \\ CrO_3, \text{acetic} \\ \text{acid, warm} \\ CrO_3\text{-pyridine}\end{array}} \quad R-\overset{\displaystyle O}{\overset{\displaystyle \|}{C}}-H$$

Primary alcohol   Aldehyde

B. Cleavage of Carbon-Carbon Double Bonds by Ozonolysis (Ozonolysis, Sec. 8.31)

Though used mostly for structure determination, ozonolysis can be used to prepare aldehydes and ketones:

$$\underset{H}{\overset{R'}{\diagdown}}C=C\underset{H}{\overset{R}{\diagup}} \quad \xrightarrow[\text{2. } H_2O, Zn]{\text{1. } O_3} \quad R-\overset{\displaystyle O}{\overset{\displaystyle \|}{C}}-H + R'-\overset{\displaystyle O}{\overset{\displaystyle \|}{C}}-H$$

C. Oxidation of Glycols by Periodic Acid (Sec. 19.18)

Certain glycols can be used in obtaining aldehydes and ketones by the periodic acid oxidation reaction:

$$R-\underset{OH}{\overset{H}{\underset{|}{\overset{|}{C}}}}-\underset{OH}{\overset{H}{\underset{|}{\overset{|}{C}}}}-R' \quad \xrightarrow{HIO_4} \quad R-\overset{\displaystyle O}{\overset{\displaystyle \|}{C}}-H + R'-\overset{\displaystyle O}{\overset{\displaystyle \|}{C}}-H$$

D. Phenolic Aldehydes from the Reimer-Tiemann Reaction (Sec. 19.8, Question 19.12)

Though most useful for the synthesis of salicylaldehyde, the Reimer-Tiemann reaction can be used to synthesize phenolic aldehydes (those containing the hydroxyl and aldehyde group on the ring):

Phenol   Salicylaldehyde

20.6   Other Methods for Preparing Aldehydes

A. Rosenmund Reduction: Preparation of Aliphatic and Aromatic Aldehydes

Another method for preparing aldehydes, one used principally in the laboratory, is **Rosenmund reduction.** It is generally difficult to convert a carboxylic acid, RCOOH, into an aldehyde, because most reducing agents (catalytic hydrogenation and other chemical methods we will discuss) convert the more reactive aldehyde group into the corresponding primary alcohol. However, Rosenmund reduction starts with acid chloride, RCOCl, which is obtained from the corresponding carboxylic acid.[1]

Carefully controlled hydrogenation of the acid chloride gives the corresponding aldehyde, often in high yield. Hydrogenation is controlled by "poisoning" the metal catalyst (usually palladium) with sulfur or barium sulfate:

$$
\underset{\text{Acid chloride}}{(Ar)R-\overset{\displaystyle O}{\overset{\|}{C}}-Cl} \xrightarrow[\text{Pd(BaSO}_4)]{H_2} \underset{\text{Aldehyde}}{(Ar)R-\overset{\displaystyle O}{\overset{\|}{C}}-H}
$$

Used for aliphatic and aromatic aldehydes

The mechanism of this particular reaction is not well established. Two examples of the Rosenmund reduction are:

$$
\underset{\substack{\text{3-Methylbutanoyl chloride}\\ \text{(isovaleryl chloride)}}}{CH_3-\overset{\displaystyle CH_3}{\overset{|}{CH}}-CH_2-\overset{\displaystyle O}{\overset{\|}{C}}-Cl} \xrightarrow[\text{Pd(BaSO}_4)]{H_2} \underset{\substack{\text{3-Methylbutanal}\\ \text{(isovaleraldehyde)}}}{CH_3-\overset{\displaystyle CH_3}{\overset{|}{CH}}-CH_2-\overset{\displaystyle O}{\overset{\|}{C}}-H} \; (+\; HCl)
$$

$$
\underset{\text{Benzoyl chloride}}{\text{C}_6\text{H}_5-\overset{\displaystyle O}{\overset{\|}{C}}-Cl} \xrightarrow[\text{Pd(BaSO}_4)]{H_2} \underset{\text{Benzaldehyde}}{\text{C}_6\text{H}_5-\overset{\displaystyle O}{\overset{\|}{C}}-H}
$$

B. Metal Hydride Reduction: Preparation of Aliphatic and Aromatic Aldehydes

Recent developments in metal hydride chemistry have produced several reagents for carrying out this same reaction. One reagent, *lithium tri-tert-butoxyaluminum hydride*, $LiAlH[OC(CH_3)_3]_3$, is a highly selective reducing agent, and as the name implies, the hydrogen attached to aluminum is a source of hydride ion. The structure of this reducing agent is:

$$
Li^{\oplus}
$$

$$
(CH_3)_3C-O-\overset{\displaystyle O-C(CH_3)_3}{\underset{\displaystyle H}{\overset{|}{\underset{|}{Al}}}}^{\ominus}-O-C(CH_3)_3
$$

Lithium tri-*tert*-butoxyaluminum hydride
Selective reducing agent

---

[1] We have not studied acids or acid chlorides, but for now it is enough to indicate the relationship between these two compounds. An acid chloride can be prepared from the corresponding carboxylic acid by allowing it to react with either phosphorus trichloride (PCl_3) or thionyl chloride ($SOCl_2$); for example:

$$
\underset{\text{Carboxylic acid}}{R-\overset{\displaystyle O}{\overset{\|}{C}}-OH \;\; or \;\; Ar-\overset{\displaystyle O}{\overset{\|}{C}}-OH} \xrightarrow[\text{SOCl}_2]{PCl_3\, or} \underset{\text{Acid chloride}}{R-\overset{\displaystyle O}{\overset{\|}{C}}-Cl \;\; or \;\; Ar-\overset{\displaystyle O}{\overset{\|}{C}}-Cl}
$$

In a sense this reaction resembles the conversion of an alcohol to an alkyl halide because the —OH group in the carboxylic acid is replaced by —Cl (a typical substitution reaction). On the other hand, one cannot use HCl and concentrated H_2SO_4 for acid chloride preparation even though they are commonly used with alcohols. Acids and acid chlorides are discussed in detail in Chaps. 22 and 23.

Recall, however, that a carboxylic acid can be prepared by oxidizing a primary alcohol or from an alkylbenzene by vigorous oxidation.

Treatment of an acid chloride with lithium tri-*tert*-butoxyaluminum hydride also produces the corresponding aldehyde·

$$
\underset{\substack{\text{Acid}\\\text{chloride}}}{(Ar)R-\overset{\overset{\textstyle O}{\|}}{C}-Cl} \xrightarrow{\text{LiAlH[OC(CH}_3)_3]_3} \underset{\text{Aldehyde}}{(Ar)R-\overset{\overset{\textstyle O}{\|}}{C}-H}
$$

Used for
aliphatic
or aromatic
aldehydes

The bulkiness of the reducing agent prevents further reduction of the aldehyde.

The mechanism of this reaction is similar to the nucleophilic substitution reactions we study for carboxylic acid derivatives in Chap. 23. For now, let us state that this reaction involves the nucleophilic substitution of hydride ion ($H{:}^{\ominus}$ on the reducing agent) for $:\!\overset{..}{Cl}\!:^{\ominus}$ on the acid chloride. We study other metal hydride reducing agents later in this chapter.

C. Oxidation of Methylbenzenes: Preparation of Aromatic Aldehydes

The previous methods can be used to prepare aliphatic and aromatic aldehydes, depending on the structure of the starting material. Another method, which is used in the industrial preparation of benzaldehyde, is the controlled oxidation of methylbenzenes. It involves the free-radical chlorination of a methylbenzene, followed by isolation of the dichloro isomer, which is subsequently hydrolyzed. The hydrolysis reaction produces the unstable *gem*-diol, which loses water to form the aldehyde.

$$
\underset{\substack{\text{Methylbenzene}\\ \textit{Must have methyl}\\ \textit{group}}}{Ar-CH_3} \xrightarrow[\substack{\text{heat or}\\\text{light}}]{Cl_2} \underset{\text{(+ other isomers)}}{Ar-\underset{\underset{\textstyle Cl}{|}}{CH}-Cl} \xrightarrow[\substack{\text{CaCO}_3,\\\text{heat}}]{H_2O} \left[\underset{\substack{\textit{gem-}\text{Diol}\\ \textit{Unstable}}}{Ar-\underset{\underset{\textstyle OH}{|}}{CH}-OH} \right] \rightleftharpoons Ar-\overset{\overset{\textstyle O}{\|}}{C}-H + H_2O
$$

The mechanism of this hydrolysis reaction was discussed in Sec. 19.8 (Question 19.12) in connection with the Reimer-Tiemann reaction.

20.7   Preparation of Ketones: Summary

We encountered several methods for preparing ketones, which are given here as a summary and review. Other methods are presented in Sec. 20.8.

A. Oxidation of Secondary Alcohols (Sec. 10.10)

One useful method for preparing ketones is the oxidation of the corresponding secondary alcohol. Chromic acid and basic potassium permanganate are usually used in the laboratory.

$$
\begin{array}{c}
\underset{\substack{\text{Secondary} \\ \text{alcohol} \\ R \text{ and } R' \\ \textit{may be alkyl} \\ \textit{or aryl}}}{
\underset{\displaystyle H}{\overset{\displaystyle OH}{R-C-R'}}
}
\quad
\left[
\begin{array}{c}
\xrightarrow[\text{H}_2\text{SO}_4,\ \text{heat}]{\text{K}_2\text{Cr}_2\text{O}_7} \\[4pt]
\xrightarrow[\text{acid, heat}]{\text{CrO}_3,\ \text{acetic}} \\[4pt]
\xrightarrow{\text{Cu}^\circ,\ 200\text{--}300^\circ} \\[10pt]
\xrightarrow[\text{H}_2\text{O, heat}]{\text{KMnO}_4,\ :\ddot{\text{O}}\text{H}^\ominus} \\[4pt]
\xrightarrow{\text{CrO}_3\text{-pyridine}}
\end{array}
\right]
\quad
\underset{\text{Ketone}}{\overset{\displaystyle O}{R-\overset{\|}{C}-R'}}
$$

B. Cleavage of Carbon-Carbon Double Bonds by Ozone (Sec. 8.31)

$$
\underset{\text{Alkene}}{\underset{R}{\overset{R'}{>}}C=C\underset{R''}{\overset{R'''}{<}}}
\xrightarrow{\text{O}_3}
\xrightarrow[\text{Zn}]{\text{H}_2\text{O}}
R-\overset{\displaystyle O}{\overset{\|}{C}}-R' + R''-\overset{\displaystyle O}{\overset{\|}{C}}-R'''
$$

C. Oxidation of 1,2-Glycols by Periodic Acid (Sec. 19.18)

$$
\underset{\text{1,2-Glycol}}{
R-\underset{\displaystyle OH}{\overset{\displaystyle R'}{C}}-\underset{\displaystyle OH}{\overset{\displaystyle R''}{C}}-R'''
}
\xrightarrow{\text{HIO}_4}
R-\overset{\displaystyle O}{\overset{\|}{C}}-R' + R''-\overset{\displaystyle O}{\overset{\|}{C}}-R'''
$$

D. Hydration of Alkynes (Sec. 11.13)

The acid-catalyzed hydration of acetylene is used in the industrial preparation of ethanal (acetaldehyde). Acetylene is the only alkyne that produces an aldehyde, however; all other alkynes yield ketones:

$$
\underset{\text{Alkyne}}{R-C\equiv C-R}
\xrightarrow[\text{H}_2\text{SO}_4,\ \text{HgSO}_4]{\text{H}_2\text{O}}
\left[
\underset{\substack{\text{Vinyl alcohol} \\ \textit{Unstable}}}{R-\underset{\displaystyle}{\overset{\displaystyle OH}{C}}=\underset{\displaystyle}{\overset{\displaystyle H}{C}}-R}
\right]
\rightleftharpoons
\underset{\text{Ketone}}{R-\overset{\displaystyle O}{\overset{\|}{C}}-CH_2-R}
$$

20.8   Other Methods for Preparing Ketones

In this section we discuss some other common methods for preparing ketones. The oxidation of a secondary alcohol is the most general method for preparing ketones (see Sec. 20.7A). However, the following methods may also be encountered, and each has some limitation.

A. Friedel-Crafts Acylation: Preparation of Aromatic Ketones

As we know, an aromatic ring is alkylated by an alkyl halide in the presence of aluminum trichloride (Friedel-Crafts *alkylation* reaction, see Sec. 14.6). A similar reaction, called **Friedel-Crafts acylation,** occurs with an acid chloride, aluminum trichloride, and an aromatic compound:

$$\text{Ar}-\text{H} + (\text{Ar})\text{R}-\overset{\overset{\textstyle O}{\|}}{\text{C}}-\text{Cl} \xrightarrow[\text{(AlCl}_3)]{\text{anhydrous}} \text{Ar}-\overset{\overset{\textstyle O}{\|}}{\text{C}}-\text{R(Ar)} + \text{HCl}$$

Acid
chloride   Aromatic
ketone

This is called *acylation* because the R—$\overset{\overset{\textstyle O}{\|}}{\text{C}}$— and Ar—$\overset{\overset{\textstyle O}{\|}}{\text{C}}$— moieties, referred to as *acyl* groups, are the substituents being introduced onto the ring. Friedel-Crafts alkylation, on the other hand, introduces an alkyl group, R—, onto the ring.

As the following two examples illustrate, either alkyl aryl ketones or diaryl ketones can be obtained from this reaction:

Benzene   Propanoyl chloride   Propiophenone
             (propionyl chloride)   (alkyl aryl ketone)

p-Methylbenzoyl   4-Methylbenzophenone
chloride   (diaryl ketone)

The mechanism of the acylation reaction appears to be analogous to that of the alkylation reaction.

Step 1. Reaction of AlCl_3 with acid chloride: formation of electrophile

$$\text{R}-\overset{\overset{\textstyle O}{\|}}{\text{C}}-\overset{\cdot\cdot}{\underset{\cdot\cdot}{\text{Cl}}}\colon + \text{AlCl}_3 \longrightarrow \text{R}-\overset{\oplus}{\text{C}}=\overset{\cdot\cdot}{\underset{\cdot\cdot}{\text{O}}} + \text{AlCl}_4^{\ominus}$$

Lewis
acid   Acylium
ion

Electrophile, E$^{\oplus}$

Aluminum trichloride serves as a Lewis acid and removes chloride from the acid chloride to produce the electrophile. (This reaction is analogous to that between an alkyl chloride and AlCl_3: $\text{R}-\text{Cl} + \text{AlCl}_3 \rightarrow \text{R}^{\oplus} + \text{AlCl}_4^{\ominus}$.) The electrophile produced is called the **acylium ion, R**—$\overset{\oplus}{\text{C}}=\overset{\cdot\cdot}{\text{O}}\colon$. This ion is especially stable because of the resonance stabilization of the positive charge by the unshared electrons on oxygen:

$$\left[\text{R}-\overset{\oplus}{\text{C}}=\overset{\cdot\cdot}{\text{O}}\colon \longleftrightarrow \text{R}-\text{C}\equiv\overset{\oplus}{\text{O}}\colon \right]$$

Carbon contains
six electrons and
is electron-deficient   More stable ion contributes to
a larger extent to resonance
hybrid; all atoms contain
octet of electrons

Acylium ion resonance structures

Chemical kinetics show that the rate of acylation depends on the concentrations of the acid chloride, the aromatic compound, and aluminum trichloride, so that all three may well be involved in the transition state for the reaction. Accordingly, it is suggested that the acylium ion and $AlCl_4^{\ominus}$ never completely dissociate before ring attack occurs (step 2). They may exist as either an ion pair or a complex of the structure:

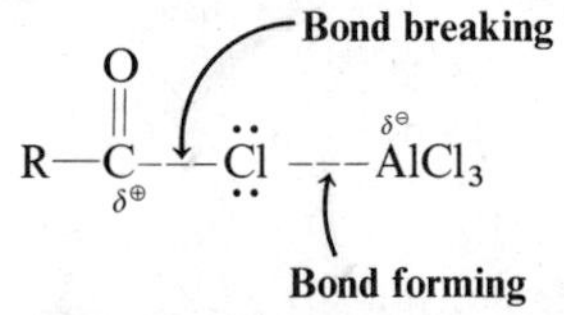

Step 2. Attack on electrophile on aromatic ring: slow reaction

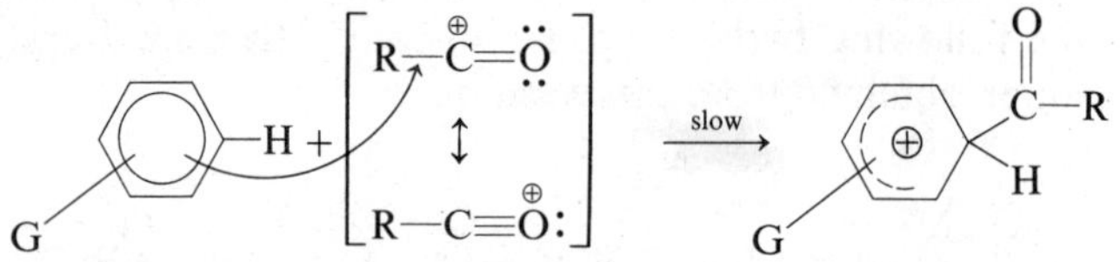

This attack is shown in simplified form although the aluminum chloride-acid chloride complex mentioned in the comments under step 1 is more likely involved in the transition state as shown here:

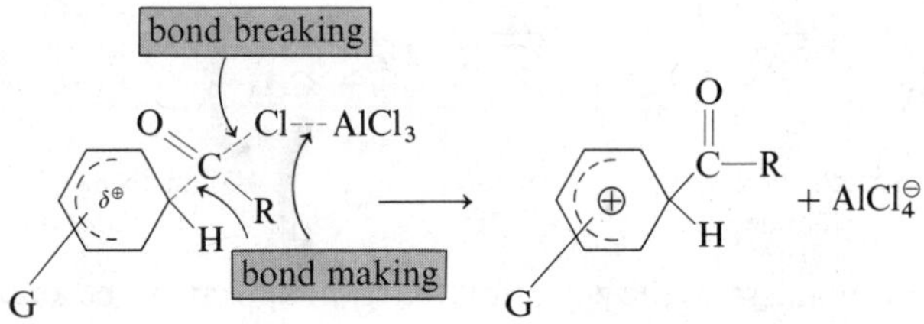

Transition state for acylation   Intermediate

Step 3. Proton loss and product formation: fast reaction

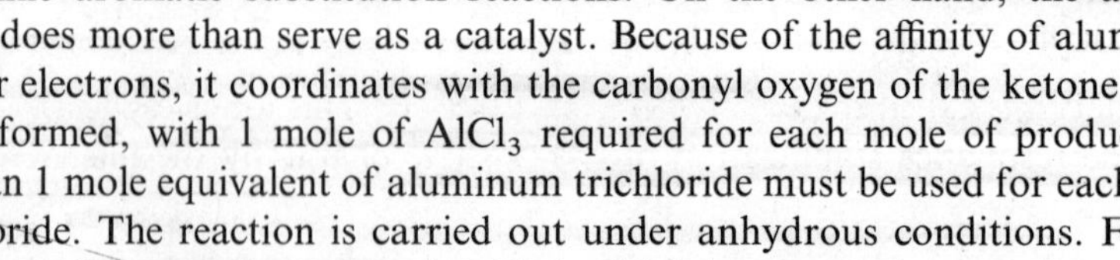

The mechanism for the Friedel-Crafts acylation is typical of that for most electrophilic aromatic substitution reactions. On the other hand, the aluminum chloride does more than serve as a catalyst. Because of the affinity of aluminum in $AlCl_3$ for electrons, it coordinates with the carbonyl oxygen of the ketone once the latter is formed, with 1 mole of $AlCl_3$ required for each mole of product. Thus, more than 1 mole equivalent of aluminum trichloride must be used for each mole of acid chloride. The reaction is carried out under anhydrous conditions. Following completion of the reaction, water is added, which breaks up the complex to liberate the ketone:

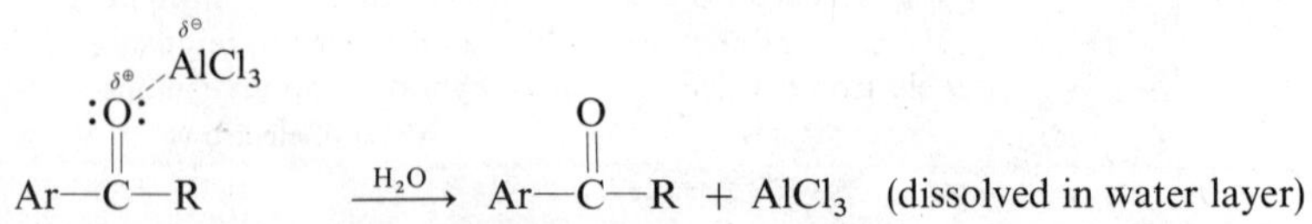

$AlCl_3$ complex of ketone   Free ketone

In conclusion, $AlCl_3$ serves as a catalyst only in the Friedel-Crafts alkylation where less than one equivalent of it is required. In acylation, however, it serves as a catalyst and is then tied up as the complex after the product is formed. Be certain you understand the distinction between the roles AlCl$_3$ plays in alkylation and acylation reactions. Note also the differences in the quantities of AlCl$_3$ that are required for these two types of Friedel-Crafts reactions.

Friedel-Crafts alkylation reactions are frequently accompanied by carbocation rearrangements before electrophilic substitution occurs (see Sec. 14.6). However, the Friedel-Crafts acylation reaction is not hampered by any rearrangement of the acylium ion. The acyl group on the aromatic ring has the same structure as that of the starting acid chloride.

$$R\text{—}X + AlCl_3 \longrightarrow R^{\oplus} \qquad \longrightarrow \quad \text{may rearrange to}$$
$$+ AlCl_3X^{\ominus} \qquad \qquad \text{more stable}$$
$$\text{carbocation}$$

Alkyl
halide

$$R\overset{\overset{\displaystyle O}{\|}}{\text{—}C}\text{—}Cl + AlCl_3 \longrightarrow \left[R\text{—}\overset{\oplus}{C}\text{=}\overset{..}{\underset{..}{O}} \longleftrightarrow R\text{—}C\text{≡}\overset{\oplus}{\underset{..}{O}} \right] + AlCl_4^{\ominus}$$

Acyl halide   Acylium ion

Does not rearrange

As in alkylation, Friedel-Crafts acylation generally cannot be carried out on an aromatic compound that contains one or more strongly deactivating *meta* directors. Yet the acid chloride may be an aromatic compound that contains one or more strongly deactivating groups. For example, 3,5-dinitrobenzophenone can be prepared from 3,5-dinitrobenzoyl chloride and benzene because electrophilic substitution is carried out on the nondeactivated benzene molecule:

3,5-Dinitrobenzoyl
chloride

3,5-Dinitrobenzophenone

Benzoyl chloride does not react with *m*-dinitrobenzene because the nitro groups deactivate the ring toward further substitution:

m-Dinitrobenzene   Benzoyl
chloride

This limitation on the Friedel-Crafts acylation reaction makes the introduction of a second acyl group onto the ring impossible. For example, the reaction between benzene and a large excess of ethanoyl chloride (acetyl chloride) and AlCl$_3$ produces only acetophenone, because once introduced, the acetyl group, —C—CH$_3$, de-
$\overset{\displaystyle \|}{\underset{\displaystyle O}{}}$

activates the ring toward further electrophilic aromatic substitution by a Friedel-Crafts reaction. However, acetophenone can be nitrated, sulfonated, and halogenated with the formation of a *meta* isomer (see Sec. 14.8).

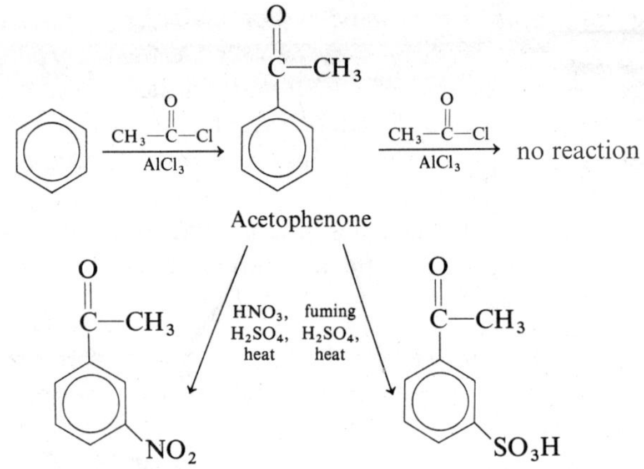

Acetophenone

m-Nitroacetophenone   *m*-Acetylbenzenesulfonic acid

B. The Organocadmium Reagent: Preparation of Aliphatic and Aromatic Ketones

Organometallic compounds were introduced in Sec. 4.2. These reagents have a common feature in the carbon-metal bond, which is very polar; the carbon portion of this bond, containing a high degree of electron density, is both basic and nucleophilic (see Sec. 4.2). Two other versatile organometallic reagents often used in organic synthesis are the **Grignard reagent** and the **organocadmium reagent.**

The Grignard reagent (which we discuss again in Sec. 21.10) is prepared by reacting an alkyl or aryl halide with magnesium metal under anhydrous conditions. The reaction is usually performed in anhydrous ethyl ether. Although alkyl chlorides,

$$(Ar)R-X + Mg\colon \xrightarrow[\text{ether}]{\text{anhydrous}} (Ar)R-MgX$$

X = Cl, Br, I

Grignard reagent
(organomagnesium
reagent)

bromides, or iodides may be used, the Grignard reagent is seldom prepared from alkyl fluorides because they are both unreactive and difficult to obtain. The Grignard reagent is one of the most versatile and useful compounds in organic chemistry.

Let us first review the structure of RMgX and, by analogy, other organometallics. Diethyl ether (often called *ether* for short) forms a complex with the Grignard reagent and has the composition $RMgX \cdot 2(CH_3CH_2)_2O$. The magnesium probably completes its octet by accepting electrons from the oxygen in ether, as follows:

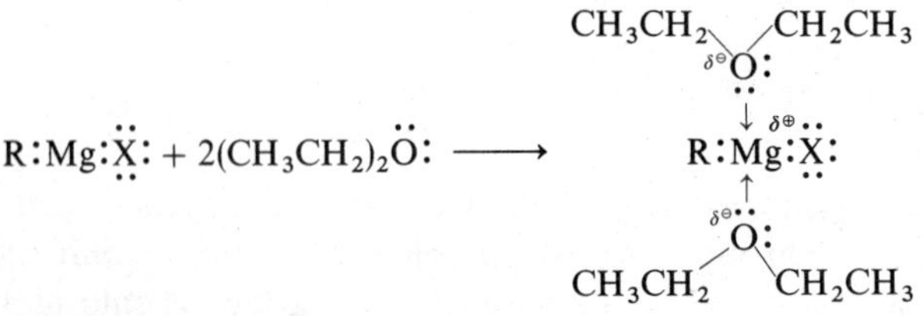

$$R\colon Mg\colon \overset{..}{\underset{..}{X}}\colon + 2(CH_3CH_2)_2\overset{..}{\underset{..}{O}}\colon \longrightarrow$$

The magnesium in RMgX is a *Lewis acid* (electron-pair acceptor) and the oxygen a *Lewis base* (electron-pair donor).

A closer look at the formation of the Grignard reagent shows that the magnesium metal, $Mg\colon$, *donates* its valence electrons to the carbon-halogen bond in R—X; oxidation and reduction occur. As far as oxidation states are concerned, the

Grignard reagent can be represented as $R \overset{\ominus}{\textbf{:}} Mg^{\overset{+2}{}} \overset{..}{\underset{..}{\textbf{:}}} X \overset{\ominus}{\textbf{:}}$. It is felt that the bonds are polar rather than completely covalent or ionic but that the magnesium-halogen bond may be mostly ionic. We might use the following structure for it:

$$\overset{\delta\ominus}{R}\text{-----}\overset{\delta\oplus}{Mg}\text{-----}\overset{\delta\ominus}{X}$$

The reactions we study involve the carbon of the Grignard reagent. This carbon, as part of a highly polar bond, behaves much as a free **carbanion, $R \overset{\ominus}{\textbf{:}}$**. Although the entire molecular complex is actually involved in each reaction, we often write this reaction and other reactions of organometallics as involving free carbanions for simplicity.

The Grignard reagent is referred to as an **organomagnesium compound** because it contains both an organic moiety (the alkyl or aryl group) and magnesium. Three specific Grignard reagents are

$$CH_3I + Mg: \xrightarrow[\text{ether}]{\text{anhydrous}} CH_3MgI$$

Iodomethane
(methyl
iodide)

Methylmagnesium
iodide

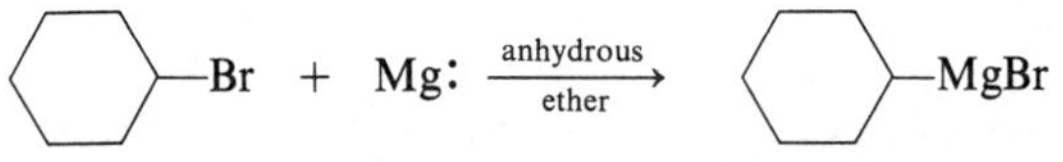

$$\langle\ \rangle\text{-Br} + Mg: \xrightarrow[\text{ether}]{\text{anhydrous}} \langle\ \rangle\text{-MgBr}$$

Bromocyclohexane   Cyclohexylmagnesium
(cyclohexyl bromide)   bromide

$$\langle\ \rangle\text{-Br} + Mg: \xrightarrow[\text{ether}]{\text{anhydrous}} \langle\ \rangle\text{-MgBr}$$

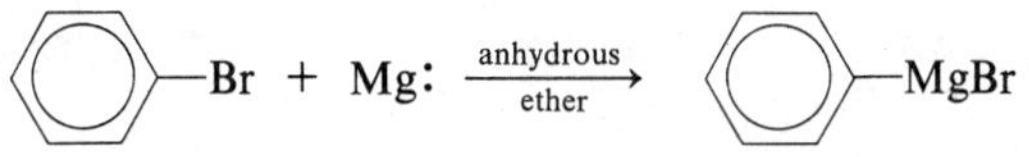

Bromobenzene   Phenylmagnesium
bromide

The Grignard reagent is fairly stable and can be kept for periods of time under anhydrous conditions. It is very reactive toward several other organic and inorganic compounds.

Among the many reactions of the Grignard reagent is its conversion to another organometallic compound, the **organocadmium reagent, R_2Cd**. When the Grignard reagent is treated with dry cadmium chloride, $CdCl_2$, a metal-metal interchange occurs; the magnesium is replaced by cadmium and the organocadmium reagent, R_2Cd, results:

$$2R\text{-}X + 2Mg \xrightarrow[\text{ether}]{\text{anhydrous}} 2R\text{-}MgX \xrightarrow{CdCl_2} R_2Cd + 2MgXCl$$

R must be   Grignard   Organocadmium
primary alkyl   reagent   reagent
or aryl
in this
application

Organocadmium reagents are useful in the synthesis of ketones. When an organocadmium reagent reacts with an acid chloride, a ketone results:

$$R_2Cd + R'\overset{O}{\overset{\|}{-C}}-Cl \xrightarrow{\text{mix}} \xrightarrow[H_2O]{H^{\oplus}} R\overset{O}{\overset{\|}{-C}}-R'$$

Organocadmium   Acid   Ketone
reagent   chloride

The organocadmium reaction is not completely understood, but the alkyl or aryl group in R_2Cd possesses some nucleophilic character and probably displaces the chlorine atom in the acid chloride by nucleophilic substitution. (See Sec. 23.3 for the mechanism of nucleophilic substitution at *acyl* carbon, which is *not* the same as that for alkyl halides.)

Two examples of the organocadmium reagent-acid chloride reaction are:

$$CH_3CH_2-\overset{\overset{\displaystyle O}{\|}}{C}-Cl + (CH_3CH_2)_2Cd \xrightarrow[\text{together}]{\text{mix}} \xrightarrow[H_2O]{H^\oplus} CH_3CH_2-\overset{\overset{\displaystyle O}{\|}}{C}-CH_2CH_3$$

Propanoyl chloride (propionyl chloride)	Diethylcadmium	3-Pentanone (diethyl ketone)

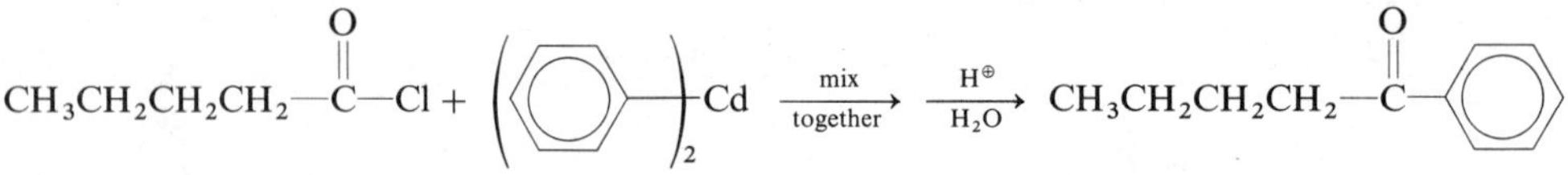

$$CH_3CH_2CH_2CH_2-\overset{\overset{\displaystyle O}{\|}}{C}-Cl + \left(\bigcirc\right)_2 Cd \xrightarrow[\text{together}]{\text{mix}} \xrightarrow[H_2O]{H^\oplus} CH_3CH_2CH_2CH_2-\overset{\overset{\displaystyle O}{\|}}{C}-\bigcirc$$

Pentanoyl chloride (valeroyl chloride)	Diphenylcadmium	1-Phenyl-1-pentanone (valerophenone or *n*-butyl phenyl ketone)

The major limitation of the organocadmium reagent is that it can be prepared *only from primary or aryl halides*. (Secondary and tertiary alkyl groups provide a reagent that, for some reason, is highly unstable and decomposes.)

Of the common organometallic compounds, the organocadmium is one of the least reactive toward many functional groups. That it reacts exclusively with acid chlorides makes it a highly useful reagent for introducing a keto group into molecules that already contain an ester (—COOR), an aldehyde (—CHO), or another ketone (—CO—) substituent. We discuss this reaction further in Sec. 21.10 F, f.

Question 20.4

Using any acid chloride, benzene, toluene, and alcohols of four or less carbon atoms, indicate suitable methods for synthesizing each of the following six compounds.

(*a*) 2-methyl-3-pentanone

(*b*) *p*-nitrobenzophenone

(*c*) *p*-methylbenzophenone

(*d*) 2-methyl-1-phenyl-1-pentanone

(*e*) $CH_3-\bigcirc-\overset{\overset{\displaystyle O}{\|}}{C}-C(CH_3)_3$

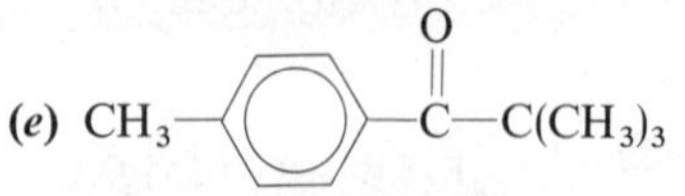

(*f*) 1,3-diphenyl-2-propanone

C. Ketones from Nitriles

As an alternative to using the organocadmium reagent, the organomagnesium reagent can also react with a nitrile to give a desired ketone directly. This is shown in the following scheme:

$$R-C\!\equiv\!\overset{\displaystyle ..}{N} + R'MgX \longrightarrow R-\underset{\underset{\displaystyle R'}{|}}{C}\!=\!\overset{\overset{\delta\ominus\quad\delta\oplus}{}}{\overset{\displaystyle ..}{N}}\text{---}MgX \xrightarrow[H^\oplus]{H_2O} R-\underset{\underset{\displaystyle R'}{|}}{C}\!=\!O$$

Ketone

The mechanism of this reaction is discussed in Sec. 23.16, where methods for the preparation of nitriles are also provided. Two additional examples of its use in the synthesis of ketones are:

$$CH_3CH_2-C\equiv N + CH_3CH_2-MgBr \xrightarrow[\substack{\text{by} \\ H^\oplus, H_2O}]{\text{followed}} CH_3CH_2-\overset{\overset{\displaystyle O}{\|}}{C}-CH_2CH_3$$

See Sec. 23.16 for additional examples of the reactions of nitriles.

20.9 Spectral Analysis of Aldehydes and Ketones

A. Infrared Spectroscopy

The carbon-oxygen double bond is easily detected in the infrared because it shows strong IR absorption in the region of 1,650 to 1,900 cm^{-1} (6.06 to 5.26 μ) due to the carbon-oxygen stretching vibration. More specifically, the carbonyl group in aldehydes and ketones commonly appears in the range of 1,690 to 1,770 cm^{-1} (5.92 to 5.88 μ), with strain-free (open chain and large cyclic) compounds appearing close to 1,700 cm^{-1} (5.88 μ). Aldehydes can be distinguished from ketones because the former give a characteristic absorption at 2,720 to 2,820 cm^{-1} (3.68 to 3.55 μ) due to stretching of the C—H bond on the aldehyde group. Four typical carbonyl stretching frequencies for different aldehydes and ketones are:

R—C=O	Ar—C=O	R—C=O	Ar—C=O
H	H	R	R
1,725 cm^{-1}	1,700 cm^{-1}	1,710 cm^{-1}	1,690 cm^{-1}
(5.78 μ)	(5.88 μ)	(5.85 μ)	(5.92 μ)

Attachment of a double bond adjacent to the aldehyde or ketone group causes the carbonyl group to absorb at lower frequencies as a result of *conjugation* (delocalization of electrons over the π bond of the double bond and the π bond of the carbonyl group). The following trends are evident:

α,β-Unsaturated ketone	α,β-Unsaturated aldehyde
1,675 cm^{-1} (5.97 μ)	1,685 cm^{-1} (5.93 μ)

Infrared absorption due to carbonyl stretching occurs at higher frequencies than does absorption caused by stretching of the carbon-carbon double bond. There is a relationship between bond strength and frequency of absorption. This can be explained in terms of the π bond strength of the C=O bond, which is considerably stronger than the π bond in C=C (see Sec. 2.7D). Most open-chain ketones absorb at ~1,700 cm^{-1} (5.88 μ), but cyclic ketones generally absorb at higher frequencies than their open-chain counterparts, as shown by the following data:

$$(CH_2)_n \quad C=O$$

Cyclopropanone: $(n = 2)$ 1,815 cm^{-1} (5.51 μ)
Cyclobutanone: $(n = 3)$ 1,775 cm^{-1} (5.63 μ)
Cyclopentanone: $(n = 4)$ 1,745 cm^{-1} (5.73 μ)
Cyclohexanone: $(n = 5)$ 1,710 cm^{-1} (5.85 μ)

The ring size of five-membered and smaller cyclic ketones is often deduced by the carbonyl absorption frequency, although it is of little use in determining ring size of six-membered and larger rings.

The IR spectra of several typical aldehydes and ketones are shown in Fig. 20.2.

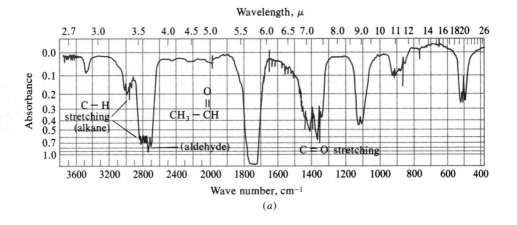

(a)

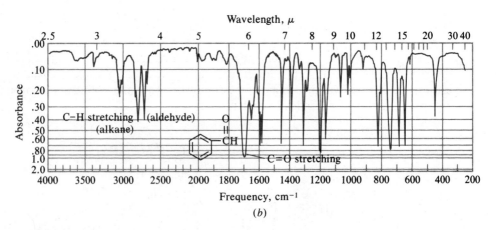

(b)

FIGURE 20.2   IR spectra of (a) acetaldehyde, (b) benzaldehyde, (c) acetone, and (d) benzophenone. [Data for (a) and (c) from *Indexes to Evaluated Infrared Reference Spectra*, copyright 1975, The Coblentz Society, Norwalk CT; data for (b) and (d) from *The Sadtler Handbook of Infrared Spectra*, W. W. Simmons (ed.), © Sadtler Research Laboratories, Division of Bio-Rad Laboratories, Inc. (1978)]

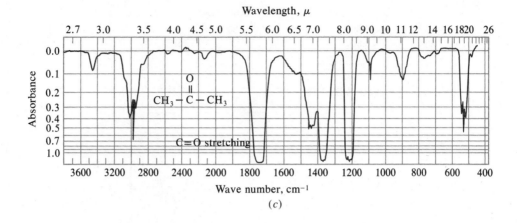

(c)

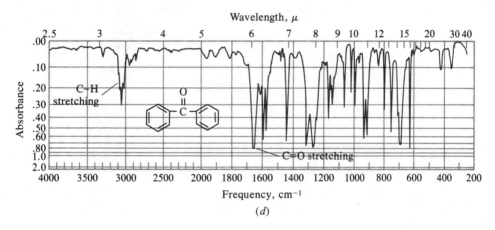

(d)

FIGURE 20.2 *(continued)*

ᵒ᷎ₘᵢₜ **B. Ultraviolet Spectroscopy**

The carbonyl group exhibits a strong primary absorption band in the region of 270 to 300 nm, with a very small extinction coefficient of about 10 to 25. This band is attributed to the $n \rightarrow \pi^*$ transition associated with this chromophore. A second, more intense band can sometimes be observed at ~ 180 nm due to $n \rightarrow \sigma^*$ transitions; like other bands in this region, it is very difficult to observe experimentally. The UV spectra of acetone and crotonaldehyde are given in Figs. 12.12 and 12.14, respectively.

When the carbonyl chromophore is part of a conjugated system, the band caused by the $n \rightarrow \pi^*$ transition is shifted to a longer wavelength. The greater the extent of conjugation, the greater the shift (see Sec. 12.7).

C. Nuclear Magnetic Resonance Spectroscopy

Aldehydes exhibit a characteristic absorption in the NMR spectrum because of the hydrogen attached to the carbonyl group (the "aldehydic" proton) in the range 9.4 to 10 δ (0 to 0.6 τ). The electron-withdrawing nature of the carbonyl group greatly deshields the proton and causes it to absorb so far downfield.

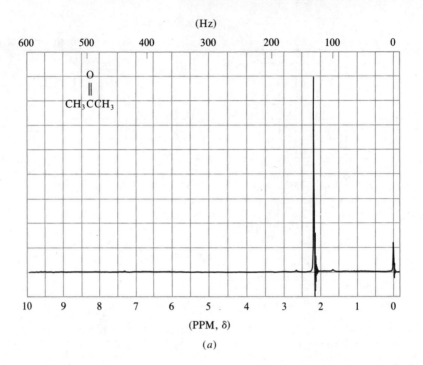

(a)

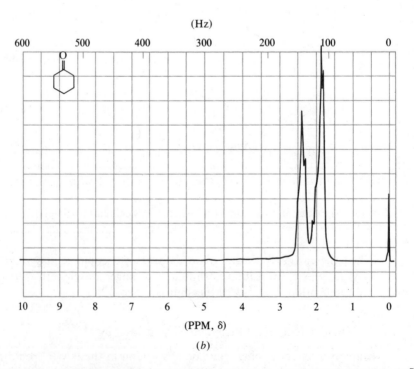

(b)

FIGURE 20.3   NMR spectra of (a) acetone, (b) cyclohexanone, and (c) propionaldehyde. [Data from *The Aldrich Library of NMR Spectra*, C. J. Pouchert and J. R. Campbell (eds.), copyright 1974, Aldrich Chemical Co., Milwaukee]

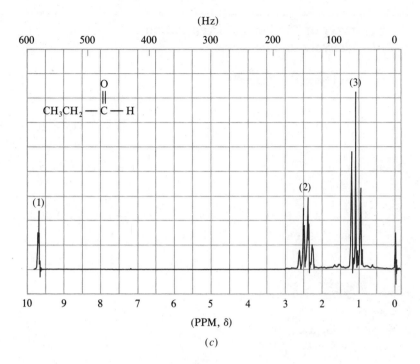

FIGURE 20.3 (*continued*)

The carbonyl group also deshields the hydrogens α and β to it.

The NMR spectra of three typical aldehydes and ketones are shown in Fig. 20.3.

omit **D. Mass Spectrometry**

The molecular ion peak of both aldehydes and ketones is generally observed; although it is very intense in the latter, it is not naturally abundant in the former.

The major point of fragmentation in both these families is between the carbonyl carbon and the adjacent carbon (ketones or aldehydes) or hydrogen (aldehydes), resulting in the formation of the acylium ion (see Sec. 20.8A):

$$
\underset{(H)\,R}{\overset{R}{\diagup}}\!C\!=\!\overset{..}{\underset{\oplus}{O}}\ \xrightarrow[\text{or}]{\substack{-R\cdot \\ -H\cdot}}\ R\!-\!C\!\equiv\!\overset{..}{\underset{\oplus}{O}}
$$

Acylium ion

Aromatic aldehydes and ketones both yield a molecular ion of high relative abundance and both produce the relatively stable aromatic acylium ion, $C_6H_5\!-\!C\!\equiv\!\overset{..}{\underset{\oplus}{O}}$

Straight-chain aldehydes or ketones that contain more than three carbon atoms undergo an intramolecular fragmentation called the McLafferty rearrangement. The fragmentation of the $P^\oplus$ of 2-pentanone provides a good example of this mode of fragmentation:

McLafferty rearrangement

20.10   Summary of Methods of Preparation of Aldehydes and Ketones

A. Oxidation of Alcohols

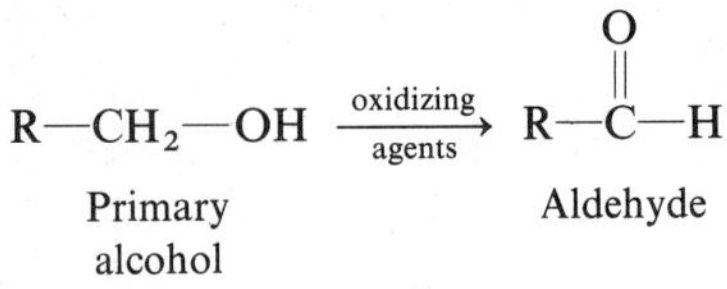

$$R-CH_2-OH \xrightarrow[\text{agents}]{\text{oxidizing}} R-\overset{\overset{\displaystyle O}{\|}}{C}-H \qquad \text{Secs. 10.10 and 20.5A}$$

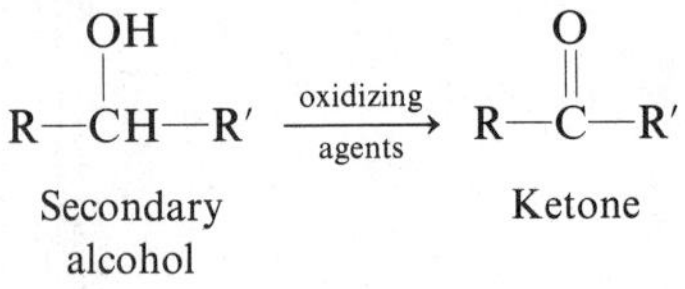

Primary   Aldehyde
alcohol

$$R-\overset{\overset{\displaystyle OH}{|}}{C}H-R' \xrightarrow[\text{agents}]{\text{oxidizing}} R-\overset{\overset{\displaystyle O}{\|}}{C}-R' \qquad \text{Secs. 10.10 and 20.7A}$$

Secondary   Ketone
alcohol

Oxidizing Agents

1. Cu°, 250–300°
2. $K_2Cr_2O_7$, H_2SO_4, warm
3. CrO_3, acetic acid, warm
4. $KMnO_4$, $OH^{\ominus}$, H_2O, heat (*ketones only*)
5. CrO_3-pyridine

B. Ozonolysis of Alkenes

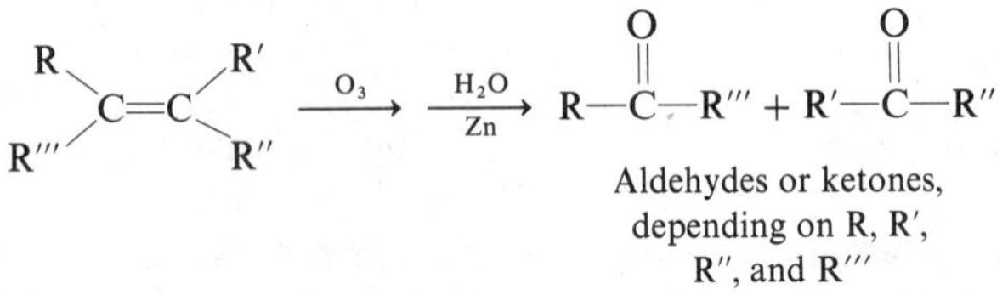

$$\underset{R'''}{\overset{R}{>}}C=C\underset{R''}{\overset{R'}{<}} \xrightarrow{O_3} \xrightarrow[\text{Zn}]{H_2O} R-\overset{\overset{\displaystyle O}{\|}}{C}-R''' + R'-\overset{\overset{\displaystyle O}{\|}}{C}-R'' \qquad \text{Secs. 8.31, 20.5B, and 20. 20.7B}$$

Aldehydes or ketones,
depending on R, R',
R'', and R'''

C. Oxidation of Glycols by Periodic Acid

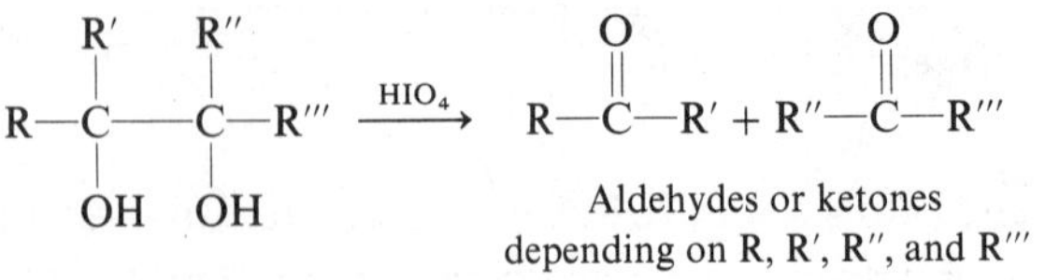

$$R-\underset{\overset{\displaystyle |}{OH}}{\overset{\overset{\displaystyle R'}{|}}{C}}-\underset{\overset{\displaystyle |}{OH}}{\overset{\overset{\displaystyle R''}{|}}{C}}-R''' \xrightarrow{HIO_4} R-\overset{\overset{\displaystyle O}{\|}}{C}-R' + R''-\overset{\overset{\displaystyle O}{\|}}{C}-R''' \qquad \text{Secs. 19.18, 20.5C, and 20.7C}$$

Aldehydes or ketones
depending on R, R', R'', and R'''

D. Phenolic Aldehydes (Reimer-Tiemann Reaction)

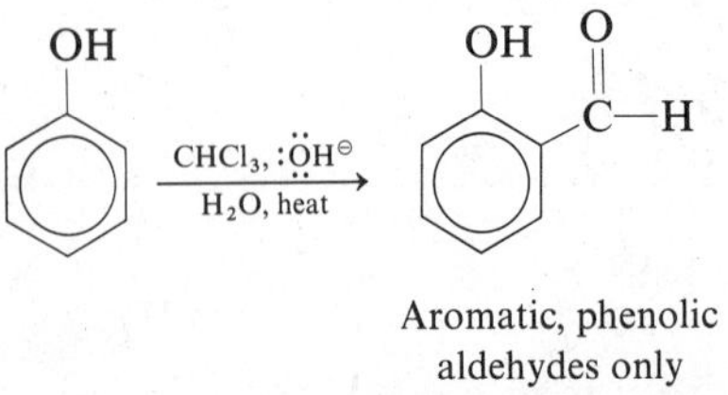

$$\text{OH} \xrightarrow[\text{H}_2\text{O, heat}]{\text{CHCl}_3, \ddot{\text{O}}\text{H}^{\ominus}} \text{OH} \ \overset{\overset{\displaystyle O}{\|}}{C}-H \qquad \text{Sec. 19.8 and 20.5D}$$

Aromatic, phenolic
aldehydes only

E. Reduction of Acid Chlorides

1. Rosenmund reduction (catalytic hydrogenation)

$$(Ar)R-\overset{\overset{\displaystyle O}{\|}}{C}-Cl \xrightarrow[Pd(BaSO_4)]{H_2} (Ar)R-\overset{\overset{\displaystyle O}{\|}}{C}-H \qquad Sec.\ 20.6A$$

Acid chloride   Aldehyde

2. Metal hydride reduction

$$(Ar)R-\overset{\overset{\displaystyle O}{\|}}{C}-Cl \xrightarrow[\text{2. }H_3O^\oplus]{\text{1. LiAlH[OC(CH}_3)_3]_3} (Ar)R-\overset{\overset{\displaystyle O}{\|}}{C}-H \qquad Sec.\ 20.6B$$

Acid chloride   Aldehyde

F. Oxidation of Methylbenzenes

$$Ar-CH_3 \xrightarrow[\text{heat or light}]{Cl_2} Ar-CHCl_2 \xrightarrow[\substack{CaCO_3, \\ heat}]{H_2O} Ar-\overset{\overset{\displaystyle O}{\|}}{C}-H \qquad Sec.\ 20.6C$$

Aromatic
aldehyde

G. Hydration of Alkynes

$$R-C\equiv C-R \xrightarrow[\substack{H_2SO_4, \\ HgSO_4}]{H_2O} R-\overset{\overset{\displaystyle O}{\|}}{C}-CH_2-R \qquad Secs.\ 11.13\ and\ 20.7D$$

Ketone
(acetaldehyde
when R = H)

H. Acylation of Aromatic Rings (Friedel-Crafts Acylation)

$$ArH\ +\ (Ar)R-\overset{\overset{\displaystyle O}{\|}}{C}-Cl \xrightarrow[(>1\ equivalent)]{AlCl_3} Ar-\overset{\overset{\displaystyle O}{\|}}{C}-R(Ar) \qquad Sec.\ 20.8A$$

ArH may *not*   Aromatic
contain a   ketone
deactivating
meta director

I. Ketones from Acid Chlorides and Organocadmium Reagent (Sec. 20.8B)

$$(Ar)R-X \xrightarrow[\substack{anhydrous \\ ether}]{Mg} (Ar)R-MgX \xrightarrow{CdCl_2} (Ar_2)R_2Cd$$

R must   Organocadmium
be primary   reagent

$$(Ar_2)R_2Cd + R'-\overset{\overset{\displaystyle O}{\|}}{C}-Cl \xrightarrow[\text{2. }H_3O^\oplus]{\text{1. mix}} (Ar)R-\overset{\overset{\displaystyle O}{\|}}{C}-R'$$

Ketone

J. Ketones from Nitriles and Organomagnesium Reagent

$$(Ar)R\text{—}MgX + (Ar)R'\text{—}C\equiv N \xrightarrow[\substack{\text{by} \\ H^{\oplus}, H_2O}]{\text{followed}} (Ar)R\text{—}\overset{\overset{\displaystyle O}{\|}}{C}\text{—}R'(Ar) \qquad \text{Secs. 20.8C and 23.16}$$

$$\underset{\text{Ketone}}{}$$

Study Questions

20.5 Give IUPAC names (except as noted) for the following compounds:

(a) $CH_3CH_2CH_2CH_2\text{—}\overset{\overset{\displaystyle O}{\|}}{C}\text{—}H$

(also give common name)

(b) [structure: 3-hydroxyphenyl group attached to $\overset{\overset{\displaystyle O}{\|}}{C}\text{—}H$, with HO on ring]

(c) $C_6H_5\text{—}\overset{\overset{\displaystyle O}{\|}}{C}\text{—}C_6H_5$

(d) $CH_3\text{—}\underset{\underset{\displaystyle CH_3}{|}}{CH}\text{—}\overset{\overset{\displaystyle O}{\|}}{C}\text{—}\underset{\underset{\displaystyle CH_3}{|}}{CH}\text{—}CH_3$

(also give common name)

(e) $CH_3CH_2\text{—}\underset{\underset{\displaystyle \bigcirc}{|}}{\overset{\overset{\displaystyle CH_3}{|}}{CH}}\text{—}CH\text{—}\overset{\overset{\displaystyle O}{\|}}{C}\text{—}H$

(f) [structure: cyclohexane-1,3-dione]

(g) $CH_3CH_2\text{—}\underset{\underset{\displaystyle CH_2}{\|}}{C}\text{—}CH_2\text{—}\overset{\overset{\displaystyle O}{\|}}{C}\text{—}CH_2CH_3$

(h) $CH_3O\text{—}CH_2\text{—}\overset{\overset{\displaystyle O}{\|}}{C}\text{—}H$

(also give common name)

(i) $CF_3\text{—}\overset{\overset{\displaystyle O}{\|}}{C}\text{—}\overset{\overset{\displaystyle O}{\|}}{C}\text{—}H$

(j) $Br\text{—}\bigcirc\text{—}\overset{\overset{\displaystyle O}{\|}}{C}\text{—}\bigcirc\text{—}NO_2$ (with Br and NO_2 substituents)

(k) [structure: bicyclic ketone (norbornanone)]

(l) $CH_3\text{—}\overset{\overset{\displaystyle O}{\|}}{C}\text{—}CH_2CH_2\text{—}OCH_3$

(m) $CH_3CH\text{—}\overset{\overset{\displaystyle O}{\|}}{C}\text{—}CHCH_2CH_3$, with CH_3 and $CH_2CH_2CH_2CH_3$ substituents below

20.6 Give structures that correspond to the following 12 names.
(a) pentanal
(b) diisobutyl ketone
(c) 3-phenyl-2-butanone
(d) isovaleraldehyde
(e) 3-methylcyclohexanone
(f) *trans*-2,3-dimethylcyclopropanone
(g) *p*-chlorobenzaldehyde
(h) 3,3-dimethoxy-2-pentanone
(i) 3-hydroxy-3-methylbutanal
(j) 2-methyl-5-*n*-butyl-3-decanone
(k) α,α,γ-tribromobutyraldehyde
(l) β-phenylvaleraldehyde

20.7 We often mention the oxidation and reduction of organic compounds, especially in this chapter on aldehydes and ketones. For each of these 12 reactions, indicate whether the carbon shown in red has undergone oxidation or reduction by computing the oxidation number of carbon in each compound (see Sec. 4.2). (You will likely become familiar with which reactions are oxidation and which are reduction simply through continued use. This exercise puts the designation on a firmer basis.)

(*a*) $C_6H_5-\overset{\overset{\displaystyle O}{\|}}{C}-CH_3 \longrightarrow C_6H_5-CH_2CH_3$

(*b*) $C_6H_5-\overset{\overset{\displaystyle O}{\|}}{C}-CH_3 \longrightarrow C_6H_5-CHOH-CH_3$

(*c*)

(*d*) $CH_3-\overset{\overset{\displaystyle O}{\|}}{C}-H \longrightarrow CH_3-\overset{\overset{\displaystyle N-OH}{\|}}{C}-H$

(*e*) $CH_3-\overset{\overset{\displaystyle O}{\|}}{C}-CH_3 \longrightarrow (CH_3)_3COH$

(*f*)

(*g*) $C_6H_5-\overset{\overset{\displaystyle O}{\|}}{C}-H \longrightarrow C_6H_5-\overset{\overset{\displaystyle O}{\|}}{C}-OH$

(*h*) $H-\overset{\overset{\displaystyle O}{\|}}{C}-H \longrightarrow CH_3OH$

(*i*) $C_6H_5-CH_3 \longrightarrow C_6H_5-\overset{\overset{\displaystyle O}{\|}}{C}-H$

(*j*) $R-\overset{\overset{\displaystyle O}{\|}}{C}-Cl \longrightarrow R-\overset{\overset{\displaystyle O}{\|}}{C}-H$

(*k*) $CH_3-\overset{\overset{\displaystyle O}{\|}}{C}-CH_3 \longrightarrow (CH_3)_2\overset{\overset{\displaystyle Cl}{|}}{C}-Cl$

20.8 Discuss the electronic structure of *acetaldehyde methyl imine*, $CH_3-CH{=}\overset{..}{N}-CH_3$, especially with regard to the double bond. Are geometric isomers possible for this compound? Why or why not? (*Hint:* Reason by analogy with the structure of the carbon-carbon double bond.)

20.9 A major fragment in the mass spectrum of each of the following three compounds has an *m/e* ratio of 105. What is this fragment? Explain why it forms so readily from each of these compounds.

20.10 Supply the missing reactants, reagents, or products, as required, to complete the following five equations. The items to be supplied are indicated by number [for example, (1)].

(*a*)

(**b**) $CH_3-\underset{\underset{Br}{|}}{CH}-\underset{\underset{Br}{|}}{CH}-CH_3 \xrightarrow[\text{ethanol}]{\text{KOH}}$ (5) $\xrightarrow{(6)}$ $CH_3-C\equiv C-CH_3$

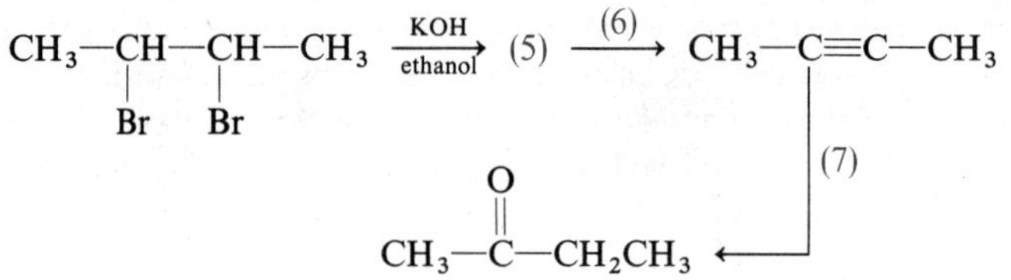

$\Big\downarrow$ (7)

$CH_3-\overset{\overset{O}{\|}}{C}-CH_2CH_3 \longleftarrow$

(**c**)

$\xrightarrow{(8)}$ (toluene, CH_3) $\xrightarrow{(9)}$ (benzaldehyde, $\overset{\overset{O}{\|}}{C}-H$) $\xrightarrow{\text{Fe, Br}_2}$ (10)

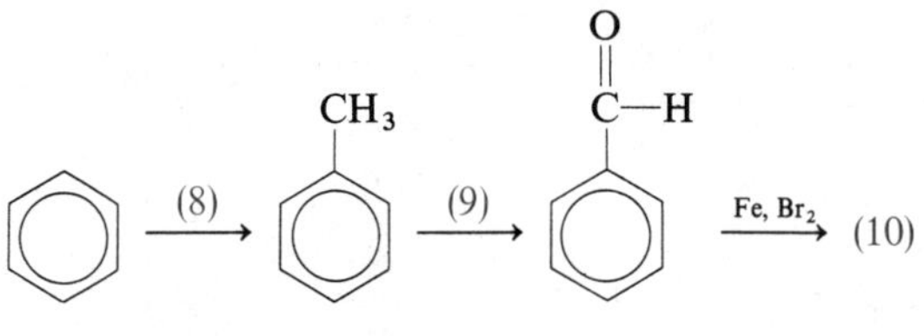

(**d**)

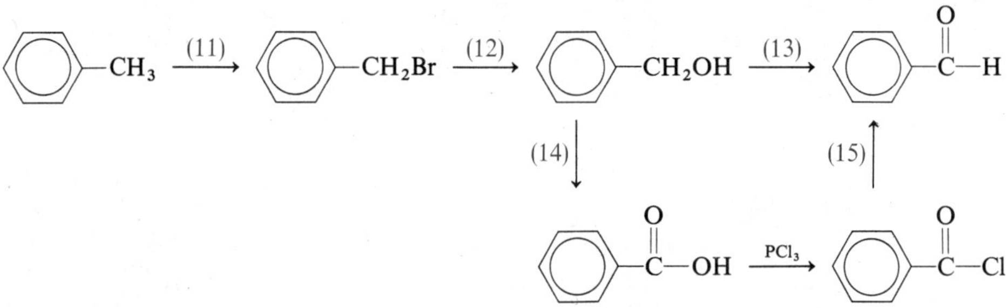

$\langle \rangle-CH_3 \xrightarrow{(11)} \langle \rangle-CH_2Br \xrightarrow{(12)} \langle \rangle-CH_2OH \xrightarrow{(13)} \langle \rangle-\overset{\overset{O}{\|}}{C}-H$

(14) $\Big\downarrow$ (15) $\Big\uparrow$

$\langle \rangle-\overset{\overset{O}{\|}}{C}-OH \xrightarrow{\text{PCl}_3} \langle \rangle-\overset{\overset{O}{\|}}{C}-Cl$

(**e**) $CH_3-\underset{\underset{CH_3}{|}}{\overset{\overset{CH_3}{|}}{C}}-CH_2-\underset{\underset{Br}{|}}{CH}-CH_3 \xrightarrow[\text{alcohol}]{\text{KOH}}$ (16) $\xrightarrow[\text{2. Zn, H}_2\text{O}]{\text{1. O}_3}$ (17)

(18) $\Big\downarrow$

$(CH_3)_3C-CH_2CH_2-C\equiv N \xleftarrow{(19)} (CH_3)_3C-CH_2CH_2-Br$

(20) $\Big\downarrow$

$(CH_3)_3C-CH_2CH_2-\underset{\underset{O}{\|}}{C}-CH_2CH_3$

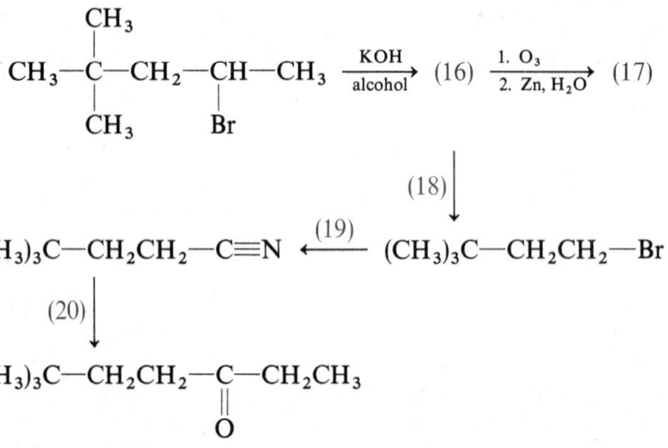

20.11 Show all the steps, conditions, and reagents required to carry out the following eight transformations using the indicated starting material. You may use any common solvents, inorganic reagents, or monofunctional organic reagents of six carbons or less.

(**a**)

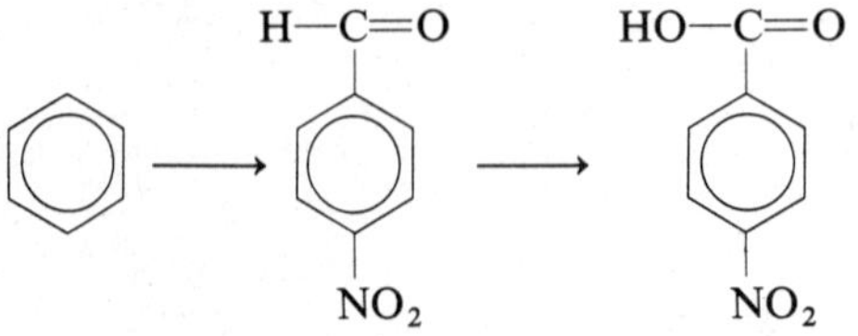

(**b**)

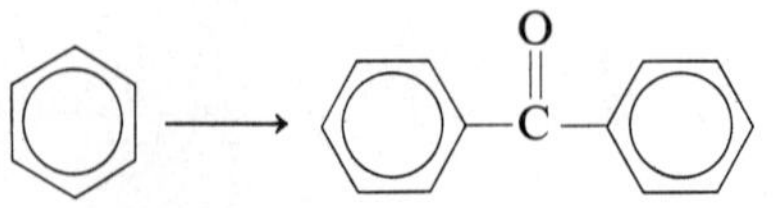

(**c**) $CH_3CH_2CH_2CH_3 \longrightarrow CH_3CH_2CH_2CH_2-\underset{\underset{O}{\|}}{C}-CH_3$

(d)

$$\text{:}\overset{\cdot\cdot}{\underset{\cdot\cdot}{O}}\text{:}^{\ominus} \text{Na}^{\oplus}$$

(e)

$$\bigcirc\!\!-\text{CH}_2\text{OH} \longrightarrow \bigcirc\!\!-\text{CH}_2\!-\!\text{C}\!\equiv\!\text{N}$$

(f)

$$\longrightarrow \overset{O}{\underset{\|}{\text{H}-\text{C}}}-(\text{CH}_2)_3-\overset{O}{\underset{\|}{\text{C}}}-\text{H}$$

(g)

$$\bigcirc\!\!-\text{CH}\!=\!\text{CH}_2 \longrightarrow \bigcirc\!\!-\text{CH}_2\!-\!\overset{O}{\underset{\|}{\text{C}}}-\text{H}$$

(h)

$$\overset{\text{Br Br}}{\underset{\text{Br Br}}{\text{CH}_3\text{CH}_2-\text{C}-\text{C}-\text{CH}_2\text{CH}_3}} \longrightarrow \text{CH}_3\text{CH}_2-\overset{}{\underset{\underset{O}{\|}}{\text{C}}}-\text{OH} \quad (2 \text{ moles})$$

20.12 The mass spectrum of 2-methylcyclopentanone gives the following fragments, among others. Explain how they result mechanistically.

$$\begin{array}{cc} \text{:O}^{\oplus} & \text{CH}_3 \\ \| & | \\ \text{C} & \text{CH}_2 \\ | & | \\ \text{HC} & \text{CH}_2 \\ \| & \cdot \\ \text{CH}_2 & \end{array}$$

20.13 (a) Do you think cyclohexanone is strain-free in the same way as cyclohexane? Why or why not? **(b)** Account for cyclopropanone forming a stable hydrate whereas most other cyclic and open-chain ketones do not. (*Hint:* Consider bond angles.)

20.14 Compound *A* is prepared by the Friedel-Crafts alkylation of benzene. *A* reacts with hot, aqueous $KMnO_4$ to produce compound *B*, $C_7H_6O_2$. *B* reacts with thionyl chloride to give *C*, C_7H_5OCl. *C* reacts in turn with dimethylcadmium followed by acid hydrolysis to give *D*, C_8H_8O. The UV and IR spectra of compound *B* as well as those of compound *D* are provided here. What are the structures of *A* to *D*?

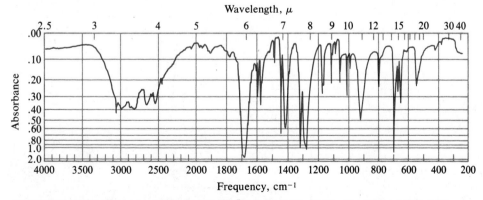

PROBLEM 20.14   IR spectrum of unknown (*B*). [Data from *The Sadtler Handbook of Infrared Spectra*, W. W. Simmons (ed.), © Sadtler Research Laboratories, Division of Bio-Rad Laboratories, Inc. (1978)]

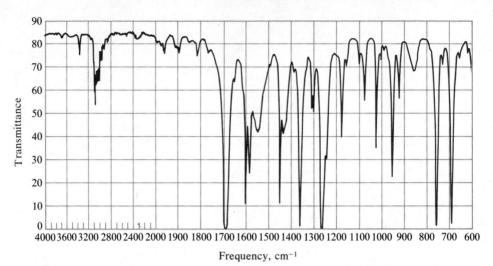

PROBLEM 20.14   IR spectrum of unknown (*D*). [Data from *Indexes to Evaluated Infrared Reference Spectra*, copyright 1975, The Coblentz Society, Norwalk CT]

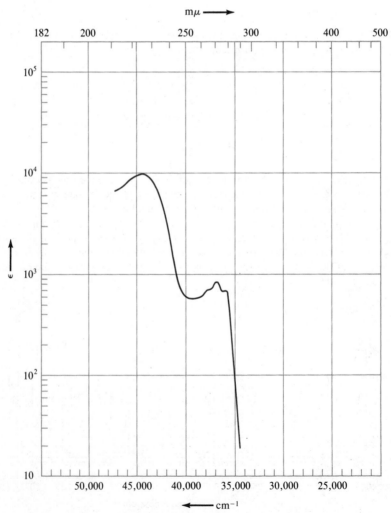

PROBLEM 20.14   UV spectrum of unknown (*B*). [Data from the *DMS UV Atlas*, 1966–1971, Plenum Press, New York]

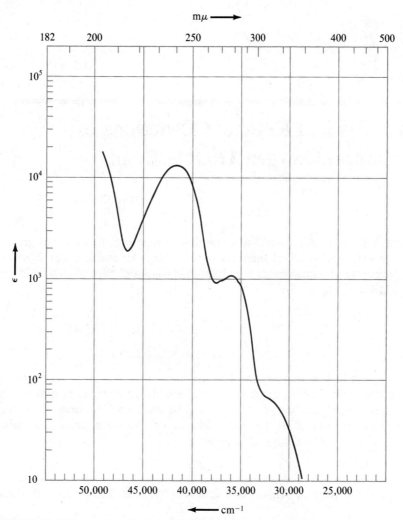

PROBLEM 20.14   UV spectrum of unknown (*D*). [Data from the *DMS UV Atlas*, 1966–1971, Plenum Press, New York]

21

Aldehydes and Ketones: Reactions of the Carbon-Oxygen Double Bond

In Chap. 20 we discussed the methods of preparing the carbonyl group, $>C=O$. In this chapter we study the reactions of aldehydes and ketones. We are concerned primarily with reactions of the carbonyl group and the mechanism of nucleophilic addition:

$$Nu-H + \quad >C=\ddot{O} \quad \longrightarrow \quad Nu-\overset{|}{\underset{|}{C}}-\ddot{\underset{..}{O}}-H$$

Nucleophilic addition

Although aldehydes and ketones are capable of other reactions, particularly reactions that result from the acidity of the hydrogens on the carbon α to the carbonyl carbon, these reactions are more easily discussed when we understand the chemistry of the carboxylic acids and their derivatives. We defer presenting these reactions until that time (see Chaps. 23 and 24).

21.1   Reactions of Aldehydes and Ketones: Relative Reactivities

The polarity of the carbonyl group in aldehydes and ketones is largely responsible for the reactions they undergo. In this section we examine the *types* of general reactions of aldehydes and ketones.

A. Nucleophilic Addition Reactions

One type of reaction typical of the carbonyl group is the addition of the reagent H—Nu across the carbon-oxygen double bond. The proton ends up on the more electronegative oxygen atom, while the nucleophile becomes attached to the electron-deficient carbonyl carbon atom:

$$\underset{\substack{R \quad\quad R' \\ \text{Aldehyde} \\ \text{or ketone}}}{\overset{\overset{\displaystyle :\overset{\delta\ominus}{\ddot{O}}:}{\underset{\delta\oplus}{\|}}}{\underset{}{C}}} \overset{\overset{\displaystyle H^{\delta\oplus}}{|}}{\underset{\delta\ominus}{}} + Nu \quad \longrightarrow \quad R-\overset{\overset{\displaystyle :\ddot{O}-H}{|}}{\underset{\underset{R'}{|}}{C}}-Nu \qquad \text{Nucleophilic addition}$$

This type of addition is called **nucleophilic addition** and is discussed in detail in the remainder of this chapter.

The **orientation of nucleophilic addition** to carbonyl compounds is always the same, proton on oxygen and nucleophile on carbon. On the other hand, recall that the orientation of electrophilic addition to an alkene depends on its structure. The proton (or electrophile) adds to the double bond to give the more stable carbocation, and then the nucleophile reacts with the carbocation to give the addition product. Nucleophilic addition to carbonyl compounds occurs without rearrangement, whereas electrophilic addition to alkenes may occur with carbocation rearrangement.

Nucleophilic Versus Electrophilic Addition

The terminology involved in the addition reactions of carbonyl compounds (nucleophilic addition) and alkenes (electrophilic addition) may appear contradictory, but it is not. The terms refer to the mechanism of addition and not the net result because H—Nu adds to both carbon-carbon and carbon-oxygen double bonds. The addition of H—Nu to alkenes is called *electrophilic addition* because the slow, rate-determining step is the reaction of the electrophilic species (often $H^{\oplus}$ or $^\delta H$-----$^\delta Nu$) with the carbon-carbon double bond; the reaction of the nucleophile with the carbocation is fast in comparison. The addition of H—Nu to carbonyl compounds is called *nucleophilic addition* because the slow, rate-determining step is the attachment of the nucleophile to the carbonyl carbon atom; the oxygen may be protonated either before or after nucleophilic addition, depending on the reaction conditions.

The precise details of addition depend largely on the reactants involved, but we can generalize by showing the major changes that occur. On collision, the nucleophile reacts with the partially positive sp^2 hybridized carbon atom of the trigonal carbonyl group, which results in the formation of the tetrahedral sp^3 hybridized carbon atom. In this process, oxygen acquires a formal negative charge. The oxygen anion (an alkoxide ion) then picks up a proton either from the solvent or from added acid, and an alcohol is produced. This sequence of events is envisioned as follows:

Aldehyde or ketone
Trigonal (120° angle),
contains sp^2 hybridized carbon

Alcohol
Tetrahedral (~109.5° angle),
contains sp^3 hybridized carbon

B. Reactions Involving Acidic α Hydrogens

Other important reactions of aldehydes and ketones involve the acidity of the hydrogens α to the carbonyl group. The electron-withdrawing properties of the partially positive carbon of the carbonyl group are partially responsible for this acidity. As we will see later, the α hydrogen can easily be removed as a proton by a suitable base, and the anion that forms undergoes some important reactions. We study these reactions in more detail in Chap. 24.

α Hydrogen acidity

Acidic α hydrogen   Carbanion

The carbanion may be written in a variety of ways and actually exists as a

resonance hybrid, stabilized by the delocalization of electrons over the C—C—O bond. It is primarily the stabilization of this intermediate that is responsible for the acidity of the hydrogens α to a carbonyl group (see Sec. 24.1).

C. Electrophilic Aromatic Substitution

The *meta*-directing and deactivating effect of the carbonyl group in electrophilic aromatic substitution reactions was discussed in Chap. 14. Now that we are acquainted with the electronic structure of the carbon-oxygen double bond, we are better able to understand the reasons for its effect in electrophilic aromatic substitution. The polarity of the carbonyl group in aldehydes and ketones means that a partially positive carbon atom is attached to the aromatic ring. This deactivates the ring by electron withdrawal and causes an incoming electrophile to occupy the *meta* position; the electron-withdrawing nature of the carbonyl group is discussed in Sec. 21.2B.

Aldehyde or ketone   *meta* Isomer

The limitation on this reaction is that *only sulfonation, nitration, and halogenation can be carried out on aromatic aldehydes and ketones.* Friedel-Crafts alkylation and acylation do not occur because the Lewis acid catalyst complexes with the carbonyl oxygen and causes the ring to become even more deactivated (see Sec. 20.6).

21.2 Nucleophilic Addition: General Mechanisms and Relative Reactivities of Aldehydes and Ketones

A. General Mechanisms of Addition

Before considering specific examples of nucleophilic addition, let us look at the details of the addition mechanism. Generally two types of conditions can be used for addition: (1) acidic and (2) neutral or basic conditions.

Addition under neutral or basic conditions usually occurs with attack of the nucleophile on the carbonyl group, transforming its planar structure to the tetrahedral structure of the product. Carbon begins to become tetrahedral in the transition state between reactant and product, and an alkoxide ion is formed. Following addition, the alkoxide ion abstracts a proton from solvent if a protic solvent is used, but if an aprotic solvent is used protonation does not occur until aqueous acid is added from an external source. We will see examples of both types of protonation, but what is important for now is that protonation occurs *after* the nucleophile adds to the carbonyl group.

General mechanism for nucleophilic addition under neutral or basic conditions:

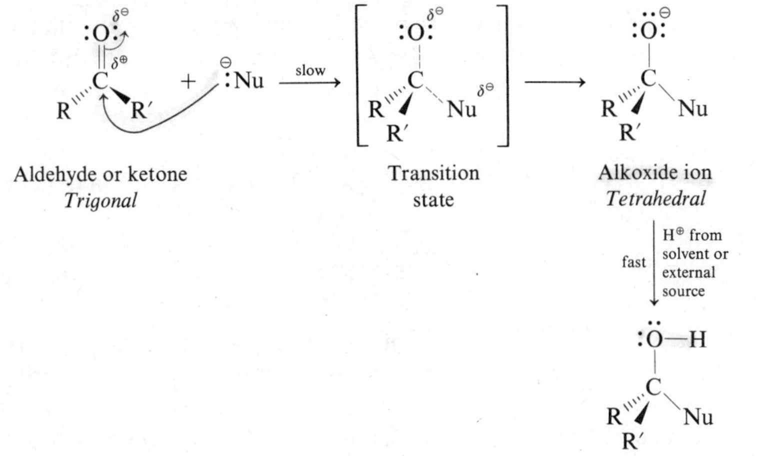
slow
Aldehyde or ketone
Trigonal
Transition
state
Alkoxide ion
Tetrahedral
fast
H⊕ from
solvent or
external
source

Addition under acidic conditions usually starts with protonation of the carbonyl oxygen (a reversible reaction). The nonbonding electrons on oxygen are especially open to attack by protons. The carbon atom in the protonated carbonyl group acquires even more positive character because of the electron-withdrawing nature of the protonated oxygen atom; the following resonance structures show this:

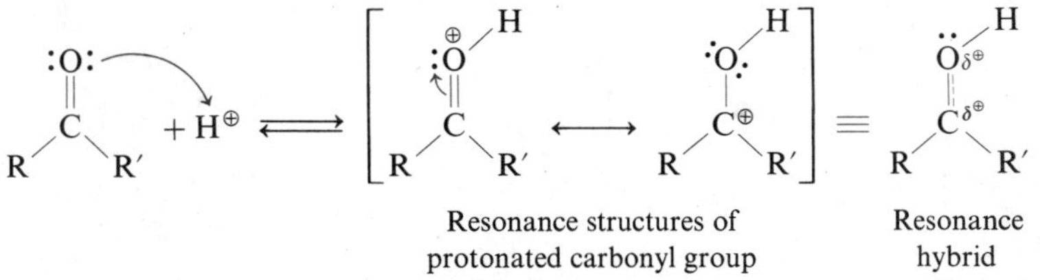
Resonance structures of
protonated carbonyl group
Resonance
hybrid

The net result of protonation is that nucleophilic addition to carbon becomes considerably easier because of the increased positive charge it bears; in other words, protonation causes carbon to gain electrophilic character, which renders it better able to react with a nucleophile. Protonation is followed by nucleophilic attack to give product. In contrast to addition under neutral or basic conditions (where protonation occurs *after* nucleophilic addition), this reaction involves protonation *before* nucleophilic addition to the carbonyl carbon atom.

General mechanism for nucleophilic addition under acidic conditions:

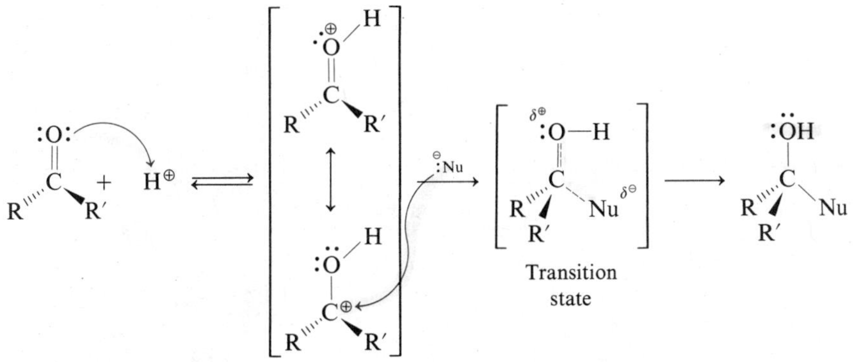
Transition
state

In some cases addition is acid-catalyzed and the proton is regenerated in the final step.

B. Relative Reactivities of Aldehydes and Ketones

The previous mechanisms provide considerable insight into the relative reactivities of various types of aldehydes and ketones in nucleophilic addition. The four trends are summarized and then justified on the basis of the mechanisms.

1. Aldehydes are more reactive than ketones.
2. Aldehydes and ketones with less bulky substituents attached to the carbonyl group are more reactive.
3. Cyclic ketones are more reactive than their noncyclic counterparts.
4. Aromatic aldehydes and ketones are less reactive than aliphatic compounds.

The effect of the bulk size of substituents on reactivity can be explained in terms of the structure of the products. We start with the planar carbonyl group. As addition occurs, the substituents are pushed closer to one another and the bond angle about carbon changes from 120° in the carbonyl group to 109.5° in the product. The larger the substituents, the more difficult it is to push them closer to one another. The *steric effect* can be seen vividly by examining structures of molecular models of two typical compounds in Fig. 21.1. Increased steric hindrance about the carbonyl group also makes attack by the nucleophile more difficult, as Fig. 21.1 illustrates. On the basis of this alone, it is not surprising that aldehydes are more reactive than ketones because hydrogen is always one of the substituents in an aldehyde.

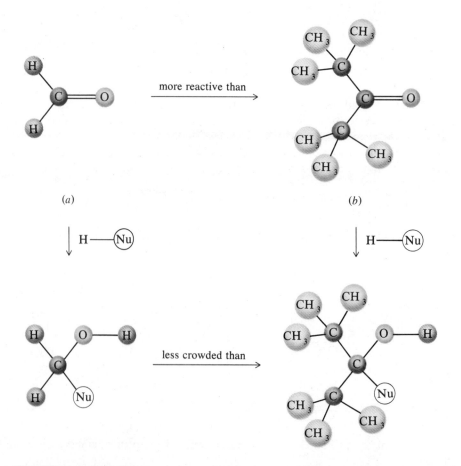

FIGURE 21.1 Sketches of molecular models showing the differences in steric crowding in the products resulting from the addition of H—Nu to (*a*) formaldehyde and (*b*) di-*tert*-butyl ketone.

$$CH_3CH_2-\overset{\overset{\displaystyle O}{\|}}{C}-H \qquad \text{more reactive than} \qquad \overset{\overset{\displaystyle O}{\|}}{\underset{CH_3 \quad CH_3}{C}}$$

Propanal
(propionaldehyde)

Propanone
(acetone)

The greater reactivity of cyclic over open-chain ketones is closely related to this concept. There is less crowding in the product, and nucleophilic attack on carbon is easier in cyclic compounds in which the carbonyl group is held rigidly in place by the ring while the substituents are held out of the way.

$$\overset{\overset{\displaystyle O}{\|}}{\underset{\underset{CH_2-CH_2}{CH_2 \quad CH_2}}{C}} \qquad \text{more reactive than} \qquad \overset{\overset{\displaystyle O}{\|}}{\underset{\underset{CH_3 \quad CH_3}{CH_2 \quad CH_2}}{C}}$$

Cyclopentanone

3-Pentanone

Electronic factors also play some role in dictating the reactivity of aldehydes and ketones. The nucleophile attacks the carbonyl carbon. If the partially positive character of that carbon atom is changed, then its affinity for a nucleophile should also be changed. As we know, alkyl groups release electrons, so attachment of alkyl groups to the carbonyl carbon should decrease its positive character and make it less susceptible to nucleophilic attack. Thus ketones should be and are less reactive than aldehydes in addition reactions:

$$\underset{H}{\overset{H}{>}}C\overset{\delta\oplus \quad \delta\ominus}{=}\ddot{\underset{\cdot\cdot}{O}} \qquad \underset{H}{\overset{R}{>}}C\overset{\delta\oplus \quad \delta\ominus}{=}\ddot{\underset{\cdot\cdot}{O}} \qquad \underset{R'}{\overset{R}{>}}C\overset{\delta\oplus \quad \delta\ominus}{=}\ddot{\underset{\cdot\cdot}{O}}$$

Decreasing positive character of carbon →

Decreasing reactivity toward Nu:⁻ →

Finally, we come to the reactivity differences between aliphatic and aromatic carbonyl compounds, that is, those that have the aromatic ring attached directly to the carbonyl carbon. Aromatic compounds undergo nucleophilic addition less rapidly than aliphatic compounds. This is explained by the unusual stability of the reactant. Even though the aryl ring is an electron-withdrawing group (by the inductive effect), it can interact with the π electrons of the carbonyl group by the resonance effect. The following resonance structures show how the carbonyl carbon is deactivated toward addition:

Best
representation

As we study the reactions in the following sections, remember that most aldehydes and ketones give identical reactions with a given reagent but that the relative reactivities stated previously generally hold.

21.3   Stereochemistry of Addition to Carbonyl Compounds

Because it is planar the carbonyl group per se is not capable of imparting chiral properties to a molecule. The reason for this is quite clear; the carbonyl group has a *plane of symmetry*:

$$R \,,,,_{''} \!\!\!\!\diagdown \atop R' \diagup C{=}O$$

Carbonyl group is achiral

On the other hand, molecules containing the carbonyl group may be chiral (that is, they may exist in enantiomeric forms) because of asymmetric centers (or other structural features) elsewhere in the molecule.

All carbonyl compounds except for symmetrically substituted ketones and formaldehyde generate a new asymmetric center when they undergo nucleophilic addition:

$$\overset{\displaystyle :O:}{\underset{}{\overset{\|}{R{-}C{-}R'}}} + H{-}Nu \longrightarrow \overset{\displaystyle :\overset{..}{O}{-}H}{\underset{\displaystyle R'}{\overset{|}{R{-}\overset{*}{C}{-}Nu}}}$$

R ≠ R'   New asymmetric center

In general, however, the product formed is optically inactive, even though the compound contains a new asymmetric center. The explanation for this is straightforward. Because the carbonyl precursor is symmetric, attack occurs with equal probability on either side of the planar species, thus producing an equimolar mixture of enantiomers. This is shown as follows for aldehydes and ketones:

Equal mixture of enantiomers;
Optically inactive

The result is an equimolar mixture of two enantiomeric products (a racemic mixture), which is *optically inactive*.

21.4   Nucleophilic Addition of Hydrogen Cyanide: Cyanohydrins

The elements of hydrogen cyanide, HCN, add to most aldehydes and many ketones to give a compound that contains both the cyano group, $-C{\equiv}\overset{..}{N}$, and the hydroxyl group on the same carbon atom; these compounds are called **cyanohydrins**:

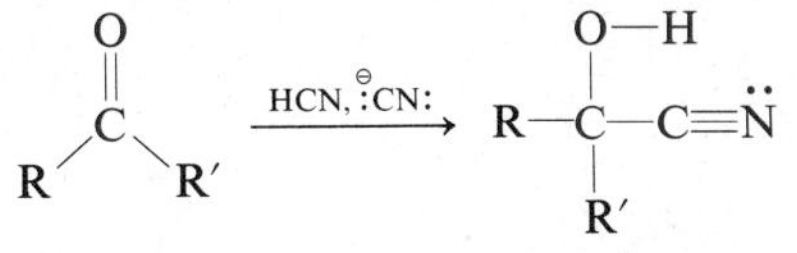

Cyanohydrin
(α-cyanoalcohol)

For example:

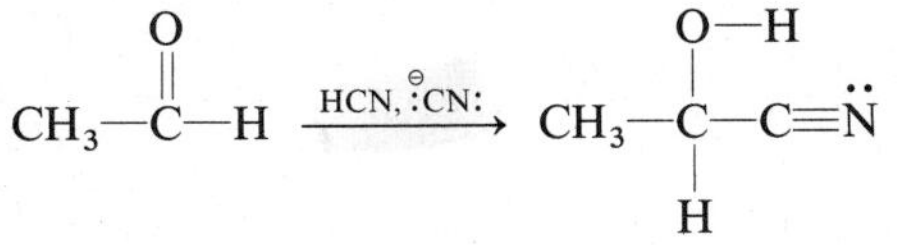

Ethanal
(acetaldehyde)

Acetaldehyde cyanohydrin
(common name)

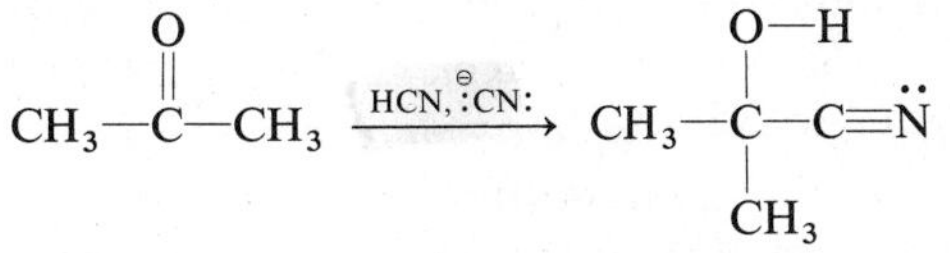

2-Propanone
(acetone)

Acetone cyanohydrin
(common name)

In principle, this addition should be favored under strongly acidic conditions, but hydrogen cyanide, HCN, is a weak acid ($K_a \approx 10^{-10}$), so there is not enough nucleophile (cyanide ion, $CN^{\ominus}$) in acidic medium for addition to occur. The addition is carried out by treating an aqueous solution of sodium cyanide with ~0.5 equivalent of mineral acid, thus generating a buffer solution that maintains the pH of the solution near 10 (that is, basic conditions). This reaction occurs via initial attack of cyanide ion (the nucleophile) on the carbonyl carbon. The resulting alkoxide ion rapidly accepts a proton from water to form the cyanohydrin and a molecule of hydroxide ion. Because cyanohydrin formation is reversible, the mechanism is shown here as a series of reversible equilibria:

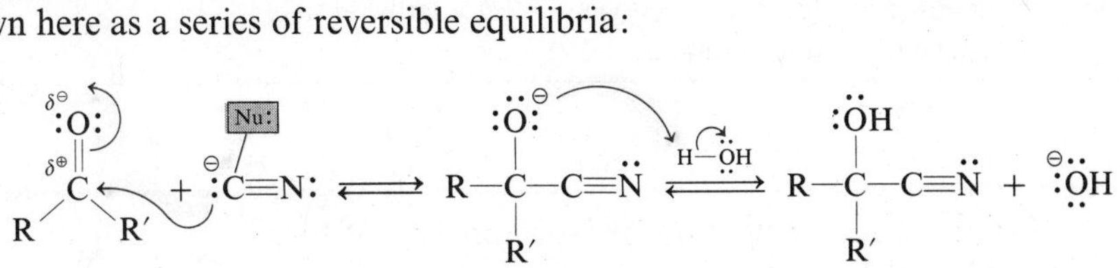

This is a base-catalyzed reaction because the hydroxide ion liberated in the formation of cyanohydrin reacts with some undissociated HCN and produces more cyanide ion, which in turn reacts with more carbonyl compound:

The experimental conditions are important. In strongly acidic solution virtually no nucleophilic cyanide ion is present, and in strongly basic solution the reverse reaction between the cyanohydrin and hydroxide ion to regenerate reactants is important.

Cyanohydrins have relatively little use on their own but they are important precursors to several useful products. The dehydration of a cyanohydrin produces an unsaturated nitrile, which can be used in polymerization reactions. Also, the acid-catalyzed hydrolysis of the nitrile group produces an α-hydroxy carboxylic acid (see Sec. 22.5 for details of the hydrolysis); for example:

$$CH_2{=}CH{-}C{\equiv}\ddot{N}$$

Cyanoethene
(acrylonitrile)

$$CH_3{-}\overset{O}{\underset{}{\overset{\|}{C}}}{-}H \xrightarrow[\underset{:CN:}{\ominus}]{HCN} CH_3{-}\overset{O-H}{\underset{H}{\overset{|}{C}}}{-}C{\equiv}\ddot{N}$$

with $\xrightarrow{H^{\oplus},\, heat}_{-H_2O}$ and $\xrightarrow{H^{\oplus},\, H_2O}$

$$CH_3{-}\overset{O-H}{\underset{H}{\overset{|}{C}}}{-}\overset{O}{\underset{}{\overset{\|}{C}}}{-}OH$$

2-Hydroxypropanoic acid
(α-hydroxypropionic acid)

Note that in the preceding examples the common name of the aldehyde or ketone addition product is provided. The common names are more routinely used in the naming of these compounds. The IUPAC system could also be used, for example, acetaldehyde cyanohydrin is also named 2-hydroxypropanenitrile (see Sec. 23.1). *In the examples throughout this chapter, if the only name provided for an addition product of an aldehyde or ketone is a common name, we will indicate it as such.*

Cyanohydrins play an important role in the synthesis of sugars, in which they are converted first into α-hydroxy acids and then into α-hydroxy aldehydes (the Kiliani-Fischer synthesis, see Sec. 29.9).

Question 21.1

Write the complete, stepwise mechanism for the conversion of acetone cyanohydrin in basic solution into acetone and cyanide ion.

Question 21.2

Give the structures of the product(s) that would be obtained by allowing the following compounds to react with a mixture of aqueous NaCN/HCN:

(*a*) benzaldehyde   (*b*) acetophenone   (*c*) diethyl ketone
(*d*) 2-butanone   (*e*) 1,4-butanedial

21.5   Nucleophilic Addition of Sodium Bisulfite: Addition Compounds

Sodium bisulfite, $NaHSO_3$, adds to most aldehydes and unhindered ketones (usually methyl ketones and cyclic ketones) in the overall reaction:

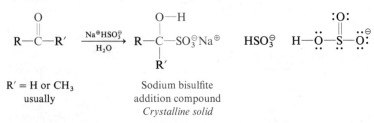

$$R{-}\overset{O}{\underset{}{\overset{\|}{C}}}{-}R' \xrightarrow[H_2O]{Na^{\oplus}HSO_3^{\ominus}} R{-}\overset{O-H}{\underset{R'}{\overset{|}{C}}}{-}SO_3^{\ominus}Na^{\oplus} \qquad HSO_3^{\ominus} \quad H{-}\ddot{\overset{..}{O}}{-}\overset{:O:}{\underset{..}{\overset{\|}{S}}}{-}\overset{..}{\overset{\ominus}{O}}:$$

R' = H or CH₃
usually

Sodium bisulfite
addition compound
Crystalline solid

omit

For example:

$$CH_3-\overset{\overset{\displaystyle O}{\|}}{C}-H \xrightarrow[H_2O]{Na^\oplus HSO_3^\ominus} CH_3-\overset{\overset{\displaystyle O-H}{|}}{\underset{\underset{\displaystyle H}{|}}{C}}-SO_3^\ominus Na^\oplus$$

Ethanal
(acetaldehyde)

Sodium bisulfite addition product
of ethanal

$$CH_3CH_2-\overset{\overset{\displaystyle O}{\|}}{C}-CH_3 \xrightarrow[H_2O]{Na^\oplus HSO_3^\ominus} CH_3CH_2-\overset{\overset{\displaystyle O-H}{|}}{\underset{\underset{\displaystyle CH_3}{|}}{C}}-SO_3^\ominus Na^\oplus$$

2-Butanone
(methyl ethyl ketone)

Sodium bisulfite addition product
of 2-butanone

It seems likely that the nucleophile which adds to the carbonyl group is the sulfite ion, $SO_3^{\ominus 2}$, formed from the dissociation of $HSO_3^\ominus$ (a weak acid). The steps in the reaction follow the general pattern for nucleophilic addition and are as follows:

$$HSO_3^\ominus \rightleftharpoons SO_3^{\ominus 2} + H^\oplus \qquad SO_3^{\ominus 2} \equiv \overset{\ominus}{\underset{..}{:}}\overset{..}{\underset{..}{O}}-\overset{\overset{\displaystyle :O:}{\|}}{S}-\overset{..}{\underset{..}{O}}\overset{\ominus}{:}$$

$$R-\overset{\overset{\displaystyle :O:}{\|}}{C}-R' + SO_3^{\ominus 2} \rightleftharpoons R-\overset{\overset{\displaystyle :O:^\ominus}{|}}{\underset{\underset{\displaystyle R'}{|}}{C}}-SO_3^\ominus \underset{H^\oplus}{\rightleftharpoons} R-\overset{\overset{\displaystyle :O-H}{|}}{\underset{\underset{\displaystyle R'}{|}}{C}}-SO_3^\ominus$$

$$\Updownarrow Na^\oplus$$

$$R-\overset{\overset{\displaystyle :O-H}{|}}{\underset{\underset{\displaystyle R'}{|}}{C}}-SO_3^\ominus Na^\oplus$$

Bisulfite addition compounds are usually crystalline solids that are soluble in water, so they can be purified by recrystallization. They are usually used in the purification of aldehydes and unhindered ketones because the water soluble bisulfite addition compounds can be separated from noncarbonyl or hindered carbonyl-containing impurities.

Bisulfite addition is reversible, and treatment of the addition compound with either strong acid or base regenerates the aldehyde or ketone:

$$R-\overset{\overset{\displaystyle O-H}{|}}{\underset{\underset{\displaystyle R'}{|}}{C}}-SO_3^\ominus Na^\oplus \xrightarrow[:\overset{..}{O}H^\ominus]{H^\oplus \text{ or}} R-\overset{\overset{\displaystyle O}{\|}}{C}-R' + SO_2 \quad \text{in acid solution} \quad \text{or } SO_3^{\ominus 2} \quad \text{in base solution}$$

Question 21.3

Indicate how the following two compounds in each pair might be separated from one another and each obtained in pure form:

(**a**) cyclohexanone and cyclohexanol   (**b**) acetaldehyde and diisopropyl ketone

21.6   Nucleophilic Addition of Alcohols: Hemiacetals and Acetals

Alcohols add to aldehydes and ketones under acidic or basic conditions, and the product(s) depends on the conditions used. We discuss the acid-catalyzed addition first and then the reaction that occurs under basic conditions. Both types of reactions are reversible.

A. Acid-Catalyzed Reaction

Under acidic conditions, an aldehyde or ketone reacts with an alcohol and gives one of two possible products depending on the reaction conditions. The general reaction is:

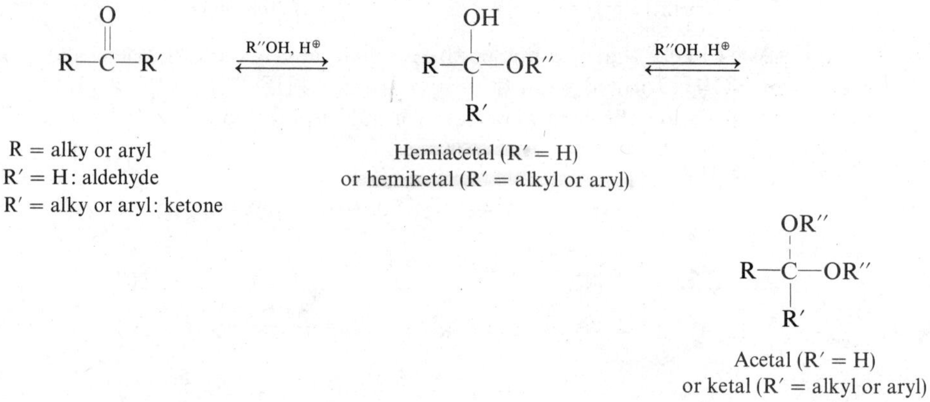

$$
\underset{\text{Aldehyde}}{\overset{\displaystyle \text{O}}{\underset{\displaystyle }{R\!-\!\overset{\|}{C}\!-\!R'}}}
\quad \underset{\longleftarrow}{\overset{R''\text{OH, H}^{\oplus}}{\rightleftharpoons}} \quad
\underset{\underset{\displaystyle R'}{|}}{\overset{\overset{\displaystyle \text{OH}}{|}}{R\!-\!\overset{|}{\underset{|}{C}}\!-\!\text{OR}''}}
\quad \underset{\longleftarrow}{\overset{R''\text{OH, H}^{\oplus}}{\rightleftharpoons}}
$$

R = alky or aryl
R′ = H: aldehyde
R′ = alky or aryl: ketone

Hemiacetal (R′ = H)
or hemiketal (R′ = alkyl or aryl)

$$
\underset{\underset{\displaystyle R'}{|}}{\overset{\overset{\displaystyle \text{OR}''}{|}}{R\!-\!\overset{|}{\underset{|}{C}}\!-\!\text{OR}''}}
$$

Acetal (R′ = H)
or ketal (R′ = alkyl or aryl)

When we start with an aldehyde, the products are **hemiacetals** and **acetals**; if a ketone is used as the starting carbonyl compound, the products are **hemiketals** and **ketals.** The terms *hemiacetal* and *hemiketal* are often used interchangeably, as are *acetal* and *ketal*. In our discussions we emphasize hemiacetals and acetals, but the same mechanisms and reactions apply for hemiketal and ketal formation.

The characteristic functional group in hemiacetals and hemiketals is the **C—O—C—OH** linkage, so they are both an ether and an alcohol. The characteristic functional group in acetals and ketals is the **C—O—C—O—C** linkage, so they are diethers.

The mechanism of hemiacetal formation is identical for hemiketal formation and is shown here:

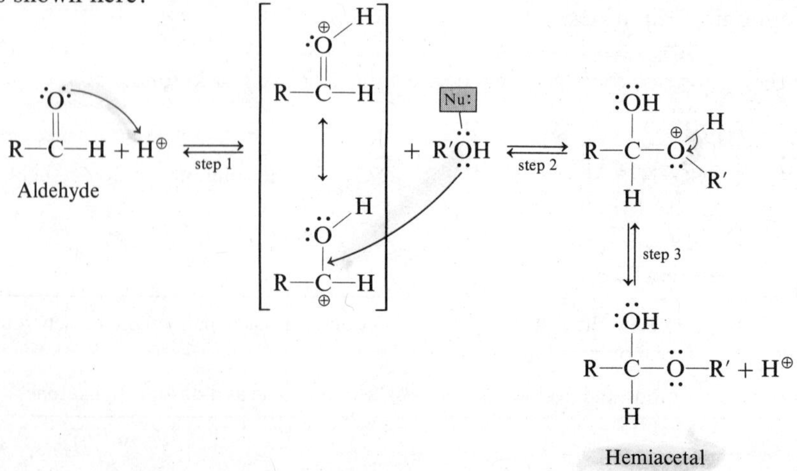

Hemiacetal

Step 1 is the protonation of the carbonyl compound, a reaction typical of oxygen containing unshared pairs of electrons. Step 2 shows the nucleophilic attack of the alcohol, R′OH, on the electron-deficient carbon atom; this is analogous to the formation of an ether via an S_N1 mechanism (see Sec. 19.4). Step 3 shows the loss of a proton from oxonium ion to form the hemiacetal; this is the familiar deprotonation reaction many oxonium ions undergo.

Once formed, the hemiacetal can react with more alcohol to form an acetal. A hemiacetal contains two oxygens, either of which can be protonated:

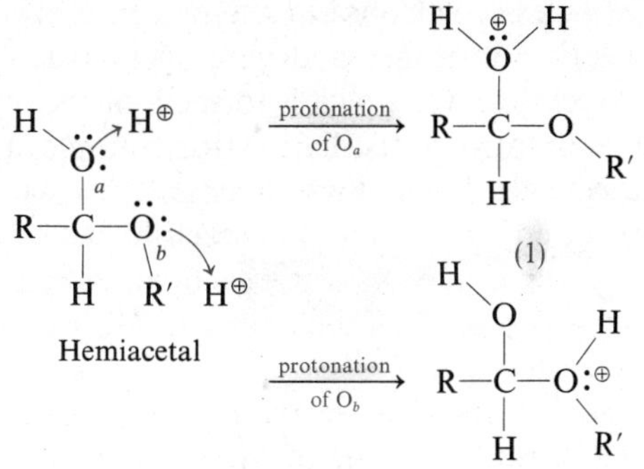

(1)

(2)

Oxonium ion (2) is the precursor to the hemiacetal (shown in the mechanism of hemiacetal formation). The formation of (2) marks the beginning of a reaction sequence in which the hemiacetal is converted back into an aldehyde and an alcohol:

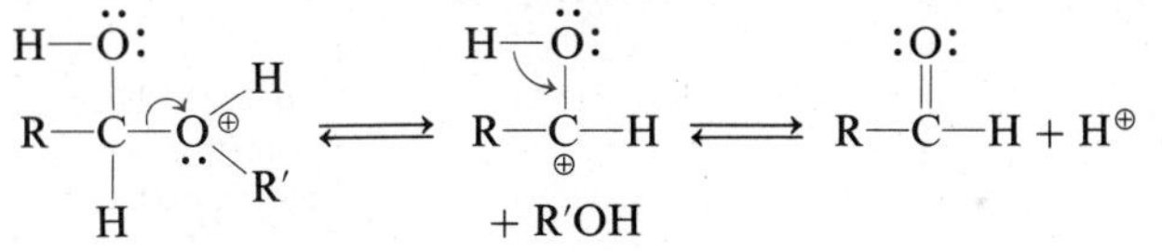

(2)

If oxonium ion (1) is formed, it can undergo other reactions that ultimately lead to an acetal. The currently accepted mechanism for this transformation is shown here:

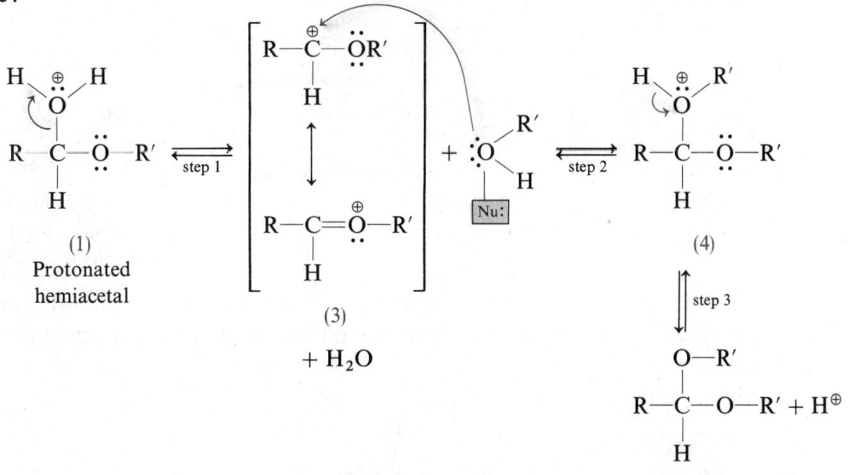

Step 1 involves the loss of a water molecule to form a carbocation that is especially stable because of the resonance interaction of the unshared pairs of electrons on oxygen with the positive charge. Step 2 is the nucleophilic attack of an alcohol molecule on cation (3) to give a new oxonium ion (4). Step 3 is the loss of a proton from oxonium ion (4) to give the final product, an acetal. Note the role of acid as a catalyst in this reaction; it is consumed in step 1 and regenerated in step 3.

We now come to an interesting and important aspect of equilibrium in hemiacetal and acetal formation. If each step in their formation is reversible, we would expect to be able to convert them back into an aldehyde or ketone and alcohol. As we will see, this is possible.

Experimentally, acetals are often formed by allowing an aldehyde or ketone to react with a large excess of alcohol in the presence of an anhydrous acid catalyst, usually dry HCl gas. Sometimes a drying agent (such as anhydrous magnesium sulfate) is added to remove water as it is formed, or the reaction vessel is equipped to remove the water by azeotropic distillation (for example, using benzene; see Sec. 10.7). These factors—excess alcohol and water removal—favor acetal (or ketal) formation because they shift the equilibrium in the direction of the acetal (or ketal).

Conversely, the **hydrolysis of acetals** is favored by the presence of excess water, with hydrogen ion again used as the catalyst. *The mechanism for the hydrolysis of acetals is just the reverse of the mechanism for their formation.*

Increasing the concentration of the reactants favors a particular product, and increasing the concentration of one of the products regenerates the reactants. The following scheme shows the reversible nature of acetal formation:

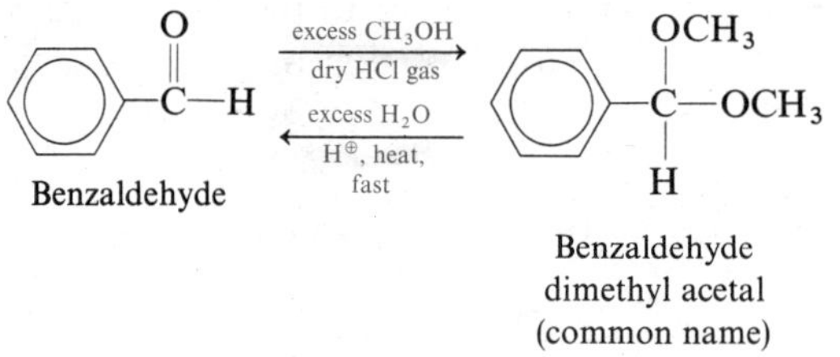

Benzaldehyde

Benzaldehyde dimethyl acetal (common name)

Acetals are often made from simple alcohols. On the other hand, ketals are seldom made from monohydroxylic alcohols, which give low yields, but instead are made from glycols. This is probably because of the lower reactivity of ketones in relation to aldehydes (see Sec. 21.2). A ketal is also readily hydrolyzed back into a ketone and glycol; for example:

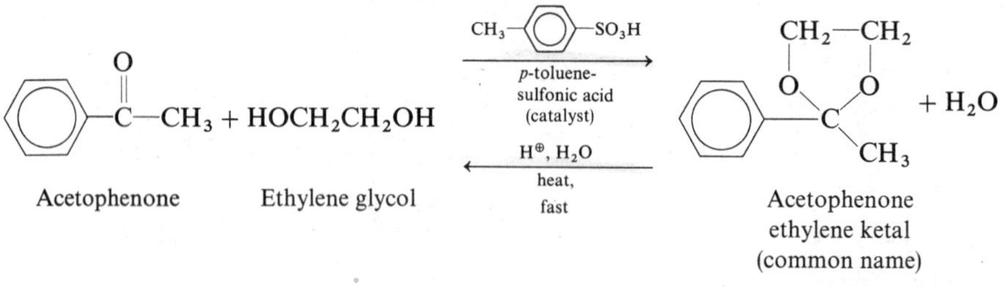

Acetophenone   Ethylene glycol

Acetophenone ethylene ketal (common name)

The first intermediate in the formation of the preceding ketal (and one that is not isolated) is the following hemiketal, which reacts internally to form the ethylene ketal:

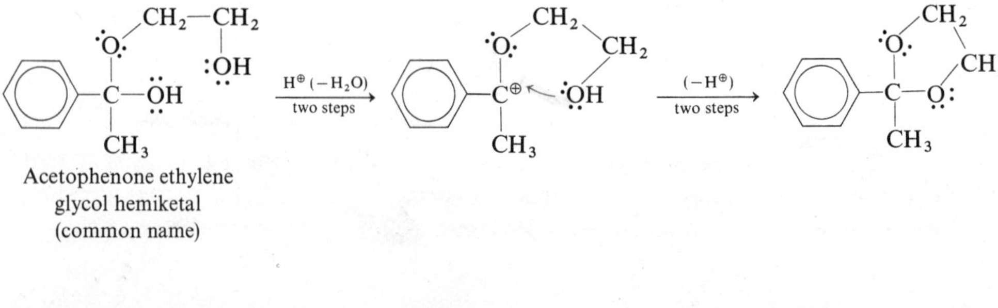

Acetophenone ethylene glycol hemiketal (common name)

It is probably surprising that the hydrolysis of acetals is fast because ethers in general are difficult to cleave and an acetal is a diether because of the

$$C—O—C—O—C$$

linkage. However, one driving force for the rapid hydrolysis of acetals is the formation of a resonance-stabilized cation.

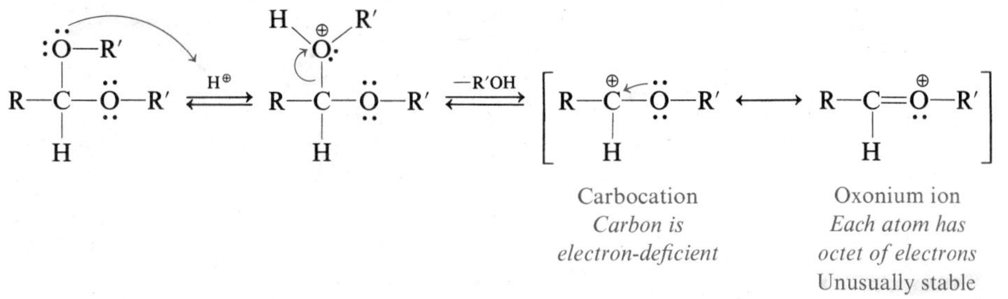

Carbocation
Carbon is
electron-deficient

Oxonium ion
Each atom has
octet of electrons
Unusually stable

In the next part of this section we see that hemiacetals and hemiketals can be formed in basic solution but that acetals and ketals cannot. Likewise, acetal or ketal hydrolysis requires acidic conditions because the initial step is the protonation of one of the "ether" linkages. Thus acetals and ketals are stable toward base and hemiacetals and hemiketals are not.

There are a myriad of uses of acetals and ketals in organic chemistry. One is in "protecting" the carbonyl group so that certain reactions of other functional groups in a molecule can be carried out without affecting the carbonyl group. (See Sec 21.4 for an example.)

Hemiacetals and hemiketals frequently occur in carbohydrates (sugars) because the latter contain hydroxyl groups and an aldehyde or ketone group in the same molecule. Five- and six-membered cyclic hemiacetals and hemiketals are often observed rather than the "free" aldehyde or ketone because of the formation of strain-free, cyclic ring systems. The following example shows this *internal* hemiacetal formation in a carbohydrate:

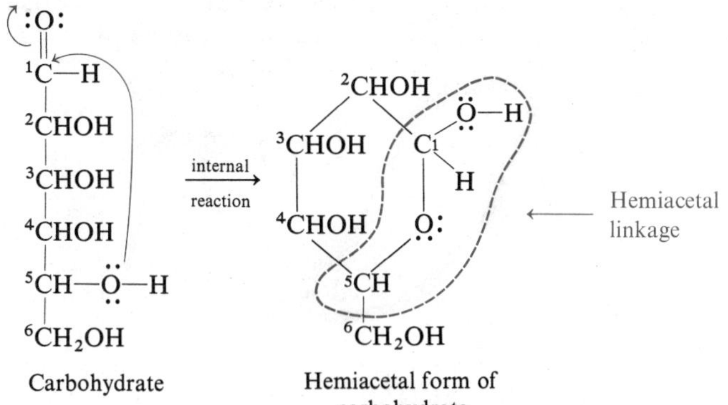

Carbohydrate

Hemiacetal form of
carbohydrate

Hemiacetal
linkage

We discuss the cyclic hemiacetals, hemiketals, acetals, and ketals of carbohydrates in Chap. 29.

Question 21.4

What carbonyl compound and alcohol would be required to synthesize the following three compounds?

(*a*) cyclohexanone diethyl ketal   (*b*) methyl ethyl ketone di-*n*-propyl ketal
(*c*) hexanal ethylene acetal

Question 21.5

Provide complete, stepwise mechanisms to account for the following reactions, giving the product where required:

(*a*) Acetone + excess ethanol in the presence of dry HCl gas

(*b*) Benzaldehyde ethylene acetal + H_2O in the presence of $H^{\oplus}$ to give benzaldehyde + ethylene glycol

(*c*)

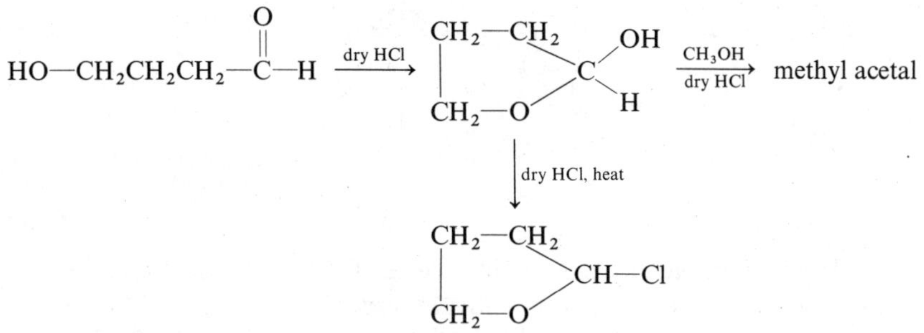

Question 21.6

The mechanism of the acid-catalyzed addition of an alcohol to a carbonyl compound starts with protonation of the carbonyl group. In many other mechanisms, however, we started with the protonated alcohol. Because an alcohol is present in this reaction, why did we not start by protonating it rather than the carbonyl group?

B. Base-Catalyzed Reaction

Hemiacetals and hemiketals can be formed in certain instances under basic conditions. In contrast, acetals and ketals require an acid-catalyzed reaction; they are stable under basic conditions and cannot be produced under such conditions.

In its most general form, the base-catalyzed reaction is as follows:

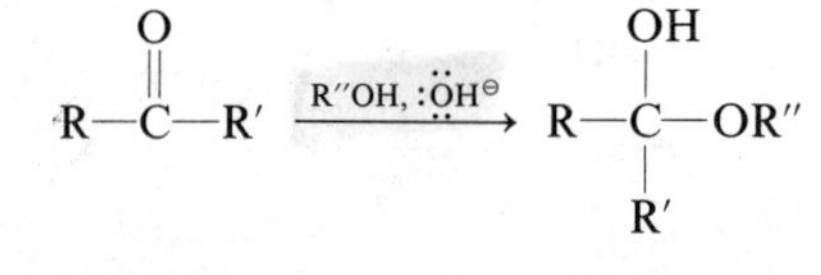

Aldehyde   Hemiacetal
or ketone   or hemiketal

For example:

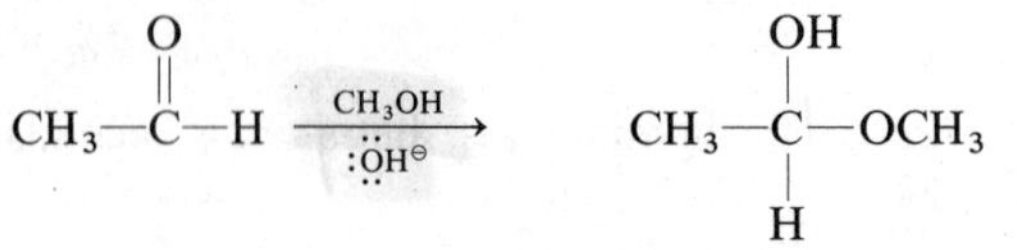

Ethanal   Acetaldehyde methyl hemiacetal
(acetaldehyde)   (common name)

The mechanism of base-catalyzed hemiacetal formation is quite similar to the nucleophilic addition of hydrogen cyanide to carbonyl compounds. The currently accepted mechanism of this reversible reaction is:

$$R''OH + :\ddot{O}H^{\ominus} \rightleftharpoons R''\ddot{O}:^{\ominus} + H_2O$$

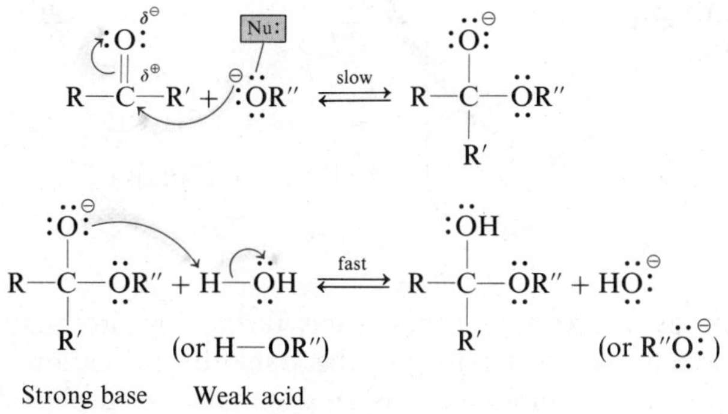

R—C—R' + :OR'' $\xrightarrow{\text{slow}}$ R—C—OR''

R—C—OR'' + H—OH $\xrightarrow{\text{fast}}$ R—C—OR'' + HO:$^{\ominus}$
   | (or H—OR'') | (or R''O:$^{\ominus}$)
   R'   R'

Strong base   Weak acid

21.7   Addition of Phenol to Formaldehyde: Condensation Polymerization Reactions

Another useful reaction of aldehydes occurs between formaldehyde and phenol, from which *resins* such as Bakelite are formed. The general reaction is:

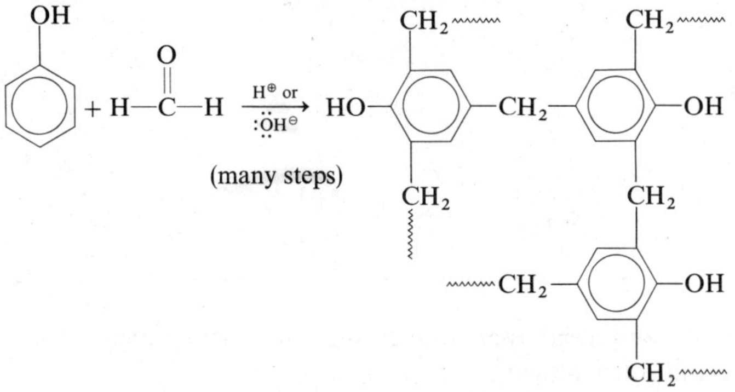

Bakelite (small segment)

The key intermediates in the reaction are *o*- and *p*-hydroxymethylphenol, which react with other molecules of formaldehyde and/or phenol to give the polymeric material, which has a high molecular weight. The Bakelite segment shown is only a small portion of the polymer.

There are several steps in the reaction, each involving principles with which we are familiar. Although the reaction can be carried out under either acidic or basic conditions, we discuss only the mechanism of the acid-catalyzed condensation. The first step is the protonation of formaldehyde, which gives the carbonyl carbon more electrophilic character. The aromatic ring in phenol is highly activated, and it undergoes electrophilic aromatic substitution with protonated formaldehyde, which serves as the electrophile:

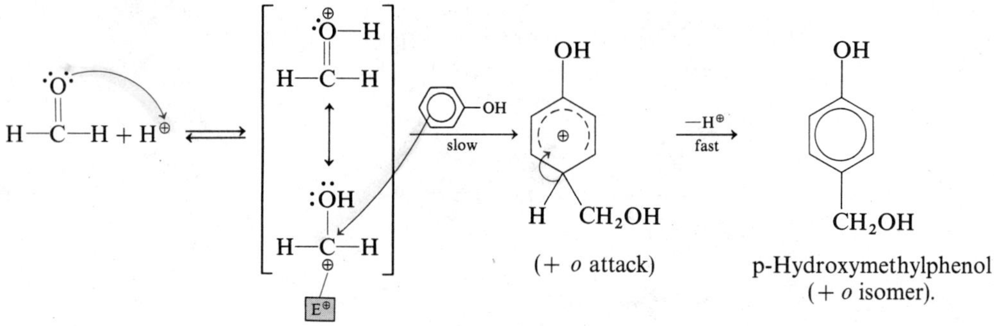

(+ o attack)

p-Hydroxymethylphenol
(+ o isomer).

The hydroxymethylphenol reacts further by protonation of the —CH₂OH group, which loses water to give the stable benzyl cation. The carbocation then reacts with another molecule of phenol (or substituted phenol) by electrophilic aromatic substitution, providing the methylene (—CH₂—) link between two phenol molecules. These reactions are shown here:

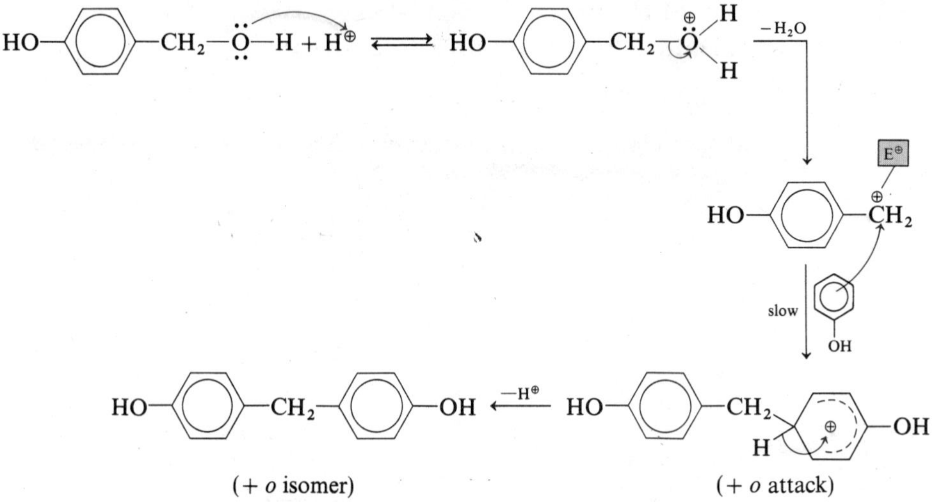

(+ o isomer)

(+ o attack)

The diaryl compound then reacts with more formaldehyde and phenol to give Bakelite and other polymeric resins.

Question 21.7

Devise a mechanism to account for the following addition reaction: (*Hint:* See Sec. 22.6.)

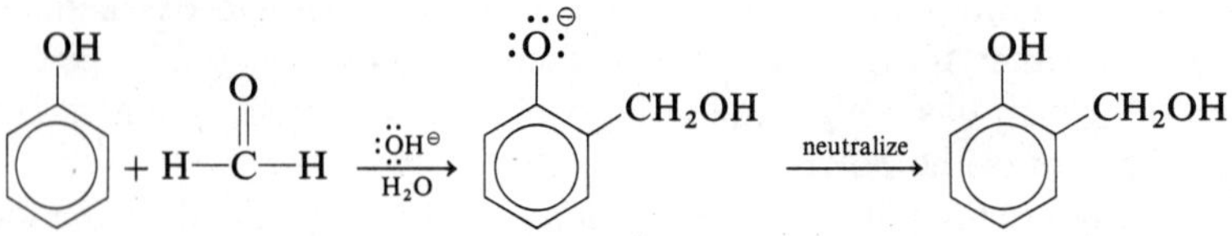

Question 21.8

When benzyl alcohol is treated with concentrated sulfuric acid, a white, gummy organic substance (containing no sulfur or oxygen) is formed. What do you think the structure of this substance might be? Propose an explanation for its formation.

21.8   Nucleophilic Addition of Ammonia and Substituted Amines

Ammonia, $\overset{..}{N}H_3$, and various substituted amines, $G—\overset{..}{N}H_2$ (where G is the variety of substituents listed under the following reaction), add across the carbon-oxygen double bond. The intermediate amino alcohol is unstable and dehydrates to give a carbon-nitrogen double bond. The general reaction is:

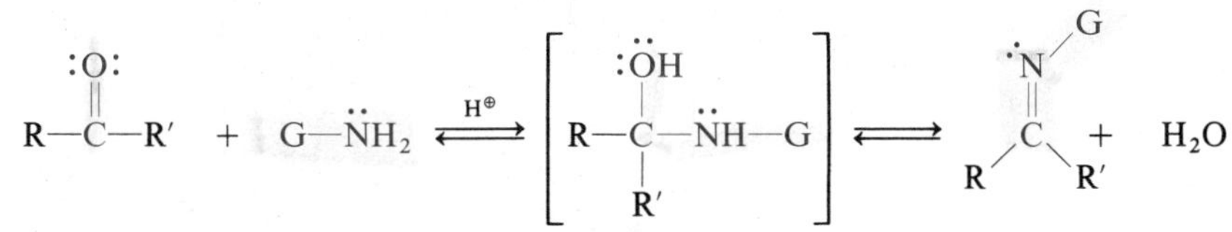

Aldehyde or ketone   Amino alcohol
                                     Unstable, not isolated

where $G = —H, —\overset{..}{O}H, —\overset{..}{N}H_2, —\overset{..}{N}HC_6H_5, —\overset{..}{N}H\overset{O}{\overset{||}{C}}\overset{..}{N}H_2$, and others

In a sense this is an **addition-elimination reaction** because the elements of water are lost after addition occurs.

The mechanism of the **addition** is viewed in the following manner:

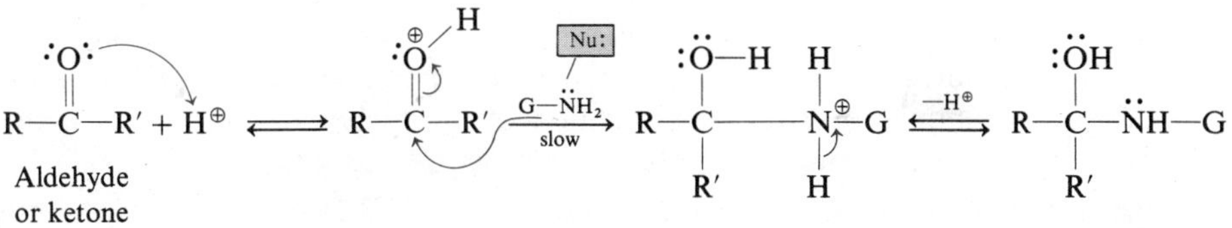

This nucleophilic addition is catalyzed by acid, which protonates the carbonyl compound and makes it more susceptible to nucleophilic attack by $G—\overset{..}{N}H_2$. However, the substituted amine is also protonated in acidic solution (by analogy with the protonation of ammonia to give the ammonium ion: $\overset{..}{N}H_3 + H^\oplus \rightleftarrows NH_4^\oplus$), and the protonated amine cannot serve as a nucleophile because it contains no unshared electron pair.

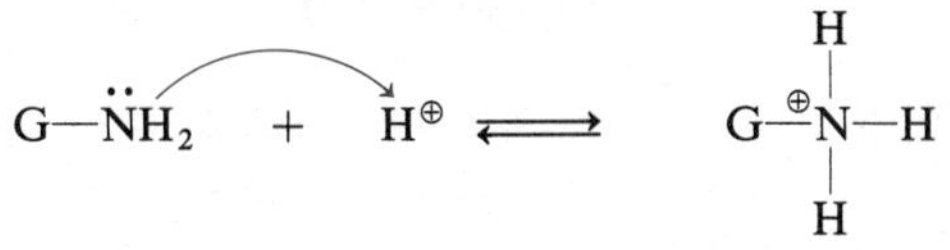

Free amine   Protonated amine
Is a nucleophile   *Contains no unshared*
(contains electron pair)   *electron pair and is not*
                                 a nucleophile

As a consequence, the hydrogen ion concentration must be controlled so that the reaction occurs at a reasonable rate (as dictated by the extent to which the carbonyl compound is protonated) and so that the substituted amine is not completely protonated (for no reaction occurs if it is). The acidity is easily controlled by buffer solutions; the exact hydrogen ion requirements differ for various carbonyl compounds and different substituted amines. Thus, there is no rule of thumb for predicting the required $[H^\oplus]$, and this must be determined experimentally; we can only indicate the necessity of carrying out the addition reactions in acidic solution.

The mechanism of the elimination reaction resembles that for the dehydration of alcohols, except that here an especially stable carbocation is formed because of the positive charge adjacent to the unshared electron pair on nitrogen:

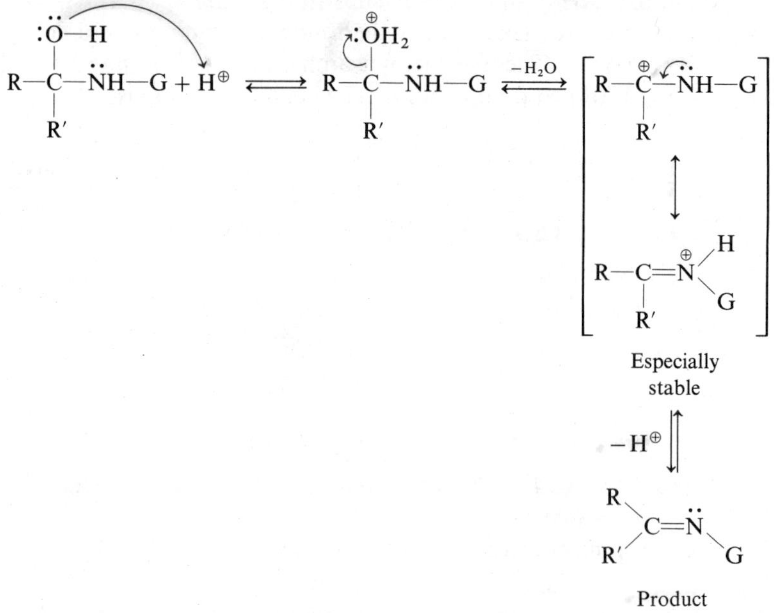

$$
\begin{array}{ccc}
\underset{\substack{|\\R'}}{\overset{\substack{:\overset{..}{O}-H\\|}}{R-C-\overset{..}{N}H-G}} + H^{\oplus} & \rightleftharpoons & \underset{\substack{|\\R'}}{\overset{\substack{\overset{\oplus}{:}\overset{..}{O}H_2\\|}}{R-C-\overset{..}{N}H-G}} \xrightarrow{-H_2O}
\end{array}
$$

$$
\left[\begin{array}{c}
\underset{\substack{|\\R'}}{R-\overset{\oplus}{C}-\overset{..}{N}H-G} \\
\updownarrow \\
\underset{\substack{|\\R'}}{R-C=\overset{\oplus}{N}\underset{G}{\overset{H}{\diagdown}}}
\end{array} \right]
$$

Especially stable

$$-H^{\oplus} \updownarrow$$

$$
\underset{R'}{\overset{R}{\diagup}}C=\overset{..}{N}\underset{G}{\diagdown}
$$

Product

An important use of the addition products of carbonyl compounds and various substituted amines is in the identification of the carbonyl compounds themselves. The principal structural change that occurs is the removal of the oxygen atom of the carbonyl group and its replacement by $=\overset{..}{N}-G$; the rest of the molecule remains intact. The addition compounds are referred to as **derivatives** of the corresponding carbonyl compound. Most derivatives are *solids*, which are easily purified and have melting points typically much higher than those of the aldehyde or ketone from which they are derived. Even a liquid carbonyl compound is usually converted to a solid derivative. Extensive tables relating the melting points of various derivatives to a given carbonyl compound have been tabulated from experimental work, so that identification of an unknown is possible if the boiling point or melting point of the original unknown and the melting points of several derivatives are known. (See also Sec. 21.16.)

The structures and common names of three typical addition products of carbonyl compounds are shown here. Semicarbazones and 2,4-dinitrophenylhydrazones are the most useful because they have higher melting points. Many oximes melt just above room temperature or are oils, which are difficult to purify and identify.

Oximes:

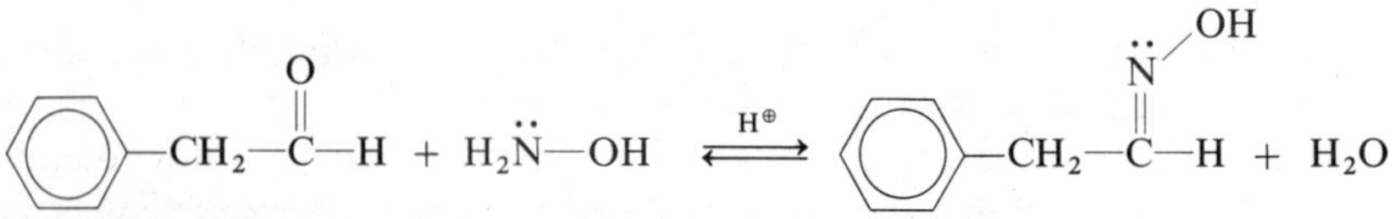

$$
\underset{\substack{\text{Phenylethanal}\\\text{(phenylacetaldehyde)}\\\textit{bp 194}°}}{\text{C}_6\text{H}_5-\text{CH}_2-\overset{\overset{\textstyle O}{\|}}{\text{C}}-\text{H}} + \underset{\text{Hydroxylamine}}{H_2\overset{..}{N}-OH} \xrightarrow{H^{\oplus}} \underset{\substack{\text{Oxime:}\\\text{phenylacetaldehyde oxime}\\\text{(common name)}\\\textit{mp 103}°}}{\text{C}_6\text{H}_5-\text{CH}_2-\overset{\overset{\textstyle \overset{..}{N}\diagup^{OH}}{\|}}{\text{C}}-\text{H}} + H_2O
$$

Semicarbazones:

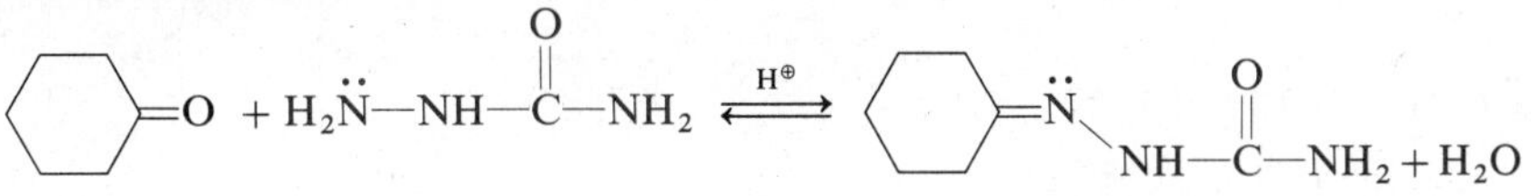

Cyclohexanone	Semicarbazide		Semicarbazone:
bp 156°			cyclohexanone
			semicarbazone
			(common name)
			mp 167°

2,4-Dinitrophenylhydrazones:

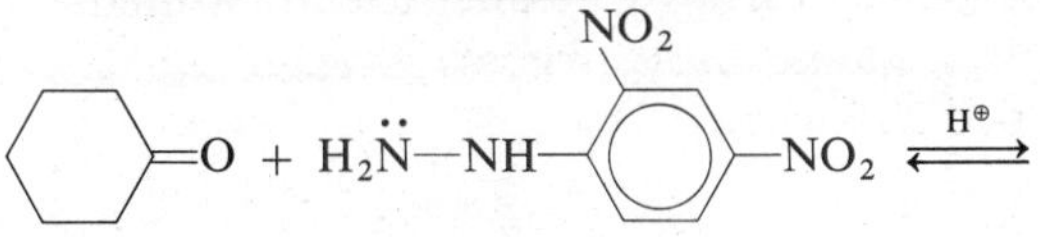

Cyclohexanone　　2,4-Dinitrophenylhydrazine
　　　　　　　　　　(abbreviated 2,4-DNPH)

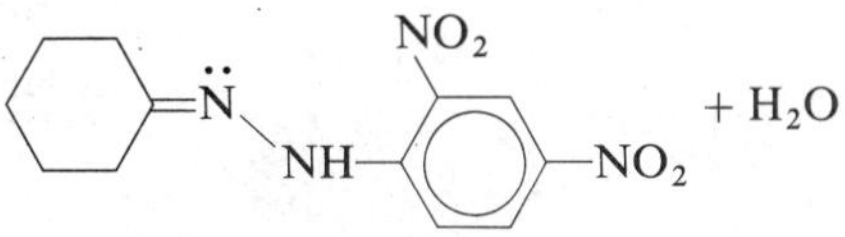

2,4-Dinitrophenylhydrazone
(abbreviated 2,4-DNP):
cyclohexanone
2,4-dinitrophenylhydrazone
(common name)
mp 162°

Other more simple ammonia derivatives also add to aldehydes and ketones, such as ammonia ($\overset{\cdot\cdot}{N}H_3$) and primary amines ($R—\overset{\cdot\cdot}{N}H_2$). The product is an **imine**, often referred to as a **Schiff's base;** for example:

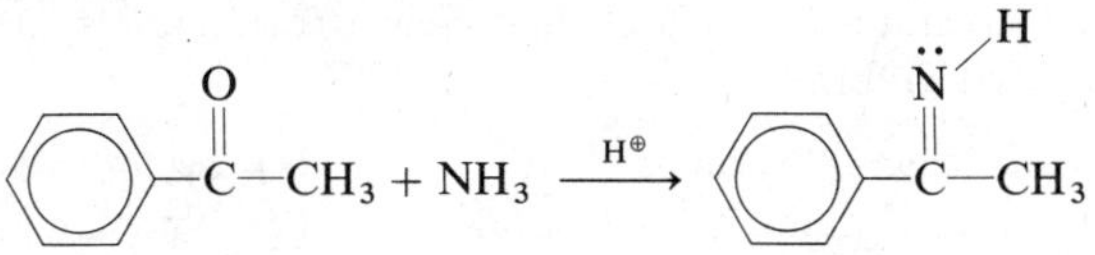

Acetophenone　　　　　　　　　　　　　　　　Acetophenone
　　　　　　　　　　　　　　　　　　　　　　imine
　　　　　　　　　　　　　　　　　　　　　　(common name)

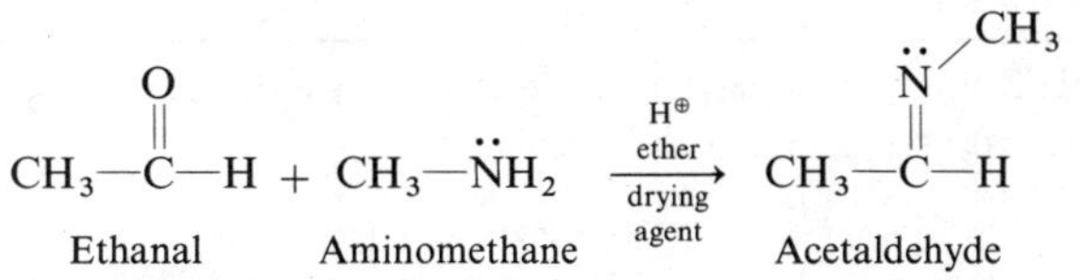

Ethanal	Aminomethane		Acetaldehyde
(acetaldehyde)	(methylamine)		methylimine
			(common name)

Schiff's bases are more reactive toward hydrolysis than are the oximes, semi-carbazones, and 2,4-dinitrophenylhydrazones.

Hydrazine, $H_2\overset{\cdot\cdot}{N}—\overset{\cdot\cdot}{N}H_2$, also adds to carbonyl compounds, although the major reaction is the condensation of 2 moles of carbonyl compound with 1 mole of hydra-zine to give an **azine.** Using 2-propanone as an example, the following reactions occur:

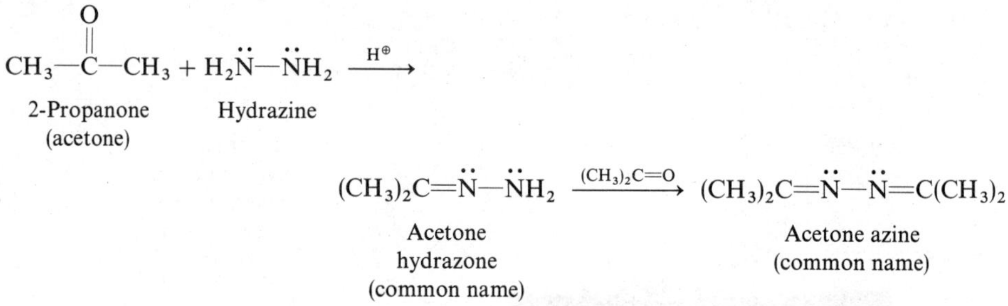

$$CH_3-\overset{\overset{\displaystyle O}{\|}}{C}-CH_3 + H_2\overset{..}{N}-\overset{..}{N}H_2 \xrightarrow{H^\oplus}$$

2-Propanone   Hydrazine
(acetone)

$$(CH_3)_2C=\overset{..}{N}-\overset{..}{N}H_2 \xrightarrow{(CH_3)_2C=O} (CH_3)_2C=\overset{..}{N}-\overset{..}{N}=C(CH_3)_2$$

Acetone   Acetone azine
hydrazone   (common name)
(common name)

Because of azine formation, substituted hydrazines are often used to ensure monocondensation. For example, semicarbazide, phenylhydrazine, and 2,4-dinitrophenylhydrazine are all substituted hydrazines:

$$H_2\overset{..}{N}-\overset{..}{N}H-H \qquad H_2\overset{..}{N}-\overset{..}{N}H-\hexagon$$

Hydrazine   Phenylhydrazine

$$H_2\overset{..}{N}-\overset{..}{N}H-\underset{NO_2}{\overset{NO_2}{\hexagon}}-NO_2 \qquad H_2\overset{..}{N}-\overset{..}{N}H-\overset{\overset{\displaystyle O}{\|}}{C}-NH_2$$

2,4-Dinitrophenylhydrazine   Semicarbazide

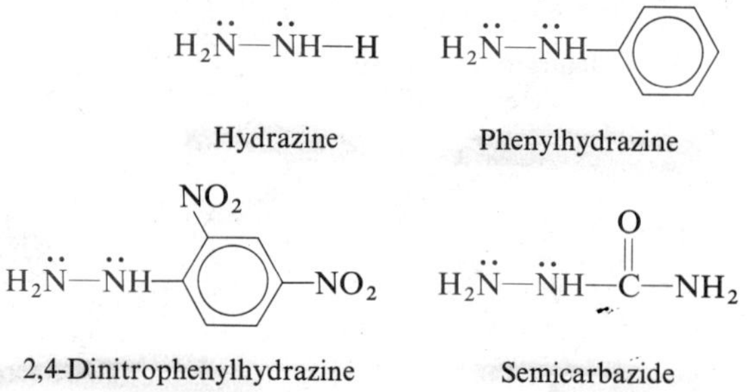

Question 21.9

Provide a complete mechanism to show how benzaldehyde imine, $C_6H_5CH=\overset{..}{N}H$, undergoes acid-catalyzed hydrolysis to form benzaldehyde and the ammonium ion.

Question 21.10

Provide structures for each of the following compounds, and indicate the reagents that would be required to form each:

(*a*) phenylacetaldehyde ethylimine   (*b*) cyclopentanone imine
(*c*) benzophenone 2,4-dinitrophenylhydrazone   (*d*) 2-pentanone semicarbazone
(*e*) cyclohexanone hydrazone   (*f*) cyclohexanone azine
(*g*) 1,4-butanedial di(phenylhydrazone)   (*h*) diisopropyl ketone oxime

Question 21.11

(*a*) Explain how propionaldehyde forms two isomeric oximes, (1) and (2), which have different physical properties and can be isolated.

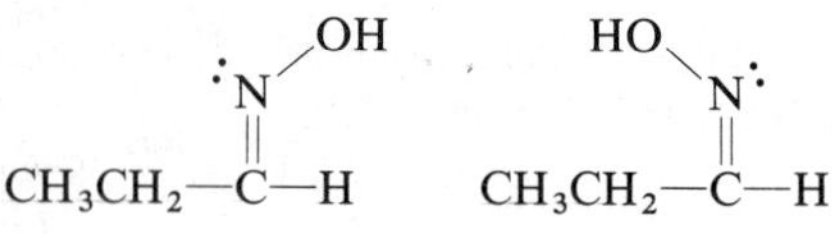

$$\underset{(1)}{CH_3CH_2-\overset{\overset{\displaystyle \overset{..}{N}\diagup OH}{\|}}{C}-H} \qquad \underset{(2)}{CH_3CH_2-\overset{\overset{\displaystyle HO\diagdown \overset{..}{N}}{\|}}{C}-H}$$

(*b*) The naming convention for the oximes of aldehydes is that the isomer with the H— and —OH on the same side of the double bond is *syn* and the isomer with H— and —OH on opposite sides is *anti*. Would you expect to find this type of isomerism from the oximes of the following compounds, and why: (1) acetone, (2) acetophenone, (3) methyl ethyl ketone, (4) formaldehyde.
(*c*) Name all the oximes from (*b*) using the *Z/E* system from Chap. 7.

21.9   Nucleophilic Addition of Water and Hydrogen Halides

We know that hydrogen halides and water add to alkenes via carbocation intermediates, but the curious student may wonder why we do not discuss the addition of these reagents to aldehydes and ketones. There is a good reason, however; these addition reactions do not occur to any appreciable extent!

The addition of water to various aldehydes and ketones is reversible, but the equilibrium lies far on the side of the carbonyl compound in most cases. Methanal (formaldehyde) exists in aqueous solution as the hydrate (see Sec. 20.4), as do compounds that contain strong electron-withdrawing groups on the carbon adjacent to the carbonyl group; trichloroacetaldehyde (called *chloral*) and hexafluoroacetone both form stable hydrates:

$$
\begin{array}{c}
\text{H} \\
\diagdown \\
\text{H}
\end{array}
\text{C=O} + \text{H}_2\text{O} \ \rightleftharpoons \
\begin{array}{c}
\text{OH} \\
| \\
\text{H}-\text{C}-\text{OH} \\
| \\
\text{H}
\end{array}
$$

Methanal   Formaldehyde hydrate
(formaldehyde)   *Not isolated*
                  (common name)

$$
\begin{array}{c}
\text{Cl} \ \ \text{O} \\
| \ \ \ || \\
\text{Cl}-\text{C}-\text{C}-\text{H} \\
| \\
\text{Cl}
\end{array}
+ \text{H}_2\text{O} \ \rightleftharpoons \
\begin{array}{c}
\text{Cl} \ \ \text{OH} \\
| \ \ \ | \\
\text{Cl}-\text{C}-\text{C}-\text{OH} \\
| \ \ \ | \\
\text{Cl} \ \ \text{H}
\end{array}
$$

Trichloroethanal   Chloral hydrate
(trichloroacetaldehyde   (common name)
or chloral)   *Isolable*

$$
\begin{array}{c}
\text{F} \ \ \text{O} \ \ \text{F} \\
| \ \ \ || \ \ | \\
\text{F}-\text{C}-\text{C}-\text{C}-\text{F} \\
| \ \ \ \ \ \ | \\
\text{F} \ \ \ \ \ \text{F}
\end{array}
+ \text{H}_2\text{O} \ \rightleftharpoons \
\begin{array}{c}
\text{F} \ \ \text{OH} \ \ \text{F} \\
| \ \ \ | \ \ \ | \\
\text{F}-\text{C}-\text{C}-\text{C}-\text{F} \\
| \ \ \ | \ \ \ | \\
\text{F} \ \ \text{OH} \ \ \text{F}
\end{array}
$$

Hexafluoropropanone   Hexafluoroacetone
(hexafluoroacetone)   hydrate
                   (common name)
                   Isolable

In general, however, hydrates (which are *gem*-diols) are unstable and not isolable:

$$
\begin{array}{c}
\text{O} \\
|| \\
\text{R}-\text{C}-\text{R}'
\end{array}
+ \text{H}_2\text{O} \ \rightleftharpoons \
\begin{array}{c}
\text{OH} \\
| \\
\text{R}-\text{C}-\text{OH} \\
| \\
\text{R}'
\end{array}
$$

Aldehyde or ketone   Unstable hydrate
                   (*gem*-diol)

That most aldehydes and ketones are in equilibrium with the corresponding hydrate in either acidic or basic solution is shown by the following results. If an aldehyde or ketone containing only a "normal" abundance of ^{18}O in the carbonyl oxygen position is allowed to stand with water enriched with oxygen 18 ($\text{H}_2{}^{18}\text{O}$) and hydroxide ion ($^{18}\text{:}\overset{..}{\text{O}}\text{H}^{\ominus}$), some ^{18}O is incorporated into the aldehyde or ketone.

The carbonyl compound is enriched with ^{18}O. The following mechanism illustrates this reversible reaction, which involves the hydrate (1) of the carbonyl compound as an intermediate:

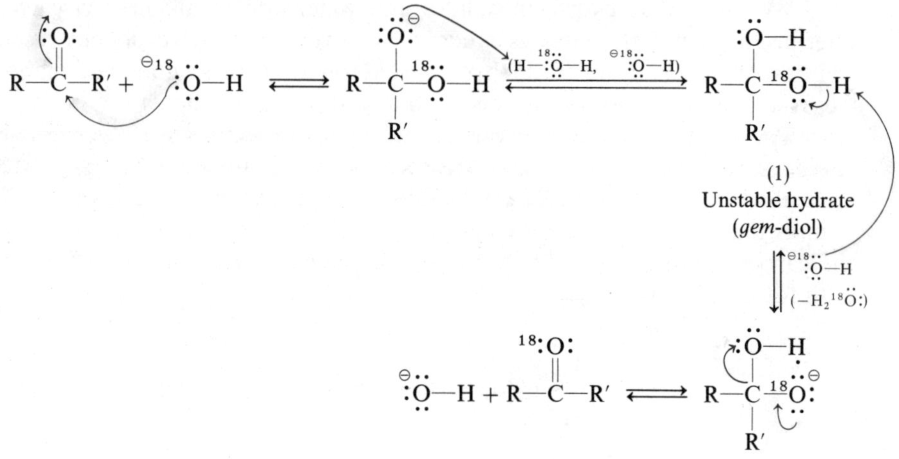

(1)
Unstable hydrate
(*gem*-diol)

The key to the exchange reaction is the tetrahedral *gem*-diol intermediate (1). Hydroxide ion can remove a proton from either of the equivalent hydroxyl groups on (1) when it is formed. Proton removal from the —OH group, loss of $^{\ominus18}:\ddot{O}H$, and regeneration of the carbonyl compound do not give any ^{18}O incorporation. Proton removal from the —^{18}OH group and subsequent loss of $:\ddot{O}H$ results in ^{18}O incorporation into the carbonyl compound. Thus, an "addition-elimination" mechanism explains the exchange reaction. Evidence similar to this is used in Chap. 23 to explain certain substitution reactions in the derivatives of the carboxylic acids.

Likewise, the addition of a hydrogen halide to a carbonyl compound is not observed. If addition does occur, the product is unstable and immediately reverts to reactants, with the equilibrium lying far to the reactant side:

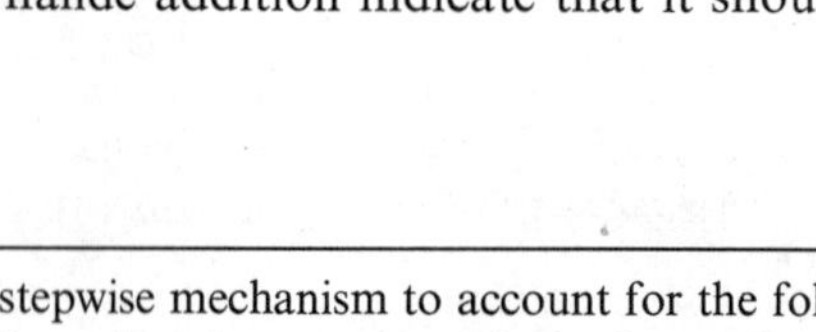

Unstable

Thus far there is no evidence for this type of equilibrium. The thermodynamics of water and hydrogen halide addition indicate that it should not be favored. (See Question 21.13.)

Question 21.12

Provide a complete, stepwise mechanism to account for the following exchange reaction, keeping in mind the type of reaction that occurs under basic conditions.

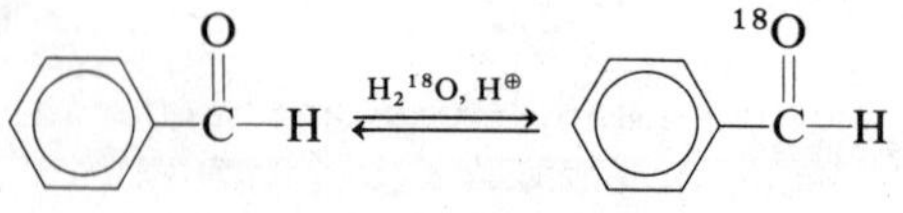

^{18}O incorporated

Question 21.13

Given the following (average) bond energies (top number in kcal, bottom number in kJ/mole),

C—H	C=O	C≡C	H—OH	H—Cl	O—H	C—OH	C—Cl	C—C
101	179	163	103	103	103	92	82	84
423	749	682	431	431	431	385	343	352

compute $\Delta H°$ for the following reactions:

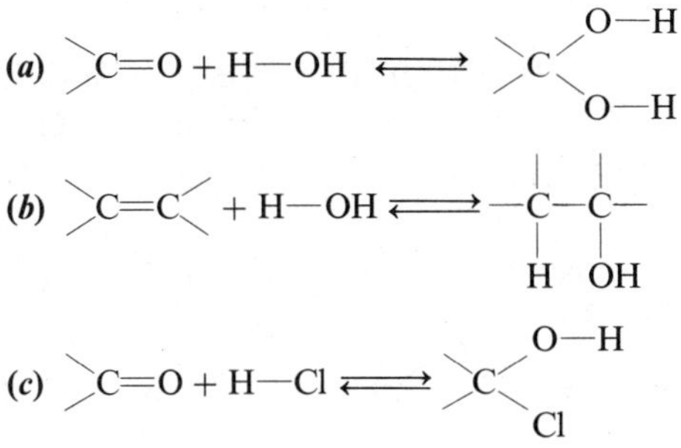

(a) $\;$C=O + H—OH $\rightleftharpoons$ $\;$C$\begin{smallmatrix} O—H \\ O—H \end{smallmatrix}$

(b) $\;$C=C$\;$ + H—OH $\rightleftharpoons$ —C—C—$\begin{smallmatrix} \;\;\; \\ H \;\; OH \end{smallmatrix}$

(c) $\;$C=O + H—Cl $\rightleftharpoons$ $\;$C$\begin{smallmatrix} O—H \\ Cl \end{smallmatrix}$

Based on your data, which reactions are favored and which are not? Why?

21.10 Nucleophilic Addition of Grignard Reagents: Versatile Method for Preparing Alcohols

The Grignard reagent, RMgX or ArMgX, is a strong base and a powerful nucleophile prepared from the corresponding alkyl or aryl halide and magnesium metal under anhydrous conditions:

$$(Ar) RX + Mg \xrightarrow[\text{ether}]{\text{anhydrous}} (Ar) RMgX$$

where R = any alkyl group or aryl group
$\qquad$ X = Cl, Br, I

Like alkyl halides (see Sec. 20.8B), aryl halides can be converted into the corresponding Grignard reagent when treated with magnesium metal. Aryl bromides and iodides react with magnesium when diethyl ether is used as the solvent, but they react more slowly than do alkyl bromides and iodides.

Aryl chlorides react with magnesium in the presence of diethyl ether to form the corresponding arylmagnesium chlorides but only with great difficulty (and sometimes not at all). However, the use of *tetrahydrofuran* (THF) in place of diethyl ether often allows Ar—MgCl to be formed readily and in good yield:

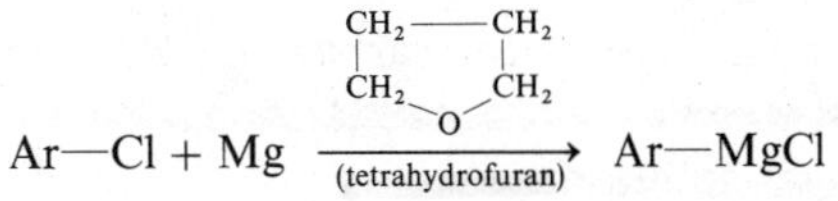

$$Ar—Cl + Mg \xrightarrow[\text{(tetrahydrofuran)}]{\begin{smallmatrix} CH_2——CH_2 \\ | \qquad | \\ CH_2 \quad CH_2 \\ \diagdown O \diagup \end{smallmatrix}} Ar—MgCl$$

The differences in reactivity between chlorides and bromides or iodides sometimes can be used to advantage. For example, when both chlorine and bromine are attached to an aromatic ring, it is possible to convert the bromine to the corresponding

Grignard reagent without affecting the chlorine:

p-Chlorobromobenzene p-Chlorophenylmagnesium
                                                bromide

Also, aryl Grignard reagents can be prepared only when certain substituents are attached to the aromatic ring; there can be no group with which a Grignard reagent might react. These limitations are discussed in more detail later in this section.

The reactions of aryl Grignard reagents are identical with those of alkyl Grignard reagents; for example, they react with water:

$$(R) \, Ar\!-\!Br \xrightarrow[\substack{anhydrous \\ ether}]{Mg} (R) \, Ar\!-\!MgBr \xrightarrow{H_2O} (R) \, Ar\!-\!H + MgBrOH$$

Several types of compounds have an acidic hydrogen that will protonate the Grignard reagent. They are all stronger acids than are alkanes, as the following data show:

Water: $H\!-\!OH \rightleftarrows H^{\oplus} + :\!\ddot{O}H^{\ominus}$ $K_a = 10^{-15}$

Alcohols: $H\!-\!OR \rightleftarrows H^{\oplus} + R\ddot{O}:^{\ominus}$ $K_a = 10^{-15}\text{--}10^{-18}$

Ammonia: $H\!-\!NH_2 \rightleftarrows H^{\oplus} + :\!\ddot{N}H_2^{\ominus}$ $K_a = 10^{-32}$

Alkanes: $R\!-\!H \rightleftarrows H^{\oplus} + R:^{\ominus}$ $K_a < 10^{-40}$

Order of decreasing acidity:

$$H_2\ddot{O}: \, > \, R\ddot{O}H \, > \, \ddot{N}H_3 \, > \, RH$$

Most   Least
acidic   acidic

Order of increasing basicity:

$$:\ddot{O}H^{\ominus} < R\ddot{O}:^{\ominus} \; < :\ddot{N}H_2^{\ominus} < R:^{\ominus}$$

Least   Most
basic   basic

The less acidic the compound, the stronger the base (often called the conjugate base) derived from that compound; for example, the weakest acid is the alkane, $R\!-\!H$, so the strongest base is the carbanion, $R:^{\ominus}$, derived from it.

The conversion of an alkyl halide to an alkane thus involves two steps, as shown by the general equation:

$$R\!-\!X + Mg: \xrightarrow[\substack{anhydrous \\ ether}]{} RMgX \xrightarrow[\substack{ROH \text{ or} \\ NH_3}]{H_2O \text{ or}} R\!-\!H + \begin{array}{l} MgX(OH) \text{ or} \\ MgX(OR) \text{ or} \\ MgX(NH_2) \end{array}$$

Alkyl   Grignard   Alkane
halide   reagent

Two specific examples are

$$\underset{\substack{\text{2-Bromopropane}\\(\text{isopropyl}\\\text{bromide})}}{\overset{\text{CH}_3}{\underset{\text{CH}_3}{\diagdown}}\text{CH—Br}} + \text{Mg:} \xrightarrow[\text{ether}]{\text{anhydrous}} \underset{\substack{\text{Isopropylmagnesium}\\\text{bromide}}}{\overset{\text{CH}_3}{\underset{\text{CH}_3}{\diagdown}}\text{CH—MgBr}} \xrightarrow{\text{H}_2\text{O}} \underset{\text{Propane}}{\overset{\text{CH}_3}{\underset{\text{CH}_3}{\diagdown}}\text{CH—H}}$$

Chlorocyclobutane
(cyclobutyl
chloride)
+ Mg: $\xrightarrow[\text{ether}]{\text{anhydrous}}$
Cyclobutylmagnesium
chloride
$\xrightarrow[\text{alcohol}]{\substack{\text{CH}_3\text{OH}\\(\text{methyl}}}$
Cyclobutane

Question 21.14

Based on our discussions about the Grignard reagent, why is it imperative that it be prepared under scrupulously anhydrous (dry) conditions? (Remember that we are not always interested in converting an alkyl halide to an alkane, so we often want to prepare the reagent and use it for another type of reaction.)

Question 21.15

Deuterium oxide, D_2O, has chemical properties similar to those of water. Draw the structure of the compound that would result from treating *tert*-butylmagnesium chloride with D_2O.

The addition of the Grignard reagent to an aldehyde or ketone *always* results in the formation of an alcohol. The anhydrous conditions required for the preparation and existence of the Grignard reagent mean that the addition reaction must be carried out in two steps. *First*, the carbonyl compound is added to the Grignard reagent and a new carbon-carbon bond is made. *Second*, the reaction is hydrolyzed by acidic water, which protonates the alkoxide ion (to give the alcohol) and dissolves the magnesium salts. The following somewhat simplified mechanism shows these reactions:

Aldehyde or ketone
(R, R′ = H, alkyl, or aryl)
+
Grignard reagent
(R″ = alkyl or aryl)
$\xrightarrow{\text{addition}}$
Alkoxide
ion

hydrolysis $\downarrow$ $\text{H}^{\oplus}$, $\text{H}_2\ddot{\text{O}}$:

Alcohol

The details of the Grignard reaction indicate that it involves 2 moles of Grignard reagent and 1 mole of carbonyl compound. Step 1 is believed to involve the formation of a complex between the carbonyl oxygen and magnesium in RMgX, which makes the carbonyl carbon more electropositive and thus more susceptible toward nucleophilic attack. Attack by another mole of RMgX (step 2) produces addition via transition state (1). The resulting intermediate (2) reacts with MgX_2 (step 3) to produce alkoxide ion (3), which remains as a magnesium salt until hydrolysis is carried out.

Step 1:

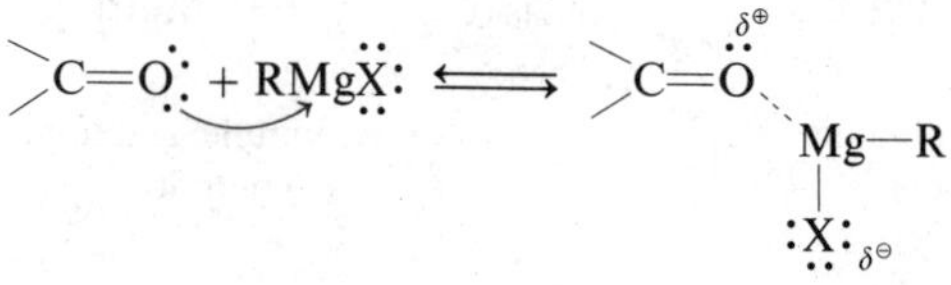

$$\text{>C=O: + RMgX:} \rightleftharpoons \text{>C=O}\overset{\delta\oplus}{\cdots}\underset{\underset{\text{:X:}_{\delta\ominus}}{|}}{\text{Mg—R}}$$

Step 2:

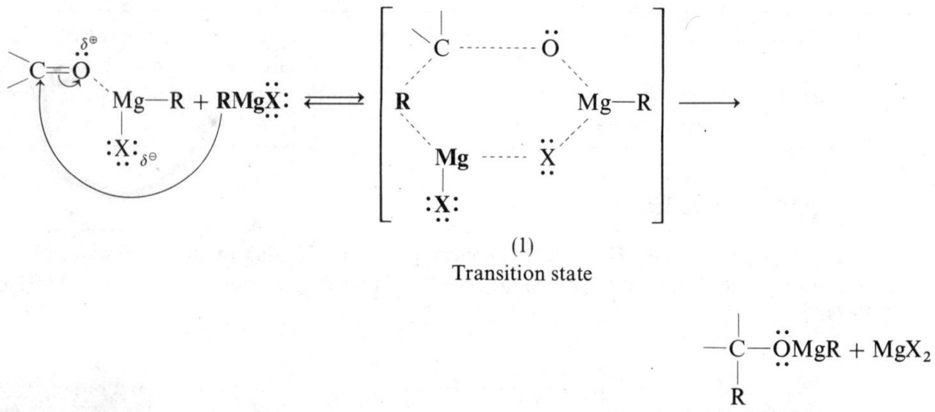

(1)
Transition state

$$\underset{\text{R}}{\overset{|}{-\text{C}}}-\ddot{\text{O}}\text{MgR} + MgX_2$$

(2)

Step 3:

$$\underset{\text{R}}{\overset{|}{-\text{C}}}-\ddot{\text{O}}\text{MgR} + MgX_2 \rightleftharpoons \underset{\text{R}}{\overset{|}{-\text{C}}}-\overset{\ominus}{\ddot{\text{O}}}\text{:}[MgX]^{\oplus} + RMgX$$

(3)

Various classes of alcohols are produced, depending on the carbonyl compound used:

Methanal (formaldehyde):

$$\text{H}-\overset{\overset{\text{O}}{\|}}{\text{C}}-\text{H} + \text{RMgX} \xrightarrow{\text{mix}} \xrightarrow{\text{H}_3\text{O:}^{\oplus}} \text{R}-\text{CH}_2\text{OH}$$

Primary alcohol

Extends carbon chain by one carbon atom

Aldehyde:

$$\text{R}'-\overset{\overset{\text{O}}{\|}}{\text{C}}-\text{H} + \text{RMgX} \xrightarrow{\text{mix}} \xrightarrow{\text{H}_3\text{O:}^{\oplus}} \text{R}'-\underset{\text{R}}{\overset{\overset{\text{OH}}{|}}{\text{C}}}-\text{H}$$

Secondary alcohol

Ketone:

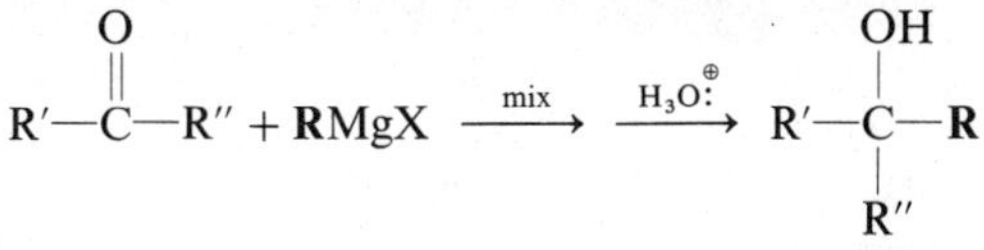

$$R'—\overset{\overset{\textstyle O}{\|}}{C}—R'' + RMgX \xrightarrow{\text{mix}} \xrightarrow{H_3O:^{\oplus}} R'—\overset{\overset{\textstyle OH}{|}}{\underset{\underset{\textstyle R''}{|}}{C}}—R$$

Tertiary
alcohol

The reaction between a carbonyl compound and the Grignard reagent is important because it (1) *forms a new carbon-carbon bond*, (2) *adds carbon atoms to a molecule*, and (3) *produces a new alcohol*, which can be transformed into a myriad of other functional groups. Three examples are shown here:

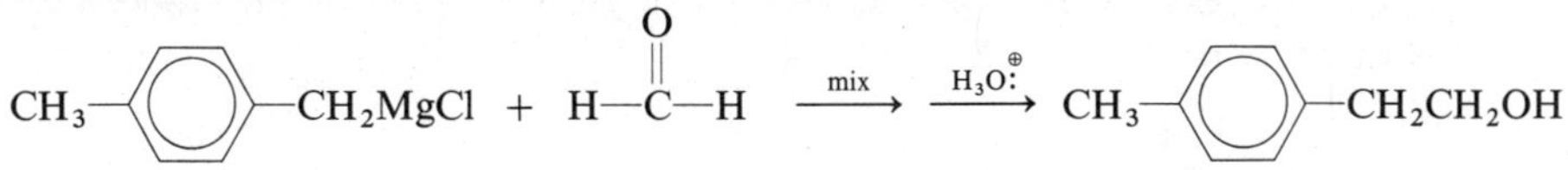

$$CH_3—\bigcirc—CH_2MgCl + H—\overset{\overset{\textstyle O}{\|}}{C}—H \xrightarrow{\text{mix}} \xrightarrow{H_3O:^{\oplus}} CH_3—\bigcirc—CH_2CH_2OH$$

| *p*-Methylbenzylmagnesium chloride | Methanal (formaldehyde) | 2-(*p*-Tolyl)ethanol *1° Alcohol* |

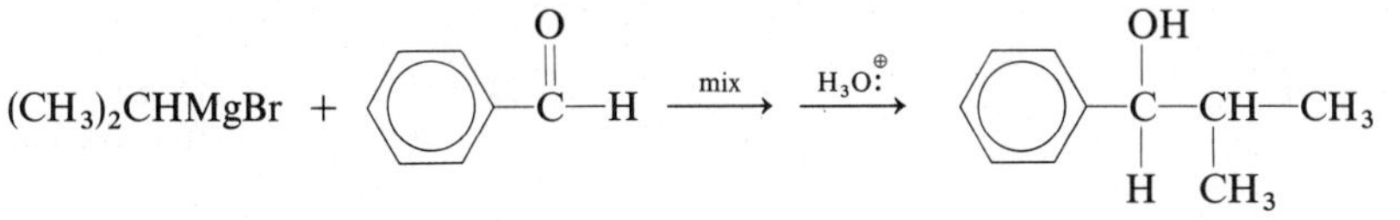

$$(CH_3)_2CHMgBr + \bigcirc—\overset{\overset{\textstyle O}{\|}}{C}—H \xrightarrow{\text{mix}} \xrightarrow{H_3O:^{\oplus}} \bigcirc—\overset{\overset{\textstyle OH}{|}}{\underset{\underset{\textstyle H}{|}}{C}}—\overset{}{\underset{\underset{\textstyle CH_3}{|}}{CH}}—CH_3$$

| Isopropylmagnesium bromide | Benzaldehyde | 2-Methyl-1-phenyl-1-propanol *2° Alcohol* |

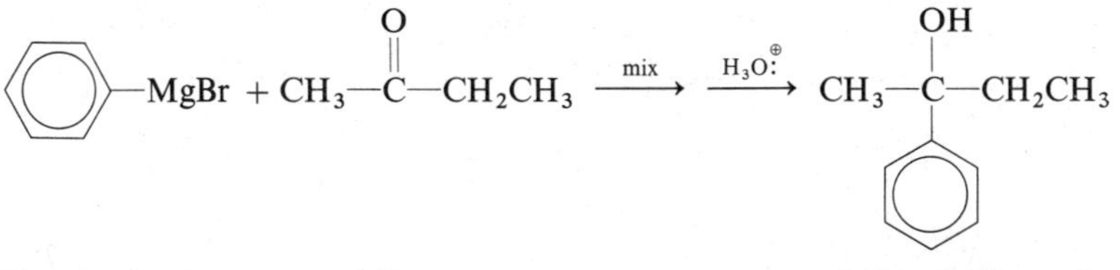

$$\bigcirc—MgBr + CH_3—\overset{\overset{\textstyle O}{\|}}{C}—CH_2CH_3 \xrightarrow{\text{mix}} \xrightarrow{H_3O:^{\oplus}} CH_3—\overset{\overset{\textstyle OH}{|}}{\underset{\underset{\textstyle \bigcirc}{|}}{C}}—CH_2CH_3$$

| Phenylmagnesium bromide | 2-Butanone | 2-Phenyl-2-butanol *3° Alcohol* |

Several limitations are placed on the use of Grignard reagents because of the various reactions they undergo with other functional groups. These limitations involve the structure of the alkyl or aryl halide from which the Grignard reagent is to be prepared and the compound with which it will be reacted.

We know, for example, that the Grignard reagent readily reacts with protons: $RMgX + \text{"}H^{\oplus}\text{"} \rightarrow R—H + Mg^{\oplus 2} + :\ddot{X}:^{\ominus}$. If the reacting molecules contain any potential source of protons, the Grignard reagent would decompose as soon as it is formed. The following functional groups are all proton donors and may *not* be present in molecules used in a Grignard preparation:

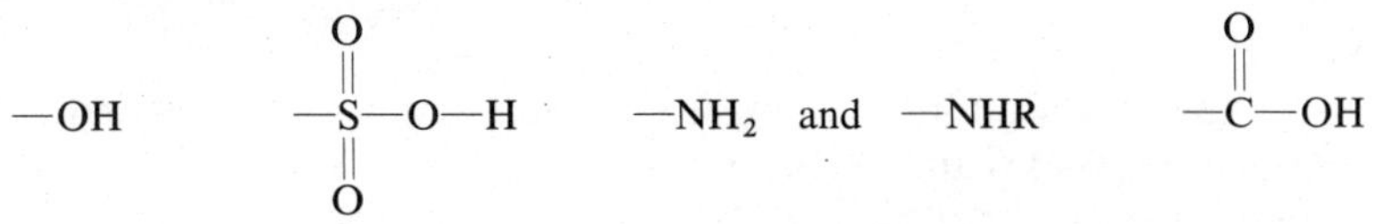

$$—OH \qquad —\overset{\overset{\textstyle O}{\|}}{\underset{\underset{\textstyle O}{\|}}{S}}—O—H \qquad —NH_2 \text{ and } —NHR \qquad —\overset{\overset{\textstyle O}{\|}}{C}—OH$$

| Alcohol or phenol | Sulfonic acid | Primary and secondary amine | Carboxylic acid |

In addition, other functional groups would *react* with the Grignard reagent (addition reaction) as soon as it is formed if they were present in the same molecule. We know that the Grignard reagent adds to aldehydes, ketones, and nitriles; later we will see that it adds to esters and acid chlorides as well. The following substituents may *not* be present in molecules used in a Grignard reaction:

$$\underset{\text{Aldehyde}}{\overset{\overset{\displaystyle O}{\|}}{-C-H}} \qquad \underset{\text{Ketone}}{\overset{\overset{\displaystyle O}{\|}}{-C-R}} \qquad \underset{\text{Ester}}{\overset{\overset{\displaystyle O}{\|}}{-C-OR}}$$

$$\underset{\text{Nitrile}}{-C\equiv N} \qquad \underset{\text{Acid chloride}}{\overset{\overset{\displaystyle O}{\|}}{-C-Cl}} \qquad \underset{\text{Acid anhydride}}{\overset{\overset{\displaystyle O}{\|}}{-C}-O-\overset{\overset{\displaystyle O}{\|}}{C}-} \qquad \text{and others}$$

It is possible to "protect" aldehyde and keto groups by acetal (ketal) formation, however, to prepare a Grignard reagent in another part of the molecule. This is illustrated by the following example:

$$CH_3\overset{\overset{\displaystyle O}{\|}}{-C}-CH_2CH_2CH_2Br \xrightarrow[\substack{\text{anhydrous}\\\text{ether}}]{Mg} CH_3\overset{\overset{\displaystyle O}{\|}}{-C}-CH_2CH_2CH_2MgBr$$

React together
(inter- and/or intramolecular)
Does not exist

but

$$CH_3\overset{\overset{\displaystyle O}{\|}}{-C}-CH_2CH_2CH_2Br \xrightarrow[\substack{H^\oplus,\text{ heat}\\\text{(remove water)}}]{HOCH_2CH_2OH} CH_3\overset{\overset{O\qquad O}{\underset{\displaystyle CH_2-CH_2}{| \quad\quad |}}}{C}-CH_2CH_2CH_2Br$$

Ketal
Protects keto group

$$\xrightarrow[\substack{\text{anhydrous}\\\text{ether}}]{Mg} CH_3\overset{\overset{O\quad O}{\underset{\displaystyle CH_2-CH_2}{| \quad |}}}{C}-CH_2CH_2CH_2MgBr \xrightarrow{\overset{\overset{\displaystyle O}{\|}}{\underset{}{\bigcirc-C-CH_3}}} \xrightarrow[\substack{\text{mild}\\\text{(hydrolyzes}\\\text{Grignard)}}]{H_3O^\oplus}$$

$$CH_3\overset{\overset{O\quad O}{\underset{\displaystyle CH_2-CH_2}{| \quad |}}}{C}-CH_2CH_2CH_2-\overset{\overset{\displaystyle OH}{|}}{\underset{\underset{\displaystyle CH_3}{|}}{C}}-\bigcirc \xrightarrow[\substack{H_2O,\text{ warm}\\\text{(hydrolyzes}\\\text{ketal, regenerates}\\\text{ketone)}}]{H^\oplus,\text{ excess}}$$

$$CH_3\overset{\overset{\displaystyle O}{\|}}{-C}-CH_2CH_2CH_2-\overset{\overset{\displaystyle OH}{|}}{\underset{\underset{\displaystyle CH_3}{|}}{C}}-\bigcirc \qquad (+ HOCH_2CH_2OH)$$

Other applications of using acetals and ketals for protecting carbonyl groups appear in Secs. 21.6 and 21.14.

21.11   Application of Grignard Synthesis of Alcohols

It is now clear that alcohols are useful in organic chemistry. The hydroxyl group, —OH, can be converted into halogen, aldehyde or carboxylic acid (from a primary alcohol), or ketone (from a secondary alcohol). Alcohols can be dehydrated and in certain cases can be used in the Friedel-Crafts alkylation reaction. A versatile way of introducing more carbon atoms into a molecule is via the Grignard addition to carbonyl compounds which gives alcohols. Let us now look at yet another application of the Grignard reagent in the synthesis of alcohols: the reaction of the Grignard reagent with epoxides.

From the viewpoint of synthetic organic chemistry, the addition of the Grignard reagent to an epoxide is a highly useful reaction, which is shown in general as follows:

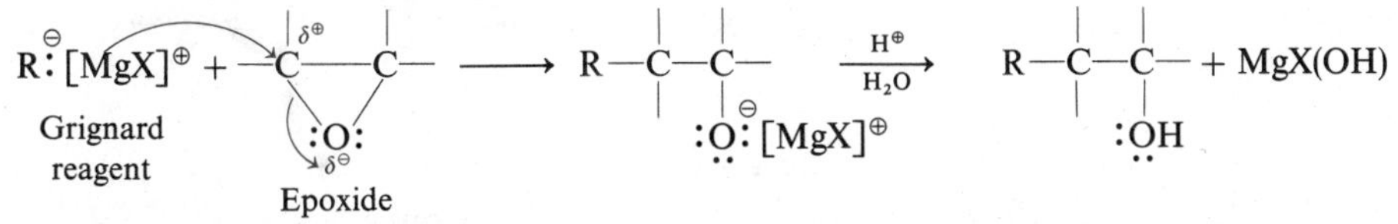

The nucleophilic nature of the Grignard reagent is responsible for this addition reaction. The carbanion attacks the electropositive carbon on the epoxide; at the same time, magnesium ion is attracted to the unshared electron pairs on the epoxide oxygen.

The Grignard reagent must be prepared under anhydrous conditions, and then the epoxide is added to it until the reaction is complete. This results in the formation of the magnesium salt of the corresponding alcohol. To obtain the free alcohol, the reaction mixture must be treated with dilute, aqueous acid, which protonates the alkoxide ion and liberates free magnesium ion and halide ion into the aqueous solution. In contrast to most other ring-opening reactions of epoxides, Grignard addition involves several steps: the Grignard reagent must be prepared, the epoxide must be added, and protonation occurs only after the addition of dilute, aqueous acid.

This reaction's great utility is that *it extends a carbon chain by precisely two carbon atoms in the case of ethylene oxide* and the new compound formed is a primary alcohol. The following two examples illustrate this:

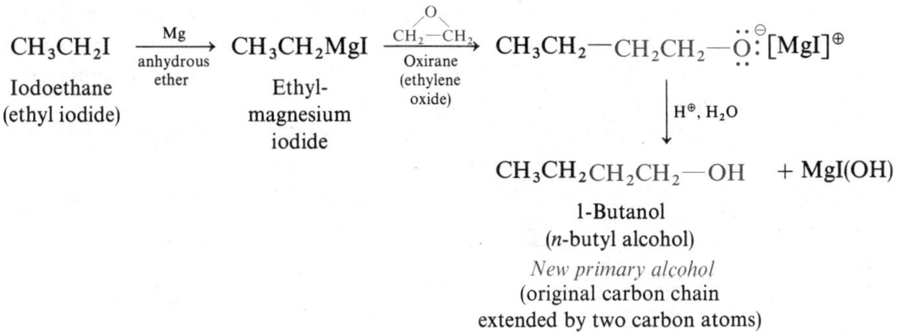

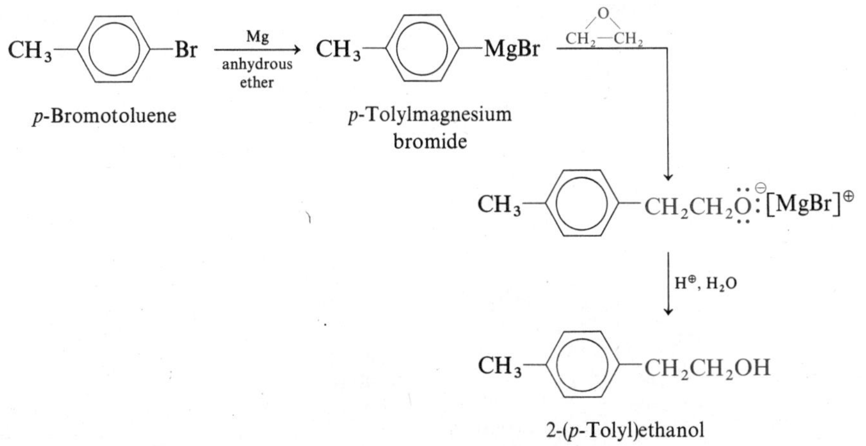

Five examples of alcohol synthesis using the Grignard reagent are given now.

Example 1

Synthesize 2-methyl-1-phenyl-1-propanol (1) from a carbonyl compound and a Grignard reagent:

$$CH_3-CH(CH_3)-CH(OH)-C_6H_5$$

(1)
2-Methyl-1-phenyl-1-propanol

As we examine this compound, we first note that it is a secondary alcohol, so we should naturally think of a reaction involving a Grignard reagent and an aldehyde (other than methanal). Two possible combinations of reagents can be used:

$(CH_3)_2CHMgBr$ + benzaldehyde $\xrightarrow{\text{mix}} \xrightarrow{H_3\ddot{O}^\oplus}$ (1)

Isopropylmagnesium bromide   Benzaldehyde   (1)

or

Phenylmagnesium bromide + $CH_3-CH(CH_3)-C(=O)-H$ $\xrightarrow{\text{mix}} \xrightarrow{H_3\ddot{O}^\oplus}$ (1)

Phenylmagnesium   2-Methylpropanal
bromide   (isobutyraldehyde)

Three factors come into play in choosing between these two routes. The first is the availability and cost of the starting materials; naturally, the least expensive reagents are preferred. Second, the least hindered carbonyl compound gives the highest yield in addition reactions. Also, the possibility of side reactions must be considered.

In this case the reaction between isopropylmagnesium bromide and benzaldehyde is preferred because the reagents are more readily available. There is some possibility of side reaction with aldehydes containing hydrogens on the carbon adjacent to the carbonyl (see Sec. 24.1), and benzaldehyde is less hindered than 2-methylpropanal.

Example 2

Synthesize 3-methyl-2-phenyl-2-butanol (2) from a carbonyl compound and a Grignard reagent:

$$CH_3-CH(CH_3)-C(OH)(C_6H_5)-CH_3$$

(2)
3-Methyl-2-phenyl-2-butanol

We first note that this is a tertiary alcohol, so we should consider reactions of a ketone with a Grignard reagent. Two possible routes merit evaluation:

$$CH_3-\underset{\underset{\displaystyle CH_3}{|}}{CH}-\underset{\underset{\displaystyle O}{\|}}{C}-CH_3 + \underset{\text{bromide}}{\underset{\text{Phenylmagnesium}}{\bigcirc\!\!\!-MgBr}} \xrightarrow{\text{mix}} \xrightarrow{H_3\ddot{O}^{\oplus}} CH_3-\underset{\underset{\displaystyle CH_3}{|}}{CH}-\underset{\underset{\displaystyle \bigcirc}{\underset{|}{C}}}{\overset{\overset{\displaystyle OH}{|}}{C}}-CH_3$$

3-Methyl-2-butanone   Phenylmagnesium
                              bromide

(2)

or

$$\underset{\text{Acetophenone}}{\bigcirc\!\!\!-\underset{\underset{\displaystyle O}{\|}}{C}-CH_3} + \underset{\underset{\text{chloride}}{\text{Isopropylmagnesium}}}{(CH_3)_2CHMgCl} \xrightarrow{\text{mix}} \xrightarrow{H_3\ddot{O}^{\oplus}} (2)$$

Here, the second route—acetophenone and isopropylmagnesium chloride—is preferred because the reagents are readily available and acetophenone is less hindered than 3-methyl-2-butanone.

Example 3

Starting with any alcohols of four carbon atoms, or less, synthesize 3-methylbutanal (3):

$$CH_3-\underset{\underset{\displaystyle CH_3}{|}}{CH}-CH_2-\underset{\underset{\displaystyle O}{\|}}{C}-H$$

(3)
3-Methylbutanal

We are to prepare a compound containing five carbon atoms, but because we have only alcohols of four carbon atoms or less available as starting materials, we must devise a method for synthesizing a five-carbon molecule. Working backward, we must ask what precursors might give the desired aldehyde. One possibility is the corresponding alcohol and another is the corresponding acid chloride:

$$\underset{\text{3-Methyl-1-butanol}}{CH_3\underset{\underset{\displaystyle CH_3}{|}}{CH}CH_2CH_2OH} \xrightarrow[\text{warm}]{H_2SO_4, K_2Cr_2O_7} (3) \qquad \text{Preferred method}$$

$$\underset{\underset{\text{chloride}}{\text{3-Methylbutanoyl}}}{CH_3\underset{\underset{\displaystyle CH_3}{|}}{CH}CH_2\underset{\underset{\displaystyle O}{\|}}{C}-Cl} \xrightarrow[\underset{\text{reduction)}}{\underset{\text{(Rosenmund}}{Pd(BaSO_4)}}]{H_2} (3)$$

Of the two, oxidation of the primary alcohol, 3-methyl-1-butanol, is preferred because the yields are better.

The next question is: How can we prepare 3-methyl-1-butanol from alcohols of four carbon atoms or less? This is a primary alcohol, and one possibility we should consider is the reaction between a Grignard reagent and methanal:

$$\underset{\underset{\text{bromide}}{\text{Isobutylmagnesium}}}{CH_3\underset{\underset{\displaystyle CH_3}{|}}{CH}CH_2MgBr} + \underset{\underset{\text{(formaldehyde)}}{\text{Methanal}}}{H-\underset{\underset{\displaystyle O}{\|}}{C}-H} \xrightarrow{\text{mix}} \xrightarrow{H_3\ddot{O}^{\oplus}} \underset{\text{3-Methyl-1-butanol}}{CH_3\underset{\underset{\displaystyle CH_3}{|}}{CH}CH_2CH_2OH}$$

The Grignard reagent and methanal are readily obtainable from the given starting materials, as shown in the entire synthesis:

$$\underset{\substack{\text{2-Methyl-1-propanol}\\\text{(isobutyl alcohol)}\\\textit{Four-carbon alcohol}}}{\overset{\overset{\displaystyle CH_3}{|}}{CH_3CHCH_2OH}} \xrightarrow{\text{PBr}_3} \underset{\substack{\text{2-Methyl-1-bromopropane}\\\text{(isobutyl bromide)}}}{\overset{\overset{\displaystyle CH_3}{|}}{CH_3CHCH_2Br}} \xrightarrow[\substack{\text{anhydrous}\\\text{ether}}]{\text{Mg}}$$

$$\underset{\substack{\text{2-Methyl-1-propylmagnesium bromide}\\\text{(isobutylmagnesium bromide)}}}{\overset{\overset{\displaystyle CH_3}{|}}{CH_3CHCH_2MgBr}}$$

and

$$\underset{\substack{\text{Methanol}\\\text{(methyl alcohol)}\\\textit{Three carbon alcohol}}}{CH_3OH} \xrightarrow[\text{warm}]{K_2Cr_2O_7,\ H_2SO_4} \underset{\substack{\text{Methanal}\\\text{(formaldehyde)}}}{\overset{\displaystyle O}{\overset{\|}{H-C-H}}}$$

Then

$$\underset{}{\overset{\overset{\displaystyle CH_3}{|}}{CH_3CHCH_2MgBr}} + \overset{\displaystyle O}{\overset{\|}{H-C-H}} \xrightarrow{\text{mix}} \xrightarrow{H_3\ddot{O}^\oplus} \underset{\text{3-Methyl-1-butanol}}{\overset{\overset{\displaystyle CH_3}{|}}{CH_3CHCH_2CH_2OH}}$$

$$\bigg\downarrow \substack{K_2Cr_2O_7,\ H_2SO_4,\\ \text{warm}}$$

$$\underset{\substack{\text{3-Methylbutanal}\\\textit{Desired product}}}{\overset{\overset{\displaystyle CH_3}{|}\quad\ \overset{\displaystyle O}{}}{CH_3CHCH_2-\overset{\|}{C}-H}}$$

As with most multistep syntheses, this approach of working backward is most fruitful until the student has more experience in devising synthetic schemes.

Example 4

Starting with alcohols of four carbon atoms or less, synthesize 3,3-dimethyl-1-butene (4):

$$\underset{\substack{(4)\\\text{3,3-Dimethyl-1-butene}}}{\overset{\overset{\displaystyle CH_3}{|}}{\underset{\underset{\displaystyle CH_3}{|}}{CH_3-C-CH=CH_2}}}$$

As before, we ask what precursors and reactions might be used to obtain the desired alkene. The following are four obvious ones:

$$\underset{\text{3,3-Dimethyl-1-bromobutane}}{\overset{\overset{\displaystyle CH_3}{|}}{\underset{\underset{\displaystyle CH_3}{|}}{CH_3-C-CH_2CH_2Br}}} \quad \text{or} \quad \underset{\text{3-Bromo-2,2-dimethylbutane}}{\overset{\overset{\displaystyle CH_3}{|}}{\underset{\underset{\displaystyle CH_3\ \ Br}{|\ \ |}}{CH_3-C----CH-CH_3}}} \xrightarrow[\text{alcohol}]{\text{KOH}} (4)$$

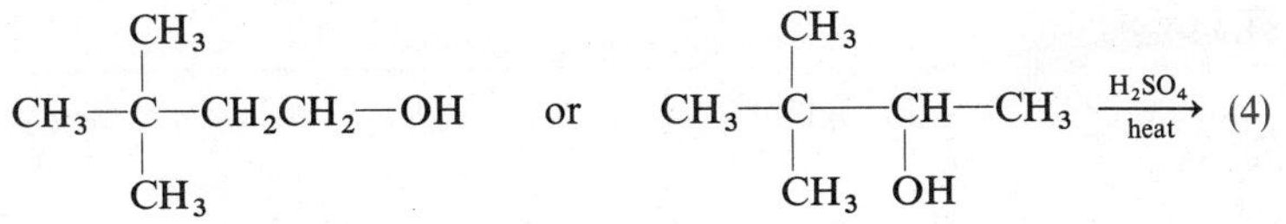

$$CH_3-\underset{\underset{CH_3}{|}}{\overset{\overset{CH_3}{|}}{C}}-CH_2CH_2-OH \quad \text{or} \quad CH_3-\underset{\underset{CH_3}{|}}{\overset{\overset{CH_3}{|}}{C}}-\underset{\underset{OH}{|}}{CH}-CH_3 \xrightarrow[\text{heat}]{H_2SO_4} \quad (4)$$

3,3-Dimethyl-1-butanol   3,3-Dimethyl-2-butanol

Of these two types of reaction—dehydrohalogenation and dehydration—the former is preferable because it is not accompanied by rearrangements. Dehydration is likely to give molecular rearrangements because of the carbocation that intervenes, thus giving a mixture of products. (*Question:* What rearrangements would the preceding alcohols undergo and what isomeric alkenes would be produced?)

Having chosen the type of reaction (dehydrohalogenation), let us now devise routes to the alkyl bromides, which, as we might expect, start with the corresponding alcohols:

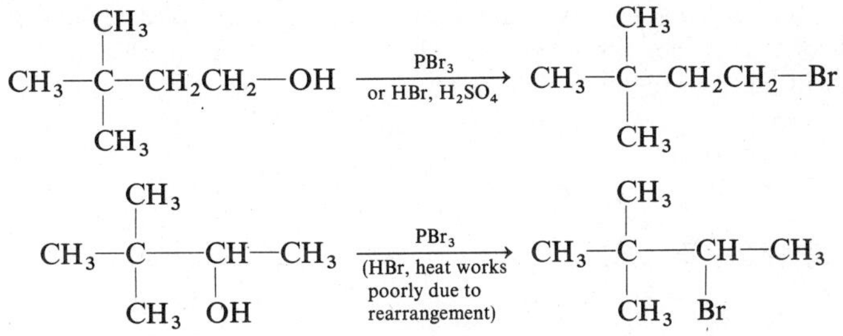

$$CH_3-\underset{\underset{CH_3}{|}}{\overset{\overset{CH_3}{|}}{C}}-CH_2CH_2-OH \xrightarrow[\text{or HBr, H}_2\text{SO}_4]{PBr_3} CH_3-\underset{\underset{CH_3}{|}}{\overset{\overset{CH_3}{|}}{C}}-CH_2CH_2-Br$$

$$CH_3-\underset{\underset{CH_3}{|}}{\overset{\overset{CH_3}{|}}{C}}-\underset{\underset{OH}{|}}{CH}-CH_3 \xrightarrow[\substack{\text{(HBr, heat works}\\ \text{poorly due to}\\ \text{rearrangement)}}]{PBr_3} CH_3-\underset{\underset{CH_3}{|}}{\overset{\overset{CH_3}{|}}{C}}-\underset{\underset{Br}{|}}{CH}-CH_3$$

The conversion of the primary alcohol to the primary alkyl halide is better (higher yield, greater purity) than that of the secondary alcohol.

Now, what route can be used to convert alcohols of four carbon atoms or less to a primary alcohol with six carbon atoms? One method, a one-step reaction, stands out from the rest:

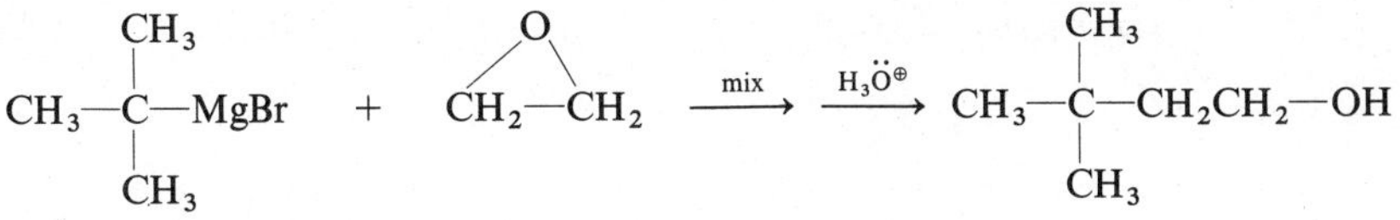

$$CH_3-\underset{\underset{CH_3}{|}}{\overset{\overset{CH_3}{|}}{C}}-MgBr \quad + \quad \overset{\overset{O}{\diagup\diagdown}}{CH_2-CH_2} \xrightarrow{\text{mix}} \xrightarrow{H_3\overset{..}{O}^{\oplus}} CH_3-\underset{\underset{CH_3}{|}}{\overset{\overset{CH_3}{|}}{C}}-CH_2CH_2-OH$$

2-Methyl-2-propylmagnesium   Oxirane   3,3-Dimethyl-1-butanol
bromide   (ethylene oxide)

(*tert*-butylmagnesium
bromide)

(*Question:* Can you outline another route to 3,3-dimethyl-1-butanol?)

The Grignard reagent and ethylene oxide are readily obtainable from alcohols of four carbon atoms or less. The following scheme shows the complete sequence of reactions leading to the desired alkene (4):

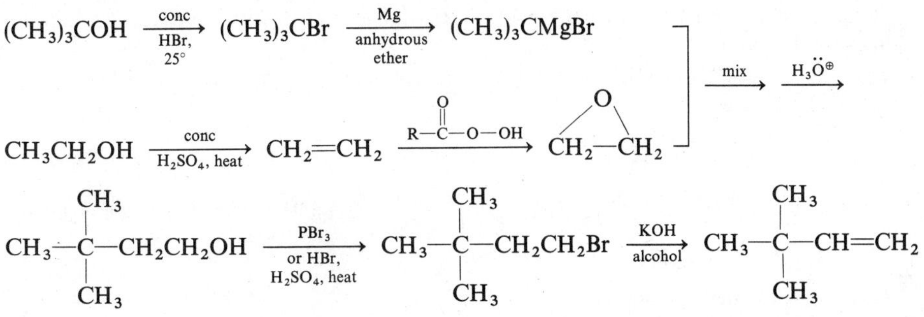

$$(CH_3)_3COH \xrightarrow[\substack{\text{HBr,}\\25°}]{\text{conc}} (CH_3)_3CBr \xrightarrow[\substack{\text{anhydrous}\\\text{ether}}]{Mg} (CH_3)_3CMgBr$$

$$CH_3CH_2OH \xrightarrow[\text{H}_2\text{SO}_4\text{, heat}]{\text{conc}} CH_2{=}CH_2 \xrightarrow{R-\overset{\overset{O}{\|}}{C}-O-OH} \overset{\overset{O}{\diagup\diagdown}}{CH_2-CH_2} \Bigg] \xrightarrow{\text{mix}} \xrightarrow{H_3\overset{..}{O}^{\oplus}}$$

$$CH_3-\underset{\underset{CH_3}{|}}{\overset{\overset{CH_3}{|}}{C}}-CH_2CH_2OH \xrightarrow[\substack{\text{or HBr,}\\\text{H}_2\text{SO}_4\text{, heat}}]{PBr_3} CH_3-\underset{\underset{CH_3}{|}}{\overset{\overset{CH_3}{|}}{C}}-CH_2CH_2Br \xrightarrow[\text{alcohol}]{KOH} CH_3-\underset{\underset{CH_3}{|}}{\overset{\overset{CH_3}{|}}{C}}-CH{=}CH_2$$

(4)
3,3-Dimethyl-1-butene
Desired product

Example 5

Starting with alcohols of four carbon atoms or less, synthesize 2-methyl-4-*tert*-butyl-4-octanol (5):

$$
\begin{array}{c}
\text{OH} \\
| \\
CH_3CH_2CH_2CH_2-\overset{}{C}-CH_2-CH-CH_3 \\
| \qquad\qquad | \\
CH_3-\overset{}{C}-CH_3 \quad CH_3 \\
| \\
CH_3
\end{array}
$$

(5)

2-Methyl-4-*tert*-butyl-4-octanol

Examination of this tertiary alcohol shows it contains 13 carbon atoms. We must devise a method for adding several carbon fragments to a simple molecule because we must start with alcohols of four carbon atoms or less. Also, one step must involve the reaction of a Grignard reagent with a ketone to produce the tertiary alcohol. Working backward, we ask what precursors might be brought together to produce (5), and the following three possibilities come to mind:

$$
\begin{array}{c}
O \quad CH_3 \\
|| \quad | \\
CH_3CH_2CH_2CH_2-C-C-CH_3 + (CH_3)_2CHCH_2MgBr \xrightarrow{\text{mix}} \xrightarrow{H_3\overset{..}{O}^{\oplus}} (5) \\
| \\
CH_3
\end{array}
$$

2,2-Dimethyl-3-heptanone   Isobutylmagnesium bromide

or

$$
\begin{array}{c}
O \quad CH_3 \\
|| \quad | \\
(CH_3)_2CHCH_2-C-C-CH_3 + CH_3CH_2CH_2CH_2MgBr \xrightarrow{\text{mix}} \xrightarrow{H_3\overset{..}{O}^{\oplus}} (5) \\
| \\
CH_3
\end{array}
$$

2,2,5-Trimethyl-3-hexanone   *n*-Butylmagnesium bromide

or

$$
\begin{array}{c}
O \\
|| \\
CH_3CH_2CH_2CH_2-C-CH_2CH(CH_3)_2 + \quad (CH_3)_3CMgBr \xrightarrow{\text{mix}} \xrightarrow{H_3\overset{..}{O}^{\oplus}} (5)
\end{array}
$$

2-Methyl-4-octanone   *tert*-Butylmagnesium bromide

Of these possible routes, the last one is preferred because the addition of a Grignard reagent to less hindered ketones gives higher yields (both ketones in the first two routes contain the bulky *tert*-butyl group).

Now, how could we prepare 2-methyl-4-octanone? One way is to oxidize the corresponding secondary alcohol:

$$
\begin{array}{c}
OH \\
| \\
CH_3CH_2CH_2CH_2-C-CH_2CH(CH_3)_2 \xrightarrow[\substack{H_2SO_4, \\ \text{heat}}]{K_2Cr_2O_7} CH_3CH_2CH_2CH_2-\overset{\overset{O}{||}}{C}-CH_2CH(CH_3)_2 \\
| \\
H
\end{array}
$$

2-Methyl-4-octanol   2-Methyl-4-octanone

Another method is the action of an organocadmium reagent on an acid chloride, but the oxidation of the alcohol is usually the method of choice because it is easier to do in the laboratory and usually gives better yields.

The synthesis of 2-methyl-4-octanol, a secondary alcohol, can be done in one of two ways:

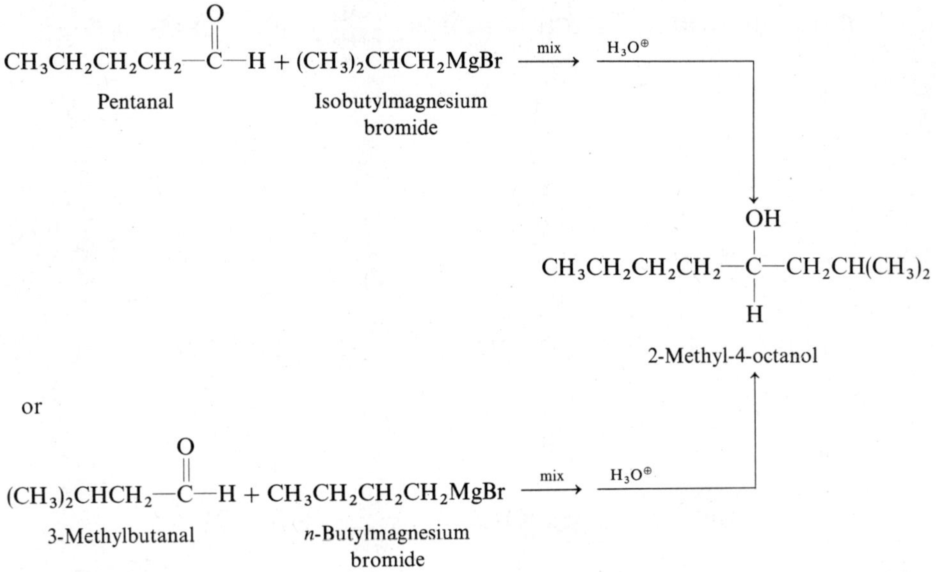

$$CH_3CH_2CH_2CH_2-\overset{\overset{\displaystyle O}{\|}}{C}-H + (CH_3)_2CHCH_2MgBr \xrightarrow{mix} \xrightarrow{H_3O^\oplus}$$

Pentanal   Isobutylmagnesium
                                       bromide

$$CH_3CH_2CH_2CH_2-\overset{\overset{\displaystyle OH}{|}}{\underset{\underset{\displaystyle H}{|}}{C}}-CH_2CH(CH_3)_2$$

2-Methyl-4-octanol

or

$$(CH_3)_2CHCH_2-\overset{\overset{\displaystyle O}{\|}}{C}-H + CH_3CH_2CH_2CH_2MgBr \xrightarrow{mix} \xrightarrow{H_3O^\oplus}$$

3-Methylbutanal   *n*-Butylmagnesium
                                   bromide

Either route is acceptable, and there is little preference for one over the other. Either route involves a *five-carbon* aldehyde, which must be prepared from alcohols of four carbon atoms or less. Using the first route as the example, pentanal can be prepared from 1-pentanol, which can be prepared from *n*-butylmagnesium bromide and formaldehyde:

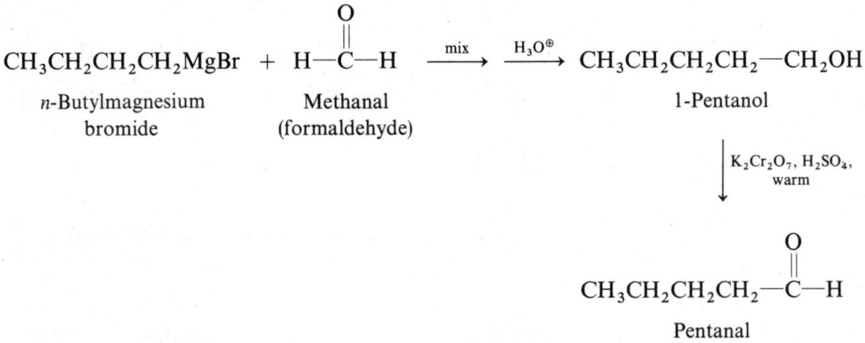

$$CH_3CH_2CH_2CH_2MgBr \; + \; H-\overset{\overset{\displaystyle O}{\|}}{C}-H \xrightarrow{mix} \xrightarrow{H_3O^\oplus} CH_3CH_2CH_2CH_2-CH_2OH$$

n-Butylmagnesium   Methanal   1-Pentanol
     bromide   (formaldehyde)

$$\Big\downarrow \; K_2Cr_2O_7, H_2SO_4, \; warm$$

$$CH_3CH_2CH_2CH_2-\overset{\overset{\displaystyle O}{\|}}{C}-H$$

Pentanal

We have now worked backward to the point where each starting compound contains four carbon atoms or less. The following scheme depicts one possible route for the complete synthesis of 2-methyl-4-*tert*-butyl-4-octanol (5) from alcohols of four carbon atoms or less:

$$CH_3CH_2CH_2CH_2OH \xrightarrow[\substack{H_2SO_4, \\ heat}]{NaBr} CH_3CH_2CH_2CH_2Br \xrightarrow[\substack{anhydrous \\ ether}]{Mg} CH_3CH_2CH_2CH_2MgBr$$

$$CH_3OH \xrightarrow[\substack{H_2SO_4, \\ warm}]{K_2Cr_2O_7} H-\overset{\overset{\displaystyle O}{\|}}{C}-H \Big\downarrow$$

$$\Big\downarrow H_3\overset{..}{O}{}^\oplus$$

$$CH_3CH_2CH_2CH_2-\overset{\overset{\displaystyle O}{\|}}{C}-H \xleftarrow[\substack{H_2SO_4, \\ warm}]{K_2Cr_2O_7} CH_3CH_2CH_2CH_2CH_2OH$$

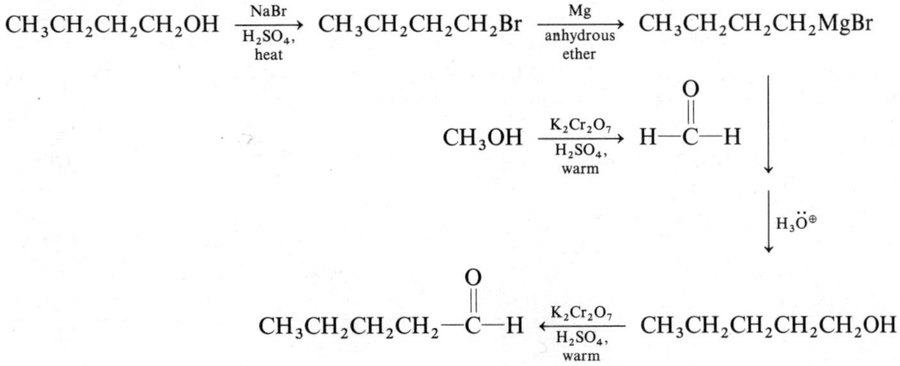

then

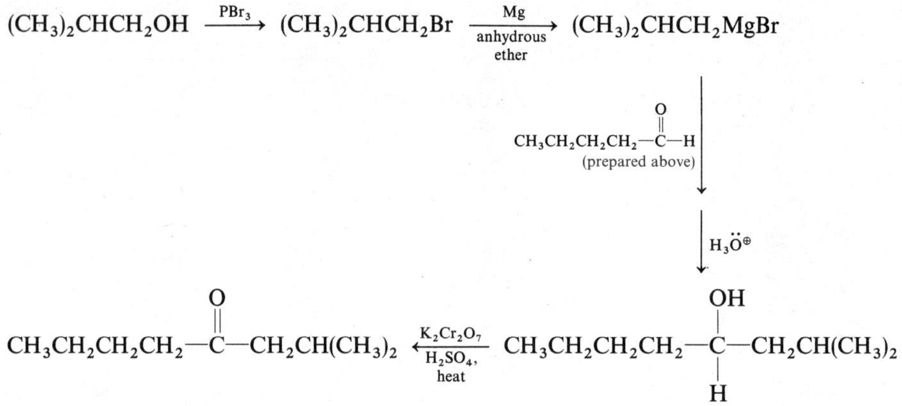

$$(CH_3)_2CHCH_2OH \xrightarrow{PBr_3} (CH_3)_2CHCH_2Br \xrightarrow[\substack{anhydrous \\ ether}]{Mg} (CH_3)_2CHCH_2MgBr$$

$$\underset{\text{(prepared above)}}{CH_3CH_2CH_2CH_2\overset{O}{\underset{\|}{-}}C-H}$$

$\downarrow H_3\overset{..}{O}{}^{\oplus}$

$$CH_3CH_2CH_2CH_2\overset{O}{\underset{\|}{-}}C-CH_2CH(CH_3)_2 \xleftarrow[\substack{H_2SO_4, \\ heat}]{K_2Cr_2O_7} CH_3CH_2CH_2CH_2\overset{\overset{OH}{|}}{\underset{\underset{H}{|}}{C}}-CH_2CH(CH_3)_2$$

and finally

$$(CH_3)_3COH \xrightarrow[\substack{(conc),\ 25°}]{HBr} (CH_3)_3CBr \xrightarrow[\substack{anhydrous \\ ether}]{Mg} (CH_3)_3CMgBr$$

$$\underset{\text{(prepared above)}}{CH_3CH_2CH_2CH_2\overset{O}{\underset{\|}{-}}C-CH_2CH(CH_3)_2}$$

$\downarrow H_3\overset{..}{O}{}^{\oplus}$

$$CH_3CH_2CH_2CH_2\overset{\overset{OH}{|}}{\underset{\underset{\underset{\underset{CH_3}{|}}{CH_3-C-CH_3}}{|}}{C}}-CH_2\overset{\overset{CH_3}{|}}{CH}-CH_3$$

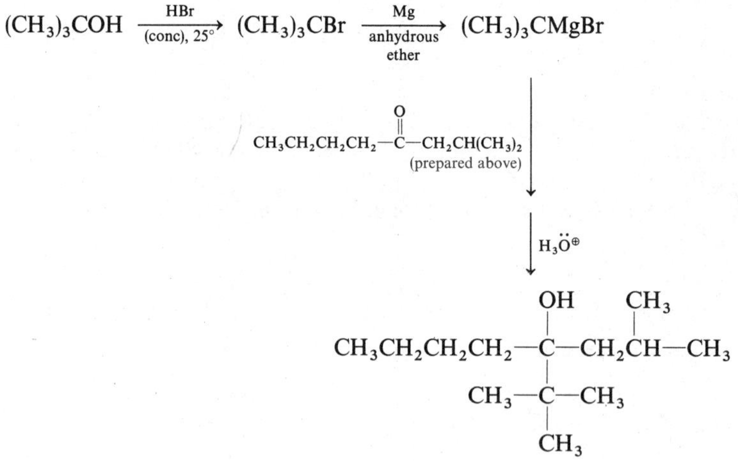

2-Methyl-4-*tert*-butyl-4-octanol

·(5)

Several of the alternative reactions discussed could have been used in place of those in the preceding scheme. There are often several acceptable routes in multistep syntheses.

Question 21.16

Starting with alcohols of four carbon atoms or less and benzene as the only sources of carbon, indicate how the following compounds can be prepared in good yield. Any needed inorganic reagents or organic solvents can be used.

(*a*) 3,3-dimethyl-1-butanol   (*b*) 2-methylpentane   (*c*) 1-bromo-3-methylbutane
(*d*) 3-methyl-1-butene   (*e*) 1-hexanal

(*f*) $(CH_3)_2CH-$⬡$-CH_2CH_2-Br$

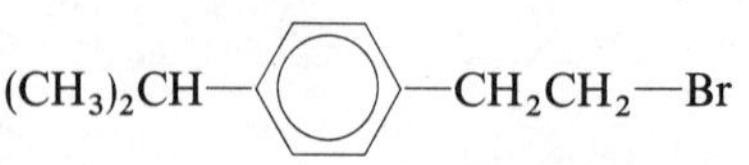

Question 21.17

Starting with alcohols containing four carbon atoms or less, synthesize each of the following compounds:

(a) 4-octanol   (b) 4-octanone   (c) *n*-propyl isopropyl ketone
(d) 2-methyl-3-hexanol (e) 3,3-diethoxyhexane (f) 4-chloro-4-*n*-propylheptane
(g) 2-methyl-1-butene   (h) 3,3-dimethyl-1-butyne

21.12   Nucleophilic Addition of Organolithium Reagents: Alcohol Preparation

Perhaps the most widely used reaction for preparing alcohols involves aldehydes and ketones and the Grignard reagent. However, another organometallic compound, the **organolithium reagent, RLi** or **ArLi,** adds to the carbonyl group in the same way as the Grignard reagent.

The organolithium reagent is usually made by allowing an alkyl or aryl halide to react with lithium metal under anhydrous conditions and at low temperatures (see Sec. 4.2):

$$RX + 2Li \xrightarrow[0-25°]{\text{anhydrous ether}} RLi + LiX$$
$$\text{Organolithium reagent}$$

and

$$ArX + 2Li \xrightarrow[0-25°]{\text{anhydrous ether}} ArLi + LiX$$
$$\text{Organolithium reagent}$$

For example:

$$CH_3I + 2Li \xrightarrow[25°]{\text{ether}} CH_3Li + LiI$$
Iodomethane   Methyllithium
(methyl iodide)

$$(CH_3)_3CBr + 2Li \xrightarrow[-10°]{\text{pentane}} (CH_3)_3CLi + LiBr$$
2-Bromo-2-methylpropane   *tert*-Butyllithium
(*tert*-butyl bromide)

$$C_6H_5Br + 2Li \xrightarrow[25°]{\text{ether}} C_6H_5Li + LiBr$$
Bromobenzene   Phenyllithium

The organolithium reagent reacts with oxygen, water, and carbon dioxide, so it is usually prepared under an inert atmosphere of nitrogen or helium gas.

The organolithium reagent is much more reactive than the corresponding Grignard reagent, and it often reacts with the ether solvent to give rearrangement products:

$$R-CH_2-O-R' + R''-Li \xrightarrow[\text{by } H_3O^\oplus]{\text{followed}} R-\underset{\underset{R'}{|}}{CH}-O-H + R''-H + Li^\oplus$$

Question 21.18

Outline a mechanism to account for the preceding rearrangement.

Organolithium compounds can be kept in ether solution for short times, but they are often prepared and kept in a hydrocarbon solvent, such as pentane or hexane, in which they are stable for a longer time.

That organolithium reagents are more reactive than Grignard reagents is attributed to the greater ionic character of the former. In other words, carbon has greater negative charge in R—Li (which may be viewed as $\overset{\ominus}{R}\!:\overset{\oplus}{Li}$) than in R—MgX.

The organolithium reagent adds to the carbonyl group as shown in the following general reaction:

$$\overset{\ominus}{R}\!:\overset{\oplus}{Li} + R'\!-\!\overset{\overset{\displaystyle :\!O\!:}{\|}}{C}\!-\!R'' \xrightarrow[\text{conditions}]{\text{anhydrous}} R'\!-\!\overset{\overset{\displaystyle :\!\overset{\ominus}{O}\!:\overset{\oplus}{Li}}{|}}{\underset{\underset{\displaystyle R}{|}}{C}}\!-\!R'' \xrightarrow{H_3\overset{..}{O}{}^{\oplus}} R'\!-\!\overset{\overset{\displaystyle :\!\overset{..}{O}\!-\!H}{|}}{\underset{\underset{\displaystyle R}{|}}{C}}\!-\!R''$$

As an illustration of the reactivity differences of Grignard and organolithium reagents, *tert*-butyllithium adds to di-*tert*-butyl ketone (a hindered ketone), but *tert*-butylmagnesium bromide does not:

$$\underset{\substack{\text{2,2,4,4-Tetramethyl-3-pentanone} \\ \text{(di-\textit{tert}-butyl ketone)}}}{(CH_3)_3C\!-\!\overset{\overset{\displaystyle O}{\|}}{C}\!-\!C(CH_3)_3}$$

$$\xrightarrow[\text{\textit{tert}-butyllithium}]{(CH_3)_3CLi} \xrightarrow{H_3\overset{..}{O}{}^{\oplus}} \underset{\substack{\text{2,2,4,4-Tetramethyl-3-\textit{tert}-butyl-3-pentanol} \\ \text{(tri-\textit{tert}-butylcarbinol)}}}{(CH_3)_3C\!-\!\overset{\overset{\displaystyle OH}{|}}{\underset{\underset{\displaystyle C(CH_3)_3}{|}}{C}}\!-\!C(CH_3)_3}$$

$$\xrightarrow[\substack{\text{\textit{tert}-butylmagnesium} \\ \text{bromide}}]{(CH_3)_3CMgBr} \xrightarrow{H_3\overset{..}{O}{}^{\oplus}} \text{no addition}$$

Question 21.19

Design a synthesis of di-*tert*-butyl ketone from neopentyl alcohol.

21.13   Reduction of Carbonyl Group to Alcohol

Another synthetic use of aldehydes and ketones is their ready reduction to the corresponding alcohol. There are two general methods for reducing aldehydes and ketones: (1) catalytic hydrogenation and (2) chemical methods. We discuss these methods in this section, but for now let us indicate the general nature of reduction:

$$\underset{\substack{\text{Aldehyde} \\ \text{or ketone}}}{R\!-\!\overset{\overset{\displaystyle O}{\|}}{C}\!-\!R'} \xrightarrow{[H]^1} \underset{\text{Alcohol}}{R\!-\!\overset{\overset{\displaystyle OH}{|}}{\underset{\underset{\displaystyle H}{|}}{C}}\!-\!R'}$$

---

[1] [H] stands for a suitable reducing agent, whereas [O] is used to symbolize an oxidizing agent (see Sec. 1.5).

Reduction of ketones gives secondary alcohols and reduction of aldehydes gives primary alcohols.

A. Catalytic Hydrogenation

The carbonyl group in aldehydes and ketones undergoes catalytic hydrogenation. The catalysts used in the reduction of carbon-carbon double and triple bonds are also effective in the addition of hydrogen to carbonyl groups; the metals often used are platinum, palladium, and nickel. In addition, copper chromite, $CuCrO_2$, is especially effective as a catalyst in the hydrogenation of the polar carbonyl group, presumably because the polar carbonyl group is effectively adsorbed on the surface of the metal oxide. Three examples of catalytic hydrogenation are:

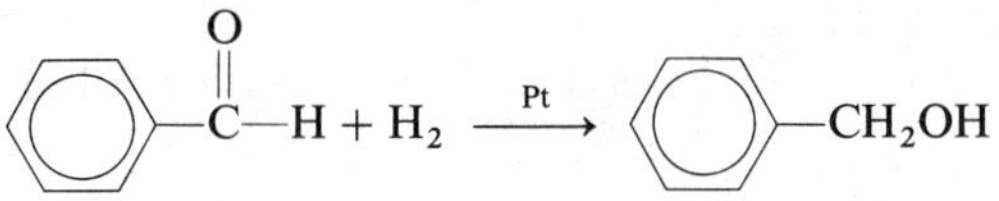

Benzaldehyde   Benzyl alcohol

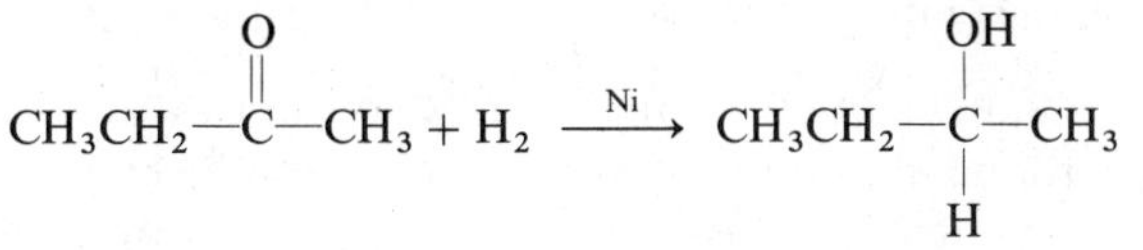

Butanone   2-Butanol

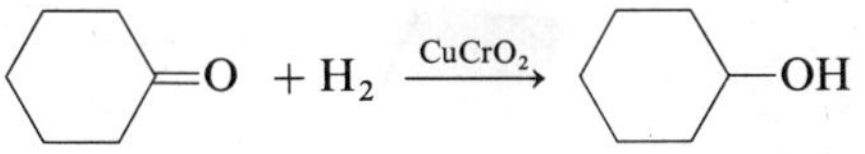

Cyclohexanone   Cyclohexanol

The mechanism for the hydrogenation of the carbon-oxygen double bond under catalytic conditions resembles greatly the mechanism for the reduction of alkenes and alkynes in Secs. 8.12 and 11.10. It too is a surface reaction where the carbonyl group "sits" on top of the metal surface containing absorbed hydrogen.

By varying the nature of the catalyst, however, it is often possible to reduce one type of bond and not the other (and vice versa) when both are in the same molecule. Reaction conditions must be worked out for each reaction. The following example illustrates how this can be done:

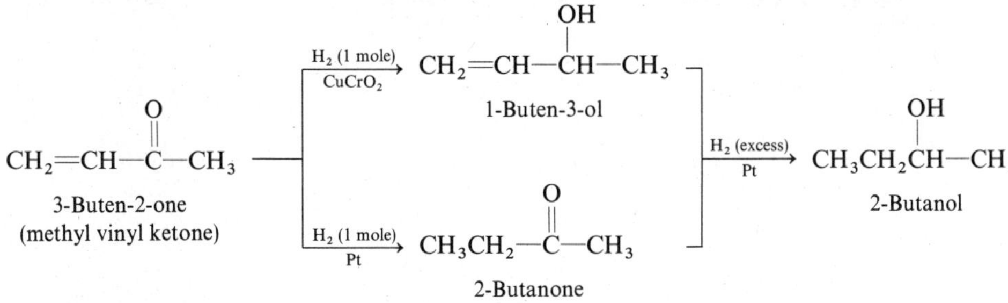

B. Chemical Reduction

Two common chemical reducing agents have been developed in recent years, and they are frequently used in a convenient laboratory method for reducing multiple bonds. These reagents, **lithium aluminum hydride, LiAlH₄** (abbreviated LAH), and

sodium borohydride, NaBH$_4$, react similarly with aldehydes and ketones, as shown by:

$$R-\overset{\overset{\displaystyle O}{\|}}{C}-R' \xrightarrow[\substack{\text{or} \\ \text{1. NaBH}_4, \;\; \text{2. H}_3\ddot{O}^{\oplus}}]{\text{1. LiAlH}_4, \;\; \text{2. H}_3\ddot{O}^{\oplus}} R-\overset{\overset{\displaystyle OH}{|}}{\underset{\underset{\displaystyle H}{|}}{C}}-R'$$

Some understanding of why these reagents behave similarly can be gained from examining their structures:

$$H-\overset{\overset{\displaystyle H}{|}}{\underset{\underset{\displaystyle H}{|}}{Al^{\ominus}}}-H \quad Li^{\oplus} \qquad H-\overset{\overset{\displaystyle H}{|}}{\underset{\underset{\displaystyle H}{|}}{B^{\ominus}}}-H \quad Na^{\oplus}$$

Lithium aluminum   Sodium borohydride
hydride (LAH)

Hydrogen is present in both compounds and provides a source of hydride ion, $H\overset{\ominus}{\colon}$ (as the names imply). This is an extension of the nucleophilic addition reaction mechanism; the nucleophile is the hydride ion, $H\overset{\ominus}{\colon}$, in this case.

Using lithium aluminum hydride as the example, let us consider the mechanism of carbonyl reduction. The first step is the nucleophilic addition of hydride ion to the carbonyl carbon. The resulting intermediate (1) still contains three hydride ions attached to aluminum, and they can add to 3 more moles of carbonyl compound:

(1)
Intermediate

In other words, a total of 4 moles of carbonyl compound react with each mole of LAH:

$$4 \overset{}{\underset{}{>}}C=\ddot{O}\colon + AlH_4^{\ominus} \xrightarrow{\text{ether}} \left[-\overset{|}{\underset{\underset{\displaystyle H}{|}}{C}}-\ddot{O}\colon \right]_4 Al^{\ominus}$$

(1)

For reasons that become clear, the addition reaction is carried out in an anhydrous solvent such as ether. The aluminum salt (1) is subsequently decomposed by addition of acidic water, which breaks the Al—O bonds and protonates oxygen; aluminum is liberated as Al$^{\oplus 3}$, which remains in aqueous solution. The lithium ion plays no role in the reaction other than to counterbalance the formal negative charge on aluminum.

$$\left[\begin{array}{c} -\overset{|}{\underset{|}{C}}-\ddot{\underset{\cdot\cdot}{O}}- \\ H \end{array}\right]_4\!\!\!-Al^{\ominus} \xrightarrow{H_3\ddot{O}^{\oplus}} 4-\overset{|}{\underset{H}{C}}-\ddot{\underset{\cdot\cdot}{O}}H + Al^{\oplus 3}$$

(1)

In summary, then, the reduction mechanism involves nucleophilic addition under neutral, anhydrous conditions, followed by hydrolysis.

Sodium borohydride reduction occurs by a mechanism analogous to that for LAH reduction.

Lithium aluminum hydride reductions must always be carried out in two steps: reaction of the carbonyl compound with LAH under anhydrous conditions *and then* hydrolysis. This is necessary because LAH reacts vigorously (almost violently) when it comes in contact with a proton source (such as water or an alcohol); hydride ion unites with a proton to form hydrogen gas: $H^{\oplus} + H\overset{\ominus}{\underset{\cdot\cdot}{:}}$ (from LAH) $\rightarrow H_2$ (gas). LAH is soluble in most ether solvents (such as ethyl ether and diglyme), which are preferred. On the other hand, sodium borohydride is a much milder reducing agent, and it reacts slowly with water or alcohol, which are often used as solvents in $NaBH_4$ reductions. Because of the relative insolubility of $NaBH_4$ in most organic solvents (except diglyme), the use of protic solvents is even more desirable. When protic solvents are used, the hydrolysis step often is not necessary.

Cyclopentanone   Cyclopentanol

$$CH_3CH_2-\overset{O}{\overset{||}{C}}-\overset{|}{\underset{|}{CH}}-CH_3 \xrightarrow[\text{2. }H_3\ddot{O}^{\oplus}]{\text{1. }NaBH_4} CH_3CH_2-\overset{OH}{\overset{|}{CH}}-\overset{|}{\underset{CH_3}{CH}}-CH_3$$

2-Methyl-3-pentanone   2-Methyl-3-pentanol

m-Bromobenzaldehyde   *m*-Bromobenzyl alcohol

The lesser reactivity of $NaBH_4$ (compared with $LiAlH_4$) can be used to advantage when there are several functional groups in the same molecule; for example:

$$CH_3-CH{=}CH-\overset{|}{\underset{H}{C}}{=}O \xrightarrow[\text{2. }H_3\ddot{O}^{\oplus}]{\text{1. }LiAlH_4} CH_3CH_2CH_2CH_2OH$$

$LiAlH_4$ reduces $C{=}C$ (conjugated with a $C{=}O$) and $C{=}O$

2-Butenal   1-Butanol
(*n*-butyl alcohol)

but

$$CH_3-CH{=}CH-\overset{|}{\underset{H}{C}}{=}O \xrightarrow[\text{2. }H_3\ddot{O}^{\oplus}]{\text{1. }NaBH_4} CH_3-CH{=}CH-CH_2OH$$

$NaBH_4$ reduces only $C{=}O$

2-Butenal   2-Buten-1-ol

Table 21.1   Effectiveness of Selected Metal Hydride Reducing Agents*

	LiAlH$_4$/Ether	NaBH$_4$/Ethanol	NaAlH$_2$(OR)$_2$	B$_2$H$_6$/THF
$-$C$=$O $\rightarrow$ $-$C$-$OH	+	+	+	+
$-$COOR' $\rightarrow$ $-$CH$_2$$-$OH	+	$-$	+†	+
$-$COOH, $-$COO$^\ominus$ $\rightarrow$ $-$CH$_2$$-$OH	+	$-$	+	+ (COOH) $-$ (COO$^\ominus$)
$-$CONR$_2$ $\rightarrow$ $-$CH$_2$NR$_2$	+	$-$	+	+
$-$C$\equiv$N $\rightarrow$ $-$CH$_2$NH$_2$	+	$-$	+ (Ar) $-$ (R)	+
$-$NO$_2$ $\rightarrow$ $-$NH$_2$	+	$-$	+	$-$
$-$COCl $\rightarrow$ $-$CH$_2$$-$OH	+†	+	+	$-$
$-$CO$-$O$-$OC$-$ $\rightarrow$ $-$CH$_2$$-$OH	+	$-$	+	
C$-$C $\rightarrow$ $-$C$-$C$-$OH with O	+	$-$		+
C$=$C $\rightarrow$ $-$C$-$C$-$ H H	$-$‡	$-$	$-$	+
$-$C$-$X, $-$C$-$OTs $\rightarrow$ $-$C$-$H	+	$-$	+	$-$
Ar$-$X $\rightarrow$ Ar$-$H	$-$	$-$	+	$-$

* A plus sign ($+$) designates a positive reaction; a minus sign ($-$), a negative reaction.
† Inverse addition at $-70°$ gives the aldehyde; reaction at higher temperature with two equivalents of hydride gives alcohols.
‡ Double bonds conjugated with carbonyls are sometimes reduced (e.g., cinnamates).

In general, NaBH$_4$ reduces only aldehydes and ketones, whereas the more powerful reducing agent LiAlH$_4$ reduces many multiply bonded functional groups (Table 21.1). Care should be taken, however, because NaBH$_4$ will often reduce C$=$C double bonds in conjugation with a C$=$O, and reaction conditions must be worked out in advance.

Question 21.20

Indicate the products that might be expected when each of the following compounds is treated with an excess of LiAlH$_4$ in ether and then hydrolyzed:

(*a*) 2,2-dimethylpropanol
(*b*) 2-octanone
(*c*) di-*tert*-butyl ketone
(*d*) 2,4-pentanedione

(*e*) cinnamaldehyde, C$_6$H$_5$CH$=$CH$-$C(=O)$-$H
(*f*)

(*g*) *p*-methylacetophenone

Question 21.21

Indicate the products that might be expected when each of the compounds listed in Question 21.20 is treated with excess NaBH$_4$ and then hydrolyzed.

Question 21.22

When acetophenone, $C_6H_5COCH_3$, is treated with $LiAlH_4$ in ether and then hydrolyzed, the product, 1-phenylethanol ($C_6H_5CHOHCH_3$), is optically inactive even though it contains an asymmetric carbon atom. Explain this result.

Question 21.23

When 1 mole of 4-hydroxy-2-butanone is treated with 1 mole of $LiAlH_4$ and the resulting reaction mixture is hydrolyzed, the product consists of an appreciable amount (~ 0.25 mole) of 4-hydroxy-2-butanone in addition to ~ 0.75 mole of 1,3-butanediol. When this identical experiment is carried out with 2-butanone, however, the product is essentially pure 2-butanol. Explain these results. (*Hint:* Consider the reaction between LAH and an alcohol.)

21.14   Carbonyl to Methylene Reductions

The conversion of a carbonyl group (aldehyde or ketone) to a methylene group ($-CH_2-$) can be accomplished by the following multistep sequence:

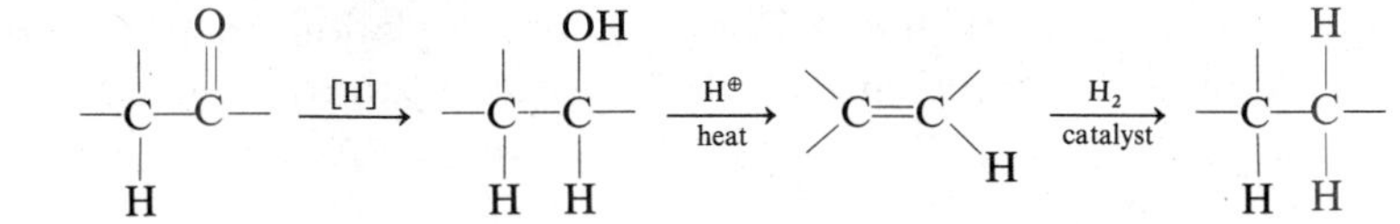

In addition to requiring these three steps, this sequence has the disadvantage that molecular rearrangements may occur at the dehydration stage. However, two common one-step methods accomplish this same transformation: (1) Clemmensen reduction and (2) Wolff-Kishner reduction.

Clemmensen reduction uses concentrated hydrochloric acid and the reducing agent *zinc amalgam*, which is zinc metal coated with elemental mercury and written as Zn(Hg). The heterogeneous mixture of reducing agent, concentrated HCl, and aldehyde or ketone is heated and decent yields (often better than 50%) of the corresponding methylene compound are obtained; for example:

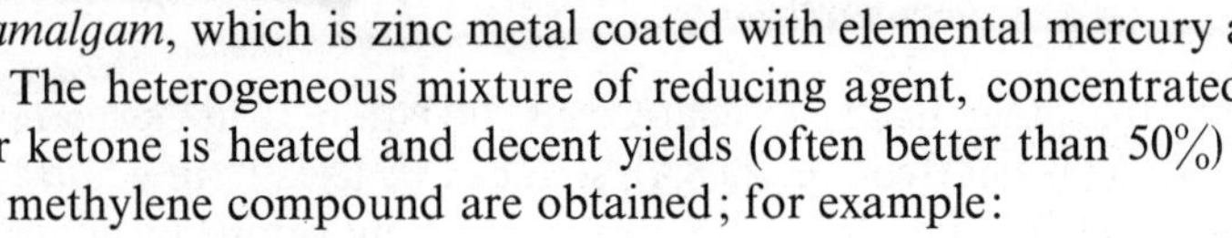

1-Phenyl-1-butanone   1-Phenylbutane
(butyrophenone)   (*n*-butylbenzene)

The mechanism has not been studied extensively, but current knowledge suggests that it involves attachment of the carbonyl group onto the zinc metal surface where reduction occurs. Alcohols are not intermediates in the reduction. Clemmensen reduction is often used on compounds that are sensitive to base.

Wolff-Kishner reduction involves mixing together hydrazine (NH_2-NH_2), the aldehyde or ketone, and potassium hydroxide in a high-boiling, protic solvent, such as ethylene glycol. The mixture is first heated at lower temperatures (110 to 130°), which forms the hydrazone of the carbonyl compound (see Sec. 21.18); the yield of hydrazone is improved by removing water by distillation as it is formed. The remaining reactants are then heated to $\sim 200°$ to decompose the hydrazone, which results in the formation of the corresponding methylene compound and nitrogen gas. In this particular method both reactions are carried out in the same reaction vessel and the intermediate hydrazone is not isolated or purified. An example of Wolff-Kishner reduction is:

$$CH_3-\underset{\underset{CH_3}{|}}{\overset{\overset{CH_3}{|}}{C}}-\overset{\overset{O}{||}}{C}-CH_3 \xrightarrow[\substack{KOH, HOCH_2CH_2OH, \\ 110-130°}]{NH_2-NH_2} H_2O + \left[CH_3-\underset{\underset{CH_3}{|}}{\overset{\overset{CH_3}{|}}{C}}-\overset{\overset{\ddot{N}\diagup^{NH_2}}{||}}{C}-CH_3 \right] \xrightarrow[\substack{HOCH_2CH_2OH, \\ 200°}]{KOH}$$

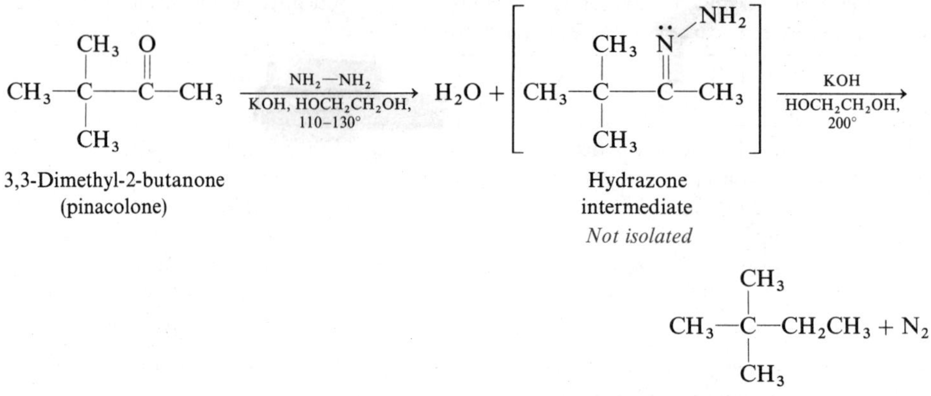

3,3-Dimethyl-2-butanone
(pinacolone)

Hydrazone
intermediate
Not isolated

$$CH_3-\underset{\underset{CH_3}{|}}{\overset{\overset{CH_3}{|}}{C}}-CH_2CH_3 + N_2$$

2,2-Dimethylbutane
(neohexane)

Wolff-Kishner reduction is often used on compounds that are sensitive to acid.

Both Clemmensen and Wolff-Kishner reductions are used in the preparation of pure alkylbenzenes. Even though Friedel-Crafts alkylation reaction introduces alkyl groups onto the aromatic ring, it is often accompanied by extensive (and often exclusive) rearrangement of the alkyl group; for example:

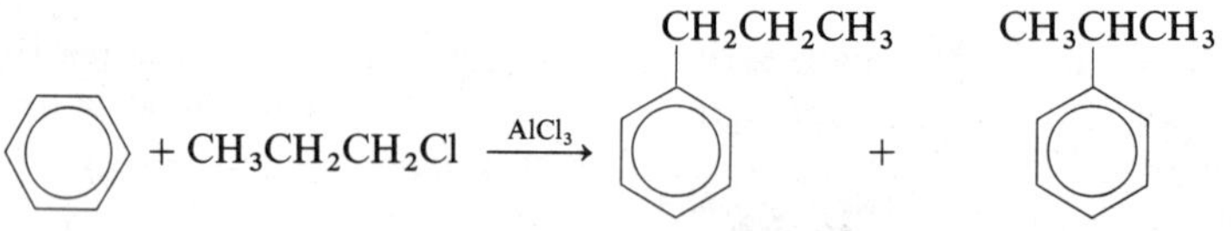

$$\bigcirc + CH_3CH_2CH_2Cl \xrightarrow{AlCl_3} \overset{CH_2CH_2CH_3}{\bigcirc} + \overset{CH_3CHCH_3}{\bigcirc}$$

1-Chloropropane
(*n*-propyl chloride)

n-Propylbenzene   Isopropylbenzene

On the other hand, Friedel-Crafts acylation is not accompanied by molecular rearrangement (see Sec. 20.8A), so it can be used to introduce the carbon chain onto the aromatic ring. Reduction of the resulting ketone by either the Clemmensen or the Wolff-Kishner method gives rise to the corresponding *n*-alkylbenzene. This is illustrated by the following reaction sequence:

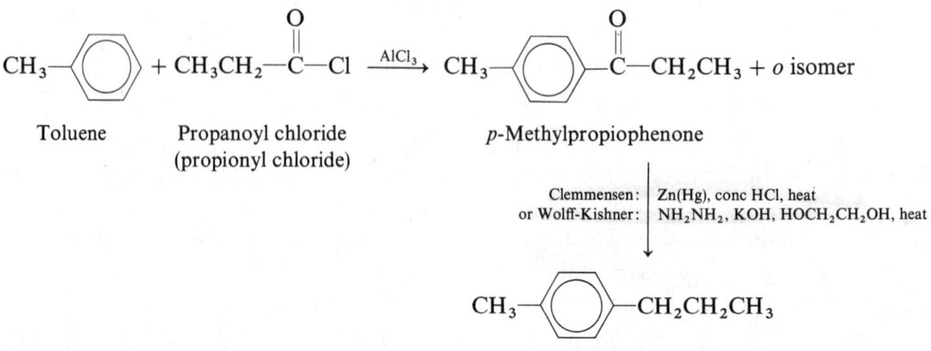

$$CH_3-\bigcirc + CH_3CH_2-\overset{\overset{O}{||}}{C}-Cl \xrightarrow{AlCl_3} CH_3-\bigcirc-\overset{\overset{O}{||}}{C}-CH_2CH_3 + o \text{ isomer}$$

Toluene   Propanoyl chloride
(propionyl chloride)

p-Methylpropiophenone

Clemmensen: │ Zn(Hg), conc HCl, heat
or Wolff-Kishner: │ NH₂NH₂, KOH, HOCH₂CH₂OH, heat

$$CH_3-\bigcirc-CH_2CH_2CH_3$$

p-(1-Propyl)toluene

Another example of a ketal protecting a carbonyl group is the following, where we want to remove the —OH group from *testosterone*, a naturally occurring steroid:

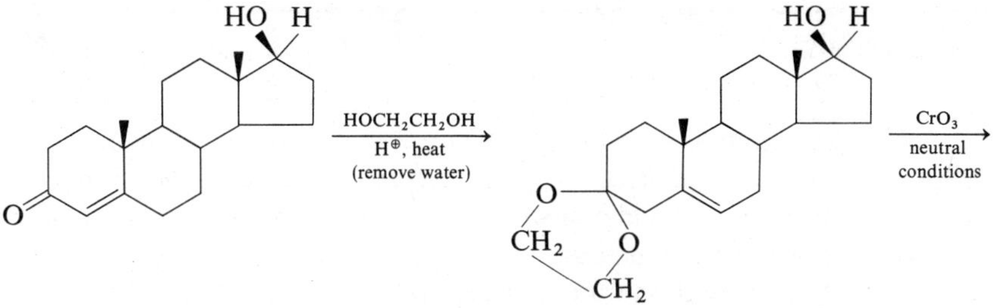

Testosterone

Ketal
Protects keto group

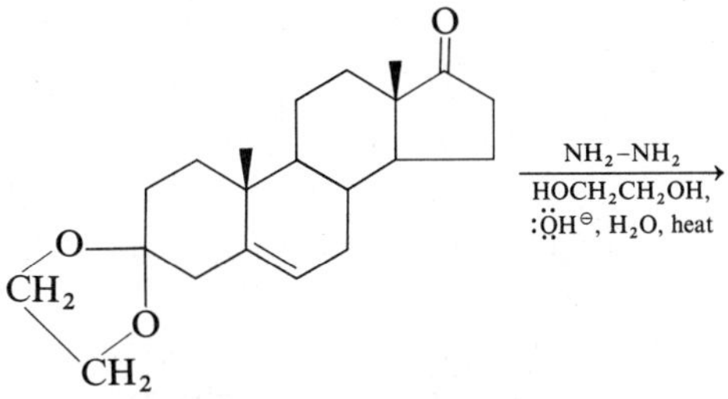

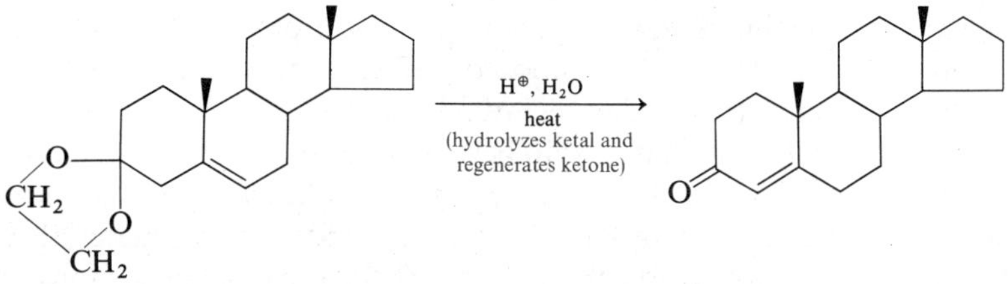

In the first step, the double bond migrates, through a rearrangement, from the A ring to the B ring (see Sec. 21.15). After the final hydrolysis, however, the bond has re-arranged and migrated back to its initial position.

Question 21.24

Refer to the preceding sequence of transformations.

(*a*) Why was the oxidation of the alcohol to the ketone carried out under neutral conditions, and what, if anything, would have happened if strongly acidic conditions (for example, sulfuric acid) had been used?

(*b*) Why was Wolff-Kishner reduction used rather than Clemmensen reduction? Briefly explain.

Question 21.25

Indicate a sequence of reactions to transform testosterone (shown earlier) into compound (1). What stereochemistry do you think would be observed at the carbon marked with an asterisk (*)? Why?

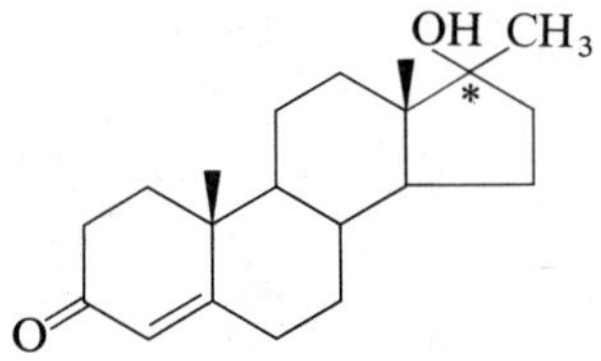

(1)

21.15   Steroids

We have encountered numerous examples of steroids in this text. The *steroid* carbon skeleton (often referred to as the *steroid nucleus*) is found in many naturally occurring molecules of biochemical interest. The steroids all contain the skeletal framework drawn here. This four-ring system is labeled in the accepted A, B, C, D convention. The conventional numbering system used in naming these compounds is also provided.

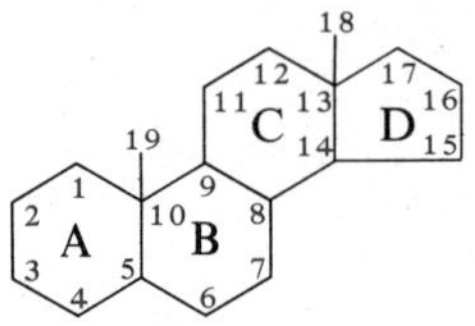

Note the methyl groups, carbons 18 and 19, which are a common feature of most steroids.

Cholesterol (see Sec. 19.15) is by far the most common steroid. It is found in most animal tissue, with the greatest amount in the tissues of the central nervous system—the brain and spinal cord. Cholesterol is also found in high concentrations in gallstones, which are composed predominantly of this substance. There is a great deal of controversy today regarding the role of cholesterol in heart disease. Results indicate a definite relationship between the concentration of cholesterol in the blood (found in the plasma) and the occurrence of heart disease (hardening of the arteries, arteriosclerosis). Cholesterol (in combination with other materials) appears to coat the arteries, causing a constriction, which in turn requires more and more pressure (hypertension) to pass the required volume of liquid through the channel. Research is still in progress regarding these issues.

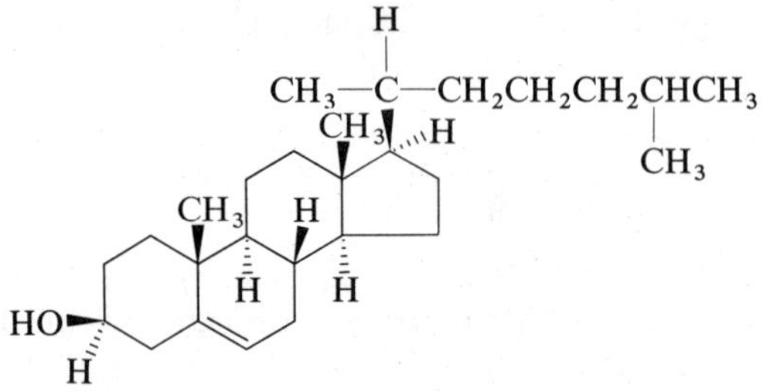

Cholesterol

Cholesterol is involved in many biochemical reactions. One of its metabolites is cholic acid, commonly referred to as bile acid (the most common of several human bile acids).

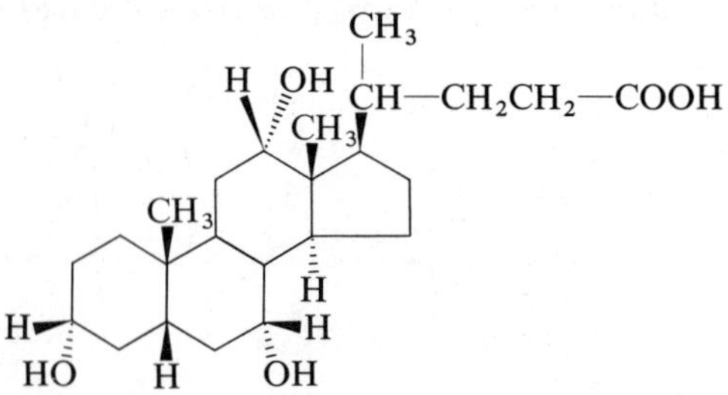

Cholic acid

Cholic acid is produced in the liver and stored in the gallbladder. These acids are released into the small intestine, as needed, to aid in the digestion of lipids.

Question 21.26

Build a framework molecular model of cholic acid (drawn above) to get an idea of the complex conformational and steric interactions in the steroid nucleus. Vary the stereochemistry of the C-5 hydrogen, which is *cis* to the C-19 methyl in the natural isomer, to determine its effect on the geometry of the molecule.

Cholesterol is also the biochemical precursor of pregnenolone in the body. Pregnenolone plays a vital role in the production of all other human sex hormones. It is converted to *androgens* (male sex hormones), *estrogens* (female sex hormones), and *progestins* (hormones involved in pregnancy):

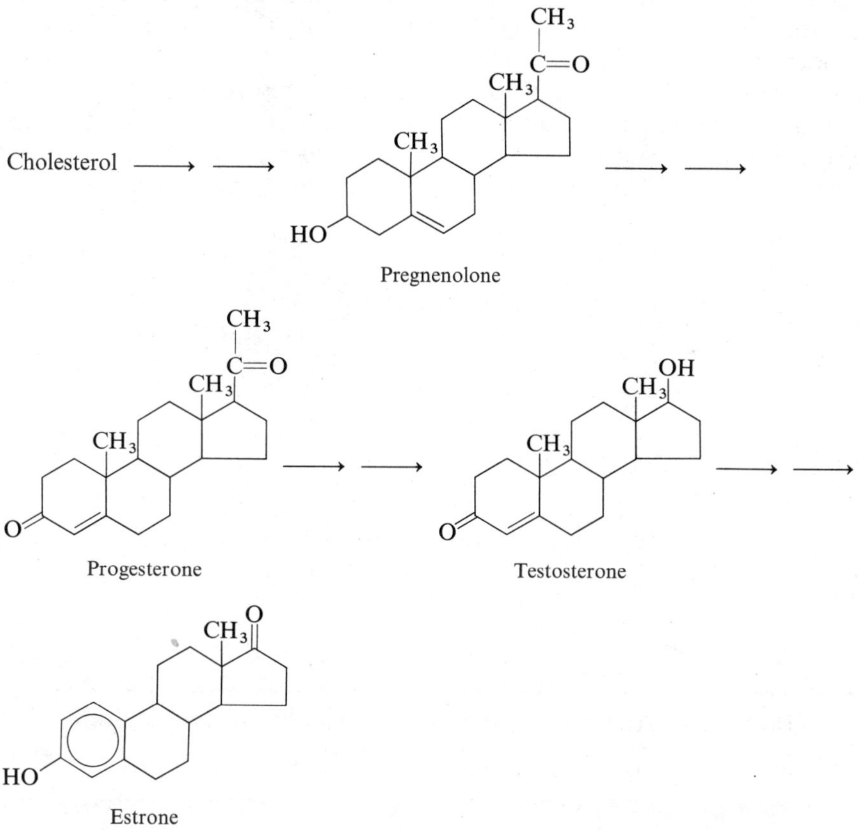

Cholesterol $\longrightarrow$ $\longrightarrow$

Pregnenolone

$\longrightarrow$ $\longrightarrow$

Progesterone

$\longrightarrow$ $\longrightarrow$

Testosterone

$\longrightarrow$ $\longrightarrow$

Estrone

Progesterone is the most important pregnancy hormone. It is produced in the ovaries by the enzymatic oxidation of cholesterol and is responsible for both the successful initiation of a pregnancy and the successful completion of pregnancy. Its initial role is to prepare the uterine mucosa (lining) for reception of a fertilized ovum. When the fertilized ovum is successfully attached to the uterine wall, the progesterone continues to be produced, aiding in the successful development of the fetus and at the same time suppressing further ovulation.

Progesterone in turn can be biochemically converted to *testosterone* (an androgen), which is found in the testes, and finally to *estrone* (an estrogen), which was first isolated from the urine of pregnant women. Both these steroids play a major role in the development of male and female characteristics and are human hormones.

Progesterone can also be converted to *cortisone* (a corticosteroid found in the adrenal gland), which is responsible for regulating a variety of metabolic processes. Cortisone

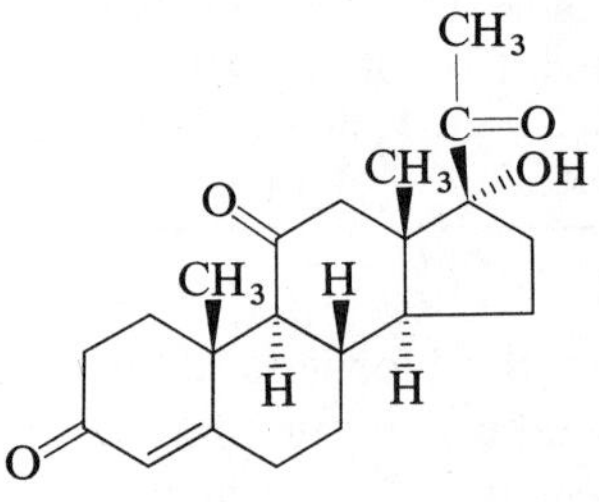

Cortisone

appears to play a principal role in the metabolism of carbohydrates, primarily the biochemical synthesis of carbohydrate from protein. Cortisone is also widely used in the treatment of rheumatoid arthritis.

The role of steroids in pregnancy had led to research into the uses of these compounds as birth control agents. Initially progesterone itself was studied in this regard, but it was found that the dose required to prevent ovulation is much too large. One successful steroid in this regard is ethynylestradiol diacetate:

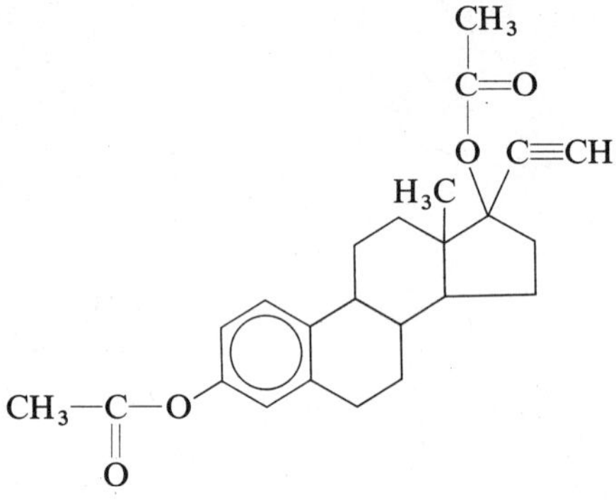

Ethynylestradiol diacetate

Lab 21.16   Qualitative Analysis: Detection of Aldehydes and Ketones

One of the common chemical methods for detecting the carbonyl group in aldehydes and ketones is to note the fast reaction they undergo with 2,4-dinitrophenylhydrazine (2,4-DNPH). That a reaction occurs is *observed* by the formation of a yellow to orange precipitate:

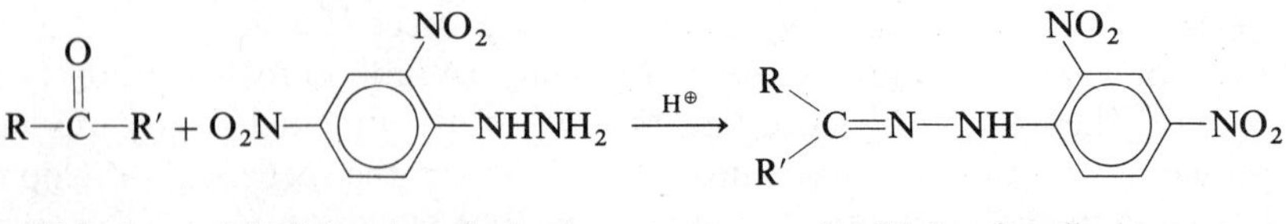

Aldehyde or ketone	2,4-Dinitrophenylhydrazine (2,4-DNPH)	2,4-Dinitrophenylhydrazone (2,4-DNP)
		Yellow to orange solid

The resulting 2,4-dinitrophenylhydrazone (2,4-DNP) can be isolated, purified, and used as a derivative for identification.

The preceding test does not distinguish between an aldehyde and a ketone, however, because it indicates only the presence of the carbonyl group. A simple chemical test to distinguish between aldehydes and ketones is based on aldehydes being readily oxidized to carboxylic acids whereas ketones are not. **Tollens' test,** which is used most frequently, involves treating an aldehyde with silver ammonia complex, $Ag(NH_3)_2^{\oplus}$, in basic solution. This *oxidizes* the aldehyde to the corresponding carboxylic acid, RCOOH, which ends up in solution as the soluble ammonium salt. The silver ion ($Ag^{\oplus}$) in the silver ammonia complex is *reduced* to silver metal, which is *observed* as a silver mirror on the side of the test tube. The reaction is:

$$\underset{\text{Aldehyde}}{R-\overset{\displaystyle O}{\overset{\|}{C}}-H} + \underset{\substack{\text{Silver} \\ \text{ammonia} \\ \text{complex}}}{Ag(NH_3)_2^{\oplus}} \longrightarrow \underset{\substack{\text{Ammonium} \\ \text{salt of} \\ \text{carboxylic acid}}}{R-\overset{\displaystyle O}{\overset{\|}{C}}-\overset{..}{\underset{..}{O}}{:}^{\ominus} NH_4^{\oplus}} + \underset{\substack{\text{Silver} \\ \text{mirror}}}{Ag^0}$$

$$\underset{\text{Ketone}}{R-\overset{\displaystyle O}{\overset{\|}{C}}-R'} + Ag(NH_3)_2^{\oplus} \longrightarrow \text{no reaction}$$

Another common test for aldehydes is **Fehling's test,** in which a basic solution of cupric tartrate complex reacts with the aldehyde. The aldehyde is again oxidized to the corresponding carboxylic acid, while copper(II) is reduced to copper(I), which precipitates as *red* cuprous oxide, Cu_2O. The reaction is:

$$\underset{\text{Aldehyde}}{R-\overset{\displaystyle O}{\overset{\|}{C}}-H} + Cu^{\oplus 2} \underset{\text{Cupric tartrate complex}}{\left[{}^{\ominus}\overset{..}{\underset{..}{O}}-\overset{\displaystyle O}{\overset{\|}{C}}-\underset{\underset{OH}{|}}{CH}-\underset{\underset{OH}{|}}{CH}-\overset{\displaystyle O}{\overset{\|}{C}}-\overset{..}{\underset{..}{O}}{:}^{\ominus} \right]} \xrightarrow{:\overset{..}{O}H^{\ominus}} \underset{\substack{\text{Salt of} \\ \text{carboxylic} \\ \text{acid}}}{R-\overset{\displaystyle O}{\overset{\|}{C}}-\overset{..}{\underset{..}{O}}{:}^{\ominus}} + \underset{\substack{\text{Cuprous} \\ \text{oxide} \\ \textit{Red} \\ \textit{precipitate}}}{Cu_2O}$$

A sensitive test for aldehydes is the **Schiff's test,** in which decolorized *Fuchsin* (a magenta dye) turns magenta on treatment with an aldehyde but *not* with a ketone.

The **iodoform test** is specific for detecting either *methyl ketones or methyl carbinols.* Upon treatment with a mixture of sodium hydroxide and iodine, organic compounds containing these specific functional groups are converted into the soluble salt of a carboxylic acid (*which contains one less carbon atom*), and a yellow precipitate of iodoform, CHI_3, is *observed.* Iodoform has a characteristic odor and a melting point of 119°, which are used to confirm its presence.

$$\underset{\substack{\text{Methyl} \\ \text{carbinol}}}{R-\underset{\underset{H}{|}}{\overset{\overset{OH}{|}}{C}}-CH_3} \text{ or } \underset{\substack{\text{Methyl} \\ \text{ketone}}}{R-\overset{\displaystyle O}{\overset{\|}{C}}-CH_3} \xrightarrow[:\overset{..}{O}H^{\ominus}, \, H_2O]{I_2} \underset{\substack{\text{Salt of} \\ \text{carboxylic} \\ \text{acid} \\ \textit{Soluble}}}{R-\overset{\displaystyle O}{\overset{\|}{C}}-\overset{..}{\underset{..}{O}}{:}^{\ominus}} + \underset{\substack{\text{Iodoform} \\ \textit{Yellow} \\ \textit{precipitate,} \\ \textit{mp 119°}}}{CHI_3(s)}$$

(R = H, alkyl or aryl)

Evidence suggests that the methyl carbinols are first oxidized to the corresponding methyl ketones, which in turn react to produce iodoform:

$$
\underset{\underset{H}{|}}{\overset{\overset{OH}{|}}{R-C-CH_3}} \xrightarrow[H_2O]{I_2, :\ddot{O}H^{\ominus}} \underset{}{\overset{\overset{O}{\|}}{R-C-CH_3}} \xrightarrow[H_2O]{I_2, :\ddot{O}H^{\ominus}} \overset{\overset{O}{\|}}{R-C-\ddot{\overset{..}{O}}:^{\ominus}} + CHI_3(s)
$$

When iodine is dissolved in aqueous base, the following reactions occur:

$$I_2 + H_2O \rightleftharpoons HI + HOI$$

$$HI + HOI + 2:\ddot{O}H^{\ominus} \rightleftharpoons :\ddot{I}:^{\ominus} + :\ddot{O}I^{\ominus} + 2H_2O$$

Iodine disproportionates (is both oxidized and reduced) to $:\ddot{I}:^{\ominus}$ and $:\ddot{O}I^{\ominus}$; $:\ddot{O}I^{\ominus}$ is called *hypoiodite* ion and iodine in it has an oxidation number of $+1$. As a result, either $I_2 + :\ddot{O}H^{\ominus}$ or $:\ddot{O}I^{\ominus} + :\ddot{O}H^{\ominus}$ are commonly specified as the reagents for the iodoform reaction.

Further details about the iodoform reaction are presented in Sec. 24.6. This reaction can also be used synthetically to remove *one carbon atom* from a molecule.

Question 21.27

Indicate simple tests to distinguish between the members of each of the following pairs of compounds. Indicate what you would *observe* as a positive test.

(*a*) benzaldehyde and acetophenone

(*b*) benzaldehyde and cyclohexane carboxaldehyde,

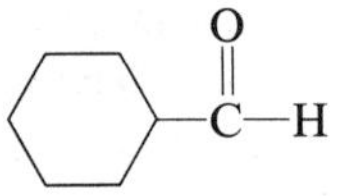

(*c*) 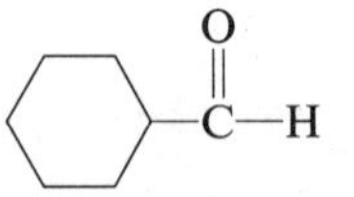   and 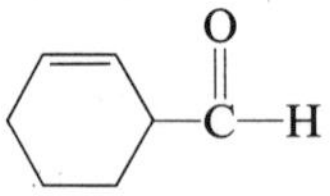   (*d*) benzaldehyde and phenol

(*e*) 2-butanone and 2-butanol   (*f*) *tert*-butyl bromide and *n*-butyl bromide
(*g*) cyclohexylbenzene and phenol   (*h*) acetophenone and propiophenone
(*i*) 2-pentanone and 3-pentanone   (*j*) 2-hexanol and 3-hexanol

Question 21.28

Which of the following 13 compounds would give a positive iodoform test?

(*a*) formaldehyde   (*b*) acetaldehyde   (*c*) acetone
(*d*) ethanol   (*e*) β-phenylethanol   (*f*) cyclohexanone
(*g*) *sec*-butyl alcohol   (*h*) isobutyl alcohol   (*i*) 2-methyl-2-hexanol
(*j*) 3-pentanol   (*k*) *tert*-butyl alcohol   (*l*) 2,4-pentanedione

(*m*) mesityl oxide $(CH_3)_2C=CH-\overset{\overset{O}{\|}}{C}-CH_3$

* 21.17   Alcohols As Key Intermediates in Aliphatic Chemistry

It is now clear how very important alcohols are in the chemistry of aliphatic compounds. Several functional groups can be obtained, either directly or indirectly, from the —OH group. As we saw, alcohols can be oxidized to aldehydes or ketones, which can react with the Grignard reagent to provide a convenient method for making new carbon-carbon bonds and thus for building complex molecules from simple ones. Rather than write out in words the usefulness of alcohols, we summarize many of their important reactions in Fig. 21.2; the reagents and conditions are intentionally left out for clarity.

Keep in mind, however, that the figure shows only *some* interrelationships. It is intended mostly as a sample but at the same time emphasizes the great utility

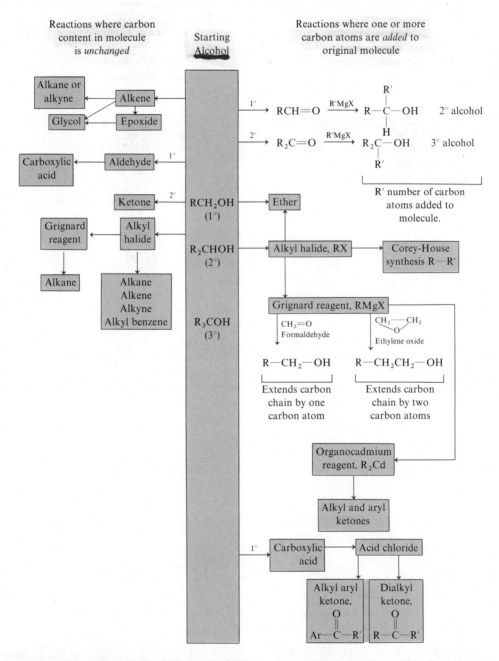

FIGURE 21.2 Chart showing some of the interrelationships between alcohols and other families of organic compounds (reagents and conditions intentionally omitted).

of alcohols. You might find it useful to construct a more complex chart of your own and insert the various reagents. The direction of many reactions changes with different reagents. For example, alcohols can be oxidized to aldehydes and ketones, and the latter can be reduced to alcohols. It might be useful to identify these cases. Many times a reaction occurs with *both* alkyl and aryl groups present, although only alkyl groups are indicated in the chart.

Devising the Synthesis of Organic Compounds

We now comment on the approach for devising synthetic routes for organic compounds. First, however, keep in mind that alcohols provide a convenient point of departure for many syntheses, and many alcohols are readily available and inexpensive.

In planning an organic synthesis, first examine the starting compound(s) and the product to determine the relationship of the carbon content in the desired compound to that in the starting material. Then thinking often follows one or more of the following paths:

1. *If the starting material and desired compound contain the same number of carbon atoms*, then consider mainly reactions that convert one functional group to another.

2. *If the carbon content increases in going from starting material to product*, then think about reactions that involve the making of carbon-carbon bonds. A summary of eight important reactions of this type follows:

(a) $RX + R'_2CuLi \longrightarrow$ R—R' Lithium dialkylcopper reagent, Sec. 4.2A

R must be methyl or   May be symmetrical
primary alkyl   or unsymmetrical

(b) $2RX \xrightarrow[\text{heat}]{\text{Na metal}}$ R—R Wurtz reaction, Sec. 4.2A

Doubles
carbon content

(c) R—$C{\equiv}\overset{\ominus}{C}{:}Na^{\oplus}$ + $R'X \longrightarrow R$—$C{\equiv}C$—R' Sec. 11.9

R′ must be methyl
or primary alkyl

(d) $RX + ArH \xrightarrow{\text{AlCl}_3}$ Ar—R Friedel-Crafts alkylation, Sec. 14.6

Rearrangement of
R— may occur

(e) $R\overset{\overset{\displaystyle O}{\|}}{-C}-Cl + ArH \xrightarrow{\text{AlCl}_3} Ar\overset{\overset{\displaystyle O}{\|}}{-C}-R$ Friedel-Crafts acylation, Sec. 20.8

(f) $R\overset{\overset{\displaystyle O}{\|}}{-C}-Cl + R'_2Cd \longrightarrow R\overset{\overset{\displaystyle O}{\|}}{-C}-R'$ Sec. 20.8

(g) R—$C{\equiv}\overset{..}{N} + R'MgX \longrightarrow R\overset{\overset{\displaystyle O}{\|}}{-C}-R'$ Sec. 20.8

(h) $RMgX +$ aldehyde or ketone $\longrightarrow$ alcohol   Sec. 21.10

May use alkyl or aryl groups; structure of product depends
on choice of reagents

3. *If the carbon content decreases in going from starting material to product*, think of reactions that break a carbon-carbon bond. The majority of the reactions we have studied all involve breaking a carbon-carbon double bond, for example:

$$R_2C{=}CR_2 \xrightarrow[\text{2. } H_2O, Zn]{\text{1. } O_3} R_2C{=}O + O{=}CR_2$$

Aldehydes and ketones

$$\Bigg\downarrow \begin{array}{l} KMnO_4, :\ddot{O}H^\ominus, \\ H_2O, \text{ heat} \end{array}$$

$$R_2C{=}O + O{=}CR_2$$

Carboxylic acids, $R\overset{\displaystyle O}{\overset{\|}{C}}{-}OH$, formed when
one R = H; CO_2 when both R's = H

Two additional example of carbon-carbon bond cleavage we have seen or will see are the cleavage of glycols by periodic acid (Sec. 20.7C) and the cleavage of methylcarbinols or methyl-carbonyls in the haloform reaction (Sec. 24.6):

$$R{-}\overset{\overset{\displaystyle R'}{\overset{|}{}}}{\underset{\underset{\displaystyle OH}{|}}{C}}{-}\overset{\overset{\displaystyle R''}{\overset{|}{}}}{\underset{\underset{\displaystyle OH}{|}}{C}}{-}R''' \xrightarrow{HIO_4} R{-}\overset{\displaystyle O}{\overset{\|}{C}}{-}R' + R''{-}\overset{\displaystyle O}{\overset{\|}{C}}{-}R'''$$

$$R{-}\overset{\displaystyle O}{\overset{\|}{C}}{-}CH_3 \xrightarrow[H_2O]{I_2, \ H\ddot{O}{:}^\ominus} R{-}\overset{\displaystyle O}{\overset{\|}{C}}{-}\ddot{O}{:}^\ominus + CHI_3(s) \quad \text{Haloform reaction}$$

As we did in solving examples of syntheses, *working backward* is usually quite important: syntheses require that you know all the reactions quite well. In preparing an alcohol it is useful to classify it as primary, secondary, or tertiary because this suggests whether an aldehyde or a ketone should be used in the Grignard reaction. The examples in Sec. 21.11 illustrate this approach. Finally, work as many synthesis problems as possible on your own; practice is highly important in becoming proficient in this area.

21.18 Summary of Reactions of Aldehydes and Ketones

A. General Types of Reactions

1. Nucleophilic addition (Sec. 21.1A)

$$R{-}\overset{\displaystyle O}{\overset{\|}{C}}{-}R' + H{-}Nu \xrightarrow[\text{or neutral conditions}]{\text{acidic, basic}} R{-}\overset{\overset{\displaystyle OH}{\overset{|}{}}}{\underset{\underset{\displaystyle R'}{|}}{C}}{-}Nu$$

Aldehyde
or ketone

Reactivities (Sec. 21.2B)

(**a**) Aldehydes > ketones
(**b**) Aliphatic > aromatic
(**c**) Less hindered > more hindered
(**d**) Cyclic > noncyclic

2. Acidity of α hydrogens (Sec. 21.1B, see also Chap. 24)

$$R{-}\overset{\displaystyle O}{\overset{\|}{C}}{-}\overset{\overset{\displaystyle}{}}{\underset{\underset{\displaystyle H}{|}}{C}}{-} \underset{\longleftarrow}{\overset{\text{base}}{\rightleftharpoons}} \left[R{-}\overset{\displaystyle :O:}{\overset{\|}{C}}{-}\underset{\ominus}{\overset{|}{\ddot{C}}}{-} \longleftrightarrow R{-}\overset{\displaystyle :\ddot{O}:^\ominus}{\overset{|}{C}}{=}\overset{|}{C}{-} \right]$$

3. Electrophilic aromatic substitution: deactivating, *meta* directors (Sec. 21.1C)

Aldehyde
or ketone

Occurs only with sulfonation, nitration, and halogenation

B. Nucleophilic Addition Reactions

1. General mechanisms (Sec. 21.2A)
2. Stereochemistry of addition (Sec. 21.3)
3. Addition of hydrogen cyanide (Sec. 21.4)

Aldehyde   Cyanohydrin
or ketone

4. Addition of sodium bisulfite (Sec. 21.5)

Aldehyde   Sodium bisulfite
or ketone   addition compound
Also works with   *Water soluble*
cyclic ketones   *crystalline compound*

5. Addition of alcohols: acid-catalyzed (Sec. 21.6A)

Aldehyde   Hemiacetal   Acetal
or ketone   or hemiketal   or ketal

6. Addition of alcohols: base-catalyzed (Sec. 21.6B)

Aldehyde   Hemiacetal
or ketone   or hemiketal *only*

7. Addition of phenol to formaldehyde: condensation polymerization (Sec. 21.7)

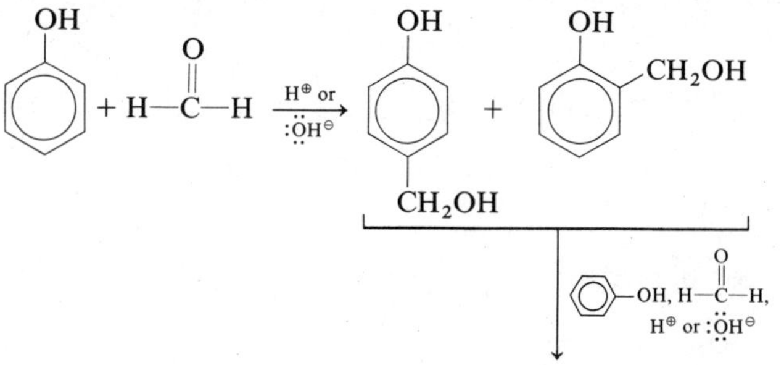
OH
O
H⊕ or
:ÖH⊖
CH₂OH
OH, H—C—H,
H⊕ or :ÖH⊖

Bakelite polymer
(a resin)

8. Addition of ammonia and substituted amines: used as derivatives for identification (Sec. 21.8)

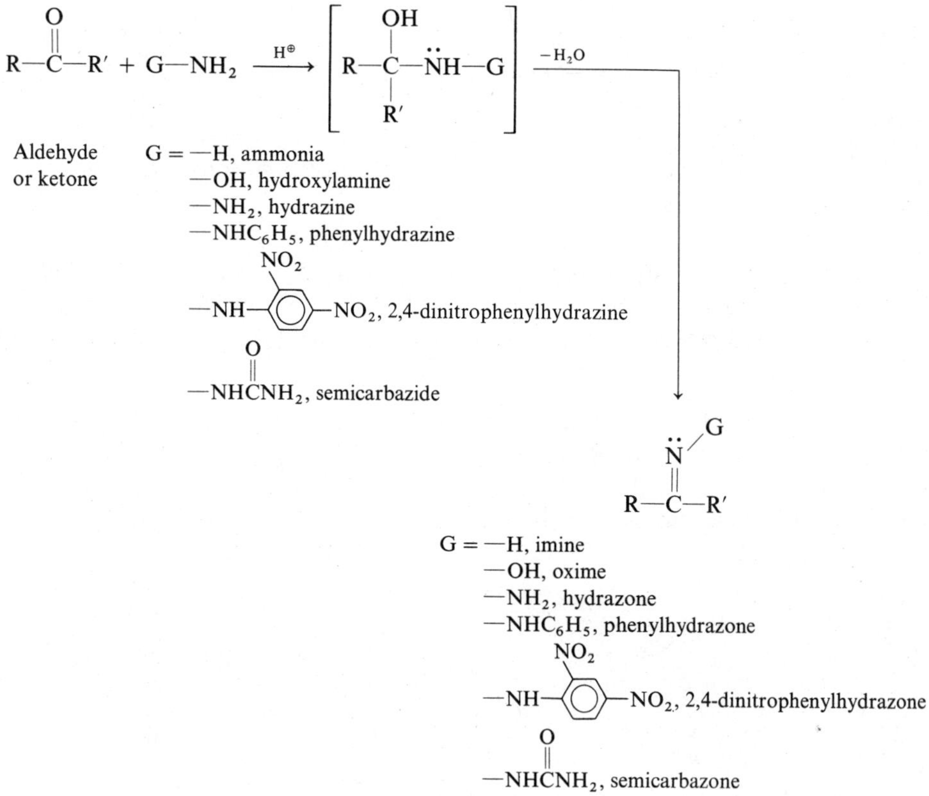
R—C—R′ + G—NH₂
H⊕
R—C—NH—G
OH
R′
−H₂O
Aldehyde
or ketone
G = —H, ammonia
—OH, hydroxylamine
—NH₂, hydrazine
—NHC₆H₅, phenylhydrazine
NO₂
—NH— —NO₂, 2,4-dinitrophenylhydrazine
—NHCNH₂, semicarbazide
G
N
R—C—R′
G = —H, imine
—OH, oxime
—NH₂, hydrazone
—NHC₆H₅, phenylhydrazone
NO₂
—NH— —NO₂, 2,4-dinitrophenylhydrazone
—NHCNH₂, semicarbazone

9. Addition of water and hydrogen halide

$$R{-}\overset{\displaystyle O}{\overset{\|}{C}}{-}R' + H_2O \rightleftharpoons R{-}\underset{\displaystyle R'}{\underset{|}{\overset{\displaystyle OH}{\overset{|}{C}}}}{-}OH \qquad \text{Sec. 21.9}$$

Hydrate
Usually unstable

Stable hydrates occur when strong electron-withdrawing substitutents (such as Cl or F) are α to carbonyl group:

$$\underset{\text{Not observed}}{R-\overset{\displaystyle O}{\overset{\|}{C}}-R' + HX \;\rightleftharpoons\; R-\overset{\displaystyle OH}{\underset{\displaystyle R'}{\overset{|}{\underset{|}{C}}}}-X} \qquad \text{Sec. 21.9}$$

10. Addition of Grignard reagent (Sec. 21.10)

$$\underset{\substack{\text{Aldehyde} \\ \text{or ketone}}}{R-\overset{\displaystyle O}{\overset{\|}{C}}-R'} + R''\colon^{\ominus}[MgX]^{\oplus} \xrightarrow[\text{ether}]{\text{anhydrous}} R-\overset{\displaystyle :\ddot{O}:^{\ominus}[MgX]^{\oplus}}{\underset{\displaystyle R'}{\overset{|}{\underset{|}{C}}}}-R'' \xrightarrow{H_3O^{\oplus}} \underset{\text{Alcohol}}{R-\overset{\displaystyle OH}{\underset{\displaystyle R'}{\overset{|}{\underset{|}{C}}}}-R''}$$

Products

Methanal (formaldehyde) → primary alcohol
Other aldehydes → secondary alcohols
Ketones → tertiary alcohols
Oxirane (ethylene oxide) → primary alcohol two carbons longer than original

Limitations

(*a*) Grignard reagent cannot be prepared from alkyl or aryl halides that contain acidic hydrogens; —OH, —COOH, —SO₃H, —NH₂ groups may not be present.
(*b*) Grignard reagent cannot be prepared from alkyl or aryl halides that contains

other multiply bonded functional groups: $-\overset{\displaystyle O}{\overset{\|}{C}}-H$, $-\overset{\displaystyle O}{\overset{\|}{C}}-R$, $-\overset{\displaystyle O}{\overset{\|}{C}}-OR$, $-C\equiv\ddot{N}$.

11. Addition of organolithium reagent (Sec. 21.12)

$$RX \xrightarrow[\substack{\text{anhydrous} \\ \text{ether}}]{Li} \underset{\substack{\text{Organolithium} \\ \text{reagent}}}{RLi}$$

$$\underset{\substack{\text{Aldehyde} \\ \text{or ketone}}}{R'-\overset{\displaystyle O}{\overset{\|}{C}}-R''} + RLi \xrightarrow[\text{ether}]{\text{anhydrous}} R-\overset{\displaystyle :\ddot{O}:^{\ominus}Li^{\oplus}}{\underset{\displaystyle R''}{\overset{|}{\underset{|}{C}}}}-R' \xrightarrow{H_3\ddot{O}^{\oplus}} R-\overset{\displaystyle OH}{\underset{\displaystyle R''}{\overset{|}{\underset{|}{C}}}}-R'$$

Works especially well with hindered ketones.

C. Hydrolysis of Acetals and Hemiacetals (reverse of addition reaction of alcohols discussed in Sec. 21.6)

$$\underset{\substack{\text{Acetal (R'} = \text{H)} \\ \text{or ketal}}}{R-\overset{\displaystyle OR''}{\underset{\displaystyle R'}{\overset{|}{\underset{|}{C}}}}-OR''} \xrightarrow{\text{excess } H_2O,\, H^{\oplus}} \underset{\substack{\text{Hemiacetal (R'} = \text{H)} \\ \text{or hemiketal}}}{R-\overset{\displaystyle OH}{\underset{\displaystyle R'}{\overset{|}{\underset{|}{C}}}}-OR''} \xrightarrow{\text{excess } H_2O,\, H^{\oplus}} \underset{\substack{\text{Aldehyde} \\ \text{or ketone}}}{R-\overset{\displaystyle O}{\overset{\|}{C}}-R'} \qquad \text{Sec. 21.6A}$$

D. Reduction Reactions

1. Carbonyl group to alcohol (Sec. 21.13)

$$R—\overset{\overset{\displaystyle O}{\|}}{C}—R' \xrightarrow[\text{agents}\equiv[H]]{\text{reducing}} R—\overset{\overset{\displaystyle OH}{|}}{\underset{\underset{\displaystyle H}{|}}{C}}—R'$$

Products

Aldehyde → primary alcohol
Ketone → secondary alcohol

Reducing agents

(*a*) Catalytic hydrogenation: H_2 and Pd, Pt, Ni, or $CuCrO_2$
(*b*) Chemical reduction: $LiAlH_4$ then $H_3O^{\oplus}$ or $NaBH_4$ then $H_3O^{\oplus}$

2. Carbonyl to methylene (Sec. 21.14)

$$R—\overset{\overset{\displaystyle O}{\|}}{C}—R' \xrightarrow[\text{agents}\equiv[H]]{\text{reducing}} R—CH_2—R'$$

Aldehyde
or ketone

Reducing agents

(*a*) Clemmensen reduction: Zn(Hg), HCl, heat
(*b*) Wolff-Kishner reduction: NH_2NH_2, $:\!\overset{..}{O}\!H^{\ominus}$, ethylene glycol, heat

E. Oxidation Reactions of Aldehydes (Sec. 21.16)

$$R—\overset{\overset{\displaystyle O}{\|}}{C}—H \begin{array}{c} \xrightarrow[\text{heat}]{K_2Cr_2O_7,\ H_2SO_4} \\ \xrightarrow[\text{heat}]{KMnO_4,\ :\overset{..}{O}H^{\ominus},\ H_2O} \\ \xrightarrow[\text{Tollens' test (Sec. 21.16)}]{Ag(NH_3)_2^{\oplus}} \end{array} R—\overset{\overset{\displaystyle O}{\|}}{C}—OH \quad \text{(or salt)}$$

Aldehyde   Carboxylic acid

Study Questions

21.29 Supply the missing reactants, reagents, and products, as required, to complete the following 26 equations. The items to be supplied are indicated by number [for example, (1)].

(*a*) Cyclopentene $\xrightarrow[\substack{\text{aqueous} \\ KMnO_4, \\ 25°}]{\text{dilute}}$ (1) $\xrightarrow{HIO_4}$ (2)

(*b*) Cyclohexanone $\xrightarrow[H^{\oplus}]{NH_2OH}$ (3)

(*c*) $CH_3—CH{=}CH—CH_2—\overset{\overset{\displaystyle O}{\|}}{C}—CH_3 \xrightarrow[\text{2. } H_3O^{\oplus}]{\text{1. } NaBH_4}$ (4)

(5) ↓

$$CH_3—\overset{\overset{\displaystyle O}{\|}}{C}—H + H—\overset{\overset{\displaystyle O}{\|}}{C}—CH_2—\overset{\overset{\displaystyle O}{\|}}{C}—CH_3$$

(6) → $CH_3CH_2CH_2CH_2—\overset{\overset{\displaystyle OH}{|}}{C}H—CH_3$

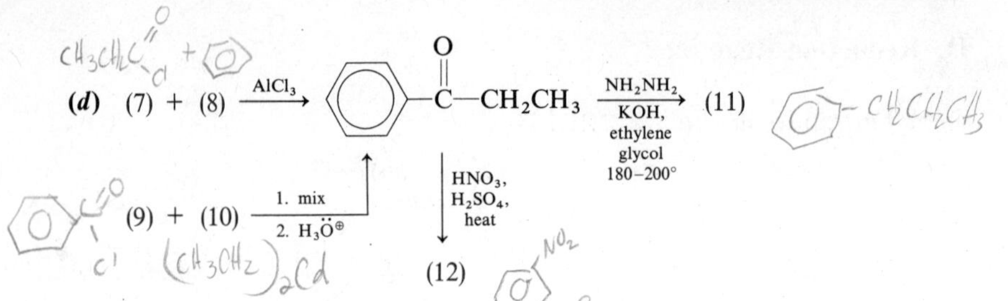

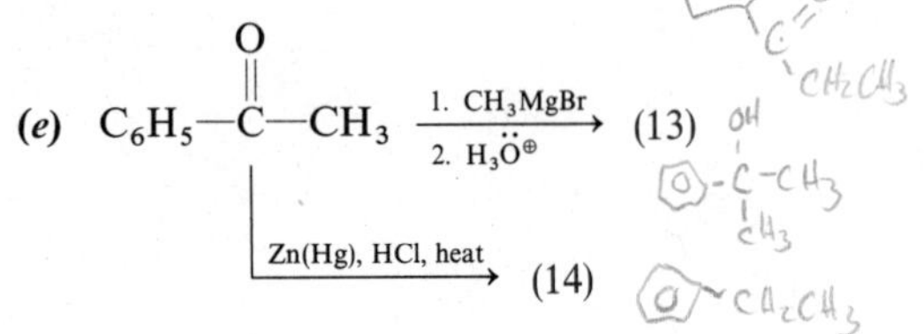

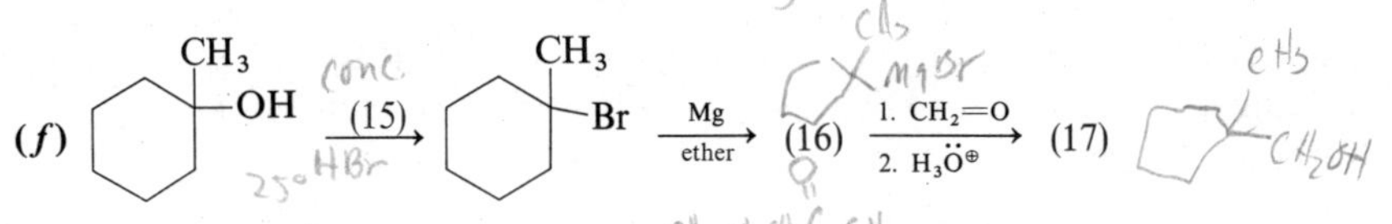

(d)   (7) + (8) $\xrightarrow{\text{AlCl}_3}$ C$_6$H$_5$—C(=O)—CH$_2$CH$_3$ $\xrightarrow[\substack{\text{KOH,}\\\text{ethylene}\\\text{glycol}\\180-200°}]{\text{NH}_2\text{NH}_2}$ (11)

(9) + (10) $\xrightarrow[\text{2. H}_3\ddot{\text{O}}^{\oplus}]{\text{1. mix}}$ ↑   (12)

$\xrightarrow[\text{heat}]{\substack{\text{HNO}_3,\\\text{H}_2\text{SO}_4,}}$ ↓

(e)   C$_6$H$_5$—C(=O)—CH$_3$ $\xrightarrow[\text{2. H}_3\ddot{\text{O}}^{\oplus}]{\text{1. CH}_3\text{MgBr}}$ (13)

$\xrightarrow{\text{Zn(Hg), HCl, heat}}$ (14)

(f)   [1-methylcyclohexan-1-ol, CH$_3$, —OH] $\xrightarrow{(15)}$ [1-methyl-1-bromocyclohexane, CH$_3$, —Br] $\xrightarrow[\text{ether}]{\text{Mg}}$ (16) $\xrightarrow[\text{2. H}_3\ddot{\text{O}}^{\oplus}]{\text{1. CH}_2=\text{O}}$ (17)

(g)   CH$_3$CH$_2$—C≡C—CH$_3$ $\xrightarrow[\text{HgSO}_4]{\text{H}_2\text{O, H}_2\text{SO}_4}$ (18) $\xrightarrow[\text{HCl, heat}]{\text{Zn(Hg)}}$ (19)

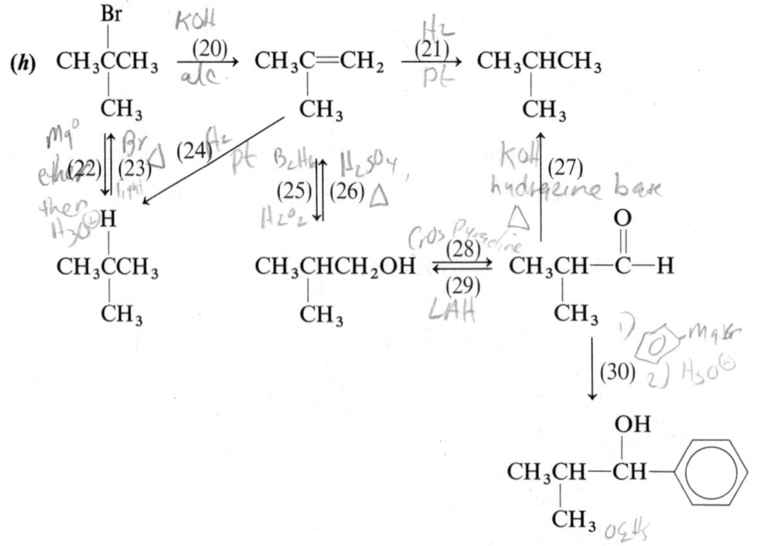

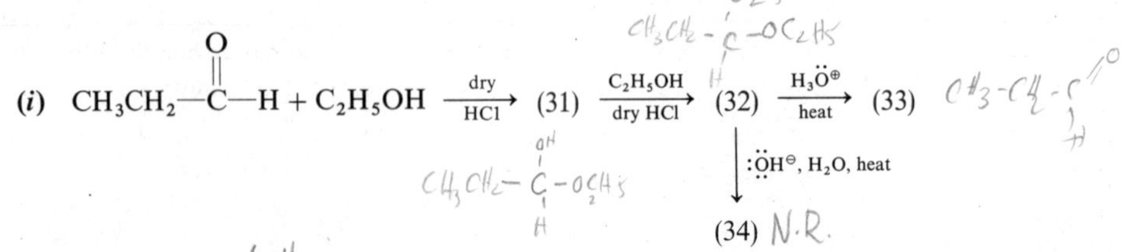

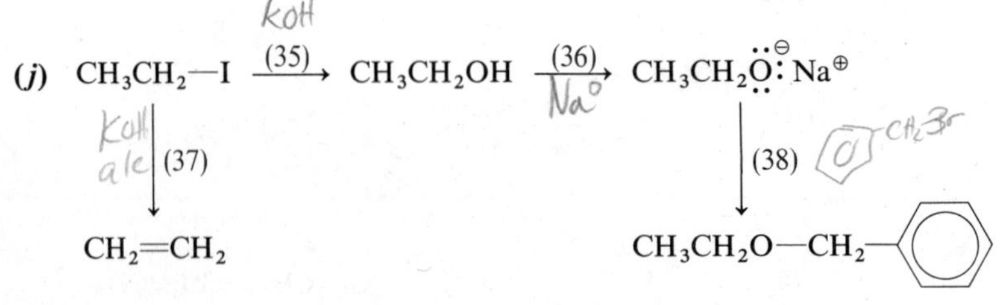

(h)   CH$_3$CCH$_3$ (with Br and CH$_3$) $\xrightarrow{(20)}$ CH$_3$C=CH$_2$ (with CH$_3$) $\xrightarrow[\text{Pt}]{(21)}$ CH$_3$CHCH$_3$ (with CH$_3$)

(22) ↕ (23)   (24)   (25) ↕ (26)   (27)

CH$_3$CCH$_3$ (with H and CH$_3$)   CH$_3$CHCH$_2$OH (with CH$_3$) $\underset{(29)}{\overset{(28)}{\rightleftharpoons}}$ CH$_3$CH—C—H (with CH$_3$ and =O)

$\downarrow$ (30)

OH
CH$_3$CH—CH—C$_6$H$_5$
CH$_3$

(i)   CH$_3$CH$_2$—C(=O)—H + C$_2$H$_5$OH $\xrightarrow[\text{HCl}]{\text{dry}}$ (31) $\xrightarrow[\text{dry HCl}]{\text{C}_2\text{H}_5\text{OH}}$ (32) $\xrightarrow[\text{heat}]{\text{H}_3\ddot{\text{O}}^{\oplus}}$ (33)

$\downarrow$ $:\ddot{\text{O}}\text{H}^{\ominus}$, H$_2$O, heat

(34)

(j)   CH$_3$CH$_2$—I $\xrightarrow{(35)}$ CH$_3$CH$_2$OH $\xrightarrow[\text{Na}]{(36)}$ CH$_3$CH$_2\ddot{\text{O}}:^{\ominus}$ Na$^{\oplus}$

(37) ↓   (38) ↓

CH$_2$=CH$_2$ CH$_3$CH$_2$O—CH$_2$—C$_6$H$_5$

(k) $CH_3CH_2-\overset{O}{\overset{\|}{C}}-CH_3$ $\xleftarrow[H_8SO_4]{\overset{H_2O}{\underset{(39)}{H_2SO_4}}}$ $CH_3CH_2C\equiv CH$ $\xleftarrow{(40)}$ CH_3CH_2-I $\xrightarrow[ether]{(41)}$ CH_3CH_2Li

[handwritten: $HC\equiv C^- Na^{\oplus}$ above (40); Li° above (41)]

1. $C_6H_5\overset{O}{\overset{\|}{C}}-H$
2. $H_3\ddot{O}^{\oplus}$

(42)

[handwritten: OH; ring$-\overset{}{\underset{H}{C}}-CH_2CH_3$]

(l) Cyclohexanone $\xrightarrow[\underset{(44)}{\overset{(43)}{}}]{}$ cyclohexanol

[handwritten: LAH above (43); CrO_3/H_2SO_4 below (44)]

(m) Ethylene $\xrightarrow{(45)}$ $\overset{O}{CH_2-CH_2}$ $\xrightarrow{(46)}$ $CH_3CH_2CH_2OH$

[handwritten: $C_6H_5CO_3H$ above (45); $CH_3MgBr/H_3O^{\oplus}$ above (46)]

(n) $CH_3CH=CH_2$ $\xrightarrow{(47)}$ $CH_3CH_2CH_2OH$ $\xrightarrow{(48)}$ $\begin{matrix} CH_3CH_2CH_2-O & & CH_3 \\ & \overset{}{\underset{}{C}} & \\ CH_3CH_2CH_2-O & & CH_3 \end{matrix}$

[handwritten: $(BH_3)_2/H_2O_2$ above (47); $(CH_3)_2C=O$, dry HCl gas below (48)]

(o) Ethylbenzene $\xrightarrow[NBS]{(49)}$ α-bromoethylbenzene

[handwritten: KOH alc. above (50)]

$\downarrow$ (50)

$C_6H_5-\overset{O}{\overset{\|}{C}}-H$ $\xleftarrow{(51)}$ $C_6H_5CH=CH_2$

[handwritten: $O_3/Zn H_2O$ above (51)]

(p) $(CH_3)_2CHCH_2MgBr + H_2C=O$ $\xrightarrow{mix}$ (52) $\xrightarrow{H_3\ddot{O}^{\oplus}}$ (53) *[handwritten: $(CH_3)_2CHCH_2CH_2OH$]*

(q) Acetone $\xrightarrow{(54)}$ $(CH_3)_2\overset{}{\underset{OH}{C}}-C(CH_3)_3$ $\xrightarrow[heat]{H^{\oplus}}$ (55) *[handwritten: $CH_3-\overset{}{\underset{CH_2}{C}}-C(CH_3)_3$]*

[handwritten: $(CH_3)_3C Li$ below (54)]

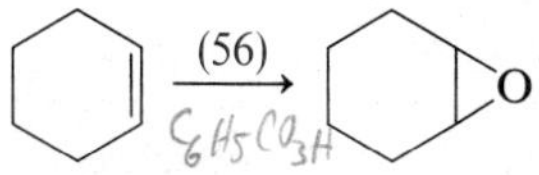

(r) [cyclohexene] $\xrightarrow{(56)}$ [epoxide O] $\xrightarrow{H_3\ddot{O}^{\oplus}}$ (57) (Indicate sterochemistry)

[handwritten: $C_6H_5CO_3H$ below (56); trans-diol structure OH + E, OH]

$\downarrow$ dilute basic KMnO$_4$, 25°

(58) (Indicate stereochemistry)

[handwritten: cis-diol structure OH OH at left]

(s) $CH_3\overset{}{\underset{CH_2CH_3}{CH}}MgBr + (59)$ $\xrightarrow[2.\ H_3\ddot{O}^{\oplus}]{1.\ mix}$ $CH_3CH_2-\overset{}{\underset{CH_2OH}{CH}}-CH_3$

[handwritten: $H_2C=O$ above (59)]

(t) [tricyclic enone structure, O=] $\xrightarrow[H_2O]{NaBH_4}$ (60)

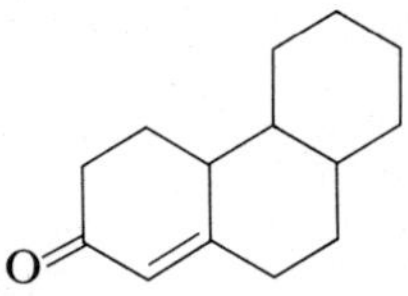

$\downarrow$ HOCHCH$_2$OH, dry HCl

(61) $\xrightarrow{H_2,\ Pt}$ (62) $\xrightarrow{H_3\ddot{O}^{\oplus}}$ (63)

(u) [decalin structure with OH] $\xrightarrow[warm]{K_2Cr_2O_7,\ H_2SO_4}$ (64)

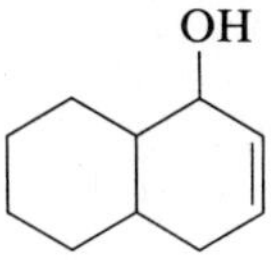

(**v**)   2-Butanone $\xrightarrow[\text{H}^{\oplus}]{\text{H}_2\text{NNH}-\overset{\overset{\text{O}}{\|}}{\text{C}}-\text{NH}_2}$ (65)

$\downarrow$ NaOH, I$_2$ $\searrow$ (67)

(66) $\underset{\text{CH}_3-\overset{\|}{\text{C}}-\text{CH}_2\text{CH}_3}{\overset{\text{N}}{\overset{\|}{}}}\overset{\text{NH}-\text{C}_6\text{H}_5}{}$

(**w**) $\text{CH}_3\text{CH}=\text{CH}-\overset{\overset{\text{O}}{\|}}{\text{C}}-\text{CH}_3$ $\xrightarrow[\text{CuCrO}_2]{\text{H}_2}$ (68)

$\downarrow$ H$_2$, Pt $\searrow$ 1. LiAlH$_4$, ether   2. H$_3\overset{..}{\text{O}}{}^{\oplus}$

(69)   (70)

(**x**)   Cyclopentanone $\xrightarrow{\text{NaHSO}_3}$ (71) $\xrightarrow{(72)}$ cyclopentanone

(**y**) $\text{CH}_2=\text{CH}-\overset{\overset{\text{O}}{\|}}{\text{C}}-\text{H}$ $\xrightarrow{(73)}$ $\text{CH}_2=\text{CH}-\overset{\overset{\text{O}\quad\text{O}}{\overline{}}}{\text{C}}-\text{H}$ $\xrightarrow[\substack{\text{H}_2\text{O},\,:\overset{..}{\text{O}}\text{H}^{\ominus}, \\ 25°}]{\text{KMnO}_4}$ (74)

$\downarrow$ H$_3\overset{..}{\text{O}}{}^{\oplus}$, warm

(75)

(**z**)   (76) $\xrightarrow{(77)}$ $\text{C}_6\text{H}_5\text{CO}\overset{..}{\underset{..}{\text{O}}}{:}^{\ominus}$ + CHI$_3$

21.30 Draw the structure of the product formed by treating each of the following carbonyl compounds with the Grignard reagent derived from *n*-butyl bromide (followed by hydrolysis).
(**a**) acetone   (**b**) formaldehyde   (**c**) benzaldehyde
(**d**) cyclopentanone   (**e**) ethylene oxide

21.31 Indicate *two* possible ways to prepare *n*-butyl alcohol from a Grignard reagent and another organic compound.

21.32 What carbonyl compound and Grignard reagent are required to prepare each of the following alcohols?
(**a**) 1-butanol   (**b**) 2-butanol   (**c**) 2-methyl-2-butanol

21.33 Show all the steps, conditions, and reagents required to carry out the following 27 transformations, using the indicated starting materials as the only source of carbon. All intermediate organic compounds must be prepared from them, except for common organic solvents (ether, CCl$_4$, and so on). All inorganic reagents are available.

(**a**)   2-Propanol + C$_1$ or C$_2$ compounds $\longrightarrow$ $\text{CH}_3\text{CH}_2\text{CH}_2-\overset{\overset{\text{OH}}{|}}{\underset{\underset{\text{CH}_3}{|}}{\text{C}}}-\text{CH}_3$

(**b**)   2-Propanol $\longrightarrow$ $\text{CH}_3\text{CH}_2\text{CH}_2-\overset{\overset{\text{OH}}{|}}{\underset{\underset{\text{CH}_3}{|}}{\text{C}}}-\text{CH}_3$

(**c**)   1-Propanol + C$_1$ or C$_2$ compounds $\longrightarrow$ $\text{CH}_3\text{CH}_2-\overset{\overset{\text{O}}{\|}}{\text{C}}-\text{CH}_2\text{CH}_3$

(**d**)   1-Propanol + C$_1$ or C$_2$ compounds $\longrightarrow$ $\text{CH}_3\text{CH}_2\text{CH}_2\text{CH}_2-\overset{\overset{\text{O}}{\|}}{\text{C}}-\text{H}$

(*e*) Cyclopentene + C_1 or C_2 compounds $\longrightarrow$ (cyclopentyl)—CH_2OH

(*f*) (methylenecyclobutane) + C_1 or C_2 compounds $\longrightarrow$ (cyclobutyl)—CH_2—CH—CH_3 with OH

(*g*) C_2 compounds $\longrightarrow$ CH_3CH_2—$\overset{\overset{\displaystyle O}{\|}}{C}$—$CH_2CH_2CH_3$

(*h*) Cyclopentene + C_2 compounds $\longrightarrow$ (cyclopentenyl)—CH—CH_3 with OH

(*i*) Cyclopentene $\longrightarrow$ (bicyclopentylidene)

(*j*) Cyclohexane + C_2 compounds $\longrightarrow$ (cyclohexyl)—CH_2CH_2OH

(*k*) 2-Propanol $\longrightarrow$ CH_3—$\overset{\overset{\displaystyle OH}{|}}{\underset{\underset{\displaystyle CH_3}{|}}{C}}$—$C\equiv N$

(*l*) 1-Bromopropane + C_2 compounds $\longrightarrow$ $CH_3\underset{\underset{\displaystyle CH_3}{|}}{CH}CH_2$—$\overset{\overset{\displaystyle O}{\|}}{C}$—$H$

(*m*) Benzaldehyde $\longrightarrow$ $C_6H_5CH_2Br$

(*n*) 2-Butanol + 1-propanol $\longrightarrow$ CH_3CH_2—$\overset{\overset{\displaystyle CH_3}{|}}{\underset{\underset{\displaystyle OH}{|}}{C}}$—$CH_2CH_2CH_3$

(*o*) Acetaldehyde $\longrightarrow$ $CH_3\overset{\overset{\displaystyle OH}{|}}{CH}$—$\overset{\overset{\displaystyle O}{\|}}{C}$—$OH$

(*p*) Benzaldehyde $\longrightarrow$ C_6H_5—$\overset{\overset{\displaystyle OCH_2C_6H_5}{|}}{CH}$—$OCH_2C_6H_5$

(*q*) 2-Butanol + C_1 or C_2 compounds $\longrightarrow$ CH_3CH_2—$\overset{\overset{\displaystyle H}{|}}{\underset{\underset{\displaystyle HO-CH-CH_3}{|}}{C}}$—$CH_3$

(*r*) Cyclopentanone $\longrightarrow$ H—$\overset{\overset{\displaystyle O}{\|}}{C}$—$CH_2CH_2CH_2$—$\overset{\overset{\displaystyle O}{\|}}{C}$—$H$

(*s*) Ethanol $\longrightarrow$ 2-butanone

(*t*) Ethyne $\longrightarrow$ 2-butanone

(*u*) $CH_3\underset{\underset{\displaystyle CH_3}{|}}{CH}CH_2$—$\overset{\overset{\displaystyle O}{\|}}{C}$—$OH$ + C_1 or C_2 compounds $\longrightarrow$ $CH_3\underset{\underset{\displaystyle CH_3}{|}}{CH}CH_2$—$\overset{\overset{\displaystyle O}{\|}}{C}$—$CH_2CH_3$

(*v*) Propanol $\longrightarrow$ 2-bromopropane

(*w*)   2-Propanone $\longrightarrow$ 1-bromopropane

(*x*)   2-Propanol + C_1 or C_2 compounds $\longrightarrow$ 2-methylpropane

(*y*)   1-Bromo-2-methylpropane $\longrightarrow (CH_3)_2CH-\overset{\displaystyle O}{\overset{\displaystyle \|}{C}}-H$

(*z*)   1-Propanol $\longrightarrow CH_3CH_2CH{=}CH_2$ (must contain no 2-butene)

(*aa*)   1-Propanol $\longrightarrow$ *cis*-2-hexene

21.34 Devise suitable synthetic methods for preparing each of the following compounds, using as the only starting materials alcohols containing four carbon atoms or less and benzene. Indicate reactants and approximate reaction conditions, but do not balance equations or give mechanisms.

(*a*)

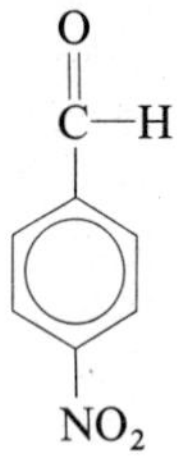

(*b*)   Br—⬡—$\overset{\displaystyle OH}{\underset{\displaystyle CH_3}{\overset{\displaystyle |}{\underset{\displaystyle |}{C}}}}$—$CH_3$

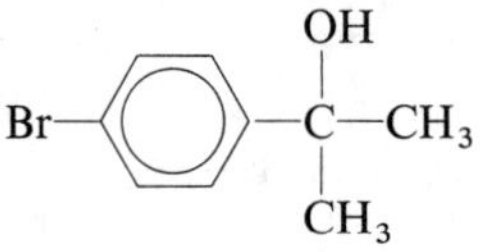

(*c*) $CH_3-\overset{\displaystyle Br}{\underset{\displaystyle CH_3}{\overset{\displaystyle |}{\underset{\displaystyle |}{C}}}}-CH_2CH_2CH_3$

(*d*)   ⬡—$\overset{\displaystyle O}{\overset{\displaystyle \|}{C}}$—$CH_2\overset{\displaystyle CH_3}{\overset{\displaystyle |}{C}}HCH_3$

(*e*)   ⬡—$CH_2CH_2\overset{\displaystyle CH_3}{\overset{\displaystyle |}{C}}HCH_3$

(*f*) $CH_3\underset{\displaystyle CH_2CH_3}{\overset{\displaystyle |}{C}}HCH_2CH_2CH_3$

(*g*) $CH_3\overset{\displaystyle OH}{\overset{\displaystyle |}{C}}HCH_2CH_2\overset{\displaystyle CH_3}{\overset{\displaystyle |}{C}}HCH_3$

(*h*) $CH_3CH_2-O-\overset{\displaystyle CH_3}{\underset{\displaystyle CH_2CH_3}{\overset{\displaystyle |}{\underset{\displaystyle |}{C}}}}-CH_2CH_3$

(*i*) $CH_2{=}C\overset{\displaystyle \diagup CH_2CH_3}{\diagdown CH_2CH_3}$

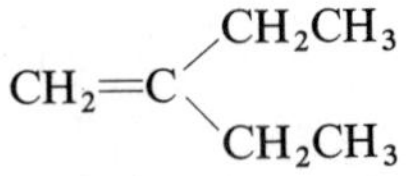

(*j*)

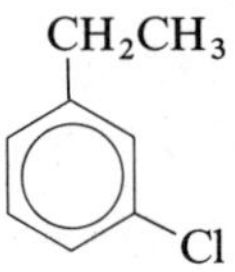

(*k*)

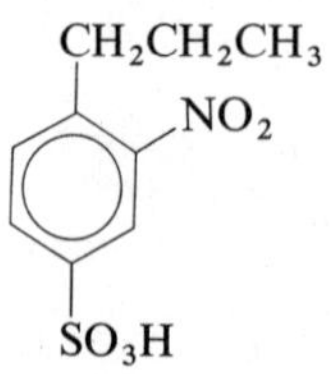

(*l*) CH_3—⬡—$\overset{\displaystyle OH}{\underset{\displaystyle CH_3}{\overset{\displaystyle |}{\underset{\displaystyle |}{C}}}}$—$CH_2CH_3$

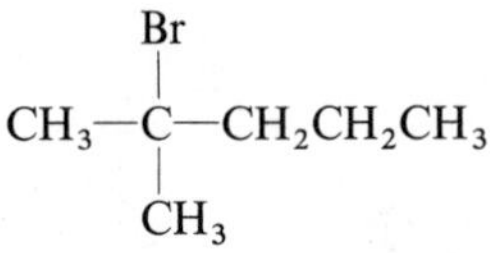

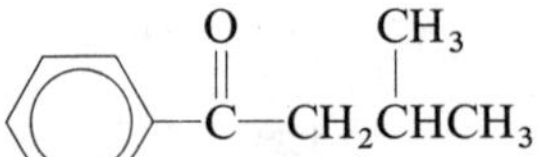

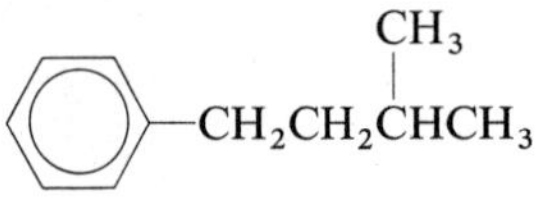

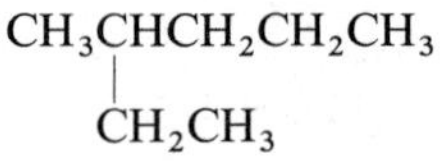

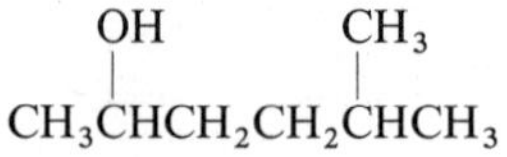

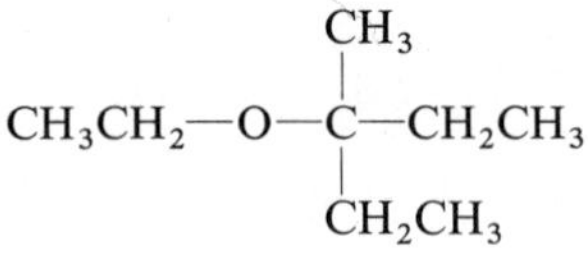

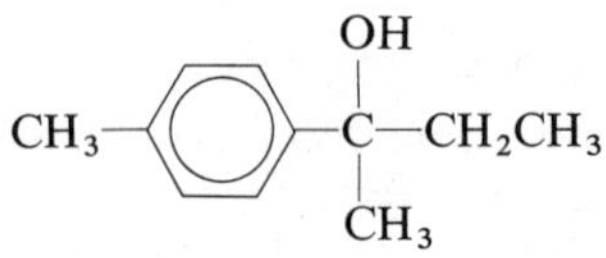

(*m*)

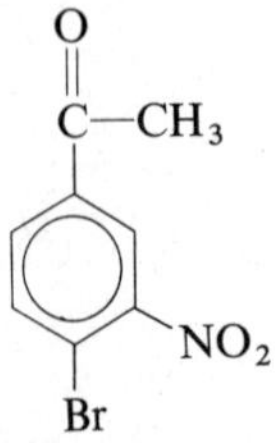

(*n*)

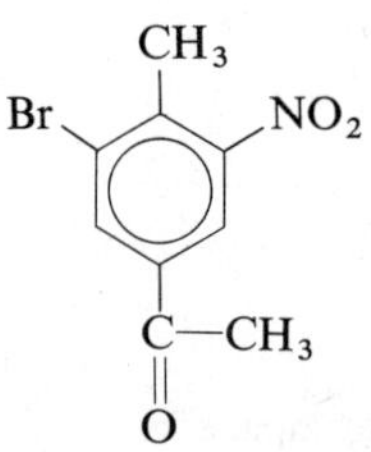

(o) [structure: benzene ring with C(=O)—CH₃, —CH₂CH₃, and NO₂ substituents]

(p) CH₃—[benzene ring]—C(=O)—[benzene ring with NO₂]

(q) $C_6H_5CH_2\overset{O}{\underset{\|}{C}}-H$

21.35 For each of the following pairs of compounds, indicate the member that would be more reactive toward the reagent indicated; give the structure of the product.

(a) formaldehyde or acetaldehyde toward sodium bisulfite

(b) acetophenone or benzophenone toward phenylhydrazine

(c) cyclohexanone or 3-hexanone toward semicarbazide

(d) acetone or diisopropyl ketone toward a mixture of sodium cyanide and hydrogen cyanide

(e) benzaldehyde or phenylacetaldehyde toward hydroxylamine

21.36 Provide complete, stepwise mechanisms to account for the following eight reactions. Show all intermediates.

(a) [structure: cyclic compound CH₂—CH₂, CH₂, C—H with OH and O] $\xrightarrow[\text{dry HCl}]{CH_3OH}$ [tetrahydrofuran ring with —OCH₃]

(b) $CH_3CH_2\overset{^{16}O}{\underset{\|}{C}}-CH_3 + H_2{}^{18}O \longrightarrow CH_3CH_2\overset{^{18}O}{\underset{\|}{C}}-CH_3 + H_2{}^{16}O$

What catalyst, if any, is needed?

(c) When *indene oxide* (1) is heated with aqueous sulfuric acid, ketone (2) forms in a high yield. Very little (if any) ketone with structure (3) is formed. In addition to rationalizing the formation of (2), explain why (3) is *not* formed.

[structure (1): indene oxide] $\xrightarrow[\text{heat}]{aq\ H_2SO_4}$ [structure (2)] No [structure (3)] is formed

(1)   (2)   (3)
          Major product

(d) [cyclobutane with C(=O)—H] $\xrightarrow[\text{heat}]{H^{\oplus}}$ [cyclopentanone]

(e) [cyclohexane with C=CH—O—CH₃] + $H_2{}^{18}O$ $\xrightarrow[\text{warm}]{H^{\oplus}}$ [cyclohexane with $\overset{^{18}O}{\underset{\|}{C}}$—H] + CH_3OH

Vinyl ether

(f) $C_6H_5CH_2CH_2CH_2\overset{O}{\underset{\|}{C}}-Cl$ $\xrightarrow{AlCl_3}$ [bicyclic ketone structure]

(g) $C_6H_5\overset{^{18}O}{\underset{\|}{C}}-H + CH_3OH$ $\xrightarrow{\text{dry HCl}}$ $C_6H_5-\overset{OCH_3}{\underset{OCH_3}{C}}-H + H_2O$

(Indicate where ${}^{18}O$ ends up)

(*h*) $(CH_3)_2C{=}O + CHCl_3 \xrightarrow[\text{H}_2\text{O}]{:\ddot{O}H^{\ominus}} CH_3{-}\overset{\overset{\displaystyle OH}{|}}{\underset{\underset{\displaystyle CH_3}{|}}{C}}{-}CCl_3$

21.37 For each of the following pairs of compounds, designate a simple chemical test to distinguish one member of the pair from the other. Choose only tests that give visible results (that is, those that are readily observed in the laboratory). Write equations for the reactions that occur and indicate what is observed.

(*a*) 3-pentanol and 3-pentanone

(*b*) ethylbenzene and acetophenone

(*c*) —C—H and cyclohexene

(*d*) $C_6H_5C{\equiv}CH$ and benzaldehyde

(*e*) $C_6H_5C{\equiv}CH$ and $C_6H_5CH{=}CH_2$

(*f*) 1-pentyne and 2-pentyne

(*g*) phenol and cyclohexanol

(*h*) phenol and 2,4,6-trimethylphenol

(*i*) 1-pentanol and $CH_3CH_2CH_2CH_2CH(OC_2H_5)_2$

(*j*) $CH_3{-}CH{=}CH{-}\overset{\overset{\displaystyle O}{\|}}{C}{-}H$ and $CH_3CH_2CH_2{-}\overset{\overset{\displaystyle O}{\|}}{C}{-}H$

(*k*) cyclohexanone and cyclohexanone ethylene ketal

(*l*) 1,2-dimethoxybutane and 2,2-dimethoxybutane

(*m*) 2-pentanone and 2-pentanol

21.38 Compound A, C_3H_7Br, reacts with magnesium metal in anhydrous ether to give compound B. B, on treatment with heavy water, D_2O, forms $CH_3{-}CHD{-}CH_3$. What is the structure of A? Write equations for the reactions that occur.

21.39 An alkyl bromide A reacts with magnesium metal in dry ether to form Grignard reagent B. Treatment of B with heavy water, D_2O, gives a monodeuterated compound C. When C is brominated (Br_2, 125°, *hv*), a new alkyl bromide D is formed, which still contains deuterium; A and D are isomeric. Treatment of Grignard reagent B with water gives 2-methylpropane. On the basis of this information, draw structures for A to D. Indicate your reasoning and write equations for the reactions that occur.

21.40 Alkyl bromide A reacts with magnesium in dry ether to give Grignard reagent B. Treatment of B with heavy water, D_2O, yields product C. When A is treated with sodium metal, a single product is formed and identified as 3,4-dimethylhexane. Write structures for A to C, and give equations for the reactions that occur.

21.41 Two compounds, A and B, are isomeric and have the molecular formula C_4H_9Cl. Both A and B react with magnesium metal in anhydrous ether to give the corresponding Grignard reagents. The Grignard reagents from A and from B, on treatment with water, both give the same alkane C, C_4H_{10}.

A, when heated with chlorine gas, gives only one dichloro product D, $C_4H_8Cl_2$. B, when heated with chlorine gas, gives only three dichloro products, each having the molecular formula $C_4H_8Cl_2$.

On the basis of the information given, draw the structures of compounds A to D. Also, draw the structures of the dichloro compounds arising from compound B. *Give your reasoning!* Indicate what the observations teach you about the structures of the lettered compounds.

21.42 (*a*) Lithium aluminum hydride, $LiAlH_4$, reacts with ethylene oxide in anhydrous ether, and on hydrolysis of the reaction mixture, ethanol is obtained. Suggest a plausible mechanism for this observation. (*b*) What product do you think would be obtained from the LAH reduction of 2,2-diphenylethylene oxide after the mixture is hydrolyzed? Explain.

21.43 (*a*) The alkynide anion, $R—C{\equiv}C\colon^{\ominus} Na^{\oplus}$, undergoes many chemical reactions that are observed for the Grignard reagent. What product do you think would be obtained from this anion and a carbonyl compound $R'_2C{=}O$ (after hydrolysis)? Explain. (*b*) Explain why 3-methyl-1-butyn-3-ol (1) is commercially available at an amazingly low price. What simple method can be used to prepare (1)?

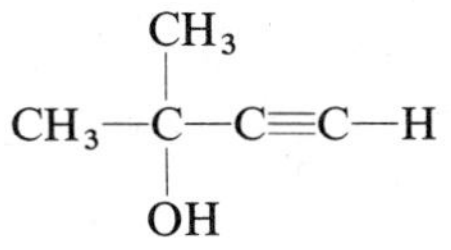

$$CH_3—\overset{\overset{\displaystyle CH_3}{|}}{\underset{\underset{\displaystyle OH}{|}}{C}}—C{\equiv}C—H$$

(1)

21.44 Lithium aluminum deuteride, $LiAlD_4$, is commercially available and reacts with carbonyl compounds in the same way that LAH does. Predict the product of each of the following reactions.

(*a*) Acetophenone + $LiAlD_4$ in ether, followed by hydrolysis with $H_3\ddot{O}^{\oplus}$

(*b*) Acetophenone + $LiAlD_4$ in ether, followed by hydrolysis with $D_3\ddot{O}^{\oplus}$

21.45 A possible route sometimes used to replace the hydroxyl group (in alcohols) by hydrogen is the sequence:

$$R—OH \xrightarrow[\text{base}]{\text{TsCl}} R—OTs \xrightarrow[\text{2. } H_3\ddot{O}^{\oplus}]{\text{1. } LiAlH_4} R—H$$

Alcohol   Tosylate

(*a*) Suggest a mechanistic explanation for the replacement of the tosylate group by hydrogen.
(*b*) Suppose optically active 1-phenylethanol is subjected to the preceding reactions, using lithium aluminum deuteride, $LiAlD_4$, in place of $LiAlH_4$. What product would be formed? What stereochemistry would be observed in the overall transformation?

21.46 In light of your answers to Study Questions 21.44 and 21.45, suggest synthetic methods for converting propiophenone, $C_6H_5—COCH_2CH_3$, into the following isotopic labeled compounds.

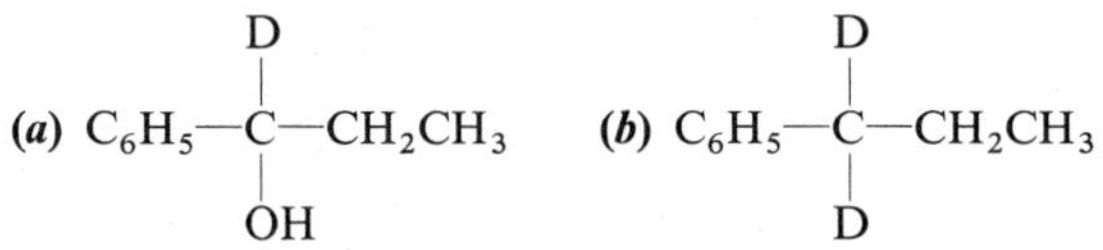

(*a*) $C_6H_5—\overset{\overset{\displaystyle D}{|}}{\underset{\underset{\displaystyle OH}{|}}{C}}—CH_2CH_3$ (*b*) $C_6H_5—\overset{\overset{\displaystyle D}{|}}{\underset{\underset{\displaystyle D}{|}}{C}}—CH_2CH_3$

21.47 In some respects an α,β-unsaturated aldehyde or ketone resembles a conjugated diene, and in other ways these two types of compounds are different. Explain the following similarities and differences, using mechanistic interpretations freely.

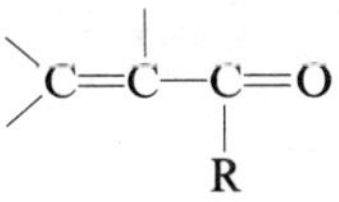

$$\overset{\diagup}{\underset{\diagup}{C}}{=}\overset{\overset{\displaystyle |}{}}{C}—\overset{\overset{\displaystyle |}{}}{\underset{\underset{\displaystyle R}{|}}{C}}{=}O$$

α,β-Unsaturated aldehyde (R = H)
or ketone (R = alkyl or aryl)

(*a*) In the base-promoted elimination reaction, the following reactions occur and are therefore similar:

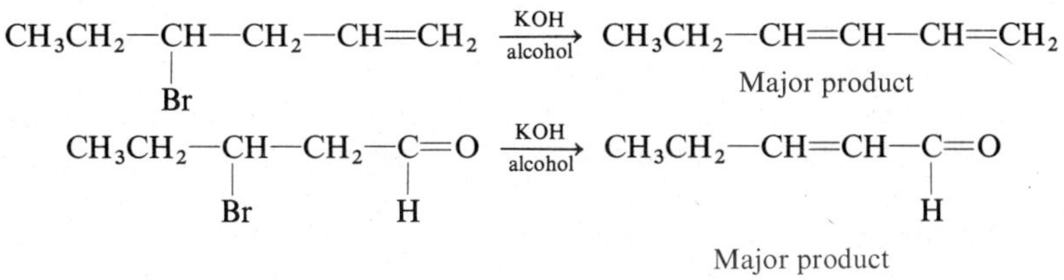

$$CH_3CH_2—\underset{\underset{\displaystyle Br}{|}}{CH}—CH_2—CH{=}CH_2 \xrightarrow[\text{alcohol}]{\text{KOH}} CH_3CH_2—CH{=}CH—CH{=}CH_2$$

Major product

$$CH_3CH_2—\underset{\underset{\displaystyle Br}{|}}{CH}—CH_2—\underset{\underset{\displaystyle H}{|}}{C}{=}O \xrightarrow[\text{alcohol}]{\text{KOH}} CH_3CH_2—CH{=}CH—\underset{\underset{\displaystyle H}{|}}{C}{=}O$$

Major product

(*b*) In the ionic addition of electrophilic reagents, the following reactions occur and are therefore different:

$$CH_3—CH{=}CH—CH{=}CH_2 \xrightarrow[CCl_4]{HBr} CH_3—\underset{\underset{H}{|}}{CH}—\underset{\underset{Br}{|}}{CH}—CH{=}CH_2 + CH_3—\underset{\underset{H}{|}}{CH}—\underset{\underset{Br}{|}}{CH}—CH{=}CH_2$$

$$CH_3—CH{=}CH—\underset{\underset{H}{|}}{C}{=}O \xrightarrow[CCl_4]{HBr} CH_3—\underset{\underset{Br}{|}}{CH}—\underset{\underset{H}{|}}{CH}—\underset{\underset{H}{|}}{C}{=}O$$

Only product

21.48 (*a*) *cis*-1,2-Cyclopentanediol reacts with acetone in the presence of dry HCl gas to give a compound with the molecular formula $C_8H_{14}O_2$. On treatment with excess acidified water, this compound regenerates the starting materials. Indicate the reactions that are occurring here. (*b*) When *cis*-1,2-cyclopentanediol is replaced by the *trans* isomer, no reaction occurs with acetone in the presence of dry HCl. Why?

21.49 Compound *A* gives a yellow to dark red precipitate on treatment with 2,4-DNPH. Reaction with $Ag(NH_3)_2^{\oplus}$ gives no visible reaction. This compound gives the following spectral data:

MS:
  $P^{\oplus}$ (molecular ion) m/e ratio $= 100 \pm 2$
NMR:
  1.8δ (6H) *bs*
  2.25δ (4H) *bs*
What is the structure of *A*?

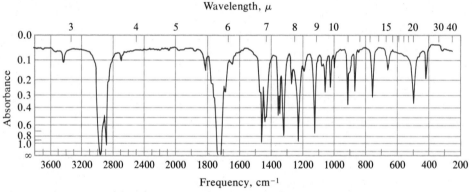

PROBLEM 21.49   IR spectrum of unknown compound (*A*). [Data from *Indexes to Evaluated Infrared Reference Spectra*, copyright 1975, The Coblentz Society, Norwalk CT]

21.50 Compound *A*, $C_{13}H_{18}$, is *optically active* and reacts readily with bromine in carbon tetrachloride to give compound *B*, $C_{13}H_{18}Br_2$. *A* is treated with ozone and then with zinc and water to give two new compounds, *C* and *D*.

  C, $C_5H_{10}O$, is *optically active* and it reacts with hydroxylamine and phenylhydrazine. *C* also gives a silver mirror with Tollens' reagent.

  D, C_8H_8O, is not optically active, but it reacts with hydroxylamine and phenylhydrazine. The NMR and IR spectra of *D* are provided here. *D*, on treatment with either hydrazine (NH_2NH_2) and base or with Zn(Hg) and hydrochloric acid, gives compound *E*, C_8H_{10}, which is identified as ethylbenzene.

(*a*) Draw the structures of compounds *A* to *E*, and briefly indicate your reasoning in reaching these answers. Be careful to indicate the source of the optical activity in those compounds that are optically active.

(*b*) Is the preceding information enough for you to provide a completely unambiguous structure for compound *A*? Briefly explain why or why not. If not, show what ambiguity exists.

(Hz)

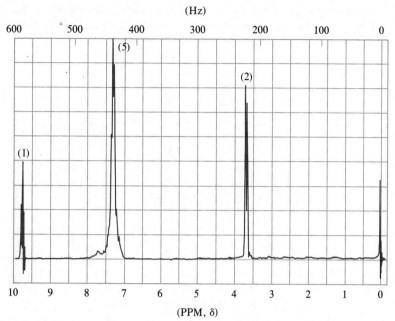

PROBLEM 21.50   NMR spectrum of compound (*D*). [Data from *The Aldrich Library of NMR Spectra*, C. R. Pouchert and J. R. Campbell (eds.), copyright 1974, Aldrich Chemical Co., Milwaukee]

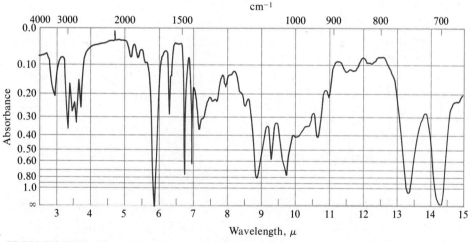

PROBLEM 21.50   IR spectrum of compound (*D*). [Data from *Indexes to Evaluated Infrared Reference Spectra*, copyright 1975, The Coblentz Society, Norwalk CT]

21.51 Compound *A*, $C_{14}H_{20}$, is an *optically active, aromatic* compound, which decolorizes bromine in carbon tetrachloride. *A*, when treated with ozone and then with zinc and water, gives two new compounds, *B* and *C*.

Compound *B*, $C_9H_{10}O$, is *optically active* and contains an *aromatic* ring. *B* gives a silver mirror when treated with Tollens' reagent, $Ag(NH_3)_2^{\oplus}$. Compound *C*, $C_5H_{10}O$, is optically inactive and unaffected by Tollens' reagent. When treated with lithium aluminum hydride and then with $H_3O^{\oplus}$, *C* gives compound *D*, $C_5H_{12}O$. When *D* is heated with concentrated H_2SO_4, it gives compound *E*, C_5H_{10}. The NMR spectrum of *D* is provided. *E*, when treated with ozone and then with zinc and water, gives *equimolar* amounts of acetaldehyde and propionaldehyde.

On the basis of these data, draw the structures of compounds *A* to *E*. Indicate the source of the optical activity in those compounds that are described here. (*Note:* Ignore the possibility of *geometric* isomers throughout the problem.)

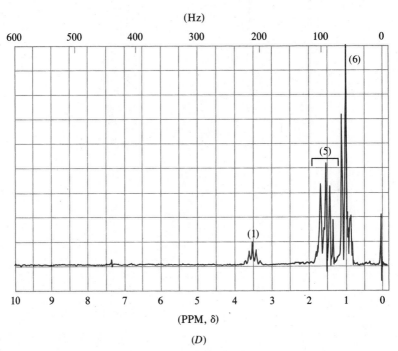

(*D*)

PROBLEM 21.50   NMR spectrum of compound (*D*). [Data from *The Aldrich Library of NMR Spectra*, C. J. Pouchert and J. R. Campbell (eds.), copyright 1974, Aldrich Chemical Co., Milwaukee]

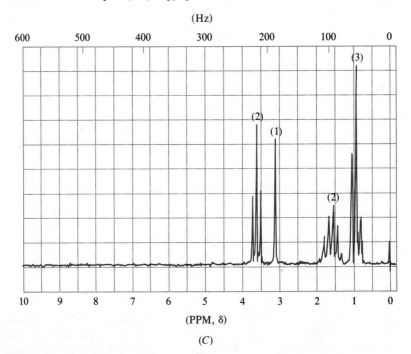

(*C*)

PROBLEM 21.52   NMR spectrum of compound (*C*). [Data from *The Aldrich Library of NMR Spectra*, C. J. Pouchert and J. R. Campbell (eds.), copyright 1974, Aldrich Chemical Co., Milwaukee]

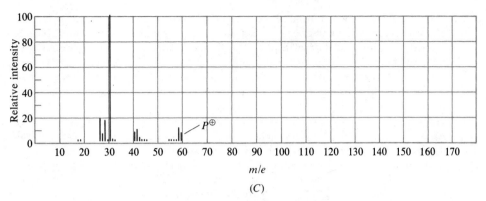

(C)

PROBLEM 21.52   Mass spectrum of compound (C). [Data from the *EPA/NIH Mass Spectral Data Base*, S. R. Heller and G. W. A. Milne (eds.), vols. 1–4, copyright 1978, U.S. Dept. of Commerce]

21.52 An unknown, neutral compound A, $C_{12}H_{26}O_2$, is *optically active*. A does not react with Br_2 in CCl_4 or 2,4-dinitrophenylhydrazine or sodium metal. A is unaffected by heating it in the presence of aqueous base. When heated with aqueous acid, however, A is converted into two new compounds, B and C; it is found that 1 mole of A yields 1 mole of B, $C_6H_{12}O$, and 2 moles of C, C_3H_8O. B is *optically active*, and it reacts with NH_2OH to give an oxime but does not give a silver mirror when treated with $Ag(NH_3)_2^{\oplus}$. B, when treated first with I_2 and NaOH (aqueous) and then neutralized with acid, gives D, $C_5H_{10}O_2$, and iodoform. The mass spectrum and NMR spectrum of C are given here. Draw the structures of compounds A to D. Indicate briefly the reasoning you used in reaching these answers, and show what each piece of data teaches you about the unknown compounds. For compounds that are optically active, indicate the source of the optical activity (e.g., the asymmetric carbon atom) with an asterisk (*).

21.53 *Optically active* compound A, $C_{10}H_{20}$, gives optically active compound B, $C_{10}H_{22}O_2$, on treatment with dilute potassium permanganate. B reacts with periodic acid, HIO_4, to yield two products, C and D. These same products can be obtained by treating A first with ozone and then with water and zinc metal. Compound C, $C_5H_{10}O$, is still optically active and it gives a positive Tollens' test. Compound D, $C_5H_{10}O$, is optically inactive and gives a positive iodoform test but a negative Tollens' test; it does react readily with 2,4-dinitrophenylhydrazine to form a yellow-orange precipitate. Clemmensen reduction of compounds C and D gives the same hydrocarbon E, C_5H_{12}. Deduce the structures of compounds A to E, and write equations for the reactions that occur.

21.54 Give structures of the lettered compounds.
(a) Compound A, $C_6H_{12}O$, is optically active. A gives a negative Tollens' test but it reacts with 2,4-dinitrophenylhydrazine.
(b) Compound B, $C_5H_{10}O$, is optically active; B gives a positive Tollens' test and reacts with 2,4-dinitrophenylhydrazine.
(c) Compound C, $C_5H_{10}O$, gives negative Tollens' and iodoform tests, but it forms an oxime.
(d) Compound D, $C_6H_{10}O$, reacts with 2,4-dinitrophenylhydrazine but not with Tollens' reagent. It also gives a negative iodoform test. Treatment of D with hydrogen gas in the presence of a platinum catalyst gives E, $C_6H_{12}O$, and E undergoes dehydration with concentrated sulfuric acid to give compound F, C_6H_{10}. F, on treatment with ozone and then with water and zinc metal, gives 1,6-hexanedial.

21.55 Compound A, with molecular formula $C_5H_{12}O$, reacts with sodium metal and gives off hydrogen gas. On heating with metallic copper, A gives B, $C_5H_{10}O$, which gives a silver mirror on reaction with Tollens' reagent. On heating with concentrated sulfuric acid, A gives C, which decolorizes dilute basic potassium permanganate. Treatment of C with ozone and then with water and zinc metal gives compound D, C_4H_8O, which does not give a silver mirror with Tollens' reagent. Deduce the structural formulas of compounds A to D, and indicate your reasoning.

21.56 Compound A, $C_9H_{11}ClO$, is aromatic, and on vigorous oxidation with basic potassium permanganate (followed by neutralization), a new aromatic compound B, $C_7H_5ClO_2$, is formed; B is a substituted benzoic acid. On treatment with bromine and ferric bromide catalyst, B produces two and only two monobromo derivatives, C and D, each having the formula $C_7H_4BrClO_2$.

On treatment with sodium metal, A produces bubbles of hydrogen gas, and controlled oxidation of A with chromic acid first gives E, C_9H_9ClO, which gives a silver mirror with Tollens' reagent; further controlled oxidation of E gives F, $C_9H_9ClO_2$, which is a carboxylic acid. When A is heated with concentrated sulfuric acid, a single compound G, C_9H_9Cl, is produced. On ozonolysis (reaction with ozone and then with water and zinc metal), G gives formaldehyde and H, which cannot be oxidized by chromic acid. On the basis of the data provided, draw the structures of compounds A to H. Explain.

21.57 Compound A, $C_{11}H_{12}$, gives a precipitate when treated with silver ammonia complex, $Ag(NH_3)_2^{\oplus}$. On catalytic hydrogenation (mild conditions), A consumes two equivalents of hydrogen gas and yields compound B, $C_{11}H_{16}$. B is also obtained by treating C, $C_{11}H_{14}O$, with zinc amalgam and concentrated hydrochloric acid. Nitration of B gives two isomeric products, D and E, and both D and E give 2-nitroterephthalic acid (1) upon treatment with hot, basic potassium permanganate. The NMR of C consists of:

0.9δ (3H) t
1.4δ (4H) m
2.8δ (2H) t
7.7δ (4H) m
9.7δ (1H) s

Provide the structures of A to E, and outline the reactions described.

(1)

2-Nitroterephthalic acid

21.58 Compound A, $C_{14}H_{20}$, is an *optically active*, *aromatic* compound, which decolorizes bromine in carbon tetrachloride. A, when treated with ozone and then with zinc and water, gives two new compounds, B and C. Compound B, $C_9H_{10}O$, is *optically active* and contains an *aromatic* ring. B gives a silver mirror when treated with Tollens' reagent, $Ag(NH_3)_2^{\oplus}$. Compound C, $C_5H_{10}O$, is optically inactive and unaffected by Tollens' reagent. C, when treated with lithium aluminum hydride and then with $H_3O^{\oplus}$, gives compound D, $C_5H_{12}O$. D, when heated with concentrated H_2SO_4, gives a *single* compound E, C_5H_{10}; no isomeric compound is formed. E, when treated with ozone and then with zinc and water, gives *equimolar* amounts of acetaldehyde and propionaldehyde. On the basis of these data, draw the structures of compounds A to E. Indicate the source of the optical activity in the compounds described. (*Note:* Ignore the possibility of *geometric* isomers throughout this problem.)

21.59 Compound A, $C_{13}H_{18}$, is *optically active* and reacts readily with bromine in carbon tetrachloride to give compound B, $C_{13}H_{18}Br_2$. A is treated with ozone and then with zinc and water to give two new compounds, C and D. C, $C_5H_{10}O$, is *optically active*, and it reacts with hydroxylamine and phenylhydrazine. C also gives a silver mirror with Tollens' reagent. D, C_8H_8O, is an optically inactive aromatic compound, and it reacts with hydroxylamine and phenylhydrazine. D does not react at all with Tollens' reagent. D, on treatment either with hydrazine (NH_2NH_2) and base or with Zn(Hg) and hydrochloric acid, gives compound E, C_8H_{10}.
(*a*) Draw the structures of compounds A to E. Indicate the source of the optical activity in those compounds that are optically active.
(*b*) Is the information given here enough for you to provide a completely unambiguous structure for compound A? Briefly explain why or why not, and if not, show what ambiguity exists.

21.60 *Optically active aromatic* compound A, $C_{13}H_{16}$, readily decolorizes Br_2 in CCl_4 but does not react with $Cu(NH_3)_2^{\oplus}$. On hydrogenation in the presence of Pt catalyst, A gives aromatic compound B, $C_{13}H_{20}$, whereas treatment of A with H_2 gas in the presence of Ni-B (Lindlar) catalyst gives compound C, $C_{13}H_{18}$. C is treated with ozone, followed by zinc and water, to yield two new compounds, D, $C_5H_{10}O$, and E, C_8H_8O. D gives a silver mirror with Tollens' reagent, and D is also *optically active*. E, an aromatic compound, is treated with $LiAlH_4$ and then with acidic water to give compound F, $C_8H_{10}O$. F readily reacts with sodium metal, and it cannot be resolved into enantiomers. F does not react with Br_2 in CCl_4, but when heated with concentrated H_2SO_4, F yields aromatic compound G, C_8H_8. G readily decolorizes Br_2 in CCl_4.

On the basis of these data, draw the structures of compounds A to G. Indicate the source of the optical activity in those compounds that are optically active. Trace the reactions given here.

22

Carboxylic Acids and Dicarboxylic Acids

In this chapter we discuss carboxylic and dicarboxylic acids. Their general structures are:

Carboxylic acid:

$$\text{(Ar—) R—}\overset{\overset{\displaystyle O}{\|}}{\text{C}}\text{—OH} \qquad \text{or} \qquad \text{(Ar—) R—COOH}$$

Structural   Condensed
formula   formula

Aliphatic dicarboxylic acid:

$$\text{HO—}\overset{\overset{\displaystyle O}{\|}}{\text{C}}\text{—}\left(\overset{|}{\underset{|}{\text{C}}}\right)_{n}\text{—}\overset{\overset{\displaystyle O}{\|}}{\text{C}}\text{—OH} \qquad \text{or} \qquad \text{HOOC—}\left(\overset{|}{\underset{|}{\text{C}}}\right)_{n}\text{—COOH}$$

Structural   Condensed
formula   formula

Aromatic dicarboxylic acid:

$$\text{Ar}\left(\overset{\overset{\displaystyle O}{\|}}{\text{C}}\text{—OH}\right)_{2} \qquad \text{or} \qquad \text{Ar(COOH)}_2$$

Structural   Condensed
formula   formula

Compounds with only hydrogen or alkyl or aryl groups attached to the carbon-oxygen double bond (that is, aldehydes or ketones) are called *carbonyl compounds*, and the C=O bond they contain is the *carbonyl group*. On the other hand, the —COOH group (sometimes written as —CO_2H) in carboxylic acids is called the **carboxyl group,** and the R—C=O bond it contains is referred to as the **acyl group.**

The name *carboxylic acid* describes this functional group in two ways. First, the word *carboxylic* is a contraction of the words *carbo*nyl and hydr*oxyl*, which are the two structural units in the —COOH group. Second, the word *acid* is appended because of the acidic properties of this family of compounds.

We concentrate on the nomenclature, physical and spectral properties, and methods of preparation of acids and diacids in this chapter. We also mention their reactions. Discussion of those reactions that produce derivatives of acids or diacids (acid chlorides, amides, esters, anhydrides, and nitriles) are deferred until Chap. 23.

22.1 Nomenclature of Carboxylic Acids

Hydrogen, an alkyl group, or an aryl group can be attached to the carboxyl group to produce the entire family of carboxylic acids. As we will see in Sec. 22.9, physical evidence indicates both a carbon-oxygen single bond and a carbon-oxygen double bond in carboxylic acids.

A. Common Names

Several names exist because of common usage in the past or because they were derived from the source of a particular acid. For example, *formic acid*, HCOOH, was obtained from ants (Latin *formica*, ant); *acetic acid*, CH_3COOH, is responsible for the sour taste in vinegar (Latin *acetum*, vinegar); and *butyric acid* produces the stench in rancid butter (Latin *butyrum*, butter). Most aliphatic carboxylic acids have a pungent odor, and the C_6 (*caproic acid*), C_8 (*caprylic acid*), and C_{10} (*capric acid*) straight-chain carboxylic acids are so named because they are found in goat fat (Latin *caper*, goat). There are many straight-chain aliphatic carboxylic acids with an *even* number of carbon atoms in naturally occurring fats and oils (see Sec. 25.1), and they are often called **fatty acids.** Table 22.1 gives the common names of the homologous series of carboxylic acids through C_{10}, as well as some common fatty acids.

Substituted carboxylic acids can be named by using the common names for the parent compounds. Greek letters are used to identify the positions, with the α position being *adjacent* to the carboxyl group. The remainder of the positions in the parent chain are identified in the following manner (carbon framework only is shown):

$$-\overset{|}{\underset{|}{C}}^{\varepsilon}-\overset{|}{\underset{|}{C}}^{\delta}-\overset{|}{\underset{|}{C}}^{\gamma}-\overset{|}{\underset{|}{C}}^{\beta}-\overset{|}{\underset{|}{C}}^{\alpha}-\overset{O}{\overset{\|}{C}}-OH$$

The following examples illustrate the use of this system:

Phenylacetic acid β-Methylbutyric acid α-Chloro-δ-phenylvaleric acid

Sometimes a compound is named as a derivative of a simple carboxylic acid:

$$CH_3-\overset{CH_3}{\underset{CH_3}{\overset{|}{\underset{|}{C}}}}-\overset{O}{\overset{\|}{C}}-OH$$

Trimethylacetic acid
(*Other common names:* α,α-dimethylpropionic acid and pivalic acid)

The Greek letter ω is used in common names to designate that the substituent is attached to the carbon atom farthest from the carboxyl group, provided it is attached to the longest continuous chain that is used for the parent name of the acid. For example, $Cl-CH_2(CH_2)_{10}COOH$ is ω-chlorolauric acid. The letter ω can be used regardless of the chain length.

TABLE 22.1   Names for Carboxylic Acids

Common Name	IUPAC Name	Structure
Formic acid	Methanoic acid	HCOOH
Acetic acid	Ethanoic acid	CH_3COOH
Propionic acid	Propanoic acid	CH_3CH_2COOH
Butyric acid	Butanoic acid	$CH_3(CH_2)_2COOH$
Valeric acid	Pentanoic acid	$CH_3(CH_2)_3COOH$
Caproic acid	Hexanoic acid	$CH_3(CH_2)_4COOH$
	Heptanoic acid	$CH_3(CH_2)_5COOH$
Caprylic acid	Octanoic acid	$CH_3(CH_2)_6COOH$
	Nonanoic acid	$CH_3(CH_2)_7COOH$
Capric acid	Decanoic acid	$CH_3(CH_2)_8COOH$
Lauric acid	Dodecanoic acid	$CH_3(CH_2)_{10}COOH$
Myristic acid	Tetradecanoic acid	$CH_3(CH_2)_{12}COOH$
Palmitic acid	Hexadecanoic acid	$CH_3(CH_2)_{14}COOH$
Stearic acid	Octadecanoic acid	$CH_3(CH_2)_{16}COOH$
Arachidic acid	Eicosanoic acid	$CH_3(CH_2)_{18}COOH$
Behenic acid	Docosanoic acid	$CH_3(CH_2)_{20}COOH$
Oleic acid	cis-9-Octadecenoic acid	$CH_3(CH_2)_7-\overset{H}{\underset{cis}{C}}=\overset{H}{C}-(CH_2)_7COOH$
Linoleic acid	cis,cis-9,12-Octadecadienoic acid	$CH_3(CH_2)_4-\overset{H}{\underset{cis}{C}}=\overset{H}{C}-CH_2-\overset{H}{\underset{cis}{C}}=\overset{H}{C}-(CH_2)_7COOH$
Linolenic acid	cis,cis,cis-9,12,15-Octadecatrienoic acid	$CH_3CH_2-\overset{H}{\underset{cis}{C}}=\overset{H}{C}-CH_2-\overset{H}{\underset{cis}{C}}=\overset{H}{C}-CH_2-\overset{H}{\underset{cis}{C}}=\overset{H}{C}-(CH_2)_7COOH$

B. IUPAC System

The IUPAC system requires that the longest continuous chain of carbon atoms that contains the carboxyl group be used as the parent structure. Such a compound is named by taking the alkane name corresponding to the number of carbon atoms in the longest chain, dropping the *-e* ending, and replacing it with **-oic acid.** The compound is numbered by assigning the carbon atom in the carboxyl group as C-1:

$$-\overset{|}{\underset{|6}{C}}-\overset{|}{\underset{|5}{C}}-\overset{|}{\underset{|4}{C}}-\overset{|}{\underset{|3}{C}}-\overset{|}{\underset{|2}{C}}-\overset{O}{\overset{\|}{\underset{1}{C}}}-OH$$

The IUPAC names of some straight-chain carboxylic acids are in Table 22.1, and the following examples illustrate the use of the system for substituted compounds:

$$CH_3-\overset{O}{\overset{\|}{C}}-OH \qquad \overset{4}{CH_3}\overset{3}{CH_2}\overset{2}{\underset{\underset{CH_3}{|}}{CH}}-\overset{O}{\overset{\|}{\underset{1}{C}}}-OH \qquad \overset{6}{CH_3}\overset{5}{\underset{\bigcirc}{CH}}\overset{4}{CH_2}\overset{3}{CH_2}\overset{2}{CH_2}-\overset{O}{\overset{\|}{\underset{1}{C}}}-OH$$

| Ethanoic | 2-Methylbutanoic | 5-Phenylhexanoic |
| acid | acid | acid |

$$\overset{CH_3}{\underset{}{\diagdown}}\overset{}{\underset{CH}{}}\overset{CH_3}{\underset{}{\diagup}}$$
$$\overset{9}{CH_3}\overset{8}{CH_2}\overset{7}{CH_2}\overset{6}{CH_2}\overset{5}{CH}\overset{4}{}\overset{3}{CH}\overset{2}{CH_2}\overset{}{CH_2}\overset{2}{\underset{\underset{Br}{|}}{CH}}-\overset{O}{\overset{\|}{\underset{1}{C}}}-OH$$

2-Bromo-5-isopropylnonanoic
acid

$$\overset{H}{\underset{\overset{6}{CH_3}\overset{5}{CH_2}}{\diagdown}}\overset{}{\underset{4}{C}}=\overset{}{\underset{3}{C}}\overset{\overset{2}{CH_2}-\overset{O}{\overset{\|}{\underset{1}{C}}}-OH}{\underset{H}{\diagup}}$$

trans-3-Hexenoic
acid

Benzoic acid is the accepted IUPAC name for C_6H_5COOH, and substituted benzoic acids are often named as derivatives of the parent compound; for example:

Benzoic
acid

p-Nitrobenzoic
acid

m-Methylbenzoic acid
(*m*-toluic acid)

Sometimes it is necessary to treat the carboxyl group as a *substituent*, as when it is attached to a cyclic ring system (other than benzene); for example:

Cyclobutanecarboxylic
acid

Cyclohexanone-2-carboxylic
acid

TABLE 22.2   Names of Dicarboxylic Acids

Common Name	IUPAC Name	Structure
Oxalic acid	Ethanedioic acid	HOOC—COOH
Malonic acid	Propanedioic acid	HOOC—CH_2—COOH
Succinic acid	Butanedioic acid	HOOC—$(CH_2)_2$—COOH
Glutaric acid	Pentanedioic acid	HOOC—$(CH_2)_3$—COOH
Adipic acid	Hexanedioic acid	HOOC—$(CH_2)_4$—COOH
Pimelic acid	Heptanedioic acid	HOOC—$(CH_2)_5$—COOH
Suberic acid	Octanedioic acid	HOOC—$(CH_2)_6$—COOH
Azelaic acid	Nonanedioic acid	HOOC—$(CH_2)_7$—COOH
Sebacic acid	Decanedioic acid	HOOC—$(CH_2)_8$—COOH
Phthalic acid	1,2-Benzenedicarboxylic acid	
Isophthalic acid	1,3-Benzenedicarboxylic acid	
Terephthalic acid	1,4-Benzenedicarboxylic acid	

22.2   Nomenclature of Dicarboxylic Acids

In this section we emphasize the system of nomenclature for dicarboxylic acids, and mention briefly the naming of compounds that contain more than two carboxyl groups.

A. Common Names

There are many straight-chain *di*carboxylic acids, and their common names are used much more frequently than their systematic names; for example:

Malonic acid   Glutaric acid

The entire series is in Table 22.2.

Substituted dicarboxylic acids can be named by lettering the carbon atoms between the two carboxyl groups with Greek letters, as shown by these examples:

α-Methylglutaric acid   α-Methyl-γ-phenyladipic acid

Aromatic compounds containing two carboxyl groups on the ring have common names, which also merit memorization:

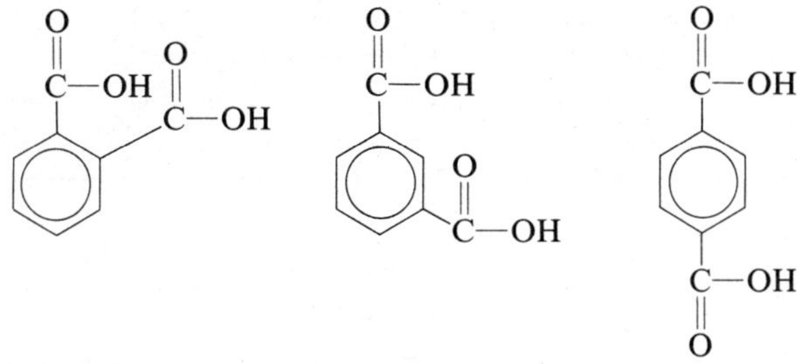

Phthalic
acid

Isophthalic
acid

Terephthalic
acid

B. IUPAC Names

Straight-chain dicarboxylic acids are named by using the longest continuous carbon chain that contains both carboxyl groups. The suffix **-dioic acid** is added to the alkane name of the parent compound (nothing is dropped from the alkane name). If substituents are present, they are indicated by position after the continuous chain is numbered starting with C-1 as the carbon atom of one of the carboxyl groups; numbering is done so the substituents have the smallest numbers. Two examples are:

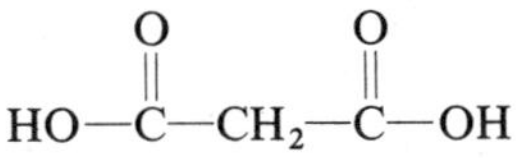

Propanedioic acid
(malonic acid)

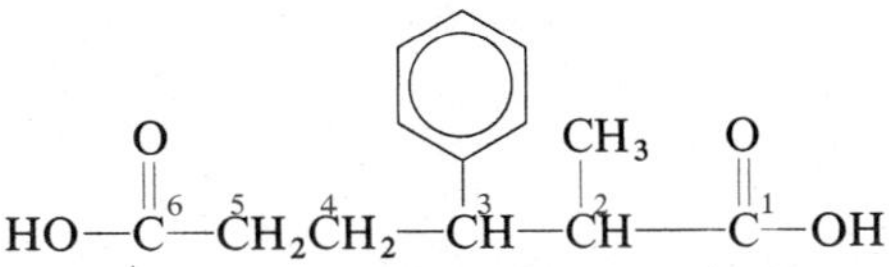

2-Methyl-3-phenylhexanedioic acid
(α-methyl-β-phenyladipic acid)

The IUPAC names of some nonsubstituted dicarboxylic acids are in Table 22.2.

As with carboxylic acids, sometimes two or more carboxyl groups must be treated as substituents; for example:

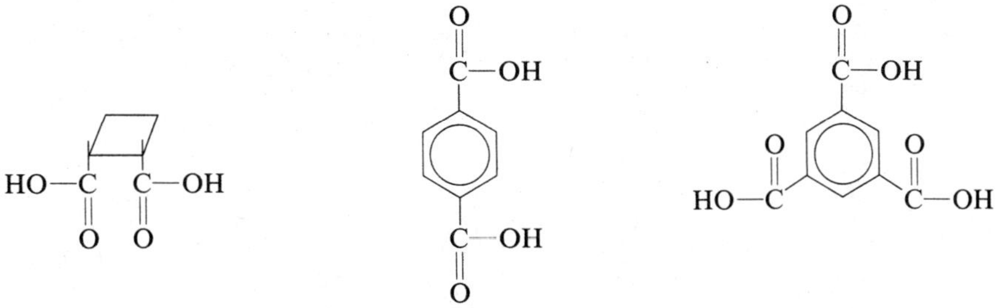

cis-1,2-Cyclobutanedicarboxylic
acid

1,4-Benzenedicarboxylic acid

1,3,5-Benzenetricarboxylic
acid

Question 22.1

Draw the possible structural isomers (ignoring stereoisomers) of the following compounds, and give two acceptable names for each (one of which must be the IUPAC name).

(*a*) all carboxylic acids having the molecular formula $C_4H_8O_2$
(*b*) all carboxylic acids having the molecular formula $C_5H_{10}O_2$
(*c*) all carboxylic acids having the molecular formula $C_6H_{12}O_2$
(*d*) all *cyclic* carboxylic acids having the molecular formula $C_5H_8O_2$
(*e*) all dicarboxylic acids having the molecular formula $C_6H_{10}O_4$

TABLE 22.3   Physical Properties of Carboxylic and Dicarboxylic Acids

Name	Formula	mp, °C	bp, °C	Solubility, g/100 g of water
Carboxylic acids:				
Formic	HCOOH	8	100.5	∞
Acetic	CH_3COOH	17	118	∞
Propionic	CH_3CH_2COOH	-22	141	∞
Butyric	$CH_3(CH_2)_2COOH$	-4.7	162.5	∞
Valeric	$CH_3(CH_2)_3COOH$	-34.5	187	3.9
Caproic	$CH_3(CH_2)_4COOH$	-1.5	205	1.2
Caprylic	$CH_3(CH_2)_6COOH$	16	237	0.7
Capric	$CH_3(CH_2)_8COOH$	31	270	0.3
Lauric	$CH_3(CH_2)_{10}COOH$	44		insoluble
Stearic	$CH_3(CH_2)_{16}COOH$	70		insoluble
Benzoic	C_6H_5COOH	122	250	0.35
o-Toluic		107.8	259	0.12
m-Toluic		112	263	0.1
p-Toluic		179	274	0.03
Dicarboxylic acids:				
Oxalic	HOOCCOOH	187		10.2
Malonic	$HOOCCH_2COOH$	135[†]		138
Succinic	$HOOCCH_2CH_2COOH$	185[‡]		6.8
Glutaric	$HOOC(CH_2)_3COOH$	98[‡]		64
Phthalic	$o\text{-}C_6H_4(COOH)_2$	207[‡]	>191*	0.7
Isophthalic	$m\text{-}C_6H_4(COOH)_2$	352		0.01
Terephthalic	$p\text{-}C_6H_4(COOH)_2$	302[§]		0.003

For the o-, m-, p-Toluic entries, the structural formula shown is a benzene ring with a COOH group and a CH_3 group.

* Decomposes on melting.
[†] Loses carbon dioxide on heating just above melting point to give acetic acid (see Sec. 22.17).
[‡] Loses water on heating just above melting point to give anhydride (see Sec. 23.12).
[§] Sublimes.

22.3   Physical Properties of Carboxylic and Dicarboxylic Acids

Table 22.3 lists the physical properties of some carboxylic and dicarboxylic acids. Comments about each family of compounds follow.

A. Carboxylic Acids

Carboxylic acids boil at higher temperatures than do alcohols, aldehydes, or ketones of similar molecular weight. For example, formic acid (mol wt 46) boils at 100.5°, whereas ethyl alcohol (mol wt 46) boils at 78° and acetaldehyde (mol wt 44) at 21°. This trend is not surprising because the carboxyl group contains both polar C=O and the —O—H groups. Molecular weight determinations show that simple carboxylic acids exist as **dimers** in both the liquid and vapor states. This is attributed to intermolecular hydrogen bonding between the C=O group of one molecule and the —O—H group of the other molecule, two such hydrogen bonds are possible:

$$
\begin{array}{ccc}
& \ddot{O}\!:\text{---}H\!-\!\ddot{O} & \\
R\!-\!C\!\!\diagup\!\!\diagdown & & \diagup\!\!\diagdown C\!-\!R \\
& :\ddot{O}\!-\!H\text{---}:\ddot{O} &
\end{array}
$$

Carboxylic acids resemble the corresponding alcohols in water solubility. Acids of four carbon atoms or less are completely soluble in water, and the solubility is most likely due to hydrogen bonding between the carboxyl group (both the C=O and the —OH groups) and water. Five-carbon acids are partially soluble, and the remainder are considered insoluble. Many acids are soluble in alcohol, which is less polar than water. An alcohol not only has the polar hydroxyl group for hydrogen bonding with the carboxyl group but also some organic character because the hydro-carbon chain attached to the —OH group has some affinity for the hydrocarbon part of a carboxylic acid.

The lower molecular weight organic acids have a sharp, sour taste (like many inorganic acids). They have strong, unpleasant odors, especially butyric, valeric, and caproic acids.

B. Dicarboxylic Acids

Table 22.3 indicates that the dicarboxylic acids are relatively high melting compounds and are not liquids at room temperature. They are fairly water-soluble because of the two carboxyl groups. Some diacids undergo special reactions (loss of carbon dioxide or water) on heating, references to these reactions are in the footnotes to the table. Because of their high melting points, diacids have little or no odor.

22.4   Methods of Preparation of Carboxylic Acids: Summary

Before considering new methods for preparing carboxylic acids, let us review some methods we encountered earlier in this text.

A. Oxidation of Primary Alcohols

One important method for preparing a carboxylic acid is by oxidation of the primary alcohol containing the same number of carbon atoms, as discussed in Sec. 10.10:

$$
R\!-\!CH_2\!-\!OH \xrightarrow[\substack{\text{or 1. }KMnO_4,\ :\!\ddot{O}H^{\ominus},\ \text{heat}\\ \text{2. }H_3O^{\oplus}}]{K_2Cr_2O_7,\ H_2SO_4,\ \text{heat}} \overset{\displaystyle O}{\overset{\|}{R\!-\!C\!-\!OH}}
$$

Primary   Carboxylic acid
alcohol   *Contains*
same number of
carbon atoms

This method is necessarily limited to the alcohols containing no other functional groups that are sensitive toward oxidation (for example, double and triple bonds). Basic permanganate oxidation must be followed by neutralization with acid ($H_3\ddot{O}^{\oplus}$) to produce the free carboxylic acid (see Sec. 22.9).

B. Oxidation of Alkylbenzenes

Aromatic rings containing an alkyl group are converted into the corresponding benzoic acid derivative on treatment with strong oxidizing agents under vigorous conditions (see Sec. 14.18):

$$\underset{\text{Alkylbenzene}}{\text{Ar}-\text{R}} \quad \xrightarrow[\substack{\text{or 1. KMnO}_4, \, :\ddot{\text{O}}\text{H}^{\ominus}, \text{ heat} \\ \text{2. H}_3\ddot{\text{O}}^{\oplus}}]{\text{K}_2\text{Cr}_2\text{O}_7, \, \text{H}_2\text{SO}_4, \text{ heat}} \quad \underset{\substack{\text{Benzoic acid} \\ \text{(or substituted benzoic acid)}}}{\text{Ar}-\overset{\displaystyle\overset{O}{\|}}{\text{C}}-\text{OH}}$$

Recall that methyl groups, as well as more complex side chains such as *n*-butyl and isopropyl groups, are oxidized so that only the carboxyl group involving the carbon attached directly to the aromatic ring remains. Each alkyl group on the ring is also oxidized, so that *di*, *tri*, and so on aromatic carboxylic acids can be obtained.

This reaction is limited to aromatic rings that are *not* highly activated (phenols or aromatic amines are completely degraded by vigorous oxidation).

C. Permanganate Oxidation of Alkenes

Certain alkenes react with basic potassium permanganate under vigorous conditions to produce carboxylic acids (see Sec. 8.32):

$$\underset{\text{H}}{\overset{\text{R}}{>}}\text{C}=\text{C}\underset{\text{H}}{\overset{\text{R}'}{<}} \quad \xrightarrow[\text{2. H}_3\ddot{\text{O}}^{\oplus}]{\text{1. KMnO}_4, \, :\ddot{\text{O}}\text{H}^{\ominus}, \text{ heat}} \quad \text{R}-\overset{\displaystyle\overset{O}{\|}}{\text{C}}-\text{OH} + \text{HO}-\overset{\displaystyle\overset{O}{\|}}{\text{C}}-\text{R}'$$

22.5   Other Common Methods for Preparing Carboxylic Acids

Ðmit

In this section we study two other common methods for preparing carboxylic acids: (1) **Grignard method** and (2) **nitrile hydrolysis.**

A. Grignard Method

One of the most useful general methods for preparing both aliphatic and aromatic carboxylic acids begins with the corresponding alkyl or aryl halide, which is first converted into the corresponding Grignard reagent and then allowed to react with carbon dioxide. On hydrolysis, a carboxylic acid *containing one more carbon atom* is obtained. The reactions are:

$$\text{R}-\text{X} + \text{Mg} \xrightarrow[\text{ether}]{\text{anhydrous}} \text{R}-\text{MgX} \xrightarrow[\text{(dry ice)}]{\text{CO}_2} \text{R}-\overset{\displaystyle\overset{O}{\|}}{\text{C}}-\text{OMgX} \xrightarrow{\text{H}_3\ddot{\text{O}}^{\oplus}} \text{R}-\overset{\displaystyle\overset{O}{\|}}{\text{C}}-\text{OH}$$

where X = Cl, Br, I

$$\underset{\substack{\text{ArCl may be used} \\ \text{with tetrahydrofuran} \\ \text{as solvent}}}{\text{Ar}-\text{Br} + \text{Mg}} \xrightarrow[\text{ether}]{\text{anhydrous}} \text{Ar}-\text{MgBr} \xrightarrow[\text{(dry ice)}]{\text{CO}_2} \text{Ar}-\overset{\displaystyle\overset{O}{\|}}{\text{C}}-\text{OMgBr} \xrightarrow{\text{H}_3\ddot{\text{O}}^{\oplus}} \text{Ar}-\overset{\displaystyle\overset{O}{\|}}{\text{C}}-\text{OH}$$

The carbon dioxide is usually supplied from dry ice, and this reaction is sometimes called *carbonation*. Experimentally, the reaction is carried out by preparing the

Grignard reagent in anhydrous ether; the solution is then poured *onto* dry ice so that a large excess of carbon dioxide is present. After the excess dry ice has evaporated, the reaction mixture is acidified with mineral acid (commonly sulfuric or hydrochloric acid) to liberate the free carboxylic acid from the magnesium salt. The carboxylic acid goes into the ether layer, while the mineral acid and other inorganic salts remain in the aqueous layer.

As the following examples illustrate, *Grignard synthesis of carboxylic acids always produces an acid with one additional carbon atom*. This method is very useful because alkyl halides, which may be primary, secondary, or tertiary, can be readily obtained from alcohols, many of which are commercially available.

$$CH_3CH_2-\underset{\underset{\displaystyle OH}{|}}{CH}-CH_3 \xrightarrow{PBr_3} CH_3CH_2-\underset{\underset{\displaystyle Br}{|}}{CH}-CH_3 \xrightarrow[\text{ether}]{\underset{\text{anhydrous}}{Mg}}$$

2-Butanol	2-Bromobutane
(*sec*-butyl alcohol)	(*sec*-butyl bromide)

$$CH_3CH_2-\underset{\underset{\displaystyle MgBr}{|}}{CH}-CH_3 \xrightarrow{CO_2} CH_3CH_2-\underset{\underset{\displaystyle CH_3}{|}}{CH}-\overset{\overset{\displaystyle O}{\|}}{C}-OMgBr \xrightarrow{H_3\overset{..}{O}^{\oplus}} CH_3CH_2-\underset{\underset{\displaystyle CH_3}{|}}{CH}-\overset{\overset{\displaystyle O}{\|}}{C}-OH$$

2-Methylbutanoic acid
(α-methylbutyric acid)

$$CH_3-\underset{\underset{\displaystyle CH_3}{|}}{\overset{\overset{\displaystyle CH_3}{|}}{C}}-OH \xrightarrow[\text{HBr}]{conc} CH_3-\underset{\underset{\displaystyle CH_3}{|}}{\overset{\overset{\displaystyle CH_3}{|}}{C}}-Br \xrightarrow[\text{ether}]{\underset{\text{anhydrous}}{Mg}} CH_3-\underset{\underset{\displaystyle CH_3}{|}}{\overset{\overset{\displaystyle CH_3}{|}}{C}}-MgBr \xrightarrow{CO_2}$$

2-Methyl-2-propanol	2-Bromo-2-methylpropane
(*tert*-butyl alcohol)	(*tert*-butyl bromide)

$$CH_3-\underset{\underset{\displaystyle CH_3}{|}}{\overset{\overset{\displaystyle CH_3}{|}}{C}}-\overset{\overset{\displaystyle O}{\|}}{C}-OMgBr \xrightarrow{H_3\overset{..}{O}^{\oplus}} CH_3-\underset{\underset{\displaystyle CH_3}{|}}{\overset{\overset{\displaystyle CH_3}{|}}{C}}-\overset{\overset{\displaystyle O}{\|}}{C}-OH$$

2,2-Dimethylpropanoic acid
(pivalic acid or trimethylacetic acid)

o-Bromotoluene → (Mg, anhydrous ether) → → (CO₂) → → (H₃O⁺) → *o*-Methylbenzoic acid (*o*-toluic acid)

The mechanism of the reaction between the Grignard reagent and carbon dioxide takes the following path:

$$R\overset{\ominus}{:}[MgX]^{\oplus} + \overset{\overset{\displaystyle \delta\ominus}{..}}{O}{=}\overset{\delta\oplus}{C}{=}\overset{\overset{\displaystyle \delta\ominus}{..}}{\underset{..}{O}} \xrightarrow[\text{ether}]{\text{anhydrous}} R-\overset{\overset{\displaystyle :O:}{\|}}{C}-\overset{..}{\underset{..}{O}}{:}^{\ominus}[MgX]^{\oplus}$$

This is a typical nucleophilic *addition* reaction of a carbon-oxygen double bond. This reaction resembles the addition of the Grignard reagent to aldehydes and ketones (see Sec. 21.10), although in carbon dioxide there are two such polar bonds in the same molecule. For reasons explained in Sec. 22.9, the reaction mixture must be hydrolyzed by mineral acid to liberate the free acid. Yields are often low because the moisture in the dry ice reacts with and destroys a portion of the Grignard reagent.

$$
\begin{array}{c}
\text{O} \\
\parallel \\
\text{R—C—}\overset{..}{\underset{..}{\text{O}}}\!:^{\ominus}[\text{MgX}]^{\oplus}
\end{array}
\xrightarrow{\text{H}_3\overset{..}{\text{O}}{}^{\oplus}}
\begin{array}{c}
\text{O} \\
\parallel \\
\text{R—C—OH}
\end{array}
+ \text{Mg}^{\oplus 2} + :\overset{..}{\underset{..}{\text{X}}}\!:^{\ominus}
$$

Question 22.2

For the reaction between the Grignard reagent and carbon dioxide, indicate the *changes* in hybridization and geometry that the carbon atom in carbon dioxide undergoes as the magnesium salt of the carboxylic acid is formed. Are there any changes in the oxidation state of that carbon atom? If so, what are they?

B. Nitrile Hydrolysis

Aliphatic and aromatic nitriles give the corresponding carboxylic acids on hydrolysis in either acidic or basic solution:

Acidic hydrolysis:

$$
\text{R—C}\!\equiv\!\overset{..}{\text{N}} \quad \text{or} \quad \text{Ar—C}\!\equiv\!\overset{..}{\text{N}}
\xrightarrow[\text{heat}]{\text{H}_3\overset{..}{\text{O}}{}^{\oplus}}
\begin{array}{c}\text{O}\\\parallel\\\text{R—C—OH}\end{array}
\quad \text{or} \quad
\begin{array}{c}\text{O}\\\parallel\\\text{Ar—C—OH}\end{array}
+ \text{NH}_4^{\oplus}
$$

Basic hydrolysis:

$$
\text{R—C}\!\equiv\!\overset{..}{\text{N}} \quad \text{or} \quad \text{Ar—C}\!\equiv\!\overset{..}{\text{N}}
\xrightarrow[\text{H}_2\text{O, heat}]{:\overset{..}{\text{O}}\text{H}^{\ominus}}
\begin{array}{c}\text{O}\\\parallel\\\text{R—C—}\overset{..}{\underset{..}{\text{O}}}\!:^{\ominus}\end{array}
\quad \text{or} \quad
\begin{array}{c}\text{O}\\\parallel\\\text{Ar—C—}\overset{..}{\underset{..}{\text{O}}}\!:^{\ominus}\end{array}
+ \overset{..}{\text{N}}\text{H}_3
$$

$$\Big\downarrow \text{H}_3\overset{..}{\text{O}}{}^{\oplus} \text{ (neutralize)}$$

$$
\begin{array}{c}\text{O}\\\parallel\\\text{R—C—OH}\end{array}
\quad \text{or} \quad
\begin{array}{c}\text{O}\\\parallel\\\text{Ar—C—OH}\end{array}
$$

We discuss the mechanism of nitrile hydrolysis in Sec. 23.16.

Aliphatic nitriles are often prepared from an alkyl halide (or alkyl tosylate, Sec. 19.6) and sodium or potassium cyanide, a reaction that requires the use of a polar solvent such as aqueous alcohol that dissolves both the alkyl halide (or tosylate) and the inorganic salt (NaCN or KCN). There are, however, certain limitations to this typical S_N2 substitution reaction. Tertiary halides and tosylates undergo predominant elimination by cyanide ion, which is a fairly strong base. The K_a of its conjugate acid, hydrogen cyanide (HCN), is very small—4×10^{-10}. The cyanide ion, therefore, is fairly effective in promoting E2 eliminations. Secondary alkyl halides and tosylates undergo considerable elimination also, so that *only primary or methyl alkyl halides and tosylates can be used to prepare nitriles in good yield.* The general equation for the reaction is:

$$
\text{R—CH}_2\!\overset{..}{\underset{..}{\text{X}}}\!:^{\ominus} + \;:\!\text{C}\!\equiv\!\text{N}:
\xrightarrow{S_N2}
\text{R—CH}_2\text{—C}\!\equiv\!\overset{..}{\text{N}} + :\overset{..}{\underset{..}{\text{X}}}\!:^{\ominus}
$$

Primary
or methyl
alkyl
halide or
tosylate   Nu: Nitrile

With tertiary (and most secondary) alkyl halides and tosylates, the elimination reaction probably occurs by an E2 mechanism:

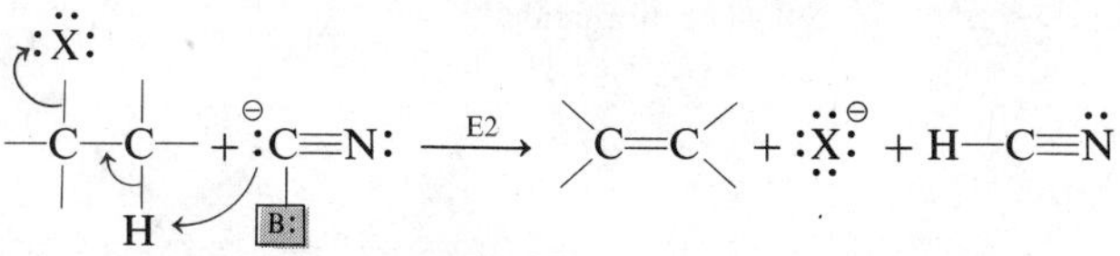

The nitrile synthesis of a carboxylic acid from an alkyl halide or tosylate also results in the introduction of *one more carbon into the carbon chain*. As the following two examples illustrate, the nitrile synthesis of an acid can originate with an alcohol containing one less carbon atom:

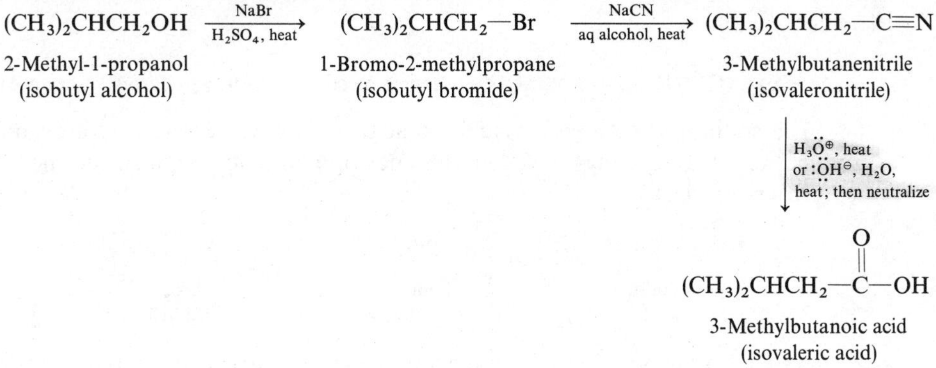

$(CH_3)_2CHCH_2OH$ $\xrightarrow[\text{H}_2\text{SO}_4, \text{ heat}]{\text{NaBr}}$ $(CH_3)_2CHCH_2-Br$ $\xrightarrow[\text{aq alcohol, heat}]{\text{NaCN}}$ $(CH_3)_2CHCH_2-C\equiv N$

2-Methyl-1-propanol   1-Bromo-2-methylpropane   3-Methylbutanenitrile
(isobutyl alcohol)   (isobutyl bromide)   (isovaleronitrile)

$\Big\downarrow$ $\text{H}_3\overset{\ddot{}}{\text{O}}{}^{\oplus}, \text{ heat}$
or $:\overset{..}{\text{O}}\text{H}^{\ominus}, \text{H}_2\text{O},$
heat; then neutralize

$(CH_3)_2CHCH_2-\overset{\displaystyle O}{\overset{\|}{C}}-OH$

3-Methylbutanoic acid
(isovaleric acid)

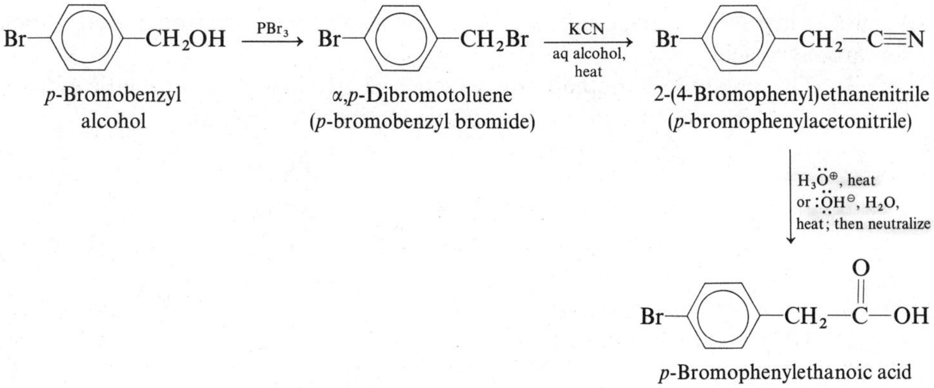

$Br-\bigcirc-CH_2OH$ $\xrightarrow{\text{PBr}_3}$ $Br-\bigcirc-CH_2Br$ $\xrightarrow[\text{aq alcohol, heat}]{\text{KCN}}$ $Br-\bigcirc-CH_2-C\equiv N$

p-Bromobenzyl   α,*p*-Dibromotoluene   2-(4-Bromophenyl)ethanenitrile
alcohol   (*p*-bromobenzyl bromide)   (*p*-bromophenylacetonitrile)

$\Big\downarrow$ $\text{H}_3\overset{\ddot{}}{\text{O}}{}^{\oplus}, \text{ heat}$
or $:\overset{..}{\text{O}}\text{H}^{\ominus}, \text{H}_2\text{O},$
heat; then neutralize

$Br-\bigcirc-CH_2-\overset{\displaystyle O}{\overset{\|}{C}}-OH$

p-Bromophenylethanoic acid

Dicarboxylic acids can also be prepared by the nitrile method, as the following two examples illustrate:

α-Halo acid
(see Sec. 22.18)

$\Big\downarrow$ $:\overset{..}{\text{O}}\text{H}^{\ominus}, \text{H}_2\text{O}, \text{ heat};$
then neutralize

$R-CH\Big\langle \begin{matrix} \overset{\displaystyle O}{\overset{\|}{C}}-OH \\ \overset{\|}{\underset{\displaystyle O}{C}}-OH \end{matrix}$

Substituted malonic acid

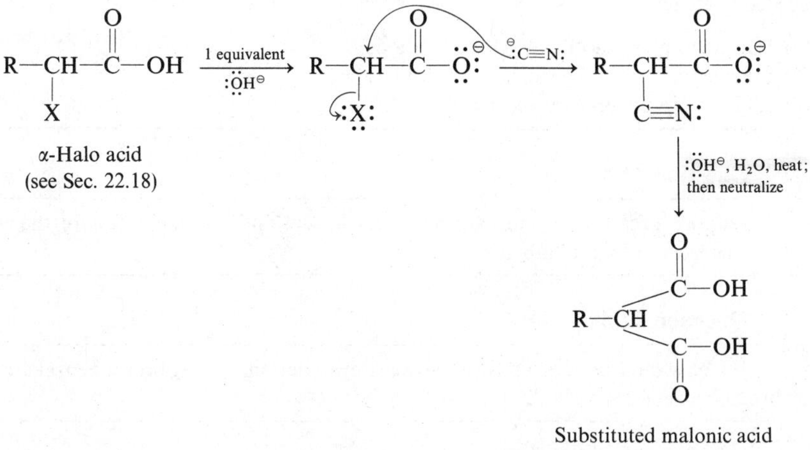

$$CH_2{=}CH_2 \xrightarrow[\text{CCl}_4]{\text{Br}_2} Br{-}CH_2{-}CH_2{-}Br \xrightarrow[\text{aq alcohol}]{\text{NaCN (2 moles)}} N{\equiv}C{-}CH_2CH_2{-}C{\equiv}N$$

<table>
<tr><td align="center">Ethene
(ethylene)</td><td align="center">1,2-Dibromoethane
(ethylene dibromide)</td><td align="center">Butanedinitrile</td></tr>
</table>

$$\Big\downarrow \text{H}_3\overset{\oplus}{\text{O}}\text{, heat}$$

$$\underset{\substack{\text{Butanedioic acid}\\ \text{(succinic acid)}}}{HO{-}\overset{\displaystyle O}{\overset{\|}{C}}{-}CH_2CH_2{-}\overset{\displaystyle O}{\overset{\|}{C}}{-}OH}$$

Aromatic nitriles are somewhat more difficult to obtain because of the great difficulty in carrying out a nucleophilic displacement reaction on an aryl halide (see Sec. 16.4):

$$Ar{-}X + \overset{\ominus}{:}C{\equiv}N: \longrightarrow \text{ no reaction (unless electron-withdrawing groups are on Ar)}$$

The sodium salt of an arylsulfonic acid undergoes reaction with cyanide to produce the corresponding nitrile, but this nucleophilic aromatic substitution reaction also gives poor yields:

$$Ar{-}SO_3H \xrightarrow[\text{(neutralize)}]{\text{NaOH}} Ar{-}SO_3^{\ominus}Na^{\oplus} \xrightarrow[\text{fuse}]{\text{solid NaCN}} Ar{-}C{\equiv}\overset{..}{N}$$

<table>
<tr><td align="center">Arylsulfonic
acid</td><td align="center">Sodium
arylsulfonate</td><td align="center">Aromatic
nitrile</td></tr>
</table>

As we learn in Sec. 26.8 one of the most general methods for preparing aromatic nitriles involves the diazonium salt. They can also be prepared from the corresponding carboxylic acid.

The use of organothallium reagents as precursors of aromatic nitriles (see Sec. 14.23) is a much easier route to this family of compounds.

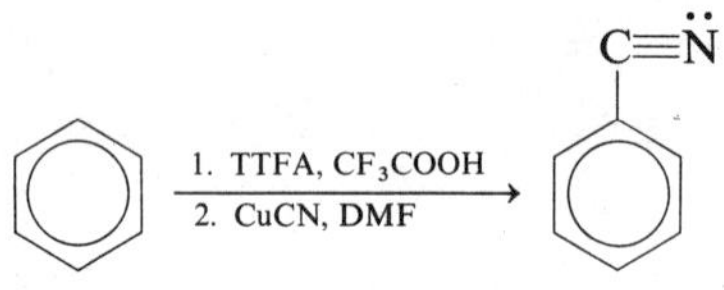

where TTFA = thallium trifluoroacetate
      DMF = dimethyl formamide

Question 22.3

Devise suitable methods for carrying out the following transformations, using any other needed reagents:

(*a*) toluene to phenylacetic acid (two methods)
(*b*) toluene to *o*-methylbenzoic acid (*o*-toluic acid)
(*c*) toluene to *p*-nitrophenylacetic acid

Question 22.4

Propose a reasonable mechanism (show all steps and intermediates) for the conversion of sodium benzenesulfonate to benzonitrile.

Question 22.5

Which would be more reactive toward cyanide ion, *p*-nitrobromobenzene or 2,4,6-trinitrobromobenzene? Why?

22.6   Special Methods for Preparing Carboxylic and Dicarboxylic Acids

Certain carboxylic and dicarboxylic acids are important industrial commodities and are prepared by special methods from readily available starting compounds. We discuss several of these compounds in this section.

Methanoic acid (formic acid), HCOOH, is usually prepared from the reaction between carbon monoxide, CO, and aqueous sodium hydroxide solution:

$$C{\equiv}O\ (gas) + NaOH \xrightarrow[\text{200–250 lb/sq in.}]{\text{160–200°}} \underset{\substack{\text{Sodium methanoate} \\ \text{(sodium formate)}}}{H-\overset{\overset{\displaystyle O}{\|}}{C}-\overset{\ominus}{\ddot{O}}{:}\ Na^{\oplus}} \xrightarrow{H_3\ddot{O}^{\oplus}} \underset{\substack{\text{Methanoic acid} \\ \text{(formic acid)}}}{H-\overset{\overset{\displaystyle O}{\|}}{C}-OH}$$

Carbon monoxide

Formic acid is less stable than the other carboxylic acids. Because it contains a hydrogen attached to the acyl group (in addition to the hydroxyl group), it possesses properties quite similar to many aldehydes. For example, it is readily oxidized to carbonic acid, which is unstable and decomposes to water and carbon dioxide. It also thermally decomposes on heating and gives carbon dioxide and hydrogen gas.

$$\underset{\text{Carbonic acid}}{\left[HO-\overset{\overset{\displaystyle O}{\|}}{C}-OH \right]} \overset{[O]}{\underset{}{\nwarrow}} \underset{\substack{ }}{H-\overset{\overset{\displaystyle O}{\|}}{C}-OH} \overset{\text{160–180°}}{\searrow} H_2 + CO_2$$

$$H_2O + CO_2$$

Ethanoic acid (acetic acid), CH_3COOH, is present to the extent of about 5% in vinegar. Acetic acid also results from the enzyme-catalyzed oxidation of the ethanol in wine, beer, and cider. One commercial method for preparing acetic acid involves the oxidation of dilute aqueous solutions of ethanol. The reaction is catalyzed by adding compounds that produce the microorganisms necessary for catalyzing the oxidation. This method often gives aqueous solutions with 12 to 14% acetic acid.

In addition to other natural sources, acetic acid is today most commonly prepared from the oxidation of ethanal (acetaldehyde). This process involves passing acetaldehyde and air over a cobalt acetate catalyst with heating:

$$\underset{\substack{\text{Ethanal} \\ \text{(acetaldehyde)}}}{CH_3-\overset{\overset{\displaystyle O}{\|}}{C}-H} \xrightarrow[\text{cobalt acetate, 70–80°}]{O_2} \underset{\substack{\text{Ethanoic acid} \\ \text{(acetic acid)}}}{CH_3-\overset{\overset{\displaystyle O}{\|}}{C}-OH}$$

Recall that acetaldehyde can be prepared from either the acid-catalyzed hydration of ethyne (acetylene) (see Sec. 11.13) or the dehydrogenation of ethanol by copper metal catalyst (see Secs. 20.5 and 10.10).

Glacial acetic acid is acetic acid that is at least 99.5% pure. It is called glacial because it melts at 16.7° and freezes in icelike crystals. Most glacial acetic acid in the United States comes from the air oxidation of acetaldehyde.

Ethanoic acid (acetic acid) is commonly used as a solvent itself and in the preparation of many esters (usually acetates), which are important solvents. Cellulose acetate, acetic anhydride, and polyvinyl acetate are other common and important acetate-containing products, and products derived from acetic acid are used in the textile, photographic, and paint industries.

Other lower molecular weight carboxylic acids are usually prepared by oxidation of either the corresponding alcohol or aldehyde. Many higher molecular weight, straight-chain acids, especially those containing an even number of carbon atoms (*the fatty acids*), are obtained from the hydrolysis of fats and oils (see Sec. 25.2).

Benzoic acid, C_6H_5COOH, is usually obtained from the vigorous oxidation of alkylbenzenes, such as toluene, which are available from petroleum.

o-Hydroxybenzoic acid (salicylic acid) (1) is obtained from the reaction between sodium phenoxide and carbon dioxide:

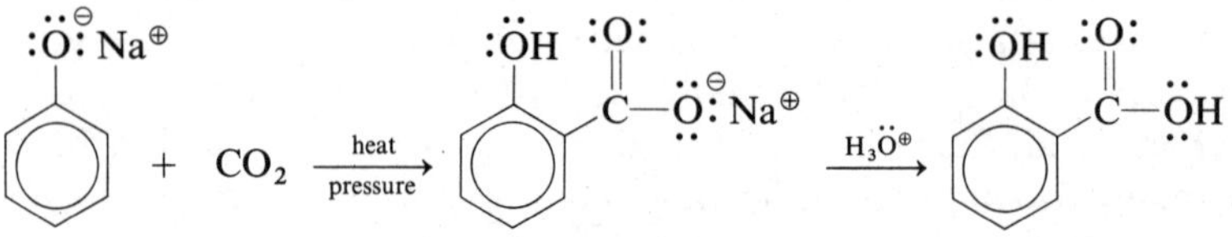

| Sodium phenoxide | Sodium salicylate | (1) Salicylic acid |

This reaction, known as the **Kolbe reaction,** is conceptually similar to the Reimer-Tiemann reaction (see Question 19.12). The negative charge in sodium phenoxide, which is distributed throughout the ring, attacks the partially positive carbon atom in carbon dioxide, followed by a proton transfer to give sodium salicylate. The mechanism for this reaction may be visualized as follows:

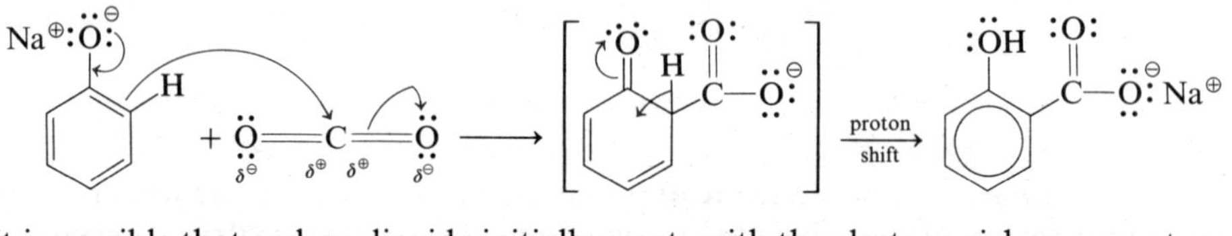

(It is possible that carbon dioxide initially reacts with the electron-rich oxygen atom of the phenoxide ion and that rearrangement by migration to the ring occurs subsequently.) Some *para* isomer is also formed, but it is readily separated from the *ortho* isomer by steam distillation.

Five commonly encountered dicarboxylic acids are obtained industrially by the reactions depicted here:

Oxalic acid:

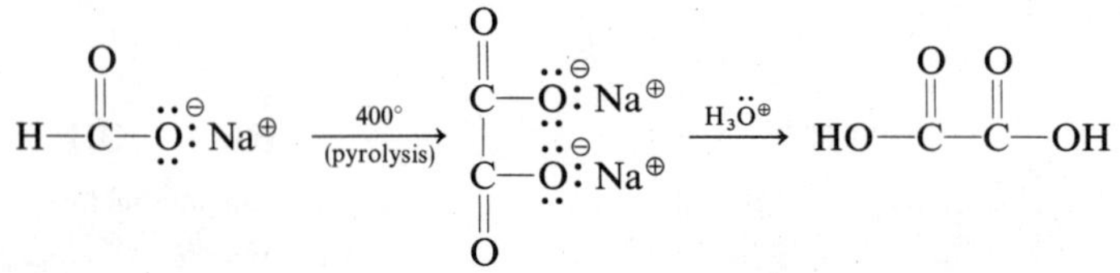

| Sodium methanoate (sodium formate) | Sodium oxalate | Ethanedioic acid (oxalic acid) |

Malonic acid:

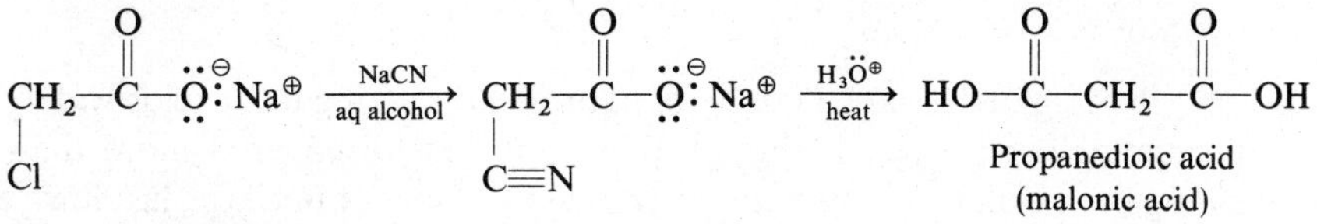

Sodium chloroethanoate
(sodium chloroacetate)

Sodium cyanoacetate

Propanedioic acid
(malonic acid)

Succinic acid:

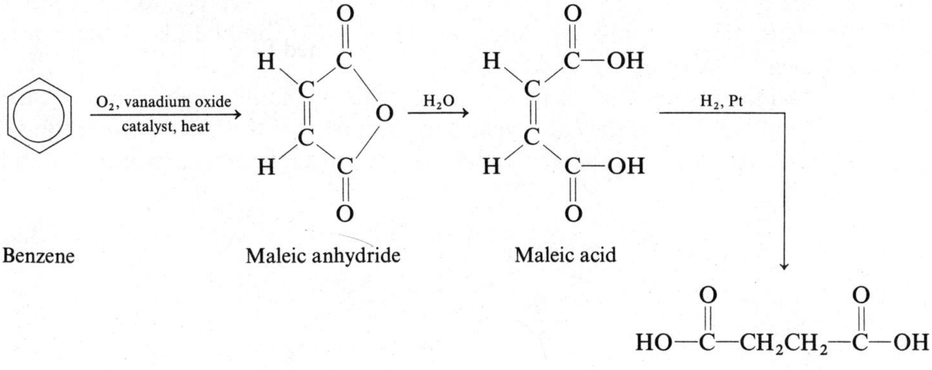

Benzene

Maleic anhydride

Maleic acid

Butanedioic
acid
(succinic acid)

Adipic acid:

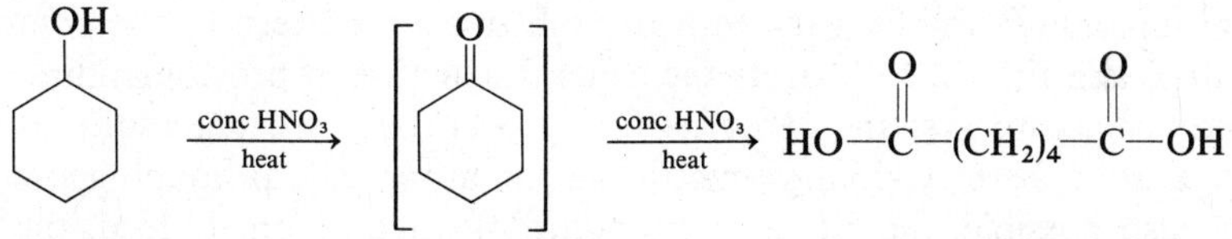

Cyclohexanol

Cyclohexanone
Not isolated

Hexanedioic acid
(adipic acid)

Phthalic acid:

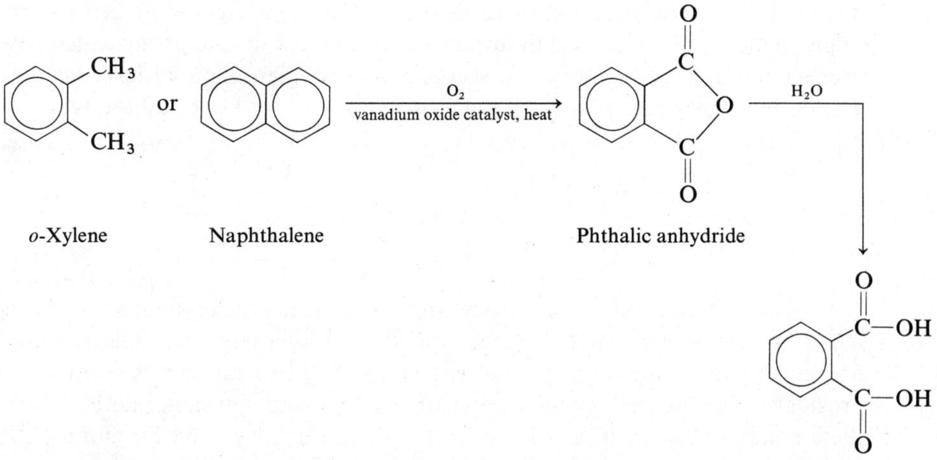

o-Xylene

Naphthalene

Phthalic anhydride

Phthalic acid

22.7   Prostaglandins

Prostaglandins are a family of naturally occurring fatty acids widely distributed in mammalian tissues. Although they were first discovered more than 50 yr ago, having been isolated from human seminal plasma, not until the last few years was the potent and wide-ranging biological activity of these compounds finally realized. Since then, many of these compounds have been found to demonstrate extremely potent biological activity in fertility control, muscle contraction and hence labor induction, regulation of blood pressure, and kidney problems; more recently they have been shown to play a major role in blood platelet aggregation. The latter point stimulated research into the potential use of these compounds in the control and treatment of heart attack and stroke.

All the prostaglandins have certain structural similarities. They are carboxylic acids, but in addition they are cyclic and composed of the basic C_{20} carbon skeleton of *prostanoic acid*. Several general classes of prostaglandins are grouped under the

Prostanoic acid

designations of A, B, E, and F among others depending on the basic carbon skeletal arrangement and the number and orientation of double bonds, hydroxyl groups, and ketone groups. A compound is given the more complete symbolism PGE_2 or $PGF_{2\alpha}$, for example, where PG stands for prostaglandin, E or F indicates the prostaglandin group, and 2 and 2α vary with the substitution pattern of the compound. The examples in Fig. 22.1 illustrate the general structure of prostaglandins and the current nomenclature system. Most naturally occurring prostaglandins have the *S* configuration at the C-15 asymmetric carbon atom and, although geometric isomerism is also possible, the 5,6-*cis* isomers are most common. Indeed, the isomers of the *R* configuration at C-15 exhibit little or no biological activity when compared with the naturally occurring *S* isomers.

The six E and F prostaglandins are derivatives of *eicosanoic acid*, as shown in Fig. 22.1. These are referred to as the *primary* prostaglandins because they are found in most cells. They can in turn be converted to a second group called *secondary* prostaglandins. The primary and secondary prostaglandins that have been isolated and/or synthesized as well as characterized number well over 20 today.

Eicosanoic acid

As is often the case when new materials with great potential in the field of medicinal chemistry are discovered, the sources of these materials are much too limited to supply the demand that is generated. The next step, which occurred in prostaglandin research, is the design of one or more synthetic methods by which these compounds can be prepared in the laboratory. By 1968 Dr. John E. Pike of Upjohn Pharmaceuticals and Dr. E. J. Corey of Harvard University had developed and published methods for the total synthesis of several naturally occurring prostaglandins that had exhibited biological activity. Their cost then was prohibitively

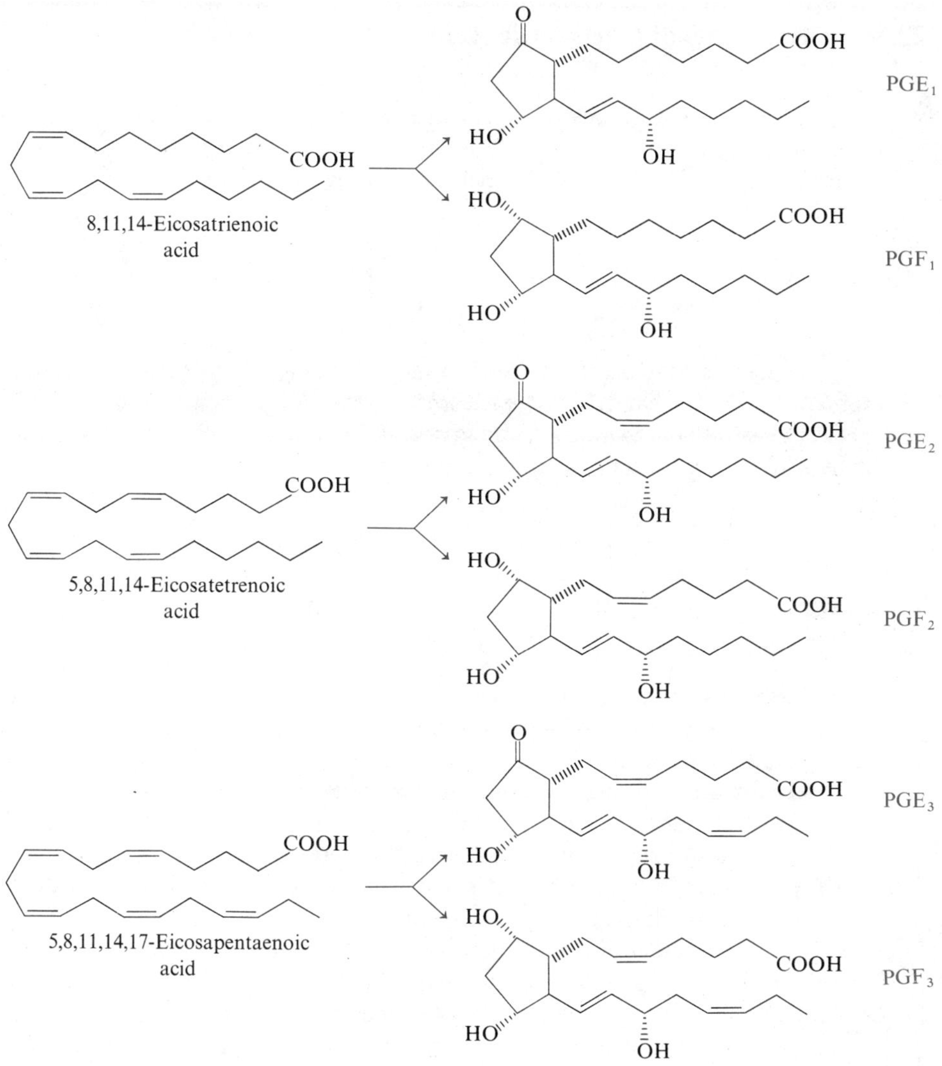

FIGURE 22.1 The six primary prostaglandins and the eicosanoic acid analogs from which they are derived.

high—approaching $1,000/g. In recent years a new and promising source of pros-taglandins and a variety of other drugs has emerged—the oceans. In 1969 a sig-nificant concentration of prostaglandins was isolated from the *gorgonian sea whip coral (Plexaura homomalla)*, which is found in high concentrations in the Caribbean off the Florida coast. Since that time, many other natural sources of these materials have been discovered. Research is under way to determine the feasibility of cultivating or farming the sea to produce these materials in the quantity needed. At the same time, newer and, more important, lower cost methods for synthesizing these com-pounds in the laboratory are emerging.

Prostaglandins have the potential of becoming the most important class of biologically active compounds ever discovered. As research progresses, their role in mammalian physiology and biochemistry is continually expanding and promises to do so for years to come.

22.8   Reactions of Carboxylic Acids

Several general types of reactions are typical of the carboxyl group. As we examine the structure of the —COOH group, we find that there are two places where bonds can be broken and made with preservation of the acyl group:

$$
\begin{array}{c}
:\!O\!: \\
\|\\
R\!-\!\overset{}{C}\!\!\!\underset{A}{\vdots}\!\!\!\overset{\cdot\cdot}{\underset{B}{O}}\!\!\!\vdots\!\!\!H
\end{array}
$$

Reactions that break the acyl-oxygen bond (bond A) belong to a family of similar reactions that involve **nucleophilic acyl substitution.** Replacement of the —OH group by other substituents gives rise to acid chlorides, amides, esters, and anhydrides. These reactions are presented in detail in Chap. 23.

Nucleophilic acyl substitution:

$$
\underset{}{\overset{O}{\underset{\|}{R\!-\!C\!-\!OH}}} \longrightarrow \quad \underset{\text{Acid chlorides}}{\overset{O}{\underset{\|}{R\!-\!C\!-\!Cl}}} \quad or \quad \underset{\text{Amides}}{\overset{O}{\underset{\|}{R\!-\!C\!-\!NH_2}}} \quad or \quad \underset{\text{Esters}}{\overset{O}{\underset{\|}{R\!-\!C\!-\!OR'}}} \quad or \quad \underset{\text{Anhydrides}}{\overset{O\qquad O}{\underset{\|\qquad\|}{R\!-\!C\!-\!O\!-\!C\!-\!R}}}
$$

Reactions that involve the oxygen-hydrogen bond (bond B) largely revolve about the **acidity of carboxylic acids.** In Sec. 22.9 we study the reasons for their acidic nature in some detail.

Certain other reactions are specific for the carboxyl group. One is the **reduction** of a carboxylic acid to the corresponding primary alcohol, and another is the **decarboxylation** of a carboxylic acid (and in certain cases of dicarboxylic acids), Carboxylic acids may also be **halogenated in the α position** to give α-halo acids. whereas the other derivatives of carboxylic acids do not so react.

In the remainder of this chapter we discuss—from a mechanistic standpoint— these various types of reactions of carboxylic acids, dicarboxylic acids, and their derivatives. We start with the reactions of the hydroxyl group and then discuss the remaining reactions. The substitution reactions of acids are deferred to Chap. 23.

22.9   Acidic Nature of Carboxylic Acids

Carboxylic acids have considerable acidic nature and are the acids of organic chemistry in much the same way that mineral acids are the acids of inorganic chemistry. We encountered several other families of organic compounds that are acidic to varying extents due to rupture of the oxygen-hydrogen bond of the —O—H group: sulfonic acids, $ArSO_2$—O—H, alcohols, RO—H, and phenols, ArO—H.

Let us start our discussion of carboxylic acid acidity by comparing the carboxyl group with the hydroxyl group in an alcohol. This is best accomplished by considering the structure of the anions formed after dissociation occurs.

Dissociation of the O—H group occurs to a much greater extent in acids than in alcohols. The reaction is reversible, and the equilibrium constants give a good indication of the relative acidities of these two families of compounds. The following equations express the equilibrium, and the K_a values given are only an order of magnitude correct.

Alcohols:

$$R-\ddot{\underset{..}{O}}-H + H_2\ddot{\underset{..}{O}}: \;\rightleftharpoons\; R-\ddot{\underset{..}{\overset{..}{O}}}:^{\ominus} + H_3\overset{..}{\underset{..}{O}}{}^{\oplus} \qquad K_a = \dfrac{[R\ddot{\underset{..}{O}}:^{\ominus}][H_3\overset{..}{\underset{..}{O}}{}^{\oplus}]}{[R\ddot{\underset{.}{O}}-H]} \approx 10^{-15}\text{--}10^{-18}$$

Carboxylic acids:

$$\underset{\substack{\big\Vert\\ }}{R-\overset{\displaystyle O}{C}}-\ddot{\underset{.}{O}}H + H_2\ddot{\underset{..}{O}}: \;\rightleftharpoons\; \underset{\substack{\big\Vert\\ }}{R-\overset{\displaystyle O}{C}}-\ddot{\underset{..}{\overset{..}{O}}}:^{\ominus} + H_3\overset{..}{\underset{..}{O}}{}^{\oplus} \qquad K_a = \dfrac{[RCO\ddot{\underset{..}{O}}:^{\ominus}][H_3\overset{..}{\underset{..}{O}}{}^{\oplus}]}{[RCO\ddot{\underset{.}{O}}H]} \approx 10^{-4}\text{--}10^{-5}$$

Clearly, however, *acids are approximately 10^{10} (10 trillion) times more acidic than alcohols!* Why this tremendous acidity difference?

Let us first look at the structure of the undissociated alcohol and the undissociated acid and their conjugate bases, the alkoxide and carboxylate anions.

When an alcohol dissociates, the alkoxide ion has no special stability or instability relative to the carboxylate ion:

$$R\ddot{\underset{..}{O}}-H \;\rightleftharpoons\; R\ddot{\underset{..}{\overset{..}{O}}}:^{\ominus} + H^{\oplus}$$

Alkoxide ion
*No special
stability or
instability*

In a carboxylic acid, however, the corresponding carboxylate ion, which is trigonal ($\sim 120°$ bond angles), has special electronic properties. Two equivalent electronic structures (*resonance structures*) may be drawn for the carboxylate anion:

$$\underset{\substack{\big\Vert\\ }}{R-\overset{\displaystyle O}{C}}\diagdown\underset{..}{\overset{..}{O}}{-}H \;\rightleftharpoons\; H^{\oplus} + \left[R-C\diagup\!\!\!\diagdown\overset{\textstyle \cdot\cdot\,O\cdot\cdot}{\underset{\cdot\cdot\,O\cdot\cdot}{}}{}^{\ominus} \;\longleftrightarrow\; R-C\diagup\!\!\!\diagdown\overset{\textstyle \cdot\cdot O:^{\ominus}}{\underset{\cdot\cdot O}{}} \right] \;\equiv\; R-C\diagup\!\!\!\diagdown\overset{\textstyle \cdot\,O\cdot^{\delta\ominus}}{\underset{\cdot\,O\cdot^{\delta\ominus}}{}}$$

Carboxylate anion   Resonance hybrid
Especially stable   *Best representation*

The electronic structure of the carboxylate anion may be best viewed in terms of the orbitals involved, which show that the negative charge is distributed over two oxygen atoms and one carbon atom; the *p* orbital on carbon can overlap with the *p* orbitals of each of the oxygen atoms (the overlap orbitals on oxygen may have some hybrid character):

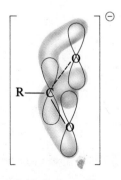

Orbital view of
carboxylate anion

X-ray studies on carboxylic acids and their salts substantiate our thinking about the structure of the anion. For example, the following bond lengths have been measured for ethanoic (formic) acid and the ethanoate (formate) ion:

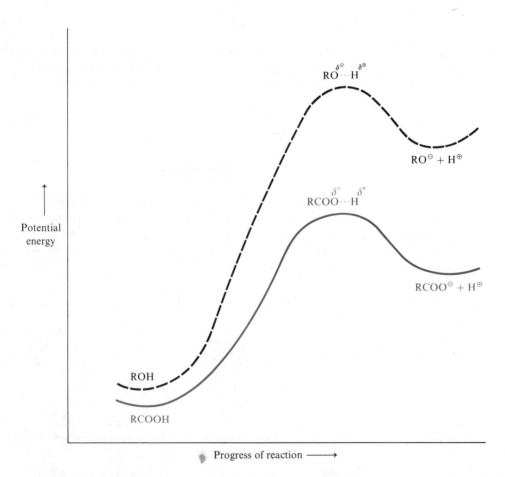

In the free acid, one bond (C=O bond, 1.23 Å) is appreciably shorter than the other (C—O bond, 1.36 Å). In the formate anion, however, both carbon-oxygen bonds have the same length (1.27 Å), which is intermediate between those for a carbon-oxygen single bond and a carbon-oxygen double bond. This is precisely what is expected on the basis of our orbital picture of the resonance-stabilized anion where the electrons are shared *equally* between carbon and the two oxygen atoms.

In summary, the carbonyl group on the carbon bearing a hydroxyl group affects the acidity of the —COOH group; the carboxylate anion that results from ionization of an acid is stabilized by resonance. And, the carboxylate anion is considerably more stable than our standard of comparison, the alkoxide ion.

FIGURE 22.2   Energy-profile diagram comparing the stabilities of an alcohol and its alkoxide ion with an acid and its anion.

There is also evidence of some resonance stabilization in the free acid, however, as shown by the structures:

$$
\left[
\begin{array}{ccc}
\overset{\cdot\cdot}{\underset{}{O}} & & \overset{\cdot\cdot}{\underset{}{O}}{}^{\ominus} \\
\| & & | \\
R-C & \longleftrightarrow & R-C \\
\diagdown\overset{\cdot\cdot}{O}-H & & \diagdown\overset{\oplus}{O}-H
\end{array}
\right]
$$

The resonance structures for the free acid are not nearly so important as they are for the carboxylate anion. A graphic display of the stability of a carboxylic acid and its anion compared with the stability of an alcohol and its anion is shown in the energy-profile diagram in Fig. 22.2.

22.10 Structure-Acidity Correlations: Aliphatic Carboxylic Acids

Our ideas about the electron-withdrawing character of the acyl group and its effect on the acidity of the O—H bond are supported by structure-acidity correlations, in which the effect on acidity by substituents elsewhere in the molecule is studied.

Let us start by considering methanoic (formic) and ethanoic (acetic) acids, with K_a values of 1.8×10^{-4} and 1.8×10^{-5}, respectively, which indicates that formic acid is a *stronger* acid than acetic acid. (The larger the K_a value, the equilibrium constant for dissociation, the more acidic the compound, because this indicates that the equilibrium lies farther on the side of the dissociated form of the acid.) Are these data in accord with what we know about the inductive effect? Yes, they are. The methyl group in acetic acid donates (releases) electrons compared with hydrogen in formic acid. A group that releases electrons to the already negatively charged carboxylate anion should make it less stable, as shown in comparing acetate ion with formate ion:

$$
CH_3 \rightarrow C\!\!\overset{\overset{\cdot\cdot}{O}}{\underset{\underset{\cdot\cdot}{O}{}^{\ominus}}{}} \qquad \text{less stable than} \qquad H-C\!\!\overset{\overset{\cdot\cdot}{O}}{\underset{\underset{\cdot\cdot}{O}{}^{\ominus}}{}}
$$

Additional methyl groups attached to the acid result in decreased acidity (smaller K_a values), as the following data indicate:

$$
\underset{}{CH_3\!-\!\overset{\overset{O}{\|}}{C}\!-\!OH} \qquad CH_3\!\rightarrow\!CH_2\!\rightarrow\!\overset{\overset{O}{\|}}{C}\!-\!OH \qquad CH_3\!\rightarrow\!\overset{\overset{CH_3}{\downarrow}}{CH}\!\rightarrow\!\overset{\overset{O}{\|}}{C}\!-\!OH \qquad CH_3\!\rightarrow\!\overset{\overset{CH_3}{\downarrow}}{\underset{\underset{CH_3}{\uparrow}}{C}}\!\longrightarrow\!\overset{\overset{O}{\|}}{C}\!-\!OH
$$

$K_a = 1.8 \times 10^{-5}$ 1.35×10^{-5} 1.44×10^{-5} 9.4×10^{-6}
Standard of comparison

The inductive effect may well account for some differences in acidity because increasing the number of electron-donating methyl groups in an acetic acid molecule should decrease the acidity. On the other hand, energy considerations (enthalpy and entropy of ionization) suggest that steric factors may also be involved. The more hindered acids, as well as the corresponding anions, are less highly solvated than the less hindered acids. Solvation plays an important role (perhaps the major role) in assisting the initial dissociation and in dictating the related acidity of a given acid. As with alcohols and phenols (see Secs. 10.9 and 17.4), the gas-phase acidities of carboxylic acids are the reverse of those in solution. This points to the importance of solvation effects in ionization as well as in stabilizing the negative charge of the carboxylate anion that is formed. But there are no hard and fast generalizations that correlate alkyl group size with acidity.

The results observed for various halogenated acids are more clear cut. Halogens are electronegative atoms and good electron-withdrawing substituents. The following data show the effect of introducing additional halogen atoms in the α position:

$$CH_3—\overset{\overset{\textstyle O}{\|}}{C}—OH \qquad Cl{\leftarrow}CH_2—\overset{\overset{\textstyle O}{\|}}{C}—OH \qquad Cl{\leftarrow}\overset{\overset{\textstyle Cl}{\uparrow}}{C}H—\overset{\overset{\textstyle O}{\|}}{C}—OH \qquad Cl{\leftarrow}\overset{\overset{\textstyle Cl}{\uparrow}}{\underset{\underset{\textstyle Cl}{\downarrow}}{C}}—\overset{\overset{\textstyle O}{\|}}{C}—OH$$

$$K_a = 1.8 \times 10^{-5} \qquad 1.4 \times 10^{-3} \qquad 3.3 \times 10^{-2} \qquad 2 \times 10^{-1}$$

Chlorine withdraws electrons most strongly from the α carbon, which in turn withdraws electrons (but to a lesser extent) from the carbonyl or carboxylate group. The negatively charged carboxylate anion, which forms on dissociation, is *stabilized* by electron withdrawal, so that α-halo acids are more acidic than acetic acid. Increasing the number of α-halo substituents drastically increases acidity.

As we might expect, changing the electron-withdrawing power of the halogen affects acidity. The decreasing electronegativity order of the halogens is F > Cl > Br > I, and as the following K_a values show, electronegativity parallels acidity fairly well.

$$X—CH_2—\overset{\overset{\textstyle O}{\|}}{C}—OH$$

X	K_a
F	2.6×10^{-3}
Cl	1.4×10^{-3}
Br	1.3×10^{-3}
I	6.7×10^{-4}

Finally, we can see how the inductive effect varies as a substituent is moved farther from the carboxyl group by considering the dissociation constants for a series of chlorobutanoic acids:

$$CH_3CH_2CH_2—\overset{\overset{\textstyle O}{\|}}{C}—OH \qquad CH_3CH_2—\underset{\underset{\textstyle Cl}{|}}{C}H—\overset{\overset{\textstyle O}{\|}}{C}—OH \qquad CH_3—\underset{\underset{\textstyle Cl}{|}}{C}H—CH_2—\overset{\overset{\textstyle O}{\|}}{C}—OH$$

$$K_a = 1.5 \times 10^{-5} \qquad 1.4 \times 10^{-3} \qquad 9.0 \times 10^{-5}$$

$$\underset{\underset{\textstyle Cl}{|}}{C}H_2CH_2CH_2—\overset{\overset{\textstyle O}{\|}}{C}—OH$$

$$2.8 \times 10^{-5}$$

The inductive effect (as it relates to the carboxyl group) decreases quite rapidly as the chlorine atom is moved farther from the carboxyl group. Indeed, 4-chloro-butanoic acid is not much more acidic than butanoic acid. Note also that 2-chloro-butanoic acid is about as acidic as chloroethanoic acid.

In summary, the inductive effect may involve electron donation or electron withdrawal, although there are many more electron-withdrawing groups than electron-donating groups. Electron donation decreases acidity by destabilizing the resulting carboxylate anion. Electron withdrawal increases acidity by stabilizing the resulting carboxylate anion. The inductive effect falls off very rapidly as the substituent is moved farther from the carboxyl group.

Question 22.6

Which acid, nitroacetic (O_2N—CH_2COOH) or acetic, is stronger? Why?

Question 22.7

In each of the following four pairs of compounds, which is the stronger acid? Briefly justify your choice.

(a) α,α-dichloropropionic acid or α,β-dichloropropionic acid
(b) propionic acid or α-hydroxypropionic acid
(c) butyric acid or isobutyric acid
(d) α-chloro-β-fluorobutyric acid or α-fluoro-β-chlorobutyric acid

22.11 Aromatic Carboxylic Acids and Structure-Acidity Correlations

The parent aromatic carboxylic acid is benzoic acid, C_6H_5COOH. Its K_a is 6.5×10^{-5} so benzoic acid is a little more acidic than most aliphatic acids (recall that $K_a = 1.8 \times 10^{-5}$ for acetic acid, 1.35×10^{-5} for propionic acid, and so on). Two factors may influence its acidity. One is the inductive effect of the phenyl group, which is known to be electron withdrawing. The other possibility is the resonance effect, whereby the phenyl group donates electrons to the resulting carboxylate anion. The following electronic structures illustrate the resonance interaction of the phenyl ring with the anion:

Benzoate anion   Resonance hybrid

Electron donation through resonance would make the benzoate anion less stable. Both effects may be operating, but experimental data suggest that the electron-withdrawing inductive effect of the phenyl group is more important.

Structure-acidity correlations for substituted benzoic acids have also been determined, and the data for selected substituents are in Table 22.4. Judging from these data, the following trends are important. Substituents that activate the aromatic ring toward electrophilic aromatic substitution (see Sec. 14.8) decrease acidity because they donate electrons through the inductive effect or the resonance effect. Substituents that deactivate the aromatic ring toward electrophilic aromatic substitution (see Sec. 14.8) increase acidity because they withdraw electrons, usually through the inductive effect. The most activating substituents in electrophilic aromatic substitution (such

TABLE 22.4   Dissociation Constants for Selected Monosubstituted Aromatic Carboxylic Acids: $ArCOOH + H_2\overset{..}{\underset{..}{O}}: \rightleftarrows Ar\overset{..}{\underset{..}{C}}\overset{..}{\underset{..}{O}}\overset{\ominus}{:} + H_3\overset{..}{O}{}^{\oplus}$

	K_a* for Substituent in		
Substituent in Ar—	*ortho* Position	*meta* Position	*para* Position
—CH_3	12.9×10^{-5}	5.8×10^{-5}	4.6×10^{-5}
—$C(CH_3)_3$	35×10^{-5}		
—Cl	115×10^{-5}	14.8×10^{-5}	10.3×10^{-5}
—NO_2	680×10^{-5}	36×10^{-5}	37×10^{-5}
—OH	110×10^{-5}	8.6×10^{-5}	2.6×10^{-5}
—OCH_3	8.4×10^{-5}	8.1×10^{-5}	5.3×10^{-5}
—NH_2	1.8×10^{-5}	1.8×10^{-5}	1.4×10^{-5}

* K_a for benzoic acid $= 6.5 \times 10^{-5}$

as —OH and —NH_2) decrease acidity the most, and the most deactivating substituents (such as —NO_2) increase acidity the most. These trends agree well with our view of acidity.

The *ortho*-substituted benzoic acids are considerably stronger than their *meta* and *para* counterparts. The reasons for this are not entirely known, although there may be a steric effect from having two substituents *ortho* to one another. For example, the resonance stabilization in aromatic systems is most effective when all the atoms can be in the same plane (that is, they are *coplanar*). Substituents in the *ortho* position reduce the degree of coplanarity of both the carboxyl group and the carboxylate anion, thereby reducing orbital overlap and the concomitant effect of the delocalization (which is a destabilizing effect in these examples). As an illustration of this effect, *o-tert*-butylbenzoic acid ($K_a = 35 \times 10^{-5}$) is a stronger acid than *o*-toluic acid ($K_a = 12.9 \times 10^{-5}$).

Another interesting comparison is *o*-hydroxybenzoic acid (salicylic acid, $K_a = 110 \times 10^{-5}$) versus *o*-methoxybenzoic acid ($K_a = 8.4 \times 10^{-5}$). The —OH and —OCH_3 substituents are somewhat similar in their activating properties in electrophilic aromatic substitution reactions, but they exert a great difference in activation in acidity. Salicylic acid is more acidic, and it is suggested that the carboxylate ion is stabilized by internal hydrogen bonding of the type shown here:

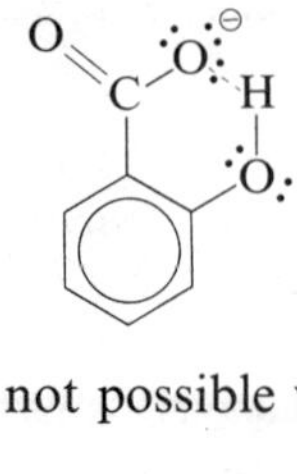

This type of hydrogen bonding is not possible with an *o*-OCH_3 group.

Question 22.8

How would you account for hydroxyacetic acid ($K_a = 1.5 \times 10^{-4}$) being stronger than acetic acid ($K_a = 1.8 \times 10^{-5}$), whereas *p*-hydroxybenzoic acid ($K_a = 2.6 \times 10^{-5}$) is weaker than benzoic acid ($K_a = 6.5 \times 10^{-5}$)?

Question 22.9

For each of the following five pairs of compounds, indicate which compound you would expect to be the stronger acid and briefly indicate your reasoning.

(*a*) *p*-acetylbenzoic acid (*p*-$CH_3COC_6H_4COOH$) or benzoic acid
(*b*) *m*-fluorobenzoic acid or *m*-bromobenzoic acid
(*c*) *p*-cyanobenzoic acid or *p*-chlorobenzoic acid
(*d*) *p*-methylbenzoic acid or *p*-trichloromethylbenzoic acid (*p*-$CCl_3C_6H_4COOH$)
(*e*) benzoic acid or phenylacetic acid

22.12   Dicarboxylic Acids and Structure-Acidity Correlations

Di- and *tri*basic inorganic acids, such as H_2CO_3, H_2S, and H_3PO_4, have two or three acid dissociation constants, and removal of the first proton is easier than removal of the second (or third) proton. The relative ease of dissociation can be seen from the magnitude of the dissociation constants. Likewise, dicarboxylic acids undergo dissociation in two steps as shown by the general equation:

$$HO-\overset{O}{\overset{\|}{C}}-(CH_2)_n-\overset{O}{\overset{\|}{C}}-OH + H_2O \overset{K_1}{\rightleftharpoons} HO-\overset{O}{\overset{\|}{C}}-(CH_2)_n-\overset{O}{\overset{\|}{C}}-O^{\ominus} + H_3O^{\oplus}$$

$$HO-\overset{O}{\overset{\|}{C}}-(CH_2)_n-\overset{O}{\overset{\|}{C}}-O^{\ominus} + H_2O \overset{K_2}{\rightleftharpoons} {}^{\ominus}O-\overset{O}{\overset{\|}{C}}-(CH_2)_n-\overset{O}{\overset{\|}{C}}-O^{\ominus} + H_3O^{\oplus}$$

Dissociation constants for some dicarboxylic acids are in Table 22.5.

TABLE 22.5   Dissociation Constants for Dicarboxylic Acids

Name	Structure	K_1	K_2
Ethanedioic acid (oxalic acid)	$HOOC-COOH$	$3{,}200 \times 10^{-5}$	5.2×10^{-5}
Propanedioic acid (malonic acid)	$HOOC-CH_2-COOH$	$1{,}300 \times 10^{-5}$	0.2×10^{-5}
Butanedioic acid (succinic acid)	$HOOC-(CH_2)_2-COOH$	6.3×10^{-5}	0.25×10^{-5}
Pentanedioic acid (glutaric acid)	$HOOC-(CH_2)_3-COOH$	4×10^{-5}	0.4×10^{-5}
cis-2-Butanedioic acid (maleic acid)	*cis*-$HOOC-CH{=}CH-COOH$	$1{,}000 \times 10^{-5}$	0.05×10^{-5}
trans-2-Butenedioic acid (fumaric acid)	*trans*-$HOOC-CH{=}CH-COOH$	100×10^{-5}	3.2×10^{-5}
1,2-Benzenedicarboxylic acid (phthalic acid)	*o*-$C_6H_4(COOH)_2$	120×10^{-5}	0.4×10^{-5}
1,3-Benzenedicarboxylic acid (isophthalic acid)	*m*-$C_6H_4(COOH)_2$	30×10^{-5}	2.5×10^{-5}
1,4-Benzenedicarboxylic acid (terephthalic acid)	*p*-$C_6H_4(COOH)_2$	30×10^{-5}	3.0×10^{-5}

There is a large difference between the first and second dissociation constants of oxalic, malonic, maleic, and phthalic acids. Two reasons have been offered for this. One explanation hinges on the electrostatic repulsion generated when two carboxylate anions are close to one another in the same molecule. An alternate explanation involves the stability gained from internal hydrogen bonding possible within the monoanion, for example:

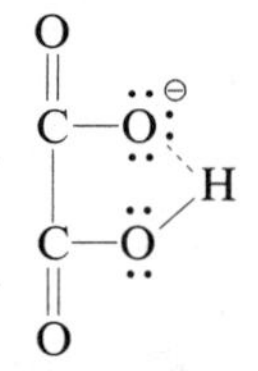

$$
\begin{array}{c}
\text{O} \\
\parallel \\
\text{C}-\ddot{\text{O}}\colon^{\ominus} \\
\mid \qquad \ddots \\
\qquad \qquad \text{H} \\
\text{C}-\ddot{\text{O}} \\
\parallel \\
\text{O}
\end{array}
$$

The effect of geometry on acidity is particularly well demonstrated by maleic and fumaric acids, where the carboxyl groups are forced close to one another in maleic acid (the *cis* isomer) but are held far apart in fumaric acid (the *trans* isomer).

Question 22.10
___
What effect do you think is responsible for the K_1 values for oxalic acid and malonic acid being considerably larger than those for succinic acid and glutaric acid?
___

Question 22.11
___
Account for the borderline of solubility for dicarboxylic acids in water occurring at six or seven carbon atoms, whereas that for monocarboxylic acids occurs at about four carbon atoms.
___

Question 22.12
___
Account for the large difference between K_1 and K_2 for phthalic acid, and explain why there is no similar difference for isophthalic and terephthalic acids even though all three compounds contain two carboxyl groups attached to an aromatic ring.
___

22.13   Salt Formation of Carboxylic and Dicarboxylic Acids

An important reaction of carboxylic and dicarboxylic acids results from their acidity. They react readily with inorganic bases, such as hydroxide ion or ammonia, to form salts. Most acids are acidic enough to react with *sodium bicarbonate* solution; this reaction is accompanied by the evolution of carbon dioxide gas. That a reaction has occurred with $NaHCO_3$ is evidenced by the formation of bubbles of CO_2 gas. The three general reactions are:

1. With sodium hydroxide

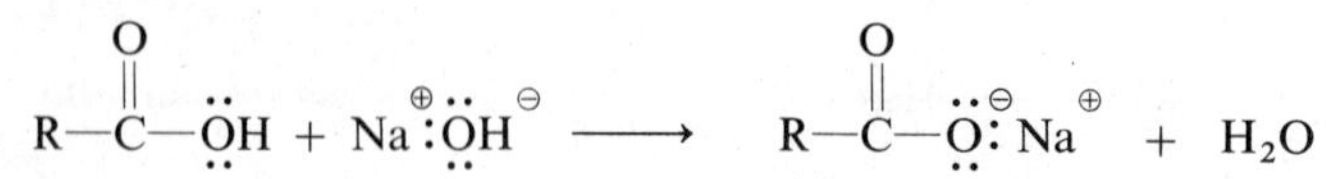

$$
\begin{array}{c}
\text{O} \\
\parallel \\
\text{R}-\text{C}-\ddot{\text{O}}\text{H} + \text{Na}^{\oplus}\ddot{\colon}\ddot{\text{O}}\text{H}^{\ominus} \longrightarrow \text{R}-\text{C}-\ddot{\text{O}}\colon^{\ominus}\,\text{Na}^{\oplus} + \text{H}_2\text{O}
\end{array}
$$

Carboxylic acid   Sodium carboxylate salt
Water-insoluble   *Water-soluble*

2. With sodium bicarbonate

$$R-\overset{\overset{\displaystyle O}{\|}}{C}-\ddot{O}H \ + NaHCO_3 \longrightarrow R-\overset{\overset{\displaystyle O}{\|}}{C}-\overset{..}{\underset{..}{O}}{:}^{\ominus}Na^{\oplus} + CO_2(gas) + H_2O$$

Water-insoluble   *Water-soluble*

3. With ammonia (ammonium hydroxide)

$$R-\overset{\overset{\displaystyle O}{\|}}{C}-\ddot{O}H \ + \ddot{N}H_3 \longrightarrow R-\overset{\overset{\displaystyle O}{\|}}{C}-\overset{..}{\underset{..}{O}}{:}^{\ominus} NH_4^{\oplus}$$

Water-insoluble   Ammonium carboxylate salt
                            Water-soluble

Although most acids with more than four carbon atoms and diacids with more than six carbon atoms are insoluble in water, the corresponding sodium, potassium, and ammonium salts are highly polar and water-soluble. Solubility in sodium bicarbonate is often a distinguishing test for carboxylic acids (see also Sec. 22.14). Solubility of carboxylic acids in sodium bicarbonate solution is often used to separate them from other organic compounds (see Sec. 22.14).

The salts of acids are named by taking either the common or IUPAC name of the acid from which they are prepared, dropping the *-ic acid* ending, and adding **-ate** as a suffix. The name of the cation (sodium, ammonium, and so on) precedes the carboxyl*ate* name; for example:

$$CH_3-\overset{\overset{\displaystyle O}{\|}}{C}-\overset{..}{\underset{..}{O}}{:}^{\ominus} NH_4^{\oplus} \qquad \overset{\overset{\displaystyle O}{\|}}{\underset{\bigcirc}{C}}-\overset{..}{\underset{..}{O}}{:}^{\ominus} Na^{\oplus} \qquad CH_3CH_2-\overset{\overset{\displaystyle O}{\|}}{\underset{\bigcirc}{CH}}-\overset{\overset{\displaystyle O}{\|}}{C}-\overset{..}{\underset{..}{O}}{:}^{\ominus} K^{\oplus}$$

Ammonium ethanoate   Sodium benzoate   Potassium 2-phenylbutanoate
(ammonium acetate)

$$Na^{\oplus} {:}\overset{\ominus}{\underset{..}{O}}-\overset{\overset{\displaystyle O}{\|}}{C}-CH_2CH_2-\overset{\overset{\displaystyle O}{\|}}{C}-\overset{..}{\underset{..}{O}}{:}^{\ominus}Na^{\oplus} \qquad \overset{\overset{\displaystyle O}{\|}}{\underset{\underset{\overset{\|}{O}}{C}-\overset{..}{\underset{..}{O}}{:}^{\ominus}K^{\oplus}}{C}}-\overset{..}{\underset{..}{O}}{:}^{\ominus}K^{\oplus}$$

Sodium butanedioate
(sodium succinate)

                            Potassium phthalate

When dicarboxylic acids are treated with one equivalent of base, only one of the two acidic hydrogens is neutralized. The following two examples illustrate the naming of the resulting compounds:

$$\overset{\overset{\displaystyle O}{\|}}{\underset{\underset{\overset{\|}{O}}{C}-\ddot{O}H}{C}}-\overset{..}{\underset{..}{O}}{:}^{\ominus}K^{\oplus} \qquad HO-\overset{\overset{\displaystyle O}{\|}}{C}-CH_2CH_2-\overset{\overset{\displaystyle O}{\|}}{C}-\overset{..}{\underset{..}{O}}{:}^{\ominus}NH_4^{\oplus}$$

                            Ammonium hydrogen succinate

Potassium hydrogen phthalate

An important adjunct to the reaction of carboxylic acids with base is the conversion of a carboxylate salt back into the free acid. This is accomplished simply by neutralizing the salt with mineral acid, which is shown as part of the reaction:

$$
\underset{\substack{\text{Carboxylic acid} \\ \textit{Water-insoluble}}}{R-\overset{\overset{\text{O}}{\|}}{C}-\overset{..}{\underset{..}{O}}H}
\;\underset{H^{\oplus}}{\overset{:\overset{..}{O}H^{\ominus}\ \text{or}\ HCO_3^{\ominus}\ \text{or}\ NH_3}{\rightleftharpoons}}\;
\underset{\substack{\text{Carboxylate anion} \\ \textit{Water-soluble}}}{R-\overset{\overset{\text{O}}{\|}}{C}-\overset{..}{\underset{..}{O}}{:}^{\ominus}}
$$

The carboxylate salts are water-soluble (unless they have very high molecular weight) when the cation is an alkali metal ($Li^{\oplus}$, $Na^{\oplus}$, $K^{\oplus}$, $Rb^{\oplus}$, or $Cs^{\oplus}$) or ammonium ($NH_4^{\oplus}$). The alkaline earth metal (Mg^{+2}, Ca^{+2}, Ba^{+2}) and most other heavy metal salts of acids are insoluble, a property that is important in the chemistry of soaps (see Sec. 25.2).

The interconversion of carboxylic acids and their salts is often encountered in organic chemistry. We indicated in discussing the preparation of a carboxylic acid by the Grignard method (in Sec. 22.5A) that the crude reaction mixture containing RCOOMgX, a carboxylate salt, must be acidified to obtain the free acid. The oxidation of alkenes, primary alcohols, or alkylbenzenes by hot, basic potassium permanganate (see Sec. 22.4) yields the water-soluble salt of a carboxylic acid. In each case, subsequent neutralization by mineral acid is required to give the free acid. In the chemistry of various derivatives of carboxylic acids discussed in Chap. 23 we frequently acidify a basic solution containing the carboxylate salt as a means of obtaining the free acid.

22.14 Acidity of Various Hydroxyl-Containing Compounds; Use of Solubility in Separations

We now draw together all the hydroxyl-containing compounds to compare their acidities and solubility properties. In the summary of solubility properties and acidities of alcohols, phenols, carboxylic acids, and sulfonic acids in Table 22.6, keep in mind that the lower molecular weight alcohols (four carbon atoms or less) and acids (six carbon atoms or less) are water-soluble.

TABLE 22.6 Solubility Properties and Acidities of Hydroxyl-Containing Compounds

Family of Compound	Approximate Acid Dissociation Constant, K_a	Special Solubility Properties
Alcohols, ROH	10^{-16}–10^{-18}	No greater solubility in base than in water; soluble in cold, concentrated sulfuric acid
Phenols, ArOH	10^{-10}	Soluble in dilute NaOH solution; not soluble in $NaHCO_3$
Carboxylic acids, RCOOH	10^{-5}	Soluble in dilute NaOH solution and in $NaHCO_3$ solution
Sulfonic acids, $ArSO_2OH$	0.1–1	Soluble in dilute NaOH and in dilute $NaHCO_3$; more soluble in water than other compounds listed

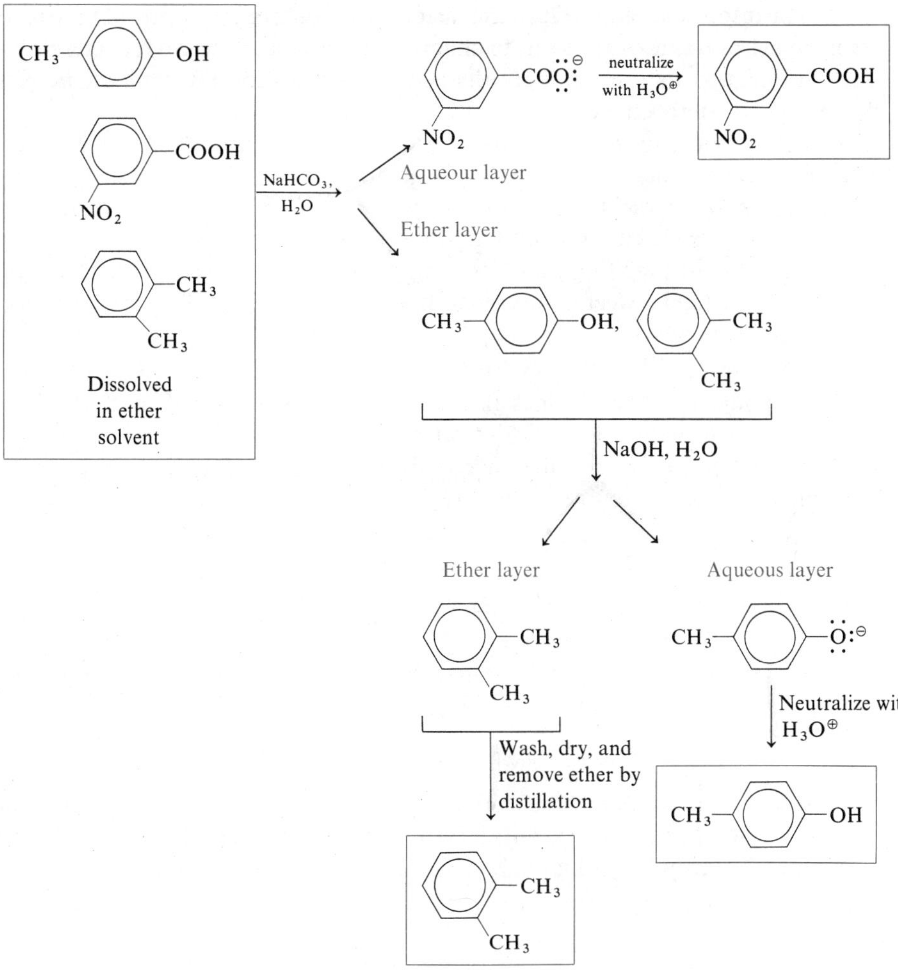

FIGURE 22.3 Scheme showing the separation of *para*-cresol, *meta*-nitrobenzoic acid, and *ortho*-xylene from one another using solubility properties.

The differing base solubilities of the various compounds can be used to separate them from one another. The scheme in Fig. 22.3 shows the separation of a mixture of *p*-cresol (*p*-methylphenol), *m*-nitrobenzoic acid, and *o*-xylene so that each compound is obtained in pure form. As is true of most organic compounds, each of these compounds is soluble in ether, which is used as the organic solvent. The separation starts by extracting the ethereal mixture with $NaHCO_3$, which dissolves only the carboxylic acid. (If the mixture were treated first with NaOH, both the acid and the phenol would be removed without separation.) After the carboxylic acid has been removed, treatment with NaOH solution removes the phenol and leaves the neutral compound (*o*-xylene) in the ether layer.

22.15 Resolution of Carboxylic Acids into Enantiomers

Many classes of organic compounds cannot be resolved directly into optically active enantiomers even though an asymmetric center is present in a given molecule. Alkanes, alkenes, alkyl halides, aldehydes, and ketones are typical compounds that cannot be resolved directly. Our familiarity with the carboxyl group and the reactions of carboxylic acids now allows us to understand how a racemic mixture of a carboxylic acid can be resolved into its optically active enantiomers. We shall draw upon our knowledge of stereochemistry from Chap. 6. (See also Sec. 26.6.)

Let us begin by recalling from Sec. 6.6 that two enantiomers have identical chemical and physical properties (with achiral probes, Sec. 6.6), so that physical methods cannot be used to separate them per se. Remember, however, that diastereomers can be separated from one another because they possess different physical properties, such as melting point, boiling point, and solubility in various solvents. What relationship between enantiomers of carboxylic acids and diastereomers allows an acid to be resolved?

From Sec. 22.13 we know that carboxylic acids form salts when they react with ammonia or other amines (for example, $R\ddot{N}H_2$, $R_2\ddot{N}H$, and $R_3\ddot{N}$). Many amines occur in nature in optically active form (for example, brucine, strychnine, and quinine) and they are called **alkaloid bases** (see Sec. 26.10D). These and other alkaloid bases have complex structures, which are discussed further in Chap. 26.

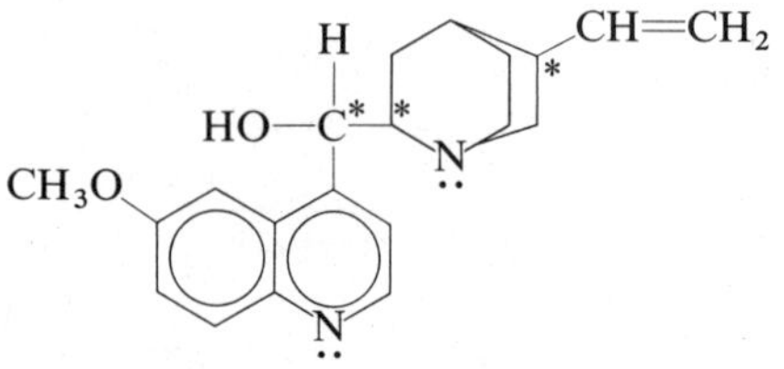

Quinine (optically pure, naturally occurring amine)

Because an alkaloid base is optically active (that is, it is one enantiomer), a mixture of two diastereomeric salts is formed when it reacts with a racemic carboxylic acid. By arbitrarily choosing the base to be the *d* enantiomer and representing it as $d\text{-}R_3\ddot{N}$, the following reaction occurs between it and an acid:

$$R'\!-\!COOH + R_3\ddot{N} \longrightarrow R'\!-\!C\overset{..}{\underset{..}{O}}\overset{\ominus}{O}\text{:}\ R_3\overset{\oplus}{N}H$$

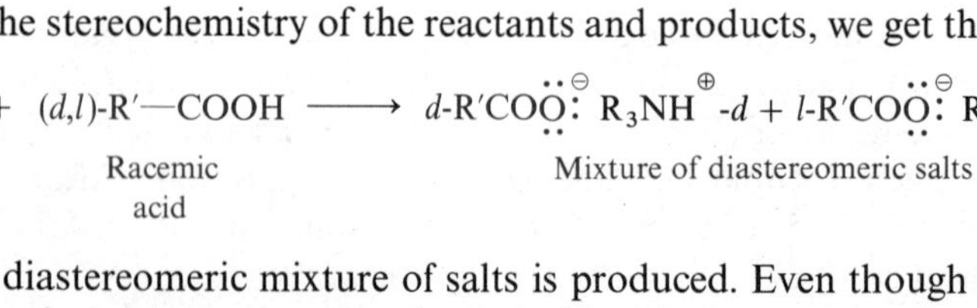

Quinine salt (formed on reaction with carboxylic acid)

Considering the stereochemistry of the reactants and products, we get the following:

$$d\text{-}R_3\ddot{N} + (d,l)\text{-}R'\!-\!COOH \longrightarrow d\text{-}R'C\overset{..\ominus}{O}O\text{:}\ R_3\overset{\oplus}{N}H\text{-}d + l\text{-}R'C\overset{..\ominus}{O}O\text{:}\ R_3\overset{\oplus}{N}H\text{-}d$$

Optically active base	Racemic acid	Mixture of diastereomeric salts

Thus, a diastereomeric mixture of salts is produced. Even though no covalent bond connects the asymmetric center in the acid with the asymmetric center in the base, each diastereomer has a unique diastereomeric crystal structure. The resulting diastereomers have different solubility properties, and an organic solvent can usually be found that permits them to be separated by *fractional crystallization*. Ideally, a solvent is chosen in which both diastereomers are soluble when the solution is heated, and then on cooling one diastereomer crystallizes out while the other remains in solution. Repetition of this fractional crystallization usually yields pure diastereomers, and because they are amine salts, they yield optically active acid (plus the amine salt) on treatment with excess mineral acid:

Mixture of diastereomeric salts $\xrightarrow[\text{crystallization}]{\text{fractional}}$ d-R′COO̤⁻ : R$_3$N̈H-d + l-R′COO̤⁻ : R$_3$N̈H-d

Pure diastereomer separated and Pure diastereomer separated

$\downarrow$ H$_3$Ö⊕ $\downarrow$ H$_3$Ö⊕

d-R′COOH l-R′COOH

Optically pure d enantiomer Optically pure l enantiomer

d-R$_3$N̈H, also found in both reactions, remains dissolved in aqueous layer

In the final hydrolysis step the optically active acid precipitates because it is water-insoluble, whereas the alkaloid base remains in the aqueous layer as the soluble protonated salt. This entire process is called **resolution,** and it is an indirect method for obtaining optically active acids. As we learn in this and the next chapter, the carboxyl group can be converted into numerous other functional groups, so acids are an important intermediate for obtaining other families of optically active compounds.

Many alkaloid bases commonly used in resolutions are quite expensive, but they can be regenerated and used again by treating the aqueous layers remaining from the acidification step with excess base:

$$d\text{-R}_3\text{NH}^\oplus + \text{:Ö̈H}^\ominus \text{ (excess)} \rightarrow d\text{-R}_3\text{N̈} + \text{H}_2\text{Ö:}$$

In principle, resolution appears simple and straightforward. In practice, however, there are many experimental problems. One must find the right combination of solvent and alkaloid base to separate the diastereomeric mixture of amine salts, and this must be done by trial and error. Sometimes fractional crystallization must be repeated numerous times before an optically pure acid is obtained because each recrystallization removes only a portion of the other diastereomer. Cases are also known where one alkaloid-solvent combination gives one enantiomer of the acid, whereas another alkaloid-solvent combination must be used to obtain the other enantiomer. Many resolutions are reported in the chemical literature in which an acid is obtained in optically active form (that is, the sample contains a preponderance of one enantiomer) but not in optically pure form (the sample contains *only* one enantiomer). Some studies require only the optically active form, whereas others require only the optically pure form.

We discuss the resolution of alcohols in Sec. 23.14 and amines in Sec. 26.6, and we continue to examine the use of stereochemistry to elucidate structure and mechanisms.

22.16 Reduction of Carboxylic Acids

In this and the next two sections we discuss several special reactions of carboxylic acids, one of the most important is the reduction of a carboxylic acid. Lithium aluminum hydride, LiAlH$_4$, is one of the few reducing agents that can convert a carboxylic acid into the corresponding primary alcohol *in one step*. The overall reaction involves first treating the acid with lithium aluminum hydride in anhydrous ether, followed by hydrolysis ($H_3\overset{..}{O}{}^{\oplus}$):

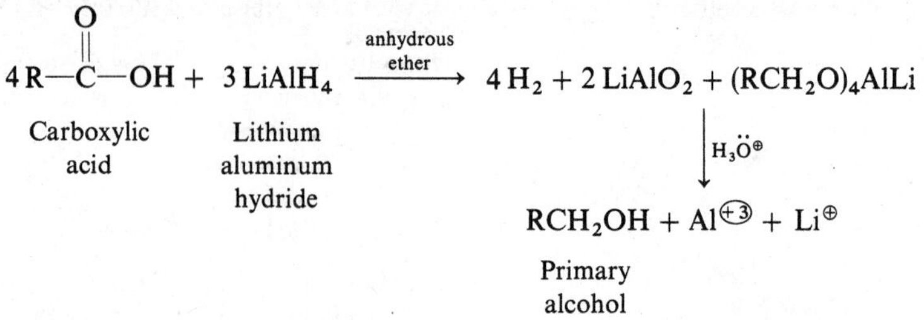

$$4\,R-\overset{\overset{\displaystyle O}{\|}}{C}-OH \;+\; 3\,LiAlH_4 \;\xrightarrow[\text{ether}]{\text{anhydrous}}\; 4\,H_2 \;+\; 2\,LiAlO_2 \;+\; (RCH_2O)_4AlLi$$

Carboxylic acid   Lithium aluminum hydride

$$\Big\downarrow {}_{H_3\overset{..}{O}{}^{\oplus}}$$

$$RCH_2OH \;+\; Al^{\textcircled{3}} \;+\; Li^{\oplus}$$

Primary alcohol

The mechanism of the $LiAlH_4$ reduction of acids is believed to involve the following four steps.

Step 1:

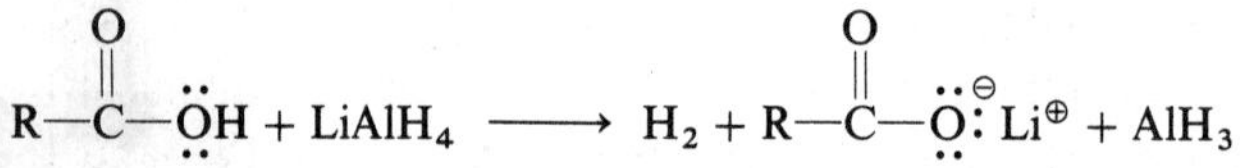

$$R-\overset{\overset{\displaystyle O}{\|}}{C}-\overset{..}{\underset{..}{O}}H \;+\; LiAlH_4 \;\longrightarrow\; H_2 \;+\; R-\overset{\overset{\displaystyle O}{\|}}{C}-\overset{..}{\underset{..}{O}}{:}^{\ominus}Li^{\oplus} \;+\; AlH_3$$

Step 2:

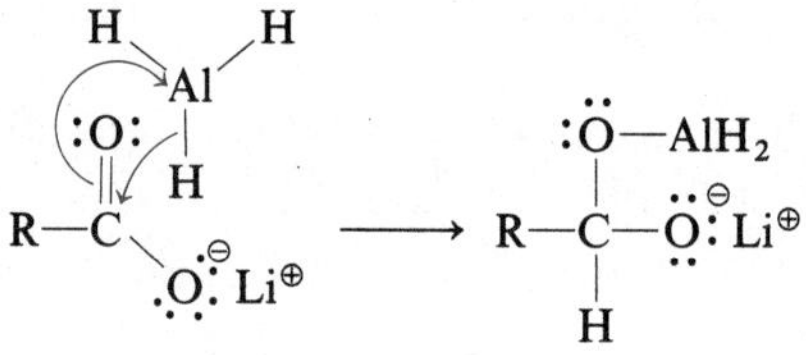

(1)

Step 3:

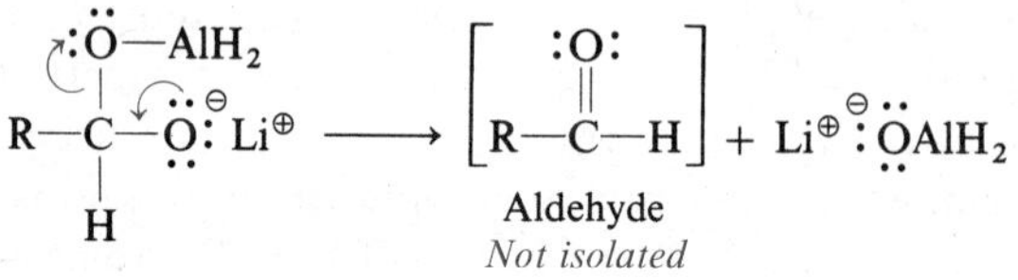

Aldehyde
Not isolated

Step 4:

$$R-\overset{\overset{\displaystyle :O:}{\|}}{C}-H \;\xrightarrow[\text{(as in Sec. 21.13)}]{LiAlH_4}\; (RCH_2O)_4Al^{\ominus}Li^{\oplus} \;\xrightarrow{H_3\overset{..}{O}{}^{\oplus}}\; RCH_2OH \;+\; Al^{\textcircled{3}} \;+\; Li^{\oplus}$$

Step 1 is the reaction between the acidic proton on the acid and a hydride ion of $LiAlH_4$, which occurs with the liberation of hydrogen gas. (This reaction destroys some of the expensive $LiAlH_4$ reagent, and for this reason acids are often converted into esters before reduction; see Sec. 23.7E.) Step 2 is the nucleophilic addition of hydride ion to the lithium salt of the acid. Step 3 shows a possible route for the decomposition of intermediate (1) formed in step 2. The resulting salt, $LiOAlH_2$, still contains two hydrides that can continue the reaction. Intermediate (1) is stabilized by the tendency that the Lewis acid, aluminium, has for gaining electrons from oxygen:

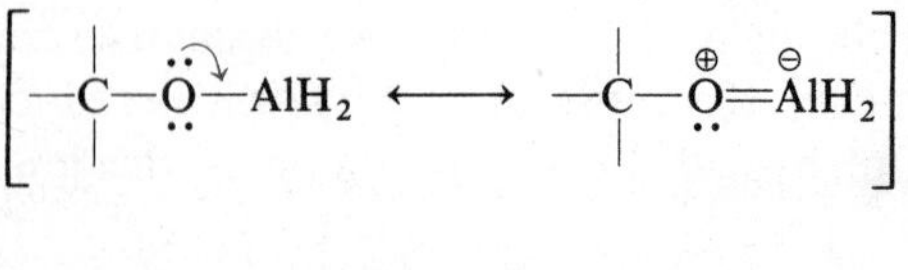

An aldehyde is believed to be formed in step 3, and once formed, it reacts rapidly with LiAlH$_4$ (step 4) to produce the salt of the primary alcohol. Acidification of the final reaction mixture yields the primary alcohol.

The following three examples illustrate the products obtained from acids and diacids:

Benzoic acid → Benzyl alcohol

Cyclohexanecarboxylic acid → Cyclohexylmethanol

Pentanedioic acid (glutaric acid) → 1,5-Pentanediol

$$HO-\overset{O}{\underset{\|}{C}}-(CH_2)_3-\overset{O}{\underset{\|}{C}}-OH \xrightarrow[\text{2. } H_3\ddot{O}^{\oplus}]{\text{1. LiAlH}_4\text{, anhydrous ether}} HOCH_2-(CH_2)_3-CH_2OH$$

Lithium aluminum hydride (LAH) reductions give excellent yields of products. Because of its relatively high cost, however, LAH is used mostly in the research laboratory; it enjoys limited industrial use where the starting compound is difficult to obtain. Although LiAlH$_4$ is one of the few reagents that reduces acids directly to alcohols, an acid can be converted to a derivative (such as an ester), which in turn can be reduced to a primary alcohol by other less expensive reagents (see Sec. 23.7E).

We know that a primary alcohol can be oxidized into a carboxylic acid, and now we have a method for converting an acid into a primary alcohol, which in turn is readily converted into other functional groups (Fig. 21.2). The reduction of an acid is of lesser importance than the oxidation of a primary alcohol into an acid. Many higher molecular weight carboxylic acids come from natural sources, especially the *fatty acids* (that is, those containing an even number of carbon atoms) obtained from fats and oils (see Sec. 25.1). Direct reduction of a fat or oil, as discussed in Sec. 25.2B, produces long-chain primary alcohols, which can be converted into other functional groups.

One limitation is that lithium aluminum hydride is such a powerful reducing agent that it also reduces other multiply bonded substituents [such as —COOR (ester), —CONH$_2$ (amide), —C≡N (nitrile), —COCl (acid chloride), —CO—O—COR (anhydride), —NO$_2$ (nitro), and the carbonyl group (aldehydes and ketones)] that may also be in a molecule.

Let us now compare the reduction of acids with the reduction of aldehydes and ketones. Aldehydes and ketones are reduced by both sodium borohydride (NaBH$_4$) and LiAlH$_4$ (see Sec. 21.13); acids are also reduced by LiAlH$_4$. For example, if an acid that contains a carbonyl group reacts with LiAlH$_4$, reduction occurs at both places:

2-Methyl-3-oxopentanoic acid → 2-Methyl-1,3-pentanediol

If we treat the same compound with $NaBH_4$, only the keto group is reduced because $NaBH_4$ is not powerful enough to reduce the carboxyl group:

$$CH_3CH_2-\overset{\overset{O}{\|}}{C}-\underset{\underset{CH_3}{|}}{CH}-\overset{\overset{O}{\|}}{C}-OH \xrightarrow[\text{2. } H_3\ddot{O}^{\oplus}]{\text{1. } NaBH_4} CH_3CH_2-\overset{\overset{OH}{|}}{CH}-\underset{\underset{CH_3}{|}}{CH}-\overset{\overset{O}{\|}}{C}-OH$$

2-Methyl-3-hydroxypentanoic acid

It is also possible to reduce the carboxyl group and not the keto group. To accomplish this, however, the keto group must first be converted to a ketal, which is formed only from an aldehyde or ketone (see Sec. 21.6). Subsequent treatment with $LiAlH_4$ reduces the carboxyl group because the ketal is unaffected by $LiAlH_4$. Finally, treatment with acidic water hydrolyzes the ketal back into the keto group and simultaneously protonates the aluminum salt of the alcohol. These reactions are outlined here:

$$CH_3CH_2-\overset{\overset{O}{\|}}{C}-\underset{\underset{CH_3}{|}}{CH}-\overset{\overset{O}{\|}}{C}-OH \xrightarrow[\text{dry HCl}]{HOCH_2CH_2OH} CH_3CH_2-\overset{\overset{O\quad O}{\diagdown\diagup}}{C}-\underset{\underset{CH_3}{|}}{CH}-\overset{\overset{O}{\|}}{C}-OH$$

Ketal

$$\xrightarrow[\text{2. } H_3\ddot{O}^{\oplus}, \text{ warm}]{\text{1. } LiAlH_4, \text{ anhydrous ether}} CH_3CH_2-\overset{\overset{O}{\|}}{C}-\underset{\underset{CH_3}{|}}{CH}-CH_2OH$$

1-Hydroxy-2-methyl-3-pentanone

22.17   Decarboxylation of Acids and Diacids

Decarboxylation reactions are relatively rare in organic chemistry because special structural features must usually be present in a molecule for decarboxylation to occur readily.

Aromatic carboxylic acids undergo decarboxylation on fusing the sodium salt of the acid with soda lime, which is a mixture of sodium hydroxide and calcium oxide. *Aliphatic acids* react similarly, but the reaction conditions must be more vigorous and are often accompanied by a large amount of decomposition; for example:

$$\text{C}_6\text{H}_5-\overset{\overset{O}{\|}}{C}-\ddot{\overset{\ominus}{\underset{..}{O}}}:Na^{\oplus} \xrightarrow[\text{fuse}]{\text{soda lime}} \text{C}_6\text{H}_6-H + Na_2CO_3$$

Sodium benzoate   Benzene

$$CH_3(CH_2)_{14}\overset{\overset{O}{\|}}{C}-\ddot{\overset{\ominus}{\underset{..}{O}}}:Na^{\oplus} \xrightarrow[\text{fuse}]{\text{soda lime}} CH_3(CH_2)_{13}CH_3 + Na_2CO_3$$

Sodium hexadecanoate   Pentadecane
(sodium palmitate)

This reaction is used in degradation studies and structure proof; it has little use in synthesis because of low yields.

Two types of *dicarboxylic acids* readily undergo decarboxylation simply on heating the free acid. **Ethanedioic acid (oxalic acid)** undergoes loss of CO_2 to give formic acid:

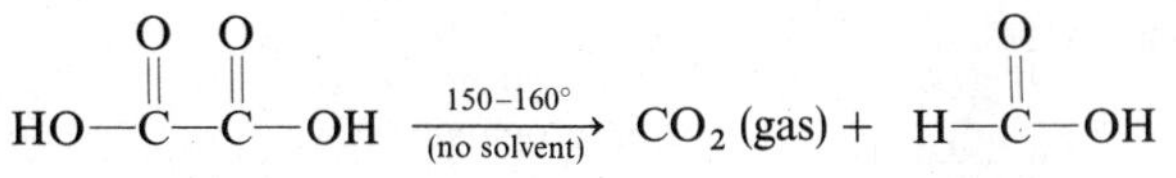

$$HO-\overset{\overset{O}{\|}}{C}-\overset{\overset{O}{\|}}{C}-OH \xrightarrow[\text{(no solvent)}]{150-160°} CO_2 \text{ (gas)} + H-\overset{\overset{O}{\|}}{C}-OH$$

Ethanedioic acid   Methanoic acid
(oxalic acid)   (formic acid)

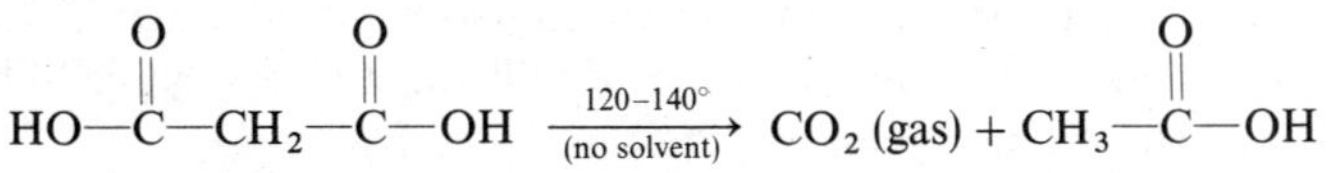

Propanedioic acid (malonic acid) and **substituted propanedioic acids** lose CO_2 even more readily and give acetic acid or a substituted acetic acid as the product; for example:

$$HO-\overset{\overset{O}{\|}}{C}-CH_2-\overset{\overset{O}{\|}}{C}-OH \xrightarrow[\text{(no solvent)}]{120-140°} CO_2 \text{ (gas)} + CH_3-\overset{\overset{O}{\|}}{C}-OH$$

Propanedioic acid   Ethanoic acid
(malonic acid)   (acetic acid)

$$HO-\overset{\overset{O}{\|}}{C}-\overset{\overset{R}{|}}{\underset{\underset{R'}{|}}{C}}-\overset{\overset{O}{\|}}{C}-OH \xrightarrow[\text{(no solvent)}]{120-140°} H-\overset{\overset{R}{|}}{\underset{\underset{R'}{|}}{C}}-\overset{\overset{O}{\|}}{C}-OH + CO_2 \text{ (gas)}$$

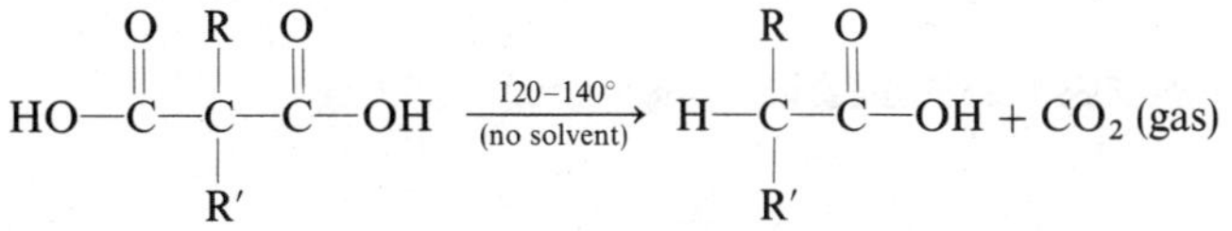

Substituted malonic acid   Substituted acid

The mechanism of the decarboxylation of malonic acid is presented in Sec. 23.12. This reaction is important in preparing certain substituted carboxylic acids by the malonic ester synthesis (see Sec. 24.20).

Other dicarboxylic acids usually undergo anhydride formation on heating (see Sec. 23.12).

22.18 α Halogenation of Carboxylic Acids

Carboxylic acids and diacids that contain one or more hydrogens α to the carboxyl group undergo *substitution* on treatment with chlorine or bromine in the presence of a trace of elemental phosphorus, which catalyzes the reaction. This reaction, known as the **Hell-Volhard-Zelinsky reaction** (often abbreviated as *HVZ reaction* for convenience), is shown here:

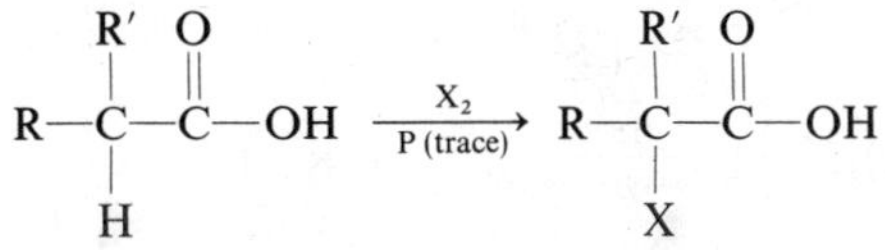

$$R-\overset{\overset{R'}{|}}{\underset{\underset{H}{|}}{C}}-\overset{\overset{O}{\|}}{C}-OH \xrightarrow[\text{P (trace)}]{X_2} R-\overset{\overset{R'}{|}}{\underset{\underset{X}{|}}{C}}-\overset{\overset{O}{\|}}{C}-OH$$

where X = Cl, Br

If more than one α hydrogen is present, it may also be substituted by halogen; for example:

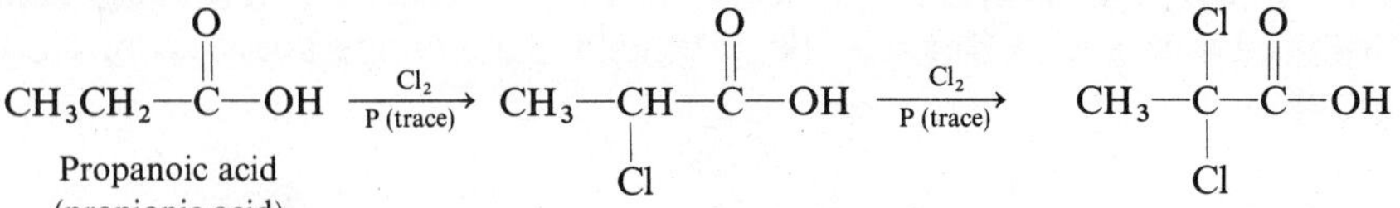

$$CH_3CH_2-\overset{O}{\overset{\|}{C}}-OH \xrightarrow[\text{P (trace)}]{Cl_2} CH_3-\underset{Cl}{\overset{}{CH}}-\overset{O}{\overset{\|}{C}}-OH \xrightarrow[\text{P (trace)}]{Cl_2} CH_3-\underset{Cl}{\overset{Cl}{C}}-\overset{O}{\overset{\|}{C}}-OH$$

Propanoic acid (propionic acid) ; 2-Chloropropanoic acid (α-chloropropionic acid) ; 2,2-Dichloropropanoic acid (α,α-dichloropropionic acid)

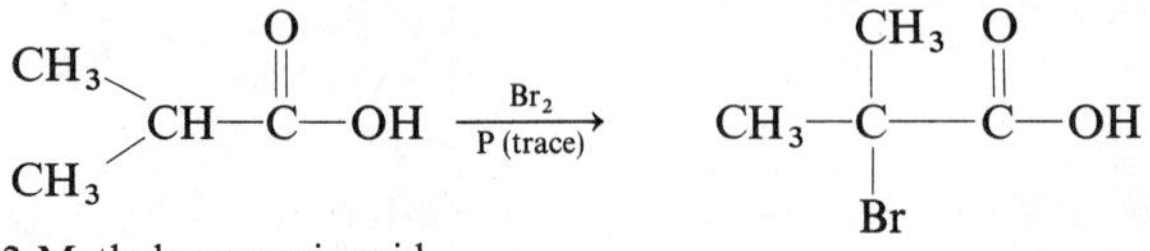

$$\underset{CH_3}{\overset{CH_3}{}}CH-\overset{O}{\overset{\|}{C}}-OH \xrightarrow[\text{P (trace)}]{Br_2} CH_3-\underset{Br}{\overset{CH_3}{C}}-\overset{O}{\overset{\|}{C}}-OH$$

2-Methylpropanoic acid

2-Bromo-2-methylpropanoic acid
Only product

The degree of substitution can be controlled experimentally, so it is possible to stop the reaction at either the mono- or disubstitution stage. The mechanism of the HVZ reaction is discussed in detail in Sec. 24.7C.

The introduction of an α halogen into a carboxylic acid gives a compound of considerable importance in synthetic organic chemistry. As the following reactions indicate, an α-halo acid can be converted into many different types of compounds.

1. Conversion into a dicarboxylic acid (Sec. 22.5B)

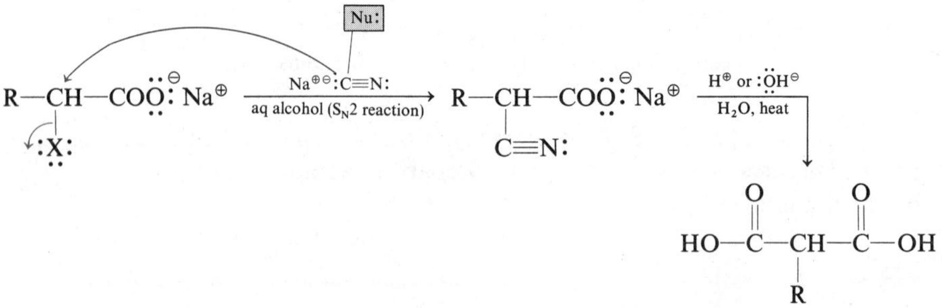

$$R-\underset{:X:}{\overset{}{CH}}-COO:^{\ominus} Na^{\oplus} \xrightarrow[\text{aq alcohol (S_N2 reaction)}]{Na^{\oplus\ominus}:C\equiv N:} R-\underset{C\equiv N:}{\overset{}{CH}}-COO:^{\ominus} Na^{\oplus} \xrightarrow[\text{H_2O, heat}]{H^{\oplus} \text{ or } :\overset{..}{O}H^{\ominus}}$$

$$HO-\overset{O}{\overset{\|}{C}}-\underset{R}{\overset{}{CH}}-\overset{O}{\overset{\|}{C}}-OH$$

Substituted malonic acid

2. Conversion into an α-hydroxy acid

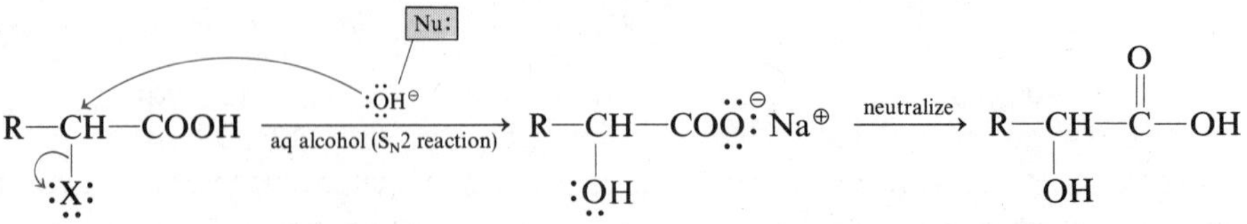

$$R-\underset{:X:}{\overset{}{CH}}-COOH \xrightarrow[\text{aq alcohol (S_N2 reaction)}]{:\overset{..}{O}H^{\ominus}} R-\underset{:OH}{\overset{}{CH}}-COO:^{\ominus} Na^{\oplus} \xrightarrow{\text{neutralize}} R-\underset{OH}{\overset{}{CH}}-\overset{O}{\overset{\|}{C}}-OH$$

α-Hydroxy acid

3. Conversion into an α-amino acid (Sec. 28.7)

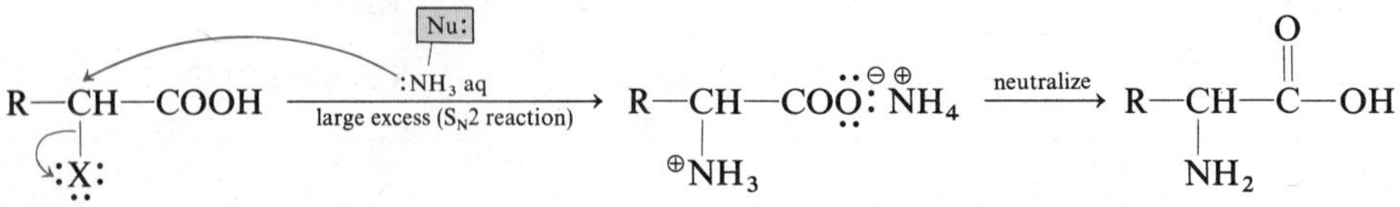

$$R-\underset{:X:}{\overset{}{CH}}-COOH \xrightarrow[\text{large excess (S_N2 reaction)}]{:NH_3 \text{ aq}} R-\underset{\oplus NH_3}{\overset{}{CH}}-COO:^{\ominus\oplus} NH_4 \xrightarrow{\text{neutralize}} R-\underset{NH_2}{\overset{}{CH}}-\overset{O}{\overset{\|}{C}}-OH$$

α-Amino acid

Review Chp 9 [alkyl halide Elim. & Subs.)

Chp 10 alcenes
to (stereoch.)

7 & 8 a/(lcene & allcadreine)

Study Chp. 11 omit 11.14
Chp 13 omit 13.4 & 13.18
Chp 14 Omit 14.23, 14.26 Bu
Chp 15 sec. 15.8 only
Chp 16

4. Conversion into an α,β-unsaturated acid

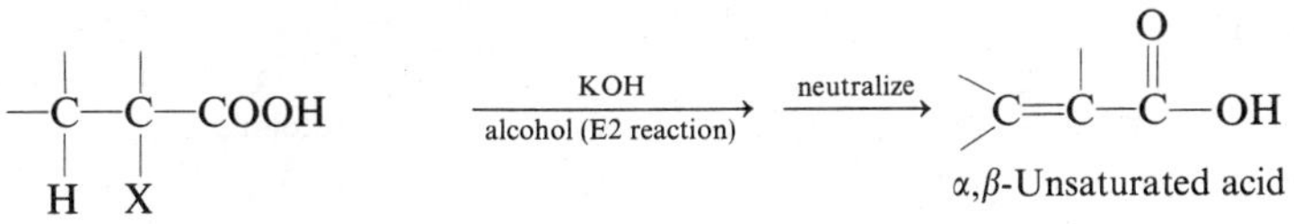

$$\underset{\underset{H \quad X}{|\quad\;|}}{-\overset{|}{C}-\overset{|}{C}-COOH} \quad \xrightarrow[\text{alcohol (E2 reaction)}]{\text{KOH}} \quad \xrightarrow{\text{neutralize}} \quad \underset{\alpha,\beta\text{-Unsaturated acid}}{\overset{\overset{\displaystyle O}{\displaystyle\|}}{C=C-C-OH}}$$

Must have at least one β hydrogen

Question 22.13

Indicate the possible α-halogenated carboxylic acids that might be obtained from treating each of the following compounds with bromine in the presence of a trace of phosphorus.

(*a*) acetic acid   (*b*) phenylacetic acid   (*c*) butyric acid
(*d*) cyclohexanecarboxylic acid   (*e*) succinic acid

Question 22.14

Indicate the principal products that would be formed when each of the monohalogenated compounds in Question 22.13 reacts with the following reagents:

(*a*) excess aqueous ammonia
(*b*) alcoholic potassium hydroxide
(*c*) excess aqueous potassium hydroxide
(*d*) sodium cyanide after being neutralized with NaOH

Question 22.15

Treatment of an α-halo acid with excess ammonia results in the formation of the ammonium salt of an α-amino acid. Explain why this is so.

22.19   Spectral Analysis of Carboxylic Acids

A. Infrared Spectroscopy

Carboxylic acids show two strong and characteristic absorptions in the infrared, one due to C=O and the other to O—H. Most carboxylic acids are dimers, and the O—H stretching frequency normally appears as a broad and very strong band between 2,500 cm^{-1} (4.0 μ) and 3,000 cm^{-1} (3.3 μ). This is caused by the O—H---O=C (hydrogen bonded) group. [Recall from Sec. 17.10 that alcohols and phenols are hydrogen bonded and that they appear in the IR region 3,000 to 3,500 cm^{-1} (3.3 to 2.86 μ) and from 3,200 to 3,600 cm^{-1} (3.125 to 2.77 μ) respectively.]

The carbon-oxygen double bond in carboxylic acids normally appears at $\sim$1,700 cm^{-1} (5.88 μ); for aliphatic acids the strong band is usually in the region 1,700 to 1,725 cm^{-1} (5.88 to 5.80 μ), and for aromatic and α,β-unsaturated acids it is typically at 1,680 to 1,700 cm^{-1} (5.95 to 5.88 μ).

Stretching due to the carbon-oxygen bond in C—OH appears at $\sim$1,250 cm^{-1} (8.0 μ), and bending of the O—H bond at $\sim$1,400 cm^{-1} (7.14 μ) and as a broad peak at $\sim$920 cm^{-1} (10.87 μ). The absorption characteristics of the carbon-oxygen single

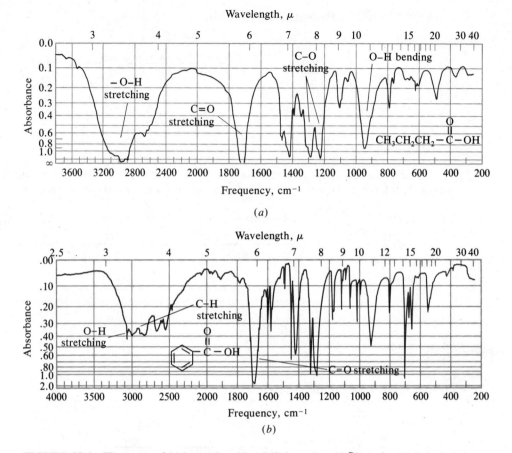

FIGURE 22.4   IR spectra of (*a*) butanoic acid and (*b*) benzoic acid. [Data for (*a*) from *Indexes to Evaluated Infrared Reference Spectra*, copyright 1975, The Coblentz Society, Norwalk CT. Data for (*b*) from *The Sadtler Handbook of Infrared Spectra*, W. W. Simmons (ed.), © Sadtler Research Laboratories, Division of Bio-Rad Laboratories, Inc. (1978)]

bond should be compared with those of alcohols and phenols (see Sec. 17.10), and ethers (see Sec. 19.19).

The IR spectra of two typical carboxylic acids are shown in Fig. 22.4.

ᵒ*miɫ*

B. Ultraviolet Spectroscopy

The UV spectra exhibited by carboxylic acids are discussed in Sec. 23.19 along with the derivatives of carboxylic acids.

C. Nuclear Magnetic Resonance Spectroscopy

Carboxylic acids are moderately strong acids (see Sec. 22.9), and as a result, the hydroxylic proton is highly deshielded and appears far downfield from TMS. In nonpolar solvents, it normally occurs as a singlet in the 10.5 to 12 δ (-2 to -0.5 τ) region. [Compare this with the hydroxylic proton of alcohols and phenols, which absorb in the NMR at 1 to 5.5 δ (4.5 to 9 τ) and at 4 to 12 δ (-2 to 6 τ), respectively.] The hydrogens α to the acyl group usually appears in the 2 to 2.6 δ (7.4 to 8 τ) region. The NMR spectrum of a typical carboxylic acid is shown in Fig. 22.5.

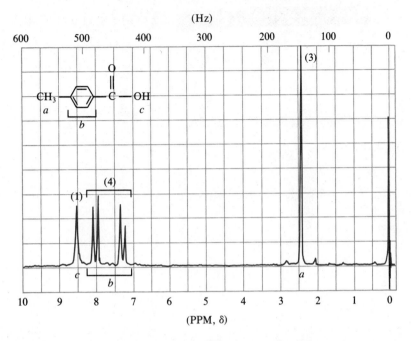

FIGURE 22.5   NMR spectrum of *p***-toluic acid.** [Data from *The Aldrich Library of NMR Spectra*, C. J. Pouchert and J. R. Campbell (eds.), copyright 1974, Aldrich Chemical Co., Milwaukee]

D. Mass Spectrometry

The molecular ion peak of a carboxylic acid, though sometimes present, at best has low natural abundance. Two modes of fragmentation readily occur:

$$
R-\overset{:\overset{.}{O}^{\cdot \oplus}}{\underset{}{C}}\overset{\frown}{O}-H \xrightarrow[\text{H\"O}\cdot]{\text{loss of}} R-C\equiv\overset{..}{\underset{\oplus}{O}} \qquad \text{Acylium ion}
$$

McLafferty rearrangement (see Sec. 20.10)

22.20   Summary of Methods of Preparation of Carboxylic Acids

A. Oxidation of Primary Alcohols

$$
(\text{Ar})\,RCH_2OH \xrightarrow[\substack{\text{or 1. KMnO}_4,\, :\overset{..}{\underset{..}{O}}H^{\ominus},\, \text{heat} \\ 2.\ H_3\overset{\oplus}{O}}]{K_2Cr_2O_7,\ H_2SO_4,\ \text{heat}} (\text{Ar})\,R-\overset{\overset{\displaystyle O}{\|}}{C}-OH
$$

B. Oxidation of Alkylbenzenes (Secs. 14.8 and 22.4B)

$$
Ar-R \xrightarrow[\substack{\text{or 1. KMnO}_4,\, :\overset{..}{\underset{..}{O}}H^{\ominus},\, \text{heat} \\ 2.\ H_3\overset{\oplus}{O}}]{K_2Cr_2O_7,\ H_2SO_4,\ \text{heat}} Ar-\overset{\overset{\displaystyle O}{\|}}{C}-OH
$$

C. Permanganate Oxidation of Alkenes (Secs. 8.32 and 22.4C)

$$(Ar)R-CH{=}CH-R'(Ar') \xrightarrow[\text{2. } H_3\overset{\oplus}{O}]{\text{1. } KMnO_4, \ :\overset{..}{O}H^{\ominus}, \text{ heat}} (Ar)R-\overset{\overset{\displaystyle O}{\|}}{C}-OH + (Ar')R'-\overset{\overset{\displaystyle O}{\|}}{C}-OH$$

D. Grignard Method (Sec. 22.5A)

1. Aliphatic acids

ROH $\xrightarrow{PX_3}$ RX $\xrightarrow[\text{anhydrous ether}]{Mg}$ RMgX $\xrightarrow[\text{(dry ice)}]{CO_2}$ R$-\overset{\overset{\displaystyle O}{\|}}{C}-\overset{..}{\underset{..}{O}}\overset{\ominus}{}[MgX]^{\oplus}$ $\xrightarrow{H_3\overset{\oplus}{O}}$

1°, 2°, or 3°
Alcohol

$$R-\overset{\overset{\displaystyle O}{\|}}{C}-OH$$

Carboxylic acid
Contains one more
carbon atom

2. Aromatic acids

Ar$-$X $\xrightarrow[\text{anhydrous ether}]{Mg}$ ArMgX $\xrightarrow[\text{(dry ice)}]{CO_2}$ ArCOO$^{\ominus}$[MgX]$^{\oplus}$ $\xrightarrow{H_3\overset{\oplus}{O}}$ Ar$-\overset{\overset{\displaystyle O}{\|}}{C}-$OH

E. Hydrolysis of Nitriles (Sec. 22.5B)

1. Aliphatic acids

ROH $\xrightarrow[\text{or Ts-Cl}]{PX_3}$ RX $\xrightarrow{:CN:^{\ominus}}$ R$-$C$\equiv$N $\xrightarrow[H^{\oplus} \text{ or } :\overset{..}{O}H^{\ominus}]{H_2O}$ R$-\overset{\overset{\displaystyle O}{\|}}{C}-$OH

Methyl, 1°   or
(and some 2°)   ROTs
alcohols

Carboxylic acid
Contains one more
carbon atom

2. Aromatic acids

Ar$-$C$\equiv$N $\xrightarrow[H^{\oplus} \text{ or } :\overset{..}{O}H^{\ominus}]{H_2O}$ Ar$-\overset{\overset{\displaystyle O}{\|}}{C}-$OH

F. Special Methods for Preparing Certain Acids (Sec. 22.6)

Salicylic acid; Kolbe reaction

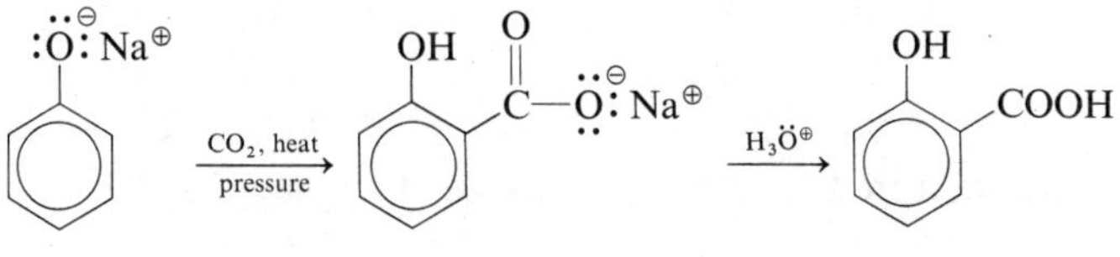

Salicylic acid

22.21 Summary of Reactions of Carboxylic Acids

A. General Reactions

1. Nucleophilic acyl substitution (Sec. 22.8 and Chap. 23)

$$R-\overset{\displaystyle \overset{O}{\|}}{C}\{O-H \xrightarrow{\text{Nu:}} R-\overset{\displaystyle \overset{O}{\|}}{C}-Nu$$

2. Acidic character (Sec. 22.9)

$$R-\overset{\displaystyle \overset{O}{\|}}{C}-O\{H + H_2O \rightleftharpoons R-\overset{\displaystyle \overset{O}{\|}}{C}-\ddot{\overset{..}{O}}\colon^{\ominus} + H_3\overset{..}{O}^{\oplus}$$

B. Acidity of Carboxylic Acids

$$R-\overset{\displaystyle \overset{O}{\|}}{C}-OH + H_2O \rightleftharpoons R-\overset{\displaystyle \overset{O}{\|}}{C}-\ddot{\overset{..}{O}}\colon^{\ominus} + H_3\overset{..}{O}^{\oplus}$$

Theory of dissociation (Sec. 22.9)

Structure-acidity correlation for aliphatic acids (Sec. 22.10)

Structure-acidity correlations for aromatic acids (Sec. 22.11)

Comparison of acids with other acidic organic compounds (Sec. 22.14)

C. Acidity of Dicarboxylic Acids (Sec. 22.12)

$$HO-\overset{\displaystyle \overset{O}{\|}}{C}-(C)_n-\overset{\displaystyle \overset{O}{\|}}{C}-OH \rightleftharpoons \overset{-H^{\oplus}}{} HO-\overset{\displaystyle \overset{O}{\|}}{C}-(C)_n-\overset{\displaystyle \overset{O}{\|}}{C}-\ddot{\overset{..}{O}}\colon^{\ominus}$$

$$\Big\updownarrow {-H^{\oplus}}$$

$$^{\ominus}\ddot{\overset{..}{O}}-\overset{\displaystyle \overset{O}{\|}}{C}-(C)_n-\overset{\displaystyle \overset{O}{\|}}{C}-\ddot{\overset{..}{O}}\colon^{\ominus}$$

D. Salt Formation of Acids and Diacids (Sec. 22.13)

$$R-\overset{\displaystyle \overset{O}{\|}}{C}-OH \xrightarrow[\text{or HCO}_3^{\ominus}]{\colon\ddot{O}H^{\ominus},\ NH_3} R-\overset{\displaystyle \overset{O}{\|}}{C}-\ddot{\overset{..}{O}}\colon^{\ominus}$$

$$\xleftarrow{H_3\ddot{O}^{\oplus}}$$

Application of salt formation to separations (Sec. 22.14)

E. Resolution of Carboxylic Acids Via Salt Formation with Alkaloid Bases (Sec. 22.15)

$$(d,l)-R-COOH \xrightarrow[\text{base}]{\text{alkaloid}} \text{salt} \xrightarrow[\text{crystallization}]{\text{fractional}} \xrightarrow{H_3\ddot{O}^{\oplus}} d-R-COOH + l-R-COOH$$

Resolved

F. Reduction of Carboxylic Acids and Diacids (Sec. 22.16)

$$R-\overset{\overset{\displaystyle O}{\|}}{C}-OH \xrightarrow[\text{2. } H_3\ddot{O}^{\oplus}]{\text{1. LiAlH}_4, \text{ anhydrous ether}} R-CH_2-OH$$

Works for diacids also

Limitations: Also reduces aldehydes, ketones, esters, amides, acid chlorides, anhydrides, nitriles, and nitro compounds.

Modifications: Protection of carbonyl group (aldehyde or ketone) by ketal formation (see Sec. 21.6) permits reduction of carboxyl group elsewhere in molecule. Hydrolysis regenerates carbonyl group.

G. Decarboxylation Reactions (Sec. 22.17)

1. Carboxylic acids

$$(Ar)R-\overset{\overset{\displaystyle O}{\|}}{C}-\overset{\ominus}{\underset{\cdot\cdot}{\ddot{O}}:Na^{\oplus}} \xrightarrow[\text{lime, fuse}]{\text{soda}} (Ar)R-H$$

2. Dicarboxylic acids

$$HO-\overset{\overset{\displaystyle O}{\|}}{C}-\overset{\overset{\displaystyle O}{\|}}{C}-OH \xrightarrow{\text{heat}} H-\overset{\overset{\displaystyle O}{\|}}{C}-OH + CO_2$$

Oxalic acid   Formic acid

$$HO-\overset{\overset{\displaystyle O}{\|}}{C}-\underset{\underset{\displaystyle R'}{|}}{\overset{\overset{\displaystyle R}{|}}{C}}-\overset{\overset{\displaystyle O}{\|}}{C}-OH \xrightarrow{\text{heat}} H-\underset{\underset{\displaystyle R'}{|}}{\overset{\overset{\displaystyle R}{|}}{C}}-\overset{\overset{\displaystyle O}{\|}}{C}-OH + CO_2$$

Substituted malonic acid   Substituted acid

H. α Halogenation of Acids: HVZ Reaction (Sec. 22.18)

$$R-\underset{\underset{\displaystyle H}{|}}{\overset{\overset{\displaystyle R'}{|}}{C}}-\overset{\overset{\displaystyle O}{\|}}{C}-OH \xrightarrow[\text{P (trace)}]{X_2} R-\underset{\underset{\displaystyle X}{|}}{\overset{\overset{\displaystyle R'}{|}}{C}}-\overset{\overset{\displaystyle O}{\|}}{C}-OH$$

where X=Cl, Br

I. Conversion of α-Halo Acids to Other Acid Derivatives

1. Dicarboxylic acids (Secs. 22.5B, 22.6, and 22.18)

$$R-\underset{\underset{\displaystyle X}{|}}{CH}-C\overset{\ominus}{\underset{\cdot\cdot}{\ddot{O}}:} \xrightarrow[\text{aq alcohol}]{:\overset{\ominus}{C}N:} R-\underset{\underset{\displaystyle C\equiv N}{|}}{CH}-C\overset{\ominus}{\underset{\cdot\cdot}{\ddot{O}}:} \xrightarrow[H^{\oplus} \text{ or } :\ddot{O}H^{\ominus}, \text{ heat}]{H_2O} HO-\overset{\overset{\displaystyle O}{\|}}{C}-\underset{\underset{\displaystyle R}{|}}{CH}-\overset{\overset{\displaystyle O}{\|}}{C}-OH$$

Substituted malonic acid

2. α-Hydroxy acids (Sec. 22.18)

$$R{-}CH{-}COOH \xrightarrow[\text{(excess)}]{:\ddot{O}H^{\ominus}} R{-}CH{-}CO\ddot{O}:^{\ominus} \xrightarrow{H_3\ddot{O}^{\oplus}} R{-}\underset{\underset{OH}{|}}{CH}{-}\overset{\overset{O}{\|}}{C}{-}OH$$

with X and OH substituents

α-Hydroxy acid

3. α-Amino acids (Secs. 22.16 and 28.7)

$$R{-}CH{-}COOH \xrightarrow[\text{(excess)}]{NH_3} R{-}CH{-}CO\ddot{O}:^{\ominus} \xrightarrow{\text{neutralize}} R{-}\underset{\underset{NH_2}{|}}{CH}{-}\overset{\overset{O}{\|}}{C}{-}OH$$

with X and NH_2 substituents

α-Amino acid

4. α,β-Unsaturated acids (Sec. 22.18)

$$\underset{\underset{H\ \ X}{|\ \ |}}{{-}C{-}C{-}}COOH \xrightarrow[\text{alcohol (E2 reaction)}]{KOH} \xrightarrow{\text{neutralize}} \overset{}{\underset{}{{>}C{=}C{-}\overset{\overset{O}{\|}}{C}{-}OH}}$$

α,β-Unsaturated acid

Study Questions

22.16 Give IUPAC names for the following 10 compounds:

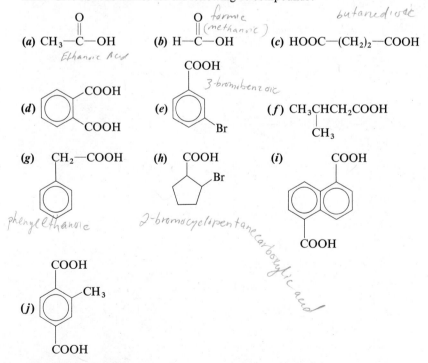

(a) $CH_3{-}\overset{\overset{O}{\|}}{C}{-}OH$ *Ethanoic Acid*

(b) $H{-}\overset{\overset{O}{\|}}{C}{-}OH$ *formic (methanoic)*

(c) $HOOC{-}(CH_2)_2{-}COOH$ *butanedioc*

(d) [benzene ring with two COOH groups ortho]

(e) [benzene ring with COOH and Br]   *3-bromobenzoic*

(f) CH_3CHCH_2COOH with CH_3

(g) $CH_2{-}COOH$ on benzene ring   *phenyl ethanoic*

(h) cyclopentane with COOH and Br   *2-bromocyclopentanecarboxylic acid*

(i) naphthalene with two COOH groups

(j) benzene ring with COOH, CH_3, and COOH

22.17 Give structures that correspond to the following names:
(a) 3-methylpentanoic acid
(b) *m*-toluic acid
(c) β-methylglutaric acid
(d) terephthalic acid

[hand-drawn benzene ring with two COOH groups]

(*e*)   nonanedioic acid
(*f*)   *trans*-1,3-cyclohexanedicarboxylic acid
(*g*)   potassium ethanoate
(*h*)   *cis*-7-bromo-4-heptenoic acid
(*i*)   2-ethyl-3-phenylpentanedioic acid
(*j*)   stearic acid

22.18 Supply the missing reactants, reagents, and products to complete the following eight equations. The items to be supplied are indicated by number [for example, (1)].

(*a*) $CH_3CH_2CH_2COOH$ $\xrightarrow[Br_2]{P}$ (1) $\xrightarrow[\text{aq alcohol}]{NaCN}$ (2) $\xrightarrow{(3)}$ $CH_3CH_2CH{-}COOH$ with $COOH$

$CH_3CH_2CH{-}\overset{O}{\overset{\|}{C}}{-}OH$ with Br

(*b*) $HO{-}\langle\text{ring}\rangle{-}COOH$ $\xrightarrow[\text{aqueous}]{NaHCO_3}$ (4) $\xrightarrow{H_3\overset{..}{O}^{\oplus}}$ (5)

(*c*)   [naphthalene with Br] $\xrightarrow[\text{ether}]{Mg}$ (6) $\xrightarrow[\text{2. } H_3\overset{..}{O}^{\oplus}]{\text{1. } CO_2}$ (7)

(*d*) $CH_3CH_2{-}\overset{O}{\overset{\|}{C}}{-}OH$ $\xrightarrow[(8)]{LAH}$ $CH_3CH_2CH_2OH$ $\xrightarrow[\Delta]{H_2SO_4 \;(9)}$ $CH_3{-}CH{=}CH_2$

$\Big\downarrow P\Big| Br_2$ (10) $\downarrow$

(11)   (13) CH_3CHCH_3 with Br

(*e*) $CH_3CHCH_2CH_3$ with OH $\xrightarrow[\substack{H_2SO_4,\\ \text{heat}}]{K_2Cr_2O_7}$ (12) $\xrightarrow{HCN}$ (14) $\xrightarrow[\text{heat}]{H^{\oplus}}$ (15)

       (12) $\uparrow NH_2OH$ (14) $\downarrow H\overset{..}{O}\overset{\ominus}{:},\ H_2O,\ \text{heat}$

   excess CH_3OH, dry $\Big|HCl$

      (17)   (16)

(*f*) $O_2N{-}\langle\text{ring}\rangle{-}CH_2CH_2CH_3$ $\xrightarrow[\substack{H\overset{..}{O}\overset{\ominus}{:},\\ \text{heat}}]{KMnO_4}$ (18) $\xrightarrow[\text{aqueous}]{H_3\overset{..}{O}^{\oplus}}$ (19)

(*g*)   [toluene CH_3] $\xrightarrow{Br_2,\ \text{light}}$ (20) $\xrightarrow{(21)}$ [ring with CH_2CN] $\xrightarrow{(22)}$ [ring with CH_2COOH]

         (23) $\searrow$ (24)

[ring with $\overset{CN}{\underset{}{CH}}{-}COOH$] $\xrightarrow[\text{aq alcohol}]{NaCN}$

   $\xrightarrow[\text{heat}]{H^{\oplus},\ H_2O}$ (25)

   heat $\downarrow$

     (26)

(h) $\xrightarrow[\text{H}_2\text{O, heat}]{\text{KMnO}_4}$ (27)

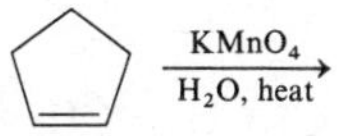

22.19 There are three isomeric cyclopropanedicarboxylic acids, A, B, and C, each having the formula $C_5H_6O_4$. Isomer A is unaffected by heat. Isomer B, on heating, loses a molecule of CO_2 to give a new compound D, $C_4H_6O_2$. Isomer C, on heating, loses a molecule of water to give a new compound E, $C_5H_4O_3$. On the basis of these data alone, write the structures of compounds A to E. Write equations for the reactions that occur. (*Note:* Throughout this problem, ignore the possibility of enantiomers.)

22.20 Answer the following questions regarding the acidity of carboxylic acids.

(a) Briefly explain why a carboxylic acid, $RCOOH$, is much more acidic than an alcohol, RCH_2OH.

(b) In each of the two compounds shown here, one of the carbon atoms is circled. In which compound would you expect the hydrogen atoms attached to the circled carbon atom to be the more acidic? Briefly explain.

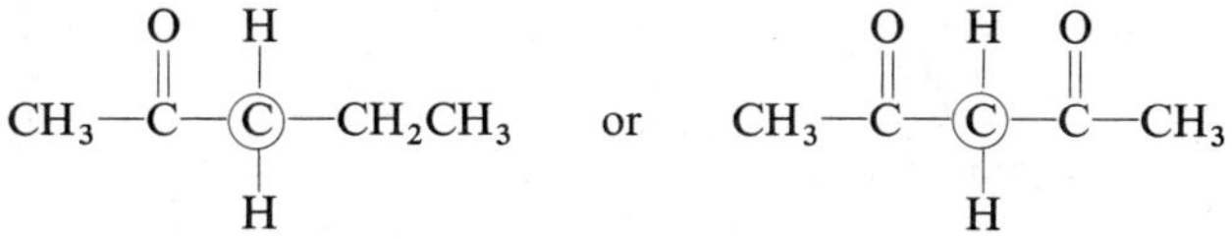

(c) Which acid in each of the following pairs is *more* acidic? Briefly explain.

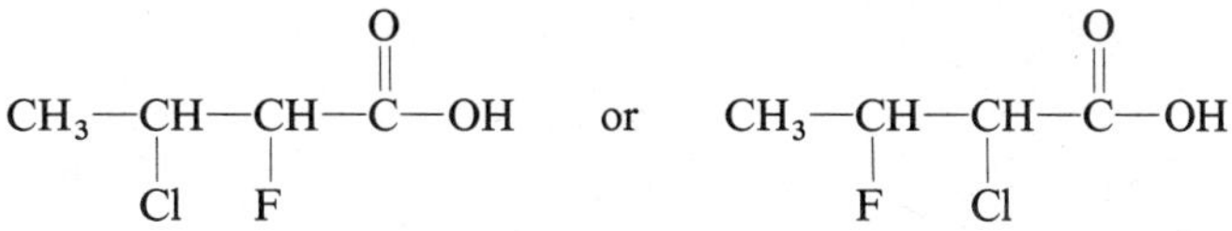

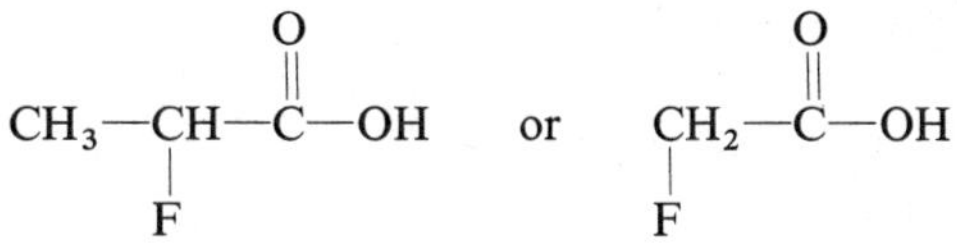

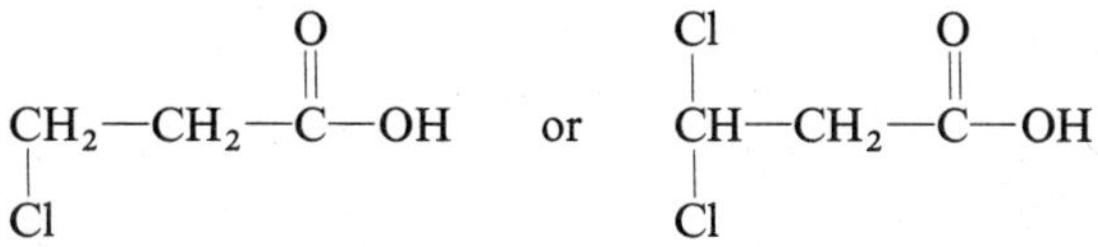

22.21 Arrange the following three groups of compounds *by letter* in order of *increasing acidity;* that is, place the least acidic compound first and the most acidic last. No reasoning is required.

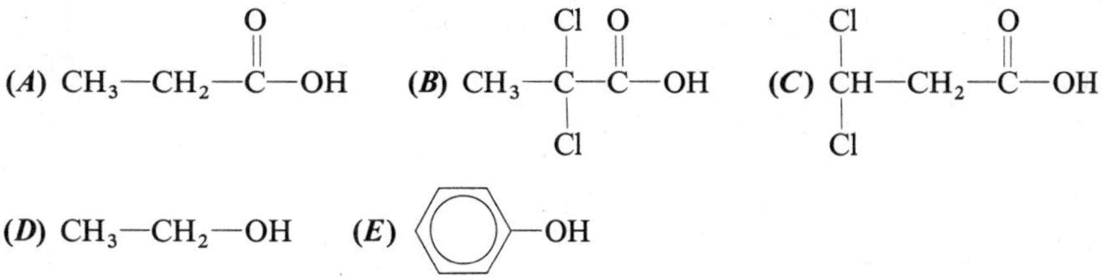

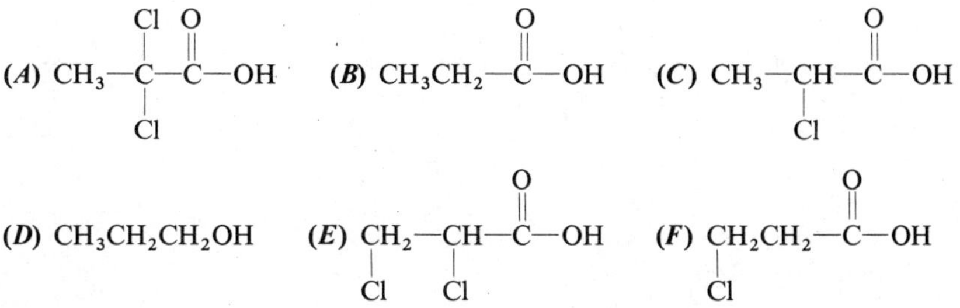

(c) (*A*) Br—CH$_2$CH$_2$—$\overset{\displaystyle O}{\overset{\|}{C}}$—OH (*B*) CH$_3$—$\underset{\underset{\displaystyle CH_3}{|}}{CH}$—$\overset{\displaystyle O}{\overset{\|}{C}}$—OH (*C*) CH$_3$—$\underset{\underset{\displaystyle Br}{|}}{CH}$—$\overset{\displaystyle O}{\overset{\|}{C}}$—OH

(*D*) CH$_3$—$\underset{\underset{\displaystyle Br}{|}}{\overset{\overset{\displaystyle Br}{|}}{C}}$—$\overset{\displaystyle O}{\overset{\|}{C}}$—OH (*E*) CH$_3CH_2$—$\overset{\displaystyle O}{\overset{\|}{C}}$—OH

22.22 Show all the steps, conditions, and reagents required to carry out the following 14 transformations, using the indicated starting materials plus any C_1 or C_2 compounds as the only sources of carbon. All intermediate organic compounds must be prepared from them, except for common organic solvents (ether, CCl_4, and so on). All inorganic reagents are available.

(*a*) Benzoic acid ⟶ (C$_6$H$_5$)—CH$_2$—$\overset{\displaystyle O}{\overset{\|}{C}}$—OH

(*b*) Ethanol ⟶ propanoic acid

(*c*) Propanoic acid ⟶ acetic acid

(*d*) Ethylene ⟶ CH$_3$CH$_2$—$\overset{\displaystyle O}{\overset{\|}{C}}$—OH

(*e*) 1-Propanol ⟶ CH$_3$—C≡C—$\overset{\displaystyle O}{\overset{\|}{C}}$—OH

(*f*) Toluene ⟶ (C$_6$H$_5$)—$\underset{\underset{\displaystyle OH}{|}}{CH}$—$\overset{\displaystyle O}{\overset{\|}{C}}$—OH

(*g*) Butyric acid ⟶ CH$_3$CH—CH$_2$—$\overset{\displaystyle O}{\overset{\|}{C}}$—OH with Br on the CH

(*h*) 1,2-Dibromocyclopentane ⟶ HO—$\overset{\displaystyle O}{\overset{\|}{C}}$—(CH$_2$)$_3$—$\overset{\displaystyle O}{\overset{\|}{C}}$—OH

(*i*) α-Methylpropionic acid ⟶ 2-methylpropanal

(*j*) Acetic acid ⟶ CH$_3$CH$_2$—O—CH$_3$

(*k*) Ethylene bromide ⟶ propanal

(*l*) Hexanoic acid ⟶ 2-hexenoic acid

(*m*) Propanoic acid ⟶ CH$_3$—$\underset{\underset{\displaystyle NH_2}{|}}{\overset{\overset{\displaystyle H}{|}}{C}}$—COOH   (alanine)

(*n*) α-Bromonaphthalene ⟶ naphthalene with $COO^{\ominus}$ Na$^{\oplus}$ substituent

22.23 How might $CH_3COCH_2CH_2COOH$ be converted to each of the following compounds in *good yield*?

(*a*) $CH_3COCH_2CH_2CHO$ (*b*) $CH_3CHOHCH_2CH_2COOH$

(*c*) $CH_3CH_2CH_2CH_2CHO$ (*d*) $CH_3CHOHCH_2CH_2CH_2OH$

(*e*) $CH_3COCH_2CH_2CH_2OH$

22.24 The IR and UV spectra of one of the following two acids are given here.

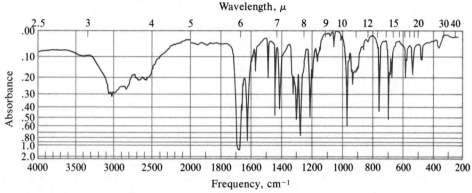

(1)   (2)

Which structure is consistent with the spectral data? Explain.

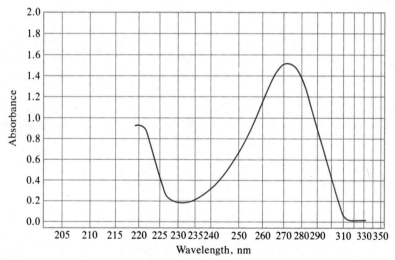

PROBLEM 22.24   IR spectrum of the unknown acid. [Data from *The Sadtler Handbook of Infrared Spectra*, W. W. Simmons (ed.), © Sadtler Research Laboratories, Division of Bio-Rad Laboratories, Inc. (1978)]

PROBLEM 22.24   UV spectrum of the unknown acid. [Data from *The Sadtler Handbook of Ultraviolet Spectra*, W. W. Simmons (ed.), © Sadtler Research Laboratories, Division of Bio-Rad Laboratories, Inc., (1979)]

22.25 Unknown *A* gives the following NMR and mass spectra. Provide a structure consistent with these observations.

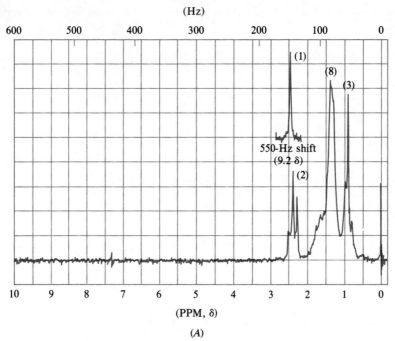

PROBLEM 22.25   NMR spectrum of unknown (*A*). [Data from *The Aldrich Library of NMR Spectra*, C. J. Pouchert and J. R. Campbell (eds.), copyright 1974, Aldrich Chemical Co., Milwaukee]

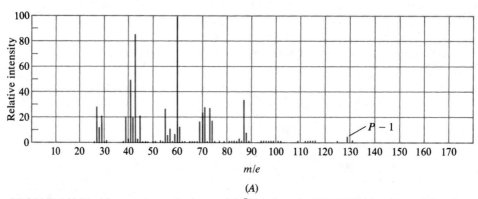

PROBLEM 22.25   Mass spectrum of unknown (*A*). [Data from the *EPA/NIH Mass Spectral Data Base*, S. R. Heller and G. W. A. Milne (eds.), vols. 1–4, copyright 1978, U.S. Dept. of Commerce]

23

Derivatives of Carboxylic and Dicarboxylic Acids

In this chapter we discuss the chemistry of various derivatives of carboxylic and dicarboxylic acids. Compounds are said to be derivatives of carboxylic acids if they have an oxygen- or nitrogen-containing substituent or a halogen atom attached to the carbon-oxygen double bond in place of the hydroxyl group of the carboxyl group, —COOH. In particular, we focus our attention on the following families of compounds:

Acid chloride:

$$
\text{(Ar—) R—}\overset{\overset{\displaystyle O}{\|}}{\text{C}}\text{—Cl} \quad \text{or} \quad \text{(Ar) RCOCl}
$$

Structural   Condensed
formula   formula

Ester:

$$
\text{(Ar—) R—}\overset{\overset{\displaystyle O}{\|}}{\text{C}}\text{—OR}' \quad \text{or} \quad \text{(Ar) RCOOR}' \ \text{or} \ \text{(Ar) ROCOR}'
$$

Structural   Condensed
formula   formula

Amide:

$$
\text{(Ar—) R—}\overset{\overset{\displaystyle O}{\|}}{\text{C}}\text{—NR}_2' \quad \text{or} \quad \text{(Ar) RCONR}_2'
$$

Structural   Condensed
formula   formula

Anhydride:

$$
\text{(Ar—) R—}\overset{\overset{\displaystyle O}{\|}}{\text{C}}\text{—O—}\overset{\overset{\displaystyle O}{\|}}{\text{C}}\text{—R (—Ar)} \quad \text{or} \quad \text{(Ar) RCOOCOR (Ar)}
$$

Structural   Condensed
formula   formula

All these compounds have the common structural feature of the carbon-oxygen double bond.

We also discuss another functional group in this chapter, the nitrile:

Nitrile:

$$(Ar—)\ R—C{\equiv}N \qquad or \qquad (Ar)\ RCN$$

Structural formula   Condensed formula

In a sense, the carboxylic acid is the parent compound because the other compounds are derived from it. There are two reasons for studying all these families of compounds as a unit: first, most of them are interconvertible and all can be derived from the corresponding carboxylic acid, and second, most of their reactions are mechanistically similar in that they usually involve nucleophilic acyl substitution.

Finally, we include dicarboxylic acids in our discussions because they undergo reactions quite similar to those of carboxylic acids. In several instances, however, we find exceptions that merit special attention and discussion.

23.1   Nomenclature

The common and IUPAC names of the derivatives of carboxylic acids and diacids are derived from the common and IUPAC names of the corresponding acids (see Sec. 22.1) and diacids (see Sec. 22.2).

A. Amides

Amides are named by changing the *-ic acid* ending of the common name of the carboxylic acid to **-amide**. When using the IUPAC system, the *-oic acid* ending of the acid is changed to **-amide**. Five examples are:

$$CH_3—\overset{\displaystyle O}{\overset{\|}{C}}—NH_2$$

Ethanamide
(acetamide)

$$CH_3—\overset{\displaystyle \bigcirc}{\underset{}{CH}}—CH_2—\overset{\displaystyle O}{\overset{\|}{C}}—NH_2$$

3-Phenylbutanamide
(*β*-phenylbutyramide)

$$CH_3—C{\equiv}C—\overset{\displaystyle O}{\overset{\|}{C}}—NH_2$$

2-Butynamide

$$Br—\bigcirc—\overset{\displaystyle O}{\overset{\|}{C}}—NH_2$$

p-Bromobenzamide

$$H_2N—\overset{\displaystyle O}{\overset{\|}{C}}—CH_2CH_2—\overset{\displaystyle O}{\overset{\|}{C}}—NH_2$$

Butanediamide
(succinamide)

The —$CONH_2$ group must be named as a substituent when it is attached to a ring other than benzene or is present in complex molecules, and then it is called **carboxamide.** If the words *carboxylic acid* appear in the name of the acid (for example, cyclobutanecarboxylic acid), they are replaced by *carboxamide* in the amide derivative; for example:

$$\underset{\square}{\overset{\displaystyle O}{\overset{\|}{C}}—NH_2}$$

Cyclobutanecarboxamide

Sometimes the nitrogen atom of an amide contains substituents other than hydrogen, as for example in RCONHR' and RCONR$_2$'. In these cases, the nitrogen substituents appear as a prefix to the regular amide name and are indicated by *N*- followed by the name of the alkyl or aryl group(s). Three examples of this nomenclature are:

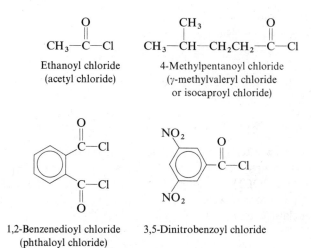

CH$_3$—C—NH—CH$_3$ CH$_3$—C—N(CH$_3$)(CH$_3$)   CH$_3$—C—N—H

N-Methylethanamide   *N,N*-Dimethylethanamide   *N*-Phenylethanamide
(*N*-methylacetamide)   (*N,N*-dimethylacetamide)   (*N*-phenylacetamide
                                              or acetanilide)

B. Acid Chlorides

Acid chlorides are named by dropping the *-ic acid* ending of either the common or IUPAC name of the corresponding carboxylic acid and adding **-yl chloride**; for example:

CH$_3$—C—Cl   CH$_3$—CH—CH$_2$CH$_2$—C—Cl

Ethanoyl chloride   4-Methylpentanoyl chloride
(acetyl chloride)   (γ-methylvaleryl chloride
                       or isocaproyl chloride)

1,2-Benzenedioyl chloride   3,5-Dinitrobenzoyl chloride
(phthaloyl chloride)

In more complicated systems if an acid is named with the words *carboxylic acid* (as in cyclobutanecarboxylic acid), the corresponding acid chloride is named by substituting **carbonyl chloride** for *carboxylic acid;* for example:

Cyclobutanecarbonyl chloride

It is possible to prepare acid bromides and iodides by reactions analogous to those used in the preparation of chlorides. However, in comparison to the chlorides, they are seldom used, and we do not discuss them at this time.

C. Esters

Esters are synthesized from two organic moieties: an acid and an alcohol or a phenol:

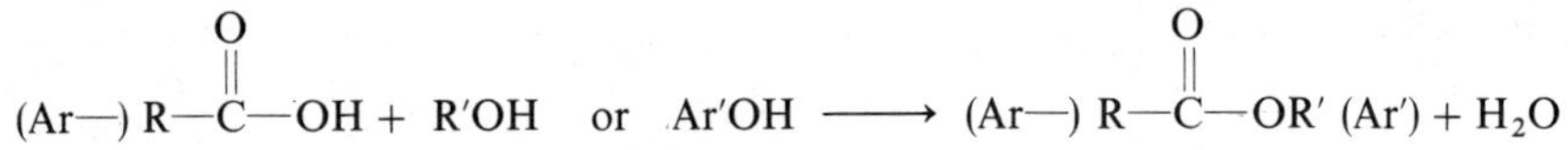

$$(Ar—) R—\overset{\overset{\displaystyle O}{\|}}{C}—OH + R'OH \quad \text{or} \quad Ar'OH \longrightarrow (Ar—) R—\overset{\overset{\displaystyle O}{\|}}{C}—OR' (Ar') + H_2O$$

Carboxylic   Alcohol   Phenol   Ester
acid

As a result, their names must reflect both the acid and the alcohol or phenol. The approach is to place first the name of the *alkyl or aryl group* that was originally in the alcohol or phenol, and follow it with the name of the carboxylic acid from which the ester was prepared, changing the *-ic acid* ending of the carboxylic acid name (IUPAC or common) to *-ate;* for example:

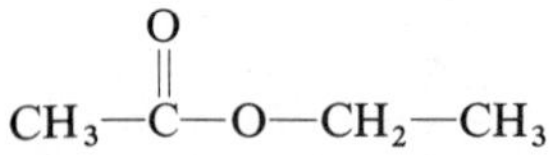

$$CH_3—\overset{\overset{\displaystyle O}{\|}}{C}—O—CH_2—CH_3 \qquad\qquad CH_3—\underset{\underset{\displaystyle CH_3}{|}}{CH}—CH_2CH_2—\overset{\overset{\displaystyle O}{\|}}{C}—O—CH_2CH_2CH_2CH_3$$

Ethyl ethanoate (derived from   Butyl 4-methylpentanoate
ethanoic acid and ethanol)
or
(ethyl acetate, derived from
acetic acid and ethyl alcohol)

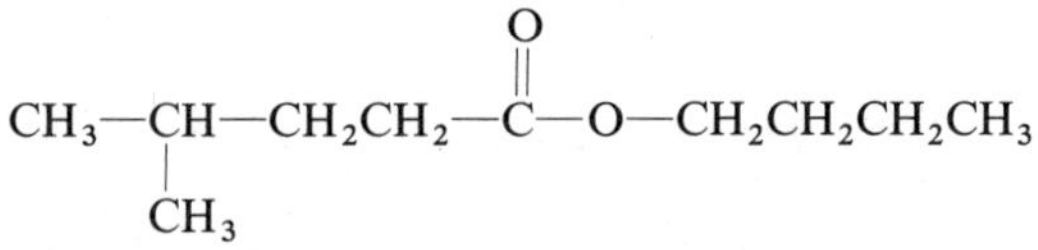

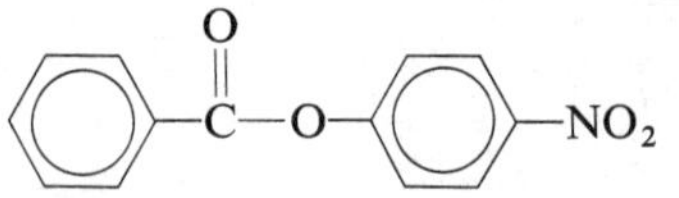

4-Nitrophenyl benzoate   Di-butyl hexanedioate
(*p*-nitrophenyl benzoate)   (di-*n*-butyl adipate)

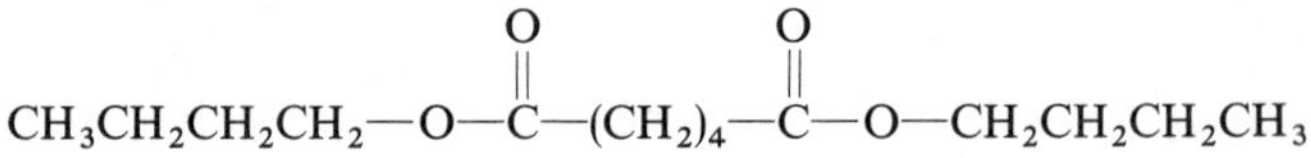

Sometimes an ester group must be designated as a substituent, and **carboalkoxy** (for alkyl groups) or **carboaryloxy** (for aryl groups) is used for this purpose. For example, —$COOCH_3$ is called *carbomethoxy*, —$COOCH_2CH_3$ is *carbethoxy* (the *o* is omitted before vowels), and —$COOC_6H_5$ is *carbophenoxy*. Two examples of this nomenclature are:

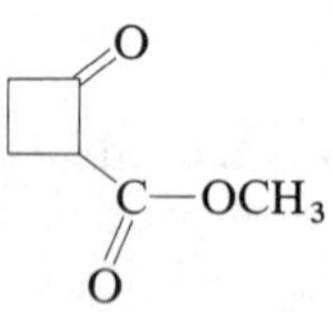

2-Carbomethoxycyclobutanone 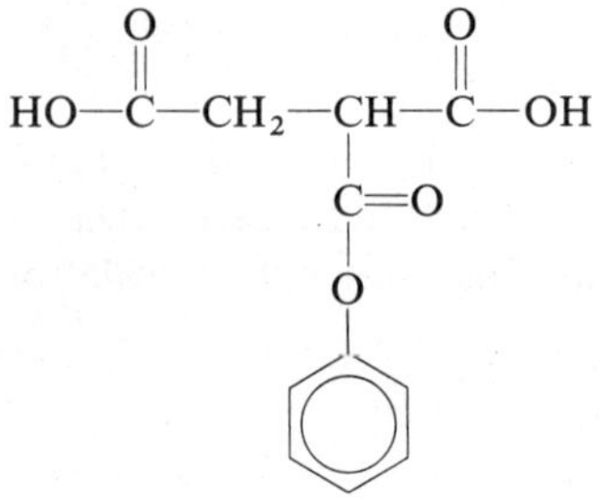   2-Carbophenoxybutanedioic acid
                                   (α-carbophenoxysuccinic acid)

D. Anhydrides

Anhydrides are formed from 2 moles of a carboxylic acid by elimination of a molecule of water. Cyclic anhydrides are formed from certain dicarboxylic acids through internal ring closure and loss of water:

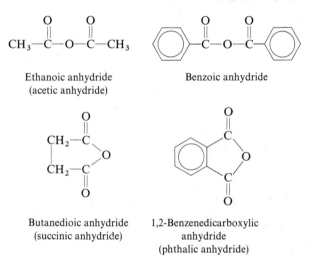

$$2 \ (Ar—) \ R—\overset{\overset{\displaystyle O}{\|}}{C}—OH \longrightarrow (Ar—) \ R—\overset{\overset{\displaystyle O}{\|}}{C}—O—\overset{\overset{\displaystyle O}{\|}}{C}—R \ (—Ar) + H_2O$$

Carboxylic acid   Anhydride

Dicarboxylic acid   Cyclic anhydride
$n = 2, 3$

As a result, anhydrides are named by taking either the common or the IUPAC name of the acid or diacid and changing the *acid* ending to **anhydride**; for example:

$$CH_3—\overset{\overset{\displaystyle O}{\|}}{C}—O—\overset{\overset{\displaystyle O}{\|}}{C}—CH_3$$

Ethanoic anhydride
(acetic anhydride)

Benzoic anhydride

Butanedioic anhydride
(succinic anhydride)

1,2-Benzenedicarboxylic
anhydride
(phthalic anhydride)

E. Nitriles

Nitriles are closely related to carboxylic acids, and they take the name of the carboxylic acid containing the same number of carbon atoms. Common names are derived by dropping the *-ic acid* ending of the acid and adding **-nitrile.** Often it is necessary to insert an *o* between the stem name and the *-nitrile* ending to aid pronunciation. In the IUPAC system, however, the suffix **-nitrile** is added to the name of the parent hydrocarbon corresponding to the number of carbon atoms in the longest chain that bears the —C≡N group; for example:

$$CH_3—C≡N \qquad CH_3—CH—C≡N \qquad N≡C—(CH_2)_3—C≡N$$

Ethanenitrile
(acetonitrile)

2-Phenylpropanenitrile
(α-phenylpropionitrile)

Pentanedinitrile
(glutaronitrile)

Common names of nitriles are used more frequently than IUPAC names.

Sometimes it is necessary to call the —C≡N group *cyano*, which is used to name compounds that bear other substituents; for example:

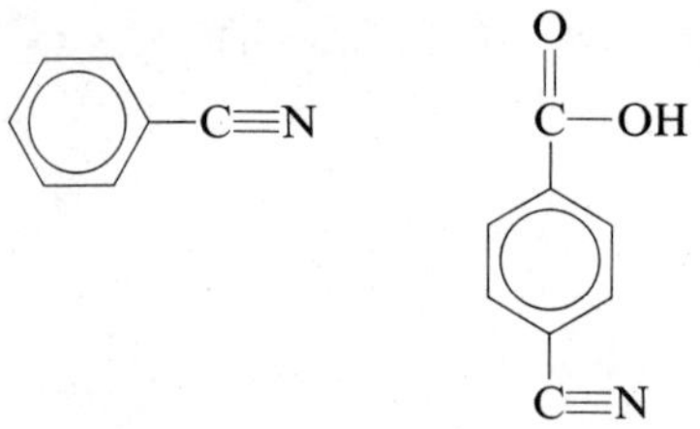

Cyanobenzene   *p*-Cyanobenzoic acid
(benzonitrile)

Naming More Complex Molecules

Some molecules contain two or more different functional groups. When only one of them is used in the name of the parent compound, the following order of functional groups determines which is the member of the parent compound. A group on the left takes precedence over one on the right.

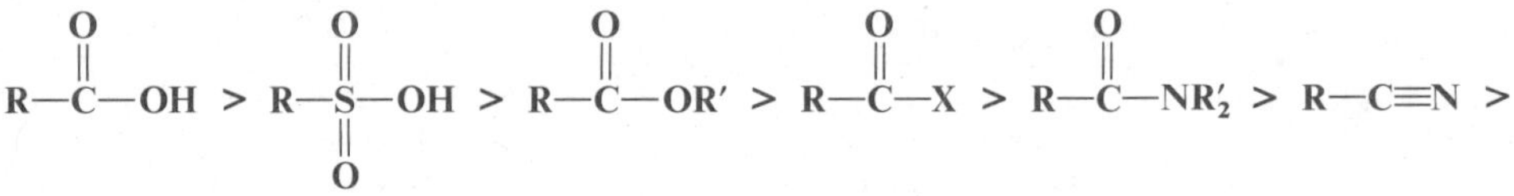

$$\underset{\substack{\text{O}}}{\text{R—C—OH}} > \underset{\substack{\text{O}\\\text{O}}}{\text{R—S—OH}} > \underset{\substack{\text{O}}}{\text{R—C—OR'}} > \underset{\substack{\text{O}}}{\text{R—C—X}} > \underset{\substack{\text{O}}}{\text{R—C—NR'}_2} > \text{R—C≡N} >$$

Carboxylic   Sulfonic   Ester   Acid   Amide   Nitrile
acid   acid   halide

**Highest
priority**

$$\underset{\substack{\text{O}}}{\text{R—C—H}} > \underset{\substack{\text{O}}}{\text{R—C—R'}} > \text{R—OH} > \text{R—NR'}_2 > \text{R—OR'}$$

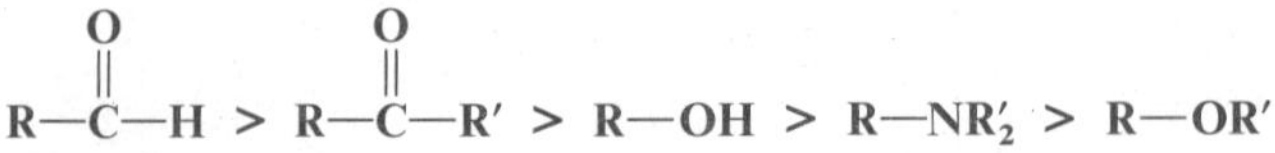

Aldehyde   Ketone   Alcohol   Amine   Ether

**Lowest
priority**

This order is not rigorously adhered to by organic chemists today. Carbon-carbon double or triple bonds are usually indicated by integrating *-en-* or *-yn-* (respectively) into the suffix, as in *-enoic acid* and *-ynamide*.

23.2   Physical Properties

Some of the physical properties and characteristics of the derivatives of carboxylic acids are discussed in this section (see Table 23.1 also).

A. Esters

Unlike the carboxylic acids from which they are derived, esters have considerably lower boiling points than acids of comparable molecular weight. They boil at

TABLE 23.1 Physical Properties of Acid Chlorides, Esters, Amides, and Anhydrides

Parent Acid	Acid Chloride		Ethyl Ester		Amide		Anhydride	
	mp, °C	bp, °C	mp, °C	bp, °C	mp, °C	bp, °C	mp, °C	bp, °C
Methanoic (formic)	Does not exist		−80	54	2	193	Does not exist	
Ethanoic (acetic)	−112	52	−84	77.1	82	222	−73	140
Propanoic (propionic)	−94	80	−74	99	80	213	−45	168
Butanoic (n-butyric)	−89	102	−93	121	116	216	−75	198
Benzoic	−1	197	−35	213	130	290	42	360
o-Toluic		213	−10	221	147			
m-Toluic	−25	218		226	97		70	
p-Toluic	−2	226		235	155		98	
Benzenedioic (phthalic)*	11			296	219		131	284

* Diamide, dichloride, and diethyl ester.

nearly the same temperature as aldehydes or ketones of similar weight. This is because esters do not contain the hydroxyl group so that hydrogen bonding, which is largely responsible for the high boiling point of acids, is not possible in esters. Esters of four carbon atoms (total) or less have appreciable water solubility and this decreases greatly with increasing carbon content.

Esters are normally pleasant, sweet-smelling compounds and are often constituents of flavorings. For example, pentyl acetate, $CH_3COOCH_2CH_2CH_2CH_2CH_3$ (often called *amyl acetate*) has a banana odor, and methyl salicylate (1), commonly called oil of wintergreen, has the aroma its name implies.

Methyl salicylate
(1)

B. Acid Chlorides

Acid chlorides boil at about the same temperature as aldehydes and ketones of similar molecular weight. The reason for this is that the hydroxyl group of the carboxylic acid is replaced, this time by chlorine, so that no intermolecular hydrogen bonding is possible. Acid chlorides have a noxious odor, especially some of the lower molecular weight ones that are volatile. This is in part because they react readily with moisture in the air and undergo hydrolysis to give hydrochloric acid—a reaction we study later.

C. Amides

As Table 23.1 indicates, most amides are solids at room temperature and thus have very high boiling points. This is because of the strong intermolecular hydrogen bonding between the N—H bond and the C=O group:

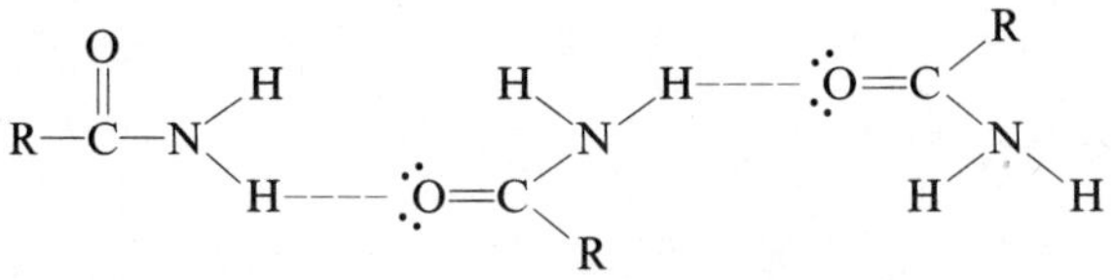

This hydrogen bonding is apparently quite strong—stronger than in carboxylic acids—as evidenced by the high melting points of most amides. Hydrogen bonding between amide linkages is partially responsible for the structure of proteins and peptides, which are naturally occurring polymers that contain numerous amide linkages (see Sec. 28.10).

Question 23.1

N,N-Dimethylacetamide (mp $-20°$, bp $165°$) is a liquid at room temperature, whereas acetamide (mp $82°$, bp $222°$) is a solid. Explain these facts, keeping in mind that the molecular weight of acetamide is considerably less than that of *N,N*-dimethylacetamide.

23.3   Interconversion of Carboxylic Acid Derivatives and Nucleophilic Acyl Substitution

Before we turn to specific reactions of carboxylic acids and their derivatives, let us look at a general picture of this type of reaction. As the following reaction shows, almost any carboxylic acid or its derivative can be converted into another derivative:

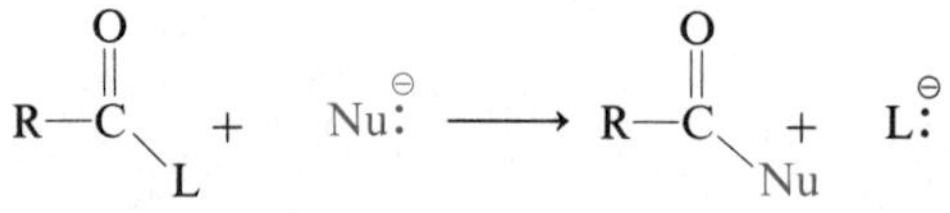

Acid or derivative	Acid or derivative

where —L and —Nu = —OH, acid
                              —X, acid halide
                              —OR′, ester
                              —NH$_2$, amide

$$\overset{\displaystyle O}{\overset{\|}{-O-C-R}},\ \text{anhydride}$$

and others

This equation does not specify what nucleophile, $\overset{\ominus}{Nu:}$, is required for the conversion or what derivative is formed. We cover all these details in the next few sections. It is important, however, to realize that *substitution* occurs, because one substituent attached to the acyl group is replaced by another one. Our intent in this section is to discuss the *general mechanism of substitution*, which is commonly referred to as **nucleophilic acyl substitution.** The mechanisms are shown here.

Nucleophilic acyl substitution in neutral or basic solution:

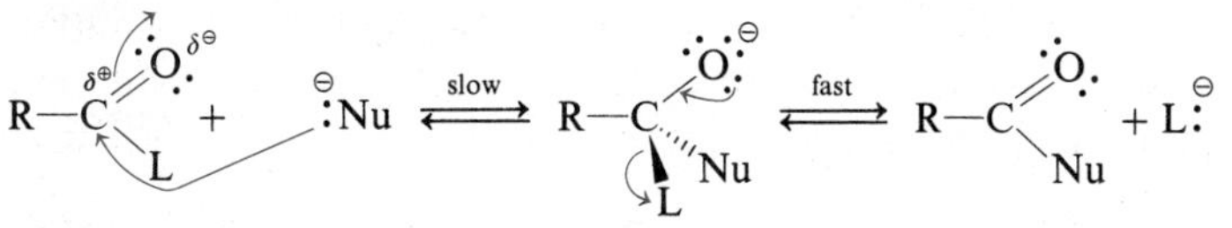

Acid or derivative *Trigonal* — Unstable intermediate *Tetrahedral* — Product *Trigonal*

Nucleophilic acyl substitution in acidic solution:

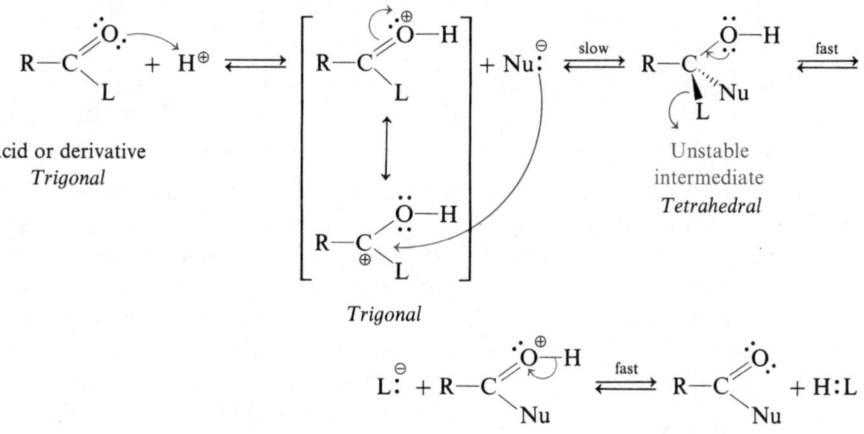

Acid or derivative *Trigonal* — *Trigonal* — Unstable intermediate *Tetrahedral* — Product *Trigonal*

These mechanisms resemble those for nucleophilic addition to aldehydes and ketones in many respects; both involve nucleophilic attack on the electropositive carbon atom of the carbon-oxygen double bond. In acidic solution both mechanisms involve the initial protonation of the oxygen of the carbon-oxygen double bond, followed by nucleophilic attack on carbon. Let us review the nucleophilic addition mechanisms (see Sec. 21.2).

Nucleophilic addition in neutral or basic solution:

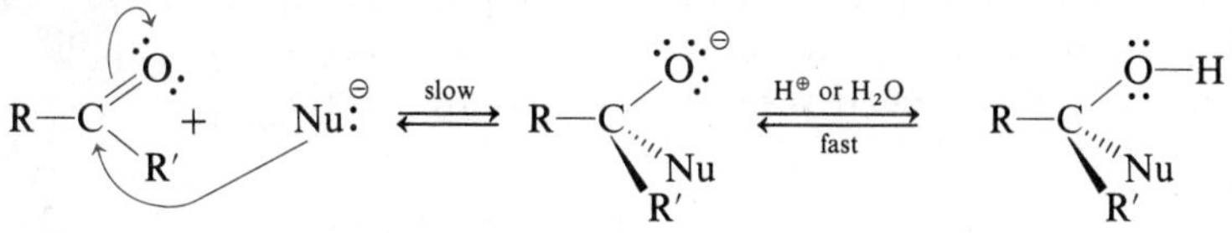

Aldehyde or ketone — Product *Tetrahedral and stable*

Nucleophilic addition in acidic solution:

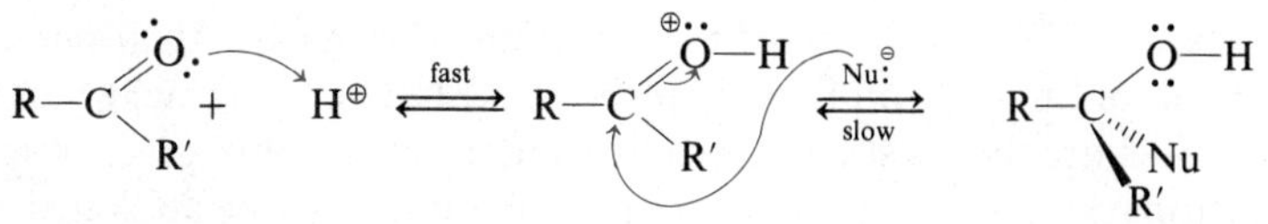

Aldehyde or ketone — Product *Tetrahedral and stable*

The steps leading to the formation of the tetrahedral structure are indeed the same for both addition and substitution reactions. The tetrahedral structures themselves are identical for both types of reactions. What then are the differences between substitution and addition? Or, put another way, why is the tetrahedral structure in nucleophilic addition stable as evidenced by the observed products, whereas it is an unstable intermediate in the reactions of carboxylic acids and their derivatives? To answer these questions, look more closely at the tetrahedral structures:

Relatively weak bond; easily broken $R-C\overset{OH}{\underset{L}{\cdots Nu}}$ Strong bond; hard to break $R-C\overset{OH}{\underset{R'}{\cdots Nu}}$

Tetrahedral intermediate   Tetrahedral product
in substitution   in addition
Unstable   *Stable*

The major difference between the preceding structures is in the **L—C** and **R′—C** bonds. In addition reactions of aldehydes and ketones, R′ is always either H or another carbon-containing group, and thus there is always a strong C—H or C—C covalent bond. These two bonds are difficult to break, and the H:$^{\ominus}$ or R:$^{\ominus}$ that would form if they did break are poor leaving groups (see Secs. 9.8 and 9.11). Once the tetrahedral structure is formed, therefore, a stable product results; the net result is *addition*. Certain addition compounds are formed reversibly, however, especially the cyanohydrins and the bisulfite addition products. With these products, the relatively weak C—Nu bond is formed, but if it breaks, reactants are *regenerated*. The C—H and C—R′ bonds are *never* broken.

In the substitution reaction, **L**—(*leaving group*) always contains some atom more electronegative than carbon (for example, *oxygen* in —OH, —OR′, —OCOR, or *nitrogen* in —NH$_2$, or *halogen* in —X) attached to the tetrahedral carbon atom. These C—L bonds are relatively weak compared with the C—H and C—R′ bonds in the aldehydes and ketones and are fairly easily broken yielding good leaving groups. When the tetrahedral intermediate forms, it decomposes and forms products because of the affinity of the electronegative element for electrons. Sometimes the oxygen- and nitrogen-containing substituents are protonated, which makes them even better leaving groups.

In summary, nucleophilic acyl substitution occurs in two steps: (1) **addition** of the nucleophile to form the *tetrahedral* intermediate and then (2) **elimination** (loss) of the leaving group to generate the *trigonal* product. Several nucleophilic acyl substitution reactions are reversible. Whether or not the equilibrium favors the formation of products depends on the specific reaction.

23.4   Preparation of Acid Chlorides

We start our discussion of the reactions of acids and their derivatives with the conversion of an acid to an acid chloride. Although other acid halides are known, the chlorides are the most common. We might expect some analogies in the reactions of the hydroxyl group in an acid and in an alcohol, and one analogy is the conversion of —OH to —Cl. Phosphorus trichloride (PCl$_3$), phosphorus pentachloride (PCl$_5$), and thionyl chloride ($SOCl_2$) are all used to convert an acid into the corresponding acid chloride. (Although HX converts an alcohol into an alkyl halide, it does *not* convert an acid into an acid halide.) The general reactions are:

$$
\underset{\substack{\text{Carboxylic} \\ \text{acid}}}{R-\overset{\overset{\displaystyle O}{\|}}{C}-OH} + \begin{array}{c} PCl_3 \\ \text{or } PCl_5 \\ \text{or } SOCl_2 \end{array} \longrightarrow \underset{\substack{\text{Acid} \\ \text{chloride}}}{R-\overset{\overset{\displaystyle O}{\|}}{C}-Cl} + \begin{array}{c} P(OH)_3 \\ \text{or } POCl_3 + HCl \\ \text{or } SO_2 + HCl \end{array}
$$

Thionyl chloride is often used because its low boiling point (79°) allows excess reagent to be removed easily. Also, the products are SO_2 and HCl, which are liberated as gases. Three specific examples are:

$$
\underset{\substack{\text{Pentanoic acid} \\ \text{(valeric acid)}}}{CH_3CH_2CH_2CH_2-\overset{\overset{\displaystyle O}{\|}}{C}-OH} \xrightarrow{PCl_3} \underset{\substack{\text{Pentanoyl chloride} \\ \text{(valeroyl chloride)}}}{CH_3CH_2CH_2CH_2-\overset{\overset{\displaystyle O}{\|}}{C}-Cl}
$$

m-Bromobenzoic
acid

m-Bromobenzoyl
chloride

$$
\underset{\substack{\text{Hexanedioic acid} \\ \text{(adipic acid)}}}{HO-\overset{\overset{\displaystyle O}{\|}}{C}-(CH_2)_4-\overset{\overset{\displaystyle O}{\|}}{C}-OH} \xrightarrow{SOCl_2} \underset{\substack{\text{Hexanedioyl chloride} \\ \text{(adipoyl chloride)}}}{Cl-\overset{\overset{\displaystyle O}{\|}}{C}-(CH_2)_4-\overset{\overset{\displaystyle O}{\|}}{C}-Cl}
$$

Question 23.2

Reasoning by analogy with the mechanism for the reaction of an alcohol with PCl_3 and $SOCl_2$ (see Sec. 10.15), suggest a plausible mechanism for similar reactions with carboxylic acids.

23.5 Reactions of Acid Chlorides

We encountered several reactions of acid chlorides, which are listed here for review. Acid chlorides react with the organocadmium reagent to give ketones (see Sec. 20.8), they undergo Friedel-Crafts acylation in the presence of aluminum trichloride to give an aryl ketone (see Sec. 20.8), and they can be reduced selectively by lithium tri-*tert*-butoxyaluminum hydride or with hydrogen gas in the presence of palladium and barium sulfate to give an aldehyde (Rosenmund reduction, see Sec. 20.6).

Acid chlorides are also very important intermediates in the preparation of carboxylic acids and derivatives of carboxylic acids, summarized as follows:

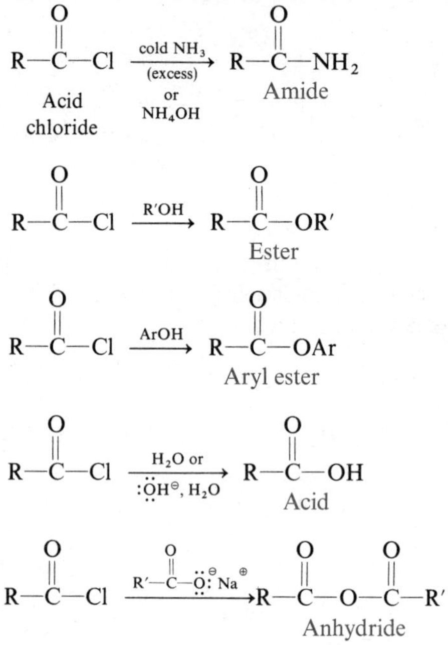

$$
\underset{\substack{\text{Acid} \\ \text{chloride}}}{R-\overset{\overset{\displaystyle O}{\|}}{C}-Cl} \xrightarrow[\substack{\text{or} \\ NH_4OH}]{\substack{\text{cold } NH_3 \\ \text{(excess)}}} \underset{\text{Amide}}{R-\overset{\overset{\displaystyle O}{\|}}{C}-NH_2}
$$

$$
R-\overset{\overset{\displaystyle O}{\|}}{C}-Cl \xrightarrow{R'OH} \underset{\text{Ester}}{R-\overset{\overset{\displaystyle O}{\|}}{C}-OR'}
$$

$$
R-\overset{\overset{\displaystyle O}{\|}}{C}-Cl \xrightarrow{ArOH} \underset{\text{Aryl ester}}{R-\overset{\overset{\displaystyle O}{\|}}{C}-OAr}
$$

$$
R-\overset{\overset{\displaystyle O}{\|}}{C}-Cl \xrightarrow[:\ddot{O}H^{\ominus},\, H_2O]{H_2O \text{ or}} \underset{\text{Acid}}{R-\overset{\overset{\displaystyle O}{\|}}{C}-OH}
$$

$$
R-\overset{\overset{\displaystyle O}{\|}}{C}-Cl \xrightarrow{R'-\overset{\overset{\displaystyle O}{\|}}{C}-\ddot{O}:^{\ominus}\, Na^{\oplus}} \underset{\text{Anhydride}}{R-\overset{\overset{\displaystyle O}{\|}}{C}-O-\overset{\overset{\displaystyle O}{\|}}{C}-R'}
$$

A. Amide Formation

When an acid chloride reacts with excess *cold, concentrated* aqueous ammonia, which is obtained as concentrated ammonium hydroxide in the laboratory, an amide is produced:

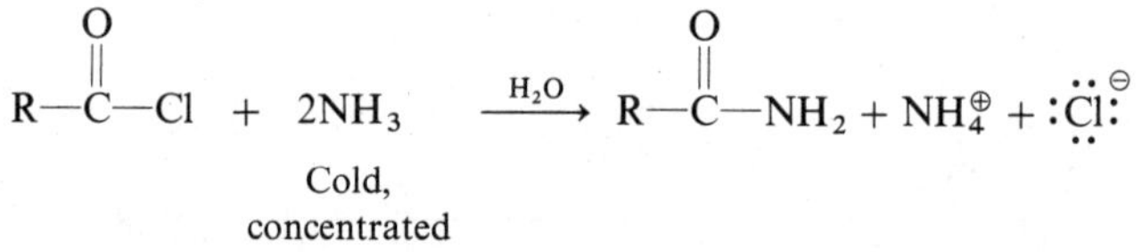

$$
\underset{\substack{\text{Cold,} \\ \text{concentrated}}}{R-\overset{\overset{\displaystyle O}{\|}}{C}-Cl} + 2NH_3 \xrightarrow{H_2O} R-\overset{\overset{\displaystyle O}{\|}}{C}-NH_2 + NH_4^{\oplus} + :\ddot{\underset{\cdot\cdot}{C}l}:^{\ominus}
$$

Mechanistically, this is a typical nucleophilic acyl substitution reaction; the steps are shown here:

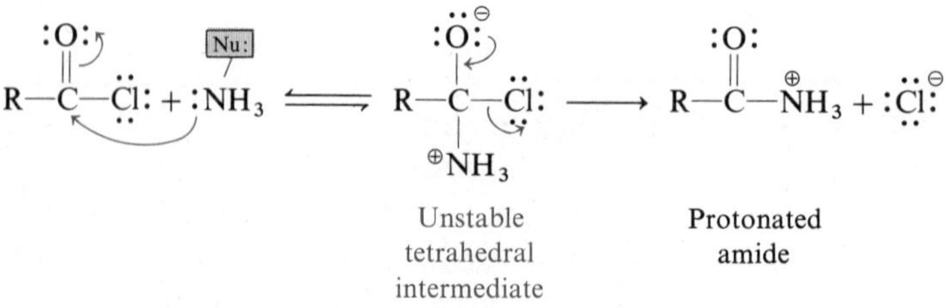

$$
R-\overset{\overset{\displaystyle :\ddot{O}:}{\|}}{C}-\ddot{\underset{\cdot\cdot}{C}l}: + :NH_3 \;\rightleftharpoons\; \underset{\substack{\text{Unstable} \\ \text{tetrahedral} \\ \text{intermediate}}}{R-\overset{\overset{\displaystyle :\ddot{O}:^{\ominus}}{|}}{\underset{\underset{\displaystyle \oplus NH_3}{|}}{C}}-\ddot{\underset{\cdot\cdot}{C}l}:} \;\longrightarrow\; \underset{\substack{\text{Protonated} \\ \text{amide}}}{R-\overset{\overset{\displaystyle :O:}{\|}}{C}-\overset{\oplus}{N}H_3} + :\ddot{\underset{\cdot\cdot}{C}l}:^{\ominus}
$$

then

$$
\underset{\underset{\displaystyle \underset{H}{|}}{\substack{\text{Acts as} \\ \text{base}}}}{R-\overset{\overset{\displaystyle O}{\|}}{C}-\overset{\overset{\displaystyle H}{|}}{\underset{}{\overset{\oplus}{N}}}-H} + :NH_3 \;\longrightarrow\; R-\overset{\overset{\displaystyle O}{\|}}{C}-\ddot{N}H_2 + NH_4^{\oplus}
$$

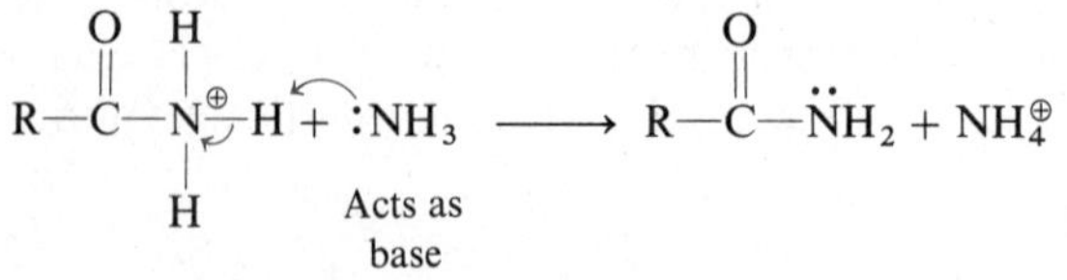

The nucleophilic ammonia molecule adds to the acyl group and forms the tetrahedral intermediate, which decomposes to give a "protonated" amide. Another molecule of ammonia acts as a base and abstracts a proton from the protonated amide to produce the free amide and an ammonium ion.

The reaction between an acid chloride and an amine is rather general because the amine can be ammonia or a primary or secondary amine (RNH_2 or R_2NH, respectively). *Tertiary amines cannot be used*. The following examples of amide formation are illustrative:

$$CH_3-\overset{\overset{O}{\|}}{C}-Cl \xrightarrow[NH_3]{cold} CH_3-\overset{\overset{O}{\|}}{C}-NH_2$$

Ethanoyl
chloride
(acetyl
chloride)

Ethanamide
(acetamide)

$$CH_3-\langle\bigcirc\rangle-\overset{\overset{O}{\|}}{C}-Cl \xrightarrow[NH_3]{cold} CH_3-\langle\bigcirc\rangle-\overset{\overset{O}{\|}}{C}-NH_2$$

p-Toluoyl
chloride

p-Toluamide

$$CH_3CH_2CH_2CH_2-\overset{\overset{O}{\|}}{C}-Cl \xrightarrow[(methylamine)]{CH_3NH_2} CH_3CH_2CH_2CH_2-\overset{\overset{O}{\|}}{C}-NHCH_3$$

Pentanoyl chloride
(valeryl chloride)

N-Methylpentanamide
(*N*-methylvaleramide)

$$\langle\bigcirc\rangle-\overset{\overset{O}{\|}}{C}-Cl \xrightarrow[(dimethylamine)]{(CH_3)_2NH} \langle\bigcirc\rangle-\overset{\overset{O}{\|}}{C}-N\overset{CH_3}{\underset{CH_3}{<}}$$

Cyclohexanecarbonyl
chloride

N,N-Dimethylcyclohexanecarboxamide

B. Ester Formation

The reaction between an acid chloride and an alcohol produces an ester. When an aromatic acid chloride (ArCOCl) is one of the reactants, some base is usually added to remove HCl as it is formed. The base is usually dilute sodium hydroxide solution, triethylamine, or pyridine (the latter two are weak organic bases, see Secs. 26.3 and 26.5), and this procedure is referred to as the **Schotten-Baumann method.** The same reactions occur when a phenol (ArOH) is used in place of an alcohol, and this is one of the best (and few) ways to prepare aryl esters. The following shows these general reactions:

$$(Ar-) R-\overset{\overset{O}{\|}}{C}-Cl + R'OH \xrightarrow{base} (Ar-) R-\overset{\overset{O}{\|}}{C}-OR' + HCl$$

$$\downarrow \text{B: (base)}$$

$$B{:}H^{\oplus} + :\overset{..}{\underset{..}{Cl}}{:}^{\ominus}$$

Both reactions occur by nucleophilic acyl substitution, as outlined:

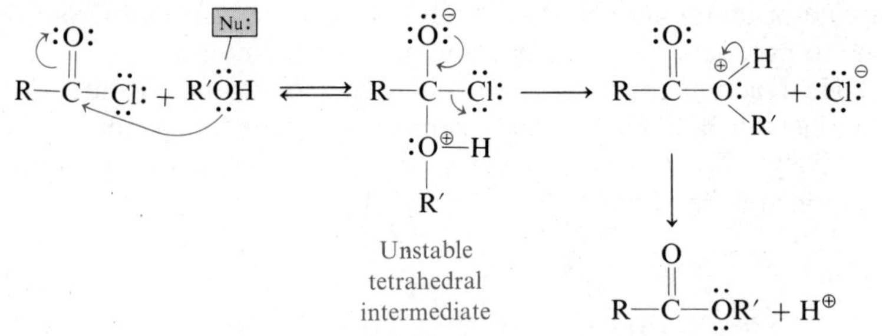

$$R-\overset{\overset{\displaystyle :O:}{\|}}{C}-\overset{\cdot\cdot}{\underset{\cdot\cdot}{Cl}}: + R'\overset{\cdot\cdot}{O}H \rightleftharpoons R-\overset{\overset{\displaystyle :\overset{\ominus}{O}:}{|}}{\underset{\underset{R'}{|}}{\underset{:O^{\oplus}-H}{C}}}-\overset{\cdot\cdot}{\underset{\cdot\cdot}{Cl}}: \longrightarrow R-\overset{\overset{\displaystyle :O:}{\|}}{C}-\overset{\oplus}{O}\overset{H}{\underset{R'}{\diagdown}} + :\overset{\cdot\cdot}{\underset{\cdot\cdot}{Cl}}:^{\ominus}$$

Unstable
tetrahedral
intermediate

$$R-\overset{\overset{\displaystyle O}{\|}}{C}-\overset{\cdot\cdot}{O}R' + H^{\oplus}$$

The following examples illustrate the use of acid chlorides in the preparation of esters:

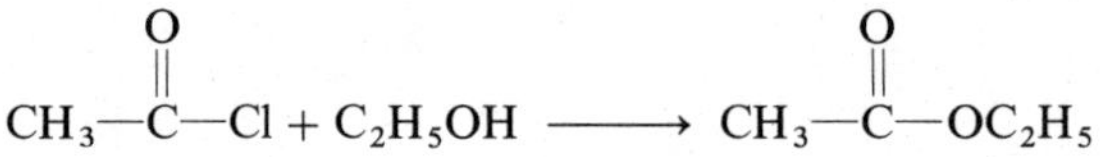

$$CH_3-\overset{\overset{\displaystyle O}{\|}}{C}-Cl + C_2H_5OH \longrightarrow CH_3-\overset{\overset{\displaystyle O}{\|}}{C}-OC_2H_5$$

Ethanoyl
chloride
(acetyl
chloride)

Ethyl ethanoate
(ethyl acetate)

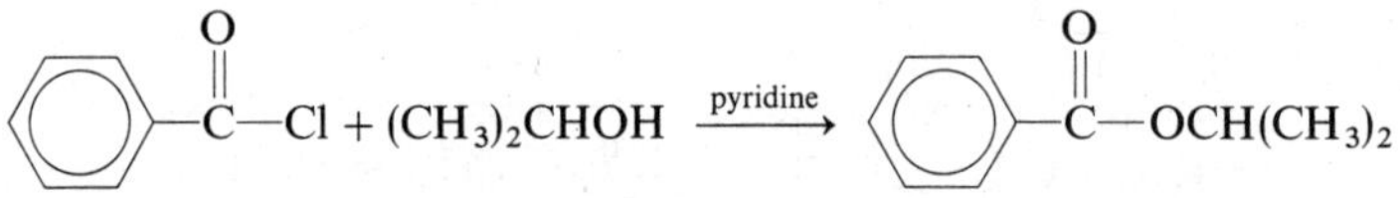

$$\text{C}_6\text{H}_5-\overset{\overset{\displaystyle O}{\|}}{C}-Cl + (CH_3)_2CHOH \xrightarrow{\text{pyridine}} \text{C}_6\text{H}_5-\overset{\overset{\displaystyle O}{\|}}{C}-OCH(CH_3)_2$$

Benzoyl
chloride

Isopropyl
alcohol

Isopropyl
benzoate

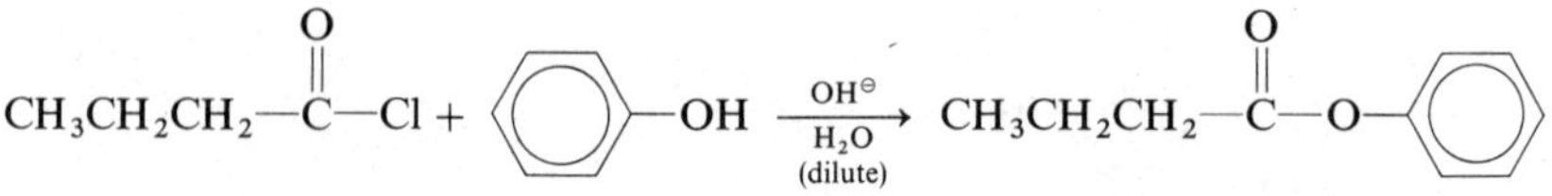

$$CH_3CH_2CH_2-\overset{\overset{\displaystyle O}{\|}}{C}-Cl + \text{C}_6\text{H}_5-OH \xrightarrow[\substack{H_2O \\ (dilute)}]{OH^{\ominus}} CH_3CH_2CH_2-\overset{\overset{\displaystyle O}{\|}}{C}-O-\text{C}_6\text{H}_5$$

Butanoyl chloride
(n-butyryl chloride)

Phenol

Phenyl butanoate
(phenyl n-butyrate)

C. Hydrolysis

Acid chlorides are hydrolyzed on heating with water, but the reaction is speeded by the addition of base (such as NaOH). The corresponding acid is formed in either case. This reaction has little use because most acid chlorides are prepared from acids. The following is the general hydrolysis reaction:

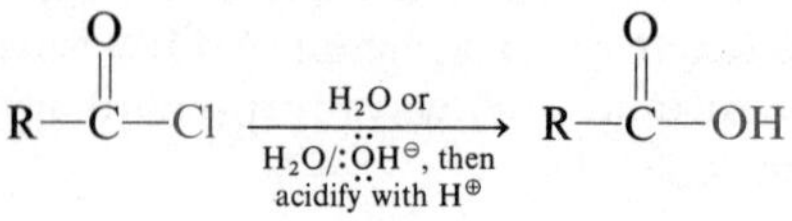

$$R-\overset{\overset{\displaystyle O}{\|}}{C}-Cl \xrightarrow[\substack{H_2O/:\overset{\cdot\cdot}{O}H^{\ominus}, \text{ then} \\ \text{acidify with } H^{\oplus}}]{H_2O \text{ or}} R-\overset{\overset{\displaystyle O}{\|}}{C}-OH$$

The following reactions are examples of hydrolysis:

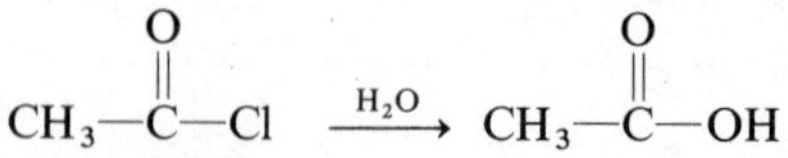

$$CH_3-\overset{\overset{\displaystyle O}{\|}}{C}-Cl \xrightarrow{H_2O} CH_3-\overset{\overset{\displaystyle O}{\|}}{C}-OH \quad \text{Very fast reaction}$$

Ethanoyl chloride
(acetyl chloride)

Ethanoic acid
(acetic acid)

The structures show: Benzoyl chloride reacting with H₂O/NaOH to form the sodium benzoate salt, which upon acidification with H⊕ gives Benzoic acid.

Benzoyl
chloride

Benzoic
acid

Aromatic acid chlorides generally react *much* slower than aliphatic compounds. The addition of hydroxide ion, which is considerably more nucleophilic than water, materially speeds the hydrolysis reaction.

The mechanism of hydrolysis in neutral or basic solution follows the general picture of nucleophilic acyl substitution, as shown by the following mechanistic steps for basic hydrolysis:

The reaction scheme shows:

$$R-\overset{\overset{\displaystyle :O:}{\|}}{C}-\ddot{C}l: + :\ddot{O}H \rightleftharpoons R-\underset{\underset{\displaystyle :\ddot{O}H}{|}}{\overset{\overset{\displaystyle :\ddot{O}:^{\ominus}}{|}}{C}}-\ddot{C}l: \longrightarrow R-\overset{\overset{\displaystyle :O:}{\|}}{C}-\ddot{O}H + :\ddot{C}l:^{\ominus}$$

Unstable
tetrahedral
intermediate

NaOH
fast

$$R-\overset{\overset{\displaystyle O}{\|}}{C}-\ddot{O}:^{\ominus} Na^{\oplus} + H_2O$$

The product from basic hydrolysis is the salt of the carboxylic acid, which must be neutralized to obtain the free acid.

Question 23.3

Reasoning by analogy with the reaction of an alcohol with an acid chloride (see Sec. 23.5B), propose a reasonable mechanism for the neutral hydrolysis of an acid chloride. Show all steps and likely intermediates.

D. Anhydride Formation

Acid chlorides react with the carboxylate anion, $R'CO\ddot{O}:^{\ominus}$, to form an acid anhydride; the general reaction is:

$$R-\overset{\overset{\displaystyle O}{\|}}{C}-Cl + R'-\overset{\overset{\displaystyle O}{\|}}{C}-\ddot{O}:^{\ominus} \xrightarrow{\text{polar} \atop \text{solvent}} R-\overset{\overset{\displaystyle O}{\|}}{C}-O-\overset{\overset{\displaystyle O}{\|}}{C}-R'$$

Anhydride

As the following examples show, this reaction can be used for preparing either symmetrical anhydrides (R = R') or unsymmetrical (mixed) anhydrides (R ≠ R'):

$$CH_3-\overset{\overset{\displaystyle O}{\|}}{C}-\ddot{O}:^{\ominus} Na^{\oplus} + CH_3-\overset{\overset{\displaystyle O}{\|}}{C}-Cl \xrightarrow{\text{polar} \atop \text{solvent}} CH_3-\overset{\overset{\displaystyle O}{\|}}{C}-O-\overset{\overset{\displaystyle O}{\|}}{C}-CH_3$$

Sodium
ethanoate

Ethanoyl
chloride

Ethanoic anhydride
Symmetrical anhydride

$$CH_3-\overset{\overset{\displaystyle O}{\|}}{C}-\ddot{O}:^{\ominus} Na^{\oplus} + \text{(benzoyl chloride)} \xrightarrow{\text{polar} \atop \text{solvent}} CH_3-\overset{\overset{\displaystyle O}{\|}}{C}-O-\overset{\overset{\displaystyle O}{\|}}{C}-\text{(phenyl)}$$

Benzoyl
chloride

Ethanoic benzoic anhydride
*Unsymmetrical (mixed)
anhydride*

In practice, more direct methods are used for synthesizing symmetrical anhydrides (see Secs. 23.11 and 23.12), so this reaction is used mostly for unsymmetrical anhydride preparation.

Question 23.4

The reaction between an acid chloride and a carboxylate anion is believed to be a typical nucleophilic acyl substitution reaction.

(*a*) What is the nucleophile?
(*b*) Propose a complete mechanism for the reaction, showing all intermediates.

Question 23.5

What products are formed when α-methylvaleryl chloride reacts with each of the following reagents?

(*a*)   water and heat
(*b*)   excess aniline, $C_6H_5NH_2$
(*c*)   *p*-bromophenol in the presence of dilute base
(*d*)   2-phenylethanol in the presence of dilute base
(*e*)   excess dilute base
(*f*)   cold, concentrated ammonia

Question 23.6

The reaction between an acid chloride and a phenol or an alcohol is often carried out in the presence of dilute aqueous base, and in both cases, the net result is the same—hydrogen chloride is removed as it is formed. An alternative mechanism for the phenol reaction is likely to occur when aqueous sodium hydroxide is used as the base. Suggest what this might be, showing all steps and intermediates.

23.6   Preparation of Esters

Esters can be prepared by several methods, one of which is the reaction between an acid chloride and an alcohol or phenol (see Sec. 23.5B). In this section we study several other methods for preparing esters.

A. Acid-Catalyzed Reaction Between Acid and Alcohol

The most important method for preparing an ester is the acid-catalyzed reaction between an acid and an alcohol, which is called **Fischer esterification.** The general equation is:

$$(Ar-)R-\overset{\overset{\displaystyle O}{\|}}{C}-OH \; + \; R'OH \; \underset{}{\overset{H^{\oplus},\;heat}{\rightleftharpoons}} \; (Ar-)R-\overset{\overset{\displaystyle O}{\|}}{C}-OR' + H_2O$$

Acid   Alcohol   Ester
                R′ may not
                be aryl

This method is particularly versatile because most alcohols are commercially available and economical. Experimentally, the reaction is carried out using a *large* excess of alcohol with a small amount of sulfuric acid as the catalyst. The alcohol often serves as the solvent for the reaction. The reaction is reversible. In later discussions we learn of the role of sulfuric acid in dictating the reaction conditions to a favorable equilibrium.

Before going further, consider some examples of esterification by this method:

$$CH_3-\overset{\overset{\displaystyle O}{\|}}{C}-OH + CH_3(CH_2)_3CH_2OH \xrightarrow[\text{heat}]{H^\oplus} CH_3-\overset{\overset{\displaystyle O}{\|}}{C}-O-(CH_2)_4CH_3 + H_2O$$

| Ethanoic acid (acetic acid) | 1-Pentanol (n-amyl alcohol) | | Pentyl ethanoate (n-amyl acetate, Banana oil) |

Salicylic acid + Methyl alcohol → Methyl salicylate *Oil of wintergreen* + H_2O

Cyclohexanecarboxylic acid + Cyclohexanol → Cyclohexyl cyclohexanecarboxylate + H_2O

Phthalic acid + $2CH_3CH_2CH_2CH_2OH$ → Di-n-butyl phthalate + $2H_2O$

The mechanism of esterification has been studied extensively. Let us look at some facts about the reaction.

a. Equilibrium: Reversibility of Esterification

The esterification reaction between ethanoic acid and ethanol has been investigated. If 1 mole of ethanoic acid and 1 mole of ethanol react and equilibrium is established, analysis of the reaction mixture reveals that ~0.33 mole each of unreacted acid and unreacted alcohol remain and that ~0.67 mole each of ester and water are produced. For this reaction, we can write the following equilibrium equation and expression:[1]

$$CH_3-\overset{\overset{\displaystyle O}{\|}}{C}-OH + C_2H_5OH \underset{}{\overset{H^\oplus}{\rightleftharpoons}} CH_3-\overset{\overset{\displaystyle O}{\|}}{C}-OC_2H_5 + H_2O$$

$$K_{\text{esterification}} = \frac{[CH_3COOC_2H_5][H_2O]}{[CH_3COOH][C_2H_5OH]}$$

[1] The equilibrium expression for esterification *must include* the concentration of water because a small amount of water is *formed* in the reaction. Contrast this to most equilibrium expressions involving water (for example, those encountered in general chemistry), where it is present in large excess and is left out of the equilibrium expression because its concentration remains effectively constant.

By inserting the equilibrium concentrations of reactants and products, $K_{esterification}$ can be computed as follows:

$$K_{esterification} = \frac{[0.67][0.67]}{[0.33][0.33]} \approx 4.1$$

Further evidence of the reversible nature of the esterification reaction comes from varying the concentration of reactants. For example, if 5 moles of ethanol and 1 mole of ethanoic acid react, ethyl ethanoate is produced in 96% yield, as compared with 67% yield obtained when 1 mole of each reactant is used. This is in accord with Le Chatelier's principle dealing with the effect of concentration on equilibrium.

The yield of ester can also be increased by removing water as it is formed. In certain reactions, the boiling points of the starting acid and alcohol and of the ester product are sufficiently high that the water can be removed by distillation, as in the preparation of di-*n*-butyl phthalate:

| Phthalic acid | *n*-Butyl alcohol | Di-*n*-butyl phthalate | Removed by distillation |

$$\overrightarrow{\frac{\text{position of equilibrium}}{\text{shifted to right}}}$$

In other cases, water can be removed by azeotropic distillation using, for example, benzene or toluene as the solvent, which forms an azeotrope with water (see Sec. 10.7 for discussion of azeotropes).

A reasonable concentration of sulfuric acid or phosphoric acid catalyst can be used in the reaction to tie up the water as the hydronium ion, $H_3\overset{..}{O}^{\oplus}$, as it is formed. This too shifts the equilibrium to favor ester formation.

Finally, if 1 mole of ethyl ethanoate and 1 mole of water react and equilibrium is reached, 0.67 mole each of unreacted water and unreacted ester remain and 0.33 mole each of ethanoic acid and ethanol are formed. *The same equilibrium mixture is obtained starting either with ethanoic acid and ethanol or with ethyl ethanoate and water.* We discuss the reverse of the esterification reaction—called *hydrolysis* (the reaction of an ester with water)—in Sec. 23.7A.

b. Acid Catalysis

If we attempted esterification by allowing the free acid to react with an alcohol, several days of heating might be required for equilibrium to be attained. Yet, the addition of a few drops of sulfuric acid causes equilibrium to be established in several hours or less. We must be prepared to include the hydrogen ion catalyst in our mechanism of esterification, even though it influences only the *rate* at which equilibrium is attained. The equilibrium concentrations of reactants and products are the same with or without added acid.

c. Chemical Kinetics

Chemical kinetic studies reveal that esterification is a third-order reaction. The rate of reaction depends on the concentrations of carboxylic acid, alcohol, and hydrogen ion:

$$\text{Rate of esterification} = k_3[\text{R—COOH}][\text{R'OH}][\text{H}^\oplus]$$

where k_3 = rate constant

This suggests that the mechanism of esterification should involve a slow, rate-determining step with a transition state that contains a hydrogen ion, a molecule of alcohol, and a molecule of carboxylic acid.

d. Isotope Tracer Studies

An additional question is: in which of the two possible ways (shown as follows) are bonds broken and made in esterification?

Breaking of acyl-oxygen bond   or   Breaking of alkyl-oxygen bond

This question was not answered until the 1930s when the use of oxygen isotopes became routine. When esterification is carried out using alcohols that contain oxygen-18 (^{18}O), the following results are obtained as a function of the general structure of the alcohol.

Primary and most secondary alcohols:

$$\text{R—C(=O)—OH} + \text{R'}^{18}\text{OH} \underset{}{\overset{\text{H}^\oplus}{\rightleftharpoons}} \text{R—C(=O)—}^{18}\text{OR'} + \text{H}_2\text{O}$$

Contains all of ^{18}O Contains no ^{18}O

Tertiary (and a few secondary) alcohols:

$$\text{R—C(=O)—OH} + \text{R'}^{18}\text{OH} \underset{}{\overset{\text{H}^\oplus}{\rightleftharpoons}} \text{R—C(=O)—}^{18}\text{OR'} + \text{H}_2{}^{18}\text{O}$$

Each contains some ^{18}O

These results show that only the acyl-oxygen bond is broken in forming esters of primary and most secondary alcohols, but that both acyl-oxygen and alkyl-oxygen bonds are broken with tertiary alcohols (and a few secondary alcohols). We must also account for these results in our mechanisms.

e. Steric Effects on Rate of Esterification

The steric effect of the substituents around the carboxyl group play a role in the *rate* of esterification. For example, the relative rates of esterification of various carboxylic acids with methanol at 40° are presented in Table 23.2 These data show that increasing the bulk size decreases the rate of reaction.

It is interesting that the position of equilibrium in esterification (that is, the yield of ester) is relatively insensitive to steric factors. It simply takes a longer time to establish equilibrium with bulky acids or alcohols.

TABLE 23.2   Rates of Esterification at 40° for Reaction

$$RCOOH + CH_3OH \xrightarrow{H^{\oplus}} RCOOCH_3 + H_2O$$

Name and Structure of Acid (RCOOH)	Rates of Reaction
CH_3COOH (ethanoic acid)	1.0
$(CH_3)_3CCOOH$ (2,2-dimethylpropanoic acid)	3.7×10^{-2}
$(CH_3CH_2)_3CCOOH$ (2,2-diethylbutanoic acid)	4.3×10^{-3}
$[(CH_3)_2CH]_2CHCOOH$ (2-isopropyl-3-methylbutanoic acid)	Too slow to measure

The mechanism of **esterification with primary and most secondary alcohols** (R′ = primary or secondary alkyl group) is presented as follows. The comments following the mechanism indicate how it accounts for the known facts about esterification.

Step 1. Protonation of acid

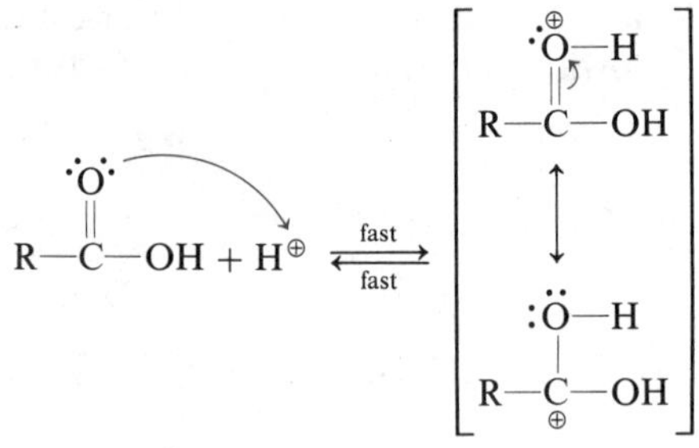

Step 2. Nucleophilic attack on protonated acid

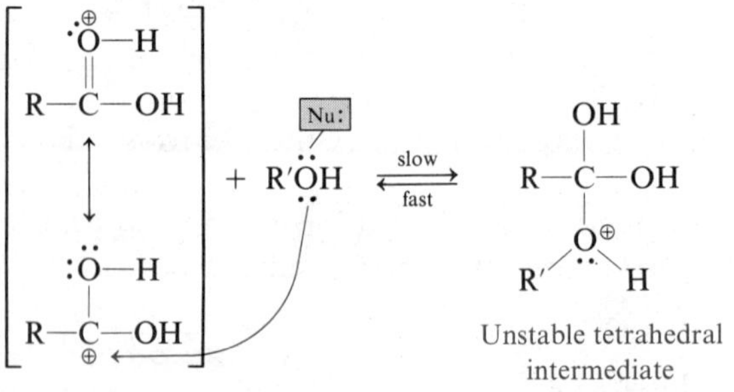

Step 3. Proton transfer

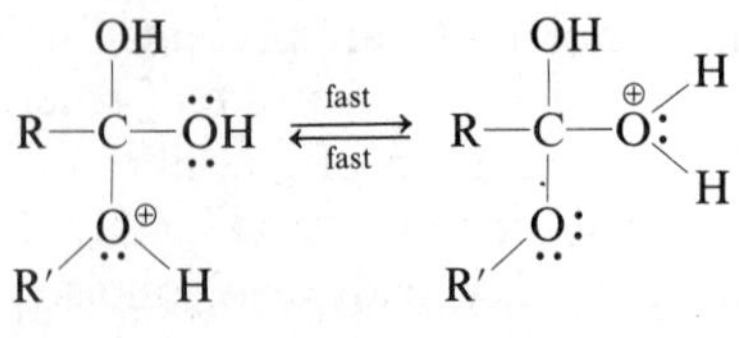

Step 4. Loss of water to form protonated ester

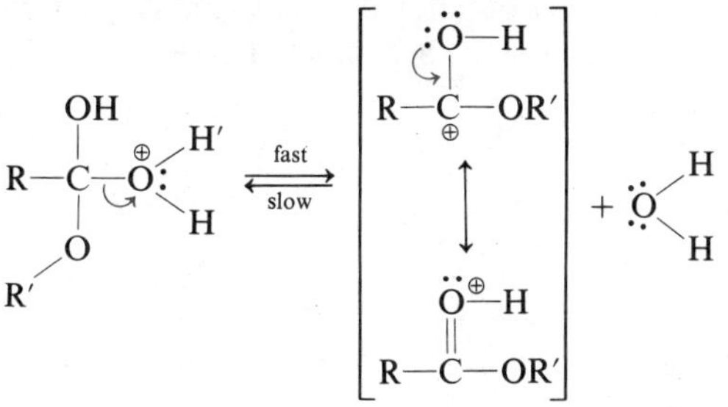

Protonated
ester

Step 5. Loss of proton to form ester

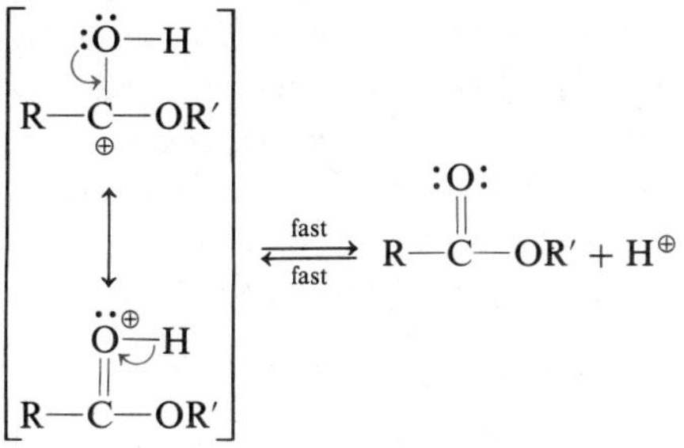

Step 1 shows the protonation of the carboxylic acid. Although both oxygens contain two unshared pairs of electrons, the carbonyl oxygen atom is more likely to be protonated because it is surrounded by a higher density of electrons (the π bond and the unshared pairs of electrons). The electronic structure of the protonated acid can be represented by two resonance structures (shown in steps 1 and 2) but is best depicted by the hybrid:

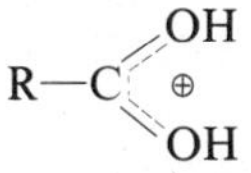

Step 2 shows the attack of the nucleophilic R′OH molecule on the protonated acid to form the unstable tetrahedral intermediate. This is the *slow, rate-determining step*. The intermediate contains one molecule each of the carboxylic acid and the alcohol and one hydrogen ion, a feature that is in accord with the chemical kinetics of esterification. The protonation in step 1 renders the acyl carbon atom more electron-deficient and thus more susceptible to nucleophilic attack by alcohol. Note also that the carbon-oxygen bond in R′—OH is unaffected; the new bond is between the alcohol oxygen and the acyl carbon atom. This view agrees with the results obtained from the ^{18}O studies.

Step 3 shows a proton shift from one oxygen atom to another, which likely occurs by a deprotonation-protonation sequence (that is, proton transfer is not intramolecular).

Step 4 shows the loss of the neutral water molecule to form the protonated ester. Either hydroxyl group can be protonated, and loss of either gives the same product— protonated ester. On the other hand, if the oxygen in R′O— is protonated and

R′—OH is the leaving group, then we have regenerated the reactants (that is, this is the reverse of the reaction shown in step 2). We can view the overall changes in steps 2, 3, and 4 as involving the exchange of an R′OH molecule for an H_2O molecule:

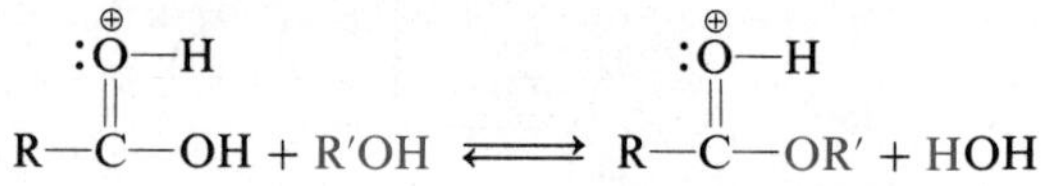

$$R\overset{\overset{\oplus}{\underset{\|}{:O-H}}}{-C}-OH + R'OH \rightleftharpoons R\overset{\overset{\oplus}{\underset{\|}{:O-H}}}{-C}-OR' + HOH$$

Step 5 shows the loss of a proton from the protonated ester produced in step 4. The desired ester is formed, and the proton that entered the reaction in step 1 is regenerated. Thus, this reaction is truly *acid-catalyzed* because the proton is consumed and then regenerated.

All steps in the esterification mechanism are shown as reversible, a feature we discuss in Sec. 23.7A concerning the hydrolysis of esters. The mechanism is consistent with all the known facts previously outlined.

The results of the ^{18}O tracer experiments with **esterification reactions involving tertiary alcohols** indicate that some ^{18}O ends up in the ester and some in the water. The ^{18}O incorporation into the ester is explained by the mechanism presented earlier for primary alcohols. However, that not all the ^{18}O ends up in the ester but some ends up in the water suggests that another mechanism must also operate.

The key to understanding these results lies in carbocation chemistry. We know that tertiary alcohols react with protons to form carbocations. The following three steps outline a plausible explanation for these results.

Step 1. Carbocation formation

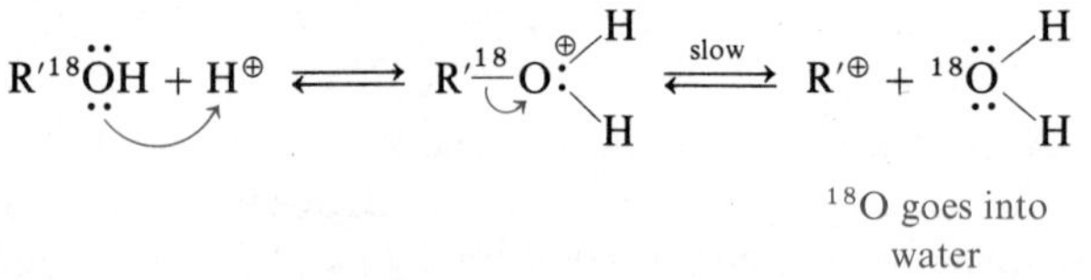

^{18}O goes into water

Step 2. Attack of carbocation on acid

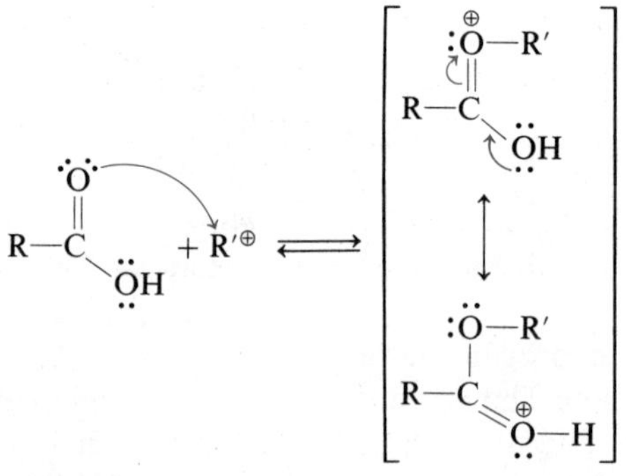

Protonated ester

Step 3. Loss of proton to give ester

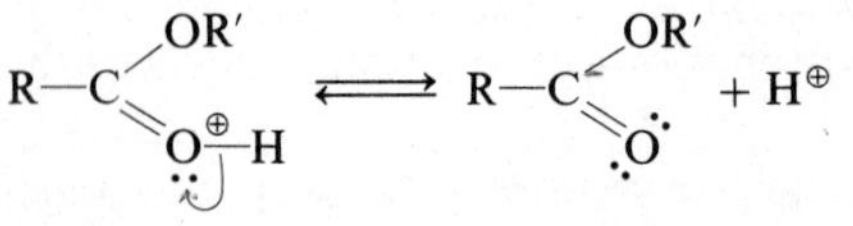

As this mechanism shows, the carbon-oxygen bond of the alcohol is broken first, and the resulting carbocation attacks the carbonyl oxygen atom (again because it has a greater electron density). The protonated ester is formed, and on loss of a proton, it gives ester that contains no ^{18}O. To account for the *partial* incorporation of ^{18}O into the ester, it is suggested that *two mechanisms—the one presented for primary alcohols and the one presented here—operate simultaneously*. The distribution of ^{18}O between ester and water is dependent on the structure of the alcohol used.

Question 23.7

An alternative possibility for esterification using primary alcohols involves nucleophilic attack of the free acid on the protonated form of the alcohol (an S_N2 mechanism):

$$R-\overset{O}{\overset{\|}{C}}-\overset{..}{\underset{..}{O}}H + R'CH_2-\overset{\oplus}{O}H_2 \xrightarrow[\text{steps}]{\text{several}} R-\overset{O}{\overset{\|}{C}}-OCH_2R' + H_2O + H^\oplus$$
$$(S_N2 \text{ reaction})$$

Do the oxygen-18 data support or refute this possibility? Explain.

Question 23.8

Predict the fate of ^{18}O when each of the following alcohols reacts with benzoic acid in the presence of sulfuric acid catalyst. Justify your answer in each case.

(*a*) $CH_3CH_2CH_2\,^{18}OH$ (*b*) $(CH_3)_3C^{18}OH$ (*c*) $C_6H_5CH_2\,^{18}OH$

(*d*) $(C_6H_5)_3C^{18}OH$ (*e*) $CH_2{=}CH-CH_2\,^{18}OH$

B. Reaction of Carboxylate Anion with Alkyl Halide

Another useful method for preparing esters is the reaction between the carboxylate anion, $RCO\overset{..}{\underset{..}{O}}{:}^{\ominus}$, and an alkyl halide:

Carboxylate anion   Alkyl halide (methyl, 1° or 2°)   Ester

This S_N2 reaction is limited to methyl, primary and secondary alkyl halides because the carboxylate anion has some basic properties and promotes elimination in tertiary halides. An example of this reaction is:

Sodium benzoate   *n*-Butyl bromide

n-Butyl benzoate

Question 23.9

Salts of carboxylic acids are water-soluble when the cation is $Na^\oplus$, $K^\oplus$, and so on, but when these soluble salts are treated with silver nitrate, the corresponding insoluble silver salt of the acid, RCOOAg, is formed. These silver salts react with methyl, primary, and secondary alkyl halides much more readily than the corresponding sodium or potassium salts. Explain this.

C. Reaction of Alcohol with Anhydride

Acid anhydrides also react with alcohols to give esters.

Open-chain anhydride:

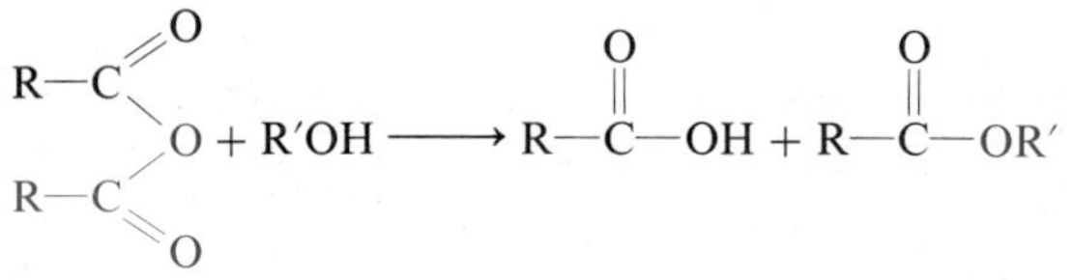

Cyclic anhydride:

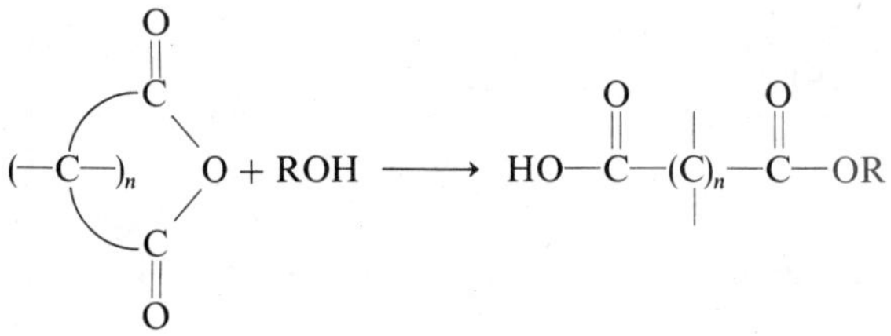

$n \geq 2$ Half acid-half
                                                          ester

A specific example of each type of reaction is given here:

Acetic anhydride   Ethyl alcohol   Ethyl acetate   Acetic acid

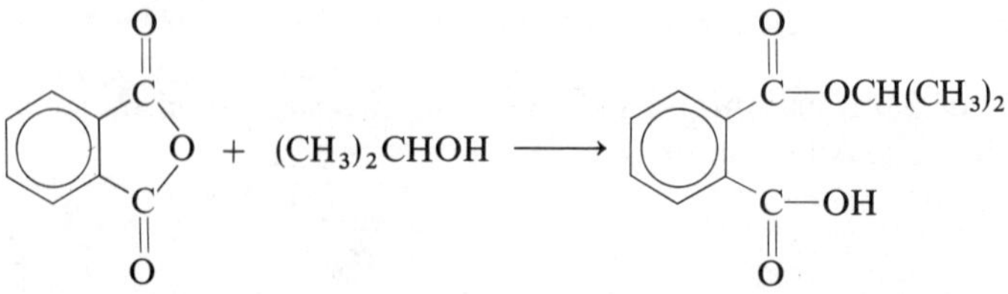

Phthalic anhydride   Isopropyl   Isopropyl hydrogen
                        alcohol   phthalate

A chief use of the reaction of an anhydride with an alcohol is in the formation of esters, especially acetates, because acetic anhydride is readily available and inexpensive. It is also used extensively in the acetylation of carbohydrates and cellulose (see Sec. 29.9). We discuss the mechanism of this reaction in Sec. 23.13A.

D. Preparation of Methyl Esters Using Diazomethane

The preparation of methyl esters is often accomplished by allowing a carboxylic acid to react with *diazomethane*, CH_2N_2, in ether solution:

$$
\underset{\text{Diazomethane}}{\overset{\displaystyle O \atop \displaystyle \|}{R-C-OH} + CH_2N_2} \xrightarrow{\text{ether}} \underset{\substack{\text{Methyl} \\ \text{ester}}}{\overset{\displaystyle O \atop \displaystyle \|}{R-C-OCH_3}} + N_2 \text{ (gas)}
$$

We discuss other reactions of diazomethane in Sec. 27.6.

The electronic structure of diazomethane is shown in the following resonance structures:

$$
\left[\overset{\oplus}{CH_2=N}\overset{\ominus}{=N:} \longleftrightarrow \overset{\ominus}{:CH_2}-\overset{\oplus}{N\equiv N:} \right]
$$

When diazomethane comes in contact with a carboxylic acid, the first step is believed to be the transfer of a proton from the acid to diazomethane. This is followed by a nucleophilic displacement reaction (S_N2 mechanism) of the resulting carboxylate anion on $CH_3N_2^{\oplus}$ to produce the methyl ester and liberate nitrogen gas:

$$
\underset{}{\overset{\displaystyle O \atop \displaystyle \|}{R-C-\overset{\cdot\cdot}{\underset{\cdot\cdot}{O}}-H}} + \left[\begin{array}{c} \overset{\oplus}{CH_2=N}\overset{\ominus}{=N:} \\ \updownarrow \\ \overset{\ominus}{:CH_2}-\overset{\oplus}{N\equiv N:} \end{array} \right] \longrightarrow \underset{}{\overset{\displaystyle O \atop \displaystyle \|}{R-C-\overset{\cdot\cdot}{\underset{\cdot\cdot}{O}}{}^{\ominus}}} + CH_3-\overset{\oplus}{N\equiv N:}
$$

then

$$
\underset{}{\overset{\displaystyle O \atop \displaystyle \|}{R-C-\overset{\text{Nu:}}{\overset{\cdot\cdot}{\underset{\cdot\cdot}{O}}{}^{\ominus}}}} + CH_3-\overset{\oplus}{N\equiv N:} \xrightarrow[\text{reaction}]{S_N2} \underset{}{\overset{\displaystyle O \atop \displaystyle \|}{R-C-\overset{\cdot\cdot}{O}CH_3}} + :N\equiv N:
$$

The protonated diazomethane molecule contains the *diazonium group*, $-\overset{\oplus}{N}\equiv N$, which is an excellent leaving group. When substitution occurs, nitrogen gas is liberated. Because the triple bond in nitrogen is one of the strongest covalent bonds known, the formation of nitrogen is a major driving force for the esterification reaction. The reaction is driven further to the right as one of the products—nitrogen gas in this case—is released. We encounter diazonium salts again in Sec. 26.8.

The yields of methyl esters are very high when diazomethane is used, so this reaction is used if the starting acid is very valuable or if it is so sensitive to acid that Fischer esterification cannot be used. The major disadvantage is that diazomethane is somewhat hazardous because it has a tendency to explode violently unless care is taken in handling it.

E. γ- and δ-Hydroxy Acids: Lactone Formation

Because hydroxy acids contain both —OH and —COOH groups, it is not surprising that certain of them form cyclic esters either spontaneously or under the influence of heat. γ- and δ-Hydroxy acids form cyclic esters which are called **lactones**:

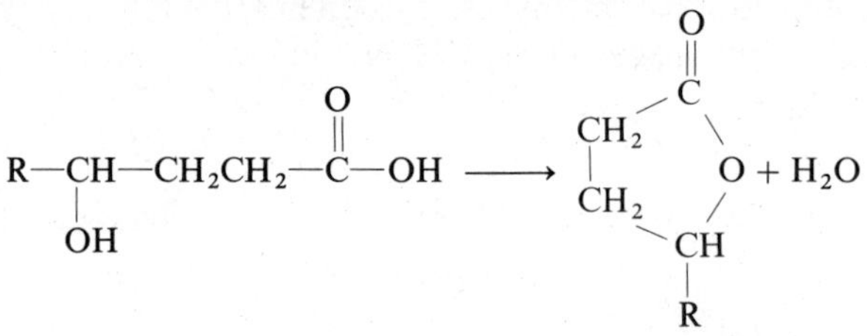

γ-Hydroxy acid

γ-Lactone
*Five-membered,
cyclic ester*

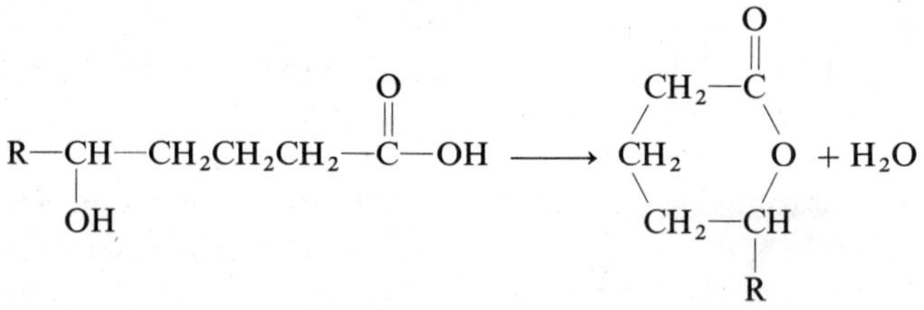

δ-Hydroxy acid

δ-Lactone
*Six-membered
cyclic ester*

The driving force for lactone formation is the well-known stability of the five- and six-membered rings that result.

Because a lactone is really an ester, it should come as no surprise that the lactone ring is opened by various nucleophilic reagents, such as water, ammonia, or an alcohol (see Sec. 23.7). However, lactone formation is so rapid that it is virtually impossible to keep the free γ- or δ-hydroxy acid per se. These hydroxy acids are generated when the corresponding lactone is treated with base, but upon acidification of the salt of the acid, the lactone is regenerated. The following illustrate this:

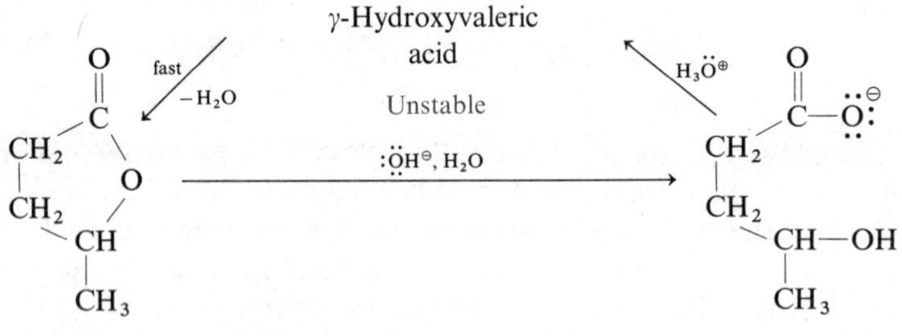

γ-Valerolactone

γ-Hydroxyvaleric acid
Unstable

γ-Hydroxyvalerate

δ-Valerolactone

δ-Hydroxyvaleric acid
Unstable

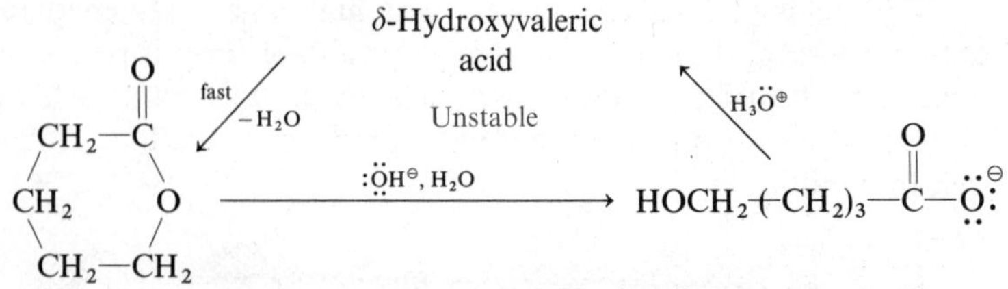

δ-Hydroxyvalerate

Note the nomenclature of the lactones, which take the parent name of the hydroxy acid; the *-ic acid* suffix is dropped and replaced by *-olactone.*

Lactones are likely to form whenever a strain-free five- or six-membered ring can be generated, sometimes even in compounds that already contain a ring; for example:

cis-3-Hydroxycyclohexane-
carboxylic acid

Five-membered γ-lactone

The *trans* isomer does not form a lactone.

Lactones are often produced internally when a carboxylate anion is generated in a molecule that contains a halogen atom. This is a typical S_N2 reaction, which requires that the geometry of the halogen atom be favorable; for example:

trans-3-Bromocyclo-
hexanecarboxylic acid

Five-membered
γ-lactone

In this case, the *cis* isomer does not react. Lactones can also be obtained from γ- and δ-halocarboxylic acids by the same type of mechanism shown earlier.

23.7 Reactions of Esters

Esters usually react by the nucleophilic acyl substitution mechanism.

A. With Water: Hydrolysis

We saw in Sec. 23.6 that the reaction between an acid and an alcohol produces an ester and water. We now study the reverse reaction, in which an ester reacts with water. Ester hydrolysis can be run under either acidic or basic conditions; the major difference is that in acid the reaction is reversible, whereas in base it is not.

The hydrolysis of esters under basic conditions is much faster than in water alone. The reaction is called *base-promoted*. The base is not a catalyst because it is actually consumed in the process of the reaction. Ester hydrolysis under basic conditions is often referred to as **saponification** (Latin *sapo*, soap), a term that originated in connection with soap-making. The importance of saponification in soaps is seen in Sec. 25.2.

Before we consider the hydrolysis mechanisms (acidic or basic conditions), let us look at some experimental evidence.

1. The use of oxygen-18 (^{18}O) as an isotopic tracer indicates that the *base-promoted* reaction generally occurs by acyl-oxygen breakage and is independent of the structure of the R and R′ groups:

$$\underset{\substack{\| \\ O}}{R-C}\!\!\!\Big\{\!^{18}OR' \xrightarrow[\text{or } H_2O]{:\ddot{O}H^\ominus} \underset{\substack{\| \\ O}}{R-C}-OH + R'^{18}OH$$

$$(\text{or } RC\ddot{O}\ddot{O}\!:^{\ominus})$$

where R′ = 1°, 2°, or 3°

2. Oxygen-18 studies show acid-catalyzed hydrolysis to be more sensitive to the structure of the alkyl group. *For primary and most secondary alkyl groups*, the acyl-oxygen bond is broken:

$$\underset{\substack{\| \\ O}}{R-C}\!\!\!\Big\{\!^{18}OR' \xrightarrow{H_3\ddot{O}^\oplus} \underset{\substack{\| \\ O}}{R-C}-OH + R'^{18}OH$$

where R′ = 1° or 2°

For most *tertiary alkyl groups*, however, considerable ^{18}O ends up in the carboxylic acid, thus indicating some alkyl-oxygen breakage:

$$\underset{\substack{\| \\ O}}{R-C}-{}^{18}O\!\!\Big\{\!R' \xrightarrow{H_3\ddot{O}^\oplus} \underset{\substack{\| \\ O}}{R-C}-{}^{18}OH + R'OH$$

where R′ = 3°

3. Chemical kinetics of the base-promoted reaction show it to be a bimolecular reaction, the rate of which depends on the concentration of both the ester and the base:

$$\text{Rate of reaction} = k_2\,[\text{ester}][:\!\ddot{O}H^\ominus]$$

where k_2 = rate constant

4. Stereochemistry of both acid-catalyzed and base-promoted reactions (*except for tertiary alkyl groups in the case of the acid-catalyzed reaction*) show that hydrolysis occurs with retention of configuration at the alkyl carbon atom; for example:

$$CH_3-\underset{\substack{\| \\ O}}{C}-O-\overset{*}{C}H-CH_3 \xrightarrow[\text{H}_2\text{O}]{H^\oplus \text{ or } :\ddot{O}H^\ominus} CH_3-\underset{\substack{\| \\ O}}{C}-OH + HO-\overset{*}{C}H-CH_3$$

$$\underset{\text{n-C$_6$H$_{13}$}}{}\qquad\qquad\qquad\qquad \underset{\text{n-C$_6$H$_{13}$}}{}$$

Optically active acetate ester   2-Octanol
*Optically active
and retained
configuration*

Now let us turn to a discussion of the actual mechanisms.

a. Acid-Catalyzed Hydrolysis

The mechanism for the hydrolysis of esters that contain primary and secondary alkyl groups is precisely the reverse of the mechanism for the acid-catalyzed esterification reaction (see Sec. 23.6A). The important difference between hydrolysis and esterification is that hydrolysis requires a large excess of water, whereas a large excess of alcohol is used in esterification (when possible) to increase the yield of ester.

For esters derived from tertiary alcohols, however, another mechanism must be invoked to account for the ^{18}O results. In a sense, this mechanism is similar to that for the esterification reaction involving tertiary alcohols (see Sec. 23.6A), but the reverse. For this type of ester, the reaction is believed to begin with the protonation of the alkyl oxygen, so that the leaving group is the relatively stable tertiary carbocation. The carbocation then reacts with water to produce the alcohol. The currently accepted mechanism for this S_N1 reaction is the following:

$$R-\overset{O}{\overset{\|}{C}}-{}^{18}\ddot{O}-R' + H^{\oplus} \rightleftharpoons R-\overset{O}{\overset{\|}{C}}-{}^{18}\overset{\overset{\displaystyle R'}{\oplus}}{\ddot{O}:} \rightleftharpoons R-\overset{O}{\overset{\|}{C}}-{}^{18}\ddot{O}H + R'^{\oplus}$$

where R' = 3°

Then

$$R'^{\oplus} + :\overset{H}{\underset{H}{\overset{\displaystyle \diagup}{O}}} \rightleftharpoons R'-\overset{\oplus}{\underset{H}{\overset{\displaystyle \diagup H}{O:}}} \rightleftharpoons R'\ddot{O}H + H^{\oplus}$$

b. Base-Promoted Hydrolysis

In proposing a mechanism for the base-promoted cleavage of esters, we must provide one consistent with the ^{18}O studies, the stereochemical results, and the chemical kinetics mentioned earlier. These facts all point to breaking the acyl-oxygen bond as shown by the following mechanism:

$$R-\overset{\overset{\displaystyle :O:}{\|}}{C}-{}^{18}\ddot{O}R' + :\ddot{O}H^{\ominus} \overset{slow}{\rightleftharpoons} \left[R-\overset{\overset{\displaystyle :\overset{\ominus}{\ddot{O}}:}{|}}{\underset{\displaystyle :OH}{C}}-{}^{18}\ddot{O}R' \right] \overset{fast}{\rightleftharpoons} R-\overset{\overset{\displaystyle :O:}{\|}}{C}-\ddot{O}H + R'^{18}\ddot{O}:^{\ominus}$$

Unstable tetrahedral
intermediate

Note that this mechanism involves the addition of the nucleophilic hydroxide ion to the acyl group, followed by decomposition of the unstable tetrahedral intermediate to the two products, RCOOH and $R'^{18}\ddot{O}:^{\ominus}$. The reaction is also bimolecular because the slow, rate-determining step is the attack of $:\ddot{O}H^{\ominus}$ on the ester; this agrees with the chemical kinetics because one molecule *each* of hydroxide ion and ester are involved in the intermediate.

When the products are formed in a fast, *nonreversible* reaction, they react further. The carboxylic acid cannot exist as such in basic solution, and it reacts either with the liberated alkoxide ion ($R'^{18}\ddot{O}:^{\ominus}$) or with the excess hydroxide ion in the reaction mixture to form the carboxylate anion and $R'^{18}OH$ or HOH:

$$R-\overset{O}{\overset{\|}{C}}-\ddot{O}H + R'^{18}\ddot{O}:^{\ominus} \text{ or } :\ddot{O}H^{\ominus} \longrightarrow R-\overset{O}{\overset{\|}{C}}-\ddot{O}:^{\ominus} + R'^{18}\ddot{O}H \text{ or } H\ddot{O}H$$

Weak acid	Strong bases	Weaker base	Weaker acids

Hence, base-promoted ester hydrolysis reactions are usually followed by neutralization with mineral acid to produce the free carboxylic acid.

Base-promoted ester hydrolyses are not reversible; once the products are formed, there is no driving force for the carboxylate anion to react with a molecule of alcohol. (*Question:* What electrostatic interactions would have to be overcome for this to occur?)

On the other hand, acid-catalyzed ester hydrolysis and acid-catalyzed esterification are reversible because they involve protonation of the acyl group, which facilitates nucleophilic attack at the carbon atom by water or alcohol.

c. Evidence for Tetrahedral Intermediate

We drew a tetrahedral intermediate in the mechanisms for nucleophilic acyl substitution reactions. We now discuss some elegant experimental work that lends strong and convincing support for this tetrahedral intermediate, which is so unstable that it has not been isolated or detected directly in the laboratory.

Isotope tracers have been used with tremendous success, but this time the carbonyl-oxygen is labeled with ^{18}O:

$$\text{(phenyl)}-\overset{\overset{\displaystyle ^{18}O}{\|}}{C}-OCH_3$$

This ester reacts partially in aqueous sodium hydroxide containing no ^{18}O enrichment, and when the unhydrolyzed ester is isolated, it is found to have lost some of the original ^{18}O. Put another way, the unhydrolyzed ester consists of a mixture of

$$\text{(phenyl)}-\overset{\overset{\displaystyle ^{18}O}{\|}}{C}-OCH_3 + \text{(phenyl)}-\overset{\overset{\displaystyle O}{\|}}{C}-OCH_3$$

Now, let us interpret this result in mechanistic terms. First, intermediate (1) is formed by nucleophilic addition of hydroxide ion to the starting ester. The resulting alkoxide ion is a strong base, and because the reaction is carried out in aqueous solution, (1) can react with water to form (2):

$$R-\overset{\overset{\displaystyle ^{18}\!:\!O:}{\|}}{C}-OR' + HO:^{\ominus} \;\rightleftharpoons\; R-\overset{\overset{\displaystyle ^{18}\!:\!O:^{\ominus}}{|}}{\underset{\displaystyle :OH}{C}}-OR' + H-\overset{..}{O}H \;\rightleftharpoons\;$$

(1)   Water
Alkoxide ion   Weak acid
Strong base

$$R-\overset{\overset{\displaystyle ^{18}\!:\!OH}{|}}{\underset{\displaystyle :OH}{C}}-OR' \quad + \quad :\overset{..}{\underset{..}{O}}H^{\ominus}$$

(2)
Tetrahedral intermediate
Contains two equivalent —OH groups

Furthermore, (2) contains *two chemically equivalent hydroxyl groups*, so that another molecule of hydroxide ion can attack either of the equivalent —OH groups in it. If the base removes the proton attached to the —^{18}OH group, intermediate (1) is re-formed, and (1) can either lose :$\overset{..}{O}H^{\ominus}$ to re-form the starting ^{18}O-labeled ester or form products. On the other hand, if the base removes the proton attached to the

—OH group in (2), new intermediate (3) is formed:

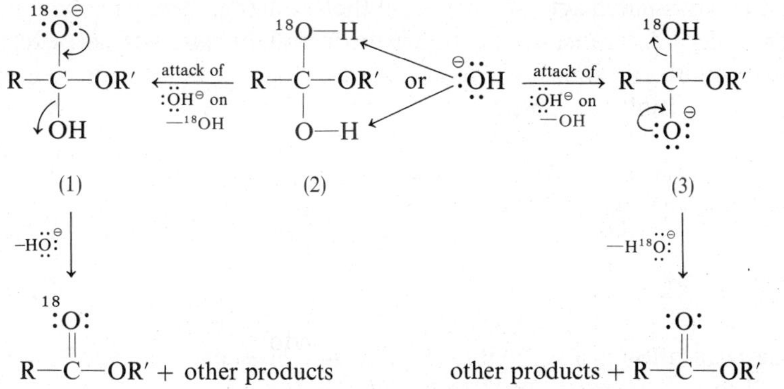

(1) (2) (3)

When (3) is produced, it can either lose $^{18}:\overset{..}{\underset{..}{O}}H^{\ominus}$ to give *ester that contains no* ^{18}O or decompose irreversibly to give products.

The two oxygen atoms in the carboxylate anion are equivalent so the same products are obtained from either (1) or (3).

This mechanism for base-catalyzed ester hydrolysis is now complete, and it explains all the known facts. The preceding experiment supports the intervention of an unstable *tetrahedral* intermediate in nucleophilic acyl substitution reactions. Reasoning by analogy, many other similar substitution reactions that are carried out in aqueous solution (or in the presence of a protic solvent) likely occur by this more complete mechanism.

Question 23.10

The acid-catalyzed hydrolysis of an ester labeled with ^{18}O in the carbonyl-oxygen position is halted before reaction is complete, and the unhydrolyzed ester is found to have lost some ^{18}O. Reasoning by analogy to the base-promoted hydrolysis, suggest a *complete* mechanism to explain these results.

B. With Other Alcohols: Transesterification

Esters react with other alcohols under acidic or basic conditions to give new esters, and this reaction is called *transesterification* because an ester is transformed into another ester:

$$R-\overset{O}{\overset{\|}{C}}-OR' + R''OH \underset{}{\overset{H^{\oplus} \text{ or } R''\overset{..}{\underset{..}{O}}\colon^{\ominus}}{\rightleftharpoons}} R-\overset{O}{\overset{\|}{C}}-OR'' + R'OH$$

Excess

The base is the alkoxide base that corresponds to R''OH.

a. Acid-Catalyzed Transesterification

This reaction can be carried out in the presence of a mineral acid such as H_2SO_4 or HCl, as shown by the example:

$$C_6H_5-\overset{O}{\overset{\|}{C}}-OCH_3 + CH_3CH_2OH \overset{H^{\oplus}}{\rightleftharpoons} C_6H_5-\overset{O}{\overset{\|}{C}}-OCH_2CH_3 + CH_3OH$$

Methyl benzoate   Large excess   Ethyl benzoate

We know that acid-catalyzed esterification involves the nucleophilic attack of an alcohol on a protonated acid and results in the "exchange" of an alcohol molecule for a water molecule. Transesterification is conceptually no different, except that one alcohol molecule (R''OH) is exchanged for a different alcohol molecule (R'OH), as shown by the partial mechanism:

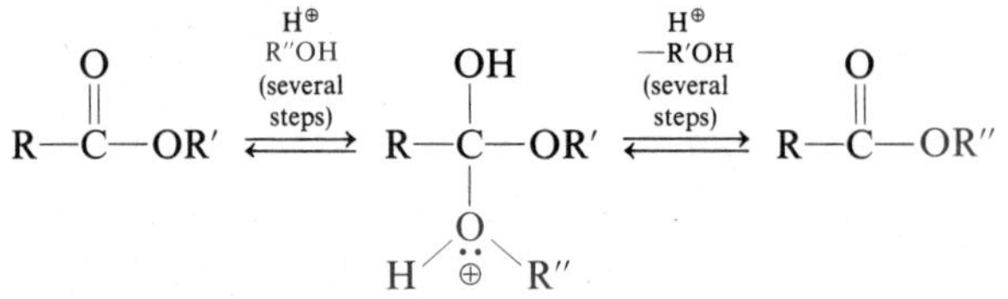

$$
\underset{\substack{\parallel \\ \text{R—C—OR'}}}{\overset{O}{}} \quad \overset{\substack{H^\oplus \\ R''OH \\ (\text{several} \\ \text{steps})}}{\rightleftharpoons} \quad \underset{\substack{| \\ \text{R—C—OR'} \\ | \\ O \\ H^{\overset{\oplus}{}} R''}}{\overset{OH}{}} \quad \overset{\substack{H^\oplus \\ -R'OH \\ (\text{several} \\ \text{steps})}}{\rightleftharpoons} \quad \underset{\substack{\parallel \\ \text{R—C—OR''}}}{\overset{O}{}}
$$

Transesterification is a reversible reaction so a large excess of alcohol (R''OH) must be used to maximize the yield of RCOOR''.

b. Base-Catalyzed Transesterification

Transesterification can also be carried out in the presence of a basic catalyst, which is usually the alkoxide base of the corresponding alcohol; for example:

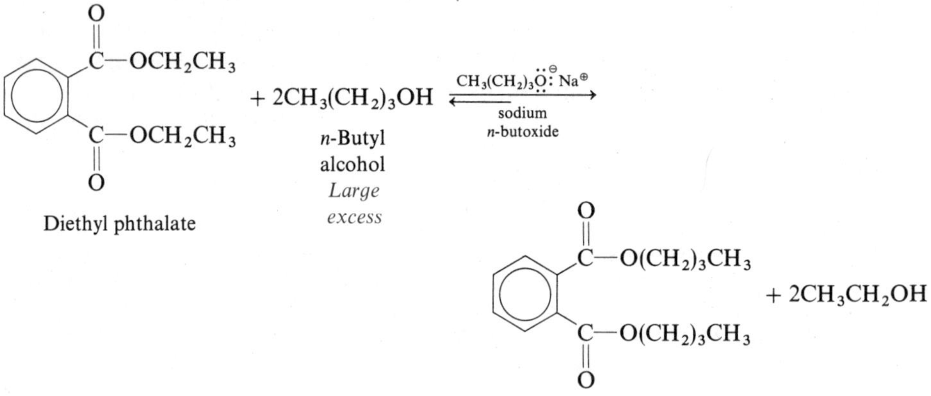

Diethyl phthalate

+ 2CH$_3$(CH$_2$)$_3$OH

n-Butyl alcohol
Large excess

$\xrightarrow[\text{sodium} \\ \textit{n}\text{-butoxide}]{\text{CH}_3\text{(CH}_2\text{)}_3\ddot{\text{O}}\colon^\ominus \text{Na}^\oplus}$

Di-*n*-butyl phthalate

+ 2CH$_3$CH$_2$OH

This is a typical nucleophilic acyl substitution reaction carried out under basic conditions.

c. Applications of Transesterification

In addition to synthetic applications in the research laboratory, transesterification is widely used in the preparation of the synthetic fabric Dacron, as shown by the following equation:

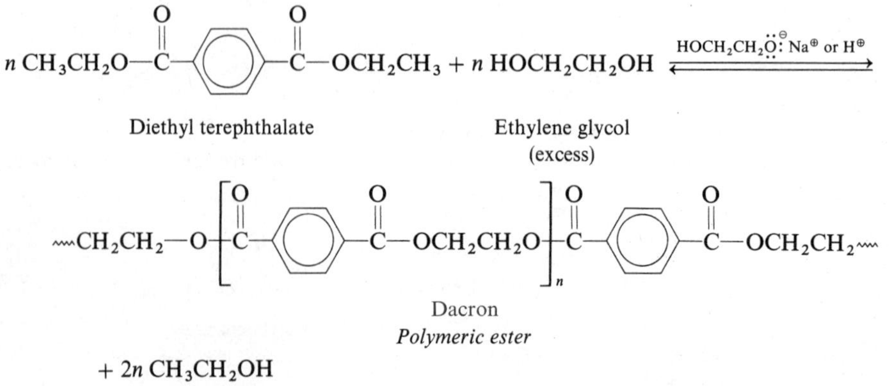

Diethyl terephthalate

Ethylene glycol
(excess)

$\xrightarrow{\text{HOCH}_2\text{CH}_2\ddot{\text{O}}\colon^\ominus \text{Na}^\oplus \text{ or } H^\oplus}$

Dacron
Polymeric ester

+ 2*n* CH$_3$CH$_2$OH

Dacron is a polyester prepared from two readily available starting materials, diethyl or dimethyl terephthalate and ethylene glycol. Either acid or base catalysis can be used. This reaction is another example of condensation polymerization (see Sec. 21.7), in which two molecules come together with the liberation of a smaller, simpler molecule (in this case, ethyl alcohol).

A polymer forms for the following reasons. The first step involves transesterification, where one end of the ethylene glycol molecule displaces an ethyl alcohol molecule:

$$CH_3CH_2O-\overset{\overset{O}{\|}}{C}-\underset{}{\bigcirc}-\overset{\overset{O}{\|}}{C}-OCH_2CH_3 + HOCH_2CH_2OH \xrightarrow[\text{or } H^{\oplus}]{HOCH_2CH_2\overset{..}{\overset{..}{O}}\overset{\ominus}{:}}$$

1 mole   1 mole

$$CH_3CH_2O-\overset{\overset{O}{\|}}{C}-\underset{}{\bigcirc}-\overset{\overset{O}{\|}}{C}-OCH_2CH_2OH \xrightarrow[\substack{\text{many} \\ \text{times}}]{\text{repeat}} \quad \text{Dacron} + CH_3CH_2OH$$

free to react further

The new molecule still contains a reactive hydroxyl group and acyl group, which can react further to give another larger and still reactive molecule. Condensation continues until the reactants are consumed.

C. With Ammonia: Ammonolysis

Esters react with ammonia to form the corresponding amide. The general equation is:

$$R-\overset{\overset{O}{\|}}{C}-OR' + NH_3 \rightleftharpoons R-\overset{\overset{O}{\|}}{C}-NH_2 + R'OH$$

For example:

$$Br-\bigcirc-\overset{\overset{O}{\|}}{C}-OCH_3 + NH_3 \longrightarrow Br-\bigcirc-\overset{\overset{O}{\|}}{C}-NH_2 + CH_3OH$$

Methyl p-bromobenzoate p-Bromobenzamide

Ammonia serves as the nucleophile in this reaction and attacks the electron-deficient acyl carbon atom, whereupon nucleophilic acyl substitution occurs.

Question 23.11

Propose a complete, stepwise mechanism to account for the ammonolysis reaction. Show all intermediates.

D. With Grignard Reagents

An ester reacts with the Grignard reagent to form an alcohol—a reaction that is a good method for preparing certain alcohols. This reaction is carried out by allowing an excess of Grignard reagent to react with an ester under anhydrous conditions, followed by hydrolysis to give an alcohol. *Two* moles of Grignard reagent are consumed per mole of ester, as shown:

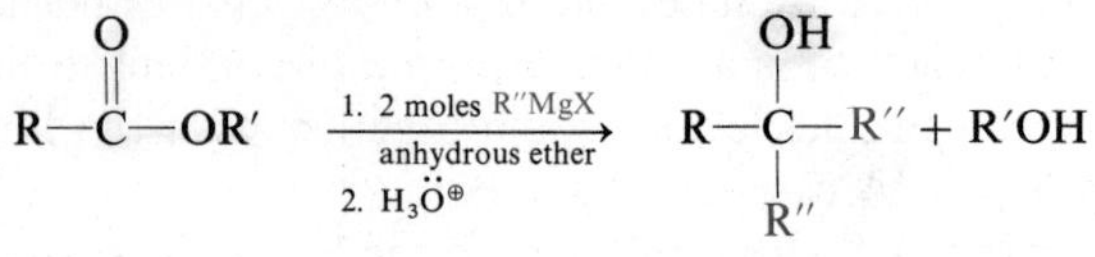

$$
\underset{\text{}}{R-\overset{\overset{\displaystyle O}{\|}}{C}-OR'} \quad \xrightarrow[\substack{\text{anhydrous ether} \\ \text{2. } H_3\overset{\cdot\cdot}{O}^{\oplus}}]{\text{1. 2 moles } R''MgX} \quad R-\overset{\overset{\displaystyle OH}{|}}{\underset{\underset{\displaystyle R''}{|}}{C}}-R'' + R'OH
$$

Alcohol
Contains two identical R'' groups

Note that the alcohol produced always contains *two identical groups* (R'') corresponding to the structure of the alkyl or aryl group in the Grignard reagent.

The following two examples illustrate the use of this reaction in synthesis. A tertiary alcohol is formed *except* when the ester is a formate ester (HCOOR') and then the product is a secondary alcohol.

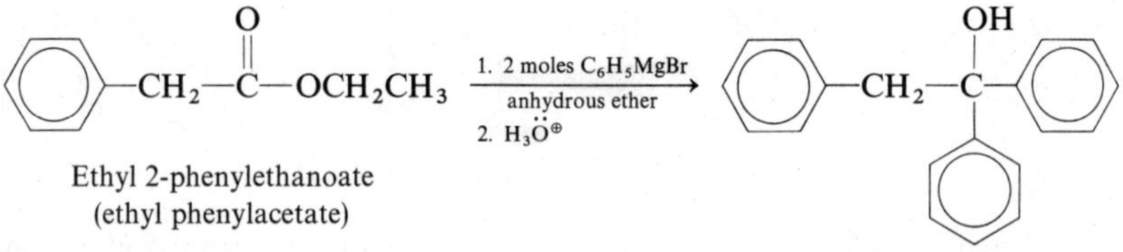

Ethyl 2-phenylethanoate
(ethyl phenylacetate)

1,1,2-Triphenylethanol
3° Alcohol

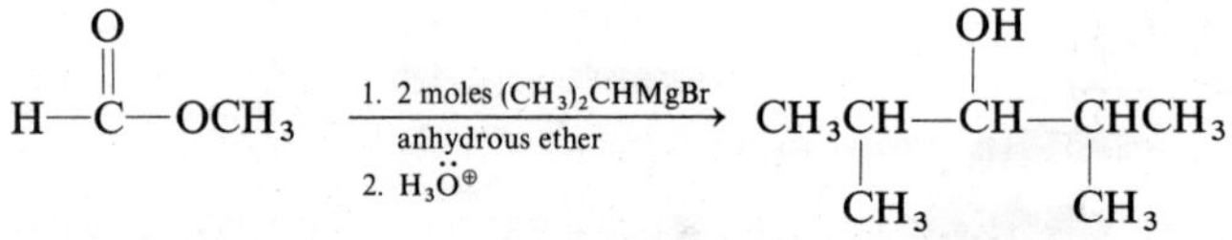

Methyl methanoate
(methyl formate)

2,4-Dimethyl-3-pentanol
2° Alcohol

The mechanism of this reaction involves the initial attack of the nucleophilic Grignard reagent on the acyl carbon atom. Nucleophilic acyl attack gives an unstable tetrahedral intermediate, which decomposes with the loss of the alkoxide ion, $R'\overset{\cdot\cdot}{\underset{\cdot\cdot}{O}}\colon^{\ominus}$ (step 1 in the following mechanism). The first stable intermediate in the reaction is a ketone, which can be isolated under certain conditions. Ketones also react with the Grignard reagent by nucleophilic addition (see Sec. 21.10), however, and on hydrolysis a tertiary alcohol is produced (step 2).

Step 1. Nucleophilic acyl substitution

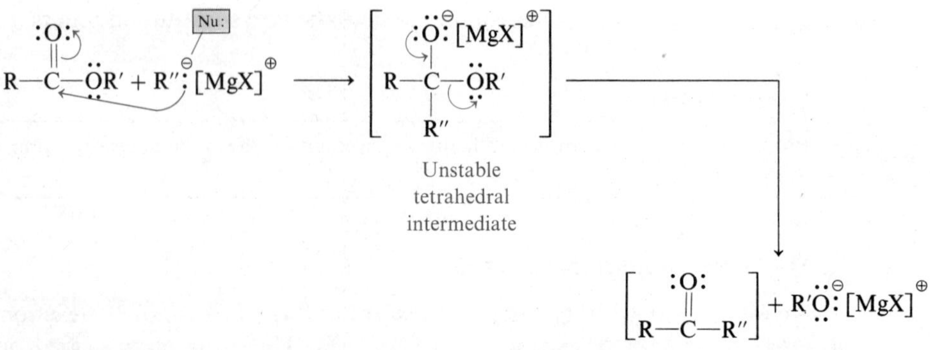

Unstable
tetrahedral
intermediate

$$
\left[\underset{R-\overset{\overset{\displaystyle :O:}{\|}}{C}-R''}{} \right] + R'\overset{\cdot\cdot}{\underset{\cdot\cdot}{O}}\colon^{\ominus}[MgX]^{\oplus}
$$

Ketone
Stable intermediate
but not isolated

Step 2. Nucleophilic addition

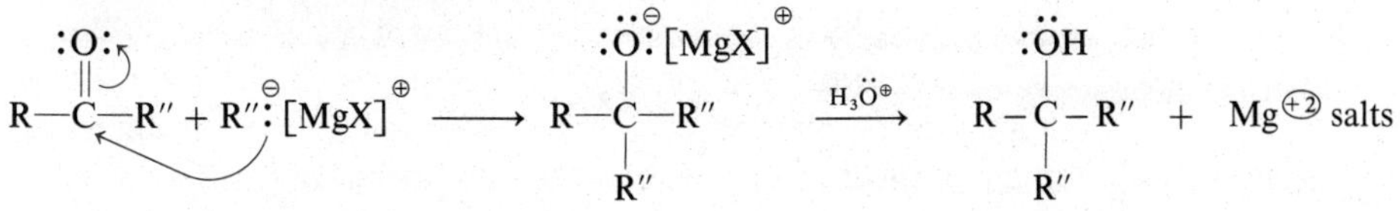

$$R\overset{\displaystyle :\overset{..}{O}:}{\underset{}{\overset{\|}{C}}}R'' + R''\overset{\ominus}{:}[MgX]^{\oplus} \longrightarrow R\overset{\displaystyle :\overset{..}{\overset{\ominus}{O}}:[MgX]^{\oplus}}{\underset{R''}{\overset{|}{C}}}R'' \xrightarrow{H_3\overset{..}{O}^{\oplus}} R\overset{\displaystyle :\overset{..}{O}H}{\underset{R''}{\overset{|}{C}}}R'' + Mg^{\oplus 2} \text{ salts}$$

Observed product

The preceding mechanism may not be completely correct, although it is plausible. For example, in some reactions a ketone is not observed as a product even when equimolar amounts of Grignard reagent and ester react. It is suggested that the following competing step occurs *without* intervention of a ketone:

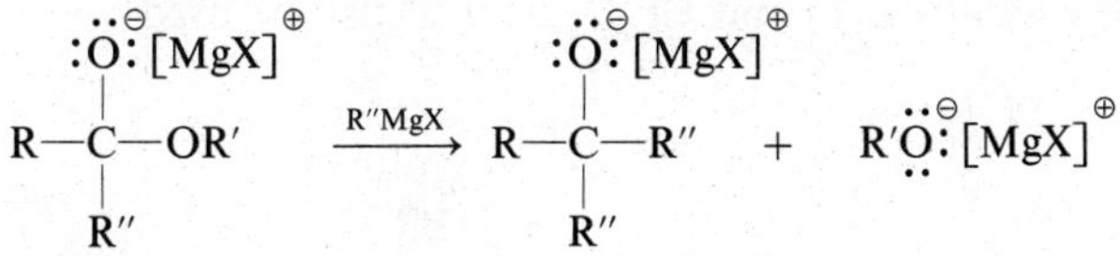

$$R\overset{\displaystyle :\overset{..}{\overset{\ominus}{O}}:[MgX]^{\oplus}}{\underset{R''}{\overset{|}{C}}}OR' \xrightarrow{R''MgX} R\overset{\displaystyle :\overset{..}{\overset{\ominus}{O}}:[MgX]^{\oplus}}{\underset{R''}{\overset{|}{C}}}R'' + R'\overset{..}{\underset{..}{O}}{}^{\ominus}[MgX]^{\oplus}$$

Unfortunately, there is little experimental evidence to support either possibility, so a single mechanism cannot be offered with certainty. Both mechanisms may operate in this reaction.

Compare the reaction of the Grignard reagent with an ester and with an aldehyde or ketone to understand the different products that are obtained.

Question 23.12

What combination of ester and Grignard reagent should be chosen so that, on hydrolysis, the following alcohols are produced?

(*a*)

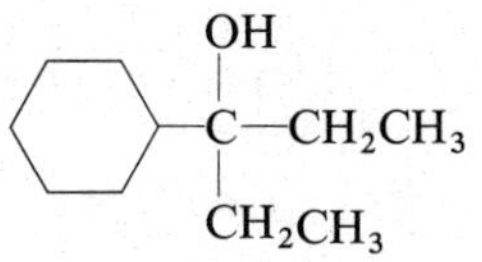

$$\text{cyclohexyl}-\overset{\displaystyle OH}{\underset{CH_2CH_3}{\overset{|}{C}}}-CH_2CH_3$$

(*b*) $(C_6H_5)_2CHOH$

(*c*) $(C_6H_5)_3COH$

(*d*)

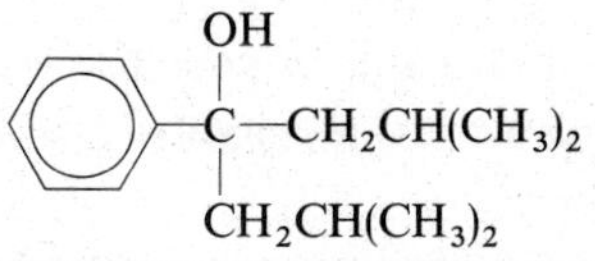

$$\text{phenyl}-\overset{\displaystyle OH}{\underset{CH_2CH(CH_3)_2}{\overset{|}{C}}}-CH_2CH(CH_3)_2$$

Question 23.13

Diethyl carbonate (1) reacts with the Grignard reagent and gives carbinol (2), which contains *three* identical substituents:

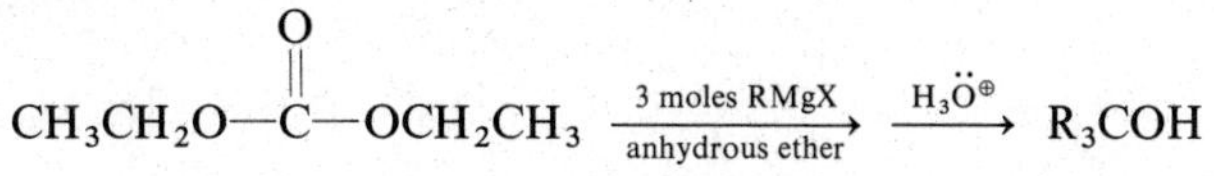

$$CH_3CH_2O\overset{\displaystyle O}{\overset{\|}{C}}OCH_2CH_3 \xrightarrow[\text{anhydrous ether}]{\text{3 moles RMgX}} \xrightarrow{H_3\overset{..}{O}^{\oplus}} R_3COH$$

(1)   (2)

(*a*) Propose a plausible mechanism for this reaction, showing likely steps and intermediates.
(*b*) Diethyl carbonate is the diester of what acid?
(*c*) Diethyl carbonate can be prepared by the following reaction:

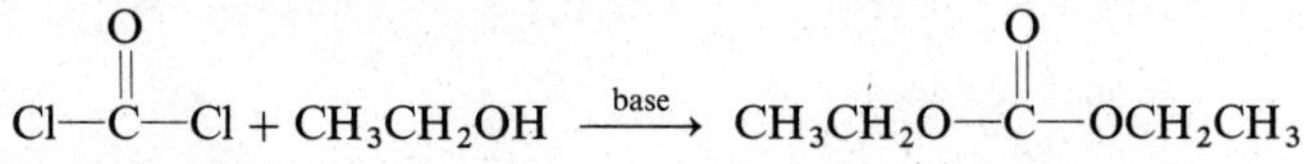

$$Cl\overset{\displaystyle O}{\overset{\|}{C}}Cl + CH_3CH_2OH \xrightarrow{\text{base}} CH_3CH_2O\overset{\displaystyle O}{\overset{\|}{C}}OCH_2CH_3$$

Phosgene   2 moles

Propose a complete, stepwise mechanism to account for this reaction.

E. With Reducing Agents: Reduction

Esters can be reduced by the addition of the elements of hydrogen. The products are always alcohols because the acid portion of the ester is converted into a primary alcohol, and the alcohol or phenol part of the ester is liberated as such. Catalytic hydrogenation and chemical reduction can be used in the following manner.

a. Catalytic Hydrogenation

The addition of hydrogen to an ester is more difficult to accomplish than is hydrogenation of an alkene, alkyne, aldehyde, or ketone. For example, the hydrogenation of esters requires a high temperature (200 to 250°) and a high pressure (2,000 to 4,000 lb/sq in.) of hydrogen gas. A special catalyst, called *copper chromite* (which consists of Cu_2O and Cr_2O_3), must also be used:

$$\underset{\substack{\text{2 moles} \\ \text{2,000–4,000 lb/sq in.}}}{R-\overset{\overset{\displaystyle O}{\|}}{C}-OR' \ + \ H_2} \ \xrightarrow[\text{200–250°,}]{Cu_2O + Cr_2O_3} \ R-CH_2-OH + R'OH$$

For example:

$$\underset{\substack{\text{Diethyl hexanedioate} \\ \text{(diethyl adipate)}}}{CH_3CH_2O-\overset{\overset{\displaystyle O}{\|}}{C}-(CH_2)_4-\overset{\overset{\displaystyle O}{\|}}{C}-OCH_2CH_3} \ \xrightarrow[\substack{\text{200–250°,} \\ \text{2,000–4,000 lb/sq in.}}]{H_2, \ Cu_2O + Cr_2O_3} \ \underset{\text{1,6-Hexanediol}}{HO-(CH_2)_6-OH} + 2CH_3CH_2OH$$

b. Chemical Reduction

Two common methods are used in the laboratory to reduce esters. One employs lithium aluminum hydride and the other uses sodium in ethanol.

On a small laboratory scale, **lithium aluminum hydride reduction** is the most convenient method; it affords high yields of product and does not attack isolated carbon-carbon double or triple bonds. The reduction is usually carried out in anhydrous ether and is followed by hydrolysis to liberate the alcohol products; for example:

n-Butyl benzoate

$$\underset{\substack{\text{Benzyl} \\ \text{alcohol}}}{\text{(phenyl)}-CH_2OH} + \underset{\text{1-Butanol}}{CH_3(CH_2)_3OH}$$

$$\underset{\substack{\text{Methyl 9-octadecenoate} \\ \text{(methyl oleate)}}}{CH_3(CH_2)_7-CH=CH-(CH_2)_7-\overset{\overset{\displaystyle O}{\|}}{C}-OCH_3} \ \xrightarrow[\text{anhydrous ether}]{LiAlH_4} \ \xrightarrow{H_3\overset{..}{O}^{\oplus}}$$

$$\underset{\substack{\text{9-Octadecen-1-ol} \\ \text{(oleyl alcohol)}}}{CH_3(CH_2)_7-CH=CH-(CH_2)_8OH} + CH_3OH$$

The mechanism of this reduction is believed to follow a path in which a hydride ion in $LiAlH_4$ replaces the alkoxide ion by way of the tetrahedral intermediate. An intermediate aldehyde probably forms but reacts so readily with $LiAlH_4$ that it is never present in any significant amount.

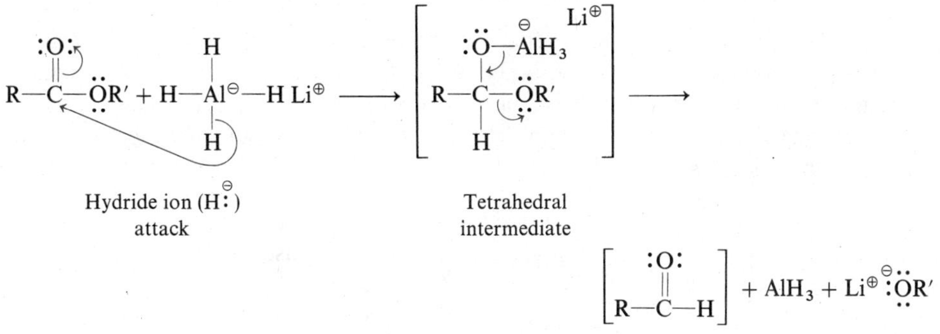

Hydride ion ($H:^{\ominus}$) attack

Tetrahedral intermediate

The reduction of an aldehyde with $LiAlH_4$ was discussed in Sec. 21.13, and the final products are the lithium and/or aluminum salts of the alkoxide ions $RCH_2\overset{..}{\underset{..}{O}}:^{\ominus}$ and $R'\overset{..}{\underset{..}{O}}:^{\ominus}$, which are protonated on hydrolysis and give RCH_2OH and $R'OH$. No other metal hydride reducing agents that we have discussed (for example, sodium borohydride) are strong enough to reduce an ester.

Sodium metal in ethanol also reduces esters. This method, called the **Bouveault-Blanc procedure,** involves adding sodium metal to a refluxing mixture of the ester and ethanol. This method does not reduce isolated carbon-carbon double or triple bonds either.

$$CH_3(CH_2)_7—CH=CH—(CH_2)_7—\overset{\overset{O}{\|}}{C}—OCH_2CH_3 \xrightarrow[CH_3CH_2OH]{Na}$$

Ethyl 9-octadecaenoate
(ethyl oleate)

$$CH_3(CH_2)_7—CH=CH—(CH_2)_8OH + CH_3CH_2OH$$

9-Octadecen-1-ol
(oleyl alcohol)

23.8 Preparation of Amides

In this section we discuss the common methods for preparing amides, $RCONH_2$, or substituted amides, $RCONHR'$ and $RCONR'_2$.

A. From Acid Chlorides and Amines

The most general laboratory method for preparing amides is the reaction that occurs between an acid chloride and an amine, as shown by the general equation:

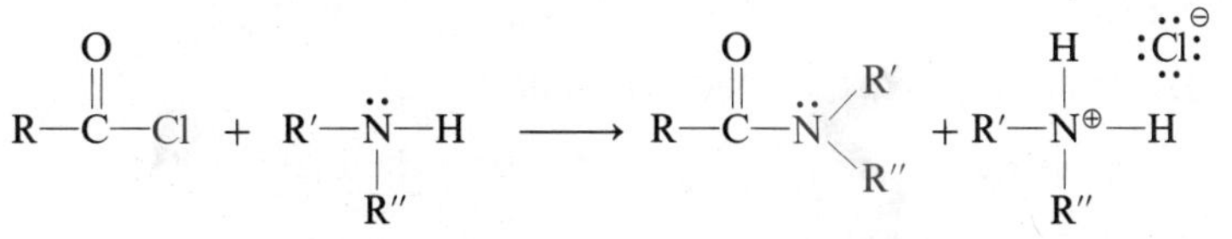

2 moles
*R' and R'' may
be alkyl, aryl, or
hydrogen*

The amine, $R'R''\ddot{N}H$, may be either ammonia ($\ddot{N}H_3$) or a primary or secondary amine; this reaction is discussed in detail in Sec. 23.5A.

B. From Acids and Ammonia

A special method for preparing amides involves the conversion of an acid to an unsubstituted amide, $RCONH_2$. When an acid is treated with an excess of ammonia, the ammonium salt of the acid forms. The ammonium salt is then heated vigorously to form the corresponding amide and a molecule of water. This sequence of steps is shown by the following:

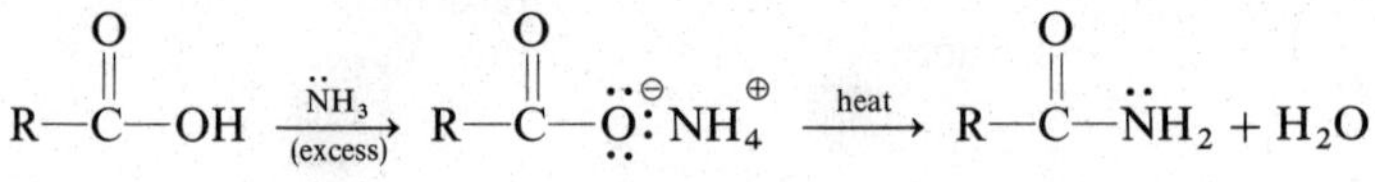

$$R\overset{\overset{\displaystyle O}{\|}}{-C}-OH \xrightarrow[\text{(excess)}]{\ddot{N}H_3} R\overset{\overset{\displaystyle O}{\|}}{-C}-\overset{\ominus}{\ddot{O}}\!:\!\overset{\oplus}{NH_4} \xrightarrow{\text{heat}} R\overset{\overset{\displaystyle O}{\|}}{-C}-\ddot{N}H_2 + H_2O$$

Ammonium
salt

For example:

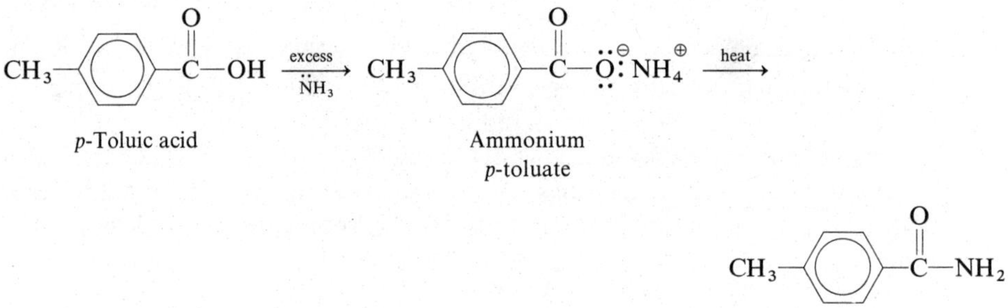

$$CH_3-\!\!\left\langle\bigcirc\right\rangle\!\!-\overset{\overset{\displaystyle O}{\|}}{C}-OH \xrightarrow[\ddot{N}H_3]{\text{excess}} CH_3-\!\!\left\langle\bigcirc\right\rangle\!\!-\overset{\overset{\displaystyle O}{\|}}{C}-\overset{\ominus}{\ddot{O}}\!:\!\overset{\oplus}{NH_4} \xrightarrow{\text{heat}}$$

p-Toluic acid   Ammonium
                                        p-toluate

$$CH_3-\!\!\left\langle\bigcirc\right\rangle\!\!-\overset{\overset{\displaystyle O}{\|}}{C}-NH_2$$

p-Toluamide

The mechanism of amide formation by this reaction is rather obscure and is not discussed here. This route for preparing amides is most frequently encountered in industrial processes and then usually only with ammonia as the amine. Yields are often low, accompanied by a great deal of decomposition because of the high reaction temperatures. The reaction between an acid chloride and an amine is usually used in the laboratory.

C. From Esters and Ammonia

As discussed in Sec. 23.7C, most esters react with ammonia to form the corresponding amide:

$$R\overset{\overset{\displaystyle O}{\|}}{-C}-OR' + \ddot{N}H_3 \rightleftharpoons R\overset{\overset{\displaystyle O}{\|}}{-C}-\ddot{N}H_2 + R'OH$$

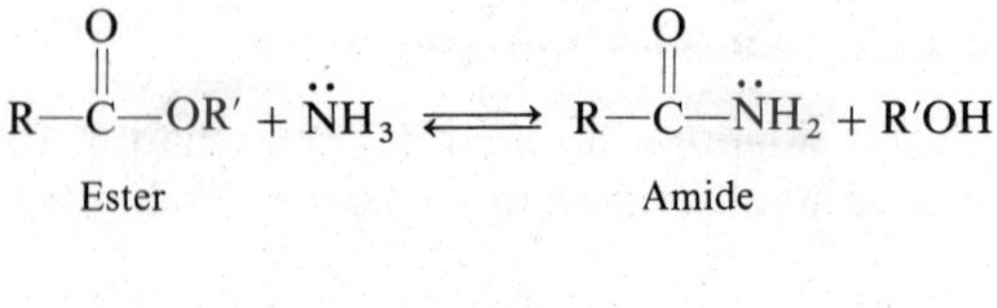

Ester   Amide

D. From Anhydrides and Ammonia

Acid anhydrides react with ammonia in the manner shown here:

Noncyclic anhydrides:

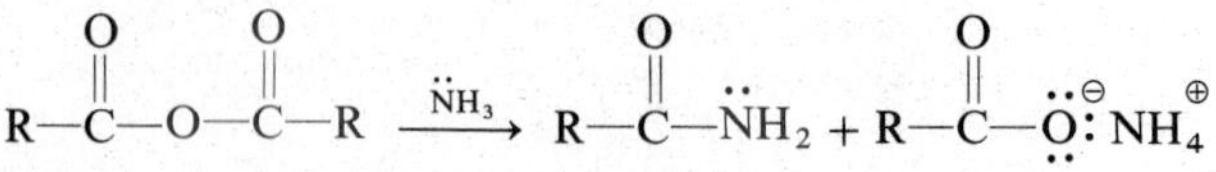

$$R\overset{\overset{\displaystyle O}{\|}}{-C}-O\overset{\overset{\displaystyle O}{\|}}{-C}-R \xrightarrow{\ddot{N}H_3} R\overset{\overset{\displaystyle O}{\|}}{-C}-\ddot{N}H_2 + R\overset{\overset{\displaystyle O}{\|}}{-C}-\overset{\ominus}{\ddot{O}}\!:\!\overset{\oplus}{NH_4}$$

Cyclic anhydrides:

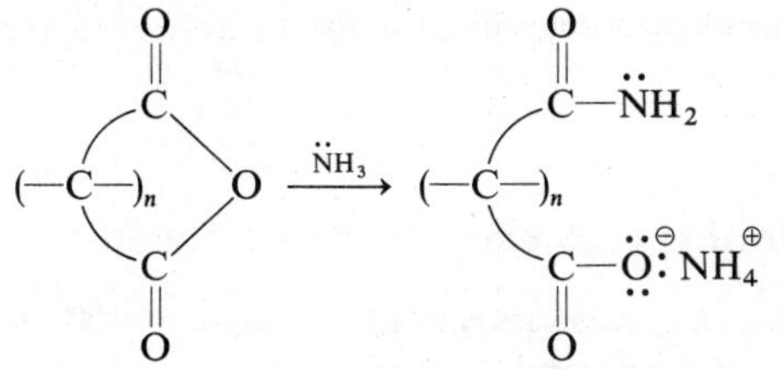

$n \geq 2$

These reactions and their mechanisms are discussed in more detail in Sec. 23.13A. This type of reaction is useful commercially but only when the anhydrides are readily available.

23.9 Reactions of Amides

We discuss some common reactions of amides in this section. Their mechanisms follow the general pattern of nucleophilic acyl substitution.

A. Hydrolysis of Amides

Amides can be hydrolyzed under either acidic or basic conditions, with the corresponding carboxylic acid or its salt being the respective products. The general reactions may be depicted as follows:

Acid-catalyzed:

$$R-\overset{\overset{\large O}{\|}}{C}-\ddot{N}H_2 \xrightarrow[\text{heat}]{H^{\oplus},\ H_2O} R-\overset{\overset{\large O}{\|}}{C}-OH + NH_4^{\oplus}$$

Base-promoted

$$R-\overset{\overset{\large O}{\|}}{C}-NH_2 \xrightarrow[\text{heat}]{:\ddot{O}H^{\ominus},\ H_2O} R-\overset{\overset{\large O}{\|}}{C}-\ddot{O}:^{\ominus} + NH_3$$

Two examples of amide hydrolysis are:

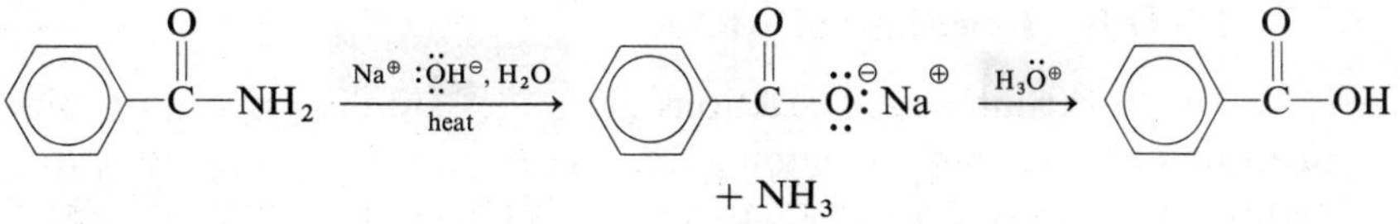

Benzamide Sodium benzoate Benzoic acid

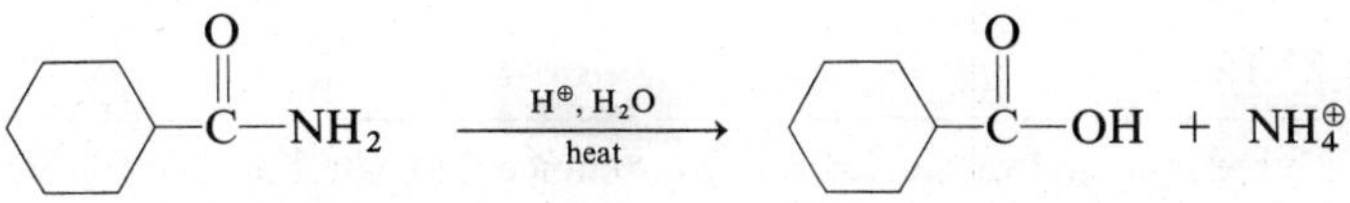

Cyclohexanecarboxamide Cyclohexanecarboxylic acid

Note that the reaction involving base must be neutralized with mineral acid to obtain the carboxylic acid.

Question 23.14

Write out detailed mechanisms for the hydrolysis of benzamide under acid and base conditions.

B. Dehydration of Amides: Nitrile Formation

Nitriles (that is, compounds containing the —C≡N group) can be prepared by allowing methyl, primary and certain secondary alkyl halides or tosylates to react with cyanide ion under S_N2 conditions (see Sec. 22.5B). The limitations of this method are that the yield is poor with secondary halides or tosylates and no nitrile is obtained with either tertiary or aryl halides or tosylates.

On the other hand, amides are converted smoothly and in good yield to the corresponding nitriles on heating with thionyl chloride ($SOCl_2$), phosphorus penta-chloride (PCl_5), phosphorus pentoxide (P_2O_5), or phosphorus oxychloride ($POCl_3$), as shown by the following equation:

$$3\,(Ar)\,R\overset{\overset{\displaystyle O}{\|}}{-C}-\ddot{N}H_2 + POCl_3 \longrightarrow 3\,(Ar)\,R-C\equiv\ddot{N} + H_3PO_4 + 3HCl$$

Phosphorus oxychloride is the reagent of choice in these reactions.

As the following sequence shows, it is often easy to obtain a nitrile group attached to a tertiary carbon atom via the dehydration of an amide:

$$(CH_3)_3COH \xrightarrow[25°]{\text{conc} \atop \text{HBr}} (CH_3)_3CBr \xrightarrow[\text{ether}]{\text{Mg} \atop \text{anhydrous}} (CH_3)_3CMgBr \xrightarrow[\text{2. } H_3O^\oplus]{\text{1. dry ice}} (CH_3)_3CCOOH$$

tert-Butyl   tert-Butyl   Pivalic acid
alcohol   bromide

1. PCl_3
2. cold NH_3

$$(CH_3)_3C-C\equiv\ddot{N} \xleftarrow[\text{heat}]{SOCl_2} (CH_3)_3C\overset{\overset{\displaystyle O}{\|}}{-C}-\ddot{N}H_2$$

Pivalonitrile   Pivalamide

These reactions show one of the relationships between acids and nitriles, in that an acid can be converted to an amide and thence to a nitrile. In Sec. 23.16 we see that nitriles can be converted to amides and carboxylic acids, which is one reason for discussing them along with derivatives of acids.

C. Other Reactions of Amides

Two other important reactions of amides are discussed later: the conversion of an amide to an amine containing the same number of carbon atoms (see Sec. 26.7A) and the conversion of an amide to a primary amine containing one less carbon atom (the Hofmann reaction, see Sec. 26.7C).

Question 23.15

Starting with benzene and/or alcohols of your choice, indicate how the following nitriles can be synthesized.

(a) $CH_3(CH_2)_4-C\equiv\ddot{N}$ (b) $CH_3CH_2-\underset{\underset{\displaystyle CH_3}{|}}{CH}-C\equiv\ddot{N}$

(c) $(CH_3CH_2)_2C-C\equiv\ddot{N}$
         |
         CH_3

(d) $CH_3-\langle\bigcirc\rangle-C\equiv\ddot{N}$

23.10   Occurrence and Use of Amides

The amide linkage is the backbone of proteins and peptides as well as a host of other naturally occurring products. We discuss these types of compounds further in Chap. 28. For now we study synthetic fibers that are polymeric amides.

Nylon was the first synthetic polymer made and is still one of the most important fibers. It is prepared by heating together adipic acid (the six-carbon, straight-chain diacid) and hexamethylene diamine (a six-carbon compound with —NH_2 groups at each end). The carboxyl groups (—COOH) and the amine groups (—NH_2) first form the corresponding ammonium salts in the same way that a monocarboxylic acid and ammonia react $(RCOOH + NH_3 \rightarrow RCO\ddot{O}:^{\ominus} NH_4^{\oplus})$. On heating, these salts lose water to form amide bonds. The following equations show the preparation of one type of nylon:

$$HO-\overset{\overset{O}{\|}}{C}-(CH_2)_4-\overset{\overset{O}{\|}}{C}-OH$$

Adipic acid
+
$$H_2N-(CH_2)_6-NH_2$$
Hexamethylenediamine

$$\longrightarrow HO-\overset{\overset{O}{\|}}{C}-(CH_2)_4-\overset{\overset{O}{\|}}{C}-\ddot{O}:^{\ominus}H_3\overset{\oplus}{N}-(CH_2)_6-NH_2$$

Ammonium salt

heat, $-H_2O$

$$HO-\overset{\overset{O}{\|}}{C}-(CH_2)_4-\overset{\overset{O}{\|}}{C}-OH \quad \xrightarrow{heat, -H_2O} \quad HO-\overset{\overset{O}{\|}}{C}-(CH_2)_4-\overset{\overset{O}{\|}}{C}-NH-(CH_2)_6-NH_2$$

$$HO-\overset{\overset{O}{\|}}{C}-(CH_2)_4-\overset{\overset{O}{\|}}{C}-NH-(CH_2)_6-NH-\overset{\overset{O}{\|}}{C}-(CH_2)_4-\overset{\overset{O}{\|}}{C}-OH$$

$H_2N-(CH_2)_6-NH_2$,
heat, $-H_2O$

$$HO-\overset{\overset{O}{\|}}{C}-(CH_2)_4-\overset{\overset{O}{\|}}{C}-NH-(CH_2)_6-NH-\overset{\overset{O}{\|}}{C}-(CH_2)_4-\overset{\overset{O}{\|}}{C}-NH-(CH_2)_6-NH_2$$

These reactions continue because each time a molecule of diacid or diamine is added, the ends of the new molecule contain a —COOH group or an —NH_2 group that reacts further with another molecule of diamine or diacid, respectively. The general structure of the nylon polymer is the following:

$$HO-\overset{\overset{O}{\|}}{C}-(CH_2)_4-\overset{\overset{O}{\|}}{C}\Big[NH-(CH_2)_6-NH-\overset{\overset{O}{\|}}{C}-(CH_2)_4-\overset{\overset{O}{\|}}{C}\Big]_n NH-(CH_2)_6-NH_2$$

Nylon 66
$n = 50-65$

Depending on the reaction conditions, n is usually 50 to 65, thus providing nylon polymers with molecular weights on the order of 12,000 to 15,000. This particular polymer is called Nylon 66 because it is composed of a six-carbon diacid and a six-carbon diamine, both of which are available industrially at very low cost (see also Sec. 26.10).

Question 23.16

Adipic acid, one of the starting materials in the preparation of Nylon 66, is available commercially from several different routes. One of these is the vigorous oxidation of cyclohexanol (see Sec. 22.6). Another more common method starts with the addition of 1 mole of chlorine to 1 mole of 1,3-butadiene in a nonpolar solvent. Based on reactions we have studied, suggest the remaining steps in the transformation of 1,3-butadiene (a readily available petroleum product) to adipic acid.

Question 23.17

Another polymer, called Nylon 6, is prepared by heating the cyclic amide caprolactam (1) in the presence of a trace of water. Noting that caprolactam has an amide linkage and given that cyclic amides of this type are especially reactive toward water,

(*a*) Give the structure of the intermediate monomer that contains both an —NH$_2$ group and a —COOH group.

(*b*) Suggest a structure for Nylon 6 that could be formed from the compound described in (*a*).

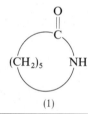

(1)

Question 23.18

Nylon 610 is prepared from hexamethylenediamine and a particular dicarboxylic acid. Give the structure and name of the diacid as well as the structure of Nylon 610.

23.11   Preparation of Anhydrides from Monocarboxylic Acids

The most important anhydride of monoacids is ethanoic anhydride (acetic anhydride), $(CH_3CO)_2O$. The most usual industrial preparation of this compound starts with ethanoic acid (acetic acid), which can be dehydrated at high temperature in the presence of a catalyst. The dehydration product is *ketene*, $CH_2=C=O$, see Sec. 27.6. Once formed, ketene reacts with another molecule of acetic acid, which adds across the reactive carbon-oxygen double bond in ketene to form an unstable enol intermediate. The enol compound rapidly tautomerizes to form acetic anhydride. We encountered keto-enol tautomerization in Sec. 11.13 and we discuss it in more detail in Chap. 24.

$$CH_3-\overset{\overset{\displaystyle O}{\|}}{C}-OH \xrightarrow[\text{AlPO}_4]{700°} CH_2=C=O + H_2O$$

Ketene

$$CH_2=C=O + CH_3-\overset{\overset{\displaystyle O}{\|}}{C}-O-H \longrightarrow \begin{bmatrix} CH_2=C\overset{\displaystyle OH}{\underset{\displaystyle O}{\diagdown}} \\ CH_3-C\overset{\displaystyle}{\underset{\displaystyle O}{\diagdown\!\!\!\parallel}} \end{bmatrix}$$

Enol
Unstable

↑↓ tautomerization

$$\begin{array}{c} CH_3-C\overset{\displaystyle O}{\diagup}\\ \qquad\diagdown O \\ CH_3-C\overset{\displaystyle}{\diagdown\!\!\!O} \end{array}$$

Ethanoic
anhydride
(acetic anhydride)

Ketene can also be prepared by the pyrolysis of acetone at 700°:

$$CH_3COCH_3 \longrightarrow CH_2=C=O + CH_4$$

Ethanoic anhydride can be used to convert other carboxylic acids into the corresponding anhydrides by allowing an excess of the low-cost ethanoic anhydride to react with a carboxylic acid. A reversible reaction results, and because ethanoic acid is formed and is the lowest boiling component in the mixture, it can be removed by fractional distillation as it is formed. The use of excess ethanoic anhydride and the removal of ethanoic acid as it is formed shift the equilibrium in the following equation to the right and increase the yield of the new anhydride.

$$CH_3-\overset{\overset{\displaystyle O}{\|}}{C}-O-\overset{\overset{\displaystyle O}{\|}}{C}-CH_3 + R-\overset{\overset{\displaystyle O}{\|}}{C}-OH \rightleftharpoons R-\overset{\overset{\displaystyle O}{\|}}{C}-O-\overset{\overset{\displaystyle O}{\|}}{C}-R + CH_3-\overset{\overset{\displaystyle O}{\|}}{C}-OH$$

| Ethanoic anhydride *Large excess* | 2 moles | 1 mole | 2 moles *Removed by continuous distillation* |

The following example illustrates the use of this method for the synthesis of symmetrical anhydrides:

$$CH_3-\overset{\overset{\displaystyle O}{\|}}{C}-O-\overset{\overset{\displaystyle O}{\|}}{C}-CH_3 + 2\,\underset{\text{Benzoic acid}}{\bigcirc\!\!-\overset{\overset{\displaystyle O}{\|}}{C}-OH} \rightleftharpoons$$

Ethanoic
anhydride
Excess

$$2\,CH_3-\overset{\overset{\displaystyle O}{\|}}{C}-OH + \bigcirc\!\!-\overset{\overset{\displaystyle O}{\|}}{C}-O-\overset{\overset{\displaystyle O}{\|}}{C}-\!\!\bigcirc$$

Remove by distillation

Benzoic
anhydride

Unsymmetrical (mixed) and symmetrical anhydrides can be prepared by the reaction between an acid chloride and the salt of a carboxylic acid, as shown by the general equation:

$$\underset{\text{Mixed}}{\underset{\text{anhydride}}{R-\overset{\displaystyle O}{\overset{\|}{C}}-\overset{..}{\underset{..}{O}}\!:\!M^{\oplus} + R'-\overset{\displaystyle O}{\overset{\|}{C}}-Cl \longrightarrow R-\overset{\displaystyle O}{\overset{\|}{C}}-O-\overset{\displaystyle O}{\overset{\|}{C}}-R' + M^{\oplus} + :\overset{..}{\underset{..}{Cl}}\!:^{\ominus}}}$$

The carboxylate anion serves as a nucleophile in this reaction and displaces chloride ion from the acid chloride via nucleophilic acyl substitution. Though not used commercially, this reaction is valuable in the laboratory where relatively small amounts of an anhydride are needed.

Question 23.19

Provide a complete, stepwise mechanism for the reaction between acetyl chloride and the acetate anion.

Question 23.20

Indicate suitable methods for preparing the following compounds, using acetic anhydride and any other desired carboxylic acids:

(*a*) butyric anhydride (give two methods)   (*b*) benzoic hexanoic anhydride

23.12   Dicarboxylic Acids: Anhydride Formation Versus Decarboxylation

Some interesting chemistry arises in dicarboxylic acids. On heating, some diacids form anhydrides, others lose a molecule of carbon dioxide (decarboxylation) to form simpler molecules, and yet others are unaffected by heat.

We now examine a sequence of aliphatic diacids where there is a "crossover" from decarboxylation to anhydride formation. Both oxalic acid and malonic acid undergo decarboxylation on heating, as shown here:

$$\underset{\substack{\text{Ethanedioic acid} \\ \text{(oxalic acid)}}}{HO-\overset{\displaystyle O}{\overset{\|}{C}}-\overset{\displaystyle O}{\overset{\|}{C}}-OH} \xrightarrow{140-160°} \underset{\substack{\text{Methanoic acid} \\ \text{(formic acid)}}}{H-\overset{\displaystyle O}{\overset{\|}{C}}-OH} + CO_2$$

$$\underset{\substack{\text{Propanedioic acid} \\ \text{(malonic acid)}}}{HO-\overset{\displaystyle O}{\overset{\|}{C}}-CH_2-\overset{\displaystyle O}{\overset{\|}{C}}-OH} \xrightarrow{140-160°} \underset{\substack{\text{Ethanoic acid} \\ \text{(acetic} \\ \text{acid)}}}{CH_3-\overset{\displaystyle O}{\overset{\|}{C}}-OH} + CO_2$$

We know from Sec. 23.11 that anhydrides form between separate carboxylic acid molecules, and because the preceding compounds contain two carboxyl groups in the same molecule, we might ask why they too do not form anhydrides. For oxalic acid and malonic acid, the answer is quite simple. The anhydride of oxalic acid would contain a three-membered ring and that of malonic acid would possess a four-membered ring (draw the structures of these cyclic anhydrides). Yet we know that three- and four-membered rings are highly unstable and so we predict anhydride formation is unlikely, as indeed it is for oxalic and malonic acid. The mechanism of decarboxylation of malonic acid has been studied, and experimental evidence suggests that a six-centered transition state is involved.

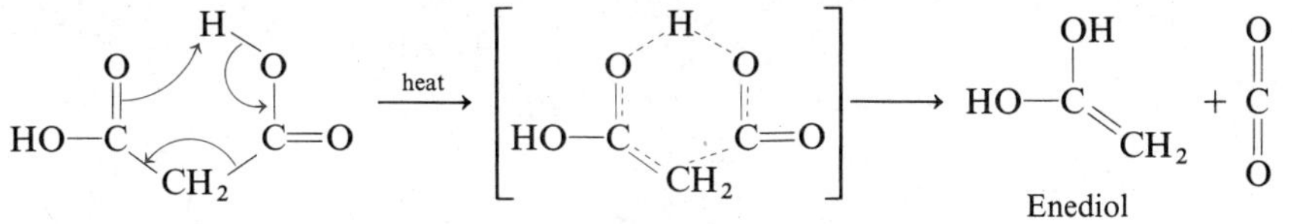

Transition state

Enediol
Unstable

An *enediol*[1] is formed initially, but like an enol (see Sec. 11.13), it is unstable and tautomerizes to yield acetic acid as the observed product:

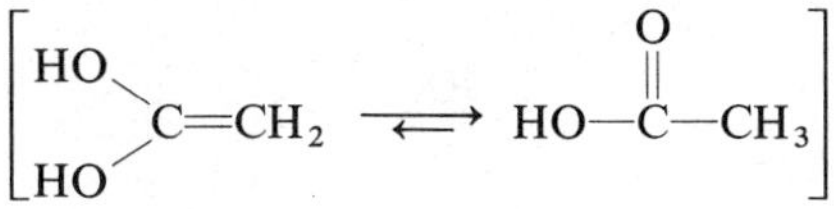

Enediol

Another driving force for decarboxylation is the formation of the very stable CO_2 molecule.

In contrast, succinic acid and glutaric acid form anhydrides when subjected to heat. Here anhydride formation is viable because stable five- and six-membered rings are formed:

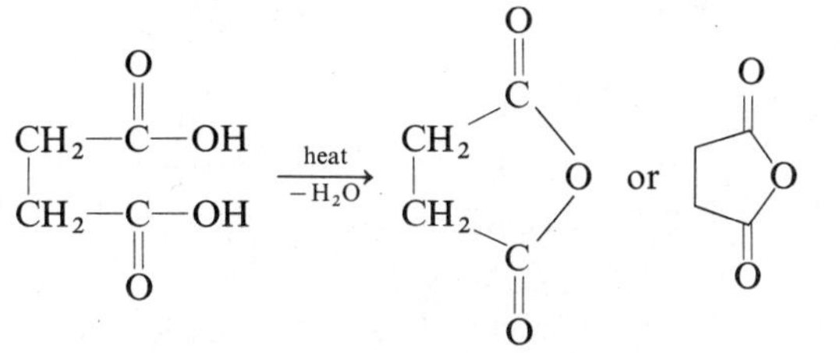

Butanedioic acid
(succinic acid)

Butanedioic anhydride
(succinic anhydride)

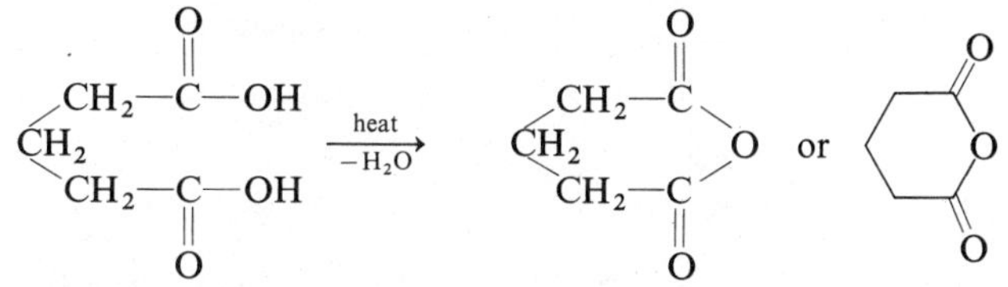

Pentanedioic
acid
(glutaric
acid)

Pentanedioic
anhydride
(glutaric anhydride)

The mechanism of anhydride formation in diacids is suggested to involve an internal proton transfer from one hydroxyl group to another, followed by nucleophilic acyl substitution.

Adipic acid (the six-carbon diacid) and other straight-chain aliphatic diacids containing more than six carbon atoms do not form anhydrides and generally do not undergo decarboxylation when heated.

Certain other diacids also form anhydrides. For example, phthalic acid readily forms phthalic anhydride on heating, whereas the other isomeric benzenedicarboxylic acids (isophthalic acid and terephthalic acid) do not.

---

[1] The term *enediol* describes this functional group, which contains an alk*ene* with two hydroxyl groups (a *diol*) attached to it. This term arises by analogy with *enol*, which names a group containing a carbon-carbon double bond and a hydroxyl group.

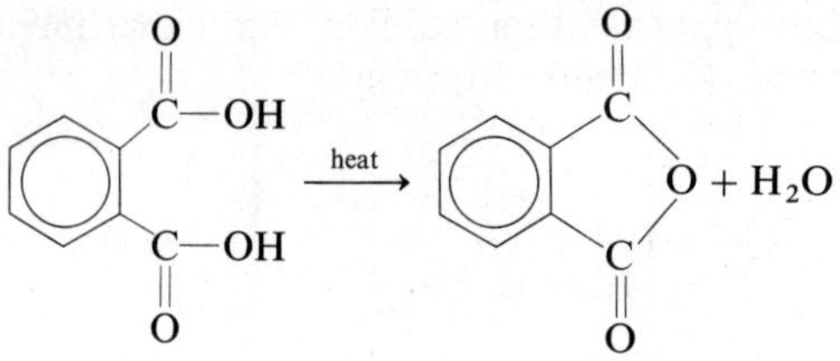

Phthalic acid   Phthalic
                                anhydride

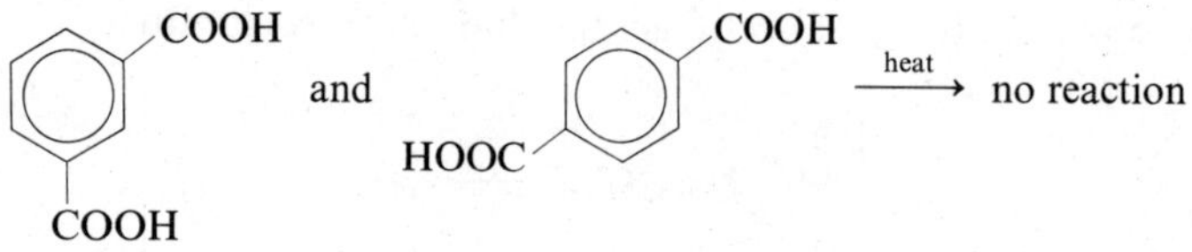

Isophthalic   Terephthalic
acid   acid

The general requirement for anhydride formation is as follows. *Whenever anhydride formation results in a five- or six-membered ring, it is likely to occur.* Anhydride formation has been used to distinguish between geometric isomers of certain cyclic dicarboxylic acids. For example, *cis*-1,2-cyclopropanedicarboxylic acid forms a stable anhydride on gentle heating, whereas the *trans* isomer does not:

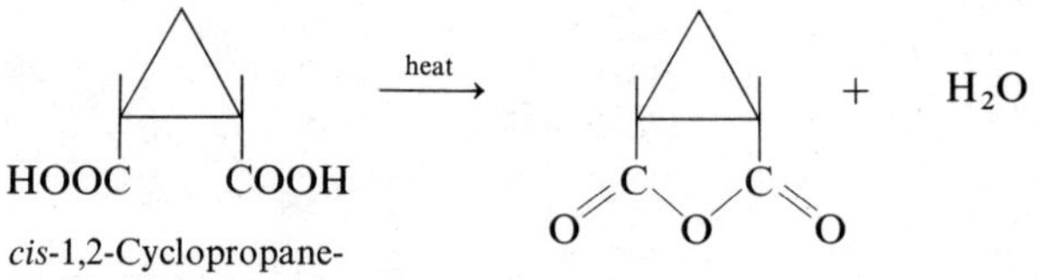

cis-1,2-Cyclopropane-
dicarboxylic acid

cis-1,2-Cyclopropane-
dicarboxylic anhydride

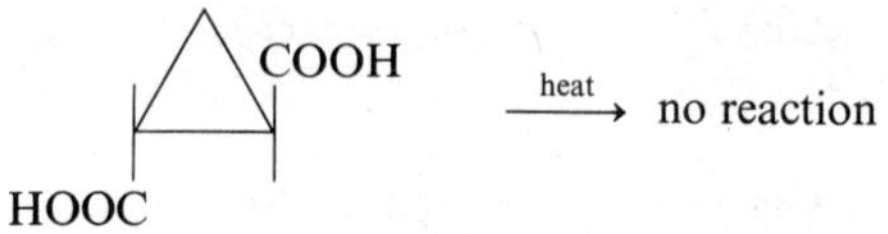

trans-1,2-Cyclopropane-
dicarboxylic acid

Certain geometric isomers of unsaturated diacids exhibit this same trend. For example, maleic acid readily forms an anhydride, whereas fumaric acid (the *trans* isomer) does not:

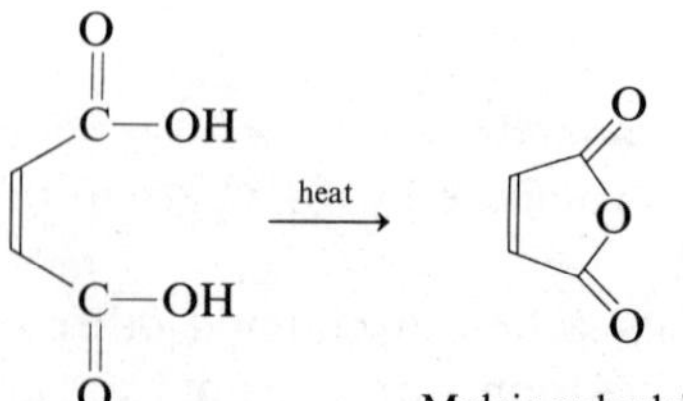

Maleic acid

Maleic anhydride

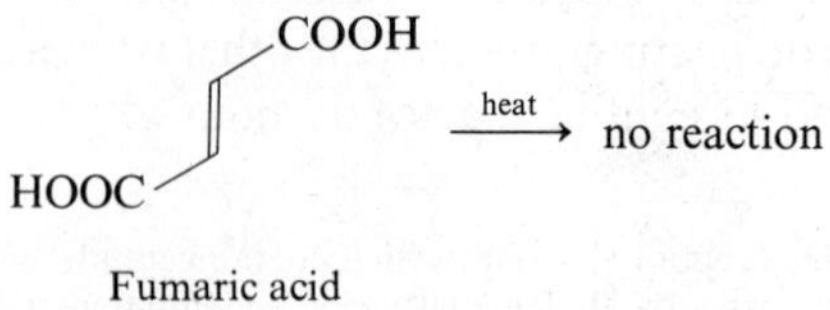

Fumaric acid

Question 23.21

Predict the effect of heat on each compound identified here.

(*a*) all the isomers (geometric and structural, but excluding optical isomers) of cyclopropanedicarboxylic acid

(*b*) all the isomers (geometric and structural, but excluding optical isomers) of cyclobutanedicarboxylic acid

(*c*)

(*d*)

(*e*) *cis* and *trans* isomers of $HOOC-CH=CH-CH_2COOH$

(*f*) *cis*- and *trans*-1,3-cyclohexanedicarboxylic acid

Question 23.22

Draw all the structural isomers of the dicarboxylic acids that have the molecular formula $C_6H_{10}O_4$, and indicate the effect of heat on each of them.

23.13 Reactions of Anhydrides

The most general type of reaction anhydrides undergo is the opening of the anhydride linkage.

A. Reaction with Nucleophilic Reagents: Water, Alcohol, and Ammonia

Water, alcohol, and ammonia are nucleophilic reagents that react with anhydrides to break the anhydride bond:

$$R-\overset{\overset{\displaystyle O}{\|}}{C}-O-\overset{\overset{\displaystyle O}{\|}}{C}-R + H:Nu \longrightarrow R-\overset{\overset{\displaystyle O}{\|}}{C}-Nu + R-\overset{\overset{\displaystyle O}{\|}}{C}-OH$$

where $HNu = H_2\ddot{O}:$, $\ddot{N}H_3$, or $R'\ddot{O}H$

These reactions all involve nucleophilic acyl substitution.

The addition of a trace of mineral acid or hydroxide ion to water increases the rate of hydrolysis appreciably, and the alcoholysis reaction is also speeded by the addition of a trace of mineral acid or alkoxide ion to the alcohol. Three typical examples of these reactions are:

$$CH_3-\overset{\overset{\displaystyle O}{\|}}{C}-O-\overset{\overset{\displaystyle O}{\|}}{C}-CH_3 + H_2O \longrightarrow CH_3-\overset{\overset{\displaystyle O}{\|}}{C}-OH$$

Acetic anhydride   2 moles

$$(CH_3CH_2CH_2C)_2O + 2NH_3 \longrightarrow CH_3CH_2CH_2-\overset{\overset{\displaystyle O}{\|}}{C}-NH_2 + CH_3CH_2CH_2-\overset{\overset{\displaystyle O}{\|}}{C}-\overset{\ominus}{\ddot{O}}: \overset{\oplus}{N}H_4$$

Butanoic anhydride (butyric anhydride)   Butanamide (butyramide)   Ammonium butanoate (ammonium butyrate)

(*Note:* In this reaction, butyramide and butyric acid are formed first, but in the presence of excess ammonia, butyric acid is converted to ammonium butyrate.)

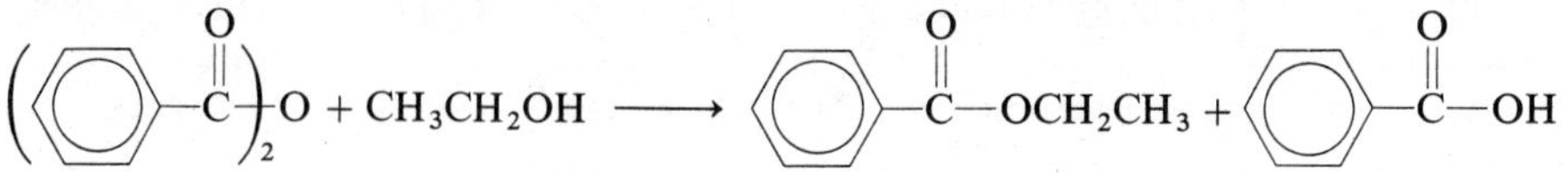

| Benzoic anhydride | Ethyl benzoate | Benzoic acid |

Cyclic anhydrides are affected similarly, except that a compound containing two carboxylic acid derivatives in the same molecule is formed rather than the two molecules that come from open-chain anhydrides. Care must be exercised in the reaction between an alcohol and an anhydride, and usually equimolar amounts of the two compounds are mixed to prevent the formation of some diester. For example, succinic anhydride reacts with ethanol to give a molecule containing both the carbethoxy (ester) and the carboxyl groups. The carboxyl group can be converted into an acid chloride, which can then be transformed into an amide, a ketone, and so on. This reaction scheme is as follows:

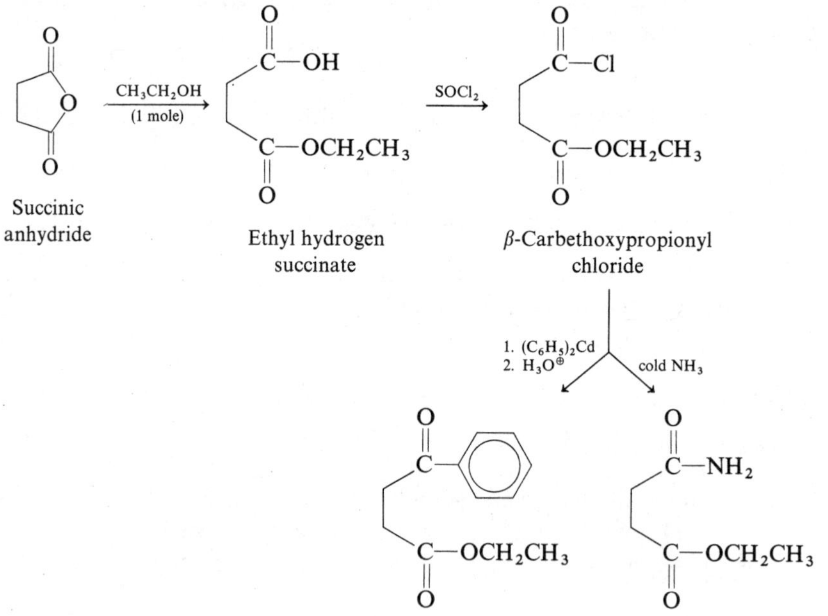

Question 23.23

Propose complete, stepwise mechanisms to show the reactions that occur when acetic anhydride reacts with each of the following reagents:

(*a*) water   (*b*) water in the presence of a trace of acid
(*c*) basic water   (*d*) benzyl alcohol   (*e*) ammonia

What products are obtained from each of these reactions? Draw structures.

Question 23.24

Give the product(s) that would be formed when each of the following compounds reacts with ethanol:

(*a*) propionic anhydride (*b*) glutaric anhydride (*c*) phthalic anhydride
(*d*) 3-nitrophthalic anhydride (*e*) α-methylsuccinic anhydride (1)

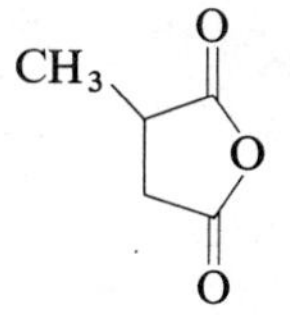

(1)

B. Use of Anhydrides in Friedel-Crafts Reactions; Ketone Formation

Anhydrides can be used in the Friedel-Crafts acylation of aromatic compounds. The general reaction is:

$$(R-\overset{O}{\overset{\|}{C}})_2O + Ar-H \xrightarrow[BF_3]{AlX_3 \text{ or}} Ar-\overset{O}{\overset{\|}{C}}-R + R-\overset{O}{\overset{\|}{C}}-OH$$

Anhydride — Aryl ketone

In principle any anhydride can be used, but in practice only commercially available anhydrides are employed. The following examples illustrate both open-chain and cyclic anhydrides. Reactions involving acetic anhydride (a process called *acetylation* because it involves the introduction of the *acetyl group*, $CH_3-\overset{O}{\overset{\|}{C}}-$) are probably the most common.

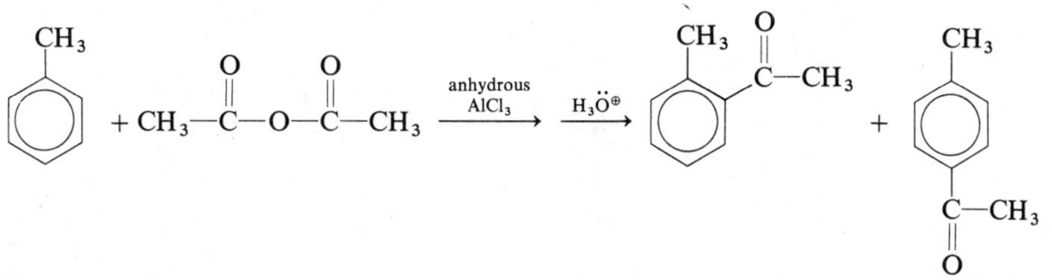

o-Methylacetophenone   *p*-Methylacetophenone

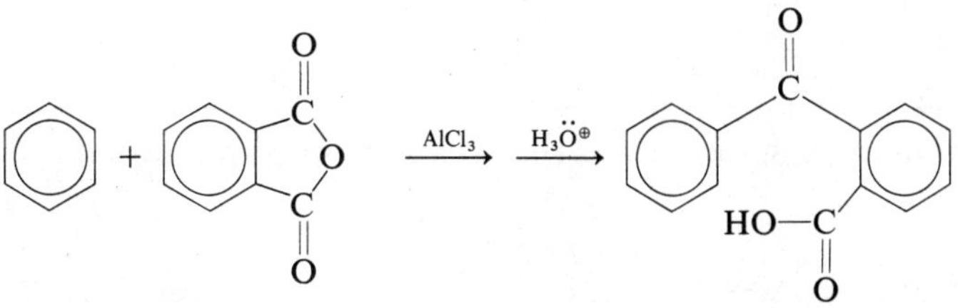

Phthalic anhydride   *o*-Benzoylbenzoic acid

Once formed, compounds such as *o*-benzoylbenzoic acid can be subjected to other reactions. For example, Wolff-Kishner or Clemmensen reduction of the carbonyl group does not affect the carboxyl group:

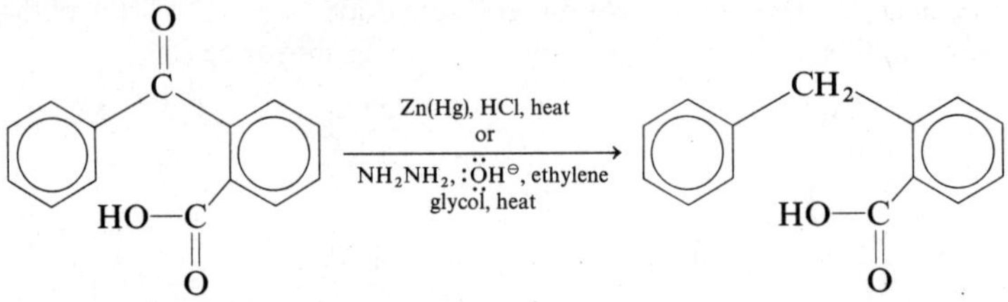

The mechanism for this Friedel-Crafts acylation reaction involves electrophilic aromatic substitution. The Lewis acid catalyst complexes with the acyl group on the anhydride, thus increasing its electrophilic character. The electrophile then attacks the aromatic ring through the usual electrophilic aromatic substitution mechanism. Finally, the reaction must be hydrolyzed to free the aluminum halide complex and produce the free aryl ketone:

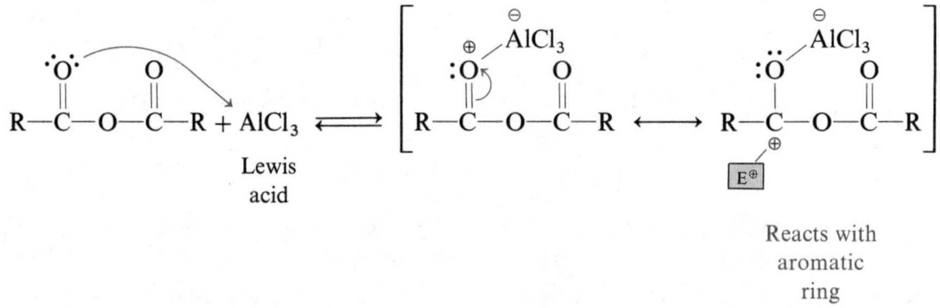

Alternatively, it is possible that the initial step in acylation involves the formation of the acylium ion and the aluminum salt of the carboxylate anion:

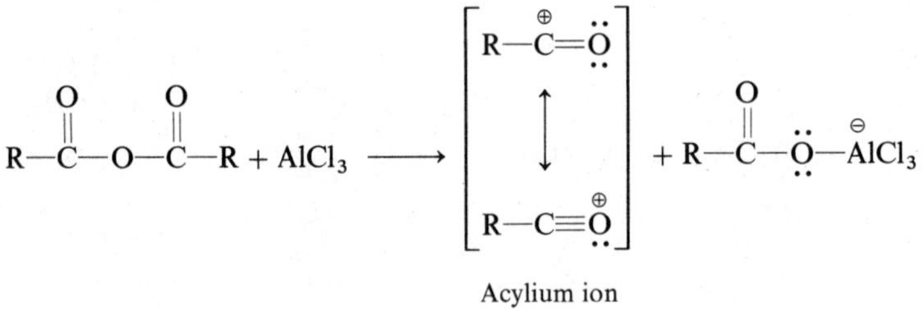

Acylium ion

The acylium ion then attacks the aromatic ring by electrophilic aromatic substitution.

23.14   Use of Anhydrides in the Resolution of Alcohols

Many classes of organic compounds cannot be resolved directly into optically active enantiomers even though an asymmetric center is present in a given molecule (see Sec. 6.7). As we saw, alcohols are highly important in organic chemistry because the hydroxyl group can be converted into a myriad of other functional groups. Optically active alcohols are also important for this reason. Alcohols cannot be resolved directly, and we learn in this section an indirect method for doing so.

We know that an alcohol reacts with a cyclic anhydride to form a half acid-half ester (see Sec. 23.13). If we start with racemic (*d,l*) alcohol, then the half acid-half ester is also racemic (a *d,l* mixture). Using phthalic anhydride as an example, the following reaction takes place:

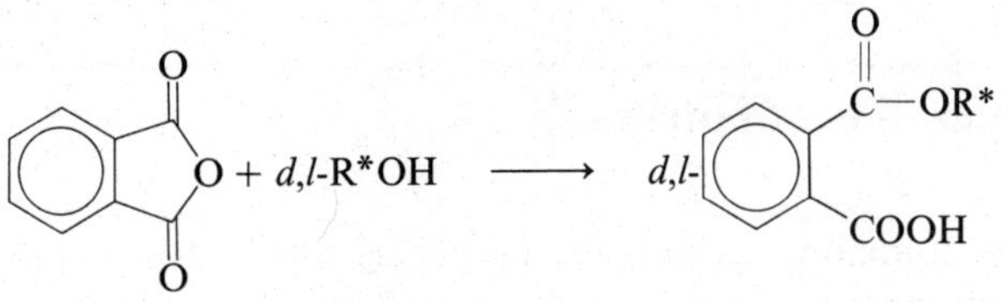

Racemic   Racemic

This reaction introduces a *handle* onto the original alcohol molecule, and the handle is the carboxyl group, —COOH. We discussed the principles of resolution as applied to carboxylic acids in Sec. 22.15, and in the present case, the same principles apply. The racemic half acid-half ester is resolved via an optically active amine to give the optically active enantiomers of half acid-half ester. The half acid-half esters are then hydrolyzed under basic conditions to give the two optically active enantiomers of the alcohol, and the hydrolysis reaction occurs with complete retention of configuration (see Sec. 23.7A).

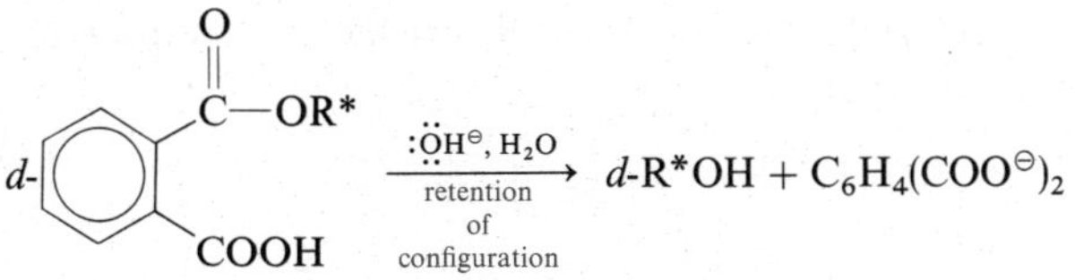

Optically   Optically
active   active

Optically   Optically
active   active

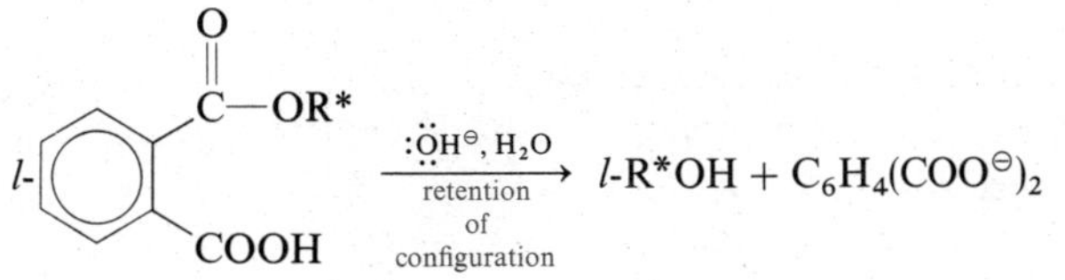

The hydroxyl group may be attached to the asymmetric carbon atom or elsewhere in the molecule.

Question 23.25

Suggest how the following five compounds can be prepared in optically active form, starting with a suitable alcohol.

(*a*) $CH_3CH_2-\overset{\overset{\displaystyle CH_3}{|}}{C}H-CH_2CH_2Br$

(*b*) $CH_3CH_2-\overset{\overset{\displaystyle CH_3}{|}}{C}H-CH_2-\overset{\overset{\displaystyle O}{||}}{C}-H$

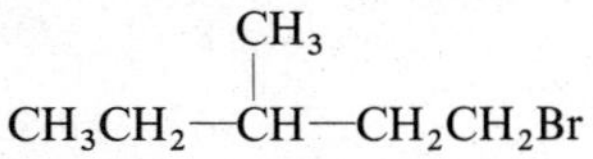

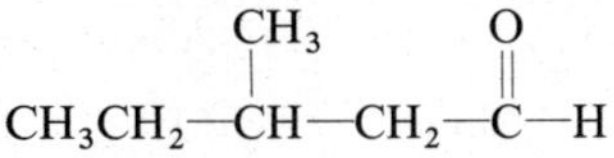

(c) CH_3CH_2—$\overset{\overset{\displaystyle CH_3}{|}}{CH}$—$CH_2CH_2$—$\overset{\overset{\displaystyle CH_3}{|}}{\underset{\underset{\displaystyle CH_3}{|}}{C}}$—OH   (d) CH_3CH_2—$\overset{\overset{\displaystyle CH_3}{|}}{CH}$—$CH_2CH_2$—⟨benzene ring⟩

(e) CH_3CH_2—$\overset{\overset{\displaystyle}{|}}{\underset{\underset{\displaystyle Br}{|}}{CH}}$—$CH_3$

23.15   Preparation of Nitriles

Three common methods for preparing nitriles were discussed in Sec. 22.5B; they are summarized here.

1. From alkyl halides or tosylates

$$R—X + \overset{\ominus}{:}C{\equiv}N: \xrightarrow{S_N2} R—C{\equiv}\ddot{N} + :\overset{\ominus}{\ddot{X}}:$$

where R = methyl or primary

2. From aryl compounds

$$Ar—SO_3^{\ominus}Na^{\oplus} + \overset{\ominus}{:}C{\equiv}N: \xrightarrow{heat} Ar—C{\equiv}\ddot{N} + SO_3^{\ominus 2} + Na^{\oplus}$$

Often low yields

The laboratory synthesis of aryl nitriles usually uses the organothallium reagent (see Sec. 22.5B):

$$Ar—H \xrightarrow[CF_3COOH]{TTFA} Ar—Tl\underset{O—\overset{\displaystyle O}{\underset{\displaystyle \|}{C}}—CF_3}{\overset{O—\overset{\displaystyle O}{\overset{\displaystyle \|}{C}}—CF_3}{}} \xrightarrow{CuCN} Ar—C{\equiv}\ddot{N}$$

3. From carboxylic acids and amides

Any nitrile can be prepared from the corresponding carboxylic acid, as discussed in Sec. 23.9.

23.16   Reactions of Nitriles

Although there are several other reactions that nitriles undergo, an important one is their hydrolysis in either acidic or basic media. The hydrolysis reaction appears to occur in two steps; the nitrile is first converted to an amide, which is then hydrolyzed to the corresponding carboxylic acid:

$$(Ar)R—C{\equiv}\ddot{N} \xrightarrow[H_2O,\ heat]{H^{\oplus}\ or\ :\ddot{O}H^{\ominus}} \left[(Ar)R—\overset{\overset{\displaystyle O}{\|}}{C}—\ddot{N}H_2 \right] \xrightarrow[H_2O]{H^{\oplus}\ or\ :\ddot{O}H^{\ominus}} (Ar)R—\overset{\overset{\displaystyle O}{\|}}{C}—OH$$

Nitrile   Amide   (or its salt)
                                    Seldom isolated   Carboxylic acid

In general, basic conditions are chosen to convert a nitrile to an acid because the soluble salt of the carboxylic acid is formed and evidence of complete reaction can be seen visually when the original heterogeneous reaction mixture becomes homogeneous. Under acidic aqueous conditions, the reaction mixture remains heterogeneous unless a polar organic solvent is used.

In basic solution, the nitrile is converted to the corresponding amide by the following sequence of steps:

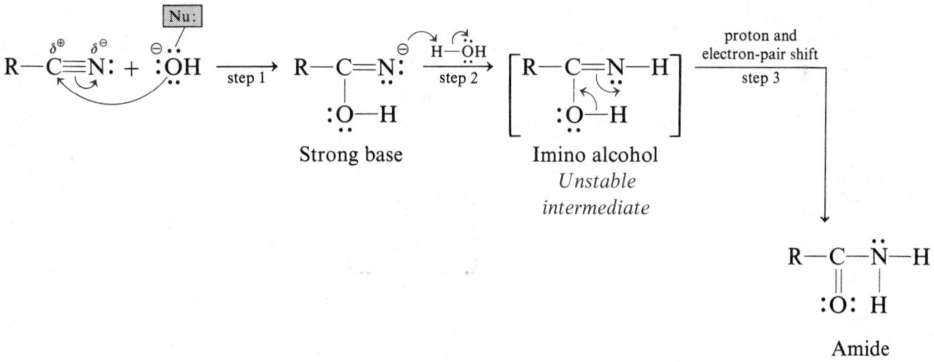

omit

Strong base

Imino alcohol
Unstable intermediate

Amide

Question 23.26

In Sec. 20.8B we learned that nitriles react with Grignard reagents to give ketones. Write the detailed mechanism of this reaction using acetonitrile and methylmagnesium bromide as the reactants.

Once the amide is formed, it is usually converted into the corresponding carboxylic acid salt by a mechanism discussed in Sec. 23.9A.

The acid hydrolysis of a nitrile first produces an amide, which is further hydrolyzed into a carboxylic acid.

Question 23.27

What is the mechanism for this reaction? Write it.

omit

23.17 Summary of Interconversions of Carboxylic Acid Derivatives

The possible interconversions among various derivatives of carboxylic acids are indicated in Fig. 23.1. Arrows are used to show the possible routes in this figure, but the reagents are omitted for clarity. Outlines of this type may be useful as a study guide, and the concerned student should construct a more comprehensive chart including other functional groups and indicating the reagents. Note the linkage of carboxylic acids and their derivatives with primary alcohols. We saw in Fig. 21.2 the importance of alcohols in leading to so many other functional groups.

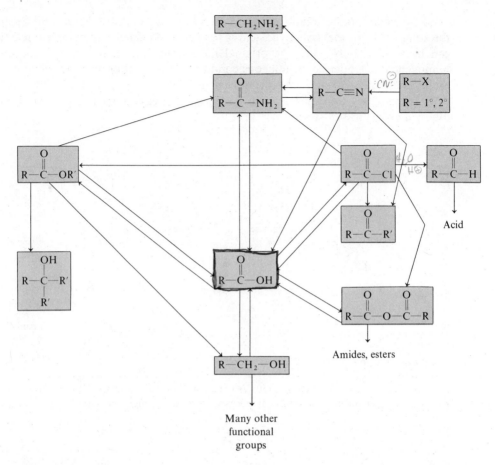

FIGURE 23.1   Chart showing the direction of possible interconversions of carboxylic acids and their derivatives.

23.18   Reactivity in Nucleophilic Acyl Substitution Reactions

In our discussions of the reactions of carboxylic acids and their derivatives, we focused our attention on mechanisms and on the interconversions for these functional groups. Yet we did not mention reactivity. In this section we discuss two aspects of reactivity: (1) relative reactivities of carboxylic acid derivatives, and (2) reactivities of carboxylic acid derivatives compared with compounds that do not contain the acyl group.

A.   Relative Reactivities of Carboxylic Acid Derivatives

The relative reactivities of the carboxylic acid derivatives are indicated in Fig. 23.2, which lists the starting acyl compounds as well as the leaving group that forms after substitution occurs. This sequence can be rationalized in three ways.

First, there is some correlation between the electronegativity of the leaving group and reactivity. For example, the more electronegative the group attached to the acyl carbon, the better it withdraws electrons from the acyl group; this increases the affinity that the acyl carbon atom has for a nucleophile. We would certainly

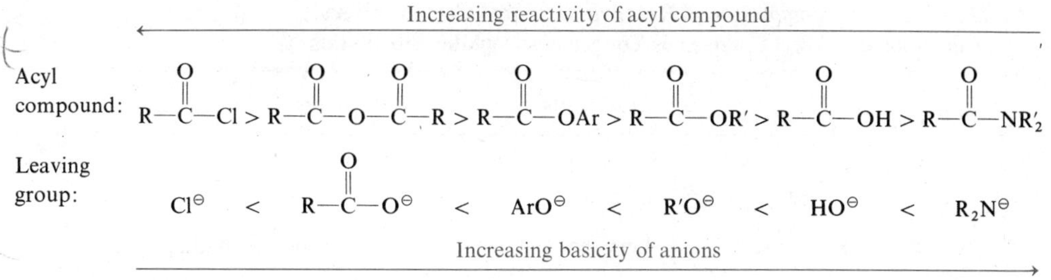

Smit

Increasing reactivity of acyl compound

Acyl compound:

$$R-\overset{\overset{\displaystyle O}{\|}}{C}-Cl > R-\overset{\overset{\displaystyle O}{\|}}{C}-O-\overset{\overset{\displaystyle O}{\|}}{C}-R > R-\overset{\overset{\displaystyle O}{\|}}{C}-OAr > R-\overset{\overset{\displaystyle O}{\|}}{C}-OR' > R-\overset{\overset{\displaystyle O}{\|}}{C}-OH > R-\overset{\overset{\displaystyle O}{\|}}{C}-NR'_2$$

Leaving group:

$$Cl^{\ominus} \quad < \quad R-\overset{\overset{\displaystyle O}{\|}}{C}-O^{\ominus} \quad < \quad ArO^{\ominus} \quad < \quad R'O^{\ominus} \quad < \quad HO^{\ominus} \quad < \quad R_2N^{\ominus}$$

Increasing basicity of anions

FIGURE 23.2   Relative reactivities of five acyl compounds toward a given nucleophile; reaction assumed to occur in basic solution.

anticipate that the more electronegative groups also leave easier because of their affinity for electrons.

Second, there is a fairly good correlation between the basicity of the leaving group and the compound's reactivity. The more basic leaving groups (ions) cause nucleophilic acyl substitution to occur less readily. Basicity is tied closely with electronegativity, which is the underlying concept in reactivity.

Finally, there is a rough correlation between reactivity and the carbon-oxygen stretching frequency of the acyl group in the IR spectra of these compounds. These frequencies are listed in Sec. 23.19A.

In Fig. 23.2, a more reactive acyl group can be converted into a less reactive functional group by use of suitable reagents.

B. Comparison of Reactivity of Nucleophilic Displacement Reactions of Acyl and Alkyl Compounds

In comparison with alkyl analogues, acyl compounds are much more reactive, and indeed some alkyl analogues are completely unreactive toward nucleophilic substitution. In the discussion that follows, we see that the factors governing reactivity are electronic and steric effects, which are more favorable in acyl compounds than in alkyl compounds. Furthermore, the arguments based on these two factors reinforce one another.

Before discussing the relative reactivities, let us examine the facts about alkyl and acyl compounds. This comparision is summarized in Table 23.3.

The steric effect must be viewed in terms of the structures of both the starting compounds and the intermediates and transition states that result from nucleophilic attack. In acyl compounds, the starting compound is always trigonal (flat), so that nucleophilic attack can occur on either side of the acyl group. In the tetrahedral alkyl compound, attack must occur from the backside in the S_N2 reaction or the leaving group must depart in the S_N1 reaction before nucleophilic attack can occur. Nucleophilic attack on the tetrahedral alkyl structure is sterically more difficult than is attack on the relatively unhindered acyl group.

The steric effect is also important in comparing the relative stabilities of the species that intervene between starting compound and final product. In acyl compounds, the intermediate is tetrahedral. On the other hand, a considerably less stable, crowded, pentacoordinate transition state intervenes in S_N2 nucleophilic substitution on alkyl compounds. This transition state is quite hindered, and carbon atoms do not generally hybridize to allow for discrete molecules possessing the pentacoordinate geometry.

TABLE 23.3   Comparison of Reactivities of Acyl Compounds and Corresponding Alkyl Compounds Toward Nucleophilic Substitution

Compounds	Comments on Relative Reactivities
Acid halide, $R-\overset{\overset{\textstyle O}{\|\|}}{C}-Cl$	*Very* reactive toward all nucleophiles, including water
Alkyl halide, $R-CH_2-Cl$	Reactive toward negatively charged nucleophiles; quite unreactive toward water
Ester, $R-\overset{\overset{\textstyle O}{\|\|}}{C}-OR'$	Reactive toward many nucleophiles in acidic or basic solution, and most reactive toward negatively charged species
Ether, $R-CH_2-OR'$	Unreactive toward most nucleophiles; strongly acidic conditions required for reaction to occur
Amide, $R-\overset{\overset{\textstyle O}{\|\|}}{C}-NH_2$	Reactive toward nucleophiles in acidic or basic solution
Amine, $R-CH_2-NH_2$	Unreactive toward all nucleophiles

The electronic effect is also important. In acyl compounds, the oxygen on the acyl group increases the electropositive character of the acyl carbon atom where nucleophilic attack occurs. In alkyl analogues, the polarity of the bond between carbon and the leaving group is not nearly as great, so nucleophilic attack is not as likely to occur on this less electropositive carbon atom.

The following equations show the electronic and steric effects in substitutions involving acyl and nonacyl compounds.

Acyl compounds:

Trigonal, sp^2 carbon
Nucleophilic attack easy

Tetrahedral, sp^3 carbon
*Formation favored,
though an unstable intermediate*

Alkyl compounds:

Tetrahedral, sp^3 carbon
Nucleophilic attack more difficult

Pentacoordinate transition state
Very unfavored

The steric and electronic effects combine to predict what is actually observed: acyl compounds are more reactive than the corresponding alkyl compounds. In summary, the electronic effect of the polar acyl group makes the acyl carbon more susceptible to nucleophilic attack. The steric effect plays a twofold role. First, nucleophilic attack on the planar acyl group is easier than is attack on a tetrahedral carbon atom, and second, the tetrahedral intermediate in acyl substitution is more stable than the pentacovalent transition state in nucleophilic alkyl substitution.

23.19 Spectral Analysis of Acyl Compounds and Nitriles

A. Infrared Spectroscopy

The carboxylic acid derivatives all exhibit a strong carbonyl stretching absorption peak in the 1,700-cm^{-1} (5.88 μ) region. Infrared spectroscopy is particularly useful in identifying compounds that contain the acyl group because the C=O stretching frequency varies with functional group. The following summarizes the usual trends:

$$\text{Acid chlorides: } R-\overset{\overset{\textstyle O}{\|}}{C}-Cl, \text{ 1,770–1,815 cm}^{-1} \text{ (5.65–5.51 } \mu)$$

$$\text{Anhydrides: } R-\overset{\overset{\textstyle O}{\|}}{C}-O-\overset{\overset{\textstyle O}{\|}}{C}-R, \text{ 1,740–1,790 cm}^{-1} \text{ (5.75–5.59 } \mu) \text{ and 1,800–}$$
1,850 cm^{-1} (5.56–5.41 μ) [two bands]

$$\text{Esters: } R-\overset{\overset{\textstyle O}{\|}}{C}-OR, \text{ 1,720–1,750 cm}^{-1} \text{ (5.81–5.71 } \mu)$$

$$\text{Amides: } R-\overset{\overset{\textstyle O}{\|}}{C}-NH_2, \text{ 1,630–1,690 cm}^{-1} \text{ (6.13–5.92 } \mu)$$

The C=O absorption is shifted in **α,β-unsaturated acyl compounds** and in compounds where an aromatic ring is attached to the acyl carbon. Resonance is presumably responsible for these shifts to lower frequencies; for example:

$$\text{Ar}-\overset{\overset{\textstyle O}{\|}}{C}-OR \quad \text{or} \quad \overset{}{\diagup}C=\overset{\overset{\textstyle O}{\|}}{\underset{|}{C}}-\overset{\overset{\textstyle O}{\|}}{C}-OR, \text{ 1,715–1,730 cm}^{-1} \text{ (5.83–5.78 } \mu)$$

(Compare with 1,720 to 1,750 cm^{-1} for saturated esters.)

Further distinctions in the IR spectrum are often possible for certain acyl compounds. For example, **carboxylic acids** exhibit a strong and broad O—H stretching frequency in the region 2,500 to 3,000 cm^{-1} (4 to 3.33 μ), which is lacking in esters. Furthermore, carboxylic acids show a C—O stretching frequency at $\sim$1,250 cm^{-1} (8 μ), whereas the C—O stretching in esters normally occurs as two relatively strong bands in the region 1,050 to 1,300 cm^{-1} (9.52 to 7.69 μ). Aldehydes and ketones show no absorption in this broad region, so they are readily distinguishable from carboxylic acids and esters.

Amides exhibit N—H stretching in the 3,050 to 3,540-cm^{-1} (3.28 to 2.82 μ) region (this may vary due to hydrogen bonding) and show another moderately strong absorption in the 1,600 to 1,645-cm^{-1} (6.25 to 6.08 μ) region caused by N—H bonding.

Nitriles show a characteristic absorption due to C≡N stretching in the region 2,210 to 2,280 cm^{-1} (4.8 to 4.4 μ), but this absorption is often variable and weak. Its presence, however, is quite indicative of the carbon-nitrogen triple bond.

The IR spectra of five typical derivatives of carboxylic acids are shown in Fig. 23.3.

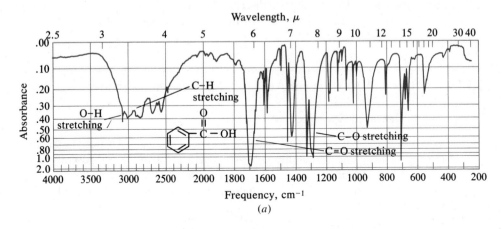

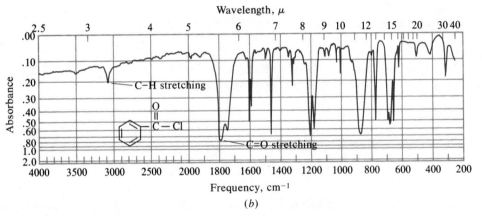

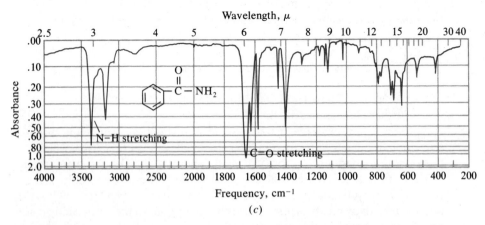

FIGURE 23.3 IR spectra of (*a*) benzoic acid, (*b*) benzoyl chloride, (*c*) benzamide, (*d*) methyl benzoate, and (*e*) benzoic anhydride. [(*a*), (*d*), and (*e*) data from *Indexes to Evaluated Infrared Reference Spectra*, copyright 1975, The Coblentz Society, Norwalk CT; (*b*) and (*c*) data from *The Sadtler Handbook of Infrared Spectra*, W. W. Simmons (ed.), © Sadtler Research Laboratories, Division of Bio-Rad Laboratories, Inc. (1978)]

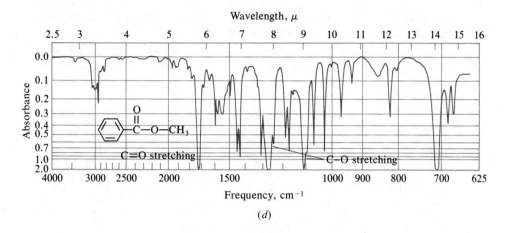

(d)

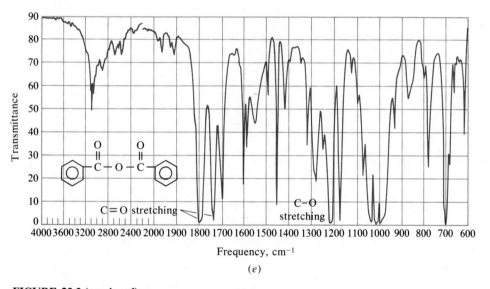

(e)

FIGURE 23.3 (*continued*)

B. Ultraviolet Spectroscopy

Contrasting acyl compounds with aldehydes and ketones, one finds that the acyl band associated with the $n \rightarrow \pi^*$ transition is shifted to much shorter wavelength. Whereas aldehydes and ketones absorb at 270 to 300 nm, the acyl group absorbs in the range of 210 to 250 nm. Like aldehydes and ketones, the extinction coefficient associated with these transitions is quite small, on the order of 40 to 60 (10 to 20 for aldehydes and ketones).

Both these general classes of chromophores, carbonyl and acyl groups, have generated intense research in electronic spectroscopy. This is in large part because of their presence in molecules of biochemical interest, for example, carbohydrates (see Chap. 29) and amino acids (see Chap. 28).

If the acyl group is part of a conjugated system, the same effect observed for carbonyls is seen here. The longer the conjugated system, the greater the $n \rightarrow \pi^*$ transition shift to longer wavelength. Refer to Sec. 23.19 and contrast the λ of absorption of the $n \rightarrow \pi^*$ band in the following examples:

$$CH_3—CH_2—\overset{\overset{\displaystyle O}{\|}}{C}—H \quad 292\ nm \quad versus \quad CH_2{=}CH—\overset{\overset{\displaystyle O}{\|}}{C}—H \quad 315\ nm$$

$$CH_2{=}CH—\overset{\overset{\displaystyle O}{\|}}{C}—OH \quad 200\ nm \quad versus \quad CH_3—CH{=}CH—CH{=}CH—\overset{\overset{\displaystyle O}{\|}}{C}—OH \quad 254\ nm$$

C. Nuclear Magnetic Resonance Spectroscopy

The hydrogens attached to nitrogen in **amides** normally appear as a broad peak (usually a "hump") in the 5 to 8 δ (2 to 5 τ) region.

Esters typically show two absorptions. The hydrogens α to the acyl group occur at 2 to 2.6 δ (7.4 to 8 τ), and those α to the alkyl oxygen occur at 3.7 to 4.1 δ (5.9 to 6.3 τ).

Acid chlorides and anhydrides also have hydrogens α to the acyl group, and if present they absorb in the 1 to 4 δ (6 to 9 τ) region.

Fig. 23.4 shows the NMR spectra of typical carboxylic acid derivatives.

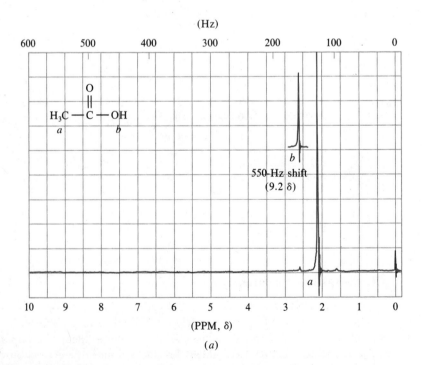

(a)

FIGURE 23.4 NMR spectra of (a) acetic acid, (b) acetyl chloride, (c) acetamide, (d) ethyl acetate, and (e) acetic anhydride. [Data from *The Aldrich Library of NMR Spectra*, C. J. Pouchert and J. R. Campbell (eds.), copyright 1974, Aldrich Chemical Co., Milwaukee]

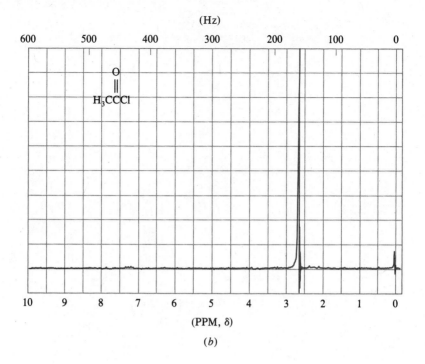

(b)

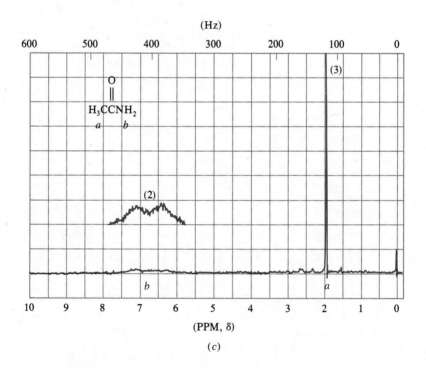

(c)

FIGURE 23.4 (*continued*)

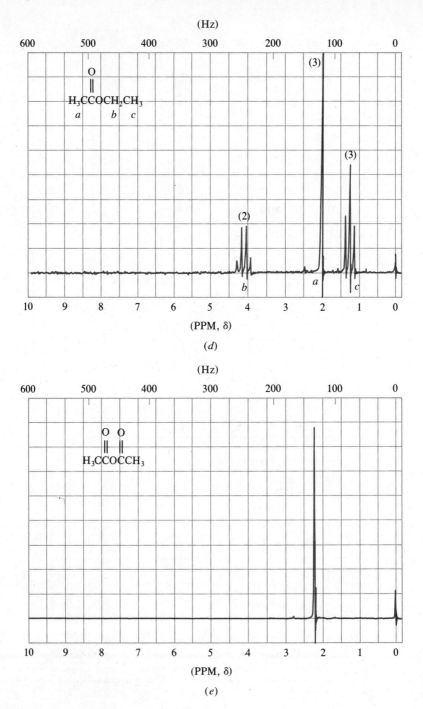

FIGURE 23.4 (*continued*)

D. Mass Spectrometry

This discussion is limited to esters. The reader is referred to the Reading References for Chap. 18 for additional information.

In contrast to acids, esters provide an abundant molecular ion peak. The two primary modes of ester fragmentation are as follows:

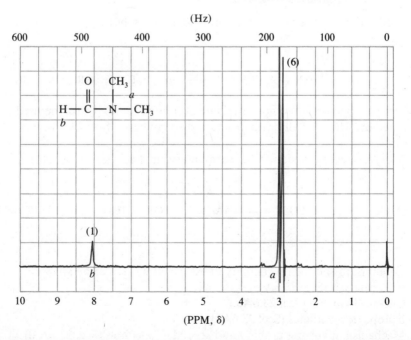

E. Spectroscopy and Structure Elucidation

The NMR spectra of *N,N*-dimethylformamide at room temperature is shown in Fig. 23.5.

Note that at room temperature the two methyl groups exhibit magnetic non-equivalence and appear as two singlets. At 130°, however, they appear as one singlet (magnetically equivalent). How can we account for this observation? The following shows what actually occurs:

The H_A and H_B protons are in magnetically (and chemically) different environments in the preceding structures. Interconversion of the two by rotation about the

FIGURE 23.5 NMR spectrum of *N,N*-dimethylformamide at room temperature. [Data from *The Aldrich Library of NMR Spectra*, C. J. Pouchert and J. R. Campbell (eds.), copyright 1974, Aldrich Chemical Co., Milwaukee]

C—N bond is slow at room temperature but fast at 130°. At room temperature the NMR "sees" two different types of methyl groups and we get two singlets. At 130° the H_A and H_B protons are interconverting environments very rapidly. The NMR sees one type of proton that is with an environment like the H_A's for part of the time and the H_B's for the remainder of the time. A singlet results for the two methyl groups that are now averaged. Thermodynamic measurements indicate that the energy barrier to rotation about the C—N amide bond is 18 kcal/mole (75.3 kJ/mole).

Question 23.28

Explain the experimental observations outlined in terms of the bonding of the amide functional group. (*Hint:* see Sec. 28.10A).

23.20   Summary of Methods of Preparation of Carboxylic Acid Derivatives

A. Preparation of Acid Chlorides

$$(Ar)\,R-\overset{\overset{\displaystyle O}{\|}}{C}-OH + \begin{matrix} PCl_3 \\ or\ PCl_5 \\ or\ SOCl_2 \end{matrix} \longrightarrow (Ar)\,R-\overset{\overset{\displaystyle O}{\|}}{C}-Cl \qquad Sec.\ 23.4$$

B. Preparation of Esters

1. Reaction between an acid chloride and an alcohol or phenol

$$(Ar)\,R-\overset{\overset{\displaystyle O}{\|}}{C}-Cl + (Ar)\,R'-OH \xrightarrow[\text{conditions}]{\text{neutral} \atop \text{or basic}} (Ar)\,R-\overset{\overset{\displaystyle O}{\|}}{C}-OR'\,(Ar) \qquad Sec.\ 23.5B$$

Schotten-Baumann method: pyridine or dilute sodium hydroxide used as base to remove HCl; usually with aromatic acid chlorides (Sec. 23.5B)

2. Acid-catalyzed reaction between an acid and an alcohol (Fischer esterification)

$$(Ar)\,R-\overset{\overset{\displaystyle O}{\|}}{C}-OH + R'OH \underset{}{\overset{H^{\oplus},\ heat}{\rightleftharpoons}} (Ar)\,R-\overset{\overset{\displaystyle O}{\|}}{C}-OR' + H_2O \qquad Secs.\ 23.1C\ and\ 23.6A$$

alcohols only,
no phenols

Equilibrium and the reversibility of esterification (Sec 23.6A,a)
Acid catalysis (Sec. 23.6A,b)
Chemical kinetics (Sec. 23.6A,c)
Isotope tracer studies (Sec. 23.6A,d)
Mechanism involving primary and secondary alcohols (Sec. 23.6A,d)
Mechanism involving tertiary (and a few secondary) alcohols (Sec. 23.6A,d)
Steric effects (Sec. 23.6A,e)

3. Reaction of carboxylate anion with an alkyl halide

$$(Ar)R-\overset{\overset{O}{\|}}{C}-\overset{..\ominus}{\underset{..}{O}:} + -\overset{|}{\underset{|}{C}}-X \xrightarrow{S_N2 \text{ reaction}} (Ar)R-\overset{\overset{O}{\|}}{C}-O-\overset{|}{\underset{|}{C}}- \qquad \text{Sec. 23.6B}$$

Halide
*Must be
methyl,
1° or 2°*

4. Reaction of an alcohol with an anhydride

$$\begin{matrix} (Ar)R-C\overset{\nwarrow O}{\diagdown} \\ O \\ (Ar)R-C\overset{\diagup O}{\diagdown} \end{matrix} + (Ar)R'OH \longrightarrow (Ar)R-\overset{\overset{O}{\|}}{C}-OH + (Ar)R-\overset{\overset{O}{\|}}{C}-OR'(Ar)$$

Secs. 23.6C, 23.13A, and 23.14

Cyclic anhydrides yield half acid-half ester products.

5. Preparation of methyl esters using diazomethane

$$(Ar)R-\overset{\overset{O}{\|}}{C}-OH + CH_2N_2 \xrightarrow{\text{ether}} (Ar)R-\overset{\overset{O}{\|}}{C}-O-CH_3 \qquad \text{Sec. 23.6D}$$

Diazomethane

C. Preparation of Amides

1. From acid chlorides and amines

$$(Ar)R-\overset{\overset{O}{\|}}{C}-Cl + R'R''NH \longrightarrow (Ar)R-\overset{\overset{O}{\|}}{C}-N\overset{\nearrow R'}{\diagdown R''} \qquad \text{Secs. 23.5A and 23.8A}$$

2 moles
*R' and R'' may be
alkyl, aryl, or hydrogen*

2. From acids and amines

$$(Ar)R-\overset{\overset{O}{\|}}{C}-OH \xrightarrow[\text{excess}]{NH_3} (Ar)R-\overset{\overset{O}{\|}}{C}-\overset{..\ominus}{\underset{..}{O}:}\overset{\oplus}{N}H_4 \xrightarrow{\text{heat}} (Ar)R-\overset{\overset{O}{\|}}{C}-NH_2 \qquad \text{Sec. 23.8B}$$

3. From esters and amines

$$(Ar)R-\overset{\overset{O}{\|}}{C}-OR' \xrightarrow{NH_3} (Ar)R-\overset{\overset{O}{\|}}{C}-NH_2 \qquad \text{Secs. 23.7C and 23.8C}$$

4. From anhydrides and amines

$$\begin{matrix} (Ar)R-C\overset{\nwarrow O}{\diagdown} \\ O \\ (Ar)R-C\overset{\diagup O}{\diagdown} \end{matrix} \longrightarrow (Ar)R-\overset{\overset{O}{\|}}{C}-NH_2 + (Ar)R-\overset{\overset{O}{\|}}{C}-\overset{..\ominus}{\underset{..}{O}:}\overset{\oplus}{N}H_4 \qquad \text{Secs. 23.8D and 23.13A}$$

Cyclic anhydrides yield half amide-half ammonium carboxylate salt products.

D. Preparation of Anhydrides

1. From monocarboxylic acids: acetic anhydride

$$
\begin{array}{c}
\underset{\substack{O \\ \parallel}}{CH_3-C-OH} \\
\\
\underset{\substack{O \\ \parallel}}{CH_3-C-CH_3}
\end{array}
\xrightarrow[\substack{700° \\ \text{pyrolysis}}]{\substack{700° \\ AlPO_4}}
CH_2{=}C{=}O
\xrightarrow{CH_3C-OH}
\begin{array}{c}
CH_3-C{\diagup}^{O} \\
\quad\diagdown_{O} \\
CH_3-C{\diagdown}_{O}
\end{array}
\qquad \text{Sec. 23.11}
$$

Ketene   Ethanoic
                  anhydride
              (acetic anhydride)

2. From monocarboxylic acids: other symmetrical anhydrides

$$
\underset{\text{2 moles}}{(Ar)\,R-\underset{\substack{\parallel \\ O}}{C}-OH}
\underset{\substack{\text{anhydride} \\ \text{(large excess)}}}{\overset{\text{acetic}}{\rightleftharpoons}}
\begin{array}{c}
(Ar)\,R-C{\diagup}^{O} \\
\qquad\diagdown_{O} \\
(Ar)\,R-C{\diagdown}_{O}
\end{array}
\qquad \text{Sec. 23.1D}
$$

3. From acid chlorides and carboxylate anions: unsymmetrical anhydrides

$$
(Ar)\,R-\underset{\substack{\parallel \\ O}}{C}-\ddot{\underset{\cdot\cdot}{O}}{:}\,M^{\oplus} + (Ar)\,R'-\underset{\substack{\parallel \\ O}}{C}-Cl
\xrightarrow[\text{solvent}]{\text{polar}}
\begin{array}{c}
R-C{\diagup}^{O} \\
\quad\diagdown_{O} \\
R'-C{\diagdown}_{O}
\end{array}
\qquad \text{Secs. 23.5D and 23.11}
$$

where R ≠ R′: mixed anhydride
    R = R′: symmetrical anhydride

4. From dicarboxylic acids

$$
\begin{array}{c}
COOH \\
| \\
(CH_2)_n \\
| \\
COOH
\end{array}
\xrightarrow[140-160°]{\text{heat}}
\begin{array}{c}
\quad C{\diagup}^{O} \\
(CH_2)_n \quad O \\
\quad C{\diagdown}_{O}
\end{array}
\qquad \text{Secs. 23.1D and 23.12}
$$

$n \geq 2$

If $n < 2$, decarboxylation occurs

E. Preparation of Nitriles

1. From alkyl halides and Tosylates

$$
R-X \; + \; {:}\overset{\ominus}{C}{\equiv}N{:} \xrightarrow{S_N2} R-C{\equiv}\ddot{N} \qquad \text{Sec. 23.15}
$$

Methyl, 1° or 2°
alkyl halide or tosylate only

2. From arylsulfonic acids

$$
Ar-SO_3^{\ominus}Na^{\oplus} + {:}\overset{\ominus}{C}{\equiv}N{:} \xrightarrow{\text{heat}} Ar-C{\equiv}\ddot{N} \qquad \text{Secs. 22.5B and 23.15}
$$

3. From carboxylic acids

$$
(Ar)\,R-\underset{\substack{\parallel \\ O}}{C}-OH \longrightarrow (Ar)\,R-\underset{\substack{\parallel \\ O}}{C}-NH_2 \xrightarrow[\substack{SOCl_2 \text{ or} \\ PCl_5 \text{ or } POCl_3 \\ \text{heat} \\ \text{dehydration}}]{P_2O_5 \text{ or}} (Ar)\,R-C{\equiv}\ddot{N} \qquad \begin{array}{l}\text{Secs. 22.5B and} \\ \text{23.15}\end{array}
$$

4. From arylthallium reagents

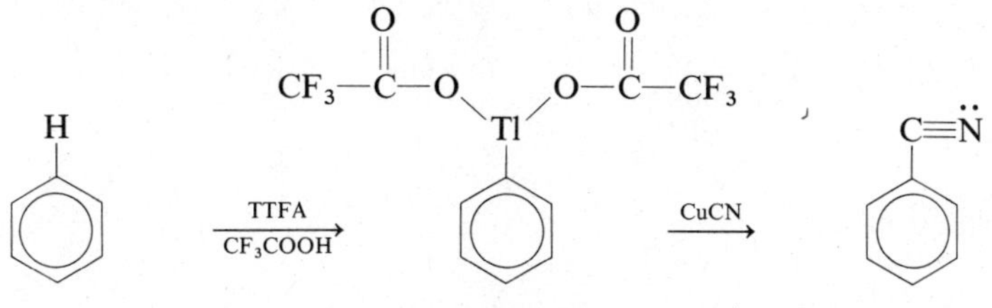

Arylthallium
reagent

Secs. 22.5B and 23.15

23.21   Summary of Reactions of Carboxylic Acid Derivatives

A. Reactions of Acid Chlorides

1. Reaction with organocadmium reagents

$$\underset{}{R-\overset{\overset{\textstyle O}{\|}}{C}-Cl} + R'_2Cd \longrightarrow \underset{\text{Ketone}}{R-\overset{\overset{\textstyle O}{\|}}{C}-R'} \quad \text{Sec. 20.8}$$

2. Friedel-Crafts acylation

$$(Ar)\,R-\overset{\overset{\textstyle O}{\|}}{C}-Cl + Ar-H \xrightarrow{AlCl_3} \underset{\text{Ketone}}{(Ar)\,R-\overset{\overset{\textstyle O}{\|}}{C}-Ar} \quad \text{Sec. 20.8}$$

3. Reduction

$$(Ar)\,R-\overset{\overset{\textstyle O}{\|}}{C}-Cl \xrightarrow[\substack{BaSO_4 \\ \text{or} \\ LiAlH[OC(CH_3)_3]_3}]{H_2,\,Pd} \underset{\text{Aldehyde}}{(Ar)\,R-\overset{\overset{\textstyle O}{\|}}{C}-H} \quad \text{Sec. 20.6}$$

4. Amide formation

$$(Ar)\,R-\overset{\overset{\textstyle O}{\|}}{C}-Cl + R'R''NH \longrightarrow (Ar)\,R-\overset{\overset{\textstyle O}{\|}}{C}-N\overset{R'}{\underset{R''}{\diagdown}} \quad \text{Secs. 23.5A and 23.8A}$$

2 moles
*R' and R'' may
be alkyl, aryl,
or hydrogen*

5. Ester formation

$$(Ar)\,R-\overset{\overset{\textstyle O}{\|}}{C}-Cl + (Ar)\,R'-OH \xrightarrow[\substack{\text{or basic} \\ \text{condition}}]{\text{neutral}} (Ar)\,R-\overset{\overset{\textstyle O}{\|}}{C}-OR'\,(Ar) \quad \text{Sec. 23.5B}$$

Schotten-Baumann method: pyridine, triethylamine, or dilute sodium hydroxide
used as base to remove HCl; usually with aromatic acid chlorides (Sec. 23.5B)

6. Hydrolysis

$$(Ar)\,R-\overset{\displaystyle O}{\overset{\|}{C}}-Cl \xrightarrow[\substack{\text{then} \\ \text{neutralize}}]{\substack{H_2O \text{ or} \\ H_2O/:\ddot{O}H^{\ominus}}} (Ar)\,R-\overset{\displaystyle O}{\overset{\|}{C}}-OH \qquad \text{Sec. 23.5C}$$

7. Anhydride formation

$$(Ar)\,R-\overset{\displaystyle O}{\overset{\|}{C}}-Cl \;+\; (Ar)\,R'-\overset{\displaystyle O}{\overset{\|}{C}}-\overset{..}{\underset{..}{O}}\!\!:^{\ominus}\;M^{\oplus} \xrightarrow[\text{solvent}]{\text{polar}} \begin{array}{c}(Ar)\,R-C\diagup\!\!\!\diagdown\!\!^O \\ \diagdown O \\ (Ar)\,R'-C\diagup\!\!\!\diagdown\!\!_O\end{array} \qquad \text{Secs. 23.5D and 23.11}$$

where R ≠ R′: mixed anhydride

R = R′: symmetrical anhydride

B. Reactions of Esters

1. Hydrolysis

$$(Ar)\,R-\overset{\displaystyle O}{\overset{\|}{C}}-OR' \xrightarrow[\substack{H\ddot{O}\!:^{\ominus}, H_2O, \\ \text{then} \\ \text{neutralization}}]{H_3\ddot{O}^{\oplus} \text{ or}} (Ar)\,R-\overset{\displaystyle O}{\overset{\|}{C}}-OH \qquad \text{Sec. 23.7A}$$

General discussion of mechanism (Sec. 23.7A)
Acid-catalyzed hydrolysis (Sec. 23.7A,a)
Base-promoted hydrolysis (Sec. 23.7A,b)
Evidence for the tetrahedral intermediate (Sec. 23.7A,c)
S_N2 mechanism (Sec. 5.16)
Glycerides (naturally occurring esters), fats, oils, soaps, detergents, and waxes
  (Chap. 25)

2. Transesterification

$$(Ar)\,R-\overset{\displaystyle O}{\overset{\|}{C}}-OR' + R''OH \xrightarrow[R''\ddot{O}\!:^{\ominus}]{H^{\oplus} \text{ or}} (Ar)\,R-\overset{\displaystyle O}{\overset{\|}{C}}-OR'' + R'OH \qquad \text{Sec. 23.7B}$$
$$\text{Excess}$$

Acid-catalyzed reaction (Sec. 23.7B,a)
Base-catalyzed reaction (Sec. 23.7B,b)
Synthetic applications: polymers (Dacron)(Sec. 23.7B,d)
Lactone formation (Sec. 23.7B,c)

3. Ammonolysis

$$(Ar)\,R-\overset{\displaystyle O}{\overset{\|}{C}}-OR' + NH_3 \longrightarrow (Ar)\,R-\overset{\displaystyle O}{\overset{\|}{C}}-NH_2 \qquad \text{Sec. 23.7C}$$

4. Reaction with Grignard reagents

$$(Ar)\,R-\overset{\displaystyle O}{\overset{\|}{C}}-OR' \xrightarrow[\substack{\text{anhydrous ether} \\ 2.\; H_3\overset{\oplus}{O}}]{1.\; 2 \text{ moles } R''MgX} (Ar)\,R-\overset{\displaystyle OH}{\underset{\displaystyle R''}{\overset{|}{\underset{|}{C}}}}-R'' \qquad \text{Sec. 23.7D}$$

Alcohol
Contains two identical R″ groups

5. Reduction

(*a*) Catalytic hydrogenation

$$(Ar)R-\overset{\overset{\displaystyle O}{\|}}{C}-OR' + H_2 \xrightarrow[\substack{200-250° \\ 2,000-4,000 \text{ lb/sq in.}}]{Cu_2O,\ Cr_2O_3} (Ar)R-CH_2-OH + R'OH \qquad \text{Sec. 23.7E,a}$$

2 moles

Reduces carbon-carbon double and triple bonds also

(*b*) Chemical reduction

$$(Ar)R-\overset{\overset{\displaystyle O}{\|}}{C}-OR' \xrightarrow[\substack{\text{anhydrous ether} \\ 2.\ H_3\overset{..}{O}^{\oplus}}]{1.\ LiAlH_4} (Ar)R-CH_2-OH + R'OH \qquad \text{Sec. 23.7E,b}$$

or

$$(Ar)R-\overset{\overset{\displaystyle O}{\|}}{C}-OR' \xrightarrow[\text{heat}]{Na,\ CH_3CH_2OH} (Ar)R-CH_2-OH + R'OH \qquad \text{Sec. 23.7E,b}$$

Bouveault-Blanc
procedure

Neither of these two methods attacks *isolated* carbon-carbon double or triple bonds.

C. Reactions of Amides

1. Hydrolysis

$$(Ar)R-\overset{\overset{\displaystyle O}{\|}}{C}-NH_2 \xrightarrow[\substack{\text{or} \\ H\overset{..}{\overset{\ominus}{O}}:,\ H_2O,\ heat}]{H^{\oplus},\ H_2O,\ heat} (Ar)R-\overset{\overset{\displaystyle O}{\|}}{C}-OH \qquad \text{Sec. 23.9A}$$

(or its salt)

Acid hydrolysis (Sec. 23.9A)
Base hydrolysis (Sec. 23.9A)

2. Dehydration: nitrile formation

$$(Ar)R-\overset{\overset{\displaystyle O}{\|}}{C}-NH_2 \xrightarrow[\substack{SOCl_2\ or \\ PCl_5\ or\ POCl_3 \\ heat}]{P_2O_5\ or} (Ar)R-C\equiv\overset{..}{N} \qquad \text{Sec. 23.9B}$$

Reduction of amides (Sec. 26.7A)
Hofmann reaction (Sec. 26.7)
Occurrence and use of amides: nylon (Sec. 23.10)

D. Reactions of Anhydrides

1. Hydrolysis, alcoholysis, ammonolysis

$$\begin{matrix} (Ar)R-C\overset{\displaystyle\nearrow O}{\underset{\displaystyle\searrow}{}} \\ \qquad\qquad O + H\text{:}Nu \\ (Ar)R-C\underset{\displaystyle\searrow O}{} \end{matrix} \longrightarrow (Ar)R-\overset{\overset{\displaystyle O}{\|}}{C}-Nu + (Ar)R-\overset{\overset{\displaystyle O}{\|}}{C}-OH \qquad \text{Sec. 23.13A}$$

where H:Nu = H_2O, NH_3, or R'OH

Resolution of alcohols (Sec. 23.14)

2. Friedel-Crafts acylation

$$(R-\overset{O}{\overset{\|}{C}})_2O + Ar-H \xrightarrow[BF_3]{AlX_3 \text{ or}} Ar-\overset{O}{\overset{\|}{C}}-R + R-\overset{O}{\overset{\|}{C}}-OH \qquad \text{Sec. 23.13B}$$

Ketone

E. Reactions of Nitriles: Hydrolysis

$$(Ar)R-C\equiv\ddot{N} \xrightarrow[H_2O, \text{ heat}]{H^\oplus \text{ or } H\ddot{O}\overset{\ominus}{:}} (Ar)R-\overset{O}{\overset{\|}{C}}-OH \qquad \text{Sec. 23.16}$$

(or its salt)

Basic hydrolysis (Sec. 23.16)
Acid hydrolysis (Sec. 23.16)
Comparison of acidic and basic hydrolysis (Sec. 23.16)

Study Questions

23.29 Give IUPAC names (except as noted) for the following 10 compounds.

(*a*) $CH_3-\overset{O}{\overset{\|}{C}}-O-\overset{O}{\overset{\|}{C}}-CH_3$

   (also give common name)

(*b*) $CH_3CH_2CH-\overset{O}{\overset{\|}{C}}-Cl$
         $\underset{CH_3}{|}$

(*c*) ⬡$-CH_2CHCH-\overset{O}{\overset{\|}{C}}-O-CH_2CH_3$
              $\underset{Cl}{|}\ \underset{CH_3}{|}$

(*d*) ⬡$-\overset{O}{\overset{\|}{C}}-\overset{..}{\underset{..}{O}}\overset{\ominus}{:}K^\oplus$

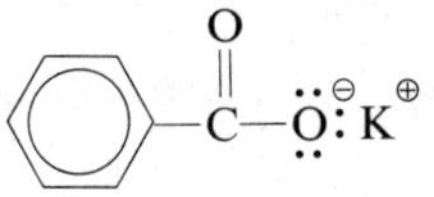

(*e*) $CH_3CH_2-\overset{O}{\overset{\|}{C}}-N\overset{CH_2CH_3}{\underset{CH_3}{<}}$

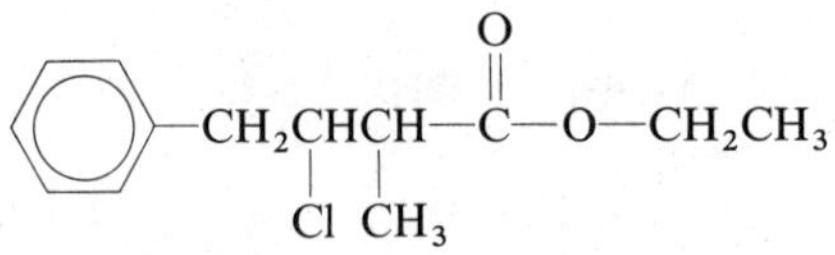

(*f*)
$\overset{\text{COO}\overset{\ominus}{:}Na^\oplus}{\underset{\text{COOH}}{⬡}}$

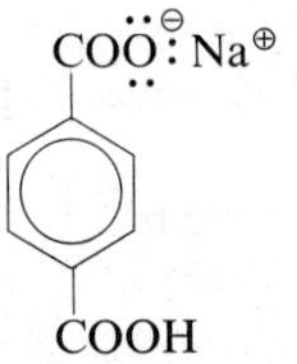

(*g*)
$\underset{CH_3}{\overset{\overset{O}{\overset{\|}{C}}-Br}{⬡}}$

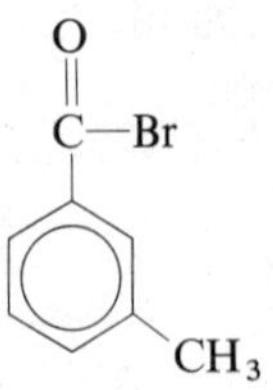

(*h*)
$\overset{CONH_2}{⬡}$

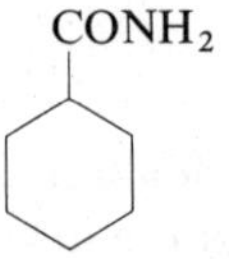

(*i*)
$\overset{COOCH_3}{\underset{COOCH_3}{⬠}}$

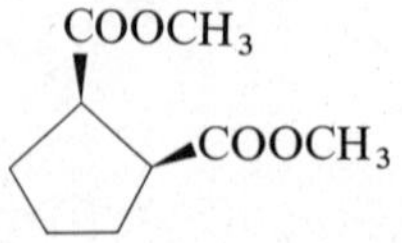

(*j*)
⬡⬡⬡$-CN$

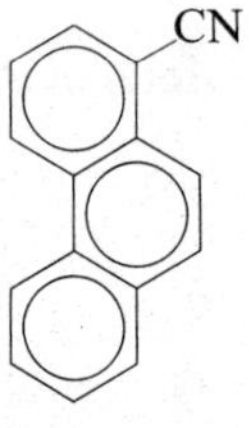

23.30 Give structures that correspond to the following names.
(*a*) methyl benzoate
(*b*) α-bromoacetamide
(*c*) diethyl heptanedioate
(*d*) *p*-toluamide

(e) *N*-ethyl-*N*-methylpropionamide   (f) 3-chloropentanoyl bromide
(g) propanoic anhydride   (h) methyl *p*-bromobenzoate
(i) β-bromobutyric acid   (j) succinic acid
(k) 2-oxocyclopentanecarboxamide

23.31 Supply the missing reactants, reagents, and products, as required, to complete the following 22 equations. The items to be supplied are indicated by number [for example, (1)].

(a)

$$\text{(succinic anhydride)} \xrightarrow[\text{CH}_3\text{CH}_2\text{OH}]{\text{1 mole}} (1) \xrightarrow{\text{PCl}_3} (2) \xrightarrow{(\text{CH}_3\text{CH}_2)_2\text{Cd}} (3)$$

(b) $\text{CH}_3\text{CH}_2\text{Br} \xrightarrow[\text{alcohol, heat}]{\text{KCN}} (4) \xrightarrow[\text{heat}]{\text{H}_3\ddot{\text{O}}^\oplus} (5) \xrightarrow{\text{PCl}_3} (6) \xrightarrow[\text{Pd + BaSO}_4]{\text{H}_2} (7)$

(c) $\text{CH}_3-\overset{\text{O}}{\overset{\|}{\text{C}}}-{}^{18}\text{OCH}_3 \xrightarrow[\text{heat}]{\text{H}^\oplus, \text{H}_2\text{O}} (8)$ Show location of ^{18}O in the products

(d)

$$\begin{array}{l} \text{CH}_2-\text{O}-\overset{\text{O}}{\overset{\|}{\text{C}}}-\text{R} \\ | \\ \text{CH}-\text{O}-\overset{\text{O}}{\overset{\|}{\text{C}}}-\text{R} \\ | \\ \text{CH}_2-\text{O}-\overset{\text{O}}{\overset{\|}{\text{C}}}-\text{R} \end{array} \xrightarrow[\text{2. H}_3\text{O}^\oplus]{\text{1. LiAlH}_4} (9)$$

$$\xrightarrow[\text{H}^\oplus, \text{heat}]{\text{large excess CH}_3\text{OH}} (10)$$

(e)

$$\text{(cyclopropane-1,1-dicarboxylic acid)} \xrightarrow{\text{strong heat}} (11)$$

(f) $\text{(benzene)} + \text{CH}_3\text{CH}_2-\overset{\text{O}}{\overset{\|}{\text{C}}}-\text{Cl} \xrightarrow{\text{AlCl}_3} (12) \xrightarrow{(13)} \text{(phenyl)}-\text{CH}_2\text{CH}_2\text{CH}_3$

(g) $\text{CH}_3-\text{CH}_2-\overset{\text{O}}{\overset{\|}{\text{C}}}-\text{O}-\text{CH}_2\text{CH}_3 \xrightarrow[\text{2. H}_3\ddot{\text{O}}^\oplus]{\text{1. 2 moles CH}_3\text{MgI}} (14)$

(h) $\text{HO}-\overset{\text{O}}{\overset{\|}{\text{C}}}-\underset{\underset{\text{CH}_3}{|}}{\text{CH}}-\overset{\text{O}}{\overset{\|}{\text{C}}}-\text{OH} \xrightarrow{\text{heat}} (15) \xrightarrow{(16)} \text{CH}_3\text{CH}_2\text{CH}_2\text{OH}$

(i)

$$\text{(acetophenone)} \xrightarrow[\text{excess :}\ddot{\text{O}}\text{H}^\ominus]{\text{excess I}_2} (17) \nearrow^{(18)} \text{(benzoyl bromide)}$$
$$\searrow_{(19)} \text{(}N,N\text{-dimethylbenzamide)}$$

(*j*)
$$\begin{array}{c} CH_3 \\ \\ CH_3 \end{array}\!\!\!CHCH_2-\overset{\overset{\displaystyle O}{\|}}{C}-OH \xrightarrow[\text{2. } H_3\ddot{O}^\oplus]{\text{1. LiAlH}_4} (20)$$

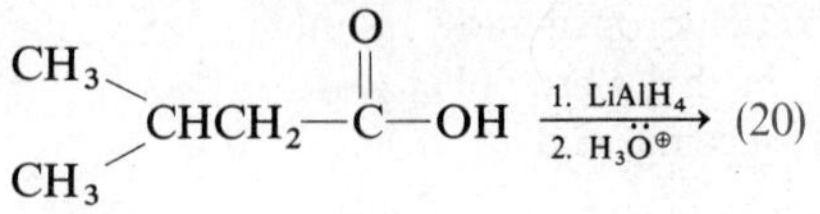

(*k*)

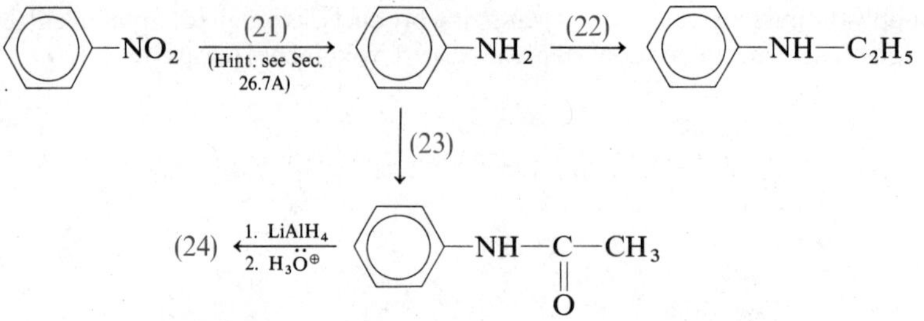

$$\text{(benzene ring)}-NO_2 \xrightarrow[\substack{\text{(Hint: see Sec.}\\ \text{26.7A)}}]{(21)} \text{(benzene ring)}-NH_2 \xrightarrow{(22)} \text{(benzene ring)}-NH-C_2H_5$$

$$\Big\downarrow (23)$$

$$(24) \xleftarrow[\text{2. } H_3\ddot{O}^\oplus]{\text{1. LiAlH}_4} \text{(benzene ring)}-NH-\overset{\overset{\displaystyle }{}}{\underset{\underset{\displaystyle O}{\|}}{C}}-CH_3$$

(*l*)
$$CH_3-\overset{\overset{\displaystyle O}{\|}}{C}-OH \xrightarrow[H^\oplus,\text{ heat}]{CH_3CH_2-^{18}O-H} (25)$$

$$\overset{CH_3}{\underset{CH_3}{CH_3-C-^{18}O-H}} \Big\downarrow{}^{H^\oplus,\text{ heat}}$$

$$(26)$$

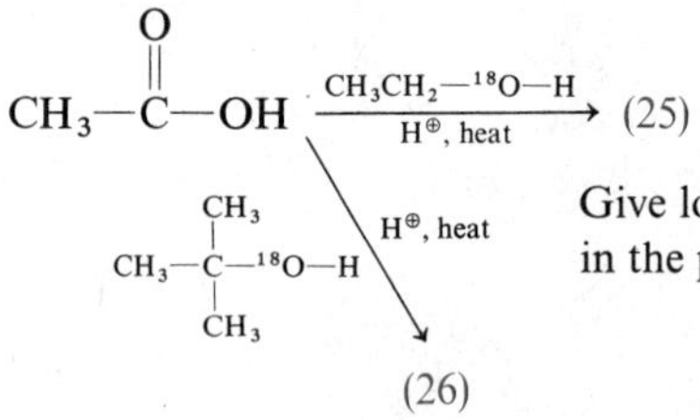

Give location of ^{18}O
in the products (25) and (26)

(*m*) $CH_3CH_2CH_2-O-CH_2CH_2CH_3 \xleftarrow{(27)} CH_3CH_2CH_2OH$

$$CH_3CH_2-\overset{\overset{\displaystyle O}{\|}}{C}-OH \nearrow{}^{(28)}$$

$$\Big\downarrow (29)$$

$$CH_3CH_2-\overset{\overset{\displaystyle O}{\|}}{C}-H \xleftarrow{(32)} CH_3CH_2-\overset{\overset{\displaystyle O}{\|}}{C}-Cl \xrightarrow{(30)} CH_3CH_2-\overset{\overset{\displaystyle O}{\|}}{C}-NH_2$$

$$\Big\downarrow (31)$$

$$CH_3CH_2-\overset{\overset{\displaystyle O}{\|}}{C}-\text{(benzene ring)}$$

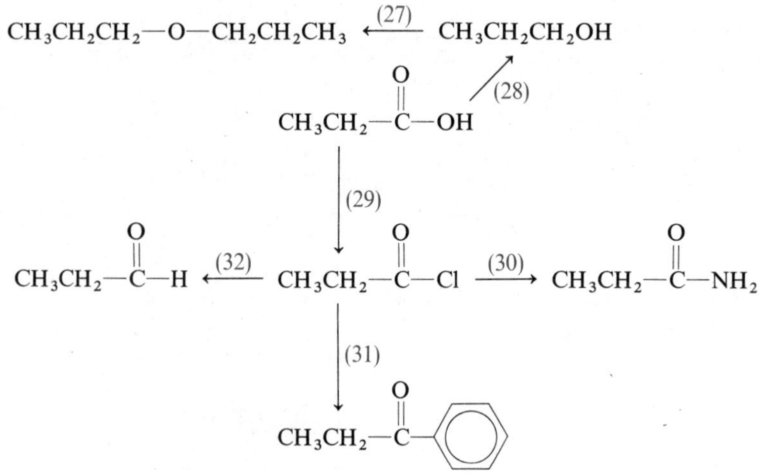

(*n*)

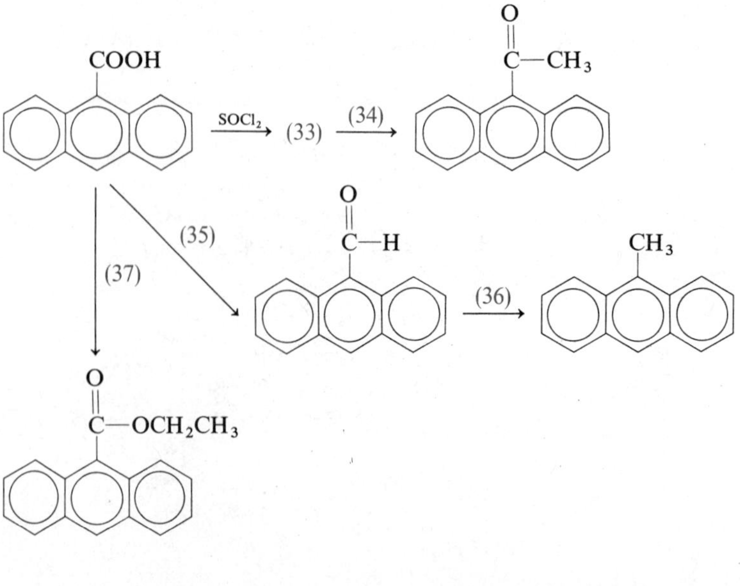

$$
(o) \quad CH_3CH_2-\overset{\displaystyle |}{\underset{\displaystyle OH}{C}}HCH_2-\overset{\displaystyle O}{\overset{\displaystyle ||}{C}}-OH \xrightarrow{\text{heat}} (38)
$$

$$
(p) \quad CH_3CHO \xrightarrow{HCN} (39) \xrightarrow[\text{heat}]{H_3\ddot{O}^{\oplus}} (40) \xrightarrow[H_2SO_4,\ \text{heat}]{CH_3OH} (41)
$$

$$
(q) \quad CH_3CH=CH_2 \xrightarrow{H_2SO_4} (42) \xrightarrow{H_2O} (43) \xrightarrow[NaOH]{I_2} (44)
$$

$$
(r) \quad CH_3CH_2Br \xrightarrow[\text{alcohol, heat}]{KCN} (45) \xrightarrow[\text{heat}]{H_3\ddot{O}^{\oplus}} (46) \xrightarrow{PCl_3} (47) \xrightarrow{H_2}_{Pd\ +\ BaSO_4} (48)
$$

$$
(s) \quad CH_3CH_2Br \xrightarrow[\text{ether}]{Mg} (49) \xrightarrow{CdCl_2} (50) \xrightarrow{CH_3COCl} (51)
$$

$$
(t) \quad CH_2=CH_2 \xrightarrow[CCl_4]{Br_2} (52) \xrightarrow[KCN]{2\ \text{moles}} (53) \xrightarrow[\text{heat}]{H_3\ddot{O}^{\oplus}} (54)
$$

$$
(u) \quad \text{Acetylene} \xrightarrow[H_2SO_4,\ H_2O]{HgSO_4} (55) \xrightarrow[\substack{K_2Cr_2O_7,\\ \text{heat}}]{H_2SO_4} (56) \xrightarrow{PCl_3} (57) \xrightarrow[\text{cold}]{NH_3} (58)
$$

$$
(v) \quad CH_3CH_2Br \xrightarrow[\text{ether}]{Mg} (59) \xrightarrow{CO_2} (60) \xrightarrow{H_3\ddot{O}^{\oplus}} (61)
$$

23.32 Indicate how the following 19 transformations may be carried out. Do not balance the equations, but indicate the necessary reagents and reaction conditions. *You must start with the indicated starting material.* You may also use *methanol, ethanol, benzene,* or *carbon dioxide* and any needed inorganic reagents, as well as common organic solvents (e.g., ether). Once a compound has been prepared from methanol or ethanol or inorganic reagents, it may be used again in subsequent reactions provided it was prepared before.

$$
(a) \quad CH_3CH_2-\overset{\displaystyle O}{\overset{\displaystyle ||}{C}}-CH_3 \longrightarrow CH_3CH_2-\overset{\displaystyle O}{\overset{\displaystyle ||}{C}}-H
$$

(b) Toluene $\longrightarrow$

$$
(c) \quad CH_3CH_2CH_2-OH \longrightarrow CH_3CH_2CH_2-\overset{\displaystyle O}{\overset{\displaystyle ||}{C}}-O-CH_2CH_3
$$

(d) $\longrightarrow CH_3-O-\overset{\displaystyle O}{\overset{\displaystyle ||}{C}}-CH_2CH_2-\overset{\displaystyle O}{\overset{\displaystyle ||}{C}}-CH_3$

(e) $-CH_2-\overset{\displaystyle O}{\overset{\displaystyle ||}{C}}-H \longrightarrow$ $-CH_2-\overset{\displaystyle O}{\overset{\displaystyle ||}{C}}-O-CH_2-CH_2-$

$$
(f) \quad CH_3CH_2CH_2-OH \longrightarrow CH_3CH_2-\overset{\displaystyle OH}{\underset{\displaystyle CH_2CH_3}{C}}-CH_2CH_3
$$

(g) $-CH_3 \longrightarrow$ $-CH_2-\overset{\displaystyle O}{\overset{\displaystyle ||}{C}}-NH_2$

(h) Benzene $\longrightarrow Br-$$-\overset{\displaystyle O}{\overset{\displaystyle ||}{C}}-O-CH_3$

(*i*) $CH_3CH_2CH_2OH \longrightarrow CH_3CH_2\underset{\underset{OH}{|}}{CH}{-}COOH$

(*j*) $(CH_3)_2CHCH{=}CH_2 \longrightarrow CH_3{-}\overset{\overset{O}{\|}}{C}{-}O{-}CH_2CH_2\underset{\underset{CH_3}{|}}{CH}CH_3$ Banana oil

(*k*) 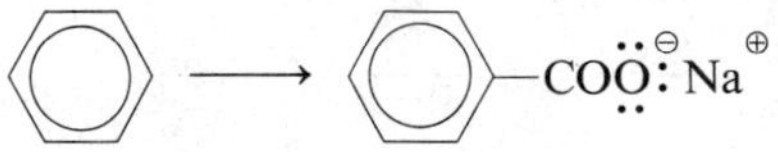   Spoilage inhibitor in food

(*l*) $CH_3{-}\overset{\overset{O}{\|}}{C}{-}OH \longrightarrow C_2H_5{-}O{-}\overset{\overset{O}{\|}}{C}{-}CH_2{-}\overset{\overset{O}{\|}}{C}{-}O{-}C_2H_5$

(*m*)

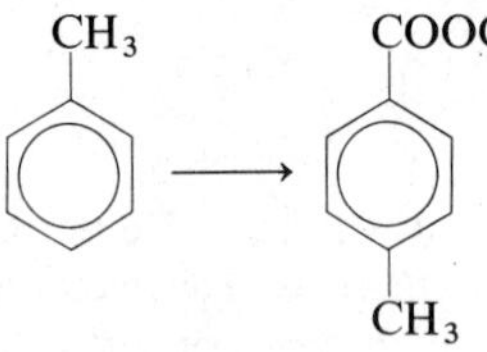

(*n*)

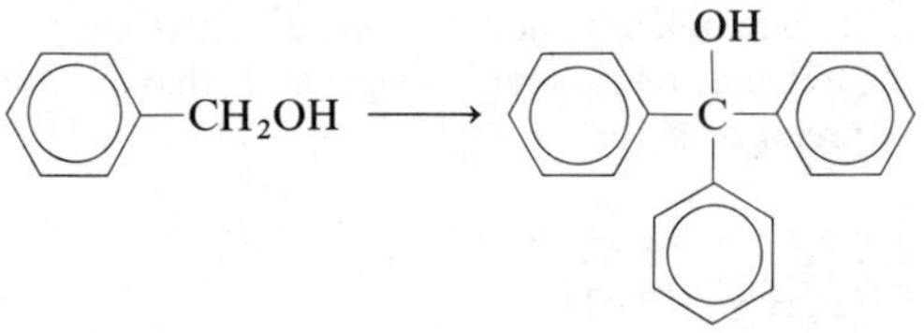

(*o*)

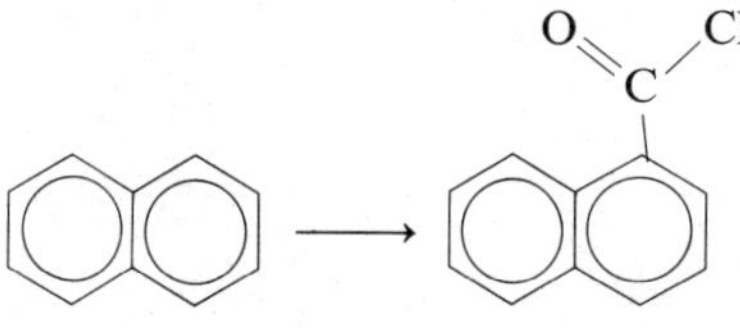

(*p*) $CH_3CH_2{-}\overset{\overset{O}{\|}}{C}{-}O{-}CH_3 \longrightarrow CH_3\underset{\underset{Br}{|}}{CH}CH_3$

(*q*) $(CH_3)_2CHCOOH \longrightarrow (CH_3)_2CHCHO$

(*r*) $\longrightarrow HOOC{-}(CH_2)_6{-}COOH$

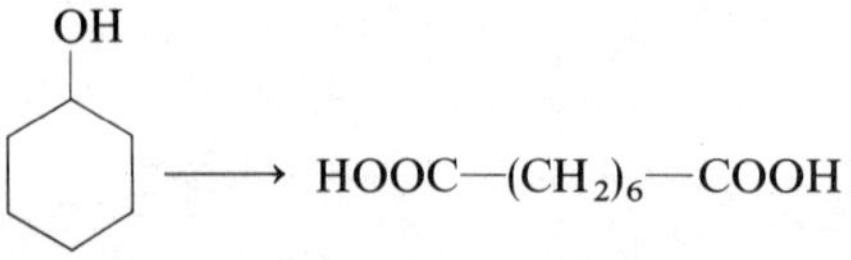

23.33 Show how each of the following transformations can be accomplished in no more than three steps, using no other organic reagents. Use any inorganic reagents and organic solvents you wish.

(*a*) Ethyl ether → propionitrile

(*b*) Ethanol → propanoic acid

(*c*) Methyl bromide → *tert*-butyl alcohol

(*d*) Ethyl propanoate → acetic acid

(*e*) Isopropyl chloride → *n*-propyl bromide

(*f*) Ethyl acetate → acetyl chloride

(*g*) 3-Methyl-2-bromobutane → 2-methyl-2-butanol

(*h*) Ethyl butanoate → butyronitrile

(*i*) Ethylene → bromoacetic acid

(*j*) Acetylene → 2-butanone

(*k*) Cyclohexene → cyclohexanol

(*l*) Ethyl bromide → 3-pentanone

(*m*) Acetylene → 1-butanol

(*n*) Ethylene → propionitrile

23.34 Based on the following information and on your knowledge of organic chemistry, provide logical chemical structures for compounds *A* to *G*.

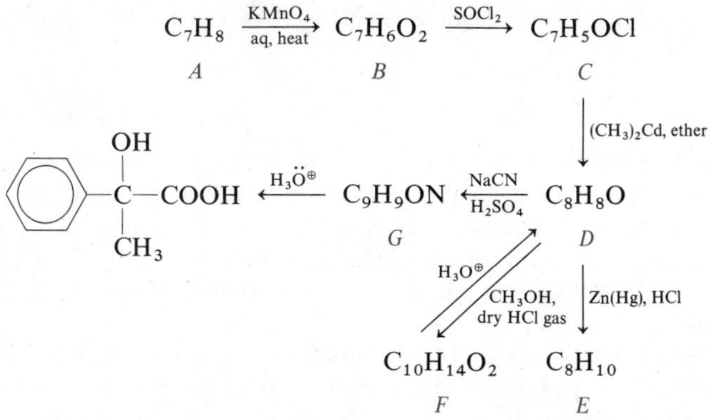

$$C_7H_8 \xrightarrow[\text{aq, heat}]{\text{KMnO}_4} C_7H_6O_2 \xrightarrow{\text{SOCl}_2} C_7H_5OCl$$

A   *B*   *C*

↓ (CH₃)₂Cd, ether

A structure with OH, C—COOH, CH₃ on a benzene ring ← $\xleftarrow{\text{H}_3\ddot{\text{O}}^\oplus}$ C₉H₉ON $\xleftarrow[\text{H}_2\text{SO}_4]{\text{NaCN}}$ C₈H₈O

G   *D*

H₃O⊕ / CH₃OH, dry HCl gas   Zn(Hg), HCl

C₁₀H₁₄O₂   C₈H₁₀

F   *E*

23.35 Draw the complete mechanism for the following transformation. Be sure to explain the expected stereochemistry of the product.

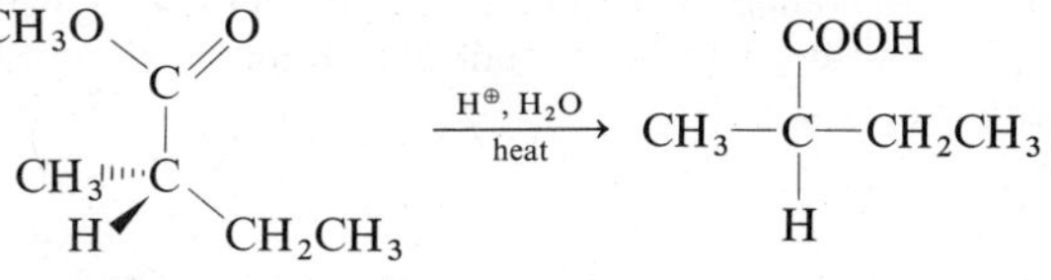

$$\xrightarrow[\text{heat}]{\text{H}^\oplus, \text{H}_2\text{O}}$$

S-Configuration

23.36 Aspirin (acetylsalicylic acid, I) is the most widely used drug in the world. People spend more than $100 million on this compound annually. Aspirin is used primarily as a pain reliever and as a means of reducing fever. APC tablets are also used widely for this purpose. APC tablets are a combination of aspirin, phenacetin (II), and caffeine.

In Excedrin, the phenacetin is replaced by acetaminophen (III). Acetaminophen is a good pain reliever and is marketed under the brand names Tylenol, Tempera, Nebs, Apamide, and Lyceta. Both phenacetin and acetaminophen are related structurally to acetanilide (IV).

Design a synthesis for each of these compounds from benzene and hydrocarbons of four carbon atoms or less. You may use any other organic or inorganic reagents you require. (*Hint:* see Sec. 26.7A.)

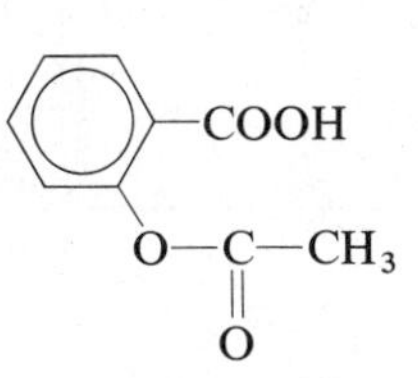

I 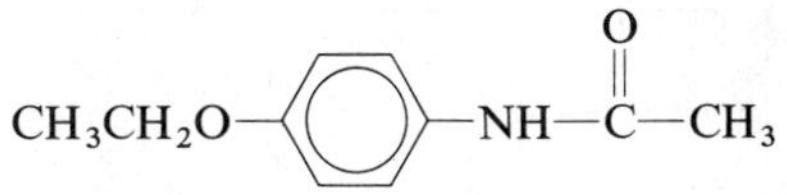   II

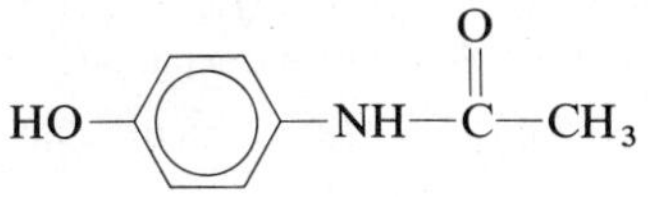

III 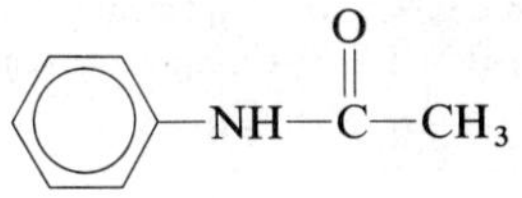   IV

23.37 Given the following three compounds:

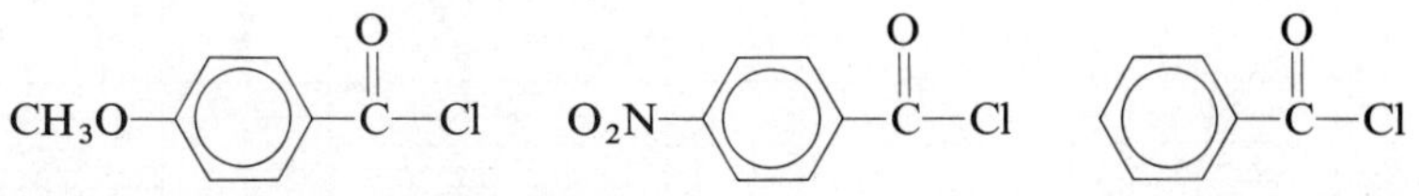

Which of these compounds would you expect to be the *most reactive toward NH₃*? Explain.

23.38 The products of the following two reactions (after each of the reaction mixtures has been neutralized) are given to the right of the arrow. The rate in both reactions is dependent on the concentrations of both reactants; that is, the rate of the reaction is dependent on both the concentration of the hydroxide *and* the respective alkyl or acyl halide.

$$CH_3CH_2Br + NaOH \longrightarrow CH_3CH_2OH + NaBr \qquad (1)$$

$$CH_3\overset{\overset{\displaystyle O}{\|}}{C}Br + NaOH \longrightarrow CH_3COOH + NaBr \qquad (2)$$

(*a*) What is the name of the mechanism type involved in reaction (1)? In reaction (2)?

(*b*) Under similar reaction conditions of concentration and temperature, reaction (2) proceeds at a *much* faster rate than reaction (1). With this in mind, draw the two mechanisms involved in reactions (1) and (2) and contrast the differences in the two mechanisms that affect the rate.

(*c*) Based on the mechanisms you drew, write a generalized rate expression for each reaction. Label each.

(*d*) In reaction (2), the oxygen involved in the carbon-oxygen double bond in the starting acyl halide is the same oxygen involved in the carbon-oxygen double bond in the acid that is formed. That is, the oxygen from the hydroxide ion does not exchange with the carbonyl oxygen to any significant extent. What model experiment could you run to prove this? Explain.

23.39 Compound A, $C_{13}H_{19}NO_2$, is soluble in cold dilute HCl. A is hydrolyzed under acidic or basic conditions to give B and C. Compound B, $C_8H_9NO_2$, is soluble in dilute HCl and NaHCO$_3$ solution. C, $C_5H_{12}O$, reacts with Na° metal. The IR and NMR spectra of C are provided here. Draw the structures of A to C.

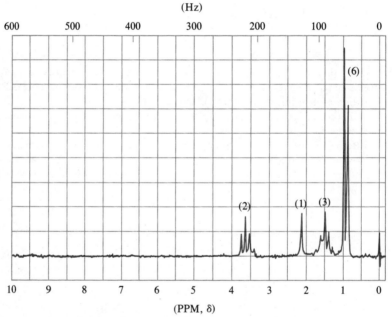

PROBLEM 23.39   NMR spectrum of unknown (*C*). [Data from *The Aldrich Library of NMR Spectra*, C. J. Pouchert and J. R. Campbell (eds.), copyright 1974, Aldrich Chemical Co., Milwaukee]

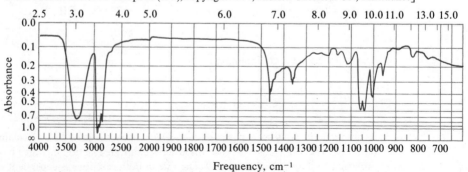

PROBLEM 23.39   IR spectrum of unknown (*C*). [Data from *Indexes to Evaluated Infrared Reference Spectra*, copyright 1975, The Coblentz Society, Norwalk CT]

23.40 Compound *A*, $C_{10}H_{20}O_2$, is a pleasant-smelling compound that is *optically active*. On hydrolysis, *A* gives two new compounds, *B* and *C*. Compound *B*, $C_4H_{10}O$, is *optically active*, and on treatment with chromic acid, *B* gives a new compound *D*, C_4H_8O. *D* is *optically inactive* and is resistant to further oxidation. *D*, when treated with methylmagnesium bromide in anhydrous ether and then with dilute aqueous acid, gives compound *E*, $C_5H_{12}O$. When treated with concentrated HBr, *E* gives compound *F*, $C_5H_{11}Br$. *F* is carried through the sequence of reactions: (1) Mg, ether, (2) CO_2, and (3) $H_3\overset{..}{O}{}^\oplus$; the final product from this sequence of reactions is *identical* with compound *C*. Compound *C* is acidic, is optically inactive, and has the formula $C_6H_{12}O_2$.

Draw the structures of compounds *A* to *F*, and indicate briefly the reasoning you used in reaching the answers. Be careful to indicate the source of optical activity in compounds that are optically active. Trace the reactions described in equation form (unbalanced).

23.41 Compound *A*, $C_{11}H_{20}O_4$, is an *optically active* neutral compound, which does not react with phenylhydrazine or with bromine in carbon tetrachloride. On hydrolysis, *A* yields three new compounds: *B*, *C*, and *D*. Compound *B* has the formula CH_4O and is optically inactive. Compound *C*, $C_5H_8O_4$, is acidic and *optically active*. On heating, *C* is converted to a neutral, optically active compound *E*, $C_5H_6O_3$. *E* reacts slowly with excess water to regenerate optically active *C*. Compound *D*, $C_5H_{12}O$, is optically *inactive*, and on treatment with chromic acid, *D* gives compound *F*, $C_5H_{10}O$. *F* gives an oxime, but it does not react with Tollens' reagent. *F* also gives a negative iodoform test.

Draw the structures of compounds *A* to *F*, and indicate briefly the reasoning you used in reaching the answers. Be careful to indicate the source of the optical activity in compounds that are optically active. (*Hint:* Start at the end of the problem and deduce the structures of *B*, *C*, and *D*; then put these structures together to give the structure of *A*.)

23.42 An unknown organic compound *A*, $C_{13}H_{16}O_4$, is *optically active*. It does not react with bromine in CCl_4. It is insoluble in water but soluble in dilute base. On hydrolysis, *A* give two new compounds *B* and *C*. *B*, $C_4H_6O_4$, is optically inactive and insoluble in water. It is, however, readily soluble in dilute base. When *B* is heated, it readily loses CO_2 to give compound *D*, $C_3H_6O_2$. *C*, $C_9H_{12}O$, is an *optically active aromatic* compound. When *C* is heated with $KMnO_4$ and base, followed by neutralization with dilute H_2SO_4, aromatic compound *E*, $C_7H_6O_2$, is obtained.

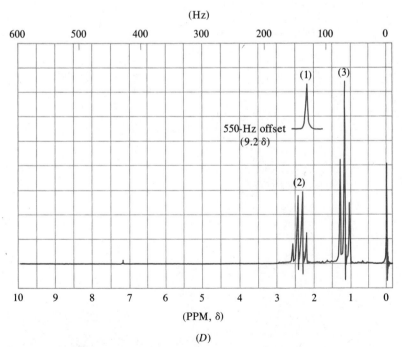

PROBLEM 23.42 NMR spectrum of unknown compound (*D*). [Data from *The Aldrich Library of NMR Spectra*, C. J. Pouchert and J. R. Campbell (eds.), copyright 1974, Aldrich Chemical Co., Milwaukee]

When C is heated with concentrated H_2SO_4, compound F, C_9H_{10}, is formed. F is optically inactive and decolorizes Br_2 in CCl_4. F, on treatment with ozone followed by zinc and water, gives formaldehyde and aromatic ketone G, C_8H_8O. The NMR spectrum of compound D is provided.

Draw the structures of compounds A to G, and indicate briefly the reasoning you used in reaching the answers. Be careful to indicate the source of the optical activity in compounds that are optically active.

23.43 Compound A, $C_{16}H_{16}O_2$, is an *optically active* neutral compound. On hydrolysis, A yields two new compounds, B and C. Compound B, $C_8H_8O_2$, is *optically inactive* and aromatic. Treatment of B with thionyl chloride gives D, C_8H_7OCl. When D reacts with an excess of cold ammonia, compound E, C_8H_9ON, is formed. E is then treated with sodium hypobromite (NaOBr) (Hofmann degradation, see Sec. 26.7C) and aqueous sodium hydroxide to produce F, C_7H_9N. F reacts rapidly with bromine water to give G, $C_7H_6Br_3N$. Compound C, $C_8H_{10}O$, is *optically active* and aromatic. C, when treated with sodium dichromate and sulfuric acid, gives compound H, C_8H_8O. H gives a yellow precipitate with 2,4-dinitrophenylhydrazine, no reaction with Tollens' reagent, and iodoform when treated with iodine in the presence of sodium hydroxide. H is *optically inactive*.

Draw the structures of compounds A to H, and indicate briefly the *reasoning* you used in reaching answers. Be careful to indicate the source of the optical activity in compounds that are optically active. Also, trace each of the reactions described in the given information.

24

Reactions of α Hydrogens in Carbonyl and β-Dicarbonyl Compounds: Condensation Reactions

We normally think of *acids* as the mineral acids of inorganic chemistry or the carboxylic acids of organic chemistry. Yet several families of organic compounds have moderately weak carbon-hydrogen bonds and may be classified as **carbon acids.** In this chapter we study the reactions of some carbon acids that involve the rupture of a carbon-hydrogen bond and the removal of hydrogen as a proton. A base is usually required to break these relatively weak carbon-hydrogen bonds, and the organic product of this reaction is called a **carbanion** (a contraction for *carb*on *anion*):

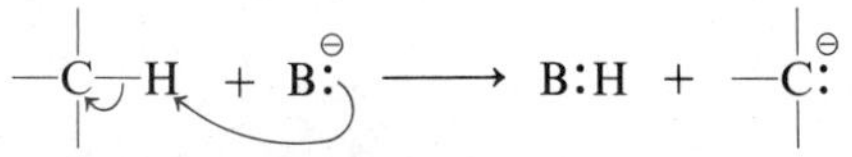

$$-\overset{|}{\underset{|}{C}}-H \; + \; B\colon^{\ominus} \longrightarrow \; B\colon H \; + \; -\overset{|}{\underset{|}{C}}\colon^{\ominus}$$

Carbon acid   Base   Carbanion

The hydrogens of most carbon-hydrogen bonds are not very acidic so that strong bases must be used to break them; however, the hydrogens attached to a carbon atom adjacent to a carbon-oxygen double bond are quite acidic compared to, for example, a C—H bond in an alkane. These hydrogens are called **α hydrogens:**

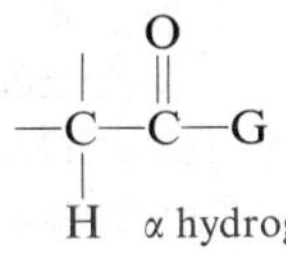

$$-\overset{|}{\underset{\underset{H}{|}}{C}}-\overset{\overset{O}{\|}}{C}-G$$

H   α hydrogen

where G = —H, —R, —Ar, —OR, and others

We devote a major portion of this chapter to reactions involving α hydrogens and the **α carbanions** that result from their abstraction by base. Many of these involve *condensation reactions* (see Sec. 24.8).

The α hydrogens in aldehydes, ketones, and esters can be abstracted by strong base. Hydrogens that are α to two carbon-oxygen double bonds are even more acidic and so are abstracted by even weaker bases; compounds bearing this type of hydrogen are examples of **β-dicarbonyl compounds:**

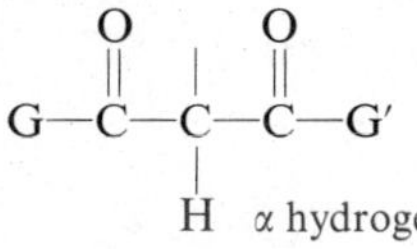

$$G-\overset{\overset{O}{\|}}{C}-\overset{|}{\underset{\underset{H}{|}}{C}}-\overset{\overset{O}{\|}}{C}-G'$$

H   α hydrogen

β-Dicarbonyl compound

where G and G′ = —H, —R, —Ar, —OR, and others

In our discussion of α hydrogens, we examine **keto-enol equilibrium** in some detail because it involves α carbanions.

Aldehydes that contain no α hydrogens cannot form an α carbanion. They undergo a special type of reaction in basic solution called the **Cannizzaro reaction.**

Carbonyl Compounds

24.1 Carbon Acids; Formation and Structure of α Carbanions

One method for comparing the acidities of compounds is to examine the equilibrium constants for the dissociation of a proton, the K_a values. Consider the following compounds and their dissociation constants:

Propane: $CH_3CH_2-CH_2-H \rightleftharpoons CH_3CH_2-CH_2{:}^{\ominus} + H^{\oplus}$ $K_a < 10^{-40}$

Propene: $CH_2{=}CH-CH_2-H \rightleftharpoons CH_2{=}CH-CH_2{:}^{\ominus} + H^{\oplus}$ $K_a \approx 10^{-36}$

Acetaldehyde: $\underset{\displaystyle H}{O{=}C}-CH_2-H \rightleftharpoons \underset{\displaystyle H}{O{=}C}-CH_2{:}^{\ominus} + H^{\oplus}$

Acetone: $\underset{\displaystyle CH_3}{O{=}C}-CH_2-H \rightleftharpoons \underset{\displaystyle CH_3}{O{=}C}-CH_2{:}^{\ominus} + H^{\oplus}$ $K_a \approx 10^{-20}$

Ethyl acetate: $\underset{\displaystyle OC_2H_5}{O{=}C}-CH_2-H \rightleftharpoons \underset{\displaystyle OC_2H_5}{O{=}C}-CH_2{:}^{\ominus} + H^{\oplus}$

In comparing these compounds, we note that acetaldehyde, acetone, and ethyl acetate are of roughly comparable acidity and they all possess a carbon-oxygen double bond, as emphasized in **red.** We are familiar with the polarity of the carbon-oxygen double bond by now, so it is not surprising that this polar group greatly eases the removal of α hydrogens as compared with the effect exerted by the nonpolar carbon-carbon double bond in propene. The electron-withdrawing properties of the carbon-oxygen double bond are quite strong. It withdraws electrons from the α carbon atom and stabilizes the carbanion that results from proton extraction.

Aqueous sodium hydroxide is basic enough to abstract an α hydrogen from an aldehyde or ketone, but the equilibrium lies far to the left and favors the neutral carbonyl compound:

Aldehydes and ketones:

$$\underset{\substack{\text{Carbon} \\ \text{acid}}}{R-\overset{\displaystyle O}{\overset{\|}{C}}-\overset{\displaystyle |}{\underset{\displaystyle H}{C}}{\to}} + \underset{\text{Base}}{{:}\overset{\ominus}{O}H} \rightleftharpoons \underset{\alpha\text{ Carbanion}}{R-\overset{\displaystyle O}{\overset{\|}{C}}-\overset{\displaystyle |}{\underset{\ominus}{C}}{:}} + H-\overset{\cdot\cdot}{\underset{\cdot\cdot}{O}}H$$

where R = H, alkyl, or aryl

The equilibrium concentration of carbanion is always very small.

Aqueous sodium hydroxide solution cannot be used to remove an α hydrogen from an ester because it would hydrolyze the ester group. However, an α carbanion can be formed in an ester by using alkoxide ion as the base in alcohol. The alkoxide ion and the alcohol must always contain the same R group as the alcohol part of ester itself. Here too equilibrium greatly favors the free ester rather than its α carbanion.

Esters:

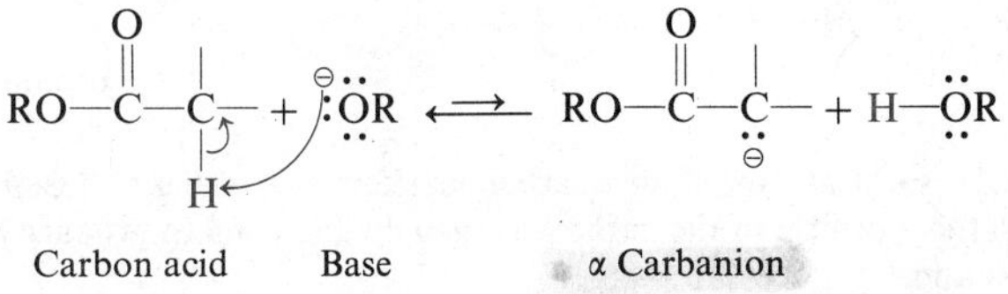

$$\underset{\text{Carbon acid}}{\overset{\displaystyle\overset{O}{\overset{\|}{}}}{RO\!-\!C\!-\!\underset{\underset{H}{|}}{C}\!-}} + \underset{\text{Base}}{:\!\overset{\ominus}{\underset{\cdot\cdot}{O}}\!R} \rightleftharpoons \underset{\text{α Carbanion}}{\overset{\displaystyle\overset{O}{\overset{\|}{}}}{RO\!-\!C\!-\!\underset{\ominus}{C}\!-}} + H\!-\!\overset{\cdot\cdot}{\underset{\cdot\cdot}{O}}R$$

Indeed, α carbanions can be formed in compounds other than aldehydes, ketones, esters, and β-dicarbonyl compounds. By proper choice of the base, an α carbanion can be generated adjacent to the acyl group in an *anhydride* (see Sec. 24.17) or to the cyano group in a *nitrile* or to the nitro group in a *nitro compound* (see Sec. 24.18). Each compound *must* of course possess an α hydrogen. On the other hand, α carbanions cannot be generated in *acid chlorides* because this functional group is so reactive toward any basic species that it undergoes nucleophilic acyl substitution in preference to carbanion formation. A *carboxylic acid* contains the very acidic proton attached to the carboxyl group that is abstracted in the presence of a base to give the carboxylate anion rather than a carbanion. Finally, the hydrogens attached to nitrogen in an *amide* are more acidic and usually more readily abstracted by base than are the α hydrogens. We discuss the acidic properties of the N—H bond in amides in Sec. 28.10A. In addition, amides undergo nucleophilic acyl substitution (for example, hydrolysis) in the presence of certain bases.

The unusual stability of an α carbanion generated adjacent to a carbon-oxygen double bond provides an additional driving force for its formation. Let us now examine the electronic structure of an α carbanion. We can write two electronic structures for it. These two resonance structures are shown here along with the hybrid structure that best represents the electron distribution in the anion:

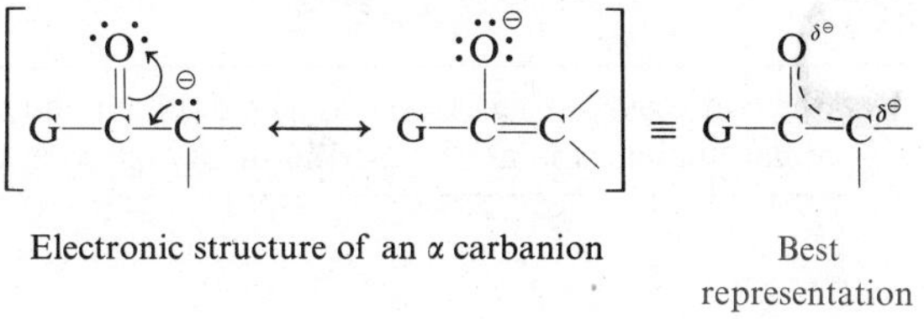

$$\left[\underset{}{G\!-\!\overset{\overset{\displaystyle\cdot\cdot O}{\|}}{C}\!\overset{\ominus}{\underset{|}{C}}\!-} \longleftrightarrow G\!-\!\overset{\overset{\displaystyle:\!\overset{\cdot\cdot}{O}:^{\ominus}}{|}}{C}\!=\!C\!\big\langle \right] \equiv G\!-\!\overset{\overset{\displaystyle O^{\delta\ominus}}{|}}{C}\!\overset{\cdot}{\cdots}\!C^{\delta\ominus}$$

Electronic structure of an α carbanion   Best
representation

where G = H, R, Ar, OR, and others

As we saw in other molecules, electron delocalization over several atoms usually results in a more stable species. Thus, the principal driving force for forming an α carbanion is the stability of the resonance-stabilized carbanion.

The orbital view of an α carbanion is the best picture of its electron distribution. The carbon atom bearing the negative charge is most likely sp^2 hybridized (having been formed from an sp^3 hybridized carbon), and the two electrons it contains overlap with the π bond of the adjacent carbon-oxygen double bond.[1] This overlap

---

[1] A rapidly inverting sp^3 hybridized species may also be possible.

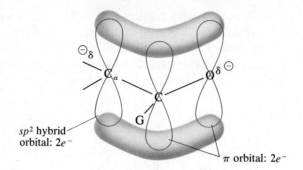

sp^2 hybrid orbital: $2e^-$

π orbital: $2e^-$

FIGURE 24.1   Orbital view of an α carbanion showing overlap of the sp^2 hybridized orbital on the α carbon with the π orbital of the carbon-oxygen double bond to produce the new π bonding orbitals of the α carbanion.

is illustrated in Fig. 24.1. Stereochemical studies on α carbanions (see Sec. 24.3) show this to be correct.

In view of its electronic structure, the term α *carbanion* seems to be a misnomer because this ion contains negative charge on both the α carbon and the oxygen atom. This is an example of an *ambident anion* (Latin *ambo*, both). The terms *ambident base* and *ambident nucleophile* are also used. It is likely that the more electronegative oxygen atom bears the greater concentration of negative charge. We use the α carbanion designation because a hydrogen on the α carbon atom must be removed by base to produce the anion.

Question 24.1

The polarity of the carbon-oxygen double bond appears to be responsible for stabilizing the α carbanion. However, the *aldehydic hydrogen* in an aldehyde (that is, the hydrogen attached directly to the carbonyl group) is not abstracted by base as shown by isotope exchange studies.

Assuming that the aldehydic hydrogen is removed, draw the structure of the resulting anion, and on the basis of its structure, explain why its removal is quite difficult.

Question 24.2

Suggest why the hydrogens α to a carbon-oxygen double bond are much more acidic than those β or γ to it. What analogies might be used in justifying your answer?

24.2   Evidence for α-Carbanion Formation: Isotopic Exchange Studies

Before going further, we might inquire about the experimental evidence for the existence of α carbanions in solution. Carbanions generated by the reaction between a base and an α hydrogen in a carbonyl-containing compound are present in *very low concentration* in solution, so they generally cannot be detected or isolated. We must therefore turn to indirect methods to prove their existence, and isotopes have been used with outstanding success in this respect.

For example, when an aldehyde or a ketone is dissolved in deuterium oxide containing sodium deuteroxide as the base, the α hydrogens are quickly replaced by deuterium:

$$R-\overset{\overset{\displaystyle O}{\|}}{C}-\underset{\underset{\displaystyle H}{|}}{C}- \quad \xrightarrow{\text{D}_2\text{O, Na}^\oplus \,:\!\ddot{\text{O}}\text{D}} \quad R-\overset{\overset{\displaystyle O}{\|}}{C}-\underset{\underset{\displaystyle D}{|}}{C}-$$

This result has been explained on the basis of carbanion formation. A carbanion is a very strong base that is capable of abstracting a proton (deuteron in this case) from the solvent. The following sequence illustrates this deuterium exchange:

$$R-\overset{\overset{\displaystyle O}{\|}}{C}\!-\!\underset{\underset{\displaystyle H}{|}}{C}- \;+\; {}^{\ominus}\!:\!\ddot{\text{O}}\text{D} \;\overset{k_2}{\rightleftharpoons}\; R-\overset{\overset{\displaystyle O}{\|}}{C}-\underset{\underset{\displaystyle \ominus}{|}}{C}- \;+\; H\!-\!\ddot{\text{O}}\text{D}$$

$$\rightarrow D-\ddot{\text{O}}\text{D} \;\text{(serves as acid)}$$

α Carbanion
Strong base

$$\Big\Updownarrow k_{2'}$$

$$R-\overset{\overset{\displaystyle O}{\|}}{C}-\underset{\underset{\displaystyle D}{|}}{C}- \;+\; {}^{\ominus}\!:\!\ddot{\text{O}}\text{D}$$

Once the carbanion is formed, it can react with H—OD by route k_2 and re-form the starting materials. However, there is usually a large excess of D_2O in the solution. The amount of hydrogen incorporated into the solvent (as HOD) is quite small, so it is much more probable that the carbanion reacts with D—OD by route $k_{2'}$ to give a carbonyl compound that contains deuterium in place of hydrogen.

Similar isotopic exchange results are obtained with esters, but they require the use of an alkoxide base ($R\ddot{\text{O}}\!:^\ominus$) in the presence of an O-deuterated solvent (ROD).

24.3 Stereochemical Evidence for Three-Dimensional Structure of Carbanions

We now examine some evidence that has been gathered to help us understand the three-dimensional structure of carbanions. We consider the structure of carbanions α to a carbon-oxygen double bond shortly but start our discussion with simple, nonstabilized carbanions.

A. Structure of Nonstabilized Carbanions

Simple, nonstabilized carbanions are considerably less stable than the corresponding α carbanions. They exist for extremely short times in the absence of any ions or molecules that stabilize them. We know, for example, that the Grignard reagent can be prepared and kept for reasonable periods of time, but this is only because magnesium salts and ether are present to help stabilize the negative charge.

In general, carbanions are not configurationally stable. For example, if we take optically active *sec*-butyl bromide and allow it to react with magnesium metal in anhydrous ether, we obtain the corresponding Grignard reagent as expected. When this Grignard reagent reacts with acetone, the product is optically inactive even though it contains an asymmetric center:

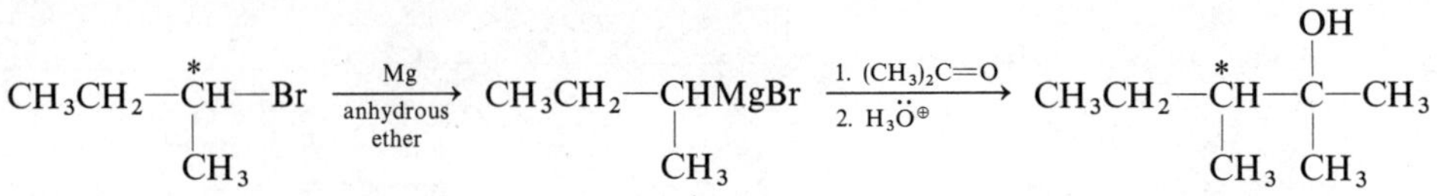

$$CH_3CH_2-\overset{*}{\underset{\underset{CH_3}{|}}{CH}}-Br \xrightarrow[\substack{\text{anhydrous} \\ \text{ether}}]{Mg} CH_3CH_2-\underset{\underset{CH_3}{|}}{CHMgBr} \xrightarrow[\text{2. } H_3\overset{..}{O}^{\oplus}]{\text{1. } (CH_3)_2C=O} CH_3CH_2-\overset{OH}{\underset{\underset{CH_3}{|}}{\overset{|}{\underset{*}{CH}}}}-\overset{|}{\underset{\underset{CH_3}{|}}{C}}-CH_3$$

2-Bromobutane
Optically active

Product
Optically inactive

That the Grignard reagent itself becomes optically inactive has been confirmed with many different Grignard reagents and other carbanions.

On the basis of extensive investigations on the structure and stereochemistry of carbanions, it appears that the best description of their three-dimensional structure is the following. The carbon bearing the unshared electron pair and the negative charge is sp^3 hybridized, and the geometry about it is **pyramidal.** It is believed that carbanions undergo a rapid interconversion between two pyramidal forms, as shown here:

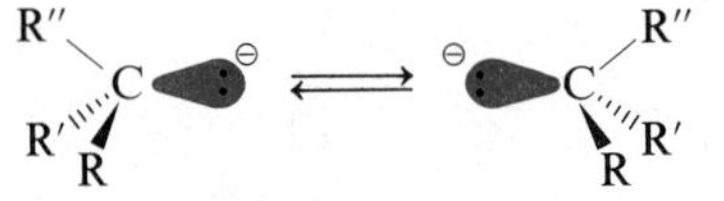

Inversion in simple, nonconjugated carbanions
Pyramidal structure

This inversion is analogous to an umbrella being blown inside out by a strong wind, except that with carbanions one pyramidal form is converted to the other (and vice versa) very rapidly. When three different R groups are attached to the carbanion, one pyramidal structure is the nonsuperimposable mirror image of the other. Thus, this inversion process results in a racemic mixture of carbanions, which is optically inactive regardless of the optical properties of the starting compound. The general reactions that follow illustrate how an optically active alkane can be racemized on treatment with strong base. Deuterium exchange is incorporated into this scheme to provide positive evidence that carbanions are involved in the racemization process.

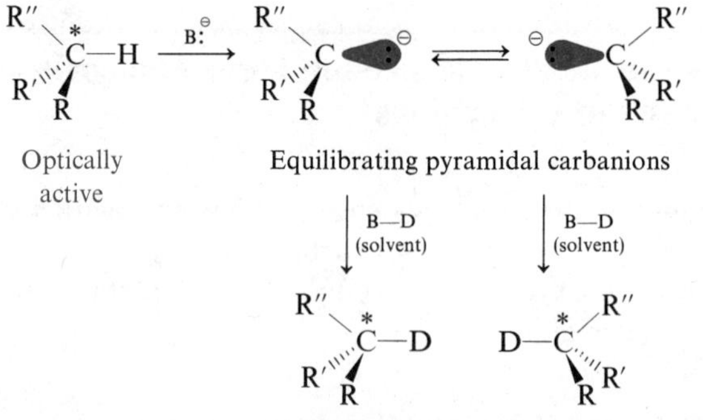

Optically
active

Equilibrating pyramidal carbanions

B—D
(solvent)

B—D
(solvent)

Mixture of enantiomers: racemic

The equilibration of the pyramidal carbanions appears to be faster than protonation (deuteration in this case), so that racemic hydrocarbon is obtained from optically active starting hydrocarbon.

Pyramidal Inversion of Carbanions

This view is simplified, but it nonetheless emphasizes the pyramidal structure of the carbanion in simple, nonconjugated compounds. The chemistry of carbanions has been the subject of extensive investigations which show that the *net* steric course of a reaction (that is, retention or inversion of configuration, or racemization) is highly dependent on the nature of the solvent. The two major variables in the properties of a solvent are *polarity* and *acidity*. An indication of solvent polarity is obtained from the dielectric constant ε, with larger dielectric constants associated with more polar solvents (see Sec. 5.14). The acidity of a solvent is indicated by its acid dissociation constant, K_a. The larger the dissociation constant, the more acidic the solvent (see Sec. 5.7). Some solvents like dimethyl sulfoxide, $(CH_3)_2SO$, are quite poor at donating protons, whereas others like methanol or ethylene glycol are very good proton donors.

The stereochemistry of the carbanion depends on the type of solvent in which it is generated, and the net steric course of a reaction is determined by the *environment* that surrounds the carbanion. If the carbanion reacts with a proton or deuteron from solvent as soon as it is formed and before it has a chance to undergo equilibration. *Asymmetric solvation* occurs and optically active product is formed. The carbanion lives longer in polar solvents that are poor proton donors, in which protonation occurs after it has a chance to be *symmetrically solvated*. A carbanion under these conditions undergoes extensive equilibration before protonation, and optically inactive product results so that racemization is observed.

This sketchy view of carbanion stereochemistry should give you an idea of the complexity of the problem. There is a detailed analysis of the effect of solvent on carbanion structure in the Reading References for this chapter.

Amines are a good model of carbanions because they are *isoelectronic* with one another (that is, they contain the same number of valence electrons about the central nitrogen or carbon). Amines are electronically neutral, whereas carbanions have a negative charge. Amines are stable compounds that can be isolated and studied easily, whereas carbanions are not. Spectral methods, such as nuclear magnetic resonance and microwave spectroscopy, show that an amine is pyramidal but that it too is in rapid equilibration with its "enantiomer."

Amines
Inversion between pyramidal forms

This type of inversion is conceptually identical with the inversion we discussed for carbanions. Spectral studies indicate *how fast* the inversion occurs in amines, and at 25° it is estimated to be *approximately* 2×10^5 times per second! The rate of inversion for carbanions is presumed to be even faster.

Further support for the pyramidal structure of amines comes from the fact that they *cannot* be obtained in optically active form (see Sec. 26.6) even though four different groups (one of which is a pair of electrons) are attached to nitrogen.

Question 24.3

Amines cannot be resolved and obtained in optically active form. On the other hand, optically active *sulfoxides*, such as *p*-carboxyphenyl methyl sulfoxide (1), have been obtained and found to be quite stable for a long time. The sulfur atom in the sulfoxide group and the nitrogen in an amine both contain an unshared pair of electrons; yet the former exists in enantiomeric forms whereas the latter does not. Suggest an explanation for this fact.

(1)

B. Structure of Stabilized α Carbanions

Stereochemical studies in conjunction with hydrogen-deuterium exchange studies demonstrate the structure of carbanions that are α to a carbon-oxygen double bond.

When optically active phenyl *sec*-butyl ketone (1) comes in contact with base in solution, the ketone loses its optical activity. In other words, optically active ketone is converted into a racemic mixture of enantiomers:

(1)
Phenyl *sec*-butyl ketone
Optically active

Optically inactive

A reaction occurs at the asymmetric center in the ketone. This observation alone strongly suggests that the α hydrogen is removed by base and that at some later time it reattaches itself to the α carbanion that was formed. Furthermore, optically active ketones that have no α hydrogens are unaffected by base; for example, optically active ketone (2) does not lose *any* optical activity on standing with base.

(2)

Isotopic exchange studies point conclusively to the formation of the α carbanion in optically active ketone (1). When phenyl *sec*-butyl ketone reacts with deuterium oxide containing sodium deuteroxide, deuterium is incorporated into the α position in place of hydrogen, and racemization still occurs:

(1)
*Optically
active*

*Optically
inactive*

It has been found that *the rate of deuterium incorporation into the ketone is exactly the same as the rate of loss of optical activity.*

Now let us interpret these results to deduce the structure of the carbanion that intervenes in the reaction. The stereochemical and isotopic exchange results point to the removal by base of the α hydrogen attached to the asymmetric center in optically active ketone (1). The resulting α carbanion reacts with deuterium, and optical activity is lost at the same rate and thus at the same time. It is conceivable that this carbanion is pyramidal and rapidly inverting, but it is much more likely that the carbanion is flat because of conjugation with the adjacent carbon-oxygen double bond as shown by the orbital view in Fig. 24.1. Attack by hydrogen (or deuterium) is *equally likely* on one side or the other of this flat delocalized carbanion. This predicts that an equal mixture of enantiomers is produced so that racemic (optically inactive) product is obtained. This is indeed what happens experimentally. Fig. 24.2 shows the stereochemistry and exchange results, as well as the structure of the α carbanion.

Other data, such as the rate of halogenation of a ketone (see Sec. 24.5), further support the formation of a carbanion α to the carbonyl group.

Similar stereochemical and exchange results are recorded for optically active esters that have an α hydrogen adjacent to the acyl group. Here also a conjugated, planar α carbanion intervenes.

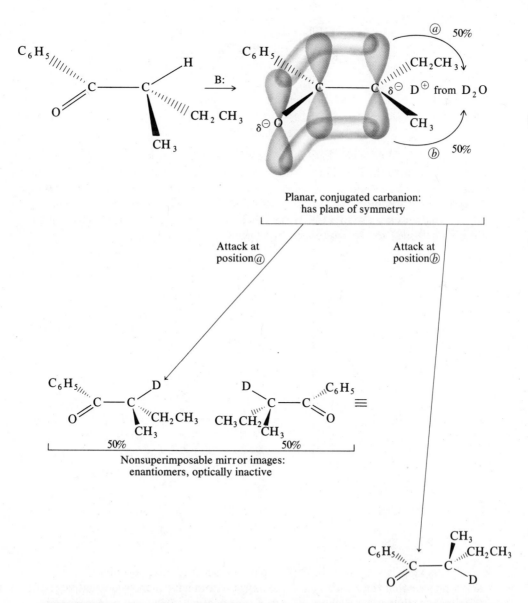

FIGURE 24.2 Stereochemical fate of the reaction of an optically active ketone with base. Deuteron (or proton) from solvent attacks either face of the planar, conjugated carbanion with equal probability to produce an equal mixture of enantiomers.

24.4   Base-Promoted Keto-enol Equilibrium

Many reactions we study in the remainder of this chapter revolve about the α hydrogens in aldehydes, ketones, esters, and anhydrides. As a prelude, let us review the details of keto-enol equilibrium or tautomerization, which exists in both basic and acidic solutions (see Sec. 11.13).

The base-promoted reactions are more commonly encountered and we discuss these first. The following equation illustrates the structural changes that can occur when a carbonyl compound containing an acidic α hydrogen comes in contact with base:

| **Keto form** | **Enol form** |
| More stable | Less stable |

Keto-enol equilibrium

The keto form is normally more stable for most compounds so that the equilibrium lies far to the left. This is shown mechanistically as follows. The α hydrogen in a carbonyl compound is abstracted by base to form a conjugated carbanion. Once formed, this carbanion can react in one of two ways: (1) a proton may become attached to the α carbon atom to regenerate the starting materials, or (2) a proton may become attached to the oxygen atom to produce the enol form:

| Keto form | Enol form |
| regenerated | produced |

This view is simplified because the resonance structures do not exist, but they do emphasize that the α carbon and the oxygen atom have increased negative charge. We illustrate that either carbon or oxygen—both of which have some negative charge and are strong bases—is capable of receiving a proton from solvent. This is another example of an ambident base (see Sec. 19.8). The oxygen is likely to bear the greater amount of negative charge, so it should preferentially abstract a proton. The enol form may be produced first (kinetic control), but because the keto form is more stable it is the ultimate product (thermodynamic control).

24.5 Base-Promoted Halogenation of Aldehydes and Ketones

Aldehydes and ketones are halogenated under basic or acidic conditions, as shown by the following equation:

$$R-\overset{\displaystyle O}{\overset{\|}{C}}-\overset{\displaystyle |}{\underset{\displaystyle H}{C}}- \xrightarrow[\text{H}_2\text{O}]{X_2, :\ddot{O}H^{\ominus}, \text{ or } H^{\oplus}} R-\overset{\displaystyle O}{\overset{\|}{C}}-\overset{\displaystyle |}{\underset{\displaystyle X}{C}}-$$

where R = H, R′, Ar
X = Cl, Br, I

As evidenced by the structural changes shown, a halogen is *substituted* for an α hydrogen. In this section we focus our attention on the reaction that occurs in *basic* solution; the reaction in acidic solution is discussed in Sec. 24.7.

The following are two known facts about the reaction.

1. If there is more than one α hydrogen present, substitution can continue until all of them are substituted by halogen; for example:

$$CH_3-\overset{O}{\overset{\|}{C}}-CH_3 \xrightarrow[\text{H}_2\text{O}]{Br_2, :\ddot{O}H^{\ominus}} CH_3-\overset{O}{\overset{\|}{C}}-CH_2-Br \xrightarrow[\text{H}_2\text{O}]{Br_2, :\ddot{O}H^{\ominus}} CH_3-\overset{O}{\overset{\|}{C}}-\overset{|}{\underset{\displaystyle Br}{C}H}-Br$$

Acetone   α-Bromoacetone   α,α-Dibromoacetone

$$\downarrow \overset{Br_2, :\ddot{O}H^{\ominus},}{\underset{H_2O}{}}$$

$$CH_3-\overset{O}{\overset{\|}{C}}-\overset{Br}{\overset{|}{\underset{\displaystyle Br}{C}}}-Br$$

α,α,α-Tribromoacetone

Proper choice of experimental conditions usually dictates the desired product. Note, however, that α-bromoacetone gives only α,α-dibromoacetone and not α,α′-dibromoacetone, $BrCH_2COCH_2Br$. We account for this later in this section.

2. The rate of halogenation depends on the concentrations of base and of aldehyde or ketone; that is,

Rate of halogenation = k_2[base][aldehyde or ketone]

where k_2 = second-order rate constant

The rate of halogenation is completely independent of the halogen concentration. Furthermore, the rate does not even depend on the nature of the halogen because chlorine, bromine, and iodine all react at the same rate for a given concentration of aldehyde or ketone and of base.

From these observations, we can suggest a plausible mechanism for halogenation in basic solution. The key to understanding the pathway of this reaction comes from chemical kinetics. That the rate of reaction depends only on the concentrations of carbonyl compound and base means that these two species must be involved in the slow, rate-determining step. The intermediate formed then reacts rapidly with halogen to produce the observed product. Our previous knowledge about the acidity of α hydrogens and their reactions with base to form a conjugated carbanion suggests the intervention of an α carbanion in this halogenation reaction. The following mechanism agrees with the chemical kinetics of the reaction.

Step 1. Formation of conjugated carbanion: *slow, rate-determining step*

(1)   (2)

Conjugated
carbanion

Step 2. Reaction of carbanion with halogen: *fast step*

(1)

Alternate step 2. Fast step

(2)

Step 1 is the formation of the stablized α carbanion, a reaction we discussed earlier in this chapter. It is the *slow, rate-determining step* and it involves a molecule of aldehyde or ketone and a molecule of base.

Once formed, the carbanion can react with halogen. This can be drawn in two ways, as shown in step 2 and alternate step 2. (The separate resonance structures of the carbanion emphasize what electrons are involved in the reaction.) Step 2 shows the attack by a negatively charged carbon on the polarizable halogen molecule to produce the α-halo carbonyl compound and halide ion, $:\ddot{X}:^{\ominus}$. Alternate step 2 shows how the unshared pairs of electrons on oxygen in the anion of the enol (or, more correct, enol*ate*) form are "donated" to the carbon-carbon double bond, which in turn attacks the positive end of the polarized halogen molecule. Both step 2 and alternate step 2 are very fast reactions, thus accounting for halogenation being independent of the nature and concentration of the halogen. The reaction of halogen with the conjugated carbanion cannot occur until it is formed in step 1, the slow, rate-determining step. Note that this is a base-promoted reaction; base is consumed in the reaction.

We can now understand the orientation of halogenation. Granted that one halogen atom is introduced into a ketone like acetone by the preceding mechanism, we might ask why a second halogen atom becomes attached to the same carbon as the first one rather than on the other side of the carbonyl group; for example:

$$CH_3-\overset{\overset{\displaystyle O}{\|}}{\underset{\alpha}{C}}-\underset{\alpha}{CH_2}-Br \xrightarrow[H_2O]{Br_2,\,:\overset{..}{O}H^\ominus} CH_3-\overset{\overset{\displaystyle O}{\|}}{C}-\underset{\underset{\displaystyle Br}{|}}{CH}-Br \quad \text{and not} \quad Br-CH_2-\overset{\overset{\displaystyle O}{\|}}{C}-CH_2-Br$$

α-Bromoacetone   α,α-Dibromoacetone   α,α'-Dibromoacetone

To answer this question, we need only to examine the bromoacetone molecule. The two hydrogens attached to the α carbon atom are more acidic than the three hydrogens attached to the α' carbon atom, because the former are on a carbon bearing two electron-withdrawing groups—the carbonyl group and the bromine atom. Thus the α hydrogens are more easily abstracted by base than are the α' hydrogens, and once the α carbanion is formed, it reacts further with bromine to produce α,α-dibromoacetone:

$$CH_3-\overset{\overset{\displaystyle O}{\|}}{\underset{\alpha'}{C}}-\underset{\alpha}{CH_2}-Br \xrightarrow{:\overset{..}{O}H^\ominus} CH_3-\overset{\overset{\displaystyle O^{\delta\ominus}}{\vdots}}{\underset{\alpha'}{C}}\!\!=\!\!\underset{\alpha}{\overset{\delta\ominus}{CH}}-Br \quad \text{and not} \quad \overset{\delta\ominus}{CH_2}\!\!=\!\!\overset{\overset{\displaystyle O^{\delta\ominus}}{\vdots}}{C}-CH_2-Br$$

More readily
formed

Another example of how structure affects the orientation of halogenation is the following:

$$CH_3-CH_2-\overset{\overset{\displaystyle O}{\|}}{C}-CH_3 \xrightarrow[H_2O]{Br_2,\,:\overset{..}{O}H^\ominus}$$

2-Butanone

$$CH_3-CH_2-\overset{\overset{\displaystyle O}{\|}}{C}-CH_2-Br \quad \text{and not} \quad CH_3-\underset{\underset{\displaystyle Br}{|}}{CH}-\overset{\overset{\displaystyle O}{\|}}{C}-CH_3$$

1-Bromo-2-butanone   3-Bromo-2-butanone

As expected, a primary carbanion is more stable than a secondary one because of the electron-releasing nature of any attached alkyl groups. Extending this to conjugated carbanions, we might anticipate that removal of a primary α hydrogen produces a more stable carbanion than would removal of a secondary α hydrogen:

$$CH_3-CH_2-\overset{\overset{\displaystyle O}{\|}}{C}-CH_3 \xrightarrow{B:^\ominus}$$

2° hydrogens   1° hydrogens

$$CH_3CH_2-\overset{\overset{\displaystyle O^{\delta\ominus}}{\vdots}}{C}\!\!=\!\!\overset{\delta\ominus}{CH_2} \quad \text{and not} \quad CH_3-\overset{\delta\ominus}{CH}\!\!=\!\!\overset{\overset{\displaystyle \delta\ominus O}{\vdots}}{C}-CH_3$$

Carbon bearing   Carbon bearing
negative charge   negative charge
is primary   is secondary
More stable   *Less stable*

Once the carbanion is formed, it reacts rapidly with halogen to yield product. The observed product confirms our prediction based on theory.

In summary, *halogenation of a ketone in basic solution occurs so that halogen ends up on the α carbon that originally contained more hydrogens.*

Question 24.4

Predict the major monochloro compound that would be formed when each of the following reacts with chlorine in the presence of aqueous base:

(*a*) acetophenone   (*b*) 1-phenyl-2-propanone   (*c*) 1-phenyl-2-butanone
(*d*) cyclohexanone   (*e*) 2-methylcyclohexanone   (*f*) *tert*-butyl methyl ketone (pinacolone)

24.6   Halogenation of Methyl Ketones and Methyl Carbinols in Basic Solution: Haloform Test

One important application of the halogenation of carbonyl compounds in basic solution occurs when a methyl group is attached to the carbon-oxygen double bond. With excess halogen and base, the following reactions occur:

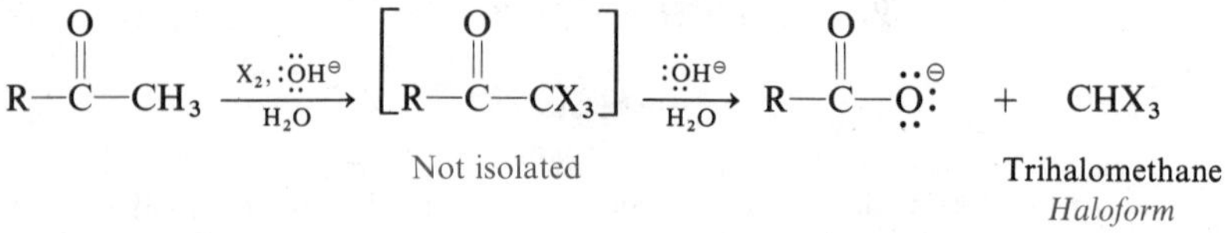

$$
\underset{\text{R}-\overset{\overset{\text{O}}{\|}}{\text{C}}-\text{CH}_3} \xrightarrow[\text{H}_2\text{O}]{\text{X}_2,\, :\ddot{\text{O}}\text{H}^\ominus} \left[\text{R}-\overset{\overset{\text{O}}{\|}}{\text{C}}-\text{CX}_3\right] \xrightarrow[\text{H}_2\text{O}]{:\ddot{\text{O}}\text{H}^\ominus} \text{R}-\overset{\overset{\text{O}}{\|}}{\text{C}}-\ddot{\text{O}}\text{:}^\ominus + \text{CHX}_3
$$

Not isolated   Trihalomethane
                                     Haloform

where R = H, R, Ar

X = Cl, Br, I

Note that a carbon-carbon bond is broken and a salt of a carboxylic acid is produced. Because a haloform (CHX_3) is formed, this reaction is generally called the **haloform test** and is a specific qualitative test for the methyl ketone group

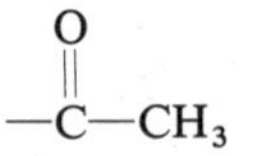

$$
\overset{\overset{\text{O}}{\|}}{-\text{C}}-\text{CH}_3
$$

(Certain other structural features in a molecule produce the same result, as we see later in this section.)

In the laboratory, excess iodine and base are usually used because the haloform produced from this reaction is *iodoform*, CHI_3. Iodoform is *observed* as a yellow precipitate, which is readily identified by its characteristic odor and yellow color. Its melting point is 119°, so a physical method can also confirm its identity. Another reason for using iodine as the halogen is that it is a solid and is more easily handled than chlorine (a toxic gas) or bromine (a liquid), which are noxious-smelling.

The Haloform Reaction

Some confusion arises in the terminology of this reaction. The term *haloform* is used because a haloform is one of the products. Alternatively, this same reaction is sometimes referred to as the sodium hypohalite reaction because halogens disproportionate in basic solution to produce the halide ion ($:\ddot{\text{X}}:^\ominus$) and the *hypohalite* ion ($:\ddot{\text{O}}:\ddot{\text{X}}:^\ominus$) in which the halogen has an oxidation state of +1; the following reaction shows this disproportionation: $X_2 + 2:\ddot{\text{O}}\text{H}^\ominus \rightarrow :\ddot{\text{X}}:^\ominus + :\ddot{\text{O}}:\ddot{\text{X}}:^\ominus + \text{H}_2\ddot{\text{O}}:$. A convenient source of hypohalite is Clorox bleach which contains sodium hypochlorite (NaOCl), an especially useful reagent in preparative reactions.

Accordingly, either of the following reagents produces the same reaction: $X_2,\, :\ddot{\text{O}}\text{H}^\ominus$ or NaOX, $:\ddot{\text{O}}\text{H}^\ominus$.

The mechanism of the haloform (hypohalite) reaction is as follows:

Step 1. α Halogenation in base

$$R-\overset{\overset{\displaystyle O}{\|}}{C}-CH_3 \xrightarrow[\text{(see Sec. 24.5)}]{\overset{\text{excess}}{X_2,\,:\ddot{O}H^{\ominus}}} R-\overset{\overset{\displaystyle O}{\|}}{C}-CX_3$$

Not isolated

Step 2. Nucleophilic addition of hydroxide ion to carbonyl group

$$R-\overset{\overset{\displaystyle :O:}{\|}}{C}-CX_3 + :\ddot{O}H^{\ominus} \longrightarrow R-\overset{\overset{\displaystyle :\ddot{O}:^{\ominus}}{|}}{\underset{\displaystyle \underset{:O-H}{|}}{C}}-CX_3$$

(1)
Tetrahedral intermediate
Unstable

Step 3. Decomposition of tetrahedral intermediate

$$R-\overset{\overset{\displaystyle :\ddot{O}:^{\ominus}}{|}}{\underset{\displaystyle \underset{:OH}{|}}{C}}-CX_3 \longrightarrow R-\overset{\overset{\displaystyle :O:}{\|}}{C}\overset{\displaystyle }{\underset{\displaystyle \ddot{O}H}{}} + \overset{\ominus}{:}CX_3$$

(1)   Trihalomethyl
                         carbanion

Step 4. Neutralization reactions

$$R-\overset{\overset{\displaystyle O}{\|}}{C}\underset{\displaystyle \ddot{O}H}{} + \left\{:\ddot{O}H^{\ominus} \text{ or } \overset{\ominus}{:}CX_3\right\} \longrightarrow R-\overset{\overset{\displaystyle O}{\|}}{C}\underset{\displaystyle \ddot{O}:^{\ominus}}{} + H_2O \text{ or } CHX_3$$

Acid   Strong
                bases

or

$$\overset{\ominus}{:}CX_3 + H-\ddot{O}H \longrightarrow CHX_3 + :\ddot{O}H^{\ominus}$$

Strong   Weak
base   acid

Step 1 is the α halogenation of an aldehyde or ketone in basic solution, the mechanism discussed in Sec. 24.5. Substitution occurs three times to give the trihalo-carbonyl compound. Step 2 shows the nucleophilic *addition* of hydroxide ion to the carbon-oxygen double bond. There is a strong driving force for addition in this case because of the three strong electron-withdrawing halogens on the α carbon atom. They increase the positive charge on carbon in the carbon-oxygen double bond and facilitate nucleophilic attack and the resulting addition. Intermediate (1) is tetrahedral and analogous to several intermediates we encountered in both nucleophilic addition (see Sec. 21.2) and substitution (see Sec. 23.3) of compounds containing the carbon-oxygen double bond. Step 3 shows the decomposition of intermediate (1) into products. That this reaction occurs may at first be surprising because we know that aldehydes and ketones undergo nucleophilic addition, whereas only carboxylic acids and their derivatives undergo substitution. We attributed these differences to the relative bond strengths of the various leaving groups in Sec. 23.3. In particular, aldehydes and ketones should not undergo substitution because it involves breaking a strong, covalent carbon-carbon bond. Why then does substitution occur with a trihalomethyl aldehyde or ketone? The three halogen atoms are also responsible for this reaction because they withdraw electrons and stabilize the resulting anion, $X_3\overset{\ominus}{C}:$. Step 4 shows various "neutralization" reactions that may occur between the carboxylic acid and the strong bases, hydroxide ion, or $:CX_3^{\ominus}$. The $:CX_3^{\ominus}$ ion may also abstract a proton from water (the solvent).

The *methylcarbinol group* also gives a positive haloform test as long as *at least one hydrogen atom is attached to the carbon bearing the hydroxyl group*:

$$
\begin{array}{c}
\text{OH} \\
| \\
\text{R—C—CH}_3 \\
| \\
\text{H}
\end{array}
$$

Methylcarbinol
Gives positive haloform reaction

where R = H, alkyl, or aryl

Secondary (and primary) methylcarbinols react with hypohalite to produce a haloform because the alcohol is first oxidized to a methyl ketone by the reagents. X_2 and $:\overset{..}{\underset{..}{O}}\overset{\ominus}{X}:$ are mild oxidizing agents which cause this oxidation (see Sec. 29.9C). The methyl ketone then reacts as described earlier to yield the haloform and carboxylate salt. This reaction sequence is:

$$
\begin{array}{c}
\text{OH} \\
| \\
\text{R}_3\text{—C—CH}_3 \\
| \\
\text{H}
\end{array}
\xrightarrow{X_2,\,:\overset{..}{O}H^{\ominus}}
\left[
\begin{array}{c}
\text{O} \\
\| \\
\text{R—C—CH}_3
\end{array}
\right]
\xrightarrow{X_2,\,:\overset{..}{O}H^{\ominus}}
$$

1° or 2°   Methyl ketone
Methylcarbinol

$$
\left[
\begin{array}{c}
\text{O} \\
\| \\
\text{R—C—CX}_3
\end{array}
\right]
\xrightarrow{:\overset{..}{O}H^{\ominus}}
\begin{array}{c}
\text{O} \\
\| \\
\text{R—C—}\overset{..}{\underset{..}{O}}:^{\ominus}
\end{array}
+ \text{CHX}_3
$$

where R = H, alkyl, or aryl

Although the haloform reaction enjoys its greatest popularity as a laboratory test for methyl ketones or methylcarbinols, it is also a convenient way to break a carbon-carbon bond and obtain an organic compound that *contains one less carbon atom*. As we know, the carboxylate salt is readily converted into a carboxylic acid on neutralization, and the carboxyl group can be transformed into numerous other functional groups. The following example illustrates this application of the haloform (hypohalite) reaction:

Acetophenone   Benzoate ion   Iodoform
                      (salt of benzoic acid)

Question 24.5

Which, if any, of the following 13 compounds would react with iodine and sodium hydroxide solution to produce iodoform? Give the expected products for the compounds that do react.

(*a*) *tert*-butyl alcohol　　(*b*) 2-pentanol　　　　(*c*) α-phenylethanol
(*d*) acetaldehyde　　　　　(*e*) formaldehyde　　　　(*f*) methyl alcohol
(*g*) ethyl alcohol　　　　　(*h*) *m*-bromoacetophenone　(*i*) phenylacetaldehyde
(*j*) 3-phenyl-2-butanol　　 (*k*) 3-phenyl-1-butanol　　(*l*) ⬜—COCH₃

(*m*) cyclohexanone.

Question 24.6

Indicate how the following compounds can be prepared from *acetophenone* in good yield using no other source of carbon. Organic solvents and inorganic reagents can be used freely.

(*a*) benzoic acid　　(*b*) benzyl benzoate　　　　　(*c*) benzamide
(*d*) toluene　　　　 (*e*) *m*-bromobenzoyl chloride　(*f*) benzaldehyde

---

24.7　Acid-Catalyzed Reactions of Carbonyl Compounds

The reactions of aldehydes, ketones, and esters in basic solution are the most prevalent and we return to these reactions shortly. We now discuss several important reactions of aldehydes and ketones in acidic solution. We focus our attention on acid-catalyzed halogenation and keto-enol equilibrium as well as the mechanism of the HVZ reaction.

A. Acid-Catalyzed Halogenation

Aldehydes and ketones that have hydrogens α to the carbonyl group also undergo halogenation under acidic conditions, as shown by the equation:

$$R-\overset{\overset{\displaystyle O}{\|}}{C}-\overset{\overset{\displaystyle |}{}}{\underset{\underset{\displaystyle H}{|}}{C}}- + X_2 \xrightarrow{H^\oplus} R-\overset{\overset{\displaystyle O}{\|}}{C}-\overset{\overset{\displaystyle |}{}}{\underset{\underset{\displaystyle X}{|}}{C}}- + H^\oplus + :\overset{..}{\underset{..}{X}}:^\ominus$$

where R = H, alkyl, or aryl

　　　X = Cl, Br, I

A convenient acid medium for the halogenation reaction is *acetic acid* because many organic compounds are soluble in it. The following two examples illustrate this reaction:

Acetophenone　　　　　　　　　　α-Bromoacetophenone

$$(CH_3)_2CH-\overset{\overset{\displaystyle O}{\|}}{C}-CH(CH_3)_2 \xrightarrow[H_2O,\ H^\oplus]{Cl_2} (CH_3)_2CH-\overset{\overset{\displaystyle O}{\|}}{C}-\overset{\overset{\displaystyle CH_3}{|}}{\underset{\underset{\displaystyle Cl}{|}}{C}}-CH_3$$

2,4-Dimethyl-3-pentanone　　　　　2-Chloro-2,4-dimethyl-3-pentanone
(diisobutyl ketone)

Acid-catalyzed halogenation is similar in many respects to the reaction that occurs in basic solution.

The following steps outline the most plausible mechanism based on the experimental evidence.

Step 1. Protonation of carbonyl compound

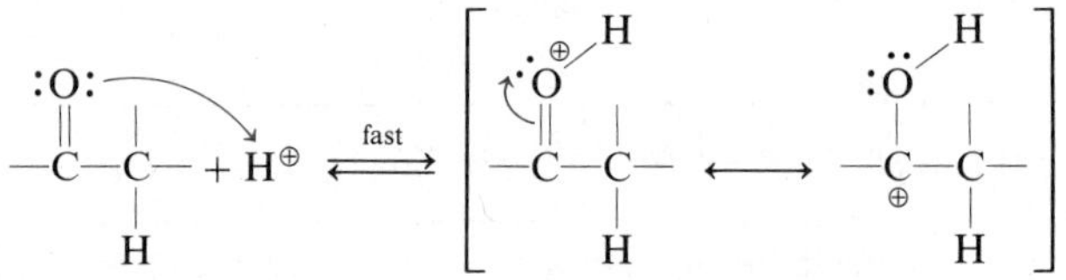

Step 2. Abstraction of α hydrogen by base

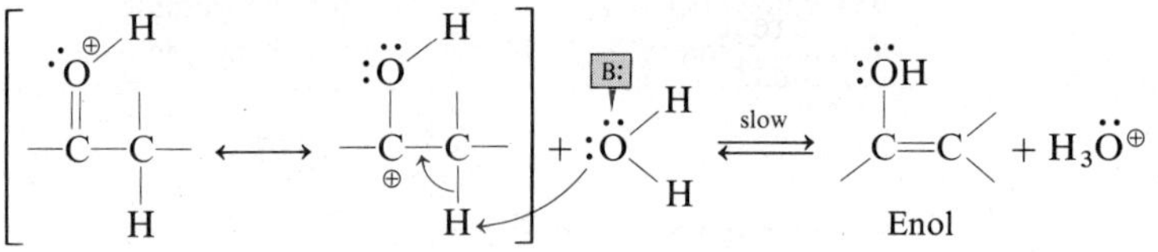

Step 3. Reaction of enol with halogen

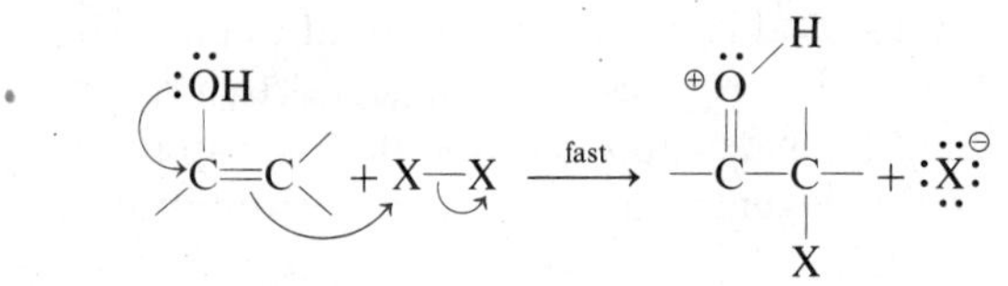

Step 4. Loss of proton

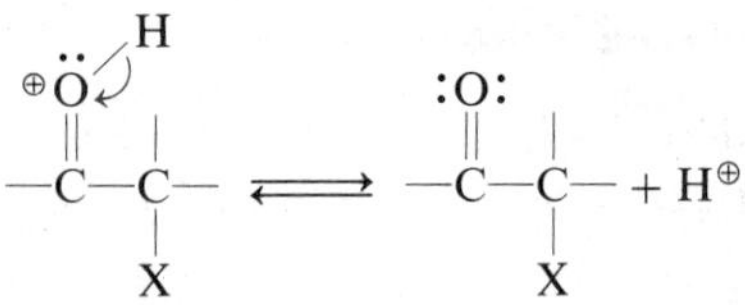

As with the base-promoted reaction, the acid-catalyzed reaction can continue until all the α hydrogens are replaced by halogen atoms.

Question 24.7

On the basis of the preceding discussion, provide a mechanism to explain each of the following observations:

(*a*) The rates of racemization and hydrogen-deuterium exchange are nearly equal when optically active *sec*-butyl phenyl ketone comes in contact with D_2SO_4 and D_2O.
(*b*) The rates of acid-catalyzed isotopic exchange and of acid-catalyzed bromination of *sec*-butyl phenyl ketone are nearly identical.

In unsymmetric ketones, there are two possible α carbon atoms at which substitution can occur. As the following example illustrates, one product predominates:

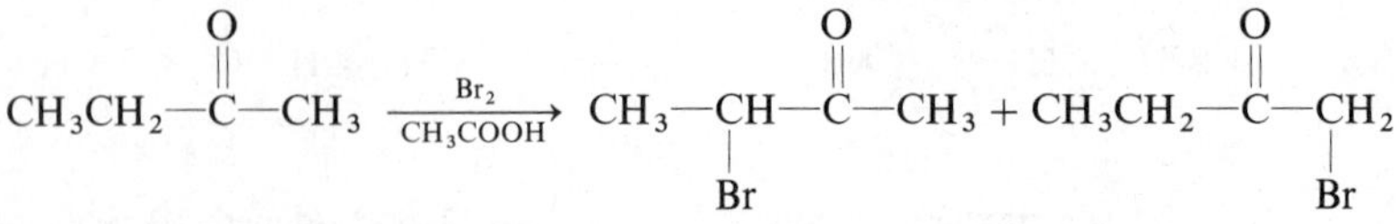

2-Butanone   3-Bromo-2-butanone   1-Bromo-2-butanone
                                Major product   *Minor product*

An examination of the structures of the two possible enols immediately explains the orientation. Recall from Sec. 7.12 that the more highly substituted carbon-carbon double bond is the more stable. Even though both enols are unstable with respect to the keto form, the more stable enol is more likely to be formed than the less stable one:

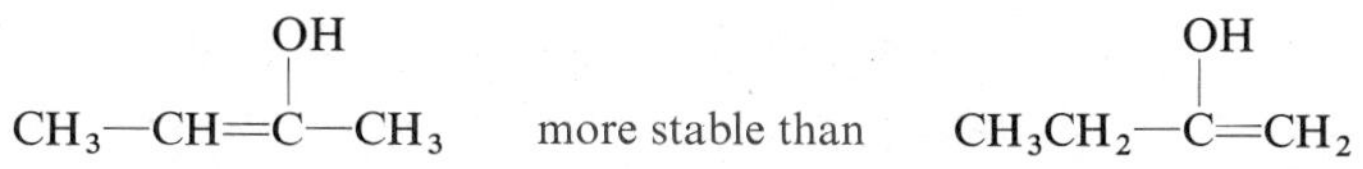

$$\underset{\substack{\text{Contains more}\\\text{highly substituted}\\\text{double bond}}}{CH_3-CH=\overset{\overset{\displaystyle OH}{|}}{C}-CH_3} \quad \text{more stable than} \quad CH_3CH_2-\overset{\overset{\displaystyle OH}{|}}{C}=CH_2$$

Once the enol forms, it reacts immediately with halogen.

In summary, *acid-catalyzed halogenation of unsymmetrical ketones occurs with the halogen becoming attached to the α carbon that originally contained fewer hydrogens.* Contrast this to the base-promoted reaction discussed in Sec. 24.5.

Question 24.8

Indicate the major monobromo product that would be obtained by allowing each of the following compounds to react with bromine under acidic conditions:

(*a*) butyraldehyde   (*b*) cyclohexanone   (*c*) 2-methylcyclohexanone
(*d*) 2-pentanone   (*e*) 3-pentanone   (*f*) 2-methyl-3-pentanone

---

B. Acid-Catalyzed Keto-enol Equilibrium

Implicit in the mechanism for the acid-catalyzed halogenation of a carbonyl compound is *keto-enol equilibrium*. The important intermediate in this equilibrium is the protonated carbonyl compound, which can lose a proton from oxygen and regenerate the carbonyl compound or lose a proton from the α carbon atom to produce the enol from:

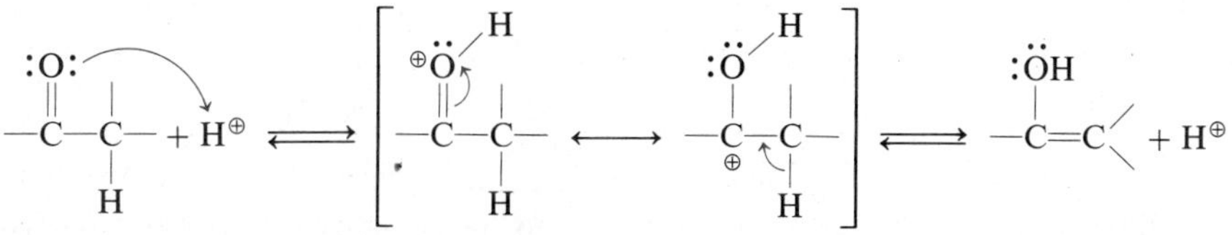

Keto form   Enol form

Usually the equilibrium overwhelmingly favors the keto form.

It is not surprising that the reverse reaction, the acid-catalyzed conversion of an enol to a keto compound, involves the protonation of the enol to produce the more stable carbocation. Protonation occurs with the positive charge placed on the carbon atom bearing the oxygen atom. The resulting carbocation is exceptionally stable because oxygen can accept the positive charge so that all atoms have their full octet of electrons; for example:

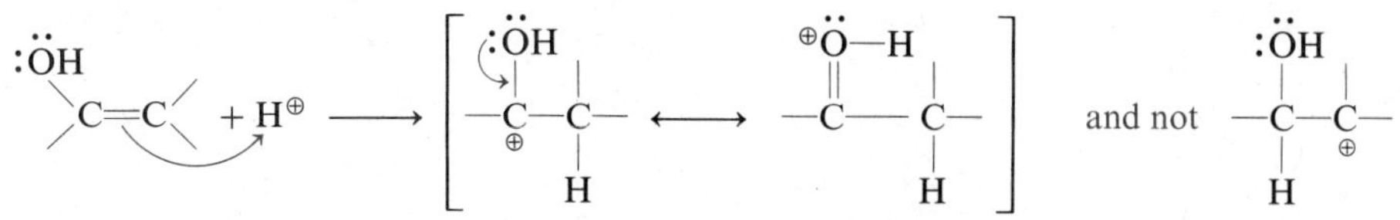

Stable
All atoms have octet of electrons

Less stable

C. Mechanism of Hell-Volhard-Zelinsky Reaction

Recall from Sec. 22.18 that the HVZ reaction is useful for introducing a halogen α to the carboxyl group:

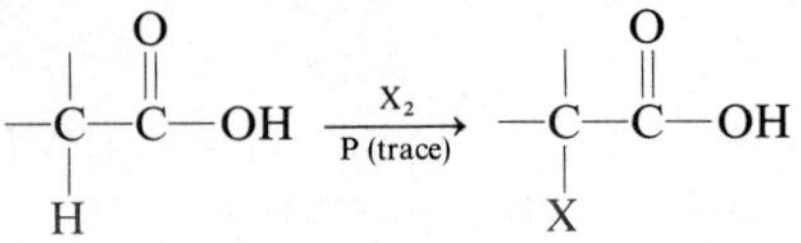

$$-\overset{\displaystyle |}{\underset{\displaystyle H}{C}}-\overset{\displaystyle O}{\overset{\|}{C}}-OH \xrightarrow[\text{P (trace)}]{X_2} -\overset{\displaystyle |}{\underset{\displaystyle X}{C}}-\overset{\displaystyle O}{\overset{\|}{C}}-OH$$

where X = Cl, Br

This reaction may not appear to involve anything remotely related to keto-enol equilibrium, but it does.

A catalytic amount of phosphorus is used in the HVZ reaction to convert some halogen into PX_3, which then reacts with the carboxylic acid to produce acid halide:

$$2P + 3X_2 \longrightarrow 2PX_3$$

$$R-\overset{\displaystyle O}{\overset{\|}{C}}-OH + PX_3 \longrightarrow R-\overset{\displaystyle O}{\overset{\|}{C}}-X + HOPX_2$$

The formation of the acid halide generates some protons that help establish keto-enol equilibrium within the acid halide. Alternatively, an acid halide can be treated directly with halogen to give the same results. The enol form of the acid halide reacts with halogen to produce the α-halo acid halide, as shown by the following reactions:

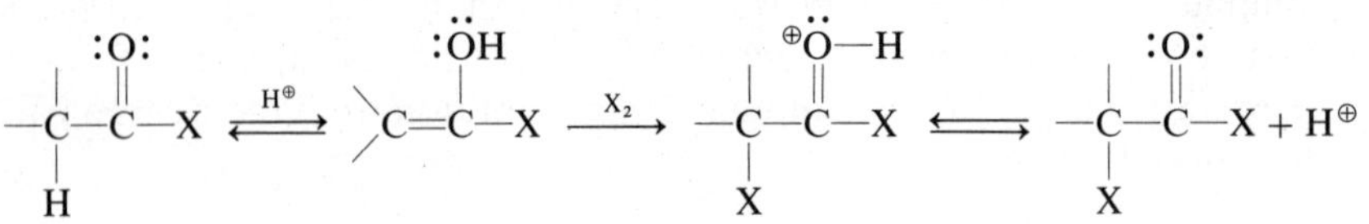

$$-\overset{\displaystyle :O:}{\underset{\displaystyle H}{\overset{\displaystyle |}{C}}}-\overset{\displaystyle |}{C}-X \underset{}{\overset{H^{\oplus}}{\rightleftharpoons}} \underset{}{\overset{:\ddot{O}H}{\underset{}{>}}}C=C-X \xrightarrow{X_2} -\overset{\displaystyle \overset{\oplus}{O}-H}{\overset{\displaystyle |}{\underset{\displaystyle X}{C}}}-\overset{\displaystyle |}{C}-X \rightleftharpoons -\overset{\displaystyle :O:}{\underset{\displaystyle X}{\overset{\displaystyle |}{C}}}-\overset{\displaystyle |}{C}-X + H^{\oplus}$$

    Acid   Enol   α-Halo acid
    halide   halide

The α-halo acid is produced by exchange with the free acid or when the reaction mixture containing the α-halo acid halide is hydrolyzed during workup, because the acid halide is exceptionally reactive toward water. The α halogen is not affected under these conditions.

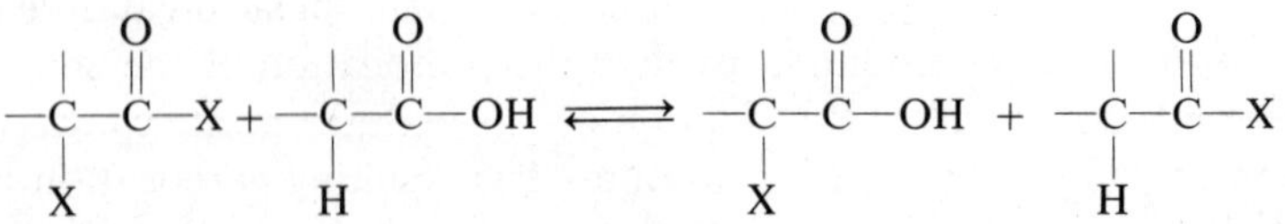

$$-\overset{\displaystyle O}{\underset{\displaystyle X}{\overset{\displaystyle |}{C}}}-\overset{\displaystyle \|}{C}-X + -\overset{\displaystyle |}{\underset{\displaystyle H}{C}}-\overset{\displaystyle O}{\overset{\|}{C}}-OH \rightleftharpoons -\overset{\displaystyle |}{\underset{\displaystyle X}{C}}-\overset{\displaystyle O}{\overset{\|}{C}}-OH + -\overset{\displaystyle |}{\underset{\displaystyle H}{C}}-\overset{\displaystyle O}{\overset{\|}{C}}-X$$

                                                             Product   Acid halide
                                                            Continues reaction

or

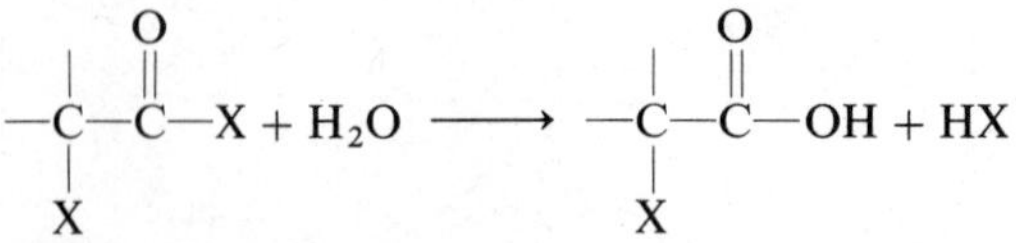

$$-\overset{\displaystyle |}{\underset{\displaystyle X}{C}}-\overset{\displaystyle O}{\overset{\|}{C}}-X + H_2O \longrightarrow -\overset{\displaystyle |}{\underset{\displaystyle X}{C}}-\overset{\displaystyle O}{\overset{\|}{C}}-OH + HX$$

                                                             Product

Condensation Reactions

24.8   Condensation of Aldehydes and Ketones: Aldol Condensation

We now return to the reactions of aldehydes and ketones in basic solution.

A. Base-Catalyzed Condensation

As we saw in Sec. 24.1, hydrogens attached to a carbon atom adjacent to the carbon-oxygen double bond are acidic and are fairly easily removed by base. If an aldehyde or ketone is treated with base in the presence of a halogen, α halogenation occurs. But what, if anything, happens when a carbonyl compound comes in contact with dilute aqueous base and no halogen is present?

In the presence of dilute base, there is a condensation reaction[1] involving two molecules of carbonyl compound. The product is a β-hydroxyaldehyde when an aldehyde is used or a β-hydroxyketone when a ketone is the carbonyl compound:

$$\underset{\substack{\text{Aldehyde}\\ \textit{2 moles}}}{-\overset{\overset{\displaystyle O}{\|}}{\underset{\underset{H}{|}}{C}}-\overset{\displaystyle\|}{C}-H} \xrightarrow[\text{H}_2\text{O}]{:\ddot{O}\text{H}^{\ominus}} \underset{\beta\text{-Hydroxyaldehyde}}{-\overset{\overset{\displaystyle OH}{|}}{\underset{\delta}{C}}-\underset{\underset{H}{|}}{\overset{|}{\underset{\beta}{C}}}-\underset{\underset{H}{|}}{\overset{|}{\underset{\alpha}{C}}}-\overset{\overset{\displaystyle O}{\|}}{C}-H}$$

$$\underset{\substack{\text{Ketone}\\ \textit{2 moles}}}{-\overset{\overset{\displaystyle O}{\|}}{\underset{\underset{H}{|}}{C}}-\overset{\displaystyle\|}{C}-R} \xrightarrow[\text{H}_2\text{O}]{:\ddot{O}\text{H}^{\ominus}} \underset{\beta\text{-Hydroxyketone}}{-\overset{\overset{\displaystyle OH}{|}}{\underset{\delta}{C}}-\underset{\underset{H}{|}}{\overset{|}{\underset{\beta}{C}}}-\underset{\underset{R}{|}}{\overset{|}{\underset{\alpha}{C}}}-\overset{\overset{\displaystyle O}{\|}}{C}-R}$$

These reactions are called **aldol condensations.** This term is not completely descriptive and indeed may be misleading. It arose because the product from an aldehyde contains both an aldehyde and an alcohol group; *aldol* is a contraction for *ald*ehyde and alco*hol*. Note that an aldol is formed and is *not* the compound undergoing condensation. The term *aldol condensation* is used widely for the condensation reactions of both aldehydes and ketones in basic solution.

To understand the aldol condensation reaction, remember that when an aldehyde or ketone comes in contact with base, the result is carbanion formation:

$$-\overset{\overset{\displaystyle :O:}{\|}}{\underset{\underset{H}{|}}{C}}-C-R + :\ddot{O}\text{H}^{\ominus} \rightleftharpoons \overset{\delta\ominus}{-}\overset{\delta\ominus}{C}\cdots\overset{\overset{\displaystyle \overset{\delta\ominus}{O}}{\|}}{\underset{R}{C}} + \text{H}_2\text{O}$$

where R = H, alkyl, or aryl

Carbon in the carbanion bears some negative charge. What do we know about the chemistry of negatively charged carbon in other reactions? For one thing, the

---

[1] A condensation reaction, in the most general sense, is the combination of two molecules with the liberation of a smaller, simpler molecule (for example, H_2O, ROH, and so on).

Grignard reagent contains carbon bearing a negative charge and this reagent *adds* to the carbonyl group (see Sec. 21.10):

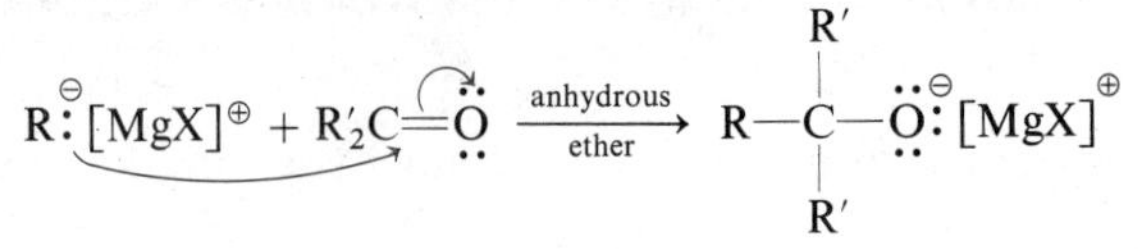

$$R:^{\ominus}[MgX]^{\oplus} + R'_2C{=}\overset{..}{\underset{..}{O}} \xrightarrow[\text{ether}]{\text{anhydrous}} R{-}\overset{\overset{\displaystyle R'}{|}}{\underset{\underset{\displaystyle R'}{|}}{C}}{-}\overset{..}{\underset{..}{O}}{:}^{\ominus}[MgX]^{\oplus}$$

By analogy, consider what a carbanion is likely to do. Keeping in mind that there is only a small concentration of carbanion and a large excess of carbonyl compound, it is reasonable to expect the carbanion to *add* to another molecule of carbonyl compound:

Step 1. Carbanion formation

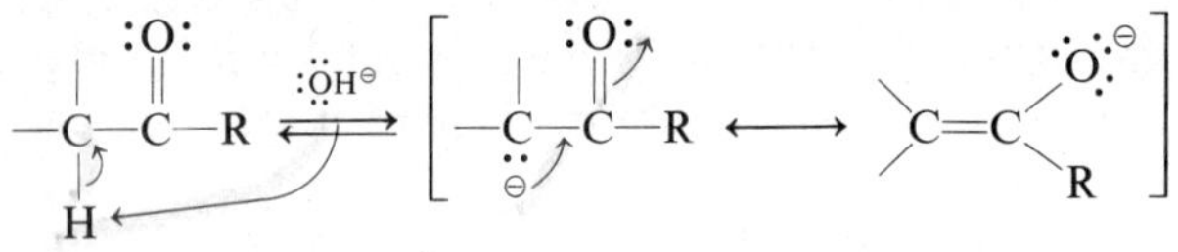

Conjugated carbanion

Step 2. Nucleophilic addition of carbanion to carbonyl group

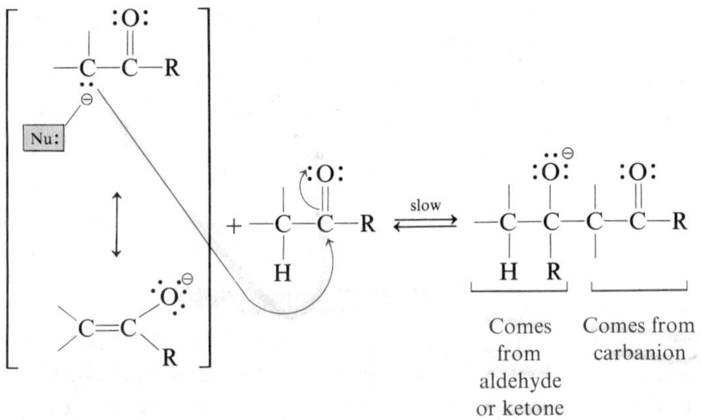

Comes from aldehyde or ketone

Comes from carbanion

Step 3. Abstraction of proton from water to form product

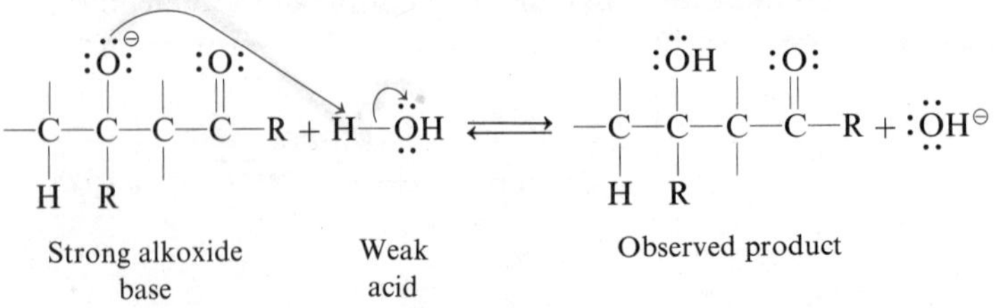

Strong alkoxide base

Weak acid

Observed product

Step 1 shows the reversible formation of the conjugated carbanion. Step 2 is the addition of the carbanion to another molecule of carbonyl compound. For emphasis, the resonance structure with negative charge on the α carbon is shown as the attacking nucleophile. Step 3 is a typical acid-base reaction, in which the strong alkoxide base abstracts a proton from water, a weak acid. There are likely to be relatively few "free" protons in basic solution, so protonation most likely involves attack of the alkoxide base on water. Three examples are:

$$CH_3-\overset{\overset{\textstyle O}{\|}}{C}-H + \underset{\underset{\textstyle H}{|}}{CH_2}-\overset{\overset{\textstyle O}{\|}}{C}-H \xrightarrow[H_2O]{:\ddot{O}H^{\ominus}} CH_3-\underset{\underset{\textstyle H}{|}}{\overset{\overset{\textstyle OH}{|}}{C}}-CH_2-\overset{\overset{\textstyle O}{\|}}{C}-H$$

Ethanal	3-Hydroxybutanal
(acetaldehyde)	(β-hydroxybutyraldehyde)

$$CH_3CH_2-\overset{\overset{\textstyle O}{\|}}{C}-H + \underset{\underset{\textstyle H}{|}}{CH_3CH}-\overset{\overset{\textstyle O}{\|}}{C}-H \xrightarrow[H_2O]{:\ddot{O}H^{\ominus}} CH_3CH_2-\underset{\underset{\textstyle H\ \ CH_3}{|}}{\overset{\overset{\textstyle OH}{|}}{C}}-CH-\overset{\overset{\textstyle O}{\|}}{C}-H$$

Propanal	3-Hydroxy-2-methylpentanal
(propionaldehyde)	β-hydroxy-α-methylvaleraldehyde

$$CH_3-\overset{\overset{\textstyle O}{\|}}{C}-CH_3 + \underset{\underset{\textstyle H}{|}}{CH_2}-\overset{\overset{\textstyle O}{\|}}{C}-CH_3 \xrightarrow[H_2O]{:\ddot{O}H^{\ominus}} CH_3-\underset{\underset{\textstyle CH_3}{|}}{\overset{\overset{\textstyle OH}{|}}{C}}-CH_2-\overset{\overset{\textstyle O}{\|}}{C}-CH_3$$

2-Propanone	4-Hydroxy-4-methyl-2-pentanone
(acetone)	(diacetone alcohol)

As the mechanism of the aldol condensation reveals, the *net* change is the addition of an α-C—H bond of one molecule across the carbon-oxygen double bond of the other molecule:

$$-\underset{\underset{\textstyle H}{|}}{\overset{\overset{\textstyle O}{\|}}{C}}-\overset{\overset{\textstyle O}{\|}}{C}-R + -\underset{\underset{\textstyle H}{|}}{\overset{\overset{\textstyle O}{\|}}{C}}-\overset{\overset{\textstyle O}{\|}}{C}-R \longrightarrow -\overset{\overset{\textstyle OH}{|}}{C}-\underset{\underset{\textstyle H\ \ R}{|}}{C}-\overset{}{C}-\overset{\overset{\textstyle O}{\|}}{C}-R$$

Not all aldol condensation reactions give an "aldol" as the final product. When an aromatic ring is attached to one side of the carbonyl group in a ketone (and this applies only for ketones), an aldol is formed initially. The resulting compound, however, loses water spontaneously to produce an α,β-unsaturated carbonyl compound. The following illustrates these reactions:

$$\text{Acetophenone} + \underset{\underset{\textstyle H}{|}}{CH_2}-\overset{\overset{\textstyle O}{\|}}{C}- \xrightarrow[H_2O]{:\ddot{O}H^{\ominus}} \left[\ \right]$$

Acetophenone	3-Hydroxy-1,3-diphenyl-1-butanone
	Unstable

1,3-Diphenyl-2-buten-1-one
Observed product

$+ H_2O$

The driving force for dehydration of the aldol is the formation of the unusually stable α,β-unsaturated carbonyl linkage, which is in conjugation with the aromatic ring attached to the carbon-carbon double bond. The orbital view of this

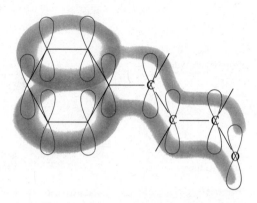

Represented by

FIGURE 24.3   Orbital view of an α,β-unsaturated ketone conjugated with an aromatic ring which produces an unusually stable system.

conjugated π system is shown in Fig. 24.3. *The aromatic ring adjacent to the hydroxyl group in the aldol is largely responsible for the facile dehydration* because it becomes part of the conjugated system in the unsaturated product. That the alcohol part of the aldol is tertiary also facilitates dehydration.

Conjugated systems merit further mention at this point. We know that diene systems are most stable when the double bonds are conjugated, as in 1,3-butadiene, $CH_2=CH—CH=CH_2$ (see Sec. 7.9). Likewise, we would expect α,β-unsaturated carbonyl compounds to be more stable than when the unsaturation is present elsewhere in the molecule; that is, conjugated dienes and α,β-unsaturated carbonyl compounds are analogous:

Conjugated diene   analogous to   α,β-Unsaturated carbonyl compound
Conjugated system

Considerable evidence supports the stability of α,β-unsaturated carbonyl compounds that do not even contain an aromatic ring. For example, 3-hydroxybutanal (obtained from the aldol condensation of acetaldehyde) undergoes facile dehydration under the influence of acid; the product is 2-butenal but not 3-butenal:

3-Hydroxybutanal   2-Butenal   3-Butenal
(crotonaldehyde)   *Not formed*
Only product

Even though this particular example obeys the Saytzeff rule for product distribution in dehydration, which states that the more highly substituted alkene is the major product (see Sec. 7.17), the formation of the conjugated system is of prime importance. Only one product is formed, but some alcohol dehydrations give two (or more) products, with the α,β-unsaturated compound formed in major amounts. We encounter more examples of dehydration when discussing the use of aldol products in synthesis in Sec. 24.10.

Question 24.9

Is the aldol condensation base-catalyzed or base-promoted? That is, does base serve as a catalyst or is it actually consumed in the reaction, as it is in the base-promoted halogenation of aldehydes and ketones? Briefly explain.

Question 24.10

Provide complete, stepwise mechanisms for each example in this section.

Question 24.11

With the aid of mechanisms, deduce the aldol products that are obtained by allowing each of the following five carbonyl compounds to come in contact with dilute aqueous base:

(*a*) 3-pentanone   (*b*) *tert*-butyl methyl ketone (pinacolone)
(*c*) phenylacetaldehyde   (*d*) α-phenylpropionaldehyde
(*e*) cyclohexanone

B. Acid-Catalyzed Condensation

Simple aldol condensations occur under *both* basic and acidic conditions. We discussed the base-catalyzed reaction, which is the most common one. We now look at the acid-catalyzed reaction using acetaldehyde as an example:

$$2\ CH_3-\overset{\overset{O}{\|}}{C}-H \xrightarrow[H_2O]{H^\oplus} \left[CH_3-\overset{\overset{OH}{|}}{CH}-CH_2-\overset{\overset{O}{\|}}{C}-H \right] \xrightarrow{H^\oplus} CH_3-CH=CH-\overset{\overset{O}{\|}}{C}-H$$

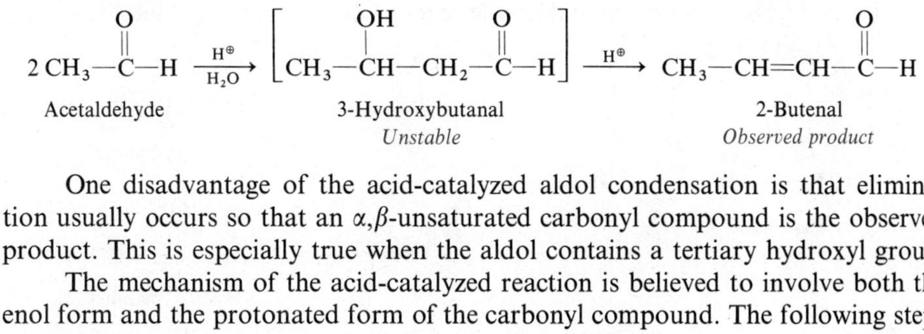

Acetaldehyde   3-Hydroxybutanal   2-Butenal
                          Unstable   *Observed product*

One disadvantage of the acid-catalyzed aldol condensation is that elimination usually occurs so that an α,β-unsaturated carbonyl compound is the observed product. This is especially true when the aldol contains a tertiary hydroxyl group.

The mechanism of the acid-catalyzed reaction is believed to involve both the enol form and the protonated form of the carbonyl compound. The following steps outline the reaction with acetaldehyde as an example:

$$\left[CH_3-\overset{\overset{\oplus \ddot{O}-H}{\|}}{C}-H \longleftrightarrow CH_3-\overset{\overset{:\ddot{O}H}{|}}{\underset{\oplus}{C}}-H \right] + CH_2=\overset{\overset{\ddot{O}H}{|}}{C}-H \xrightarrow{\text{step 1}}$$

Protonated acetaldehyde   Enol of
                                          acetaldehyde

$$\left[CH_3-\overset{\overset{OH}{|}}{CH}-CH_2-\overset{\overset{:\ddot{O}H}{|}}{\underset{\oplus}{C}}-H \longleftrightarrow CH_3-\overset{\overset{OH}{|}}{CH}-CH_2-\overset{\overset{\oplus\ddot{O}-H}{\|}}{C}-H \right] \xrightarrow{\text{step 2}}$$

$$CH_3-\overset{\overset{OH}{|}}{CH}-CH_2-\overset{\overset{:O:}{\|}}{C}-H + H^\oplus$$

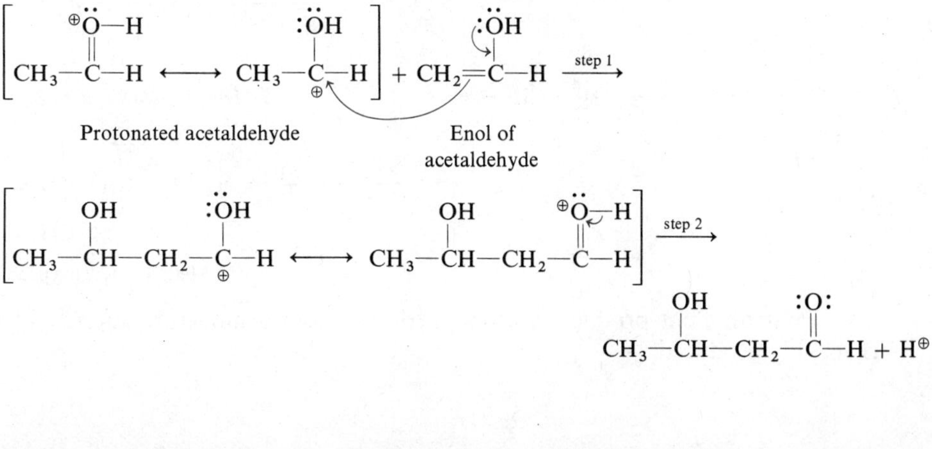

The β-hydroxyaldehyde thus formed then undergoes acid-catalyzed dehydration of the alcohol to give crotonaldehyde (2-butenal):

$$CH_3-\overset{\underset{|}{OH}}{CH}-CH_2-\overset{\underset{||}{O}}{C}-H \xrightarrow[-H_2O]{H^\oplus} CH_3-CH=CH-\overset{\underset{||}{O}}{C}-H$$

3-Hydroxybutanal   2-Butenal

(aldol)   (crotonaldehyde)

Question 24.12

Indicate complete, stepwise mechanisms to show the formation of the aldol product from the acid-catalyzed condensation of the following compounds:

(*a*) acetone   (*b*) acetophenone

24.9 Crossed Aldol Condensations

In Sec. 24.8 we considered only the self-condensation of an aldehyde or ketone containing one or more α hydrogens. We now consider reactions between two different carbonyl-containing compounds in basic solution. Because they involve different molecules, these reactions are called **crossed** or **mixed** aldol condensations.

A. Reaction of Two Different Aldehydes That Have α Hydrogens

Generally speaking, the aldol condensation reaction involving two different aldehydes having α hydrogens is of little utility because a mixture of products is obtained. This is illustrated for the reactions that occur when equimolar mixtures of acetaldehyde and propionaldehyde react in the presence of dilute base:

Four different products are obtained, although a moderate amount of each is actually produced.

B. Crossed Aldol Condensations Between Aldehydes and Ketones

Crossed aldol condensation reactions involving aldehydes and ketones that have α hydrogens can be carried out experimentally so that certain cross condensation products are not obtained. These are sometimes called **Claisen-Schmidt reactions.** For example, in the crossed reaction of aldehyde (1) and ketone (2), the self-condensation of the ketone (3) and the condensation resulting from the addition of the α hydrogen of the aldehyde to the ketone (4) do not occur to any extent:

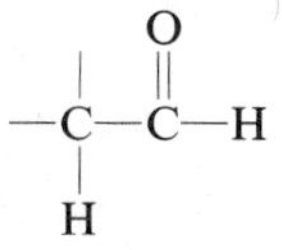

(1) 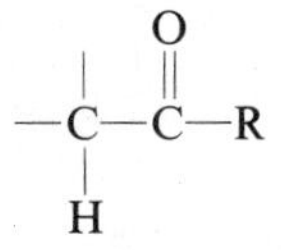   (2)

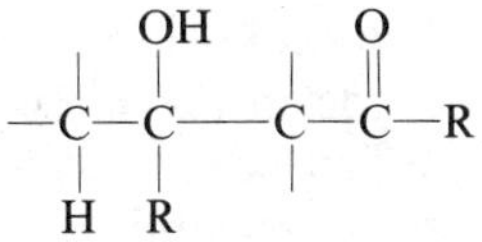

(3)   and   (4)

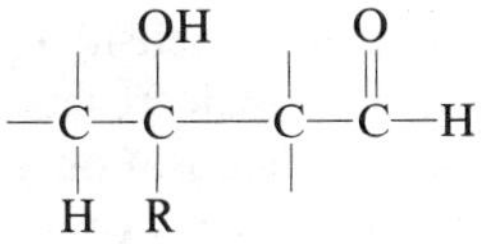

Not formed to any extent

The self-condensation of the aldehyde product (5) can be minimized by slowly adding the aldehyde to a mixture containing an excess of the ketone and the base.

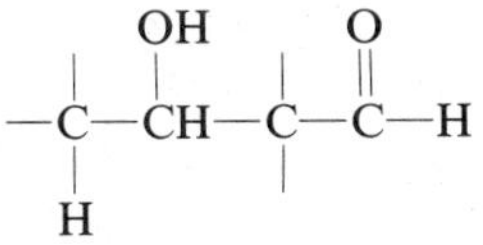

(5) 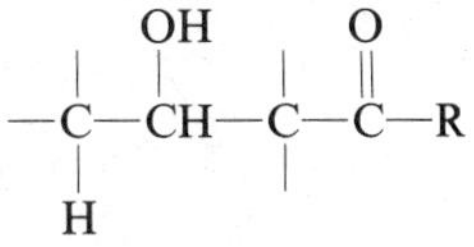   (6)

Major product

The remaining crossed condensation product (6) arises from the addition of the ketone α hydrogen across the carbonyl group in the aldehyde, and it is produced in good yield.

When unsymmetrical ketones are used in base-catalyzed reactions, preferential abstraction of an α hydrogen from the methyl group rather than from the methylene group occurs in the compound $R—CH_2—CO—CH_3$ (see Sec. 24.5).

C. Crossed Aldol Condensations Involving Aldehydes or Ketones That Have No α Hydrogens

Crossed aldol condensations of an aldehyde or ketone having α hydrogens and one having no α hydrogens are usually more successful because fewer side reactions occur. The compound with no α hydrogens cannot undergo self-condensation, and self-condensation of the other carbonyl compound can be minimized by controlling the experimental conditions. The following two examples illustrate this reaction:

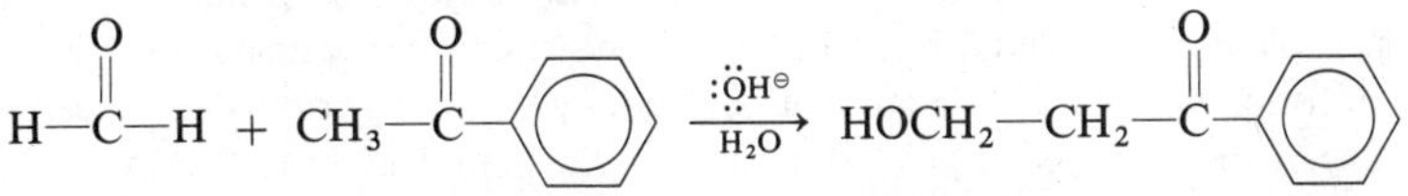

Formaldehyde   Acetophenone   3-Hydroxy-1-phenyl-1-propanone

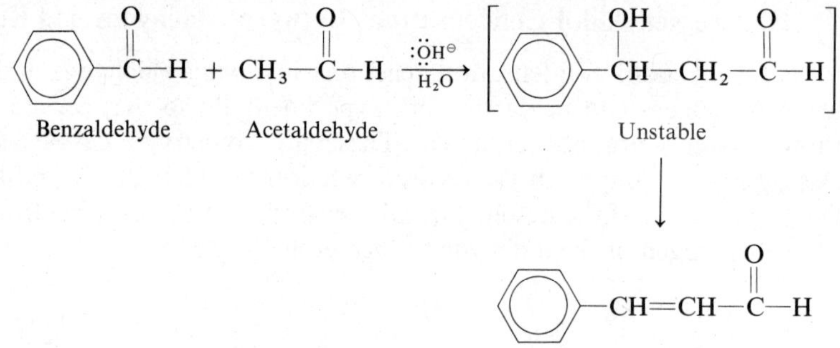

Benzaldehyde   Acetaldehyde   Unstable

3-Phenylpropenal
(cinnamaldehyde[1])

The mechanisms for these reactions are exactly the same as those for the self-condensation reaction in basic solution (see Sec. 24.8A).

An interesting extension of the crossed aldol condensation occurs when a large excess of formaldehyde reacts with acetaldehyde in the presence of base. As expected, crossed condensation occurs, but the first condensation product still contains two acidic α hydrogens. Condensation occurs until all the α hydrogens in acetaldehyde have reacted.

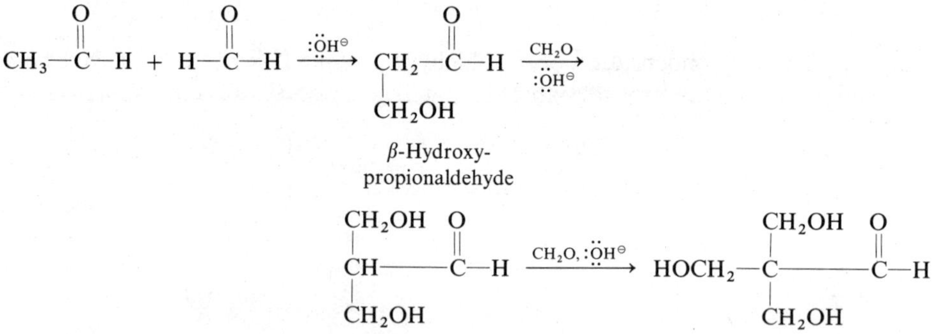

β-Hydroxy-
propionaldehyde

α-Hydroxymethyl-
β-hydroxypropionaldehyde

α,α-Di(hydroxymethyl)-
β-hydroxypropionaldehyde

This reaction is of industrial importance because the final product can be reduced to give pentaerythritol, $C(CH_2OH)_4$, which is often used in the manufacture of dishwasher soaps (see Sec. 25.2). The reduction of the aldehyde group can be done in a number of ways, but as we see in Sec. 24.11, the Cannizzaro reaction is the most convenient and most widely used.

Question 24.13

Give the major product(s) for the condensation of the following compounds under basic conditions:

(*a*) acetophenone + acetaldehyde
(*b*) propionaldehyde + acetophenone
(*c*) propionaldehyde + formaldehyde (large excess)
(*d*) acetophenone + benzaldehyde

Question 24.14

The base-promoted reaction between acetaldehyde and a large excess of formaldehyde produces $(HOCH_2)_3CCHO$, but when the same reaction is carried out using benzaldehyde in place of formaldehyde, only 1 mole of benzaldehyde is incorporated into the product. Draw the structure of the anticipated product and briefly explain this observation.

---

[1] Cinnamaldehyde is a common name for this compound because its odor resembles that of cinnamon.

24.10   Applications of Aldol Condensations to Synthesis

Aldol condensations enjoy great use as a synthetic tool for the organic chemist because they are another method for making a carbon-carbon bond. Aldehydes and ketones are readily prepared from available alcohols or are commercially available. In this section we see some important transformations of aldol products to other functional groups.

The self-condensation of aldehydes or ketones yields a product in which *the carbon content is doubled.* If a synthesis starts with an aldehyde or ketone and if the desired product contains twice as many carbon atoms as the starting material, an aldol condensation may be involved. Consider the example of converting 1-butanol into

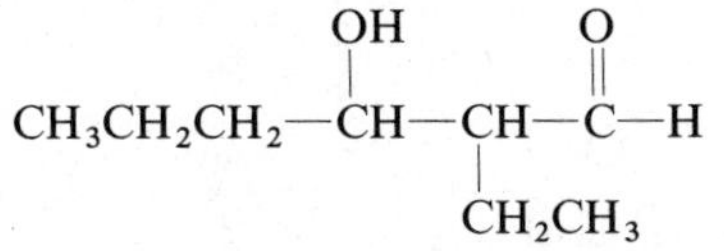

$$CH_3CH_2CH_2-\overset{\overset{\displaystyle OH}{|}}{CH}-\overset{\overset{\displaystyle}{|}}{\underset{\underset{\displaystyle CH_2CH_3}{|}}{CH}}-\overset{\overset{\displaystyle O}{\|}}{C}-H$$

Observe that the product contains twice as many carbon atoms (eight) as 1-butanol (four). Furthermore, the product is a β-hydroxyaldehyde, a functional-group arrangement that is produced in the aldol condensation reaction. In this particularly simple example, the following steps are required:

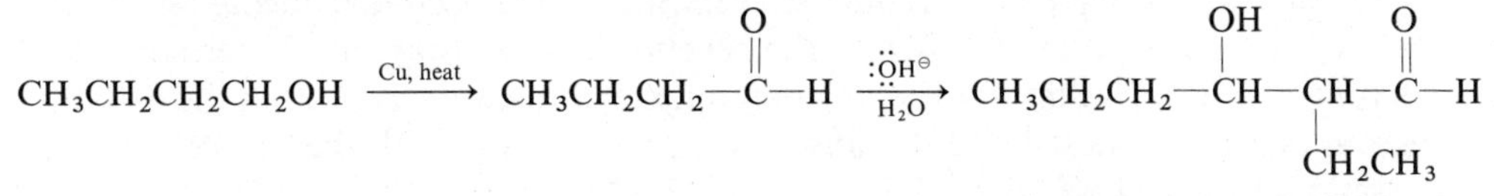

$$CH_3CH_2CH_2CH_2OH \xrightarrow{Cu,\ heat} CH_3CH_2CH_2-\overset{\overset{\displaystyle O}{\|}}{C}-H \xrightarrow[H_2O]{\overset{\cdot\cdot}{O}H^{\ominus}} CH_3CH_2CH_2-\overset{\overset{\displaystyle OH}{|}}{CH}-\overset{\underset{\underset{\displaystyle CH_2CH_3}{|}}{}}{CH}-\overset{\overset{\displaystyle O}{\|}}{C}-H$$

1-Butanol   2 moles   3-Hydroxy-2-ethylhexanal

Aldol products can be converted to other products by different chemical reactions. Industrially, aldol condensations are often used in the synthesis of alcohols, diols, and acids. The following scheme shows how the aldol derived from propanal can be converted into other compounds. Pay particular attention to how other functional groups are introduced into a molecule.

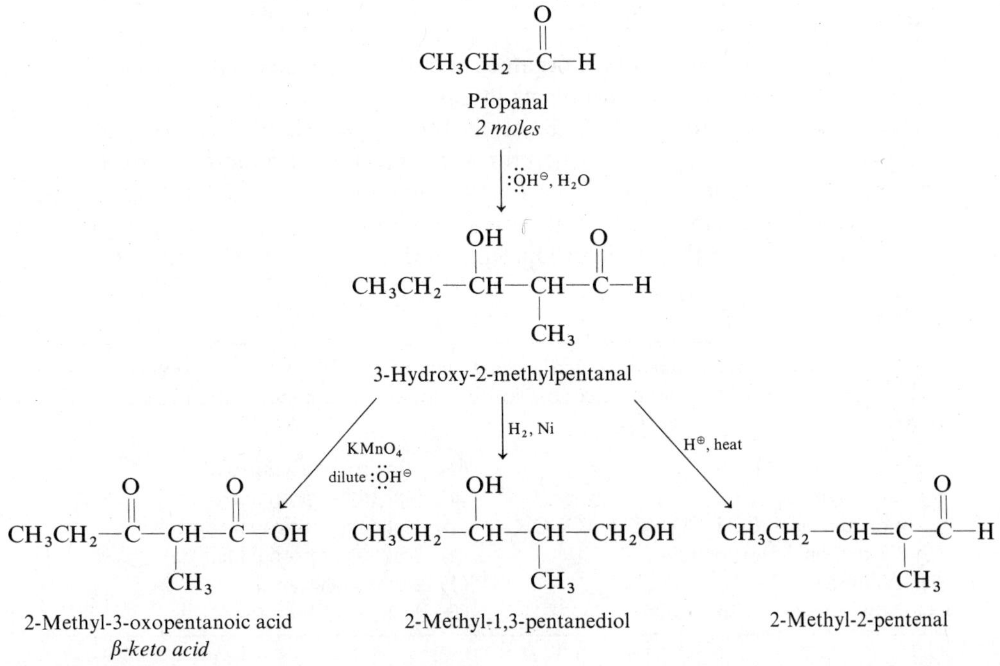

$$CH_3CH_2-\overset{\overset{\displaystyle O}{\|}}{C}-H$$

Propanal
2 moles

$$\Big\downarrow \overset{\cdot\cdot}{:}OH^{\ominus},\ H_2O$$

$$CH_3CH_2-\overset{\overset{\displaystyle OH}{|}}{CH}-\overset{\underset{\underset{\displaystyle CH_3}{|}}{}}{CH}-\overset{\overset{\displaystyle O}{\|}}{C}-H$$

3-Hydroxy-2-methylpentanal

KMnO₄ path: dilute $:\overset{\cdot\cdot}{O}H^{\ominus}$

H_2, Ni

$H^{\oplus}$, heat

$$CH_3CH_2-\overset{\overset{\displaystyle O}{\|}}{C}-\overset{\underset{\underset{\displaystyle CH_3}{|}}{}}{CH}-\overset{\overset{\displaystyle O}{\|}}{C}-OH$$

$$CH_3CH_2-\overset{\overset{\displaystyle OH}{|}}{CH}-\overset{\underset{\underset{\displaystyle CH_3}{|}}{}}{CH}-CH_2OH$$

$$CH_3CH_2-CH{=}\overset{\underset{\underset{\displaystyle CH_3}{|}}{}}{C}-\overset{\overset{\displaystyle O}{\|}}{C}-H$$

2-Methyl-3-oxopentanoic acid
β-keto acid

2-Methyl-1,3-pentanediol

2-Methyl-2-pentenal

The oxidation of 3-hydroxy-2-methylpentanal to 2-methyl-3-oxopentanoic acid requires a relatively mild oxidizing agent and slightly basic solution, because dehydration would occur if the oxidation were carried out in acidic solution (for example, chromic acid).

The acid-catalyzed dehydration of an aldol product is a very useful method for obtaining an *α,β-unsaturated aldehyde or ketone*, and the unsaturated aldehyde can in turn be converted into an *α,β-unsaturated acid or ester*. We discuss some important reactions of these unsaturated compounds in Sec. 24.23.

The α,β-unsaturated aldehyde, 2-methyl-2-pentenal, can be converted into either a saturated alcohol or an unsaturated alcohol depending on the reagents used:

$$\text{CH}_3\text{CH}_2\text{—CH}\text{=}\underset{\underset{\text{CH}_3}{|}}{\text{C}}\text{—}\overset{\overset{\text{O}}{\|}}{\text{C}}\text{—H}$$

2-Methyl-2-pentenal

$\xrightarrow{\text{H}_2,\text{ Ni}}$ CH$_3$CH$_2$CH$_2$—CH—CH$_2$OH

　　　　　　　　　　　　　　|
　　　　　　　　　　　　　CH$_3$

2-Methyl-1-pentanol

$\xrightarrow[\text{2. H}_3\ddot{\text{O}}^{\oplus}]{\text{1. NaBH}_4}$ CH$_3$CH$_2$—CH=C—CH$_2$OH

　　　　　　　　　　　　　　　　|
　　　　　　　　　　　　　　　CH$_3$

2-Methyl-2-penten-1-ol

The reduction of the aldehyde group to the primary alcohol without affecting the conjugated double bond requires a very mild and selective reducing agent such as sodium borohydride (NaBH$_4$). The reaction is not always 100% selective. If a stronger reducing agent like lithium aluminum hydride were used, both the carbonyl group and the double bond (because it is in conjugation with the carbonyl group) would be reduced (see Sec. 21.13).

If the α,β-unsaturated compound is an ester, then it is also possible to reduce the double bond without affecting the carbonyl portion of the molecule. The reaction is also difficult and yields are often low:

$$\text{CH}_3\text{—CH}\text{=}\text{CH}\text{—}\overset{\overset{\text{O}}{\|}}{\text{C}}\text{—OCH}_3 \xrightarrow[\text{CH}_3\text{CH}_2\text{OH}]{\text{NaBH}_4} \text{CH}_3\text{—CH}_2\text{—CH}_2\text{—}\overset{\overset{\text{O}}{\|}}{\text{C}}\text{—OCH}_3$$

Methyl 2-butenoate　　　　　　　　　　　　　　　Methyl butanoate

As with the rest of synthetic organic chemistry, it is essential to work problems. In addition to the preceding problems involving aldol condensations, two features that are most difficult to recognize are: (1) the building up of compounds using crossed aldol condensations where the structures of the products are more complex so that care must be used to select the right carbonyl compounds (that is, the carbon content is usually not doubled in a crossed aldol reaction), and (2) the transformation of the functional groups in the aldol product into other functional groups.

Question 24.15

Outline suitable methods for synthesizing each of the following compounds. Start with any needed aldehyde, ketone, or alcohol, and use at least one condensation reaction in each synthesis.

(*a*) 2-methylpentanal
(*b*) 3-hydroxybutyric acid
(*c*) 2-methylpentane
(*d*) 3-phenyl-1-propanol
(*e*) cinnamic acid, C$_6$H$_5$CH=CH—COOH
(*f*) 1-phenyl-1,3-propanediol
(*g*) 1,3-diphenyl-1-propanol
(*h*) 2-butenal (crotonaldehyde)
(*i*) 1,3-diphenyl-2-buten-1-one
(*j*) 2-ethyl-1-hexanol
(*k*) 2-(hydroxymethyl)propanal
(*l*) 4-phenyl-2-butanol

24.11   Reaction of Aldehydes That Contain No α Hydrogens: Cannizzaro Reaction

We may expect aldehydes with no α hydrogens to be unreactive in basic solution in the absence of any other reagents (halogens or other aldehydes and ketones); yet in concentrated aqueous base such aldehydes undergo an interesting and often useful reaction. Two moles of such an aldehyde react to produce 1 mole of the corresponding primary alcohol and 1 mole of the salt of the corresponding carboxylic acid:

$$
\underset{\substack{\text{No α hydrogens} \\ \text{2 moles}}}{R_3C-\overset{\displaystyle O}{\overset{\|}{C}}-H} \xrightarrow[\text{H}_2\text{O}]{\text{conc} :\ddot{O}H^{\ominus}} \underset{\text{1° Alcohol}}{R_3C-CH_2OH} + \underset{\substack{\text{Salt of} \\ \text{carboxylic} \\ \text{acid}}}{R_3C-C\ddot{O}\ddot{O}^{\ominus}}
$$

For example, formaldehyde is converted into methyl alcohol and sodium formate on treatment with *concentrated* sodium hydroxide solution:

$$
\underset{\substack{\text{Methanal} \\ \text{(formaldehyde)} \\ \text{2 moles}}}{H-\overset{\displaystyle O}{\overset{\|}{C}}-H} \xrightarrow[\text{H}_2\text{O, heat}]{\text{conc NaOH}} \underset{\substack{\text{Methanol} \\ \text{(methyl alcohol)}}}{CH_3OH} \quad \underset{\substack{\text{Sodium} \\ \text{methanoate} \\ \text{(sodium formate)}}}{H-\overset{\displaystyle O}{\overset{\|}{C}}-\ddot{O}^{\ominus}\ \overset{\oplus}{Na}}
$$

This reaction is called the **Cannizzaro reaction.** The aldehyde is both oxidized (to the salt of formic acid) and reduced (to methyl alcohol), a general reaction that is sometimes called *auto-oxidation* or *disproportionation*. The mechanism of the Cannizzaro reaction is as follows.

Step 1. Nucleophilic addition of hydroxide ion to carbonyl group

$$
R_3C-\overset{\displaystyle :O:}{\overset{\|}{C}}-H + \overset{\ominus\,..}{:}OH \quad \underset{}{\overset{\text{fast}}{\rightleftharpoons}} \quad R_3C-\overset{\displaystyle :\ddot{O}:^{\ominus}}{\overset{|}{\underset{|}{C}}}-H
$$

$$\underset{:\ddot{O}H}{}$$

(1)
Unstable tetrahedral intermediate

Step 2. Intermolecular hydride ion transfer

$$
R_3C-\overset{\displaystyle :\ddot{O}:^{\ominus}}{\underset{:\ddot{O}H}{\overset{|}{\underset{|}{C}}-H}} + H-\overset{\displaystyle :O:}{\overset{\|}{C}}-CR_3 \xrightarrow{\text{slow}} R_3C-\overset{\displaystyle :O:}{\overset{\|}{C}}-\ddot{O}H + H-\overset{\displaystyle :\ddot{O}:^{\ominus}}{\underset{H}{\overset{|}{\underset{|}{C}}}-CR_3}
$$

Step 3. Acid-base reactions yielding observed products

$$
R_3C-\overset{\displaystyle :O:}{\overset{\|}{C}}-\ddot{O}H + NaOH \longrightarrow R_3C-\overset{\displaystyle O}{\overset{\|}{C}}-\ddot{O}:^{\ominus}\ Na^{\oplus} + H_2O
$$

$$
R_3C-CH_2\ddot{O}:^{\ominus} + H_2O \rightleftharpoons R_3C-CH_2\ddot{O}H + :\ddot{O}H^{\ominus}
$$

The Cannizzaro reaction is used in the industrial synthesis of pentaerythritol (see Sec. 24.9), a synthetic intermediate in the manufacture of many dishwasher detergents. The reactions are summarized as follows:

$$
\underset{\substack{\text{Acetaldehyde}}}{CH_3-\overset{\overset{O}{\|}}{C}-H} \xrightarrow[\substack{\text{crossed aldol}\\ \text{3 times}}]{H-\overset{\overset{O}{\|}}{C}-H,\ H\ddot{O}{:}^{\ominus}} \underset{\substack{\alpha,\alpha\text{-Di(hydroxymethyl)-}\\ \beta\text{-hydroxypropionaldehyde}}}{HOCH_2-\underset{\underset{CH_2OH}{|}}{\overset{\overset{CH_2OH}{|}}{C}}-\overset{\overset{O}{\|}}{C}-H} \xrightarrow[\text{Cannizzaro}]{H-\overset{\overset{O}{\|}}{C}-H,\ H\ddot{O}{:}^{\ominus}} \underset{\substack{\text{Pentaerythritol}}}{HOCH_2-\underset{\underset{CH_2OH}{|}}{\overset{\overset{CH_2OH}{|}}{C}}-CH_2OH}
$$

Question 24.16

Provide complete, stepwise mechanisms for the following four reactions, and give the structure(s) of the organic products.

(*a*) benzaldehyde + *conc* NaOH
(*b*) benzaldehyde + formaldehyde + *conc* NaOH
(*c*) propionaldehyde + *excess* formaldehyde + *conc* NaOH
(*d*) 1,5-pentanedial + *conc* NaOH

Question 24.17

The addition of hydroxide ion to the carbonyl group in an aldehyde that has no α hydrogens is reversible in the Cannizzaro reaction. Suggest an experimental method for proving this fact.

24.12   Alkylation of Ketones: Enamine Synthesis

We have concentrated on the base-promoted condensation reactions of a carbonyl compound either with itself or with another carbonyl compound. Our knowledge about nucleophilic substitution at a saturated carbon in alkyl halides (that is, S_N2 reactions; see Sec. 5.16) suggests that carbanions derived from carbonyl compounds should undergo a similar type of reaction. The conjugated carbanion should serve as the nucleophile, with the result being alkylation of the α carbon atom:

$$
R-\overset{\overset{O}{\|}}{C}-\underset{\underset{H}{|}}{\overset{|}{C}}-\ +\ B{:}^{\ominus} \ \rightleftharpoons\ B{:}H\ +\ R-\overset{\overset{{:}O{:}}{\|}}{C}-\underset{Nu{:}}{\overset{|}{C}}{:}^{\ominus} \xrightarrow[(S_N2\ \text{reaction})]{R'-CH_2-X} R-\overset{\overset{O}{\|}}{C}-\underset{\underset{CH_2R'}{|}}{\overset{|}{C}}-
$$

Carbonyl compounds, especially ketones, are indeed alkylated in accordance with these expectations. Strong bases must usually be used to insure a reasonable concentration of carbanion. Potassium *tert*-butoxide, $(CH_3)_3C\ddot{O}{:}^{\ominus}\ K^{\oplus}$, and sodium amide (*sodamide*), $NaNH_2$, are often used, but they have the disadvantage of causing most secondary and tertiary alkyl halides to undergo dehydrohalogenation. In practice, methyl halides and allyl halides ($CH_2{=}CH-CH_2-X$) work best because they are not subject to elimination.

Three disadvantages to this alkylation reaction are:

1. Aldehydes and ketones are subject to self-condensation in the presence of strong base (see Sec. 24.5).

2. In the presence of strong base, two possible carbanions are generated from an unsymmetrical ketone, thus giving two different products; for example:

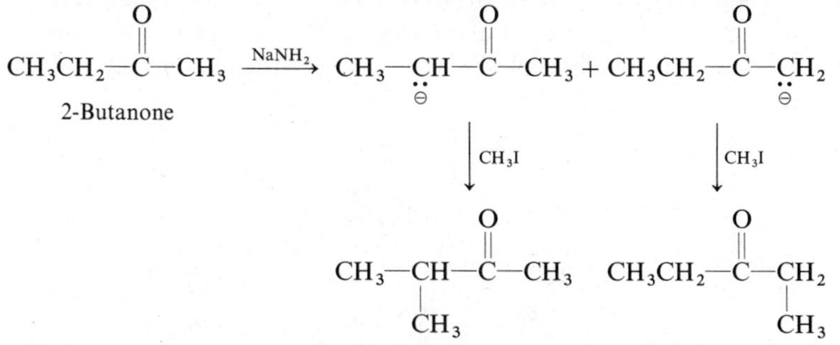

$$CH_3CH_2-\overset{\overset{\textstyle O}{\|}}{C}-CH_3 \xrightarrow{NaNH_2} CH_3-\underset{\ominus}{\overset{\cdot\cdot}{C}H}-\overset{\overset{\textstyle O}{\|}}{C}-CH_3 + CH_3CH_2-\overset{\overset{\textstyle O}{\|}}{C}-\underset{\ominus}{\overset{\cdot\cdot}{C}H_2}$$

2-Butanone

(left branch) ↓ CH₃I   (right branch) ↓ CH₃I

$$CH_3-\underset{\underset{\textstyle CH_3}{|}}{CH}-\overset{\overset{\textstyle O}{\|}}{C}-CH_3 \qquad CH_3CH_2-\overset{\overset{\textstyle O}{\|}}{C}-\underset{\underset{\textstyle CH_3}{|}}{CH_2}$$

3-Methyl-2-butanone   3-Pentanone

3. Polyalkylation often occurs to give a mixture of products that may be quite difficult to separate because their physical properties are nearly identical. For example, the alkylation of cyclohexanone using RX and sodamide gives the following products:

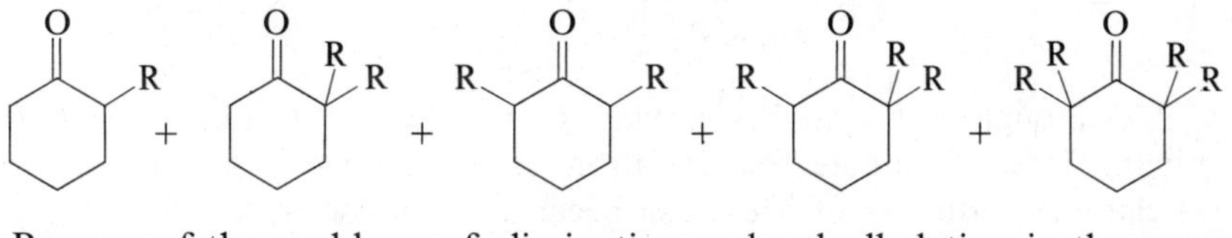

Because of the problems of elimination and polyalkylation in the preceding procedure, a novel method for alkylation was discovered and developed for controlling the alkylation of α carbon atoms. The carbonyl compound first reacts with a secondary amine, $R_2\overset{\cdot\cdot}{N}H$, which undergoes nucleophilic *addition* to the carbonyl group followed by loss of water:

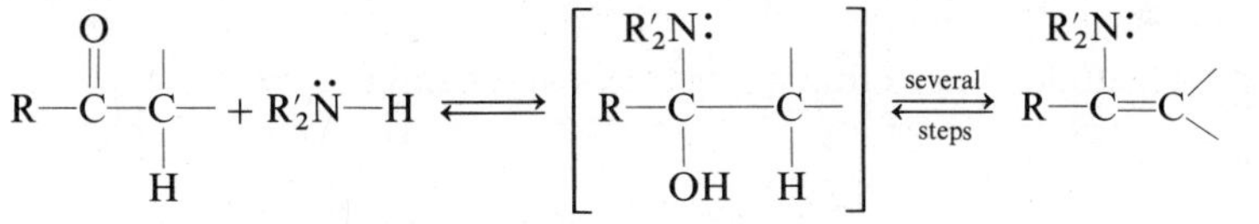

$$R-\overset{\overset{\textstyle O}{\|}}{C}-\underset{\underset{\textstyle H}{|}}{C}- + R_2'\overset{\cdot\cdot}{N}-H \rightleftharpoons \left[R-\underset{\underset{\textstyle OH}{|}}{\overset{\overset{\textstyle R_2'N:}{|}}{C}}-\underset{\underset{\textstyle H}{|}}{C}- \right] \underset{steps}{\overset{several}{\rightleftharpoons}} R-\overset{\overset{\textstyle R_2'N:}{|}}{C}=C\diagdown$$

2°   (1)   (2)

Amine   Unstable   Enamine

The mechanism for the addition of the secondary amine to the carbonyl group is the same as that for ammonia or a primary amine (see Sec. 21.8). Most amine additions are acid-catalyzed, and the acidity must be carefully controlled. There is only one way for amino alcohol intermediate (1) to lose water and that is by forming the carbon-carbon double bond. (*Question:* Why is this? The addition of ammonia or a primary amine to a carbonyl group produces a compound having the $\diagup C{=}N{-}R$ linkage.) The product of addition-dehydration (2) is called an **enamine,** where *en-* refers to the double bond (alk*ene*) and *amine* to the $R_2'N-$ group attached to it.

The enamine is a nucleophile and can undergo nucleophilic substitution with alkyl halides. The carbon-carbon double bond interacts with the unshared electron pair on nitrogen; this delocalization is represented by the resonance structures:

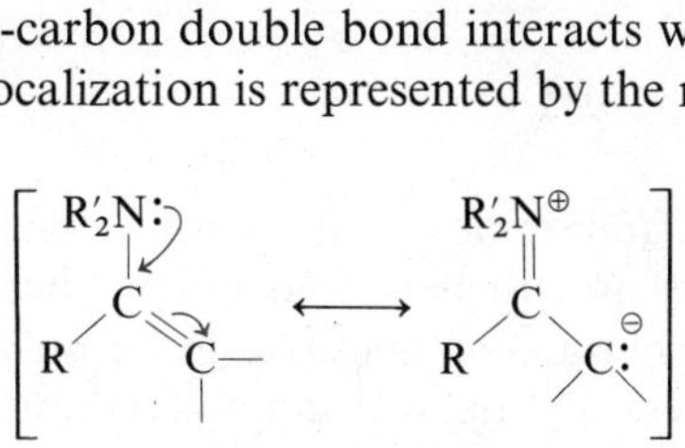

In principle, this same type of delocalization is involved in the electrophilic aromatic substitution of aryl amines (see Sec. 14.11). The added stability of the enamine grouping provides some driving force for dehydration of the amino alcohol (1).

The enamine undergoes alkylation at the two places bearing negative charge (it is an ambident nucleophile): (1) at nitrogen **(N alkylation)** and (2) at carbon **(C alkylation).**

N Alkylation:

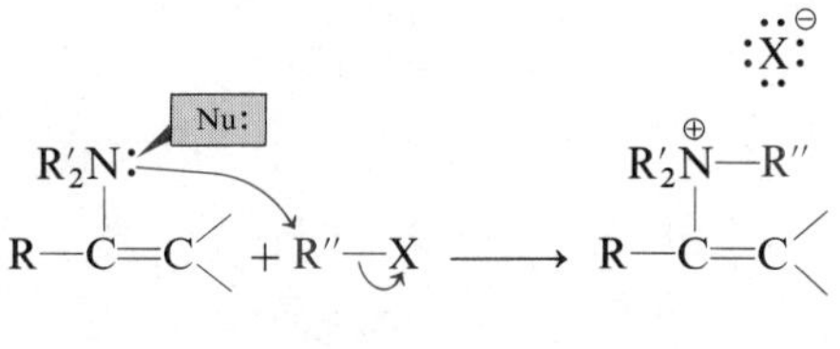

C Alkylation:

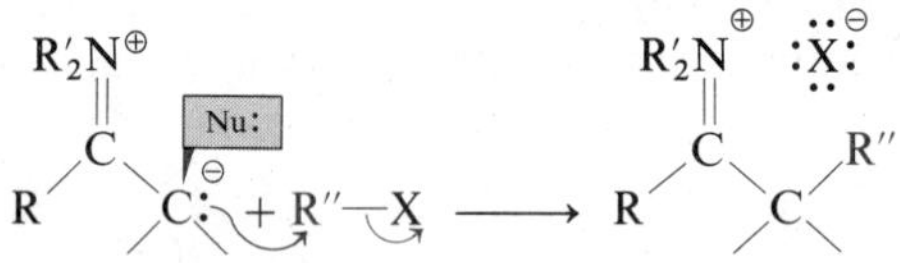

Although both C and N alkylation are observed, C alkylation is often predominant. This is fortunate, for alkylation on nitrogen has little synthetic value. The C alkylation product is of great use because it can be hydrolyzed into a carbonyl compound, which is alkylated in the α position:

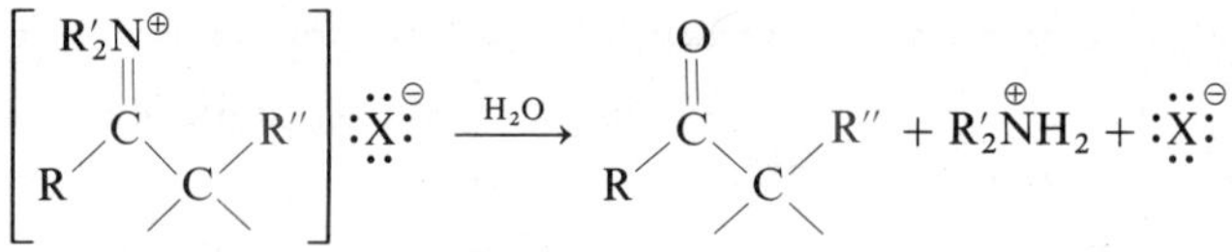

The following example illustrates an enamine synthesis:

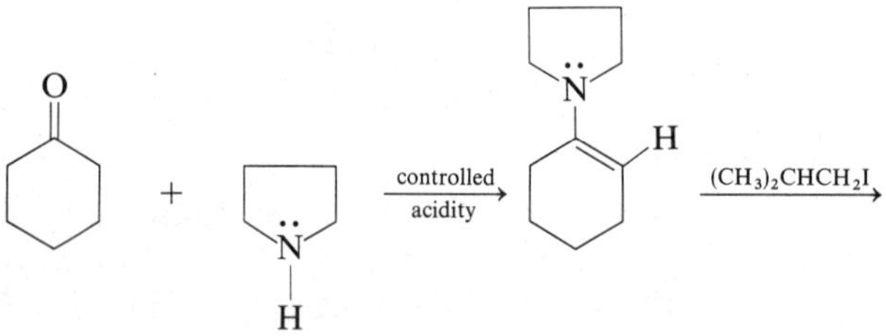

Cyclohexanone   Pyrrolidine

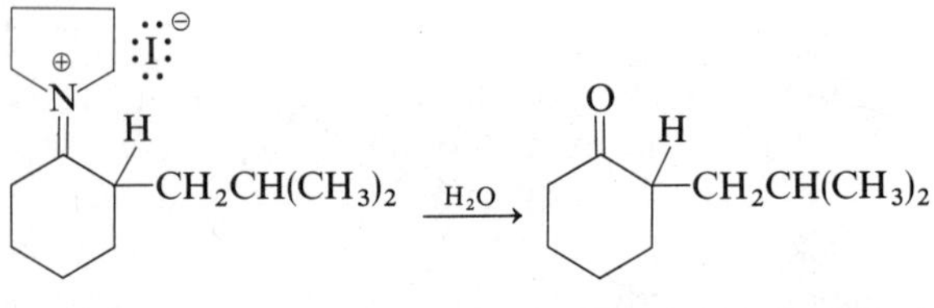

2-Isobutylcyclohexanone

The secondary amine most often used in enamine synthesis is pyrrolidine. This and other amines are discussed in detail in Chap. 26.

The enamine is not a *good* nucleophile, so only reactive halides such as alkyl iodides, benzyl halides, allyl halides, α-haloketones (XCH_2COR), and α-haloethers

(XCH$_2$OR) can be used. This method permits the use of many primary alkyl iodides because elimination is not important.

Question 24.18

Suggest routes for accomplishing the following syntheses, using any other needed organic reagents.

(*a*) acetone → hexamethylacetone

(*b*)

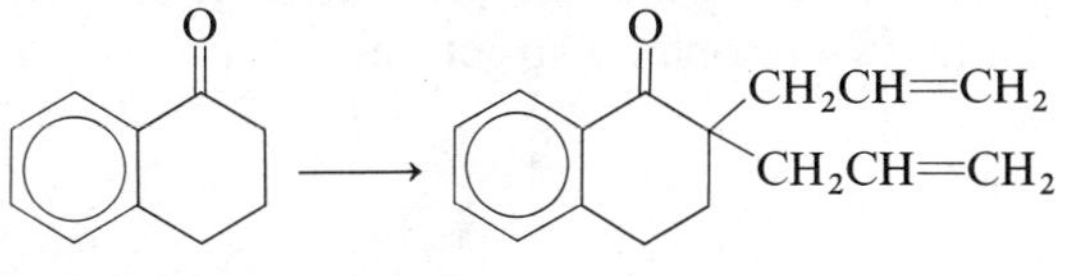

(*c*) butanal → 2-ethylbutanal
(*d*) cyclopentanone → 2-benzylcyclopentanone

Question 24.19

Propose a complete, stepwise mechanism to show the hydrolysis of an enamine. Show all intermediates. Keep in mind that the positively charged nitrogen atom withdraws electrons from carbon:

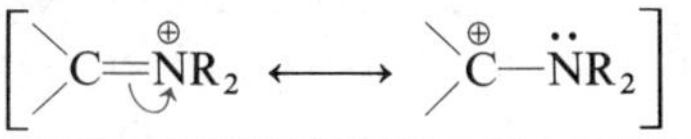

Question 24.20

Propose an explanation for 2-methylcyclohexanone reacting with pyrrolidine to give the following enamine mixture:

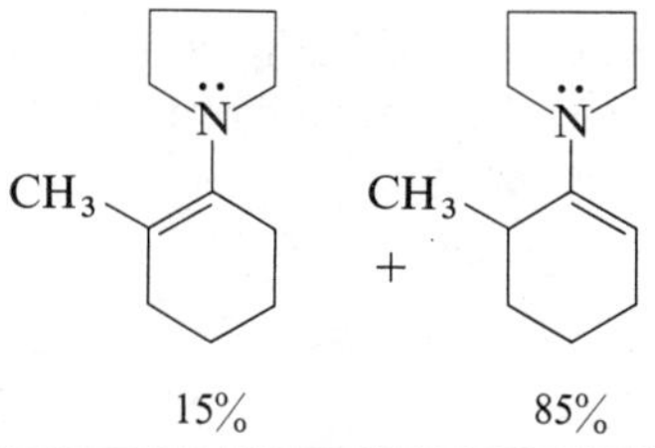

15%   85%

24.13   Wittig Reaction: Use of Carbanions in Organic Synthesis

The Wittig reaction is used primarily for the conversion of a carbonyl compound to an alkene. The overall result, illustrated here, is the substitution of a methylene group, **CH$_2$**, for the oxygen of the carbonyl group. The reaction derives its name from Professor George Wittig, who was the first to work with these extremely versatile reagents.

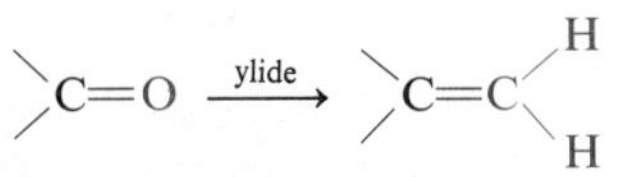

The transformation is accomplished by treatment of the carbonyl compound with an interesting class of carbanion reagents called *ylides*. Ylides are compounds of the following general structure:

$$\overset{\mid}{\underset{\ominus}{C}}\!\!:-\overset{\mid}{\underset{\oplus}{G}}$$

where G = appropriately substituted N, S, or P

Ylides are carbanionic species in which the carbanion is attached to a hetero-atom (sulfur, phosphorus, or nitrogen normally), which generally bears a positive charge. Although the dipolar form of the ylide was illustrated, ylides are actually resonance hybrids of the following two resonance structures:

$$\left[-\overset{\mid}{\underset{\underset{\ominus}{\cdot\cdot}}{C}}\!\!\overset{}{\frown}\overset{\mid}{\underset{\oplus}{G}} \longleftrightarrow \overset{}{\diagdown}C\!\!=\!\!\overset{\mid}{G} \right] \equiv -\overset{\mid}{\underset{\delta\ominus}{C}}\!\!=\!\!=\!\!\overset{\mid}{\underset{\delta\oplus}{G}}$$

Three examples of ylides frequently encountered in organic synthesis are:

$$R-\overset{\overset{\displaystyle R}{\mid}}{\underset{\underset{\displaystyle R}{\mid}}{P}}\!\!=\!\!C\!\!\diagup^{R'}_{\diagdown R''} \qquad R-\overset{\overset{\displaystyle O}{\parallel}}{\underset{\underset{\displaystyle R}{\mid}}{S}}\!\!=\!\!C\!\!\diagup^{R'}_{\diagdown R''} \qquad R-\overset{\overset{\displaystyle R}{\mid}}{\underset{\cdot\cdot}{S}}\!\!=\!\!C\!\!\diagup^{R'}_{\diagdown R''}$$

Depending on the ylide used, R' and R'' may be a large variety of substituents including, but not limited to, H, R, Ar, and COOR. It is not only possible, therefore, to transform a carbonyl oxygen into a methylene group, but by careful choice of the ylide, one can also selectively synthesize a wide variety of substituted olefins.

$$\diagup^{\diagdown}C\!\!=\!\!O \xrightarrow{\text{ylide}} \diagup^{\diagdown}C\!\!=\!\!C\!\!\diagup^{R}_{\diagdown R''}$$

where R' and R'' = H, R, Ar, COOR, and others

Although a large number of both sulfur and phosphorus ylides are used routinely in organic synthesis, the remainder of this discussion is concerned with phosphorus ylides of the following type:

$$\text{Ph}\!-\!\!\overset{\mid}{P}\!\!=\!\!C\!\!\diagup^{R}_{\diagdown H} \qquad \text{abbreviated (Ph)}_3\text{P}\!\!=\!\!\text{CHR}$$

These ylides are prepared from triphenylphosphine through this reaction sequence:

$$\text{Ph}_3\text{P}\!: \quad + \quad \text{RCH}_2\!\!-\!\!X \xrightarrow[\text{reaction}]{S_N2} \text{Ph}_3\overset{\oplus}{P}\!\!-\!\!\text{CH}_2\text{R} \\ \underset{\cdot\cdot}{:}\overset{\cdot\cdot}{X}\overset{\ominus}{:}$$

Triphenylphosphine   1° Alkyl halide   Phosphonium salt

$$\downarrow \text{base}$$

$$\left[\text{Ph}_3\text{P}\!\!=\!\!\text{CHR} \longleftrightarrow \text{Ph}_3\overset{\oplus}{P}\!\!-\!\!\overset{\cdot\cdot}{\text{CHR}} \right]$$

Ylide

The first step involves a nucleophilic displacement (S_N2) reaction to produce the phosphonium salt. The hydrogens adjacent to the positively charged phosphorus are reasonably acidic and may be abstracted with strong base to produce the desired ylide. Many bases may be used in this regard, but alkyllithium reagents (for example, *n*-butyllithium in anhydrous tetrahydrofuran solvent) and sodium alkoxides (for example, $(CH_3)_3CONa$ in dimethylformamide) are the usual ones. Ylides are not very stable and are normally prepared as needed followed by immediate reaction with the desired carbonyl compound.

Three specific examples of the use of ylides to synthesize alkenes are given here:

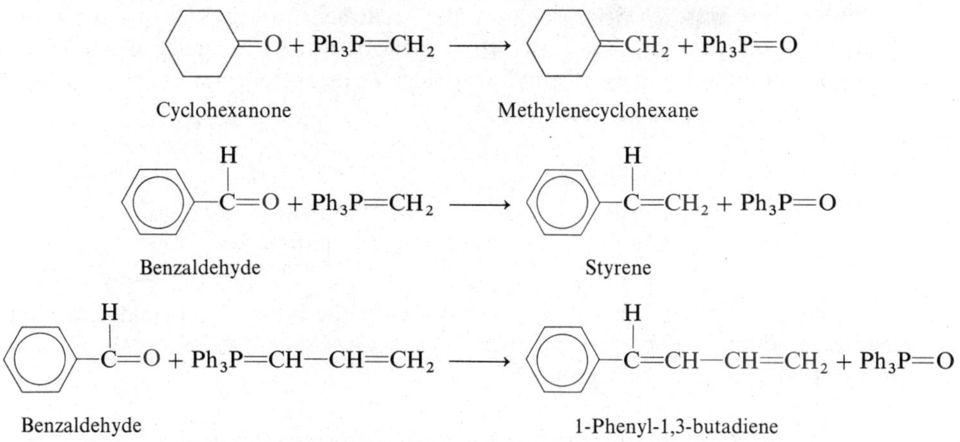

Cyclohexanone   Methylenecyclohexane

Benzaldehyde   Styrene

Benzaldehyde   1-Phenyl-1,3-butadiene

The reaction of the ylide with the carbonyl compound to produce an alkene occurs through the following reaction sequence:

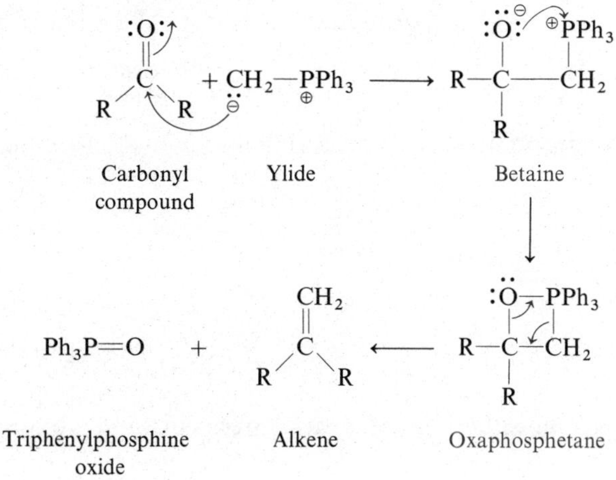

Carbonyl   Ylide   Betaine
compound

Triphenylphosphine   Alkene   Oxaphosphetane
oxide

The first step is a nucleophilic addition on the carbonyl compound to produce the dipolar intermediate called a *betaine*. The *betaine*, in either one or two steps, eliminates the triphenylphosphine oxide to yield the desired alkene. The reaction probably proceeds through two steps by way of the *oxaphosphetane* intermediate. The alkene obtained is a mixture of the *cis* and *trans* isomers.

24.14   Self-Condensation of Esters: Claisen Condensation

In view of our knowledge about the self-condensation reactions of aldehydes and ketones, we might anticipate that esters behave analogously. Indeed, treatment of an ester containing α hydrogens with a strong alkoxide base results in self-condensation of the ester:

$$-\overset{\displaystyle \text{O}}{\underset{\displaystyle \text{H}}{\overset{\|}{\underset{|}{\text{C}}}}}-\overset{\text{O}}{\overset{\|}{\text{C}}}-\text{OR} \xrightarrow[\text{ROH}]{\text{RÖ}^{\ominus}} -\overset{\text{O}}{\overset{\|}{\text{C}}}-\underset{\beta}{\overset{\text{O}}{\underset{\text{H}}{\overset{\|}{\text{C}}}}}-\underset{\alpha}{\overset{\text{O}}{\overset{\|}{\text{C}}}}-\text{OR}$$

　　　　　2 moles　　　　　　　　　β Keto ester

This reaction, called the **Claisen ester condensation,** produces a family of compounds called **β keto esters.** For example, treatment of ethyl acetate with sodium ethoxide produces the β keto ester ethyl 3-oxobutanoate, which is usually referred to by the common names of ethyl acetoacetate or acetoacetic ester:

$$\text{CH}_3-\overset{\text{O}}{\overset{\|}{\text{C}}}-\text{OCH}_2\text{CH}_3 \xrightarrow[\text{CH}_3\text{CH}_2\text{OH}]{\text{CH}_3\text{CH}_2\overset{..}{\text{O}}{}^{\ominus}\ \text{Na}^{\oplus}} \text{CH}_3-\overset{\text{O}}{\overset{\|}{\text{C}}}-\text{CH}_2-\overset{\text{O}}{\overset{\|}{\text{C}}}-\text{OCH}_2\text{CH}_3$$

Ethyl acetate　　　　　　　　Ethyl 3-oxobutanoate
2 moles　　　　　　　　　　　(ethyl acetoacetate
　　　　　　　　　　　　　　or acetoacetic ester)

In this reaction, what might we expect on the basis of our understanding of similar reactions? The steps involved in Claisen condensation are outlined here, using ethyl acetate as an example.

Step 1.　Removal of α hydrogen from ester by strong base

$$\text{CH}_3\text{CH}_2\overset{..}{\underset{..}{\text{O}}}{:}^{\ominus} + \overset{\displaystyle \overset{B:}{}}{\text{CH}_2}-\overset{\text{O}}{\overset{\|}{\underset{\text{H}}{\text{C}}}}-\text{OCH}_2\text{CH}_3 \rightleftharpoons {:}\text{CH}_2-\overset{\text{O}}{\overset{\|}{\text{C}}}-\text{OCH}_2\text{CH}_3 + \text{CH}_3\text{CH}_2\overset{..}{\text{O}}\text{H}$$

　　　　　　　　　　　　　　　　　　　　　　Conjugated
　　　　　　　　　　　　　　　　　　　　　　carbanion

Step 2.　Nucleophilic addition of carbanion to another molecule of ester

$$\text{CH}_3-\overset{\overset{\displaystyle :\text{O}:}{}}{\text{C}}-\text{OCH}_2\text{CH}_3 + {:}\text{CH}_2-\overset{\text{O}}{\overset{\|}{\text{C}}}-\text{OCH}_2\text{CH}_3 \rightleftharpoons \text{CH}_3-\overset{:\overset{..}{\text{O}}:^{\ominus}}{\underset{\text{OCH}_2\text{CH}_3}{\overset{|}{\underset{|}{\text{C}}}}}-\text{CH}_2-\overset{\text{O}}{\overset{\|}{\text{C}}}-\text{OCH}_2\text{CH}_3$$

　　　　　　　　　　　　　　　　　　　　　　　　　　　　　Tetrahedral intermediate
　　　　　　　　　　　　　　　　　　　　　　　　　　　　　Unstable

Step 3.　Decomposition of tetrahedral intermediate to give product

$$\text{CH}_3-\overset{:\overset{..}{\text{O}}:^{\ominus}}{\underset{:\overset{..}{\text{O}}\text{CH}_2\text{CH}_3}{\overset{|}{\underset{|}{\text{C}}}}}-\text{CH}_2-\overset{\text{O}}{\overset{\|}{\text{C}}}-\text{OCH}_2\text{CH}_3 \rightleftharpoons \text{CH}_3-\overset{\text{O}}{\overset{\|}{\text{C}}}-\text{CH}_2-\overset{\text{O}}{\overset{\|}{\text{C}}}-\text{OCH}_2\text{CH}_3 + \text{CH}_3\text{CH}_2\overset{..}{\underset{..}{\text{O}}}{:}^{\ominus}\ \text{Na}^{\oplus}$$

The main difference between the self-condensations of aldehydes and ketones and of esters is that in the former a stable tetrahedral intermediate is formed (protonation of which gives the aldol product), whereas with esters an unstable tetrahedral intermediate is formed. Step 3 shows the decomposition of this unstable intermediate to give the β keto ester. In other words, steps 2 and 3 constitute the addition-elimination process we encountered so often in Chap. 23 in the reactions of carboxylic acid derivatives.

The picture is not quite complete, however. The ethyl acetoacetate product contains *two hydrogens* that are α to two carbon-oxygen double bonds: the carbonyl group and the ester group. We saw the profound effect one carbon-oxygen double bond has on the acidity of α hydrogens, so we might expect that α hydrogens adjacent to two similar groups are even more acidic. This is indeed the case because the β keto ester (ethyl acetoacetate) reacts further with ethoxide ion according to the following equation:

$$CH_3-\overset{\overset{\displaystyle O}{\|}}{C}-CH_2-\overset{\overset{\displaystyle O}{\|}}{C}-OCH_2CH_3 + CH_3CH_2\overset{..}{\overset{..}{O}}{:}^{\ominus} \; Na^{\oplus} \; \rightleftharpoons$$

α to two
C=O bonds
Very acidic

$$CH_3-\overset{\overset{\displaystyle O}{\|}}{C}-\underset{\underset{\ominus \; Na^{\oplus}}{}}{CH}-\overset{\overset{\displaystyle O}{\|}}{C}-OCH_2CH_3 + CH_3CH_2\overset{..}{O}H$$

Sodium ethyl acetoacetate

The reaction mixture in the Claisen condensation consists of ethoxide ion and the anion of ethyl acetoacetate. In the laboratory, this reaction is followed by neutralization with dilute mineral acid, which protonates the anions that are present and liberates the keto ester:

$$CH_3-\overset{\overset{\displaystyle O}{\|}}{C}-\underset{\underset{\ominus}{}}{CH}-\overset{\overset{\displaystyle O}{\|}}{C}-OCH_2CH_3 \xrightarrow{\text{dilute } H_3\overset{..}{O}{}^{\oplus}} CH_3-\overset{\overset{\displaystyle O}{\|}}{C}-CH_2-\overset{\overset{\displaystyle O}{\|}}{C}-OCH_2CH_3$$

Ethyl acetoacetate

The formation of the anion from a β keto ester is the driving force for this particular reaction to occur in good yield. Whereas the formation of the anion from ethyl acetate (step 1 of the mechanism) and the nucleophilic acyl substitution of this anion on another molecule of ethyl acetate (steps 2 and 3) involve reversible reactions with equilibria that lie moderately far to the left, the equilibrium in the formation of the anion of ethyl acetoacetate lies to the right because of the relative stability of the anion. This has the net effect of driving the Claisen condensation to completion.

Put another way, the anion from ethyl acetoacetate is more stable than the anion of ethyl acetate because of the greater delocalization of negative charge throughout the ion (as compared with ethyl acetate), which is shown by the resonance structures:

Anion of ethyl acetoacetate
Resonance hybrid

We encounter other examples of stabilized anions with similar structures in Secs. 24.20 and 24.21.

The important role played by the formation of the anion of a β keto ester is supported by the observation that an ester with only *one α hydrogen* gives a very poor yield of the condensation product because it cannot form a stabilized anion; for example:

$$(CH_3)_2\overset{\displaystyle O}{\underset{\displaystyle H}{\overset{\displaystyle \|}{C}}}-C-OCH_2CH_3 \;\underset{}{\overset{CH_3CH_2\ddot{O}:^{\ominus} Na^{\oplus}}{\rightleftharpoons}}\; (CH_3)_2CH-\overset{O}{\overset{\|}{C}}-\underset{CH_3\,CH_3}{C}-\overset{O}{\overset{\|}{C}}-OCH_2CH_3$$

<table>
<tr><td>Ethyl isobutyrate</td><td>Ethyl 2,2,4-trimethyl-3-oxopentanoate</td></tr>
<tr><td>*2 moles*</td><td>*Low yield*</td></tr>
</table>

Finally, the stabilized anion that is formed probably prevents further condensation of the product. This is because the keto ester forms a negatively charged anion, and the likelihood of the anion of ethyl acetate coming near the keto ester anion is not great due to electrostatic repulsions.

Note that ester condensation involves nucleophilic acyl substitution, whereas aldol condensation involves nucleophilic addition to the carbonyl group in an aldehyde or ketone.

Although alkoxide ion is often used as the base in Claisen condensation, other methods improve the yield of the keto ester. For example, continuous removal of a volatile product (such as alcohol formed from the alkoxide base) shifts the equilibrium to favor the keto ester. Other bases that are stronger than alkoxide can also be used. Sodium hydride, NaH, is sometimes used with good results because the hydride ion (H:$^{\ominus}$) it contains is a very strong base. The hydride ion reacts with an acidic α hydrogen to form hydrogen gas, so carbanion formation is irreversible with this base:

$$-\underset{H}{\overset{|}{\underset{|}{C}}}-COOR + NaH \longrightarrow \left[-\underset{:\ominus}{\overset{|}{C}}-COOR\right] Na^{\oplus} + H_2\uparrow$$

We encounter several important reactions of β keto esters in Sec. 24.20, one of which is their hydrolysis into a β keto acid.

Question 24.21

In the Claisen ester condensation of ethyl acetate, sodium ethoxide is used as the base. What would be the disadvantage of using sodium methoxide as the base (even though it is strong enough to cause condensation)? Explain briefly.

Question 24.22

Provide complete, stepwise mechanisms for the base-promoted Claisen condensation of the following esters. Show the structure of each product.

(*a*) ethyl propionate   (*b*) ethyl phenylacetate   (*c*) ethyl butyrate

Question 24.23

Although the Claisen ester condensation usually occurs between 2 moles of the same ester, reactions are known in which two ester groups in the same molecule undergo condensation to form cyclic compounds. This type of reaction is usually called the **Dieckmann condensation.**

(*a*) Propose a mechanism for the following reaction, showing all steps and intermediates:

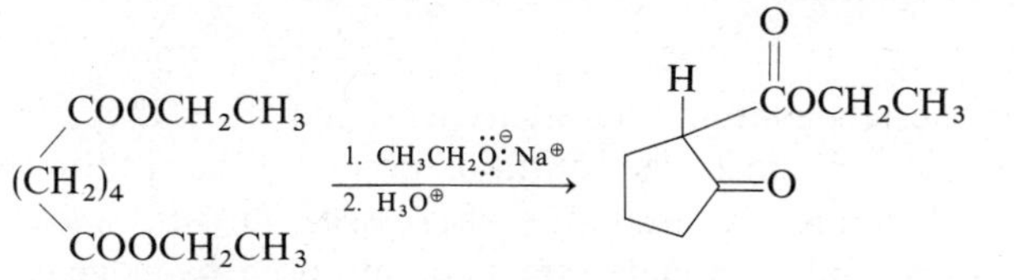

$$\underset{\text{Diethyl adipate}}{(CH_2)_4 \Big\langle \begin{matrix} COOCH_2CH_3 \\ COOCH_2CH_3 \end{matrix}} \quad \xrightarrow[\text{2. } H_3O^\oplus]{\text{1. } CH_3CH_2\ddot{\underset{\cdot\cdot}{O}}{:}^{\ominus} Na^\oplus} \quad \underset{\text{2-Carbethoxycyclopentanone}}{}$$

(*b*) Propose a mechanism for the Dieckmann condensation of the following ester, and on the basis of your mechanism, deduce the structure of the product:

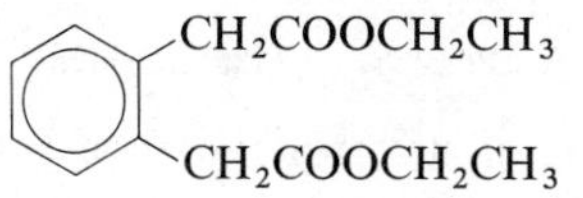

$$\underset{}{\bigcirc\!\!\!\!\!\bigcirc}\Big\langle \begin{matrix} CH_2COOCH_2CH_3 \\ CH_2COOCH_2CH_3 \end{matrix}$$

(*c*) When the following diester reacts with sodium ethoxide, two possible products can be formed. Show the mechanism for the formation of each, as well as their structures.

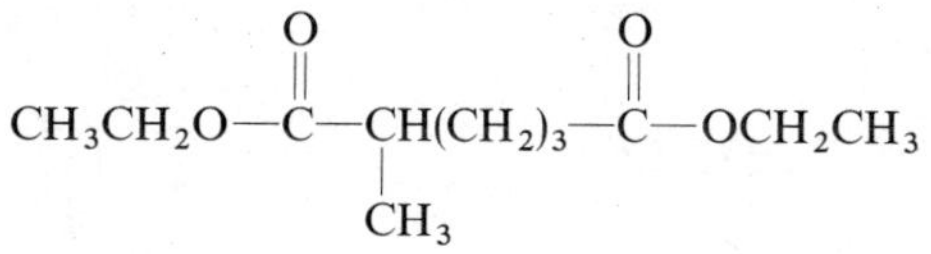

$$CH_3CH_2O-\overset{O}{\overset{\|}{C}}-\underset{\underset{CH_3}{|}}{CH}(CH_2)_3-\overset{O}{\overset{\|}{C}}-OCH_2CH_3$$

Which product do you think is likely to be formed in greatest amount, and why? (*Hint:* Evaluate relative carbanion stabilities.)

24.15 Crossed Condensation of Esters

Two different esters can be condensed (*crossed Claisen condensation*) in the same way that two different carbonyl compounds are in the aldol condensation. Often, however, to provide high yields one ester in a crossed ester condensation contains no α hydrogen atoms, as shown in the following example:

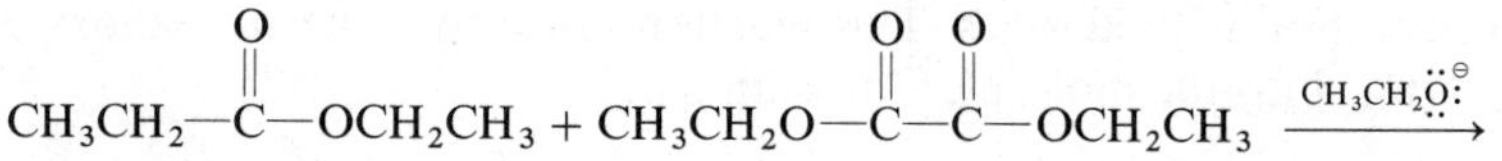

$$\underset{\text{Ethyl propionate}}{CH_3CH_2-\overset{O}{\overset{\|}{C}}-OCH_2CH_3} + \underset{\text{Diethyl oxalate}}{CH_3CH_2O-\overset{O}{\overset{\|}{C}}-\overset{O}{\overset{\|}{C}}-OCH_2CH_3} \xrightarrow{CH_3CH_2\ddot{\underset{\cdot\cdot}{O}}{:}^{\ominus}}$$

$$\underset{\text{Ethyl α-ethoxalylpropionate}}{CH_3-\underset{\underset{\overset{\|}{O}}{\underset{O=C}{|}}}{CH}-\overset{O}{\overset{\|}{C}}-OCH_2CH_3}$$

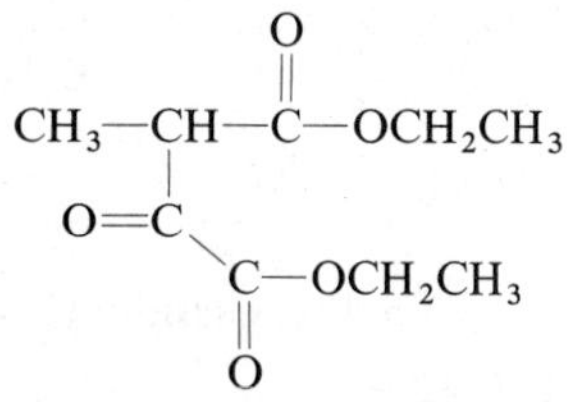

with $O=C-OCH_2CH_3$ shown bonded

Ethyl α-ethoxalylpropionate

Question 24.24

For the preceding reaction, provide a complete mechanism. What side product(s) are formed? Show how they are formed.

24.16   Condensation of Esters with Ketones

Other esters contain no α hydrogens, and when they are used in condensation reactions, nucleophilic acyl substitution occurs at the carbon-oxygen double bond in the ester. These are also examples of *crossed Claisen condensations*. Formate ester, oxalate esters, and carbonate esters contain no α hydrogens and are often used in such condensations:

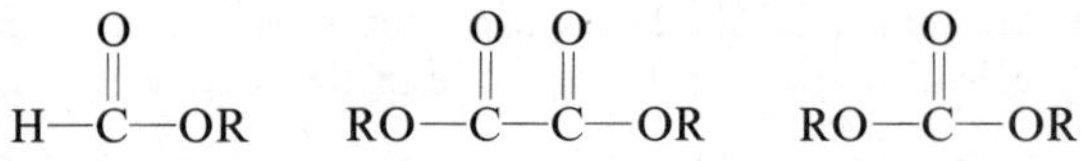

Formate ester   Oxalate ester   Carbonate ester

The condensation of a ketone with a formate ester introduces the *formyl* group, —CHO, into the molecule, a process sometimes called *formylation;* for example (see Sec. 24.24):

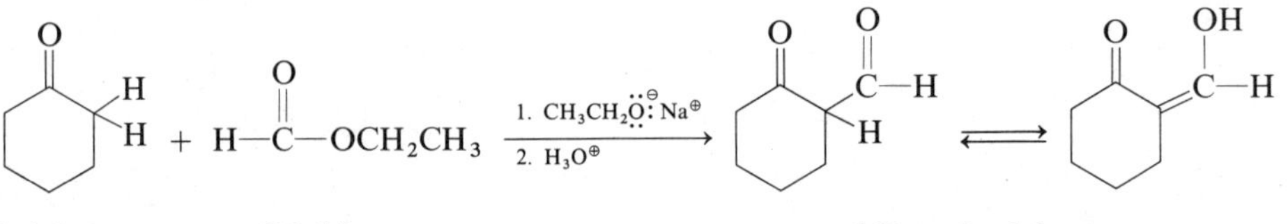

Cyclohexanone   Ethyl formate   2-Formylcyclohexanone
Tautomers

In like manner, the base-promoted reaction between a ketone and an oxalate ester incorporates the oxalyl group, —COCOOR, into the position α to the carbonyl group in the ketone; this process is called *oxalylation* and is illustrated in the following example:

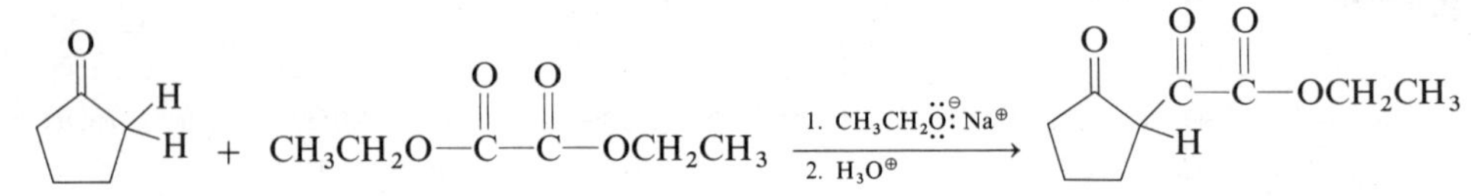

Cyclopentanone   Diethyl oxalate   2-Ethoxalylcyclopentanone

Diethyl carbonate, $(CH_3CH_2O)_2C=O$, contains no α hydrogens and also undergoes condensation with ketones. This reaction introduces the carbethoxy group, —COOCH$_2$CH$_3$, into the molecule; for example:

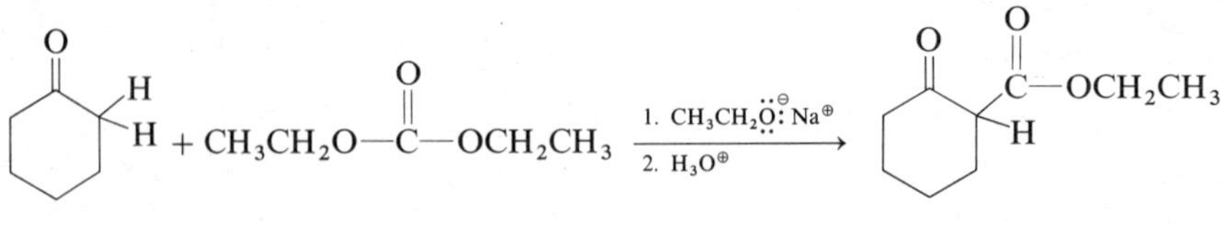

Diethyl carbonate   2-Carbethoxycyclohexanone

Diethyl carbonate is less reactive than the formate or the oxalate esters.

Question 24.25

Provide complete mechanisms for the preceding two reactions shown.

Question 24.26

Give major condensation products for the following reactions:

(*a*) acetone + diethyl oxalate in presence of sodium ethoxide

(b) acetophenone + ethyl formate in presence of sodium ethoxide
(c) ethyl phenylacetate + diethyl oxalate in presence of sodium ethoxide
(d) acetone + diethyl carbonate in presence of sodium ethoxide

Question 24.27

The formate ester and the oxalate diester are much more reactive in condensation reactions than is the carbonate diester. Suggest an explanation for these facts.

24.17 Condensation Reactions of Anhydrides: Perkin Condensation

Anhydrides also undergo condensation reactions in the presence of base. The base is usually the carboxylate salt corresponding to the acid in the anhydride instead of the alkoxide or hydroxide bases we used in previous condensations. This type of reaction is called the **Perkin condensation** and is illustrated by the following example:

Benzaldehyde   Acetic anhydride   (1) Unstable, not isolated

CH₃COOH + Cinnamic acid (2)

This reaction involves the initial addition of the α carbon in acetic anhydride to the carbon-oxygen double bond in benzaldehyde. The mechanism for this addition is analogous to that in aldol condensation. Note, however, that fairly vigorous reaction conditions are required (175 to 200°) probably because acetate ion is a weak base (compared with hydroxide or alkoxide ion). It should be no surprise that the α hydrogens in an anhydride are acidic because they are adjacent to the carbonyl group. The initial product of addition is an unstable intermediate (1) that cannot be isolated because it spontaneously loses water to form unsaturated compound (2). [Recall from Sec. 24.9 that compounds similar to (1) are readily dehydrated because they yield the more stable conjugated system.] On treatment with water, (2) is converted into cinnamic acid (3-phenylpropenoic acid) because the anhydride linkage is hydrolyzed. This is an important commercial preparation for cinnamic acid, which can be hydrogenated under mild conditions to hydrocinnamic acid (3-phenylpropanoic acid) or converted into other related compounds.

Perkin condensation occurs only with aromatic aldehydes because they have no α hydrogens and cannot undergo the aldol condensation.

Question 24.28

Outline the steps involved in the reaction of benzaldehyde, acetic anhydride, and sodium acetate to produce unstable intermediate (1).

Question 24.29

Explain briefly why hydroxide ion or alkoxide ion cannot be used as the base in Perkin condensation. Do you think sodium hydride, NaH, could be used? Why or why not?

Question 24.30

Give the important organic product that would result from each of the following reactions, assuming that hydrolysis is also carried out at the end of the reaction:

(*a*) benzaldehyde + propionic anhydride + sodium propionate (175°)
(*b*) *p*-phenylbenzaldehyde + acetic anhydride + sodium acetate (175°)

Question 24.31

Starting with any aromatic aldehyde of your choosing and using any needed organic or inorganic reagents, indicate how the following compounds can be synthesized in good yield:

(*a*) 2-methyl-3-phenylpropanoic acid   (*b*) 2-methyl-3-phenyl-1-propanol
(*c*) 3-phenylpropanamide   (*d*) phenylpropiolic acid ($C_6H_5C\equiv C-COOH$)

24.18   Cyano and Nitro Groups as Acidifying Substituents

The **cyano group** ($-C\equiv N$) and the **nitro group** ($-NO_2$) also withdraw electrons and cause α hydrogens to be somewhat acidic. The resulting α carbanions are stabilized by resonance as shown by the following reactions:

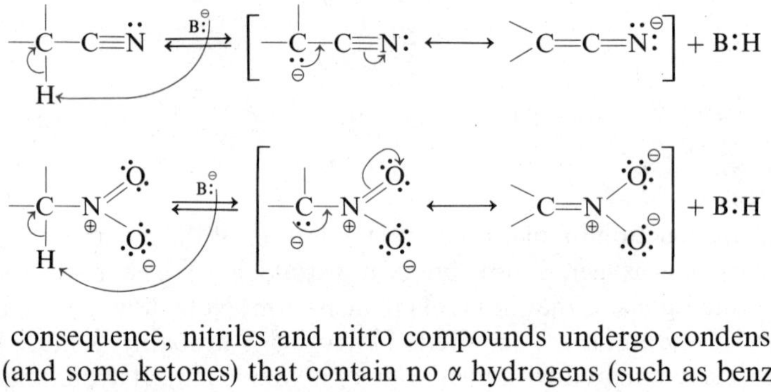

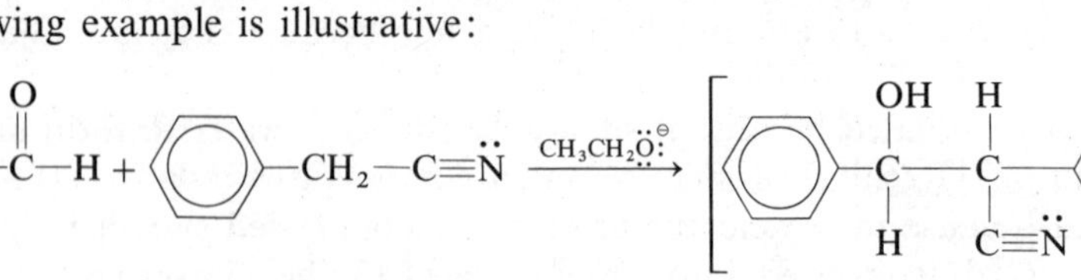

As a consequence, nitriles and nitro compounds undergo condensation with aldehydes (and some ketones) that contain no α hydrogens (such as benzaldehyde). The following example is illustrative:

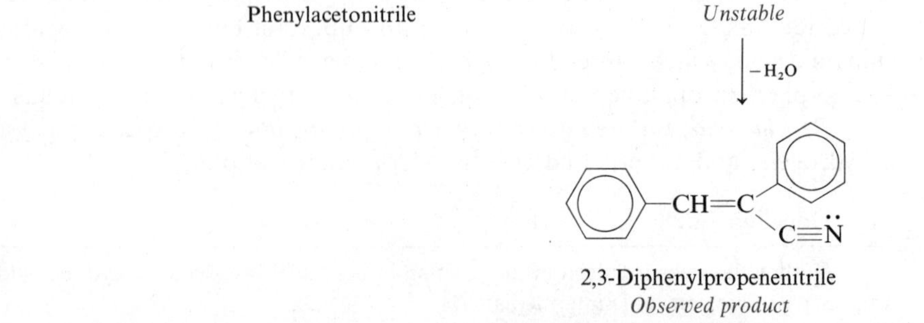

Phenylacetonitrile   *Unstable*

2,3-Diphenylpropenenitrile
Observed product

Question 24.32

(*a*) Propose a mechanism for the following reaction:

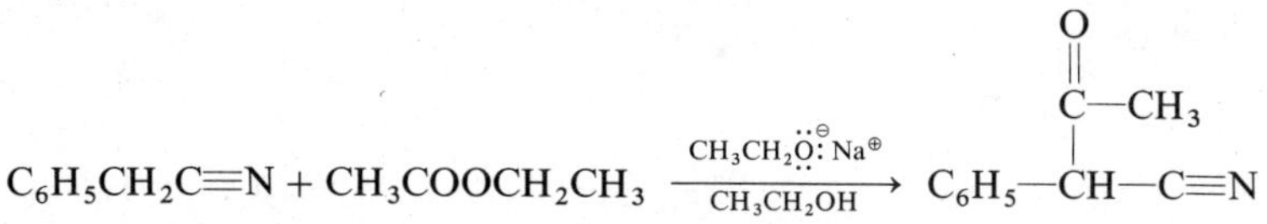

$$C_6H_5CH_2C{\equiv}N + CH_3COOCH_2CH_3 \xrightarrow[\text{CH}_3\text{CH}_2\text{OH}]{\text{CH}_3\text{CH}_2\ddot{\text{O}}{:}^{\ominus}\text{ Na}^{\oplus}} C_6H_5{-}\underset{\displaystyle \overset{\big|}{\underset{\displaystyle\overset{\|}{\text{C}}-\text{CH}_3}{\text{O}}}}{\text{CH}}{-}C{\equiv}N$$

(*b*) To what other *type* of reaction is this analogous?

(*c*) What side product(s) would you expect from this reaction? Provide a mechanism for their formation.

β-Dicarbonyl Compounds

We now turn to compounds that contain two carbon-oxygen double bonds *β* to one another, and we focus our attention on the hydrogens that are *α* to *both* double bonds:

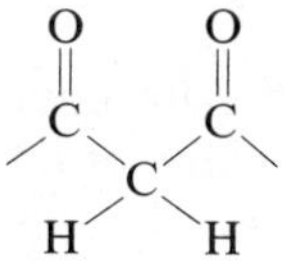

$$\underset{\substack{\displaystyle \\ \text{β-Dicarbonyl} \\ \text{compound}}}{\overset{\displaystyle \overset{\text{O}}{\|}\qquad\overset{\text{O}}{\|}}{\underset{\displaystyle \;\;\underset{\displaystyle \underset{\text{H}\quad\;\;\text{H}}{\text{C}}}{\;}\;\;}{\text{C}\qquad\quad\text{C}}}}$$

We saw many examples of compounds that have acidic *α* hydrogens because they are *α* to one carbon-oxygen double bond, and we expect hydrogens that are *α* to two such bonds to be even more acidic because of the electron-withdrawing influence of the second carbonyl group. As we saw in Sec. 24.14, this is indeed the case.

Although there are various possible combinations of aldehydes, ketones, and esters that can be classed as *β*-dicarbonyl compounds, we concentrate on *β*-diketones, *β* keto esters, and *β*-diesters with the following general structures:

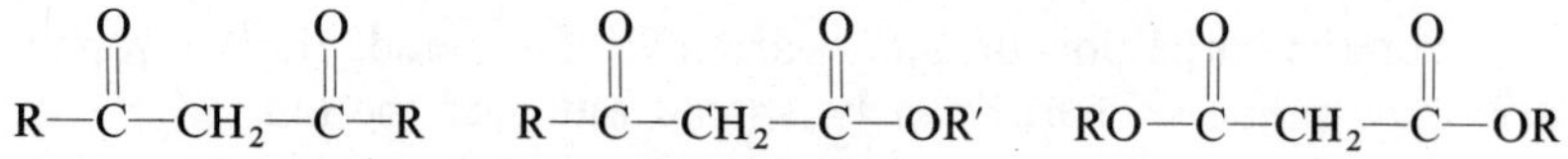

$$\underset{\text{β-Diketone}}{R{-}\overset{\overset{\text{O}}{\|}}{C}{-}CH_2{-}\overset{\overset{\text{O}}{\|}}{C}{-}R} \qquad \underset{\text{β Keto ester}}{R{-}\overset{\overset{\text{O}}{\|}}{C}{-}CH_2{-}\overset{\overset{\text{O}}{\|}}{C}{-}OR'} \qquad \underset{\text{β-Diester}}{RO{-}\overset{\overset{\text{O}}{\|}}{C}{-}CH_2{-}\overset{\overset{\text{O}}{\|}}{C}{-}OR}$$

24.19   *β*-Dicarbonyl Compounds As Acids: Formation and Structure of Carbanions

Dissociation constants of acids are a handy reference for relative acidities, and this concept can be applied to *β*-dicarbonyl compounds. The following data compare the acidity of carbonyl compounds and esters with that of various *β*-dicarbonyl compounds:

Acetone, acetaldehyde, and ethyl acetate:

$$K_a \approx 10^{-20} \text{ (see Sec. 24.1)}$$

Diethyl malonate (malonic ester):

$$CH_3CH_2O-\overset{\overset{O}{\|}}{C}-CH_2-\overset{\overset{O}{\|}}{C}-OCH_2CH_3 \rightleftharpoons$$

$$CH_3CH_2O-\overset{\overset{O}{\|}}{C}-\underset{\underset{\ominus}{\overset{..}{}}}{CH}-\overset{\overset{O}{\|}}{C}-OCH_2CH_3 + H^\oplus \qquad K_a = 5 \times 10^{-14}$$

Ethyl acetoacetate:

$$CH_3-\overset{\overset{O}{\|}}{C}-CH_2-\overset{\overset{O}{\|}}{C}-OCH_2CH_3 \rightleftharpoons$$

$$CH_3-\overset{\overset{O}{\|}}{C}-\underset{\underset{\ominus}{\overset{..}{}}}{CH}-\overset{\overset{O}{\|}}{C}-OCH_2CH_3 + H^\oplus \qquad K_a = 2.1 \times 10^{-11}$$

2,4-Pentanedione:

$$CH_3-\overset{\overset{O}{\|}}{C}-CH_2-\overset{\overset{O}{\|}}{C}-CH_3 \rightleftharpoons CH_3-\overset{\overset{O}{\|}}{C}-\underset{\underset{\ominus}{\overset{..}{}}}{CH}-\overset{\overset{O}{\|}}{C}-CH_3 + H^\oplus \qquad K_a = 1 \times 10^{-9}$$

These data show that the introduction of a second carbon-oxygen double bond markedly increases the acidity of the hydrogens α to both C=O bonds. What are the reasons for this increase?

The carbanion that forms on abstraction of an α hydrogen is much more stable in β-dicarbonyl compounds than in simple aldehydes, ketones, and esters. Let us compare these two general types of anions. For the conjugated carbanion of a monocarbonyl compound, two resonance structures show the delocalization of negative charge in the anion:

Resonance
hybrid

For the carbanion of a β-dicarbonyl compound, three resonance structures show how negative charge can be spread out over the ion:

Resonance
hybrid

(The preceding structures are shown for a general β-dicarbonyl compound, which may be any of the ones we mentioned previously.)

We know that stability normally increases with increased delocalization. We thus expect a carbanion with negative charge spread out over three atoms (as in the anion of a β-dicarbonyl compound) to be more stable than one with charge delocalized over two atoms (as in the anion of an aldehyde, ketone, or ester).

In summary, theory and experimental fact are in accord in predicting the increased acidity of β-dicarbonyl compounds over monocarbonyl compounds.

Question 24.33

Nitromethane, CH_3NO_2, has a dissociation constant of 6.1×10^{-11}. Which compound would you expect to be more acidic, $CH_3COCH_2COCH_3$ or $CH_3COCH_2NO_2$? Why?

Question 24.34

Draw the resonance structures of the anion derived from $CH_3COCH_2NO_2$.

Question 24.35

In each of the following four pairs of compounds, which would you expect to be more acidic and why?

(*a*) $CH_3COCH(CH_3)COCH_3$ or $CH_3COCH_2COCH_3$

(*b*) $CH_3COCH_2NO_2$ or $CH_2(NO_2)_2$

(*c*) $CH_3COCH_2COOCH_2CH_3$ or

(*d*) $CH_3COCH_2COCH_3$ or $CH_3COCHBrCOCH_3$

24.20 Malonic Ester Synthesis: Alkylation Reactions

A sequence of steps that starts with malonic ester (usually diethyl malonate) and produces a mono- or disubstituted acetic acid derivative is often referred to as the **malonic ester synthesis.** The overall conversion is the following:

$$RO-\overset{\overset{O}{\|}}{C}-CH_2-\overset{\overset{O}{\|}}{C}-OR \xrightarrow[\text{steps}]{\text{several}} R'-\underset{\underset{R''}{|}}{CH}-\overset{\overset{O}{\|}}{C}-OH$$

Malonic ester   Mono- or disubstituted
                                  acetic acid derivative

The two key steps in malonic ester synthesis are (1) alkylation of the malonic ester and (2) hydrolysis and decarboxylation of the substituted malonic ester. Let us begin by looking at the alkylation reaction.

Diethyl malonate, which is synthesized from readily available starting materials,[1] is acidic so that it is converted into the corresponding anion on treatment

---

[1] Diethyl malonate can be synthesized from acetic acid by the following sequence:

$$CH_3COOH \xrightarrow[\text{HVZ reaction}]{Cl_2,\ P\ (trace)} ClCH_2COOH \xrightarrow[\text{with }:\ddot{O}H^{\ominus}]{\text{neutralize}} ClCH_2CO\ddot{O}{:}^{\ominus}\ Na^{\oplus} \xrightarrow{{:}CN{:}^{\ominus}}$$

$$N{\equiv}C-CH_2-CO\ddot{O}{:}^{\ominus}\ Na^{\oplus} \xrightarrow[H_2O]{:\ddot{O}H^{\ominus}} Na^{\oplus}\ {:}\ddot{O}OC-CH_2-CO\ddot{O}{:}^{\ominus}\ Na^{\oplus} \xrightarrow[H^{\oplus},\ \text{warm}]{CH_3CH_2OH}$$

$$CH_3CH_2O-\overset{\overset{O}{\|}}{C}-CH_2-\overset{\overset{O}{\|}}{C}-OCH_2CH_3$$

The nitrile, $N{\equiv}C-CH_2-CO\ddot{O}{:}^{\ominus}\ Na^{\oplus}$, can also be treated with ethanol and acid to give diethyl malonate directly; this is often done commerically.

with sodium ethoxide in ethanol solvent:

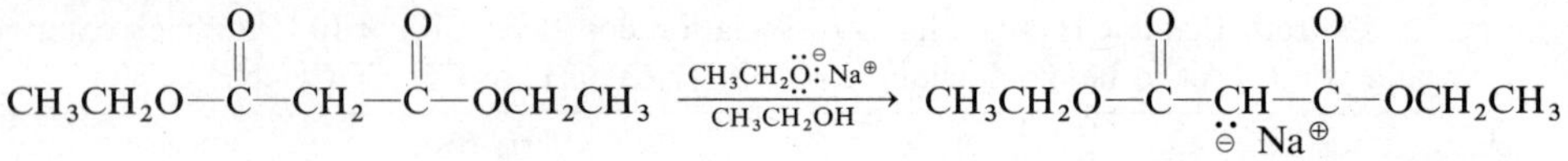

$$CH_3CH_2O-\overset{\overset{O}{\|}}{C}-CH_2-\overset{\overset{O}{\|}}{C}-OCH_2CH_3 \xrightarrow[\text{CH}_3\text{CH}_2\text{OH}]{\text{CH}_3\text{CH}_2\ddot{\text{O}}\text{:}^\ominus \text{ Na}^\oplus} CH_3CH_2O-\overset{\overset{O}{\|}}{C}-\underset{\underset{\ominus \text{ Na}^\oplus}{\ddot{}}}{CH}-\overset{\overset{O}{\|}}{C}-OCH_2CH_3$$

Sodium ethoxide used as the base converts the malonic ester into its anion. (Sodium hydride, NaH, is sometimes used for anion formation, although an aprotic solvent such as benzene or ether must be used.)

The anion of diethyl malonate can serve as a nucleophile and displace halide from an alkyl halide via an S_N2 reaction, and the net effect is C alkylation. This process is illustrated using methyl iodide as the alkyl halide:

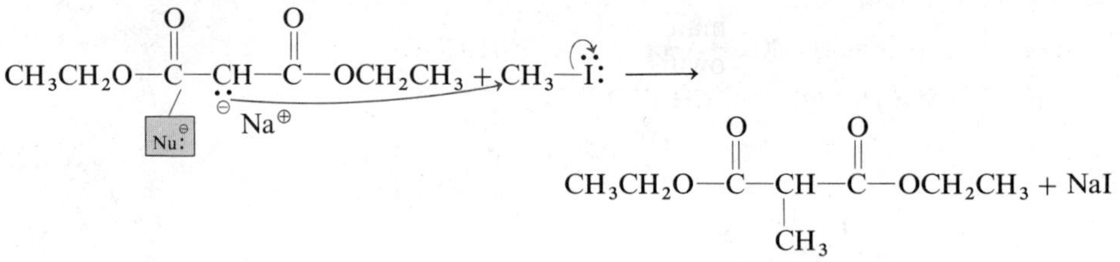

$$CH_3CH_2O-\overset{\overset{O}{\|}}{C}-\underset{\underset{\boxed{Nu\text{:}}\ \ \ddot{}\ Na^\oplus}{}}{CH}-\overset{\overset{O}{\|}}{C}-OCH_2CH_3 + CH_3-\overset{\curvearrowright}{\ddot{I}}\text{:} \longrightarrow$$

$$CH_3CH_2O-\overset{\overset{O}{\|}}{C}-\underset{\underset{CH_3}{|}}{CH}-\overset{\overset{O}{\|}}{C}-OCH_2CH_3 + NaI$$

Diethyl methylmalonate

This is a useful reaction in which a new carbon-carbon bond is made by alkylation.

C versus O Alkylation

An interesting question arises as to why alkylation occurs on carbon (C alkylation) rather than on oxygen (O alkylation), because as we saw, some of the negative charge of the anion is on both oxygen and carbon:

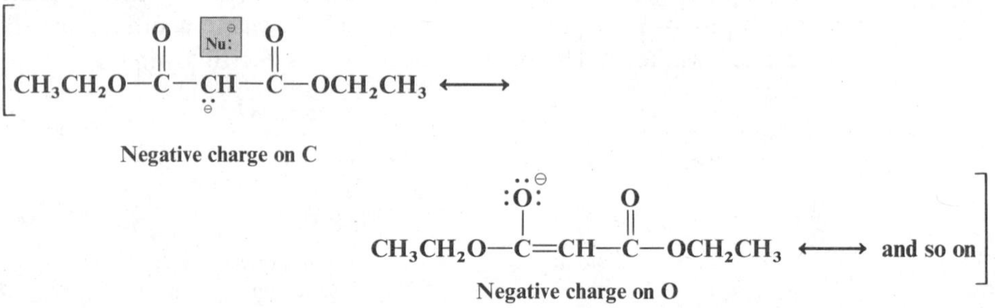

$$\left[CH_3CH_2O-\overset{\overset{O}{\|}}{C}-\underset{\underset{\ominus}{\ddot{}}}{\overset{\boxed{Nu\text{:}^\ominus}}{CH}}-\overset{\overset{O}{\|}}{C}-OCH_2CH_3 \longleftrightarrow \right.$$

Negative charge on C

$$\left. CH_3CH_2O-\overset{\overset{:\ddot{O}\text{:}^\ominus}{|}}{C}=CH-\overset{\overset{O}{\|}}{C}-OCH_2CH_3 \longleftrightarrow \text{ and so on} \right]$$

Negative charge on O

It is reasonable to expect that alkylation occurs at both negative carbon and negative oxygen. Indeed, alkylation does occur at both places, but C alkylation predominates. It may well be that carbon bears the greater amount of negative charge or that negative carbon is a better nucleophile than negative oxygen. Variations in experimental parameters, such as solvent and temperature, produce different amounts of C versus O alkylation, the details of which are not discussed here.

As we might expect, the same limitations that apply to other S_N2 reactions apply to malonic ester synthesis: *Substitution works well with methyl and primary alkyl halides, it gives acceptable results with many secondary alkyl halides, and it fails completely with tertiary halides* (with which elimination is the major reaction).

Note also that the introduction of one alkyl group into diethyl malonate produces a compound that still contains one hydrogen α to both acyl groups. Further treatment of the monoalkylation product with base and then with either the same or a different alkyl halide yields a disubstituted malonic ester. The introduction of a second methyl group into diethyl methylmalonate is shown here:

$$CH_3CH_2O-\overset{\overset{O}{\|}}{C}-\underset{\underset{CH_3}{|}}{CH}-\overset{\overset{O}{\|}}{C}-OCH_2CH_3 \xrightarrow[CH_3CH_2OH]{CH_3CH_2\overset{..}{\overset{..}{O}}:^{\ominus} Na^{\oplus}}$$

Diethyl methylmalonate

$$CH_3CH_2O-\overset{\overset{O}{\|}}{C}-\underset{\underset{CH_3}{|}}{\overset{\ominus}{\underset{..}{C}}}-\overset{\overset{O}{\|}}{C}-OCH_2CH_3 \xrightarrow[\substack{S_N2 \\ reaction}]{CH_3-I} CH_3CH_2O-\overset{\overset{O}{\|}}{C}-\underset{\underset{CH_3}{|}}{\overset{\overset{CH_3}{|}}{C}}-\overset{\overset{O}{\|}}{C}-OCH_2CH_3$$

Diethyl dimethylmalonate

The alkylation of malonic ester is frequently indicated in a somewhat condensed form, without showing the intermediate carbanion. The following sequence illustrates this procedure:

$$CH_3CH_2O-\overset{\overset{O}{\|}}{C}-CH_2-\overset{\overset{O}{\|}}{C}-OCH_2CH_3 \xrightarrow[\substack{2. R-X}]{1. \ CH_3CH_2\overset{..}{\overset{..}{O}}:^{\ominus} Na^{\oplus}, CH_3CH_2OH}$$

$$CH_3CH_2O-\overset{\overset{O}{\|}}{C}-\underset{\underset{R}{|}}{CH}-\overset{\overset{O}{\|}}{C}-OCH_2CH_3 \xrightarrow[\substack{2. R'-X}]{1. \ CH_3CH_2\overset{..}{\overset{..}{O}}:^{\ominus} Na^{\oplus}, CH_3CH_2OH}$$

$$CH_3CH_2O-\overset{\overset{O}{\|}}{C}-\underset{\underset{R}{|}}{\overset{\overset{R'}{|}}{C}}-\overset{\overset{O}{\|}}{C}-OCH_2CH_3$$

Even when we want to introduce two identical R groups, two separate steps are generally used. When introducing two different alkyl groups into diethyl malonate, the larger or more bulky one is usually put on first.

A. Hydrolysis and Decarboxylation

After alkylation or dialkylation, the substituted malonic ester is hydrolyzed to the corresponding malonic acid, which is decarboxylated by heat. Hydrolysis is usually accomplished in aqueous base, followed by neutralization to yield a substituted malonic acid. Recall from Sec. 22.17 that malonic acid and substituted malonic acids lose carbon dioxide on heating to produce acetic acid or a substituted acetic acid, the mechanism of which was discussed in Sec. 24.21.

Using general equations, the following sequences of hydrolysis and decarboxylation illustrate these reactions:

Malonic ester:

$$RO-\overset{\overset{O}{\|}}{C}-CH_2-\overset{\overset{O}{\|}}{C}-OR \xrightarrow[\substack{H_2O, \\ heat}]{:\overset{..}{O}H^{\ominus}} \xrightarrow[\text{with } H^{\oplus}]{\text{neutralize}} HO-\overset{\overset{O}{\|}}{C}-CH_2-\overset{\overset{O}{\|}}{C}-OH$$

$$\downarrow heat$$

$$CH_3-\overset{\overset{O}{\|}}{C}-OH + CO_2$$

Acetic
acid

Monosubstituted malonic ester:

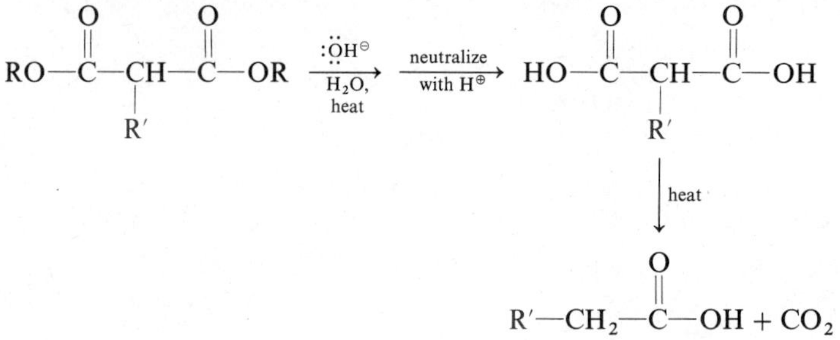

Monosubstituted
acetic acid

The reaction extends the R′ group by *two carbon atoms* and places the carboxyl group at the end of the chain.

Disubstituted malonic ester:

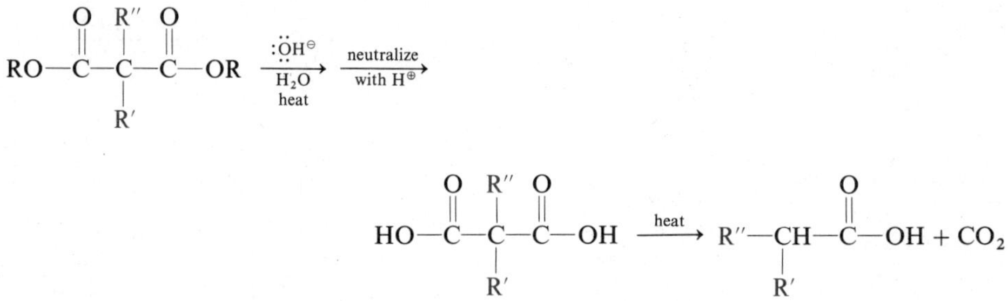

Disubstituted
acetic acid

B. Malonic Ester Syntheses

The applications of malonic ester syntheses are wide and varied, and indeed they even extend to the preparation of various barbituric acids (see Question 24.54). We first look at the synthesis of 2-benzylhexanoic acid (1) from diethyl malonate and other organic reagents:

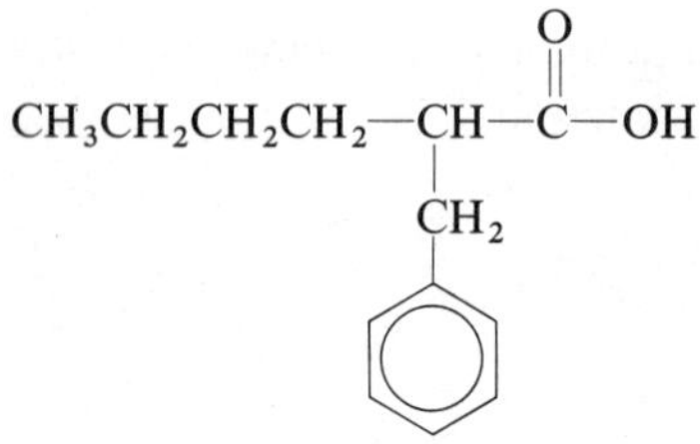

(1)

We start by examining the structure of the desired compound and note that it can be considered as a derivative of acetic acid in which the *n*-butyl group and the benzyl group are the substituents:

$$CH_3CH_2CH_2CH_2 \!-\!\overset{\displaystyle \overset{O}{\|}}{\underset{\underset{\displaystyle \underset{\displaystyle \bigcirc}{CH_2}}{CH}}{CH}} \!-\! C \!-\! OH$$

n-Butyl group   Benzyl group

With this in mind, we set out to introduce these two substituents into diethyl malonate, followed by hydrolysis of the disubstituted malonate ester and decarboxylation of the disubstituted malonic acid. These steps are outlined here:

$$CH_3CH_2O \!-\! \overset{O}{\overset{\|}{C}} \!-\! CH_2 \!-\! \overset{O}{\overset{\|}{C}} \!-\! OCH_2CH_3 \quad \xrightarrow[\text{2. } C_6H_5CH_2Br]{\substack{\text{1. } CH_3CH_2\ddot{O}{:}^{\ominus} \, Na^{\oplus}, \\ CH_3CH_2OH}}$$

$$CH_3CH_2O \!-\! \overset{O}{\overset{\|}{C}} \!-\! \underset{\underset{\displaystyle CH_2C_6H_5}{|}}{CH} \!-\! \overset{O}{\overset{\|}{C}} \!-\! OCH_2CH_3 \quad \xrightarrow[\text{2. } CH_3CH_2CH_2CH_2Br]{\substack{\text{1. } CH_3CH_2\ddot{O}{:}^{\ominus} \, Na^{\oplus}, \\ CH_3CH_2OH}}$$

$$CH_3CH_2O \!-\! \overset{O}{\overset{\|}{C}} \!-\! \underset{\underset{\displaystyle CH_2 \!-\! C_6H_5}{|}}{\overset{\overset{\displaystyle CH_2CH_2CH_2CH_3}{|}}{C}} \!-\! \overset{O}{\overset{\|}{C}} \!-\! OCH_2CH_3 \quad \xrightarrow[\text{heat}]{:\ddot{O}H^{\ominus}, \, H_2O} \quad \xrightarrow{\substack{\text{neutralize} \\ \text{with } H^{\oplus}}}$$

$$HO \!-\! \overset{O}{\overset{\|}{C}} \!-\! \underset{\underset{\displaystyle CH_2 \!-\! C_6H_5}{|}}{\overset{\overset{\displaystyle CH_2CH_2CH_2CH_3}{|}}{C}} \!-\! \overset{O}{\overset{\|}{C}} \!-\! OH \quad \xrightarrow{\text{heat}} \quad CH_3CH_2CH_2CH_2 \!-\! \underset{\underset{\displaystyle \underset{\displaystyle \bigcirc}{CH_2}}{|}}{CH} \!-\! \overset{O}{\overset{\|}{C}} \!-\! OH$$

Note that the more bulky benzyl group is put into the malonic ester first, followed by the less bulky *n*-butyl group.

Other variations in malonic ester synthesis sometimes use alkyl dihalides of the general structure X—$(CH_2)_n$—X. For example, cyclobutanecarboxylic acid is often prepared from diethyl malonate and 1,3-dibromopropane by the following route:

$$CH_2(COOCH_2CH_3)_2 \quad \xrightarrow[\text{2. } BrCH_2CH_2CH_2Br \, (1 \text{ mole})]{\substack{\text{1. } CH_3CH_2\ddot{O}{:}^{\ominus} \, Na^{\oplus} \, (1 \text{ mole}), \\ CH_3CH_2OH}} \quad BrCH_2CH_2CH_2 \!-\! CH(COOCH_2CH_3)_2$$

1 mole

$$\xrightarrow[\text{CH}_3CH_2OH]{CH_3CH_2\ddot{O}{:}^{\ominus} \, Na^{\oplus} \, (1 \text{ mole})} \quad \underset{\underset{\displaystyle CH_2 \!-\! CH_2}{|}}{\overset{\overset{\displaystyle COOCH_2CH_3}{|}}{\underset{\displaystyle CH_2 \!-\! \underset{}{C}}{C}}} \!-\! COOCH_2CH_3 \quad \xrightarrow[\text{and decarboxylation}]{\text{hydrolysis}} \quad \overset{H}{\square} \!-\! COOH$$

Internal displacement   Cyclobutanecarboxylic
reaction   acid

This ring closure is rather novel in that it occurs internally. Variation in experimental details can be used to displace the remaining bromo group by another malonic ester molecule, thus providing a pathway to the dicarboxylic acid family. This sequence is:

$$BrCH_2CH_2CH_2—CH(COOCH_2CH_3)_2 + Na^{\oplus} : \overset{\ominus}{C}H(COOCH_2CH_3)_2 \longrightarrow$$

Prepared earlier

$$\Big\uparrow \quad CH_3CH_2\overset{..}{\underset{..}{O}} : {}^{\ominus} Na^{\oplus},$$
$$CH_3CH_2OH$$

$$CH_2(COOCH_2CH_3)_2$$

$$(CH_3CH_2OOC)_2CH—CH_2CH_2CH_2—CH(COOCH_2CH_3)_2 \xrightarrow{\text{hydrolysis}}$$

Tetraester

$$(HOOC)_2CH—CH_2CH_2CH_2—CH(COOH)_2 \xrightarrow{\text{heat}}$$

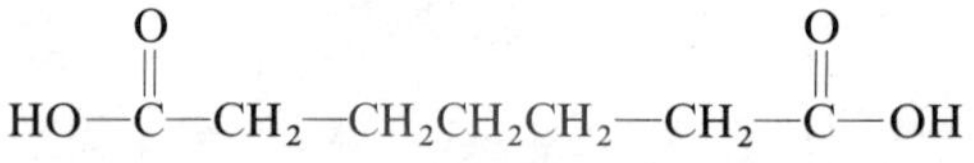

Tetraacid

$$HO—\overset{\overset{\displaystyle O}{\|}}{C}—CH_2—CH_2CH_2CH_2—CH_2—\overset{\overset{\displaystyle O}{\|}}{C}—OH$$

Heptanedioic acid

The substituted malonic ester or the substituted acetic acid derived from it can also be reduced to the corresponding diol or alcohol, respectively, and the carboxyl group can be transformed into various other functional groups.

Question 24.36

In the alkylation of diethyl malonate by an alkyl halide, RX, using sodium ethoxide in ethanol, a small amount of dialkylation is usually observed. Provide a mechanistic explanation for this observation. (*Hint:* Assume all the ethoxide is consumed by the diethyl malonate in forming the anion, so that ethoxide should not be used in your mechanism. However, what other "basic" species is present when alkylation is, say, 50% complete?)

24.21   Acetoacetic Ester Synthesis: Alkylation

Acetoacetic ester (for example, ethyl acetoacetate) is alkylated in the same manner as malonic ester. The hydrogens α to both carbon-oxygen double bonds are acidic and readily removed by bases such as alkoxide or sodium hydride. The overall steps here also involve C alkylation, followed by hydrolysis and decarboxylation, and they are shown here:

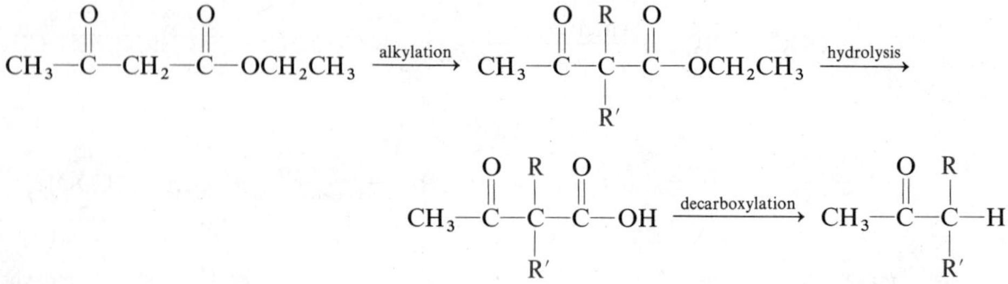

$$CH_3—\overset{\overset{\displaystyle O}{\|}}{C}—CH_2—\overset{\overset{\displaystyle O}{\|}}{C}—OCH_2CH_3 \xrightarrow{\text{alkylation}} CH_3—\overset{\overset{\displaystyle O}{\|}}{C}—\overset{\overset{\displaystyle R}{|}}{\underset{\underset{\displaystyle R'}{|}}{C}}—\overset{\overset{\displaystyle O}{\|}}{C}—OCH_2CH_3 \xrightarrow{\text{hydrolysis}}$$

$$CH_3—\overset{\overset{\displaystyle O}{\|}}{C}—\overset{\overset{\displaystyle R}{|}}{\underset{\underset{\displaystyle R'}{|}}{C}}—\overset{\overset{\displaystyle O}{\|}}{C}—OH \xrightarrow{\text{decarboxylation}} CH_3—\overset{\overset{\displaystyle O}{\|}}{C}—\overset{\overset{\displaystyle R}{|}}{\underset{\underset{\displaystyle R'}{|}}{C}}—H$$

Disubstituted
acetone derivative

As this sequence illustrates, the acetoacetic ester method provides a ketone that can be considered as a derivative of acetone. One or two substituents (R and R′) can be attached to the same α carbon atom, depending on the mode of synthesis.

Before looking at examples of the acetoacetic ester synthesis, let us review the synthesis of ethyl acetoacetic ester, which is accomplished via Claisen ester condensation of ethyl acetate (see Sec. 24.14):

$$CH_3\!-\!\overset{\overset{O}{\|}}{C}\!-\!OCH_2CH_3 \xrightarrow[CH_3CH_2OH]{CH_3CH_2\ddot{O}\!:^{\ominus} Na^{\oplus}} CH_3\!-\!\overset{\overset{O}{\|}}{C}\!-\!CH_2\!-\!\overset{\overset{O}{\|}}{C}\!-\!OCH_2CH_3$$

2 moles   Ethyl acetoacetate

Other variations in Claisen condensation permit the synthesis of more complex β keto esters; for example:

$$R\!-\!CH_2\!-\!\overset{\overset{O}{\|}}{C}\!-\!OR' \xrightarrow[ROH]{R'\ddot{O}\!:^{\ominus} Na^{\oplus}} R\!-\!CH_2\!-\!\overset{\overset{O}{\|}}{C}\!-\!\underset{\underset{R}{|}}{CH}\!-\!\overset{\overset{O}{\|}}{C}\!-\!OR'$$

2 moles

Let us start by presenting the general acetoacetic ester synthesis sequence:

Alkylation:

$$CH_3\!-\!\overset{\overset{O}{\|}}{C}\!-\!CH_2\!-\!\overset{\overset{O}{\|}}{C}\!-\!OCH_2CH_3 \xrightarrow[CH_3CH_2OH]{CH_3CH_2\ddot{O}\!:^{\ominus} Na^{\oplus}} CH_3\!-\!\overset{\overset{O}{\|}}{C}\!-\!\underset{\underset{Na^{\oplus}\;\ddot{\;}^{\ominus}}{}}{CH}\!-\!\overset{\overset{O}{\|}}{C}\!-\!OCH_2CH_3$$

$$\downarrow R\!-\!X\ (S_N2\ \text{reaction})$$

$$CH_3\!-\!\overset{\overset{O}{\|}}{C}\!-\!\underset{\underset{R}{|}}{\overset{\overset{R'}{|}}{C}}\!-\!\overset{\overset{O}{\|}}{C}\!-\!OCH_2CH_3 \xleftarrow[\substack{2.\ R-X}]{\substack{1.\ CH_3CH_2\ddot{O}\!:^{\ominus} Na^{\oplus},\\ CH_3CH_2OH}} CH_3\overset{\overset{O}{\|}}{C}\!-\!\underset{\underset{R}{|}}{CH}\!-\!\overset{\overset{O}{\|}}{C}\!-\!OCH_2CH_3$$

Hydrolysis and decarboxylation:

$$CH_3\!-\!\overset{\overset{O}{\|}}{C}\!-\!\underset{\underset{R}{|}}{\overset{\overset{R'}{|}}{C}}\!-\!\overset{\overset{O}{\|}}{C}\!-\!OCH_2CH_3 \xrightarrow[\substack{2.\ \text{neutralize}\\ \text{with } H^{\oplus}}]{1.\ :\ddot{O}H^{\ominus},\ H_2O,\ \text{heat}} CH_3\!-\!\overset{\overset{O}{\|}}{C}\!-\!\underset{\underset{R}{|}}{\overset{\overset{R'}{|}}{C}}\!-\!\overset{\overset{O}{\|}}{C}\!-\!OH$$

$$\downarrow \text{heat}$$

$$CH_3\!-\!\overset{\overset{O}{\|}}{C}\!-\!\underset{\underset{R}{|}}{\overset{\overset{R'}{|}}{C}}\!-\!H + CO_2$$

The following specific example illustrates the synthesis of 3-methyl-2-octanone (1) from ethyl acetoacetate and other organic or inorganic reagents.

$$CH_3CH_2CH_2CH_2CH_2\!-\!\underset{\underset{CH_3}{|}}{CH}\!-\!\overset{\overset{O}{\|}}{C}\!-\!CH_3$$

(1)

We first examine the structure of the desired compound and note that it can be considered as an α,α-disubstituted acetone:

$$CH_3CH_2CH_2CH_2CH_2 \boxed{CH-\overset{\overset{\displaystyle O}{\|}}{C}-CH_3}$$
$$\underset{CH_3}{|}$$

n-Pentyl group   Methyl group

Because the acetoacetic ester synthesis produces this type of compound on alkylation, hydrolysis, and decarboxylation, we need to introduce a methyl and an *n*-pentyl group as substituents. The complete synthesis is as follows:

$$CH_3-\overset{\overset{\displaystyle O}{\|}}{C}-CH_2-\overset{\overset{\displaystyle O}{\|}}{C}-OCH_2CH_3 \xrightarrow[\text{2. } CH_3CH_2CH_2CH_2CH_2Br]{\text{1. } CH_3CH_2\overset{\ominus}{\ddot{O}}: Na^{\oplus}, CH_3CH_2OH}$$

Ethyl acetoacetate

$$CH_3-\overset{\overset{\displaystyle O}{\|}}{C}-\underset{\underset{\displaystyle CH_2CH_2CH_2CH_2CH_3}{|}}{CH}-\overset{\overset{\displaystyle O}{\|}}{C}-OCH_2CH_3 \xrightarrow[\text{2. } CH_3I]{\substack{\text{1. } CH_3CH_2\overset{\ominus}{\ddot{O}}: Na^{\oplus}, \\ CH_3CH_2OH}}$$

$$CH_3-\overset{\overset{\displaystyle O}{\|}}{C}-\underset{\underset{\displaystyle CH_2CH_2CH_2CH_2CH_3}{|}}{\overset{\overset{\displaystyle CH_3}{|}}{C}}-\overset{\overset{\displaystyle O}{\|}}{C}-OCH_2CH_3 \xrightarrow[\substack{\text{2. neutralize} \\ \text{with } H^{\oplus}}]{\text{1. } :\ddot{O}H^{\ominus}, H_2O, \text{ heat}} CH_3-\overset{\overset{\displaystyle O}{\|}}{C}-\underset{\underset{\displaystyle CH_2CH_2CH_2CH_2CH_3}{|}}{\overset{\overset{\displaystyle CH_3}{|}}{C}}-\overset{\overset{\displaystyle O}{\|}}{C}-OH \xrightarrow{\text{heat}}$$

$$CH_3-\overset{\overset{\displaystyle O}{\|}}{C}-\underset{\underset{\displaystyle CH_2CH_2CH_2CH_2CH_3}{|}}{\overset{\overset{\displaystyle CH_3}{|}}{C}}-H$$

As with the malonic ester synthesis, the carbonyl group and/or the ester group can be converted into other functional groups by choosing suitable reagents.

The alkylation and hydrolysis reactions are completely analogous to those studied in Sec. 24.20B on malonic ester syntheses. However, some additional comment needs to be made about the decarboxylation of the β keto acid formed by hydrolysis. We studied the decarboxylation of malonic acid, but this type of reaction is indeed more general: *Carboxylic acids that have a carbon-oxygen double bond one carbon atom removed from the carboxyl group undergo loss of carbon dioxide (decarboxylation) on heating:*

Cyclic   Enol   Keto form
transition state

where G = —OH (diacid)
        —R (β keto acid)
        —H (β aldehyde acid)

In this mechanism a cyclic six-membered transition state is involved. The stable carbon dioxide molecule is formed (and is a major driving force for the reaction) and is expelled as a gas. The enol form is produced initially but is transformed into the corresponding keto form via enol-keto equilibrium.

Question 24.37

Cyanoacetic acid, $N{\equiv}C{-}CH_2{-}COOH$, undergoes decarboxylation on heating in much the same way as malonic acid and acetoacetic acid, and the product is acetonitrile, $CH_3{-}C{\equiv}\ddot{N}$. Provide a mechanism for this transformation, given an equilibrium of the type $H{-}\ddot{N}{=}C{=}CH_2 \rightleftarrows \ddot{N}{\equiv}C{-}CH_2{-}H$.

Question 24.38

Starting with ethyl acetoacetate and using any other needed reagents, indicate how the following compounds could be prepared in good yield.

(*a*) 2-pentanone   (*b*) 3-ethyl-2-pentanone   (*c*) 3-ethylpentane
(*d*) 3-ethyl-2-pentanol   (*e*) diethylmalonic acid, $(CH_3CH_2)_2C(COOH)_2$
(*f*) 2,5-hexanedione (*Hint:* CH_3COCH_2Br is useful in this synthesis)
(*g*) 2,6-hexanedione (*Hint:* two moles of ethyl acetoacetate are used in the *best* method for synthesizing this compound)

Question 24.39

2-Carbethoxycyclohexanone (2) is a β keto ester even though a cyclic system is involved:

(2)

(*a*) Suggest how (2) can be prepared from cyclohexanone and any other needed organic reagents.
(*b*) Provide a method of converting (2) into 2-methylcyclohexanone. Note that there is no direct way of converting the $-COOCH_2CH_3$ group into a methyl group, so another route must be used.
(*c*) How might (2) be converted into compound (3) using any other needed reagents?

(3)

24.22   Acylation of Malonic Ester and Acetoacetic Ester

Secs. 24.20 and 24.21 show how malonic ester and acetoacetic ester can be *alkylated.* We now see by analogy with the many other nucleophilic acyl substitution reactions of acid chlorides, that they can be used to *acylate* either malonic ester or acetoacetic ester.

The main difficulty in carrying out these reactions is in using conditions so that the acid chloride is not destroyed before acylation occurs. For example, if ethoxide ion in ethanol were used to form the anion of either malonic ester or acetoacetic ester, the acid chloride would react with ethanol before it reacts with the carbanion. This problem can be overcome by allowing the malonic ester or acetoacetic ester to react with sodium metal or sodium hydride with ether or benzene as the solvent. Both Na metal and NaH react with the acidic hydrogens to liberate hydrogen gas, and the sodium salt of the malonic ester or acetoacetic ester remains in the organic solvent as an insoluble precipitate:

$$
\left[\begin{array}{c} CH_3-\overset{O}{\overset{\|}{C}}-CH_2-\overset{O}{\overset{\|}{C}}-OR \\ \\ or \\ \\ RO-\overset{O}{\overset{\|}{C}}-CH_2-\overset{O}{\overset{\|}{C}}-OR \end{array} \right] \xrightarrow[\substack{\text{benzene, or} \\ \text{ether}}]{\substack{\text{Na metal} \\ \text{or NaH,}}} \left[\begin{array}{c} CH_3-\overset{O}{\overset{\|}{C}}-\underset{\overset{\ominus}{}\overset{.}{\underset{.}{C}}H}{\overset{}{C}}\,Na^{\oplus}-\overset{O}{\overset{\|}{C}}-OR \\ \\ or \\ \\ RO-\overset{O}{\overset{\|}{C}}-\underset{\overset{\ominus}{}\overset{.}{\underset{.}{C}}H}{\overset{}{C}}\,Na^{\oplus}-\overset{O}{\overset{\|}{C}}-OR \end{array} \right] + H_2
$$

The addition of an acid chloride to the suspension of one of the preceding carbanions results in acylation, as shown by the following specific example:

$$
CH_3-\overset{O}{\overset{\|}{C}}-\underset{\underset{Na^{\oplus}}{\overset{\ominus}{}}}{\overset{..}{C}H}-\overset{O}{\overset{\|}{C}}-OCH_2CH_3 + CH_3-\overset{O}{\overset{\|}{C}}-Cl \xrightarrow{\text{benzene}} CH_3-\overset{O}{\overset{\|}{C}}-\underset{\underset{}{\overset{|}{C}=O}}{\overset{|}{C}H}-\overset{O}{\overset{\|}{C}}-OCH_2CH_3
$$
$$
 \underset{CH_3}{O=C}
$$

| Ethyl | Acetyl | Ethyl acetylacetoacetate |
| sodioacetoacetate | chloride | *Tricarbonyl compound* |

24.23 α,β-Unsaturated Carbonyl Compounds: Structure and Nucleophilic Addition Reactions

In this section we study briefly the structure of α,β-unsaturated carbonyl compounds (aldehydes, ketones, and esters), followed by a discussion of orientation or addition of nucleophilic reagents to these types of compounds. In particular, we examine the addition of carbanions to α,β-unsaturated carbonyl compounds, a topic that is particularly germane to this chapter on carbanion chemistry.

A. Structure of α,β-Unsaturated Carbonyl Compounds

We learned about the unusual stability of α,β-unsaturated carbonyl compounds in Sec. 24.8; β-hydroxy aldehydes and ketones readily lose water to produce a compound with the double bond adjacent to the carbon-oxygen double bond. The orbital view of an α,β-unsaturated carbonyl compound (Fig. 24.3) best shows the overlap of atomic *p* orbitals to produce the resonance-stabilized system.

We can also draw the following resonance structures to show electron distribution in an α,β-unsaturated carbonyl compound:

$$
\left[\overset{\diagdown}{\underset{|}{C}}{=}\overset{\diagdown}{\underset{|}{C}}{-}\overset{\diagdown}{C}{=}\overset{..}{\underset{..}{O}}{:} \longleftrightarrow \overset{\diagdown}{\underset{|}{C}}{=}\overset{\diagdown}{\underset{|}{C}}{-}\overset{\oplus}{C}{-}\overset{..}{\underset{..}{O}}{:}^{\ominus} \longleftrightarrow {-}\overset{|}{\underset{\oplus}{C}}{-}\overset{|}{\underset{|}{C}}{=}\overset{}{C}{-}\overset{..}{\underset{..}{O}}{:}^{\ominus} \right] \equiv \overset{\diagdown}{\underset{|}{C}}{\overset{\delta\oplus}{=}}\overset{}{\underset{|}{C}}{\overset{\delta\oplus}{=}}\overset{}{C}{\overset{\delta\ominus}{=}}\overset{}{O}
$$

$$(1)(2)\text{Resonance hybrid}$$

Structure (1) shows the polarity of the carbon-oxygen double bond, a feature with which we are quite familiar. Note that this structure places a positive charge *adjacent* to the carbon-carbon double bond, and we know from our studies on the allyl cation in Sec. 13.6C that this results in the positive charge drawing electrons away from the adjacent π bond. Structure (2) shows this electron delocalization. Indeed, the electron distribution in an α,β-unsaturated carbonyl compound may be shown as a resonance hybrid in which a partially positive charge is placed on both the carbon atom of the carbonyl group and the β carbon atom. In the following discussion, we see the importance of this electron delocalization in determining the mode of addition to α,β-unsaturated carbonyl compounds.

B. Nucleophilic Addition to α,β-Unsaturated Carbonyl Compounds

Many polar reagents add to the carbon-oxygen double bond in aldehydes and ketones (see Chap. 21). In the case of certain reagents such as the Grignard reagent, the addition is unaffected by *isolated* carbon-carbon double bonds in the molecule. When the double bond is adjacent to the carbonyl group (that is, in α,β-unsaturated carbonyl compounds), however, more than one addition product is obtained. Let us first look at a general picture of nucleophilic addition to this type of compound.

The addition of a polar reagent, H—Nu, to an α,β-unsaturated carbonyl compound may occur in one of the following two ways:

1. 1,2-Addition: addition to the carbon-oxygen double bond.

We consider reactions in which the nucleophile adds first, followed by protonation:

α,β-Unsaturated
carbonyl compound

1,2-Addition

This is the same mechanism used for nucleophilic addition to an aldehyde or ketone that we studied in Sec. 21.2, and it occurs as though the carbon-carbon double bond were not present. The α,β-unsaturated carbonyl compound is numbered as shown previously, and addition to the carbon-oxygen double bond is called **1,2-addition.**

In a simple aldehyde or ketone, the carbonyl carbon atom is partially positive. Placement of the carbon-carbon double bond adjacent to the carbonyl group serves to partially "absorb" some positive charge of the carbonyl carbon and place it on atom 4. This decreases the electron-deficient character of the carbonyl carbon and thus reduces the ease with which nucleophiles attack it. In other words, the carbon-oxygen double bond in a α,β-unsaturated carbonyl compound is less reactive toward nucleophilic addition than is a simple aldehyde or ketone.

2. 1,4-Addition or conjugate addition.

Again considering addition of the nucleophile first, followed by protonation, we have the following sequence of reactions:

Resonance-stabilized α carbanion
Especially stable

The driving force for addition is in part the formation of the resonance-stabilized α carbanion because addition across the carbon-carbon bond in the reverse direction provides a carbanion that is not especially stable:

$$Nu\!:^{\ominus} + \quad \overset{\displaystyle \diagup}{\underset{\displaystyle \diagdown}{C}}=\overset{\displaystyle |}{C}-C=O \quad \xrightarrow{\quad\times\quad} \quad -\overset{\displaystyle |}{\underset{\displaystyle \ominus}{C}}-\overset{\displaystyle \overset{Nu}{|}}{C}-C=O$$

Does not form

Once the nucleophile has added to the carbon-carbon double bond, the carbanion can be protonated either on carbon bearing negative charge or on oxygen bearing negative charge:

$$-\overset{|}{\underset{Nu}{C}}-\overset{|}{\underset{\ominus}{C}}-C=O + H^{\oplus} \longrightarrow -\overset{|}{\underset{Nu}{C}}-\overset{|}{\underset{H}{C}}-C=O \qquad \text{C protonation}$$

$$-\overset{|}{\underset{Nu}{C}}-\overset{|}{C}=C-\overset{\ominus}{\underset{..}{O}}: + H^{\oplus} \longrightarrow -\overset{|}{\underset{Nu}{C}}-\overset{|}{C}=C-O-H \qquad \text{O protonation}$$

1,4-Addition

The product arising from O protonation contains the elements of HNu, with H on oxygen (1 position) and Nu on carbon (4 position). For this reason, addition involving the carbon-carbon double bond in an α,β-unsaturated carbonyl compound is often referred to as **1,4-addition** or **conjugate addition**; the latter term arises because the original unsaturated carbonyl compound is conjugated.

In most cases, however, the 1,4-addition product yields an enol, which is generally unstable and is usually transformed to the more stable keto form via keto-enol equilibrium:

$$-\overset{|}{\underset{Nu}{C}}-\overset{|}{C}=C-O-H \rightleftharpoons -\overset{|}{\underset{Nu}{C}}-\overset{|}{\underset{H}{C}}-C=O$$

<table>
<tr><td align="center">Enol form</td><td align="center">Keto form</td></tr>
</table>

Thus either C protonation or O protonation ultimately results in the same product, namely one that has the nucleophile (Nu) attached to atom 4 and the proton attached to atom 3 in the conjugated molecule.

In a sense, the terminology with respect to numbers is confusing because the addition is really 3,4 addition. Keep in mind that the term 1,4 addition arises for the reasons given earlier.

Question 24.40

In considering resonance structures for an α,β-unsaturated carbonyl compound, the following electronic structure was not included:

$$-\overset{|}{\underset{\ominus}{C}}-C=C-\overset{\oplus}{\underset{..}{O}}:$$

Even though this is a possible structure, suggest why it is not an important resonance contributor.

C. Conjugate Addition of Carbanions to α,β-Unsaturated Carbonyl Compounds

We now examine the addition of carbanions derived from different sources to α,β-unsaturated carbonyl compounds. The general mechanisms presented earlier apply.

a. Addition of Grignard Reagents

Several factors are involved in the addition of the Grignard reagent to an α,β-unsaturated carbonyl compound, and we present only the general trends here. There is always competition between 1,2- and 1,4-addition, but there are some useful rules of thumb for predicting which mode of addition predominates. The most important factors involved are the size of R in the Grignard reagent, RMgX, and the size of the substituent attached to the carbonyl group in the α,β-unsaturated carbonyl compound. 1,2-Addition is favored by small groups in both cases, whereas 1,4-addition predominates when both groups are quite large. Let us look at some examples.

The aldehyde group is more reactive toward nucleophilic addition than is the keto group, so α,β-unsaturated aldehydes react mostly by 1,2-addition:

$$R-CH=CH-C \underset{\|}{\overset{:\ddot{O}:}{}}-H + R' :^{\ominus} [MgX]^{\oplus} \xrightarrow{\text{ether}} R-CH=CH-\underset{R'}{\overset{:\ddot{O}:^{\ominus}[MgX]^{\oplus}}{\underset{|}{\overset{|}{C}}}}-H \xrightarrow{H_3\ddot{O}^{\oplus}}$$

$$R-CH=CH-\underset{R'}{\overset{OH}{\underset{|}{\overset{|}{C}}}}-H$$

1,2-Addition
Major product

When the Grignard reagent is bulky, however, as in the case of *tert*-butyl-magnesium bromide, 1,4- or conjugate addition is the major mode of addition.

α,β-Unsaturated ketones have less tendency to undergo 1,2-addition because of the decreased reactivity of the keto group toward nucleophilic addition. The amount of 1,2-addition is decreased even more when the substituent attached to the carbonyl group is bulky.

The presence of substituents in the 4 position in a conjugated system decreases the amount of 1,4-addition product because they also hinder the attack of the Grignard reagent at the 4 position; for example:

$$R:^{\ominus} [MgX]^{\oplus} + R_2''C=CH-\overset{O}{\overset{\|}{C}}-R' \xrightarrow{\text{ether}} \xrightarrow{H_3\ddot{O}^{\oplus}} R_2''C-\underset{R\ H}{\overset{O}{\underset{|}{C}H-\overset{\|}{C}}}-R' + R_2''C=CH-\underset{R}{\overset{OH}{\underset{|}{C}}}-R'$$

R'' = bulky group   1,4-Addition   1,2-Addition
Minor product   *Major product*

Question 24.41

Indicate the possible addition products as well as the one formed in *major* amounts that might be expected when equimolar amounts of organometallic reagent and unsaturated carbonyl compound react in anhydrous ether (followed by hydrolysis with aqueous acid) in each case.

(a) crotonaldehyde (CH_3—CH=CH—CH=O) + methylmagnesium bromide
(b) acrolein (CH_2=CH—CH=O) + ethylmagnesium iodide
(c) methyl vinyl ketone (CH_2=CH—CO—CH_3) + phenylmagnesium bromide
(d) methyl vinyl ketone + phenyllithium (C_6H_5Li)

(e) $(CH_3)_2C$=CH—$\overset{\displaystyle O}{\overset{\|}{C}}$—$CH_3$ + CH_3MgI

(f) $(CH_3)_2C$=CH—$\overset{\displaystyle O}{\overset{\|}{C}}$—$C(CH_3)_3$ + CH_3MgI

Question 24.42

Account mechanistically for the addition of a Grignard reagent to an α,β-unsaturated ketone producing a 1,2- and/or a 1,4-addition product but no diaddition product as shown in the following equation, even though a large excess of Grignard reagent is used.

$$\text{>C=C—C=O + RMgX} \xrightarrow[\text{2. } H_3O^\oplus]{\text{1. ether}} \text{—C—C—C—OH}$$

$$\underset{\text{R \quad H \quad R}}{}$$

2 moles   Not formed

---

b. Addition of Conjugated Carbanions: Michael Addition Reaction

Like negatively charged carbon in the Grignard reagent, certain other carbanions also add to α,β-unsaturated carbonyl compounds. In particular, β-dicarbonyl compounds have very acidic hydrogens that are easily abstracted and form unusually stable conjugated carbanions. This nucleophilic addition reaction, called **Michael addition,** is viewed generally as follows:

$$R-\overset{O}{\overset{\|}{C}}-\overset{}{\underset{H}{C}}-\overset{O}{\overset{\|}{C}}-R' \xrightarrow{\text{base}} R-\overset{O}{\overset{\|}{C}}-\overset{\ominus}{\underset{\cdot\cdot}{C}}-\overset{O}{\overset{\|}{C}}-R' \xrightarrow{\hspace{2cm}}$$

β-Dicarbonyl
compound

$$\underset{\overset{\|}{O}}{\overset{\displaystyle \overset{O}{\overset{\|}{C}-R'}}{-C-\overset{}{C}-C=C-\overset{\cdot\cdot}{\underset{\cdot\cdot}{O}}{:}^{\ominus}}}\underset{\overset{}{\underset{C-R}{}}}{} \xrightarrow[\text{solvent and}]{\text{protonation by}} \underset{\overset{}{\underset{tautomerization}{}}}{}$$

The Michael addition reaction is used widely in synthetic organic chemistry as a method for making a new carbon-carbon bond and producing compounds that are key intermediates in the synthesis of new molecules. Two of the most common β-dicarbonyl compounds are ethyl acetoacetate and diethyl malonate. The α hydrogens in ethyl cyanoacetate, N≡C—CH_2—$COOCH_2CH_3$, are also quite acidic so this compound is useful in a Michael reaction.

The following example illustrates the use of Michael additions to unsaturated carbonyl compounds:

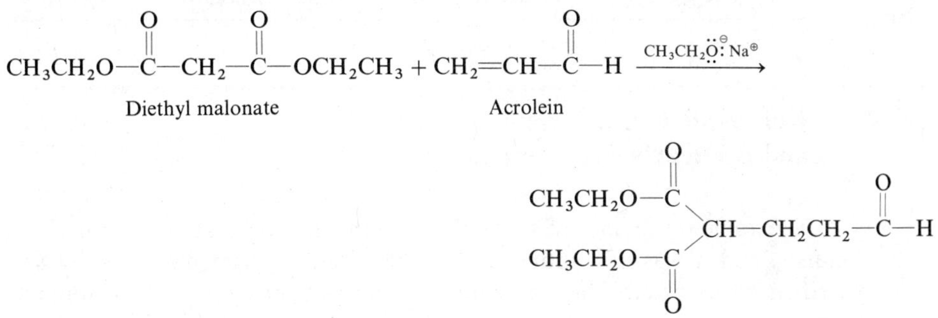

$$CH_3CH_2O-\overset{\overset{\displaystyle O}{\|}}{C}-CH_2-\overset{\overset{\displaystyle O}{\|}}{C}-OCH_2CH_3 + CH_2=CH-\overset{\overset{\displaystyle O}{\|}}{C}-H \xrightarrow{CH_3CH_2\overset{\ominus}{\ddot{O}:} Na^{\oplus}}$$

Diethyl malonate ⠀⠀⠀⠀⠀⠀ Acrolein

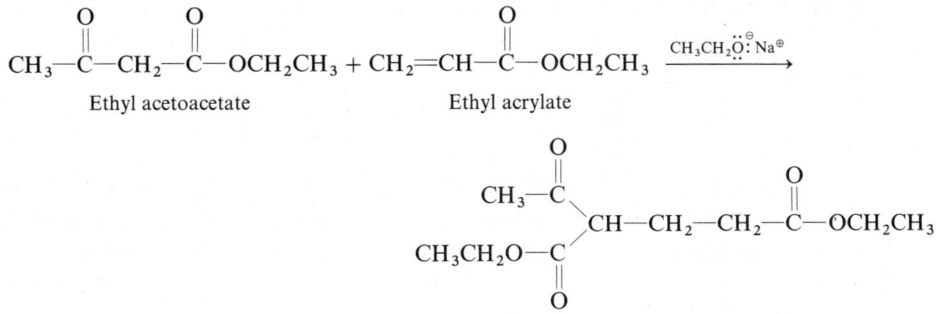

$$\begin{array}{c} CH_3CH_2O-\overset{\overset{\displaystyle O}{\|}}{C} \\ \diagdown \\ \diagup \\ CH_3CH_2O-\overset{\underset{\displaystyle O}{\|}}{C} \end{array} CH-CH_2CH_2-\overset{\overset{\displaystyle O}{\|}}{C}-H$$

$$CH_3-\overset{\overset{\displaystyle O}{\|}}{C}-CH_2-\overset{\overset{\displaystyle O}{\|}}{C}-OCH_2CH_3 + CH_2=CH-\overset{\overset{\displaystyle O}{\|}}{C}-OCH_2CH_3 \xrightarrow{CH_3CH_2\overset{\ominus}{\ddot{O}:} Na^{\oplus}}$$

Ethyl acetoacetate ⠀⠀⠀⠀⠀⠀ Ethyl acrylate

$$\begin{array}{c} CH_3-\overset{\overset{\displaystyle O}{\|}}{C} \\ \diagdown \\ \diagup \\ CH_3CH_2O-\overset{\underset{\displaystyle O}{\|}}{C} \end{array} CH-CH_2-CH_2-\overset{\overset{\displaystyle O}{\|}}{C}-OCH_2CH_3$$

Question 24.43

Draw the resonance structures and resonance hybrid of the carbanion derived from abstraction of an α hydrogen from ethyl cyanoacetate.

Question 24.44

The conjugate addition of water to an α,β-unsaturated carbonyl compound produces a β-hydroxy carbonyl compound, but on prolonged reaction the products shown in the following equation are sometimes observed:

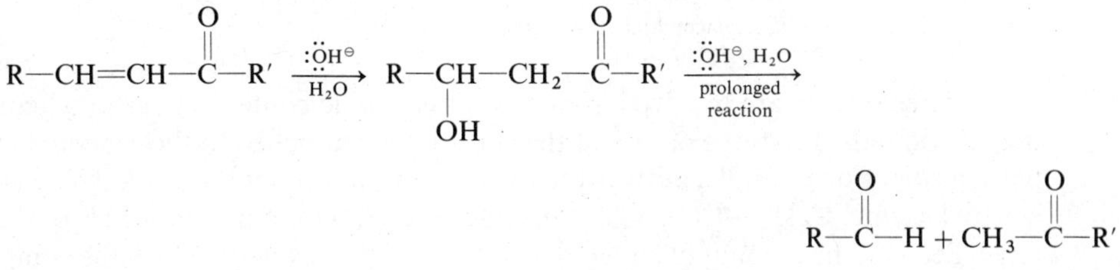

$$R-CH=CH-\overset{\overset{\displaystyle O}{\|}}{C}-R' \xrightarrow[H_2O]{:\ddot{O}H^{\ominus}} R-\underset{\underset{\displaystyle OH}{|}}{CH}-CH_2-\overset{\overset{\displaystyle O}{\|}}{C}-R' \xrightarrow[\substack{prolonged \\ reaction}]{:\ddot{O}H^{\ominus}, H_2O}$$

$$R-\overset{\overset{\displaystyle O}{\|}}{C}-H + CH_3-\overset{\overset{\displaystyle O}{\|}}{C}-R'$$

(*a*) Provide a complete mechanism showing the addition of water to give the β-hydroxy carbonyl compound.
(*b*) Suggest a mechanistic pathway to explain the formation of the two products of prolonged reaction.

Question 24.45

Give the structures of the Michael addition products for each of the following five reactions:

(*a*) ethyl cinnamate ($C_6H_5CH=CHCOOCH_2CH_3$) + ethyl acetoacetate
(*b*) methyl vinyl ketone ($CH_2=CH-CO-CH_3$) + ethyl cyanoacetate

    (c) α-methyl acrolein $[CH_2\!\!=\!\!C(CH_3)CHO]$ + diethyl malonate
    (d) benzalacetophenone $(C_6H_5CH\!\!=\!\!CH\!-\!COC_6H_5)$ + ethyl acetoacetate
    (e) mesityl oxide $[(CH_3)_2C\!\!=\!\!CHCOCH_3]$ + diethyl methylmalonate

24.24   Keto-enol Tautomerism in Carbonyl and β-Dicarbonyl Compounds

    The apparent duality of structures for ethyl acetoacetate was a puzzle to early chemists, and it was not until the late 1800s that experimental results were interpreted in terms of ethyl acetoacetate existing as a *mixture* of two isomeric compounds: the *keto form* and the *enol form*. The keto and enol forms are called **tautomers** of one another, and they are discrete, isolable isomers.

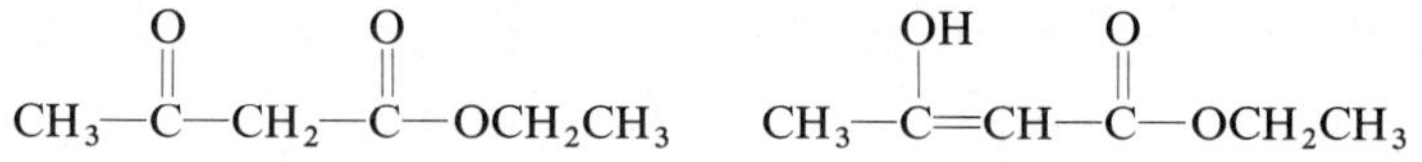

          Keto form   Enol form

    Since the original discovery of tautomers, methods have been developed for accurately determining the amount of enol in carbonyl-containing compounds. For example, pure acetone contains about $10^{-4}\%$ of the enol form, whereas β-dicarbonyl compounds contain appreciably more enol. The following reactions are typical examples:

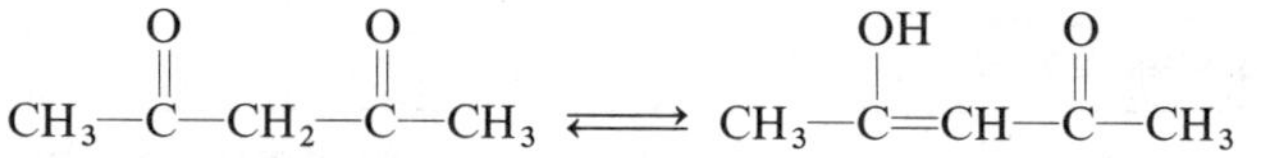

       2,4-Pentanedione   *70% enol*
        (acetylacetone)
          30% keto

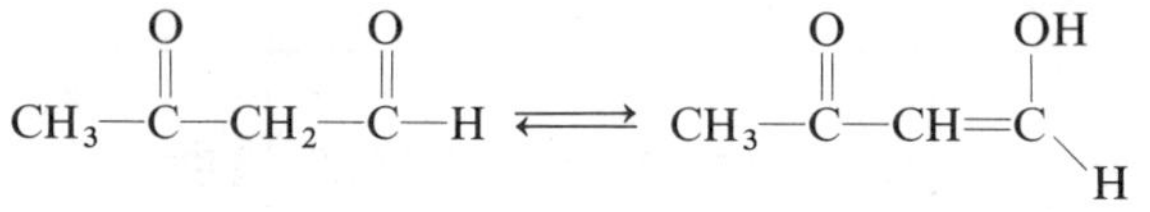

       3-Oxobutanal   *98% enol*
      (acetylacetaldehyde)
        2% keto

    Three factors are involved in dictating the enolic content of carbonyl-containing compounds. First, the nature of the substituents attached to the carbonyl groups plays a role. For example, ethyl acetoacetate contains about 8% enol, whereas ethyl benzoylacetate, $C_6H_5\!-\!CO\!-\!CH_2\!-\!COOCH_2CH_3$, contains about 23% of the enol form. Second, the amount of enol is dependent on the solvent in which the compound is dissolved. Concentration and temperature of the mixture are other variables. Third, the greatest driving force for enol formation in β-dicarbonyl compounds is the conversion of a nonconjugated form (the keto form) into a conjugated form (the enol form). The enol form is further stabilized by the formation of a six-membered ring, which involves hydrogen bonding of the following nature:

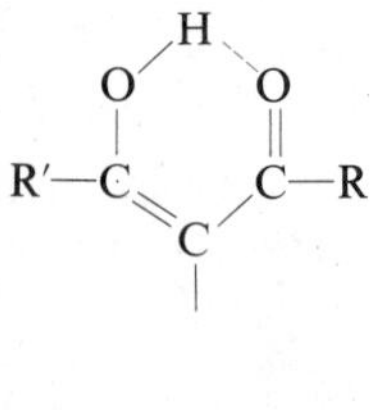

This internal hydrogen bonding is favored in nonpolar solvents where the enol content is greatest, but in polar solvents (such as ethanol) external hydrogen bonding with solvent becomes more important and the enol content decreases.

24.25 Summary of Reactions of α Hydrogens in Carbonyl and β-Dicarbonyl Compounds

A. Aldehydes and Ketones

1 Base-promoted halogenation of aldehydes and ketones

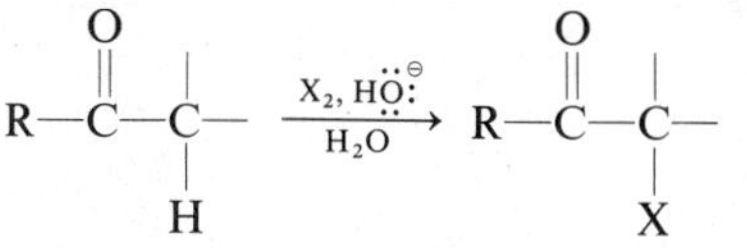

where X = Cl, Br, I

The haloform test:

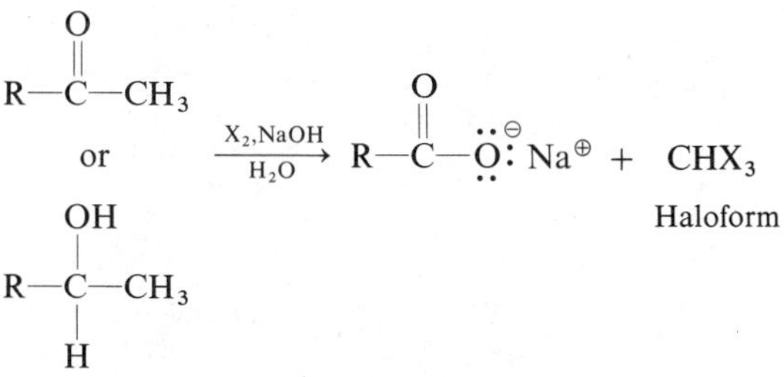

where X = Cl, Br, I

A yellow precipitate is obtained when iodine is used in the reaction.

2. Acid-catalyzed halogenation of aldehydes and ketones

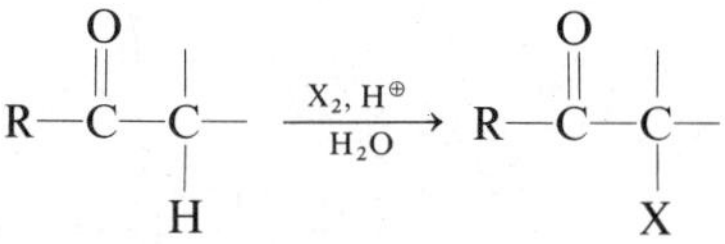

where X = Cl, Br, I

3. Hell-Volhard-Zelinsky (HVZ) reaction

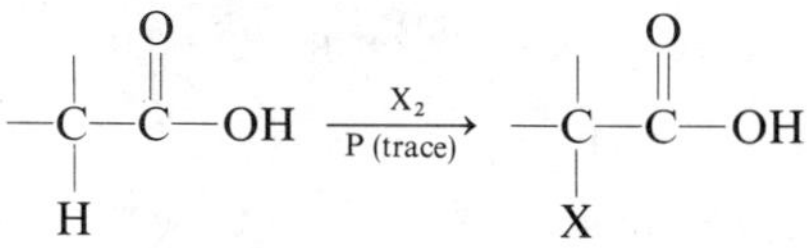

where X = Cl, Br

4. Aldol condensation of aldehydes and ketones

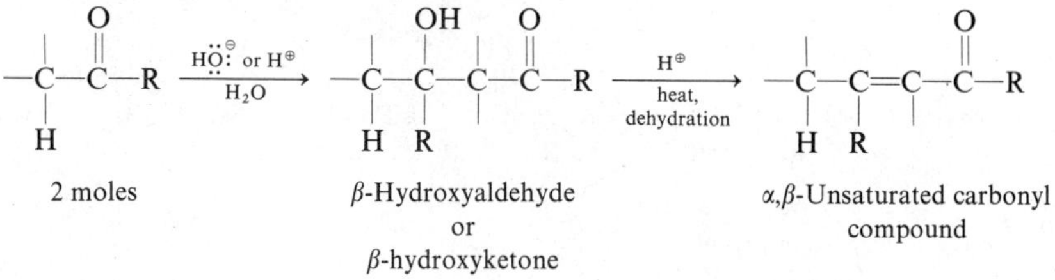

2 moles   β-Hydroxyaldehyde   α,β-Unsaturated carbonyl
            or   compound
         β-hydroxyketone

Crossed aldol condensations (Sec. 24.9)
Reactions of two different aldehydes (Sec. 24.9A)
Reactions between aldehydes and ketones (Sec. 24.9B, C)
Applications to synthesis (Sec. 24.10)

5. Reaction of aldehydes with no α hydrogens: Cannizzaro reaction

$$R_3C-\overset{\overset{\displaystyle O}{\|}}{C}-H \xrightarrow[\substack{H_2O}]{\substack{conc \\ NaOH}} R_3C-CH_2OH + R_3C-CO\overset{..}{\underset{..}{O}}\text{:}^{\ominus} Na^{\oplus} \quad \text{Sec. 24.11}$$

No α   1° Alcohol   Salt of
hydrogens   carboxylic
2 moles   acid

6. Alkylation of ketones and enamines

$$R-\overset{\overset{\displaystyle O}{\|}}{C}-\overset{\overset{\displaystyle |}{}}{\underset{\underset{\displaystyle H}{|}}{C}}- \xrightarrow[\substack{2.\ R'-CH_2-X \\ (S_N2\ reaction)}]{\substack{1.\ (CH_3)_3C\overset{..}{\underset{..}{O}}\text{:}^{\ominus}K^{\oplus} \\ or\ NaNH_2}} R-\overset{\overset{\displaystyle O}{\|}}{C}-\overset{\overset{\displaystyle |}{}}{\underset{\underset{\displaystyle CH_2R'}{|}}{C}}- \quad \text{Sec. 24.12}$$

Enamines in synthesis:

$$R-\overset{\overset{\displaystyle O}{\|}}{C}-\overset{\overset{\displaystyle |}{}}{\underset{\underset{\displaystyle H}{|}}{C}}- + R'_2N-H \rightleftharpoons R-\overset{\overset{\displaystyle R'_2N\text{:}}{|}}{C}=C\diagup \quad \text{Sec. 24.12}$$

2°   Enamine
Amine

R alkylation:

$$\overset{\overset{\displaystyle R'_2N\text{:}}{|}}{R-C}=C\diagup \quad \xrightarrow{R''-X}$$

C alkylation $\longrightarrow R-\overset{\overset{\displaystyle :\overset{..}{\underset{..}{X}}\text{:}^{\ominus}}{\overset{\displaystyle R'_2\overset{\oplus}{N}}{\|}}}{C}-\overset{\overset{\displaystyle |}{}}{\underset{\underset{\displaystyle R''}{|}}{C}}-$

N alkylation $\longrightarrow R-\overset{\overset{\displaystyle R'_2\overset{\oplus}{N}-R''\text{:}\overset{..}{\underset{..}{X}}\text{:}^{\ominus}}{|}}{C}=C\diagup$

$$R-\overset{\overset{\displaystyle :\overset{..}{\underset{..}{X}}\text{:}^{\ominus}}{\overset{\displaystyle R'_2\overset{\oplus}{N}}{\|}}}{C}-\overset{\overset{\displaystyle |}{}}{\underset{\underset{\displaystyle R''}{|}}{C}}- \xrightarrow{H_2O} R-\overset{\overset{\displaystyle O}{\|}}{C}-\overset{\overset{\displaystyle |}{}}{\underset{\underset{\displaystyle R''}{|}}{C}}-$$

C-Alkylation
product

7. Wittig reaction

$$Ph_3\overset{\oplus}{P}-\overset{..}{\underset{\ominus}{C}}\diagup + \overset{\overset{\displaystyle O}{\|}}{\underset{\underset{\displaystyle R\quad R'}{}}{C}} \longrightarrow \overset{\overset{\displaystyle \diagup C\diagdown}{\|}}{\underset{\underset{\displaystyle R\quad R'}{}}{C}} + Ph_3PO \quad \text{Sec. 24.13}$$

Ylide   Aldehyde
or ketone

B. Esters

Base-catalyzed condensation of esters: Claisen condensation

$$\underset{\underset{H}{\overset{\overset{\displaystyle O}{\|}}{-C}}}{-C}-OR \quad \xrightarrow[ROH]{R\ddot{O}\colon^{\ominus}} \quad \underset{\underset{H}{\overset{\overset{\displaystyle O}{\|}}{-C}}}{-C}-\underset{\overset{\displaystyle O}{\|}}{C}-OR \qquad \text{Sec. 24.14}$$

2 moles   β Keto ester

Intramolecular condensation: the Dieckmann condensation (Sec. 24.14)
Alternative bases (Sec. 24.14)
Crossed condensations (Sec. 24.15)
Condensations of esters and ketones (Sec. 24.16)

C. Anhydrides and Other Compounds Containing Acidic α Hydrogens

1. Condensation reaction of anhydrides: Perkin condensation

$$Ar-\overset{\overset{\displaystyle O}{\|}}{C}-H + CH_3-\overset{\overset{\displaystyle O}{\|}}{C}-O-\overset{\overset{\displaystyle O}{\|}}{C}-CH_3 \xrightarrow[\text{heat}]{CH_3C\ddot{O}\ddot{O}\colon^{\ominus}Na^{\oplus}} \left[Ar-\underset{\overset{\displaystyle H}{|}}{\overset{\overset{\displaystyle OH}{|}}{C}}-CH_2-\overset{\overset{\displaystyle O}{\|}}{C}-O-\overset{\overset{\displaystyle O}{\|}}{C}-CH_3 \right] \longrightarrow$$

Other anhydrides
may be used

$$Ar-CH=CH-\overset{\overset{\displaystyle O}{\|}}{C}-O-\overset{\overset{\displaystyle O}{\|}}{C}-CH_3 \xrightarrow{H_2O} Ar-CH=CH-\overset{\overset{\displaystyle O}{\|}}{C}-OH \qquad \text{Sec. 24.17}$$

Cinnamic acid

Cyano and nitro groups as acidifying agents (Sec. 24.18)
2. General

α Carbanions: formation and structure (Sec. 24.1)
Evidence for carbanion formation (Sec. 24.2)
  Isotope exchange studies (Sec. 24.3)
  Stereochemical evidence (Sec. 24.3)
  Nonstabilized carbanions (Sec. 24.3A)
  Stabilized carbanions (Sec. 24.3B)
Summary of reactive intermediates (Sec. 2.8)
Base-catalyzed keto-enol equilibria (Sec. 24.4)
Acid-catalyzed keto-enol equilibria (Sec. 24.7)

D. β-Dicarbonyl Compounds

1. Malonic ester synthesis

$$RO-\overset{\overset{\displaystyle O}{\|}}{C}-CH_2-\overset{\overset{\displaystyle O}{\|}}{C}-OR \xrightarrow[\substack{2.\ R'-X \\ (S_N2\ \text{reaction})}]{1.\ \text{base}} RO-\overset{\overset{\displaystyle O}{\|}}{C}-\underset{\overset{\displaystyle R'}{|}}{CH}-\overset{\overset{\displaystyle O}{\|}}{C}-OR \xrightarrow[2.\ R''-X]{1.\ \text{base}}$$

$$RO-\overset{\overset{\displaystyle O}{\|}}{C}-\underset{\overset{\displaystyle R'}{|}}{\overset{\overset{\displaystyle R''}{|}}{C}}-\overset{\overset{\displaystyle O}{\|}}{C}-OR \xrightarrow[\substack{2.\ \text{heat} \\ (\text{decarboxylation})}]{1.\ H^{\oplus},\ H_2O} H-\underset{\overset{\displaystyle R'}{|}}{\overset{\overset{\displaystyle R''}{|}}{C}}-\overset{\overset{\displaystyle O}{\|}}{C}-OH \qquad \text{Sec. 24.20}$$

Substituted
acetic acid

2. Acetoacetic ester synthesis

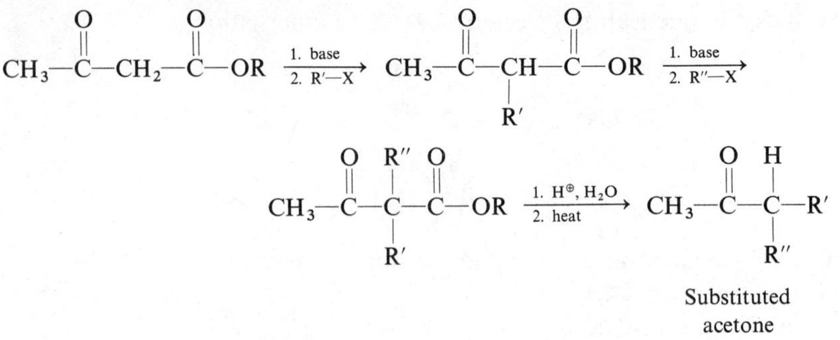

$$CH_3-\overset{O}{\underset{\|}{C}}-CH_2-\overset{O}{\underset{\|}{C}}-OR \xrightarrow[\text{2. R'—X}]{\text{1. base}} CH_3-\overset{O}{\underset{\|}{C}}-\underset{\underset{R'}{|}}{CH}-\overset{O}{\underset{\|}{C}}-OR \xrightarrow[\text{2. R''—X}]{\text{1. base}}$$

$$CH_3-\overset{O}{\underset{\|}{C}}-\underset{\underset{R'}{|}}{\overset{\overset{R''}{|}}{C}}-\overset{O}{\underset{\|}{C}}-OR \xrightarrow[\text{2. heat}]{\text{1. H}^{\oplus}\text{, H}_2\text{O}} CH_3-\overset{O}{\underset{\|}{C}}-\underset{\underset{R''}{|}}{\overset{\overset{H}{|}}{C}}-R' \qquad \text{Sec. 24.21}$$

Substituted
acetone

Acylation of malonic ester and acetoacetic ester (Sec. 24.22)

E. α,β-Unsaturated Carbonyl Compounds

1. Conjugate addition of Grignard reagents

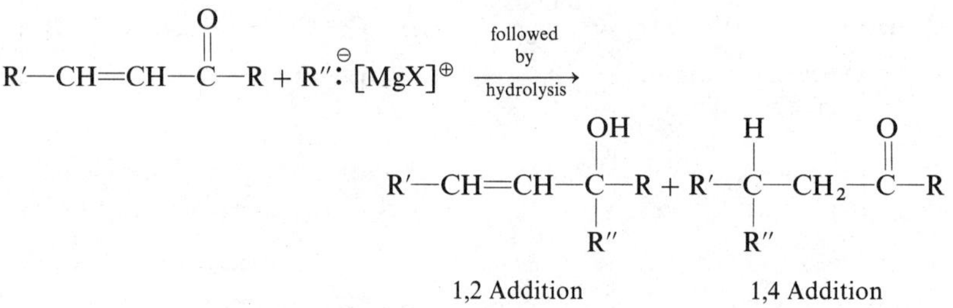

$$R'-CH=CH-\overset{O}{\underset{\|}{C}}-R + R''\overset{\ominus}{:}[MgX]^{\oplus} \xrightarrow[\text{hydrolysis}]{\substack{\text{followed} \\ \text{by}}}$$

$$R'-CH=CH-\underset{\underset{R''}{|}}{\overset{\overset{OH}{|}}{C}}-R + R'-\underset{\underset{R''}{|}}{\overset{\overset{H}{|}}{C}}-CH_2-\overset{O}{\underset{\|}{C}}-R \qquad \text{Sec. 24.23C}$$

1,2 Addition   1,4 Addition

Conjugate addition of other carbon nucleophiles: Michael reaction (Sec. 24.23C)

2. General

Structure of α,β-unsaturated carbonyl compounds (Sec. 24.23A)
General picture of nucleophilic addition: 1,2 versus 1,4 addition (Sec. 24.23B)
Michael addition (Sec. 24.23C)

Study Questions

24.46 Give IUPAC names for the following 11 compounds.

(*a*) $CH_3CH_2-\overset{O}{\underset{\|}{C}}-CH_2-Br$

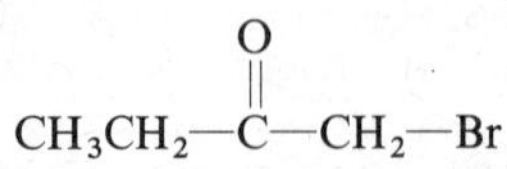

(*b*) $CH_3CH_2\underset{\underset{OH}{|}}{CH}CH_2-\overset{O}{\underset{\|}{C}}-H$

(*c*) $CH_3-\text{⟨◯⟩}-CH=CH-\overset{O}{\underset{\|}{C}}-H$

(*d*) cyclopentanone with $\overset{O}{\underset{\|}{C}}-NH_2$ substituent

(*e*) cyclohexanone with CH_3 and CH_2CH_3 substituents

(*f*) $\text{⟨◯⟩}-CH=CH-COOCH_3$

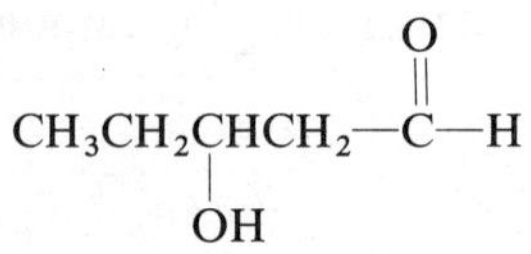

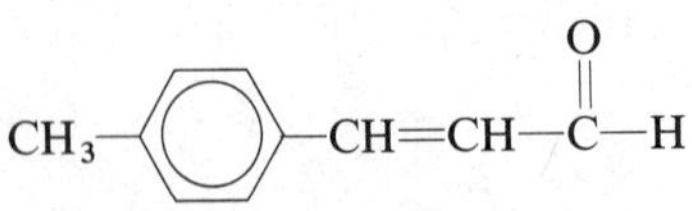

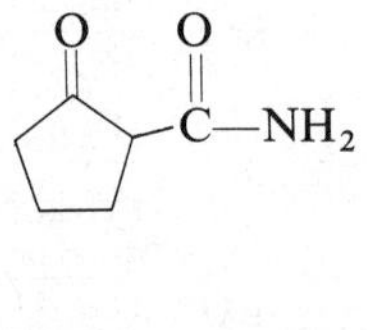

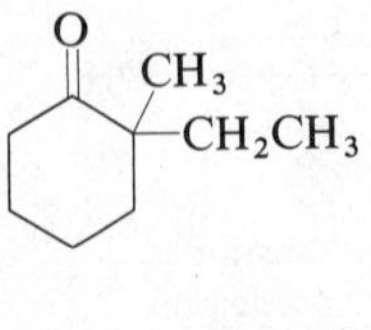

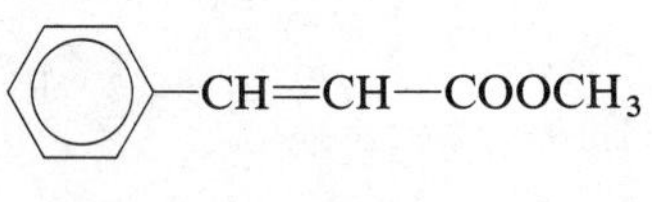

(g) 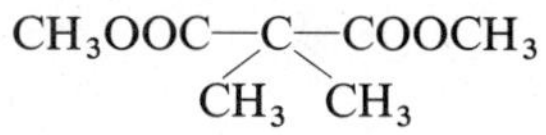 CH_3OOC—$\underset{\underset{CH_3}{|}}{\overset{\overset{CH_3}{|}}{C}}$—$COOCH_3$

(h) 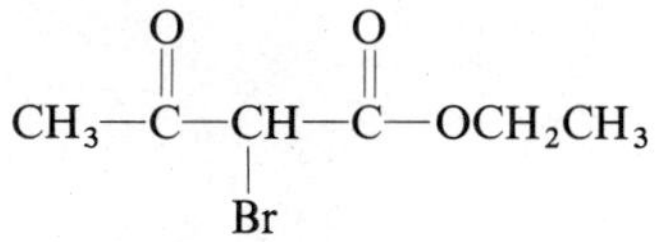 CH_3—$\overset{\overset{O}{||}}{C}$—$\underset{\underset{Br}{|}}{CH}$—$\overset{\overset{O}{||}}{C}$—$OCH_2CH_3$

(i)

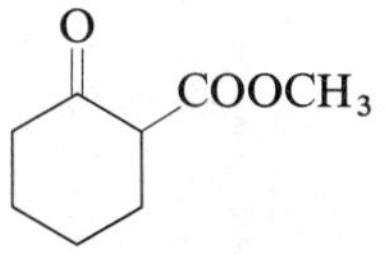

COOCH_3

(j) 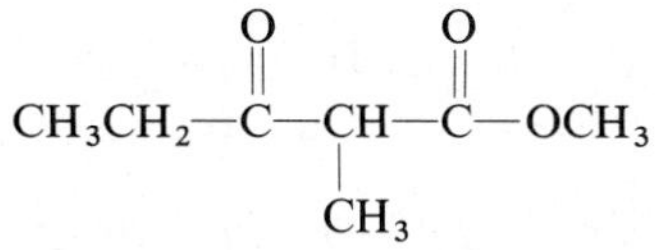 CH_3CH_2—$\overset{\overset{O}{||}}{C}$—$\underset{\underset{CH_3}{|}}{CH}$—$\overset{\overset{O}{||}}{C}$—$OCH_3$

(k) 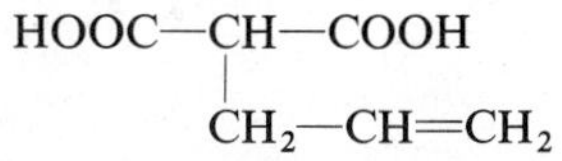 $HOOC$—$\underset{\underset{CH_2-CH=CH_2}{|}}{CH}$—$COOH$

24.47 Give structures that correspond to the following names.

(a) 5-hydroxy-4-methyl-5-ethyl-3-heptanone   (b) α-bromoacetophenone
(c) crotonaldehyde (2-butenal)   (d) 3-oxobutanal
(e) dimethyl succinate   (f) ethyl α-methylacetoacetate
(g) 2-carboxycyclohexanol   (h) acrolein

24.48 Supply the missing reactants, reagents, and products, as required, to complete the following 20 reactions. The items to be supplied are indicated by number [for example, (1)].

(a) 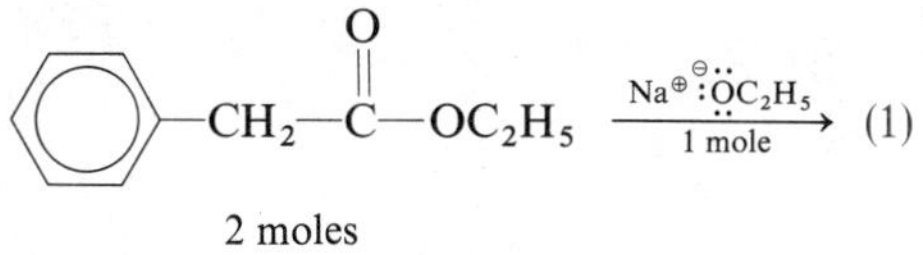 (⬡)—CH_2—$\overset{\overset{O}{||}}{C}$—$OC_2H_5$ $\xrightarrow[\text{1 mole}]{Na^{\oplus}:\ddot{O}C_2H_5}$ (1)

2 moles

(b) 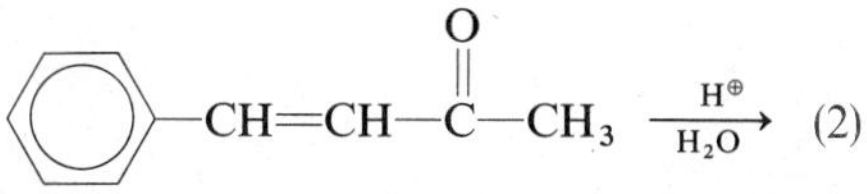 (⬡)—$CH=CH$—$\overset{\overset{O}{||}}{C}$—$CH_3$ $\xrightarrow[H_2O]{H^{\oplus}}$ (2)

(c) CH_3—$\overset{\overset{O}{||}}{C}$—$OCH_2CH_3$ $\xrightarrow{(3)}$ CH_3—$\overset{\overset{O}{||}}{C}$—$CH_2$—$\overset{\overset{O}{||}}{C}$—$OCH_2CH_3$

2 moles

$\downarrow$ 1. $NaOCH_2CH_3$
  2. $CH_3CH_2CH_2Br$

(4)

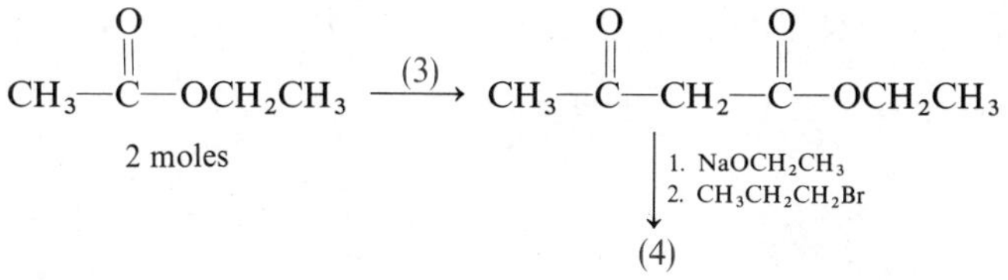

(d) 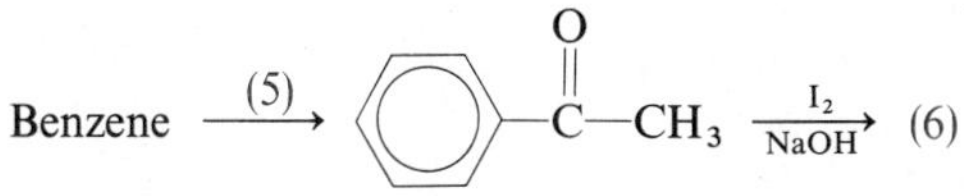 Benzene $\xrightarrow{(5)}$ (⬡)—$\overset{\overset{O}{||}}{C}$—$CH_3$ $\xrightarrow[NaOH]{I_2}$ (6)

(e) 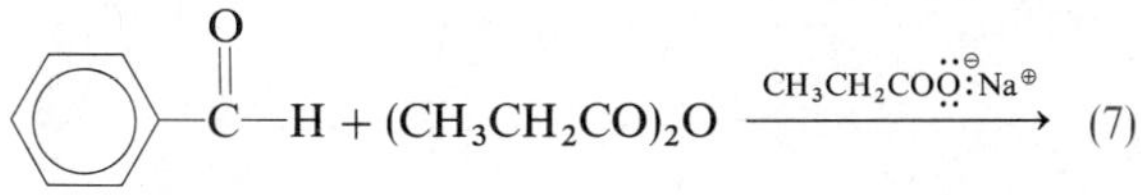 (⬡)—$\overset{\overset{O}{||}}{C}$—$H$ + $(CH_3CH_2CO)_2O$ $\xrightarrow{CH_3CH_2CO\ddot{O}:^{\ominus}Na^{\oplus}}$ (7)

(f) 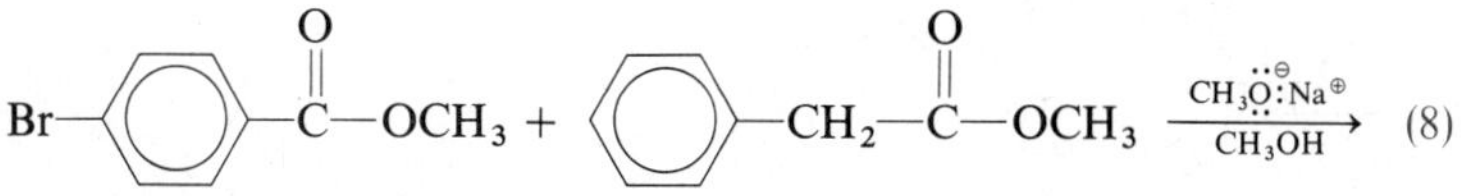 Br—(⬡)—$\overset{\overset{O}{||}}{C}$—$OCH_3$ + (⬡)—CH_2—$\overset{\overset{O}{||}}{C}$—$OCH_3$ $\xrightarrow[CH_3OH]{CH_3\ddot{O}:^{\ominus}Na^{\oplus}}$ (8)

(g)

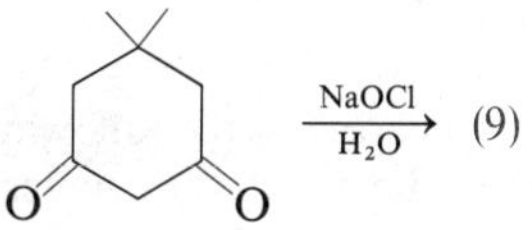

$\xrightarrow[H_2O]{NaOCl}$ (9)

(h) $CH_3CH_2\overset{\overset{O}{||}}{C}CH_3$ $\xrightarrow[H_2O]{H\ddot{O}:^{\ominus}}$ (10)

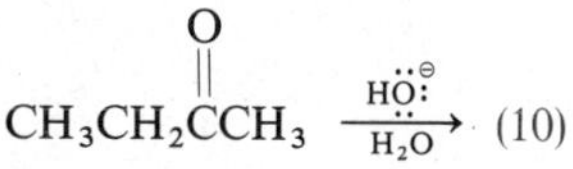

(*i*) $\text{C}_6\text{H}_5\text{–C(=O)–H} \xrightarrow[\text{H}_2\text{O}]{\text{conc HO}^{\ominus}} (11)$

(*j*) $\text{C}_6\text{H}_5\text{–C(=O)–H} \xrightarrow[\text{CN}^{\ominus}]{\text{HCN}} (12) \xrightarrow[\text{heat}]{\text{H}_3\text{O}^{\oplus}} (13)$

(*k*) $(\text{CH}_3)_2\text{CH–CH}_2\text{–C}_6\text{H}_5 \xleftarrow{(14)} (\text{CH}_3)_2\text{CH–C(=O)–C}_6\text{H}_5 \xrightarrow[\substack{:\text{OH}^{\ominus}, \\ \text{H}_2\text{O}}]{\text{Br}_2} (15)$

(*l*) o-C$_6$H$_4$(CH$_2$COOCH$_2$CH$_3$)$_2$ $\xrightarrow{(16)}$ indanone–COOCH$_2$CH$_3$ $\xrightarrow{\text{heat}} (17)$

$\xrightarrow[\text{conc}]{\text{HO:}^{\ominus}} (18)$

(*m*) $\text{CH}_3\text{–C(=O)–OH} \xrightarrow{(19)} \underset{\underset{\text{Br}}{|}}{\text{CH}_2}\text{–C(=O)–OH} \xrightarrow[\text{NH}_3]{\text{excess}} (20)$

(*n*) C$_6$H$_{11}$–C(=O)–CH$_3$ $\xrightarrow[\text{I}_2]{\text{NaOH}} (21)$

(*o*) $\text{CH}_3\text{COOH} \xrightarrow[\text{H}_2\text{SO}_4, \text{ heat}]{\text{CH}_3\text{CH}_2\text{OH}} (22) \xrightarrow{\text{NaOCH}_2\text{CH}_3} (23) \xrightarrow[\text{H}_2\text{O, heat}]{\text{H}_2\text{SO}_4} (24) \xrightarrow{\text{heat}} (25)$

(*p*) $\text{CH}_3\text{CH}_2\text{COOH} \xrightarrow[\text{H}_2\text{SO}_4, \text{ heat}]{\text{CH}_3\text{CH}_2\text{OH}} (26) \xrightarrow{\text{NaOCH}_2\text{CH}_3} (27) \xrightarrow[\text{2. (CH}_3)_2\text{CHCH}_2\text{Br}]{\text{1. NaOCH}_2\text{CH}_3} (28)$

(*q*) $(\text{CH}_3)_2\text{CHCH}_2\text{COOH} \xrightarrow[\text{Cl}_2]{\text{P}} (29) \xrightarrow[\text{alcohol}]{\text{NaOH}} (30) \xrightarrow[\text{heat}]{\text{KMnO}_4} (31)$

(*r*) C$_6$H$_5$–C(=O)–CH$_3$ $\xrightarrow[\text{H}_2\text{O}]{\text{HO}^{\ominus}} (32)$

(*s*) $\text{C}_6\text{H}_5\text{–C(=O)–}\underset{\underset{\text{CH}_3}{|}}{\text{CH}}\text{–C(=O)–OCH}_3 \xrightarrow[\text{HO:}^{\ominus}]{\text{H}_2\text{O}} (33)$

(*t*) cyclopentanone $\xrightarrow{(34)}$ enamine(piperidine) $\xrightarrow{(35)}$ 2-methylcyclopentanone

24.49 Show all the steps, conditions, and reagents required to carry out the following transformations from the indicated starting materials. Start with the indicated starting material and use any other monofunctional organic reagent you wish. All organic solvents (ether, CCl$_4$, and so on) and all inorganic reagents are available.

(*a*) $\text{CH}_3\text{CH}_2\text{CH}_2\text{COOH} \longrightarrow \text{CH}_3\text{CH}_2\underset{\underset{\text{COOCH}_3}{|}}{\text{CH}}\text{COOCH}_3$

(b) $CH_3COOH \longrightarrow$ ⬡$-CH_2CH_2COOH$

(c) Adipic acid $\longrightarrow$ ⬠ with CH_3 (*Hint:* see Sec. 24.14)

(d) $CH_3\overset{O}{\underset{\|}{C}}-H \longrightarrow CH_3\underset{\underset{OH}{|}}{CH}CH_2CH_2-OH$

(e) $CH_3CH_2CH_2CH_2-OH \longrightarrow CH_3CH_2CH_2CH_2\underset{\underset{CH_2CH_3}{|}}{CH}CH_2-OH$

(f) ⬡ with CH_3 $\longrightarrow$ ⬡$-\underset{\underset{Br}{|}}{CH}-\underset{\underset{Br}{|}}{CH}-\overset{O}{\underset{\|}{C}}-CH_3$

(g) $CH_3CH_2-\overset{O}{\underset{\|}{C}}-H \longrightarrow CH_3CH_2-\overset{O}{\underset{\|}{C}}-\underset{\underset{CH_3}{|}}{CH}-\overset{O}{\underset{\|}{C}}-OH$

(h) $CH_3-\overset{O}{\underset{\|}{C}}-CH_3 \longrightarrow CH_3-\overset{O}{\underset{\|}{C}}-CH_2CH_2-$⬡

(i) $CH_3-\overset{O}{\underset{\|}{C}}-OCH_3 \longrightarrow CH_3CH_2CH_2-COOCH_3$

(j) $CH_3CH_2CH_2OH \longrightarrow CH_3CH_2CH=\underset{\underset{CH_3}{|}}{C}-\overset{O}{\underset{\|}{C}}-H$

(k) $CH_3CH_2-\overset{O}{\underset{\|}{C}}-OH \longrightarrow CH_3-\underset{\underset{CH_3}{|}}{\overset{\overset{OH}{|}}{CH}}-CH-\overset{O}{\underset{\|}{C}}-OH$

24.50 Indicate how you might prepare the following 12 compounds, starting from *either diethyl malonate or ethyl acetoacetate* and any other organic or inorganic reagents you wish. Give reaction conditions and reagents, but do not balance the equations.

(a) $CH_3CH_2CH_2\underset{\underset{CH_3}{|}}{CH}CH_2CH_2CH_3$

(b) $CH_3\underset{\underset{CH_3}{|}}{CH}\overset{\overset{OH}{|}}{CH}CH_2CH_2CH_3$

(c) ⬡$-CH_2CH_2COOH$

(d) $CH_3CH_2CH_2\overset{O}{\underset{\|}{C}}-OH$

(e) $\underset{\underset{CH_2CH(CH_3)_2}{|}}{CH_3CH_2CH}-\overset{\overset{O}{||}}{C}-OCH_2CH_3$

(f) $HO-\overset{\overset{O}{||}}{C}-(CH_2)_5-\overset{\overset{O}{||}}{C}-OH$

(g) $CH_3COCH(CH_3)CH_2CH_3$

(h) $(CH_3CH_2)_2CHCH_2OH$

(i) $(CH_3CH_2)_3CH$

(j) $HOOC(CH_2)_5COOH$

(k) $(HOCH_2)_2C(CH_3)_2$

(l) $\underset{\underset{CH_3}{|}}{CH_3CH_2CHCH_2CH_2COOH}$

24.51 Provide reasonable mechanisms for the following reactions. Be sure to show *all* steps and indicate all equilibria where important. Also, show all intermediates.

(a) $CH_3-CH_2-\overset{\overset{O}{||}}{C}-OCH_2CH_3 \xrightarrow[heat]{CH_3CH_2\overset{\ominus}{\ddot{O}:}\ Na^{\oplus}}$ condensation

Give structure of product in addition to the mechanism leading to its formation.

(b) (benzaldehyde) $\overset{\overset{O}{||}}{C}-H \xrightarrow[heat]{conc\ NaOH}$ (benzoate) $\overset{\overset{O}{||}}{C}-\overset{\ominus}{\ddot{O}:}\ Na^{\oplus} +$ (benzyl) $-CH_2OH$

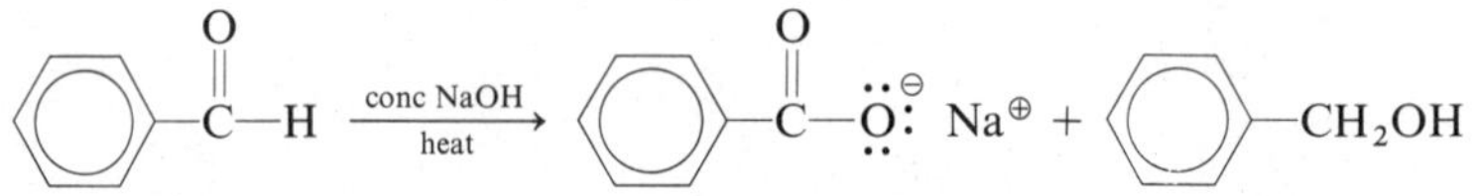

2 moles → 1 mole + 1 mole

(c) $\underset{\underset{CH_2OH}{|}}{\overset{\overset{CH_2OH}{|}}{CH_3-C}}-\overset{O}{\underset{H}{\overset{||}{C}}} + H_2C=O \xrightarrow[heat]{aq\ conc\ NaOH} \underset{\underset{CH_2OH}{|}}{\overset{\overset{CH_2OH}{|}}{CH_3-C}}-CH_2OH + H-\overset{\overset{O}{||}}{C}-\overset{\ominus}{\ddot{O}:}\ Na^{\oplus}$

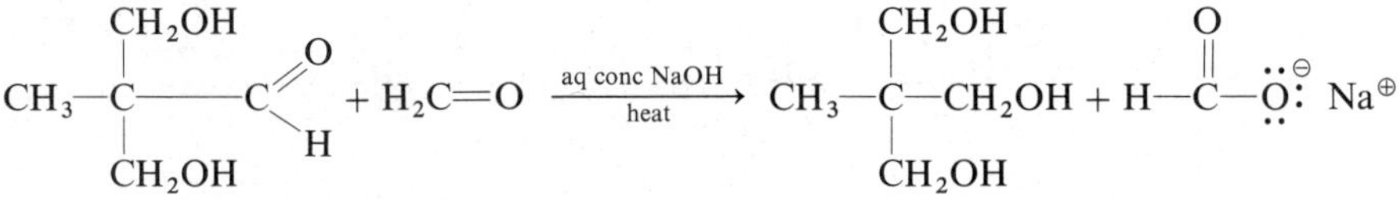

(d) $CH_3-\overset{\overset{O}{||}}{C}-CH_3 \xrightarrow[H_2O]{dilute\ :\ddot{O}H^{\ominus}} CH_3-\overset{\overset{O}{||}}{C}-CH_2-\underset{\underset{CH_3}{|}}{\overset{\overset{OH}{|}}{C}}-CH_3$

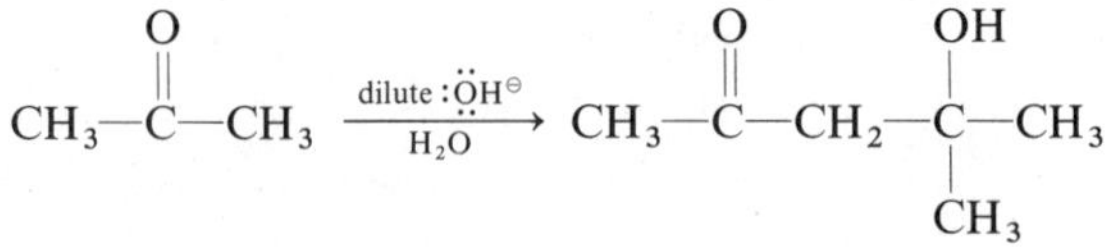

(e) $C_6H_5-\overset{\overset{O}{||}}{C}-CHBr_2 \xrightarrow{Br_2,\ :\ddot{O}H^{\ominus}} C_6H_5-\overset{\overset{O}{||}}{C}-\overset{\ominus}{\ddot{O}:}\ Na^{\oplus} + CHBr_3$

(f) $CH_3CH_2O-\underset{\underset{CH_2-CH_2}{|}}{\overset{\overset{O}{||}}{\underset{CH_2}{C}}}\quad\underset{}{\overset{\overset{O}{||}}{\underset{CH_2}{C}}}-OCH_2CH_3 \xrightarrow{Na^{\oplus}\ :\ddot{O}CH_2CH_3} \underset{CH_2-CH_2}{\overset{C}{\underset{CH_2}{}}}\quad\underset{}{\overset{}{CH}}-\overset{\overset{O}{||}}{C}-OCH_2CH_3$

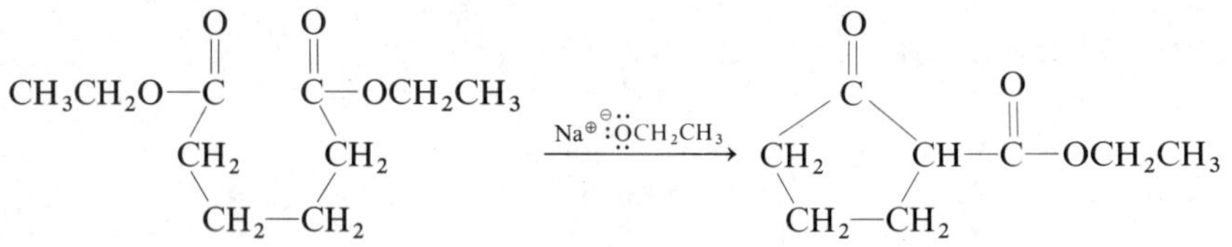

(g) $CH_3CH_2\overset{\overset{O}{||}}{C}-H \xrightarrow[HO:^{\ominus}]{Br_2}$

Give structure of *all* products in addition to the mechanism leading to their formation.

24.52 Match the following λ_{max} of absorption values with the given compounds: 232, 262, and 278 nm.

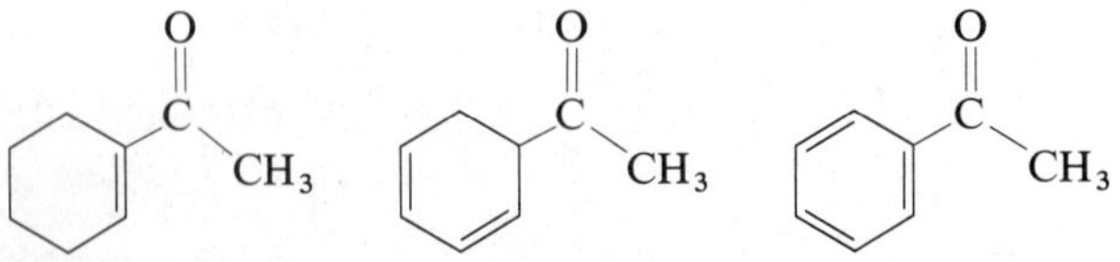

24.53 The IR spectrum, and NMR spectrum of compound *A* are shown. The mass spectrum exhibits a $P^{\oplus}$ peak at $127 \pm 2\ m/e$ units. Provide a structure consistent with the observed data. Completely interpret all spectra.

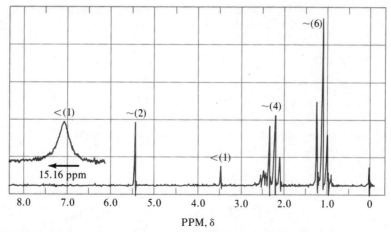

PROBLEM 24.53   NMR spectrum of unknown (*A*). [Data from the *Sadtler Standard Reference Spectra (NMR)*, © Sadtler Research Laboratories, Division of Bio-Rad Laboratories, Inc. (1965–1974)]

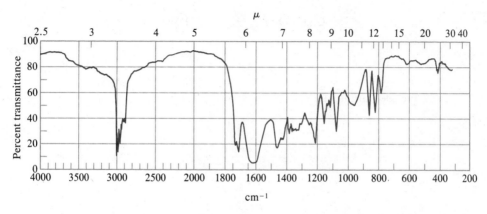

PROBLEM 24.53   IR spectrum of unknown (*A*). [Data from *The Sadtler Handbook of Infrared Spectra*, W. W. Simmons (ed.), © Sadtler Research Laboratories, Division of Bio-Rad Laboratories, Inc. (1978)]

24.54 Barbiturates, commonly known by the the slang term "downers," are derivatives of barbituric acid (1). Barbituric acid is obtained from urea and diethyl malonate as follows:

$$H_2NCNH_2 + CH_3CH_2O-CCH_2C-OCH_2CH_3 \xrightarrow[CH_3CH_2OH]{CH_3CH_2\overset{..}{\underset{..}{O}}:^{\ominus} Na^{\oplus}}$$

(1)

Starting with diethyl malonate, how might you synthesize barbital (2), a strong sedative marketed as sleeping pills:

(2)

Barbituric acid has a pK_a of approximately 4, whereas barbital has a pK_a of about 8. Explain.

24.55 Explain the following observation.

How would the UV, IR, and NMR spectra of the two products differ?

24.56 Provide structures for compounds *A* to *C*.

(*Hint:* The controlled acidity step involves the *dialkylation* of an enamine.)

24.57 Provide structures for compounds *A* to *C*.

24.58 Given the following information, provide structures for compounds *A* to *D*.

24.59 Design a mechanism to explain the following observation:

Sulfur ylide

The following two compounds are also produced during the course of the reaction:

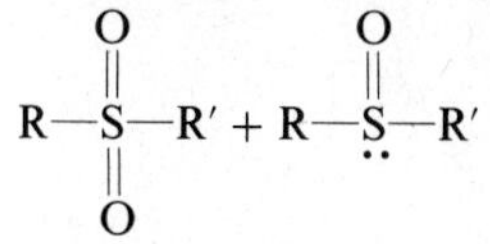

$$R-\overset{\displaystyle O}{\underset{\displaystyle O}{\overset{\displaystyle \|}{\underset{\displaystyle \|}{S}}}}-R' + R-\overset{\displaystyle O}{\underset{\displaystyle \cdot\cdot}{\overset{\displaystyle \|}{S}}}-R'$$

24.60 Would the following electronic representation of a nitrogen ylide be correct? Why or why not?

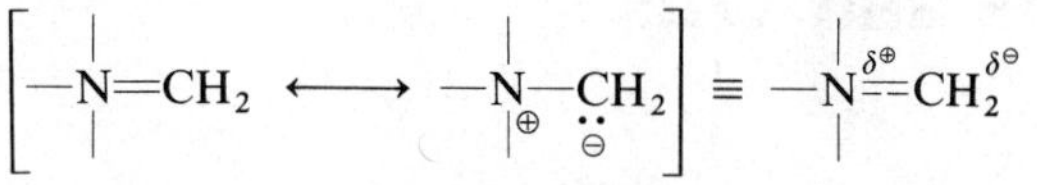

$$\left[-\overset{|}{\underset{|}{N}}=CH_2 \longleftrightarrow -\overset{|}{\underset{|}{\overset{\oplus}{N}}}-\overset{\cdot\cdot}{\underset{\ominus}{CH_2}} \right] \equiv -\overset{|}{\underset{|}{N}}\overset{\delta\oplus}{=\!\!=}CH_2^{\delta\ominus}$$

24.61 Given the following information, deduce the structures of compounds A to G.

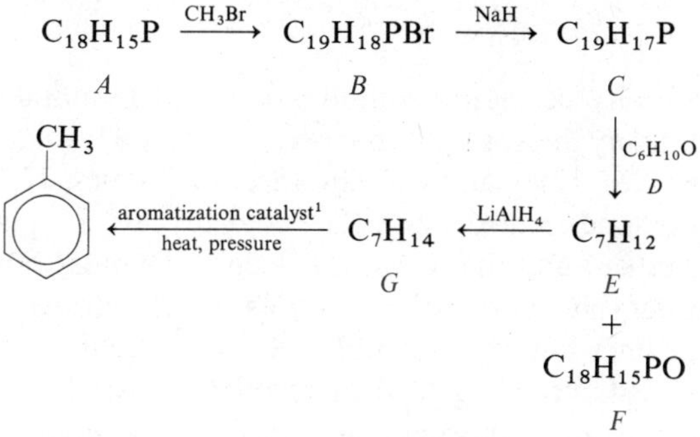

$$C_{18}H_{15}P \xrightarrow{\;CH_3Br\;} C_{19}H_{18}PBr \xrightarrow{\;NaH\;} C_{19}H_{17}P$$
$$\qquad\quad A \qquad\qquad\qquad\quad B \qquad\qquad\qquad\quad C$$

CH₃

$$\underset{\overset{\text{aromatization catalyst}^{1}}{\text{heat, pressure}}}{\longleftarrow} C_7H_{14} \xleftarrow{\;LiAlH_4\;} C_7H_{12}$$

$C_{19}H_{17}P \xrightarrow{C_6H_{10}O}$ D

G $\qquad\qquad\qquad E$

$$+$$

$$C_{18}H_{15}PO$$
$$F$$

The NMR spectrum of compound A exhibits only one peak, a multiplet (which actually appears to be a doublet), centered at 7.28 δ.

24.62 Z-9-Tricosene is the sex pheromone of the common American house fly. Structure elucidation of the compound, which was isolated from the cuticle and feces of the female fly, gave the structure drawn here. Outline a synthesis of this compound from organic starting materials containing no more than six carbon atoms.

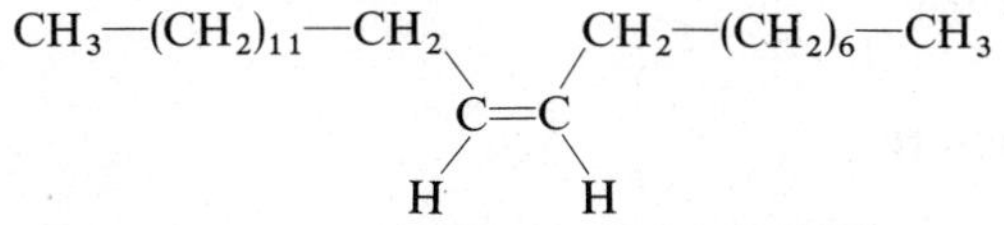

$$\underset{H}{\overset{CH_3-(CH_2)_{11}-CH_2}{\diagdown}}C=C\underset{H}{\overset{CH_2-(CH_2)_6-CH_3}{\diagup}}$$

Z-9-Tricosene

---

[1] Many cyclic compounds undergo dehydrogenation (loss of hydrogen) and aromatization when heated under pressure with catalysts such as Pd or Se. For example:

25

Lipids: Fats and Oils

Lipids are naturally occurring compounds found in animals and plants which are important in many biological processes. Fats, oil, waxes, steroids (Sec. 21.15), prostaglandins (Sec. 22.7), and vitamins (Sec. 19.3, Ques. 27.17) belong to the general family of compounds called **lipids** (Greek *lipos*, fat). Lipids are insoluble in water but soluble in ether and other organic solvents of similar polarity. Natural products that are considerably more polar, such as carbohydrates, proteins, peptides, and alkaloids, are somewhat water-soluble, which distinguishes them from lipids. Lipids are often classified according to their reactivity toward base. Fats, oils, and waxes react with aqueous base and undergo hydrolysis, whereas steroids and other lipids often do not (see Sec. 21.15). In this chapter we are concerned mainly with lipids that react with base, namely fats, oils, and waxes. The relationship of fats and oils to soaps, detergents, paints, waxes, and biological membranes is also discussed.

25.1 Glycerides

A. Structure

Fats and oils have the general structure shown here and are called **glycerides.** These compounds are *triesters* derived from glycerol (a trihydroxy compound) and three carboxylic acids:

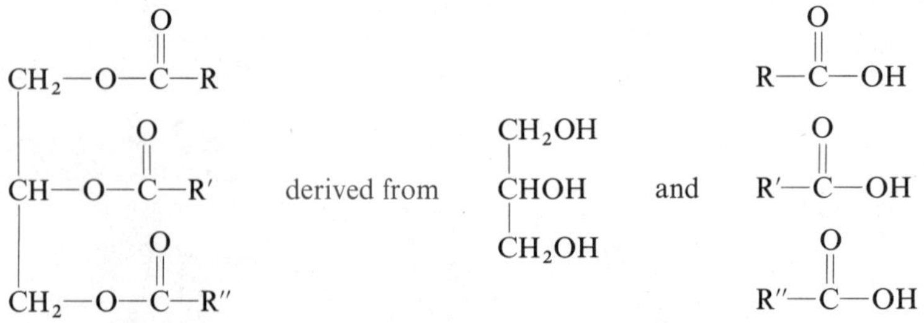

$$
\begin{array}{ccc}
\underset{\substack{| \\ }}{CH_2\!-\!O\!-\!\overset{\overset{O}{\|}}{C}\!-\!R} & & R\!-\!\overset{\overset{O}{\|}}{C}\!-\!OH \\
\underset{\substack{| \\ }}{CH\!-\!O\!-\!\overset{\overset{O}{\|}}{C}\!-\!R'} \quad \text{derived from} \quad \underset{\substack{| \\ CH_2OH}}{\overset{CH_2OH}{\underset{|}{CHOH}}} \quad \text{and} \quad R'\!-\!\overset{\overset{O}{\|}}{C}\!-\!OH \\
CH_2\!-\!O\!-\!\overset{\overset{O}{\|}}{C}\!-\!R'' & & R''\!-\!\overset{\overset{O}{\|}}{C}\!-\!OH
\end{array}
$$

| Glyceride (fat or oil) *Triester* | 1,2,3-Propanetriol (glycerine) | |

Glyceride
(fat or oil)
Triester

1,2,3-Propanetriol
(glycerine)

Fats and **oils** are both triglycerides. The main difference between them is that a fat is a solid at room temperature whereas an oil is a liquid under the same conditions. Fats usually come from animals and oils are generally obtained from plants. Therefore, we often use the terms *animal fats* and *vegetable oils* (see Sec. 25.3).

More general observations about fats and oils concern the carboxylic acid portions of the ester. Some of the acids are saturated whereas others are unsaturated, and most are straight-chain compounds. In those that are unsaturated, the double bonds are usually *cis* and are very seldom *trans*. Most of the acids contain an *even* number of carbon atoms, usually from 12 to 20. For this reason most acids containing six of more carbon atoms in even multiples are called **fatty acids.**

Some new fatty acids discovered recently contain a highly strained *cyclopropene* ring, for example:

$$CH_3(CH_2)_7 \quad (CH_2)_7COOH \qquad CH_3(CH_2)_7 \quad (CH_2)_6COOH$$

Sterculic acid · · · · · · · · · · · · · · Malvalic acid

The seed fat of the Java olive (*Sterculia foetida*) contains more than 70% *sterculic acid*, which is a C_{19} acid.

Some important fats and oils are listed in Table 25.1 The composition of the carboxylic acids that each contains is somewhat approximate, and the glycerides have widely varying compositions depending on their source. There are several reasons for variations in composition. For example, an animal fat's composition is dependent on the age, sex, and diet of the animal, and the composition of a vegetable oil depends on the climate (temperature and amount of sunlight and rain) and the soil where it grew.

The existence of *cis* and *trans* isomers in the acids contained in glycerides is interesting, and very little is known about why double bonds appear where they do. Because glycerides are synthesized in living systems, there is some rationale for finding acids that contain an even number of carbon atoms (see Reading References for Chap. 2 in the appendix). The location of the double bonds and the existence of predominantly *cis* geometry are probably caused by the enzymes involved in the biosynthesis of glycerides.

As we saw in Sec. 8.15 the *trans* isomers of alkenes are generally more stable thermodynamically than the *cis* isomers. Fatty acids containing the *trans* double bond generally melt at higher temperatures than those containing the *cis* bond. Thus, a *cis* acid (for example, oleic acid) is converted to the *trans* acid on heating with sulfur or selenium until equilibrium is established; for example, equilibration of oleic acid gives about 67% of elaidic acid (*trans* isomer). This procedure can be used to convert a lower melting *cis* acid into the higher melting *trans* acid, so that certain oils can be transformed into fats.

Finally, there is a rather even distribution of carboxylic acids within a given glyceride sample. For example, if a sample of glyceride contains about one-third of one carboxylic acid, then each glyceride molecule of that type usually contains one molecule of that acid rather than some glyceride molecules containing two molecules of the acid with others containing none. Glyceride molecules generally do not contain three identical acids unless the source of the glyceride contains more than 67% of that particular acid.

B. Nomenclature

The following common system is used to name glycerides. The three positions on the glycerine molecule are labeled α, β, and γ. Simple triglycerides are named by

TABLE 25.1   Composition of Fatty Acids in Various Fats and Oils

Fat or Oil	Carboxylic Acids Present (Weight %)						
	Lauric	Myristic	Palmitic	Stearic	Oleic	Linoleic	Others*
Vegetable fats:							
Palm		1–3	35–40	3–6	38–40	5–11	
Coconut	45–51	17–20	4–10	1–5	5–8	0–2	Caprylic, 5–10; capric, 5–11
Animal fats:							
Butter (cow milk)	3–5	7–11	23–26	10–13	30–40	4–6	Butyric, 3–4; caproic, 2; caprylic, 1; capric, 2–3; palmitoleic,[†] 5
Human milk	5–7	8–14	22–25	8–10	30–35	4–8	Capric, 1–3; palmitoleic,[†] 3–4; arachidic, 1
Lard (hog fat)		1–2	28–30	15–22	41–52	6–8	Palmitoleic,[†] 1–3
Tallow (beef fat)		2–3	24–32	14–32	35–48	2–7	Palmitoleic,[†] 1–3
Vegetable oils:							
Olive		0–1	6–15	1–4	69–85	4–12	
Peanut			6–9	2–6	50–70	13–26	Arachidic, 2–5; tetracosanoic,[‡] 1–5
Castor			0–1		0–9	3–7	Ricinoleic,[§] 80–92
Corn		0–2	8–10	3–5	43–50	34–42	Palmitoleic,[†] 1–2
Cottonseed		1–3	19–24	1–2	23–31	40–50	Palmitoleic,[†] 0–2; arachidic, 1
Soybean		0–1	7–10	2–4	21–31	50–62	Linolenic, 4–8
Tung			4–5	0–2	4–16	1–10	Eleostearic,[¶] 74–88
Linseed			4–7	3–5	9–30	3–40	Linolenic, 25–55
Animal oils:							
Lard oil			22–26	15–17	45–55	8–10	
Whale		4–6	11–18	2–4	33–38		Palmitoleic,[†] 13–18; 11–20 unsaturated C_{20} acids and 6–10 C_{22} unsaturated acids

---

* See the following footnotes for the structures of certain of these acids; the structures of those without footnotes are in Table 22.1.

[†] Palmitoleic acid: $CH_3(CH_2)_5—CH=CH—(CH_2)_7COOH$

[‡] Tetracosanoic acid: $CH_3(CH_2)_{22}COOH$

[§] Ricinoleic acid: $CH_3(CH_2)_5—\underset{\underset{OH}{|}}{CH}—CH_2—\underset{cis}{CH=CH}—(CH_2)_7COOH$

[¶] Eleostearic acid: $CH_3(CH_2)_3—\underset{trans}{CH=CH}—\underset{trans}{CH=CH}—\underset{cis}{CH=CH}—(CH_2)_7—COOH$

dropping the *-ic acid* ending and adding the suffix **-in**; for example:

$$\overset{\alpha}{C}H_2-O-\overset{\overset{\textstyle O}{\|}}{C}-(CH_2)_{10}-CH_3$$

$$\overset{\beta}{C}HOH$$

$$\overset{\gamma}{C}H_2OH$$

$$CH_2-O-\overset{\overset{\textstyle O}{\|}}{C}-(CH_2)_{10}-CH_3$$

$$CH-O-\overset{\overset{\textstyle O}{\|}}{C}-(CH_2)_{10}-CH_3$$

$$CH_3-O-\overset{\overset{\textstyle O}{\|}}{C}-(CH_2)_{10}-CH_3$$

α-Monolaurin $\qquad\qquad$ Trilaurin

When there are different acids in a glyceride, the acid that appears last is named as described earlier. The name(s) of the other acids precede it, but the *-ic acid* ending of the acid is dropped and replaced by **-o**; for example:

$$CH_2-O-\overset{\overset{\textstyle O}{\|}}{C}-(CH_2)_{14}-CH_3$$

$$CHOH$$

$$CH_2-O-\overset{\overset{\textstyle O}{\|}}{C}-(CH_2)_7-CH=CH-(CH_2)_7CH_3$$

cis

α-Palmito-γ-olein

or

α-oleo-γ-palmitin

The names of the acids can be placed in any order; there are no firm rules governing this aspect of glyceride nomenclature.

25.2 Soaps and Detergents

A. Soaps

The lore of soap-making dates back to the times of the ancient Romans. Originally it involved the hydrolysis of a fat or oil with aqueous alkali, which was usually a mixture of potassium carbonate and potassium hydroxide obtained from the leaching of wood ashes with water.

Today soap is obtained in much the same way from the base-catalyzed hydrolysis (*saponification*, see Sec. 23.7) of animal or vegetable fats or oils, which means that it is derived from natural sources (plants or animals). It is not surprising that treatment of a fat or oil (a triester) with excess base hydrolyzes each ester grouping. This reaction produces glycerol and a mixture of salts of carboxylic acids.

The carbon content of the fatty acid salts governs the solubility of a soap. The lower molecular weight acid salts (up to C_{12}) have greater solubility in water and give a lather containing large bubbles. The higher molecular weight acid salts (C_{14} to C_{20}) are much less soluble in water and give a lather with fine bubbles. The potassium salts of the acids are more soluble in water than the corresponding sodium salts.

The question of how soap works can be answered by considering the functional groups in soap and how they interact with oil and water. The long, continuous side chains of carbon atoms in a soap molecule resemble an alkane and they dissolve other nonpolar compounds such as oils and greases. The adage "like dissolves like" applies here. The large nonpolar hydrocarbon part of the molecule is

$$CH_2-O-\overset{\displaystyle O}{\overset{\|}{C}}-R$$

$$CH-O-\overset{\displaystyle O}{\overset{\|}{C}}-R' \quad \xrightarrow[\substack{H_2O, \\ heat}]{M^{\oplus}:\overset{..}{\overset{..}{O}}H^{\ominus}} \quad \begin{array}{l} CH_2-OH \\ | \\ CH-OH \quad + \\ | \\ CH_2-OH \end{array}$$

$$CH_2-O-\overset{\displaystyle O}{\overset{\|}{C}}-R''$$

Fat or oil
Triglyceride

Glycerol

$$R-\overset{\displaystyle O}{\overset{\|}{C}}-\overset{..}{\underset{..}{O}}{:}^{\ominus} M^{\oplus} \; + \; R'-\overset{\displaystyle O}{\overset{\|}{C}}-\overset{..}{\underset{..}{O}}{:}^{\ominus} M^{\oplus} \; + \; R''-\overset{\displaystyle O}{\overset{\|}{C}}-\overset{..}{\underset{..}{O}}{:}^{\ominus} M^{\oplus}$$

Soap
*Mixture of
carboxylic.acid
salts*

where $M^{\oplus} = Na^{\oplus}$ or $K^{\oplus}$

described as **hydrophobic,** which means repelled by water. The highly polar carb-oxylate end of the molecule is called **hydrophilic,** which means water-soluble. When soap is dissolved in water, the carboxylate end actually dissolves. The hydrocarbon part is repelled by the water molecules so that a thin film of soap is formed on the surface of the aqueous layer, with the hydrocarbon chains protruding outward. This greatly lowers the surface tension of the water. When soap solution comes in con-tact with oil or grease, the hydrocarbon part dissolves in the oil or grease but the polar carboxylate group remains dissolved in water. When particles of oil or grease are surrounded by soap molecules, the resulting "units" formed are called *micelles* (see Sec. 25.5). A simplified view of this phenomenon is shown in Fig. 25.1. Keep in mind that this view is highly simplified because it does not show hydrogen bond-ing and solvation. These micelles repel one another because they are surrounded on the surface by the negatively charged carboxylate ions. Mechanical action (for

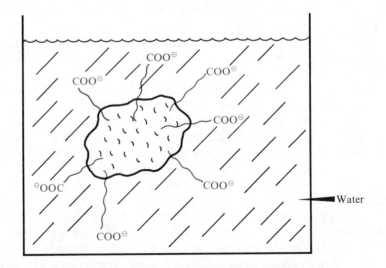

FIGURE 25.1   Simplified view of the action of soap on oil particles in aqueous solution. The wiggly line represents the long, continuous carbon chain of each soap molecule.

example, scrubbing or tumbling in a washing machine) causes oil or grease to be broken into small droplets so that relatively small micelles are formed with soap solution. Careful examination of this solution shows that it is an **emulsion** containing suspended micelles.

We considered the action of soap on grease and oil, yet we know that soap is also effective in removing inert dirt particles. Most dirt is held onto cloth by a thin film of oil or grease, which is removed by soap. Soap also reduces the surface tension of water, which in combination with mechanical action facilitates the removal of dirt from cloth.

The major disadvantage of soap arises when it is used in *hard water*, which is water containing a high content of certain metallic cations such as calcium (Ca^{+2}), magnesium (Mg^{+2}), barium (Ba^{+2}), and iron (Fe^{+2} and Fe^{+3}). These cations convert the soluble sodium or potassium salts of the carboxylic acids into insoluble carboxylate salts:

$$2\ CH_3(CH_2)_{16}COO{:}^{\ominus}\ Na^{\oplus}\ +\ Mg^{+2}\ \longrightarrow\ [CH_3(CH_2)_{16}COO{:}^{\ominus}\ ]_2Mg^{+2}\ +\ 2\ Na^{\oplus}$$

Sodium stearate   Magnesium stearate
Water-soluble salt   *Water-insoluble salt*

These water-insoluble salts are responsible for the "ring" around the bathtub that occurs in regions where the water is hard. The precipitation of soap as an insoluble salt renders it ineffective as a cleansing agent. This problem has been overcome partially by the development of synthetic detergents, which are discussed in Sec. 14.14B.

B. Uses of Glycerides in Synthetic Chemistry

Glycerides are a useful and cheap source of fatty acids as well as the corresponding alcohols. For example, acid or base hydrolysis of a glyceride yields the carboxylic acids directly:

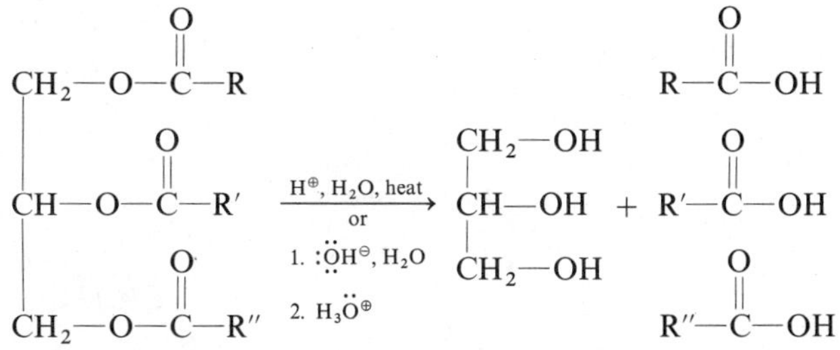

Glyceride   Glycerol   Mixture of
                                                  carboxylic acids

Fractional distillation can be used to obtain each carboxylic acid in a fairly pure state. A glyceride can also be transesterified by simple alcohols (for example, methanol or ethanol) to provide a mixture of methyl or ethyl esters, which can be separated by fractional distillation and then converted to other carboxylic acid derivatives.

The acids or esters obtained by the preceding methods can also be reduced to the corresponding long-chain alcohols. However, methods are available for converting glycerides directly into these alcohols by reducing them with common reducing agents. Lithium aluminum hydride is the usual choice in the laboratory, but either catalytic hydrogenation or sodium in ethanol is used industrially.

Because catalytic hydrogenation of glycerides also reduces any carbon-carbon double bonds that may be present (see Sec. 8.12), sodium in ethanol (the **Beauvault-Blanc** process, see Sec. 23.7E) is used to avoid this side reaction. As we will see, the

$$
\begin{array}{l}
\text{CH}_2\text{—O—}\overset{\displaystyle O}{\overset{\|}{\text{C}}}\text{—R} \\[2mm]
\text{CH—O—}\overset{\displaystyle O}{\overset{\|}{\text{C}}}\text{—R}' \\[2mm]
\text{CH}_2\text{—O—}\overset{\displaystyle O}{\overset{\|}{\text{C}}}\text{—R}''
\end{array}
\quad
\xrightarrow[\substack{\text{H}_2,\ \text{Cu}_2\text{O} + \text{Cr}_2\text{O}_3, \\ \text{heat, pressure or} \\ \text{Na, CH}_3\text{CH}_2\text{OH}}]{\substack{\text{1. LiAlH}_4,\ 2.\ \text{H}_3\ddot{\text{O}}^\oplus \\ \text{or}}}
\quad
\begin{array}{ll}
\text{CH}_2\text{—OH} & \text{RCH}_2\text{OH} \\[2mm]
\text{CH—OH} & + \text{R}'\text{CH}_2\text{OH} \\[2mm]
\text{CH}_2\text{—OH} & \text{R}''\text{CH}_2\text{OH}
\end{array}
$$

Glyceride   Glycerol   Long-chain
                                                      primary
                                                      alcohols

conversion of glycerides to long, continuous-chain alcohols is important because the latter can then be converted into detergents.

C. Detergents

The term **detergent** is general and refers to any cleansing agent (thus, soaps fall under this broad definition), but it is most commonly associated with synthetic detergents. Numerous detergents have been synthesized in recent years. They are important because they do not form precipitates with the divalent metallic ions in hard water. There are several general classes of detergents (cationic, anionic, and nonionic), but we focus our attention on the anionic ones.

Anionic detergents have the same general feature of soaps; they contain a highly polar, negatively charged (in this case $-\text{SO}_3^\ominus$) group and a long hydrocarbon chain that dissolves oil and grease. The starting materials for many detergents are long-chain, saturated C_{12} to C_{18} alcohols such as those available from the reduction of naturally occurring fats and oils by the methods discussed earlier. The following shows the preparation of a typical detergent from an alcohol:

$$
\text{CH}_3(\text{CH}_2)_n\text{CH}_2\text{OH} \xrightarrow[\text{H}_2\text{SO}_4]{\text{cold, conc}} \text{CH}_3(\text{CH}_2)_n\text{CH}_2\text{—O—}\overset{\displaystyle O}{\underset{\displaystyle O}{\overset{\|}{\underset{\|}{\text{S}}}}}\text{—OH} \xrightarrow[\text{with NaOH}]{\text{neutralize}}
$$

$n = 10, 12, 14, 16$ Alkyl hydrogen
                               sulfate

$$
\text{CH}_3(\text{CH}_2)_n\text{CH}_2\text{—O—}\overset{\displaystyle O}{\underset{\displaystyle O}{\overset{\|}{\underset{\|}{\text{S}}}}}\text{—}\overset{..}{\underset{..}{\text{O}}}\overset{\ominus}{:}\ \text{Na}^\oplus
$$

Sodium alkyl sulfate
Detergent

The sodium alkylbenzenesulfonates are another widely used class of anionic detergent, and they are prepared as follows:

$$
\text{Benzene} \xrightarrow[\text{alkylation}]{\text{Friedel-Crafts}} \text{Ar—R} \xrightarrow{\text{H}_2\text{SO}_4} \underset{R}{\overset{\text{SO}_3\text{H}}{\bigcirc}} \xrightarrow[\text{with NaOH}]{\text{neutralize}} \underset{R}{\overset{\text{SO}_3^\ominus \text{Na}^\oplus}{\bigcirc}}
$$

Sodium alkylbenzenesulfonate

Alkylation occurs by the Friedel-Crafts reaction of an alkene, alkyl halide, or alcohol on benzene (see Secs. 14.6 and 14.7). The sulfonation of the arene yields a mixture of monosubstituted arylsulfonic acids, so a detergent is a mixture of compounds.

The alkylbenzenesulfonates were originally prepared by alkylation with tetra-propylene (see Sec. 14.14B) because of the ready availability and low cost of this alkene. However, it was found that detergents with highly branched side chains were not being degraded in sewage treatment plants and septic tanks but instead were discharged into streams, where they killed wildlife and even made their way back into drinking water systems. This kind of detergent is referred to as being "hard" or *nonbiodegradable*. The discovery that straight-chain alkyl groups are degraded by microorganisms in septic tanks and sewage treatment systems helped solve the pollu-tion problem. Since 1966, most detergents are prepared with the aromatic ring attached to various positions along a straight carbon chain to produce compounds that are "soft" or *biodegradable*. The starting compounds for the alkylation reaction are straight-chain alcohols obtained from the reduction of fats and oils or alkenes and alkyl halides from petroleum fractions. Although the preparation of biodegrad-able detergents is more costly than that of the nonbiodegradable variety, the benefits of decreased pollution and preservation of natural resources are far more important.

Soaps (containing the carboxylate group) and detergents (containing the sulfate or sulfonate group) are useful because they are water-soluble and possess a long carbon chain in which oil and grease are soluble. Yet most soaps and detergents produce foam (soap suds) on stirring or in washing machines. This makes them undesirable for use in dishwashers or other places where suds are not wanted. The development of nonionic detergents, such as the one shown here, has helped overcome this problem:

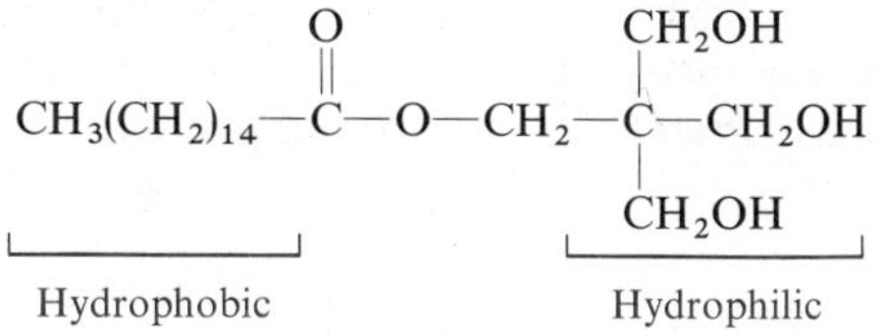

$$CH_3(CH_2)_{14}-\overset{\overset{\displaystyle O}{\|}}{C}-O-CH_2-\overset{\overset{\displaystyle CH_2OH}{|}}{\underset{\underset{\displaystyle CH_2OH}{|}}{C}}-CH_2OH$$

$$\underbrace{\hspace{4cm}}_{\text{Hydrophobic}} \qquad \underbrace{\hspace{4cm}}_{\text{Hydrophilic}}$$

Pentaerythrityl palmitate

The three hydroxyl groups increase the compound's water solubility greatly, and the long carbon chain dissolves oil and grease. Compounds of this type are biodegradable.

25.3   Unsaturated Glycerides

Unsaturated glycerides (those with one or more carbon-carbon double bonds in the side chain) have several special uses because of their unsaturation.

Terms such as *polyunsaturated*, *saturated*, and *hydrogenated* fats and oils are familiar to us from advertising. Oils are usually highly unsaturated, and vegetable oils can be transformed into solid fats industrially by catalytic hydrogenation. This process, called *hardening*, involves the low-pressure hydrogenation of an oil in the presence of a metal catalyst:

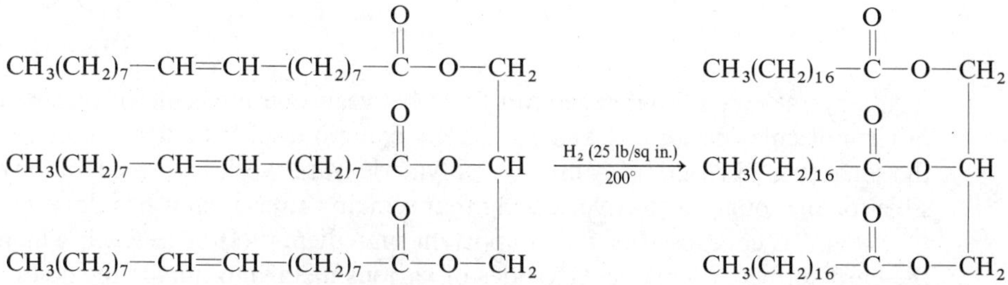

$$CH_3(CH_2)_7-CH\!=\!CH-(CH_2)_7-\overset{\overset{\displaystyle O}{\|}}{C}-O-CH_2$$

$$CH_3(CH_2)_7-CH\!=\!CH-(CH_2)_7-\overset{\overset{\displaystyle O}{\|}}{C}-O-CH \quad\xrightarrow[200°]{H_2 \text{ (25 lb/sq in.)}}\quad CH_3(CH_2)_{16}-\overset{\overset{\displaystyle O}{\|}}{C}-O-CH$$

$$CH_3(CH_2)_7-CH\!=\!CH-(CH_2)_7-\overset{\overset{\displaystyle O}{\|}}{C}-O-CH_2$$

$$CH_3(CH_2)_{16}-\overset{\overset{\displaystyle O}{\|}}{C}-O-CH_2$$

$$CH_3(CH_2)_{16}-\overset{\overset{\displaystyle O}{\|}}{C}-O-CH_2$$

Triolein   Tristearin
Oil   *Fat*

The moderately mild reaction conditions reduce only the carbon-carbon double bonds and not the ester linkages, because the latter require more vigorous conditions for reduction (see Sec. 23.7E). In practice, hydrogenation conditions are carefully controlled so that an oil is only partially hydrogenated (that is, some unsaturated linkages are still present), because complete hydrogenation gives a compound that is hard like animal fat or tallow. Partial hydrogenation of a vegetable oil (for example, corn, soybean, cottonseed, or safflower seed oil) produces oleomargarine or cooking fats. The abundance of inexpensive vegetable oils accounts for the moderately low cost of margarine and cooking fats. Recent studies suggest the possible relationship between saturated fats and deposits of cholesterol in the arteries, so vegetable oils (especially safflower seed oil) are increasingly popular for cooking.

A. Drying Oils and Paints

Unsaturated oils are frequently encountered in paints and varnishes and are responsible for the formation of hard, protective coatings on various surfaces. They are often (and incorrectly) called *drying oils,* a result of a chemical reaction that occurs between the oil and oxygen in the air. Linseed oil is used in paint because it is cheap and is especially reactive in the hardening process. When linseed oil comes in contact with air, peroxide linkages are formed between linseed oil molecules, a reaction analogous to the vulcanization of rubber by sulfur (see Sec. 8.35). We know from Sec. 8.30 that allylic hydrogens are especially reactive, and here we see that peroxide linkages are formed between the allylic carbon atoms of different molecules to form a polymeric material in the drying process. A simplified view of this "cross linking" is:

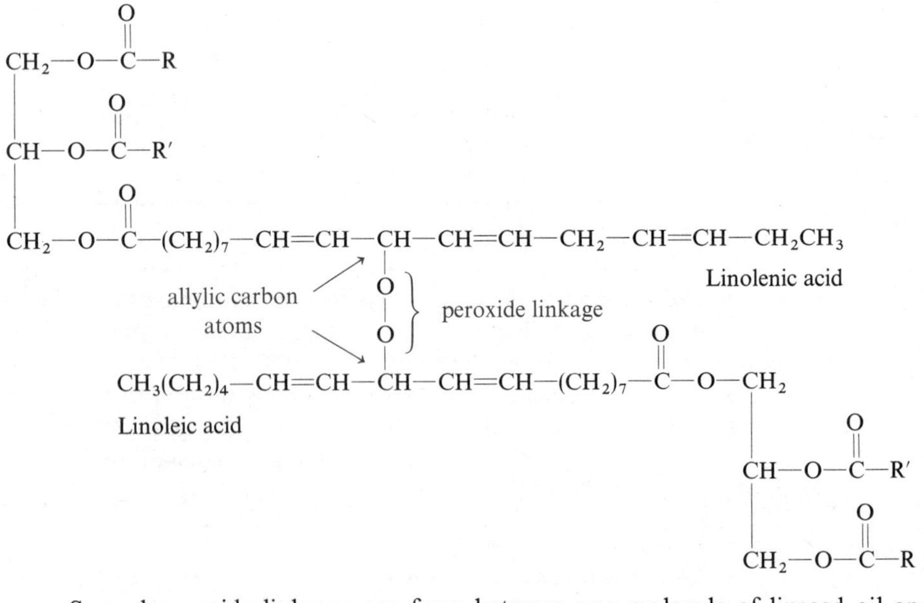

Several peroxide linkages can form between one molecule of linseed oil and other molecules because R and R′ usually contain double bonds that can undergo the same type of reaction. A massive polymeric framework is formed and is responsible for the tough protective coating that remains after a paint has "dried."

Paint is composed of four important ingredients: (1) linseed oil, which forms the hard surface; (2) pigments (oxides of various metals) to impart the color; (3) cobalt, manganese, or lead salts of carboxylic acids, which are added because these metallic ions are especially good in catalyzing the drying process; and (4) a volatile hydrocarbon solvent (such as turpentine or petroleum fractions), which makes the paint thin enough to allow even application to a surface. Thus, paint is indeed a heterogeneous mixture of compounds. After paint is applied, the volatile hydrocarbon solvent evaporates and leaves the linseed oil containing the color pigments and metal catalyst to come in contact with the oxygen in air and form the hard finish. When paint is kept away from a source of oxygen and is in an inert solvent, drying does not occur.

B. Rancid Compounds

Fats and oils that develop a disagreeable odor are called *rancid* compounds. Two important types of chemical reactions—oxidation and hydrolysis—are responsible for the odor.

For example, butter becomes rancid because of hydrolysis. It contains many lower molecular weight fatty acids, such as butyric acid and caproic acid, which have strong odors. These acids are liberated when butter warms in humid air; enzymes (called *lipases*) help catalyze the hydrolysis reaction. Butter can be prevented from becoming rancid by being kept cold and covered in the refrigerator.

Oxidation is commonly observed in unsaturated glycerides, and although the precise details of the reaction are unknown, it is believed to involve the formation of intermediate peroxides from the attack of oxygen in the air on the reactive allylic hydrogens. Subsequent decomposition of the peroxides may involve cleavage of the carbon-carbon double bonds and the liberation of lower molecular weight acids that have strong odors. This oxidation reaction is slowed by the addition of inhibitors (called *antioxidants*) to foods containing glycerides. The inhibitors, such as vitamin E and ascorbic acid, react with oxygen more rapidly than do the glycerides.

25.4   Waxes

Waxes are derived from different sources and have different chemical compositions. *Paraffin wax*, for example, is composed of a mixture of solid hydrocarbons (usually straight-chain compounds). On the other hand, *Carbowax* is a polyether that is prepared synthetically (see Sec. 19.13B). More generally, however, waxes are naturally occurring products that are monoesters derived from long-chain carboxylic acids containing 16 or more carbon atoms and long-chain alcohols containing one hydroxyl group (*monohydric*) and 16 or more carbon atoms. It should come as no surprise that each of these carbon chains contains an *even* number of carbon atoms because they are formed from fatty acids in plants. Like fats and oils, natural waxes are often mixtures of esters and may contain other impurities. For example, *beeswax* is obtained from honeycomb and contains mostly C_{26} and C_{28} acids and C_{28} to C_{32} alcohols, with typical structures being $CH_3(CH_2)_{24}COO(CH_2)_{27}CH_3$ and $CH_3(CH_2)_{26}COO(CH_2)_{27}CH_3$. *Carnauba wax* is valuable and is obtained from the leaves of the Brazilian palm where it is a coating. A typical molecule has the structure $CH_3(CH_2)_{30}COO(CH_2)_{33}CH_3$. This is a relatively high melting wax (mp 80 to 87°), and because of its comparative hardness and resistance to water, carnauba wax is widely used as an ingredient in car waxes, floor polishes, and carbon paper coatings. *Spermaceti wax* comes from the head of the sperm whale and is mainly cetyl palmitate, $CH_3(CH_2)_{14}COO(CH_2)_{15}CH_3$.

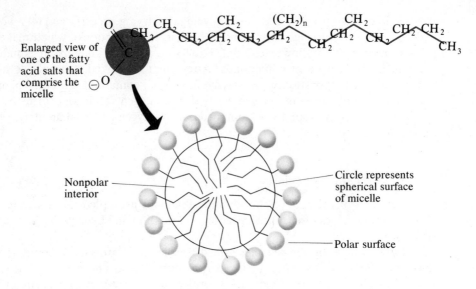

Enlarged view of one of the fatty acid salts that comprise the micelle

Nonpolar interior

Circle represents spherical surface of micelle

Polar surface

FIGURE 25.2   Micelle cluster showing the highly polar hydrophilic surface and the nonpolar hydrophobic interior.

25.5   Phospholipids: Micelles and Membranes

When a soap or detergent is dissolved in water, the result is an emulsion made up of clusters of soap molecules called *micelles*. Micelle clusters are spherical (Fig. 25.2) and are composed of a highly hydrophilic surface with a hydrophobic interior. This structure results from the ordered arrangement of the individual soap (or detergent) molecules with the carboxylate groups along the surface in contact with the polar solvent (water) and the long hydrocarbon tails pointing inward toward the center of the micelle sphere. Micelles have recently been used as biomimetic models (models for a naturally occurring biochemical system) for plant and animal cells. In particular, the micelle bears a strong resemblance both structurally and in chemical and physical behavior to the membrane that comprises the outer surface and principal structural material of many cells.

Cell membranes also have a hydrophilic outer surface and a hydrophobic interior. The molecules that constitute this membrane resemble the long-chain fatty acid salts we discussed. They are ordered with their long hydrocarbon portion aligned inward and the polar groups associated with these molecules pointing toward the outer surface of the membrane (Fig. 25.3).

This membrane acts as the barrier between the organic material of the cell interior and the aqueous environment that surrounds it. In addition, the membrane is a filtering device, allowing the passage of materials such as nutrients and wastes into and out of the interior of the cell while selectively blocking the passage of others. The similarity between the micelle and cell membrane is evident.

Let us now look in more detail at the individual molecules that constitute the cell membrane. Like the long-chain fatty acid salts that make up the micelle, the membrane is constructed of lipids. In particular, it is composed of a class of lipids called **phospholipids**; their general structure is shown in Fig. 25.4.

Phospholipids are triesters of glycerol in which one fatty acid side chain is replaced by an unsymmetric diester of phosphoric acid. Half this diester is attached to

Cell
Membrane

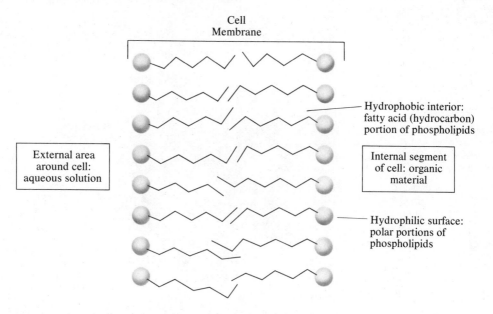

Hydrophobic interior:
fatty acid (hydrocarbon)
portion of phospholipids

External area
around cell:
aqueous solution

Internal segment
of cell: organic
material

Hydrophilic surface:
polar portions of
phospholipids

FIGURE 25.3 **Schematic representation of a cross section of a cell membrane.**

the glycerol molecule while the other half may be attached to a variety of groups forming several classes of phospholipids; among these are lecithins, cephalins, and phosphatidylserines (Fig. 25.4).

A more detailed view of the membrane wall is shown in Fig. 25.5.

The picture of the cell membrane we presented here is greatly simplified. The membrane actually is composed of phospholipids plus a variety of proteins, enzymes, and carbohydrates depending on the source of the membrane. The relative percentages, arrangements, and functions of these components also vary with the source of the material. In addition to the outer surface membrane, many cells may have internal membranes involved in specialized actions and functions.

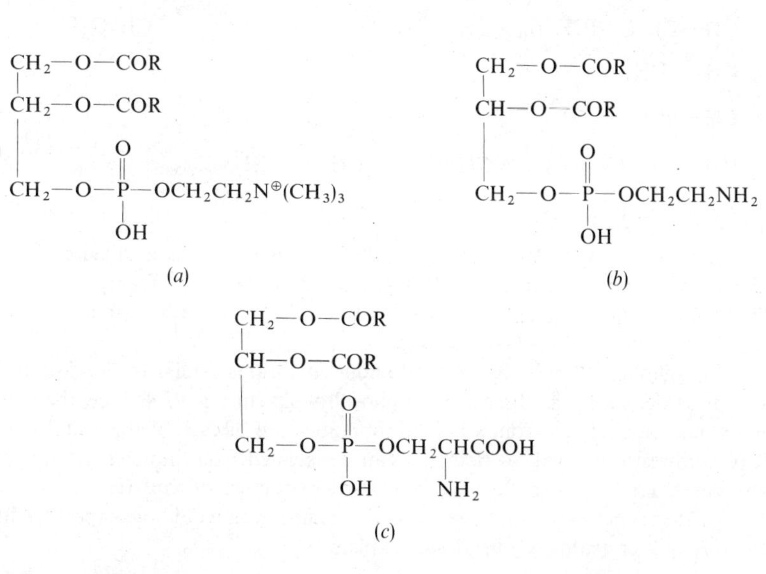

$$CH_2-O-COR$$
$$CH_2-O-COR$$
$$CH_2-O-\overset{\overset{O}{\parallel}}{\underset{OH}{P}}-OCH_2CH_2N^{\oplus}(CH_3)_3$$

(a)

$$CH_2-O-COR$$
$$CH-O-COR$$
$$CH_2-O-\overset{\overset{O}{\parallel}}{\underset{OH}{P}}-OCH_2CH_2NH_2$$

(b)

$$CH_2-O-COR$$
$$CH-O-COR$$
$$CH_2-O-\overset{\overset{O}{\parallel}}{\underset{OH}{P}}-OCH_2\underset{NH_2}{CHCOOH}$$

(c)

FIGURE 25.4 **General structure of several classes of phospholipids: (a) lecithins, (b) cephalins, and (c) phosphatidylserines.**

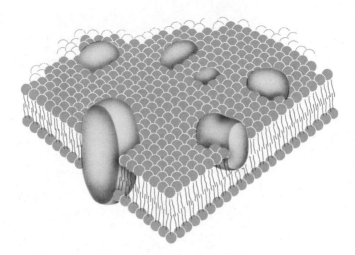

FIGURE 25.5   Detailed view of the cell membrane and the phospholipids and proteins that comprise it. (Based on S. J. Singer and G. L. Nicolson, "The Fluid Mosaic Model of the Structure, *Science*, **vol. 175, pp. 720–731, Fig. 3, Feb. 18, 1972.)**

Study Questions

25.1 Name each of the following glycerides.

(*a*)
$$\begin{array}{l} CH_2-O-\overset{\displaystyle O}{\overset{\|}{C}}-(CH_2)_{10}-CH_3 \\ CH-OH \\ CH-O-\overset{\displaystyle O}{\overset{\|}{C}}-(CH_2)_{10}-CH_3 \end{array}$$

(*b*)
$$\begin{array}{l} CH_2-O-\overset{\displaystyle O}{\overset{\|}{C}}-(CH_2)_{14}-CH_3 \\ CH-O-\overset{\displaystyle O}{\overset{\|}{C}}-(CH_2)_{14}-CH_3 \\ CH_2OH \end{array}$$

(*c*)
$$\begin{array}{l} CH_2-OH \\ CH-OH \\ CH_2-O-\overset{\displaystyle}{\underset{\displaystyle O}{\overset{\|}{C}}}-(CH_2)_7-CH{=}CH-(CH_2)_7-CH_3 \end{array}$$

25.2 Explain why some glycerides are called *fats* and others are called *oils*.

25.3 Using balanced chemical equations, show the conversion of a glyceride molecule to a soap.

25.4 Using chemical equations, show the conversion of a glyceride to a detergent. What advantage does a detergent have over a soap?

25.5 Phospholipids can be hydrolyzed under conditions similar to those used in the hydrolysis of a glyceride. Using the three general phospholipids in Fig. 25.4, show the hydrolysis products that would be obtained from a typical saponification of each of these molecules.

25.6 Trimyristin is a triglyceride that can be isolated from nutmeg. Hydrolysis of trimyristin with base gives only myristic acid. What is the structure of trimyristin?

25.7 Of the six compounds shown here, *circle* those you would *not* expect to find in a naturally occurring glyceride (fat or oil). Briefly explain.

$$CH_3(CH_2)_5COOH \qquad CH_3(CH_2)_{10}COOH \qquad CH_3(CH_2)_{11}COOH$$
$$CH_3(CH_2)_{16}COOH \qquad CH_3(CH_2)_{17}COOH \qquad CH_3(CH_2)_{18}COOH$$

25.8 Show how each carboxylic acid in Study Question 25.7 can be synthesized from alkanes of six carbons or less. You may assume any common solvent, organic reagent, or inorganic reagent is available.

25.9 How many stereoisomers are possible for each of the following compounds?

(*a*) sterculic acid   (*b*) palmitoleic acid   (*c*) eleostearic acid

(*d*) oleic acid   (*e*) linoleic acid

25.10 What is the index of hydrogen deficiency (see Sec. 8.14) for the triglyceride composed of each of the five carboxylic acids in Study Question 25.9?

25.11 A certain glyceride (fat) is analyzed and contains 2 moles of stearic acid, $CH_3(CH_2)_{16}COOH$, and 1 mole of palmitic acid, $CH_3(CH_2)_{14}COOH$, per mole of glyceride. On the basis of this information only, draw the possible structure(s) for the glyceride.

25.12 What mechanism is involved when a glyceride is hydrolyzed by aqueous base? Draw this complete mechanism for a typical monoglyceride.

26

Amines

The last principal functional group to be studied is the amino group, $-NH_2$, which is found in the family of compounds called **amines**:

$$(Ar) \ R-NH_2$$

Amine

The amines are an extremely important class of compounds. The amino function is found in many naturally occurring compounds (alkaloids, nucleic acids, proteins, and amino acids), compounds of medicinal interest (amphetamines, antibiotics, and local anesthetics), industrial compounds (nylons and dyes), and food additives (sweeteners). Each of these classes is discussed in this chapter.

The **heterocyclic compounds,** cyclic compounds that contain some atom besides carbon as part of their ring skeleton, are also discussed. The chemical properties of the better-known members of this family are presented. The amino group, which is found in the compounds listed in the previous paragraph, is most often part of a heterocyclic ring system.

26.1   Structure and Nomenclature

The $-NH_2$ group is designated by the prefix *amino*. Attachment of this group to a carbon atom results in the family of amines, $(Ar)R-NH_2$.

Amines may be classified further according to the number of alkyl or aryl groups attached to the nitrogen. For example, RNH_2 is a *primary* (1°) amine, R_2NH is a *secondary* (2°) amine, R_3N is a *tertiary* (3°) amine, and $ArNH_2$ is an *aromatic* amine. As a result of their basicity, amines also form stable, tetravalent species of the type $R_4N^\oplus$, where R can be an alkyl or aryl group or a proton. Species of this type are called *quaternary* (4°) ammonium salts. We discuss these salts in more detail in Sec. 26.3. *Anilinium* is used in place of *ammonium* for the quaternary salts of aromatic amines that contain the aniline skeleton ($C_6H_5-NR_3^\oplus$).

A. Common Names

The rules for the common names of amines are simple though limited. These names are derived from the various alkyl or aryl groups attached to the nitrogen in the amines. Each alkyl group is listed alphabetically or by size (smallest to largest) in one continuous word followed by the suffix *amine*. The following examples illustrate:

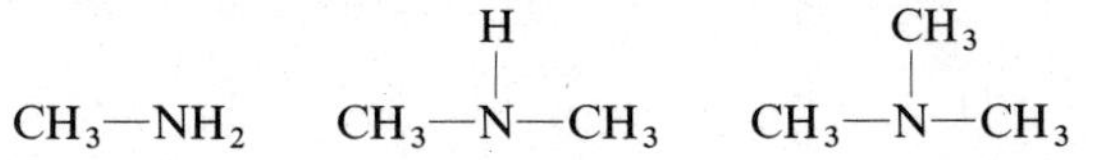

$$CH_3—NH_2 \qquad CH_3—\overset{\overset{\displaystyle H}{|}}{N}—CH_3 \qquad CH_3—\overset{\overset{\displaystyle CH_3}{|}}{N}—CH_3$$

Methylamine   Dimethylamine   Trimethylamine

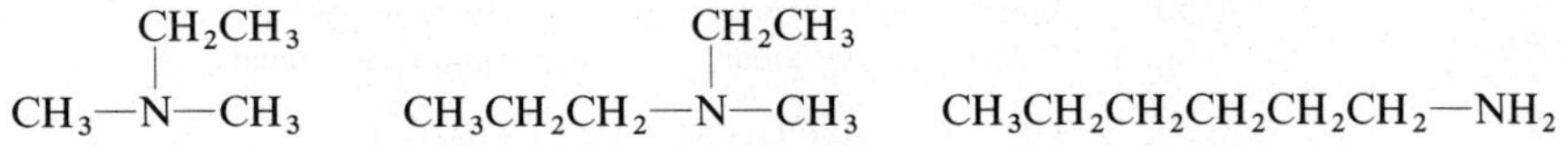

$$CH_3—\overset{\overset{\displaystyle CH_2CH_3}{|}}{N}—CH_3 \qquad CH_3CH_2CH_2—\overset{\overset{\displaystyle CH_2CH_3}{|}}{N}—CH_3 \qquad CH_3CH_2CH_2CH_2CH_2CH_2—NH_2$$

Ethyldimethylamine   Ethylmethyl-*n*-propylamine   *n*-Hexylamine

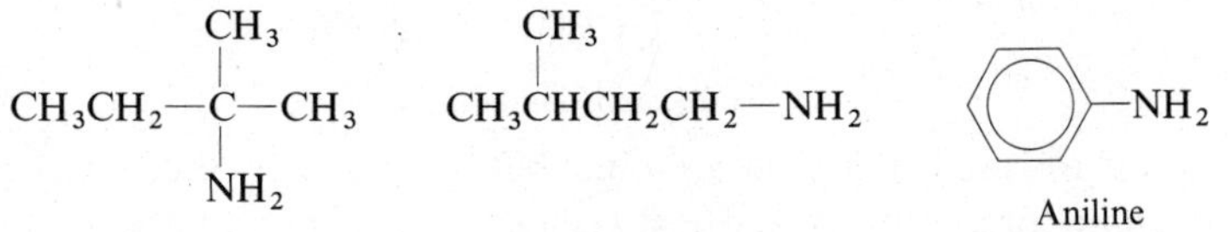

$$CH_3CH_2—\overset{\overset{\displaystyle CH_3}{|}}{\underset{\underset{\displaystyle NH_2}{|}}{C}}—CH_3 \qquad CH_3\overset{\overset{\displaystyle CH_3}{|}}{C}HCH_2CH_2—NH_2 \qquad \langle\!\!\bigcirc\!\!\rangle—NH_2$$

tert-Pentylamine   Isopentylamine   Aniline
(not aminobenzene)

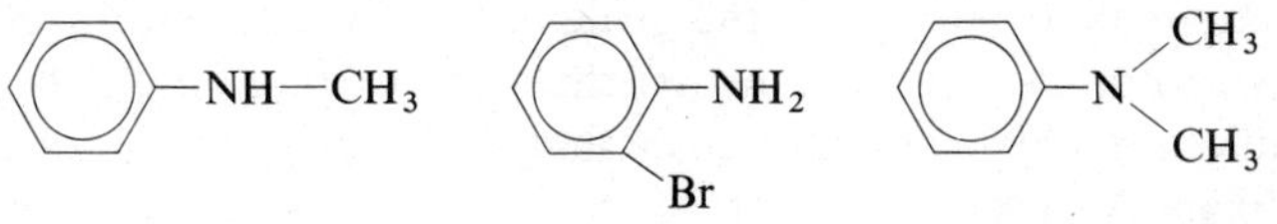

$$\langle\!\!\bigcirc\!\!\rangle—NH—CH_3 \qquad \langle\!\!\bigcirc\!\!\rangle\overset{\displaystyle —NH_2}{\underset{\displaystyle Br}{|}} \qquad \langle\!\!\bigcirc\!\!\rangle—N\overset{\displaystyle \diagup CH_3}{\diagdown CH_3}$$

N-Methylaniline   *o*-Bromoaniline   *N,N*-Dimethylaniline

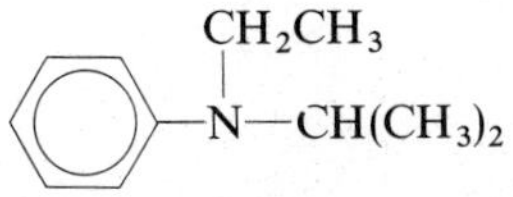

$$\langle\!\!\bigcirc\!\!\rangle—\overset{\overset{\displaystyle CH_2CH_3}{|}}{N}—CH(CH_3)_2$$

N-Ethyl-*N*-isopropylaniline

Note the use of the prefix *N*- to designate a substituent on the nitrogen. The main problem with the system, as with the common system in general, is the limited availability of common names for the alkyl groups.

B. IUPAC Names

The rules for applying the IUPAC nomenclature system to amines are as follows:

1. The longest continuous chain of carbon atoms to which the —NH$_2$ group is bonded is the parent compound. The alkane name corresponding to this chain of carbon atoms is used as the suffix of the amine name. The alkane name is preceded by *amino* and a number designating the position of the amino group on the chain:

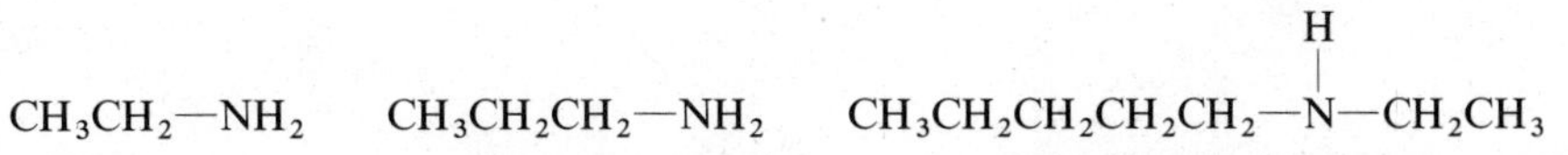

$$CH_3CH_2—NH_2 \qquad CH_3CH_2CH_2—NH_2 \qquad CH_3CH_2CH_2CH_2CH_2—\overset{\overset{\displaystyle H}{|}}{N}—CH_2CH_3$$

Aminoethane   1-Aminopropane   *N*-Ethyl-1-aminopentane

2. As the complexity of the amine increases, all substituents are named and numbered in the usual way. The amino group or substituted amino group becomes one more substituent. Note that in these cases the names of the *substituents on the amino group* precede the name *amino*. For example, *p*-dimethylaminobromobenzene means there is an aminobromobenzene with two methyls on the amino group.

$$CH_3 \underset{N}{\diagdown} CH_3$$

p-Dimethylaminobromobenzene
(*p*-bromo-*N*,*N*-dimethylaniline)

$$CH_3CHCH_2CH_2\overset{O}{\overset{\|}{-}C-}OH$$
$$\underset{NH_2}{|}$$

4-Aminopentanoic acid
(γ-aminovaleric acid)

$$CH_3CH_2CH_2CH_2CH_2CHCH_2CH_2-OH$$
$$\underset{CH_3}{\overset{N}{\diagup}}\diagdown CH_2CH_3$$

3-(Ethylmethylamino)-1-octanol

Parentheses are used when there could be ambiguity in the naming if they were omitted. In the preceding example, it is clear that the amino group at position 3 has a methyl and an ethyl attached to it. If the parentheses were not used, for example, in 3-ethylmethylamino-1-octanol, the structure could contain an ethyl group on the number 3 carbon, a methylamino on the same carbon, and a 1-octanol backbone.

3. Several amines have special names that are also approved for use in the IUPAC system. Examples are:

o-Toluidine   *p*-Anisidine

The naming of the salts of the amines is presented in Sec. 26.3.

Question 26.1

Name each of the following compounds using the IUPAC convention:

(*a*) $$CH_3CHCH_2CH_2CH_2CH_2CH_2CH_2CH_2\overset{O}{\overset{\|}{C}}-OCH_3$$
$$\underset{CH_3 \quad H}{\overset{N}{\diagup}\diagdown}$$

(*b*) $$CH_3 \diagdown \underset{CH_3}{\diagup} N - \bigcirc - N \underset{\diagdown CH_3}{\diagup CH_3}$$

(*c*) $$CH_3 \diagdown \underset{N}{\diagup} CH_2CH_3$$

(*d*) $$H \diagdown \underset{N}{\diagup} CH_2CH_2CH_3$$
$$\underset{CH_2}{|}$$

26.2   Physical Properties of Amines

The electronegativity of the nitrogen in the amines results in a highly polar N—H bond. This polarity is reflected in the physical properties of these compounds. In particular, boiling points and solubilities can be explained qualitatively with this concept.

A comparison of the boiling points of amines and alcohols versus alkanes and ethers of comparable molecular weight provides the following information:

$CH_3CH_2CH_3$ $CH_3—O—CH_3$ $CH_3—NH—CH_3$ $CH_3CH_2—OH$ $H—O—H$

bp $-42.2°$	bp $-24°$	bp $7.5°$	bp $78°$	bp $100°$
mol wt 44	mol wt 46	mol wt 45	mol wt 46	mol wt 18

The greater polarity of the O—H bond is reflected in the boiling points of ethanol and water, both of which are much higher than that of either dimethyl ether or dimethylamine. The amine, in contrast, boils at a significantly higher temperature than the ether or the alkane. The polarity of the N—H bond is responsible for this. This trend is also observed in the solubilities of these compounds. Whereas ethanol is infinitely soluble in water and dimethylamine is also very soluble, dimethyl ether is essentially insoluble in water.

The effect of hydrogen bonding is quite evident from Table 26.1. Note particularly that the lower molecular weight amines are readily soluble in water. This

TABLE 26.1 Physical Properties of Selected Amines

Name	Structure	mp, °C	bp, °C	Density at 20°, g/cc
Methylamine	CH_3NH_2	-93	-7	0.699
Ethylamine	$CH_3CH_2NH_2$	-81	17	0.689
n-Propylamine	$CH_3CH_2CH_2NH_2$	-83	49	0.719
Isopropylamine	$(CH_3)_2CHNH_2$	-101	33	
n-Butylamine	$CH_3CH_2CH_2CH_2NH_2$	-50	77	0.740
Isobutylamine	$(CH_3)_2CHCH_2NH_2$	-86	68	
sec-Butylamine	$CH_3CH_2CH(NH_2)CH_3$	-104	63	
tert-Butylamine	$(CH_3)_3CNH_2$	-68	45	0.696
Dimethylamine	$(CH_3)_2NH$	-96	7	0.680
Diethylamine	$(CH_3CH_2)_2NH$	-42	56	0.711
Di-n-propylamine	$(CH_3CH_2CH_2)_2NH$	-40	111	0.738
Diisopropylamine	$[(CH_3)_2CH]_2NH$	-61	84	0.717
Di-n-butylamine	$(CH_3CH_2CH_2CH_2)_2NH$	-59	159	0.767
Trimethylamine	$(CH_3)_3N$	-117	3.5	0.662
Triethylamine	$(CH_3CH_2)_3N$	-115	90	0.728
Tri-n-propylamine	$(CH_3CH_2CH_2)_3N$	-94	156	0.757
Ethylmethylamine	$CH_3CH_2NHCH_3$		37	
Cyclohexylamine	$cyclo\text{-}C_6H_{11}NH_2$		134	
Benzylamine	$C_6H_5CH_2NH_2$		185	
Allylamine	$CH_2\text{=}CH—CH_2NH_2$		53	0.761
Aniline	$C_6H_5NH_2$	-6	184	1.022
N-Methylaniline	$C_6H_5NHCH_3$	-57	196	0.989
N,N-Dimethylaniline	$C_6H_5N(CH_3)_2$	2	194	0.956
o-Toluidine	$o\text{-}CH_3—C_6H_4—NH_2$	-16	200	0.998
m-Toluidine	$m\text{-}CH_3—C_6H_4—NH_2$	-30	203	0.989
p-Toluidine	$p\text{-}CH_3—C_6H_4—NH_2$	45	201	0.962
Diphenylamine	$(C_6H_5)_2NH$	54	302	1.160
Triphenylamine	$(C_6H_5)_3N$	126	365	0.774

$$\text{IIIII}H-\underset{\underset{H}{|}}{\overset{\overset{H}{|}}{N}}:\text{IIIII}H-\underset{\underset{H}{|}}{\overset{\overset{H}{|}}{\underset{..}{N}}}-H\text{IIIII}:\underset{\underset{H}{|}}{\overset{\overset{H}{|}}{N}}-H\text{IIIII}$$

Hydrogen bonding in amines

effect decreases as the size of the hydrocarbon portion of the amine increases, rendering the amine more nonpolar. For example, *n*-propylamine is infinitely soluble in water; diphenylamine is insoluble. In addition to a decrease in solubility and an increase in boiling point as the molecular weight increases, both solubility in water and boiling point decrease as branching of the amine increases. This too would be predicted. As the branching increases, the steric bulk increases, making it much more difficult for the molecules to approach each other and form hydrogen bonds. Also, as the amine goes from 1° to 2° to 3°, the number of nitrogen-hydrogen bonds decreases, also resulting in fewer hydrogen bonds. The decrease in boiling points is due to the decrease in both hydrogen bonding and van der Waals interactions as the branching increases.

Question 26.2

Compare the boiling points of the isomeric butylamines in Table 26.1. Explain the experimental data.

26.3   Basic Properties of Amines: Formation of Ammonium Salts

The nitrogen atom of an amine contains a nonbonding pair of electrons that it can donate in a Lewis base manner. If the electron pair is donated to a proton, then the amine fits the definition of a Lowry-Brønsted base as well as a Lewis base.

$$H-\underset{\underset{H}{|}}{\overset{\overset{H}{|}}{N}}: + HX \longrightarrow H-\underset{\underset{H}{|}}{\overset{\overset{H}{|}}{\overset{\oplus}{N}}}H \quad X^{\ominus}$$

Amine   *Acid*   Ammonium ion
Base

Amines are fairly strong bases ($K_b \approx 1 \times 10^{-4}$ to 1×10^{-5}). Because they are strong bases, they react readily with any acid, even a weak acid like water. The reaction of ammonia with water has a *base dissociation constant*, K_b, of approximately 10^{-5}.

$$\overset{..}{N}H_3 + H_2O \rightleftharpoons NH_4^{\oplus} + OH^{\ominus}$$

The concentration of ammonium hydroxide is quite small. The K_b for this reaction is:

$$K_{equil}[H_2O] = K_b = \frac{[NH_4^{\oplus}][OH^{\ominus}]}{[NH_3]} = 1.8 \times 10^{-5}$$

and

$$pK_b = -\log K_b = 4.74$$

The product of the equilibrium constant (K_{equil}) and the constant (nonvarying) concentration of water (55.5 M) is the K_b. The pK_b values of selected amines are in Table 26.2.

When a strong acid is used, the equilibrium is shifted markedly to the right. For example, when ammonia reacts with hydrochloric acid, the equilibrium is shifted quantitatively in the direction of the ammonium chloride product:

$$\ddot{N}H_3 + HCl \rightleftharpoons NH_4^{\oplus}Cl^{\ominus}$$

Ammonium
chloride

When naming the salt of an amine, take the complete common name of the amine and change the suffix *amine* to *ammonium*. The suffix *ammonium* is followed by a space and then the name of the counter ion (anion) associated with the salt. In the preceding example, the unsubstituted $NH_4^{\oplus}$ cation is the *ammonium ion*, so

TABLE 26.2 K_b's of Selected Amines

Amine	K_b	pK_b
Ammonia	1.8×10^{-5}	4.74
Methylamine	4.3×10^{-4}	3.36
Ethylamine	4.4×10^{-4}	3.36
n-Propylamine	4.7×10^{-4}	3.32
Isopropylamine	4.0×10^{-4}	3.40
n-Butylamine	4.8×10^{-4}	3.32
Dimethylamine	5.3×10^{-4}	3.28
Diethylamine	9.8×10^{-4}	3.01
Di-*n*-propylamine	10.0×10^{-4}	3.00
Trimethylamine	5.5×10^{-5}	4.26
Triethylamine	5.7×10^{-4}	3.24
Tri-*n*-propylamine	4.5×10^{-4}	3.35
Cyclohexylamine	4.7×10^{-4}	3.33
Benzylamine	2.0×10^{-5}	4.67
Aniline	4.0×10^{-10}	9.40
N-Methylaniline	6.1×10^{-10}	9.21
N,N-Dimethylaniline	11.6×10^{-10}	8.94
p-Toluidine	1.2×10^{-9}	8.92
p-Fluoroaniline	4.4×10^{-10}	9.36
p-Chloroaniline	1×10^{-10}	10.00
p-Bromoaniline	7×10^{-11}	10.15
p-Iodoaniline	6×10^{-11}	10.22
p-Anisidine	2×10^{-9}	8.70
p-Nitroaniline	1×10^{-13}	13.00

the complete name of the salt is *ammonium chloride*. See the following examples:

$$\begin{array}{cc}
\overset{\displaystyle CH_2CH_3}{\underset{\displaystyle H \quad I^{\ominus}}{CH_3CH_2 - \overset{\oplus}{N} - CH_2CH_3}} & \overset{\displaystyle CH_3}{\underset{\displaystyle H \quad HO^{\ominus}}{CH_3CH_2 - \overset{\oplus}{N} - CH_3}}
\end{array}$$

Triethylammonium iodide   Ethyldimethylammonium hydroxide

In instances in which the amine salt is derived from aniline or a substituted aniline (for example, *p*-bromoaniline or *p*-nitroaniline), the suffix *aniline* is changed to *anilinium* and the name of the anion is added; for example,

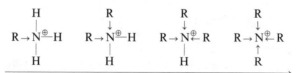

Anilinium   *p*-Bromoanilinium
hydrogen sulfate   chloride

Aromatic amines are less basic than aliphatic amines. Why? To answer we can apply the same logic we used in discussing the acidity of alcohols, phenols, and carboxylic acids (see Secs. 10.9, 17.4, and 22.9, respectively). By comparing the stabilities of the resulting conjugate acids, a qualitative order of relative basicity can be derived. The stabilities of the conjugate acids can be explained in terms of the inductive and/or resonance effects of the groups attached to the basic nitrogen. Compare the following series of alkyl amines:

$$\overset{\displaystyle H}{\underset{\displaystyle H}{R \rightarrow \overset{\oplus}{N} - H}} \qquad \overset{\displaystyle R\downarrow}{\underset{\displaystyle H}{R \rightarrow \overset{\oplus}{N} - H}} \qquad \overset{\displaystyle R\downarrow}{\underset{\displaystyle H}{R \rightarrow \overset{\oplus}{N} \leftarrow R}} \qquad \overset{\displaystyle R\downarrow}{\underset{\displaystyle R\uparrow}{R \rightarrow \overset{\oplus}{N} \leftarrow R}}$$

$\xrightarrow{\hspace{8cm}}$

Increasing ammonium ion stability; increasing ease of formation

The electron-donating effect of the attached alkyl groups on the nitrogen increases the concentration of electrons on the nitrogen (increases the electron density), making the positive charge more stable. Tertiary amines should be more basic than secondary amines, secondary more basic than primary, and primary more basic than ammonia. In reality, as the number of groups attached to the nitrogen increases, the basicity changes very little. Secondary amines are approximately as basic as primary amines, but tertiary are actually a little less basic. This may be because of steric crowding around the nitrogen, making it more difficult for protonation to occur. It may also be that solvation effects are crucial in determining basicity. Research is still going on to answer these questions.

$$\overset{\displaystyle R}{\underset{\displaystyle H}{H - N:}} \approx \overset{\displaystyle R}{\underset{\displaystyle H}{R - N:}} > \overset{\displaystyle R}{\underset{\displaystyle R}{R - N:}} > \overset{\displaystyle H}{\underset{\displaystyle H}{H - N:}}$$

1°   2°   3°   Ammonia

$\xleftarrow{\hspace{8cm}}$

Actual increasing base strength

For example,

$$\begin{array}{ccc}
\overset{\displaystyle H}{\underset{\displaystyle pK_b = 3.36}{CH_3-\overset{|}{\underset{\cdot\cdot}{N}}-H}} & \overset{\displaystyle CH_3}{\underset{\displaystyle pK_b = 3.28}{CH_3-\overset{|}{\underset{\cdot\cdot}{N}}-H}} & \overset{\displaystyle CH_3}{\underset{\displaystyle pK_b = 4.26}{CH_3-\overset{|}{N}-CH_3}}
\end{array}$$

$$\begin{array}{ccc}
\overset{\displaystyle H}{\underset{\displaystyle pK_b = 3.36}{CH_3CH_2-\overset{|}{\underset{\cdot\cdot}{N}}-H}} & \overset{\displaystyle CH_2CH_3}{\underset{\displaystyle pK_b = 3.01}{CH_3CH_2-\overset{|}{\underset{\cdot\cdot}{N}}-H}} & \overset{\displaystyle CH_2CH_3}{\underset{\displaystyle pK_b = 3.24}{CH_3CH_2-\overset{|}{N}-CH_2CH_3}}
\end{array}$$

$$\begin{array}{ccc}
\overset{\displaystyle H}{\underset{\displaystyle pK_b = 3.32}{CH_3CH_2CH_2-\overset{|}{\underset{\cdot\cdot}{N}}-H}} & \overset{\displaystyle CH_2CH_2CH_3}{\underset{\displaystyle pK_b = 3.00}{CH_3CH_2CH_2-\overset{|}{\underset{\cdot\cdot}{N}}-H}} & \overset{\displaystyle CH_2CH_2CH_3}{\underset{\displaystyle pK_b = 3.35}{CH_3CH_2CH_2-\overset{|}{N}-CH_2CH_2CH_3}}
\end{array}$$

When comparing aromatic amines, the resonance effect of any attached group must also be taken into account:

G = electron-donating group such as R—, Ar—	Aniline: standard	G = electron-withdrawing group, such as —NO_2, —X, —NR_3
G makes cation more stable due to donating electrons to positive charge	*No special stability or instability*	*G makes cation less stable due to affinity of G for electrons*

The G group can act as an electron-donating or -releasing group by the inductive and/or resonance effect. Consider the following examples. [*Note:* The pK_b values are for the free amines, although they are written in their cationic (conjugate acid) form.]

$$\begin{array}{ccc}
\underset{\displaystyle pK_b = 13}{NO_2-\!\!\bigcirc\!\!-\overset{\oplus}{N}H_3} & \underset{\displaystyle pK_b = 9.4}{H-\!\!\bigcirc\!\!-\overset{\oplus}{N}H_3} & \underset{\displaystyle pK_b = 8.9}{CH_3-\!\!\bigcirc\!\!-\overset{\oplus}{N}H_3}
\end{array}$$

As the electron-releasing, inductive effect of the attached group increases, the positive charge of the ion (in this case the anilinium ion) becomes more stable. *p*-Toluidine is a stronger base than aniline, and aniline is a stronger base than *p*-nitroaniline. When the attached groups can donate or withdraw electrons through delocalization of electron density (the resonance effect), then both effects must be taken into account in any discussion of the acidity or basicity of the compound. In the case of *p*-nitroaniline, both effects reinforce each other; the inductive and resonance effects remove electron density from the positive nitrogen center.

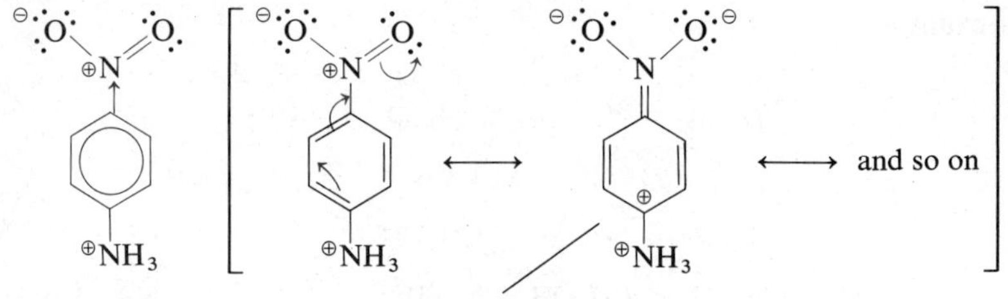

Inductive effect of nitro group destabilizes cation

Especially unstable due to like charges adjacent to each other

Resonance effect of nitro group destabilizes cation

With a methoxy substituent, however, the inductive effect of the oxygen withdraws electron density from the positive center, but the resonance effect involving the nonbonding electrons on oxygen works in the opposite direction to donate electron density to the center. Knowing that methoxy and hydroxy are strong activating groups in electrophilic aromatic substitution (see Sec. 14.11), the resonance effect should predominate. A methoxy group has a net effect of releasing electrons. The pK_b of *p*-anisidine is 8.7; it is a stronger base than *p*-toluidine. The importance of the resonance contribution is also reflected in the pK_b values of the *p*-haloanilines, which are on the order of 10, slightly weaker than aniline but significantly stronger than *p*-nitroaniline. This too is a result of the resonance effect.

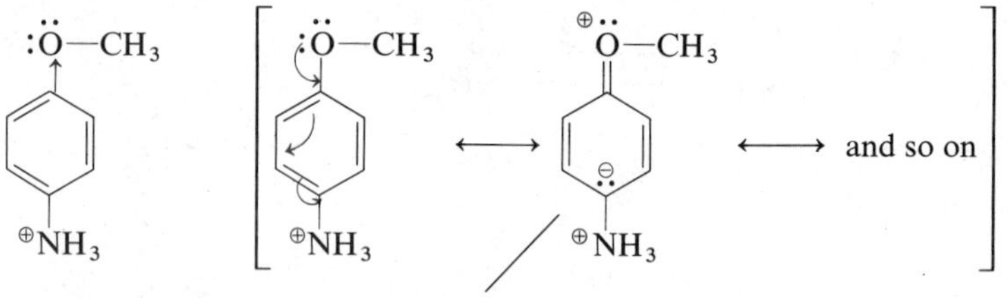

Inductive effect of methoxy group destabilizes cation

Especially stable due to unlike charges adjacent to each other

Resonance effect of methoxy group stabilizes cation

Note that all the aromatic amines are weaker bases than the aliphatic amines. We can explain this by the contribution of the attached groups, the aromatic ring (or substituted aromatic ring) versus an alkyl group. Contrasting methylamine and aniline, the following reactions can be written:

$$CH_3\text{—}\overset{..}{N}H_2 + H_2O \rightleftharpoons CH_3\overset{\oplus}{N}H_3 + :\overset{..}{\underset{..}{O}}H^{\ominus} \qquad pK_b = 3.36$$

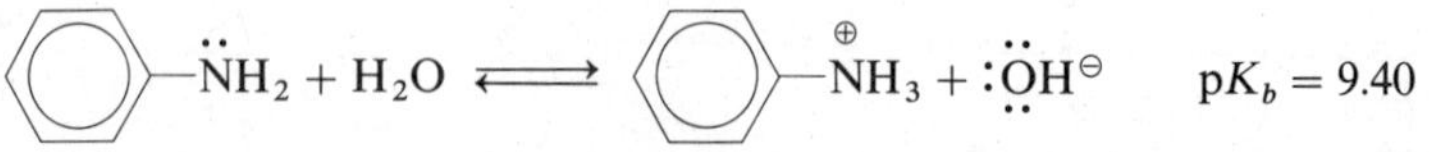

$$\text{⬡}\text{—}\overset{..}{N}H_2 + H_2O \rightleftharpoons \text{⬡}\text{—}\overset{\oplus}{N}H_3 + :\overset{..}{\underset{..}{O}}H^{\ominus} \qquad pK_b = 9.40$$

Aniline is a weaker base by six powers of 10. This is what we would predict. The methyl group is an electron-releasing group by the inductive effect, whereas the benzene ring is an electron-withdrawing group by a combination of the inductive and resonance effects.

Alkyl group donates electron and stabilizes cation

Phenyl group has affinity for electrons and destabilizes cation

The methylammonium ion is stabilized to a much greater extent than the anilinium ion. Even when an electron-releasing group is on the aromatic ring, as in the case of *p*-anisidine, the effect of this group is small in contrast to that of a directly attached aryl group.

Question 26.3

If a base has a pK_b of 9.4 (aniline, for example), what pK_a is associated with its conjugate acid? What are the actual dissociation (equilibrium) constants (K_{equil}) for each of these compounds?

Question 26.4

The acid dissociation constant (K_a) of ammonia is 9.8×10^{-37}.

(*a*) Write the equilibrium reaction that this K_a measures.
(*b*) Does this reaction favor products or reactants? Explain.
(*c*) In light of your answer in (*b*), how could the K_a be determined? [*Hint:* Write equations for the K_a, K_b, and autodissociation (protonation-deprotonation) constant K_D for ammonia. The answer lies in the equations.]

Question 26.5

Compare the basicities of the *p*-haloanilines in Table 26.2. Explain them in terms of the electronic effects involved.

Question 26.6

In Secs. 10.9 and 17.4 the relative acidities of alcohols and phenols were discussed. Using an analogous argument, contrast the difference in the basicities of ammonia and aniline by comparing the relative stabilities of the free base and conjugate acid. Pay particular attention to any resonance structures that may be contributing to this stabilization.

26.4   Solubility Properties of Amines

The basic property of amines provides a convenient and efficient method by which they can be separated from many other organic compounds and purified. For example, suppose we had a mixture of aniline and *p*-xylene dissolved in methylene chloride. It is quite easy to separate these two compounds. By extracting the organic solution with dilute acid (for example, 2 *N* HCl), the aniline is converted to the anilinium ion, a highly polar, water-soluble salt, which enters the aqueous layer. The xylene is unaffected by the acid and remains in the organic layer. Separation of the two layers separates the two compounds. Adding base to the aqueous layer until it is strongly basic gives the free amine, which can be purified. The following scheme outlines this separation:

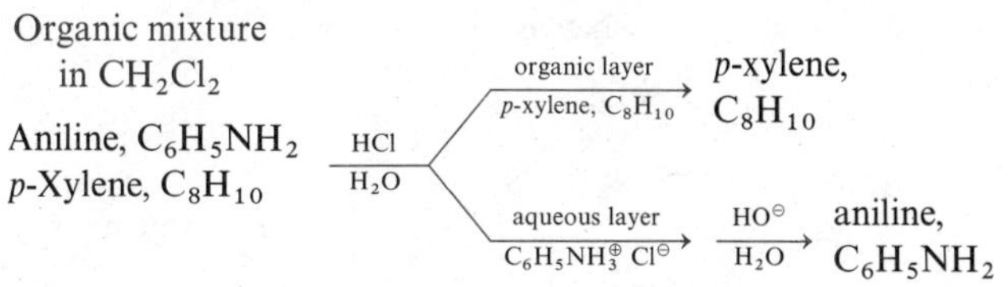

Organic mixture
in CH_2Cl_2

Aniline, $C_6H_5NH_2$ $\xrightarrow[\text{H}_2\text{O}]{\text{HCl}}$
p-Xylene, C_8H_{10}

organic layer
p-xylene, C_8H_{10} →   *p*-xylene, C_8H_{10}

aqueous layer
$C_6H_5NH_3^{\oplus}\ Cl^{\ominus}$ $\xrightarrow[\text{H}_2\text{O}]{\text{HO}^{\ominus}}$ aniline, $C_6H_5NH_2$

Question 26.7

In general, aliphatic amines are much more water-soluble than aromatic amines. Explain this fact.

26.5 Heterocyclic Nitrogen Compounds

Heterocyclic compounds were introduced in previous chapters. The oxygen heterocyclics (cyclic ethers), sulfur heterocyclics, and phosphorus heterocyclics were all presented in varying detail (see Secs. 19.1 and 19.10).

The nitrogen heterocyclics are the most common and important heterocyclic compounds. They are found in many naturally occurring compounds (amino acids and proteins, DNA and RNA, and alkaloids, for example), compounds of medicinal interest (amphetamines and antibiotics, for example), polymers (nylons), dyes, and food additives (such as sweeteners).

A. Nomenclature

Like all families of compounds we encountered, the nitrogen heterocyclic compounds can be named in two different ways. A system of common names is well established, and there is also an IUPAC convention for each member of this series. The IUPAC convention for the saturated oxygen, sulfur, and phosphorus heterocyclics is presented in Sec. 19.1. The IUPAC names for saturated nitrogen heterocyclics are as follows:

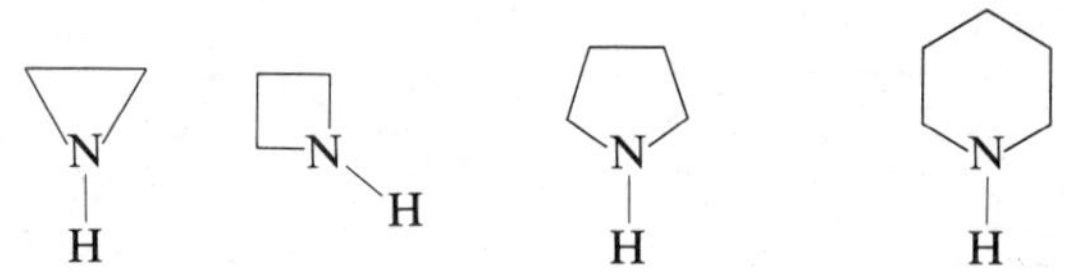

| Aziridine | Azetidine | Azolidine (pyrrolidine) | Perhydroazine (piperidine) |

Aziridine and azetidine are routinely used, but pyrrolidine and piperidine are used more commonly for five- and six-membered rings.

The remaining common nitrogen heterocyclic ring systems are as follows. The name most often encountered for each compound is provided.

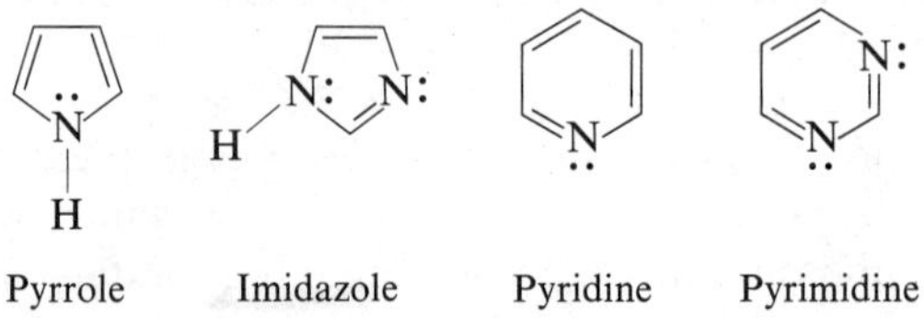

| Pyrrole | Imidazole | Pyridine | Pyrimidine |

Several other important nitrogen heterocyclics are composed of more than one ring:

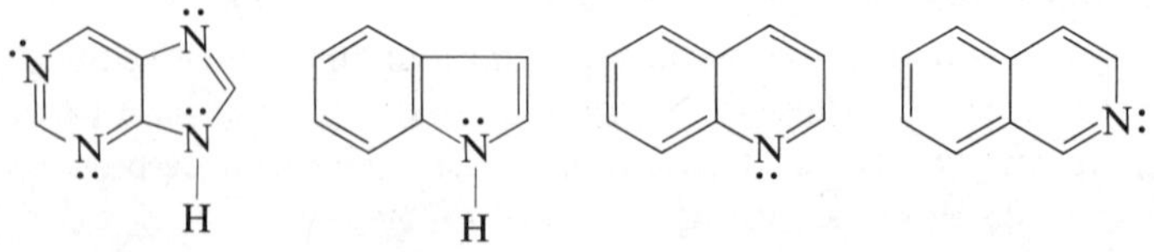

| Purine | Indole | Quinoline | Isoquinoline |

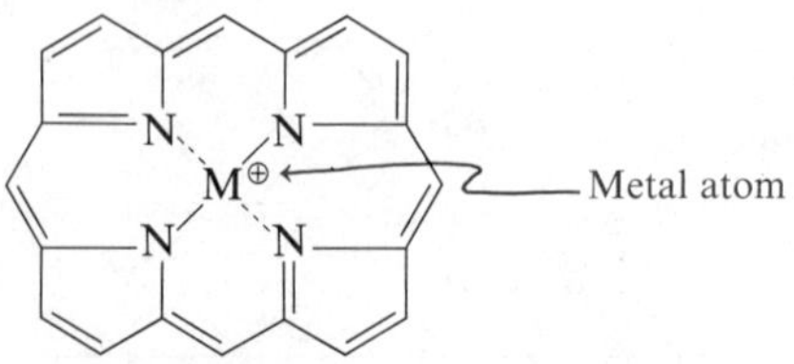

Metal atom

Porphyrin

The pyridine ring is found in vitamin B_6, for example:

Vitamin B_6
Nitrogen heterocycle

The purine and pyrimidine ring systems are found in the nucleic acids DNA and RNA (see Sec. 29.8), the porphyrin ring in hemoglobin and myoglobin (two blood proteins, see Sec. 28.14) and chlorophyll (a plant pigment, see Sec. 29.2), and the indole ring of quinoline and isoquinoline in the family of compounds named *alkaloids* (see Sec. 26.11D). The naming is often cumbersome. In all but the simplest of these compounds (which were provided earlier and which the conscientious student should commit to memory), the most advantageous way is to use the literature. Determine the parent skeleton to be used in the naming, and then consult a reference (the *Handbook of Chemistry and Physics* is a good source) for the convention. In particular, each heterocyclic ring system has a designated numbering sequence that is accepted internationally and must be adhered to. For example:

Main skeletal feature: Quinoline ring system
Numbering:

Substituents: Methyl at position 2 (2-methyl), nitro at position 6 (6-nitro)
Total name: 2-Methyl-6-nitroquinoline

With simple, one-ring heterocyclics, the numbering always begins with the heteroatom and continues in the direction that gives the lowest numbers to any attached substituents, for example:

3-Methyloxane
(3-methyltetrahydropyran)

When there are heteroatoms in the ring system, the numbering begins with one of them and proceeds to the second heteroatom to give it the lowest possible number, for example:

4-Bromo-1,3-dioxane
(not 2-bromo-1,5-dioxane)

Other accepted numberings for the common ring systems are:

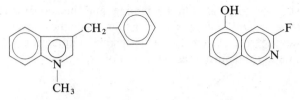

(*Note:* Numbering does not
begin on the nitrogen)

Two examples are:

<table>
</table>

3-Benzyl-*N*-methylindole   3-Fluoro-5-hydroxyisoquinoline

B. Basicity; Salt Formation

Many cyclic amines also contain nonbonding electron pairs on nitrogen. The compounds would be predicted to be Lewis bases and the prediction is correct. Cyclic saturated amines are as basic as their acyclic counterparts (Table 26.3).

The aromatic heterocyclic amines, however, are very weak bases. This too is predicted. Recall the orbital view of pyridine and pyrrole in Sec. 13.7C. In many of these cyclic compounds, the nonbonding electrons on the heteroatom are actually

TABLE 26.3 K_b's of Selected Nitrogen Heterocyclic Compounds

Compound	K_b	pK_b
Aliphatic:		
Pyrrolidine	1.3×10^{-3}	2.88
Piperidine	0.6×10^{-3}	2.79
Aromatic:		
Pyrrole	2.5×10^{-14}	13.60
Imidazole	9.9×10^{-8}	7.00
Pyridine	1.7×10^{-9}	8.77
Pyrimidine	5×10^{-12}	11.30
Quinoline	6.3×10^{-10}	9.20
Purine	2×10^{-12}	11.70

part of the bonding orbitals of the molecule. For example, in pyrrole these electrons are part of the aromatic sextet of this species and are not available to react with an electrophilic acid. The K_b of pyrrole is thus very low, 2.5×10^{-14}.

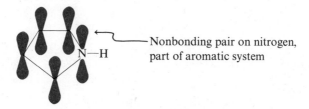

Orbital picture of pyrrole

The bonding in pyridine, pyrimidine, and quinoline, on the other hand, does not require the nonbonding nitrogen lone pair to make up the required aromatic system. The K_b values of these compounds are several powers of 10 larger than that of pyrrole.

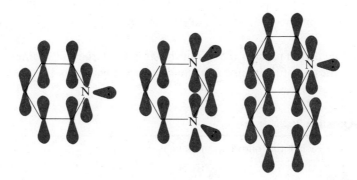

Orbital pictures of pyridine, pyrimidine, and quinoline

In the preceding pictures, in addition to a pair of electrons which is part of the aromatic system, all three have a pair of nonbonding electrons on nitrogen in an sp^2 orbital that is orthogonal (does not overlap) with the π aromatic system.

Finally, imidazole and purine contain more than one heteroatom and they exhibit both effects:

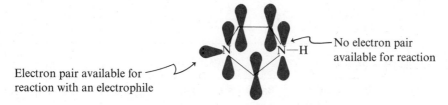

Orbital picture of imidazole

The decreased basicity in the aromatic heterocyclic amines, which contain nonbonding electrons available for sharing, versus the nonaromatic nonheterocyclic amines can also be explained by viewing the orbitals involved in the bonding in each compound. In aliphatic cyclic amines the nonbonding electrons are in an sp^3 orbital, whereas in the aromatic cyclic amines they are in a nitrogen sp^2 orbital. The sp^2 orbital has 33% s character, whereas the sp^3 orbital has only 25% s character. The electrons in the sp^2 orbital should be closer to and more tightly held by the nucleus. They are not so available for sharing and the compounds are weaker bases.

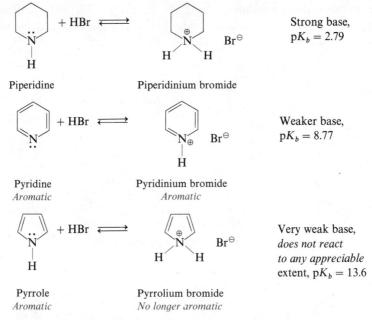

Note that in naming the salt of the simple heterocyclic amines, we simply drop the -*e* at the end of the amine name and replace it with -*ium* followed by the name of the anion.

Question 26.8

Name each of the following compounds.

(*a*)   (*b*)   (*c*)   (*d*)

Question 26.9

Quinuclidine is drawn here. Based on the compounds presented thus far, what would you predict its K_b to be? Explain.

Quinuclidine

Question 26.10

The pK_b values for ammonia and methylamine are 4.74 and 3.36, respectively. The two amino acids (see Sec. 28.1), glycine and alanine, have pK_b's of 11.63 and 11.64, respectively. Why is the effect of the methyl group so small in the latter pair as compared with in the amines?

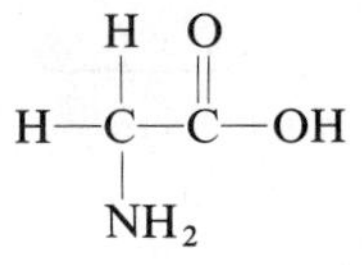

Glycine 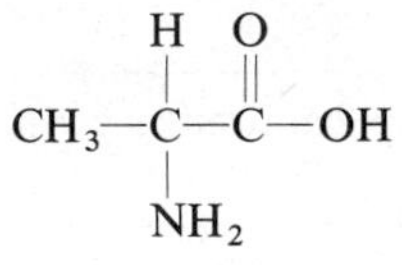   Alanine

26.6   Stereochemistry of Nitrogen

Although amines are sp^3 hybridized and pyramidal, they cannot generally be resolved into their two enantiomeric forms. This is because of rapid interconversion of the two enantiomers, called *pyramidal inversion* (see Secs. 6.19 and 24.3). The interconversion has been measured experimentally and occurs on the order of a million times per second.

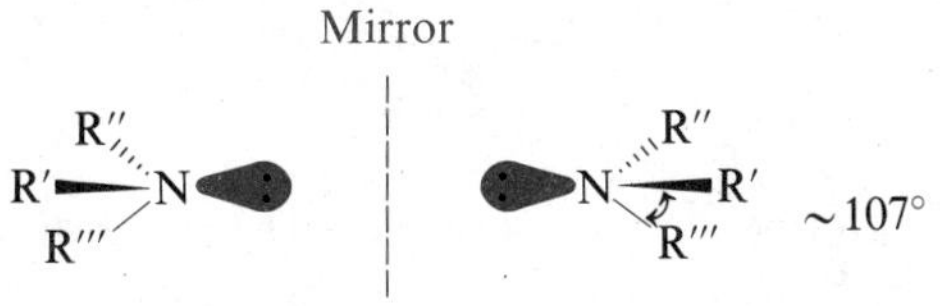

Nonsuperimposable mirror images of tertiary nitrogen compound: nonresolvable enantiomers

If the amine nitrogen is part of a rigid cyclic system or if it is in the form of one of its ammonium salts, then pyramidal inversion is either impossible or extremely slow at room temperature. In such cases it is possible to isolate or observe the two enantiomeric forms of the amine. In the first two examples that follow the amine nitrogen is one of two asymmetric centers in the two molecules. *Troger's base* is an example of a resolvable amine in which only the nitrogen atoms are asymmetric; that is, there are no other asymmetric centers.

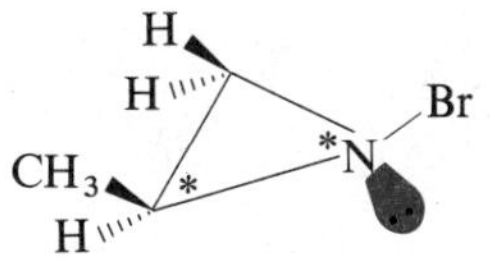

An aziridine 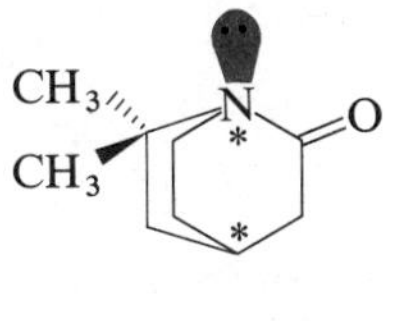   A quinuclidone

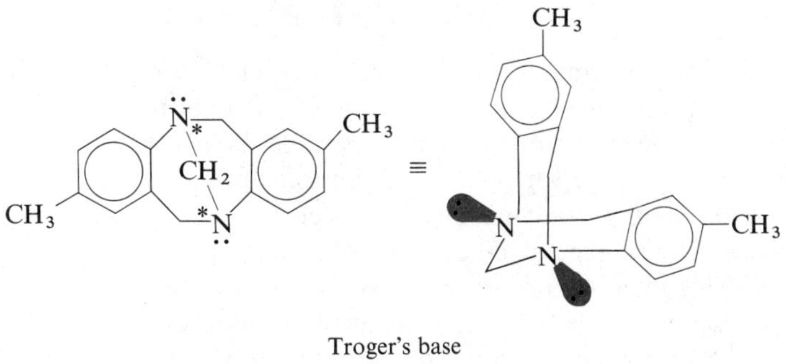

Troger's base

Question 26.11

Draw the other enantiomer of the aziridine and the quinuclidone.

Question 26.12

How many optical isomers are actually possible based on the aziridine and quinuclidone skeletons drawn?

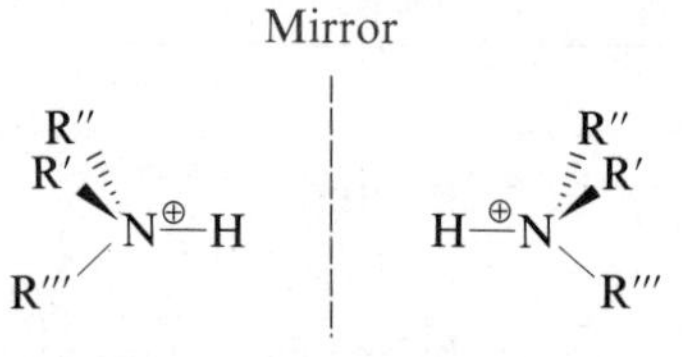

Mirror

Two enantiomers of quaternary nitrogen compound: resolvable enantiomers

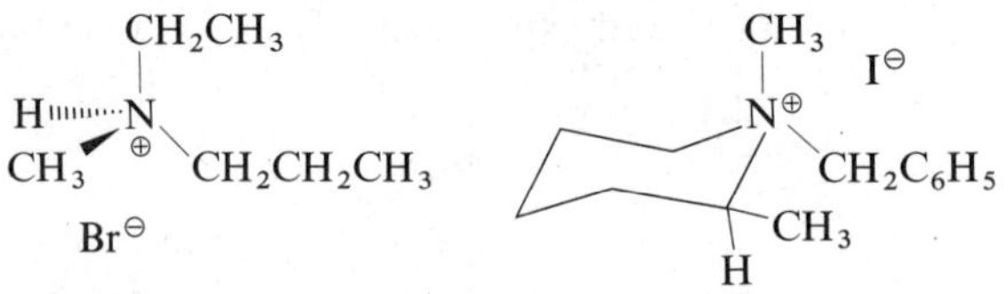

Amine salts that can be resolved into two enantiomeric forms

Question 26.13

Give the complete IUPAC name (including stereochemistry) for each of the two preceding amine salts.

Amines are extremely useful resolving agents. In particular, they are often used to resolve carboxylic acids through their ammonium salts. This method was presented in detail in Sec. 22.15.

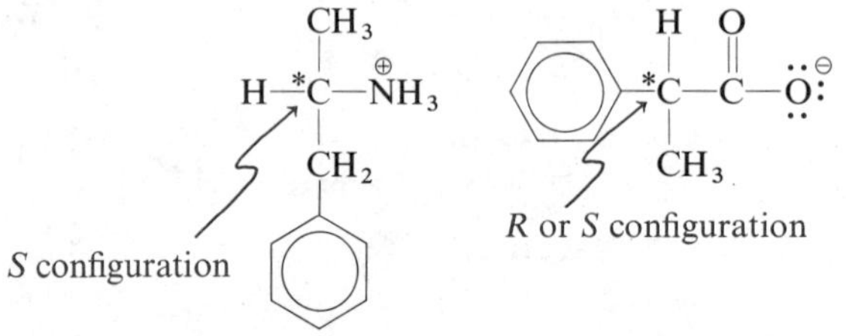

S configuration

R or S configuration

Pair of diastereomeric amine salts: separable and provide direct route to resolved enantiomeric carboxylic acids

The same technique can be used in reverse to resolve amines. For example, the optically pure carboxylic acids in Fig. 26.1 may be reacted with a racemic mixture of an amine, leading to a pair of separable diastereomeric salts. The salts, once separated, can be converted back to the free amines (Figs. 26.1 and 26.2).

Another important geometric aspect of amines that relates directly to their stereochemistry is the partial double-bond character of the carbon-nitrogen bond in amides. This phenomenon is discussed in detail in Sec. 28.10. Although the Z and E isomers that result from the restricted rotation about this bond are constantly interconverting rapidly at room temperature, the separate forms can be observed by lowering the temperature and thereby slowing the interconversion. The effect of this geometric constraint is perhaps most evident in the natural macromolecules, the proteins. For example, the helical arrangement of the α helix is due in large part to the double-bond nature of the amides that constitute its backbone (see Sec. 28.10).

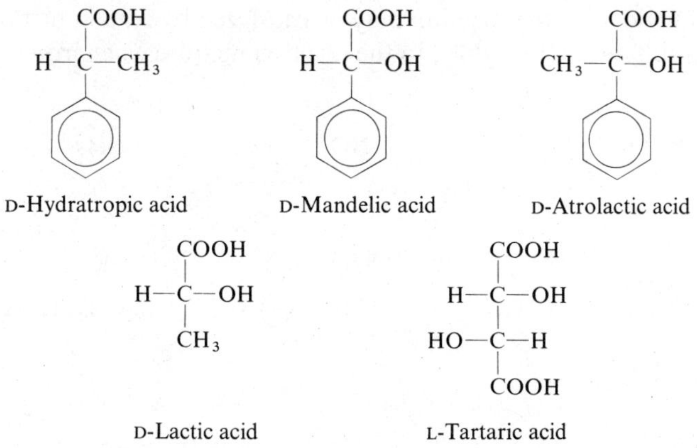

FIGURE 26.1 Five of the optically active carboxylic acids used in the resolution of amines and other compounds.

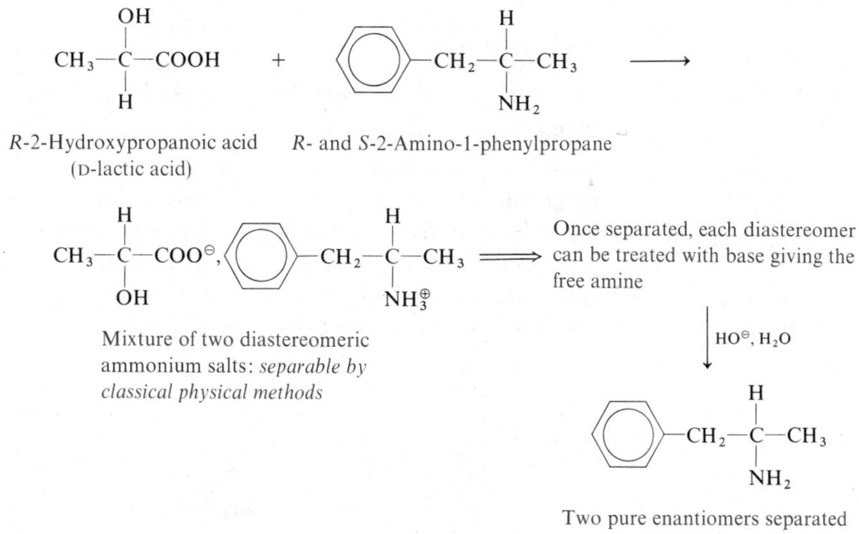

FIGURE 26.2 Resolution of amines by way of their diastereomeric ammonium salts.

26.7 Methods of Preparation of Amines

The reactions used in the preparation of amines are summarized here.

A. Reductions

a. Reduction of Nitro Compounds

Aryl nitro compounds are synthesized by the nitration of benzene or a benzene analogue; alkyl nitro compounds are prepared by the photolytic free-radical substitution of alkanes with nitric acid. (These reactions are presented in Secs. 14.3 and 4.8, respectively.)

Nitro compounds can be reduced to the corresponding amines directly. The reaction is usually accomplished by using a metal in acid. Iron, tin, and zinc are all used. In addition, the reaction may be catalyzed by certain metal salts; for example, ferrous sulfate is often added to the reaction mixture when iron is the reducing agent:

o-Nitrotoluene   o-Toluidine

$$CH_3CH_2CH_2\underset{\underset{NO_2}{|}}{CH}CH_3 \xrightarrow[H_2SO_4]{Sn} CH_3CH_2CH_2\underset{\underset{NH_2}{|}}{CH}CH_3$$

2-Nitropentane   2-Aminopentane

In the strongly acidic solution in which the reduction is carried out, the amine is protonated and exists as the ammonium salt. On workup the solution is made basic to give the free amine, which is then isolated and purified, but the reaction is written as going from nitro compound to amine in one step for simplification.

$$R\overset{\oplus}{-}NH_3 \ HSO_4^{\ominus} \xrightarrow[H_2O]{H\overset{..}{\overset{..}{O}}:^{\ominus}} R-NH_2$$

Ammonium   (solution   Free
salt   made basic)   amine

Ammonium sulfides are also used in the reduction of aromatic nitro compounds, particularly those bearing more than one nitro group. In such cases it is often possible to reduce one nitro group selectively. $(NH_4)_2S$ (prepared from ammonia and hydrogen sulfide) and NaHS are most commonly used. In most cases it is not possible to predict which nitro group will be reduced, although once determined it is reproducible. Attention must also be paid to the amount of reducing agent added in these reactions to ensure that only one nitro group will be reduced.

p-Dinitrobenzene   p-Nitroaniline

2,4-Dinitrotoluene   4-Amino-2-nitrotoluene

Hydride reducing agents, such as $LiAlH_4$ or $NaBH_4$, do not work well and are seldom used, although reduction with hydrogen and a catalyst may be used.

b.   Reduction of Amides

Amides are prepared from the corresponding carboxylic acid or acid derivative (see Secs. 22.19 and 23.8).

Reduction of these compounds requires vigorous conditions and only strong reducing agents provide good yields. $LiAlH_4$ works quite well, but $NaBH_4$ gives very low yields unless specialized cobalt catalysts are added. Also, diborane in tetrahydrofuran (B_2H_6/THF) and catalytic hydrogenation both work quite well.

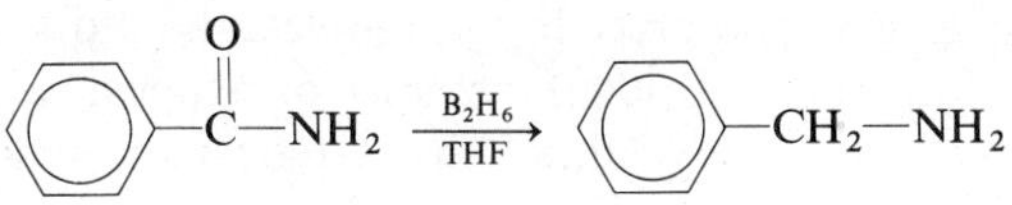

Benzamide   Benzylamine

Acetamide   Aminoethane

The reaction is reminiscent of Clemmensen or Wolff-Kishner reduction, in which the carbonyl group ($-\overset{|}{C}=O$) is reduced to a methylene group ($-CH_2-$).

c. Reduction of Nitriles

Alkyl nitriles are prepared by allowing primary and certain secondary alkyl halides or tosylates to react with cyanide ion under S_N2 conditions (see Sec. 23.15); aryl nitriles are synthesized by means of organothallium reagents (see Sec. 23.15), via diazonium salts (see Sec. 26.9A), or from nucleophilic aromatic substitution of sodium arylsulfonates (see Sec. 23.15). The reduction of nitriles again requires fairly vigorous conditions. $LiAlH_4$/ether, B_2H_6/THF, and catalytic hydrogenation provide the best yields:

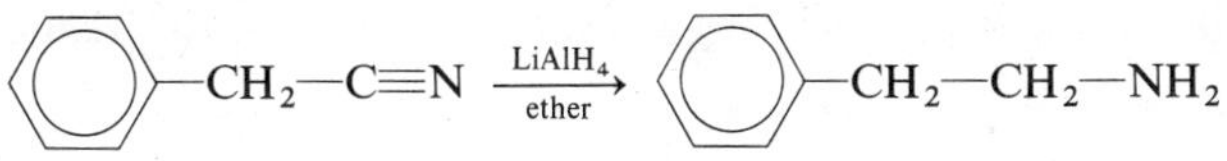

α-Phenylacetonitrile   2-Phenyl-1-aminoethane

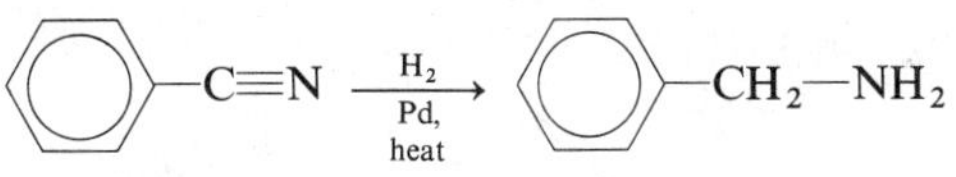

Benzonitrile   Benzylamine

As with amides, $NaBH_4$ may be used, but yields are acceptable only if specialized cobalt catalysts are added.

d. Reduction of Unsaturated Nitrogen Compounds

The reaction of aldehydes and ketones with ammonia derivatives ($G-NH_2$) results in a wide variety of derivatives (see Sec. 21.8). Among these are *imines* and *oximes*. All these derivatives contain a carbon-nitrogen double bond. In the case of the imine and oxime, reduction of this bond results in the formation of an amine:

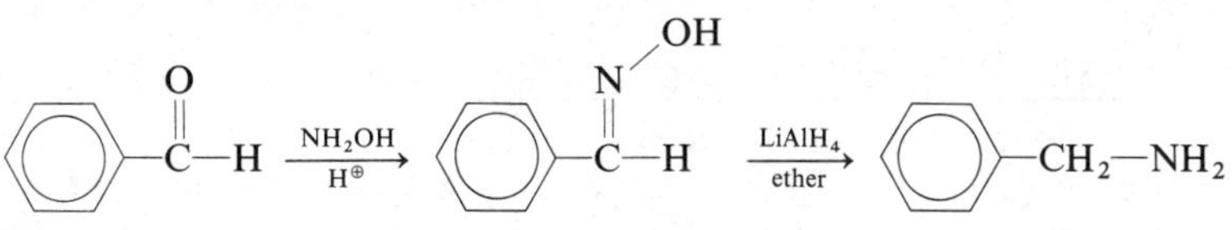

Benzaldehyde   Benzaldehyde   Benzylamine
                               oxime

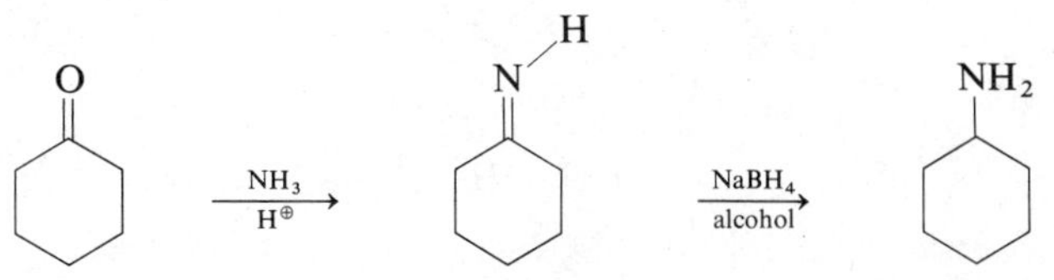

Cyclohexanone   Cyclohexanone   Aminocyclohexane
                                imine   (cyclohexylamine)
                          Unstable, not isolated

As illustrated in the preceding examples, the reduction is usually carried out with lithium aluminum hydride (for imines or oximes), sodium borohydride (imines only), or borane (imines only), although catalytic hydrogenation also works quite well (imines or oximes).

Question 26.14

Propose mechanisms for each of the following reductions:

(*a*) benzamide to benzylamine with $LiAlH_4$/ether
(*b*) cyclohexanone imine to cyclohexylamine with $NaBH_4$/CH_3CH_2OH

Imines are unstable and seldom isolated. Because of this, a shortcut often applied to the reaction sequence for imines is shown as follows:

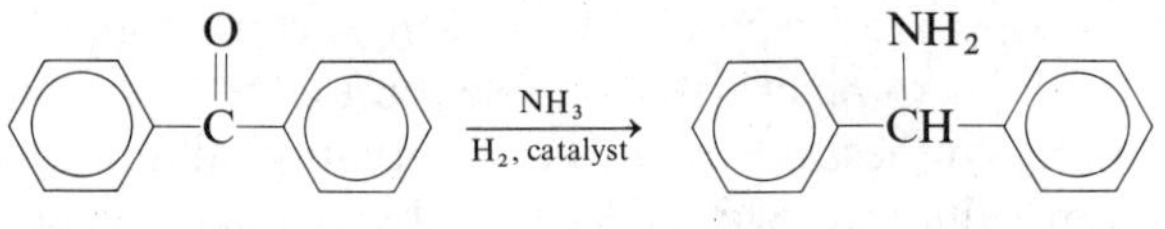

Benzophenone   1,1-Diphenylaminomethane

As an alternative to preparing the imine and then reducing it, it is possible to prepare the imine in the presence of hydrogen and a catalyst. As soon as the imine is formed, it is quickly reduced to the amine. The entire reaction sequence is called **reductive amination.**

Question 26.15

Explain the following observation mechanistically.

How could you control the reaction to give product 1 as the major product? Product 2 as the major product?

The reaction can be extended successfully to substituted imines, which in turn provide secondary amines on reduction. A list of common hydride reducing agents and a summary of the functional groups reduced by each of them are in Table 21.1.

Question 26.16

The **Eschweiler-Clark reaction** is another example of reductive amination. In this reaction, the carbonyl compound is formaldehyde, which is mixed with formic acid and a 1° or 2° amine. The formic acid acts as a hydride source and is oxidized to carbon dioxide. The overall reaction is as follows:

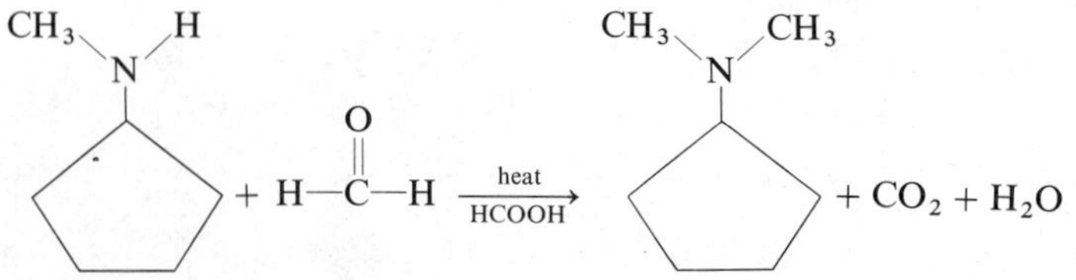

Outline a mechanism for this transformation.

Question 26.17

Complete each of the following syntheses.

(*a*) benzene to *p*-bromoaniline
(*b*) toluene to *p*-(methylamino)benzoic acid
(*c*) bromobenzene to benzonitrile
(*d*) benzonitrile to benzamide
(*e*) benzoic acid to 1,1-diphenylaminomethane
(*f*) benzyl alcohol to *N,N*-diphenylbenzylamine

B. Substitution Reactions

a. Nucleophilic Substitution by Ammonia and Ammonia Analogues

The reaction of ammonia with primary (or some secondary) alkyl halides under S_N2 conditions (in this case, with excess nucleophile and heat and pressure as needed) yields a primary alkyl amine:

$$H_3N: \; + \; CH_3-I: \xrightarrow{S_N2} H_3N^{\oplus}-CH_3 \quad :I:^{\ominus} \xrightarrow[H_2O]{HO:^{\ominus}} H_2N-CH_3$$

1° Alkyl halide
or
methyl halide
Methylamine

A problem with the actual application of this reaction is that it involves a series of competing reactions:

$$NH_3 \xrightarrow{CH_3I} NH_2CH_3 \xrightarrow{CH_3I} NH(CH_3)_2 \xrightarrow{CH_3I}$$

1°
Product
2°
Product

$$N(CH_3)_3 \xrightarrow{CH_3I} \overset{\oplus}{N}(CH_3)_4 \; :I:^{\ominus}$$

3°
Product
4°
Product

If one starts with ammonia, then a primary, secondary, tertiary, or even quaternary product is possible. By using a large excess of the amine, it is possible to preferentially alkylate only once, but yields are often low because of the competing reactions. The reaction is most useful if it is applied to the complete alkylation of an amine to give either a tertiary or perhaps a quaternary nitrogen compound. In the examples that follow, aniline is converted to *N,N*-diethylaniline. This compound is bulky and a poor nucleophile. The reaction stops at the 3° amine. Piperidine, on the other hand, is a better nucleophile and methyl iodide is more susceptible to attack so the 4° salt results. Additional examples of the synthetic utility of this reaction can be found in Sec. 26.9C.

Aniline $\xrightarrow[\text{2. HO:}^{\ominus}, H_2O]{\text{1. } CH_3CH_2I \text{ excess}}$ *N,N*-Diethylaniline

Piperidine $\xrightarrow[\text{2. HO:}^{\ominus}, H_2O]{\text{1. } CH_3I \text{ excess}}$ *N,N*-Dimethylpiperidinium iodide

The reaction works well with an α-halo acid and results in the formation of an amino acid. This reaction is discussed in Sec. 28.7.

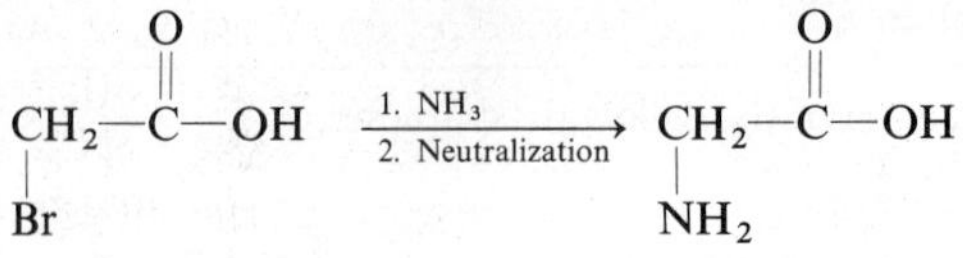

$$\underset{\substack{| \\ \text{Br} \\ \text{α-Bromoacetic acid}}}{CH_2-\overset{\overset{\displaystyle O}{\|}}{C}-OH} \xrightarrow[\text{2. Neutralization}]{\text{1. NH}_3} \underset{\substack{| \\ NH_2 \\ \text{Glycine} \\ \text{(an amino acid)}}}{CH_2-\overset{\overset{\displaystyle O}{\|}}{C}-OH}$$

Question 26.18

(*a* Whereas amines are strong bases and good nucleophiles, the conjugate acids (salts) of amines are not. Explain.

(*b*) When methylamine reacts with methyl iodide, dimethylammonium iodide results. Dimethylammonium iodide is a salt. Keeping in mind your answer to (*a*), why do primary amines, such as methylamine, react with excess alkylating agents, such as methyl iodide, to produce quaternary salts?

$$CH_3-\overset{..}{N}H_2 + CH_3-I \text{ (excess)} \longrightarrow \underset{\substack{| \\ CH_3}}{\overset{\overset{\displaystyle CH_3}{|}}{CH_3-\overset{\oplus}{N}-CH_3}} \quad :\overset{..}{\underset{..}{I}}:^{\ominus}$$

Tetramethylammonium
iodide
Quaternary salt

b.   Gabriel Synthesis

In contrast to ammonia and its derivatives, potassium phthalimide provides an efficient and versatile route to primary amines. Phthalimide is prepared by the reaction of phthalic acid with ammonia followed by heating. Treatment of phthalimide with KOH/alcohol gives potassium phthalimide, a good nucleophile. The first step in the **Gabriel synthesis** is nucleophilic substitution by the potassium phthalimide on an alkyl halide. The *N*-alkylated phthalimide can be isolated or directly hydrolyzed to the primary amine.

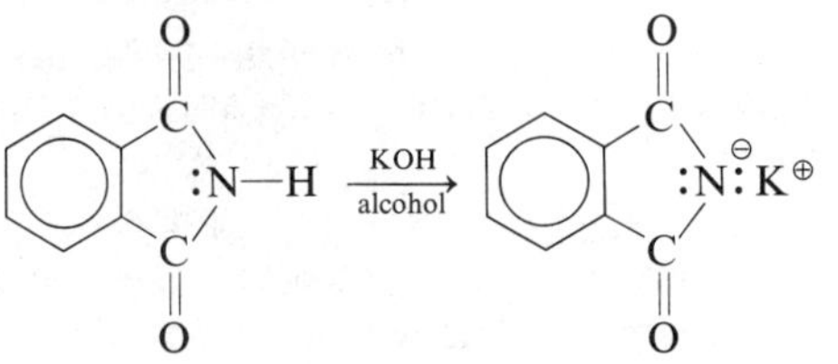

Phthalimide   Potassium phthalimide

Methyl or
1° alkyl halide

$$\underset{\substack{| \\ O \\ \text{1° Amine} \\ \text{(methylamine)}}}{} + CH_3-\overset{..}{N}H_2$$

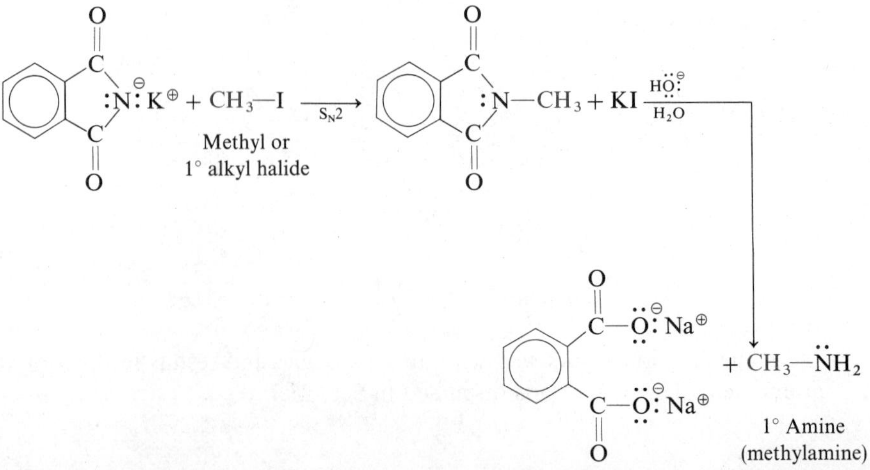

In all the reactions discussed in Secs. 26.7B,a and 26.7B,b, tosylates can be used in place of alkyl halides. This provides greater versatility to this synthetic pathway. Further examples are:

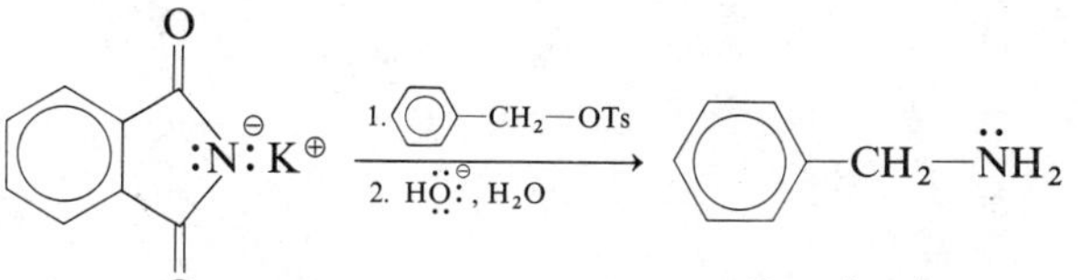

Benzyl amine

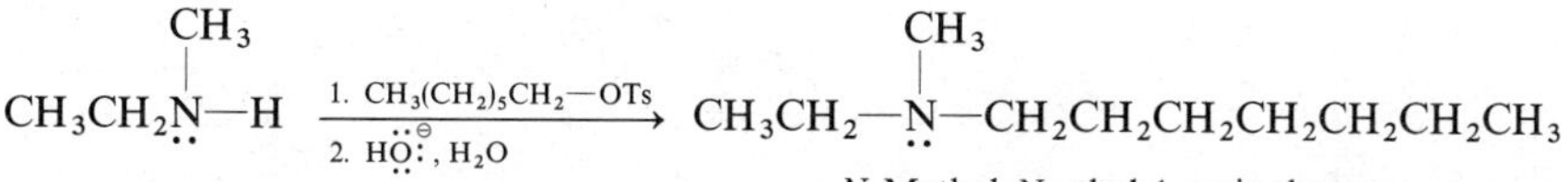

N-Methyl-N-ethyl-1-aminoheptane

The Gabriel synthesis is also used to prepare amino acids through reaction with an α-halomalonic ester. This reaction is presented in Sec. 28.7.

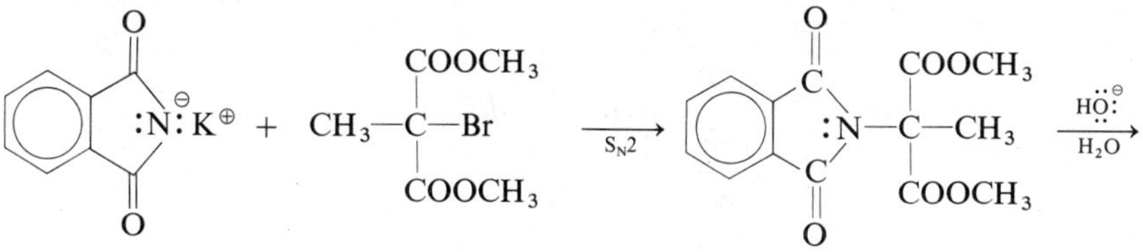

α-Bromomalonic ester

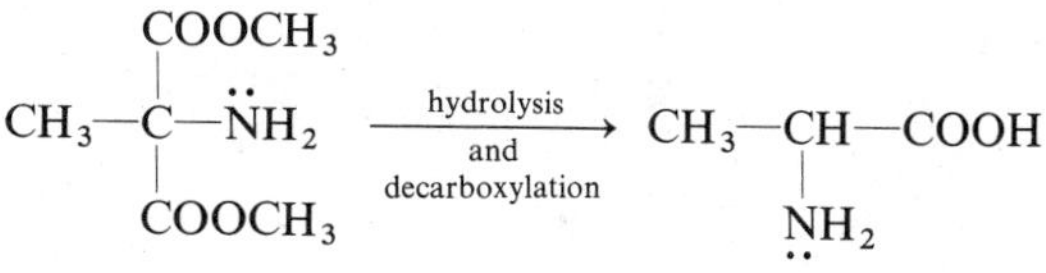

Alanine
(an amino acid)

Question 26.19

Phthalimide is prepared most conveniently by heating the diammonium salt of phthalic acid (ammonium phthalate) to high temperatures. The products are phthalimide, water, and ammonia. Is this reaction consistent with what we know about intramolecular cyclizations? Explain. (*Hint:* See Secs. 23.7B,c and 23.12 on lactones, anhydrides, and intramolecular condensation reactions.)

Question 26.20

Phthalimide is a fairly strong acid with a pK_a of approximately 9.6. This is about the same acid strength as phenol or the ammonium ion. Explain.

C. Rearrangements; Hofmann Reaction

A primary amide can be transformed directly to an amine that contains one less carbon atom by treatment with hypohalite. Either sodium hypobromite ($Br_2/NaOH$) or sodium hypochlorite ($Cl_2/NaOH$) is used:

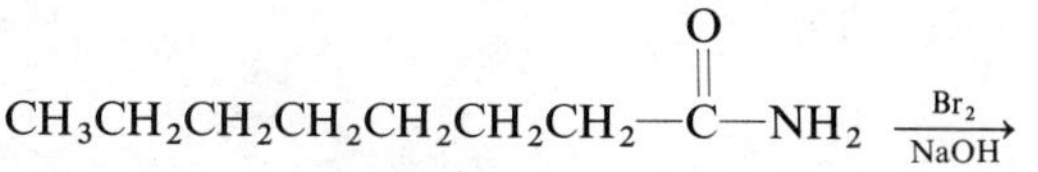

$$CH_3CH_2CH_2CH_2CH_2CH_2CH_2-\overset{\overset{\displaystyle O}{\|}}{C}-NH_2 \xrightarrow[\text{NaOH}]{\text{Br}_2}$$

Octanamide

$$CH_3CH_2CH_2CH_2CH_2CH_2CH_2-NH_2 + CO_3^{(-2)}$$

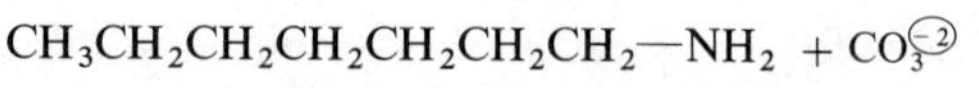

1-Aminoheptane

The reaction goes through an isocyanate, as shown by the following mechanism:

Step 1

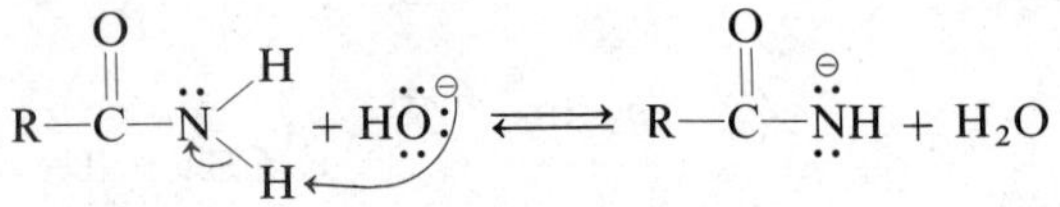

$$R-\overset{\overset{O}{\|}}{C}-\overset{\overset{..}{N}}{\underset{H}{}}\overset{H}{} + HO\colon^{\ominus} \rightleftharpoons R-\overset{\overset{O}{\|}}{C}-\overset{\ominus}{\overset{..}{N}}H + H_2O$$

Step 2

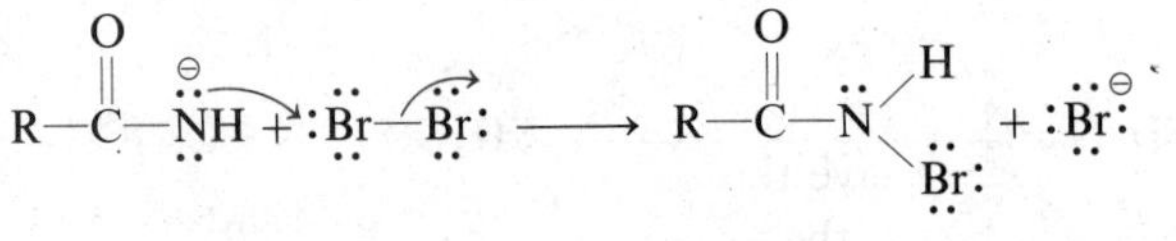

$$R-\overset{\overset{O}{\|}}{C}-\overset{\ominus}{\overset{..}{N}}H + \colon\!\overset{..}{\underset{..}{Br}}-\overset{..}{\underset{..}{Br}}\colon \longrightarrow R-\overset{\overset{O}{\|}}{C}-\overset{..}{N}\underset{\overset{..}{\underset{..}{Br}}\colon}{\overset{H}{}} + \colon\!\overset{..}{\underset{..}{Br}}\colon^{\ominus}$$

N-Bromoamide

Step 3

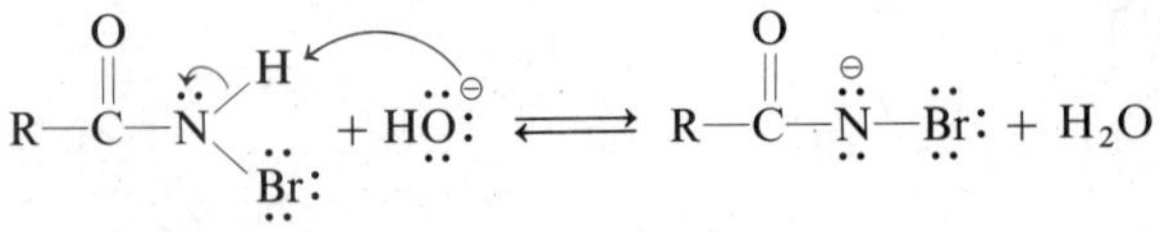

$$R-\overset{\overset{O}{\|}}{C}-\overset{..}{N}\underset{\overset{..}{Br}\colon}{\overset{H}{}} + HO\colon^{\ominus} \rightleftharpoons R-\overset{\overset{O}{\|}}{C}-\overset{\ominus}{\overset{..}{N}}-\overset{..}{\underset{..}{Br}}\colon + H_2O$$

Step 4

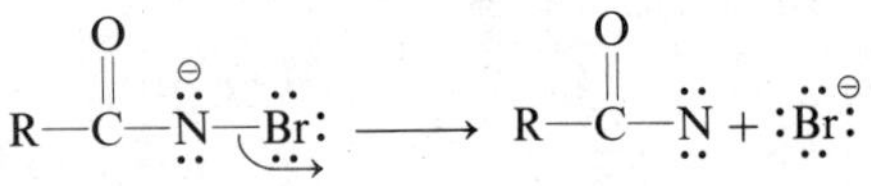

$$R-\overset{\overset{O}{\|}}{C}-\overset{\ominus}{\overset{..}{N}}-\overset{..}{\underset{..}{Br}}\colon \longrightarrow R-\overset{\overset{O}{\|}}{C}-\overset{..}{\underset{..}{N}} + \colon\!\overset{..}{\underset{..}{Br}}\colon^{\ominus}$$

Nitrene

Step 5

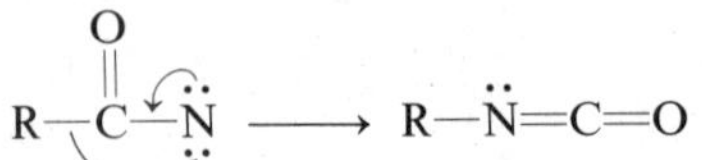

$$R-\overset{\overset{O}{\|}}{C}-\overset{..}{\underset{..}{N}} \longrightarrow R-\overset{..}{N}=C=O$$

Isocyanate

Step 6

$$H_2\overset{..}{O}\colon + R-\overset{..}{N}=C=O \rightarrow \rightarrow R-\overset{..}{\underset{H}{N}}-\overset{\overset{O}{\|}}{C}-OH$$

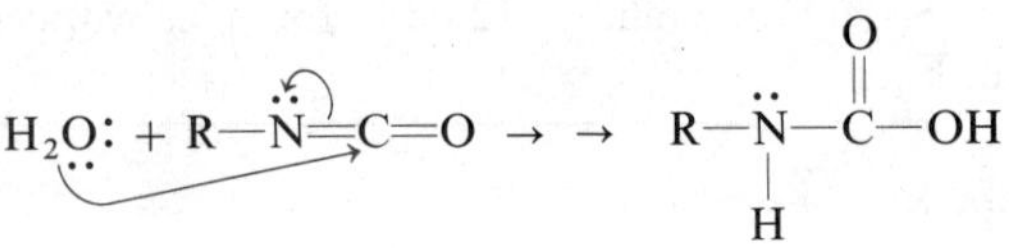

Unstable carbamic acid

Step 7

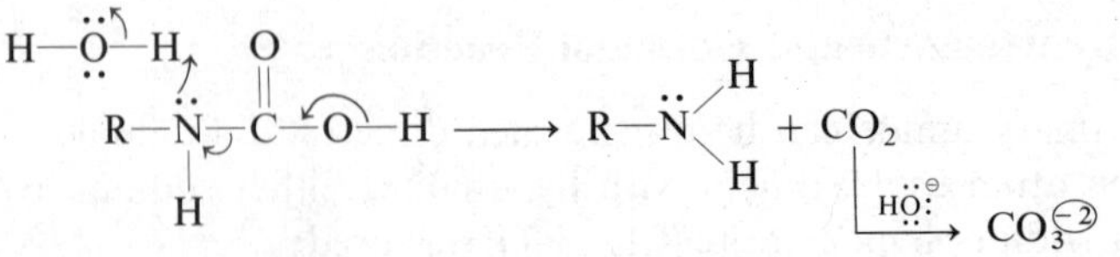

$$\overset{H-\overset{..}{O}-H}{\underset{}{}} \quad R-\overset{..}{\underset{H}{N}}-\overset{\overset{O}{\|}}{C}-O-H \longrightarrow R-\overset{..}{N}\overset{H}{\underset{H}{}} + CO_2 \overset{HO\colon^{\ominus}}{\longrightarrow} CO_3^{\ominus 2}$$

Amine

In step 1 the amide acts as an acid, a proton is abstracted by base, and the following resonance-stabilized *amide anion* results:

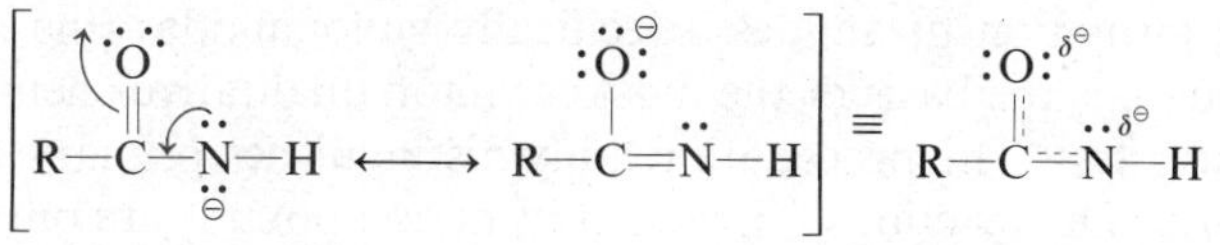

$$\left[\begin{array}{ccc} \overset{\displaystyle \overset{\uparrow}{:\!\ddot{O}}\,}{R-C\overset{\curvearrowleft}{\;}\underset{\underset{\ominus}{\ddot{N}}}{\;}-H} & \longleftrightarrow & \overset{:\ddot{O}:\,^{\ominus}}{R-C=\overset{\ddot{}}{N}-H} \end{array} \right] \equiv \overset{:O:^{\delta\ominus}}{R-C\!\!=\!\!\overset{\;\;\delta\ominus}{\ddot{N}}\!-H}$$

The equilibrium highly favors the reverse direction, but the concentration of the amide anion allows for a reaction at a sufficiently rapid rate with bromine to provide *N*-bromoamide (step 2). The amide hydrogen in *N*-bromoamide is more acidic than that in the original amide because of the inductive effect of bromine. Base abstracts this proton to give the *N*-bromoamide anion (step 3). Elimination of bromide from the anion produces the highly reactive *nitrene* intermediate (step 4). The nitrene, like the carbene (see Sec. 27.6), is capable of insertion reactions. Rearrangement and insertion of the nitrene give the *isocyanate* (step 5). Hydrolysis and decarboxylation of the isocyanate provide the amine product (steps 6 and 7).

Two additional examples of the Hofman reaction are:

$$\underset{\text{3-Methylbutanamide}}{\overset{\displaystyle \overset{CH_3}{|}\qquad\qquad \overset{O}{\|}}{CH_3CH-CH_2-C-NH_2}} \quad \xrightarrow[\underset{\ddot{}}{H\ddot{O}:}]{Cl_2} \quad \underset{\text{2-Methyl-1-aminopropane}}{\overset{\displaystyle \overset{CH_3}{|}}{CH_3CH-CH_2-NH_2}}$$

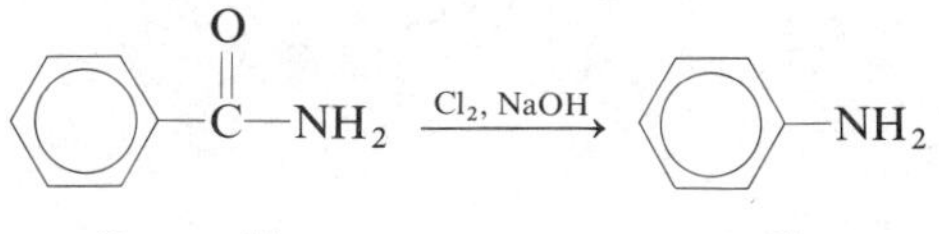

$$\underset{\text{Benzamide}}{\overset{\displaystyle \overset{O}{\|}}{\bigcirc\!\!-C-NH_2}} \quad \xrightarrow{Cl_2,\ NaOH} \quad \underset{\text{Aniline}}{\bigcirc\!\!-NH_2}$$

Question 26.21

Complete the following four reactions by supplying any missing reagents.

(*a*) toluene to aniline   (*b*) toluene to benzylamine
(*c*) toluene to β-phenethylamine (2-phenyl-1-aminoethane)
(*d*) toluene to *N*-(*p*-nitrophenyl)benzylamine

26.8   Reactions of Amines Covered in Past Chapters

Many reactions of amines were covered in previous chapters. These are summarized here, followed by several new reactions of this family.

A. Preparation of Amides; The Hinsberg Test

Amides can be prepared from carboxylic acids directly or via more reactive carboxylic acid derivatives. These reactions are illustrated here and discussed in Sec. 23.8.

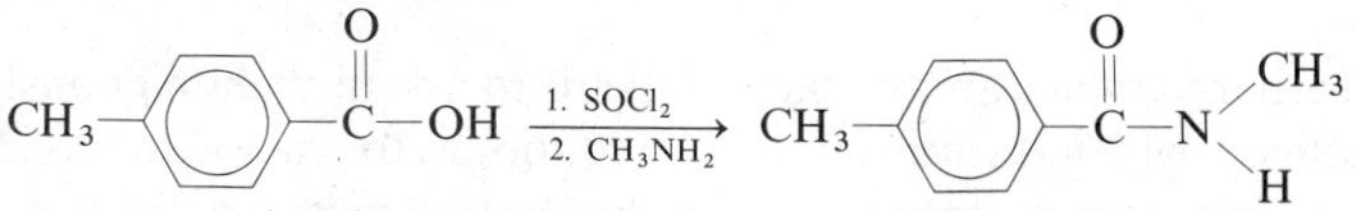

$$\underset{\text{p-Toluic acid}}{CH_3-\bigcirc\!\!-\overset{\displaystyle \overset{O}{\|}}{C}-OH} \quad \xrightarrow[\text{2. } CH_3NH_2]{\text{1. } SOCl_2} \quad \underset{\text{N-Methyl-p-toluamide}}{CH_3-\bigcirc\!\!-\overset{\displaystyle \overset{O}{\|}}{C}-\overset{CH_3}{\underset{H}{N}}}$$

$$\underset{\text{Hexanoic acid}}{CH_3CH_2CH_2CH_2CH_2-\overset{\displaystyle \overset{O}{\|}}{C}-OH} \quad \xrightarrow[\text{2. heat}]{\text{1. } NH_3\ \text{(excess)}} \quad \underset{\text{Hexanamide}}{CH_3CH_2CH_2CH_2CH_2-\overset{\displaystyle \overset{O}{\|}}{C}-NH_2}$$

The formation of amides, specifically sulfonamides (the amides of benzene-sulfonic acid), is the basis of the most common qualitative chemical test for amines, the **Hinsberg test.** The procedure readily distinguishes primary, secondary, and ter-tiary amines. The reactions for each class of compounds are outlined here.

The amine reacts with benzenesulfonyl chloride in basic medium. In all three reactions, the products are sulfonamides of benzenesulfonic acid.

Benzenesulfonyl
chloride

(1)

(2)

(3)

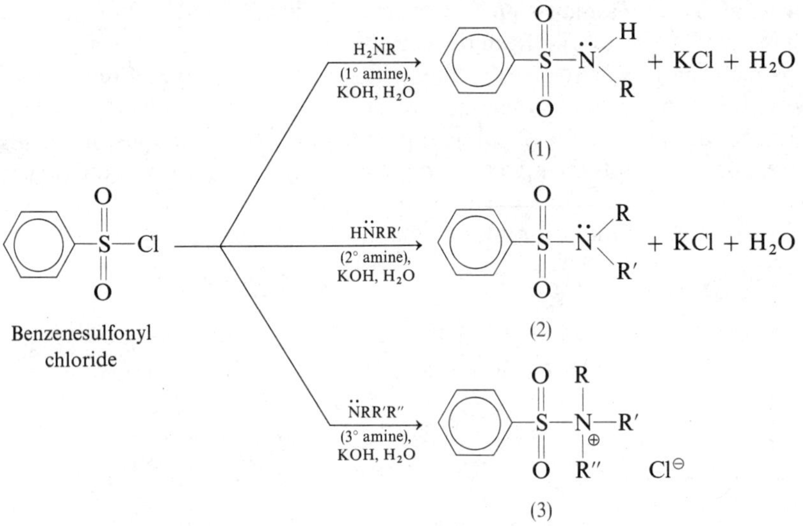

Benzenesulfonamides

The benzenesulfonamide of the primary amine continues to react with KOH to give a water-soluble salt of the sulfonamide. This solution, when acidified, becomes heterogeneous as the water-insoluble sulfonamide precipitates.

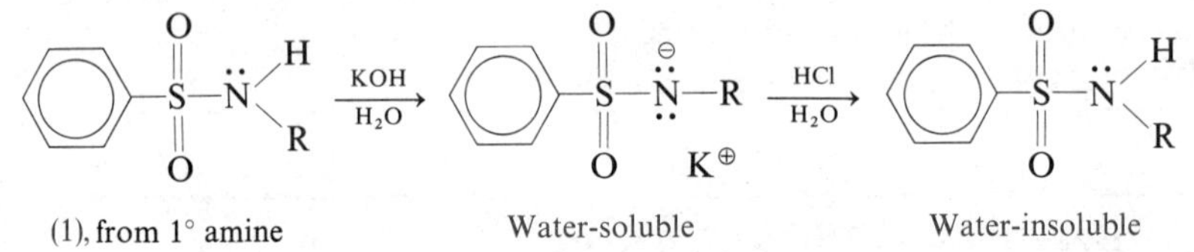

(1), from 1° amine

Water-soluble

Water-insoluble

The secondary amine, in contrast, immediately gives a water-insoluble sulfon-amide, which precipitates out of solution. The sulfonamide is incapable of further reaction with the base to give a salt.

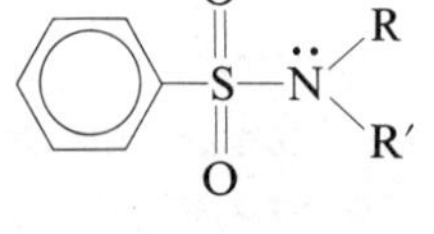

(2), from 2° amine
Water-insoluble

Tertiary amines are the most difficult to assign unambiguously. The salt of the sulfonamide, which is formed initially, continues to react with base to give the potas-sium salt of benzenesulfonic acid and the starting material (the tertiary amine) again. The solution is therefore heterogeneous, as it is with secondary amines, because of the water-insoluble tertiary amine. Acidification of this solution produces a homoge-neous solution with the formation of a water-soluble ammonium salt. Acidification of the solution of the sulfonamide of a secondary amine results in no change in the solubility of the sulfonamide, with the solution remaining heterogeneous.

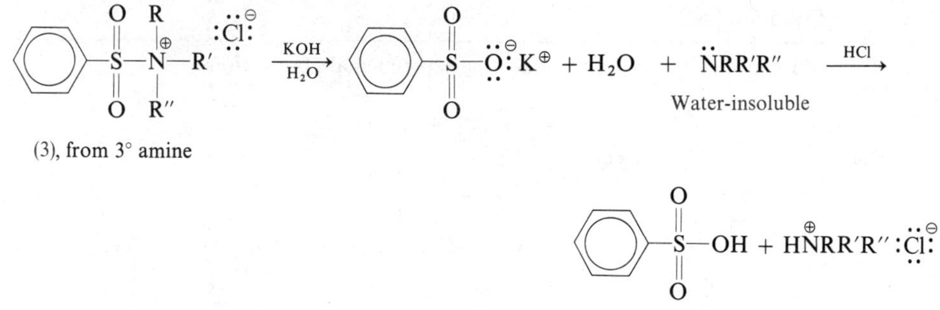

(3), from 3° amine

Water-soluble

Question 26.22

The sulfonamide salt intermediate of a tertiary amine, compound (3), is capable of another reaction under the conditions of the Hinsberg reaction:

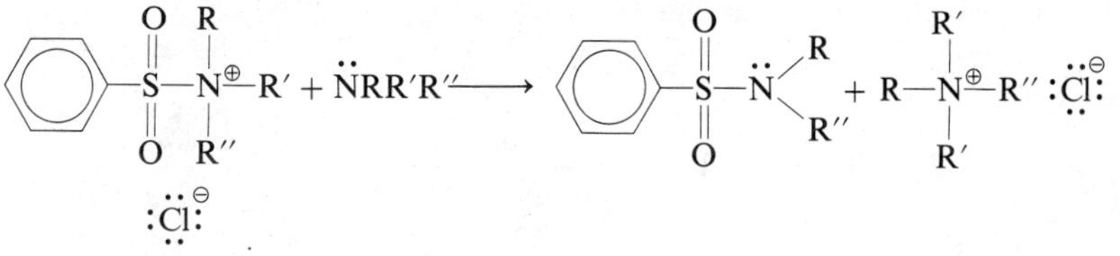

Explain this reaction mechanistically. What would you do experimentally to minimize this side reaction?

B. Preparation of Enamines

Enamines are prepared by the reaction of a secondary amine with a carbonyl compound (see Sec. 24.12):

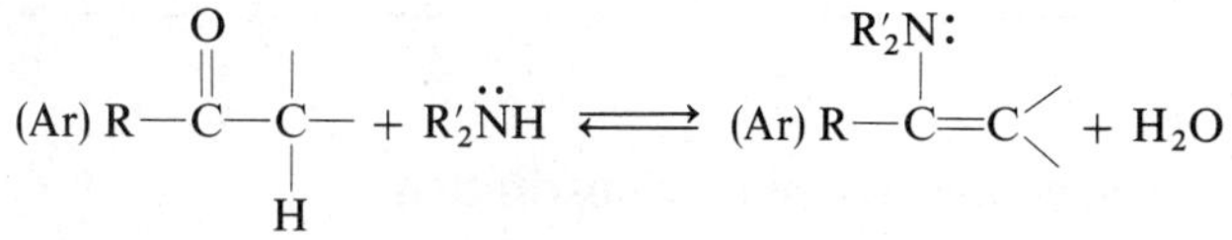

| Carbonyl compound | 2° Amine | Enamine |

The enamine is a weak nucleophile that can react only with reactive alkylating agents such as alkyl iodides, allyl halides, benzyl halides, α-haloketones, and α-haloethers. Enamines are very useful synthetic tools because they give only mono-alkylation of the initial carbonyl compound. Eliminations do not occur with the use of enamines in condensation reactions.

An example is:

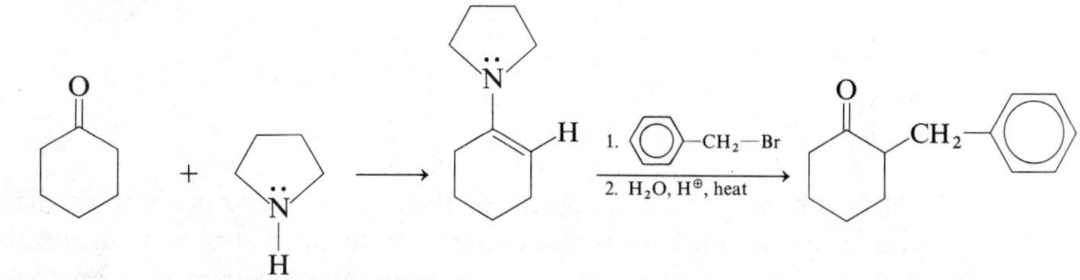

Cyclohexanone   Pyrrolidine   2-Benzylcyclohexanone

Question 26.23

Mechanistically explain the following experimental observations.

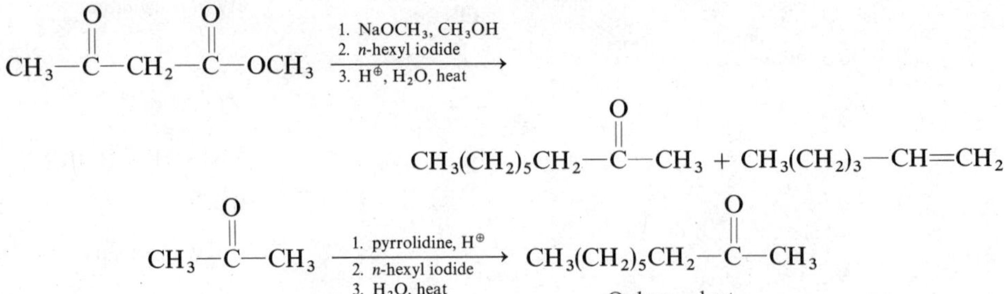

$$CH_3-\overset{\overset{\displaystyle O}{\|}}{C}-CH_2-\overset{\overset{\displaystyle O}{\|}}{C}-OCH_3 \xrightarrow[\text{3. H}^\oplus\text{, H}_2\text{O, heat}]{\begin{array}{l}\text{1. NaOCH}_3\text{, CH}_3\text{OH}\\\text{2. }n\text{-hexyl iodide}\end{array}}$$

$$CH_3(CH_2)_5CH_2-\overset{\overset{\displaystyle O}{\|}}{C}-CH_3 + CH_3(CH_2)_3-CH=CH_2$$

$$CH_3-\overset{\overset{\displaystyle O}{\|}}{C}-CH_3 \xrightarrow[\begin{array}{l}\text{2. }n\text{-hexyl iodide}\\\text{3. H}_2\text{O, heat}\end{array}]{\text{1. pyrrolidine, H}^\oplus} CH_3(CH_2)_5CH_2-\overset{\overset{\displaystyle O}{\|}}{C}-CH_3$$

Only product

Question 26.24

Explain the following reaction:

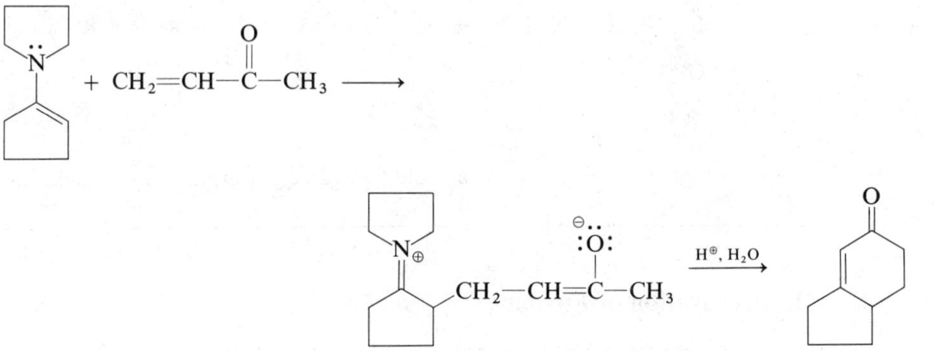

Not isolated   Isolated product

C. Electrophilic Aromatic Substitution

The effect of the amino group ($-\overset{..}{N}H_2$) or substituted amino group ($-\overset{..}{N}R_2$) on electrophilic aromatic substitution is discussed in Sec. 14.11. The amino group is a very strong activating group; it is also an *ortho,para* director.

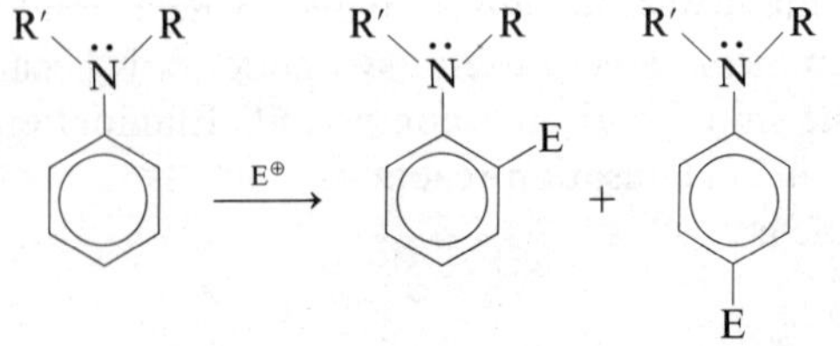

ortho and *para* Substitution

where R and R′ = alkyl, aryl, or hydrogen

The attachment of an electrophile at the *ortho* or *para* position to the amino group on an aromatic ring results in an intermediate ion that is resonance-stabilized. The resonance stabilization involves the nonbonding electron pair on the amino group (see Sec. 14.11).

Although nitrogen, like oxygen, withdraws electrons inductively, it donates electrons through resonance. The resonance effect is the more important of the two.

The amine group is susceptible to a variety of reactions including oxidation and alkylation (see Secs. 26.7 and 26.8B,b). To block these side reactions in an organic synthesis, the amine is often converted to an amide, which can then be hydrolyzed back to the free amine whenever necessary. Substituted acetamides are most commonly used for this purpose. In addition to acting as a protecting group, the amide group is a weaker activating group than is a free amino group. As a result, electrophilic aromatic substitution reactions are often much easier to control in a desired direction if this substitution is made. The following examples illustrate this point:

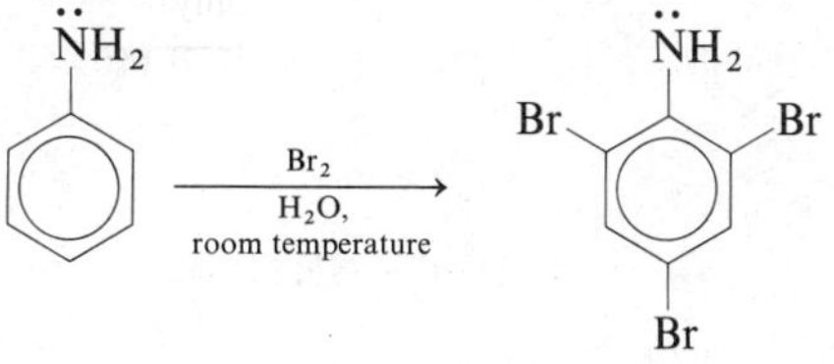

2,4,6-Tribromoaniline
Major product

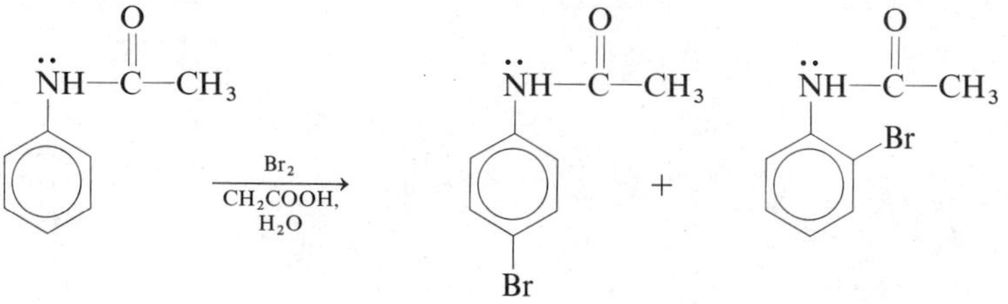

o- and *p*-Bromoacetanilide

>90%   *<10%*

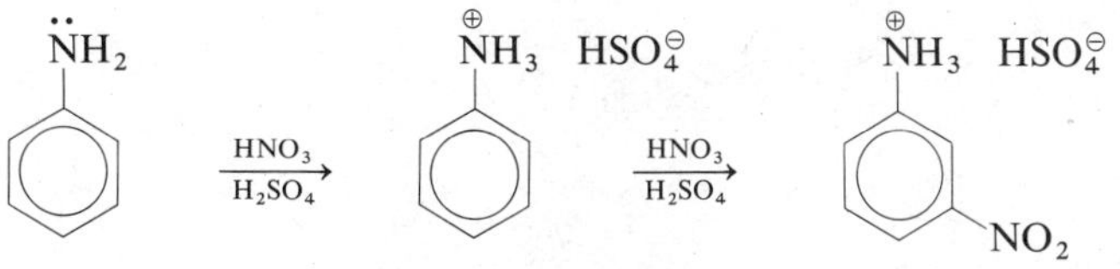

Anilinium
hydrogen sulfate
$\oplus NH_3$ *group is*
meta director

Major
product

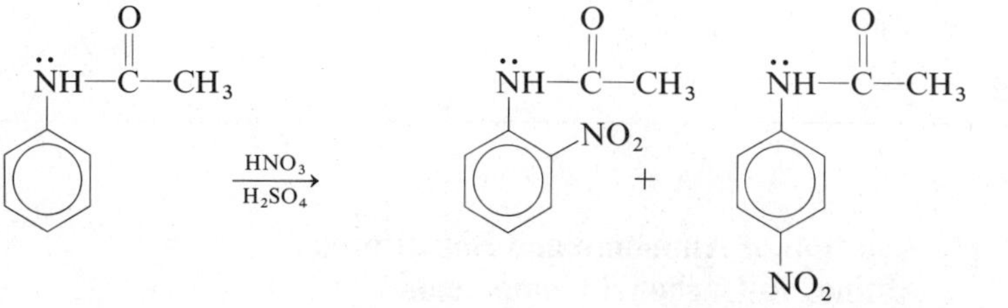

<10%   *>90%*

Note the distribution of the *o-* and *p*-bromoacetanilide. The acetamido group directs electrophiles almost exclusively (more than 90%) to the *para* position in most electrophilic aromatic substitution reactions.

The synthesis of sulfanilamide provides another example of the use of the acetamide group in synthesis.

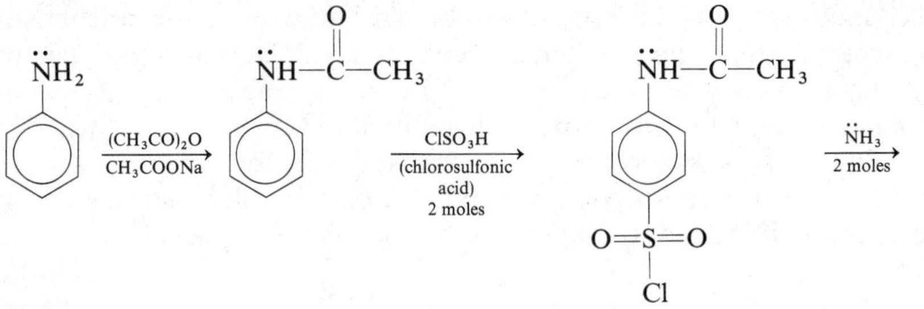

Aniline   Acetanalide   Major product

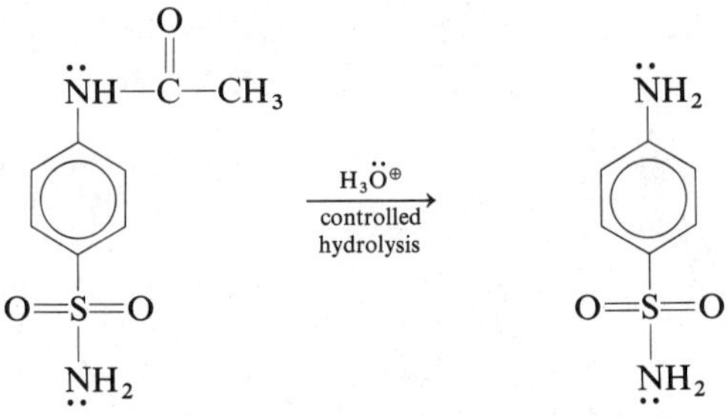

p-Aminobenzenesulfonamide
[sulfanilamide: an antibacterial
(*see Sec. 26.11B,a*)]

Question 26.25

How can you explain the following experimental observations?

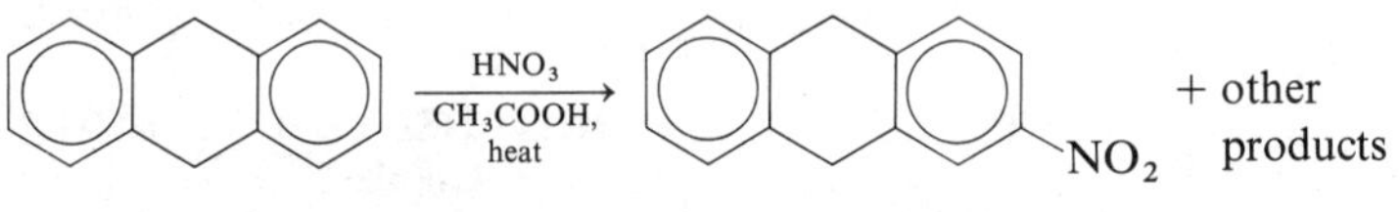

+ other products

79%

but

+ other products

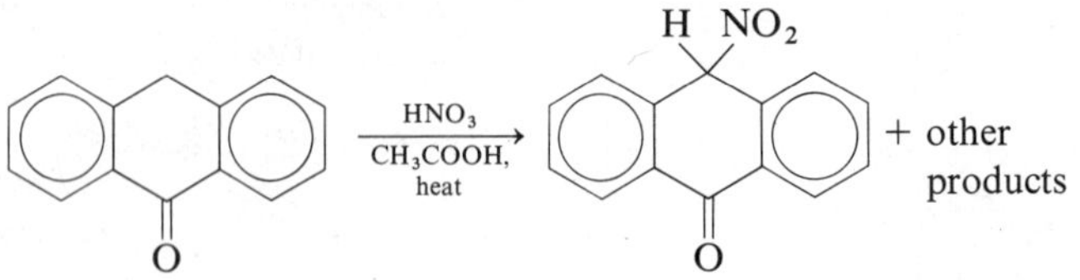

67%

D. Addition of Ammonia and Substituted Amines to Carbonyl Compounds

Ammonia, NH_3, and derivatives of ammonia, $G—NH_2$, react with carbonyl compounds by adding across the carbon-oxygen double bond (see Sec. 21.8). The intermediate amino alcohol is unstable and dehydrates to give a carbon-nitrogen double bond:

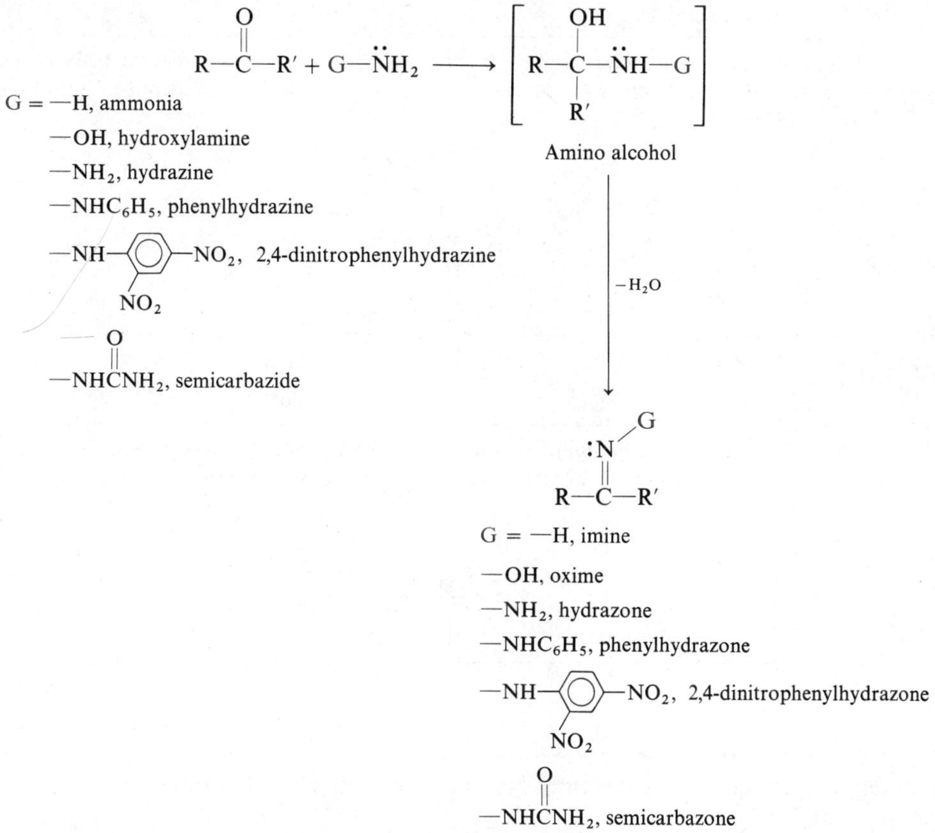

$$R-\overset{\overset{O}{\|}}{C}-R' + G-\overset{\cdot\cdot}{N}H_2 \longrightarrow \left[R-\overset{\overset{OH}{|}}{\underset{R'}{C}}-\overset{\cdot\cdot}{N}H-G \right]$$

G = —H, ammonia

—OH, hydroxylamine

—NH$_2$, hydrazine

—NHC$_6$H$_5$, phenylhydrazine

—NH—⬡—NO$_2$, 2,4-dinitrophenylhydrazine
  |
  NO$_2$

—NH$\overset{\overset{O}{\|}}{C}NH_2$, semicarbazide

Amino alcohol

$\downarrow$ $-H_2O$

$$R-\overset{\overset{\displaystyle :N^{\nearrow G}}{\|}}{C}-R'$$

G = —H, imine

—OH, oxime

—NH$_2$, hydrazone

—NHC$_6$H$_5$, phenylhydrazone

—NH—⬡—NO$_2$, 2,4-dinitrophenylhydrazone
  |
  NO$_2$

—NH$\overset{\overset{O}{\|}}{C}NH_2$, semicarbazone

The primary use of these reactions is as a method of purification and identification through the formation of a derivative; for example:

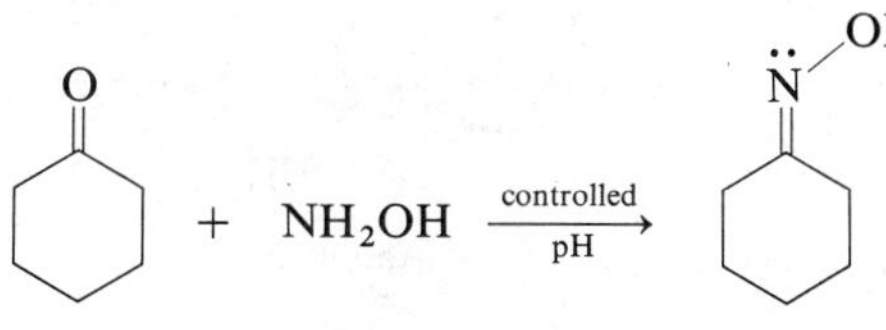

Cyclohexanone   +   NH$_2$OH $\xrightarrow{\text{controlled pH}}$ Cyclohexanone oxime

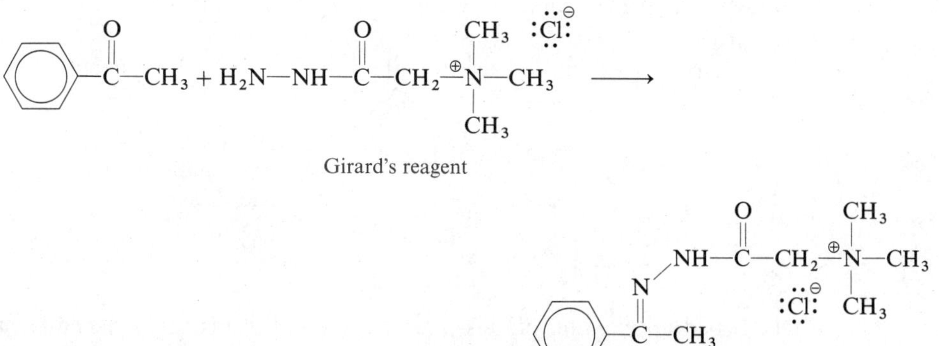

Girard's reagent

Hydrazone of acetophenone

Question 26.26

Another reagent often used to prepare hydrazones is Girard's reagent P. What property of these two reagents makes them so useful in the purification and identification of carbonyl compounds?

$$\text{:Cl:}^{\ominus} \quad \overset{O}{\overset{\|}{CH_2-C-NH-NH_2}}$$

Girard's reagent P

Question 26.27

The reaction of 2-butanone with hydroxylamine produces two isomers. The reduction of either of these isomers with $LiAlH_4$/ether gives *sec*-butylamine. Give the structures of the two isomers.

$$\underset{CH_3 \quad CH_2CH_3}{\overset{O}{\overset{\|}{C}}} \xrightarrow[H^\oplus]{NH_2OH} \quad \begin{array}{c} \text{two isomers of molecular} \\ \text{formula } C_4H_9NO \end{array}$$

26.9   Reactions of Amines Not Covered in Past Chapters

A. Nitrosation

The reaction of **primary amines** with nitrous acid ($NaNO_2/H_3O^\oplus$) results in the formation of a diazonium salt:

$$(Ar)\,R-\overset{..}{N}\overset{H}{\underset{H}{\diagdown}} \xrightarrow[HX]{NaNO_2} (Ar)\,R-\overset{\oplus}{N}\equiv\overset{..}{N}X^{\ominus}$$

1° Amine   Diazonium salt

The reaction must be carried out at low temperatures because of the relative instability of these salts. The salts themselves are seldom isolated and are often explosive when dry. The reactions are generally carried out in baths of dry ice and acetone ($-78°$) or salt and water ($-15°$).

The reaction is of limited use for aliphatic amines, where decomposition is often the result:

$$R-\overset{\oplus}{N}\equiv\overset{..}{N} \xrightarrow[\substack{(S_N1 \text{ or } S_N2) \\ \text{pathway}}]{-N_2} R^\oplus \longrightarrow \begin{array}{c} \text{decomposition and} \\ \text{rearrangement} \\ \text{products} \end{array}$$

Aliphatic   Carbocation
diazonium   (if S_N1)
cation

One broad application of this reaction to synthesis is the **Tiffeneu-Demyanov reaction,** which involves the diazotization of an aminomethyl group attached to the carbinol carbon of a ring. The result is a cylic ketone that is one carbon larger than the starting cyclic system:

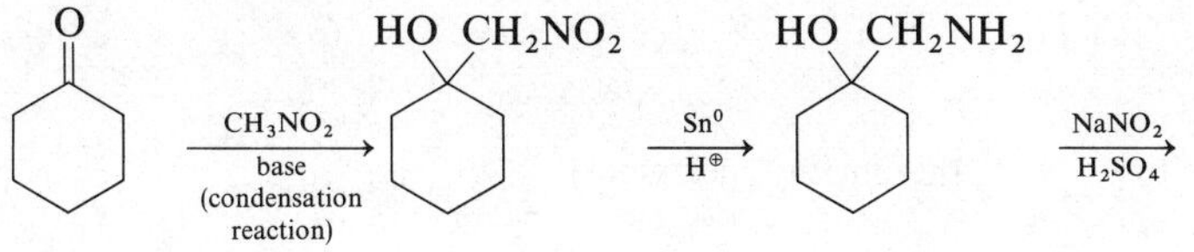

Cyclohexanone

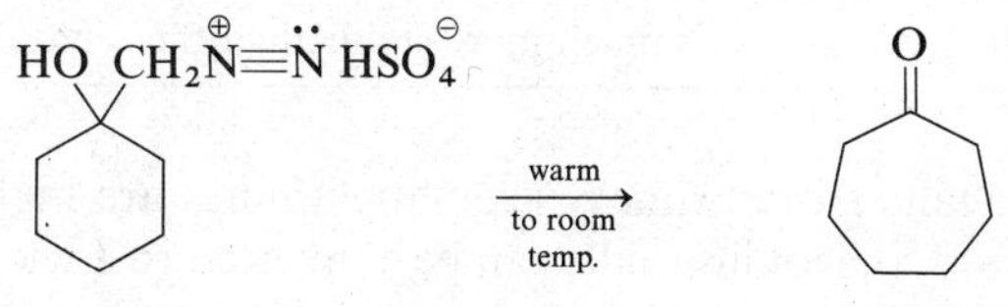

warm
to room
temp.

Cycloheptanone
*Product contains one more carbon
in ring than starting material*

In these ring expansions a carbocation is generated that is in close proximity to the migrating carbon:

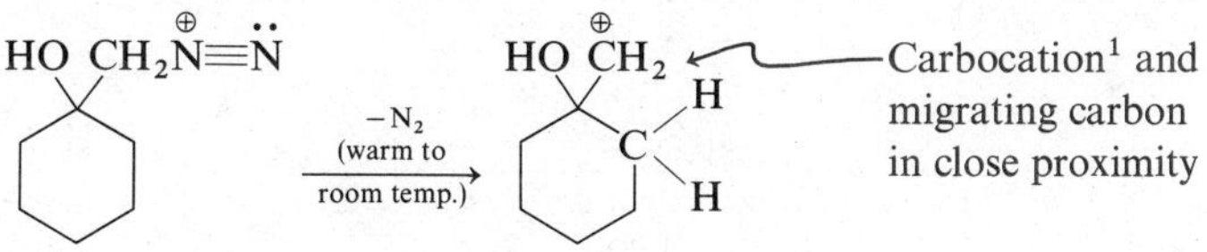

−N₂
(warm to
room temp.)

Carbocation[1] and
migrating carbon
in close proximity

Migration also results in a resonance-stabilized carbocation:

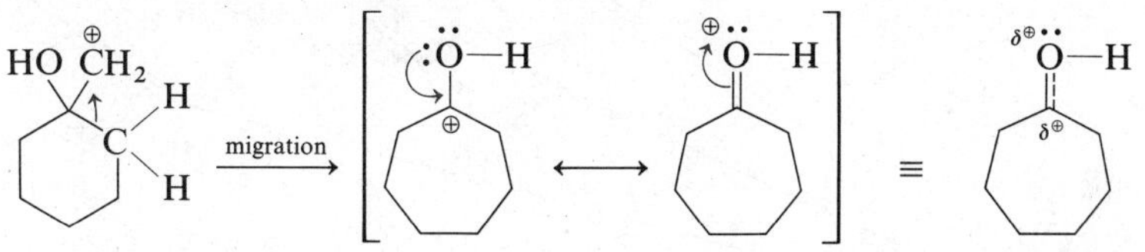

migration

Resonance-stabilized
carbocation

The carbocation intermediate then loses a proton to give the desired product:

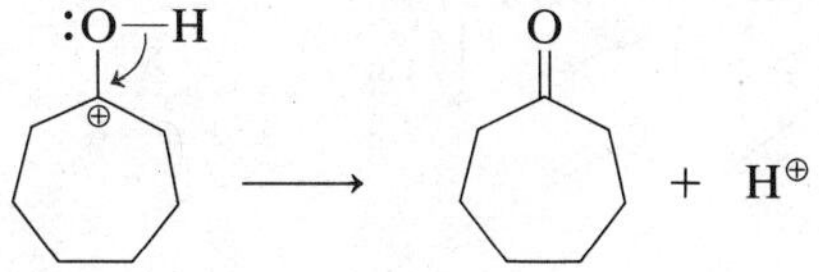

+ $H^{\oplus}$

Cycloheptanone
Product

Question 26.28

Write the complete mechanism for each step in the preceding reaction in which cyclo-hexanone is converted to the nitro intermediate.

Question 26.29

Complete each of the following syntheses.

(*a*) cyclopentene to cyclohexanol   (*b*) cyclopentene to cycloheptane

---

[1] See the Study Hint, "Primary Carbocations," Sec. 7.15.

(c) cyclohexene to 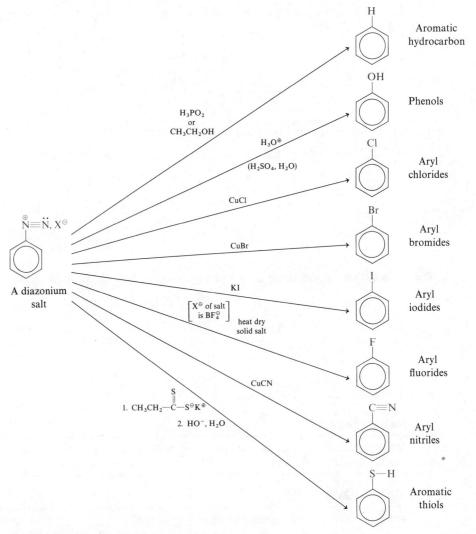 (hexamethyleneamine)

Question 26.30

The benzenediazonium ion is much more stable than the *n*-hexyldiazonium ion. Explain.

Diazonium salts from primary aromatic amines are highly versatile synthetic tools. As we will see, diazonium salts can be converted to a wide spectrum of organic compounds through a series of simple and quick reactions. These are summarized here.

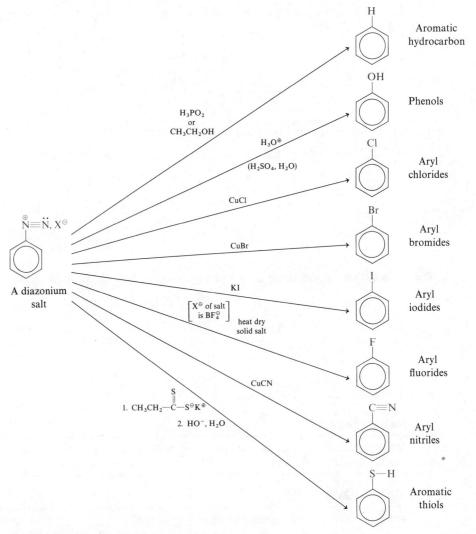

FIGURE 26.3   Reactions of diazonium salts.

The diazotization reaction is carried out in aqueous acid at low temperature, usually around 0°. Once the amine is converted to the diazonium salt, the second step in the reaction sequence, for example the addition of CuCl, is carried out in situ. The reaction mixture is then allowed to warm to room temperature. A normal isolation and purification sequence yields the desired product.

Reaction conditions vary slightly from experiment to experiment, but the preceding experimental conditions are fairly representative of the type of conditions used.

When cuprous chloride, cuprous bromide, or cuprous cyanide is used to prepare the corresponding **aryl halide** or **aryl nitrile,** the reaction is called the **Sandmeyer reaction.** Aryl iodide is prepared through the use of potassium iodide. The fluorides, however, are synthesized somewhat differently. When the diazonium salt is prepared, the acid used is fluoroboric acid (HBF_4) rather than the more common sulfuric or hydrochloric acid. The dry fluoroborate salt on heating gives the aryl fluoride directly; this is the **Schiemann reaction.** The possible danger of handling dry diazonium salt requires extreme caution in using this method.

m-Dinitrobenzene	*m*-Nitroaniline	*m*-Nitrobenzenediazonium chloride

m-Nitrobenzonitrile	*m*-Nitrobenzoic acid

p-Toluidine	*p*-Tolyldiazonium fluoroborate	*p*-Fluorotoluene

A diazonium salt can be converted to an **aromatic hydrocarbon,** that is, replacement of $-N_2^{\oplus}$ by $-H$, on treatment with ethanol or hypophosphorus acid (H_3PO_2) followed by heating:

m-Nitrobenzenediazonium hydrogen sulfate	Nitrobenzene

Acidification of the reaction mixture of the diazonium salt with aqueous sulfuric acid followed by heating gives the corresponding **phenol;** —OH replaces —N$_2^{\oplus}$:

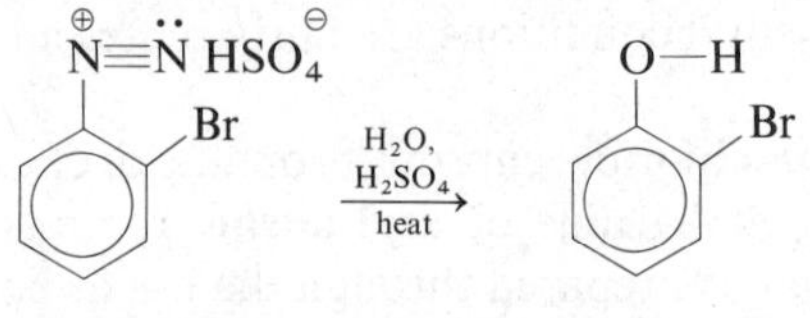

o-Bromobenzenediazonium
hydrogen sulfate

o-Bromophenol

Thiols are prepared from diazonium salts in a somewhat different manner. The diazonium salt reacts with potassium ethyl xanthate (see Sec. 26.8B,c), which gives an aryl xanthate intermediate. Hydrolysis of this xanthate (H$\ddot{\text{O}}$:$^{\ominus}$, H$_2$O), which is really a diester (half thioester and half oxygen ester), results in the thiol product:

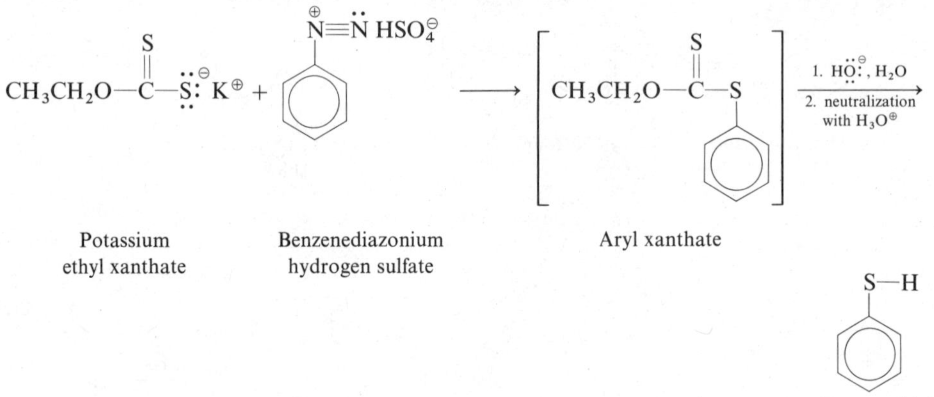

Potassium
ethyl xanthate

Benzenediazonium
hydrogen sulfate

Aryl xanthate

Benzenethiol

The reaction involves nucleophilic aromatic substitution by the xanthate on the aromatic ring:

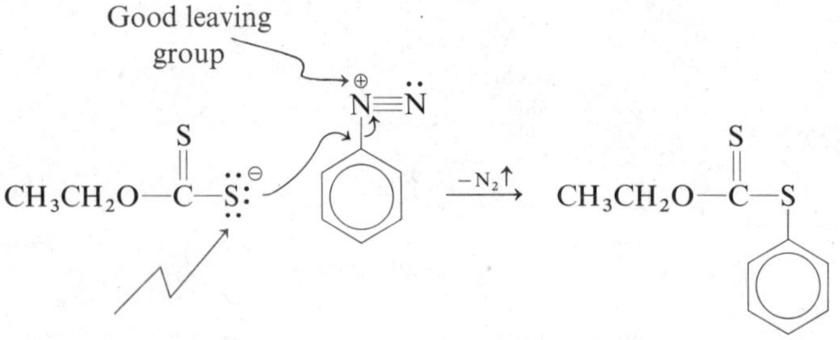

Good leaving
group

Nucleophile

Aryl xanthate intermediate

Question 26.31

Xanthates are prepared by the reaction of alcohols with carbon disulfide in base (KOH). Write the complete mechanism of this reaction.

Question 26.32

The aryl xanthate intermediate in the preceding reaction sequence is converted to the thiol by basic hydrolysis. Describe this reaction mechanistically.

Question 26.33

Complete each of the following syntheses.

(*a*) benzene to 1,3,5-tribromobenzene   (*b*) toluene to *p*-iodobenzoic acid
(*c*) nitrobenzene to *m*-bromophenol   (*d*) nitrobenzene to *N*,*N*-dimethylbenzamide

There are alternate routes to many of the compounds prepared through diazotization. Bromination and chlorination may be carried out directly with iron (see Sec. 14.4) and iodination and nitrile formation can be accomplished through the organothallium reaction (see Secs. 14.23 and 23.15), to name a few. Nevertheless the diazotization reaction is one of the easiest and most convenient routes to many of these compounds. One's arsenal of synthetic preparations can never be too extensive.

Another reaction of diazonium compounds is the *coupling reaction*. This is particularly important in the synthesis of dyes, discussed in Sec. 26.11A,b.

Before proceeding to the secondary amines, let's look at the mechanism of diazotization.

Step 1

$$2\,HNO_2 \;\rightleftharpoons\; H_2O + N_2O_3 \equiv \ddot{\underset{..}{O}} = \ddot{N} - \ddot{\underset{..}{O}} - \ddot{N} = \ddot{\underset{..}{O}}$$

Nitrous acid anhydride

Step 2

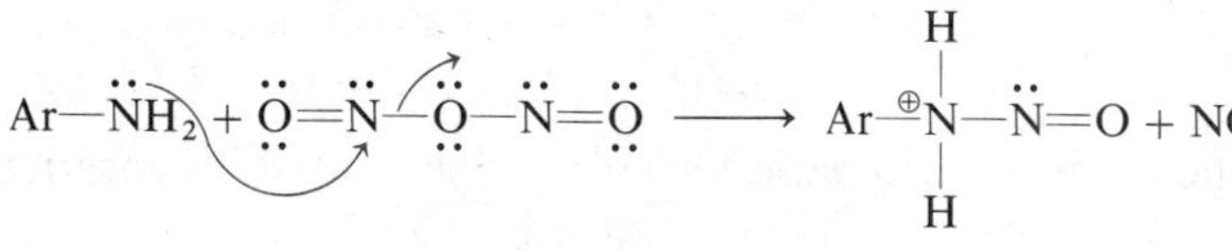

Protonated
N-nitrosoamine

Steps 3 and 4

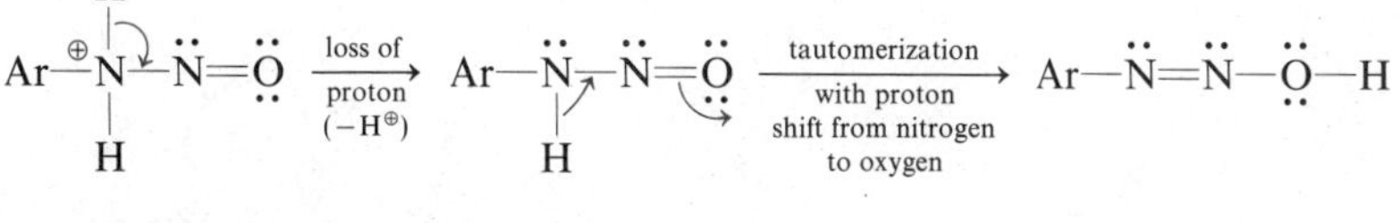

Steps 5 and 6

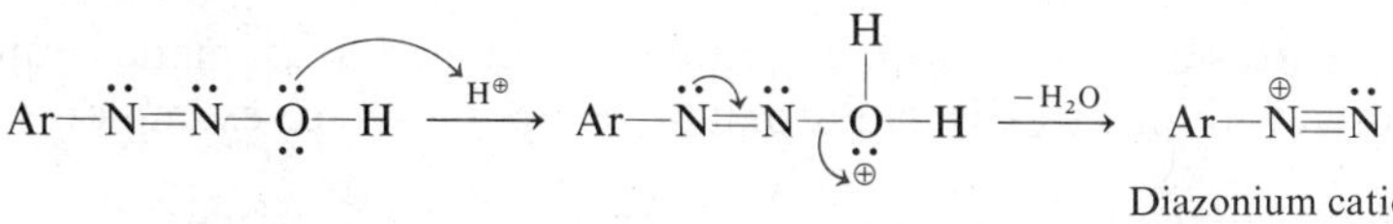

Diazonium cation

Step 1. The nitrous acid is in equilibrium with a small amount of its anhydride (N_2O_3), which is the actual attacking species.

Steps 2 and 3. The aromatic primary amine reacts with the anhydride to give a protonated form of the *N*-nitroso amine. Loss of a proton in step 3 gives the free *N*-nitroso amine of the starting amine.

Step 4. The amine tautomerizes in a reversible reaction analogous to a keto-enol equilibrium.

Steps 5 and 6. The enol tautomer of the amine *N*-nitroso amine protonates followed by loss of water to give the diazonium salt.

Step 2 may appear anomalous. Can there really be attack by the amine in the acid solution? Isn't the amine completely in the form of its conjugate acid, the anilinium salt? The answer to both questions is yes. Although essentially all the amine is protonated in the acid, the aromatic amines are weak enough bases that there is also a small yet finite concentration of the free amine. It is this free base that acts as the nucleophile in step 2.

Secondary amines react with nitrous acid to give *N*-nitroso amines:

2° Amine   N-Nitroso amine

These compounds have interesting environmental implications. They have been shown to be carcinogens (cancer-causing chemicals) and are discussed in Sec. 26.11C.

Tertiary amines do not react under nitrosation conditions. Any reaction that does occur usually leads to a variety of decomposition products.

3° Amine

no reaction or decomposition products

If, however, an aromatic tertiary amine is used, nitrosation of the ring often results:

o-Nitroso-*N*,*N*-dimethylaniline   *p*-Nitroso-*N*,*N*-dimethylaniline

Nitrosation of an aromatic ring is not limited to aromatic amines but may also be used with other activated aromatic compounds; for example:

o-Nitrosophenol   *p*-Nitrosophenol

Question 26.34

In the IR spectrum of *p*-nitrosophenol, an absorption for both carbonyl ($\diagdown C = O$) and hydroxyl (—O—H) is present. All other expected peaks are also found. How can you explain this observation?

The mechanism by which nitrosation is accomplished is normal electrophilic aromatic substitution. The electrophile in these cases is the *nitrosonium* ion, $NO^\oplus$. Actually the nitrosonium ion may be the reactive species in many diazotization reactions of primary amines.

$$N_2O_3 \equiv \ddot{O} = \ddot{N} - \ddot{O} - \ddot{N} = \ddot{O} \longrightarrow NO_2^\ominus + NO^\oplus$$

Nitrous acid   Nitrosonium
anhydride   ion

Other nitrosating reagents may also be used, for example nitrosonium hexafluorophosphate, $NO^\oplus PF_6^\ominus$.

B. Oxidation

Amines can exist in many different stable oxidation states; six of these are:

$$-\overset{|}{\underset{|}{N}}{}^\oplus \qquad -\overset{|}{\underset{\cdot\cdot}{N}}- \qquad -\overset{|}{\underset{\cdot\cdot}{N}}-O-H \qquad -\overset{|}{\underset{\underset{\cdot\cdot}{:O:}}{N}}{}^\oplus_\ominus \qquad -\ddot{N} = \ddot{N}- \qquad -\overset{\oplus}{N} \equiv \ddot{N}$$

Ammonium Amine Hydroxylamine Amine   Azo Diazonium
ion   oxide compounds cation

We discussed all but amine oxides and azo compounds. The azo compounds are discussed with the dyes in Sec. 26.11A,b, and we discuss the amine oxides shortly.

Although primary, secondary, and tertiary amines can all be oxidized with common oxidizing agents (for example, H_2O_2, $ArCO_3H$), the primary and secondary amines give several products. The reaction has very limited synthetic use. With tertiary amines, however, the amine oxide results:

$$R - \overset{\overset{\textstyle R}{|}}{\underset{\underset{\textstyle R}{|}}{N}}: \quad \xrightarrow[H_2O]{H_2O_2} \quad R - \overset{\overset{\textstyle R}{|}}{\underset{\underset{\textstyle R}{|}}{N}}{}^\oplus - \ddot{\underset{\cdot\cdot}{O}}{}^\ominus$$

3° Amine   Amine oxide

The amine oxides are of interest as intermediates in elucidating the structure of amines; examples are:

$$CH_3CH_2CH_2CH_2CH_2 - \overset{\overset{\textstyle CH_3}{|}}{\underset{\underset{\textstyle CH_3}{|}}{N}}{}^\oplus - \ddot{\underset{\cdot\cdot}{O}}{}^\ominus \qquad\qquad CH_3 - \overset{\overset{\textstyle CH_2CH_3}{|}}{\underset{\underset{\underset{\cdot\cdot}{:O:}}{\ominus}}{N}}{}^\oplus - CH_3$$

N,N-Dimethylpentylamine   Dimethylethylamine
N-oxide   *N*-oxide

Question 26.35

What stereochemistry would the amine oxides be predicted to exhibit?

C. Eliminations: Hofmann and Cope

Two reactions of amines that are highly useful in elucidating the structure of these compounds involve eliminations. These are the Hofmann elimination and the Cope elimination.

We first discuss **Hofmann elimination.** If one heats a quaternary ammonium hydroxide, elimination results. The overall reaction, given in the following example, produces an alkene and a tertiary amine:

$$CH_3-\overset{\overset{\displaystyle CH_3}{|}}{\underset{\underset{\displaystyle CH_3\ \ H\overset{..}{\underset{..}{O}}{:}^{\ominus}}{|}}{N^{\oplus}}}-CH_2-CH_3 \xrightarrow{\text{heat}} CH_3-\overset{\overset{\displaystyle CH_3}{|}}{\underset{\underset{\displaystyle CH_3}{|}}{N}}{:}\ + CH_2{=}CH_2 + H_2O$$

The mechanism involved is the familiar E2 elimination; hydroxide is the base and the tertiary amine is the leaving group:

E2 elimination of
quaternary ammonium hydroxide

The reaction would be of limited use except that any amine can be converted to a quaternary ammonium ion quite readily (see Sec. 26.7). The most straightforward method for this is the reaction of amine with an excess of an alkylating agent, which for this sequence is usually methyl iodide.

The quaternary ammonium iodide is converted to the quaternary ammonium hydroxide by reaction with silver oxide (Ag_2O). The insoluble silver iodide produced can be filtered off:

$$R-\overset{\overset{\displaystyle R}{|}}{\underset{\underset{\displaystyle R\ \ \ {:}\overset{..}{\underset{..}{I}}{:}^{\ominus}}{|}}{N^{\oplus}}}-R \xrightarrow[\text{H}_2\text{O}]{\text{Ag}_2\text{O}} R-\overset{\overset{\displaystyle R}{|}}{\underset{\underset{\displaystyle R\ \ \ H\overset{..}{\underset{..}{O}}{:}^{\ominus}}{|}}{N^{\oplus}}}-R + AgI\downarrow$$

Once the quaternary ammonium hydroxide is in hand, all that is required is heat to effect the elimination.

In certain compounds it is possible to continue the process—alkylation with methyl iodide, anion exchange with silver oxide, and elimination with heat—to give a second or perhaps a third elimination.

$$CH_3-CH_2-\overset{\overset{\displaystyle CH_2-CH_3}{|}}{\underset{\underset{\displaystyle CH_3-CH_2\ \ H\overset{..}{\underset{..}{O}}{:}^{\ominus}}{|}}{N^{\oplus}}}-CH_3 \xrightarrow{\text{heat}} CH_3-CH_2-\overset{\overset{\displaystyle CH_2-CH_3}{|}}{\underset{..}{N}}-CH_3$$
$$+ CH_2{=}CH_2 + H_2O \quad\xrightarrow[\substack{2.\ \text{Ag}_2\text{O} \\ 3.\ \text{heat}}]{\substack{1.\ \text{CH}_3\text{I,} \\ \text{excess}}}$$

$$CH_3-CH_2-\overset{\overset{\displaystyle CH_3}{|}}{\underset{..}{N}}-CH_3 \xrightarrow[\substack{2.\ \text{Ag}_2\text{O} \\ 3.\ \text{heat}}]{\substack{1.\ \text{CH}_3\text{I, excess}}} CH_3-\overset{\overset{\displaystyle CH_3}{|}}{\underset{..}{N}}-CH_3$$
$$+ CH_2{=}CH_2 + H_2O \qquad\qquad + CH_2{=}CH_2 + H_2O$$

If one can isolate and determine the products obtained from each elimination, one can work backward to determine the structure of the starting amine.

Consider the following natural product, for example. An alkaloid obtained from the roots of a certain variety of yam gives as one of its degradation products compound (1), $C_8H_{13}NO$. Hofmann exhaustive methylation and elimination give compounds (2) and (3):

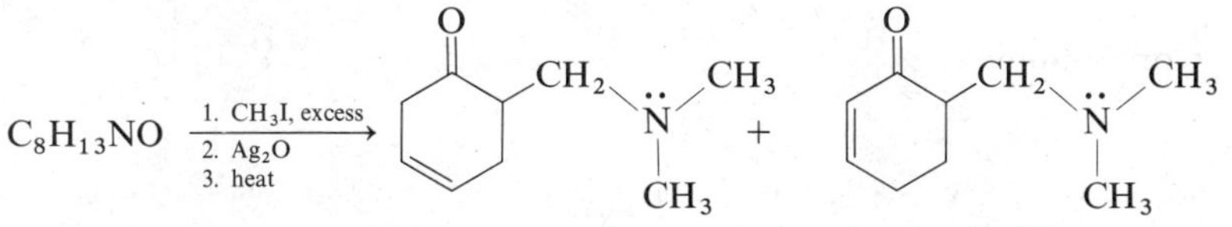

(1)   (2)   (3)

It is evident from these results that compound (1) must be the following:

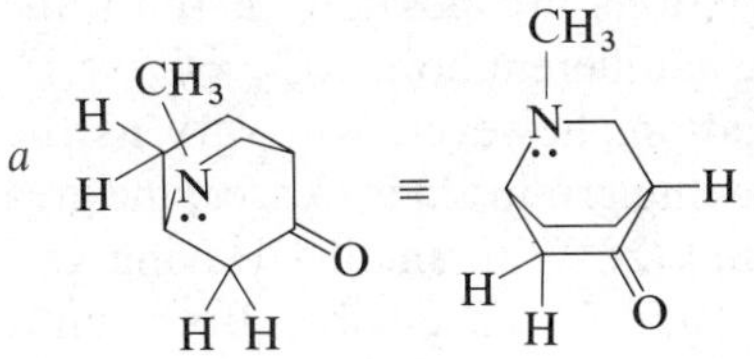

(1)

Elimination of the hydrogens at position b gives (3), whereas elimination of those at a gives (2).

Question 26.36

In the preceding reaction predict the major product. Briefly explain.

Experimentally, this E2 elimination presents an interesting and apparent dichotomy. Consider the following examples:

Both reactions go through E2 elimination, yet in one the major product is the more highly substituted, more stable alkene (the *Saytzeff product*, see Sec. 7.12); whereas in the other we get the opposite product distribution (the *Hofmann product*). It appears that the predominant force governing the orientation in these reactions is steric.

Consider what we know about E2 eliminations experimentally:

1. The reaction occurs in one step; it is a concerted reaction.
2. The two groups that are being eliminated are usually *anti* to each other.
3. With alkyl halides, elimination (KOH/alcohol) gives the expected, more stable, thermodynamic product—the more highly substituted alkene.
4. With larger leaving groups, for example $\overset{\cdot\cdot}{N}R_3$, one gets the less highly substituted alkene as the major product.

If we look at the conformation of the substrate (the substituted 2-pentyl group) just before it undergoes elimination, we can explain these observations.

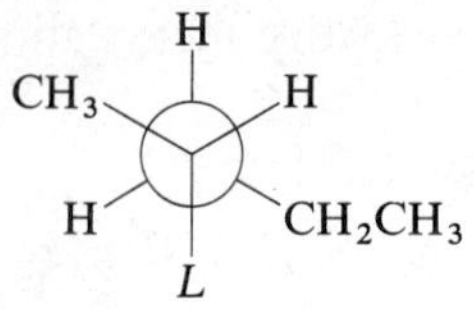

(1)
Most stable conformation for
anti-elimination along C_2—C_3
bond

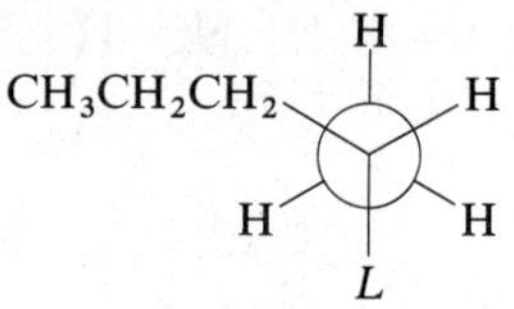

(2)
Most stable conformation for
anti-elimination along C_1—C_2
bond

Both conformations are possible in the same molecule at the same instant since we are looking at different bonds (C_2—C_3 vs. C_1—C_2). The relative amount of a particular conformation, however, will vary as the size of the substituent varies. As the size of the leaving group (L) increases, the greater is the difference in stabilities of conformers (1) and (2). With smaller leaving groups like halogen, the difference in energy between (1) and (2) is negligible. Both conformations are present in solution and most elimination goes through (1) to give the thermodynamic product. As L gets larger (for example, $R_3\overset{..}{N}$), conformation (1) becomes much less stable so that there is a great deal more of (2) in solution, with the majority of the elimination occurring to give the Hofmann product.

To further substantiate this argument, all the following compounds that have large leaving groups go in the direction one would predict for Hofmann elimination.

Elimination Products

G—	1-Pentene	2-Pentene
Br—	31%	69% (Saytzeff)
$(CH_3)_3\overset{\oplus}{N}$—	98%	2% (Hofmann)
$(CH_3)_2\overset{\oplus}{S}$—	87%	13% (Hofmann)
TsO—	48%	52% (Saytzeff/Hofmann)

Question 26.37

Predict the major product in each of the following reactions.

(*a*) CH_3—$\underset{\underset{H}{|}}{\overset{\overset{CH_3}{|}}{C}}$—$\underset{\underset{Br}{|}}{CH}$—$CH_3$ $\xrightarrow[\text{alcohol}]{\text{KOH}}$

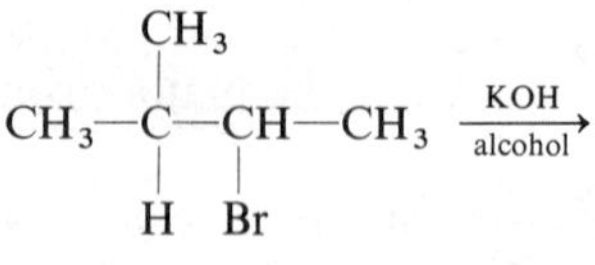

(*b*) CH_3—$\underset{\underset{H}{|}}{\overset{\overset{CH_3}{|}}{C}}$—$\underset{\underset{OH}{|}}{CH}$—$CH_3$ $\xrightarrow[\text{heat}]{H_2SO_4}$

(*c*) CH_3—$\underset{\underset{H}{|}}{\overset{\overset{CH_3}{|}}{C}}$—$\underset{\underset{NH_2}{|}}{CH}$—$CH_3$ $\xrightarrow[\substack{2.\ Ag_2O \\ 3.\ heat}]{1.\ CH_3I,\ excess}$

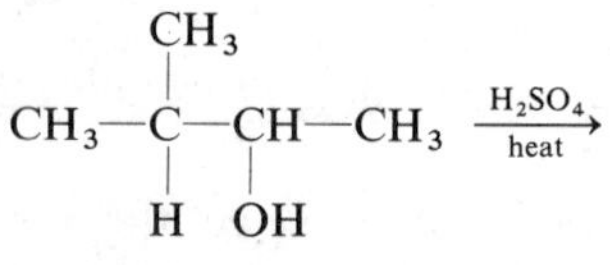

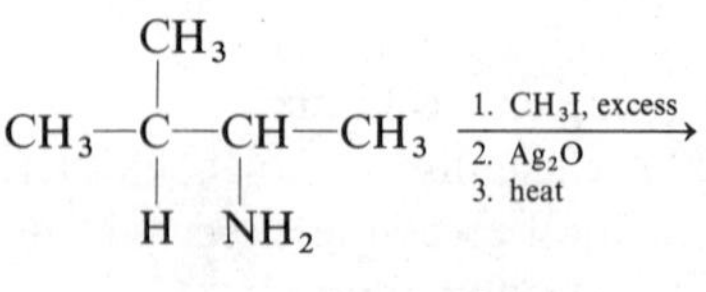

(*d*)

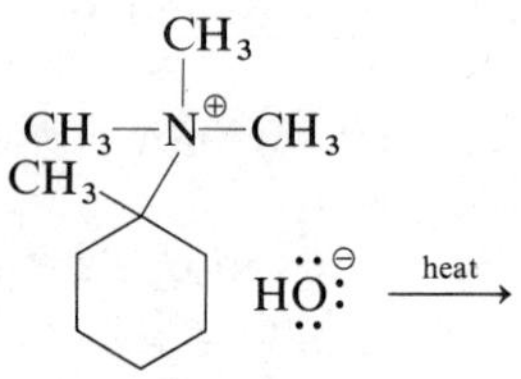

(*e*)

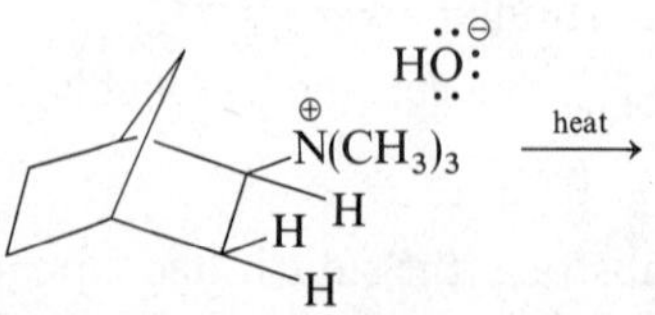

(*f*) *trans*-1-(*N,N*-Dimethylamino)-2-methylcyclohexane $\xrightarrow[\substack{2.\ Ag_2O \\ 3.\ heat}]{1.\ CH_3I,\ excess}$

A second elimination reaction that can be used in elucidating the structure of amines is the **Cope elimination** of amine oxides:

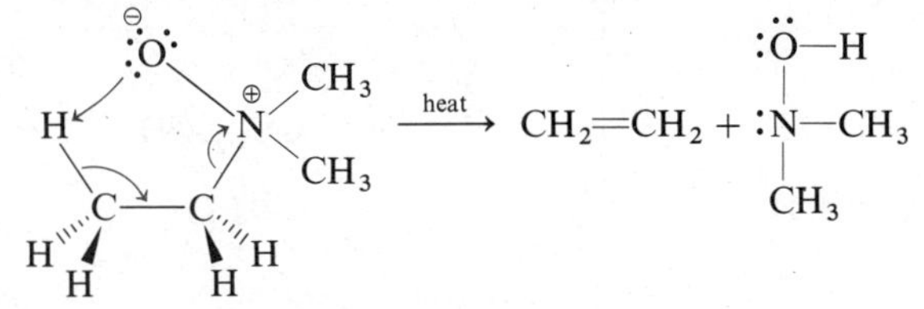

Amine oxide

The reaction probably proceeds through an intramolecular E2 mechanism but in this instance through *syn* elimination. Proof of the required *syn* orientation comes in part from the following experimental observation:

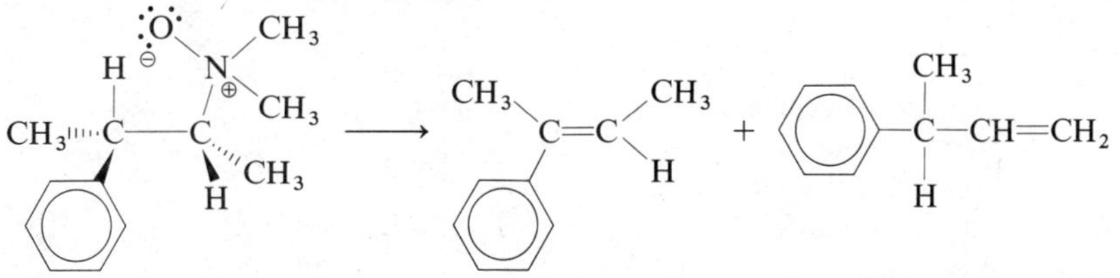

cis-2-Phenyl-2-butene
93%
No *trans*-2-phenyl-2-butene
is formed

7%

Cope elimination can also be used in synthetic preparations. For example, it provides an interesting route to 3,4-dimethylenecyclopentanol, which would be very difficult to synthesize otherwise.

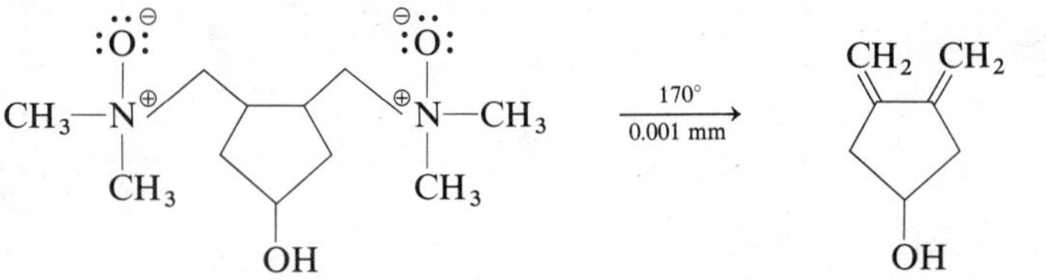

3,4-Dimethylenecyclopentanol
70%
Very Unstable

Following are examples of reactions that are analogous to the Cope elimination.

Sulfoxide pyrolysis:

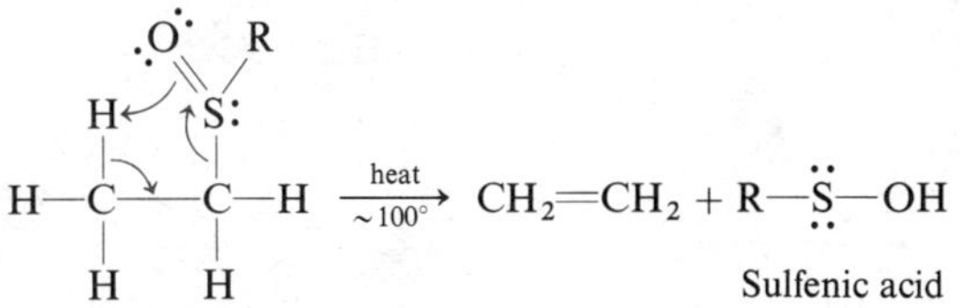

Sulfenic acid

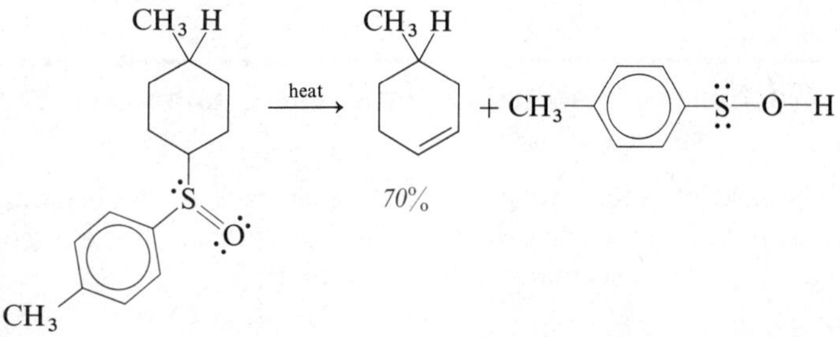

70%

Selenoxide pyrolysis:

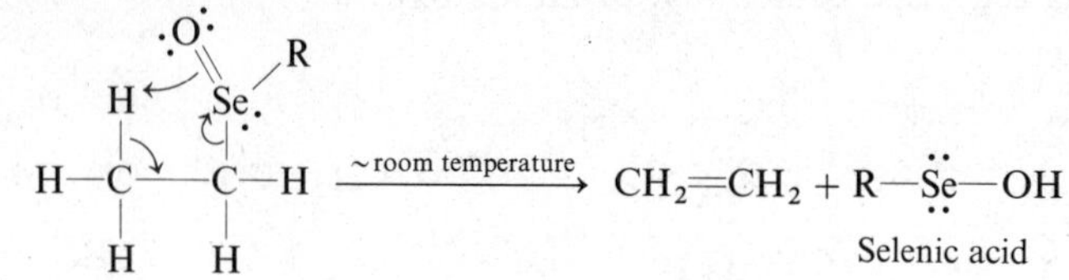

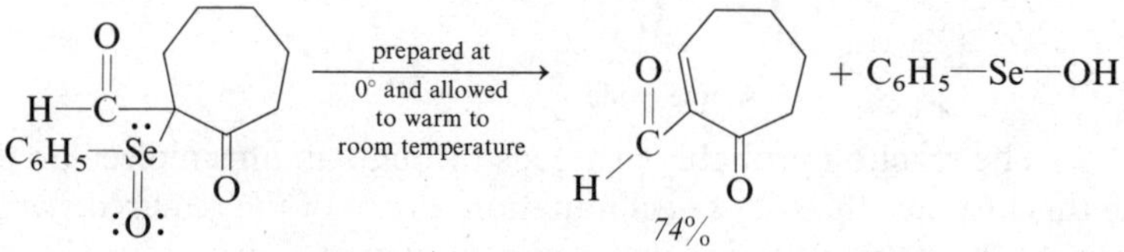

$$H-C-C-H \xrightarrow{\sim \text{room temperature}} CH_2{=}CH_2 + R-\overset{..}{Se}-OH$$

Selenic acid

prepared at
0° and allowed
to warm to
room temperature

74%

Ester pyrolysis:

Note: a six-membered
transition state as
opposed to five-membered
transition state in
prior examples

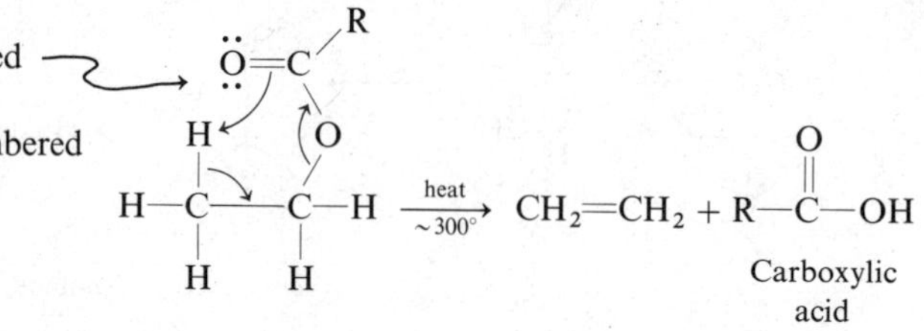

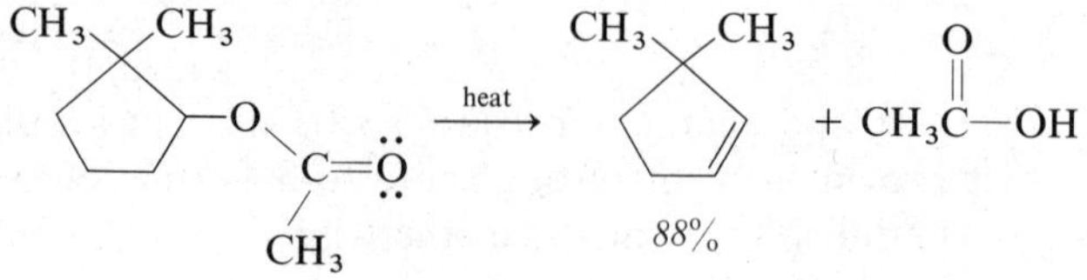

$$\xrightarrow[\sim 300°]{\text{heat}} CH_2{=}CH_2 + R-\overset{O}{\overset{\|}{C}}-OH$$

Carboxylic
acid

$$\xrightarrow{\text{heat}}$$

88%

Xanthate pyrolysis:

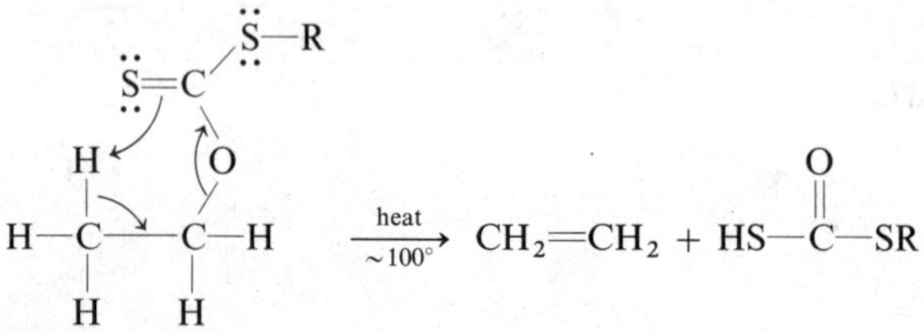

$$\xrightarrow[\sim 100°]{\text{heat}} CH_2{=}CH_2 + HS-\overset{O}{\overset{\|}{C}}-SR$$

$$\xrightarrow{\text{heat}}$$

75%

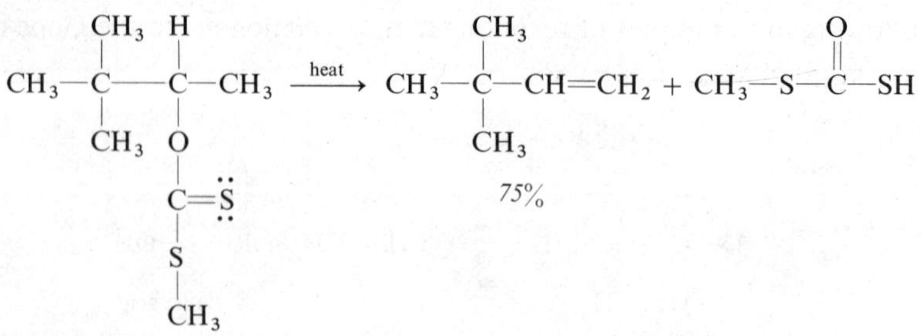

26.10   Reactions of Selected Heterocyclic Compounds

The field of heterocyclic chemistry goes back to the beginnings of the science known as organic chemistry. This is particularly true of the nitrogen heterocyclics, which are present in so many natural products.

We now attempt to scratch the surface of this vast subject. To illustrate the chemistry of the nitrogen heterocyclics, we look at the reactions of pyridine and pyrrole.

Pyridine and pyrrole are both aromatic compounds (see Sec. 13.7C) and undergo electrophilic aromatic substitution reactions. Whereas pyridine is quite unreactive under the normal conditions of electrophilic aromatic substitution, pyrrole is very reactive. The former behaves similarly to nitrobenzene and does not undergo Friedel-Crafts reactions or diazonium coupling reactions; the latter is as reactive as aniline and undergoes all these reactions quite easily. Several of these reactions are shown in Fig. 26.3. Note that the reagents are specific to each compound and differ from the conditions used with benzene and other aromatics that do not contain a heteroatom.

The major product is illustrated in each reaction. Pyridine substitutes preferentially at the 3 position, whereas pyrrole substitutes at the 2 position.

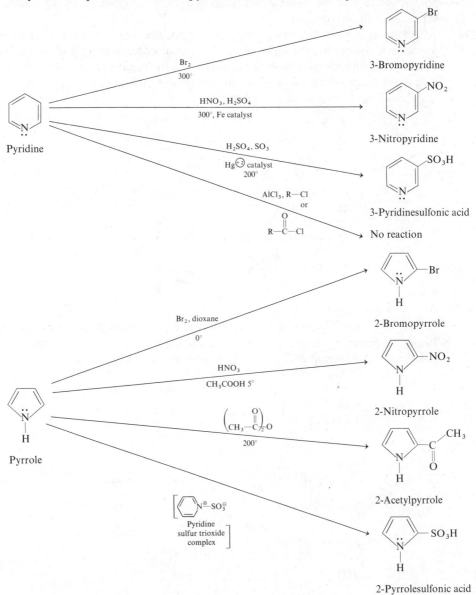

FIGURE 26.4   Reactions of Pyridine and Pyrrole

Question 26.38

(*a*) Draw all the resonance structures that would result from electrophilic attack at the 2, 3, and 4 positions of pyridine and explain why the 3 position is preferred.
(*b*) Do the same for the 2 and 3 positions of pyrrole.
(*c*) From these resonance structures, explain the reactivity of pyrrole versus pyridine.

In addition to the differences in the stabilities of the intermediates resulting from attack at the various positions of these two rings, the Lewis base nature of the heterocyclic nitrogen must also be taken into account in explaining reactivity differences. This base can easily react with any electrophile to give the cyclic ammonium ion, a cation that is not susceptible to electrophilic attack. We saw in Sec. 26.5 that pyridine is a stronger base than pyrrole by several orders of magnitude. Pyridine is complexed to a much larger extent than is pyrrole, as reflected in the different reactivities of these two compounds.

Pyrrole is a weak base and a weak nucleophile (see Sec. 26.5). Pyrrole is also a weak acid, however, with a pK_a of ~15 (the same order of magnitude as methanol). It can react with alkali base such as potassium hydroxide or sodium hydroxide to give the corresponding alkali metal salt. These salts are quite basic (pK_b of approximately −1) and are good nucleophiles.

Pyrrole as an acid, base, and nucleophile:

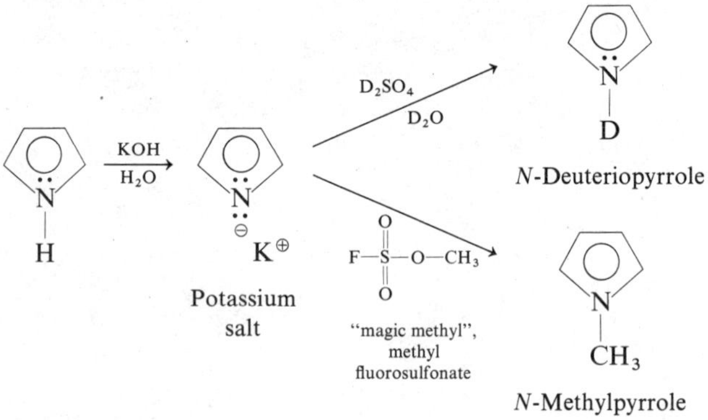
KOH
H2O
K⊕
Potassium salt
D2SO4
D2O
N-Deuteriopyrrole
F—S—O—CH3
"magic methyl", methyl fluorosulfonate
N-Methylpyrrole

Pyridine, in contrast, is a good base and a good nucleophile but is not acidic. It has no acidic hydrogen and the nonbonding electron pair on nitrogen is not involved in the aromatic character as it is in pyrrole (see Sec. 26.5).

Pyridine as a base:

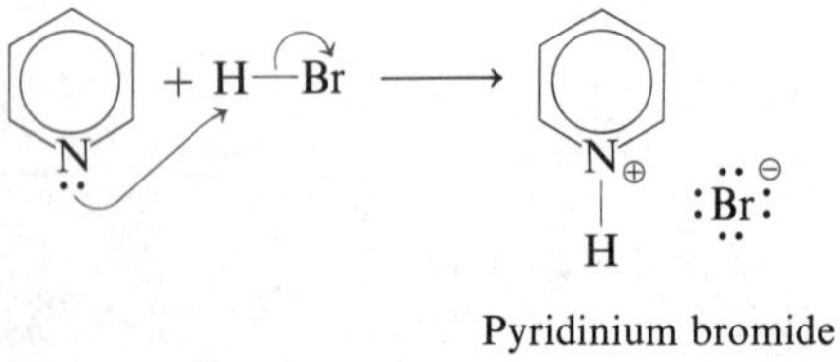
+ H—Br
:Br:
Pyridinium bromide

Pyridine as a nucleophile:

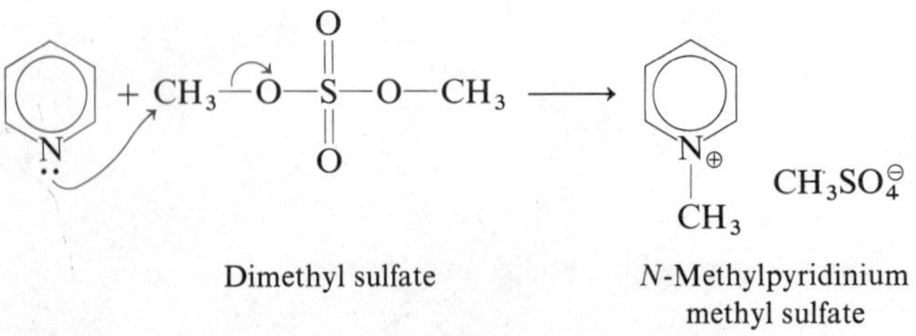
+ CH3—O—S—O—CH3
CH3SO4⊖
Dimethyl sulfate
N-Methylpyridinium methyl sulfate

Pyridine as a 3° amine:

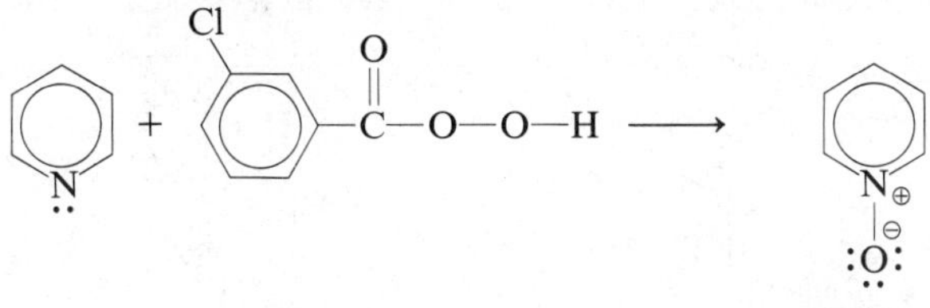

| m-Chloroperbenzoic acid | Pyridine N-oxide |

Pyridine, as a result of its electronic configuration, is also capable of undergoing nucleophilic substitution reactions. This is illustrated here. The nucleophilic attack occurs at positions 2 and 4, with 2 being the more common site.

Step 1

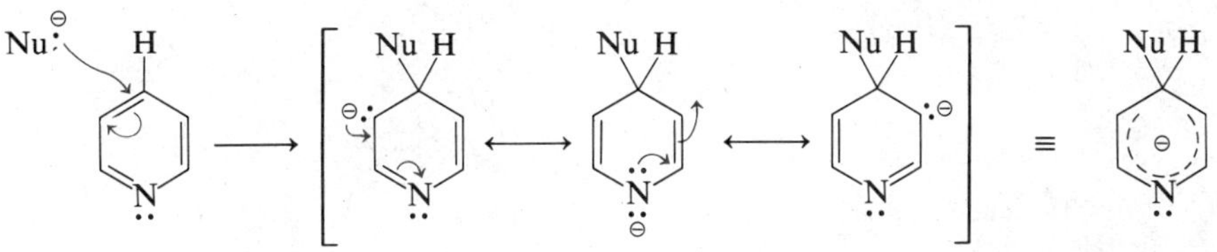

Resonance-stabilized intermediate

Step 2

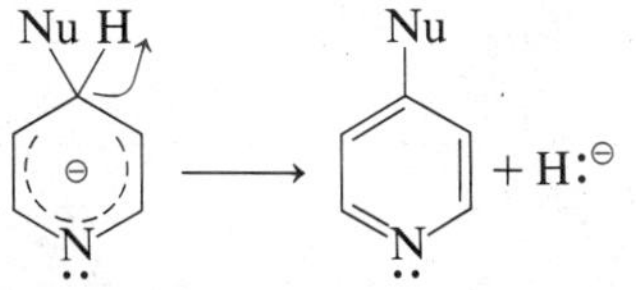

Examples are:

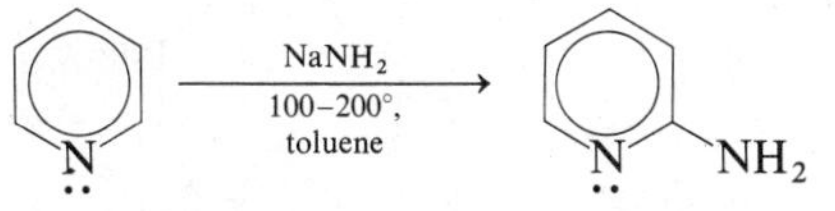

2-Aminopyridine

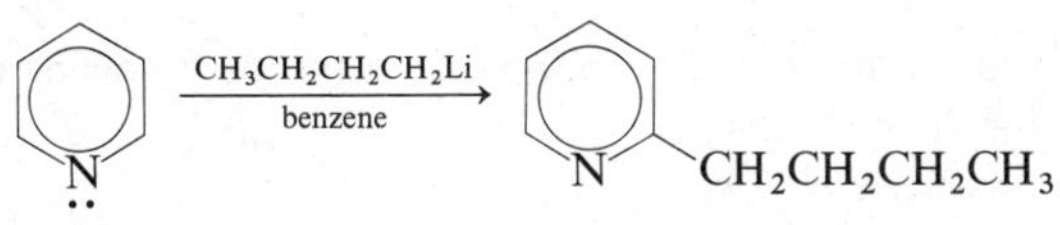

2-n-Butylpyridine

A similar reaction occurs with pyridinium ions, which results in a dihydropyridine skeleton:

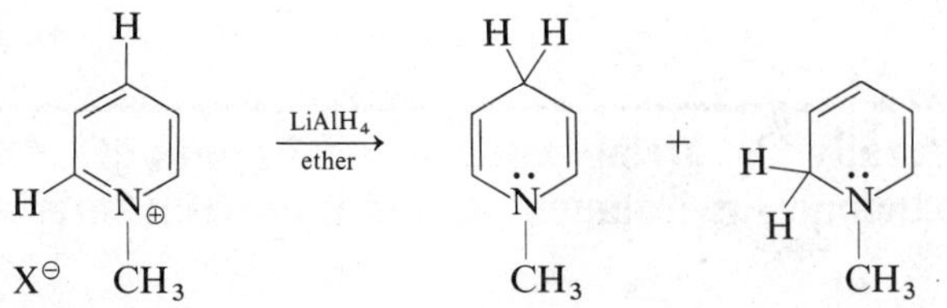

| N-Methylpyridinium salt | N-Methyl-1,4-dihydropyridine | N-Methyl-1,2-dihydropyridine |

This type of reaction also has biological significance. It is the basis of the NADH-NAD$^{\oplus}$ oxidation-reduction half-reaction (see Sec. 29.10):

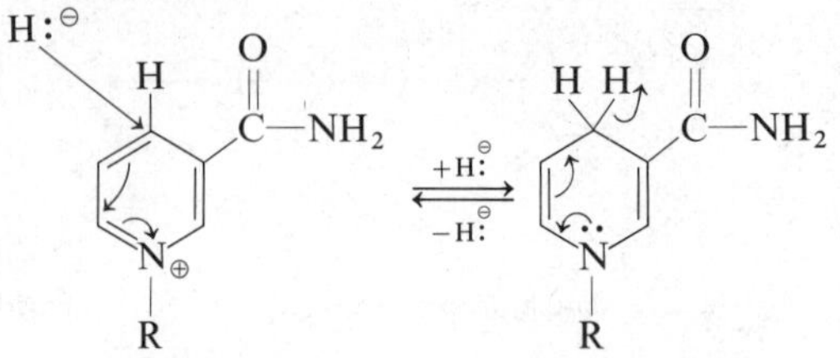

NAD$^{\oplus}$ NADH

Question 26.39

Based on the preceding reactions and your understanding of chemistry, determine the products in each of the following reactions.

(a) $\xrightarrow{\text{CH}_3\text{I}}$ (b) Thiolane $\xrightarrow[\text{H}_2\text{O}]{\text{H}_2\text{O}_2}$

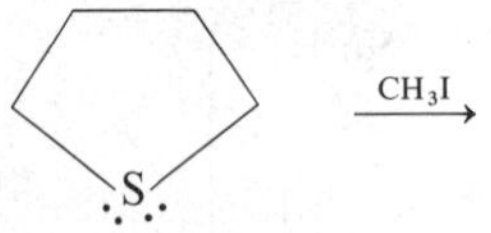

Thiolane
(tetrahydrothiophene)

(c) $\xrightarrow{\text{(CH}_3)_2\text{SO}_4}$ (d) HS$\overset{\frown}{\qquad}$ $\xrightarrow[\text{H}_2\text{O}]{\text{NaOH}}$

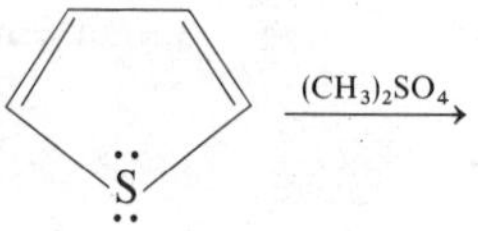

Thiophene   Butanethiol

Question 26.40

(a) Draw all the stereoisomers of each of the following; label enantiomers and diastereomers.

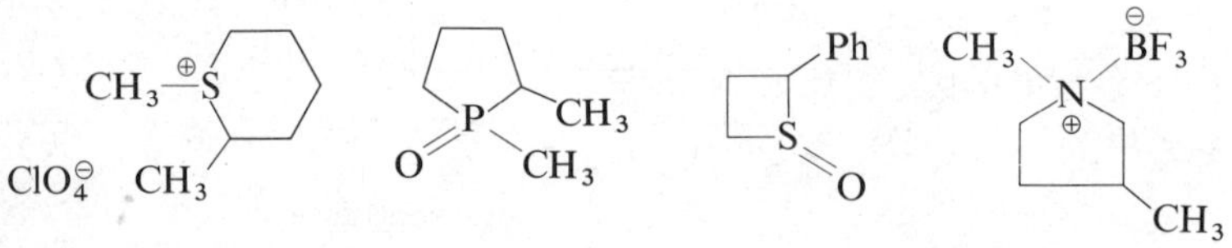

(b) How could you synthesize these compounds from the following starting materials? Ignore stereochemistry for the synthesis.

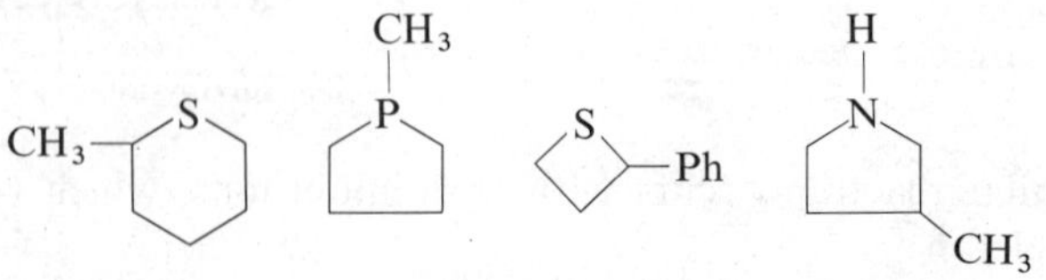

26.11   Naturally Occurring Amines and Amines of Medicinal, Environmental, and Industrial Interest

Amines play a role in every facet of our daily lives. We now summarize some common members of this family.

A. Industrial Amines

Two important classes of compounds that originate from amines come primarily from the industrial sector; these are the polyamides (nylons) and dyes.

a. Nylons

Nylons were discussed in Sec. 23.10. The structure of Nylon 66 is as follows:

$$\sim\sim\sim\overset{\displaystyle O}{\overset{\|}{C}}-(CH_2)_4-\overset{\displaystyle O}{\overset{\|}{C}}-[NH-(CH_2)_6-NH-\overset{\displaystyle O}{\overset{\|}{C}}-(CH_2)_4-\overset{\displaystyle O}{\overset{\|}{C}}-]_nNH-(CH_2)_6-NH\sim\sim\sim$$

Nylon 66 ($n = 50$–65)
Polyamide

b. Dyes

Dyes may seem to be an obscure group of compounds in this category, but actually they play a role in many aspects of our day-to-day lives. Dyes are used to color foodstuffs (*butter yellow*, Fig. 26.4), in the clothing and textile industries (*fast red A*), in paint pigments (*hansa yellow G*), as indicators (*methyl orange*), in photography (*brilliant yellow*), and in medicinal compounds (*trypan blue*). (See the Reading References for this chapter for additional readings on this subject.)

Many dyes contain the *azo group*, $-\overset{..}{N}=\overset{..}{N}-$ which is prepared most directly from diazonium salts. This is done by reacting the diazonium salt with another aromatic compound, a reaction called a *coupling reaction*. The diazonium salt is a cation and hence an electrophile. It is a weak electrophile and gives good yields only

Diazonium cation
Electrophile

with rings that are activated. The aromatic ring to be substituted is therefore usually a phenol or aniline analogue. The reaction goes through a normal electrophilic aromatic substitution, with the predominant substitution occurring at the *para* position (*ortho* if the *para* position is blocked by a substituent).

Activated ring,
G = —OR or —NR$_2$ (R = R, Ar, H)
usually

Azo compound

The rate of the coupling is dependent on the pH of the solution. There are two equilibria of interest, one involving the diazonium salt and its hydrolysis products and the other involving the aromatic compound undergoing substitution and its

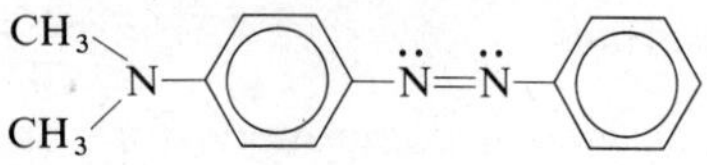

Butter yellow
(food coloring)

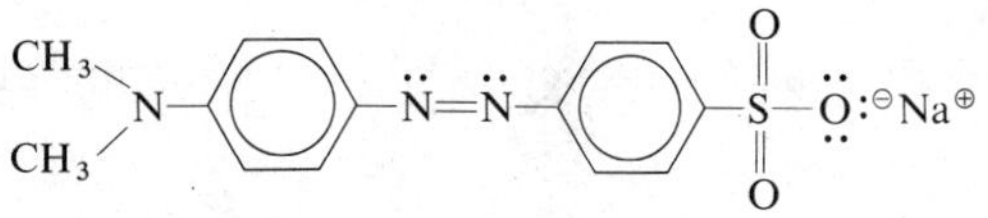

Methyl orange
(indicator)

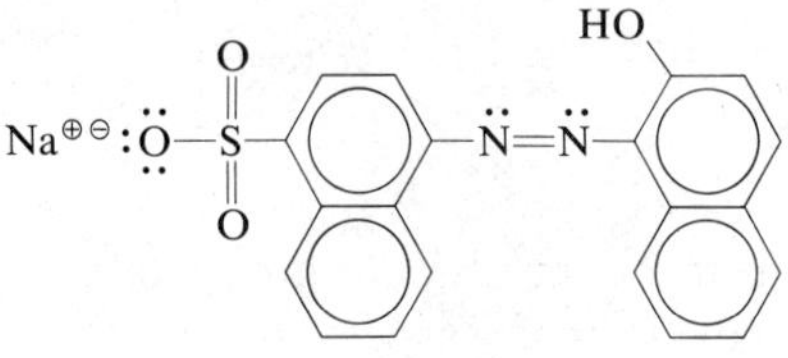

Fast red A
(textile dye)

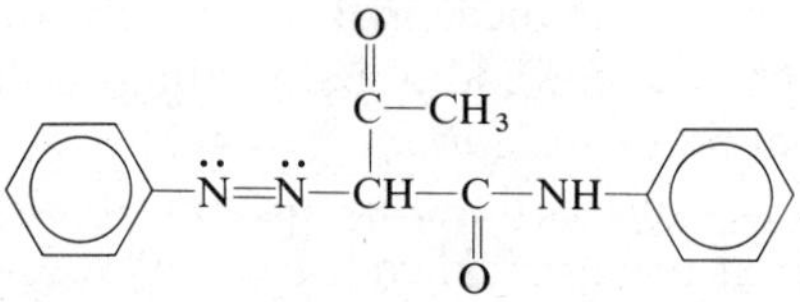

Hansa yellow
(paint pigment)

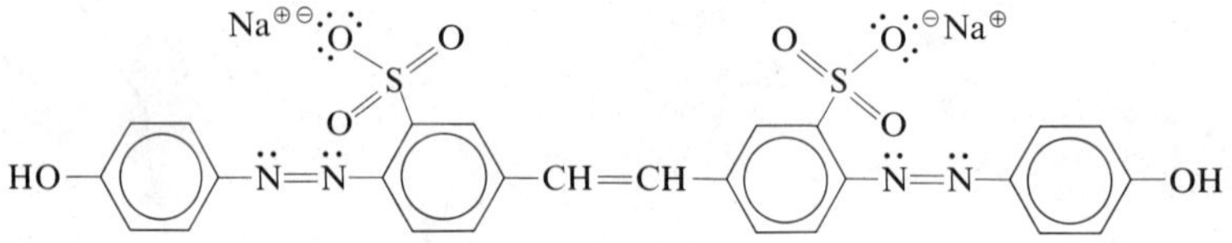

Brilliant yellow
(printing color)

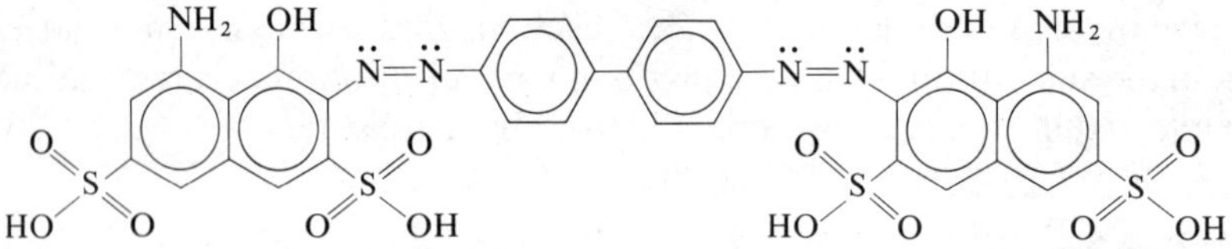

Trypan blue
(histological stain)

FIGURE 26.5   Six examples of azo dyes.

conjugate base or acid:

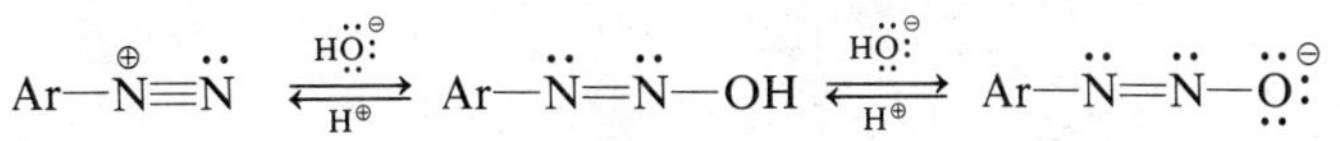

Diazonium salt
Electrophile

Not an electrophile

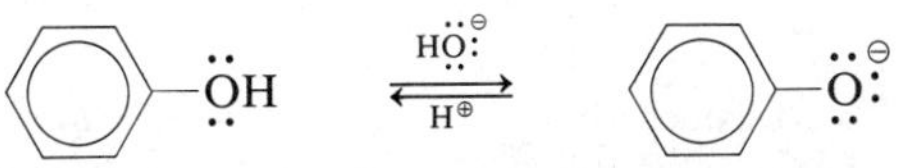

Not an electrophile

Phenol

Phenoxide ion
More reactive toward electrophilic aromatic substitution than phenol

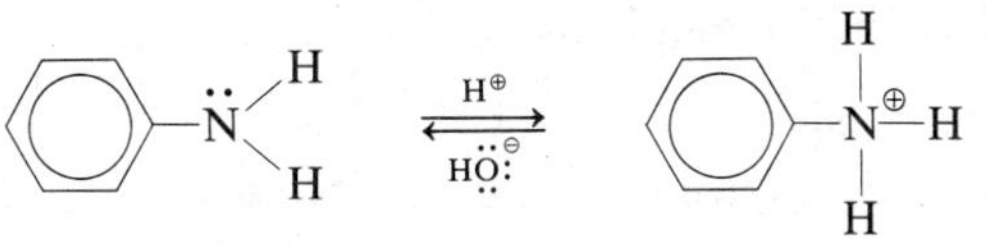

Aniline
More reactive toward electrophilic aromatic substitution than the anilinium ion

Anilinium ion

For phenols, then, a pH that maximizes the concentrations of the diazonium cation and the phenoxide anion is most favorable (pH >7, <10), whereas for anilines a pH that is slightly more acidic favors the diazonium cation and the neutral aniline (pH >5, <7).

Question 26.41

Write the complete mechanism for the hydrolysis of a diazonium salt with base. Explain why attack occurs at the terminal nitrogen rather than at the nitrogen attached to the alkyl or aryl group.

Question 26.42

When *p*-dimethylaminobenzenediazonium salt reacts with phenol, two diastereomeric azo compounds result. What are they?

The naming of the simpler azo compounds is illustrated in the following three examples:

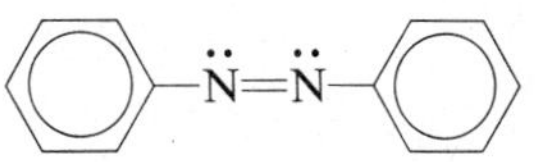

Azobenzene

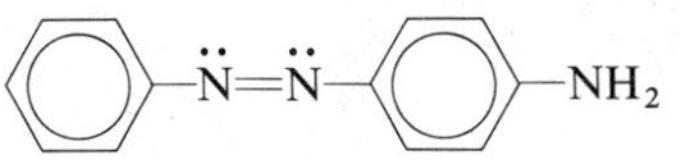

p-Aminoazobenzene

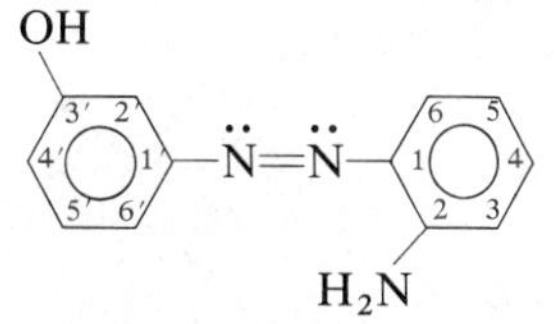

2-Amino-3′-hydroxyazobenzene

The geometry of the azo function is actually not linear as depicted here. The nitrogen-nitrogen double bond is between two sp^2 hybridized nitrogen atoms. There are, therefore, two geometric isomers possible for each compound:

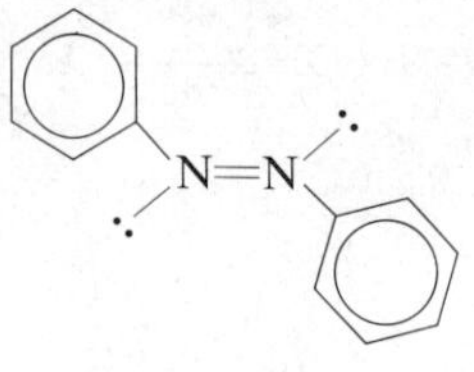

anti-Azobenzene
(*trans* or *E* may also
be used)

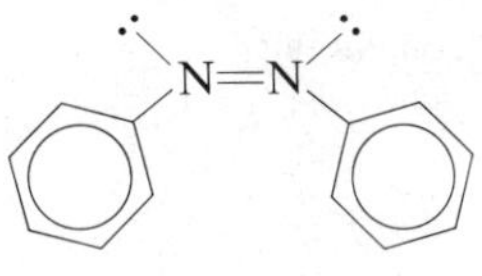

syn-Azobenzene
(*cis* or *Z* may also be used)

The barrier to interconverting these two isomers is ∼ 15 to 20 kcal/mole.

B. Medicinal Chemicals

a. Sulfa Drugs

The sulfa drugs are a class of antibiotics related to sulfanilamide:

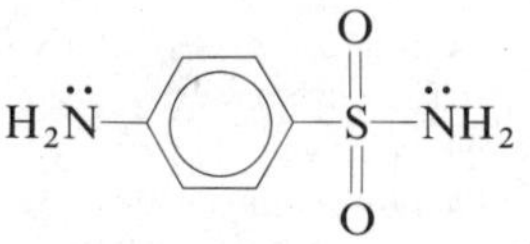

Sulfanilamide
(*p*-aminobenzenesulfonamide)

Antibiotics work to either destroy or inhibit the activity of a parasitic micro-organism, such as a fungus or bacterium. Other common antibiotics are tetracyclines, penicillins, cephalosporins, and polypeptide antibiotics (see Sec. 28.4).

Sulfanilamide was the first medicinal agent to be discovered and used. Various substituted forms of sulfanilamide have also been developed and may be more active against certain classes of bacteria; for example:

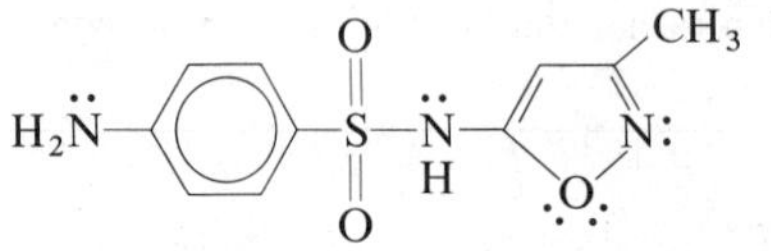

Isooxazole
ring

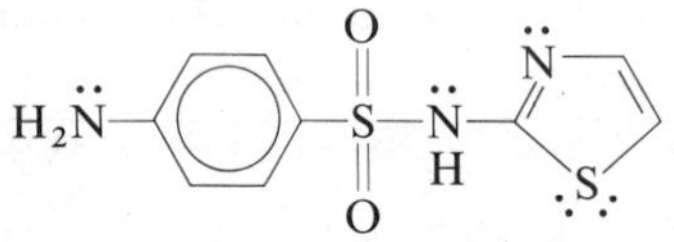

Thiazole
ring

Numerous other substitution patterns of sulfanilamide have been tried and are summarized here:

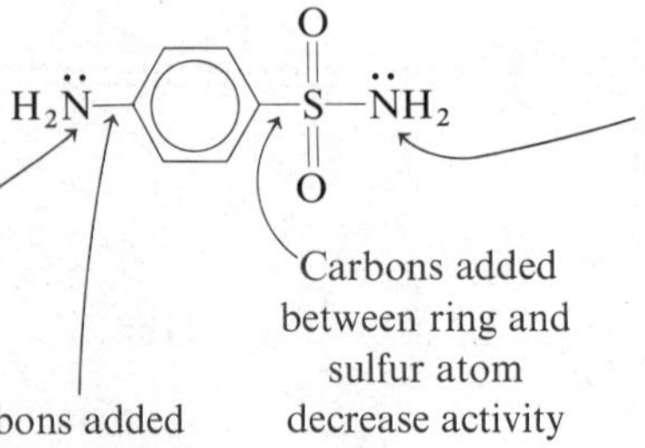

Substituents on
amide nitrogen
can increase
activity

Substituents on
amine nitrogen
decrease activity

Carbons added
between ring and
sulfur atom
decrease activity

Carbons added
between ring and
nitrogen have no
real effect

The synthesis of sulfanilamide itself is outlined in Sec. 26.8A,c.

Sulfanilamide destroys microorganisms by acting as an antagonist; that is, it substitutes for one of the chemicals required for the normal development and metabolism of microorganisms, thus short-circuiting their development. With bacteria, sulfanilamide (and its analogues) is an antagonist of folic acid, a compound required by the bacteria for development. Actually they are antagonists of *p*-aminobenzoic acid, which is used in folic acid synthesis.

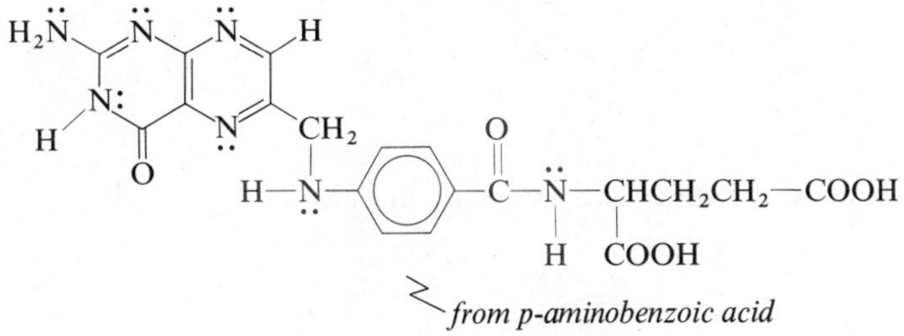

from *p*-aminobenzoic acid

Folic acid

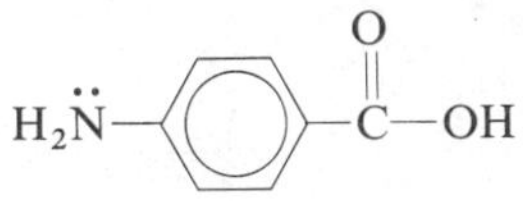

p-Aminobenzoic acid

Both folic acid and sulfanilamide contain the amino group attached to an aromatic ring. In nucleic acid (DNA) synthesis, as the bacteria require folic acid, sulfanilamide substitutes. The normal DNA synthesis is interrupted as is cell development. The result is eventual destruction of the bacteria.

b. Amphetamines

Barbiturates (downers) were introduced in Chap. 24 (see Question 24.54). The amphetamines (uppers) act in the reverse manner; they stimulate. These compounds are widely used to treat psychiatric disorders such as depression, in weight reduction programs, to increase alertness, and in the treatment of other central nervous system disorders. Two common amphetamines are:

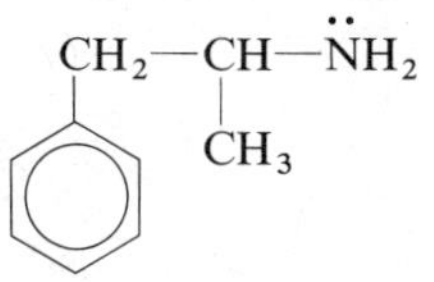

Amphetamine
(2-amino-1-phenylpropane)
Benzedrine

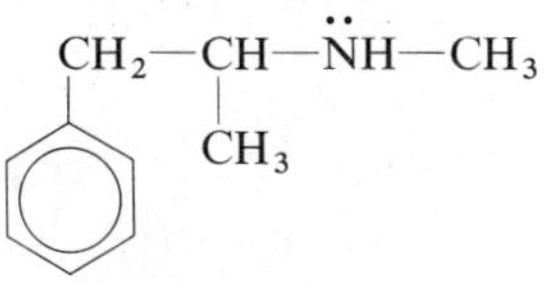

Methamphetamine
(2-methylamino-1-phenylpropane)
Methedrine

c. Local Anesthetics

Local anesthetics took medicine and dentistry from what could be called the *dark ages* to modern times. Until the advent of anesthetics, the medical treatment was often worse than the disease. More people probably died from the trauma of an operation than from the operation itself.

Local anesthetics work on the peripheral nervous system. Their mode of action is not completely known. They appear to function through a variety of different pathways, including (1) interfering with the transmission of nerve impulses and (2) changing the permeability of cell membranes, thereby interfering with ion flow and muscular action.

The classical local anesthetic is cocaine (see also Sec. 26.11D).

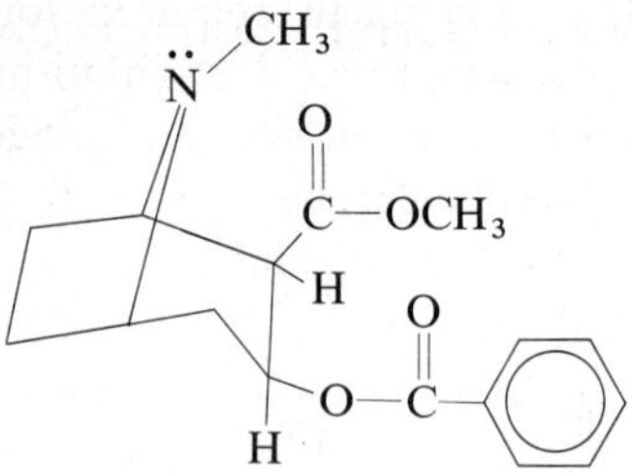

Cocaine

Benzocaine, novocaine, and lidocaine are also widely used:

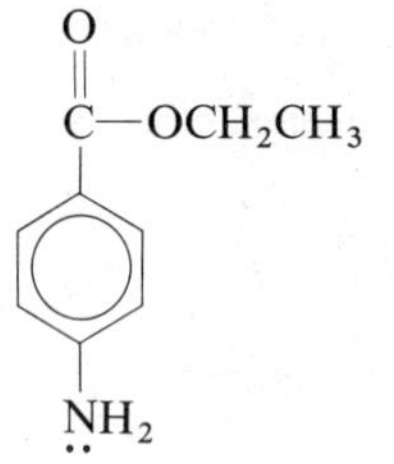

Benzocaine

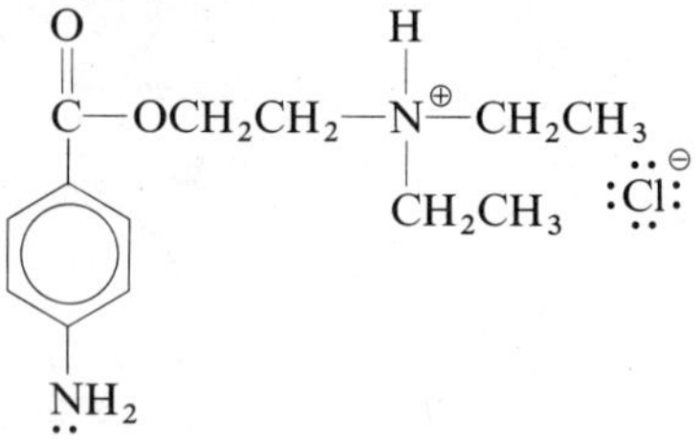

Novocaine

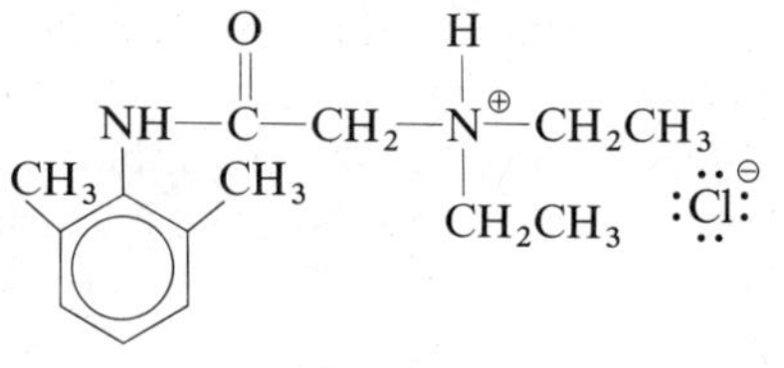

Lidocaine

d. Pain Relievers

The pain relievers most people use on a day-to-day basis are usually nitrogen compounds. Aspirin, phenacetin, and acetaminophen, in combination with other materials such as caffein, buffering agents, and antihistamines, make up just about all the aspirin and aspirin-type compounds on the market.

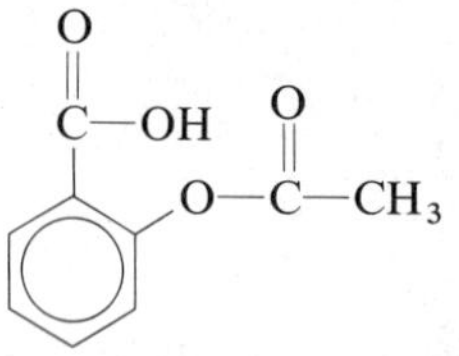

Aspirin

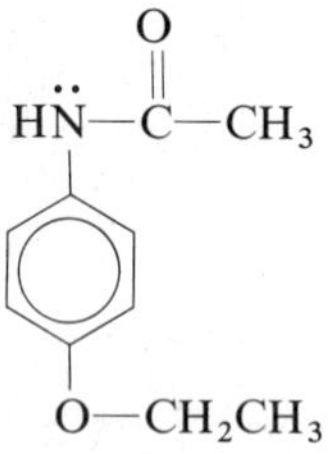

Phenacetin

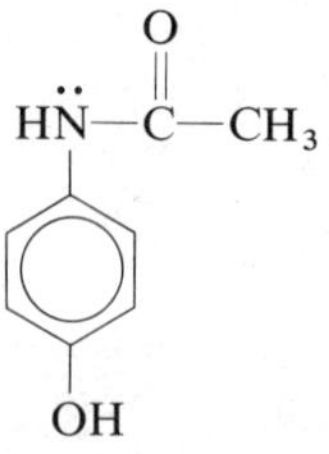

Acetaminophen

Demarol, a stronger (prescription) pain reliever, is also a nitrogen compound, in this case a nitrogen heterocyclic:

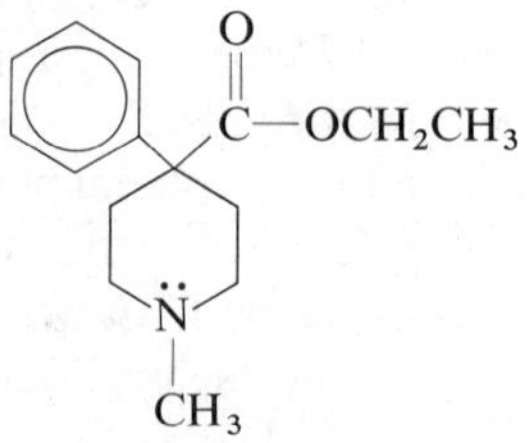

Demarol

Demarol is related in structure to cocaine (see Sec. 26.11D) and is a good example of a drug that was designed from a known model (in this case cocaine).

e. Sweeteners

Some people must reduce or completely cut out their sugar intake. Diabetics, for example, are in the latter category, whereas obese people fall into the former. Over the years artificial sweeteners have been developed and marketed as sugar substitutes. They are sweet but do not contain any sucrose (see Sec. 29.12A). Saccharin and cyclamate are the more famous sugar substitutes:

Saccharin   Calcium cyclamate

Both these nitrogen-containing compounds are amides of sulfonic acid (sulfonamides) and are thought to be possible carcinogens (see Sec. 26.11C). These allegations may or may not be true. Research is ongoing to develop and test alternate sugar substitutes.

C. Carcinogens

Most nitrogen-containing functional groups have been linked to cancer in animals and in some cases in humans. Five are listed here:

Aromatic amines: for example, aniline, α- and β-naphthylamine, and 4,4′-diaminobiphenyl

N-Nitroso compounds: for example, N-nitroso-N-methylaniline

Hydrazines: for example, dimethylhydrazine

Azo compounds: for example methyl orange

Aliphatic amines: for example, cyclohexylamine (a metabolite of cyclamate)

Little is known about the mode of action of these carcinogens, but from the small amount of available information they all appear to react in a similar fashion. They all appear to bind to the growing DNA chain in cell metabolism by alkylation of one of the numerous nucleophilic nitrogens on the chain (see Secs. 29.8 and 19.15). Again, as with the sulfa drugs, DNA synthesis is short-circuited. In this case this works to the detriment of humans because it results in the formation and development of nonnormal (cancerous) cells.

These compounds are found in every aspect of our daily lives. Sodium nitrite, which can react with other amines we ingest to form *N*-nitroso compounds, is used in the preservation of most red and pink meats. *N*-nitroso compounds are found in pork products, particularly bacon (the crisper the bacon, the higher the concentration). Artificial sweeteners, medicinal compounds, food additives, and much of what we eat contain nitrogen compounds.

D. Alkaloids

The name *alkaloid*, which literally means alkali-like, has its origin in the chemical behavior of these compounds. Chemically, they behave like alkali bases (sodium

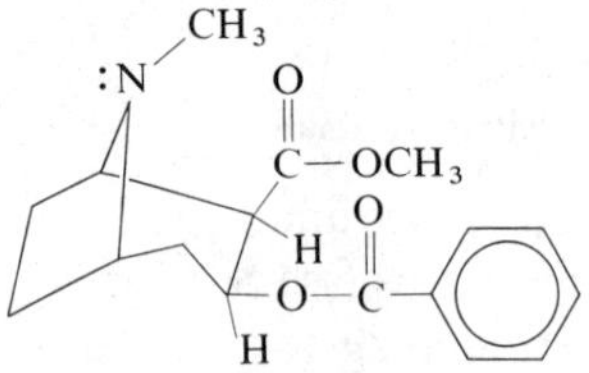

Cocaine
(tropane ring skeleton)

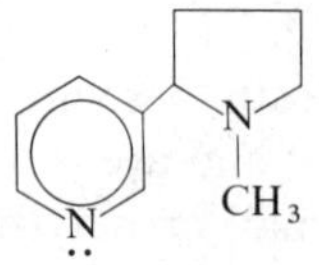

Nicotine
(pyridine and pyrrolidine
ring skeleton)

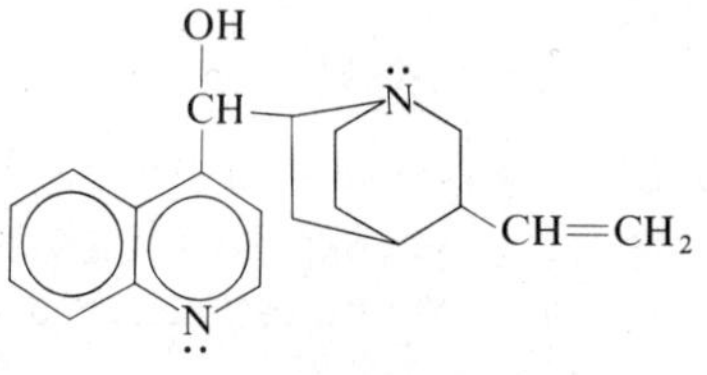

Quinine
(quinoline ring skeleton)

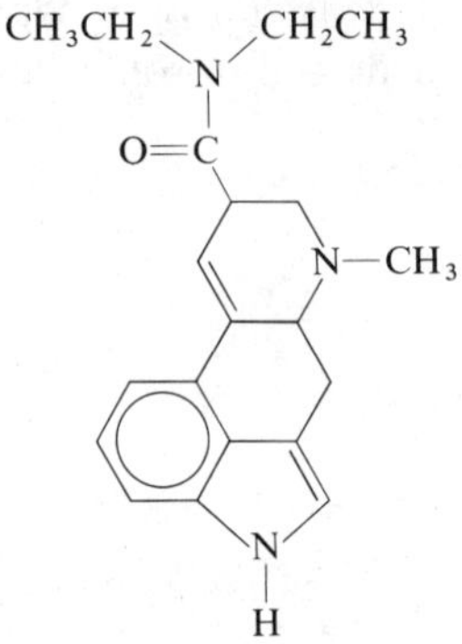

Lysergic acid diethylamide
(LSD)
(indole ring skeleton)

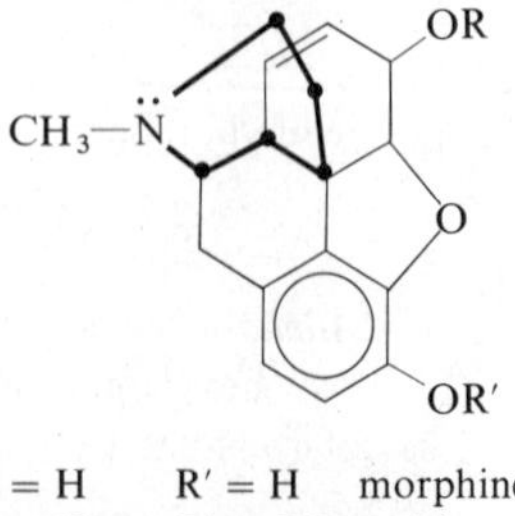

R = H   R′ = H   morphine
R = H   R′ = CH₃ codeine
R = OAc R′ = OAc heroin
(piperidine ring skeleton)

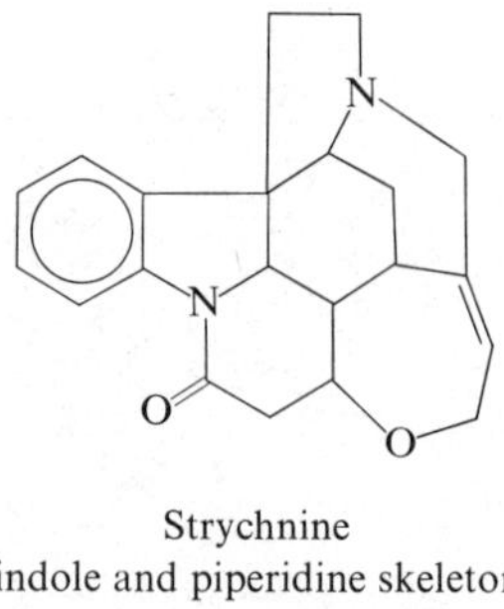

Strychnine
(indole and piperidine skeleton)

FIGURE 26.5 Six examples of alkaloids.

hydroxide, for example). Alkaloids, by definition, originate in plants. They all contain one or more nitrogen heterocyclic rings and are classified based on the complexity of their structure. Examples of several classes of alkaloids are in Fig. 26.5.

The alkaloids are used in a variety of roles: as medicines (cocaine is a local anesthetic), as pain relievers (morphine and codeine), and in optical resolutions (quinine, which is also an antimalarial agent, and strychnine). See the Reading References for this chapter in the appendix for more information on this family of nitrogen heterocyclics.

26.12 Spectral Analysis of Amines

A. Infrared Spectroscopy

The N—H bond of a primary or secondary amine absorbs in the 3,200 to 3,500-cm^{-1} (3.12 to 2.86 μ) region of the IR spectrum. The absorption of the nitrogen-hydrogen stretching frequency of a secondary amine consists of one very weak band, whereas that of a primary amine consists of two bands [an asymmetric stretching band (the stronger of the two) and a symmetric stretching band]. Tertiary amines contain no N—H bonds and exhibit no absorption in this region of the spectrum.

As with the alcohols (see Sec. 17.10), hydrogen bonding in amines shifts the N—H stretching band to lower frequencies. The shifts are not as large because of the weaker hydrogen bonding of amines as compared with that of alcohols (approximately 250 cm^{-1} for amines versus 300 cm^{-1} for alcohols).

In addition to the N—H stretching vibrations, there are absorptions caused by the carbon-nitrogen (C—N) stretching vibration at 1,030 to 1,230 cm^{-1} (8.13 to 9.71 μ) and two strong absorptions between 1,560 to 1,640 cm^{-1} (6.10 to 6.41 μ) and 650 to 900 cm^{-1} (11.11 to 15.38 μ) due to the N—H bending or wagging vibration.

The IR spectra of aniline, N-methylaniline, and N,N-dimethylaniline are in Fig. 26.6.

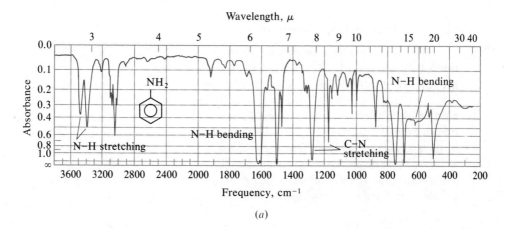

(a)

FIGURE 26.7 IR spectra of (a) aniline, (b) N-methylaniline, and (c) N,N-dimethylaniline. [Data from *Indexes to Evaluated Infrared Reference Spectra*, copyright 1975, The Coblentz Society, Norwalk CT]

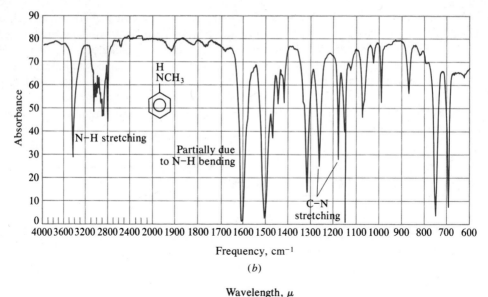

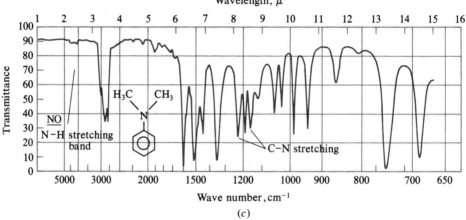

FIGURE 26.6 (continued)

B. Ultraviolet Spectroscopy

The amino group is an auxochromic group. It exhibits no appreciable absorption of its own in the UV portion of the electromagnetic spectrum, but it has a pronounced effect on the wavelength of absorption of any chromophore to which it is attached. The $n \rightarrow \sigma^*$ transitions for the aliphatic amines are found in the region below 230 nm and have an extinction coefficient on the order of 10^3. The aromatic amines absorb in the 230 to 300-nm region, a result of the $\pi \rightarrow \pi^*$ transitions of the aromatic ring. Note that these absorptions are shifted to longer wavelengths as a result of the amino group being attached to the ring. Benzene has its principal absorption centered at 203 nm. The extinction coefficients associated with the transitions of the aromatic ring are on the order of 10^4.

C. Nuclear Magnetic Resonance Spectroscopy

The NMR spectra of amines are very much like those of alcohols. The N—H absorption of an aliphatic amine is normally in the 0.5 to 3.0-δ range, whereas the absorption for aromatic amines is at 3.0 to 5.0 δ. As a result of hydrogen bonding in the primary and secondary amines, the chemical shift of the N—H proton varies markedly with solvent, concentration, and temperature (see Sec. 17.10). This is analogous to the alcohols. Again as with the alcohols, it is possible to differentiate an N—H proton by allowing it to undergo deuterium exchange with D_2O. The N—H absorption is easily picked out by its disappearance as the isotopes exchange, with the concomitant appearance of the HOD peak of the exchanged water. See Fig. 26.7 for examples of NMR spectra of amines.

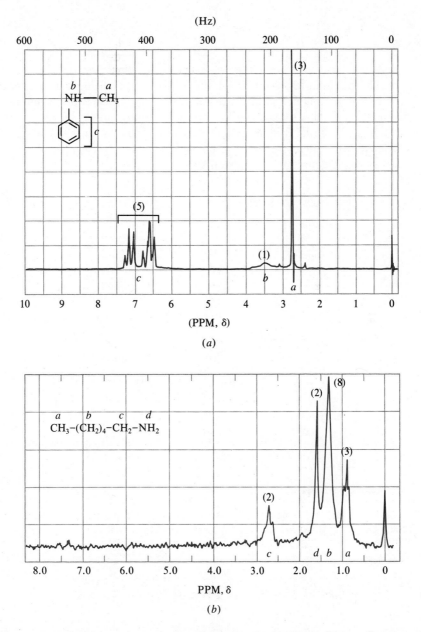

FIGURE 26.8   NMR spectra of (*a*) *N*-**methylaniline and** (*b*) **1-aminohexane.** [(*a*) **Data from** *The Aldrich Library of NMR Spectra*, **C. J. Pouchert and J. R. Campbell (eds.), copyright 1974, Aldrich Chemical Co., Milwaukee;** (*b*) **data from the** *Sadtler Standard Reference Spectra* (*NMR*), © **Sadtler Research Laboratories, Division of Bio-Rad Laboratories, Inc. (1965–1974)]**

D. Mass Spectrometry

Amines do give molecular ion fragments but they are usually quite weak (as a result of facile cleavage, which is described in the next paragraph) and often difficult to see.

Amines fragment in a predictable and reproducible pattern. When possible, the amine fragments between the carbons α and β to the amino function:

$$-\overset{|}{\underset{|}{C}}-\overset{|}{\underset{|}{C}}-\ddot{N} \xleftarrow{\hspace{0.3cm}} \xrightarrow{-e^{\ominus}} \quad -\overset{|}{\underset{|}{C^{\beta}}}-\overset{}{\overset{\oplus}{\underset{\alpha}{C_{\alpha}}}-\overset{\oplus}{\underset{|}{N}}} \xrightarrow[\text{fragmentation}]{C_{\alpha}-C_{\beta}} \quad >C=\overset{\oplus}{N}<$$

$$P^{\oplus}$$

$$+$$

$$-\overset{|}{\underset{|}{C}}\cdot$$

The resulting fragment has an m/e ratio of 30 if the amine is primary and has a methylene group (CH_2) at the α carbon. If the α carbon is branched, the principal fragment has an m/e ratio equal to 29 ($-CH=NH_2$) plus the weight of the branch.

After the initial fragmentation occurs, the ion continues to cleave into smaller and smaller fragments, giving the normal series of peaks expected for a straight-chain aliphatic compound; for example:

$$CH_3CH_2CH_2CH_2CH_2-\ddot{N}\overset{H}{\underset{H}{\big<}} \longrightarrow CH_3CH_2CH_2\overset{\beta}{C}H_2\overset{\alpha}{C}H_2-\overset{\oplus}{N}\overset{H}{\underset{H}{\big<}} \quad (P^{\oplus})$$

$$C_{\alpha}-C_{\beta} \Big| \text{cleavage}$$

$$CH_3CH_2CH_2CH_2^{\cdot} + CH_2=\overset{\oplus}{N}\overset{H}{\underset{H}{\big<}}$$

$$\Big|$$

fragments to give

$$CH_3^{\cdot}, CH_3CH_2^{\cdot}, CH_3CH_2CH_2^{\cdot} \text{ and others}$$

Question 26.43

2-Aminopentane gives the following mass spectral fragmentation pattern: m/e 87, 44, 43, 29, 26, 15, and some smaller fragments. Explain this mechanistically.

Question 26.44

2-Methyl-2-aminopentane gives a strong mass spectral peak at m/e 86 as one of its principal peaks. Draw this fragment and explain mechanistically how it is formed.

26.13   Summary of Methods of Preparation of Amines

1. Reduction of nitro compounds (Sec. 26.7A,a)

$$\underset{\text{Nitro compound}}{(Ar)\,R-NO_2} \xrightarrow[H_3\ddot{O}^{\oplus}]{\text{Sn or Fe}} \underset{\text{Amine}}{(Ar)\,R-\ddot{N}H_2}$$

Ammonium sulfide [for example, $(NH_4)_2S$] and catalytic hydrogenation (H_2/catalyst) may also be used.

2. Reduction of amides (Secs. 23.9 and 26.7A,b)

$$\text{(Ar) R}-\overset{\overset{\displaystyle O}{\|}}{C}-NH_2 \xrightarrow[\text{ether}]{LiAlH_4} \text{(Ar) R}-\overset{\overset{\displaystyle H}{|}}{\underset{\underset{\displaystyle H}{|}}{C}}-NH_2$$

Amide   Amine

B_2H_6/tetrahydrofuran or H_2 catalyst also may be used.

3. Reduction of nitriles (Secs. 23.16 and 26.7A,c)

$$\text{(Ar) R}-C{\equiv}N \xrightarrow[\text{catalyst}]{H_2} \text{(Ar) R}-\overset{\overset{\displaystyle H}{|}}{\underset{\underset{\displaystyle H}{|}}{C}}-NH_2$$

Nitrile   Amine

$LiAlH_4$/ether or B_2H_6/tetrahydrofuran also works well.

4. Reduction of unsaturated nitrogen compounds (Secs. 26.7A,d and 26.8A,d)

$$\underset{\underset{\displaystyle R \qquad R'}{\overset{\displaystyle |}{\underset{\displaystyle C}{}}}}{\overset{\displaystyle N{\diagdown}G}{\overset{\displaystyle \|}{}}} \xrightarrow[\substack{\text{or} \\ LiAlH_4, \\ \text{ether}}]{\substack{H_2 \\ \text{catalyst}}} \underset{\underset{\displaystyle R'}{\overset{\displaystyle |}{}}}{\overset{\overset{\displaystyle H{\diagdown}\,{\diagup}G}{\displaystyle N}}{R-C-H}}$$

Imine or oxime   Amine

where R and R′ = alkyl, aryl, or hydrogen

$$-G = -H, -R, -Ar, -OH \text{ (imines or oximes)}$$

Reductive amination, in which the carbonyl compound is converted to the imine and catalytically reduced to the amine all in one step, is also possible. $NaBH_4$/alcohol and B_2H_6/tetrahydrofuran can also be used for imines but not oximes.

5. Nucleophilic substitution by ammonia and ammonia analogs (Sec. 26.7B,a)

$$\overset{..}{N}H_3 \xrightarrow[\substack{(S_N2) \\ \text{step 1}}]{R-X} \underset{\underset{\displaystyle H}{|}}{\overset{\overset{\displaystyle H}{|}}{R-\overset{\oplus}{N}-H}} \;\; X^{\ominus} \xrightarrow[\substack{H_2O \\ \text{step 2}}]{\overset{..}{H}\overset{..}{O}{:}^{\ominus}} R-\overset{..}{N}H_2$$

         1° Amine

$$\xrightarrow[\substack{\text{and 2} \\ \text{again}}]{\text{steps 1}} \underset{..}{\overset{\overset{\displaystyle R'}{|}}{R-N-H}} \xrightarrow[\substack{\text{and 2} \\ \text{again}}]{\text{steps 1}} \underset{..}{\overset{\overset{\displaystyle R'}{|}}{R-N-R''}} \xrightarrow[\substack{\text{and 2} \\ \text{again}}]{\text{steps 1}} \underset{\underset{\displaystyle R'''}{|}}{\overset{\overset{\displaystyle R'}{|}}{R-\overset{\oplus}{N}-R''}} \;\; X^{\ominus}$$

2° Amine   3° Amine   Ammonium salt

Works best for the preparation of 3° amines or ammonium salts. α-Halo acids give *amino acids*.

6. **Gabriel synthesis** (Sec. 26.7B,b)

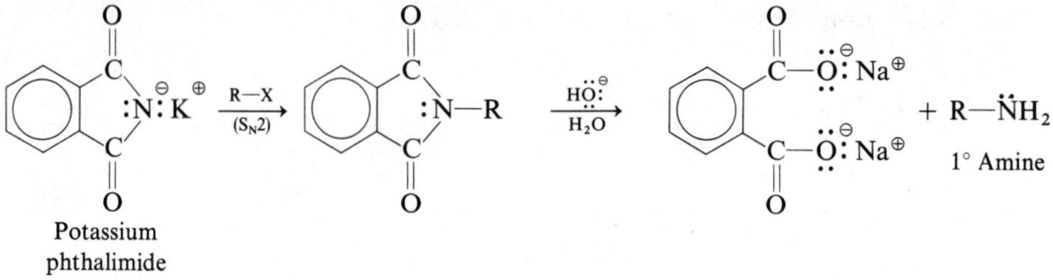
R—X
(S$_N$2)
HO:⊖
H$_2$O
Na⊕
Na⊕
+ R—NH$_2$
1° Amine
Potassium
phthalimide

May be used to synthesize *amino acids.*

7. **Hofmann reaction** (Sec. 26.7C)

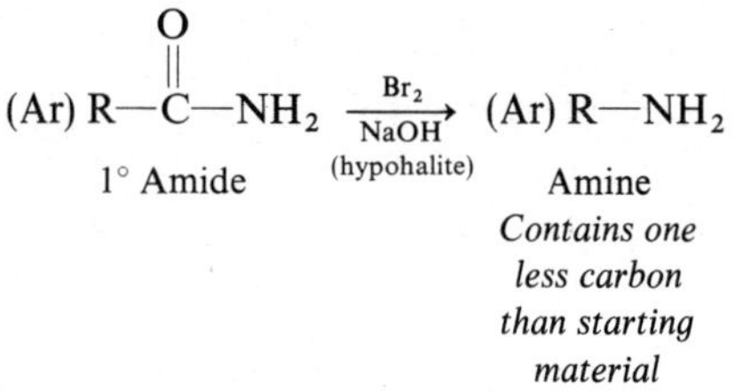
(Ar) R—C—NH$_2$
Br$_2$
NaOH
(hypohalite)
(Ar) R—NH$_2$
1° Amide
Amine
Contains one
less carbon
than starting
material

26.14 Summary of Reactions of Amines

1. Preparation of amides (Secs. 23.8 and 26.8A,a)

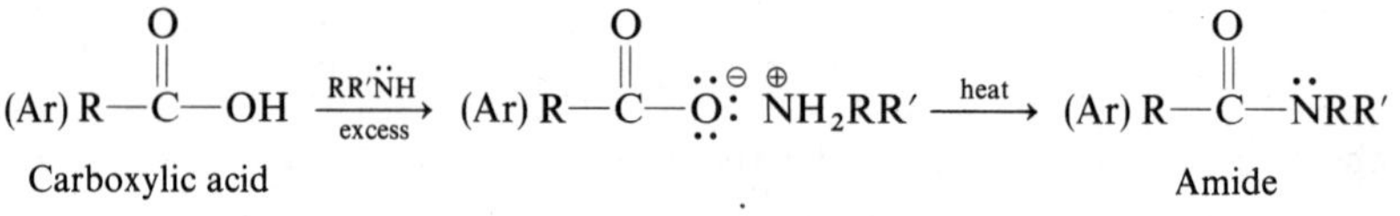
(Ar) R—C—OH
RR′NH
excess
(Ar) R—C—O: NH$_2$RR′
heat
(Ar) R—C—NRR′
Carboxylic acid
Amide

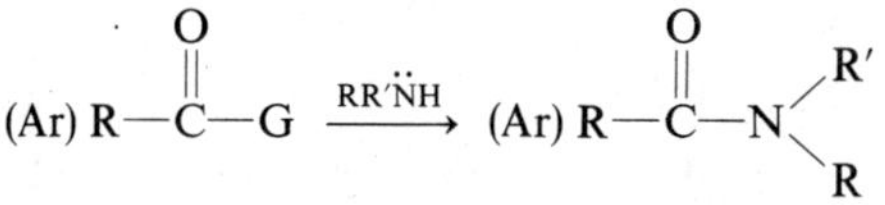
(Ar) R—C—G
RR′NH
(Ar) R—C—N
R′
R

where —G = —X (Cl, Br), —OR′, —O—C(=O)—R (Ar)

Hinsberg test with benzenesulfonyl chloride for 1°, 2°, or 3° amines

2. Preparation of enamines (Secs. 24.12 and 26.8A,b)

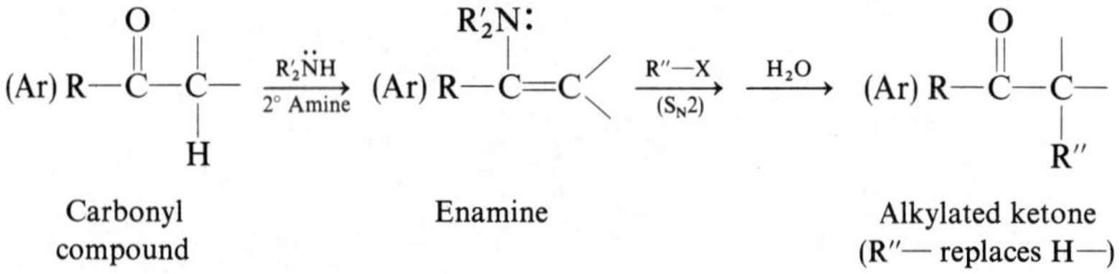
(Ar) R—C—C—
H
R′$_2$NH
2° Amine
R′$_2$N:
(Ar) R—C=C
R″—X
(S$_N$2)
H$_2$O
(Ar) R—C—C—
R″
Carbonyl
compound
Enamine
Alkylated ketone
(R″— replaces H—)

N-Alkylation is also possible.

3. Electrophilic aromatic substitution (Secs. 14.11 and 26.8A,c)

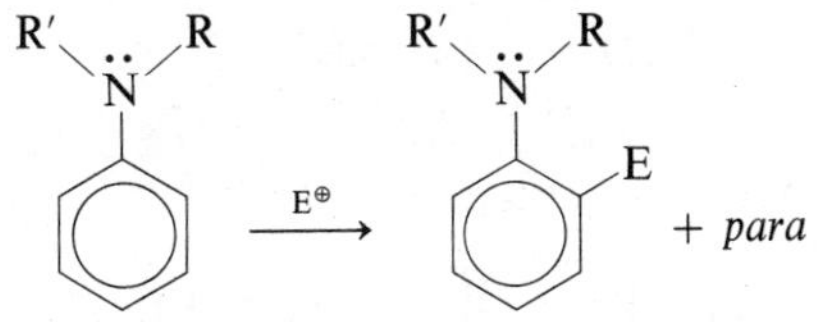
R'
R
N
E⊕
R'
R
N
E
+ *para*

Strong activating group,
ortho,para director

where R and R' = alkyl, aryl, or hydrogen

4. Addition of ammonia and substituted amines to carbonyl compounds (Secs. 21.8 and 26.8A,d)

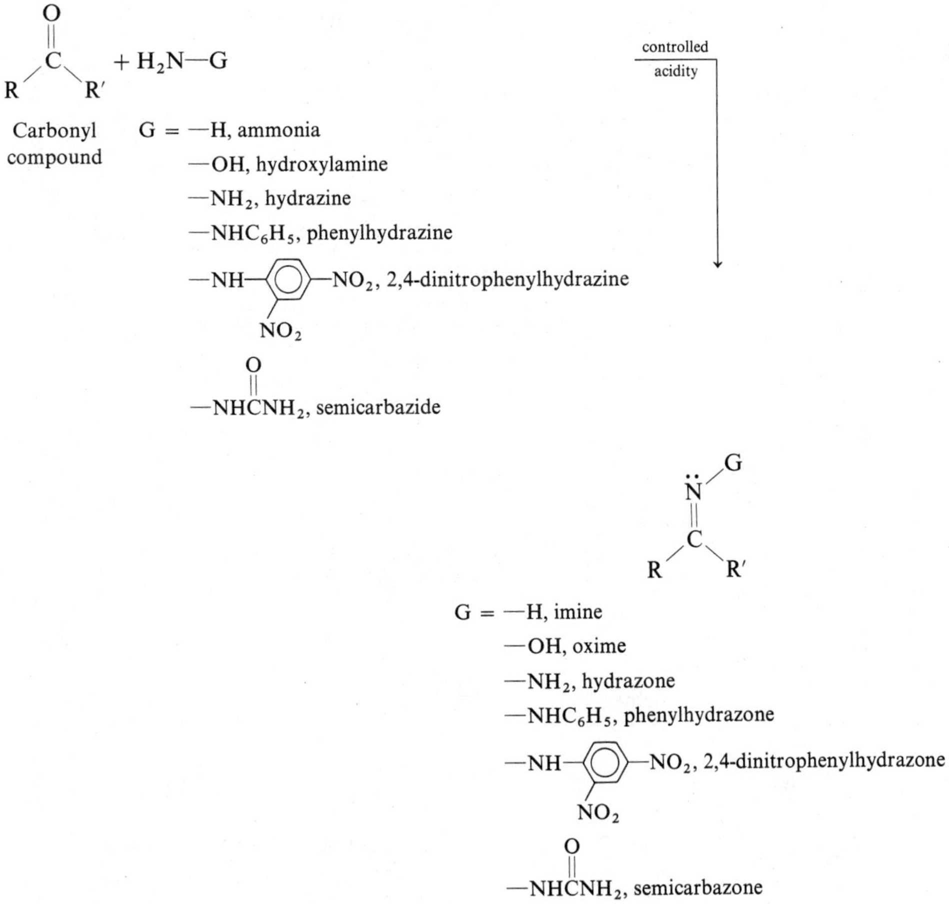
O
C
R
R'
+ H₂N—G
controlled
acidity
Carbonyl
compound
G = —H, ammonia
—OH, hydroxylamine
—NH₂, hydrazine
—NHC₆H₅, phenylhydrazine
—NH— —NO₂, 2,4-dinitrophenylhydrazine
NO₂
O
—NHCNH₂, semicarbazide
G
N
C
R
R'
G = —H, imine
—OH, oxime
—NH₂, hydrazone
—NHC₆H₅, phenylhydrazone
—NH— —NO₂, 2,4-dinitrophenylhydrazone
NO₂
O
—NHCNH₂, semicarbazone

5. Nitrosation (Sec. 26.8B,a)

$$\underset{\text{1° Amine}}{(\text{Ar})\,\text{R—NH}_2} \xrightarrow[\text{HX}]{\text{NaNO}_2} \underset{\text{Diazonium salt}}{(\text{Ar})\,\text{R—}\overset{\oplus}{\text{N}}\equiv\ddot{\text{N}}\ \text{X}^{\ominus}}$$

Aliphatic diazonium salts usually lead to decomposition products. One case where they are versatile synthetic intermediates is the Tiffeneu-Demyanov ring expansion:

$$
\underset{(CH_2)_n}{\overset{HO}{\diagdown}}\underset{}{\overset{CH_2NH_2}{\diagup}} \quad \xrightarrow[\substack{2. \text{ warm to} \\ \text{room temp.}}]{1. \text{ NaNO}_2,\ HX} \quad \underset{(CH_2)_{n+1}}{\overset{O}{\diagup}}
$$

Aromatic diazonium salts can be converted to a variety of functional groups:

$$
\underset{X^{\ominus}}{Ar\overset{\oplus}{-}N\equiv\overset{..}{N}}
$$

$\xrightarrow[\substack{\text{or} \\ CH_3CH_2OH}]{H_3PO_2}$	Ar—H
$\xrightarrow[(H_2SO_4,\ H_2O)]{H_3O^{\oplus}}$	Ar—OH
$\xrightarrow[CuBr]{CuCl\ or}$	Ar—X
$\xrightarrow{KI}$	Ar—I
$\xrightarrow[\substack{\text{heat, dry} \\ \text{solid salt}}]{[X^{\ominus} \text{ of salt is } BF_4^{\ominus}]}$	Ar—F
$\xrightarrow{CuCN}$	Ar—CN
$\xrightarrow[2.\ H\overset{..}{\overset{..}{O}}:^{\ominus},\ H_2O]{1.\ CH_3CH_2O-\overset{\overset{S}{\|}}{C}-\overset{..}{\underset{..}{S}}:^{\ominus}K^{\oplus}}$	Ar—SH
$\xrightarrow[\substack{\text{activated} \\ \text{ring}}]{Ar-G}$	$Ar-\overset{..}{N}=\overset{..}{N}-Ar-G$

Azo compound

Dyes (Sec. 26.11A,b)
Carcinogens (Sec. 26.11C)

$$
\underset{\text{2° Amine}}{R-\overset{\overset{R'}{\|}}{\underset{..}{N}}-H} \quad \xrightarrow[HX]{NaNO_2} \quad \underset{\substack{\textit{N}\text{-Nitroso} \\ \text{amine} \\ \textit{Carcinogen}}}{R-\overset{\overset{R'}{\|}}{N}:\ \underset{\overset{\|}{O}}{\overset{\|}{N}}:}
$$

$$
\underset{\text{3° Amine}}{R-\overset{\overset{R'}{\|}}{\underset{..}{N}}-R''} \quad \xrightarrow[HX]{NaNO_2} \quad \text{no reaction or decomposition products}
$$

Will give nitroso aromatic compounds if the 3° amine contains an aromatic ring:

$$
\underset{}{\overset{\overset{R\diagdown\ \ \diagup R'}{\overset{..}{N}}}{\bigcirc}} \quad \xrightarrow[HX]{NaNO_2} \quad \underset{\text{Nitrosoaniline}}{\overset{\overset{R\diagdown\ \ \diagup R'}{\overset{..}{N}}}{\bigcirc}\overset{..}{N}=O} \quad +\ \textit{para}
$$

6. Oxidation (Sec. 26.8B,b)

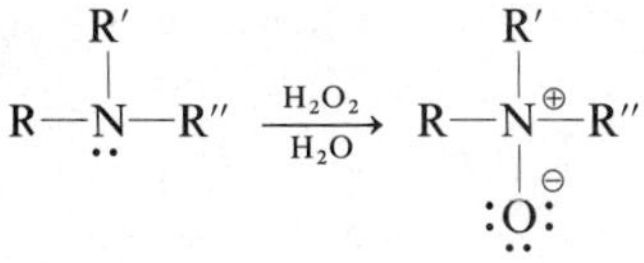

$$R\overset{R'}{\underset{\cdot\cdot}{-N-}}R'' \xrightarrow[H_2O]{H_2O_2} R\overset{R'}{\underset{\underset{:\overset{\cdot\cdot}{O}:}{\overset{\ominus}{|}}}{-N^{\oplus}-}}R''$$

3° Amine   Amine oxide

1° and 2° Amines give several products. See the Cope reaction.

7. Hofmann elimination (Sec. 26.8B,c)

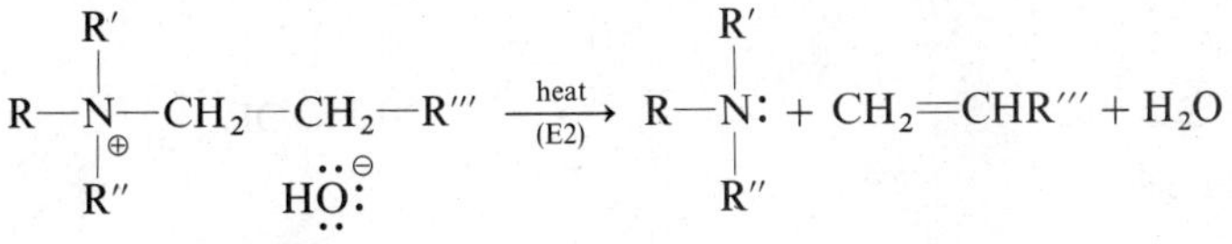

$$R\overset{R'}{\underset{\underset{R''}{\overset{|}{N^{\oplus}}}}{-N-}}CH_2-CH_2-R''' \xrightarrow[(E2)]{heat} R\overset{R'}{\underset{R''}{-N:}} + CH_2{=}CHR''' + H_2O$$

$$HO:^{\ominus}$$

Quaternary   Alkene
ammonium hydroxide
Must contain
β hydrogens

Used primarily in the structure elucidation of cyclic amines. *Hofmann versus Sayt-*
zeff elimination

8. Cope elimination (Sec. 26.8B,c)

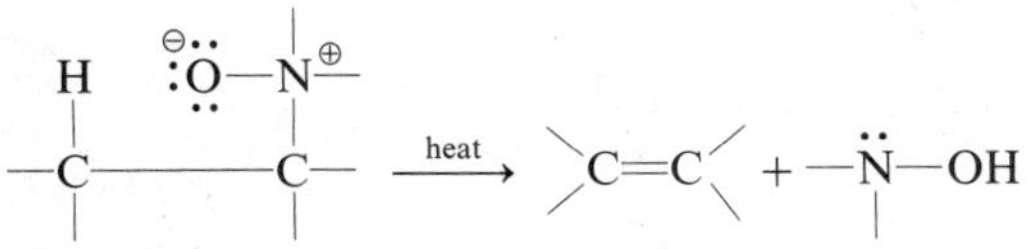

$$\overset{H \quad \overset{\ominus\cdot\cdot}{:O}-\overset{\oplus}{N}{-}}{\underset{|}{-C}{-}{-}{-}{-}\underset{|}{C}-} \xrightarrow{heat} {>}C{=}C{<} + {-}\overset{\cdot\cdot}{N}{-}OH$$

Amine   Alkene
oxide

Reactions analogous to Cope elimination: sulfoxide pyrolysis, selenoxide pyrolysis,
ester pyrolysis, and xanthate pyrolysis.

9. *Heterocyclic compounds*

Nomenclature and structure (Secs. 19.1B and 26.5A)
Basicity (Sec. 26.5B)
Stereochemistry (Sec. 26.6)
Reactions (Sec. 26.10)

Study Questions

26.45 Name each of the following 16 compounds.

(a)
$$CH_2CH_2CH_2CH_2\overset{O}{\overset{\|}{-C}}-O-CH_2CH_2CH_3$$
$$\underset{H}{\overset{N:}{\diagup}}\underset{CH_2CH_3}{\diagdown}$$

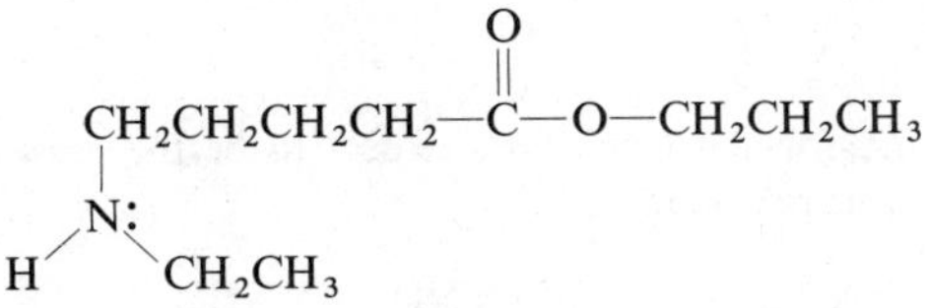

(b)

(c) $CH_3-\langle\bigcirc\rangle-\overset{\cdot\cdot}{N}{=}\overset{\cdot\cdot}{N}-\langle\bigcirc\rangle-O-CH_3$

(d) $\langle\bigcirc\rangle-CH_2-CH_2-\overset{\cdot\cdot}{N}\overset{CH_3}{\underset{CH_3}{\diagup}}$

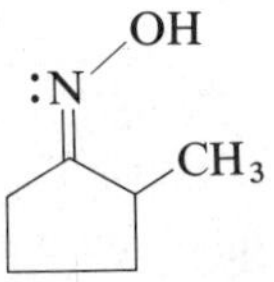

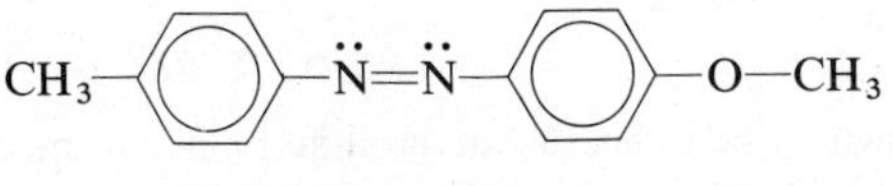

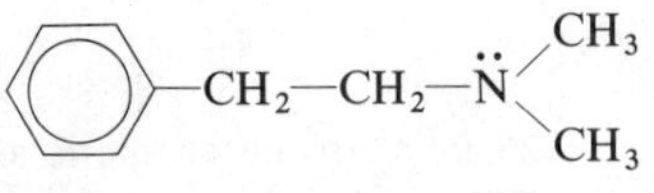

(e)

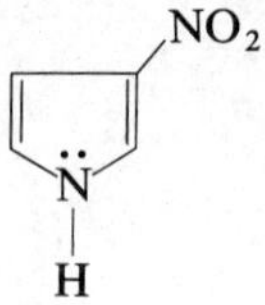

(f)

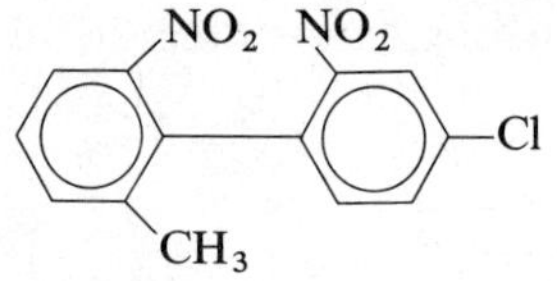

(g)

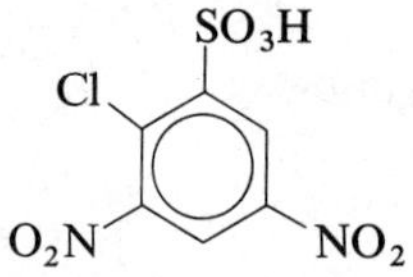

(h)

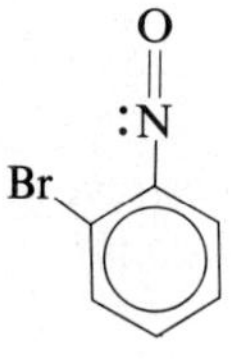

(i)

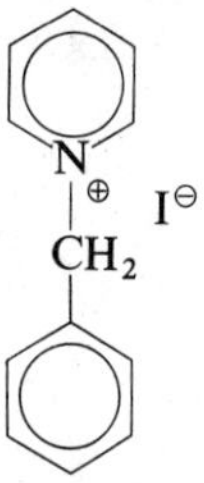

(j)

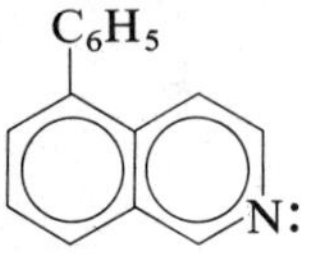

(k)

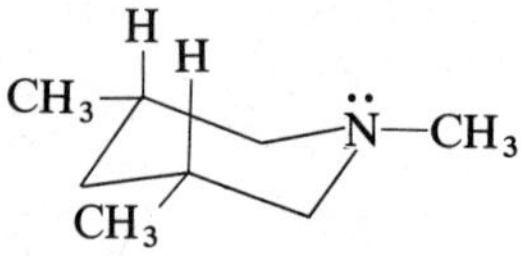

(l)

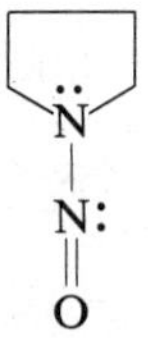

(m)

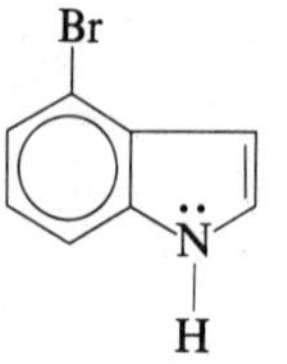

(n)

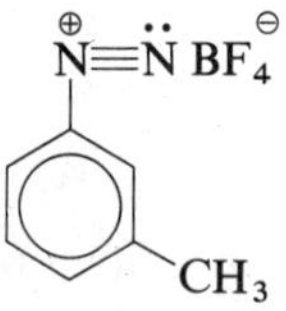

(o)

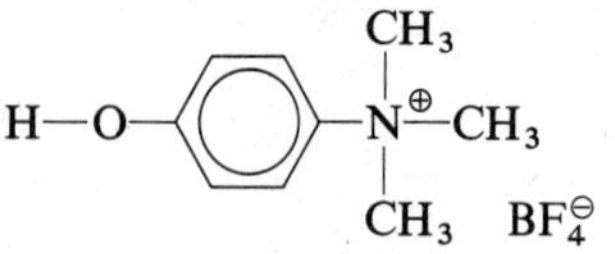

(p)

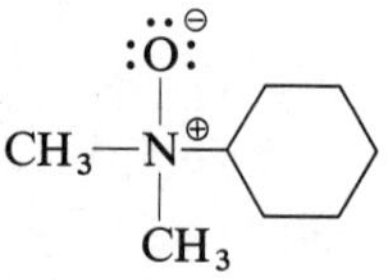

26.46 Arrange the following compounds *by letter* in order of *increasing basicity;* that is, list the least basic compound first and the most basic compound last.

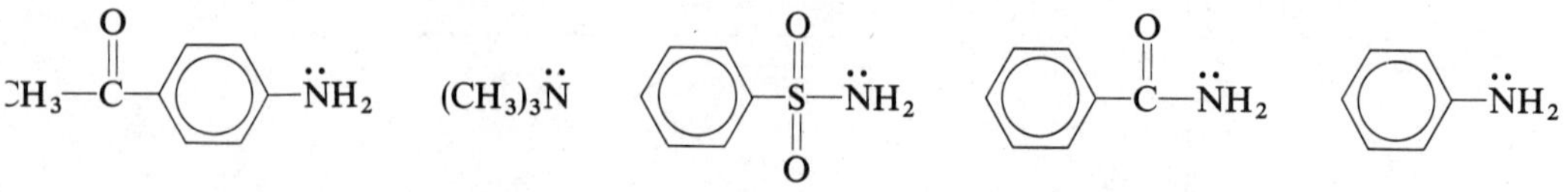

A   *B*   *C*   *D*   *E*

26.47 Of the following four compounds, which would you expect to be the *most basic* and which the *least basic*? Briefly explain your reasoning.

$$Br-\ddot{N}H_2 \qquad CH_3-\ddot{N}-CH_3 \qquad CH_3-\ddot{N}H_2 \qquad \ddot{N}H_3$$
$$\phantom{CH_3-\ddot{N}-CH_3aaaaaa}\underset{CH_3}{|}$$

26.48 A certain aromatic amine is known to be either *N*-ethylaniline or *N,N*-dimethylaniline. When the amine is shaken with benzenesulfonyl chloride and an excess of aqueous potassium hydroxide, a heterogeneous mixture results. The organic layer is removed and found to be completely soluble in dilute hydrochloric acid. On the basis of this information, what is the amine? Give your reasoning and write equations for any and all reactions that occur.

26.49 When the molecule shown here is subjected to carefully controlled, base-catalyzed hydrolysis ($:\ddot{O}H^{\ominus}/H_2O$), one of the two functional groups in it is preferentially hydrolyzed. Which group is hydrolyzed first? Provide a complete, stepwise mechanism to show its hydrolysis (show all intermediates) and then give at least *two* reasons for its preferential hydrolysis. (*Note:* The same group is preferentially hydrolyzed under acidic conditions as well.)

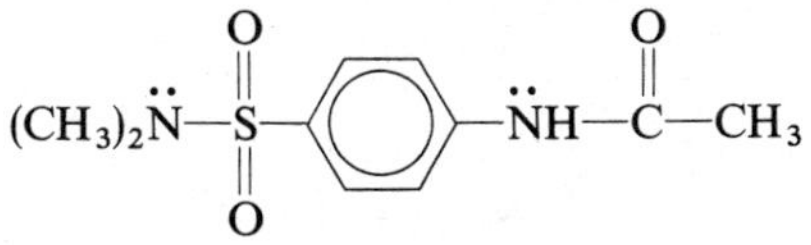

26.50 Supply the missing reactants and products, as required, to complete the following 21 equations. The items to be inserted are indicated by number [for example, (1)]. If no reaction occurs, so indicate.

(*a*)

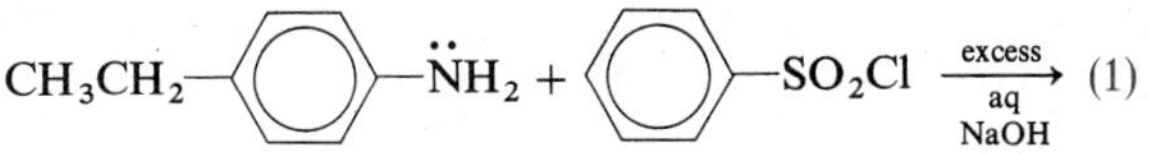

(*b*)

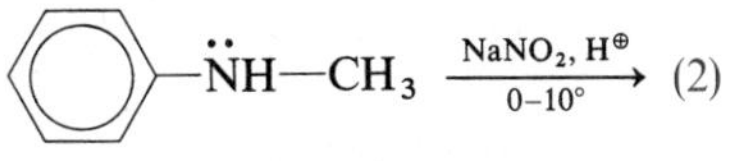

(*c*)

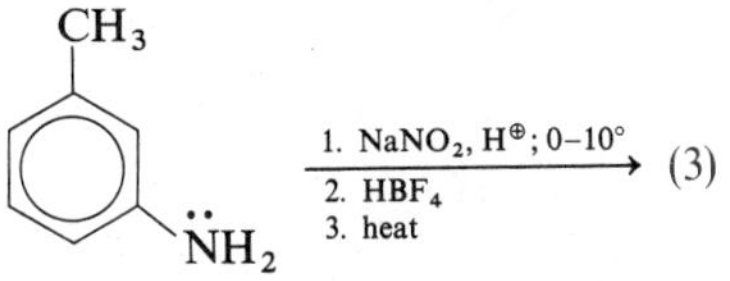

(*d*)

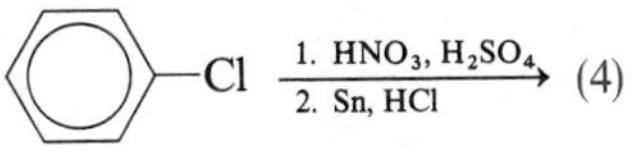

(*e*)

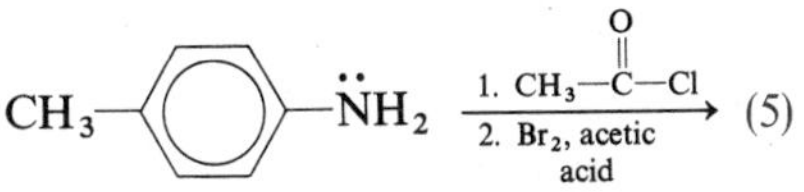

(*f*)

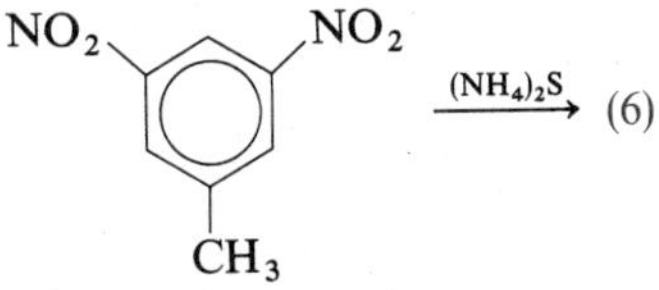

(*g*) $O_2N-\langle\rangle-\ddot{N}(CH_3)_2$. $\xrightarrow[\substack{2.\ H^{\oplus},\ NaNO_2,\\0-10°}]{1.\ Sn,\ HCl}$ (7)

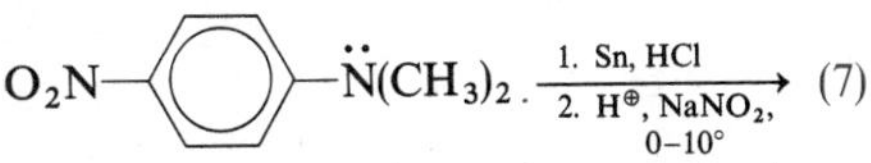

(*h*) $H_2\ddot{N}-\langle\rangle-SO_3H$ $\xrightarrow[\text{2. phenol}]{1.\ H^{\oplus},\ NaNO_2,\ 0-10°}$ (8)

(*i*)

(*j*) $CH_3CH_2CH_2\ddot{N}H_2$ $\xleftarrow{(10)}$ CH_3CH_2Cl $\xrightarrow{(11)}$ $CH_3CH_2\ddot{N}H_2$

$CH_3\ddot{N}H_2$ $\xleftarrow{(14)}$ $CH_3\overset{O}{\overset{\|}{C}}-\ddot{N}H_2$ $\quad\overset{(13)}{\nearrow}\quad\overset{(12)}{\Big\downarrow}$ $CH_3CH_2\ddot{N}HCH_3$

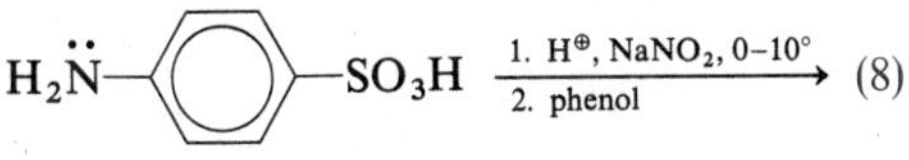

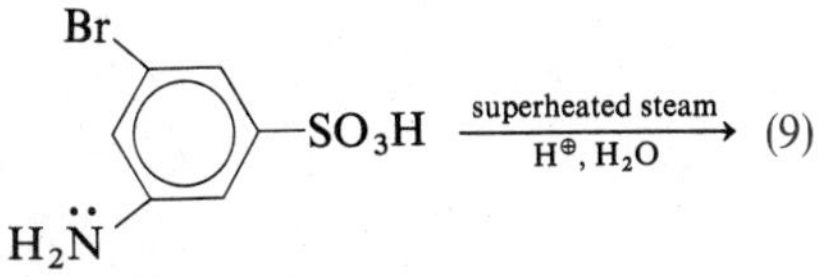

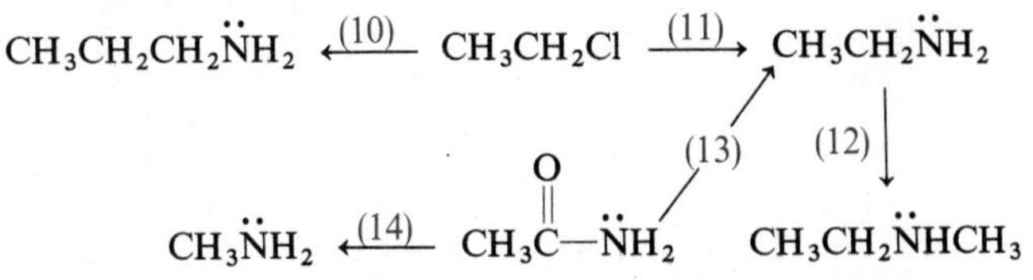

(**k**) $\underset{}{\text{C}_6\text{H}_5}\overset{..}{\text{N}}\overset{\text{H}}{\underset{\text{CH}_3}{\big|}}$ $\xrightarrow[\text{0–10°}]{\text{H}^{\oplus},\ \text{NaNO}_2}$ (15)

(**l**) [benzene] $\xrightarrow{(16)}$ [C$_6$H$_5$NO$_2$] $\xrightarrow{(17)}$ [C$_6$H$_5\overset{..}{\text{N}}\text{H}_2$] $\xrightarrow{(18)}$ [C$_6$H$_5$—$\overset{\oplus}{\text{N}}$≡$\overset{..}{\text{N}}$ HSO$_4^{\ominus}$]

$\xrightarrow{(19)}$ [Br—C$_6$H$_5$]

$\xrightarrow{(20)}$ [I—C$_6$H$_5$]

$\xrightarrow{(21)}$ [H—C$_6$H$_5$]

$\xrightarrow{(22)}$ [OH—C$_6$H$_5$]

$\xrightarrow{(23)}$ [C$_6$H$_5$—C≡$\overset{..}{\text{N}}$]

$\xrightarrow{(24)}$ [C$_6$H$_5$—S—CH$_3$]

(**m**) [pyridine] $\xrightarrow[\text{H}_2\text{O}]{\text{HCl}}$ (25)

$\xrightarrow{(26)}$ [N-ethylpyridinium, $\overset{\oplus}{\text{N}}$—CH$_2CH_3$]   CH$_3CH_2SO_4^{\ominus}$ $\xrightarrow[\text{2. H}_3\overset{..}{\text{O}}{}^{\oplus}]{\text{1. LiAlH}_4\ \text{ether}}$ (27)

(**n**) [C$_6$H$_5$—CH$_2\overset{..}{\text{N}}$H$_2$] $\xleftarrow{(28)}$ [C$_6$H$_5$—CH$_2$—$\overset{\overset{\text{O}}{\|}}{\text{C}}$—NH$_2$]

$\xrightarrow[\text{2. H}_3\overset{..}{\text{O}}{}^{\oplus}]{\text{1. LiAlH}_4}$ (29)

(**o**) [phthalimide, N—H] $\xrightarrow[\text{alcohol}]{\text{KOH}}$ (30) $\xrightarrow{\text{CH}_3\text{—}\overset{\overset{\text{Cl}}{|}}{\text{CH}}\text{—COOCH}_2\text{CH}_3}$ (31) $\xrightarrow[\substack{\text{H}_2\text{O,}\\ \text{heat}}]{:\overset{..}{\text{O}}\text{H}^{\ominus}}$ (32)

(p)

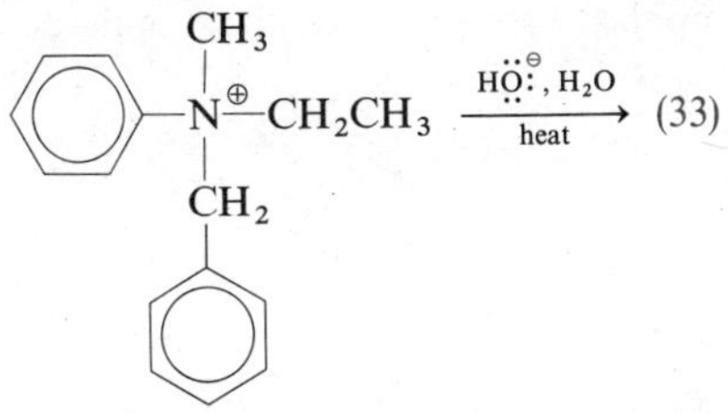

$$\underset{\overset{|}{CH_2}}{\overset{\overset{CH_3}{|}}{\underset{|}{N}}} \overset{\oplus}{-}CH_2CH_3 \xrightarrow[\text{heat}]{H\ddot{O}:^{\ominus},\ H_2O} \quad (33)$$

(q) Benzene $\xrightarrow{(34)}$ $\langle\bigcirc\rangle$—$\ddot{N}$=$\ddot{N}$—$\langle\bigcirc\rangle$—$\ddot{N}(CH_3)_2$

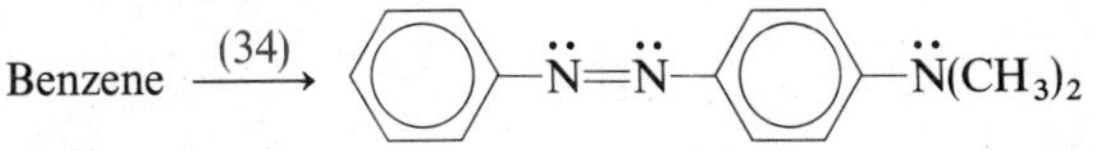

(r)

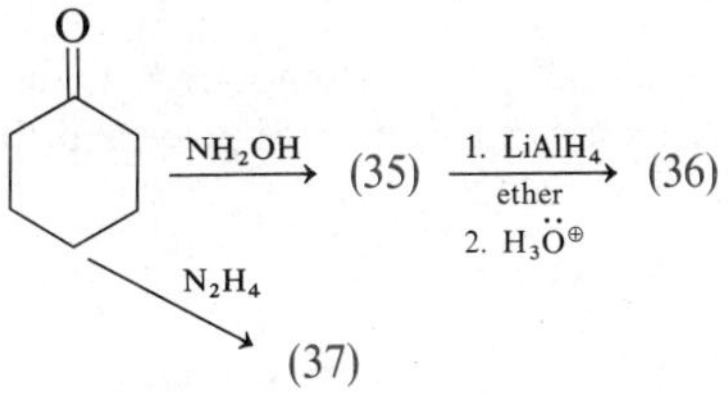

cyclohexanone $\xrightarrow{NH_2OH}$ (35) $\xrightarrow[\text{ether}]{1.\ LiAlH_4}$ (36)

$\xrightarrow{N_2H_4}$ (37)

(s) CH_3—$\langle\bigcirc\rangle$—CH_2CH_3 $\xrightarrow[\substack{\text{liquid}\\NH_3}]{:\ddot{N}H_2^{\ominus}}$ (38)

with Br substituent

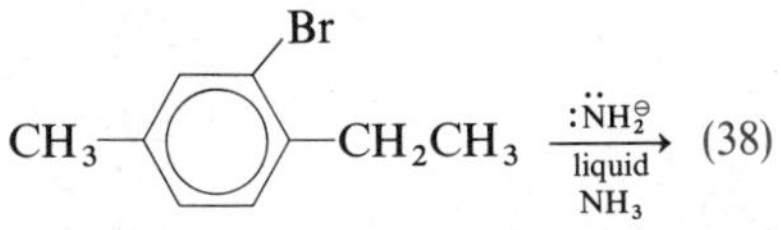

(t)

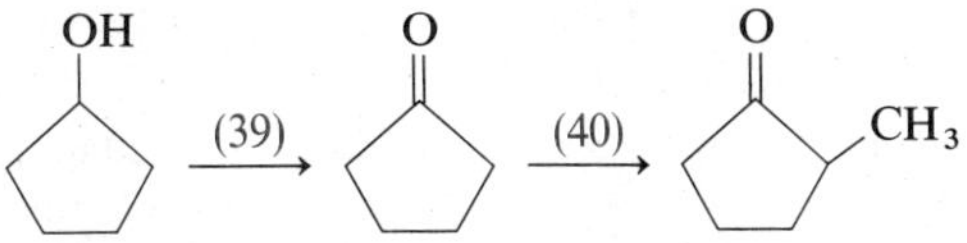

OH-cyclopentane $\xrightarrow{(39)}$ cyclopentanone $\xrightarrow{(40)}$ methyl cyclopentanone

(u) $\langle\bigcirc\rangle$—$\ddot{N}H_2$ $\xrightarrow{(41)}$ O_2N—$\langle\bigcirc\rangle$—$\ddot{N}H_2$ $\xrightarrow{(42)}$ O_2N—$\langle\bigcirc\rangle$—CN

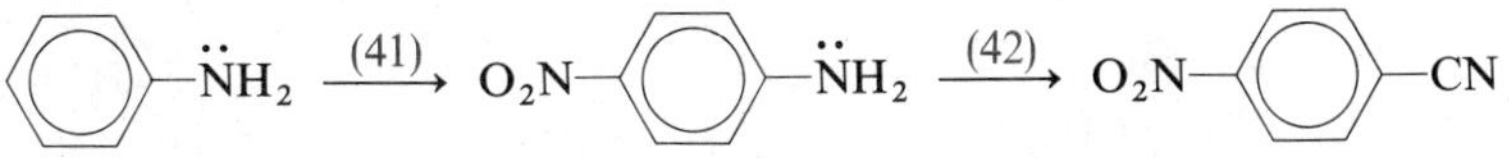

26.51 Novocaine, a widely used local anesthetic, has the following structure:

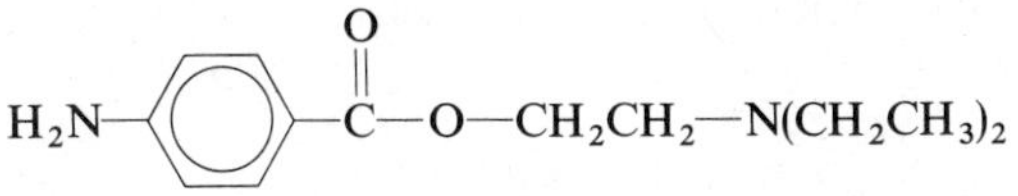

$$H_2N-\langle\bigcirc\rangle-\overset{\overset{O}{\|}}{C}-O-CH_2CH_2-N(CH_2CH_3)_2$$

Design a logical laboratory synthesis for Novocaine beginning with benzene and diethylamine and using any other C_1 or C_2 compounds you wish. In addition, you may use any inorganic reagents.

26.52 For the following reaction

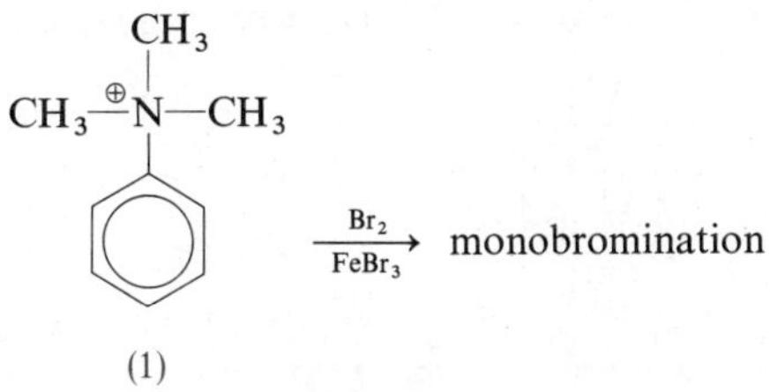

$$CH_3-\overset{\overset{CH_3}{|}}{\underset{|}{N}}\overset{\oplus}{-}CH_3$$

on benzene ring $\xrightarrow[\text{FeBr}_3]{Br_2}$ monobromination

(1)

consider the possibility of attachment by $Br^{\oplus}$ at the *meta* and at the *para* position of compound 1.

(*a*) Write the resonance structures for the intermediate carbocations that result when $Br^{\oplus}$ attacks at the *meta* position of compound (1).

(*b*) Write the resonance structures for the intermediate carbocations that result when $Br^{\oplus}$ attacks at the *para* position of compound (1).

(*c*) On the basis of the resonance structures you drew in (*a*) and (*b*), deduce whether the —$N(CH_3)_3^{\oplus}$ group is an *ortho,para* director or a *meta* director. Briefly explain.

26.53 Codeine is a monomethyl ether of morphine. Show how morphine could be methylated at one —OH only to give codeine.

HO ... O ... N—CH₃ ... HO

→

CH₃—O ... O ... N—CH₃ ... H—O

Morphine   Codeine

26.54 Epoxy resins are well-known, two-component adhesives. Compound *A*, the hardener, and compound *B*, the adhesive, are mixed, whereupon a reaction takes place and the glue sets or hardens. What is the structure of the hard resin formed?

$$CH_2{-}CH{-}CH_2{-}(O{-}C_6H_4{-}\underset{CH_3}{\overset{CH_3}{C}}{-}O{-}CH_2{-}\underset{OH}{CH}{-}CH_2{-})_n O{-}CH_2{-}CH{-}CH_2$$

B

$$H_2\ddot{N}{-}CH_2CH_2{-}\ddot{N}H{-}CH_2CH_2{-}\ddot{N}H_2$$

A

26.55 *L*-DOPA (*L*-dihydroxyphenylalanine) is one of the intermediates involved in the synthesis of adrenalin (epinephrine) in the body.

$$CH_2{-}CH{-}\overset{O}{\overset{\|}{C}}{-}OH$$
$$\underset{NH_2}{}$$
HO ... OH

enzymatic action →

$$\overset{OH}{CH}{-}CH_2{-}N\overset{H}{\underset{CH_3}{}}$$
HO ... OH

L-DOPA   Epinephrine

The IR spectra of both these compounds are shown here. Which spectrum goes with which compound? Briefly explain.

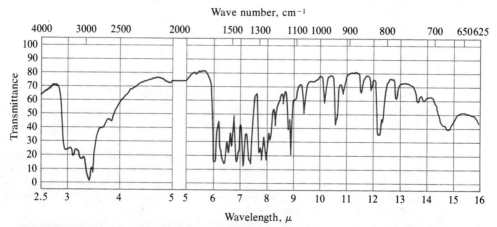

PROBLEM 26.55   IR spectrum (I). [Data from *The Aldrich Library of Infrared Spectra*, C. J. Pouchert (ed.), copyright 1975, Aldrich Chemical Company, Milwaukee]

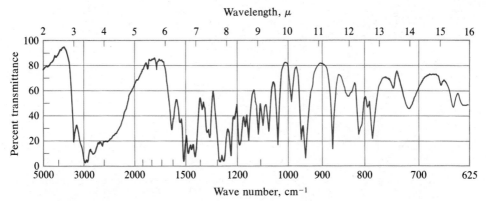

Wavelength, μ

PROBLEM 26.55 IR spectrum (II). [Data from *Indexes to Evaluated Infrared Reference Spectra*, copyright 1975, The Coblentz Society, Norwalk CT]

26.56 Complete each of the following 10 syntheses. In addition to the indicated starting material, you can use any inorganic compound, organic solvent, or *monofunctional* organic compound.

(*a*) Toluene ⟶ [2-iodotoluene structure, CH₃ and I on benzene ring]

(*b*) [benzene] ⟶ [Ph–N̈=N̈–C₆H₄–OH]

(*c*) [nitrobenzene, NO₂] ⟶ HO–[C₆H₄]–CH₃

(*d*) [C₆H₅–NO₂] ⟶ [3-iodonitrobenzene, NO₂ and I on ring]

(*e*) [acetanilide: HN̈–C(=O)–CH₃ on benzene] ⟶ [4-chlorobenzoic acid: C(=O)–OH on ring with Cl]

(*f*) [aniline, N̈H₂] ⟶ H₂N̈–[C₆H₄]–S(=O)₂–N̈H–[C₆H₅]

(*g*) [anthracene, three fused rings] ⟶ [9-fluoroanthracene, F substituent]

(*h*) CH₃CH₂CH₂CH₂—OH ⟶ CH₂CH₂CH₂—N̈H₂

(*i*)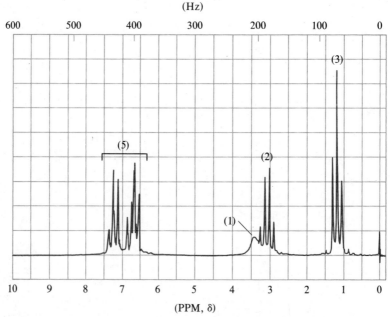

$\underset{\text{Br}}{\text{aniline}}$—$NH_2$

(*j*)

26.57 An aromatic compound *A*, $C_8H_{11}N$, is examined in the laboratory to elucidate its structure. The following observations are made:

Test 1: *A* is soluble in dilute hydrochloric acid but insoluble in sodium hydroxide solution.
Test 2: Treatment of *A* with sodium hydroxide and benzenesulfonyl chloride, $C_6H_5SO_2Cl$, results in the formation of a heterogeneous mixture. The NMR spectrum of *A* is given.
Test 3: *A* is treated with acetic anhydride to give *B*, $C_{10}H_{13}ON$. *B* is insoluble in dilute acid or dilute base at room temperature; heating *B* in dilute acid or base regenerates *A*.
Test 4: When *B* is heated with concentrated nitric acid, a single product *C*, $C_{10}H_{12}O_3N_2$, is formed in excellent yield.

On the basis of these observations, draw the structures of compounds *A* to *C*. Write an equation (not balanced) for *each* reaction that occurs, and indicate clearly what *each* test teaches you about the structures of the unknown compounds.

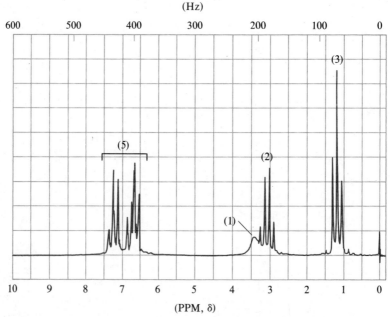

PROBLEM 26.57   NMR spectrum of unknown (*A*). [Data from *The Aldrich Library of NMR Spectra*, C. J. Pouchert and J. R. Campbell (eds.), copyright 1974, Aldrich Chemical Co., Milwaukee]

26.58 PCP, 1-(1-phenylcyclohexyl)piperidine, is one of the most widely abused drugs on the American street today. Deduce its structure from the following experimental data:
(*a*) PCP gives a $P^{\oplus}$ peak in the mass spectrometer of $m/e\ 243 \pm 1$.
(*b*) Hofmann exhaustive methylation and elimination (two times) give 1,4-pentadiene and compound *A*, $C_{14}H_{21}N$.
(*c*) Hofmann exhaustive methylation and elimination of *A* give trimethylamine and *B*, $C_{12}H_{14}$.

(*d*) *B* can be synthesized from cyclohexanone by reacting first with phenylmagnesium bromide and then with $H_3\overset{..}{O}{}^{\oplus}$ and then H_2SO_4 (conc) and heat.

26.59 Compound *A*, $C_8H_{13}NO$, an alkaloid, reacts with hydrogen and a catalyst to give *B*, $C_8H_{15}NO$. *B* reacts with thionyl chloride to give *C*, $C_8H_{14}NCl$, which undergoes reductive dehalogenation (H_2/Pt) to give *D*, $C_8H_{15}N$. Hofmann exhaustive methylation of *D* gives *E*.

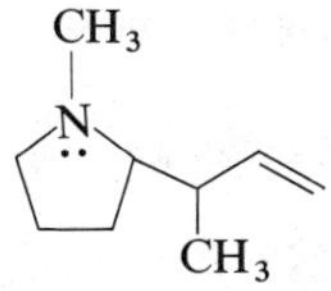

E

Give structures for *A* to *D*.

26.60 Explain the following experimental observations mechanistically.

(*a*)

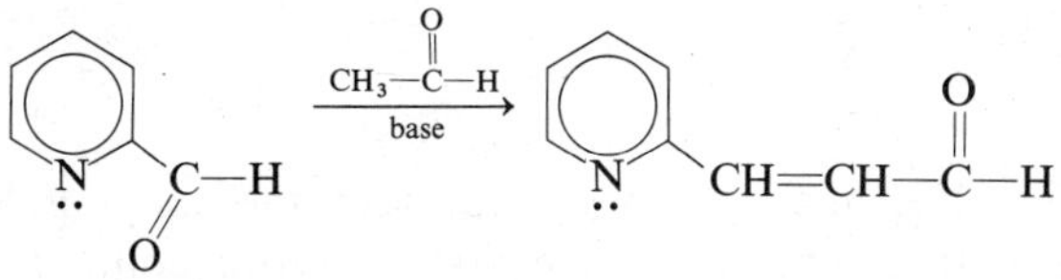

(*b*)

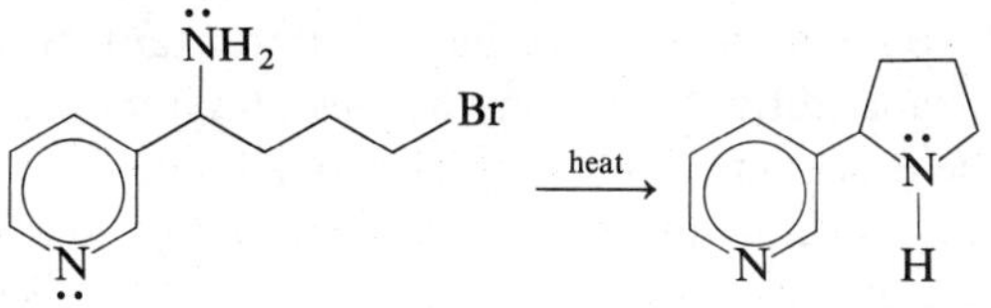

Nicotine

(*c*)

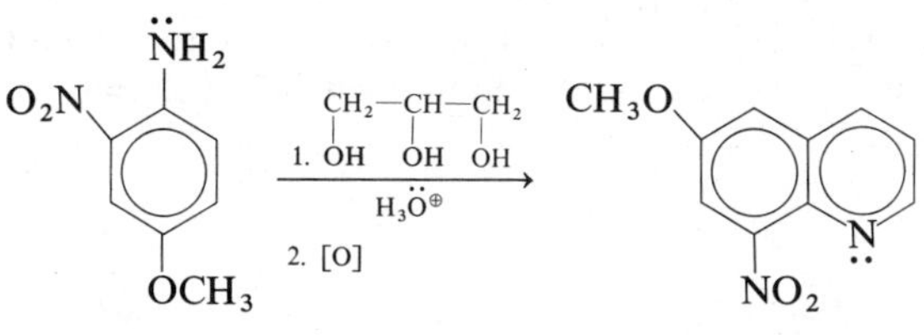

Quinoline
Skraup synthesis

(*d*)

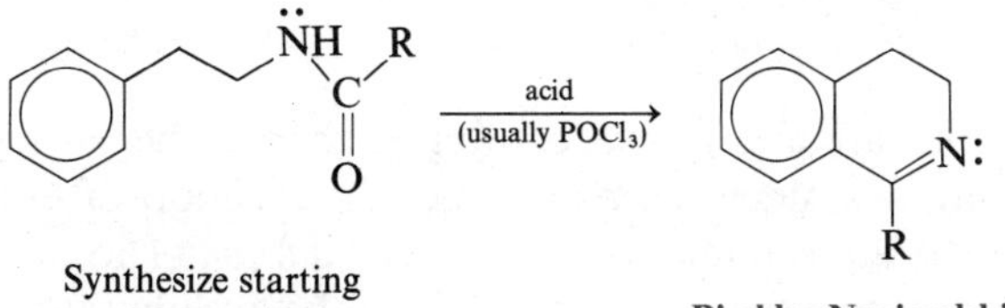

Synthesize starting
material from benzene

**Bischler-Napieralski
reaction**

(*e*)

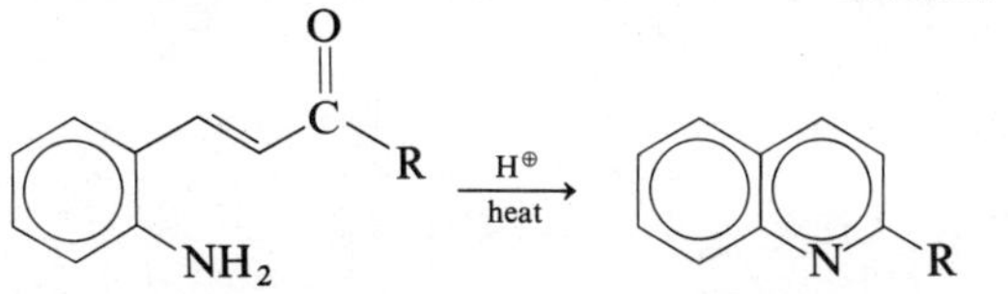

Synthesize starting
material from benzene

**Friedlander quinoline
synthesis**

27

Cycloaddition Reactions:
Woodward-Hoffmann Rules

Cycloaddition reactions involve, in the most general case, the reaction of two π-electron systems, and result in the overall conversion of two π bonds into two σ bonds. They provide a direct route to a wide variety of important compounds. We discuss several common cycloaddition reactions in this chapter including the **Diels-Alder reaction,** perhaps the most versatile reaction of this type.

In this chapter, we also see several interesting applications of molecular orbital theory to the study of reaction mechanisms in organic chemistry. We see, for example, how MO theory can be used to predict whether or not various reactions will indeed occur and, if so, by what pathway. We learn how the stereochemistry of these reactions can be explained and predicted and see how this can be of use to the synthetic organic chemist.

Synthetic Applications

27.1   Cycloaddition Reactions

Cycloaddition reactions are classified according to the number of π electrons in each of the two reacting π systems. For example, the reaction of ethene with ethene to give cyclobutane is an example of *2 + 2 cycloaddition*. The two π bonds of each ethene molecule are converted to two σ bonds in the cyclobutane molecule. The reaction occurs under the influence of heat or light.

2 + 2 Cycloaddition:

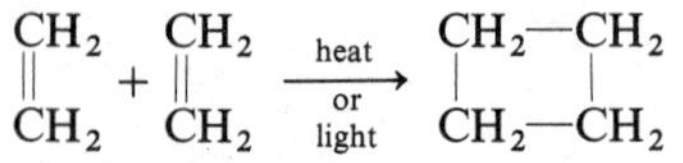

$$
\begin{array}{c}
CH_2 \\
\parallel \\
CH_2
\end{array}
+
\begin{array}{c}
CH_2 \\
\parallel \\
CH_2
\end{array}
\xrightarrow[\substack{or \\ light}]{heat}
\begin{array}{c}
CH_2\!-\!CH_2 \\
| \qquad | \\
CH_2\!-\!CH_2
\end{array}
$$

The most common cycloaddition reaction is the Diels-Alder reaction, a *4 + 2 cycloaddition*. This reaction was introduced in Secs. 8.29 and 11.15. The two-π-electron system usually comes from an alkene or alkyne and is called the *dienophile*. The four π electron system usually comes from a *diene*. The reaction is shown here in its simplest form.

4 + 2 Cycloaddition:

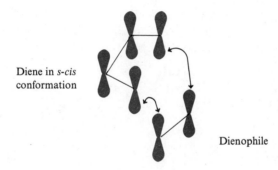

| Diene | Dienophile | Diels-Alder product |

Although there are other types of cycloaddition reactions, we focus primarily on the preceding two classes.

27.2   Diels-Alder Reaction: 4 + 2 Cycloaddition

The Diels-Alder reaction between a hydrocarbon diene and a dienophile requires both heat and pressure to achieve good yields. When the rate of this reaction is followed kinetically, this rate expression is obtained:

$$\text{Rate} = k_2[\text{diene}][\text{dienophile}]$$

where k_2 = rate constant

Thus, the transition state must involve one molecule each of the diene and dienophile. Further, the reaction rate is relatively unaffected by catalysts and solvents. Experiments indicate that radicals and carbocations are not involved in the mechanistic pathway.

The currently accepted mechanism consistent with these observations involves *simultaneous bond breaking and bond making* in the transition state. As shown in three dimensions, the transition state probably resembles the following:

Transition
state

It seems reasonable that the p orbitals in the diene and the dienophile overlap before reaction can occur, so the diene must be in the *s-cis* conformation:

Diene in *s-cis*
conformation

Dienophile

A simplified way to represent the transition state is shown here, but remember that it does *not* adequately represent the overlap of orbitals:

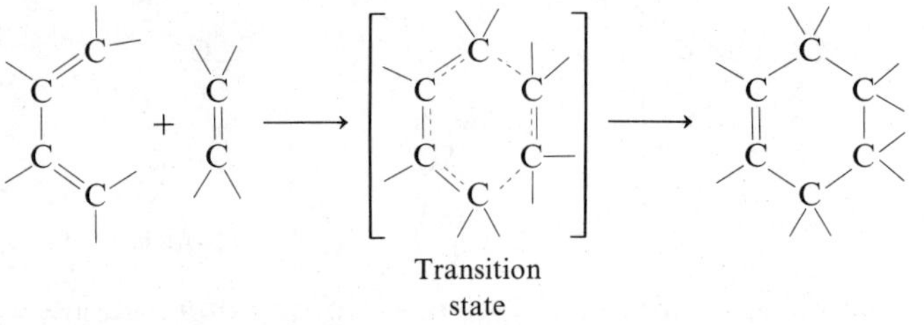

Transition
state

Note that the transition state has little free-radical character and is not polar because positive and negative charges are not generated. To keep track of the electrons with arrows, the following designation is often used:

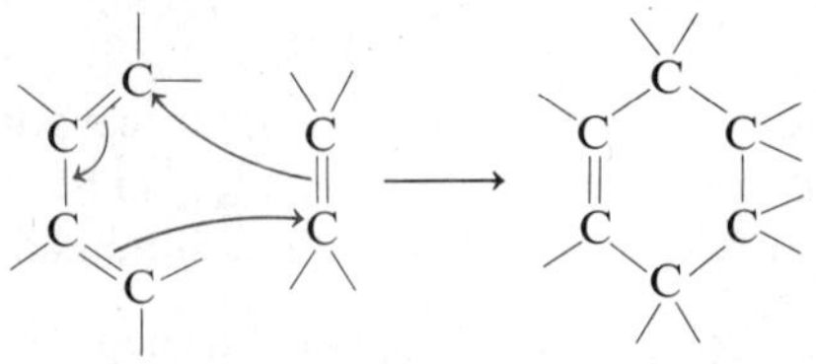

Many different dienes and dienophiles are used to afford good yields of adducts. The presence of electron-donating groups (such as alkyl or alkoxy, —O—R) on the diene or of electron-withdrawing groups (such as cyano, —C≡N, and carbonyl, $>$C=O) on the dienophile seems to improve yields and accelerate the rate of reaction. With these substituents present, the Diels-Alder reaction is often exothermic. When ethene is used as the dienophile, however, heat must be supplied.

Other interesting aspects of the reaction support the cyclic transition state. As pointed out, the diene must be capable of existing in the *s-cis* conformation, which is the geometry required for the formation of the *cis* double bond in the product. If the diene is *s-trans*, it would lead to a *trans* double bond, which cannot be formed in a six-membered ring. Using 1,3-butadiene as an example, we have

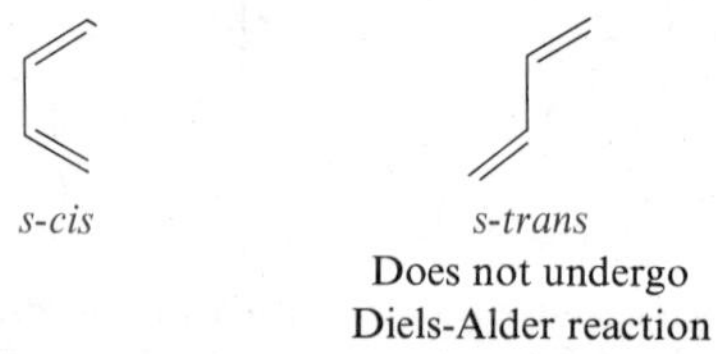

s-cis

s-trans
Does not undergo
Diels-Alder reaction

The Diels-Alder reaction is stereospecific because of the strict orientation required in the transition state. For example, the reaction between *trans,trans*-2,4-hexadiene and a dienophile yields exclusively *cis*-3,6-dimethylcyclohexene (see Sec. 27.13 for more details):

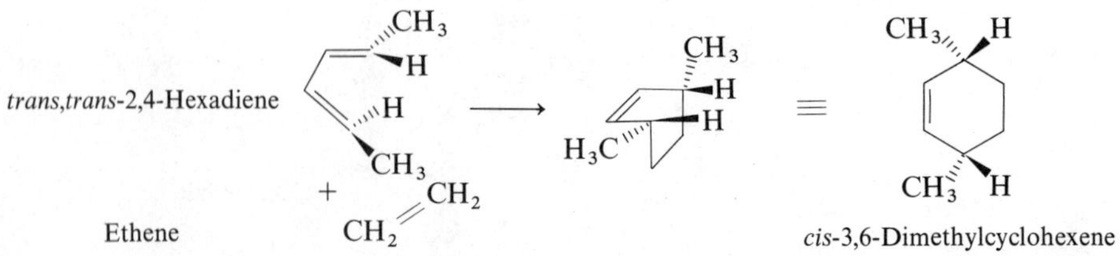

trans,trans-2,4-Hexadiene

Ethene

cis-3,6-Dimethylcyclohexene

Relatively few side reactions accompany the Diels-Alder reaction, which is often quantitative. The single most important side reaction is the dimerization of the diene in which one molecule of diene serves as the "diene" and one double bond of the other molecule serves as the "dienophile." This side reaction is more important when a poor dienophile, such as ethene, is used. A typical side reaction involving 1,3-butadiene is

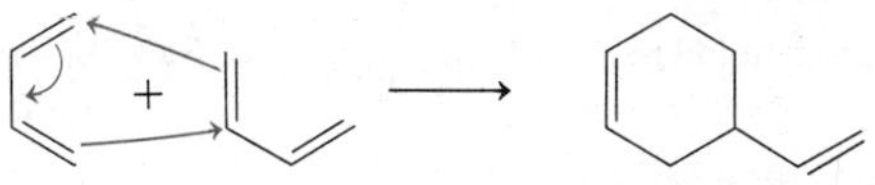

4-Ethenylcyclohexene
(4-vinylcyclohexene)

Question 27.1

Explain why 1-methyl-1,3-cyclohexadiene (1) undergoes a Diels-Alder reaction whereas 3-methylenecyclohexene (2) does not.

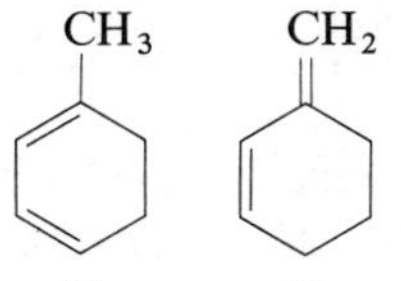

CH_3 CH_2

(1)   (2)

Question 27.2

Predict the stereochemistry of the product from the reaction between ethene and *cis,trans*-2,4-hexadiene.

Question 27.3

Suggest why *cis,cis*-2,4-hexadiene does not undergo a Diels-Alder reaction.

Question 27.4

The Diels-Alder reaction is stereospecific and *trans,trans*-2,4-hexadiene reacts with ethene to give *cis*-3,6-dimethylcyclohexene (see above). Suppose 1,3-butadiene reacted with a dienophile such as *cis*-1,2-dichloroethene or *trans*-1,2-dichloroethene. What stereochemistry would you expect from each of these dienophiles? Why?

27.3 Bicyclic Compounds: Application of Diels-Alder Reaction

In Sec. 27.2 we presented reactions of open-chain dienes and dienophiles to form cyclohexene derivatives. Modification of the diene structure produces some interesting and often complex cyclic compounds.

One of the simplest cyclic dienes is cyclopentadiene, which reacts with ethene at elevated temperature ($\sim 200°$) and pressure:

| Cyclopentadiene | Ethene | Bicyclo[2.2.1]-2-heptene (norbornene) |

1,3-Cyclohexadiene also reacts with ethene to give a Diels-Alder adduct:

1,3-Cyclohexadiene   Bicyclo[2.2.2]-2-octene

Cyclopentadiene is so reactive that it readily dimerizes at room temperature to produce dicyclopentadiene; however, this reaction is reversible, and at 170° the dimer is "cracked" and converted back into monomer. The dimerization reaction is similar to that of butadiene with itself, which was discussed in Sec. 27.2.

Cyclopentadiene   Cyclopentadiene dimer

Question 27.5

Draw structures corresponding to the following names of bicyclic compounds:

(*a*) bicyclo[3.2.2]-2-nonene   (*b*) 1-methylbicyclo[2.2.2]octane
(*c*) 1,3-dimethylbicyclo[1.1.0]butane

Question 27.6

Name the following compounds by the IUPAC system.

27.4 Diels-Alder Reaction Using α,β-Unsaturated Carbonyl Compounds

In Chaps. 23 and 24 we learned that many reactions (such as Aldol, Perkin, and Claisen reactions) yield α,β-unsaturated carbonyl compounds. We now use these α,β-unsaturated carbonyl compounds in the Diels-Alder reaction. In doing so, we see that using an α,β-unsaturated carbonyl compound as the dienophile provides a good synthetic approach for many new types of compounds.

Certain dienes or dienophiles react to give a product in which there is no question about the stereochemistry of the adduct because only one isomer is formed:

| 2-Methyl-1,3-butadiene (isoprene) | Maleic anhydride | 4-Methyl-1,2,3,6-tetrahydrophthalic anhydride |

| 1,3-Butadiene | Diethyl butynedioate (diethyl acetylenedicarboxylate) | Diethyl 3,6-dihydrophthalate |

| 2,3-Dimethyl-1,3-butadiene | Propenal | 3,4-Dimethyl-1,2,5,6-tetrahydrobenzaldehyde |

Adducts such as these can be dehydrogenated (aromatized) on heating with sulfur, selenium, or palladium to produce the corresponding aromatic compounds. This is a highly useful synthetic aspect of the Diels-Alder reaction; for example:

| Prepared earlier | 4-Methylphthalic anhydride |

In reactions involving cyclic dienes, certain dienophiles, such as 1,1-disubstituted alkenes, tetrasubstituted alkenes, and disubstituted alkynes, react to give only one possible adduct; for example:

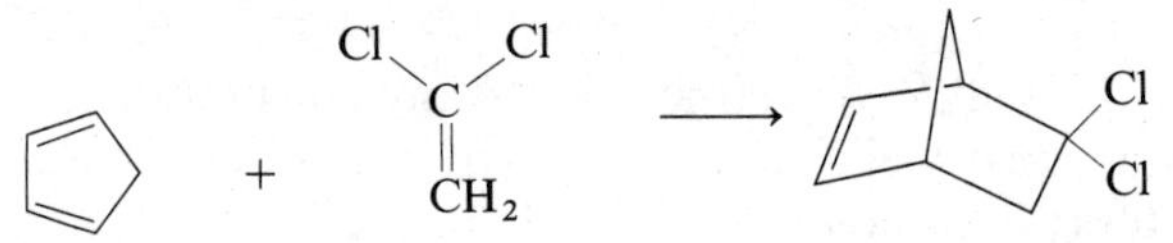

Cyclopentadiene   1,1-Dichloroethene

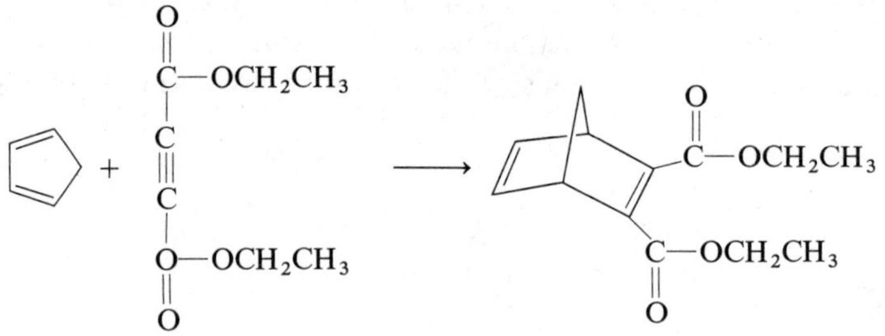

Diethyl butynedioate

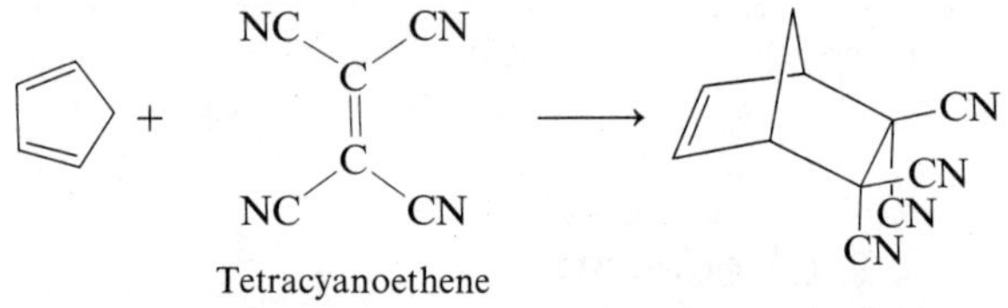

Tetracyanoethene

One virtue of the Diels-Alder reaction is that cycloaddition occurs with known stereospecificity. The addition of a dienophile to a diene involves the overlap of π bonds, and when the dienophile contains more than one double bond (as in α,β-unsaturated carbonyl compounds), the second double bond appears to prefer being near the diene even though this bond does not enter directly into the reaction. In short, *the stereochemistry of the Diels-Alder reaction is dictated by maximum overlap of as many π bonds as possible of both the dienophile and the diene* (see Sec. 27.13).

Let us translate these principles into practice by considering the reaction of cyclopentadiene with itself. The following mode of cycloaddition must occur:

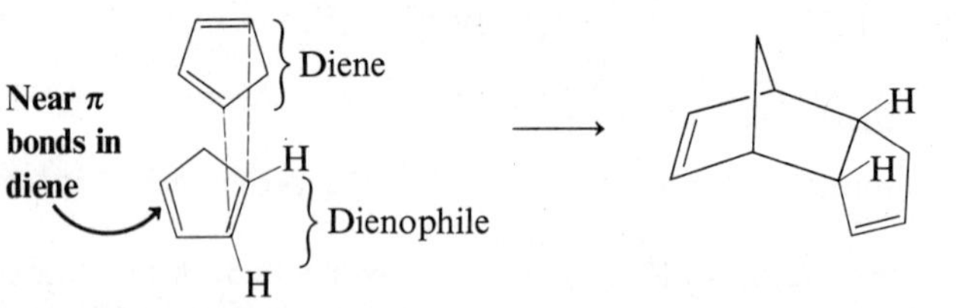

Near π bonds in diene   } Diene   } Dienophile

Cyclopentadiene   *endo*-Dicyclopentadiene
2 moles   *Major product*

The two molecules of cyclopentadiene "sit" on top of one another to maximize overlap of the π bonds. The product that forms has the original dienophile molecule folded underneath with the two hydrogens sticking outward, as shown in the preceding diagram. This is called the *endo* isomer, which is the observed product.

The alternative mode of addition is shown as follows, with the second double bond in the dienophile pointed in the opposite direction from the double bonds in the diene:

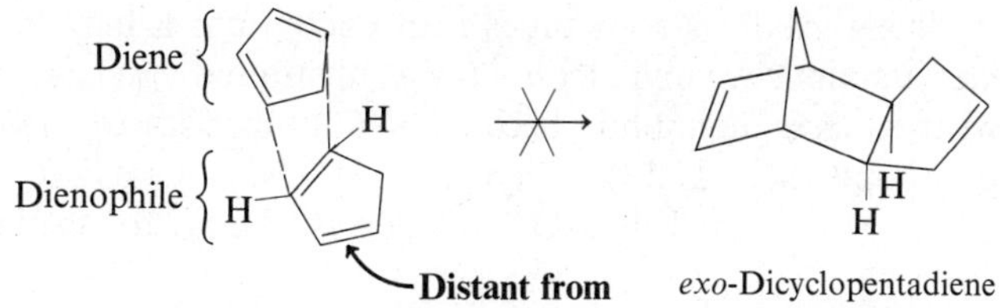

Diene

Dienophile

Distant from
π bonds in
diene

exo-Dicyclopentadiene
Minor product

This mode of addition does not occur to any appreciable extent; only minor amounts of this *exo* compound are observed.

The terms *exo-* (Greek *exo*, out) and *endo* (Greek *endon*, within) designate the stereochemistry of the products. The *exo* substituents protrude outward from the bicyclic ring structure and the *endo* substituents are folded back under it.

The following examples illustrate two additional applications of the Diels–Alder reaction:

Cyclopentadiene

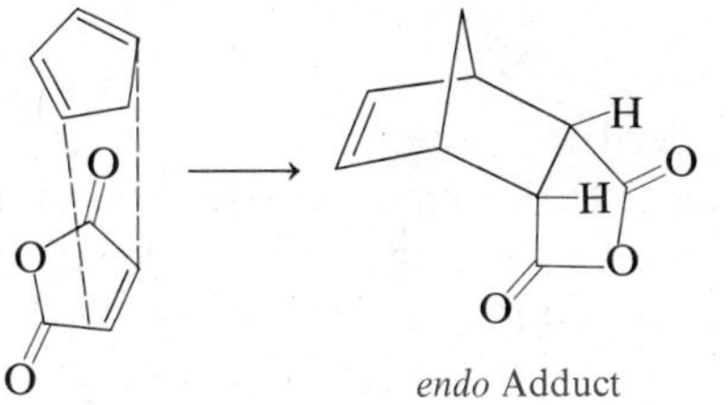

Maleic anhydride

endo Adduct

Comment: Carbon-oxygen double bonds are folded under cyclopentadiene to maximize π overlap

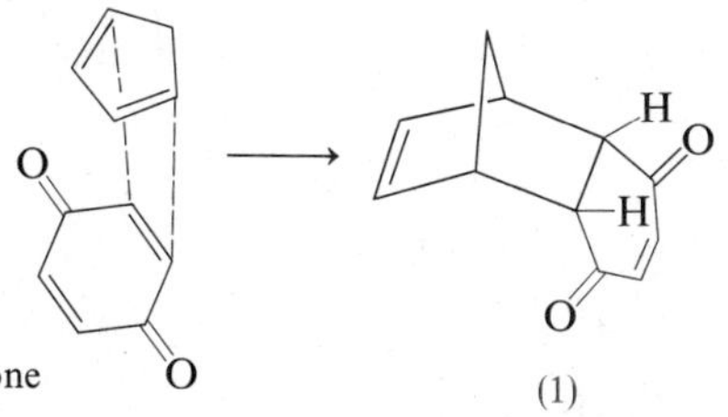

1,4-Benzoquinone

(1)
endo Adduct

The monoadduct (1) from the reaction between 1,4-benzoquinone and cyclopentadiene can then undergo a second Diels-Alder reaction with another molecule of cyclopentadiene. The principles of stereochemistry are still followed.

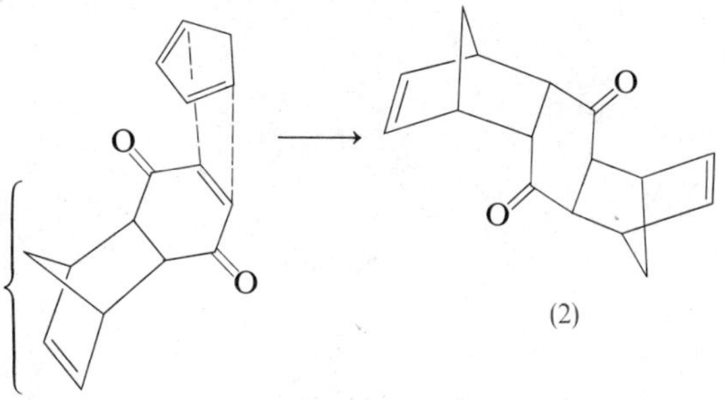

Adduct

(1)

(2)

The diadduct (2) can be redrawn in the following way to emphasize its structure:

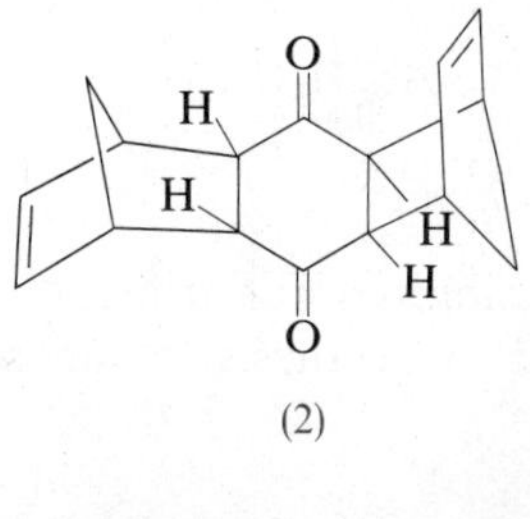

(2)

A hydroxy group is sometimes introduced into a bicyclic compound with a Diels-Alder reaction. A vinyl alcohol (or a substituted vinyl alcohol) is unstable and cannot be used as a dienophile because it tautomerizes to the corresponding carbonyl compound (Sec. 11.13). Vinyl acetate, CH_3—COO—CH=CH$_2$, is stable, however, and can be used in a Diels-Alder synthesis; for example:

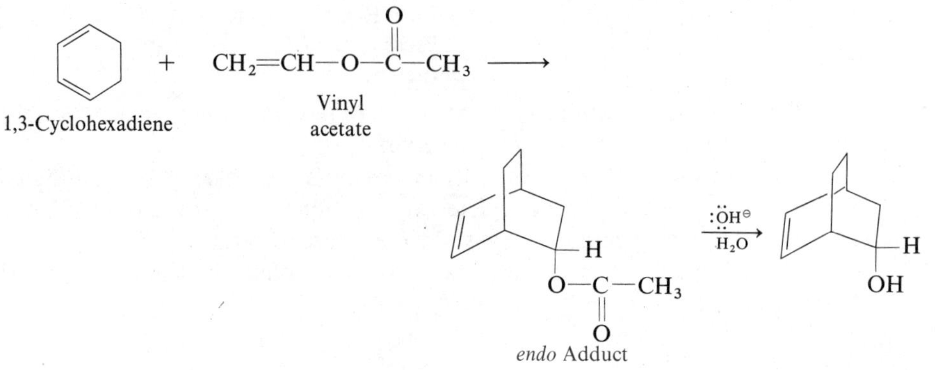

1,3-Cyclohexadiene

Vinyl acetate

endo Adduct

Thus far we emphasized the generation of carbon skeletons through the use of the Diels-Alder reaction. Yet this reaction also gives cyclic compounds that contain atoms other than carbon in the ring. This is in keeping with the prime requirement for Diels-Alder, in that a compound containing two (or more) alternate double bonds (the "diene") reacts with a compound containing one (or more) double bond (the "dienophile"). The following three examples illustrate methods for synthesizing compounds that contain oxygen and nitrogen in the ring structure.

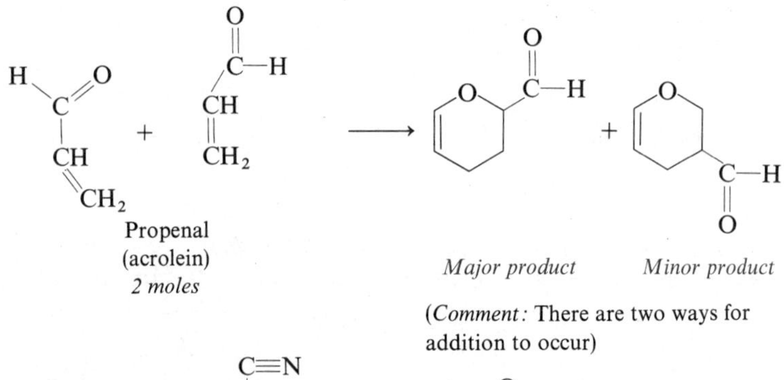

Propenal (acrolein)
2 moles

Major product   *Minor product*

(*Comment:* There are two ways for addition to occur)

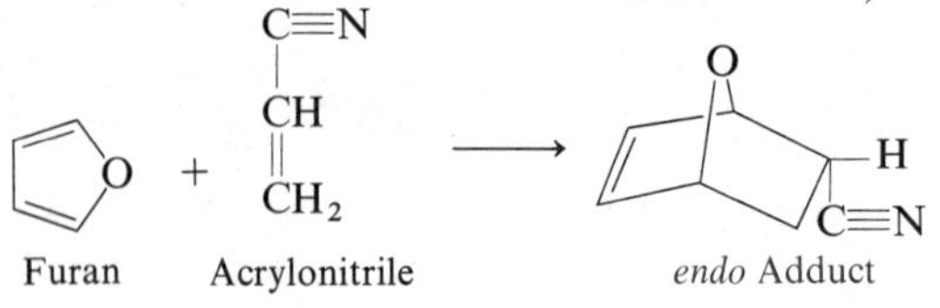

Furan   Acrylonitrile   *endo* Adduct

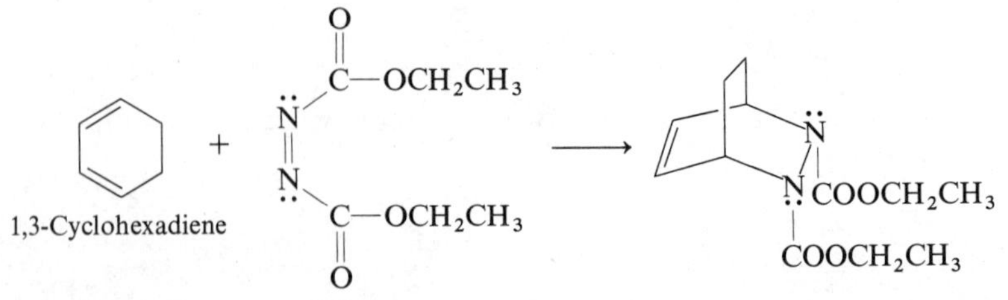

1,3-Cyclohexadiene

The number of possibilities for using the Diels-Alder reaction in making carbon-carbon bonds and in synthesis is almost unlimited.

Question 27.7

Provide the structures of the Diels-Alder adducts that would be formed in each of the following situations (monoadducts only). Be certain to indicate stereochemistry.

(*a*) 1,3-cyclohexadiene + maleic anhydride   (*b*) cyclopentadiene + methyl vinyl ketone
(*c*) 1,3-butadiene + methyl acrylate, $CH_2\!=\!CH\!-\!COOCH_3$
(*d*) isoprene + diethyl fumarate, *trans*-$CH_3CH_2OOC\!-\!CH\!=\!CH\!-\!COOCH_2CH_3$
(*e*) 2-methoxy-1,3-butadiene + *p*-benzoquinone
(*f*) cyclopentadiene + *cis*-benzalacetophenone, *cis*-$C_6H_5\!-\!CH\!=\!CH\!-\!CO\!-\!C_6H_5$
(*g*) 1,3-butadiene + nitroethene, $CH_2\!=\!CH\!-\!NO_2$

27.5   2 + 2 Cycloaddition Reactions

The 2 + 2 cycloaddition of alkenes is not as general a reaction as the 4 + 2 cycloaddition. Yields are often low and mixtures of products are often obtained. Three examples of successful 2 + 2 cycloadditions are:

The following mechanisms are proposed.

Concerted mechanism:

Radical mechanism:

The 2 + 2 cycloaddition mechanism is discussed in more detail in Sec. 27.12.

27.6   Addition of Methylene to Alkenes: Cyclopropanes

A reaction also classified as 2 + 2 cycloaddition involves the addition of the reactive intermediate **methylene** (also called **carbene**) to alkenes (see Sec. 8.29). Before continuing our discussion of cycloaddition reactions, we look at the synthetic applications of the methylene addition reaction.

Methylene is formed by subjecting diazomethane, CH_2N_2, or ketene, CH_2CO, to irradiation by UV light or by thermal decomposition (pyrolysis):

$$CH_2{=}\overset{\oplus}{N}{=}\overset{\ominus}{\underset{\cdot\cdot}{N}}: \quad \xrightarrow{\textit{hv or heat}} \quad :CH_2 + N_2$$

Diazomethane   Methylene

$$CH_2{=}C{=}\overset{\cdot\cdot}{\underset{\cdot\cdot}{O}} \quad \xrightarrow{\textit{hv or heat}} \quad :CH_2 + CO$$

Ketene   Methylene

The driving force for these reactions is the formation of the very stable nitrogen and carbon monoxide molecules. The existence of methylene has been shown spectroscopically, and chemical reactions support its existence.

The methylene molecule is highly reactive, and when produced by photochemical decomposition it appears to be in the **singlet** state (see Sec. 2.8). This species undergoes **insertion** reactions in which the elements of CH_2 are inserted **between** each carbon-hydrogen bond:

Singlet insertion:

$$-\overset{|}{\underset{|}{C}}{-}H + :CH_2 \longrightarrow -\overset{|}{\underset{|}{C}}{-}CH_2{-}H \equiv -\overset{|}{\underset{|}{C}}{-}CH_3$$

It also undergoes **addition** reactions to carbon-carbon double bonds to form cyclopropanes:

Singlet addition:

$$\overset{}{\underset{}{\diagup}}C{=}C\overset{}{\underset{}{\diagdown}} + :CH_2 \longrightarrow -\overset{|}{C}\underset{\underset{CH_2}{\diagdown\diagup}}{}\overset{|}{C}-$$

For example, the photochemical decomposition of diazomethane in the presence of 2-methylpropene yields the following:

$$CH_2{=}C\overset{CH_3}{\underset{CH_3}{\diagup\diagdown}} \xrightarrow[\textit{hv}]{CH_2=N=N} CH_3{-}CH{=}C\overset{CH_3}{\underset{CH_3}{\diagup\diagdown}} + CH_2{=}C\overset{CH_2{-}CH_3}{\underset{CH_3}{\diagup\diagdown}} + CH_2{-}C\overset{CH_3}{\underset{CH_3}{\diagup\diagdown}}$$

Insertion   Addition

*Reactions carried out in the liquid phase favor singlet methylene, whereas those in the gas-phase favor **triplet** methylene.* Other typical changes include varying the wavelength of light used in irradiation (photolysis), changing the source of methylene, and adding various types of gases in gas-phase reactions. The chemistry of methylene (and of carbenes in general) has been investigated extensively during the past years, and we do not attempt to delve into specific reaction conditions. However, let us compare the reactions of singlet and triplet methylene.

Singlet methylene undergoes both insertion and addition. However, it is so reactive that it exhibits no selectivity in the bonds it attacks; that is, it attacks all carbon-hydrogen bonds and adds to all carbon-carbon double bonds. Furthermore, the addition of singlet methylene is *stereospecific*, as the following general example illustrates. The reaction occurs in a one-step, concerted manner, which accounts for the observed stereospecificity (see Sec. 6.16).

The degree of stereospecificity of the addition reaction can be altered by changing experimental conditions, thus indicating that the conversion from the singlet state to the triplet state can occur.

cis-Alkene   cis-Cyclopropane

It is believed that triplet methylene is less reactive and more selective. It also undergoes insertion and addition, but the mechanisms are different from the reactions of singlet methylene. These differences are attributed to the radical character of the triplet methylene. We know from studies on alkanes that radicals can abstract a hydrogen and form a new radical. Because triplet methylene is a diradical, it is not unreasonable that it abstracts a hydrogen from carbon, followed by the subsequent recombination of the two radicals.

Triplet insertion:

Abstraction   Recombination

The addition reaction between triplet methylene and a carbon-carbon double bond also appears to occur in two steps. The first is the attack of the electron-deficient methylene on the electron-rich double bond. This breaks the double bond and forms a new diradical, which then can undergo isomerization by rotation about the carbon-carbon single bond. On recombination of the radicals, both *cis*- and *trans*-cyclopropane isomers are obtained.

Triplet addition:

cis-Alkene   Diradical

cis-Cyclopropane   trans-Cyclopropane

The addition of carbenes to double bonds is not limited to alkenes because aromatic rings and other ring systems are also attacked.

It is reasonable that the electron-deficient carbene, $R_2C:$, which has only six valence electrons about carbon, seeks electrons in the same way that radicals and carbocations do. In their reactions, carbenes obtain electrons in two ways: (1) they attack the electron cloud in carbon-hydrogen σ bonds and undergo *insertion* reactions, and (2) they attack the electron cloud of π bonds and undergo *addition* reactions. In the latter case the reaction may be classified as $2 + 2$ cycloaddition. Two electrons from the π bond of the alkene and two electrons from the orbitals of the carbene react to form two new σ bonds, and cyclopropanes result.

Question 27.8

The photolysis of diazomethane in liquid pentane gives 47% *n*-hexane, 36% 2-methyl-pentane, and 17% 3-methylpentane.

(*a*) Account for the formation of these products.
(*b*) Using probabilities alone (based on the number of each type of hydrogen in *n*-pentane that gives each product), compute the product distribution assuming methylene insertion is completely random.
(*c*) Does the data support or refute the claim that methylene insertion is random [that is, that the product distribution does not depend on the type of hydrogen (1°, 2°, or 3°) attacked]? Explain.

Question 27.9

Give the products that would be expected when singlet methylene reacts with the following compounds:

(*a*) *n*-butane   (*b*) cyclohexane
(*c*) *cis*-2-pentene   (*d*) 1-deuteriopropane (CH_2D—CH_2—CH_3)

Question 27.10

When diazomethane is photolyzed in liquid *cis*-2-butene, *cis*-1,2-dimethylcyclopropane is obtained as the only cyclic product. When this same reaction is carried out in the gas phase and with low alkene concentration, a mixture of *cis*- and *trans*-1,2-dimethylcyclopropane is obtained. Which type of methylene must be involved in each of these reactions? Why?

Woodward-Hoffmann Rules

27.7 Utility of Wave Mechanics and Molecular Orbital Theory

Many properties and much of the behavior of many chemical systems can be explained (or at least clarified) through the application of wave mechanics. Properties such as bond strengths, dipole moments, bond lengths, bond multiplicity, stability, resonance, and reactivity, to name a few, can all be explained by wave mechanics. In recent years, a major application is the prediction and explanation of organic reaction mechanisms.

27.8 Conservation of Orbital Symmetry

In 1965 the first in a long, remarkable series of papers on the application of MO theory to the study of organic reaction mechanisms was published by the late Dr. R. B. Woodward of Harvard University and Dr. R. Hoffmann of Cornell University. Although many reactions were known to proceed with a high degree of stereo-specificity and selectivity and although detailed mechanisms for many of these reactions were routinely written, there was no formal explanation of *why* the reactions occurred in this manner.

Woodward and Hoffmann[1] applied MO theory to these reactions and formulated a general set of rules to explain and even predict all concerted chemical reactions. The entire theory they developed can be summarized in a few words from one of their papers:

> Orbital symmetry is conserved in concerted reactions.

Let us analyze this quote. We have a good idea what an orbital (atomic or molecular) is. In addition, in Chap. 6 we introduced the idea of grouping molecules according to their *symmetry* properties. Molecules of the same symmetry group exhibit certain similar chemical characteristics or behavior. We are familiar with the term *concerted*, which we used repeatedly in this text. Examples of reactions that proceed in a concerted fashion include the S_N2, E2, Diels-Alder, and Cope elimination. What does the term mean with reference to reaction mechanisms? *Concerted reactions are those that occur in one step.* That is, all bonds are broken and formed at the same time such that the rearrangement of the electrons occurs in one continuous (concerted) step. The energy required for these transformations is normally supplied in the form of heat (*thermal*) or light (*photochemical*).

With this brief review, we can now analyze the previous quote. It is possible to derive the MO's associated with a given reactant and product in some theoretical reaction. It is also possible to classify these MO's in terms of their symmetry. Finally, if a reaction proceeds in a concerted fashion,[2] the orbitals associated with the reactant must be capable of being transformed into the orbitals of the product with an overall preservation of orbital symmetry.

In a qualitative way at least, we have been making this statement throughout this text. When we first encountered the σ bond, we saw that it is formed by the overlap of two neighboring atomic orbitals so that the MO formed is cylindrically symmetric along the bond axis joining the two nuclei. We introduced the idea of the phase of orbitals. For two atomic orbitals to form a σ bond, overlap must be possible geometrically. Additionally, the orbitals must be in phase with each other; otherwise the overlap results in the formation of an antibonding orbital. For bonding to occur, the overlapping atomic orbitals (which have a symmetry associated with them) must generate MO's of the same symmetry—for example, the plus portion of a p orbital must overlap with the plus portion of another p orbital. This is just another way of saying that *reactions must proceed to give maximum bonding* (maximum overlap) in the transition state and hence provide the greatest stability of the resulting product (see Sec. 2.6).

27.9   MO's: Mode of Attack

There are only two ways in which orbitals can react. These are called *suprafacial attack* in which the orbitals undergoing a change lie on the same face of the molecules

---

[1] The application of MO theory to organic reaction mechanisms was a well-established method of investigation in 1965. Building on the ideas of such men as Fukui (frontier orbitals), Dewar, and others, Woodward and Hoffmann provided a rigorously stated set of rules applicable to any concerted organic reaction and by which those reactions could be predicted and explained.

[2] Although the examples of concerted reactions we discuss in this chapter occur in one step, multistep reactions contain several concerted steps, each of which must abide by these rules.

that are reacting, and *antarafacial attack* in which overlap of the reacting orbitals or product orbitals occurs on opposite faces of the reacting molecules. Illustrations of both modes for the bonding π orbital of ethene reacting with atomic *s* orbitals are shown here (see Sec. 2.15):

π Orbitals:

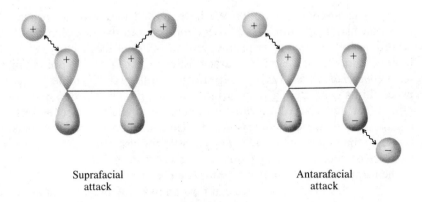

Suprafacial
attack

Antarafacial
attack

Another method used to designate this mode of attack for ethene is:

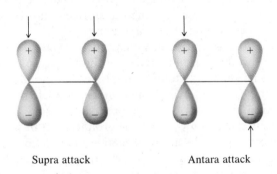

Supra attack   Antara attack

In this example the arrows show incoming orbitals that are about to interact with and perhaps react with the ethene orbitals.

We focus on reactions involving π bonds (π orbitals). The principles put forward, however, can be applied to all molecular and atomic orbitals.

27.10   Molecular Orbital Symmetry

We are familiar with two of the symmetry elements used by chemists to classify compounds into *symmetry groups*. These are the σ plane (C_s) and the axis of symmetry (C_n). (Refer to Sec. 6.4 for more detail.) Any molecule can be categorized into a given group depending on the number and types of C_n axes and σ planes it contains. Furthermore, molecules of a given group exhibit similar chemical characteristics or behavior. Knowing what symmetry group a molecule belongs to gives us insight into its behavior and properties. The investigation of compounds with reference to their symmetry groups is called *group theory*. The consequences of these applications are far-reaching (see General Textbooks with Reading References in the appendix).

In many of the following examples, MO's are assigned to a symmetry group. For further discussions of these assignments, see the Reading References for this chapter in the appendix.

Question 27.11

For each of the following molecules, label each C_n axis and σ plane that exists.

(*a*) ☐   (*b*) ☐   (*c*) CH$_2$—CH$_2$ (antibonding orbital)

27.11   Applying MO Symmetry Rules to Organic Reaction Mechanisms

In applying the Woodward-Hoffmann symmetry rules to a given organic reaction, the following four steps must be followed.

1. The MO's of the reactant(s) and product(s) must be determined. The relative energy and symmetry of each must be known. Generally, the orbitals of interest are the highest occupied (containing electrons) MO's of the reactant(s) and the lowest unoccupied (containing no electrons) MO's of the product(s). All other orbitals can be ignored for all practical purposes.

2. The orbitals associated with the reactant molecules are next envisioned as approaching each other in a predefined mode, for example, parallel or perpendicular to each other.

3. As the orbitals overlap, electrons flow and the orbitals of the reactant(s) are envisioned as undergoing a transformation through some transition state into the orbitals of the product(s)

4. The process is described with a *correlation diagram* (Fig. 27.4). If the conversion of the reactant orbital(s) into the product orbital(s) is favored in terms of energy *and* if orbital symmetry is conserved in the process, the reaction is said to be *symmetry-allowed* and will generally occur under normal thermal conditions. If either of these two stipulations is not met, the reaction is called *symmetry-forbidden* and generally will not occur under normal thermal conditions.[1]

27.12   Cycloaddition Reactions Revisited

Cycloaddition reactions are classified according to the number of π electrons in each of the two reacting π systems. For example, in the Diels-Alder reaction, a dieneophile containing two π electrons reacts with a diene containing four π electrons. The reaction is called 4 + 2 cycloaddition.

---

[1] A reaction labeled symmetry-forbidden cannot occur through the pathway under discussion (the reaction is not favored in terms of energy). This does not rule out that the reaction can occur easily through a different path that fulfills steps 1 to 4.

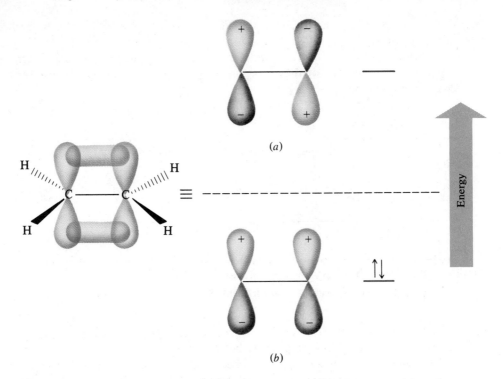

FIGURE 27.1   (*a*) Antibonding and (*b*) bonding orbitals associated with the π MO's of ethene.

An example of 2 + 2 cycloaddition involves the conversion of two molecules of ethene into one molecule of cyclobutane:

$$
\begin{array}{ccc}
CH_2 & CH_2 & CH_2{-}CH_2 \\
\| & \| & | \quad | \\
CH_2 & CH_2 & CH_2{-}CH_2
\end{array}
\quad \xrightarrow{\text{heat}} \quad
$$

π bonds broken σ bonds formed

Applying the Woodward-Hoffmann rules (1 to 4 in Sec. 27.11), we get the following information:

1. In the preceding 2 + 2 cycloaddition, the two π orbitals of the two ethene molecules are converted into two σ (bond) orbitals in cyclobutane. *These are the only orbitals of interest in this reaction.* The π orbital of each ethene is formed from the overlap of two atomic p orbitals. Two atomic orbitals always combine to yield two MO's. In the case of ethene, these are the bonding and antibonding orbitals in Fig. 27.1. Note that the two π electrons are paired and are both in the bonding orbital. As cyclobutane is formed, two new σ (bond) orbitals are formed. These bonds are derived from the four original p orbitals of the two ethene molecules. These four atomic p orbitals combine to form the four MO's in Fig. 27.2. Note that all four σ electrons are in bonding orbitals.

There are other bonds and MO's associated with both ethene and cyclobutane. For example, in ethene we omitted all the σ orbitals in our correlation diagram. This is typical of the Woodward-Hoffmann approach to MO theory. One is concerned only with those orbitals (bonds) that are broken or formed in a given chemical reaction.

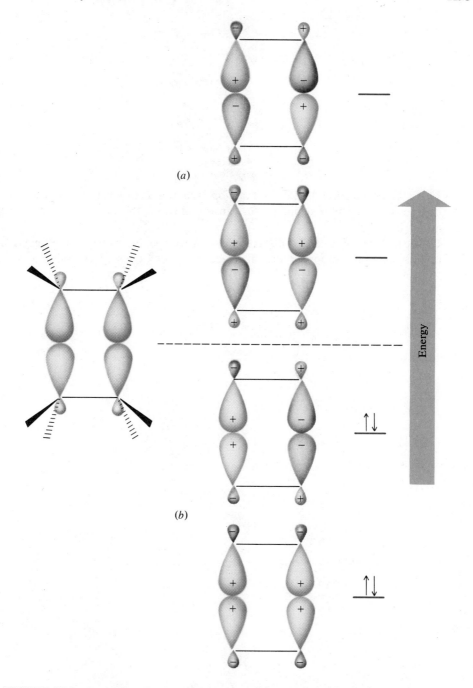

FIGURE 27.2   (a) Antibonding and (b) bonding orbitals associated with two carbon-carbon σ MO's in cyclobutane. The two bonds are on opposite sides of the ring.

2. How does the reaction occur? One can envision, as an example,[2] the two ethene molecules approaching each other with a parallel arrangement of the molecular

[2] At this juncture in the Woodward-Hoffmann approach, a reaction orientation must be chosen arbitrarily (presumably based on chemical logic) and the results for that particular orientation determined. Other possible orientations can be tested subsequently.

planes:

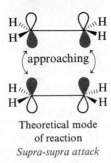

Theoretical mode
of reaction
Supra-supra attack

3. As the orbitals overlap, if the reaction is favorable in terms of energy, electrons flow and the reactant orbitals are transformed into the product orbitals. This is depicted in Fig. 27.3.

4. The picture in Fig. 27.3 is not complete. As we saw, the π orbital of ethene is made up of a bonding component and an antibonding component. Depending on whether

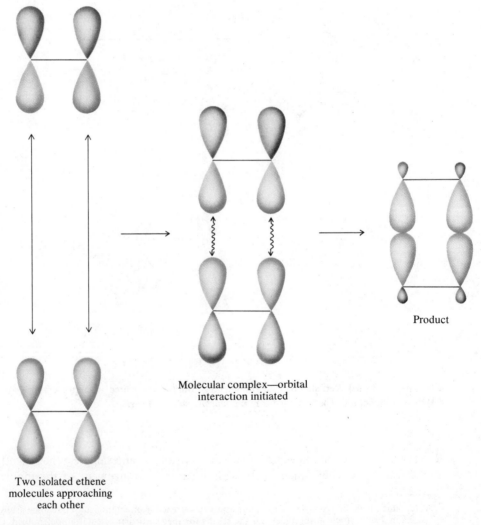

Product

Molecular complex—orbital
interaction initiated

Two isolated ethene
molecules approaching
each other

FIGURE 27.3 Generalized picture of the reaction of two ethene molecules to produce cyclobutane.

the bonding orbitals and antibonding orbitals are involved, there are four transition states of formation, all with different energies: bonding orbital reaction with bonding orbital, antibonding with antibonding, bonding with antibonding, and antibonding with bonding. These in turn yield the four MO's associated with cyclobutane. The complete correlation diagram for this reaction is in Fig. 27.4.

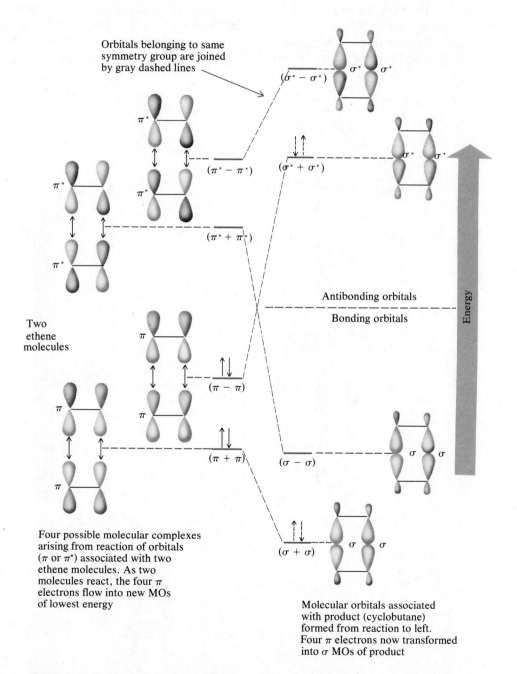

Orbitals belonging to same symmetry group are joined by gray dashed lines

Two ethene molecules

Four possible molecular complexes arising from reaction of orbitals (π or π^*) associated with two ethene molecules. As two molecules react, the four π electrons flow into new MOs of lowest energy

Molecular orbitals associated with product (cyclobutane) formed from reaction to left. Four π electrons now transformed into σ MOs of product

FIGURE 27.4 Correlation diagram for the symmetry-forbidden thermal 2 + 2 cycloaddition of two ethene molecules to give cyclobutane.

We can see from the diagram that, because of symmetry constraints, as the two ethene molecules react, the two π electrons in the $(\pi + \pi)$ orbital state flow into the $(\sigma + \sigma)$ state. The remaining two π electrons, however, flow from the $(\pi - \pi)$ state into the $(\sigma^* + \sigma^*)$ state rather than to the next lowest $(\sigma - \sigma)$ state. For symmetry to be conserved in the process, two electrons flow to bonding orbitals while the remaining two electrons flow to an antibonding orbital. The reaction therefore has a high energy of activation and will not generally proceed under normal thermal conditions. The thermal 2 + 2 cycloaddition of ethene is said to be symmetry-forbidden.

How then can we get this reaction to go? Experimentally, it is known that ethene reacts very easily under photochemical conditions to produce cyclobutane. Why does light allow the reaction to proceed so easily, whereas the same reaction with heat for all practical purposes does not occur? A look at Fig. 27.5 explains this.

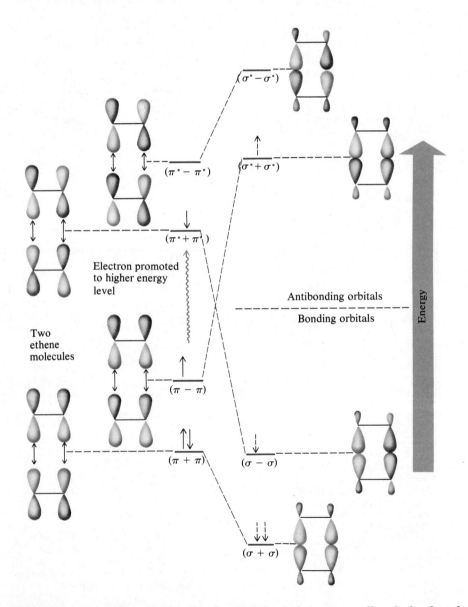

FIGURE 27.5   Correlation diagram for the photochemical (symmetry-allowed) 2 + 2 cycloaddition of ethene to produce cyclobutane.

When ethene is irradiated with light, energy is absorbed and an electron is excited to one of the higher orbitals associated with the interacting ethene molecules; the ($\pi^* + \pi^*$) orbital. Still adhering to the symmetry constraints, as the orbitals rearrange the electron flow is allowed in terms of energy. We end up with three electrons in bonding orbitals and only one in an antibonding orbital. The reaction is symmetry-allowed photochemically.

As illustrated in the previous example, using the semitheoretical, semiempirical Woodward-Hoffmann approach of the study of concerted reactions, it is possible to rationalize a given experimental observation and even predict whether a given process will be allowed or forbidden. The applications are limitless.

There are, however, five major drawbacks to the technique, rendering it tedious and often difficult to apply.

1. All wave equations associated with a given set of reactants and products must be known.
2. The relative energy of each orbital must be known.
3. The symmetry group of the reactant(s), product(s), and often the molecular complexes must be determined.
4. The orientation of the reacting species may be varied, requiring further calculations.
5. Other rules often must be applied that we have not discussed.

Despite these obstacles, the development of the Woodward-Hoffmann rules is a major scientific achievement of the past decade. The method is applied to virtually all postulated concerted reaction mechanisms, providing unquestionable evidence either for or against a given pathway.

27.13 Generalized Selection Rules for Pericyclic Reactions

As pointed out in Sec. 27.12, the application of the Woodward-Hoffmann rules to chemical systems is often a formidable task. In the development of the rules, however, Drs. Woodward and Hoffmann noted the close similarity among various concerted pericyclic reactions. Pericyclic reactions (Greek *peri*, around; *kyklos*, ring) involve bond breaking or bond forming in the synthesis or degradation of a ring. Woodward and Hoffmann found that certain reactions could be categorized as symmetry-allowed or symmetry-forbidden based on a general rule applicable to all concerted pericyclic changes. The rule, paraphrased from their text on the subject, is:

A ground state pericyclic change is symmetry-allowed when the total number of ($4q + 2$) *supra* and ($4q$) *antara* components is odd.

Let us put this in a more understandable form.

In any pericyclic reaction, orbitals are interacting; each orbital (σ, π, s, p, and so on) is called a *component*. We can therefore have components that are π orbitals, σ orbitals, or atomic orbitals. Each orbital in turn has a given number of electrons associated with it. These electrons may total ($4q$) ($0, 4, 8, 12, \ldots$) or $4q + 2$ ($2, 6, 10, 14, \ldots$), where q is any integer. Finally, as these orbitals react with one another, they may do so in two ways, suprafacially or antarafacially. The following examples illustrate how each of these points can be characterized symbolically. Note that in applying the general selection rule we are not concerned with the phase of the components undergoing reaction. This has been taken into account in deriving the rule.

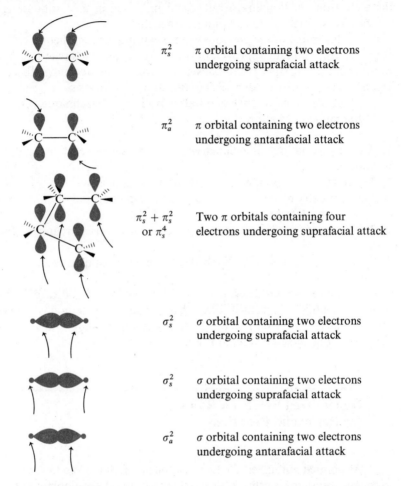

π_s^2 — π orbital containing two electrons undergoing suprafacial attack

π_a^2 — π orbital containing two electrons undergoing antarafacial attack

$\pi_s^2 + \pi_s^2$ or π_s^4 — Two π orbitals containing four electrons undergoing suprafacial attack

σ_s^2 — σ orbital containing two electrons undergoing suprafacial attack

σ_s^2 — σ orbital containing two electrons undergoing suprafacial attack

σ_a^2 — σ orbital containing two electrons undergoing antarafacial attack

In applying the general selection rule, each component is characterized symbolically and then assigned a value of zero or unity based on the number of electrons involved and the mode of attack. Referring to the original quote, if the component contains $(4q + 2)$ electrons and is involved in a suprafacial attack, it is assigned unity. If the component contains $4q$ electrons and is involved in an antarafacial attack, unity is also assigned. If, however, the component contains $(4q + 2)$ electrons and is involved in antarafacial attack or $4q$ electrons with suprafacial attack, the component is assigned a value of zero. Note the following examples:

$\pi_s^2 = 1$ Two electrons, $(4q + 2)$ electron system, *supra attack*
$\pi_a^2 = 0$ Two electrons, $(4q + 2)$ electron system, *antara attack*
$\sigma_a^2 = 0$ Two electrons, $(4q + 2)$ electron system, *antara attack*
$\pi_s^4 = 0$ Four electrons, $4q$ electron system, *supra attack*

When all components involved in a given reaction are assigned a value, the components are summed. If the sum is an *odd* number, the reaction is ground state (thermally allowed). If the result is *even*, a photochemical process is allowed but a thermal process is symmetry-forbidden.

Let us now apply the general selection rule to several reactions we encountered previously.

A. 2 + 2 Cycloadditions

$(\pi_s^2 + \pi_s^2)$ Cycloaddition of two ethene molecules

The components involved in the cycloaddition are assigned a value of unity. Because $1 + 1 = 2$, an even number, the reaction is thermally symmetry-forbidden. This is the same result obtained by the more intense treatment of this system in Sec. 27.12. The reaction is, therefore, photochemically allowed.

Before proceeding to another reaction, let us discuss the following second possible reaction pathway for 2 + 2 cycloaddition of ethene:

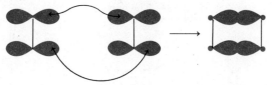

$(\pi_s^2 + \pi_a^2)$ Cycloaddition of two ethene molecules

It would appear from the general selection rule $(1 + 0 = 1, \text{odd})$ that there is a thermally allowed pathway by which two ethene molecules can undergo cycloaddition to produce cyclobutane. Why then is this not normally observed experimentally, and is this a contradiction of the results obtained by applying the rule to the $(\pi_s^2 + \pi_s^2)$ case?

The problem is in the system itself. For the rule to be applied, the reaction must proceed in a concerted fashion. This may or may not be the case. Although a given symmetry-allowed pathway exists, it does not mean that the reaction actually proceeds by that pathway. In many cases reactive intermediates, for example free radicals, may actually be involved and the reaction does not proceed in a concerted fashion.

A second problem involves the geometry of the reacting species. As in the previous 2 + 2 cycloaddition, there is often more than one allowable pathway by which a given reaction may proceed. The geometric constraints of the reacting systems, in terms of both the orbitals undergoing reaction and the nonreacting bonds that we generally tend to ignore in the Woodward-Hoffmann approach, must be taken into account. In the case of the $(\pi_s^2 + \pi_a^2)$ process, the two molecules could undergo such a process only by approaching each other in a perpendicular manner:

$(\pi_s^2 + \pi_a^2)$ Reaction pathway for two ethene molecules

The resulting orbital overlap would be small compared with the $(\pi_s^2 + \pi_s^2)$ photochemical case discussed previously. In addition, for overlap to occur at all, the

π orbitals of the ethene molecules would have to twist to some extent; this would generally require energy consumption by the system. There is added steric strain associated with this perpendicular approach. Our chemical intuition, therefore, would lead us to predict an easy $(\pi_s^2 + \pi_s^2)$ photochemical 2 + 2 cycloaddition and a possible but difficult $(\pi_s^2 + \pi_a^2)$ thermal 2 + 2 cycloaddition. Examples of both reaction pathways are known.

B. 4 + 2 Cycloaddition (Diels-Adler Reaction)

$(\pi_s^4 + \pi_s^2)$ Cycloaddition of ethene and 1,3-butadiene

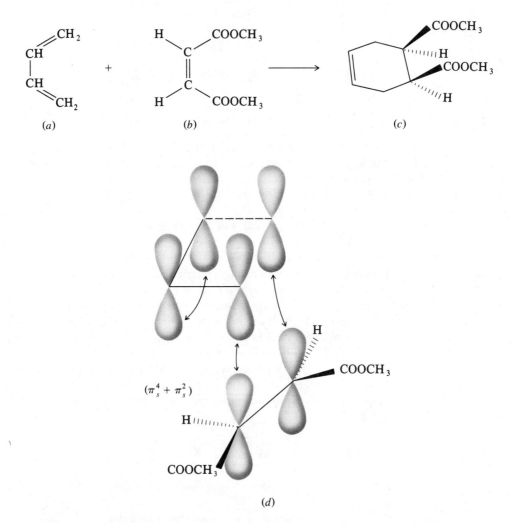

FIGURE 27.6   Diels-Alder reaction between (a) 1,3-butadiene and (b) cis-dimethyl fumarate to produce (c) cis-4,5-dicarbomethoxycyclohexene. In (c), no trans isomer is formed; (d) is an orbital picture of the reaction patheway.

The reaction is predicted to be and is thermally and geometrically allowed. Symbolically, it can be treated in either of the following two ways:

$$\pi_s^4 + \pi_s^2: \qquad 0 + 1 \qquad = 1 \qquad \text{odd; allowed}$$
$$\pi_s^2 + \pi_s^2 + \pi_s^2: \qquad 1 + 1 + 1 = 3 \qquad \text{odd; allowed}$$

We can also deduce from this allowed pathway a preferred mode of attack of the two π systems. The reaction is predicted to occur through the *syn* addition of the two π systems. One would, therefore, predict a stereospecific product distribution, which is observed experimentally. See the reactions in Figs. 27.6 and 27.7.

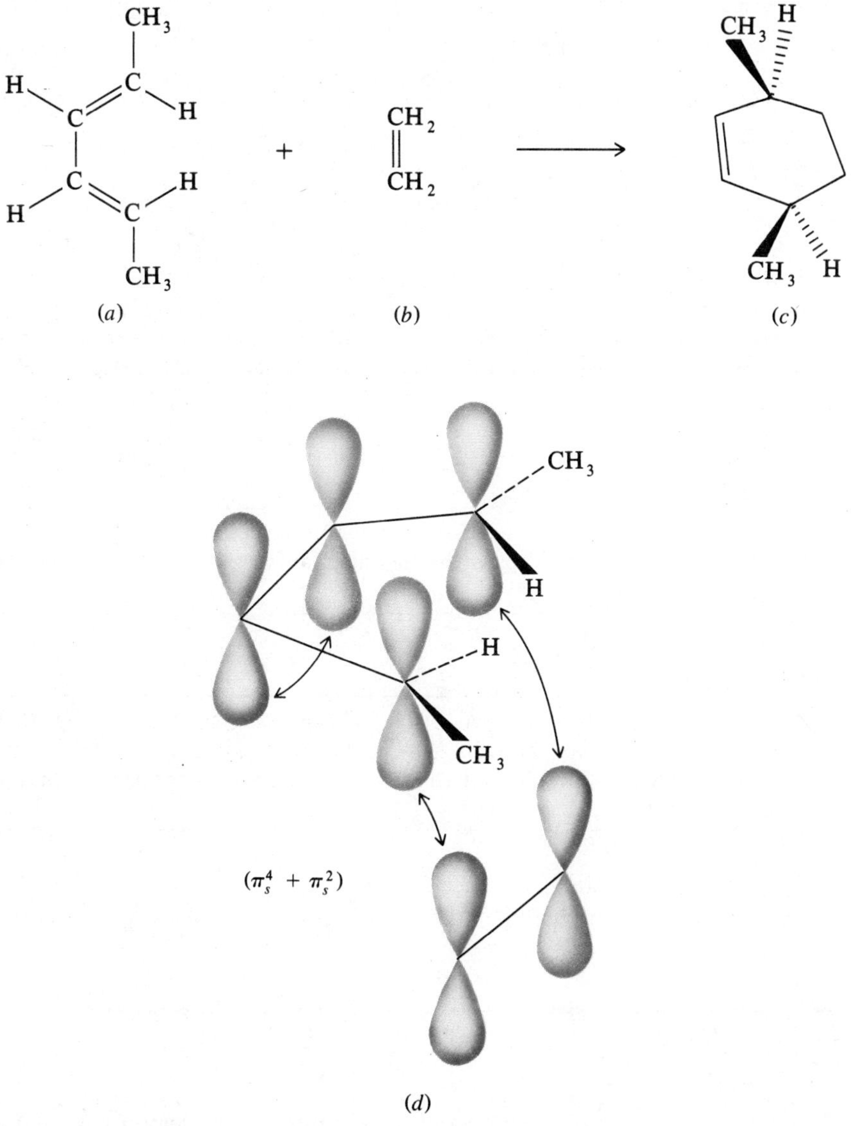

(a)   (b)   (c)

$(\pi_s^4 + \pi_s^2)$

(d)

FIGURE 27.7   Diels-Alder reaction between (a) *trans, trans*-2,4-hexadiene and (b) ethene to produce (c) *cis*-3,6-dimethycyclohexene. In (c), no *trans* isomer is formed; (d) is an orbital picture of the reaction pathway.

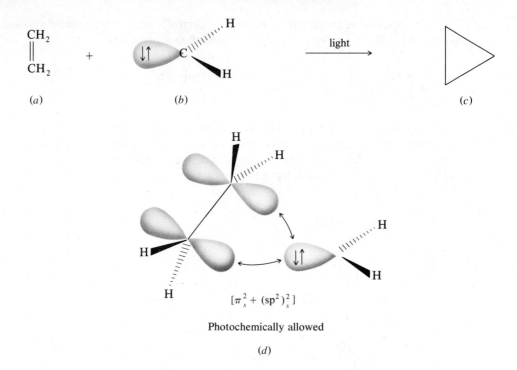

$$[\pi_s^2 + (sp^2)_s^2]$$

Photochemically allowed

(d)

FIGURE 27.8   Reaction of (a) ethene with (b) singlet carbene to give (c) cyclopropane. The reaction is photochemically allowed, as shown in (d), the orbital picture of the reaction pathway. When triplet carbene adds to a double bond, the reaction occurs in two steps (see Sec. 27.6) and is nonstereospecific.

Question 27.12

Treating each of the p orbitals in 1,3-butadiene and ethene as an atomic orbital, derive the equation that symbolically depicts the thermally allowed Diels-Alder reaction to produce cyclohexene.

Question 27.13

The $(\pi_a^4 + \pi_a^2)$ Diels-Alder reaction $(1 + 0 = 1$ odd; allowed) would appear to be a thermally allowed process, yet it does not occur. Why not? How would one be able to determine whether or not this particular pathway is involved in a given cycloaddition (for example, the cycloaddition of *trans*-2-butene and 1,3-butadiene)?

See Figs. 27.8 and 27.9 for additional examples of the use of the Woodward-Hoffmann rules in predicting and explaining reaction pathways.

27.14   Summary

We attempted to give a greater insight into and understanding of MO theory. We tried to present a concise and clear picture of how organic chemists use MO theory in their investigations and what type of information one is able to determine using this method. The reactions presented merely scratch the surface. There are

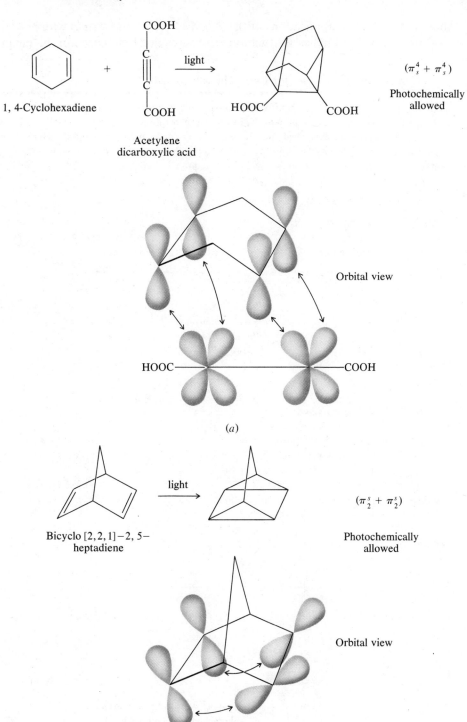

1, 4-Cyclohexadiene

Acetylene dicarboxylic acid

$(\pi_s^4 + \pi_s^4)$

Photochemically allowed

Orbital view

(a)

Bicyclo [2, 2, 1] – 2, 5 – heptadiene

light

$(\pi_2^s + \pi_2^s)$

Photochemically allowed

Orbital view

(b)

FIGURE 27.9 *(a)* and *(b)* **Examples of cycloaddition reactions.**

thousands of known applications of the Woodward-Hoffmann rules with new ones published daily. For further information see the Reading References for this chapter.

General Selection Rules for Pericyclic Reactions

The general selection rule is often difficult to apply to a given chemical system, particularly if one has little experience in this area of chemistry. The general selection rule is actually a summary of many rules that were developed for working with various pericyclic reactions. Here we give more detailed rules that may be easier to apply to these chemical reactions. They may be of some help to the student in solving the questions for this chapter.

Cycloaddition Reactions

Total Number of π Electrons in $(m + n)$ Cycloaddition*	Thermally Allowed	Photochemically Allowed
4q	$m_{supra} + n_{antara}$ $m_{antara} + n_{supra}$	$m_{supra} + n_{supra}$ $m_{antara} + n_{antara}$
4q + 2	$m_{supra} + n_{supra}$ $m_{antara} + n_{antara}$	$m_{supra} + n_{antara}$ $m_{antara} + n_{supra}$

* For example, six electrons in a $(4 + 2)$ cycloaddition: $m = 4$, $n = 2$, or vice versa.

An important point is that although the preceding rules predict a reaction to be allowed, this does not mean that it necessarily occurs. Each system must be analyzed individually, particularly as to the geometric constraints necessary for the reaction to proceed.

There are similar selection rules for all classes of concerted reactions. We introduced only one class in this chapter—cycloaddition reactions. Additional examples are found in the Reading References for this chapter in the Appendix.

Selection Rules Versus General Selection Rule

$\pi_s^2 + \pi_s^2 + \pi_s^2 + \pi_s^2 = 4$, even; **photochemically allowed**

Through application of the general selection rule, the reaction is predicted to be allowed photochemically. Applying the selection rules for cycloaddition reactions, we find that $m = 4$ and $n = 4$ and $m + n = 8$. Because 8 is a $4q$ number, the reaction may proceed thermally with one of the reacting π systems undergoing antarafacial attack and the other suprafacial attack. Geometrically, this is clearly impossible. Because of geometric constraints, the reaction must undergo *supra-supra* attack. The table tells us this is allowed photochemically in agreement with the results from the general selection rule.

Study Questions

27.14 Label the following as examples of suprafacial and/or antarafacial reactions.

(a)

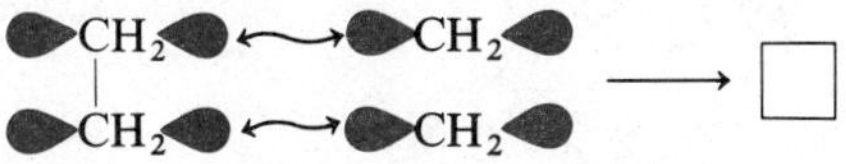

(b)

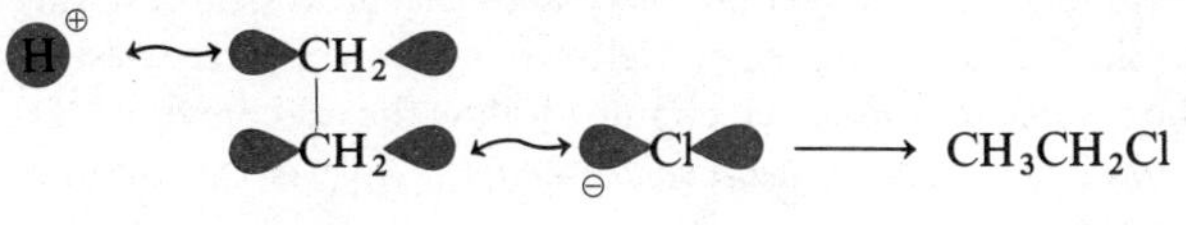

(c)

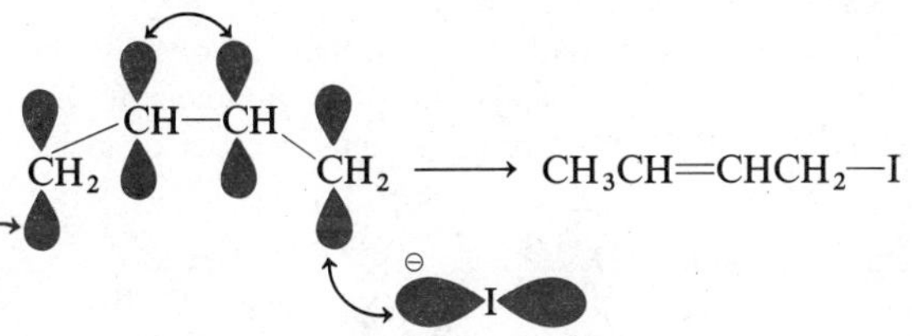

27.15 Each of the following six reactions occurs under either thermal or photochemical conditions. Answer the following questions as they pertain to each reaction.

1. Is the indicated reaction run under thermal or photochemical conditions?
2. Explain the observed product by presenting a clear discussion of the mechanistic pathway involved.
3. Give the general selection rule notation to symbolize each reaction.

(a) 2

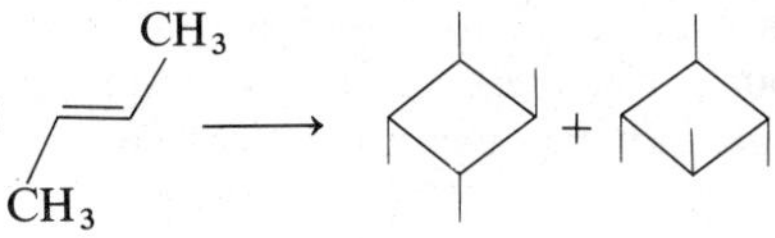

(b) 2 *cis*-2-Butene

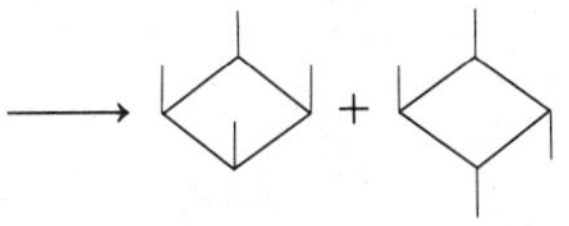

(c)

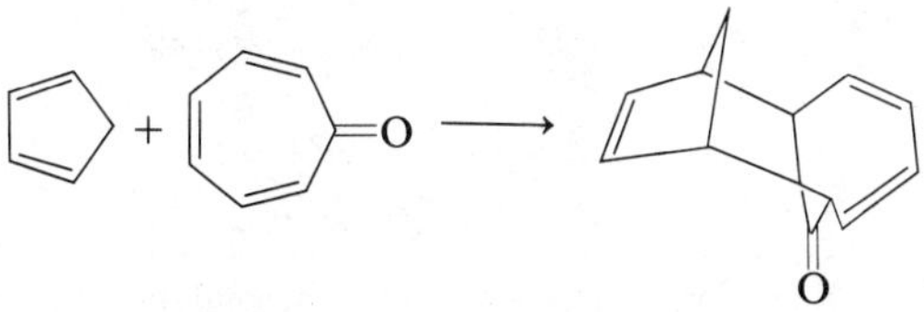

(d)

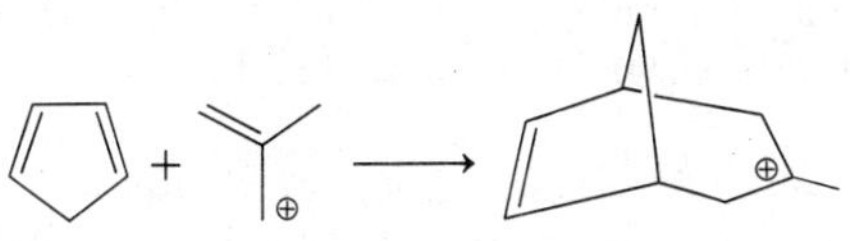

(e)

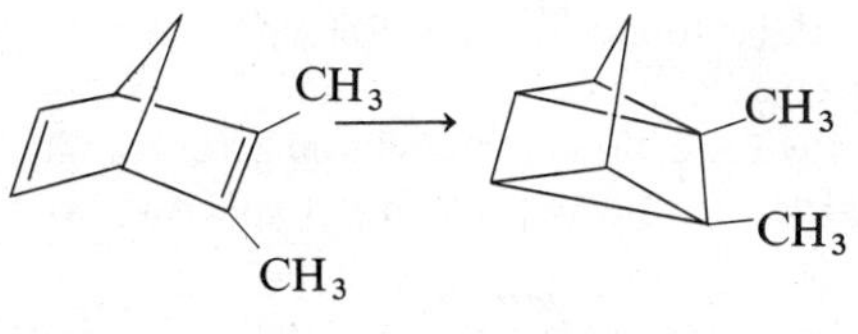

(*f*) 2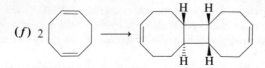

27.16 We have been concerned solely with intermolecular cycloaddition reactions. It is possible for cycloaddition reactions to occur in an intramolecular fashion also. For example, the thermal cyclization of 1,3-butadiene to cyclobutene is a well-known process:

Reactions of this type (intramolecular cycloadditions of a π-electron system resulting in cyclization and a decrease of two in the total number of π electrons involved in the system) are called *electrocyclic reactions*. The reverse reactions, which must follow the exact reverse of the pathway associated with the forward reactions, are called *cycloreversions*. Answer the following questions as they pertain to the preceding electrocyclic reaction.

(*a*) Which bond(s) of the reactant and which of the product are involved in the interconversion?

   The MO's associated with the π-electron system of 1,3-butadiene were described in Question 13.19; those of interest for cyclobutene are given here in order of increasing energy:

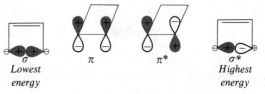

(*b*) Construct the partial MO correlation diagram associated with the preceding cyclization. The diagram should include the reactant orbitals, the order of increasing energy (see Fig. 27.4 for example), and the product orbitals in a similar fashion. Bonding and antibonding orbitals should be labeled appropriately.

   If one studies this diagram it should be evident that for the orbitals of the reactant to be converted to those of the product, the π orbitals of the reactant species must rotate to interact and eventually form product orbitals. This can occur in two ways. Recall that when orbitals of like phase overlap, bonding occurs.

Conrotatory   Disrotatory
rotation   rotation

(*c*) Using all the preceding information, draw the complete MO correlation diagrams for the following transformations: (1) a thermal conrotatory process, (2) a thermal disrotatory process, (3) a photochemical conrotatory process, and (4) a photochemical disrotatory process.

(*d*) Which of the preceding reaction pathways are symmetry-allowed and which are symmetry-forbidden?

(*e*) What product will the thermal intramolecular cycloaddition of *cis,trans*-2,4-hexadiene produce? What product(s) will the photochemical reaction produce?

   The student is referred to the General Textbook Reading References for this chapter in the appendix for more detail concerning electrocyclic reactions.

27.17 Vitamin D, drawn here, is produced by the action of sunlight on precursor (1). Precursor (2) is formed initially and is then converted to vitamin D thermally. Explain the conversion of (1) to (2) mechanistically.

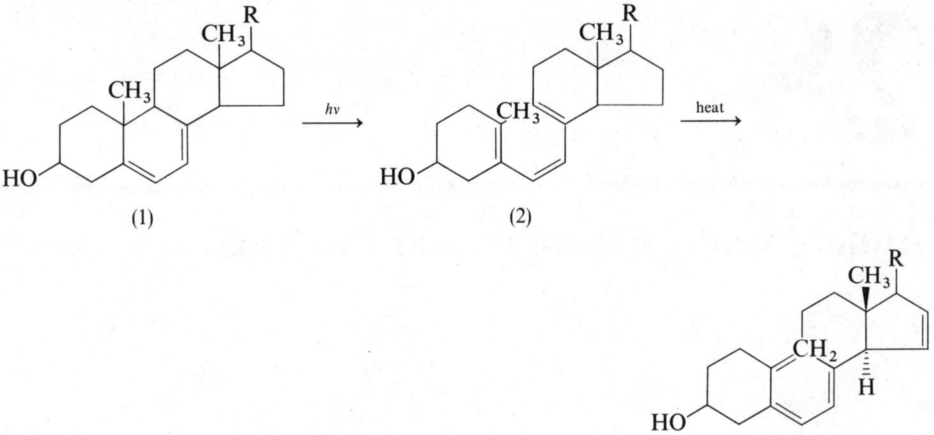

(1)   (2)

Vitamin D

where R = $CH_3CH-CH=CH-CH-CH(CH_3)_2$
       | |
             CH_3

trans

27.18 Explain, through use of the general selection rule, each of the following observations.

(*a*)

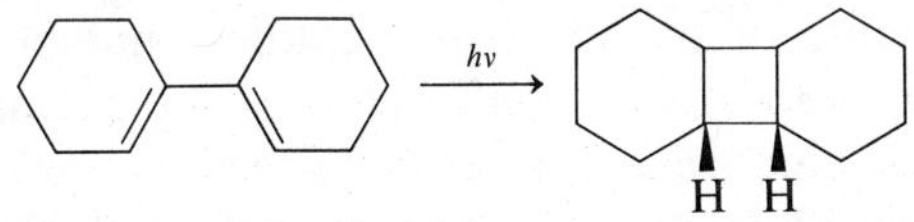

(*b*)

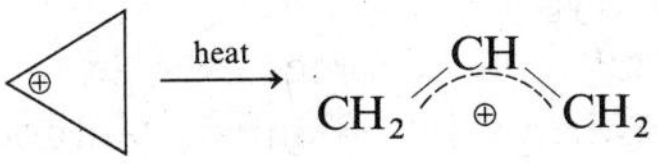

(*c*)

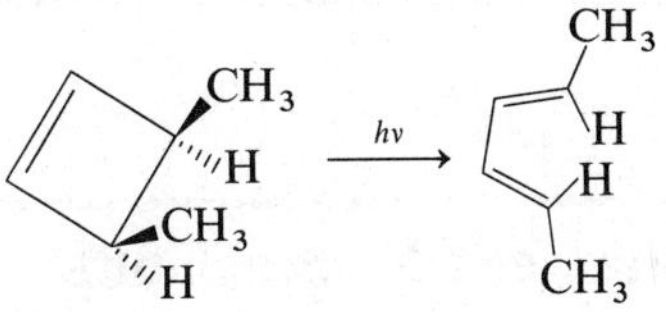

(*d*)

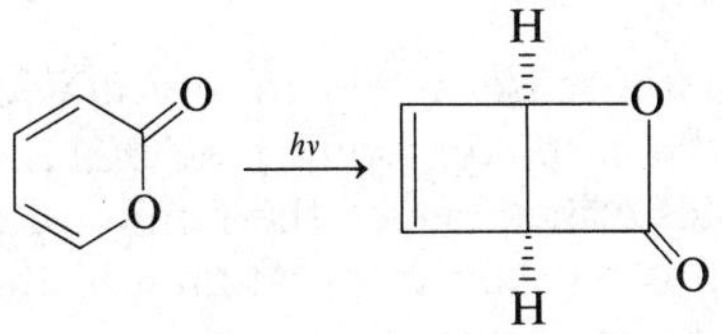

(*e*)

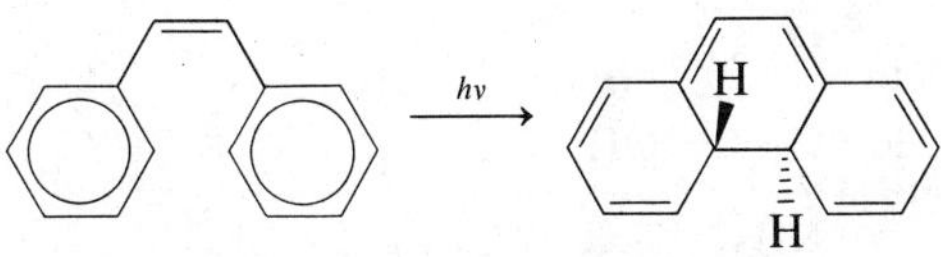

28

Amino Acids, Peptides, and Proteins

This chapter deals with the chemistry of amino acids and the natural polymers, peptides and proteins, which are composed of amino acid units (sometimes referred to as **residues**). In particular, we present the chemistry of the α-amino acids, those in which the amino group (—NH$_2$) and the carboxyl group (—COOH) are both attached to the same carbon atom.

Beginning with a short introduction to the chemistry and physical properties of the α-amino acids, we review some methods used in synthesizing these compounds. A discussion of the structure of amino acids, peptides, and proteins is included, paying particular attention to the overall conformations of these macromolecular species. How and why these conformations result as well as the role they play in the chemistry of the protein or peptide are presented through the use of three examples—insulin, keratins, and chymotrypsin.

Structure elucidation is extremely difficult and becomes even harder with increasing size and complexity of the molecules being investigated. Methods used in elucidating the structure of proteins and peptides are summarized and examples of their application are provided.

28.1   α-Amino Acids: Structure, Properties, and Nomenclature

The term **amino acid,** in its most general sense, refers to any molecule that contains both an amino group (—NH$_2$) and a carboxyl group (—COOH). By convention, however, the term generally denotes a member of the family of α-amino acids—molecules in which the amino group is on the carbon atom α to the carboxyl carbon. The general structure of most α-amino acids is:

$$R-\underset{\underset{\displaystyle NH_2}{|}}{\overset{\overset{\displaystyle H}{|}}{C}}-\overset{\overset{\displaystyle O}{||}}{C}-OH$$

Although more than 100 α-amino acids have been isolated, only approximately 20 are indigenous to the plant or animal matter in which they are found. The remainder result from biochemical transformations (metabolism). Twenty-one common amino acids have the preceding general structure and hence differ only in the structure of the R group attached to the α carbon. Indeed, the amino acids are most commonly classified according to the R group making up the molecule and the chemistry associated with and influenced by this group. These 21 naturally occurring amino acids are listed in Table 28.1 along with the three-letter abbreviation commonly used for each. Note the diverse possibilities for the R group, which ranges from a completely aliphatic or aromatic moiety (e.g., alanine or phenylalanine, respectively) to such diverse groups as hydroxyl, thiol, sulfide, amide, and acid. We discuss the chemistry of these groups in more detail in subsequent sections. For now, it is sufficient to note that some are hydrophilic and polar, whereas others are quite nonpolar and

TABLE 28.1   α-Amino Acids

R Group	Name*	Symbol†	pK_I	pK_{II}	pK_{III}
Neutral:					
H—	Glycine	gly	2.34	9.6	
CH₃—	Alanine	ala	2.35	9.69	
(CH₃)₂CH—	*Valine*	val	2.32	9.62	
(CH₃)₂CHCH₂—	*Leucine*	leu	2.36	9.60	
CH₃CH₂CH— | CH₃	*Isoleucine*	ile	2.36	9.68	
—CH₂—	*Phenylalanine*	phe	1.83	9.13	
	Proline‡	pro	1.99	10.60	
	Tryptophan	trp	2.38	9.39	
H₂N̈—C—CH₂— ‖ O	Asparagine	asn	2.02	8.8	
H₂N̈—C—CH₂CH₂— ‖ O	Glutamine	gln	2.17	9.13	
Basic:					
H₃⊕NCH₂CH₂CH₂CH₂—	*Lysine*	lys	2.18	8.95	10.53
H₂⊕N⟍ C—NHCH₂CH₂CH₂— H₂N̈⁄	Arginine	arg	2.17	9.04	12.48
	Histidine	his	1.82	9.17	6.00

TABLE 28.1   (*continued*)

R Group	Name*	Symbol[†]	pK_I	pK_{II}	pK_{III}
Acidic:					
$HOOCCH_2—$	Aspartic acid	asp	2.09	9.82	3.86
$HOOCCH_2CH_2—$	Glutamic acid	glu	2.19	9.67	4.25
Miscellaneous:					
$HSCH_2—$	Cysteine	cys	1.71	8.91	8.50
$—CH_2SSCH_2—$	Cystine[§]	cys-cys	1.65	7.86	
			2.26	9.85	
$CH_3SCH_2CH_2—$	*Methionine*	met	2.28	9.21	
$HOCH_2—$	Serine	ser	2.21	9.15	
$HOCH—$ $\quad\vert$ $\quad CH_3$	*Threonine*	thr	2.09	9.10	
$HO—\langle\bigcirc\rangle—CH_2—$	Tyrosine	tyr	2.20	9.11	10.07

* All the essential amino acids necessary for life processes in humans yet *not* synthesized by the body are in *italic*.
[†] Classified by the acidity-basicity of the R group.
[‡] The α-amino group is in the side chain.
[§] A diamino acid.

hence hydrophobic. As we will see, the polarity as well as the overall size and shape of a given R group influence its function in biologically active systems such as proteins and enzymes.

28.2   Chemical and Physical Properties of α-Amino Acids

Amino acids are amphoteric compounds; they contain both an amino group, which can exist in a cationic form as the ammonium ion ($—\overset{\oplus}{N}H_3$), and an acidic carboxyl group, which can easily exist as the carboxylate anion ($—CO\overset{\cdot\cdot}{O}\overset{\ominus}{:}$). Although in the solid crystalline state all amino acids exist in stable, high-melting, zwitterionic forms, this is not true in aqueous solution. This is most easily seen by studying the various equilibrium forms possible in acidic, basic, and neutral solutions, as summarized in the following equations:

$$R—CH—CO\overset{\cdot\cdot}{O}\overset{\ominus}{:} \underset{H\overset{\cdot\cdot}{O}:^{\ominus}}{\overset{H^{\oplus}}{\rightleftharpoons}} R—CH—CO\overset{\cdot\cdot}{O}\overset{\ominus}{:} \underset{H\overset{\cdot\cdot}{O}:^{\ominus}}{\overset{H^{\oplus}}{\rightleftharpoons}} R—CH—COOH$$

$$\underset{:NH_2}{\quad} \qquad \underset{\underset{\oplus}{NH_3}}{\quad} \qquad \underset{\underset{\oplus}{NH_3}}{\quad}$$

Anionic   II   Zwitterionic   I   Cationic

Note that in strongly acidic solutions all the amino acids exist in cationic forms, whereas in strongly basic solutions the equilibrium is shifted quantitatively toward the anionic form. At intermediate pH ranges each of the two equilibria of interest has an associated ionization constant and therefore has all three species (anionic, zwitterionic, and cationic) present in solution in various percentages. The equilibrium constants of interest are:[1]

$$K_{\mathrm{I}} = \frac{\left[\begin{array}{c} R-CH-COO^{\ominus} \\ \underset{\oplus}{\mid} \\ NH_3 \end{array}\right]}{\left[\begin{array}{c} R-CH-COOH \\ \underset{\oplus}{\mid} \\ NH_3 \end{array}\right]} = \frac{[\text{zwitterionic}]}{[\text{cationic}]}$$

$$K_{\mathrm{II}} = \frac{\left[\begin{array}{c} R-CH-COO^{\ominus} \\ \mid \\ NH_2 \end{array}\right]}{\left[\begin{array}{c} R-CH-COO^{\ominus} \\ \underset{\oplus}{\mid} \\ NH_3 \end{array}\right]} = \frac{[\text{anionic}]}{[\text{zwitterionic}]}$$

Varying the pH of the solution allows us to shift the equilibria and concentrations of the various species of interest. The pH at which a distribution results such that the concentration of the zwitterionic form is at a maximum and the concentrations of both the anionic and cationic forms are identical is called the **isoelectric point** and has a fixed value for a given amino acid. The position of the isoelectric point depends on the magnitude of the two equilibrium constants and hence on the acidity of the ammonium grouping, the basicity of the carboxylate grouping, and the nature of the R group and its influence on these two equilibria.

The effect of the R group on the isoelectric point results in a wide range of pH values for the equilibrium associated with each individual amino acid. These effects are summarized in the following list.

1. Neutral R groups: The isoelectric points of amino acids with hydrocarbon, amide, and other neutral side chains are all in the same range, pH = 5 to 6. Hence, in cells (where the pH range of the cellular fluid is in the range of 6 to 7) these species are very close to their isoelectric points.

Glycine	pH = 5.97 at the isoelectric point
Alanine	6.02
Valine	5.97
Leucine	5.98
Isoleucine	6.02
Phenylalanine	5.48
Proline	6.20
Tryptophan	5.88
Asparagine	5.41
Glutamine	5.70

2. Basic R groups: If an amino acid contains a basic group as part of its side chain, this normally results in a decrease in the overall acidity of the amino acid and a concurrent shift of the isoelectric point to higher pH. In the cellular fluid these species exist predominantly in their cationic form.

---

[1] Note that the equilibrium constants are written so that the more acidic member of each pair is the denominator and its conjugate base (the salt of the acid) is the numerator.

Lysine	9.74
Arginine	10.76
Histidine	7.59

3. Acidic R groups: The presence of an acidic R group (generally a carboxyl group) on the side chain of an amino acid results in an overall increase in the acidity of these species and shifts the isoelectric point to lower pH. These species are predominantly in their anionic form in the cellular fluid.

| Aspartic acid | 2.88 |
| Glutamic acid | 3.22 |

4. Sulfides (RSR), thiols (RSH), and alcohols or phenols on the side chains of amino acids are all neutral or very weakly acidic and can be classified in group 1 (see above) with pH ranges at the isoelectric points of 5 to 6.5.

Cysteine	5.02
Cystine	5.06
Methionine	5.06
Serine	5.68
Threonine	5.60
Tyrosine	5.67

The acidity of the amino acids can be rationalized in a Lowry-Brønsted sense (proton donor and proton acceptor) and described with reasonable accuracy by the Henderson-Hasselbach equation:

$$pH = pK + \log \frac{[\text{salt of acid}]}{[\text{acid}]}$$

Because there are two or more K_a values of interest in a given amino acid, there are separate equations for the various equilibria involved. For example, the titration curve for leucine (Fig. 28.1) consists of two parts (one associated with pK_I of leucine and the second with pK_{II}), each of which can be described by the Henderson-Hasselbach equation. The pH at the isoelectric point can be calculated as follows: $pH = \frac{1}{2}(pK_I + pK_{II})$. Given pK_I and pK_{II} for an amino acid and knowing the pH of the solution, one can calculate the concentrations of the anionic, cationic, and zwitterionic forms in solution.

If the amino acid contains a third acidic moiety as part of its side chain, a third K_a (K_{III}) is involved in the overall acidity of the amino acid. It can be shown that if the moiety is itself acidic (for example, asp, $R = -CH_2COOH$), the pH at the isoelectric point is described by the equation $pH = \frac{1}{2}(pK_I + pK_{III})$. If there is a basic side chain (for example, lys, $R = -CH_2CH_2CH_2CH_2NH_2$), the pH at the isoelectric point is $pH = \frac{1}{2}(pK_{II} + pK_{III})$.

Consider the following example: 0.1 mole of leucine (Fig. 28.1) is dissolved in 100 ml of water at pH 4. What are the concentrations of the zwitterionic (Z), cationic (C), and anionic (A) forms of the amino acid in solution?

There are two equilibria of interest:

1. $pH = pK_I + \log \dfrac{[\text{zwitterionic}]}{[\text{cationic}]}$

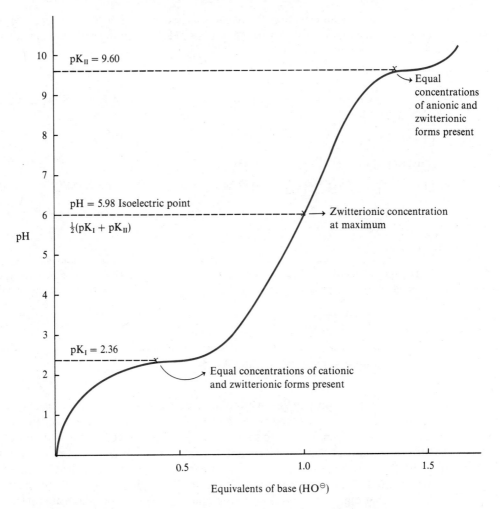

FIGURE 28.1   Titration curve obtained on titrating leucine (leu) with sodium hydroxide. pK_I, pK_{II}, and isoelectric points are labeled.

2. $pH = pK_{II} + \log \dfrac{[\text{anionic}]}{[\text{zwitterionic}]}$

Therefore,

3. from equation 1: $4 = 2.36 + \log \dfrac{[Z]}{[C]}$ $\qquad \log \dfrac{[Z]}{[C]} = 1.64$

4. from equation 2: $4 = 9.60 + \log \dfrac{[A]}{[Z]}$ $\qquad \log \dfrac{[A]}{[Z]} = -5.6$

and

5. from equation 3: $\dfrac{[Z]}{[C]} = 43.7$

6. from equation 4: $\dfrac{[Z]}{[A]} = 399{,}033.2$

It is evident from equation 6 that *the concentration of A is negligible*, $[A] \simeq 0$ when compared with those of $[Z]$ and $[C]$ at this pH. Continuing with the calculation, one gets

7. $[Z] = 43.7[C]$

and because we started with a 1 M solution of Z (0.1 mole of Z in 100 ml of water), we also know that

8. $[Z] = [1] - [C]$

Substituting $([1] - [C])$ for Z in equation 7 gives

9. $[1] - [C] = 43.7[C]$, $[C] = 0.022$ M   and $[Z] = 0.978$ M

   and $[A] \simeq 0$ M

Question 28.1

The pK_a's of histidine are 1.82, 9.17, and 6.0 for the C_α-carboxyl, C_α-ammonium, and histidine side chains, respectively. Calculate the ratio of neutral to positively charged histidine at pH 7.4.

28.3   Stereochemistry of α-Amino Acids

α-Amino acids (except glycine) contain a structural feature we encountered throughout this text—an asymmetric carbon atom. α-Amino acids are, therefore, chiral molecules capable of exhibiting optical activity. Indeed, most α-amino acids isolated from natural sources do exhibit optical activity. With very few exceptions, all naturally occurring α-amino acids have the same absolute configuration. We find them to be predominantly of the *S* configuration. Throughout the literature on amino acids (and other molecules of biochemical importance), however, there are references to D and L configurations of these species (see Table 28.2 for examples). What do these symbols refer to?

TABLE 28.2 Absolute Configurations and Specific Rotations of Selected α-Amino Acids

Amino Acid	Absolute Configuration*	Specific Rotation†
Alanine	L	+1.8
Phenylalanine	L	−34.5
Leucine	L	−11.0
Isoleucine	L	+12.4
Proline	L	−86.2
Tryptophan	L	−33.7
Lysine	L	+13.5

* Common form found in nature.
† $[\alpha]_D^{25}$ (H_2O).

The D and L system of classification of absolute configurations was developed and used for approximately half a century before the *R* and *S* system preferred today. It is an alternative method of structurally classifying optical isomers. The D and L system relies on a difficult and at times impossible chemical assignment of the absolute configurations of one or more compounds. Why then do we present and discuss this system in organic chemistry today? The problem is similar to that of the common versus IUPAC system of nomenclature. The D and L system is an old, widely established system that is still in common use, particularly in biochemical and biological literature. It is, therefore, necessary to understand both it and the newer *R* and *S* system. The basis for the D and L system is simple. A model compound that exists in two enantiomeric forms is chosen and the enantiomers are assigned the D and L configurations.[1] All other molecules that can exist as enantiomers are related structurally to these two models and assigned a corresponding configuration.

The compound chosen as the reference model in this system is glyceraldehyde. Its two enantiomers with their assigned absolute configurations are:

L-Glyceraldehyde D-Glyceraldehyde
S-Glyceraldehyde R-Glyceraldehyde

One can see from the following three examples that the system functions quite well.

L-Glyceraldehyde   L-Lactic acid   L-Alanine

Question 28.2

Assign the absolute configuration as *R* or *S* and D or L for each of the following:

(*a*) HOCH₂—C—NH₂   (*b*)   (*c*)

---

[1] The assignment is not so arbitrary as it may seem. In reality, the model compounds (and only the model compounds) were assigned configurations based on the direction in which they rotated plane-polarized light, dextrorotatory = D and levorotatory = L.

28.4   Amino Acids, Peptides, and Proteins

The reaction of an amine (primary or secondary) with an appropriately activated carboxylic acid derivative (acid chloride, ester, or anhydride, for example) produces an amide (see Secs. 23.8 and 26.8). The reactions involve nucleophilic attack of the amine at the acyl carbon of the activated carboxylic acid derivative, resulting in nucleophilic acyl substitution:

$$
\underset{\substack{\text{Activated carboxylic}\\\text{acid derivative}}}{R-\overset{\overset{\textstyle O}{\|}}{C}-G} \;+\; \underset{\substack{1° \text{ or } 2°\\\text{Amine}}}{R'-\overset{R''}{\underset{\cdot\cdot}{N}}H} \;\longrightarrow\; \underset{\text{Amide}}{R-\overset{\overset{\textstyle O}{\|}}{C}-\underset{R'}{\overset{R''}{N}\!:}} \;+\; HG
$$

where $-G = -Cl, -OR, -O-\overset{\overset{\textstyle O}{\|}}{C}-R$, and others

The same reaction is possible for two amino acids; one amino acid behaves as the carboxylic acid and the second behaves as the amine. The generalized reaction is:

$$
\underset{R}{H_2N-\overset{H}{\underset{|}{C}}-\overset{\overset{\textstyle O}{\|}}{C}-OH} + \underset{R'}{H_2N-\overset{H}{\underset{|}{C}}-\overset{\overset{\textstyle O}{\|}}{C}-OH} \xrightarrow[-H_2O]{\text{dehydrating agent}} H_2N-\overset{H}{\underset{R}{C}}-\overset{\overset{\textstyle O}{\|}}{C}-NH-\overset{H}{\underset{R'}{C}}-\overset{\overset{\textstyle O}{\|}}{C}-OH
$$

Amide bond

Although we wrote the reaction as occurring in one step involving a coupling of the two amino acids with a subsequent loss of water, we see in Sec. 28.5 that the conditions necessary for a successful reaction are often very complex.

The coupling of two amino acids through a reaction of the preceding type results in the formation of a new "amide bond" between the acyl carbon of one amino acid and the amino group of the second amino acid. The resulting compound still contains an amine end and a carboxylic acid end and is capable of further coupling reactions with other monomeric amino acids or polyamino acids, for example:

$$
\underset{\substack{\text{Amino acid}}}{H_2N-\overset{H}{\underset{R''}{C}}-\overset{\overset{\textstyle O}{\|}}{C}-OH} + \underset{\substack{\text{Diamino acid}\\\text{(dipeptide)}}}{H_2N-\overset{H}{\underset{R}{C}}-\overset{\overset{\textstyle O}{\|}}{C}-NH-\overset{H}{\underset{R'}{C}}-\overset{\overset{\textstyle O}{\|}}{C}-OH} \xrightarrow[\substack{\text{dehydrating}\\\text{agent}}]{-H_2O}
$$

$$
\underset{\substack{\text{Triamino acid}\\\text{(tripeptide)}}}{H_2N-\overset{H}{\underset{R''}{C}}-\overset{\overset{\textstyle O}{\|}}{C}-NH-\overset{H}{\underset{R}{C}}-\overset{\overset{\textstyle O}{\|}}{C}-NH-\overset{H}{\underset{R'}{C}}-\overset{\overset{\textstyle O}{\|}}{C}-OH}
$$

or

$$H_2N-\underset{\underset{R}{|}}{\overset{\overset{H}{|}}{C}}-\overset{\overset{O}{||}}{C}-NH-\underset{\underset{R'}{|}}{\overset{\overset{H}{|}}{C}}-\overset{\overset{O}{||}}{C}-OH + H_2N-\underset{\underset{R}{|}}{\overset{\overset{H}{|}}{C}}-\overset{\overset{O}{||}}{C}-NH-\underset{\underset{R'}{|}}{\overset{\overset{H}{|}}{C}}-\overset{\overset{O}{||}}{C}-OH \quad \xrightarrow[\substack{\text{dehydrating} \\ \text{agent}}]{-H_2O}$$

Diamino acid   Diamino acid
(dipeptide)   (dipeptide)

$$H_2N-\underset{\underset{R}{|}}{\overset{\overset{H}{|}}{C}}-\overset{\overset{O}{||}}{C}-NH-\underset{\underset{R'}{|}}{\overset{\overset{H}{|}}{C}}-\overset{\overset{O}{||}}{C}-NH-\underset{\underset{R}{|}}{\overset{\overset{H}{|}}{C}}-\overset{\overset{O}{||}}{C}-NH-\underset{\underset{R'}{|}}{\overset{\overset{H}{|}}{C}}-\overset{\overset{O}{||}}{C}-OH$$

Tetraamino acid
(tetrapeptide)

The polymeric compounds that result from these reactions are called **peptides** and the amide bond that joins them is called a **peptide bond** or *peptide link*. If the chain is composed of two amino acid units, it is called a *dipeptide;* if three amino acids are involved, it is a *tripeptide*. The naming system continues to *tetrapeptide*, *pentapeptide*, and so on, as the number of amino acids in the chain increases. The general term *polypeptide* designates any peptide composed of three or more monomeric amino acid units. As the polypeptide chains increase in length, the molecular weight also increases. Polypeptides of molecular weight greater than approximately 5,000 (consisting of roughly 50 or more amino acids) are called **proteins.**

Traditionally, the amine end of the peptide is written to the left (as one faces the page and is called the *N-terminal end*, whereas the carboxylic acid end, called the *C-terminal end*, is written to the right. This may seem trivial on first inspection, but the following example illustrates the problem one encounters if this convention is not followed. The dipeptide that results from the coupling of alanine and glycine is drawn as follows. Note that there are two possible dipeptides that could result from this coupling—ala-gly and gly-ala:

$$H_2N-\underset{\underset{H}{|}}{\overset{\overset{H}{|}}{C}}-\overset{\overset{O}{||}}{C}-NH-\underset{\underset{CH_3}{|}}{\overset{\overset{H}{|}}{C}}-\overset{\overset{O}{||}}{C}-OH \qquad H_2N-\underset{\underset{CH_3}{|}}{\overset{\overset{H}{|}}{C}}-\overset{\overset{O}{||}}{C}-NH-\underset{\underset{H}{|}}{\overset{\overset{H}{|}}{C}}-\overset{\overset{O}{||}}{C}-OH$$

gly-ala   ala-gly

Realizing that the N-terminal end goes to the left and the C-terminal end to the right, we have no problem in assigning the correct structure to each symbol. Without this convention, however, the abbreviated nomenclature system would be ambiguous.

To illustrate this method of naming peptides, the structure (both complete and abbreviated) of the hormone lysine vasopressin is in Fig. 28.2. Lysine vasopressin is a nonapeptide excreted by the pituitary gland. It is used clinically as a hypertensive agent and was the first naturally occurring hormonal peptide synthesized in the laboratory. The Nobel Prize in chemistry in 1955 was awarded to V. du Vigneaud for this excellent work.

Among the most exciting recently discovered or synthesized peptides are the **endorphins.** Endorphins are a small group of peptides with morphinelike activity; that is, they are extremely potent analgesics (pain killers). Though the mechanism of their action is yet to be elucidated completely, they function primarily by blocking

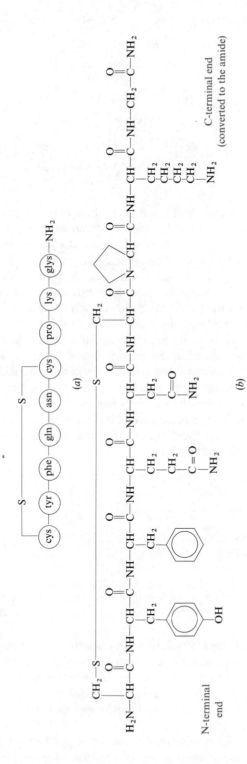

FIGURE 28.2 Lysine vasopressin: (*a*) abbreviated notation, (*b*) complete formula with the disulfide bridge between the two *cys* shown. The —S—S— bond in both (*a*) and (*b*) is not drawn to scale but is elongated to allow the peptide to be drawn linearly.

nerve transmission in the body. The endorphins are a subclass of another family of peptides—the **enkephalins.** The enkephalins differ from the endorphins primarily in their size; the enkephalins are much lower molecular weight peptides. The primary structure of a typical enkephalin is:

Methionine enkephalin

Much research is under way on all phases of the chemistry of these compounds: structure elucidation, mode of action, and, perhaps most important, synthesis of the parent compounds and analogues of the parent compounds.

Question 28.3

How many tripeptides can be formed that contain one of each of the three amino acids gly, ala, and phe? Draw complete structural formulas of each.

28.5   Synthesis of Peptides

The actual synthesis of a peptide is much more complicated than we indicated. The two species involved in the coupling reaction, whether they are free amino acids or peptides, often possess a wide variety of chemically reactive substituents, including carboxyl groups (—COOH), amino groups (—NH$_2$), thiol groups (—SH), hydroxy groups (—OH), and many others. The possibility of a large number of side reactions competing with the formation of the peptide link is a real concern. To avoid these side reactions, which result in lower yields and increased cost, a number of chemical blocking (or protecting) and activating groups have been developed. The sequence for the synthesis of a dipeptide is outlined in Fig. 28.3.

The blocking or protecting groups render a particular functional group unreactive under the coupling conditions used. At the same time, the protecting group must be capable of being removed under relatively mild conditions that do not disrupt the remaining portions of the molecule. The activating groups serve the reverse function. They make the portion of the molecule to which they are attached highly reactive, allowing for easy and quantitative coupling. The general sequence outlined in Fig. 28.3 is as follows.

1. The *N-terminal* end of one of the amino acids (or peptides) to be coupled is protected—Ⓟ.
2. The *C-terminal* end of the second amino acid (or peptide) is protected—Ⓟ.
3. If there are functional groups on the side chains (R or R′) of either of the two species being coupled, they are also protected with specific protecting groups.
4. The two amino acids (or peptides) are then coupled (activating groups are used if necessary), forming the protected dipeptide.

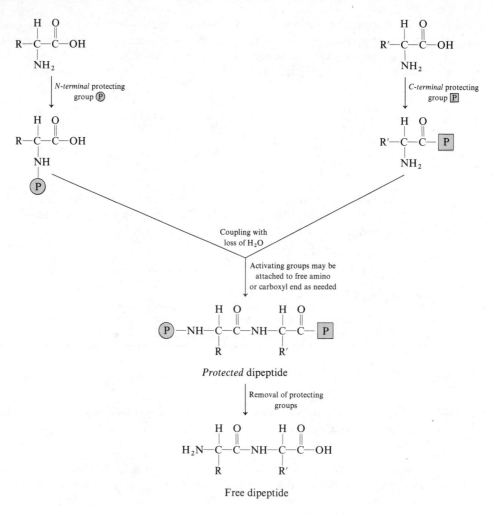

FIGURE 28.3 **Schematic representation of the general sequence used in the coupling of two amino acids or peptides.**

5. All protecting groups are removed if the synthesis is complete, or specific ones are removed if the new species is to be coupled further. That is, the *N-terminal* or *C-terminal* end may now be deprotected and coupled with another amino acid or peptide through the same sequence outlined earlier. The process is continued until the desired peptide is obtained.

Examples of *N-terminal* and *C-terminal* protecting and activating groups are in Tables 28.3 and 28.4.

In addition to the reagents in Tables 28.3 and 28.4, which are used in a multistep reaction sequence—protection, activation, coupling, deprotection—several reagents have been developed that allow both coupling and activation to be carried out in a single step. The most versatile of these is dicyclohexylcarbodiimide (DCC), which is an excellent dehydrating agent. DCC reacts with the free carboxyl end of one amino acid or peptide that is to be coupled, forming an activated ester intermediate. The intermediate then reacts in situ with the second amino acid or peptide to provide

TABLE 28.3 [*] **Protecting Groups**

Side Group	Blocking Group	Product*	Deprotection
N-terminal	$CH_2O-CO-Cl$ Carbobenzoxy chloride	$CH_2O-CO-NH-$(AA) Carbamate	$2\,M$ HBr, room temperature or Na, liquid NH_3
	$(CH_3)_3CO-CO-Cl$ t-Butoxycarbonyl chloride	$(CH_3)_3CO-CO-NH-$(AA) Carbamate	HCl in CH_3COOH, cold or CF_3COOH
	Phthalic anhydride	Imide (N—AA)	NH_2-NH_2
C-terminal	$R-OH$ ($R = CH_3-$, CH_3CH_2-, benzyl alcohols)	(AA)$-C(O)-O-R$ Ester	Dilute base or H_2O in dioxane, or H_2, Pd for benzyl
Side chain (AA)$-NH_2^{**}$	$Cl-S(O_2)-$ CH_3 P-Toluenesulfonyl chloride	(AA)$-NH-S(O_2)-$ CH_3 Sulfonamide	Na, liquid NH_3

TABLE 28.3 *(continued)*

Side Group	Blocking Group	Product*	Deprotection
(AA)—S—H	Cl—CH₂—⬡ Benzyl chloride	(AA)—S—CH₂—⬡ Benzyl sulfide	Na, liquid NH₃
(AA)—O—H	Cl—CH₂—⬡ Benzyl chloride	(AA)—O—CH₂—⬡ Benzyl ether	H₂, Pd
(AA)—COOH***	HO—CH₂—⬡ Benzyl alcohol	(AA)—C(=O)—O—CH₂—⬡ Benzyl ester	H₂, Pd
(AA)—(imidazole)N—H	Cl—CH₂—⬡ Benzyl chloride	(AA)—(imidazole)N—CH₂—⬡	Na, liquid NH₃
(AA)—CH₂—N(H)—C(=NH)—NH₂	Acid, H⊕	(AA)—CH₂—NH—C(=NH)—NH₃⊕	Neutralization

* The symbol (AA) designates a general amino acid. The portion of the amino acid we are interested in is drawn as branching off the circle, for example, (AA)—NH₂, (AA)—COOH, (AA)—OH.

** Carbobenzoxy chloride and *t*-butoxycarbonyl chloride may also be used. See N-terminal protecting groups.

*** Other esters may also be used. See C-terminal protecting groups.

TABLE 28.4 C-Terminal Activating Groups*

Activated $\overset{\dagger}{\widehat{(AA)}}$

Acyl azides

Active esters

$-R = -p\text{-}C_6H_4NO_2, -p\text{-}S-C_6H_4NO_2$ (a thioester),

$-CH_2CN$, and others

Active acyl compounds

$-X = -\overset{O}{\overset{\|}{C}}-R, -O-\overset{O}{\overset{\|}{P}}(OR)_2$, and others

* Although it is possible to activate either the C-terminal or N-terminal end of the amino acid or peptide, activation of the C-terminal end is by far more common. Examples of N-terminal activation can be found in the Reading References for this chapter.

† The activated portion of the amino acid is shown in red.

the new peptide, joined through the normal peptide link, and dicyclohexyl urea:

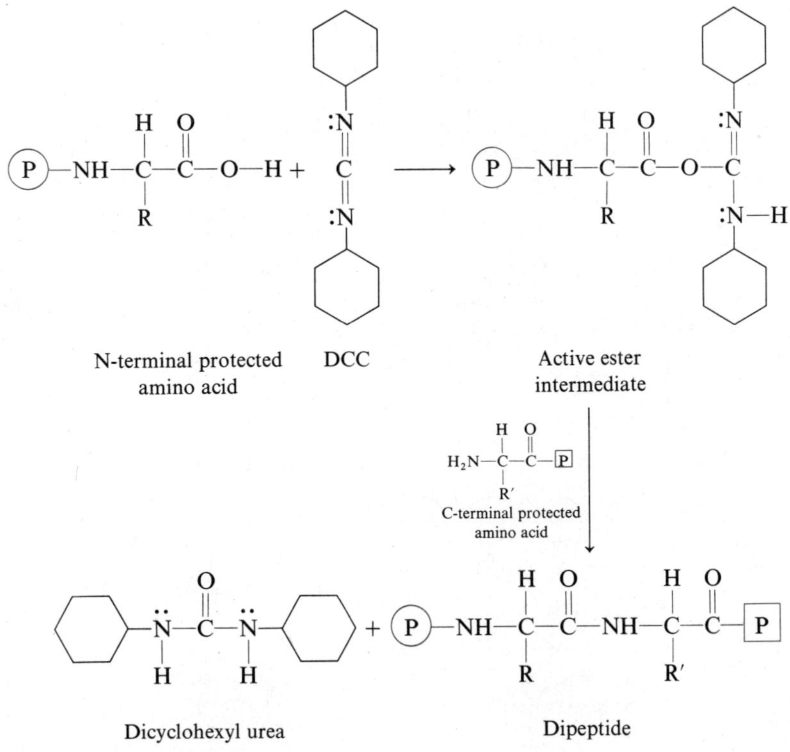

N-terminal protected amino acid   DCC   Active ester intermediate

C-terminal protected amino acid

Dicyclohexyl urea   Dipeptide

The use of the protecting and activating groups is illustrated in the following synthetic sequence—the final steps in one possible route to the total synthesis of lysine vasopressin, a nonapeptide we encountered in Fig. 28.2. The protecting groups are set off from the rest of the molecule in red.

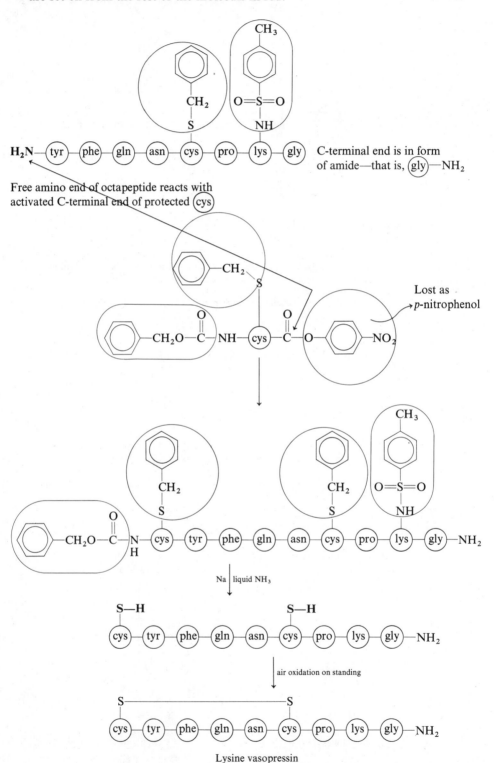

Lysine vasopressin

The free N-terminal end of the octapeptide precursor readily reacts with the activated C-terminal end of the protected cysteine. The reactive portions of the octapeptide, the—SH group of cysteine and the—NH_2 group of lysine, are protected as the benzyl sulfide and the sulfonamide, respectively (Table 28.3). In addition, the—SH of the cysteine coupled to the octapeptide is blocked as the benzyl sulfide, whereas the N-terminal end is protected as the carbobenzoxycarbonyl and the C-terminal end is activated as the p-nitrobenzoate ester.

The resulting nonapeptide is subsequently treated with Na in liquid NH_3, which removes the four remaining protecting groups to give the free nonapeptide. On standing, the free nonapeptide undergoes air oxidation to give the —S—S— bridge and lysine vasopressin.

Question 28.4

Using the preceding generalized approach—Ⓟ and ☐P☐ for protecting groups and Ⓐ for activating groups—outline a general synthesis that can be used to produce the tripeptide val-his-ser.

Question 28.5

The C-terminal activating groups have several esters including the p-nitrophenyl ester and the p-nitrobenzenethiol ester, NO_2—⟨◯⟩—O— and NO_2—⟨◯⟩—S—, respectively. What role does the nitro group play in activating these esters?

28.6   Automated Peptide Synthesis

A main problem associated with the synthesis of peptides is obtaining the product in a pure form. The problem is magnified as the length of the peptide increases. With each coupling of a new amino acid residue or peptide to the growing peptide chain, some percentage of the precursor is left unreacted. To assume that 100% coupling occurs at each step would be naïve. It is necessary, therefore, to wash each intermediate completely before proceeding to the next coupling reaction. All unreacted materials, which could compete with the desired peptide in subsequent steps, must be removed. Washing should also be done after each protection, activation, or deprotection sequence—again to avoid side reactions and a decrease in purity. In principle an exhaustive washing technique should be adequate, and it often is. However, the yield obtained on large peptides is also affected by these many purifications; yields of larger peptides are often less than 1% after months of work.

A method that has been used successfully in an attempt to alleviate this problem is called the *solid-support method* of peptide synthesis. In this method, the peptide being synthesized is attached at one end to a polymeric material that is insoluble in the solvents used in the synthesis. The use of the solid polymer with the desired peptide chain attached allows numerous washings and reactions without loss of material due to the mechanical and dilution problems associated with a completely liquid system. The polymeric product is washed and simply filtered to remove solvents. The method is summarized in Fig. 28.4.

A polymeric resin support (the material is normally a chloromethylated styrene or styrene-divinylbenzene copolymer) reacts with the free C-terminal end of a desired

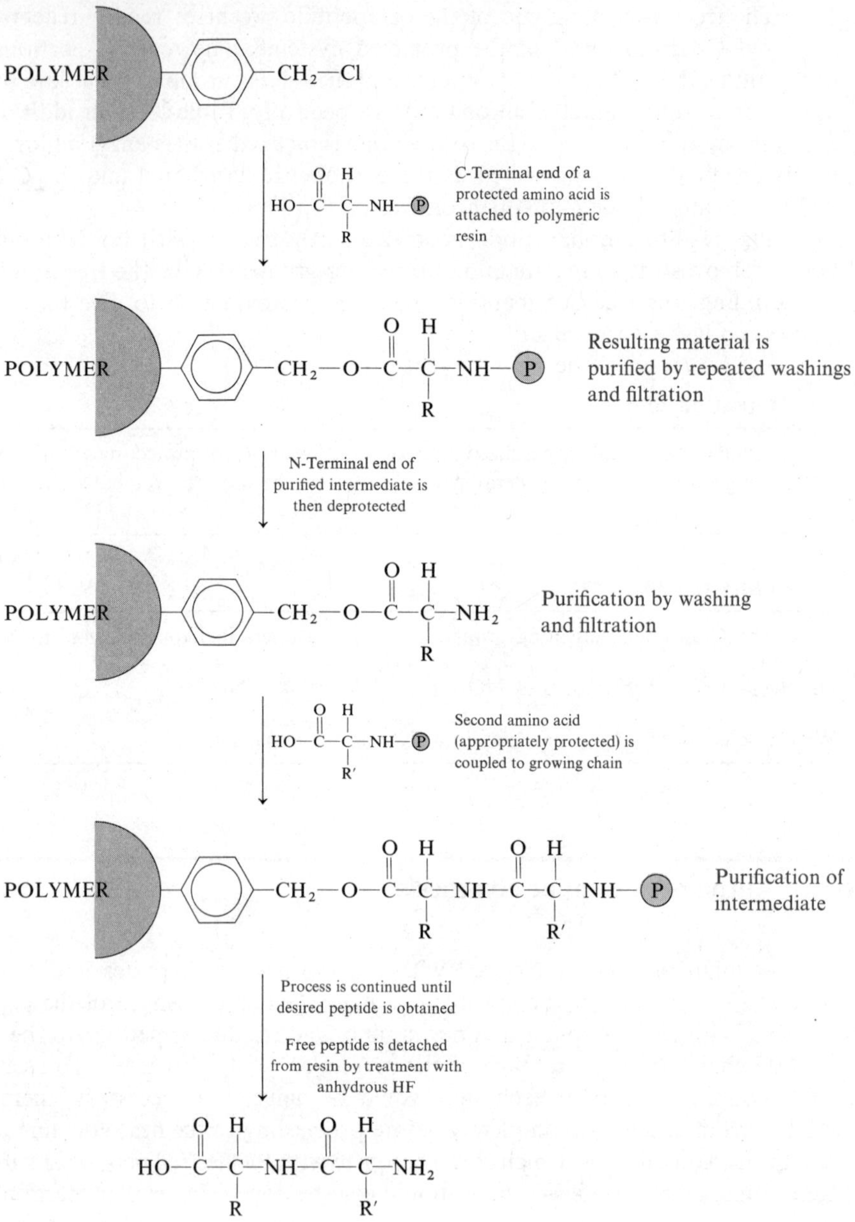

FIGURE 28.4   Solid-phase peptide synthesis.

amino acid. Following washing to remove any unreacted amino acid, the N-terminal end of the attached amino acid is deprotected, purified, and subsequently reacted with an activated C-terminal end of another amino acid. The DCC method of coupling is used most frequently in this step. The intermediate is washed, deprotected, washed again, and finally reacted again with a third amino acid. This process is continued until the desired peptide is finally obtained. The peptide is next detached from the polymeric resin by treatment with anhydrous hydrogen fluoride. All side-chain protecting groups are also removed at this time. The peptides prepared by this method are reasonably pure and may be purified further by normal chromatographic methods if necessary.

The entire solid-phase synthetic process has taken a giant step forward through the efforts of Professor R. Bruce Merrifield of the Rockefeller University in New York. Merrifield is responsible for the complete automation of the solid-phase system. Using a player-piano, drumlike device connected to reservoirs of washing solvents, blocking groups, activating groups, and amino acids through a series of selector valves (Fig. 28.5), Merrifield programs the addition, washing, and draining sequence needed in the synthesis of a given peptide so that the entire process, once initiated, can continue to completion on its own. The reaction vessel containing the solid support is divided into two parts separated by a fritted glass disk and is connected to a shaker. After a reagent is added to the solid support and reacted with shaking, a suction is applied and any unreacted materials pass through the fritted disks to the waste reservoirs. The next reagent is then added and the process continued.

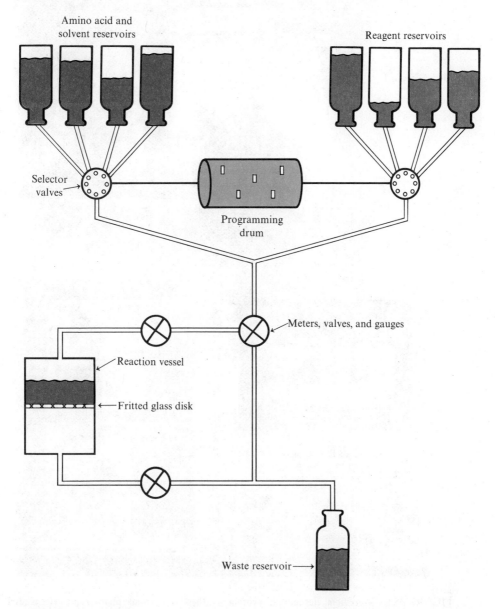

FIGURE 28.5 **Merrifield automated protein synthesizer—schematic diagram.**

Materials that are removed from the reaction vessel are cycled to a series of waste reservoirs and are eventually recycled and purified for future use. The entire apparatus is diagrammed in Fig. 28.5 and an actual picture of Merrifield's original device is in Fig. 28.6.

The automated system was first used successfully to synthesize the nonapeptide brandykinin in 1965. The entire process took approximately one week and resulted in an overall yield of about 68%. The feat was remarkable as compared with the

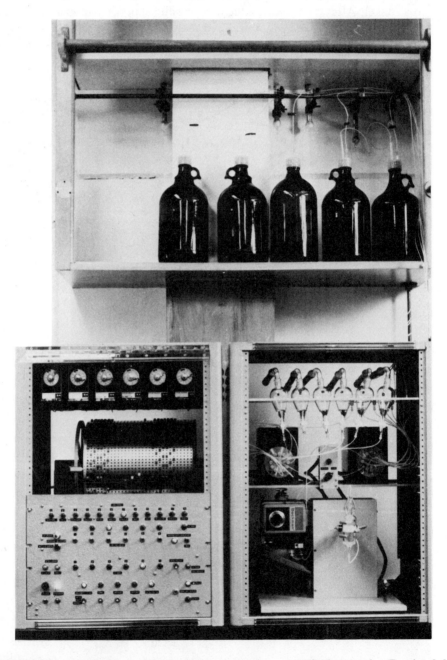

FIGURE 28.6 Merrifield automated protein synthesizer—actual photograph. (Reprinted by permission of Dr. R. Bruce Merrifield, Department of Chemistry, Rockefeller University)

usual methods used at that time. Peptide synthesis had entered the realm of modern synthetic organic chemistry. The versatility of the technique was further demonstrated by the Merrifield group four years later in the total synthesis of ribonuclease, a protein containing 124 amino acid residues.

28.7 Synthesis of Amino Acids

The synthesis of a peptide or protein requires amino acids as starting materials. The next logical question is: How does one obtain amino acids?

The human body can obtain most (12 of the essential 20) of the amino acids it requires through a variety of metabolic pathways starting from either carbohydrates (see Chap. 29) or lipids (see Chap. 25). The nitrogen required for the conversion is available in a variety of forms, often ammonium ion ($NH_4^\oplus$), which in turn originates from the degradation of protein ingested by organisms. The conversion of α-ketoglutarate to L-glutamate by the action of the enzyme NADH is an example of such a process (see Sec. 29.10A):

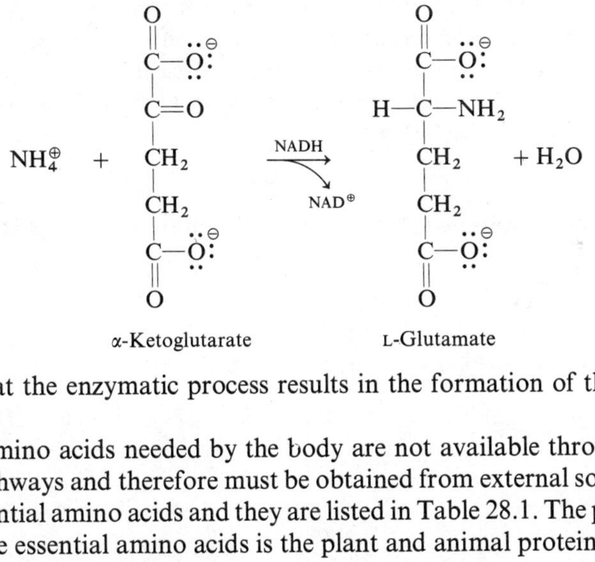

$$NH_4^\oplus \;+\; \text{α-Ketoglutarate} \xrightarrow[NAD^\oplus]{NADH} \text{L-Glutamate} \;+\; H_2O$$

Note that the enzymatic process results in the formation of the L-amino acid exclusively.

Many amino acids needed by the body are not available through any normal metabolic pathways and therefore must be obtained from external sources. These are called the essential amino acids and they are listed in Table 28.1. The primary external source of these essential amino acids is the plant and animal proteins that constitute the main portion of the human diet.

Amino acids can be synthesized in the laboratory by methods that we have seen or that are closely related to one or more synthetic processes we have studied. These are listed here:

A. From carboxylic acids—Hell-Volhard-Zelinsky reaction

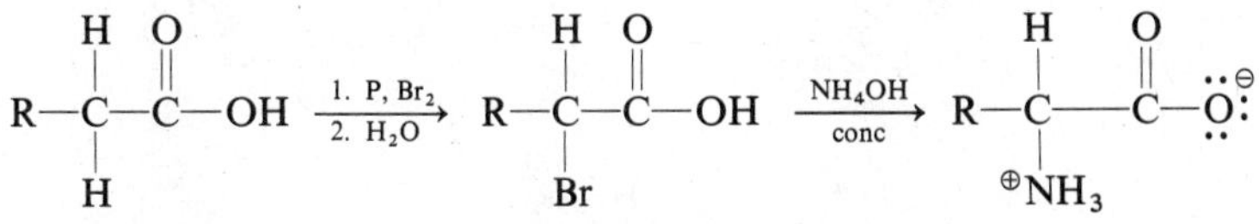

$$\underset{\substack{\text{Carboxylic}\\ \text{acid}}}{R-\overset{H}{\underset{H}{C}}-\overset{O}{C}-OH} \xrightarrow[\text{2. } H_2O]{\text{1. P, Br}_2} \underset{\substack{\text{α-Bromocarboxylic}\\ \text{acid}}}{R-\overset{H}{\underset{Br}{C}}-\overset{O}{C}-OH} \xrightarrow[\text{conc}]{NH_4OH} \underset{\text{Amino acid}}{R-\overset{H}{\underset{\oplus NH_3}{C}}-\overset{O}{C}-O^\oplus}$$

Carboxylic acids are readily converted to the corresponding α-bromocarboxylic acid by the reaction of bromine in the presence of a trace of phosphorus (see Sec.

22.18). Treatment of the α-bromo acid with a concentrated solution of aqueous ammonia results in the formation of the amino acid desired.

B. From aldehydes or ketones—Strecker synthesis

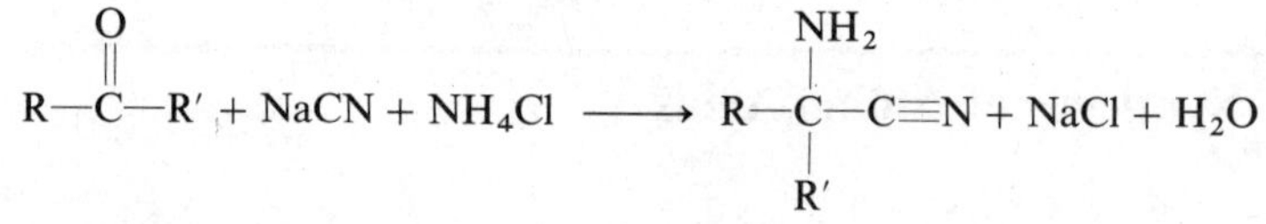

$$R-\overset{\overset{O}{\|}}{C}-R' + NaCN + NH_4Cl \longrightarrow R-\overset{\overset{NH_2}{|}}{\underset{R'}{C}}-C\equiv N + NaCl + H_2O$$

Aldehyde or   α-Amino
ketone   nitrile

$$R-\overset{\overset{NH_2}{|}}{\underset{R'}{C}}-C\equiv N \xrightarrow[H_2O]{H^\oplus} R-\overset{\overset{NH_2}{|}}{\underset{R'}{C}}-\overset{\overset{O}{\|}}{C}-OH \equiv R-\overset{\overset{R'}{|}}{\underset{\oplus NH_3}{C}}-\overset{\overset{O}{\|}}{C}-\overset{..}{\underset{..}{O}}{:}^\ominus$$

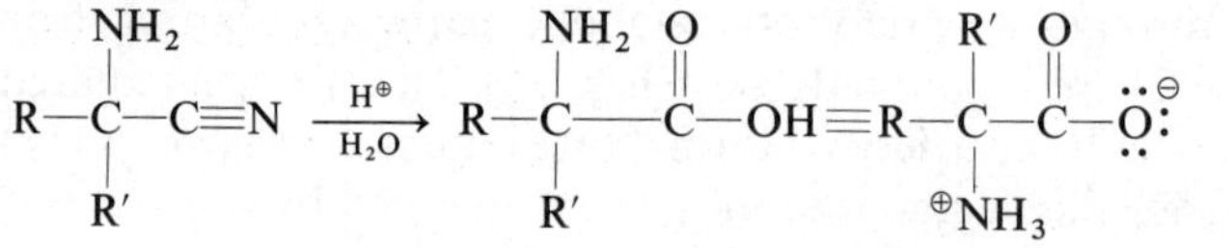

Amino acid

The reaction of an aldehyde or ketone with sodium cyanide in the presence of ammonium chloride results in the formation of a α-amino nitrile. The amino nitrile is often isolable and can be hydrolyzed easily into the corresponding amino acid. The reaction is similar to the formation of cyanohydrins from aldehydes or ketones discussed in Sec. 21.4.

C. From malonic ester—phthalimidomalonic ester synthesis

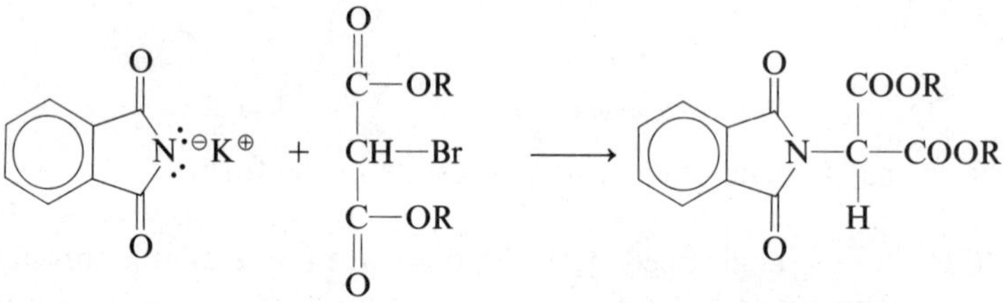

Potassium   α-Bromomalonic   N-Phthalimidomalonic ester
phthalimide   ester

1. base
2. R—X,
   R = R—CH₂—X or
   CH₂—COOR
   X
   usually

$$\xrightarrow[\text{(decarboxylation)}]{\overset{H_2O,\ heat}{H^\oplus,}}$$

$$R-\overset{\overset{H}{|}}{\underset{\oplus NH_3}{C}}-\overset{\overset{O}{\|}}{C}-\overset{..}{\underset{..}{O}}{:}^\ominus$$

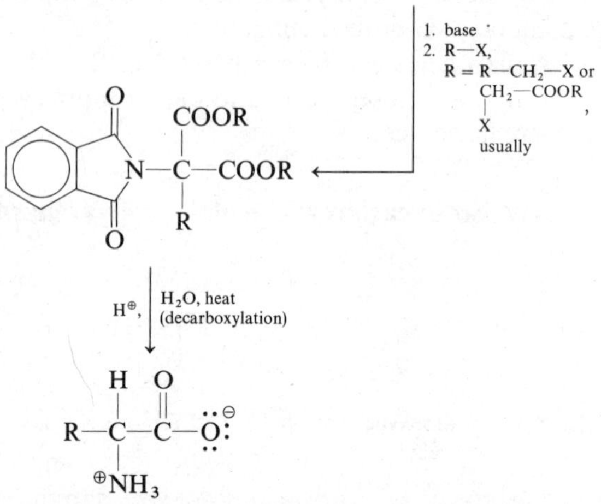

Amino acid

The versatility of potassium phthalimide in the synthesis of primary amines (Gabriel synthesis) was discussed in Sec. 26.7. The preceding reaction sequence is extended to the synthesis of amino acids. It involves the reaction of potassium phthalimide with α-bromomalonic ester, easily obtained by treatment of malonic ester with bromine in carbon tetrachloride. The resulting substituted phthalimide may then be alkylated as in malonic ester synthesis (see Sec. 24.20) to provide a wide variety of R groups on what was the α carbon of the original malonic ester. Acid hydrolysis and decarboxylation (see Sec. 24.20) provide the desired amino acid.

Examples of syntheses using these three methods are:

$$CH_3-\overset{\overset{\displaystyle O}{\|}}{C}-H \xrightarrow[NH_4Cl]{NaCN} \xrightarrow[H_2O]{H^\oplus} CH_3-\overset{\overset{\displaystyle H}{|}}{\underset{\underset{\displaystyle \oplus NH_3}{|}}{C}}-\overset{\overset{\displaystyle O}{\|}}{C}-\overset{..}{\underset{..}{O}}{:}^\ominus$$

Acetaldehyde

Alanine
72% Yield

$$\text{Diethyl } N\text{-phthalimidomalonate} \xrightarrow[\text{2.} \underset{}{\bigcirc}-CH_2-Cl]{\text{1. NaOCH}_2CH_3} \xrightarrow[\text{heat}]{H^\oplus, H_2O} \bigcirc-CH_2-\overset{\overset{\displaystyle H}{|}}{\underset{\underset{\displaystyle \oplus NH_3}{|}}{C}}-\overset{\overset{\displaystyle O}{\|}}{C}-\overset{..}{\underset{..}{O}}{:}^\ominus$$

Diethyl *N*-phthalimidomalonate

Phenylalanine
90% Yield

$$CH_3-CH_2-\overset{\overset{\displaystyle O}{\|}}{C}-OH \xrightarrow[P]{Br_2} CH_3\underset{\underset{\displaystyle Br}{|}}{CH}-\overset{\overset{\displaystyle O}{\|}}{C}-OH \xrightarrow[\text{aq}]{\overset{\text{excess}}{\underset{}{NH_3}}} CH_3-\overset{\overset{\displaystyle H}{|}}{\underset{\underset{\displaystyle \oplus NH_3}{|}}{C}}-\overset{\overset{\displaystyle O}{\|}}{C}-\overset{..}{\underset{..}{O}}{:}^\ominus$$

Propionic acid

Alanine
70% Yield

Question 28.6

Show how one might synthesize each of the following amino acids from readily available starting materials of six carbons or less.

(*a*) leu   (*b*) phe   (*c*) val

28.8   Synthesis of Proteins

Proteins are very large peptides with molecular weights usually greater than 5,000. They are found in the cells of all living organisms, each performing a specific act or set of functions necessary to the survival of the cell. Indeed, thousands of these natural polymers can be found in a given cell of most organisms.

How are proteins synthesized? The answer to this question was presented in part earlier in the chapter. It is possible, for example, to use classical methods to synthesize a protein from its component amino acids. The task is monumental, however, and success eluded the synthetic chemist until recent times. The automated method of peptide and protein synthesis developed by the Merrifield group rendered the dream possible. Ribonuclease, found in mammals, was the first protein to be successfully synthesized with this method. The synthesis required more than 350 different chemical reactions and resulted in a 17% overall recovery. The huge number of manipulations involved and the relatively high yield even after all these manipulations provide undeniable proof of the versatility of the technique.

The synthesis or manufacture of proteins in the body occurs in the cytoplasm of the cell—the intracellular fluid surrounding the nucleus of a cell and containing dissolved **DNA** *(deoxyribonucleic acid)*, various types of **RNA** *(ribonucleic acid)*, and a variety of proteins and other materials. The DNA, which is found primarily in the nucleus of the cell, is the storage center of all hereditary knowledge associated with a given organism. The structures of DNA and RNA are very similar, but a comprehensive discussion of their structure and action is best left to a course in biochemistry. The remainder of this section is devoted to a qualitative review of their mode of action, which is shown schematically in Figs. 28.7 and 28.8.

The DNA in the nucleus of the cell is composed of many segments called *genes*. Each segment of the DNA chain contains the hereditary code (blueprint) needed for the complete synthesis of a given protein. The information from a gene is chemically transcribed in the nucleus onto a new, smaller molecule called messenger RNA (mRNA). The mRNA is actually synthesized in the nucleus, with the DNA acting as a template and an enzyme called RNA polymerase joining the various segments of the mRNA chain together. The mRNA, once formed, can migrate through the nuclear membrane surrounding the nucleus and enter the cytoplasm of

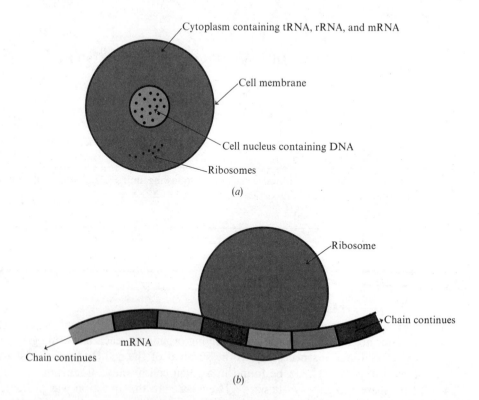

(a)

(b)

FIGURE 28.7 (*a*) Schematic representation of a cell showing relative sizes and positions of nucleus, cytoplasm, membrane, and ribosomes. (*b*) Expanded view of the *r*RNA:*m*RNA couple. Segments shown in various shades illustrate the portions of the chain containing the information to be transferred to the *t*RNA molecules, telling them which amino acid to add to the growing chain.

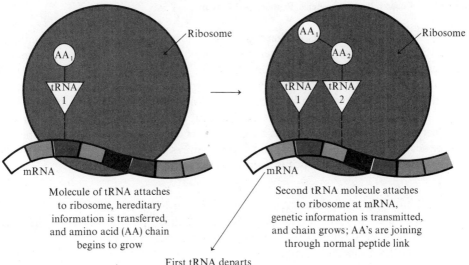

Molecule of tRNA attaches
to ribosome, hereditary
information is transferred,
and amino acid (AA) chain
begins to grow

Second tRNA molecule attaches
to ribosome at mRNA,
genetic information is transmitted,
and chain grows; AA's are joining
through normal peptide link

First tRNA departs

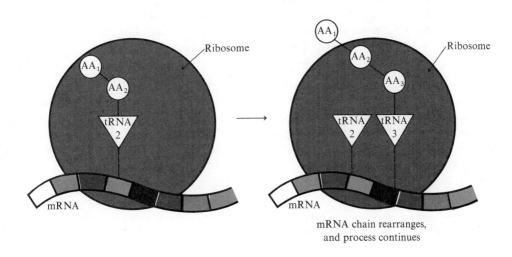

mRNA chain rearranges,
and process continues

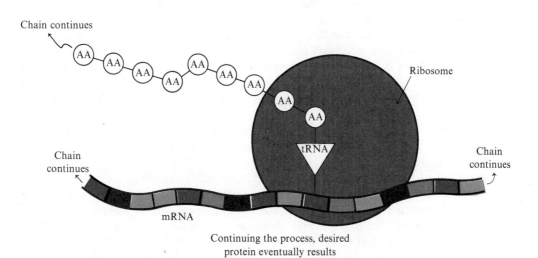

Continuing the process, desired
protein eventually results

FIGURE 28.8 Natural synthesis of proteins.

the cell. In the cytoplasm, there is another form of RNA called ribosomal RNA (rRNA) and it comprises the major portion of the ribosomes in the cytoplasm. The remainder of the ribosomes is comprised of a variety of proteins. The mRNA, after entering the cytoplasm, becomes attached to one or more ribosomes, and it is at these new ribosomes that protein synthesis actually occurs.

The rRNA:mRNA couple contains all the information necessary to synthesize a protein and also provides a reaction surface at which the actual synthesis may be accomplished. The synthesis, however, requires yet a third form of RNA—transfer RNA (tRNA). The sequence is as follows (Fig. 28.8):

1. A molecule of tRNA attaches to the rRNA:mRNA couple.
2. The hereditary information in mRNA is transmitted to tRNA, resulting in the attachment of an amino acid to the rRNA:mRNA couple.
3. The tRNA molecule departs.
4. The mRNA rearranges in such a way as to provide the chemical information necessary for the next step at the site of the ongoing synthesis.
5. A second molecule of tRNA now attaches itself to the ribosome, the genetic information is transmitted, and a second amino acid is attached to the chain.
6. Steps 3 to 5 are repeated until the desired protein or polypeptide is complete.

Protein synthesis is a constantly ongoing process in cells. As proteins are metabolized they must be replaced, or if a larger concentration of a specific protein is needed by the cells for a given function, it can be synthesized at that time. Compared with the synthetic protein-making machines of humans, the protein-making machines of the cells are much quicker and more efficient, often completing the synthesis of a desired protein in seconds.

28.9   Structure of Proteins

Proteins may be divided into three primary classes: (1) **structural proteins**, making up the major portion of all structural units of plants and animals (for example, hair, fur, skin, feathers, muscle), (2) proteins involved in the regulation of the many biochemical reactions necessary to the existence of the organism **(regulator proteins)**, and (3) proteins involved in transport (foodstuffs, waste, and so on), the **transport proteins.**

A second broad classification system is based on the overall physical shape of proteins. According to this scheme, proteins may be divided into two classes: (1) **fibrous** and (2) **globular.** Considering the protein in terms of its bulk geometry only, one could liken it to a strand of spaghetti; the strand is a long, continuous, non-branched chain of amino acids. Fibrous proteins (which are water-insoluble) are made up of these chains entwined in what could be described as a side-by-side, parallel fashion, whereas globular proteins (which are water-soluble) have a large degree of twisting and folding in each chain and appear to be more spherical and rounded—hence the name *globular*. Examples of naturally occurring proteins classified as to function and origin are in Table 28.5.

A more detailed description of a protein (or polypeptide) is often necessary. Due to the complexity of these molecules, a fairly rigorous classification system has

TABLE 28.5   Proteins: Function and Origin

Function	Example*	Origin and Description
Structural proteins	*Keratins*	Make up most protective coatings of animals, for example, skin, feathers, hair, nails, hoofs, claws, silk, and wool.
	Collagen	Comprises majority of all connective tissue in animals, for example, bones, cartilage, and tendons.
Regulator proteins:		
Enzymes	*Chymotrypsin*	Excreted by the pancreas, capable of cleaving polypeptide chains, involved in digestive process.
	Lysozyme	Found in a variety of natural sources (a high concentration in egg whites), capable of cleaving polysaccharide chains, involved in digestion.
	Ribonuclease	Also excreted by the pancreas, this cleavage enzyme cleaves polynucleotide chains—RNA.
Hormonal proteins	*Insulin*	Excreted by the pancreas, required for normal metabolism of glucose.
	Vasopressin	Excreted by the pituitary gland, assists in regulation of blood pressure.
	Bradykinin	Found in blood plasma, also involved in regulation of blood pressure.
Transport proteins	Hemoglobin	Found in erythrocytes (red blood cells) of most animals, responsible for the transport of oxygen from the lungs to the cells and for the removal of waste carbon dioxide from the cells, which it returns to the lungs.
	Myoglobin	Found in muscle tissue, is responsible for binding oxygen, which it receives from the hemoglobin, and storing it until needed by the cells.

* Proteins in *italic* are discussed in detail in the next few sections.

been developed that divides molecular geometry into its component parts. This system of protein structure is summarized here.

Primary structure: The primary structure associated with a protein is concerned with all the covalent bonding in the protein chain or between protein chains. The primary structure is the amino acid sequence of the protein.

Secondary structure: As the protein chain grows, many nonbonding types of interactions are possible between the various functional groups present. These interactions result in a variety of conformational folds in the chain. These conformations are called the α helix, β sheet, and π helix, to name a few. Each designation refers to a specific ordered arrangement of the amino acids in the protein chain. The various portions of the chain that are, for example, in an α or β form, are part of the secondary structure, and they result from nonbonding interactions in the chain.

Tertiary structure: The dividing line between secondary and tertiary structure is fuzzy at best. Discussions of tertiary structure are primarily concerned with the gross overall folding of the protein chain. The chain is again viewed as a piece of twine, but the amino acids involved and the portions of the chain that are either helical or sheetlike are of no real concern. Only the gross intertwining is of interest. Again, it is primarily the nonbonding interactions that are responsible for the geometry, but covalent bonding in the form of —S—S— bridges also plays a role.

Quaternary structure: Also primarily the result of nonbonding interactions, quaternary structure is concerned with these interactions as they result in gross folding patterns and arrangements between two or more protein chains.

The use of these terms and the system itself becomes much clearer as we apply them to real examples in the next sections.

28.10 Factors Influencing Protein Structure

A variety of factors taken together yield the overall molecular geometry associated with a given protein. Each factor is discussed in turn in this section.

A. Peptide Link

At the backbone of the protein chain is the peptide link, which is the amide bond that attaches one amino acid to another. The peptide link plays a major role in the geometry of proteins. This pronounced effect is due in large part to the bond between the acyl carbon and the amide nitrogen in an amide having a great deal of double-bond character. For example, the bond is stronger than a normal carbon-nitrogen single bond, it is shorter than a carbon-nitrogen bond, and restricted rotation (*cis/trans* integrity) is observed about the bond. The term *restricted rotation* describes a rate of rotation that is, with respect to the amide carbon-nitrogen bond in this case, much slower than would be predicted if a pure single bond existed between these two atoms. The partial double-bond character of the carbon-nitrogen bond explains this observation. Indeed, through the use of a variety of instruments (NMR for example), it is possible to observe the difference between the two groups attached to an amide nitrogen, for example the two methyl groups in *N,N*-dimethyl-formamide shown here:

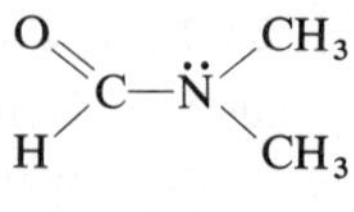

$$\begin{array}{c} O \\ \diagdown \\ C-\ddot{N} \\ \diagup \diagdown \\ H \end{array} \begin{array}{c} CH_3 \\ \diagup \\ \\ \diagdown \\ CH_3 \end{array}$$

N,N-Dimethylformamide

That these two groups can be distinguished instrumentally (see Sec. 23.19E) indicates that they are in different chemical environments and are not identical. A partial double bond would explain this observation. The *cis/trans* geometry that results would place one group (methyl in this example) *cis* to the acyl carbon and the second group *trans* to that carbon.

Why do amides exhibit partial double-bond character? Because of the proximity of the acyl carbon oxygen to the amide nitrogen, stabilization of the amide itself is possible through resonance overlap of the carbon-oxygen double bond with the nonbonding electrons on the nitrogen. Resonance hybrids of the following forms

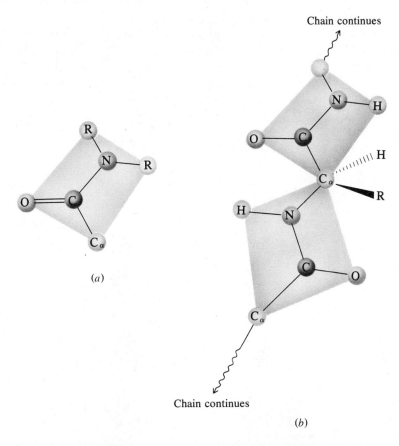

FIGURE 28.9 (*a*) Coplanar arrangement of atoms in an amide because of restricted rotation about the carbon-nitrogen bond. (*b*) Partial peptide chain showing the peptide links and the coplanar arrangement of atoms that result. The resulting planes may have a variety of conformational arrangements with respect to each other.

result:

Resonance-stabilized
amide

Partial double bond

The resulting amide links may be drawn in a simplified view as a coplanar arrangement of atoms of the type in Fig. 28.9.

Question 28.7

Draw the complete orbital picture (in three dimensions) of the peptide link.

Joining a large number of amino acids through peptide links results in a series of coplanar arrangements of atoms in space. These arrangements are conformationally biased to provide the most stable arrangement of atoms by minimizing repulsions and maximizing attractions between the various functional groups. The

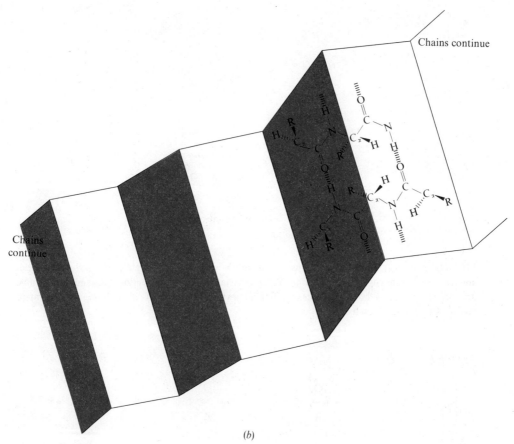

FIGURE 28.10   (a) Two-dimensional geometry associated with the β-sheet structure of proteins with hydrogen bonds indicated. (b) Three-dimensional arrangement of the β-sheet structure.

two most common arrangements in proteins are the **α helix** and the **β sheet** (Figs. 28.10 to 28.12).

B. α Helix

As shown in Fig. 28.10, the peptide links that bind a protein together result in a long series of coplanar arrangements of atoms. Each plane is free to rotate about the C_α through which the planes are connected, but, as in all molecular arrangements, certain preferred conformations result in the most stable orientations of the

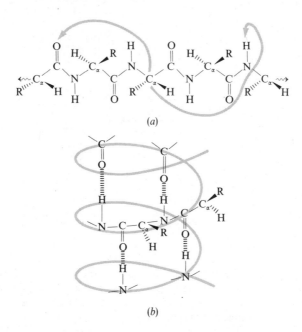

(a)

(b)

FIGURE 28.11 (*a*) Two-dimensional view of the α helix. The arrow indicates the hydrogen bonding between the carbonyl oxygen and amide hydrogen, the distance between which (13 atoms) equals one turn in the helix. (*b*) Three-dimensional view of the α helix partially illustrating the hydrogen bonding that holds the helix in the observed conformation.

FIGURE 28.12 Space filling model of a typical polypeptide α helix. *Left:* protons omitted for clarity; *right:* protons added.

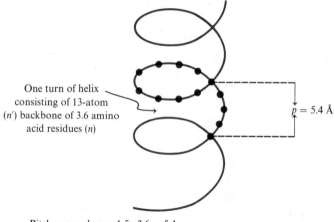

One turn of helix
consisting of 13-atom
(n') backbone of 3.6 amino
acid residues (n)

$p = 5.4$ Å

Pitch $= p = d \cdot n = 1.5 \cdot 3.6 = 5.4$

FIGURE 28.13 Typical α helix, with structural characteristics indicated. Note that the helix above is in a right-handed screw sense. This is true of all α helices composed of L-amino acids. D-amino acids (because of the stereochemical constraints of their geometry) provide a left-handed helix that is the mirror image of the one above.

molecule. One of the more common of these, found in a variety of naturally occurring polymeric systems, is the α helix.

Other helical arrangements of atoms are found in nature, and a method of describing them has been formalized. The following six features are of interest.

1. Is the helix a right-handed screw or a left-handed screw? A right-handed screw if turned clockwise around an axis through its center moves away from you. A left-handed screw moves toward you (Fig. 28.13).
2. The number of amino acids that are in the skeletal backbone of the protein and are present in a complete turn of the helix $= n$.
3. The distance (in Å usually) along the helix axis traversed by one amino acid residue $= d$.
4. The pitch of the helix is defined as $p = d \cdot n$.
5. The total number of atoms in the skeletal backbone that constitute one complete turn of the helix $= n'$.

These units are used to characterize a typical α helix in Fig. 28.13.

The primary force responsible for holding the α helix as well as the β sheet in these preferred conformations (and also the other naturally occurring conformations we do not discuss) is a nonbonding interaction we are familiar with—the *hydrogen bond*. These bonds are illustrated in Figs. 28.10 and 28.11.

A helical arrangement in a polymer, because of its coiled nature and intramolecular hydrogen bonding, is flexible yet at the same time reasonably strong.

C. β Sheet

A linear parallel arrangement of the repeating planar amide units of a protein results in a conformation called the β-pleated sheet, which is illustrated in Fig. 28.10.

The sheet, like the helix, is held in its conformation by numerous hydrogen bonds between carbonyl oxygens and amide hydrogens. The β sheet is often referred to as the antiparallel pleated sheet because any two adjacent protein chains run in opposite directions, one in a head-to-tail direction and the other in a tail-to-head direction. This too is illustrated in Fig. 28.10. Note that the upper chain is in the following sequence proceeding from left to right—C_α—CO—NH—whereas the lower chain follows the sequence—NH—CO—C_α.

The β sheets are the strongest protein conformations, and although they are somewhat flexible, they are resistant to stretching.

D. Triple Helix

The final conformational arrangement presented here is the **triple helix.** The triple helix (Fig. 28.14) is an arrangement of protein chains that provides more strength than the α helix and more strength and flexibility than the β sheet. The triple helix results from the intertwining of three α helices into a semirigid cylindrical triple helix. The three helices are themselves held together in the normal way through a large number of hydrogen bonds and disulfide linkages (—S—S—). The triple helix normally has an average length of approximately 3,000 Å and a diameter of approximately 15 Å.

The triple-helix arrangement is often referred to as a *protofibril*. A second arrangement of the triple helix found in a variety of naturally occurring proteins is the *microfibril*, which is composed of 11 protofibrils in a cylindrical arrangement.

E. Nonbonding Interactions

All the previous conformational arrangements result mainly from two molecular structural features—the peptide link and the hydrogen bond. The hydrogen

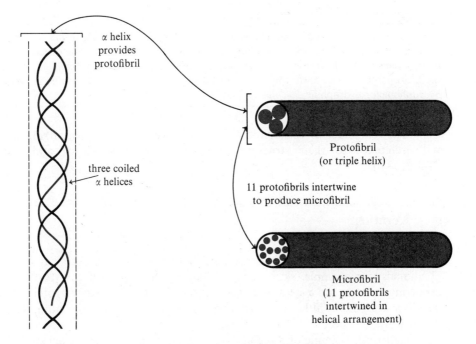

FIGURE 28.14   The α helix, triple helix (also called *protofibril*), and microfibril.

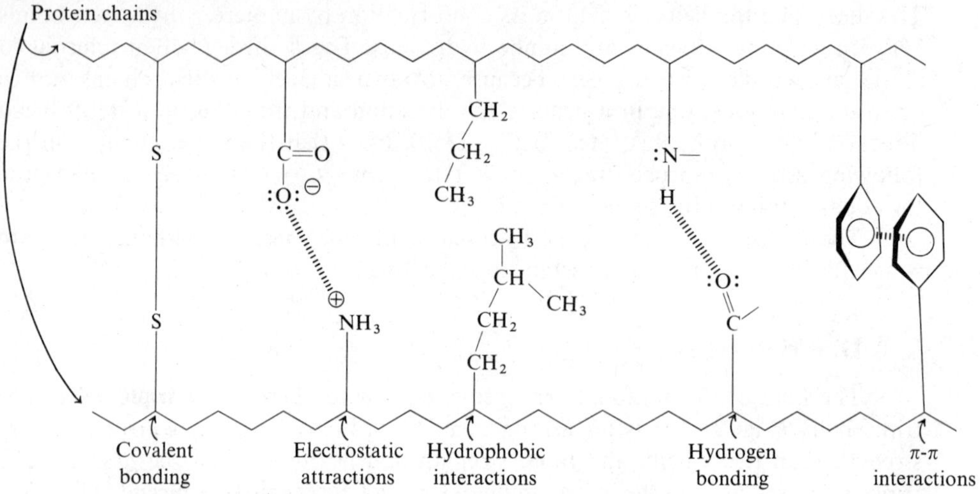

FIGURE 28.15 Nonbonding interactions present in and partially responsible for protein conformations. The protein chains may be one chain folded over or two or more chains (e.g., in the protofibril) that are in close proximity to one another.

bond is only one of many nonbonding interactions possible in conformational analyses of proteins. Many others, for example hydrophobic interactions, electrostatic interactions between ionic groups, and other polar dipole-dipole interactions, are also possible depending on the types of groups in the protein and their proximity to each other. Covalent bonding, particularly the disulfide link (—S—S—), also plays an important role in conformational geometry. These interactions are summarized in Fig. 28.15.

Question 28.8

Hydrophobic (nonbonding) interactions play a very important role in the conformation, structure, and function of proteins. Globular proteins, in their intertwining, orient the hydrophilic groups of the chain toward the surface, thus accounting for their solubility in water, and the hydrophobic groups toward the interior of the sphere. Classify the following as containing hydrophobic or hydrophilic side chains: phe, gly, his, val, ile, lys.

The remainder of this chapter discusses several naturally occurring proteins with reference to their structure (using the terms described) and function (Table 28.5).

28.11   Keratins

Keratins are strong, fibrous proteins that constitute most protective coatings of animals, for example, skin, hair, or feathers (see Table 28.5). Most naturally occurring keratins are α keratins, meaning they contain primarily protein in an α-helical arrangement, although other conformations are known including the antiparallel pleated sheet.

Hair is an example of a readily available keratin that exists exclusively in the α-helix form. The basic α-helix structural conformation provides the hair with its

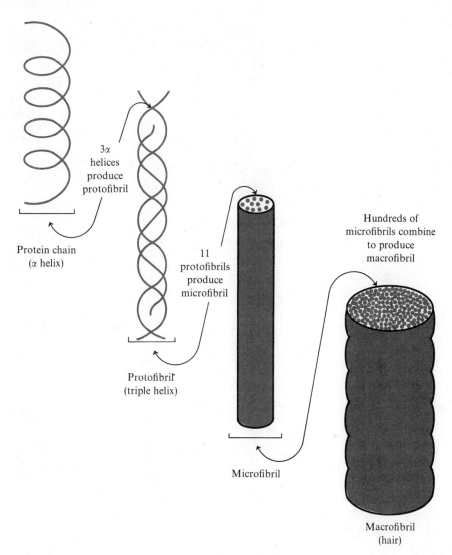

3α
helices
produce
protofibril

Protein chain
(α helix)

11
protofibrils
produce
microfibril

Protofibril
(triple helix)

Hundreds of
microfibrils combine
to produce
macrofibril

Microfibril

Macrofibril
(hair)

FIGURE 28.16 **Macrofibril that comprises a hair fiber.**

strength, resilience, and flexibility. Hairs are actually composed of large units called *macrofibrils*, illustrated in Fig. 28.16.

Along with the hydrogen bonds, many S—S bridges between cysteines are also responsible in large part for holding the helices together. In wool, which is also composed of protein primarily in an α helix, it is postulated that an enzyme excreted by a moth breaks these S—S bridges and is at least in part responsible for the breakdown of the fiber.

28.12   Insulin

The earliest regulator protein to have its total amino acid sequence determined was insulin. Insulin is a hormonal protein excreted by the pancreas, which is required for the normal metabolism of glucose (blood sugar). As illustrated in Fig.

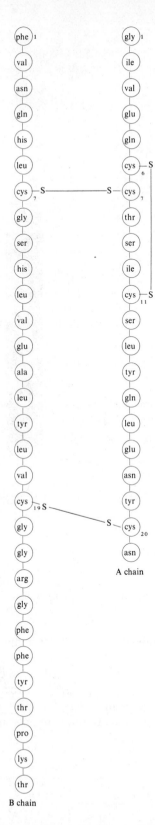

FIGURE 28.17 Primary structure of human insulin—A (21 amino acid residues) and B (30 amino acid residues) chain labeled and all S—S bridges labeled. The molecular weight of the protein is approximately 6,000.

28.17, the primary structure of insulin consists of two polypeptide chains joined by two intermolecular covalent bonds through the disulfide bridges of the cysteines.

The role of insulin in the body is both well known and a mystery. For example, it is well documented that insulin is involved in maintaining the correct level of blood glucose in the body. But how this is actually accomplished is still debated. Insulin is also required for the normal synthesis of fatty acids in the liver and for the uptake and consumption of oxygen in the muscles, but again there is a great deal of debate as to how insulin is involved in these processes. One common belief is diagrammed schematically here:

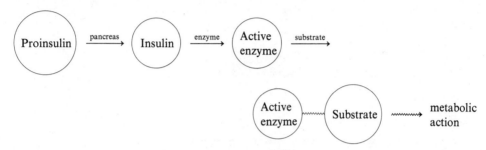

According to this theory, insulin is excreted by the pancreas, enters the bloodstream, and is carried to various parts of the body where it is required. On arrival, the insulin, when needed, reacts with and activates a given enzyme, which in turn reacts with a substrate to effect one of the actions listed earlier. A consequence of this mechanism is that the insulin, as it is involved in various metabolic roles in the body, behaves independently in each role. The second belief diagrammed here says that it is the insulin itself that is the activating species that reacts with the substrate to promote the desired reactions. Based on this reasoning, no other secondary species (for example, an enzyme) is needed.

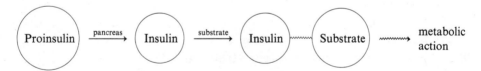

Actually, it is not insulin but rather a precursor called *proinsulin* (Fig. 28.18) that is excreted by the pancreas. The proinsulin is then acted on enzymatically to provide insulin.

When the body loses its ability to produce sufficient insulin for normal glucose metabolism, a condition known as *diabetes mellitus* results. The condition is treated by the daily administration of insulin through one or more injections directly into the bloodstream. The hormone, once injected, acts very rapidly to metabolize the glucose in the blood. This sometimes results in too rapid a decrease in the level of blood glucose and a large swing in the opposite direction to a condition of low blood glucose and a comalike state known as *insulin shock*.

Insulin must be taken intravenously because it is unable to pass through the digestive tract in a stable form when taken orally. Proteolytic enzymes (proteases) that cleave protein chains are present throughout the digestive system and would destroy the insulin before it could arrive at its destination, for example, the liver.

Several hypoglycemic agents (agents that lower the levels of blood glucose) on the market may be taken orally. None of these, however, is an insulin analogue.

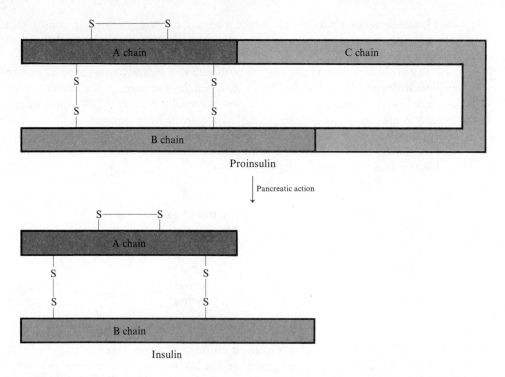

Proinsulin

Insulin

FIGURE 28.18 Production of insulin from proinsulin in the pancreas. The proinsulin, like the insulin, contains the A and B chains intact (including the three S—S bridges). The A and B chains of proinsulin are joined by a third C chain that contains 33 amino acid residues; this third chain is cleaved by protolytic enzyme action in the pancreas.

Although such analogues have been synthesized and tested, none has yet proved satisfactory.

28.13 Determination of Protein Structure

The gross three-dimensional conformations associated with proteins are primarily studied by instrumental methods, for example, X-ray crystallography. Also, many wet chemical methods have been developed and provide an indirect yet clear picture of the total primary structure (covalent) of the protein and may even lend credence to the instrumental data with regard to the secondary, tertiary, and quaternary structure of the protein.

A. Total Amino Acid Analysis

The hydrolysis of all amide bonds in a polypeptide or protein (6 N HCl at 110° for 24 hr is sufficient) yields all the individual amino acids that make up the polymer. Separation and analysis of these amino acids by chromatographic methods provide the complete amino acid composition of the protein of interest, including the relative amounts of each. For example, if the total hydrolysis of a pentapeptide yields the following information, ala:gly:phe:val:ser, 1:1:1:1:1, we can say unequivocally that the pentapeptide is composed of the five different amino acids listed, but we can say little else. Statistically, there could be 5! (five factorial) possible combinations of these amino acids, which would yield 120 different possible pentapeptides. To determine the total primary structure, much more information is required. How this information is obtained is discussed in the next few sections.

B. Partial Hydrolysis of Proteins and Peptides

If a protein or peptide is hydrolyzed under somewhat less rigorous conditions (for example 25°, 3 N HCl, 12 hr), it is possible for incomplete hydrolysis of the peptide (or protein) to result. The partial hydrolysis of the previous pentapeptide could yield the following observations as one example: ala-gly:phe-val:ser. With these pieces of the puzzle there would be only 3! possible combinations of these three segments to give six possible isomers. We still do not know the complete covalent structure, but we made a great deal of progress in going from 120 to 6 possibilities.

C. N-Terminal Analysis

In any peptide (other than cyclic peptides) there is an N-terminal end and a C-terminal end. Reagents have been developed that will react selectively with one end of the chain to the complete exclusion of the other.

Sanger reaction:

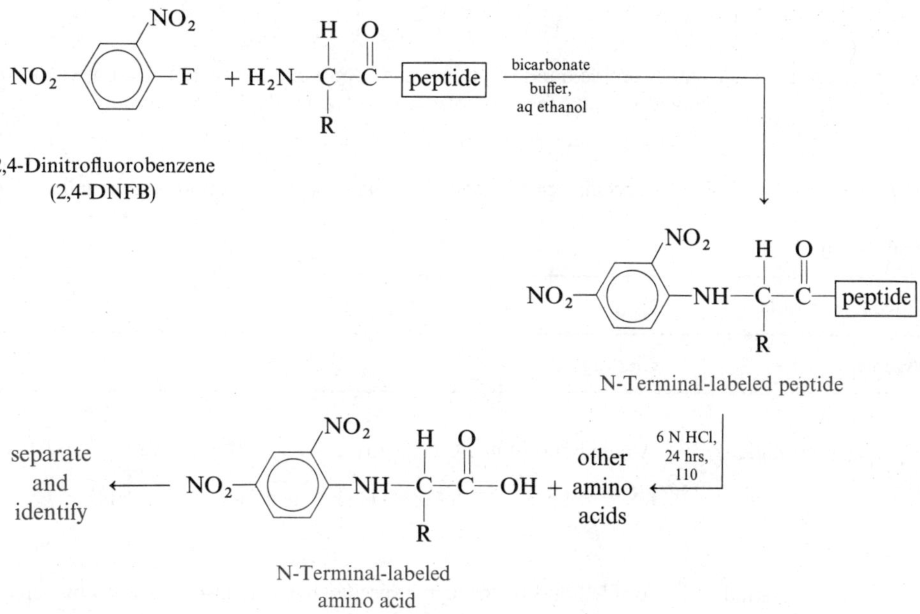

2,4-Dinitrofluorobenzene (2,4-DNFB) reacts with the peptide in ethanol that has been buffered with bicarbonate to provide a slightly basic solution. A nucleophilic aromatic substitution reaction (see Sec. 16.5) occurs and the peptide labeled at the N-terminal end results. Total hydrolysis of the peptide now provides all the free amino acids, but only one of these has a 2,4-dinitrophenyl (2,4-DNP) group attached. The labeled amino acid can be identified using paper chromatography, for example.

TABLE 28.6 Specific Protein Cleavage Agents

Exopeptidases:* Enzymes that Cleave the Terminal Residues of the Protein Chain.

Agent	Cleavage Point	
Carboxypeptidase	C-Terminal amino acid	peptide—C(=O)┼NH—CH(R)—C(=O)—OH
Leucine aminopeptidase	N-Terminal amino acid	H₂N—CH(R)—C(=O)┼NH—peptide

Endopeptidases: Enzymes that Cleave the Peptide Within the Chain But Not at the End Residues.

Agent	Cleavage Point	
Trypsin	Amide bond of *lys* or *arg*	peptide—CH(R')—C(=O)┼NH—CH(R)—C(=O)—OH

$$[R'— = H_2N(CH_2)_4— \text{ or } H_2N—\overset{\oplus NH_2}{\underset{}{C}}—NH—(CH_2)_3—]$$

Chymotrypsin — Primarily amide bond of amino acids containing aromatic R groups: *tyr*, *trp*, *phe*. Others also react, for example, *leu*, *met*, *asn*, *gln*, but much more slowly.

$$[R'— = p\text{-HO}—C_6H_4—CH_2—, C_6H_5—CH_2—, \text{ or } (C_8H_6N)—CH_2—]$$

Pepsin, subtilisin, and papain — Not very specific; give a general breakdown of the peptide into smaller fragments.

Chemical Agents

Agent	Cleavage Point	
Cyanogen bromide (BrCN)	Amide bond of *met* specifically	peptide—CH(R')—C(=O)┼NH—CH(R)—C(=O)—OH

$$[R'— = CH_3—S—CH_2CH_2—]$$

N-Bromosuccinamide — Amide bond of *trp* or *tyr* specifically; may also cleave *his* but much more slowly.

$$[R'— = p\text{-HO}—C_6H_4—CH_2— \text{ or } (C_8H_6N)—CH_2—]$$

* Enzymes are classified by function as follows: *catenases*: any chain-cutting enzyme: *proteases*: a protein-cutting enzyme (a specific subclass of catenases); and *peptidases*: an enzyme that acts on peptides (also a subclass of catenase).

D. C-Terminal Analysis

Treatment of a protein with hydrazine (N_2H_4) for several days at approximately 90°C results in the complete cleavage of each amide bond. The amino acids other than the C-terminal amino acid form hydrazides, whereas the C-terminal amino acid ends up in the free acid form. The free acid can easily be isolated and identified by several methods, including chromatography.

Reaction with hydrazine:

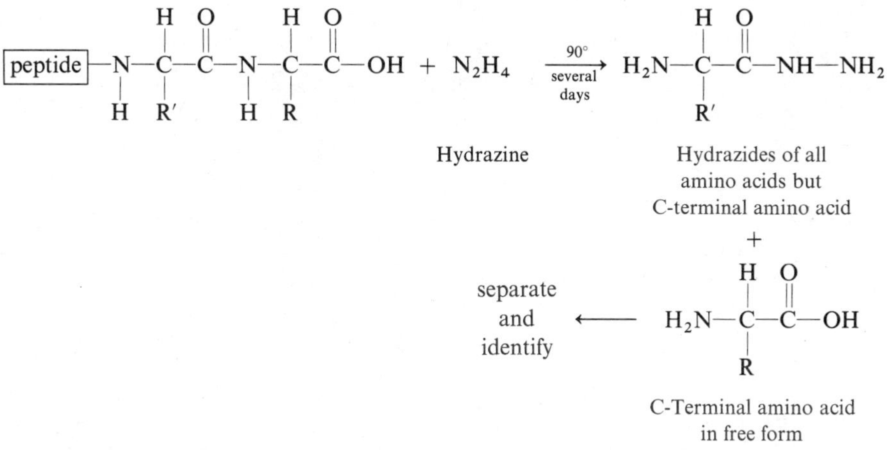

E. Specific Cleavage Agents

Several agents, both enzymatic and chemical, are capable of cleaving the peptide chain at selected positions. Knowing the position of cleavage, one can identify the resulting fragments and work backward to determine part of the peptide structure. Several reagents are listed in Table 28.6.

Returning to the original problem of determining the total primary structure of the unknown pentapeptide, we have the following information:

1. Total amino acid analysis: ala:gly:phe:val:ser/1:1:1:1:1
2. Partial hydrolysis: ala-gly, phe-val, ser
3. C-terminal analysis (reaction with hydrazine): gly
4. N-terminal analysis (reaction with 2,4-DNFB): ser
5. Total primary structure: ser-phe-val-ala-gly

28.14 Prosthetic Groups

Enzymes are biological catalysts involved in a variety of reactions in the body requiring highly specific reagents that can perform remarkably complex functions. Enzymes may be broadly classified into two groups: (1) those that are composed of protein totally and (2) those that are composed of both a protein and a nonprotein portion. The nonprotein portion of the enzyme in the second category is referred to as the **prosthetic group.** The prosthetic group either may be covalently bonded to the protein portion of the enzyme or may be attached to the protein through the

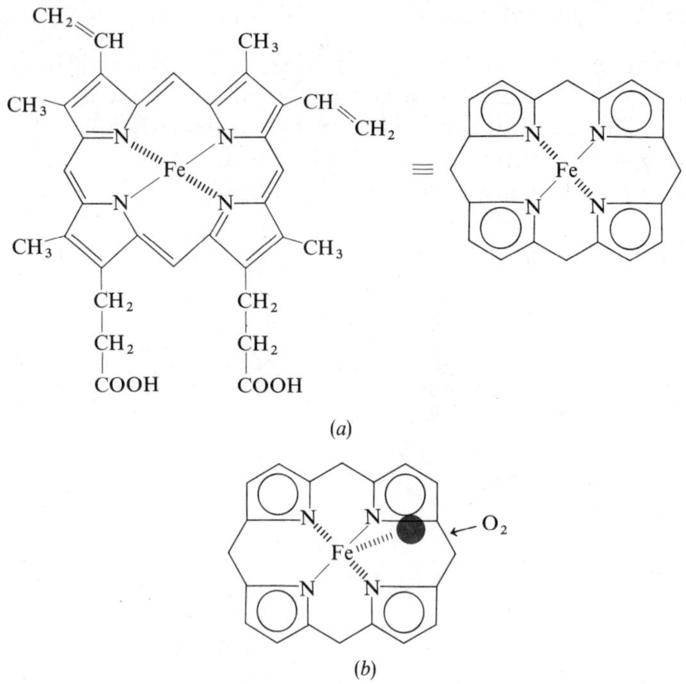

(a)

(b)

FIGURE 28.19 (*a*) *Left:* **Heme group (prosthetic group) of hemoglobin and myoglobin.** *Right:* **Abbreviated notation for the heme group.** (*b*) **Schematic representation of the binding of oxygen to the heme group.**

many nonbonding interactions discussed previously. The prosthetic group is often quite complex. Examples of proteins that contain prosthetic groups are *nucleoproteins*, which contain nucleic acid segments (see Sec. 29.8), *glycoproteins* and *mucoproteins*, which are composed of large portions of carbohydrates (see Chap. 29), and *chromoproteins*, which contain light-absorbing prosthetic groups such as the heme group (Fig. 28.19). The heme group in the transport proteins hemoglobin and myoglobin is responsible for binding oxygen and transporting oxygen from the lungs to the muscles where it is needed.

28.15   Chymotrypsin

Chymotrypsin is the most intensely studied enzyme and as such is perhaps the best understood in terms of its mode of action—the mechanism by which it reacts. Chymotrypsin is a globular catenase—a chain-cutting enzyme—consisting of several chains, containing 245 amino acid residues, and linked by several cys-cys bridges. More precisely, it is a protease or peptidase; it cuts protein and peptide chains. In an attempt to better understand the mode of action of this enzyme, several model experiments were designed and undertaken. The results of these along with other known data about chymotrypsin are summarized here.

1. Chymotrypsin cleaves peptide bonds by hydrolyzing the peptide link—the amide bond.
2. Chymotrypsin reacts preferentially with tyr, trp, and phe amino acids that contain an aromatic side chain.

3. Chymotrypsin is also capable of hydrolyzing other moieties, for example esters, amides, and anhydrides.

4. With points 1 to 3 in mind, *p*-nitrophenyl acetate, an ester containing an aromatic portion, was reacted with chymotrypsin. Rapid hydrolysis of the ester occurred with the concomitant formation of *p*-nitrophenol and acetic acid:

p-Nitrophenyl acetate	Acetic acid	*p*-Nitrophenol

The rates of formation of *p*-nitrophenol and acetic acid are quite different, however. The *p*-nitrophenol forms very rapidly, whereas the acetic acid is generated more slowly over a much longer time. The results indicate a two-step reaction sequence commonly observed in enzyme catalysis. According to this scheme, the enzyme (*E*) and substrate (*S*) react to form an intermediate *enzyme-substrate complex* (*E-S*). This complex may then undergo a reverse reaction or may react further to produce product (*P*) and the free enzyme. This sequence is summarized as:

Enzyme + substrate ⇌ enzyme-substrate → enzyme + product

The observed reaction of *p*-nitrophenyl acetate fits this scheme very nicely. The ester reacts with chymotrypsin in the first (reversible) step to produce *p*-nitrophenol and the *E-S* of chymotrypsin and the acyl portion of the original ester:

The *E-S* continues to react through the second step, which is much slower, to produce acetic acid:

The same two-step sequence and rate formation of products are observed with other model compounds.

5. If the reaction in 4 is carried out in the presence of an excess of the substrate (*p*-nitrophenyl acetate in this case), the *E-S* can actually be isolated in a pure crystalline form.

6. Reaction of the enzyme with diisopropyl fluorophosphate followed by complete amino acid analysis shows that the fluorophosphate group attaches to the *ser* residue at position 195 in the protein chain. Furthermore, after reaction with diisopropyl fluorophosphate, the enzyme loses all its catalytic activity. These observations illustrate that *ser* 195 is necessary to the normal action of the enzyme.

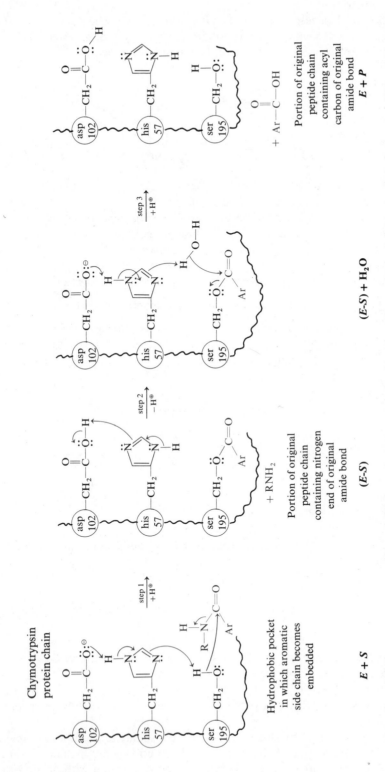

FIGURE 28.20 Step 1 illustrates the nucleophilic attack of the *ser* 195 residue on one of the amide bonds of the polypeptide—*E* + *S*. The polypeptide is held in position through the hydrophobic interactions between the aromatic side chain of the polypeptide and the hydrophobic pocket formed by the amino acids of the enzyme. In step 2, one segment of the polypeptide chain is lost and the second portion becomes attached to *ser* 195 as ester—the (*E-S*). In step 3, the enzyme-substrate complex reacts further with water, and the reverse electron flow *asp* 102 through *his* 57 occurs to give the free enzyme and the hydrolyzed portion of the polypeptide chain—*E* + *P*. Protons are lost and gained (shifted) throughout the reaction sequence, probably through the intervention of several molecules of water.

Protein (ser 195)—CH$_2$—OH + F—P(=O)—OCH(CH$_3$)$_2$ with OCH(CH$_3$)$_2$

Diisopropyl
fluorophosphate

Protein (ser 195)—CH$_2$—O—P(=O)—OCH(CH$_3$)$_2$ with OCH(CH$_3$)$_2$

ser-195 Position blocked
Deactivation of enzyme

7. The *his* residue at position 57 in the protein chain is part of the active site—point of enzyme reaction—of the enzyme. Specific methylation or protonation of this residue causes deactivation and complete loss of activity of the enzyme:

(his 57)—CH$_2$—(imidazole) $\xrightarrow{\text{H}^{\oplus}}$ (his 57)—CH$_2$—(imidazolium)

his-57 Position protonated
Deactivation of enzyme

8. Similar experiments, which are not discussed here, show that *asp* 102 is also involved in the action of chymotrypsin.

The complete mechanism of the action of chymotrypsin as it is understood to date is diagrammed in Figs. 28.20 to 28.22.

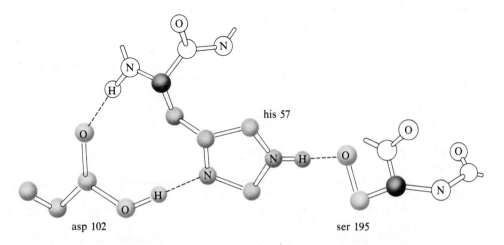

his 57

asp 102   ser 195

FIGURE 28.21 The active site of chymotrypsin. [From *The Structure and Action of Proteins* by R. E. Dickerson and I. Geis. Benjamin/Cummings, Menlo Park CA, Publishers. Copyright 1969 by Dickerson and Geis.]

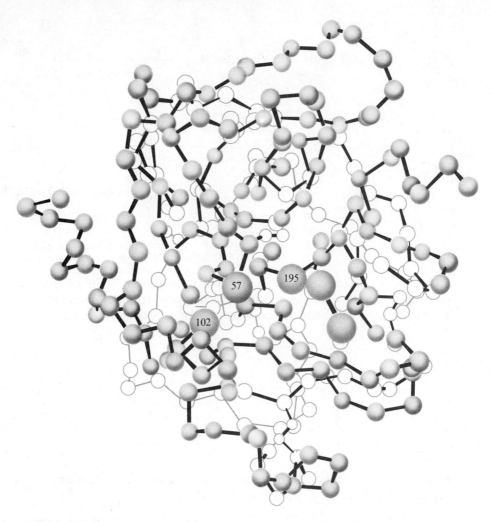

FIGURE 28.22   Total structure of chymotrypsin—the active site is illustrated in red. [From *The Structure and Action of Proteins* by R. E. Dickerson and I. Geis. Benjamin/Cummings, Menlo Park CA, Publishers, Copyright 1969 by Dickerson and Geis.]

Study Questions

28.9 Briefly define (or describe) the following terms as they pertain to protein chemistry.

(*a*) primary protein structure   (*b*) secondary protein structure
(*c*) tertiary protein structure   (*d*) quaternary protein structure
(*e*) proteins versus peptides—how they differ   (*f*) peptide link
(*g*) fibrous versus globular proteins—how they differ
(*h*) α helix versus β pleated sheet—how they differ
(*i*) protein synthesis

28.10 (*a*) Draw the dipolar ion (zwitterion) of alanine, $CH_3CH(NH_2)COOH$.

(*b*) Draw the structure of alanine as it would exist in strongly acidic solution.

(*c*) Draw the structure of alanine as it would exist in strongly basic solution.

(*d*) Define briefly the term *isoelectric point*.

28.11 Show in equation form the *complete equilibrium* that occurs when serine

$$H_2N-CH-COOH$$
$$|$$
$$CH_2OH$$

is dissolved in water. Be certain to consider possible reaction(s) of serine with water. Of all the possible species that might be present at equilibrium, which would be expected to predominate at the *isoelectric* point?

28.12 Aspartic acid has pK_a values of approximately 2.1, 3.86, and 9.82 for the C_α—COOH group, —CH_2—COOH group, and —$NH_3^\oplus$ group, respectively. Draw the titration curve for aspartic acid. Write equilibria corresponding to each ionization. Label each inflection point on your titration curve and indicate the predominant species in solution at each inflection point. Calculate the isoelectric point.

28.13 A typical mechanistic sequence for an enzyme-catalyzed reaction is:

$$E + S \xrightarrow{k_2} E\text{-}S \xrightarrow{k_1} E + F$$

Draw a potential energy versus time reaction coordinate for this generalized reaction. Label the positions of all starting materials, intermediates, and products. Assume k_2 is much slower than k_1. Label the rate-determining step.

28.14 The biosynthetic origin of insulin in the body, as discussed in the chapter, probably involves proinsulin as the precursor. Treatment of this precursor with insulinase (a pancreatic enzyme) results in the formation of insulin. Until this protein and enzyme were both isolated, one school of thought favored a biosynthetic sequence involving the combination of the two chains of insulin (the A and B chains) to produce the hormone itself. Discuss the problems associated with this synthetic route if it occurred in the body.

28.15 Using the protecting groups in Table 28.3 and the activating groups in Table 28.4, synthesize each of the following peptides from the amino acids that constitute it. You may use any organic or inorganic reagents you want. Show all steps in the synthesis and all intermediate compounds.

(*a*) *gly-gly-ala*

(*b*)

(*c*)

(*d*) *ala-cys-gly-cys-ala*

28.16 The synthesis of carbobenzoxycarbonylisoleucyl-serine methyl ester has recently been reported. A solution of serine methyl ester hydrochloride was made basic with triethylamine, and benzyloxycarbonylisoleucine-*p*-nitrophenyl ester was added. After 1 hr the solution was neutralized and a crystalline product was isolated. Write the chemical reaction that occurred. Give the structures of all reactants and products.

This dipeptide was also prepared by coupling the two preceding protected amino acids using DCC as a condensing reagent. Write the chemical reaction again including the structure of all reactants and products.

28.17 Cyanogen bromide (BrCN) is a chemical reagent for the selective cleavage of peptide chains at the amide link of methionine.

Polypeptide chain   Two fragments of original chain

The mechanism of the reaction proceeds through the steps summarized here:

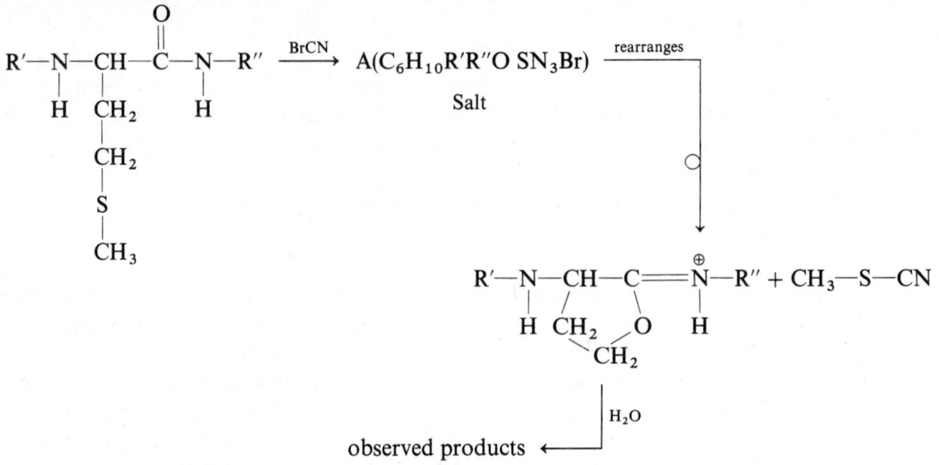

Given this information, draw the complete stepwise mechanism in going from the original polypeptide to the two polypeptide fragments.

28.18 Cation and anion exchange resins are often used for purification and separation in amino acid chemistry. These resins are polymeric materials that are used as the stationary phase of a normal liquid-solid chromatography system. The mobile phase is normally an aqueous solution with varying pH. The polymer itself acts as a backbone on which are attached ionic functional groups capable of complexing with the anionic or cationic forms of the materials being purified and separated. A typical anionic resin, for example, would have the following formula— $R—SO_3^{\ominus}H^{\oplus}$. Cationic materials passed through this resin would complex with the sulfonate group to differing degrees, hence the varying rates of elution, purification, and separation. The pH at the isoelectric point for each of the following amino acids is given in parentheses: glycine (pH 5.97), threonine (pH 6.53), leucine (pH 5.98), and lysine (pH 9.74). If a solution containing these four amino acids dissolved in an aqueous buffer solution at pH 3.0 was applied to a cation exchange resin and the column was eluted with the pH 3.0 buffer, in what order would the amino acids be eluted?

28.19 A certain hexapeptide (*H*) is isolated from natural sources and analyzed, with the following results obtained:

(*a*) Complete hydrolysis of *H* yields the following amino acids per mole of *H*: 1 mole glycine (*gly*), 1 mole serine (*ser*), 2 moles alanine (*ala*), 1 mole aspartic acid (*asp*), and 1 mole proline (*pro*).

(*b*) *H* is *unaffected* by the enzyme leucine aminopeptidase.

(*c*) *H* is *unaffected* by the enzyme carboxypeptidase.

(*d*) *H* *does not react* with 2,4-dinitrofluorobenzene in the presence of dilute base.

(*e*) In a large number of partial hydrolysis reactions the following dipeptides are identified:

ala-asp   *ala-pro*   *asp-ser*

pro-gly   *ser-ala*   *gly-ala*

On the basis of these data, what is the amino acid sequence in the unknown hexapeptide (*H*)? [*Hint: Think carefully* about the implications derived from observations (*b*) to (*d*).]

28.20 Heptapeptide (*H*) is analyzed to elucidate its structure, and the following observations are made:

(*a*) Complete acid hydrolysis of *H* gives the following amino acids per mole of *H*: 2 moles aspartic acid (*asp*), 2 moles histidine (*his*), 1 mole glycine (*gly*), 1 mole serine (*ser*), and 1 mole valine (*val*).

(*b*) Treatment of *H* with 2,4-dinitrofluorobenzene (DNFB), followed by complete hydrolysis, gives 2,4-DNP-*asp*.

(*c*) Treatment of *H* with carboxypeptidase gives aspartic acid (*asp*) and a hexapeptide.

(*d*) Partial hydrolysis of *H* gives *val-his-gly*, *his-gly-his*, and *asp-ser-val* as some tripeptide products. What is the amino acid sequence in heptapeptide *H*?

28.21 A pentapeptide (*P*) is found to contain each of the following amino acids: *ala, gly, val, cys, ser*.

(*a*) On partial hydrolysis, the following fragments are obtained:

gly-val   cys-ser   ala   val-cys

(*b*) Reaction of the peptide with 2,4-dinitrofluorobenzene (under slightly basic conditions), followed by complete hydrolysis, yields *ala* labeled with 2,4-DNP, 2,4-DNP-*ala*.
Draw the complete primary structure of the pentapeptide.

28.22 A certain peptide has the following formula:

arg,cys,glu,gly$_2$,leu,phe$_2$,tyr,val

where the commas indicate an unknown sequence of amino acids and the subscripts indicate the numbers of each different amino acid per mole of peptide. On partial hydrolysis, the peptide yields the following tripeptides:

val-cys-gly + *gly-phe-phe* + *glu-arg-gly* + *tyr-leu-val* + *gly-glu-arg*

Work out the primary sequence of the amino acid residues in the unknown peptide.

28.23 Pentapeptide (*P*) is analyzed to elucidate its structure. The following observations are made:

(*a*) Total hydrolysis of *P* (in acidic solution) gives 1 mole of *each* of the following amino acids per mole of *P*: aspartic acid (*asp*), glutamic acid (*glu*), histidine (*his*), phenylalanine (*phe*), and glycine (*gly*).

(*b*) Treatment of *P* with 2,4-dinitrofluorobenzene (2,4-DNFB), followed by partial hydrolysis, gives 2,4-DNP-*gly* and 2,4-DNP-*gly-his*.

(*c*) Treatment of *P* with carboxypeptidase gives phenylalanine (*phe*) and a tetrapeptide.

(*d*) Partial hydrolysis of *P* gives *gly-his* and *glu-phe* as two dipeptide products.
What is the amino acid sequence in pentapeptide *P*?

28.24 An octapeptide (*O*) is analyzed to elucidate its structure, and the following observations are made:

(*a*) Breakdown of the peptide with pepsin gives the following fragments among others:

ala-cys-gly   phe-cys-met   gly-ala   ala-phe

(*b*) Reaction of the peptide with hydrazine at 90° for four days followed by separation of the component amino acid residues indicates the presence of *val*.

(*c*) Treatment of the peptide with 2,4-DNFB, followed by hydrolysis, yields an *ala* labeled with a 2,4-DNP group.

(*d*) Complete amino acid analysis yields the following amino acids in the stoichiometric quantities indicated.

ala,cys$_2$,gly,met,phe,val

What is the structure of the peptide?

28.25 The following observations are obtained from several experimental degradations of hexapeptide (*H*) with a variety of reagents.

(*a*) A total amino acid analysis followed by paper chromatography of the products indicate that the peptide is composed of six different amino acids.

(*b*) Reaction of *H* with 2,4-DNFB, followed by the normal workup, indicates an *asp* end residue.

(*c*) Hydrazinolysis of *H* gives a *tyr* residue on analysis.

(*d*) Hydrazinolysis of the remaining fragment of *H* after a quantitative one-step attack by the enzyme carboxypeptidase shows a *gly* end residue on the fragment chain.

(*e*) Two very different molecular weight peptides are produced on the reaction of *H* with trypsin. The reaction of each chain with 2,4-DNFB shows end residues of *asp* and *met*, respectively.

(*f*) Cyanogen bromide reacts with *H* to yield two peptides. Reaction of these with 2,4-DNFB shows end residues of *asp* and *ala*.

(*g*) A one-step reaction of *H* with the enzyme leucine aminopeptidase, followed by analysis with the remaining chain by the Sanger reaction, indicates an end residue of *arg*.
What is the complete primary structure of *H*?

29

Carbohydrates

The term *carbohydrate* is used to describe any compound that is composed of primarily carbon, hydrogen, and oxygen and has the molecular formula $C_x(H_2O)_y$. These compounds are found in many naturally occurring sources, such as milk, cereals, bread, sugarcane, sugar beets, and all fruits. Carbohydrates are a main source of energy for both plants and animals.

In this chapter we learn about the structures of sugars (or saccharides), names applied to the simple carbohydrates. We see how sugars can be interconverted, and several examples are provided of how one elucidates structure in carbohydrate chemistry. The final sections discuss some of the biopolymers—cellulose, amylose, and amylopectin—that are composed of monomeric sugar units.

29.1   Carbohydrates

The term **carbohydrate** literally means that the compound is a hydrate of carbon. This is also indicated by the manner in which the general formula is written. Most naturally occurring carbohydrates contain elements other than carbon, hydrogen, and oxygen. In particular, nitrogen, phosphorus, and sulfur are often encountered. The names **sugars** and **saccharides** are synonyms and apply to the simple **carbohydrates.** All three names are used interchangeably.

Another broad definition often encountered when reading about carbohydrates is based on the functional groups in the carbohydrate molecule. By these definitions, carbohydrates are either *polyhydroxyaldehydes* or *polyhydroxyketones*, compounds containing a large number of hydroxyl groups (—OH) and either an aldehyde or ketone carbonyl function:

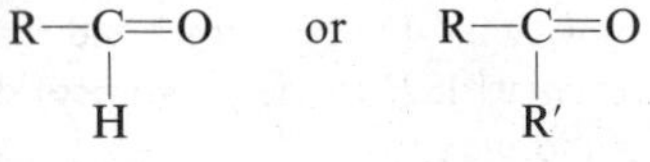

$$R-\underset{\underset{H}{|}}{C}=O \quad \text{or} \quad R-\underset{\underset{R'}{|}}{C}=O$$

For most purposes this definition is quite satisfactory, although as we will see, it is not entirely accurate. Three carbohydrates are:

$$\begin{array}{ccc}
\text{CH}{=}\text{O} & \text{CH}{=}\text{O} & \text{CH}_2{-}\text{OH} \\
| & | & | \\
\text{CH}{-}\text{OH} & \text{CH}{-}\text{OH} & \text{C}{=}\text{O} \\
| & | & | \\
\text{CH}{-}\text{OH} & \text{CH}{-}\text{OH} & \text{CH}{-}\text{OH} \\
| & | & | \\
\text{CH}{-}\text{OH} & \text{CH}{-}\text{OH} & \text{CH}{-}\text{OH} \\
| & | & | \\
\text{CH}{-}\text{OH} & \text{CH}_2{-}\text{OH} & \text{CH}{-}\text{OH} \\
| & & | \\
\text{CH}_2{-}\text{OH} & & \text{CH}_2{-}\text{OH}
\end{array}$$

Aldose or hexose or aldohexose	Aldose or pentose or aldopentose	Ketose or hexose or ketohexose

Note that carbohydrates may be classified by two broad systems. The first refers to the presence of either the aldehyde or ketone function in the sugar. If a ketone is present, the sugar is called a *ketose*, whereas the term *aldose* is applied to sugars that contain an aldehyde. These names are general and tell us little about the structure of the carbohydrate. A slightly more detailed naming system includes the number of carbon atoms in the main skeletal chain of the sugar. A three-carbon chain is referred to as a *triose*, four carbons a *tetrose*, five a *pentose*, six a *hexose*, and so on. This system tells the size of the carbohydrate chain but contains no information about the functionality in the sugar. A contraction of the two names, as illustrated in the preceding examples, provides a reasonable amount of information concerning the structure of the carbohydrate. For example, a *ketopentose* is a carbohydrate five carbons long with a ketone function and, by definition, a large number of hydroxyl groups. We return to the structure of carbohydrates in Sec. 29.4.

29.2   Photosynthesis

The photosynthetic process is described by the following stoichiometric equation:

$$6CO_2 + 6H_2O \xrightarrow[\text{light}]{\text{chlorophyll}} C_6H_{12}O_6 + 6O_2$$

Glucose
Carbohydrate

The process involves the conversion of carbon dioxide and water to oxygen and the carbohydrate glucose ($C_6H_{12}O_6$, a hexose). The reaction occurs in green plants and many microorganisms such as algae. The process of photosynthesis, though seemingly straightforward according to the equation, actually involves a large number of complicated enzyme-catalyzed biosynthetic reactions. A thorough discussion of these reactions is best left to a biochemistry course. However, a central step that is essential to the process involves the green-plant pigment chlorophyll. Chlorophyll absorbs light provided by the sun. The light energy is converted to chemical energy and provides the driving force for the oxidation and reduction processes that result in photosynthesis. The structure of chlorophyll is shown in Fig. 29.1. Note the similarity between this compound and the heme group of hemoglobin and myoglobin in Sec. 28.14. This highly unsaturated molecule contains many conjugated double bonds and would be expected to absorb light in the visible as well as the UV portion of the electromagnetic spectrum (see Sec. 12.7D), which is observed (Fig. 29.1).

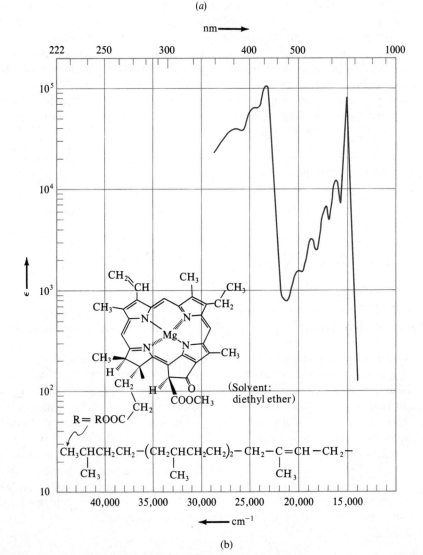

$$-R' = -CH_2CH_2-\overset{\displaystyle O}{\overset{\|}{C}}-O-CH_2CH=CCH_2CH_2CH_2CHCH_2CH_2CH_2CHCH_2CH_2CH_2CHCH_3$$
$$CH_3CH_3CH_3CH_3$$

$$-R'' = -CH_3 \text{ in chlorophyll } a \text{ and } -\overset{\displaystyle O}{\overset{\|}{C}}-H \text{ in chlorophyll } b$$

(a)

(b)

FIGURE 29.1 (a) Chlorophyll a and b. (b) Absorption spectrum of chlorophyll b. [Data from the *DMS UV Atlas*, 1966–1971, Plenum Press, New York.]

29.3   Glucose Metabolism: Biosynthetic Pathways

The main metabolic pathways involving glucose in plants and animals are summarized in Fig. 29.2.

The glucose produced by the photosynthetic process in plants is partially consumed to produce the energy necessary for the normal metabolism of the plant. The remainder of the glucose is converted to *cellulose*, the main structural material of the plant (see Sec. 29.12), and plant starch, the reserve food supply of the plant (see Sec. 29.12). When plants are consumed by animals, the glucose, in the form of cellulose, plant starch, free glucose, and others, is metabolized by the animal. A variety of enzymatic processes are involved in this biosynthesis and are also summarized in Fig. 29.2. As with the plants, part of the glucose is consumed to produce the immediate energy needed by the organism and part is converted to *glycogen* (animal starch), which is stored in the liver and serves as a ready food supply for the animal (see Sec. 29.12). The glycogen can be converted back to free glucose, which can then be oxidized to produce energy as needed. In addition to the glycogen, another portion of the glucose is converted to animal fat through what is known as fatty acid synthesis. The animal fat also serves as a reserve store of energy for the organism (see Chap. 25). Finally, the glucose can be enzymatically converted to a variety of other carbohydrates (which in turn are used in nucleic acid synthesis, see Sec. 29.8), amino acids (used in protein synthesis, see Sec. 28.8), and a variety of hormones.

29.4   Stereochemistry of Carbohydrates

An examination of the structure of carbohydrates reveals molecules that contain several asymmetric carbon atoms as in the following aldohexose:

$$
\begin{array}{l}
H \\
\;\;\;\;\searrow C{=}O \\
\;\;\;\;\;\;| \\
H\sim \overset{*}{C}\sim OH \\
\;\;\;\;\;\;| \\
H\sim \overset{*}{C}\sim OH \\
\;\;\;\;\;\;| \\
H\sim \overset{*}{C}\sim OH \\
\;\;\;\;\;\;| \\
H\sim \overset{*}{C}\sim OH \\
\;\;\;\;\;\;| \\
\;\;\;\;CH_2{-}OH
\end{array}
$$

The curved bonds ($\sim$) designate a center of unspecified stereochemistry. There are four asymmetric carbon atoms in the aldohexose, which means there is a maximum of 16 optical isomers (enantiomers or diastereomers of each other) possible. This number is obtained by the 2^n rule presented in Sec. 6.13, where $n = 4$. Of the 16 possible optical isomers of the aldohexose, one, glucose, is the most common found naturally (Fig. 29.3). The only other two aldohexoses found naturally are galactose and mannose.

Before proceeding with a discussion of the stereochemistry of the aldohexoses and other sugars, a review of terms is in order.

All chiral molecules exist as stereoisomers. If two molecules are stereoisomers, they must, by definition, bear an enantiomeric or a diastereomeric relationship to each other. If they are enantiomers, one member of the pair rotates plane-polarized light in the ($+$) direction, whereas the other rotates the plane of light in the ($-$)

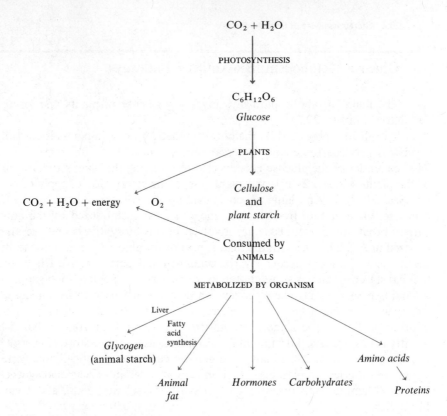

FIGURE 29.2 Glucose metabolism.

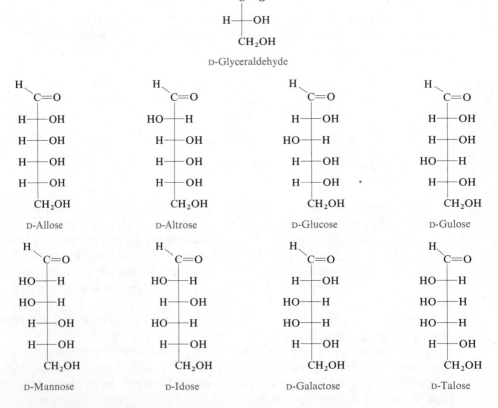

FIGURE 29.3 D-Enantiomers of the aldohexoses.

direction. In classifying carbohydrates according to the configuration they exhibit, the D/L system[1] introduced in Sec. 28.3 is used.

Assignment of D or L Configuration to Carbohydrates

1. The carbohydrate considered is drawn in two dimensions using the Fischer projection method of depicting asymmetric centers (see Sec. 6.10). One of the aldohexoses in Fig. 29.3, D-glucose, is used as the example:

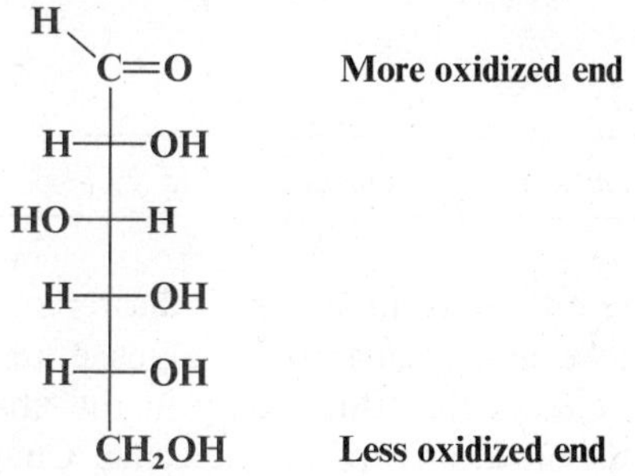

$$
\begin{array}{ccc}
\text{H} \\
\quad\diagdown \\
\qquad \text{C}=\text{O} & \qquad \textbf{More oxidized end} \\
\text{H} \!-\!\!-\!\!-\! \text{OH} \\
\text{HO} \!-\!\!-\!\!-\! \text{H} \\
\text{H} \!-\!\!-\!\!-\! \text{OH} \\
\text{H} \!-\!\!-\!\!-\! \text{OH} \\
\text{CH}_2\text{OH} & \qquad \textbf{Less oxidized end}
\end{array}
$$

D-Glucose $(C_6H_{12}O_6)$

2. The carbohydrate is *always* drawn as illustrated in the preceding example, with the most oxidized carbon end pointing to the top of the page and the remaining molecule in a linear form stretching below it.

3. The terminal asymmetric carbon at the bottom of the chain in the carbohydrate is then compared with the asymmetric carbon of D- and L-glyceraldehyde (the standard reference compounds for the

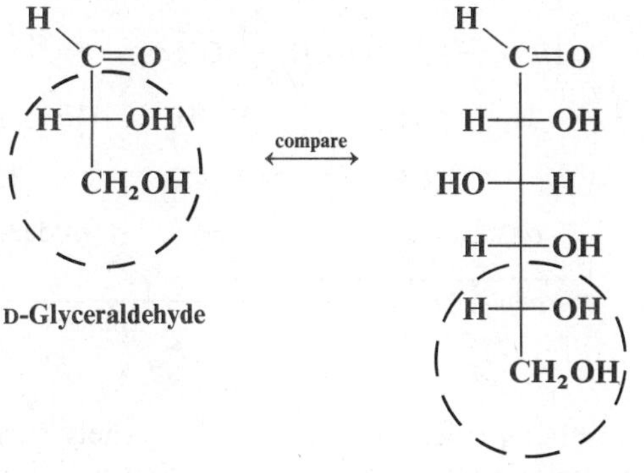

D-Glyceraldehyde   compare   D-Glucose

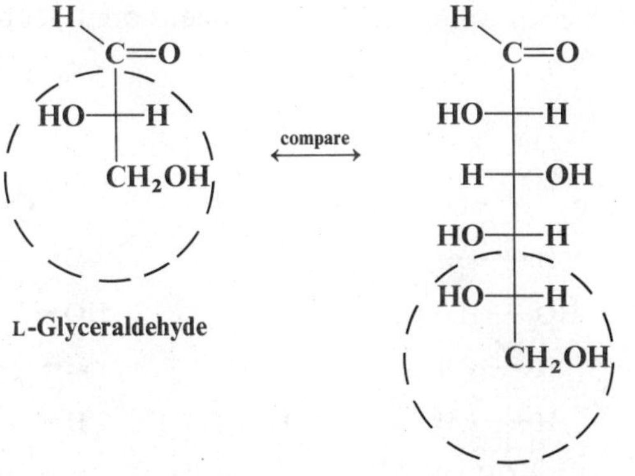

L-Glyceraldehyde   compare   L-Glucose

---

[1] The R/S (Cahn-Ingold-Prelog) nomenclature system may also be employed but is seldom used in carbohydrate chemistry. The D/L convention is used throughout this chapter. Assign the absolute configuration as R or S for each asymmetric center in D-glucose (Fig. 29.4).

D/L system of absolute configuration, see Sec. 28.3). If the end carbon bears the same relative configuration as the end carbon of D-glyceraldehyde, the sugar is classified as a D sugar. If it is arranged as in L-glyceraldehyde, the sugar is classified as L.

The naturally occurring aldohexoses, glucose, galactose, and mannose, all have the D configuration. These three compounds are drawn here. The D enantiomers of the aldohexoses are illustrated in Fig. 29.3.

Question 29.1

Draw the L enantiomer of each sugar in Fig. 29.3.

Note that these three isomers are all diastereomers of each other. In addition, D-mannose/D-glucose and D-galactose/D-glucose are **epimers** of one another. *Epimers* are optical isomers that differ only in the absolute configuration of one of the asymmetric carbon atoms in the molecule. Changing the configuration at the epimeric carbon converts one epimer into the other.

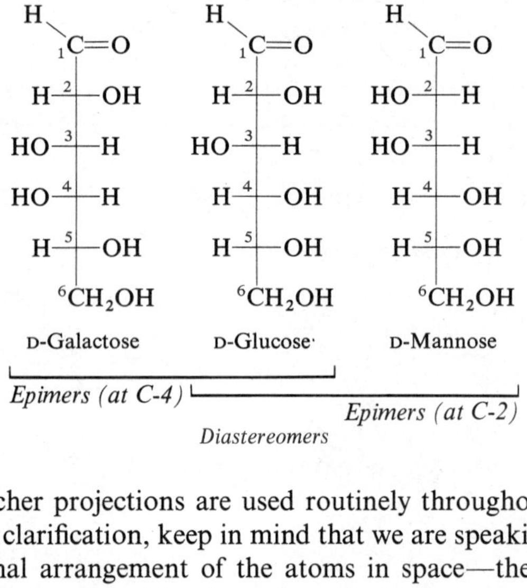

D-Galactose D-Glucose D-Mannose

Epimers (at C-4) *Epimers (at C-2)*

Diastereomers

Although Fischer projections are used routinely throughout this chapter for both simplicity and clarification, keep in mind that we are speaking of differences in the three-dimensional arrangement of the atoms in space—these compounds are optical isomers of each other. The three-dimensional geometry of D-glucose is illustrated in Fig. 29.4.

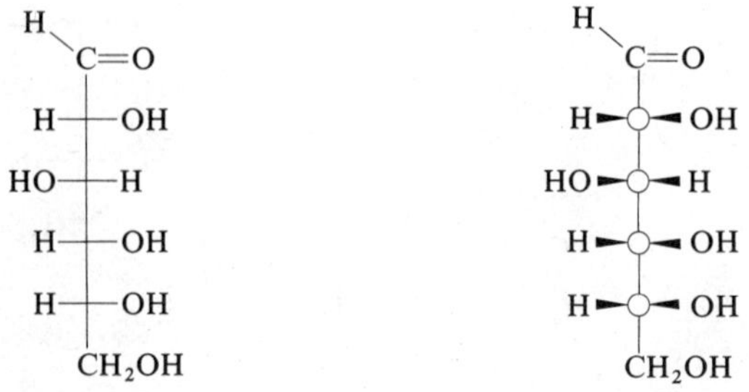

(*a*) (*b*)

FIGURE 29.4 D-Glucose: (*a*) Fischer projection, (*b*) three-dimensional representation.

29.5   D-Glucose: Mutarotation

Of the three naturally occurring aldohexoses (D-glucose, D-galactose, and D-mannose), D-glucose is by far the most common, being found in a variety of compounds including cellulose, starch, and proteins. Experimentally, however, several observations indicate that D-glucose exists in two isomeric forms. They are both isomers of D-glucose (as opposed to any being the enantiomeric L isomer) but they exhibit different physical and chemical properties. For example, when D-glucose is recrystallized from water, depending on the temperature at which the recrystallization occurs, the two isomers of D-glucose may be isolated. One isomer has a sharp melting point of 146° and a specific rotation ($[\alpha]_D^{25}$) of $+112°$. The other isomer has a melting point of 150° and a specific rotation of $+18.7°$. In addition, if either isomer is obtained in pure form and dissolved in water and allowed to sit for a time, the glucose has a specific rotation of $+52.7°$. More information is needed before a clear picture of the structure of the D-glucose isomers can be elucidated. These results are summarized in Fig. 29.5.

To understand the reasons for these observations and statements, a review of several classes of compounds introduced in Chap. 21 is in order—*acetals*, *ketals*, *hemiacetals*, and *hemiketals*.

Acetals and ketals or hemiacetals and hemiketals are formed from the reaction of aldehydes or ketones with alcohols or glycols under acid-catalyzed or in certain instances base-catalyzed conditions. These reactions were discussed in Sec. 21.6 and are summarized in Fig. 29.6.

Under acid-catalyzed conditions either the acetal or hemiacetal (or ketal or hemiketal) derivative can be synthesized. (Base-catalyzed conditions yield the hemiacetal or hemiketal only.) Considering the acid-catalyzed reaction, the conversions of a carbonyl compound (take an aldehyde for the sake of this discussion) to a hemiacetal and subsequently to an acetal are both reversible. The reaction favors the hemiacetal/acetal only if there is a large excess of alcohol and anhydrous acid is used as a catalyst. The removal of water as it forms is normally essential to a complete reaction (see Sec. 21.6).

Unlike the hemiacetal and hemiketal, which are fairly labile and unstable compounds, carbohydrate hemiacetals and hemiketals are quite stable. Indeed, most

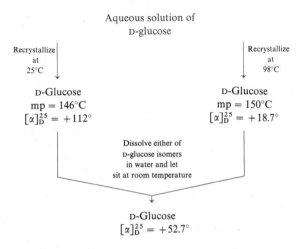

FIGURE 29.5   Isomers of D-glucose.

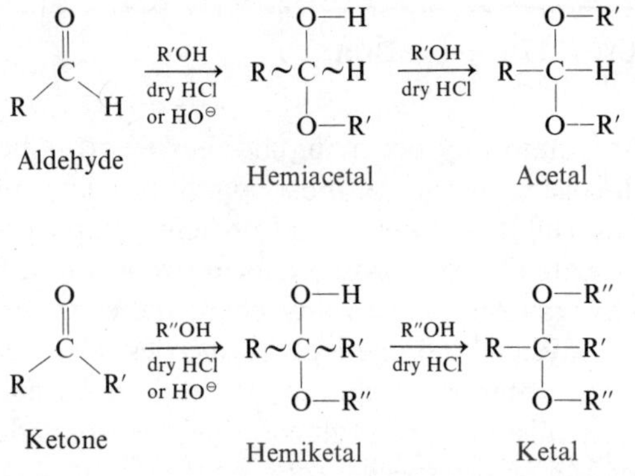

FIGURE 29.6 Acetals and ketals, hemiacetals and hemiketals. The formation of the hemiacetal (or hemiketal) linkage results from the bonding of an —OH oxygen on the carbohydrate skelton to the aldehyde (or ketone) carbonyl. Attack on either side of the carbonyl carbon is possible so that the new asymmetric carbon which results may be of either configuration. This is indicated by the curved bonds. The two resulting diastereomers (of the R configuration and the S configuration at carbon 1, respectively) are called *anomers* (see Sec. 29.6).

carbohydrates exist preferentially in their hemiacetal or hemiketal forms. This equilibrium is shown here in a general sense:

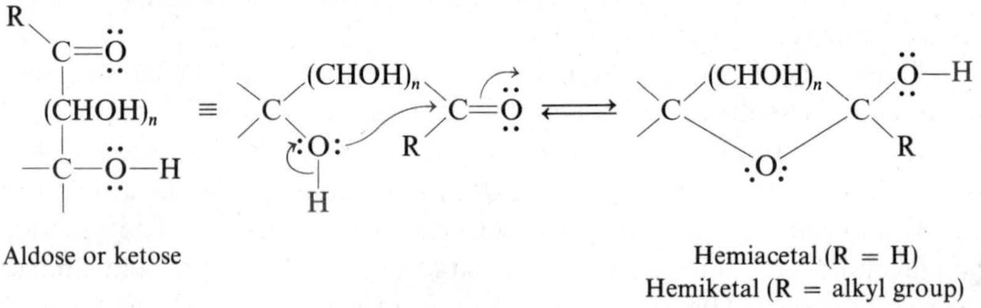

Aldose or ketose

Hemiacetal (R = H)
Hemiketal (R = alkyl group)

The open-chain form of the carbohydrate can undergo an *intramolecular* reaction of the preceding type and form a cyclic carbohydrate that contains a hemiacetal or hemiketal group. This is an example of ring-chain tautomerism, which was introduced in Sec. 11.13. The reaction applied to D-glucose is:

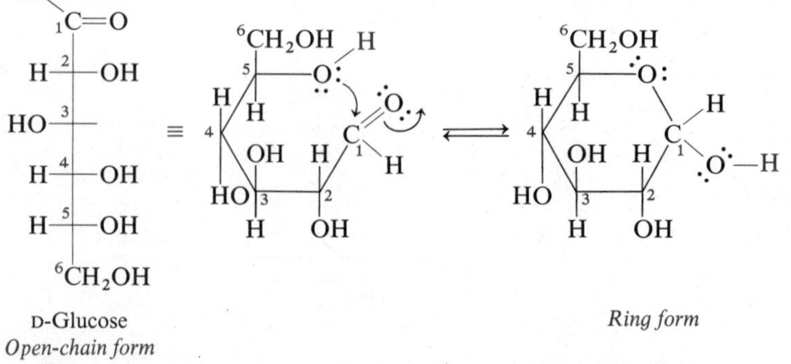

D-Glucose
Open-chain form

Ring form

Although a three-, four-, five-, six-, or seven-membered ring is possible (using the hydroxyl oxygens of carbons 2, 3, 4, 5, or 6, respectively), D-glucose is found principally in the six-membered ring form. Thermodynamically, the six-membered ring form is favored. In general, the formation of five- and six-membered rings in any intramolecular reaction is energetically (thermodynamically) favored over larger or smaller rings. Contrasting the five- and six-membered rings, studies show that the former is often formed more quickly, but if allowed to equilibrate, the latter is the more stable product. This is another example of *kinetic versus thermodynamic control* of reactions (see Sec. 8.11).

In yet another nomenclature classification system for carbohydrates, the carbohydrate name is based on the size of the ring involved in the cyclic hemiacetal or hemiketal. The naming relates to the two cyclic ethers first studied in Sec. 19.1: furan and pyran.

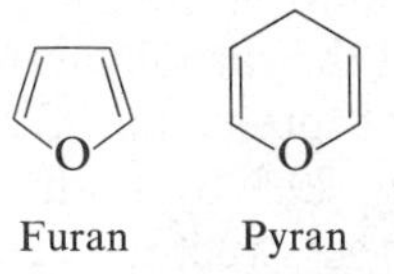

Furan   Pyran

Carbohydrates containing a five-membered cyclic hemiacetal or hemiketal form as their predominant tautomer are classified as *furanoses*, whereas those that have a six-membered ring structure are *pyranoses* (Fig. 29.7). Like the aldose or ketose system or any other broad classification systems presented in this chapter, little information is gained by knowing that a given carbohydrate is, for example, a furanose. Yet each piece of information makes the complete structure of the carbohydrate a little clearer. Prefacing the furanose or pyranose name with the name of the open-chain carbohydrate from which it is derived is a better and more complete method of naming these compounds. The carbohydrate name is shortened by

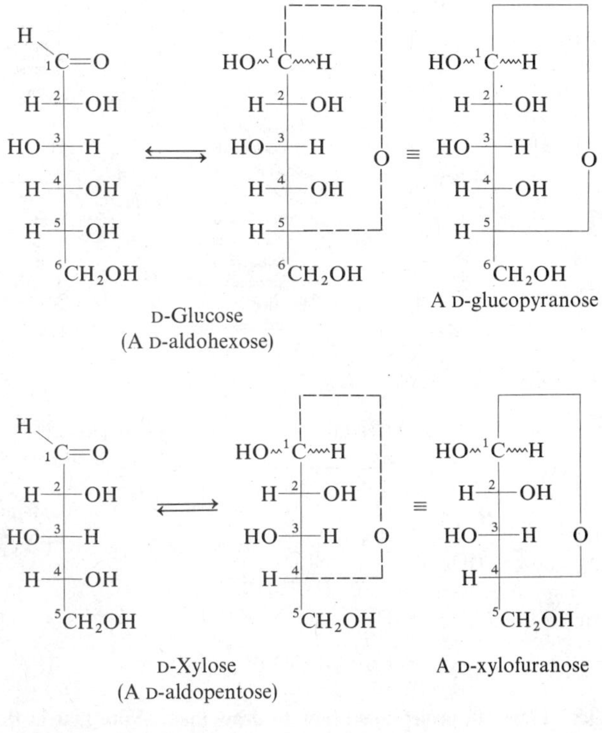

D-Glucose
(A D-aldohexose)

A D-glucopyranose

D-Xylose
(A D-aldopentose)

A D-xylofuranose

FIGURE 29.7   Fischer projection formulas: how to draw them.

the omission of the -*se* ending and the resulting prefix is attached to the furanose or pyranose name. The configuration of the parent carbohydrate should also be included. For example, the cyclic hemiacetal form of D-glucose is D-glucopyranose. Other examples are in Figs. 29.7 to 29.9. See also Fig. 29.10.

Most aldohexoses (Fig. 29.9, for example) exist almost exclusively in the pyranose form with perhaps a very small percentage of the furanose form in equilibrium with it. The percentage of the open-chain form involved in the equilibrium is considered negligible.

It is possible to draw the carbohydrates (open-chain or cyclic forms) in a variety of ways. Figure 29.4 illustrates the use of the Fischer projection and the dash-wedge representations. In addition to these methods, Haworth structures are used routinely for cyclic forms of the carbohydrates. The previous ring form of D-glucose is represented as a Haworth projection. Several examples of furanoses and pyranoses are illustrated in Fig. 29.9 by both Fischer and Haworth projections. The methods for constructing these structures are provided in Figs. 29.7 and 29.8.

Returning to the original question concerning the structure of D-glucose, we can now explain the experimental data. Realizing that D-glucose exists preferentially in the cyclic form drawn earlier, it is easy to see that D-glucose exists in two diastereomeric forms, which are called *anomeric forms*. The term *anomer* is applied to two

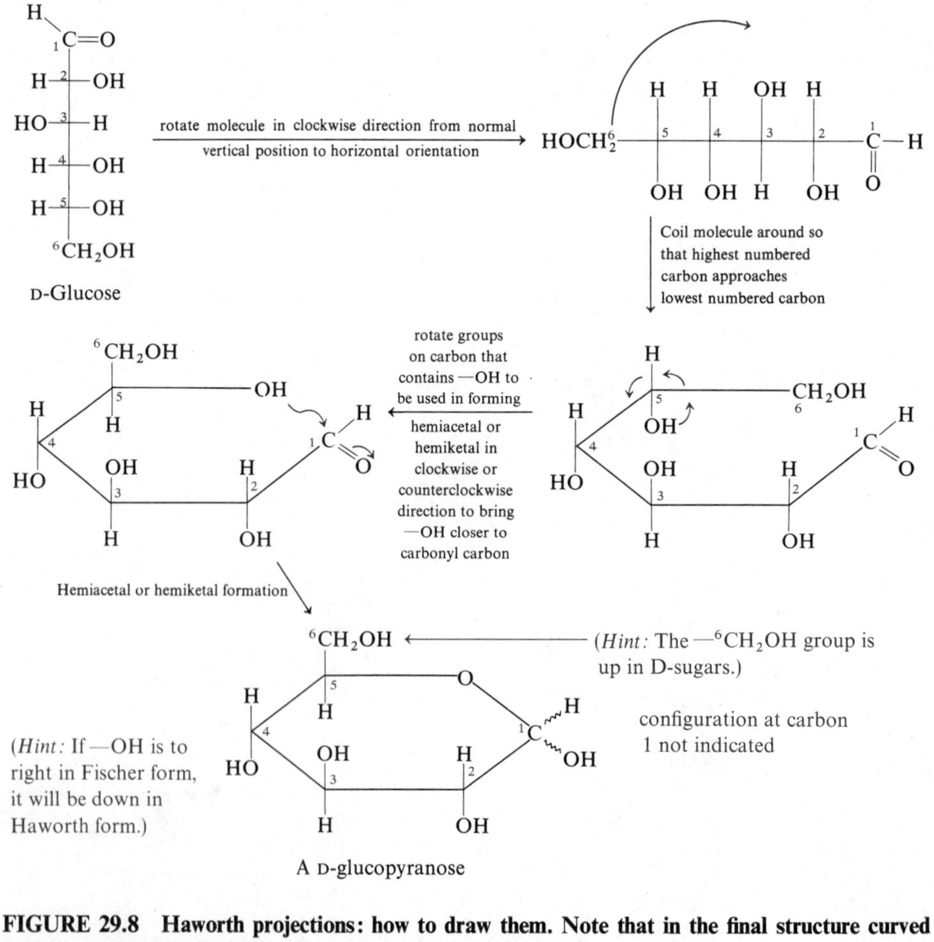

FIGURE 29.8   Haworth projections: how to draw them. Note that in the final structure curved bonds are used to designate a carbon of unknown configuration.

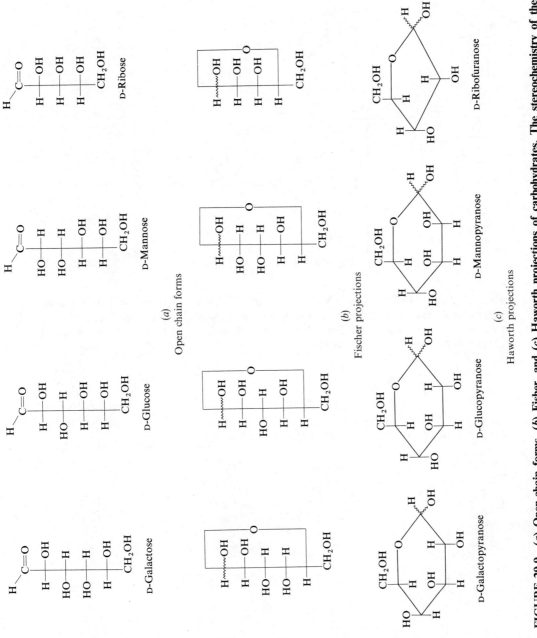

FIGURE 29.9 (*a*) Open-chain forms, (*b*) Fisher, and (*c*) Haworth projections of carbohydrates. The stereochemistry of the hemiacetal carbon in each Haworth projection is not designated—the hydroxyl can be either *cis* or *trans* to the —CH₂OH group.

1271

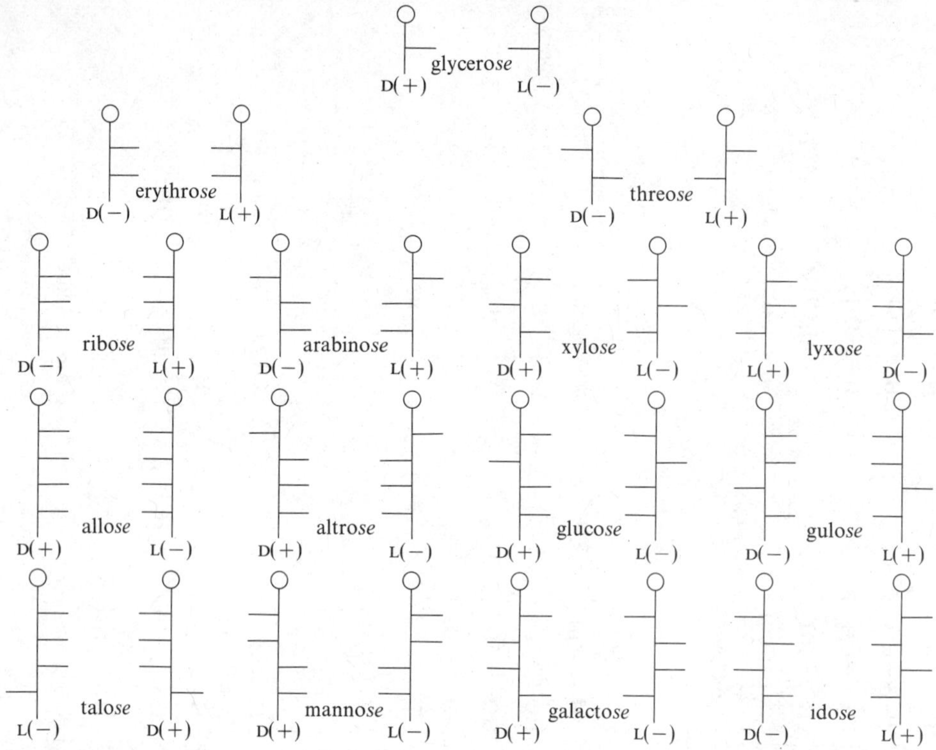

FIGURE 29.10   Skeletal framework of each of the commonly encountered carbohydrates, along with the prefix associated with each given structure. The name of the sugar is given by adding the suffix *se* to each prefix provided. The aldehyde carbon is indicated by a circle, and a —CH$_2$OH group is on the bottom end of the skeleton.

carbohydrates that differ in their configuration at carbon 1, the most highly oxidized carbon of the open-chain form (the hemiacetal or hemiketal carbon of the cyclic form). The two anomeric forms of glucose are:

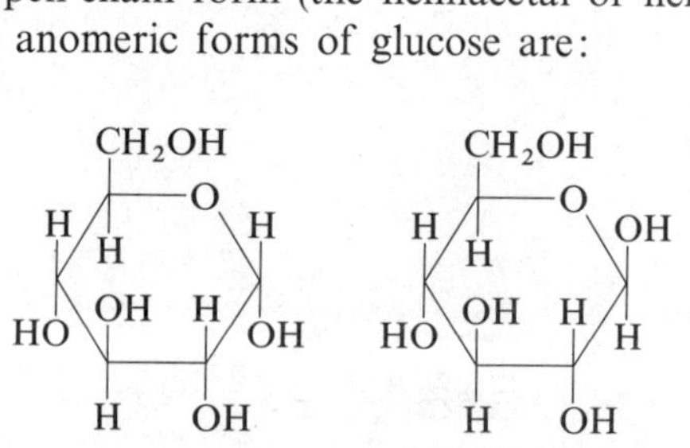

α-D-Glucose
mp 146°

β-D-Glucose
mp 150°

Haworth projections

Note the difference in the configurations of the two anomers (α and β) at position 1 in the carbon skeleton of the carbohydrate. In the β form, the hydroxyl on carbon 1 is *cis* to the —CH$_2$OH group of carbon 5, whereas in the α form the two groups are *trans*. This is true for most simple carbohydrates. The designation of α and β is not so arbitrary as it may seem. The system used in this determination is summarized here.

1. Draw the furanose or pyranose using the Fischer projection system.

2. Compare the configurations of the anomeric carbon (carbon 1) with the configuration of the highest numbered asymmetric carbon in the carbohydrate skeleton (carbon 5 in the two preceding examples).

3. If the groups on the two carbons in the Fischer projection are *cis* to each other, the anomer is called the α anomer. If the groups are *trans*, the β designation is used.[1]

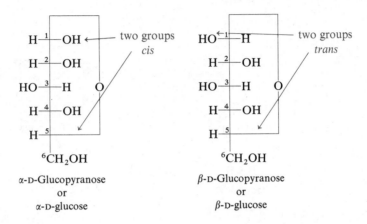

α-D-Glucopyranose
or
α-D-glucose

β-D-Glucopyranose
or
β-D-glucose

Several other examples of this system are in Fig. 29.11.

The two forms can interconvert when dissolved in water. This most likely results through an equilibration involving the open-chain form. The mechanism of the interconversion is the reverse of the hemiacetal formation—the hydrolysis of a hemiacetal (for example, the α form) and the recyclization of the open-chain aldohexose to the β anomer in the normal way:

$$
\begin{array}{ccc}
\alpha\text{-D-Glucose} & \underset{\text{cyclization}}{\overset{\text{hydrolysis}}{\rightleftharpoons}} & \text{D-Glucose chain form} & \underset{\text{hydrolysis}}{\overset{\text{cyclization}}{\rightleftharpoons}} & \beta\text{-D-Glucose}
\end{array}
$$

α-D-Glucose

mp 146°C
$[\alpha]_D^{25}$ +112°

β-D-Glucose

mp 150°C
$[\alpha]_D^{25}$ +18.7°

D-Glucose
*Aldohexose
Chain form*

The composition of the equilibrium mixture of the two D-glucopyranoses at room temperature is approximately 36% α anomer and 64% β anomer (with a negligible amount of the open-chain form). This mixture, attainable by starting with either anomer, produces the D-glucose with a specific rotation of +52.7°.

---

[1] *Cis* and *trans*, as used in this system, refer to the *relative geometries of the two groups in the Fischer projection formula*. They are not used to designate a particular geometry (that is, axial or equatorial) in the ring itself.

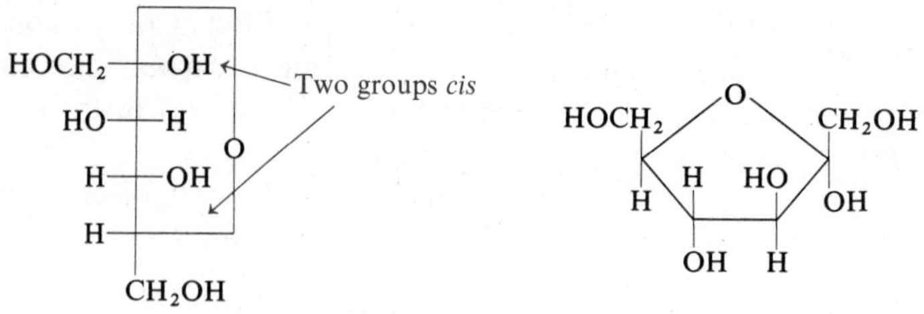

α-D-Fructofuranose

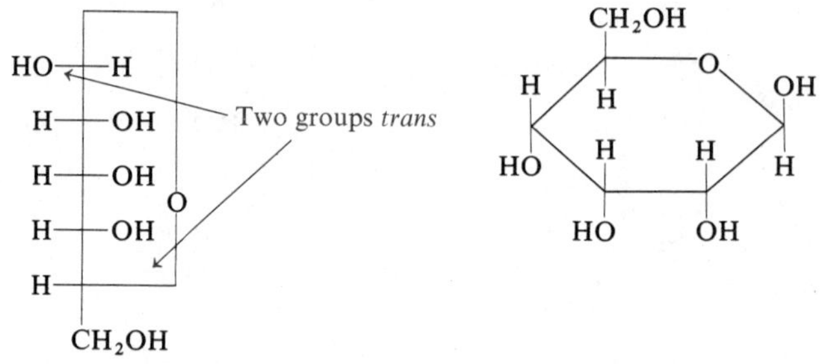

β-D-Allopyranose

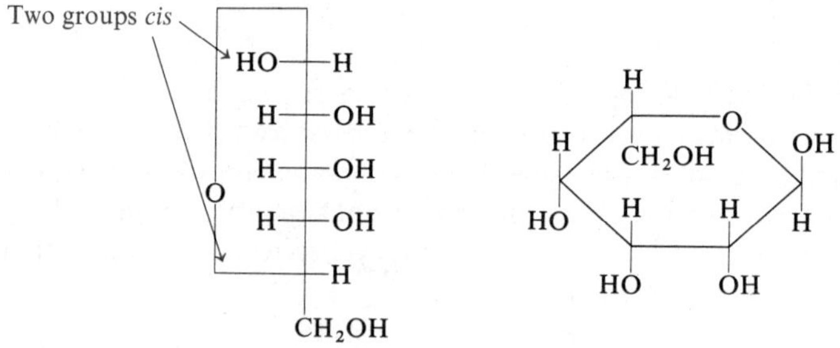

α-L-Talopyranose

FIGURE 29.11 α **and** β **Designations using Fischer projections. The** *cis* **and** *trans* **designations refer only to the apparent orientation of the groups in the Fischer projection.**

The original configuration assignment was based on the magnitude and sign of the specific rotation of a given carbohydrate. For the D isomers the α form was the more highly dextrorotatory, and for the L isomers the more highly dextrorotatory form was designated as β. Time has shown the two conventions to give the same results for the simple carbohydrates.

Question 29.2

Draw the structures of the anomers of D-(+)-talose (pyranose forms only). Give *both* Fischer projection and Haworth structures for each anomer. Also, name each anomer and indicate which would have the higher positive rotation.

Question 29.3

Name *completely* the compound (+)-I shown here; note that it has a positive rotation. Draw Haworth *and* Fischer formulas for the anomer of the compound and name it as completely as possible.

(+)-I

Question 29.4

Draw the structure of the enantiomer of D-(−)-idose using a noncyclic Fischer projection formula. Give a complete name for the enantiomer.

Question 29.5

Draw the Haworth structure for β-D-fructofuranose.

Question 29.6

Suppose the two ends of D-(+)-allose are interconverted (that is, the —CH$_2$OH group is converted to the —CHO group and the —CHO group is converted to the —CH$_2$OH group). Draw the structure of the resulting sugar. Give a complete name for the new compound, being careful to indicate its family and sign of rotation.

29.6   D-Glucose Conformational Analysis

The Haworth and Fischer projection formulas, though extremely useful, do not provide a complete picture of the three-dimensional geometry of the carbohydrate. The three-dimensional geometry can be of the utmost importance in explaining differences in reactivity between two anomers or differences in the stability of two anomers.

We limit the remainder of this section to a discussion of the pyranoses. Extension to the furanoses is straightforward.

It is not surprising that the pyranoses exist preferentially in chair conformations analogous to the six-membered cyclohexane ring. As with cyclohexane (see Sec. 3.14), the boat form and other less stable conformational forms are in rapid equilibrium with the chair form, but for all practical purposes the majority of the molecules in any sample are almost exclusively in the chair conformation. Unless one is specifically concerned with one or more of the less favored conformations, they are generally ignored.

In determining the stability of the six-membered ring chair conformation of a pyran skeleton, two primary factors must be taken into account.

1. As in all examples involving six-membered rings, the more stable ring conformation is that conformation in which the majority of the substituents are in equatorial positions. This orientation minimizes the steric repulsions (1,3-interactions, see Sec. 3.18) found for axial substituents.
2. When the six-membered ring contains a heterocyclic oxygen, a second important effect, referred to as the *anomeric effect*, must be considered. The anomeric effect derives its name from the anomeric carbon or, more precisely, the hydroxyl group on the anomeric carbon. Electrostatic repulsions between the nonbonding electrons

of the hydroxyl group on this carbon and the two nonbonded pairs of electrons on the ring oxygen are assumed responsible for the preferred orientation of these two groups as illustrated:

Preferred orientation

The two preferred conformations for α-D-glucose and β-D-glucose are:

α-D-Glucose   β-D-Glucose

Because 64% of the equilibrium mixture of the two anomers of glucose consists of the β form, in this example the anomeric effect plays a secondary role to the steric difference of having the anomeric hydroxyl axial versus equatorial. An example in which the anomeric effect does manifest itself is in the mannopyranoses in which the α form predominates over the β anomer:

α-D-Mannose   β-D-Mannose

Question 29.7

For *one* of the two isomeric forms of D-galactose, draw two possible chair conformations. Indicate the conformation you would expect to be more stable and briefly explain the basis of your choice.

Question 29.8

Draw the Haworth structures for α-D-(+)-altropyranose and β-D-(+)-altropyranose. Identify which is which.

29.7   Acetals and Ketals of Carbohydrates

A variety of reagents are available to the synthetic chemist that allow for the conversion of the hemiacetal or hemiketal form of a carbohydrate to the full acetal or ketal (and related forms). Some common ones are listed here:

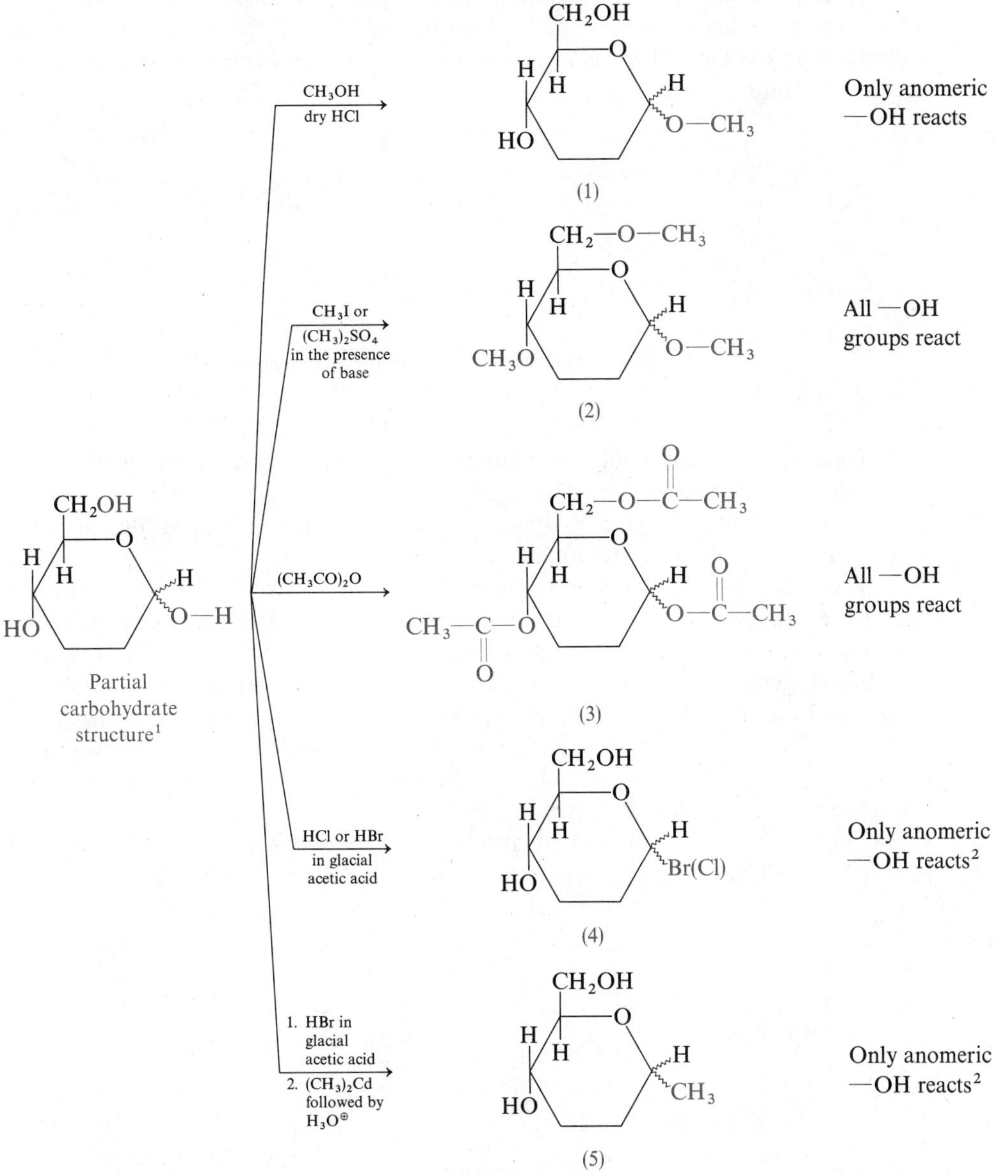

CH_2OH

H | H

HO

CH_3OH
dry HCl →

$O—CH_3$

(1)

Only anomeric
—OH reacts

$CH_2—O—CH_3$

CH_3O

CH_3I or
$(CH_3)_2SO_4$
in the presence
of base →

$O—CH_3$

(2)

All —OH
groups react

CH_2OH

H | H

HO

H
$O—H$

Partial
carbohydrate
structure[1]

$(CH_3CO)_2O$ →

$CH_2—O—\overset{\displaystyle O}{\overset{\|}{C}}—CH_3$

$CH_3—\overset{\displaystyle O}{\underset{\|}{C}}—O$

$O—\overset{\displaystyle O}{\overset{\|}{C}}—CH_3$

(3)

All —OH
groups react

CH_2OH

H | H

HO

H
Br(Cl)

HCl or HBr
in glacial
acetic acid →

(4)

Only anomeric
—OH reacts[2]

CH_2OH

H | H

HO

H
CH_3

1. HBr in
glacial
acetic acid
2. $(CH_3)_2Cd$
followed by
$H_3O^\oplus$ →

(5)

Only anomeric
—OH reacts[2]

Compounds of the general form of (1) and (2), that is, those that contain an —OR group at the anomeric center, are called *glycosides*.

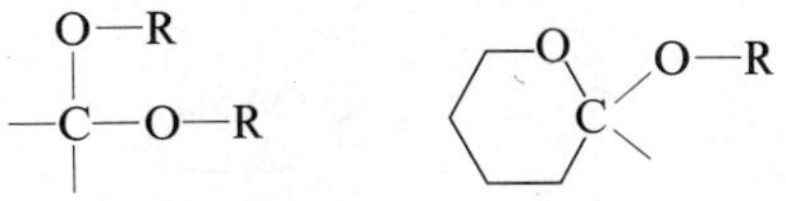

$$-\overset{\displaystyle O—R}{\underset{\displaystyle |}{C}}—O—R$$

Acetal or ketal group

$$\overset{O}{\diagup}\overset{O—R}{C}$$

Glycoside

---

[1] Note that curved bonds are used to designate carbons of unknown configuration or samples in which either configuration (α or β) may be present.

[2] All the hydroxyl groups except the anomeric hydroxyl of the carbohydrate are protected as the acetates before reaction with the indicated reagents. The acetates are converted to the free alcohol functions subsequent to the indicated reaction. See the examples on the next pages.

When referring to a specific glycoside, the -*side* suffix is added to the carbohydrate name (Fig. 29.10); for example, galactoside, glucoside, xyloside, and fructoside. The glycoside name is itself prefixed by the name of the alkyl substituent on the anomeric oxygen, which is itself not part of the main ring. The following examples illustrate this system:

HOCH$_2$

HOCH$_2$

Methyl α-D-glucoside
or
methyl α-D-glucopyranoside

Methyl β-D-alloside
or
methyl β-D-allopyranoside

The resulting compounds are of immense use in investigations concerning such questions as conformation, ring size, ring reactivity, and ring stability. They may also be used as intermediates in synthesizing nucleosides or nucleotides or other carbohydrate-containing compounds.

Through a careful selection of reagents and the sequence of reacting the carbohydrate with each of these reagents, it is possible to synthesize a wide selection of substituted saccharides. The reaction of D-glucose with acetic anhydride (or other acylating reagents such as acetyl chloride) results in the acetylation of all the hydroxyl groups and the formation of D-glucose pentaacetate, analogous to the earlier compound (3). Treatment of the pentaacetate with dilute HCl results in hydrolysis of the acetate function on the anomeric carbon exclusively. The pentaacetates may in turn be treated with either HBr or HCl dissolved in glacial acetic acid, which results in the cleavage of the acetate group on the anomeric carbon exclusively and the replacement of this group with either —Br or —Cl [see structure (4)].

HOCH$_2$

$\xrightarrow{(CH_3CO)_2O}$

AcOCH$_2$

$\xrightarrow[\substack{glacial \\ acetic \\ acid}]{HX}$

D-Glucose

D-Glucose
pentaacetate

AcOCH$_2$

$\xrightarrow[\substack{2. \ H_3O^\oplus \\ dilute}]{1. \ (CH_3)_2Cd}$

AcOCH$_2$

where Ac— = $CH_3-\overset{\overset{\textstyle O}{\|}}{C}-$

Subsequent reaction of the halogenated carbohydrate with dimethylcadmium (followed by hydrolysis with dilute acid) places a methyl substituent on the anomeric carbon [see structure (5)].

The halogenated saccharides are also extremely important intermediates in the synthesis of nucleosides and nucleotides, the nucleic acid monomers. The general synthetic approach is:

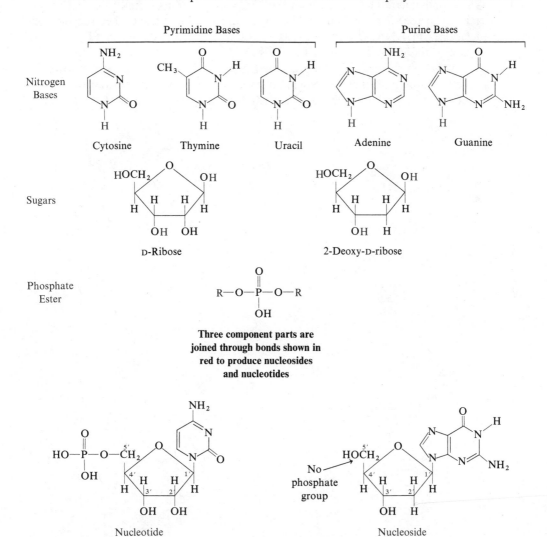

29.8 Nucleic Acids

The nucleic acids, DNA (deoxyribonucleic acid) and RNA (ribonucleic acid), were introduced in Chap. 28. These acids are immense biopolymers involved in the

FIGURE 29.12 General structure of nucleosides and nucleotides.

transfer of genetic information from the cell to the ribosomes, where it is transcribed and used in protein synthesis. This process is schematically illustrated in Fig. 28.8. The DNA in the cell nucleus and a variety of forms of RNA (transfer RNA, messenger RNA, and ribosomal RNA) are involved in the synthetic sequence. The synthesis occurs at a ribosome, which is part nucleic acid and part protein and is classified as a *nucleoprotein.*

The nucleic acids are composed of many monomeric units called nucleotides, which are themselves composed of a carbohydrate portion, a nitrogen heterocyclic base portion, and a phosphate ester grouping. Removal of the phosphate ester converts the nucleotide to the corresponding nucleoside.

The carbohydrate portion of nucleotides or nucleosides is generally D-ribose, which is one of the primary monomeric units of RNA, and 2-deoxy-D-ribose (in which a hydroxyl at the 2 position of D-ribose has been replaced by a hydrogen atom), found in the DNA polymer skeleton. The nitrogen base is attached at the 1 position in both sugars and is always of the β configuration. The nucleotides are joined by phosphate ester linkages from the 3 position of one sugar to the 5 position of a second sugar. These are called the 3′ and 5′ positions to indicate that we are referring to the ring positions of the sugar and not the nitrogen heterocycle. The nucleotides in DNA and the various forms of RNA are shown in Fig. 29.12; a portion of the DNA chain is in Fig. 29.13.

Although the primary structure of DNA (see Sec. 28.8) has been known for a long time, the total three-dimensional structure (secondary and tertiary structure) of these macromolecules eluded scientists for many years. In 1953 Watson and

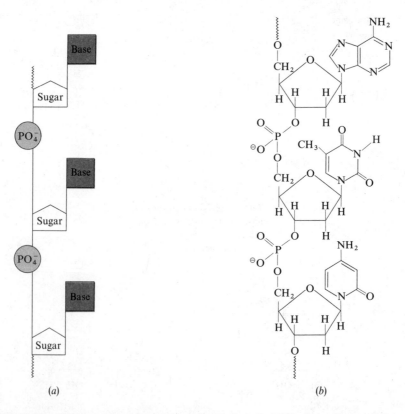

(a)   (b)

FIGURE 29.13   (*a*) **Schematic representation of a portion of a DNA chain.** (*b*) **Detailed representation of a portion of the nucleic acid chain of DNA.**

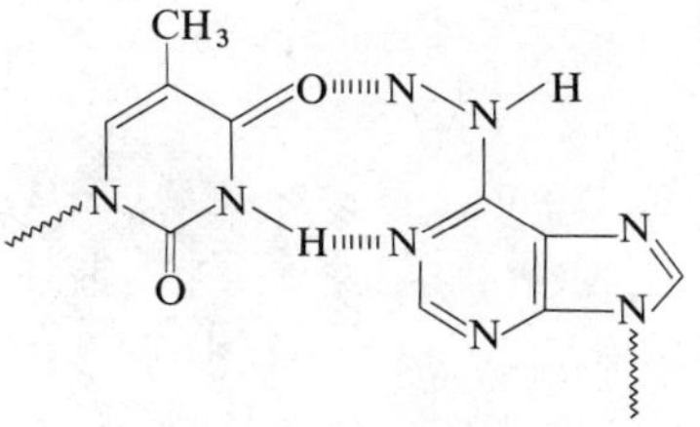

Thymine—Adenine

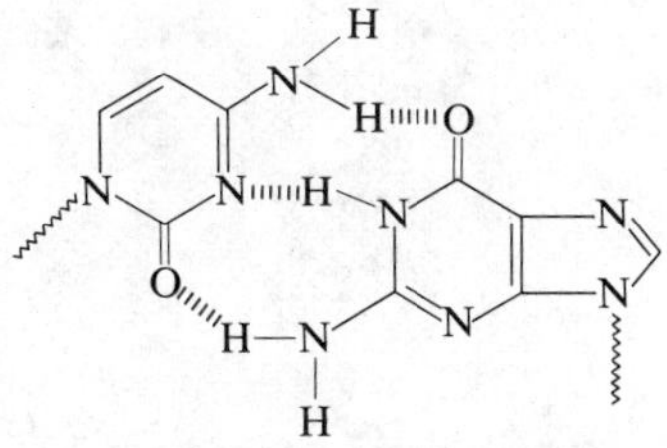

Cytosine—Guanine

FIGURE 29.14 Base pairing in DNA. Hydrogen bonds are indicated by color dashed lines.

Crick postulated the now-famous *double helix*. Their conclusion was based on years of work and an immense amount of data covering all aspects of the chemistry of the DNA species. Working with these data and highly accurate models (scaled to replicate the real molecules), Watson and Crick were able to elucidate that two DNA chains are intimately linked together (paired) to form an α-helical species (see Sec. 29.8). They also proved that this helix must be right-handed (a consequence of the D configurations of the sugars involved) and that the primary force involved in this binding is hydrogen bonds, which are present in large numbers between the various nitrogen bases along the backbone of the chain. A consequence of this model, and also of the real system, is that the bases join in certain geometrically allowed pairs and only in these pairs. Thymine and adenine form such a pair as do cytosine and guanine. The hydrogen bonding within these pairs is illustrated in Fig. 29.14.

A complete picture of the DNA double helix is in Fig. 29.15.

The RNA macromolecules, unlike the DNA, exist in single chains, some of which are fibrous whereas others are globular.

Fig. 29.16 illustrates the actual synthesis of a nucleoside from the appropriately substituted carbohydrate and activated nitrogen base.

Question 29.9

(*a*) Using the structures of adenine cytosine shown here, draw two units of a DNA (deoxyribonucleic acid) molecule that incorporates these two bases in it and shows only the *primary* structure.

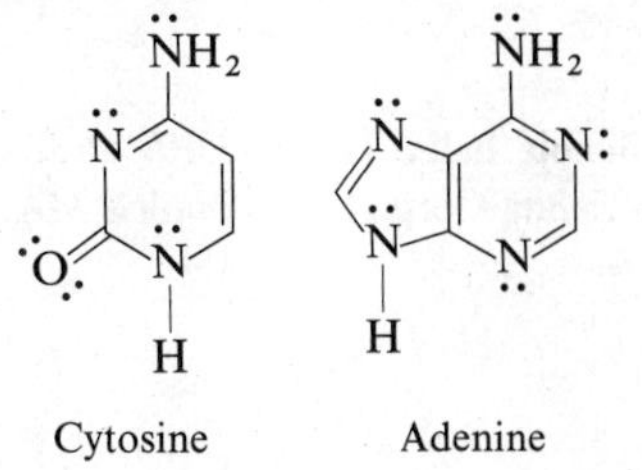

Cytosine   Adenine

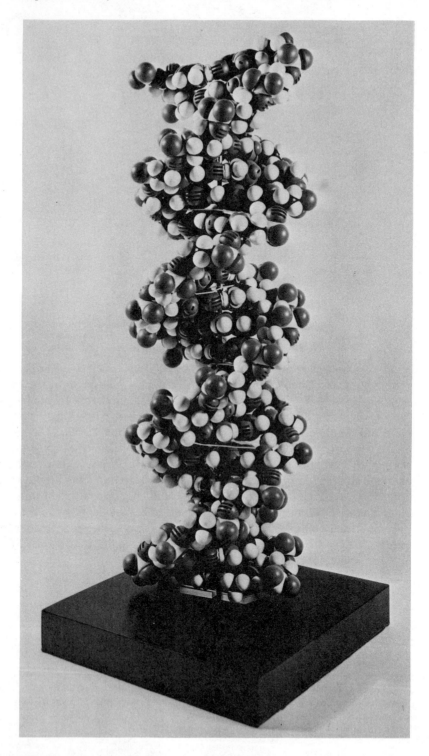

FIGURE 29.15   DNA double helix. [From *CPK Precision Molecular Models Catalog*, 1976, p. 5, by permission of the Ealing Corp., South Natick MA.]

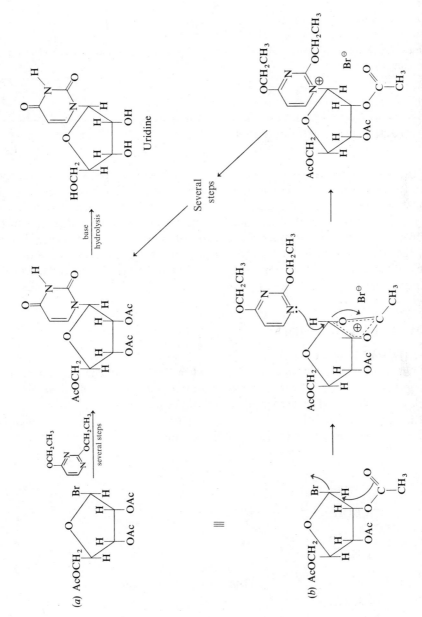

FIGURE 29.16 Synthesis of the nucleoside uridine. (*a*) Overall reaction. (*b*) Partial mechanism of the reaction explaining the observed retention of relative configuration in going from the bromosugar to the desired product.

(b) Pure cytosine (e.g., not attached to another molecule) possesses the following structure. Based on principles studied thus far, explain why you would expect this structure to be preferred (and more stable) over that shown in (a).

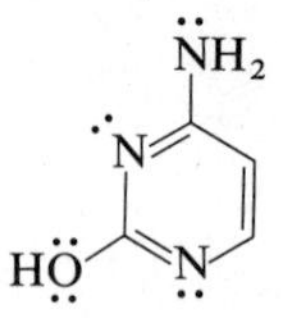

(c) Draw the complete orbital picture of the compound in (b) and show all the π electrons and nonbonding electrons.

29.9   Reactions of Carbohydrates

In this section we give a representative selection of the types of reactions carbohydrates undergo. Carbohydrate chemistry is one of the oldest and most well established fields of organic chemistry—an area in which an immense amount of data has been generated. The reactions presented here merely scratch the surface and provide a "taste" of the type of chemistry of interest to carbohydrate chemists. Also, the synthesis and isolation and purification of a carbohydrate are often extremely difficult tasks. These compounds tend to form syrups (oils or tars) rather than crystalline compounds, and are difficult to purify.

A. Oxidation Reactions

a. Tollens' Reagent

The Tollens' reagent was first discussed in Sec. 21.16. The reaction involves treating any aldose or α-hydroxyketose with a basic solution of silver ammonia complex, $Ag(NH_3)_2^{\oplus}$. The aldose or α-hydroxyketose is oxidized to a variety of oxidation products, and the silver ion ($Ag^{\oplus}$) is reduced to silver metal, which precipitates as a silver mirror on the inside of the test tube in which the reaction is carried out.

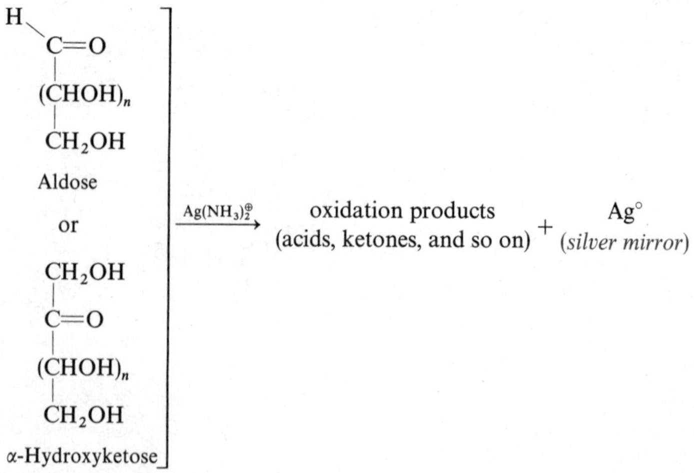

Most sugars exist preferentially in their cyclic tautomeric forms, although the acyclic forms are depicted in this and following examples for clarity.

The oxidation reaction is successful only if the carbohydrate is in the form of its hemiacetal or hemi-(α-hydroxy) ketal. If the sugar is in the full acetal or ketal form, the reaction does not occur because of the inherent stability and nonreactivity of acetals and ketals (see Sec. 21.6). Sugars that give a positive test with Tollens' reagent are called **reducing sugars;** those that give a negative test are **nonreducing sugars.**

A major problem in using this reagent is the strong alkaline solution in which the reaction is carried out. As discussed in Chaps. 23 and 24, hydrogens on a carbon α to a carbonyl carbon are reasonably acidic. This acidity is used to the benefit of the synthetic organic chemist in the aldol reaction and other related reactions. This same acidity manifests itself in the carbohydrates (both aldoses and ketoses). When an aldose, for example, is treated with Tollens' reagent, the major oxidation product from the reaction is the corresponding **aldonic acid** (a carbohydrate in which the carbonyl has been oxidized to a carboxyl group). If this product is isolated, however, there are two stereoisomers that differ in configuration about the asymmetric carbon next to the carbonyl carbon (α to the carbonyl carbon); two different diastereomers of the aldonic acid differing in configuration about only one carbon atom are produced. Diastereomers of this type are called *epimers* (Sec. 29.4), and the process of converting one stereoisomer into a pair of epimers is *epimerization*. The mechanism involved in this epimerization of aldoses is illustrated here:

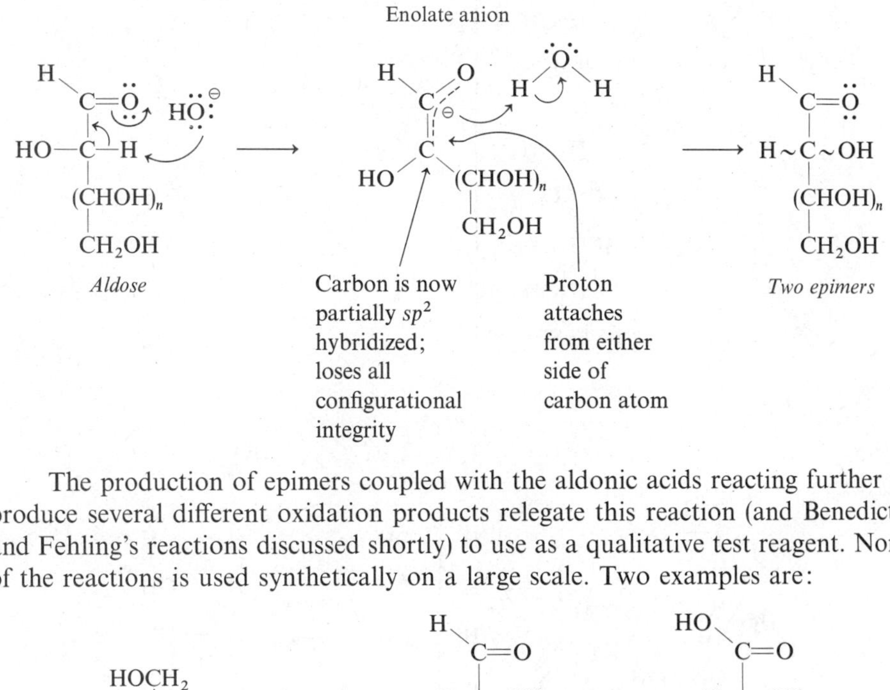

Enolate anion

Aldose

Carbon is now partially sp^2 hybridized; loses all configurational integrity

Proton attaches from either side of carbon atom

Two epimers

The production of epimers coupled with the aldonic acids reacting further to produce several different oxidation products relegate this reaction (and Benedict's and Fehling's reactions discussed shortly) to use as a qualitative test reagent. None of the reactions is used synthetically on a large scale. Two examples are:

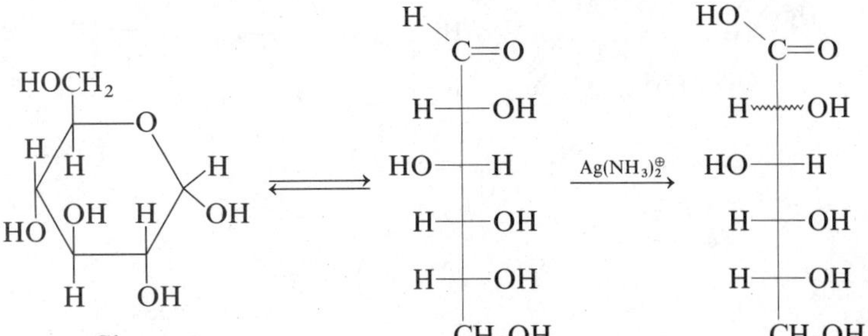

D-Glucose
Aldose

$Ag(NH_3)_2^{\ominus}$

D-Gluconic acid
and D-Mannonic acid
aldonic acids
Epimeric

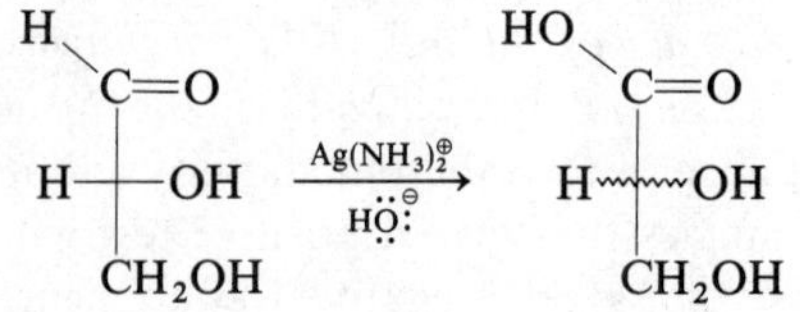

D-Glyceraldehyde

Glyceric acid
Racemic mixture

b. Benedict's and Fehling's Reagents

Benedict's reagent (cupric citrate in base solution)[1] and Fehling's reagent (cupric tartrate in base solution),[1] also introduced in Sec. 21.16, are analogous to Tollens' reagent in their reaction with carbohydrates. Upon treatment with these reagents, reducing sugars undergo oxidation to a mixture of various products, and the metal ion, Cu^{+2} in this case, is reduced to Cu^{+}, which precipitates as *red* cuprous oxide, Cu_2O:

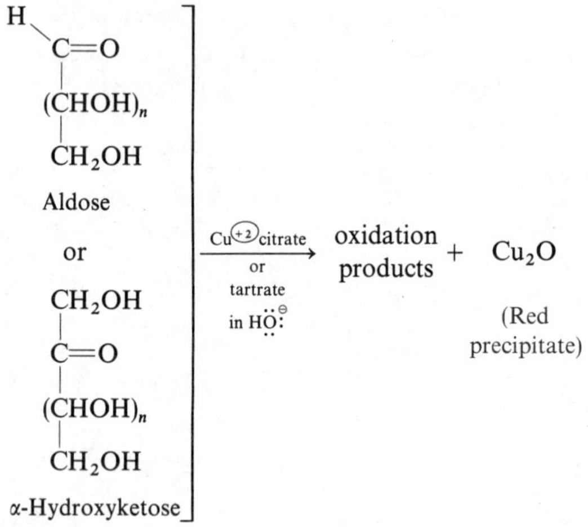

Aldose

or

α-Hydroxyketose

oxidation products + Cu_2O

(Red precipitate)

For example:

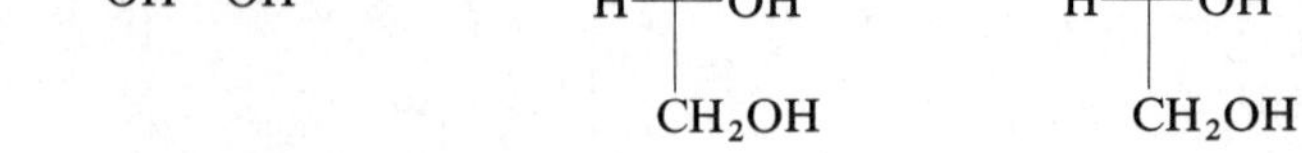

D-Allose
Aldose

Epimeric aldonic acids

---

[1] Tartrate = $^{\ominus}:\!\ddot{O}OC\!-\!CH\!-\!CH\!-\!CO\ddot{O}:^{\ominus}$; citrate = $^{\ominus}:\!\ddot{O}OCCH_2\!-\!\underset{\underset{\textstyle CO\ddot{O}:^{\ominus}}{|}}{\overset{\overset{\textstyle OH}{|}}{C}}\!-\!CH_2CO\ddot{O}:^{\ominus}$.

Question 29.10

Explain (show all intermediates) why D-(−)-tagatose (1) gives a silver mirror when treated with Tollens' reagent even though it does not contain an aldehyde group. Give the structure(s) of the sugar molecule(s) that remain after the silver mirror is formed.

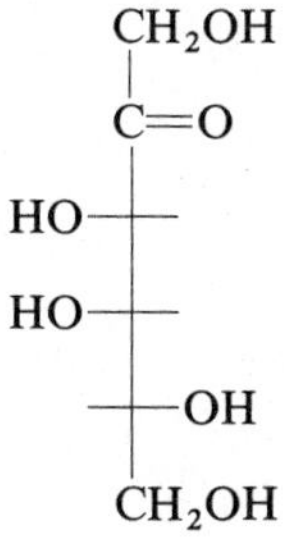

$$
\begin{array}{c}
CH_2OH \\
| \\
C=O \\
HO-\!\!-\!\!| \\
HO-\!\!-\!\!| \\
|-\!\!-OH \\
CH_2OH
\end{array}
$$

(1)

Question 29.11

Which of the following would be classified as reducing sugars?

(a) HOCH₂ ... (pyranose ring structure)

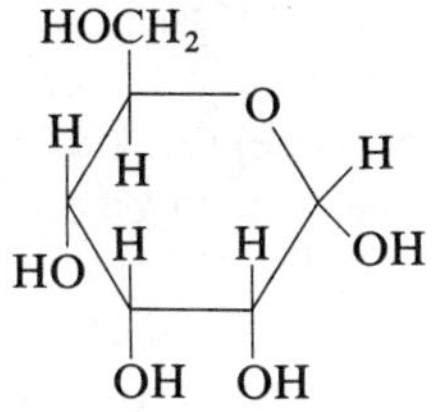

(b)
$$
\begin{array}{c}
CH_3O-\!\!-H \\
H-\!\!-OH \\
H-\!\!-OH \\
H-\!\!- \\
CH_2OCH_3
\end{array}
$$

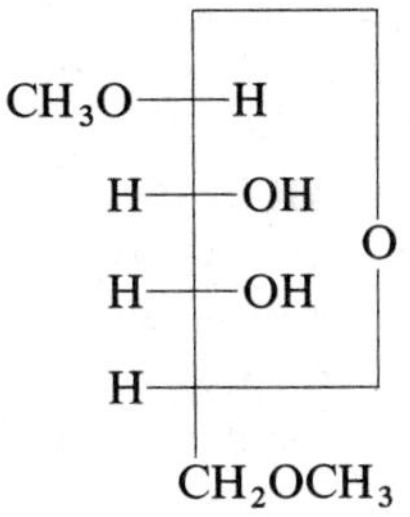

(c)
$$
\begin{array}{c}
H-\!\!-OH \\
H-\!\!-OH \\
H-\!\!-OH \\
H-\!\!- \\
CH_2OH
\end{array}
$$

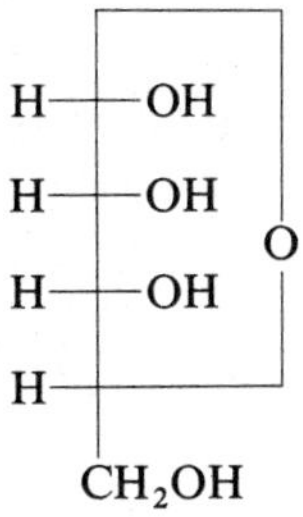

(d)
$$
CH_3C-O-CH_2 \text{ ... (acetylated pyranose with } O-C-CH_3 \text{ groups)}
$$

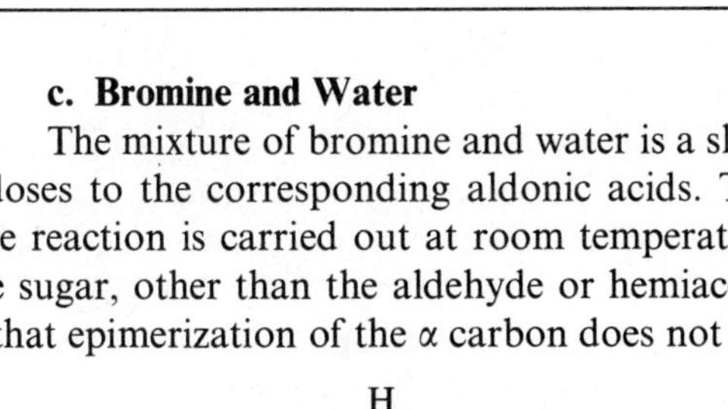

(e)
$$
\begin{array}{c}
H \\ \ \ \ C=O \\
H-\!\!-OH \\
CH_2OH
\end{array}
$$

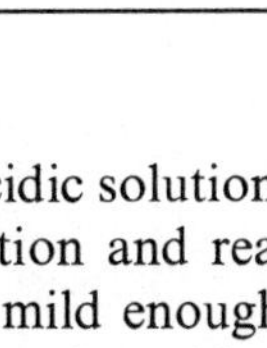

c. Bromine and Water

The mixture of bromine and water is a slightly acidic solution that can oxidize aldoses to the corresponding aldonic acids. The solution and reaction conditions (the reaction is carried out at room temperature) are mild enough that the rest of the sugar, other than the aldehyde or hemiacetal group, is stable. Most important is that epimerization of the α carbon does not occur.

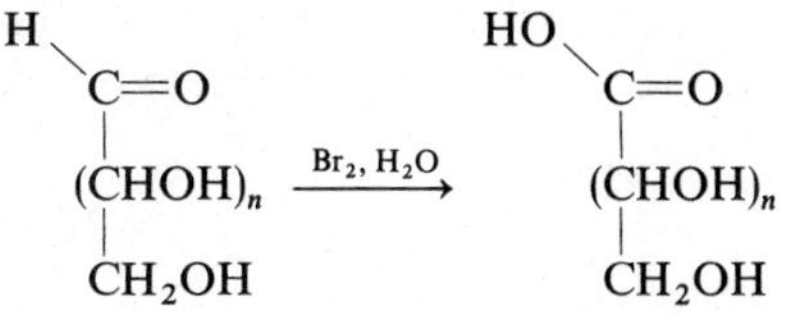

$$
\begin{array}{c}
H \\ \ \ C=O \\
(CHOH)_n \\
CH_2OH
\end{array}
\quad \xrightarrow{Br_2,\ H_2O} \quad
\begin{array}{c}
HO \\ \ \ C=O \\
(CHOH)_n \\
CH_2OH
\end{array}
$$

Aldose   Aldonic acid

The β isomer of D-glucose (the isomer containing the equatorial hydroxyl on the anomeric carbon) for example, is the reactive species. Any of the α isomer present must equilibrate and rearrange to this isomer before the oxidation can occur.

The partial mechanism is:

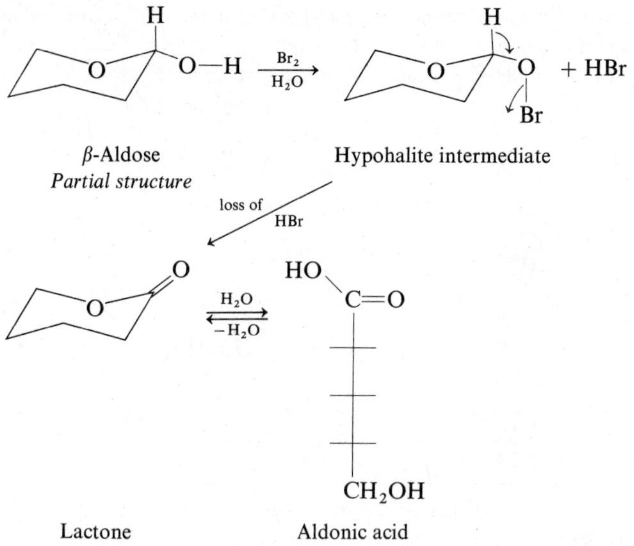

β-Aldose
Partial structure

Hypohalite intermediate

Lactone   Aldonic acid

Reaction of the hemiacetal hydroxyl group with bromine in water results in an intermediate containing the hypohalite (—OBr, hypobromite) functionality. This intermediate probably forms with approximately equal ease from either the α or β anomer. The second step, however, requires *anti* elimination of HBr (see Sec. 7.11). The β anomer can readily align the groups being eliminated in this *anti*-orientation. The α anomer would have difficulty doing so because of the steric crowding of the 1,3-interactions common to axially substituted six-membered rings. For example:

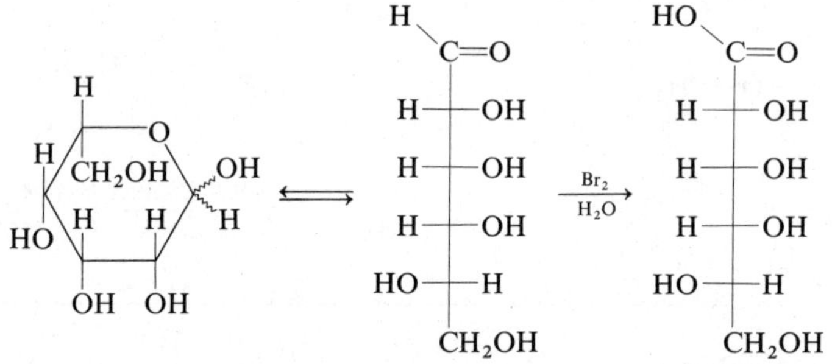

L-Talose
Aldose

L-Talonic acid
Aldonic acid
No Epimerization

d. Nitric Acid

A much stronger oxidizing agent than bromine and water, nitric acid oxidizes aldoses to the corresponding **aldaric acids,** in which the two ends of the carbohydrate, the aldehyde end and the —CH$_2$OH end, are both oxidized to carboxylic acids.

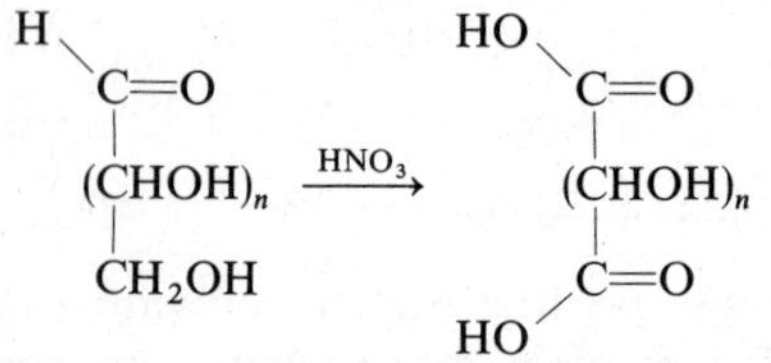

Aldose   Aldaric acid

For example:

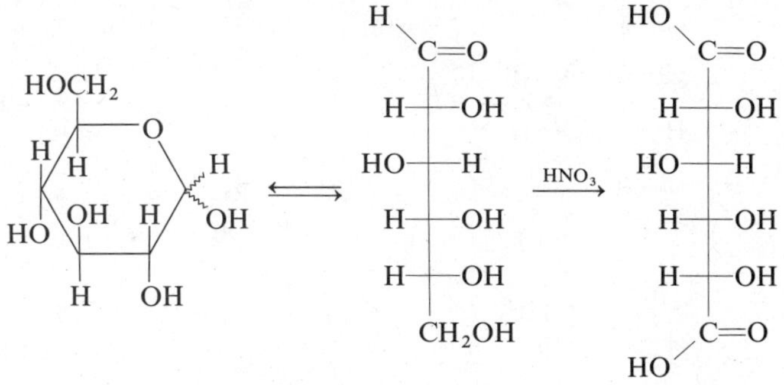

D-Glucose
Aldose

D-Glucaric acid
Aldaric acid

e. Periodic Acid Oxidation

When two groups that are capable of being oxidized are present on adjacent carbon atoms in the carbohydrate skeleton, treatment of the carbohydrate with periodic acid, HIO_4, cleaves the carbon-carbon bond of the carbohydrate skeleton between the carbons containing these two oxidizable functions (see Sec. 19.18):

Cleavage of this
carbon-carbon bond results

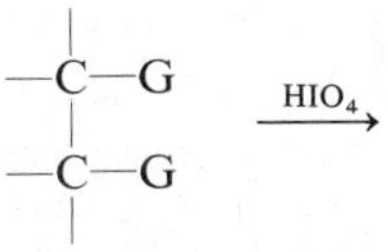

$$\begin{matrix} -C-G \\ -C-G \end{matrix} \xrightarrow{\ HIO_4\ }$$

aldehydes
ketones
carboxylic acids
carbon dioxide

Two oxidizable
groups (C—G)
vicinal to
each other

Oxidation products

If one of the groups $\left(-\overset{|}{\underset{|}{C}}-G\right)$ is

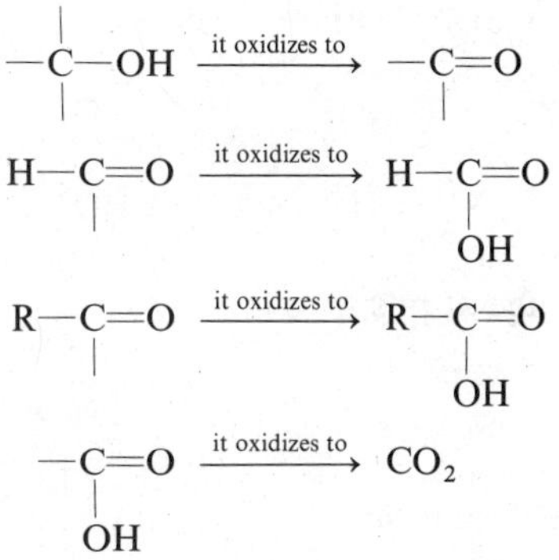

$$-\overset{|}{\underset{|}{C}}-OH \xrightarrow{\text{it oxidizes to}} -C=O$$

$$H-\overset{|}{C}=O \xrightarrow{\text{it oxidizes to}} H-\underset{OH}{\overset{|}{C}}=O$$

$$R-\overset{|}{C}=O \xrightarrow{\text{it oxidizes to}} R-\underset{OH}{\overset{|}{C}}=O$$

$$-\underset{OH}{C}=O \xrightarrow{\text{it oxidizes to}} CO_2$$

Between two other
oxidizable groups

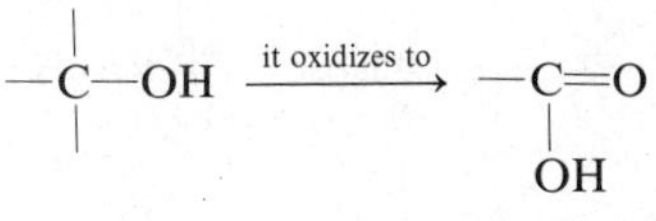

$$-\overset{|}{\underset{|}{C}}-OH \xrightarrow{\text{it oxidizes to}} -\underset{OH}{C}=O$$

Between two other
oxidizable groups

$$-\overset{|}{C}=O \xrightarrow{\text{it oxidizes to}} CO_2$$

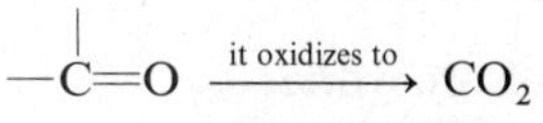

respectively. The partial mechanism for two hydroxyl groups is:

$$\begin{array}{c}
\text{R} \\
\text{H}-\overset{|}{\text{C}}-\text{O}-\text{H} \\
\text{H}-\overset{|}{\text{C}}-\text{O}-\text{H} \\
\text{R}
\end{array} + \text{IO}_4^{\ominus} \longrightarrow
\begin{array}{c}
\text{R} \\
\text{H}-\overset{|}{\text{C}}-\text{O} \\
\text{H}-\overset{|}{\text{C}}-\text{O} \\
\text{R}
\end{array}
\begin{array}{c}
\text{O} \\
\overset{|}{\text{I}}-\overset{\ominus}{\ddot{\text{O}}}\text{:} + \text{H}_2\text{O} \\
\text{O}
\end{array}$$

$$2 \quad \begin{array}{c} \text{H} \\ \diagdown \\ \underset{\text{R}}{\diagup}\text{C}=\text{O} \end{array}$$

$$+$$

$$\text{IO}_3^{\ominus}$$

For example:

D-Glucose

Cleavage at positions 1 and 2:

Cleavage at positions 3 and 4:

Cleavage at position 5:

$$\begin{array}{c} \text{H} \\ \diagdown \\ \underset{\text{HO}}{\diagup}\text{C}=\text{O} \end{array} + \begin{array}{c} \text{H} \\ \diagdown \\ \underset{\text{HO}}{\diagup}\text{C}=\text{O} \end{array}$$

Overall reaction:

$$\text{D-Glucose} \xrightarrow[\text{H}_3\overset{\oplus}{\ddot{\text{O}}}]{\text{NaIO}_4} 5\ \text{H}-\overset{\overset{\text{O}}{\|}}{\text{C}}-\text{OH} + \text{H}-\overset{\overset{\text{O}}{\|}}{\text{C}}-\text{H}$$

B. Reduction Reactions

a. Sodium Borohydride

An aqueous solution of sodium borohydride, $NaBH_4$, a reagent often used in the reduction of carbonyl compounds, easily reduces the carbonyl group of an aldose or ketose to the corresponding polyhydroxyalkane (**alditol**):

Aldose → Alditol (NaBH₄, H₂O)

Ketose → Epimeric alditols (NaBH₄, H₂O)

b. Hydrogen and Metal

Hydrogen and platinum (Pt) or other metal catalysts such as nickel (Ni) may also be used to reduce the carbonyl group of an aldose or ketose to the alditol. The generalized reaction is the same as that illustrated for sodium borohydride.

c. Sodium Amalgam

Another reagent capable of accomplishing the preceding reduction is sodium amalgam [Na(Hg)]. A trace of water is necessary to act as the proton source in the oxidation-reduction processes that result; for example:

D-Glucose (Aldose) ⇌ ... —[H]→ D-Glucitol (Alditol)

D-Fructose (Ketose) ⇌ ... —[H]→ D-Glucitol + D-Mannitol (Alditols)

where $[H] = H_2$, metal; $NaBH_4$, H_2O; $Na(Hg)$, H_2O (trace)

C. Chain Lengthening and Shortening Reactions

a. Chain Lengthening Kiliani-Fischer Synthesis

In Secs. 22.4 and 23.16 the versatility of using nitriles as carboxylic acid and diacid precursors was illustrated. The conversion is accompanied by hydrolysis of the nitrile under either acidic or basic conditions. This same reaction may be used to increase the length of a carbohydrate chain by one carbon atom. Concurrent with the chain-lengthening process, the aldose or ketose that is reacted is converted to the aldonic acid. The reaction proceeds through the formation of a cyanohydrin from the carbonyl compound (by the addition of HCN-NaCN, H_2O). This reaction proceeds through a nucleophilic addition mechanism (see Sec. 21.2). When an aldose or ketose is the carbonyl compound being reacted, the carbonyl carbon becomes an asymmetric carbon atom on conversion to the cyanohydrin. Due to the easy attack from either side of the carbonyl group, a pair of enantiomers are formed, resulting in an epimeric mixture of two cyanohydrins. These may be hydrolyzed to the corresponding aldonic acids:

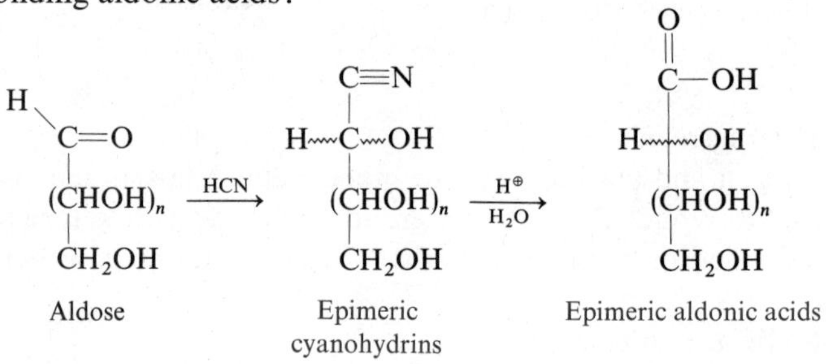

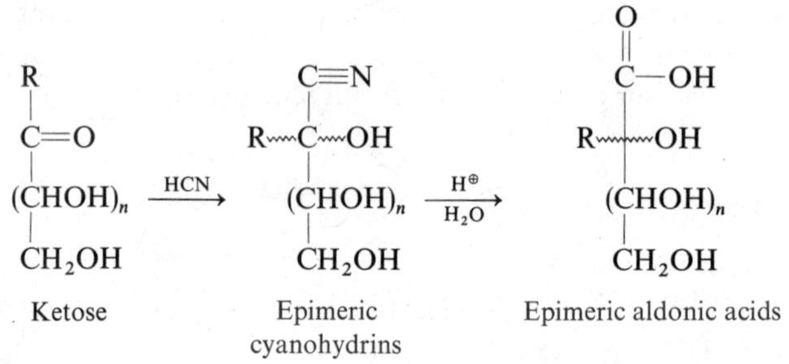

Aldose   Epimeric cyanohydrins   Epimeric aldonic acids

Ketose   Epimeric cyanohydrins   Epimeric aldonic acids

For example:

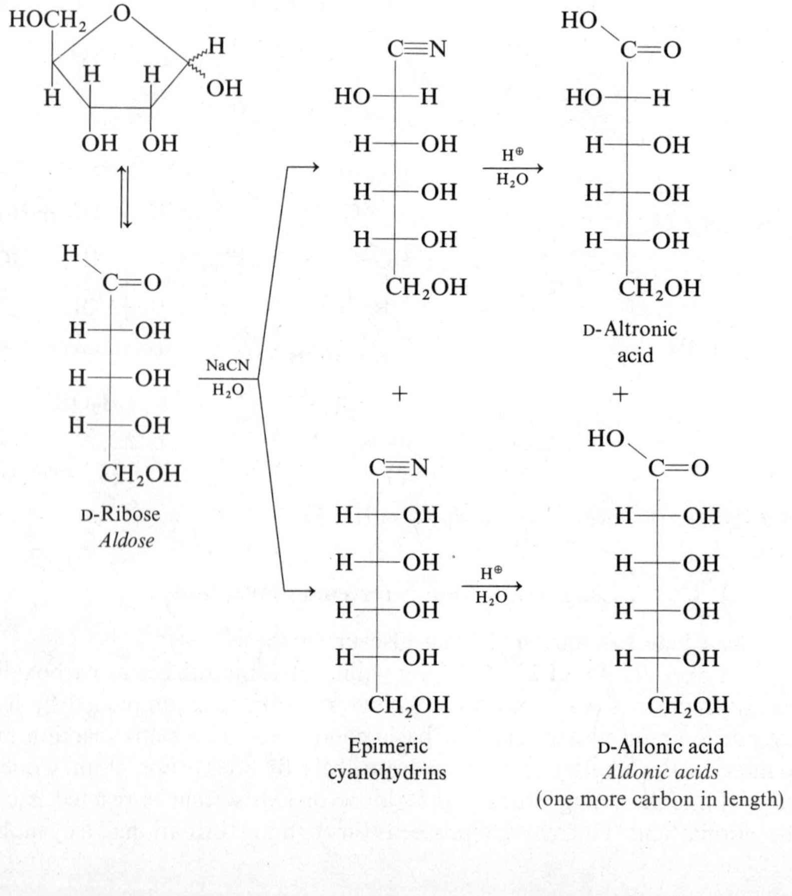

D-Ribose
Aldose

Epimeric cyanohydrins

D-Altronic acid

+

D-Allonic acid
Aldonic acids
(one more carbon in length)

Actually it is not the aldonic acid that is isolated after the hydrolysis, but the γ-lactone. These lactones are in turn reduced [with Na(Hg) and a trace of water] to the aldose, which is one carbon longer than the original aldose:

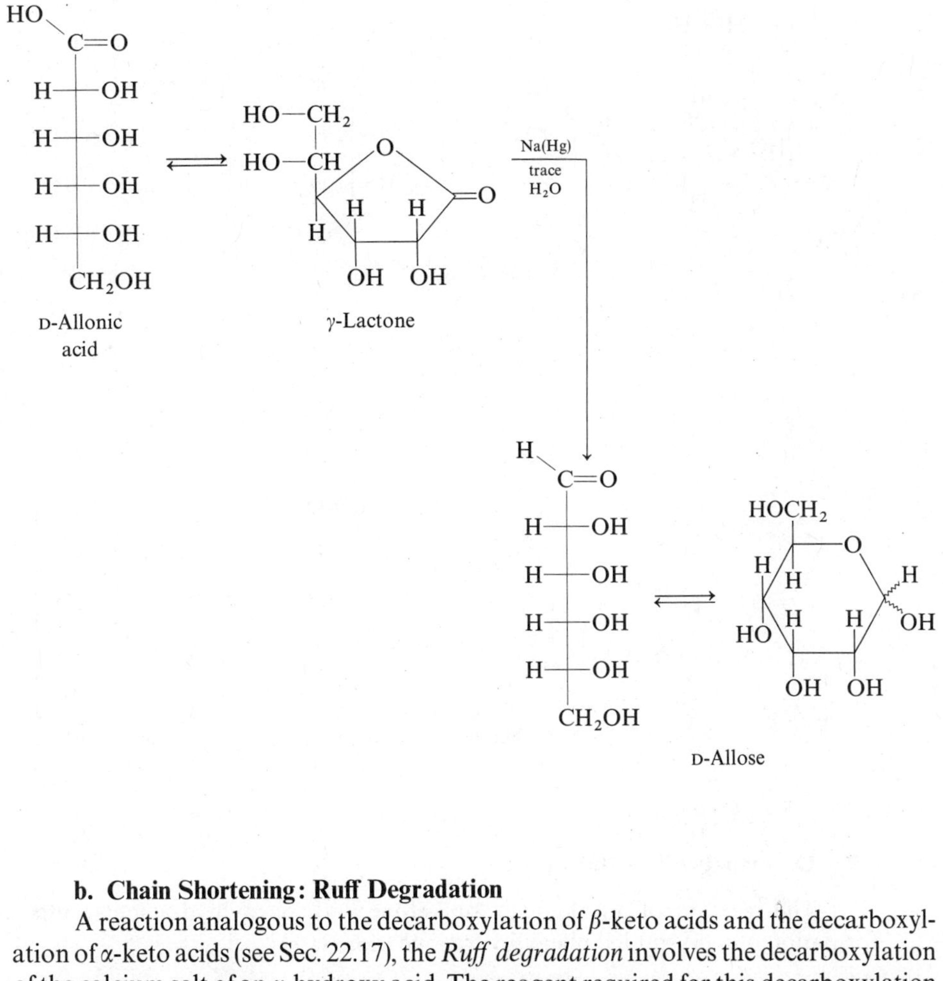

D-Allonic acid γ-Lactone

D-Allose

b. Chain Shortening: Ruff Degradation

A reaction analogous to the decarboxylation of β-keto acids and the decarboxylation of α-keto acids (see Sec. 22.17), the *Ruff degradation* involves the decarboxylation of the calcium salt of an α-hydroxy acid. The reagent required for this decarboxylation is hydrogen peroxide (H_2O_2) in which ferric ions (Fe^{+3}) are present; the hydrogen peroxide/ferric ion solution is commonly called *Fenton's reagent*. The reaction, which occurs in one step, involves decarboxylation of carbon 1 of an aldonic acid and the concurrent oxidation of carbon 2 of the chain. Loss of carbon dioxide results in the formation of an aldose with one less carbon than the aldonic acid from which it was prepared:

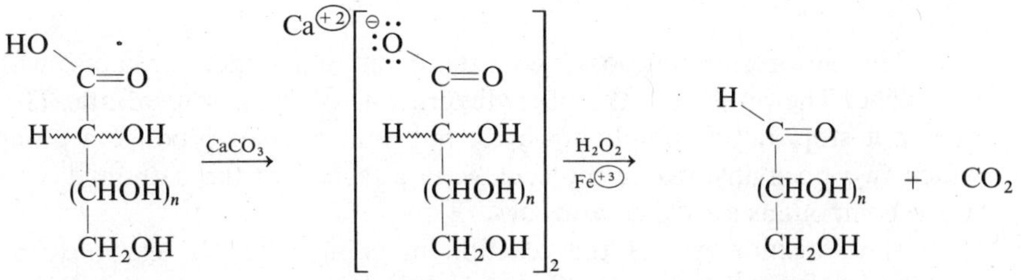

Aldonic acid   Calcium salt   Aldose
                                                (one less carbon in length)

For example:

D-Mannose   D-Mannonic acid

D-Arabinose   Calcium salt

D. Osazone Formation

The reaction of an aldehyde or ketone with phenylhydrazine results in the formation of a crystalline phenylhydrazone (Sec. 21.8):

Carbonyl   Phenylhydrazone
compound

The same reaction is possible on treatment of an aldose or ketose with phenyl-hydrazine. The product is the phenylhydrazone of the carbohydrate. The reaction does not stop at this point, however, but continues to produce a compound in which two phenylhydrazine molecules are attached to the carbohydrate skeleton. These compounds are called **osazones.**

The stoichiometry of the reaction in going from the carbohydrate to the osazone results in the consumption of 1 mole of sugar for every 3 moles of phenylhydrazine. A mechanism consistent with these findings is given in Fig. 29.17. An example is:

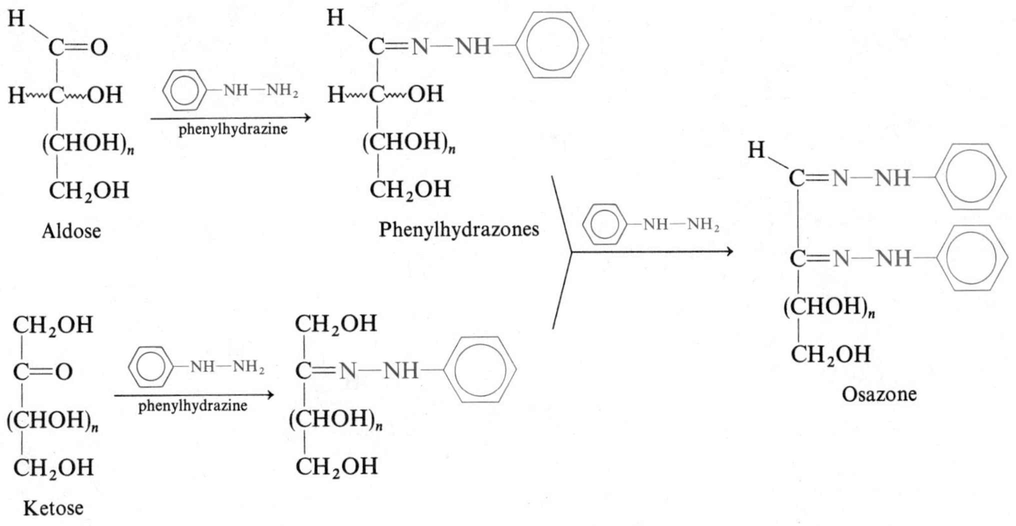

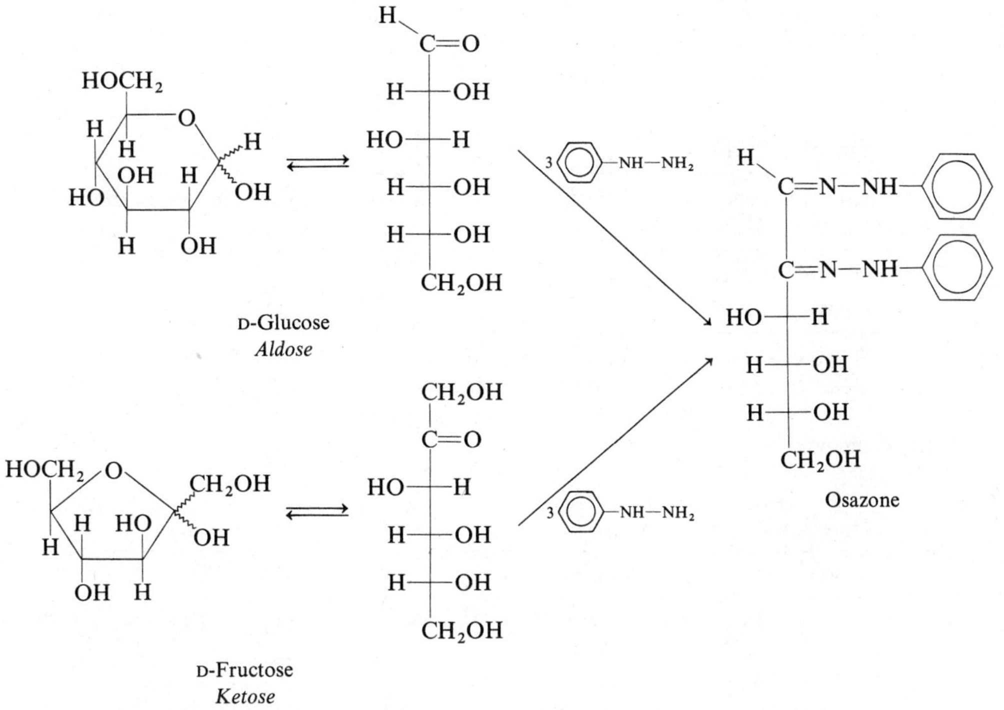

Question 29.12

Suppose (D)-glyceraldehyde is subjected to Kiliani-Fischer synthesis. Draw the structure(s) of the new compound(s) produced. Completely name each, giving family and sign of rotation.

Question 29.13

A certain aldohexose, *A*, gives the same osazone as D-(+)-allose. Write the structures of the osazone and *A*, and briefly explain your reasoning.

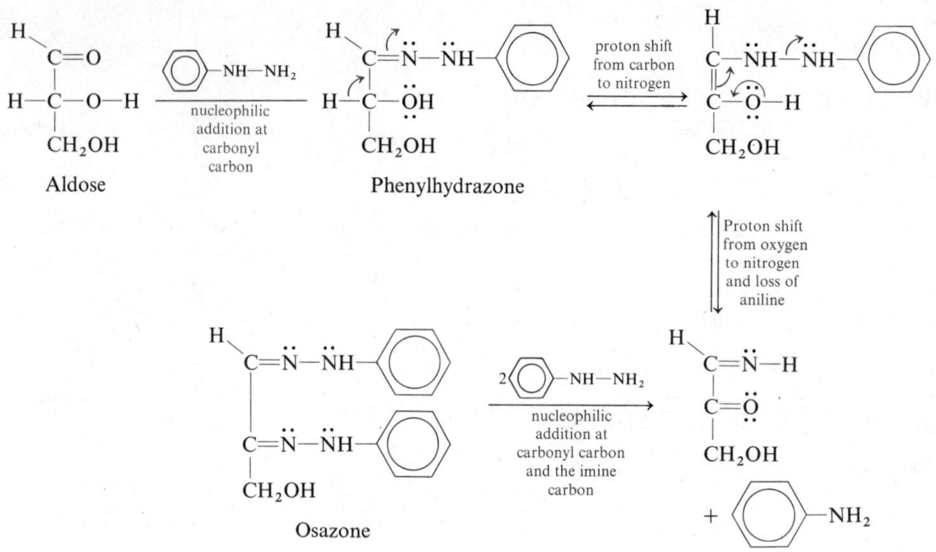

FIGURE 29.17   Partial mechanism for the formation of osazones from aldoses. The mechanism accounts for the observed stoichiometry of one sugar to three phenylhydrazines.

Question 29.14

Draw the noncyclic Fischer projection structure and name the compound formed by a single Ruff degradation of D-(+)-altrose.

Question 29.15

Suppose you have an unknown aldopentose B. On reduction with hydrogen gas (Pt or Ni catalyst), B gives an optically inactive pentaalcohol. On Ruff degradation, B gives an aldotetrose that is identified as D-threose. What is the structure of B? To what family (D or L) does this compound belong? Briefly explain.

Question 29.16

Suppose D-(−)-lyxose is subjected to Kiliani-Fischer synthesis.

(a) Using noncyclic Fischer projection formulas, draw the structure(s) of the new compound(s) produced.

(b) Suppose the new compound(s) in (a) is(are) oxidized with nitric acid. Draw the structure(s) of the diacid(s) produced and indicate whether each is optically active or inactive. Briefly explain, and use Fischer projection formulas.

29.10   Biochemical Reactions Involving Carbohydrates

A glance at Fig. 29.2 gives an idea of the role and importance of carbohydrates to biochemical pathways. Several specific examples that better illustrate their function are provided in this section. Note that enzymes play a fundamental role in

these reactions. Indeed, all the reaction types discussed in this section may be accomplished through the action of various enzymes.

A. Reduction

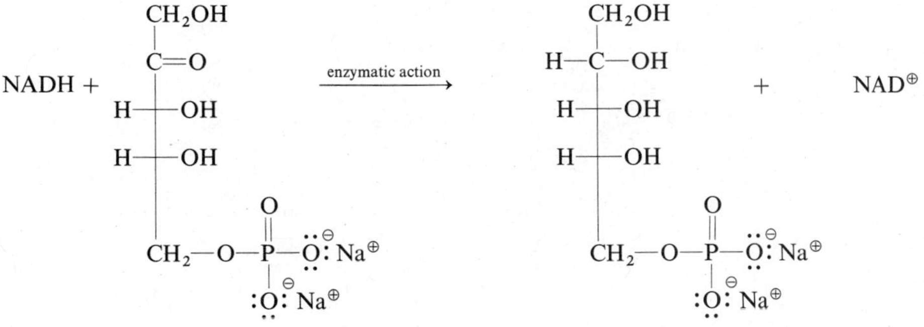

$$\text{D-Ribulose 5-phosphate} \qquad\qquad \text{D-Ribitol 5-phosphate}$$

Enzymatic action on D-ribulose 5-phosphate in conjunction with the action of
the reduced form of nicotinamide adenine dinucleotide (NADH) results in the
reduction of the carbonyl group to the corresponding alcohol. At the same time the
carbonyl is reduced, the dinucleotide is oxidized from NADH to $NAD^{\oplus}$:

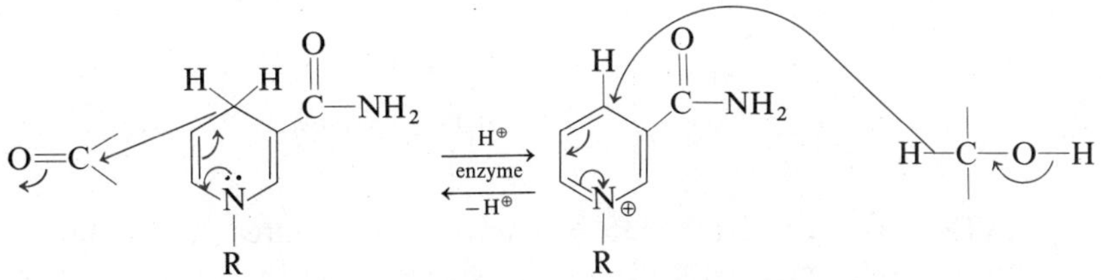

Note the high degree of stereochemical integrity in the enzymatic process—only
one of the two possible epimeric products is produced.

$NAD^{\oplus}$ is a coenzyme—a nonprotein *required* for the activity of a given enzyme.
A coenzyme may be viewed as a subclass of prosthetic groups (see Sec. 28.14); whereas
a prosthetic group is normally very tightly bound to the protein, a coenzyme need
not be. $NAD^{\oplus}$ is one of the most important coenzymes in biochemical transformations. The complete structure of $NAD^{\oplus}$ is as follows; the active site is indicated:

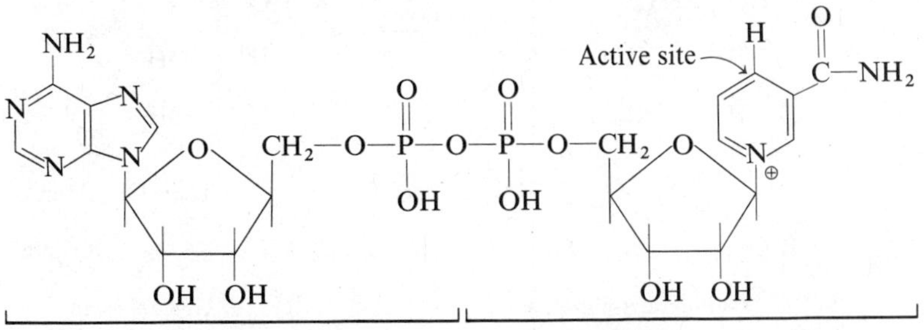

$$\underbrace{\text{Adenine nucleotide}}\qquad\qquad \underbrace{\text{Nicotinamide nucleotide}}$$

Nicotinamide adenine dinucleotide $(NAD^{\oplus})$

The preceding shorthand structure is sufficient for our discussions.

A second extremely important biochemical molecule similar in structure to NAD$^{\oplus}$ is ATP, adenosine triphosphate. ATP is the body's main energy storehouse. Hydrolysis of the phosphate links is exothermic—energy is released ($\Delta H^{\circ} = -9$ kcal/mole; 38 kJ/mole):

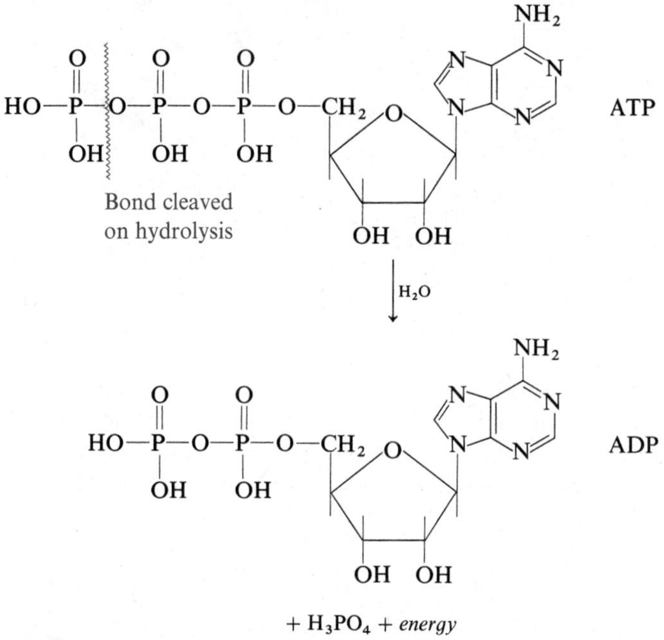

ATP is converted to ADP whenever it is required by the body. The ATP consumed is reformed in the glycolysis cycle (Fig. 29.18).

B. Oxidation

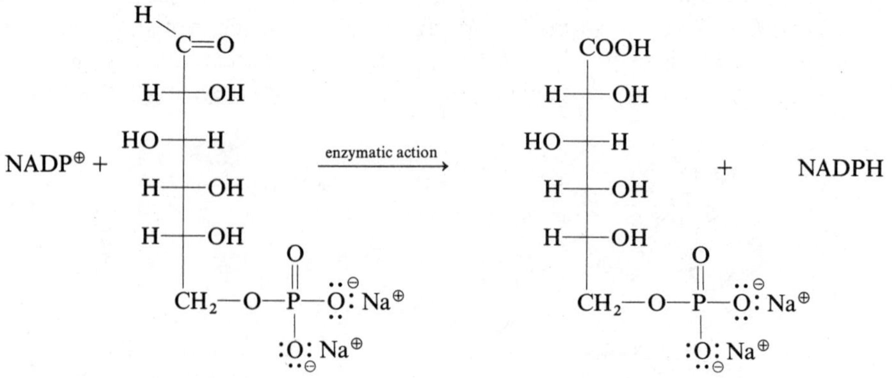

D-Glucose 6-phosphate

D-Gluconic acid 6-phosphate

In this example, as the aldose is oxidized to the aldonic acid, nicotinamide adenine dinucleotide phosphate is reduced—NADP$^{\oplus}$ → NAPDH:

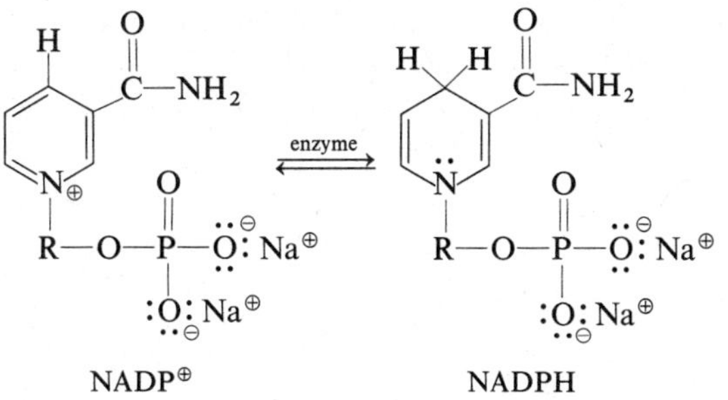

NADP⊕   NADPH

C. Glycolysis

Glycolysis is the term for the metabolic pathway by which glucose is converted to lactic acid. It is through this degradation that many organisms convert glucose into the energy required for their existence. The partial schematic of the intermediate involved in the anaerobic glycolysis process is presented in Fig. 29.18.

Question 29.17

Write an equation to show the conversion of D-glucose to D-glucose 6-phosphate, which involves ATP. Show only the starting materials and products; do not provide mechanisms.

29.11   Structure Elucidation in Carbohydrate Chemistry

A. The Configuration of Glucose

In the late 1800s, through a series of *chemical correlations*, Emil Fischer determined the configuration of D-glucose. The chemistry and logic applied to the solution of this problem were remarkable. There is not sufficient time to do justice to this marvelous example of classical organic chemistry. The enterprising student might find it fascinating to review this work. The original work, as published by Emil Fischer (and for which he was awarded the Nobel Prize), is in *Berichte*, vol. 24, pp. 1836 and 2683 (1891). A simplified proof, which is easier to follow, is in the *Journal of Chemical Education*, vol. 46, p. 55 (1969), in an article by Dr. Y. Z. Frohwein of the Hebrew University. The salient features of the Fischer proof are summarized here.

It was known that (+)-glucose was an aldohexose, and therefore one of 16 possible stereoisomers (aldohexoses contain 4 asymmetric centers—see Sec. 29.4— and exist as 2^4 possible stereoisomers). Eight of these stereoisomers are of the D configuration and eight of the L configuration. Fischer realized that there were no methods available to differentiate one enantiomer from another. He arbitrarily *assumed* that (+)-glucose was of the D configuration. His task now was to determine which of the eight D aldohexoses actually were (+)-glucose. These are drawn here using the skeletal drawings first presented in Fig. 29.10:

Possible structures of (+)-glucose:

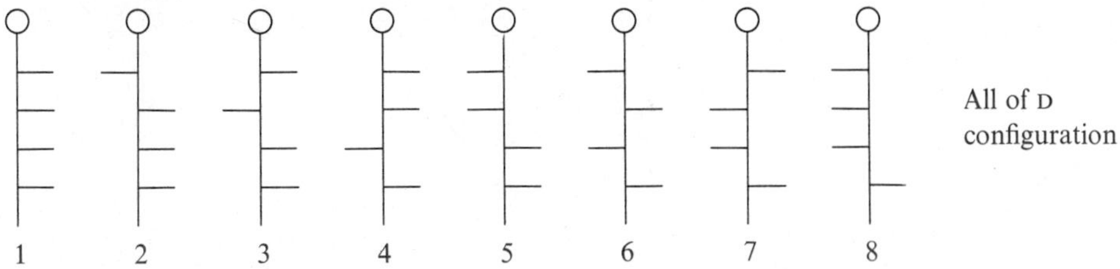

All of D
configuration

1   2   3   4   5   6   7   8

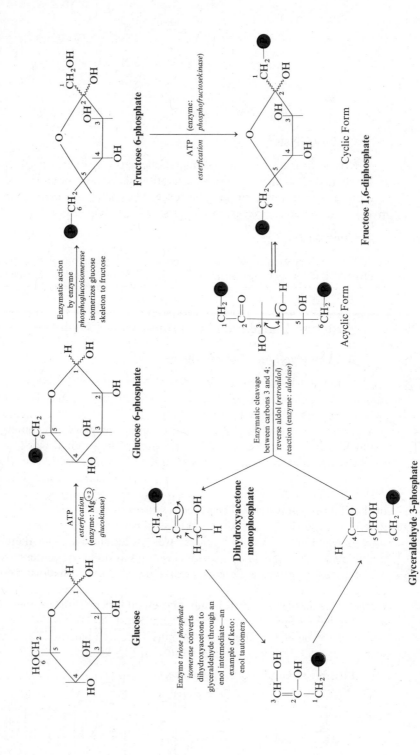

Glucose

Glucose 6-phosphate

Fructose 6-phosphate

ATP
esterfication
(enzyme: Mg^{+2}
glucokinase)

Enzymatic action
by enzyme
phosphoglucoisomerase
isomerizes glucose
skeleton to fructose

ATP
esterfication
(enzyme:
phosphofructosekinase)

Fructose 1,6-diphosphate

Cyclic Form

Acyclic Form

Enzymatic cleavage
between carbons 3 and 4;
reverse aldol (*retroaldol*)
reaction (enzyme: *aldolase*)

**Dihydroxyacetone
monophosphate**

Enzyme *triose phosphate
isomerase* converts
dihydroxyacetone to
glyceraldehyde through an
enol intermediate—an
example of keto:
enol tautomers

Glyceraldehyde 3-phosphate

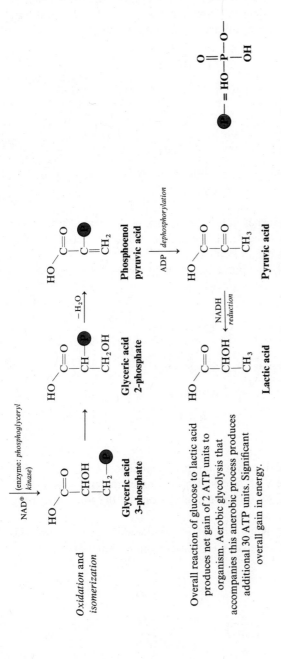

FIGURE 29.18 Partial schematic of the intermediates involved in glycolysis under anaerobic conditions. Many reactions listed are reversible but are shown as proceeding only in the forward direction for simplification. For more information about glycolysis, see Reading References for this chapter in the Appendix.

1301

From this basic assumption, Fischer proceeded as follows:

1. The reaction of (+)-glucose with concentrated nitric acid gives an optically active aldaric acid (Sec. 29.9A,d). This eliminates 1 and 7 as possibilities for the aldaric acids that would result from their oxidation are meso compounds.

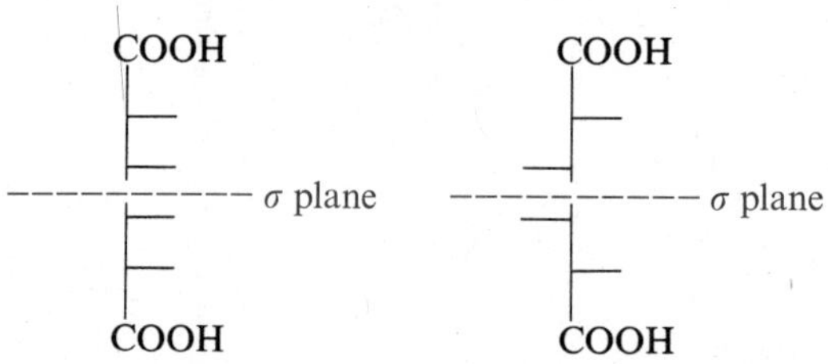

Aldaric acid from 1   Aldaric acid from 7

Meso compounds

2. The reaction of (+)-mannose, another aldohexose, with concentrated nitric acid also yields an optically active aldaric acid. (+)-Mannose cannot be 1 or 7 for the same reasons presented for (+)-glucose in item (1).

3. The reaction of (+)-glucose or (+)-mannose with phenylhydrazine gives the same osazone (Sec. 29.9D).

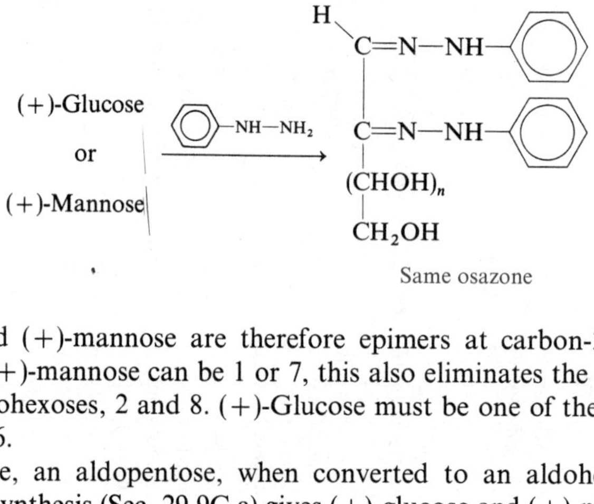

Same osazone

(+)-Glucose and (+)-mannose are therefore epimers at carbon-2. Since neither (+)-glucose or (+)-mannose can be 1 or 7, this also eliminates the epimers (at C-2) of these two aldohexoses, 2 and 8. (+)-Glucose must be one of the four remaining structures: 3 to 6.

4. (−)-Arabinose, an aldopentose, when converted to an aldohexose using the Kiliani-Fischer synthesis (Sec. 29.9C,a) gives (+)-glucose and (+)-mannose.

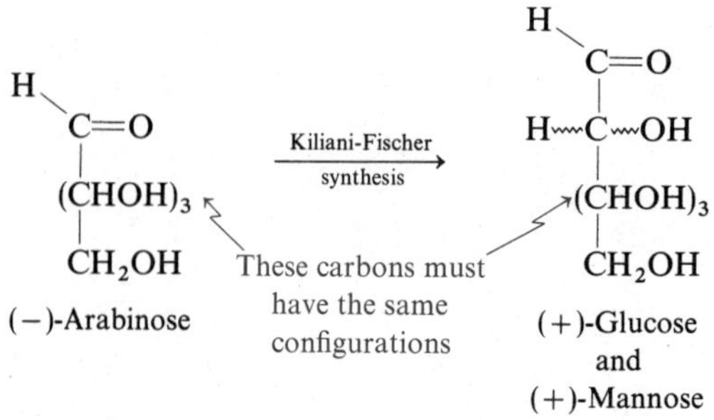

The configuration of carbons 3 to 5 in (+)-glucose must therefore be the same as carbons 2 to 4 in (−)-arabinose. From structures 3 to 6 we see that there are only two possibilities for the configuration of (−)-arabinose, 9 or 10:

Possible configurations of (−)-arabinose at carbons 2, 3, and 4:

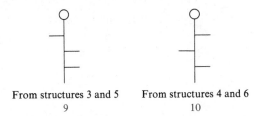

From structures 3 and 5   From structures 4 and 6
9   10

5. (−)-Arabinose gives an optically active aldaric acid when oxidized with nitric acid. This eliminates structure 10, which would give a meso compound on oxidation. (−)-Arabinose is structure 9.

6. If (−)-arabinose is structure 9, then (+)-glucose must be either structure 3 or 5. (+)-Mannose is the other epimer.

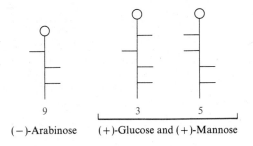

9   3   5

(−)-Arabinose   (+)-Glucose and (+)-Mannose

7. (+)-Gulose and (+)-glucose give the same aldaric acid when oxidized with nitric acid.[1]

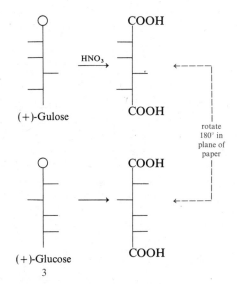

(+)-Glucose is therefore structure 3 and (+)-Mannose is structure 5.

Because Fischer had to make an assumption originally that (+)-glucose is of the D configuration, his final structure could still have been wrong. That is, the same arguments applied here can be equally well applied if one assumes that (+)-glucose is of the L configuration. If this were true, then all the assignments made would be incorrect; the correct assignments would be the mirror images of each of

---

[1] (+)-Gulose was synthesized by Fischer for this investigation.

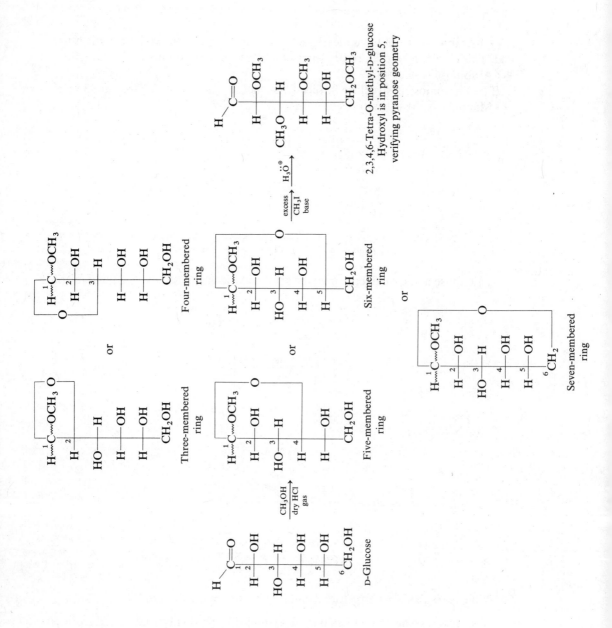

2,3,4,6-Tetra-O-methyl-D-glucose
Hydroxyl is in position 5,
verifying pyranose geometry

Four-membered ring

Three-membered ring

Six-membered ring

Five-membered ring

Seven-membered ring

or

or

or

D-Glucose

the compounds drawn. X-ray crystallographic analysis has shown Fischer's original assignment to be correct: ($+$)-Glucose is of the D configuration.

B. The Pyranose Geometry of Glucose

All the reactions discussed in the past few sections may be used to elucidate various aspects of the structures of carbohydrates.

Although we have drawn the cyclic hemiacetal form of glucose as a pyranose (six-membered ring) structure, no direct proof of this geometry has been given. Using two reactions discussed in Sec. 29.9, it is now possible to prove that the pyranose geometry is indeed correct. The reaction of methyl D-glucoside in base with either methyl iodide (CH_3I) or dimethyl sulfate ($CH_3OSO_2OCH_3$) results in the pentamethyl analogue of the glucoside. The carbons bearing the methoxy groups are all ether carbons except for the anomeric carbon, which is an acetal carbon. Hydrolysis of the resultant derivative with aqueous acid results in cleavage of the acetal and formation of the tetramethyl analogue of D-glucose. Ring size may now be determined by finding the position of the hydroxyl group on the carbon skeleton (which may be accomplished in several ways) and relating this to the original starting material. This process is illustrated on page 1304.

Additional evidence for the pyranose geometry comes from the cleavage of methyl D-glucoside with periodic acid. The dialdehyde shown here could only result from the pyranose form of D-glucose:

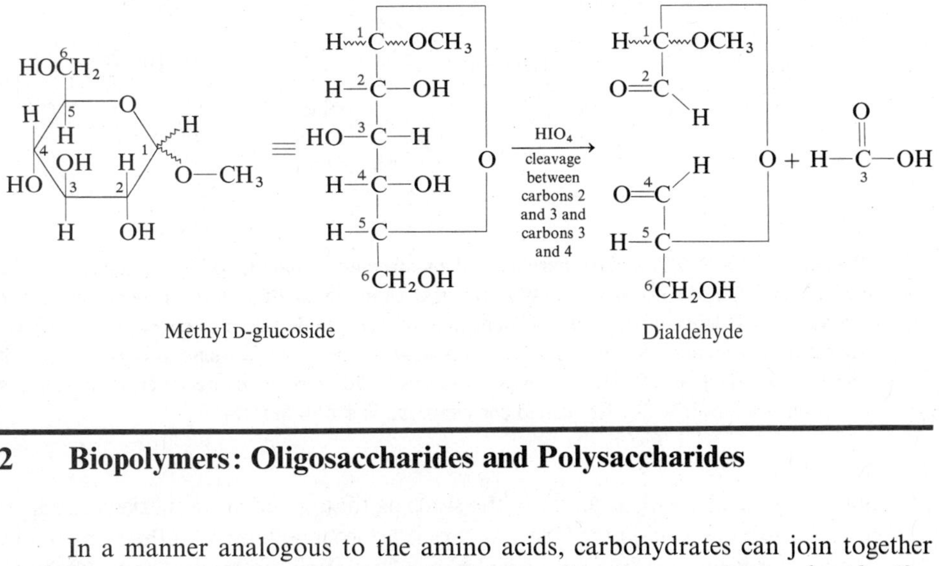

Methyl D-glucoside   Dialdehyde

29.12   Biopolymers: Oligosaccharides and Polysaccharides

In a manner analogous to the amino acids, carbohydrates can join together to form polymeric units. These polymers are referred to as *oligosaccharides* (for polymeric units consisting of two to approximately eight monomeric saccharide units) or *polysaccharides* (for polymeric units consisting of more than eight monomeric saccharide units). Small oligosaccharides are called *disaccharides* (two monomeric units), *trisaccharides* (three monomeric units), and so on.

A. Disaccharides

Disaccharides consist of two saccharide units joined by either an acetal (or ketal) or hemiacetal (or hemiketal) linkage. The chemistry of the oligosaccharides

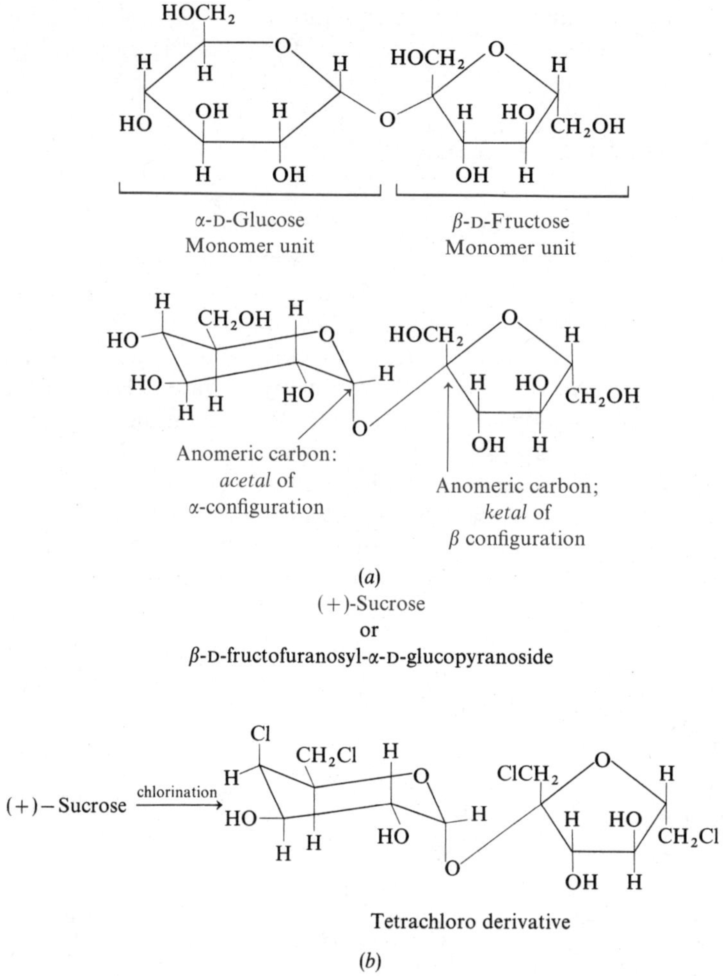

α-D-Glucose
Monomer unit

β-D-Fructose
Monomer unit

Anomeric carbon:
acetal of
α-configuration

Anomeric carbon;
ketal of
β configuration

(a)

(+)-Sucrose
or
β-D-fructofuranosyl-α-D-glucopyranoside

(+)-Sucrose $\xrightarrow{\text{chlorination}}$

Tetrachloro derivative

(b)

FIGURE 29.19   (*a*) **This disaccharide has the molecular formula $C_{12}H_{22}O_{11}$ and is a nonreducing sugar. It is obtained naturally, mainly from sugar cane (about 20% sucrose) and sugar beets (about 15% sucrose). It is primarily used as a sweetener and as the main starting material in the fermentation process that produces ethanol. It recently was reported that the chlorination of sucrose produces a tetrachloro derivative 500 times as sweet as sucrose. This derivative sweetener, still being tested to determine whether it is safe for human consumption, is drawn in (*b*).**

and polysaccharides is essentially the same as that of the monomeric carbohydrates. Structure elucidation in the larger sugars is approached in much the same manner as in the simple sugars, but the complexity of these polymers makes the overall task much more formidable. Several disaccharides are illustrated in Figs. 29.19 to 29.21.

B. Structure Elucidation

With (+)-lactose as an example, we now present a typical reaction sequence used in elucidating the structure of carbohydrates (disaccharides in this case). The method is identical with that used earlier in determining the ring size of glucose. The reactions are illustrated in Fig. 29.22.

Treatment of (+)-lactose with Tollens' reagent gives a positive silver mirror test, indicating the presence of a hemiacetal or hemiketal group. Oxidation of

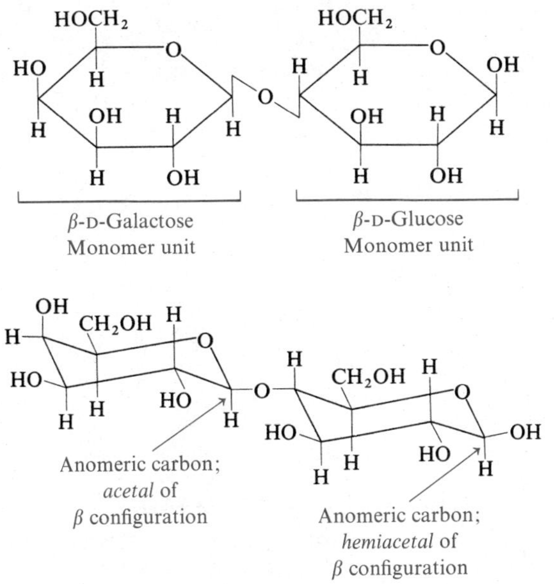

β-D-Galactose
Monomer unit

β-D-Glucose
Monomer unit

Anomeric carbon;
acetal of
β configuration

Anomeric carbon;
hemiacetal of
β configuration

(*a*)

(+)-Lactose
or
4-O-(β-D-galactopyranosyl)-β-D-glucopyranose

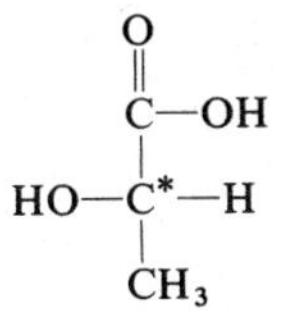

Lactic acid
(*b*)

FIGURE 29.20 (*a*) **This disaccharide has the molecular formula $C_{12}H_{22}O_{11}$ and is a reducing sugar. It is found naturally in the milk of humans and other mammals (comprising 5 to 10% of the milk) and is commonly called milk sugar. Lactose is used in the food industry in the production of many modified milk products. Bacterial action is often responsible for the spoilage of milk and milk products. When the lactose is acted upon by certain bacteria, one by-product is lactic acid, drawn in (*b*), which causes the bitter taste and odor of spoiled milk.**

lactose with bromine and water yields the corresponding carboxylic acid. Characterization of this compound shows the position of the hemiacetal functionality.

The reaction of lactose with methanol and acid (CH_3OH, $H^{\oplus}$) results in a methyl glycoside, which upon exhaustive methylation yields an octamethyl intermediate, which upon hydrolysis yields 2,3,4,6-tetra-O-methyl-D-galactose and 2,3,6-tri-O-methyl-D-glucose. Identification of these fragments unequivocally proves that the two sugar monomers are joined at the anomeric carbon of one and the carbon 4 of the other and that the glucose ring (as opposed to the galactose ring) contains the reducible hemiacetal group.

Performing the same reaction sequence on the carboxylic acid intermediate obtained by the action of bromine water provides two fragments, 2,3,5,6-tetra-O-methyl-D-gluconic acid and 2,3,4,6-tetra-O-methyl-D-galactose. The structure of the

HOCH₂ structures:

α-D-Glucose
Monomeric
unit

β-D-Glucose
Monomeric
unit

Anomeric carbon; *acetal*
of α configuration

Anomeric carbon;
hemiacetal of
β configuration

(+)-Maltose
or
4-O-(α-D-glucopyranosyl)-β-D-glucopyranose

FIGURE 29.21 **(+)-Maltose or 4-O-(α-D-glucopyranosyl)-β-D-glucopyranose. This has the molecular formula C₁₂H₂₂O₁₁ and is a reducing sugar. It is obtained principally from starch by enzyme degradation. It is used as a commercial food supplement and as an alternative sweetener and also in the fermentation process that produces beer.**

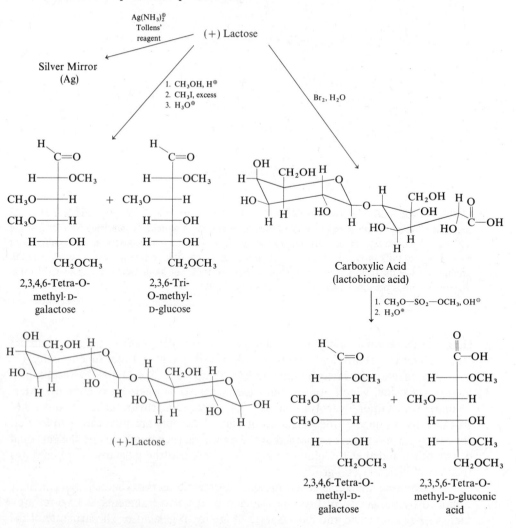

Ag(NH₃)₂⊕
Tollens'
reagent

(+) Lactose

Silver Mirror
(Ag)

1. CH₃OH, H⊕
2. CH₃I, excess
3. H₃O⊕

Br₂, H₂O

2,3,4,6-Tetra-O-
methyl-D-
galactose

2,3,6-Tri-
O-methyl-
D-glucose

Carboxylic Acid
(lactobionic acid)

1. CH₃O—SO₂—OCH₃, OH⊖
2. H₃O⊕

(+)-Lactose

2,3,4,6-Tetra-O-
methyl-D-
galactose

2,3,5,6-Tetra-O-
methyl-D-gluconic
acid

FIGURE 29.22 **Structure elucidation of (+)-lactose.**

1308

isolated gluconic acid shows that the position of attachment of the hemiacetal ring form to the second ring is carbon 4 of the hemiacetal ring. The glucose fragment indicates a pyranoside attached to the second ring through the oxygen of its anomeric carbon in the form of an acetal link.

The structure provided in Fig. 29.20 is consistent with all these results.

C. Oligosaccharides and Polysaccharides

Several examples of polymers composed solely of D-glucose units are in Figs. 29.23 and 29.24.

Question 29.18

Draw the structure of a nonreducing disaccharide derived from β-D-fructofuranose. (It contains only β linkages.)

Question 29.19

Consider the compound (−)-I shown here and then answer the questions that follow.

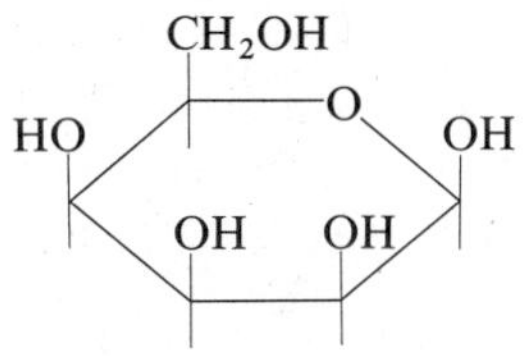

(−)-I

(a) Draw the Fischer structure for this compound and provide the *complete* name for it.
(b) Is (−)-I a reducing sugar or a nonreducing sugar?
(c) Suppose (−)-I is joined with itself in such a way as to give a nonreducing disaccharide. Using Haworth structures, draw the structure of the resulting compound. Assume the configurations at *all* carbon atoms in (−)-I remain unchanged in forming the disaccharide.

TABLE 29.1 Naturally Occurring Polysaccharides

Polysaccharide	Source	Monomeric Sugar Units
Amylopectin	Starch	D-Glucose
Amylose	Starch	D-Glucose
Cellulose	Plants (structural fiber)	D-Glucose
Chitin	Animals (structural fiber)	2-Acetamido-D-glucose*
Floridean starch	Red algae	D-Glucose
Galactan	Beef lung tissue	D-Galactose
Glycogen	Animals (muscles and liver)	D-Glucose
Gum tragacanth	Plant resin	L-Arabinose and D-galactose
Hyaluronic acid	Animals (connective tissue)	D-Glucose and 2-acetamido-D-glucose
Inulin	Artichokes	D-Fructose
Laminaran	Seaweed	D-Glucose
Mannan	Ivory nuts	D-Mannose
Yeast mannan	Yeast	D-Mannose
Xylan	Plants	D-Xylose

$$* \; \text{Acetamido} = CH_3\overset{\displaystyle O}{\overset{\|}{C}}NH—$$

Cellulose

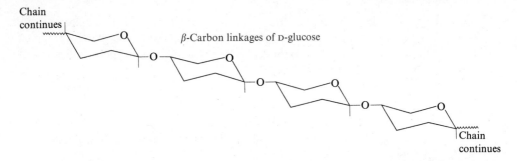

β-Carbon linkages of D-glucose

Chain continues

Chain continues

FIGURE 29.23   Cellulose. This polysaccharide is composed solely of D-glucose units that have been linked in a 1,4 manner (carbon 1 on one ring to carbon 4 of the next ring). All the glucose units are of the β configuration. Cellulose's molecular weight ranges from 100,000 to 1,000,000. Cotton is probably the richest source of cellulose, containing more than 90% cellulose by weight. Most other plants also contain a relatively high percentage of cellulose, generally ranging from 10 to 15%. Cellulose is a very important industrial chemical; for example, it is used in the manufacture of paper goods, paints, explosives, and rayon. Cellulose is one of the most abundant compounds on earth.

Amylose

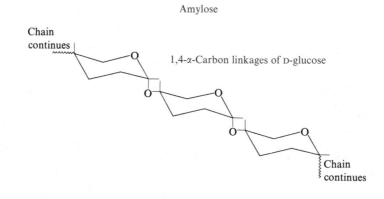

1,4-α-Carbon linkages of D-glucose

Chain continues

Chain continues

Amylopectin

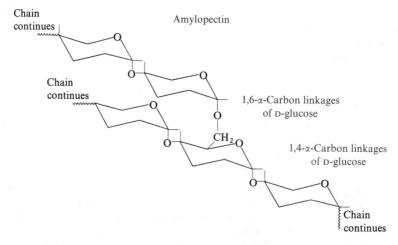

1,6-α-Carbon linkages of D-glucose

1,4-α-Carbon linkages of D-glucose

Chain continues

Chain continues

Chain continues

FIGURE 29.24   Starch. This is a mixture of two polysaccharides, amylose and amylopectin, which are present in roughly a 1:3 ratio in plants. Both these polysaccharides contain only one type of monomer: D-glucose. In both polysaccharides the D-glucose units are joined in an α manner. However, amylopectin differs from amylose in that the former is a highly branched sugar, whereas the latter consists of a linear group of monomeric glucose units. Starch provides a ready storehouse of glucose to plants whenever needed and is also used in the manufacture of foodstuffs, paste, and glues and in the clothes cleaning industry.

Table 29.1 lists several other naturally occurring polymeric carbohydrates according to source and composition.

Study Questions

29.20 Using examples where possible, define each of the following 20 terms.

(*a*) monosaccharide	(*b*) oligosaccharide	(*c*) polysaccharide	(*d*) aldose
(*e*) ketose	(*f*) tetrose	(*g*) ketopentose	(*h*) pyranose
(*i*) furanose	(*j*) Haworth structures	(*k*) Fischer structures	(*l*) anomers
(*m*) epimers	(*n*) glycoside	(*o*) glucoside	(*p*) aldonic acid
(*q*) aldaric acid	(*r*) alditol	(*s*) mutarotation	(*t*) sugar

29.21 Give the structures of the compounds produced in each of the following reactions. When there is more than one product, be sure to indicate all possible products.

(*a*)

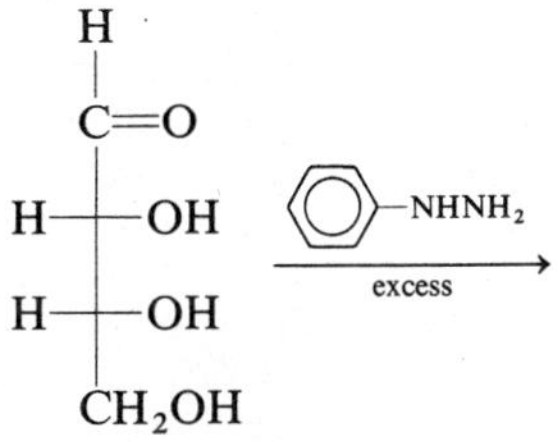

$$
\begin{array}{c}
\text{H} \\
|\\
\text{C}=\text{O} \\
\text{H}-\!\!\!-\text{OH} \\
\text{H}-\!\!\!-\text{OH} \\
\text{CH}_2\text{OH}
\end{array}
\xrightarrow[\text{excess}]{\langle\!\bigcirc\!\rangle-\text{NHNH}_2}
$$

(*b*)

$$
\begin{array}{c}
\text{H} \\
|\\
\text{C}=\text{O} \\
\text{HO}-\!\!\!-\text{H} \\
\text{H}-\!\!\!-\text{OH} \\
\text{CH}_2\text{OH}
\end{array}
\xrightarrow[\text{synthesis}]{\text{Kiliani-Fischer}}
$$

(*c*)

$$
\begin{array}{c}
\text{CH}_2\text{OH} \\
\text{HO}-\!\!\!-\text{H} \\
\text{C}=\text{O} \\
|\\
\text{H}
\end{array}
\xrightarrow{\text{Br}_2/\text{H}_2\text{O}}
$$

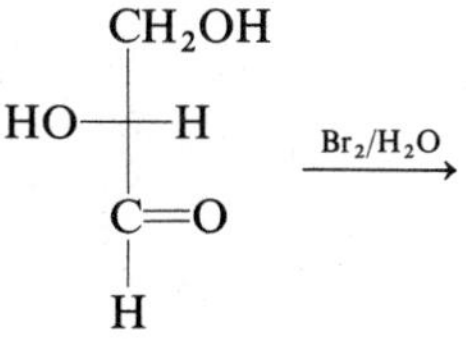

(*d*)

$$
\begin{array}{c}
\text{H}-\text{C}-\text{OCH}_3 \\
\text{H}-\!\!\!-\text{OH} \\
\text{H}-\!\!\!-\text{OH} \\
\text{HO}-\!\!\!-\text{H} \\
\text{H}-\!\!\!- \\
\text{H}-\!\!\!-\text{OH} \\
\text{CH}_2\text{OH}
\end{array}
\xrightarrow{\text{HIO}_4}
$$

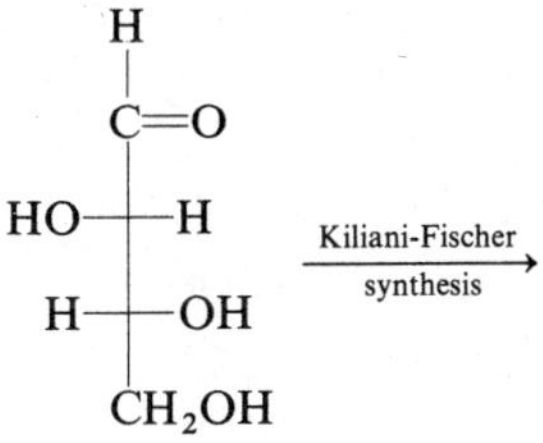

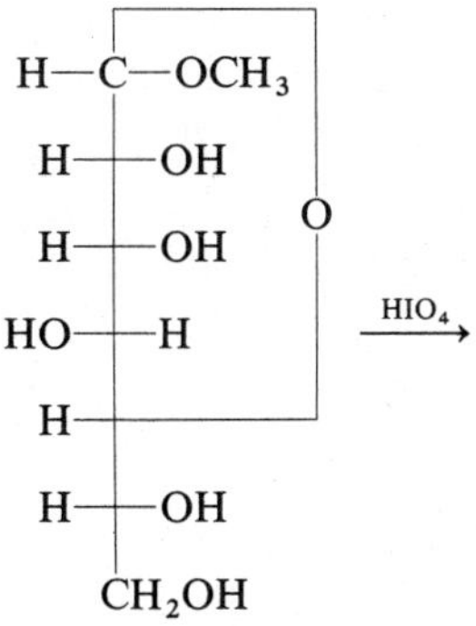

(*e*) L-(−)-Glyceraldehyde $\xrightarrow{\text{HCN}}$

(*f*) D-(+)-Glucose $\xrightarrow[\text{H}_2\text{O}]{\text{Br}_2}$ $\xrightarrow[(-\text{H}_2\text{O})]{\text{heat}}$ Lactone formation

(*g*) D-(+)-Allose $\xrightarrow[\text{Ni}]{\text{H}_2}$ $\xrightarrow[\text{oxidation}]{\text{mild}}$ monoacid

(*h*) D-(+)-Galactose $\xrightarrow[\text{Ni}]{\text{H}_2}$ $\xrightarrow[\text{oxidation}]{\text{mild}}$ monocarboxylic acid (Fischer structure)

(*i*) D-(−)-Arabinose $\xrightarrow[\text{excess}]{\text{C}_6\text{H}_5-\text{NH}-\text{NH}_2}$

(*j*)

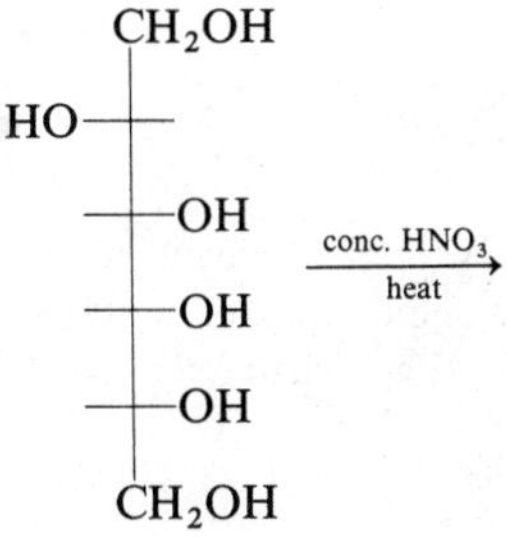

$$
\begin{array}{c}
\text{CH}_2\text{OH} \\
\text{HO}-\!\!\!- \\
-\!\!\!-\text{OH} \\
-\!\!\!-\text{OH} \\
-\!\!\!-\text{OH} \\
\text{CH}_2\text{OH}
\end{array}
\xrightarrow[\text{heat}]{\text{conc. HNO}_3}
$$

(k) β-D-(−)-Fructofuranose $\xrightarrow[\text{HCl gas}]{\text{CH}_3\text{OH}}$

(l)

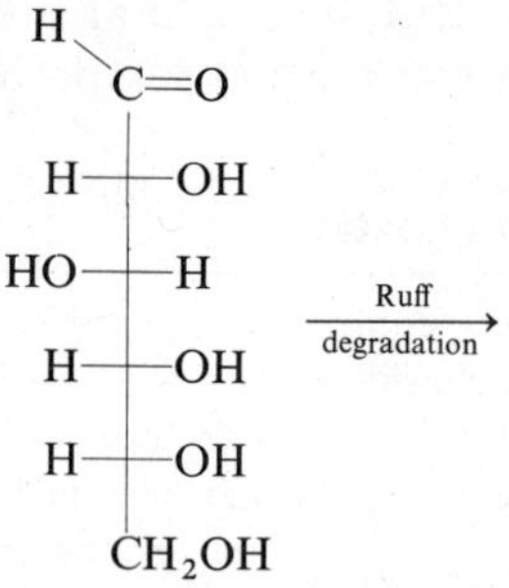

$$
\begin{array}{c}
\text{H} \\
\quad \diagdown \\
\quad\quad \text{C}=\text{O} \\
\text{H} \!-\!\!-\! \text{OH} \\
\text{HO} \!-\!\!-\! \text{H} \\
\text{H} \!-\!\!-\! \text{OH} \\
\text{H} \!-\!\!-\! \text{OH} \\
\text{CH}_2\text{OH}
\end{array}
\xrightarrow[\text{degradation}]{\text{Ruff}}
$$

(m) D-(+)-Glucose $\xrightarrow[\substack{\text{citrate,} \\ \overset{\cdot\cdot}{\text{HO}}:^{\ominus}}]{\text{Cu}^{+2}}$

(n) D-(+)-Mannose $\xrightarrow{\text{Ag(NH}_3)_2^{\oplus}}$

(o) D-(+)-Glucose $\xrightarrow{\text{NaBH}_4}$

29.22 Indicate whether each of the following would be classified as a reducing or nonreducing sugar.

(a)

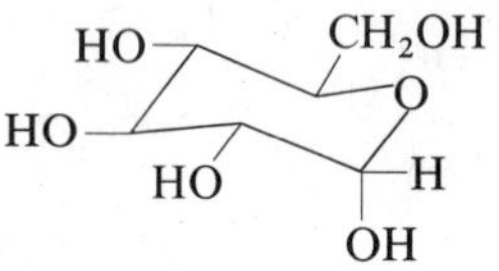

(b)

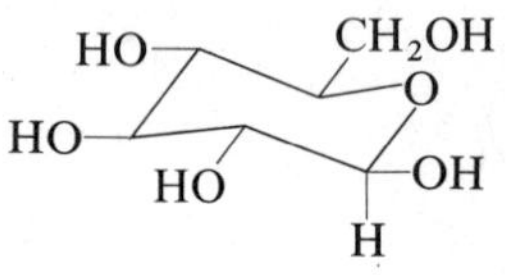

(c)

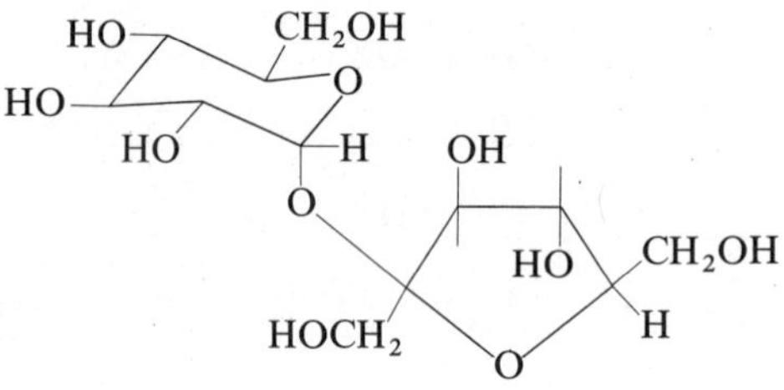

(d)

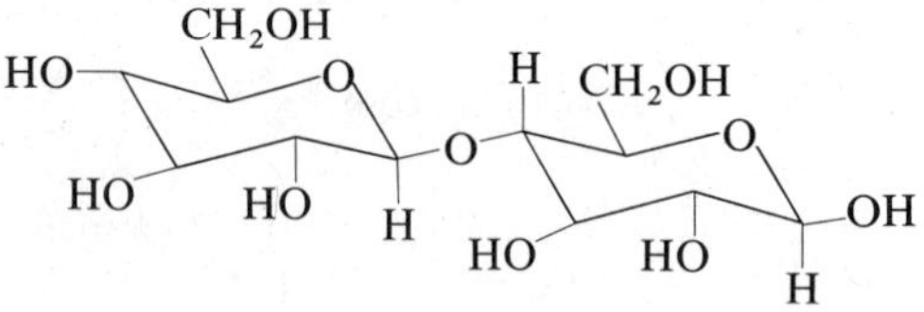

(e)

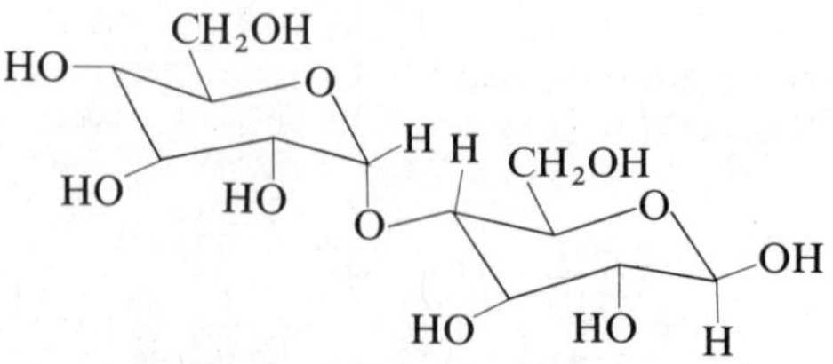

29.23 Identify the stereochemical relationship between the two members in each of the following four pairs of compounds by writing beside each pair *as many* of the following descriptive terms as may apply. (Each term may be used more than once, and more than one term may be needed for each pair.)

Epimers
Enantiomers
Diastereomers
Structural isomers
Geometric isomers
Identical compounds

Under *each* pair of structures, indicate whether they are optically active or optically inactive.

(a)

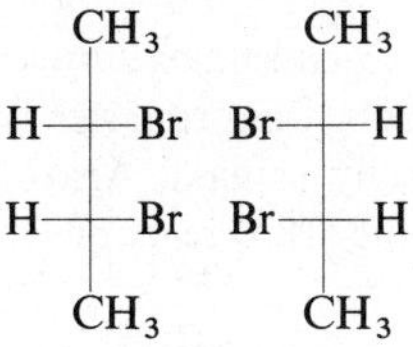

```
    CH₃        CH₃
H──Br   Br──H
H──Br   Br──H
    CH₃        CH₃
```

(b)

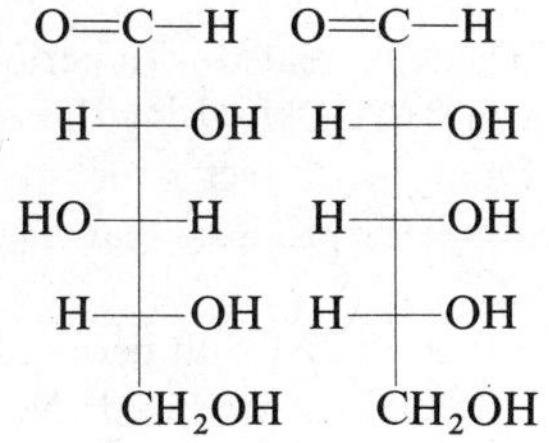

```
O=C─H   O=C─H
H──OH   H──OH
HO──H   H──OH
H──OH   H──OH
  CH₂OH    CH₂OH
```

(c)

```
O=C─OH      CH₃
H──Cl    H──Cl
H──Cl    H──Cl
   CH₃     O=C─OH
```

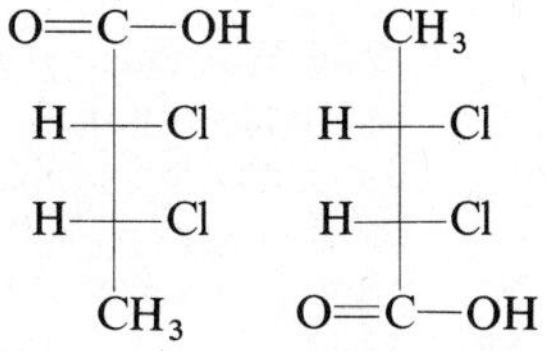

(d)

```
    Cl              Cl
     C               C
H⋮ ⟩CH₃     H⋮ ⟩CH₂CH₃
  CH₂CH₃         CH₃
```

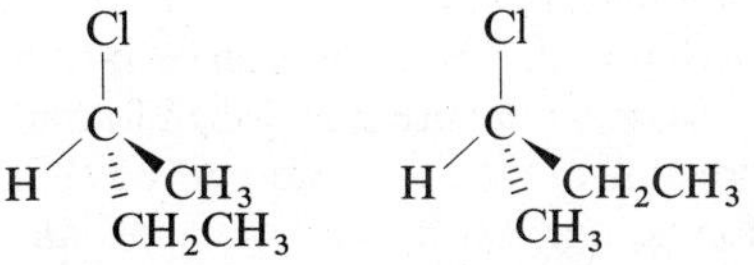

29.24 Complete the following table by filling in the blanks. If a question does not apply to a particular structure *as it is drawn*, write Not Applicable as the answer.

Structure 1:
```
CH₂OH
C=O
H──OH
H──OH
HO──H
CH₂OH
```

Structure 2:
```
HO──H
H──OH
HO──H   O
HO──H
H──
CH₂OH
```

Structure 3:
```
H──OCH₃
HO──H
H──OH   O
H──OH
H──
CH₂OH
```

	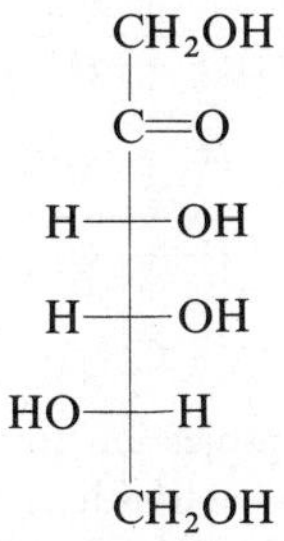 Structure 1	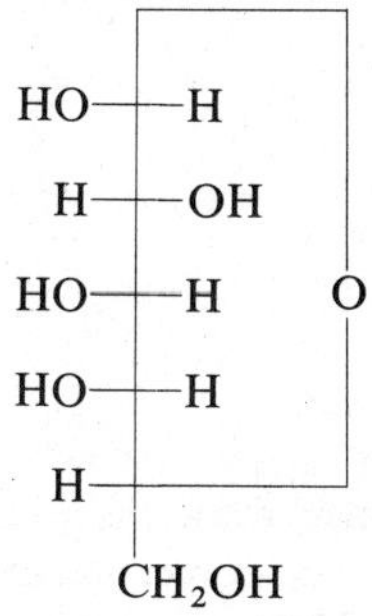 Structure 2	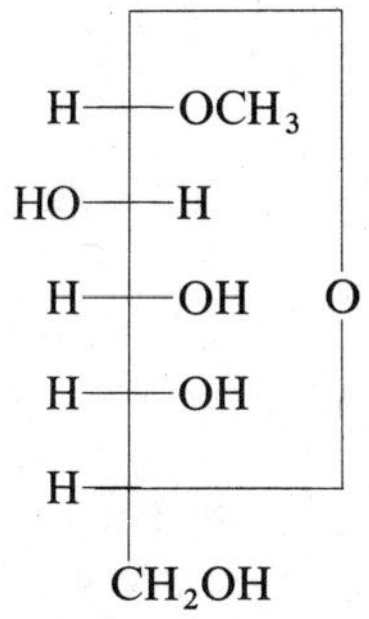 Structure 3
Is a reducing sugar? (yes or no)			
Undergoes mutarotation? (yes or no)			
Is a D sugar or L sugar? (state which)			
Is an α anomer or a β anomer? (state which)			

29.25 Propose a reasonable mechanism (show all steps and intermediates) to show how methyl α-D-(+)-glucopyranoside, *A*, is hydrolyzed by acidic water to D-glucopyranose, *B*. Based on your mechanism, what stereochemistry would you expect to find at the anomeric carbon atom (C-1) at the very *instant* hydrolysis occurs (that is, before the mutarotation occurs)? (*Note:* In answering this, you may use the abbreviated structure in *B*.)

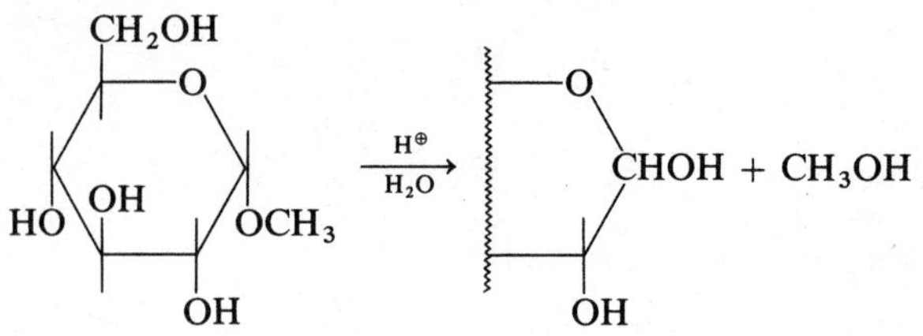

$$A \qquad\qquad B$$

29.26 When methyl α-L-arabinoside is treated with periodic acid, it consumes 2 moles of HIO_4 to produce 1 mole of formic acid, HCOOH. Using this information, what ring size is indicated for methyl α-L-arabinoside? Draw its correct structure and briefly explain. Also name it as completely as possible, and show all the products that result from the HIO_4 reaction. Use Fischer projection formulas throughout.

29.27 When methyl β-D-riboside is treated with periodic acid, 2 moles of HIO_4 are consumed and 1 mole of formic acid, HCOOH, is produced. No formaldehyde is observed. Based on this information, is methyl β-D-riboside in the furanose or the pyranose form? Briefly give your reasoning, and write the equation for the reaction that occurs.

29.28 Suppose a nonreducing disaccharide, $C_{12}H_{22}O_{11}$, gives on hydrolysis D-(+)-allose and D-(+)-talose. Draw a possible structure for the disaccharide, using Haworth formulas.

29.29 The Fischer projection formula for the structure of D-(+)-mannose is shown here. D-(+)-Mannose has been isolated in two isomeric forms, one having $[\alpha]_D^{25} = +29°$ and the other $[\alpha]_D^{25} = -17°$. Use this information in answering the following questions. (*Hint:* The α form of virtually all D sugars and glycosides exhibits the higher positive specfic rotation.)

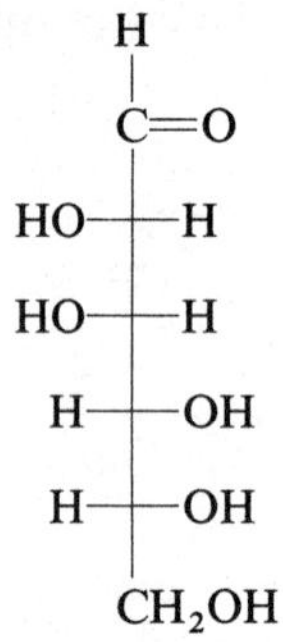

(*a*) For *each* isomer, draw the Fischer projection formula and the Haworth structure. Give correct names for each isomer and identify each with its correct rotation. (Consider only pyranose forms.)

(*b*) For *one* of the two isomeric forms of D-(+)-mannose, draw the two possible chair conformations. Indicate the conformation you would expect to be more stable and indicate the basis for your choice.

(*c*) Suppose the isomer having $[\alpha]_D^{25} = +29°$ is dissolved in water. What change in rotation, if any, would be observed in this solution with time? (Answer only in a qualitative sense.) Briefly explain why changes in rotation, if any, occur.

29.30 Draw Haworth structures for the following typical disaccharides, each of which has the molecular formula $C_{12}H_{22}O_{11}$. On hydrolysis, each disaccharide yields equimolar amounts of D-glucose and L-glucose. (*Note:* In answering this question, there *may* be several possible correct structures; your job is to show one correct structure that is consistent with the facts given.)

(*a*) Sugar *A* does not reduce Fehling's solution and *A* is *optically inactive*.

(*b*) Sugar *B* does *not* reduce Fehling's solution, and *B* is *optically active*.

(*c*) Sugar *C* reduces Fehling's solution.

(*d*) Which, if any, of the hydrolysis mixtures from sugars *A* to *C* are *optically active*? Briefly explain.

29.31 Suppose sugars (1) and (2) are reduced with hydrogen gas in the presence of a catalyst (for example, Ni or Pt). From both (1) and (2), compounds with the *general* formula (3) are formed. (Note that no stereochemistry is implied in (3), as shown.)

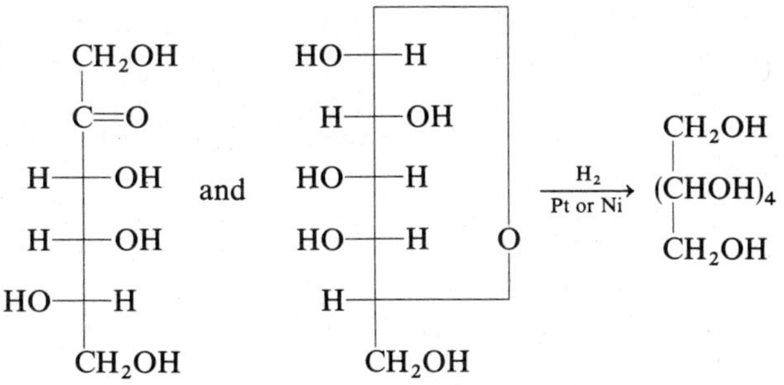

The reaction scheme at top:

```
   CH₂OH              HO——H
    |
   C=O                H——OH                    CH₂OH
                                                |
 H——OH    and        HO——H          H₂        (CHOH)₄
                                    ——————→      |
 H——OH               HO——H         Pt or Ni    CH₂OH
                                        O
HO——H                HO——H
    |
   CH₂OH             H——
                      |
                     CH₂OH

   (1)                 (2)                       (3)
```

(a) Draw the structure(s) of all stereoisomer(s) formed from (1) and (2). Show which come from (1) and which from (2). Indicate whether each product [each has the *general* formula (3)] is optically active or optically inactive. Briefly give your reasoning.

(b) Draw the structures of the osazones formed when (1) and (2) are treated with phenylhydrazine. Indicate which comes from (1) and which from (2).

29.32 Two aldoses, *A* and *B*, and ketose *C* all give the same osazone *D* on treatment with excess phenylhydrazine. A mixture of *A* and *B* can be produced by subjecting L-(+)-ribose to Kiliani-Fischer synthesis.

(a) On the basis of these data alone, suggest structure(s) for *A* to *D* as far as you possibly can. Explain your reasoning.

(b) Is the preceding information sufficient to allow you to deduce *unique* structures for *A* to *D*? If not, indicate what ambiguity exists.

29.33 A certain polysaccharide, *P*, is analyzed as follows:

(a) Complete hydrolysis of *P* gives only D-(+)-allose. Partial hydrolysis of *P* gives, in addition to other products, a disaccharide, *A*. *A* is hydrolyzed to 2 moles of D-(+)-allose by the enzyme *emulsin*, an enzyme that hydrolyzes only β-glycoside linkages.

(b) When *P* is methylated (dimethyl sulfate and NaOH) and then completely hydrolyzed, 0.2% of compound *B* and 99.8% of compound *C* are produced. Using this information, deduce the

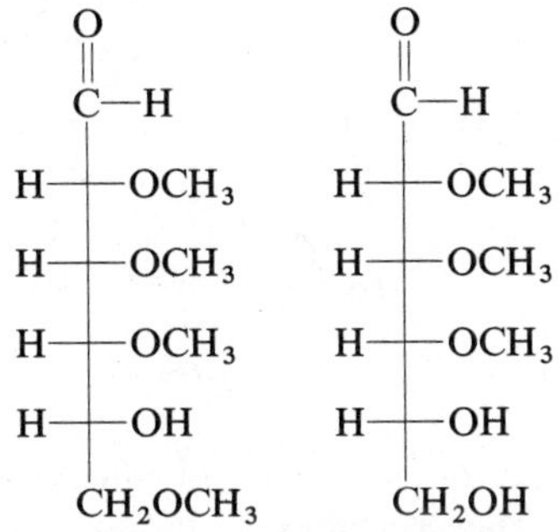

```
      O                      O
      ‖                      ‖
      C—H                    C—H
 H——OCH₃                H——OCH₃
 H——OCH₃                H——OCH₃
 H——OCH₃                H——OCH₃
 H——OH                  H——OH
   CH₂OCH₃                CH₂OH

    B, 0.2%                C, 99.8%
```

structure of *P*. Be sure to indicate *about* how many allose molecules there are in a molecule of *P*. Also, show the ring size of allose in *P*. Tell what information allows you to deduce the structure you have drawn. Use Haworth structures throughout. *Give your reasoning.*

29.34 Two aldohexoses, *A* and *B*, belong to the D family and both give the same osazone, *C*. On nitric acid oxidation, *A* gives optically active diacid *D*, and *B* gives optically active diacid *E*. On Ruff degradation, *A* gives aldopentose *F*, which in turn gives an optically inactive diacid *G* on oxidation with concentrated nitric acid. *A*, after being subjected to Kiliani-Fischer synthesis, gives two aldoheptoses, both of which give optically active diacids on oxidation with nitric acid (the two diacids are not identical). On Kiliani-Fischer synthesis, *B* also gives two aldoheptoses; one gives an optically active diacid and the other gives an optically inactive diacid. On the basis of these data, deduce the structures of compounds *A* to *G*. Use open-chain Fischer projection formulas throughout, and briefly indicate your reasoning.

29.35 Consider the following facts about a *disacccharide A*, and then, using Haworth structures, draw the structures for compounds *A* to *C*. Be certain to indicate the ring sizes.

(*a*) *A* is a reducing sugar.

(*b*) *A* can be hydrolyzed to give equimolar amounts of D-ribose and D-galactose.

(*c*) After the reaction sequence:

$$A \xrightarrow[\text{HCl}]{\text{CH}_3\text{OH}} B \xrightarrow[\text{NaOH}]{\text{CH}_3\text{O}-\text{SO}_2-\text{OCH}_3} C \xrightarrow[\text{heat}]{\text{H}^{\oplus},\ \text{H}_2\text{O}} (1)\ \text{and}\ (2)$$

A yields products (1) and (2).

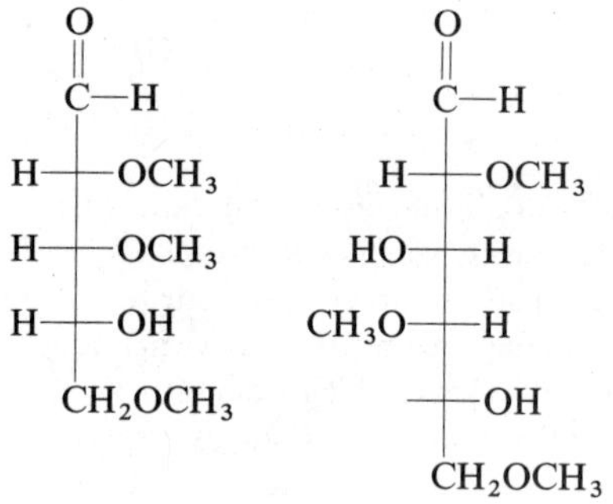

Structure (1):
```
        O
        ||
        C—H
  H —|— OCH3
  H —|— OCH3
  H —|— OH
      CH2OCH3
```

Structure (2):
```
        O
        ||
        C—H
  H  —|— OCH3
  HO —|— H
  CH3O —|— H
       —|— OH
      CH2OCH3
```

(1)   (2)

29.36 Compound *A*, $C_{12}H_{22}O_{11}$, is optically active and is a reducing sugar that forms an osazone. *A* can be hydrolyzed by dilute acid or by the enzyme *maltase* to give only D-(+)-fructose and D-(+)-allose. Methylation of *A*, using dimethyl sulfate and sodium hydroxide, gives compound *B*; *B* contains eight methyl groups and is not a reducing sugar. Hydrolysis of *B* under acidic conditions gives compounds *C* and *D*:

Structure C:
```
        H
        |
        C=O
  H —|— OCH3
  H —|— OCH3
  H —|— OCH3
  H —|— OH
      CH2OCH3
```

Structure D:
```
      CH2OCH3
        |
        C=O
  CH3O —|— H
  H   —|— OCH3
  H   —|— OH
      CH2OH
```

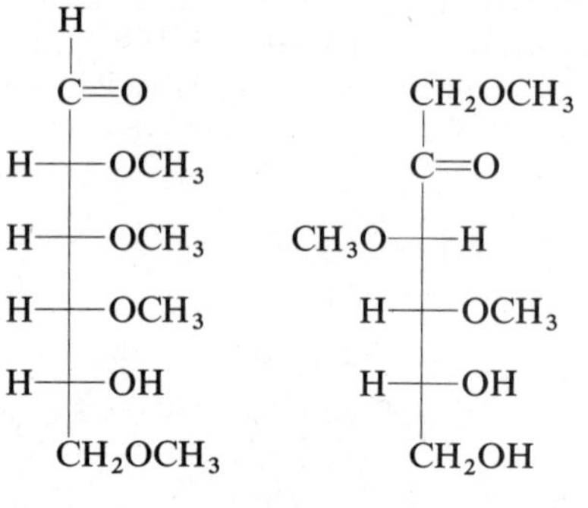

C   *D*

(*a*) Assuming that the fructose part of the molecule exists in its furanose form in *A*, write the structures of *A* and *B*. Use Haworth projection formulas and *give your reasoning*.

(*b*) What uncertainties, if any, are in the structure you propose for *A*? (That is, does the information given allow you to make an unequivocal structure assignment? If not, indicate what added data are needed.)

29.37 A new disaccharide *A* was recently isolated from natural sources, and some information leading to its structure is as follows:

(*a*) *A* gives a brick-red precipitate when treated with Fehling's solution.

(*b*) Hydrolysis of *A* gives D-(−)-arabinose and D-(+)-glucose. Enzyme hydrolysis suggests that only α linkages are present in *A*.

(*c*) *A* reacts readily with bromine water to give compound *B*, which is a monocarboxylic acid.

(*d*) *B*, on methylation (using dimethyl sulfate and NaOH) followed by hydrolysis, gives compounds *C* and *D* with the structures shown here.

On the basis of the information given, draw the structures of *A* and *B*. Be careful to show the correct ring size of the D-arabinose part of the disaccharide, and clearly state the reasons for your choice. *Use Haworth structures.*

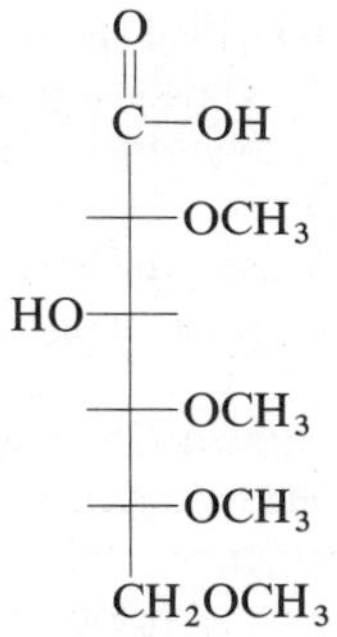

$$
\begin{array}{c}
\overset{\displaystyle O}{\overset{\displaystyle \|}{C}}-OH \\
\hline\;-OCH_3 \\
HO-\;\; \\
\hline\;-OCH_3 \\
\hline\;-OCH_3 \\
CH_2OCH_3 \\
C
\end{array}
\qquad
\begin{array}{c}
\overset{\displaystyle O}{\overset{\displaystyle \|}{C}}-H \\
CH_3O-\;\; \\
\hline\;-OCH_3 \\
\hline\;-OH \\
CH_2OCH_3 \\
D
\end{array}
$$

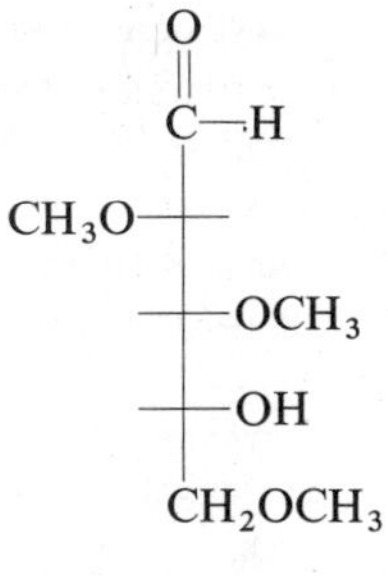

29.38 An unlabeled bottle contains only one of the following aldohexoses in pure form.

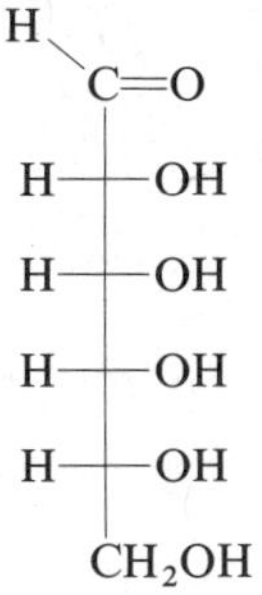

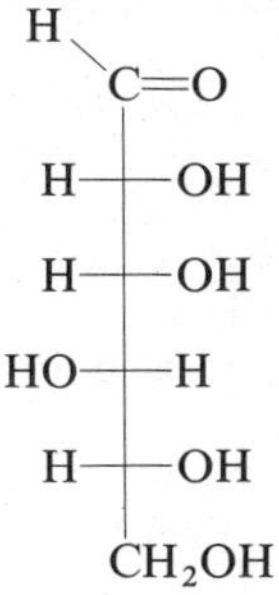

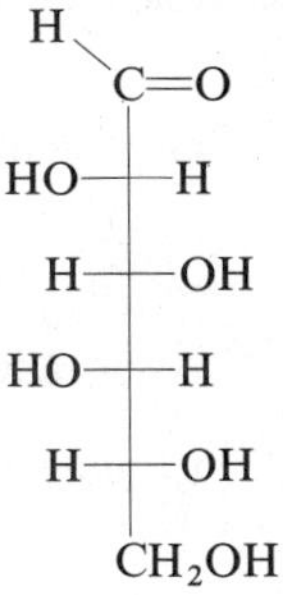

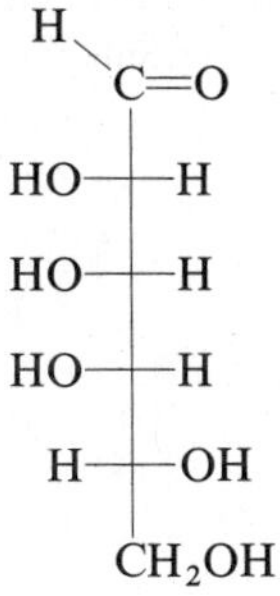

$$
\begin{array}{c}
\overset{H}{\underset{}{\diagdown}}C=O \\
H-\!\!-OH \\
H-\!\!-OH \\
H-\!\!-OH \\
H-\!\!-OH \\
CH_2OH \\
1
\end{array}
\quad
\begin{array}{c}
\overset{H}{\underset{}{\diagdown}}C=O \\
H-\!\!-OH \\
H-\!\!-OH \\
HO-\!\!-H \\
H-\!\!-OH \\
CH_2OH \\
2
\end{array}
\quad
\begin{array}{c}
\overset{H}{\underset{}{\diagdown}}C=O \\
HO-\!\!-H \\
H-\!\!-OH \\
HO-\!\!-H \\
H-\!\!-OH \\
CH_2OH \\
3
\end{array}
\quad
\begin{array}{c}
\overset{H}{\underset{}{\diagdown}}C=O \\
HO-\!\!-H \\
HO-\!\!-H \\
HO-\!\!-H \\
H-\!\!-OH \\
CH_2OH \\
4
\end{array}
$$

To identify the unknown sugar (called compound A), some tests are carried out and the following results are obtained.

Test 1: Oxidation of A with nitric acid gives an *optically active* diacid B.

Test 2: Ruff degradation of A gives aldopentose C, and on treatment with nitric acid, C gives *optically active* diacid D.

Test 3: When A is subjected to Kiliani-Fischer synthesis, two new aldoheptoses, E and F, are formed. On oxidation with nitric acid, E gives an *optically active* diacid G, and F gives an *optically inactive* diacid H.

On the basis of these data, draw the structures of compounds A to H. Give your reasoning, and show in a logical and organized fashion how you can eliminate the possible aldohexose structures one by one until you are left with an unambiguous structure for compound A—the unknown sugar.

29.39 A naturally occurring disaccharide A, $C_{12}H_{22}O_{11}$, exhibits mutarotation and gives a silver mirror with Tollens' reagent. Hydrolysis of A with aqueous acid or by the enzyme *emulsin* (see Ques. 29.33) yields equimolar amounts of D-(+)-galactose and D-(−)-fructose. Treatment of A with bromine and water gives a monocarboxylic acid B. B is methylated (using dimethyl sulfate and sodium hydroxide) and then hydrolyzed to give compounds C and D with the structures shown here.

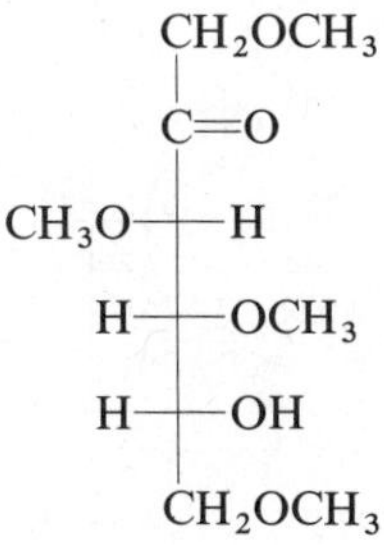

$$
\begin{array}{c}
CH_2OCH_3 \\
|\;\; \\
C=O \\
CH_3O-\!\!-H \\
H-\!\!-OCH_3 \\
H-\!\!-OH \\
CH_2OCH_3 \\
C
\end{array}
\qquad
\begin{array}{c}
COOH \\
H-\!\!-OCH_3 \\
CH_3O-\!\!-H \\
CH_3O-\!\!-H \\
H-\!\!-OCH_3 \\
CH_2OH \\
D
\end{array}
$$

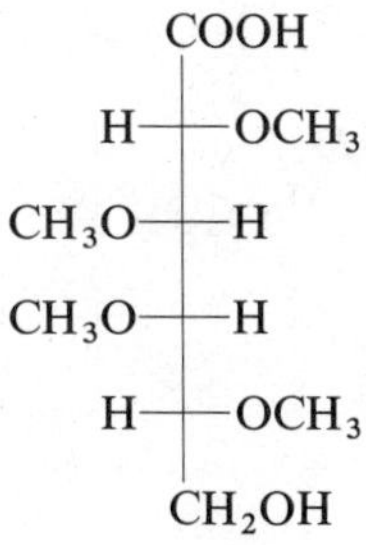

Using these data, deduce the structures of compounds A and B. Assume fructose is in the furanose form and galactose is in the pyranose form. Use Haworth structures. Give your reasoning.

29.40 An unknown aldopentose A is subjected to a series of chemical reactions to elucidate its structure.

(a) Ruff degradation converts A to an aldotetrose, which is identified as (+)-threose.

(b) A is subjected to Kiliani-Fischer synthesis, whereupon two new compounds B and C are formed.

(c) When compound B is treated with concentrated HNO_3, the resulting diacid D is *optically active*. On the other hand, C yields *optically inactive* diacid E on similar treatment. On the basis of these data alone, deduce the structure of the unknown aldopentose A. Also, draw the structures of B to E. To what family (D or L) does compound A belong, and why?

29.41 Compound A is an aldopentose, which on oxidation with nitric acid gives an optically active diacid B. A is subjected to two successive Ruff degradations to yield (−)-glyceraldehyde. A is subjected to Kiliani-Fischer synthesis to give two aldohexoses, C and D. Catalytic hydrogenation of C and D gives optically active hexaalcohols E and F. Aldohexoses C and D are examined further as outlined:

$$C \xrightarrow[\text{synthesis}]{\text{Kiliani-Fischer}} \text{aldoheptose, } G \quad + \quad \text{aldoheptose, } H$$

$$\big\downarrow HNO_3 \qquad\qquad\qquad \big\downarrow HNO_3$$

optically active diacid   optically active diacid

$$D \xrightarrow[\text{synthesis}]{\text{Kiliani-Fischer}} \text{aldoheptose, } I \quad + \quad \text{aldoheptose, } J$$

$$\big\downarrow HNO_3 \qquad\qquad\qquad \big\downarrow HNO_3$$

optically active diacid   optically inactive diacid

Using *noncyclic Fischer projection formulas*, draw the structures of compounds A to J. Indicate your reasoning. To what family does A belong, and why?

29.42 A certain trisaccharide gives, on hydrolysis, three different hexoses, A, B, and C. Treatment of the hydrolysis mixture with excess phenylhydrazine gives one and only one osazone D. (That is, each monosaccharide gives the same osazone D.) One hexose, call it A, is isolated and studied further as follows.

Hexose A gives an optically active diacid E on treatment with concentrated HNO_3. Hexose A is subjected to Ruff degradation and yields pentose F, which gives optically active diacid G on treatment with concentrated HNO_3. Pentose F is degraded once more (Ruff method) to give a tetrose, which is identified as (−)-threose.

Draw the structures for compounds A to G, using *skeletal formulas*. To what family do hexoses A to C belong, and why? Indicate your reasoning. (*Note:* Do not attempt to show the structure of the original trisaccharide.)

29.43 A new trisaccharide A, called "nonsensose," was recently "discovered." A is a nonreducing sugar, which on complete hydrolysis gives 2 moles of D-glucose and 1 mole of D-talose. Partial hydrolysis of A gives two *different* disaccharides, B and C. B is a nonreducing sugar that contains

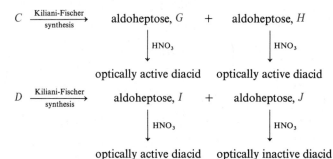

D-glucose and D-talose, whereas *C* is a reducing sugar that also contains D-glucose and D-talose. Enzyme hydrolysis indicates that only β linkages are involved. *A* is methylated completely (dimethyl sulfate and NaOH) and then hydrolyzed to give 2 moles of *D* and 1 mole of *E*, with the structures shown.

On the basis of the information given, deduce the structures of compounds *A* to *C*. Use Haworth structures throughout, and be sure to show proper ring sizes.

29.44 Polysaccharide *A* is analyzed as follows:

(*a*) Complete hydrolysis of *A* gives only D-(+)-galactose.

(*b*) Partial hydrolysis of *A* gives, in addition to other products, a disaccharide *B*. *B* is hydrolyzed to 2 moles of D-(+)-galactose.

(*c*) When *A* is methylated (using dimethyl sulfate and sodium hydroxide) and then completely hydrolyzed, 2% of compound *C* and 98% of compound *D* are produced. The structures of *C* and *D* are shown here.

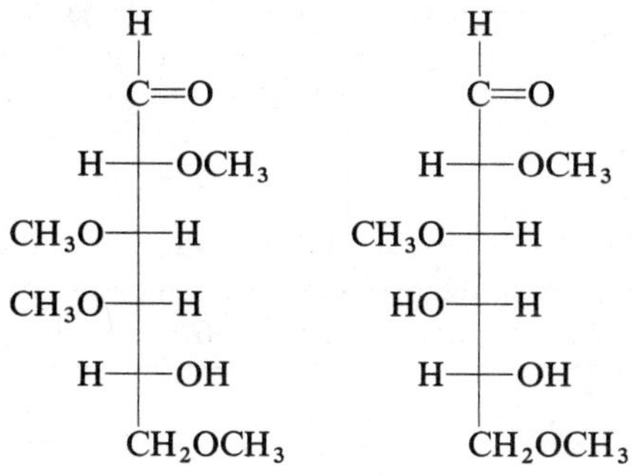

$$
\begin{array}{cc}
\text{H} & \text{H} \\
| & | \\
\text{C=O} & \text{C=O} \\
\text{H}-\text{OCH}_3 & \text{H}-\text{OCH}_3 \\
\text{CH}_3\text{O}-\text{H} & \text{CH}_3\text{O}-\text{H} \\
\text{CH}_3\text{O}-\text{H} & \text{HO}-\text{H} \\
\text{H}-\text{OH} & \text{H}-\text{OH} \\
\text{CH}_2\text{OCH}_3 & \text{CH}_2\text{OCH}_3 \\
C,\ 2\% & D,\ 98\%
\end{array}
$$

Using this information, deduce the structures of *A* and *B*. Assume galactose is present in the pyranose form. Be sure to indicate *about* how many galactose molecules there are per molecule of *A*. Use Haworth structures, and give your reasoning.

29.45 A nonreducing sugar, *A*, has the formula $C_{18}H_{32}O_{16}$. Complete hydrolysis of *A* gives D-glucose, D-galactose, and D-fructose. Treatment of *A* with the enzyme β-glucosidase (which hydrolyzes only β-glucoside linkages) gives D-glucose and a disaccharide *B*. *B* is a nonreducing sugar, and its high optical rotation indicates that only α linkages are present in it. *A* is methylated (using dimethyl sulfate and sodium hydroxide) and then hydrolyzed to give *C*, *D*, and *E*, with the structures shown here.

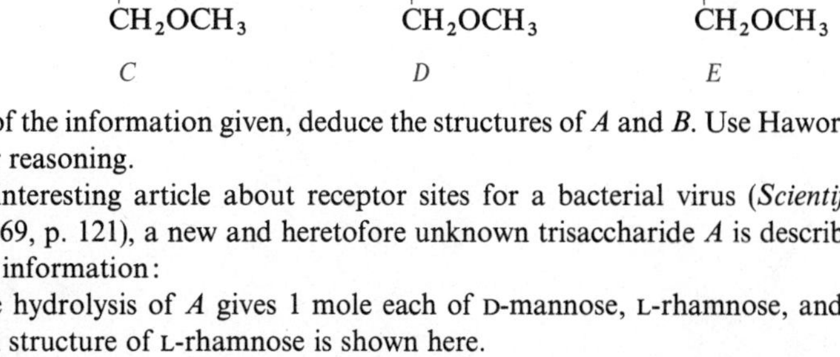

$$
\begin{array}{ccc}
\text{O=C}-\text{H} & \text{CH}_2\text{OH} & \text{O=C}-\text{H} \\
\text{H}-\text{OCH}_3 & \text{C=O} & \text{H}-\text{OCH}_3 \\
\text{CH}_3\text{O}-\text{H} & \text{CH}_3\text{O}-\text{H} & \text{CH}_3\text{O}-\text{H} \\
\text{CH}_3\text{O}-\text{H} & \text{H}-\text{OCH}_3 & \text{H}-\text{OCH}_3 \\
\text{H}-\text{OH} & \text{H}-\text{OH} & \text{H}-\text{OH} \\
\text{CH}_2\text{OCH}_3 & \text{CH}_2\text{OCH}_3 & \text{CH}_2\text{OCH}_3 \\
C & D & E
\end{array}
$$

On the basis of the information given, deduce the structures of *A* and *B*. Use Haworth structures, and give your reasoning.

29.46 In an interesting article about receptor sites for a bacterial virus (*Scientific American*, November 1969, p. 121), a new and heretofore unknown trisaccharide *A* is described. Consider the following information:

(*a*) Complete hydrolysis of *A* gives 1 mole each of D-mannose, L-rhamnose, and D-galactose. The Haworth structure of L-rhamnose is shown here.

(*b*) *A* is a reducing sugar, and partial hydrolysis of *A* gives a disaccharide *B* as one product. *B* is also a reducing sugar, which is hydrolyzed by the enzyme α-mannosidase (an enzyme that cleaves only α-mannose linkages) to give D-mannose and L-rhamnose.

(c) *A*, on treatment with the enzyme α-rhamnosidase (an enzyme that cleaves only α-rhamnoside linkages), gives *B* and D-galactose.

(d) Oxidation of *A* with bromine and water gives a monocarboxylic acid *C*. *C* is methylated with dimethyl sulfate and base and then hydrolyzed to give the three sugars, *D*, *E*, and *F*, shown here.

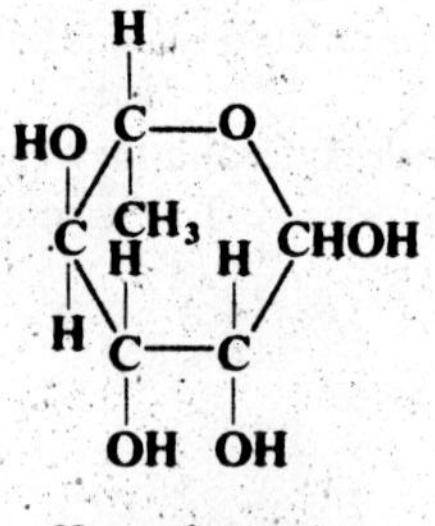

Haworth structure
of L-rhamnose

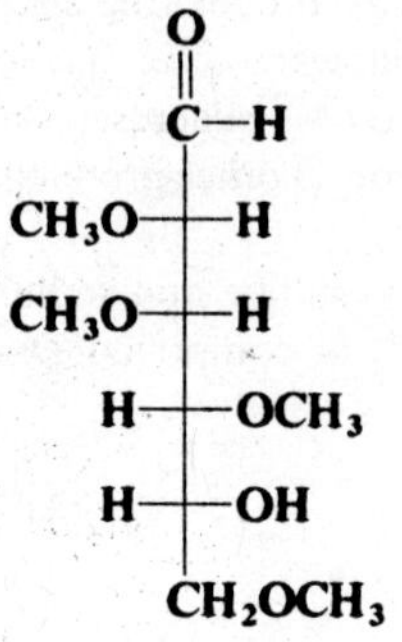

D

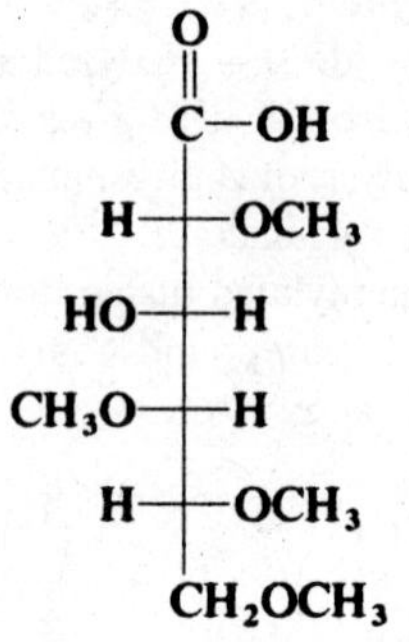

$$\begin{array}{c} C\!-\!H \\ \| \\ O \end{array}$$

CH$_3$O—H
CH$_3$O—H
H—OCH$_3$
H—OH
CH$_2$OCH$_3$

E

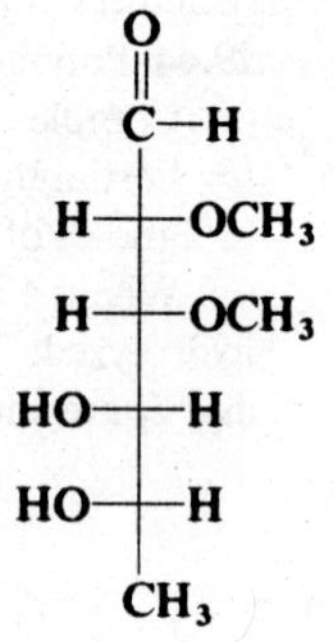

$$\begin{array}{c} C\!-\!OH \\ \| \\ O \end{array}$$

H—OCH$_3$
HO—H
CH$_3$O—H
H—OCH$_3$
CH$_2$OCH$_3$

F

$$\begin{array}{c} C\!-\!H \\ \| \\ O \end{array}$$

H—OCH$_3$
H—OCH$_3$
HO—H
HO—H
CH$_3$

On the basis of the information given, deduce the structures of compounds *A* and *B*. Assume L-rhamnose exists in the pyranose form, and note that structure *F* is derived from L-rhamnose. Use Haworth structures throughout. *Give your reasoning!*

Appendix: Reading References

Chapter 1

The first and third references are comprehensive general chemistry texts that are useful for reviewing empirical formula, molecular formula, molecular weight determinations, and other topics.

C. W. Keenan, J. H. Wood, and D. C. Kleinfelter, *General College Chemistry*, 6th ed., Harper & Row, New York, 1979.

M. J. Sienko and R. A. Plane, *Chemical Principles and Applications*, 1st ed., McGraw-Hill, New York, 1979.

W. L. Masterton and E. J. Slowinski, *Chemical Principles*, 4th ed., Saunders, Philadelphia, 1977.

Chapter 2

Books

W. L. Jorgensen and L. Salem, *The Organic Chemist's Book of Orbitals*, Academic Press, New York, 1973.

A. Liberles, *Introduction to Molecular Orbital Theory*, Holt, New York, 1966.

L. Pauling, *The Nature of the Chemical Bond*, 3d ed., Cornell University Press, Ithaca NY, 1960.

W. B. Smith, *Molecular Orbital Methods in Organic Chemistry HMO and PMO, An Introduction*, Marcel Dekker, New York, 1974.

Short Articles

W. T. Bordass and J. W. Linnett, "A New Way of Presenting Atomic Orbitals," *J. Chem. Ed.*, vol. 47, pp. 672–675, 1970.

I. Cohen and J. DelBene, "Hybrid Orbitals in Molecular Orbital Theory," *J. Chem. Ed.*, vol. 46, p. 487, 1969.

O. Reinmuth, "Some Aspects of Organic Molecules and Their Behavior, I. Electronegativity," *J. Chem. Ed.*, vol. 34, pp. 272–275, 1957.

Chapter 3

Isomers

H. R. Henze and C. M. Blair, "The Number of Isomeric Hydrocarbons of the Methane Series," *J. Am. Chem. Soc.*, vol. 53, p. 3077, 1931.

Nomenclature

E. A. Evieux, "The Geneva Congress on Organic Nomenclature, 1892," *J. Chem. Ed.*, vol. 31, p. 326, 1954.

C. D. Hurd, "The General Philosophy of Organic Nomenclature," *J. Chem. Ed.*, vol. 38, p. 43, 1961.

"Definitive Rules for Nomenclature of Organic Chemistry, sec. A; Hydrocarbons, sec. B; Heterocyclic Systems," *J. Am. Chem. Soc.*, vol. 82, p. 5517, 1960.

A. M. Patterson, "Definitive Report on the Reform of the Nomenclature of Organic Chemistry," *J. Am. Chem. Soc.*, vol. 55, p. 3905, 1933.

———, L. T. Capell, and M. A. Magill, "Nomenclature of Organic Compounds," *Chem. Abstr.*, vol. 39, p. 5875, 1945.

J. G. Traynham, *Organic Nomenclature: A Programmed Introduction*, Prentice-Hall, Englewood Cliffs NJ, 1966.

Bonding and Cyclic Compounds

W. A. Bernett, "A Unified Theory of Bonding of Cyclopropanes," *J. Chem. Ed.*, vol. 44, p. 17, 1967.

L. N. Ferguson, "Alicyclic Chemistry: The Playground of Organic Chemists," *J. Chem. Ed.*, vol. 46, p. 404, 1969.

———, "Ring Strain and Reactivity of Alicycles," *J. Chem. Ed.*, vol. 47, p. 46, 1970.

Molecular Models

W. R. Brode and C. E. Boord, "Molecular Models in the Elementary Organic Laboratory, I," *J. Chem. Ed.*, vol. 9, p. 1774, 1932.

Q. R. Petersen, "Some Reflections on the Use and Abuse of Molecular Models, *J. Chem. Ed.*, vol. 47, p. 24, 1970.

F. Vogtle and E. Goldschmitt, "Simple and Inexpensive Orbital Lobe Molecular Models," *J. Chem. Ed.*, vol. 51, p. 350, 1974.

Stereochemistry and Conformational Analysis

W. G. Dauben and K. S. Pitzer, "Conformational Analysis," in *Steric Effects in Organic Chemistry*, M. S. Newman (ed.), Wiley, New York, 1956.

J. D. Dunitz, "The Two Forms of Cyclohexane," *J. Chem. Ed.*, vol. 47, p. 488, 1970.

E. L. Eliel, *Stereochemistry of Carbon Compounds*, McGraw-Hill, New York, 1962.

———, N. L. Allinger, S. J. Angyal, and G. A. Morrison, *Conformational Analysis*, Wiley, New York, 1965.

———, *Elements of Stereochemistry*, Wiley, New York, 1969.

J. P. Idoux, "Conformational Analysis in Chemical Reactivity," *J. Chem. Ed.*, vol. 44, p. 495, 1967.

E. Juaristi, "The Attractive and Repulsive Gauche Effects," *J. Chem. Ed.*, vol. 56, p. 438, 1979.

C. A. Kingsbury, "Conformations of Substituted Ethanes," *J. Chem. Ed.*, vol. 56, p. 431, 1979.

K. Mislow, *Introduction to Stereochemistry*, Benjamin/Cummings, Menlo Park CA 1965.

Chapter 4

Books

E. S. Gould, *Mechanism and Structure in Organic Chemistry*, Holt, New York, 1959.

J. Hine, *Physical Organic Chemistry*, 2d ed., pp. 402–483, McGraw-Hill, New York, 1962.

A. Liberles, *Introduction to Theoretical Organic Chemistry*, Macmillan, New York, 1968.

W. A. Pryor, *Introduction to Free Radical Chemistry*, Prentice-Hall, Englewood Cliffs NJ, 1965.
———, *Free Radicals*, McGraw-Hill, New York, 1966.
C. Walling, *Free Radicals in Solution*, Wiley, New York, 1957.

Short Articles

D. F. Larder, "Historical Aspects of the Tetrahedron in Chemistry," *J. Chem. Ed.*, vol. 44, p. 661, 1967.
M. K. Kaloustian, "The Electrostatic Dimension in Conformational Analysis," *J. Chem. Ed.*, vol. 51, p. 777, 1974.
O. Reinmuth, "Some Aspects of Organic Molecules and Their Behavior, II. Bond Energies," *J. Chem. Ed.*, vol. 34, p. 318, 1957.
F. D. Rossini, "Hydrocarbons in Petroleum," *J. Chem. Ed.*, vol. 37, p. 554, 1960.
L. Schmerling, "The Mechanisms of the Reactions of Aliphatic Hydrocarbons," *J. Chem. Ed.*, vol. 28, p. 562, 1951.
R. H. Shoemaker, E. L. d'Ouville, and R. F. Marscher, "Recent Advances in Petroleum Refining," *J. Chem. Ed.*, vol. 32, p. 30, 1955.
C. J. M. Stirling, "Closure of Three-Membered Rings," *J. Chem. Ed.*, vol. 50, p. 855, 1973.

Chapter 5

Books

F. A. Carey and R. J. Sundberg, "Advanced Organic Chemistry." part A, *Structure and Mechanisms*, Plenum, New York, 1977.
E. Gould, *Mechanism and Structure in Organic Chemistry*, Holt, New York, 1959.
J. M. Harris and C. C. Wamser, *Fundamentals of Organic Reaction Mechanisms*, Wiley, New York, 1976.
T. H. Lowry and K. S. Richardson, *Mechanism and Theory in Organic Chemistry*, 2nd. ed., Harper & Row, New York, to be published.
J. March, *Advanced Organic Chemistry: Reactions, Mechanisms, and Structures*, 2d ed., McGraw-Hill, New York, 1977.

Short Articles

C. L. Deasey, "The Walden Inversion in Nucleophilic Aliphatic Substitution Reactions," *J. Chem. Ed.*, vol. 22, p. 455, 1962.
J. O. Edwards, "Bimolecular Nucleophilic Displacement Reactions," *J. Chem. Ed.*, vol. 45, p. 386, 1968.
M. S. Paur, "Nucleophilic Reactivities of the Halide Ions," *J. Chem. Ed.*, vol. 47, p. 473, 1970.

Chapter 6

Books

E. L. Eliel, *Stereochemistry of Carbon Compounds*, McGraw-Hill, New York, 1962.
———, N. L. Allinger, S. J. Angyal, and G. A. Morrison, *Conformational Analysis*, Wiley, New York, 1965.
K. Mislow, *Introduction to Stereochemistry*, Benjamin/Cummings, Menlo Park CA, 1965.
G. Natta and M. Farina, *Stereochemistry*, Harper & Row, New York, 1972.
G. W. Wheland, *Advanced Organic Chemistry*, 3d ed., chaps. 2, 6–9, Wiley, New York, 1960.

Short Articles

M. J. H. Van't Hoff, "Sur les Formules de Structure," *Bull. Soc. Chim. Fr.*, vol. 23, p. 293, 1875.

J. M. Bijovet, "Determination of the Absolute Configuration of Optical Antipodes," *Endeavor*, vol. 14, p. 71, 1955.

R. L. Bent, "Aspects of Isomerism and Mesomerism, III. Stereoisomerism," *J. Chem. Ed.*, vol. 30, p. 328, 1953.

J. L. Carlos, Jr., "Molecular Symmetry and Optical Inactivity," *J. Chem. Ed.*, vol. 45, p. 248, 1968.

O. Cori, "Complimentary Rules to Define R and S Configuration," *J. Chem. Ed.*, vol. 49, p. 461, 1972.

J. Figueras, Jr., "Stereochemistry of Simple Ring Systems," *J. Chem. Ed.*, vol 28, p. 134, 1951.

J. E. Garvin, "Inexpensive Polarimeter for Demonstrations and Student Use," *J. Chem. Ed.*, vol. 37, p. 515, 1960.

E. J. Gill, "A Demonstration of Optical Activity with an Eskimo Yo-Yo," *J. Chem. Ed.*, vol. 38, p. 263, 1961.

A. F. Holleman, "My Reminiscences of van't Hoff," *J. Chem. Ed.*, vol. 29, p. 379, 1952.

R. S. Kahn, "An Introduction to the Sequence Rule," *J. Chem. Ed.*, vol. 41, p. 116, 1964.

H. W. Moseley, "Pasteur: The Chemist," *J. Chem. Ed.*, vol. 5, p. 50, 1928.

D. F. Mowery, Jr., "The Cause of Optical Inactivity," *J. Chem. Ed.*, vol. 29, p. 138, 1952.

———, "Criteria for Optical Activity," *J. Chem. Ed.*, vol. 46, p. 269, 1969.

S. E. Murov and M. Pickering, "The Odor of Optical Isomers," *J. Chem. Ed.*, vol. 50, p. 74, 1973.

W. K. Noyce, "Stereoisomerism of Carbon Compounds," *J. Chem. Ed.*, vol. 38, p. 23, 1961.

H. B. Thompson, "The Criterion for Optical Isomerism," *J. Chem. Ed.*, vol. 37, p. 530, 1960.

F. T. Williams, "Resolution by the Method for Racemic Modification," *J. Chem. Ed.*, vol. 39, p. 211, 1962.

Chapter 7

Books

J. Hine, *Physical Organic Chemistry*, 2d ed., chap. 8, McGraw-Hill, New York, 1962.

C. K. Ingold, *Structure and Mechanism in Organic Chemistry*, chaps. 7 and 8, Cornell University Press, Ithaca NY 1953.

J. March, *Advanced Organic Chemistry: Reactions, Mechanisms, and Structure*, 2d ed., McGraw-Hill, New York, 1977.

Short Articles

D. Bethell and V. Gold. "The Structure of Carbonium Ions," *Quart. Rev. (London)*, vol. 12, p. 173, 1958.

A. Liberles, A. Greenberg, and J. E. Eilers, "Attractive Steric Effects," *J. Chem. Ed.*, vol. 50, p. 676, 1973.

Chapter 8

Books

M. J. S. Dewar, *Hyperconjugation*, Ronald Press, New York, 1962.

J. Hine, *Physical Organic Chemistry*, 2d ed., chap. 9, McGraw-Hill, New York, 1962.

H. L. Holmes, "The Diels-Alder Reaction—Ethylenic and Acetylenic Dienophiles," in *Organic Reactions*, vol. 4, pp. 60–173, Roger Adams (ed.), Wiley, New York, 1948.

Short Articles

M. C. Caserio, "Reaction Mechanisms in Organic Chemistry," *J. Chem. Ed.*, vol. 48, p. 782, 1971.

W. R. Dolbier, Jr., "Research Summaries for Teaching Organic Chemistry: Electrophilic Addition to Alkenes," *J. Chem. Ed.*, vol. 46, p. 342, 1969.

D. G. M. Diaper, "Ozonolysis," *J. Chem. Ed.*, vol. 44, p. 354, 1967.

N. Isenberg and M. Grdinic, "A Modern Look at Markovnikov's Rule and the Peroxide Effect," *J. Chem. Ed.*, vol. 46, p. 601, 1969.

G. Jones, "The Markovnikov Rule," *J. Chem. Ed.*, vol. 38, p. 297, 1961.

H. M. Leicaster, "Vladimir Vasilevich Markovnikov," *J. Chem. Ed.*, vol. 18, p. 53, 1941.

G. Natta, "Precisely Constructed Polymers," *Scientific American*, August 1961.

———, "Polymerization," *Scientific American*, September 1967.

W. S. Partridge and E. R. Schiertz, "Otto Wallach: The First Organizer of the Terpenes," *J. Chem. Ed.*, vol. 24, p. 106, 1947.

D. Seyferth, "Phenyl (trihalomethyl) Mercury Compounds: Exceptionally Versatile Dihalocarbene Precursors," *Acct. Chem. Res.*, vol. 5, p. 65, 1972.

J. G. Traynham, "The Bromonium Ion," *J. Chem. Ed.*, vol. 40, p. 392, 1963.

F. X. Werber and D. F. Hoeg, "How to Fix Polymer Molecules in Space," *Chem. and Eng. News*, vol. 37, Mar. 23, 1959.

Chapter 9

Books

C. A. Bunton, *Nucleophilic Substitution at a Saturated Carbon Atom*, Elsevier, New York, 1963.

J. Hine, *Physical Organic Chemistry*, 2d ed., chaps. 6 and 7, McGraw-Hill, New York, 1962.

C. K. Ingold, *Structure and Mechanism in Organic Chemistry*, chap. 7, Cornell University Press, Ithaca NY, 1953.

J. March, *Advanced Organic Chemistry: Reactions, Mechanisms, and Structure*, 2d ed., McGraw-Hill, New York, 1977.

A. Streitwieser, Jr., *Solvolytic Displacement Reactions*, McGraw-Hill, New York, 1962.

E. R. Thornton, *Solvolysis Mechanisms*, Ronald Press, New York, 1964.

Short Articles

F. G. Bordwell, "How Common Are Base-Initiated 1,2-Eliminations?", *Acct. Chem. Res.*, vol. 5, p. 374, 1972.

C. L. Deasy, "The Walden Inversion in Nucleophilic Aliphatic Substitution Reactions," *J. Chem. Ed.*, vol. 22, p. 455, 1962.

J. O. Edwards, "Bimolecular Nucleophilic Displacement Reactions," *J. Chem. Ed.*, vol. 45, p. 386, 1968.

M. S. Paur, "Nucleophilic Reactivities of the Halide Ions," *J. Chem. Ed.*, vol. 47, p. 473, 1970.

W. H. Saunders, Jr., "Distinguishing Between Concerted and Nonconcerted Eliminations," *Acct. Chem. Res.*, vol 9, p. 19, 1976.

Chapter 10

Books

L. B. Clapp. *The Chemistry of the OH Group*, chaps. 1 and 2, Prentice-Hall, Englewood Cliffs NJ, 1967.

S. Patai (ed.), *The Chemistry of the Hydroxyl Group*, parts 1 and 2, Wiley, New York, 1971.

R. Stewart, *Oxidation Mechanisms*, Benjamin/Cummings, Menlo Park CA, 1964.

Short Articles

L. N. Ferguson, "Hydrogen Bonding and the Physical Properties of Substances," *J. Chem. Ed.*, vol. 33, p. 267, 1956.

L. A. Greenberg, "Alcohol in the Body," *Scientific American*, December 1963.

Chapter 11

Books

W. Franke, W. Ziegenbein, and H. Meister, "The Formation of the Acetylenic Bonds," in *Newer Methods of Preparative Organic Chemistry*, vol. 3, pp. 425–450, W. Foerest (ed.), Academic Press, New York, 1964.

H. L. Holmes, "The Diels-Alder Reaction—Ethylenic and Acetylenic Dienophiles," in *Organic Reactions*, vol. 4, p. 60, Roger Adams (ed.), Wiley, New York, 1948.

C. K. Ingold, *Structure and Mechanism in Organic Chemistry*, p. 645, Cornell University Press, Ithaca NY, 1953.

J. A. Nieuwland and R. R. Vogt, *The Chemistry of Acetylene*, Reinhold, New York, 1945.

T. L. Jacobs, "The Synthesis of Acetylenes," in *Organic Reactions*, vol. 5, p. 1, Roger Adams (ed.), Wiley, New York, 1949.

R. A. Raphael, *Acetylenic Compounds in Organic Synthesis*, Wiley, New York, 1955.

Short Articles

P. D. Bartlett, "Mechanisms of Cycloadditions," *Quart. Rev. (London)*, vol. 24, p. 473, 1970.

G. C. Bond, "Mechanism of Catalytic Hydrogenation and Related Reactions," *Quart. Rev. (London)*, vol. 8, p. 279, 1954.

W. E. Hanford and D. L. Fuller, "Acetylene Chemistry," *Ind. Eng. Chem.*, vol. 40, p. 1171, 1948.

Chapter 12

Books

The following contain discussion on more than one type of spectroscopy, as noted.

J. R. Dyer, *Applications of Absorption Spectroscopy of Organic Compounds*, Prentice-Hall, Englewood Cliffs NJ, 1965 (IR, UV, NMR)

I. Fleming and D. H. Williams, *Spectroscopic Methods in Organic Chemistry*, McGraw-Hill, New York, 1966 (IR, UV, NMR, MS)

J. B. Lambert, H. F. Shurvell, L. Verbit, R. G. Cooks, and G. H. Stout, *Organic Structural Analysis*, Macmillan, New York, 1976 (IR, UV, NMR, MS)

R. M. Silverstein, G. C. Bassler, and T. C. Morrill, *Spectrometric Identification of Organic Compounds*, 3d ed., Wiley, New York, 1974 (IR, UV, NMR, MS)

IR Spectroscopy

L. J. Bellamy, *The Infrared Spectra of Complex Molecules*, Meuthen, London, 1958.

R. T. Conley, *Infrared Spectroscopy*, Allyn and Bacon, Boston, 1966.

A. D. Cross, *Introduction to Practical Infrared Spectroscopy*, 2d ed., Butterworth, London, 1964.

K. Nakanishi, *Infrared Absorption Spectroscopy*, Holden-Day, San Francisco, 1962.

UV Spectroscopy

A. E. Gillam and E. S. Stern, *Electronic Absorption Spectroscopy*, 2d ed., Arnold, London, 1957.

C. N. R. Rao, *Ultraviolet and Visible Spectroscopy*, Butterworth, London, 1961.

A. I. Scott, *Interpretation of the Ultraviolet Spectra of Natural Products*, Pergamon Press, Oxford, 1964.

Catalogs of Spectra

Aldrich Library of Infrared Spectra, 2d ed., C. J. Pouchert (ed.), Aldrich Chemical Co., Milwaukee.
Sadtler Standard Spectra, Midget Edition, vols. 1–26, Sadtler Research Laboratories.

Chapter 13

Books

J. R. Dyer, *Applications of Absorption Spectroscopy of Organic Compounds*, pp. 81–82, Prentice-Hall, Englewood Cliffs NJ, 1965.

E. S. Gould, *Mechanism and Structure in Organic Chemistry*, pp. 412–415, Holt, New York, 1959.

J. March, *Advanced Organic Chemistry: Reactions, Mechanisms, and Structures*, 2d ed., chap. 2, McGraw-Hill, New York, 1977.

Short Articles

A. Gero, "The Concept of Resonance in Elementary Organic Chemistry," *J. Chem. Ed.*, vol. 29, p. 82, 1952.

——, "Kekulé's Theory of Aromaticity," *J. Chem. Ed.*, vol. 31, p. 201, 1954.

L. C. Newell, "Faraday's Discovery of Benzene," *J. Chem. Ed.*, vol. 3, p. 1248, 1962.

C. R. Noller, "A Physical Picture of Covalent Bonding and Resonance in Organic Chemistry," *J. Chem. Ed.*, vol. 27, p. 504, 1950.

M. D. Saltzman, "Benzene and the Triumph of the Octet Theory," *J. Chem. Ed.*, vol. 51, p. 498, 1974.

L. J. Schadd and B. A. Hess, Jr., "Hückel Theory and Aromaticity," *J. Chem. Ed.*, vol. 51, p. 640, 1974.

E. E. Van Tamelen, "Benzene—The Story of Its Formulas," *Chem.*, vol. 38, p. 6, 1965.

A. Willemart, "Charles Friedel," *J. Chem. Ed.*, vol. 26, p. 2, 1949.

Chapter 14

Books

E. S. Gould, *Mechanism and Structure in Organic Chemistry*, pp. 412–452, Holt, New York, 1959.

J. Hine, *Physical Organic Chemistry*, 2d ed., chap. 16, McGraw-Hill, New York, 1960.

E. K. Ingold, *Structure and Mechanism in Organic Chemistry*, chap. 6, Cornell University Press, Ithaca NY, 1953.

P. B. D. de la Mare and J. H. Ridd, *Aromatic Substitution—Nitration and Halogenation*, Academic Press, New York, 1959.

L. M. Stocks and H. C. Brown, *Advances in Physical Organic Chemistry*, vol. 1, pp. 62–80, V. Gold (ed.), Academic Press, New York, 1963.

Short Articles

P. D. Bartlett, "The Chemical Properties of the Methyl Group," *J. Chem. Ed.*, vol. 30, p. 22, 1953.

H. Duewell, "Aromatic Substitution," *J. Chem. Ed.*, vol. 42, p. 138, 1966.

L. N. Ferguson, "The Orientation and Mechanism of Electrophilic Aromatic Substitution," *J. Chem. Ed.*, vol. 32, p. 42, 1955.

M. Gomberg, "An Instance of Trivalent Carbon: Triphenylmethyl," *J. Am. Chem. Soc.*, vol. 22, p. 757, 1900.

F. L. Lambert, "Substituent Effects in the Benzene Ring. A Demonstration," *J. Chem. Ed.*, vol. 35, p. 342, 1958.

K. L. Marsi and S. H. Wilen, "Friedel-Crafts Alkylation," *J. Chem. Ed.*, vol. 40, p. 214, 1963.

H. Meislich, "Teaching Aromatic Substitution," *J. Chem. Ed.*, vol. 44, p. 153, 1967.

R. M. Roberts, "Friedel-Crafts Chemistry," *Chem. Eng. News*, p. 96, Jan. 25, 1965.

L. M. Stock, "The Origin of the Inductive Effect," *J. Chem. Ed.*, vol. 49, p. 400, 1972.

Y. P. Varshni, "Directive Influence of Substituents in the Benzene Ring," *J. Chem. Ed.*, vol. 30, p. 342, 1953.

Chapter 15

Books

The following contain discussion on more than one type of spectroscopy, as noted.

J. R. Dyer, *Applications of Absorption Spectroscopy of Organic Compounds*, Prentice-Hall, Englewood Cliffs NJ, 1965 (IR, UV, NMR)

I. Fleming and D. H. Williams, *Spectroscopic Methods in Organic Chemistry*, McGraw-Hill, New York, 1966 (IR, UV, NMR, MS)

J. B. Lambert, H. F. Shurvell, L. Verbit, R. G. Cooks, and G. H. Stout, *Organic Structural Analysis*, Macmillan, New York, 1976 (IR, UV, NMR, MS)

R. M. Silverstein, G. C. Bassler, and T. C. Morrill, *Spectrometric Identification of Organic Compounds*, 3d ed., Wiley, New York, 1974 (IR, UV, NMR, MS)

Nuclear Magnetic Resonance Spectroscopy

J. W. Akitt, *NMR and Chemistry*, Halsted Press, New York, 1973.

A. Ault and G. O. Dudek, *NMR—An Introduction to Proton Nuclear Magnetic Resonance Spectroscopy*, Holden-Day, San Francisco, 1976.

L. M. Jackman, *Applications of Nuclear Magnetic Resonance Spectroscopy in Organic Chemistry*, Pergamon Press, New York, 1959.

—— and S. Sternhell, *Applications of Nuclear Magnetic Resonance Spectroscopy in Organic Chemistry*, 2d ed., Pergamon Press, New York, 1966.

J. D. Roberts, *Nuclear Magnetic Resonance, Applications to Organic Chemistry*, McGraw-Hill, New York, 1959.

——, *An Introduction to the Analysis of Spin-Spin Splitting in High-Resolution Nuclear Magnetic Resonance Spectroscopy*, Benjamin/Cummings, Menlo Park CA, 1962.

Catalogs of Spectra

Sadtler Research Laboratories, Inc., *Sadtler NMR Spectra*, all vols.

Varian Associates, *High Resolution NMR Spectra Catalogue*, vol. 1, 1962, vol. 2, 1963.

The Aldrich Library of NMR Spectra, vols. 1–11, 1974.

Chapter 16

Books

J. Hine, *Physical Organic Chemistry*, 2d ed., chap. 17, McGraw-Hill, New York, 1962.

R. W. Hoffman, *Dehydrobenzene and Cycloalkynes*, Academic Press, New York, 1967.

Short Articles

J. F. Bunnett and R. E. Zahler, "Aromatic Nucleophilic Substitution Reactions," *Chem. Rev.*, vol. 49, p. 273, 1951.

——, "The Chemistry of Benzyne," *J. Chem. Ed.*, vol. 38, p. 278, 1961.

———, "The Remarkable Reactivity of Aryl Halides with Nucleophiles," *J. Chem. Ed.*, vol. 51, p. 312, 1974.

J. O. Edwards, "Bimolecular Nucleophilic Displacement Reactions," *J. Chem. Ed.*, vol. 45, p. 386, 1968.

Chapter 17

Books

W. J. Hickinbottom, "Nuclear Hydroxy Derivatives of Benzene and Its Homologs. Phenols," in *Chemistry of Carbon Compounds*, vol. IIIA, pp. 413–486, E. H. Rodd (ed.), Elsevier, New York, 1954.

J. A. Monick, *Alcohols—Their Chemistry, Properties, and Manufacture*, Rheinhold, New York, 1968.

S. Patai (ed.), *The Chemistry of the Hydroxyl Group*, parts 1 and 2, Wiley, New York, 1971.

Chapter 18

Books

The following contain discussion on more than one type of spectroscopy, as noted.

J. R. Dyer, *Applications of Absorption Spectroscopy of Organic Compounds*, Prentice-Hall, Englewood Cliffs NJ, 1965 (IR, UV, NMR)

I. Fleming and D. H. Williams, *Spectroscopic Methods in Organic Chemistry*, McGraw-Hill, New York, 1966 (IR, UV, NMR, MS)

J. B. Lambert, H. F. Shurvell, L. Verbit, R. G. Cooks, and G. H. Stout, *Organic Structural Analysis*, Macmillan, New York, 1976 (IR, UV, NMR, MS)

R. M. Silverstein, G. C. Bassler, and T. C. Morrill, *Spectrometric Identification of Organic Compounds*, 3d ed., Wiley, New York, 1974 (IR, UV, NMR, MS)

Mass Spectroscopy

J. H. Benyon, *Mass Spectrometry and Its Applications to Organic Chemistry*, Elsevier, Amsterdam, 1960.

K. Biemann, *Mass Spectrometry*, McGraw-Hill, New York, 1962.

H. Budzikiewicz, C. Djerassi, and D. H. Williams, *Interpretation of Mass Spectra of Organic Compounds*, Holden-Day, San Francisco, 1964.

F. W. McLafferty, *Interpretation of Mass Spectra*, 2d ed., Benjamin/Cummings, Menlo Park CA, 1973.

Problem Books Involving Spectroscopy

In the past few years many paperbacks have appeared that contain problems which involve the identification of unknown compounds. Molecular formula and molecular weight are usually provided, along with spectral information from IR, UV, NMR, and/or MS. The problems in these books cover a cross section of functional groups, many of which you have not yet studied. However, you may find several of these books useful at a later time.

R. H. Shapiro, *Spectral Exercises in Structural Determination of Organic Compounds*, Holt, New York, 1969.

R. M. Silverstein and G. C. Bassler, *Spectrometric Identification of Organic Compounds*, 3d ed., Wiley, New York, 1974 (IR, UV, NMR, MS)

B. Trost, *Problems in Spectroscopy*, Benjamin/Cummings, Menlo Park CA, 1969.

D. H. Williams and I. Fleming, *Spectroscopic Problems in Organic Chemistry*, McGraw-Hill, New York, 1967.

Chapter 19

Books

S. Patai (ed.), *The Chemistry of the Ether Linkage*, Wiley, New York, 1967.
H. Perst, *Oxonium Ions in Organic Chemistry*, Academic Press, New York, 1971.

Short Articles

H. K. Beecher, "Anesthesia," *Scientific American*, January 1957.
E. L. Jackson, "Periodic Acid Oxidation," in *Organic Reactions*, vol. 2, Roger Adams (ed.), Wiley, New York, 1948.
D. Swern, "Epoxidation and Hydroxylation of Ethylenic Compounds with Organic Peracids," in *Organic Reactions*, vol. 7, Roger Adams (ed.), Wiley, New York, 1948.

Chapter 20

Books

C. D. Gutsche, *The Chemistry of Carbonyl Compounds*, chaps. 1–4, Prentice-Hall, Englewood Cliffs NJ, 1967.

Short Articles

W. G. Brown, "Reductions by Lithium Aluminum Hydride," in *Organic Reactions*, vol. 6, Roger Adams (ed.), Wiley, New York, 1951.
A. Crease and P. Legzdins, "The Carbonyl Ligand As a Hard and Soft Base," *J. Chem. Ed.*, vol. 52, p. 499, 1975.
D. A. Shirley, "The Synthesis of Ketones from Acid Chlorides and Organometallic Compounds of Magnesium, Zinc and Cadmium," in *Organic Reactions*, vol. 8, Roger Adams (ed.), Wiley, New York, 1951.
A. L. Wilds, "Reduction with Aluminum Alkoxides [the Meerwein-Pondorf-Verley reaction]," in *Organic Reactions*, vol. 2, Roger Adams (ed.), Wiley, New York, 1944.

Chapter 21

Books

H. O. House, *Modern Synthetic Reactions*, 2d ed., Benjamin/Cummings, Menlo Park CA, 1972.

Short Articles

G. C. Bond, "Mechanism of Catalytic Hydrogenation and Related Reactions," *Quart. Rev.*, vol. 8, p. 279, 1954.
C. Djerassi, "The Oppenauer Oxidation," in *Organic Reactions*, vol. 6, Roger Adams (ed.), Wiley, New York, 1951.
E. J. Goller, "Stereochemistry of Carbonyl Addition Reactions," *J. Chem. Ed.*, vol. 52, p. 499, 1975.

C. Hurd, "Hemiacetals, Aldals, and Hemialdals," *J. Chem. Ed.*, vol. 43, p. 527, 1966.
D. Todd, "The Wolff-Kishner Reduction," in *Organic Reactions*, vol. 4, Roger Adams (ed.), Wiley, New York, 1948.

Chapter 22

Books

R. P. Bell, *The Proton in Chemistry*, Cornell University Press, Ithaca NY, 1959.
J. Hine, *Physical Organic Chemistry*, 2d ed., McGraw-Hill-Hill, New York, 1962.
C. A. VanderWerf, *Acids, Bases, and the Chemistry of the Covalent Bond*, Reinhold, New York, 1961.

Short Articles

D. Davidson, "Acids and Bases in Organic Chemistry," *J. Chem. Ed.*, vol. 19, p. 154, 1942.
W. J. Hickinbottom, "Monocarboxylic Acids of the Benzene Series," in *Chemistry of Carbon Compounds*, vol. IIIA, pp. 541–601, E. H. Rodd (ed.), Elsevier, New York, 1954.
A. Magliulo, "Prostaglandins," *J. Chem. Ed.*, vol. 50, p. 602, 1973.
D. A. Nugteren, D. A. VanDorp, S. Bergstrom, M. Hamberg, and B. Samuelsson, "Absolute Configuration of the Prostaglandins," *Nature*, vol. 212, p. 38, 1966.

Chapter 23

Books

E. S. Gould, *Mechanism and Structure in Organic Chemistry*, chap. 13, Holt, New York, 1959.
J. March, *Advanced Organic Chemistry: Reactions, Mechanisms, and Structure*, 2d ed., McGraw-Hill, New York, 1977.
S. Patai, (ed.), *The Chemistry of Carboxylic Acids and Esters*, Wiley, New York, 1969.

Chapter 24

Books

A. G. Cook, *Enamines: Synthesis, Structure, and Reactivity*, Dekker, New York, 1969.
C. D. Gutsche, *The Chemistry of Carbonyl Compounds*, Prentice-Hall, Englewood Cliffs NJ, 1969.
H. O. House, *Modern Synthetic Reactions*, 2d ed., chaps. 8–11, Benjamin/Cummings, Menlo Park CA, 1972.
R. E. Ireland, *Organic Synthesis*, Prentice-Hall, Englewood Cliffs NJ, 1969.
J. March, Advanced Organic Chemistry: *Reactions, Mechanisms, and Structure*, 2d ed., McGraw-Hill, New York, 1977.

Short Articles

E. Adams, "Barbiturates," *Scientific American*, p. 60, January 1968.
L. B. Hendry, "Insect Pheromones: Diet Related?", *Science*, vol. 192, p. 143, 1976.
J. Jacobus, "End Group Transfers: Mechanism of the Cannizzaro Reaction," *J. Chem. Ed.*, vol. 49, p. 349, 1972.
J. A. Katzenellenbogen, "Insect Pheromone Synthesis: New Methodology," *Science*, vol. 194, p. 139, 1976.

B. P. Mundy, "Alkylations in Organic Chemistry," *J. Chem. Ed.*, vol. 49, p. 91, 1972.

E. O. Wilson, "Pheromones," *Scientific American*, vol. 208, p. 100, 1963.

Chapter 25

Books

A. L. Lehninger, *Biochemistry*, 2d ed., chap. 23, Worth, New York, 1975.

F. J. Reithel, *Concepts in Biochemistry*, chap. 30, McGraw-Hill, New York, 1967.

Short Articles

A. J. Koning, "Analysis of Egg Lipids," *J. Chem. Ed.*, vol. 51, p. 48, 1974.

D. R. Paulson, J. R. Saranto, and W. A. Forman, "The Fatty Acid Composition of Edible Oils and Fats," *J. Chem. Ed.*, vol. 51, p. 406, 1974.

I. A. Wolff, "Seed Lipids," *Science*, vol. 154, p. 1140, 1966.

Chapter 26

Books

J. ApSimon (ed.), *The Total Synthesis of Natural Products*, vols. 1 and 2, Wiley, New York, 1973.

K. W. Bentley, *The Alkaloids*, Wiley, New York, 1957.

L. Fieser and M. Fieser, *Organic Chemistry*, 3d ed., D. C. Health, Lexington MA, 1956.

L. A. Paquette, *Principles of Modern Heterocyclic Chemistry*, Benjamin/Cummings, Menlo Park CA, 1968.

H. Zollinger, *Azo and Diazo Chemistry*, Wiley, New York, 1961.

Short Articles

L. N. Ferguson, "Cancer: How Can Chemists Help?", *J. Chem. Ed.*, vol. 52, p. 688, 1975.

A. R. Katritzky, "Heterocycles," *Chem. Eng. News*, p. 80, Apr. 13, 1971.

R. A. Moss, "Deamination Chemistry," *Chem. Eng. News*, p. 28, Nov. 22, 1971.

P. Rademacher and H. Gilde, "Chemical Carcinogens," *J. Chem. Ed.*, vol. 53, p. 757, 1976.

"Synthetic Drugs Used and Abused," *Chem. Eng. News*, p. 26, Nov. 2, 1970.

Chapter 27

Books

R. E. Lehr and A. P. Marchand, *Orbital Symmetry: A Problem Solving Approach*, Academic Press, New York, 1972.

T. H. Lowry and K. S. Richardson, *Mechanism and Theory in Organic Chemistry*, 2nd ed., Harper & Row, New York, to be published.

R. B. Woodward and R. Hoffmann, *The Conservation of Orbital Symmetry*, Academic Press, New York, 1970.

Short Articles

J. E. Baldwin, A. H. Andrist, and R. K. Pinschmidt, Jr., "Orbital-Symmetry-Disallowed Energetically Concerted Reactions," *Acct. Chem. Res.*, vol. 5, p. 402, 1972.

M. C. Caserio, "Reaction Mechanisms in Organic Chemistry," *J. Chem. Ed.*, vol. 48, p. 782, 1971.

K. N. Houck, "The Frontier Molecular Orbital Theory of Cycloaddition Reactions," *Accounts of Chemical Research*, vol. 8, p. 361, 1975.

R. G. Pearson, "Symmetry Rules for Chemical Reactions," *Chemistry in Britain*, vol. 12, p. 160, 1976.

J. J. Vollmer and K. L. Servis, "Woodward-Hoffmann Rules: Cycloaddition Reaction," *J. Chem. Ed.*, vol. 47, p. 491, 1970.

H. E. Zimmerman, "MO Following: The Molecular Orbital Counterpart of Electron Pushing," *Acct. Chem. Res.*, vol. 5, p. 393, 1972.

Chapter 28

Books

R. Barker, *Organic Chemistry of Biological Molecules*, Prentice-Hall, Englewood Cliffs NJ, 1971.

R. C. Bohinski, *Modern Concepts in Biochemistry*, Allyn and Bacon, Boston, 1973.

R. E. Dickerson and I. Geis, *The Structure and Action of Proteins*, Harper & Row, New York, 1969.

D. T. Elmore, *Peptides and Proteins*, Cambridge University Press, London, 1969.

A. L. Lehninger, *Biochemistry*, 2d ed., Worth, New York, 1975.

Organic Chemistry of Life, from *Readings from Scientific American*, W. H. Freeman, San Francisco, 1973.

Short Articles

A. Ault, "An Introduction to Enzyme Kinetics," *J. Chem. Ed.*, vol. 51, p. 381, 1974.

D. N. Buchanan and R. W. Kleinman, "Peptide Hydrolysis and Amino Acid Analysis," *J. Chem. Ed.*, vol. 53, 255, 1976.

L. C. Dickson, "Metal Replaced Hemoproteins," *J. Chem. Ed.*, vol. 53, p. 381, 1976.

H. M. L. Dieteren and A. P. H. Schouteten, "Preparation of the Proteins by Microorganisms," *J. Chem. Ed.*, vol. 47, p. 663, 1970.

J. R. Knox, "Protein Molecular Weight by X-Ray Diffraction," *J. Chem. Ed.*, vol. 49, p. 476, 1972.

D. C. Neckers, "Solid Phase Synthesis," *J. Chem. Ed.*, vol. 52, p. 695, 1975.

W. G. Nigh, "A Kinetic Investigation of an Enzyme Catalyzed Reaction," *J. Chem. Ed.*, vol. 53, p. 668, 1976.

R. Olby, "The Macromolecular Concept and the Origin of Molecular Biology," *J. Chem. Ed.*, vol. 47, p. 168, 1970.

I. Smith, M. J. Smith, and L. Roberts, "Models for Tertiary Structures; Myoglobin and Lysozyme," *J. Chem. Ed.*, vol. 47, p. 303, 1970.

A. G. Splittgerber, K. Mitchell, G. Dahle, M. Puffer, and K. Blomquist, "The Kinetics and Inhibition of the Enzyme Methemoglobin Reductase," *J. Chem. Ed.*, vol. 52, p. 680, 1975.

T. Vedvick and M. Coates, "Hemoglobin: A Simple Backbone Type of Molecular Model," *J. Chem. Ed.*, vol. 48, 537, 1971.

Chapter 29

Books

R. Barker, *Organic Chemistry of Biological Molecules*, Prentice-Hall, Englewood Cliffs NJ, 1971.

A. L. Lehninger, *Biochemistry*, 2d ed., Worth, New York, 1975.

Organic Chemistry of Life, from *Readings from Scientific American*, W. H. Freeman, San Francisco, 1973.

J. M. Stanek, J. Kocourek, and J. Pacak, *The Monosaccharides*, Academic Press, New York, 1963.

R. L. Whistler and C. L. Smart, *Polysaccharide Chemistry*, Academic Press, New York, 1953.

Short Articles

P. F. Blackmore, J. F. Williams, and M. G. Clark, "Biological Asymmetry of Glycerol," *J. Chem. Ed.*, vol. 50, 555, 1973.

Y. Z. Frohwein, "A Simplified Proof of the Constitution and the Configuration of D-Glucose," *J. Chem. Ed.*, vol. 46, p. 55, 1969.

F. Gabrielli, "Gluconeogenesis: A Teaching Pathway," *J. Chem. Ed.*, vol. 53, p. 86, 1976.

M. M. Green, G. Blackenhorn, and H. Hart, "Which Starch Fraction Is Water-Soluble, Amylose or Amylopectin?", *J. Chem. Ed.*, vol. 52, p. 729, 1975.

W. Guild, Jr., "Theory of Sweet Taste," *J. Chem. Ed.*, vol. 49, p. 171, 1972.

A. M. Lesk, "Progress in Our Understanding of the Optical Properties of Nucleic Acids," *J. Chem. Ed.*, vol. 46, p. 821, 1969.

R. B. Martin, "Configurational and Conformational Relations Among Sugars," *J. Chem. Ed.*, vol. 56, p. 641, 1979.

Index

Boldface numbers represent pages on which physical constants can be found. Items in red type represent material that is highlighted in the text.

Absolute configuration, 201
Absorption, relation to color, 494
Acetaldehyde (*See also* Ethanal), 431, 787, **791**
   acidity of, 1018
   discussion of, 794
   IR spectrum of, 806
Acetaldehyde cyanohydrin, preparation of, 825
Acetaldehyde methyl hemiacetal, preparation of, 832
Acetaldehyde methylimine, 813
Acetaldehyde methylimine, formation of, 837
Acetals, 828
   formation of, 828
   hydrolysis of, 830
   in carbohydrates, 1277
   mechanism of formation, 829
   mechanism of hydrolysis, 831
Acetamide, 940, **945**
   from acetyl chloride, 951
   NMR of, 998
Acetamido group, 1133
   directing effect of, 1133
p-Acetamidotoluene, from aniline, 587
Acetaminophen, 1013, 1158
Acetanilide, 941
Acetic acid (*See also* Ethanoic acid), 891, 892, **896**
   discussion of, 903
   from acetaldehyde, 903
   from malonic acid, 925
   NMR of, 998
Acetic anhydride (*See also* Ethanoic anhydride), 943, **945**, 981
   NMR of, 998
Acetoacetic ester (*See also* Ethyl acetoacetate), 1054
   acylation of, 1071
   ketones from, 1068
   synthesis, 1068
Acetone (*See also* Propanone), 788, **791**
   acidity of, 1018

aldol condensation of, 1039
   discussion of, 794
   IR spectrum of, 806
   NMR spectrum of, 808
   UV spectrum of, 512
Acetone azine, formation of, 838
Acetone cyanohydrin, preparation of, 825
Acetone hydrazone, formation of, 838
Acetonitrile, 176, 943
Acetophenone (*See also* Methyl phenyl ketone), 789, **791**
   aldol condensation of, 1039
Acetophenone ethylene ketal, preparation of, 830
Acetophenone ethylene glycol hemiketal, 830
Acetophenone imine, formation of, 837
Acetyl chloride, 941, **945**
   NMR of, 998
Acetyl group, 801, 987
Acetylacetaldehyde, 1078
Acetylacetone, 1078
Acetylation, 987
m-Acetylbenzesulfonic acid, from acetophenone, 802
Acetylene (*See also* Ethyne), 37
   dimerization of, 481
   MO's of, 465
   NMR chemical shifts in, 636
   preparation of, 467
Acetylide ion, 470
Acetylide ion, from acetylene, 470
2-Acetylpyrrole, 1149
Acetylsalicylic acid (*See also* Aspirin), 1013
Acid chlorides, acids from, 952
   aldehydes from, 796
   amides from, 950, 975
   anhydrides from, 953
   esters from, 951
   from carboxylic acids, 949
   hydrogenation of, 796
   infrared spectroscopy of, 997
   NMR of, 998

nomenclature of, 941
   physical properties of, 945
   preparation of, 948
   reaction with amines, 1129
   reactions of, 949
   reduction by lithium tri-*tert*-butoxyaluminum hydride, 797
   reduction of, 796
   table of, 945
Acidity constants. *See* Dissociation constants
Acidity, as a function of *s* character, 471
   of alcohols, 429
   of alkynyl anions, 471
   of β-dicarbonyl compounds, 1055, 1061
   of β-keto esters, 1055
   of barbitol, 1088
   of barbituric acid, 1088
   of carboxylic acids, 908
   of phenols, 700
   of various hydroxyl-containing compounds, 918
Acids. *See* Carboxylic acids
Acids and Bases, 166
Acrolein (*See also* Propenal), **791**, 1077
Acrylonitrile, from ethyne, 481
   preparation of, 826
Activated complex, 136, 148
Activating groups, in electrophilic substitution, 570
   in peptide syntheses, 1225
Acyl azides, 1225
Acyl compounds. *See* Carboxylic acid derivatives or specific group
Acyl group, 890
Acylation, of malonic ester and acetoacetic ester, 1071
Acylium ion, 799
   from anhydrides, 988
   resonance structures of, 799
   in mass spectroscopy, 809
Addition reactions, of aromatic compounds, 610

Addition, of free radicals to alkenes, 304
1,2-Addition, to α,β-unsaturated carbonyl compounds, 1073
1,2-Addition, to conjugated dienes, 305
1,4-Addition, to α,β-unsaturated carbonyl compounds, 1073
1,4-Addition, to conjugated dienes, 305
1,4-Addition, in Diels-Alder reaction, 336
Adenine, 1279
Adenine nucleotide, 1297
Adenosine diphosphate (ADP), 441, 1300
Adenosine monophosphate (AMP), 441
Adenosine triphosphate (ATP), 441, 1300
Apidic acid (*See also* Hexanedioic acid), 894, 896
  synthesis of, 905
Adipoyl chloride, synthesis of, 949
ADP, 441, 1300
Adrenaline, 228
Air pollution, 140
Alanine, 1211
  enantiomers of, 200
Alcohol dehydration, product distribution in, 277
  reactivity in, 277
  transition state in, 273
Alcohol, denatured, 426
Alcoholic silver nitrate, classification of alkyl halides, 399
Alcohols, acidity of, 428
  addition to alkynes, 481
  alkyl halides from, 749
  application of Grignard synthesis of, 846
  as key intermediates in aliphatic chemistry, 868
  carbinol names of, 410
  carboxylic acids from, 899
  charts showing reactions of, 869
  classification of, 409
  classification tests for, 453
  common names of, 410
  dehydration of, 270
  ease of dehydration, 270
  elimination versus substitution, 8 449
  esters from, 954
  ethers from, 742
  from carbonyl compounds, 841
  from carbonyl compounds and Grignard reagents, 844
  from carboxylic acids. 921
  from epoxides and Grignard reagents, 847
  from esters and Grignard reagents, 972
  from fatty acids, 923
  from glycerides, 1096
  from reduction of carbonyl compounds, 856
  from reduction of esters, 974
  general reactions of, 427
  in alkylation of benzene, 566
  in transesterification, 969
  IR spectroscopy of, 709
  IUPAC names of, 411
  mechanism for reaction of 2° and 3° with HX, 446
  mechanism for reaction with HX, 444
  mechanism of dehydration, 271
  mechanism of oxidation of, 435
  mechanism of reaction with PX_3 and PX_5, 451
  mechanism of reaction with thionyl chloride, 452
  NMR spectroscopy of, 711
  nomenclature of, 409
  of industrial importance, 425
  oxidation of, 430
  oxidation of secondary, 797
  physical properties of, 412
  pK_a's of, 430
  preparation by hydroboration-oxidation, 420
  primary, oxidation of, 431
  primary, oxidation to carboxylic acids, 433
  properties of, 413
  reaction with hydrogen halides, 442
  reaction with phosphorus trihalides, 450
  reaction with thionyl chloride, 452
  reaction with acids to give esters, 438
  reaction with metal, 428
  reactivity toward metals, 429
  resolution of, using anhydrides, 988
  secondary, oxidation of, 433
  solubility in water, 414
  spectral analysis of, 709
  summary of methods of preparation, 415, 455
  summary of reactions of, 456
  table of, 412
  tertiary, oxidation of, 434
  tosylates from, 748
Alcoholysis, in S_N1 reactions, 396
Aldaric acids, from aldoses, 1288
Aldehydes, addition of hydrogen cyanide to, 825
  aldol condensation of, 1037
  aldol condensations with nitriles, 1060
  aldol condensations with nitro compounds, 1060
  alkanes from, 861
  α-amino acids from, 1232
  α-carbanions from, 1018
  common names of, 787
  condensation with anhydrides, 1059
  crossed aldol condensations in, 1043
  derivatives of, 836
  dipole moments of, 793
  from acid chlorides, 796
  from alkenes, 795
  from glycols, 795
  from oxidation of primary alcohols, 431
  from primary alcohols, 795
  halogenation of, 1027
  hydrogenation of, 857
  infrared spectroscopy of, 805
  IUPAC names of, 787
  mass spectrometry of, 809
  NMR of, 807
  nomenclature of, 787
  nucleophilic addition reactions of, 818
  oxidation by Tollens' reagent, 867
  physical properties of, 790
  preparation of aromatic, 797
  qualitative tests for, 866
  reduction by lithium aluminum hydride, 858
  reduction by sodium borohydride, 858
  reduction of, 856
  resonance in aromatic aldehydes, 823
  spectral analysis of, 805
  summary of preparation of, 794, 810
  summary of reactions of, 871
  table of, 791
  UV spectroscopy of, 807
Aldehydic hydrogen, 1020
Alder, Kurt, 335
Alditols, from aldoses, 1291
Alditols, in carbohydrates, 1291
Aldohexoses, D-enantiomers of, 1264

Aldol condensations, 1037
  acid-catalyzed, 1041
  applications to synthesis, 1045
  base-catalyzed, 1037
  crossed, 1042
  dehydration, 1040
  mechanism of, in acid, 1041
  mechanism of, in base, 1038
  reverse of, in carbohydrates, 1298
  using nitriles, 1060
  using nitro compounds, 1060
Aldonic acids, from aldoses, 1287, 1292
  from ketoses, 1292
  in carbohydrates, 1285
D-Aldopentose, 1269
Aldoses, 1261
  aldaric acids from, 1288
  alditols from, 1291
  aldonic acids from, 1287
Aldrin, 689
Alkadienes (See also Dienes), 254
  addition reactions of, 304
  classification of, 255
  hydrogenation of, 313
  preparation of, 283
  summary of preparation of, 284
Alkaloid bases, 920
Alkaloids, 1160
Alkane family, names of, 82
Alkane halogenation, product distribution in, 160
Alkane nomenclature, examples of, 86
Alkanes, addition to alkenes, 329
  boiling points and van der Waals forces, 92
  classification of structure, 159
  combustion of, 138
  common names, 90
  density of, 95
  from alkenes by hydroboration, 423
  from alkyl halides, 842
  from alkynes, 475
  from carbonyl compounds, 861
  general formula, 78
  halogenation of, 130
  homologous series, 78
  industrial sources of, 121
  IUPAC nomenclature, 83
  melting and boiling points and densities, 93
  melting points, 94
  nitration of, 137

physical properties, 92
  pK$_a$ of, 470
  preparation by hydrogenation reactions, 127
  preparation by zinc metal reduction, 127
  preparation of by coupling reactions, 124
  solubility properties of, 95
  summary of methods of preparation, 142
  summary of reactions, 143
  table of, 93
Alkenes, addition of alkenes to, 327
  addition of bromine to, 318
  addition of halogens to, 317
  addition of hydrogen halides to, 290
  addition of hydrogen to, 310
  addition of hypohalous acid to, 326
  addition of methylene to, 336, 1187
  addition of sulfuric acid to, 300
  addition of water to, 296
  alkylation of benzene by, 567
  common names, 247
  conjugation with aromatic rings, 598
  conversion to alkanes by hydroboration, 423
  conversion to alkynes, 468
  cyclopropanes from, 1187
  decreasing stability, 264
  from alcohols, 270
  from alkyl halides, 261
  from alkynes, 475
  from amine oxides, 1147
  from amines, 1144
  from carbonyl compounds, 1051
  from esters, 1148
  from quaternary ammonium salts, 1144
  from selenoxides, 1148
  from sulfoxides, 1147
  from vicinal dihalides, 282
  from xanthates, 1148
  general formula, 243
  hydration of, 296
  hydroboration of, 417
  hydroxylation of, 333
  ionic addition reactions, 289
  IUPAC nomenclature, 248
  nomenclature of, 247
  oxidation of permanganate, 342
  ozonolysis of, 339

physical properties of, 252
  pK$_a$ of, 470
  preparation by dehydrogenation, 259
  preparation of dehydrohalogenation, 261
  properties of, 252
  qualitative analysis of, 344
  qualitative test for, 334
  reaction with alkanes, 329
  relative reactivities in addition reactions, 324
  stability from heats of hydrogenation, 315
  stability order, 317
  substitution reactions of, 336
  summary of addition reactions, 355
  summary of cleavage reactions, 357
  summary of preparation of, 283
  summary of substitution reactions, 357
  table of, 252
Alkenyl anion, structure of, 471
Alkenylbenzenes, addition of ionic reagents to, 599
  conjugation in, 598
  orbital view of conjugation in, 599
Alkenyl groups, names of, 248
Alkyl dihydrogen phosphate, 441
Alkoxide ion, in ether formation, 744
Alkoxide salts, formation of, 428
  naming of, 429
  reaction with water, 430
Alkoxy alcohols, from epoxides, 759
Alkoxy radical, 303
Alkyl anion, structure of, 471
Alkyl aryl ethers (See also Ethers), cleavage of, 755
Alkyl p-bromobenzenesulfonate, 750
Alkyl compounds, comparison of reactivities with carboxylic acid derivatives, 993
Alkyl groups, naming of, 84
  theory of orientation in electrophilic substitution, 575
Alkyl halides (See also Dihalides)
  amines from, 1125
  carboxylic acids from, 898
  classification of, 399
  conversion to alkanes, 842
  dehydrohalogenation of (See also Dehydrohalogenation). 261

Alkyl halides—*Continued*
 from alcohols, 442, 749
 from alcohols using PX₃ and
  PX₅, 450
 Grignard reagents from, 802
 in Friedel-Crafts reactions,
  563
 IR of, 690
 lactones from, 965
 nitriles from, 900
 organolithium reagents from,
  855
 physical properties of, 366
 spectral analysis of, 689
 summary of methods of preparation, 365
 summary of reactions, 401
 table of, 367
 UV of, 690
Alkyl hydrogen sulfate, 439,
  440, 1096
Alkyl migration, 279
Alkyl *p*-nitrobenzenesulfonate,
  750
Alkyl *p*-toluenesulfonate (*See
  also* Tosylates), 748
Alkyl shift, 279
1,2-Alkyl shift, example of, 281
 transition state in, 279
1,2-Alkyl shifts, 278
Alkyl sulfate, 440
Alkylation reactions, 329
 rearrangements in, 330
C-Alkylation, in enamine synthesis, 1050
C-Alkylation, in ether formation,
  752
N-Alkylation, in enamine synthesis, 1050
O-Alkylation, in ether formation,
  752
Alkylation, of alkenes by alkanes, 329
 of benzene (*See also* Friedel-
  Crafts alkylation), 563
 of benzene, using alcohols,
  566
 of diethyl malonate, 1063
 of ethyl acetoacetate, 1068
 of ketones, 1048
 of phenol, 706
Alkylbenzene ion radical, in MS,
  724
Alkylbenzenes, electrophilic substitution on, 587
 from alcohols, 566
 from benzene, 563
 side-chain halogenation of, 594
 side-chain oxidation of, 593
Alkylboron compounds, 417

Alkyllithium reagent, 124
Alkynes, addition of alcohols to,
  481
 addition of diborane to, 480
 addition of dienes to, 480
 addition of halogens to, 477
 addition of hydrogen cyanide
  to, 481
 addition of hydrogen halides
  to, 477
 addition of water to, 478
 as carbon acids, 470
 cis-alkenes from, 476
 cleavage reactions of, 482
 common names of, 465
 conversion to alkanes, 475
 conversion to alkenes, 475
 formation of heavy metal salts
  from, 472
 from alkenes, 468
 from heavy metal salts, 473
 from tetrahalides, 468
 from vicinal dihalides, 468
 general formula of, 464
 hydration of, 479
 hydroboration-reduction reactions of, 480
 hydrogenation of, 475, 476
 ketones from, 798
 oxidation of, 482
 physical properties of, 467
 pKₐ of, 470
 preparation by substitution reactions, 469
 reaction with base, 470
 reaction with cuprous ammonia complex, 472
 reaction with silver ammonia
  complex, 472
 reactions of ozone with, 482
 reactions of permanganate
  with, 482
 reactions with dienes, 480
 stereospecific hydrogenation
  of, 476
 structure of, 464
 substitution reactions on, 473
 summary of methods of preparation, 482
 summary of reactions, 483
 synthesis of, 473
 systematic nomenclature of,
  465
 table of, 466
 trans-alkenes from, 476
 use in Diels-Alder reactions,
  480
Alkynyl anions, 469
 as bases, 471, 474
 as nucleophiles, 473

 reaction with acids, 471
 structure of, 471
Allene, 255
D-Allonic acid, 1292
β-D-Allopyranose, 1274
D-Allose, 1264
Allyl alcohol, 409, **412**
Allyl bromide (*See also* 3-
  Bromopropene), 248, **367**
Allyl cations, 308
 from conjugated dienes, 307
 in nucleophilic substitution,
  389
 MO's of, 533
 resonance structures of, 307,
  533
Allyl chloride (*See also* 3-Chloro-
  propene), 336, **367**, 389
Allyl ether, **739**
Allyl group, 248
Allyl halides, reactivity in nu-
  cleophilic substitution, 389
Allyl iodide (*See also* 3-Iodo-
  propene), **367**
Ally phenyl ether, 752
Allyl radicals, in catalytic hy-
  drogenation, 313
 in substitution reactions of al-
  kenes, 337
 resonance in, 338
 resonance structures of, 533
 stability of, 339
 structure of, 337
Allylamine, **1107**
Allylbenzene, 549, **550**
Allylic hydrogens, 338
 ease of abstraction of, 338
o-Allylphenol, 752
D-Altronic acid, 1292
D-Altrose, 1264
Aluminum alkyls, 347
Aluminum trihalide, catalyst, in
  Friedel-Crafts alkylation,
  564
Ambident anion, 1020
Ambident base, 1020
Ambident nucleophile, 1020,
  1050
Amide anion, resonance structures of, 1129
Amide bond, in proteins and
  peptides, 1218
Amides, amines from, 1122
 coplanar structure of, 1239
 evidence for restricted rotation
  using NMR, 1001
 from acid chlorides, 975
 from anhydrides, 976, 985
 from carboxylic acids, 976
 from esters, 971, 976

from nitriles, 978
Hofmann degradation of, 1127
hydrogen bonding in, 946
hydrolysis of, 977
in proteins and peptides, 1218
infrared spectroscopy of, 997
NMR of, 998
nomenclature of, 940
physical properties of, 946
preparation from acid chlo-
    rides, 950
reactions of, 977
reduction of, 1122
resonance structures of, 1239
table of, 945
Amination, reductive, 1124
Amine oxide pyrolysis, 1147
Amine oxides, alkenes from,
    1147
from tertiary amines, 1143
Amines, addition to carbonyl
    compounds, 835, 1134
alkenes from, 1144
aromatic electrophilic substitu-
    tion on, 1132
aromatic, protecting groups in,
    1133
as resolving agents, 1120
basic properties of, 1108
basicity, discussion, 1110
classification of, 1104
classification of, using Hins-
    berg test, 1130
common names of, 1104
enamines from, 1131
from alkyl halides, 1125
from amides, 1122
from Gabriel synthesis, 1126
from Hofmann degradation of
    amides, 1127
from ketones, 1124
from nitriles, 1123
from nitro compounds, 1121
from oximes, 1123
hydrogen bonding in, 1108
IR spectroscopy of, 1161
IUPAC names of, 1105
K_b's of, 1109
mass spectrometry of, 1164
methods of preparation, 1121
NMR of, 1162
oxidation of, 1143
physical properties of, 1106
pK_b's of, 1109, 1111
primary, nitrosation of, 1136
pyramidal inversion in, 1023
reaction with acid, 1109
reaction with acid chlorides,
    975, 1129
resolution of, 1120

secondary, nitrosation of, 1142
secondary, reaction with ni-
    trous acid, 1142
solubility properties of, 1113
spectral analysis of, 1161
stereochemistry of, 1119
stereoisomers of, 237
structure of, 1104
sulfonamides from, 1130
summary of methods of prepa-
    ration, 1164
summary of reactions of, 1166
table of, 1107
tertiary, amine oxides from,
    1143
tertiary, nitrosation of, 1142
tertiary, reaction with nitrous
    acid, 1142
UV spectroscopy of, 1162
α-Amino acids, anionic form of,
    1212
cationic form of, 1212
chemical properties of, 1212
equilibrium in, 1212
from aldehydes, 1232
from α-halo acids, 926, 1125
from carboxylic acids, 1231
from ketones, 1232
from malonic ester, 1232
from Strecker synthesis, 1232
Henderson-Hasselbach equa-
    tion, 1214
isoelectric point in, 1213
L-configuration of, 1216
nomenclature of, 1211
phthalimidomalonic ester syn-
    thesis of, 1232
stereochemistry of, 1216
structure of, 1210
synthesis of, 1231
table of, 1211
titration curve for, 1215
zwitterions in, 1212
Amino alcohols, 835
dehydration of, 836
Amino group, 1105
α-Amino nitrile, 1232
2-Amino-3'-hydroxyazobenzene,
    1155
2-Amino-1-phenylpropane, 1157
4-Amino-2-nitrotoluene, 1122
p-Aminoazobenzene, 1155
p-Aminobenzenesulfonamide,
    1134, 1156
p-Aminobenzoic acid, 1157
Aminocyclohexane, 1123
Aminoethane, 1105, 1123
1-Aminoheptane, 1127
1-Aminohexane, NMR spectrum
    of, 1163

Aminomethane, 837
2-Aminopentane, 1122
4-Aminopentanoic acid, 1106
o-Aminophenol, **697**
m-Aminophenol, **697**
p-Aminophenol, **697**
oxidation of, 707
1-Aminopropane, 1105
2-Aminopyridine, 1151
γ-Aminovaleric acid, 1106
Ammonia, **414**
addition to carbonyl com-
    pounds, 835, 837
mechanism of addition to car-
    bonyl compounds, 835
structure of, 47
Ammonium acetate, 917
Ammonium butanoate, 985
Ammonium butyrate, 985
Ammonium cyanate, 1
Ammonium ethanoate, 917
Ammonium hydrogen succinate,
    917
Ammonium p-toluoate, 976
Ammonium salts, formation of,
    1108
nomenclature of, 1110
stereoisomers of, 237
Ammonolysis, 397
AMP, 441
Amphetamines, 1157
(n-)Amyl acetate, 945
synthesis of, 956
(n-)Amyl alcohol, 410
tert-Amyl alcohol, 410
(n-)Amyl ether, **739**
Amylopectin, 1309, 1310
Amylose, 1309, 1310
Androgens, 865
Anesthetics, 1157
Angle of rotation, α, 186
Anhydrides, acids from, 985
amides from, 976, 985
condensation with aldehydes,
    1059
cyclic, 943
esters from, 962, 985
from acid chlorides, 953
from carboxylic acids, 980
from dicarboxylic acids, 982
in resolution of alcohols, 988
infrared spectroscopy of, 997
ketones from, 987
NMR of, 998
nomenclature of, 942
reaction with alcohols, 985
reaction with ammonia, 985
reaction with water, 985
reactions of, 985
table of, 945

Aniline, 546, 1105, **1107**
  from bromobenzene, 683
  from organothallium reagent,
    603
  Ir spectrum of, 1161
  resonance effect in elec-
    trophilic substitution, 582
Anilinium hydrogen sulfate,
  1110, 1133
Anilinium ions, resonance structures of, 1112
Animal fats, 1090
Anions, ambident, 1020
p-Anisidine, 1106
Anisole, 546, 734, **739**
  IR spectrum of, 773
[14]-Annulene, 540, 690
  NMR chemical shift in, 691
[16]-Annulene, 541
[18]-Annulene, 541
  NMR chemical shifts in, 638
Annulenes, nomenclature of,
  690
Anomeric effect, in carbohydrates, 1275
Anomers, in carbohydrates,
  1270
Antara attack, 1192
Antarafacial attack, 1192
Anthracene, 552
  mass spectrum of, 728
  orbital picture of, 552
  reaction with benzyne, 687
  UV spectrum of, 608
9,10-Anthraquinone, 552
anti-Addition, of bromine to alkenes, 320
anti-Conformation, 73
anti-Elimination, 263
  evidence for in dehydroha-
    logenation, 269
  in Hofmann elimination,
    1146
  in dehydrohalogenation, evi-
    dence for, 269
anti-Hydroxylation, from epoxides, 760
anti-Markovnikov addition, 302
  in hydroboration, 419
Antiaromatic, 535
Antiaromatic systems, 545
  summary of, 540
Antibonding orbital, 56
Antifreeze, 427
Antioxidants, 1099
Apocamphyl chloride, 391
Aprotic solvents, 175, 387
Ar-, Defined, 585
D-Arabinose, 1294
Arabinose, 1272
Arachidic acid, 892

Arginine, 1211
Aromatic compounds (*See also*
    Aromatic hydrocarbons), hy-
    drogenation of, 592
  IR spectroscopy of, 605
  NMR spectra of, 650
  physical properties of, 549
  qualitative tests for, 604
  spectral analysis of, 605
  summary of reactions of, 609
  ultraviolet spectroscopy of,
    607
Aromatic hydrocarbons (*See
    also* Aromatic compounds),
    from diazonium salts, 1138
  polyhalogenated, 688
  polynuclear, 551
  source of, 550
  table of, 550
  UV absorptions of, 608
Aromatic ketones, from acid
  chlorides, 799
Aromatic systems, summary of,
  540
Aromatic thiols, from diazonium
  salts, 1138
Aromaticity and molecular orbital theory, 542
Aromaticity and ring currents, in
  NMR, 637
Aromaticity, defined, 534
  evidence for using NMR, 691
  of anthracene, 552
  of cycloheptatrienyl cation,
    536
  of cyclopentadienyl anion,
    535
  of furan, 539
  of naphthalene, 552
  of phenanthrene, 552
  of polynuclear aromatic hydro-
    carbons, 552
  of pyridine, 538
  of pyrrole, 538
  of thiophene, 539
  requirements for, 534
  summary of, 539
  use of Hückel's $(4n + 2)$ π-
    electron rule, 534
Aromatization, 551
  of cyclic compounds, 1089
  of cycloalkenes, 1183
Arrows In Mechanisms, 276
Arrows, single-headed, 134
Arrows, Use Of In Mechanisms,
  125, 134
Aryl bromides, from diazonium
  salts, 1138
Aryl chlorides, from diazonium
  salts, 1138
Aryl ethers. *See* ethers

Aryl fluorides, from diazonium
  salts, 1138
Aryl halides, carboxylic acids
    from, 898
  dehydrohalogenation of, 683
  electrophilic substitution on,
    583
  elimination-addition reactions
    of, 685
  from diazonium salts, 1138
  Grignard reagents from, 802
  IR of, 690
  nomenclature of, 545
  nucleophilic substitution on,
    672
  organolithium reagents from,
    855
  physical properties of, 671
  preparation of, 561, 603, 670
  resonance structures of, 671
  spectral analysis of, 689
  structure of, 670
  summary of reactions, 691
  table of, 671
Aryl iodides, from diazonium
  salts, 1138
Aryl nitriles, from diazonium
  salts, 1138
Aryl xanthate, 1140
Arylsulfonic acids, conversion to
    phenols, 699
  electrophilic substitution on,
    588
  nitriles from, 902
Asparagine, 1211
Aspartic acid, 1212
Aspirin, 1013, 1158
Asymmetric, 191
Asymmetric carbon atom, 191
  generation of by free-radical
    halogenation, 211
Asymmetric germanium com-
    pounds, stereochemistry,
    238
  other atoms, 238
Asymmetric induction, 230
Asymmetric molecules, no chiral
    centers, 196
  one asymmetric carbon, 195
Asymmetric solvation, 1023
Asymmetrical stretching, 495
Asymmetry, 194
Atactic polymers, 347
Atomic and molecular structure,
    historical overview, 17
Atomic orbital structure, $1s$, $2s$,
    $2p_x$, 54
  $2p$, $3d$, 55
Atomic orbital theory, 54
Atomic orbital, $2p$, 21
  $2s$, 21

Atomic orbitals, geometry of, 20
  linear combination of, 56
ATP, 441, 1300
D-Atrolactic acid, 1121
Aufbau principle, 23
Auto-oxidation, 1047
Auxochrome, 510
Avogadro's number, 493
Axial hydrogens, 102
Azelaic acid, 894
Azeotrope, defined, 426
Azeotropes, in acetal formation, 830
Azeotropic mixture, ethanol-water, 426
Azetidine, 1114
Azine, 837
Aziridine, 1114
Azo compound, 1153
Azo dyes, 1154
Azo group, 1153
Azobenzene, 1155
anti-Azobenzene, 1156
syn-Azobenzene, 1156
Azolidine, 1114

Baeyer strain theory, 98
Baeyer test, 334, 344
  for alkenes, 334
Bakelite, 833
  mechanism of formation of, 834
Banana oil, 945
Barbituates, synthesis of, 1087
Barbituric acid, 1087
Base dissociation constants. See Dissociation constants
Bases, ambident, 1020
Basicity, of heterocyclic nitrogen compounds, 1116
Bathochromic shift, 510
Beauvault-Blanc process, 1095
Beer-Lambert law, 503
Behenic acid, 892
Benedict's reagent, in carbohydrates, 1286
Benzal chloride, from toluene, 595
Benzaldehyde, 546, 788, **791**
  from benzoyl chloride, 796
  IR spectrum of, 806
Benzaldehyde dimethyl acetal, preparation of, 830
Benzaldehyde oxime, 1123
Benzamide, **945**
  IR spectrum of, 993
Benzedrine, 1157
Benzene, **550**
  addition of chlorine to, 593

alkylation by alcohols, 566
alkylation by alkenes, 567
alkylation of, 563
alkylation with tetrapropylene, 588
bromination of, 561
chlorination of, 561
delocalization energy of, 525, 529
diagram showing delocalization energy of, 530
disubstitution of, 589
electronic structure of, 525
electrophilic substitution reactions of, 556
from sodium benzoate, 924
halogenation of, 561
heat of combustion of, 530
heats of hydrogenation of, 529
hydrogenation of, 128, 524
Kekulé structures of, 523
mass spectrum of, 728
mechanism of nitration of, 558
MO's of, 543
MO's of the π system of, 527
molecular orbit (MO) theory of, 526, 542
monosubstitution reactions of, 557
nitration of, 558
NMR chemical shifts in, 636
orbital picture of, 525
π clouds in, 525
possible structures of, 522
reactivity and properties of, 523
resonance hybrids, 529
resonance in, 525, 528
resonance structures in nitration of, 559
resonance structures of, 528, 531
structure of, 525
structure proof of, 523
sulfonation of, 562
transition states in nitration of, 560
UV spectrum of, 607
Benzene derivatives, nomenclature of, 545
Benzenediamide, **945**
Benzenediazonium hydrogen sulfate, 1140
1,4-Benzenedicarboxylic acid, 895
1,2-Benzenedicarboxylic anhydride, 943
Benzenedioic acid, **896**
1,2-Benzenedioyl chloride, 941
Benzenedioyl chloride, **945**

Benzenehexachloride (BHC), 593, 688
Benzenesulfonamides, 1130
Benzenesulfonic acid, 546
  conversion to phenol, 699
  electrophilic substitution, 577
  from benzene, 562
Benzenesulfonyl chloride, 1130
1,3,5-Benzenetricarboxylic acid, 895
Benzoate anion, resonance structures of, 913
Benzocaine, 1158
Benzoic acid, 546, 893, **896**
  from acetophenone, 1032
  from benzoyl chloride, 953
  from n-butylbenzene, 594
  from toluene, 593
  IR spectrum of, 928, 993
  reduction of, 923
Benzoic anhydride, 943, 981
  IR spectrum of, 993
Benzonitrile, 944, 1123
  from organothallium reagent, 603
Benzophenone (See also Diphenyl ketone), 789, **791**
  IR spectrum of, 806
Benzopyrene, 553
p-Benzoquinone, 707
1,4-Benzoquinone, 707
Benzotrichloride, from toluene, 595
Benzoyl chloride, **945**
  IR spectrum of, 993
Benzoyl peroxide, 136
o-Benzoylbenzoic acid, 987
Benzyl alcohol, **412**
  from benzaldehyde, 857
  from benzoic acid, 923
  mass spectrum of, 726
Benzyl amine, 1127
Benzyl cations, in electrophilic addition to alkenylbenzenes, 600
  relative stability of, 601
  resonance structures of, 601
  structure of, 601
Benzyl chloride, 594
Benzyl group, 547
Benzyl radicals, 596
  delocalization in, 597
  formation of, 596
  relative stability of, 596
  structure of, 597
Benzylamine, **1107**, 1123
2-Benzylcyclohexanone, 1131
2-Benzylhexanoic acid, 1066
Benzylic hydrogens, 595
3-Benzyl-N-methylindole, 1116

Benzyne, 684
  evidence for mechanism in formation of, 685
  formation of, 683
  mechanism for formation of, 684
  orbital view of, 687
  structure of, 687
Betaine, 1053
BHC (*See also* Benzenehexachloride), 688
Bicyclic compounds, application of Diels-Alder reaction, 1181
Bicyclic compounds, nomenclature of, 390
Bicycloalkyl halides, reactivity in nucleophilic substitution, 390
Bicyclo [2.2.1] heptane, 391
Bicyclo [2.2.1]-2-heptene, 391, 1182
Bicyclo [2.2.2]-2-octene, 1182
Bicycloundecapentaene, 540
Bimolecular reaction, 178
  in nucleophilic aromatic substitution, 674
  in S$_N$2 reaction, 369
Biodegradable, 1097
Biodegradable detergent, 588
Biopolymers, oligosaccharides and polysacchrides, 1305
Biot, Jean-baptiste, 188
Biphenyl 549, **550**
Birth control compounds, 866
Bisabolene, 331
Boat conformation, 103
Bohr atom, 17
Boiling point, 7
Boiling points, use to determine hydrogen bonding, 414
Bond dissociation energies, table of, 152
Bond dissociation energy, 25, 151
Bond length, 24
Bond Polarity and Reactive Intermediates, 169
Bond strength, 151
Bonding orbital, 56
Borane, 416
Borane-tetrahydrofuran complex, 416
Borane-ether complex, 417
Boron, hybridization in, 46
  structure of, 46
Boron trifluoride, 417
  structure of, 47
Boron trifluoride etherate, 417
Bouveault-Blanc procedure, 975

Branched-chain molecules, 71
Bradykinin, 1237
Bridgehead carbon atoms, 390
Brilliant Yellow, 1154
Bromination, mechanism of bromine addition to cycloalkenes, 320
  of benzene, 561
  of methane, energy profile diagram of, 155
Bromine in carbon tetrachloride, 344
Bromine water, oxidation of carbohydrates, 1287
  reaction with phenol, 705
  reaction with phenoxide ion, 706
  test for phenols, 708
Bromine, addition to alkenes, 318
Bromine/carbon tetrachloride, test for alkenes, 317
1-Bromo-2-butanone, 1029, 1034
3-Bromo-2-butanone, 1029, 1034
1-Bromo-2-butene, 305
2-Bromo-2-butene, 468
3-Bromo-1-butene, 305
2-Bromo-3-chlorobutane, 230
(2R,3R)-2-Bromo-3-chlorobutane, 230
(2S.3R)-2-Bromo-3-chlorobutane, 230
1-Bromo-1-chloroethane, enantiomers of, 213
trans-1-Bromo-2-chloroethene, 249
E-2-Bromo-1-chloro-1-iodopropene, 250
1-Bromo-2-chloropropane, from propene, 319
1-Bromo-1,2-dichloropropane, stereoisomers of, 215
p-Bromo-N,N-dimethylaniline, 1106
3-Bromo-2,2-dimethylbutane, 850
4-Bromo-1,3-dioxane, 1115
2-Bromo-3-hexene, 306
4-Bromo-2-hexene, 306
1-Bromo-2-iodoethane, from ethene, 319
2-Bromo-5-isopropylnonanoic acid, 893
trans-4-Bromo-4-methylcyclohexanol, 406
1-Bromo-2-methylpropane (*See also* Isobutyl bromide), 133, **367**, 475
2-Bromo-2-methylpropane (*See also* tert-Butyl bromide), 133, **367**

2-Bromo-2-methylpropanoic acid, synthesis of, 926
2-Bromo-5-nitrotoluene, 548
1-Bromo-1-phenylethane, from ethylbenzene, 595
1-Bromo-2-propanol, from propene, 326
3-Bromo-1-propene, 752
p-Bromoacetanilide, 1133
α-Bromoacetone, 1027
α-Bromoacetophenone, 1033
N-Bromoamides, 1128
o-Bromoaniline, 1105
p-Bromoanilinium chloride, 1110
o-Bromobenzaldehyde, 788
m-Bromobenzaldehyde, 859
p-Bromobenzamide, 940, 971
Bromobenzene, 545, **671**
  from benzene, 561
o-Bromobenzoic acid, 548
m-Bromobenzoyl chloride, synthesis of, 949
m-Bromobenzyl alcohol, from *m*-bromobenzaldehyde, 859
p-Bromobenzyl bromide, 901
1-Bromobutane (*See also* n-Butyl bromide), 133, **367**
  from 1-butanol, 442, 450
1-Bromobutane-1-*d*, 755
2-Bromobutane (*See also* sec-Butyl bromide), **367**
  dehydrohalogenation of, 265
  enantiomers of, 191
  from 1-butene, 291
  from 2-butanol, 450
S-2-Bromobutane, 210
Bromochlorofluoromethane, enantiomers of, 191
R-Bromochlorofluoromethane, 203
S-Bromochlorofluoromethane, 203
Bromocyclobutane, 96
Bromocyclohexane (*See also* Cyclohexyl bromide), **367**
  conformations of, 657
  from cyclohexanol, 450
  IR spectrum of, 502
trans-3-Bromocyclohexanecarboxylic acid, 965
trans-4-Bromocyclohexanol, 406
3-Bromocyclohexene, from cyclohexene, 337
Bromocyclopentane, from cyclopentanol, 442
Bromoethane (*See also* Ethyl bromide), **367**
  from ethene, 290
Bromoethene, **671**
  addition of HBr to, 301

β-Bromoethyl alcohol, 410
α-Bromoethylbenzene, 595
1-Bromohexane (See also n-Hexyl bromide), 367
α-Bromomalonic ester, 1127
Bromomethane (See also Methyl bromide), 367
from methane, 130
NMR chemical shifts in, 633
m-Bromonitrobenzene, 547
from nitrobenzene, 586
Bromonium ion, 320
as an electrophile, 561
in halogenation of benzene, 561
2-Bromooctane, from 2-octanol using tosylates, 749
1-Bromopentane (See also n-Pentyl bromide), 127, 367
o-Bromophenol, 704
m-Bromophenol, 695
p-Bromophenol, 697, 704
2-(4-Bromophenyl)ethanenitrile, 901
p-Bromophenylacetonitrile, 901
p-Bromophenylethanoic acid, synthesis of, 901
1-Bromopropane (See also n-Propyl bromide), 367
from cyclopropane, 294
from propane, 132
from propene, 302
2-Bromopropane (See also Isopropyl bromide), 367
from propane, 132
from propene, 302
2-Bromopropene, from propyne, 478
3-Bromopropene (See also Allyl bromide), 248, 367
3-Bromopyridine, 1149
2-Bromopyrrole, 1149
N-Bromosuccinimide, 337
α-Bromotoluene, 547
Brønsted-Lowry theory, 167
Brosylates, 750
Brosylate group, 386
Buna-S, 349
1,2-Butadiene, 255
1,3-Butadiene, 255, 258
addition of chlorine to, 324
addition of HBr to, 305
addition of hydrogen to, 313
addition to, 305
from butane, 283
from ethyne, 481
preparation of, 283
structure of, 257
Butanal (See also n-Butyraldehyde), 791
Butanamide, 945, 985

(n-)Butane, 72, 93
conformations, 72
cracking of, 283
from 1,3-butadiene, 313
from 1-butene, 423
from 2-butene, 311
from 1-butyne, 475
from 2-butyne, 475
from cyclobutane, 312
halogenation of, 133
structural isomers of, 71
Butanediamide, 940
Butanedinitrile, 902
Butanedioic acid, 894
synthesis of, 902
Butanedioic anhydride, 943
1,2-Butanediol, 769
1,4-Butanediol, 283, 737, 769
2,3-Butanediol, from 2-butene, 335
Butanethiol, 1152
Butanoic acid, 896
Butanoic anhydride, 985
1-Butanol (See also n-Butyl alcohol), 412
dehydration of, 271
from 1-butene, 420
n-butyl ether from, 744
mechanism of dehydration of, 278
synthesis of, 847
1-Butanol-1-d, 755
2-Butanol (See also sec-Butyl alcohol), 412
dehydration of, 271
enantiomers of, 200
from 2-butanone, 857
racemization of, 299
Butanone (See also Methyl ethyl ketone), 789, 791, 857
from 2-butanol, 433
sodium bisulfite addition product, 827
Butanoyl chloride, 945
1-Buten-3-ol, 857
2-Buten-1-ol, 859
1-Buten-3-one (See also Methyl vinyl ketone), 791
3-Buten-2-one, 789, 857
1-Buten-3-yne, 481
2-Butenal (See also Crotonaldehyde), 791, 1040
reduction of, 859
3-Butenal, 1040
1-Butene, 244, 252, 258, 316
from 1-bromobutane, 430
mechanism of HBr addition to, 293
preparation of, 261
2-Butene, 244
geometric isomers of, 245

isomerization of, 246
cis-2-Butene, 245, 252, 316
trans-2-Butene, 245, 252, 316
E-2-Butene, 250
Z-2-Butene, 250
cis-2-Butene oxide, 737
1-Butene-3-yne, from ethyne, 481
2-Butenyl group, 249
tert-Butoxycarbonyl chloride, 1223
Butter, 1092
Butter Yellow, 1154
Butyl 4-methylpentanoate, 942
Di-n-butyl adipate, 942
(n-)Butyl alcohol (See also 1-Butanol), 412
sec-Butyl alcohol (See also 2-Butanol), 270, 412
tert-Butyl alcohol (See also 2-Methyl-2-propanol), 91, 270, 412
n-Butyl benzoate, 974
synthesis of, 961
n-Butyl bromide (See also 1-Bromobutane), 367
sec-Butyl bromide (See also 2-Bromobutane), 91, 367
tert-Butyl bromide (See also 2-Bromo-2-methylpropane), 367
n-Butyl tert-butyl ether, synthesis of, 746
tert-Butyl cation, 164
n-Butyl chloride (See also 1-Chlorobutane), 90, 367
sec-Butyl chloride (See also 2-Chlorobutane), 367
tert-Butyl chloride (See also 2-Chloro-2-methylpropane), 91, 367
(n-)Butyl ether, 734, 739
from 1-butanol, 744
IR spectrum of, 773
synthesis of, 749
(n-)Butyl group, 84
sec-Butyl group, 84
tert-Butyl group, 84
Di-butyl hexanedioate, 942
n-Butyl iodide (See also 1-Iodobutane), 367
sec-Butyl iodide (See also 2-Iodobutane), 367
tert-Butyl iodide (See also 2-Iodo-2-methylpropane), 367
Di-tert-butyl ketone, 856
(n-)Butyl phenyl ether, synthesis of, 747
Butyl phenyl ketone, 788
Di-n-butyl phthalate, 970
synthesis of, 955

(*n*-)Butyl tosylate, 749

cis-4-*tert*-Butyl-1-chlorocyclo-
hexane, 269

tert-Butylamine, **1107**

Di-*n*-butylamine, **1107**

n-Butylamine, **1107**

sec-Butylamine, **1107**

(*n*-)Butylbenzene, **550**
oxidation of, 594

tert-Butylbenzene, 545, **550**
from benzene and *tert*-butyl
alcohol, 566
nitration of, 572

Tri-*n*-butylboron, 417

Tri-*tert*-butylcarbinol, 856

Butylene (*See also* Butene), 247

tert-Butyllithium, 855

(*n*-)Butylmagnesium bromide,
853

5-*sec*-Butylnonane, 87

2-*n*-Butylpyridine, 1151

2-Butynamide, 940

1-Butyne, **466**

2-Butyne, **466**, 469
addition of HX to, 478
from 2-butene, 468
halogenation of, 477
IR spectrum of, 501

(*n*-)Butyraldehyde (*See also* Butanal), 787, **791**

Butyramide, 985

(*n*-)Butyramide, **945**

(*n*-)Butyric acid, 891, 892, **896**

Butyric anhydride, 985

Butyrophenone, 861

Butyryl chloride, 945

C- versus O-Alkylation, in Malonic Ester Synthesis, 1064

Cahn-Ingold-Prelog system of
configuration, 201

Cahn, R. S., 201

Calcium carbide, 467

Calcium cyclamate, 1159

Camphor, 351

Cancer, from polynuclear aromatic hydrocarbons, 768

Cannizzaro reaction, 1047
mechanism of, 1047

Capric acid, 891, 892, **896**

Caproaldehyde (*See also* Hexanal), **791**

Caproic acid, 891, 892, **896**

Caprolactam, 980

Caprylic acid, 891, 892, **896**

Carbamate, 1223

Carbamic acid, 1128

Carbanions, addition to α, β-unsaturated carbonyl com-
pounds, 1075

alkynyl anions, 470
from aldehydes and ketones,
819
from carbon acids, 1017
general discussion of, 166
hydrolysis of, 842
in Grignard reagents, 803
in nucleophilic aromatic substitution, 677
in organolithium reagents,
855
in Wittig reaction, 1051
pyramidal inversion in, 1022
reaction with water, 842
stability of, 166
stabilization by electron-
withdrawing substituents,
679
structure of, 43
structure of nonstabilized,
1021

α-Carbanions, 1017
evidence for 1020
from carbonyl compounds,
1018
from esters, 1019
in aldol condensations, 1037
in conjugate addition reactions, 1073
in keto-enol equilibrium, 1020
orbital view of, 1020
structure of 820, 1019, 1025

Carbene (*See also* Methylene),
44, 1187
structure of, 44

Carbethoxy group, 942

2-Carbethoxycyclohexanone,
1058, 1071

2-Carbethoxycyclopentanone,
1057

β-Carbethoxypropionamide, 986

β-Carbethoxypropionyl chloride,
986

β-Carbethoxypropiophenone,
986

Carbinol names, of alcohols,
410

Carboalkoxy group, 942

Carboaryloxy group, 942

Carbobenzoxy chloride, 1223

Carbocation character, 322
in hydroboration, 418

Carbocation rearrangements, 278
in S$_N$1 reactions, 382

Carbocation stabilities, 301

Carbocation stability, in electrophilic substitution, 573

Carbocations in Dehydration of
Alcohols, 274

Carbocations versus carbonium
ions, 163

Carbocations, acylium ion, 799
addition of nucleophiles to,
292, 301
addition to alkenes, 328
benzyl cations, 601
ease of formation, 170
evidence for structure of, 391
from alkenes, 292
from alkyl halides in Friedel-
Crafts reactions, 563
from diazonium salts, 1136
from oxonium ions, 273
general discussion, 163
geometry of, 298
in acetal formation, 829
in alcohol dehydration, 272
in aldol condensations, 1041
in ether formation, 743
in hydration of alkenes, 297
in keto-enol equilibrium, 1035
in natural products, 331
in reactions of alcohols with
HX, 447
in S$_E$2 reactions, 563
in S$_N$1 reactions, 379
loss of protons from, 273
rearrangement of, 278
rearrangements in biochemical
systems, 767
rearrangements in electrophilic
addition to alkenes, 295
rearrangements in Friedel-
Crafts alkylation, 565
resonance-stabilized, 1137
stability due to resonance effect, 581
stability of, 165
stabilized, 601
stereochemical evidence for
structure of, 299
structure of, 43, 274
summary, 332

Carbohydrates, α-designation of,
1273
acetals and ketals of, 1276
addition of HCN to, 1292
alditols in, 1291
aldohexoses, enantiomers of,
1264
aldonic acids in, 1285
anomeric effect in, 1275
anomers in, 1270
β-designation of, 1273
Benedict's reagent in, 1286
biochemical reactions, 1296
bromine and water oxidation
of, 1287
conformational analysis of,
1275

Carbohydrates, D- or L-Configuration of, 1265

Carbohydrates, D-family of, 1265
diastereomers in, 1270
disaccharides, 1305
disaccharides, structure elucidation, 1306
epimerization in, 1285
epimers in, 1266, 1285
equilibrium in, 1273
Fehling's reagent in, 1286
Fischer projection formulas, drawing of, 1269
furanoses in, 1269
glucose, structure proof of, 1301
glycosides of, 1277, 1278
Haworth structures of, 1270
Haworth structures, drawing of, 1270
hemiacetals and hemiketals, reactions of, 1277
hemiacetals in, 1267, 1268
hemiketals in, 1268
in nucleic acids, 1280
keto-enol equilibrium in, 1285
Kiliani-Fischer synthesis in, 1291
kinetic vs thermodynamic control in, 1269
L-family of, 1265
lactones in, 1293
molecular formula of, 1260
mutarotation in D-glucose, 1267
nitric acid oxidation of, 1288
nonreducing, 1285
osazone formation, 1294
oxidation reactions of, 1284
periodic acid oxidation of, 1289
photosynthesis of, 1261
pyranoses in, 1269
reactions of, 1284
reducing, 1285
reduction by NADH, 1297
reduction reactions of, 1290
ring size in, 1269
ring size proof, of glucose, 1304
ring-chain tautomerism in, 1268
Ruff degradation of, 1293
skeletal framework of, 1272
stereochemistry of, 1263
Tollens' test for, 1284
Carbomethoxy group, 942
2-Carbomethoxycyclobutanone, 942
Carbon acids, acidity of, 1017
Carbon anions (See also Carbanions), 43

Carbon Atoms, Types of, 85
Carbon tetrachloride (See also Tetrachloromethane), 53
Carbon, classification of (1°, 2°, 3°, 4°), 123, 159
β-Carbon, defined, 373
α-Carbon, defined, 373
Carbon, hydridization in, 28
primary, 123
quarternary, 123
secondary, 123
tertiary, 123
types of, 123
Carbon-13 magnetic resonance, discussion of, 660
example of, 660
Carbon-13, in NMR, 660
Carbon-14, use in benzyne mechanism, 686
Carbon-carbon double bond, 33
Carbon-carbon double bond, in alkenes, 243
Carbon-carbon triple bonds, physical properties of, 37
structure of, 37, 464
Carbon-oxygen double bonds, physical properties of, 41
polarity in, 40
preparation of, 786
reactions of, 818
structure of, 40
Carbonate ester, 1058
Carbonate ion, resonance structures of, 532
structure of, 532
Carbonation, 898
Carbonium ion, 164
Carbonium Ions versus Carbocations, 163
Carbonium ions (See Carbocations), 45
Carbonyl chloride group, 941
Carbonyl compounds, α-carbanions from 1018
acid-catalyzed aldol condensations of, 1041
acid-catalyzed reactions of, 1033
addition of alcohol to, 828
addition of amines to, 1134
addition of ammonia and amines to, 835
addition of Grignard reagents to, 841
addition of hydrogen cyanide to, 825
addition of sodium bisulfite to, 826
alcohols from, using Grignard reagents, 844
alkenes from, 1051

conjugated, reduction of, 859
derivatives of, oximes, 836
electrophilic substitution on, 820
hydration of, 839
hydrogenation of, 857
mechanism of acid-catalyzed halogenation, 1034
mechanism of addition of ammonia and amines, 835
mechanism of addition of Grignard reagents to, 843
reaction with organolithium reagents, 856
reduction by lithium aluminum hydride, 858
reduction by sodium borohydride, 858
reduction to methylene, 861
stereochemistry of nucleophilic addition to, 824
Carbonyl group, structure of, 40, 786
Carbonyl groups, reduction of, 856
Carbophenoxy group, 942
2-Carbophenoxybutanedioic acid, 942
α-Carbophenoxysuccinic acid, 942
Carbowax, 1099
Carbowaxes, 763
Carboxamide group, 940
Carboxyl group, 890
Carboxylate anions, as nucleophiles, 953
esters from, 961
in nucleophilic substitution, 965
reaction with acid, 918
resonance structures of, 909
structure of, 909
Carboxylate salts, formation of, 916
Carboxylate salts, nomenclature of, 917
Carboxylic acid derivatives, interconversion of, 946
nomenclature of, 940
relative reactivities of, 992, 993
spectral analysis of, 996
Carboxylic Acid Derivatives, Summary of Interconversions, 991
Carboxylic acid derivatives, comparison of reactivities with alkyl compounds, 993
summary of methods of preparation, 1002
summary of reactions of, 1005

Carboxylic acids, 890
 α-amino acids from, 1231
 acid chlorides from, 949
 acidity of, 908
 α-halogenation of, 925
 amides from, 976
 anhydrides from, 980
 aromatic, acidity of, 913
 aromatic, K_a values of, 914
 decarboxylation of, 924
 derivatives of, 939
 factors affecting acidity of, 911
 from acid chlorides, 952
 from alcohols, 899
 from aldehydes, 867
 from alkenes, 898
 from alkyl halides, 898
 from alkylbenzenes, 593, 898
 from amides, 977
 from anhydrides, 985
 from aryl halides, 898
 from carboxylate anions, 918
 from esters, 965
 from glycerides, 1095
 from Grignard reagents, 898
 from nitrile hydrolysis, 900
 from oxidation of primary alcohols, 433
 from primary alcohols, 897
 hydrogen bonding in, 897
 infrared spectroscopy of, 927,
  997
 IUPAC system of nomenclature, 893
 K_a values of, 912
 lactones, from hydroxy acids,
  963
 limitations in reduction of, 923
 malonic ester synthesis of,
  1065
 mass spectrometry of, 929
 mechanism of formation from
  Grignard reagents, 899
 mechanism of reduction of,
  922
 methyl esters from, 963
 NMR of, 928
 physical properties of, 896
 reaction with ammonia, 917
 reaction with sodium bicarbonate, 917
 reaction with sodium hydroxide, 917
 reactions of, 908
 reduction of, 921
 resolution of enantiomers of,
  919
 resonance structures of, 911
 salt formation of, 916
 spectral analysis of, 927

 structure of, 910
 summary of preparation of,
  897, 929
 summary of reactions of, 931
 table of, 892, 896
 ultraviolet spectroscopy of,
  928
 using malonic ester synthesis,
  1065
Carboxypeptidase, 1250
p-Carboxyphenyl methyl sulfox-
  ide, 1023
Carcinogens, 1159
Carnauba wax, 1099
β-Carotene, 352
 UV absorption in, 511
 UV spectrum of, 504
Carvacrol, 696
Carvone, 229, 351
Castor oil, 1092
Catalysts, 6
Catalytic cracking 122, 139, 260
Catalytic hydrogenation (See
  also Hydrogenation)
 stereochemistry of, 312
Catalytic isomerization, 139
Catalytic reforming, 122, 139
Catechol, 696, 697
Cell membrane, 1101
Cells, schematic representation
  of, 1234
Cellulose, 1309, 1310
Cephalins, 1101
Ceric nitrate test, for phenols,
  708
Cetyl palmitate, 1099
Chain reaction, 134
Chain-initiation, 135
Chain-propagation, 135
Chain-termination, 135
Chair conformation, 101
Chemical Reactions, Philosophy
  of Studying, 123
Chemical shift and structure cor-
  relations, in NMR, 638
Chemical shifts and structure, in
  NMR, 632
Chemical shifts, examples of, in
  NMR, 633, 636
 in NMR, 628
 table of, in NMR, 634
Chirality, 194
Chitin, 1310
Chloral hydrate, 410
 formation of, 839
Chlordane, 689
Chlorination of methane, energy
  profile diagram of, 154
Chlorination, addition to ben-
  zene, 593

 of alkenes, 323
 of benzene, 561
 of side chains in alkylbenzenes, 594
Chlorine, addition to alkenes,
  323
Chloroacetone, 791
Chloroacetone (See also
  Chloropropanone), 791
Chlorobenzene, 545, 671
 electrophilic substitution on,
  583
 from benzene, 561
 nitration of, 589
2-Chloro-1,3-butadiene, 348
1-Chlorobutane (See also n-Butyl chloride), 133, 367
 from 1-butanol, 443
2-Chlorobutane (See also sec-
  Butyl chloride), 87, 133, 367
R-2-Chlorobutane, 204
S-2-Chlorobutane, 204
(−)-1-Chloro-2-methylbutane, 207
2-Chloro-2-methylbutane, 295
2-Chloro-2-methylbutane, from
  3-methyl-2-butanol, 750
3-Chloro-2-methylbutane, 295
4-Chloro-1-butanol, tetrahydrofuran from, 751
1-Chloro-2-butene, 389
2-Chloro-2-butene, from 2-bu-
  tyne, 478
3-Chloro-1-butene, 389
1-Chloro-2,3-dimethylbutane,
  137
2-Chloro-2,3-dimethylbutane,
  137
2-Chloro-2,4-dimethyl-3-penta-
  none, 1033
2-Chloro-2-iodobutane, from
  propyne, 478
1-Chloro-5-methyl-2-oxaheptane,
  735
1-Chloro-2-methylpropane (See
  also Isobutyl chloride), 133,
  367
2-Chloro-3-methylbutane, from
  3-methyl-2-butanol, 750
3-Chloro-3-methylpentane, dehydrohalogenation of, 265
2-Chloro-2-methylpropane (See
  also tert-Butyl chloride),
  133, 367
 from methylpropene, 291
4-Chloro-1-pentene, 305
1-Chloro-1-phenylethane, 380,
  549
 from ethylbenzene, 595
2-Chloro-1-phenylethane, from
  ethylbenzene, 595

α-Chloro-δ-phenylvaleric acid, 891
3-Chloro-1-propene, 389
  from propene, 336
Chlorocyclohexane (*See also* Cyclohexyl chloride), 130, 367
  from cyclohexane, 130
  from cyclohexene, 290
Chlorocyclopentane, 231
Chlorocyclopropane, 233
Chlorodiphenylmethane, 549
Chloroethane (*See also* Ethyl chloride), 367
2-Chloroethanol, from ethene, 326, 752
Chloroethene, 247, 248, **671**
  from ethene, 338
α-Chloroethylbenzene, 595
β-Chloroethylbenzene, 595
Chloroethylene, 247, 248, **671**
Chloroform (*See also* Trichloromethane), 53
1-Chlorohexane (*See also* n-Hexyl chloride), **367**
p-Chloroiodobenzene, 547
ω-Chlorolauric acid, 891
Chloromethane (*See also* Methyl chloride), 53, 131, **367**
  NMR chemical shifts in, 633
  resonance structures of, 531
Chloromethylbenzene, from toluene, 594
o-Chloronitrobenzene, from chlorobenzene, 589
m-Chloronitrobenzene, from nitrobenzene, 589
p-Chloronitrobenzene, from chlorobenzene, 589
1-Chloropentane (*See also* n-Pentyl chloride), 210, **367**
2-Chloropentane, 127, 210
  dehydrohalogenation of, 265
3-Chloropentane, 210
m-Chloroperoxybenzoic acid, 756
o-Chlorophenol, **697**, 704
m-chlorophenol, **697**
p-Chlorophenol, **697**, 704
p-Chlorophenoxyethane, 734
p-Chlorophenylmagnesium bromide, 842
Chlorophyll, adsorption spectrum of, 1262
  structure of, 1262
  UV spectrum of, 1262
Chloroprene, 348
1-Chloropropane (*See also* n-Propyl chloride), **367**
  from propane, 132

2-Chloropropane (*See also* Isopropyl chloride), **367**
  from propane, 132
2-Chloropropanoic acid, synthesis of, 926
Chloropropanone (*See also* Chloroacetone), **791**
1-Chloropropene, 248
3-Chloropropene (*See also* Allyl chloride), **367**
α-Chloropropionic acid, synthesis of, 926
Chlorosulfite ester, 452
α-Chlorotoluene, 594
o-Chlorotoluene, **671**
  dehydrohalogenation of, 684
m-Chlorotoluene, **671**
  dehydrohalogenation of, 683
p-Chlorotoluene, **671**
  dehydrohalogenation of, 684
Cholesterol, 228, 766, 864
Cholic acid, 864
Chromatography, 6
Chromic acid, 435
  use in oxidation of alcohols, 432
Chromic acid in acetone, as a classification test for alcohols, 454
Chromic acid oxidation, mechanism of, 435
Chromium trioxide, use in oxidation of alcohols, 432
Chromophore, 509
Chromoproteins, 1252
Chymotrypsin, 1237, 1250
  action of, 1253
  active site of, 1255
  discussion of, 1252
  mechanism of hydrolysis by, 1254
  total structure of, 1256
D/L-Configuration, of carbohydrates, 1265
Cinnamaldehyde, 1044
Cinnamic acid, 1046, 1059
cis-trans Isomers in cycloalkanes, 114
Citrate ion, 1286
Claisen condensations, 1053
  crossed, 1057
  mechanism of, 1054
Claisen-Schmidt reactions, 1043
Clemmensen reduction, 861, 988
Cloves, 741
CMR. *See* Carbon-13 magnetic resonance
Coal gas, 551
Coal oil, 551
Cocaine, 1158, 1160

Coconut oil, 1092
Codeine, 1160
Coenzyme, 1297
Coke, 467
Collagen, 1237
Color, relation to absorption, 494
Common names, 81
Concerted reaction, dehydrohalogenation, 263
  in alcohols to alkyl halides, 445
  in hydroboration, 418
  in S$_N$2 reaction, 370
Concerted reactions, 1191
Condensation polymerization, formation of bakelite, 833
  polyesters, 970, 979
  in proteins and peptides, 1218
Condensed structure, 32
D-Configuration, in carbohydrates, 1265
L-Configuration, in carbohydrates, 1265
Conformational analysis, 101
  in D-glucose, 1275
  of cyclohexane by NMR, 659
  using NMR, 657
Conformational isomers, 185
Conformations, 68
  eclipsed, 68
  NMR studies of, 655
  nonequivalence in enantiomers, in NMR, 655
  skew, 68
  staggered, 68
Conformers, 68
  in NMR, 656
  NMR spectra of, 656
Conjugate addition, to α, β-unsaturated carbonyl compounds, 1073
Conjugated dienes (*See also* Dienes, conjugated), 255, 256
  addition to, 305
Conjugated systems, UV absorption frequences of, 512
  UV spectra of, 510
Conjugation, in alkenylbenzenes, 598
Conrotatory rotation, 1208
Conservation of orbital symmetry, 1190
Constitutional isomers, 71
Continuous-chain molecules, 71
Cope elimination, 1147
Coplanar, 914
Coplanar arrangement, of amide in proteins, 1239

Copolymer, 345
Copper chromate, as a catalyst, 857
Corey, E. J., 906
Corey-House synthesis, mechanism of, 180
Corn oil, 1092
Correlation diagrams, in Woodward-Hoffmann rules, 1193
Cortisone, 866
Cottonseed oil, 1092
Coupling constants, due to spin-spin splitting, in NMR, 644
examples of, in NMR, 648
Coupling reaction, using diazonium cations, 1153
Covalent bond, 3, 26
Cps, 493
Cracking, 260
o-Cresol, **697**
m-Cresol, **697**
p-Cresol, 548, 696, **697**
Crick, 1281
Crossed aldol condensations, 1042
Crossed Claisen condensations, 1057
Crotonaldehyde (See also 2-Butenal), **719**, 1040
UV spectrum of, 514
Crystallization, 6
Cumene, **550**
Cumulated diene, 255
Cuprous alkynide, 472
Cuprous ammonia complex, reaction with terminal alkynes, 472
Cyanide ion, as a base, 901
as a nucleophilie, 900
Cyano group, 944
Cyanoacetic acid, 1071
α-Cyanoalcohol, 825
Cyanobenzene, 944
p-Cyanobenzoic acid, 944
Cyanoethene (acrylonitrile) preparation of, 826
Cyanogen bromide, 1250
Cyanohydrins, preparation of, 825
mechanism of formation, 825
Cyclamates, 1159
Cycles per second, 493
Cyclic ethers. See Ethers
Cycloaddition reactions, 1178
Cycloaddition reactions, Diels-Alder reaction, 335
Cycloaddition reactions, Woodward-Hoffmann rules in, 1193
2+2 Cycloaddition reactions, 1178, 1187

correlation diagram of symmetry-forbidden, 1197
correlation diagram of symmetry-allowed, 1198
of ethene, 1194
photochemical, 1198
selection rules for, 1201
thermal, 1197
4+2 Cycloadditions, Diels-Alder reaction, 1202
Cycloalkanes, 95
cis-trans isomerism in, 113
heats of combustion and strain energy, 107
nomenclature of, 96
industrial sources of, 121
melting and boiling points and densities of, 97
physical properties of, 97
preparation by cyclization reactions, 126
restricted rotation in, 114
stability, evidence for, 106
structure of, 95
summary of methods of preparation, 142
summary of reactions, 143
table of, 97
Cycloalkenes, 251
using Diels-Alder reaction, 1179
Cyclobutadiene, 540
Cyclobutane, 96, **97**
Cyclobutane, addition reactions of, 358
antibonding orbitals, (MO's) in, 1195
bonding orbitals, (MO's) in, 1195
conformations of, 100
from cycloaddition of ethene, 1194
from cyclobutyl chloride, 843
hydrogenation of, 312
MO's of, 1195
structure of, 98
Cyclobutanecarbonyl chloride, 941
Cyclobutanecarboxamide, 940
Cyclobutanecarboxylic acid, 893, 1067
cis-1,2-Cyclobutanedicarboxylic acid, 895
Cyclobutene, 251
Cyclobutyl bromide, 96
Cyclobutylmagnesium chloride, 843
Cycloheptane, **97**, 105
conformation of, 106
Cycloheptanone, 1137
Cycloheptatriene, 537, 602

Cycloheptatrienyl anion, 540
Cycloheptatrienyl cation, 540
from cycloheptatriene, 602
resonance structures of, 537
Cycloheptatrienyl fluoroborate, from cycloheptatriene, 537
Cycloheptatrienyl iodide, 537
1,3-Cyclohexadiene, 255, 258 524, 1182
1,4-Cyclohexadiene, 255, 524
preparation of, 480
Cyclohexane carboxaldehyde, 868
Cyclohexane Derivatives, Drawing Conformational Structures, 112
Cyclohexane, 96, **97**
boat conformation of, 103
chair conformation of, 101
conformational analysis studies by NMR, 659
Newman projections of, 103
spin-spin coupling for, 656
structure of, 100
Cyclohexanecarboxylic acid, reduction of, 923
cis-1,2-Cyclohexanediol, **769**
trans-1,2-Cyclohexanediol, **769**
1,4-Cyclohexanediol, 287
cis-1,4-Cyclohexanediol, 406
1,4-Cyclohexanedione, 789
Cyclohexanes, 1,3-diaxial interactions in monsubstituted, 108
conformational analysis for disubstituted, 116
conformational analysis of monosubstituted, 108
conformations of polysubstituted, 117
energy differences in monosubstituted, 111
Cyclohexanol, **412**
adipic acid from, 905
dehydration of, 271
from cyclohexanone, 857
Cyclohexanone, **791**
from cyclohexanol, 433
NMR spectrum of, 808
Cyclohexanone 2-4-dinitrophenylhydrazone, formation of, 837
Cyclohexanone imine, 1123
Cyclohexanone oxime, 1135
Cyclohexanone semicarbazone, formation of, 837
Cyclohexanone-2-carboxylic acid, 893
1,3,5-Cyclohexatriene, 524
Cyclohexene, 251, 258
cyclohexene oxide from, 757

INDEX

NMR chemical shifts in, 636
Cyclohexene chlorohydrin, from cyclohexene, 752
Cyclohexene oxide, 737
from cyclohexene, 752, 757
from cyclohexene chlorohydrin, 752
Cyclohexyl alcohol, 410, 412
Cyclohexyl bromide (*See also* Bromocyclohexane), 367
Cyclohexyl chloride (*See also* Chlorocyclohexane), 130, 367
Cyclohexyl cyclohexanecarboxylate, synthesis of, 955
Cyclohexyl iodide (*See also* Iodocyclohexane), 367
Cyclohexylamine, 1107, 1123, 1159
2-Cyclohexylethanol, from cyclohexylethene, 420
3-Cyclohexylhexane, 96
Cyclohexylmagnesium bromide, 803
Cyclohexylmethanol, from cyclohexanecarboxylic acid, 923
Cyclononane, conformation of, 106
Cyclononatetraenyl anion, 540
Cyclooctane, 97
conformation of, 106
Cyclooctatetraene, 540
NMR chemical shifts in, 638
Cyclooctatraenyl dianion, 540
Cycloparaffins, 129
1,3-Cyclopentadiene, 255, 535, 1182
Cyclopentadiene dimer, 1182
Cyclopentadienyl anion, 540
from cyclopentadiene, 535
MO's of, 543
orbital picture of, 536
resonance structures of, 535
structure of, 535
Cyclopentadienyl cation, 540
in mass spectrometry, 724
Cyclopentane, 96, 97
structure of, 99, 101
cis-1,2-Cyclopentanediol, 769
from cyclopentene, 333
trans-1,2-Cyclopentanediol, 737, 769
from cyclopentene oxide
Cyclopentanol, 412, 442
from cyclopentanone, 859
from cyclopentene, 297
Cyclopentene, 251
cyclopentene oxide from, 757
Cyclopentene glycol, 333
Cyclopentene oxide, from cyclopentene, 757

trans-1,2-cyclopentanediol from, 760
Cyclopentyl alcohol, 412, 442
Cyclopentyl bromide, 442
mass spectrum of, 726
Cyclopentyl mercaptan, 461
Cyclopentyl nitrile, 461
Cyclopentylmethanol, from 4-methylcyclopentene, 424
Cyclophanes, NMR chemical shifts in, 664
Cyclopropane, 96, 97
addition of bromine to, 318
hydrogenation of, 312
orbital view of, 99
structure of, 98
summary of addition reactions, 358
cis-1,2-Cyclopropanedicarboxylic acid, 984
cis-1,2-Cyclopropanedicarboxylic anhydride, preparation of, 984
Cyclopropanes, addition of HX to, 294
from alkenes, 336, 1187
mechanism of HX addition to, 295
Cyclopropenyl anion, 540
MO's of, 545
Cyclopropenyl cation, 540
MO's of, 544
Cycloreversions, 1208
p-Cymene, 550
Cysteine, 1212
Cystine, 1212
Cystosine, 1279

2,4-D (*See also* 2,4-Dichlorophenoxyacetic acid), 689, 741
d-Form, 186
D-Family, of carbohydrates, 1265
Dacron, synthesis of, 970
DCC (*See also* Dicyclohexylcarbodiimide), 1222
DDT (*See also* Dichlorodiphenyltrichloroethane), 688
Deactivating groups, in electrophilic substitution, 570
DeBroglie theory, 18
Decanal, 788
(n-)Decane, 93
Decanedioic acid, 894
Decanoic acid, 892
1-Decanol (*See also* n-Decyl alcohol), 412
Decarboxylation, of β-keto acids, 1069

(n-)Decyl alcohol (*See also* 1-Decanol), 412
1-Decyne, 466
Degree of unsaturation, 314
Dehalogenation, transition state in, 282
Dehydration and hydration, compared, 298
Dehydration of alcohols, mechanism of, 271
Dehydration reactions, reactivity in, 276
Dehydration, in peptide syntheses, 1225
of aldol condensation products, 1040
of amino alcohols, 836
product distribution in, 277
Dehydrobenzene. See Benzyne
Dehydrogenation, 259
Dehydrogenations, of alkanes, 260
of cyclic compounds, 1089
preparation of alkenes, 259
Dehydrohalogenation, anti-elimination in, 267
mechanism of, 262
preparation of alkenes, 261
product distribution in, 265
reactivity in, 266
stereochemistry of, 266, 269
transition state in, 264
Delocalization energy, in benzene, 525, 529
Delocalization of electrons by resonance, 308
Delta (δ) scale, in NMR, 628
Delta minus (δ⁻) definition, 28
Delta plus (δ⁺), definition, 28
Demarol, 1158
Denatured alcohol, 426
2-Deoxy-D-ribose, 1279
Deoxyribonucleic acid (DNA), 1234, 1279
Derivatives, in compound identification, 344
in identification of carbonyl compounds, 836
Deshielding, in NMR, 632
Detergents, 588
discussion of, 1096
from glycerides, 1096
synthesis of, 1096
Deuterated solvents, in NMR, 629
3-Deuterio-2-butanol, from 2-butene, 422
Deuteriobenzene, from benzene, 563
N-Deuteriopyrrole, 1150
Deuteriosulfuric acid, 563

Deuterium exchange, in α-carbanions, 1021
Deuterium isotope effects, 436
Deuterium oxide, use in NMR, 712
Deuterium, in NMR, 658, 659
Deuterochloroform, 629
Developer, in photography, 707
Dextrorotatory, 186
Diacetone alcohol, 1039
Dialkyl ethers. See Ethers
Dialkyl hydrogen phosphate, 441
4,4'-Diaminobiphenyl, 1159
Diaryl ethers (See also Ethers)
  cleavage of, 755
Diastereomeric salts, 920
Diastereomers, 185, 217
  definition of, 217
  in carbohydrates, 1270
  physical properties of, 224
  properties of, 224
  use in resolution of carboxylic acids, 920
1,3-Diaxial interaction, 108
Diazomethane, 336, 963, 1188
  in esterification, 963
  photolysis of, 1188
Diazonium cation, 1136
  as an electrophile, 1153
Diazonium group, 963
Diazonium salts, aromatic hydrocarbons from, 1138
  aryl bromides from, 1138
  aryl chlorides from, 1138
  aryl halides from, 1138
  aryl iodides from, 1138
  aryl nitriles from, 1138
  aryl thiols from, 1138
  carboxylic acids from, 1139
  decomposition of, 1136
  electrophilic substitution by, 1153
  fluoroborates from, 1139
  formation of, 1136
  from primary amines, 1136
  mechanism of formation of, 1141
  phenols from, 1138
  reduction of, using hypophosphorus acid, 1139
  ring expansions in, 1136
  structure of, 1136
  thiols from, 1140
Dibenzopyrene, 553
Diborane, 416
  preparation of, 417
  reduction of amides by, 1123
α,α-Dibromoacetone, 1027
2,3-Dibromobutane, 468
  from 2-butene, 323
  stereoisomers of, 219

meso-2,3-Dibromobutane, dehydrobromination of, 267
d,l-2,3-Dibromobutane, dehydrobromination of, 268
3,4-Dibromocyclobutene, 555
2,3-Dibromo-2-butene, from 2-butyne, 477
trans-1,2-Dibromocyclopentane, from cyclopentene, 319
trans-1,3-Dibromocyclopentane, 115
cis-1,3-Dibromocyclopentane, 115
1,2-Dibromoethane, 139, 902
  from ethene, 319
cis-1,2-Dibromoethene, 249, **254**
trans-1,2-Dibromoethene, **254**
1,6-Dibromohexane, 126
R-1,2-Dibromo-3-methylbutane, 204
2,4-Dibromonitrobenzene, 548
2,4-Dibromo-2'-nitrobenzophenone, 790
1,2-Dibromo-1-phenylethane, NMR spectrum of, 655
1,2-Dibromopropane, from propene, 318, 319
1,3-Dibromopropane, from cyclopropane, 318
2,2-Dibromopropane, from propyne, 478
α,p-Dibromotoulene, 901
α,β-Dicarbonyl compounds, cleavage by periodic acid, 771
β-Dicarbonyl compounds, 1017, 1061
  acidity of, 1055
  addition of α,β-unsaturated carbonyl compound to, 1076
  hydrogen bonding in, 1078
  keto-enol equilibrium in, 1078
  resonance structures of, 1062
Dicarboxylic acids, 890
  anhydrides from, 982
  common names of, 894
  decarboxylation of, 924, 982
  derivatives of, 939
  dissociation constants of, 915
  IUPAC names of, 895
  K_a values of, 915
  nomenclature of, 894
  physical properties of, 896
  salt formation of, 916
  tables of, 894, 896
3,4-Dichloro-1-butene, from 1,3-butadiene, 324
1,4-Dichloro-2-butene, from 1,3-butadiene, 324
1,2-Dichloro-2-methylbutane, 207

1,3-Dichloroallene, enantiomers of, 196
m-Dichlorobenzene, 547, **671**
p-Dichlorobenzene, 547, **671**
o-Dichlorobenzene, 547, **671**
Dichlorocarbene, 753
trans-1,2-Dichlorocyclohexane, enantiomers of, 235, 236
cis-1,2-Dichlorocyclohexane, stereoisomers of, 235, 236
trans-1,2-Dichlorocyclopentane, from chlorocyclopentane, 231
cis-1,2-Dichlorocyclopentane, from chlorocyclopentane, 231
1,1-Dichlorocyclopropane, 113
1,2-Dichlorocyclopropane, 113
  stereoisomers of, 233
trans-1,2-Dichlorocyclopropane, 114
  enantiomers of, 233
cis-1,2-Dichlorocyclopropane, 114, 233
  stereoisomers of, 234
Dichlorodiphenyltrichloroethane (DDT), 688
1,2-Dichloroethane, from ethene, 318
trans-1,2-Dichloroethene, **254**
cis-1,2-Dichloroethene, **254**
1,1-Dichloroethylene, polymerization of, 345
Dichloromethane (See also Methylene chloride), 53, 131
  NMR chemical shifts in, 633
2,4-Dichloropentane, 305
2,4-Dichlorophenoxyacetic acid (2,4-D), 689
2,4-Dichlorophenoxyethanoic acid, 741
1,2-Dichloropropane, from propene, 336
1,3-Dichloropropane, 126
  NMR chemical shifts in, 633
1,3-Dichloropropane, NMR spectrum of, 628, 631
2,2-Dichloropropanoic acid, synthesis of, 926
α,α-Dichloropropionic acid, synthesis of, 926
α,α-Dichlorotoluene, 595
Dicyclohexylcarbodiimide, 1222
Dicyclohexyl urea, 1225
endo-Dicyclopentadiene, 1184
exo-Dicyclopentadiene, 1185
Dieckmann condensation, 1056
Dieldrin, 689
Dielectric constant, 175
Dielectric constants of certain solvents, 176

Diels, Otto, 335
Diels-Adler reaction, cycloalkenes from, 335, 1179
4+2 cycloaddition, 1202
4+2 cycloadditions, 1202
bicyclic compounds from,
1181
in trapping benzyne, 687
mechanism of, 1179
stereochemistry of, 1180
using α,β-unsaturated carbonyl
compounds, 1183
Woodward-Hoffmann rules
for, 1202
use of alkynes in, 480
s-cis-Diene, 256
s-trans-Diene, 256
Dienes, addition of alkynes to,
480
addition reactions of, 304
chlorination of, 324
conjugated, structure of, 256
cumulated, 255
hydrogenation of, 313
in Diels-Alder reaction, 336,
1181
isolated, 255
mechanism of polymerization,
349
polymerization of, 348
product distribution of addition to, 308
reactions with alkynes, 480
structure of conjugated, 256
summary of addition reactions,
357
Dienophile, 335, 1184
Dienophiles, in Diels-Alder reaction, 336
β-Diesters, 1061
1,2-Diethoxyethane, 734
Diethyl acetylenedicarboxylate,
1183
Diethyl benzenedioate, **945**
Diethyl butynedioate, 1183
Diethyl carbonate, 973
Diethyl 3,6-dihydrophthalate,
1183
Diethyl dimethylmalonate, 1065
Diethyl ketone (*See also* 3-Pen-
tanone), **791**
Diethyl malonate, acidity of,
1062
alkylation of, 1063
barbiturates from, 1087
synthesis of, 1063
3,4-Diethyl-2-methyl-*cis*-2-hep-
tene, 249
Diethyl methylmalonate, 1064
Diethyl oxalate, 1057
Diethyl phthalate, **945**, 970

Diethyl terephthalate, 970
Diethylamine, **1107**
N,N-Diethylaniline, 1125
Diethylcadmium, 804
Diethylene glycol, 759
Diethylene glycol dimethyl
ether, 734
Diglyme, 734, 760
Dihalides (*vincinal*), alkenes
from, 282
dehalogenation of, 282
Dihedral angle and spin-spin
splitting, in NMR, 656
3,5-Dihydroxy-1-(1-pentyl)ben-
zenes, 742
Dihydroxyacetone monophosphate, 1298
α,α-Di(hydroxymethyl)-β-
hydroxypropionaldehyde,
1044
L-Dihydroxyphenylalanine (L-
DOPA), 1174
cis-1,2-Diiodoethene, **254**
trans-1,2-Diiodoethene, **254**
Diisopropylamine, **1107**
Diisopropyl fluorophosphate,
1255
β-Diketones, 1061
Dimer, 328
Dimerization, 328
mechanism of alkene addition
to alkene, 328
of alkenes, 328
of cyclopentadiene, 1182
of dienes, 1182
N,N-Dimethylacetamide, 941
Dimethylacetylene, 465
Dimethylamine, 1105, **1107**
p-Dimethylaminobromobenzene,
1106
3,4-Dimethylaniline, 548
N,N-Dimethylaniline, 1105, **1107**
IR spectrum of, 1161
3,3-Dimethyl-1-bromobutane,
850
2.3-Dimethyl-1,3-butadiene, 255,
258, 1183
preparation of, 283
2,2-Dimethylbutane, **77**
from 3,3-dimethyl-2-butanone,
862
2,3-Dimethylbutane, **77**
from methylpropane and ethene, 329
2,3-Dimethyl-1,4-butanediol, 411
2,3-Dimethyl-2,3-butanediol, 283
3,3-Dimethyl-1-butanol, 851
3,3-Dimethyl-2-butanol, 851
3,3-Dimethyl-2-butanone, reduction of, 862
3,3-Dimethyl-1-butene, 316

synthesis of, 850
2,3-Dimethyl-2-butene, 316
3,3-Dimethyl-1-butyne, **466**
Dimethylcarbinol, 411
Dimethylcarbonium ion, 164
2,2-Dimethyl-1-chloropropane,
564
cis-1,3-Dimethylcyclobutane,
115
trans-Dimethylcyclobutane,
115
stereochemistry of, 234
cis-1,2-Dimethylcyclohexane,
conformations of, 116
from 1,2-dimethylcyclohexane,
311
trans-1,2-Dimethylcyclohexane,
116
conformations of, 116
1,4-Dimethylcyclohexane, from
p-xylene, 592
cis-1,4-Dimethylcyclohexane,
115
trans-1,4-Dimethylcyclohexane,
115
N,N-Dimethylcyclohexanecar-
boxamide, from cyclohex-
anecarbonyl chloride, 951
2,2-Dimethylcyclohexanone,
789
1,2-Dimethylcyclohexene, 311
1,3-Dimethylcyclohexene, 251
cis-3,6-Dimethylcyclohexene,
1180
1,5-Dimethyl-1,5-cyclooctadiene,
364
1,2-Dimethylenecyclohexane,
251
3,4-Dimethylenecyclopentanol,
1147
N,N-Dimethylethanamide, 941
Dimethyl ether, 734
Dimethylethylamine N-oxide,
1143
N,N-Dimethylformamide, 176
NMR spectrum of, 1001
4,5-Dimethyl-3,5-heptadien-1-
yne, 466
3,3-Dimethylheptane, mass spectrum of, 725
2,2-Dimethyl-3-heptanone, 852
Dimethylhydrazine, 1159
3,4-Dimethyl-2.5-dihydro-
benzaldehyde, 1183
Dimethyl ketone (*See also* Acetone), 788
3,7-Dimethyl-4-nonyne, 465
2,2-Dimethyl-3-oxaheptane, synthesis of, 746
2,2-Dimethyloxirane, 737
cis-2,3-Dimethyloxirane, 737

2,3-Dimethyl-3-pentanol, dehydration of, 278
2,4-Dimethyl-3-pentanol, 972
3,4-Dimethyl-1-pentene, 248
N,N-Dimethylpentylamine N-oxide, 1143
2,4-Dimethylphenol, oxidation of, 707
2,3-Dimethyl-2-phenylpentane, 546
N,N-Dimethylpiperidinium iodide, 1125
Dimethylpropane, 77
2,2-Dimethylpropanoic acid, preparation of, 899
2,2-Dimethyl-1-propanol, 412
  from 1-iodo-2,2-dimethylpropane, 382
α,α-Dimethylpropionic acid, 891
6,6-Dimethyl-2-quinuclidone, 237
Dimethyl sulfate, 751
Dimethyl sulfoxide, 176
Dimethyl-para-tolylsulfonium tetrafluoroborate, Carbon-13 NMR spectrum, 660
α,β-Dimethylvaleraldehyde, 787
m-Dinitrobenzene, 1139
  from nitrobenzene, 586
3,5-Dinitrobenzophenone chloride, preparation of, 801
3,5-Dinitrobenzoyl, 801, 941
2,4-Dinitrochlorobenzene, 673
2,4-Dinitrofluorobenzene, 1249
2,4-Dinitrophenol, 697
  from 2,4-dinitrochlorobenzene, 673
3,4-Dinitrophenol, 695
2,4-Dinitrophenylhydrazine, 837
  addition to carbonyl compounds, 837
2,4-Dinitrophenylhydrazones, 837, 866
gem-Diol, 409, 737
Diols, 283
1,4-Dioxacyclohexane, 735
1,4-Dioxane, 735, 739
Dioxane, 735
3,6-Dioxo-1,4-cyclohexadiene, 790
Dipeptide, 1218
Diphenylacetylene, 550
Diphenylamine, 1107
1,3-Diphenyl-2-buten-1-one. 1039
Diphenylcadmium, 804
Diphenylcarbitol, 760
trans-1,2-Diphenylethene, 549
Diphenyl ether, 734, 739
Diphenyl ketone (See also Benzophenone), 791

Diphenylmethane, 550
Diphenylmethanol, 412
Diphenylmethyl alcohol (diphenylcarbinol), 412
2,3-Diphenylpropenenitrile, 1060
tris-(Dipivaloylmethanato)-praseodymium, 667
Dipole, moments, derivation of, 52
  of 1,2-dihaloethenes, 253
  of aldehydes, 793
  of alkenes, 253
  of ketones, 793
Dipole-dipole interaction, in aldehydes and ketones, 792
  in alkanes, 92
Di-n-propyl ketone (See also 4-Heptanone), 791
Diradical, 44
Disaccharides, 1305
  structure elucidation of, 1306
Disproportionation, 1047
Disrotatory rotation, 1208
Dissociation constant,
  of diethyl malonate, 1062
  of ethyl acetoacetate, 1062
  of 2,4-pentanedione, 1062
Dissociation constants, of aliphatic carboxylic acids, 912
  of amines, 1108
  of aromatic carboxylic acids, 914
  of carbonyl compounds, 1018
  of dicarboxylic aids, 915
  of heterocyclic nitrogen compounds, 1116
Dissymmetry, 194
Distillation, 6
Disubstitution reactions, of aromatic compounds, 610
Disubstitution, in aromatic compounds, 568
Diterpenes, 350, 351
Divinyl ether, 740
DNA (See also Deoxyribonucleic acid), 1279
  base pairing in, 1281
  double helix, 1282
  hydrogen bonding in, 1281
  structure of, 1280
2,4-DNP. See 2,4-Dinitrophenylhydrazones
2,4-DNPH. See 2,4-Dinitrophenylhydrazine
(n-)Docosane, 93
Docosanoic acid, 892
(n-)Dodecane, 93
Dodecanoic acid, 892
L-DOPA (L-Dihydroxyphenylalanine), 1174
Double bond. See type of bond

Double helix, in DNA, 1282
Double resonance, in NMR, 660
Dow process, 673
  ether formation in, 748
Downfield chemical shift, in NMR, 632
Du Vigneaud, V., 1219
Durene, 550
Dyes, 1153

E, represents electrophile, 169
E1 reaction, defined (See also Elimination), 275
  in dehydration of alcohols, 275
E1 reactions, competition with S_N1 reactions, 394
E1 versus S_N1 reactions, 394
E1_{CB} reaction, defined, 685,
E2 reaction, defined (See also Elimination), 262
  preparation of alkenes, 262
E2 reactions, competition with S_N2 reactions, 392
E2 reactions, effect on bases in, 393
E2 versus S_N2 reactions, 392
Eclipsed conformation, 67, 68
(n-)Eicosane, 93
Eicosanoic acid, 892, 906
5,8,11,14,17-Eicosapentaenoic acid, 907
5,8,11,14-Eicosatetrenoic acid, 907
8,11,14-Eicosatrieonic acid, 907
Electrocyclic reactions, 1208
Electromagnetic radiation, 492
Electromagnetic radiation spectrum, regions of, 492
Electromagnetic spectrum, 492
Electron configuration, 23
Electron-donation, effect on amine basicity, 1111
Electron-donation, in carboxylic acid acidity, 911
Electron-releasing groups, in electrophilic substitution, 573
Electron-withdrawal, effect on amine basicity, 1111
  effect on carboxylic acid acidity, 912
Electron-withdrawing groups, chemical shift, in NMR, 632
  in electrophilic substitution, 573
  in nucleophilic aromatic substitution, 678
Electronegativities, table of, 51
Electronegativity, 50
Electronegativity, effect on carboxylic acid acidity, 912

Electronic effects, in S$_N$1 reactions, 383
Electronic integration, in NMR, 631
Electronic structure, 23
Electronic structure in the periodic table, 24
Electronic transitions, UV spectroscopy, 503
Electrophile, definition of, 169
Electrophilic addition, of alkenes, 290
  benzyl cations in, 600
  to alkenylbenzenes, 600
  to conjugated dienes, 306
  rearrangements in, 295
  reactivity of alkenes in addition reactions, 325
Electrophilic aromatic substitution. See Electrophilic substitution.
Electrophilic substitution, alkylation of benzene, 563
  carbocation stability in, 573
  compared with electrophilic addition, 557
  comparison with nucleophilic aromatic substitution, 681
  disubstitution reactions, 589
  disubstitution, discussion of, 568
  effect of aryl substituents, 585
  energy diagram for nitration, 560
  examples of synthesis in, 590
  Friedel-Crafts acylation, 799
  general discussion of, 556
  general factors in reactivity on, 573
  general mechanism of, 557
  halogenation, 561
  inductive effect in, 574
  inductive and resonance effects in, 585
  intermediates in, 557, 559
  limitations on, 585
  nitration, 558
  nitrosation, 1142
  of aromatic amines, 1132
  on aldehydes and ketones, 820
  on aryl halides, 583
  on phenols, 704
  on pyridine, 1149
  on pyrrole, 1149
  orientation and reactivity in, 570
  orientation and reactivity in synthesis, 590
  polysubstitution reactions, 590
  product distribution in disubstitution, 569

resonance effects in, 579
resonance structures in, 575, 577, 580, 583
steric effect in, 572
sulfonation, 562
synthesis examples in, 590
theory of orientation with alkylbenzenes, 575
theory of orientation with aryl halides, 583
theory of orientation with *meta* directors, 577
theory of orientation with *ortho, para* directors, 580
transition states and intermediates in, 576
using alcohols, 566
using alkenes, 567
writing resonance structures for, 579
Electrophilic versus Nucleophilic Addition, 819
Element effect, in nucleophilic aromatic substitution, 675
Elemental analysis, 7, 344
Eleostearic acid, 1092
Elimination and substitution, 398
Elimination reactions, by alkynyl anions, 474
  in preparation of alkynes, 468
Elimination, competition with ether formation, 743
  competition with substitution, 392, 394
  control of, 398
  in amine oxides, 1147
  in esters, 1148
  in ether synthesis, 746
  in selenoxides, 1148
  in sulfoxides, 1147
  in xanthates, 1148
  of quaternary ammonium hydroxides, 1144
  summary of factors in, 402
  using cyanide ion as a base, 901
Elimination-addition, in benzyne formation and reactions, 685
Empirical formulas, 8
Emulsin, 1315
Emulsion, 1095
Enamine synthesis, 1048
Enamines, from amines, 1131
  preparation of, 1049, 1131
  resonance structures of, 1049
Enantiomers, 185, 195
  criteria for existence of, 191
  definition of, 190
  discovery of, 188
  nonequivalence of protons in NMR, 655

of glucose, 1265
properties of, 197
resolution of, 200
structures of, 190
endo-Isomer, 1184
Endocyclic double bond, 251
Endopeptidases, 1250
Endorphins, 1219
Endothermic reaction, 147
Energy barrier to free rotation, 69
Energy barrier to rotation, 68
Energy of activation, 148
  effect of, in HBr addition to 1,3-butadiene, 310
Energy of electromagnetic radiation, 492
Energy states of hydrogen nucleus, in NMR, 624
Energy states of molecules, diagram of, 505
Energy states, in electronic transitions, 494
Enkephalins, 1221
Enolate, 1028
Entgegen, 250
Enthalpy, 147
Entropy, 147
Epimerization, 1285
Epimers, 1266, 1285
Epoxidation, mechanism of using peroxyacids, 757
  of alkenes, 756
Epoxides, acidic versus basic conditions in ring opening, 763
  halohydrins from, 759
  mechanism of opening by acidic reagents, 758
  mechanism of ring opening by basic reagents, 762
  nomenclature of, 736
  opening of, 757
  orientation in ring opening, 763
  preparation of, 756
  reaction with Grignard reagents, 847
  ring opening by acidic reagents, 757
  ring opening by basic reagents, 761
  ring opening in biochemical systems, 765
  strain in, 757
  summary of methods of preparation, 777
  summary of reactions, 778
Epoxy resins, 1174
1,2-Epoxy-2-methylpropane, 737, 763

1,2-Epoxycyclohexane, 737
Epoxyethane, 737
1,2-Epoxyhexane, from 1-hexene, 757
cis-2,3-Epoxybutane, 737
1,2-Epoxypropane, 737
Equatorial hydrogens, 102
Equilibrium constants, 168
Equilibrium constants and free energy, 111, 148
Erg, 493
Erythrose, 1272
Eschweiler-Clark reaction, 1124
Ester pyrolysis, 1148
Esterification, equilibrium in, 955
    isotope tracer studies in, 957
    kinetics of, 957
    mechanism of, 955–961
    rates of, 958
    steric effects in, 957
Esters, ammonolysis of, 971
    condensation of, 1053
    condensation with ketones, 1058
    crossed condensation of, 1057
    cyclic. See Lactones
    from acid chlorides, 951
    from acids and alcohols, 954
    from alcohols, 438
    from amides, 976
    from anhydrides, 962, 985
    from carboxylate anions and alkyl halides, 961
    hydrogenation of, 974
    hydrolysis of, 965
    in glycerides, 1090
    infrared spectroscopy of, 997
    mass spectrometry of, 1000
    methyl, using diazomethane, 963
    NMR of, 998
    nomenclature of, 942
    of carboxylic acids, 439
    of nitric acid, 440
    of phosphoric acid, 441
    of sulfuric acid, 439
    physical properties of, 944
    preparation of, 954
    reactions of, 965
    reactions with Grignard reagents, 971
    reduction by sodium in ethanol, 975
    reduction of, 974
    stereochemistry of hydrolysis, 966
    table of, 945
    transesterification of, 969
Estrogens, 865
Estrone, 865

Ethanal (See also Acetaldehyde), 788, **791**
    aldol condensation of, 1039
    from ethanol, 431
    from ethyne, 478
    sodium bisulfite addition product, 827
    trimer of, 794
Ethanamide (See also Acetamide), 940, **945**
    from ethanoyl chloride, 951
Ethane, **93**
    conformational analysis in, 66
    from ethene, 311
    structure of, 33, 65
Ethanedioic acid (See also Oxalic acid), 894, **896**
1,2-Ethanediol (See also Ethylene glycol), 426, 737, 759, **769**
    from ethene, 333
Ethanenitrile (See also Acetonitrile), 943
Ethanoic acid (See also Acetic acid), 892, 893, **896**
    from ethanol, 903
    from ethyne, 478
Ethanoic anhydride (See also Acetic anhydride), 943, 981
    from ethanoyl chloride, 953
Ethanoic benzoic anhydride, 953
Ethanol (See also Ethyl alcohol), **412**
    dehydration of, 271
    discussion of, 425
    from ethene, 300
    high-resolution NMR spectrum of, 712
    in beverages, 425
    IR spectrum of, 502, 710
    NMR spectrum of, 661, 711
Ethanolysis, 397
Ethanoyl chloride (See also Acetyl chloride), 941, **945**
Ethene (See also Ethylene), **252**, 316
    antibonding MO's of, 1194
    bonding MO's of, 1194
    2+2 cycloaddition of, 1194
    cyclobutane from, 1194
    ethylene oxide from, 752, 756
    MO's of, 1194
    polymerization of, 345
    preparation of, 261
    structure of, 33, 243
Ethenyl group, 248
4-Ethenylcyclohexene, 1181
Ethenylcyclopropane, 248
Ether, in Grignard reagents, 802
Ether. See Ethyl ether

Ethers (cyclic), by the Williamson method, 751
Ethers, basic properties of, 740
    cleavage mechanism of, 754
    cleavage of, 753
    cleavage by organolithium reagents, 855
    common names of, 734
    control of formation of, 743
    cyclic, 735
    cyclic ether synthesis, 751
    dipole moment of, 738
    elimination in competition with formation of, 743
    elimination in synthesis of, 746
    from alkynes, 481
    hydrogen bonding of in water, 739
    in Grignard reactions, 740
    IR spectroscopy of, 773
    IUPAC nomenclature of, 734
    mass spectroscopy of, 774
    mechanism of formation of from alcohols, 742
    NMR of, 774
    nomenclature of cyclic, 735
    preparation of alcohols, 742
    properties of, 738
    reaction with acid, 740
    reaction with BF_3, 740
    spectral analysis of, 773
    structure of, 733, 738
    summary of methods of preparation, 776
    summary of reactions, 777
    symmetrical, 743
    synthesis by Williamson method, 744
    synthesis of using dimethyl sulfate, 751
    table of, 739
    use in hydroboration reactions, 740
    UV spectroscopy, 773
    Williamson method of synthesis, 744
2-Ethoxalylcyclopentanone. 1058
1-Ethoxybutane, synthesis of, 745
2-Ethoxyethanol, from ethylene oxide, 762
2-Ethoxy-2-methyl-1-propanol, 763
1-Ethoxy-2-methyl-2-propanol, 764
Ethyl 2-phenylethanoate. 972
Ethyl acetate (See also Ethyl ethanoate), **945**
    acidity of, 1018

from acetyl chloride, 952
NMR of, 998
synthesis of, 962
Ethyl acetoacetate, 1054
acidity of, 1062
anion of, 1055
ketones from, 1068
synthesis of, 1069
Ethyl acetoacetate anion, resonance structures of, 1055
Ethyl acrylate, 1077
Ethyl alcohol (*See also* Ethanol), **412**
Ethyl benzoate, **945**
Ethyl benzoylacetate, 1078
Ethyl bromide (*See also* Bromoethane), **367**
mass spectrum of, 723
Ethyl butanoate, **945**
Ethyl *n*-butyl ether, 745
Ethyl *n*-butyrate, **945**
Ethyl cation, 164
Ethyl chloride (*See also* Chloroethane), **367**
Ethyl cyanoacetate, 1076
Ethyl ethanoate (*See also* Ethyl acetate), 942, **945**
from ethanoyl chloride, 952
Ethyl ether, **739**
discussion of, 740
Ethyl α-ethoxalylpropionate, 1057
Ethyl formate, **945**, 1058
Ethyl gasoline, 139
Ethyl hydrogen succinate, 986
Ethyl hydrogen sulfate, 300
Ethyl iodide (*See also* Iodoethane), **367**
Ethyl methanoate, **945**
Ethyl phenyl ether, **739**
cleavage of, 755
Ethyl phenyl ketone (*See also* Propiophenone), **791**
Ethyl phenylacetate, 972
Ethyl propanoate, **945**
Ethyl propionate, **945**
Ethyl *o*-toluoate, **945**
Ethyl *m*-toluoate, **945**
Ethyl *p*-toluoate, **945**
N-Ethyl-1-aminopentane, 1105
Ethyl 3-oxobutanoate, 1054
Ethyl-*n*-butylacetylene, 465
N-Ethyl-N-isopropylaniline, 1105
Ethylamine, **1107**
Ethylbenzene, 545, **550**
halogenation of side chain in, 595
mass spectrum of, 731
nitration of, 572
Ethylcyclohexane, 126

Ethyldiisopropylcarbinol, 411
Ethyldimethylamine, 1105
Ethyldimethylammonium hydroxide, 1110
Ethylene chloride, 326
Ethylene chlorohydrin, 326, 752
Ethylene glycol, 333, 426, 737, **769**
from ethylene oxide, 759, 762
synthesis of, 427
Ethylene glycol dimethyl ether, 734
Ethylene glycoxide ion, 762
Ethylene oxide, 735, 737
2-ethoxyethanol from, 762
2-phenoxyethanol from, 762
ethylene glycol from, 759, 762
from 2-chloroethanol, 752
from ethene, 752, 756
polymerization of, 763
Ethylene. *See* Ethene
Ethyllithium, 125
Ethylmethylamine, **1107**
3-(Ethylmethylamino)-1-octanol, 1106
Ethylmethyl-*n*-propylamine, 1105
o-Ethylnitrobenzene, 547
2-Ethylpentanal, 788
p-Ethyltoluene, 548
α-Ethylvaleraldehyde, 788
Ethyne (*See also* Acetylene), **466**
structure of, 37
Ethynylestradiol diacetate, 866
Eugenol, 741
Exhaustive methylation, 1145
Exo compounds, 1185
Exocyclic double bond, 251
Exopeptidases, 1250
Exothermic reaction, 147
External magnetic field, in NMR, 625
Extraction, 6

Farnesol, 331
Fast Red A, 1154
Fats, fatty acids in, 1092
from oils, 1097
properties of, 1091
structure of, 1090
table of, 1092
Fatty acids, 891, 1091
alcohols from, 923
from glycerides, 1095
in fats and oils, 1092
Fehling's reagent, in carbohydrates, 1286
Fehling's test, 867
Fenton's reagent, 1293
Ferric chloride test, for phenols, 708

Ferric halide, in halogenation of benzene, 561
Ferrocene, preparation of, 536
structure of, 536
Fibrous proteins, 1236
Fingerprint region, in IR spectra, 500
Fischer esterification, 954
Fischer Projection Formulas, 212
Fischer projection formulas, 211, 212
how to draw, 212
of D-glucose, 1266
one asymmetric carbon, 212
Fischer, Emil, 1301
Fischer-Tropsch synthesis, 122, 793
Fixer, in photography, 707
Flagpole hydrogens, 104
Flame test, 2
Flash Cards, 342
Flipping of spin states, in NMR, 625
Floridean starch, 1310
Fluorine molecule, bonding in, 27
Fluorine, in NMR, 659
Fluorobenzene, **671**
from organothallium reagent, 603
Fluoroboric acid, 602
3-Fluoro-5-hydroxyisoquinoline, 1116
Fluoromethane, NMR chemical shifts in, 633
3-Fluoropentane, from 3-pentanol, 749
p-Fluorophenol, **697**
p-Fluorotoluene, 1139
Folic acid, 1157
Formal charge, 4
effect in electrophilic substitution, 570
Formaldehyde (*See also* Methanal), 431, 787, **791**
condensation with phenol (*See also* Bakelite), 833
discussion of, 793
hydrate, 839
Formamide, 945
Formate ester, 1058
Formic acid, 891, 892, **896**
from oxalic acid, 925
synthesis of, 903
Formyl chloride, **945**
Formyl group, 787
Formylation, 1058
4-Formylbenzenesulfonic acid, 787

2-Formylcyclohexanone, 1058
β-Formylcyclohexene, 787
Fractional crystallization, use in resolution of enantiomers, 920
Fragmentation, in MS, 723
Free energy, 111, 147
Free energy and equilibrium constants, 111, 148
Free radicals, 157
  ease of formation, 159
  selectivity of, 159
  stability of, 157
  stability and the inductive effect, 163
  stereochemistry of, 208, 209
  structure of, 209
  structure-reactivity correlations, 162
Free radical addition, of hydrogen bromide to alkenes, 303
Free radical chain reaction, 134
  in HBr addition to alkenes, 303
Free radical chlorination, 208, 211
Free radical halogenation, 134
  evidence for, 206
  of alkylbenzenes, 594
Free rotation about single bonds, 66
Freon 12, 142
Freons, 142
Frequency (ν), 492
Frequency, conversion to wavelength, 500
Frequency, relation to mass, 496
Frequency, relation to wavelength, 492
Friedel-Crafts acylation, 798
  limitations in, 801
  mechanism of, 799
  preparation of ketones, 798
  using anhydrides, 987
Friedel-Crafts alkylation, 563
  catalysts in, 564
  in synthesis of detergents, 1096
  limitations on, 586
  mechanism of, 563
  of phenol, 706
  rearrangements in, 565
Friedel-Crafts Alkylation, Rearrangements, 565
α-D-Fructofuranose, 1274
β-D-Fructofuranosyl-α-D-glucopyranoside, 1306
Fructose-6-phosphate, 1298
Fuels, 139
Fumaric acid, 984

Functional group determination, use of IR in, 495
Functional groups, 11
  listing of, 12
Furan, orbital picture of, 539
  structure of, 539
Furanoses, in carbohydrates, 1269
Fused-ring systems, 551

Gabriel synthesis, 1126
Galactan, 1310
D-Galactopyranose, 1271
4-O-(β-D-Galactopyransoyl)-β-D-glucopyranose, 1307
D-Galactose, 1264
Gammexane, 115, 593
Gasoline, 139
Gauche conformations, 74
gem-Diols, 410
General selection rules, versus selection rules, 1206
Geneva system, 81
Geneva system of nomenclature, 81
Geometric isomers, 185
  in alkenes, 246
  in cycloalkanes, 114
Geraniol, 351
Gibbs standard free energy, 111
Girard's Reagent, 1135
Girard's Reagent P, 1136
Glacial acetic acid, 904
Globular proteins, 1236
D-Glucaric acid, 1289
D-Gluconic acid, 1285
D-Glucopyranose, 1269, 1271
  Haworth structure of, 1270
α-D-Glucopyranose, 1273
β-D-Glucopyranose, 1273
4-O-(D-Glucopyranosyl)-β-D-glucopyranose, 1308
D-Glucose, 1264
  conformational analysis in, 1275
  enantiomers of, 1265
  equilibrium in, 1273
  Fischer projection of, 1266
  general discussion, 1263
  glycolysis of, 1299
  metabolism of, 1264
  mutarotation in, 1267
  open-chain form, 1268
  pentaacetate, 1278
  pyranose geometry of, 1305
  ring form, 1268
  ring size proof, 1304
  structure proof of, 1301
  three-dimensional representation of, 1266

α-D-Glucose, conformation of, 1276
  structure of, 1272
β-D-Glucose, conformation of, 1276
  structure of, 1272
Glucose-6-phosphate, 1298
L-Glutamate, 1231
Glutamic acid, 1212
Glutamine, 1211
Glutaric acid, 894
Glutaric anhydride, preparation of, 983
Glutaronitrile, 943
D-Glyceraldehyde, 1264
Glyceraldehyde-3-phosphate, 1298
Glyceric acid, 1286
Glyceric acid 2-phosphate, 1299
Glyceric acid 3-phosphate, 1299
Glycerides, alcohols from, 1096
  carboxylic acids from, 1095
  fatty acids from, 1095
  in synthetic chemistry, 1095
  nomenclature of, 1091
  occurrence of, 1091
  soaps from, 1094
  structure of, 1090
  unsaturated, 1097
Glycerine, 426, 1090
Glycerol, 426, 737, **769**
Glycercose, 1272
Glycine, 1211
Glycogen, 1263, 1310
Glycols, 737
  cleavage by periodic acid, 771
  from alkenes, 332
  nomenclature of, 737
  properties of, 769
  reactions of, 770
  summary of methods of preparation, 770
  summary of reactions, 779
  table of, 769
Glycolysis, 1301
  mechanism of, 1299
Glycoproteins, 1252
Glycosides, in carbohydrates, 1277, 1278
Glyme, 734, 760
GR-S, 349
Grignard reactions, use of ketals in, 846
Grignard reagents, 802
  addition to α,β unsaturated carbonyl compounds, 1075
  carboxylic acids from, 898
  compared with organolithium reagents, 856
  hydrolysis of, 842
  limitations on use of, 845

mechanism of addition to carbonyl compounds, 843
mechanism, of formation of carboxylic acids from, 899
preparation of, 802
reactions with esters, 971
stereochemistry of, 1022
structure of, 802
Grignard synthesis, of alcohols, 846
Group theory, 1192
Guanine, 1279
D-Glucose, 1264
Gum tragacanth, 1310
Gyromagnetic ratio, in NMR, 624

Half acid-half ester, 962
Half-chair form of cyclohexane, 105
α-Halo acids
α-amino acids from, 926, 1125
α-hydroxy acids from, 926
α,β-unsaturated acids from, 927
malonic acids from, 901, 926
Halobenzenes (See also Aryl Halides), electrophilic substitution on, 588
α-Haloethylbenzene, 596
β-Haloethylbenzene, 596
Haloform Reaction, 1030
Haloform test, 867
discussion of, 1030
mechanism of, 1031
Halogenation, of aldehydes and ketones, base-promoted, 1027
of alkanes, by sulfuryl chloride, 136
of alkenes, 317
of alkynes, 477
of benzene, 561
of benzene, mechanism of, 561
of carbonyl compounds in acid, 1033
of carbonyl compounds, mechanism of, 1027
of carbonyl compounds, orientation in, 1029
of methane, mechanism, 134
of phenols, 704
of side chains in alkyl benzenes, 594
Halogens, addition to alkenes, 317
Halohydrins, 326
from alkenes, 326
from epoxides, 759
Hansa Yellow, 1154

Hard water, 1095
Hardening, 1097
Haworth projections. See Haworth structures
Haworth structures, in carbohydrates, 1270
Heats of combustion, 106
alkanes, 138
benzene, 530
for geometric isomers of alkenes, 252
Heats of hydrogenation, alkene stability, 315
alkenes and alkadienes, 258
of alkenes, 316
use in benzene structure, 529
use for diene structure, 257
Heats of reaction, 147
calculation of, 153
α-Helix, in proteins, 1240
structure of, 1241
Hell-Volhard-Zelinsky reaction, 925
mechanism of, 1036
Heme group, 1252
Hemiacetals (See also Hemiketals), 828
acid-catalyzed formation of, 828
base-catalyzed formation of, 832
formation of, 828
in carbohydrates, 831, 1267, 1268
mechanism of base-catalyzed formation of, 833
mechanism of formation, 828
Hemiketals (See also Hemiacetals), 828
in carbohydrates, 1268
Hemimellitene, 550
Hemoglobin, 1237, 1252
Henderson-Hasselbach equation, for α-amino acids, 1214
(n-)Heneicosane, 93
(n-)Heptacontane, 93
(n-)Heptadecane, 93
(n-)Heptaldehyde (See also Heptanal), 791
Heptanal (See also n-Heptaldehyde), 791
(n-)Heptane, 93
structural isomers of, 79
Heptanedioic acid, 894, 1068
Heptanoic acid, 892
1-Heptanol, 412
1-Heptanol (See also n-Heptyl alcohol), 412
4-Heptanone (See also Di-n-Propyl ketone), 791
1-Heptene, 252

trans-2-Heptene, 249
cis-3-Heptene, from 3-heptyne, 480
(n-)Heptyl alcohol (See also 1-Heptanol), 412
1-Heptyne, 466
3-Heptyne, addition of diborane to, 480
Heroin, 1160
Hertz, in NMR, 626
Heterocyclic Compounds, IUPAC Nomenclature of, 736
Heterocyclic compounds, nomenclature of, 1114
reactions of, 1148
structure of, 537
Heterocyclic nitrogen compounds, 1114
K_b's and pK_b's of, 1116
reaction with acids, 1118
Heterolytic bond cleavage, 164
1,2,3,4,5,6-Hexachlorocyclohexane, 115
Hexachlorophene, 688, 696
(n-)Hexacontane, 93
(n-)Hexadecane, 93
Hexadecanoic acid, 892
Hexadeuteriobenzene, from benzene, 563
Hexadeuteroacetone, 629
Hexadeutero dimethyl sulfoxide, 629
1,5-Hexadiene, 258
2,4-Hexadiene, 306
addition of HBr to, 306
cis,cis-2,4-Hexadiene, 256
cis,trans-2,4-Hexadiene, 256
trans,trans-2,4-Hexadiene, 256, 1180
E,E-2,4-Hexadiene, 256
E,Z-2,4-Hexadiene, 256
Z,Z-2,4-Hexadiene, 256
1,4-Hexadiyne, 465
Hexaethylbenzene, 550
Hexafluororacetone, 839
Hexafluoroacetone hydrate, formation of, 839
Hexafluoropropanone, 839
Hexamethylbenzene, 550
Hexamethyleneamine, 1138
Hexamethylenediamine, 979
Hexamethylphosphoramide, 176
Hexanal (See also Caproaldehyde), 791
(n-)Hexane, 77, 93
from 3-hexene, 424
IR spectrum of, 501
structural isomers of, 77
Hexanedioic acid, 894
1,6-Hexanediol, 974

Hexanedioyl chloride, synthesis of, 949
Hexanoic acid, 892
1-Hexanol (*See also* *n*-Hexyl alcohol), **412**
  from 1-hexene, 424
2-Hexanone, 250, **791**
3-Hexanone, **791**
1,3,5-Hexatriene, 255
4-Hexen-1-yne, 466
1-Hexene, **252**, 258, 316
  1,2-epoxyhexane from, 757
  IR spectrum of, 501
trans-3-Hexenoic acid, 893
Hexose, 1261
(*n*-)Hexyl alcohol (*See also* 1-Hexanol), **412**
(*n*-)Hexyl bromide (*See also* 1-Bromohexane), **367**
(*n*-)Hexyl chloride (*See also* 1-Chlorohexane), **367**
(*n*-)Hexyl iodide (*See also* 1-Iodohexane), **367**
(*n*-)Hexylamine, 1105
Hexylresorcinol, 696
1-Hexyne, **466**, 473
2-Hexyne, **466**
3-Hexyne, **466**, 473
Hinsberg test, 1130
Histidine, 1211
Hoffmann, R., 1190
Hofmann elimination, 1144
  compared to Saytzeff elimination, 1146
  in structure determination, 1145
  orientation in, 1145
Hofmann reaction, 1127
  degradation of amides, 1127
  mechanism of, 1128
  rearrangements in, 1128
Homolog, 79
Homologous series, 78, 79
Homolytic cleavage, 153
Hormones, 865
Hückel's $(4n + 2)$ rule, 534
Hund's rule, 23
HVZ reaction (*See* Hell-Volhard-Zelinsky reaction), 925
Hyaluronic acid, 1309
Hybridization, 49
Hydration and dehydration, compared, 298
Hydration of alkenes, 296, 297
  of alkynes, 478, 479
  of carbonyl compounds, 839
D-Hydratropic acid, 1121
Hydrazine, addition to carbonyl compounds, 837

  use in Wolff-Kishner reduction, 861
Hydrazones, formation of, 838
Hydride abstraction, by carbocations on alkanes, 330
  of cycloheptatriene, 602
Hydride ion, 5
  as a nucleophile in reductions, 858
  in Cannizzaro reaction, 1047
  in lithium aluminum hydride, 859
  in sodium borohydride, 859
Hydride migration, 279
Hydride reducing agents, table of, 860
Hydride shift, 279
1,2-Hydride shift, example of, 280
  transition state in, 279
1,2-Hydride shifts, 278
Hydride transfer, from triarylmethanes, 605
  in oxidation of alcohols, 438
Hydroboration, isomerization in, 423
  mechanism of, 418
  of alkenes, 416
  orientation of addition in, 418
  transition state in, 418
Hydroboration-oxidation, 420
  mechanism of oxidation step, 421
  stereochemistry of, 420
Hydroboration-reduction, of alkynes, 480
Hydrocarbons, acidity of, 470
  classification of, 64
  definition of, 64
Hydroforming, 551
α-Hydrogen, acidity, 819
  in carbonyl and β-dicarbonyl compounds, summary of reactions, 1079
Hydrogen atom, 5
Hydrogen bonding, comparison with other elements, 414
  in alcohols, 413
  in amides, 946
  in amines, 1108
  in β-dicarbonyl compounds, 1078
  in carboxylic acids, 897
  in DNA, 1281
  in phenols, 697
  in proteins, 1242
  in salicyclic acid, 914
  internal in phenols, 698
  of aldehydes and ketones in water, 792

  of ethers in water, 739
  with fluorine, 413
  with nitrogen, 413
  with oxygen, 413
Hydrogen bromide, **414**
Hydrogen bromide addition to alkenes, *anti*-Markovnikov, 302
Hydrogen chloride, **414**
Hydrogen cyanide, acidity of, 825
  addition to alkynes, 481
  addition to carbohydrates, 1292
  addition to carbonyl compounds, 825
Hydrogen deficiency, index of, 314
Hydrogen fluoride, **414**
  bonding in, 28
  resonance structures of, 531
Hydrogen halides, addition to alkenes, 290
Hydrogen iodide, **414**
Hydrogen molecule, bonding, 26
Hydogen peroxide, 421
Hydrogen selenide, **412**
Hydrogen sulfide, **414**
Hydrogen telluride, **414**
Hydrogen, classification of, (1°, 2°, 3°), 123
  classification of type, 159
  primary, 123
  secondary, 123
  tertiary, 123
  types of, 123
Hydrogenation, in structure determination, 314
  mechanism in alkyne reactions, 476
  of α,β-unsaturated carbonyl compounds, 1046
  of acid chlorides, 796
  of aldehydes, 857
  of alkenes, 310
  of alkynes, 475
  of aromatic compounds, 592
  of benzene, 524
  of carbonyl compounds, 857
  of conjugated ketones, 857
  of dienes, 313
  of esters, 974
  of fats, 1097
  of ketones, 857
  of nitro compounds, 1121
  selective in unsaturated ketones, 857
  stereochemistry of, 311
α-Hydrogens, in carbonyl compounds, 1017

reactions of, 819
Hydrolysis, 397
  by chymotrypsin, 1253
  in S_N1 reactions, 395
  of acid chlorides, 952
  of amides, 977
  of esters, 965
  of glycerides, 1095
  of Grignard reagents, 842
  of nitriles to give carboxylic
    acids, 900
Hydronium ion, 272
Hydroperoxy anion, 421
Hydrophilic, 1094
Hydrophobic, 1094
Hydroquinone, 696, **697**
α-Hydroxy acids, from α-halo
    acids, 926
δ-Hydroxy acid, 964
γ-Hydroxy acid, 964
α-Hydroxy aldehydes, cleavage
    by periodic acid, 771
α-Hydroxy ketones, cleavage by
    periodic acid, 771
3-Hydroxy-1,3-diphenyl-1-bu-
    tanone, 1039
3-Hydroxy-2- ethylhexanal, 1045
1-Hydroxy-2-methyl-3-pen-
    tanone, synthesis of, 924
3-Hydroxy-2-methylpentanal,
    1039, 1045
4-Hydroxy-4-methyl-2-pen-
    tanone, 1039
3-Hydroxy-1-phenyl-1-
    propanone, 1043
o-Hydroxybenzaldehyde (*See
    also* Salicylaldehyde), 788,
    791
  from phenol, 753
p-Hydroxybenzaldehyde, **791**
o-Hydroxybenzenesulfonic acid,
    695, 705
p-Hydroxybenzenesulfonic acid,
    705
3-Hydroxybutanal, 1039, 1042
  dehydration of, 1040
β-Hydroxybutyraldehyde, 1039
cis-3-Hydroxycyclohexanecar-
    boxylic acid, 965
Hydroxyl radical, 138
Hydroxylamine, 836
  addition to carbonyl com-
    pounds, 836
Hydroxylation, mechanism of,
    334
  of alkenes, 333
  of alkenes by osmium tetrox-
    ide, 333
  of alkenes by potassium
    permanganate, 333

stereochemistry of, 334
α-Hydroxymethyl-β-hydroxy-
    propionaldehyde, 1044
p-Hydroxymethylphenol, prepa-
    ration of, 834
3-Hydroxypentanal, 1042
2-Hydroxypropanoic acid, prepa-
    ration of, 826
β-Hydroxypropionaldehyde,
    1044
α-Hydroxypropionic acid, prepa-
    ration of, 826
p-Hydroxytoluene, 548
δ-Hydroxyvalerate, 964
γ-Hydroxyvalerate, 964
δ-Hydroxyvaleric acid, 964
γ-Hydroxyvaleric acid, 964
Hyperchromic effect, 510
Hyperconjugation, 533
  in *tert*-butyl cation, 534
Hypochromic effect, 510
Hypohalite, 1127
Hypohalite reaction, 868
Hypohalite test, 1030
Hypohalous acid, addition to al-
    kenes, 326
  from halogens and water, 327
Hypophosphorus acid, 1139
Hypsochromic shift, 510
Hz, 493

D-Idose, 1264
Imidazole, 1114
  orbital picture of, 1117
Imide, 1223
Imine, 837
Imines, reduction of, 1123
in situ, defined, 442
Index of hydrogen deficiency,
    314
Indole, 1114
Inductive effects, 163
  electron-donation in elec-
    trophilic substitution, 575
  electron-withdrawal in elec-
    trophilic substitution, 577
  in electrophilic substitution,
    574
Infrared spectra, 495
Infrared spectroscopy (*See also*
    IR), 491
  introduction to, 495
  of acid chlorides, 997
  of alcohols, 709
  of aldehydes, 805
  of alkyl halides, 690
  of amides, 997
  of amines, 1161
  of anhydrides, 997

of aromatic compounds, 605
of aryl halides, 690
of carboxylic acid derivatives,
    996
of carboxylic acids, 927, 997
of esters, 997
of ethers, 773
of ketones, 805
of nitriles, 997
of phenols, 709
Ingold, Sir C. K., 201
Insulin, 1237
  discussion of, 1245
  from proinsulin, 1248
  primary structure of, 1246
Intermediate, 136
Intermolecular hydrogen bond-
    ing, in alcohols, 413
Internal hydrogen bonding, 698
Interorbital angle, 98
Intramolecular hydrogen bond-
    ing, 698
Introduction, 1
Inulin, 1310
Inversion of configuration, 199
  definition of, 371
  in reaction of alcohols with
    HBr, 446
  in S_N2 reaction, 370
Inversion of relative configura-
    tion, 210
1-Iodo-2,2-dimethylpropane, 382
1-Iodo-2-methylpropane (*See
    also* Isobutyl iodide), **367**
2-Iodo-2-methylpropane (*See
    also tert*-Butyl iodide), **367**
Iodobenzene, **671**
  from organothallium reagent,
    603
1-Iodobutane (*See also n*-Butyl
    iodide), **367**
2-Iodobutane (*See also sec*-
    Butyl iodide), **367**
  from 2-butene, 291
  from methylcyclopropane, 294
R-2-Iodobutane, 210
Iodocyclohexane (*See also*
    Cyclohexyl iodide), **367**
Iodoethane (*See also* Ethyl
    iodide), **367**
  from ethanol, 450
Iodoform, 867
Iodoform test, 867
1-Iodohexane (*See also n*-Hexyl
    iodide), **367**
Iodomethane (*See also* Methyl
    iodide), **367**
  NMR chemical shifts in, 633
1-Iodopentane (*See also
    n*-Pentyl iodide), **367**

2-Iodopentane, from 2-pentanol, 443

p-Iodophenol, **697**

1-Iodopropane (See also n-Propyl iodide), **367**
from 1-propanol, 443

2-Iodopropane (See also Isopropyl iodide), **367**

3-Iodopropene (See also Allyl iodide), **367**

Ion pairs, in S$_N$1 reaction, 453
in halogenation of benzene, 562

Ion radical, in MS, 721

IR absorption and structure, correlations between, 500

IR absorption frequencies, table of, 497

IR spectra, interpretation of, 502

IR spectrophotometer, diagram of, 499
operation of, 498

IR spectroscopy (See also Infrared spectroscopy), 495

IR spectroscopy and structure, 500

Isobutane (See also 2-methylpropane or Methylpropane), 72, 90

Isobutyl alcohol (See also 2-Methyl-1-propanol), **412**

Isobutyl bromide (See also 1-Bromo-2-methylpropane), **367**

Isobutyl chloride (See also 1-Chloro-2-methylpropane), 91, **367**

Isobutyl group, 84

Isobutyl iodide, (See also 1-Iodo-2-methylpropane), **367**

Isobutylamine, **1107**

2-Isobutylcyclohexanone, 1050

Isobutylene. See 2-Methylpropene

Isobutylene glycol, 737

Isobutylene oxide, 737

Isobutyraldehyde (See also 2-Methylpropanal), **791**

Isocaproyl chloride, 941

Isooctane, 139

Isocyanate, 1128

Isodurene, **550**

Isoelectric point, for α-amino acids, 1213

Isoeugenol, 741

Isohexane, 90

Isohexyl bromide, 91

Isohexyl group, 85

Isolated diene, 255

Isoleucine, 1211

Isomerization, in hydroboration, 423

Isomers, classification of, 185

cis,trans-Isomers, in alkenes, 245
in cycloalkanes, 114

Isooxazole ring, 1156

Isopentane, **77**

Isopentenyl pyrophosphate, 350

Isopentyl alcohol (See also 3-Methyl-1-butanol), 410, **412**

Isopentyl group, 85

Isopentyl pyrophosphate, 768

Isopentylamine, 1105

Isophthalic acid, 894, **896**, 897, 984

Isoprene, 348
UV spectrum of, 513

Isoprene rule, 350

Isopropenyl group, 248

Isopropyl alcohol, (See also 2-Propanol), **412**
discussion of, 426

Isopropyl benzoate, from benzoyl chloride, 952

Isopropyl bromide (See also 2-Bromopropane), **367**
conformations of, 658

Isopropyl cation, 164

Isopropyl chloride (See also 2-Chloropropane), **367**

Isopropyl ether, **739**

Isopropyl group, 84

Isopropyl hydrogen chromate, 435

Isopropyl hydrogen phthalate, synthesis of, 962

Isopropyl hydrogen sulfate, 300

Isopropyl iodide (See also 2-Iodopropane), **367**

Isopropylamine, **1107**

Isopropylbenzene, from propene and benzene, 567
nitration of, 572

Isopropylmagnesium bromide, 843

o-Isopropylphenol, 706

p-Isopropylphenol, 706

p-Isopropyltoluene, oxidation of, 594

Isoquinoline, 1114

Isotactic polymers, 347

Isotope effects, energy diagram for, 437

Isotope Effects, Kinetic, 436

Isotope effects, primary, 436
primary hydrogen, 436
secondary, 436

Isotopes, carbon-13 in NMR, 659
carbon-14 in benzyne formation, 686
deuterium in NMR, 658
deuterium, in α-carbanion formation, 1021
in mass spectrometry, 722
natural abundance of, 722
oxygen-18, in ester hydrolysis, 966
oxygen-18, in esterification, 957
oxygen-18, in hydration of carbonyl compounds, 840

Isotopic exchange studies, α-carbanions, 1020

Isovaleraldehyde (See also 3-Methylbutanal), **791**

IUPAC system of nomenclature, 81

J (coupling constants), in NMR, 645

J. J. Thompson atom, 18

K$_a$, acid dissociation constants, 168

K$_b$, base dissociation constants, 1108

Kekulé structures, of benzene, 523

Kekulé, August, 523

Kepone, 689

Keratins, 1237
discussion of, 1244

Ketals (See also Acetals), 828
as protecting groups, 863, 924
in carbohydrates, 1277
use as protecting groups in Grignard reactions, 846

Ketene, 981, 1188

β-Keto acids, decarboxylation of, 1069

β-Keto esters, 1061
acidity of, 1055
using Claisen condensations, 1054

Keto-enol equilibrium, acid-catalyzed, 1035
base-promoted, 1026
in β-dicarbonyl compounds, 1078
in carbohydrates, 1285
in conjugate addition reactions, 1074
in decarboxylation of β-keto acids, 1070

Keto-enol tautomerism, (See also Keto-enol equilibrium), 479
in hydration of alkynes, 479

α-Ketoglutarate, 1231
Ketones (*See also* Carbonyl compounds) α-amino acids from, 1232
α-carbanions from, 1018
addition of hydrogen cyanide to, 825
aldol condensation of, 1037
alkanes from, 861
alkylation of, by enamine synthesis, 1048
amines from, 1124
common names of, 788
condensation with nitro compounds, 1137
crossed aldol condensations of, 1043
crossed condensation with esters, 1058
derivatives of, 836
dipole moments of, 793
from alkenes, 798
from alkynes, 798
from anhydrides, 987
from ethyl acetoacetate, 1068
from glycols, 798
from nitriles, 804
from secondary alcohols, 433, 797
haloform test for, 1030
halogenation of, 1027
hydrogenation of, 857
infrared spectroscopy of, 805
IUPAC names of, 789
mass spectrometry of, 809
NMR of, 807
nomenclature of, 788
nucleophilic addition reactions of, 818
physical properties of, 790
preparation of aromatic, 798
preparation using organocadmium reagent, 802
qualitative tests for, 866
reaction with organolithium reagents, 856
reduction by lithium aluminum hydride, 858
reduction by sodium borohydride, 858
reduction of, 856
resonance in aromatic ketones, 823
spectral analysis of, 805
summary of preparation, 797
summary of preparation of, 810
summary of reactions, 871
table of, 791
UV spectroscopy of, 807
Ketopentose, 1261

Ketose, 1261
Ketoses, as reducing sugars, 1286
Kiliani-Fischer synthesis, 1291
Kinetic control, in HBr addition to 1,3-butadiene, 309
Kinetic Isotope Effects, 436
Kinetic versus thermodynamic control, in carbohydrates, 1269
in HBr addition to 1,3-butadiene, 309
Kinetically controlled reactions, 169
Kinetics, 174
Kinetics versus thermodynamics, 169
Kolbe reaction, 904
Körner method, of structure determination, 616

L-Configuration, of α-amino acids, 1216
L-Form, 186
L, represents leaving group, 172
L-Family of carbohydrates, 1265
L-Lactic acid, 1121
Lactic acid, 1299
enantiomers of, 196
Fischer projection formulas of enantiomers of, 214
Lactobionic acid, 1308
γ-Lactone, 964
δ-Lactone, 964
Lactones, from hydroxy acids, 963
in carbohydrates, 1293
Lactose, 1307, 1308
LAH. *See* Lithium aluminum hydride
Laminaran, 1310
Lanosterol, 352
from squalene, 766
Lard, 1092
Lard oil, 1092
Lauric acid, 892, **896**
Lauryl alcohol, 440
Lauryl hydrogen sulfate, 440
LCAO, 56
Leaving groups, effect in S_N2 reactions, 378
in S_N1 reactions, 385
in S_N2 reactions, 385
Lebel, 189
Lecithins, 1101
Leucine, 1211
Leucine aminopeptidase, 1250
Levorotatory, 186
Lewis acid-base reaction, in halogenation of benzene, 561

Lewis acid-base theory, 167
Lewis dot structure, 32
Lewis electron structures, 3
Lidocaine, 1158
Light, interaction with matter, 195
Limestone, 467
Limonene, 350
Lindane, 115, 593, 688
Lindlar's catalyst, 475
Line structure, 32
Linear combination of atomic orbitals (LCAO's), 56
Linoleic acid, 892
Linolenic acid, 892
Linseed oil, 1092
in paints, 1098
Lipases, 1099
Lipids, 1090
Lithium aluminum deuteride, 883
Lithium aluminum hydride, comparison with sodium borohydride, 859
ester reduction, 974
mechanism of carbonyl reduction by, 858
mechanism of reduction of acids, 922
reaction with acid, 859
reduction of amides, 1122
reduction of carbonyl compounds by, 858
reduction of carboxylic acids by, 922
structure of, 858
Lithium dialkylcopper reaction, 353
Lithium dialkylcopper reagent, 124
Lithium dicyclohexylcopper, 126
Lithium diethylcopper, 125
Lithium *tri-tert*-butoxyaluminum hydride, 796
Lowry-Brønsted acid-base reaction, in alcohol dehydration, 273
LSD, 1160
Lucas reagent, 443
as a classification test for alcohols, 454
Lycopene, 352
UV absorption in, 513
UV spectrum of, 515
Lysergic acid diethylamide (LSD), 1160
Lysine, 1211
Lysine vasopressin, structure of, 1220
Lysozyme, 1237
Lyxose, 1272

Magic methyl, 1150
Magnesium ethoxide, 428
Magnesium stearate, 1095
Magnetic and spin properties, in NMR, 624
Magnetic orientations and spin-spin splitting, in NMR, 639
Maleic acid, 905, 984
Maleic anhydride, 905
preparation of, 984
Malonic acid, 894, **896**
synthesis of, 905
Malonic acids, decarboxylation of, 925
decarboxylation of in malonic ester synthesis, 1065
from α-halo acids, 901, 926
from nitriles, 901
Malonic ester syntheses, examples of, 1066
Malonic ester synthesis, 1063
carboxylic acids from, 1065
Malonic ester (*See also* Diethyl malonate) α-amino acids from, 1232
acylation of, 1071
Maltase, 1316
Maltose, 1308
Malvalic acid, 1091
D-Mandelic acid, 1121
Mannan, 1310
D-Mannonic acid, 1294
D-Mannopyranose, 1271
D-Mannose, 1264
α-D-Mannose, conformation of, 1276
β-D-Mannose, conformation of, 1276
Margarine, 1098
Marijuana, 741
Markovnikov addition, to alkenes, 301
Markovinikov's rule, defined, 301
in addition of HX to alkynes, 477
in addition of water to alkynes, 478
Mass spectra, interpretation of, 723
Mass spectrometer, diagram of, 722
operation of, 722
Mass spectrometry (MS), 721
alkylbenzene ion radical in, 724
cyclopentadienyl cation in, 724
discussion of, 721
examples of, 724–728
fragmentation in, 723

interpretation of mass spectra in, 723
ion radical in, 721
mass/charge (*m/e*) in, 721
of aldehydes, 809
of amines, 1164
of carboxylic acids, 929
of esters, 1000
of ethers, 774
of ketones, 809
parent ion in, 721
parent ion radical in, 721
toluene ion radical in, 724
tropylium cation in, 724
McLafferty rearrangements in, 809
Mass, relation to frequency, 496
Mass/charge (*m/e*), in MS, 722
Matter, interaction with light, 195
McKillop, A., 603
McLafferty rearrangement, in mass spectroscopy, 809
Mechanism, definition of, 125
Mechanisms, Use of Arrows in, 276
Megahertz, in NMR, 626
Melting point, 7
Membrane wall, 1101
Membranes, 1100
2-Menthene, 270
3-Menthene, 270
Menthyl chloride, 270
Merrifield automated protein synthesizer, 1230
Merrifield, R. Bruce, 1229
Mesityl oxide (*See also* 4-Methyl-3-penten-2-one), **791**
Mesitylene, **550**
meso Compound, 220
meso Compounds, optical property of, 220
Messenger RNA, 1234
meta Director, defined, 569
meta Directors, 570
orientation theory in, 577
meta, Use in nomenclature, 547
Metal alkoxide, 428
Methamphetamine, 1157
Methanal (*See also* Formaldehyde), 788, **791**
from methanol, 431
hydration of, 793
polymers of, 794
Methanamide (*See also* Formamide), **945**
Methane, **93, 414**
antibonding MO's in, 60
bonding MO's in, 60
halogenation, mechanism, 134
hybridization, 28

NMR chemical shifts in, 633
physical properties, 65
resonance structures of, 532
sources of, 65
structure of, 28
Methanoic acid (*See also* Acetic acid), 892, **896**
synthesis of, 903
Methanol (*See also* Methyl alcohol), 270, **412**
discussion of, 425
Methanoyl chloride, **945**
Methedrine, 1157
Methionine, 1212
2-Methoxybutane, 734
Methoxychlor, 689, 741
1-Methoxy-1-phenylhexane, 734
p-Methoxytoluene, NMR spectrum of, 651
Methyl alcohol (*See also* Methanol), 270, **412**
Methyl β-D-allopyranoside, 1278
Methyl β-D-alloside, 1278
Methyl anion, 5
structure of, 43
Methyl benzoate, IR spectrum of, 993
Methyl bromide (*See also* Bromomethane), **367**
Methyl *p*-bromobenzoate, 971
Methyl butanoate, 1046
Methyl 2-butenoate, 1046
Methyl *sec*-butyl ether, 734
Methyl *tert*-butyl ether, cleavage of, 754
Methyl carbinol, 867
Methyl carbinols, haloform test for, 1030
tests for, 867
Methyl carbitol, 759
Methyl cation, 5, 164
Methyl cation, structure of, 43
Methyl chloride (*See also* Chloromethane), 53, **367**
Methyl ether, 734, **739**
Methyl ethyl ether, **739**
Methyl ethyl ketone (*See also* Butanone), 788, **791**
Methyl fluorosulfonate, 1150
Methyl α-D-glucopyranoside, 1278
Methyl α-D-glucoside, 1278
Methyl iodide (*See also* Iodomethane) 90, **367**
Methyl ketone, 867
tests for, 867
Methyl methanoate, 972
Methyl 9-octadecenoate, 974
Methyl oleate, 974
Methyl Orange, 1154
Methyl phenyl ether, 734, **739**

Methyl phenyl ketone (*See also* Acetophenone), **791**

Methyl *n*-propyl ketone (*See also* 2-Pentanone), **791**

Methyl radical, 5
structure, 41

Methyl salicylate, 945
synthesis of, 955

Methyl vinyl ether, 734

Methyl vinyl ketone (*See also* 1-Buten-3-one), 788, **791**

2-Methyl-1-aminopropane, 1129

2-Methyl-1,4-benzoquinone, from 2,4-dimethylphenol, 707

2-Methyl-N-bromoaziridine, 237

2-Methyl-1-bromopropane, 850

2-Methyl-1,3-butadiene, 258, 348

2-Methyl-2-butanol, from 2-methyl-2-butene, 297

3-Methyl-1-butanol (*See also* Isopentyl alcohol), **412**, 850

2-Methyl-2-butanol (*See also* *tert*-Pentyl alcohol), **412**

3-Methyl-2-butanol, 2-chloro-2-methylbutane from, 750
2-chloro-3-methylbutane from, 750
from 2-methyl-2-butene, 420

3-Methyl-2-butanone, 849, 1049

2-Methyl-1-butene, **252**

3-Methyl-1-butene, **252**, 295, 316
addition of HX to, 295

2-Methyl-2-butene, 316

2-Methyl-4-*tert*-butyl-4-octanol, synthesis of, 852

3-Methyl-2-butyl tosylate, 750

3-Methyl-1-butyne, **466**

2-Methyl-2-chlorobutane, from 2-methyl-2-butanol, 443

1-Methyl-1,3-cyclohexadiene, 1181

1-Methyl-1,4-cyclohexanediol, 406

N-Methyl-N-ethyl-1-aminoheptane, 1127

6-Methyl-3-heptanol, 411

N-Methyl-1,2-dihydropyridine, 1151

N-Methyl-1,4-dihydropyridine, 1151

2-Methyl-3-hydroxybutanal, 1042

2-Methyl-3-hydroxypentanal, 1042

2-Methyl-3-hydroxypentanoic acid, synthesis of, 924

2-Methyl-6-nitroquinoline, 1115

2-Methyl-4-nonyne, from ethyne, 475

2-Methyl-4-octanol, 852

2-Methyl-4-octanone, 852

3-Methyl-2-octanone, 1069

2-Methyl-5-oxohexanal, 790

2-Methyl-3-oxopentanoic acid, 923, 1045

2-Methyl-1,3-pentanediol, 1045
synthesis of, 923

2-Methyl-1-pentanol, 1046

2-Methyl-3-pentanol, from 2-methyl-3-pentanone, 859

4-Methyl-1-pentene, from propene, 328

4-Methyl-2-pentene, from propene, 328

2-Methyl-2-penten-1-ol, 1046

2-Methyl-2-pentenal, 1045, 1046

4-Methyl-3-penten-2-one (*See also* Mesityl oxide), **791**

α-Methyl-γ-phenyladipic acid, 894

3-Methyl-3-phenylbutanal, 788

2-Methyl-2-phenylbutane, from benzene, 564

2-Methyl-3-phenylbutane, from 2-methyl-2-phenylbutane, 565

3-Methyl-2-phenyl-2-butanol, synthesis of, 848

β-Methyl-β-phenylbutyraldehyde, 788

2-Methyl-2-phenyl-4-heptanone, 789

2-Methyl-3-phenylhexanedioic acid, 895

2-Methyl-4-phenylphenol, 548

2-Methyl-1-phenyl-1-propanol, preparation of, 845, 848

2-Methyl-1,2-propanediol, 737

2-Methyl-1-propanol, **412**, 850

2-Methyl-1-propanol (*See also* Isobutyl alcohol), **412**

2-Methyl-2-propanol, 270, **412**
dehydration of, 271

2-Methyl-2-propanol (*See also* *tert*-Butyl alcohol), **412**

2-Methyl-1-propenyl group, 249

2-Methyl-1-propylmagnesium bromide, 850

N-Methylacetamide, 941

o-Methylacetophenone, 987

p-Methylacetophenone, 987

Methylacetylene, 465

Methylamine, 1105, **1107**

2-Methylamino-1-phenylpropane, 1157

N-Methylaniline, 1105, **1107**
IR spectrum of, 1161
NMR spectrum of, 1163

Methylbenzenes, oxidation of, 797

m-Methylbenzoic acid, 893

4-Methylbenzophenone, preparation of, 799

p-Methylbenzoyl chloride, 799

p-Methylbenzylmagnesium chloride, 845

2-Methylbicyclo[2.2.1]-2.5-heptadiene, preparation of, 480

2-Methylbicyclo[2.2.2]octane, 391

5-Methylbicyclo[4.2.0]-2-octene, 391

3-Methylbutanal (*See also* Isovaleraldehyde), **791**
from 3-methylbutanoyl chloride, 796
synthesis of, 849

3-Methylbutanamide, 1129

Methylbutane, 77

3-Methylbutanenitrile, 901

2-Methylbutanoic acid, 893
preparation of, 899

3-Methylbutanoic acid, synthesis of, 901

3-Methylbutanoyl chloride, 796, 849

β-Methylbutyraldehyde, 787

β-Methylbutyric acid, 891

Methylcarbonium ion, 164

Methylcyclohexane, 97
conformational equilibrium in, 112
conformations of, 109
from toluene, 592

1-Methylcyclohexene, 363
NMR chemical shifts in, 636

Methylcyclopentane, 97

trans-2-Methylcyclopentanol, from 1-methylcyclopentene, 421

cis-3-Methylcyclopentanol, from *trans*-3-bromomethylcyclopentane, 372

3-Methylcyclopentene, 251

4-Methylcyclopentene, 424

Methylene, 336
addition to alkenes, 1187
from diazomethane, 1188
singlet, 44, 1188
singlet addition of, 1188
singlet insertion of, 1188
singlet, and Woodward-Hoffmann rules, 1204
stereochemistry of singlet addition, 1188
structure of, 44
triplet, 44, 1188
triplet addition of, 1189
triplet insertion of, 1189

Methylene choride (*See also* Dichloromethane), 53

Methylene group, 251
Methylene groups, from carbonyl groups, 861
Methylenecyclobutane, 251
Methylenecyclohexane, 1053
Methylenecyclohexane oxide, 765
3-Methylenecyclohexene, 1181
N-Methylethanamide, 941
1-Methylethenyl group, 248
α-Methylglutaric acid, 894
Methyllithium, 855
Methylmagnesium iodide, 803
3-Methyloxane, 1115
Methyloxirane, 737
N-Methylpentanamide, from pentanoyl chloride, 951
2-Methylpentane, 77, 86
  from propene, 328
3-Methylpentane, 77
4-Methylpentanoyl chloride, 941
p-Methylphenol, 548
4-Methylphthalic anhydride, 1183
2-Methylpropanal (See also Isobutyraldehyde), 791, 848
  from 2-methyl-1-propanol, 432
Methylpropane, (See also Isobutane) 72, 86
  halogenation of, 133
2-Methylpropanoic acid, 926
Methylpropene (See also Isobutylene), 244, 252, 316
  mechanism of HCl addition to, 293
p-Methylpropiophenone, reduction of, 862
N-Methylpyridinium salt, 1151
N-Methylpyrrole, 1150
α-Methylstyrene (See also 2-Phenylpropene), 549, 598
α-Methylsuccinic anhydride, 987
3-Methyltetrahydropyran, 1115
N-Methylvaleramide, from valeryl chloride, 951
γ-Methylvaleryl chloride, 941
Micelles, 1094, 1100
Michael addition reactions, 1076
Microfibril, 1243
Microns (μ), 492
Microwave spectroscopy, 494
1,2-Migration, 279
Millimicrons (mμ), 492
Mirror images, 189
Molar extinction coefficient, in UV spectra, 503
Molecular (bond) orbitals, 24
Molecular formulas, 9
  use in structure determination by hydrogenation, 314
Molecular Models, Use Of, 76

Molecular orbital (MO) theory, 56
  theory, aromaticity, 542
  theory, of benzene, 542
  Woodward-Hoffmann rules and, 1190
Molecular orbital symmetry, Woodward-Hoffmann rules, 1192
Molecular orbitals (MO's) in ethene, 61
  in ethyne, 61
  of benzene, 543
  of cyclopentadienyl anion, 543
  of cyclopropenyl anion, 545
  of cyclopropenyl cation, 544
Molecular orbitals, for π bonds, 59
  for σ bonds, 59
  σ versus π, 59
  in hydrogen, 57
  in methane, 31
Molecular structure, 24
Molecular weights, 9
Molozonide, 339
Monochromatic light, 185
α-Monolaurin, 1093
Monomolecular reaction, in S_N1 reaction, 369
Monosubstitution reactions, of aromatic compounds, 609
Monosubstitution, of benzene, 556
Monoterpenes, 350, 351
Morphine, 1160
MS. See Mass spectrometry
Mucoproteins, 1252
Multiplicity in spin-spin splitting, in NMR, 644
Multistep Synthesis, 352
Multistep Synthesis, Hints, for Doing, 870
Mutarotation, in D-glucose, 1267
Myoglobin 1237, 1252
Myrcene, 350
Myristic acid, 892

($N + 1$) rule, in NMR, 644
NAD$^{\oplus}$, 1152
NADH, 1152
Names, Common versus IUPAC, 91
Nanometers, (nm), 492
NADP, 1300
NADPH, 1300
Naphthalene, mass spectrum of, 728, 726
Naphthalene, orbital picture of, 552
α-Naphthalenesulfonic acid, 553
1-Naphthalenesulfonic acid, 553

Naphthenes, 121
α-Naphthylamine, 1159
β-Naphthylamine, 1159
Natta, Giulio, 347
Natural rubber, 348
NBS, 337
Neohexane, 91
Neohexyl chloride, 91
Neomenthyl chloride, 270
Neopentane, 77, 91
  NMR chemical shifts in, 633
Neopentyl alcohol (See also 2,2-Dimethyl-1-propanol), 412
Neopentyl chloride, 91, 564
Neopentyl iodide, 382
Neoprene, 348
Nerolidol, 351
Net dipole moment (See also Dipole moment), 52
Newman projection formulas, 66
  drawing of, 68
Nicol prisms, 185
Nicotinamide adenine dinucleotide (NAD$^{\oplus}$), 1297
Nicotinamide adenine dinucleotide phosphate (NADP), 1300
Nicotinamide nucleotide, 1297
Nicotine, 1160
Nitrate ester, 441
Nitration of benzene, resonance structures in, 559
Nitration, of phenol, 705
Nitrene, 1128
Nitric oxide, 140
Nitriles, amines from, 1123
  aryl, from diazonium salts, 1138
  carboxylic acids from, 900
  from alkyl halides, 900
  from amides, 978
  from arylsulfonic acids, 902
  from tosylates, 900
  in aldol condensations, 1060
  infrared spectroscopy of, 997
  ketones from, 804
  malonic acids from, 901
  nomenclature of, 943
  preparation of, 990
  preparation, using organothallium reagents, 902
  reaction of Grignard reagents with, 804
  reactions of, 990
  reduction of, 1123
Nitro compounds, amines from, 1121
  condensation with ketones, 1137
  in aldol condensations, 1060
  selective reduction of, to give amines, 1122

m-Nitroacetophenone, from acetophenone, 802
m-Nitroaniline, 548, 1139
p-Nitroaniline, 1122
Nitrobenzene, 545
chlorination of, 589
dinitration of, 568
electrophilic substitution on, 586
from benzene, 558
m-Nitrobenzenediazonium chloride, 1139
m-Nitrobenzenediazonium hydrogen sulfate, 1139
m-Nitrobenzenesulfonic acid, from nitrobenzene, 586
m-Nitrobenzoic acid, 1139
from *m*-nitrotoluene, 593
p-Nitrobenzoic acid, 893
m-Nitrobenzonitrile, 1139
2-Nitrobutane, 654
NMR spectrum of, 654
o-Nitrochlorobenzene, **671**
m-Nitrochlorobenzene, **671**
p-Nitrochlorobenzene, **671**, 673
NMR spectrum of, 652
Nitroethane, 138
Nitrogen, hybridization in, 46
stereochemistry of, 1119
Nitroglycerine, 440
Nitromethane, 137, 176, 1063
Nitrone, 173
Nitronium ion, 559
as an electrophile, 559
from nitric acid, 559
Nitronium perchlorate, 559
2-Nitropentane, 1122
9-Nitrophenanthrene, 553
o-Nitrophenol, **697**, 705
m-Nitrophenol, **697**
p-Nitrophenol, **697**, 705
from *p*-nitrochlorobenzene, 673
p-Nitrophenyl benzoate, 942
4-Nitrophenyl benzoate, 942
1-Nitropropane, NMR spectrum of, 647
2-Nitropropane, NMR spectrum of, 646
3-Nitropyridine, 1149
2-Nitropyrrole, 1149
Nitrosation, of primary amines, 1136
of secondary amines, 1142
of tertiary aromatic amines, 1142
N-Nitroso amines, from secondary amines, 1142
Nitroso group, 693
o-Nitroso-N,N-dimethylaniline, 1142

p-Nitroso-N,N-dimethylaniline, 1142
N-Nitroso-N-methylaniline, 1159
Nitrosonium hexafluorophosphate, 1143
Nitrosonium ion, 1143
o-Nitrosophenol, 1142
p-Nitrosophenol, 1142
Nitroterephtalic acid, 888
o-Nitrotoluene, 1122
m-Nitrotoluene, oxidation of, 593
Nitrous acid anhydride, 1141
Nitrous acid, use in forming diazonium salts, 1136
No-bond resonance, 534
Nomenclature, Naming Complex Molecules, 944
Nomenclature, of *α*-amino acids, 1211
of acid chlorides, 941
of aldehydes, 787
of alkadienes, 254
of alkenes, 247
of alkynes, 465
of amides, 940
of amines, 1104
of ammonium salts, 1110
of anhydrides, 942
of annulenes, 690
of aryl halides, 545
of benzene derivatives, 545
of carboxylate salts, 917
of carboxylic acid derivatives, 940
of carboxylic acids, 891
of dicarboxylic acids, 894
of dienes, 254
of disubstituted benzenes, 547
of esters, 942
of ethers, 734
of glycerides, 1091
of glycols, 737
of heterocyclic nitrogen compounds, 1114
of ketones, 788
of monosubstituted benzenes, 545
of nitriles, 943
of phenols, 695
of side chains in alkylbenzenes, 595
of steroids, 864
of tri- and polysubstituted benzenes, 548
(*n*-)Nonadecane, **93**
(*n*-)Nonane, **93**
Nonanedioic acid, 894
Nonanoic acid, 892

Nonaromaticity, evidence for using NMR, 691
Nonbiodegradable, 1097
Nonequivalence of protons, in NMR, 630, 652
Nonpolar bonds, 52
Nonpolar solvent, 175
Nonreducing sugars, 1285
(*n*-)Nonyl bromide, 90
1-Nonyne, **466**
Norbornane, 390
Norbornene, 1182
Nosylate, 750
Novocaine, 1158, 1173
Nu, represents nucleophile, 169
Nuclear magnetic moment, in NMR, 623
Nuclear magnetic resonance (NMR) spectroscopy, 623
Nuclear magnetic resonance spectrometer, diagram of, 626
Nuclear magnetic resonance (NMR), aromatic compounds, spectra of, 650
aromaticity and ring currents, 637
atoms detected in, 623
atoms other than hydrogen in, 658
carbon-13 in, 660
chemical shift and structure correlations, 638
chemical shifts and structure in, 632
chemical shifts in, 628
chemical shifts in, table of, 634
chemical shifts, examples of, 633, 636
conformations and nonequivalence in enantiomers, 655
conformers in, 656
coupling constants due to spin-spin splitting, 644
coupling constants, examples of, 648
delta (δ) scale in, 628
deshielding in, 632
deuterated solvents, 629
deuterium in, 658
dihedral angles and spin-spin splitting, 656
downfield chemical shift in, 632
electronic integration in, 631
electron-withdrawing groups and chemical shift in, 632
energy states of hydrogen nucleus in, 624
external magnetic field in, 625

Nuclear magnetic resonance
  (NMR)—*Continued*
flipping of spin states in, 625
fluorine in, 659
gyromagnetic ratio in, 624
Hertz, 626
interpretation of spectra from, 629
interpretation of spectra, a summary, 648
J (coupling constants), 645
kinds of protons in, 629
magnetic and spin properties in, 624
Megahertz, 626
multiplicity in spin-spin splitting, 644
N + 1 rule, 644
nonequivalence of protons in, 630, 652
nuclear magnetic moment in, 623
of acid chlorides, 998
of alcohols, 711
of aldehydes, 807
of alkyl halides, 690
of amides, 998
of amines, 1162
of anhydrides, 998
of aryl halides, 690
of carboxylic acids, 928
of esters, 998
of ethers, 774
of ketones, 807
of phenols, 711
operation of, 626
parts per million (*ppm*) in, 629
peak integration in, 631
precession in, 623
proton counting in, 630
proton equivalence in, 629
proton exchange in, 661, 712
proton ratios in, 631
resonance, 625
ring currents and aromaticity, 637
ring currents and deshielding in, 636
ring currents and shielding in, 636
ring currents in, 636
ring currents in acetylene, 637
ring currents in benzene, 637
shielding in, 632
shift reagents in, 775
spectra in, 627
spin-spin coupling, 638
spin-spin splitting, 638
spin-spin splitting theory, 639-644
splitting as a function of pro-

tons on adjacent carbons, 639
splitting patterns, idealized, 649
structure correlations with chemical shift, 638
tau (τ) scale in, 629
tetramethylsilane as standard, 627
theory of, 624
upfield chemical shift in, 632
upfield in, 627
Nucleic acids, 1279
Nucleophile, definition of, 125, 169
Nucleophiles, addition to carbonyl compounds, 818
ambident, 1020
basicity of, 393
effect of, in S$_N$2 reaction, 375
reactivity of, in S$_N$2 reactions, 377
Nucleophilic acyl substitution, 946
compared with nucleophilic addition, 947
comparison with nucleophilic aliphatic substitution, 993
general discussion of, 948
general mechanism of, 947
in acidic solution, 947
in neutral or basic solution, 947
reactivity in, 992
tetrahedral intermediates in, 968
Nucleophilic addition, compared with nucleophilic substitution, 947
electronic factors in, 823
general mechanism on carbonyl compounds, 820
in enamine synthesis, 1049
mechanism of Grignard addition, 843
mechanism of sodium bisulfite addition, 827
of alcohols, 828
of aldehydes, 818
of ammonia and substituted amines, 835
of Grignard reagents to carbon dioxide, 899
of Grignard reagents to carbonyl compounds, 842
of hydride ion to carbonyl compounds, 858
of hydrogen cyanide, 824
of ketones, 818
of organolithium reagents to carbonyl compounds, 855

of sodium bisulfite, 826
of water and hydrogen halides to carbonyl compounds, 839
orientation of, 819
relative reactivities of aldehydes and ketones, 822
to α,β-unsaturated carbonyl compounds, 1072
under acidic conditions, 821
under neutral or basic conditions, 821
Nucleophilic aliphatic substitution, 172, 368
amine preparation using, 1125
chemical kinetics of, 368
comparison with nucleophilic acyl substitution, 993
comparison with nucleophilic aromatic substitution, 682
energy profile diagrams for, 179
epoxide ring opening, 761
ether cleavage, 754
ether synthesis, 745
formation of ethers from alcohols, 742
general reaction, 368
in coupling reactions, 125
in acetoacetic ester synthesis, 1068
in malonic ester synthesis, 1064
in nucleoside synthesis, 1279
internal reactions, 751
introduction of, 177
nitriles using, 900
S$_N$1, 379
S$_N$2, 177
stereochemistry of, 209
summary of factors in, 401
using carboxylate anions, 953, 965
Nucleophilic aromatic substitution, by xanthate ions, 1140
comparison with electrophilic aromatic substitution, 681
comparison with nucleophilic aliphatic substitution, 682
effect of substituents on, 678
energy diagram for, 676
ether formation by, 747
in protein structure determination, 1249
mechanism of, 674
on aryl halides, 672
on pyridine, 1151
reactivity in, 673
structure-reactivity correlations in, 676
transition states and intermediates in, 676

Nucleophilic backside attack, 370

Nucleophilic substitution, alkyl versus acyl reactivity, 995

Nucleophilic versus Electrophilic Addition, 819

Nucleophilicity, 178

Nucleophilicity, Factors That Favor, 377

Nucleophilicity, in S_N2 reactions, 375

Nucleoproteins, 1252, 1280

Nucleosides, general structure of, 1279

Nucleotides, general structure of, 1279

Nutmeg, 741

Nylon, 1153

Nylon 66, 979

Observed rotation, $\alpha_{observed}$, 186

cis-9,12-Octadecadienoic acid, 892

(n-)Octadecane, **93**

Octadecanoic acid, 892

cis,cis,cis-9,12,15-Octadecatrienoic acid, 892

9-Octadecen-1-ol, 974

cis-9-Octadecenoic acid, 892

Octanamide, 1127

(n-)Octane, **93**

Octane numbers of hydrocarbons, 139

Octanedioic acid, 894

Octanoic acid, 892

1-Octanol, **412**

2-Octanol, from 2-octyl brosylate, 386

1-Octanol (See also n-Octyl alcohol), **412**

1-Octene, **252**

(n-)Octyl alcohol (See also 1-Octanol), 410, **412**

2-Octyl brosylate, 386

2-Octyl tosylate, from 2-octanol, 749

1-Octyne, **466**

Oil of wintergreen, 955

Oils, fats from, 1097

fatty acids in, 1092

properties of, 1091

structure of, 1090

table of, 1092

Olah, G., 669

Olefiant gas, 243

Oleic acid, 892

α-Oleo-γ-palmitin, 1093

Oleyl alcohol, 974

Oligosaccharides, 1305, 1310

Olive oil, 1092

Olivetol, 742

Optical activity, 184, 194

causes of, 198

Optically active, definition of, 186

Optically inactive, definition of, 186

Optically pure enantiomers, 200

Orbital symmetry, 1191

conservation of, 1190

Orbital, antibonding, 56

bonding, 56

Organic chemistry, applications of, 14

study of, 13

Organic Chemistry, Studying of, 15

Organic compounds, classification of, 10

Organic Compounds, Devising The Synthesis Of, 870

Organic compounds, identification of, 6

Organic versus inorganic chemistry, bonding, 3

physical properties, 2

reactivity, 5

Organocadmium reagents, 802

from Grignard reagents, 803

preparation of, 803

use in carbohydrates, 1278

Organolithium reagents, from alkyl and aryl halides, 855

preparation of, 855

stability of, 855

structure of, 856

compared with Grignard reagents, 856

Organomagnesium compounds, 802

Organometallic compounds, lithium dialkylcopper reagent, 123

Grignard reagent, 802

organocadmium reagents, 802

organolithium reagents, 123, 855

organomagnesium compounds, 803

organothallium reagent, 603

Organothallium reagent, preparation of, 603

reactions of, 603

use of in synthesis, 604

Orientation, in electrophilic substitution, 569

Orientation and reactivity, in synthesis using electrophilic substitution, 590

Orlon, 346

from acrylonitrile, 481

ortho, para Directors, defined, 569, 570

orientation theory of, 575, 580

ortho, use in nomenclature, 547

Orthophosphorus acid, 450

Osazone formation, in carbohydrates, 1294

Osazones, mechanism of formation, in carbohydrates, 1296

Osmium tetroxide, 333

7-Oxabicyclo[4.1.0]heptane, 737

Oxalate ester, 1058

Oxalic acid, 894, **896**

decarboxylation of, 925

synthesis of, 904

Oxalylation, 1058

Oxaphosphetane, 1053

Oxidation and Reduction, 128

Oxidation States of Organic Compounds, 128

Oxidation, of alkylboron compounds, 421

of amines, 1143

of carbohydrates, 1284

of methylbenzenes, 797

of side chains in alkylbenzenes, 593

Oximes, 836

amines from, 1123

reduction of, 1123

Oxirane (See also Ethylene oxide), 736, 737, 756

3-Oxobutanal, 1078

Oxonium ion, 272

Oxonium ions, from alcohols, 444

Oxygen atom, 140

Oxygen, hybridization in, 46

Oxygen-18, in esterification mechanism, 957

in hydration of carbonyl compounds, 840

incorporation in carbonyl compounds, 840

use in ester hydrolysis, 966

Ozone, 140, 339

formation of, 341

resonance structures of, 341

structure of, 341

Ozone and the environment, 141

Ozonide, 339

Ozonolysis, mechanism of, 341

of alkenes, 339

of alkynes, 482

P-2 catalyst, 475

PAH. See Polynuclear aromatic hydrocarbons

Pain relievers, 1158

Paints, discussion of, 1098
Palm oil, 1092
Palmitic acid, 892
α-Palmito-γ-olein, 1093
Palmitoleic acid, 1092
PAN, 141
Papain, 1250
para, use in nomenclature, 547
Paraffin wax, 1099
Paraffins, 129
Paraformaldehyde, 794
Paraldehyde, 794
Parent ion radical, in MS, 721
Parent ion, in MS, 721
Parts per million (*ppm*), in NMR, 629
Pasteur, Louis, 188
Pauli Exclusion Principle, 23
PBB (*See also* Polybromo-biphenyl), 688
PCB (*See also* Polychloro-biphenyl), 688
PCP, 1176
Peak integration, in NMR, 631
Peanut oil, 1092
Penicillin, 228
(*n*-)Pentacontane, **93**
(*n*-)Pentacosane, **93**
(*n*-)Pentadecane, **93**
Pentadecane, from sodium pal-mitate, 924
2,4-Pentadien-1-ol, 411
1,3-Pentadiene, 255, 258
1,4-Pentadiene, 255, 258, 305
addition of, HCl to, 305
Pentaerythritol, **769**, 1044, 1048
Pentaerythrityl palmitate, 1097
Pentamethylbenzene, **550**
Pentanal (*See also n*-Valeral-dehyde), 788, **791**, 853
(*n*-)Pentane, **77, 93**
Pentane, structural isomers of, 76, 77
Pentanedial, from cyclopentene, 340
Pentanedinitrile, 943
2,4-Pentanedione, 789, 1078
acidity of, 1062
Pentanoic acid, 892
1-Pentanol (*See also n*-Pentyl al-cohol), **412**
3-Pentanol, 3-fluoropentane from, 749
2-Pentanone (*See also* Methyl *n*-propyl ketone), **791**
3-Pentanone (*See also* Diethyl ketone), 789, **791**, 1049
preparation of, 804
Pentanoyl chloride, synthesis of, 949
Pentapeptide, 1219

(*n*-)Pentatriacotane, **93**
4-Penten-1-ol, 411
1-Penten-4-yne, 466
2-Pentenal, 788
1-Pentene, **252**, 258, 316
cis-2-Pentene, **252**
from 2-pentyne, 476
trans-2-Pentene, **252**
from 2-pentyne, 476
Pentose, 1261
Pentyl acetate, 945
(*n*-)Pentyl alcohol (*See also* 1-Pentanol), **412**
tert-Pentyl alcohol (*See also* 2-Methyl-2-butanol), 410, **412**
(*n*-)Pentyl bromide (*See also* 1-Bromopentane), **367**
(*n*-)Pentyl chloride (*See also* 1-Chloropentane), **367**
Pentyl ethanoate, synthesis of, 955
(*n*-)Pentyl ether, **739**
(*n*-)Pentyl iodide (*See also* 1-Iodopentane), **367**
3-Pentyl tosylate, from 3-pen-tanol, 749
tert-Pentylamine, 1105
1-Pentyne, **466**
2-Pentyne, **466**, 465
Pepsin, 1250
Peptide bond, 1219
Peptide link, 1238
Peptide syntheses, automated, 1227
general approach, 1228
solid-phase, 1228
C-terminal activating groups in, 1225
general method of, 1222
Lysine Vasopressin as ex-ample of, 1226
Peptides (*See also* Proteins), 1219
amides in, 1218
C-terminal end in, 1219
N-terminal end in, 1219
partial hydrolysis of, 1249
synthesis of, 1221
Peracids. *See* Peroxyacids
Perhydroazine, 1114
Pericyclic reactions, 1199
Pericyclic Reactions, General Selection Rules For, 1206
Pericyclic reactions, selection rules for, 1199
Periodic acid, 770
α,β-dicarbonyls cleavage by, 771
α-hydroxy aldehyde cleavage by, 771
α-hydroxy ketone cleavage by, 771

carbohydrates, oxidation of, 1289
glycol cleavage by, 771
Perkin condensation, 1059
Peroxy acyl nitrate (PAN), 141
Peroxyacids, 756
in epoxide syntheses, 757
Peroxybenzoic acid, 756
Petroleum, 121, 139
fractional distillation of, 122
Phenacetin, 1013, 1158
Phenanthrene, 552
orbital picture of, 552
UV spectrum of, 608
9,10-Phenanthroquinone, 552
α-Phenethyl alcohol, 380
α-Phenethyl bromide, 595
α-Phenethyl chloride, 380, 549, 595
β-Phenethyl chloride, 595
α-Phenethyl halide, 596
β-Phenethyl halide, 596
Phenetole, **739**
Phenol, 546, **697**
(Friedel-Crafts) alkylation of, 706
condensation with for-maldehyde (*See also* Bakelite), 833
electrophilic substitution, 579
from benzenesulfonic acid, 699
from chlorobenzene, 673
IR spectrum of, 710
K_a of, 700
nitration of, 705
resonance effect in elec-trophilic substitution, 579
salicylic acid from, 904
sulfonation of, 705
Phenols, acid dissociation con-stants of, 697
as acids, 700
bromine water test for, 708
ceric nitrate test for, 708
effect of substituents on acid-ity of, 701
electrophilic aromatic substitu-tion reactions on, 704
ferric chloride test for, 708
from aromatic amines, 699
from arylsulfonic acids, 699
from diazonium salts, 1138
from halobenzenes, 699
from organothallium reagents, 700
halogenation of, 704
hydrogen bonding in, 697
IR spectroscopy of, 709
monochlorination of, 704
nitrosation of, 1142

NMR spectroscopy of, 711
nomenclature of, 695
oxidation of, 706
physical properties of, 696
preparation of, from aryl halides, 673
qualitative tests for, 708
reaction with base, 702
separation from neutral compounds, 703
solubility test for, 708
spectral analysis of, 709
structure of, 695
summary of methods of preparation, 713
summary of reactions of, 713
table of, 697
UV spectroscopy of, 709
Phenoxide ion, in ether synthesis, 745
reaction with bromine water, 706
resonance structures of, 701, 753
Phenoxide ions, alkylation of, 752
C-alkylation versus O-alkylation, 752
reactions of, 706
1-Phenoxybutane, synthesis of, 747
2-Phenoxyethanol, from ethylene oxide, 762
Phenyl butanoate, from butanoyl chloride, 952
Phenyl ether, 734
Phenyl group, structure of, 546
Phenyl n-butyrate, from n-butyryl chloride, 952
Phenyl sec-butyl ketone, 1024
2-Phenyl-1-aminoethane, 1123
1-Phenyl-1,3-butadiene, 1053
1-Phenyl-2-bromobutane, dehydrohalogenation of, 598
2-Phenyl-2,3-butanediol, 772
1-Phenyl-2-butanol, dehydration of, 598
2-Phenyl-2-butanol, preparation of, 845
1-Phenyl-1-butanone, 861
1-Phenyl-1-butene, from 1-phenyl-2-bromobutane, 598
1-Phenyl-1-butene, from 1-phenyl-2-butanol, 598
1-Phenyl-1-pentanone, preparation of, 804
1-Phenyl-1-propanol, from 1-phenyl-1-propene, 600
1-Phenyl-1-propene, hydration of, 600
cis-2-Phenyl-2-butene, 1147

Phenylacetaldehyde, 787
Phenylacetaldehyde oxime, formation of, 836
N-Phenylacetamide, 941
Phenylacetic acid, 891
Phenylacetonitrile, 1060
α-Phenylacetonitrile, 1123
Phenylacetylene, 549, **550**
Phenylalanine, 1211
3-Phenylbutanamide, 940
1-Phenylbutane, from 1-phenyl-1-butanone, 861
β-Phenylbutyramide, 940
1-(1-Phenylcyclohexyl) piperidine (PCP), 1176
N-Phenylethanamide, 941
1-Phenylethanol, from 1-chloro-1-phenylethane, 380
Phenylethene (See also Styrene), 549, 598
5-Phenylhexanoic acid, 893
Phenylhydrazine, 838
osazones from, in carbohydrates, 1294
Phenyllithium, 855
Phenylmagnesium bromide, 803, 845
Phenylmethanol, **412**
o-Phenylphenol, 696
p-Phenylphenol, 695
2-Phenylpropanenitrile, 943
3-Phenylpropenal, 1044
2-Phenylpropene (See also α-Methylstyrene), 549, 598
3-Phenylpropene, 549
3-Phenylpropenoic acid, 1059
α-Phenylpropionitrile, 943
Phosgene, 973
Phosphatidylserines, 1101
Phosphine, **414**
Phosphine oxides, stereoisomers of, 238
Phosphines, stereoisomers of, 238
Phosphoenol pyruvic acid, 1299
Phospholane, 736
Phospholipids, 1100
Phosphonium salt, 1052
Phosphorus oxyhalide, 450
Phosphorus pentabromide, 450
Phosphorus pentachloride, 450
Phosphorus pentahalide, 450
Phosphorus trihalide, 450
Phosphorus triiodide, 450
Phosphorus ylides, 1052
Photochemical smog formation, 140
Photography, chemistry of, 707
Photon, 184
Photosynthesis of carbohydrates, 1261

Phthalamide, **945**
Phthalic acid, 894, 895, **896**
from naphthalene, 905
from o-xylene, 593, 905
synthesis of, 905
Phthalic anhydride, **945**, 1223
preparation of, 984
Phthalimide, 1126
Phthaloyl chloride, 941, **945**
Pi (π) bond, 33
physical properties of, 35
structure of, 35
in benzene, 525
Pi cloud, in benzene, 525
Picric acid, 673
Picryl chloride, 673
Pike, John E., 906
Pimelic acid, 894
Pinacolone, 862
Piperidine, 1114
Piperidinium bromide, 1118
Pivalamide, 978
Pivalic acid, 891, 978
Pivalonitrile, 978
pK_a, 168
of alkanes, 470
of alkenes, 470
of alkynes, 470
pK_b, defined, 1109
Planck's constant, 19, 184, 493
Plane of symmetry, 221
Plane-polarized light, 185, 184
Platforming, 551
Polar bonds, 52
Polar solvent, 175
Polarimeter, 185
schematic representation of, 187
Polarity, 50
Polarizability, effect in S_N2 reactions, 376
Polarized light, properties of, 198
Pollution, 139
Poly(ethylene glycol) polymer, from ethylene oxide, 763
Polyacrylonitrile, 346
Polyamides, Nylon 66, 979
Polybromobiphenyl (PBB), 688
Polychlorobiphenyl (PCB), 688
Polychloroprene, 348
Polydichloroethylene, 345
Polyesters, Dacron, 970
Polyethylene, from ethylene, 345
Polyhalogenated hydrocarbons, 688
Polyhydroxyaldehydes, 1260
Polyhydroxyketones, 1260
Polyisoprene, 348
Polymer, 345

Polymerization, addition, 345
condensation, in proteins and peptides, 1218
condensation, polyamides, 979
mechansim of cationic, 346
mechanism of free-radical, 346
types of, 345
Polymers, atactic, 347
isotactic, 347
syndiotactic, 347
Polynuclear aromatic hydrocarbons (PAHs), 551
and cancer, 768
UV spectroscopy of, 607
Polypeptide, 1219
Polysaccharides, 1305, 1310
Polytetrafluoroethylene, 346
Polyunsaturated, 1097
Polyunsaturated fats, 1097
Polyvinyl chloride, 346
Porphyrin, 1114
Potassium *tert*-butoxide, 428
Potassium ethyl xanthate, 1140
Potassium hydrogen phthalate, 917
Potassium phthalate, 917
Potassium phthalimide, 1126
Pr(DPM)³, 776
Precession, in NMR, 623
Pregnenolone, 865
Prehnitene, **550**
Prelog, V., 201
Primary carbon, 123
Primary hydrogen, 123
Primary isotope effect, 436
Principle of Microscopic Reversibility, 297
Product distribution, effect of substituents in electrophilic substitution, 571
in electrophilic substitution, 569
Progesterone, 865
Progestins, 865
Proinsulin, 1247
Proline, 1211
Propadiene, 255, 258
Propanal (*See also* Propionaldehdye), **791**
aldol condensation of, 1039
from 1-propanol, 432
Propanamide, **945**
Propane, **93**
conformations of, 71
from 2-bromopropane, 843
from cyclopropane, 312
halogenation of, 132
NMR chemical shifts in, 633
product distribution in chlorination of, 162
structure of, 71

Propanedioic acid, 894, 895
1,2-Propanediol, 426, **769**
from propene, 333
1,3-Propanediol, **769**
1,2,3-Propanetriol, 426, 737, **769**, 1090
Propanone (*See also* Acetone), 789, **791**
Propanoyl chloride, **945**
Propenal (*See also* Acrolein), **791**, 1183
Propene (*See also* Propylene), **252**, 316
preparation of, 261
stereospecific polymerization of, 347
structure of, 244
use in alkylation of benzene, 567
2-Propen-1-ol, **412**
2-Propenyl group, 248
Propionaldehyde (*See also* Propanal), 787, 791
NMR spectrum of, 808
Propionamide, **945**
Propionic acid, 892, **896**
Propionyl chloride, **945**
Propiophenone (*See also* Ethyl phenyl ketone), 789, **791**
preparation of, 799
3-Propoxypentane, 734
(*n*-)Propyl alcohol (*See also* 1-Propanol), **412**
(*n*-)Propyl bromide (*See also* 1-Bromopropane), **367**
(*n*-)Propyl chloride (*See also* 1-Chloropropane), **367**
(*n*-)Propyl ether, **739**
(*n*-)Propyl group, **84**
(*n*-)Propyl iodide (*See also* 1-Iodopropane), **367**
4-Propyl-2-nonene, 248
(*n*-)Propylamine, **1107**
Di-*n*-propylamine, **1107**
Tri-*n*-propylamine, **1107**
Propylbenzene, **550**
Propylene (*See also* Propene), 244, 247
Propylene bromide, 326
Propylene bromohydrin, 326
Propylene chloride, 336
Propylene glycol, 333, 426, **769**

Propylene oxide, 737
Propyne, **466**
addition of HBr to, 478
Prostaglandins, 906
Prostanoic acid, 906
Prosthetic group, 1252
Prosthetic groups, 1251
Protecting groups, in protein synthesis, 1223
Protein structure, C-terminal analysis in, 1251
determination of, 1248
enzyme cleavage in, 1249
factors influencing, 1238
Sanger reaction in, 1249
specific cleavage agents in, 1251
Protein syntheses, protecting group removal, 1223
protecting groups in, 1223
Proteins (*See also* Peptides), 1219
α-helix in, 1240
amides in, 1218
β-sheet structure, 1240
β-sheets in, 1242
coplanar structure of amide bond, 1239
covalent bonding in, 1244
disulfide link in, 1244
electrostatic attractions, 1244
examples of, 1237
fibrous, 1236
globular, 1236
hydrogen bonding in, 1242, 1244
hydrophobic interactions, 1244
keratins, 1244
N-terminal analysis in, 1249
natural synthesis of, 1235
nonbonding interactions in, 1243
partial hydrolysis of, 1249
peptide link in, 1238
π-π interactions in, 1244
primary structure, 1237
quaternary structure, 1238
regulator, 1236
secondary structure, 1237
structural, 1236
structure of, 1236
structure of α-helix, 1241
synthesis of, 1233
tertiary structure, 1237
total amino acid analysis of, 1248
transport, 1236
triple helix in, 1243
Protic solvent, 175, 387
Protofibril, 1243
Proton counting, in NMR, 630

Proton equivalence, in NMR, 629
Proton Exchange, in NMR, 712
Proton exchange, in NMR, 661, 712
use of deuterium oxide in, 712
Proton magnetic resonance (PMR). *See* Nuclear magnetic resonance
Proton ratios, in NMR, 631
Protons, 274
Protons, equivalent versus nonequivalent in NMR, 652
Protons, kinds, of, in NMR, 629
Prussian Blue, 8
Pseudo first order reaction, 272
Pseudocumene, **550**
Purine, 1114
Purine bases, 1279
PVC, 346
Pyramidal carbanions, 1022
Pyramidal inversion in amines, 1023
Pyramidal Inversion of Carbanions, 1022
Pyranoses, in carbohydrates, 1269
Pyrene, 553
Pyrene-epoxide, 768
Pyridine, 540, 1114
Pyridine N-oxide, 1151
Pyridine, as a base, 748, 1150
as a nucleophile, 1150
electrophilic substitution on, 1149
nucleophilic substitution, 1151
orbital picture of, 538, 1117
structure of, 538
3-Pyridinesulfonic acid, 1149
Pyridinium bromide, 1118
Pyridinium ion, 748
reduction of, 1151
Pyrimidine, 1114
orbital picture of, 1117
Pyrimidine bases, 1279
Pyrolysis, 260
Pyrrole, 1114
as a base, 1150
as a nucleophile, 1150
as an acid, 1150
electrophilic substitution on, 1149
orbital picture of, 538, 1117
structure of, 538
2-Pyrrolesulfonic acid, 1149
Pyrrolidine, 1050, 1114
Pyrrolium bromide, 1118
Pyruvic acid, 1299

Qualitative analysis, alkynes, 482

classification of alcohols, 453
classification of alkyl halides, 399
detection of aldehydes and ketones, 866
detection of aromatic rings, 604
detection of phenols, 708
for alkenes, 317, 334
Quanta, 492
Quantum mechanical atom, 18
Quaternary carbon, 123
Quaternary ammonium hydroxides, preparation of, 1144
Quaternary ammonium salts, stereoisomers of, 237
Quinine, 920, 1160
as a resolving agent, 920
Quinoline, 1114
orbital picture of, 1117
Quinone (*See also* Benzoquinone), 707
Quinuclidine, 1118

R/S system, of configuration, 201
two asymmetric carbons, 219, 223
R-configuration, 201
R group, definition of, 124
Racemic mixture, 200
Racemization, 207
in free-radical halogenation, 207
in S_N1 reactions, 381
of optically active ketones, 1024
Radicals, addition to alkenes, 303
ease of formation, 170
stability of, 163
Rancid compounds, 1099
Rate law, 174
Rate-determining step in reactions, 150
Reaction intermediate versus transition state, 149
Reaction mechanisms, catalysts in, 175
intermediates in, 172
introduction to, 145
kinetics in, 173
product distribution in, 171
solvent effects in, 175
structure-reactivity effects, 177
study of, 171
substituent effects in, 177
Reaction Outline, Chaps 1–8, 343
Reactions, Studying of, 342

Reactive intermediates, structures of, 41
Reactive Intermediates, Summary of Structures, 45
Reactive intermediates, type of, 41
Reactivity, in electrophilic substitution, 569
Reading References, 1321
Rearrangements, in synthesis of alkyl halides from alcohols, 448
Reasoning by Analogy, 50
Reciprocal centimeters (cm^{-1}), 499
Reducing sugars, 1285
Reduction, of α,β-unsaturated carbonyl compounds, 1046
of alkylboron compounds, 424
of carbohydrates, 1290
of carbohydrates, by NADH, 1297
of carbonyl groups, 856
of glycerides, 1096
of imines, 1123
of nitro compounds, 1121
of oximes, 1123
Reductive amination, 1124
Reforming, 551
Regulator proteins, 1236
Reimer-Tiemann reaction, 753
Resolution, 921
Resolution, of alcohols, 988
of carboxylic acids into enantiomers, 919
of enantiomers, 200
using amines, 1120
Resonance effect, defined, 581
orbital view of, 581
in amine basicity, 1112
in electrophilic substitution, 579
in electrophilic substitution on aryl halides, 584
Resonance energy. *See* Delocalization energy
Resonance hybrid, 307
Resonance structures, 307
defined, 307
drawing of, 530
in allyl cations, 307
in aryl aldehydes and ketones, 823
in benzene, 528
in electrophilic substitution, 579, 580
in nucleophilic aromatic substitution, 677, 678
of α-carbanions, 1019
of acylium ion, 799
of amide ion, 1129

Resonance structures—*Continued*
of amides, 1239
of anilinium ions, 1112
of aryl halides, 671
of benzoate anion, 913
of benzyl cations, 601
of benzyl radicals, 597
of carboxylate anions, 909
of enamines, 1049
of ozone, 341
of phenoxide ion, 701
of protonated epoxides, 764
of vinyl halides, 671
of ylides, 1052
Resonance, in allyl cations, 307
in NMR, 624
Resorcinol, 696, **697**
Restricted rotation in alkenes, 246
Retention of configuration, in hydroboration-oxidation, 422
in reduction of alkylboron compounds, 424
L-Rhamnose, 1320
D-Ribitol 5-phosphate, 1297
D-Ribofuranose, 1271
Ribonuclease, 1237
Ribonucleic acid, 1234, 1279
Ribose, 1272
D-Ribulose 5-phosphate, 1297
Ricinoleic acid, 1092
Ring closure, 126
involving carbocations, 331
Ring currents and aromaticity, in NMR, 637
Ring currents and deshielding, in NMR, 636
Ring currents and shielding, in NMR, 636
Ring currents in acetylene, in NMR, 637
Ring currents in benzene, in NMR, 637
Ring currents, in NMR, 636
Ring-chain tautomerism, in carbohydrates, 1268
RNA (*See also* Ribonucleic acid), 1234, 1279
Roberts, John D., 683
Rosenmund reduction, 795
Rubber, 348
vulcanization of, 349
Rubbing alcohol, 426
Ruff degradation of carbohydrates, 1293

s-Atomic orbital, 20
s-Character, Percentage of, 39

s-Character, relationship to acidity, 471
S-Configuration, 201
Saccharides. *See* Carbohydrates
Saccharin, 1159
Safrole, 741
Salicylaldehyde (*See also o*-Hydroxybenzaldehyde), 788, **791**
from phenol, 795
Salicylic acid, from phenol, 904
hydrogen bonding in, 914
Sandmeyer reaction, 1139
Sanger reaction, 1249
Saponification, 965, 1093
Saran, 345
Sassafras, 741
Saturated fats, 1097
Sawhorse structures, 66
Saytzeff, Alexander, 264
Saytzeff elimination, compared to Hofmann elimination, 1146
Saytzeff rule, 264
defined, 264
in alkenylbenzenes, 598
in dehydration reactions, 277
in dehydrohalogenation reactions, 264
in Hofmann elimination, 1145
Schiemann reaction, 1139
Schiff's base, 837
Schiff's test, 867
Schotten-Baumann method, 951
Schrödinger wave mechanics, 20
S_E2, defined, 557
Sebacic acid, 894
Secondary carbon, 123
Secondary hydrogen, 123
Secondary isotope effects, 436
Selection Rules, General, for Pericyclic Reactions, 1206
Selection rules, for 2+2 cycloaddition reactions, 1201
for pericyclic reactions, 1199
versus general selection rule, 1206
Selenic acid, 1148
Selenoxide pyrolysis, 1148
β-Selinene, 351
Semicarbazide, 837
addition to carbonyl compounds, 837
Semicarbazones, 837
Separations, based on acidity, 919
Sequence rules for R,S-configuration, 201
Serine, 1212
Sesquiterpenes, 350, 351
Shielding, in NMR, 632

1,2-Shift, 279
Shift reagents, in NMR, 775
SI units, conversion table, 25
Sigma (σ) bonds, 26
Sigma (σ) orbital, 27
Silane, **414**
Silicon compounds, stereochemistry of, 238
Silver alkynide, 472
Silver ammonia complex, reaction with terminal alkynes, 472
Singlet methylene, 1188
sterochemistry of addition, 1188
Skew conformations, 68
S_N1 and S_N2 reactions, comparison of, 385
summary of factors in, 385
S_N1 reactions, defined (*See also* Nucleophilic aliphatic substitution), 379
allyl halides in, 389
competition with E1 reactions, 394
effect of leaving group in, 385
effect of nucleophile concentration in, 388
effect of nucleophile in, 385
effect of solvent in, 386
effect of structure in, 382
electronic effects in, 383
energy-profile diagram for, 380
in reaction of alcohols with HX, 446
kinetics of, 369
mechanism of, 379
racemization in, 381
rates of reaction in, 383
rearrangements in, 382
silver ion catalysis in, 384
solvation in, 384
stereochemistry of, 380
S_N1 versus E1 reactions, 394
S_N2 reactions, competition with E2 reactions, 392
effect of α-substitution on, 373
effect of bases in, 393
effect of β-substitution in, 375
effect of leaving group on, 378
effect of nucleophile concentration in, 388
effect of nucleophiles on, 375
effect of solvent in, 387
effect of steric factors in, 373
effect of structure on, 372
in synthesis of alkynes, 474
kinetics of, 369
mechanism of, 370
order of nucleophilic reactivity in, 377

order of reactivity in, 374
relative rates of reaction, 373
stereochemistry of, 209, 371
transition states in, 210
S_N2 versus E2 reactions, 392
S_Ni mechanism, 453
Soaps, action of, 1094
discussion of, 1093
from glycerides, 1094
Sodamide (*See also* Sodium amide), 468
Sodium 4,4-dimethyl-4-silapentane sulfonate, 664
Sodium acetylide, 475
Sodium alkyl sulfate, 1096
Sodium amide, 468
preparation of, 470
use in dehydrohalogenation of aryl halides, 683
use of in elimination reactions, 468
Sodium arylsulfonate, 902
Sodium benzoate, 917
Sodium bisulfite addition compounds, 826
hydrolysis of, 827
mechanism of addition to carbonyl compounds, 827
Sodium borohydride, 858
comparison with lithium aluminum hydride, 859
mechanism of carbonyl reduction by, 858
reduction of carbonyl compounds by, 858
selective reduction of conjugated aldehydes, 859
structure of, 858
Sodium ethyl acetoacetate, 1055
Sodium fusion, 7
Sodium hexadecanoate, 924
Sodium hydride, 1056
Sodium iodide in acetone, classification of alkyl halides, 400
Sodium lauryl sulfate, 440
Sodium metal, reduction of esters, 975
Sodium methoxide, 428
Sodium methylate, 430
Sodium palmitate, 924
Sodium stearate, 1095
Sodium succinate, 917
Solubility properties, in separation of amines, 1113
Solvation, effect in S_N2 reactions, 376
in S_N1 reactions, 384
Solvent, effect of, in S_N1 reactions, 386
in S_N2 reactions, **387**

Solvolysis, 397
in S_N1 reactions, 396
Soybean oil, 1092
sp Hybridization of carbon, 37
sp^2 Hybridization of carbon, 34
sp^3 Hybridization of carbon, 29, 30
Specific rotation, $[\alpha]$, 187
calculation of, 188
Spectrograms, 491
Spectrometers, 491
Spectrometry, 491
Spectrophotometers, 491
Spectroscopy, 491
Spectroscopy, infrared (IR), 491
Spectroscopy, nuclear magnetic resonance (NMR), 623
Spectroscopy, types of, 491
Spectroscopy, ultraviolet (UV), 491
Spectrums, 491
Speed of light, 493
Spermaceti wax, 1099
Spin decoupling, in NMR, 660
Spin-spin coupling, in NMR, 638
Spin-spin splitting, in NMR, 638–644
Splitting as a function of protons on adjacent carbons, in NMR, 639
Splitting patterns, idealized, in NMR, 649
Squalene, 352, 766
Squalene 2,3-epoxide, 767
Staggered conformation, 67, 68
Stearic acid, 892, **896**
Sterculic acid, 1091
Stereochemistry, Important Terms in, 192
Stereochemistry, 184
alkyl halides from alcohols, 749
of α-amino acids, 1216
of addition to carbonyl compounds, 824
of addition of chlorine to alkenes, 324
of amines, 1119
of bromine addition to alkenes, 319, 323
of α-carbanions, 1021
of carbocations, 299
of carbohydrates, 1263
of Cope elimination, 147
of dehydrohalogenation, 266
of Diels-Alder reaction, 1180
of epoxide hydrolysis, 760
of ester hydrolysis, 966
of Grignard reagents, 1022
of hydoxylation of cycloalkenes, 333

of hydroboration of alkynes, 480
of hydroboration-oxidation, 422
of hydrogenation of alkynes, 476
of hydroxylation of alkenes, 335
of polymerization, 347
of reaction of alcohols with HBr, 446
of reactions involving asymmetric carbon atom, 206
of reactions not involving asymmetric carbon atom, 206
of reactions that generate asymmetric carbon atom, 210
of reactions that generate second asymmetric center, 229
of S_N1 reaction, 380
of S_N2 reaction, 371
of S_Ni reaction, 453
of singlet methylene addition, 1188
of tosylate formation, 748
Stereoisomerism, 184
Stereoisomers, 185
classification of, 185
determining the number of, 223
how to draw, 217
in cyclic compounds, 232
in cyclohexane derivatives, 235
involving isotopes, 238
of amines, 237
of ammonium salts, 237
of germanium compounds, 238
of isotopically-labeled compounds, 238
of phosphine oxides, 238
of phosphines, 238
of silicon compounds, 238
of sulfonium salts, 238
of sulfoxides, 238
one asymmetric carbon, 195
two analogous asymmetric carbons, 219
two asymmetric carbons, 215
two different asymmetric centers, 215
Stereoselective reactions, 230
Stereospecific addition, of bromine to cycloalkenes, 321
Stereospecific polymerization, 347
Stereospecific reaction, defined, 322

Steric effect, in electrophilic sub-
    stitution, 572
  in nucleophilic addition to car-
    bonyl compounds, 822
Steric factors, in S_N2 reactions,
    373
Steric repulsion, 75
Steroid nucleus, 864
Steroids, 864
Steroids, nomenclature of, 864
cis-Stilbene, **550**
trans-Stilbene, 549, **550**
Structural isomerism, 10
Structural isomers, 71, 185
  alkanes, 78
  system for writing, 79
Structural proteins, 1236
Structure correlations with
    chemical shift, in NMR,
    638
Structure, using IR, 500
Structure, using UV, 509
Strychnine, 1160
Styrene (*See also* Phenylethene),
    549, **550**, 598, 1053
  from ethylbenzene, 598
Suberic acid, 894
Substitution and elimination, 398
Substitution versus Elimination
    of Alcohols, 449
Substitution, competition with
    elimination, 392, 394
  control of, 398
α-Substitution, defined, 373
β-Substitution, defined, 374
Subtilisin, 1250
Succinamide, 940
Succinic acid, 894, **896**
Succinic acid, synthesis of, 902,
    905
Succinic anhydride, 943
  preparation of, 983
Succinimide, 337
Sucrose, 1306
Sugars. *See* Carbohydrates.
Sulfa drugs, 1156
Sulfanilamide, 1134, 1156
  synthesis of, 1134
Sulfenic acid, 1147
Sulfonamides, acidity of, 1130
  from amines, 1130
  solubility in base, 1130
Sulfonates, As Leaving Groups,
    750
Sulfonation, of benzene, 562
  of benzene, mechanism of, 562
  of phenols, 705
Sulfonium salts, stereoisomers
    of, 238
Sulfoxide pyrolysis, 1147
Sulfoxides, stereoisomers of, 238

Sulfur trioxide, as an elec-
    trophile, 562
  structure of, 562
Sulfur ylides, 1052, 1088
Sulfuryl chloride, 136
Supra attack, 1192
Supra-supra attack, in Wood-
    ward-Hoffmann rules, 1196
Suprafacial attack, 1191
Sweeteners, 1159
Symmetric solvation, 1023
Symmetrical in-plane bending
    (scissoring), 495
Symmetrical reagents, 290
Symmetrical stretching, 495
Symmetry groups, 1192
Symmetry, axis of, 193
  center of, 194
  orbital, 1191
  plane of, 193
Symmetry-allowed 2+2 cycload-
    dition, 1198
Symmetry-allowed reactions, in
    Woodward-Hoffmann rules,
    1193
Symmetry-forbidden 2+2 cy-
    cloaddition, 1197
Symmetry-forbidden reactions,
    in Woodward-Hoffmann
    rules, 1193
syn-Addition, in chlorination of
    alkenes, 324
  in hydroboration, 421
  in hydrogenation of alkenes,
    311
  in hydroxylation reactions,
    334
  of hydrogen to alkynes, 476
syn-Elimination, 1147
syn-Hydrogenation, of alkynes,
    476
Syndiotactic polymers, 347
Syntheses, examples of multistep
    involving alkenes, 354
Syntheses, Multistep, Hints For
    Doing, 870
Synthesis, Multistep, 352

2,4,5-T (*See also* 2,4,5-
    Trichlorophenoxyacetic acid),
    689
Tallow, 1092
L-Talonic acid, 1288
α-L-Talopyranose, 1274
D-Talose, 1264
(−)-Tartaric acid, 227
(+)-Tartaric acid, 227
L-Tartaric acid, 1121
meso-Tartaric acid, 227
Tartrate ion, 1286

Tau (τ) scale, in NMR, 629
Tautomerism, 479
  ring-chain in carbohydrates,
    479, 1268
Tautomers, 479
Taylor, Edward C., 603
Teflon, 346
Terephthalic acid, 894, 895, **896**,
    984
  from p-isopropyltoluene, 594
Terminal alkynes (*See also*
    Alkynes), 470
Terpenes, 350
Terpenoid family, 331
Tertiary carbon, 123
Tertiary hydrogen, 123
Testosterone, 863
2,3,4,6-Tetra-O-methyl-D-galac-
    tose, 1308
2,3,5,6-Tetra-O-methyl-D-
    gluconic acid, 1308
1,2,3,4-Tetrabromobenzene, 548
2,2,3,3-Tetrabromobutane, 469
2,2,3,3-Tetrabromobutane, from
    2-butyne, 477
1,2,4,5-Tetrabromocyclohexane,
    524
1,2,3,4-Tetrabromocyclohexane,
    524
1,2,3,4-Tetrachlorobutane, from
    1,3-butadiene, 324
Tetrachloromethane (*See also*
    Carbon tetrachloride), 53,
    131
(n-)Tetracontane, **93**
(n-)Tetracosane, **93**
Tetracosanoic acid, 1092
Tetracyanoethene, 1184
(n-)Tetradecane, **93**
Tetradecanoic acid, 892
Tetraethyl lead, 139
Tetrafluoroethylene, 247, 346
Tetrahedral carbon, 30
Tetrahedral carbon atom, evi-
    dence for, 192
Tetrahedral intermediates, evi-
    dence for in nucleophilic
    acyl substitution, 968
  in haloform test, 1031
Tetrahydrocannabinol (THC),
    741
Tetrahydrofuran (THF), 735,
    739
  discussion of, 741
  in Grignard reagent formation,
    841
  synthesis of, 751
Tetrahydrothiophene, 1152
Tetramer, 328
2,2,4,4-Tetramethyl-3-pentanone,
    856

2,2,4,4-Tetramethyl-3-*tert*-butyl-3-pentanol, synthesis of, 856
Tetramethylammonium iodide, 1126
Tetramethylene glycol, 737
Tetramethylethylene, 247
Tetramethylsilane as standard, in NMR, 627
Tetrapeptide, 1219
Tetrapropylene, 588
Tetraterpenes, 350, 352
Tetrose, 1261
Thallium trifluoroacetate (TTFA), 602
  preparation of, 603
Thallium, in organic synthesis, 602
THC, 741
Thermodynamic control, in HBr addition to 1,3-butadiene, 309
Thermodynamically controlled reactions, 169
Thermodynamics versus kinetics, 169
Thermodynamics, discussion of, 147
Thiazole ring, 1156
Thietane, 736
Thiolane, 1152
Thiols, aryl, from diazonium salts, 1138
Thionyl chloride, 452
Thiophene, 1152
  structure of, 539
Three-dimensional structures, drawing of, 32
Threonine, 1212
Threose, 1272
Thymine, 1279
Thymol, 696
Tiffeneu-Demyanov reaction, 1136
Titanium tetrachloride, 347
Tollens' reagent, in carbohydrates, 1284
Tollens' test, 867
o-Tolualdehyde, **791**
m-Tolualdehyde, **791**
p-Tolualdehyde, **791**
o-Toluamide, **945**
m-Toluamide, **945**
p-Toluamide, **945**, 976
  from *p*-toluoyl chloride, 951
Toluene, 546, **550**
  electrophilic attack in, 576
  electrophilic substitution, 575
  ion radical in MS, 724
  mass spectrum of, 724
  nitration of, 568
  NMR chemical shifts in, 636

NMR spectrum of, 650
oxidation of, 593
side-chain halogenation of, 594
p-Toluenesulfonic acid, 830
p-Toluenesulfonyl chloride, 748
o-Toluic acid, **896**
  preparation of, 899
m-Toluic acid, 893, **896**
p-Toluic acid, **896**
  NMR spectrum of, 929
o-Toluidine, 1106, **1107**, 1122
m-Toluidine, **1107**
p-Toluidine, 548, **1107**
  from aniline, 587
o-Toluoyl chloride, **945**
m-Toluoyl chloride, **945**
p-Toluoyl chloride, **945**
Toluquinone, 707
p-Tolyldiazonium fluoroborate, 1139
2-*p*-Tolylethanol, preparation of, 845
2-(4-Tolyl)ethanol, synthesis of, 847
Torsional energy, 69
Tosylates, 748
  as leaving groups, 748
  formation of from alcohols, 748
  nitriles from, 900
Transesterification, 969
  applications of, 970
Transition state, 136, 148
  in decarboxylation reactions, 1070
  in hydroboration of alkenes, 424
  in S_N2 reaction, 370
Transition states, in nitration of benzene, 560
Transition state versus reaction intermediate, 149
Transmittance, percent, 499
Transport proteins, 1236
2,3,6-Tri-O-methyl-D-glucose, 1308
(*n*-)Triacontane, **93**
Trialkyl phosphate, 441
α,α,α-Tribromoacetone, 1027
2,4,6-Tribromoaniline, 1133
2,4,6-Tribromophenol, **697**
  from phenol, 705
Trichloroacetaldehyde, 839
1,1,2-Trichlorocyclopropane, enantiomers of, 233
1,1,1-Trichloro-2,2-di(*p*-methoxyphenyl)ethane, 741
Trichloroethanal, 839
1,1,2-Trichloroethane, NMR spectrum of, 646
Trichloromethane (*See also*

Chloroform), 53, 131
  NMR chemical shifts in, 633
2,4,6-Trichlorophenol, **697**
2,4,5-Trichlorophenoxyacetic acid (2,4,5-T), 689
α,α,α-Trichlorotoluene, 595
(*n*-)Tricosane, **93**
Z-9-Tricosene, 1089
(*n*-)Tridecane, **93**
Triethylamine, **1107**
Triethylammonium iodide, 1110
1,3,5-Triethylbenzene, **550**
Triethylboron, 417
2,2,2-Trifluoroethanol, NMR chemical shifts in, 659
Trifluoromethane, 53
3,3,3-Trifluoropropene, 302
Trigonal geometry, 34
Trilaurin, 1093
Trimer, 328
2,2,5-Trimethyl-3-hexanone, 852
Trimethylacetic acid, 891
Trimethylamine, 1105, **1107**
Trimethylcarbonium ion, 164
Trimethylene oxide, 735
2,2,4-Trimethylhexane, 87
2,4,6-Trinitrochlorobenzene, 673
Trinitroglycerine, 440
2,4,6-Trinitrophenol, **697**
  from 2,4,6-trinitrochlorobenzene, 673
Triolein, 1097
Triose, 1261
Trioxane, 794
2,5,8-Trioxanonane, 735
Tripeptide, 1218
Triphenylamine, **1107**
1,1,2-Triphenylethanol, 972
Triphenylmethane, **550**
Triphenylmethanol, **412**, 602
Triphenylmethyl alcohol (triphenylcarbinol), **412**
Triphenylmethyl fluoroborate, 602
  hydride abstractions using, 602
Triphenylphosphine, 1052
Triple bond, 37
Triple helix, in proteins, 1243
Triplet addition, of methylene, 1189
Triplet insertion, of methylene, 1189
Triplet methylene, 1188
  addition of, 1189
  insertion of, 1189
Triptycene, from benzyne and anthracene, 687
Trisaccharides, 1305
Tristearin, 1097

INDEX

Trisubstitution reactions, of aromatic compounds, 610
Trisubstitution, in benzene, 590
Triterpenes, 350, 351
Tritium, 498, 620
Tröger's base, 237, 1119
Tropolone, 621
Tropylium cation, 540
in mass spectrometry, 724
Tropylium ion (*See also* Cycloheptatrienyl cation), 537
Trypan Blue, 1154
Trypsin, 1250
Tryptophan, 1211
TsCl (*See also* Tosyl chloride and *p*-Toluenesulfonyl chloride), 748
TTFA. *See* Thallium trifluoroacetate
Tung oil, 1092
Twist form of cyclohexane, 105
Two-step reactions, 150
Tyrosine, 1212

Ultraviolet spectroscopy (*See also* UV), 502
of aldehydes, 807
of alkyl halides, 690
of amines, 1162
of aromatic compounds, 607
of carboxylic acid derivatives, 997
of carboxylic acids, 928, 997
of ethers, 773
of ketones, 807
of phenols, 709
of vinyl halides, 690
(*n*-)Undecane, **93**
α,β-Unsaturated acids, from α-halo acids, 927
α,β-Unsaturated carbonyl compounds, addition of β-dicarbonyl compounds to, 1076
addition of Grignard reagents to, 1075
from aldol condensations, 1039
in Diels-Alder reaction, 1183
orbital view of, 1040
reduction of, 1046
resonance structures of, 1072
structure, 1072
Unsaturated hydrocarbons, 243
Unsaturation, degree of, 314
Unsymmetrical ethers. *See* Ethers
Unsymmetrical reagents, 290
Upfield chemical shift, in NMR, 632

Upfield, in NMR, 627
Uracil, 1279
Urea, 1
Uridine, 1283
synthesis of, 1283
UV. *See* Ultraviolet spectroscopy and UV spectra
UV absorption frequencies, of conjugated compounds, 512
table of, 507
UV spectra and structure, correlation between, 509
UV spectra, effect of conjugation on, 509
effect of π bonds on, 506
effect of σ bonds on, 505
effect of structure on, 505
effect of unshared electron pairs on, 506
of double bonds, 506
table of absorption frequencies, 507
UV spectral measurements, 503
UV spectroscopy and structure, 509
UV spectroscopy, of conjugated systems, 510
of polynuclear aromatic hydrocarbons, 607
prediction of absorption maximum in, 511
structure identification in, 513
terminology of, 509
UV transitions, diagram of, 506

(*n*-)Valeraldehyde (*See also* Pentanal), 787, **791**
Valeric acid, 892, **896**
γ-Valerolactone, 964
δ-Valerolactone, 964
Valeroyl chloride, synthesis of, 949
Valine, 1211
Van't Hoff, 189
Vanilla, 741
Vanillin, 741
Vasopressin, 1237
Vegetable oils, 1090
vic-Diols, 410
Vicinal dihalides, 282, 317
dehydrohalogenation of, 468
Vicinal diols, 737
Vinyl alcohols, 409
from alkynes, 478
Vinyl bromide, **671**
Vinyl chloride, 247, 248, 338, **671**
Vinyl cyanide, 481

Vinyl cyclopropane, 248
Vinyl ether, **739**
Vinyl group, 248, 338
Vinyl halides, physical properties of, 671
resonance structures of, 671
structure of, 670
UV of, 690
Vinyl radical, 338
Vinylacetylene, 481
Vinylbenzene, 549
Vinylboron compounds, from alkynes, 480
4-Vinylcyclohexene, 1181
Vinylic hydrogens, 338
Vitamin A$_1$, 315, 351
Vitamin B$_6$, 1115
Vitamin D, 1209
Vitamin E, 741
Vulcanization, 349

Wallach, O., 350
Water, **414**
addition of. *See* Hydration
dipole moment of, 52
loss of. *See* Dehydration
structure of, 48
Watson, 1281
Watson and Crick, 1281
Wave function, atomic orbitals, 20
Wave function, molecular orbitals, 20
Wave functions, 21
Wave numbers, in IR spectroscopy, 496
Wavelength (λ), 492
conversion to frequency, 500
in IR spectroscopy, 496
relation to frequency, 492
Waves, properties of, 19
Whale oil, 1092
Williamson ether synthesis, 744
variations in, 748
Wittig reaction, 1051
mechanism of, 1053
Wittig, George, 1051
Wolff-Kishner reduction, 861, 988
Woodward, R. B., 510, 1190
Woodward-Hoffmann rules, 1190
and singlet methylene, 1204
application to organic reaction mechanisms, 1193
correlation diagrams in, 1193
for Diels-Alder reaction, 1202
for pericyclic reactions, 1199

MO theory in, 1193
summary of, 1204
supra-supra attack in, 1196
symmetry-allowed reactions, 1193
symmetry-forbidden reactions, 1193
Wurtz reaction, 126, 353

Xanthate pyrolysis, 1148
Xylan, 1310
m-Xylene, 548, **550**
IR spectrum of, 606

o-Xylene, 548, **550**
from toluene, 587
IR spectrum of, 606
oxidation of, 593
p-Xylene, 548, **550**
from toluene, 587
IR spectrum of, 606
D-Xylofuranose, 1269
Xylose, 1272
D-Xylose, 1269

Yeast Mannan, 1310
Ylides, 1051

formation of, 1052
structure of, 1052

Z and E, nomenclature, 250
Z/E system of nomenclature, 250
Ziegler, Karl, 347
Ziegler-Natta catalysts, 347
Zinc amalgam, 861
Zingiberene, 351
Zusammen, 250
Zwitterions, in α-amino acids, 1212

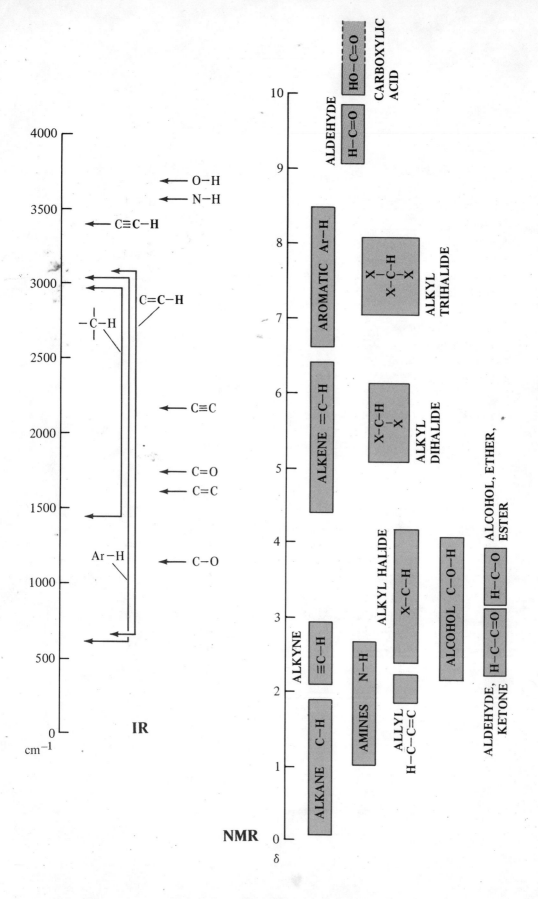